Proceedings of the 10th International Symposium on Cavitation (CAV2018)

May 14-16, 2018
Baltimore, Maryland

I0894727

© 2018, The American Society of Mechanical Engineers (ASME), 2 Park Avenue, New York, NY 10016, USA (www.asme.org)

All rights reserved. Printed in the United States of America. Except as permitted under the United States Copyright Act of 1976, no part of this publication may be reproduced or distributed in any form or by any means, or stored in a database or retrieval system, without the prior written permission of the publisher.

INFORMATION CONTAINED IN THIS WORK HAS BEEN OBTAINED BY THE AMERICAN SOCIETY OF MECHANICAL ENGINEERS FROM SOURCES BELIEVED TO BE RELIABLE. HOWEVER, NEITHER ASME NOR ITS AUTHORS OR EDITORS GUARANTEE THE ACCURACY OR COMPLETENESS OF ANY INFORMATION PUBLISHED IN THIS WORK. NEITHER ASME NOR ITS AUTHORS AND EDITORS SHALL BE RESPONSIBLE FOR ANY ERRORS, OMISSIONS, OR DAMAGES ARISING OUT OF THE USE OF THIS INFORMATION. THE WORK IS PUBLISHED WITH THE UNDERSTANDING THAT ASME AND ITS AUTHORS AND EDITORS ARE SUPPLYING INFORMATION BUT ARE NOT ATTEMPTING TO RENDER ENGINEERING OR OTHER PROFESSIONAL SERVICES. IF SUCH ENGINEERING OR PROFESSIONAL SERVICES ARE REQUIRED, THE ASSISTANCE OF AN APPROPRIATE PROFESSIONAL SHOULD BE SOUGHT.

ASME shall not be responsible for statements or opinions advanced in papers or . . . printed in its publications (B7.1.3). Statement from the Bylaws.

For authorization to photocopy material for internal or personal use under those circumstances not falling within the fair use provisions of the Copyright Act, contact the Copyright Clearance Center (CCC), 222 Rosewood Drive, Danvers, MA 01923, tel: 978-750-8400, www.copyright.com.

Requests for special permission or bulk reproduction should be addressed to the ASME Publishing Department, or submitted online at https://www.asme.org/shop/books/book-proposals/permissions

ASME Press books are available at special quantity discounts to use as premiums or for use in corporate training programs. For more information, contact Special Sales at CustomerCare@asme.org

ISBN: 978-0-7918-6185-1

Table of Contents

Proceedings of the 10th Symposium on Cavitation (CAV2018)

CAV2018 was the 10th international symposium on cavitation. It is the world's leading event committed to cavitation where leading scientists and industry experts converge and discuss all phases of cavitation, both fundamental and applied. CAV2018 provides a unique opportunity to discuss the latest advances in theoretical modelling, numerical simulation and experimentation related to the cavitation phenomenon. It was hosted by the Johns Hopkins University and took place at the Renaissance Baltimore Harborplace Hotel in Baltimore, Maryland, USA. The proceedings contains over 200 papers, and is available for public access on The ASME Digital Collection, as well as for purchase in print-book format.

6DOF motion and Cavity Dynamics of a Ventilated Super-cavitating Vehicle with Control Fins

[1]Sungtaek Park; [1]Jeonghwa Seo; [1]Shin Hyung Rhee*; [2]Sangdon Lee;

[1]Seoul National University, Seoul, Korea; [2]NEXTfoam Co., Ltd., Seoul, Korea

Abstract

Flow fields around a super-cavitating vehicle with control fins in ventilated cavity and its six-degrees-of-freedom (6DOF) motion were investigated by computational fluid dynamics (CFD) analysis. SNUFOAM, a CFD program based on an open-source CFD toolkit and specialized for naval hydrodynamics, was used in this study. A mixture fluid-based cavitation model and k - ω SST model were employed for the analysis of turbulent multiphase flow with liquid water, vapor, and non-condensable gas. Convection and diffusion terms were discretized by the second order upwind and central differencing scheme, respectively. The geometry of interest was a super-cavitating vehicle consisted of a cone-shaped nose, cylindrical main body, and four control fins near the tail. A cavitator and ventilation holes for injecting non-condensable gas were located at the nose of the vehicle. To accommodate the motion of deploying fins, a layering mesh technique was employed, while a mesh morphing technique was used for the motion of the vehicle body. Two different control fin operations, which are deploying and deflecting fins, were applied and the effects of injecting non-condensable gas were also examined.

Keywords: Super-cavitation; 6DOF motion; CFD; Ventilated cavitation

Introduction

Super-cavitating underwater vehicles can significantly reduce frictional drag, unlike conventional ones, by generating cavity around the body and achieve high speed. Control fins of a super-cavitating vehicle are partially immersed in the cavity, thus interaction of cavity dynamics and force on the fins should be considered to analyze the maneuverability of the super-cavitating vehicle [1, 2].

For initiating and maintain super-cavitation, cavitation number should be kept sufficiently low. It is tough to achieve and maintain the status by natural cavitation only, and therefore ventilated cavitation by injection of non-condensable gas has been also studied [3]. In those studies, due to extreme difficulty in model testing, CFD analysis is the most feasible method to analyze the cavity dynamics and maneuverability of the ventilated super-cavitation vehicle. The present study aims to introduce CFD methods for a ventilated super-cavitation vehicle and analyze 6DOF motion of the vehicle with fin operation.

Problem Description

The test model was an axisymmetric super-cavitating vehicle. Figure 1 shows the model geometry with a cavitator and six injector holes. Four control fins were attached near the tail. The length (L) and diameter (D) of the model were 2.1 m and 0.145 m, respectively. The span of control fins was 0.128 m and the chord at the fin root was 0.0196 m. Two conditions of control fin operation, i.e., deploying fins after launching and deflecting fins for maneuvering, were tested. The vehicle advance speed when deploying control fins was 20 m/s. Fins were initially lain in slits on the body and deployed within 0.5 sec after the cavitation surrounding the vehicle was fully developed. The case of deflecting fins assumes that the cavity is fully developed at the design speed 50 m/s. The deflection angle was 1.5°. For both cases, the injection volume of non-condensable gas from ventilation holes was 0.069 m^3/s. No-injection condition was also tested in the fin deflection case.

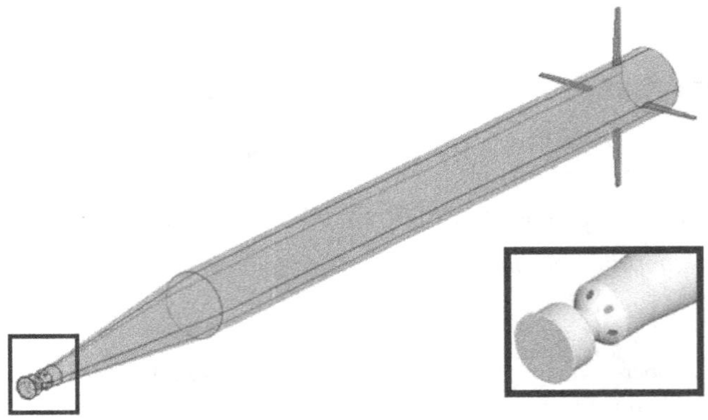

Figure 1 Design of the super-cavitating vehicle

Governing Equations and Numerical Methods

The governing equations of the numerical simulation were the continuity equation and unsteady Reynolds-averaged Navier-Stokes equation. The presence of three phases (water liquid, vapor, and non-condensable gas) was represented by volume fraction.

Computations were carried out using SNUFOAM, which was developed based on an open-source CFD toolkit, OpenFOAM [4]. SNUFOAM employed a cell-centered finite-volume method based on a multi-dimensional linear reconstruction scheme that permitted the use of computational elements (cells) with arbitrary polyhedral cell topology. The solution gradients at cell centers were obtained by applying Green-Gauss theorem based on a node-based quadrature. For discretization of convection terms and diffusion terms in filtered momentum equations, we adopted a quadratic upwind differencing (QUICK) scheme and the second order central-differencing scheme, respectively. The system of discretized governing equations was solved using point-implicit Gauss-Seidel relaxation along with algebraic multi-grid method to accelerate solution convergence. For turbulence closure, the k-ω SST model [5] was used.

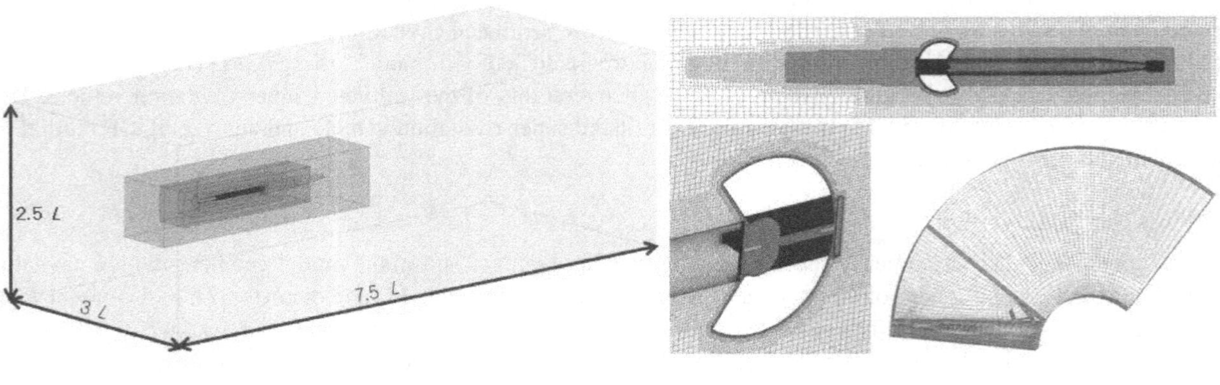

Figure 2 Computational domain and mesh around the vehicle and control fins

Figure 2 shows the computational domain and grids. The upstream inlet and downstream exit boundary was located at 1.5 L forward and 5 L aft from the model, respectively. The width of the computational domain was 3 L and height was 2.5 L. To enable deploying motion of control fins, the layering mesh method was applied. The layering mesh method is effective to analyze a motion of large displacement along a trajectory. Morphing mesh technique was used to analyze 6DOF motion of the vehicle body with small displacement and arbitrary direction. The dynamic mesh zone that morphing was allowed was within 0.75 L from the vehicle body.

Results and Discussion

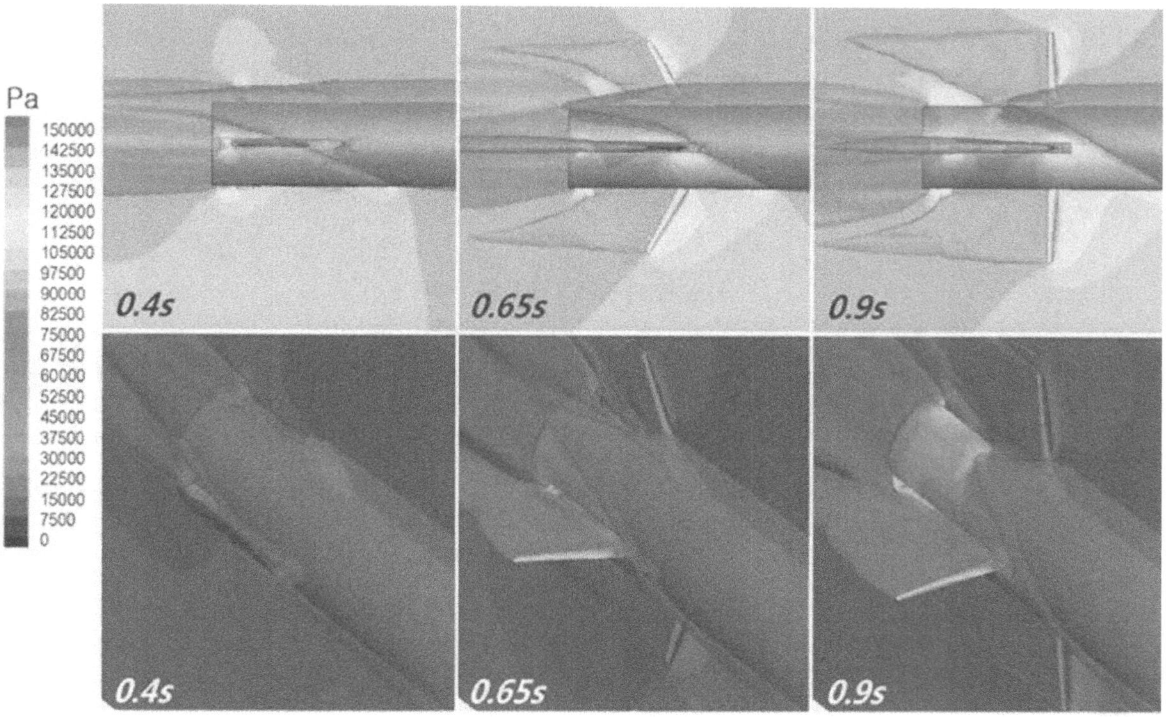

Figure 3 Pressure contours and the boundary of the cavity at deploying control fins in fully developed super-cavitation

Figure 3 shows pressure distribution and cavity shape around the vehicle and control fins during the deploying phase. The cavity was generated from the nose with small local cavities were observed behind the fins and body. Owing to the buoyancy of the non-condensable gas, the cavity on the upper side was larger than that on the lower side. At 0.3 s, the super-cavitation was fully developed and the control fins began to deploy. Drag force on control fins increased as they were unfolded. At 0.8 s, the control fins were completely deployed and the drag force was converged (see Figure 4). The asymmetry of the cavity resulted in imbalance of drag force on the control fins. Drag force on the top fin was smaller than that on the bottom fin, but showed fluctuation after 0.8 s, due to tail slap on the cavity boundary.

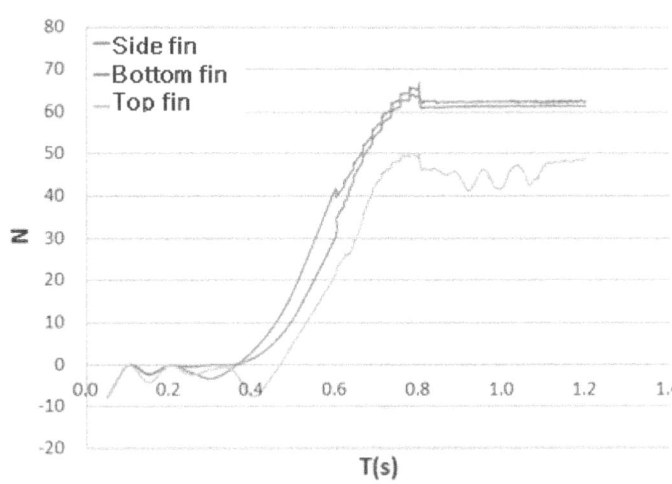

Figure 4 Drag on the control fins at deployment

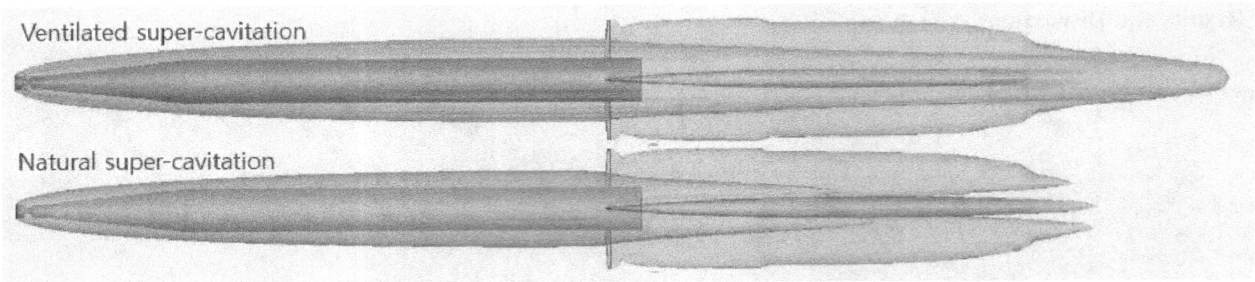

Figure 5 Cavity around the super-cavitation vehicle

Figure 5 shows fully developed cavity around the test model before fin deflection at advance speed of 50 m/s. With the ventilated super-cavitation case, the cavity was thicker than that with the natural case, resulting in different immersed depth of the fins. In both of cases, the gravity caused a bias of cavity toward the upward direction, but the ventilation case with non-condensable gas injection showed more severe bias, and the resultant pitching moment on control fins was generated.

For maneuvering of the super-cavitating vehicle, top and bottom fins were deflected by 1.5° to cause yaw motion. Figure 6 shows time history of roll and yaw motion of the vehicle. Deflection of control fins began at 0 s. At 0.1 s, yaw motion fluctuated as the tail slapped the cavity boundary, but increased gradually from 0.2 s.

Roll motion of the vehicle was also observed, because of different immersed depth of the top and bottom fins. Lift on the bottom fin was larger than that on the top fin, and the imbalance of lift resulted in roll moment on the vehicle body caused roll motion. The ventilated cavitation case, where the imbalance was larger, showed greater change of roll motion.

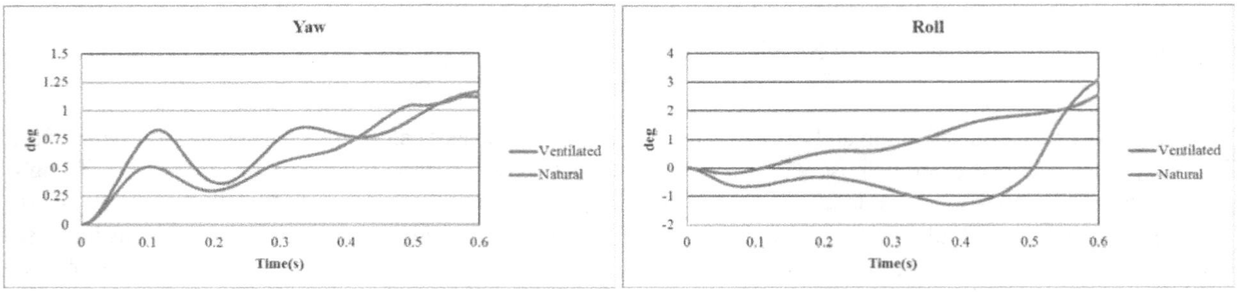

Figure 6 Time history of the motion of the super-cavitating vehicle

Conclusion

Mesh morphing technique in a multi-phase flow CFD solver was applied to analyze flow fields around a super-cavitating vehicle with control fins deployment and deflection. The ventilated cavitation was asymmetric and skewed upward. Tail slap and fluctuation of drag on the top fin were observed. For fin deflection, effects of non-condensable gas injection were examined. Ventilated super-cavitation case showed less fluctuation in yaw motion, but greater change in roll motion than natural super-cavitation.

References

[1] Lin, G., Balachandran, B., Abed, E. H. (2007). *Nonlinear dynamics and bifurcations of a supercavitating vehicle.* IEEE Journal of Oceanic Engineering. 32(4).

[2] Lin, G., Balachandran, B., Abed, E. H. (2008). *Dynamics and control of supercavitating vehicles.* Journal of Dynamic Systems, Measurement, and Control. 130(2).

[3] Lindau, J. W., Kunz, R. F., Boger, D. A., Stinebring, D. R., Gibeling, H. J. (2002). *High Reynolds number, unsteady, multiphase CFD modeling of cavitating flows.* Journal of Fluids Engineering. 124(3).

[4] Jasak, H. (2009). *OpenFOAM: Open source CFD in research and industry.* International Journal of Naval Architecture and Ocean Engineering. 1(2).

[5] Menter, F. R. (1994). Two-equation eddy-viscosity turbulence models for engineering applications. AIAA Journal. 32(8).

© 2018 ASME

10th International Symposium on Cavitation - CAV2018
Baltimore, Maryland, USA, May 14 – 16, 2018 CAV18-05002

A Computational Assessment of Gas Jets in a Bubbly Co-Flow

[1]Melissa Fronzeo*, [1]Michael Kinzel

[1]*The Pennsylvania State University, University Park, PA, USA*

Abstract

In this effort, Computational Fluid Dynamics (CFD) is used to investigate the dynamics of gas jetted into an internal pipe flow. Specifically, we aim to understand the resulting cavity and resulting interaction modes that drive the gas to mix with the exterior flow. In order to vary the bubbly flow, the pipe fluid is treated as a weighted mixture of air and water with the volume fraction of air being a controlled variable. The velocity of the external flow and the mass flow rate of the air jetted in the flow are also varied to obtain a global understanding of the variation. These variations indicate the occurrence of several flow regimes of the primary cavity forming at the jet exit, as the jet gas interacts with the bulk pipe flow. Results of the weighted air/water mixture are compared to previous results involving only water as the bulk pipe flow. The overall results provide insight into the various flow regimes that occur.

Introduction

In the present effort, we consider the primary breakup mechanism in a cavity that initiates gas mixing within a pipe co-flow. The diagram in Figure 1 indicates the scenario being investigated, which indicates flow through a pipe with a sting-mounted body that is used to inject and mix gas into the pipe flow. In general, gas jets expelled into a gaseous flow are well understood. However, the mechanisms of multiphase gas jets expelled into a liquid are less understood than single phase jets, with the majority of the studies relating to gas jetted into a quiescent liquid. When it comes to gas jets expelled into a liquid co-flow, or even bubbly liquid, the field of study is even smaller. Despite this, understanding the dynamics of the jet-liquid interactions are important to industrial uses such as underwater cutting and chemical mixing [1]. In this effort, a parametric study of this multiphase flow system is investigated using CFD, which has been previously used and validated in regards to similar multiphase jets applications [2-4].

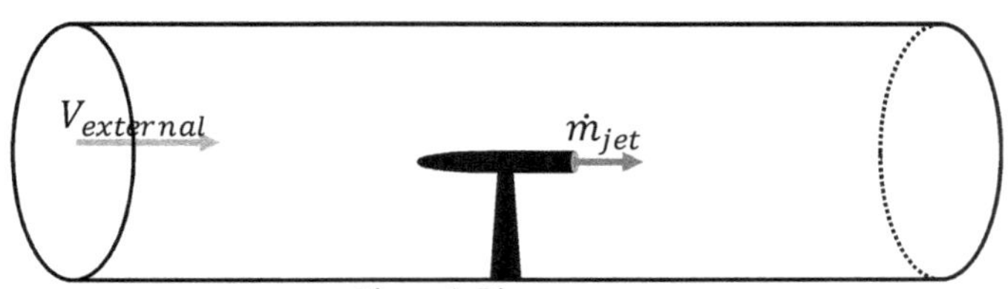

Figure 1: Pipe geometry

In previous work from the authors, the variation of the freestream liquid velocity and the mass-flow rate of the gaseous jet were investigated [2,3]. This data was then used to identify different regimes based on the gas-liquid interactions, and subsequently create a regime map of these interactions. Results indicated that the interaction involves at least five-different dynamic modes. In those studies, pure water was used as the freestream liquid, hence, the conditions do not include free-stream gas content that is expected to occur with recirculation or when placing injection ports in parallel. The present effort expands on the previous work in order to develop a more generalized understanding of submerged jets by also considering variations in the freestream liquid properties. Specifically, the liquid is varied by using a weighted mixture of air and water. Such mixture variation, controlled by a volume fraction parameter, leads to density, compressibility, and viscosity changes in the freestream flow. For these various free-stream conditions, parametric evaluations of freestream velocity and jet-mass-flow rate are still evaluated creating a three-dimensional set of data. In these efforts, previously used regime analysis methods are applied to identify the flow regimes for each data set. The result is a three-dimensional matrix of data with parametric variations in the freestream velocity, jet mass-flow rate, and free-stream gaseous volume fraction.

Methods

The present work was performed within the commercial code, Star-CCM+ [5]. The physical modeling for the gas jet and varying freestream liquid was previously used and validated for gas jetted into water [2]. The modeling was specifically validated against the experimental data found in Weiland's PhD dissertation [1]. All of the simulations are conducted using 2D axisymmetric modeling, with a time-resolving flow model. The previous work conducted on air jetted into water classified several multiphase regimes of the gas-liquid interactions, with infographics in Figure 2:

- Mode 1: Shedding Jet: Large bubble ejections; gas is attached to the nozzle (Figure 2(e)).

*Corresponding Author, Melissa Fronzeo: maf5524@psu.edu

- Mode 2: Toroidal Cavity: Small gaseous cavity; liquid enters cavity in constant stream, ejections of gas pockets, often in toroidal vortex form. (Liquid entering the cavity indicated by dotted lines in Figure 2(a)).
- Mode 3: Pulsating Cavity: Medium gaseous cavity; large interface waves; liquid enters cavity in small pockets (indicated by dotted lines and circles in Figure 2(b)).
- Mode 4: Stable Cavity: Large gaseous cavity; little to no liquid enters cavity; small interface waves (Figure 2(c)).
- Mode 5: Over-Ventilated Cavity: Small gaseous cavity; large bubble ejections from end of cavity (Figure 2(d)).

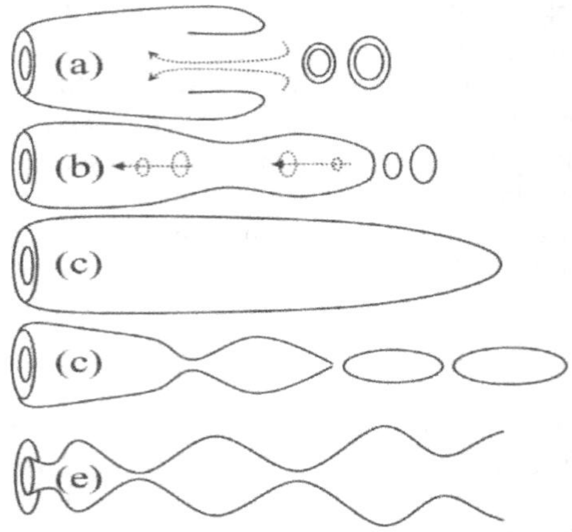

Figure 2: Cartoons of each mode; (a) Toroidal Cavity, (b) Pulsating Cavity, (c) Stable Cavity, (d) Over-Ventilated Cavity, (e) Shedding Jet.

Simulation examples of each of these regimes are shown in Figure 3. In the images, the grayscale lines indicate the volume fraction of the gas (with inside of white being all gas, outside of the black lines being all liquid) and the contours tend to outline the predicted mixture regions of the gaseous cavity. The color contour indicates Mach number, with dark blue being low Mach, and the light blue/cyan color showing the jet (high Mach). The top images are of the time-averaged simulation, while the bottom images are a snapshot of the unsteady simulation.

These regimes were identified using visual inspection; videos of the unsteady simulation, as well as images of the time-averaged solutions were compared for sixty-four separate cases (varied freestream velocity and mass-flow rate of the gas). The following figures show an example of this process, using a toroidal vortex cavity case (Figure 4). The regimes are then assigned an integer value associated with the regime (Figure 5), and then plotted on a grid to indicate the flow-regime map.

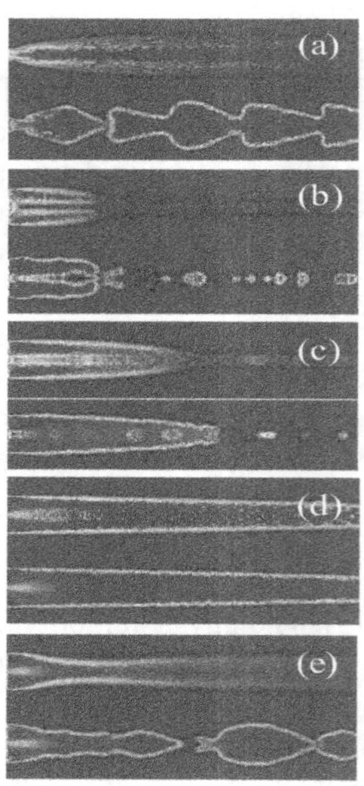

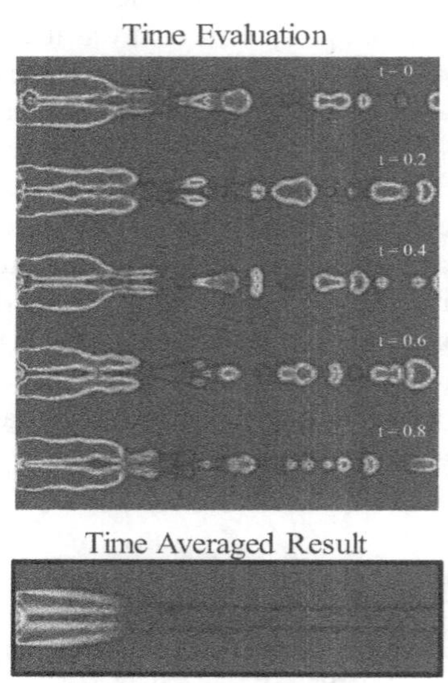

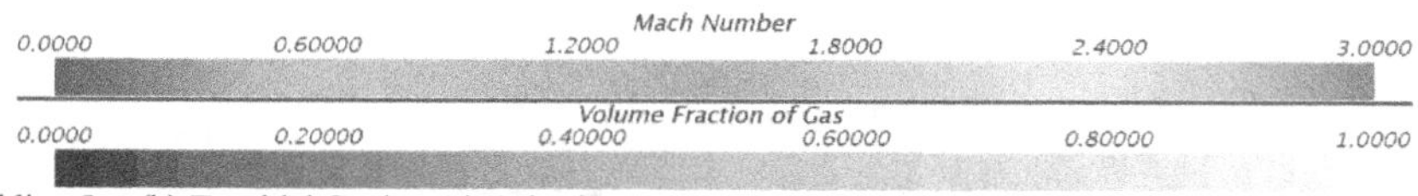

Figure 3: (a) Shedding Jet, (b) Toroidal Cavity, (c) Pulsating Cavity, (d) Stable Cavity, (e) Over-Ventilated Cavity

Figure 4: Time Sequence (top) and time-averaged solution (bottom) of a Toroidal Cavity

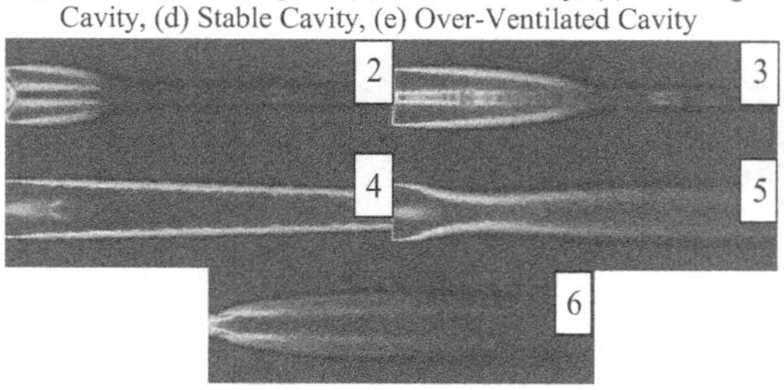

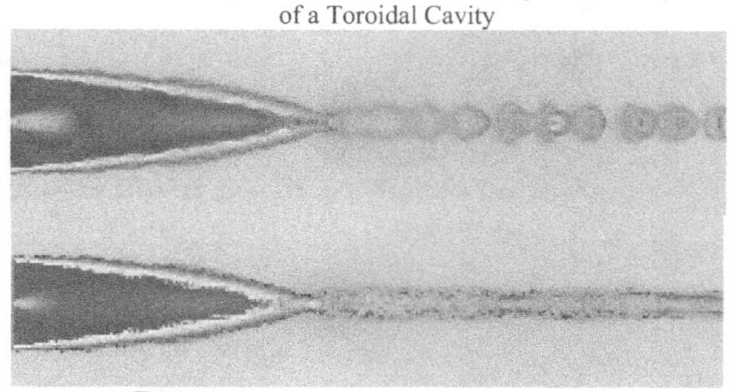

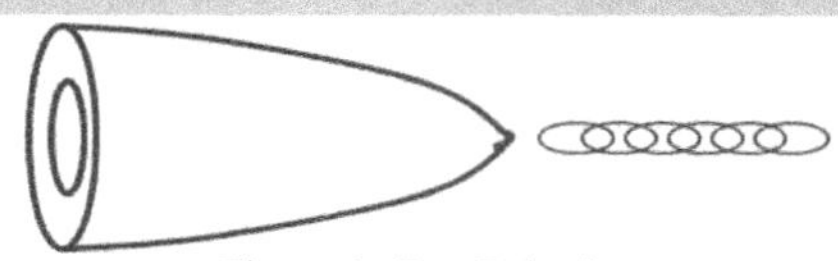

Figure 5: Numeric numbering of regimes

Figure 6: Fast Pulsating

These numeric values are used to create a regime map over a grid of input parameters. Regime maps such as these have been used before to evaluate multiphase flows, such as gas and liquid in a pipe. These maps provide additional insight into controlling parameters between regime, and are a predictive aid in future work [6-7]. The original cases were taken over a sweep of freestream velocity values (1, 5, 10, 15, 25, 30, 40, and 50 m/s), and a sweep of mass-flow rate of the jet values (0.0025, 0.0040, 0.0050, 0.0060, 0.0075, 0.0080, 0.0100, and 0.0125 kg/s). In this work, we expand on the previous wok with additional cases that vary the free-stream volume fraction of air. Hence, the original data set used only water as the freestream liquid. The new data will use a weighted air/water mixture, with volume fractions of air equal to 0.25, 0.50, and 0.75. The same variations of freestream velocity as well as mass-flow rate of the jet are used.

Results

As the volume fraction of air in the freestream is increased, the first major change is the appearance of a new regime. When air is mixed into the water, the higher freestream velocity cases begin to form the regime shown in Figure 6. This regime, which will be called the Fast Pulsating Cavity, is characterized by a small cavity with no visible waves on the interface, but a constant stream of small gaseous ejections from the cavity tip. Comparing over a reduced sets of data for the volume fractions of 0.25, 0.50, and 0.75, it appears that this regime would fit into the numeric value of 1, for plotting and transitional purposes. Reduced data sets (only 16 cases; 4 mass-flow rates and 4 freestream velocities) are shown in Figure 7, Figure 8, and Figure 9.

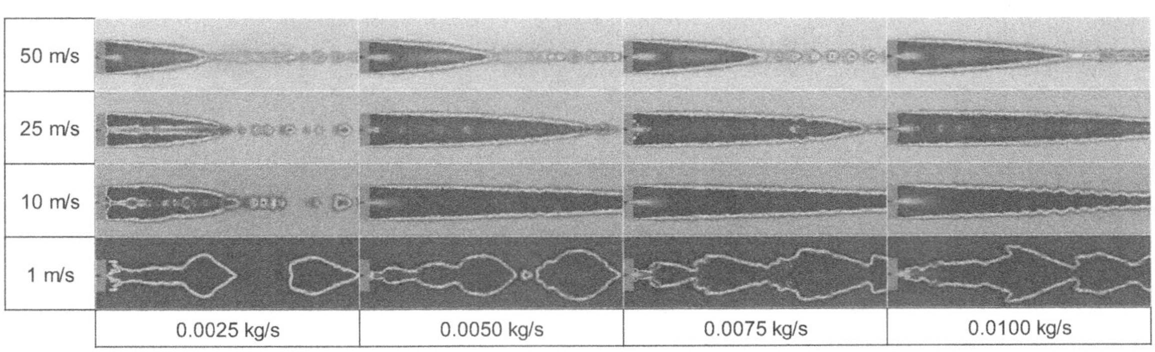

Figure 7: Reduced Data Set, Volume Fraction of Air = 0.25

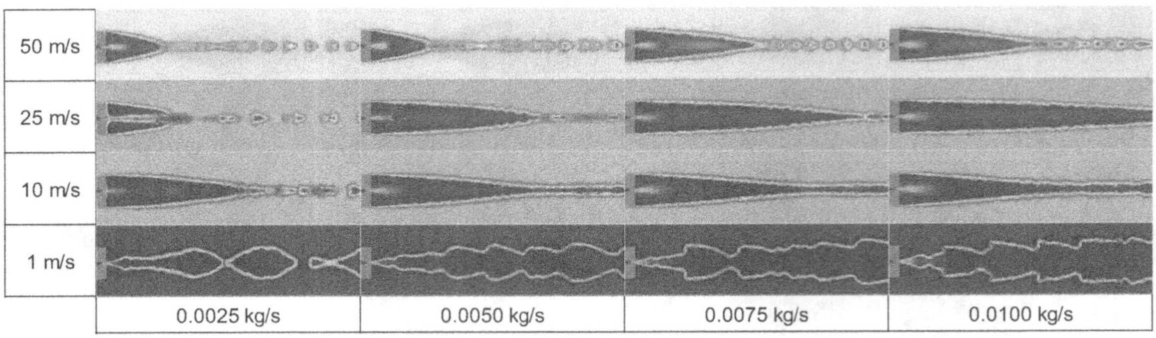

Figure 8: Reduced Data Set, Volume Fraction of Air = 0.50

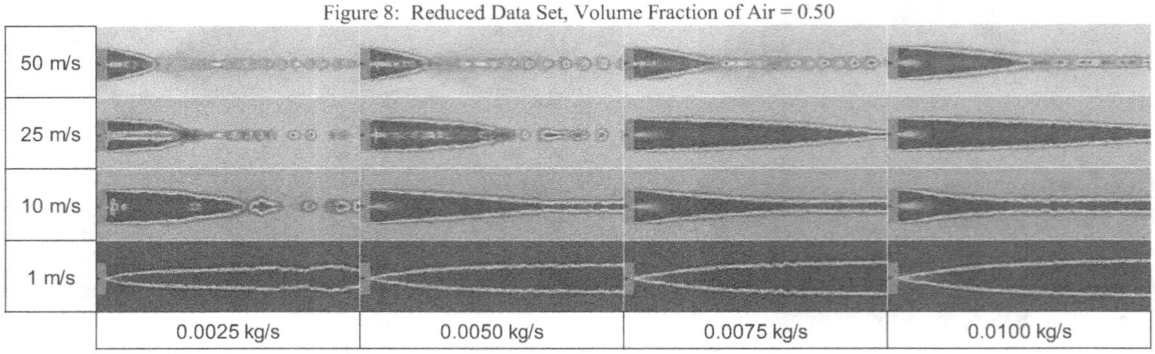

Figure 9: Reduced Data Set, Volume Fraction of Air = 0.75

The resulting regime maps were created by plotting the associated integer for reach regime values over a two-dimensional grid of freestream velocity and mass-flow rate for each volume fraction, interpolating between given values. The limits between each regime (and intermediate zones) are then smoothed by applying curve fits. An interpolated and smoothed plot are shown in Figure 10, with the rest of the smoothed plots shown in Figure 11. Displaying the information in this map format shows how flow is affected by increasing the volume fraction of air in the freestream. When the volume fraction is increased from 0.00 to 0.25, the fast pulsating regime appears. Similar to the shedding jet, it appears to be almost completely controlled by freestream velocity, creating an upper cap on the freestream velocity values. In addition, the over-ventilated and stable regimes expand. When the volume fraction is raised to 0.50, the fast pulsating regime appears at lower freestream velocities, and the over-ventilated and stable cavity regimes expand further, showing that the intermediate regime bands become compressed (fewer instances of Toroidal and Pulsating cavities). These changes are depicted in Figure 11.

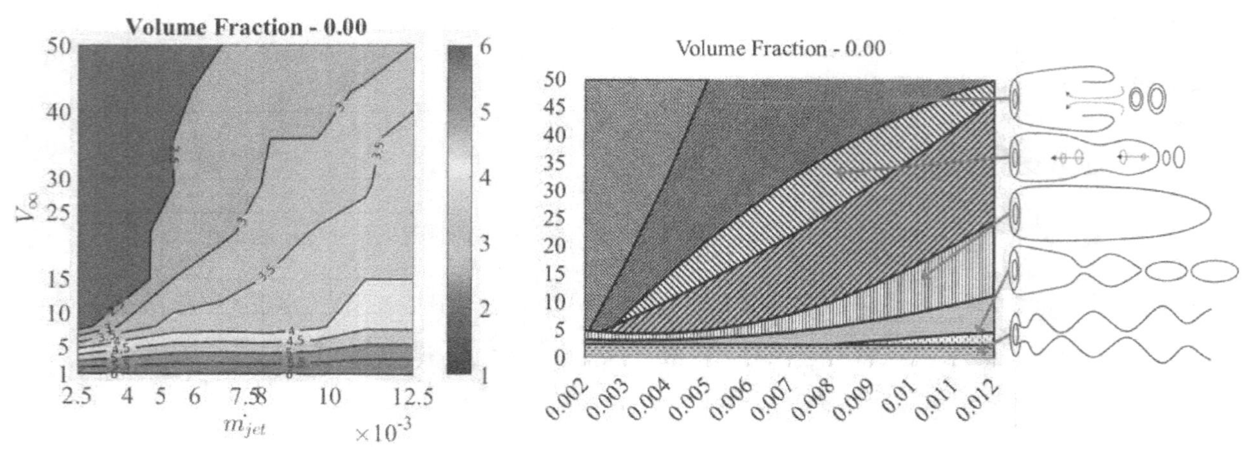

Figure 10: Original regime map (left) and smoothed regime map (right).

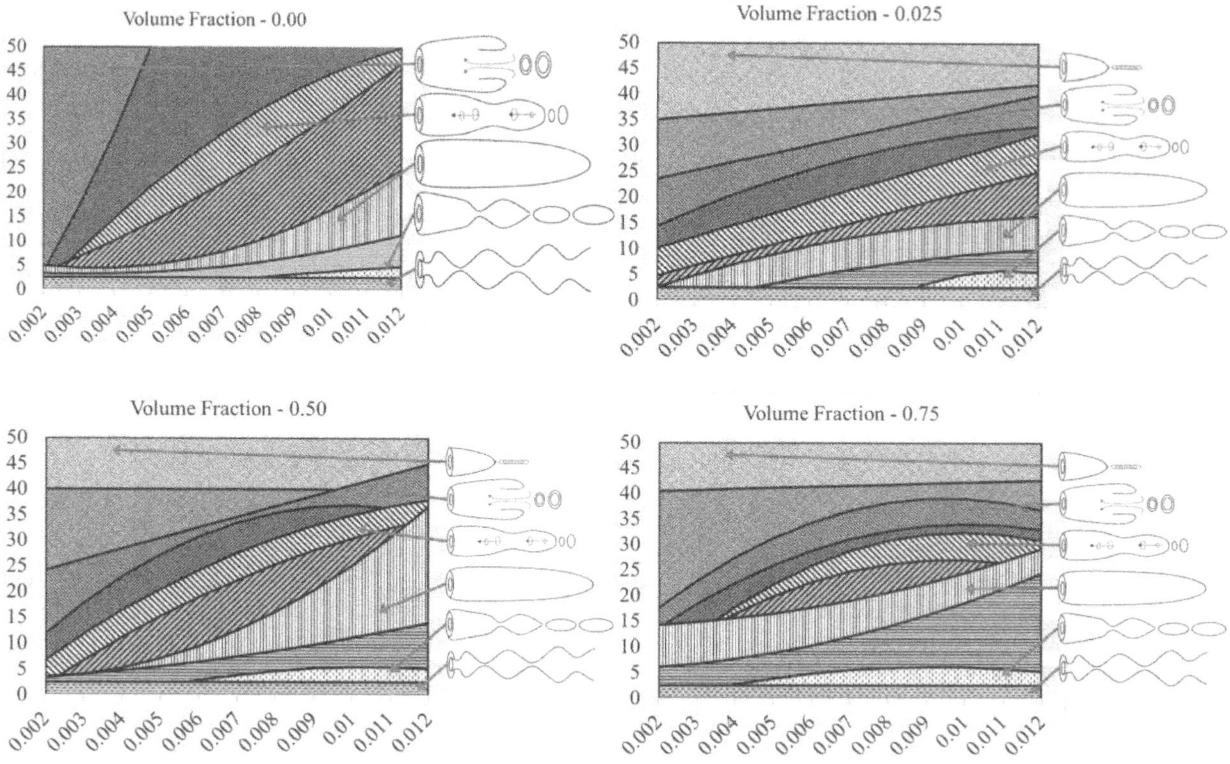

Figure 11: Smoothed regime maps of volume fractions 0.00 (top left) 0.25 (top right), 0.50 (bottom left), and 0.75 (bottom right).

Conclusion

The present simulations evaluated the initial breakup mechanisms for a gas jet vented into a liquid pipe flow with parametric variations in the free-stream gaseous air content, the freestream velocity, and the gaseous mass-flow rate. The overall results indicate a rather complicated regime map that is still presently being evaluated. When the new simulations are compared to an only water co-flow, it is found that an additional regime occurs which seems to dominate when the free stream is high, appearing as the gas content of the freestream is increased. These datasets can be further analyzed to in the hopes of identifying two non-dimensional parameters which can collapse the all of the data to a single map. This can better generalize the dynamics found in high speed multiphase mixing of a gas into a freestream liquid, or a liquid/gas mixture.

References

[1] Weiland, C. J., 2009, "Characteristics of the high speed gas-liquid interface," Ph.D. thesis, Virginia Tech, Blacksburg

[2] Fronzeo, Melissa, Michael Kinzel, and Jules Lindau "A Flow Regime Map for Gas Jets in a Bubbly Flow" *Proceedings of the 91st Fluid Engineering Division Summer Meeting FEDSM* July 2017.

[3] Fronzeo, M. A., 2017, "A Computational Study of Gaseous Jets Submerged in a Liquid Co-Flow", M.S. thesis, The Pennsylvania State Univeristy, State College

[4] Kinzel, Michael P., et al. "A numerical assessment of the interaction of a supercavitating flow with a gas jet." *Ocean Engineering* 136 (2017): 304-313.

[5] Siemens, "StarCCM+," 2017.

[6] Açikgöz, M., F. Franca, and R. T. Lahey. "An experimental study of three-phase flow regimes." International Journal of Multiphase Flow 18.3 (1992): 327-336.

[7] Taitel, Yemada, and A. E. Dukler. "A model for predicting flow regime transitions in horizontal and near horizontal gas-liquid flow." AIChE Journal 22.1 (1976): 47-55.

© 2018 ASME

A level-set based free-surface tracking method for the simulation of bubble collapse and jetting in generalized Newtonian fluids

[1,2]Ratnesh K. Shukla*; [2]Arpit Tiwari; [2,3]Jonathan B. Freund

[1]*Department of Mechanical Engineering, Indian Institute of Science, Bangalore, Karnataka, INDIA*

[2]*Department of Mechanical Science & Engineering, University of Illinois at Urbana-Champaign, Urbana, IL, USA*

[3]*Department of Aerospace Engineering, University of Illinois at Urbana-Champaign, Urbana, IL, USA*

Abstract

Predictive simulations of aspherical bubble collapse and jetting are known to be particularly challenging. Under the influence of an external pressure perturbation, bubbles can also undergo prolonged oscillations with potentially large variations in their size and shape. Simulation techniques must robustly represent these large variations while maintaining accuracy. The challenge of nonlinear rheology introduces additional challenges. To do this, we design a new level-set-based free-surface tracking technique. The bubble contents are assumed to be inviscid and spatially homogeneous, governed by a polytropic gas equation of state. A least-squares methodology enforces the normal pressure jump and shear-free boundary conditions precisely at the sharp bubble-liquid interface. Constraints maintain a divergence-free velocity. Numerical tests on spherically symmetric configurations on rectangular meshes demonstrate that the method accurately predicts the bubble response to impulsive and time-periodic pressure perturbations in a generalized Newtonian fluid over multiple growth and collapse cycles. Asymmetries remain small throughout the course of computations. Demonstrations of aspherical bubble dynamics and jetting next to a rigid wall illustrate the formulation's ability to represent complex topological features, including interface breakup in shear-thinning/thickening fluids. These simulations are used to study the influence of shear-thinning and shear-thickening rheology on cavitation damage in tissue-mimicking soft materials. Simulations of growth and collapse of a spherically symmetric vapor bubble that fully incorporate the phase change dynamics, illustrate the flexibility and robustness of our implementation.

Keywords: cavitation; bubble collapse; sharp interface method

Introduction

Cavitation bubble collapse plays an important role in a variety of biomedical techniques that involve interaction of shock or acoustic waves with tissue or blood [1-3]. In applications such as shock-wave and burst-wave lithotripsy, cavitation bubble collapse is thought to be principally responsible for both kidney stone comminution and undesirable collateral damage caused to the surrounding tissue. Similarly, microcavitation induced by the interaction of the tensile portion of the blast waves with brain tissue has been proposed as a likely cause for blast-wave-induced traumatic brain injury. Detailed understanding of the aspherical bubble collapse and jetting process in tissue-mimicking soft materials can aid design of damage-resistant shields that mitigate or suppress microcavitation-induced tissue injury and can help optimize biomedical techniques such as lithotripsy.

The bubble collapse process is intense and rapid with the entire process, including the formation of high-speed jet, typically lasting for only tens of microseconds for micron sized bubbles [1]. The violence of the collapse and jetting process cause high local stresses and strain rates. The mechanical properties of tissues differ vastly from those of simple fluids and under these extreme conditions the nonlinear rheology of the tissue material is expected to play an important role in the collapse, jetting, and injury [3]. The influence of complex rheology on the collapse and jetting dynamics is thus of significance and motivates our simulation-based analysis of aspherical bubble dynamics in shear-thinning and shear-thickening generalized Newtonian fluids.

*Corresponding Author, Ratnesh K. Shukla: ratnesh@iisc.ac.in

The nonlinearity in the constitutive relationship presents additional computational challenges in simulations of bubble growth and collapse in a rheologically complex fluid medium. To accurately account for the nonlinear rheology of a generalized Newtonian fluid medium, simulation methods are most faithful to the mechanics they represent if they ensure exact continuity of the normal and tangential stresses across the bubble surface. Enforcing this condition is particularly challenging in the commonly employed fully compressible formulations that integrate shock and interface capturing methodologies with techniques that ensure correct transmission of the shock and acoustic waves across the bubble-liquid interface [4-11]. Resolution of rheology-dependent sharp flow features, such as boundary layers, within a fully compressible formulation tends to be prohibitively expensive. This is due to the restrictions on spatiotemporal resolutions that must overcome the numerical dissipation fundamental to effective shock and interface capturing techniques. Important for motivating our formulation, for many relevant physiological conditions, the acoustic perturbations are relatively weak so that a general compressible treatment of the nearly incompressible liquid phase is overly descriptive, inefficient, and potentially inaccurate. To overcome the aforementioned challenges and limitations of a compressible multifluid treatment we develop a level-set-based free-surface tracking technique that fully incorporates the important complex rheology of the generalized Newtonian fluid medium but with approximations suited to our condition. In the subsequent sections, we describe our modeling and simulation approach, and demonstrate simulations of a near-wall bubble pair collapse in a shear-thinning fluid and a spherically symmetric collapse with inclusion of mass transfer across the interface.

Free-surface model for bubble dynamics

Typical cavitation-induced tissue damage scenarios involve interactions of micron-sized bubbles with relatively weak acoustic perturbations with a wavelength significantly larger than the bubble size. The bubble density and viscosity are typically orders of magnitude lower than those of the surrounding fluid. In such situations, spatial variations of the state variables in the bubble interior are negligible compared to those in the surrounding medium, which itself is very nearly incompressible. We therefore consider a modeling framework in which the bubble contents are assumed to be inviscid and spatially homogeneous, as in the well-established Rayleigh-Plesset model for spherical bubble dynamics [12-16]. The spatially uniform bubble density ρ_B is related to its uniform pressure P_B via a polytropic gas equation of state of the form $P_B = P_{B0} (\rho_B/\rho_{B0})^{\gamma}$, where γ is the polytropic exponent and subscript '0' denotes initial bubble pressure and density. The complex rheology of the incompressible medium surrounding the bubble is accounted for through a generalized Newtonian fluid model for the shear-dependent viscosity. Thus, within this relatively simplified framework, the shear-free bubble-medium interface simply serves as a boundary for the incompressible fluid medium. The principle novelty of the numerical discretization is enforcement of the kinematic and dynamic conditions at the interface to ensure continuity of the tangential and normal stresses across the material interface.

Numerical discretization

The method is built on a standard second-order accurate spatial discretization of the incompressible Navier-Stokes equations for a generalized Newtonian fluid. We employ a fully staggered arrangement of the primitive variables, taking advantage of its well-understood advantageous properties. The bubble boundary is tracked with an efficient narrow-band level-set technique, which relies on high-order advection and frequent reinitialization of a signed distance function [17]. A least-squares extrapolation method enforces the normal pressure jump and shear-free boundary conditions at the sharp bubble-medium interface. To minimize conservation errors in the transport of the level-set function, constraints are built into the extrapolation procedure for computing a divergence-free velocity field. We employ a second-order accurate explicit Runge-Kutta projection method for time advancement of the flow field and the bubble-medium interface.

Demonstration simulations

As a first demonstration example, we simulate axisymmetric growth and collapse of a bubble pair in a shear thinning medium next to a solid wall. The initially equisized gas ($\gamma = 1.4$) bubbles are located at non-dimensional stand-off distances of 2.33 and 5.0 (with the initial bubble radius R_0 as the characteristic length scale) from a left solid bound-

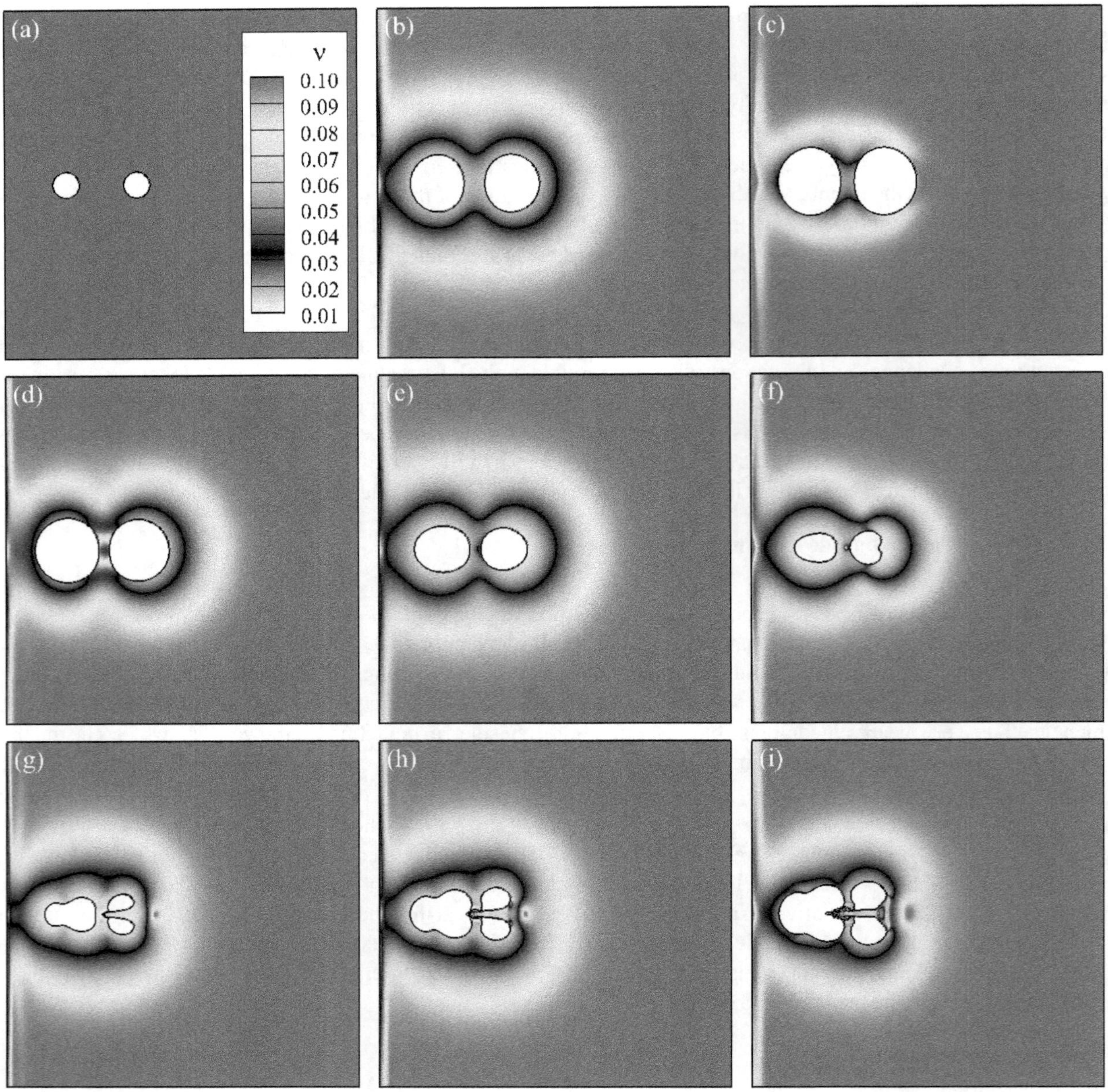

Figure 1: Generalized Newtonian viscosity (colors) and the shear-free bubble-liquid interface (solid black line) as a function of time for near-wall growth and collapse of a bubble pair in a shear-thinning fluid medium.

ary as depicted in figure 1(a). The initial bubble pressure P_{B0} is ten times the far-field pressure P_∞. No-slip and symmetry boundary conditions are applied at the left solid wall and lower boundary, while at the remaining two boundaries the far-field pressure P_∞ is prescribed along with a normal extrapolation for the velocity components. The shear-thinning is modeled with the commonly used Carreau expression [18]:

$$\nu(\dot{\gamma}) = \nu_\infty + \frac{\nu_0 - \nu_\infty}{\left(1 + \lambda^2 \dot{\gamma}^2\right)^{N/2}}$$

where limit viscosities are $\nu_0 = 0.1$ and $\nu_\infty = 0.01$, with the exponent $N = 0.8$. In this model, $\dot{\gamma}$ and λ denote the non-dimensional shear rate and a material constant, respectively.

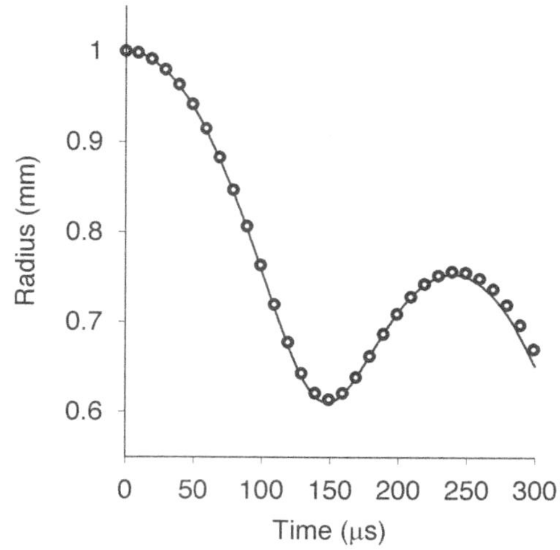

Figure 2: Radius histories of a spherically-symmetric vapor bubble collapse under an initial pressure ratio of 2 obtained from semi-analytical solution (line) and detailed axisymmetric simulation (symbols).

Figure 1 shows the evolution of the bubble-medium interface and the generalized Newtonian viscosity for a typical simulation on a uniform mesh with a spatial resolution that corresponds to about 19 nodes across the initial bubble radius R_0. The initial expansion of the bubble pair induces a shear-driven reduction in the generalized Newtonian viscosity around the bubble and in the boundary layer on the solid boundary. The reduction in the local viscosity, owing to an aspherical expansion of the two bubbles, eventually results in a jet formation and jetting impact (spherical to toroidal shape transition) of the bubble that is located further away from the solid boundary. However, this shear-thinning-induced local reduction in viscosity is less pronounced in the vicinity of the solid wall and therefore jet formation and therefore jetting impact is completely suppressed for the bubble located next to the solid boundary.

In our second example, we demonstrate the flexibility of our method to include phase change phenomenon. As a first step, we implemented a simplified from of the Hertz-Knudsen-Langmuir expression (see Hauke *et al.* [19] for details) for non-equilibrium mass transfer across the interface. The primary simplification we made here is that the saturation pressure in the mass transfer expression is assumed to be constant throughout the simulation, equal to the initial bubble pressure. For demonstration, we simulated spherically symmetric collapse of a vapor ($\gamma = 1.3$) bubble under an initial pressure ratio of 2. The bubble is assumed to be filled with saturated vapor initially at 50 kPa pressure; the mass transfer accommodation coefficient is set to 0.01. Figure 2 shows excellent agreement between the radius histories obtained from our axisymmetric simulation and the semi-analytical solution of a generalized Rayleigh-Plesset equation.

Summary

In summary, a sharp-interface Eulerian method that employs least-squares extrapolation in simulations of growth and collapse of multiple bubbles in an incompressible, generalized Newtonian fluid was presented. Axisymmetric simulations of near-wall growth and collapse of a bubble pair in a shear-thinning surrounding medium were conducted to demonstrate the method. In the examples considered, shear-thinning promotes jetting in aspherical bubble collapse through a reduction in the local generalized Newtonian viscosity. We also demonstrated the ability of our method to handle phase transition.

This work received support from the National Institutes of Health (NIH) NIDDK Grant No. P01-DK043881.

References

[1] Lauterborn, W., and Kurz, T. (2010). *Physics of bubble oscillations*, Reports on Progress in Physics, 73, 106501.

[2] Loske, A. M. (2017). *Medical and biomedical applications of shockwaves*. Springer.

[3] Brujan, E.-A. (2010). *Cavitation in non-Newtonian fluids*. Springer.

[4] Johnsen, E., and Colonius, T. (2006). *Implementation of WENO schemes incompressible multicomponent flow problems*. Journal of Computational Physics, 219, 715 - 732.

[5] Johnsen, E., and Colonius, T. (2009). *Numerical simulations of non-spherical bubble collapse*. Journal of Fluid Mechanics, 629, 231 - 262.

[6] Shukla, R. K., Pantano, C., and Freund, J. B. (2010). *An interface capturing method for simulation of multiphase compressible flows*. Journal of Computational Physics, 229, 7411 - 7439.

[7] Tiwari, A., Freund, J. B., and Pantano, C. (2013). *A diffuse interface model with immiscibility preservation.* Journal of Computational Physics, 252, 290 – 309.

[8] Coralic, V., and Colonius, T. (2014). *Finite-volume WENO scheme for viscous compressible multicomponent flows*. Journal of Computational Physics, 274, 95 – 121.

[9] Shukla, R. K. (2014). *Nonlinear preconditioning for efficient and accurate interface capturing in simulation of multicomponent compressible flows.* Journal of Computational Physics, 276, 508 – 540.

[10] Tiwari, A., Pantano, C., and Freund, J. B. (2015). *Growth-and-collapse dynamics of small bubble clusters near a wall.* Journal of Fluid Mechanics, 775, 1 - 23.

[11] Saurel, R. and Pantano, C. (2018) *Diffuse-interface capturing methods for compressible two-phase flows.* Annual Review of Fluid Mechanics, 50, 105 – 130.

[12] Plesset, M. S. and Prosperetti, A. (1977). *Bubble dynamics and cavitation.* Annual Review of Fluid Mechanics, 9, 145 - 185.

[13] Blake, J. R. and Gibson, D. C. (1987). *Cavitation bubbles near boundaries*. Annual Review of Fluid Mechanics, 19, 99 - 123.

[14] Popinet, S. and Zaleski, S. (2002). *Bubble collapse near a solid boundary: a numerical study of the influence of viscosity*. Journal of Fluid Mechanics, 464, 137 - 163.

[15] Sussman, M. (2003). *A second order coupled level set and volume-of-fluid method for computing growth and collapse of vapor bubbles.* Journal of Computational Physics, 187, 110 – 136.

[16] Can, E. and Prosperetti, A. (2012). *A level set method for vapor bubble dynamics*. Journal of Computational Physics, 231(4), 1533 – 1552.

[17] Hartmann, D., Meinke, M. and Schröder, W. (2008). *Differential equation based constrained reinitialization for level set methods*. Journal of Computational Physics, 227, 6821 - 6845.

[18] Carreau, P. J., De Kee, D.C.R. and Chhabra, R.P. (1997) *Rheology of Polymeric Systems*. Hanser, Cincinnati.

[19] Hauke, G., Fuster, D., and Dopazo, C. (2007). *Dynamics of a single cavitating and reacting bubble.* Physical Review E, 75(6), 066310.

© 2018 ASME

A method for triggering surface modes by bubble coalescence

[1]Sarah Cleve*; [2]Matthieu Guédra; [1]Cyril Mauger; [2]Claude Inserra; [1]Philippe Blanc-Benon

[1]*Univ Lyon, Ecole Centrale de Lyon, INSA de Lyon, CNRS, LMFA UMR 5509, F-69134 Ecully Cedex, France;*
[2]*Univ Lyon, Université Lyon 1, INSERM, LabTAU, F-69003 Lyon, France*

Abstract

Acoustically driven microbubbles find numerous applications in the biomedical field (targeted drug delivery or blood-brain barrier opening for instance) and in engineering (surface cleaning or sonochemistry). In most cases only spherically oscillating bubbles are considered in these techniques. However, bubbles exhibiting surface modes can lead to additional effects (enhanced micromixing amongst others), and are scarcely investigated (mainly due to positional and growing surface instabilities). We present an experimental setup which allows the triggering and control of surface modes of free bubbles. As opposed to earlier studies which are based on short-time exposure to ultrasound waves [1] or an amplitude modulated pressure field [2], we obtain bubbles with steady-state surface modes. The experimental procedure consists of the following steps: (i) An acoustic field is induced in a water tank by a transducer. The driving frequency is chosen as to create an acoustic standing wave inside the tank. (ii) A bubble is nucleated by a short focused laser pulse. Our bubbles are all below resonance size and are hence driven towards a pressure antinode due to primary Bjerknes forces. (iii) By nucleating a second bubble, the two bubbles will coalesce due to primary Bjerknes forces and mutual attraction by secondary Bjerknes forces. The coalescence leads to a single, initially deformed bubble. By choosing the appropriate set of acoustic pressure and bubble size, a desired surface mode can be triggered. Additionally, we observed that the trajectory of the coalescing bubbles will define the orientation of the axis of symmetry. In sum, coalescence is a fast method to induce surface modes in free acoustically driven bubbles.

Keywords: microbubbles; shape modes; surface modes; bubble coalescence

Introduction

Shape modes of microbubbles have mainly been investigated theoretically. First theoretical studies were based on uncoupled mode theory [3,4]. These models allow in particular predicting the appearance of parametrically excited surface modes of order n when the driving amplitude exceeds a certain pressure threshold p_n^{thresh}. Experimentally, the onset surface modes can be observed by exposing bubbles to short ultrasound bursts [1].
Coupling between different modes has been introduced with studies on energy transfer between the spherical mode and one or two shape modes [5,6]. Later, models were extended to include all energy transfer between translational, spherical and surface modes [7,8]. These models allow taking into account the saturation of the shape mode growth. They are hence able to describe stead-state dynamics of a bubble with surface modes. Experimental validations of these models have been effected [2] using amplitude-modulated sound fields.
The above cited experimental methods of exposure to short ultrasound bursts and the amplitude-modulated sound fields allow the triggering and subsequent analysis of temporal dynamics. Steady-state oscillations of surface modes are however difficult to obtain as their stability has to be assured. We propose bubble coalescence as a means of triggering surface modes.

Experimental setup

The experimental setup is presented in Figure 1. It consists of a 8cm-edge cubic tank filled with water. On the bottom area, an ultrasonic plane transducer (SinapTec®, diameter of the active area 35mm) is connected to it in order to create an acoustic field inside the tank. The driving frequency is set to 31.25kHz and the voltage amplitude of the transducer can be varied between 0 and 10V. No gain amplifier is used as the estimated acoustic pressure in the tank for 5V applied voltage is ~30kPa, a sufficient value to generate surface modes for nearly all our experimentally studied bubbles. Backlight illumination with a continuous light source (LED) and a CMOS camera (Vision Research® V12.1) equipped with a 12x objective lens (Navitar®) allow capturing the bubble dynamics. A best compromise between frame size and frame rate is found to be 256x256 pixels and 67.065 kHz. Single bubbles are nucleated by focused laser pulses (λ= 532nm, New Wave Solo III, 6 ns pulse duration). All nucleated as well as coalesced bubbles are smaller than 90µm. To capture coalescence, a first bubble is trapped in the acoustic field, then a second bubble will be nucleated and encounter the first one. An exemplary result is shown in Figure 2. In order to simplify the manipulation during the experiments, the position of the water tank can be adjusted by millimeter screws in all three directions.

*Corresponding Author, Sarah Cleve: sarah.cleve@ec-lyon.fr

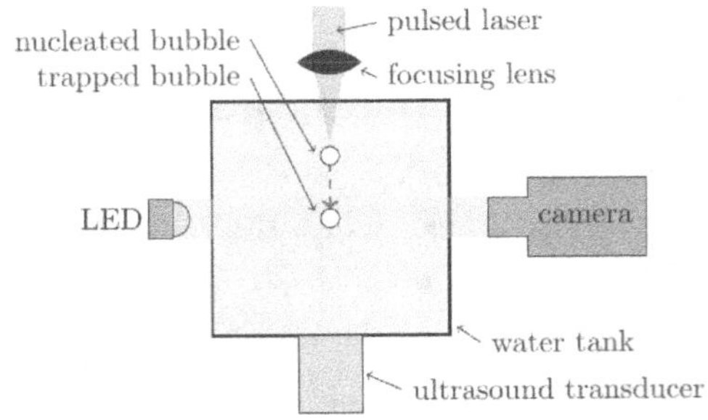

Figure 1: Schematic presentation of the experimental setup.

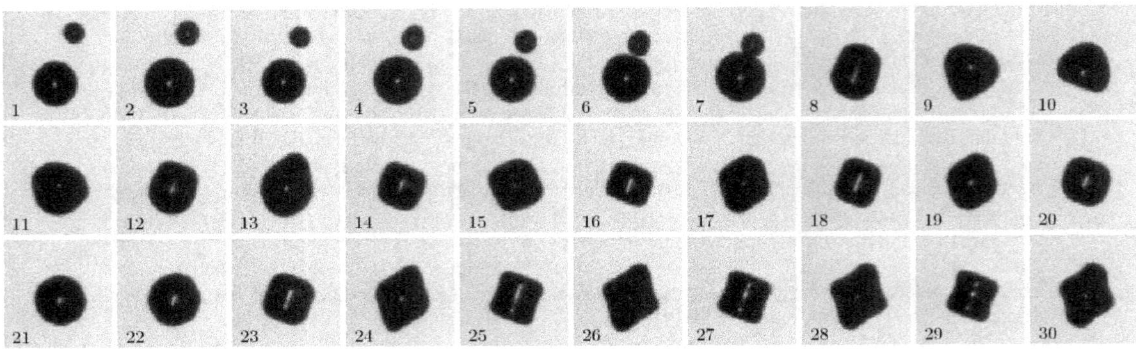

Figure 2: Exemplary series of 30 consecutive snapshots of two coalescing bubbles resulting in a shape mode 4. The images are recorded at time intervals of 14.9μs. The initial bubble radii are 52.5μm and 24.2μm respectively, the acoustic pressure has been evaluated to be 24.5kPa.

A microbubble in an acoustic field

A gas bubble in water, exposed to an acoustic field will be influenced by several forces. The bubble oscillations caused by the pressure fluctuations has been approached by different theoretical investigation, one of them being the Keller-Miksis model [12]. Fitting the Keller-Miksis model to the experimentally obtained bubble radius allows deducing the acoustic pressure on the bubble if there is no direct pressure measurement.

The bubble position is defined by the sum of forces acting on the bubble. Those are mainly the force of buoyancy and the primary Bjerknes forces. Primary Bjerknes forces are the net forces $\langle F \rangle = -\langle V\ \nabla p \rangle$ induced by the pressure gradients ∇p of the acoustic field onto the bubble volume V [13]. In a relatively weak acoustic field (hence omitting large bubble oscillations leading to translational instability and so-called dancing motion [14]), bubbles smaller than resonant size oscillate in-phase with the acoustic field, bubbles larger than resonant size out-of-phase. Consequently, the former are driven towards a pressure antinode, the latter are driven towards a pressure node. According to Minnaert's theory [15], the resonant radius corresponding to 31.25kHz is R_{res}=104μm. All our studied bubbles are hence smaller than resonant size and driven towards a pressure antinode. The bubble will find a stable position closely above a pressure antinode to equal out the buoyancy. Hence, the exact position depends on the bubble size. Here, the importance to have a rapidly adjustable tank position becomes obvious.

As soon as a second bubble is nucleated, there are also mutual forces between the two bubbles, called secondary Bjerknes forces. For bubbles oscillating in-phase with one another, secondary Bjerknes forces cause attraction between them. These forces will therefore generally cause the first bubble to leave its equilibrium location before encountering the second one. The coalesced bubble will later find a new equilibrium location slightly beneath the original one.

Stability of surface modes

There are two main difficulties, when we wish to obtain bubbles with stable surface modes. They are (i) the triggering of the surface mode and (ii) the stability of the surface mode as well as the appearance of surface modes alone without any translational instability. No studies in literature present a method which overcomes all of those difficulties. Exposing bubbles to short ultrasound bursts (see e.g. [1]) allows triggering surface modes. Also, due to the short pulse duration the shape mode amplitude and the possible appearance of translational instabilities can be limited. However, no steady-state regime can be obtained. Another existing method is to drive bubbles with an amplitude-modulated ultrasound field (see e.g. [2]), where the driving frequency is modulated with a low modulation frequency. This allows periodic triggering of non-spherical oscillations and their observation during short time intervals (before their extinction) that will obviously limit their growth. However, this technique does not allow for a steady-state regime. It is also possible to place a bubble in an acoustic field and rise the acoustic pressure until surface modes appear. However, without an additional triggering, bubbles often rest in a metastable spherical mode before abruptly instabilities take over. With very careful tuning, stable shape modes can be obtained in this way. But as this method is laborious and does not always succeed, it is desirable to find another method to trigger surface modes. Bubble coalescence can be used to do so.

Conditions for triggering surface modes

A surface mode can be triggered if the acoustic pressure exceeds the respective pressure threshold. This threshold $p_n^{thresh}(f_{osc},R_o)$ of a mode of order n depends on the driving frequency f_{osc} and the bubble radius at rest R_o [4]. Numerical calculations of the stability regions for a fixed driving frequency f_{osc}=31.25kHz are given in Figure 3. For a chosen set of parameters, the triggered mode can hence be predicted. For example, for a coalesced bubble with a radius of 70μm at 10kPa we expect to obtain a mode 3, whereas for the same bubble at 5kPa we expect to obtain a purely spherically oscillating bubble. Figure 4 shows three exemplary series of bubble coalescence with similar sized bubbles but different acoustic pressures resulting respectively in a spherical mode, a mode 2 and sustained rebound. Sustained rebound of bubbles has been observed in some cases (see Figure 4c)). The exact conditions under which the bubbles do not coalesce have however not yet been studied in detail. In literature, one of the discussed parameters is the Weber number [9-11].

The map in Figure 3 allows choosing the correct parameters in order to trigger a desired mode. We can slightly influence the size of the nucleated bubble by tuning the power of the pulsed laser and we can grow the trapped bubble by multiple coalescences. The acoustic pressure amplitude is directly related to the transducer voltage. In our study, a linear relation between the pressure and the voltage has been found in preliminary experiments. During these experiments, a spherically oscillating bubble has been driven at different transducer voltages. By fitting a Keller-Miksis model [12] to the experimentally obtained radii, the respective pressures can be deduced.

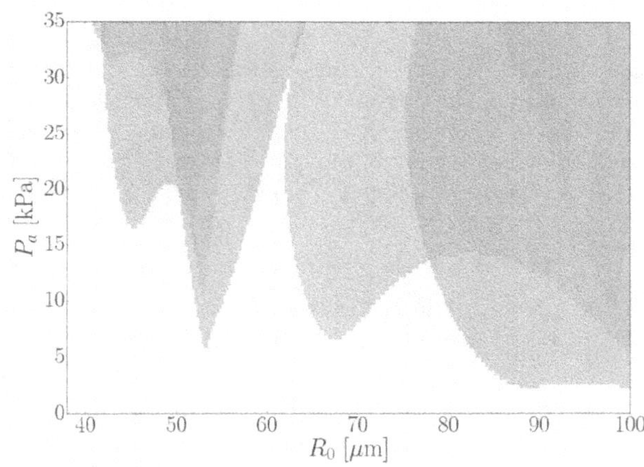

Figure 3: Radius-pressure map for a driving frequency f_{osc}=31.25kHz giving the stability boundaries and indicating the theoretically expected mode for bubble radii R_0=[40μm 100μm] and driving pressures P_a=[0kPa 35kPa]. White – spherical mode, blue – mode 2, green – mode 3, red – mode 4. Theory based on [4].

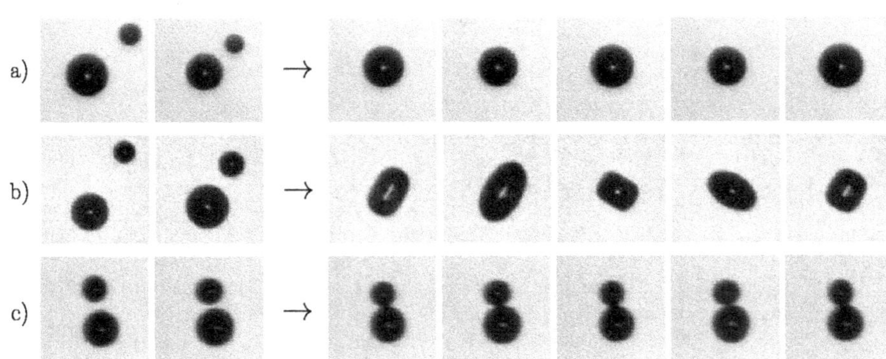

Figure 4: Three examples of similar sized bubbles at different acoustic pressures resulting in different shape modes: a) 40.9μm and a 21.8μm radius bubble at 27kPa coalescing to a spherically oscillating bubble b) a 40.8μm and a 24.3μm bubble 28kPa resulting in a shape mode 2 c) a 36.9μm and a 25.9μm radius bubble at 13kPa resulting in sustained rebound. For all series the first two snapshots are taken at about 0.3s and 0.1s before the encounter of the two bubbles, the following images are five consecutive snapshots taken at about 1s after the encounter.

The axis of symmetry

Nearly all experimental and theoretical studies on shape modes of bubbles are based on the expansion of the bubble interface as a sum of Legendre Polynomials (meaning zonal harmonics amongst the basis of spherical harmonics). This approach supposes axisymmetry which seems to be fairly correct for a large number of observed bubbles. However, for experimental purposes, the definition of the axis of symmetry is crucial for a correct post-processing of the bubble shape. The easiest way to post-process without any additional correction and without loss of information requires the axis of symmetry to be in the plane of the camera.

Earlier studies have shown that the axis of symmetry can be defined by the direction of traveling ultrasound waves [1]. However, in the present setup no such possibility is given as we work with a standing wave without any preferential direction. To achieve the definition of the axis of symmetry, our experiments show that coalescence is a means to influence it. It is defined by the direction of approach between the two bubbles. An example is demonstrated in Figure 5. Earlier classical studies on bubble coalescence (e.g. [10]) already let foresee these hypothesis by presenting coalescing ellipsoidal bubbles respecting the here formalized symmetry axis. With the present study we are able to formalize this and in particular to validate its extension to surface modes of higher order.

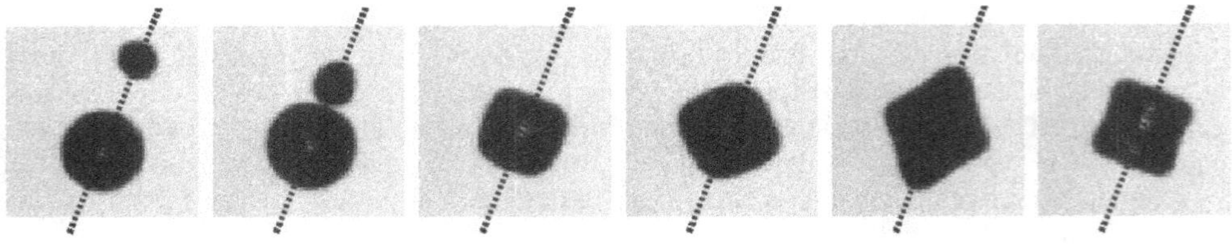

Fig 5: Demonstration of the axis of symmetry for some chosen snapshots of Figure 3.

Conclusion

We present bubble coalescence as a reliable method to trigger surface modes. As opposed to earlier methods such as short exposition to ultrasound pulses or modulated pressure fields, the coalesced bubbles can remain in a steady-state regime. With the definition of the axis of symmetry, it is possible to completely control the appearing shape mode, including the shape mode number and the orientation of the bubble. In conclusion bubble coalescence is a fast method to trigger surface modes employable for further investigation of bubble dynamics.

Acknowledgements

This work was performed within the framework of the Labex *CeLyA* of the *Université de Lyon*, within the programme *'Investissements d'Avenir'* (ANR-10-LABX-0060/ANR-11-IDEX-0007) operated by the French National Research Agency (ANR).

References

[1] Versluis M., Goertz D. E., Palanchon P., Heitman I. L., van der Meer S. M., Dollet B., de Jong N., Lohse D. (2010). *Microbubble shape oscillations excited through ultrasonic parametric driving.* Physical Review E. 82(2): 026321.

[2] Guédra M., Inserra C., Mauger C., Gilles B. (2016). *Experimental evidence of nonlinear mode coupling between spherical and nonspherical oscillations of microbubbles.* Physical Review E. 94(5): 053115.

[3] Francescutto A., Nabergoj R. (1978). *Pulsation amplitude threshold for surface waves on oscillating bubbles.* Acta Acustica united with Acustica, 41(3): 215.

[4] Brenner M. P., Lohse D., Dupont T. F. (1995). *Bubble shape oscillations and the onset of sonoluminescence.* Physical Review Letters. 75: 954.

[5] Feng Z. C., Leal L. G. (1994). *Bifurcation and chaos in shape and volume oscillations of a periodically driven bubble with two-to-one internal resonance.* Journal of Fluid Mechanics, 266: 209.

[6] Mei C. C., Zhou X. (1991). *Parametric resonance of a spherical bubble.* Journal of Fluid Mechanics. 229: 29.

[7] Doinikov A. A. (2004). *Translational motion of a bubble undergoing shape oscillations.* Journal of Fluid Mechanics. 501: 1.

[8] Shaw S. J. (2006). *Translation and oscillation of a bubble under axisymmetric deformation.* Physics of Fluids. 18(7): 072104.

[9] Duineveld P. C. (1996). *The influence of an applied sound field on bubble coalescence.* The Journal of the Acoustical Society of America, 99(1): 622.

[10] Postema M., Marmottant P., Lancée C. T., Hilgenfeldt S., de Jong N. (2004). *Ultrasound-induced microbubble coalescence.* Ultrasound in Medicine & Biology, 30(10): 1337.

[11] Jiao J., He Y., Yasui K., Kentish S. E., Ashokkumar M., Manasseh R., Lee J. (2105) *Influence of acoustic pressure and bubble sizes on the coalescence of two contacting bubbles in an acoustic field.* Ultrasonics Sonochemistry, 22: 70.

[12] Keller J. B., Miksis M. (1933). *Bubble oscillations of large amplitude.* The Journal of the Acoustical Society of America, 68: 628.

[13] Leighton T. (2012). The acoustic bubble. Academic press.

[14] Mettin, R., Doinikov A. A. (2009). *Translational instability of a spherical bubble in a standing ultrasound wave.* Applied Acoustics. 70(10): 1330.

[15] Minnaert M. (1933). *Xvi. On musical air-bubbles and the sounds of running water.* The London, Edinburgh, and Dublin Philosophical Magazine and Journal of Science. 16: 235.

© 2018 ASME

A Numerical Study on the Effects of Cavitation Number on Cavitating Mixing Layer of Liquefied Natural Gas (LNG) behind a Flat Plate Splitter

Saeed Rahbarimanesh; Joshua Brinkerhoff*

The University of British Columbia, Kelowna, BC, Canada

Abstract

The mutual interaction of shear layer instabilities and phase change in a two-dimensional, unsteady cryogenic cavitating mixing layer is investigated using numerical simulation. The model is developed based on homogeneous equilibrium mixture (HEM) approach in a density-based framework to compute the temperature-dependent cavitation field for liquefied natural gas (LNG). The mixing layer is simulated for vorticity-thickness Reynolds numbers of 44 to 215 and cavitation numbers of 0.1 to 1.1. At the lowest cavitation numbers, steady vapor cavities are initiated on the splitter plate, followed by roll-up of the separated shear layer via the well-known Kelvin-Helmholtz mode. Unsteady, shear-driven cavitation then occurs as vapor cavities nucleate and grow from the low-pressure cores in the rolled-up vortices. As the Reynolds number and cavitation number are varied, thermal effects and baroclinic vorticity production are found to have significant impacts on the mutual interaction of shear-layer instabilities and unsteady cavitation processes.

Keywords: cryogenic cavitation model; unsteady cavitation; LNG; shear layer instability

Introduction

Pressure modulated liquid-to-vapor phase changes (cavitation) is a challenging topic of research in engineering and is a key issue in turbomachinery design. For instance, positive-displacement fuel pumps are significantly degraded by the presence of even a small volumetric percentage of vapor, leading to significant decay in output head and loss of machine efficiency [1]. The complexity of the cavitation phenomenon becomes even more pronounced in the case of cryogenic fluids such as liquefied natural gas. Phase change process in cryogenic fluid systems, where the operating temperature is typically close to the critical temperature of the working fluid, occurs with exchanging substantial thermal effects and abrupt changes in fluid properties, which alter the vapor phase fraction characteristics compared to isothermal fluids. In particular, the cryogenic phase change process involves not only mass transfer effects due to evaporation/condensation, but also heat transfer corresponding to the latent heat of vaporization of liquid phase. Such a behavior, combined with lower liquid to vapor density ratios, causes cryogenic cavitation to experience altered vaporization and condensation rates compared to isothermal cavity structures [2].

In dealing with cavitation on turbomachinery installations two major patterns are frequently observed: 1) attached sheet cavitation, such as steady cavities on the suction side of rotary inducer blades, and 2) unsteady shear-flow cavitation, such as separated cavitating flows in the trailing edge region of rotor blades or tip-clearance cavitation observed in pumps [2,3]. Attached sheet cavity structures, normally initiated as a developing boundary layer on the wall in the case of thermo-insensitive fluids such as water, is often followed by unsteady vortex cavitation which is caused by shedding of vortical structures into the flow downstream of vapor cavities [1]. Continuation of the vortex cavity shedding causes the formation of large-scale vapor clouds downstream of the flow which might be accompanied by re-entrant jets in the near-wall regions. The detached cloud structures consist of bubble-filled large-scale coherent vortices [1,3,4]. In the case of cryogenic cavitation, the vortex-cavity interaction mechanisms can occur simultaneously with temperature-induced phase change (boiling) and thermal gradients that reshape the cavity structures. For instance, temperature decreases around cavity spots due to the exchange of latent heat during vaporization cause these structures to expand and grow. As a result of this temperature drop, saturated vapor pressure in the surrounding liquid around the cavity spots is reduced and delays the expansion rate of cavities compared to the thermo-insensitive cavitation [5]. The physics in cryogenic cavitating flows is thus heavily influenced by interactions between phase instabilities, thermal interactions, and shear layer instabilities, leading to a complex interplay between coherent vortical structures, phase change process, and thermodynamic effects. Moreover, compared with isothermal cavitation, these interactions are further amplified by local temperature fluctuations added to other sources of unsteadiness such as density and pressure variations [6].

*Corresponding Author, Joshua Brinkerhoff: joshua.brinkerhoff@ubc.ca

A number of experimental and numerical studies on unsteady cavitation, mainly focusing on the effect of fluid property variations in isothermal cavitating shear layers of water, have been reported in literatures [3,7,8]. Among these studies, Wang and Ostoja-Starzewski [1] developed a single fluid model with an incorporated LES scheme, using a fifth-order polynomial relationship between density coefficient ratio and pressure coefficient, for weakly-compressible cavitating fluids. Time-dependent sheet/cloud cavitation structures around a NACA0015 airfoil in a range of operating conditions were simulated to illustrate the non-linear interaction between macroscopic vortex motion and microscopic bubble dynamics. More recently, Iyer and Ceccio [9] used particle image velocimetry-laser induced fluorescence experiments to investigate the influence of developed cavitation on a turbulent shear layer flow of water with different range of void fractions. They suggested that the overall formation, growth mechanism, and convection of the primary and secondary vortical structures of the shear layer are not influenced significantly by the presence of the vapor phase. However, an increase of turbulent fluctuations in the center of the cavitating shear layer is observed compared to a baseline non-cavitating case. It is also reported that cavitation has a strong effect on Reynolds stress tensor anisotropy especially in downstream regions where the cross-stream velocity fluctuations and the Reynolds stresses of the cavitating shear layer are reduced in comparison with the non-cavitating case.

Due to lack of knowledge concerning the detailed mechanisms of shear layer instabilities and their correlation with cryogenic phase change processes, the objective of the present research is to explore the nature of these interactions through fundamental numerical simulations of a cryogenic cavitating mixing layer of LNG. The resulting numerical database on the detailed dynamic behavior of LNG cavitating shear flows will aid in characterizing the unsteady interactions of phase instabilities and shear layer instabilities in turbulent two-phase flows at cryogenic conditions. In particular, the present study assesses the influence of cavitation number on the unsteady cryogenic cavitation.

Computational Model

The current study uses a cryogenic cavitation model developed by the authors [10], which is built on the currently-available *cavitatingFoam* solver in the open-source OpenFOAM® libraries. The developed solver employs the homogeneous equilibrium mixture (HEM) approach with a barotropic equation of state in a density-based framework. Baroclinic and thermal effects on the flow field are accounted by solving an enthalpy-based energy conservation equation in conjunction with thermophysical property models that accurately captures the properties of LNG at cryogenic temperatures for a broad range of pressures. In addition, the saturation vapor pressure is made dependent on temperature by using the saturation pressure relationship from the Clausius-Clapeyron equation [4]. An extended merged PISO-SIMPLE algorithm is used to iteratively couple the resulting temperature field to the pressure, velocity, and vapor fraction fields.

Computational Setup

Following the experimental study of Aeschlimann et al. [3], a two-dimensional cavitating mixing layer of LNG behind a flat splitter plate is simulated using the developed cryogenic cavitation solver. The splitter plate has a thickness of 6 mm with a rounded edge of 0.2 mm radius, which is placed in a 300×80 mm^2 rectangular domain (figure 1). The domain is meshed using 525,700 non-uniform structured 2D elements. The velocity above (U_1) and below (U_2) the splitter is 15.8 m/s and 3.5 m/s, respectively, and three cavitation numbers 0.21, 0.65, and 1 are studied; the cavitation numbers are defined in terms of the low-speed flow pressure at the inlet, correspondingly giving total pressure ratios of 1.094, 1.064, and 1.04 between the high- and low -speed sides. Outlet static pressure is fixed at 4.13 bar. Initial flow field contains only liquid methane (LNG) with operating temperature of 132 K.

Results and Discussion

The temporal evolution of vapor phase fraction and temperature fields for LNG cavitating mixing layer with operating cavitation numbers of 0.21 and 0.65 at time instant t = 120 s is depicted in figure 2. Figure 2(a) and (c) show that a steady attached vapor cavity forms in the upper splitter plate boundary layer. Downstream of the splitter plate, the near-wall vorticity quickly rolls-up into discrete vortices owing to the development of Kelvin-Helmholtz instability in the highly-strained braid regions [4,11-13]. Pairing of these counter-rotating vortices occurs within a few wavelengths downstream of the splitter plate, leading to larger-scale structures with twice the streamwise period; this secondary instability is due to the non-uniform acceleration of the primarily-shed structures by the non-

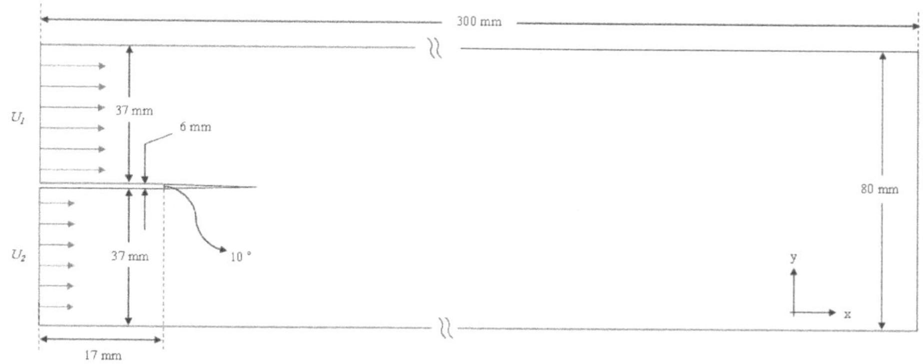

Figure 1: Computational domain

uniform velocity field. A subsequent tertiary pairing also occurs at the lowest cavitation number (σ = 0.21) by the same mechanism. These observations of the vortical structures are based on the observed coincidence of coherent vortices and vapor cavities. Vapor cavities nucleate in the core of the coherent vortices as the local static pressure reduces due to locally high rotation rate [3,4]. As the secondary and tertiary vortex pairing occurs, the vapor cavities also grow into unsteady cavitation clouds. The growth of unsteady cavitation clouds is sustained by the continual shedding of the shear layer vortices which induce continuous pressure and density fluctuations affecting the size of the cavitating cores [14].

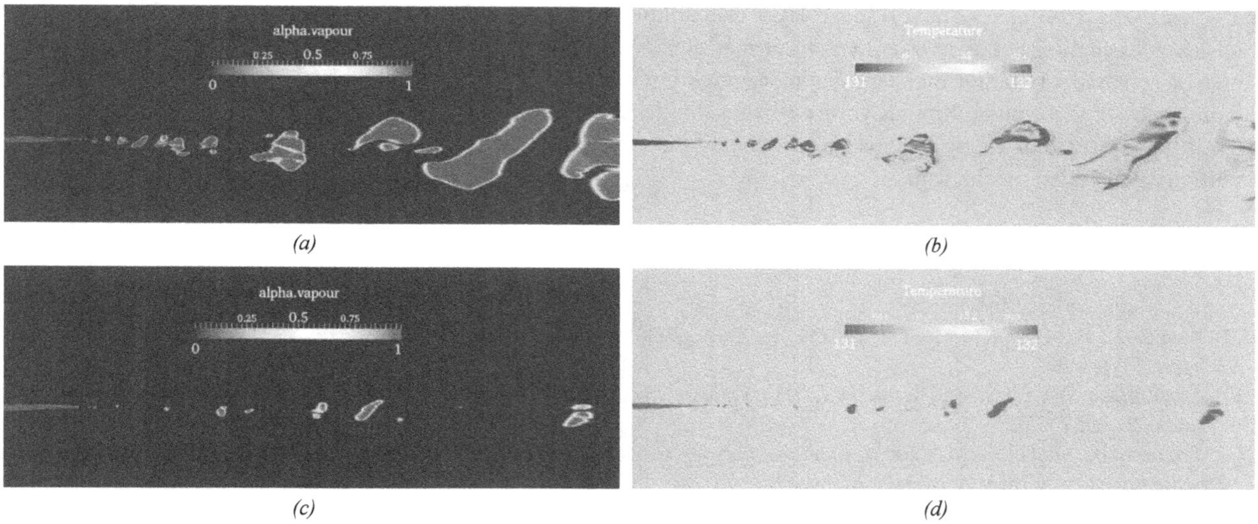

Figure 2: Temporal evolution of cavitating mixing layer of LNG at t = 120 s: vapor phase fraction (α) field at cavitation number (a) σ = 0.21 and (c) σ = 0.65; temperature field at cavitation number (b) σ = 0.21 and (d) σ = 0.65.

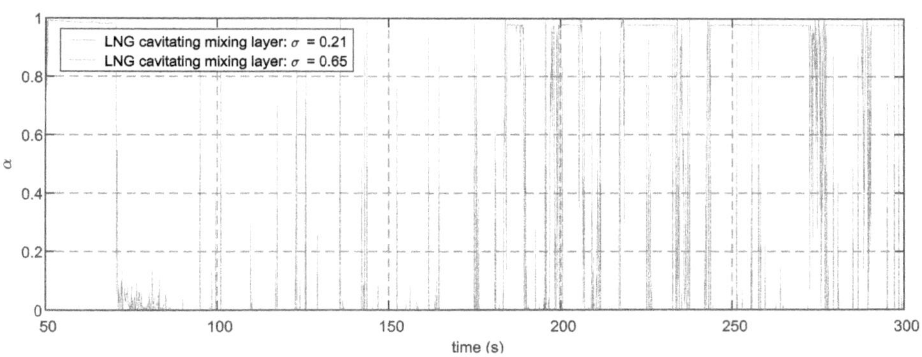

Figure 3: Time history of vapor fraction field (α) of cavitating mixing layer of LNG at the probe placed at x = 1 mm and y = 0 mm close to the splitter trailing edge; the vapor field fluctuation is shown for two cavitation numbers of 0.21 and 0.65.

In the temperature field (figure 2(b)-(d)), a slight temperature change of ±1.8 K is observed with high temperature gradients along the liquid/vapor interfaces and the frontal side of cavities. This small variation significantly influences the evolution of the vapor cavities. As vapor spots are generated and grow, the latent heat of vaporization is absorbed from the surrounding liquid causing a temperature drop around the vapor area. Because of the direct connection between temperature and saturation pressure, the drop in temperature decreases the saturation pressure, resulting in an overall delay in the inception and growth rate of the vapor spots [4,10]. This behavior diminishes at higher cavitation numbers, as shown in figure 2(c)-(d), which yields smaller vapor cavities. This suggests that reducing the cavitation number enhances the growth of vapor cavities by increasing the amplification rate of shear-layer instability [3,4,14]. Such a behavior is illustrated in figure 3 by plotting the temporal variation of vapor fraction field at a point 1 mm downstream of the splitter plate (x = 1 mm, y = 0 mm) for cavitation numbers of σ = 0.21 and 0.65. As seen, successive vaporization and collapsing of cavities at lower cavitation number adds additional unsteadiness to the flow and thus fluctuations in the local temperature. The vapor cavities are more strongly affected by temperature variations at lower cavitation numbers. Evaluation of the compressible vorticity transport equation (not shown) indicates that baroclinic vorticity production terms, introduced as a result of temperature dependence of the density field, enhances vorticity production at phase interfaces, leading to higher amplification rates of shear-layer instability of these cavities [4,10].

Conclusion

A cryogenic cavitation solver is used to simulate a cavitating mixing layer of LNG behind a flat splitter plate, aimed at identifying the interaction of shear layer instabilities and phase change in cryogenic fluids. The growth of unsteady cavitation is linked to coherent vortices generated following the roll-up of the shear layer past the splitter plate. For small cavitation numbers, vapor cavities nucleate at the center of the coherent vortices and evolve into large-scale cavity clouds through vortex pairing due to secondary shear layer instabilities. Non-isothermal effects modify the growth of vapor cavities by influencing the shear-layer instability process through introducing baroclinic vorticity generation mechanisms.

References

[1] Wang, G., Ostoja-Starzewski, M. (2007). *Large Eddy Simulation of a Sheet/Cloud Cavitation on a NACA0015 Hydrofoil.* Applied Mathematical Modeling. 31.

[2] Zhang, X. B., Qiu, L. M., Gao, Y., Zhang, X. J. (2008). *Computational Fluid Dynamic Study on Cavitation in Liquid Nitrogen.* Cryogenics. 48.

[3] Aeschlimann, V., Barre, S., Djeridi, H. (2011). *Velocity Field Analysis in an Experimental Cavitating Mixing Layer.* Physics of Fluids. 23. 055105.

[4] Franc, J. P., Michel, J. M. (2005). *Fundamentals of Cavitation,* in: Fluid Mechanics and Its Applications Series (76), Grenoble Sciences: Kluwer Academic Publishers.

[5] Niiyama, K., Hasegawa, S., Tsuda, S., et al. (2009). *Thermodynamic Effects on Cryogenic Cavitating Flow in an Orifice.* Proceedings of the 7th International Symposium on Cavitation (CAV2009).

[6] Hosangadi, A., Ahuja, V. (2005). *Numerical Study of Cavitation in Cryogenic Fluids.* Journal of Fluids Engineering. 127.

[7] Hassan, W., Legoupil, S., Chambellan, D., Barre, S. (2008). *Dynamic Localization of Vapor Fraction in Turbopump Inducers by X-ray Tomography.* IEEE Transactions on Nuclear Science. 55(1).

[8] Laberteaux, K. R., Ceccio, S. L. (2001). *Partial Cavity Flows. Part 1: Cavities Forming on Models without Spanwise Variation.* Journal of Fluid Mechanics. 431(1).

[9] Iyer, C. O., Ceccio, S. L., Prosperetti, A. (2002). *The Influence of Developed Cavitation on the Flow of a Turbulent Shear Layer.* Physics of Fluids. 14. 3414.

[10] Rahbarimanesh, S., Brinkerhoff, J., Huang, J. (2017 - in press). *Development and Validation of a Homogeneous Flow Model for Simulating Cavitation in Cryogenic Fluids.* Applied Mathematical Modeling. https://doi.org/10.1016/j.apm.2017.12.004.

[11] Lasheras, J. C., Cho, J. S., Maxworthy, T. (1986). *On the Origin and Evolution of Streamwise Vortical Structures in a Plane, Free Shear Layer.* Journal of Fluid Mechanics. 172.

[12] Brinkerhoff, J. R., Yaras, M. I. (2014). *Numerical Investigation of the Generation and Growth of Coherent Flow Structures in a Triggered Turbulent Spot.* Journal of Fluid Mechanics. 759.

[13] Bernal, L., Roshko, A. (1986). *Streamwise Vortex Structure in Plane Mixing Layers.* Journal of Fluid Mechanics. 170.

[14] Belahadji, B., Franc, J. P., Michel, J. M. (1995). *Cavitation in the Rotational Structures of a Turbulent Wake.* Journal of Fluid Mechanics. 287.

© 2018 ASME

A Parametric Study of Bubble Cloud Dynamics near a Wall in an Acoustic Field

Chao-Tsung Hsiao*; Jingsen Ma; Georges L. Chahine

DYNAFLOW, INC., Jessup, MD, USA

Abstract

A parametric study is conducted to investigate bubble cloud dynamics near a rigid wall when excited by a sinusoidal pressure field. It is shown that a preferred driving frequency which can incur the strongest collective bubble behavior and result in the highest pressure impact on the nearby wall exists for a given bubble cloud initial condition. Such preferred driving frequency is strongly dependent on the driving amplitude but independent on the initial ambient pressure. The preferred driving frequency decrease as the driving amplitude is increased and approaches the natural frequency of bubble cloud predicted by linear theory when the driving amplitude is very small.

Keywords: Cloud Cavitation, Bubble Dynamics, Multiphase Flow

Introduction

The collapse of a cloud of microbubbles near a rigid boundary is known to be as one of the most destructive forms of cavitation due collective bubble dynamics resulting in high pressure generation during successive collapses and rebounds. Cloud cavitation can be observed in hydrodynamic applications such as on rotating propellers or hydrofoils[1]–[5] as well as in high intensity acoustic fields such as in ultrasonic devices, Shock Wave Lithotripsy (SWL) for kidney stone fragmentation [6] and High Intensity Focused Ultrasound (HIFU) for tumor ablation [7]. Our previous studies [3], [4] have shown that the bubbles in the cloud, collapse in a cascading fashion with the bubbles farthest from the wall and cloud center collapsing first and those closest to the wall and center collapsing last. This results in a pressure wave moving inward and toward the wall. In order to contribute to the understanding of the physics involved in the complex interaction between the many bubbles and the imposed pressure field, and to predict the loading on a nearby object, we have applied our coupled Eulerian-Lagrangian two-phase flow modeling [8], [9] to an initially spherical bubble cloud to study the effects of the excitation amplitude and frequency. The advantage of the Eulerian-Lagrangian numerical approach is that it is a multiscale approach and captures key characteristics of both the bubbles' dynamics and the overall cloud dynamics. It also enables one to account for the often-neglected slip velocity between the bubbles and the liquid, which results in micro-streaming with the bubbles migrating significantly when the driving pressure amplitude is increased. In [9] we also considered the effects of the initial bubble radii, the bubble distribution, and the cloud distance from the wall and found that the strongest collective bubble behavior occurs at a preferred driving frequency for a given initial cloud condition. At this preferred driving frequency, pressure peaks orders of magnitudes higher than the excitation pressure, result from bubble interaction. However, our previous studies mainly focused on a high ambient pressure. In this study we will study the effect of the ambient pressure on the bubble cloud dynamics. In addition, the driving pressure amplitude and frequency are varied for different ambient pressures to investigate their effects on the bubble cloud dynamics and the resulting pressure at the wall as the bubble cloud collapses.

Numerical Method

The two-phase mixture in the bubble cloud is treated as a continuum with the continuity and momentum equations expressed as follows:

$$\frac{\partial \rho_m}{\partial t} + \nabla \cdot (\rho_m \mathbf{u}) = 0, \qquad \rho_m \frac{D\mathbf{u}}{Dt} = -\nabla p + \mu_m \nabla^2 \mathbf{u}, \tag{1}$$

where ρ_m, μ_m, $\mathbf{u}$, and p are respectively the mixture density, dynamic viscosity, velocity, and pressure. ρ_m and μ_m are related to the liquid and gas properties and to the gas volume fraction, α, by:

$$\rho_m = (1-\alpha)\rho_l + \alpha\rho_g, \qquad \mu_m = (1-\alpha)\mu_l + \alpha\mu_g. \tag{2}$$

Eq. (1) is coupled with equations describing the dynamics of the discrete individual bubbles in the cloud. Knowing at each instant all bubble radii and locations provides α (thus ρ_m and μ_m) as a function of space and time and substitutes for the need of a mixture equation of state to close the system of equations. Each bubble is treated as a source, which represents volume change, and a dipole to represent translation. For each bubble, the equivalent spherical radius, $R(t)$, is obtained using a modified Keller-Herring equation [5], which accounts for the mixture compressibility and non-uniform pressure field:

$$\left(1-\frac{\dot{R}}{c_m}\right)R\ddot{R}+\frac{3}{2}\left(1-\frac{\dot{R}}{3c_m}\right)\dot{R}=\frac{\mathbf{u}_s^2}{4}+\frac{1}{\rho_m}\left(1+\frac{\dot{R}}{c_m}+\frac{R}{c_m}\frac{d}{dt}\right)\left[p_v+p_g-p_{enc}-\frac{2\gamma}{R}-4\mu_m\frac{\dot{R}}{R}\right], \quad (3)$$

where c_m is the local sound speed in the mixture, p_v is the liquid vapor pressure, p_g is the bubble gas pressure, and γ is the surface tension. The term $\mathbf{u}_s^2/4$, accounts for the pressure resulting from the slip velocity, $\mathbf{u}_s=\mathbf{u}_{enc}-\mathbf{u}_b$, between the host medium velocity, $\mathbf{u}_{enc}$, and the bubble velocity, $\mathbf{u}_b$, with [6]. p_{enc} and $\mathbf{u}_{enc}$ are the encountered pressures and velocities averaged over the bubble surface to account for local non-uniform flow. The bubble trajectory is obtained using the following equation of motion [7]:

$$\frac{d\mathbf{u}_b}{dt}=\left(\frac{\rho_l}{\rho_g}\right)\left[\frac{3}{8R}C_D\,|\,\mathbf{u}_s\,|\,\mathbf{u}_s+\frac{1}{2}\left(\frac{d\mathbf{u}_{enc}}{dt}-\frac{d\mathbf{u}_b}{dt}\right)-\frac{\nabla p}{\rho_l}+\frac{(\rho_b-\rho_l)}{\rho_l}g+\frac{3C_L}{4\pi}\frac{\sqrt{\nu}}{R}\frac{\mathbf{u}_s\times\boldsymbol{\Omega}}{\sqrt{|\boldsymbol{\Omega}|}}+\frac{3\dot{R}}{2R}\mathbf{u}_s\right], \quad (4)$$

where C_L is a lift coefficient, $\boldsymbol{\omega}$ is the local vorticity, and C_D is a drag coefficient. The last term is the Bjerknes force due to coupling between bubble volume rate and bubble motion.

Numerical Simulations and Discussion

We consider an initially spherical bubble cloud with a radius, A_0, (Fig. 1), driven by a sinusoidal pressure, $P(t)=P_0\left[1-\xi\sin(2\pi ft)\right]$, where P_0 is the initial and average ambient pressure, $\xi=P_{amp}/\,P_0$ is the normalized pressure oscillation amplitude, and f is the frequency. The cloud center is initially at a distance X_0 from a rigid wall and is composed of small bubbles of initial radii R_0. The bubbles are randomly distributed within A_0, resulting in a quasi-uniform initial α_0 within the cloud and all bubbles are initially at equilibrium with the pressure P_0. We can identify in this problem two groups of parameters: one for the imposed pressure field (driving pressure and frequency) and one for the bubble cloud characteristics (cloud and bubble sizes and the void fraction).

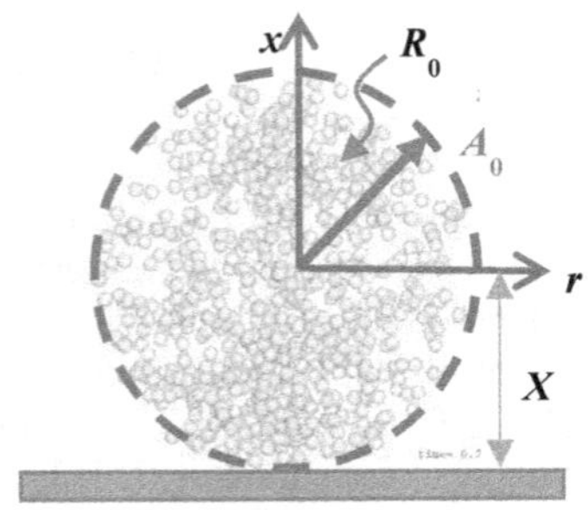

Fig. 1: Schematic of the problem of the dynamics of a bubble cloud near a rigid wall.

Fig. 2 shows an example time sequence of the bubble cloud response during the first cycle of oscillation for $A_0=X_0=1.5$ mm, $R_0=50$ µm, and $\alpha_0=5\%$, driven by the imposed pressure $P_0=1$ atm, $\zeta=0.9$ and $f=7$kHz. The color contours indicate the pressure inside each bubble and the corresponding pressure loading at the wall. It is seen that the bubbles in the cloud grow first then collapse in a cascading fashion starting with the bubbles at the cloud top (farthest from the wall) collapsing first, and finally those on the bottom (closest to the wall) collapsing last and resulting a high pressure loading at the wall.

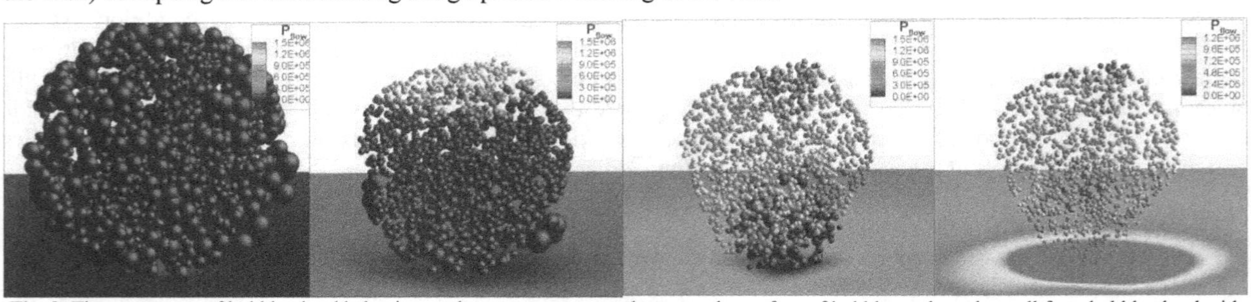

Fig. 2: Time sequence of bubble cloud behaviors and pressure contours shown on the surface of bubbles and nearby wall for a bubble cloud with the initial conditions, $R_0=50$ µm, $\alpha_0=5\%$, and $A_0=X_0=1.5$ mm, driven by a sinusoidal pressure field with $P_0=1$ atm, $\zeta=0.9$, and $f=7$ kHz.

Our previous studies [8], [9] have shown that for a given bubble cloud initial geometric condition, the pressure loading at the wall when the bubble cloud collapses is highly dependent on the driving pressure amplitude and frequency. A preferred driving frequency was found to result in a maximum pressure loading for a given ζ. However, the previous studies focused on a high initial pressure, $P_0=10$ atm, and thus it would be interesting to investigate if the same trend exists for different initial pressures.

To start with, Fig. 3 compares the effect of the driving frequency on the pressure loading at the wall centre when $P_0=1$ atm (which is a more practical case for ultrasonic applications than the previously studied $P_0=10$ atm) and $\zeta=0.9$. It is seen that the highest collective effects appear here close to 7 kHz. This is compared with the results of

P_0 = 10 atm as shown in Fig. 4. It can be seen that only a small shift in the peak driving frequency (7 kHz vs. 8 kHz) can be observed between these two initial ambient pressures even though the resulting maximum collapse pressure is more than five times higher for P_0 = 10 atm. This implies that the preferred driving frequency may be little dependent on the initial ambient pressure. We are considering next a larger range of pressure changes to understand this behaviour.

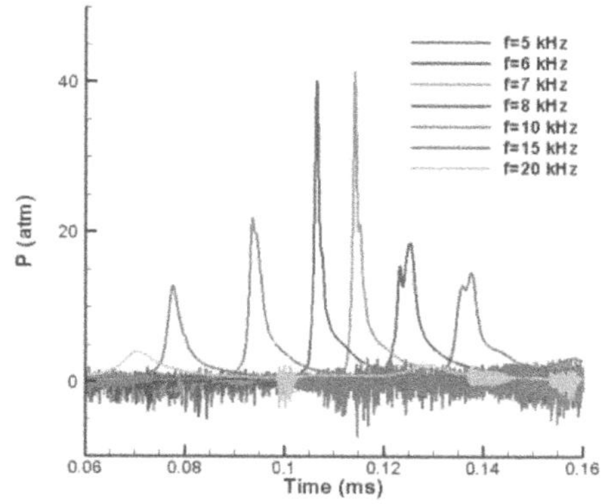

Fig. 3: Pressure versus time at the wall center created by bubble clouds for different driving frequency for P_0=1 atm, ζ=0.9, R_0=50μm, α_0=5%, and A_0=X_0=1.5 mm.

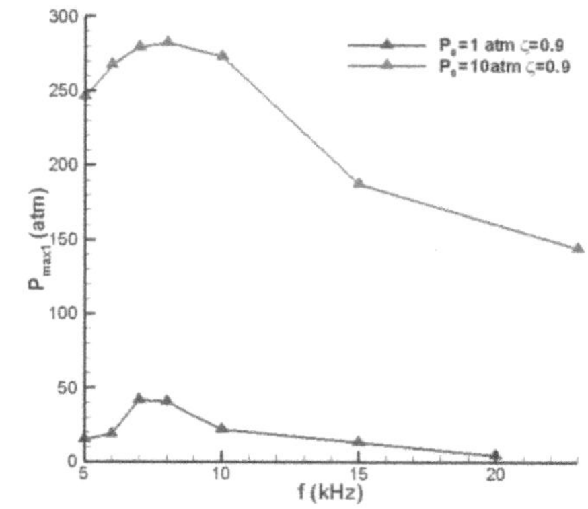

Fig. 4: Effect of initial ambient pressure on $P_{\max 1}$ vs f, for ζ=0.9, R_0=50μm, α_0=5%, and A_0=X_0=1.5 mm.

Fig. 5 shows for the same driving frequency, f = 8 kHz, how the peak pressure, $P_{\max 1}$, varies with P_0. It is seen that the maximum pressure loading at the wall increases as the initial ambient pressure is increased indicating stronger bubble collapses at the higher pressures. This, however, reaches a plateau at the largest initial ambient pressures considered.

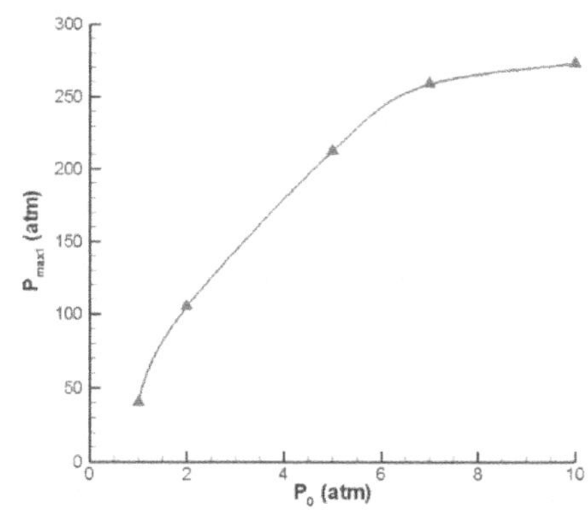

Fig. 5: Effect of initial ambient pressure on the maximum pressure loading at the wall center for f=8kHz, ζ=0.9, R_0=50μm, α_0=5%, and A_0=X_0=1.5 mm

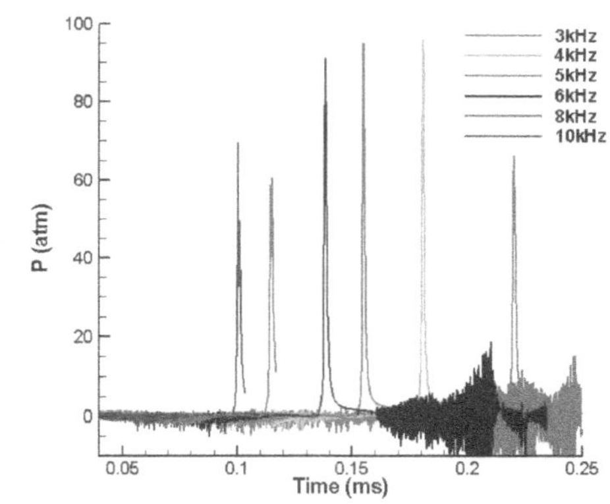

Fig. 6: Pressure versus time at the wall center created by bubble clouds for different driving frequency for P_0=1 atm, ζ=1.25, R_0=50μm, α_0=5%, and A_0=X_0=1.5 mm.

Another set of cases at P_0 = 1 atm and ζ=1.25 is simulated and compared to ζ=0.9 for different acoustic driving frequencies, to examine the effect of the driving amplitude, ζ, on the peak amplitude at the wall. It is seen from Fig. 7 that the preferred driving frequency, $f_{\max}$, is about 4 kHz for ζ=1.25, while it was 8 kHz for ζ=0.9. This indicates a strong influence of ζ. The effects of the driving amplitude on $P_{\max 1}$ vs f is also seen in Fig. 7 It can be seen that the peak value of the pressure at the wall, $P_{\max 1}$, increases and $f_{\max}$ decreases as ζ is increased. It is important to note that the preferred driving frequency, $f_{\max}$, highlighted here is different from the natural frequency of a bubble cloud derived from linear bubble dynamics theory by [10]:

$$f_{cloud} = f_0 \left(1 + \frac{12}{\pi^2} \frac{A_0^2}{R_0^2} \frac{\alpha_0}{1-\alpha_0} \right)^{-1/2} \quad \text{with} \quad f_0 = \frac{1}{2\pi} \sqrt{\frac{P_{g0}}{\rho R_0^2} \left(3 - \frac{2\gamma}{P_{g0} R_0} \right)}, \tag{5}$$

which gives a value of f_{cloud} = 23 kHz for the current case. Further expanding the range of ζ will allow one to investigate such dependency. Fig. 8 shows the dependency of $f_{\max}$ on ζ. It can be seen that when ζ is less than 0.2, $f_{\max}$ is compatible with f_{cloud}. For ζ > 0.2, $f_{\max}$ deviates significantly from f_{cloud}. $f_{\max}$ decreases almost linearly with increasing ξ until ξ >1 where the f_{max} seems to reach a plateau. A plausible explanation for this saturation is the appearance of inertial cavitation for ξ >1 ,where the bubble expansion and collapse are controlled by inertia of the fluid [11].

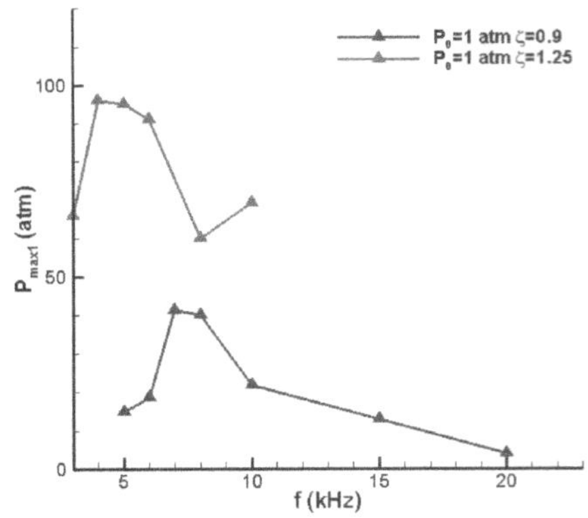

Fig. 7: Effect of driving pressure amplitude on P_{max1} vs f, for P_0=1 atm, R_0=50μm, α_0=5%, and A_0=X_0=1.5 mm

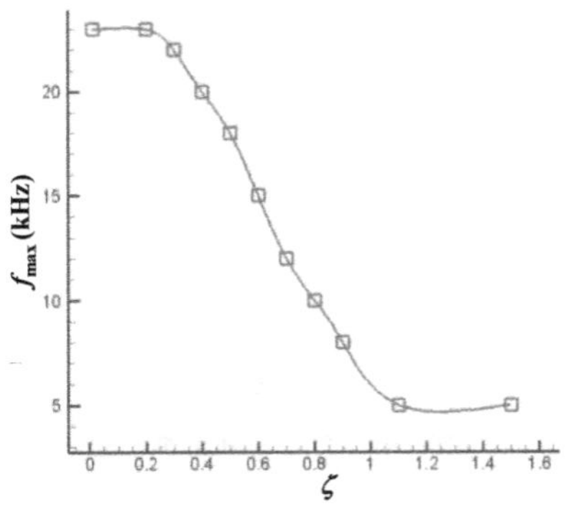

Fig. 8: driving frequency, for different relative driving pressure amplitudes, b) bubble cloud resonance frequency versus ξ, for P_0=1 atm, R_0=50μm, α_0=5%, and A_0=X_0=1.5 mm.

Conclusions

The dynamics of a bubble cloud subjected to a sinusoidal pressure field near a rigid wall is numerically studied using an Eulerian/Lagrangian two-phase flow model. Very strong pressures are generated at the wall during the cloud collapse at a resonance driving frequency. The study shows that the magnitude of pressure at the wall highly depends on the ambient pressure but not the resonance frequency. The study also shows that both the magnitude of the pressure and the resonance frequency are strongly dependent on the amplitude of the driving pressure. The magnitude of the pressure increases as the driving pressure is increased. The resonance frequency decreases as the driving amplitude is increased and approaches the natural frequency of the bubble cloud at small amplitude oscillations.

Acknowledgements

This work was funded by ONR under Contract No. N00014-15-C-5029 monitored by Dr. Ki-Han Kim.

References

[1] L. van Wijngaarden, "On the collective collapse of a large number of cavitation bubbles in water," in *11th Int. Congr. of Appl. Mech*, 1964.

[2] I. Hansson, V. Kedrinskii, and K. Mørch, "On the dynamics of cavity clusters," *J. Phys. D. Appl. Phys.*, vol. 15, no. 9, p. 1725, 1982.

[3] G. L. Chahine, "Cloud Cavitation: Theory," in *Proceedings of the 14th Symposium on Naval Hydrodynamics; Ann Arbor Michigan National Academy*, 1983, pp. 165–195.

[4] G. L. Chahine and R. Duraiswami, "Dynamical Interactions ina Multi-Bubble Cloud," *ASME J. Fluids Eng.*, vol. 114, no. 4, pp. 680–686, 1992.

[5] G. L. Chahine, C.-. T. Hsiao, and R. Raju, "Scaling of Cavitation Bubble Cloud Dynamics on Propellers," in *Advanced Experimental and Numerical Techniques for Cavitation Erosion Prediction*, vol. 160, K.-H. Kim, G. L. Chahine, J.-P. Franc, and A. Karimi, Eds. Spinger, Berlin, 2014, pp. 345–373.

[6] E. A. Brujan, T. Ikeda, K. Yoshinaka, and Y. Matsumoto, "The final stage of the collapse of a cloud of bubbles close to a rigid boundary," *Ultrason. Sonochem.*, vol. 18, no. 1, pp. 59–64, 2011.

[7] B. C. Tran, T. L. Hall, J. B. Fowlkes, and C. A. Cain, "Microbubble-enhanced cavitation for noninvasive ultrasound surgery," *IEEE Trans. Ultrason. Ferroelectr. Freq. Control*, vol. 50, no. 10, pp. 1296–1304, Oct. 2003.

[8] J. Ma, C.-T. Hsiao, and G. L. Chahine, "Euler-Lagrange Simulations of Bubble Cloud Dynamics Near a Wall," *ASME J. Fluids Eng.*, vol. 137, no. 4, pp. 041301–10, 2015.

[9] J. Ma, C.-T. Hsiao, and G. L. Chahine, "Numerical Study of Bubble Cloud Dynamics near a Rigid Wall," *Ultrason. Sonochemisty*, vol. 40, pp. 944–954, 2018.

[10] C. E. E. Brennen, *Cavitation and bubble dynamics*, vol. 9, no. 1. New York: Oxford University Press, 1995.

[11] T. J. Matula, "Inertial cavitation and single-bubble sonoluminescence," *Philos. Trans. R. Soc. A Math. Phys. Eng. Sci.*, vol. 357, no. 1751, pp. 225–249, 1999.

© 2018 ASME

A Reduced Order Gas Pressure Law for Single Acoustic Cavitation Bubbles

[1]Can F. Delale*; [2]Şenay Pasinlioğlu

[1]*MEF University, Istanbul, Turkey;* [2]*Istanbul Technical University, Istanbul, Turkey*

Abstract

An acoustic cavitation model for noncondensable gas/vapor bubbles that couples spherical bubble dynamics by the Keller-Miksis equation to the Plesset-Zwick equation is constructed by accounting for phase change, but neglecting the mass diffusion of the noncondensable gas. Results obtained for acoustically driven air/water-vapor cavitation bubbles using two different acoustic pressure signals and variable fluid properties show reasonable agreement with the reduced order model of Preston et al.

Keywords: acoustic cavitation, reduced order gas pressure law, bubble dynamics.

Introduction

Acoustic cavitation has aplications in sonochemistry [1], sonoluminescence [2,3] and medical ultasound [4,5] where bubbles of a few microns size grow and collapse to sizes that vary by orders of magnitude, especially at ultrasonic frequencies. In this case the pressure and temperature inside the bubble can show variations by orders of magnitude. Many complex acoustic cavitation models are constructed [6-12] for the numerical simulation of the temperature and pressure distributions under acoustically driven frequencies. However, when the complexity of the models and the computational time required for their simulation are taken into account, the need for simplified expressions for the gas pressure and temperature inside the bubble is obvious. This need becomes more important in hydrodynamic cavitation. In this investigation the heat conduction through the bubble is considered by the solution of Prosperetti's equation [8] in the uniform pressure approximation by a novel iterative method [13]. This iterative method leads to the desired reduced order gas pressure law exhibiting power law dependence on the bubble wall temperature and bubble radius, with the polytropic index depending on the isentropic exponent of the gas and on a parameter which is a function of the Peclet number. Moreover, it is shown that this reduced order gas pressure law reduces to the classical isothermal and adiabatic laws in the appropriate limits of the parameter. The bubble wall temperature entering this reduced order gas pressure law is obtained from the Plesset-Zwick solution [14]. Using this reduced order gas pressure law, an acoustic cavitation model for noncondensable gas/vapor bubbles that couples spherical bubble dynamics by the Keller-Miksis equation to the Plesset-Zwick equation is constructed by accounting for phase change, but neglecting the mass diffusion of the noncondensable gas. Results obtained for acoustically driven air/water-vapor cavitation bubbles using two different acoustic pressure signals and variable fluid properties show reasonable agreement with the reduced order model of Preston et al. [10] for a suitable value of the parameter.

An Acoustic Cavitation Model Using A Novel Reduced Order Gas Pressure Law

We consider the thermal behavior of a single spherical gas bubble surrounded by a liquid in the uniform pressure appproximation. The temperature distribution inside the bubble is then given by the Prosperetti equation [8, 13,15]

$$\frac{p}{T}\left\{\frac{\partial T}{\partial t}+\frac{1}{(Pe)pR^2}\left[\lambda(T)\frac{\partial T}{\partial y}-y\left(\frac{\partial T}{\partial y}\right)_{y=1^-}\right]\frac{\partial T}{\partial y}\right\}=\frac{(\gamma-1)}{\gamma}\frac{dp}{dt}+\frac{1}{(Pe)R^2y^2}\frac{\partial}{\partial y}\left[y^2\lambda(T)\frac{\partial T}{\partial y}\right] \tag{1}$$

where T is the temperature within the bubble, p is the uniform gas pressure inside the bubble, R is the instantaneous bubble radius, $\lambda(T)$ is the temperature dependent thermal conductivity of the gas, γ is the isentropic exponent of the gas, y is the normalized radial coordinate normalized and t is the time, all normalized with respect to some reference quantities [15]. In eq. (1) the Peclet number Pe is defined by

$$Pe=\frac{\gamma p_0' R_0'^2}{(\gamma-1)\lambda_R' T_0'\Theta'} \tag{2}$$

*Corresponding Author, Can F. Delale: delalec@mef.edu.tr

where p'_0 and T'_0 are, respectively, the bulk liquid equilibrium pressure and temperature with Θ' denoting a characteristic time, λ'_R denoting the thermal conductivity of the gas at the bubble wall temperature and R'_0 denoting the initial equilibrium bubble radius. The gas pressure within the bubble in the uniform pressure approximation is then obtained by solving the differential equation

$$\frac{dp}{dt}=\frac{3\gamma}{R}\left[\frac{1}{(Pe)R}\left(\frac{\partial T}{\partial y}\right)_{y=1^-}-p\frac{dR}{dt}\right]. \tag{3}$$

The system of the coupled differential eqs. (1) and (3) is then solved subject to the initial and boundary conditions

$$p(0)=p_i \quad ; \quad T(t=0,y)=1 \quad ; \quad T(t,y=1^-)=T_R(t) \quad ; \quad (\partial T/\partial y)_{y=0}=0 . \tag{4}$$

By expanding the temperature distribution near the bubble wall in the form

$$T(t,y)=T_R(t)+\left(\frac{\partial T}{\partial y}\right)_{y=1^-}(y-1^-)+\frac{1}{2!}\left(\frac{\partial^2 T}{\partial y^2}\right)_{y=1^-}(y-1^-)^2+O[(y-1^-)^3] \tag{5}$$

and by neglecting the temperature dependence of the gas thermal conductivity, substitution of eq. (5) into eq. (1) and taking the limit as $y\rightarrow 1^-$ yield the expression

$$c(t)=\left(\frac{\partial T}{\partial y}\right)_{y=1^-}=\frac{(Pe)R^2}{2}\left[\frac{p}{T_R}\frac{dT_R}{dt}-\frac{(\gamma-1)}{\gamma}\frac{dp}{dt}\right]-\frac{1}{2}\left(\frac{\partial^2 T}{\partial y^2}\right)_{y=1^-} \tag{6}$$

for the temperature gradient at the bubble wall. Utilizing eq. (6), we can obtain successive iterative approximations for the gas pressure from the solution of eq. (3). The first approximation, which neglects the second order radial derivative, is already discussed in [13], but this approximation does not produce dependence on the Peclet number and is thus inconvenient for all Peclet numbers. In the second approximation the second order radial derivative is taken to be proportional to the radial temperature gradient at the bubble wall, with the proportionality factor f exhibiting Peclet number dependence as well as weak dependence on time scale. Under this assumption substitution of eq. (6) into eq. (3), upon exact integration, leads to the novel reduced order gas pressure law in the form

$$p=p_i\left[\frac{(T_R)^{1/2(1+f)}}{R}\right]^{3\Gamma} \tag{7}$$

where Γ is a polytropic index given by

$$\Gamma=\frac{2\gamma(1+f)}{(3\gamma-1+2f)} \tag{8}$$

with f denoting a parameter depending on the Peclet number. It can further be shown that the novel reduced order gas pressure law given by eqs. (7) and (8) reduces to the isothermal law when γ=1 (or f=1/2) and T_R=1, and to the classical adiabatic law as $f\rightarrow\infty$. Using this reduced order gas pressure law, we can construct an acoustic cavitation model for bubbly liquids containing noncondensable gas/vapor bubbles. Assuming that the noncondensable gas/vapor bubbles consist of ideal gases and using the reduced order gas pressure law for the noncondensable gas, the total normalized mixture pressure p_b inside the bubble can be written in the form

$$p_b=p_v+p_g=p_{v,sat}(T_R)+p_{g0}\left[\frac{(T_R)^{1/2(1+f)}}{R}\right]^{3\Gamma} \tag{9}$$

where $p_{v,sat}(T_R)$ is the normalized saturation pressure of the vapor at the bubble wall and p_{g0} is the normalized initial equilibrium gas pressure. For the bubble wall temperature T_R, by assuming that most of the latent heat of condensation at the bubble wall is transferred to the liquid side since the thermal conductivity of the noncondensable gas/vapor mixture is much smaller than that of the liquid, we use the Plesset-Zwick solution in the form [14]

$$T_R(t) = 1 - B\int_0^t \frac{L(\xi)\rho_{v,sat}(\xi)R^2(\xi)(dR/d\xi)}{[\int_\xi^t R^4(\tau)d\tau]^{1/2}}d\xi \tag{10}$$

where $L=L'/L_0'$ and $\rho_{v,sat}=\rho'_{v,sat}/\rho'_{v0}$ with L' and $\rho'_{v,sat}$ denoting, respectively, the latent heat of vaporization of the liquid and the saturated vapor density at the bubble wall temperature and where 0 denotes the properties at the bulk liquid temperature. The constant B in eq. (10) is defined by

$$B = \left(\frac{L_0'}{T_0'c'_{p\ell 0}}\right)\left(\frac{\rho'_{v0}}{\rho'_{\ell 0}}\right)\frac{R_0'}{\left(\pi\,\Theta'\alpha'_{\ell 0}\right)^{1/2}} \tag{11}$$

with $c_p'_{\ell 0}$, $\rho'_{\ell 0}$ and $\alpha'_{\ell 0}$ denoting, respectively, the specific heat of the liquid, the density of the liquid and the thermal diffusivity of the liquid at the bulk liquid temperature. Using the reduced order gas pressure law for the partial pressure of the noncondensible gas, the spherical dynamics of the bubble can be characterized by the Keller-Miksis equation which takes the normalized form

$$\begin{aligned}&\left[1-MR+\frac{4M}{R(\mathrm{Re})}\right]RR+\frac{3}{2}\left[1-\frac{M}{3}R-\frac{8M}{3R(\mathrm{Re})}\right]R^2+\left[\frac{4(1+MR)}{\mathrm{Re}}-\frac{2M}{We}\right]\frac{R}{R}\\&+2(1+MR)\left\{\frac{1}{R(We)}-\frac{1}{We_0}\left[\frac{(T_R)^{1/2(1+f)}}{R}\right]^{3\Gamma}\right\}+\frac{(1+MR)}{2}\left\{\sigma-\sigma_0\left[\frac{(T_R)^{1/2(1+f)}}{R}\right]^{3\Gamma}\right\}\\&+\frac{(1+MR)C_p}{2}-MR\frac{d}{dt}\left\{\left(\frac{2}{We_0}+\frac{\sigma_0}{2}\right)\left[\frac{(T_R)^{1/2(1+f)}}{R}\right]^{3\Gamma}-\frac{C_p}{2}\right\}=0\end{aligned} \tag{12}$$

where the variable cavitation number σ, the variable Reynolds number Re and the variable Weber number We are defined by

$$\sigma = \frac{p_0'-p'_{v,sat}}{\frac{1}{2}\rho'_{\ell 0}U_0'^2}, \qquad \mathrm{Re} = \frac{\rho'_{\ell 0}U_0'R_0'}{\mu_\ell'}, \qquad \text{and} \qquad We = \frac{\rho'_{\ell 0}U_0'^2R_0'}{S'} \tag{13}$$

with S' denoting the surface tension, μ'_ℓ denoting the liquid dynamic viscosity (both to be evaluated at the bubble wall temperature), $U'_0 = R'_0/\Theta'$ denoting a characteristic speed and with subscript 0 denoting the initial value of the variable properties. The Mach number is given by $M=U'_0/c'_{\ell 0}$ with $c'_{\ell 0}$ denoting the speed of sound in the liquid, and the pressure coefficient entering the normalized Keller-Miksis equation (12) is defined as

$$C_p(t) = -\frac{p_A'(t)}{\frac{1}{2}\rho'_{\ell 0}U_0'^2} \tag{14}$$

where $p'_A(t)$ is the deriving acoustic pressure. Equations (10) and (12) form the basic equations of the proposed acoustic cavitation model using the novel reduced order gas pressure law. For the numerical solution of the initial value problem of the Keller-Miksis equation (12) for spherical bubble dynamics, we use the Runge-Kutta-Fehler method with adaptive time step size together with the Plesset-Zwick solution (10) for the bubble wall temperature evaluated by Simpson's 3/8-numerical integration scheme iteratively. We now apply the proposed acoustic cavitation model to water-vapor/air bubbles in water. We choose two different acoustic deriving pressure signals with pressure coefficients given by [10,16]

$$C_p(t) = -0.25\left[1-\cos\left(\frac{2\pi t}{500}\right)\right] \;;\qquad 0<t<500 \tag{15}$$

and

$$C_p(t) = -0.662\exp\left[-\left(\frac{t-130}{43.5}\right)^2\right]; \qquad 0 < t < 300. \tag{16}$$

The bulk water temperature is chosen to be T_0'=20°C. In this case the saturated vapor pressure is p'_{v0}=0.0234 bar, the surface tension is S_0'=0.071 N/m and the dynamic viscosity of water is $\mu'_{\ell 0}=10^{-3}$ kg/m-s at the bulk liquid temperature. For the acoustic driving pressure signal [16] given by eq. (15) , the initial cavitation number is σ = 0.492, the initial bubble radius is R_0'=100 μm and the characteristic time corresponding to the acoustic deriving pressure is taken as $\Theta' = 10^{-5}$ s which yield an initial Reynolds number *Re=1000* and an initial Weber number *We=137*. Figures 1(a) and (1b) show, respectively, the variations of the normalized bubble radius and of the normalized bubble wall temperature with time for different values of the parameter *f* (*f*=0.5, 1.0 and 10.0) using the present acoustic cavitation model. The time variation of the bubble radius shows reduction in the bubble oscillation amplitudes as the parameter *f* increases from 0.5 to 10.0 exhibiting typical behavior from near-isothermal case (low Peclet number) to near-adiabatic case (high Peclet number). The bubble wall temperature variations also show reduction in magnitude as the parameter increases from 0.5 to 10.0. For the acoustic driving pressure signal [10] given by eq. (16) , the initial cavitation number is σ = 0.654, the initial bubble radius is R_0'=100 μm and the characteristic time corresponding to the acoustic deriving pressure is taken as $\Theta' = 2.3 \text{ X } 10^{-6}$ s which yield an initial Reynolds number *Re=695* and an initial Weber number *We=167*. Figures 2(a) and (2b) show, respectively, the variations of the normalized bubble radius and of the normalized bubble wall temperature with time for different values of the parameter *f* (*f* = 0.5, 1.0 and 10.0) for this case using the present acoustic cavitation model. Similar behavior in the reduction of the bubble oscillation amplitudes and in the bubble wall temperature is observed as the parameter *f* increases from 0.5 to 10.0. In particular, the results for *f*=0.5 are in agreement with those given by Preston et al. [10] using their reduced order constant transfer model without mass diffusion where they employ the Rayleigh-Plesset equation instead of the Keller-Miksis equation. Finally, under the same conditions of Figure 2, we investigate the effect of variable fluid properties using the proposed acoustic cavitation model. Figures 3(a) and 3(b) show, respectively, the variations of the normalized bubble radius and of the normalized bubble wall temperature with time for variable fluid properties. Considerable reductions in both buble oscillatiion amplitudes and in bubble wall temperatures are observed as compared to the case of constant fluid properties. Moreover, a fast relaxation to equilibrium of the bubble radius and of the bubble wall temperature is observed depending on the value of the parameter *f*.

(a) (b

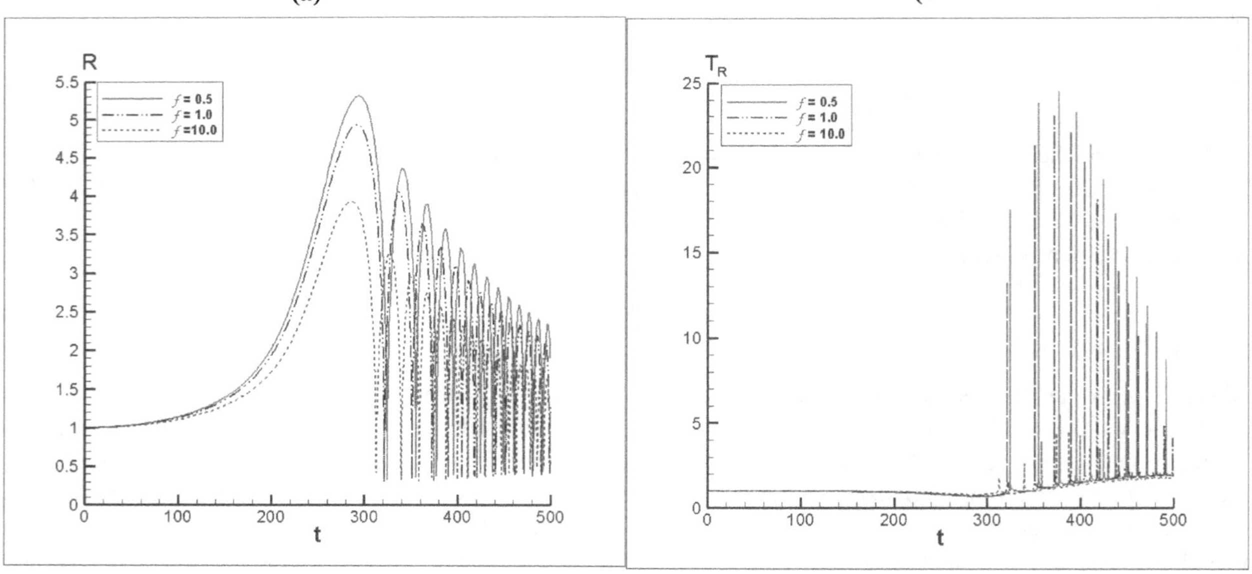

Figure 1: The temporal evolution of (a) the normalized bubble radius, (b) the normalized bubble wall temperature, driven by the acoustic pressure given by eq. (15) for an air-water vapor bubble in water with σ = 0.492, We=137, Re=1000 for different values of *f* (0.5, 1.0 and 10.0) obtained by the proposed acoustic cavitation model using constant fluid properties.

(a) (b)

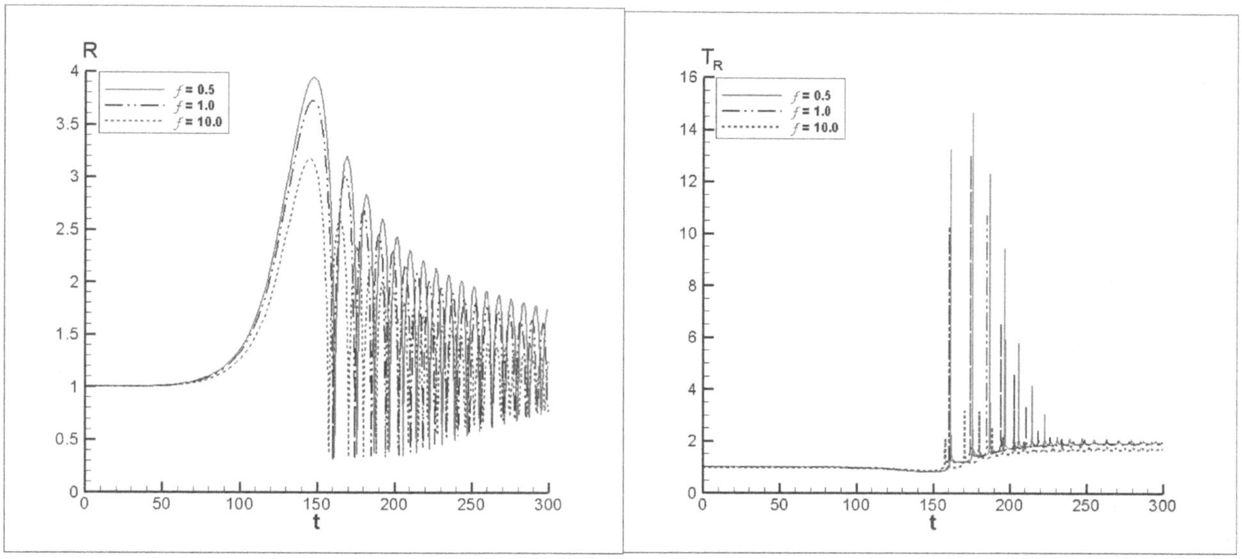

Figure 2: The temporal evolution of (a) the normalized bubble radius, (b) the normalized bubble wall temperature, driven by the acoustic pressure given by eq. (16) for an air-water vapor bubble in water with $\sigma = 0.654$, We=167, Re=695 for different values of f (0.5, 1.0 and 10.0) obtained by the proposed acoustic cavitation model using constant fluid properties.

(a) (b)

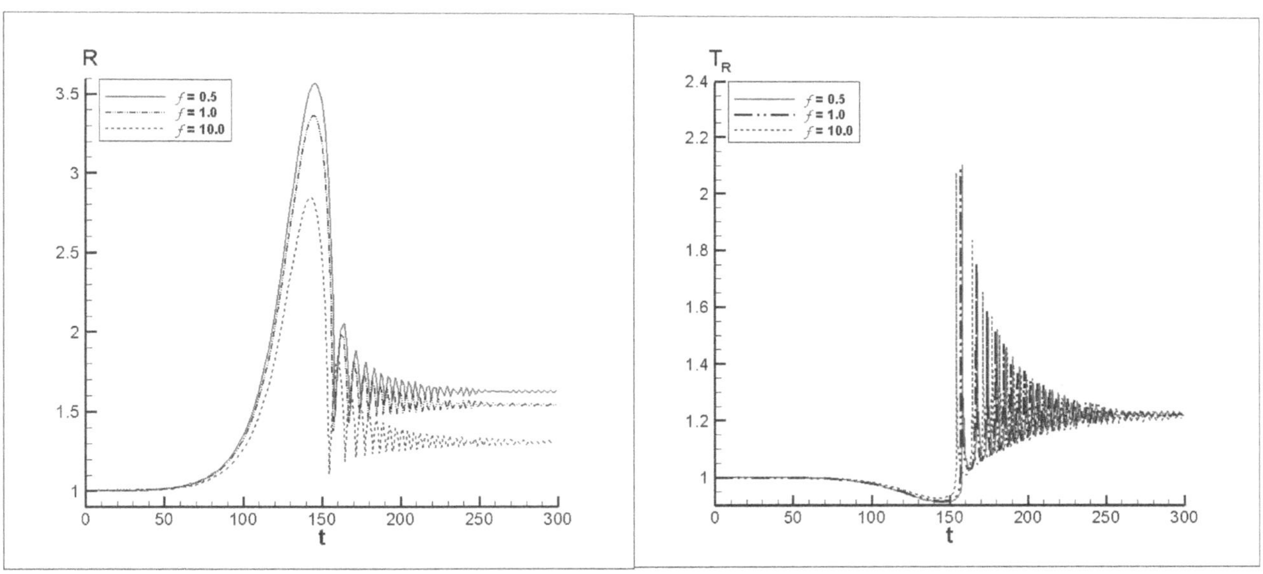

Figure 3: The temporal evolution of **(a)** the normalized bubble radius, **(b)** the normalized bubble wall temperature, driven by the acoustic pressure given by eq. (16) for an air-water vapor bubble in water with initial values $\sigma = 0.654$, We=167, Re=695 for different values of f (0.5, 1.0 and 10.0) obtained by the proposed acoustic cavitation model using variable fluid properties.

Conclusion

The thermal behavior of gas and gas/vapor bubbles is investigated by considering th energy balance between a single spherical bubble and the surrounding liquid, neglecting the effect of gas diffusion to the liquid. For gas bubbles we solve the well-known coupled differential equations for the gas pressure and temperature inside the bubble in the uniform pressure approximation by using an iterative technique for approximating the second order radial derivative of the temperature at the bubble wall. In this way we obtain a novel reduced order gas pressure law which exhibits power law dependence on the bubble wall temperature and bubble radius, with the polytropic index depending on the isentropic exponent of the gas and on a parameter, which is a function of the Peclet number and of the characteristic time scale. Moreover, it is shown that this parameter is an increasing function of the Peclet number where the reduced order gas pressure law reduces to the classical isothermal law when this parameter takes the value 1/2 and to the classical adiabatic law when this parameter tends to infinity. The bubble wall temperature entering this reduced order gas pressure law is obtained from the Plesset-Zwick solution of the temperature distribution of the liquid side in the thin boundary layer approximation. Using this reduced order gas pressure law, a reduced order acoustic cavitation model for noncondensable gas/vapor bubbles that couples spherical bubble dynamics by the Keller-Miksis equation to the Plesset-Zwick equation is constructed by accounting for phase change, but neglecting the mass diffusion of the noncondensable gas. Results obtained for acoustically driven air/water vapor cavitation bubbles using two different acoustic pressure signals and variable fluid properties show reasonable agreement with the reduced order model of Preston et al. [10] for a suitable value of the parameter. It remains to correlate the model parameter with Peclet number and characteristic time scale (driving frequency) using the results of the DNS of the original PDEs over a wide range of Peclet numbers and driving frequencies.

Acknowledgement
This work is, in part, supported by the Scientific and Technological Research Foundation of Turkey (TÜBİTAK) under grant number 117M072.

References

[1] Suslick, K.S. (1990). *Sonochemistry*. Science, 247, 1439-1445.
[2] Gaitan, D.F., Crum, L.A., Church, C.C., Roy, R.A. (1992) *Sonoluminescence and Bubble Dynamics for a Single, Stable, Cavitation Bubble*. Journal of Acoustical Society of America. 91, 3166-3183.
[3] Crum, L.A. (1994) *Sonoluminescence*. Physics Today. 47(9), 22-29.
[4] Ferrara, K., Polard, R., Borden, M. (2007) *Ultrasound Microbubble Contrast Agents: Fundamental Applications to Gene and Drug Delivery*. Annual Review of Biomedical Engineering. 9, 415-447.
[5] Qin, S., Caskey, C.F., Ferrara, K.W. (2009) *Ultrasound Contrast Microbubbles in Imaging and Therapy: Physical Principles and Engineering*. Physics in Medicine and Biology. 54, R27-R57.
[6] Nigmatulin, R.I., Khabeev, N.S., Nagiev, F.B. (1981) *Dynamics, Heat and Mass Transfer of Vapour-Gas Bubbles in a Liquid*. International Journal of. Heat and Mass Transfer. 24, 1033-1044.
[7] Matsumoto, Y., Takemura, F. (1994) *Influence of Internal Phenomena on Gas Bubble Motion*. JSME International Journal Series B, 37, 288-296.
[8] Prosperetti, A. (1991) *The Thermal Behavior of Oscillating Gas Bubbles*. Journal of Fluid Mechanics. 222, 587-615.
[9] Toegel, R., Gompf, B., Pecha, R., Lohse, D. (2000) *Does Water Vapor Prevent Upscaling Sonoluminescence?* Physical Review Letters. 85, 3165-3168.
[10] Preston, A. T., Colonius, T., Brennen, C. E. (2007) *A Reduced-Order Model of Diffusive Effects on the Dynamisc of Bubbles*. Physics of Fluids. 19, Article No. 123302.
[11] Stricker, L., Prosperetti, A., Lohse, D. (2011) *Validation of an Approximate Model for the Thermal Behavior in Acoustically Driven Bubbles*. Journal of Acoustical Society of America, 130, 3243-3251.
[12] Kreider, W., Crum, L.A., Bailey, M., Sapozhnikov, O.A. (2011) *A Reduced-Order, Single-Bubble Cavitation Model with Applications to Ultrasound*. Journal of Acoustical Society of America. 130(5), 3511-3530.
[13] Delale, C. F., Pasinlioğlu, Ş. (2015) *First Iterative Solution of the Thermal Behaviour of Acoustic Cavitation Bubbles in the Uniform Pressure Approximation*. Journal of Physics: Conference Series. 656, Article No. 012016.
[14] Plesset, M. S., Zwick, S. A. (1952) *A Nonsteady Heat Diffusion Problem with Spherical Symmetry*. Journal of Applied Physics. 23, 95-98.
[15] Delale, C. F. (2002) *Thermal Damping in Cavitating Nozzle Flows*. Journal of Fluids Engineering. 124, 969-976.
[16] Wang, Y.C. (1996). *Shock Waves in Bubbly Cavitating Flows*. PhD Thesis, California Institute of Technology.

© 2018 ASME

Acoustic Cavitation as Process Intensifier: A Phenomenological Study

[1]Ramamurthy Nagarajan*; [1]Srivalli Hariharan;

[1]*Department of Chemical Engineering, Indian Institute of Technology-Madras, Chennai-600036, India*

Abstract

When an acoustic field is coupled to a liquid and transient cavitation is induced, it can significantly intensify ongoing processes, making them run faster, cheaper, better. Surface cleaning is a classic example that has been investigated in various studies over the years. Sonochemical reactors are increasingly being deployed in industry. Cavitational enhancement of mass transfer phenomena such as diffusion, dissolution and leaching has been quantified in literature, and its effect on heat transfer from fluids to adjoining solids has been reported as well. Nano-particle synthesis by "sono-fragmentation" is widely practised as a top-down alternative to bottom-up methods of nano-particle production with tons-per-day throughput. Cavitation-induced destratification has been investigated earlier in this laboratory as a method to keep cryogenic fuels well-mixed during storage on the ground. More recently, cavitation has been evaluated as a mechanism to remove ash, sulfur and alkalis from coal prior to combustion, thereby greatly mitigating the downstream propensity for slagging, fouling, corrosion and erosion. Practical applications abound, as the additional energy expenditure associated with the acoustic field is easily compensated by savings in process time and productivity enhancement.

Unification of these cavitation-induced effects in multiple domains is a worthwhile exercise in order that the effect may be fully understood and optimized. In particular, ultrasonic frequency is a tunable parameter that can have dramatic effects on the outcome. The transition from the cavitation regime to acoustic streaming (as frequency is increased from the ultrasonic to the megasonic regime) is one that needs to be characterized and managed with care. In many applications, acoustic streaming can provide a synergistic effect when deployed in conjunction with cavitation, as in, for example, a dual-frequency system that can operate in the high (> 100 kHz) and low (<100 kHz) rage. Alternatively, an intermediate frequency such 300 kHz which combines both cavitational and streaming features may be employed as well.

The definition of a "process intensification factor" and assessment of its dependence on ultrasonic frequency for various thermodynamic and transport phenomena constitutes a valuable contribution for practitioners, and this has been attempted in this paper. Based on the frequency dependence established, processes may be categorized with respect to their sensitivity to acoustic cavitation. This categorization enables optimum application of cavitational fields to achieve desired results with minimum expenditure of energy, and mitigation of undesirable side effects.

Keywords: Cavitation, Ultrasound, Process Intensification, Sono-fragmentation, Coal beneficiation, De-sulphurisation

Introduction

Application of ultrasound on liquids such as water leads to two types of physical mechanisms: acoustic cavitation and acoustic streaming. The former refers to the implosion of bubbles resulting in a shock wave that travels in all directions, while the latter is described by a flow that is unidirectional in nature along the normal to the transducer in the ultrasonic equipment. When a medium is exposed to sound waves of frequency above 18 kHz [1], it results in the propagation of oscillating compression and rarefaction cycles. When a local pressure gradient develops due to difference in ambient pressure and vapor pressure of the liquid, bubbles are formed [2]. These bubbles grow in size till they can no longer sustain themselves, leading to a violent bubble collapse which causes release of shock waves that traverse the medium. This is associated with a local increase of temperature and pressure, often reported to be up to 5000K and 1000 atm., respectively.

When the ambient pressure of a liquid is reduced adequately, it can boil without heating. This principle is used in an ultrasonic tank where the reduction of local pressure results in formation and implosion of bubbles, leading to release of shock waves throughout the medium [1]. Cavitation can be stable (occurs at lower intensities where the bubbles oscillate about their mean position) or transient (occurs at higher intensities where the bubbles grow and collapse violently). As rarefaction and compression occur in alternating cycles, bubbles form and grow during the former and collapse in the latter cycles. As the frequency increases, the number of bubbles formed increases correspondingly [3]. However, the size of the bubble decreases with increasing frequency and leads to reduced intensity of cavitation as the frequency increases. This explains the pronounced effect of cavitation intensity at lower frequencies.

Cavitation intensity depends on a number of factors [4]. While the force of cavitation is inversely proportional to the cube of frequency, other properties of the medium such as surface tension, vapour pressure, temperature, viscosity and density play a key role as well. For example, altering the temperature results in change in ambient pressure and the gradient developed becomes higher leading to higher cavitation effects. Similarly, lowering the viscosity leads to a better response of the liquid to bubble growth and collapse, and increases cavitation effects in turn. Each of these properties can be altered depending on the medium used for bubble growth and collapse.

The objective of this work is to study the "intensification" effect of ultrasound-assisted processes. Although ultrasound has been used for several decades in diverse fields to improve processes [5], the pace of applications has accelerated in recent times. Cavitation can have two types of effects on a chemical reaction: it can either accelerate the reaction, or promote reactions which would not have occurred otherwise. The latter happens due to the dissociation of sonicated water into hydrogen ions and hydroxyl radicals which, in turn, leads to multiple back reactions and formation of new products. The mechanical effects associated with cavitation are also of significance in a number of systems. The formation and collapse of bubbles at a surface leads to a sudden inflow of liquid at the surface, and since this inflow occurs at high local temperature, pressure and velocity, it is an important mechanism in ultrasonic surface cleaning applications. Cavitation has also found abundant applications in wastewater treatment, rupture of cells, synthesis of nano-particles, etc. While some processes may be highly sensitive to cavitation, some might need a combination of cavitation and acoustic streaming to demonstrate intensification effects which are noteworthy. A consolidated understanding of the role that cavitation plays in process-intensification in a variety of heterogeneous systems would be useful in identifying system-specific parameters which can be tuned for optimization.

Cavitation as a process-intensifier: Applications

a. Heat-Transfer enhancement in furnace tubes

Dhanalakshmi et al. [6] studied the effect of ultrasound-assisted heat transfer enhancement in furnace tubes. This work focused on the heat transfer characteristics of water flowing in a tube exposed to heat in the presence and absence of ultrasound. It was found that the critical parameter in the process is the ratio of characteristic ultrasonic field velocity and prevailing flow velocity. Though heat transfer enhancement due to ultrasonic exposure depended very much on the flow conditions, i.e., the Reynolds number, heat transfer enhancement rate (HTER) at 20 kHz frequency increased till a Reynolds number of 500 was reached; it decreased as Reynolds number increased from 500 to 1200 and showed no significant change thereafter. This study was done with 3 different frequencies – 20 kHz, 33 kHz and a combination of 20 and 33 kHz. While 20 kHz showed much better HTER compared to 33 kHz, 20+33 kHz was slightly better than 20 kHz alone. This illustrates that while 20 kHz alone is a first-order effect, using a combination of frequencies introduces a second-order effect. They concluded that the cavitation technique for heat transfer enhancement is ineffective for higher flow rates, but showed considerable improvement for controlled laminar flow regimes.

b. Degradation of Methyl Violet

High-intensity, low-frequency ultrasound is widely used to augment sono-chemical reactions. The sono-chemical hotspots formed due to cavitational implosion lead to disintegration of water molecules into H^+ and OH^- radicals. These recombine to form hydrogen peroxide, molecular hydrogen and a number of other radicals and species. Methyl violet dye is released as an industrial waste and is extremely harmful to plant and aquatic life owing to its highly toxic characteristics. The main aim of this study was to boost decolorization of methyl violet using ultrasound

[6]. The experiments were conducted in both tank-type and probe-type sonicators, and utilizing frequencies ranging from 20 kHz to 1 MHz. The main steps involved preparation of Fenton's reagent, mixing with dye dissolved in water and irradiating with ultrasound for 150 minutes. The results showed that in the presence of Fenton's reagent, ultrasonic irradiation of the contaminated sample led to near-complete removal of dye from the water. The effect of frequency is interesting to note because as the frequency increases, decolorization of water is faster, and reaches near-completion although the increase in rate is not significant when the frequency exceeds 430 kHz. At higher frequencies, the cavitation effect decreases and is dominated by acoustic streaming, the large density of bubbles leads to higher gas-liquid interfacial area and hence greater contact with OH radicals, resulting in higher reaction rates. The rate of decolorization also increased with increasing power.

c. Sono-fragmentation of coal

This study [6], focused on the breakage of coal particles suspended in a solution. Frequencies spanning from low (<100 kHz), intermediate (100-400 kHz) and high (> 400 kHz) were studied. The ground coal was mixed with three different reagents (hydrogen peroxide, acetic acid and ethanol) separately and sonicated at the chosen power and different frequencies. The results obtained showed that particle breakage was higher at lower frequencies, while particle agglomeration set in at higher frequencies. A combined effect of both the phenomena was observed at intermediate frequencies. Particle breakage was found to be a function of both frequency and time. Longer time of sonication, up to 5 minutes, led to reduced particle size, and particle breakage gradually decreased between 5 to 10 minutes of ultrasonic exposure. Particle size reduction with respect to frequency followed the order: 25 > dual > 132 >192 >470 >1000 kHz.

d. Destratification

Thermal stratification refers to the variation in temperature at different levels of a stored fluid. This is a challenged posed by cryogenic liquids due to boil off losses associated in handling, storage and transport. The work reported in [7] studies the potential effect of ultrasound on thermal destratification. Experiments were initially conducted in a storage tank with water and later extended to cryogenic liquids. This was done by heating the top layer of a tank containing water while the bottom layer remained at room temperature. This system was irradiated with ultrasound and the temperature was measured at all levels creating a complete temperature profile along the height of the vessel at different points of time. A range of frequencies starting from 25 kHz to 470 kHz were used to study the destratification effect. Of all the frequencies chosen, the highest frequency of 470 kHz provided maximum breakage of thermal layer and destratification. When the power input and the storage pressure was increased, 470 kHz system showed drastic increase in stratification Index compared to other lower frequencies. However, dual frequency mode of operation such as 58/192 kHz, which combines the effect of cavitation and streaming, provides better mixing as opposed to using a high frequency tank in isolation.

e. Effervescent atomization

When an effervescent gas is mixed in with a liquid inside a chamber in such a way that bubbles are formed inside a nozzle, it is known as "effervescent atomization". This is used in spraying of fuel as liquid droplets in the combustor and leads to instantaneous formation of vapor. The objective of this study [8] was to produce very fine droplets from an effervescent atomizer, with the assistance of ultrasound. Ultrasound was used to break the bubbles formed by the aerator and distribute their size homogeneously. Experiments were carried out with 20 kHz frequency and showed that ultrasound helped in breaking up of larger sized bubbles into smaller ones. At lower power (of less than 40% input power), bubbles undergo shape change without fragmentation. Although acoustic streaming plays a role in elongation of bubble in the direction of the streaming force, the main mechanism involved in this process of bubble break up is cavitation, and the spray obtained is qualitatively superior when compared to traditional methods employed for effervescent atomization.

f. Coal beneficiation studies

The high percentage of sulphur and ash content found in Indian coals make them suitable for techniques such as ultrasonic coal wash. This method, employed at different frequencies to remove ash and sulphur from Indian coals, forms the main basis of the study reported in [9, 10]. Along with different frequencies, both aqueous- and reagent-based ultrasonic coal washing methods were explored. 20 kHz, 25 kHz, 58/192 kHz dual-mode and 430 kHz frequencies were used for the study. While the 20 kHz frequency was transmitted via a probe-type sonicator, the rest of the equipment was tank-type. For de-ashing experiments, the sieved coal was sonicated for 15 minutes and subjected to three-level decantation using a fabricated settling column. Ash, being heavier, settles faster than coal. Three layers are then separated and analyzed for ash content. The de-sulphurisation experiments involved sonication of the sample, filtration, washing, drying and analysis as their main steps. For de-ashing, the lowest frequency gave the best size reduction owing to cavitational bubble collapse and subsequent breakage of coal particles. Higher frequencies showed lesser rupture of particles due to streaming being the dominant mechanism in such systems. However, ash removal percentage was optimum at 132 kHz and this behavior may be attributed to the simultaneous presence of streaming and cavitation producing a balanced effect. For de-sulphurisation of coal, the lowest frequency systems showed excellent contact of oxidizing agents produced by ultrasound with fine coal particles, leading to higher removal of pyritic and organic sulphur compared to the high-frequency systems. The rate of removal was, however, more than 90% in all cases, proving the viability of ultrasonic coal wash as an option for de-sulphurisation of coal.

g. Alkali leaching of coal

The alkali content in coal has been a cause of concern with regard to deposition of fly-ash on heat transfer surfaces. The sodium and potassium salts leave the char and react with calcium, magnesium and alumino-silicates leading to the formation of a sticky layer. This sticky layer aids in the deposition of tiny fly-ash particles on the surface of boilers, on super heater tubes and elsewhere in the system. Ultrasound has been effectively used in removal of alkali from coal. Extensive studies have been undertaken to characterize ultrasound-assisted alkali leaching of coal. This technique involves preparing a solution of ground coal, chosen reagent and water, followed by sonication for a short period of time. Analysis is done using ICP. While preliminary studies [11] involved a combination of low- (25 kHz) and high-frequency (430 kHz) systems, further studies done on low-frequency cavitation systems involved 25 kHz, 40 kHz, 58 kHz and 68 kHz. Though the results were satisfactory for potassium leaching, removal of sodium was low. This was followed up with further leaching using a range of ultrasonic and megasonic frequencies, namely 58 kHz, 132 kHz, 58/132 kHz dual, 360 kHz and 1 MHz. The results showed the presence of different types of alkali salts in coal, with only the water-soluble salts being removed. The reaction is diffusion-controlled as observed by fitting the data to a "shrinking core" model. 360 kHz worked best for removal of both sodium and potassium owing to the right mix of cavitation and streaming. While cavitation aids in breakage of coal matrix and particle fragmentation, streaming accounts for efficient leaching of alkali that occurs subsequently.

Cavitation as a process-intensifier: Mechanistic Understanding

The following Table 1 summarizes the observed intensification effects of ultrasonic fields on various physico-chemical phenomena:

Table 1: Consolidated representation of process intensification under the influence of ultrasound

No.	Process	Procedure	Frequency	Observations
a	Heat-Transfer enhancement in furnace tubes	• To compare of heat transfer characteristics of water flowing in a tube exposed to heat in the presence and absence of ultrasound.	20,33, 20+33 kHz	• Order of HTER was as follows: 33 kHz > 20+33 kHz combined > 20 kHz. • While 20 kHz alone was a first-order effect, combination of 20 kHz and 33 kHz was found to have a second-order

		• Critical parameter found to be ratio of ultrasonic flow velocity and prevailing flow velocity.		effect.
b	Degradation of Methyl Violet	• To boost decolorization of methyl violet using ultrasound. • Preparing Fenton's reagent, mixing with dye and irradiating with ultrasound were the main steps involved.	25, 132, 430 kHz, 1 MHz	• Decolorization increases with increasing frequency and reaches near completion at 430 kHz. • At higher frequencies, though cavitation effect is reduced, there is a higher density of bubbles leading to better contact with OH radicals and higher reaction rates.
c	Sono-fragmentation of coal	• To mix ground coal with three different reagents and sonicate at the chosen frequency. • Results were analyzed quantitatively using a laser particle counter and qualitatively in SEM.	25, 132, 58+192, 190, 470 kHz, 1 MHz	• Lower frequencies show effective particle breakage while particle agglomeration happens at higher frequencies. • Particle size reduction with respect to frequency (kHz) followed the order: 25 > 58+192 > 132 >192 >470 >1000.
d	Destratification	• To study the effect of ultrasound on thermal destratification. • A tank filled with water is heated at the top, while the bottom is maintained at room temperature. The system is exposed to ultrasound and the temperature profile is obtained.	25, 40, 58, 68, 132, 192, 58+192, 172+192, 470 kHz	• 470 kHz provided maximum breakage of thermal layer and destratification. • When the power input and the storage pressure was increased, 470 kHz system showed a dramatic increase in Stratification Index. • Dual frequencies, which combine effect of cavitation and streaming, provided better mixing compared to high or low frequencies.
e	Effervescent atomization	• To produce very fine droplets from an effervescent atomizer, with the assistance of ultrasound • Ultrasound was used to break the bubbles formed by the aerator and distribute them homogeneously	20 kHz	• At lower frequency, ultrasound helped in breaking up of larger sized bubbles into smaller ones. • The main mechanism involved in bubble break-up is cavitation, and the spray obtained is qualitatively superior.
f	Coal beneficiation studies	• To study aqueous and reagent-based ultrasonic coal wash • For de-ashing, the sieved coal is subjected to sonication followed by a three-level decantation. • De-sulphurisation was done by sonication, followed by titration for analysis	20, 25, 58+192, 430 kHz	• For de-ashing, the lowest frequency gave the best size reduction owing to cavitational bubble collapse and subsequent breakage of coal particles. • For de-sulphurisation of coal, the low-frequency systems showed excellent contact of oxidizing agents produced by ultrasound.
g	Alkali leaching of coal	• To sonicate ground and sieved coal for a fixed period of time after mixing with reagent. • Samples are tested using ICP.	25, 40, 58, 68, 132, 58+132, 360 kHz, 1 MHz	• Low frequency ultrasound alone was not effective for alkali leaching. • When a medium frequency system such as 360 kHz is used, combined effect of cavitation and streaming produces best results for removal of

				both sodium and potassium from coal.

When an acoustic field is coupled to a liquid medium, two phenomena are induced--- cavitation and acoustic streaming—and two effects are created—physical and chemical. The interaction between these can be quite complex. However, some general insights emerge from a cause-and-effect analysis:

- When first-order physical effects are pronounced-- for example, in particle breakage and in aerosolization of liquids-- low-frequency cavitational effects are dominant..
- However, when second-order physical effects, such as diffusion and leaching, the combined effect of cavitation and acoustic streaming is superior to that produced by either mechanism acting alone.
- Convection and mixing may be intensified by either cavitation or acoustic streaming, since they operate by different but equally effective mechanisms. While cavitational fields literally "throw" parcels of fluid around, streaming induces a unidirectional flow of the fluid away from the transducer, which can set-up convection rolls that assist redistribution.
- Higher frequencies lead to greater density of smaller-sized bubbles, thus increasing net effective interfacial area. Chemical reactions that rely upon surface contact between reacting species and radicals are therefore maximally enhanced at such frequencies at the higher end of the ultrasonic spectrum (or lower end of the megasonic range).
- When particle agglomeration is desirable, such as in separation of contaminants from effluents, higher frequencies are ideal for promoting it, while lower frequencies will actually break-up agglomerates.
- Acoustic field effects on transport phenomena—momentum, mass and heat—are very similar when the dominant mechanisms are the same.
- In the case of chemical reactions, while the localized high-temperatures and high-pressures associated with cavitation can increase local rates, frequency and length of contact between reacting species can be more directly affected by streaming effects. Thus, a frequency-blended field may be more appropriate.
- In general, streaming enhancement factors scale as square of frequency, while cavitational-enhancement effects scale as $1/f^3$. Both scale linearly with input power (amplitude) until a lower threshold value at which the field collapses. When both effects are exerting a combined influence, the frequency exponent will range from +2 to -3, depending on which influence is dominant. When multiple frequencies are used in series, as opposed to being integrated into the same tank or same stage, a cleaner separation of effects can be achieved, and this is the recommended mode of operation.

References

[1] **Suslick, K.S.,** (1988), *Ultrasound: Its chemical, physical and biological effects*, VCH: New York
[2] **Ginsburg. E, Kinsley.M.D, and Quitral. A**. (1998), *The power of ultrasound.*, Adm. Radiol. J., 17(5),17–20
[3] **Suslick, K.S., Y. Didenko, M. M. Fang, T. Hyeon, K. J. Kolbeck, W. B. McNamara, M. M. Mdleleni** and **M. Wong** (1990) *Acoustic Cavitation and Its Chemical Consequences,* Phil. Trans. R. Soc. London A, **357**, 335 – 353
[4] **Vetrimurugan. M**., (2007) "*Investigation of High-frequency, High-Intensity ultrasound for de-stratification of liquids stored in insulated containers*," Ph.D thesis, Indian Institute of Technology, Madras
[5] **Laborde, J. L., Hita, A., Caltagirone, J. P., and Gerard, A., (2000)** *Fluid Dynamics Phenomena Induced by Power Ultrasound*, Ultrasonics, 38, 297-300
[6] **Dhanalakshmi NP**., (2013), *Process intensification via power ultrasound - unifying principles.* Ph D Thesis, Indian Institute of Technology, Madras
[7] **Jagannathan, T.K**., (2010), *Mechanistic modeling, visualization and applications of flow induced by high-intensity, high-frequency acoustic fields,* Ph D Thesis, Indian Institute of Technology, Madras
[8] **Jagannathan, T.K., Nagarajan, R., and Ramamurthi, K.,** (2011), Effect *of ultrasound on bubble breakup within the mixing chamber of an effervescent atomizer*, Chemical Engineering and Processing, 50, 305-315
[9] **Ambedkar. B, Nagarajan. R, and Jayanti. S**, (2011), *Ultrasonic coal-wash for de-sulfurization*, Ultrason. Sonochem., 18(3), 718–726
[10] **Ambedkar. B, Chintala T.N, Nagarajan. R, and Jayanti, S.,** (2011) *Feasibility of using ultrasound-assisted process for sulphur and ash removal from coal*, Chem. Eng. Process Intensification, 50(3), 236–246
[11] **Srivalli. H, Nirmal. L, and Nagarajan. R**, (2015), *Effect of cavitation on removal of alkali elements from coal,* vol. 012107. (dx.doi.org/ 10.1088/1742-6596/656/1/012107)

© 2018 ASME

Air-Water Ring in the Vaneless Gap of a Reversible Pump-Turbine Operating in Condenser Mode

Elena Vagnoni*[1], Loïc Andolfatto[1], Renaud Guillaume[2], Pierre Leroy[2] and François Avellan[1]

[1] EPFL, Laboratory for Hydraulic Machines, Lausanne, Switzerland

[2]GE Renewable Energy, Grenoble, France

Abstract

The present research aims to experimentally study the dynamic characteristics of a rotating bubbly flow in the vaneless gap of a reversible pump-turbine operating in condenser mode by performing high speed image acquisition. Image processing is developed to detect and track the bubbles in the flow for estimating their size and shape and for measuring the velocity profiles. The influence of the gauge pressure is investigated by performing the experiments at five gauge pressure conditions between 2 bar and 17 bar.

The experimental investigation evidences the formation of an air-water ring in the vaneless gap between the impeller blades and the closed guide vanes. The air-water ring is characterized by a rotating bubbly flow due to the air-water mixing. Bubbles have an oblate spheroid shape whose size and velocity depend on the gauge pressure. The velocity is mainly tangential and a periodic fluctuation of both velocity components is recorded owing to the geometrical shape of the guide vanes walls.

Keywords: Two-phase flows; pump-turbine; condenser mode.

Introduction

Hydraulic pump-turbine can be required to operate in condenser mode to compensate the grid instability caused by the penetration of intermittent renewable energy sources [1]. In this particular operating mode, the guide vanes of the pump-turbine are closed and pressurized air is injected in the draft tube cone to reduce the resisting torque by maintaining the water level below the impeller. Furthermore, cooling water is injected through the labyrinth seals.

The sloshing motion of a free surface, turbulent waves causing the formation and entrainment of bubbles diffusing in the water volume and the formation of droplets interfering with the impeller are the main two-phase flow phenomena which have been reported in literature for a turbine operating in condenser mode [2][3][4]. Another expected phenomenon is the onset of an air-water ring in the vaneless gap between the impeller and the closed guide vanes due to the interaction of the water discharge coming from the labyrinth cooling system and water droplets which are generated by the oscillation of the air-water interface. The water droplets entrained by the impeller are expelled from the outer periphery of the impeller by centrifugal force and it forms a rotating air-water ring in the vaneless gap. The rotating flow interacts with the impeller causing power losses. In addition, it is a suitable place for bubble formations and the mass transfer of air into water.

The air-water ring requires further investigations to understand the mechanism of interaction with the impeller and the air mass transfer. To accomplish this goal, the present paper seeks the geometric and dynamic characterization of the rotating air-water ring by performing experimental investigations to estimate the bubbles size and the velocity fields of the flow as a function of the gauge pressure.

Experimental set-up

The experiments are performed on a reduced scale physical model of a high head reversible pump-turbine. A sketch of the main components of the model is illustrated in Figure 1. The model includes the spiral case, 20 guide vanes, 9 blades impeller and the draft tube. The vaneless gap between the impeller and the guide vanes has a ring shape and it measures $0.06 \times D_e$ in height h and $0.09 \times D_e$ in maximum width w, where D_e is the high pressure diameter of the impeller. A generator connected to the impeller regulates the rotational speed. The test rig is operated in closed loop configuration and is equipped with an axial pump which generates the specific hydraulic energy. To operate in condenser mode, the guide vanes are closed and a system of air injection is installed in the draft tube cone. Water is

*Corresponding Author, Elena Vagnoni: elena.vagnoni@epfl.ch

injected through the labyrinth seals for cooling purpose. The Froude similarity is fulfilled by defining the densimetric Froude number as follows:

$$Fr_d = \sqrt{\frac{\rho_a}{\rho_w}}\frac{ND_e}{\sqrt{gD_e}} \quad (1)$$

where N is the rotational speed of the impeller, ρ_a and ρ_w are the air and water density, respectively. The air density is computed from the ideal gas law by measuring the static air pressure in the cone of the draft tube and the temperature. This similarity law has been introduced by Tanaka [2]. The visualization of the water ring in the vaneless gap between the guide vanes and the impeller is performed through two flat optical windows to allow the illumination and the optical access of the vaneless gap and guide vanes as illustrated in Figure 1. The air-water ring visualization is performed with high speed camera, at image acquisition rates of 4'000 frames per second for 8 s. A 95 % uniform LED screen is installed as a back light to enhance the contrast between gaseous and liquid phases.

Tests are performed at five gauge pressure conditions p measured in the cone of the draft tube in air and at $Fr_d = 0.5$. The machines operates in pump rotating direction for all the investigated operating points.

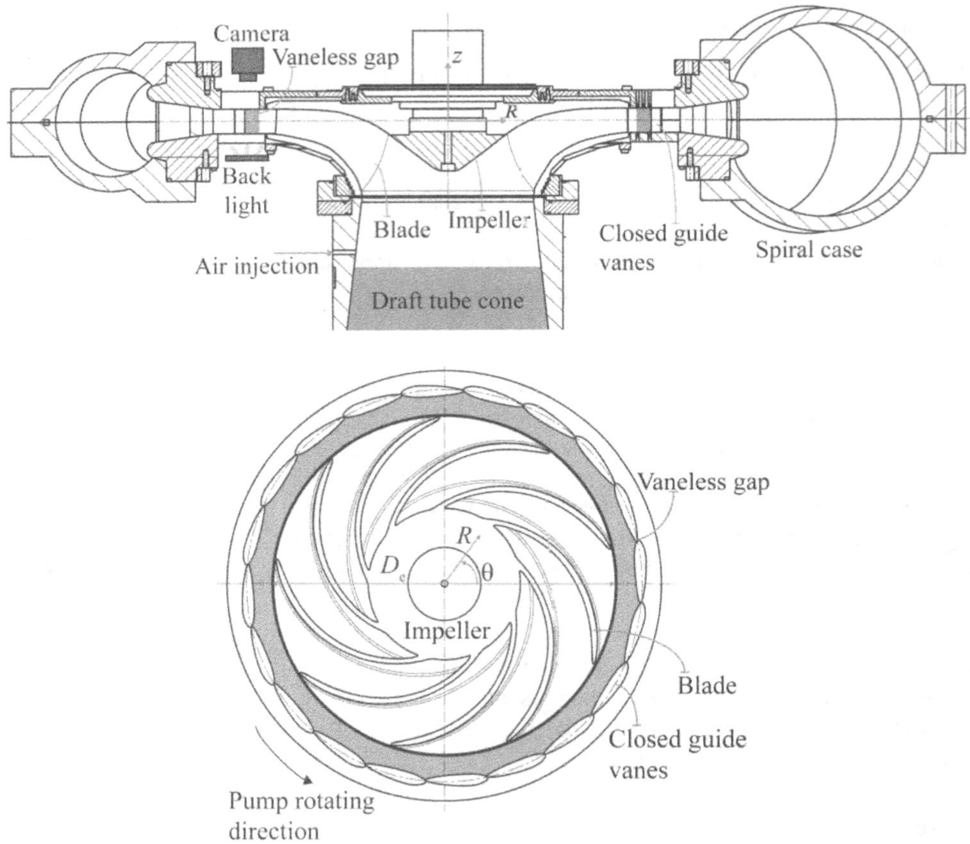

Figure 1. Test facility layout with the instrumentation (right). Top view of the vaneless gap (left).

Image processing method

An image processing method is developed to determine the bubble size and velocity profiles by detecting and tracking the bubbles in the recorded images [5]. As a first step, the images are trimmed to select the region of interest and each pixel of the image is identified by a polar coordinate which corresponds to the polar coordinates system presented in

Figure 1 and Figure 2a). In a second step, the Otsu filter [6] is applied to identify the edge of the bubbles in contrast with the bright background, as shown in Figure 2b). The patterns of interest are the single bubbles represented by the small black areas surrounded by white liquid film. The black areas at the bottom and top of the window correspond to the casing of the machine and they should not be considered as a pattern. Bubbles clusters which cannot be individually distinguished also are discarded. Since the size of one bubble is clearly different on the top and on the bottom black areas and on the clustering of many bubbles, only the features with a perimeter smaller then a fixed value are selected, as illustrated in Figure 2c). On the other side, clusters of few tiny bubbles can be considered as a single feature by the image processing method. In the previous step, each feature is determined by the perimeter, major and minor axis, and the position on the image. The bubbles are assumed to be oblate spheroid with minor semi-axis a and major semi-axis b. This hypothesis has been validated in literature [7]. For each bubble the equivalent diameter is estimated as follows:

$$d_{eq} = \sqrt[3]{8b^2 a} \tag{2}$$

In the final step, the method links the successive positions of each feature in the following images. The trajectories of the bubbles are identified and the velocity vectors are computed for each bubble as shown in Figure 2d). To compute the radial and angular velocity profiles of both velocity components, the velocities of the bubbles are averaged depending on the bubble position.

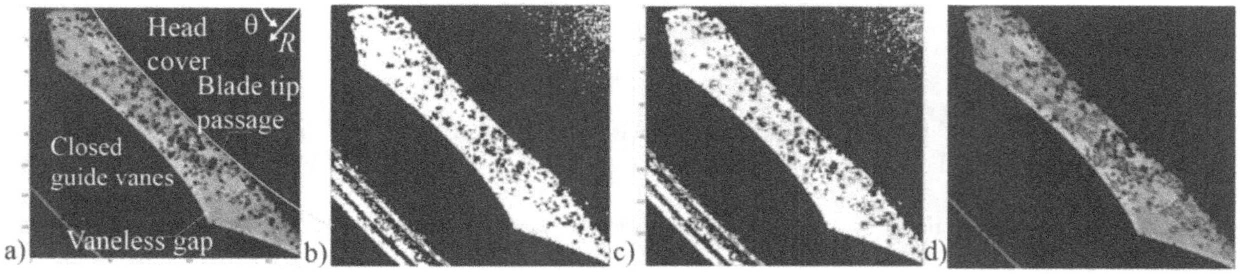

Figure 2. Steps of the image processing method.

Results

Size of bubbles

The high speed visualization enables the observation of the air-water ring in the vaneless gap between the impeller and the guide vanes of a pump-turbine. Images of the air-water ring and the mean equivalent diameter of the bubbles, made non dimensional by dividing for the vaneless gap width w, are presented in Figure 3 as a function of the gauge pressure. The boxes extend from the lower to upper quartile values of the data, with a line at the median and a point at the mean value. The whiskers extend from the box to the maximum and minimum values to show the range of the data. It is observed that the boxes and the whiskers extension increases by increasing the gauge pressure. The mean size of the bubbles increases by increasing the gauge pressure and as a consequence, there are more bubbles of different sizes which explains the distribution of the data on a wider range of values.

a)

b)

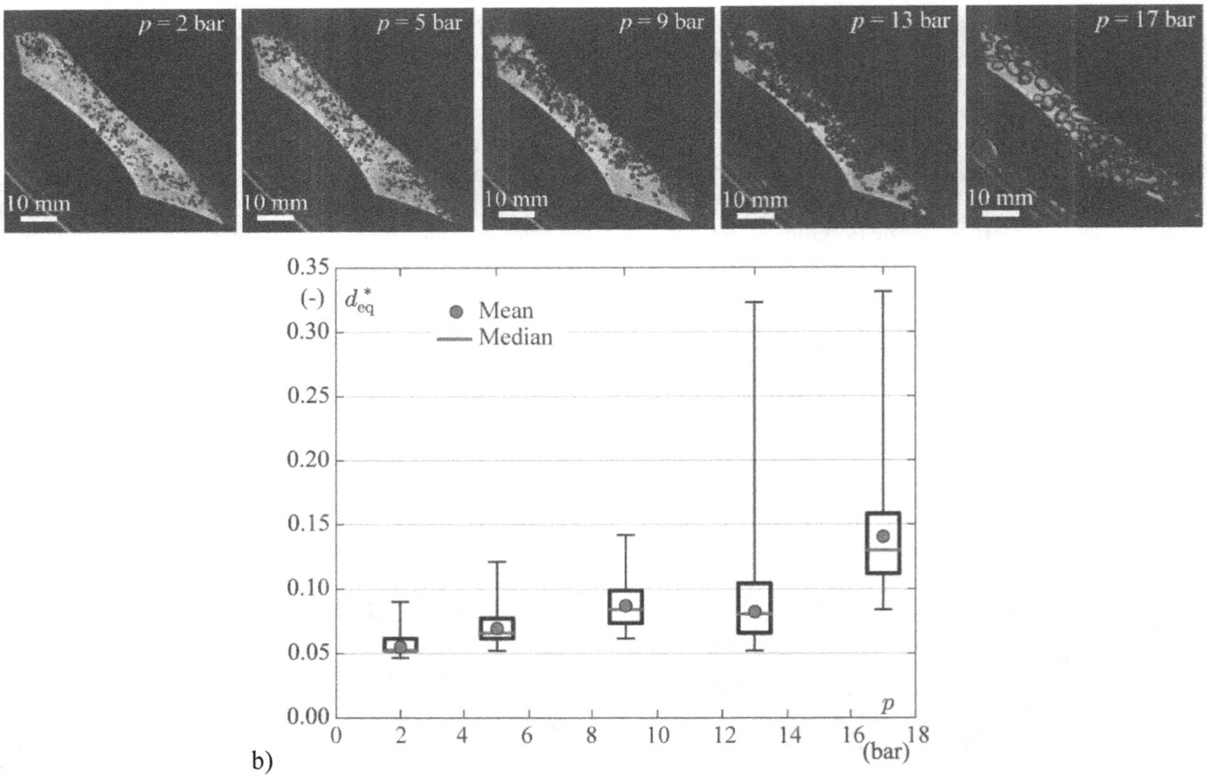

Figure 3. Images of the air water ring (a) and equivalent diameter distribution statistics (b) for the five investigated gauge pressure conditions.

Velocity profiles

The velocity profiles of both tangential and radial components are illustrated in Figure 4. The bubbles velocity is mainly tangential and is slower than the tip of the blades. An influence of the gauge pressure on the tangential velocity divided by the peripheral velocity is recorded. The tangential velocity of the flow at p = 17 bar is lower than the one recorded at the other pressure conditions. This can be explained by the larger size of the bubbles: the formation of clusters of bubbles is observed and they slow down the flow velocity. At p = 13 bar a lower number of bubbles could be tracked due to the higher void fraction. The bubbles at the radius closer to the tip could not be distinguished by the image processing method. Moreover, this introduces a higher standard deviation on both tangential and radial velocity components.

The angular distributions of the tangential and radial velocity components of the bubbles of the air-water ring are presented in Figure 5(a) and (b), respectively at p = 5 bar. The bubbles follow the geometrical shape of the guide vanes by causing a fluctuation of the flow velocity in the vaneless gap which can justify the value of the standard deviation observed in Figure 4 of both velocity components profiles. The tangential velocity decreases in correspondence of the enlargements of the vaneless gap due to the guide vanes shape. The radial velocity component negatively increases in the narrow section of the vaneless gap due to the guide vanes shape and it positively increases in the corresponding enlargement.

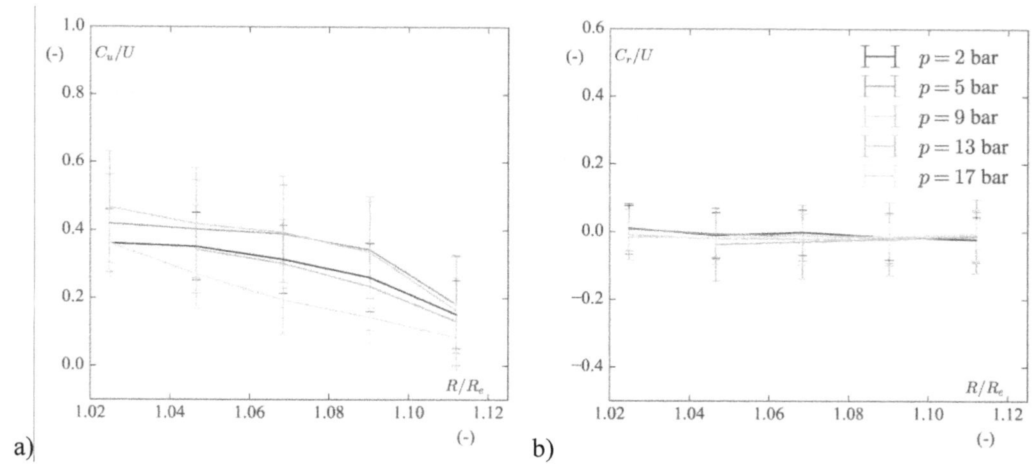

Figure 4. Velocity profile of tangential (a) and radial (b) velocity for the five investigated gauge pressure conditions.

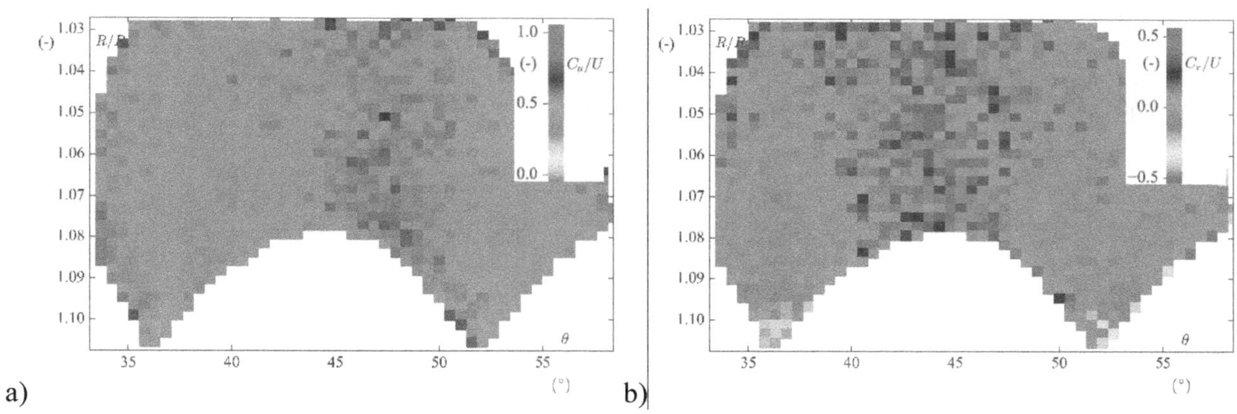

Figure 5. Angular distribution of the tangential (a) and radial (b) velocity components of the bubbles in the air-water ring for p = 5 bar.

Conclusion

In the present work, the dynamic characteristics of the air-water ring in the vaneless gap of a pump-turbine operating in condenser mode are investigated.

High speed visualization and image processing are performed to measure the size and the velocity profiles of the bubbly flow in the air-water ring. Both size and velocity of the bubbles depend on the gauge pressure. The higher the pressure, the larger is the equivalent diameter of the bubbles. At $p = 17$ bar the size of the bubbles causes the formation of clusters which slow down the velocity of the flow. The measured velocity profiles showed that the flow is mainly tangential in the radial section of the vaneless gap and the fluctuation of both velocity component is due to the geometrical shape of the guide vanes wall.

This study contributes to the understanding of the dynamic characteristics rotating air-water ring in the vaneless gap of a pump-turbine operated in condenser mode. In a further study, we envisage to estimate the parameters of the mass transfer of air in water to compute the air losses through the vaneless gap.

Acknowledgements

The research leading to the results published in this paper has received funding from GE Renewable Energy France. The authors would like to thank the team of GE Renewable Energy France for their collaboration and technical support, in particular Yann Viguen Heratchian and Pierre-Yves Lowys.

References

[1] I. Rychkov, *Reactive power control services based on a generator operating as a synchronous condenser*, Power Technology and Engineering 46, pp. 405-409, (2013).

[2] H. Tanaka, K. Matsumoto, K. Yamamoto, *Sloshing motion of the depressed water in the draft tube in dewatered operation of high head pump-turbines*, XVII IAHR Symposium on hydraulic machines and cavitation, Beijing, Chine (1994).

[3] Rossi, G. and Zanetti, V., *Starting in air and synchronous condenser operation of pump-turbines - Model research*, 9th IAHR-SHMEC Symposium, vol. 2, pp. 337-352,Fort Collins (USA), (1978).

[4] O. Ceravola, M. Fanelli, B. Lazzaro, *The behavior of the free level below the impeller of Francis turbines and pump-turbine in operation as synchronous condenser*, Proceedings of X IAHR Symposium on hydraulic machines and cavitation, Tokyo, Japan 1 (1980) 765-775.

[5] S. van der Walt, J. Schonberger, J. Nunez-Iglesias, F. Boulogne, J. Warner, N. Yager, E. Gouillart, T. Yu, *scikit-image: image processing in python*, 350 PeerJ 2:e453 doi:https://doi.org/10.7717/peerj.453.

[6] R. Gonzales, R. Woods, *Digital image processing*, Pearson Education International, 2008.

[7] D. Colombet, D. Legendre, A. Cockx, P. Guiraudand, F. Risso, C. Daniel, S. Galinat, *Experimental study of mass transfer in a dense bubble swarm*, Chemical Engineering Science 66 (2011) 3432-3440.

© 2018 ASME

An Equation of State Tabulation Approach for Injectors with Non-Condensable Gases: Development and Analysis

Mathis Bode[a,*], Sutharsan Satcunanathan[b], Kazuki Maeda[c], Tim Colonius[d], Heinz Pitsch[a]

[a]*Institute for Combustion Technology, RWTH Aachen University, 52056 Aachen, Germany*
[b]*Institute for Aerodynamics, RWTH Aachen University, 52056 Aachen, Germany*
[c]*Mechanical Engineering, University of Washington, Seattle, WA 98195, USA*
[d]*Department of Mechanical Engineering, California Institute of Technology, Pasadena, CA 91125, USA*

Abstract

In this work, a general equation of state (EOS) tabulation method is presented, which allows arbitrary combinations of EOSs in different phases and can be used with single-phase flow solvers by adding one additional transport equation for the total partial density of all non-condensable gases. The new tabulation method assumes instantaneous equilibrium for all phase change processes and uses Legendre transformation to construct the convex hull of the energy surface. Newton iterations are applied to improve the accuracy within the tabulation step as well as of the data retrieved at runtime. A high-order 5-equation multiphase solver with stiffened-gas equations as EOS for all phases and with the ability to use different time scales for the relaxation processes between liquid and vapor phase is used to discuss the full equilibrium assumption of the tabulation approach. Furthermore, results using different EOSs for the tabulation are compared. The implication of choosing a stiffened-gas equation or a cubic EOS, such as the Peng-Robinson equation for the vapor phase, on the saturation quantities is discussed. A nozzle simulation performed under typical gasoline direct injection (GDI) conditions is finally used to demonstrate the advantages of the new tabulation method and to evaluate additional computational cost.

Keywords: Equation of State; Homogeneous Equilibrium Model; Cavitation; Non-Condensable Gases; Injector Flows; Homogeneous Relaxation Model

Introduction

The performance of direct injection engines strongly depends on the injection process of the liquid fuel. This process is influenced by phase change, including cavitation inside the nozzle. For an accurate prediction of the nozzle flow, non-condensable gases have been found to be important, resulting in a flow configuration with three different phases: liquid fuel, fuel vaporl, and non-condensable gases [1, 2]. Small amounts of non-condensable gases are typically dissolved in the liquid fuel, but they can also penetrate into the nozzle from the low-pressure chamber. This phenomenon is known as hydraulic flip.

For the described multiphase systems, N-equation models have become popular in recent years [3, 4, 5, 6, 7, 8]. The so-called 7-equation model, which yields conservation equations for mass, momentum, and energy for a liquid and a gas phase as well as an additional gas volume fraction transport equation, has been successfully reduced to 6-, 5-, and 4-equation systems by assuming kinetic, mechanic, and thermodynamic equilibrium. Also, extensions to more than two phases and systems with phase change have been presented. However, the application of these models is both expensive due to the additional equations and complex due to the requirement of additional boundary conditions compared to single-phase solvers. Also the closure of the extra source terms, for example in the context of large-eddy simulations (LESs), is unsolved.

Due to their flexibility and simple utilization in flow solvers originally designed for single-phase flows, equation of state (EOS) tabulation methods are an alternative approach. However, for nozzle flows, existing tabulation methods suffer from one of the following issues: Either they cannot account for non-condensable gases or they are unable to

*Corresponding Author, Mathis Bode: m.bode@itv.rwth-aachen.de

use different analytical EOSs for liquid fuel and fuel vapor, which is important for accurately predicting both fuel phases and phase changes simultaneously. This is adressed in this work by developing and analyzing a 3-equation homogeneous equilibrium model (HEM) for 3-phase systems. It emerges from the 7-equation model by assuming kinematic, mechanic, thermodynamic, and chemical equilibrium and requires only the solution of the well-known single-phase Navier-Stokes equations. The complexity of the model is hidden in the construction of a convex EOS that covers multiple phases. Since Maxwell constructions are not able to deal with arbitrary EOSs [9], Legendre transformation (LT) is used to construct the convex hull of the energy surface and recover convexity [10]. This is crucial for a wide validity of the model. The resulting thermodynamic properties are tabulated and a subsequent table lookup method is employed for incorporation of the HEM into a 3-equation flow solver. The accuracy of both, the tabulation step and the data retrieved, is improved by Newton iterations.

3-Phase Homogeneous Equilibrium Model

The EOS alters the flow equations through the flux function. In pure phases, these are known to be smooth convex functions. When phase transfer effects are included in the EOS, the flux function becomes non-smooth allowing new wave structures. A single EOS that attempts to cover the pure liquid, the pure vapor, the pure gas, and the mixture region exhibits a non-physical region of negative speed of sound, where convexity constraints from the 2nd law of thermodynamics are violated. Thus, the construction of a convex, i. e. thermodynamically admissible, equilibrium EOS out of convex single-fluid EOSs is one main challenge for tabulated EOS of 3-phase systems. One way to achieve the convexification is by strictly enforcing the equilibrium conditions. Denoting liquid by the subscript 'l', the corresponding vapor by 'v', the consolidation of all non-condensable gases by 'o', and mass fractions by Y, these can be summarized as

Temperature equilibrium (T)	$T = T_\mathrm{l} = T_\mathrm{v} = T_\mathrm{o}$	(1a)
Pressure equilibrium (p)	$p_\mathrm{l} = p_\mathrm{v} + p_\mathrm{o}$	(1b)
Chemical equilibrium (μ)	$\mu_\mathrm{l} = \mu_\mathrm{v}$	(1c)
Specific volume (v)	$v = (1 - Y_\mathrm{o} - Y_\mathrm{v})v_\mathrm{l} + Y_\mathrm{v} v_\mathrm{v}$	(1d)
Specific internal energy (e)	$e = (1 - Y_\mathrm{o} - Y_\mathrm{v})e_\mathrm{l} + Y_\mathrm{v} e_\mathrm{v} + Y_\mathrm{o} e_\mathrm{o}$	(1e)
Miscibility condition	$v_\mathrm{v} Y_\mathrm{v} = v_\mathrm{o} Y_\mathrm{o}$	(1f)

Alternatively, the convexification can be done by means of LT. Due to the higher flexibility, LT is used for this work as explained in the next subsection.
Besides the thermodynamic validity of the 3-phase HEM, accuracy is an important property of the model. This is especially true for the mixture region, where strong gradients occur across the phase boundaries. The developed HEM allows to flexibly combine exact, analytical expressions in the pure fluid regions and looked up values in the mixture region, which can be improved by Newton iterations on runtime using tabulated initial values and locally converging optimized Newton schemes. This hybrid combination allows high accuracy at moderate computational cost with a maximum of flexibility. Details are given in Subsection 'Hybrid Approach'.

Convexification
In this work, LT is used for the convexification and is denoted for any function $\tilde{f}$ as

$$\tilde{f}^* : \xi \in \mathbb{R}^n \to \mathbb{R}; \xi \mapsto \sup_{x \in \mathbb{R}^n} \left\{ \langle \xi, x \rangle - \tilde{f}(x) \right\} \tag{2}$$

with $\langle \cdot, \cdot \rangle$ as scalar product on $\mathbb{R}^n$. Applying LT twice results in the biconjugate $\tilde{f}^{**}$, which is in general the convex hull of $\tilde{f}$. The construction of the convex hull of the energy surface is complicated by the non-intuitive set of variables v, s, e (with s as entropy) given by the fundamental relations [10], since it makes the choice of an appropriate domain for the LT difficult. Especially the edges of the domain might be located in regions, where the respective EOS loses its validity, leading to steep gradients, which have major impact on the dual space since $\xi \propto \nabla e$. This was further emphasized by computations, which have shown that the choice of the dual sets Ω_{ξ^v} and Ω_{ξ^s} is crucial for the accuracy of the resulting table. If the dual set is chosen to be equally spaced between $\min(\xi_i)$ and $\max(\xi_i)$, this might distort the physically interesting regions by decreasing the resolution. To overcome these problems in this work, the easier measurable quantities temperature and pressure are used to cluster grid points in regions of interest, which is possible due to $(\xi^v, \xi^s)^T \propto \nabla e(v, s) = (-p, T)^T$. I.e. the negative of the pressure is dual to specific volume and temperature is dual to entropy. Consequently, the EOS-table is built as follows: Given a temperature and pressure range $\Omega_{T,p} = [T_{\min}, T_{\max}] \times [p_{\min}, p_{\max}]$, the sets Ω_{ξ^v} and Ω_{ξ^s} are chosen such that

$$n_{T,\mathrm{rel}} = \left\{ |\Omega^T_{\xi^v}| / |\Omega_{\xi^v}| \mid T_{\min} \le \xi^v_i \le T_{\max}, \forall \xi^v_i \in \Omega^T_{\xi^v}, \Omega^T_{\xi^v} \subseteq \Omega_{\xi^v} \right\} \tag{3}$$

and

$$n_{p,\mathrm{rel}} = \left\{ |\Omega^p_{\xi^s}| / |\Omega_{\xi^s}| \mid -p_{\max} \le \xi^s_i \le -p_{\min}, \forall \xi^s_i \in \Omega^p_{\xi^s}, \Omega^p_{\xi^s} \subseteq \Omega_{\xi^s} \right\} \tag{4}$$

are greater than some prescribed values and $\xi^v_1 \le \min_j \xi^{v,j}_1, \xi^v_{N_{\xi^v}} \ge \max_j \xi^{v,j}_{N_{\xi^v}}$ and $\xi^s_1 \le \min_i \xi^{s,i}_1, \xi^s_{N_{\xi^s}} \ge \max_i \xi^{s,i}_{N_{\xi^s}}$ with n as number of points for the specified dimension for the tabulation. The superscripts j and i indicate their belonging to the sets $\Omega^j_{\xi^v}$ and $\Omega^i_{\xi^v}$ and the latter condition ensures that the domain of e^{**} matches with that of e. Consequently, in total $n_{T,\mathrm{rel}} \cdot n_{p,\mathrm{rel}}$ of the points will be located in $\Omega_{T,p}$. This can even be improved, if the distribution density of the slopes additionally enters the determination of the dual sets, which is done in this work. Furthermore, to account for the stiff liquid phase it is crucial to have high resolution for small v to capture the steep gradients. Therefore, a logarithmic scale for the v-axis is chosen, while the s-axis is kept linear during tabulation.

Hybrid Approach

The physics of 3-phase systems considerably differs from 2-phase systems. For example, due to the Gibbs' phase rule, which gives degrees-of-freedom $\hat{f} = 2$, the well-known thermodynamic characteristics of phase equilibrium ($v_{\mathrm{l,sat}}(v_{\mathrm{v,sat}})$, $T_{\mathrm{sat}}(v_{\mathrm{v,sat}})$, $p_{\mathrm{sat}}(v_{\mathrm{v,sat}})$, ... with 'sat' denoting saturation) cannot be employed and $v_{\mathrm{v,sat}}$ is not suited to parametrize the boundary of the mixture region, although it is most suited for 2-phase systems [9]. The isotherms in the p-v-space are no longer horizontal lines and the phase boundary is not a 1D set of points.

System 1 summarizes the equilibrium conditions for 3-phase systems, which are the starting point of the HEM. It can be seen that for computing the thermodynamic state for given v, e, and Y_o, iterating on v_v and T circumvents the necessity to solve implicit relations. Since this set of nonlinear equations cannot be solved explicitly, Newton iteration methods are used. More than one possible combination of two equations, on which to perform the iteration, exists where condition Equation 1a is implicitly enforced by using one global temperature T and therefore drops out. There still remain $(4-1)! = 6$ combinations, when the miscibility condition is not included. It has been observed that the choice of the Newton scheme (NS) for NS[p, μ], NS[v, e], and NS[p, e] strongly affects the convergence for different regions in the v-e-Y_o-space and thus the performance of the tabulation method. Therefore, the optimal choice has been also precomputed for this work and stored as part of the EOS-table in order to save computational cost. Furthermore, the domain of convergence was extended by dynamically adapting the relaxation parameter of the Newton iteration based on physical considerations. This was found to be especially important for very low volume fractions of the non-condensable gases, where the system of equations is ill-conditioned since the physics changes considerably from 3-phase to 2-phase systems rendering the system very stiff.

Results and Discussion

Comparison of Stiffened-Gas/Stiffened-Gas with Stiffened-Gas/Peng-Robinson

The choice of the EOSs of the different phases in a multiphase simulation is crucial. It is well-known that cubic EOS such as the Peng-Robinson EOS (PR-EOS) have problems to accurately represent liquids due to their small compressibility. This combined with the simple usage of stiffened-gas EOSs (SG-EOSs) and the limited possibility of combining arbitrary EOSs have made the usage of SG-EOSs for both the liquid and the vapor phase popular. The developed HEM allows to study the accuracy of various EOS combinations. For that, EOS coefficients were fitted [11, 12] and compared to experimental data from the National Institute of Standards Technology (NIST) database. The comparison is shown in Figure 1. The liquid parameters are not influenced by the coupled vapor EOS and only the slightly different equilibrium states result in negligible differences for the saturated liquid quantities. Thus, all

EOS combinations with SG-EOS for the liquid phase share almost the same graph and for the sake of readability only one representative graph labeled 'SG' is plotted. The plots cover a fairly large temperature range nearly up to the critical point. Since SG-EOS does not support this point, 'SG/PR' does not give meaningful results when approaching the critical point either. For p_{sat}, L_{v} and all vapor quantities except for the speed of sound, 'SG/SG' shows good agreement within a limited temperature range, but then starts to gradually deviate from the experimental curves for higher temperatures. Overall, the quality of the fit for SG-EOS depends strongly on the chosen temperature interval for the parameter determination. The saturated liquid speed of sound reveals one main shortcoming of SG-EOS, which is that complex molecular interactions are neglected. Even the monotonicity is miss-predicted. This can be ascribed to the expression for the speed of sound, which is a pure function of the temperature with $\frac{\partial c^2}{\partial T} = C_{\mathrm{v}}\gamma(\gamma - 1) > 0$ (with c as speed of sound, C_{v} as specific heat capacity at constant volume, and γ as heat capacity ratio) and thus unable to predict the tendency of decreasing speed of sound with temperature.

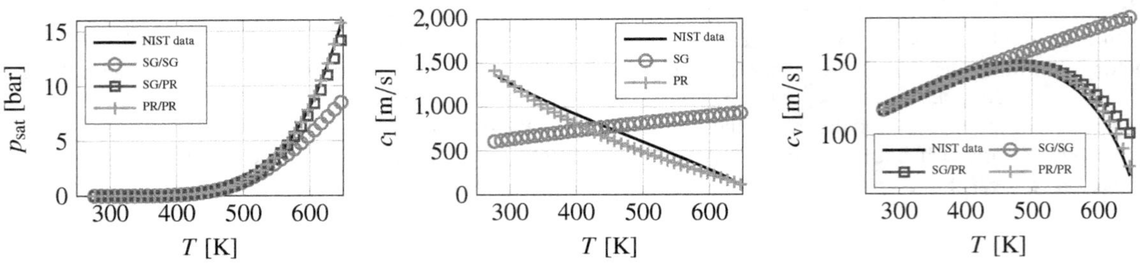

(a) Saturation pressure (b) Liquid sound speed (c) Vapor sound speed

Figure 1: Dodecane at saturation in the temperature interval T = [273 K, 647 K].

1D Validation

To evaluate the 3-equation HEM for multiphase applications, a two-phase shock tube ($Y_{\mathrm{o}} = 0$) with dodecane and the conditions proposed by Saurel et al. [3] was computed. For a better comparability with other N-equtation models, SG-EOS is used for both phases and the results are compared to results computed with a 5-equation model with complete relaxation [8]. Due to different model assumptions, two points needs to be kept in mind during the analysis: First, according to Gibbs' phase rule, each pure phase in the mixture is determined by one state variable and exactly one combination of T, p in the initial conditions (ICs) will allow the system to be in a mixture equilibrium state. In contrast, in the relaxation models, phase transfer can be dynamically controlled, for instance to suppress phase transfer at metastable states. Consequently, these models can handle non-equilibrium ICs and the developed HEM cannot. Second, even in pure phase regions, a small fraction of the respective other (actually non-present) phase is usually included in N-equation model simulations for numerical reasons. The proposed model is able to deal with real pure phases without a need for a lower threshold.

In Figure 2, the results at $t = 0.473$ ms are compared to those of the 5-equation model with infinite relaxation and the same ICs, number of cells, and CFL number. From left to right, an expansion wave in the pure liquid, a contact discontinuity separating the liquid-vapor mixture from the pure vapor, and a shock in the pure vapor can be seen. The pressure plateau belongs to the state at the phase boundary with $p_{\mathrm{sat}}(T = 606\,\mathrm{K}) = 11.7$ bar, where the isentrope exhibits a kink, resulting in a discontinuity in the speed of sound. This leads to the observed wave splitting into the fast-running expansion wave in the pure liquid followed by the continuation of the expansion in the mixture region by the low speed wave, propagating with the mixture speed of sound. The left-running expansion wave (including the evaporation front) propagates with the wave speed of the acoustic mode $u - c$. The speed of sound of the 3-equation model over the length of the tube is depicted in Figure 2c. It has two discontinuities corresponding to the saturated liquid at $x \approx 0.21$ m and saturated vapor at the contact discontinuity.

The results of both models perfectly coincide and also agree with data in the literature [3, 5, 7]. All wave velocities have been predicted equally. This concludes that the wave structure of the 5-equation model with full relaxation (i. e. including stiff source terms) corresponds to that of the 3-equation equilibrium model. That is, by full relaxation in the mixture region, the consequences on the expansion wave coming from the discontinuity of the speed of sound in the 3-equation model, but missing in the 5-equation model, are restored.

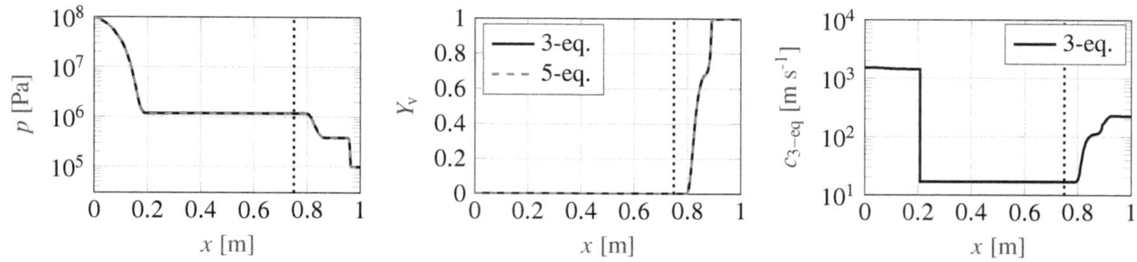

(a) Pressure (b) Vapor mass fraction (c) Speed of sound

Figure 2: Dodecane liquid-vapor shock tube results at t = 0.473 ms. The location of the initial discontinuity is highlighted by a dotted line. The number of cells is 1200 and the time step was specified by CFL = 0 .5. Comparison between the 3-equation HEM and the 5-equation model with infinite relaxation (two left plots) shows perfect agreement. The speed of sound of the 3-equation model exhibits two discontinuities, namely at the right end of the rarefaction wave and at the contact discontinuity (right plot).

2D Model Evaluation

The key assumption of the developed 3-equation HEM is that full equilibrium is reached infinitely fast. An alternative approach is represented by so-called homogeneous relaxation models (HRM) [13]. They were implemented in the context of an Eulerian mixture model by Schmidt et al. [14] for simulations of cavitating nozzle flows in order to include non-equilibrium effects by introducing a time scale θ_0, at which the local vapor mass fraction relaxes towards its equilibrium value based on an empirical correlation, which was fitted for w ater. A comparison of results from the HRM to experimental results from X-ray imaging showed good agreement [15]. However, the evaluation of the time scales is completely empirical and difficult because typically no experiments for real fuels exist. The impact of different choices of the relaxation time scale in terms of θ_0 for the shock tube case considered in the previous subsection is shown in Figure 3a. The range was chosen comparable to that by Saha et al. [16] and the effect of the relaxation time scale parameter can be clearly seen. The larger the relaxation time parameter, the closer are the results to the full equilibrium solution. Even though the effect of the relaxation time parameter is large for the shock tube case, the question of what the physically correct solution is remains. Since it is not possible to answer that question due to the lack of experimental data, it is not possible to give the HRM nor the HEM an edge over the other. Instead, the sensitivity with respect to the relaxation time scale parameter on the results for a more realistic dodecane 2D test case representing typical gasoline direct injection (GDI) conditions as given in Bode et al. [2] is studied here. Results of this test case during early time steps, which should be more sensitive to the relaxation time parameter than time steps during steady-state, are shown in Figure 3b and Figure 3c. It can be seen that vapor is formed at the nozzle inlet edge, which moves towards the orifice. Since the orifice is still locked by liquid, non-condensable gases from outside have not penetrated into the nozzle yet and hydraulic flip has not o ccurred. Comparing both plots, the differences in the gas volume fraction inside the nozzle are negligible. This was also found to be true for all values of C = [0.01, 100] considered here. Thus, the application of the developed 3-equation HEM for complex nozzle simulations seems reasonable.

Besides the simple usage of the developed 3-equation HEM due to its compatibility with single-phase solvers, the computational cost is another advantage. For the shown 2D test case, the computational cost of the 3-equation HEM simulations was about 46% smaller than that of the 5-equation model.

Conclusions

A 3-equation HEM is presented in this work. It was developed for injector flows with non-condensable gases and allows arbitrary combinations of EOSs, which was shown to be important. Improvements of the convexification step and a novel tabulation method in the framework of HEM were summarized. Optimizations with respect to computational

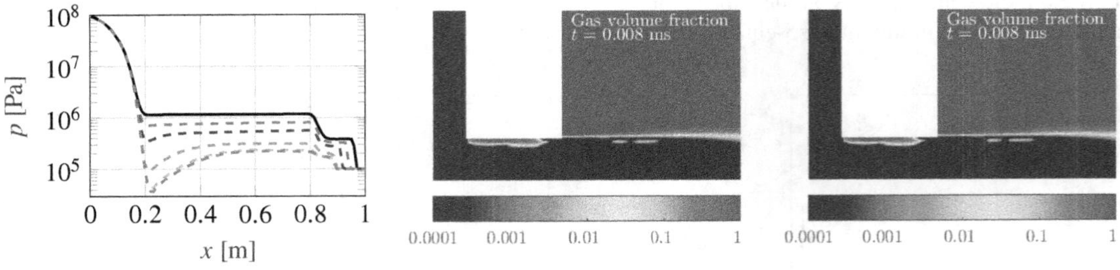

(a) Shock tube (b) 3-eq.: HEM (c) 5-eq.: HRM θ_0

Figure 3: Shock tube results at $t = 0.473$ ms computed with the 5-equation model with infinitely fast relaxation (black, solid line) and various chemical relaxation time scales $C\theta_0$: red - $C = 100$, blue - $C = 10$, cyan - $C = 1$, orange - $C = 0.1$, green - $C = 0.01$, all dotted (left plot). Gas vapor fraction results of the 2D test case with a nozzle height of 0.065 mm and a nozzle length of 0.140 mm computed with the 3-equation HEM and a 5-equation HRM with θ_0 using structured, uniform meshes at dimensionless time $t = 0.008$ (two right plots).

cost were explained. The HEM was found to match results of a 5-equation model with full relaxation and phase change and the reasonability of the equilibrium assumption was shown in the context of a GDI test case.

Acknowledgements

The authors gratefully acknowledge funding by the Cluster of Excellence "Tailor-Made Fuels from Biomass" and Honda R&D.

References

[1] M. Battistoni, D. Duke, A. B. Swantek, F. Z. Tilocco, C. F. Powell, S. Som, Effects of noncondensable gas on cavitating nozzles, Atomization and Sprays 25 (6) (2015) 453–483.

[2] M. Bode, T. Falkenstein, M. Davidovic, H. Pitsch, H. Taniguchi, K. Murayama, T. Arima, S. Moon, J. Wang, A. Arioka, Effects of cavitation and hydraulic flip in 3-hole gdi injectors, SAE International Journal of Fuels and Lubricants 10 (2) (2017) 380–393.

[3] R. Saurel, F. Petitpas, R. Abgrall, Modelling phase transition in metastable liquids: application to cavitating and flashing flows, Journal of Fluid Mechanics 607 (2008) 313–350.

[4] R. Saurel, F. Petitpas, R. A. Berry, Simple and efficient relaxation methods for interfaces separating compressible fluids, cavitating flows and shocks in multiphase mixtures, Journal of Computational Physics 228 (2009) 1678–1712.

[5] A. Zein, M. Hantke, G. Warnecke, Modeling phase transition for compressible two-phase flows applied to metastable liquids, Journal of Computational Physics 229 (8) (2010) 2964–2998.

[6] Y. Wang, L. Qiu, R. D. Reitz, R. Diwakar, Simulating cavitating liquid jets using a compressible and equilibrium two-phase flow solver, International Journal of Multiphase Flow 63 (2014) 52–67.

[7] M. Pelanti, K.-M. Shyue, A mixture-energy-consistent six-equation two-phase numerical model for fluids with interfaces, cavitation and evaporation waves, Journal of Computational Physics 259 (2014) 331–357.

[8] M. Bode, F. vom Lehn, H. Pitsch, Numerical investigation of the effect of dissolved non-condensable gases on hydraulic flip in cavitating nozzles, in: 10th International Cavitation Symposium, Baltimore, USA, 2018.

[9] A. Voss, W. Dahmen, Exact riemann solution for the euler equations with nonconvex and nonsmooth equation of state, Tech. rep., Fakultät für Mathematik, Informatik und Naturwissenschaften (2005).

[10] P. Helluy, H. Mathis, Pressure laws and fast legendre transform, Mathematical Models and Methods in Applied Sciences 21 (04) (2011) 745–775.

[11] O. Le Métayer, J. Massoni, R. Saurel, Élaboration des lois d'état d'un liquide et de sa vapeur pour les modèles d'écoulements diphasiques, International journal of thermal sciences 43 (3) (2004) 265–276.

[12] O. Le Métayer, R. Saurel, The noble-abel stiffened-gas equation of state, Physics of Fluids 28 (4) (2016) 046102.

[13] P. Downar-Zapolski, Z. Bilicki, L. Bolle, J. Franco, The non-equilibrium relaxation model for one-dimensional flashing liquid flow, International Journal of Multiphase Flow 22 (3) (1996) 473–483.

[14] D. P. Schmidt, S. Gopalakrishnan, H. Jasak, Multi-dimensional simulation of thermal non-equilibrium channel flow, International Journal of Multiphase Flow 36 (2010) 284–292.

[15] D. Duke, A. Swantek, Z. Tilocco, A. Kastengren, K. Fezzaa, K. Neroorkar, M. Moulai, C. Powell, D. Schmidt, X-ray imaging of cavitation in diesel injectors, SAE International Journal of Engines 7 (2) (2014) 1003–1016.

[16] K. Saha, S. Som, M. Battistoni, Investigation of homogeneous relaxation model parameters and their implications for gasoline injectors, Atomization and Sprays 27 (4) (2017) 345–365.

© 2018 ASME

10th International Symposium on Cavitation - CAV2018
Baltimore, Maryland, USA, May 14 – 16, 2018 CAV18-05012

An Evaluation of CFD Cavitation Models using Streamline Data

Michael P. Kinzel[1], Jules W. Lindau[1], and Robert F. Kunz[2]

The Pennsylvania State University, University Park, PA, USA
[1] *Applied Research Laboratory* [2] *Department of Mechanical Engineering*

Abstract
In the present work, finite-rate cavitation models common to multi-phase computational fluid dynamics (CFD) are evaluated. These evaluations are based along streamlines extracted from benchmarked CFD. The results include comparative studies of: (1) full Rayleigh-Plesset equation (RPE), (2) simplifications to the RPE, and (3) other cavitation models. Additionally, using this approach, numerical uncertainty is evaluated at a much higher level than previously considered. In the context of developed cavitation, the present assessments elucidate similarities and differences between cavitation models and suggest mesh requirements are more demanding the normally considered.

Introduction

Cavitation models used in CFD stem from a range of physical assumptions and can be roughly categorized into thermodynamic state or bubble dynamics modeling. The present work focuses on the latter, which spans full bubble dynamics and finite-rate cavitation models. Both approaches model rates of nuclei growth and bubble collapse. Finite-rate cavitation models are common practice for multiphase CFD model formulations. Examples of widely used forms include those developed by Merkle *et al.*[1], Kunz *et al.* [2], Sauer and Schnerr [3], Singhal *et al.*[4], Zwart *et al.*[5] and others. One distinction of the models of Singhal, Sauer, and Zwart, with respect to the comparatively ad-hoc models of Merkle and Kunz, is that the former are approximate forms to the RPE. In other words, many finite-rate models for cavity growth and collapse are based on a simplified RPE. Despite their more physical grounding, such approximate RPE models have not displayed more accurate results.

In the context of developed cavitation, the present effort aims to improve the understanding of cavitation models. The analysis is based on benchmarked cavitating CFD results for the cavitating flow over an axisymmetric head form of varying cavitation numbers ($\sigma = \frac{p_\infty - p_{sat}}{0.5\rho_l V_\infty^2}$). Rather than evaluating the fully-coupled system of equations on the entire domain, the process is simplified to an ordinary differential equation (ODE) modeling nuclei growth and collapse along streamlines extracted from the benchmarked CFD solutions. Thus, the CFD pressure fields on the streamlines are used as time-varying forcing functions for the cavitation model formulated as an ODE. Such an approach has the advantage of evaluating and comparing cavitation models in the context of a validated CFD flow field, while avoiding complications and costs associated with the full domain and equation set. Such efforts aim to better understand various models and computational mesh requirements for predicting cavitation using multiphase CFD.

Methods

Cavitation Modeling

Finite-rate cavitation models are often cast into a vapor mass conservation equation form that uses source terms to model gas formation and destruction processes. Such a vapor-mass conservation equation is given by [1]:

$$\frac{\partial \rho_v \alpha_v}{\partial t} + \frac{\partial \rho_v \alpha_v u_i}{\partial x_i} = \dot{S}^+ - \dot{S}^- \tag{1}$$

Here, ρ_v is the vapor density, α_v is the vapor volume fraction, and the source terms, $\dot{S}^+$and $\dot{S}^-$, model bubble growth and collapse, respectively. When combined with a Navier-Stokes-based solver, via fluid properties, this yields a well-established cavitation model approach for CFD [1, 2, 4, 3]. As previously

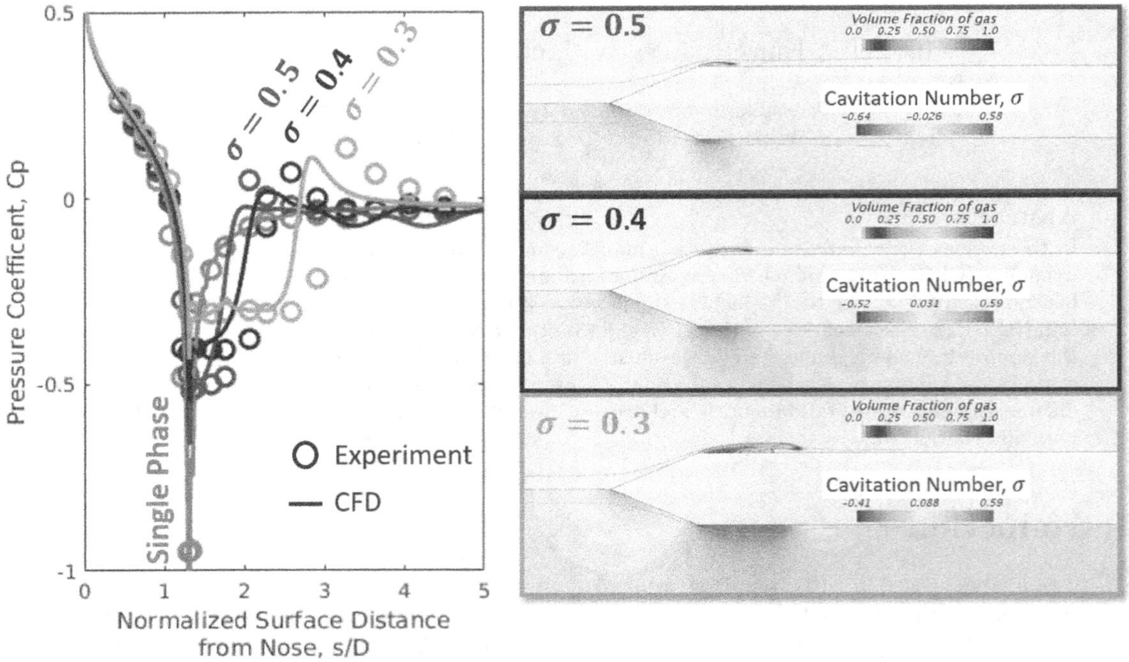

Figure 1: Benchmark comparing CFD to Rouse and Mcknown experiments measuring cavitation over a conical-shaped head form. The left plot indicates C_P predictions (lines) versus experimental measurements (symbols). The contour plots to the left are flow visualizations (upper plots are vapor volume fractions, lower plots are C_P).

mentioned, several cavitation models are formed using an approximate RPE. The full RPE, in an Eulerian reference frame consistent with CFD models, is given as

$$R\frac{D^2R}{Dt^2}+\frac{3}{2}\left(\frac{DR}{Dt}\right)^2+\frac{4\nu_L}{R}\frac{DR}{Dt}+\frac{2S}{\rho_L R}=\frac{p_v-p}{\rho_L}+\frac{p_{G_0}}{\rho_L}\left(\frac{R_0}{R}\right)^{3\gamma}. \tag{2}$$

Here we discuss the model of Singhal *et al.* [4], with the turbulence terms neglected, providing source terms that approximate the RPE as

$$\dot{S}^+ = F_{vap}\frac{\rho_v\rho_l\alpha_l}{\sigma}\sqrt{\frac{2}{3}\frac{max(p-p_{sat},0)}{\rho_l}} \tag{3}$$

and

$$\dot{S}^- = F_{cond}\frac{\rho_v\rho_l\alpha_v}{\sigma}\sqrt{\frac{2}{3}\frac{max(p_{sat}-p,0)}{\rho_l}}. \tag{4}$$

Note that the model clearly displays the Rayleigh approximation to the RPE, i.e. $\frac{dR}{dt}=\sqrt{\frac{2}{3}\frac{p-p_{sat}}{\rho_l}}$. The remaining constants and variables indicate how the model deviates from the Rayleigh approximation (i.e., empirical corrections). Similar to the Singhal model [4], the most of the ubiquitous finite-rate cavitation models may be cast algebraically into a similar form and can be solved as a first-order ODE in time[6]. The Singhal model is applied in the baseline CFD solution to generate streamlines for ODE integration of the Singhal, Sauer, Kunz, and Rayleigh models.

CFD Model

The basis of this effort rely on benchmarked CFD results obtained using an implementation of the Singhal model [6] into Star-CCM+ 12.06 [7]. The model relies on the segregated-flow solver in Star-CCM+ that

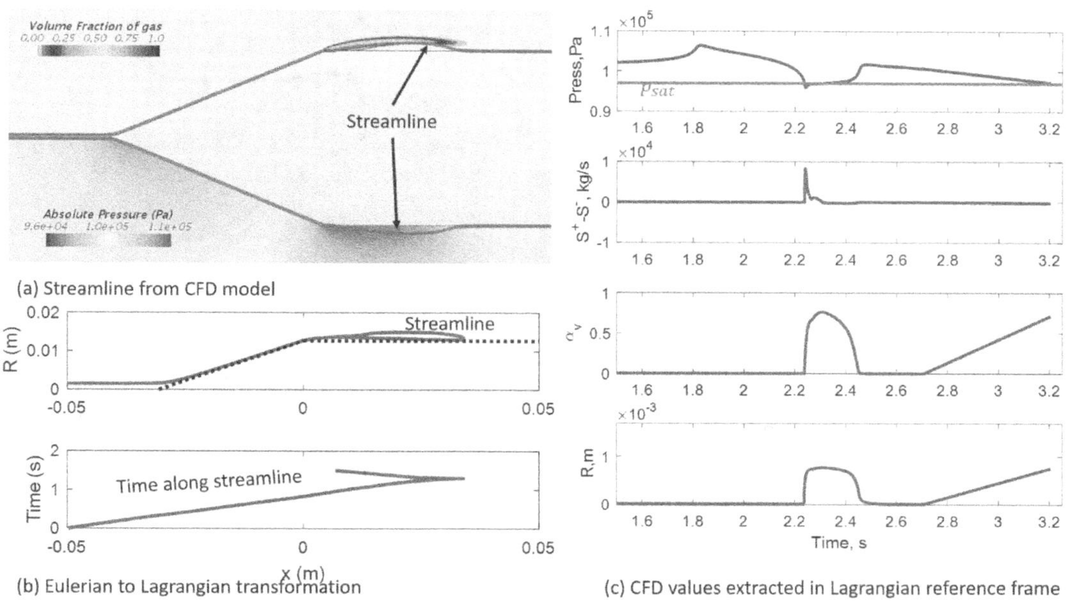

Figure 2: Description of the conversion to a Lagrangian reference frame. The plots display the streamline extraction process to convert to ODE evaluations of cavitation models.

conserves global mass, momentum, and species mass. The numerical scheme is formally second-order accurate in space and the simulations are run using a steady solution approach. A Reynolds Averaged Navier-Stokes (RANS) turbulence model is used based on the $k-\epsilon$ turbulence model in the context of the models discussed in Kinzel *et al.*[8]. Note that the present analyses are expected to be independent of turbulence model and findings should extend to any turbulence model choice.

The physical properties used in this modeling effort are as follows. The model is based on an incompressible fluids with a liquid water phase ($\rho_l = 1000kg/m^3$,$\mu_l = 0.001Pa-s$) and a gaseous phase(with properties similar to air, $\rho_v = 1.2kg/m^3$, $\mu_v = 1.85 \times 10^{-5}Pa-s$). In addition, in the context of the unified model presented in Kinzel et. al [6], the model uses a free stream nuclei with a site density of $400,000m^{-3}$ and a radius, R_0, of $10\mu m$. Lastly, the saturated vapor pressure, p_{sat}, is specified as $p_\infty - \sigma 0.5\rho_l V_\infty^2$ where p_∞ is 1 atmosphere. These properties define the physical inputs relevant to the presented results.

Results from the aforementioned CFD model are compared to experimental measurements of surface pressure from a cavitating flow over a conical-shaped head form from Rouse and Mcknown [9]. The benchmarking exercise is summarized in Fig. 1. In the left plot, are comparisons of the pressure coefficients, i.e.,$C_P = \frac{p-p_\infty}{0.5\rho_l V_\infty^2}$, measured (symbols) to the CFD predictions (lines) at various cavitation numbers. The overall correlation between CFD and experiment is reasonable, hence, extracting streamlines and pressure data are a reasonable starting point to assess cavitation models.

Streamline Analyses

Enabling cavitation-model evaluation using ODE solutions demands that the CFD streamline data are converted to the Lagrangian reference frame. The process demands temporal data on the nuclei as the transport along the streamline, hence, time along the streamline must be computed as

$$t_p = \int_0^p \frac{ds}{|V_p|}. \tag{5}$$

Here, p is a point on the streamline at time t_p that is relative to the streamline start time. Additionally, the velocity magnitude, $|V_p|$, and a differential distance ds, i.e., the differential distance along the

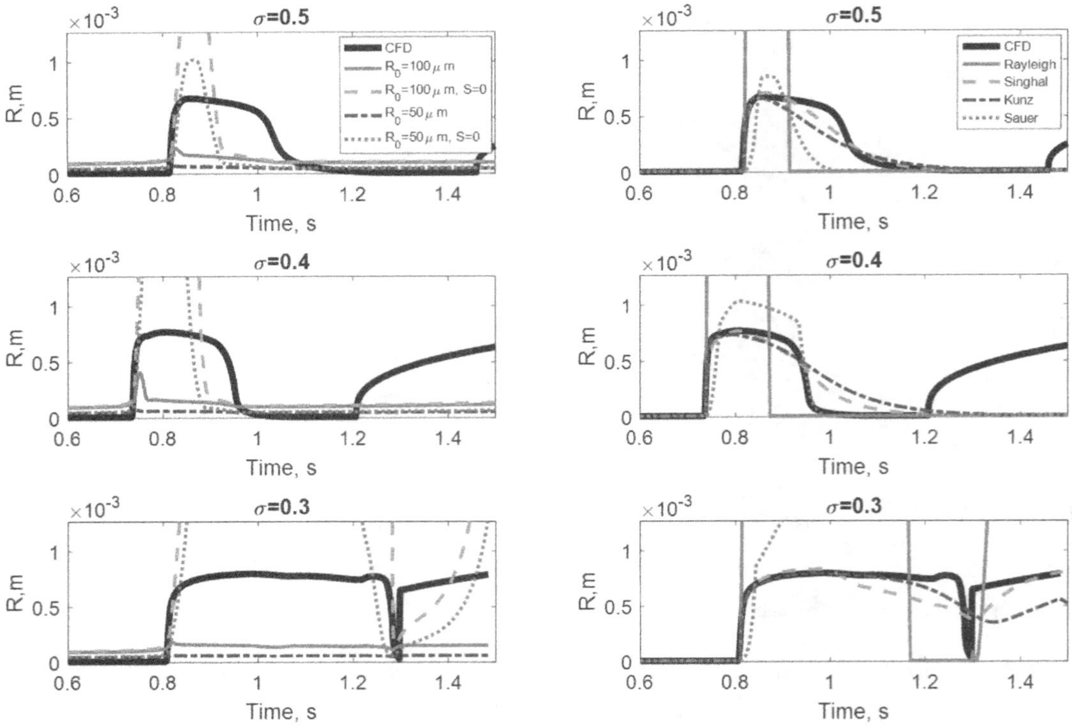

(a) Evaluation of RPE (b) Cavitation-model comparisons

Figure 3: Evaluation of the various cavitation models along a streamline. Part (a) indicates the RPE solutions whereas part m(b) compares results from various finite-rate models.

streamline, are required. Sample results and extraction of the data are depicted in Fig. 2. The pressure and velocity are extracted along the streamline providing the necessary data for the transformation and the ODE forcing function (i.e., the pressure-time curve indicated the upper-right plot in Fig. 2). Such a pressure history drives the cavitation model, which can be compared to the CFD cavitation model, vapor content, and effective radius in the second through forth plots to the right of Fig. 2. Note that the CFD radius is computed using an approximation of: $R = \left(\frac{\alpha_v}{4/3\pi N_b}\right)^{1/3}$.

Results and Discussion

We now compare various cavitation models to the RPE and evaluate mesh dependencies associated with cavity gas content. Results are expected to yield insight into cavitation models, their performance, and paths to improvement.

RPE Solution

First, consider the application of the RPE in the context of these developed cavitation cases. Results are plotted in Fig. 3 (a) for three different σ values and are compared to CFD Singhal-model results. Note that these RPE results assume free-stream nuclei having radii of $R_0 = 50$ or $100\mu m$, which are both evaluated with and without surface tension. Results indicate that the full RPE, with surface tension and exceedingly large nuclei $10\times$ larger than assumed in the CFD model, do not correspond to the benchmarked CFD. Specifically, the RPE predicts that nuclei growth is minimal when passing into an already ruptured cavity void. Alternatively, and similar to finite-rate model formulations, neglecting surface tension indicates that the RPE predicts cavitation similar to the CFD. Such results are not intended to suggest that full RPE is an invalid model. It does, however, suggest that the RPE is not directly applicable in the present analysis. We hypothesize that the full RPE may not be directly applicable to developed cavitation due to it not considering vapor addition.

Cavitation Model Comparison

Now consider comparing various cavitation models along these streamlines. This may be considered an extension of the efforts from Kinzel *et al.* [8] that compared cavitation models using artificial pressure distributions. In Fig. 3 (b), are comparisons of results from the (1) Rayleigh, (2) Singhal, (3) Kunz, and (4) Sauer models for three σ values. Recall that the reference CFD result was generated with the Singhal model.

First notice the discrepancies between the CFD- (CFD) and ODE-Singhal (Singhal) models in Fig. 3 (b). The observed discrepancies are expected to be a result of the ODE-model treating cavitation processes as a pure advection process, which neglects the vapor diffusion, limiters, and other numerical factors present in the CFD. In general, the model results are comparable suggesting verifying that the streamline method can yield insight.

Model-to-model comparisons in Fig. 3 (b) suggests good agreeable between the cavitation models. Note that, for each model, the cavitation constants are consistent across all σ values. The results form the Singhal- and Kunz-ODE (Kunz) models correlate well with the Kunz model constants set to $C_{evap} = 2.0$, $C_{e,ref}$, and $C_{cond,1} = 1.25C_{c,ref}$. This is a slight modification from the analysis[6] suggesting $C_{evap} = 1.15$, $C_{e,ref}$, and $C_{cond,1} = 10.0C_{c,ref}$. The Singhal and Kunz models do deviate from the Sauer model, and when compared to the RPE, the the Sauer model better correlates with the RPE model without surface tension. Note that the Sauer model constants deviated significantly from those predicted in Kinzel *et al.*[6]. Specifically, the present Sauer-model results use $C_{evap} = \frac{\rho_l}{40{,}000\alpha_{v,0}}C_{e,ref}$, and $C_{cond,2} = 0.0065\frac{\rho_l^2}{\rho_v}C_{c,ref}$ as compared to the previous analysis[6] suggesting $C_{evap} = 1.15C_{e,ref}$, and $C_{cond,2} = \frac{\rho_l^2}{\rho_v}C_{c,ref}$. This indicates that the previous analysis[6] should be refined for the Sauer model. Lastly, the Rayleigh model indicates much faster evaporation and condensation rates as the model indicates it models too rapid of growth. Overall, the general character of the Singhal and Kunz models are very similar, some deviation is observed with the Sauer model, and the Rayleigh model indicates a much more violent cavitation process.

Effect of Mesh

The next analyses indicates that cavitation prediction has a strong mesh resolution dependency that is exacerbated with lower σ. The analysis focuses on the Singhal-ODE solution and only focuses on cavitation generation (which does not necessarily correlate to loads). In Fig. 4 (a), the results with different time-step sizes are compared with respect to refining from a Δt of unity which corresponds to the CFD mesh (depicted in Fig. 4 (b)). Such a mesh is considered to be a standard cavitation resolution. Refined resolutions are obtained through ODE integration to shorter times steps. In evaluating Fig. 4 (a), it is clear that cavity generation is highly dependent on the resolution. Such an observation indirectly relates to CFD mesh resolution suggesting CFD meshes may need to be significantly refined to evaluate cavitation dynamics.

Summary

In this work, flow-field data extracted on streamlines from a CFD solution are used to evaluate cavitation models. In comparing CFD results to results from such streamline analyses, a reasonable correlation was observed indicating the usefulness of the approach and greatly simplifying model comparisons. When cavitation models are analyzed, the present results suggest strong similarity in various cavitation models. However, there was one exception. The full RPE with surface tension deviated significantly indicating that, for the given pressure fields, nuclei could not effectively supply gas volume for these developed cavitation conditions. This may suggest that nuclei expansion is not an important process for developed cavitation. In addition, we also observed a significant sensitivity in the development of a cavity associated with the computational mesh, suggesting that numerical uncertainty is large cavitation models. Such insights are useful in that they provide insight to improve cavitation models.

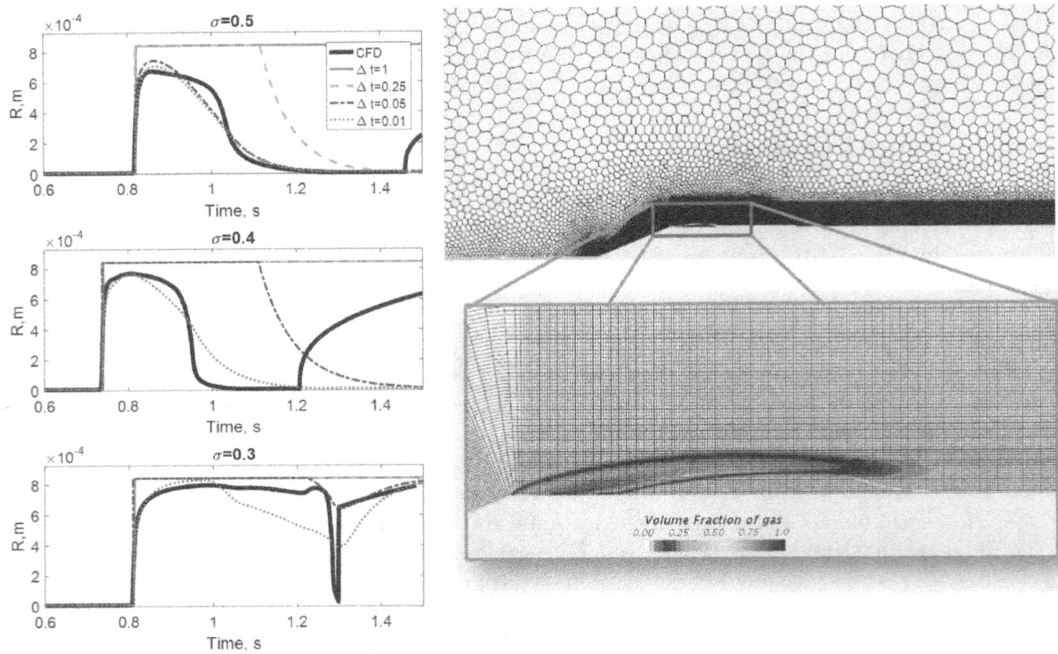

(a) Effective mesh sensitivity (b) Computational mesh and detail near cavity

Figure 4: Evaluation of the discretization sensitivity. Discretization sensitivities are presented in part (a) where a Δt of unity corresponds to the CFD computational mesh used provided in part (b).

References

[1] Charles L Merkle, J Feng, and Phillip E O Buelow. Computational modeling of the dynamics of sheet cavitation. In *3rd International symposium on cavitation, Grenoble, France*, volume 2, pages 47–54, 1998.

[2] Robert F. Kunz, David A. Boger, David R. Stinebring, Thomas S. Chyczewski, Jules W. Lindau, Howard J. Gibeling, Sankaran Venkateswaran, and T. R. Govindan. Preconditioned Navier-Stokes method for two-phase flows with application to cavitation prediction. *Computers & Fluids*, 29(8):849–875, 2000.

[3] Jürgen Sauer and G. H. Schnerr. Development of a new cavitation model based on bubble dynamics. *Zeitschrift für Angewandte Mathematik und Mechanik*, 81(S3):561–562, 2001.

[4] Ashok K Singhal, Mahesh M Athavale, Huiying Li, and Yu Jiang. Mathematical Basis and Validation of the Full Cavitation Model. *Journal of Fluids Engineering*, 124(3):617–624, 8 2002.

[5] Philip J Zwart, Andrew G Gerber, and Thabet Belamri. A Two-Phase Flow Model for Predicting Cavitation Dynamics. In *ICMF 2004 International Conference on Multiphase Flow*, 2004.

[6] Michael P Kinzel, Jules W Lindau, and Robert Kunz. A Unified Model for Cavitation. In *FEDSM17 2017 ASME*, Waikoloa, Hawaii, USA, 0. ASME.

[7] Siemens. Star CCM+ 12.06, 2017.

[8] M P Kinzel, R F Kunz, and J W Lindau. An Assessment of CFD Cavitation Models Using Bubble Growth Theory and Bubble Transport Modeling. In *International Symposium on Transport Phenomena and Dynamics of Rotating Machinery (ISROMAC 2017)*, Maui, Hawaii, USA, 2017.

[9] Hunter Rouse and John McNown. Cavitation and pressure distribution: head forms at zero angle of yaw. *University of Iowa Studies in Engineering*, 1 1948.

© 2018 ASME

An Experimental Investigation of Artificial Supercavitation with Variation of the Body Shape

[1]Byoung-Kwon Ahn*; [1]So-Won Jeong; [1]Sang-Tae Park

[1]*Dept. of Naval Architecture and Ocean Engineering, Chungnam National University, Daejon, Korea*

Abstract

In this study, artificial super-cavitating flow around a three-dimensional body has been investigated. Experiments were conducted in a cavitation tunnel of the Chungnam National University that has a special equipment to remove injected air for carrying out ventilation tests. Behaviors and dimensions of the artificial supercavity depending on the shape configuration of the body were examined. Pressures inside the supercavity were measured by using an absolute pressure transducer and the cavity formation was observed by using high-speed visualizations. The results show that several parameters defining the body geometry and the combination of the cavitator have a significant influence on the development of the supercavity.

Keywords: cavitation tunnel experiment; cavitator; natural supercavitation; artificial supercavitation

Introduction

In recent years, supercavitation has again attracted interest and attention for practical advantages in drag reduction of underwater vehicles. As a submerged object travels at very high speeds and a vaporous cavity naturally grows to cover the entire body. Supercavitation can also be formed by injecting gas into the low pressure regions behind the front part of a body, cavitator, which is termed as a ventilated or an artificial supercavity. Both types of supercavities appear to be alike at the same cavitation number conditions. However, the main difference between the two is the effect of gravity. Artificial supercavitation can be generated at relatively lower speeds than the natural case for given conditions, thereby making Froude number effects significant. Most previous experimental researches on artificial supercavitating flows were performed for the case of cavitators alone [1, 2, 3, 4, 5]. In this work, we investigated the effects of the combination of body and cavitator on the supercavity dimension and development. Here, three different cavitators including disk, cone and ellipsoidal shape and also three different fore body angles were considered.

Experimental apparatus and test models

A vaporous cavity is naturally formed on the body by increasing the flow speed, or by decreasing the ambient pressure. Natural cavitation is governed by the parameter referred to as the cavitation number, $\sigma_n = (P_\infty - P_v)/\left(\frac{1}{2}\rho U_\infty^2\right)$. As the cavitation number is gradually decreased, a partial cavity grows longer and transitions into a supercavity. Based on the same principle, a supercavity can also be attained by injecting gas into the low pressure regions and making cavities to envelope the body. Artificially created supercavities mainly depend on the amount of injected air and the pressure inside the cavity. Especially, the influence of gravity causes the cavity to deform and its tail to go up. The phenomenon of artificial supercavitation is characterized by the artificial cavitation number, $\sigma_c = (P_\infty - P_c)/\left(\frac{1}{2}\rho U_\infty^2\right)$, and other parameters such as air entrainment coefficient, $C_q = \dot{Q}/\left(U_\infty d_c^2\right)$, and Froude number, $F_n = U_\infty / \sqrt{g d_c}$. The key parameter of artificial supercavitation is pressure inside the cavity, which is determined by injected gas rate. Contrary to the case of unbounded flow, when the supercavity is generated in a flow confined by the walls in tunnel experiments, blockage effects should be considered. Figure 1 shows the CNU cavitation tunnel employed in this experimental work. The test section is 100 mm square and 1,400 mm long and maximum flow speed is 20 m/s. Cavitators and the overall shape of the test model are also shown in Figure 1. Disk type including two different shapes, cone and ellipsoidal type cavitators can be combined with different fore body parts. Dimensions of the test models

*Corresponding Author, Byoung-Kwon Ahn: bkahn@cnu.ac.kr

named M-I, M-II and M-II are shown in Table 1. Here, d_c, S_L, B_D, B_L are cavitator diameter, fore body (shoulder) length, body diameter and body length respectively.

The air is supplied through a pipe inside the body from the outside of the tunnel, and it is injected through 1.5 mm four holes behind the cavitator. Using an absolute pressure sensor, we measured the pressure inside the cavity at 30mm apart from the cavitator. We also recorded the growth process of the supercavity by using a high-speed camera placed at the side of the test section.

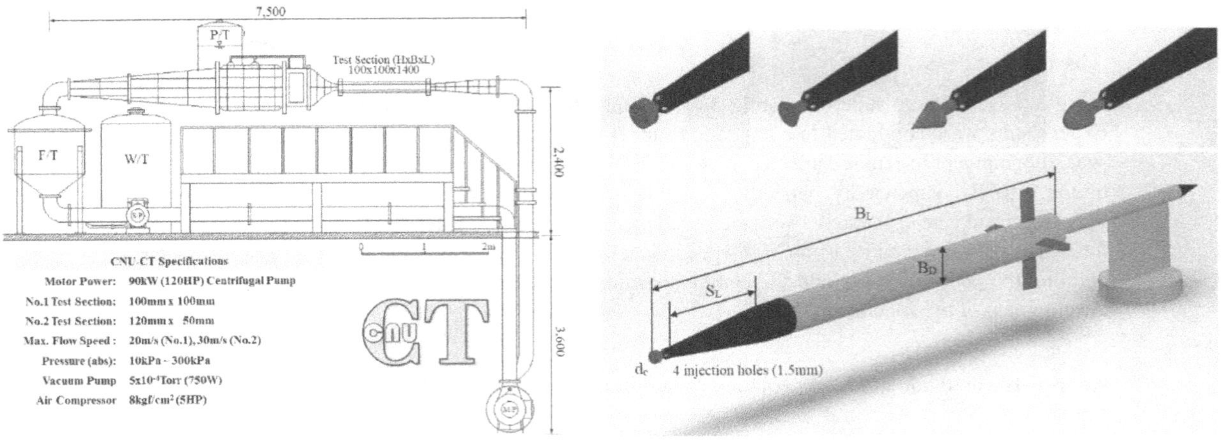

Figure 1. CNU cavitation tunnel (left) and test model and cavitators (right)

Table 1. Specifications of the test models

Model name	d_c [mm]	S_L [mm] (angle)	B_D [mm]	B_L [mm]	B_D /d_c	S_L /d_c
M-I	8.6	82.5 (6.1°)	25	326	2.9	9.5
M-II	9.4	160 (3.1°)	25	326	2.7	17.0
M-III	10.5	180 (2.9°)	25	326	2.4	17.1

Results and discussions

Figure 2 shows a typical formation of the artificial supercavity occurring from the test model, M-I and M-II, with respect to different air entrainment coefficients, C_q. Initially a short foamy cavity appears and grows as injected air increases. This foamy cavity is closed by an unstable wake flow of the cavitator and the rear part of the cavity is re-entrained by the outer flow. At a critical value for the air entrainment coefficient, the cavity varies rapidly from a foamy cavity to a clear cavity. As the injection rate increases, the cavitation number reaches to a minimum value the clear supercavity occurs and maintains the shape even if the injection rate decreased. This mechanism of the hysteresis effect has been reviewed by many experts [3, 6, 7]. Figure 3 shows gravitational effects on M-III model according to the Froude number. The supercavity is more straightforward as the Froude number increases by an increase in velocity. It is easy to figure out this effect by observing the location of the closure of the supercavity. Figure 4 shows the dimension of the supercavity generated in each model. Comparing with the case without body, it shows that the cavity length is greatly influenced by the presence of the body, and the smaller the angle of the fore body the larger cavity. In Figure 5 (left), cavitation numbers are plotted against Froude numbers for each model. As the Froude number increases the cavitation number continually decreases until it reaches the minimum value dictated by blockage. Figure 5 (right) shows non-dimensional cavity length according to the fore-body angle, which is corresponding to the results of Figure 4. Figure 6 shows high-speed images of the development of the supercavity from the disk and cone type cavitator for 0.08 second. Supercavity grows faster in cone type cavitator rather than that of the disk type cavitator, but the cavity dimension is larger in the disk type cavitator.

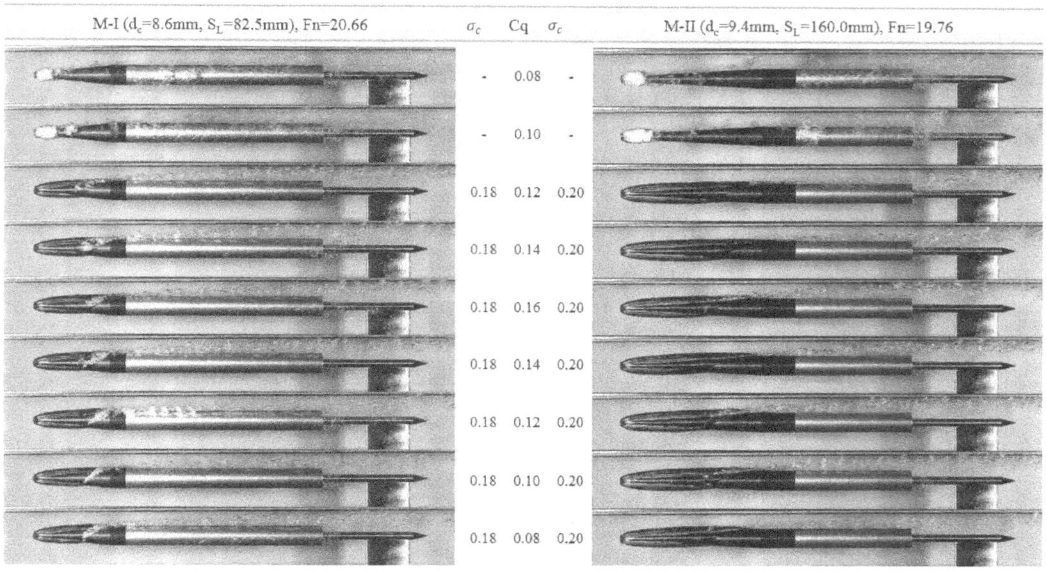

Figure 2. Supercavity formation corresponding to M-I model (left) and M-II model (right)

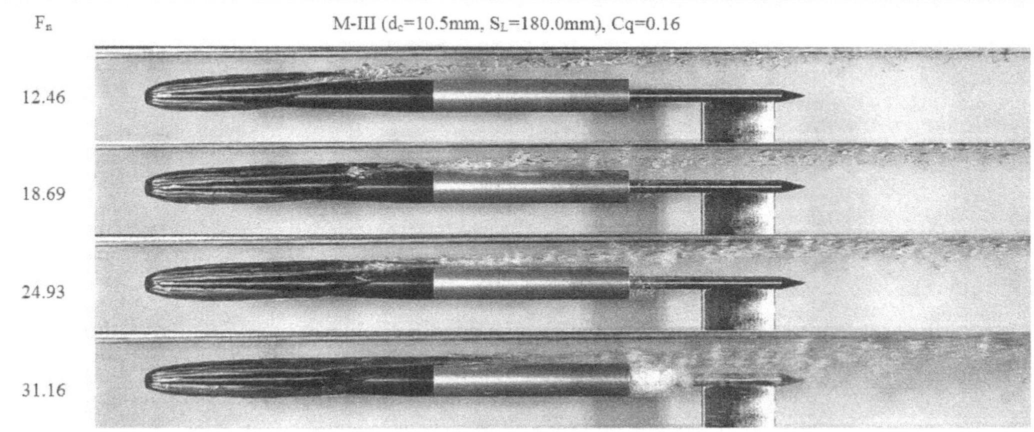

Figure 3. Gravity effects of M-III model

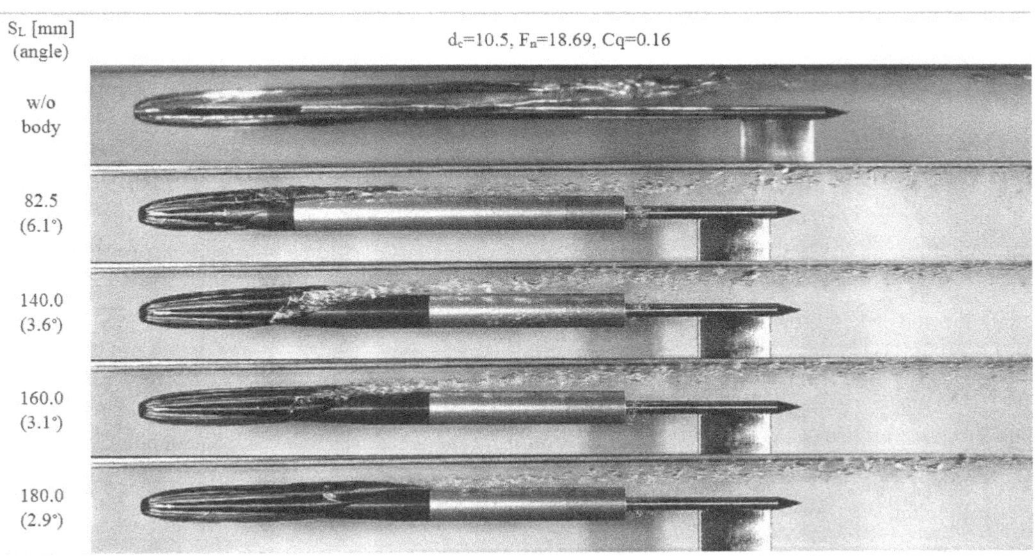

Figure 4. Cavity dimensions according to the fore body angle

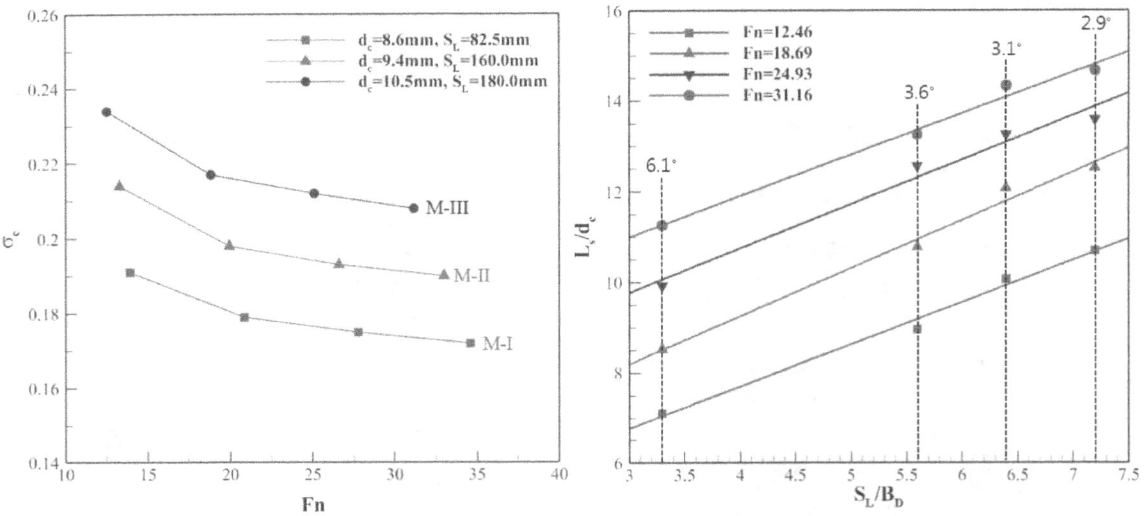

Figure 5. Cavitation number against Froude number for each model (left) and non-dimensional supercavity length according to the fore body angle (right)

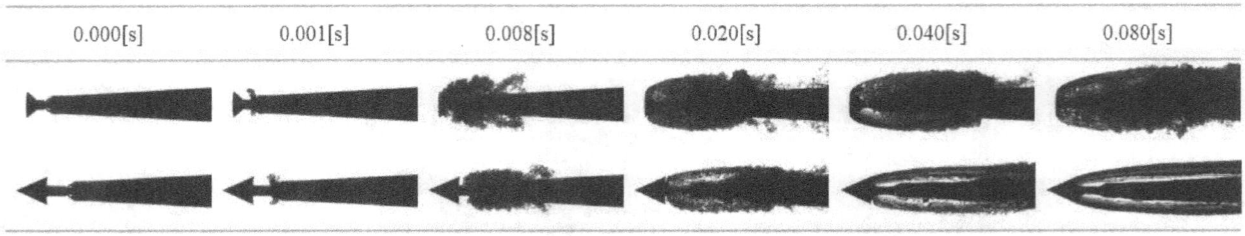

Figure 6. Comparison of the supercavity formation of the disk type (top) and cone type cavitator (bottom) for 0.08 second

Concluding remarks

On the basis of the previous experiment of artificial supercavitation in a disk type cavitator without the body, in this study we investigated the effects of the body behind the cavitator on the supercavity dimension. Three models with different body shape combinations were tested in a cavitation tunnel. Typical characteristics of the artificial supercavity such as the gravity effect and hysteresis effect were observed for each model. Using high-speed visualizations, supercavity dimensions were measured according to the fore body angle, and also the formation process of the supercavity in different cavitators was examined. In conclusion, it was found that the combination of the fore body shape greatly affects the development of the supercavity.

References

[1] Ahn, B.-K., Jeong, S.-W., Kim, J.-H., Shao, S., Hong, J. and Arndt, R. E. A. (2017). *An experimental investigation of artificial supercavitation generated by air injection behind disk-shaped cavitator,* Int. Journal of Naval Architecture and Ocean Engineering, 9, pp. 227-237.

[2] Karn, A., Arndt, R. E. A., & Hong, J. (2015). *Dependence of supercavity closure upon flow unsteadiness*. Experimental Thermal and Fluid Science, 68, pp. 493-498.

[3] Kawakami, E., and R.E.A. Arndt. (2011). *Investigation of the Behavior of Ventilated Supercavities*. Journal of Fluids Engineering, 133(9), pp. 1-11.

[4] Knapp, R.T., Daily, J.W., and Hammitt, F.G. (1979). *Cavitation*. University of Iowa, pp.188-201.

[5] Self, M., and Ripken, J.F. (1955). *Steady-state Cavity Studies in a Free-jet Water Tunnel*. 47, St. Anthony Falls Hydr. Laboratory.

[6] Semenenko, V.N. (2001). *Artificial Supercavitation, Physics and Calculation. RTO AVT Lecture Series on Supercavitating Flows*, Brussels, Belgium.

[7] Spurk, J.H. (2002). *On the Gas Loss from Ventilated Supercavities*. Acta Mechanica, 155, pp. 125-135.

© 2018 ASME

An Experimental Study of Cavity Shedding Mechanisms for Unsteady Cloud Cavitation

[1,2]Qin Wu*; [2]Guoyu Wang; [3]Mohamed Farhat; [2]Biao Huang; [1]Shuliang Cao

[1]Tsinghua University, Beijing, 100084, China; [2]Beijing Institute of Technology, Beijing, 100081, China;

[3]Ecole Polytechnique Fédérale de Lausanne, 1007 Lausanne, Switzerland

Abstract

The objective of this work is to shed light on different mechanisms associated to the cloud cavitation instability and corresponding vibration characteristics. Results are presented for a modified NACA66 hydrofoil tested in the EPFL high-speed cavitation tunnel. The high-speed digital camera and the single point Laser Doppler Vibrometer are applied to analyze the unsteady cavitating flow structures and corresponding structural vibration characteristics. For the partial cloud cavitation, it occurs when the maximum cavity length is less than 1.0c with Strouhal number of 0.3-0.4. As for the trailing cloud cavitation, the Strouhal number decreases to 0.1-0.15 with the cavity extending to the downstream of the hydrofoil, accompanied with intense vibration and noise. Compared with the cavity shedding dynamics in the partial cloud cavitation, it has a more complex cavitating flow patterns for the trailing cloud cavitation. When the shedding cloud cavity reaches to the rear part of the hydrofoil, it collapses and induces a strong shock wave, which moves backward. As a result, the main cavity shrinks and may even entirely disappears on the suction side. After that, the cavity begins to form and develop again in the next period. Hence, a significant reduction of the shedding frequency and much larger vibration velocities can be observed for the trailing cloud cavitation.

Keywords: cloud cavitation; cavity shedding; flow-induced vibration; experimental study

Introduction

Cavitation generally occurs when the local fluid pressure reduces to the saturated vapor pressure and consequently the gas filled or gas and vapor filled bubbles are formed. For a long time, a large number of experiments [1-4] and numerical simulations [5-7] have been conducted to study the cavitating flow structures and dynamics on hydrofoils. As mentioned in the previous researches, the temporal periodicity may occur as a result of the unsteady break-off and periodical shedding of the large cloud cavities, which will generate significant pressure fluctuations and aggravates the flow-induced vibration of structures, so that it can lead to many problems and become a primary concern for the hydraulic machine and marine engineering [8-10]. Arndt et al. [11-12] investigated the instability of partial cavitation noting the cavity patterns, maximum amplitude in lift and noise characteristics. They found that the vertical extent of the cavitating flow is less at lower angle of attack. Leroux et al. [13-14] observed two distinct cavity self-oscillation dynamics and found that the re-entrant jet was mainly responsible for the cavity breakdown and the strong pressure wave generated by the cloud cavity collapse is responsible for the instabilities. Seo and Lele [15] investigated the partial cavitating flow and cloud cavitating flow with focus on the pressure wave generated by collapse of cloud cavity. The results showed that the pressure pulse generated by the collapse of bubble cloud and the flow-blockage effect caused by a large cloud cavity are responsible for the shifting of cloud cavitation dynamics.

The objective of this work is to analyze different mechanisms associated to the cloud cavitation instability, in particular, to further explore the stable partial cloud cavitating and the unstable trailing cloud cavitating flow regimes. In this paper, the experimental setup and the measuring techniques are first presented. Then the typical vibration characteristics along with corresponding cavitation patterns are given. Finally, the conclusions are summarized in the last section.

Experimental setup and methods

The experimental studies are conducted in the high-speed cavitation tunnel in EPFL [16], as shown in Fig. 1. The test section is 0.7m (length)×0.07m (width)×0.19m (height), with the free stream turbulence intensity around 2%. As sketched in Fig. 2, the test hydrofoil is a modified-NACA66 made of stainless steel, which has a uniform cross-section of shape with chord length c=0.1m, the span b=0.15m and a maximum thickness-to-chord ratio of 12%. The flow conditions are set based on the Reynolds number $Re=U_\infty c/\nu=1.0\times10^6$, where U_∞ is the free stream velocity, ν is the dynamic viscosity of the liquid, and the cavitation number $\sigma=(p_\infty-p_v)/(0.5\rho_1 U_\infty^2)$, where p_∞ is the pressure in the test section, p_v is the saturated vapor pressure at ambient temperature, ρ_l is the liquid density.

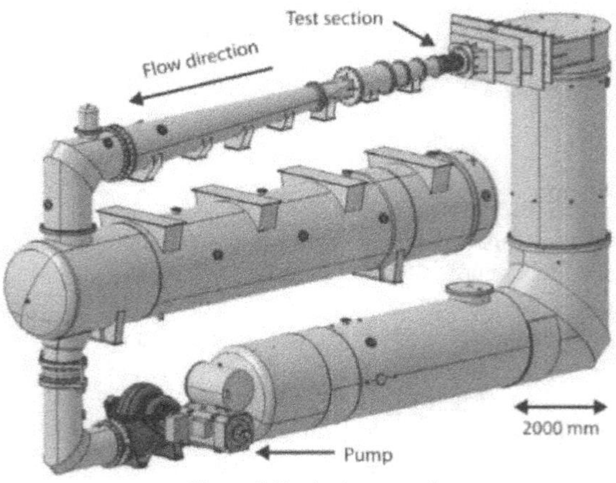

Figure 1 Cavitation tunnel

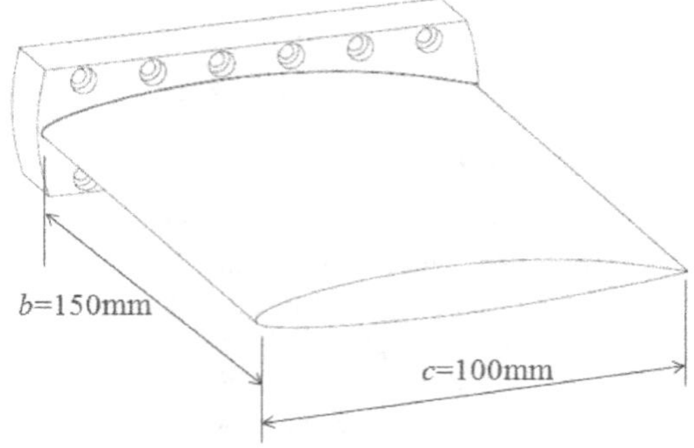

Figure 2 NACA66 hydrofoil

The cavitation patterns are visulized by a Photron Fastcam SA1.1 digital camera from the front (704×400 pixels) and top (576×480 pixels) of the section at a sampling rate of 20000 frames per second. The flow-induced vibration is measured with a laser Doppler vibrometer, Polytec PDV100. The sample frequency of 5kHz is adopted in the present work and the smallest possible measurement range is used to optimize the signal-to-noise ratio. The measurement point is focused on the pressure side of the hydrofoil at the position of x/c = 0.75 and z/b = 0.6, where the vibration signal is strong enough without interfere of the cavity formed on the suction side and wake region of the hydrofoil. The data acquisition system is controlled by a NI-Labview programming and a sampling frequency of 5kHz is adopted in this work. For frequency spectral analysis, Fast Fourier Transform is applied to 30 equal segments of the time signals with 50% overlap between the adjacent segments, followed by averaging of the spectrum, so the time and frequency resolutions are 0.0002s and 0.5Hz respectively.

Results and discussions

As to the regime of the cloud cavitation, the unsteady breakdown and shedding of the cavity occurs violently, which may lead to strong dynamic instabilities. The visualization of two typical cavitation development, trailing cloud cavitation (α=6°, σ=1.2, $\sigma/2(\alpha-\alpha_0)$=4.58) and partial cloud cavitation (α=8°, σ=1.5, $\sigma/2(\alpha-\alpha_0)$=4.52), is shown in Figs. 3 and 4. Figure 5 shows the evolution of the cavity lengths in two consecutive cycles, which are extracted from the recorded images. The cavity growth rate is changing through the whole cavitation developing process. The periodical formation and shedding of large-scale cloud cavity can be captured in both cases, but compared to the cavity patterns for the partial cloud cavitation, the cavity shedding process for the trailing cloud cavitation is more complex and characterized by a lower frequency.

For the trailing cloud cavitation, at first, the partial sheet cavity forms and develops on the suction side of the hydrofoil, along with formation of the re-entrant jet at the rear end of the cavity, as shown in Figs. 3(a)-(b), which causes the oscillating of the cavity length in the range of 0.2c, as shown in Fig. 5(a). When the cavity develops to 0.9c-1.0c, the trailing end of the cavity becomes unstable and the latter part of the cavity from 0.5c to 0.9c rolls into cloud cavity cluster due to the interaction between the re-entrant jet and the cavity interface, cutting the cavity length into 0.5c, as shown in Figs. 3(c)-(d) and Fig. 5(a). With the shedding of the cloud cavity cluster, the residual attached cavity continue growing with a larger cavity growth rate, as shown in Figs. 3(d)-(e) and Fig. 5(a), because the cavity is lifted away from the near surface by the re-entrant jet and the frictional resistance reduces. As the shedding cloud cavity moves downstream of the hydrofoil, it collapses due to the pressure gradient between the flow field and the inner part of the cavity, as shown in Fig. 3(e). Almost at the same time, the residual cavity stops to grow and on the contrary, it shrinks dramatically and even disappears entirely on the suction side, as shown in Figs. 3(f)-(h), corresponding to the sudden decrease of cavity length seen in Fig. 5(a).

As for the partial cloud cavitation, the partial sheet cavity develops from the leading edge of the foil to the position of x/c=0.8, as shown in Figs. 4(a)-(f) and Fig. 5(b). During the cavity developing process, the re-entrant jet forms at the rear end of the cavity and moves from both side of the trailing edge of the cavity towards the center of the leading edge of the cavity. The collision of these two bundles of upward flow cuts off the downstream part of cavity and induces the shedding of the cloud cavity, generating a wedge-shape hole at the rear part of the remaining sheet cavity, as shown in Fig. 4(g).

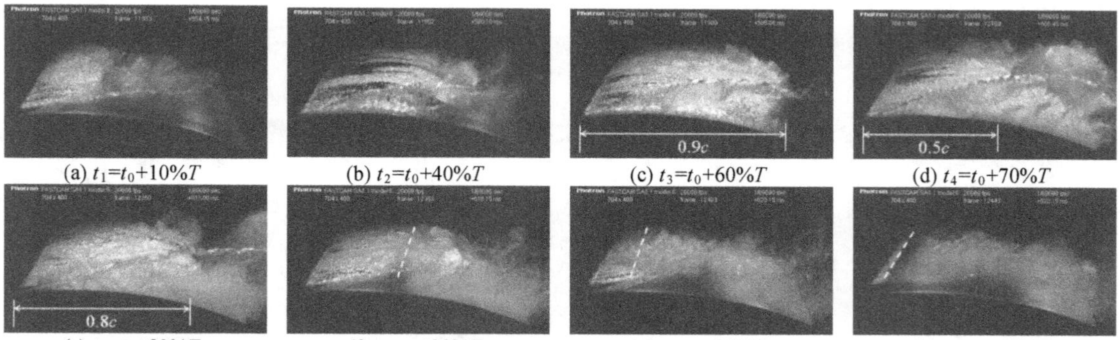

(a) $t_1=t_0+10\%T$ (b) $t_2=t_0+40\%T$ (c) $t_3=t_0+60\%T$ (d) $t_4=t_0+70\%T$

(e) $t_5=t_0+80\%T$ (f) $t_6=t_0+90\%T$ (g) $t_7=t_0+92\%T$ (h) $t_8=t_0+95\%T$

Figure 3 Observations of the cavitation pattern (trailing cloud cavitation)

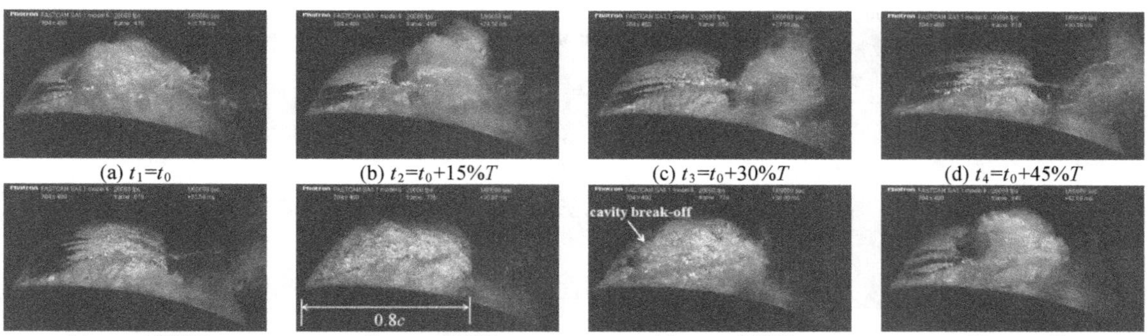

(a) $t_1=t_0$ (b) $t_2=t_0+15\%T$ (c) $t_3=t_0+30\%T$ (d) $t_4=t_0+45\%T$

(e) $t_5=t_0+60\%T$ (f) $t_6=t_0+80\%T$ (g) $t_7=t_0+90\%T$ (h) $t_8=t_0+T$

Figure 4 Observations of the cavitation pattern (partial cloud cavitation)

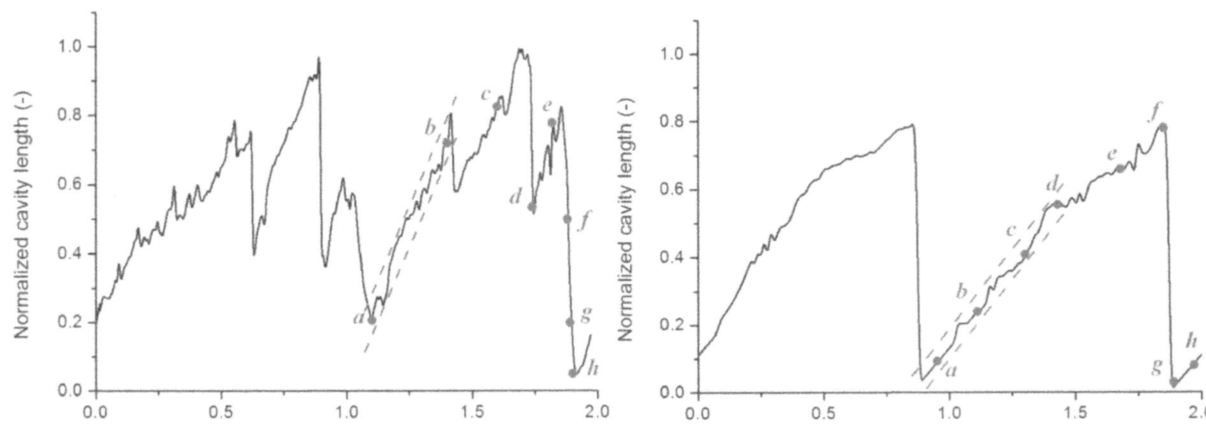

(a) trailing cloud cavitation (b) partial cloud cavitation

Figure 5 Evolution of the cavity lengths

Figure 6 shows the time evolution of the vibration velocity for the cases with trailing cloud cavitation and partial cloud cavitation. Hibert transform is applied to determine the envelop of the vibration signals. The vibration signals for both cases exhibit marked periodic variation, but the vibration amplitude for the trailing cloud cavitation is much larger. Also noted in Fig. 6, for the trailing cloud cavitation, each period of oscillation consists of distinct two parts, small amplitude fluctuations accompanied with the increase of the envelop of the vibration signals and large amplitude fluctuations together with the decrease of the vibration signals envelop curve. While for the partial cloud cavitation, the vibration velocity goes up and down periodically with small amplitude fluctuations. Based on the Hilbert transform, the spectra for these two cavitating flow regimes is shown in Fig. 7. The dominant frequencies for the trailing cloud cavitation and partial cloud cavitation are 14Hz and 44Hz respectively.

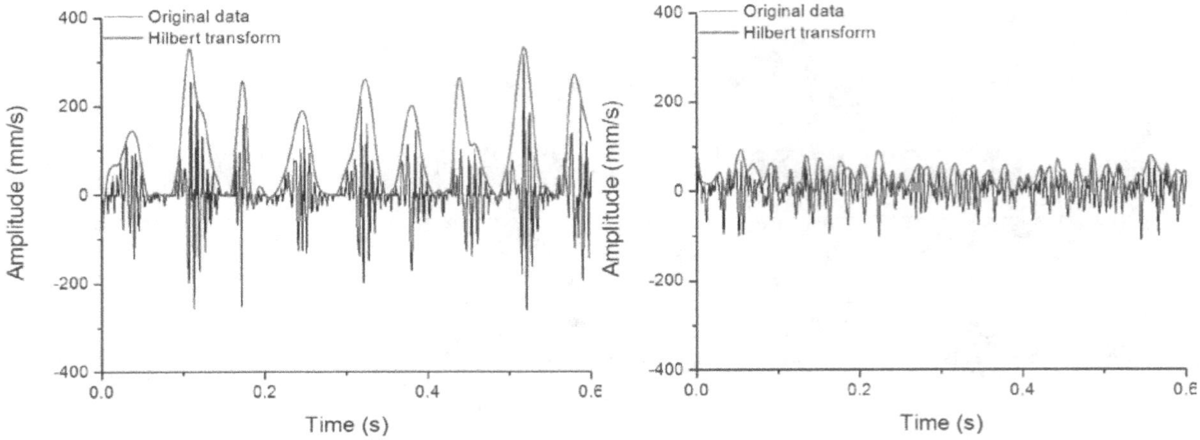

(a) trailing cloud cavitation (b) partial cloud cavitation

Figure 6 Time evolution of the vibration velocities

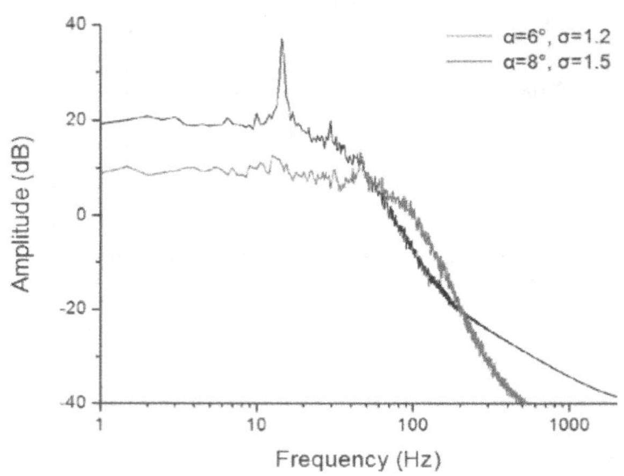

Figure 7 Spectrum analysis of the vibration velocities

Comparing the shedding process occurred in both cases, it can be seen that there are two competing mechanism acting on the cavity shedding. One is the re-entrant jet due to the pressure gradient at the rear end of the cavity, as seen in Fig. 3(b) and Fig. 4(d). The other is the shock wave induced by the collapse of the cloud cavity, as seen in Figs. 3(f)-(h) and Figs. 4(d)-(e). When the maximum cavity length is larger than 1.0*c*, the collapse of the cavity occurs in the downstream of the hydrofoil and the pressure difference between the suction side and pressure side of the hydrofoil aggravates the pressure gradient between the inner and outer region of the cavity, so that the shock wave induced by the collapse dominates the evolution of the cavity structures, which is corresponding to the larger amplitude fluctuation of the vibration velocities shown in Fig. 5.

Conclusion

In this paper, experiments on a modified NACA66 hydrofoil in cloud cavitating flow have been conducted based on high speed photography and laser vibrometry. Two typical cavity patterns and the corresponding vibration characteristics are analyzed.

For the partial cloud cavitation, with the maximum cavity length less than 1.0 and the Strouhal number of 0.3-0.4, the partial sheet cavity forms and develops on the suction side of the hydrofoil with the formation of the re-entrant jet at the rear end of the cavity, as the re-entrant jet flowing toward the leading edge of the hydrofoil, the cavity is lifted away from the near surface and a large-scale cloud cavity breaks up with the main cavity and sheds downstream, accompanied with the growth of the residual cavity on the suction side of the hydrofoil.

For the trailing cloud cavitation, the cavity extends to the downstream of the hydrofoil and the Strouhal number decreases to 0.1-0.15 accompanied with intense vibration and noise. A more complex cavitating flow pattern is observed. When the shedding cloud cavity reaches to the rear part of the hydrofoil, it collapses and induces a strong shock wave. As a result, the main cavity shrinks and may even entirely disappears on the suction side. Compared with the partial cloud cavity shedding dynamics (St=0.3-0.4), a significant reduction of the shedding frequency can be observed, which results to a much lower Strouhal number (St=0.1-0.12), and larger vibration velocities were obtained.

Acknowledgments

The authors gratefully acknowledge the support by the National Postdoctoral Program for Innovative Talents (Grant No: BX201700126), the China Postdoctoral Science Foundation (Grant No: 2017M620043), the National Natural Science Foundation of China (Grant No: 51679005) and the National Natural Science Foundation of Beijing (Grant No: 3172029).

References

[1] Wang, G. Y., Senocak, I., Shyy, W., Ikohagi, T., Cao, S. L. (2001) *Dynamics of attached turbulent cavitating flows.* Progress in Aerospace Sciences. 37.

[2] Foeth, E. J., Van Doorne, C. W. H., Van Terwisga, T., Wieneke, B. (2006) *Time resolved PIV and flow visualization of 3D sheet cavitation.* Experiments in Fluids. 40.

[3] Ausoni, P., Farhat, M., Escaler, X., Egusquiza, E., Avellan, F. (2007) *Cavitation Influence on von Kármán Vortex Shedding and Induced Hydrofoil Vibrations.* Journal of Fluids Engineering. 129.

[4] Wu, Q., Wang, Y., Wang, G. (2017) *Experimental investigation of cavitating flow-induced vibration of hydrofoils.* Ocean Engineering. 144(1).

[5] Coutier-Delgosha, O., Reboud, J. L., Dellanoy, Y. (2003) *Numerical simulation of the unsteady behaviour of cavitating flows.* International Journal of Numerical Methods in Fluids. 42.

[6] Huang, B., Zhao, Y., Wang, G. Y. (2014) *Large eddy simulation of turbulent vortex-cavitation interactions in transient sheet/cloud cavitating flows.* Computers & Fluids. 92.

[7] Kubota, A., Kato, H., Yamaguchi, H. (1992) *A new modelling of cavitating flows: a numerical study of unsteady cavitation on a hydrofoil section.* Journal of Fluid Mechanics. 240

[8] Arndt, R. E. A. (1981) *Cavitation in fluid machinery and hydraulic structures.* Annual Review of Fluid Mechanics. 3.

[9] Luo, X., Ji, B., Tsujimoto, Y. (2016) *A review of cavitation in hydraulic machinery.* Journal of Hydrodynamics. 28.

[10] Wu, Q., Wang, G., Huang, B., Gao, Y. (2015) *Experimental and numerical investigation of hydroelastic response of a flexible hydrofoil in cavitating flow.* International Journal of Multiphase Flow. 74.

[11] Arndt, R. E. A. (2012) *Some remarks on hydrofoil cavitation.* Journal of Hydrodynamics. 24(3).

[12] Arndt, R. E. A., Song, C. C. S., Kjeldsen, M., Keller, A. (2000) *Instability of partial cavitation: A numerical/experimental approach.* Proceedings of 23rd Symposium for Naval Hydrodynamics.

[13] Leroux, J. B., Coutier-Delgosha, O., Astolfi, J. A. (2005) *A joint experimental and numerical study of mechanisms associated to instability of partial cavitation on two-dimensional hydrofoil.* Physics of Fluids. 17.

[14] Leroux, J. B., Astolfi, J. A., Billard, J. Y. (2004) *An experimental study of unsteady partial cavitation.* Journal of Fluids Engineering. 126.

[15] Seo, J. H., Lele, S. K. (2009) *Numerical investigation of cloud cavitation and cavitation noise on a hydrofoil section.* Proceedings of 7th Symposium of Cavitation. CAV2009-0062.

[16] Avellan, F., Henry, P., Rhyming, L. (1987) *A new high speed cavitation tunnel for cavitation studies in hydaulic machinery* ASME Fluids Engineering Division FED. 57.

© 2018 ASME

An Experimental Study of Unsteady Behaviour of Cavity Flow Over a 2-D Wall-Mounted Fence

[1]Luka Barbaca*; [1]Bryce W. Pearce; [1]Paul A. Brandner; [2]Harish Ganesh; [2,3]Steven L. Ceccio

[1]*Australian Maritime College, University of Tasmania, Launceston, Tasmania, Australia*

[2]*Department of Mechanical Engineering, University of Michigan, Ann Arbor, MI, USA*

[3]*Department of Naval Architecture and Marine Engineering, University of Michigan, Ann Arbor, MI, USA*

Abstract

The unsteady behaviour of ventilated and natural cavity flows over a 2-D wall-mounted fence is investigated for fixed length cavities with varying free-stream velocity using still imaging, X-ray densitometry and dynamic surface pressure measurement in two experimental facilities. Cavities in both ventilated and natural flows were found to have a re-entrant jet closure, but not to exhibit large-scale oscillations, rather irregular small-scale shedding at the cavity closure. Small-scale cavity break-up was associated with a high-frequency broad-band peak in the wall pressure spectra, found to be governed by the overlying turbulent boundary layer characteristics, similar to observations from single-phase flow over a forward-facing step. A low-frequency peak reflecting the oscillations in size of the re-entrant jet region, analogous to 'flapping' motion in single-phase flow, was found to be modulated by gravity effects (i.e. a Froude number dependence).

Keywords: cavitation; ventilation; wall-mounted fence

Introduction

Flow over a geometric discontinuity immersed in a turbulent boundary layer will result with flow separation. Cavitation may occur in a separated flow region as the local pressure approaches vapour pressure. As the cavitation number is reduced the gaseous bubble/cavity grows and eventually surpasses the original length of the separated flow region [6]. Depending on the location of the cavity closure, i.e. if the closure is on the wall of the cavitating body or rather downstream, cavities are classified as either 'partial' or 'super-cavities'. Cavities of similar appearance and behaviour, termed 'ventilated' can be formed artificially if sufficient gas (typically air) is injected into the separated flow region [11]. It is the condition of the cavity closure on the wall that is responsible for the emergence of significant unsteadiness in flows featuring partial cavitation. The closure region of an unsteady partial cavity is found to generally exhibit a re-entrant jet behaviour.

In a seminal study Callenaere et al. [5] were able to map various cavitation regimes, among which they differentiated two regimes with the established re-entrant jet flow based on the re-entrant jet to cavity thickness ratio. In the case of the re-entrant jet being much thinner than the cavity, the classical cloud cavitation instability characterized by the re-entrant jet propagation up to the cavity leading edge and periodic shedding of large-scale bubbly structures is observed. In the second regime, labelled as 'thin non-auto-oscillating cavities with periodic re-entrant jet', the re-entrant jet thickness is relatively large compared to the cavity thickness, promoting interaction between the jet/cavity and cavity/water interfaces. The small-scale instability resulting from this interaction induces cavity break-up at many random points, leading to irregular shedding of a large number of small-scale structures and a globally quasi-steady cavity appearance. Due to the less severe effect on the flow this second regime type has received little attention in comparison with cloud cavitation.

Until the recent numerical and experimental studies reported by the authors [12, 2, 1, 3], cavity flow over a wall-mounted fence has not been reported on in the open literature. To date the mean characteristics and global behaviour of the cavity flow over a fence immersed in an oncoming wall boundary layer of variable thickness for a range of free-stream conditions has been reported on. The next step towards a better understanding of this canonical flow with a nominally zero stream-wise pressure gradient is to analyse unsteady behaviour.

Within the present study the unsteady behaviour of ventilated and natural cavity flow over a 2-D wall-mounted fence is experimentally investigated in two testing facilities. Large-scale experiments are performed in the University of Tasmania water tunnel with a 600 × 600 mm square test section, whilst smaller-scale experiments are performed in

*Corresponding Author, Luka Barbaca: Luka.Barbaca@utas.edu.au

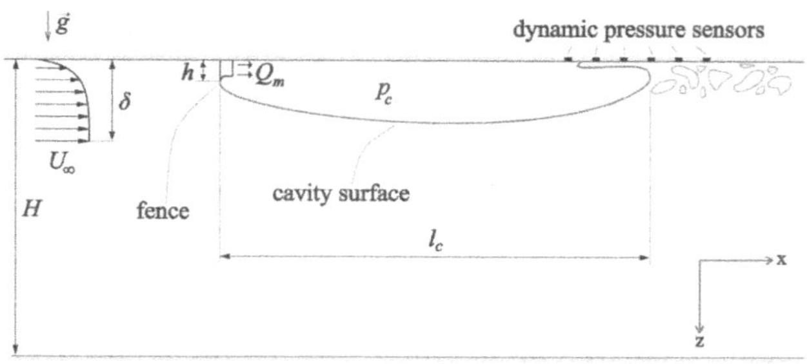

Figure 1: Sketch of a wall mounted fence immersed in the oncoming wall boundary layer with a cavity detaching from the sharp fence tip. The location along the test-section ceiling centre-line of an array of six dynamic pressure sensors in the cavity closure region is indicated.

the University of Michigan 9-inch water tunnel with the test section size reduced to 76.2 × 76.2 mm. These experimental setups are developed to utilize the complimentary capabilities from each facility, with X-ray densitometry measurements from the small-scale experiment and long-period pressure measurements from an array of six dynamic pressure sensors from the large-scale tests. The scope of work is to examine the influence of free-stream Reynolds and Froude numbers on the cavity closure dynamics and the mechanisms of gas entrainment into the main flow for a fixed cavity length to fence height ratio.

1 Experimental Overview

A schematic of cavity flow over a wall-mounted fence representative of the tests in both facilities is shown in figure 1. A fence of height h is immersed in the upstream wall boundary layer of thickness δ. In the case of a ventilated cavity, air is supplied to the wake region of the resulting bluff body flow with a mass flow rate Q_m. Alternatively, for $Q_m = 0$, a natural cavity may be formed due to phase change when the pressure in the wake of the fence, $p = p_c$, reduces to vapour pressure, p_v. Irrespective of the type of formation (i.e. natural or ventilated), the cavity detaches from the sharp fence tip and exhibits a re-entrant jet closure region.
The flow dependence on the free-stream velocity has been quantified using fence height based Reynolds ($Re_h = U_\infty h/\nu$, where U_∞ is the reference free-stream velocity and ν is the kinematic viscosity of the water) and Froude numbers ($Fr_h = U_\infty/\sqrt{gh}$, where g is the gravitational acceleration). For the characterization of unsteady behaviour of the flow a fence height based Strouhal number is used, $St_h = fh/U_\infty$, where f is the frequency of the unsteady phenomenon of interest.
The large-scale experiments were performed in the University of Tasmania variable pressure water tunnel. Detailed description of the facility and the instrumentation used can be found in Barbaca et al. [3]. The fence model incorporating an internal air supply channel is machined from a stainless steel bar, with a manifold for air distribution attached to the downstream face. The fence is 599 mm wide × 10 mm high, with a sharp tip machined on the front face to ensure a stable cavity detachment. Air is supplied through the fore-body of the fence to the manifold, from which it is distributed into the wake of the fence through 40 × ϕ2 mm equi-spaced stream-wise outlets. The fence is located 230 mm from the test section entrance, with the oncoming wall boundary layer thickness at this stream-wise position being 15 mm [4], giving a boundary layer thickness to fence height ratio of $\delta/h = 1.5$.
An array of six flush-mounted 8.1 mm diameter FUTEK PFT510 dynamic pressure sensors installed within the test-section ceiling window was used for the measurements of unsteady pressure within the cavity closure region. The most upstream sensor was located 730 mm downstream of the fence tip, with each subsequent sensor positioned 60 mm further downstream along the test-section horizontal centre-line. The sensor range is 15 bar absolute, with a rated output of $1-2$ mV/V (estimated output uncertainty of 0.8%) and natural frequency of 6 kHz. A 5V excitation and signal amplification is provided through an Entran MSC6 signal conditioning box, with data acquired simultaneously using a National Instruments PXIe-4497 card. High-resolution front-lit still images were captured using a Nikon D810E camera with a Nikon Nikkor f/2.8D 24mm lens.
Natural and ventilated cavities of nominally fixed length, $l_c/h \approx 90$, were investigated for a range of free-stream speeds. Tests were conducted over the full operational range, $2 \leq U_\infty \leq 12$ m/s, resulting in Reynolds and Froude numbers varying between $2 \times 10^4 \leq Re_h \leq 1.2 \times 10^5$ and $6.4 \leq Fr \leq 38.3$, respectively. All tests were conducted with a dissolved O_2 content between 2 and 6 ppm.

The small-scale experiments were performed at the University of Michigan 9-Inch Water Tunnel. Detailed description of the facility and the instrumentation can be found in Ganesh et al. [7]. To achieve flow similarity between the two facilities it was decided to have the same δ/h. The fence model spanning the test section width is machined from a 6 mm brass bar and is mounted inside a recess in the acrylic ceiling window. The fence is 75.2 mm wide, with the in-flow height of $h = 1.33$ mm. A sharp tip is machined on the fence front tip to ensure a stable cavity detachment. A gap between the fence body and the recess walls serves as a plenum for air injection which is distributed into the fence wake through a full-span slanted slot between the angled downstream face of the fence and the ceiling. The fence tip is located 115.6 mm downstream of the entrance to reduced test section. The oncoming wall boundary layer thickness at this position is ≈ 2 mm, resulting in $\delta/h \approx 1.5$.
A cinematographic X-ray densitometry system was used to measure the span-wise averaged spatial distribution of the void fraction for the ventilated and natural cavity flows over the fence. A complete description of the X-ray setup and data reduction process is provided by Mäkiharju et al. [10]. The X-ray images were acquired at a sample rate of 1 kHz for ≈ 0.8 s. Ventilated and natural cavities of nominally fixed length, $l_c/h \approx 60$, were investigated for a constant free-stream speed, $U_\infty = 10$ m/s ($Re_h = 1.33 \times 10^4$, $Fr_h = 87.6$). All tests were conducted with a dissolved O_2 content between 2 and 6 ppm.

2 Results

A typical topology of a fully developed ventilated cavity observed in the large-scale test is shown in figure 2 for $Re_h = 1 \times 10^5$. The cavity has a stable detachment from the fence tip and two distinct regions are evident along the cavity length. The region closer to the fence is air/vapour filled with a relatively sharp cavity/water interface and a transparent appearance. The rear or closure of the cavity is dominated by re-entrant jet flow and is observed as an opaque region. From the observation that the re-entrant jet does not propagate all the way upstream to the fence, but is contained to the closure region only, the flow can be classified within the ‘thin cavities’ regime as distinguished by Callenaere et al. [5]. The opaque cavity surface appearance is a result of interaction between a relatively thick re-entrant flow region and the cavity/water interface, inducing irregular/chaotic cavity surface break-up. The surface break-up triggers the shedding of small-scale bubbly structures into the cavity wake with consequent entrainment into the main flow. This small-scale pinch-off at the cavity closure results in a relatively small variation in the cavity length and the cavity can be considered as non-auto-oscillating.
The main difference between the ventilated and natural cavity flows is observed in the wake region. The wake of a ventilated cavity consists of a dense mixture of shed structures of variable scale. In the case of a natural cavity the vaporous structures condense within a short distance from the closure and only a population of micro-bubbles containing incondensable gas remains in the far wake.
In figure 3 instantaneous void fraction distribution extracted from X-ray imaging, along with the root-mean-square deviation (RMSD) of the X-ray sequence, is shown for the ventilated (a) and natural (b) cavity flows from the small-scale test for $l_c/h \approx 60$ and $Re_h = 1.33 \times 10^4$. Note that the scale in X-ray images is exaggerated (times two) in the vertical direction. From the instantaneous X-ray frames a difference in the mean void fraction distribution can be observed between the ventilated and natural cases. While the upstream part of the cavity has a similar maximum void fraction value in both cases, a more significant difference can be seen in the closure. Firstly, the mean α (void fraction) value is noticeably lower in the naturally cavitating flow across the whole re-entrant jet affected region and secondly, the re-entrant jet itself has a much higher liquid content when compared with a seemingly more bubbly jet present for the

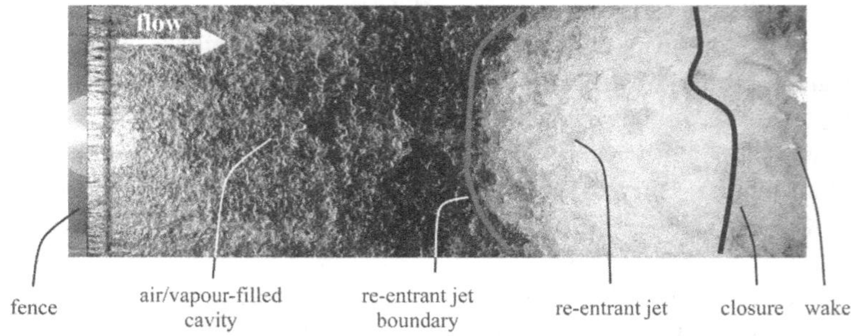

Figure 2: Typical topology of a cavity observed in the large-scale experiments for $Re_h = 1 \times 10^5$. Cavity has an air/vapour filled region present closer to the fence and a re-entrant jet type closure, and can be classified within ‘thin cavity’ regime as specified by Callenaere et al. [5].

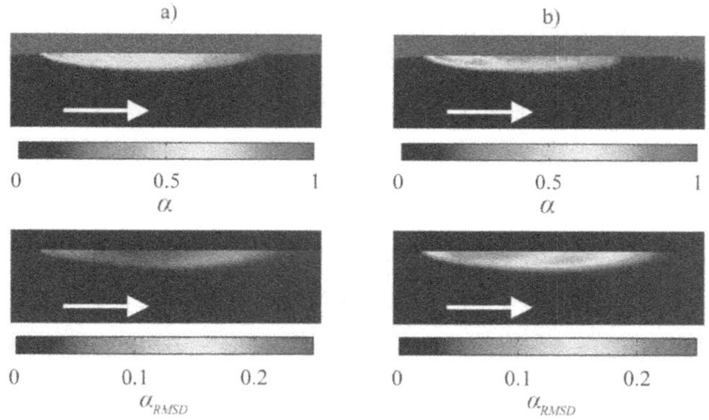

Figure 3: Topologies of ventilated (a) and natural (b) cavity flow from the small-scale experiment for $Re_h = 1.33 \times 10^4$ shown as instantaneous frames extracted from X-ray imaging. Additionally, the root-mean-square deviation (RMSD) of the void fraction (α) from the X-ray sequences is presented for both cases. Note that the scale in the X-ray images is shown exaggerated (times two) in the vertical direction.

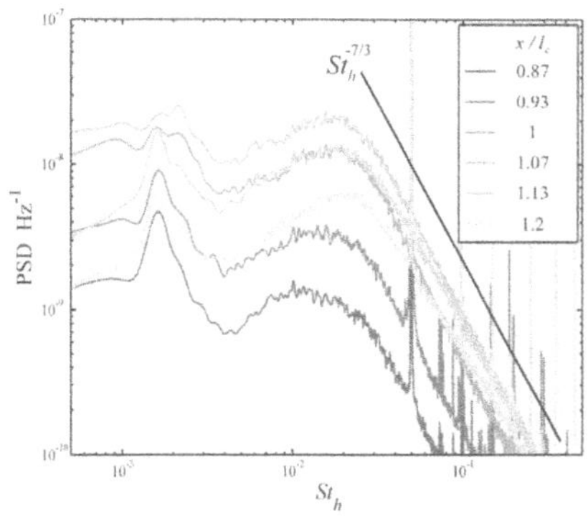

Figure 4: Wall pressure power spectra from six dynamic pressure sensors located on the test-section ceiling centre-line for ventilated flow at $Re_h = 1 \times 10^5$. A black line indicates the power spectra decay following a power law behaviour proportional to $St_h^{-7/3}$.

ventilated flow. From the plots of void fraction RMSD it is observable that the void fraction distribution undergoes much more pronounced oscillations (approximately a 100 % increase in RMSD value) for natural cavitation.

The power spectral densities (PSDs) of long sampled pressure signals from an array of six pressure sensors located in the cavity closure region were analysed using the Welch method. The pressure sensors are positioned with the nominal position of cavity closure approximately aligned with the third most upstream sensor. The time-series were recorded for 1000 s at 1 kHz. In figure 4 ventilated flow pressure spectra from each sensor is given for a constant $Re_h = 1 \times 10^5$. Spectral content observed for each sensor is characterized by two broad-band peaks and a high-frequency roll-off that follows a power law behaviour ($\propto St_h^{-7/3}$). Interestingly, this behaviour in two-phase flow is similar to that observed in the single-phase flow over a forward-facing step [8]. Although, the similarity between the single-phase and cavitating flow is remarkable, a significant difference exists. For two-phase flow peaks in pressure fluctuations are observed at $St_h \approx 0.002$ and $St_h \approx 0.02$, whereas in single-phase flow these peaks occur an order of magnitude higher. The low-frequency peak can be associated with the oscillations of the re-entrant jet length (analogous to 'flapping' motion in single-phase flow), while the high-frequency peak can be related to the cavity break-up induced by the large-scale structures contained within the overlaying turbulent boundary layer. The observed power law like high-frequency power spectra decay characterizes the spectral content of pressure fluctuations within a turbulent boundary layer in a single-phase flow, which further supports the premise that cavity closure break-up is dominated by the characteristics of the overlaying boundary layer.

In figure 5 wall pressure spectra from the sensor nominally aligned with the cavity closure is given for ventilated cavity case for the full range of Re_h investigated. From figure 5a it can be seen that if the PSD is plotted against the Strouhal number defined using a constant length scale (i.e. fence height), the high frequency peak will collapse to a constant $St_h \approx 0.02$ value, leading to a conclusion that the frequency of turbulent shedding is proportional to free-

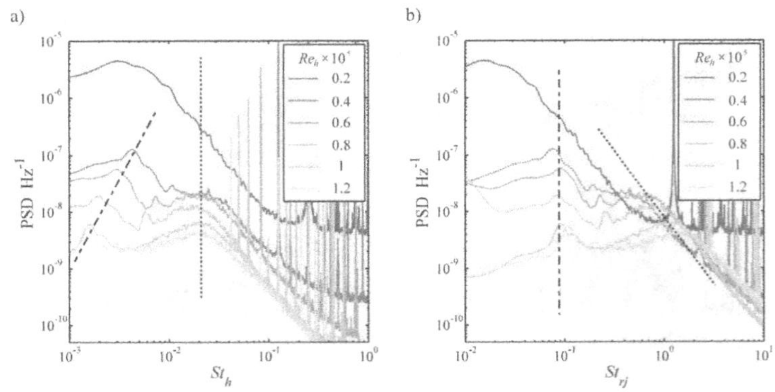

Figure 5: Wall pressure power spectra from the dynamic pressure sensor nominally aligned with the cavity closure. Data is presented for ventilated flow with the frequency non-dimensionalised as St_h (a) and St_{rj} (b). The dash-dot lines are added to the plots to indicate the low-frequency peaks, while the dotted lines denote the high-frequency peaks.

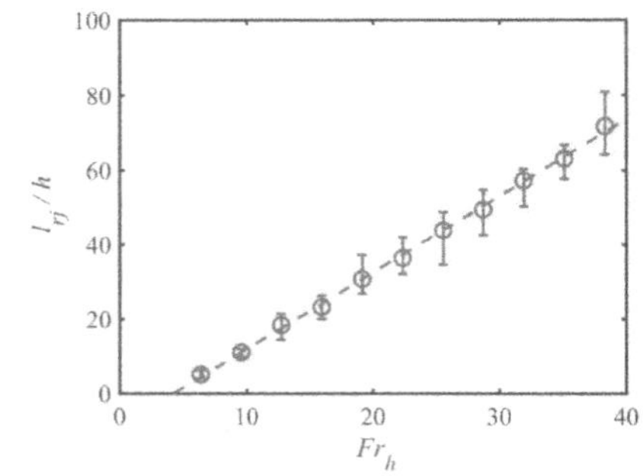

Figure 6: Dependence of non-dimensionalised re-entrant jet length (l_{rj}/h) on Fr_h for ventilated flow.

stream velocity. The high-frequency power law like decay is not affected by change in Re_h, which is consistent with the reasoning that the cavity shedding is controlled by the overlying turbulent boundary layer. If the amount of energy contained within the low- and high-frequency peaks is compared across the range of Re_h, it is observable that peaks have approximately equal value at higher Re_h, but the low-frequency oscillation become increasingly more dominant with decreasing Re_h.

The location of the low-frequency peak does not collapse with St_h. As previously noted, the low frequency peak is associated with the large-scale re-entrant jet oscillation and it may be useful to alternatively define a re-entrant jet length based Strouhal number, $St_{rj} = fl_{rj}/U_\infty$. A plot showing the dependence of re-entrant jet length, non-dimensionalised by fence height (l_{rj}/h), on Fr_h is given for ventilated flow in figure 6. The maximum and minimum recorded length are represented with error-bars for each investigated condition. A linear dependence of l_{rj}/h on Fr_h is evident. The justification for designation of Fr_h as a dimensionless parameter governing the re-entrant jet length can be found if the studied flow is considered as analogous to the recirculating flow associated with the undular jump phenomenon in super-critical open-channel flow. As presented in experimental work by Hager et al. [9], the length of recirculating zone (termed 'roller length') has a dependence on a fixed length scale based Fr which can be closely approximated with a linear fit function.

When the pressure spectra is plotted against the St_{rj} (figure 5b) it can be seen that the low frequency peak collapses to a constant St_{rj} value of about 0.1. The collapse of low-frequency peak to a nominally constant St_{rj} in two-phase flow is in contrast to the observations from single-phase flow, where the 'flapping' oscillation frequency occurs at a constant St_h. This difference can be attributed to the variation in re-entrant jet length (i.e. recirculating zone length) in the two-phase flow, in comparison to a nominally constant length of recirculating zone observed in single-phase flow for $Re_h > 8.5 \times 10^3$ [8]. On this basis, it may be concluded that the influence of gravity, arising from the presence of two phases with substantially different densities, in the recirculating region modulates its dynamic behaviour.

Conclusion

The topology and unsteady behaviour of ventilated and natural cavity flows over a fixed height 2-D wall-mounted fence were investigated for fixed length cavities using still imaging, X-ray densitometry and dynamic surface pressure measurements in two experimental facilities. Based on the observations of cavity topology and shedding mechanisms, the examined flow is classified within the 'thin non-auto-oscillating cavity' regime [5]. Two main features contributing to cavity unsteady behaviour were found, irregular small-scale shedding of gaseous/vaporous structures at the cavity closure and larger-scale oscillations in size of the re-entrant jet affected region. From the spectral analysis of wall pressure fluctuations, the small-scale shedding is associated with a high-frequency broad-band peak and a power law decay ($\propto St^{-7/3}$), analogous to that observed in a single-phase flow over a forward-facing step. The location of this peak collapsed for $St_h \approx 0.02$, showing that the cavity break-up is invariant of Re_h and Fr_h, and governed by the overlying turbulent boundary layer characteristics only. The re-entrant jet oscillation was observed as a low-frequency peak in wall pressure spectra, analogous to the 'flapping' instability of the recirculation zone in single-phase flow. The low-frequency peak was found to collapse for a Strouhal number based on the re-entrant jet length (of about 0.1). The re-entrant length jet was found to have a linear dependence on free-stream velocity, analogous to the recirculation zone length in gravity dominated super-critical open-channel flow. This similarity indicates that the low-frequency oscillation of the re-entrant jet is influenced by gravity (i.e. influence of buoyancy due to inclusion of the gaseous/vaporous phase of much lower density), which is in contrast with the observations from single-phase flow where it was found to be independent of free-stream velocity.

Acknowledgements

The authors wish to acknowledge the assistance of Mr Robert Wrigley and Mr Steven Kent in carrying out the experiments in Univeristy of Tasmania cavitation tunnel, and Mr Kent Pruss and Miss Juliana Wu in carrying out the experiments in University of Michigan cavitation tunnel. The authors wish to acknowledge the financial support received through University of Tasmania Conference and Research Travel Funding Scheme for travel to University of Michigan and the support of Australian Maritime College.

References

[1] Barbaca, L., Pearce, B. W., and Brandner, P. A. (2017a). *Experimental study of ventilated cavity flow over a 3-D wall-mounted fence.* International Journal of Multiphase Flow, 97:10–22.

[2] Barbaca, L., Pearce, B. W., and Brandner, P. A. (2017b). *Numerical analysis of ventilated cavity flow over a 2-D wall mounted fence.* Ocean Engineering, 141:143–153.

[3] Barbaca, L., Pearce, B. W., and Brandner, P. A. (2018). *An experimental study of cavity flow over a 2-D wall-mounted fence in a variable boundary layer* (under revision). International Journal of Multiphase Flow.

[4] Belle, A., Brandner, P. A., Pearce, B. W., de Graaf, K. L., and Clarke, D. B. (2016). *Artificial thickening and thinning of cavitation tunnel boundary layers.* Experimental Thermal and Fluid Science, 78:75–89.

[5] Callenaere, M., Franc, J. P., Michel, J. M., and Riondet, M. (2001). *The cavitation instability induced by the development of a re-entrant jet.* Journal of Fluid Mechanics, 444:223–256.

[6] Franc, J. and Michel, J. M. (2004). *Fundamentals of Cavitation*, volume 76 of Fluid Mechanics and Its Applications. Kluwer Academic Publishers, Dordrecht.

[7] Ganesh, H., Mäkiharju, S. A., and Ceccio, S. L. (2016). *Bubbly shock propagation as a mechanism for sheet-to-cloud transition of partial cavities.* Journal of Fluid Mechanics, 802:37–78.

[8] Graziani, A., Lippert, M., Uystepruyst, D., and Keirsbulck, L. (2017). *Scaling and flow dependencies over forward-facing steps.* International Journal of Heat and Fluid Flow, 67(Part A):220–229.

[9] Hager, W. H., Bremen, R., and Kawagoshi, N. (1990). *Classical hydraulic jump: length of roller.* Journal of Hydraulic Research, 28(5):591–608.

[10] Mäkiharju, S. A., Gabillet, C., Paik, B. G., Chang, N. A., Perlin, M., and Ceccio, S. L. (2013). *Time-resolved two-dimensional X-ray densitometry of a two-phase flow downstream of a ventilated cavity.* Experiments in Fluids, 54(7).

[11] May, A. (1975). *Water Entry and the Cavity-Running Behaviour of Missiles.* Report 75-2, NAVSEA Hydroballistics Advisory Comitee.

[12] Pearce, B. and Brandner, P. (2014). *Inviscid cavity flow over a wall-mounted fence.* Ocean Engineering, 80:13–24.

© 2018 ASME

An experimental study on the cavity evolution of a continuous entry of sphere through a viscous fluid into water

[1]Sun Tiezhi*;[1]Wang Heng;[1,2,3]Zhang Guiyong; [1,2,3]Zong Zhi; [1]Li Haitao

[1]Dalian University of Technology, Dalian, Liaoning Province, China;

[2]State Key Laboratory of Structural Analysis for Industrial Equipment, Dalian, Liaoning Province, China;

[3]Collaborative Innovation Center for Advanced Ship and Deep-Sea Exploration, Shanghai, China

Abstract

In the present paper, cavity evolution of a continuous entry of sphere through dimethicone into water was studied experimentally in DLUT (Dalian University of Technology) Marine Hydrodynamics Laboratory. High-speed video was used to capture the movement sequences of the sphere and cavity evolution. Firstly, we conducted a vertical entry case of the sphere dropped from 0.7m above the free surface. After the entire sphere submerged under the free surface, the splash crown and subsurface air cavity began to form. It was important to note that a continuous cavity appeared at the interface of the dimethicone and water. Furthermore, the influence of various parameters, specifically the thickness of the dimethicone layer, were studied. One of the most interesting results was that relationship was found between the thickness of the dimethicone layer and the pitch off time. The interaction between the sphere and the cavity was also noted. These would help to understand the water entry problem with the change in fluid properties.

Keywords: water entry; dimethicone; cavity evolution; pitch off

Introduction

It is well known that considerable efforts have been devoted to addressing the water entry problems due to its wide applications including industry, sport, biology and military. The water entry of objects present a rich physical phenomenon, address the flow and evolution mechanism of water entry cavity is a goal that researchers have long been pursuing.

The earliest experimental study of water entry problems can be found in Worthington and Cole[1][2], they used single-spark photography to investigate cavity shape through the vertical entry of a sphere into the water. In the 1950s, May[3] was the first to describe this buckling instability and found the surface seal phenomena. He described it as 'ribbon-like filaments' appearing just prior to surface seal. With the developments in high-speed imaging technology, recent studies have attributed to better and further understanding of transient cavity dynamics. Duez et *al.*[4] investigated the effect of liquid properties and sphere wettability on the threshold velocity of cavity entrainment. They demonstrated that the wettability of the impacting body was a key factor in determining the degree of splashing. Their novel understanding provided a new perspective for impacts on the free surface. In addition to these studies, other notable works pertaining to water entry problems. Shi *et al.*[5] analyzed supercavitation induced by high-speed water entry of projectiles at the velocity of 342 m/s and found the 'pull-away' of the cavity beneath the water surface. Gekle *et al.*[6] found the unique effect of capillary ripples on cavity dynamics and two asymptotic regimes separated by discrete jumps for the scaled pinch-off depths. Zhao *et al.*[7] used high-speed photography technology investigated an experimental study of an axisymmetric slender body. They found that the impact between the tail of slender body and supercavity wall resulted in the rotation of the slender body.

Spheres are used as the most simple and representative objects to investigate water entry problems. However, usually, this canonical shape impacting on the free surface does not yield simple hydrodynamic characteristics[8]. Truscott *et al.* [8] experimentally investigated the complex hydrodynamics of water entry by a spinning sphere for low Froude numbers. They found the cavity dynamics and splash are highly altered for the spinning case compared to a sphere without spin. To address the physical characteristics involved in the evolution of cavity, Yan *et al.*[9] conducted an experiment of water entry with freely dropping spheres. Jeffrey et al. conducted theoretical and

*Corresponding Author, Tiezhi Sun: suntiezhi@dlut.edu.cn

experimental investigation on the vertical impact of low-density spheres on a water surface. The sphere dynamics and the effect of its deceleration on the shape of the resulting air cavity were analyzed systematically. More recently, Marston *et al.* [10]presented new observations by an experimental investigation of the sealing phenomena and crown splash observed during the impact of spheres onto quiescent liquid pools, and they varied the ambient pressure, liquid properties and sphere diameter.

However, little studies have been carried out on the effect of variable liquid properties such as surface tension and density during the process of the cavity formation, evolution and jet. In order to address these characteristics, in this paper, we present new phenomena from an experimental investigation of the cavity dynamics during the continuous entry of sphere through high viscous liquid into water.

Experimental setup

A schematic of the experimental setup used in present study is shown in figure 1. We used a billiard ball to perform the experiment. The ball's diameter d is 5.72cm, the mass is 0.18kg, and the surface roughness is P1000. The dimensions of this reinforced glass tank is 0.8m×1.0m×1.5m (width × depth × height), and a plexiglass tube is vertically mounted above the tank. In the experiments, there is a layer of high viscous dimethicone on the top of the water, whose density is 941kg/m^3, and dynamic viscosity is 0.01Pa s. High-speed video was used to record the movement sequences of the sphere and cavity evolution. The frame rate is 1000 fps with an image resolution of 1280×800 pixels and the exposure time is 0.001s. Due to the higher frame rate and shorter exposure time, the intensity of illumination needs to be increased appropriately to ensure the clarity of the images in this experiment. Thus, two Starison CE-1500WS spotlights were set up on both sides of the high-speed camera to ensure the quality of the captured images. After a series of image processing, the cavity formation, development, pitch-off and jet can be obtained.

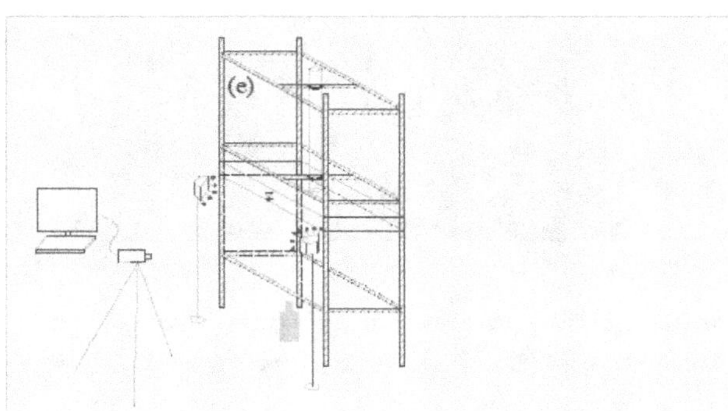

Figure 1. Schematic of the experimental setup used in the present study (Figure is not to scale)

Results and discussion

To address the special phenomenon in a practical evolution process using high-speed visualization, we measured the pinch-off time and captured the cavity shape evolution under various parameters: the thickness of dimethicone is 0mm, 5mm, 10mm, 15mm, 20mm, 25mm and 30mm respectively.

Figure 2 shows the sequence of images depicting the splash and air cavity formed in the wake of the sphere impacting the dimethicone and water. Herein, the thickness of dimethicone is $h = 10$ mm, the initial impact velocity is $V_0 = 3.7$ m/s and the Froude number is $Fr = V_0 / \sqrt{gD} = 4.95$, where g is the acceleration of gravity, D is the diameter of the sphere, respectively. It is apparent that an initial jet of fluid forms as the sphere impacts the free surface and continues to extend outwards as the sphere descends into the dimethicone. At t=7ms, the sphere has passed the layer of dimethicone and into the water. From t=7ms to t=50ms, we can see that the jet transitions from outwards to upwards growth and the splash crown appears near the free surface. The subsurface air cavity grows outwards and elongates vertically as the sphere descends. From t=80ms to t=100ms, the subsurface air cavity is still open to the atmosphere. A low pressure induced by the air flows into the splash, which results in driving the splash curtain inwards. Meanwhile, the cavity below the free surface shrinks gradually. At t=105ms, the pitch off phenomenon occurs and the subsurface air cavity is splited? ? into two distinct cavities. After pinch off, distinct jets of fluid eject away from the point of pinch-off in opposite directions. One of the most interesting results for the cavity evolution of a continuous entry of sphere through viscous liquid into water is that the characteristics of the buckling instability. In the experiments by Marston *et al.*[10], they found that the crown has collapsed in on itself with a significantly reduced neck radius in the upper region and the crown wall appears more-or-less vertical in perfluorohexane. However, interestingly, the expansion and contraction of the crown radius appear in the present study. In other words, the shape of the crown wall like a 'drum' (as shown in Figure 2(d)).

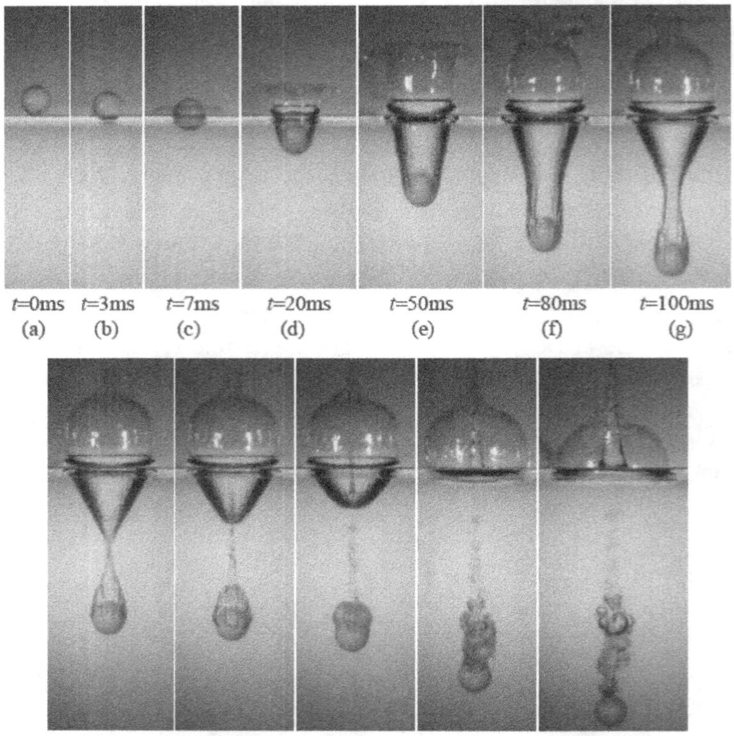

t=0ms *t*=3ms *t*=7ms *t*=20ms *t*=50ms *t*=80ms *t*=100ms
(a) (b) (c) (d) (e) (f) (g)

t=105ms *t*=110ms *t*=120ms *t*=160ms *t*=200ms
(h) (i) (j) (k) (m)

Figure 2. A sequence of images depicting the splash and air cavity formed in the wake of the sphere impacting the dimethicone and water, the initial impact velocity is $V_0 = 3.7$ m/s, the Froude number is $Fr = 4.95$

Figure 3 presents the images at the moments of pinch off for the seven thicknesses of dimethicone, where the moment of pinch-off is taken when the subsurface air cavity has completely necked down. All the impact velocity is the same, where $V_0 = 3.7$. We can see that the depth of pinch-off distances almost are fundamentally similar. Moreover, we note that the thickness of the dimethicone appears to have no effect on cavity pinch-off time. For all the cases, the pinch-off time is 105ms.

h=0mm (105ms) *h*=5mm(105ms) *h*=10mm(105ms) *h*=15mm(105ms) *h*=20mm(105ms) *h*=25mm(105ms) *h*=30mm(105ms)
(a) (b) (c) (d) (e) (f) (g)

Figure 3. A comparison of the formed cavity at the moment of pinch-off with increasing the thickness (left to right) of dimethicone ($V_0 = 3.7$ m/s, $Fr = 4.95$), the first frame is 0mm, subsequent frames are 5mm apart.

Conclusion

In this paper, we experimentally investigated the well-known crown splash during a continuous entry of sphere through a viscous fluid into water. Using high-speed videography, we have provided a detailed cavity evolution. One of the most interesting results in the present study is that the shape of the crown wall looks like a 'drum'. Furthermore, the pinch-off time is dependent on the thickness of dimethicone above the water at a certain drop height of the sphere.

Acknowledgement

This work was supported by the National Natural Science Foundation of China (51709042, 51679037, 51639003, 51579042), the Recruitment Program of Global Young Experts (D1007001) and the Fundamental Research Funds for the Central Universities (DUT16RC(3)085, DUT17ZD311, DUT16ZD218).

References

[1] Worthington, A. M., & Cole, R. S. (1897). *Impact with a liquid surface studied by the aid of instantaneous photography.* Philosophical Transactions of the Royal Society of London. 189, 137-148.

[2] Worthington, A. M., & Cole, R. S. (1900). *Impact with a liquid surface studied by the aid of instantaneous photography. paper II.* Philosophical Transactions of the Royal Society of London. 194, 175-199.

[3] May, A. (1952). *Vertical entry of missiles into water.* Journal of Applied Physics. 23(12): 1362-1372.

[4] Duez, C., Ybert, C., Clanet, C., & Bocquet, L. (2007). *Making a splash with water repellency.* Nature Physics. 3(3), 180-183.

[5] Shi, H. H., Itoh, M., & Takami, T. (2000). *Optical observation of the supercavitation induced by high-speed water entry.* Journal of Fluids Engineering, 122(4), 806-810.

[6] Gekle, S., Van, d. B. A., Bergmann, R., Van, d. M. D., & Lohse, D. (2008). *Noncontinuous Froude number scaling for the closure depth of a cylindrical cavity.* Physical Review Letters, 100(8), 084502.

[7] Zhao, C. G., Wang, C., Wei, Y. J., & Zhang, X. S. (2015). *An experimental study on characteristics of cavitation and ballistic of axisymmetric slender body underwater movement.* Journal of Physics: Conference Series. IOP Publishing, 656,012175.

[8] Truscott, T. T., & Techet, A. H. (2009). *Water entry of spinning spheres. Journal of Fluid Mechanics.* 625(625), 135-165.

[9] Yan, H., Liu, Y., Kominiarczuk, J., & Yue, D. K. P. (2009). *Cavity dynamics in water entry at low Froude numbers.* Journal of Fluid Mechanics. 641(641), 441-461.

[10] Marston, J. O., Truscott, T. T., Speirs, N. B., Mansoor, M. M., & Thoroddsen, S. T. (2016). *Crown sealing and buckling instability during water entry of spheres.* Journal of Fluid Mechanics. 794, 506-529.

© 2018 ASME

10th International Symposium on Cavitation - CAV2018
Baltimore, Maryland, USA, May 14 – 16, 2018

An Improved Tip Vortex Cavitation Model for Propeller-Rudder Interaction

[1]Naz Yilmaz*; [1]Mehmet Atlar, [1]Patrick A. Fitzsimmons

[1]*University of Strathclyde, Glasgow, UK*

Abstract

The paper starts with the computational modelling of the tip vortex cavitation in uniform flow conditions with an isolated propeller in detail and provides experimental validation. It then moves onto further modelling to include the effect of non-uniform flow and the presence of a rudder placed in the propeller slipstream. The propeller-rudder arrangement of the Newcastle University research vessel, The Princess Royal, and associated experimental data were used for Experimental Fluid Dynamics (EFD) analysis to validate the modelling. The cavitation simulations were conducted using commercial CFD software, Star CCM+. A new meshing technique, which utilizes a Mesh Adaptive Refinement approach for Cavitation Simulations (MARCS), recently developed by the authors, has been applied successfully to simulate the tip vortex cavitation, particularly to trace its extension up to the rudder in the propeller slipstream. The comparison of the CFD and EFD methods for the isolated propeller in cavitation tunnel conditions showed very good agreement in terms of the thrust and torque coefficients of the propeller as well as the sheet and tip vortex cavitation patterns observed. The cavitation simulations have been extended for the same propeller by using the new mesh refinement approach to include the effect of the hull wake and the presence of the rudder. Although the latter simulations fall short of the EFD results and hence they are still under development, the paper presents the developments and results so far to achieve the ultimate aim of this study, i.e. computational modelling of cavitating tip vortices of a propeller interacting with a rudder.

Keywords: Tip Vortex Cavitation, EFD, CFD, Propeller-Rudder Interaction

Introduction

Sheet cavitation of marine propellers can be predicted accurately with existing methods such as lifting surface, BEM and even more accurately with computational fluid dynamics methods (CFD) due to better modelling of the physics of the flow, thanks to new developments in computational power and technology. However, computational modelling of the cavitating tip vortices of a propeller has its challenges, in particular for extending these vortices from the blades up to a rudder in the wake of a ship hull and interacting with the rudder. Even sheet cavitation predictions sometimes give unstable results due to the lack of accurate tip vortex cavitation modelling [1].

There are many numerical and computational studies to predict tip vortex cavitation in literature [2, 3, 4 and 5], using especially CFD methods in which RANS based models for tip vortex cavitation simulations of marine propellers are preferred [6]. Although the RANS model is recognized as a reliable method for sheet cavitation simulations, further studies are still required particularly for accurate predictions of tip vortex cavitation [6]. In contrast to the RANS model, scale-resolving simulations can model small-scale motions and resolve the large scales of turbulence. Within this context, there are two popular approaches for scale-resolving simulations which are known to be Detached Eddy Simulations (DES) and Large Eddy Simulations (LES) models. Recently, the turbulence models based on these two approaches have been preferred widely by the CFD community for simulating complex physical phenomenon such as cavitation and especially for the tip vortex type. These two approaches are also implemented in the commercial CFD software, Star CCM+ [7] which is used in this study as described below, where the LES model is preferred for simulating the tip vortex cavitation.

Detailed and different grid generation techniques have been investigated in literature with regards to bubble dynamics, cavitating bubble diameter and cavitation inception phenomenon. Although, each bubble cannot be modelled, followed and tracked in space using the Rayleigh-Plesset model, which is implemented in the Star CCM+ code, cavitation phenomena have been investigated by creatively pushing the limits of mesh generation capabilities of the software for capturing tip vortex cavitation. In terms of bubble dynamics, Rayleigh published the first analysis

*Corresponding Author, Naz Yilmaz: naz.gorener@strath.ac.uk

about cavitation and bubble dynamics results [8]. After that, Rayleigh-Plesset equation was presented by Plesset and Prosperetti including bubble growth and collapse phenomena by neglecting bubble-bubble interaction and assuming the bubble was spherical in shape [9]. Based on the Rayleigh-Plesset equation, Schneer and Sauer developed a new volume of fraction (VOF) method for time dependent growth and collapse of cavitating bubbles [10]. The Schneer-Sauer cavitation model, neglecting surface tension, viscous effects and bubble growth acceleration, is implemented in the Star CCM+ Code and hence also used for cavitation simulations in this study [7].

Recently, numerical modeling of the tip vortex cavitation phenomenon has been the focal point by some researchers using CFD methods and commercial codes by creating mesh refinement regions around a propeller's tip area for capturing cavity bubbles in the slipstream including the authors, e.g. [12, 13]. However, if the mesh is generated using a larger than required surface size in the tip vortices region, tip vortex cavitation cannot be captured in the propeller slipstream. On this basis there is a need for studies to investigate the cavitating bubble radius for determining surface size in the mesh refinement region based on some useful experimental data e.g. [11]. Within this framework, the authors have developed a new mesh refinement technique, which utilizes a Mesh Adaptive Refinement approach for Cavitation Simulations (MARCS), based on the relationship between the surface size of the generated mesh and the cavity bubble radius [12]. In this paper, the MARCS approach has been applied to the propeller-rudder arrangement of the Newcastle University research vessel, the Princess Royal by taking advantage of the experimental data which were generated by the authors in their previous work [17] and the results are discussed.

Numerical Approach

For cavitation simulation, the Volume of Fluid (VOF) method was used in defining two states of the fluid (water and vapour) in flow domain. DES and LES turbulence models were preferred for the reasons explained in the introduction.

The Schnerr-Sauer cavitation model used in this study is based on a reduced Rayleigh–Plesset (RP) equation within Star CCM+. It neglects the influence of surface tension, bubble growth acceleration, viscous effects and the bubble-bubble interaction but includes scaling of the bubble growth and collapse rates for both single-component and multi-component materials. Using this cavitation model, the cavitation bubble growth rate can be calculated as follows,

$$\left(\frac{dR}{dt}\right)^2 = \frac{2}{3}\left(\frac{p_{sat} - p_\infty}{\rho_l}\right)$$

Where p_{sat} is the saturation pressure, p_∞ is the pressure of the liquid and ρ_l is the liquid density.

The cavitation number (rotational speed, σ_n) is also defined as,

$$\sigma_n = \frac{p - p_{sat}}{0.5\rho_l (nD)^2}$$

The new adaptive mesh refinement approach (MARCS) proposed by the authors is used to enhance the capture of tip vortex cavitation in a propeller slipstream. In MARCS, the adaptive mesh refinement was created only in the region where the tip vortex cavitation may occur. At the beginning of this application, the upper limit of absolute pressure in the solution was determined by creating a threshold region in Star CCM+. In such cavitation simulations, the volume fraction of vapour shows the volume where the absolute pressure is below the saturation pressure of water, thus identifying the cavitating volume. A threshold region was created by increasing the saturation pressure from the default saturation pressure, 3169 [Pa] used by Star CCM+ to a higher value, 17,000 [Pa] thus generating, the pink region shown in Figure 1. This artifice provides an indication of the volumetric trajectory on which to generate a fine mesh for accurately capturing the pressure-drop correctly and tracking the cavity bubbles in the propeller slipstream.

In applying this idea on a generated mesh, two field functions were created in order to prepare a table including coordinates and the surface size of the new refined mesh for generating the adaptive mesh. These were needed in order to extend the tip vortex cavitation, much as shown in Figure 1. One of the two field functions, called "Cell Width", specifies the one dimension of each cubic cell. Whereas the other field function termed "Refinement" is defined for creating a refinement table which includes coordinates of each cell in x, y and z directions and the surface size of the new mesh to be used while an adaptive mesh is being generated. The "Refinement" field function represents each mesh cell, where the absolute pressure below 17000 [Pa] and above 3169 [Pa], is sub-divided by three in the three dimensions. The upper limit of absolute pressure was defined by creating a threshold (17000 [Pa]) and checked visually as shown in Figure 1. The lowest limit of the pressure was determined by the saturation pressure (3169 [Pa]). Figure 1 shows, respectively, the threshold below 17000 [Pa] and mesh generated using the new adaptive mesh refinement approach (MARCS).

The most important part of the MARCS approach is to determine the surface size of the each cubic cell in the tip vortex region for capturing cavity bubbles in the slipstream. Kuiper [11] investigated and measured cavitation inception, including tip vortex cavitation, using a model scale propeller (Propeller V) at J-values of 0.3, 0.4 and 0.5. Additionally, he defined an experimental relationship between the cavitating core radius (a_c) and cavitation index (σ_n) by preparing equations and graphs in this range of J values. In his study, the minimum radius of each bubble (a_i) was consistently found to be about 0.25 mm on a 250 mm diameter propeller at the cavitation inception stage. According to these investigations, the core radius (a_c) always tends to go to the minimum core radius (a_i). On the basis of Kuiper's study, similar relations between bubble radius, mesh size and simulating tip vortex cavitation are determined within the new adaptive mesh approach. The mesh size was always required to be maintained below 0.25 mm (it is approximately 0.22 mm for this case) for capturing the tip vortex cavitation structure in the propeller slipstream. Using a mesh size larger than 0.25 mm, the tip vortex cavitation could not be simulated as extended as shown in Figure 2.

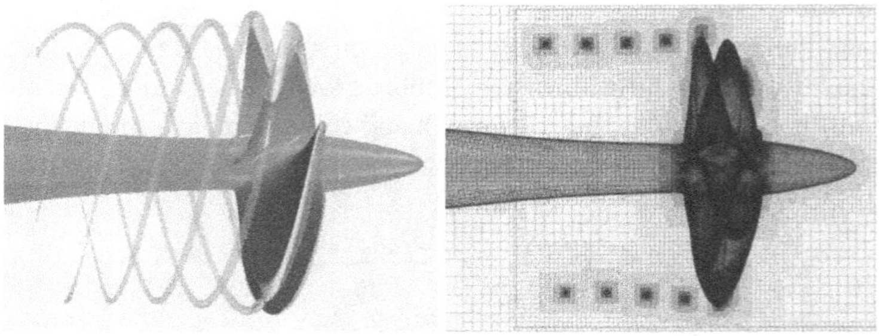

Figure 1. Adaptive Mesh Refinement (MARCS)

Results

Cavitation phenomena, particularly the cavitating tip vortex, have been investigated using EFD and CFD methods for the Princess Royal Propeller. This propeller has been recommended recently by the ITTC [14] as a benchmark propeller for test ranging from open water tests to noise measurement studies.

In 2017, this propeller was tested in the cavitation tunnel of the Shanghai Jiao Tong University (SJTU) as a part of a collaborative study between the University of Strathclyde (UoS) and SJTU. The model propeller was manufactured using a 3.41 scale factor, resulting in model scale diameter of 0.22 m for SJTU model propeller. The main particulars of the propeller and the CAD model of it were supplied by the UoS [15]. The cavitation test matrix was composed according to the test conditions that have been prepared and published before in literature as a part of a round-robin test campaign [16]. Accordingly, 12 different conditions were determined with 2 different J (0.4 & 0.5) and 3 different σ_V values (13.9, 8.1 and 4.5-5.5) using J and K_T similarities. CFD simulations of each condition were conducted simultaneously to allow validation of the computational model for an isolated propeller. Figure 2 shows the comparisons between EFD and CFD results for only one condition at J=0.4 and σ_V=8.1 (Condition 2) where the strongest tip vortex cavitation was observed and with sheet cavitation rolling up to the tip vortex. This condition was selected for comparison with CFD results due to the behavior of the strong tip vortex cavitation. The CFD predictions of the cavitation patterns were created using a volume of vapour (VOF) value of 0.1. The comparison results showed good agreement in terms of propeller performance coefficients -deviation is 2.5% and 5% for K_T and K_Q respectively- as well as the sheet and tip vortex cavitation patterns as shown in Figure 2.

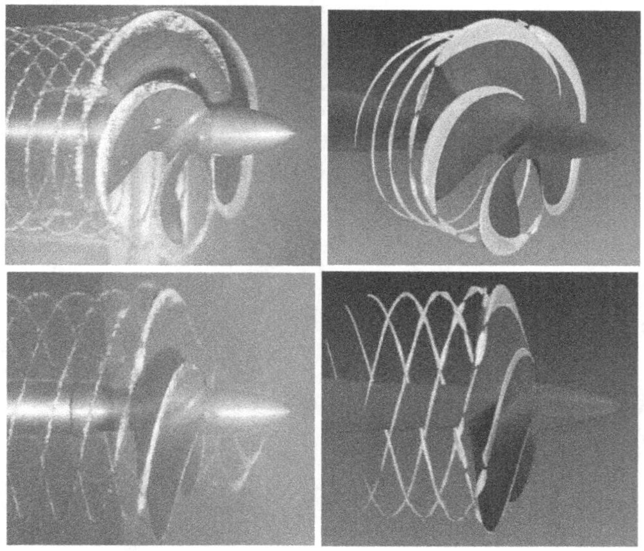

Figure 2. Cavitation comparisons between EFD and CFD Results (Left; EFD, Right;CFD)

Following this achievement in simulating the tip vortex cavitation of the Princess Royal Propeller in uniform flow, the study was extended to simulations in non-uniform flow in the presence of a rudder in order to investigate propeller-rudder interaction with cavitating tip vortices. Thus the Princess Royal Propeller with the rudder arrangement in non-uniform hull wake was simulated using the same mesh refinement approach (MARCS) to allow the tip vortices to interact with the rudder. While the sliding mesh method was used for the case of an isolated propeller, the "overset mesh" method was preferred in order to eliminate interface problems between the rotating and stationary domains for the propeller-rudder interaction simulations.

Figure 3 shows cavitation pattern comparisons between the EFD and CFD results including sheet and tip vortex cavitation. The EFD results were obtained from cavitation tests conducted in the Emerson Cavitation Tunnel, University of Newcastle, as presented in the literature [17]. It can be observed that although the same mesh refinement approach was used for the propeller-rudder interaction case, the CFD prediction of the tip vortex cavitation could not be extended as far as in the isolated propeller simulations (Figure 2 and 3) and hence currently presenting a discrepancy compare to the EFD results.

Bearing in mind the differences in physical conditions for the isolated propeller and the propeller-rudder arrangement cases, which include shaft inclination, non-uniform flow (wake screen) and the presence of a rudder for the latter case, the discrepancy between the two simulation cases can be investigated using Figure 4. In this figure, CFD simulations at 35000, 60000 and 70000 iterations are presented (from Left to Right) while the simulation was running and the generated mesh is rotating with the propeller. At 35000 iterations, a new mesh was generated using the MARCS method as described above. During the simulation, other images from 60000 and 70000 iterations were captured as seen in Figure 4. It can be observed that the generated mesh at 35000 iterations cannot match with the tip vortices that are produced at subsequent time steps in the simulation (i.e. 60000 and 70000 iterations) for the same shape at 35000 iterations. While the vortices were produced for each blade position and each time step and the propeller is rotating with inclined shaft, in non-uniform flow and presence of a rudder, the mesh generated at the beginning was no longer suitable during the subsequent solution time. Thus, tip vortex cavitation cannot be extended for propeller-rudder interaction simulation as much as in the isolated propeller simulations. Although the latter simulations are still under development during the preparation of this paper, so far the tip vortex cavitation simulations have been encouraging for evaluating the propeller performance in cavitating conditions interacting with the rudder in its slipstream.

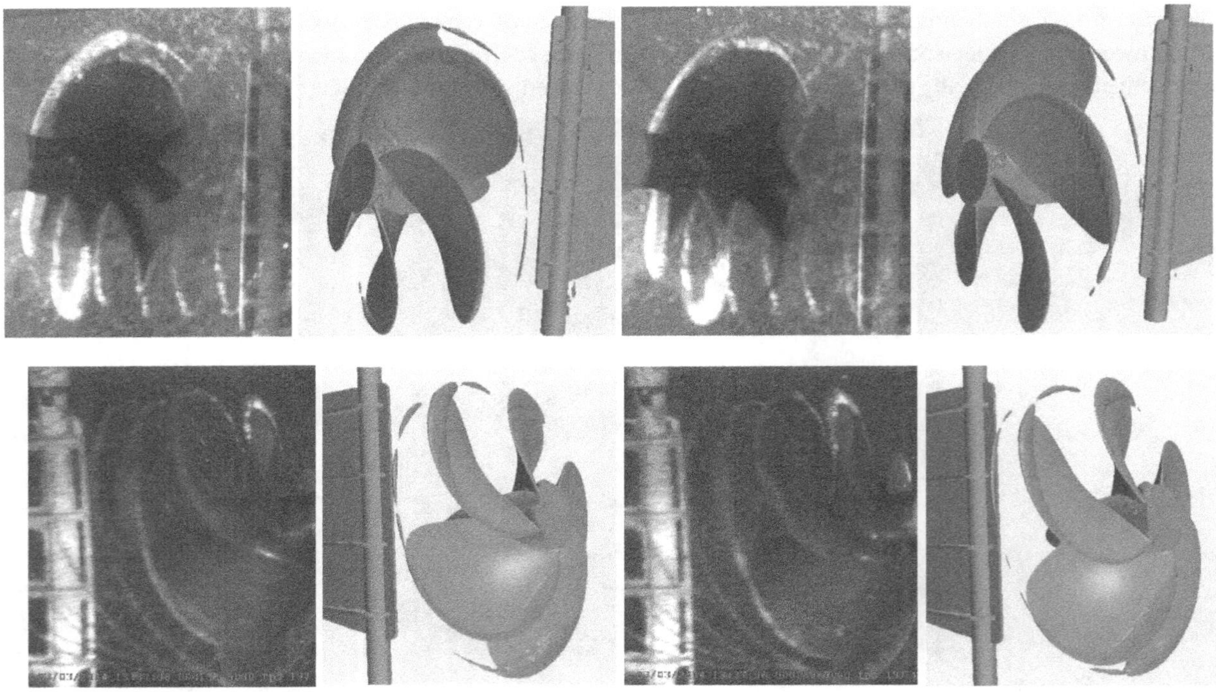

Figure 3. Cavitation comparisons between EFD and CFD Results (Left; EFD, Right;CFD)

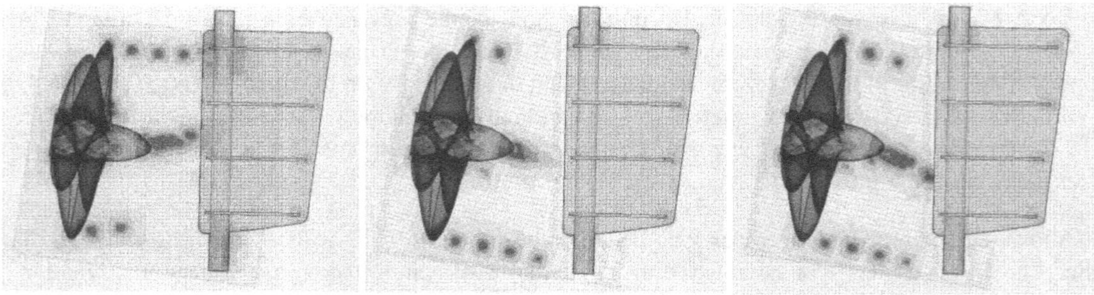

Figure 4. Adaptive Mesh Refienement (MARCS) during solution time (From Left to Right; Mesh from 35000, 60000 and 70000 iterations)

Conclusion

This paper has presented a new and efficient Mesh Adaption and Refinement approach for Cavitation Simulation (MARCS) of marine propellers, particularly for tip vortex cavitation. Successful simulations on an isolated propeller were achieved for the extension of the cavitating tip vortices well downstream of the propeller plane using the new approach such simulations. Further, the Princess Royal propeller was simulated using the same method in non-uniform flow and in the presence of a rudder. The simulation results so far for the latter case present a discrepancy compared to the EFD results and hence require further development in order to extend the tip vortex cavitation through the rudder. It is hoped, in the near future, to achieve further improved simulations of the propeller, rudder and hull flow interaction leading to more accurate prediction of the performance of propellers in cavitating conditions both at model and full-scale which is the ultimate aim of this study.

Acknowledgments

The principal author of this paper is sponsored by the Turkish Ministry of Education during this study. For CFD simulations, the access provided to High Performance Computing for the West of Scotland (Archie-West) through

EPSRC grant no. EP/K000586/1 is gratefully acknowledged. The authors also thank to Prof. Chen-Jun Yang providing opportunity to the principle author for attending the cavitation tunnel tests in SJTU cavitation tunnel.

References

[1] Fine, N. E. 1992, Nonlinear Analysis of Cavitating Propellers in Nonuniform Flow, PhD thesis, Department of Ocean Engineering, MIT.

[2] Lee, H., 2002, Modeling of Unsteady Wake Alignment and Developed Tip Vortex Cavitation, PhD Dissertation, University of Texas at Austin.

[3] Hsiao, C. T., Pauley, L. L., 1998, Numerical Computation of Tip Vortex Flow Generated by a Marine Propeller, 1998 ASME Fluids Engineering Division Summer Meeting FEDSM'98 June 21 - 25, 1998 Washington D.C.

[4] Lee, H., Kinnas, S. A., 2004, Application of a Boundary Element Method in the Prediction of Unsteady Blade Sheet and Developed Tip Vortex Cavitation on Marine Propellers, Journal of Ship Research, Vol. 48, No. 1, March 2004, pp. 15–30.

[5] Hsiao, J. T., Chahine. G.L., 2008, Scaling of Tip Vortex Cavitation Inception for a Marine Open Propeller, 27th Symposium on Naval Hydrodynamics Seoul, Korea.

[6] Gaggero, S., Tani, G., Viviani, M. and Conti, F., 2014, A Study on The Numerical Prediction of Propellers Cavitating Tip Vortex, Ocean Engineering 92(2014)137–161.

[7] Star CCM+ User Guide, 2018

[8] Rayleigh, Lord, 1917, Phi/os. Mag. 34: 94-98

[9] Plesset, M. S., Prosperetti, A., 1977, Bubble Dynamics and Cavitation, Ann. Rev. Fluid Mech. 1977. 9: 145-85

[10] Schneer, G. H., Sauer, J., 2001, Physical and Numerical Modeling of Unsteady Cavitation Dynamics, ICMF-2001, 4th International Conference on Multiphase Flow, New Orleans, USA, May 27-June 1, 2001

[11] Kuiper G., 1981, Cavitation Inception on Ship Propeller Models, Wageningen.

[12] Yilmaz, N., Khorasanchi, M., Atlar, M., 2017, An Investigation into Computational Modelling of Cavitation in a Propeller's Slipstream, Fifth International Symposium on Marine Propulsion smp'17, Espoo, Finland, June 2017

[13] Lloyd, T., Vaz, G., Rijpkema, D., Reverberi, A., 2017, Computational fluid dynamics prediction of marine propeller cavitation including solution verification, Fifth International Symposium on Marine Propulsors smp'17, Espoo, June 2017

[14] ITTC Report, 2017, Specialist Committee on Hydrodynamic Noise Final Report and Recommendations to the 28th ITTC.

[15] Atlar M., Aktas B., Sampson R., Seo K.C., Viola I.M., Fitzsimmons P., Fetherstonhaug C., 2013, A Multi-Purpose Marine Science & Technology Research Vessel For Full Scale Observations And Measurements, 3rd International Conference on Advanced Model Measurement Technologies for the Marine Industry, Jan.

[16] Tani, G., Aktas, B., Viviani, M., Atlar, M., 2017, Two medium size cavitation tunnel hydro-acoustic benchmark experiment comparisons as part of a round robin test campaign, Ocean Engineering 138 (2017) 179–207.

[17] Aktas, B., Atlar, M., Turkmen, S., Shi, W., Sampson, R., Korkut, E., Fitzsimmons, P. Propeller cavitation noise investigations of a research vessel using medium size cavitation tunnel tests and full scale trials, Ocean Engineering Volume 120, 1 July 2016, Pages 122-135.

© 2018 ASME

An Investigation of Tip-Vortex Turbulence Structure using Large-Eddy Simulation

[1]David Hanson*; [1]Michael Kinzel; [1]Kyle Sinding; [1]Michael Krane

[1]*The Pennsylvania State University, University Park, PA, USA*

Abstract

In this effort, the tip vortex behind a rectangular plan-form lifting fin with a cavitation-relevant airfoil section (NACA-66 mod) is studied using Large Eddy Simulation (LES) at a free-stream Reynolds number of 500,000. Elementary flow quantities (lift, drag, velocities) are compared with measurements made in the Garfield Thomas Water Tunnel (GTWT) 12-inch circular test-section. Good agreement is seen with the measured lift and drag. The characteristic 45-degree lag between the mean strain rate and the Reynolds stress components in the plane of the tip vortex is observed. The calculated flow field provides insight into the turbulent structure of the tip vortex for this fin.

Keywords: tip-vortex, turbulence, computational fluid dynamics

Introduction

Cavitation within tip-vortices, *i.e.*, Tip-Vortex Cavitation (TVC), is an important aspect relevant to the design of propellers for underwater vehicles [1]. One specific adverse aspect is erosion associated when a cavitating vortex impacts downstream surfaces. Understanding the complicated turbulent flow-field in the region of TVC inception is necessary in order to develop Computational Fluid Dynamics (CFD) models to predict TVC.

In the present literature, the authors found a limited database of detailed turbulent structure information within tip vortices. The turbulence, however, directly informs the overall tip-vortex evolution and the development of TVC. The cavitation behavior within a tip vortex is informed by the turbulent pressure and velocity fluctuations. The challenges involved in measuring experimentally both the time-varying pressure and velocity fields within a tip vortex are considerable. As such, the established understanding of the correlation between these quantities is limited. The present work involves a numerical study using Large-Eddy Simulation (LES) of a tip-vortex flow around a rectangular fin to gain detailed insight into the resulting turbulent flow-field. The tip-vortex flow geometry in question was recently studied experimentally by Sinding and Krane [2].

For the present work, the authors will interrogate the unsteady LES flow-field solution to extract the fine details of the turbulent structure. LES will, thus, be used to develop a thorough understanding of Reynolds-Averaged Navier-Stokes (RANS) turbulence model failure in tip-vortex flows.

Description of the Flow

The fin geometry considered in this work was studied experimentally by Sinding and Krane [2]. The chord is 4 inch (0.1016 meter) and the aspect ratio is two. The fin has an NACA-66(mod) thickness distribution [3] wrapped around an NACA 4-digit camber line. The maximum camber in the chord-wise direction is $0.0192 \times$ chord, located at 50% chord. The camber is uniform along the span and was chosen such that at zero incidence the lift coefficient is 0.4 with a span-wise lift distribution that is approximately elliptical. A similar fin was also studied experimentally by Hanson [4]; however, the fin at that time had a geometry with a rounded tip. Sinding and Krane modified the fin model to have a flat tip in order to simplify the boundary-layer detachment as the flow enters the tip vortex. Because the shape of the fin tip is expected to influence strongly the tip-vortex dynamics, this paper focuses only on the flat-tipped fin studied by Sinding and Krane. The physical fin model is shown in the 12-inch (0.3048 meter) circular test-section water tunnel at Penn State's Applied Research Laboratory in Figure 1.

The present computational model includes the fin geometry and the circular water tunnel section. The free-stream speed is 15 ft/s (4.57 m/s), yielding a moderate Reynolds number of $\sim 5 \times 10^5$. The tunnel is modeled in order for the calculated lift and drag to be directly comparable with the experimental measurements without correcting for the blockage or the presence of walls. The present work uses a geometry and computational mesh corresponding to that presented in Figure 2.

*Corresponding Author, David Hanson: drh5013@psu.edu

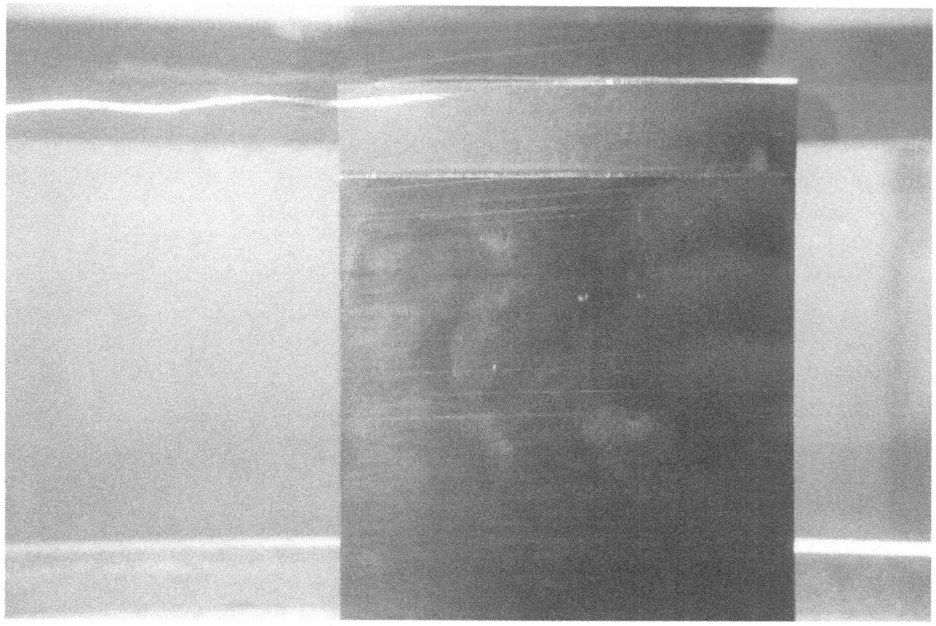

Figure 1: Side view of the fin in the water tunnel. TVC is visible near the top of the figure downstream of the tip. Flow is right-to-left.

Flow Solver

In the present work, the discretized LES equations of fluid motion are solved in the context of the commercial CFD suite Star-CCM+ [5]. Star-CCM+ is developed and maintained by CD-Adapco, a division of Siemens Corporation. Star-CCM+ is capable of many different spatial and temporal solution methods; only those relevant to the present work are discussed here.

The spatial- and temporal-discretization schemes used are formally second-order accurate. Pressure-velocity coupling in the present incompressible flow is accomplished at each time step in a segregated manner via the SIMPLE algorithm, wherein a discrete pressure-Poisson equation provides an update to the solution of the discretized momentum conservation equations within each linear sub-iteration. All solution variables are located at the cell centers; facial fluxes are approximated via a Total-Variation Bounding (TVB) central-difference scheme.

Equations of Fluid Motion

The subgrid-filtered equations of mass- and momentum-conservation for LES of an incompressible flow are written as

$$\frac{\partial \overline{u}_i}{\partial x_i} = 0 \qquad (1)$$

and

$$\frac{\partial \overline{u}_i}{\partial t} + \frac{\partial}{\partial x_j}\left(\overline{u}_i \overline{u}_j + \delta\overline{p} - \overline{T}_{ij}\right) = \frac{\partial \tau_{ij}^{r}}{\partial x_j} \qquad (2)$$

where the overline in this context refers to variables that have undergone the subgrid filter.

Formally, the residual stress tensor $\tau_{ij}^{r} = \overline{u_i u_j} - \overline{u}_i \overline{u}_j$ represents contributions from three types of interactions: interactions among resolved scales, interactions among the unresolved scales, and cross-interactions between resolved and unresolved scales. These are typically called the Leonard stress, the Reynolds stress, and the Clark stress, respectively. In practice, however, this term is modeled via a Boussinesq approximation, $\tau_{ij}^{r} = 2\left(\nu_{t,SGS} S_{ij} - \frac{\delta_{ij} k_{SGS}}{3}\right)$, where k_{SGS} is the subgrid-scale turbulent kinetic energy. The subgrid turbulent viscosity, $\nu_{t,SGS}$, is given by a simple function evaluation in terms of the local strain rate and the mesh size. The Wall-Adapting Large Eddy (WALE) constitutive model is used to define $\nu_{t,SGS}$ in the present work [6].

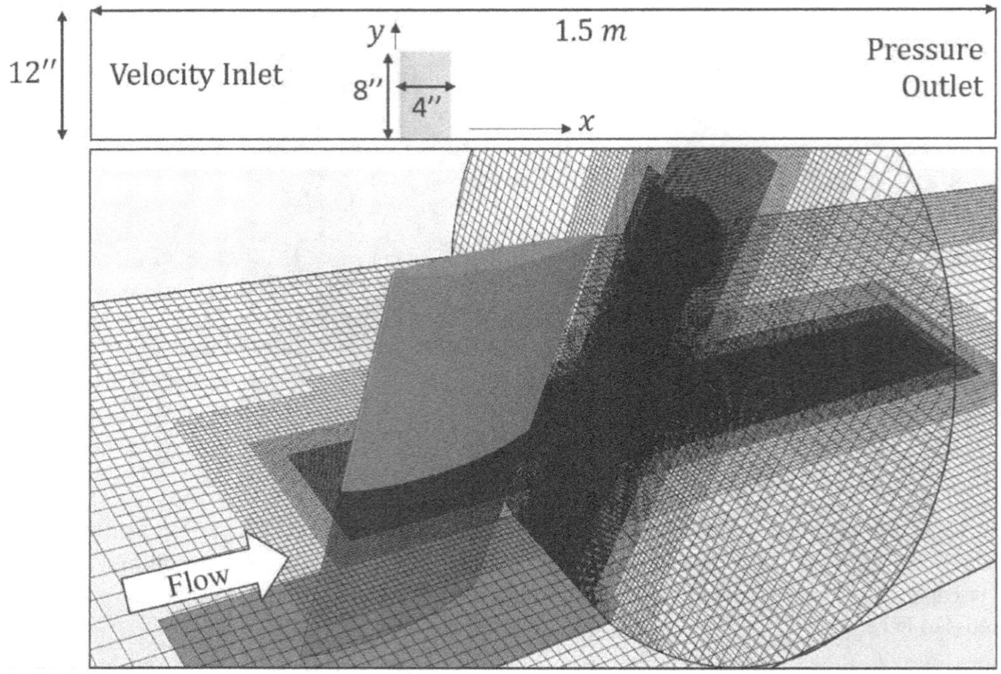

Figure 2: Top: *geometry of the flow over the finite-span fin in the 12-inch (0.3048 meter) circular test section of the water-tunnel. The z-direction points into the page and completes the right-handed set. The origin is coincident with where the fin leading edge meets the tunnel wall.* Bottom: *the refinement regions overlaid upon the fin geometry.*

Discretization

The flow equations of motion are discretized and solved on an unstructured trimmed Cartesian mesh with prism layers. This mesh type was chosen because it allows the best compromise between the competing goals of minimizing the numerical dissipation, minimizing the total cell count required to resolve the geometry, and minimizing the mesh generation time. Within the high-resolution turbulence-resolving region, the mesh is uniform and rectilinear with the exception of the thin near-wall prism layer. Such a uniform mesh reduces the importance of numerical diffusion and errors due to the gradient reconstruction.

The computational mesh used in the present study has approximately 70×10^6 cells. As mentioned by Georgiadis *et al.* [7], the spatial resolution of the mesh used for LES is typically determined by the maximum limit of the computing resources available. That is also the case here. Rather than resolving the viscous sublayer, the lowest region of the boundary layer at the surface of the fin is approximated using a blended wall function based on Reichardt's law. The dimensionless distance, $y^+ = \Delta y\, u^*/\nu$, from the wall to the nearest cell center, is between 20 and 100 on the fin surface. In outer variables, this yields a wall-normal cell size of $\frac{\Delta y}{\delta} \approx 0.02$, consistent with the recommendation of Larsson *et al.* [8] for LES. Because this is a relatively simple wing geometry, without strong pressure gradients or separated flow regions, use of wall functions is not expected to be detrimental to the solution quality. The linear size of wall cells in the wall-parallel direction is 0.25 mm, or x^+, $z^+ \sim 50$. The near-wall region of mesh refinement is shown in Figure 2. The present mesh provides approximately 40 grid points across the vortex core diameter within the region of highest refinement. Using the computed solution, the Taylor microscale, λ, is approximated and compared to the computational cell size in the Results section.

The time step is chosen such that the convective Courant number is less than unity in almost all computational cells. For the present flow, a time step of 25 microseconds satisfies this requirement. Ten linear iterations are performed within each non-linear time step. The solution was run for a total simulation time of 0.55 seconds; this corresponds to a simulation time of approximately 24 convective time scales. Time-averages for the various quantities discussed in the Results section were recorded for approximately the last quarter second, or half the total simulation time, in order to avoid contaminating the averages with flow transients.

Turbulence Quantities

A goal of this study is to provide quantitative information regarding the detailed turbulent structure within the tip vortex as it undergoes the wrap-up process. Toward this end, every term both in the Reynolds Stress (RS) equation and in the much simpler Turbulent Kinetic Energy (TKE) equation is recorded at every fourth time step on two planes (at one-quarter and one chord downstream of the trailing edge) across the tip vortex. Furthermore, time-averaged values of every term in the TKE equation are recorded at every cell in the domain.

In the context of a Cartesian coordinate system, the transport equation for each component of the RS tensor for an incompressible Newtonian flow is:

$$\frac{\partial}{\partial t}\left(\overline{u_i'u_j'}\right) + \bar{u}_i \frac{\partial}{\partial x_i}\left(\overline{u_j'u_k'}\right) = P_{RS} + DS_{RS} + PDC_{RS} + DF_{RS,1} + DF_{RS,2} + DF_{RS,3} \quad 3$$

where the symbols on the right side of the equation represent tensors that express turbulence production, pseudo-dissipation, the correlation between pressure and strain, and three types of diffusion, respectively. The forms for these terms used in the present work are summarized in Table 1. Each term in this table is a symmetric second order tensor and therefore has six independent components that must be recorded.

P	$-\overline{u_j'u_k'}\frac{\partial \bar{u}_j}{\partial x_i} - \overline{u_i'u_j'}\frac{\partial \bar{u}_k}{\partial x_i}$
DS	$-2\nu\overline{\frac{\partial u_k'}{\partial x_i}\frac{\partial u_j'}{\partial x_i}}$
PSC	$\overline{\frac{p'}{\rho}\left(\frac{\partial u_j'}{\partial x_k} + \frac{\partial u_k'}{\partial x_j}\right)}$
DF_1	$\frac{\partial}{\partial x_i}\left(-\overline{u_i'u_j'u_k'}\right)$
DF_2	$\nu\frac{\partial^2}{\partial x_i \partial x_i}\left(\overline{u_j'u_k'}\right)$
DF_3	$\frac{\partial}{\partial x_i}\left(-\overline{\frac{p'}{\rho}\left(\delta_{ij}u_k' + \delta_{ij}u_j'\right)}\right)$

P_{TKE}	$-\overline{u_i'u_j'}\frac{\partial \bar{u}_j}{\partial x_i}$
ϵ	$\nu\overline{s_{ij}'s_{ij}'}$
$DF_{TKE,1,pressure}$	$\frac{\partial}{\partial x_i}\left(-\frac{\overline{u_i'p'}}{\rho}\right)$
$DF_{TKE,1}$	$\frac{\partial}{\partial x_i}\left(-\overline{u_i'u_j'u_j'}\right)$
$DF_{TKE,2}$	$\nu\frac{\partial^2 k}{\partial x_i x_i}$
$DF_{TKE,3}$	$\nu\frac{\partial^2}{\partial x_i x_i}\left(\overline{u_j'u_j'}\right)$

Table 1: Meaning of terms in the RS equation (left) and the TKE equation (right)

A common simplification of the RS tensor equation is to assume that the dynamics of turbulence is controlled by the trace of the RS tensor, interpreted as the TKE, $k = \frac{1}{2}\left(\overline{u'^2} + \overline{v'^2} + \overline{w'^2}\right)$. This reduces the six tensor RS equations to a scalar equation for k:

$$\frac{\partial k}{\partial t} = \overline{u}_i \frac{\partial k}{\partial x_i} = P_{TKE} - \epsilon + DF_{TKE,1,pressure} + DF_{TKE,1} + DF_{TKE,2} + DF_{TKE,3} \qquad 4$$

where the right-hand-side symbols represent turbulence production, dissipation, and four types of diffusion, respectively. These are also summarized in Table 1.

The approach taken for the present work is to track the instantaneous and mean values of the five field variables for the TKE equation and the instantaneous values for the further 36 variables for the RS equation. This is in addition to seven field variables for RS and TKE. All instantaneous variables in the RS equation were exported every fourth time step on two planes normal to the tip vortex located 0.25 × chord and 1 × chord downstream of the trailing edge, respectively; time-averaged quantities for the RS equation are therefore available only on these planes.

Validation

The lift- and drag-coefficients, *i.e.*, C_L and C_D, respectively, computed for the present flow are compared with experimental measurements made by Hanson for this fin [4] in Figure 3. Recall that the fin used in those experimental measurements had a rounded geometry in contrast to the CFD model. It is not anticipated that the tip geometry greatly alters C_L and C_D. Agreement between the present CFD and experimental measurements is quite close.

TKE spectra are shown for two different point probes within the tip vortex in Figure 3. The velocity point probes are located on the inboard and outboard edges of the tip vortex, respectively, 1 inch (0.0254 meter) downstream of the trailing edge. The expected $f^{-5/3}$ decay of the kinetic energy with respect to frequency is observed at higher frequencies at both locations.

In the present work, λ is approximated from the flow solution as $\lambda \sim \sqrt{10\nu k/\epsilon}$. The ratio of this length scale to the computational cell linear size, h, is shown in Figure 4 on a planar slice through the tip vortex at an axial location one chord downstream of the trailing edge. It is seen that the present mesh has approximately five cells per λ within the wake and at the edge of the vortex core.

Elementary Turbulence Quantities

For the sake of brevity, only one elementary turbulence quantity is mentioned here. The rightmost image in Figure 4 compares the $y - z$ component of the mean strain rate tensor, S_{ij}, with the corresponding component of the specific Reynolds stress tensor, τ_{ij}. Chow [9] observed that the Reynolds shear stress in the plane of the tip vortex tends to lag the corresponding component of the mean strain rate by 45°. It is worth noting that RANS turbulence models that use a linear Boussinesq closure expression are unable to produce Reynolds stresses that are not aligned with the mean strain rate. Several authors (see, for example, Churchfield and Blaisdell [10]) have proposed that this weakness in the linear Boussinesq approximation makes a key contribution to RANS models diffusing tip vortices much more quickly than is seen in experiments. The present simulation shows the expected 45° offset between the lobes of S_{23} and τ_{23} once the vortex rollup is complete approximately one chord downstream of the trailing edge.

Future Work

The remaining terms in the TKE equation are to be extracted from the flow solution and explored in detail. This will include determining the dominant terms in the turbulent energy budget as the vortex evolves downstream of the trailing edge.

Furthermore, the pressure variance, $(p')^2_{rms}$, will be compared with a common approximate expression for the pressure fluctuations, $p' \sim 0.39\rho k$. This approximation is appropriate for isotropic, homogeneous turbulence; however, it will be seen that it does not adequately capture the character of the turbulent pressure fluctuations within the highly ordered, anisotropic vortex core. This has implications for RANS-based cavitation models.

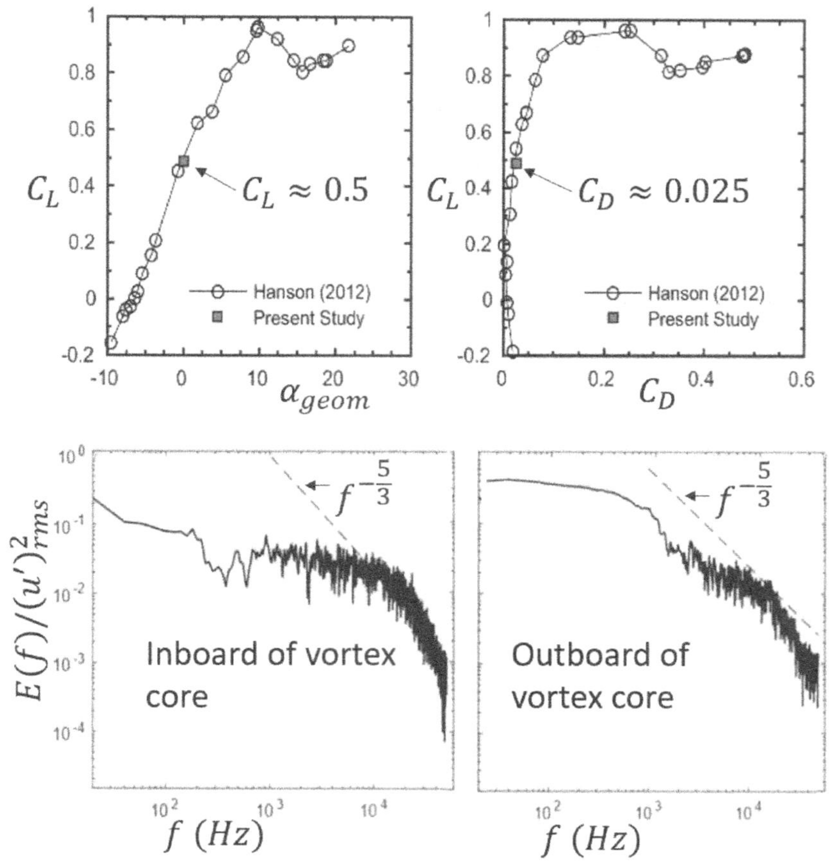

Figure 3: Top Row: *C_L from the present study compared with the measurements of Hanson [4]. C_D measurements from that study are compared with the present CFD in the rightmost figure.* Bottom Row: *spectra of turbulent velocity fluctuations at two locations in the flow: immediately inboard (left) and outboard (right) of the vortex core; both probes are located one inch downstream of the trailing edge. The specific kinetic energy in the flow, $E = \frac{1}{2}u^2$, is made dimensionless by the local velocity variance, u'^2_{rms}.*

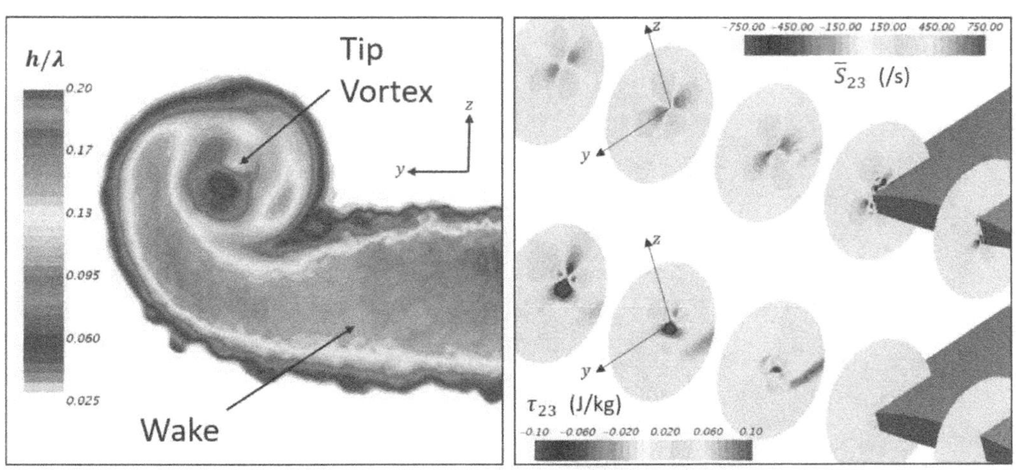

Figure 4: Ratio of cell linear size, h, to the approximate λ. Right: comparison of the mean shear rate component $\bar{S}_{23}$ with the corresponding Reynolds stress component τ_{23}.

References

[1] Barnes W McCormick, Jr. *A Study of the Minimum Pressure in a Trailing Vortex System.* Doctoral dissertation, Pennsylvania State University, 1954.

[2] K Sinding and M Krane. "Tip vortex core pressure estimates derived from velocity field measurements." In *APS Division of Fluid Dynamics Meeting Abstracts*, 2016.

[3] T Brockett. "Minimum pressure envelopes for modified NACA-66 sections with NACA A= 0.8 camber and BuShips type 1 and type 2 sections." Technical report, David Taylor Model Basin, 1966.

[4] David Hanson. *Cavitation Inception for Non-Elliptically Loaded Fins*. Master's thesis, The Pennsylvania State University, 2012.

[5] Siemens. Star CCM+ v 12.04 User Manual, 2017.

[6] F. Nicoud and F. Ducros. "Subgrid-scale stress modelling based on the square of the velocity gradient tensor." *Flow, Turbulence and Combustion,* 62(3):183–200, 1999.

[7] Nicholas J Georgiadis, Donald P Rizzetta, and Christer Fureby. "Large-Eddy Simulation: Current Capabilities,

[8] Recommended Practices, and Future Research." *AIAA Journal*, 48(8):1772–1784, 2010.

[9] Johan Larsson, Soshi Kawai, Julien Bodart, and Ivan Bermejo-Moreno. "Large eddy simulation with modeled wall-stress: recent progress and future directions." *Bulletin of the JSME Mechanical Engineering Reviews*, 3(1), 2016.

[10] Jim Chow, Greg Zilliac, and Peter Bradshaw. "Turbulence Measurements in the Near Field of a Wingtip Vortex." 1997.

[11] Matthew J Churchfield and Gregory A Blaisdell. Numerical Simulations of a Wingtip Vortex in the Near Field. *Journal of Aircraft*, 46(1):230–243, 2009.

© 2018 ASME

10th International Symposium on Cavitation - CAV2018
Baltimore, Maryland, USA, May 14 – 16, 2018

CAV18-05019

Analysis of the Finite Mass Transfer Models in the Numerical Simulation of Bubbly Flows

Ebrahim Ghahramani*; Rickard E. Bensow;

Chalmers University of Technology, Gothenburg, Sweden

Abstract

In this study, the effect of empirical constants on the performance of the finite mass transfer models as well as some numerical issues of the model are investigated through the numerical analysis of the Rayleigh problem and the collapse of a bubble cluster. Also, by implementing the exact bubble radius profile into the model, it is shown that the pressure field can be very well estimated by the model.

Keywords: cavitation modeling; finite mass transfer model; bubble collapse

Introduction

A common numerical method for simulation of cavitating flows in large scale applications (e.g. marine propellers) is the homogeneous mixture modeling based on the finite mass transfer assumption. In this approach, the multiphase fluid is considered as a homogeneous mixture and a scalar transport equation is solved to find the liquid-vapor interface based on the interface-capturing scheme and the mass transfer between the phases is modeled using empirical terms. While these models are used to study turbulent sheet and cloud cavity structures with reasonable accuracy, they are limited in resolving small scale structures such as cavitation nuclei and bubbles. One source of such a limitation is the simplification behind the derivation of the mass transfer terms. In fact, most of the popular models are based on a simplified form of the well-known Rayleigh-Plesset equation in which the second-order derivative term is ignored. As cavity inertia becomes more important in the last steps of its collapse, such a simplification can decrease model accuracy in capturing bubble collapse and rebound. Further, some empirical coefficients are introduced to the mass transfer term which usually need to be tuned for different applications.

In this paper, the capability of the finite mass transfer model in the prediction of cavitating flows is investigated through a study of the effect of the empirical coefficients; also some numerical issues in pressure estimation are discussed. The Schnerr-Sauer model [1] is used to obtain mass transfer rate and two test cases are chosen for this study. The first case is the collapse of a vapor bubble under a far-field pressure higher than the characteristic vapor pressure. The second problem is the collapse of a cluster of bubbles in which the collapse of each bubble is influenced by the dynamics of surrounding structures. Then, a different approach is used to find the exact single bubble interface, which help to clarify one aspects of the mass transfer model limitation in appropriate estimation of pressure field of the cavitating flows.

Method

In this approach, the general continuity and Navier-Stokes equations are solved to calculate the main flow field and the interface capturing scheme based on the Volume of Fluid (VOF) concept is used to calculate the liquid-vapor interface. Therefore, a scalar transport equation for the liquid volume fraction, α, is solved and the liquid-vapor mass transfer at the interface is considered as a source term to this equation as

$$\frac{\partial \alpha}{\partial t} + \frac{\partial (u_i \alpha)}{\partial x_i} = \frac{\dot{m}}{\rho_l}, \tag{1}$$

where u_i is the velocity vector and ρ_l is the liquid density. There are various empirical models in literature to estimate the mass transfer rate, $\dot{m}$. In this study, the Schnerr-Sauer model [1] of OpenFOAM is used in which the condensation and vaporization rates are given by

$$\dot{m}_c = C_c \alpha (1-\alpha) \frac{3\rho_l \rho_v}{\rho_m R_B} \sqrt{\frac{2}{3\rho_l |p - p_{sat}|}} \max(p - p_{sat}, 0), \tag{2a}$$

$$\dot{m}_v = C_v \alpha (1 + \alpha_{Nuc} - \alpha) \frac{3\rho_l \rho_v}{\rho_m R_B} \sqrt{\frac{2}{3\rho_l |p - p_{sat}|}} \min(p - p_{sat}, 0). \tag{2b}$$

*Corresponding Author, Ebrahim Ghahramani: ebrahim.ghahramani@chalmers.se where R_B and α_{Nuc} are the generic radius and volume fraction of bubble nuclei which can be considered as empirical parameters. ρ_v and ρ_m are the vapor and mixture density, respectively and p_{sat} is the saturation pressure of the fluid. Finally, C_c and C_v are the empirical condensation and vaporization coefficients implemented in OpenFOAM, respectively. R_B and ρ_m are functions of liquid volume fraction and are updated at each time step. To simplify the analysis, all other parameters are kept constant and only condensation and vaporization coefficients are modified. In this study, the liquid and vapor densities are 1000 kg/m^3 0.01389 kg/m^3, saturation pressure is 2330 kg/ms^2 and α_{Nuc} is set to $5*10^{-5}$ by considering 10^8 nuclei/m^3 with initial diameter of 0.1 mm.

The mixture properties (including density) are updated at each time step based on the liquid volume fraction at each cell. However, pure liquid and vapor are assumed to be incompressible. The governing equations are solved in the open source C++ package OpenFOAM using *interPhaseChangeFoam* solver.

Result

To investigate performance of the model, the simple problem of *Rayleigh bubble collapse* is studied first. For this problem, the analytical solution for evolution of bubble radius as well as its surrounding pressure are available in literature [2]. Here, the collapse of a vapor bubble in an infinite medium with atmospheric pressure is simulated and the effects of viscosity, non-condensable gas, and surface tension are ignored. The initial bubble radius is 0.4 mm and the vapor pressure is assumed to be 2340 Pa. Considering the spherical symmetry of the problem, only one cell layer of one eighth of the domain is simulated, using corresponding symmetry and wedge boundary conditions. The far-field is located 0.5 m from the bubble center. A polar grid is used to discretize the domain and the initial bubble is resolved with 20 radial cells (i.e. $\Delta_r/R_0 = 0.05$, around the bubble). In figure 1, the generated grid and the initial Laplacian pressure field is depicted. The solution time step is $5*10^{-9}$ s.

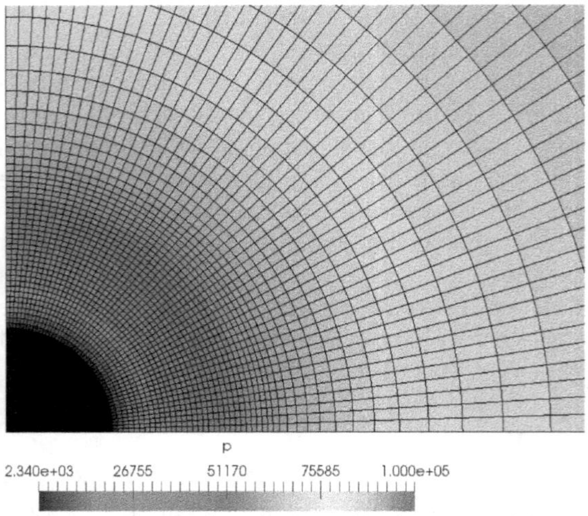

Figure 1: Domain discretization and initial pressure field

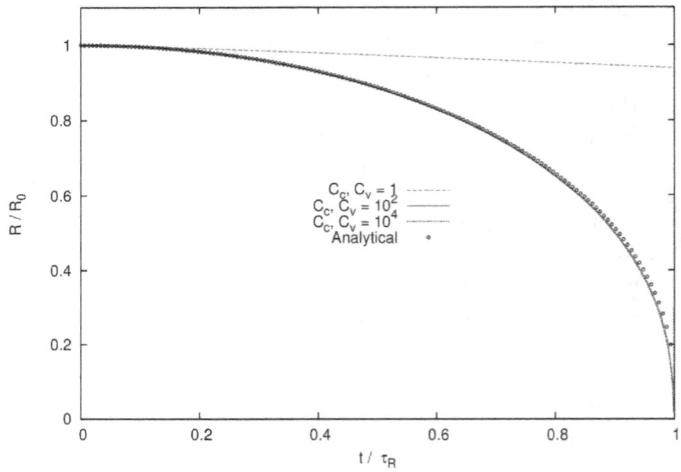

Figure 2: Temporal evolution of bubble radius

In figure 2, the temporal evolution of bubble radius with different mass transfer coefficients is compared with the analytical solution. The solution time and bubble radius are normalized with Rayleigh collapse time (τ_R) and initial radius (R_0), respectively. According to the figure, for small coefficients, the finite mass transfer model is incapable in estimation of bubble radius profile and the empirical constants should be larger than a minimum value. However, to have a better understanding of the model behavior, the calculated pressure profile should be investigated as well. In figure 3, the estimated pressure profiles in the radial direction around the bubble interface are compared to analytical data at three instances of the last stages of bubble collapse. In this figure the radial distance from the bubble center (R) is normalized with bubble initial radius (R_0) and the vertical axes shows the normalized pressure profile, given by $\Pi = (p - p_\infty)/(p_\infty - p_v)$. According to the figure, the pressure field is highly dependent on the mass transfer coefficients. In fact, while both moderate ($C = 10^2$) and large coefficients ($C = 10^4$) predict similar radius profiles, their pressure estimations are considerably different. For the lower coefficient, the estimated pressure inside the bubble is larger than the vapor pressure ($\Pi > -1$). Also, some numerical pressure pulses are emitted from bubble interface which makes the pressure values at $t / \tau_R = 0.921$ to be larger than the corresponding values at a later time ($t / \tau_R = 0.948$). If we increase the mass transfer coefficient to $C = 10^4$ these numerical pulses have almost disappeared from the solution and the pressure inside the bubble is estimated correctly. However, the pressure peaks around the interface is underestimated and the pressure lines are shifted a little as compared to analytical data. It can be shown that if the coefficients are increased further, the solution does not change. Therefore, to have an appropriate prediction of single bubble collapse, the coefficients should be set large enough. This conclusion is in agreement with results of Schenke and van Terwisga [3] using a different mass transfer model. It should be emphasized that increasing the model coefficients can cause numerical instability in the solution and special convergence measures should be applied to the solver to guarantee a converged and stable solution. To further investigate the model performance, the problem is solved with a coarser grid in which the initial bubble is discretized with 12 cells (i.e. $\Delta_r/R_0 = 0.083$). In figure 4, the calculated pressure profiles with empirical coefficients of 10^4 are compared with the corresponding ones of the fine grid. It is seen that, even with the high mass transfer coefficients, considerable numerical pressure pulses are existed in the solution when the grid resolution is not fine enough. This is an important point, since in typical engineering problems, the small cavity structures are not discretized with very fine grids.

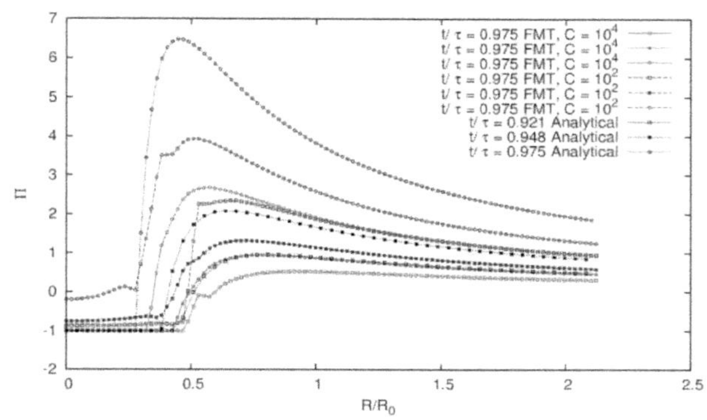

Figure 3: Pressure estimation of the finite mass transfer model with different empirical coefficients

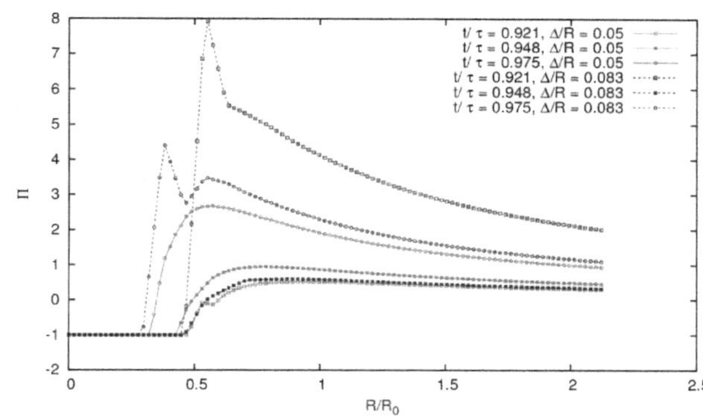

Figure 4: Pressure estimation of the finite mass transfer model with different domain discretization; C = 10^4

To study the mass transfer coefficient effect for more complex problems, the collapse of a cluster of bubbles is solved and the obtained solution is compared to the results of Schmidt et al. [4], where a compressible solver using thermodynamic equilibrium equation of state for the liquid, mixture, and vapor. In this problem, a cluster of 125 bubbles over a flat wall is exposed to an external pressure of 40 bar at infinity and the collapse resulting pressure impacts on a bottom wall are measured. The bubble cluster is in a small cube with an edge length of 20 mm. This cube is discretized uniformly by 55^3 cells. The farfield boundaries are located on the edges of a larger domain with 4*4*2 m^3 dimensions and the solution time step is $3.9*10^{-8}$ sec. For further description of the problem, the reader is referred to [4]. Here, the collapse of each bubble can affect the pressure field around the surrounding bubbles. In figure 5 the obtained pressure impacts on the bottom wall are compared with the equilibrium model results for different mass transfer coefficients. From the equilibrium model result, it is seen that, due to the collapse of different bubbles, some pressure pulses are imposed on the wall which can be seen as local peaks in the plot. For the finite mass transfer model, however, the results are highly dependent on the mass transfer coefficient. If the coefficients are low, no pressure pulse is seen from individual bubble collapse and the wall pressure impact increases smoothly to the maximum value which corresponds to the final violent collapse and after the collapse it decreases smoothly. When the coefficients are increased to moderate values (C = 10^2), some pressure pulses are imposed on the wall and the maximum pressure value is estimated much larger than the corresponding value of compressible equilibrium model. When the coefficients are increased further to high values (C = 10^4), the peak pressure estimation is decreased again, and the local peaks of the profile are changed both in value and position. Also, according to the previous findings of single bubble collapse, it is not assured which of these local peaks are physical and which of them are only spurious numerical pulses.

To investigate possible sources of this inconsistency in the finite mass transfer approach, the simple Rayleigh collapse problem is considered again. For this problem, it is possible to have the exact bubble radius evolution from the solution of Rayleigh-Plesset equation. Then this exact profile of the cavity interface can be used investigate the model capability in prediction of collapse pressure. In other words, a modified finite mass transfer approach is used in which the exact volume fraction distribution is available from the Rayleigh-Plesset equation and the mass transfer model is only used to obtain the continuity equation source term to calculate the pressure field. The obtained pressure profile is compared to analytical data in figure 6, for both coarse and fine grids. In this case, the mass transfer coefficients are set to high values (C = 10^4). According to the fine grid solution, the pressure field can be very well estimated by the mass transfer model if the coefficients are high enough and the pressure peaks and their locations are well-captures without any shift in the profile. Also, from the coarse grid solution, it is concluded that, even with less accurate discretization, the model has an acceptable accuracy and no numerical pressure pulse is generated in the domain. It should be noticed that the previously mentioned numerical instability problems for the model do not exist in the new solutions. In summary, the mass transfer model can predict the pressure profile with reasonable accuracy and without any numerical issue, provided that the coefficients are high enough. In other words, one major source of the mass transfer model inconsistencies in prediction of cavitation problems is in the solution of the scalar transport equation of the liquid volume fraction. Inheriting the interface capturing scheme nature, this equation needs very fine grid and special solution criterion to sufficiently resolve the cavity interface and avoiding numerical pulses. Also, in order to have more stable solution and diagonal dominancy of the coefficient matrix this equation is usually discretized as

$$\frac{\partial \alpha}{\partial t} + \frac{\partial (u_i \alpha)}{\partial x_i} = \frac{\dot{m}}{\rho_l} + \alpha \left(\frac{\partial u_i}{\partial x_i} - \frac{\partial u_i}{\partial x_i} \right). \tag{4}$$

While theoretically, the two velocity divergence terms should be identical, one term is replaced by the continuity equation source term and the other term is calculated from the velocity field. Since the continuity equation and the volume fraction transport equations (and its corresponding source term) are not solved simultaneously for each solution iteration, these two terms are not necessarily equal and it can cause some numerical error in the solution of the volume fraction transport equation.

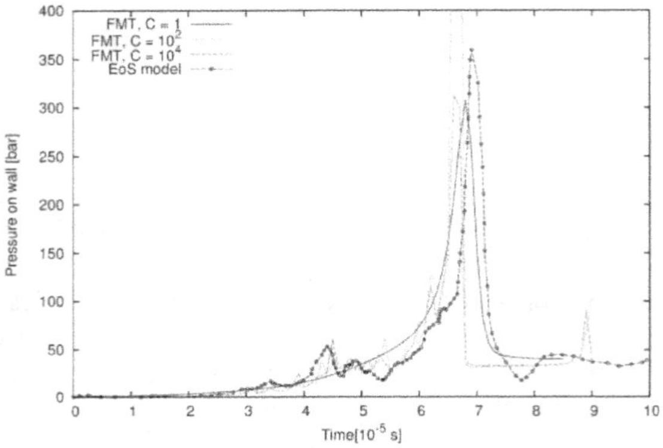

Figure 5: Pressure impact estimation of the finite mass transfer model with different coefficients for the collapse of a bubble cluster

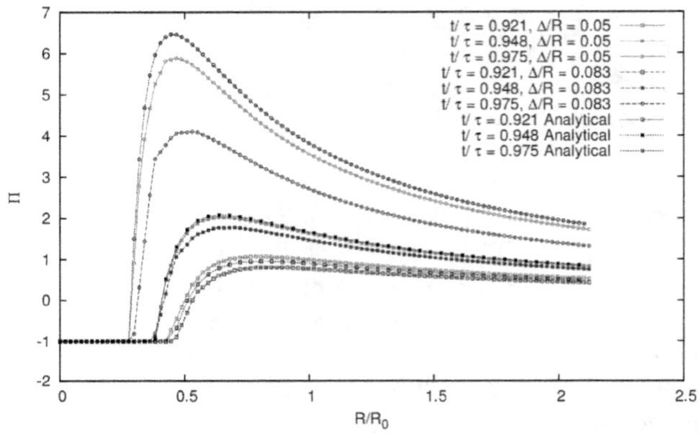

Figure 6: Estimated pressure profile of the finite mass transfer model with the exact solution of the bubble interface

Conclusion

From the solution of the single bubble collapse, it is concluded that for resolving the collapse event the empirical coefficients of the mass transfer model should be large enough and very fine grids are needed. Otherwise, the bubble collapse may have numerical delay for small coefficients or numerical pulses may happen. This can be problematic for the more complex cases when the collapse of each bubble may influence the dynamics of other structures. It is shown that a major source of this issue is the numerical problems in solving of vapor transport equation.

Acknowledgement

This work is funded through the H2020 project CaFE, a Marie Skłodowska-Curie Action Innovative Training Network project, grant number 642536. The computations were performed on resources at Chalmers Centre for Computational Sciences and Engineering (C3SE) provided by the Swedish National Infrastructure for Computing (SNIC). We thank Dr. Steffen Schmidt (TU Munich) for sharing the bubble cluster test case data as well as the inspiring discussion on the topic. We further greatly appreciate the interesting discussion about the finite mass transfer models with Sören Schenke (TU Delft), who also provided us with the mesh for the bubble cluster test case.

References

[1] Schnerr, G. H. and Sauer, J. (2001). *Physical and Numerical Modeling of Unsteady Cavitation Dynamics.* International Conference on Multiphase Flow, New Orleans

[2] Franc, J.-P. and Michel, J.-M. (2006). *Fundamentals of Cavitation.* Springer Science & Business Media, vol. 76

[3] Schenke, S., and van Terwisga, T.J.C. (2017). *Simulating Compressibility in Cavitating Flows with an Incompressible Mass Transfer Flow Solver*. Proceedings of the 5th International Symposium on Marine Propulsors, Finland
[4] Schmidt, S.J., Mihatsch, M., Thalhamer, M. and Adams, N.A. (2011). *Assessment of the prediction apability f a thermodynamic cavitation model for the collapse characteristics of a vapour-bubble cloud*. IMRC 3rd nternational Cavitation Forum, Warwick, UK

© 2018 ASME

Application of Cavitation Based Micro-Bubbles to Recover Neutrally Buoyant Oil Droplets

[1]Greg Loraine*; [1]Georges Chahine;

[1] *DYNAFLOW, INC., Jessup, MD 20794, USA*

Abstract

Small neutrally buoyant oil droplets are difficult to remove from water under any condition, but this difficulty is compounded in oil spills in open water. While floating oil can be collected using several existing methods, there are currently few practical methods for recovering neutrally buoyant oil dispersed in the water column. In this work, a large number of cavitation generated microbubbles, small enough to attach to oil droplets and large enough to help the droplets rise to the surface in a timely manner were effectively used for oil recovery.

Bubble generators using special nozzle designs induced cavitation in submerged water jets creating cavitation, bubble shearing and break up, and generated masses of poly-dispersed bubbles, ranging in size from a few microns to 300 microns. The generators used vortex cavitation initiated in swirl chambers. The bubble sizes and numbers produced were measured by means of image analysis of high speed videos. The effects of the various operating conditions on the bubble size distributions were investigated. The recovery of three types of crude oils from water under various salinities and bubble concentrations was measured.

Keywords: microbubbles, oil flotation, cavitating jets

Introduction

In the 2010 Deepwater Horizon oil spill in the Gulf of Mexico, only 15-20% of the total oil released was estimated to have reached the surface to form the primary oil slick, an additional 25% was dissolved into the water, 24% was lost as gas or through volatilization, and 31% was trapped in the water column as small neutrally buoyant oil droplets [1]. There are existing technologies for removing floating oil from the water surface, such as booms, skimmers, and chemical dispersants [2]-[9]. However, there are few available methods for removing and recovering neutrally buoyant submerged oil within the subsurface water column.

In this project, air flotation by means of microbubble injection with DYNASWIRL® bubble generators was developed to remove neutrally buoyant oil droplets from the water column. While the mechanism of the attachment of gas bubbles to particles or droplets in a host liquid is quite complex, it is generally accepted that the removal efficiency of this process is highly dependent on the relative sizes of the bubbles used and the particles to be removed [11]-[12]. Hence there is a significant advantage to using micro-bubbles in air flotation processes for enhanced removal of dispersed oil and emulsions and energy efficiency. A disadvantage of this is that the rise time of very small bubbles can be very long. However, if a range of bubble sizes is present small bubbles of the same size range as the oil droplets can attach to them and form oil-bubble complexes [11]. The larger bubbles, which rise faster, will then entrain the bubble-oil complexes and lift them to the surface for collection.

The DYNASWIRL® cavitating jets bubble generators create microbubble plumes with diameters ranging from a few microns to a few hundreds of microns (depending on operating conditions) to lift neutrally buoyant oil droplets to the surface where they were extracted using conventional oil recovery schemes.

Results

The bubble generators developed in this project (DYNASWIRL® Bubble Generators) are based on cavitating jet nozzles, which have the capability to produce both very small bubbles < 10 μm, and larger bubbles, 50 – 300 μm diameters, at the same time. The cavitation in the jets is obtained with the help of specially designed nozzles, which induce cavitation in vortical structures of the submerged water jets. In this project we selected the swirling cavitating jet, DYNASWIRL®, which achieves cavitation at high cavitation numbers (i.e. low jet velocities or low pump pressures) through a swirling flow inside the nozzle. Swirl is achieved passively by means of internal tangential injection slots, resulting in a central vortex. The DYNASWIRL® can generate cavitation at low pressures (only a fraction of the pressure required to induce cavitation using a conventional nozzle) and high flow rates, thus providing a more economical option for the deployment of bubble generators in this practical application. In addition, the DYNASWIRL® has the advantage of generating very large cavity surface areas (the vortex core) that starts inside the nozzle assembly and extend in front of the nozzle orifice (Figure 1). Breakup of this cavity into bubbles at the nozzle exit results in a large number of very fine bubbles. Also, additional air can be injected in the liquid flow or the vortex core to further increase the generated bubble numbers and sizes.

Figure 1. Cavitation Core from DYNASWIRL® nozzle (left hand photo). Large (1,350 gallon) test tank with 2 bubble generators (center hand photo). Close-up of the generators at the bottom of the tank (right hand photo).

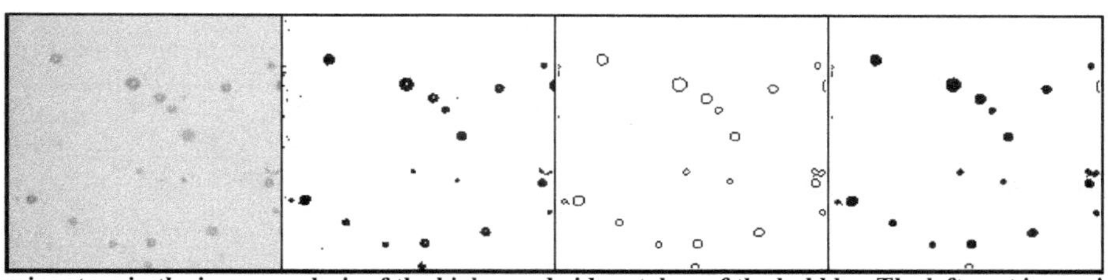

Figure 2. Successive steps in the image analysis of the high speed videos taken of the bubbles. The leftmost image is the original image and the next is the binary converted image. The edges were detected in the third image and in the last the internal areas were filled.

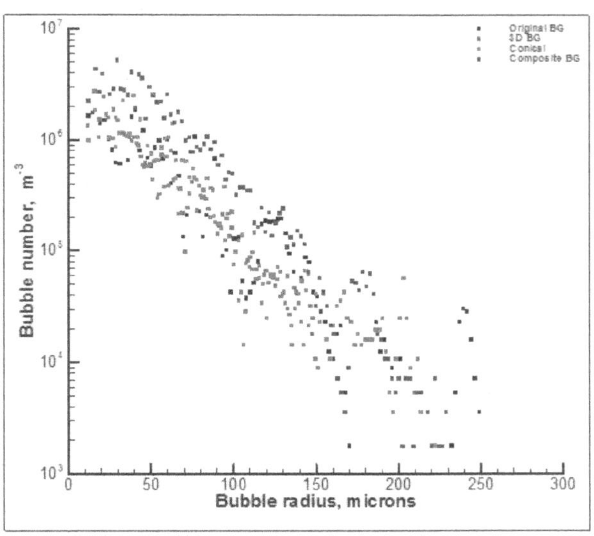

Figure 3. Bubble size distribution from nine different DYNASWIRL® bubble generators. Pump pressure 15 psi, water flow rate 26.5lpm, and air flow rate 0.46 lpm.

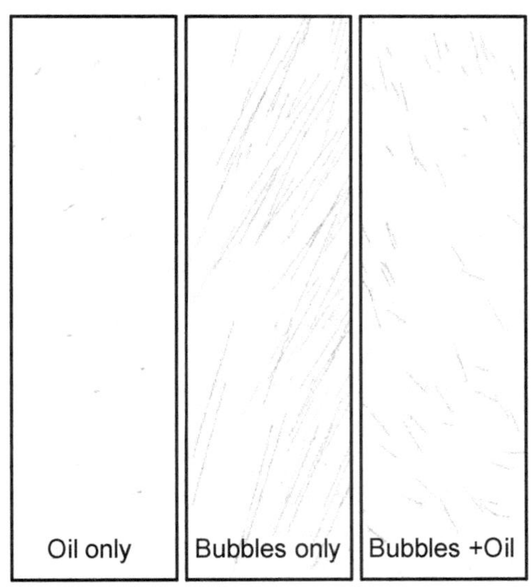

Figure 4. Trajectories in water of ANS oil droplets (left), Bubbles (center), and bubbles and oil (right) in a 35 ppt saline water. These were obtained from 60 frames of a video taken at 300 frames per second (size of image 38 mm x 11 mm).,

Bubble size distributions generated by the DYNASWIRL® nozzles under different operating conditions such as pressure drop across the nozzle, and concentrations of air injected into the core or through the slots were measured using an optical photography method. High speed videos of the bubble clouds produced were captured and processed using the image analysis software, ImageJ. Figure 3 shows the bubble size distribution of nine bubble generators operated at 15 psi, a water flow rate of 26.5 lpm, and an air injection rate of 0.46 lpm.

Neutrally buoyant Anadarko, ANS, and HOOPS crude oil droplets were also generated by using a cavitating jet to produce an oil water emulsion in a large drum and then injecting the mixture into the test tank. This resulted in oil droplets of diameters ranging from 20 µm to 150 µm. Figure 3 shows the particles trajectories from image analysis of a) almost neutrally buoyant ANS oil droplets, b) bubbles generated by DYNASWIRL® and c) a mixture of bubbles and oil particles. These were obtained from high speed photography in the 6 foot cube 1,350 L tank (Figure 1, middle picture) in salt water with a salinity of 35 ppt. As the traces show, the oil particles, being small (20 µm to 150 µm) and nearly neutrally buoyant, have short trace lengths (corresponding to rise speeds between 0.3 and 5 mm/s) and the trajectory picture is almost empty. On the other hand, the movies taken with the bubbles generated by the DYNASWIRL® in the clean water before dispersing the oil, show relatively very long traces corresponding to rise speeds between 3 and 5 cm/s. When the DYNASWIRL® bubbles are generated in the water polluted with dispersed ANS crude oil the rise speed is much larger than with the oil droplets alone and a little lower than the bubbles in clean water (~ 0.3 to 1.5 cm/s).

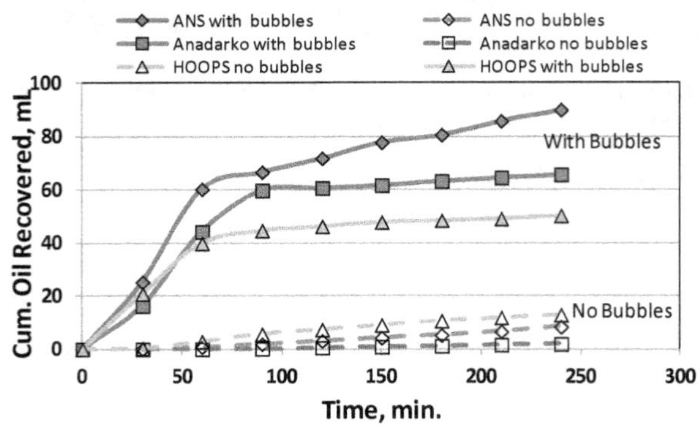

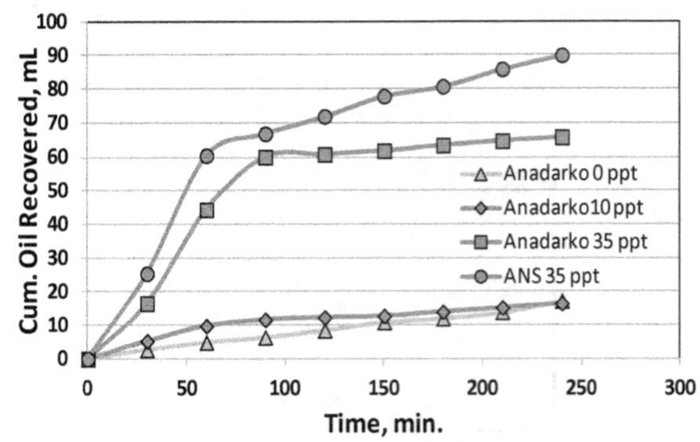

Figure 5. (Left) Comparison of cumulative oil recovery during laboratory experiments with (solid lines) and without (dashed lines) microbubble injection for three types of crude oil: Anadarko, ANS, and HOOP all at ~0.002%. (Right) Oil recovery of Anadarko at three salinity levels and ANS at 35 ppt salinity.

The oil recovery experiments were run using three types of crude oil (Anadarko, HOOPS, and ANS) by adding bubbles only for the first 20 minutes of the experiment after which bubble addition stopped. The reasoning was that it takes a long time for the very small bubble to rise to the surface, so for a 240 minutes test we cannot see the effect of the bubbles inserted after the 20 minutes. The oil that reached the surface was recovered using a floating tube skimmer that ran continuously to collect the recovered oil, and as shown in the figures, the oil continued to be collected at a relatively high rate well after the bubble injection stopped.

The recovery of the three types of crude oil with and without the addition of bubbles is shown in Figure 4. In all three cases, the recovery of crude oil was much higher when microbubbles were injected into the oil plume. However, HOOPS oil seemed to be the hardest of the three to collect and ANS the easiest. This difference among different oil types could be attributed to the composition differences among different oil types that affect the balance of coalescence and repulsion between oil droplets and bubbles. Also, the differences in the properties between the different oils will change the efficiency of the surface skimmer used to collect the oil. However, we did not observe any visual differences in the operation of the skimmer.

	Vol. Oil Added, mL	Initial Oil Concentration, mL/L	Final Oil Concentration, mL/L	Vol. Oil Recovered, mL	% Oil Recovered
Anadarko- Gravity only	97.4	0.0191	0.0183	3.9	4 %
Anadarko - Flotation	99.8	0.0196	0.0067	65.7	66 %
ANS – Gravity only	84	0.0165	0.0148	8.4	10 %
ANS - Flotation	118	0.0232	0.0055	81.4	74 %
HOOPS – Gravity only	87.4	0.0171	0.0148	12.9	14%
HOOPS Flotation	107	0.0211	0.0112	50.3	46%

Table 1. Summary of overall oil removal in the 1,350 gallon tank.

The operation of the bubble generators and their oil recovery performance were tested for fresh water and for two additional water salinity levels: 10 and 35 parts per thousands (ppt), corresponding to estuary and seawater salinities. The oil recoveries increased with salinity and were significantly higher at 35 ppt.

Conclusion

The removal of neutrally buoyant oil droplets by means of air bubble flotation was investigated. The bubbles were generated using cavitating jet nozzles as bubble generators (DYNASWIRL® bubble generators). High speed videos were recorded and image analysis software were used to characterize the sizes, numbers, and rise speeds of bubbles produced under varying operating conditions. The bubbles produced were in the range of 20 μm to 300 μm.

The rise speed of ANS oil droplets, bubbles, and oil droplets during flotation in water with 35 ppt salinity were measured using high speed videography. The rise speeds for the oil droplets were between 0.0147 mm/s and 0.16 mm/s. The rise rates for the air-oil mixture was between 0.3 mm/s and 1.6 mm/s, while the bubble rise speeds were between 30 mm/s and 50 mm/s.

Acknowledgements

This work was funded by the US Coast Guard Research and Development Center under Award No. HSCG32-15-C-R00005, monitored by Mr. Alexander Balsley.

References

[1] Ryerson, T., Camilli, R., Kessler, J., Kujawinski, E., Reddy, A., Valentine, D., Atlas, E., Blake, D., de Gouw, J., Meinardi, S., Parrish, D., Peischl, J., Seewald, J., Warneke, C. (2012). *Chemical data quantify Deepwater Horizon hydrocarbon flow rate and environmental distribution.* Proc. Natl. Acad. Sci, 109.

[2] Fitzpatrick, M., Tebeau, P., Hansen, K. (2013). *Development of Bottom Oil Recovery Systems – Final Project Report.* Department of Homeland Security, CG-D-09-13.

[3] Hansen, K. (2014). *Submerged Oil Response.* Proceedings US Coast Guard, Winter 2013-2014, www.uscg.mil/proceedings.

[4] Mitchel, J., Christopherson, S., Whipple, F. (1994). *Mechanical Protection Guidelines.* NOAA Operational Manual.

[5] Min Choi, H., Rinn, M. (1992). *Natural Sorbents in Oil Spill Cleanup.* Environmental Science Technology. 26(4).

[6] USCG, NOASS, US EPA, & CDC Protocol Document, August, 2000. *Special Monitoring of Applied Response Technologies (SMART).*

[7] Liu, Z., Liu, J., Zhu, Q., Wu, W. (2012). *The weathering of oil after the Deepwater Horizon oil spill: insights from the chemical composition of the oil from the sea surface, salt marshes and sediments.* Environ. Res. Letts., 7.

[8] Dubinsky, E., Conrad, M., Charkraborty, R., Bill, M., Borglin, S., Hollibaugh, J., Mason, O., Piceno, Y., Reid, F., Stringfellow, W., Tom, L., Hazen, T., Andersen, G. (2013). *Succession of Hydrocarbon-Degrading Bacteria in the Aftermath of the Deepwater Horizon Oil Spill in the Gulf of Mexico*, Environ. Sci. Tech., 47(19).

[9] American Petroleum Institute (1999). *A decision-maker's guide to dispersants – a review of theory and operational requirements*, A Standard, API Publ 4692, March.

[10] US National Research Council, 2005. *Oil Spill Dispersants: Efficacy & Effects.*

[11] Wang, L., Shammas, N., Selke, W., Aulenbach, D. (2010) *Vol 12 Handbook of Environmental Engineering Flotation Technology*, Humana Press, New York.

[12] Lee, D., Bateman, W., Owens, N. (2007). *Efficiency of Oil/Water Separation Controlled by Gas Bubble Size and Fluid Dynamics within the Separation Vessel*, 17th Annual Produced Water Society Seminar, League City, TX.

© 2018 ASME

10th International Symposium on Cavitation - CAV2018
Baltimore, Maryland, USA, May 14 – 16, 2018 CAV18-05021

Application of Homogeneous and Inhomogeneous Two-Phase Models to a Cavitating Tip Leakage Vortex on a NACA0009 Hydrofoil

[1]Jonas Wack*; [1]Stefan Riedelbauch

[1]University of Stuttgart, Germany

Abstract

Two-phase modeling plays an important role for the correct prediction of cavitation phenomena. However, the computational effort grows with increasing complexity of the model. Consequently, complex models should only be applied if they lead to a significant improvement of the results. In the scope of this study, homogeneous and inhomogeneous models are applied to a cavitating tip leakage vortex on a NACA0009 hydrofoil at 10° incidence angle. Two different cavitation model constants from literature are applied for the homogeneous model. For the inhomogeneous approach, the impact of bubble size diameter is investigated. Furthermore, a constant drag coefficient 0.44 is compared to Schiller Naumann drag model. The results indicate that for the presented case, the inhomogeneous model does not lead to a significant improvement of the simulation results, which can be explained by a low Stokes number due to small bubble diameters. All investigated setups show similar results, which underline that the homogeneous model is sufficient to resolve the present cloud cavitation as well as the cavitating tip separation and tip leakage vortex.

Keywords: cavitation; two-fluid model; tip leakage vortex

Introduction

Owing to the integration of other renewable energies into the electrical grid, hydro power plants are more and more operated at off-design conditions. This comes across with secondary flow phenomena and consequences like cavitation have to be considered. In a water turbine, several different forms of cavitation can occur and coexist at the same time. For instance, in a Kaplan turbine a cavitating tip leakage vortex, sheet or cloud cavitation on the runner blades and a cavitating draft tube vortex rope can occur simultaneously. The occurrence of different forms of cavitation makes it challenging to select a suitable two-phase modeling approach.

For cavitation simulations it is common practice to use a model – called homogeneous model – that treats the two phases as one mixture. However, cavitation simulations on a Francis turbine at deep part load conditions showed shortcomings to correctly reproduce the simulated cavitation appearance compared to measurements [1]. This raises the question whether the more complex inhomogeneous two-phase modeling approach, also called two-fluid model, can increase accuracy of simulation results. In the scope of this study, the two approaches are applied to a NACA0009 hydrofoil that is facing a cavitating tip leakage vortex, which is generated by the flow through the gap between hydrofoil side wall and cavitation tunnel wall. This test case has been measured in detail by Dreyer et al. [2, 3, 4]. Numerical studies have been performed by Decaix et al. [5] who showed that on a coarse mesh the cavitation volume is significantly underestimated.

Numerical modeling

The commercial code Ansys CFX version 17.2 is used for CFD simulations. In the first step, single phase simulations are performed for a mesh study and the selection of suitable turbulence models. For the two-phase simulations, the homogeneous and inhomogeneous models are applied as multiphase modeling approach.

The more complex inhomogeneous model consists of one set of governing equations (mass and momentum conservation) per phase, which results in a significantly increased computational effort compared to single phase simulations. This modeling approach allows separate velocity fields for liquid and vapor phase, which is the cause of interfacial forces like drag force. If the relative motion between the phases is neglected, the mixture can be treated as a pseudo-fluid. This results in the homogeneous model, where only one set of governing equations needs to be solved for the mixture phase. Additionally, a transport equation for the volume fraction α has to be solved. All in all, the computational effort of the homogeneous model is significantly lower compared to the inhomogeneous model.

*Corresponding Author, Jonas Wack: jonas.wack@ihs.uni-stuttgart.de

Both models have in common that liquid and vapor phase share a common pressure field. For more detailed information about the two-phase modeling approaches it has to be referred to literature (e.g. [6, 7]).
The mass transfer between the phases is considered by the Zwart cavitation model [8], which contains model constants F_e and F_c for evaporation and condensation. Contrary to recommendations by Zwart et al. [8] to set F_e=50 and F_c=0.01, Morgut et al. [9] used the constants F_e=300 and F_c=0.03 that were determined by an optimization on a NACA66(mod) test case with sheet cavitation. In this study both suggested model constants are applied.
For RANS simulations, the SST turbulence model is chosen, which is widely used for CFD applications. As second model the SBES model [10] is selected, that can operate in scale resolving simulation (SRS) mode. The idea of SRS is to have a hybrid RANS-LES model that resolves large eddies away from the wall, while for the wall boundary layers a RANS model is used [11].

Computational setup

Extensive measurements have been performed for the NACA0009 hydrofoil in the high speed cavitation tunnel of the EPFL [2, 3, 4]. The hydrofoil has an initial chord of c_0=0.11 m, a maximum thickness h=9.9 mm and a span of 0.15 m. It is truncated to a chord of c=0.1 m and the foil tip on the pressure side of the blade is rounded with a radius of 1 mm to reduce the cavitation in the clearance [3]. This radius is considered in the simulation model, but to facilitate the meshing, a small radius of 0.1 mm remains on the suction side of the blade. Stereo PIV measurements of all velocity components in three different measurement planes have been carried out for non-cavitating conditions. The planes (see figure 1) are located downstream of the hydrofoil at z/c=1, 1.2 and 1.5.

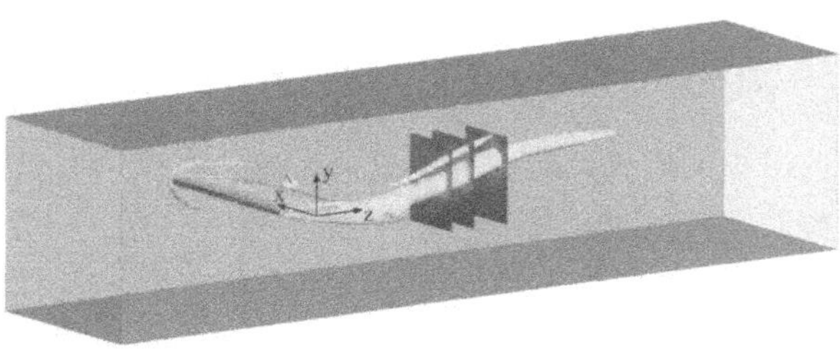

Figure 1(a): Measurement planes and visualization of the tip leakage vortex with an isosurface of the velocity invariant Q.

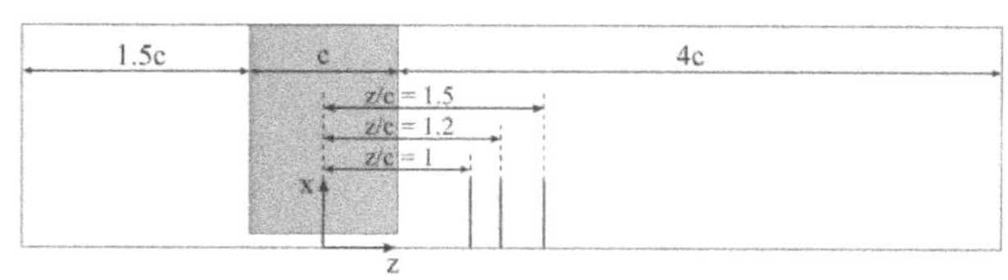

Figure 1(b): Top view on the main subdomain.

For all simulations a configuration is chosen with 10° incidence angle and a normalized tip clearance $\tau=\frac{gap}{h}=1$. As inlet boundary condition a block profile with a velocity w_∞=9.78 m/s and a turbulence intensity of 5% is set. This corresponds to the measurement with w_∞=10 m/s and takes into account the boundary layer close to the wall [3]. At the outlet, the static pressure is prescribed to obtain a pressure at the inlet p_∞=1 bar. To ensure that the boundary conditions do not affect the solution, the inlet boundary condition is set 3.5 chords upstream and the outlet boundary condition 12 chords downstream of the hydrofoil. The simulation domain is divided into three subdomains to reduce the number of cells. While the main subdomain (see figure 1(b)) around the hydrofoil has a fine resolution to resolve the tip leakage vortex, the other two subdomains consist of a coarser grid.
For all unsteady simulations, a second order backward Euler scheme is applied for temporal discretization and a time step of 10^{-5} s is prescribed, which corresponds to a RMS Courant number of 0.35. The selection of spatial discretization depends on the turbulence model. A high resolution scheme is prescribed for SST simulations and a bounded central differencing scheme is used for the SBES model.

Mesh study and turbulence model selection

A mesh study is carried out for the selection of a suitable grid. Steady state single phase simulations are performed using the SST turbulence model. In the first step, the global number of cells is varied, while the resolution of the boundary layer is kept constant, which results in an average y^+-value of 35. To quantify the differences between the grids, u-component of the velocity is compared to measurement results along a line that is chosen to pass the region of highest velocity gradient from the measurements. The results (see figure 2(a)) indicate that the smallest meshes 4M and 11M still show mesh dependencies, while the other grids have only minor differences in the velocity distribution. Nevertheless, there is still a huge deviation to the measurements that were performed at non-cavitating conditions.
While for all other variants the mesh is globally refined, grid 30M is derived from variant 26M by locally refining the grid in the region of the tip leakage vortex. In simulations made on a Francis turbine with cavitating inter-blade

vortices it was identified that mesh resolution in the vortex core region is of great importance to resolve the pressure minimum [12]. Thus, further investigations are based on grid 30M.

After the specification of the global cell distribution, meshes with different y^+-value are compared. In this step, only the resolution of the boundary layer of the wing and channel wall is varied, while the distribution in the rest of the domains remains constant. The velocity distribution of u is displayed in figure 2(b). It can be observed that the y^+-value has a significant impact on the simulation results. With decreasing y^+-value, simulation results approach to the experiment but the steep gradient from the measurements is still underestimated. At y^+=1, mesh independent results are observed, as a further refinement of the wall resolution has no relevant impact on the solution. Due to mesh independency and the best agreement with the measurements for the velocity distribution, the mesh with y^+=1 is selected for further investigations. Owing to the refinement in the boundary layers, this mesh consists in total of 46M cells.

Unsteady single-phase simulations are performed for the selection of the turbulence model. The results are displayed in figure 2(c). The SST simulation results are almost identical with the steady state simulations from the mesh study. With the more advanced SBES model, a significant improvement can be observed, which underlines that the strong swirling flow caused by the vortex needs a more complex turbulence modelling. Consequently, for the two-phase simulations the SBES model is applied.

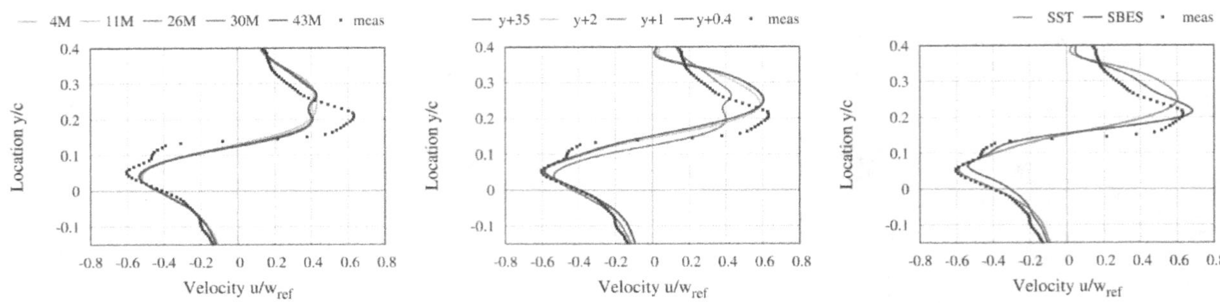

(a) Variation of global mesh size. (b) Variation of mesh resolution. (c) Comparison of turbulence models.

Figure 2: Distribution of velocity u along line in y-direction at measurement plane z/c=1. The line is located at x/c=0.12.

Numerical results

The main focus in this study lies on the comparison of the homogeneous and inhomogeneous two-phase modeling approach. Two different configurations of cavitation model constants are prescribed for the homogeneous model. The simulations with the inhomogeneous model differ in the treatment of the dispersed vapor phase. Two simulations are performed with monodisperse settings, by prescribing a constant bubble diameter of the vapor phase of 0.01 mm and 0.1 mm, respectively. Furthermore, one simulation is performed with a polydispersed treatment based on the homogeneous MUSIG approach with ten size groups, a maximum bubble diameter of 1 mm and a minimum bubble diameter of 0.001 mm. The bubble diameters are selected based on the investigations by Maeda et al. [13] that state that the bubble radius in cavitation clouds is in the range between 0.001 and 0.1 mm.

A constant drag coefficient C_D=0.44 is prescribed for all the described simulations with the inhomogeneous model. Furthermore, one simulation is carried out with the Schiller Naumann drag model and a monodispersed vapor phase with 0.1 mm bubble diameter to investigate the impact of drag model. Other interfacial forces like lift force are neglected in this study.

In figure 3, a side view on the cavitation regions is displayed. On the left side an instantaneous snapshot is shown and on the right side the time-averaged results over 20,000 time steps are presented. The cavitation volume in the simulation results is displayed by isosurfaces with three different thresholds of the vapor volume fraction α_v: 0.5, 0.2 and 0.05. The use of three thresholds is intended to facilitate the assessment of cavitation occurrence. There are three different forms of cavitation present for the investigated NACA0009 hydrofoil. At the suction side of the leading edge, cloud cavitation can be observed due to the high incidence angle. Secondly, in the gap it comes to cavitation due to the formation of a tip separation vortex. Finally, a cavitating tip leakage vortex occurs.

A comparison of the simulations with the homogeneous model indicates that for the averaged quantities, the cavitation volume is slightly bigger with the cavitation model constants from Morgut et al. [9]. The simulation results with the inhomogeneous model indicate a slight increase in vapor volume going from bubble diameter

0.01 mm to 0.1 mm to a polydispersed treatment. The change in the drag model does not show any clear change in the results. A comparison between results from homogeneous and inhomogeneous model does not show any significant difference that would justify the additional computational effort that is caused by the inhomogeneous model. The similar results between the different two-phase modeling approaches can be explained by the Stokes number *St*:

$$St = \frac{\tau_V}{\tau_F} = \frac{\frac{\rho_v D^2}{18\mu_l}}{\tau_F}$$

The symbol τ_V denotes for the momentum response time and depends on the density of the vapor phase ρ_v, the bubble diameter D and the dynamic viscosity of the liquid phase μ_l. For $St \ll 1$, which corresponds to small bubble diameters, the response time of the vapor bubble is very fast compared to the characteristic time scale of the flow τ_F. The consequence is that the bubble has enough time to adapt to changes in the velocity field of the liquid phase, which results in a velocity equilibrium between the two phases [14]. As a result, the inhomogeneous model does not have any benefit in that case. On the other hand for big bubble diameters the two-fluid model might be beneficial, that is why it is often applied to bubble column simulations.

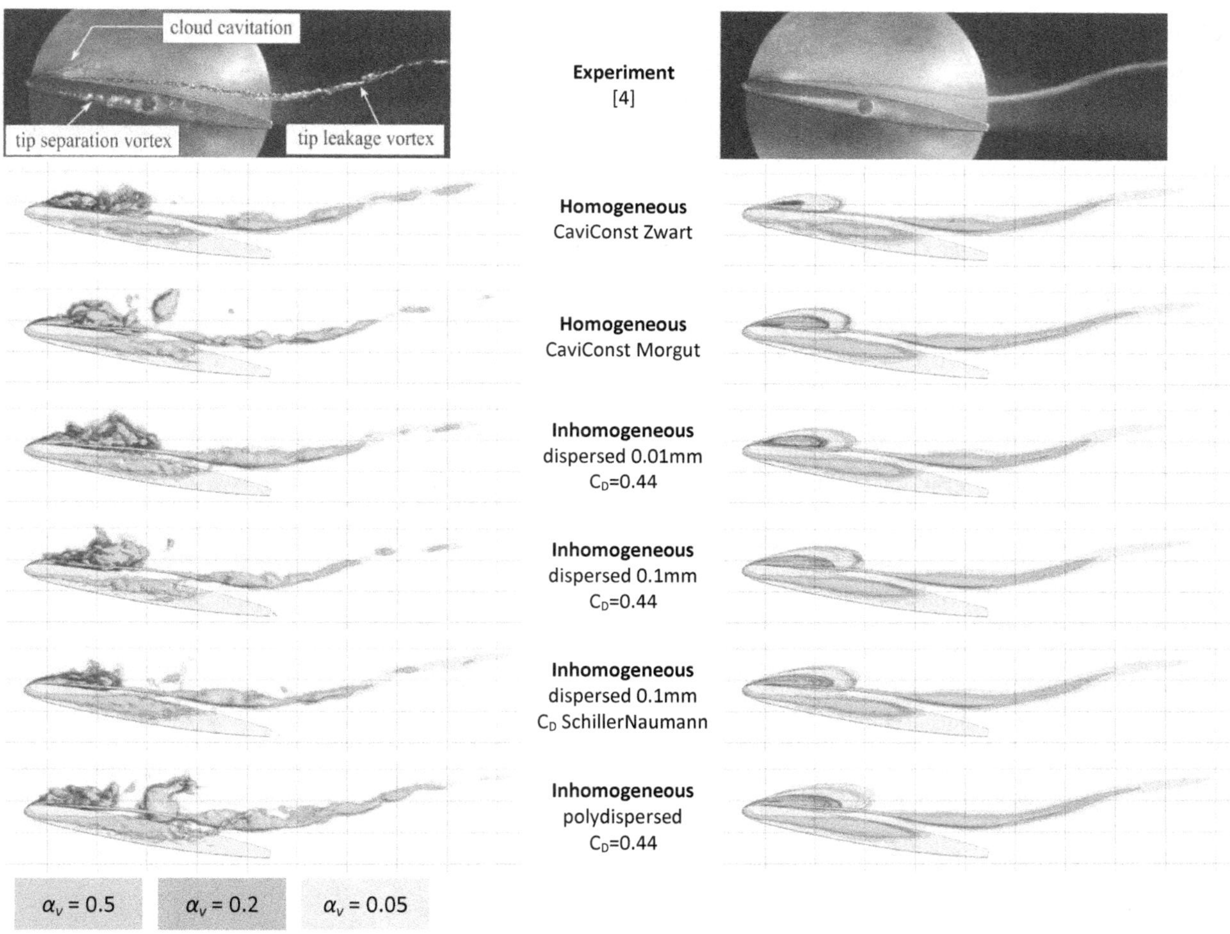

(a) Instantaneous snapshot. (b) Time-averaged result over 20,000 time steps.

Figure 3: Visualization of the cavitation volume by three isosurfaces of the vapor volume fraction with different thresholds.

All in all, it can be stated that all simulation results show a reasonable agreement with the measurement. Deviations can be found for the tip separation vortex, where the cavitation volume is overestimated in the region close to the leading edge, and in the region where the tip leakage vortex unites with the tip separation vortex. There, as well the cavitation volume is overestimated compared to the experiment. The deviations between the different simulation setups are relatively small. For this case it is more important to have a fine mesh resolution, especially to resolve the pressure minimum in the vortex core of the tip leakage vortex.

Conclusion

The selection of a suitable two-phase modeling approach is mandatory for accurate simulation results. For the presented investigations of the cavitating flow around a NACA0009 hydrofoil at 10° incidence angle, the inhomogeneous model does not show significant improvements compared to the simpler homogeneous model. The reason is that due to the small size of the vapor bubbles it comes to a strong coupling between the liquid and vapor phase, which results in velocity equilibrium.

For the selected case, it is more important to use fine grids that resolve the pressure minimum in the core of the tip leakage vortex. This is essential as the pressure level is linked with the mass transfer due to cavitation.

Acknowledgements

The authors would like to thank the HLRS Stuttgart for support and supply of computation time. Furthermore, the authors would like to thank the EPFL for providing the experimental data.

References

[1] Wack, J., Riedelbauch, S., Yamamoto, K., Avellan, F. (2016). *Two-phase flow simulations of the inter-blade vortices in a Francis turbine.* Proceedings of the 9th International Conference on Multiphase Flow, Florence, Italy.

[2] Dreyer, M., Decaix, J., Münch-Alligné, C., Farhat, M. (2014). *Mind the gap - tip leakage vortex in axial turbines.* IOP Conference Series: Earth and Environmental Science 22(5).

[3] Dreyer, M., Decaix, J., Münch-Alligné, C., Farhat, M. (2014). *Mind the gap: a new insight into the tip leakage vortex using stereo-PIV.* Experiments in Fluids. 55(11).

[4] Dreyer, M. (2015). *Mind The Gap: Tip Leakage Vortex Dynamics and Cavitation in Axial Turbines.* Ph.D. Thesis, EPFL Lausanne.

[5] Decaix, J., Balarac, G., Dreyer, M., Farhat, M., Münch, C. (2015). *RANS computations of tip vortex cavitation.* Journal of Physics: Conference Series. 656(1).

[6] Yeoh, G.H., Tu, J. (2010). *Computational Techniques for Multi-Phase Flows.* Elsevier.

[7] Ishii, M., Hibiki, T. (2010). *Thermo-Fluid Dynamics of Two-Phase Flow.* Springer.

[8] Zwart, P.J., Gerber, A.G., Belamri, T. (2004). *A Two-Phase Flow Model for Predicting Cavitation Dynamics.* Proceedings of the 5th International Conference on Multiphase Flow, Yokohama, Japan.

[9] Morgut, M., Nobile, E., Bilus, I. (2011). *Comparison of mass transfer models for the numerical prediction of sheet cavitation around a hydrofoil.* International Journal of Multiphase Flow. 37(6).

[10] Menter, F. (2016). *Stress-Blended Eddy Simulation (SBES) – A new Paradigm in hybrid RANS-LES Modeling.* Proceedings of the 6th HRLM Symposium, Strasbourg, France.

[11] Menter, F., Schütze, J., Gritskevich, M. (2012). *Global vs. Zonal Approaches in Hybrid RANS-LES Turbulence Modelling.* Progress in Hybrid RANS-LES Modeling. Springer.

[12] Wack, J., Riedelbauch, S. (2015). *Numerical simulations of the cavitation phenomena in a Francis turbine at deep part load conditions.* Journal of Physics: Conference Series. 656(1).

[13] Maeda, M., Yamaguchi, H., Kato, H. (1991). *Laser holography measurement of bubble population in cavitation cloud on a foil section.* Proceedings of the 1st ASME-JSME Fluids Engineering Conference, Portland, USA.

[14] Crowe, C.T., Schwarzkopf, J.D., Sommerfeld, M., Tsuji, Y. (2011). *Multiphase Flows with Droplets and Particles.* CRC Press.

© 2018 ASME

Application of the Stochastic Field Method to Cavitating Flows in an Injection Nozzle

[1]Martin Raquet*; [1]Andreas Class; [2]Wilfried Edelbauer;

[1] *Karlsruhe Institute of Technology, Karlsruhe, Germany;* [2] *AVL List GmbH, Graz, Austria*

Abstract

In this paper a fully Eulerian Monte Carlo method called 'Stochastic Field Method' (SFM) is applied on cavitating flows in an injection nozzle from the automotive industry. After the successful application to the field of combustion [1], the Stochastic Field Method is now applied to two phase flow. In the presence of numerous bubbles the method provides advantages in computational speed and statistical convergence compared to the classical Lagrangian-particle Monte Carlo methods. Modeling random processes which are omnipresent in combustion and cavitation requires careful treatment: Unphysical results are obtained, if in a nonlinear process a randomly distributed variable is merely approximated by its average. Thus the whole distribution must be considered. A probability density function (PDF) is used to describe the probabilistic behavior. The corresponding evolution and transport in space is governed by the PDF transport equation. Instead of solving the high dimensional PDF transport equation directly it is less complex to discretely approximate the PDF by random samples. In contrast to other stochastic methods that usually consider these samples as Lagrangian particles, the Stochastic Field Method describes them as fluctuating Eulerian fields. This point of view allows PDF description in a fully Eulerian framework, thus omitting the cumbersome coupling of Eulerian and Lagrangian solvers. The very first application of the Stochastic Field Method to a multi-phase flow was published by our working group [2]. Here, stochastic partial differential equations (SPDEs) describe samples of the disperse phase volume fraction as Eulerian fields. The collective of all samples approximates the probability density function at a considered location. Within the present paper the implementation to a commercial code is discussed. Application to an industrial case namely a cavitating flow within an injection nozzle is presented. Collecting samples of several time steps allows the SFM to visualize the volume fraction's or bubble size's distribution within individual computational cells. When applying the Stochastic Field Method to multiphase problems a number of physical aspects have to be considered which require careful analysis. These difficulties are pointed out and discussed.

Keywords: stochastic, multi-phase, stochastic field method, cavitation, nozzle, Monte Carlo, Eulerian

Introduction

Random behavior can be observed experimentally in both, combustion and cavitation problems. When including this behavior in numerical simulations the consideration of rare events is advantageous in the presence of nonlinear processes. Stochastic models yield an attractive possibility to observe these fluctuations in numerical investigations. The direct calculation of the PDF is associated with exhaustive numerical costs. Instead of considering the PDF, it is usually approximated by samples with a Monte Carlo method to reduce effort. Methods that treat these samples as Lagrangian particles are well known, where virtual particles are added to the flow and observed in a Lagrangian point of view. The flow itself is investigated in a Eulerian frame.

Basics of the Stochastic Field Method

The Stochastic Field Method regards samples approximating the PDF in a fully Eulerian description. Instead of Lagrangian particles Eulerian fields of concentrations are considered. Several Lagrangian particles and their corresponding stochastic field are presented in Figure 1 for an exemplary peculiarity. Here, the stochastic field delivers for every computational cell one concentration which is interpreted as a sample to approximate the PDF. Note that Figure 1 only represents one peculiarity. Note that, in the presence of fluctuations (e.g. in a turbulent flow) the Lagrangian particles are not located at deterministic positions. Analogously, the corresponding stochastic field exhibits statistical fluctuations. To include several peculiarities into the numerical investigation, several stochastic fields are calculated. Several possibilities are shown in Figure 2. To approximate the PDF within a specific cell, the concentrations from the stochastic fields are exploited. For illustration, the concentrations from eight stochastic fields are marked green in Figure 3 ('samples'). Building the histogram of these samples clearly approximates the corresponding PDF. Considering samples using stochastic fields offers several advantages: The fully Eulerian description makes the computationally expensive coupling of two solver architectures obsolete, which is the case for

Lagrangian particles. For a statistically reliable description of the PDF a sufficient number of samples is necessary. When using Lagrangian particles regions of dilute sample concentration or high sample concentration can occur. To overcome this, the methods of "clustering" and "cloning" were developed, see e.g. [3], but require additional computational time. The Stochastic Field Method delivers within every computational cell the same number of samples which avoids the need for methods like "clustering" and "cloning": It automatically delivers a predefined number of stochastic fields. In general the usage of 8, 16 [4] or 12 and 18 [5] fields is a good compromise between accuracy and computational effort.

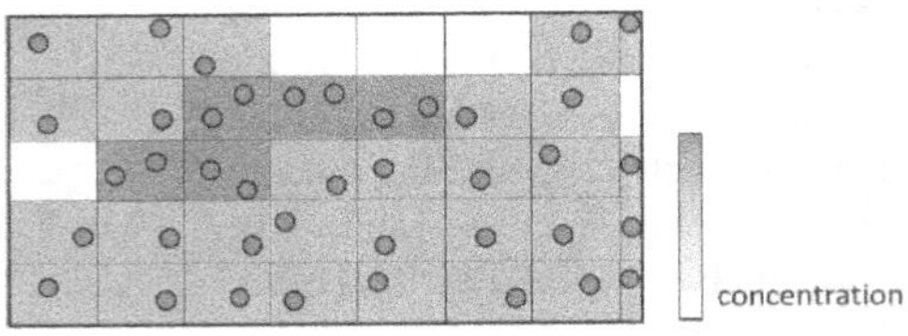

Figure 1: Lagrangian particles and corresponding Stochastic Field.

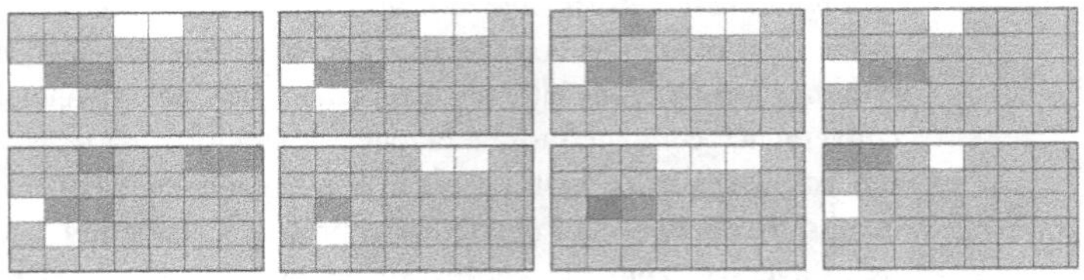

Figure 2: Several Stochastic Fields each representing one peculiarity and differing by the influence of the random term.

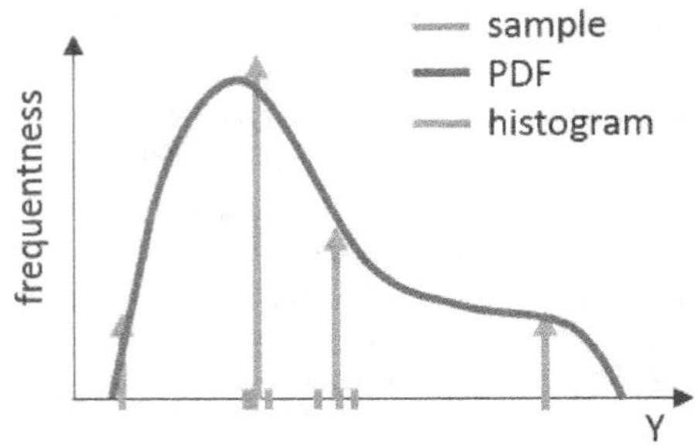

Figure 3: the histogram of the samples approximates the PDF.

Stochastic Field Method in the field of combustion

The stochastic partial differential equation describing a stochastic field was derived by Valiño [1] and Sabel'nikov and Soulard [6] in the field of combustion. It goes back to the joint probability density function (JPDF) of a scalar c and a velocity U derived by Pope [7]. This derivation builds on the conservation equation of a scalar c which reads

$$\rho\frac{\partial c}{\partial t}+\rho U_i\frac{\partial c}{\partial x_i}=-\frac{\partial J}{\partial x_i}+\rho S \tag{1}$$

where J represents the diffusive mass flux. With Fick's first law this can be expressed

$$J=-D\frac{\partial c}{\partial x_i}$$

with the molecular diffusion coefficient D, the fluid density ρ and the source term S. The mean scalar transport equation reads [8]

$$\frac{\partial \bar{c}}{\partial t}+\bar{U}\frac{\partial \bar{c}}{\partial x_j}=\frac{\partial}{\partial x_i}\left[(D+D_t)\frac{\partial \bar{c}}{\partial x_i}\right]+\bar{S} \tag{2}$$

with the turbulent diffusion coefficient D_t. Valiño [1] derived the transport equation of the stochastic fields (eq. (3)). Sabel'nikov's and Soulard's work [6] presents a less restrictive derivation, which results in the same equation, but with vanishing molecular diffusivity D. In many practical examples the molecular diffusivity is small compared to the turbulent diffusivity D_t. Consequently, both results are considered to be equivalent.

$$\frac{\partial Y}{\partial t}+\widetilde{U}_j\frac{\partial Y}{\partial x_j}+\sqrt{2D_t}\dot{W}_j\frac{\partial Y}{\partial x_j}-\frac{1}{\langle\rho\rangle}\frac{\partial}{\partial x_j}\left(\langle\rho\rangle(D+D_t)\frac{\partial Y}{\partial x_j}\right)=M(Y;x,t)+S(Y) \tag{3}$$

In equation (3) Y is one sample of the concentration of a species and W is a Wiener Process where the Ito interpretation was used. M represents the micromixing term which can be modeled, e.g. by the Interaction by Exchange with the Mean (IEM) model [9]. In the PDF transport equation of the species concentration the turbulent advection is modelled by the gradient diffusion assumption, which creates the 'diffusion-like' term (second derivative) in the transport equation of the corresponding stochastic field equation (see [6] for details). Several Stochastic fields are calculated differing by the influence of the Wiener process term approximating the PDF; depending on the influence of the Wiener process, the individual samples may have bigger or smaller values. For the calculation of the PDF's expectation Pope's [7] definition of the ensemble average of a random variable $\phi = \phi(\underline{x_0}, t_0)$ is used

$$\langle\phi\rangle_N \equiv \frac{1}{N}\sum_{n=1}^{N}\phi^{(n)}. \qquad (4)$$

In his example $\phi^{(n)}$ is the nth 'experiment' from N nominally identical 'apparatus' and $\langle\phi\rangle_N$ estimates $\langle\phi\rangle$ since

$$\mathbf{E}(\langle\phi\rangle_N) = \langle\phi\rangle \qquad (5)$$

for any N. Further Pope states that 'for large N, the Central Limit Theorem reveals the relationship between $\langle\phi\rangle_N$ and $\langle\phi\rangle$' [7]. Analogously this definition (eq. (4)) can be used to calculate the expected value $\langle Y\rangle_N$ from the samples $Y^{(n)}$ approximating the PDF

$$\langle Y\rangle_N \equiv \frac{1}{N}\sum_{n=1}^{N}Y^{(n)}. \qquad (6)$$

Ensemble averaging stochastic fields in the corresponding transport equation (eq. (3)) yields the transport equation of the expected value. With vanishing Wiener process [10] and vanishing micromixing term it matches the mean concentration from the averaged scalar transport equation (cmp. eq. (2)).

Application to the field of two phase flow

Two phase flows are often described by using the two fluid model [11], [12]. The authors present the average balance equation of mass of phase k, the so called volume fraction equation. In both descriptions the volume fraction equation reads

$$\frac{\partial \alpha_k \overline{\overline{\rho_k}}}{\partial t} + \nabla\cdot\left(\alpha_k\overline{\overline{\rho_k}}\widehat{\boldsymbol{U}_k}\right) = S_k, \qquad (7)$$

with the volume fraction[1] of phase k defined as

$$\alpha_k = \overline{X_k}, \qquad (8)$$

with the component indicator function of a realization r [12]

$$X_k(\boldsymbol{x}, t; r) = \begin{cases} 1, \text{if } x \in k \text{ in realization } r \\ 0, \text{otherwise} \end{cases} \qquad (9)$$

and the average defined for a function F_k [11]

$$\overline{F_k}(\boldsymbol{x}_0; t_0) = \lim_{\delta\to 0}\frac{1}{\Delta t}\int_{[\Delta t]_T} F_k(x_0; t)dt, \qquad (10)$$

where δ depicts the interface thickness. The phase average of a function F_k, in equation (7) applied e.g. on the density, is defined as [11]

$$\overline{\overline{F_k}} = \frac{\overline{F_k}}{\alpha_k}. \qquad (11)$$

The density weighted average of a function F_k, in equation (7) applied e.g. on the velocity, is defined as [11]

$$\widehat{F_k} = \frac{\overline{\rho_k F_k}}{\overline{\rho_k}}. \qquad (12)$$

Applying the Stochastic Field Method to two phase flows where the stochastic fields are formulated for the scalar volume fraction α reads

[1] Mostly α_k is referred to as the volume fraction of phase k. In fact, it is the probability of phase k occurring at location $\boldsymbol{x}$ at time t [13].

$$\frac{\partial \alpha}{\partial t} + \widehat{U_j} \frac{\partial \alpha}{\partial x_j} + R = S(\alpha). \quad (13)$$

Here, R represents a stochastic term in Ito interpretation. Note that even though we distinguish several fields we use the common average velocity field from the two fluid model. This equation requires some comments:

1. For a stationary and incompressible flow ($\nabla \widehat{\boldsymbol{U}_k} = 0$) the average balance equation (eq. (7)) can be divided by $\overline{\overline{\rho_k}}$ and simplified:
$$\frac{\partial \alpha_k}{\partial t} + \widehat{\boldsymbol{U}_k} \nabla \alpha_k = \frac{S_k}{\overline{\overline{\rho_k}}} = \widetilde{S_k} \quad (14)$$
It is evident that the PDF's expectation should be identical to the average volume fraction from eq. (14). Analogously, the transport equation of the PDF's expectation should be identical to the average balance equation (eq. (14)).
Applying Pope's definition (4) to calculate the expected volume fraction $\langle\alpha\rangle_N$ from the samples $\alpha^{(n)}$ approximating the PDF yields
$$\langle\alpha\rangle_N \equiv \frac{1}{N} \sum_{n=1}^{N} \alpha^{(n)}. \quad (15)$$
The transport equation of the expected volume fraction can be derived by averaging the equation of the stochastic field (eq. (13)) over an ensemble of N stochastic fields with the operator $\langle\cdot\rangle_N$
$$\langle\frac{\partial \alpha}{\partial t}\rangle_N + \langle\widehat{U_j} \frac{\partial \alpha}{\partial x_j}\rangle_N + \langle R\rangle_N = \langle S(\alpha)\rangle_N. \quad (16)$$
The averages of the first two and the last term yield (cmp. [7])
$$\langle\frac{\partial \alpha}{\partial t}\rangle_N = \frac{\partial\langle\alpha\rangle_N}{\partial t} \qquad \langle\widehat{U_j} \frac{\partial \alpha}{\partial x_j}\rangle_N = \widehat{U_j} \frac{\partial\langle\alpha\rangle_N}{\partial x_j} \qquad \langle S(\alpha)\rangle_N = \overline{S(\alpha)}. \quad (17)$$
Since R represents a stochastic term in Ito interpretation its average vanishes (see [13] for details). Consequently the transport equation of the expected volume fraction corresponds to the average balance equation (14). An essential difference between the original stochastic field formulation taken from the field of combustion (eq. (3)) and the proposed formulation for the two fluid model is the exclusion of a 'diffusion-like' term. If the 'diffusion-like' term was retained it would appear in the transport equation of the expected volume fraction rendering equation (16) and equation (14) inconsistent.
2. The micro-mixing term M in equation (3) represents the molecular mixing on the smallest scales (cmp. eq. (1)). Since no small-scale mixing is included in the local instant formulation of the continuity equation for each phase [11], there is no reason for including micro-mixing into the stochastic partial differential equation. This was also confirmed by the group of Oevermann [14].
3. In analogy to the average balance equation (eq. (7)) the convection term in equation (13) contains the density averaged velocity $\widehat{U_j}$.
4. The formulation of the stochastic term R should be formulated problem dependent. The formulation which applies the analogy to the Stochastic Field Method in combustion (eq. (3)) is conceivable. However, the Ito interpretation should be used to retain the equality of the ensemble average over stochastic fields (eq. (16)) and the average balance equation (eq. (14)).

The stochastic fields represent samples approximating the PDF of volume fractions. Since the two fluid model needs for further calculations an average volume fraction, samples are averaged after each iteration step. Nevertheless, nonlinear processes depending on the volume fraction can use the PDF approximated by samples, and this avoids the modelling error which is evident in models based on one average value.

Application to a cavitation example from automotive industry

The Stochastic Field Method is implemented into the commercial CFD Code AVL FIRE™ and applied to a 2D as well as a 3D flow in an injection nozzle where cavitation occurs. The resulting average volume fraction of the disperse phase is shown in Figure 4. For a specific cell the average volume fraction and the volume fractions from the stochastic fields can be plotted over the time.

Figure 4: average volume fraction of the disperse phase in a 2D injection nozzle. The positions for the distributions in Figure 7 are marked with A, B and C.

Figure 5 compares the development of the stochastic fields for a cell within the cavitation area. The stochastic fields representing the possible peculiarities are clearly influenced by the random term. They are perturbed around the averaged field. Using the Stochastic Field Method allows a consideration of peculiarities that greatly differ from the averaged value (big difference between stochastic field and averaged field in Figure 5). This might have strong influence on the simulation result when strong nonlinear processes are present. This influence is ignored in methods dealing only with one average value.

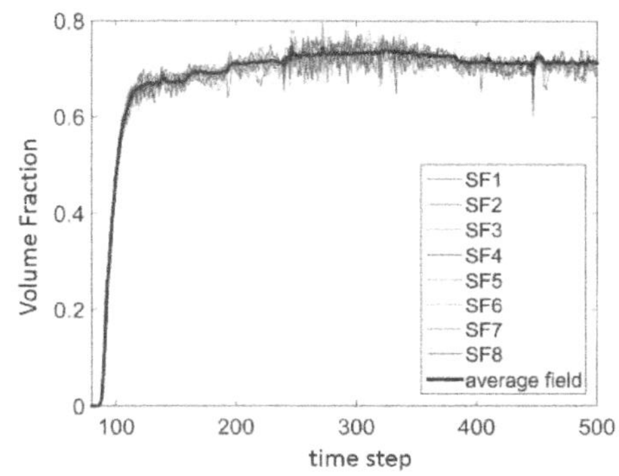

Figure 5: volume fraction of the stochastic fields and the average field for a cell within the cavitation zone.

Figure 6 shows simulation results for a 3D injection nozzle with moving needle. Due to cavitation effects the disperse vapor phase is generated at the nozzle entrance on the left-hand side. Its average volume fraction is plotted in Figure 6 (top). In addition two stochastic fields are plotted (bottom). Taking a closer look, the stochastic differences can be observed.

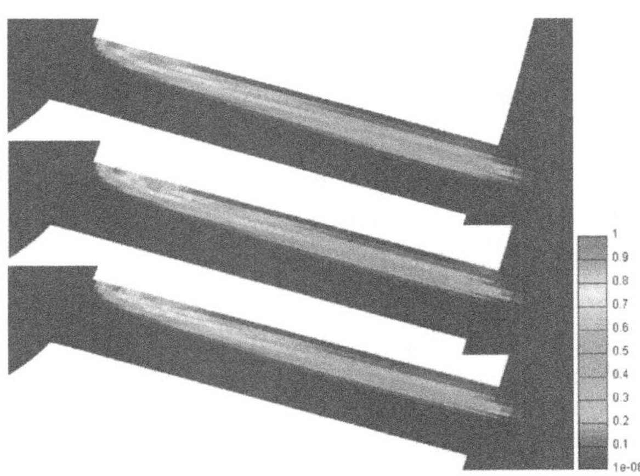

Figure 6: volume fraction of the dispersed phase; top: mean, bottom: two stochastic fields

Collecting values occurring in the stochastic fields over several time steps allows the construction of a histogram. It gives access to distributions of volume fraction within different regions. The distribution of the volume fraction of the disperse phase is given in Figure 7 for a cell outside, at the edge or within the cavitation sheet, respectively. Starting from a region outside the cavitation sheet (Figure 7a) the distribution is shifted towards higher volume fractions on the way to the center of the cavitation sheet. The distribution of the volume fraction can be used for the derivation of other distributions, such as the distribution of bubble size.

Conclusion

Originating from the field of combustion the Stochastic Field Method was applied to cavitating multi-phase flow. It was pointed out that the SPDE describing the stochastic fields cannot contain a 'diffusion-like' term nor a micro-mixing term to be consistent with the two fluid model. The Stochastic Field Method was applied to a cavitating flow within an injection nozzle. Distributions of volume fractions were shown.

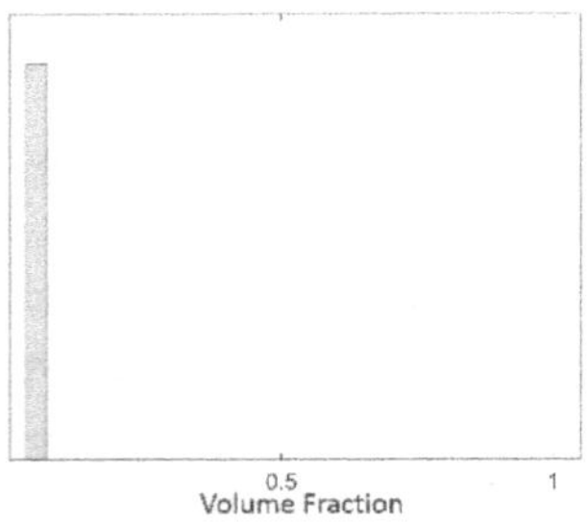

Figure 7a: Distribution of volume fraction for a cell outside the cavitation sheet (marked in Figure 4 with A). The stochastic fields only contain low volume fractions

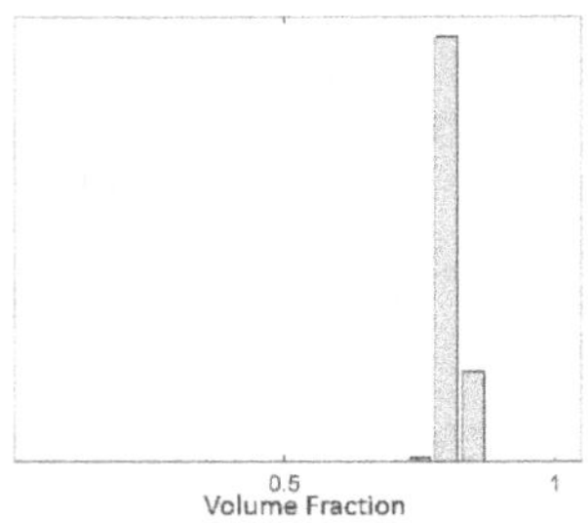

Figure 7b: Distribution of volume fraction for a cell at the edge of the cavitation sheet (marked in Figure 4 with B).

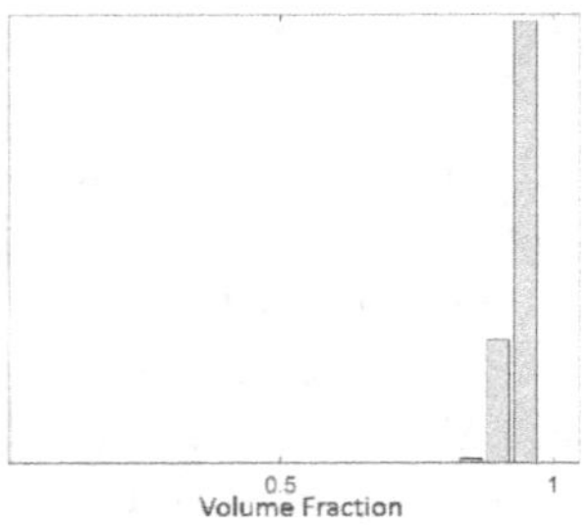

Figure 7c: Distribution of volume fraction for a cell within the cavitation sheet (marked in Figure 4 with C). The stochastic fields contain volume fractions with values close to 1.

Acknowledgment: This work was funded by AVL List GmbH and AREVA Nuclear Professional School.

References

[1] L. Valiño, "A Field Monte Carlo Formulation for Calculating the Probability Density Function of a Single Scalar in a Turbulent Flow," *Flow, Turbulence and Combustion,* pp. 157-172, 09 10 1998.

[2] J. Dumond, A. Class and F. Magagnato, "Stochastic-Field cavitation model," *Physics of Fluids,* 25 07 2013.

[3] Wang and Pope, "Lagrangian investigation of local extinction, re-ignition and auto-ignition in turbulent flames," *Combustion Theory and Modelling,* pp. 857-882, October 2008.

[4] W. Jones and S. Navarro-Martinez, "Large eddy simulation of autoignition with a subgrid probability density function method," *Combustion and Flame 150,* pp. 170-187, March 2007.

[5] A. Garmory and R. E. B. a. E. Mastorakos, "Simulation of the evolution of aircraft exhaust plumes including detailed chemistry and segregation," *JOURNAL OF GEOPHYSICAL RESEARCH,* April 2007.

[6] V. Sabel'nikov and O. Soulard, "Rapidly decorrelating velocity-field model as a tool for solving one-point Fokker-Planck equations for probability density functions of turbulent reactive scalars," *Physical Review E 72,* 06 07 2005.

[7] Pope and S. B., "PDF Methods for turbulent reactive flows," *Progr. Energy Combust. Sci.,* 1985.

[8] E. Gavi, D. L. Marchisio and A. A. Barresi, "CFD modelling and scale-up of Confined Impinging Jet Reactors," *Chemical Engineering Science,* p. 2228 – 2241, 2007.

[9] J. Devillon and J. Villermaux, *Encyclopedia of Fluid Mechanics,* 1986.

[10] H. H. G. Schöner, "A systematic elimination procedure for ito stochastic differential equation and the adiabatic approximation," *Z. Phys. B - Condensed Matter,* pp. 89-103, 2 March 1987.

[11] M. Ishii and T. Hibiki, Thermo-Fluid Dynamics of Two-Phase Flow, New York: Springer, 2011.

[12] D. A. Drew and S. L. Passman, Theory of Multicomponent Fluids, New York: Springer, 1999.

[13] C. W. Gardiner, Handbook of Stochastic Methods, Berlin: Springer-Verlag, 1985.

[14] M. Oevermann, *private conversation,* Chalmers University of Technology, June 28, 2017.

[15] D. A. Drew, "A turbulent dispersion model for particles or bubbles," *Journal of Engineering Mathematics,* pp. 259-274, 21 May 2001.

© 2018 ASME

10th International Symposium on Cavitation - CAV2018
Baltimore, Maryland, USA, May 14 – 16, 2018

CAV18-05023

Approach to Quantify Acoustic Cavitation in Absolute Physical Units

[1]Claudio Zanelli*; [2]Nicolas Candia; [1]Petrie Yam; [1]Sam Howard;

[1]Onda Corporation, 1290 Hammerwood Ave., Sunnyvale, CA, USA; [2]Universidad de Santiago de Chile, Chile

Abstract

In the study of acoustic cavitation it is desirable to quantify the acoustic signals in physical units to allow comparison among different conditions regardless of the equipment used. Motivated by cavitation present in cleaning environments, we have developed a method to measure the acoustics of cavitation in a way that is independent of the instrument used. The proposed method is based on having a point-like pressure sensor (hydrophone) calibrated in magnitude and phase over a broad frequency spectrum (well beyond the driving field frequency), capturing a limited sample of the raw time domain signal (in V), and converting the sample to frequency domain. The spectral decomposition of the raw signal is then used to deconvolve the sensor response to generate the true spectrum of the detected pressure field (in Pa). In the frequency domain, the corrected spectrum reveals the direct field as a single frequency signal, the stable cavitation by the sub-harmonics, harmonics and ultra-harmonics, and transient cavitation in the form of broadband noise above the electronic noise. By selectively integrating the various frequency components it is possible to know the acoustic pressure, in RMS units of pressure (Pa), for the direct field, stable cavitation, and transient cavitation. We present practical examples of measurements under various cleaning conditions where one can observe the effects of driving power, location relative to the source, or distribution of objects to be cleaned in the tank. It is believed that this or other method based on true, absolute pressure measurements will allow more progress in the research of cavitation by providing a means to exchange and compare quantitative data.

Keywords: Acoustic Cavitation, Stable Cavitation, Transient Cavitation, Measurement

Introduction

Acoustic cavitation is widely used in industry due to its effects on solid surfaces[1]. In cleaning processes the transient (or inertial) cavitation is often associated with deep cleaning or damage to the surface or structures on it, due to the violent collapse of bubbles, while stable cavitation provides a more gentle cleaning effect due to microstreaming. Distinct techniques to measure and detect the acoustic cavitation have been used [1,2]. While various kinds of hydrophones have been used to measure the acoustic field, and the cavitation activity has been evaluated in the frequency domain, the absence of real voltage to pressure calibration only allows measuring in electrical units (in V), which is not physically meaningful because it is dependent on the sensor used.

The proposed method to quantify the stable and transient cavitation levels is based on spectrally analyzing the acoustic emissions with a hydrophone that is small relative to the dominating wavelength, rugged enough to survive the environment, and calibrated in magnitude and phase over a broad frequency spectrum well beyond the direct field frequency. The signal is converted to the frequency domain (in V) which is then used to deconvolve the sensor response to generate the true spectrum of the detected pressure field (in Pa). A sample pressure spectrum can be seen below (Fig. 1). The primary peak colored in blue represents the direct field frequency (F_0) and pressure (P_0), the yellow peaks corresponding to harmonics and subharmonics of the direct field represent stable cavitation pressure (P_s), and the red colored region represents the signal from the bubble collapse or transient cavitation pressure (P_t) [3]. Uniquely quantifying both P_s and P_t is integral to monitoring and controlling an ultrasonic cleaning processes.

It should be noted that the acoustic field in a cleaning tank is not static, as the direct field is subject to reflections off the walls and the air above, and variations in the surface shape change the reflections and therefore the patterns of interference between the reflected waves. To obtain meaningful measurements one must take readings over an extended time window that incorporates enough of the fluctuations in order to obtain an average reading. The time of observation, as in other processes that include randomness, needs to be judiciously determined depending on the characteristics of the environment. The cleaning process, after all, relies on the integrated effect of the pressures on the particles, over the time used for cleaning.

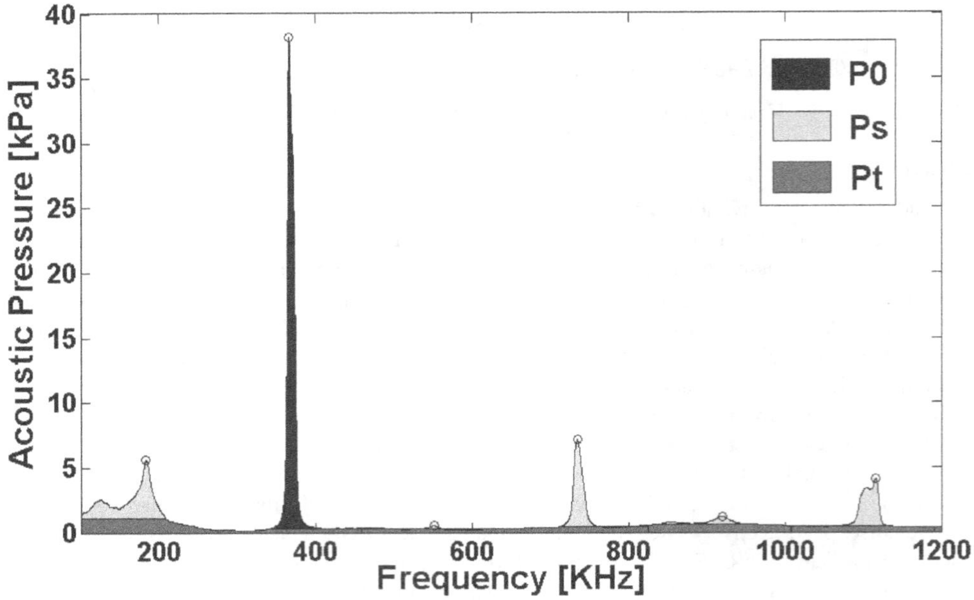

Figure 1. Acoustic pressure spectrum example. The blue area represents the direct field pressure (P_0) generated by the driving source at its fundamental frequency (F_0). The yellow areas represent harmonics, sub-harmonic and ultra-harmonics associated with stable acoustic cavitation pressure (P_s). The red area represents the broadband noise generated from the collapse of bubbles, or transient cavitation pressure (P_t).

Materials and methods

A hydrophone (HCT-0310, Onda Corporation) was used to acquire the electrical signal in a cavitation environment. The hydrophone sensing area is about 1 mm in size, appropriate for direct fields up to 1 MHz. The hydrophone is connected to a cavitation meter (MCT-2000, Onda Corporation) which has an electronic bandwidth of 10 MHz and acquires the acoustic pressure spectrum to determine measurement parameters including P_0, P_s, P_t and F_0. Figure 2 shows the process steps used to obtain the cavitation acoustic parameters.

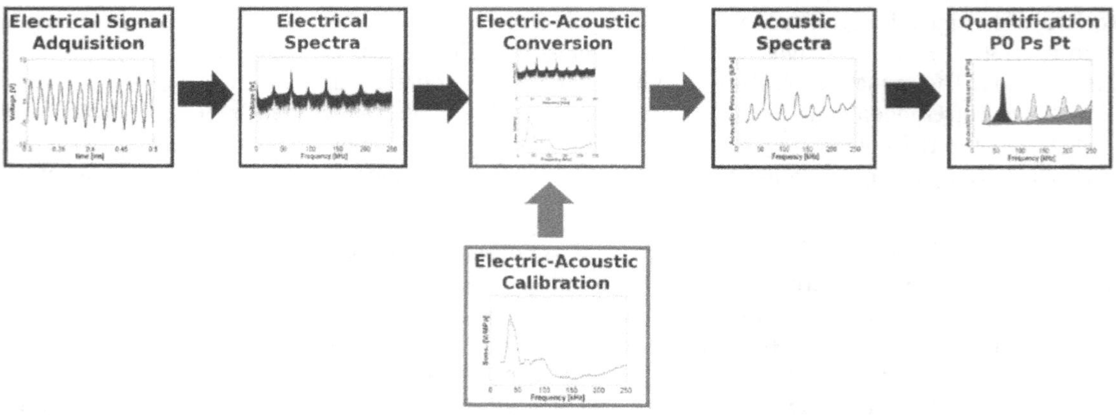

Figure 2. Processing steps to quantify the cavitation acoustic parameters.

Results and examples

We present examples of measurements under various cleaning conditions in Figure 3. Fig. 3(a) shows the cavitation pressure for several driven frequencies (36, 78, 118, 132, 172, 215 and 265 [kHz]) using approximately the same power density. In Figure 3(b), the cavitation pressure is measured at varying electrical power levels for a 78 [kHz] tank.

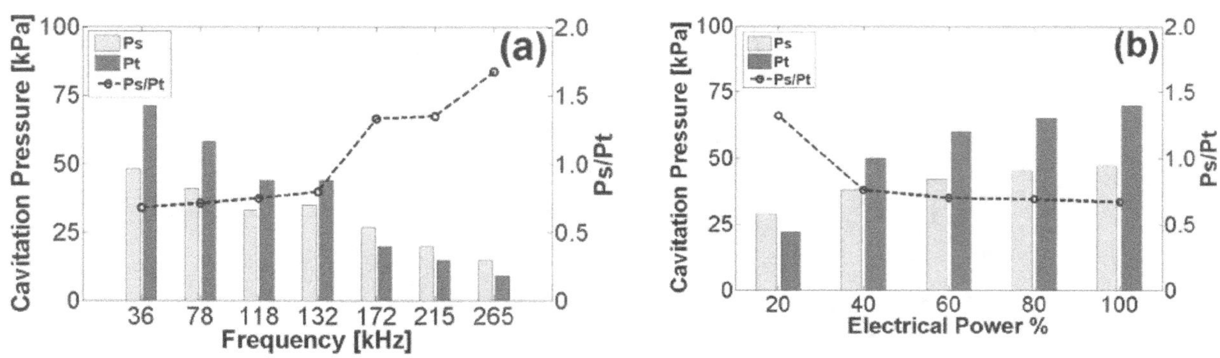

Figure 3. (a) Stable and transient cavitation comparison at different frequency. (b) Stable and transient cavitation readings in a 78 KHz tank as a function of the electric power.

To investigate the uniformity of the acoustic field in a 160x80mm, 40 KHz cleaning tank, the direct field pressure was mapped at intervals of 5 mm, creating a mesh of 17x33 points (Figure 4). This is a quantitative equivalent of the aluminum foil test [4].

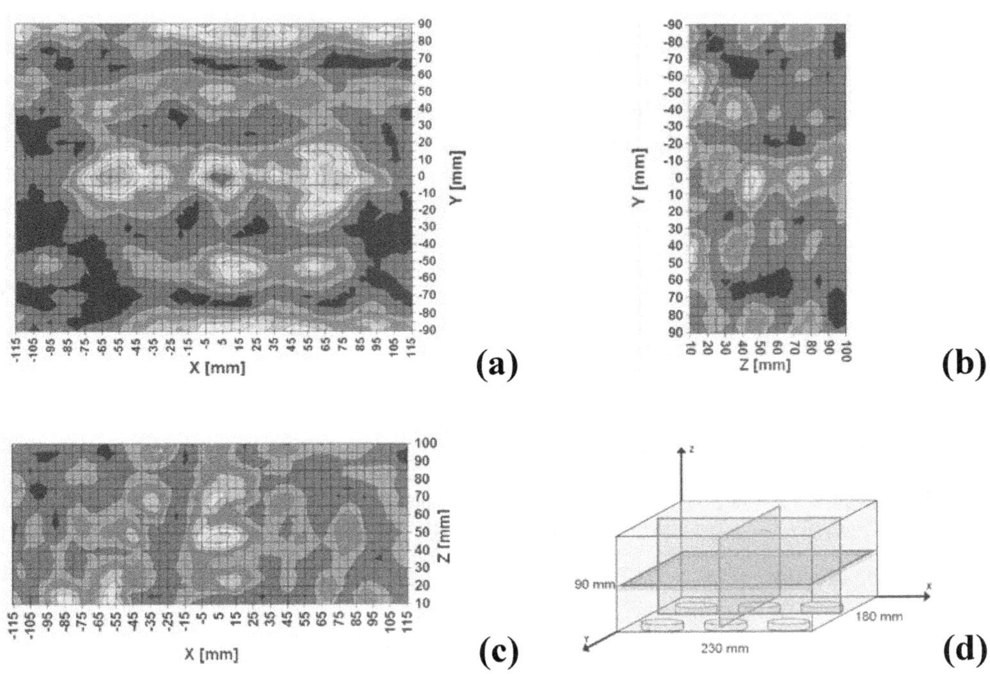

Figure 4. (a) Normalized P_0 uniformity mapping for XY plane, (b) YZ plane and (c) XZ plane and (d) Schematic of cleaning tank showing scanned planes.

Effect of the presence of objects in a tank

To understand the effect of the presence of parts in the acoustic field, the mapped cavitation pressure distribution in 40 kHz cleaning tank that is used to clean disks in cassettes is shown in Figure 5. The map is made in the plane next to where a disk is located, in the tank with water but otherwise empty (a), then adding an empty cassette (b) and cassette with 14 discs (c).

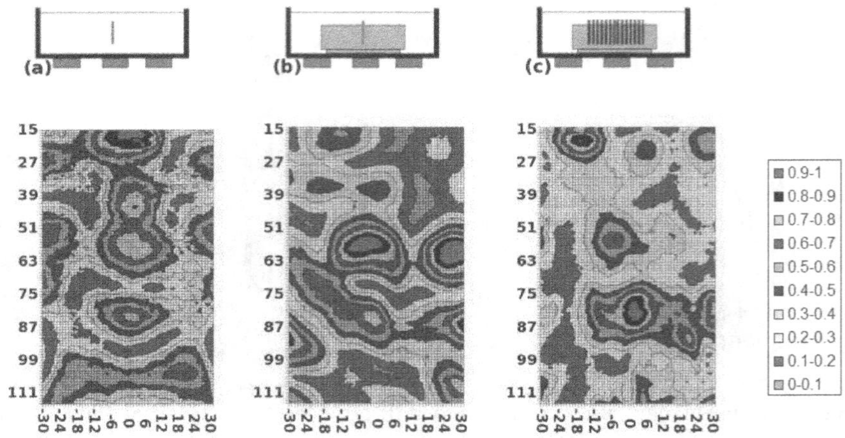

Figure 5. Loading effects on acoustic uniformity 40 kHz batch system (a) empty tank, (b) added an empty cassette and (c) added 14 disks to the cassette.

The drastically different patterns over the same plane shows that the effect of the presence of cassette and disks must be taken into account when evaluating the cleaning performance.

Conclusion

The ability to measure cavitation in its separate components and in absolute units has been demonstrated, and we propose that this or similar methods be employed in quantifying cavitation acoustics. The key elements are a durable sensor that is smaller than $\lambda/2$ (for isotropy) and calibrated (in V/Pa) over a wide range of frequencies. The range should extend from below the driving frequency to a few MHz, as transient cavitation has been observed in this range [3,5]. Processing must be done on time samples that are long enough to smooth over fluctuations of the acoustic field (usually of the order of 1 second) and short enough to provide timely information (in the order of few seconds).

References

[1] Verhaagen B., Fernández D. (2016). *Measuring cavitation and its cleaning effect.* Ultrasonics Sonochemistry 29, 619–628

[2] Sarno D., Hodnett M., Lian Wang, Zeqiri B. (2017). *An objective comparison of commercially-available cavitation meters.* Ultrasonics Sonochemistry 34, 354–364

[3] Frohly J., Labouret S., Bruneel C., Looten-Baquet I., Torguet R. (2000). *Ultrasonic cavitation monitoring by acoustic noise power measurement.* J. Acoust. Soc. Am. 108 (5), 2012-2020

[4] *Health Technical Memorandum 01-05: Decontamination in primary care dental practice.* UK Department of Health, Leeds, 2013.

[5] Song J.H., Johansen K., Prentice P. (2016). *An analysis of the acoustic cavitation noise spectrum : The role of periodic shock waves.* J. Acoust. Soc. Am. 140 (4) October 2016.

© 2018 ASME

Assessment of flow aggressiveness at an ultrasonic horn cavitation erosion test device by PVDF pressure measurements and 3D flow simulations

[1]Simon A. Paepenmöller*; [1]Johannes Kuhlmann; [1]Martin Blume; [1]Romuald Skoda

[1] *Chair of Hydraulic Fluid Machinery, Ruhr-Universität Bochum, Universitätsstr. 150, 44801 Bochum, Germany*

Abstract

Transient cavitation induced loads are investigated in a flow region subject to severe cavitation erosion at an ultrasonic horn test facility operated in indirect mode, both experimentally and via CFD. A variation of gap width and sensor position is used to vary flow aggressiveness. The experimental technique utilizes piezoelectric sensors based on Polyvinylidene fluordide (PVDF). Sensors are employed at the stationary specimen in close proximity to the oscillating horn tip and wall load spectra are obtained. Complementary CFD studies emulate the experimental sensor. Flow aggressiveness decreases when the gap is widened and increases towards the stationary specimen center. Results from experiment and simulation show good quantitative agreement.

Keywords: cavitation erosion, wall load spectrum, PVDF sensor, CFD, barotropic equation of state

1. Introduction

Ultrasonic horns are widely used to characterize the resistance of materials to cavitation erosion. The test facility employs a horn oscillating in longitudinal mode at high frequency to cause generation and collapse of voids below the oscillating tip. We employ the indirect method, in which a stationary erosion sample is placed in close proximity to the tip. The gap width is an important parameter that influences erosion pattern, mass loss rate and incubation time of the erosion sample [1–3].

To gain better understanding of the erosion process, load measurements have been performed at ultrasonic horn facilities by different groups (see e.g., [4–8]). Wall load spectra are obtained by performing peak detection on the force signal, either as postprocessing or during runtime through hardware (i.e. a peak hold circuit with finite reset interval). Derived histograms are typically presented in cumulative form, with the occurrence rate normalized to measurement time and sensitive area of the sensor. Detected events show exponential distribution. This behavior is shared with other cavitation test facilities (see e.g., [9, 10]).

Within the present study, a variation of gap width and sensor position is used to adjust flow aggressiveness. Wall load spectra are determined from the experiment as well as from numerical simulation: the experiment utilizes piezoelectric PVDF-foil based sensors, which are reproduced within the simulation to reflect sensor position, size and spatial integrating behavior. Peak detection and derivation of load collectives is performed on both, signals from experiment and simulation with identical method and the resulting statistics are compared. The numerical approach is especially attractive, since the experiment's physical limit of spatial resolution (i.e. possible sensor miniaturization) is multiple orders of magnitude larger than the spatial limit of simulation (i.e. grid resolution). While high temporal resolution can be achieved in both, experiment and simulation, the latter is solely bounded by computational cost.

Aim of the present study is to evaluate CFD as a complimentary tool for flow aggressiveness assessment. For this purpose, experimental results are carefully examined in terms of reliability and repeatability. Quality of the simulation is assured through a grid refinement study and evaluation of temporal convergence of the obtained load collectives.

2. Experimental configuration

The test facility consists of a commercially available ultrasonic horn from Hielscher Ultrasonics (UIP2000hd) operated at 19.8 kHz driving frequency and 40.9μm amplitude (peak-to-peak). The horn tip is immersed in distilled water at equilibrium air saturation, kept at 20°C by an actively controlled indirect cooling circuit, as shown in Fig. 1. Gap width between horn tip and the sensor equipped stationary specimen is adjusted with a 10μm precision leveling device.

*Corresponding Author, Simon A. Paepenmöller: simon.paepenmoeller@rub.de

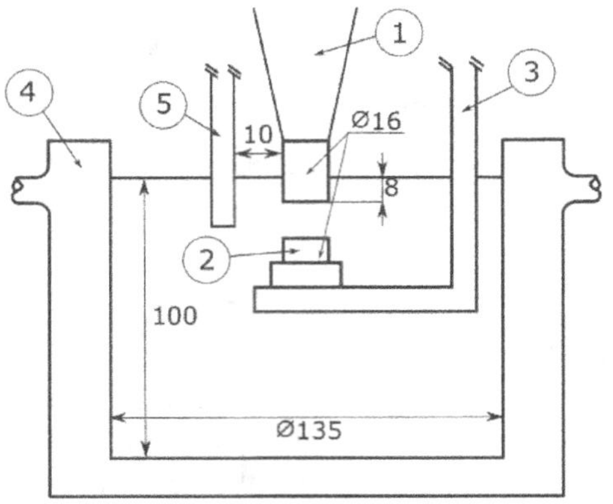

Fig. 1 Test facility with dimensions in mm.

2.1. Sensor design, calibration and signal processing

To enable temporal resolution of cavitation events, the sensors utilized follow a minimum thickness doctrine without attenuating backing layer. This design leads to maximized resonance frequency and bandwidth of the instrument. Preceding stages of instrument development have been published [11, 12] and the final design is depicted in Fig. 2. Upper and lower pole connect to a fully differential charge amplifier.

A piezoelectric PVDF foil element of 5mm diameter and 30µm thickness is laminated between an UV-cured resin as top layer, and a bottom layer made from Polyethylene terephthalate (PET). Electrodes are printed in a screen printing technique and the top layer erosion resistance is improved through a single layer of Polyimide (Kapton) adhesive tape. The overall package height is 295±25µm.

Fig. 2 Sensor with electrodes on the left and sensitive area on the right.

The sensor outlines are milled into a stationary specimen and the foils are mounted flush with cyanoacrylate glue. The two investigated radial positions are shown in Fig. 3, where the sensitive area is depicted in dark grey.

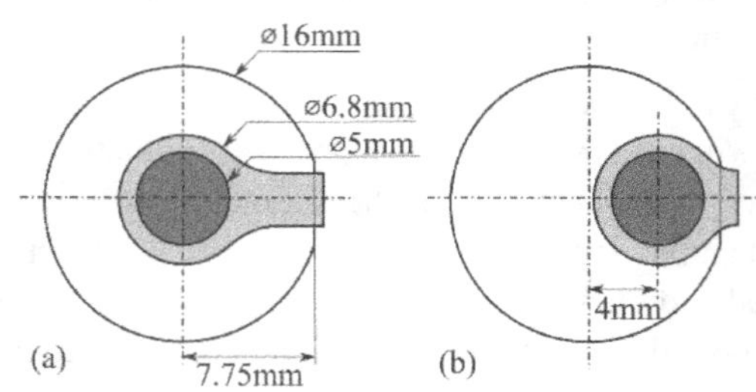

Fig. 3 Top view on stationary specimen with centric (a) and excentric (b) sensor mounting positions.

Calibration is performed with the ball drop technique, where a steel ball is dropped unto the surface from a predefined height and the rebound is measured through long exposure imaging. Contact time and amplitude are utilized to calculate the time averaged force during contact from momentum and energy conservation considerations. The calibration constant of the sensor is determined from 10 repetitions at approximately 40 N loading and 26 µs contact time. Repeatability of the mean calibration constant is about ±5%.

The voltage signal is digitized with a 12 bit deep memory high-speed oscilloscope at 15.63 MHz sampling rate for a continuous interval of at least 7 s. As the sensor surface is not loaded uniformly, we are limited to the quantification of integral force across the sensor surface. Peak detection is performed with an algorithm utilizing a minimum amplitude threshold of 10 N, a peak prominence value of 10 N and a minimum peak distance of 0.26µs. It is performed on the raw signals without smoothing or band pass filter operations. Reasonable lower bounds for threshold and prominence depend on the signal to noise ratio, while introducing a minimum peak distance has shown to improve detection quality and increase robustness. Negligible influence of the parameters has been assured through variation.

Sensors are subject to severe erosion and subsequent destruction or alteration of characteristics. We ensured reliability of the results by limiting usage time of each sensor to 8 s and applying them in a one-way manner. We performed reproduction at every operating point with a new sensor and ensured unaltered characteristics through recalibration after each run. We further ensured that the final load collectives are not subject to systematic temporal effects by splitting the force signal into 0.5 s chunks and deriving statistics from the pieces: if the collectives obtained in this manner show only slight variation around a stable mean, we assume that no degenerative effects are present.

3. Numerical method and numerical sensor

For the numerical part we perform explicit 3rd order Runge-Kutta time integration of the Euler equations for an inviscious fluid, coupled with an isentropic barotropic equation of state, in finite volume discretization with flux-corrected, conservative mesh deformation. A homogenous mixture of liquid and vapor within each cell is assumed. The method has previously been implemented into our in-house code 'Hydrub' and applied to study erosion sensitive wall zones and erosion mechanisms at ultrasonic horns [13]. By counting events exceeding threshold values for pressure and void collapse rate, erosion indicators and a map of dimensionless erosion probability are obtained. Details on the numerical scheme and the assessment of erosion sensitive wall zones by erosion probability can be found in Mottyll and Skoda [13]. Results within the present publication are generated with our in-house OpenFOAM implementation 'HydrubFOAM' [14]. For density, the Minmod scheme [15] is employed, while for velocity the GammaV scheme [16] is utilized to provide 2nd order accuracy in space.

The numeric domain mimics the experimental geometry depicted in Fig. 1. Simulation time step depends on the acoustic CFL number and therefore depends on the smallest cell within the domain: it is typically about 13 ns for basic mesh settings and 6.4 ns for refined meshes.

Cell faces of the stationary specimen top wall (i.e. numerical domain boundary faces), are grouped to a numeric sensor in accordance with the experimental sensor's position and size. Through integration of numerical cell face center pressure across the numerical sensor, a force signal is obtained in analogy to the experiment.

For the excentric position, 5 numerical sensors are deployed along the circumference and their signals are concatenated in order to provide virtually extended evaluation time. The algorithm and parameters used to derive the load collectives are identical for both numerical and experimental input data.

3.1. Grid quality and grid independency

Utilized grids are generated block structured with combined H- and O-grid topology. The grid within the gap and its immediate surrounding is optimized for minimal distortion and volume change. This is a necessary prerequisite for the erosion indicator methodology (as described in [13]). Cell volumes within the entire gap differ by ±12 % from average. Grids for all gap widths investigated share a constant cell height of 50 µm (basic mesh) respectively 25 µm (refined mesh) throughout the cavitation zone. To reduce computational effort, hanging nodes and volume expansion are employed outside the cavitation zone, i.e. the beaker geometry. The total number of cells is approximately 110k for coarse and 430k for refined grids.

Previous employment of a closely related numerical technique at hydrodynamic cavitation test facility has revealed strong mesh dependency of the numerical sensor results [17]. Our grid study proves only minor influence of the grid resolution, as depicted in Fig. 4, where collapse force rates are shown, i.e. the normalized cumulated rate of peaks exceeding a certain force. Loading of the stationary specimen is dominated by subharmonic collapse, i.e. the collapse of large vapor structures which form over a multiple of the horn driving period [13]. The weak grid dependence of numerical sensor signals presented here is attributed to the large sensor size: while numerical pressure in a single cell strongly depends on grid size, cell-face area weighted integration across a large number of cells to obtain a sensor force diminishes grid dependency. Preliminary simulations have shown, that a much smaller numerical sensor exhibits significant grid dependence, also at ultrasonic horn test cases.

All subsequently presented results have been obtained on the coarser grid.

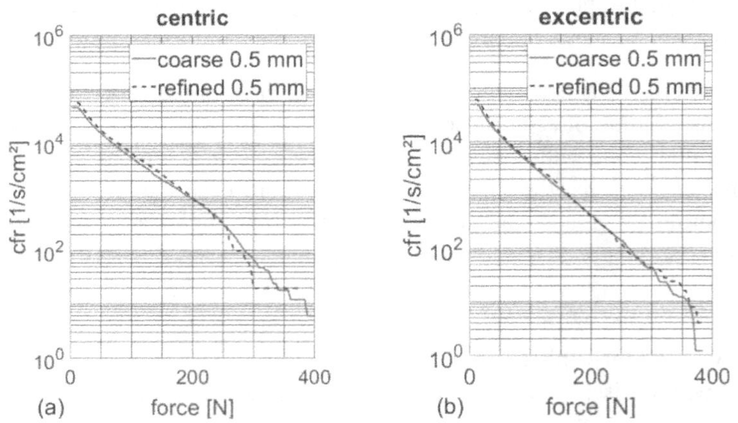

Fig. 4 Load collectives from centric (a) and excentric (b) positioned sensors for coarse and refined meshes at simulation time 0.75 s (coarse) and 0.26 s (refined). The excentric signal measurement time is extended through concatenation of 5 virtual probes.

3.2. Temporal convergence

When deriving load collectives from successively prolonged experimental data, results become indistinguishable after approximately 3 s. The computational effort of the simulation does not allow for such a period to be evaluated, but sufficient convergence in the numerical study is achieved faster, typically after approximately 0.5 s. All subsequently presented results are ensured to be temporally converged.

4. Results

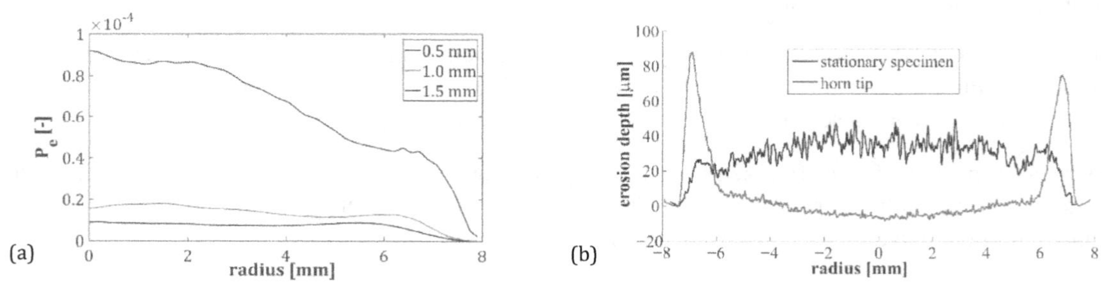

Fig. 5 (a): Radial erosion probability (P_E) distribution on the stationary specimen from simulation for varying gap widths and (b) erosion depths as measured for 0.5mm gap width (Part b reprinted from [13]).

The erosion indicator methodology predicts increasing erosion probability towards the center of the stationary specimen for 0.5 mm gap width, as shown in Fig. 5 (a): this is in good agreement with practical experience as well as topographic measurements, as reprinted in Fig. 5 (b) from [13], where the erosion depth for the stationary specimen and horn tip is shown.

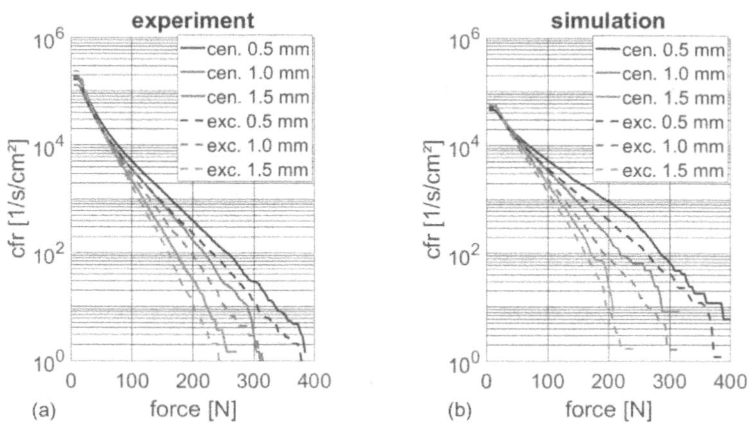

Fig. 6 Load collectives in dependence of gap width and radial position from experiment (a) and simulation (b).

Widening of the gap leads to reduced flow aggressiveness, manifesting itself in a decrease of erosion probability (see Fig. 5 a) and a decrease of the slope of the load collectives, as presented in Fig. 6. This result is confirmed by erosion tests performed at our laboratory, which show prolonged incubation time and reduced steady state mass loss rate for increased gap width (exemplarily data given for 1.4404 steel in Table 1). The flow aggressiveness measured in excentric position is decreased in comparison to the centric position, as seen in Fig. 6 (a). Simulation results reproduce this behavior, depicted in Fig. 6 (b).

Table 1 Erosion testing results of polished 1.4404 specimen in distilled water @20°C, 41µm amplitude and 19.8kHz frequency

Gap width [mm]	**0.5**	**1.0**	**1.5**
Incubation time [s]	8600	8800	12600
Steady state erosion rate [µg/min]	48	32	18

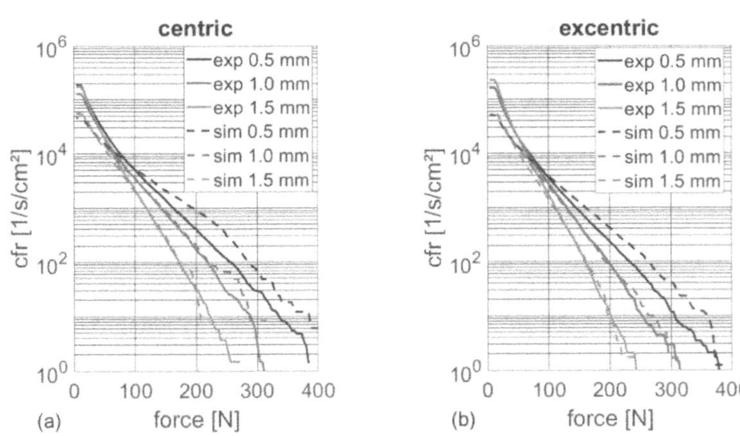

Fig. 7 Load collectives from experiment and simulation in dependence of gap width for centric (a) and excentric (b) position.

Quantitative comparison of the load collectives from experiment and simulation is given in Fig. 7. The total number of events (i.e. the vertical axis intercept) in the simulation is lower than in the experiment. For gap widths of 1.0 mm and 1.5 mm the characteristic slope is reproduced very well. For 0.5 mm there is an offset of the collective. We assume that the poorer match at 0.5 mm is due to neglect of non-condensable gas in the simulation. This assumption is further supported by preliminary measurements, which reveal a stagnation of flow aggressiveness for gaps below 0.5 mm, while the simulation results indicate increasing flow aggressiveness. Of course this hypothesis is yet speculative and demands further investigation.

5. Conclusion

The numerical method has proven capability to correctly account for the influence of gap width and radial position. Although the vertical axis intercept (i.e. the total number of events) differs between experimental and numerical study, the operating conditions can be correctly assessed by the load collective slope. Extrapolation of the linear part of the simulation collectives can be used to compensate the intercept mismatch. Forces predicted from the simulation are in good quantitative agreement with experimental values and we therefore conclude CFD to be established as a complimentary method to impact load measurement.

Further work will concern smaller gap widths, which might require extending the numerical model to incorporate effects such as the presence of non-condensable gas. Future investigation should establish a link between assessed flow aggressiveness and material response, i.e. validation of an erosion model based on localized impact load distributions.

6. References

[1] Endo, K., Okada, T., Nakashima, M. (1967). *A Study of Erosion Between Two Parallel Surfaces Oscillating at Close Proximity in Liquids.* J. Lub. Tech.; 89: 229.

[2] Kikuchi, K., Hammitt, F. G. (1985). *Effect of separation distance on cavitation erosion of vibratory and stationary specimens in a vibratory facility.* Wear; 102: 211–25.

[3] Matsumara, M., Okumoto, S., Saga, Y. (1982). *Mechanism of Damage in the Vibratory Test with Stationary Specimen.* Bulletin of JSME; 25: 898–905.

[4] Okada, T., Iwai, Y., Awazu, K. (1989). *A study of cavitation bubble collapse pressures and erosion part 1:* A method for measurement of collapse pressures. Wear; 133: 219–32.

[5] Hattori, S., Mori, H., Okada, T. (1998). *Quantitative Evaluation of Cavitation Erosion.* J. Fluids Eng.; 120: 179.

[6] Arndt, R. E. A., Paul, S., Ellis, C. R. (1997). *Application of Piezoelectric Film in Cavitation Research.* Journal of Hydraulic Engineering; 123: 539–48.

[7] Lee, M. K., Hong, S. M., Kim, G. H., Kim, K. H., Rhee, C. K., Kim, W. W. (2006). *Numerical correlation of the cavitation bubble collapse load and frequency with the pitting damage of flame quenched Cu–9Al–4.5Ni–4.5Fe alloy.* Materials Science and Engineering: A; 425: 15–21.

[8] Singh, S., Choi, J.-K., Chahine, G. L. (2013). *Characterization of Cavitation Fields From Measured Pressure Signals of Cavitating Jets and Ultrasonic Horns.* J. Fluids Eng.; 135: 91302.

[9] Franc, J.-P., Riondet, M., Karimi, A., Chahine, G. L. (2011). *Impact Load Measurements in an Erosive Cavitating Flow.* J. Fluids Eng.; 133: 121301.

[10] De, M. K., Hammitt, F. G. (1982). *New Method for Monitoring and Correlating Cavitation Noise to Erosion Capability.* J. Fluids Eng.; 104: 434.

[11] Lang, S., Dimitrov, M., Pelz, P. F. (2012). *Spatial and Temporal High Resolution Measurement of Bubble Impacts. In:* Ohl C-D, editor. CAV 2012: Proceedings of the 8th International Symposium on Cavitation, 13 - 16 August 2012, Singapore. Singapore: Research Publishing Services.

[12] Dimitrov, M. (2015). *Räumlich auflösende, dynamische Druckmesssysteme mit piezoelektrischen Foliensensoren (In German)*. Aachen: Shaker.

[13] Mottyll, S., Skoda, R. (2016). *Numerical 3D flow simulation of ultrasonic horns with attached cavitation structures and assessment of flow aggressiveness and cavitation erosion sensitive wall zones.* Ultrason Sonochem; 31: 570–89.

[14] Deimel, C., Skoda, R. (2014). *Implementation of an explicit density-based solver for the simulation of cavitating flows in OpenFOAM.* 9th OpenFOAM Workshop Zagreb.

[15] Harten, A. (1983). *High resolution schemes for hyperbolic conservation laws.* Journal of Computational Physics; 49: 357–93.

[16] Jasak, H., Weller, H. G., Gosman, A. D. (1999). *High resolution NVD differencing scheme for arbitrarily unstructured meshes.* Int. J. Numer. Meth. Fluids; 31: 431–49.

[17] Mihatsch, M. S., Schmidt, S. J., Adams, N. A. (2015). *Cavitation erosion prediction based on analysis of flow dynamics and impact load spectra.* Physics of Fluids; 27: 103302.

© 2018 ASME

Assessment of Remote Cavitation Detection Methods with Flow Visualization in a Full Scale Francis Turbine

[1]Xavier Escaler*; [2]Ingrid K Vilberg; [3]Jarle V Ekanger; [3]Hakon H Francke; [3]Morten Kjeldsen

[1]*Universitat Politècnica de Catalunya, Barcelona, Spain;* [2]*Norwegian University of Science and Technology, Trondheim, Norway;* [3]*Flow Design Bureau AS, Stavanger, Norway*

Abstract

This paper describes the experimental investigations carried out in the Francis turbine at Svorka power plant operated by Statkraft in Norway. The unit, with a head of 260 m, can deliver a maximum output load of 25 MW. The rated flow rate is 11 m^3/s and the machine rotates at 600 rpm. The turbine runner shows cavitation pitting on the suction side of the blades but some blades present more erosion than others. Moreover, preliminary studies based on remote monitoring of vibrations and acoustic emissions in this particular unit have predicted risk of erosion at high loads and the presence of a draft tube swirl affecting the cavity dynamics. In order to assess the sensitivity of these methods and the validity of the predictions, several acrylic-glass windows have been installed on the draft tube wall to visualize the runner outlet flow during operation. A high speed camera has been used to record the flow field during the tests with rates up to 5000 frames per second. A cavitation detection system has been installed comprising three high-frequency uniaxial integrated electronics piezoelectric (IEPE)-type accelerometers and an acoustic emission sensor, mounted in the turbine guide bearing pedestal and a guide vane arm. In particular, a series of measurements at different operation conditions have been carried out to correlate the simultaneous camera observations with the acceleration and acoustic emission overall levels in high frequency bands. The preliminary analysis of the camera records permits to certify the existence of erosive blade cavitation with the closure region close to the eroded areas at high loads. It can be seen that cavitation appears only in some blades and that it presents different cavity sizes for the same operation condition. As the load increases towards maximum powers, both the number of blades with cavitation and the size of the cavities grow. Moreover, the overall vibration levels also rise as expected.

Keywords: Francis turbine, cavitation erosion, high speed flow visualization, cavitation detection, vibration, acoustic emission

Introduction

Hydraulic turbine manufacturers and operators have to face with the drawbacks of suffering cavitation erosion when the units operate outside the best efficiency point. Currently, most of the machines require regular visual inspections to detect the extent of the erosion in order to replace the removed mass of material and to avoid a catastrophic failure. Obviously, this type of planned preventive maintenance is costly because it requires long machine shutdown. A most efficient procedure should be based on the predictive condition monitoring. For that, a remote measurement equipment is installed in the turbine and the adequate signal processing techniques are applied continuously without stopping the turbine.

In this direction, Flow Design Bureau AS is developing a cavitation erosion system to detect and predict the risk of erosion in hydraulic turbines with the help of UPC and NTNU. For that, a series of experiments have already been carried out in the Svorka power plant located near Surnadal in Norway. The plant is owned by Statkraft AS and Svorka Energi AS. The vertical shaft unit is a Francis turbine with a head of 260 m, rotating at 600 rpm and providing a maximum output power of about 25 MW. The runner has 15 blades and the wicket gate has 20 guide vanes. The turbine specific speed is 0.585 and the maximum turbine performance is reached around 21 MW.

This machine is of interest because it suffers aggressive erosion on the suction side of the runner blades close to the trailing edge as it can be seen in the photograph presented in figure 1. This type of erosion is caused by cloud cavitation detaching from the blade leading edge [1]. The methods used to predict cavitation are based on induced high frequency vibrations and acoustic emissions measured on the turbine guide bearing and the guide vanes [2].

*Corresponding Author, Xavier Escaler: xavier.escaler@upc.edu

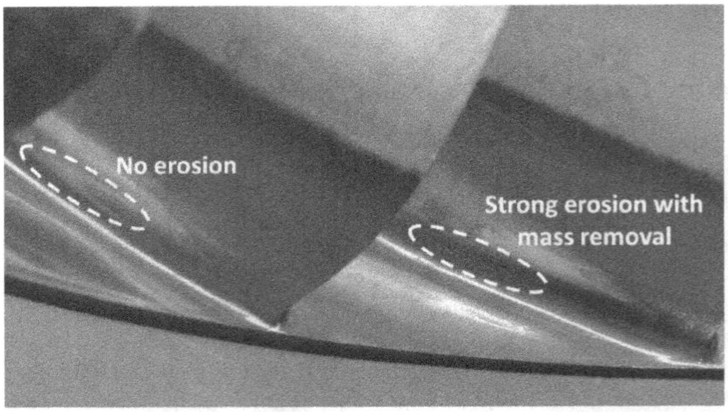

Figure 1. Photograph of the runner showing one blade with erosion and the contiguous one without erosion.

A preliminary investigation [3] has already demonstrated the feasibility of the installed measuring system and the applied signal processing techniques to monitor the activity of runner blade cavitation for unit operation above the best efficiency point (BEP). The results are consistent with the hypothesis that the highest risk of erosion would occur at full load. At this operating condition, the main hydrodynamic frequencies would force the shedding of cloud cavities, thus increasing their aggressiveness. Moreover, a draft tube instability modulating the runner blade cavitation dynamic behavior with periodic character has also been detected above BEP with maximum pressure pulsation levels at full load. Obviously, such detailed predictions could not be directly verified because the flow field inside the runner and draft tube cannot be observed in industrial environments such as any actual hydro power plant.

To the authors' knowledge and unfortunately, no literature reporting flow visualizations of blade cavitation in full scale prototypes can be found. Only in reduced scale model tests with transparent draft tube walls some limited research has been carried out to validate the cavitation remote monitoring techniques [4] which is not sufficient to validate the current methods in real machines. The typical design of the test rigs differs in many aspects from the full scale units. For instance, the signal transmission paths, the cavitation dynamic behavior and the level of aggressiveness might present significant differences that are not well known specially operating outside the BEP.

Consequently, it has been decided to design, construct and mount several transparent acrylic-glass (PMMA) windows around the draft tube wall that allow to visualize the cavitation flow exiting a Francis runner during operation. The final goal of the experimental investigation is to evaluate and validate the erosive cavitation remote measurements and procedures that permit to predict erosive cavitation and to estimate its aggressiveness.

Flow visualization setup

In order to have visual access to runner areas with detected cavitation erosion, three pipe-stubs were added to the draft tube into which 0.2 m diameter acrylic-glass bolts were fitted. In addition, a 0.42 m diameter acrylic-glass bolt was produced to replace the manhole cover. All the bolts surfaces were machined in such a fashion that allowed direct line-of-sight towards the erosion zones as indicated by the lines plotted in the outline presented on the left of figure 2. At the time of writing, the acrylic-glasses have been fitted twice in the pipe-stubs and the manhole. More specifically, they have been exposed to water between 2 and 3 days total, without presenting any observation of cracking or discoloration.

As it could be detected in a series of preliminary tests, the axis of the draft tube presents a gas-filled core for all the operation conditions as shown on the right of figure 3. The size of this cavitating vortex grows as the unit operates far from the BEP and obviously blocks the vision of the runner. Consequently, the camera had to be pointed away from the draft-tube center and given the current window positions the visible runner blades were observed when passing in front of the spiral casing end.

The high-speed camera used, model Photron® FASTCAM SA5, allows a frame-rate up to 7000 frames per second (f.p.s.) while still maintaining maximum resolution (1024x1024 pixels). During unit operation, the flow field was recorded with frame rates of 4000 and 5000 f.p.s., which correspond to a runner motion between frames of 0.9° and 0.72°, respectively, given the turbine rotating speed (10 Hz). As the field of view exceeds 24°, more than the angular separation between two consecutive blades could be recorded. The camera mounted a 50 mm f 1.8 lens with the Nikon f-mount and it was focused on the blades eroded areas through one of the two smaller bolts located at the draft tube top. Professional photographic lights were used that lighted through the manhole bolt and through the other smaller bolt located at the top.

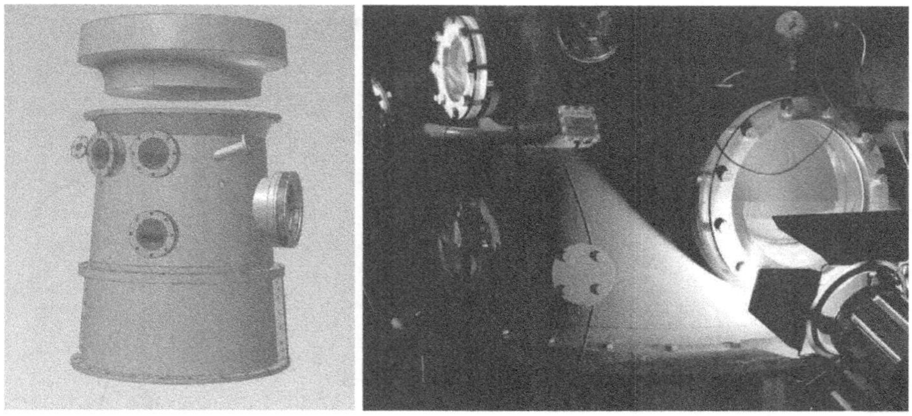

Figure 2. (a) Outline of the acrylic-glass windows mounted on the draft tube with the adequate vision angle to see the blades; (b) photograph showing the flow visualization tests.

Figure 3. Vision of the runner from the manhole with the machine stopped (a) and with the machine operating at 19 MW (b).

Vibration and acoustic emission measurement

A cavitation detection system was installed that consisted of three high-frequency uniaxial integrated electronics piezoelectric (IEPE)-type accelerometers (AC) and an acoustic emission sensor (AE) mounted in the positions indicated in figure 4. Two of the accelerometers separated 90°, TGB1 and TGB2, were mounted on the turbine guide bearing pedestal using studs glued to its base that in turn screwed onto the sensor. Their axes were orientated in radial direction relative to the shaft axis. The other accelerometer, GV, was mounted at the top of a guide vane shaft using similar studs. Its axis was parallel to the shaft axis. The acoustic emission sensor, TGB, was also mounted on the guide bearing pedestal using a spring-loaded magnetic holder. Silicone grease was used between the sensor surface and the base surface. A preamplifier, supplied by a constant voltage source, was used to condition its signal.

All data recording was made using National Instruments (NI) hardware in conjunction with LabView® software. Signals were recorded using an NI cDAQ-9178 chassis. The accelerometers were conditioned and digitized using an NI 9234 IEPE module and the AE sensor using an NI 9215 AI [V] module. Accelerometer signals were recorded at 51200 samples per second (S/s) and the AE sensor at 250 kS/s. Several raw data time segments, each of 20 seconds duration, were recorded at each steady operation condition.

High speed observations and cavitation detection results

Based on the previous results [3] it was decided to concentrate the measurements and visual records in operating conditions ranging from the BEP (21 MW) up to full load (25 MW) in steps of 1 MW. The inspection of the runner blades just before the tests had shown that some blades were more eroded than the rest. In fact, only some blades presented strong mass removal and others did not show any significant mark. For example, such differences can be clearly seen in figure 1 where two consecutive blades are shown.

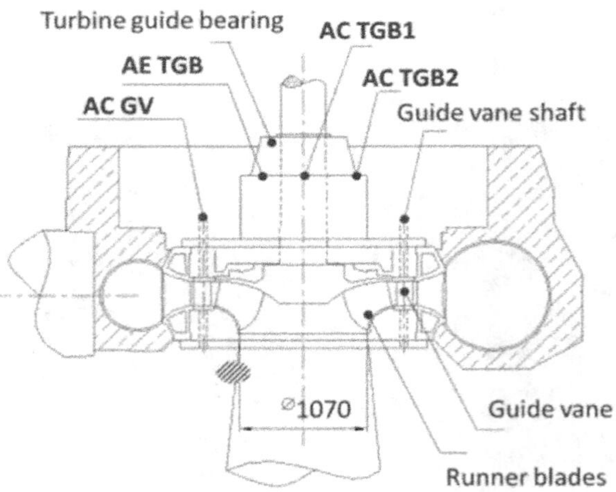

Figure 4. Location of the acceleration and acoustic emission sensors on the turbine guide bearing pedestal and guide vane shaft.

After analyzing the video records and identifying the 15 blades, it has been possible to compare their cavitation conditions among them for all the operation conditions. As it can be seen in figure 5, not all the blades present suction side cloud cavitation. For instance, blade number 6 appears not to suffer cavitation at all for any power output. On the contrary, blade number 14 is free of cavitation at BEP but develops cavitation for the rest of higher output loads.

The global results regarding the presence of cloud cavitation for all the blades and operation conditions are summarized in table 1. Here it must be reminded that the field of view only corresponds to a narrow zone located next to the end of the spiral casing and no information about the evolution of the cavitation conditions as the runner blades span the whole circumference is available.

Firstly, it can be noted that only blade number 10 shows cavitation for all the loads including the BEP (21 MW). Then, cavitation starts at 22 MW for blades number 5, 13 and 14. At 24 MW, blades number 4, 6, 11 and 12 do not show cavitation yet. Operation condition at full load (25 MW) requires a special comment. Due to the significant increase of bubbles inside the flow, all the images appear excessively blurry and it is not possible to identify the presence of any type of large scale cavities. This unexpected behavior might be due to a deficient turbine setting point that reduces the available Net Positive Head (NPSHA) especially at full load.

	#1	#2	#3	#4	#5	#6	#7	#8	#9	#10	#11	#12	#13	#14	#15
21MW	-	-	-	-	-	-	-	-	-	Y	-	-	-	-	-
22MW	-	-	-	-	Y	-	-	-	-	Y	-	-	Y	Y	-
23MW	Y	-	-	-	Y	-	-	-	-	Y	-	-	Y	Y	-
24MW	Y	Y	Y	-	Y	-	Y	Y	Y	Y	-	-	Y	Y	Y
25MW	?	?	?	?	?	?	?	?	?	?	?	?	?	?	?

Table 1. Summary of the observations regarding the presence of blade suction side cavitation on all the 15 blades for operation conditions from 21MW up to full load at 25 MW. [Y] means that cavitation is detected, [-] means that it is not detected and [?] means that visualization information is not clear enough to distinguish its presence.

A more detailed observation of the flow field at 24 MW shown in figure 6 permits to clearly see the large-scale types of cavitation taking place when blades number 10 and 13 pass in front of the vision angle. As already commented, the closure region of the cloud cavitation on the blade suction side is observed as well as other forms of cavitation such as vortex shedding cavitation at the blade trailing edge and draft tube vortex rope cavitation from the runner hub.

The overall RMS vibration acceleration and acoustic emission levels in the frequency bands from 18k to 20k Hz and from 95k to 97k Hz, respectively, have been plotted as a function of output load in figure 7. These results correspond to the average values from measurements taken in two different days. The data show that the levels increase with output load which corresponds to the increase of the number of blades with cavitation and the cavity sizes. The values in both guide bearing accelerometers are quite similar meanwhile the values in the guide vane are higher. The trends from the two accelerometers and the acoustic emission sensor located on the guide bearing are analogous but the trend observed in the guide vane accelerometer differs slightly.

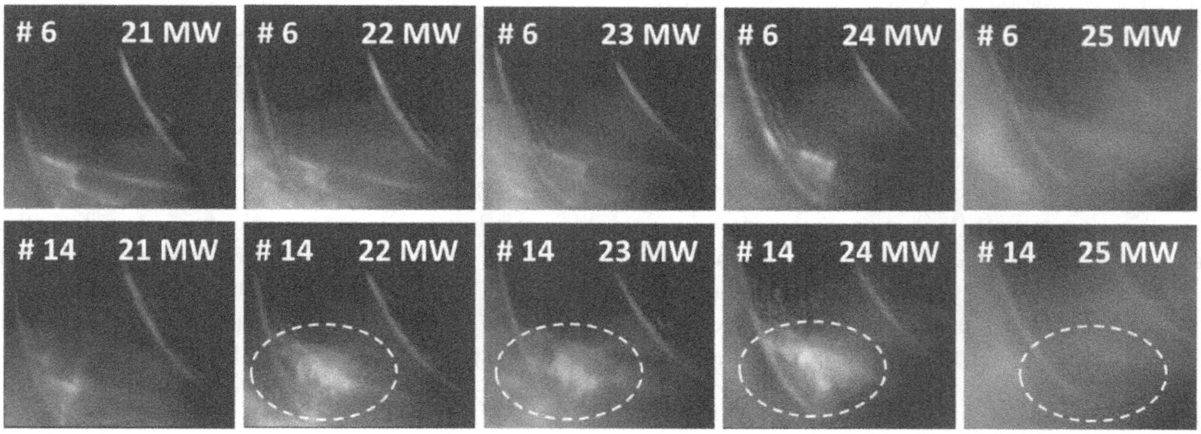

Figure 5. Flow visualization on blades number 6 (upper row) and number 14 (lower row) for operation conditions ranging from 21 MW up to full load at 25 MW. Suction side blade cavitation is observed for blade number 14 starting at 22 MW meanwhile blade number 6 does not exhibit cavitation at any operation condition.

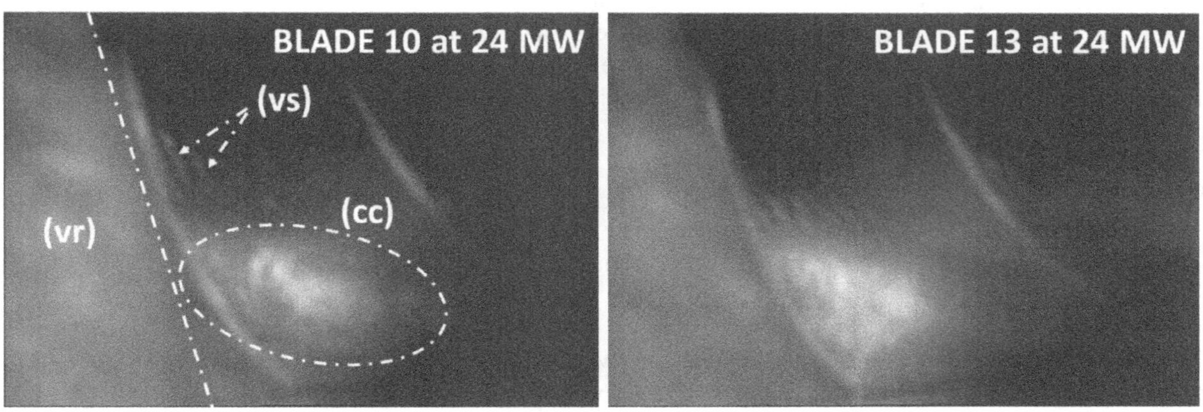

Figure 6. Zoomed view of the flow field at 24 MW when blades 10 (left) and 13 (right) pass in front of the vision angle that permits to identify cloud cavitation on the blade suction side (cc), vortex shedding cavitation at the blade trailing edge (vs) and draft tube vortex rope cavitation from the runner hub (vr).

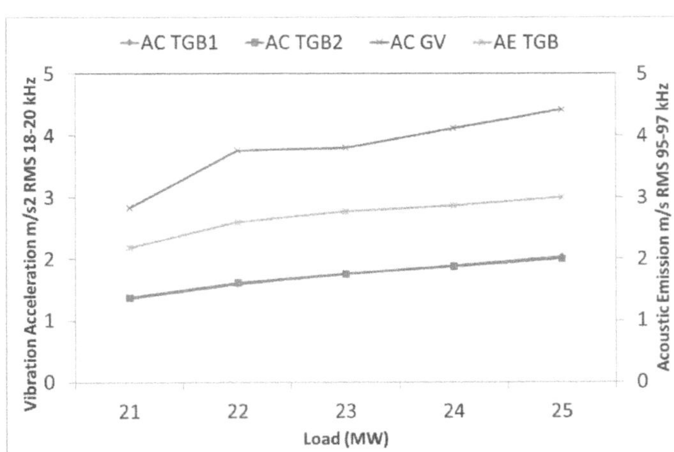

Figure 7. Overall RMS vibration acceleration and acoustic emission levels in the frequency bands [18k-20k Hz] and [95k-97k Hz] respectively, as a function of output load.

Conclusion

This work demonstrates the feasibility of installing visual access points on full-scale Francis turbine draft tubes to investigate cavitation phenomena inside the runner. The main objective is focused on answering at which turbine operation condition blade erosive cavitation takes place and how many runner blades are attacked. Analysis of video records taken with a high-speed camera has allowed identifying and quantifying runner cavitation from BEP up to full load. It has been observed that cloud cavitation on the blade suction side is responsible for the observed erosion. Surprisingly, cavitation is present in some blades even at BEP. As the load increases towards maximum powers, the number of blades with cavitation increases and the size of the cavities grows. Cloud cavitation seems to vanish at full load but it might be masked by the high bubble content of the main flow provoking images that are blurry. Based on overall measurements of cavitation erosion intensity obtained from induced noise and acoustic emissions with methods presented in previous works, a good correlation with the visual observations has been obtained. The remote cavitation measurements also suggest growing cavitation intensity from BEP towards full load, the latter with maximum aggressiveness, which is in agreement with the observations. Another significant confirmation is that the cavitation clouds are present only at selected runner blades, thus explaining the presence of erosion only at particular blades.

Acknowledgements

Ingrid K. Vilberg, an employee at Flow Design Bureau AS, is in parallel enrolled in the PhD program at NTNU. For this PhD related work, FDB receives partial funding from the Industrial PhD program at the Research Council of Norway and from the regional research funding program "RFF Vest". PhD Bjørn W. Solemslie at NTNU provided initial guidance and "tips-and-tricks" for the use of the high speed camera.
Statkraft AS and Svorka Energi AS generously make the Svorka turbine unit available for research activities. In addition, services minded and dedicated staff at all levels within the two organizations contribute to the results achieved during testing. Statkraft's Erik J. Wiborg is especially acknowledged for the fact that the visual access points were installed.

References

[1] Avellan, F. (2004). *Introduction to cavitation in hydraulic machinery*. Proceedings of the 6th Int. Conf. on Hydraulic Machinery and Hydrodynamics, Timisoara, Romania.
[2] Escaler, X., Egusquiza, E., Farhat, M., Avellan, F. and Coussirat, M. (2006). *Detection of cavitation in hydraulic turbines*. Mech. Syst. Signal Proc. 20(4).
[3] Escaler, X., Ekanger, J. V., Francke, H. H., Kjeldsen, M. and Nielsen, T. K. (2015). *Detection of draft tube Surge and Erosive Blade Cavitation in a Full-Scale Francis Turbine*. J. Fluids Eng. 137(1).
[4] Escaler, X., Farhat, M., Ausoni, P., Egusquiza, E. and Avellan, F. (2006). *Cavitation monitoring of hydroturbines: tests in a Francis turbine model.* Proceedings of the 6th Int. Symp. on Cavitation (CAV2006), Wageningen, The Netherlands.

© 2018 ASME

Asymmetrical Propagation of Powerful Sound in Double-Layer Liquid

Weizhong Chen*; Xun Wang;

The Key Laboratory of Modern Acoustics, MOE, Institution of Acoustics, Nanjing University, Nanjing 210093, China

Abstract

The asymmetry sound propagation has been observed experimentally in coupling double-layer of liquids with different cavitation thresholds. When the sound wave propagates from the high threshold liquid to the low one, it is possible that both two liquids can avoid the cavitation under suitable driving pressure. However, the driving pressure will cavitate the low-threshold liquid though the high-threshold liquid keeps uncavitated when the sound propagates oppositely. The additional attenuation due to cavitation makes the distribution of the sound field different between two opposite propagations, leading to the asymmetrical sound propagation. The observations can be reproduced by the numerical simulation in framework of the modified two-phase fluid theory.

Keywords: asymmetric transmission, frequency spectrum broadening, sound pressure

Introduction

The electron diode, invented by Fleming more than one hundred years ago, is widely applied in modern circuit to revolutionize our life dramatically. Besides, it is the first component of realizing one-direction propagation of energy. It has encouraged people to investigate the asymmetric transmission of other kinds of energy. Recently, with the development of numerical calculation, people return to the research of asymmetric transmission of energy, and have made significant progress. Li *et al.* numerically simulated the coupled Frenkel-Kontorova lattice model and find the microphenomenon of it deviates from the Fourier thermal conductivity law.[1] The asymmetric energy transmission phenomenon which is named as heat rectification was reported. After that, investigation on heat rectification by equivalent LC transmission circuit also supports the phenomenon of heat rectification.[2] The asymmetric transmission of acoustic energy, namely the acoustic-diode, is also proposed and verified by experiment during the same period.[3] The asymmetry transmission systems mentioned above are usually carefully designed. During our research on the propagation of sound wave in cavitating liquid, we find the asymmetric transmission also exists in a simple double-layer liquid system when powerful ultrasound transmits through it. Further more, the direction of sound energy rectification can be reversed under different driving sound pressure.

Model and mechanism

We consider a simple harmonic sound wave propagating from one end with driving pressure amplitude P_{dr} to the other end with output pressure amplitude p_{out} in a 1-dimensional two layers of liquids coupled by a sound passing membrane. Both thicknesses of the left and right liquids are a. For the transmission from the left liquid to the right one (L-R transmission), $P_{dr}=P_L$, and $p_{out}=p_R$ (Fig. 1). Otherwise $P_{dr}=P_R$, and $p_{out}=p_L$ for the R-L transmission. Let Q_L and Q_R to be the cavitation threshold pressures of the left and the right liquids, and $Q_L<Q_R$ without loss of generality. The asymmetry can exist between the L-R and R-L transmissions. If $Q_L>P_{dr}\geq Q_R$, it is possible that there is no cavitation in both left and right liquids for the L-R transmission. At first there should be no cavitation in the left liquid due to $P_{dr}<Q_L$. The sound pressure amplitude at the interface between two liquids, $p(0)$, has been decayed by the left liquid, so that $p(0)<P_{dr}$. If $p(0)$ is further lower than the threshold pressure of the right liquid Q_R, then the right liquid can not be cavitated too. Therefore, there is no any cavitation to be taken place for the L-R transmission. On the contrary, for the R-L transmission the right liquid will be cavitated because of $P_{dr}\geq Q_R$. As is well known, the cavitation will yield an additional attenuation called cavitation screening,[5] leading to that the R-L transmission is more difficult than the L-R transmission. In other words, the sound wave is rectified in this system.

Asymmetrical sound transmission system

For verifying the mechanism of asymmetrical transmission, we design the experimental system as shown in Fig. 2. The cross section of the rectangle glass container is $0.09\times0.09\text{m}^2$ and each layer length of the host liquids a is 0.1 m. Both end faces of the container are stainless steel plates of 0.5 mm in thickness, and two ultrasound transducers (40

*Corresponding Author, Weizhong Chen: wzchen@nju.edu.cn

kHz) are bonded to the outer sides of two end faces. A waterproof sound passing membrane is positioned at the middle of container to separate the left- and right-liquids. In our experiment, the host liquids are 30 vol% alcohol solution (AS(30%)), AS(20%) and water, and their cavitation threshold pressures are 0.4 bar, 0.5 bar and 0.9 bar respectively. Acoustic absorbing material is attached to the internal lateral face of the container to eliminate the reflection of sound wave as much as possible. During the experiment the liquids are cycled continuously so that the properties of the host liquids keep unchanged. The temperature of host liquids is maintained at 20±2°C.

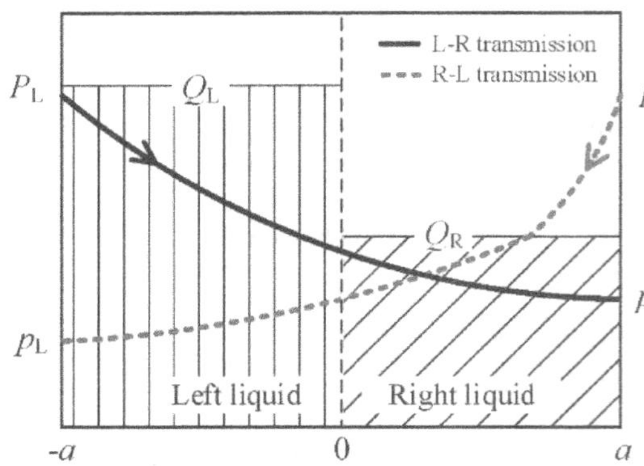

FIG. 1: Asymmetrical Propagation in double –layer cavitating liquids

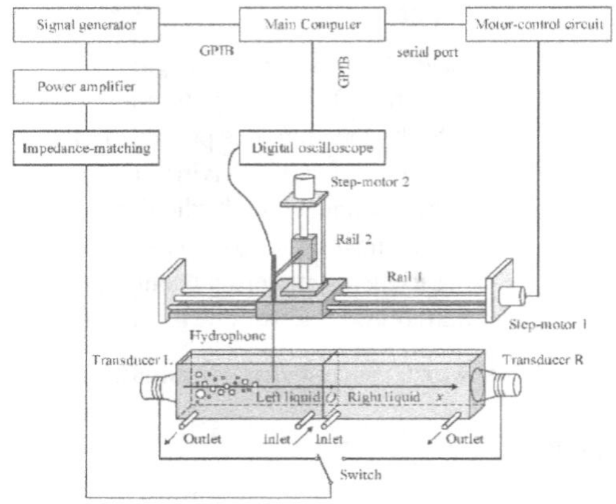

FIG. 2 Schematic diagram of the experimental system

A sinusoidal signal at about 40 kHz is output by a function generator (Agilent 28335A, US), then amplified by a power amplifier (B&K 2716, Denmark) to drive the transducer L or R after impedance matching. The L-R transmission with $P_{dr}=P_L$ and $p_{out}=p_R$, or the R-L transmission with $P_{dr}=P_R$ and $p_{out}=p_L$ is chosen by a switch. A programmed scanning system has been constructed for quickly and precisely measuring the sound field in the container. There are two orthogonal guide rails parallel and perpendicular to the container in the system, respectively. Two slide blocks are rode on them respectively (Fig. 2). A needle hydrophone(RESON, TC4013, Denmark) is held at the vertical slide block to measure the sound pressure in the container. Both two blocks are driven by step-motors via the motor-control circuit at the spatial resolution 0.05 mm. The output of the hydrophone is fed to a digital oscilloscope (Agilent Infiniium 54810, US). The function generator, motor-control circuit, and oscilloscope are programmably controlled by a main computer through its GPIB and serial port. The amplitude of fundamental pressure is obtained by FFT calculation and calibration based on the time-domain data acquired by the oscilloscope. The measuring step of the horizontal block is 5 mm. Every amplitude data has been binned over itself and its left and right adjacent measurements to smooth the spatial jumping due to cavitation.

Experimental observations

In our experiment the right liquid has been set as AS(30%) whose threshold pressure is the lowest among the host liquids, that is, Q_R=0.4 bar, while the left liquid can be chosen one of other host liquids.

At the beginning, the water with the threshold pressure Q_L=0.9 bar is chosen as the left liquid, and P_{dr} is set to be 0.6 bar, so $Q_L>P_{dr}>Q_R$, the same case mentioned in section II. Figure 3(a) shows the distribution of sound pressure amplitude $p(x)$ in this system, which consists with our prediction in Section II (see Fig.1). Indeed, hardly any cavitation bubbles have been observed in both two liquids in the L-R transmission. Nevertheless, in the R-L transmission some bubbles are clearly visible in the right liquid. As a result, the sound propagation is asymmetric between the L-R and R-L transmissions, that is, $p_R>p_L$.

Then, we set P_{dr}=1.8 bar, so $P_{dr}>Q_L>Q_R$. The distribution of $p(x)$ is plotted in Fig. 3(b) where the asymmetry between the L-R and R-L transmissions is still observed. $p(x)$ almost maintains higher than Q_L and Q_R in the L-R transmission, whereas only keeps higher than Q_R in the right liquid in the R-L transmission. Furthermore, $p_R<p_L$. It is a new

mechanism of the asymmetric transmission and differs from that described above or in Section II. We believe larger scope of cavitation leads to greater attenuation, leading to $p_R<p_L$. These experiments show the direction of the rectification is dependent on P_{dr} with respect to the cavitation threshold pressures.

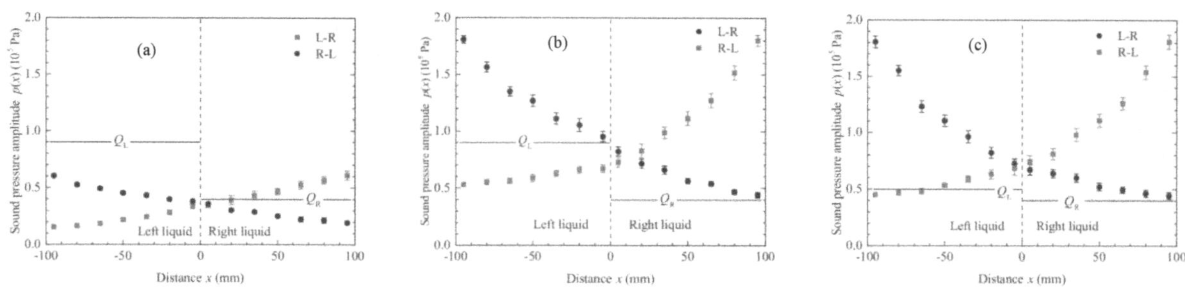

FIG. 3 Distributions of sound pressure amplitude in (a) water-AS(30%) driven by P_{dr}=0.6 bar, (b) water-AS(30%) driven by P_{dr}=1.8 bar, and (c) AS(20%)-AS(30%) driven by P_{dr}=1.8 bar, respectively.

Finally, we note that the system will loss the asymmetry if the gap between Q_L and Q_R vanishes. To verify it, we set the left liquid as AS(20%) whose cavitation threshold is 0.5 bar, near to that of the right liquid, and P_{dr}=1.8 bar. The observation of $p(x)$ in the L-R and R-L transmissions shows that output sound pressures p_R and p_L are approximately identical (see Fig. 3(c)). We can conclude that a large threshold gap is necessary for realizing the remarkable asymmetry in sound propagation.

Theory and simulation

The sound wave propagating in cavitating liquid can be described by the Helmholtz equation[6]

$$\nabla^2 p + k_m^2 p = 0, \tag{1}$$

where p is the sound pressure amplitude and k_m is the complex wave number expressed as:

$$k_m = \frac{\omega}{c_m} - \mathrm{i}\alpha_\eta, \tag{2}$$

with c_m and α_η being the acoustic velocity and attenuation coefficient in bubbly liquid respectively. Both c_m and α_η are tedious functions of the volume fraction of the two-phase liquid $\beta(x)$.[6] For our experimental condition, we assume the volume fraction $\beta(x)$ is proportional to the local sound pressure $p(x)$ exceeding the cavitation threshold $Q(x)$, that is,

$$\beta(x) = \begin{cases} 0 & p(x) < Q(x) \\ B[p(x) - Q(x)] & p(x) \geq Q(x) \end{cases}, \tag{3}$$

where B is a model parameter.

In the theoretical model of Ref. [6], the authors only take account of the attenuation caused by cavitation bubbles. In our situation the bulk attenuation of the host liquids is not ignorable, so that an additional attenuation coefficient α_0 has been inserted into α_η [7], that is, $\alpha_0 + \alpha_\eta \rightarrow \alpha_\eta$.

By solving Eq.(1) the sound pressure distribution can be acquired. For our double-layer liquid, water-AS(30%), the cavitation threshold function $Q(x)$ can be expressed as

$$Q(x) = \begin{cases} Q_L & -a \leq x \leq 0 \\ Q_R & 0 < x \leq a \end{cases}, \tag{4}$$

where a=0.1 m, Q_L=0.9 bar, and Q_R=0.4 bar, respectively. Parameters used in numerical simulation are the same as those in Ref. [6]. The bulk attenuation coefficient α_0 is set as 4.0 m^{-1} and B=2.0×10^{-8} Pa^{-1}.

Figure 4 shows the calculated distributions of the sound pressure amplitude in the water-AS(30%) driven by 0.4 bar, 0.6 bar, 1.8 bar and 1.4 bar, respectively. It is obvious that there is no any asymmetric transmission to be appeared in the double-layer liquid if the driving sound pressure amplitude is lower than both Blake threshold pressures Q_L and Q_R (see Fig. 4(a)). When P_{dr} rises, P_{dr}=0.6 bar for example, the distribution of the sound pressure (Fig. 4(b)) shows that the system undergoes the asymmetric transmission as Fig. 3(a). If P_{dr}=1.8 bar, the distribution (Fig. 4(c)) reproduces the experimental asymmetric transmission of Fig. 3(b).

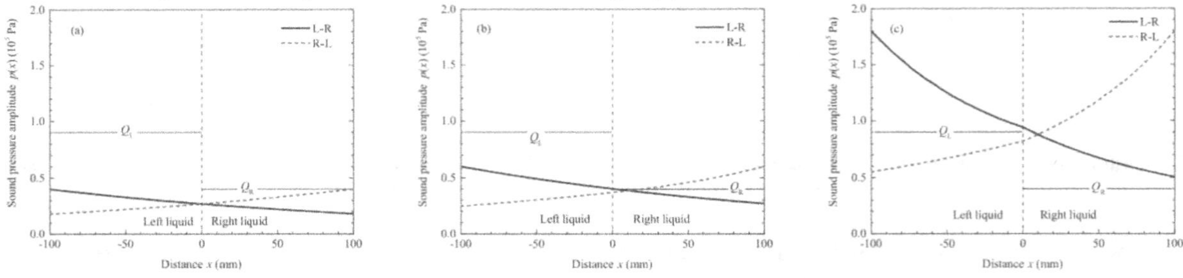

FIG. 4 Calculated distributions of the sound pressure amplitude in water-AS(30%) driven by (a) P_{dr}=0.4 bar, (b) P_{dr}=0.6 bar, and (c) P_{dr}=1.8 bar, respectively.

Conclusion and discussion

We propose a model which is composed of two liquids with different cavitation thresholds to realize the asymmetric transmission of sound wave. Experimental results show there exists two kinds of rectifying mechanism of sound wave, and the direction and efficiency of rectification is related to the driving sound pressure and the cavitation threshold pressures of the two liquids. The dependency of the rectification effect on these factors is analyzed in more details according to numerical simulation. In the future, we will explore the method to improve the asymmetry and perfect the theory.

Acknowledgments This work was supported by the National Natural Science Foundation of China (Nos.11334005 and 11574150).

References

[1] B. W. Li, L. Wang, G. Casati. (2004). *Thermal diode: Rectification of heat flux.* Phys. Rev. Lett. 93, 184301.

[2] F. Tao, W. Z. Chen, *et al.* (2001). *Asymmetric energy flux in a transmission line based on frequency multiplication.* Phys. Rev. E, 83, 056605.

[3] B. Liang, B. Yuan, J. C. Cheng.(2009). *Acoustic Diode: Rectification of Acoustic Energy Flux in One-Dimensional Systems.* Phys. Rev. Lett. 103, 104301.

[4] Y. Wang, M. Yao. (2013). *Realization of cavitation fields based on the acoustic resonance modes in an immersion-type sonochemical reactor.* Ultrasonics Sonochemistry. 20, 565.

[5] D. Sarno, M. Hodnett, L. Wang, B. Zeqiri. (2017). *An objective comparison of commerically-available cavitation meters.* Ultrasonics Sonochemistry. 34, 354.

[6] R. Jamshidi, B. Pohl, U.A. Peuker, G. Brenner. (2012). *Numerical investigation of sonochemical reactors considering the effect of inhomogeneous bubble clouds on ultrasonic wave propagation.* Chem. Eng. J. 189, 364.

[7] P. R. Gogate, V. S. Sutkar, A. B. Pandit. (2011). *Sonochemical reactors: important design and scale up considerations with a special emphasis on heterogeneous systems.* Chem. Eng. J. 166, 1066.

© 2018 ASME

10th International Symposium on Cavitation - CAV2018
Baltimore, Maryland, USA, May 14 – 16, 2018

CAV18-05027

Aviation Fuel Cavitation in a CD Nozzle: A Quantitative Experimental Characterization

[1]Michael Waldrop*; [1]Flint Thomas**

[1]*Institute for Flow Physics and Control, University of Notre Dame, Notre Dame, IN, USA*

Abstract

This work describes experiments aimed at clarifying several physical aspects of aviation fuel cavitation. The experiments are performed in a simple two-dimensional converging-diverging (CD) nozzle geometry with JP-8 fuel replaced by dodecane, a single component surrogate and its primary component by weight. Experiments are focused on gaseous cavitation with air micro-bubbles injected at the nozzle inlet. Carefully controlled initial void fraction leads to bubble growth in the bulk of the fluid and quasi-1D flow patterns downstream. These experiments seek to characterize the flow patterns with particular focus on propagating shock waves that form due to collapse of gaseous bubble clouds. Experimental documentation is made through arrays of static and unsteady pressure sensors, high-speed video and a volumetric flow rate meter. The results are analyzed to obtain choked mass flow rate, distributions of mean streamwise pressure, shock propagation speed, local shock passage frequency, and static pressure jump across the bubbly shock.

Keywords: Cavitation, Bubbly Shocks, Shockwaves, Bubbly Flows, Aviation Fuel

Introduction and Motivation

The occurrence of cavitation in an aircraft fuel system can lead to unexpected degradation in system performance and damage to fuel system components. JP-8 aircraft fuel has been shown to be prone to cavitation in piston fuel pumps and other system components which motivates this experimental examination. Several previous studies of cavitation in jet fuel have been made in order to more fully understand its behavior. Notable among these is the work of Dunn et al.[1] which examines cavitating, choked CD nozzle flows of JP-8 fuel, dodecane and water in comparison to predictions of a barotropic cavitation model. Their results show that, for flows with or without inlet air injection, a standing shock exists in the diverging portion of the nozzle for water and a significantly milder shock (over larger streamwise region) exists for dodecane. Meanwhile, JP-8 exhibits no shock behavior but instead shows a more gradual pressure rise from its minimum near the nozzle throat to the outlet pressure. The shock location changes with variation in back pressure as well as changes in initial void fraction for the air-injected case. Dunn et al.[1] attribute the differences between the choked, cavitating CD nozzle flows of water, dodecane and JP-8 to differences in their respective fluid properties, primarily surface tension and viscosity. These results are intriguing but do not encapsulate the full range of behaviors predicted by numerical modelling. For example, the work of Preston et al.[2] describes two additional cavitation regimes for flows with controlled initial void fraction. These are an unsteady regime with propagating shock waves in the diverging portion of the nozzle and a steady but underexpanded regime at extremely low back pressure. Experimental exploration of these regimes has been minimal in water, jet fuel or any other fluid. Their study in JP-8 is greatly aided by replacing the fuel with dodecane, its largest component by weight[3], as a single-component surrogate. JP-8 contains more than 200 hydrocarbons[3] in varying ratios from sample to sample. As a result, properties of the fuel are often not repeatable and the reliability of dodecane is preferable for scientific study. For this reason, the experiments described herein utilize dodecane-air mixtures to examine gaseous cavitation in aviation fuel. A simple, effectively two-dimensional CD nozzle geometry provides a pressure drop sufficient to induce cavitation in these mixtures and permits thorough study of propagating bubbly shock waves in the resulting flows.

Experimental Apparatus

A conceptual schematic of the experimental facility is shown in Figure 1(a). Flow is driven through a vertically oriented, Plexiglas™ CD nozzle by pressure differential between an upstream tank open to atmosphere and an evacuated downstream vacuum tank. The volumetric flow rate of liquid entering the nozzle is measured with Dwyer's™ TVF-02 Industrial Flow Meter. An infusion pump supplies air at either of two distinct rates to a pair of

[+]Work funded by Honeywell Aerospace
*Michael Waldrop: mwaldrop@nd.edu
**Flint Thomas: fthomas@nd.edu

33 gauge, blunt tip hypodermic needles at the nozzle inlet as shown schematically in Figure 1(b). The first infusion rate (0.429 mL/s) equates to an initial void fraction (α_0) of ~1.1% for dodecane in choked flow while the second rate (0.176 mL/s) gives $\alpha_0 \sim 0.47\%$ under the same conditions. Figure 1(c) is a conceptual schematic of the high-speed imaging process. The test section is backlit from a distance of 1.83 m with a theater lamp. A translucent plastic diffuser is placed immediately behind the test section while the lens of a Photron FASTCAM™ SA1.1 high-speed camera is positioned 0.6 m away. This technique causes the liquid phase to appear bright in the resulting images while the gas/vapor phase is dark. Average pressure measurement is performed with a streamwise array of 40 static pressure taps evenly spaced along the centerline of the nozzle and a set of four Setra™ Model 209 vacuum pressure transducers. Unsteady pressure measurement is achieved with four Measurement Specialties™ EPIH-37B absolute pressure transducers that may be placed at any point along the nozzle centerline. The pressure taps for average pressure measurements and the set of unsteady pressure transducers are each mounted in one several interchangeable, Plexiglas™ face plates that act as one of the CD nozzle walls. Another face plate, free of any pressure measurement apparatus but marked with fiduciary indicators every 1.27 cm along the nozzle length is used for high-speed imaging. With this setup shock propagation speed can be accurately determined.

The symmetric CD nozzle used for this investigation (shown in Figure 2) has a rectangular cross section, length (L) of 0.127 m and constant depth (d) of 0.1588 cm. The converging portion is 2.54 cm in length while the diverging portion is 9.525 cm long. The intervening 0.635 cm section consists of a constant area throat with height of 0.1588 cm. The nozzle inlet and outlet are each 1.905 cm in height. Both the converging and diverging section contours are defined by fifth order polynomials with zero first and second derivative end conditions.

Results and Analysis

Experimental characterization of gaseous dodecane cavitation begins with measurements of mean streamwise static pressure distribution and mass flow rates for a broad range of nozzle pressure ratios. These results are presented in Figure 2 which shows the mean pressure at all forty static tap locations across the indicated range of nozzle back pressure to inlet pressure (P_B/P_0) ratios. Curves with $P_B/P_0 < 0.5$ are indicative of choked nozzle flow as all curves follow the same path through the nozzle throat and measured mass flow rate is identical for each case ($\dot{m}_c$=0.0261 kg/s, see Table 1). These curves also show a sharp rise in static pressure over a small spatial region, consistent with the presence of a normal shock in the diverging section. As described later, each of these cases also exhibits unsteady shock waves propagating downstream in high-speed video imaging. Higher pressure ratio cases ($0.58 < P_B/P_0 < 0.8$) do not follow the same path through the throat and show a variable mass flow rate indicating unchoked flow.

The time mean pressure distributions are useful in describing the broad properties of the flow but provide little insight into the inherently unsteady mechanisms associated with shock formation and propagation. Such phenomena are more easily studied through the use of high-speed video. High-speed imaging is first used to study the properties of air micro bubbles injected at the nozzle inlet as depicted in Figure 3. Mean bubble injection frequency is measured via the number of bubbles passing a fiduciary marker over the duration of a high-speed video image sequence. The diameter of an individual injected bubble is then obtained by counting the number of pixels across its widest dimension and using the number of pixels between fiduciary markers to convert to physical units. An average of the first 20 bubbles to cross the fiduciary marker is then recorded as the mean initial bubble diameter. Injection frequencies and mean initial bubble diameters for air injected into dodecane may be found in Table 2. As shown in the sample image in Figure 3, injected bubbles initiate cavitation in the fluid bulk near the nozzle throat which results in clouds of gaseous bubbles downstream. These clouds are found to undergo collapse and emit planar shock waves which propagate downstream from the collapse location. High-speed images of these shocks are examined for emission frequency, average propagation speed across the nozzle and formation location utilizing the time derivative of pixel saturation between subsequent images. As previously discussed, regions with high pixel saturation are associated with the liquid phase while darker pixels indicate a gaseous void. These voids tend to reduce in size under the pressure jump associated with a passing shock wave, meaning regions with a positive time derivative of pixel saturation between subsequent frames indicate a passing shock. After using a spatial Butterworth filter to remove small scale phenomena and applying a color scale to the derivative images, the shock waves appear as blue regions and may be tracked between images as shown in Figure 4. This, along with the fiduciary markers in the high-speed images, permits measurement of shock speed, emission frequency and emission location.

Histograms of shock emission frequency, average shock speed and emission location have been obtained from the video imaging and may be found in Figure 5. Shocks are emitted across a fairly narrow streamwise band with $0.45 \leq (x/L)_E \leq 0.55$. This coincides with the region of sharp increases in mean static pressure for the choked flow cases shown in Figure 2. Emission frequency and propagation speed show some evidence of bimodality but confirmation requires acquisition of additional samples in order to refine the statistics. Unsteady pressure measurements provide significantly more data, allowing for verification of shock emission frequency results as well as determination of a typical shock pressure profile at a given x/L location. With unsteady pressure sensors placed at "Unsteady Pressure Tap Array" locations 1,3,4 and 5 (see Figure 2) 20 seconds of pressure time-series data sampled at 102.4 kHz yields fully converged shock statistics and a well-defined typical shock pressure distribution shape for each sensor. The arrival of a shock is determined by a sudden jump in pressure above a specified threshold for a specified duration (span). Threshold and span values are unique to each sensor and each sampling run. They are determined by comparison to a visual count of shock events in the raw pressure signal (number of shock events in first 20,000 samples). Once threshold and span are determined, frequency of shocks and their average pressure signature may be obtained.

Figure 6 shows histograms of time between shock arrivals (Figure 6(a)) and shock arrival frequency (Figure 6(b)) at Sensor 4 (x/L=0.9) in a dodecane-air flow with P_B/P_0 =0.36 and α_0 = 1.1%. They are normalized by the average time between bubble injections and average frequency of bubble injections, respectively (see Table 1). These confirm the bimodal nature of the shock emissions hinted at by the derivative image technique and show that each of these modes occurs at a harmonic of the bubble injection frequency. This is a strong indicator that shock emission in this flow is directly driven by air micro bubble injection.

Figure 7 shows typical shock pressure profiles obtained by averaging all the shock events identified in the raw pressure signal from each sensor. Figure 7(a) demonstrates the reduction in pressure jump magnitude of a typical shock as it propagates downstream. A 36.6 kPa jump in pressure occurs at x/L=0.5 while 12.6 kPa, 9.5 kPa and 8.8 kPa pressure jumps occur at x/L=0.8, 0.9 and 1.0 respectively. Figure 7(a) also demonstrates that pre and post-shock pressures are significantly lower at x/L = 0.5 (12.6 kPa pre-shock and 17.9 kPa post-shock) than x/L = 0.8, 0.9 and 1.0. These pressures are effectively identical between the sensors farther downstream (~26 kPa pre-shock and ~29 kPa post shock). This is consistent with the essentially constant average pressure in this region for the P_B/P_0=0.36 case shown in Figure 2. Figures 7(b) and 7(c) show the typical pressure signal from three different data runs normalized by their respective maximum values. The resulting curves are effectively identical and confirm the repeatability of the typical shock shape at a given streamwise location.

Figures & Tables:

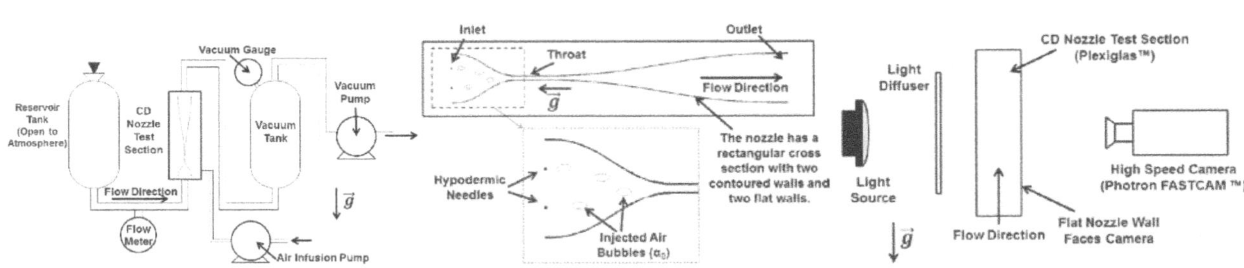

(a) (b) (c)

Figure 1: (a) Cavitation facility, (b) micro-bubble injection method, (c) high-speed imaging technique

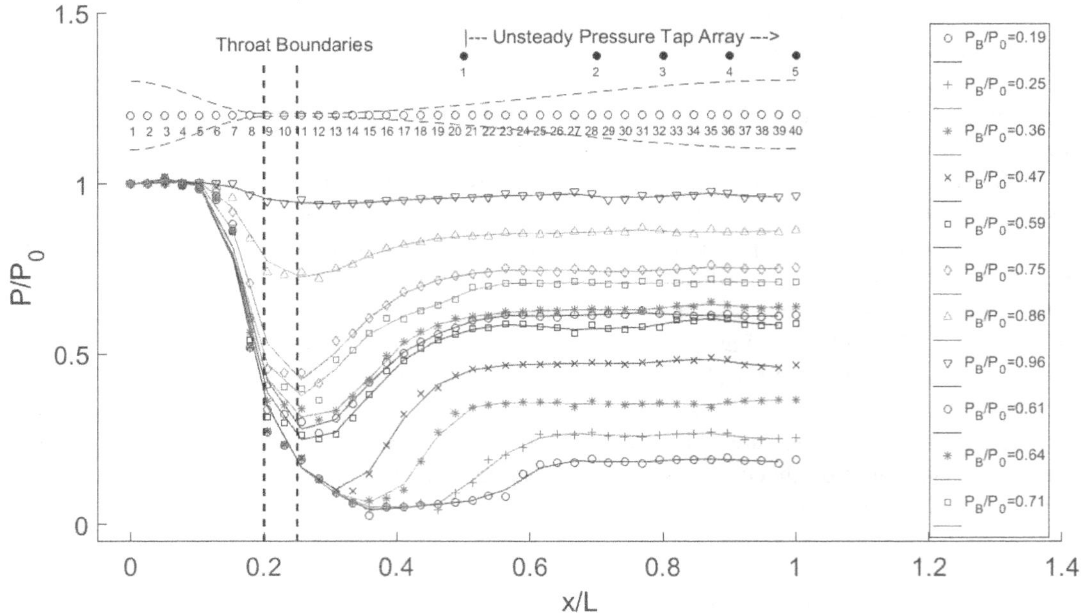

Figure 2: Average pressure distribution for gaseous cavitation of dodecane and injected air (α_0 = 1.1%) in a CD nozzle at various back pressure ratios with nozzle profile and pressure tap configurations

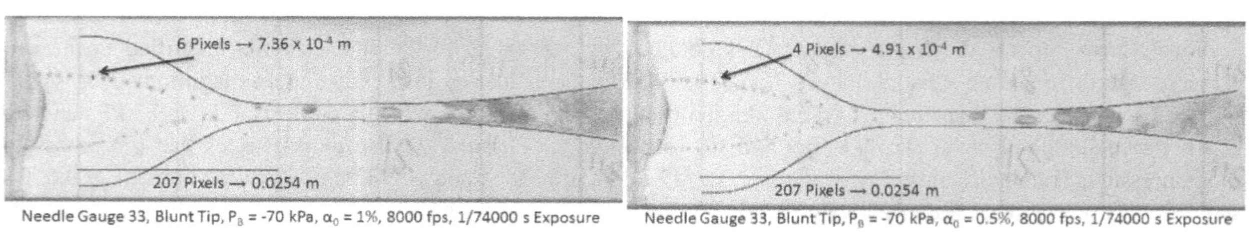

(a) (b)

Figure 3: Sample high-speed video images of air micro-bubble injection frequency and diameter for (a) α_0 = 1.1% and (b) α_0 = 0.47%

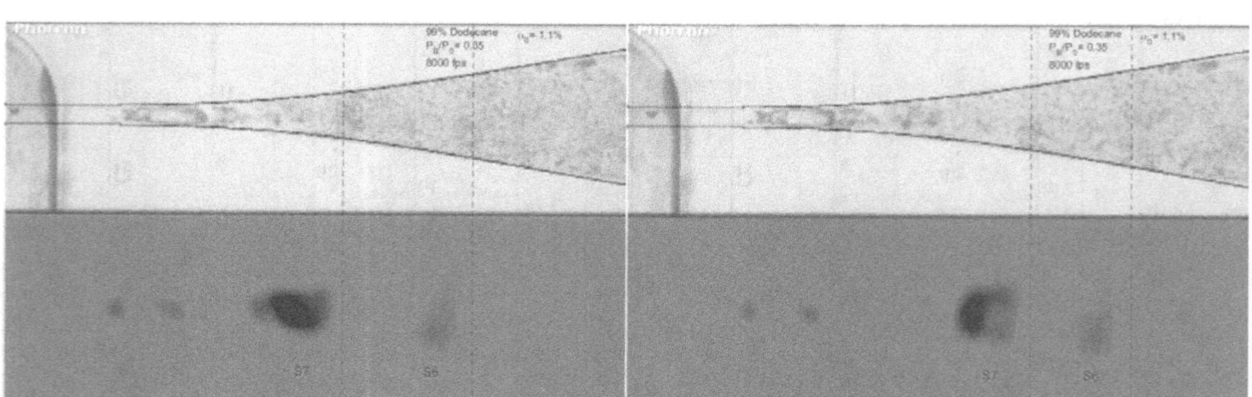

Figure 4: Shock propagation due to gaseous cavitation in micro air bubble injected dodecane flow through a CD nozzle (P_B/P_0 =0.36 and α_0 = 1.1%). (Blue regions are indicative of propagating shock waves.)

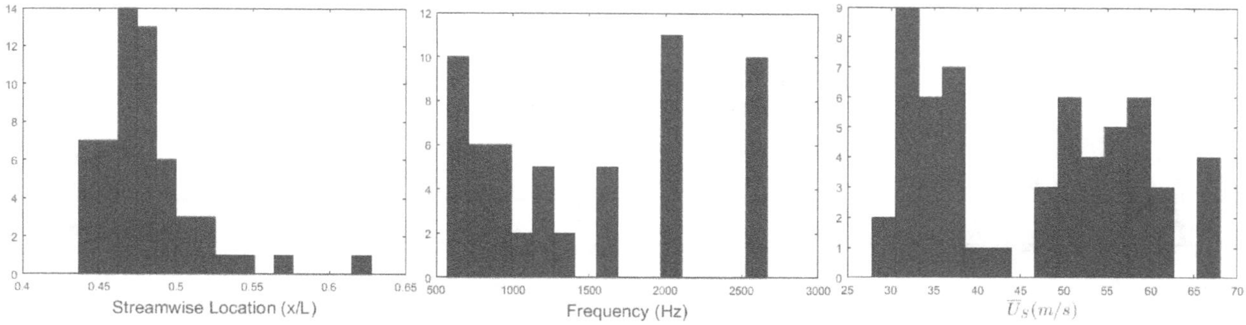

(a) (b) (c)

Figure 5: High-speed video derived histograms of shock waves due to gaseous cavitation in micro air bubble injected dodecane flow through a CD nozzle (P_B/P_0 =0.36 and α_0 = 1.1%): (a) emission location, (b) emission frequency and (c) propagation speed

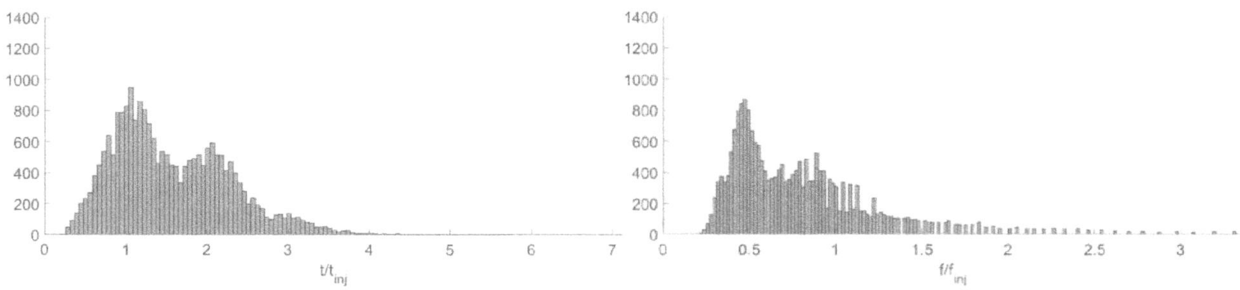

(a) (b)

Figure 6: Unsteady pressure derived histograms of shocks due to gaseous cavitation in micro air bubble injected dodecane flow through a CD nozzle (P_B/P_0 =0.36 and α_0 = 1.1%): (a) time between shocks normalized by average time between bubble injections, (b) shock emission frequency normalized by average bubble injection frequency

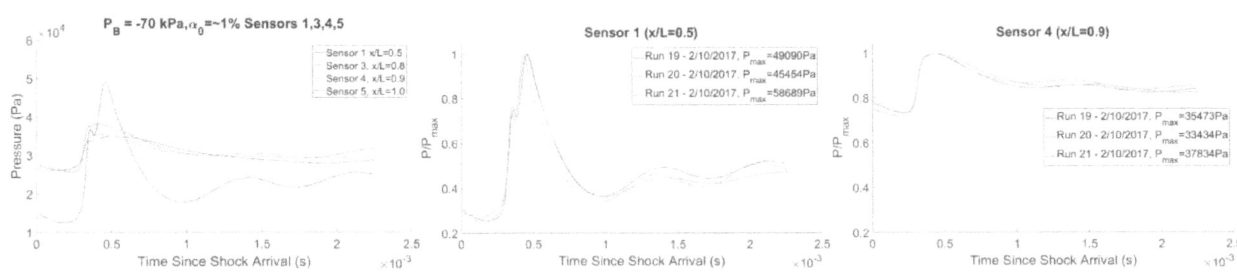

(a) (b) (c)

Figure 7: Gaseous cavitation induced shock pressure profiles in micro air bubble injected dodecane flow through a CD nozzle (P_B/P_0 =0.36 and α_0 = 1.1%): (a) various x/L locations for a single sampling run, multiple sampling runs normalized to local max shock pressure at (b) x/L= 0.5 and (c) x/L= 0.9

P_B/P_0	α_0	Mass Flow Rate (kg/s)	Bubble Count (Over Full Video)	Injection Frequency (Hz)	Average Initial Bubble Diameter (Pixels)	Average Initial Bubble Diameter (m)
0.19	1.1%	0.0261	127	1227	6.7	7.4×10^{-4}
0.25	1.1%	0.0261	126	1217	6.6	7.4×10^{-4}
0.36	1.1%	0.0261	123	1188	6.35	7.1×10^{-4}
0.47	1.1%	0.0261	133	1285	6	6.68×10^{-4}

Table 1: Inlet bubble properties and critical flow properties for gaseous cavitation of dodecane and injected air in a CD nozzle

Conclusion

Gaseous cavitation in aviation fuel is examined through the use of dodecane-air mixtures in flows through a converging-diverging nozzle where dodecane serves as a single component fuel surrogate. Cavitation induced choked flow occurs in the nozzle for back pressure ratios of 0.5 or less with a choked mass flow rate of 0.0261 kg/s and choked throat pressure ratio of 0.23. High-speed video and unsteady pressure measurements show that for gaseous cavitation due to injection of air micro bubbles at the CD nozzle inlet (1.1 % initial void fraction, 0.36 back pressure ratio) there is a bimodal distribution of shock emission frequency and propagation speed in the diverging portion of the nozzle. Shock emission location is fairly localized ($0.45 \leq (x/L)_E \leq 0.55$) for this case. Unsteady pressure data shows the two modes of shock emission frequency occur at harmonics of the air micro bubble injection frequency. This is a strong indicator that bubbly shock emission in the nozzle's diverging portion is driven by bubble injection at the nozzle inlet. This data is also used to characterize the pressure signature of a typical shock wave as it passes each pressure sensor. The magnitude of pressure jump across the shock decreases as it propagates downstream (36.6 kPa at x/L=0.5, ~10 kPa closer to the outlet) and the background pressures pre and post-shock are significantly lower at x/L=0.5 (12.6 kPa and17.9 kPa respectively) than at sensors downstream. Pre-shock pressure and post-shock pressure are greater in magnitude and effectively unchanged between sensors at x/L = 0.8, 0.9 and 1.0 (~26 kPa and ~29 kPa respectively). Typical shock shapes are also shown to be consistent between sampling runs for a given streamwise location when normalized by their respective maximum pressures. These findings demonstrate that an understanding of the distribution of extant bubbles in a fuel system and the effect of the flow geometry on individual bubbles can be used to predict the frequency and magnitude of shocks downstream of a cavitation event. This knowledge could be used to significantly mitigate any damage cavitation might cause to fuel system components.

References

[1] Dunn, P., Thomas, F., Davis, M., & Dorofeeva, I. (2010). *Experimental Characterization of Aviation-Fuel Cavitation.* Physics of Fluids. 22,117102.

[2] Preston, A. T., Colonius, T., & Brennen, C. E. (2002). *A Numerical Investigation of Unsteady Bubbly Cavitating Nozzle Flows.* Physics of Fluids, 14(1), 300-311.

[3] *Handbook of Aviation Fuel Properties*, 2004, 3rd Ed., Coordinating Research Council, Inc.

© 2018 ASME

Behavior of bubble induced by fiber-type laser for TUL near soft wall with deformability

[1]Yasuhiro Sugimoto*; [2]Daichi Nagata; [1]Keiichi Sato;

[1]Kanazawa Institute of Technology, Ishikawa, JAPAN
[2]Guraduate School of Kanazawa Institute of Technology, Ishikawa, JAPAN

Abstract

A fiber-type laser induced bubble makes an important role on laser lithotripsy. This treatment can be affected by the circumference condition such as ureter tissue and a stone. In this study, we observed behavior of the bubble in a narrow space between two walls with different deformability. According to the present experimental results, a bubble symmetrically divides into two parts and collapses in the case of two walls with same hardness. The soft wall is deformed with bubble growth and collapse. In the case of walls with different hardness materials, the bubble which shows characteristic behavior asymmetrically divides, moves and collapses even if the bubble is formed at the center of two walls. The bubble moves toward wall direction during multiple collapses. The larger bubble collapse on the hard wall, that is, the impact is predicted to be larger on the rigid wall. It is found that bubble collapse pattern is greatly affected even by slight wall deformation.

Keywords: Laser-induced bubble; Bubble behavior; Bubble collapse position; Soft wall behavior; Ho:YAG laser.

Introduction

Transurethral ureterolithotripsy (TUL) is effectively carried out by means of pulsed Holmium: YAG laser [1]-[2]. A single bubble is formed at a fiber tip when a laser with high absorption rate irradiates into water through an optical fiber. The bubble shows characteristic behavior when it grows and collapses. Shock waves generated at the bubble collapse act on the stone. The bubble collapse and shock wave behavior play an important role on the lithotripsy.
We have studied the behavior of a bubble induced by the fiber type laser [3]-[4]. The bubble behavior was observed by a high speed video camera simultaneously with impulse force measurement using a hand-made sensor for various laser irradiation conditions. As a result, in the case of a bubble generated near a solid wall, the bubble collapses not at the fiber tip but on the wall and causes the maximum impact on wall surface at a certain distance from the wall to the fiber tip. It is pointed out that there is an optimum laser irradiation condition on the wall impact for TUL.
In general, this laser surgery is carried out in a narrow space with a soft material such as kidney and ureter tissue. There are some studies on behavior near an elastic wall [5], near soft and rigid boundaries [6], between two rigid walls [7], and with stone motion [8]. It is important to understand the bubble behavior and their impact mechanism in narrow space that simulates in vivo condition because there is a possibility to be injured by its impact on the tissue.
In the present study, we observe behavior of bubble in a narrow space composed of two plates with deformability.

Experimental apparatus and procedure

Experimental setup is shown in Fig. 1. Laser irradiates from a clinical laser lithotripter (Ho:YAG, $\lambda = 2.06$ μm, Pulse duration $\tau = 0.25$ ms, Lumenis, VersaPulse Select 30W) through an optical fiber ($d = 0.2$ mm, Boston Scientific Japan, Flexiva 200) to a test section between two walls inside a vessel. The two walls which consist of rigid (made of acrylic resin) and/or soft (made of silicone rubber with durometer A hardness of 2, 5 and 8) materials are kept $W = 5$ mm apart. Hardness of silicone rubber is soft in order by A2, A5 and A8 (A2 is almost equivalent to human skin). The optical fiber is installed in the center between two walls. A bubble size is controlled by the laser power. The bubble behavior is observed by a high-speed video camera (Photron, SA5, Spatial resolution: 0.05 mm/pixel) triggered by an output of hydrophone (B&K, 8103). The bubble is illuminated by a light source (NPI, PCS-MH375RC) with a flat light guide (NPI, PLG-B100X) placed behind the test section.

*Corresponding Author, Yasuhiro Sugimoto: y-sugi@neptune.kanazawa-it.ac.jp

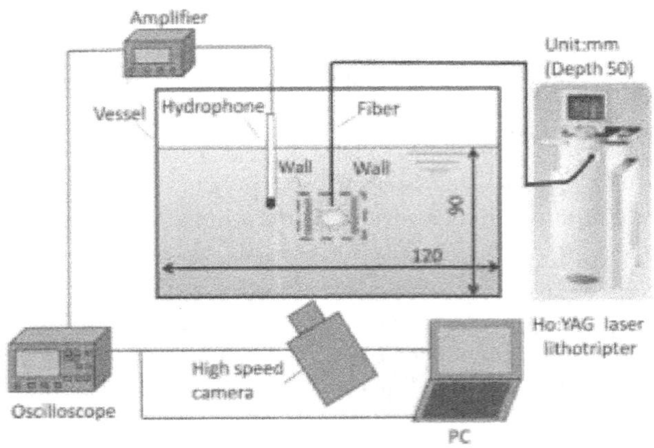

(a) Experimental setup

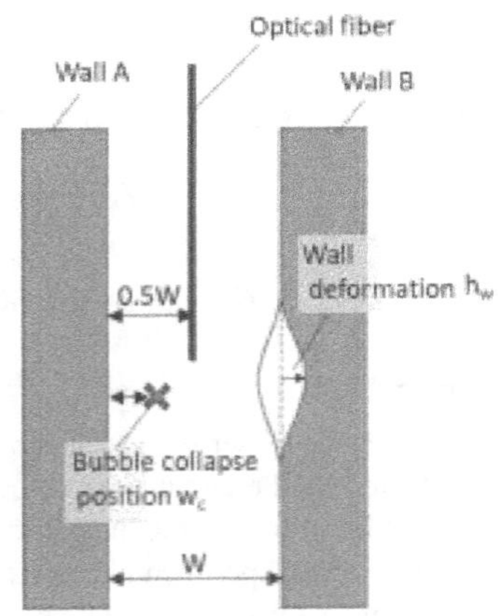

(b) Details of test section

Fig.1 Experimental setup and test section

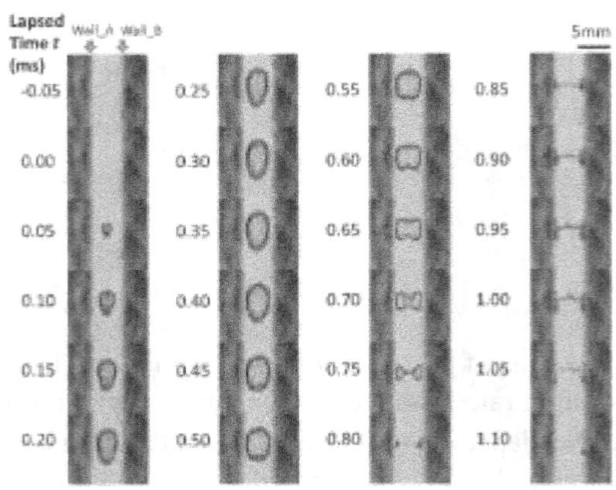

(a) Acrylic resin walls

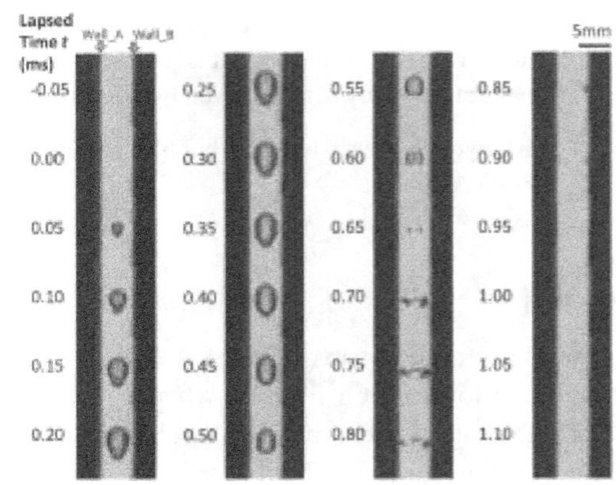

(b) Sillicone rubber (A8) walls

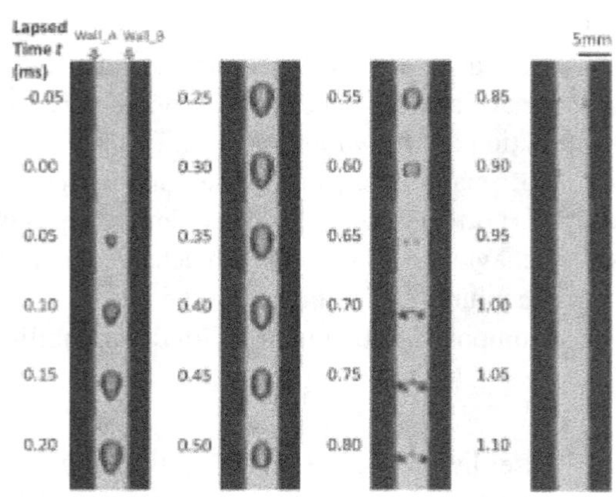

(c) Sillicone rubber (A5) walls

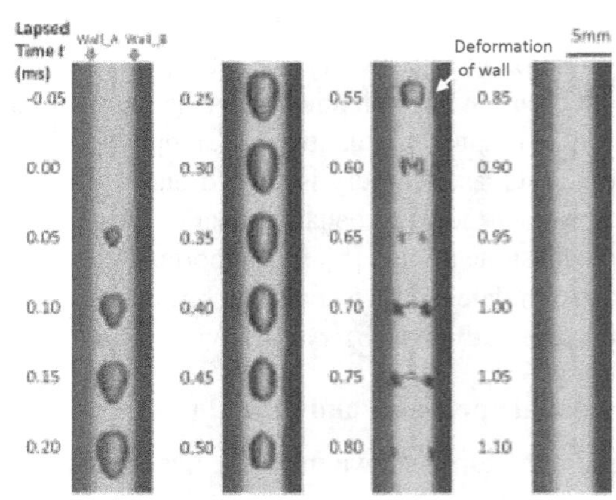

(d) Sillicone rubber (A2) walls

E = 1.0 J, d = 0.2 mm, T_w = 297 K, β = 7.8 mg/L, f_s = 100,000 fps

Fig.2 Behavior of bubble between two walls with same hardness

Bubble behavior between two walls with the same hardness

Figure 2 shows behavior of bubbles between two walls with the same hardness observed by a high-speed video camera. The lapsed time t = 0 corresponds to that at bubble formation. A bubble symmetrically divides into two parts and collapses irrespective of the wall hardness. In addition, the divided bubbles rebound, collapse several times and moves towards the wall. Figure 3 shows a time series of projected areas of bubbles. The bubble grows larger between two rigid (acrylic resin) walls than between two soft (deformable silicone rubber) walls. The collapse time for rigid walls becomes longer than that for soft walls. Figure 4 shows a time series of deformation on soft wall. The wall dents and bulges with bubble growth and collapse. The life time of bubbles can be affected by such slight deformation of the wall surface as shown in Fig. 3. Figure 5 shows a bubble collapse position w_c (distance from wall A) for various bubble sizes. Figure 5(a) shows the collapse position for the primary bubble and (b) for the rebound bubble, respectively. Bubble collapse positions tend to distribute around the fiber tip in the case of smaller bubble size. The bubble further moves toward the wall side with increase in maximum bubble (equivalent) radius R_{max} (where R_{max} is calculated by the projected bubble area). Furthermore, the collapse positions of the rebound bubbles are closer to the wall side than that of the primary bubbles.

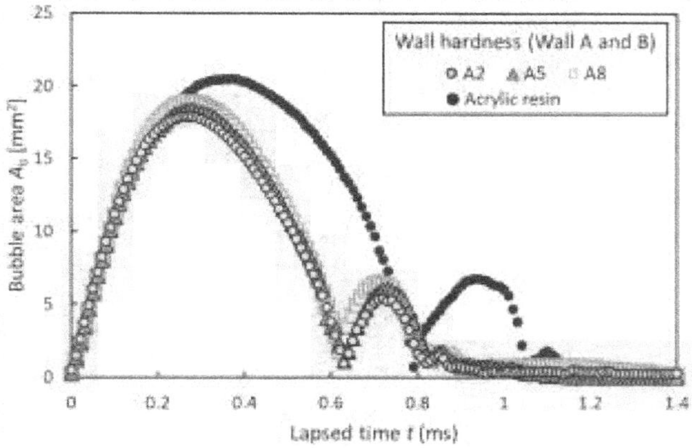

E = 1.0 J, d = 0.2 mm, T_w = 297 K, β = 7.4-7.8 mg/L
Fig.3 A time series of projected bubble area between walls with same hardness

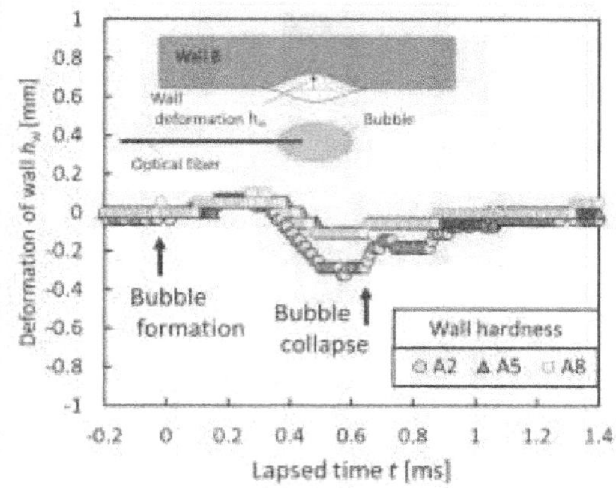

E = 1.0 J, d = 0.2 mm, T_w = 296 K, β = 7.4-7.8 mg/L
Fig.4 Deformation of soft wall between walls with same hardness

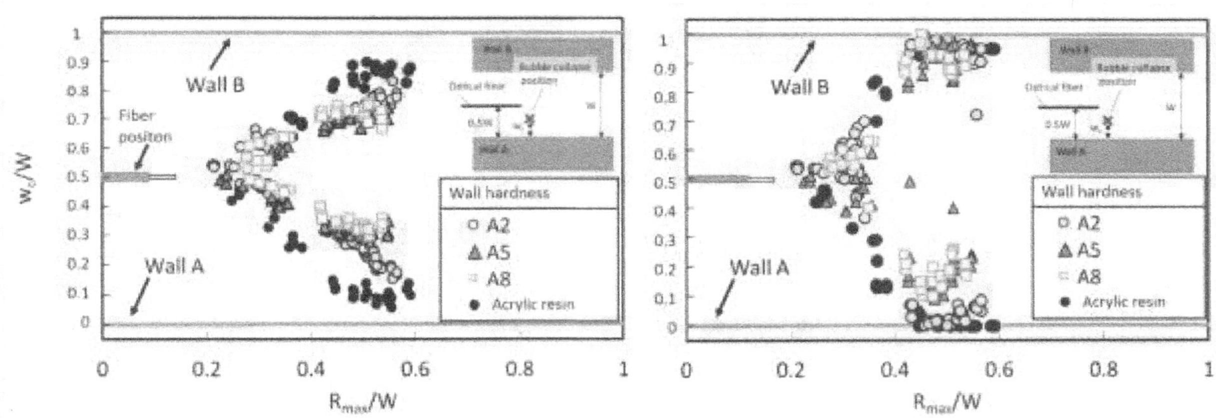

(a) Primary bubble collapse (1st bubble collapse) (b) Rebound bubble collapse (2nd bubble collapse)
d = 0.2 mm, T_w = 297 K, β = 7.4-7.8 mg/L
Fig.5 Bubble collapse positions between two walls with same hardness

Bubble behavior between two walls with different hardness

In the case of a rigid wall and a soft wall with deformability, the bubble is asymmetrically divided even though the bubble forms at the center of two walls as shown in Fig. 6 (a) - (c). Then the larger one moves toward the rigid wall and collapses on it. This characteristic bubble collapse behavior near wall can lead to high impact [9]. The impact on the wall surface can be predicted to be larger on the rigid wall. Figure 7 shows a time series of projected bubble area. In the case of soft wall in the present study, the changes of projected bubble area show almost same tendency regardless of the wall hardness. The collapse time tends to become longer than that of walls with the same hardness as shown in Fig. 3. Figure 8 shows a time series of deformation of soft wall. The wall surface dents when the bubble grows, and it bulges so as to be sucked by the bubble deformation during bubble collapse though the deformation is small. Figure 9(a) shows the collapse position for the primary bubble and (b) for the rebound bubble in various bubble size conditions, respectively. The bubble with small radius (around R_{max}/W = 0.2) tends to collapse at the rigid wall side from the center. With increase in bubble size ($R_{max}/W > 0.3$) as shown in Fig. 9(a), the bubble collapse positions shift to both wall sides in any bubble size and soft wall hardness though the distribution is biased toward the hard wall side. Collapse positions of the rebound bubble are located closer to the wall side than those of the primary bubble, maintaining the asymmetric distribution as shown in Fig. 9(b). The bubble collapse pattern and position are greatly affected by wall deformation irrespective of bubble size. The bubble collapses near/on rigid wall without collapse on the soft wall around $R_{max}/W = 0.4$. This means that effective impact can act on rigid stone without damage of soft tissue.

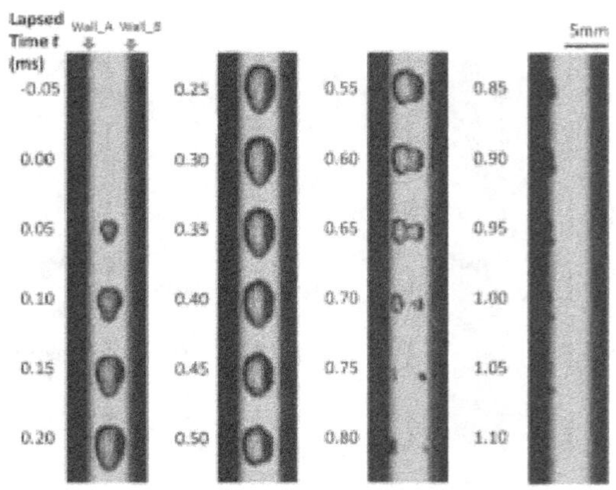

(a) Acrylic and silicone rubber (A8) walls

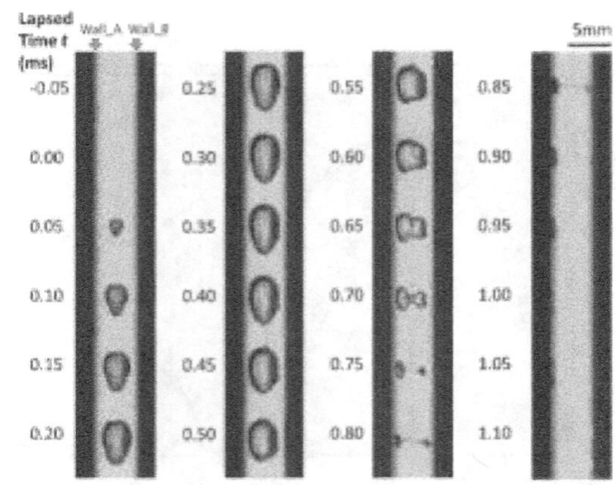

(b) Acrylic and silicone rubber (A5) walls

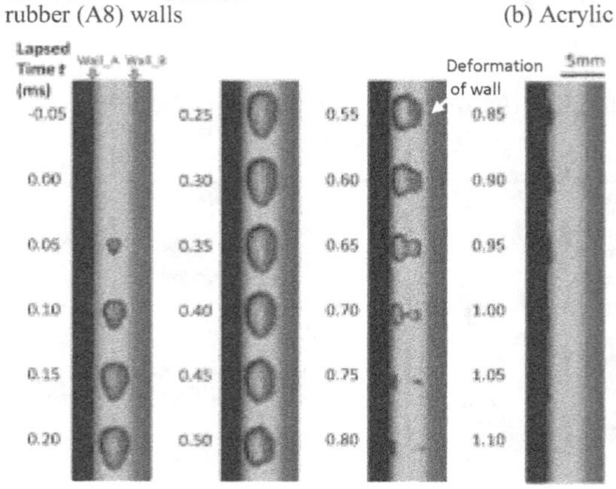

(c) Acrylic and silicone rubber (A2) walls

$E = 1.0$ J, $d = 0.2$ mm, $T_w = 296$ K, $\beta = 7.4$ mg/L, $f_s = 100{,}000$ fps

Fig.6 Bubble behavior between two walls with different hardness

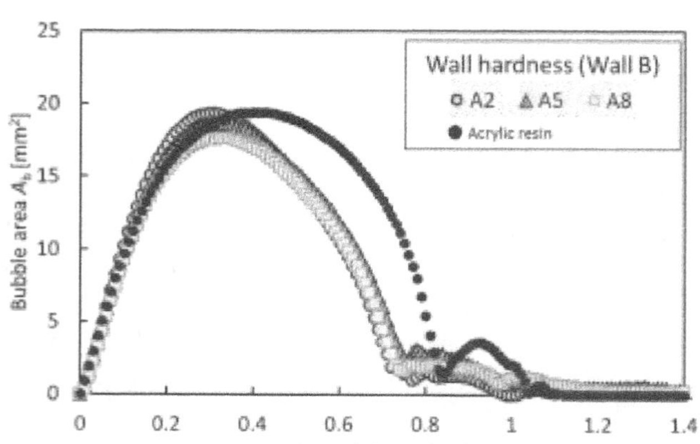

$E = 1.0$ J, $d = 0.2$ mm, $T_w = 297$ K, $\beta = 7.4$-7.8 mg/L

Fig.7 A time series of projected bubble area between two walls with different hardness

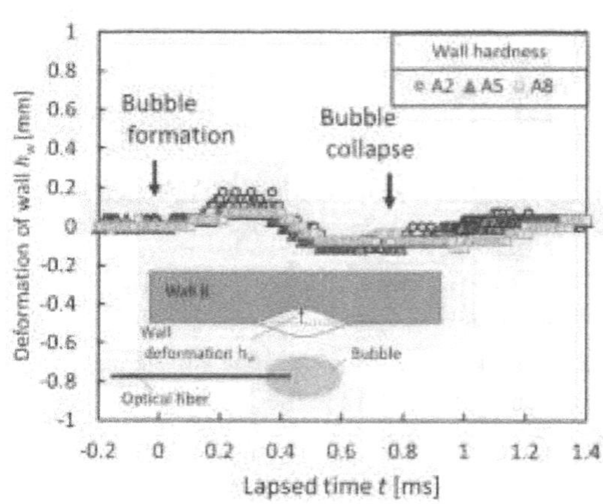

$E = 1.0$ J, $d = 0.2$ mm, $T_w = 296$ K, $\beta = 7.4$-7.8 mg/L

Fig.8 Deformation of soft wall between two walls with different hardness

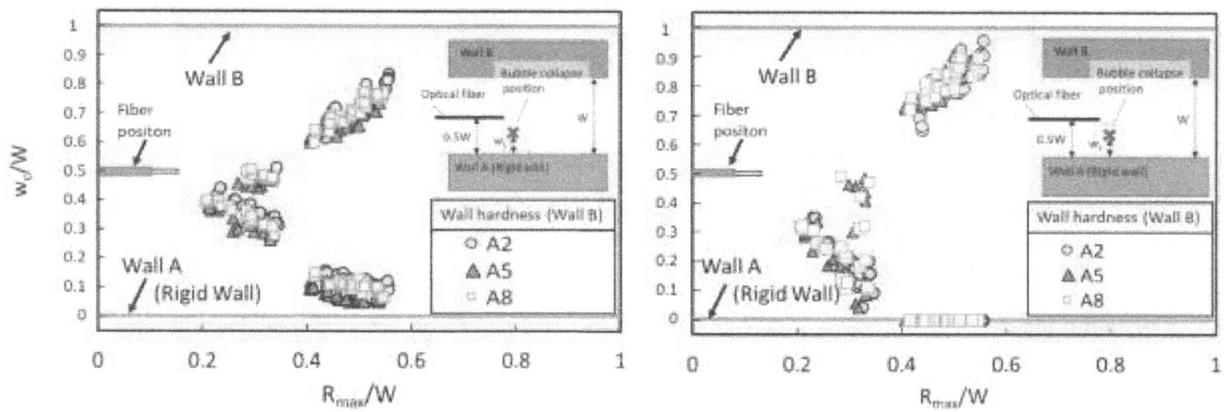

(a) Primary bubble collapse (1st bubble collapse) (b) Rebound bubble collapse (2nd bubble collapse)
$d = 0.2$ mm, $T_w = 297$ K, $\beta = 7.4$-7.8 mg/L
Fig.9 Bubble collapse positions between two walls with different hardness

Conclusion

In order to simulate in vivo condition for TUL, we observed the bubble behavior in a narrow space between two walls with various deformability. In the case of two walls with the same hardness, a bubble symmetrically divides into two parts and collapses. The soft wall deforms due to bubble behavior such as growth and collapse. The bubble collapse position moves closer to wall during multiple bubble collapses. In the case of walls with different hardness, the bubble asymmetrically divides, moves and collapses even if the bubble begins to form at the center between two walls. The bubble collapse pattern is greatly affected even by slight wall deformation. It can be pointed out that bubble size and wall deformability play an important role in safety issues of in vivo treatment.

Acknowledgements

The authors thank to Dr. M. Moriyama, Professor of Kanazawa Medical University, and Boston Scientific Japan co., ltd. for advice on medical view point and providing a laser equipment. A part of this work was supported by JSPS KAKENHI Grant Number JP16K06092.

References

[1] Marks A. J., Qiu J., Milner T. E., Chan K. F. and Teichman, J. M. H. (2011). *Laser lithotripsy physics, Urinary tract stone disease.* Springer. pp.301-309.

[2] Zhong, P. (2013). *Shock wave lithotripsy.* Bubble dynamics & shock saves. Shockwaves 8. Springer. 291-338 (ed, Delale, C. F.).

[3] Sugimoto Y., Yamanishi Y., Sato K. and Moriyama M. (2015). *Measurement of bubble behavior and impact on solid wall induced by fiber-holmium:YAG laser.* J. Flow Control, Measurement & Visualization. 3. 135-143.

[4] Yamanishi Y., Sugimoto Y., Sato K. (2015). *Various Conditions for Impulse Force on Solid Wall by a Fiber-type Laser Induced Bubble.* International Workshop on Cavitation Peening and Related Phenomena. 12th Int. Conf. Flow Dynamics (ICFD2015). Sendai.

[5] Brujan, E. A., Nahen, K., Schmidt, P. and Vogel, A. (2001). *Dynamics of laser-induced cavitation bubbles near an elastic boundary.* J. Fluid Mech. 433. 251-281.

[6] Kobayashi, K., Kodama, T., and H.Takahira. (2011). *Shock wave–bubble interaction near soft and rigid boundaries during lithotripsy: numerical analysis by the improved ghost fluid method.* Physics in Medicine & Biology. 56(19). 6421–6440.

[7] Ogasawara, T., Tsubota, N., Seki, H., Shigaki, Y. and Takahira, H. (2015). *Experimental and numerical investigations of the bubble collapse at the center between rigid walls.* J. Physics. Conf. Ser. 656. 012031.

[8] Mohammadzadeh, M., Mercado, J. M. and Ohl, C-D. (2015). *Bubble Dynamics in Laser Lithotripsy.* J. Physics. Conf. Ser. 656. 012004.

[9] Tomita, Y. and Shima, A. (1986). *Mechanisms of impulsive pressure generation and damage pit formation by bubble collapse.* J. Fluid Mech. 169. 535-564.

© 2018 ASME

Behaviors of Free Bubble Nuclei in Inception of Sheet Cavitation

[1]Wakana Tsuru*; [1]Satoshi Watanabe; [1]Shin-ichi Tsuda

[1]*Kyushu University, Fukuoka, JAPAN*

Abstract

Inception of sheet cavitation is known to be a complex phenomenon depending upon the working flow conditions, especially on the boundary layer flow and the water quality. Because of this, it seems very difficult to simulate the sheet cavity inception by CFD using homogeneous cavitation model. In the present study, experimental observation of inception of sheet cavitation from a single free nucleus in a two-dimensional convergent-divergent nozzle is conducted under well degassed condition with few free nuclei. Then numerical simulation solving the equation of motion of bubble nuclei with Rayleigh-Plesset equation is conducted to see the effect of slippage of bubble nuclei from liquid flow on their trajectories. The flow field obtained by the single phase CFD is used as a background liquid flow. Finally, using the obtained bubble slip velocity, the spatial distribution of bubble nuclei density is estimated.

Keywords: sheet cavitation, inception, free bubble nuclei, slip velocity

Introduction

Inception of sheet cavitation is known to be a complex phenomenon depending upon the working flow conditions. Many researches have been devoted to that in external flows around bodies such as hydrofoils and head forms since old times (for ex. [1]-[3]). Typically, when the laminar separation bubble is formed leading to the transition to turbulent flow, the sheet cavity forms not from the minimum pressure point but from the downstream separation bubble even though the pressure there is apparently larger than the minimum pressure [4]. This fact indicates that the behavior of free bubble nuclei is important for the formation of sheet cavity; it is necessary to have nuclei at the incipient point in sufficient time duration, but the nuclei seem to hardly approach to the minimum pressure point or hardly stay there sufficient time for their substantial growth.

The objective of this study is to understand nuclei behaviors in the process of sheet cavitation inception. The visual observation for free bubble nuclei near the throat wall of a two-dimensional convergent-divergent nozzle is conducted. Degassed water is used as working fluid. The general mechanism of attached sheet cavitation is briefly discussed. Then, numerical simulation tracking the motion of bubble nuclei in the liquid flow obtained by CFD is conducted to make clear the effect of slippage between bubble nuclei and liquid flow on their trajectories. The effect of the bubble growth is also investigated by solving Rayleigh-Plesset equation along with the equation of motion. Finally, the possible spatial distribution of bubble nuclei density is estimated by considering the conservation of bubble nuclei density with the above obtained bubble slip velocity.

Experimental observation

Observations of sheet cavitation inception from free bubble nuclei is conducted in a two-dimensional convergent-divergent nozzle as shown in Figure 1 [5]. The size of test section is rather small, 20mm x 20mm, and the throat height is h=10mm. The maximum velocity at the nozzle throat is U_{throat} =10m/s, resulting in the maximum Reynolds number is about 10^5. In order to observe the behavior of individual nucleus, it is important to reduce the nuclei density in the liquid. Therefore, multi-layered membrane filter with nominal pore diameter of 1μm is installed in the test tunnel and the sufficient deaeration is carried out to reduce the dissolved gas content in water. The behaviors of nuclei are observed at the throat by a high-speed camera, from which the inception process is discussed.

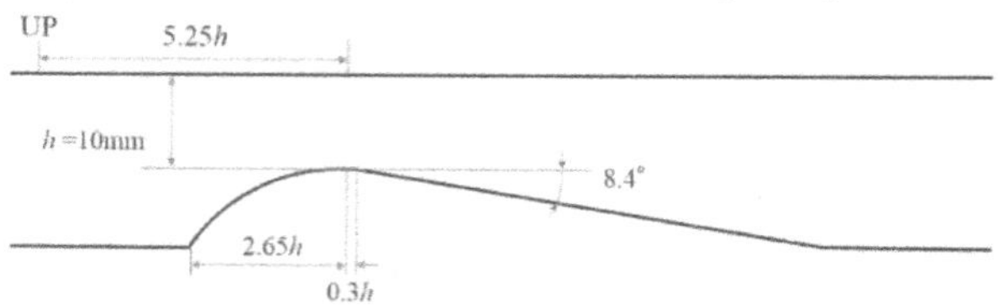

Figure 1 Dimensions of test nozzle. Flow direction is from left to right. Width is 20mm.

*Corresponding Author, Wakana Tsuru: tsuru.w.123@kyudai.jp

Figure 2 shows the typical inception process of sheet cavitation. Two series of snap shots are presented. Note that the two movies were taken separately, therefore the recorded bubbles are different ones, whereas these two movies give typical examples for sheet cavitation inception. From right-hand-side ones, it can be seen that a small bubble nucleus with the diameter less than 200μm in the top figure approaches to the nozzle wall, leading the sheet cavitation inception. From left-hand-side pictures, a large bubble attaches to the wall downstream of the throat, leaves the tail of itself on the wall and collapses downstream. In many cases of the observation results, it is confirmed that a free nucleus flowing near the wall at the throat region grows to a bubble. Then, the bubble attaches to the wall, and the tails of the bubble left on the wall develop to patch-like cavity, which can grow to sheet cavity. Although the static pressure takes the minimum at the throat, the locations of patch-like cavities and of leading edge of sheet cavity are slightly downstream.

From this movement of bubbles, it can be said that the free bubble nuclei tend to gather near the wall, as can be easily imagined by considering the light weight of bubbles with given pressure gradient, which would result in the larger number density of the bubble nuclei near the wall than in the free stream. However, the bubble nuclei itself does not grow to sheet cavitation, but would supply smaller portion onto the wall as a source of sheet cavitation.

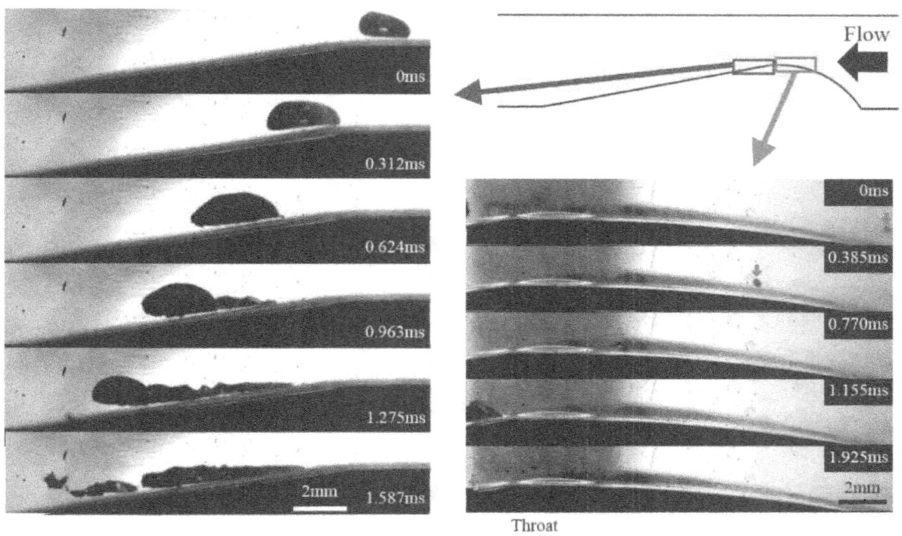

Figure 2 Typical sheet cavitation inception process from free bubble nuclei. Note that two series of movies (left and right) have been taken separately, tracking the different bubble nuclei.

Numerical analysis

In order to analyze the motion of bubble nuclei, series of numerical simulation is conducted. Firstly, steady-state single-phase CFD is carried out to obtain the liquid flow field. The commercial CFD solver, ANSYS-CFX ver. 16.2 is used. Two-dimensional model is employed for simplicity. Shear Stress Transport model is employed as a turbulence model. The number of mesh nodes is around 105,000 and the normalized wall distance y^+ is less than 0.1.

Once the liquid flow field is obtained, the bubble nucleus is tracked in Lagrange way by solving the following equation of motion of bubble nucleus.

$$\frac{4}{3}\pi r^3 \rho_b \frac{d\boldsymbol{v}_b}{dt} = \boldsymbol{F}_g + \boldsymbol{F}_p + \boldsymbol{F}_l + \boldsymbol{F}_d + \boldsymbol{F}_a + \boldsymbol{F}_r + \boldsymbol{F}_h + \boldsymbol{F}_w + \boldsymbol{F}_{other}$$

where r denotes the radius of bubble nucleus, and ρ_b and $\boldsymbol{v}_b$ are density of gas and moving velocity of bubble nucleus respectively. The forces from $\boldsymbol{F}_g$ to $\boldsymbol{F}_{other}$ in RHS are the buoyancy force[6], the pressure gradient force[6], the lift force[7], the drag force[8], the added mass force[9][10], the force due to bubble growth[6], the history force, the force due to near wall[10] and the other forces respectively, which are determined from the corresponding literature. History force and other forces are neglected. The bubble tracking is carried out by time-integrating the above equation using the fourth order Runge-Kutta method until the bubble nucleus collides to the wall.

Along with solving the above equation of motion of bubble nucleus, Rayleigh-Plesset equation is solved for the growth of the bubble, which is expressed as

$$r\frac{d^2r}{dt^2}+\frac{3}{2}\left(\frac{dr}{dt}\right)^2=\frac{1}{\rho_l}\left(p_v-p+p_g-\frac{4\mu_l}{r}\frac{DR}{Dt}-\frac{2S}{r}\right)$$

where ρ_l is the density of liquid, p_v and p_g is the vapor pressure and the pressure of non-condensable gas, μ_l is the viscosity of liquid and S is the surface tension. The pressure p is the local pressure which is taken from the result of single phase CFD. It is assumed that the non-condensable gas in the bubble nucleus obeys the isothermal law. Considering that the bubble is in a dynamic equilibrium condition at the inlet of numerical domain, the initial gas pressure p_{g0} can be determined with the given initial bubble radius r_0 and the given upstream pressure p_0 calculated from the assumed incipient cavitation number of $\sigma = 2(p_0 - p_v)/\rho_l U_{throat}{}^2$=0.9.

Figure 3(a) shows the examples of bubble trajectories with the initial bubble radius of r_0=50μm (top) and 5μm (bottom). The throat velocity is U_{throat}=10m/s. Flow direction is from left to right. In this figure, the equation of bubble motion is only solved with the constant nucleus radius. The color of trajectories indicates the magnitude of slip velocity between liquid and bubble $|\Delta\boldsymbol{v}|$. The grey colored lines show the liquid streamlines. From the figure, we can see the difference of bubble trajectories and liquid streamlines, which is more remarkable with the larger nucleus radius. Looking closely at the bottom corner of beginning of converging section (in dashed oval), so called "bubble screening effect" [2] due to the stagnant pressure can be recognized; the bubble nuclei go apart from the wall. Despite of this "screening effect," the bubble nuclei approaches again to the wall near the throat as can be seen in the enlarged figure. This motion is also observed in the experiment as shown in the right-hand side snapshots of Figure 2. Finally, many of bubbles rapidly approach and collide to the wall.

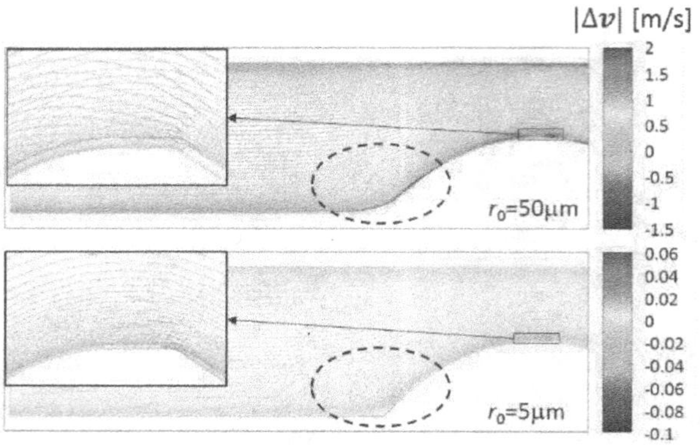

Figure 3(a) Trajectories of bubble nuclei with constant nuclei radius. The color indicates the magnitude of slip velocity.

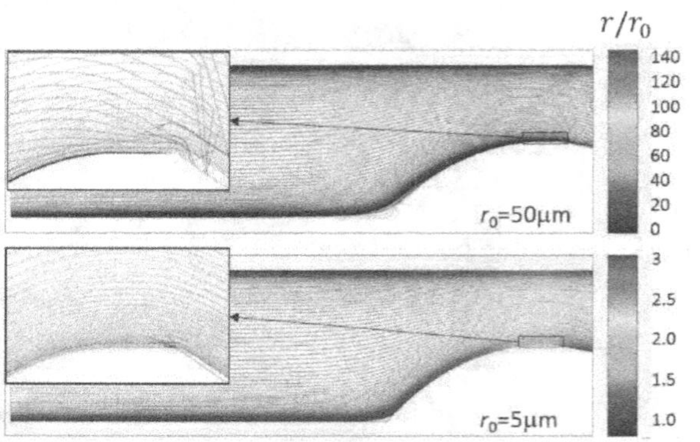

Figure 3(b) Trajectories of bubble nuclei with solving bubble growth equation. The color indicates the normalized bubble radius.

Figure 3(b) shows the examples of bubble trajectories with solving the bubble growth equation. The color of trajectories indicates the radius of bubble normalized by the initial bubble radius, r/r_0. By comparing this results with the previous one, it can been seen that the bubble trajectories are not very different in the regions far from the bottom nozzle wall. When the bubble nuclei approach to the wall roughly to the order of milli to sub-millimeter, the deviations is very remarkable, and it is seen that some bubble nuclei bounce violently. Since the spherical bubble has been assumed not only in Rayleigh-Plesset equation but also in the equation of motion, the obtained behaviors of bubbles very close to the wall seem to be inappropriate. Further modeling should be necessary to well reproduce the actual motion of bubble nuclei in near wall regions.

Although the bubble behaviors in near-wall region are somewhat strange, it seems to be still valuable to discuss the spatial distribution of the number density of bubbles due to the slippage between the liquid flow and the bubble velocity. By assuming the constant number density of bubble nuclei n_0 upstream, the spatial distribution of bubble nuclei density n can be calculated from the bubble and liquid velocity distributions obtained above. Neglecting the bubble growth for simplicity, the following equation is obtained from the conversation law of number density of bubble nuclei.

$$\frac{D_g}{Dt}\left(\frac{n}{n_0}\right)=-\frac{n}{n_0}\nabla\cdot\boldsymbol{v}_g \quad or \quad \frac{D_g}{Dt}\ln\left(\frac{n}{n_0}\right)=-\nabla\cdot\boldsymbol{v}_g$$

where D_g/Dt is the time derivative along with bubble motion. After time-integrating this equation along the bubble trajectory like in Figure 3(a), and interpolating them to the mesh used in CFD, the distribution of number density is obtained as shown in Figure 4.

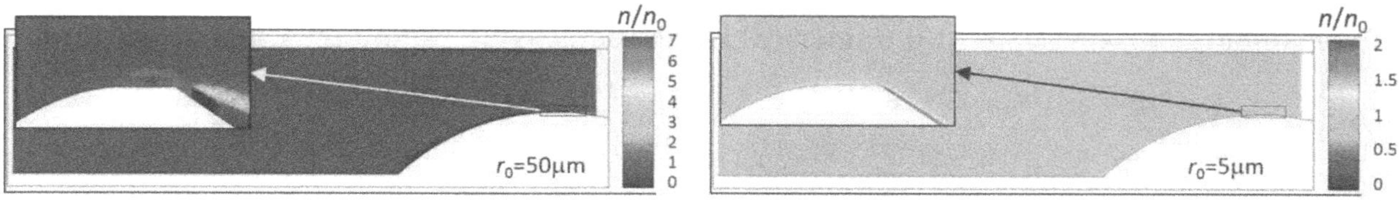

Figure 4 Calculated spatial distribution of nuclei number density for the nuclei radius of r_0=50μm (left) and r_0=5μm (right).

Since the near-wall motions of bubble nuclei are not well solved as mentioned before, the number densities near the wall (within wall distance less than $2r_0$) have been calculated by the extrapolation considering its normal gradient to the wall is zero. The accumulation of the bubble nuclei near the wall can be seen from upstream of the nozzle throat in the case with r_0=5μm (right). On the other hand, in the case with r_0=50μm (left), the largest number density can be seen downstream of the nozzle throat and a little far from the wall. However, more accurate treatment of bubble motions are necessary for further discussion on the bubble density distribution and the sheet cavitation inception.

Conclusion

In the present study, the visual observation of sheet cavitation inception from a free bubble nuclei in a two-dimensional convergent-divergent nozzle is conducted to obtain some general overview of inception mechanism of sheet cavitation. In many cases, it is observed that bubble nuclei in free stream approach to the wall and some portion is attached and left on the wall, resulting in a patch-like cavity as a source of sheet cavitation. Numerical simulation of bubble motion is conducted, and it is found that the effect of slip velocity of bubble nuclei to liquid flow is significant and the trajectories of bubble nuclei is similar to those observed in experiments. The number density of bubble nuclei is found to be larger near the throat due to the slippage effect, whereas the further study is needed especially on establishment of more sophisticated model for bubble nuclei moving near the wall.

This work was partly supported by JSPS KAKENHI Grant Number JP17J04154. Numerical simulation was mainly carried out using the computer facilities at Research Institute for Information Technology, Kyushu University.

References

[1] Knapp, R. T., Hollander, A. (1948). *Laboratory investigations of the mechanism of cavitation*. ASME J. Fluids Eng. 70, pp. 419-433.

[2] Kodama, Y., Take, N., Tamiya, S., Kato, H. (1981). *The effect of nuclei on the inception of bubble and sheet cavitation on axisymmetric bodies*. ASME J. Fluids Eng. 103, pp. 557-563.

[3] Rijsbergen, M. (2016). *A Review of Sheet Cavitation Inception Mechanisms,* Proc. ISROMAC2016. 356.

[4] Arakeri, H., Acosta, A. J. (1976). *Cavitation inception observations on axisymmetric bodies at supercritical Reynolds numbers*. J. Ship Research. 20, pp. 40-50.

[5] Tsuru, W., Konishi, T., Watanabe, S., Tsuda, S. (2017). *Observation of inception of sheet cavitation from free nuclei*. J. Thermal Science. 26(3), pp. 223-228.

[6] Shams, E., Finn, J., Apte, S. V. (2011). *A numerical scheme for Euler-Lagrange simulation of bubbly flows in complex systems*. Int. J. Numerical Method of Fluids. 67, pp. 1865-1898.

[7] Antal, S. P., Lahey Jr, R. T., Flaherty, J. E. (1991). Analysis of phase distribution in fully developed laminar bubbly two-phase flow. Int. J. Multiphase Flow. 17(5), pp. 635-52.

[8] Tomiyama, A., Sou, A., Minagawa, H., Sakaguchi, T. (1991). *Numerical analysis of a single bubble with VOF method.* Trans. JSME, Ser. B (in Japanese). 57, pp. 2167-2173.

[9] Brennen, C. E. (1995). *Cavitation and bubble dynamics*. Oxford University Press.

[10] Sanada, T., Sato, A., Shirota, M., Watanabe, M. (2009). *Motion and coalescence of a pair of bubbles rising side by side*. Chemical Engineering Science. 64(11), pp. 2659-2671.

© 2018 ASME

Bubble dynamics and High Intensity Focused Ultrasound: experimental observations and numerical simulations using Boundary Element Method

[1]Siew-Wan Ohl; [1]Evert Klaseboer; [2]Boo Cheong Khoo*

[1]*Institute of High Performance Computing, A*STAR, Singapore;* [2]*Dept of Mechanical Engineering, University of Singapore.*

Abstract

We present an experimental study of High Intensity Focused Ultrasound (HIFU) using a parabolic shaped transducer and high speed photography (Ohl et al, 2015). The transducer (Sonic Concept Inc.) has a diameter of 60 mm, and a resonant frequency of 250kHz. When it is driven between 120 to 150 Volt peak-to-peak, concentrated rings of bubbles are formed on top of the transducer. The distance between the rings is about 3 mm, which is half the wavelength of the sound wave produced by the transducer at 250 kHz. The bubbles within the rings are not stable. They move or jump between the rings, or coalesce, or float toward the free surface. To stabilise the rings, we use a reflector which is made of stainless steel, and have the same shape and dimensions as the transducer. At lower driving voltages (80 to 100 Vpp), streams of bubbles are seen moving towards the free surface. This indicates that no standing waves are present. When driven in a burst mode (500 cycles), small bubble clouds are formed. We captured the nucleation and expansion of these clouds using the high speed photography and a long distance microscopic lens. In an attempt to understand some of these observed phenomena, we employed numerical simulations based on the Boundary Element Method (BEM). The Boundary Element Method is an established numerical method for the simulation of bubble dynamics (Blake et al, 1986, 1987). Recently Klaseboer et al (2012) and Sun et al (2014) have improved the numerical solution methodology by eliminating the singularities in the Boundary Integral equation. By using an analytical function, and subtracting this from the original Boundary Integral Method in a specific manner, the method is desingularized, with the added advantage of eliminating the solid angle. This breakthrough significantly reduces the computational complexity and cost, making the use of higher order elements becomes straight forward.The same analytics is applied then to develop a three-dimensional BEM code to solve the Helmholtz equation. The code is used to simulate the ultrasound field generated by the HIFU transducer which is used in the experiments previously mentioned. From the simulation, the focused ultrasound field is clearly described.

Keywords: High Intensity Focused Ultrasound, HIFU, Boundary Element Method, numerical simulation.

1. Introduction

High Intensity Focused Ultrasound was first used in research in the 1930s (Grützmacher, 1936). It is generated typically either with a ceramic or piezo transducer, which has a bowl or a cylindrical shape, or by an array of piezo transducers which are controlled to generate ultrasound with a certain phase differences between them. The HIFU system is able to focus the ultrasound to a certain volume where high pressure or high ultrasound intensity is achieved.

HIFU offer the possibility of non-invasiveness where medical treatments can be administered extra-corporeally. The focus region is heated up, and tissues are lysed or emulsified. This thermal effect is the main mechanism behind the use of HIFU for the treatment of prostate cancer (Poissonnier et al., 2007) and the removal of kidney or liver tumours (Illing et al., 2005). There are new treatments being explored especially in the area of brain surgery, for example to treat essential tremors (Elias et al., 2013).

Another important mechanism offers by HIFU is the generation of cavitation bubbles in the focus region. We have explored using the violently oscillating and collapsing bubbles for the treatment of root canal infection (Shrestha et al., 2009) or the removal or biofilm (Iqbal et al., 2013). These bubbles can be potentially used for localized and timed drug delivery (Dromi et al., 2007), and the non-thermal lysis of tissue via histotripsy (Kim et al., 2014).

Since it is difficult to visualize or tract HIFU generated bubbles in medical treatment, we have attempted to understand the bubble sound waves interaction via experiments and numerical simulations. We will discuss our

*Corresponding Author, Boo Cheong Khoo: mpekbc@nus.edu.sg

experimental observations in the next section. For numerical simulation, we use the Boundary Element Method which is effective and efficient in modeling the change in bubble shapes by allowing meshing only on the bubble surface instead of the whole simulation domain (for example the water tank). From our previous simulations, we found that the bubbles are very sensitive to the intensity and frequency of sound waves (Fong et al., 2008), and also the collapse of the bubbles are dependent on the properties of the nearby tissues (Fong et al., 2006).

2. HIFU Experiments

2.1 Experimental setup

The experimental setup (see Fig. 1(a)) consists of the following items: a custom-made bowl-shaped transducer (Sonic Concept Inc.) which is 60 mm in diameter and has a resonant frequency of 250 kHz; an acrylic tank which is about 20x20x25 cm^3 in dimension; a linear amplifier (AG1021 LF Amplifier/Generator, T&C Power Conversion); and a signal generator (33220A 20 MHz Function/Arbitrary Waveform Generator, Agilent Technologies). The signal generator controls the input signal, which is typically a 5 mV peak-to-peak sinusoidal wave. Its output is channeled into the input of the linear amplifier, which amplifies the voltage to between 80 and 200 Vpp. The linear amplifier is connected to the transducer, which is submerged in the tank, which is filled with tap water. The transducer has been customized to be waterproof.

2.2. Experimental observations

At the resonant frequency of the transducer, i.e. 250 kHz, bubbles are seen to form in the water tank. At high driving voltage (150 to 200 Vpp), rings of bubbles are formed on top of the transducer as seen in Fig. 1(b). The rings can be stabilized with a reflector which is placed on top (below the free surface). The reflector has the same shape and dimensions as the transducer. Note that these rings are observed without the use of the reflector as well. Just that they are less stable and therefore more difficult to image. The wavelength of the standing wave in the tank is calculated as:

$$\lambda = \frac{c}{f}, \qquad (1)$$

where c is the speed of sound in water and f is the frequency of the wave. Given that c is 1500 ms^{-1}, and f is 250 kHz, λ is 0.006 m or 6 mm. The distance between the rings in Fig. 1(b) is about 3 mm. Thus it is verified that the bubbles are trapped at the pressure antinodes of the standing wave. These bubbles are not completely stationary; over time, they coalesce and form bigger bubbles, which will eventually 'escape' from the ring due to buoyancy.

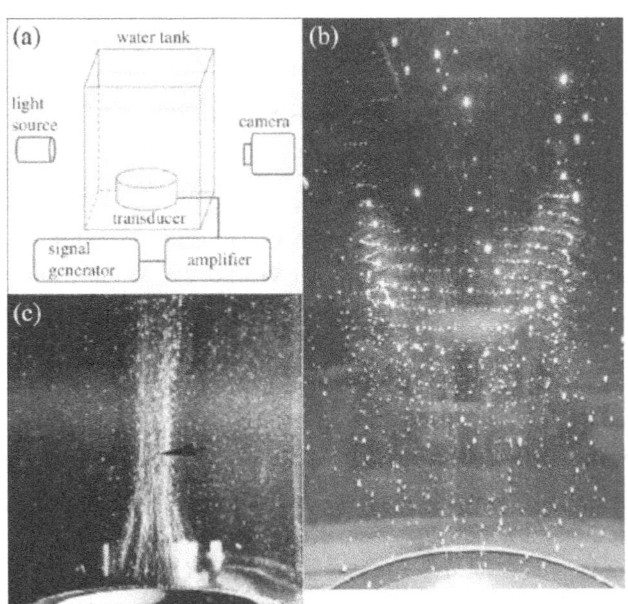

Figure 1 (a) Experimental setup. The transducer is driven by an amplifier, which has an input waveform from the signal generator. The bubbles are generated on top of the transducer in a water tank, and imaging is done with a camera system. (b) Bubble rings on top of the transducer, which is driven at 250 kHz and 200 Vpp. (c) At a lower driving voltage of 80 Vpp, a stream of bubbles is seen to move upwards from the transducer surface.

3. Desingularised Boundary Element Method

The Boundary Element Method is used to solve the potential function that defines the fluid domain. Let us first investigate the more familiar Laplace equation and show the principle of desingularization before moving on to the Helmholtz equation for sound wave simulations in the next section. Interested reader is referred to textbooks on BEM, for example Becker (1992).

We can define a potential $\phi = \phi(\vec{x})$ which satisfies the Laplace equation as:

$$\nabla^2 \phi = 0. \quad (2)$$

This equation is equivalent to an equation with only boundary integrals as:

$$c(\vec{x}_0)\phi(\vec{x}_0) + \int_S \phi \frac{\partial G}{\partial n} dS = \int_S G \frac{\partial \phi}{\partial n} dS, \qquad (3)$$

where $G = 1/|\vec{x} - \vec{x}_0|$, and it is known as the 3D free space Green function. Equation (3) is referred to as the boundary element (or integral) method. It is noted that both G and $\partial G/\partial n$ are singular with $1/|\vec{x} - \vec{x}_0|$, because $(\vec{x} - \vec{x}_0)$ is perpendicular to the normal vector n when $\vec{x}$ approaches $\vec{x}_0$. In the standard BEM treatment, special quadrature formulae have to be developed for calculations involving the singular integrals

In our desingularized BEM, we choose another known function, say χ, which satisfies the Laplace equation. Then:

$$c(\vec{x}_0)\chi(\vec{x}_0) + \int_S \chi \frac{\partial G}{\partial n} dS = \int_S G \frac{\partial \chi}{\partial n} dS. \qquad (4)$$

Also we require the following limits to be satisfied:

$$\lim_{\vec{x} \to \vec{x}_0} \chi(\vec{x}) = \phi(\vec{x}_0), \qquad (5)$$

$$\lim_{\vec{x} \to \vec{x}_0} \frac{\partial \chi(\vec{x})}{\partial n} = \frac{\partial \phi(\vec{x}_0)}{\partial n}. \qquad (6)$$

Subtracting Equation (4) from Equation (3), the solid angle c is eliminated. Thus

$$\int_S (\phi - \chi) \frac{\partial G}{\partial n} dS = \int_S G \left(\frac{\partial \phi}{\partial n} - \frac{\partial \chi}{\partial n} \right) dS. \qquad (7)$$

Equation (7) is a fully desingularized BEM. There are no singularities that require special treatment in the calculation. The only remaining problem is to choose a suitable function χ. One of the possibilities is to use a combination of a constant and a linear function:

$$\chi(\vec{x}) = \phi(\vec{x}_0) + \frac{\partial \phi(\vec{x}_0)}{\partial n} \vec{n}(\vec{x}_0) \cdot (\vec{x} - \vec{x}_0), \qquad (8)$$

$$\frac{\partial \chi(\vec{x})}{\partial n} = \vec{n}(\vec{x}) \cdot \nabla \chi(\vec{x}) = \frac{\partial \phi(\vec{x}_0)}{\partial n} \vec{n}(\vec{x}) \cdot \vec{n}(\vec{x}_0). \qquad (9)$$

Then Equation (7) becomes:

$$\int_S \left[\phi(\vec{x}) - \phi(\vec{x}_0) - \frac{\partial \phi(\vec{x}_0)}{\partial n} (\vec{x} - \vec{x}_0) \right] \frac{\partial G}{\partial n} dS = \int_S G \left[\frac{\partial \phi(\vec{x})}{\partial n} - \frac{\partial \phi(\vec{x}_0)}{\partial n} \vec{n}(\vec{x}) \cdot \vec{n}(\vec{x}_0) \right] dS. \qquad (10)$$

Equation (10) is our desingularized boundary integral method. Interested readers are referred to Klaseboer et al. (2012) for further proof that Equation (7) is indeed non-singular. If the problem under consideration is an external one, then the contribution of the surface at infinity will give an extra term $4 * \pi * \phi(x_0)$ on the left hand side of Equation (10).

3.1 Simulation of HIFU ultrasound field

The desingularized BEM method can be applied to solve the Helmholtz equation as well as presented in Sun et al. (2014). Briefly, the wave equation in the frequency domain is represented as:

$$\nabla^2 \phi(\mathbf{\vec{x}}) + k^2 \phi(\vec{x}) = 0\,, \quad (11)$$

where k is the wave number. The variable ϕ is now a complex variable. It is clear that if ϕ is a solution, then $\phi e^{i\alpha}$ is also a solution with α as a constant. The corresponding Green function is then $G^k = e^{ikr}/r$, with $r = |\vec{x} - \vec{x}_0|$. After some manipulation, it is shown in Sun et al. (2014) that one of the possible solutions is:

$$\chi(\vec{x}) = \phi(\vec{x}_0)\cos y + \frac{1}{k}\frac{\partial\phi(\vec{x}_0)}{\partial n}\sin y, \quad (12)$$

with

$$\frac{\partial\chi(\vec{x})}{\partial n} = \left[-k\phi(\vec{x}_0)\sin y + \frac{\partial\phi(\vec{x}_0)}{\partial n}\cos y\right] n(\vec{x}_0)\cdot n(\vec{x}),$$

where

$$y = k\vec{n}(\vec{x}_0)\cdot(\vec{x}-\vec{x}_0).$$

Basically Equation (12) is the solution for two standing waves in which one has nodes and the other antinodes at $\vec{x}_0$. Fig. 2 shows three-dimensional simulation from a fully desingularized Boundary Element Method code. The transducer is bowl-shaped and is vibrating with displacement in the upward and downward directions. The instantaneous potentials of the acoustic field in the surrounding water are shown in Fig. 2(a). High pressure and low pressure regions are clearly seen. In Fig. 2(b) the magnitude of the pressure field is plotted instead. There is a clear region of focus on top of the transducer. Note that in this simulation the transducer is located in an infinite volume of water. There is no reflection from any walls or free surface. Therefore there is no standing wave.

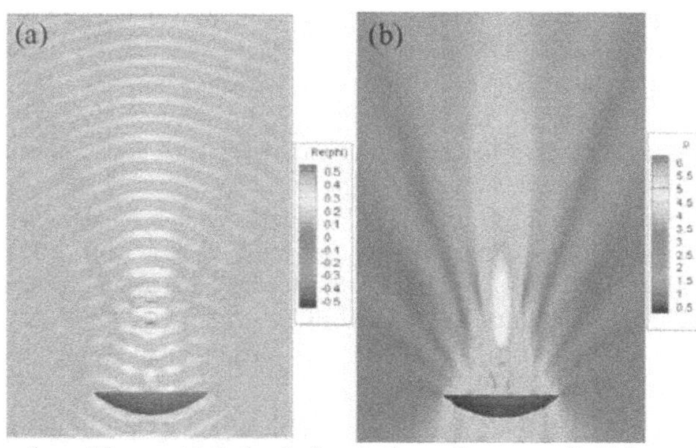

Fig. 2 Three dimensional Boundary Element Method simulation of a wave field in an infinite body of water. (a) The transducer is oscillating up and down. The color code denotes the real part of the instantenous potential in the water. (b) The absolute pressure distribution on top of the transducer is shown. It is noticed that there is a smooth and clear region of focus.

Conclusion

From our HIFU experiment, it is found that HIFU generates cavitation bubbles in its focused region. These bubbles are not stable. They coalesce and move in the sound field. When a standing wave is formed, the bubbles are trapped in the pressure anti-nodes of the sound field. When pulsed ultrasound is used, or when the HIFU is driven with low voltage, the standing wave is not formed and streams of bubbles are seen to travel from the surface of the transducer upwards.

We have also presented a new development in our Boundary Element Method code. A desingularized Boundary Element Method is used to simulate the sound field around a HIFU transducer. It is noted that with the desingularization the complexity in the coding of the Boundary Element Method is greatly reduced because no special treatment is needed for resolving singularities. Also the efficiency of this desingularized method is sufficient to allow the simulation of quadratic elements to solve the Helmholtz equation within a few minutes (for examples see the results in Fig. 2). Thus high resolution and accuracy acoustic simulations can now be done with the desingularized BEM.

Acknowledgement

The authors acknowledge the contributions from the following students from the National University of Singapore in the HIFU experiments: Nigel Chong Jia Wei, and Mohammed Haiqal Haqim Bin Mohd Rahim.

References

Becker, A. A. (1992). The boundary element method in engineering: a complete course. McGraw-Hill Companies.

Blake, J. R., Taib, B. B., & Doherty, G. (1986). Transient cavities near boundaries. Part 1. Rigid boundary. Journal of Fluid Mechanics, 170, 479-497.

Blake, J. R., Taib, B. B., & Doherty, G. (1987). Transient cavities near boundaries Part 2. Free surface. Journal of Fluid Mechanics, 181, 197-212.

Dromi, S., Frenkel, V., Luk, A., Traughber, B., Angstadt, M., Bur, M., ... & Wood, B. J. (2007). Pulsed-high intensity focused ultrasound and low temperature–sensitive liposomes for enhanced targeted drug delivery and antitumor effect. Clinical Cancer Research, 13(9), 2722-2727.

Elias, W. J., Huss, D., Voss, T., Loomba, J., Khaled, M., Zadicario, E., ... & Druzgal, J. (2013). A pilot study of focused ultrasound thalamotomy for essential tremor. New England Journal of Medicine, 369(7), 640-648.

Fong, S. W., Klaseboer, E., Turangan, C. K., Khoo, B. C., & Hung, K. C. (2006). Numerical analysis of a gas bubble near bio-materials in an ultrasound field. Ultrasound in Medicine & Biology, 32(6), 925-942.

Fong, S. W., Klaseboer, E., & Khoo, B. C. (2008). Interaction of microbubbles with high intensity pulsed ultrasound. The Journal of the Acoustical Society of America, 123(3), 1784-1793.

Grützmacher, J. (1936). Ultraakustischer Richtstrahler. *Z. Tech. Phys*, *17*, 166.

Illing, R. O., Kennedy, J. E., Wu, F., Ter Haar, G. R., Protheroe, A. S., Friend, P. J., ... & Middleton, M. R. (2005). The safety and feasibility of extracorporeal high-intensity focused ultrasound (HIFU) for the treatment of liver and kidney tumours in a Western population. British Journal of Cancer, 93(8), 890-895.

Iqbal, K., Ohl, S. W., Khoo, B. C., Neo, J., & Fawzy, A. S. (2013). Effect of high-intensity focused ultrasound on Enterococcus faecalis planktonic suspensions and biofilms. Ultrasound in Medicine & Biology, 39(5), 825-833.

Kim, Y., Hall, T. L., Xu, Z., & Cain, C. A. (2014). Transcranial histotripsy therapy: A feasibility study. IEEE Transactions on Ultrasonics, Ferroelectrics, and Frequency Control, 61(4), 582-593.

Klaseboer, E., Sun, Q., & Chan, D. Y. (2012). Non-singular boundary integral methods for fluid mechanics applications. Journal of Fluid Mechanics, 696, 468-478.

Ohl, S. W., Klaseboer, E., & Khoo, B. C. (2015). Bubbles with shock waves and ultrasound: a review. Interface Focus, 5(5), 20150019.

Poissonnier, L., Chapelon, J. Y., Rouviere, O., Curiel, L., Bouvier, R., Martin, X., ... & Gelet, A. (2007). Control of prostate cancer by transrectal HIFU in 227 patients. European Urology, 51(2), 381-387.

Shrestha, A., Fong, S. W., Khoo, B. C., & Kishen, A. (2009). Delivery of antibacterial nanoparticles into dentinal tubules using high-intensity focused ultrasound. Journal of Endodontics, 35(7), 1028-1033.

Sun, Q., Klaseboer, E., Khoo, B. C., & Chan, D. Y. (2014). A robust and non-singular formulation of the boundary integral method for the potential problem. Engineering Analysis with Boundary Elements, 43, 117-123.

© 2018 ASME

Bubble-bubble interactions and wall pressures produced by the collapse of a bubble pair near a rigid surface

Shahaboddin A. Beig*; Eric Johnsen

University of Michigan, Ann Arbor, MI, USA

Abstract

Cavitation occurs in a variety of applications ranging from naval structures to biomedical ultrasound. One important consequence of cavitation is structural damage to neighboring surfaces following repeated inertial collapse of vapor bubbles. Although the mechanical loading produced by the collapse of a single bubble has been widely investigated, less is known about how the presence of a second bubble affects the loading. In such a problem, the bubble-bubble interactions modify the dynamics, e.g., by increasing the non-sphericity of the bubbles and amplifying/hindering the collapse intensity depending on the flow parameters. Here, we quantify the effects of bubble-bubble interactions on the bubble dynamics, as well as the pressures produced by the collapse of a bubble pair near a rigid surface. We perform high-resolution simulations of this problem by solving the three-dimensional compressible Navier-Stokes equations for gas/liquid flows. The results are used to investigate the non-spherical bubble dynamics and characterize the pressure fields based on the relevant parameters entering the problem: stand-off distance from the wall surface, the angle, and the distance between the two bubbles.

Keywords: bubble dynamics; bubble-bubble interactions; re-entrant jets.

Introduction

In most applications, cavitation-induced damage is caused by the collapse of bubble clouds containing a large number of bubbles. Experimental observations reveal that the collapse of these clouds may emit intense pressures, and induce substantial structural damage [1,2]. Hansson et al. (1982) investigated the collapse of a hemispherical bubble cloud close to a rigid wall and illustrated that the collapse takes place in an inward fashion [3]. The subsequent inward-propagating shock wave concentrates the collapse energy in the cloud center, leading to the generation of high pressure regions. When a dense bubble cloud collapses near a rigid boundary, the combination of the interactions between the bubble cloud and the neighboring wall, and the interactions among the collapsing bubbles gives rise to bubble asymmetry, and the formation of re-entrant jets [4,5]. In such flows, bubble-bubble interactions may affect the overall dynamics, and must be accounted for to predict impact load, and subsequent cavitation erosion.

Owing to the complexity of these nonlinear flows, theoretical approaches are challenging. On the other hand, diagnosing these flows experimentally is not trivial, because of the wide range of spatial and temporal scales, difficult optical access, and lack of accuracy in measuring devices. Numerically resolving the full bubble cloud dynamics is not feasible at the present time. Therefore, these obstacles have triggered the development of simplified homogeneous-mixture models that are typically based on spherical bubble dynamics [6,7].These models neglect the non-spherical effects of the collapse and tend to overestimate the produced pressures at the collapse; in a numerical study, Tiwari et al. (2015) simulated the collapse of a hemispherical cluster containing 50 bubbles, and measured peak pressures for a variety of configurations over an order of magnitude lower than the values predicted by reduced cloud models [5].

To characterize bubble-bubble interactions and quantify their effects on the non-sphericity of the collapse and the resulting pressures, we carry out highly-resolved three-dimensional simulations of bubble pairs collapsing near rigid surfaces. We perform a parametric study and quantify the bubble morphologies and collapse non-sphericity, examine the radially propagating shocks, and report the pressures measured along the wall surface. By investigating interactions among the bubbles and the boundary and their effects on bubble dynamics, this study will provide knowledge necessary to develop reduced cloud models that takes the non-spherical effects of the collapse into account, and can be used to perform more realistic simulations.

Numerical method

A solution-adaptive central/discontinuity-capturing approach is employed to solve the compressible Navier-Stokes equations for a binary gas-liquid system [8]. In this problem, two bubbles with identical initial volume collapse in the vicinity of a rigid surface. The problem setup is shown in Figure 1. The bubble, located at distance H_o from the wall, is called the "primary bubble". The non-dimensional parameter $\delta_o = H_o/R_o$ defines the initial proximity of the primary bubble to the wall. The "secondary bubble"' is located at distance D_o from the primary bubble, such that the non-dimensional parameter $\gamma_o = D_o/R_o$ sets the distance between the two bubbles, and the connecting line

*Corresponding Author, Shahaboddin A. Beig: alahyari@umich.edu

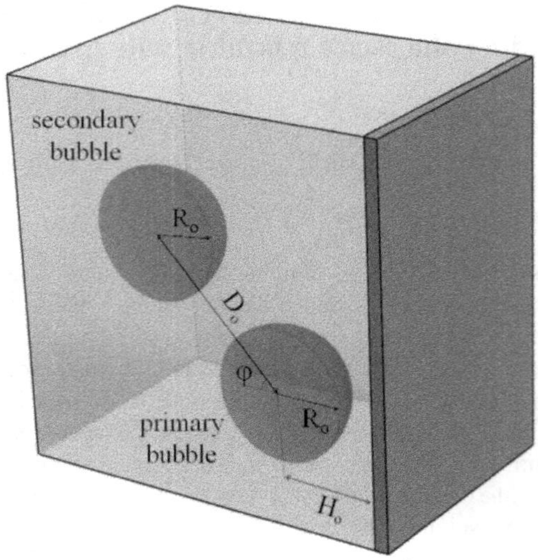

Figure 1: Schematic of the problem setup.

makes the angle ϕ with the horizontal. Thus, δ_o, γ_o, and ϕ are the geometrical variables, determining the bubbles' configuration at the beginning of the collapse, and the driving pressure is set to $p_\infty = 5$ MPa. The problem has a plane of symmetry (the middle plane), so we simulate only half of the domain, with symmetric boundary condition along the corresponding plane. Non-reflecting boundary conditions are applied along the remaining boundaries, except for the no-slip, adiabatic wall. To study the effects of geometry on the bubble dynamics, the following parameter ranges are considered: $1.1 \geq \delta_o \geq 3.0$, $1.1 \geq \gamma_o \geq 3.0$q, and $0^o \geq \phi \geq 90^o$. The total number of computational cells varies between 1 to 2 billion, depending on the geometrical configuration of the bubbles, such that there are 192 cells per initial bubble radius for all cases. The theoretical Rayleigh collapse time $t_c = 0.915R_o\sqrt{\rho_l/\Delta p}$, which is $t_c = 1.3$ μs for this problem, is used to non-dimensionalize time.

Results and discussion

Figure 2 shows contours of density gradient magnitude, and pressure at different times during the collapse, where $\delta_o = 1.5$, $\gamma_o = 2.5$, and $\phi = 45^o$. As the collapse starts, the bubbles release radially propagating rarefaction waves, which later interact with each other and the rigid wall. When the rarefaction impinges upon the bubble interface, a rarefaction is transmitted into the bubble, while a compression wave is reflected back due to the impedance mismatch. These continual interactions result in a zigzag wave pattern in the domain, which further accelerates the bubbles' interface non-uniformly ($t = 0.11$). The formation of a re-entrant jet towards the wall when a single bubble collapses near a rigid wall is thoroughly discussed in literature. A re-entrant jet is also observed in the collapse of bubble pairs in a free field, if occurring in-phase, and with bubbles relatively similar in size; in such a case, the jets accelerate towards each other. In this problem, unlike single bubble collapse near a rigid wall or the collapse of bubble pairs in a free field, the jet is directed neither towards the wall nor the other bubble, but in-between. Here, the jet forms within the secondary bubble first ($t = 1.12$). Upon the impact of the re-entrant jet onto the distal side of the secondary bubble, a water-hammer shock is generated, which, combined with the shock wave from the bubble implosion, creates a high-pressure region. The bubble then takes the form of a vortex ring migrating in the direction of the jet angle. The re-entrant jet is observed to form within the primary bubble, tilted slightly away from the wall and towards the secondary bubble. Ultimately, the jet hits the distal side of the primary bubble, and creates a water-hammer shock. At this time, the shock from the collapse of the secondary bubble has reached the primary bubble and compresses it even more, resulting in a stronger collapse of the primary bubble, and thus producing high pressure regions ($t = 1.25$). Upon the impact of the shock emitted from the collapse of the secondary bubble, followed by the shock from the primary bubble collapse, high pressures are measured along the wall. The shock waves reflect back and eventually interact with the vortex rings again ($t = 1.41$). Although the major events during the collapse (e.g., jet formation, shock propagation, and vortex ring migration) are similar in single-bubble and twin-bubble problems, adding an extra bubble to the flow creates a far more complicated dynamics.

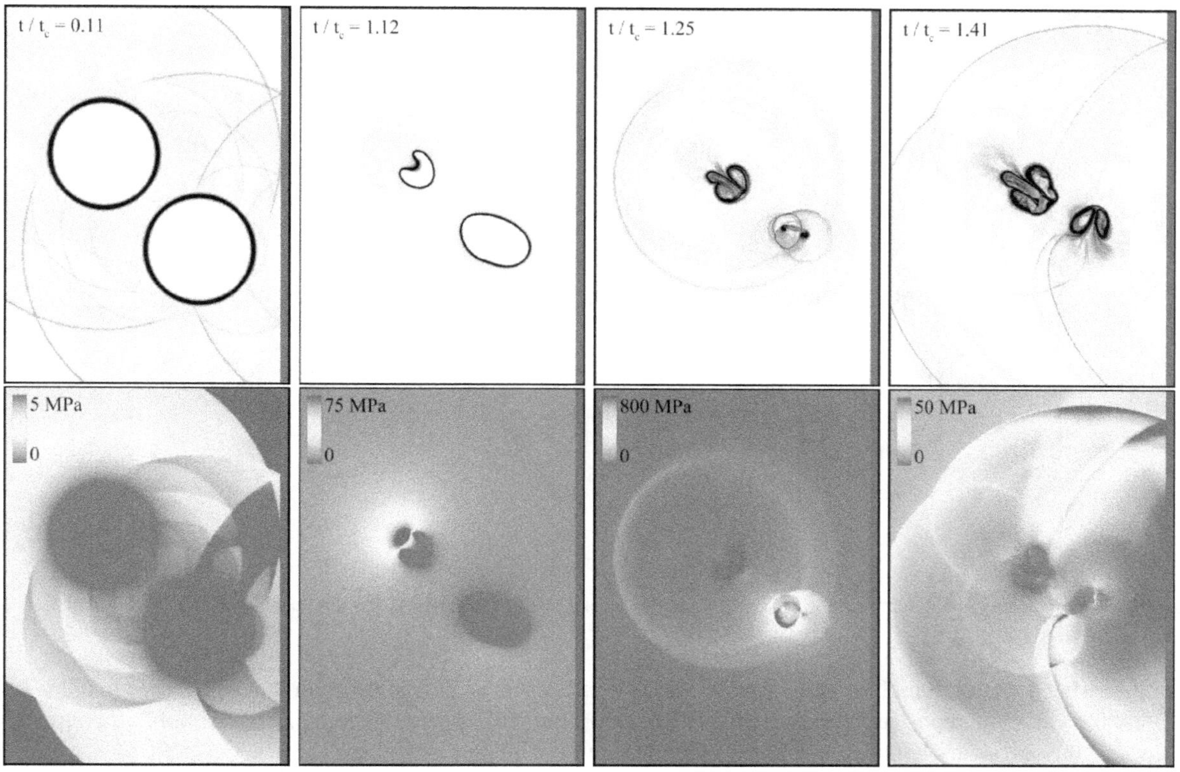

Figure 2: Collapse of a bubble pair near a rigid surface with $\delta_o = 1.5$, $\gamma_o = 2.5$, $\phi = 45^o$, and $p_\infty = 5$ MPa; top: density gradient magnitude; bottom: pressure contours.

The collapse of a bubble pair near a rigid surface combines two problems in which re-entrant jets form: collapse of a bubble next to another bubble and collapse of a bubble near a wall. Our numerical simulations show that the jets are formed within both primary and secondary bubbles, though with distorted shapes and geometry-dependent angles. According to the simulations, the secondary bubble collapses prior to the primary bubble for all configurations, except for $\phi = 90^o$, in which case the two bubbles collapse at the same time. Although the two mechanisms inducing jet formation promote different angles, they are both on the right side of the secondary bubble. Thus, it is anticipated that, the jet points towards the bottom-right and that the vortex ring convects in that same direction. Figure 3 (top) shows the jet morphology within the secondary bubble immediately before impact at the top. The jet points in the direction perpendicular to the wall ($\alpha_2 = 0$) at $\phi = 0^o$, where the secondary bubble is located directly behind the primary bubble. By increasing the angle ϕ, the jet of the secondary bubble starts to turn away from the horizontal, and reaches the maximum absolute value of $\alpha_2 = 49^o$, at $\phi = 90^o$.

Unlike the secondary bubble, the mechanisms leading to jet formation (presence of another bubble and of a wall) lie on opposite sides of the primary bubble, which can lead to a more non-spherical behavior and eventually complicate the dynamics further. Figure 3 (bottom) qualitatively illustrates the primary bubble shape, and the jet(s) formation, just before the impact. According to the simulations, two re-entrant jets are observed to form within the primary bubble at $\phi = 0^o$; one with angle 0°, and the other one with angle 180° in the exact opposite direction. At $\phi = 45^o$, "jet 1" is about to hit the distal side, while "jet 2" is still developing. However, for $\phi = 90^o$, only one single jet at angle $\phi = 49^o$ is observed. By increasing ϕ from 0° to 90°, the jets start deforming, and ultimately at $\phi = 56^o$, double-jetting no longer occurs and a single jet is observed. The emergence of double jetting affects the dynamics, the generated shock waves, and pressure loadings produced at the collapse; thus, the role of initial configuration (e.g., the arrangement of the bubbles and their proximity to the rigid wall) on the collapse non-sphericity and overall dynamics is non-negligible.

The wall pressure p_{mw} is a quantity of interest for erosion [10,11]. Figure 4 shows how this quantity depends on the angle. The maximum wall pressure behaves in a non-monotonic fashion: it initially increases and reaches a local maximum as the angle is increased; thereafter, it starts to decrease to a minimum value at $\phi = 75^o$, and increases

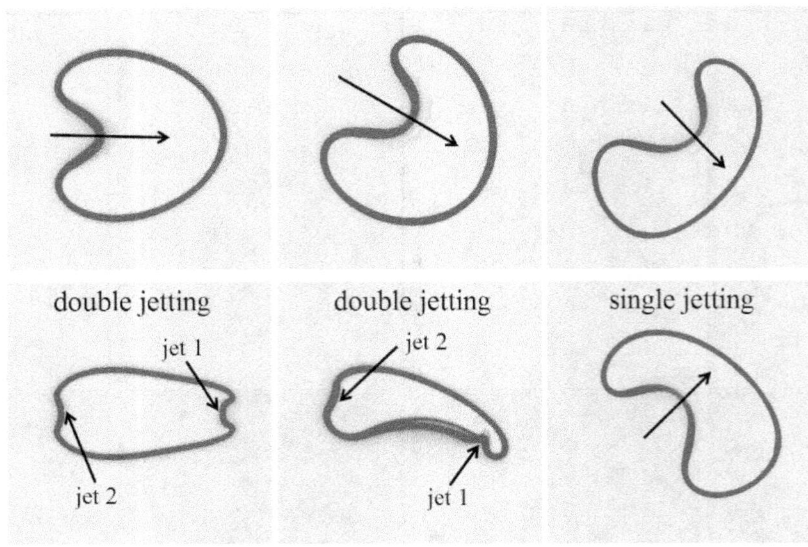

Figure 3: Jet morphology at different angles $\phi = 0^o, 45^o, 90^o$ for secondary (top) and primary (bottom) bubbles with $\delta_o = 1.5$, and $\gamma_o = 2.5$.

again thereafter to reach it highest value at $\phi = 90^o$. For all cases, p_{mw} is lower than that of a single bubble, except for $\phi = 90^o$ where the reported value is slightly higher than the pressure produced by a single bubble. In general, the presence of the secondary bubble lowers the maximum pressure along the wall surface.

Conclusion

In the collapse of bubble clouds, the inter-bubble interactions result in a complex non-spherical behavior and affect the collapse properties further. To quantify the effects of these interactions on bubble dynamics, we simulated the collapse of a pair of bubbles near a rigid wall. We showed that if a second bubble is placed in the vicinity of the original bubble, the collapse becomes far more complicated. In fact, strong bubble-bubble interactions dramatically increase the non-sphericity of the bubbles, compared to the single bubble collapse. These interactions may amplify or reduce the pressures produced at the collapse, depending on the initial configuration of the problem. We showed that the maximum pressure along the wall behaves non-monotonically in the presence of the secondary bubble with varying angle. Distorted re-entrant jets were observed to form within both bubbles, and it was shown that depending on the geometrical configuration, two jets were detected penetrating the primary bubble, an event we call "double jetting". Unlike single bubble case, the dynamics depend on at least three additional parameters (inter-bubble distance, relative size and angle), in addition to the stand-off distance and driving pressure. The results show that the dependence on even one of these parameters is complicated; determining the dependence on all parameters will be challenging. From this observation, it follows that the value of such high-resolution simulations to understand or predict the collapse of many bubbles (clouds) is questionable because there are too many parameters. As a result, modeling the behavior of clouds of bubbles in the context of erosion may require alternative approaches.

Acknowledgement

This work was supported by ONR grant N00014-12-1-0751 under Dr. Ki-Han Kim and used resources from the Blue Waters, supported by the National Science Foundation (awards OCI-0725070 and ACI-1238993) and the state of Illinois.

References

[1] Kubota, A., and Kato, H., Yamaguchi, H., and Maeda, M. (1989). *Unsteady structure measurement of cloud cavitation on a foil section using conditional sampling technique*. J. Fluids Eng., **111**, 204-210.

[2] Ceccio, S. L., and Brennen, C. E., (1991). *Observations of the dynamics and acoustics of travelling bubble cavitation*. J. Fluid Mech., **233**, 633-660.

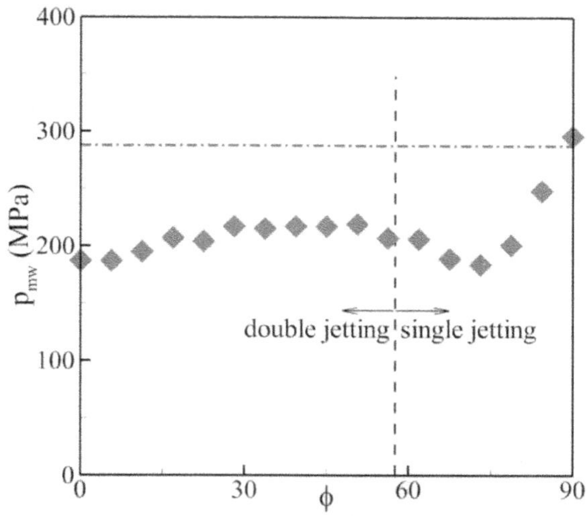

Figure 4: Maximum wall pressure as a function of the initial angle ϕ, with $\delta_o = 1.5$, $\gamma_o = 2.5$, and $p_\infty = 5$ MPa; red dashed-dotted line represents the collapse time in single bubble case.

[3] Hansson, I., Kedrinskii, V., and Morch, K. A. (1982). *On the dynamics of cavity clusters*. J. Phys. D: Appl. Phys., **15**, 1725-1734.

[4] Bremond, N., and Arora, M., and Ohl, C. D., and Lohse, D. (2006). *Controlled multibubble surface cavitation*. Phys. Rev. Lett., **96**, 224501.

[5] Tiwari, A., Pantano, C., and Freund, J. B. (2015). *Growth-and-collapse dynamics of small bubble clusters near a wall*. J. Fluid Mech., **775**, 123.

[6] Van Wijngaarden, L. (1968). *On the equations of motion for mixtures of liquid and gas bubbles*. J. Fluid Mech., **33**, 465474.

[7] Fuster, D., and Colonius, T. (2011). *Modelling bubble clusters in compressible liquids*. J. Fluid Mech., **688**, 352389.

[8] Beig, S. A., and Johnsen, E. (2015). *Maintaining interface equilibrium conditions in compressible multiphase flows using interface capturing*. J. Comput. Phys., **302**, 548566.

[9] Beig, S. A. (2018). *A computational study of the inertial collapse of gas bubbles near a rigid surface*. PhD thesis, University of Michigan.

[10] Philipp, A., and Lauterborn, W. (1998). *Cavitation erosion by single laser-produced bubbles*. J. Fluid Mech., **361**, 75116.

[11] Johnsen, E., and Colonius, T. (2009). *Numerical simulations of non-spherical bubble collapse*. J. Fluid Mech., **629**, 231264.

© 2018 ASME

Cavitating Flow in a Model Diesel Injector Return Valve

[1]Alberto Bonifacio; [1]Russel Lockett*, [2]Richard Price

[1]Department of Mechanical Engineering & Aeronautics, City, University of London, London, UK

[2]Shell Global Solutions, London, UK

Abstract

In modern magnetic solenoid actuated diesel injection equipment, the injection cycle process is regulated by the action of the ball spill valve. This connects the high-pressure region of the needle valve and the low-pressure region of the fuel tank. The precision in timing and hydrodynamic forces determines the amount injected and the amount of the fuel returned back for the following cycle. The sequence of inlet and outlet throttle, through which the high pressure diesel flows whilst the solenoid is actuated, are expected to enhance cavitation due to their microscopic size. In the long-term this may cause the accumulation of damaging internal deposits. Two acrylic models of the spill valve reproducing the geometrical features with different outlet throttle diameters were constructed to allow an optical access of three different fuels – a paraffinic rich model diesel, a 95% - 5% hexadecane-octane mixture and a 80% - 20% hexadecane-octane mixture. Variation of the upstream pressure up to 40 bar and downstream pressures up to 10 bar and manual control of the ball valve lift height were the key elements controlling and producing the flow conditions. Experimental results show the pronounced effect of the upstream pressure on the characterisation of cavitation inception occurring at the throttle entrance. It was concluded that there is a linear relationship between upstream and downstream pressures that control cavitation inception. Moreover the ball valve height plays a significant role in cavitation inception at the beginning of the ball valve lift, where higher upstream-to-downstream pressure ratios were necessary to initiate cavitation. The highest-pressure ratio was obtained with the highest fuel viscosity and largest outlet nozzle diameter, and it consistently decreased as the viscosity and diameter decreased. No cavitation was observed in the inlet throttle passage, but the change in pressure ratio between the two acrylic blocks suggests a change in pressure in the intermediate region; thus it is possible to control the cavitation occurring in the region above the injector needle in spill valves. A high-speed, high-resolution imaging system was utilised to acquire images for inception, partially and fully cavitated flows. The results obtained are presented.

Keywords: spill valve; injector; cavitation

Introduction

A significant body of research involving the chemistry underlying the formation of soot in diesel pumps and injectors, has been published. The presence and significance of cavitation has been noted in various experiments, which may have been responsible for the alteration of diesel fuel properties during operation [1]. The high diesel fuel pressures occurring in injectors (up to 2000 bar), along with the small passages, cause an increase in the dynamic pressure, large pressure gradients and local decreases in static pressure. Locally, the static pressure may fall below the saturated vapour pressure, resulting in local boiling (hydrodynamic cavitation). As the pressure recovers, the previously generated vapour pockets collapse due to the work done by the surrounding liquid, thereby generating large vapour pressures and temperatures during collapse. The collapse of the bubbles, cloud or sheet generates a series of reactions that reflect pyrolytic degradation and dehydrogenation; eventually small particles are formed, that are carbonaceous in structure, and have a mixed amorphous and graphitic-like carbon composition [2].

Ultrasonic cavitation has proved to be resourceful in understanding the correlation between cavitation and soot formation. Suslick *et al.* [3] demonstrated that *n*-alkanes can be broken down to lower alkanes and alkenes by sonication with similar characteristics to high-temperature pyrolysis. Price and McCollom followed this with an assessment of testing diesel fuel stability using ultrasonic excitation [4, 5].

Apart from the metered injection side of a diesel injector (where forms of cavitation were identified in the recirculation zones of the injector [6]), there are other areas throughout the common rail mechanism that present similar flow characteristics. One of these is the metered side of the high pressure fuel that is returned to the low pressure fuel tank. Cavitation may occur across both return valves in injectors and return valves in high-pressure pumps.

An acrylic model of a modern diesel injector return valve was designed and manufactured in order to study the occurrence of cavitation through the low pressure side at low upstream pressures for different fuels.

*Corresponding Author, Russel Lockett: R.D.Lockett@city.ac.uk

Experimental setup

A customized mechanical flow rig developed by Lockett *et al.* [7] was employed to investigate the cavitating flow occurring inside an acrylic replica model diesel injector return valve. The rig and model valve assembly is shown in Figure 1. Initially, the high pressure nitrogen cylinder was employed to fill the three high pressure cylinders (10.5l) with the fuel sample to be tested – as shown with the red arrows. Once the fuel cylinders were filled and the storage tank had been depressurised, the nitrogen cylinder was employed to pressurise the fuel in the three bottles to the required pressure for the experiment. Hence, the blue arrows in the diagram indicate the higher pressure line directions all the way to the acrylic model, where the flow cavitated as it passed through the microscopic channels of the return valve, with the downstream pressure regulated manually with a needle valve. Pressure gauges were located upstream and downstream of the acrylic model in order to control the pressure ratio occurring before and after the nozzles comprising the return valve. The upstream pressure gauge was set by manually adjusting the nitrogen bottle pressure regulator supplying nitrogen to the fuel bottles, while the downstream pressure gauge was regulated manually to a maximum pressure of 11.0 bar. The advantage of this arrangement was that it enabled manual control of the pressure difference established between the inlet and the outlet of the model return valve, hence permitting the alteration of the location and the intensity of the cavitating flow occurring across the valve.

A scaled side and isometric view of the acrylic model and housing is shown in Figure 2. The dimensions of the acrylic block were 400mm x 400mm x 300mm. The side passage represented the high pressure inlet of the fuel flow, with a Ø10mm bore, the lower channel represented the intermediate valve control chamber that was sealed to avoid leaking, and the upper passage represented the outlet of the fuel flow, with a Ø20mm bore. The functionality of the return valve was dependent on the micro passages indicated in Figure 2, comprised of the inlet throttle on the side connecting the inlet passage to the valve control chamber, and the stepped outlet throttle connecting the valve control chamber to the outlet passage. The inlet throttle was manufactured with Ø210μm, while the outlet throttle with a Ø225μm and a step of Ø450μm. The two throttles make it possible to separate the fuel lines reaching the nozzle holes at high pressure and the fuel lines returning to the fuel tank at low pressure.

Experimental Methodology

A stainless steel needle sealed the outlet throttle to allow the fuel entering to gather in the valve control chamber after passing through the inlet throttle. A rotation of the valve control elevated the stainless steel needle (shown in Figure 2); every fifth of a rotation corresponded to a 50μm needle lift. Hence, the needle is raised from its seal position to allow the fuel to flow through the outlet passage.

The upstream pressure was set by the nitrogen bottle regulator valve, beginning from as low as 10bar absolute pressure, and increased in 5 bar steps. Ball valves attached to the high pressure pipe were employed to control the fuel flow to the model return valve assembly. Once the pressurised fuel entered the acrylic model through the inlet passage, it filled the passages of the model return valve all the way to the sealed needle at the outlet throttle. When the needle was lifted (needle lift settings were 50μm and 100μm), the pressurised fuel escaped through the outlet passage and returned to the ambient pressure fuel storage tank, via the downstream pressure gauge and needle valve.

A Photron APX-RS video camera operating at 15 kHz and 3.64x optical magnification captured white light scattering obtained from the fuel flow passing through the acrylic throttle passages. This camera-lens-model arrangement achieved a spatial resolution of 5.5μm/pixel. As stated earlier, the downstream fuel pressure was adjusted to determine the onset of cavitation (inception), and the intensity of the nozzle cavitation through control of the nozzle flow velocity.

The experiment was conducted using three different fuels, a paraffinic rich model diesel (distillation profile and saturated vapour pressure similar to diesel), and mixtures of n-hexadecane and n-octane in 95:5 v/v and 80:20 v/v proportions. For each fuel and upstream pressure setting, 25 separate readings of downstream pressures for cavitation inception were obtained for a fixed needle lift. The incipient cavitation pressure ratio was recorded *'from below'* meaning that the downstream pressure was increased until the cavitation disappeared, or *'from above'* meaning that the downstream pressure was reduced until the cavitation appeared. The mean and standard deviation for each set of measurements was then determined. Table 1 summarises the properties of the different fuels.

Results & Discussion

The results showed that the fuel flow would always cavitate in the same position (at the entrance to the outlet throttle). The inlet throttle on the other hand never presented any form of cavitation, regardless of the pressure ratios, fuel types or acrylic model.

Figure 3 shows single-shot images obtained of cavitation inception (bottom of image) and fully developed cavitating flow for the paraffinic rich model diesel at 40 bar upstream and 50 μm needle lift taken at 15,000 frames/s and an exposure of 10 μs. The cavitating flow shown in

Figure 3 is characterized by regions of white light scattering obtained from discontinuous liquid/vapour surfaces formed in the multi-phase flow. The pressure measurements 'from below' were found to be more repeatable compared to the ones measured 'from above' due to flow-induced hysteresis, i.e. less stable cavitation inception when reducing the downstream pressure 'from above'. This effect was found to produce larger standard deviations in the measurement of critical downstream pressure for lower upstream pressures.

Readings for downstream pressure, i.e. pressure at the exit of the outlet throttle, were taken for a set of upstream pressures ranging from 10 to 40 bar – due to equipment limitation – for both 50 and 100μm needle lift setting, using the three different fuel types. Plots are shown in Figure 4, where the needle lift in the return valve is seen to have a significant effect on the pressure ratio producing cavitation inception. Comparing the two plots in Figure 4, the pressure ratio $\frac{P_u}{P_d}$ is larger for the 50μm lift setting, reducing for 100μm, and then remaining invariable thereafter. This phenomenon is consistent for the different fuels, meaning that the lower the needle lift, the larger is the pressure ratio required to produce cavitation inception. This occurs due to the low needle lift producing a pressure gradient across the annulus formed between the needle and the diverging conical passage, thereby reducing the pressure gradient forming across the nozzle holes.

A noteworthy finding is obtained from the variation of fuel type. From Table 1, it is possible to determine the cavitation number for the different fuels, using the following definition:[8]

$$C_N = \frac{P_u - P_d}{P_d - P_v} \qquad \text{Equation 1}$$

where P_u is the upstream pressure, P_d is the downstream pressure and P_v is the vapour pressure. Cavitation inception obtained from the paraffinic rich model diesel occurs at the largest cavitation number and lowest saturation vapour pressure when compared with the 2 hexadecane-octane mixtures (as shown in Figure 5), therefore indicating a reduced inclination to cavitate compared to mixture 2 (95:5 v/v C_{16}:C_8), which in turn shows a reduced inclination to cavitate compared with mixture 1 (80:20 v/v C_{16}:C_8), which exhibits cavitation inception at lower critical cavitation number and the largest saturated vapour pressure.

Figures & Tables:

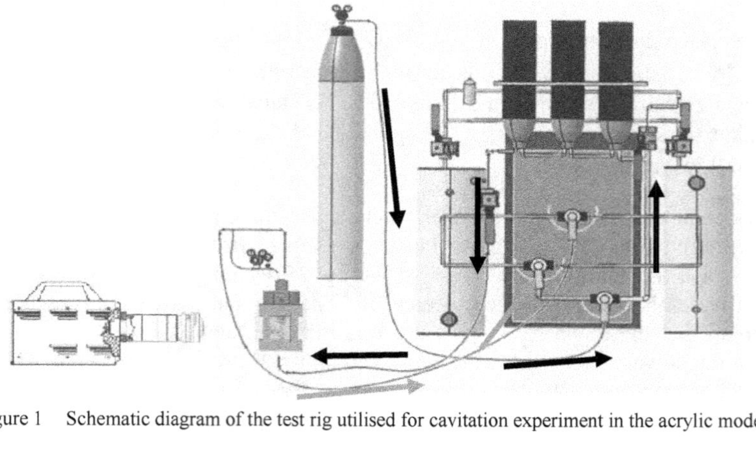

Figure 1 Schematic diagram of the test rig utilised for cavitation experiment in the acrylic model

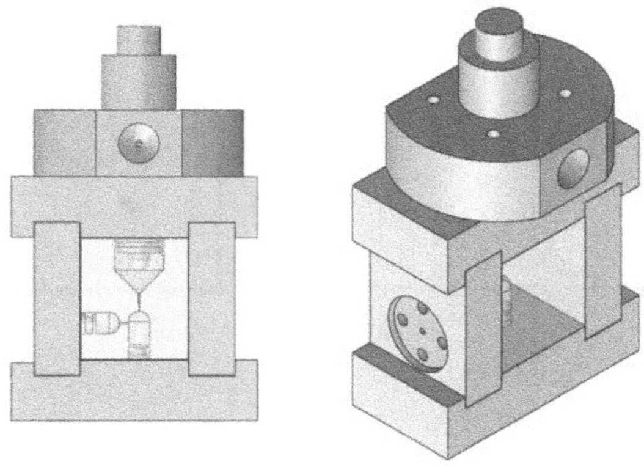

Figure 2 Front and isometric view of the acrylic model return valve assembly

Fuel	Saturated Vapour Pressure (Pa)	Viscosity (Ns/m^2)
Paraffinic Model Diesel	~ 100	0.00153
Mixture 1 (80% C_{16}, 20% C_8)	470	0.0027
Mixture 2 (95% C_{16}, 5% C_8)	150	0.00323

Table 1 Fuel properties

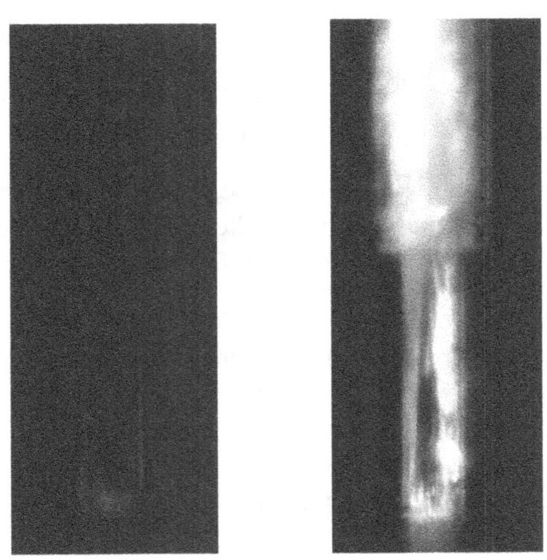

Figure 3 Incipient (left) and fully cavitated flow (right) through outlet throttle with GTL, 40bar and 50μm needle lift

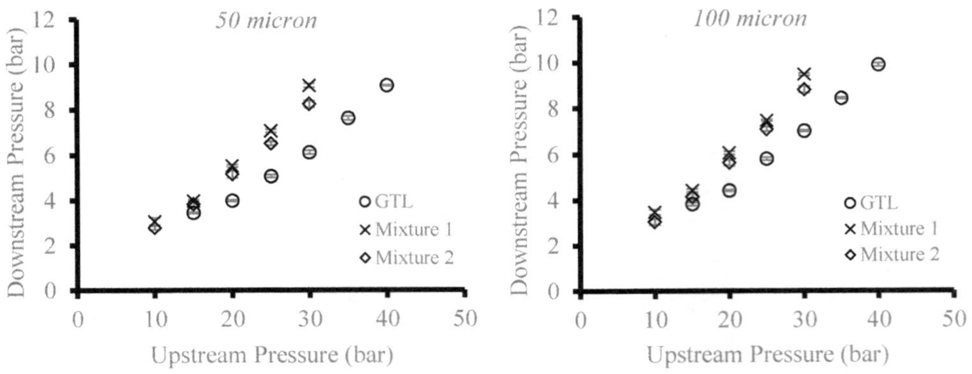

Figure 4 Upstream/Downstream pressure ratios for 50μm and 100μm needle lift

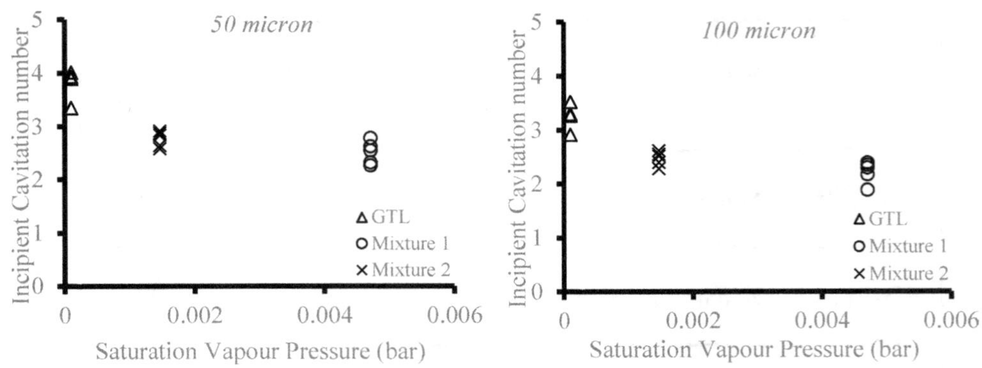

Figure 5 Incipient cavitation number for 50μm and 100μm needle lift

Conclusions

The acrylic model diesel return valve assembly enabled the determination of the flow characteristics arising from the flow developing in the pressure control system. The conclusions may be summarised as follows:

1. Cavitating flow was observed to begin at the entrance to the low pressure throttle, independently of fuel type and cavitation number (above the critical cavitation number for cavitation inception).

2. Cavitation inception was observed to depend on fuel type, cavitation number, saturated vapour pressure, and needle lift. Cavitation inception in the paraffinic model diesel occurred at the largest cavitation numbers and smallest saturated vapour pressures, as expected. This was followed by mixture 2, and then followed in turn by mixture 1, again as expected.

3. Consistently larger pressure ratios were required to produce cavitation inception for lower needle lift. This occurs because the low needle lift produces a pressure gradient in the annulus around the needle, thereby reducing the pressure gradient developing across the nozzles. Hence a larger pressure difference across the nozzles is required to produce cavitation inception. This effect is significant for the cavitating flow developing in real diesel injector return valves during needle actuation.

References

[1]Lockett, R.D. and Jeshani, M. (2013). *An experimental investigation into the effect of hydrodynamic cavitation on diesel.* International Journal of Engine Research. 14(6).

[2]Price, R.J. et al. (2015). *Understanding the impact of cavitation on hydrocarbons in the middle distillate range*. Fuel. 156.

[3]Suslick, KS, Gawienowski, JJ, Schubert, PF, Wang, HH. (1983). *Alkane sonochemistry*. Journal of Physical Chemistry. 87(13).

[4]Price, G.J. and McCollom, M. (1995). *The effect of high-intensity ultrasound on diesel fuels*. Ultrasonics sonochemistry. 2(2).

[5]Price, G.J. and Smith, P.F. (1993). *Ultrasonic degradation of polymer solutions: 2. The effect of temperature, ultrasound intensity and dissolved gases on polystyrene in toluene*. Polymer. 34(19).

[6]Andriotis, A. et al. (2008). *Vortex flow and cavitation in diesel injector nozzles*. Journal of Fluid Mechanics. 610.

[7]Lockett, R. et al. (2015). *Cavitation inception in immersed jet shear flows.* Journal of Physics: Conference Series. 656.

[8]Arcoumanis, C. et al. (1999). *Investigation of cavitation in a vertical multi-hole injector*. SAE Technical Papers. 108(3).

© 2018 ASME

10th International Symposium on Cavitation - CAV2018
Baltimore, Maryland, USA, May 14 – 16, 2018

CAV18-05033

Cavitating Flows of Varying Temperature Liquid Nitrogen in Converging-diverging Nozzle

Tairan Chen, Wendong Liang, Mindi Zhang, Biao Huang, Guoyu Wang*

School of Mechanical Engineering, Beijing Institute of Technology, Beijing, RP China

Abstracts

The objective of this paper is to investigate the unsteady cavitating flows of varying temperature liquid nitrogen. Liquid nitrogen cavitating flows in a converging-diverging (C-D) nozzle are carried out with the temperature range from 68 K to 86 K, the pressure in the tanks is within the range of 0.03 MPa~0.30 MPa. The experimental results show that there are two typical cavitation dynamics in varying temperature liquid nitrogen, namely quasi-isothermal mode and thermo-sensitive mode. As the free-stream temperature increases, the cavity area increases to its maximum around the transition temperature and then significantly decreases. Further analysis indicates that the liquid/vapor density ratio D dominates the change of the cavitation dynamics when the temperature is below the transition temperature. The cavity tends to be mushier (smaller bubbles) and longer with the increasing temperature. However, the thermodynamic effects on cavitation, which could suppress the development of the cavitating flow, dominate the change of the cavitation dynamics when the temperature is above the transition temperature. The transition temperature of cavitating flows in liquid nitrogen is between 76 K~78 K, and the transition temperature slightly increases with the increasing cavitation number.

Key words: Cavitating flows; Liquid nitrogen; Cavitation dynamics; Transition temperature

Introduction

Cavitation is a phase-change phenomenon that occurs in the liquid when the local static pressure drops below the saturated vapor pressure, this phenomenon in room temperature water is usually assumed to be an isothermal process [1,2]. However, the isothermal hypothesis is no longer valid for thermo-fluids. The temperature-dependent physical properties of thermo-fluids are thermosensitive. Furthermore, the significant temperature drop ΔT during the cavitation process changes the local physical properties, and then cavitation dynamics is affected [3,4]. Cavitation with thermodynamic effects was firstly investigated by Stahl and Stepanoff [5]. They conducted a B-factor method to estimate the thermodynamic effects during phase-change process based on a simple heat balance between these two phases. Holl et al. [6] conducted comprehensive experiments of cavitation in water and Freon 113 under different free-stream conditions with four test models. They developed a semiempirical entrainment theory to correlate the measured temperature drop ΔT. Brennen [1] developed another appropriate thermodynamic parameter named Σ by incorporating the thermodynamic effect into the Rayleigh-Plesset equation for cavitation bubble dynamics. The above parameters propose a criterion to evaluate the strength of thermodynamic effects.

To directly investigate cavitation dynamics of cavitating flows in thermo-fluids, Hord [7,8] measured the temperature and pressure data around the cavity region in liquid nitrogen and liquid hydrogen under a lot of free-stream conditions, which is the most comprehensive experimental data so far. However, it lacks of unsteady

* Corresponding Author, Guoyu Wang: wangguoyu@bit.edu.cn

visualization and data because of the technological limitation. Recently, Franc et al. [3] investigated the thermodynamic effects on a cavitating inducer in R-114, they found that the development of blade cavitation was suppressed at higher reference temperature due to thermodynamic effects. Cervone et al. [9] carried out experiments in varying temperature water around a NACA 0015 hydrofoil to investigate the thermodynamic effects on cavitation dynamics. They found that the cavity length increased with the increasing free-stream temperature. Ohira et al. [10] experimentally investigated the cavitation instability of subcooled liquid nitrogen in two types of converging-diverging circular nozzle. As the temperature decreases, the cavitation mode changed from continuous mode to intermittent mode, and the intermediate mode was approximately around 76 K. Kelly and Segal [11] investigated the cavitating flows of fluoroketone in a wide range of velocities, cavitation numbers, and temperatures. They pointed out the transition from classical to thermo-sensitive cavitation for fluoroketone taken place around T_∞=323 K. Zhu et al. [12] experimentally investigated the dynamic characteristics of liquid nitrogen cavitating flow through a venturi tube. The spatial changes of the temperature, pressure as well as the flow rate, together with the visual observation of unsteady cavitation behaviors were presented.

Although the cavitation dynamics in thermos-fluids have been deeply investigated in the past decades, the transition from quasi-isothermal to thermo-sensitive cavitation and the specific thermodynamic effects on cavitation dynamics in a wide range of temperatures are still not well understood. The objectives of this work are to present the change of cavitation dynamics in varying temperature liquid nitrogen.

Experimental apparatus and conditions

The experimental apparatus for liquid nitrogen cavitating flows in a converging-diverging (C-D) nozzle is shown in Figure 1(a). The apparatus essentially consisted of a run tank (capacity 40 L), the test nozzle and a catch tank (capacity 60 L). All of them are thermally protected with vacuum insulation. The temperature of liquid nitrogen is increased by increasing the pressure in both tanks by closing all the valves. And it is reduced from the atmospheric boiling point by simultaneously reducing the pressure in both tanks by using vacuum pumps. The pressure in both tanks can also be quickly increased by charging high-purity nitrogen gas. At the beginning of the experiment, the liquid nitrogen is flowed back and forth several times between these two tanks to ensure that the temperatures of the liquid nitrogen and the apparatus are steady. In the experiments, after pressurizing the liquid nitrogen in the run tank with high-purity nitrogen gas, the control valve is quickly opened to allow the liquid nitrogen to flow through the turbine flow meter and the test section to the catch tank. The test section is equipped with an assembled converging-diverging (C-D) square nozzle with a stainless steel body, Teflon bottom, and quartz glass side windows. As shown in Figure 1(b), the test section is a C-D nozzle with an overall length of 258 mm, the inner width of 5 mm, inner heights of 20 mm on both upstream side and downstream side. The actual height of throat is 1.58 mm (the design value is 1.5 mm), which is measured by a series of needle gauges at room temperature. Considering the thermal expansion at 77 K (the thermal expansion coefficients of stainless steel and Teflon are $1.7\times10^{\wedge-5}$ (m/m K) and $8.6\times10^{\wedge-5}$ (m/m K), respectively), the change difference due to thermal expansion is 0.39 mm. And hence, the actual throat height at 77 K is approximately 1.97 mm. The key area of the C-D nozzle is equipped with visualization windows of quartz glass with the length of 70 mm and height of 7 mm. As previously indicated, the temperatures of the liquid nitrogen in the upstream and downstream of the C-D nozzle, the run tank and the catch tank are measured by Cernox thermometers (Lake Shore Cryotronics, Inc., accuracy ±16 mK at 77 K). The pressures in both tanks, upstream and downstream, as well as the diverging section, of the C-D nozzle are measured by absolute pressure sensors (Helm Instrument Co., Inc., range 0~0.6Mpa, accuracy ±0.1% FS). A turbine flowmeter (Hoffer flow controls, Inc., range 1.3~13.2 L/min, accuracy ±0.25%) is fixed at the upstream of C-D nozzle, to measure the volumetric flow rate. At the same time, cavitating flows are captured for total 1 s via a high-speed video camera (Redlake Co., Inc.) with the resolution of 496×128 pixels at a frame rate of 12000 fps. The spatial resolution of the images is 0.0625 mm/pixel. To sufficiently illuminate the cavitating flow, a LED lamp (Larson Electronics LLC, 160 W-21,600 Lumens) is used and positioned at the back of text section. Experiments are carried out with varying temperature liquid nitrogen from 68 K to 86 K, the pressure in the tanks is set within the range of 0.03 MPa ~0.30 MPa. Assuming the liquid nitrogen is an incompressible fluid, throat velocity (U_{throat}) is calculated by the volumetric flow rate based on the continuity equation. Considering the choke effect of the cavity near the throat, the cross-section area for liquid nitrogen is estimated by the visualization.

(a) Experimental apparatus (b) Test section (C-D nozzle)

Figure 1 Experimental apparatus and test section for cavitating flows of liquid nitrogen

Results and discussion

The cavitation number *σ*, which is defined in Eq. (1), is used to characterize the state of the cavitating flow.

$$\sigma = \frac{p_{up} - p_v(T_{throat})}{p_{up} - p_{down}} \tag{1}$$

Here, p_{up} is the upstream static pressure of the C-D nozzle, p_{down} is downstream static pressure, and p_v is the saturated vapor pressure at throat temperature T_{throat}. The average of the upstream temperature T_{up} and downstream temperature T_{down} are used to indicate the throat temperature (T_{throat}) of liquid nitrogen.

An example of typical cavitating flow observed in the experiments is presented in Figure 2, which includes the original visualization, the schematic interpretations drawn and the processed cavity area via an in-house feature-recognition code [13]. For every moment, the original image is firstly changed into the grayscale image. Next, the grayscale value of each pixel is changed to be 255 (white) if its original grayscale value is above 100 (clean the background). And then, the wall in the image is cut by using a non-cavitation image. Finally, the cavity area is calculated by counting the numbers of the pixel whose grayscale value is not 255. The time evolution of the cavity area over a period of 1 s (12000 images) is processed for each test condition. And then, the cavity scale, the development of cavitation and the frequency analysis could be quantitatively characterized.

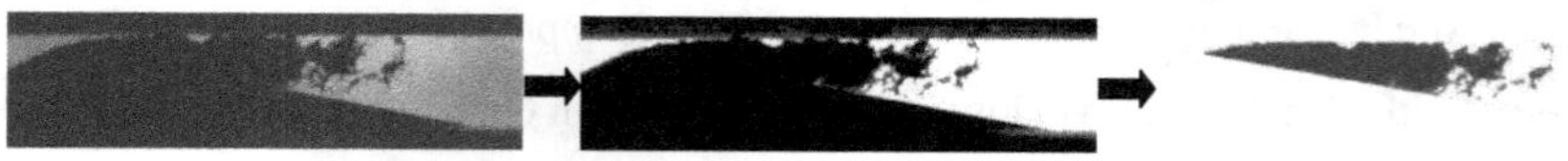

Figure 2 Typical flow visualization and schematic interpretation of cavity area (black shadow is the cavity)

(σ=1.41, T_{throat}=73.06 K, U_{throat}≈25.8 m/s)

Figure 3(a) illustrates the relationships among time-averaged cavity area, cavitation number and temperature in the throat. The cavitation conditions are indicated by solid dots (•) and the non-cavitation conditions are indicated by stars (*). The contour is drawn by the experimental data of cavitation conditions. At any given temperature, the non-cavitation flow changes to cavitating flow as the cavitation number is reduced. The blue boundary approximately indicates the incipient cavitation number versus temperature, and the incipient cavitation number decreases with the increasing temperature because of the increasing saturated vapor pressure. The lower red boundary indicates that the visualization window has been fully filled with cavity. It is worth noting that the cavity area increases and then decreases with the increasing temperature under the same cavitation number. The cavity area increases to its maximum value when the temperature is around 78 K. This behavior is quite different from the cavitation behaviors with strong thermodynamic effects in liquid nitrogen we used to know, while it is similar to the change of cavitation dynamics in varying temperature water and fluoroketone [4,9,11,14]. The change tendency in Figure 3(a) is not convincing because the throat velocities of these test conditions are different. And hence, cases with similar throat velocity (U_{throat}≈23.1 m/s) are selected and illustrated in Figure 3(b). It is found the change tendency is almost the same as that in Figure 3(a). As the temperature increases, the cavity area increases to its maximum at the transition temperature (76 K~78 K) and then decreases under the same cavitation number. It is also found that the transition temperature slightly increases with the increasing cavitation number.

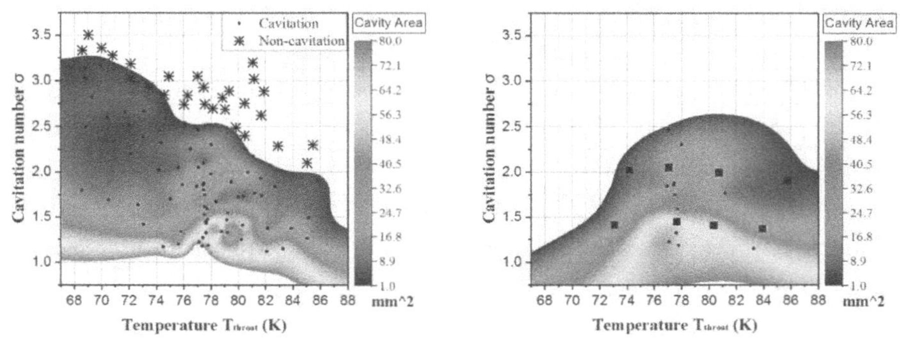

(a) All test conditions (b) Throat velocity U_{throat}≈23.1 m/s

Figure 3 Distributions of time-averaged cavity area versus throat temperature and cavitation number

To further show the behavior change of cavitating flow in varying temperature liquid nitrogen. Typical of attached and shedding cavity along the diverging section for these eight cases (black square ■ in Figure 3(b)) under the same cavitation numbers and similar throat velocity are illustrated in Figure 4. It is shown the length of attached cavity for cavitation number σ=1.41 is approximatively 10 mm, while the length for cavitation number σ=1.98 is only approximatively 5 mm. Since the attached and shedding cavities appear as shadows, the darker is the color, the greater is the fraction of the vapor phase. It is found the liquid-vapor interface of shedding cavity is clear for T_{throat}=73.06 K and T_{throat}=74.16 K, where cavity is observed to disappear immediately after shedding. As the throat temperature increases, the shedding cavity turns to be mushier and frothier. Bubbles inside and around the shedding cavity turn to be smaller and collapse more slowly. It is because the liquid/vapor density ratio D is relatively large when the temperature is close to its triple point. Meanwhile, the local changed temperature due to thermodynamic effects is small. Assuming cavitation occurs in a finite volume, there is only one nucleation site in this finite volume, and all the liquid in the finite volume evaporates. The radius of a single spherical cavitation bubble can be conducted from the mass conservation, ($r^{3}4\pi/3=V_v=V_l\rho_l/\rho_v=V_lD$). For cavitating flow, large density ratio D means large radius bubbles forming the cavity, and hence, the interface of the cavity is clear As the throat temperature increases, bubbles inside the cavity turns to be smaller because of the reduced density ratio D, the smaller bubbles tend to move more quickly, separate and coalesce more easily, so that the cavity turns to be mushier and frothier as observed in Figure 4.

The increase of cavity area is because the thermodynamic effects on cavitation, which could suppress the development of cavitation, are relatively slight when the temperature is close to its triple point. Meanwhile, the smaller bubbles inside the cavity tend to collapse more slowly and move more quickly with the increasing temperature. However, the thermodynamic effects significantly enhance with the increasing temperature. The influence of density ratio D is less than the thermodynamic effects when the temperature is above the transition temperature. And then, the cavity area significantly decreases with the increasing temperature. The transition means the cavitation dynamics transits from the quasi-isothermal mode to the thermo-sensitive mode. It is found that the transition temperature slightly increases with the increasing cavitation number in Figure 3(b). This is because thermodynamic effects are reduced due to less mass transfer from liquid to vapor under higher cavitation number.

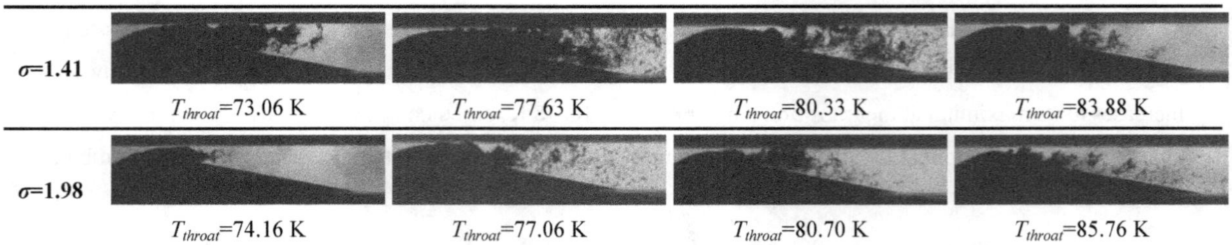

Figure 4 Cavitating flows of liquid nitrogen in converging-diverging nozzle under different flow conditions (U_{throat}≈23.1 m/s)

Conclusion

Unsteady cavitating flows of liquid nitrogen under a wide range of free-stream conditions were experimentally investigated. The cavity structure, pressure and temperature, as well as the volumetric flow rate are captured by a simultaneous sampling system. The experimental results show that there are two typical cavitation dynamics in varying temperature liquid nitrogen under similar flow conditions, namely quasi-isothermal mode and thermo-sensitive mode. As the free-stream temperature increases, the cavity area increases to its maximum around the transition temperature and then significantly decreases. Further analysis indicates that the liquid/vapor density ratio D dominates the change of the cavitation dynamics when temperature is below the transition temperature. The cavity tends to be mushier (smaller bubbles) and longer with the increasing temperature. However, the thermodynamic effects on cavitation, which could suppress the development of the cavitating flow, dominate the change of the cavitation dynamics when temperature is above the transition temperature. The transition temperature of cavitating flows in liquid nitrogen is between 76 K~78 K, and the transition temperature slightly increases with the increasing cavitation number.

Acknowledgment

This work was supported by the National Natural Science Foundation of China under Grant Nos. 51479002 and 91752105. The authors also gratefully acknowledge the support by the China Scholarship Council (No. 201706030018) and Graduate Technological Innovation Project of Beijing Institute of Technology (No. 2017CX10016).

References

[1] Brennen, C.E. (1995). *Cavitation and bubble dynamics*. Oxford University Press.

[2] Wang, G., Senocak, I., Shyy, W., Ikohagi, T., Cao, S. (2001). *Dynamics of attached turbulent cavitating flows*. Prog Aerosp Sci. 37(6).

[3] Franc, J.P., Rebattet, C., Coulon, A. (2004). *An experimental investigation of thermal effects in a cavitating inducer*. J.

Fluids Eng. 126(5).

[4] Chen, T., Huang, B., Wang, G. (2016). *Numerical study of cavitating flows in a wide range of water temperatures with special emphasis on two typical cavitation dynamics*. Int. J. Heat Mass Tran. 101.

[5] Stahl, H.A., Stepanoff, A.J. (1956). *Thermodynamic aspects of cavitation in centrifugal pumps*. J. Fluids Eng. 78.

[6] Holl, J.W., Billet, M.L., Weir, D.S. (1975). *Thermodynamic effects on developed cavitation*. J. Fluids Eng. 97(4).

[7] Hord, J. (1973a). *Cavitation in liquid cryogens II-hydrofoils*. NASA CR-2156.

[8] Hord, J. (1973b). *Cavitation in liquid cryogens III-Ogives*. NASA CR-2242.

[9] Cervone, A., Bramanti, C., Rapposelli, E., d'Agostino, L. (2006). *Thermal cavitation experiments on a NACA 0015 hydrofoil*. J. Fluids Eng. 128(2).

[10] Ohira, K., Nakayama, T., Nagai, T. (2012). *Cavitation flow instability of subcooled liquid nitrogen in converging-diverging nozzles*. Cryogenics. 52(1).

[11] Kelly, S., Segal, C. (2013). *Characteristics of Thermal Cavitation on a Two-Dimensional Hydrofoil*. J. Propul. Power. 29(2).

[12] Zhu, J., Xie, H., Feng, K., Zhang, X., Si, M. (2017). *Unsteady cavitation characteristics of liquid nitrogen flows through venturi tube*. Int. J. Heat Mass Tran. 112.

[13] Huang, B., Young, Y.L., Wang, G., Shyy, W. (2013). *Combined experimental and computational investigation of unsteady structure of sheet/cloud cavitation*. J. Fluids Eng. 135(7).

[14] Chen, T., Huang, B., Wang, G., Zhang, H., Wang, Y. (2017). *Numerical investigation of thermo-sensitive cavitating flows in a wide range of free-stream temperatures and velocities in fluoroketone*. Int. J. Heat Mass Tran. 112.

© 2018 ASME

Cavitating structures at inception in turbulent shear flow

Karuna Agarwal; Omri Ram; Joseph Katz*

Department of Mechanical Engineering, The Johns Hopkins University, Baltimore, Maryland 21218

Abstract

Cavitation inception in the near field of turbulent shear layers occurs along secondary quasi-streamwise vortices developing between the primary spanwise vortices. To observe these incepting structures and understand the flow conditions under which they form, the present paper compares the flow statistics to results of high speed imaging performed in the shear layer behind a backward facing step. Simultaneous top and side imaging captures the location and spatial arrangement of the cavitating structures at varying cavitation indices. Two-dimensional PIV measurements characterize the mean velocity and Reynolds stress profiles, for Reynolds numbers in the separating boundary layer of Re_τ = 807, 1504, 2345. Being concentrated within the quasi-streamwise strings, the cavitation events are preferentially located in the near field of the shear layer, where the unsteady flow is dominated by coherent structures, upstream of the peaks in turbulence level. They move more upstream with increasing Re and decreasing pressure. Furthermore, the probability of cavitation increases by 3-5 times when the velocity is increased from 10.5 to 16 m/s. These indicate that the pressure fluctuation peaks causing the cavitation do not follow the same trends as the Reynolds stresses which collapse when scaled by the freestream velocity.

Keywords: cavitation inception; turbulent shear layer; flow structure visualization

1. Introduction

Cavitation, being one of the fundamental phenomena that affect the efficiency, noise, and life expectancy of hydrodynamic devices, has been studied extensively throughout the last century. However, much of the basic mechanisms that lead to the inception of cavitation, the growth rate of cavities and their coupling with the surrounding flow, is yet to be understood; especially for cases involving turbulent shear flows. Previous studies, e.g. [1-4], have shown that the cavitation inception index typically increases with Reynolds number. Furthermore, while the inception process is intermittent, depending on the distributions of nuclei and low-pressure sites, cavitation appears in quasi streamwise vortices at higher ambient pressures compared to events occurring in the larger spanwise vortices. Hence, the elongated vortices developing between the spanwise structures must have lower core pressure in spite of their much lower strength. The location and the number of these axial vortices vary, but they migrate upstream with increasing Reynolds number and have preferred spanwise locations [5]. Little is known about the location and strength of these structures, hence, data that would enable prediction of cavitation inception in them is still lacking. More recent work [6,7] on turbulence-cavitation interactions have focused at developed stages of cavitation. The aim of this study is to characterize the structure of early phases of cavitation in a shear layer developing downstream of a backward facing step and relate it to the local flow and turbulence fields. It is the first step in an effort to relate the pressure field and nuclei distributions on cavitation inception.

Experimental set-up

The experiments have been conducted in a water tunnel facility described in [8]. It has a 63.5×50.8 mm test section and a maximum velocity of 18 m/s. The flow is driven by a pair of centrifugal pumps located 5 m below the test section and passes through a 1000-liter buffer tank before entering the settling chamber leading to the test section. The geometry used to generate a free shear layer downstream of a backward-facing step is illustrated in figure 1. The step height is 10 mm, and the expansion ratio is 1.19. The shape of the nozzle is a fifth-order polynomial to ensure that the curvature is nearly zero at both ends to prevent undesirable pressure gradients. Grooves machined at bottom of the entrance to the test section are used for tripping the boundary layer, ensuring that all the present tests are performed when the separating boundary layer is turbulent.
Two-dimensional PIV is used for characterizing the flow field without cavitation, from the boundary layer all the way to the reattachment region. The flow is seeded with 2 μm silver-coated glass particles, the light sheet is generated

*Corresponding Author, Joseph Katz: katz@jhu.edu

using a Quantel Evergreen Nd:YAG laser, images are recorded by a 6600×4400 pixels Imperx BB640 camera equipped with 105 mm Nikon Micro lens for slightly overlapping 36.3×24.2 mm (x×y) sample areas (Figure 1). Cross-correlation analysis using the Davis software package with 32x32 pixels interrogation areas, with 75% overlap between windows, generates 827×552 vector arrays with spacing of 43 μm. About 2000 realizations are used to obtain the mean flow and Reynolds stress statistics. Experiments have been performed at three free-stream velocities, 5.3, 10.5 and 16 m/s.

The setup used for high speed imaging of the cavitating structures is also illustrated in figure 1. Two synchronized orthogonal views, side and top views, are used for capturing all three dimensions of the cavitation bubbles. The images are recorded at 3931.8 fps for 2.56 seconds at the resolution of 1730x730 using pco.Dimax cameras. Two halogen lamps are used for back-lighting the test section, enabling an exposure time of 3.98 μs. The dissolved oxygen content is kept close to 4.5ppm, representing 70-120% of the saturation level in the test section, as measured using an optical dissolved oxygen sensor (FireStingO2). The mean pressure is measured at the inlet to the test section, and the pressure above the step is inferred using the Bernoulli's equation.

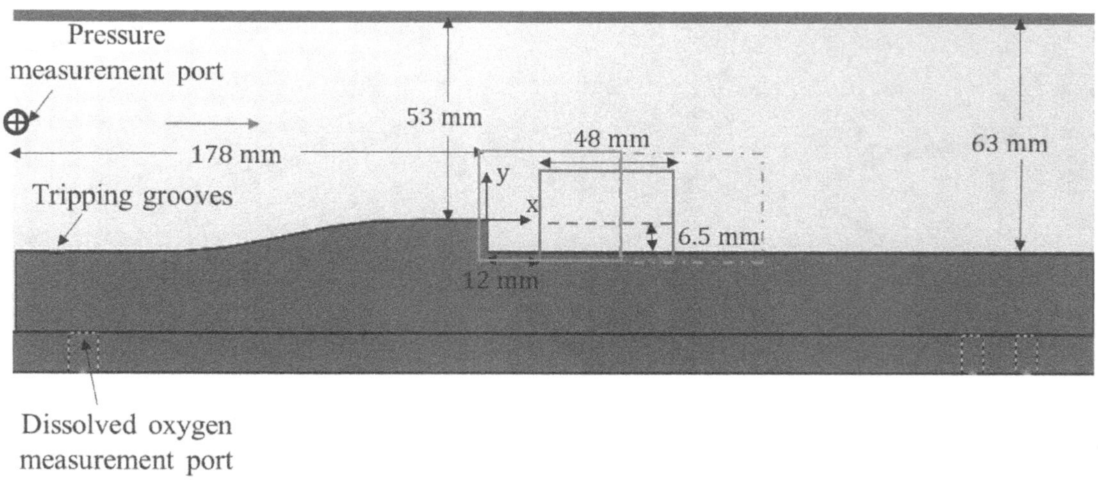

Figure 1: Setup for 2D PIV and visualizations experiments. The fields of view for PIV are marked in green, field of view of the side camera blue and the focal plane of the top camera in red.

2. Results and discussion:

2.1. 2D PIV results

A normalized plot of the boundary layer profile at three free-stream velocities is presented in figure 2(a). The wall shear velocity $u_\tau = \sqrt{\tau_{wall}/\rho}$ is estimated by least square fitting to the velocity profile in the log layer, and the boundary layer height δ is determined based on the elevation of peak velocity. The resulting $Re_\tau = u_\tau \delta/\nu$ are 807, 1504 and 2345 for free stream velocities U of 5.3 m/s, 10.5 m/s and 16 m/s, respectively. The reattachment length x_r determined based on the point where the zero velocity line in the shear layer reaches the surface, is found to decrease with increasing velocity, from 61.9, to 55.3 mm, and to 53.1 mm. As demonstrated in Figure 2(b) the mean velocity profiles collapse after normalizing the horizontal axis by is x_r and the vertical axis by h (step height). The sample profile in Figure 2(c) show that the two normal and one shear Reynolds stress components also nearly collapse using the same length scales and the freestream velocity. This figure also indicates that the freestream turbulence intensity for all cases falls below 1%. For reference, Figure 2(d) displays sample contour plots of $\overline{u_x u_y}/U$and $\overline{u_y u_y}/U$. The slight mismatch occurring at x~33mm corresponds due to the stitching of two PIV data sets. As is evident, $\overline{u_y u_y}/U$, which in boundary layers correlates well with the pressure fluctuations [9], increase with x/x_r, peaking at 0.95, far downstream of the cavitation inception area described below.

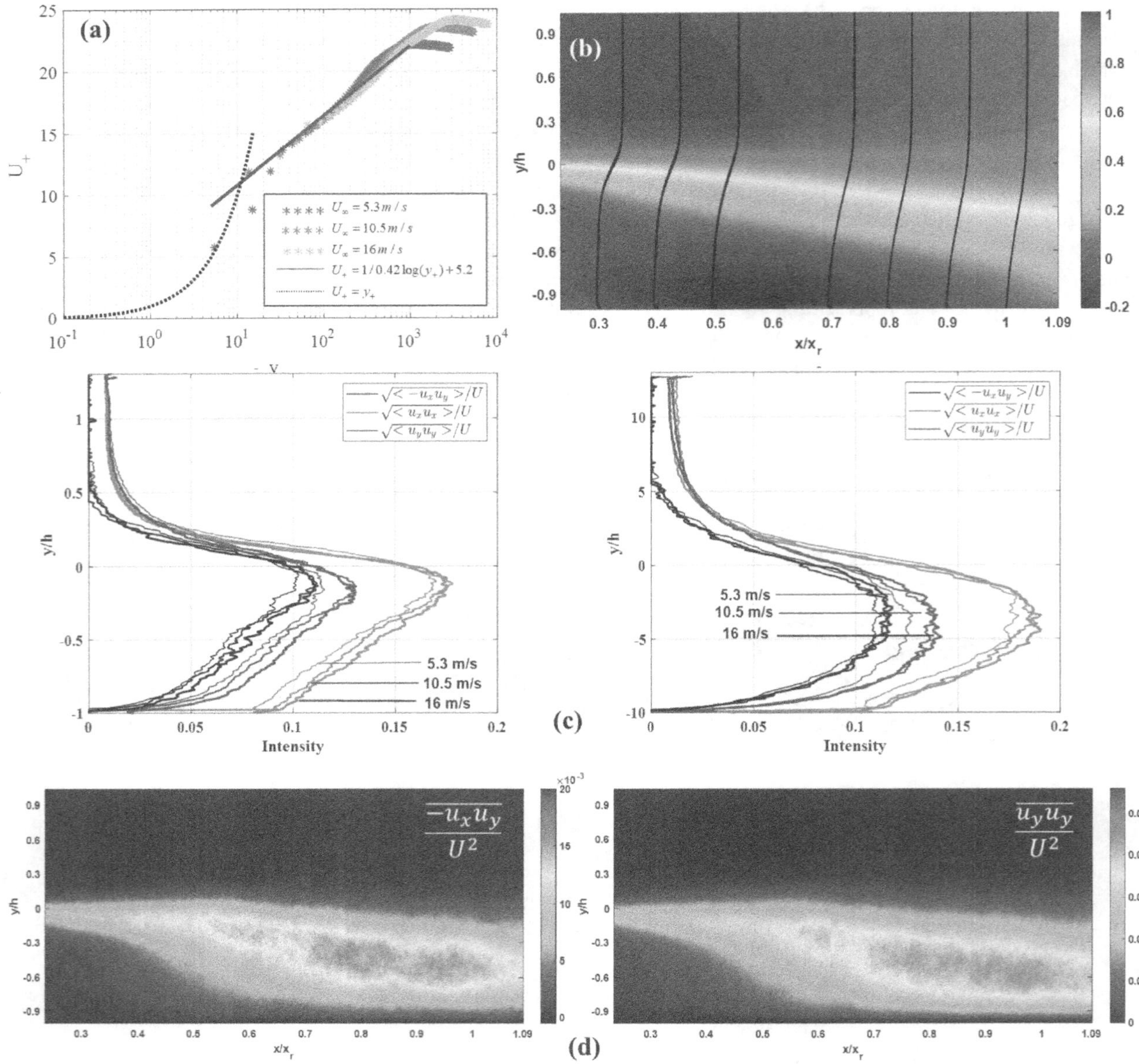

Figure 2 (a) The scaled velocity profile of the separating turbulent boundary layer and the fitted log function. (b) Average velocity profile scaled with U and x_r. (c) Turbulent fluctuation quantities scaled with free-stream velocity at $x/x_r = 0.5$ (left) and $x/x_r = 0.7$ (right). (d) Mean square wall-normal fluctuations and Reynolds stress.

2.2. Cavitation imaging

Figure 3 presents a set of sample snapshots showing the time evolution of a cavitation event where a free stream bubble (circled on the leftmost image), rapidly expand into an elongated quasi-streamwise cavity. Between the first two exposures, the bubble grows to nearly its full length (>5 mm) in less than 250 μs, indicating the present frame rate is too slow to capture the cavitation inception. Moreover, the entire life span of the cavitation event is about 2ms, during which the bubble grows along inclines the streamwise direction with small changes in its radius, and then fragments into small bubbles, before disappearing (at least in the present magnification). The relationship between these phenomena and the transient dynamics of secondary vortices developing between the primary spanwise structures in the shear layer is yet to be well understood. Interestingly, the length of these cavities does not seem to change significantly with cavitation index and freestream velocity.

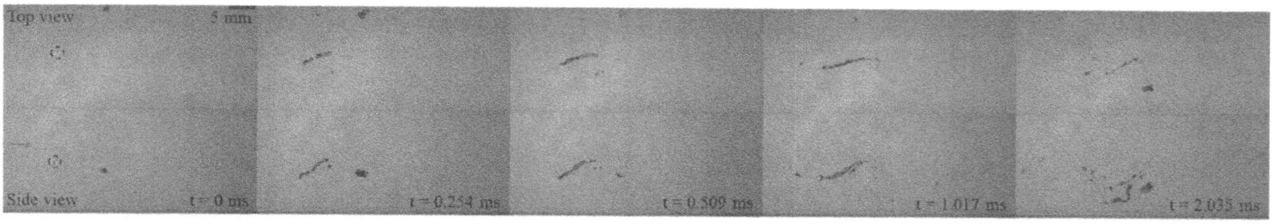

Figure 3: Time evolution of a cavitating bubble seen in two views. The nuclei marked in t = 0 ms, explodes along the streamwise direction in 0.254 ms and collapses by t = 2.035 ms. The freestream velocity is 10.5 m/s and the cavitation index 0.39.

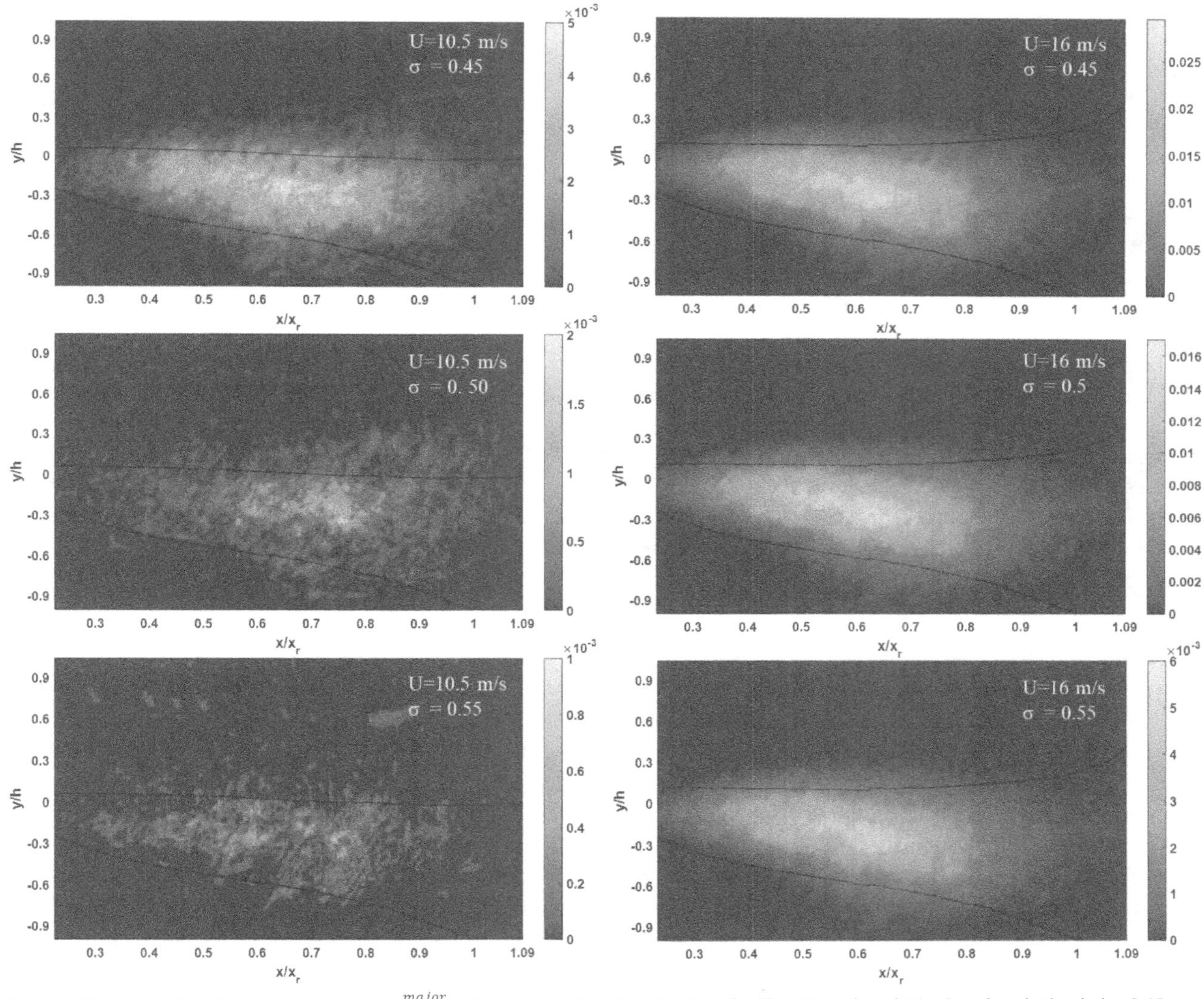

Figure 4: The probability of an elongated bubble ($\frac{major}{minor} = 2$) to be at a given location for velocities 10.5 m/s and 16 m/s and cavitation index 0.45, 0.5 and 0.55. The black lines represents 0 and $0.9U$ average velocity contours.

As a first step to characterize the behavior of these cavities, six set of 10000 images each of early phases of cavitation have been examined to determine the preferred location of the elongated bubbles. Only bubbles larger than 157 μm are accounted for. If the analysis is performed after imposing a ratio of major to minor axes of more than 2.0, the results are not significantly different, indicating that overwhelmingly, most of the bubbles are elongated structures. The time averaged 2D projected signatures of these bubbles are normalized with the total number of frames to represent the probability of having cavitation at a certain location. The results are mapped in Figure 4 (note the different scales) for two velocities and three cavitation indices $\sigma = 2(P - P_v)/\rho U^2$, where P is the mean pressure above the step, and P_v is the vapor pressure at the water temperature. As is evident, cavitation is most likely to take place at 0.5<x/x_r <0.75. As indicated by the vertically averaged probability plots, for both velocities, the most probable site moves upstream with increasing velocity (Figure 5). The axial locations of these probability peaks also do not coincide with those of the peaks in Reynolds stresses.

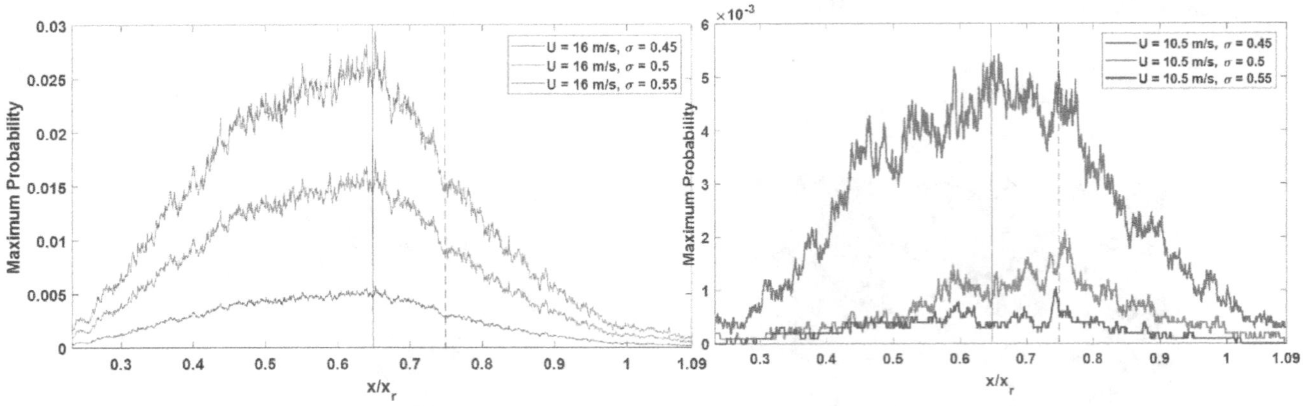

Figure 5. Vertically collapsed maximum probability plots for the six cases, higher velocity case on the left. The solid black line indicates the location of maximum probability for 16 m/s velocity and the dashed line for 10.5 m/s.

As expected, the likelihood of cavitation occurrence increases with decreasing cavitation index. However, there is 3-5 times increase in the probability of cavitation inception as the velocity is increased from 10.5 to 16 m/s for the same cavitation index. This trend contrasts with the collapse of Reynolds stresses under the same conditions, further indicating that the pressure fluctuation *peaks* causing the cavitation do not follow the same trends as the Reynolds stresses. Finally, spatially averaging the probability at 0.45 <x/x_r <0.8 shows that its magnitude increases faster than the increase in cavitation index. These findings provide motivation for measuring the scaling trends of the pressure field within the quasi streamwise vortices in the shear layer.

3. Conclusion

The present study compares the scaling trends of the most likely location of early phases cavitation to those of the Reynolds stresses in the shear layer downstream of a backward facing step. The cavitation is most likely to take place at 0.45<x/x_r <0.75, moving upstream with decreasing cavitation index and increasing velocity. Unlike the Reynolds stresses, which collapse when scaled using the free stream velocity, the likelihood of cavitation increases by 3-5 times as the velocity increases from 10.5 to 16 m/s. Hence, the pressure fluctuation *peaks* associated with the quasi streamwise structures causing the early cavitation do not follow the same trends as the turbulence in the shear layer. These findings provide the motivation for a systematic study of the pressure field generated by primary and secondary vortices in the shear layer. Fortunately, recent advances in technique for determining the instantaneous pressure distributions in turbulent flows based on data obtained from time-resolved three dimensional velocity measurements makes such a task feasible [10,11].

References

[1] Katz, J. (1984). *Cavitation phenomena within regions of flow separation*. Journal of Fluid Mechanics, 140.
[2] Katz, J., & O'hern, T. J. (1986). *Cavitation in large scale shear flows*. Journal of fluids engineering. 108(3).
[3] O'hern, T. J. (1990). *An experimental investigation of turbulent shear flow cavitation*. Journal of fluid mechanics, 215.
[4] Gopalan, S., Katz, J., & Knio, O. (1999). *The flow structure in the near field of jets and its effect on cavitation inception*. Journal of fluid mechanics, 398.
[5] Bernal, L. P., & Roshko, A. (1986). *Streamwise vortex structure in plane mixing layers*. Journal of Fluid Mechanics, 170.
[6] Iyer, C. O., & Ceccio, S. L. (2002). *The influence of developed cavitation on the flow of a turbulent shear layer*. Physics of fluids, 14(10).

[7] Aeschlimann, V., Prothin, S., Barre, S., & Djeridi, H. (2012). *High speed visualizations of the cavitating vortices of 2D mixing layer*. European Journal of Mechanics-B/Fluids, 31.
[8] Gopalan, S., & Katz, J. (2000). *Flow structure and modeling issues in the closure region of attached cavitation.* Physics of fluids, 12(4).
[9] Joshi, P., Liu, X., & Katz, J. (2014). *Effect of mean and fluctuating pressure gradients on boundary layer turbulence.* Journal of Fluid Mechanics, 748.
[10] Van Oudheusden, B. W. (2013). *PIV-based pressure measurement*. Measurement Science and Technology, 24(3).
[11] Wang, Zhang and Katz (2018). *GPU-based, parallel-line, omni-directional integration of the acceleration field to obtain the 3D pressure distribution.* 10th International symposium on cavitation.

© 2018 ASME

Cavitation bubble collapse and wall shear stress generated in a narrow gap

[1]Silvestre Roberto Gonzalez-Avila; [2]Anne Charlotte van Blokland; [1]Qingyun Zeng; [1,3]Claus-Dieter Ohl

[1]Nanyang Technological University, Singapore; [2]University of Oxford, United Kingdom;
[3]Otto-von-Guericke University of Magdeburg, Germany.

Abstract

Cavitation bubble dynamics in narrow gaps confined by rigid boundaries greatly differs from a liquid half-space. There exists a regime, where the bubble is created on one side of the gap and collapses on the opposite side [1]. Additionally, the jet velocities may be considerably larger. The oscillating bubble and the impacting jets form complex boundary layer flows with greatly increased wall shear stress. The observations have implications for cleaning applications in narrow spaces. The experimental results show that the shear flow generated is active much longer after the collapse of the bubble. The results show approximately a 3-fold increase of the shear stress as compared to a liquid half space. We also present a space-time graph from numerical simulations. Additionally, the fluid dynamics is studied with the motion of particles attached to a substrate and the effect of shear on adherent biological cells.

Keywords: Wall shear stress; bubble dynamics; particle removal

Introduction

Cavitation near boundaries generates large forces on boundaries through normal and tangentially acting stresses. Viscosity is responsible to transmit tangential surface forces, which are particularly important for cleaning and biological cell applications. After studying the flow in greater detail [2, 3] we became interested on the flow occurring between two parallel plates separated by a thin gap [1]. We revealed that cavitation bubbles created might migrate from the location where they are created to the opposite plate. Thus the aforementioned forces occur at some distance from the bubble creation. This could lead to applications of cleaning in thin gaps such as in stacks of membranes for water treatment or to induce forces on cells while preventing a physical contact of the bubble generator to the cells.

Besides the translational movement we have seen considerably higher velocities of the jet using inviscid boundary element simulations. This suggests that the wall shear stresses could also be higher in a thin gap as compared to a similar sized bubble in a semi-infinite liquid. To test this hypothesis we measure the wall shear stress and additionally conduct viscous computational fluid mechanics simulations.

Then we present results on cleaning and cell removal/drug delivery to show the advantages of the geometry. All bubbles presented here are created with a spark discharge using a simple high-voltage sparker.

Bubble Dynamics

The experimental setup consists of a piezoelectric igniter connected to two electrodes where the cavitation bubbles are created, an oscilloscope (LeCroy HRO 64zi, 2 GHz maximum sampling rate), and a high-speed camera (Photron SAX2). The electrodes on a printed circuit board (PCB) are made of copper by an etching method as described by [4]; the thickness of the wires is ~125 μm and the separation between them is ~ 150 μm although these dimensions vary ~ ±20 μm due to the etching process. The piezoelectric igniter starts the events; it creates a spark between the electrodes, which creates a hemispherical cavitation bubble originating from the PCB. To measures the wall shear stress a hot film anemometer 55R46 (Dantec Dynamics, 10 kHz bandwidth) is flush mounted in the boundaries, see figure 1a. The hot film anemometer and the electrodes were submerged in a cubic container of 75 mm in side and mounted on a 3-axis stage; in this way the relative position of the electrodes and the wall shear stress sensor can be adjusted accurately. The rising signal detected by the anemometer starts the data acquisition process, the oscilloscope triggers then the high-speed camera. The relevant experimental parameters are depicted in figure 1b; H is the distance between the walls, R_x the maximum radius attained by the bubble in the horizontal direction and δ is the horizontal distance between the center of the bubble and the hot film sensor.

*Corresponding Author, S. Roberto Gonzalez-Avila: roberto_glez83@hotmail.com

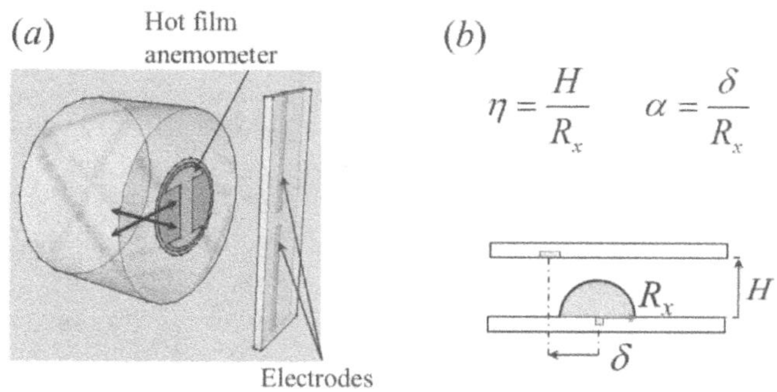

Figure 1. (a) Schematic diagram of the experimental setup (b) experimental parameters

The experiments are complimented with a compressible volume of fluid simulation using OpenFOAM. The simulation domain is axisymmetric (radial size is 5 mm, height 0.75 mm). An initial grid with 100x15 cells is refined recursively five times leading to a cell size of about 1.5 μm where the bubble is located. The simulation omits the nucleation dynamics and starts with an initially spherical gas bubble located at a distance of 250 μm above the wall with a radius of 50 μm and initial pressure of $p_g(t = 0) = 1250$ bar. The solver accounts for surface tension and liquid viscosity. A detailed discussion of the techniques and limits is provided in [5].

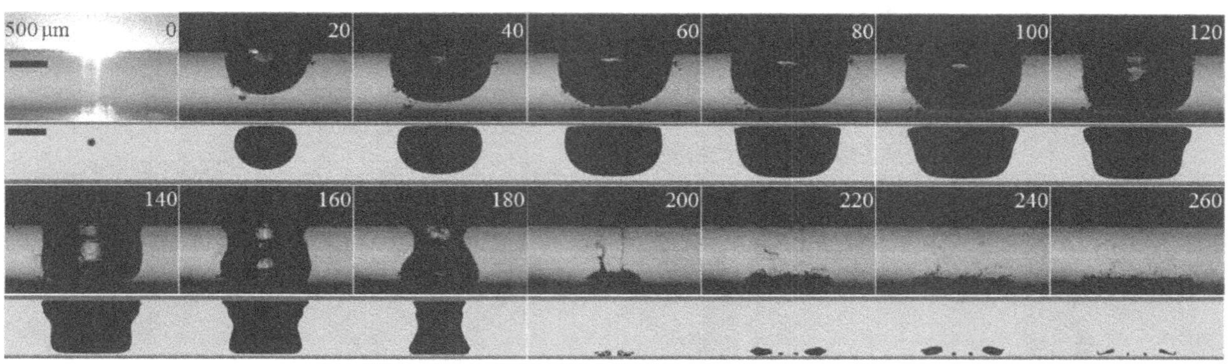

Figure 2. Comparison of experiment and simulation for a bubble of R_x = 770 μm and h = 750 μm. The upper row is the high-speed images, the lower row is the simulation result. The upper boundary is the nucleated wall and the lower boundary is the opposite wall. Time is in μs.

Figure 2 presents a bubble nucleated in a regime where the migration to the opposite wall takes place. The figure also depicts a comparison between experiment and a simulated bubble dynamics for R_x = 770 μm and h = 750 μm. The interval between frames is 20 μs for both sequences. At t = 0 the simulated bubble expands from a small spherical volume while the bubble in the experiment is nucleated on the surface. Despite the simplifications introduced in the simulations a rather good agreement of the bubble shape is obtained already after t = 20 μs. The bubbles in the experiment and the simulation reach their maximum size at $t \sim$ 100 μs with a cap shape. During the early stage of the collapse, the middle part of the bubble collapses faster than the other parts and an hourglass shape bubble is formed. However, after t = 160 μs the part closer to the wall where the bubble is nucleated shrinks faster and a jet is formed directed to the opposite wall. The jet penetrates the bubble wall and impacts on the opposite surface creating a strong radial shear flow. The toroidal bubble reaches its minimum size at around t = 200 μs and rebounds to a much smaller maximum size than that in the first cycle at around t = 240 μs. The overall bubble dynamics is nicely reproduced by the simulations. We therefore discuss the simulated wall shear stress first before we look into the measured data.

Wall Shear Stress

The wall shear stress is defined as

$$\tau = \mu \left. \frac{du_r}{dy} \right|_{y=0} \tag{1}$$

where μ is the liquid's viscosity, u_r the flow velocity component parallel to the wall and y is the distance to the boundary. Figure 3 presents the spatio-temporal wall shear stress map on both walls for a bubble of R_x = 770 µm and h = 750 µm. We take a logarithm base 10 of the shear stress value due to its large range. The stress directed away from the axis of symmetry is colored in red, while the negative values in blue color represents the stress towards the axis of symmetry. The high stresses occur approximately when the bubble reaches its minimum size at t ~ 200 µs on both walls , about 20 ~ kPa on the nucleated wall and 100 ~ kPa on the opposite wall. The stress is positive in the whole domain during most of the expansion, $0 \leq t \leq 50$ µs, on both walls due to the rapid expansion of the bubble. At the later stage of the expansion and the early period of the following shrinkage, the stress is still positive within the domain between the bubble wall and the boundary and becomes negative outside the bubble domain because of the incoming flow, due to the deformation and collapsing of the bubble. Note that for the above period, the positive stress decreases with time while the negative increases with time, and a large area with almost zero stress occurs on the nucleated wall before the formation of the jet. High stresses are found at t ~ 195 µs near the axis of symmetry (r = 0), which come from the impact and spread of the liquid jet which penetrates through the bubble.

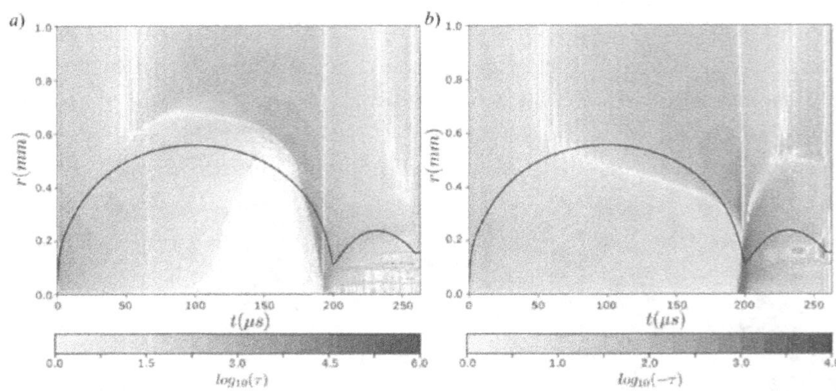

Figure 3. Wall shear stress for R_x = 770 µm, h = 750 µm. (a) on the wall the bubble is nucleated; (b) on the opposite wall. The black curve is the bubble equivalent radius. The wall shear stress is in Pa; positive values represent flow along the positive radial direction, negative towards the axis of symmetry.

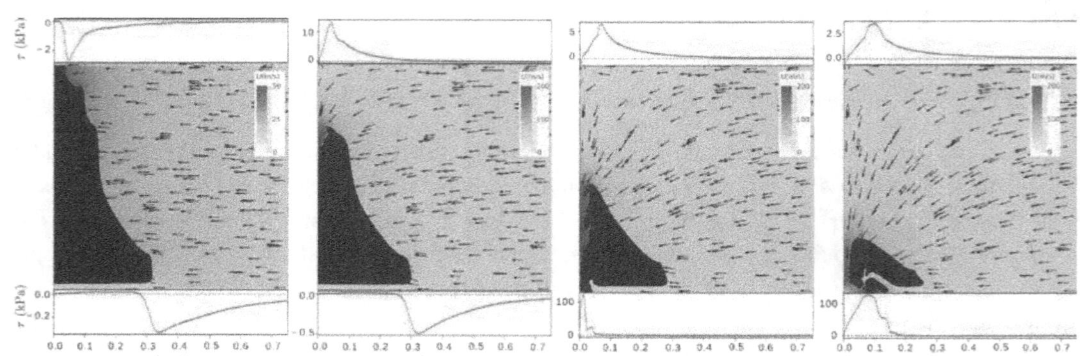

(a) 192 µs (b) 194 µs (c) 196 µs (d) 198 µs

Figure 4. The jet formation and wall shear stress distribution on the upper and lower walls for a bubble of R_x = 770 µm, h = 750 µm. The bubble is nucleated on the top wall. The black region is the bubble. The plots above and below the flow fields are the wall shear stress distribution along the radial directions on each wall. The arrows show the direction of the liquid flow. Negative wall shear stress values represent flow towards the axis of symmetry while positive values represent flow in the positive radial direction.

Figure 4 portrays this jet formation process, as well as the wall shear stress distribution on both walls. Before the jet is formed, the stresses are negative in the regime outside the bubble and the negative peaks are located near the outside boundary of the hourglass shape bubble with a lowest value of about -2.5 kPa on the nucleated wall and -0.5 kPa on the opposite wall. When the jet starts to be formed a high positive stress of 12 kPa appears on the nucleated wall (t = 194 µs), which is due to the impact and spread of the recoiled flow of the initial jet. However, this recoiled flow is much weaker than the jet-induced flow thus the shear stress is damped quickly, see figure 4b-d. The jet accelerates due to the collapsing bubble and reaches approximate 200 m/s when it impacts the lower bubble wall.

We now discuss the experimentally recorded wall shear stress, shown in figure 5a. Each curve in the figure depicts the wall shear stress for a different distances from the bubble, α. In these tests the channel height, H, is kept fixed, as shown in the inset of figure 5b, and the wall shear stress, τ, is evaluated when the offset parameter is varied in the range $0.1 \leq \alpha \leq 2.1$; three curves of increasing δ/R_x values are shown. The width of the shaded boxes represents the collapse time of the cavitation bubble, from maximum expansion to its minimum volume; the height of each box represents the increase in shear stress during the bubble's collapse time. At $\delta/R_x = 0.1$ the τ vs t exhibits a maximum value at $t = 261$ μs and a second maximum at $t = 701$ μs; the first maximum is produced by the re-expansion of the cavitation bubble; we believe the second maximum may be due to the vortex generated after the bubble's collapse. This curve also displays a third and small peak at $t = 3.1$ ms but we cannot tell from the video why the curve is distorted. The maximum τ value seen in the curves of δ/R_x =0.8 and $\delta/R_x = 1.9$, at $t = 381$ μs and at $t = 1.2$ ms, respectively occur much later after the bubble's collapse, thus this peak is more like due to the vortex generated; also note that the peak of the curve of $\delta/R_x = 1.9$ is ~ 1% of that of the curve of $\delta/R_x = 0.1$. Figure 5b portrays the summary of the peak τ value recorded at the different α values tested. The maximum τ values recorded decrease very rapidly, thus the maximum stress generated 2 radii away is ~ 1% of the value at recorded at $\delta/R_x = 0.1$. This is in agreement with previous studies that show that the effect of cavitation bubbles is very localized [3,4,6].

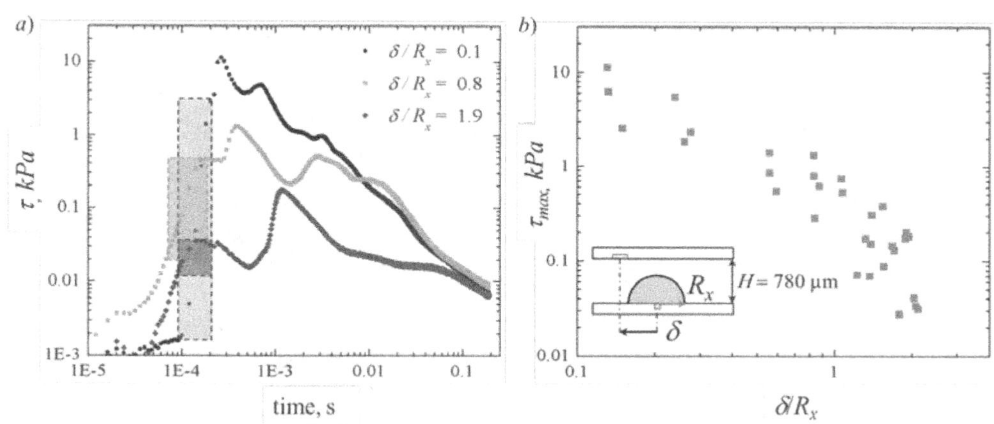

Figure 5. (a) Wall shear stress vs time for different lateral positions; (b) τ_{max} vs δ/R_x; H is kept constant. $R_x = 730 \pm 50$ μm (average of 33 tests).

Experimental and numerical results

As was shown in figure 2 the overall bubble dynamics is well reproduced by the simulations. However, there is a significant difference in the measured and simulated peak shear stress values. We believe that the peak shear stress detected is underestimated due to the low bandwidth of the anemometer, yet our results show that the peak measured shear stress value is ~ 3 times larger than that previously reported for a bubble of similar size created next to a single boundary [3]. The simulations are especially useful to interpret the complex flow generated at the bubble collapse (i.e., see figure 4a-d) which is too fast to be resolved even by the high-speed camera. In addition to that, the wall shear stress distribution in space and as a function of time is available from the simulations, a large number of experiments would be needed to create such a map due to the point wise nature of the hot film probe. In the next subsections, we indicate two potential applications where this shear flow can be utilized: cleaning in a narrow gap and drug delivery into adherent cells.

Cleaning in a narrow gap

Fluorescent polystyrene particles with a diameter of 15 μm are prepared by letting droplets with suspended particles evaporate on a heated glass substrate. This leads to a characteristic distribution of weakly bonded particles on an annular ring, see first frame in figure 6, which remains even after the substrate is re-suspended in water. In the cleaning experiments, the substrate with the attached particles forms the lower boundary in the gap. The bubble is created on the upper plate by activating the high voltage source. Figure 6 depicts the motion of the fluorescent particles exposed to a bubble following a very similar dynamics as in figure 2 (η =1.0). The six frames in figure 6 depict the particles before (t = -28 μs), during ($0 \leq t \leq 222$ μs) and after (t = 138 ms) the cavitation event. The bubble is created from a spark discharge visible in figure 6 (t = 0) with a bright region at the center. A first motion of the particles is visible during the expansion phase t = 56 μs where they are dragged outwards. During the bubble shrinkage the inward flow drags the particles towards the center, see arrows at t = 195 μs. Interestingly while the expansion of the bubble accelerates the particles purely radially there is a circumferential instability of the inward flow visible at time t = 195 μs. Once the bubble collapses a fast jet is created which reverses the particle motion as indicated with the arrows at t = 222 μs. This back and forth motion leads to complex boundary layer flow dispersing the particles such that the initial annular ring shape is lost, see figure 6 at t = 138 ms. Also, the particles have been removed from the field of view either by moving out of focus or being transported out of the observation area through the residual flow generated by the bubble.

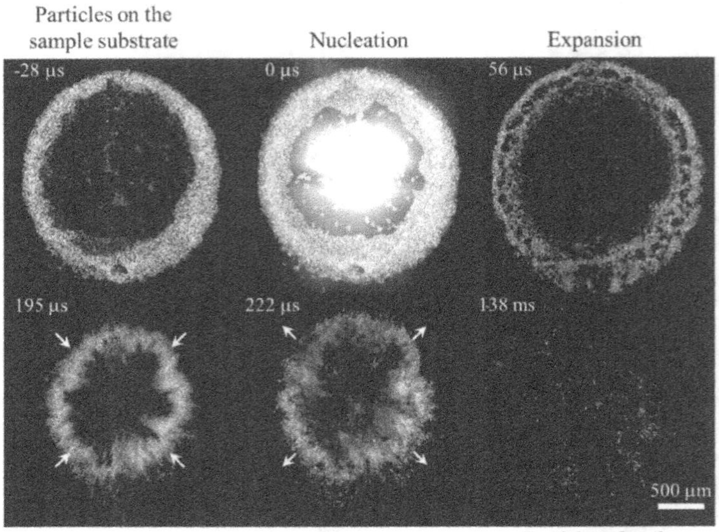

Figure 6. Fluorescent polystyrene particles removed by a spark-generated cavitation bubble, η = 1.0.

Molecule transport into cells

It is known that biological cells with a plasma membrane can be temporarily opened for the transport of large molecules through shear stress. The flow responsible for the particle motion in figure 6 demonstrates that although the cavitation is created as some distance from the substrate strong shearing flow can be create. We explore if it is feasible to induce pores in cell membranes. This geometry has the advantage that the cells do not need to be in contact with the device creating the flow. Additionally, the strength of the flow may be adjusted by changing the height of the gap H as shown in figure 7a.

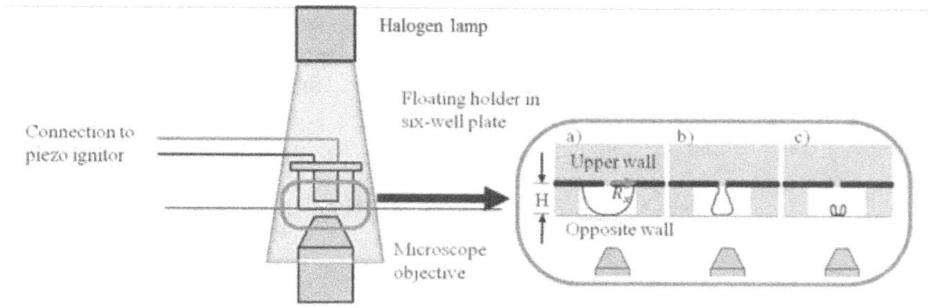

Figure 7. Sketch of the device and experimental setup to study drug delivery into cells. The electrodes are on a separate device, which is placed onto a petri-dish with the adherent cells. The cells and bubble dynamics is recorded from below though an inverted microscope.

For the experiments we use colon carcinoma cells (RKO). An added complexity is the electrical discharge to generate the bubbles. As the cell medium is conductive we add a thin layer of low viscosity silicon oil (5cSt) on top of the cell medium. The microscope objective is focused on the substrate with the cells, thus the bubble initially growing on the upper wall with electrodes is out of focus, see figure 8a, t = 3.7 μs. After bubble expansion it collapses onto the opposite wall at t = 119 μs with a torus shape due to the jet that pierces through it. The size of this jet area is comparable to the size where molecular uptake is taking place, see figure 8b.

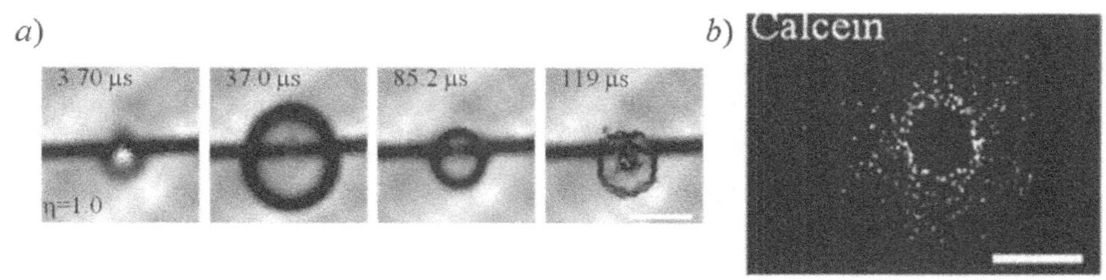

Figure 8. Fluorescent uptake of Calcein in the RKO cells for $\eta = 1.0 \pm 0.1$. (a) high-speed images of the bubble dynamics; (b) fluorescent image of cells taken up the non-membrane permeant dye Calcein. The scale bar is 500 μm.

Conclusion

Bubble collapse between two walls can produce wall shear stresses of up to 10kPa near the site where the bubble collapses; this is likely a lower bound of the shear stress reached due to the low bandwidth of the hot film probe of 10kHz. Nevertheless, the maximum shear stress measured is ~ 3 times larger than the values found in bubbles of similar size collapsing near a single wall [3]. The flow generated by the cavitation bubbles remains long after the cavitation event ends. However, a large number of tests are required to obtain a shear stress distribution on a solid wall. Due to this, numerical simulations are presented. A space-time diagram depicts the shear stress attained by a single cavitation event for specific η value. Also, we have shown that a very simple device, with no moving walls can be used to generate such a high shear flow; in addition to that, two potential applications are demonstrated, the removal of particles attached to a substrate and molecule transport though the cell membrane. The flow generated by a migrating bubble can also be used to produce fast microjets; this topic has received significant attention recently [7,8,9] due to their potential to deliver drug without needles.

References

[1] Gonzalez-Avila, S.R., Klaseboer, E., Khoo, B.C., Ohl, C.D. (2011). *Cavitation bubble dynamics in a liquid gap of variable height*. J. Fluid Mech., 682.

[2] Ohl, C-D., Arora, M., Dikink, R., de Jong, N., Versluis, M., Delius, M., Lohse, D. (2006). *Sonoporation from Jetting Cavitation Bubbles*. Biophysi J. 91, 4285–429.

[3] Dijkink, R., Ohl, C-D. (2008). *Measurement of cavitation induced wall shear stress*. Appl Phys Lett, 93, 254107.

[4] Gonzalez-Avila, S. R., Huang X., Quinto-Su, P. A., Wu, T., Ohl, C-D. (2011). *Motion of micrometer sized spherical particles exposed to a transient radial flow: attraction, repulsion, and rotation*. Phys Rev Lett, 107, 074503.

[5] Zeng, Q., Gonzalez Avila, S-R., Dijkink, R., Gavaises, M., Ohl C-D (2017). *Wall shear stress from setting cavitation bubbles*. Submitted to the J. Fluid Mech.

[6] Kim, W., Kim, T. H., Choi, J., Kim, H. Y. (2009). *Mechanism of particle removal by megasonic waves*. Appl. Phys. Lett. 94, 081908.
[7] Karri, B., Gonzalez-A., S. R., Loke, Y. C., O'Shea, S. J., Klaseboer, E., Khoo, B. C., Ohl, C-D. (2012). *High-speed jetting and spray formation from bubble collapse*. Phys. Rev. E 85, 015303.
[8] Tagawa, Y., Oudalov, N., El Ghalbzouri, A., Sun, C., Lohse, D. (2013). *Needle-free injection into skin and soft matter with highly focused microjets*. Lab Chip 13, 1357–1363.
[9] Gonzalez-Avila, S. R., Song, C., Ohl, C-D. (2015). *Fast transient microjets induced by hemispherical cavitation bubbles*. J. Fluid Mech., 767, 31-51.

© 2018 ASME

Cavitation Bubble Collapse Monitoring by Acoustic Emission in Laboratory Testing

[1]Markku Ylönen*; [1]Pentti Saarenrinne; [1]Juha Miettinen; [2]Jean-Pierre Franc; [3]Marc Fivel

[1]Tampere University of Technology, 33720 Tampere, Finland; [2]Université Grenoble Alpes, CNRS, Grenoble INP, LEGI, 38000 Grenoble, France; [3]Université Grenoble Alpes, CNRS, Grenoble INP, SIMAP, 38000 Grenoble, France

Abstract

In order to investigate the potential of the acoustic emission technique in predicting cavitation erosion, laboratory tests were conducted in a high-speed cavitation tunnel. One face of a cylindrical stainless steel sample was subjected to an annular cavitation field created by the PREVERO cavitation tunnel [1]. Acoustic emission was measured from the back surface of the sample in order to detect impacts caused by cavitation bubble or cloud collapses. Cavitation aggressiveness was varied by changing the operating parameters of the cavitation tunnel. Two different operating points were compared. Collapsing cavitation bubbles lead to impacts towards the sample surface and they induce elastic waves in the material. A resonance type acoustic emission sensor with a resonance frequency of 160 kHz captured these waves during the cavitation tests. The acoustic emission waveform was measured with a sampling frequency of 5 MHz. The sensor was mounted behind the sample using a wave-guide that maintained a transfer path for the elastic waves to travel from the impacted surface to the sensor. The elastic waves reaching the sensor were observed as distinguishable bursts in the acoustic emission waveform. Acoustic emission from cavitation impacts was estimated to be about 100 times stronger than acoustic emission from other sources, such as hydrodynamic events or machine vibration. This means that the signal was almost entirely induced by cavitation. The bursts contain multiple reflections that attenuate in time and that have a frequency content corresponding to the sensor frequency response. The bursts attenuate quickly enough not to overlap, as the cavitation events occur with a large enough temporal separation. The hypothesis in this study is that the maximum amplitude of the acoustic emission event voltage correlates with the strength of the cavitation bubble collapse impacting the surface. Voltage peak value counting was applied to the acoustic emission waveform data. As the bursts contain multiple amplitude peaks due to sensor resonance, an envelope function was fitted to the waveform for peak counting. Using this method, each counted voltage peak value is expected to correspond to a single cavitation impact event. The pulse distribution shows an exponential decrease with a decreasing voltage peak value rate as the peak voltage increases. This compares well with earlier studies, such as [2] and [3], where an exponential distribution of bubble collapse amplitudes was found. The results of this study prove acoustic emission as a direct and non-intrusive method that can be used to monitor cavitation impacts from outside of the cavitation field.

Keywords: cavitation impact detection; acoustic emission; cavitation intensity;

Introduction

The impact load induced by a cavitation bubble collapsing near a solid boundary has been studied by multiple methods, both experimental and computational. In a cavitating flow, the impact load determines material damage in a boundary caused by a single bubble or bubble cloud collapse. Another important factor in material damage is the impact frequency, as cavitation erosion tends to be a cumulative process [4-6]. The impact distribution that combines the impact frequency and amplitude of the cavitation impact loads is essential in determining the cavitation intensity of a flow.

Cavitation impact loads have been measured by various methods. Franc et al. [2] measured the impact loads in the PREVERO cavitation tunnel with conventional pressure sensors flush-mounted in the cavitation closure region. Hujer et al. [3] used PVDF pressure sensors, also flush-mounted in the same tunnel. Both of them observed exponential impact distributions. Hattori et al. [7] studied the impact pressures in an ASTM G-32 vibratory device and Okada et al. [8] used the same device to calibrate pressure sensors for a Venturi nozzle test. In the vibratory test, the impact distribution also follows an exponential law. Franc et al. [9] also observed the pits formed by cavitation impacts and they observed an exponential distribution in the pit size distribution. The pit shape factor or the ratio between pit depth and pit diameter increases with increasing cavitation intensity, meaning that larger impacts lead to deeper pits [10].

Several authors have studied acoustic emission (AE) as a method to detect and characterize cavitation and cavitation erosion. Boorsma and Fitzsimmons [11] created a cavitation monitoring method for ship rudders and propellers. Yongyong and Zaiyang [12] connected the AE-event energy to mass loss in an ASTM-G32 vibratory cavitation apparatus. Schmidt et al [13; 14] worked on a cavitation detection system based in AE on a prototype Kaplan turbine. They discovered that with properly placed sensors, cavitation leads to increase in AE root mean squared voltage value, event energy and fluctuation of both. Van Rijsbergen et al. [15] found that acoustic emission sensors in direct contact

*Corresponding Author, Markku Ylönen: markku.ylonen@tut.fi

with a hydrofoil capture bubble collapses near the foil surface and with the cavitation impact towards the foil, but not those that occur in the flow far from the foil. These studies encourage further development of cavitation monitoring by AE.

In this study, the same cavitation tunnel as used by Franc et al. [2] and Hujer et al. [3] was fitted with acoustic emission sensors. The difference between previous studies with the tunnel in question is that the AE sensors are placed outside of the liquid flow by placing them in the backside of the sample. Voltage peak values corresponding to individual cavitation events were classified by their quantity and voltage. The goal was to produce similar distributions for two different operating conditions from the acoustic emission voltage peak values and to correlate these distributions with those of previous works.

Experiments

The experiments were carried out in the PREVERO high-speed cavitation tunnel in LEGI laboratory. The tunnel circulates water through a radially diverging test section. Water comes in to the test section through a Ø 16 mm inlet nozzle and the flow stagnates in the middle of the test section where the sample is located and continues to diverge radially in a 2.5 mm thick channel. The samples are 20 mm thick cylindrical disks with a 100 mm diameter and placed so that the sample center is in the middle of the test section. Cavitation inception is located in the beginning of the radial section and cavitation closure is located between 21 and 26 mm radius of the disk when operating with the typical cavitation number $\sigma = 0.9$. Cavitation number in PREVERO is defined in equation 1:

$$\cdot = \frac{P_d - P_v}{P_u - P_d} \tag{1}$$

where P_d is the pressure after the test section, P_u is the pressure before the test section and P_v is the liquid vapor pressure.

The acoustic emission sensors were fitted to the sample using a waveguide that is fixed with a screw thread to the sample. Figure 1(a) shows the sample disk flush mounted to the sample holder and the AE sensors fitted to the sample from behind. Figure 1(b) presents the test section without the sample and the sample holder. The sample and the holder were fastened to the test section so that from the inlet nozzle, the flow radially diverges to a 2.5 mm thick channel.

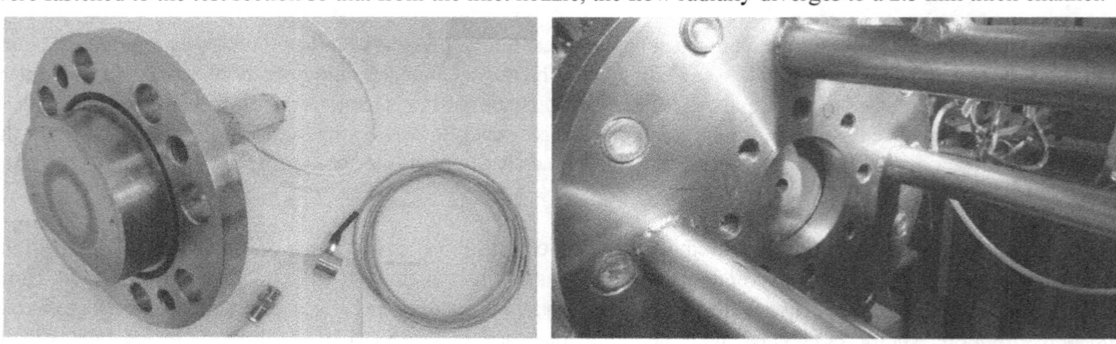

Figures 1(a) and 1(b). 1(a): Sample mounted to the sample holder and fitted with an AE sensor and a waveguide. 1(b): PREVERO test section opened and with the inlet nozzle in the middle. The sample is flush mounted so that it forms a part of the other wall in the test section.

Cavitation inception occurs at the outlet of the cylindrical nozzle, as the cross section area of the flow drops to 62.5 % of that of the inlet nozzle. Cavitation closure occurs further downstream as the cross section area and therefore the static pressure of the flow increase. Tunnel downstream tank is pressurized by nitrogen and upstream section by a pump linked to a frequency transformer. Downstream pressure varies from ambient pressure to around 3 MPa and the maximum upstream pressure is 4 MPa. The cavitation tunnel is presented in more detail in [1; 4].

The acoustic emission setup was a PAC PCI-2 two channel acquisition card fitted with PAC R15 and D9203b sensors and 20/40/60 preamplifiers. The R15 sensor is a resonance type sensor with a resonance frequency of 160 kHz. A band-pass filter from 100 kHz to 400 kHz was used. The D9203b sensor is a broadband sensor with a frequency range from 100 kHz to 900 kHz. In this study, only the data acquired with the R15 sensor were analyzed. The samples were made of a stainless steel used in Francis turbines and the waveguides were made of steel.

In this study, two samples were mirror polished with successive diamond pastes and a colloidal silica suspension. One sample was subjected to cavitation erosion for 2 minutes with a 4 MPa upstream pressure and the other for 6 minutes with a 2 MPa upstream pressure. The 4 MPa and the 2 MPa upstream pressures correspond to 89.4 m/s and 63.2 m/s cavity reference velocities, respectively. Cavity reference velocity is defined as the velocity in the test section where pressure is assumed as vapor pressure. A more detailed explanation of cavity reference velocity is in [4]. Pits with this type of exposure covered roughly 10 % of the surface of the cavitation closure area. The cavitation impacts were thus assumed to be hitting virgin material most of the time. Acoustic emission waveforms were acquired with a sampling frequency of 5 MHz for the full duration of the tests.

Peak value distribution by an envelope function

Impacts in the cavitation sample surface were observed in the AE waveform as quickly rising voltage peaks that diminish exponentially. The AE waveforms were analyzed in time-voltage space. Wave propagation in the sample and the waveguide may affect the waveforms, but this effect is not considered in detail in this study. The frequency content of each cavitation burst or AE-event resembles that of the sensor frequency response, meaning that the impacts provoke sensor resonance. This means that the events are mostly short duration impacts compared to the sensor time scales. It is assumed that each voltage peak value in the waveform, with its resonance effects, corresponds to one cavitation impact towards the sample surface. The length of a cavitation bubble collapse is in the range of some microseconds up to some tens of microseconds [2; 16]. Through waveform observations, impact overlapping seems not to be significant.

With the assumption that each measured maximum in the voltage peak value corresponds to a single cavitation event, there is a need to filter out the sensor resonance effects when peak counting is applied. In most cases in this study, the acoustic emission waveform contains more or less isolated events with breaks between them. Each event has a distinct maximum or sometimes two or more maximums. To negate the resonance effects, an envelope function is fitted to the waveform. As the signal is approximately symmetric around zero volts, the absolute value of the signal is calculated. After this, the envelope function was fitted to the waveform. The envelope was calculated by spline interpolation over local maximums. The minimum distance between local peaks was set to 16 µs, which is about five times the distance between peaks resulting from sensor resonance. This value was found to be suitable through trial and error method. Figure 2 presents a typical AE waveform sample fitted with an envelope function.

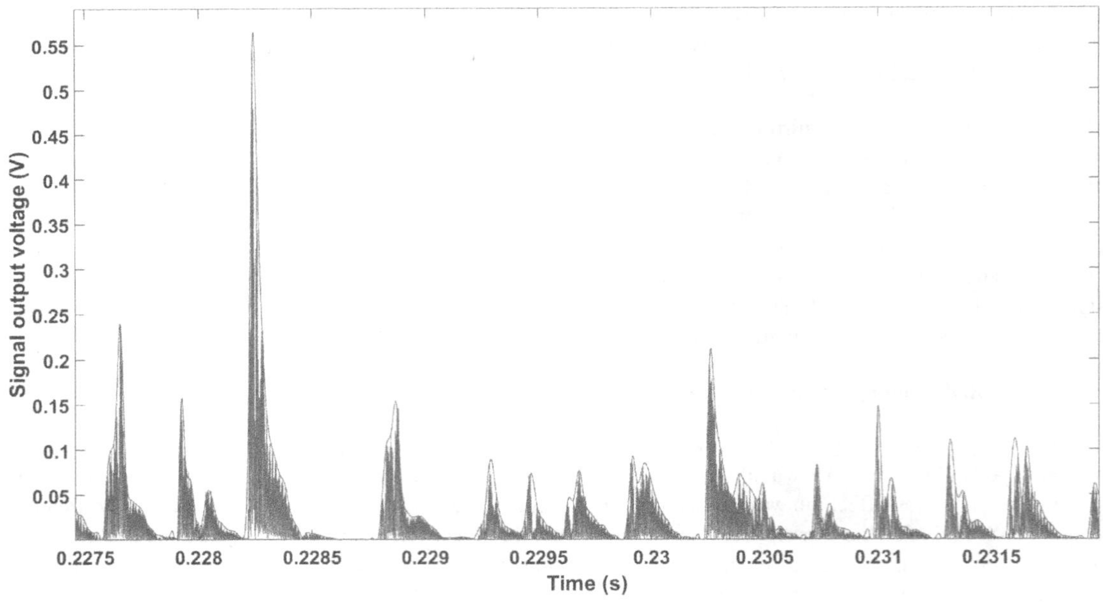

Figure 2. An extract of the AE waveform and the envelope function fitted to it.

As observed in figure 2, the envelope follows the waveform, properly addressing the maximum values of each event. If the event contains two distinct peaks, it is assumed that two events are overlapping and both are taken into account. In the area outside the events, the envelope function may create false peaks due to noise and the formulation of the function. This effect is negated in the results, as events falling below a certain threshold are considered as either static noise or bubble collapses too weak to be erosive.

Results

The peak value distributions for both the 2 MPa and the 4 MPa upstream pressure tests are presented in figure 3. The distributions are presented as cumulative so that each rate value represents the rate of voltage peak values larger than the corresponding voltage. The bin size for peak counting was 0.02 V. The rate was expressed in [3] and [4] in counts per second per area, with the area being the sensor active area. In this study, the sensor captures all events occurring in the sample, so the active area cannot be properly defined.

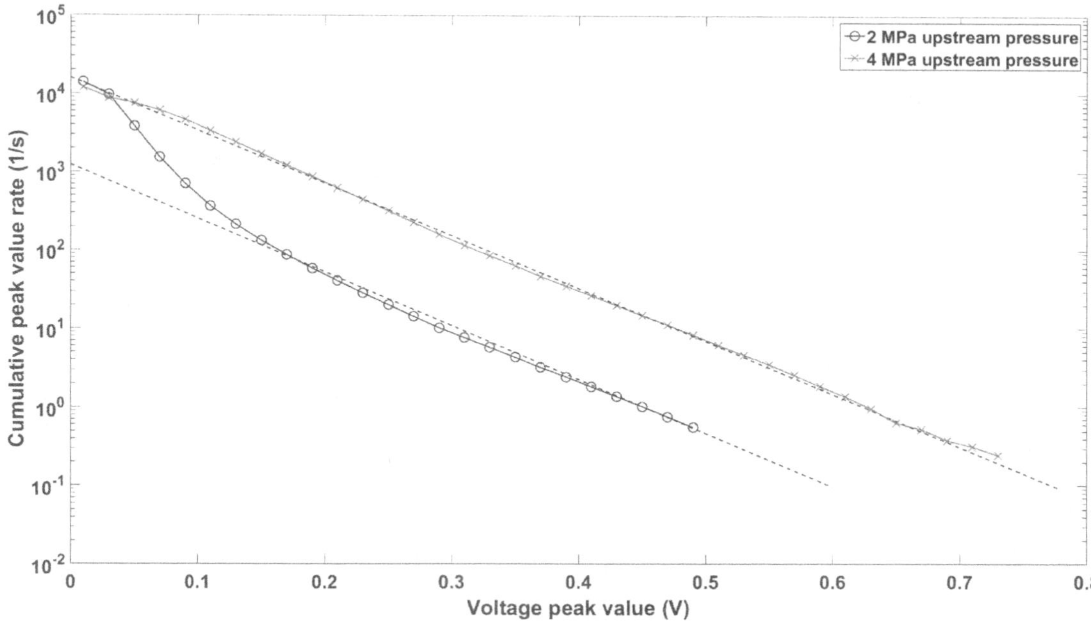

Figure 3. Cumulative peak rate vs. voltage peak value in linear-logarithmic scale. A linear fit was applied to the linear part of the curve.

The cumulative peak value distribution has a linear and a non-linear part in the linear-logarithmic scale. It was assumed that the non-linear part consists of static noise and insignificantly small bubble collapses. This study focuses only on the linear part of the distributions, corresponding to the expected exponential behavior of the cavitation impacts. The exponential law is presented in equation 2:

$$\dot{N} = \dot{N}_0 e^{-\frac{U}{U_0}} \tag{2}$$

where $\dot{N}$ is the peak rate, $\dot{N}_0$ is the reference peak rate, U is the voltage peak value and U_0 is the reference voltage. The reference values $\dot{N}_0$ and U_0 are presented in table 1.

Upstream pressure/cavity reference velocity	**Reference peak rate $\dot{N}_0$**	**Reference voltage U_0**
2 MPa / 63.2 m/s	1232 1/s	0.063 V
4 MPa / 89.4 m/s	15958 1/s	0.065 V

Table 1. Cumulative distribution reference values

The linear parts of the cumulative distributions in figure 3 are essentially parallel. The reference voltages U_0 are calculated from the slope of the linear fits and are thus almost equal. The reference peak rate $\dot{N}_0$ follows the cavity reference velocity V with a relation of $\dot{N}_0{\sim}V^{7.4}$. As the slope of the linear fit in linear-logarithmic scale does not change with the change of operating point, it means that the ratio between cumulative impact rates remains constant and independent of voltage peak value.

In this study, the connection between AE voltage peak values and the impact magnitude or impact damage is not addressed. Cavitation pitting in a virgin sample tends to produce pit sizes following a same type of exponential distribution as results in this and previous studies such as [9]. This fact is encouraging and shows that the results presented in this study somehow represent the physical phenomenon of cavitation pitting.

Conclusion

A method to monitor and characterize cavitation impacts by acoustic emission was presented. Acoustic emission was measured for two different operating conditions and envelope functions were fitted to the resulting waveforms. Voltage peak values were counted from the fitted envelope functions and cumulative distributions were calculated. Reference peak rates and reference voltages were calculated for both operating points. The reference voltage was about the same for both operating points and the reference peak rate had a relation of $\dot{N}_0 \sim V^{7.4}$.

Using the presented method, the cumulative distributions for acoustic emission voltage peak rate value may be calculated for any operating condition by calculating the reference peak rate $\bar{A}_0$ and reference voltage U_0 and then applying the relation between reference peak rates, if the distribution for one point is known. To validate the results, multiple operating points should be tested. With the assumption that the relation holds as stated, the acoustic emission distributions can be used in estimating cavitation pitting in the cavitation tunnel. The advantage of the presented method is that it is non-intrusive, as it does not require direct access to the flow. Future studies will address these issues in more detail. These results show the potential of using acoustic emission in cavitation detection and characterization in laboratory testing and eventually in actual hydro machines.

Acknowledgements

The authors would like to thank Tekes (the Finnish Funding agency for Technology and Innovation), Fortum Power and Heat Oy, Sandvik Mining and Construction Oy, Valtra Oy and Teollisuuden Voima Oyj for funding the research and for providing technical support.

References

[1] *The PREVERO cavitation erosion tunnel*. Web page. Available (accessed 13/12/2017): http://www.legi.grenoble-inp.fr/web/spip.php?article1265&lang=fr.

[2] Franc, J., Riondet, M., Karimi, A., Chahine, G. (2013). *Impact Load Measurements in an Erosive Cavitating Flow*. Journal of Fluids Engineering, 133(12).

[3] Hujer, J., Carrat, J., Müller, M., Riondet, M. (2015). *Impact load measurements with a PVDF pressure sensor in an erosive cavitating flow*. Journal of Physics: Conference Series, 656(012051).

[4] Franc, J. (2009), *Incubation Time and Cavitation Erosion Rate of Work-Hardening Materials*. Journal of Fluids Engineering, 131(2).

[5] Franc, J., Chahine, G., Karimi, A. (2014). *Pitting and Incubation period*. in: K. Kim, G.L. Chahine, J. Franc, A. Karimi (ed.), 10.1007/978-94-017-8539-6. Springer Dordrecht Heidelberg, pp. 37-69.

[6] Berchiche, N., Franc, J., Michel, J. (2002). *A Cavitation Erosion Model for Ductile Materials*. Journal of Fluids Engineering. 124(3).

[7] Hattori, S., Hirose, T., Sugiyama, K. (2010). *Prediction method for cavitation erosion based on measurement of bubble collapse impact loads*. Wear. 269(7-8).

[8] Okada, T., Iwai, Y., Hattori, S., Tanimura, N. (1995). *Relation between impact load and the damage produced by cavitation bubble collapse*. Wear. 184(2).

[9] Franc, J., Riondet, M., Karimi, A., Chahine, G. (2012). *Material and velocity effects on cavitation erosion pitting*. Wear. 274-275.

[10] Roy, S.C., Franc, J., Fivel, M. (2015). *Cavitation erosion: Using the material as a pressure sensor*. Journal of Applied Physics. 118(16).

[11] Boorsma, A., Fitzsimmons, P. (2009). *Quantification of Cavitation Impacts with Acoustic Emissions Techniques*. Proceedings of the 7th International Symposium on Cavitation (CAV).

[12] Yongyong, H., Zaiyang, S. (2012). *Experimental Research on Cavitation Erosion Detection Based on Acoustic Emission Technique*. 30th European Conference on Acoustic Emission Testing & 7th International Conference on Acoustic Emission.

[13] Schmidt, H., Kirschner, O., Riedelbauch, S., Necker, J., Kopf, E., Rieg, M., Arantes, G., Wessiak, M., Mayrhuber, J. (2014). *Influence of the vibro-acoustic sensor position on cavitation detection in a Kaplan turbine*. IOP Conference Series: Earth and Environmental Science. 22.

[14] Schmidt, H., Kirschner, O., Riedelbauch, S. (2014). *Cavitation measurements on a pump-turbine model*. Journal of Physics: Conference Series. 656(1).

[15] van Rijsbergen, M., Foeth, E., Fitzsimmons, P., Boorsma, A. (2012). *High-Speed Video Observations and Acoustic-Impact Measurements on a NACA 0015 Foil*. 8th International Symposium on Cavitation (CAV).

[16] Kim, K. Chahine, G., Franc, J., Karimi, A. (2014). *Advanced Experimental and Numerical Techniques for Cavitation Erosion Prediction*. 1st ed. Springer.

© 2018 ASME

10th International Symposium on Cavitation - CAV2018
Baltimore, Maryland, USA, May 14 – 16, 2018

CAV18-05038

Cavitation CFD Prediction for NACA0015 Hydrofoil Flow Considering Boundary Layer Characteristics

[1]Motohiko Nohmi*; [1]Tomoki Tsuneda; [1]Byungjin An; [1]Takayuki Suzuki; [2]Satoshi Watanabe; [2]Shin-ichi Tsuda

[1]EBARA Corporation, Fujisawa-shi, Kanagawa, Japan; [2]Kyushu University, Motooka Nishi-ku, Fukuoka, Japan

Abstract

The accuracy of cavitation CFD is not sufficient for the flow around very simple geometries such as single hydrofoil. The lift force and the cavity length are underestimated particularly for the high angle of attack cases. One of the possible reasons of the discrepancy is the singular characteristic of sheet cavitation in the boundary layers that is not taken into account in the recent cavitation models. It was observed that the sheet cavitation is not initiated in the laminar boundary layers. The inception and growth of the sheet cavitation occurs at the reattachment point of the laminar boundary layer separation or at the transition point of the laminar boundary layer to the turbulent boundary layer. These characteristics were reported in many literatures however they are not well modeled for cavitation CFD. In this paper the laminar boundary layer zone is identified for NACA0015 hydrofoil from the high speed video observation. Then, cavitation CFD is carried out by deactivating the cavitation model at the estimated laminar boundary layer zone. This modified cavitation CFD results are discussed and compared to conventional cavitation CFD results and measurement results.

Keywords: cavitation CFD; NACA0015 hydrofoil; boundary layer

Introduction

Cavitation CFD prediction functions are built in many commercial codes currently, and are frequently used in the development and design of turbomachinery such as pumps, hydraulic turbines and the ships propellers. On the other hand, whether the analysis accuracy of them is sufficient or not remains a matter of debate. In Japan, the effect of the cavitation coefficient on the lift force of a hydrofoil was analyzed with various kinds of cavitation CFD codes and the prediction performance was evaluated in the "Industry-University Collaborative Research Project on Numerical Predictions of Cavitating Flows in Hydraulic Machinery" conducted in Turbomachinery Society of Japan from 2009 to 2011. As a result, it was found that none of these codes gave sufficient analysis accuracy, and the reliability of cavitation CFD came into question.[(1)] Responding to this situation, several groups began studies after the project to track down the cause and discuss improvement measures.[(2-5)] As part of them, this study also conducts an analysis of the NACA0015 hydrofoil that was analyzed in the above project, considering the characteristics of cavitation in a boundary layer that was not considered sufficiently in the existing analysis, and evaluates its effect on analysis accuracy.

Target of the Analysis

The target of the analysis is the flow around the NACA0015 hydrofoil installed in a cavitation tunnel. The result of an experiment conducted in Marine Propeller Cavitation Tunnel at The University of Tokyo is used as the verification data[(1)]. The analytical domain is determined according to this experiment. The shape of the cross section of the measuring part in the tunnel is 600 mm in length and 150 mm in width. The chord length of the NACA0015 hydrofoil is 150 mm and the width 150 mm. In this study the analysis is conducted under the conditions of the inlet flow velocity = 8 m/s (constant) and the angles of attack = 6° and 8°. The analytical domain ranges up to 5C upstream from the center of the chord and down to 5C downstream from it, where C is the length of the chord. For details of other analysis conditions, see the reference [(1)]. The analysis code used in this study is a commercial cord, ANSYS-FLUENT. In a two-dimensional steady analysis, the number of grids is about 90,000, the turbulent model is SST-k-ω model, and the cavitation model is the Zwart-Gerber-Belami model[(6)].

In this study, the analysis result is arranged using the lift coefficient, C_l, with the cavitation coefficient, σ, and the static pressure coefficient, C_p, defined below:

*Corresponding Author, Motohiko Nohmi: nohmi.motohiko@ebara.com

$$\sigma = \frac{p_\infty - p_V}{\frac{1}{2}\rho_l u_\infty^2} \quad (1) \qquad C_p = \frac{p - p_\infty}{\frac{1}{2}\rho_l u_\infty^2} \quad (2) \qquad C_l = \frac{F_l}{\frac{1}{2}S\rho_l u_\infty^2} \quad (3)$$

where p is static pressure, p_∞ and u_∞ are the average static pressure of flow and the flow velocity at the inlet of the analytical domain respectively, p_V is saturated vapor pressure, ρ_l is the density of single-phase water, F_l is lift force, and S is the area of blade.

Characteristics of Cavitation in a Boundary Layer

The scale of the microscopic behavior of bubbles those form cavitation is extremely small compared to the representative dimensions of pump impellers, hydraulic turbine runners, and the blades of a ship propellers. It is therefore difficult to directly analyze the overall flow of fluid machinery using grids with a size that can track or capture the surface phenomena of cavitation bubbles. Thus, most of the current cavitation CFD use a so-called cavitation model, in which grids that can resolve the flow around the blades of fluid machinery are used, and cavitation phenomena with a sub-grid scale is modelled using simplified equations. Many kinds of cavitation models have been proposed. Here, as a representative example, the model by Chen and Heister shown in Eq. (4) is introduced which was proposed in an early stage and therefore simple, and whose physical meaning is easy to understand.(7) The Zwart-Gerber-Belami model mentioned above is one of the models developed based on the model by Chen and Heister.

$$\frac{D\rho}{Dt} = c(p - p_V) \quad (4)$$

where ρ is the local average density of the mixed fluid of gas (vapor phase) and liquid phase, and p is the static pressure defined locally in the mixed fluid. In this equation it is assumed that when there is a difference between the local static pressure and the saturated vapor pressure in a mixed fluid of water and vapor, the progress of phase change depends on the magnitude of the difference, with 'c' being the empirical constant that determines the change rate. If the local static pressure equals to the saturated vapor pressure, phase change does not progress anymore, resulting in an equilibrium state with a constant mixed-fluid density. In other words, it is assumed in this model that a water-vapor system has a nature of proceeding to an equilibrium state. This is an easy to understand concept, but it has to be noted that there are many cases familiar to us where an equilibrium state is not easily reached such as supercooling and supersaturation.

As a result of a literature research in this study, it became apparent that many cases were reported in actual cavitation where an equilibrium state was not reached in a short time and a nonequilibrium state continued. For example, it was reported by Huang and Peterson that the occurrence of cavitation on a hydrofoil was limited to the location of separation-reattachment of a laminar boundary layer or the location of transit from a laminar boundary layer to a turbulent boundary layer. Therefore, they point out that the characteristics of Reynolds numbers and boundary layers differ among blades with different sizes even if their shape is identical, and as a result, the positions of the occurrence of cavitation also greatly differ among them, leading to the difference between small-scale model tests and actual blade tests.(8) Kuiper reported a similar result, showing that cavitation did not occur when a laminar boundary layer remained on a hydrofoil even if the absolute static pressure was negative, meaning that a tensile force existed in water.(9) Kato et al. briefly summarized in his book the occurrence conditions and types of cavitation as a flowchart, which showed a result similar to the above ones.(10) From the above results, it is anticipated that a cavitation analysis of hydrofoils that does not appropriately consider the characteristics of cavitation in a boundary layer does not give sufficient accuracy. Phase change of liquid water does not occur in a laminar boundary layer, and cavitation occurs only after fluctuations by turbulent flow transition. It resembles the bumping phenomena occurring by stimulation after a supersaturation (superheated) state is reached by gentle heating or rapid heating of still water.

On the basis of the above discussion, an important subject of this study is to predict with high accuracy the separation-reattachment of a laminar boundary layer on a hydrofoil or the transition from a laminar boundary layer to a turbulent boundary layer. An option to achieve this is to conduct a high-accuracy three-dimensional turbulent analysis such as LES and identify, using that result, the location where a laminar boundary layer disappears. However, whether LES should be adopted or not is a continuous matter at present because an LES analysis requires considerable computational resources/loads, and to what degree the result of a single-phase LES analysis can be applied to the occurrence of cavitation is not clear in consideration of the possibility that cavitation itself may change the characteristics of a boundary layer. In this study, therefore, we decided to obtain it from the image of the experimental result as a simpler and more reliable method. By observing the high speed motion pictures of the occurrence of cavitation on the NACA0015 blade at angls of attack of 6° and 8°, it was found that the front edge location of sheet cavitation did not change very much when the cavitation coefficient changed. It was also found that the front edge location of sheet cavitation was not influenced very much by the unsteady fluctuation of cavitation on the blade. The front edge location of sheet cavitation was measured based on the image, and it was 7.8 % of the chord length downstream from the forward edge of the blade for the angle 6°, while 5.5 % of the chord length downstream for the angle 8°. In this study, using these values, and on the assumption that a laminar boundary layer is kept from the stagnation point at the leading edge of the hydrofoil blade to the location of these values although this is considerably an artificial assumption, the cavitation model is deactivated upstream of these points and it is applied downstream of them. In the next chapter, the case this method is adopted is compared with the case the cavitation model in a standard manner is used.

Results and Discussion

Fig. 1 shows the comparison between the experimental value of distribution of the static pressure coefficient, C_P, on the blade under the no-cavitation condition, and the calculated value by an incompressible analysis not considering cavitation for angls of attack of 6° and 8°. As seen in Fig. 1, there is close agreement between the experimental value and the calculated value. The validity of the incompressible analysis, which is the start point of the cavitation analysis, was ascertained.

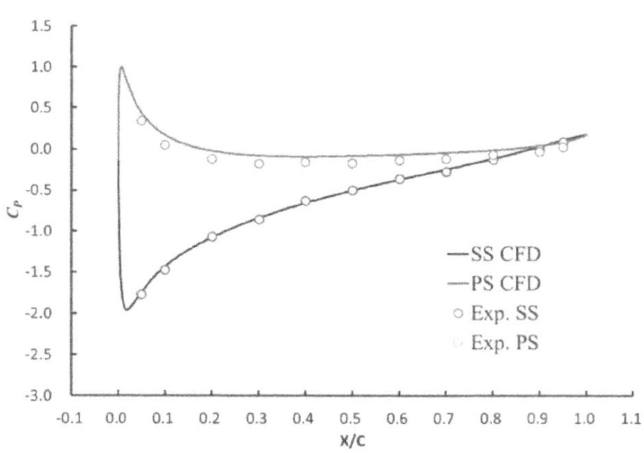

Figure 1(a) Angle of Attack 6 deg

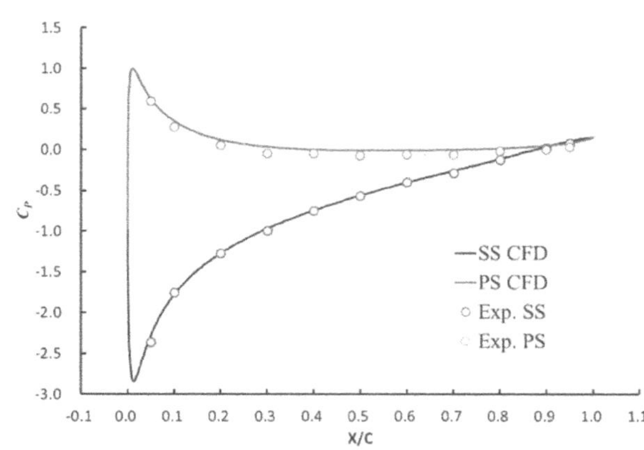

Figure 1 (b) Angle of Attack 8 deg

Figure 1 Static Pressure Distribution without Cavitation

Fig. 2 shows the σ-C_l diagram for angles of attack of 6° and 8°. For 6°, when a laminar boundary layer is considered, the point from which the lift force begins to decrease moves toward the lower side of σ, that is, the direction coming close to the experimental result, compared to the calculation result using the standard cavitation model. However, the trend that the lift force increases and then decrease in the low-σ region is not predicted in the calculation. For 8°, when a laminar boundary layer is considered, the point from which the lift force begins to decrease moves toward the lower side of σ compared to the calculation result using the standard cavitation model. However, the trend that the lift force does not decrease until σ is about 1.2, which is seen in the experimental result, is not predicted. The trend that the lift force decreases as σ decreases is common in two types of calculation results, which is a considerable difference compared to the experimental result.

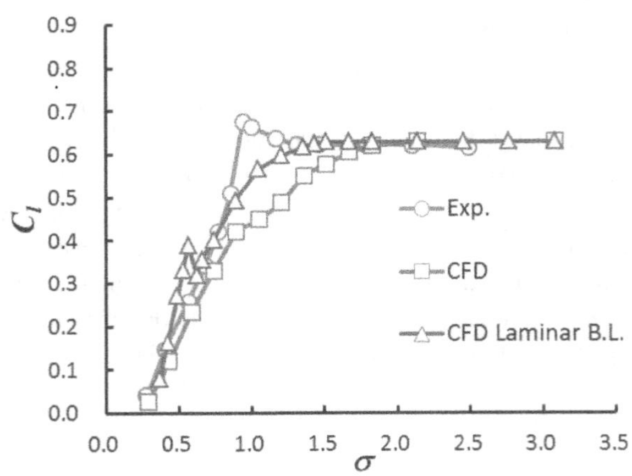

Figure 2(a) Angle of Attack 6 deg

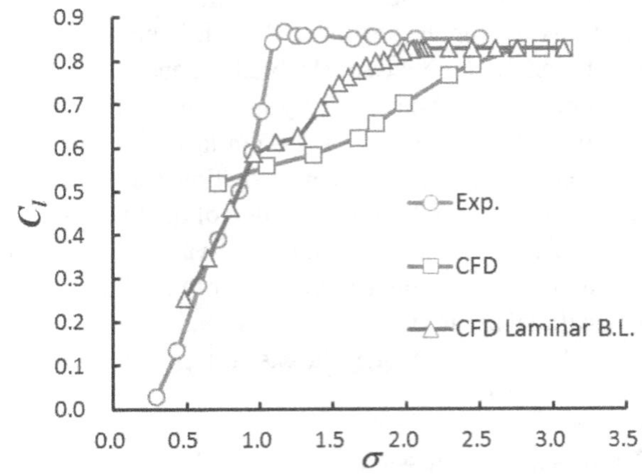

Figure2 (b) Angle of Attack 8 deg

Figure 2 Cavitation Coefficient and Lift Coefficient Characteristics

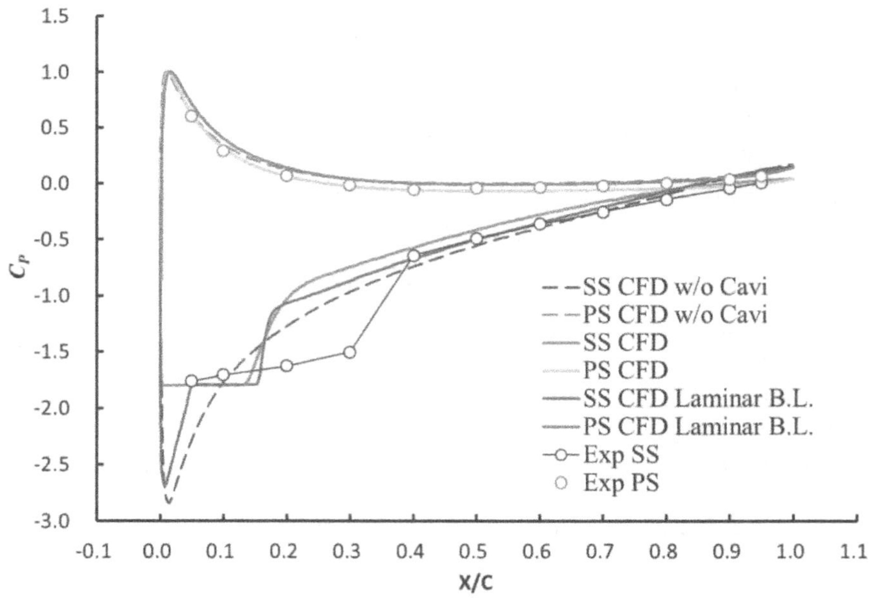

Figure 3 Static Pressure Distribution at σ=1.8

Figure 4(a) Conventional Cavi. Model

Figure 4(b) Cavi. Model with Laminar B.L.

Figure 4 Stream Lines and Void Fraction Contour at σ=1.8

Fig. 3 shows the comparison of the calculation result and the experimental result for the static pressure distribution on a blade surface for $\sigma = 1.8$. Fig. 4 shows the stream lines around the blade and the void fraction distribution. It is seen in Fig. 3 that there is good agreement between the experimental result and the calculation result for the static pressure distribution on the pressure surface even when cavitation exists. When a laminar boundary layer is considered, the peak of the lowest static pressure upstream from the point the laminar boundary layer ends becomes lower than the saturated vapor pressure, but is higher than the peak of negative pressure of the analysis result without cavitation. On the other hand, when the standard cavitation model is used, the saturated vapor pressure is nearly equal to the lower limit of the static pressure. The validity of the pressure being lower than the saturated vapor pressure cannot be verified as there is no static pressure measuring point at the relevant position. In the experiment, it is not easy to obtain reliable data because absolute negative pressure cannot be measured easily and a pressure measuring hole itself may become a cavitation generation source. In Fig. 3, the length of cavitation is predicted to be smaller compared to the experiment. This tendency is also pointed out in a past study.(1) It is also seen in Fig.4 that the position of the front edge of sheet cavitation moves downstream in the analysis considering a laminar boundary layer compared to the standard model. In both results the stream line is separated from the front edge of sheet cavitation.

Conclusion

As a cavitation fluid analysis of the NACA0015 hydrofoil, calculation was conducted considering the nature that cavitation did not occur in a laminar boundary layer. Although some improvement was seen in the calculation result, the prediction accuracy was not sufficient yet. Overall improvement considering cavitation models, turbulent models, computational schemes, analysis grids, three-dimensionality, unsteadiness, etc., is needed.

Acknowledgments

This analysis was made with huge efforts by Mr. Naoki Kodama. We would like to thank his contribution.

References

[1] Kato C., (2011). *"Industry-University Collaborative Project on Numerical Predictions of Cavitating Flows in Hydraulic Machinery -Part 1: Benchmark test on cavitating hydrofoils-"*, Proc. AJK2011-FED,, AJK2011-06084.
[2] Tsuda, Tani and Yamanishi, (2012). *"A Construction of a Multi-Process Cavitation Model for Cold Water (Implementation of Inception/Collapse and Coalescence/Break-up)"* (in Japanese), Transactions of the JSME (B) Vol. 78, No. 790 (2012-6), pp. 1165-1180.
[3] Suzuki and Kato, (2013). "Essential Problems Associated with Numerical Simulation of Cavitating Flows Based on Homogeneous Models" (in Japanese), Transactions of the JSME (B) Vol. 79, No. 808 (2013-12), pp. 2743-2758.
[4] Watanabe S., Tsuru W., Yamamoto Y. and Tsuda S, (2016). *"Cavitation simulation using a new simple homogeneous cavitation model"*, ISROMAC 2016.
[5] Iga, Matsuura and Tateishi, (2016). *"Consideration of Numerical Model of Cavitation in Hydraulic Oil Flow"* (in Japanese), Symposium on Cavitation (18th).
[6] Zwart P. J., Gerber A. G. and Belamri T., (2004). "A Two-Phase Flow Model for Predicting Cavitation Dynamics", Proc. ICMF Fifth International Conference on Multiphase Flow, see the following website:
http://www.afs.enea.it/project/neptunius/docs/fluent/html/th/node343.htm
[7] Chen Y., Heister S. D. (1995). *"Two-phase modeling of cavitated flows"*, Computers and Fluids 24, 7, pp. 799–809.
[8] Huang T. T., Peterson F.B., (1976). *"Influence of Viscous Effects on Model/Full-Scale Cavitation Scaling"*, Journal of Ship Research, Vol. 20, No. 4, pp. 215-223.
[9] Kuiper G., (2010). *"Cavitation in Ship Propulsion"*, Chapter-7, pp. 47-79.
https://ocw.tudelft.nl/wp-content/uploads/Chapter-7_02.pdf
[10] Hiroharu Kato (Editor),(2016). "Shinban Cavitation Kiso To Saishin No Shinpo ; Foundations of Cavitation and Recent Progress (New Edition) (in Japanese), Morikita Publishing, pp. 76-79.

© 2018 ASME

Cavitation Dynamics at Sub-Millimeter Scale

[1]Kunpeng Long, [2]Matevž Dular, [1,3]Olivier Coutier-Delgosha

[1]*Arts et Métiers ParisTech, Lille, France*

[2]*University of Ljubljana, Slovenia*

[3]*Virginia Tech, Kevin T. Crofton Aerospace and Ocean Engineering Dept., Blacksburg, VA, USA*

Abstract

This paper is focused on the dynamics of cloud cavitation at Sub-millimeter scales, between 0.1 and 10 mm. A simple flow configuration consisting in a small diameter jet impacting a wall at 90° and then flowing radially between two plates is used. Cavitation is obtained in the gap between the two plates, due to the local flow acceleration. The scale of the setup was changed by varying separately the gap between the two plates, and the nozzle diameter. Observations were performed from the transparent bottom plate, using a high-speed camera and intense illumination, for various Reynolds and cavitation numbers. The analysis is focused on the variations of the cavity length and the Strouhal number based on the characteristic frequency of the flow instability.

Keywords: Small Scale, cavitating flow, instability, flow imaging

1.Introduction

In the last decades, small-scale hydraulic systems have developed, due for example to new applications requiring small size underwater machines such as pipeline maintenance (Kim 2008) and urban underground or ocean engineering (Marani et al. 2009). In the medical area as well, the use of micro medical robots provides more secure, more accurate, and less invasive techniques for surgery (Kim et al. 2004). These small devices often involve high speed flows and/or low-pressure operation, which favors the development of cavitation. Although cavitating flows have been studied for a long time for various naval, nuclear or medical applications (to name only a few), there is almost no analysis in the literature of the scale effect on their structure and dynamics. However, the occurrence of scale effects on cavitation has long been known and was already mentioned by Ackeret in 1930. Numerous studies of scaling laws in cavitating flow were performed, but they all dealt with the problem of cavitation occurrence, while the case of fully developed cavitating flow was not considered (Holl *et al.* 1972; Arakeri & Acosta 1973; Gates & Billet 1980; Billet & Holl 1981; Ooi 1985; Amromin 2002). In addition, most of the studies have focused on large dimensions, typically between the cm and the meter, while Dular et al. (2012) have recently investigated the influence of lower scale (down to 3 mm) on cloud cavitation. In the later study, significant effects of the small size are reported, especially a damping of the unsteady effects at the smallest scales.

The objective of the present study is to extend this previous work to smaller scales, by analyzing the cavitation behavior for characteristic dimensions between 100 μm and 10 mm. Investigation of the flow at such small scale dramatically increases the complexity of the setup, especially to manufacture test sections with reliable dimensions, and the difficulty of investigation. Therefore, a very simple device, based on cavitation between two flat plates, has been used in the present study. Flow imaging with a high-speed camera has enabled to characterize various behaviors, according to the cavitation number, the Reynolds number and the scale of the setup.

2.Experiments Design and Setup

Cavitation tests were conducted in two small-scale cavitation tunnels at the University of Ljubljana (Slovenia) and Arts et Metiers ParisTech (Lille, France). The latter is schematically represented in figure 1.

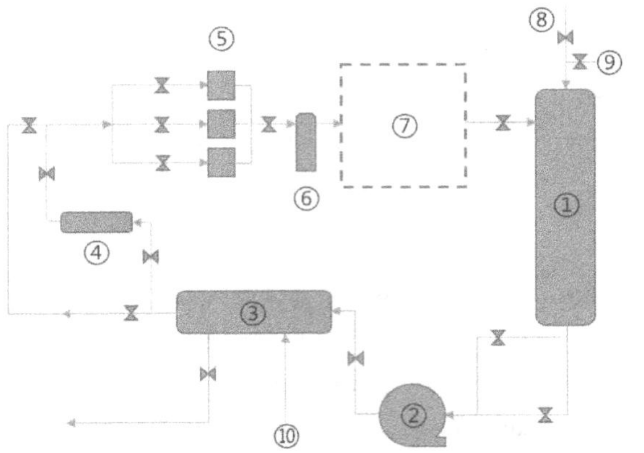

① Water tank
② Circulation pump
③ Heat exchanger
④ Heater
⑤ Flow meters
⑥ Filter
⑦ Test section
⑧ Water inlet
⑨ Vacuum pump / compressor
⑩ Cooling water

Figure 1. Cavitation tunnel at Arts et Metiers ParisTech

Circulation of water is obtained with a Salmson Multi HE 403 pump with variable rotation speed to set the flow rate. At the pump delivery, a tank completely filled with the circulation water is used for water cooling in order to maintain a constant temperature. The volume flow rate is measured a Bürkert type 8032 (uncertainty 0.15 l/min after in situ calibration) turbine flow meter for flow rates lower than 10 l/min and with a Bürkert type 8045 (2 % uncertainty) flow meter for higher flow rates. Temperature is obtained with a type K thermocouple, which is directly in contact with the circulation water. Upstream from the test section, a second tank partially filled with water is used to filter the flow rate and/or periodical pressure fluctuations due to the passage of the pump blades. The reference pressure is measured 30 mm upstream from the test section with a Rosemount 3051 pressure sensor. The uncertainty of the measurements was of the order of 10 mbar. The partially filled tank is connected to a compressor and a vacuum pump, which enables to vary the pressure in this tank between 0.1 bar and 3.5 bar, and thus to adjust the pressure in the test section.

Cavitation is obtained with a radial jet impacting a flat plate, as shown in Fig.2. The flow enters the test section at high velocity through a nozzle with a diameter D drilled in the upper plate. After it impact the bottom plate, a radial outward cavitating flow, between the bottom and upper plates is obtained. The gap between the two plates is denoted H. Cavitation is initiated slightly downstream from the edge of the inlet nozzle, and it collapses abruptly as the fluid moves radially outwards, since the pressure re-increases in the radial direction.

Figure 2. test section (a) scheme of the flow passage, (b) Top view of the bottom plate, (c) upper plate

A total of 13 geometrical configurations have been investigated, with a nozzle diameter varying between 1 and 10 mm and various gaps between the two plates, as indicated in table 1. Note that two experimental campaigns were performed. In the data set#1 (cases A to F in table 1), a D = 3 mm diameter hole was mainly considered, with H varying between 0.1 mm and 1 mm. Two additional configurations with a bigger 10 mm hole diameter (H = 3 mm) and a smaller 1 mm diameter (H = 0.33 mm) were also studied to investigate the effect of the hole size. As reported in Section 3 hereafter, an abrupt change of cavitation behavior was detected between H = 0.3 mm and H = 0.7 mm, so a second experimental campaign was setup to investigate the transition between these two behaviors. In this 2nd set of experiments (cases G to M in table 1), holes of 3 mm and 6 mm diameter have been considered, for various values of H between 0.3 and 0.9 mm.

Geometry	*D*	*H*	*D/H*	*Data Set*
A	1	0.33	3.0	1
B	3	0.1	30.0	1
C	3	0.33	9.1	1
D	3	0.7	4.3	1
E	3	1	3.0	1
F	10	3.33	3.0	1
G	3	0.3	10.0	2
H	3	0.5	6.0	2
I	3	0.7	4.3	2
J	3	0.9	3.3	2
K	6	0.5	12.0	2
L	6	0.7	8.6	2
M	6	0.9	6.7	2

Table 1. Investigated geometrical configurations

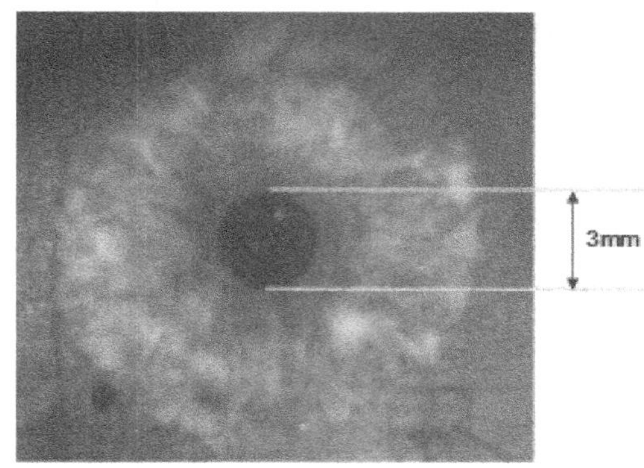

Figure 3. Example of cavitation in case J

A MotionBLITZ EoSens mini1 high speed camera is used for flow visualization from the bottom of the test section. Figure 3 presents the typical cavitation pattern obtained between the two plates. The maximum resolution of the camera is 1280×1024 pixels at 2000 fps, while it was used for the present tests at a higher frequency of 20K to 50K fps (depending on the frequency of cloud cavitation) with a typical resolution of 512×100 pixels. For that purpose, only a small band of cavitation was recorded at high frequency, instead of the whole axisymmetric cavitation pattern (see figure 4). In the second campaign, an optimization of the image quality was conducted, by testing 3 optics (Nikon AF nikkor 50mm, Nikon AF-S DX nikkor 40mm, and Nikon micro nikkor 105mm) combined extension rings and various continuous illumination sources. As a result, the contrast and the resolution of the image could be significantly improved for all data sets #2 (figure 4).

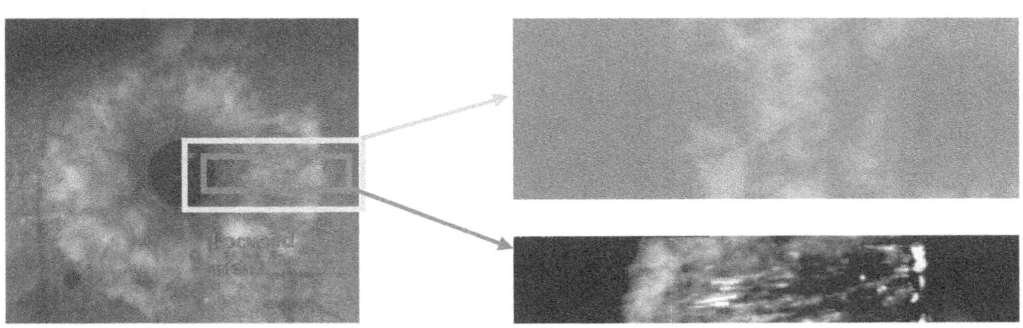

Figure 4. cavitation images (a) whole pattern, (b) zoom for high frequency acquisitions in data sets #1, (c) zoom in data sets #2

Various pressure and velocity conditions were investigated for each configuration, by varying the Reynolds number $Re = \frac{\rho v \cdot H}{\mu}$ and the cavitation number $\sigma = \frac{P - P_{vap}}{\frac{1}{2}\rho v^2}$.

Where v is the flow velocity between the plates at the nozzle radius and P is the inlet pressure measured upstream from the injection nozzle. Post-processing of the images was performed with MATLAB. It basically consists in analyzing the grey level variations in the images to extract the mean cavity length L and the cavity oscillation frequency f. The cavity length is defined as the distance from the edge of the nozzle (location of cavitation inception) to the downstream end of the attached (non-fluctuating) cavitation area. For frequency analysis, FFTs of the mean grey level in 10 windows of interest (size 10 × 10 pixels by default) are calculated, as shown in figure 5a. These 10 FFTs are eventually superimposed as shown in figure 5b, to get the main and secondary frequencies. In case the results were unclear, the windows of interest were adjusted on a case by case basis to increase the intensity of the max. peaks in the FFTs.

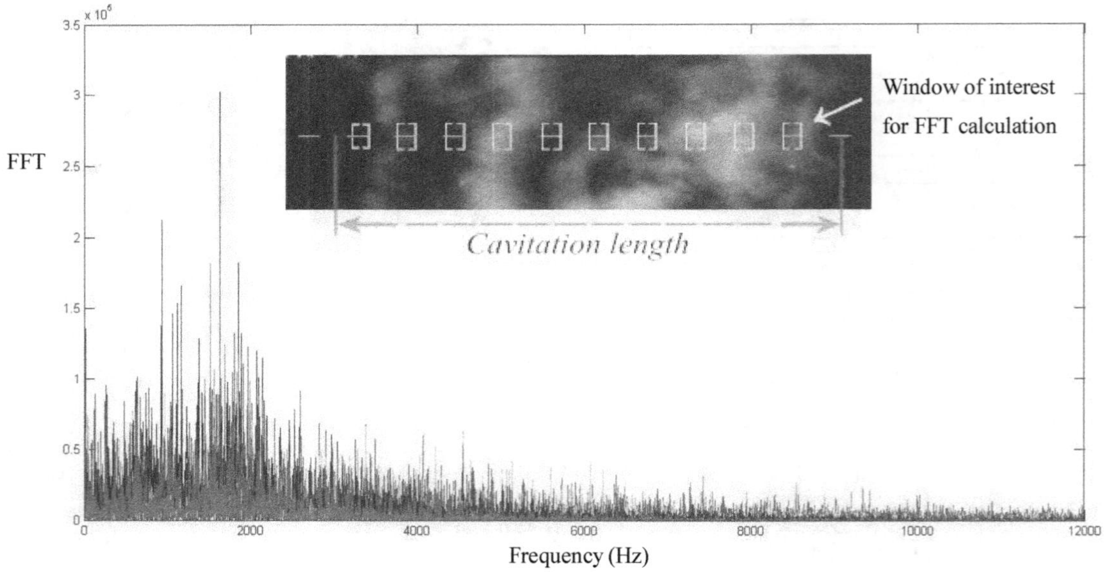

Figure 5. (a) Windows of interest for FFT calculation, (b) Superimposed 10 FFTs

3.Experimental results and discussion

The general cavitation behavior is similar in data sets 1 & 2. The axisymmetric cavitation cloud shows a periodic expansion and shrink in all experimental conditions (gap, diameter, and volume flow rate). In the results presented here, the Strouhal number $Str = \frac{f \times L}{v}$ is drawn according to $L_{norm} = \frac{L}{H}$. In figure 6, the results for data sets #1 (cases B to E) are drawn.

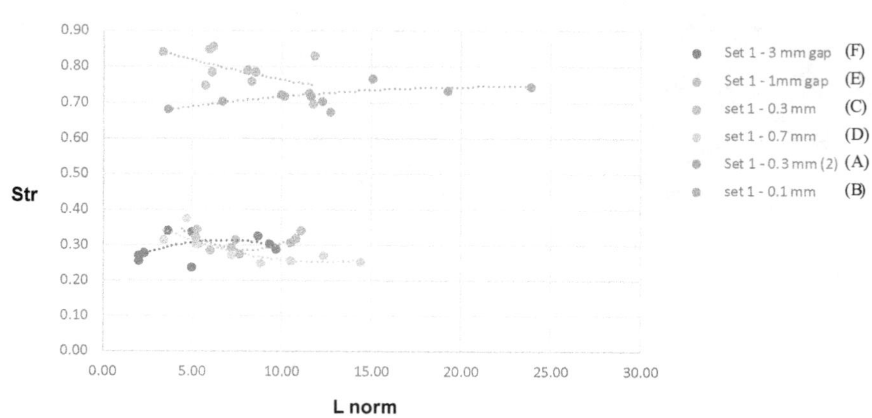

Figure 6. Variation of Str according to L_{norm} for the data sets #1

As mentioned previously, an abrupt modification of the flow behavior is obtained when H is decreased: for H higher or equal to 0.7 mm, the standard Strouhal number close to 0.3 is obtained, as usually reported in all studies of cloud cavitation at larger scale. So, it suggests that the flow dynamics is not influenced by the downscale of the geometry in such case. Conversely, when H is decreased down to 0.3 or lower, the Strouhal number becomes much higher, in the range 0.65 – 0.85. This significant and sudden change has motivated the second experimental campaign, where the objective was specially to investigate the transition from H = 0.7 mm down to 0.3 mm, and to determine which changes in the flow features could be responsible for this different unstable behavior.

Surprisingly, the results presented in figure 8 for data sets #2 look very different from the previous ones. Most of the values of the Strouhal number fall in the range 0.2 – 0.4 (for all geometrical configurations). No increase of the Strouhal number is observed when H is reduced down to 0.3 mm. Conversely, the Strouhal number increases up to 1.4 for very short sheet cavities in the specific case J where H = 0.9 mm and D = 3 mm. Generally, Str increases significantly for small values of L_{norm}. This specific behavior is related to the oscillation of very short cavities, where the characteristic frequency is hardly detectable, so it should be considered circumspectly here.

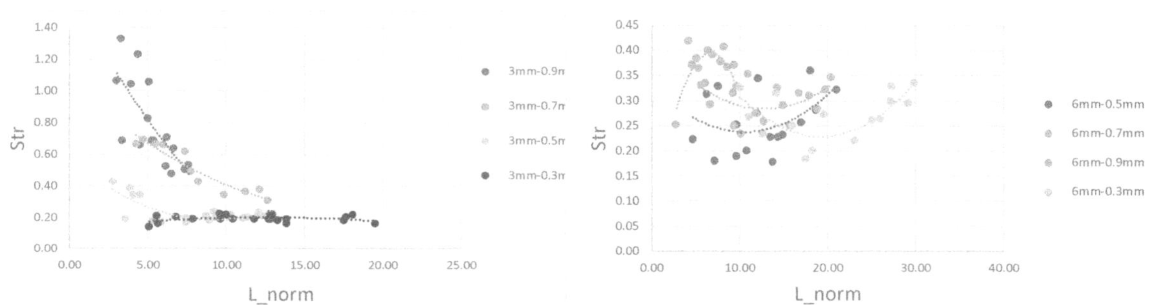

Figure 7. Strouhal number Str according to L_{norm} for cases G to M (data sets #2). In the caption, the first number refers to D and the second to H.

This difference between the two sets of data is highlighted in figure 8, which combines the results of both campaigns for the hole diameter D = 3 mm. The reason for this discrepancy is still currently investigated and the final results will be presented at the CAV 2018 conference. One possible explanation is related on some differences in the experimental process, between the two campaigns: in the first one, the Reynolds number was maintained constant and the cavity size was changed by varying the pressure with a vacuum pump and a compressor. In the second campaign, no pump was used, so the cavity size depends only on the flow velocity. A third experimental campaign is planned where a broader range of H and D values will be investigated.

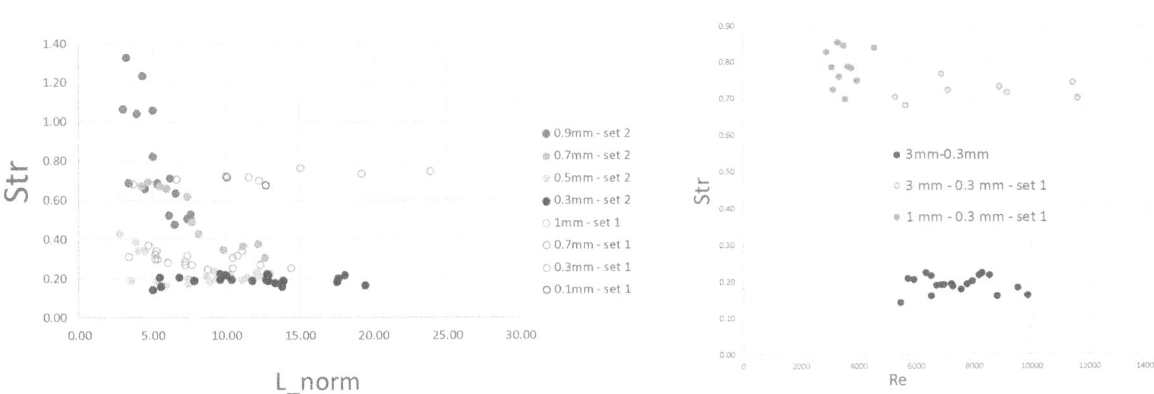

Figure 8. (a) comparison of the Strouhal number for the two sets of data and D = 3 mm, (b) Influence of the Reynolds number on the results for D = 3 mm and H = 0.3 mm

References

J. Ackeret (1930), Experimentelle und theoretische Untersuchungen über Hohlraumbildung (Kavitation) im Wasser. *Forschung im Ingenieurwesen* **1**(2), 63-72.

E. Amromin (2002) Scale Effect of Cavitation Inception on a 2D Eppler Hydrofoil. *J. Fluids Eng.* **124**(1), 186-193.

V.H. Arakeri, A.J. Acosta (1973), Viscous Effects in Inception of Cavitation on Axisymmetric Bodies. *J. Fluids Eng.* **95** (4), 519-526.

M.L. Billet, W. Holl (1981), Scale Effects on Various Types of Limited Cavitation. *J. Fluids Eng.* **103**(3), 405-414.

M. Dular, I. Khlifa, S. Fuzier, M. Adama Maiga, O. Coutier-Delgosha (2012), Scale effect on unsteady cloud cavitation, *Exp. Fluids* **53**, 1233- 1250

E.M. Gates, M.L. Billet (1980), Cavitation Nuclei and Inception. IAHR Symposium, Tokyo, Japan.

J.W. Holl, R.E.A. Arndt, M.L. Billet (1972), Limited Cavitation and the Related Scale Effects Problem. The 2nd Int. JSME Symp. on Fluid Machinery and Fluidics, Tokyo, Japan.

B. Kim, Y. Jeong, T. Kim, J.O. Park, S. Song (2004), Micro capsule type robot, US Patent 6,719,684

J. H. Kim (2008), Design of a fully autonomous mobile pipeline exploration robot (FAMPER)

G. Marani, S. K. Choi, J. Yuh (2009), Underwater autonomous manipulation for intervention missions AUVs, *Ocean Engineering* **36**(1), 15-23

K.K. Ooi (1985), Scale effects on cavitation inception in submerged water jets: a new look. *J. Fluid Mech.* **151**, 367-390.

 © 2018 ASME

10th International Symposium on Cavitation - CAV2018
Baltimore, Maryland, USA, May 14 – 16, 2018 CAV18-05040

Cavitation Dynamics in Wakes Behind Bluff Bodies

[1,2]Lisa Deijlen; [1]Anubhav Bhatt; [1]Juliana Wu; [1]Harish Ganesh*; [1]Steven Ceccio

[1]*University of Michigan, Ann Arbor, MI, USA;* [2]*University of Twente, Enschede, Netherlands*

Abstract

Properties of wakes behind bluff bodies such as shedding frequency, vortex spacing, and drag change with the presence of cavitation. As extent of cavitation in the wake increases from inception to an attached super-cavity, wake shedding Strouhal number increases, reaching a peak value before decreasing as the wake forms a super cavity. The physical mechanism responsible for this observed change in shedding dynamics is yet to be fully understood. In the current study, we employed time resolved X-ray densitometry and high-speed videography to study the cavitation dynamics in the wake of a triangular, nominally two-dimensional wedge in a re-circulating water tunnel to understand the underlying mechanisms responsible for cavity formation and shedding. Mach numbers of the bubbly mixture, obtained based on base pressure and void fraction measurements, revealed that the flow transition from subsonic to supersonic coincides with the increase in Strouhal number suggesting the importance of compressibility effects on the wake dynamics.

Keywords: cavitation; bluff bodies; Strouhal number

Introduction

Wakes of bluff bodies placed in a steady inflow experience flow induced unsteadiness characterized by the shedding of alternate vortices, termed von Kármán vortex street. Characteristics of such flows has been studied extensively in the past, and it has been found that the wake properties such as vortex shedding frequency and form drag depend on the Reynolds number and blockage effects [1], [2], and [3]. In the case of a liquid flow capable of experiencing cavitation in the shed vortex cores, vortex shedding frequencies at given Reynolds number depend on the extent of cavitation, dictated by the mean pressure level.

Cavitating wakes produced by bluff bodies have been studied in the past by [4] and [5]. It was observed that the shedding frequency, characterized by the Strouhal number, with decreasing cavitation number, increases from the non-cavitating shedding frequency, attaining a peak value, and decrease again. More recently, [6] studied the effect of cavitation for sharp bluff bodies, and found that vortex spacing also changed with the extent of cavitation. However, despite thorough studies, no satisfactory mechanism responsible for the observed behavior has been proposed.

Cavitating flows with high vapor content can have low speed of sound values as discussed in [7] and [8]. Recently, [9], found that propagating bubbly shocks act as a primary mechanism of shedding in partial cavities. It was found that the compressibility of the bubbly mixture can play an important role in dictating the vapor-liquid flow dynamics. In the present study, the influence of the mixture compressibility on the observed wake cavitation dynamics is explored by measuring the void fraction flow fields, obtained using time-resolved X-ray densitometry. With the help of time synchronous unsteady base pressure measurements, and time-averaged base static pressure measurements, the role of mixture speed of sound is being tried to understand.

Experimental setup

Experiments were carried out at the University of Michigan 9-Inch re-circulating water tunnel. The tunnel has a 6:1 round contraction leading into a test section with a diameter of 23 cm. The test section then transitions to a square cross-section that is 21 cm by 21 cm. The flow velocity U_o in the tunnel test section can be varied from 0 to 18 m/s and the static inlet pressure p_o from near vacuum to 200 kPa absolute pressure. In the present experiments, the test section was further reduced in cross-sectional area to a conduit that had a 7.6 cm by 7.6 cm cross-section. This was done to reduce the baseline X-ray attenuation produced by the non-cavitating flow. A schematic of the test-section, wedge model, hydrophone, and measurement locations are shown in Figure 1. The set-up and the high-speed imaging is discussed in more detail in [9] and [10]. Details of the X-ray densitometry system can be found in [11].

*Corresponding Author, Harish Ganesh: gharish@umich.edu

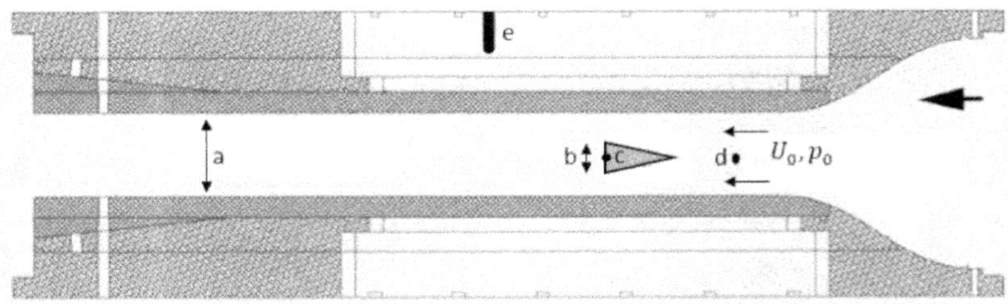

Figure 1: Schematic overview test section with (a) cross section of 7.6 cm by 7.6 cm, (b) wedge base height of 1.9 cm, (c, d) pressure transducers and (e) a hydrophone.

The static inlet pressure p_o was measured at (d) in Figure 1 using an absolute pressure transducer and was varied between 15 kPa to 100kPa. The flow velocity U_o was acquired by measuring the pressure drop across the contraction, upstream after a honey comb and point (d), and was set to 6 m/s. This gives an inlet cavitation number σ_o, Equation (1), in the range of 0.7 to 6.5.

$$\sigma_o = \frac{p_o - p_v}{(1/2)\rho_L U_o^2} \tag{1}$$

With the vapor pressure $p_v = 3.169$ kPa and the liquid density $\rho_L = 1000\ \mathrm{kg/m^3}$. The wedge has an apex angle of 15 degrees and a base height of 1.9 cm. The static pressure at the wedge base p_b and the dynamic pressure at the base of the wedge were measured using a static and a dynamic pressure transducer respectively at point (c) in Figure 1. To improve the visualization of the vortices, a mixture of milk and alcohol was injected through a single port at point (c) in Figure 1.

Results

Cavitation events in the wake were observed for $\sigma_o < 4.8$. The flow characteristics stay fairly constant for $2.5 < \sigma_o < 4.8$. However, for inlet cavitation numbers smaller than 2.5 changes in the flow are visible. Figure 2 shows the Strouhal number, Equation (2), with the frequency f obtained from the dynamic base pressure measurements and the wedge height D as function of inlet cavitation number σ_o. Similar to [4] and [5], a peak value in Strouhal number is observed. The inlet cavitation numbers were further examined using both high-speed cinematography and X-ray densitometry. Figure 3 (a)-(h), shows snapshots of the wake at four different cavitation numbers.

$$St = \frac{fD}{U_o} \tag{2}$$

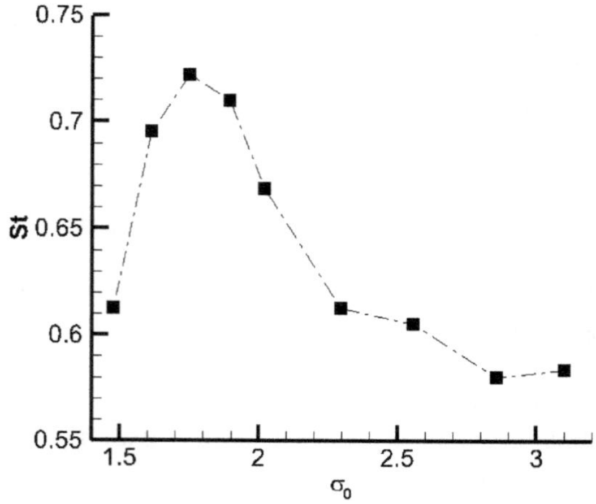

Figure 2: Strouhal number St as function of inlet cavitation number σ_o.

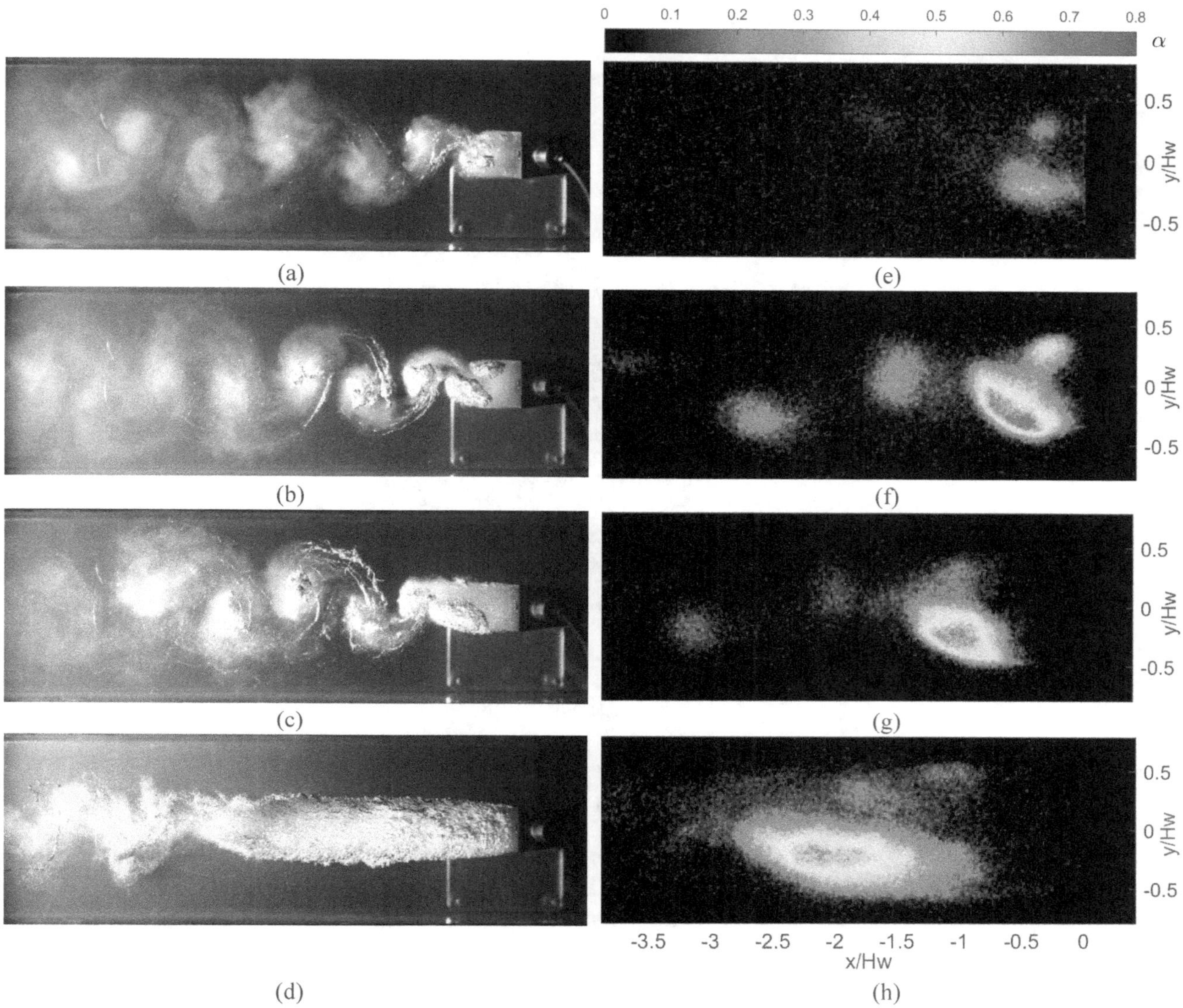

Figure 3: Instantaneous high-speed (a)-(d) and X-ray densitometry spanwise averaged void fraction field (e)-(h) snapshots of cavitation in the wake for different cavitation numbers, σ_o= 2.6 (a) and (e), σ_o= 1.9 (b) and (f), σ_o= 1.6 (c) and (g), and σ_o= 1.25 (d) and (h).

From Figure 3(a)-(d) it can be seen that the extent of cavitation increases with decreasing cavitation number, as expected. Examination of the void fraction flow fields in Figure 3(e)-(h) indicate that the instantaneous void fractions also increase in size, suggesting the presence of more vapor. In regions with such void fractions, the speed of sound of the mixture will drop significantly. The void fraction values α can be used to calculate the values for the speed of sound c and Mach number M using Equation (3), obtained from [7], and (4) respectively.

$$c = \sqrt{\frac{1}{\alpha(1-\alpha)} \frac{\left(k(p-p_v)\right)}{\rho_L}} \tag{3}$$

$$M = \frac{U_0}{c(1-\lambda)} \tag{4}$$

With the liquid density ρ_L, the polytrophic constant $k = 1.3$, the local static pressure p, the vapor pressure p_v and the blockage of the wedge in the test section $\lambda = 1/4$. In the wake region, the speed of sound c can be estimated with

Equation (3) in the wake region using the average static pressure at the base of the wedge p_b, at point (c) in Figure 1, as approximation for the local static pressure p. The Mach number M can then be estimated using Equation (4). Using this procedure, Figure 4(a)-(d) shows the instantaneous Mach number field, corresponding to the void fraction fields in Figure 3 (e)-(h). Hot spots, colors yellow to red in Figure 4, indicate $M > 1$ and thus the presence of supersonic flow.

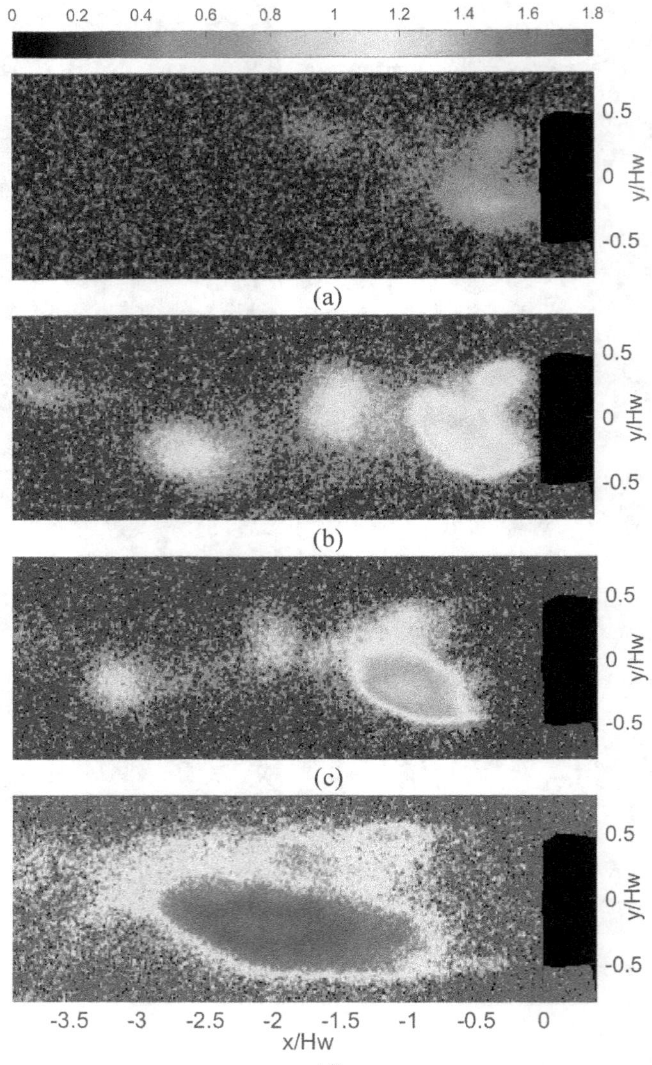

(d)

Figure 4: Instantaneous Mach number fields in the wake at cavitation numbers, σ_o= 2.6 (a), σ_o= 1.9 (b), σ_o= 1.6 (c) and σ_o= 1.25 (d).

The maximum Mach number that occurred within the time series was calculated. Figure 5 shows the Strouhal number based on base pressure measurements and the maximum Mach number in the flow field within time as function of inlet cavitation number σ_o. The Strouhal number increase occurs close to the point at which the average Mach number exceeds unity, suggesting the importance of compressibility effects in the observed change in the wake dynamics.

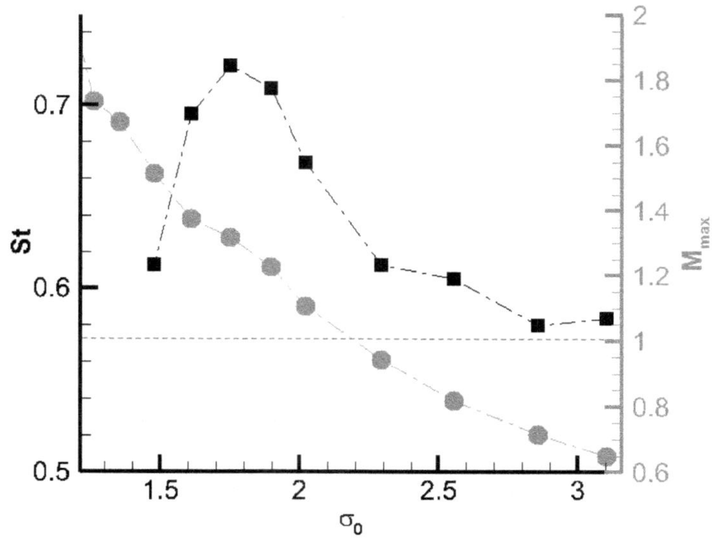

Figure 5: Strouhal number St (black) and Mach number M (red) as function of inlet cavitation number σ_o.

Acknowledgements

This work was funded by the Office of Naval Research under grant N00014-14- 1-0292, Dr. Ki-Han Kim program manager.

References

[1] Roshko, A. (1955). *On the wake and drag of bluff bodies*. Journal of the Aeronautical Sciences. 22(2), pp. 124-132.
[2] Saffman, P.G., Schatzman, J.C. (1982). *An inviscid model for the vortex-street wake*. Journal of Fluid Mechanics. 122, pp. 467-486.
[3] Gerrard, J.H. (1966). *The mechanics of the formation region of vortices behind bluff bodies*. Journal of Fluid Mechanics. 25(2), pp. 401-413.
[4] Young, J.O., Holl J.W. (1966). *Effects of cavitation on periodic wakes behind symmetric wedges*. Journal of Basic Engineering. 88(1), pp. 163-176.
[5] Ramamurthy, A.S., Bhaskaran, P. (1977). *Constrained Flow Past Cavitating Bluff Bodies*. Journal of Fluids Engineering. 99(4), pp. 717-726.
[6] Belahadji, B., Franc, J.P., Michel, J.M. (1995). *Cavitation in the rotational structures of a turbulent wake*. Journal of Fluid Mechanics. 287, pp. 383-403.
[7] Brennen, C.E. (2013). *Cavitation and bubble dynamics*. Cambridge University Press
[8] Crespo, A., (1969). *Sound and Shock Waves in Liquids Containing Bubbles*. Physics of Fluids. 12(11), pp. 2274-2282.
[9] Ganesh, H., Mäkiharju, S.A., Ceccio, S.L. (2016). *Bubbly shock propagation as a mechanism for sheet-to-cloud transition of partial cavities*. Journal of Fluid Mechanics. 802, pp. 37-78.
[10] Ganesh, H. (2015). *Bubbly Shock Propagation as a Cause of Sheet to Cloud Transition of Partial Cavitation and Stationary Cavitation Bubbles Forming on a Delta Wing Vortex*. Ph.D Thesis, University of Michigan.
[11] Mäkiharju, S.A., Gabillet, C., Paik, B.-G., Chang, N.A., Perlin, M., Ceccio, S.L (2013). *Time resolved two dimensional X-Ray densitometry of a two phase flow downstream of a ventilated cavity*. Experiments in Fluids. 54(7), pp. 1-21.

© 2018 ASME

Cavitation dynamics on laser-textured surfaces

[1]Martin Petkovšek *; [2]Matej Hočevar; [1]Peter Gregorčič;

[1]Faculty of Mechanical Engineering, University of Ljubljana, Ljubljana, Slovenia; [2]Institute of Metals and Technology, Ljubljana, Slovenia

Abstract

This study examines hydrodynamic cavitation behavior on laser-textured surfaces. Cavitation dynamics was investigated on stainless steel cylinders with 10 mm in diameter. Their surfaces were modified by direct laser texturing (DLT) with a nanosecond fiber laser (wavelength of 1060 nm). A highly-polished cylinder was used as a reference. Cavitation characteristics behind circular cylinders were monitored simultaneously by high-speed visualization and high frequency pressure transducer. Comparison at similar cavitation numbers proves, that cavitation characteristics differ significantly between different micro-structured surfaces. In some cases, cavitation extent decrease in comparison with reference – a highly polished cylinder.

Keywords: cavitation; laser texturing; fluid dynamics; surface engineering

Introduction

Cavitation as a physical phenomenon is relatively well known, and has been studied intensively since the end of 19th century on behalf of avoiding, preventing or predicting cavitation on turbine machinery. Nowadays cavitation is not considered exclusively as a negative effect of a local pressure drop in a liquid, but it can be also used as an alternative technology in various processes, such as water treatment [1]. Nevertheless, when one studies cavitation phenomena to avoid or to use them, it is preferable to know them in details. Most of the study usually considerate how basic body geometry influences on cavitation characteristics. Also scale effects on cavitation dynamics [2] have been studied, while no study with direct-laser-textured (DLT) surfaces has been performed yet to observe cavitation characteristics. The main aim of this contribution is to study, how surface properties influence on cavitation dynamics. Therefore, we used 10 mm long stainless steel cylinders with diameter of 10 mm. Their surfaces were modified by (DLT) with a nanosecond fiber laser. Several recent studies [3,4,5] have shown that DLT can be used for surface functionalization, since it influences on surface topography and also on surface wettability - immediately after DLT, the textured surfaces become superhydrophilic and after some time the transition to superhydrophobic state may occurs.

Experimental set-up

Experiments were performed in a closed loop cavitation tunnel (Figure 1a) at the Faculty of Mechanical Engineering, University of Ljubljana. For water circulation in cavitation tunnel a 4.5 kW pump is installed, which enables the variation of rotation frequency via frequency controller to control the flow rate. The water flows through the upstream reservoir into the test section and through the downstream reservoir back into the pump. The main purpose of the two reservoirs is to eliminate the pump pulsation and to avoid potential small bubbles to enter into the test section. The upstream reservoir can be also used to raise the fluid temperature with an electric heater installed, while downstream reservoir enables a cooling system with separated cooling loop, connected to the tap water system. Two valves are installed upstream and downstream of the test section, which enables the additional control of the flow rate and easy and fast disconnection of the test section from the main loop. The flowrate is monitored by an electromagnetic flowmeter ABB WaterMaster. Temperature of the fluid is monitored by Pt100, installed into the downstream reservoir, while pressure is monitored at the test section by ABB absolute and differential pressure transducers. The test section (Figure 1b) is made out of transparent acrylic glass, which enables to observe cavitation phenomena from different angles. Cavitation was observed on the 10 mm long stainless steel cylinders with diameter of 10 mm and modified surfaces by direct laser texturing, with a nanosecond fiber laser. The cylinders were installed horizontally, perpendicular to the flow direction.

*Corresponding Author, Martin Petkovšek: martin.petkovsek@fs.uni-lj.si

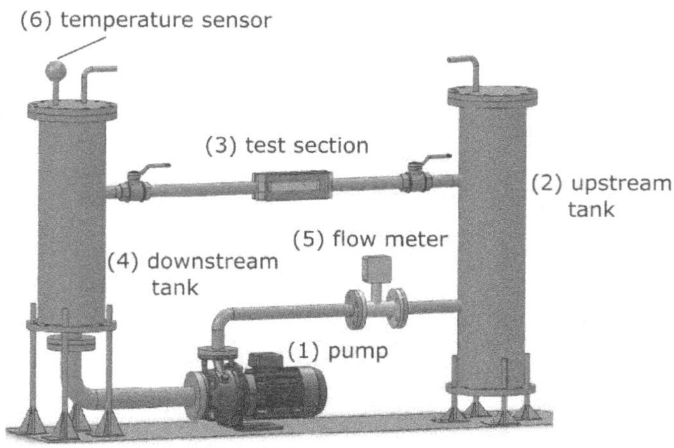

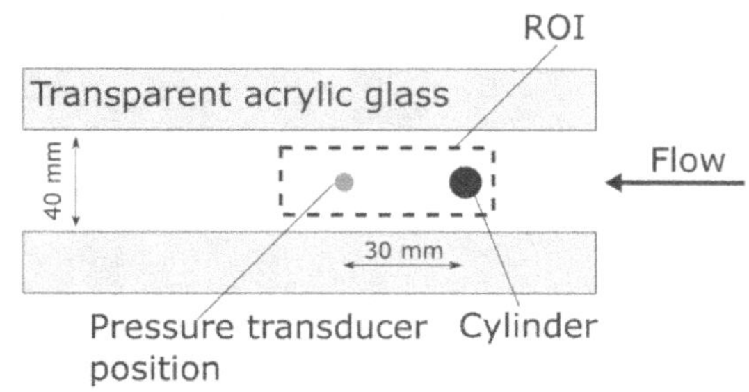

Figure 1: Cavitation tunnel (left - a) and test section (right – b).

Experiments were performed sequentially under the same operating conditions (Table 1) for all five specimens-cylinders, labeled S1-S5. The flowrates were chosen based on visual perception, where the initial flowrate determines initial cavitation on the reference specimen. Flowrates, inlet pressures and flow temperatures were monitored and controlled, to ensure the same cavitation numbers for specific measurement points, by all specimens.

Measurement point	Flowrate [L/min]	Pressure [kPa]	cav. number [ ]
A	163	94	2.2
B	182	91	1.7
C	197	92	1.5
D	212	93	1.3
E	231	92	1.1

Table 1: Operating conditions.

Cavitation characteristics behind circular cylinders were monitored simultaneously by high-speed visualization and high frequency pressure transducer. Visualization was performed by high-speed camera Fastec HiSpec4 at 15,000 fps and resolution of 464×164 pixels. High power LED illumination allowed to set shutter time down to 10μs, at medium opened aperture. Pressure oscillations were measured by PCB 113B28 pressure transducer at 1MHz via DAQ NI USB 6351 measurement card.

Cylinder surface texturing

As a reference sample we used a highly-polished surface (S1). Additionally, we prepared 4 different DLT samples (S2-S5). Their surfaces were textured by using a fiber laser (SPI Lasers, Ltd., G4, SP-020P-A-HS-S-A-Y), radiating pulses with a wavelength of 1060 nm, an average power of 10 W, pulse frequency of 50 kHz, and pulse duration of 28 ns. The spot size in a focal position equaled 0.03 mm. To process the whole surface of cylinders, they were mounted on a rotational motorized stage with an angular resolution of 0.005°. For all DLT-samples (with exception of S3) the laser beam was delivered in lines, parallel to symmetrical axis of the cylinder with scanning velocity of 300 mm/s. SEM micrographs of all samples after DLT are shown in Figure 2. In case of highly porosity surface (S2), the distance between two successive scanning lines, so called scan line separation Δy, equaled 25 μm. The micro-channels in S4 and S5 are 100 μm wide and were performed by engraving at four different angles (0°, 18°, 45°, 72°) and Δy = 10 μm. On sample S4, the micro-channels are separated by 200 μm, while on S5 this distance equals 500 μm. In sample's surface S3 we drilled micro-holes with diameter of 40 μm, that are separated by 200 μm Here, each hole was drilled by 1,000 pulses.

After DLT, all textured samples were superhydrophilic in a saturated Wenzel regime (a static contact angle equaled 0°), similar as in Ref. [5].

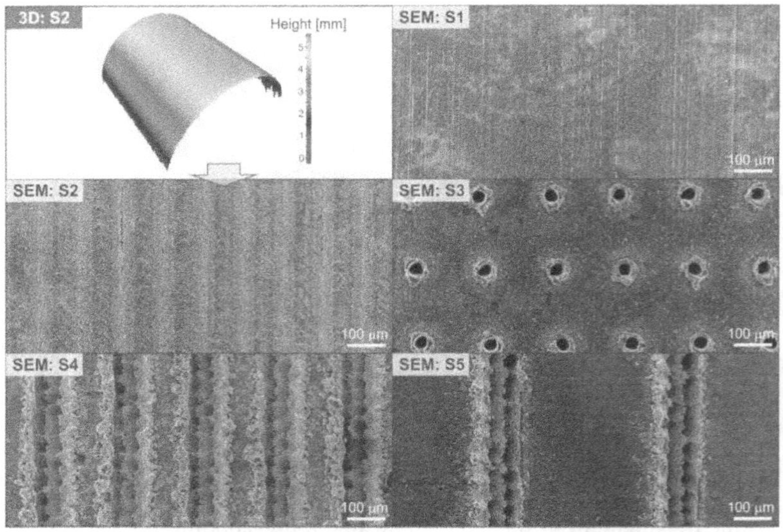

Figure 2: 3D scan and SEM micrographs of DLT surfaces.

Results

Figure 3 presents averaged cavitation extents behind the circular cylinder, while Figure 4 presents standard deviation, calculated based on visualization, for five different measurement points and five different specimens. One can see that cavitation grows with decreasing the cavitation number by all specimens. Notice that the results in Figures 3 and 4 are presented at different color bar scales for individual measurement point, due to better visual perception. Each measurement point has the same color bar scale for all five specimens. The flow is from right to left.
It is clearly visible, that cavitation differs between five specimens, by the same cavitation number (Figures 3 and 4). By S2 cavitation first appears at lower cavitation number in comparison with reference (S1). It is also clearly demonstrated, that by S3 cavitation appears in different shape as by S1, i.e., it is more compressed and centered behind the cylinder. By S4 and S5 cavitation extent is more significant in comparison with the reference, but differs in shape.

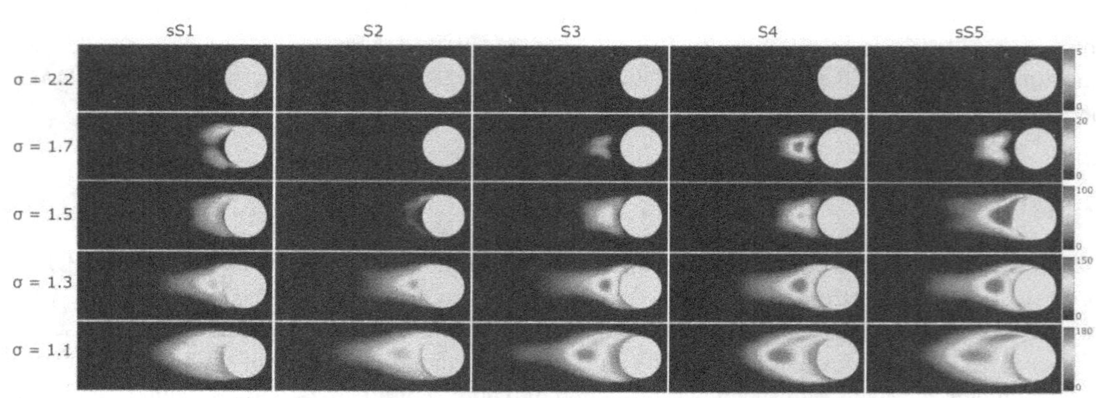

Figure 3: Averaged cavitation extents, based for a period of time of 0.67 s.

Observing standard deviations reveals that in case of S2, no cavitation dynamics is taking place till cavitation number decreases to 1.5. It is demonstrated that in all other cases (S1, S3, S4, and S5) strong cavitation dynamic is present from initial cavitation state. Comparing cavitation dynamics at cavitation numbers 2.2 and 1.7 shows, that in case of laser-textured specimens, strong cavitation dynamic is present more at the center, behind the cylinder, while in case of polished specimen (S1) dynamics are present more at the upper and lower side of the cylinder.

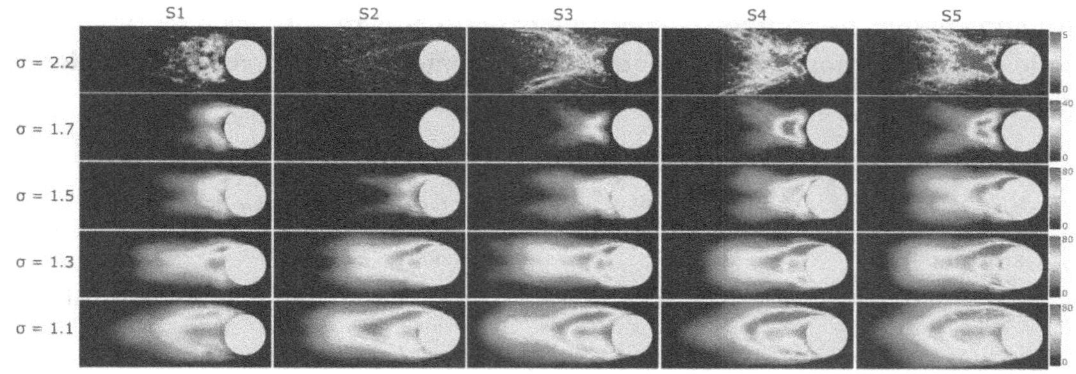

Figure 4: Calculated standard deviation, based on visualization for a period of time of 0.67 s.

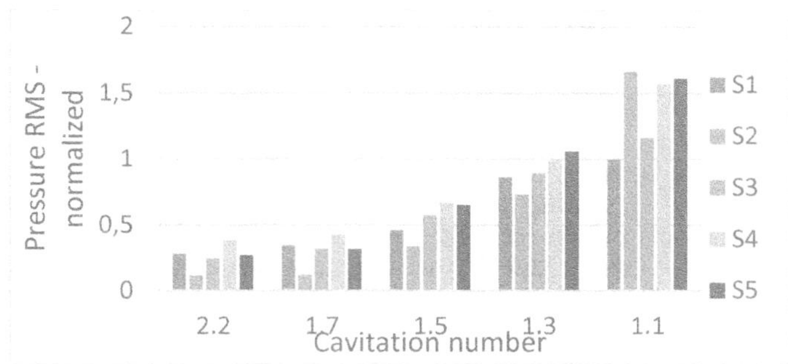

Figure 5: Pressure RMS values, normalized on S1.

Figure 5 presents RMS pressure values for all measurement points, by all specimens. Values are normalized with the S1 value at cav. number 1.1. It is shown, that in case of S2, pressure pulsations are weaker than in case of S1, which corresponds with results from visualization. Pressure pulsation by S3 are very similar in comparison with S1, while in case of S4 and S5 pressure pulsations are stronger.

Conclusion

This study clearly demonstrates that cavitation on the same basic geometry (circular cylinder, 10 mm in diameter) can be controlled by surface engineering, using nanosecond fiber-laser texturing. If appropriate laser-texturing parameters are chosen, cavitation can be decreased or initial cavitation can be shifted to lower cavitation numbers. The presented results, therefore, reveal that direct laser texturing on the exposed surfaces is offering a potential of reducing the risk of cavitation to appear or eventually reducing the cavitation extent and consequently the cavitation erosion.

Acknowledgements

The authors acknowledge the financial support from the Slovenian Research Agency (research core funding No. P2-0401, P2-0392, P2-0132 and project Z2-8188), and the support of SPI Lasers Ltd. by loan of their fiber laser within the research project *Surface functionalization by nanosecond fiber laser texturing (nsFLaT)*.

References

[1] Dular, M., Griessle-Bulc, T., Gutierrez-Aguirre, I., Heath, E., Kosjek, T., Krivograd Klemenčič, A., Oder, M., Petkovšek, M., Rački, N., Ravnikar, M., Šarc, A., Širok, B., Zupanc, M., Žitnik, M., Kompare, B. (2016). *Use of hydrodynamic cavitation in (waste)water treatment. Ultrasonic Sonochemistry.* 29.

[2] Dular, M., Khlifa, I., Fuzier, S., Adama Maiga, M., Coutier-Delgosha, O. (2012). *Scale effect on unsteady cloud cavitation.* Exp. Fluids. 53.

[3] Ta, D.V., Dunn, A., Wasley, T.J., Kay, R.W., Stringer, J., Smith, P.J., Connaughton, C., Shephard, J.D. (2015). *Nanosecond laser textured superhydrophobic metallic surfaces and their chemical sensing applications*. Appl. Surf. Sci., 357.

[4] Trdan, U., Hočevar, M., Gregorčič, P. (2017). *Transition from superhydrophilic to superhydrophobic state of laser textured stainless steel surface and its effect on corrosion resistance*. Corros. Sci., 123.

[5] Gregorčič, P., Šetina-Batič, B., Hočevar, M. (2017). *Controlling the stainless steel surface wettability by nanosecond direct laser texturing at high fluences*. Appl. Phys. A 123.

© 2018 ASME

Cavitation in engine lubricants: visualisation experiments in both a single ring test rig and a single cylinder motored diesel engine to complement on the theoretical modeling of cavitation

[1]Polychronis Dellis*

[1]ASPETE, School of Mechanical Engineering Educators, Athens, Greece

Abstract

The importance of studying cavitation lies in the fact that it can cause tremendous damage to fluid handling machinery. Cavitation generation is largely dependent on pressure fluctuation in the turbulent flow period as well as on the presence of air bubble nuclei.

The shape and dynamic behavior of oil film in a piston-ring pack after oil film rupture are greatly influenced by revolution speed. After dead center, negative pressure involving tension arises before oil film rupture occurs. With the generation of internal cavities tension is released but negative pressure is still maintained in those cavities.

This study is focused on the presentation of cavitation imaging. Images are both taken from a simplified single-ring test rig which is purpose built so that numerous uncertainties regarding the complex tribological phenomena between the piston-ring and cylinder wall to be omitted and a single cylinder diesel engine which was modified to accommodate quartz window sections, to take images of the piston-rings and liner interface. The experimental data give an insight of the rheological phenomena occurring and try to complement the lack of experimental data on this subject. Eventually, experimental results will be used on how best to model the cavitation at the interface of the piston-ring and the cylinder wall.

Keywords: piston-ring lubrication; simulating test rig; cavitation in piston-rings; engine visualization experiments

Introduction

The formation of cavities and their subsequent disposition affects the pressure generated in the continuous thin lubricated film and hence, any integrated quantities such as the load capacity of bearings and piston-rings. With a convergent-divergent geometry in a liquid film, the occurrence of gaseous cavitation is a natural consequence under all circumstances. The mechanisms responsible for cavitation are:

1) Reverse squeezing in a converging-diverging geometry

2) Shear thinning of the liquid by modifying film thickness

3) Reciprocating speed, varying load and temperature (variable film thickness)

In reciprocating piston-rings, superambient pressures are generated in the converging wedge and subambient or negative pressures at the diverging wedge of the piston-ring. In general, two forms of film breakdown can exist: the first depends on the balance between viscous shear and surface tension forces and the second is due to vaporous cavitation which may occur if the first type of rupture allows the formation of sufficiently low pressures [1].

Body

For the visualization of cavitation two test rigs were used: a simplified single-ring test rig which incorporates a steady piston-ring section and a reciprocating flat liner and a single cylinder Lister-Petter PHW1 engine that was used for motoring tests. In the case of the single-ring test rig, a glass liner specimen was used for the visualization experiments (Figure 1(a)) and the same concept was used for the engine experiments with specially mounted quartz windows (Figures 1(b) and 1(c)). The onset and development of cavitation in the lubrication film present throughout the stroke between the piston-ring and the cylinder liner in the test rig was visualized by a charge-coupled device (CCD) camera and monitored by a miniature pressure sensor. Various transient cavitation patterns were identified

*Corresponding Author, Polychronis Dellis: pasd@city.ac.uk

during the stroke, ranging from developing fern-shaped cavities to string cavities extending gradually up to the trailing edge of the piston-ring. Both the onset and the development of these cavities seem to be affected by piston speed and load; at higher speeds cavities appear later in the stroke and are larger in size, while at higher loads they appear earlier, are more numerous and, thus, smaller in size [2].

In the series of experimental data the oil film was visualized as it was seen on the surface of the piston ring. For the single-ring test rig experiments, different test conditions were used (speed, load, temperature) and pictures were taken for different pulses of the shaft encoder and with different magnification lenses in order to capture the cavity development and the growth of specific cavities. In the published study for the single-ring test rig, some very interesting images were presented of the cavity shapes on the surface of the piston-ring specimen (Figure 2 – initiation of cavitation and Figures 3 (a) – (e) transition between different cavitating conditions). It was concluded that the lubricant film thickness has a major influence on the cavitation behavior, which in turn, affects the shape of the hydrodynamic pressure profile in the entrainment direction and consequently the ring-pack's load carrying capacity. The use of a silicon chip ultra miniature transducer has provided both qualitative and quantitative pressure data, which proved to support the flow visualisation results, while its robustness and good repeatability offer promise for its application to the piston-ring pack of production engines [2].

Cavitation which was evident at the idealized test rig [1], [2], was also going to be observed in engine experiments and conclusions were going to be drawn regarding the phenomenon of cavitation and how can it be characterized with the addition of oil film pressure measurements [1], [3]. For the purpose of visualisation, pockets had to be machined in the sides of the engine block enabling a large viewing area for the imaging process. The pockets were machined with spark erosion technique (Figures 1(b) and 1(c)).

Engine visualization experimental data were also acquired and a direct comparison of the visualization results from both test rigs is attempted. There is evidence of the string cavities that appear on the surface of the barrel-faced top compression ring of the Lister PHW1 engine. The research approach to understanding cavitation, has followed the outcome of the visualisation research study of the single-ring test rig. The idealised single-ring test rig is being used as a large scale model in contrast to the real sized engine test rig. In the case of the single-ring test rig, the ring was covered in the majority of the stroke with string cavities. During flow reversal, the cavitation is initiated by fern-shaped cavities which grow further, to form up bigger ferns and fissure type cavities. Later on in the stroke they develop to strings on the surface of the piston-ring and they grow so that they will be reaching its trailing edge [2], as mentioned above.

In the case of the engine, however, the stages of cavitation inception and initial development are not clear because the size of the visualisation window has hindered imaging of the oil flow at the reversal points close to the dead centres. For comparison purposes a cavitation factor was introduced that compares the two experimental test rigs. This coefficient is the ratio of the width or the length of the string cavities divided to the piston ring width [1], [3].

Figure 4 (a), shows the most probable cavitation development in the engine during the stroke according to the images taken. The stages derived, are postulated from engine imaging and the simulation test rig results visualisation data. Figure 4 (a) is complemented with the cavitation imaging from the engine (Figures 4 (b) and 5 (a) and (b)). Figures 4 (b), 5 (a) and 5 (b) present visualization data acquired from the engine test rig with a CCD camera with different magnification lenses and distances from TDC, to focus on the findings from these experiments. Engine experiments showed unclear appearance of string cavities. The identification of oil droplets in general can be directly compared to the research that was presented by Toyota [4]. Evidence of oil mist was unclear in the images.

The schematic in Figure 4 (a) also combines the probable oil film pressure for the cavities in every stage respectively. As in the case of the single-ring test rig, the cavitation inception is being accompanied by negative or sub-atmospheric pressures. In the converging wedge of the piston-ring, the pressure is the combustion chamber pressure and at the diverging wedge the pressure reading would be that of the second land. Later on in the stroke (Figures 4 (a) - (B) – irregular fern growth) the ferns grow, and they are combined with sub-atmospheric oil film pressures at the diverging wedge of the piston-ring. Again, the pressure reading at the diverging wedge ends to the pressure measurement of the second land. The string cavities (Figure 4 (a) – Visualised Cavitation Stage (C)) can reach the trailing edge of the piston-ring or they can be cut midway. The pressure reading at the diverging wedge is that of the second land. Cavitation factors were calculated for both the single-ring test rig and the engine. They showed that the engine cavitation factor is approximately 70% greater than the respective factor calculated for the single-ring test rig [1], [3]. The residence time of the cavities was also calculated but this consideration is under some serious risk.

Figures & Tables:

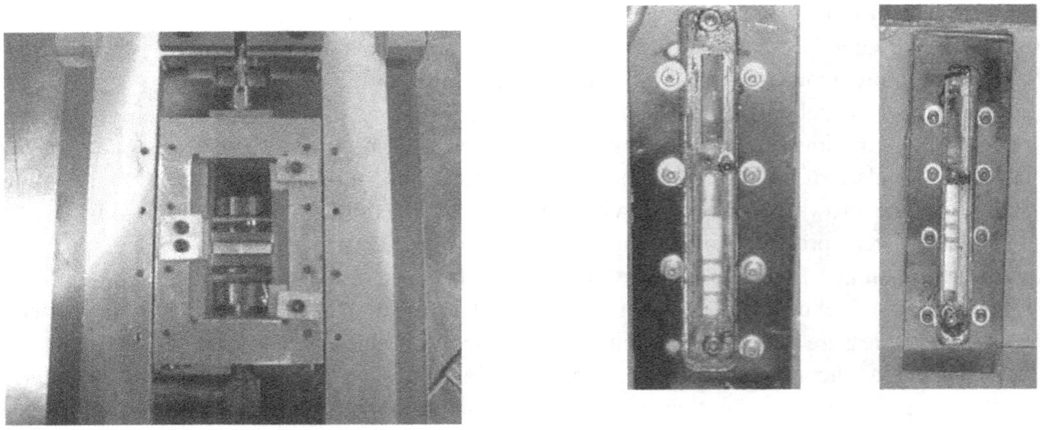

(a) (b) (c)

Figure 1. (a) The glass liner in the single-ring test rig and (b)-(c) Upper and lower window, (b) anti-thrust side window visualisation area (c) front side of the Lister-Petter block [1], [3].

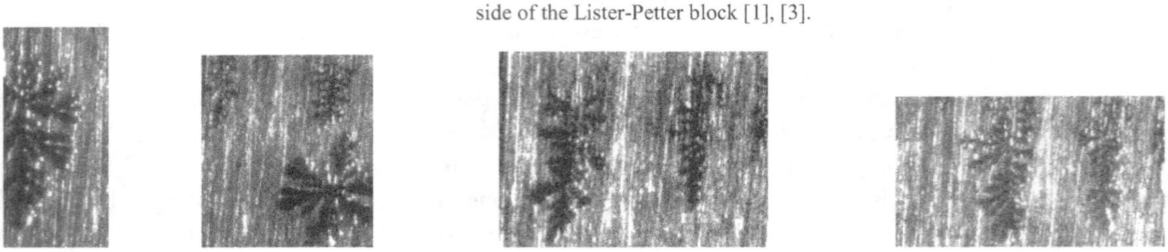

10.98° CA 11.16° CA 11.34° CA 11.52° CA

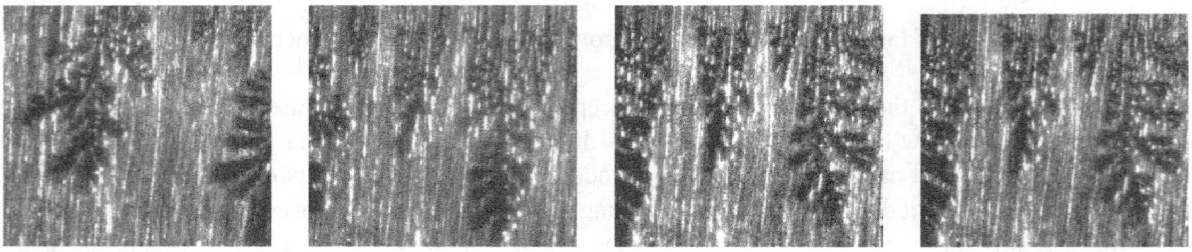

11.70° CA 11.88° CA 12.06° CA 12.24° CA

Figure 2. Fern growth captured using different magnification systems CF-4 magnification with extension lens (8.7x), 700 rpm, 22 mm stroke, 2843 N/m at different crank angles [2]

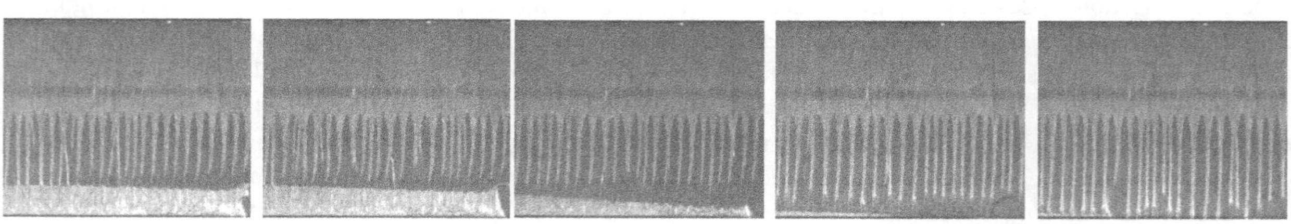

(a) 82° CA (b) 84° CA (c) 86° CA (d) 88° CA (e) 90° CA

Figure 3 (a) – (e). Visualisation of transition between different cavitating conditions [2]

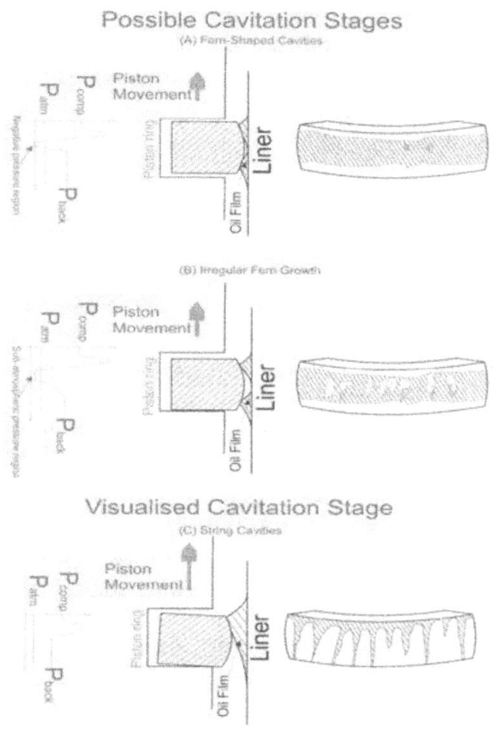

(a)

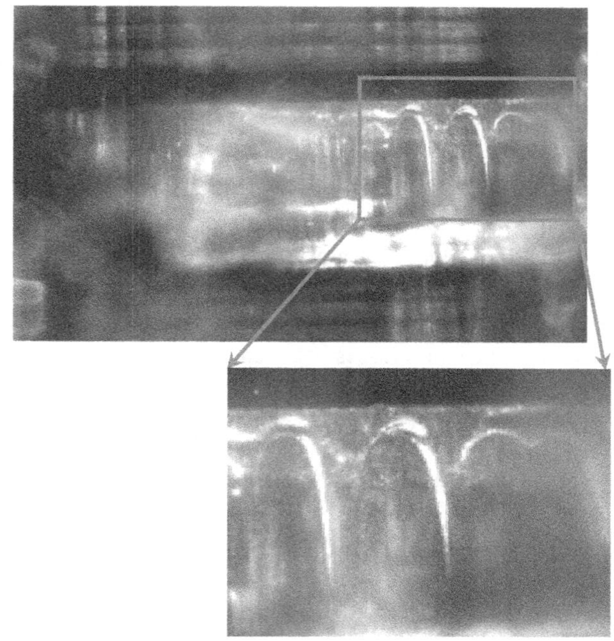

(b)

Figure 4. (a) Possible cavitation stages on the Lister-Petter engine and (b) string cavities on the surface of the piston-ring [1], [3]

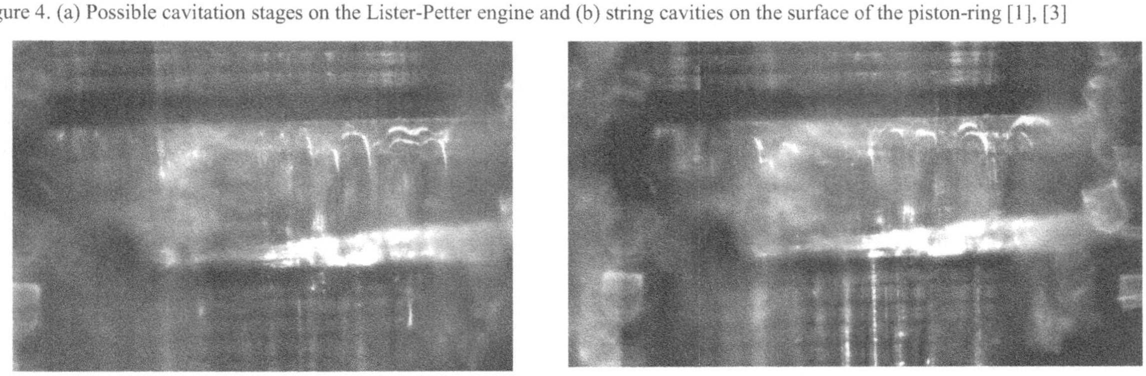

(a)

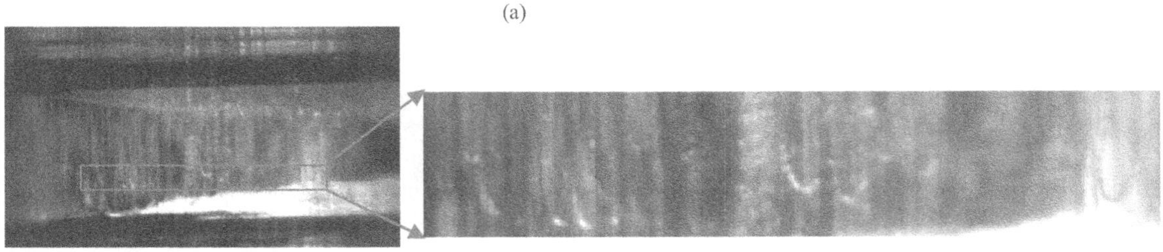

(b)

Figure 5. Images taken from engine visualization experiments showing string cavities (a) 576 °CA exhaust stroke camera at 11.17 cm from top edge and (b) 475.20° CA expansion stroke camera at 9.17 cm from top edge [1]

Conclusion

The obtained insight into the onset and development of cavitation in the piston-ring assembly will encourage further efforts in modelling cavitation under a much wider range of operating conditions than those examined here. It seems that a combined theoretical and experimental approach may eventually lead to a level of understanding of piston-ring lubrication which can directly influence engine design and, through this, engine performance, economy and exhaust emissions [5].

Experimental data fail to resolve clearly which boundary conditions are appropriate or whether perhaps different boundary conditions are suited to different conditions of load, speed, temperature and degree of lubricant starvation [5]. The effect of the range of different cavitation models on the tribological behavior of the single piston-ring is characterized by hydrodynamic pressure profiles, lubricant film boundaries, lubricant film thickness, lubricant flow rate and friction losses between the piston-ring and the cylinder wall.

The research approach to understanding cavitation that was identified in engine imaging has followed the outcome of the visualisation research study of the single-ring test rig. The idealised single-ring test rig is being used as a large scale model in contrast to the real sized engine test rig. In the case of the single-ring test rig, the ring was covered in the majority of the stroke with string cavities. During flow reversal, the cavitation is initiated by fern-shaped cavities which grow further, to form up bigger ferns and fissure type cavities. Later on in the stroke they develop to string cavities on the surface of the piston-ring and they grow so that they will be reaching its trailing edge [1], [2].

In the case of the engine, however, the stages of cavitation inception and initial development are not clear because the size of visualisation window has hindered imaging of the oil flow at the reversal points close to the dead centres. On one hand, the quartz sectioned liner provided cavitation images that resemble the ones recorded for the single - ring test rig, on another, the overall size and their span compared to the stroke length is not adequate to "scan" the whole of the engine stroke length. In the visualisation parametric study for the single-ring test rig, it was evident that as speed increases, the fern cavities appear later on in the stroke and in the case of the load increase, the opposite happens. For the motoring tests, it was not possible to have a good picture of the squeeze film effect and how does cavitation initiation change in the case of speed and load increase. Although speed was maintained in quite low levels its inception was not captured. For higher speed imaging, where the cavities are formed later on in the stroke, still no clear signs were identified. The visualised cavity shapes are not uniform. It should be pointed out, though, that for the case of the anti-thrust and front side windows, for the same speed and load, different cavitation initiation points in the stroke should be considered, as the load on the ring changes along the circumference due to piston tilt. Engine experiments showed, for the motoring tests, that top compression ring operates under serious conditions of lubricant starvation.

References

[1] Dellis P. (2005). *Aspects of lubrication in piston cylinder assemblies*, PhD thesis, Imperial College London, Department of Mechanical Engineering, London.

[2] Dellis, P. and Arcoumanis, C. (2004). *Cavitation Development in the Lubricant Film of a Reciprocating Piston – Ring Assembly*, Proceedings of the Institution of Mechanical Engineers Part J: Journal of Engineering Tribology, Vol 218 (3), 157-171 doi: 10.1243/1350650041323340.

[3] Dellis, P. (2017). *Aspects of Lubrication in a Reciprocating Single-ring Test Rig and Further Implementation to Engine Applications*. Tribology in Industry, 39, 260-269, doi: 10.24874/ti.2017.39.02.14.

[4] Inagaki, H., Saito, A., Murakami, M. and Konomi, T. (1995). *Development of Two-Dimensional Oil Film Thickness Distribution Measuring System*, SAE Paper 952346.

[5] Priest M., Dowson D. and Taylor C. M. (2000). *Theoretical Modelling of Cavitation in Piston-Ring Lubrication*, Proceedings of the Institution of Mechanical Engineers Vol. 214, Part C.

© 2018 ASME

Cavitation in Fractal Geometry Orifices

[1]Pavel Rudolf*; [1]Dávid Kubina; [1]Martin Hudec; [1]Ondřej Hochman; [1]František Pochylý

[1]*Brno University of Technology, Faculty of Mechanical Engineering, Viktor Kaplan Department of Fluid Engineering, Brno, Czech Republic*

Abstract

An experimental research of three types of orifices was carried out in a hydraulic circuit. One-hole orifice and two fractal-shaped orifices (von Koch snowflake, Apollonian gasket) were tested. Fractal-shaped orifices had lower pressure loss in non-cavitating regime. For cavitating conditions, von Koch snowflake orifice maintained the reduced pressure loss, while pressure loss curve within Apollonian gasket orifice gradually aligned with that of one-hole orifice. Dynamics of the cavitating flow behind the orifices featured distinct peak for one-hole orifice connected with shedding of one strong coherent vortical structure, while pressure amplitudes for fractal-shaped orifices (especially von Koch snowflake orifice) were significantly lower. This conclusion is attributed to multiscale behavior of the flow behind the fractal-shaped orifices.

Keywords: cavitation; orifice; fractal; pressure loss; noise; von Koch snowflake; Apollonian gasket

Introduction

Orifices are frequently used in many industries for flow measurement (thermal and nuclear power plants, water pipelines), flow restriction (process and chemical plants), but also in multiphase flow reactors to enhance mixing [1, 2]. Their advantage is very simple design and manufacturing. In many situations, in severe conditions, cavitation occurs. This is very often connected with significantly increased level of noise and vibrations [1].

Orifices can also serve as a simple model to study cavitation in valves. Standard orifices have one circular hole with prescribed thickness/diameter ratio and shape of the orifice edge.

In our previous research [3, 4], it was found out that pressure amplitudes downstream of both one-hole and multihole orifices are lower than those generated by Venturi tube.

Fractal geometry

While one-hole orifices are mostly used in practice, investigations have also been carried out on multihole orifices [2, 4]. It was proved that they can achieve lower pressure loss across the orifice. Both one and multihole orifices are manufactured with sharp and smooth edges.

However in nature most of the objects are characterized by fractal shape (leaves, feathers, roots, etc.) By definition fractal objects are those, where each part has the same statistical character as the whole. It means that some geometrical pattern is replicated on different scales and self-similarity is established. Attention to this feature of natural objects was especially brought by Mandelbrot [5]. Until recently there was no practical utilization of fractal shapes in fluid mechanics and technologies related to fluids handling. Seoud and Vassilicos [6] performed extensive investigations on turbulence generated by fractal objects and wakes behind the fractal shaped objects. They concluded that fractal shaped objects provide unique homogeneous isotropic decaying turbulence and might be used for passive multiscale control of the turbulent flow. Kearney [7] suggested practical application of fractals for improvement of the flow processes to control formation of the fluid structures for example in sugar industry. Aly et al [8] and Manshoor et al [9] applied fractal shapes to orifice plate edges. Aly et al designed and experimentally tested in wind tunnel fractal-shaped orifices based on von Koch snowflake (up to the fourth generation of fractality). Manshoor et al experimentally tested in hydraulic circuit an orifice with several sets of holes with fractally scaled diameters. All these investigations have shown improved performance of fractal-shaped orifices in terms of pressure loss compared to one-hole orifice. Later a CFD study was performed by Elsaey et al [10] using the shapes from [8].

However no information is provided in above mentioned papers about the dynamics of the flow through fractal-shaped orifices (i.e. frequency of the dominant pressure amplitudes) and about the cavitation within these orifices.

*Corresponding Author, Pavel Rudolf: rudolf@fme.vutbr.cz

Design of fractal shaped orifices

One-hole orifice with hole diameter 24 mm placed in a pipe with inner diameter 50 mm is reference case for the present study. Two types of fractal-shaped orifices are designed with the same flow cross-sectional area as one-hole orifice. The first type is based on von Koch snowflake. Derivation of this shape is illustrated in figure 1. It is formed by gradual dividing of a line into three parts and forming a triangle above the middle part. The fractal dimension of such object is 1.2619. Less known fractal shape, which is axisymmetric, is Apollonian gasket formed by so called Sode's circles. This object is made of a set of holes with scaled diameters.

In our study the third generation of von Koch snowflake was manufactured from plastic material by rapid prototyping (see figures 2, 4). Apollonian gasket was made by boring holes into plastic material plate (see figures 3, 4).

The minimum length scales for both fractal-shaped orifices, which are based on the shortest edge or smallest diameter of the hole respectively and maximum length scales which are based on largest gap within the orifice are summarized in table 1.

One-hole orifice was manufactured from stainless steel plate. The diameter/thickness ratio for one-hole orifice and von Koch snowflake orifice is 0,46, for Apollonian gasket orifice ranges from 0.96 to 0.09.

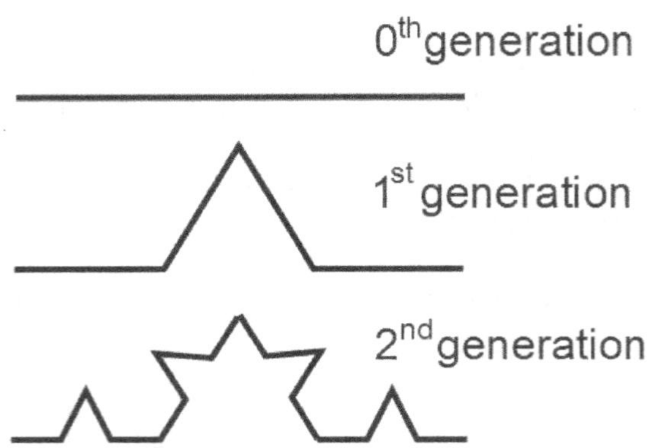

Figure 1. Fractal shape generation for von Koch curve

	von Koch snowflake	Apollonian gasket
L_{min} (mm)	30.67	11.50
L_{max} (mm)	2.95	1.00

Table 1. Length scales in fractal-shaped orifices

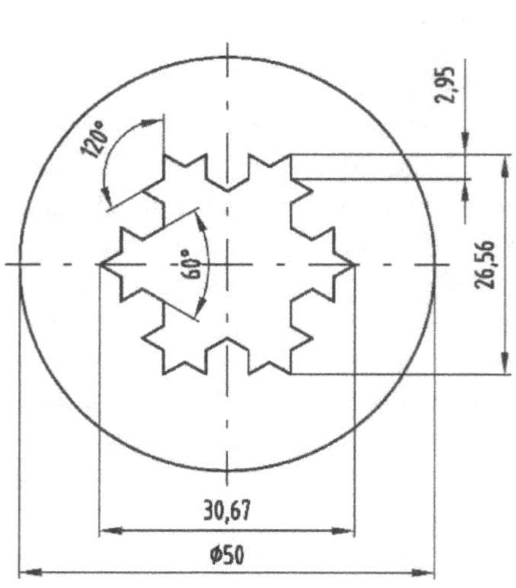

Figure 2. Dimensions of von Koch snowflake orifice

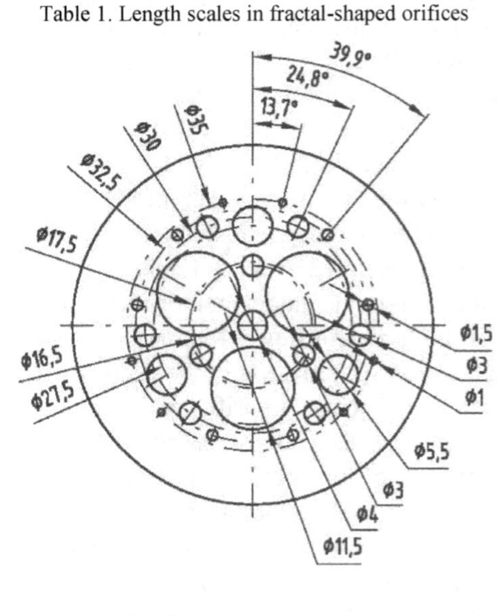

Figure 3. Dimensions of Apollonian gasket orifice

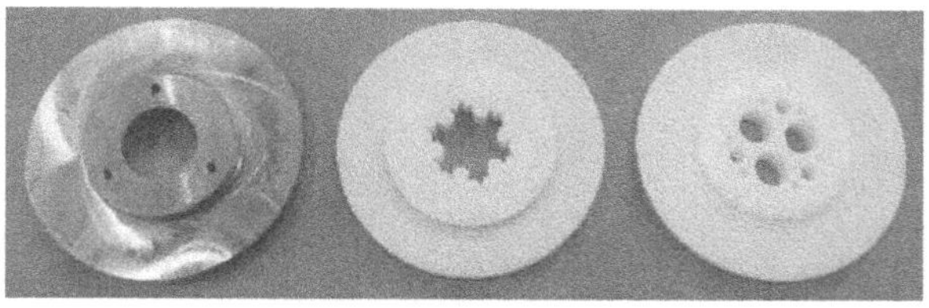

Figure 4. Dimensions of Apollonian gasket orifice

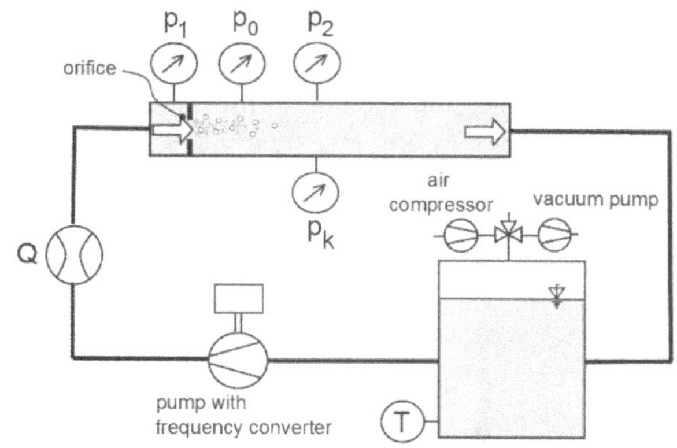

Figure 5. Scheme of the hydraulic circuit

Experiment

An experimental circuit was set up according to figure 5. Flow measured using induction flow meter, pressures were recorded in three positions, just ahead of the orifice and 150 mm (3 pipe diameters) and 500 mm (10 pipe diameters) downstream. Pressure sensors p_0 to p_2 were of tensometric type (DMP331, BD Sensors), sensor p_k is of piezoelectric type (Kistler).Pressure loss coefficient ζ and cavitation number σ were determined from pressure sensors p_1 and p_2. Vapor pressure p_{vapor} was computed from Clasius-Clapeyron equation as function of temperature. Velocity is define by bulk velocity in the orifice constriction.

$$\zeta = \frac{p_1 - p_2}{\rho \frac{v^2}{2}} \tag{1}$$

$$\sigma = \frac{p_2 - p_{vapor}}{\rho \frac{v^2}{2}} \tag{2}$$

Results in form pressure loss coefficient curves for both non-cavitating and cavitating regimes are plotted in figure 6. Curves in figure 6 also depict that cavitation inception for von Koch orifice is postponed compared to two other orifices. Loss reduction in non-cavitating regime for Apollonian gasket orifice is almost 25%. It is interesting to note that loss curve for Apollonian gasket orifice sticks to one-hole orifice curve with increasing intensity of cavitation. Pressure loss reduction is higher than measurement uncertainty.

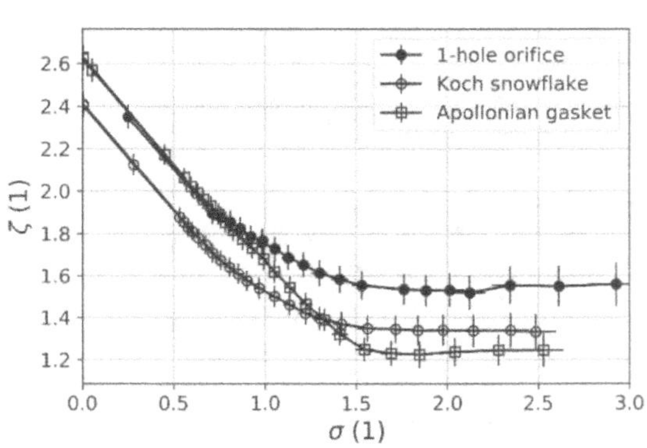

Figure 6. Loss coefficient curves

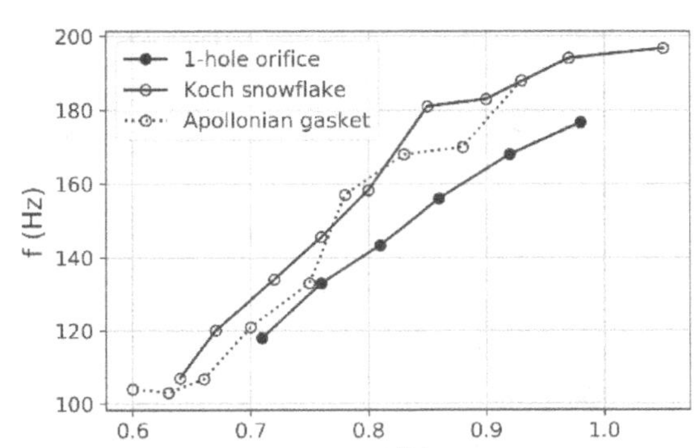

Figure 7. Frequencies of the dominant pressure peaks

Amplitude frequency spectra for cavitating regime were evaluated from wall pressures measured by piezoelectric sensor p_k. The frequency values corresponding to dominant peaks are plotted in figure 7. It is seen that those peaks are shifted from 10 to 20 Hz higher for fractal-shaped orifices compared to one-hole orifice. Pressure amplitudes are significantly higher for one-hole orifice as can be seen from comparison with von Koch snowflake orifice for cavitation number $\sigma = 0.72$, see figures 8, 9.

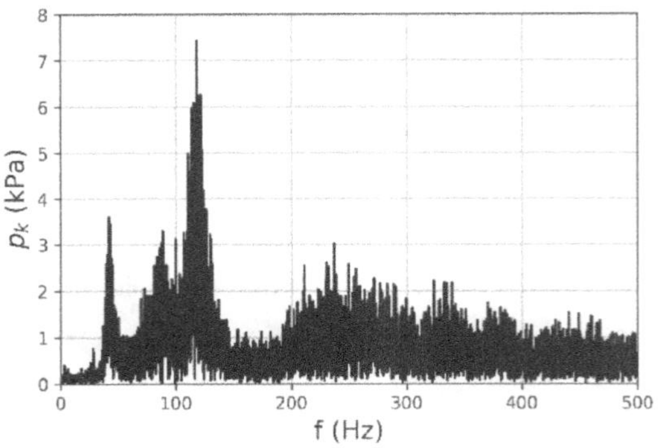

Figure 8. Amplitude frequency spectrum for one-hole orifice

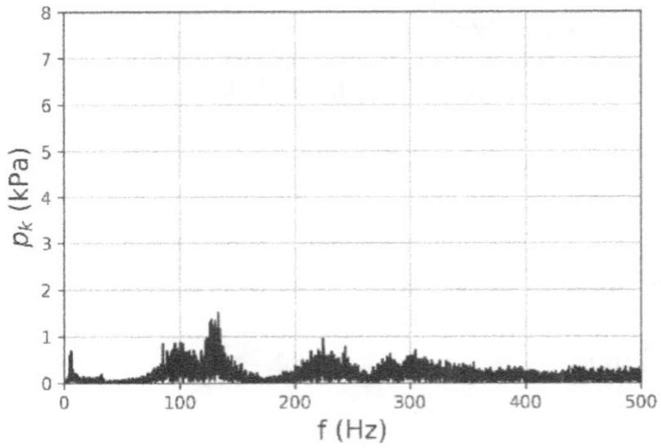

Figure 9. Amplitude frequency spectrum for von Koch snowflake orifice

Different behavior of the orifices is best seen for the supercavitation regime (figure 10). Well pronounced jet appears for one-hole orifice, while for both fractal-shaped orifices the jet is spread over the cross-section. For von Koch snowflake the spreading occurs just behind the orifice, for Apollonian gasket a bit further downstream.

Figure 10. Supercavitation regime visualization (from top: one-hole orifice, von Koch snowflake, Apollonian gasket)

Conclusion

Experimental research of three orifice types with same cross-sectional flow area was carried out: one-hole, von Koch snowflake, Apollonian gasket. Following conclusions can be drawn:

- fractal shaped orifices reduce pressure in the non-cavitating regime compared to one-hole orifice of the same cross-section. The reduction is more significant for Apollonian gasket shaped orifice.
- when cavitation occurs von Koch snowflake orifice maintains the reduced pressure loss, compared to one-hole orifice, over whole operating range from initial cavitation to supercavitation
- frequencies associated with the most dominant peak in pressure spectrum are higher for both fractal shaped orifices compared to one-hole orifice
- amplitudes of wall pressures within the cavitating regime are substantially higher for one-hole orifice than for fractal-shaped orifices. Consequently level of noise (especially in von Koch orifice) is lower for fractal-shaped orifices.

All conclusions are connected with flow pattern generated by the respective orifice plates. One-hole orifice induces a strong vertical structure (vortex ring) associated with the length scale of the hole, i.e. its diameter. On the other hand fractal shaped orifices generate vortices on a whole range of scales, energy is distributed over a range length scales. Fractal shaped orifices might be used as flow constriction devices in situations, where low operating noise is required. Fractal shapes might also be used in control valves or polyjet sleeve valves [11] to achieve lower noise and faster decay of the kinetic energy. Application for fast, yet simple, mixing (combustion, chemical processes) is apparent.

References

[1] Testud, P., Moussou, P., Hirschberg, A., Aurégan, Y. (2005). *Cavitating Orifice: Flow regime transitions and low frequency sound production*. In Proceedings of PVP2005. (pp. 437-445).

[2] Sivakumar, M., Pandit, A.B. (2002). *Wastewater treatment – a novel energy efficient hydrodynamic cavitational technique*. Ultrasonics Sonochemistry. 25(7).

[3] Rudolf, P., Hudec, M., Gríger, M., Štefan, D. (2014). *Characterization of the cavitating flow in converging-diverging nozzle based on experimental investigations*. EPJ Web of Conferences. 67 (2014).

[4] Rudolf, P.; Kubina, D.; Hudec, M.; Kozák, J.; Maršálek, B.; Maršálková, E.; Pochylý, F. (2017) *Experimental investigation of hydrodynamic cavitation through orifices of different geometries*. EPJ Web of Conferences. (2017).

[5] Mandelbrot, B. (1977). *Fractal geometry of nature*.

[6] Seoud, E.R., Vassilicos, S.J. (2007). *Dissipation and decay of fractal-generated turbulence*. Physics of Fluids. 19(10).

[7] Kearney, M. (1997). *Engineered fractal cascades for fluid control applications*. Proc. of Fractals in engineering.

[8] Aly, A.A.E, Chong, A., Nicolleau, F., Beck, S. (2010). *Experimental study of the pressure drop after fractal-shaped orifices in turbulent pipe flows*. Exp. Thermal and Fluid Sciences. 34(2010).

[9] Manshoor, B., Nicolleau, F., Beck, S. (2011). *The fractal flow conditioner for orifice plate flow meters*. Flow Measurement and Instrumentation. 22 (2011).

[10] Elsaey, A., Aly, A.A.E., Foaud, M. (2014). *CFD simulation of fractal-shaped orifices for flow measurement improvement*. Flow Measurement and Instrumentation. 36 (2014).

[11] Burgi, P.H. (1977). Hydraulic Tests and Development of Multijet Sleeve Valves. Report No REC-ERC-77-14. US Dept. of Interior.

Acknowledgement

Czech Science Foundation is gratefully acknowledged for support of this research under project No 16-18316S „Principles and mechanisms causing microorganism elimination by hydrodynamic cavitation ".

© 2018 ASME

Cavitation in Sudden Gap Expansion as a Model for Synovial Joint Cavitation

[1]Monica S. Li; [1,2]Marc Krüger-Sprengel; [1]Simo A. Mäkiharju*;

[1]University of California, Berkeley, CA, USA; [2]Technical University of Munich, Germany

Abstract

The sound of cracking joints, more commonly knuckles, is thought to be due to the cavitation and subsequent bubble collapse in the synovial fluid of the joint. This fluid is shear-thinning and rheopectic, confined within a synovial membrane sac, and surrounded by complex curved surfaces comprised of cartilage on bone. A simplified model potentially capturing the dominant physics of this problem is cavitation between axisymmetric parallel disks in an infinite domain. Two analytical models are found for this problem: the Stefan-Reynolds equation and Kuzma's derivation. Analytical solutions for flow and pressure distributions are found using these simplified models. The disks are suddenly accelerated from an initial separation of a sub-mm gap, and the pressure at the center decreases to values well below the Blake-threshold, leading to cavitation. The initial work reported in this paper utilized Newtonian fluid, namely water, and is focused on comparing the observed and expected cavitation dynamics.

Keywords: cavitation; biofluid

Introduction

It is widely believed that fluid mechanics, namely cavitation, is responsible for popping or cracking sound of synovial joints (Unsworth *et al.* 1971; Kawchuk *et al.* 2015; Dowson *et al.* 1970; Watson *et al.* 1989). This cracking sound occurs when bones are pulled apart axially, resulting in formations of gas or vapor bubble within the joint. Other joint movement can also produce sounds, but may be due to tendons or ligaments moving across bone. Cracking is studied in metacarpal joints, commonly known as knuckles (Unsworth *et al.* 1971; Kawchuk *et al.* 2015), but the same mechanism also occurs in knee, hip, and spine joints. It is also assumed that understanding more about joint geometries and synovial fluid composition may lead to early detection of joint disease, such as arthritis (Unsworth *et al.* 1971; Cooke *et al.* 1978; Fam *et al.* 1978). Synovial fluid reduces the friction between the cartilage covering the bone, as surfaces move across each other, and transports nutrients, waste (Balazs *et al.* 1967; Cooke *et al.* 1978). In a healthy subject, synovial fluid is opaque in color, similar to blood serum, and shear-thinning (Balazs *et al.* 1967; Fam *et al.* 2007). Furthermore, long polymer chains (hyaluronan) in the fluid and attached to the cartilage give the fluid rheopectic properties (Cooke *et al.* 1978; Rwei *et al.* 2007).

Unsworth *et al.* (1971) setup an apparatus to pull the index finger of human subjects and recorded the bone separation of the joint using x-ray exposures. They determined that cavitation collapse is responsible for the cracking noise, and observed that about 15 min after the crack the bone separation distance returned to normal. This is presumed to be following Henry's Law, as the gases redissolved into the fluid (Kimbrough 1999). Gas content was on average 15 percent at room conditions, with 80 percent of the gas being CO_2. In Unsworth *et al.* (1971) a joint simulator was used to study the radius of curvature of the two joint surfaces, and it was found that the synovial fluid composition and joint geometry are the two main factors determining whether a joint will crack. Kawchuk *et al.* (2015), similarly to Unsworth *et al.* (1971), pulled the index finger of subjects, but used Magnetic Resonance Imaging (MRI) to image the joint. At a frame rate of ~3Hz, they were able to record a bubble in the synovial fluid at the time of crack. This bubble was reported to disappear in unspecified time once the axial force was no longer applied. The cracking sound was attributed to the bubble inception, or tribonucleation, rather than its collapse (Kawchuk *et al.* 2015).

The apparent lack of full agreement and thorough fundamental fluid mechanical explanation motivated the present study. Beginning with a canonical geometry to analyze and experiment with, we seek to investigate the mechanisms that produces the sound in more detail, and attempt to glean useful physical insight about joint cracking grounded solidly in fluid mechanics.

Theoretical and Experimental Model

The real joint is quite complex, with the cartilage surface being porous and having non-symmetric shape, and the synovial fluid is non-Newtonian confined within a membrane. So, we begin our investigation on a simplified model of this flow. While curved geometries and non-Newtonian fluids will eventually be utilized, the first geometry we study is the canonical form of two flat disks in an infinite fluid domain, shown in Figure 1(a).

*Corresponding Author, Simo A. Mäkiharju: makiharju@berkeley.edu

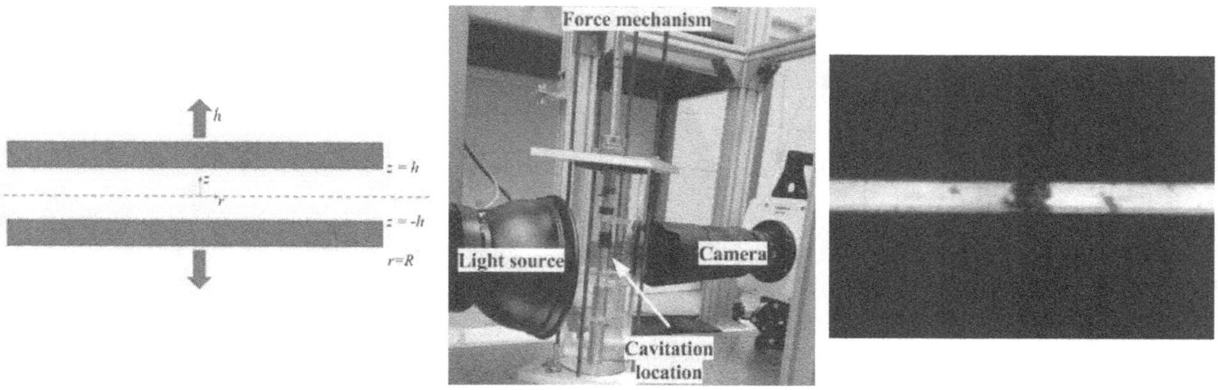

(a) (b) (c)

Figure 1. a) Coordinate system for pressure calculations. At time t=0, top rod is moved up while the bottom is stationary, and is modeled as symmetric about the z-axis, b) Experimental setup with force actuation mechanism, test section, light and high-speed video camera, c) Sample image from high-speed video capturing a large cavitation event during rapid gap expansion, image width $w = 0.7d_{rod}$.

Experimental data is obtained, and we utilize an analytical model to predict the pressure distribution, and more importantly, the minimum pressure achieved at the center of the discs. We calculate the minimum pressure at the center of the discs based on boundary conditions and the prevailing flow conditions extracted from experimental footage.

Theory

The available mathematical models, as summarized by Engmann *et al.* (2005), are analyzed, and the two most suitable models are discussed here. The first mathematical model is based on a squeeze flow derivation by Stefan and Reynolds. This formulation has been adapted for non-Newtonian fluids with a power-law shear-stress expression by Scott (1931), and re-derived as an explicit pressure formulation by Leider (1974) (Equation 4). While this derivation is applicable to non-Newtonian fluids, it fails to properly account for inertial effects. To account for the effects of fluid inertia in separating plates, Jackson (1962) derived solution with corrections from Kuzma (1967), which results in a pressure distribution given by Equation 9.

The formulation by Leider is based on an integral mass balance of the fluid contained between the disc faces and the radial component of the Cauchy momentum equations. One assumes incompressible flow and quasi-steady-state flow conditions, allowing the local and instantaneous description of the flow to be steady-state between two parallel plates with negligible inertial terms and normal stresses. $R >> h$ and u_r is the dominant velocity component. This derivation starts from the integral mass balance of the fluid between the two rod faces

$$\int_0^h v_r \, \mathrm{d}z = -\frac{r\dot{h}}{2} \, . \tag{1}$$

And, the radial component of the Cauchy momentum equations

$$\begin{gathered}\frac{\partial u_r}{\partial t} + u_r\frac{\partial u_r}{\partial r} + \frac{u_\theta}{r}\frac{\partial u_r}{\partial \theta} + u_z\frac{\partial u_r}{\partial z} - \frac{u_\theta^2}{r} = \\ -\frac{1}{\rho}\frac{\partial p}{\partial r} + \frac{1}{r\rho}\frac{\partial (r\tau_{rr})}{\partial r} + \frac{1}{r\rho}\frac{\partial \tau_{\theta r}}{\partial \theta} + \frac{1}{\rho}\frac{\partial \tau_{zr}}{\partial z} - \frac{\tau_{\theta\theta}}{r\rho} + g_r \, .\end{gathered} \tag{2}$$

With $\dot{h}$ representing the velocity of the upper rod surface, neglecting body forces g_r, all inertial terms, the normal stresses involving τ_{rr} and $\tau_{\theta\theta}$, and the shear stress $\tau_{\theta r}$ due to axial symmetry and using a rheological formulation for shear stress τ_{rz} for power-law fluids is given by

$$\tau_{rz} = -m\left|\frac{\partial u_r}{\partial z}\right|^{n-1}\frac{\partial u_r}{\partial z} = m\left(-\frac{\partial u_r}{\partial z}\right)^n \, . \tag{3}$$

Where m is the flow consistency index and n the flow behavior index (Leider 1974). With Equations 1-3, we obtain the explicit formulation for the pressure difference according to Leider (1974). The pressure difference for a power-law fluid according to Leider can then be derived through substitution of the shear stress into Equation 2 and after rearranging and integrating the result can subsequently be substituted into Equation 1, which when integrated yields

$$p - p_{\mathrm{atm}} = \frac{(-\dot{h})^n}{h^{2n+1}} \left(\frac{2n+1}{2n}\right)^n \frac{mR^{n+1}}{n+1} \left[1 - \left(\frac{r}{R}\right)^{n+1}\right]. \tag{4}$$

Kuzma's formulation is based on the continuity equation in differential form and the momentum equation in the radial direction. He assumed incompressible flow, $R >> h$, and u_r as the dominant velocity component with negligible azimuthal and vertical pressure gradients resulting in the pressure depending only on r and t. The starting point is the differential form of the continuity equation

$$\frac{1}{r}\frac{\partial(ru_r)}{r} + \frac{\partial u_z}{z} = 0, \tag{5}$$

and momentum equation for Newtonian fluid flow in this geometry

$$\rho\frac{\partial u_{r0}}{\partial t} + \rho u_{r0}\frac{\partial u_{r0}}{\partial r} + \rho u_{z0}\frac{\partial u_{r0}}{\partial z} = -\frac{\partial p}{\partial r} + \mu\left(\frac{\partial^2 u_r}{\partial z^2} + \frac{1}{r}\frac{\partial u_{r0}}{\partial r} - \frac{u_{r0}}{r^2} + \frac{\partial^2 u_{r0}}{\partial r^2}\right). \tag{6}$$

Obtaining the initial guesses for the velocity components from the lubrication assumption

$$u_{r0} = \frac{3r\dot{h}}{h^3}(z^2 - hz), \qquad u_{z0} = -\frac{\dot{h}}{h^3}(2z^3 - 3hz^2) \tag{7}$$

and substituting into Equation 6, this can be rearranged and integrated yielding an expression for u_r. Subsequently, this can be substituted into the integral continuity equation

$$\int_0^h u_r \,\mathrm{d}z = -\frac{r\dot{h}}{2} \tag{8}$$

Equation 8 can then be rearranged to yield an explicit formulation for the pressure gradient. Assuming a constant ambient pressure and no-slip conditions at the surfaces, integration yields the pressure difference for Newtonian flow according to Kuzma as

$$p - p_{\mathrm{atm}} = \frac{r^2 - R^2}{2}\left(\frac{6\mu\dot{h}}{h^3} + \frac{3\rho\ddot{h}}{5h} - \frac{15\rho\dot{h}^2}{14h^2}\right). \tag{9}$$

It has been noted that inertial effects become significant for high $Re = \rho h\dot{h}/\mu$, with h being the surface separation, $\dot{h}$ the upper surface velocity, ρ the density and μ the viscosity (Jackson 1962). While the lubrication assumption inherent in the solution is only satisfactory for small Reynolds numbers, Kuzma's experiments show close agreement up to Reynolds numbers of $Re = 60$ (Kuzma 1967). That is, while the model proposed by Kuzma relies on assumptions for the velocity profile, which are no longer wholly satisfactory for higher Reynolds numbers, it shows unexpectedly good agreement with experimental results (Kuzma 1967). While Leider's model is also applicable to non-Newtonian flows, it assumes steady-state conditions and neglects inertial effects, thus making it less suitable for the problem of rapidly separating discs with high accelerations and high Reynolds number flows. Hence, while the present paper only discusses the previous models, we are deriving a model suitable to both non-Newtonian flows and high Reynolds number flows with large accelerations.

Experimental Setup

The experimental setup mirrors the theoretical model, with two axisymmetric discs that rapidly separate. The working fluid for reported experiments was water nominally at 20°C and, for these initial experiments, presumed saturated at 1 atm. A clear 76mm diameter acrylic tube is sandwiched between two aluminum plates, with the seal made watertight using a rubber gasket. In the tube are two 16mm diameter stainless steel shafts with polished ends. One shaft is fixed to the bottom plate, and the other is free to move axially. A repeatable force is transferred through a lever mechanism from free-falling weights dropped from a chosen height to the top shaft by an 80/20 structure. The setup is backlit with 200 Watt LED light, and images are recorded at 110,000fps with a Vision Research Phantom v1210 high-speed camera (Figure 1b).

Results

High speed videos were recorded at 110,000fps for each experiment, and sample time series are shown in Figure 2, which shows a zoomed out view (a) and a higher resolution image sequence (b) of a typical experiment. We observe bubble formation and slight growth at the instant of calculated minimum pressure. The bubbles continue to grow more rapidly until reaching a maximum radius followed by rapid collapse, as expected. During collapse and rebound at the center, we can see secondary bubbles forming at other points in the domain in 2(a), presumably due to the shock-wave created by the initial bubble collapse.

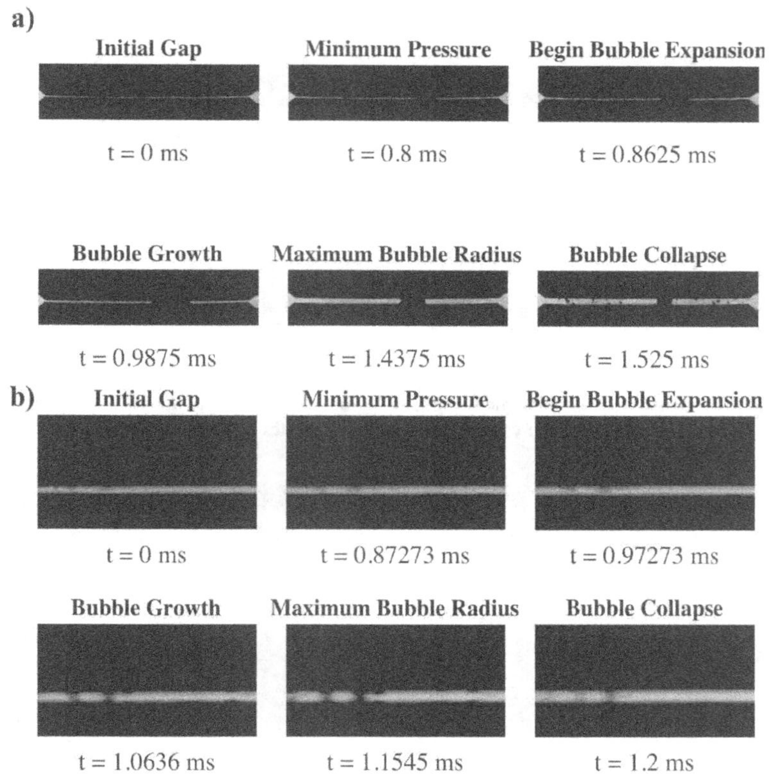

Figure 2: a) Time sequence of an experiment, image width $w = d_{rod}$, b) Time sequence of zoomed in experiment, image width $w = 0.3d_{rod}$

The experiment shown in the time-series displayed in Figure 2(b) is analysed using image tracking algorithms. We analyse the position, velocity and acceleration of the upper rod surface, with results displayed in Figure 3. The Reynolds number, relative pressures and cavitation numbers, based on both Kuzma's as well as Leider's analytical models, are presented in Figure 4. When comparing Figure 2(b) to the displacement in Figure 3(a) we can see that the minimum pressure occurs just after the initial movement of the rod. At this moment in time the gap height is still close to the initial value. While the velocity is still low, the acceleration has reached $a \approx 600\ \frac{m}{s^2}$. This highlights the relative importance of the gap height and inertial terms, categorized by the acceleration versus the rod velocity.

In Figure 4, which compares the relative pressures based on Kuzma's and Leider's analytical models, the cavitation number, σ, is defined as

$$\sigma = \frac{p_{center} - p_c}{\frac{1}{2}\rho \dot{h}_{max}^2} \tag{10}$$

Where p_c is the Blake-threshold based on an initial bubble radius of 1 μm and the surface tension and vapor pressure of water at 1 atm and 21°C (Brenner 2005). With p_{center} as the pressure in the radial and vertical center of the fluid domain and $\dot{h}_{max}$ as the maximum velocity of the upper rod during the experiment.

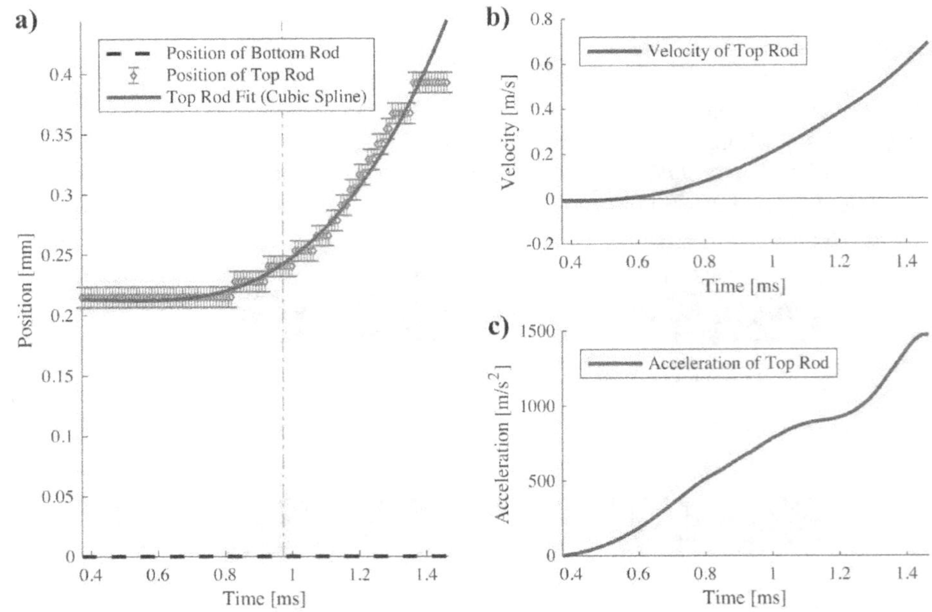

Figure 3: a) Position of the upper and lower rod, measurement points and spline fit, b) Velocity of upper rod, based on spline fit, c) Acceleration of upper rod based on spline fit. The vertical dash-dot line indicates moment cavitation was first observed.

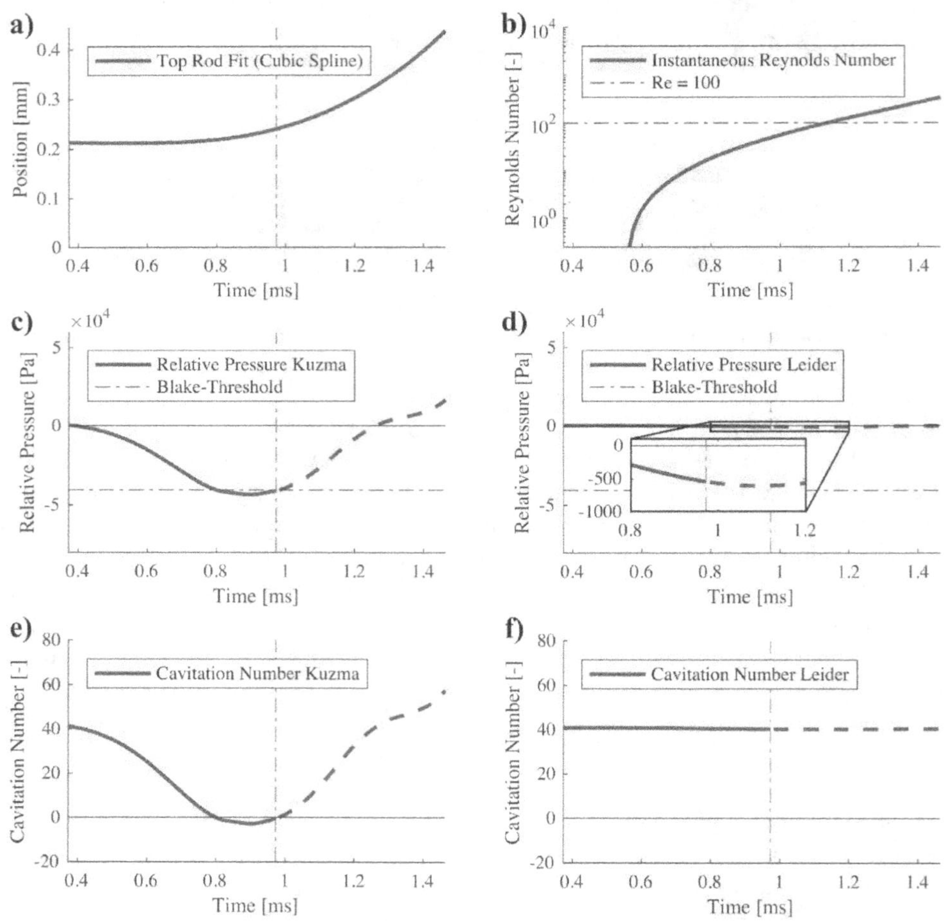

Figure 4: a) Fitted position of upper rod, b) Reynolds number, c) Relative pressure and Blake-Threshold Kuzma model, d) Relative Pressure and Blake-Threshold Leider model, e) Cavitation number Kuzma model, f) Cavitation number Leider model. The vertical dash-dot line indicates moment cavitation was first observed.

Comparing both Leider and Kuzma models highlights the disadvantages of neglecting inertial terms for a case such as ours. While Equation 9 predicts a cavitation event to occur slightly before cavitation is evident in the high speed video, the pressure change Equation 4 is orders of magnitude lower and does not predict cavitation. We should also note that Equation 4 assumes steady state flow conditions, and while Figure 4(d) indicates that this is not a valid assumption for our experiment, Engman *et al.* (2005) have discussed this model's ability to produce satisfactory results for compression cases with low Reynolds numbers. However, once cavitation occurs, the underlying assumptions regarding incompressibility and constant density are no longer valid. Therefore, the equations can no longer expected to correctly describe the flow field in the gap after the occurrence of initial cavitation (indicated by blue vertical line and dashed relative pressure and cavitation number lines in Figures 3 and 4). The dynamic pressure transducer used for next stage of the experiments will shed light on the pressures experienced during the cavitation event and the rebound of the cavitation bubbles.

Conclusion

A simplified experiment was designed to study of cavitation in joints. Adapted and rederived analytical solutions predicted the pressure minima with varying success, and results show that inertial terms cannot be neglected for the given canonical geometry and large accelerations. The experimental setup also allowed for repeatable measurements and yielded data consistent with prediction of the beginning of the cavitation event using Kuzma's model.

In the next stage the authors will incorporate direct dynamic pressure measurements, vary dissolved gas contents, and transition to study shear thinning fluids with adapted mathematical modeling. The effect of confinement due to a flexible membrane that mimics the synovial sac and curved interfaces reminiscent of the ball-and-socket geometry of a real joint will also be studied, if resources enable this. As it has been speculated (Unsworth 1971) that cracking joints can cause arthritis *via* mechanisms similar to that damaging ship propellers as vapor bubbles collapse near the surface, an investigation into joint cavitation from a fluid mechanics perspective may give us insight into how cracking affects joints. It has also been proposed that the cracking sound could be used as an initial diagnostic tool to reveal information about an individual's synovial fluid composition and joint topology (Cooke *et al.* 1978). While evaluating the validity of these is beyond the present and planned scope of study, the fundamental understanding from these laboratory experiments and theory may offer guidance for medical applications.

References

[1] Balazs, E. A., Watson, D., Duff, I. F,. and Roseman, S. (1967). *Hyaluronic acid in synovial fluid. I. Molecular parameters of hyaluronic acid in normal and arthritic human fluids*. Arthritis & Rheumatism, 10, pp. 357–376

[2] Brennen, C. E. (2005). *Fundamentals of multiphase flow*. Cambridge university press.

[3] Cooke A. F., Dowson D., and Wright V. (1978). *The Rheology of Synovial Fluid and Some Potential Synthetic Lubricants for Degenerate Synovial Joints.* Engineering in Medicine 7.2, pp. 66–72

[4] Dowson D., Unsworth A., and Wright V. (1970). *Analysis of 'Boosted Lubrication' in Human Joints.*Journal of Mechanical Engineering Science 12.5, pp. 364–369

[5] Engmann, J., Servais, C., and Burbidge, A. S. (2005). *Squeeze flow theory and applications to rheometry: A review*. Journal of Non-Newtonian Fluid Mechanics 132.1, pp. 1–27.

[6] Fam, H., Bryant, J.T., and Kontopoulou, M. (2007). *Rheological properties of synovial fluids.* Biorheology 44, pp. 59–74

[7] Jackson, J. D. (1962). *A study of squeezing flow*. Applied Scientific Research 11.1, pp. 148–152.

[8] Kawchuk, G., Fryer, J., Jaremko, J., Zeng, H., Rowe, L., and Thompson, R. (2015). *Real Time Visualization of Joint Cavitation.* PLOS ONE

[9] Kimbrough D. R. (1999). *Henry's Law and Noisy Knuckles.* Journal of Chemical Education 76 (11), p. 1509.

[10] Kuzma, D. C. (1967). *Fluid inertia effects in squeeze films*. Applied Scientific Research 18.1, pp. 15–20.

[11] Leider, P. J., Bird, R. B. (1974). *Squeezing Flow between Parallel Disks. I. Theoretical Analysis*. Transactions of the Institution of the Rubber Industry 13.4, pp. 336–341.

[12] Rwei S. P., Chen S. W., Mao C. F., and Fang H. W. (2008). *Viscoelasticity and wearability of hyaluronate solutions.* Biochemical Engineering Journal 40.2, pp. 211-217

[13] Scott, J. R. (1931). *Theory and Application of the Parallel-Plate Plastimeter*. Trans. of Inst. of Rubber Ind. 7, pp. 169–186.

[14] Unsworth, A., Dowson, D., and Wright, V. (1971). *Cracking Joints: A bioengineering study of cavitation in the metacarpophalangeal joint.* Annals of the Rheumatic Diseases 30, pp. 348–358.

[15] Watson, P., Kernohan, W. G., and Mollan, R. A. B., (1989). *A study of the cracking sounds from the metacarpophalangeal joint.* Proceedings of the Institution of Mechanical Engineers, Part H: Journal of Engineering in Medicine 203.2, pp. 109–118

© 2018 ASME

Cavitation Induction by Projectile Impacting on a Water Jet

[1]E. Stavropoulos Vasilakis*; [1]P. Koukouvinis; [2]M. Farhat; [1]M. Gavaises;

[1]*School of Mathematics, Computer Science and Engineering, CITY, University of London, UK;*

[2]*Laboratory of Hydraulic Machines, EPFL, CH*

Abstract

Following the work of Field et al. [4], who experimentally visualised cavity formation and shock propagation in impacted liquids at high velocities, the present study focuses on the simulation of the high velocity impact of a solid projectile on a water jet. The undeformable solid projectile is modelled through a direct forcing Immersed Boundary Method. The simulation is carried out using an explicit density based compressible solver, developed by Kyriazis et al. [6], which employs a two-phase flow model and includes phase change. This study gives a better insight on the phenomena following the impact of solids on liquids, including shock propagation and vapour formation, and demonstrates the capabilities of the presented Immersed Boundary Method to handle compressible cavitating flows.

Keywords: Immersed Boundary Method; Cavitation; liquid solid impact

Introduction

High velocity liquid impacts on solids consist a broaden category of physical problems of high interest, as they take place in many engineering applications and affect drastically the performance of machines operating in high velocities and under high pressures, as they are linked to cavitation formation and erosion development and eventual damage of the solid surfaces. Such applications may refer to erosion induced in hydraulic machines [4] or by rain drops on airplanes and wind turbine blades [1]. Field [3] [4] along with Lesser [7] and Bowden and Brunton [2], have conducted many theoretical and experimental studies on the mechanisms of liquid impacts on solids or liquids, and have established a solid understanding of the phenomena taking place in such cases.

Field et al. in their recent paper [4], summarise the experiments conducted on liquid to liquid and liquid to solid impacts, focusing on the complicated shock structures and the high-speed jetting on the area of impact. According to experimental data, that are in accordance with the theoretical work of Lesser, when a liquid impacts a solid surface, compressibility effects take place near the impact area that lead to high velocity jetting along the solid surface and pressure shock waves are generated that get detached and travel away from contact region. When these waves interact with the free surface of the liquid, during detachment or reflection, systems of relief waves are formed that produce enough tensile stress to initiate cavitation.

The current study focuses on the simulation of a high velocity impact of a solid projectile on a water jet following the experiments of Field et al. [4], in order to investigate the shock structures and the cavitation induction. This simulation poses a significant challenge regarding the modelling of the solid projectile motion. To tackle this issue, an Immersed Boundary (IB) Method is used to model the movement of a solid boundary over the fixed grid of the domain, avoiding any complexities arising from re-meshing or moving meshing techniques. The IB methods, first introduced by Peskin [11] model internal boundaries by either changing the computational stencil near the solid or adding source terms in the equations. The later approach is used in this study, where forcing source terms are added in the momentum equations of an explicit density based in-house solver, developed by Kyriazis [6] on OpenFOAM platform. This solver is able to simulate cavitating flows, where the local Mach number has a wide range, in addition to the advection of immiscible compressible gas. The case of the projectile impact on a water jet serves as an interesting and challenging demonstration of the capabilities of the developed numerical tools.

Numerical Method

The Immersed Boundary (IB) method used in this study, falls in the category of direct forcing methods [8], where the presence of the boundary is taken into account as solid forcing by introducing a source term in the momentum equation. Similar method has been used by Mochel et at. [9] in high velocity compressible turbulent flows as well as by Koukouvins et al. [5] to simulate the cavitating flow resulting from the high-speed closure of the claw of the pistol shrimp. To the best of the authors' knowledge, apart from the method developed by Koukouvinis, the only IB

*Corresponding Author, E. Stavropoulos-Vasilakis evangelos.stavropoulos-vasilakis@city.ac.uk

methodology applied on cavitating flows, such as Diesel injector flows with closing needle, is presented by Orley et al. [10], which consists of a cut cell methodology. The IB method used in the presented study provides important advantages regarding geometrical and topological manipulations, compared to cut-cell methods, and thus results in less time-consuming computations.

The immersed solid boundary is represented by a surface mesh while a mask is used for the computational grid cells enclosed by this surface. This mask representing the solid area, corresponds to the solid volume fraction of the cells and is calculated as the ratio of the cell volume covered by the IB surface over the total cell volume. This calculation is carried out as the average of normal distance between all the vertices of the cell and the nearest IB surface face.

The forcing source term is calculated as the difference of the fluid velocity from the IB solid velocity, and therefore tends to impose no-slip condition on the IB cells. The source term is then multiplied with the IB mask to localise the IB forcing, as in Equation 1.

$$\vec{f}_{IB} = \alpha_{IBMask} \cdot \frac{\Delta U}{\Delta t} = \alpha_{IBMask} \cdot \frac{U_{fluid} - U_{IB}}{\Delta t}, \alpha_{IBMask} \in [0,1]$$

Equation 1 Immersed Boundary forcing term.

The IB forcing term is introduced in the compressible Navier-Stokes equations that are solved using a density based, two-phase, explicit solver, developed by Kyriazis et al. [6]. This solver is based on the homogeneous mixture approach to account for liquid, liquid-vapour and gaseous phases, that are in mechanical and thermodynamic equilibrium. Phase change is considered between liquid and vapour phases, using the linear barotropic law to compute the mixture density. A transport equation is solved for the advection of the gas, considering the gas mass fraction, which is modelled as isothermal ideal gas.

The solver is making use of a special hybrid flux calculation, to tackle issues arising when density based solvers are used for three phase cases with great variations of speed of sound. The hybrid flux is based on Primitive Variable Riemann Solver and Mach consistent flux [6]. The reader is referred to [6] where Kyriazis et al. present the solver in detail.

Results

The current study examines the impact of a projectile, with a diameter of 9mm, at horizontal speed of 210m/s, on a water jet, of 25mm in diameter, which flows vertically with a velocity around 1m/s. The ambient air is initially stationary. The Mach number with respect of the surrounding air is 0.6 and with respect to the impacted water 0.14. The pressure of the domain is considered equal to 1bar and the reference temperature set to 300K. As the projectile velocity is relatively high, turbulence modelling was not used as the boundary layers are expected to be very thin and therefore their influence could be neglected. The domain is discretized by a cylindrical mesh with 128 cells along the diameter of the water jet.

When the projectile impacts on the circular water surface, a peak in pressure is observed and a shock envelop is formed (as seen in Figure 1a) that while the projectile penetrates the free surface gets detached from the solid front surface and starts to travel towards the opposite free surface of the water jet where it gets reflected. The water region near the solid gets compressed and pushed out along the projectile front surface. The pressure waves structures favour cavitation inception.

As the shock wave detaches, low pressure regions form that follow the initial high-pressure envelope. In addition, as the shockwave travels on the free surface, it gets reflected and relief waves form that start travelling towards the center of the water jet. The low-pressure region expands following the impact shock and values lower than the vapour saturation pressure are initially observed next to the free surface, inducing cavitation induced Figure 1b. Two cavities form in both sides of the water jet, that start expanding mainly on the periphery of the water jet following the high-pressure shock envelope, until the opposite free surface where they merge into one cavity that expands vertically and shrinks horizontally Figure 1c. This cavity then breaks up into two almost symmetrical cavities over and under the middle horizontal plane and collapse, rebound and collapse again before completely vanishing Figure 1 d-f.

During this expansion of the cavity, the pressure waves that follow the initial high-pressure envelope, get reflected on the vapour surface resulting in a very transient and perturbed pressure distribution, visible in Figure 1b,c. As the cavity collapses, two low pressure waves get released and travel in the jet inwards and get reflected on the front free surface of the jet, resulting in very low-pressure values and cavitation induction in almost symmetrical positions over and under the projectile, near the front free surface Figure 1e,f.

These results are in qualitative agreement with the experimental data of Field et al. [4], where the reflection of the initial impact shock wave on the opposite free surface of the jet, creates multiple vapour cavitaties that act as reflective surfaces for upcoming waves. Comparison between the side view experimental shadowgraphy images and the magnitude of the density's gradient from the simulation, plotted on the vertical and horizontal middle plane, as seen in Figure 2, indicates that the shock and the cavitation regions are predicted correctly, as well as the bulk dynamics of the deformation of the free surface. A striking difference is the splashing on the entry point, that seems not captured in the simulations. This may be a result of either the IB approach used or the inadequate spacial refinement near the entry point in combination with the diffusive nature of the gas mass fraction transport equation.

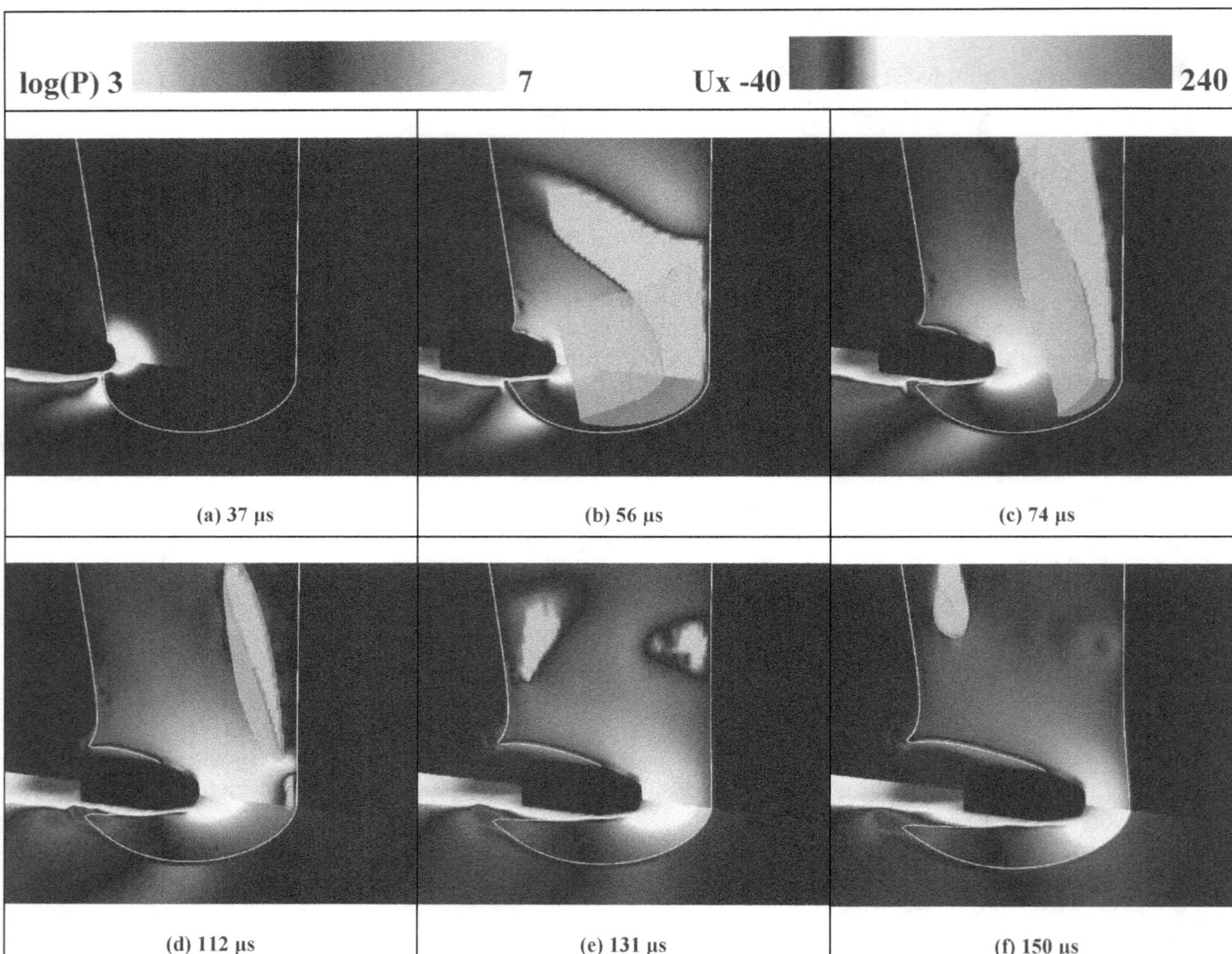

Figure 1 Projectile impact on water jet. Pressure (vertical plane) and longitudinal velocity component (horizontal plane) contours on different time steps. In addition, 0.1% of vapour volume fraction represented by iso-surface and red contour line on planes, and 50% gas volume fraction in white contour line to represent liquid-mixture/gas interface.

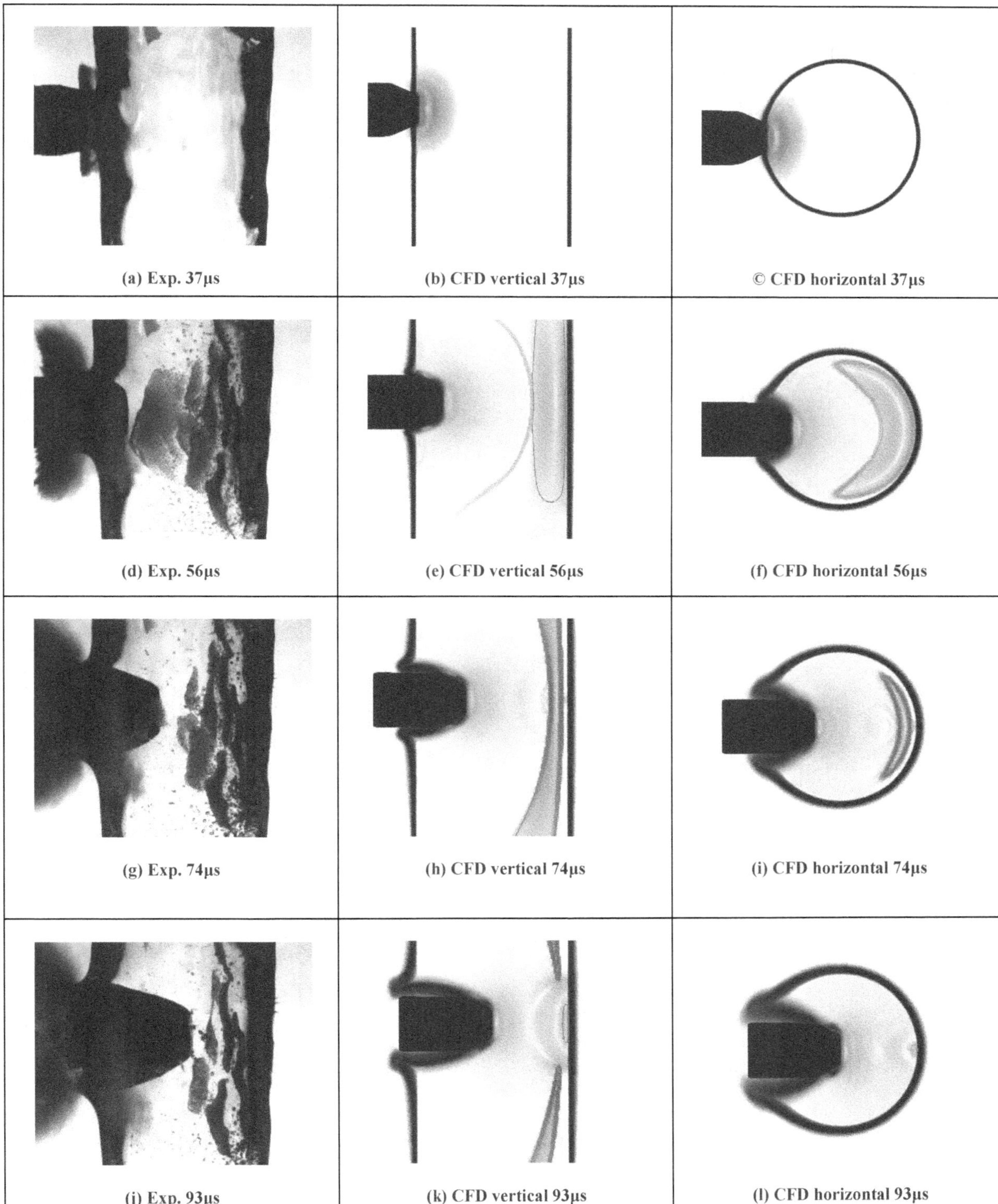

Figure 2 Experimental data from [4] (left) and numerical results (middle, right) for the impact of projectile on water jet, with Uimpact=210m/s. For the CFD results, the logarithm of the magnitude of the gradient of density is plotted on the vertical (middle) and horizontal (right) middle plane on grayscale, along with the 0.1% contour for vapour volume fraction, represented by a red line.

Conclusion

The current study examines the solid projectile impact on a water jet and assess the use of direct forcing Immersed Boundary Method in conjunction with an explicit density based compressible two-phase solver that accounts for cavitation. The physics of the case studied was accurately captured and the simulation results are in qualitative agreement with experiments presented by Field et al. [4]. Pressure shock waves, relief waves and cavitation induction, are present and numerically calculated. A more detailed description of the vapour cavity is provided that is in accordance with the analysis of the experimental data. However, liquid splashing and the anticipated pressure peak values where not observed. Although experimental pressure measurements do not exist for the specific problem, theoretical analysis and experimental data on simpler cases indicate that pressure values exceed water hammer pressures. The Immersed Boundary method used proved not good enough to capture such phenomena and shows room for improvement.

Acknowledgments

This work was carried out in the framework of CaFE project, which has received funding from the European Union Horizon 2020 Research and Innovation programme, with Grant Agreement No 642536.

References

1. Serge Abrate (2016). *Soft impacts on aerospace structures*. Progress in Aerospace Sciences, 81(Supplement C):1–17, Dynamic Loading Aspects of Composite Materials.

2. F. P. Bowden and J. H. Brunton (1961). *The Deformation of Solids by Liquid Impact at Supersonic Speeds*. Proceedings of the Royal Society of London. Series A, Mathematical and Physical Sciences, 263(1315):433–450

3. J. E. Field, M. B. Lesser, and J. P. Dear (1985). *Studies of Two-Dimensional Liquid-Wedge Impact and Their Relevance to Liquid-Drop Impact Problems*. Proceedings of the Royal Society of London. Series A, Mathematical and Physical Sciences, 401(1821):225–249

4. J.E. Field, J.-J. Camus, M. Tinguely, D. Obreschkow, and M. Farhat (2012). *Cavitation in impacted drops and jets and the effect on erosion damage thresholds*. Wear, 290–291:154–160

5. P. Koukouvinis, C. Bruecker, and M. Gavaises (2017). *Unveiling the physical mechanism behind pistol shrimp cavitation*. Scientific Reports, (1). Licensed under a Creative Commons Attribution 4.0 International License.

6. N. Kyriazis, P. Koukouvinis, and M. Gavaises. *Modelling cavitation during droplet impact on solid surfaces*. Computer Methods in Applied Mechanics and Engineering. Under Review.

7. M. B. Lesser (1981). *Analytic solutions of liquid-drop impact problems*. Proceedings of the Royal Society of London. Series A, Mathematical and Physical Sciences, 377(1770):289–308.

8. Rajat Mittal and Gianluca Iaccarino (2005). *Immersed Boundary Methods*. Annual Review of Fluid Mechanics, 37(1):239–261.

9. Loïc Mochel, Pierre-Élie Weiss, and Sébastien Deck (2014). *Zonal Immersed Boundary Conditions: Application to a High-Reynolds-Number Afterbody Flow*. AIAA Journal, 52(12):2782–2794.

10. Felix Örley, Vito Pasquariello, Stefan Hickel, and Nikolaus A. Adams (2015). *Cut-element based immersed boundary method for moving geometries in compressible liquid flows with cavitation*. J. Comput. Physics, 283:1–22.

11. Charles S Peskin (1977). *Numerical analysis of blood flow in the heart*. Journal of Computational Physics, 25(3):220–252.

© 2018 ASME

10th International Symposium on Cavitation - CAV2018
Baltimore, Maryland, USA, May 14 – 16, 2018

Cavitation modelling using real-fluid equation of state

[1]Songzhi Yang[*1,2], Chaouki Habchi[*1,2], Ping Yi[*1,2], Rafael Lugo[1]

[1]IFP Energies nouvelles, 1 et 4 avenue de Bois-Préau, 92852 Rueil-Malmaison, France

[2]Institut Carnot IFPEN Transports Energie

Abstract

The aim of this work is to improve the accuracy of cavitating flow simulations by considering real-fuel mixture thermodynamics, like gas (air) solubility and physical property deviations from ideality, and their effects on phase change. In this work, a new fully compressible two-phase six-equations model coupled with Peng and Robinson [1] real-fluid cubic equation of state (PR-EOS) is described. Assuming thermodynamic equilibrium with negligible capillarity, this diffuse interface model has been implemented in the IFP-C3D solver [3], which already includes the Gibbs Energy Relaxation Method (GERM) cavitation model [10]. One-dimensional academic test cases (cavitation-tube, shock-tube) are performed firstly to validate the real-fluid PR-EOS implementation. Comparisons of the numerical results with previously published works are also carried out. Then, the models have been applied to simulate the cavitation phenomenon inside a single-hole nozzle. With phase equilibrium theory, the simulation has proved that increased dissolved gas may facilitate gaseous cavitation and restrict the developing of vaporous cavitation. The effect of dissolved gas on fuel properties has also been presented and discussed using the suggested real fluid model. Overall, the two fluid model combined with real equation of state is able to reveal more physical details about the cavitation process.

Keywords: two fluid model; real fluid; dissolved gas; cavitation

Introduction

Many strategies like high injection pressure and smaller nozzle orifice have been adopted to improve the Diesel engine combustion efficiency and reduce emissions, like particle issues and NOx. However, much remains to be done to achieve the objectives in this area because of an insufficient understanding of the various physical phenomena governing the fuel injection and the mixing with gas before combustion. With the increasing injection pressure and decreasing nozzle diameter, the cavitation phenomenon become more and more extrusive during fuel injection in modern diesel engine. Many researchers have contributed to better understanding this phenomenon [2, 7, 9, 10, 12–14, 17]. Among various research strategies, numerical modeling is still one of the most efficient tools. In this study, a multicomponent and fully compressible two-fluid six equation model is proposed using the Peng and Robinson [1] real-fluid cubic equation of state (PR-EOS). This model is referred to below as 6EQ-PR model. The main aim of this work is to investigate the effect of non-condensable gas on the cavitation phenomenon using the 6EQ-PR model. Previous researchers like Duke et al., [6, 7, 14] have demonstrated the effect of non-condensable gas on cavitation using a standardized fuel and also degassed fuel through X-ray radiography experiments. Then Battistoni et al., [2] further studied these effects by numerical modeling with Homogeneous Relaxation Model (HRM). In both of the previous studies, it has been found that the void content appeared in the central part of the orifice has decreased significantly as the initial amount of gas nuclei is decreased. Due to the fact that pressure is above the saturated pressure in the central part of the nozzle, the cavitation cloud taking place there has been identified as gaseous cavitation, thus confirming the analysis of Habchi, et al., [4] on the existence of two kinds of cavitation, namely gaseous cavitation and vaporous cavitation. Habchi, et al., [10, 11] have also detected gaseous cavitation in a typical Diesel injector using a two-fluid seven equations diffused interface model. Besides, Mori et al. [16] has confirmed that the mixing of non-condensable gas can induce the cavitation to incept at higher pressure exceeding pure material saturated pressure. Li et al., [15] also proved experimentally that the tensile strength of water has decreased with the increase of dissolved gas like nitrogen, oxygen at given temperature. The reduced tensile strength thereby facilitates the formation of cavitation. All these previous researches have provided valuable insight into current research. To know more details about the effect of non-condensable gas on cavitation, unlike previous research, the present study has adopted a diffuse interface two-fluid model combined with a multicomponent thermodynamics' equilibrium solver based on Peng and Robinson [1] cubic equation of state (PR-EOS). Indeed, using this real-fluid EOS gives us the possibility to consider dissolved gas (like Nitrogen) in the

*Corresponding Author, Songzhi Yang:songzhi.yang@ifpen.fr

liquid in addition to the possible presence of gas nuclei bubbles. The main difference with previous models and studies is the way of dealing with gaseous cavitation and vaporous cavitation. In current study, both kinds of cavitation are resulting from the thermodynamics equilibrium solver. Due to the limited space here, the equations of the newly developed flow and thermodynamics' model solvers cannot be described in this article. More explanations it can be found in [18]. This paper has been organized as following: firstly, one dimensional (1D) test cases results including a double expansion tube and a shock tube are presented. Next, the numerical results obtained using two-dimensional (2D) simulations of real-size cavitating nozzle are discussed. Finally, conclusions section summarize the main findings of this article.

1. Results and discussion

1.1. 1D Double expansion tube

This test case consists of 1 m long tube filled with liquid water (H_2O) at atmospheric pressure and with temperature 293K. The initial mass fraction of gas nitrogen (N2) is 1e-5, and for water is 0.99999. The initial discontinuity is set at 0.5 m. At this location, the velocity (1 m/s) in the left side has been specified the same but opposite to the right side. The computational time is 3.2ms and the numerical results have been compared with the results from Chiapolino et al.,[5] model in which Stiffened Gas (SG) EoS has been chosen. As shown in Fig.1, the profiles of pressure, temperature and velocity are in a good agreement. Nevertheless, the mass fraction of water vapor induced by the expansion wave has presented some difference which are due to the different initializations of the simulations as shown in Fig .1(d). Indeed, in this study, the flow has been assumed to be in saturated state or phase equilibrium state since the beginning of the simulation, which seems to be not the case for the Chiapolino et al. [5] simulations.

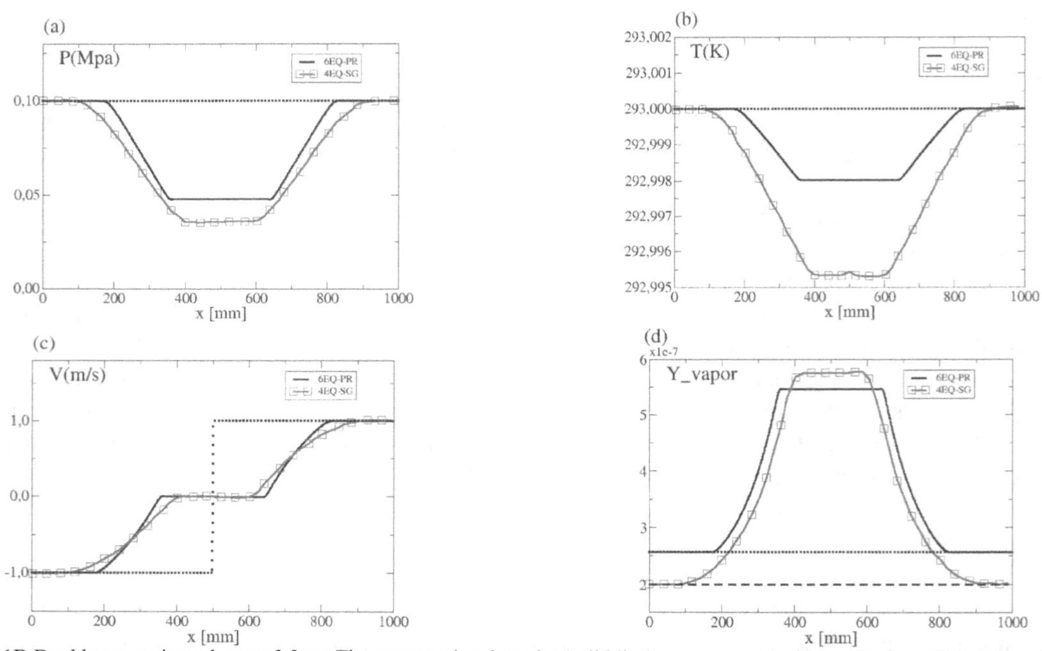

Fig 1. 1D Double expansion tube at t=3.2ms. The computational results (solid line) are compared with results from Chiapolino et al. [5] (Symbols). The dashed lines are the initial conditions. The present computations were conducted with 1000 cells and a CFL number equals to 0.2.

1.2. Shock Tube

The shock tube is 1 meter long and with the initial discontinuity at 0.5m. The pressure ratio is 2 (2 bar in the left and 1 bar in the right) and with a temperature of 293K throughout the tube. The N2 mass fraction is 0.98, and 0.02 for water. The results shown in Fig.2 are at 1ms. A good agreement has been achieved when comparing with Chiapolino et al. [5]. As mentioned above, the difference of initial water vapor fraction has resulted in some discrepancies in the final amount of vapor mass fraction. But, the spreading trend of compression wave and rarefaction wave is very similar, particularly for the comparison of pressure and velocity, as shown in Fig .2(a, c). However, the difference of initial vapor mass fraction has induced a significantly higher temperature in the right side (Fig .2(b)) along with a stronger evaporation rate (Fig .2(d)).

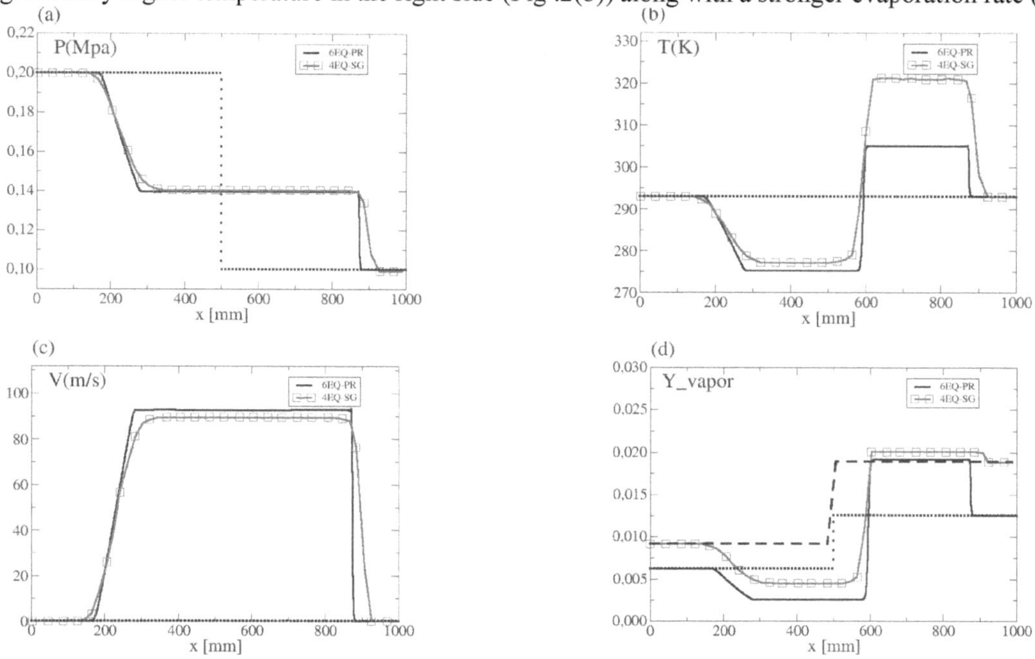

Fig 2. 1D shock tube at t=1ms. The computational results(solid line) are compared with results from Chiapolino et al. [5] (Symbols). The dashed lines are the initial conditions. The computations were conducted with 1000 cells and a CFL number equals to 0.2

1.3. Nozzle Cavitation

In this section, the suggested 6EQ-PR model has been used to simulate the cavitation phenomenon in a real size 2D nozzle. The fuel used here is gasoline calibration fluid (Viscor 16BR). Relevant fuel properties can be found in ref [2]. The non-condensable gas adopted is N_2. Firstly, the phase equilibrium analysis about Viscor and N_2 system is presented. Then, some results and discussion of the cavitation modelling in the nozzle are reported.

1.3.1 Vapor liquid equilibrium analysis of (Viscor - N_2) system

To better understand the effect of dissolved gas on cavitation, it is essential to investigate the thermodynamic equilibrium behavior of Viscor and N_2. The main method used for thermodynamics equilibrium analysis at given temperature and pressure is isothermal flash computation (TP flash) [13]. Firstly, the vapor-liquid equilibrium computation of (n-heptane - N_2) system is conducted and compared with experimental data [8]. A very good agreement has been obtained (Fig. 3(a)) except in the critical point zone where PR-EoS is known to fail. Then the phase equilibrium computation is switched back to (Viscor - N_2) system. One important parameter in thermodynamics theory is the feed vapor mole fraction, usually denoted "psi" (Ψ). In the current (Viscor- N_2) system, this parameter corresponds to the overall amount of vapor which includes the vaporized fuel and undissolved gaseous N_2. As shown in Fig. 3(b), at any given T and P, with the increase of N_2 molar fraction in the feed, Ψ increases from negative to positive values. This imply the well-known phenomenon that adding a high amount of N_2 to the system leads to a saturated system state that will be more prone to nucleation and promote cavitation. Recent research has confirmed this phenomenon with experiments [12], in which they have attributed the strengthening of cavitation to the intensifying of cavitating nuclei coming from the dissolved gas. Therefore, the dissolved gas nuclei can help decrease the energy needed to form a bubble and reduce the tensile strength of fluid. As a matter of fact, the negative Ψ in Fig. 3(b) means no vapor is generated in the flow and all the N_2 present in the system is fully dissolved inside the liquid. In this case, the fluid is in a single liquid phase state. In addition, phase transition (i.e. cavitation inception) pressure increases with N_2 concentration in the feed, as shown in Fig. 3(b). For high N_2 concentration in the feed (i.e. standard liquid fuels), cavitation may appear in region at relatively high-pressures. This kind of cavitation has been called "gaseous cavitation", as the incepted nucleus in this condition contains mostly undissolved N_2. Inversely, for low N_2 concentration in the feed (i.e. degassed liquid fuels), cavitation will appear only in relatively low-pressure regions. This kind of cavitation has been called "vaporous cavitation", as the incepted nucleus in this condition contain mostly fuel-vapor. This thermodynamics analysis corroborates the Habchi's numerical findings [9].

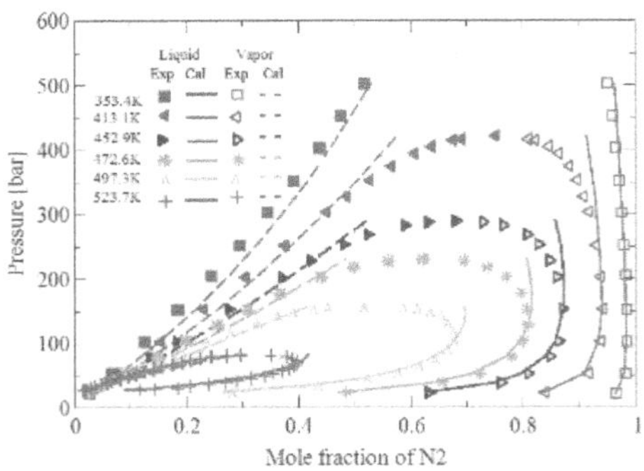

(a)

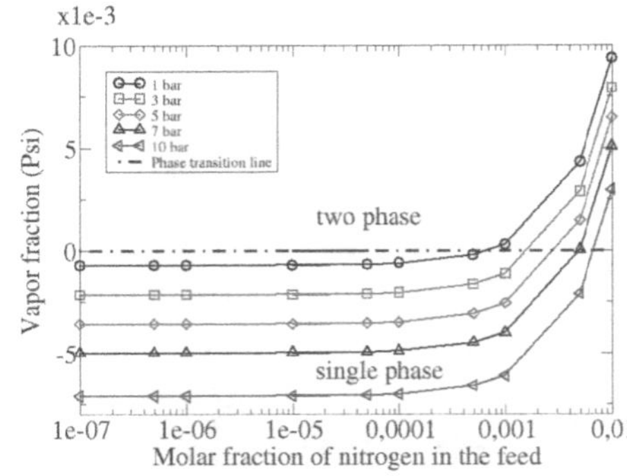

(b)

Fig 3. (a) Phase equilibrium results compared to n-heptane experimental data[8]; (b) Evolution of vapor fraction with the molar number of nitrogen in the feed at T=293 K and P=1-10 bar

1.3.2 Results discussion about the cavitation modeling

The 2D cavitating nozzle used here is axisymmetric with sharp inlet edge and a diameter of 500 μm and a length of 2.5mm. Detailed descriptions can be found in [6]. For computational efficiency reason, 1/8 of the original geometry configuration has been simulated, as shown in Fig. 4. Mesh is refined around the orifice zone and the minimum resolution of grid reaches 6 μm and the total cell number counts 25982.The nozzle outlet is submerged according to the original experimental condition [6]. The inlet and outlet are set with pressure boundary conditions with the 10 bar and 1 bar, respectively.

Due to the fact that the objective of this study is to investigate the effect of dissolved nitrogen on cavitation, a two-phase condition has been selected as initial state. The initial pressure for the overall flow field has been set the same as the inlet pressure (10 bar). Initially, the fluid has been assumed at equilibrium state with an initial mass fraction of dissolved gas (N_2) in liquid phase around 2e-3 corresponding to a gas volume fraction equaling to 1e-2(Table 1).

Table 1 Test cases initial conditions

T=293 K, P=10 bar				
Case No	Feed(z_{N2})	Vapor fraction	Y_dissolved_n2	Gas volume fraction(α_g)
1	8.08e-3	1e-3	2.01e-3	1e-2

z denotes the molar fraction of nitrogen in the feed

The modeling results shown below are at t=1.4ms. As shown in Fig. 5, the gas volume fraction (denoted alphagaz in in Fig. 5(a)) can be decomposed into two contributions: the gaseous cavitation (Fig. 5(b)) and the vaporous cavitation in Fig. 5(c). Consistent with our thermodynamics analysis reported above, one may see in this latter Figure that vaporous cavitation appears mostly near the orifice sharp edge, where the pressure is the lowest (see Fig. 6(a and d)). In contrast, gaseous cavitation is present throughout the computational domain. In addition, vaporous cavitation is very weak (see palette maximum value in Fig. 5(c)) relatively to gaseous cavitation (see Fig. 6(b)). This means that the overall gas volume fraction (Fig. 5(a)) is mostly due to the expansion of the initial nucleus volume fraction (1e-2 in this case). This expansion is nearly an isothermal process as the temperature is nearly constant. Indeed, the maximum temperature deviation is more or less equal to 0.2K in the orifice (see Fig. 6(d)).

In addition, these two kinds of cavitation are mutually restricted. To further confirm this phenomenon, the developing process of cavitation in the inlet corner has been investigated using our phase equilibrium solver. The molar fraction of nitrogen is set the same with the initial value 8.08e-3. As seen in Fig.6 (c), with the pressure lowering down to 0.5 bar, the evaporation ratio of fuel becomes much higher which further proves the vaporous cavitation is sensitive to the fast pressure drop. Dissolved nitrogen concentration also becomes higher with pressure increasing. However, the increasing of dissolved gas has seen the reduction of molar fraction of vaporous fuel (Fig.6 (c)) and a slow increment of gaseous nitrogen. This may imply that the dissolved gas is favorable to the formation of gaseous cavitation instead of the vaporous cavitation. Besides, for completeness of this real-fluid study, the influence of dissolved gas on the density and heat capacity have also been shown in Fig.6 (e, f). Overall, the liquid density has not been through significant variation with the addition of dissolved gas. In contrast, liquid heat capacity shows a clear change especially inside the cavitation pocket. The change might be caused by the small temperature variation, shown in Fig. 6(d).

As seen from the void fraction distribution (Fig. (5)), the cavitation appeared in the corner and center of inlet orifice qualitatively agreed with Duke et al. [6, 7] X-ray experiments. But the cavity cloud appearing in the center of orifice close to the outlet has not matched the experimental results well [2]. The reason may come from the relatively high initial molar fraction of nitrogen z_{N2} in this case (Table 1). Indeed, in previous research [2], a value equaling to 7e-5 has been used. But, thermodynamically speaking, such z_{N2} low value corresponds to a single-phase state as can be seen in Fig. 3(b). Therefore, future endeavors are going to tackle these issues.

Fig 4. Single-hole configuration and mesh refining with total 25982 cells and minimum grid resolution of 6 μm.

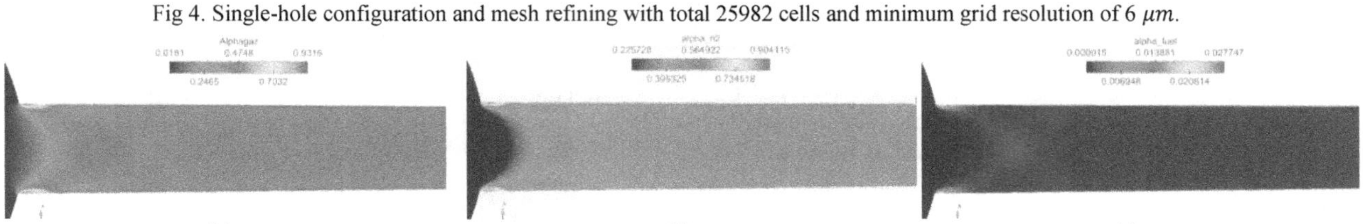

(a) (b) (c)

Fig 5. Results of the 2D cavitating nozzle (case 1)

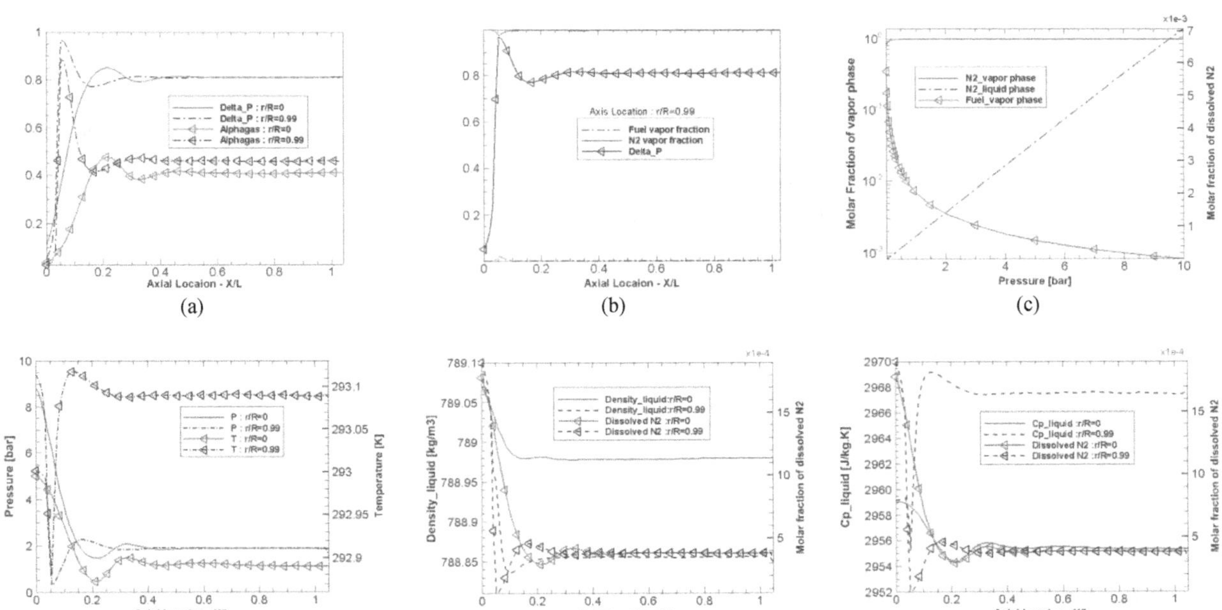

(a) (b) (c)

(d) (e) (f)

Fig 6. (a) Variation of void fraction with Delta_P; (b) Variation of vapor molar fraction of fuel and nitrogen with Delta_P at r/R=0.99; (c) Variation of vapor molar fraction of fuel and nitrogen with pressure at T=293K and $z_{N2} = 8e-3$; (d) Variation of pressure and temperature; (e, f) Variation of density and heat capacity with dissolved nitrogen. Delta_P $= \frac{(10\text{ bar}-P)}{10\text{ bar}}$.

2. Conclusion

A fully compressible two fluid model based on phase equilibrium theory with real fluid equation of state has been described and first validations have been shown and discussed in current study. One dimensional test cases including shock tube and double expansion tube have been conducted and validated with previous published results. Then, the model was used to simulate the cavitation phenomenon inside a single-hole nozzle. With phase equilibrium theory, the simulation has proved the effect of dissolved gas on cavitation. On the one hand, it has been shown that the gaseous cavitation is intensified with the increasing dissolved gas. On the other hand, the vaporous cavitation is restricted if there is more dissolved gas is in the initial feed. In addition, the model also proves that the dissolved gas concentration can become much higher with increasing pressure and some physical properties like density and heat capacity can be impacted by the dissolved gas in the fluid. Finally, the two-fluid model combined with a real-fluid phase equilibrium solver has been shown to have more advantages in revealing the cavitation physics details than previously published cavitation models using barotropic or incompressible liquid assumptions. However, more endeavors are needed to tackle real-fluid phase-transition issues when the initial fluid is in single phase state.

Acknowledgement

This project has received funding from the European Union Horizon 2020 Research and Innovation program. Grant Agreement No 675528 for the IPPAD project.

References

[1] Abudour, A. M., Mohammad, S. A., Robinson, R. L., and Gasem, K. A. 2012. Volume-translated Peng–Robinson equation of state for saturated and single-phase liquid densities. *Fluid Phase Equilibria* 335, 74–87.

[2] Battistoni, M., Duke, D. J., Swantek, A. B., Tilocco, F. Z., Powell, C. F., and Som, S. 2015. nozzles. *Atomiz Spr* 25, 6, 453–483.

[3] Bohbot, J., Gillet, N., and Benkenida, A. 2009. IFP-C3D. An Unstructured Parallel Solver for Reactive Compressible Gas Flow with Spray. *Oil & Gas Science and Technology - Rev. IFP* 64, 3, 309–335.

[4] Chawki Habchi. 2013. *A Gibbs free Energy Relaxation Model for Cavitation Simulation in Diesel injectors. DOI*=10.13140/2.1.4615.7761.

[5] Chiapolino, A., Boivin, P., and Saurel, R. 2017. A simple phase transition relaxation solver for liquid-vapor flows. *Int. J. Numer. Meth. Fluids* 83, 7, 583–605.

[6] Duke, D. J., Kastengren, A. L., Swantek, A. B., Matusik, K. E., and Powell, C. F. 2016. X-ray fluorescence measurements of dissolved gas and cavitation. *Exp Fluids* 57, 10, 538.

[7] Duke, D. J., Schmidt, D. P., Neroorkar, K., Kastengren, A. L., and Powell, C. F. 2013. High-resolution large eddy simulations of cavitating gasoline–ethanol blends. *International Journal of Engine Research* 14, 6, 578–589.

[8] García-Sánchez, F., Eliosa-Jiménez, G., Silva-Oliver, G., and Godínez-Silva, A. 2007. High-pressure (vapor+liquid) equilibria in the (nitrogen+n-heptane) system. *The Journal of Chemical Thermodynamics* 39, 6, 893–905.

[9] Gonçalves, F. M., Castier, M., and Araújo, O. Q. F. 2007. Dynamic simulation of flash drums using rigorous physical property calculations. *Braz. J. Chem. Eng.* 24, 2, 277–286.

[10] Habchi, C. 2015. A Gibbs Energy Relaxation (GERM) model for cavitation simulation. *Atomiz Spr* 25, 4, 317–334.

[11] Habchi, C., Bohbot, J., Schmid, A., and Herrmann, K. 2015. A comprehensive Two-Fluid Model for Cavitation and Primary Atomization Modelling of liquid jets - Application to a large marine Diesel injector. *J. Phys.: Conf. Ser.* 656, 12084.

[12] Habchi, C., Dumont, N., and Simonin, O. 2008. Multidimensional simulation of cavitating flows in diesel injectors by a Homogeneous Mixture Modelling approach. *Atomiz Spr* 18, 2, 129–162.

[13] J.B. Moreau, O. Simonin, C. Habchi, Ed. 2004. A numerical study of cavitation influence on diesel jet atomization. DOI 10.13140/2.1.2813.5369

[14] Kastengren, A. L., Tilocco, F. Z., Duke, D. J., Powell, C. F., Zhang, X., and Moon, S. 2014. Time-resolved x-ray radiography of sprays from Engine Combustion Network SPRAY A diesel injectors. *Atomiz Spr* 24, 3, 251–272.

[15] Li, B., Gu, Y., and Chen, M. 2017. An experimental study on the cavitation of water with dissolved gases. *Exp Fluids* 58, 12, 1065.

[16] Mori, Y., Hijikata, K., and Nagatani, T. 1976. Effect of dissolved gas on bubble nucleation. *International Journal of Heat and Mass Transfer* 19, 10, 1153–1159.

[17] Roth, H., Gavaises, M., and Arcoumanis, C. 2002. Cavitation Initiation, Its Development and Link with Flow Turbulence in Diesel Injector Nozzles. In . SAE Technical Paper Series. SAE International400 Commonwealth Drive, Warrendale, PA, United States. DOI=10.4271/2002-01-0214.

[18] Songzhi Yang, Chaouki Habchi, and Rafael Lugo. 2017. Subcritical and supercritical flow modeling using real-fluid Equation of State. A basic study of contact interface transport. *DOI*=10.13140/RG.2.2.34176.33289.

© 2018 ASME

Cavitation nuclei and tensile strength of water

K.A. Mørch

Technical University of Denmark, DK-2800 Kongens Lyngby

Abstract

The paper presents a discussion of the possibility of stabilization of cavitation nuclei of very high tensile strength in pure water by liquefaction of gas molecules at the water-gas interface of a gas bubble approaching dissolution. Likewise, the formation of such nuclei by molecular clustering of gas molecules already in solution in the water is considered. Finally, we discuss the origin of skin-stabilized gas bubbles, in particular at solid-water interfaces. Skin-stabilized gas bubbles in diffusion balance can explain most experimental results for water of low to moderate tensile strength, including effects of pressurization and de-pressurization.

Keywords: cavitation nuclei, tensile strength of water, liquefaction of gas, interfacial water

Problem analysis

Any gas bubble in water may serve as cavitation nucleus if exposed to sufficient tensile stress, but precise knowledge of cavitation nuclei in engineering systems is quite limited. Their tensile strength varies from values close to zero in plain water to values of almost 30 MPa, obtainable in highly purified and degassed water using focused ultrasonic wave-packets[1], and it comes up to the theoretical values for H_2O of the order of 140 MPa, when strained in inclusions of quartz[2]. Normally water contains dissolved gas, as well as particles and contaminating agents, which are decisive for its tensile strength. Further, in experiments the techniques used for measurement of tensile strength influence the results achieved because cavitation nuclei are not invariant.

A gas bubble in perfectly pure water is unstable and dissolves due to the Laplace force[3]. However, when the bubble shrinks to nanometer size and the gas pressure inside the bubble becomes very high, an increasing fraction of the closely packed gas molecules, usually di-atomic molecules, left in the bubble are exposed to bonding to the water molecules at the bubble boundary, and they exchange their momentum with them. The interfacial gas expectedly liquefies with a dense structure at a much higher temperature than the gas remaining in the bubble core, and forms a very small stable cavitation nucleus already at the temperatures of liquid cold water.

Likewise, when at a free surface a gas molecule has entered into liquid water it in principle constitutes a cavitation nucleus. Locally it prevents the water molecules from setting up hydrogen bonds between themselves in the space occupied by the gas molecule, but it is a nucleus too small to reduce notably the tensile strength of ideally pure liquid H_2O. However, when gas molecules in the water meet over time, they stick together because clustering reduces the potential energy of the system, the surface area of the water bounding the cluster being smaller than the sum of areas bounding the individual gas molecules in solution. Such a cluster may grow until it is about to shift into a gaseous state - a balance is achieved.

It is very difficult to remove all gas molecules from water. Therefore, stable nuclei of liquefied gas may explain why theoretical values of tensile strength of water are difficult to achieve experimentally, even when the water is highly purified, and degassed as well as possible.

The water we meet in our daily life, e.g. seawater and tap water, contains gas in solution, it is usually contaminated by alien substances, and it carries particles - and particles as well as bounding walls are known to harbor cavitation nuclei. A skin of contaminating components may stabilize free gas bubbles of size in the micrometer range[4,5], hereby allowing them to be in gas diffusion balance at the prevailing pressure[6], and causes a low tensile strength of such water. Likewise, we may assume that interfacial gas bubbles on solid surfaces can be stabilized by a skin[7], and it makes them comparable to free skin-stabilized gas bubbles.

Liquefaction of gas

Gases are characterized by the molecules moving randomly between each other, their mean free path between collisions depending on the spatial density of the molecules, and their kinetic energy being given by the temperature. The pressure p, the specific volume v, and the temperature T are connected by the equation of state.

Likewise, in the *liquid state* the molecules are moving randomly between each other, but here they are kept together by bonds, shifting between immediate neighbors, while the molecules move around thermally.

In the *solid state* the thermal energy is insufficient to break the intermolecular bonds, and the molecules vibrate thermally around equilibrium positions of a solid structure, very often a crystalline structure.

When a gas is pressurized beyond its critical level P_{crit}, then further pressure increase affects the packing of the molecules only very little. If also the temperature is beyond the critical level T_{crit} the thermal motion of the molecules prevents intermolecular bonds from being established, and the medium is kept together only by the external forcing, and though the molecules are densely packed, it maintains a gaseous character. However, in a phase diagram the boundary between the gaseous and the liquid states vanishes. For nitrogen this happens at T_{crit} = 126.2 K and P_{crit} = 3.4 MPa, and for oxygen at T_{crit} = 154.6 K and P_{crit} = 5.0 MPa, i.e. at temperatures way below the freezing point of water. However, these values of T_{crit} are valid for molecules of the gas interacting mutually, but when in water a gas bubble gradually goes in solution an increasing fraction of the remaining gas molecules interact with the dangling hydrogen bonds of the water molecules at the bubble surface. The bubble shrinkage makes the Laplace force increase, and eventually it eliminates the mean free path between molecular collisions in the gas, and translational energy shifts into rotational and vibrational energy. Within the electric near field of the dangling bonds at the bubble surface the gas liquefies with the dense structure, expectedly already at temperatures much higher than T_{crit}, while at larger distance it remains in a gaseous state.

The intermolecular force between two molecules is governed primarily by their electric fields, and the potential energy of their interaction

$$\boldsymbol{E_p \propto 1/r^{\,m+n-1}}, \qquad (1)$$

where $\boldsymbol{r}$ is the distance between the two m and n polar molecules considered. Thus, the dipole field of a water molecule has a notably larger range and strength than the quadrupole field of a di-atomic gas molecule. Therefore, water-gas interaction is much stronger than gas-gas interaction. Further, at a water-gas interface one of the four hydrogen bonds of each water molecule is dangling and acts as a monopole, while the other ones are engaged with neighboring water molecules inside the liquid water. The dangling bonds are alternatively positive and negative, making the interface electrically neutral. At a bubble surface the electric field therefore depends on the distance between neighboring oppositely charged dangling hydrogen bonds, a distance that may be estimated from the size of a "surface cell" of liquid water. Though on a molecular scale the liquid water surface is shifting and unstable, we may from the hexagonally structured surface cell of ice use a size estimate of ~ 0.5 nm. We assume the gas-water interaction at the bubble surface to cause liquefaction in a densely populated shell of gas molecules of thickness up to maybe a few times 0.5 nm, even at room temperature. When the Laplace pressure in the bubble reaches or exceeds the critical pressure P_{crit}, diffusion of gas out of the bubble is blocked, even if a gaseous kernel is left inside the droplet. Continuum calculations for a spherical gaseous cavitation nucleus of tensile strength 26 MPa, as measured[1], lead to a bubble radius of ~2 nm and a Laplace pressure of ~70 Mpa, i.e. way above P_{crit}, but calculations should be molecular dynamics simulations.

Particles/solid boundaries and contamination by alien substances

In the engineering world, water is very often close to being saturated by atmospheric air in solution, and in addition, it usually contains alien non-soluble substances that may form a stabilizing skin on gas bubble nuclei[4,5]. Further, water normally carries particles, and solid surfaces in general bound liquid flows. Removing, to the extent possible in 1950, all contamination and particles, and most of the gas from distilled water in a glass tube, Briggs[8] obtained a maximum of tensile strength of 27.7 MPa at about 10 °C. The tensile strength was found to decrease not only at rising temperature, as to be expected, but also when the temperature was reduced towards 0 °C. The same temperature dependence was found by Keller[9] for tap water used in a water tunnel at different degrees of gas saturation, though here the values of tensile strength were about 2 orders of size smaller, Figure 4a,b. Thus, contamination and gas content can be taken responsible for the level of tensile strength, but the basic physics behind the temperature dependence seems the same for water qualities ranging from plain water in a flow, to highly cleaned water of low gas content contained in a glass tube. In contrast, for ideally purified H_2O, theory predicts that the tensile strength reaches a maximum value of the order of 140 MPa at 0 °C, Figure 4c.

Low values of tensile strength may be explained by assuming free gas bubbles of appropriate size to be stabilized by a skin that allows diffusion balance at the prevailing pressure[6]. Such skin-stabilized free gas bubbles may be generated as in Johnson and Cooke's experiments[4], or naturally in the ocean by entrainment of air by breaking waves, or by biological activity. However, other experiments, such as those of Briggs and Keller, reveal that very often solid surfaces, either those of particles in the bulk of water, or solid surfaces bounding the water, carry the nuclei responsible for cavitation inception[7,10-11]. How can such interfacial bubbles arise?

The tensile strength of water goes up when the gas content is reduced, and it also rises when the particle size is reduced. However, with particles < 0.2 µm the gas content loses influence [13]. Likewise, very smooth solid spheres of small radius cause a higher tensile strength than larger ones, while surface roughness reduces their tensile strength[10,11]. This indicates that surface curvature on a local scale is a decisive parameter. Already in early studies, a molecular approach to cavitation inception seemed required[14,15].

Molecular dynamics simulations of water at 300 K[16] have shown that at an atomically planar Pt-surface the H_2O molecules bond to the Pt atoms, so that the interfacial layer of water has the solid Pt-structure. The next layer also has an orderly structure, but it is liquid, while at 300 K the third layer is influenced only faintly by the Pt-surface. The strength of the weakest bonding determines the tensile strength of such a system. At temperature rise the thermal motion of the H_2O molecules increasingly weakens the molecular bonds, and the tensile strength decreases. Conversely, at lowering of the temperature towards the freezing point the tensile strength increases - but only until more water layers than the interfacial one begins to solidify - but with the structure of ice-1h. Thus, the very interfacial water layer misfits the developing ice-1h structure of the next layers, and the bonds are weakened. They break if the misfit is too large, causing local dislocations in the interfacial water, and gas molecules may enter the interface.This expectedly leads to the drop of tensile strength observed when the temperature drops below 10 °C[8,9(17)]. Thus, solidification of interfacial water layers seems decisive, and it explains why, *in the absence* of solid surfaces, the tensile strength for H_2O continues rising when the temperature drops towards 0 °C, as found using focused ultrasonic wave packets for measurements in ultra-purified water that carries no particles and sees no solid surfaces in the region of tensile stress[1].

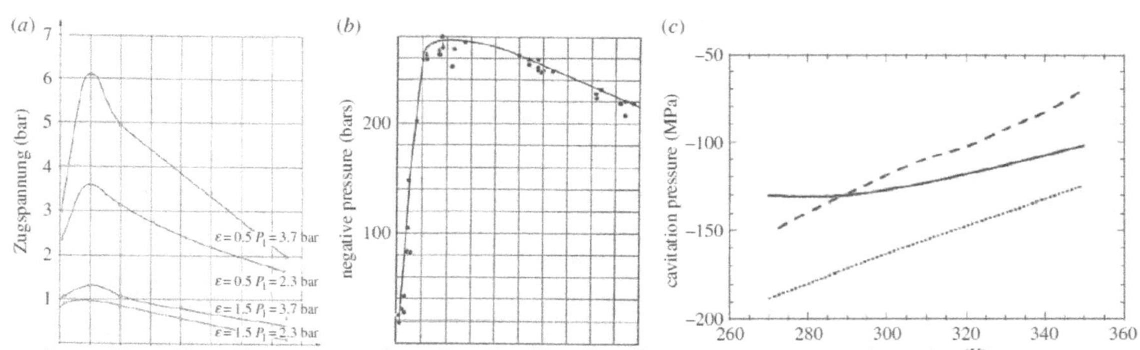

Figure 1. Measured tensile strength, *TS*, of tap water versus temperature (a) by Keller [9] at system pressure P_1 and degree of gas saturation ε, and (b) for highly purified water by Briggs [8], and (c) theoretical cavitation pressures ($p_{cav} = p_v - TS$): dotted line, thin wall approximation: solid line, density functional theory (Speedy): dashed line, TIP5P equation of state. (Reproduced from [17]).

We notice that at the Pt-H_2O interface, the water layer next to the one directly Pt-bonded has a 'surface' energy, because the solid-like layer of Pt-bonded water molecules do not bond optimally with the liquid-like neighboring layer. Expectedly, the solid-water bonding found at Pt characterizes the bonding at solid surfaces generally. And it suggests that at convex solid surface elements, the interfacial tension in the water tends to maintain the liquid-solid contact, while at concave surface elements it tends to weaken the coherence of the interfacial water, hereby promoting the formation of voids at concave surface elements. A reduction of the tensile strength is to be expected at solid-water interfaces when the curvature shifts from convex to concave shape – a result in accordance with experimental observations.

Perspectives

Cavitation nuclei of liquefied gas may explain some unresolved questions in relation to specific experimental observations. Thus, we can expect that at given pressure, oxygen liquefies easier than nitrogen, which leads to a higher oxygen/nitrogen-ratio of the air dissolved in water than in the atmosphere. Likewise, many air molecules, absorbed in the water at water-gas interfaces, coalesce within the space of water as liquefied nano-droplets until they reach critical size, or exceed it – if so, they shift into the gaseous state, and subsequently they shrink to critical size by diffusion of gas molecules back into the bounding water. This implicates that in time, nano-droplets of different sizes will be present, uniformly distributed in the liquid space. When exposed to critical tensile stress, the largest ones cause homogeneous cavitation, the nuclei forming a field of essentially hexagonal closed-packed structure.

Finally, when a liquefied gas droplet meets a submerged solid surface it is adsorbed, and large ones shift into very flat, gaseous surface nanobubbles, bubbles intensely studied by atomic force microscopy. However, when gas molecules inside such a bubble move towards the locus of bubble attachment, the space becomes very narrow and on their way the gas molecules interact effectively with the dangling bonds at the water surface. They lose their translatory kinetic energy, and the molecular density increases – they accommodate, and at the locus of attachment, the gas molecules liquefy even though the pressure is low. This liquefaction may explain the observed pinning of surface nanobubbles, Figure 2.

Figure 2. Gas bubbles, which shrink by diffusion of gas, liquefied gas droplets that grow by coalescence, and a surface naobubble, pinned by liquefied gas.

Summary

In the handling of problems of physics, we are accustomed to considering bulk parameters of solids, liquids and gases, but actually, at surfaces the physics often deviates from that in the bulk. An important example is catalysis, which itself constitutes a whole field of science, but also e.g. the surface of aluminum melts at a lower temperature than the bulk[18]. By analogy, it seems relevant to consider the water-gas interface, and here the water influences the interfacial gas layer. And gas molecules in solution in water expectedly tend to form nano-clusters of liquefied gas by coalescence. They may grow to a maximum radius determined by the range of the electric near field of the dangling bonds of the water molecules bounding the cluster.

Considering water at a hydrophilic solid surface, the very interfacial water layer is solidified with a structure as the solid, and when approaching the freezing point of bulk water from about 10 °C the adjacent layer expectedly also approaches solidification, but with an ice-1h structure. Normally this structure misfits the solid structure, and therefore, locally the water structure is weakened, hereby easing gas molecules or liquefied droplets to enter the interfacial structure – a plausible mechanism of initial formation of flat, interfacial nanobubbles, but they may grow by merging with liquefied gas droplets reaching the interface. Surface nanobubbles of contact diameters up to the 100-200 nm range are well known, and they have been studied intensively by atomic force microscopy (AFM)[19]. They have a relatively low tensile strength, and liquefaction of gas, as well as surface skin, may explain their origin and stability. We notice that using highly purified water, the generation of surface nanobubbles requires sophisticated techniques of water handling, while with lack of focus on purification they form readily[20] – exactly as cavitation nuclei! And experiments with free gas bubbles[4,5] as well as with interfacial gas bubbles[7] are well interpreted from the assumption of the bubbles being stabilized by a skin that allows diffusion balance across the bubble surface. Thus, Fox and Herzfeld's original idea of a skin that stabilizes gas bubbles[21] seems central for cavitation nuclei – but the skin has to be permeable to gas! Such skin also explains the Keller scaling law of incipient cavitation[22]. However, studies of skin on cavitation nuclei, as well as studies of the molecular aspects of interfaces discussed in this paper, are required to lift models and theories of cavitation nuclei into well-established knowledge.

Acknowledgement

The Otto Mønsted Foundation is acknowledged for grant No. 18-81-0005, financing the author's participation in CAV2018.

References

1. Herbert E, Balibar S, Caupin F. 2006. Cavitation pressure in water. Phys. Rev. E 74, 041603. (doi:10.1103/PhysRevE.74.041603)

2. Zheng Q, Durben DJ, Wolf GH. 1991. Liquids at large negative pressures: Water at the homogeneous cavitation limit. Science 254, 829-832.

3. Epstein PS, Plesset MS. 1950. On the stability of gas bubbles in liquid-gas solutions. J. Chem. Phys. 18, 1505–1509. (doi:10.1063/1.1747520)

4. Johnson BD, Cooke RC. 1981. Generation of stabilized microbubbles in seawater. Science 213, 209–211. (doi:10.1126/science.213.4504.209)

5. Yount DE. 1979 Skins of varying permeability: a stabilization mechanism for gas cavitation nuclei. J. Acoust. Soc. Am. 65, 1429–1439. (doi:10.1121/1.382930)

6. Mørch KA. 2007. Reflections on cavitation nuclei in water. Phys. Fluids 19, 072104. (doi:10.1063/1.2747210)5.

7. Andersen A, Mørch KA. 2015. Cavitation nuclei in water exposed to transient pressures. J. Fluid Mech. 771, 424–448. (doi:10.1017/jfm.2015.185)

8. Briggs LJ. 1950 Limiting negative pressure in water. J. Appl. Phys. 21, 721–722. (doi:10.1063/1.1699741)

9. Keller AP. 1980 Schlussbericht über das Forschungsvorhaben 'Beginnende Kavitation, Zugspannungen in Flüssigkeiten'. Technische Universität München.

10. Marschall HB, Mørch KA, Keller AP, Kjeldsen M. 2003 Cavitation inception by almost spherical solid particles in water. Phys. Fluids 15, 545–553. (doi:10.1063/1.1535940)

11. Arora M, Ohl C-D, Mørch KA. 2004 Cavitation inception on microparticles: a self-propelled particle accelerator. Phys. Rev. Lett. 92, 174501. (doi:10.

12. Harvey EN, Barnes DK, McElroy WD, Whiteley AH, Peace DC, Cooper KW. 1944a Bubble formation in animals I. J. Cell. Comp. Physiol. 24, 1–22. (doi:10.1002/jcp.1030240102)

13. Greenspan M, Tschiegg CE. 1967 Radiation-induced acoustic cavitation; apparatus and some results. J. Res. Natl Bur.Stand. C 71C, 299–312. (doi:10.6028/jres.071C.024)

14. Drost-Hansen W. 1971 Structure of water near solid interfaces, In "Chemistry and Physics of Interfaces II", Am.Chem.Soc. Publ., 207-241.

15.Mørch KA, Song, JP. 1992 Cavitation nuclei at solid-liquid interfaces. "Cavitation", Proc. of the Institution of Mechanical Engineers, C453/059, 1-7

16. Xia X, Perera L, Essman U, Berkowitz ML 1995 Surface Science 335, 401-415

17. Mørch KA. 2015 Cavitation inception from bubble nuclei. Interface Focus 20150006. http://dx.doi.org/10.1098/rsfs.2015.0006

18. Stoltze P, Nørskov JK, Landman U. 1988 Disordering and Melting of Aluminum Surfaces. Physical Review Letters 61,1988, 440-443.

19. Lohse D, Zhang X (张雪花). 2015 Surface nanobubbles and nanodroplets. Reviews of Modern Physics 87, 981-1035.

20. Holmberg M, Kühle A, Garnæs J, Mørch KA, Boisen A. 2003 Nanobubble Trouble on Gold Surfaces. Langmuir 2003, *19,* 10510-1051

21. Fox FE, Herzfeld KF. 1954 Gas bubbles with organic skin as cavitation nuclei. J. Acoust. Soc. Am. 26, 984–989.

22. Mørch KA, Keller AP, Liu S, Zuo Z, Ren Z. The physics behind the empirical Keller scaling law of incipient cavitation. Manuscript in preparation.

© 2018 ASME

Cavitation of Water Confined in Hydrophobic Nanoporous Materials

[1,2]Antonio Tinti*; [2]Alberto Giacomello; [2]Carlo Massimo Casciola;

[1]*Max-Planck-Institut für Intelligente Systeme, Stuttgart, Germany;* [2]*Sapienza Università di Roma, Rome, Italy*

Abstract

Hydrophobic nanoporous materials immersed in water are emerging as promising means to store or dissipate energy [1-3]. In such applications, surface energy is accumulated when the water pressure is increased causing liquid intrusion inside the pores and can be subsequently released by decreasing pressure and triggering cavitation inside the pores. Given the extreme confinement, which for zeolites and metal organic frameworks can be below the nanometer, the phenomena of liquid intrusion and cavitation are expected to significantly deviate from the classical macroscopic laws: cavitation can happen at pressures larger than tens of MPa.

In this contribution, we study via molecular dynamics the nucleation of a vapor cavity inside nanometer-sized hydrophobic pores and the opposite process of liquid intrusion. By employing advanced rare-event simulation techniques in order to tackle the long timescales typical of vapor nucleation, we obtain molecular-level insights into nanoconfined cavitation (and liquid intrusion) avoiding simulation artifacts [4]. The simulation campaign reveals deviations from the macroscopic Kelvin-Laplace law for liquid intrusion in a capillary and a significant increase of the cavitation rate as compared to the predictions of the classical nucleation theory. Furthermore, the behavior of nanoporous materials as molecular springs or as vibration dampers is critically discussed and related to their physical characteristics.

Keywords: Cavitation, Nanoscale, Molecular Dynamics, Molecular Springs, Confinement

Introduction

Cavitation in extreme confinement is responsible for the innovative energetic application of heterogeneous lyophobic systems (HLS), in which a nanoporous material is paired with a non-wetting liquid: liquid-vapor transition within the pores determines the accumulation of surface energy, owing to the immense surface area density of porous materials. During the intrusion phase, the liquid is forced by large liquid pressures inside the pores accumulating surface energy; during the extrusion phase, nanoconfined cavitation is triggered by the decreased pressure. Depending on their characteristics, HLSs are able either to efficiently dissipate or to store mechanical energy. Energy dissipation is related to the difference in the cavitation (extrusion) and intrusion pressures. If this difference is minimal, it is possible to exploit such systems as efficient energy storage devices. A number of such systems were tested experimentally in previous works [1-3,5].

The behavior of these systems, associated with the intrusion and extrusion of water inside the nanopores, challenges the present understanding based on macroscopic theories and classical nucleation theory (CNT) because of the peculiar behavior of water in extreme confinement. In this contribution we show how advanced molecular dynamics techniques can be used to investigate these non-classical effects, which are often difficult to probe directly via experiments, owing to the nanometric size and fast timescales that characterize the phenomenon. A state-of-the-art rare-event technique is used to observe, without artifacts, the thermally-activated, microscopic mechanism of water intrusion and cavitation in the pores. In addition, *in silico* intrusion-extrusion experiments are presented, showing how, as the size of the nanopore decreases, the intrusion-extrusion hysteresis is suppressed, making these materials ideal candidates for mechanical energy storage. Our simulations were able to measure the existence of non-classical effects, i.e., phenomena that are not accounted for in classical nucleation theory (CNT). In particular the cavitation free-energy barrier is reduced as compared to CNT, accounting for an accelerated cavitation process, and the pressure of liquid intrusion is increased due to nanoscale confinement. A dependence of the intrusion/extrusion hysteresis on the pore size is also exposed by the present simulations. The reported effects are able to provide a microscopic interpretation of several experimental results on nanoporous materials.

*Corresponding Author, Antonio Tinti: atinti@is.mpg.de

Cavitation at the Nanoscale

Experimental studies of confined cavitation often result in the measurement of the nucleation rates at given experimental conditions [2]. The exponential dependence of the nucleation rates on the thermodynamic properties (pressure, surface tension, contact angles, etc.) makes these experiments an exceptional probe for elusive quantities such as the line tension or the curvature dependence of the surface tension [6]. In these cases, a nucleation theory, providing a theoretical expression for the rates, is needed in order to extract these quantities from the experimental measurements of the nucleation rates.

Simple continuum theories based on the classical nucleation theory (CNT) rely on a number of approximations. These simplifications [4,7], which are often reasonable at a micrometric or mesoscale level, tend to break down when the system size falls below a few nanometers. The original CNT [8-10] can be extended to cope with heterogeneous and confined cases [7,11], by studying the energetics of a cavitation bubble growth as a function of the bubble volume.

More sophisticated approaches, based on atomistic simulations can also be adopted. In particular, molecular dynamics (MD), coupled with state-of-the-art rare events methods such as the string in collective variables [12,13], is able to provide an unbiased picture of nanoconfined cavitation including all genuine nanoscale phenomena, such as thermal fluctuations, packing and line effects, etc. A detailed comparison of MD and extended CNT techniques for cavitation in cylindrical confinement is presented in [14].

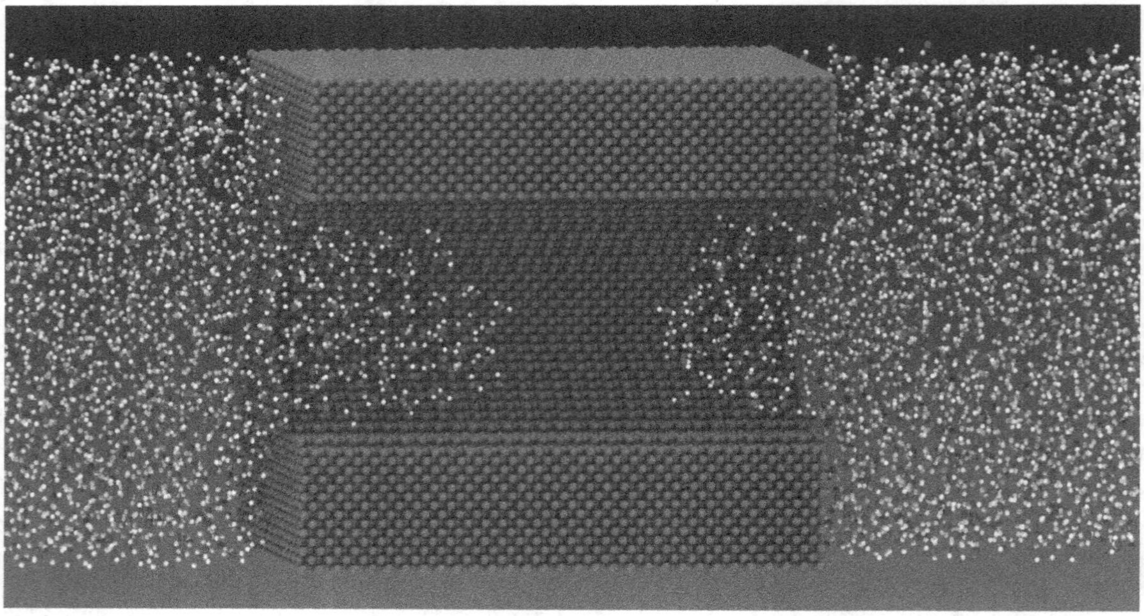

Fig. 1) Snapshot of the atomistic system during a string method calculation; a single nanopore, excavated from a hydrophobic Lennard-Jones crystal, is surrounded by ca. 13000 TIP4P/2005 molecules. In order to drive the string method we adopt, as a set of collective variables, the coarse-grained density field of water, computed on a grid of 4×4×10 cubic cells, whose sides measure ca. 7Å. The liquid pressure is kept constant, around liquid-vapor coexistence, using two solid walls acting as pistons (not shown in figure).

Simulation Methods

Atomistic simulations have been deployed in order to study cavitation inside a cylindrical nanopore with an unprecedented spatial and temporal resolution. These simulations combine molecular dynamics with the string method in collective variables [12] in order to simulate the formation of a vapor bubble inside a cylindrical hydrophobic nanopore (confined cavitation) and compute the related free energy. The pore, having a diameter of 2.6 nm, is excavated from a crystalline material whose atoms interact with the oxygen atoms of the water molecules via a modified Lennard-Jones potential. The strength of the attractive term is tuned in order to render the desired contact angle of 119° on a flat surface, as measured in an independent simulation. The TIP4P/2005 water model [15] was adopted in the simulations, as it reproduces the main features of water (and in particular the surface tension) at the temperature of our interest (300 K). The string method uses a coarse grained density field as the collective variable. For further details on the simulations we refer to [16] and for the string method in collective variables to the original article [12]. A snapshot of the atomistic system is presented in Fig. 1. The main result of the rare event atomistic simulation of cavitation inside the model hydrophobic pore is the free-energy profile shown in Fig. 2, in which the volume of the vapor bubble is used as the progress variable for the cavitation process. This profile will be discussed in the next section. Another, simpler simulative protocol is used in order to study the intrusion and extrusion process of water in a hydrophobic pore, which, however, does not yield the detailed information on the free energy. By varying the pressure in discrete steps, and letting the system equilibrate for a sufficient time, it is possible to mimic *in silico* an intrusion-extrusion experiment. This technique, similar to the one used in [17], has been adopted in order to measure the intrusion and extrusion pressures for a smaller hydrophobic pore with a diameter of 1.2 nm excavated from the same crystalline material.

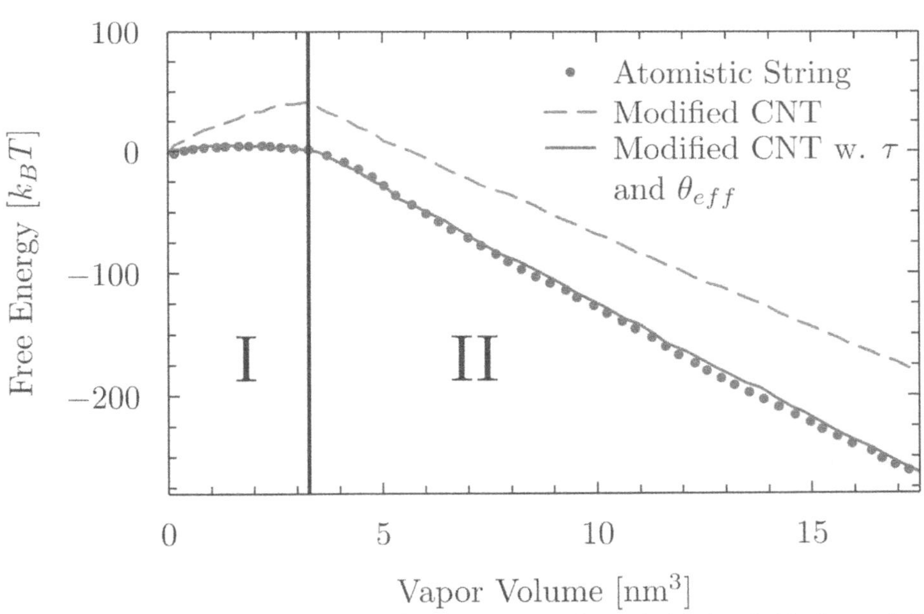

Fig. 2) Landau free energy profiles (T = 300 K, ΔP = 0) obtained via the atomistic string method (red symbols), via extended CNT (green line), and via extended CNT with the addition of terms proportional to the triple line and an effective contact angle (light blue line). We can distinguish between two regions, indicated in figure as I and II. The first one is associated with the cavitation of a critical bubble, whereas in the second region we observe the sliding of two symmetric liquid-vapor menisci. "Modified CNT" free-energy profiles are obtained by evaluating the free energy functional $\Omega = \Delta P V_v + \gamma_{lv}(A_{lv} + \cos(\theta_Y) A_{sv})$ along the path returned by the atomistic string calculations. The green line shows how it is possible to correct these predictions for nanoscale effects by adding an effective "line tension" (i.e., a term proportional to the triple line) $\tau = -1.1 \cdot 10{-}11$ N along with the adoption of an effective contact angle $\theta_{eff} = 129^\circ$ in the same free energy functional.

Discussion

Figure 2 shows a comparison of the free-energy profiles computed via the atomistic string method and via the extended CNT [11,14], revealing the occurrence of a first non-classical effect: the atomistic cavitation free-energy barrier are reduced six-fold as compared to the CNT predictions. This is consistent with the experimental observations by Charlaix and coworkers [2] on MCM-41 who reported extrusion at pressures as large as 10 MPa – the nanoporous material which inspired our model MD system. The low cavitation free-energy barrier computed by atomistic simulations indeed accounts for the observation of cavitation even at positive pressures within a timescale compatible with the experimental one.

A second important nanoscale effect is related to the intrusion pressures. Using a procedure explained in detail elsewhere [16], it is possible to add a pressure term, linear in the volume, to the atomistic free energy, in order to render the cavitation free-energy profile at different thermodynamical conditions (This term mimics the ΔPV_v contribution in the classical expression $\Omega = \Delta PV_v + \gamma_{lv}(A_{lv} + \cos\theta_Y A_{sv})$ for the free energy of the system).

By approaching the conditions that result in the vanishing of the intrusion and extrusion barriers it is therefore possible to obtain an estimate for the intrusion and extrusion pressures which determine the "spinodal" conditions. For the 2.6 nm pore, object of our string calculations these spinodal intrusion and extrusion pressures amount respectively to ΔP^{sp}_{int} = 85 MPa and ΔP^{sp}_{ext} = −22 MPa, defining a large hysteresis loop. The intrusion pressure is much higher than the 54 MPa predicted by the classic Kelvin-Laplace equation, showing a significant increase due to nanoscale confinement. A third nanoscale effect which results from the atomistic simulations is related to the suppression of this large hysteresis loop when considering smaller pore radii. The *in silico* intrusion-extrusion experiments for the pore with 1.2 nm diameter, presented in Fig. 3 reveal a drastic decrease of the intrusion-extrusion hysteresis and thus of the dissipation of mechanical energy associated with each intrusion-extrusion cycle. The ~20 MPa hysteresis observed in simulations is further reduced in real-world scenarios in which a macroscopic time is allowed for the event, which allow thermal fluctuations to activate both processes. This trend of hysteresis with the pore size is consistent with the experimental observations on zeolites and metal-organic frameworks available in the literature [1,3,18-21]. For a more quantitative discussion of these nanoscale effects, along with an in-depth comparison with individual experimental results, we refer the interested reader to [16].

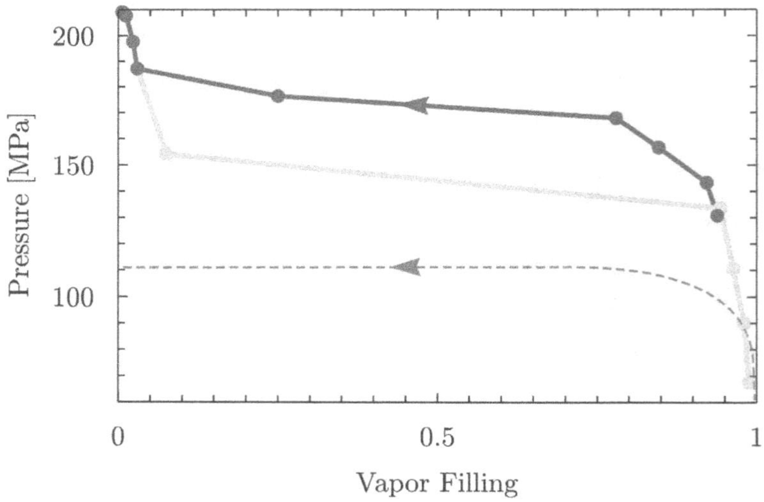

Fig. 3) Intrusion/extrusion cycle for a D = 1.2 nm nanopore showing a reduced hysteresis. The yellow and red curves result from an in silico MD experiment in which the pressure is changed in discrete steps of 12.5 MPa and the system is then allowed to stabilize for at least 3 ns; The initial steps in which pressure is varied are discarded when computing the average value for the liquid filling of the nanopore. The blue dashed line represents the outcome of the same conceptual experiment according to the predictions of classical nucleation theories.

Conclusion

In this contribution we provide a microscopic picture of cavitation in extreme hydrophobic confinement –a scenario which is relevant for energy system based on nanoporous materials. These extreme conditions result in deviations from the macroscopic theories, which are responsible for the technologically interesting phenomenology associated with intrusion and extrusion of a non-wetting liquid inside lyophobic nanoporous materials. These non-classical effects include reduced cavitation free-energy barriers, allowing for the occurrence of spontaneous cavitation even at positive pressures, deviations from the Kelvin-Laplace prediction for the intrusion pressure, and the suppression of intrusion/extrusion hysteresis for smaller nanometer-sized pores. The simulative effort presented in this contribution reveals how such deviations from the macroscopic theory of cavitation due to extreme confinement are associated with controllable physical features allowing for the design of innovative devices of technological interest for the efficient, inexpensive, and green storage or dissipation of mechanical energy.

References

[1] Eroshenko, V., Regis, R.-C., Soulard, M., and Patarin, J. (2001) *Energetics: a new field of applications for hydrophobic zeolites*. J. Am. Chem. Soc., 123, 8129–8130.
[2] Guillemot, L., Biben, T., Galarneau, A., Vigier, G., and Charlaix, E. (2012) *Activated drying in hydrophobic nanopores and the line tension of water*. Proc. Natl. Acad. Sci. USA, 109, 19557–19562.
[3] Grosu, Y., Li, M., Peng, Y.-L., Luo, D., Li, D., Faik, A., Nedelec, J.-M., and Grolier, J.-P. (2016) *A Highly Stable Nonhysteretic {Cu2 (tebpz) MOF+ water} Molecular Spring*. ChemPhysChem, 17, 3359–3364.
[4] Meloni, S., Giacomello, A., and Casciola, C. M. (2016) *Focus Article: Theoretical aspects of vapor/gas nucleation at structured surfaces*. J. Chem. Phys., 145, 211802.
[5] Cailliez, F., Trzpit, M., Soulard, M., Demachy, I., Boutin, A., Patarin, J., and Fuchs, A. H. (2008) *Thermodynamics of water intrusion in nanoporous hydrophobic solids*. Physical Chemistry Chemical Physics, 10, 4817–4826.
[6] Bruot, N. and Caupin, F. (2016) *Curvature dependence of the liquid-vapor surface tension beyond the Tolman approximation*. Phys. Rev. Lett., 116, 056102.
[7] Giacomello, A., Chinappi, M., Meloni, S., and Casciola, C. M. (2012) *Metastable Wetting on Superhydrophobic Surfaces: Continuum and Atomistic Views of the Cassie-Baxter–Wenzel Transition*. Phys. Rev. Lett., 109, 226102.
[8] Volmer, M. (1939) *Kinetik der Phasenbildung*. Verlag Theodor Steinkopff, Dresden
[9] Frenkel, Y. I. (1955) *Kinetic theory of liquids*. Dover Publications, New York
[10] Debenedetti, P. G. (1996) *Metastable liquids: concepts and principles*. Princeton University Press.
[11] Lefevre, B., Saugey, A., Barrat, J.-L., Bocquet, L., Charlaix, E., Gobin, P.-F., and Vigier, G. (2004) *Intrusion and extrusion of water in hydrophobic mesopores*. J. Chem. Phys., 120, 4927– 4938.
[12] Maragliano, L., Fischer, A., Vanden-Eijnden, E., and Ciccotti, G. (2006) *String method in collective variables: Minimum free energy paths and isocommittor surfaces*. J. Chem. Phys., 125, 024106–024106.
[13] E, W., Ren, W., and Vanden-Eijnden, E. (2007) *Simplified and improved string method for computing the minimum energy paths in barrier-crossing events*. J. Chem. Phys., 126, 164103.
[14] Lisi, E., Tinti, A., and Giacomello, A. (2017) *Intrusion and extrusion of water in hydrophobic nanopores*. Proceedings of the XXIII Conference of the Italian Association of Theoretical and Applied Mechanics.
[15] Vega, C. and De Miguel, E. (2007) Surface tension of the most popular models of water by using the test-area simulation method. J. Chem. Phys., 126, 154707.
[16] Tinti, A., Giacomello, A., Grosu, Y., and Casciola, C. M. (2017) *Intrusion and extrusion of water in hydrophobic nanopores.* Proceedings of the National Academy of Sciences, 114, E10266– E10273.
[17] Amabili, M., Giacomello, A., Meloni, S., and Casciola, C. M. (2016) *Intrusion and extrusion of a liquid on nanostructured surfaces*. Journal of Physics: Condensed Matter, 29, 014003.
[18] Eroshenko, V., Regis, R.-C., Soulard, M., and Patarin, J. (2002) *The heterogeneous systems 'water-hydrophobic zeolites': new molecular springs*. C. R. Phys., 3, 111–119.
[19] Trzpit, M., Soulard, M., and Patarin, J. (2007) *The pure silica chabazite: a high volume molecular spring at low pressure for energy storage*. Chem. Lett., 36, 980–981.
[20] Trzpit, M., Rigolet, S., Paillaud, J.-L., Marichal, C., Soulard, M., and Patarin, J. (2008) *Pure silica chabazite molecular spring: A structural study on water intrusion- extrusion processes.* J. Phys. Chem. B, 112, 7257–7266.
[21] Grosu, Y., Renaudin, G., Eroshenko, V., Nedelec, J.-M., and Grolier, J.-P. (2015) *Synergetic effect of temperature and pressure on energetic and structural characteristics of {ZIF-8+ water} molecular spring.* Nanoscale, 7, 8803–8810.

© 2018 ASME

10th International Symposium on Cavitation - CAV2018
Baltimore, Maryland, USA, May 14 – 16, 2018

CAV18-05049

CFD Analysis of Propeller Tip Vortex Cavitation in Ship Wake Fields

[1]Keun Woo Shin*; [2]Poul Andersen

[1]*MAN Diesel & Turbo, Frederikshavn, Denmark;* [2]*Technical University of Denmark (DTU), Kgs.Lyngby, Denmark*

Abstract

TVC on marine propellers in behind-hull conditions is a challenging problem for numerical predictions, because it is deformed in turbulent viscous interactions with hull wake, propeller-induced flows and rudder disturbance. DES is made for cavitating flows on a propeller and a rudder of a military vessel with a 3.5° shaft inclination in non-uniform hull wake. While a cavitation tunnel test shows TVC extending over the rudder, only a short extent of TVC is reproduced in DES on the initial grid. After an adaptive grid refinement based on the Q-criterion, a larger extent of TVC is simulated and the first- and second-order pressure pulses on two points above the propeller are closer to the experimental measurements. The effects of hull wake, propeller-induced flow, rudder disturbance and shaft inclination on TVC are analyzed by looking into the deformation of the helical trajectory identified by a Q-criterion iso-surface.

Keywords: tip vortex cavitation; marine propeller; CFD; DES; hull wake; adaptive grid

Introduction

Tip vortex cavitation (TVC) on marine propellers is increasingly a subject of concern in propeller design, because it is responsible for broadband hull vibration and rudder erosion in many cases [1, 2]. As the environmental impact of the underwater radiated noise from cavitating propellers is highlighted, the need for predicting TVC behavior is increased, because TVC is often the first cavitation type shown at the lowest inception speed and is propagated far downstream from the aftship. The intensity of TVC can be reduced by lowering the blade tip loading, but abrupt tip unloading can trigger erosive cloud cavitation and stable TVC can be a safe way to let sheet cavitation depart from the blade surface without violent collapse [3].

The prediction of TVC behavior in actual propeller operations behind a hull is a complex flow problem, as the helical trajectory of TVC is deformed by hull wake, propeller-induced flow and rudder interaction. Hybrid methods consisting of a viscous flow solver for the hull flow, a potential flow solver for the propeller flow and an empirical model for TVC are efficient for predicting high-order pressure pulses induced by TVC, but have limitations in taking interactive dynamics into account [4]. Extensive TVC on an open-water propeller is simulated by a viscous flow solver with a multiphase model [5], but the simulation in behind-hull conditions is still challenging due to high computational effort required for resolving tip vortex flows. In this research work, detached-eddy simulations (DES) are made for TVC on a propeller and a rudder of a military vessel with a 3.5° shaft inclination in non-uniform hull wake. A cavitation tunnel test conducted with a complete hull model at SSPA shows that TVC is extended over the rudder for about one blade revolution in the loading condition of the maximum engine power.

CFD setup

(a) (b) (c)

Figure 1. (a) Experimental setup and (b) initial & (c) refined grids of computational model

*Corresponding Author, Keun Woo Shin: keun.shin@man.eu

The improved delayed DES solver of the commercial CFD software StarCCM+ is adopted with the curvature-corrected k-ω SST turbulence model. Cavitation is modelled by using an Eulerian multiphase approach of the volume-of-fluid method and a vapor transport equation with an interphase mass-transfer model based on the asymptotic Rayleigh-Plesset equation. Instead of including the hull model, the hull wake is modelled as a propeller inflow by using the inlet flow and momentum sources. DES results for unsteady sheet and cloud cavitation on another propeller with hull wake modeling have been validated against experimental results [6].

The propeller model is 4-bladed with a diameter of D=230 mm and a high area ratio of A_E/A_O=0.76, where A_E is the expanded blade area and A_O is the propeller disk area. Trimmed hexahedral mesh is generated around the propeller and rudder with prismatic mesh resulting mostly in a dimensionless first-cell height of y^+<2 on the wall surface in a cylindrical fluid domain. The grid is initially refined to a cell size of Δx=0.2 mm along the blade edges and at the outer-radii region around the tip. The propeller rotation is modelled by the rigid body motion and the sliding mesh in an inner cylindrical domain around the propeller.

The propeller shaft is aligned with the axis of the cylindrical domain and instead, the rudder is rotated around the blade reference line perpendicular to the ship center plane for taking into account the shaft 3.5° inclined downward at the aft end in the computational model, so that the axial component of the hull wake applied to the inlet boundary may be less disturbed on the way to the propeller plane. The radial and tangential wake components are modelled by momentum sources applied 0.6·D upstream from the propeller plane. The strength of the momentum sources are adjusted by numerical tests in the same CFD setup excluding the propeller model and the velocity field on the propeller plane extracted from the final wake modeling test is compared with experimental wake measurements in figure 2. Since the hull wake is from a single-screw ship, it is symmetric with respect to the centerline.

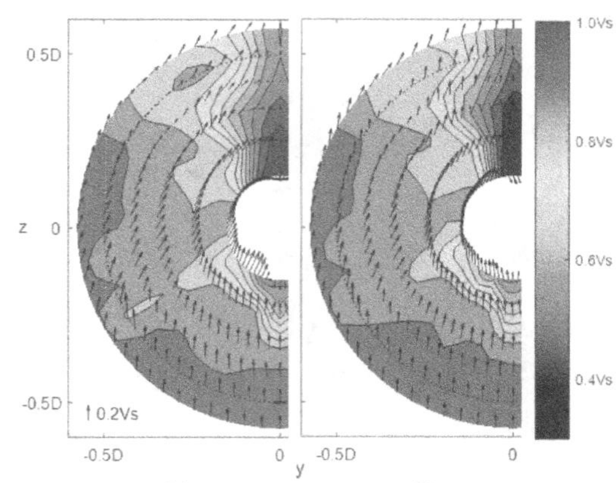

(a) (b)

Figure 2. Hull wake: (a) experimental measurement, (b) CFD modelling

An unsteady simulation is made on the initial grid starting with a time-step Δt corresponding to 6° propeller rotation per Δt and gradually reducing it to 0.5° rotation per Δt. The condition showing the largest extent of TVC in the cavitation tunnel test is considered in the simulation: ship speed V_S=4.5 m/s, propeller speed N=23.2 rps and cavitation number σ=5.42, where $\sigma=(P_\infty-P_V)/(0.5\cdot\rho\cdot V_A^2)$, P_∞ is the static pressure on the propeller shaft axis and V_A is the propeller advance speed. The propeller is on a controllable-pitch hub and the propeller pitch is increased for the maximum loading condition.

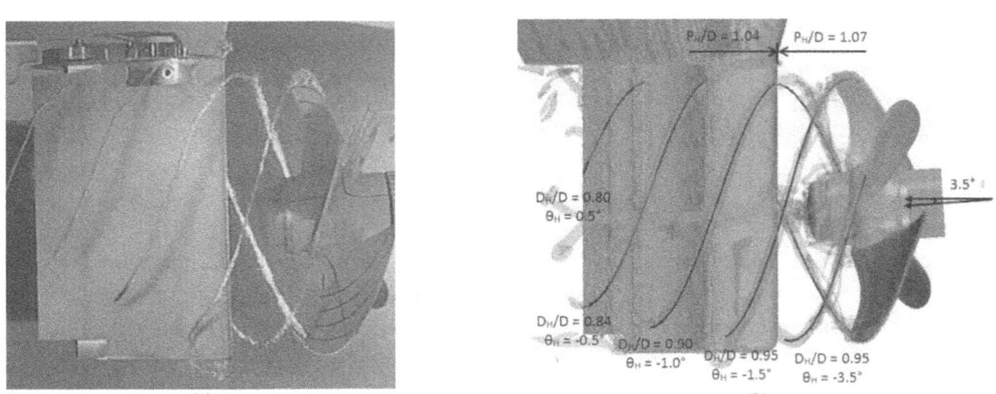

(a) (b)

Figure 3. (a) TVC in experiment and (b) iso-surface of Q-criterion=100,000 in DES with CAD helix curves

The helical trajectory of TVC identified by the iso-surface of a calibrated Q-criterion value shows a good agreement with that in the experiment until reaching the rudder. After TVC is disturbed by the rudder, the pitch P_H and diameter D_H of the helical trajectory in DES differ from those in the experiment. An overset grid is defined in a region over the rudder downstream from the rotating domain. As shown in figure 1(c), adaptive grid refinement of Δx=0.2 mm along the helical trajectory is applied to the overset grid region and the rotating domain. The overset grid rotates at the same speed as the propeller. Another simulation is made on the refined grid.

CFD result for TVC

In figure 4, the cavity interface visualized by the iso-surface of 10% vapor fraction in DES is compared with the experiment result. The comparison shows a good agreement in the leading-edge sheet cavitation at blade angles of φ=0 – 30°, but it is formed from more inner radii at φ=270 – 330° in DES, where φ=0° indicates the 12 o'clock blade position.

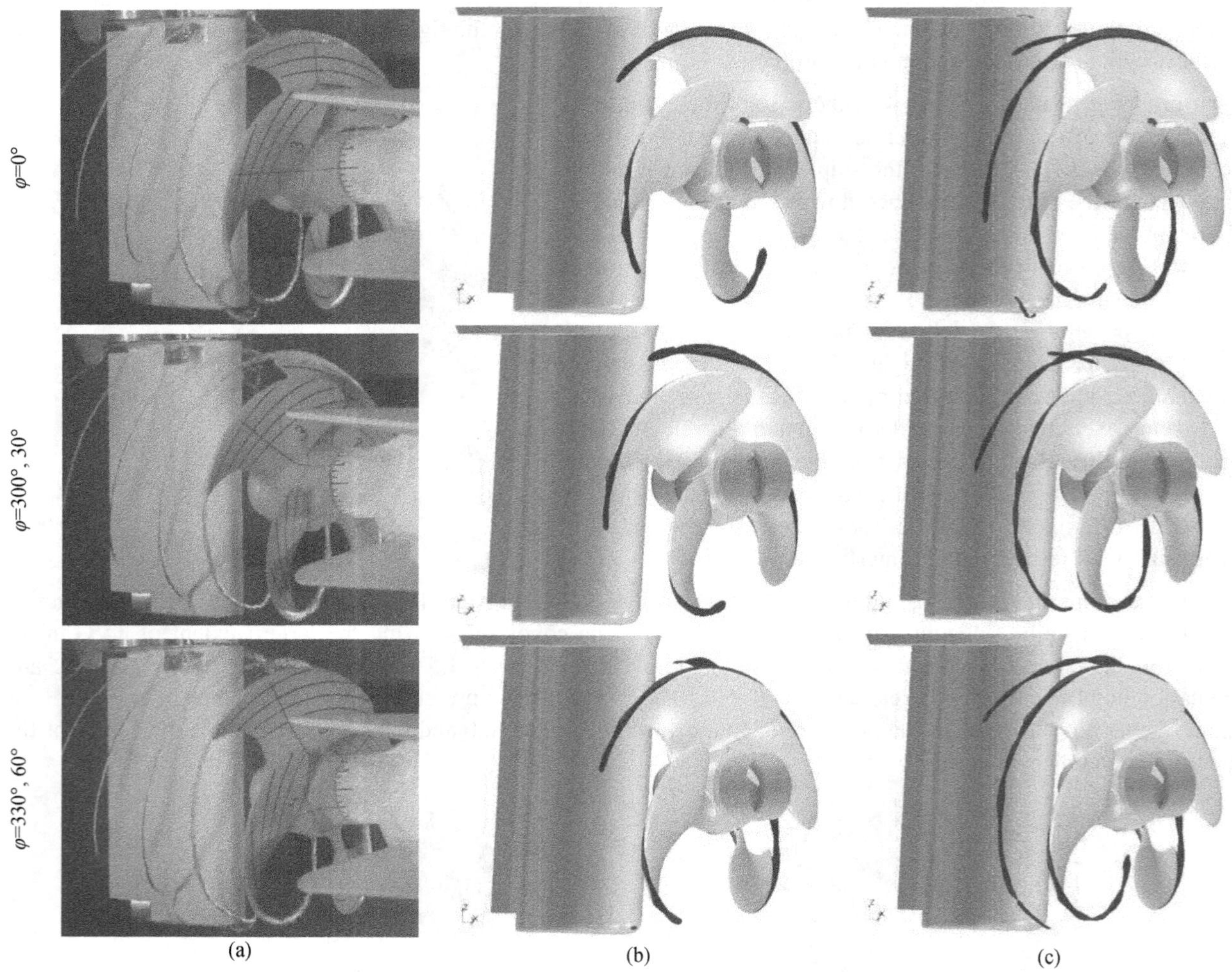

Figure 4. Cavitating flows on propeller and rudder in (a) experiment and DES on (b) initial and (c) refined grids

In the experiment, TVC is formed with a spiral structure from the roll-up of the trailing vortex sheet at the beginning and it is developed to be a concentrated vortex with a round cross section after about one quarter revolution. DES on the initial grid shows only a short extent of TVC with a round cross section, because the fine vortex core is not sufficiently resolved. After the grid refinement, TVC is more extended with a pronounced spiral structure, but it is still shorter than in the experiment. As TVC is distorted by the rudder disturbance, some parts of TVC can deviate from the refined grid region and so TVC from the blade at φ=300 – 330° is more extended than at φ=0 – 30°, because the grid refinement is made once for a certain momentary trajectory. The grid refinement to Δx=0.2 mm may not be enough to resolve the concentrated TVC core tapering downstream.

In figure 3(b), the deformation of TVC is analyzed by fitting CAD helix curves on the Q-criterion iso-surface. The initial pitch ratio of P_H/D=1.07 is reduced to P_H/D=1.04 by the rudder disturbance after a quarter of a revolution, which is contrasted to a 3% increase of the initial pitch further downstream shown in experiments on other propellers in uniform inflow with a straight shaft and no appendage [7]. P_H in the open-water experiments is about 5% higher than estimations from the actuator disc formula $P_H/D=J\cdot(1+(1+C_{Th})^{0.5})/2$, where J is the advance coefficient and C_{Th} is the thrust loading coefficient. If there is no rudder, the downstream pitch can be increased to $P_H/D=1.10$, which is 11% higher than $P_H/D=0.99$ from the actuator disc formula, because the propeller loading considered in DES corresponds to the maximum engine power, which is 1.67 times higher than the engine power of the design condition.

The initial diameter ratio of $D_H/D=0.95$ is gradually decreased to $D_H/D=0.8$ due to the radially inward propeller-induced flow. It is smaller than $D_H/D=0.83$ in the open-water experiments [7] due to the stronger flow acceleration in the high loading condition. The contraction of TVC is larger than the slipstream diameter reduction estimated to be about 4% by a formula derived from potential-flow calculations without considering rudder effects [8]. The angle θ_H of the progressing direction is 3.5° downwards along the shaft axis for an initial quarter of a revolution and it is gradually tilted upwards to $\theta_H=0.5°$ by the upward flow of the hull wake.

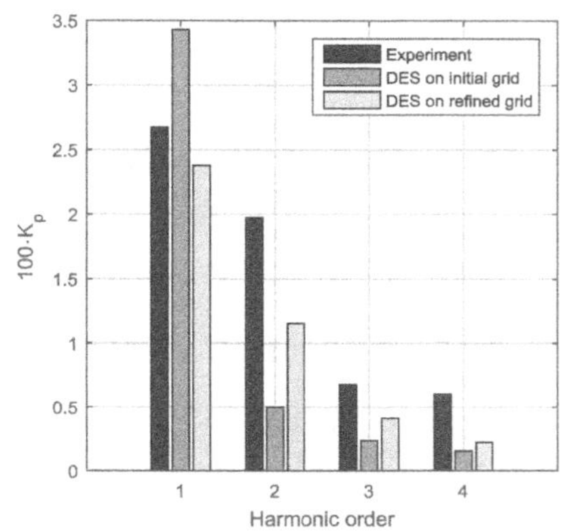

(a)

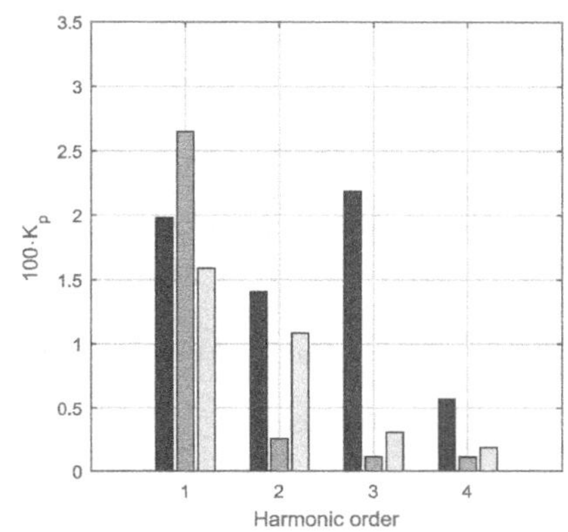

(b)

Figure 5. Pressure pulses on two points above the propeller 0.51·R downstream from the propeller plane and 0.51·R to (a) port and (b) starboard sides from the centerline

In figure 5, dimensionless pressure pulses K_P on two points above the propeller at harmonics of the blade passing frequency are compared between the experiment and DES, where $K_P=\acute{P}/(\rho\cdot N^2\cdot D^2)$ and $\acute{P}$ is the pressure amplitude. The two points are located 0.51·R from the centerline to both sides, 0.51·R downstream from the propeller plane and 1.58·R above the shaft axis. While the experimental measurements are made on the hull surface, the pressure is probed in the middle of the fluid domain without a wall boundary in DES and so the pressure pulse in DES is increased by a multiplication factor of 2 as per [8].

(a)

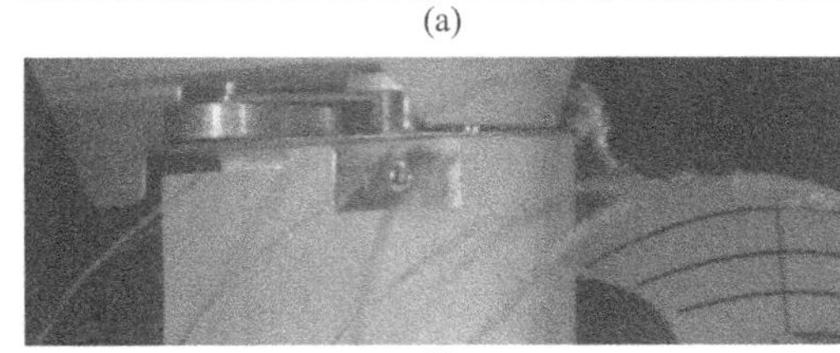

(b)

Figure 6. Collapsing of TVC at the gap between the rudder and the headbox in experiment at (a) φ=350° and (b) φ=0°

After the grid refinement, the first- and second-order pressure pulses are closer to the experimental measurements. Although the third- and fourth-order pressure pulses are increased in DES on the refined grid, these are still lower than the experimental ones. In the experiment, the third-order pressure pulse is higher than the first- and second-order ones on the starboard-side point (See figure 5 (b)), which may be related to the TVC collapse at the gap between the rudder and the headbox shown in figure 6. As TVC in DES is not extended to the rudder, the high-order pressure pulses are underestimated.

Conclusion

The adaptive grid refinement based on the Q-criterion with the overset grid over the rudder is effective in resolving TVC in DES. As TVC is better resolved by the grid refinement, the first- and second-order pressure pulses on two points above the propeller are closer to the experimental result. But the underestimation of high-order pressure pulses is significant, probably because the TVC collapse at the rudder headbox is not reproduced well. Repetitive grid refinements to smaller cell sizes of $\Delta x \leq 0.2$ mm may improve the accuracy to reproduce a larger extent of TVC.

The vortex flow visualization in DES is a useful tool for analyzing the deformed trajectory of TVC in the non-uniform hull wake field. The pitch of the TVC trajectory is reduced by the rudder disturbance in contrast with the pitch increase downstream in the open-water condition. The initial pitch is larger and the downstream radius is smaller due to the higher propeller loading than for moderately loaded propellers. The upward tilt of the TVC progress is pronounced by the upward flow of the hull wake and the downward inclination of the shaft axis.

Acknowledgements

The authors would like to express their gratitude for the support granted by the Danish Maritime Fund (Den Danske Maritime Fond).

References

[1] van Terwisga, T., Fitzsimmons, P. A., Ziru, L. and Foeth, E. J. (2009) *Proc. CAV2009* (Ann Arbor, MI, USA)
[2] Arndt, R., Pennings, P., Bosschers, J. and van Terwisga, T. (2015) *Interface Focus* 5: 20150025
[3] Shin, K. W. and Andersen, P. (2015) *Proc. CAV2015* (Lausanne, Switzeland)
[4] Berger, S., Gosda, R., Scharf, M., Klose, R., Greitsch, L. and Abdel-Maksoud, M. (2016) *Proc. 31st Symp. on Naval Hydrodynamics* (Monterey, CA, USA)
[5] Viitanen, V. M., Hynninen, A., Lubke, L., Klose, R., Tanttari, J., Sipila, T. and Siikonen, T. (2017) *Proc. SMP'17* (Espoo, Finland)
[6] Shin, K. W., Regener, P. B. and Andersen, P. (2015) *Proc. SMP'15* (Austin, TX, USA)
[7] Kerwin, J. E. (1976) *A deformed wake model for marine propellers* (Cambridge, MA, USA: MIT)
[8] Bertram, V. (2000) *Practical ship hydrodynamics* (Woburn, MA, USA: Butterworth-Heinemann)
[9] Hasuike, N., Yamasaki, S. and Ando, J. (2011) *Proc. MARINE 2011* (Lisbon, Portugal)

© 2018 ASME

CFD Simulation of Thermodynamic Effect Using a Homogeneous Cavitation Model Based on Method of Moments

[1]Shin-ichi Tsuda*; Satoshi Watanabe

[1]*Department of Mechanical Engineering, Kyushu University, Fukuoka, Japan*

Abstract

Recently, we have constructed a new homogeneous cavitation model, which includes various elementary processes in cavitation bubbles on the basis of a method of moments. We already confirmed that CFD simulation with this model can predict lift coefficient of a hydrofoil in cavitating water flow better than some conventional methods while the temperature decrease by thermodynamic effect in liquid nitrogen as a cryogenic fluid is underestimated. In this paper, we re-applied the cavitation model to a CFD simulation around the same target, cavitating liquid nitrogen flow around NACA16-012 hydrofoil. As a result, the temperature decrease by the thermodynamic effect is fairly well reproduced compared with the previous result when we arranged the model parameter of the growth/shrinkage while a limitation of homogeneous model and that of turbulence model are still indicated.

Keywords: cavitation; thermodynamic effect; NACA16-012 hydrofoil

Introduction

Cryogenic fluids such as liquid hydrogen or liquid oxygen have been used as a liquid rocket propellant, and it is well known that the suction performance of turbo-pump inducer is better in cryogenic fluids than in cold water due to thermodynamic effect. Thermodynamic effect is caused by temperature decrease inside a cavity region that arises from the latent heat transfer across the interface of cavity, and it brings about depression of bubble growth. To better understand the suction performance of cavitating cryogenic inducers, we should take account of the temperature changes due to thermodynamic effect and Computational Fluid Dynamics (CFD) is a powerful tool for this purpose. However, accurate evaluation of the temperature depression affecting the cavitating flow is even now one of the difficult tasks although some CFD studies have been carried out [1-3].

Recently, we have constructed a new homogeneous cavitation model calling "Multi-process cavitation model", which includes various elementary processes in cavitation bubbles based on a method of moments. We already confirmed that CFD simulation with this model can predict lift coefficient of a hydrofoil in cavitating water flow better than some conventional methods [4] while we could not successfully reproduce the thermodynamic effect in cryogenic fluids. In this paper, we retried the application of the present model for a cryogenic fluid of liquid nitrogen around NACA16-012 hydrofoil (angle of attack is 8 deg.) whose experiment was conducted by Niiyama et al [5].

Multi-process cavitation model

The multi-process model was constructed based on the moment method. In this method, "ith moment M_i" is defined using the size distribution function $f(R, t)$ as follows.

$$M_i = \int R^i f(R,t)dR,\ i \in \{0,1,2,3\}, \tag{1}$$

where R is the bubble radius and t is the time. Taking the time derivative for the quantity M_i based on a simple approximation [4], we can finally derive the relationship below,

$$\frac{DM_i}{Dt} = iM_{i-1}G(\overline{R}) + S_{p,i}. \tag{2}$$

Here, $\overline{R}$ is the mean bubble radius defined as M_1/M_0, $G(\overline{R})$ is the bubble growth/shrinkage rate, and $S_{p,i}$ reflects the change of the number of bubbles, which occurs with inception/collapse and coalescence/break-up. For each i (i=0, 1, 2, 3), the equation set is shown in Fig. 1 [4].

*Corresponding Author, Shin-ichi Tsuda: tsudashin@mech.kyushu-u.ac.jp

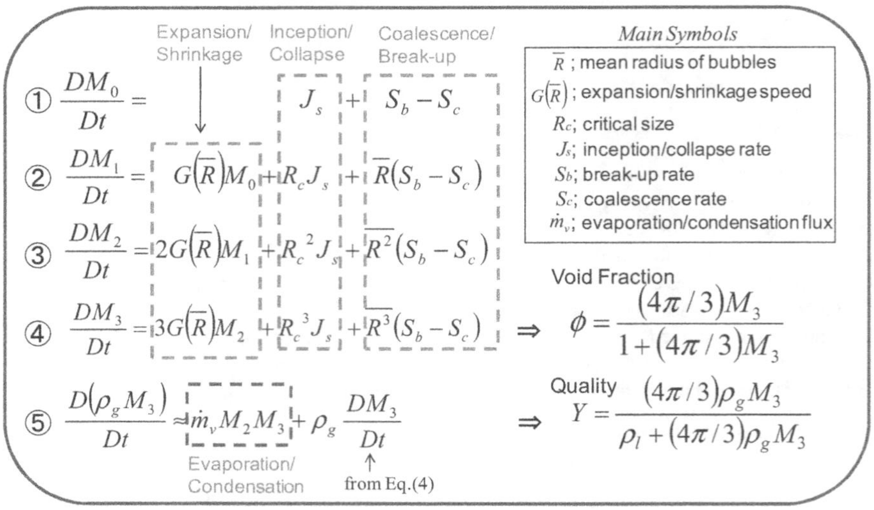

Fig. 1 Equation set of multi-process cavitation model

As shown in Fig. 1, main elementary processes such as growth/shrinkage, inception/collapse, and coalescence/breakup are reflected in each source term. The fifth equation in Fig. 1 is an extended equation of Eq. (2) reflecting the evaporation/condensation, which is the source of thermodynamic effect. The specific expression of each elementary process is described in detail in Ref. [4].

Numerical method and validation target

"Multi-process cavitation model" was implemented to a C-CUP (Cubic interpolation propagation - Combined Unified Procedure) flow solver [1] where compressibility of liquid-vapor two-phase flow can be incorporated with Baldwin-Lomax turbulence model as a zero-equation model. Using the solver, we simulated cavitating flow of liquid nitrogen around the hydrofoil of NACA16-012. The experimental set up by Niiyama et al. [5] in which five thermo-sensor to detect the temperature depression by thermodynamic effect and the present computational grid are shown in Fig. 2. In computation, each total cell number is around 30,000, and the flow velocity was given at the inlet boundary while the pressure was given at the outlet in the computational domain.

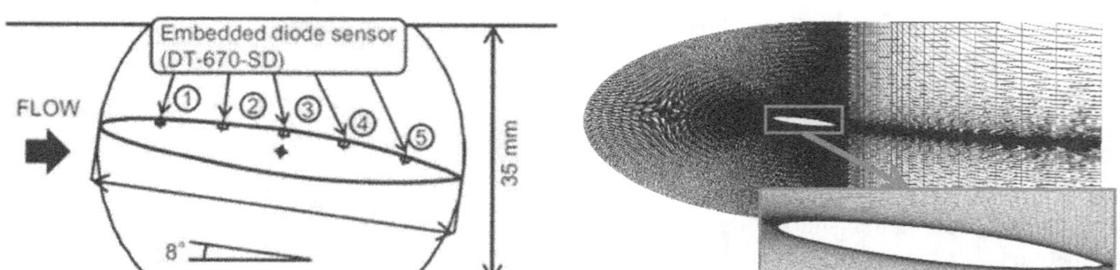

Fig. 2 Experimental set of thermos-sensor [5] and the computational grid for NACA16-012

Results and discussion

In advance, we investigated the influence of the model parameters of Multi-process cavitation model, and confirmed that the most sensitive model parameter is the coefficient of Growth/Shrinkage (GS) process, C_g, and the second one is that of Evaporation/Condensation (EC) process, C_{ec}. Each model for the two elementary processes is as follows, in which the expression for GS process is slightly different from that in Ref. [4].

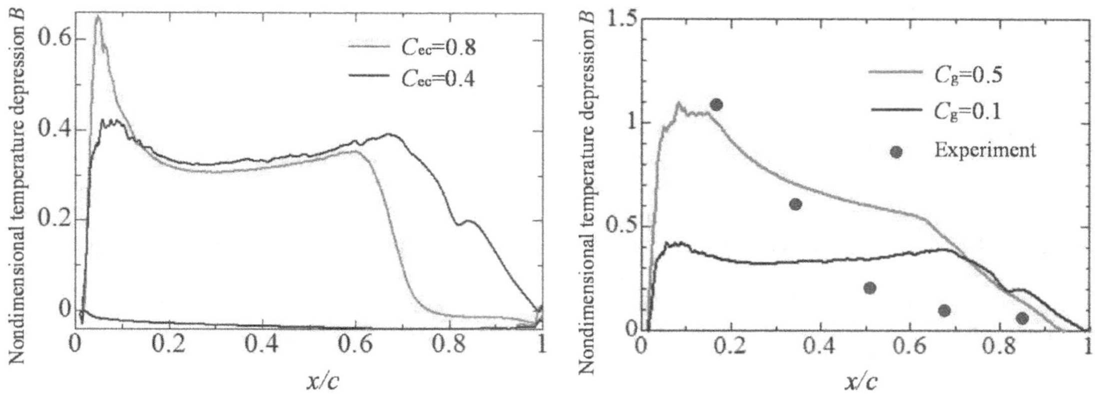

Fig. 2 Influence of the model parameters at $\sigma = 0.37$ (left: influence of C_{ec}, right: influence of C_g)

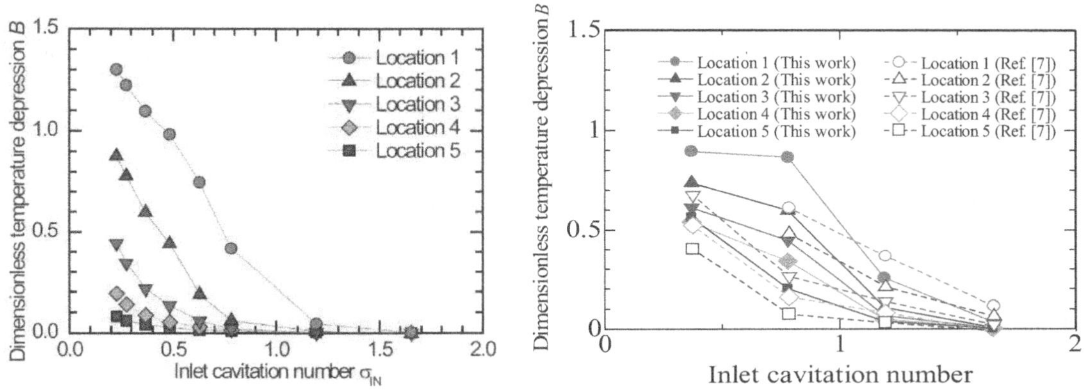

Fig. 3 Dimensionless temperature depression against cavitation number (left: experiment, right: present calculation)

$$G(\overline{R}) = C_g \sqrt{\frac{2}{3}\frac{p_v - p_l - 2\sigma_s/\overline{R}}{\rho_l}} \text{ (for GS process)}, \quad \dot{m}_v = \frac{2C_{ec}}{2 - C_{ec}} \frac{p_{s,in} - p_v}{\sqrt{2\pi R_v T}} \text{ (for EC process)} \tag{3}$$

The former is a well-employed simplified Rayleigh-Plesset model [3], and the latter is a classical evaporation/condensation model by Schrage [6]. Here, p_v, p_l, ρ_l σ_s, R_v, T are saturation vapor pressure, liquid pressure (given in CFD simulation), liquid density, surface tension, gas constant, and temperature, respectively. $p_{s,in}$ is the pressure inside bubble, which is given by Kelvin equation [6].

Figure 2 is a result showing an example of the influence of the two parameters (cavitation number σ is 0.37), and the experimental result by Niiyama et al. [5] is plotted in the right figure. Here, the vertical dimensionless temperature decrease B is defined as the following form.

$$B_N = \frac{T_{in} - T_N}{T^*} \quad (N = 1,2,3,4,5) \text{ , where } T^* = \frac{\rho_v(T_{in})L(T_{in})}{\rho_l(T_{in})C_{p,L}(T_{in})}. \tag{4}$$

Here, T_{in} is the inlet temperature, L is latent heat, $C_{p,L}$ is liquid specific heat at constant pressure, ρ is density (the subscript l and v mean liquid and vapor), respectively. The index N of B corresponds to each thermos-sensor in Fig. 1. As shown, the previous model parameter set ($C_g = 0.1$, $C_{ec} = 0.4$) [4] clearly underestimates the temperature depression at the leading edge side while just modification of C_{ec} does not reproduce the temperature distribution at the trailing edge side. In this work, we set the value $C_g = 0.5$ without changing C_{ec} (C_{ec} is kept to be 0.4).

Figure 3 compares the temperature depression at the five points along the chord direction in some cavitation numbers between experiment and this work. Also, the computational result by Takeda et al. [7] who captures the liquid-vapor interface with a method coupling VOF and Level-Set is plotted in the left side in Fig. 3. Note that the legend 'Location 1-5' in Fig. 3 corresponds to each temperature sensor 1-5 in Fig. 1. As shown, the present results using the new parameter set qualitatively reproduce the temperature depression in the experiment, which is different from the simulation using the old parameter set [4]. It suggests that the model parameter of C_g is very important for the evaluation of thermodynamic effect. In addition, our results apparently show closer result against the experiment compared with the previous CFD result by Takeda et al. [7]. However, we still have two deviations from the measurement. One is that, at the higher cavitation number, the present work overestimates the temperature depression because our simulation based on a homogeneous cavitating flow precedes the cavitation inception. In the homogeneous cavitating flow, bubble nuclei whose quantity is given according to water quality [4] are also present on the every wall surface imposed non-slip condition, and in this case, the inception starts at the suction peak of negative pressure surface. This overestimation will be common to homogeneous medium models. On the other hand, at the lower cavitation number, the present simulation even underestimates the temperature depression near the leading edge while it overestimates the depression near the trailing edge side. The reason would be in inaccurate evaluation of turbulent flow rather than the cavitation model because the present simulation employs a simple turbulence model among zero-equation models. Therefore, the revision of the flow field itself is still necessary.

Conclusion

We have applied "Multi-process cavitation model", in which the basic elementary processes in cavitation are taken into account based on the moment method, to the cryogenic cavitation around NACA16-012. As a result, the temperature decrease by the thermodynamic effect is evaluated when the model parameter of the growth/shrinkage rather than that of the evaporation/condensation is arranged. However, a limitation of homogeneous model and that of turbulence model are still present, absolutely indicating the improvement of the two factors.

Acknowledgment

This study was partially supported by Ebara Co. Ltd., paying special thanks to Dr. Motohiko Nohmi (Ebara Co. Ltd.).

References

[1] Tani, N. and Nagashima, T., 2002, Numerical Analysis of Cryogenic Cavitating Flow on Hydrofoil –Comparison between Water and Cryogenic Fluids–, Proceedings of 4th International Symposium on Launcher Technology, #27 (CD-ROM)

[2] Hosangadi, A. and Ahuja, V., 2005, Numerical Study of Cavitation in Cryogenic Fluids, Trans. ASME J. Fluids Eng., Vol. 127, pp.267-281

[3] Tsuda, S., Tani, N., Yamanishi, N., Development and Validation of a Reduced Critical Radius Model for Cryogenic Cavitation, 2012, Trans. ASME J. Fluids Eng., Vol. 134, 051301

[4] Tsuda, S. and Watanabe, S., Application of Multi-Process Cavitation Model for Cavitating Flow in Cold Water and in Liquid Nitrogen around a Hydrofoil, Proc. ASME/JSME/KSME 2015 Joint Fluids Engineering Conf., Paper No. AJKFluids2015-05532

[5] Niiyama, K., Yosihida, Y., Hasegawa, S., Watanabe, M., and Oike, M., 2012, Experimental Investigation of thermodynamic effect on cavitation in liquid Nitrogen, Proc. 8th International Symposium on Cavitation, No. 231

[6] Fujikawa, S., Yano, T., Watanabe, M., 2011, Vapor-Liquid Interfaces, Bubbles and Droplets, Springer

[7] Takeda, N., Himeno, T., Umemura, Y., Watanabe, T., 2014, Numerical Simulation of Cavitation Considering Thermodynamic Effect by Direct Interface Capturing Approach, Proc. 17th Symposium on Cavitation (Japan), Paper No. 0004

© 2018 ASME

Circular acoustic microstreaming flows due to pulsating microbubbles near a membrane

Nima Mobadersany, Kausik Sarkar*

The George Washington University, Washington, D.C., USA

Abstract

In this study, the microstreaming velocity near a plane wall due to microbubble pulsation has been studied. Microbubbles are assumed to pulsate spherically near the wall under the excitation of low amplitude ultrasound. It has been shown an axisymmetric vortex is generated near the wall inducing shear stress on the wall. The shear stress can help facilitate the uptake of drugs and genes in to desired cells through the process called sonoporation.

Keywords: microstreaming; ultrasound; sonoporation; microbubble

Introduction

It is well known that in addition to the sinusoidal movement in fluid particles, sound can generate steady vortices in which particles circulate steadily [1, 2]. These vortex motions have been observed experimentally near oscillating gas bubbles resting on a solid surface[3]. The streaming motion of the bubbles exerts shear stress on the boundary. The microstreaming shear due to the pulsation of gas bubbles has been shown as the main mechanism of hemolysis of the red blood cells [4]. It has been shown that microstreaming shear stress can be exploited in thrombolysis, drug delivery and gene therapy. Microstreaming can rupture and increase the permeability of cell membranes (sonoporation) in biological tissues for better passage of therapeutic agents across the vascular barrier and cell membranes [5-7]. The early theory for microstreaming velocity field and the resultant shear stress has been proposed by Nyborg [2]. Levin and Bjorno studied the maximum shear stress due to hemispherical gas microbubbles resting on biological cells using Nyborg's theory [8]. Doinikov applied Nyborg's theory to find the shear stress on the plane wall induced by the spherical oscillation of contrast microbubble assuming the bubble is detached from the wall [9]. Here, we study the shear stress and the streamlines due to microstreaming near a plane rigid wall generated by spherical pulsation of microbubble numerically.

Governing equation

The early theoretical work on microstreaming has been studied by Nyborg [1, 2]. Nyborg decomposed the velocity, pressure and density of fluid particles around an oscillatory body into a first order and a second order approximations, where the first order approximation is the sinusoidally oscillating velocity and the second order approximation is the streaming velocity. Taking temporal average of the Navier Stokes equation and assuming streaming velocity is steady and solenoidal, he found the following stokes type equation governing the streaming flow by keeping the second order approximation terms:

$$\mu\nabla^2\vec{V_2} - \nabla p_2 = \rho_0 < \vec{V_1}.\nabla\vec{V_1} >_t \quad , \tag{1}$$

where $\vec{V_2}$ and p_2 are the streaming velocity and the pressure (the second order velocity and pressure), $\vec{V_1}$ is the sinusoidal velocity vector (first order approximation of the velocity) and $<>_t$ is the average in time. By solving equation (1), the radial component of the streaming velocity u_2 for a flow over a plane surface is found out as a function of the local radial irrotational velocity [2]:

$$u_2 = (\omega)^{-1}\left(0.5\frac{\partial(u_\phi)^2}{\partial r}(u_\alpha - u_\beta) - \frac{(u_\phi)^2}{r}u_\beta\right), \tag{2}$$

where,

*Corresponding Author, Kausik Sarkar: sarkar@gwu.edu

$$u_\alpha = \frac{1}{4}e^{-2\beta z} + e^{-\beta z}\sin(\beta z) - \frac{1}{4},$$
$$u_\beta = \frac{1}{2}\beta z e^{-\beta z}(\cos(\beta z) - \sin(\beta z)) - e^{-\beta z}\sin(\beta z) - \frac{1}{2}e^{-\beta z}\cos(\beta z) + \frac{1}{2}$$

$\beta^{-1} = \delta = \sqrt{\frac{2\nu}{\omega}}$ is the acoustic boundary layer. ν is fluid kinematic viscosity, ω is the excitation angular frequency and u_ϕ is the amplitude of the local radial irrotational velocity.

The vertical component of the streaming velocity w_2 is solved by using incompressible mass conservation (equation(3)), while the no slip condition on the rigid wall is satisfied:

$$\frac{1}{r}\left(\frac{\partial(ru_2)}{\partial r}\right) + \left(\frac{\partial w_2}{\partial z}\right) = 0. \tag{3}$$

We study the circulatory flow close to the plane rigid wall due to the spherical pulsation of microbubbles. Figure 1 shows the schematic of the problem. The effect of the rigid wall has been accounted for by an image bubble.

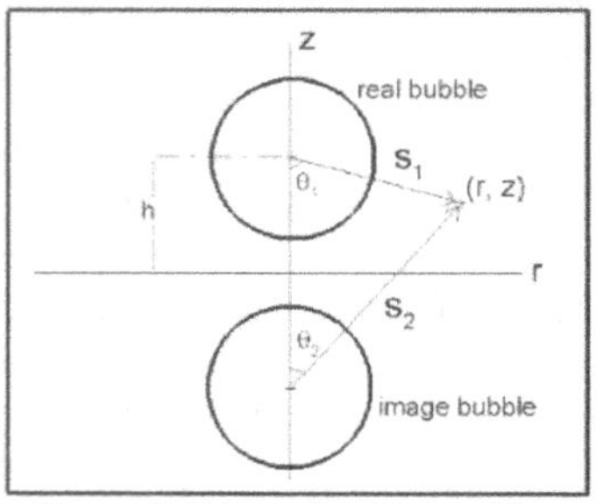

Figure 1. Schematic of the problem

As mentioned earlier, the streaming velocity can be found by having the local irrotational velocity distribution. The velocity potential of the fluid surrounding the microbubble is:

$$\phi = \left(\frac{1}{S_1} + \frac{1}{S_2}\right)\dot{R}R^2 \tag{4}$$

Where ϕ is the velocity potential, S_1 and S_2 are the distance from the center of the real and image bubble to a location in the fluid, $\dot{R}$ and R are the velocity and instant radius of the microbubble.

Equation (4) gives the irrotational local velocity $\nabla\phi$.

Numerical results

In this study, the microbubble has been excited with an ultrasound wave at excitation frequency of $f_{ex} = 1.5MHz$ and excitation pressure of $P_{ex} = 100KPa$. The initial radius of the microbubble is $R_0 = 3\mu m$. Figure 2 shows the radial pulsation of the microbubble with non-dimensional time (with respect to ambient pressure, density and initial bubble radius) located at $h = 2R_0$. As it is shown in the steady region when the transient pulsation has been subsided, the maximum pulsation amplitude of the microbubble is $0.17R_0$.

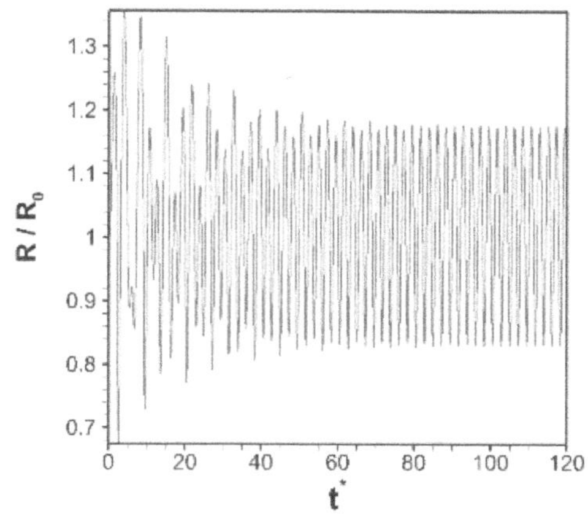

Figure 2. The radial oscillation of the free bubble when

$$P_{ex} = 100kPa, f_{ex} = 1.5MHz, h = 2R_0$$

Figure 3 shows the streamlines due to microstreaming near the wall for the condition described in figure 2. An axisymmetric vortex is generated near the wall with the length of $L = 1.41R_0$ corresponding to the location where the direction of external driving force of streaming flow changes. Inside the vortex $(0 < r < 1.41R_0)$, the flow near the wall is directed inward, while it is directed radially outward beyond $r = 1.41R_0$. Figure 4 shows the corresponding shear stress on the wall. The sign of the shear stress changes at $r = 1.41R_0$ corresponding to the location where the direction of the streaming flow near the wall changes. The shear stress induced by microstreaming can perforate the cell membranes and facilitate the uptake of drugs and large molecules in to desired tissue.

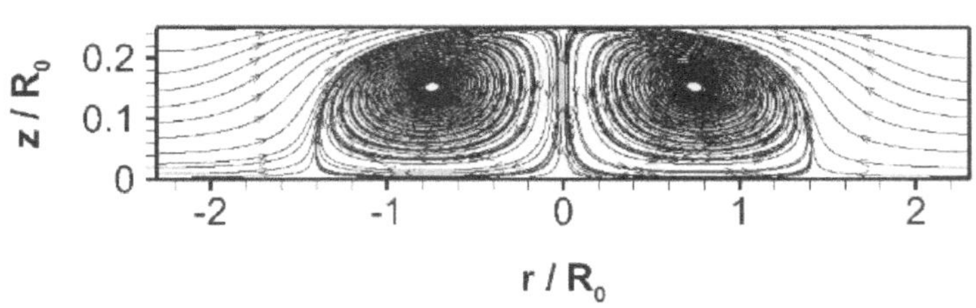

Figure 3. Streamlines of microstreaming flow near the plane rigid wall due to spherical pulsation of microbubble when

$$P_{ex} = 100kPa, f_{ex} = 1.5MHz, h = 2R_0$$

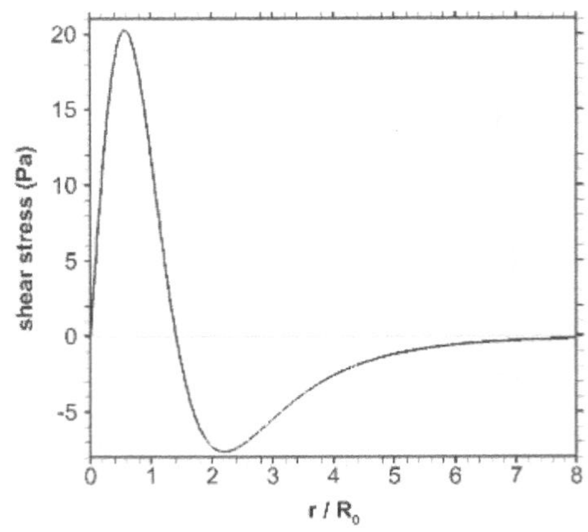

Figure 4. Shear stress on the wall due to microstreaming when $P_{ex} = 100kPa, f_{ex} = 1.5MHz, h = 2R_0$

Conclusion

In this study, the microstreaming flow near the plane rigid wall has been studied. A microbubble is assumed to pulsate spherically near the wall under the excitation of low amplitude ultrasound. It has been shown that a radial ring vortex is generated near the wall. It gives rise to a shear stress on the wall causing sonoporation which may help drug delivery.

References

[1] *W. L. Nyborg, "Acoustic streaming due to attenuated plane waves," The journal of the acoustical society of America, vol. 25, pp. 68-75, 1953.*

[2] *W. L. Nyborg, "Acoustic streaming near a boundary," The Journal of the Acoustical Society of America, vol. 30, pp. 329-339, 1958.*

[3] *J. Kolb and W. L. Nyborg, "Small-scale acoustic streaming in liquids," The Journal of the Acoustical Society of America, vol. 28, pp. 1237-1242, 1956.*

[4] *J. A. Rooney, "Hemolysis near an ultrasonically pulsating gas bubble," Science, vol. 169, pp. 869-871, 1970.*

[5] *P. Marmottant and S. Hilgenfeldt, "Controlled vesicle deformation and lysis by single oscillating bubbles," Nature, vol. 423, pp. 153-156, 2003.*

[6] *Z. Fan, R. E. Kumon, and C. X. Deng, "Mechanisms of microbubble-facilitated sonoporation for drug and gene delivery," Therapeutic delivery, vol. 5, pp. 467-486, 2014.*

[7] *A. Pommella, N. J. Brooks, J. M. Seddon, and V. Garbin, "Selective flow-induced vesicle rupture to sort by membrane mechanical properties," Scientific reports, vol. 5, 2015.*

[8] *P. A. Lewin and L. Bjo/rno/, "Acoustically induced shear stresses in the vicinity of microbubbles in tissue," The Journal of the Acoustical Society of America, vol. 71, pp. 728-734, 1982.*

[9] *A. A. Doinikov and A. Bouakaz, "Theoretical investigation of shear stress generated by a contrast microbubble on the cell membrane as a mechanism for sonoporation," The Journal of the Acoustical Society of America, vol. 128, pp. 11-19, 2010.*

© 2018 ASME

Collapse and jet formation of ultrasound contrast microbubbles near a membrane for sonoporation

Nima Mobadersany, Kausik Sarkar*

The George Washington University, Washington, D.C., USA

Abstract
Contrast agents, in the presence of ultrasound, can help facilitate the uptake of drugs and genes into desired cells through the process called sonoporation. Sonoporation is the temporarily rupture of cell membranes in the presence of ultrasound. In this work, we studied the behaviour of a contrast agent near a rigid wall in the presence of ultrasound using boundary integral equation method. A contrast microbubble forms a high-velocity microjet at the last stage of the collapse phase. The microjet and the adjacent surrounding fluid impinges on the membrane and spreads radially, generating high shear stress resulting in the rupture and perforation of the cell membrane. The encapsulation of the microbubble can be assumed as an interface with an infinitesimal thickness. There are several models to simulate the interface. In this study, the encapsulation is modelled using strain-softening interfacial elastic model called the exponential elasticity model (EEM).

Keywords: sonoporation; ultrasound; microjet; Boundary integral equation; microbubble collapse

Introduction
Ultrasound waves are pressure waves capable of transporting energy inside the body at precise locations as they are absorbed very little by tissues. Their non-invasive, safe and painless transmission through the skin is the reason for their wide use in drug delivery and gene therapy applications [1]. Microbubbles in the presence of ultrasound facilitates the transport of drugs through a process called sonoporation. Sonoporation is the acoustically induced temporary rupture of membranes. With a scanning electron microscope, it has been shown that multiple surface pores are being created on the cell membrane exposed to ultrasound [2]. The induction of sonoporation using targeted microbubbles have been studied both in vitro and vivo [3-5]. Microbubbles were initially developed for enhancing the contrast of the image inside the body due to their high echogenicity. They consist of a gas core encapsulated by a layer of proteins or lipids to prevent them against premature dissolution. The shell reduces the surface tension and stabilizes them. At high amplitude ultrasound excitations, microbubbles implode after a few cycles. If the collapsing microbubbles are in the vicinity of a solid boundary, they generate a high-velocity liquid microjet toward the wall. The impingement of microjet along with its high velocity creates high shear stresses in the cell membrane resulting in sonoporation. This consequently will facilitate uptake of drugs into the desired site [6-8]. The study of non-spherical bubble collapse shows that the dynamics of the jet strongly depends on the distance of the bubble relative to the wall [9]. It is also shown that in addition to the jet speed, jet size can also have a determining factor on the impacting power of collapsing bubble [10]. During the formation of jet, there exists very high pressure fluid on top of the jet pushing the bubble toward the wall [11, 12]. In this work, we do numerical study on the formation of microjet from contrast agent microbubbles in the vicinity of a solid wall to better understand the mechanism of sonoporation. We have assumed the cell membrane as a solid wall for simplification. We have used a zero-thickness interfacial rheology models to simulate the encapsulation of the contrast agent microbubbles as the encapsulation is practically an interface [13]. The boundary integral equation method has been used for the numerical study.

Mathematical equations
In this study, the flow around the bubble is assumed incompressible, irrotatational and inviscid. The surrounding flow is governed by Laplace equation. It is solved using a boundary integral equation method obtained using following Green's integral formula:

$$c(p)\phi(p)+\int_S \phi(q)\left\{\frac{\partial}{\partial n}\left(\frac{1}{|p-q|}+\frac{1}{|p-\hat{q}|}\right)\right\}ds=\int_S \frac{\partial}{\partial n}(\phi(q))\left(\frac{1}{|p-q|}+\frac{1}{|p-\hat{q}|}\right)ds \; . \tag{1}$$

*Corresponding Author, Kausik Sarkar: sarkar@gwu.edu

In equation(1), s is the boundary of the fluid domain which includes the bubble surface and its image; ϕ is the velocity potential and $\frac{\partial \phi}{\partial n}$ is the normal velocity; p is a point in the fluid domain, or on the boundary; q is a point on the fluid boundary and $q^{\wedge}$ is the image point of q across the plane z=0 where the cell membrane is located. $c(p)$ is a coefficient dependent on the location of the point p. The image bubble helps satisfy the impermeability condition on the cell membrane. Figure 1 shows the schematic of the problem. The bubble surface is discretized into M cubic spline elements. The velocity potential and the normal velocity are assumed to be constant on each element and are located in the middle of the elements. The normal velocity on the boundary of the fluid domain is indicated by n and is directed outward (directed toward the bubble center). The vertical axis is indicated by z, while r is the radial axis. The problem is assumed axisymmetric.

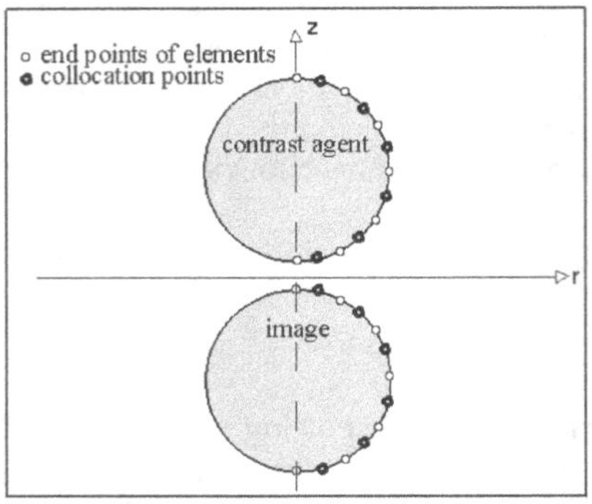

Figure 1. Schematic of the problem

Equation (1) gives a set of linear equations for evaluating $\frac{\partial}{\partial n}(\phi)$ on each segment on the bubble surface. To find the velocity potential in the next time steps, the normalized form of the unsteady Bernoulli equation is applied for the flow along the surface of the contrast agent:

$$\frac{D\phi^*}{Dt^*} = \frac{1}{2}\left|\nabla^*\phi^*\right|^2 + \frac{(P_\infty - P_{bw})}{P_{atm}} , \tag{2}$$

where ϕ^*, t^* and ∇^* are the non-dimensional velocity potential, time and gradient respectively (with respect to ambient pressure, density and initial radius). P_{bw} is the pressure in the fluid on the bubble surface. P_∞ is the pressure in the far field including the static ambient pressure (here atmospheric pressure) and excitation ultrasound pressure:

$$P_\infty = P_{atm} - P_{ex}\sin(\omega t) . \tag{3}$$

P_{ex} and ω are the excitation pressure and excitation radial frequency due to the ultrasound wave. The gas inside the contrast agent is assumed to be an ideal gas that follows a polytropic law:

$$P_g = P_{g_0}\left(\frac{V_0}{V}\right)^{\kappa} . \tag{4}$$

P_g is the gas pressure inside the microbubble when its volume is V. P_{g_0} is the initial gas pressure inside microbubble when the bubble is in its initial volume of V_0, and κ is the polytropic constant.

Modelling the encapsulation of the microbubble

The pressure inside and outside the microbubble are related by the normal stress balance. The effect of the encapsulation is considered on the normal stress balance:

$$P_g - P_{bw} = \left(\sigma_{eff} + \kappa_s \left(\nabla_s \cdot V\right)\right)\left(\nabla_s \cdot n\right) . \tag{5}$$

In equation(5), σ_{eff} and κ_s are the effective surface tension and dilatation viscosity of the contrast agent due to the encapsulation respectively, $\nabla_s \cdot V$ is the surface divergence of velocity, and $\nabla_s \cdot n$ is the curvature of the contrast agent[14]. Applying the dilatation viscosity term in the normal jump condition gives rise to instability which could not be removed by smoothing. Therefore in the current research, we neglect the dilatation viscosity for simplicity, and we will study its effect in future studies. Various different interfacial rheology models $\sigma(R)$ have been proposed. In the present study we are using eexponential elasticity model (EEM) [15] to calculate the effective surface tension.

$$\begin{cases} \sigma_{eff}\left(R\right) = \sigma_0 + \beta E^s, & \text{where } E^s = E_0^s \exp(-\alpha^s \beta) \text{ and } \beta = (\dfrac{R^2}{R_E^2} - 1), \\ \sigma_{eff}\left(R\right) = 0, & for \qquad \sigma\left(R\right) < 0 \end{cases} \tag{6}$$

$\sigma_{eff}(R)$ is the effective interfacial tension, E^s is the elasticity of the shell, E_0^s is elasticity constant and β is the area fraction. R_E is the equilibrium radius of the contrast agent. Equilibrium radius equals to the radius of the bubble where elastic stress is zero.

$$R_E = R_0[1 + (\frac{1 - \sqrt{1 + 4\gamma_0 \alpha^s / E_0^s}}{2\alpha^s})]^{-1/2} . \tag{7}$$

Results and discussion

In this study, we are using Sonazoid contrast agent as an example. Accordingly, the bubble radius is 1.6 μm. The excitation frequency is assumed to be 2 MHz same as the average damped resonance frequency of Sonazoid. Figure 2 shows the contrast microbubble and the surrounding velocity and pressure when $h^* = 4.0, P_{ex}^* = 5.0, f = 2MHz$. h^* is the non-dimensional (with respect to the initial radius) initial distance of the bubble center from the cell membrane, P_{ex}^* is the non-dimensional (with respect to the atmospheric pressure) excitation pressure and f is the excitation frequency. Note that the magnitude of velocity vectors in the figure is relative to other particles in the same figure, and they do not show the actual magnitude of fluid particles velocity. Figure 2(a) shows the contrast agent at the early stages of the growing phase of the bubble with a higher pressure inside. Figure 2(b) shows the contrast microbubble when it reaches the maximum size at $t^* = 2.1888$ where t^* is the non-dimensional time. Bubble expands nearly spherically to reach the maximum volume. During the last stage of the collapse phase, the microbubble forms a jet directed toward the membrane. Figure 2(c) shows the beginning of the collapse phase when the bubble starts to form a jet at $t^* = 3.2697$. There exists a high-pressure region on top of the contrast agent pushing the high velocity jet [16]. Figure 2(d) shows the last stage of the collapse phase of the contrast agent at $t^* = 3.3214$. The collapse time from the start of the jet formation to the end of the collapse is very short.

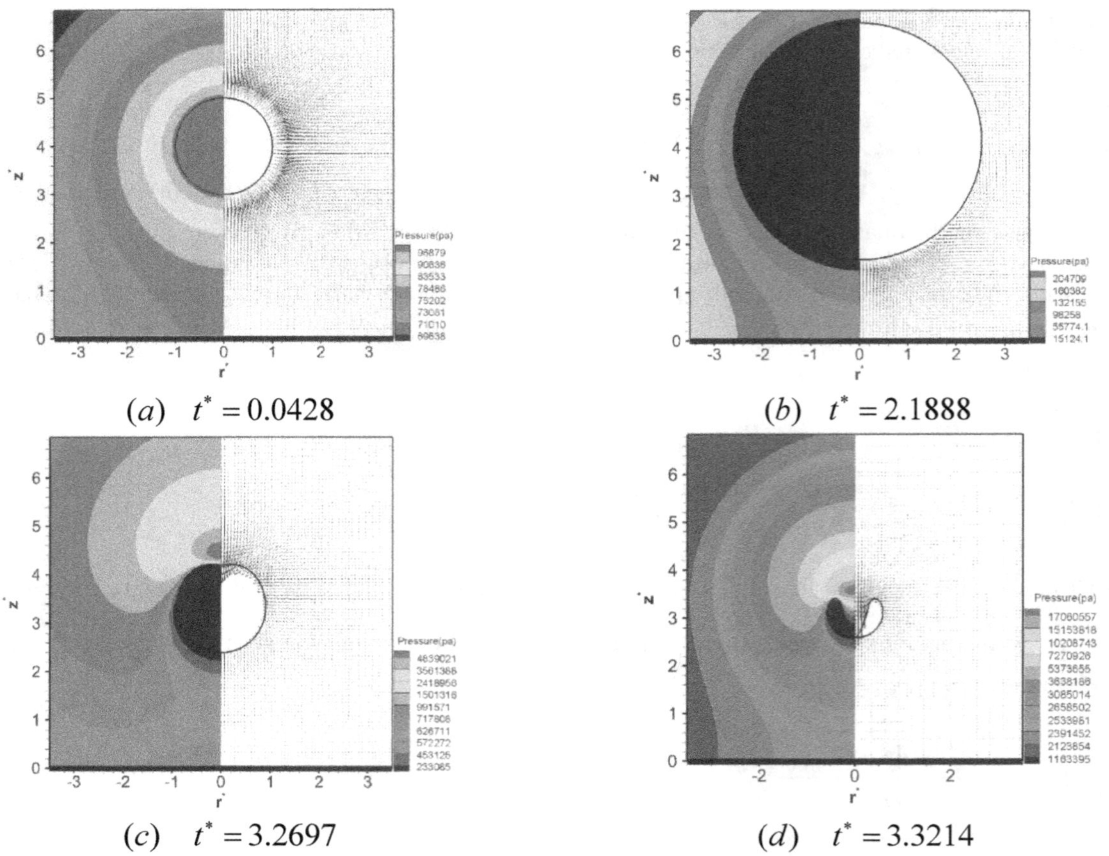

Figure 2. Velocity and pressure surrounding the contrast agent with respect to non-dimensional time using EEM model for the bubble encapsulation when $h^* = 4.0, P_{ex}^* = 5.0$

Figures 2(c-d) show that the microjet and the adjacent surrounding fluid are moving toward the cell membrane with a very high velocity. This high velocity fluid impinges the membrane and spreads radially along it. This will generate a high velocity gradient on the tissue and therefore a shear stress resulting in the rupture and perforation of cell membrane. To calculate the shear stress generated on the wall, we consider the flow field induced from the jet of the contrast agent spreading outward radially along the tissue [17]. We assume a jet flow impacting vertically on the tissue with a constant velocity equal to the averaged vertical velocity of the elements on the jet at each time. Glauret wall jet flow has been used to estimate the shear on the tissue [18] to compute the wall shear stress τ :

$$\tau = \rho\nu\left(\frac{\partial u}{\partial y}\right)_{y=0} = \rho\left(\frac{125F^3}{216\nu x^{11}}\right)^{1/4}. \tag{8}$$

The constant F is the momentum flux of the incoming jet:

$$F = \frac{1}{128}U_{jet}^3 d_{jet}^4 \ . \tag{9}$$

U_{jet} is the average jet velocity, and d_{jet} is the jet diameter. Figure 3 shows the shear stress along the wall due to the impingement of the jet.

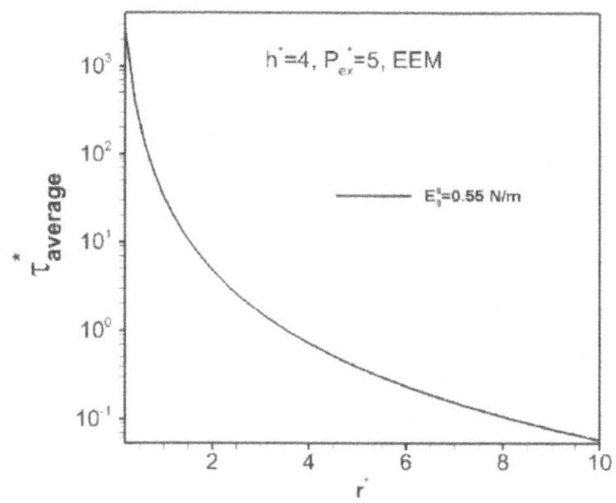

Figure 3. Corresponding shear stress on the rigid boundary with respect to non-dimensional time when $h^* = 4, \mathrm{P}_{ex}^* = 5$ using EEM model for the encapsulation

Figures 4 (a-c) show the non-dimensional volume, volume centroid and jet tip velocity of the contrast microbubble at the conditions shown in figure 2. As shown in figures 4(a), the maximum growth of the contrast microbubble decreases with the increase of shell elasticity, as the higher elasticity makes the contrast agent stiffer. Figure 4(b) shows the motion of volume centroid. Microbubbles with higher shell elasticity experiences less excursion towards the cell membrane. The life time of the contrast agent decreases with the increase of shell elasticity. Therefore, contrast agent with higher shell elasticity, has less time to move toward the cell membrane. Figure 4(c) shows that the velocity of the jet tip is higher with higher shell elasticity. In contrast to coated bubbles, free bubbles experience more expansion, more translation toward the cell membrane and less jet velocity.

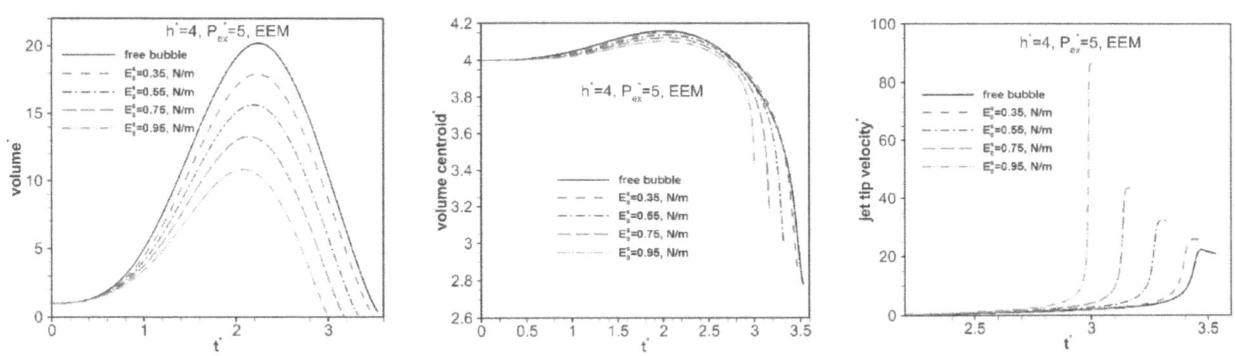

(a) (b) (c)

Figure 4. Non-dimensional volume, non-dimensional volume centroid and non-dimensional jet tip velocity of microbubble with respect to non-dimensional time when $h^* = 4.0, P_{ex}^{\ *} = 5.0$ using EEM model for the encapsulation

Conclusion

The behaviour of contrast microbubble near a rigid wall in the presence of ultrasound have been studied using boundary integral equation method. During the last stage of the collapse, the contrast agent forms a high-velocity microjet towards the wall. The microjet impinges the wall with high velocity and exerts a high shear stress. The shear stress induced by the jet may lead to sonoporation. An exponential elasticity model (EEM) has been used to simulate the interfacial rheology of the microbubble encapsulation.

References

[1] W. G. Pitt, G. A. Husseini, and B. J. Staples, "Ultrasonic Drug Delivery – A General Review," *Expert opinion on drug delivery,* vol. 1, pp. 37-56, 2004.

[2] K. Tachibana, T. Uchida, K. Ogawa, N. Yamashita, and K. Tamura, "Induction of cell-membrane porosity by ultrasound," *Lancet,* vol. 353, pp. 1409-1409, Apr 24 1999.

[3] K. Kooiman, M. Foppen-Harteveld, A. F. van der Steen, and N. de Jong, "Sonoporation of endothelial cells by vibrating targeted microbubbles," *Journal of controlled release,* vol. 154, pp. 35-41, 2011.

[4] J.-M. Escoffre, A. Zeghimi, A. Novell, and A. Bouakaz, "In-vivo gene delivery by sonoporation: recent progress and prospects," *Current gene therapy,* vol. 13, pp. 2-14, 2013.

[5] H. J. Lee, Y. I. Yoon, and Y. J. Bae, "Theragnostic ultrasound using microbubbles in the treatment of prostate cancer," *Ultrasonography,* vol. 35, p. 309, 2016.

[6] M. Lokhandwalla and B. Sturtevant, "Mechanical haemolysis in shock wave lithotripsy (SWL): I. Analysis of cell deformation due to SWL flow-fields," *Physics in Medicine and Biology,* vol. 46, pp. 413-437, Feb 2001.

[7] J. Sundaram, B. R. Mellein, and S. Mitragotri, "An experimental and theoretical analysis of ultrasound-induced permeabilization of cell membranes," *Biophysical Journal,* vol. 84, pp. 3087-3101, May 2003.

[8] C. M. H. Newman and T. Bettinger, "Gene therapy progress and prospects: Ultrasound for gene transfer," *Gene Therapy,* vol. 14, pp. 465-475, Mar 2007.

[9] M. Calvisi, J. Iloreta, and A. Szeri, "Dynamics of bubbles near a rigid surface subjected to a lithotripter shock wave. Part 2. Reflected shock intensifies non-spherical cavitation collapse," *Journal of Fluid Mechanics,* vol. 616, pp. 63-97, 2008.

[10] S. W. Fong, E. Klaseboer, and B. C. Khoo, "Interaction of microbubbles with high intensity pulsed ultrasound," *Journal of the Acoustical Society of America,* vol. 123, pp. 1784-1793, Mar 2008.

[11] M. T. Shervani-Tabar and N. Mobadersany, "Numerical study on the hydrodynamic behavior of the dielectric fluid around an electrical discharge generated bubble in EDM," *Theoretical and Computational Fluid Dynamics,* vol. 27, pp. 701-719, Sep 2013.

[12] M. T. Shervani-Tabar and N. Mobadersany, "Numerical study of the dielectric liquid around an electrical discharge generated vapor bubble in ultrasonic assisted EDM," *Ultrasonics,* vol. 53, pp. 943-955, Jul 2013.

[13] D. Chatterjee and K. Sarkar, "A Newtonian rheological model for the interface of microbubble contrast agents," *Ultrasound in medicine & biology,* vol. 29, pp. 1749-1757, 2003.

[14] L. Scriven, "Dynamics of a fluid interface equation of motion for Newtonian surface fluids," *Chemical Engineering Science,* vol. 12, pp. 98-108, 1960.

[15] S. Paul, A. Katiyar, K. Sarkar, D. Chatterjee, W. T. Shi, and F. Forsberg, "Material characterization of the encapsulation of an ultrasound contrast microbubble and its subharmonic response: Strain-softening interfacial elasticity model," *Journal of the Acoustical Society of America,* vol. 127, pp. 3846-3857, Jun 2010.

[16] M. T. Shervani-Tabar, N. Mobadersany, S. M. S. Mahmoudi, and A. Rezaee-Barmi, "Velocity field and pressure distribution around a collapsing cavitation bubble during necking and splitting," *Journal of Engineering Mathematics,* vol. 71, pp. 349-366, Dec 2011.

[17] C. D. Ohl, M. Arora, R. Ikink, N. de Jong, M. Versluis, M. Delius, *et al.*, "Sonoporation from jetting cavitation bubbles," *Biophys J,* vol. 91, pp. 4285-95, Dec 1 2006.

[18] M. B. Glauert, "The wall jet," *Journal of Fluid Mechanics,* vol. 1, pp. 625-643, 1956.

© 2018 ASME

Comparative Assessment of a Barotropic Model and a Void Fraction Transport Model for Numerically Predicting Steady Sheet Cavitation

[1]Jeremy Nahon*; [1]Mehrdad Zangeneh; [2]Motohiko Nohmi; [2]Hiroyoshi Watanabe

[1]University College London,UK; [2]EBARA Corportion, Fujisawa, Japan

Abstract

The Tohoku-Ebara barotropic model for cavitating flow is evaluated against the popular Zwart transport model. Computations are run for a 2D hydrofoil cascade and aim to predict the steady state sheet cavitation. Two numerical approaches are applied: a specifically developed in-house solver, implemented with the Tohoku-Ebara State Law; and the ANSYS-Fluent commercial suite which includes the Zwart transport terms. The features of interest are those directly associated to cavitation or vital to blade design, i.e. water vapour volume fraction and blade pressure.

Keywords: hydrofoil cascade, sheet cavitation, homogeneous mixture, cavitation modelling, barotropic state law, void fraction transport

Introduction

In hydraulic machinery, cavitation is responsible for the breakdown of hydraulic power transformed by the rotating element, material erosion and vibration. It is naturally a critical factor in the operation of hydraulic machinery as it affects the range, noise level and life expectancy of the mechanical system. Being able to accurately predict the character of sheet cavitation is, therefore, an essential step in the design process.

To that end, both homogenous and non-homogeneous mixture models have been developed. The latter multiphase approach has been applied to attached cavity flows by *Chen and Heister* [1] and *Deshpande et al.* [2] but is unable to capture detachment features without a costly computational load. The more widespread homogeneous models treat the cavitating medium as a single phase of varying density. The multitude of techniques constructed to compute the density value can be split into two major categories: i. pressure-density coupling; ii. void fraction transport equation [3].

In this study, instances emerging from both classes of cavitation models, i.e. the Tohoku-Ebara State Law (pressure-density coupling) developed by *Nohmi et al.* [4] and the *Zwart-Berger-Belamri* model (void fraction transport) [5], are assessed and compared. Analysis is carried out on the most fundamental turbomachinery element: the 2D hydrofoil cascade, which constitutes an ideal starting point for enhancements towards full turbomachinery design. The scope of the work presented here is limited to the computation of steady state attached cavity flows. The features of interest are those directly associated to cavitation or vital to blade design, i.e. water vapour volume fraction and blade pressure. Of particular importance, is the behaviour at cavitation closure as it has a direct impact on the erosive aggressiveness and vibration levels.

An In-house CFD solver has been purposefully developed to incorporate the Tohoku-Ebara (TE) model. It uses the low dissipation upwind SLAU scheme for flux discretisation [6] and is augmented with preconditioning and multigrid for computational efficiency. Results using the Zwart model are obtained from the commercial ANSYS-Fluent solver which has it set as its default configuration.

Tohoku-Ebara Barotropic Model

For pressure-density coupling methods in general, the mixture density is directly related to the pressure by a barotropic law of the form $\rho = f(p)$. Usually, three states, the pure liquid water, pure water vapour and transition state, are considered separately and carry their own properties. On the other hand, the TE model takes its inspiration from the equation of state built by *Iga et al.* from Tohoku University [7] which merges the transitional and pure liquid states by combining the Tammann and Ideal Gas laws. It is further augmented by introducing the mass fraction of air Y to arrive at

$$\rho = \begin{cases} \dfrac{(p+p_c)(p-p_v)}{(1-Y)(p-p_v)K(T+T_c) - Y\rho_v K(T+T_0)R_aT + Y(p+p_c)R_aT}, & p > p_v \\ \dfrac{p}{(YR_a + (1-Y)R_v)T}, & p \leq p_v \end{cases}$$

*Corresponding Author, Jeremy Nahon: j.nahon@ucl.ac.uk

The definition for each of the model variables are given in Table 1 along with their corresponding values. The void fraction α is evaluated by comparing the density ratios such that:

$$\alpha_v = \frac{\rho_l - \rho}{\rho_l - \rho_v - \rho_a}$$

Zwart-Gerber-Belamri Transport Model

For void fraction transport models, the mixture density is not directly given by the pressure but deduced from the relation $\rho = \rho_l\alpha_l + \rho_v\alpha_v$ where ρ_l and ρ_v are constant saturation values. The evaluation of α_l is achieved by adding to the Euler or NS equations the mass transport equation of the form

$$\frac{\partial \rho_l \alpha_l}{\partial t} + \frac{\partial \rho_l \alpha_l u_j}{\partial x_j} = \dot{m}^+ + \dot{m}^-$$

The different transport models are concerned with providing a definition for the condensation term $\dot{m}^+$ and evaporation term $\dot{m}^-$. In the method proposed by Zwart et al., which is itself an update of the Kubota model [8], the nuclei density is assumed constant with bubble growth and shrink rates obtained from the simplified form of the Rayleigh-Plesset equation for single bubble dynamics [5]. The condensation and evaporation terms are given by

$$\dot{m}^- = -C_e \frac{3\alpha_{nuc}(1-\alpha_v)\rho_v}{R_B}\left[\frac{2}{3}\frac{p_v - p}{\rho_l}\right]^{1/2} \text{ if } \qquad p < p_v$$

$$\dot{m}^+ = C_c \frac{3\alpha_v\rho_v}{R_B}\left[\frac{2}{3}\frac{p - p_v}{\rho_l}\right]^{1/2} \text{ if } \qquad p > p_v$$

where α_{nuc} is the nucleation site volume fraction, R_B is the nucleation site radius, C_e and C_c are two calibration coefficients for the evaporation and condensation rates respectively. In this study, the values recommended by Zwart et al. are chosen because they have been shown to perform for a variety of cases: $\alpha_{nuc} = 5 \times 10^{-4}$, $R_B = 10^{-6}\ m$, $C_e = 50$, $C_c = 0.01$ [8].

Numerical Methods

To assess the performance of the TE model, a novel CFD solver for cavitating flow has been specifically developed. The scheme is designed to solve the steady state 2D isothermal compressible Euler equations. In conservative form and including the preconditioning matrix Γ, the governing equations are expressed as

$$\Gamma \frac{\partial}{\partial t} \iiint_V \boldsymbol{Q}\, dV + \oint_S \boldsymbol{F}\, dS = 0$$

where $\boldsymbol{Q}$, the primitive variable vector, and $\boldsymbol{F}$, the inviscid flux vector, are given by:

$$\boldsymbol{Q} = \begin{pmatrix} p \\ u \\ v \end{pmatrix} \qquad \boldsymbol{F} = \begin{pmatrix} \rho V \\ \rho V u + p n_x \\ \rho V v + p n_y \end{pmatrix} \qquad \Gamma = \begin{pmatrix} \beta & 0 & 0 \\ \beta u & \rho & 0 \\ \beta v & 0 & \rho \end{pmatrix}$$

Here, $V = u n_x + v n_y$ is the contravariant velocity taken at each control volume face with normal vector $\boldsymbol{n} = (n_x, n_y)$, and $\beta = U_r^{-2}$ is the preconditioning variable responsible for clustering the acoustic wave speeds. The reference velocity U_r is derived by combining the multidimensional eigenvalue analysis laid out by *Turkel* [9] and the conservative flux formulation proposed by *Weiss & Smith* [10]. This allows the amount of preconditioning to be weighted according to local mesh topology and offers far higher convergence and robustness gains.

Flux discretization follows the upwinded *SLAU* scheme developed by *Shima & Kitamura* [6]. It was selected for its sharp shock resolution and applicability to arbitrary equations of state. Face values are constructed using a constant piecewise approximation. To further enhance robustness and convergence rate, an adapted multigrid technique was implemented. Its specificities are mainly contained within the restriction and prolongation operators which must be able to accommodate the coexistence of hypersonic flow ($M > 10$ in water vapour regions) and low speed flow ($M \approx 0.1$ in liquid water regions), as well as the high pressure and density gradient located at cavitation closure. To that end, techniques employed for hypersonic flow problems such as upwinded residual smoothing and a shock weighting operator have been built into the solver [11] [12].

For the Zwart transport model, numerical solutions are provided by ANSYS-Fluent. Solver parameters are configured for inviscid flow and pseudo-transient time marching.

Hydrofoil Cascade

The hydrofoil profile was extracted from the geometry of a specifically designed axial flow pump impeller. The pump's features are: purposefully large impeller diameter to minimise flow variations in the radial direction; number of blades set to obtain a solidity ratio (*chord/pitch*) close to *1.5* so that $chord = 0.1358\ m$ and $pitch = 0.09053\ m$; and consistent blade profile in the radial direction. In our 2D stationary analysis, a single channel is made up of 32 and 160 cells in the tangential and meridional directions respectively in an H-type grid configuration (see Figure 1).

Boundary Conditions

At design point, the meridional component of velocity at inlet has a fixed value $u_\infty = 10\ ms^{-1}$ and flow enters the domain at an angle $\gamma = 46.321°$. The boundary conditions are set such that the velocity/mass flow condition is ensured. For the In-house TE solver, total pressure and flow angle are defined at the inlet, while mass flow rate is specified at the outlet. Because ANSYS-Fluent does not allow mass flow boundary conditions in 2D configurations, velocity module and angle are defined at inlet, and the static pressure is set at outlet. The cavitation number $\sigma = (p_\infty - p_v)/(0.5\rho_\infty|V_\infty|^2)$ is adjusted by tuning either the inlet total pressure for the In-house TE solver, or the outlet static pressure in ANSYS-Fluent.

Results and Discussion

Preliminary numerical experiments are carried out in non-cavitating conditions for both TE and Zwart models. The treatment of the pure liquid phase is significantly different: in the case of the TE State Law, which is applied for all $p > p_v$, the density is variable producing an inherently compressible fluid; for the method proposed by *Zwart et al.*, on the other hand, mixture density is constant at $\rho = \rho_l$ as $\alpha_v \to 0$. The case is run using both approaches at cavitation number $\sigma = 0.891$ ($p_\infty = 200\ kPa$). The blade C_p and density distributions presented in Figure 2 show that the variable density computed by the TE model does not significantly affect the static pressure which coincides well with the result provided by the Zwart-Fluent solver. The discrepancy in density is balanced out by differences in fluid velocity.

To assess the cavitation prediction performance, the hydrofoil cascade is run for gradually decreasing inlet pressure values. The lowest pressure corresponds to the transition point between steady sheet cavitation and the onset of vapour break-off phenomena which cannot be accurately computed without unsteady analysis. Non-dimensional pressure coefficient results are shown in Figure 3. In the solutions produced by both models, the cavitation region is characterised by the flattening of the suction side pressure curve and a sharp rise at closure. There are, however, significant differences. For instance, the gradient at closure shock is higher in the case of the TE model than the Zwart model. The cause for this is the larger counter-streamwise spread of the cavity predicted by the TE State Law. As a result, the deceleration at closure is stronger giving rise to a more pronounced static pressure gain.

More importantly, there is a clear discrepancy in the onset of cavitation: for the Zwart model results, water vapour starts to appear around $0.68 > \sigma > 0.66$, whereas with the TE model we observe cavitation at $\sigma = 0.75$. This disparity also affects cavity size, e.g. at $\sigma = 0.642$ the location of the vapour region closure is at $x/chord \approx 0.3$ for the Zwart results and at $x/chord \approx 0.4$ for the TE results.

The amount of turning ΔV_y generated by the cascade is measured by the difference in the tangential component of the mass flow averaged velocity between inlet and outlet, i.e. $\Delta V_y = |\bar{V}_y^{out} - \bar{V}_y^{in}|$. Like lift, it is used to evaluate the cascade's work output but provides advantageous insight when applied to turbomachinery design. Its value has been shown to be proportional to blade loading so, naturally, the effect of cavitation appears in the downgrading of the ΔV_y output (see Figure 4). The mechanism responsible for the breakdown is the expansion of the low velocity wake region as the cavity grows. As expected, the early onset of vapour formation predicted by the TE model is observed in the ΔV_y curve with a decrease initiated at $\sigma = 0.75$ instead of $\sigma = 0.66$ for the Zwart model.

Figures & Tables:

Coefficients	Definition	Value
P_c	TE pressure coefficient	$1.945 \times 10^9\ Pa$
P_v	Saturation vapour pressure	$2339.2\ Pa$
Y	Mass fraction of air	10^{-5}
K	Liquid constant	$472.3\ m^2s^{-2}K^{-1}$

T	Mixture temperature	$293.15\ K$
T_c	TE temperature coefficient	$3.837 \times 10^3\ K$
ρ_v	Saturation vapour density	$0.0173\ kg\ m^{-3}$
R_a	Ideal gas constant for air	$287.1\ m^2 s^{-2} K^{-1}$
R_v	Ideal gas constant for vapour	$461\ m^2 s^{-2} K^{-1}$

Table 1: Definition and values for the TE model coefficients.

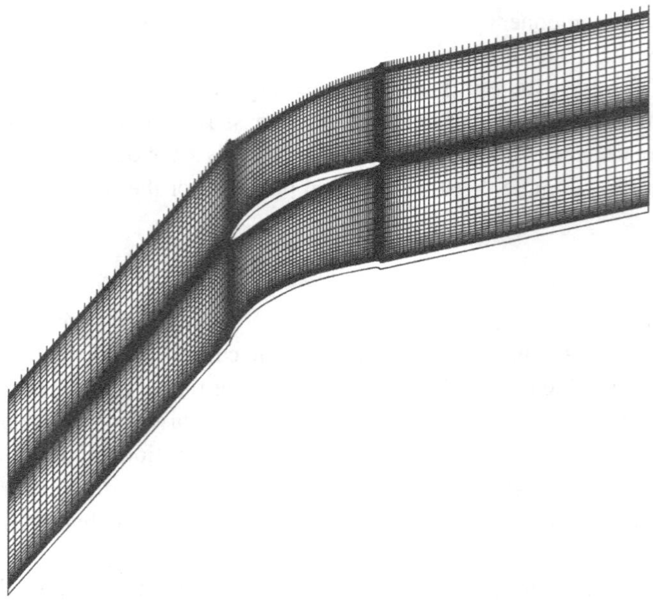

Figure 1: Hydrofoil cascade geometry; two instances of the computational domain – i.e. channel enclosed by blade surfaces - are shown here.

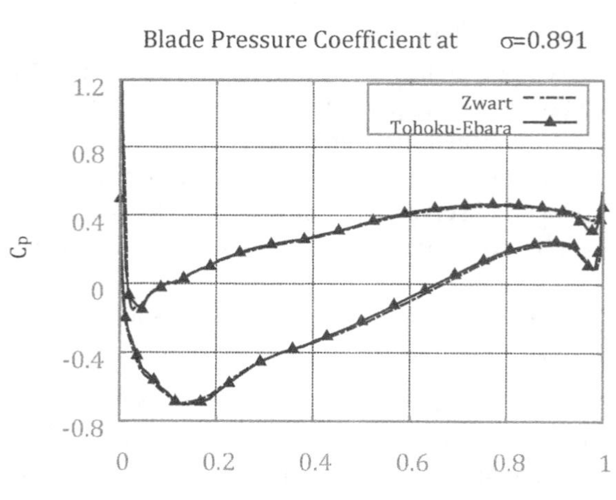

(a) Pressure Coefficient

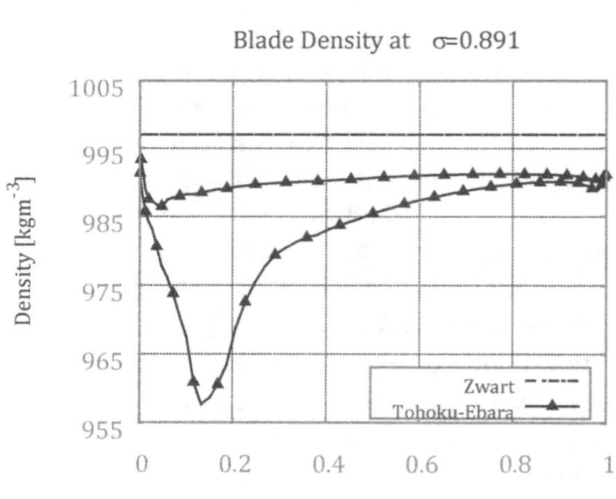

(b) Density

Figure 2: Pressure (a) and density (b) results taken at blade surfaces in non-cavitating conditions at $\sigma = 0.891$.

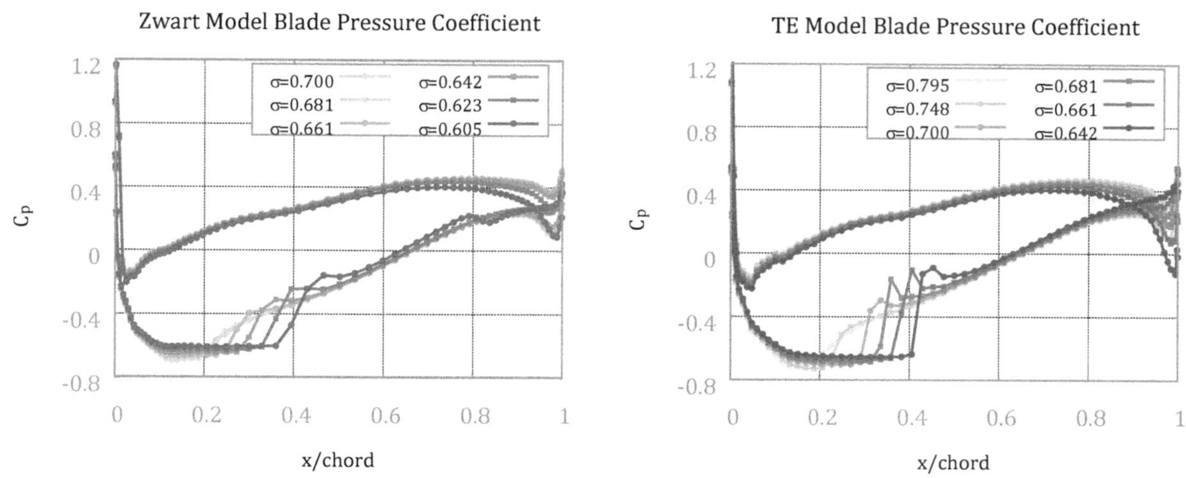

(a) Zwart Model (b) TE Model

Figure 3:Comparison of blade pressures obtained from the Zwart (a) and TE (b) models for a range of cavitation numbers.

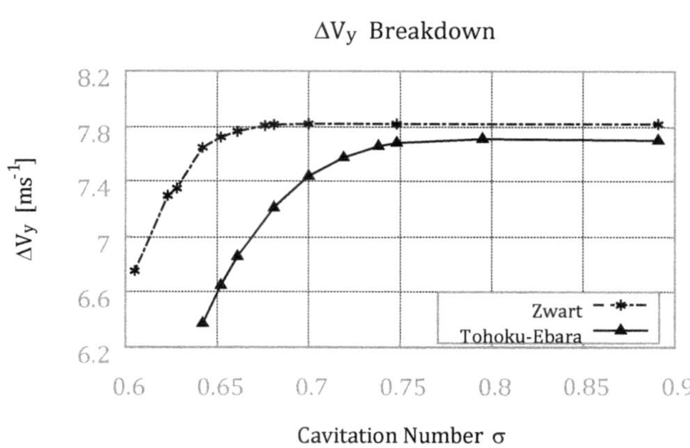

Figure 4: Comparative breakdown of the turning ΔV_y generated by the cascade in terms of the cavitation number.

Conclusion

By analysing the numerical results obtained for the hydrofoil cascade following both Zwart and TE approaches, we have been able to pick up the following discrepancies in the rendering of the sheet cavity:

- Sharper pressure rise at cavitation closure for the TE solution, caused by the wider extent of the cavity in the counter-streamwise direction which amplifies passage blockage.
- With the TE model, cavitation is initiated at higher cavitation numbers leading to an early breakdown of the cascade's work output.

Experimental assessment of the hydrofoil cascade under the same conditions is to be carried out to properly assess the accuracy of each model. Results will be presented in a follow-up paper.

References

[1] Y. Chen and S. D. Heister, "A Numerical Treatment for Attached Cavitation," *Journal of FLuids Engineering,* vol. 116, p. 613, 1994.

[2] M. Deshpande, "Numerical Modeling of the Thermodynamic Effects of Cavitation," *Journal of FLuids Engineering,* vol. 119, p. 420, 1997.

[3] M. Morgut, E. Nobile and I. Bilus, "Comparison of mass transfer models for the numerical prediction of sheet cavitation around a hydrofoil," *International Journal of Multiphase Flow,* vol. 37, no. 6, pp. 620-626, 2011.

[4] M. Nohmi, A. Goto, Y. Iga and T. Ikohagi, "Cavitation CFD in a centrifugal pump," in *Fifth International Symposium on Cavitation* , Osaka, Japan, 2003.

[5] P. J. Zwart, A. G. Gerber and T. Belamri, "A two-phase flow model for predicting cavitation dynamics," in *Fifth International Conference on Multiphase Flow*, Yokohama, Japan, 2004.

[6] E. Shima and K. Kitamura, "Parameter-free simple low-dissipation AUSM-family scheme for all speeds," *AIAA Journal,* vol. 49, pp. 1693-1709, 2011.

[7] Y. Iga, M. Nohmi, A. Goto, B. R. Shin and T. Ikohagi, "Numerical Study of Sheet Cavitation Breakoff Phenomenon on a Cascade Hydrofoil," *Journal of Fluids Engineering,* vol. 125, p. 643, 2003.

[8] A. Kubota, H. Kato and H. Yamaguchi, "A New Modelling of Cavitationg Flows: a Numerical Study of Unsteady Cavitation on a Hydrofoil Section," *Journal of Fluid Mechanics,* vol. 240, 1992.

[9] A. Ducoin, B. Huang and Y. L. Young, "Numerical modeling of unsteady cavitating flows around a stationary hydrofoil," *International Journal of |rotating Machinery,* vol. 2012, 2012.

[10] E. Turkel, "Preconditioned methods for solving the incompressible and low speed compressible equations," *Journal of computational physics,* vol. 72, no. 2, pp. 277-298, 1987.

[11] J. M. Weiss and W. A. Smith, "Precondtioning applied to variable and constant density flows," *AIAA Journal,* vol. 33, no. 11, pp. 2050-2057, 1995.

[12] F. Grasso and M. Marini, "Solutions of hypersonic vicous flows with total variation diminishing multigrid techniques," *Computers and Fluids,* vol. 24, no. 5, pp. 571-592, 1995.

[13] R. Radespiel and R. C. Swanson, "Progress with multigrid schemes for hypersonic flow problems," *Journal of Computational Physics,* vol. 116, pp. 103-122, 1995.

© 2018 ASME

Comparison of various turbulence models applied to a twisted hydrofoil

[1,2]Tingyun Yin; [2]Giorgio Pavesi, [1]Ji Pei*, [1]Shouqi Yuan, [1]Nana Adu Daniel
[1]*National Research Center of Pumps, Jiangsu University, Zhenjiang 212013, China;*
[2]*Department of Industrial Engineering, University of Padova, Padova 35131, Italy*

Abstract
In the paper, various turbulence models are compared and different modifications employed to investigate more in deep to what extent the RANS are applicable to simulate the unsteady cavitating flow around a twisted foil. The predicted vapor shedding frequency and lift force coefficient are compared with the experimental data. Moreover, the dynamics behavior induced by the cavitation is discussed in detail. The results show that frequency predicted by density correction based model (DCM) SST is much closer to the experimental data, but the model fails to predict the second shedding vapor. In general, DCM RNG shows a better prediction ability. Firstly, a vapor cloud is sheared toward downstream due to the effect of main flow and re-entrant flow. After that, the second shedding vapor catches up with the first one and becomes the main vapor. The oscillation of the lift coefficient changes consistently with the vapor fluctuations and when a second vapor shedding is highlighted, a second peak value occurs in the time history of lift coefficient.
Keywords: turbulence model; cavitation; twisted hydrofoil; unsteady shedding

1. Introduction

Cavitation, considered as a complex three-dimensional, two phases flows phenomena, always influences design, operation range and reliability of hydraulic machines[1-2]. Recent increases in available computational resources and improvements in turbulence modelling capabilities means with the correct tools and knowledge, cavitation can now be predicted reliably and accurately. Current research is concerned with cavitation simulations of Delft twisted hydrofoil using ANSYS CFX. Various turbulence models are tested and the results are compared with the experimental data obtained by Foeth et al. [3] in a cavitation tunnel. Moreover, the dynamics behavior induced by the cavitation will be discussed in detail.

2. Mathematical model

In present multiphase flow simulation approach, any physical variable is assumed to be consistent on each phase component. The mixture made up of liquid and vapor phases are regarded as a kind of one-fluid compressible medium, where no slip between the phases is considered. The density and viscosity of mixture are expressed by the vapor volume fraction α_v,

$$\mu = \mu_v \alpha_v + \mu_l \left(1-\alpha_v\right), \tag{1}$$

$$\rho = \rho_v \alpha_v + \rho_l \left(1-\alpha_l\right). \tag{2}$$

In the current study, k-ε, RNG k-ε, SST k-ω model and PANS model [4] are compared. To account the large density changes caused by cavitation, Delgosha et al. [5] have considered the compressibility of vapor-liquid mixture phases and proposed a density correction based model (DCM), which effectively reduces the mixture turbulent viscosity. However, in this method, the expression and the constants of k and ε equations are unchanged. The modified turbulent viscosity μ_t is defined as follows:

$$\mu_t = f\left(\rho_m\right) C_\mu \frac{k^2}{\varepsilon}, \tag{3}$$

$$f\left(\rho_m\right) = \rho_v + \left(\rho_m - \rho_v\right)^n \Big/ \left(\rho_l - \rho_v\right)^{n-1}, \tag{4}$$

For cavitating flow, the use of $f(\rho_m)$ decreases the turbulent viscosity in flow fields with high vapor volume fraction. However, for non-cavitating liquid flows, the formula of μ_t follows the original form. Here, n is equal to 10.

Since Johansen et al. [6] formulated a filter-based model (FBM), which combines the standard k-ε model with large eddy simulation, many researchers have applied it to cavitation simulation and extended k-ε to RNG k-ε model. In Johansen's approach, the expression and the constants of k and ε equations are unchangeable; but the formula for turbulent viscosity μ_t is computed by a filtering procedure:

$$\mu_t = Fil \rho_m C_\mu \frac{k^2}{\varepsilon}, \tag{5}$$

Corresponding author: Ji Pei. jpei@ujs.edu.cn

$$Fil = Min\left[1, C_3 \Delta \varepsilon k^{3/2}\right], \tag{6}$$

where Fil is the filter function with $C_3 = 1$. To ensure the filter process in the present study, the filter size is chosen to be not less than the largest grid scale adopted in the calculation namely, $\Delta \geq f_{grid}*\max((\Delta x \Delta y \Delta z)^{1/3})$, Δx, Δy and Δz which are the length of the grid in the three coordinate directions. f_{grid} is the constant, which is equal to 1.05. Considering compressibility of cavitating flow, a mixed filter-based model (DCM FBM) is presented, which has been discussed by Zhang et al. [7]. The modified turbulent viscosity μ_t is defined by equation (5) and in equation (6) the constant 1 is replaced by $f(n)$, which is defined as follow:

$$f(n) = \frac{\rho_v + \alpha_l^n (\rho_l - \rho_v)}{\rho_v + \alpha_l (\rho_l - \rho_v)} \tag{7}$$

Table 1 Turbulence model used in the present research.

Modified method	k-ε	RNG k-ε	SST k-ω	PANS
DCM	√	√	√	
FBM	√	√		$f_k = 0.2$ $f_\varepsilon = 1$
DCMFBM	√	√		

As listed in table 1, three modifications are applied to k-ε, RNG k-ε, SST k-ω model. The cavitation process is governed by the mass transfer equation for the conservation of the vapor volume fraction, which is defined as

$$\frac{\partial(\rho_v \alpha_v)}{\partial t} + \frac{\partial}{\partial x_j}\left(\rho_v \alpha_v u_j\right) = \dot{m} = \dot{m}^+ - \dot{m}^- \tag{8}$$

The source terms for the specific mass transfer rate corresponding to the vaporization ($\dot{m}^+$) and condensation ($\dot{m}^-$) are given by

$$\dot{m}^+ = F_{evap} \frac{3\alpha_{nuc}(1-\alpha_v)\rho_v}{R_b} \sqrt{\frac{2}{3} \frac{Max(p_v - p, 0)}{\rho_l}} \tag{9}$$

$$\dot{m}^- = F_{cond} \frac{3\alpha_v \rho_v}{R_b} \sqrt{\frac{2}{3} \frac{Max(p - p_v, 0)}{\rho_l}} \tag{10}$$

where F_{evap} and F_{cond} are empirical coefficients for the vaporization and the condensation processes with recommended values of 50 and 0.01[8]. α_{nuc} has the value of 5×10^{-4} and R_b is the radius of typical bubble size with the value of 1×10^{-6}.

3. Geometry

The models, tested on the Delft, is the Twist-11 hydrofoil shown in Fig. 1. The hydrofoil is a wing having a rectangular platform of a NACA0009 section with varying attack angles from 0° at the side section to 11° in the mid-section. The chord length of the foil is c = 0.15 m and the span length is 0.3 m. The attack angle of the entire hydrofoil was -2°. A standard right-hand coordinate system with the X-axis (red mark) in flow direction, the Y-axis (green mark) in span wise direction and the Z-axis (blue mark) directed upwards are used as reference coordinate systems.

Fig. 1. Three-dimensional twisted hydrofoil.

The computational domain is shown in Fig. 2. The hydrofoil was located in a channel with height 2c, a length of 2c upstream of the leading edge, a length of 5c downstream of the leading edge and a width of 2c. The flow simulations discretized only half of the hydrofoil due to its geometric symmetry.

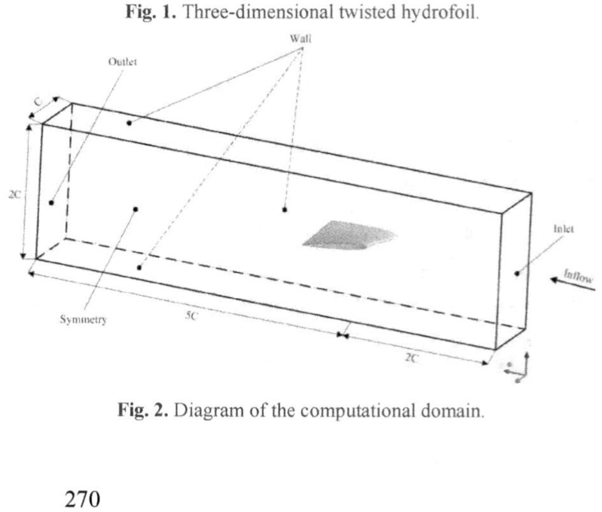

Fig. 2. Diagram of the computational domain.

4. Mesh

An O-H type grid was generated for the domain with sufficient refinement towards the foil surface, as shown in Fig. 3. To better resolve the 3D cavity structure, the mesh along the span wise direction was carefully checked in non-cavitation condition with three node numbers. The investigation was performed by monitoring the minimum and maximum pressures around hydrofoil surface and the values of the lift C_l and drag C_d coefficients.

From the results shown in Table 2, the differences can be neglected between the medium and fine resolution meshes. Thus, the medium resolution mesh with about 3.3 million nodes was selected as the final grid.

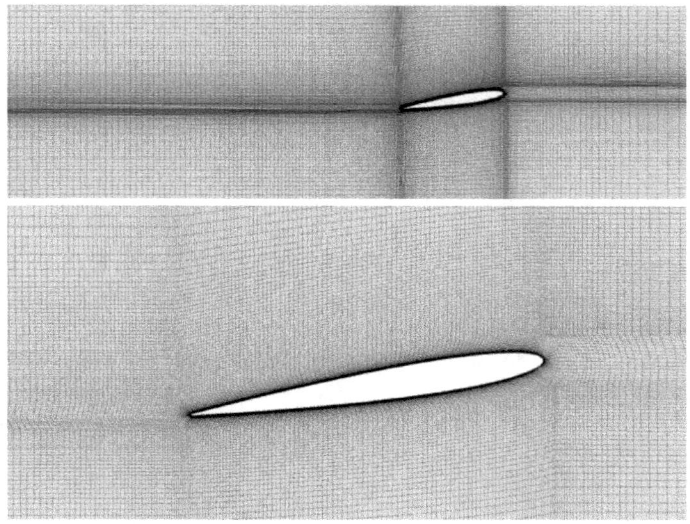

Fig. 3. Meshing in the computational domain (80nodes).

Table 2 Steady results of non-cavitation flow.

Span wise nodes	Grid number	p_{min}	p_{max}	C_l	C_d	Angle °	Determinant 2*2*2
40	1,628,400	-54,463.7	54,860.8	0.4244	0.01549	36	0.721
80	3,256,800	-54,679.1	54,847.3	0.4246	0.01546	36	0.754
120	4,885,200	-54,715.6	54,860.9	0.4246	0.01546	36	0.766

5. Boundary Conditions

The inflow velocity was set to V_{in} = 6.97 m/s. The static pressure at the outlet plane of the domain, i.e. p_{out}, was assigned according to the cavitation number 1.07. The mid-plane was a symmetry plane. The hydrofoil surface was a non-slip wall while the tunnel walls were used free slip walls. To accelerate convergence, *k-ε* model was used to simulate non-cavitating flow. 20 iterations in each time step was fixed and the iterations stopped when the RMS residual was less than 10^{-4} within each time step. The time step was chosen equal to 1×10^{-4}s through the simulation.

6. Results and discussion

6.1 Transient vapor behavior

The total vapor volume, V_{cav}, is a convenient parameter to evaluate the transient behavior of cavitating flows. The total vapor volume calculated at each time step indicates that the vapor volume variation due to the cavity shedding from the twisted hydrofoil is periodic (Fig. 4). It is found that unmodified RANS models fail to predict the unsteady cavitation characteristic. Moreover, DCM has no effect on improving the prediction capability of *k-ε* model. No significant differences are highlighted on the variation of the vapor volume among all modified models. In Fig. 5, the vapor fraction iso-surfaces α_v = 0.1 together with pressure colored foil surface are plotted and compared with the top view pictures taken by Foeth. All modified models are able to simulate the process of vapor shedding.

The sheet cavity reaches maximum length and has a convex shape at the rear in instant I. It can be found that only FBM *k-ε* (Fig. 5.I.b) and DCM FBM RNG (Fig. 5.I.d) model fail to capture the convex shape. The pressure distribution, on the hydrofoil surface highlights a strong adverse pressure gradient at the ending of the cavity that forces the re-entrant flow into the vapor

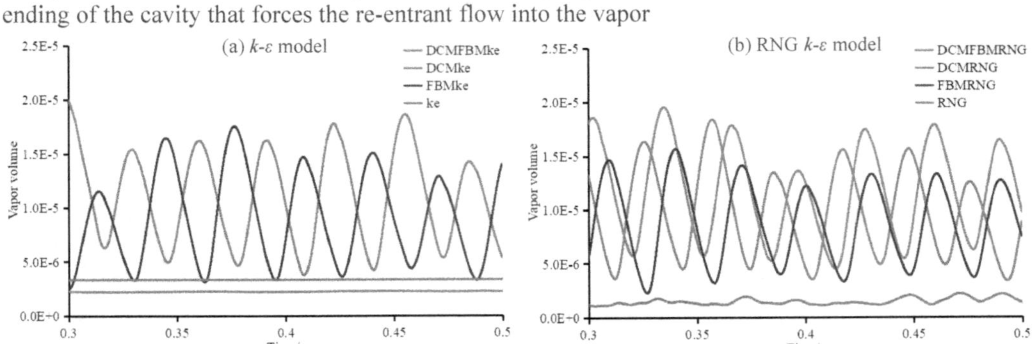

Fig. 4.a Time history of vapor volume.

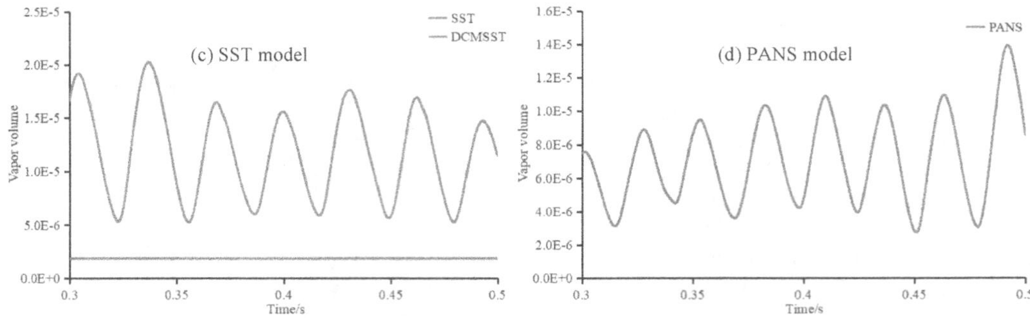

Fig. 4.b Time history of vapor volume.

region. This re-entrant flow moves upstream along the suction surface of the hydrofoil and causes the primary shedding of the cavity (Fig. 5.II). After that, the shedding cavity develops into a second-vapor structure, as shown in Fig .5.III. Not all the modified models showed a good prediction of this structure, for example, PANS (Fig. 5.III.c) and FBM RNG (Fig. 5.III.f) model.

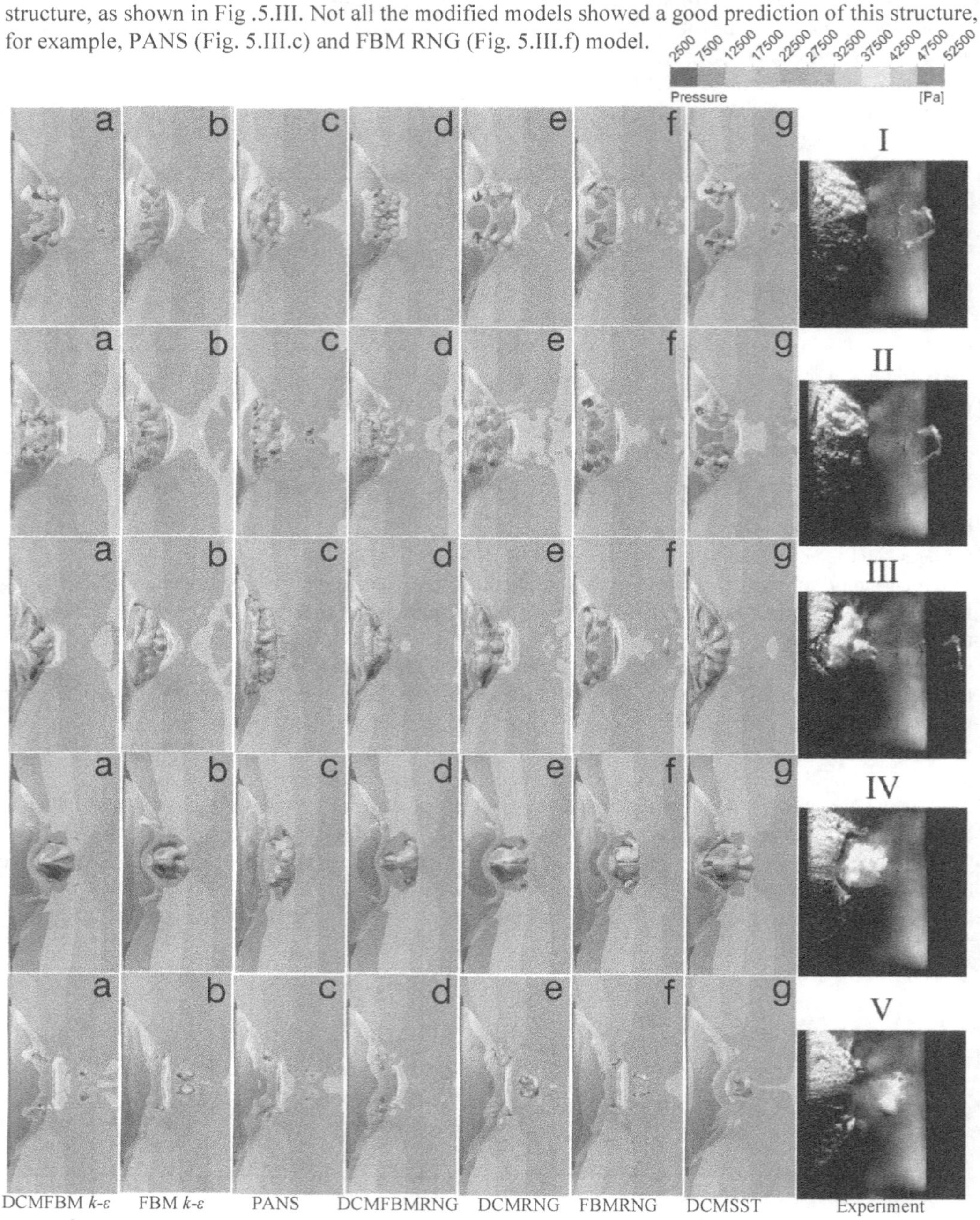

Fig. 5 Cavitation patterns during one cavity shedding cycle.

Table 3 Shedding frequency predicted by various model.

Turbulence model	Cal.	Exp.(Hz)	Error (%)
RNG *k-ε*	-	32.2	-
FBMRNG *k-ε*	33.90	32.2	5.28
DCMRNG *k-ε*	31.29	32.2	2.83
DCMFBMRNG *k-ε*	34.60	32.2	7.45
k-ε	-	32.2	-
FBM *k-ε*	33.22	32.2	3.17
DCM *k-ε*	-	32.2	-
DCMFBM *k-ε*	33.90	32.2	5.28
SST	-	32.2	-
DCMSST	32.36	32.2	0.50
PANS	34.97	32.2	8.59

Table 4 Time-averaged lift coefficients predicted by various model.

Turbulence model	Cal.	Exp.	Error (%)
RNG *k-ε*	-	0.5167	-
FBMRNG *k-ε*	0.4053	0.5167	21.56
DCMRNG *k-ε*	0.4384	0.5167	15.15
DCMFBMRNG *k-ε*	0.4090	0.5167	20.84
k-ε	-	0.5167	-
FBM *k-ε*	0.4158	0.5167	19.53
DCM *k-ε*	-	0.5167	-
DCMFBM *k-ε*	0.4180	0.5167	19.10
SST	-	0.5167	-
DCMSST	0.4375	0.5167	15.32
PANS	0.3904	0.5167	24.45

The shedding cavity quickly changes from a smooth pocket vapor into a highly turbulent vapor cloud, as shown in Fig. 5.IV. Moreover, a new sheet cavity grows from the leading edge of the hydrofoil with a concave shape of the closure line, while the shedding vapor become more turbulent and is entrained downstream by the main flow. After that, the sheet cavity grows slowly and the shedding vapor cloud quickly shrinks and finally collapses in Fig. 5.V.

Only DCM SST model fails to predict the secondary shedding structure (Fig. 5.V.g). The predicted shedding frequencies are listed in Table 3. It can be found that frequency predicted by DCM SST is closer to the experimental data, but other modified models show very good results as well.

6.2 Lift force coefficient

The predicted lift force coefficients over several cycles are show in Fig 6 and the time averaged values are listed in Table 4. Among all modified models, DCM RNG result is closer to the measurement data. The lift force coefficient evolution using DCM RNG model is compared with the cavity volume in several cycles in Fig. 7. The cavity volume reaches its maximum value contextually with the maximum lift coefficient, which is marked by pink circles. Likewise, when cavity volume drops to the minimum, force coefficient reaches the minimum values, which is marked by a red triangle. When a second vapor shedding occurs, a secondary lift coefficient peak, marked by a blue rectangle, is highlighted.

6.3 Second-vapor structure development

The comparison carried out in the sections 6.1 and 6.2 highlighted that DCM RNG model is able to predict the cavitation flow field more accurately. Consistently, the development of second-vapor structure is discussed using DCM RNG model. Many researchers (Ji et al[9], Bensow[10], Park et al[11]) have successfully captured this phenomenon, but few details were reported.

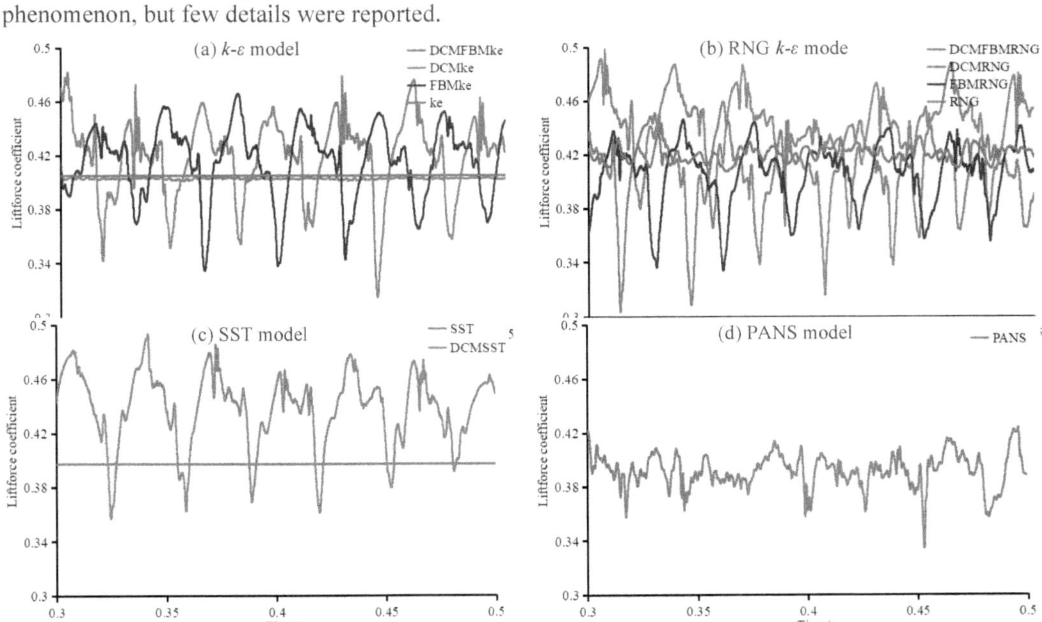

Fig. 6. Time history of lift force coefficient.

Fig 8 shows the vapor volume fraction and velocity vector in the mid-plane. A vapor cloud is sheared towards the downstream due to the effect of main flow and re-entrant flow. This vapor is called first vapor and moves along the foil surface slowly. During this process, the attached vapor in the leading edge is ready to be shed, as shown in instant 2. After the primary shedding occurs nearby the leading edge, a strong attached flow develops quickly. Together with re-entrant flow, this strong flow shears off another vapor, which is called the second vapor. The second vapor moves away from the foil surface and rotates quickly towards the downstream, as shown in instant 3. After that, the second shedding vapor catches up with the first one and becomes a main vapor. Moreover, the leading vapor begins to grow up and reproduces this process, as shown in instant 4-5.

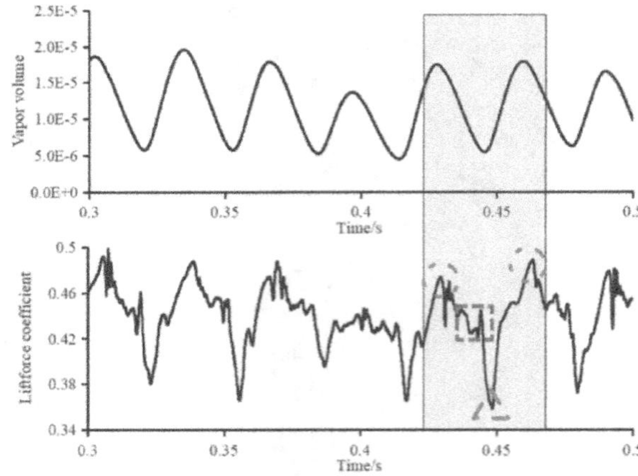

Fig. 7. The lift and drag coefficients in contrast to the cavity volume fluctuation in several cycles.

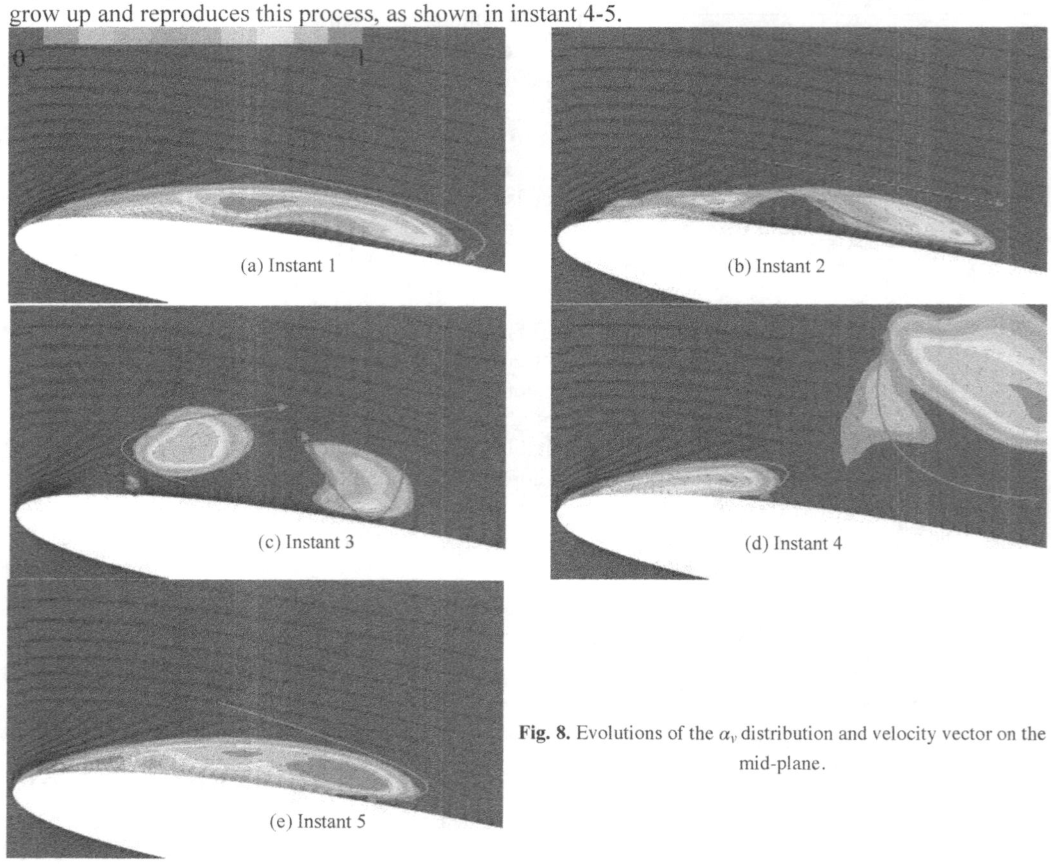

Fig. 8. Evolutions of the α_v distribution and velocity vector on the mid-plane.

7. Conclusions

Several turbulence models and different modifications were used to simulate the unsteady cavitating flow around a twisted foil in this work. The results can be summarized in the following conclusions:

1. Frequency predicted by DCM SST is much closer to the experimental data, but it fails to predict the second shedding vapor. In general, DCM RNG show a better prediction ability.
2. The oscillation of the force coefficient has some distinct features. The force changes consistently with the vapor fluctuates but a secondary peak value occurs in the time history of lift coefficient when there is a second vapor shedding.
3. A second-vapor structure was observed in the leading edge. Firstly, A vapor cloud is sheared towards the downstream due to the effect of main flow and re-entrant flow. After that, the second shedding vapor catches up with the first one and becomes a main vapor.

Acknowledgement

This study is supported by National Natural Science Foundation of China (Grant No. 51409123), Natural Science Foundation of Jiangsu Province (Grant No. BK20140554), Qing Lan Project, Training Project for Young Core Teacher of Jiangsu University. The author also gratefully acknowledges the financial support provided by the China Scholarship Council (No. 201708320237).

References

1. Kumar P, Saini R P. Study of cavitation in hydro turbines—A review, (2010), Renewable and Sustainable Energy Reviews, 2010, 14(1): 374-383. DOI: 10.1016/j.rser.2009.07.024.
2. Luo X, Bin J I, Tsujimoto Y. (2016). A review of cavitation in hydraulic machinery. Journal of Hydrodynamics, Ser. B, 2016, 28(3): 335-358. DOI: 10.1016/S1001-6058(16)60638-8.
3. Foeth E J, van Terwisga T, van Doorne C. (2008) On the collapse structure of an attached cavity on a three-dimensional hydrofoil. Journal of Fluids Engineering, 2008, 130(7): 071303. DOI:10.1115/1.2928345.
4. Ji B, Luo X, Wu Y, et al. (2013) Numerical analysis of unsteady cavitating turbulent flow and shedding horse-shoe vortex structure around a twisted hydrofoil. International Journal of Multiphase Flow, 2013, 51: 33-43. DOI: 10.1016/j.ijmultiphaseflow.2012.11.008.
5. Coutier-Delgosha O, Fortes-Patella R, Reboud J L. (2003) Evaluation of the turbulence model influence on the numerical simulations of unsteady cavitation. Journal of Fluids Engineering, 2003, 125(1): 38-45. DOI:10.1115/1.1524584.
6. Johansen S T, Wu J, Shyy W. (2004) Filter-based unsteady RANS computations. International Journal of Heat and Fluid Flow, 2004, 25(1): 10-21. DOI: 10.1016/j.ijheatfluidflow.2003.10.005.
7. Zhang D, Shi L, Zhao R, et al. (2017) Study on unsteady tip leakage vortex cavitation in an axial-flow pump using an improved filter-based model. Journal of Mechanical Science and Technology, 2017, 31(2): 659-667. DOI: 10.1007/s12206-017-0118-0.
8. Zwart P J, Gerber A G, Belamri T. (2004) A two-phase flow model for predicting cavitation dynamics. Fifth International Conference on Multiphase Flow, Yokohama, Japan. 2004, 152.
9. Ji B, Luo X, Arndt R E A, et al. (2014) Numerical simulation of three dimensional cavitation shedding dynamics with special emphasis on cavitation–vortex interaction[J]. Ocean Engineering, 2014, 87: 64-77. DOI: 10.1016/j.oceaneng.2014.05.005.
10. Bensow R E. (2011) Simulation of the unsteady cavitation on the the Delft Twist11 foil using RANS, DES and LES. Second International Symposium on Marine Propulsors, Hamburg, Germany. 2011.
11. Park S, Rhee S H. (2012) Computational analysis of unsteady cavitation on a three-dimensional twisted hydrofoil. Fiftieth AIAA Aerospace Sciences Meeting including the New Horizons Forum and Aerospace Exposition. 2012: 1100. DOI: 10.2514/6.2012-1100.

© 2018 ASME

Computational Modeling of Dynamic Planing Forces

[1]Christopher Smith *; [1]Jules Lindau; [1]Sheri Martinelli; [1]Colin Begg

[1]Applied Research Lab at Penn State University, State College, PA, USA

Abstract

Waid and Kermeen completed static experiments using cylinders planing on free liquid surfaces. Their data represents the foundation for characterization of forces imparted on a planing cylinder. Several authors have derived simplified or analytic formulations of the impact forces on a cylindrical body interacting with a curved free surface. This article presents a Computational Fluid Dynamics (CFD) based approach to modeling such dynamic forces. A curved free surface is generated in a CFD water tunnel by ventilating a gas filled wake aft of a curved deflector. The computational planing cylinder is then repeatedly plunged into the curved free surface. The CFD force record is used to reveal the dynamically added mass of the cylinder. Results indicate that the modeling approach taken accurately replicates the complex physics of the dynamic motion of the cylinder interacting with the surface. When added mass of the dynamic interaction is significant, a phase difference appears between the cylinder position and force time histories. Phase offsets between cylinder position and force have been observed across different plunging frequencies, suggesting a significant added mass effect. With the success of this approach (obtaining dynamic forces based on CFD), reduced-order models of the added mass can be created.

Keywords: planing, dynamic forces

Introduction

The dynamics of bodies interacting with free surfaces has been extensively studied in the past in the context of surface craft hulls planing on flat, or undulating surfaces [1] [2] [3] [4] (and many others). These models are for prismatic hulls in flat water or oncoming perpendicular waves, not for hull impacts or for lateral curvature of the free surface. There exists comparatively little literature on the forces imparted to a body on a laterally curved free surface. The work of Waid and Kermeen [5] represents the most extensive experimental investigation into the steady forces on a cylindrical body planing on a curved free surface, but lacks data regarding dynamic effects. Yen et al. [6] conducted experiments that were in some ways similar to the Waid and Kermeen work. In addition they developed a quasi-steady model for the planing forces. Vasin and Pareyshev [7] presented a two dimensional potential flow model for the impact forces of a circular object inside a circular ventilated wake. It would be a relatively simple task to combine Vasin and Paryshev's potential flow model with strip theory to derive an impact load. While this approach could be insightful, it is still not a fully three dimensional approach. Kulkarni and Pratap [8] and Rand et al. [9] both use an added mass estimate combined with strip theory to predict the additional loading on the body impacting a laterally curved free surface. To the authors' knowledge there exists no computational or experimental attempts specifically oriented toward assessing the dynamic component of the impact loads on a cylindrical body on a laterally curved free surface.

This paper reports on a high-fidelity Computational Fluid Dynamics (CFD) approach for assessing the forces during the impact of a cylindrical body on a curved free surface. The modeling approach is a fully three dimensional CFD simulation of the cylinder impact. This allows extraction of relevant information about the impact that is typically difficult to acquire from an experiment, such as wetted area of the cylinder. In order to better facilitate usage of the data collected from the computations, a reduced order modeling technique [10] is employed to isolate the dynamic component of impact loads. It is found that the dynamic component of the impact force can be approximated using a linear force coefficient model, but further investigation is required to understand the extent to which that model is valid.

Computational Approach

The computational fluid dynamics software, StarCCM+ [11], was employed to compute the physical quantities of interest. Numerics are second order accurate in time and space, and care was taken to ensure asymptotically convergent results. Solutions were based on a finite volume discretization with an incompressible, locally-homogeneous

*Corresponding Author, Christopher Smith, css27@arl.psu.edu

multiphase approach and the volume of fluid, HRIC scheme [12]. The resulting mean equations of motion (continuity, water volume conservation, and momentum) are cast in a Reynolds averaged form and closed with a Spalart-Allmaras Improved Delayed Detached Eddy Simulation [13] turbulence model within StarCCM+. Local convective Courant numbers were verified to fall within the bounds prescribed for best use of the HRIC scheme. An overset mesh strategy (also second order accurate) was employed to facilitate motion of the CFD test article relative to the CFD tunnel test section.

The computational domain was setup to mimic an experimental setup in a 0.3048 meter water tunnel. The two half-cylinders under test were 0.0762 meters and 0.0889 meters in diameter. Each cylinder had a 2° angle of attack relative to the incoming free stream velocity, which was perpendicular to the domain gravity vector. Figure 1 shows the computational domain used in the CFD simulations. A round deflector centered at the top dead center of the tunnel approximately 0.3 m from the leading edge of the cylinder, was used to generate a gas filled wake, bounded by a curved free surface. The cylinder was plunged into the water to the same maximum immersion depth at various frequencies. A free stream velocity of 8.53 m/s entered the domain. The average radius of the curved free surface, defined from the top dead center of the computational domain was 0.115 m.

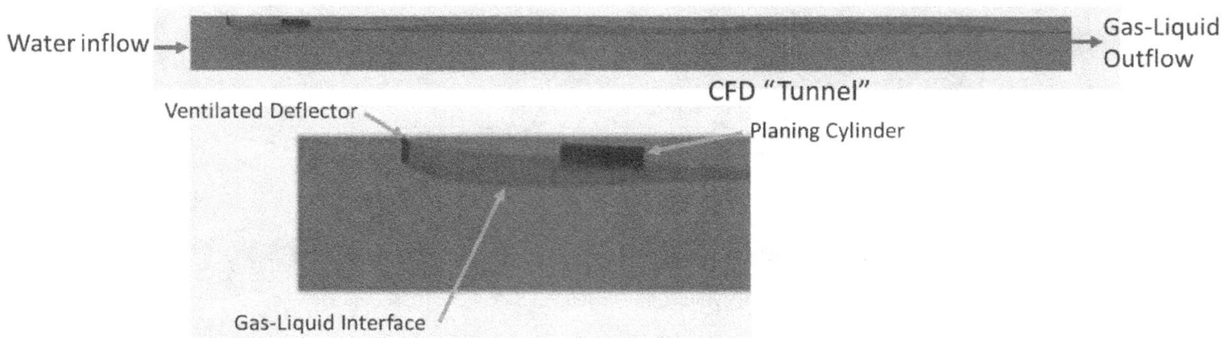

Figure 1 - Depiction of the computational domain.

Results

The data collected from each CFD simulation includes time histories of the cylinder drag, lift, cylinder vertical position, cylinder vertical velocity and the wetted area of the cylinder. Figure *2* shows the cylinder partially immersed in the curved free surface and highlights some of the terms used in the following discussions and analysis. The cylinder immersion, h, is defined as the vertical distance traveled by the cylinder after contact with the curved free surface. The immersion rate, $\dot{h}$, is time rate of change of the immersion. An approximation of the wetted length, L, of the cylinder is deduced from the wetted surface area of the cylinder using Equation 1 [14], where S is the wetted area of the cylinder, R is the cylinder radius. Equation 1 was derived from a planar slice of a cylinder, which differs from the intersection of a cylinder with a curved surface, but the error incurred by using the approximation is expected to be small. The lift and drag directions are shown in Figure *2*; the lift direction is parallel to the positive y-axis of the domain, while the drag direction is parallel to the positive x-axis of the domain. The cylinder coordinate system is also shown in Figure *2*. The cylinder normal force, Z, and cylinder axial force, X, can be derived from the lift and drag time histories via Equation 2.

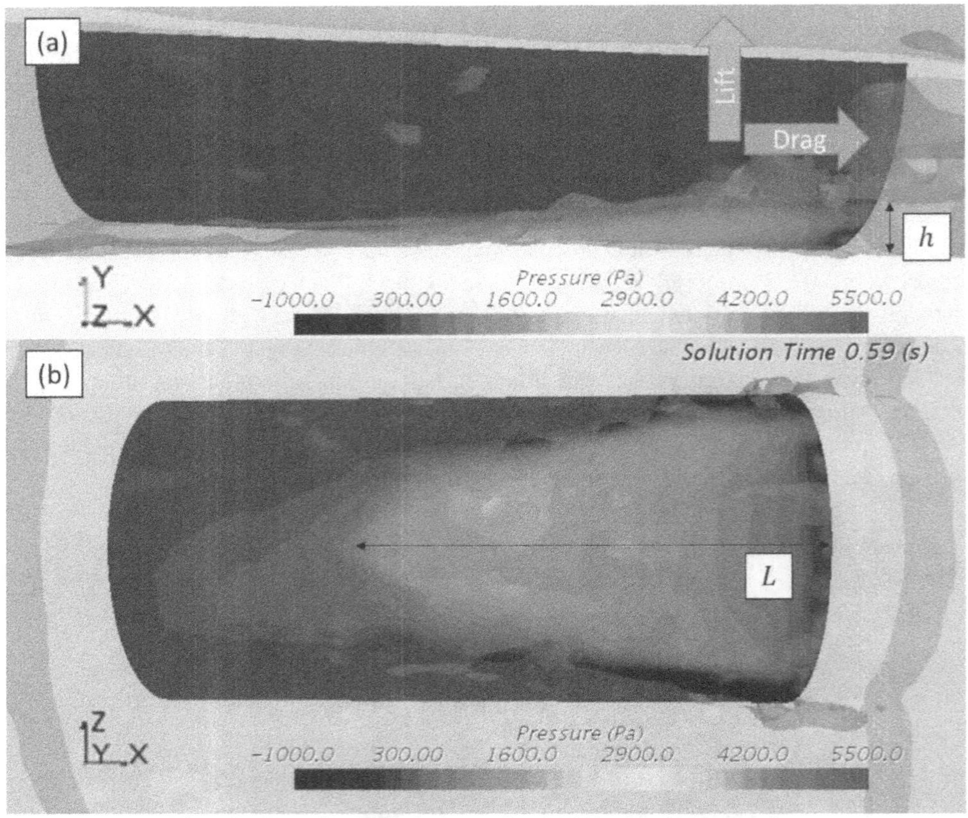

Figure 2 - Example of cylinder immersion. (a) Side view (b) Bottom view

$$L = \frac{Sh}{2R\left[\sqrt{2hR - h^2} + (h - R)\left(\frac{1}{2}\pi \tan^{-1}\frac{h - R}{\sqrt{2hR - h^2}}\right)\right]} \tag{1}$$

$$\begin{aligned} X &= -D\ \cos 2° + L \sin 2° \\ Z &= -L\ \cos 2° - D \sin 2° \end{aligned} \tag{2}$$

$$\begin{aligned} X' &= \frac{X}{\frac{1}{2}\rho V_\infty^2 R^2} \\ Z' &= \frac{Z}{\frac{1}{2}\rho V_\infty^2 R^2} \end{aligned} \tag{3}$$

The vertical and axial forces on the cylinder are non-dimensionalized using Equation 3. The time histories of non-dimensional cylinder normal force and immersion for several vertical oscillation frequencies are shown in Figure 3 (a)-(d) In Figure 3 (a)-(d) a negative immersion means that the cylinder is above the water surface. The dynamic nature of the impact forces is apparent from the phase offset between the force time history and the immersion time history, highlighted in Figure 3 (d).

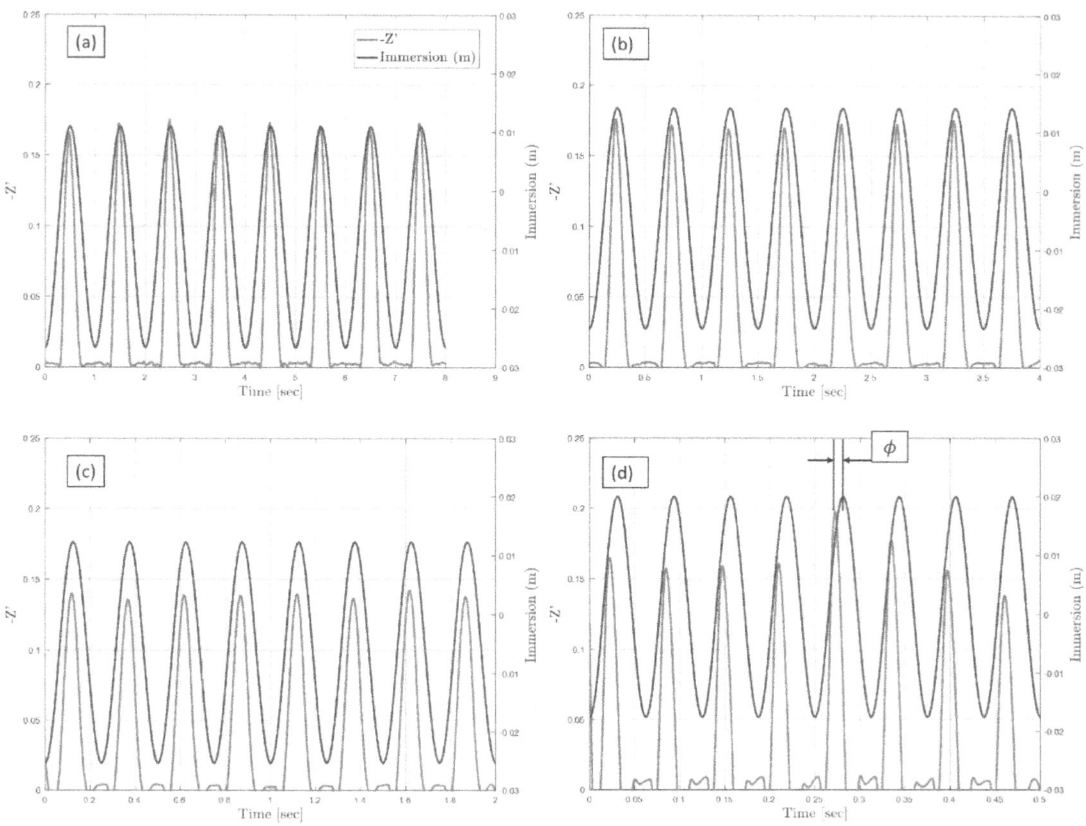

Figure 3 - Time histories of the non-dimensional vertical force on the cylinder and the immersion. (a) 1 Hz (b) 2 Hz (c) 4 Hz (d) 16 Hz.

To further demonstrate the presence of a dynamic component of the planing forces during an impact, the methodology of Yen et al. [6] was employed and compared to the force time histories from the computations. Figure 4 shows that there is a distinct difference in the force magnitudes, along with a phase difference between the quasi-steady approach implied using Yen et al.'s model and the unsteady computations. In Figure 4 the computational lift data was non-dimensionalized according to Yen et al. so as to provide a direct comparison. Also, the large excursions of the lift coefficient calculated using the model of Yen et al. are due to vanishingly small wetted lengths on the cylinder as it is enters and emerges from the water. A favorable comparison of the dynamic computational data as the cylinder is exiting the free surface and the quasi-steady approach suggests that the computational model is correctly predicting the forces present on the cylinder during the course of the impact events.

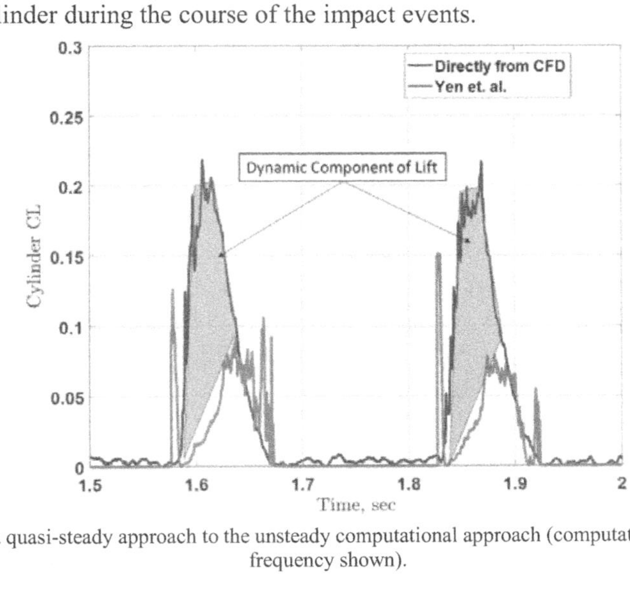

Figure 4 - Comparison against a quasi-steady approach to the unsteady computational approach (computational data from the 4Hz oscillation frequency shown).

Analysis

A well-accepted method for modeling the apparent mass of an accelerating, fully submerged, body is provided by Goodman [10]. Following Goodman, the immersion and immersion rate time histories are given by Equations 4 and 5, in which h_o is the amplitude of the immersion half-sine wave and ω is the oscillation frequency in radians per second. The vertical time history of the force can be reduced to components that are in-phase and out-of-phase with the immersion time history, as shown in Equation 6, here we use the subscripts *"in"* and *"out"* to denote the in-phase and out-of-phase components, respectively. The in-phase and out-of-phase components of the force time history then can be derived from the force and immersion time histories calculated by the CFD via Equations 7 and 8, in which Z_o is the amplitude of the force half-sine wave and ϕ is the phase difference between the immersion and force time histories in radians. Thus, the part of the force time history that is out-of-phase with the immersion time history represents the dynamic portion of the cylinder loading, $Z_{\dot{h}}$. It is proposed to model this dynamic loading as a linear coefficient derived from the change in Z_{out} with respect to the reduced frequency of oscillation given by Equation 9, where $\dot{h}_o' = \frac{h_o}{\omega}\frac{\omega R}{V_\infty}$. Using this type of model the dynamic portion of the impact loading could be given by $Z_{dynamic} = Z_{\dot{h}}' \, \dot{h} \, \frac{1}{2}\rho V_\infty^2 R^2$.

$$h = h_o \sin \omega t \tag{4}$$

$$\dot{h} = h_o \omega \cos \omega t \tag{5}$$

$$Z = Z_{in} \sin \omega t + Z_{out} \cos \omega t \tag{6}$$

$$Z_{in} = Z_o \cos \phi \tag{7}$$

$$Z_{out} = -Z_o \sin \phi \tag{8}$$

$$-\frac{\partial Z_{out}}{\partial \dot{h}_o'} = Z_{\dot{h}} \tag{9}$$

Figure 5 shows the in-phase and out-of-phase components of the force time history as a function of the non-dimensional reduced frequency. A linear, zero-intercept, fit has been applied to the out-of-phase data for both cylinder diameters. The non-dimensional immersion rate coefficient for the 0.0762 meter diameter cylinder is $Z_{\dot{h}}' = -0.58989$ and for the 0.0899 meter diameter cylinder $Z_{\dot{h}}' = -0.2963$, as provided in the figure. The negative sign of the coefficient indicates that the force due to the impact opposes the motion, which is physically correct. The data suggests that a linear approximation would provide a reasonable prediction of the additional loading imparted to the cylinder from the impact with the water. The data also implies that the assumption of a zero-intercept for the linear coefficient is sensible. The results for the 0.0889 m diameter cylinder show that the dynamic impact planing forces are dependent upon the ratio of the body radius to the radius of the curved wake. However, additional simulations must be executed in order to provide additional data at lower reduced frequency, to explore more fully the linear assumption of the model, and more deeply explore the dependency of the immersion rate coefficient on other variables such as the cylinder-to-wake radius ratio, Froude number and cylinder angle of attack relative to the gas-filled wake centerline.

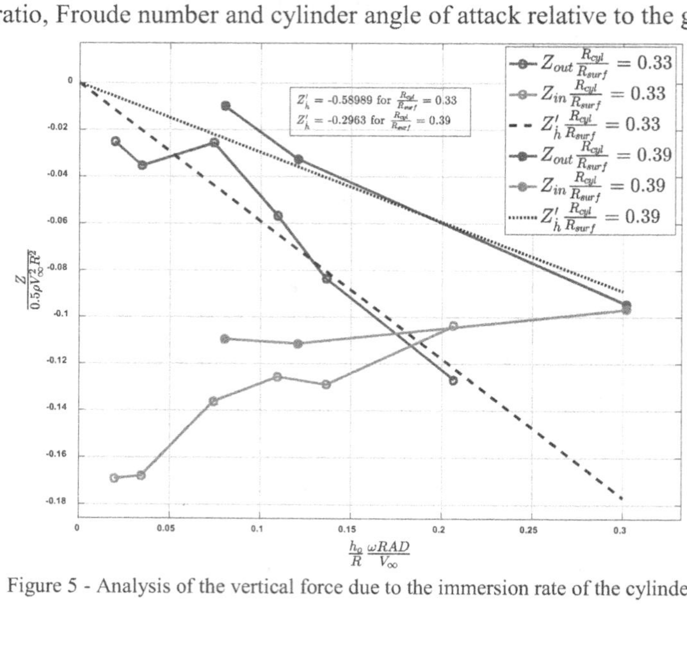

Figure 5 - Analysis of the vertical force due to the immersion rate of the cylinder.

Conclusion

A fully three dimensional CFD approach has been employed to assess the additional loading on a cylindrical body impacting a curved free surface due to the dynamic nature of the interaction. A reduced order model has been proposed. Future work includes performing additional CFD simulations to fully explore the validity of the proposed reduced order model, to explore other parameters which could affect the magnitude of the immersion rate coefficient, and collection of experimental data for validation purposes.

Acknowledgement

This material is based upon research supported by the U.S. Office of Naval Research under award number N00014-16-1-2222.

References

[1] R. Algarin and O. Tascon, "Analysis of Dynamic Stability of Planing Craft on the Vertical Plane," *Ship Science and Technology,* vol. 8, no. 15, pp. 35-43, 2014.

[2] D. L. Blount and L. T. Codega, "Dynamic Stability of Planing Boats," *Marine Technology,* vol. 29, no. 1, pp. 4-12, 1992.

[3] J. D. Hicks, A. W. Troesch and C. Jiang, "Simulation and Nolinear Dynamics Analysis of Planing Hulls," *Journal of Offshore Mechanics and Arctic Engineering,* vol. 117, pp. 38-45, 1995.

[4] E. E. Zarnick, "A Nonlinear Mathematical Model of Motions of a Planing Boat in Regular Waves," David W. Taylor Naval Ship Research and Development Center, Bethesda, MD, 1978.

[5] R. L. Waid and R. W. Kermeen, "Forces on Cylinders Planing on Flat and Curved Surfaces in Cavitating and Noncavitating Flow," Hydrodynamics Laboratory, California Institue of Technology, Pasadena, California, 1957.

[6] T. Yen, M. Morabito, L. Imas, J. Dzielski and R. Datla, "Investigation of Cylinder Planing on a Flat Free Surface," in *11th International Conference on Fast Sea Transportation*, Honolulu, Hawaii, 2011.

[7] A. D. Vasin and E. V. Paryshev, "Immersion of a Cylinder in a Fluid Through a Cylindrical Free Surface," *Fluid Dynamics,* vol. 36, no. 2, pp. 169-177, 2001.

[8] S. S. Kulkarni and R. Pratap, "Studies on the Dynamics of a Supercavitating Projectile," *Applied Mathematical Modeling,* vol. 24, pp. 113-129, 1999.

[9] R. Rand, R. Pratap, D. Ramani, J. Cipolla and I. Kirschner, "Impact Dynamics of a Supercavitating Underwater Projectile," in *1997 ASME Design Engineering Technical Conference*, Sacramento, CA, 1997.

[10] A. Goodman, "Experimental Techniques and Methods of Analysis Used in Submerged Body Research," in *Third Symposium on Naval Hydrodynamics*, Scheveningen, Netherlands, 1960.

[11] CD-adapco, *STAR-CCM+ v. 12.06 User Guide, Siemens PLM Software,* Granite Park One 5800 Granite Parkway Suite 600, Plano, TX 75024: Siemens PLM Software, 2017.

[12] I. Demirdzic, S. Muzaferija, M. Peric, H. Schreck and V. Seidl, "Computation of Flows with Free Surfaces," *Scientific Computing in Chemical Engineering II,* pp. 360-367, 1999.

[13] M. S. Gritskevich, A. V. Garbaruck and F. R. Menter, "A Comprehensive Study of Improved Delayed Detached Eddy Simulation with Wall Functions," *Flow, Turbulence and Combustion,* vol. 98, no. 2, pp. 461-479, 2017.

[14] E. W. Weinstein, ""Cylindrical Wedge." From MathWorld - A Wolfram Web Resource," 2017. [Online]. Available: http://mathworld.wolfram.com/CylindricalWedge.html. [Accessed 2017].

© 2018 ASME

Coupling a Numerical Optimization Technique with a Panel Method or a Vortex Lattice Method to Design Cavitating Propellers in Non-Uniform Inflows

Weikang Du* and Spyros Kinnas

Ocean Engineering Group, Department of Civil, Architecture and Environmental Engineering, University of Texas at Austin, Texas, TX 78712, USA

Abstract

In this study, a nonlinear optimization method, which is coupled with either a panel method or a vortex lattice method, is used to design open propellers in uniform, circumferentially averaged or non-uniform inflow. A B-spline geometry with 4×4 control points is used to ensure that the propeller blade is accurately defined with fewer parameters. The optimization objective is to maximize the efficiency of the propeller while satisfying the given propeller thrust, and different cavity area or pressure constraints are applied. The influence of those constraints are studied, and propeller geometries are designed in different cases. The optimal efficiency as a function of the thrust coefficients are compared with those from other references, and the optimal circulations from this method are compared with those predicted from the lifting line optimization theory. It is shown that this method satisfies the optimization objectives and can be used in the practice of designing cavitating propellers.

Keyword: optimization; panel method; vortex lattice method; cavitating propeller design

Introduction

The preliminary design of marine propellers has been gradually developing in the past. Generally, the methods used can be categorized into two types [Kerwin and Hadler, 2010]. The first one involves using systematic series of propellers, whose open water characteristics are already known, either from experimental tests or from computational fluid dynamic simulations [Yeo et al., 2014]. A widely known example is by using the Wageningen B-screw series [van Lammeren et al., 1969]. The second type of methods do not require and are not limited to the known propeller series, but are based on the optimal radial distribution of circulation. These methods need the circumferentially averaged inflow, and are more adapted to the nominal wake field downstream of the ship hull.

An alternative approach can be seen as an improvement of the second type of methods. Mishima and Kinnas [1997] developed a numerical optimization method which is coupled with a Vortex Lattice Method (VLM) to design the cavitating propellers in the non-uniform inflow. A major advantage of this method is that it considers the circumferential variance of the inflow, which is important in predicting the periodic pressure fluctuations and the vibrations on the ship hull caused by the propeller in unsteady non-axisymmetric inflows. This method uses a B-spline to represent the blade geometry, and the optimization objective and constraints (thrust, torque, cavitation area, etc.) are approximated via second-order Taylor expansions. This method allows a constrained amount of cavitation in the design and thus can be applied to design propellers working under a moderate or even high loading. A numerical code called CAVOPT-3D (3-Dimensional CAVitating propeller blade OPTimizaiton) was developed and later improved by Griffin and Kinnas [1998] to include the minimum pressure constraints and a one-variable quadratic skew optimization.

In the Ocean Engineering Group of University of Texas at Austin, the prediction of flows around propellers has been studied intensively in the past decades. The two major codes are MPUF-3A, which uses the VLM, and PROPCAV (PROPeller CAVitation), which uses the Boundary Element Method (BEM, more commonly known as the panel method). Some major improvements in MPUF-3A include: the thickness loading coupling scheme by Kinnas [1992], the unsteady wake alignment scheme and the non-linear terms in pressure evaluation by He [2010], etc. In PROPCAV, Kinnas et al. [2007] coupled the panel method with a boundary layer solver XFOIL [Drela, 1989] and developed the viscous/inviscid interaction method; Tian and Kinnas [2012] introduced a pseudo-unsteady wake alignment scheme

*Corresponding Author, Weikang Du: allendu1988@utexas.edu

(full wake alignment scheme, or FWA) to improve the accuracy of the predicted propeller performance, especially at low advance ratios.

In this paper, the nonlinear optimization method is coupled with the newest version of MPUF-3A and PROPCAV to design cavitating propellers. Different constraints will be applied, and various inflows, including uniform, circumferentially averaged and non-uniform inflow will be used. The efficiency and optimal circulation from the current method will be compared with those from other references [Kerwin and Hadler, 2010] or the lifting line optimization method [Menéndez Arán and Kinnas, 2014].

Methodology

The vortex lattice method and the panel method

In the VLM, line vortices and sources are placed on the mean camber surface of the blade and the trailing wake, and the strengths of those singularities are determined by solving the potential flow which satisfies the kinematic boundary condition and dynamic boundary condition. The resultant velocity $\overrightarrow{V}$ can be written as the summation of the individual contributions as:

$$\overrightarrow{n}\cdot\overrightarrow{V}=\overrightarrow{n}\cdot\left(\overrightarrow{V_\Gamma}+\overrightarrow{V_Q}+\overrightarrow{V_I}+\overrightarrow{V_B}+\overrightarrow{V_O}\right)=0, \tag{1}$$

where $\overrightarrow{V_\Gamma}$ is the velocity induced by the line vortices on the key blade and wake, $\overrightarrow{V_Q}$ the velocity induced by the cavity sources on the key blade and wake, $\overrightarrow{V_I}$ the inflow velocity, $\overrightarrow{V_B}$ the velocity induced by the line sources, which represents the thickness of the blade, and $\overrightarrow{V_O}$ the induced velocity by the other blades and wakes.

In the panel method, the total flow velocity $\overrightarrow{q}$ is decomposed into an incoming flow $\overrightarrow{q_{in}}$ and a propeller-induced flow $\overrightarrow{U_{ind}}$ (the perturbation velocity),

$$\overrightarrow{q}=\overrightarrow{q_{in}}+\overrightarrow{U_{ind}}. \tag{2}$$

$\overrightarrow{U_{ind}}$ can be treated as the potential flow which is governed by the Laplace's equation

$$\overrightarrow{U_{ind}}=\nabla\phi,\quad \nabla^2\phi=0, \tag{3}$$

where ϕ is the perturbation potential.

The Kutta condition is required which means the velocity must remain finite

$$|\nabla\phi|<\infty \quad \text{at the trailing edge.} \tag{4}$$

By using the Green's identity, (3) can be written in the boundary integrated form:

$$2\pi\phi_p=\iint_{S_H}\left[\phi_p\frac{\partial G(p,p')}{\partial \mathbf{n}_{p'}}-G(p,p')\frac{\partial\phi_p}{\partial\mathbf{n}_{p'}}\right]\mathrm{d}S+\iint_{S_W}\Delta\phi_W\frac{\partial G(p,p')}{\partial\mathbf{n}_{p'}}\mathrm{d}S, \tag{5}$$

where S_H is the surface of the hydrofoil or propeller blade, duct, and hub, S_W the surface of the trailing wake, G the Green's function, which is defined as $1/R(p,p')$ in 3D and $2lnR(p,p')$ in 2D, and R the distance between the two points p and p'.

More details of the VLM and the panel method can be found in Lee [1979] and Lee [2002] respectively.

The B-splines propeller blade geometry

The camber of the propeller blade geometry is represented as a B-spline surface, as shown in figure 1. 13 parameters are chosen to determine the movement of each vortex at each optimization iteration to specify the improved blade geometry. The advantage of using B-spline geometry is that much fewer variables are needed in the optimization (compared with the traditional geometry definition by the distribution of chord length, skew, rake, pitch etc.), while each point on the blade surface is uniquely specified. The non-uniform B-spline surface is defined as

$$P(u,w)=[x(u,w),y(u,w),z(u,w)]=\sum_{i=0}^{N_u-1}\sum_{j=0}^{N_w-1}d_{ij}N_{i,4}(u)N_{j,4}(w), \tag{6}$$

where d_{ij} are the B-spline control points, $N_{i,4}(u)$ and $N_{j,4}(w)$ B-spline basis of order 4, u and v parameters for B-splines, and N_u and N_w are number of B-spline control points in u-direction and w-direction.

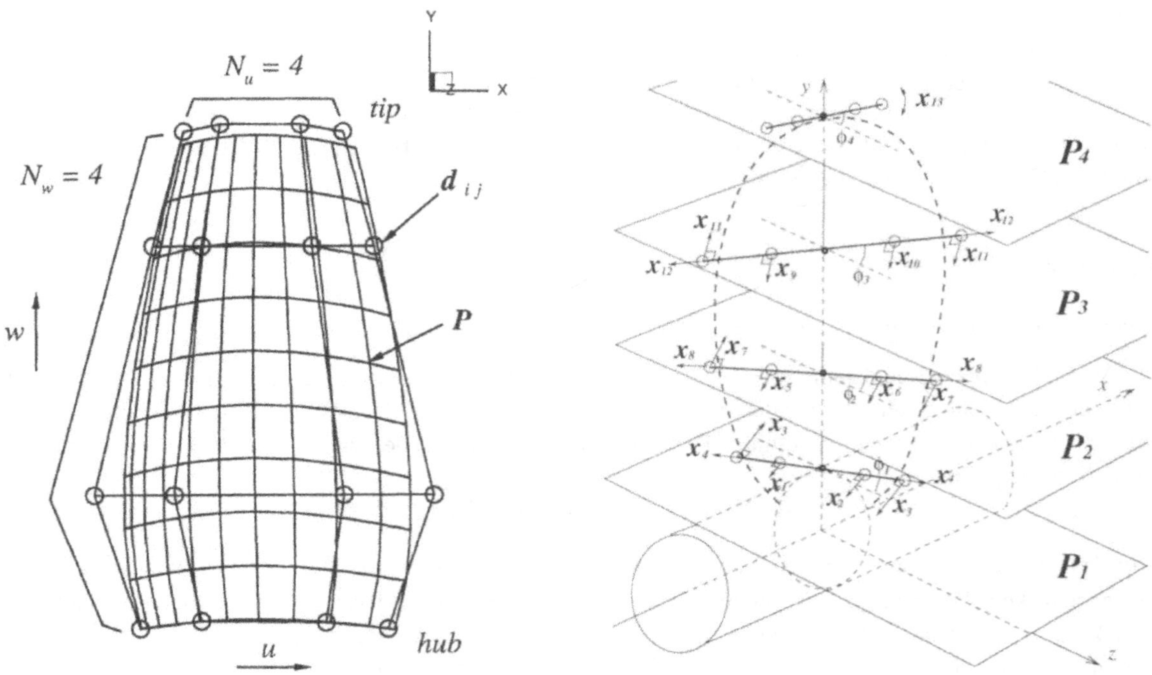

(a) cubic B-spline camber surface with 4 × 4 polygon net (b) initial B-spline vertices and the vertex movements

Figure 1: The B-spline blade geometry from Mishima and Kinnas [1997]

The numerical optimization method

The general constrained optimization problem is defined as

$$\begin{aligned} &minimize && f(x) && && (7)\\ &subject\ to && g_i(x) \leq 0 && i = 1,2,\ldots,m && (8)\\ & && h_x(x) = 0 && i = 1,2,\ldots,l && (9) \end{aligned}$$

where $f(x)$ is the optimization objective function, $g_i(x)$ the inequality constraints and $h_i(x)$ the equality constraints. In the present method, the design objective is to minimize the torque for a given thrust so that the propeller can achieve the highest efficiency. The equality constraint function is the thrust coefficient and the inequality constraints can be the maximum allowable back/face cavity area (CA), the maximum allowable blade rate cavity volume velocity harmonics, the maximum allowable cavity length at the tip and the minimum difference between the lowest pressure and the vapor pressure, shown as

$$\sigma - \text{MAX}(-Cp) \geq BTOL \tag{10}$$

where σ is the cavity number, defined as

$$\sigma = \frac{P_0 - P_v}{\frac{\rho}{2} n^2 D^2} \tag{11}$$

where P_0 is the pressure far upstream, P_v the vapor pressure, ρ the fluid density, n the propeller rotational frequency and D the propeller diameter. The pressure coefficient is defined as

$$C_p = \frac{P - P_0}{\frac{\rho}{2} n^2 D^2} \tag{12}$$

where P is the pressure on the propeller surface.

The objective function and the constraint functions are approximated as second-order Taylor expansions. More details of this method can be found in Mishima and Kinnas [1996] and Mishima and Kinnas [1997].

Results and discussion

Case 1. Uniform inflow, unconstrained

This method is first applied to design a 5-blade cavitating propeller subject to uniform inflow, without any constraints. The hub radius over the blade radius is 0.3. The target KT is 0.35, and the cavity number is 3.5. The advance ratio equals 1.0, which is defined as

$$J_s = \frac{V_s}{nD}, \tag{13}$$

where V_s is the ship speed. The Froude number, defined as

$$Fr = \frac{n^2 D}{g}, \tag{14}$$

where g is the acceleration of gravity, is taken as infinitely large. A large Froude number means that the cavitation is not influenced by the gravity.

The optimization process converges in about 500 iterations, as shown in figure 2c. Since there are no constraints on the optimization, there is some cavitation in the design, as shown in figure 2b. The propeller geometry and the optimal distribution of the pitch (P), chord length (c) and maximum camber (f), are shown in figure 2a and figure 2d.

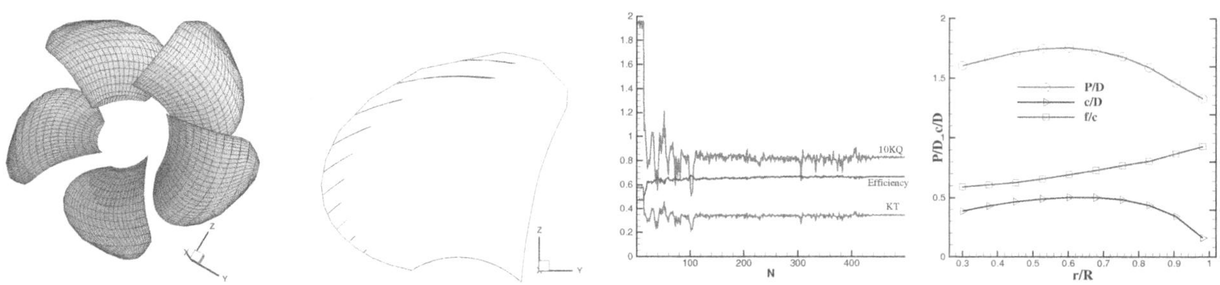

(a) blade geometry (b) cavity shape (c) convergence history (d) distribution of P, c and f

Figure 2: A 5-blade cavitating propeller is designed in uniform inflow without any constraints

Case 2. Circumferentially averaged inflow, constrained as no cavitation

In the inflow shown in figure 3a, there a low axial velocity region where blade angle equals 0 (12 o'clock), which can be considered as the influence of the ship hull boundary layer. The axial velocity is normalized by the ship speed. In this case, the inflow is circumferentially averaged, and the target KT is 0.5. The constraint is chosen so that the lowest pressure is greater or equal to the vapor pressure (i.e., setting $BTOL = 0$ in equation 10). All the other conditions are the same as in *Case 1*. These design conditions and constraint lead to a square tip propeller blade without any cavitation, as shown in figure 3b. However, some information of the non-uniform inflow is missing in averaging, especially when the variation in the circumferential direction is big. Consequently, even the optimal design satisfies the no cavitation constraint, there might be some cavitation at some blade angles in the non-uniform inflow, especially where the axial velocity is low, as shown in figure 3c and figure 3d.

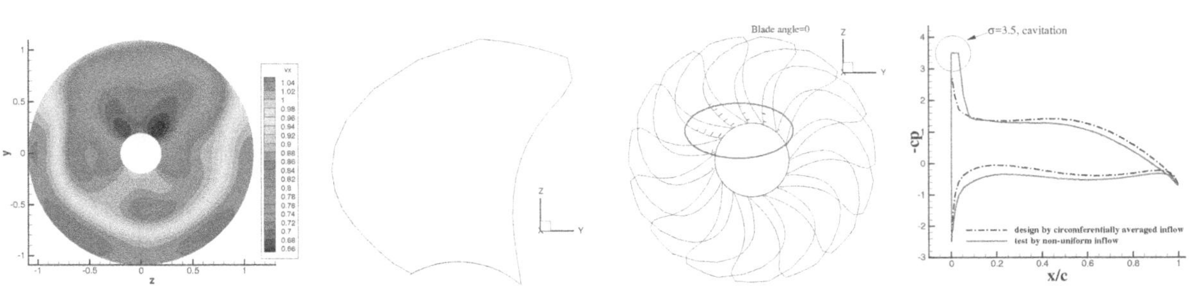

(a) non-uniform inflow (contour shows the axial velocity) (b) no cavitation in circumferentially averaged inflow (c) cavitation in non-uniform inflow where axial velocity is low (d) cavitation pressure at blade angle equals 0 near the hub

Figure 3: Propeller blade designed with no-cavitation constraint in the circumferentially averaged inflow might still show some undesirable cavitation in non-uniform inflow where the axial velocity is low.

Case 3. Non-uniform inflow with constraint on the cavity area

In this case, the cavitating propeller is designed in non-uniform inflow, which is one of the advantages of the presented method. The cavity number is 2.5, and the Froude number is 5.0. All the other design conditions are the same as in *Case 1*, except that the inflow in figure 3a is used (non-uniform, without being circumferentially averaged). The constraint is to limit the cavity area (non-dimensionalized by the blade area) as 0.2 or 0.3, as shown in figure 4a. If the maximum allowable cavity area is 0.2, the efficiency of the optimal propeller is 71.2%, while if the maximum cavity area is 0.3, the efficiency is 72.4%. Those efficiencies are close, however, the case with a smaller maximum cavity area will have wider blade near the hub, as shown in figure 4b. The pitch distributions of those two cases are similar.

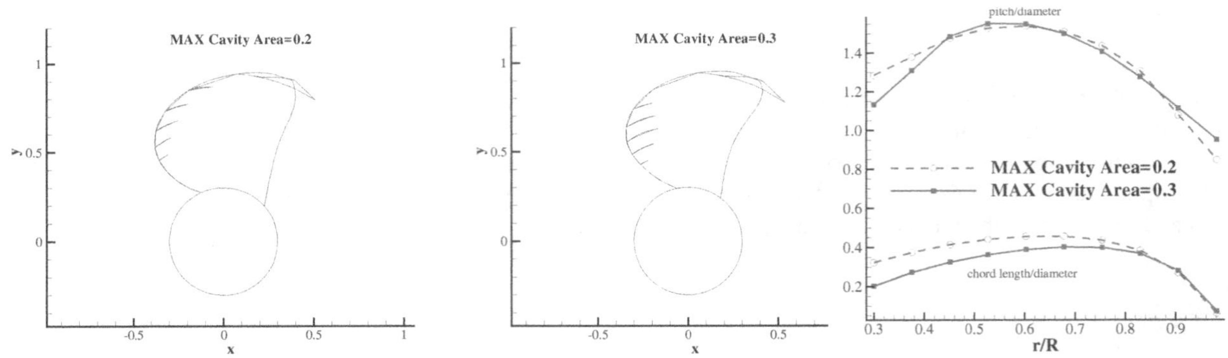

(a) different constraints on the cavity area will lead to different designs (b) distribution of pitch and chord length

Figure 4: Wider blade near the hub is needed in the case with more strict constraint on cavity area.

Optimal efficiency and circulations compared with other references or methods

For a given advance ratio, Kerwin and Hadler [2010] showed the ideal efficiency from the actuator disk model in uniform inflow, with the effect of swirl included ($J_s = 1.0$). These efficiencies are compared with those of the optimal propeller blades designed by the current method, both in the uniform inflow and non-uniform inflow, as shown in figure 5. The efficiency curve from the optimal blades are lower than those from the actuator disk model, as expected. The circulations from the optimal propeller agree well with the results from the lifting line optimization method [Menéndez Arán and Kinnas, 2014], as shown in figure 6.

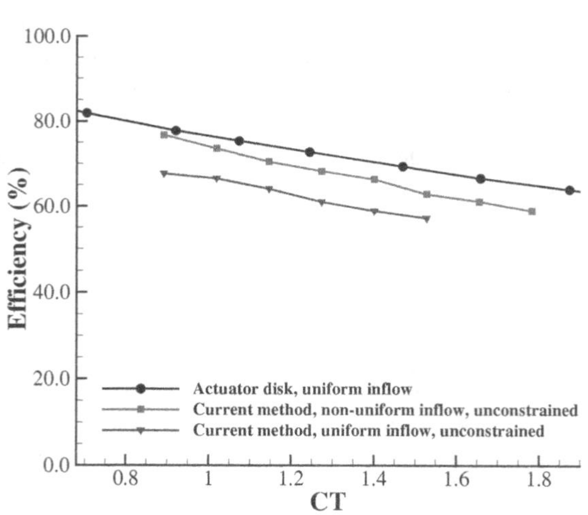

Figure 5: The efficiencies from the current method are compared with the results from the actuator disk model ($J_s = 1.0$). The thrust coefficient CT is defined as $CT = \frac{Thrust}{\frac{1}{2}\rho\pi V_s^2 R^2}$.

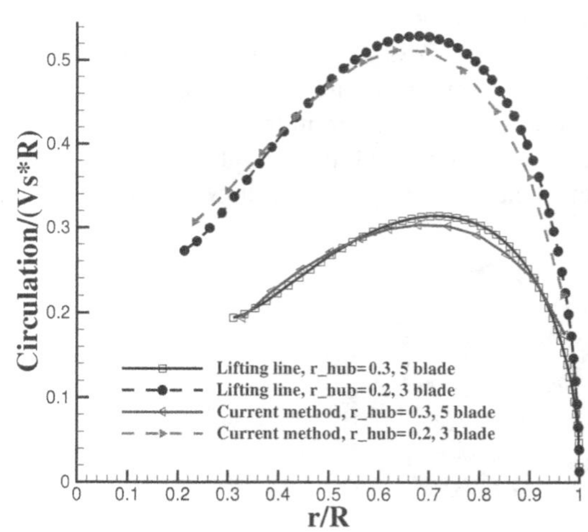

Figure 6: The optimal circulations from the current method agree well with the results from the lifting line optimization method for two cases with different hub radius and number of blades.

Conclusion

In this paper, a numerical optimization method, which is coupled with a vortex lattice method or a panel method, is applied to design cavitating propellers in uniform, circumferentially averaged or non-uniform inflows. Outputs of this method include the 3-dimensional propeller blade geometries, together with the optimal distribution of the chord length, maximum camber, pitch and skewness. Results show that the design objectives (minimize the torque for given thrust) and constraints (maximum cavity area, minimum pressure, etc.) can be satisfied. It is also shown that with a smaller allowable cavity area, the optimal propeller has a wider blade near the hub. The efficiencies as a function of the thrust are compared with the ideal value from the actuator disk model, and a reasonable trend is observed. The circulations of the optimal propeller blades agree well with those predicted from the lifting line optimization method. Finally, this numerical technique can be a useful and reliable tool in designing marine propellers with controlled amount of cavitation.

Acknowledgement

Support for this research was provided by the U.S. Office of Naval Research (Grant Number N00014-14-1-0303 and N00014-18-1-2276; Dr. Ki-Han Kim) partly by Phases VII and VIII of the Consortium on Cavitation Performance of High Speed Propulsors.

References

Mark Drela. XFOIL: An analysis and design system for low Reynolds number airfoils. In *Low Reynolds number aerodynamics*, pages 1–12. Springer, 1989.

Paul E Griffin and Spyros A Kinnas. A design method for high-speed propulsor blades. *Journal of Fluids Engineering*, 120(3):556–562, 1998.

Lei He. *Numerical simulation of unsteady rotor/stator interaction and application to propeller/rudder combination (also UT-OE Report 10-05).* PhD thesis, The University of Texas at Austin, Austin, TX, 2010.

Justin E Kerwin and Jacques B Hadler. *Principles of Naval Architecture Series: Propulsion.* The Society of Naval Architects and Marine Engineers (SNAME), 2010.

Spyros A Kinnas. A general theory for the coupling between thickness and loading for wings and propellers. *Journal of Ship Research*, 36:59–68, 1992.

Spyros A Kinnas, Hanseong Lee, Hong Sun, and Lei He. Performance prediction of single or multi-component propulsors using coupled viscous/inviscid methods. In *Proc. 10th Int. Symposium on the Practical Design of Ships and other Floating Structures, Houston, TX*, 2007.

Chang-Sup Lee. *Predicton of steady and unsteady performance of marine propellers with or without cavitation by numerical lifting-surface theory.* PhD thesis, Massachusetts Institute of Technology, 1979.

Hanseong Lee. *Modeling of unsteady wake alignment and developed tip vortex cavitation.* PhD thesis, The University of Texas at Austin, Austin, TX, 2002.

David H Menéndez Arán and Spyros A Kinnas. On fully aligned lifting line model for propellers: An assessment of betz condition. *Journal of Ship Research*, 58(3):130–145, 2014.

Shigenori Mishima and Spyros A Kinnas. A numerical optimization technique applied to the design of two-dimensional cavitating hydrofoil sections. *Journal of Ship Research*, 40:28–38, 1996.

Shigenori Mishima and Spyros A Kinnas. Application of a numerical optimization technique to the design of cavitating propellers in nonuniform flow. *Journal of Ship Research*, 41:93–107, 1997.

Ye Tian and Spyros A Kinnas. A wake model for the prediction of propeller performance at low advance ratios. *Int. J. Rotating Machinery*, 2012, 2012.

WPA van Lammeren, J D van Manen, and MWC Oosterveld. The Wageningen B-screw series. *Trans. SNAME*, 77, 1969.

Kiam B Yeo, Wai H Choong, and Wen Y Hau. Wageningen-B marine propeller performance characterization through CFD. *Journal of Applied Sciences*, 14(11):1215–1219, 2014.

© 2018 ASME

10th International Symposium on Cavitation - CAV2018
Baltimore, Maryland, USA, May 14 – 16, 2018 CAV18-05058

Details of the collapse of a cavitation bubble next to a flat, rigid wall

[1,2]Christiane Lechner*, [1]Max Koch, [1]Werner Lauterborn, [1]Robert Mettin

[1]*Georg-August-University Göttingen, Göttingen, Germany;* [2]*TU Wien, Vienna, Austria*

Abstract

A detailed numerical study of a single laser-induced cavitation bubble that is generated next to a flat, solid wall under normal ambient conditions is presented. For the numerical simulation a compressible two-phase solver of the finite volume software package OpenFOAM has been adapted for our purpose. Bubbles with an initial normalized distance γ from the wall in the range $\gamma \in [1.02, 1.5]$ are simulated. The investigation focuses on the dynamics of the liquid-on-liquid impact of the micro-jet, the splitting of the resulting torus bubble into several ring-like bubbles and their subsequent collapse. Details of the dynamics of the bubble at and after jet-impact, including the formation of an annular sheet-like "nano-jet" are presented. Pressure waves emanating from jet-impact and collapse are described. The influence of the initial distance from the wall on the above described phenomena is investigated. In particular, jet velocities and the maximum pressure at jet-impact are given for this range of normalized distances.

Keywords: bubble dynamics, microjet, pressure waves, volume of fluid method

Introduction

The interest in bubbles near boundaries stems from their destructive action on any wall material however hard (cavitation erosion), later from their use in cleaning surfaces (bubble cleaning). Nevertheless numerical studies on the behavior of bubbles near boundaries are scarce owing to the size of the problem. Large scale differences in space (mm to sub-µm) and time (ms to ns) call for large meshes with small cells and large time intervals with small time steps. It is only recently that computers and code can handle this bubble dynamics problem in compressible liquids in reasonable computer time (e.g. [1–6]). Here, a model of a laser-induced bubble near a solid wall at different normalized initial distances γ in the range 1.02 to 1.5 is studied with respect to the main axial jet formation and the thereby induced phenomenon of an annular nano-jet.

Model

A standard bubble model for a cold liquid (a liquid far from its boiling point) is used [7]. The bubble is filled with a small amount of non-condensable gas (air). The (low) vapour pressure is neglected as are thermodynamic effects, gravity and mass exchange through the interface. Compressibility of the sourrounding liquid (water) is included via the Tait equation of state to allow for pressure waves and the formation of weak shock waves. Surface tension is neglected here. Viscosity of the liquid and gas is included.

The two fluids liquid and gas are modeled as a single compressible medium with density $\rho = \alpha_l \rho_l + (1 - \alpha_l)\rho_g$, pressure p and velocity field U satisfying the continuity (1) and Navier-Stokes equations (2). The liquid volume fraction α_l is used to distinguish between liquid ($\alpha_l = 1$) and gas ($\alpha_l = 0$). α_l is evolved by means of the continuity equation of the liquid (3).

$$\frac{\partial \rho}{\partial t} + \nabla \cdot (\rho \mathrm{U}) = 0, \tag{1}$$

$$\frac{\partial (\rho \mathrm{U})}{\partial t} + \nabla \cdot (\rho \mathrm{U} \otimes \mathrm{U}) = -\nabla p + \nabla \cdot \mathbb{T}, \tag{2}$$

$$\frac{\partial (\alpha_l \rho_l)}{\partial t} + \nabla \cdot (\alpha_l \rho_l \mathrm{U}) = 0. \tag{3}$$

∇ denotes the gradient, $\nabla \cdot$ is the divergence and $\otimes$ the tensorial product. $\mathbb{T}$ is the viscous stress tensor of a Newtonian fluid:

$$\mathbb{T} := \mu \left(\nabla \mathrm{U} + (\nabla \mathrm{U})^T - \frac{2}{3}(\nabla \cdot \mathrm{U})\mathbb{I} \right), \tag{4}$$

*Corresponding Author, Christiane Lechner: christiane.lechner@tuwien.ac.at

The dynamic viscosities μ_l of the liquid and μ_g of the gas are taken to be constant ($\mu_l = 1.002 \times 10^{-3}\,\mathrm{kg/(sm)}$, $\mu_g = 1.7 \times 10^{-5}\,\mathrm{kg/(sm)}$). The equations of motion are closed by the equations of state for the gas and the liquid. For the gas in the bubble, the change of state is assumed to be adiabatic.

$$p\rho_{\mathrm{g}}^{-\gamma_{\mathrm{g}}} = \mathrm{const}\,, \tag{5}$$

with $\gamma_{\mathrm{g}} = 1.4$ (as valid for air) the ratio of the specific heats of the gas. For the liquid, the Tait equation of state for water is used.

$$p(\rho_l) = (p_\infty + B)\left(\frac{\rho_l}{\rho_\infty}\right)^{n_{\mathrm{T}}} - B\,, \tag{6}$$

with p_∞ the atmospheric pressure, ρ_∞ the equilibrium density, the Tait exponent $n_{\mathrm{T}} = 7.15$ and the Tait pressure $B = 304.6\,\mathrm{MPa}$.

The numerical implementation is based on the two-phase solver compressibleInterFoam of the software package OpenFOAM [8]. Eqs. (1) – (3) are rewritten as evolution equations for p, U and α_l and are discretized with the finite volume method (FVM). Numerical diffusion of the interface is counteracted by an interface compression term added to Eq. (3). The latter is solved explicitly in a number of sub-cycles per time step using a scheme that guarantees boundedness of α_l between 0 and 1. Eqs. (1) and (2) are solved iteratively with a PISO like algorithm. For details on the numerical implementation see the description in [3]. There also grid convergence tests and a validation of the code by comparing the bubble shape during expansion, collapse and first rebound with experimental data for a laser generated bubble with $\gamma = 1.42$ are given.

Simulations are carried out in axial symmetry, i.e. on a wedge geometry with an opening angle α taken to be $\alpha = 2^\circ$ here. The grid is Cartesian with uniform grid spacing $\Delta x = \Delta y = R_{\mathrm{max}}/500$ in a rectangular region that completely covers the bubble. This resolution has been chosen such that the finer details of the shape dynamics during jet impact are resolved. Furthermore, robust quantities of the bubble dynamics, as e.g. the maximum jet velocity, have converged at this resolution. For evaluation of the maximum pressure at jet impact in Fig. 2 below, as well as for Figs. 3, 4 the grid is refined to $\Delta x = \Delta y = R_{\mathrm{max}}/1000$. Outside the rectangular domain the grid spacing increases with a progression factor of 1.09 up to a distance of $80\,R_{\mathrm{max}}$ from the bubble center. Further layers of cells with a progression factor of 1.2 are added to damp pressure reflexions at the outer boundary.

Boundary conditions at the wall are no-slip for the velocity and zero normal gradient both for the pressure and the volume fraction α_l. At the (artificial) outer boundary of the computational domain pressure and velocity satisfy an approximate non-reflecting boundary condition with linear relaxation to the atmospheric pressure p_∞ repectively zero velocity at some distance. The volume fraction α_l has a zero normal gradient boundary condition.

The time-step is chosen such that the maximum Courant number of the flow is ≤ 0.1. During the initial expansion phase of the bubble as well as after jet impact the maximum acoustic Courant number is well below 1.

We start the evolution of the bubble shortly after formation. A spherical bubble with an initial radius of $R_{\mathrm{init}} = 20\,\mu\mathrm{m}$ is placed at various distances D_{init} from the wall, such that the dimensionless parameter $\gamma := D_{\mathrm{init}}/R_{\mathrm{max}}$ is in the interval $\gamma \in [1.02, 1.5]$. Initially the bubble is put under an internal pressure $p_i = 1.1 \times 10^4$ bar such that it would expand to a maximum radius $R_{\mathrm{max}} = 500\,\mu\mathrm{m}$ when placed in an unbounded liquid. The initial internal pressure results from adiabatic compression of the gas from an equilibrium radius of $R_{\mathrm{n,init}} = 184.1\,\mu\mathrm{m}$ to R_{init}. However, it is known that with this amount of non-condensable gas the collapse will be delayed and mitigated by the gas compression and the maximum rebound radius will be larger than is experimentally found. We therefore remove some portion of the gas during the expansion phase such that the maximum rebound radius in an unbounded liquid matches experimental data [9]. The final amount of gas corresponds to an equilibrium radius $R_{\mathrm{n,final}} = 64\,\mu\mathrm{m}$.

Results

After generation the bubble expands to a maximum equivalent radius $R_{\mathrm{max,equiv}}$ that is slightly smaller than R_{max} but approaches R_{max} for larger γ. Subsequently the bubble collapses driven by the ambient pressure of 1 bar. A pressure zone develops above the bubble on the side opposite from the solid wall which drives an axial liquid jet. The jet impacts on the opposite bubble wall. The time that elapses between maximum bubble volume and jet impact first increases with γ in the interval $[1.02, 1.1]$ and then decreases with γ. The numerical values of these times normalized by the individual collapse times are in good agreement with those in [10] (Fig. 14) obtained from an incompressible simulation for a bubble with constant vapour pressure driven by an ambient pressure of 1 bar.

Figure 1 shows the velocity of the jet tip $v_{\mathrm{jet}} := |\mathrm{v}_{\mathrm{jet}}|$ shortly before impact as a function of the stand-off parameter γ. The velocity increases with the initial distance from the wall. These values again are in good agreement with [10] (Fig. 15) for these values of γ. Figure 1 also gives the impact velocity $v_{\text{impact}} := |v_{\text{jet}} - v_{\text{lbw}}|$, with v_{lbw} the velocity of the lower bubble wall. As the jet impact occurs in a later stage of bubble collapse with increasing γ, the impact velocity also increases with γ.

The jet has a rounded tip that impacts on the opposite, curved bubble wall. Figure 1 gives the curvature radii of the jet tip and the lower bubble wall as a function of γ. The curvature radius of the jet tip increases from approximately 18 µm at $\gamma = 1.02$ to approximately 36 µm at $\gamma = 1.5$. The curvature radius of the lower bubble wall is around 800 µm at $\gamma = 1.02$ where the solid boundary has a strong influence on the lower bubble wall. It sharply drops to approx. 242 µm at $\gamma = 1.1$ and then slowly decreases to 122 µm at $\gamma = 1.5$.

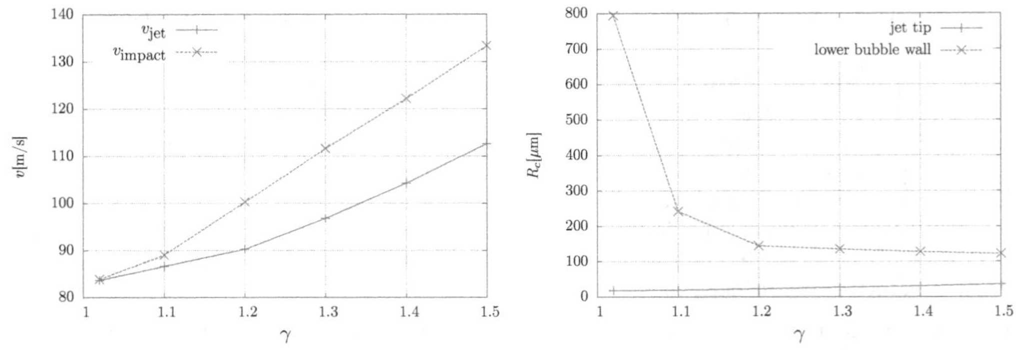

Figure 1: Left: jet velocity and impact velocity. Right: curvature radii of the jet tip and the lower bubble wall.

The maximum pressure at the axis of symmetry at jet impact is shown in Fig. 2 as function of γ. It increases with γ from approximately 200 bar at $\gamma = 1.02$ to 600 bar at $\gamma = 1.5$. Figure 2 also gives the water hammer pressure, which is an estimate for the impact pressure obtained from 1D considerations. The relation for the impact liquid-on-liquid is

$$p_{\text{wh}} = \frac{1}{2}\rho_\infty c_\infty v_{\text{impact}}. \tag{7}$$

The maximum pressure at the axis extracted from the numerical simulation is smaller than the water hammer pressure by a factor of approximately 1.7 - 3, the difference decreasing with γ. This might be mainly due to insufficient numerical resolution. The high pressure is generated in a small region $\sim R_{\text{c,jet}} v_{\text{impact}}/c_\infty$ and lasts for a very short time interval $\sim 3R_{\text{c,jet}} v_{\text{impact}}/(2c_\infty^2)$. For the jet curvature radii, $O(10\,\mu\text{m})$, and velocities, $O(100$ m/s), as determined here this corresponds to a spatial region with a radius of $O(1\,\mu\text{m})$ and a duration of $O(1$ ns). Both are not properly resolved with $\Delta x = 0.5\,\mu\text{m}$ and a time-step of $\Delta t \simeq 0.3$ ns around jet impact.

Figure 3 shows the pressure field of a bubble with $\gamma = 1.2$ from jet impact to the first collapse. The first row gives an overview for selected times, the second row presents an enlargment (for partly different times). At $t = 102.45\,\mu\text{s}$,

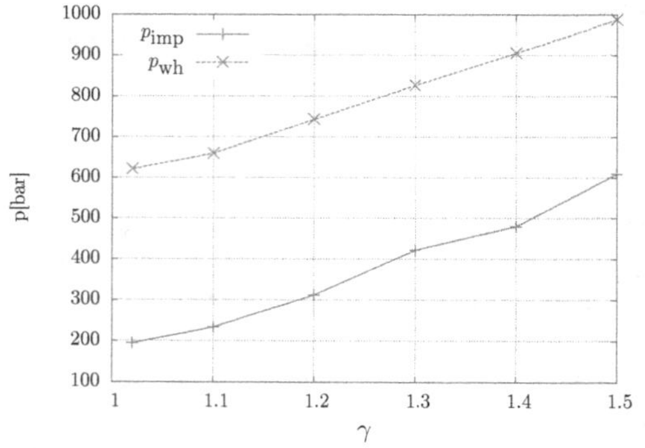

Figure 2: Maximum pressure at the axis at jet impact taken from a numerical simulation with a grid spacing $\Delta x = R_{\max}/1000$ and the water hammer pressure Eq. (7).

shortly before jet impact, the region of high pressure above the bubble is visible. At $t = 102.48\,\mu s$ the jet has impacted the opposite bubble wall changing the topology of the bubble from simply connected to torus form. The high pressure in the impact region is visible. At $t = 102.5\,\mu s$ the impact pressure wave is visible spreading in nearly spherical form. At $t = 102.6\,\mu m$ the further propagation of the pressure wave with a tension tail can be seen in the funnel of the torus bubble. On the frames from $t = 102.6\,\mu s$ to $t = 103\,\mu s$ a small annular liquid jet is visible, which is called "nano-jet" here, to distinguish it from the primary axial jet that commonly is termed "micro-jet". The nano-jet gets unstable and sheds (torus) droplets into the bubble, that impact the opposite bubble wall. $t = 103.7\,\mu s$ finally shows that the torus bubble has split into several small bubbles, that collapse individually emitting a series of shock waves.

The velocity field is shown in Figure 4. Shortly before jet impact, in the first frame, the jet with a velocity of approximately 90 m/s pushes the gas between jet and lower bubble wall into the main region of the bubble. The continuous impact of the jet pushes the gas further into the bubble, forming an expelled gas shock, visible at $t = 102.6\,\mu s$. The nano-jet and small (torus) droplets are visible on the frames $t = 102.6\,\mu s$ to $t = 103.4\,\mu s$. Several small torus bubbles are split from the main bubble subsequently in frames $t = 103\,\mu s$ to $t = 103.6\,\mu s$. For a more detailed description of the pressure and velocity field of a bubble with similar behaviour ($\gamma = 1.36$) see [5].

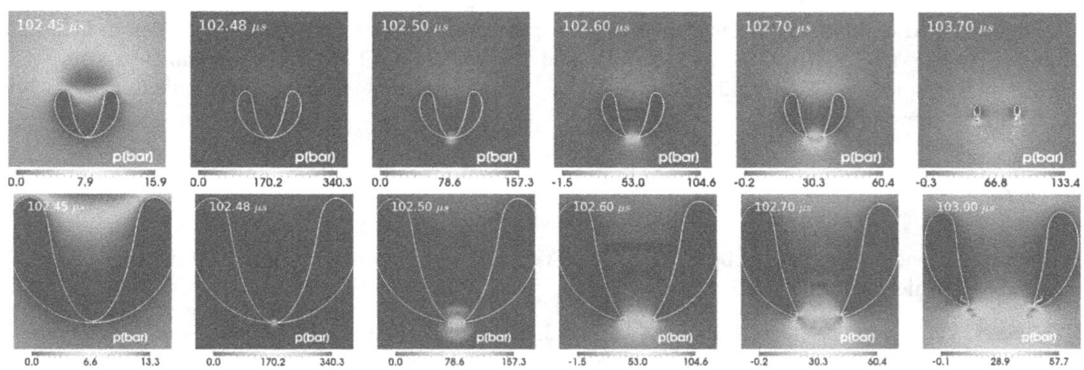

Figure 3: Pressure field of a bubble with $\gamma = 1.2$ from jet impact to the collapse of several torus bubbles. First row: frame size $800 \times 800\,\mu m$, the solid wall is located at the lower border of each frame. Second row: frame size $300 \times 300\,\mu m$

Discussion

The evolution of laser generated bubbles with a dimensionless distance of $\gamma \in [1.02, 1.5]$ from a solid wall has been modeled and simulated from expansion to the first collapse. Bubbles in this range show a similar behaviour in that the axial micro jet impacts well before the final stage of the collapse, which occurs as individual collapses of several torus bubbles. The nearby solid boundary has a pronounced influence on the lower bubble wall for $\gamma \lesssim 1.1$ which manifests itself here by the small curvature of the lower bubble wall at jet impact. The jet velocity given here for a gas filled bubble is in very good agreement with the values in [10] obtained from a boundary integral method for a bubble with constant vapour pressure driven by an ambient pressure of 1 bar. If the bubble contains a larger amount of gas, the jet velocity gets smaller [6]. For larger γ this effect is more pronounced.

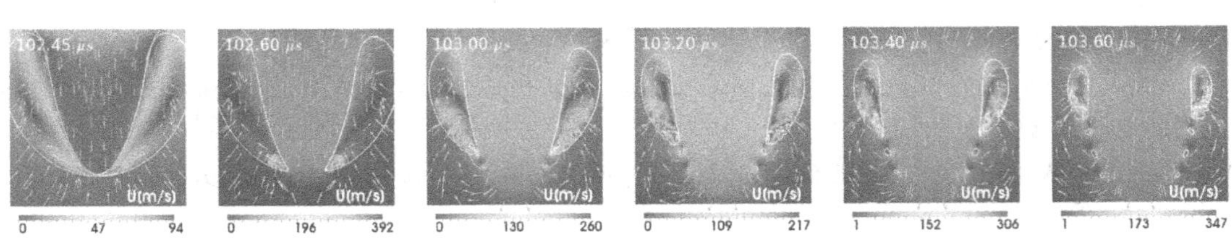

Figure 4: Velocity field of a bubble with $\gamma = 1.2$ from jet impact to first collapse. Frame size $300 \times 300\,\mu m$.

The maximum pressure at jet impact is given, which grows with the distance from the wall, corresponding to the larger impact velocity. The values obtained from the present numerical simulation are lower than the water hammer pressure. This mainly is attributed to insufficient numerical resolution (for this particular effect). Another contribution might come from the gas in the bubble which decelerates the jet before impact.

The numerical simulation reveals details on the bubble dynamics around jet impact that are not, or not easily, accessible in the experiment. The tension wave traveling upwards in the funnel of the torus bubble after jet impact supports the explanation of the counter jet observed in experiments as secondary cavitation [11]. The formation of the sheet-like annular "nano-jet" after jet impact also has been described in numerical work on shock-induced bubble collapse [12], and has been observed experimentally for larger cavitation bubbles close to a free surface [13].

Future work will include an extension of the γ-range both to values smaller than unity and to larger values as well as comparisons with experimental data.

Acknowledgements

The authors thank the Cavitation Bubble Dynamics Group at the Institute for many inspiring discussions. The work was supported in part by the Deutsche Forschungsgemeinschaft (German Science Foundation) under grant number ME 1645/8-1. C.L. thanks H. C. Kuhlmann and the Institute of Fluid Mechanics and Heat Transfer, TU Wien, for their hospitality. Computational resources on the cae cluster of Zentraler Informatikdienst, TU Wien, are gratefully acknowledged.

References

[1] Johnsen, E. and Colonius, T. (2009). *Numerical simulations of non-spherical bubble collapse.* Journal of Fluid Mechanics, 629:231 – 262.

[2] Hsiao, C.-T., Jayaprakash, A., Kapahi, A., Choi, J.-K., and Chahine, G. L. (2014). *Modelling of material pitting from cavitation bubble collapse.* Journal of Fluid Mechanics, 755:142–175.

[3] Koch, M., Lechner, C., Reuter, F., Köhler, K., Mettin, R., and Lauterborn, W. (2016). *Numerical modeling of laser generated cavitation bubbles with the finite volume and volume of fluid method, using OpenFOAM.* Computers & Fluids, 126:71 – 90.

[4] Koukouvinis, P., Gavaises, M., Supponen, O., and Farhat, M. (2016). *Numerical simulation of a collapsing bubble subject to gravity.* Physics of Fluids, 28:032110.

[5] Lechner, C., Koch, M., Lauterborn, W., and Mettin, R. (2017). *Pressure and tension waves from bubble collapse near a solid boundary: A numerical approach.* Journal of the Acoustical Society of America, 142:3649–3659.

[6] Lauterborn, W., Lechner, C., Koch, M., and Mettin, R. (2018). *Bubble models and real bubbles: Rayleigh and energy-deposit cases in a Tait-compressible liquid.* IMA Journal of Applied Mathematics. Accepted.

[7] Brennen, C. (1995). *Cavitation and Bubble Dynamics.* Cambridge University Press, New York.

[8] The OpenFOAM(R) Extend Project. http://sourceforge.net/projects/openfoam-extend/.

[9] Akhatov, I., Lindau, O., Topolnikov, A., Mettin, R., Vakhitova, N., and Lauterborn, W. (2001). *Collapse and rebound of laser-induced cavitation bubbles.* Physics of Fluids, 13:2805 – 2819.

[10] Supponen, O., Obreschkow, D., Tinguely, M., Kobel, P., Dorsaz, N., and Farhat, M. (2016). *Scaling laws for jets of single cavitation bubbles.* Journal of Fluid Mechanics, 802:263–293.

[11] Lindau, O. and Lauterborn, W. (2003). *Cinematographic observation of the collapse and rebound of a laser-produced cavitation bubble near a wall.* Journal of Fluid Mechanics, 479:327–348.

[12] Hawker, N. A. and Ventikos, Y. (2012). *Interaction of a strong shockwave with a gas bubble in a liquid medium.* Journal of Fluid Mechanics, 701:59 – 97.

[13] Supponen, O., Obreschkow, D., Kobel, P., and Farhat, M. (2015). *Detailed jet dynamics in a collapsing bubble.* Journal of Physics: Conference Series, 656:012038.

© 2018 ASME

Detection and Level Estimation of Cavitation in Hydraulic Turbines with Convolutional Neural Networks

[1]Andreas Look*; [1]Oliver Kirschner; [1]Stefan Riedelbauch; [2]Jörg Necker

[1]*University of Stuttgart, Institute of Fluid Mechanics and Hydraulic Machinery, Stuttgart, Germany*

[2]*VOITH Hydro, Heidenheim, Germany*

Abstract

In this paper a method for detecting and furthermore estimating the intensity of cavitation occurrences in hydraulic turbines is presented. The method relies on analyzing high frequency signals with a convolutional neural network (CNN). The CNN is trained in an adversarial manner in order to get more robust results. After successful training the obtained network is modified in such a way, that it is possible to obtain estimations of the intensity. For evaluation purposes a separate dataset is investigated.

Keywords: turbines; monitoring; machine learning

Introduction

Cavitation in hydraulic turbines can cause damage, and mostly occurs when turbines operate close to the limits of the operating range. Often it is not clearly known whether the machinery is currently operating under cavitation conditions or not. It would be advantageous to detect cavitation while the turbines are running, thus the operator can take actions in order to avoid it. Since visual detection of cavitation in a hydraulic power plant is not an option, acoustic event detection offers an alternative. Existing methods [1, 2, 3] rely on analyzing acoustics signals by designing statistical or handcrafted features, e.g. kurtosis. In this work a method is proposed, which does not rely on handcrafted features, i.e. the features will be automatically designed by the algorithm. Furthermore the method shows only little dependency on the sensor location. In the second part of this paper the method will be extended in order to estimate the cavitation intensity level.

Data Acquisition and Preprocessing

The radiated cavitation noise is acquired by acoustic emission sensors. The sensors used in this work have an operating range from 0.1MHz up to 1MHz. Working in such a high frequency range comes with the advantage of mostly noise free signals, i.e. no inference from bearings or other mechanical parts. The signals are divided into smaller time intervals and spectrograms are generated. Since spectrograms have high dynamics, dynamic range compression (DCR) [4] is used in order to visualize acoustic events with low energy content. DCR corresponds to the elementwise application of function 1

$$f(x) = \log(1 + C * x) \tag{1}$$

where C is a variable hyperparameter. A high value of C will emphasize more frequencies with lower energy content, though the spectrogram becomes more noisy. After DCR the spectrogram is wrapped by a filter bank, which consists of several overlapping triangular windows. This is used for dimensional reduction in order to facilitate the training process of the CNN, while keeping the most important information. Similar preprocessing is done for obtaining Mel-frequency cepstral coefficients, which are used in speech recognition applications. In the final preprocessing step the spectrograms are normalized. In the process of this work data from several different model and prototype turbines, as well as different turbine types, was collected. Furthermore every turbine was equipped with multiple sensors at diverse positions.

Detecting Cavitation

In order to detect cavitation a CNN is used. CNNs are typically used for image recognition tasks [5], but have also proven useful for acoustic event detection. A CNN basically consists of several learnable filters, followed by a multilayer perceptron. The learning process is done by minimizing a specified loss function via gradient descent.

*Andreas Look: andreas.look@ihs.uni-stuttgart.de

The network used in this work is inspired by the Visual Geometry Group (VGG)-architecture [6]. This means several alternating convolutional and max pool layers followed by a multilayer perceptron (MLP). Convolutional layers are able to augment the input picture in such a way, that specific shapes or structures become more visible. Max pool layers are interpretable as an infinitely strong prior, that a feature should be present, regardless of its position in the input data. Combining these two layers it is possible to create good and representable features. Based on these features the following MLP is capable to detect cavitation occurrences. For any machine learning algorithm and especially artificial neural networks (ANN) it is crucial to avoid overfitting. Any ANN is capable to obtain a training error of 0% even at completely randomly labeled data [7]. In order to avoid overfitting and to obtain good generalization capability dropout [8] and early stopping is used. Dropout means randomly setting a specified fraction of neuron inputs to 0, and thus to perturb the training process.
Using these overfitting techniques it is possible to obtain a CNN with a detection accuracy of 94,2%. The detection accuracy was evaluated on separate turbines, which have not been part of the training dataset, and multiple sensor positions. The test dataset had a 50/50 split between measurements with and without cavitation.
In order to further increase generalization capability an auxiliary classifier generative adversarial network (AC-GAN) [9] is built. In this network topology two networks compete against each other. The first network (generator, *G*) has the task to synthesize samples, which are ideally not distinguishable from real samples. Furthermore it is possible to control the class of the output of *G* by adding an additional input. The second network (discriminator, *D*) has the task to distinguish between real and fake samples, and also to predict their class (cavitation present or not?).

$$L_S = E[\log P(S = real|X_{real})] + E[\log P(S = fake|X_{fake})] \tag{2}$$

$$L_C = E[\log P(C = c|X_{real})] + E[\log P(C = c|X_{fake})] \tag{3}$$

$$I(X|Y) = \sum_{i,j} (x_{i,j} \log \frac{x_{i,j}}{y_{i,j}} - x_{i,j} + y_{i,j}) \tag{4}$$

The source loss L_S corresponds to the log likelihood of the correct source (real?) and L_C to the log likelihood of the correct class (cavitation?). The Kullback–Leibler divergence $I(X|Y)$ is a measurement of similarity between two different spectrograms, being commonly used in denoising tasks [10]. The generator *G* is than trained to maximize $L_C + I(X|Y) - L_S$ and *D* to maximize $L_C + L_S$.
Essentially *G* is used as a data augmentation method, in order to get more diverse training samples. After successful training of the whole network, *D* can be used as a predictor in order to detect cavitation. Using this method it is possible to further raise the accuracy to 98,2%.

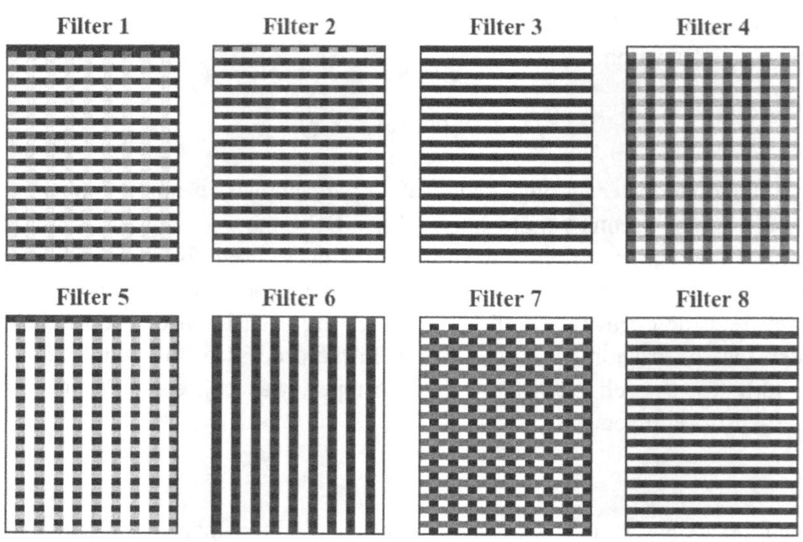

Figure 1: Maximum response inputs for convolutional filters

Figure 1 exemplarily visualizes eight learned convolutional filters in the first convolutional layer. The images have been created by modifying the input image in such a way, that it maximizes the activation of each corresponding filter. Since ANNs are fully differentiable, these images were created by the hill climbing technique. The filters maximally respond to simple geometric forms, like vertical lines, horizontal lines, or checkerboard patterns. Filter in higher layers tend to more intensively respond to inputs, with more detailed textures. Although these forms seem to be very simple, they are able to create good features, when they are stacked.

Level estimation

In order to estimate the intensity of the detected cavitation an additional network is built. Since there is less data available with reliable cavitation intensity estimation, transfer learning [5] is used for building a network for intensity estimation. In transfer learning, learned rules/features from one task shall be transferred for solving another, but still similar, problem. In this case the predictor, which was obtained for detecting cavitation, is very powerful, so the learned features principally can also be used for other cavitation related problems. For this application the convolutional part of D is used only, the following part (MLP) is removed. Instead, a new and randomly initialized MLP is used instead of the old MLP. The new MLP is practically a clone of the old MLP, but with only one output neuron (intensity). During the training process the convolutional part is not allowed to be changed, in order to avoid overfitting.

The intensity is said to be a number $\in [0,9]$, whereas 0 corresponds to no cavitation, and 9 to maximum cavitation. In order to force the network to predict numbers in the given range a stretched sigmoid function is used as the activation function for the output neuron. In order to increase generalization capacity the training data is simply augmented by making the target labels (intensity) noisy. Therefore a random number $\in U(0,1)$ is added to target labels with intensity less than five and a random number $\in U(-1,0)$ is added to target labels with intensity greater than five. Furthermore aggressive dropout (50%) is used for maximum input variance.

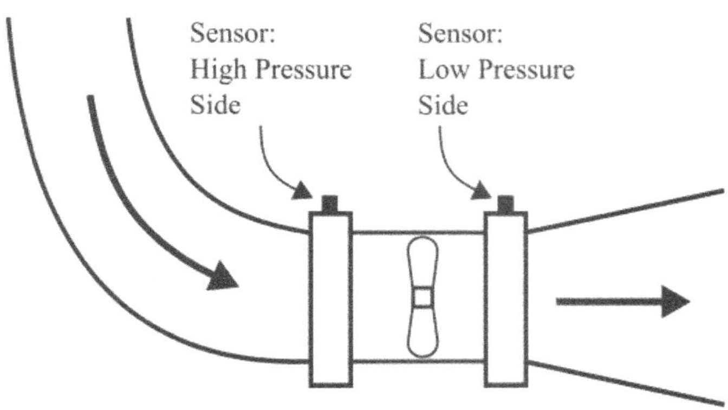

Figure 3: Principle sketch of investigated turbine

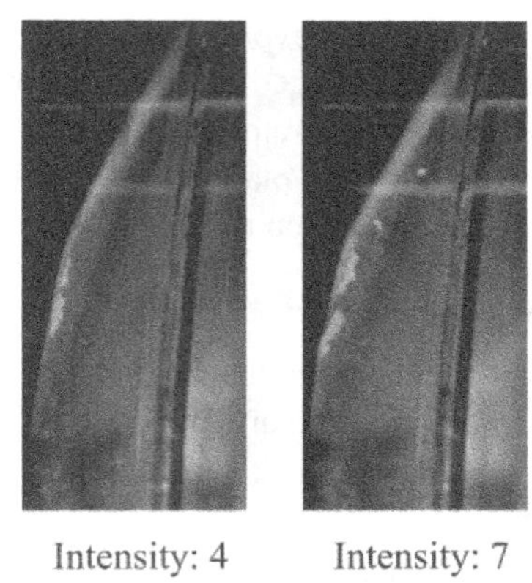

Figure 2: Intensity ranking

The algorithm for intensity estimation was evaluated on a model bulb turbine, which was not part of the training dataset. Figure 2 shows a principle sketch of the test turbine with two mounted sensors. One sensor was mounted on the high pressure side upstream the runner, the other one was mounted downstream the runner on the low pressure side. Two different measurement series have been conducted with different rotational speed. The first/second series has started with no cavitation and went on with increasing cavitation intensity on the suction/pressure side. Both series have been recorded by a high speed camera. Figure 3 shows photographs of cavitation with two different intensity levels.

Figure 4 shows the predicted and visually classified intensity. For both measurement series the RMSE is between 0.6 and 1.2 depending on the input channel and measurement. Although the signals have been normalized, the sensor near the cavitation occurrence tends to estimate higher cavitation intensity compared to the sensor, which is further away. There are different signal-to-noise-ratios, which cannot be easily adjusted to different positions by normalizing the signal. Nevertheless both sensor show, despite their different positions, the same tendency and predict intensity values in approximately the same range. The absolute values of 0 and 9 are very unlikely to be predicted, since they have never been target labels, caused by the used data augmentation technique. Therefore the predicted intensity seems to float around 0.5 when there is no cavitation present.

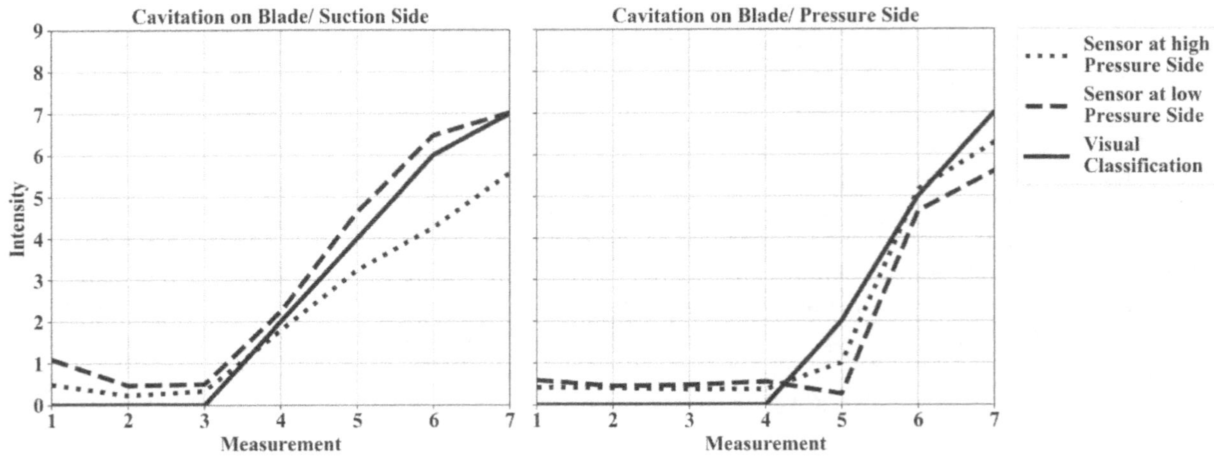

Figure 4: Estimated vs. real Intensity

Conclusion and Outlook

In this work it was shown that convolutional neural networks, which have been trained with spectrograms, are able to detect cavitation in hydraulic turbines with high accuracy. In order to increase generalization capability of the trained network, an adversarial training method has been used. Using this method it was possible to further boost the detection accuracy beyond 98%. In the second part of this work the network, which was trained for detecting cavitation, was modified in order to estimate the intensity of the cavitation. For this purpose the old network was used as a feature generator, while a new network is used for estimating the intensity based on these features.
In further work it is planned to correlate the cavitation intensity with the erosive potential. This work will be accompanied by erosion tests.

Acknowledgement

The research leading to the results presented in this paper is part of a highly appreciated common research project of the University of Stuttgart and Voith Hydro Holding GmbH.

References

[1] Schmidt, H., Kirschner, O., Riedelbauch, S. (2014). *Influence of the vibro-acoustic sensor position on cavitation.* Proceedings of the 27th. Symposium on Hydraulic Machinery and Systems.
[2] Escaler, X., Egusquiza, E., Farhat, M., Avellan, F., Coussirat, M. (2006). *Detection of cavitation in hydraulic turbines.* Mechanical systems and signal processing. 20(4)
[3] Gregg, S., Steele, J., Van Bossuyt, D. (2017). *Machine learning: A tool for predicting cavitation erosion rates on turbine runners.* Hydro Review. 36(3)
[4] Lukic, Y., Vogt, C., Durr, O., Stadelmann, T. (2016). *Speaker identification and clustering using convolutional neural networks* IEEE International Conference on Acoustic, Speech and Signal Processing.
[5] Goodfellow, I., Bengio, Y., Courville, A. (2016). *Deeplearning.* MIT Press.
[6] Simonyan, K., Zisserman, A. (2015). *Very Deep Convolutional Neural Networks for Large-Scale Image Recognition.* International Conference on Learning Representations.
[7] Zhang, C., Bengio, S., Hardt, M., Recht, B., Oriol, V. (2017). *Understanding deep learning requires rethinking generalization.* International Conference on Learning Representations.
[8] Srivastava, N., Hinton, G., Krizhevsky, A., Sutskever, I., Salakhutdinov, R. (2014). *Dropout: A Simple Way to Prevent Neural Networks from Overfitting.* Journal of Machine Learning Research
[9] Odena, A., Olah, C., Shlens, J. (2016). *Conditional image synthesis with auxiliary classifier GANs.* Arxiv Preprint
[10] Wilson, K., Raj, B., Smaragdis, P., Divakaran, A. (2008). *Speech denoising using nonnegative matrix factorization with priors.* IEEE International Conference on Acoustic, Speech and Signal Processing.

© 2018 ASME

Direct Numerical Simulations of Gas-Liquid Flows

[1]Gretar Tryggvason*; [1]Jiacai Lu; [2]Ming Ma

[1]Johns Hopkins University, Baltimore, MD, USA; [2]University of Notre Dame, Notre Dame, IN, USA

Abstract

Recent progress in direct numerical simulations (DNS) of gas-liquid flows is discussed. We start by reviewing briefly DNS of cavitating and non-cavitating flows and then address two advanced topics: How to use results for bubbly flows to help generate improved models for the large-scale or average flow, and simulations of flows undergoing massive topology changes. For the first topic we have started to experiment with the use of machine learning methods to extract complex correlations from the DNS data and for the second we are exploring how to diagnose the flow and describe its structure.

Keywords: Direct numerical simulations; turbulence; machine learning; topology changes

Introduction

Direct numerical simulations (DNS) of multiphase flows, where every continuum length and time scale is resolved for a flow spanning a large range of spatial and temporal scales, have advanced significantly over the last decade and it is now possible to simulate hundreds of bubbles in turbulent flows for a sufficiently long time so that meaningful statistical description of the flow can be obtained. Although a few DNS studies have been done for cavitating flows, most progress has been made in the context of incompressible fluids. DNS have provided significant new insight into the dynamics of such flows and the influence of bubble deformability, void fraction and surfactants, on the overall flow structure.

Cavitation can be simulated in two ways. The most complete approach is to treat it as constant temperature boiling. Significant progress has been made in simulations of boiling, although nucleate boiling still poses some challenges (see [1], for recent progress). The other approach is to simply specify the pressure inside the bubble when solving the Navier-Stokes equations. The pressure in the liquid is solved to enforce incompressibility but the pressure inside the bubbles is set to either the vapor pressure or a pressure based on the bubble volume for gas bubbles. This approach has been used to simulate the shock propagation through bubbly flows for both two- and three-dimensional flows and assuming both vapor and gas bubbles [2,3], and by [4] who compared the results for three-dimensional flow for gas bubbles with predictions by models for the average flow based on the Rayleigh-Plesset equation. The results were also used to develop an improved model. Overall the results suggest that this simplified model (assuming a constant pressure in the bubbles) works relatively well and captures the change of size and shape of bubbles as the pressure changes. A simulation of slightly more complex flow, but for two-dimensions only, can be found in [5]. This approach is a relatively straightforward extension of simulations of flows with incompressible bubbles and it should, in particular, be possible to simulate cavitation bubbles in turbulent flows. For other examples of simulations of the collapse of cavitation bubbles in inviscid compressible flows see [6], which describes what was, at the time, the largest fluid dynamics simulation ever done in terms of number of grid points used to resolve the flow field.

Studies of incompressible bubbles in turbulent flows include a demonstration of the importance of deformability for bubble induced drag reduction in turbulent flows [6], and several investigations of turbulent bubbly flows in vertical channels (see [8,9], for example). Although most of those simulations are still at modest Reynolds numbers, a few investigators have started to explore how the data can be used to help improve two-fluid models for the average flow [10,11]. Others have initiated work on using DNS data to help develop large eddy simulation (LES) models for multiphase flows [12-14].

*Corresponding Author, Gretar Tryggvason: gtryggv1@jhu.edu

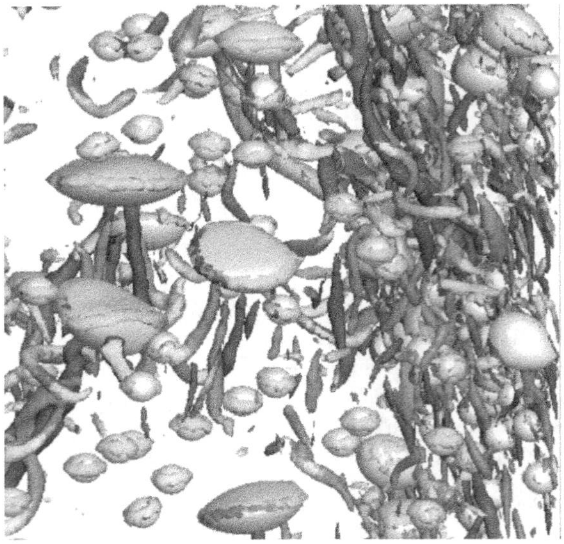

Figure 1. Details of the flow (bubbles and vorticity) at one time from a simulation of hundreds of bubbles in a turbulent channel flow.

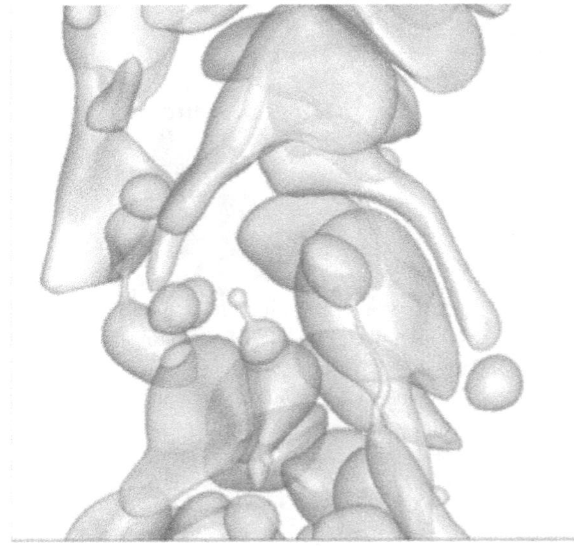

Figure 2. The interface between gas and liquid at one time from a simulation of two-fluid flow undergoing coalescence and breakup.

The most significant outcome of DNS studies is, of course, the enormous amount of data that such simulations produce and the question of how to use this data to assist with the generation of reduced order models of one sort or another is emerging as a major challenge and we discuss briefly how machine learning can be used to extract closure terms for reduced order or averaged models of two-phase flows. The second challenge is that many gas-liquid flows are very complex and the phase boundary forms a dynamic structure that cannot be described as bubbles and drops. We have recently started to examine such flows and how they can be characterized.

Numerical Approaches

Most simulations of multiphase flows rely on the so called one-fluid or one-field formulation of the governing equations, where one set of equations is solved for the whole flow field, using a stationary grid. The different fluids or phases are identified by an index function and the various material properties are set based of the index function. Forces and other effects concentrated on the interface are included using delta functions smoothed onto the fixed grid. This allows the use of flow solvers similar to those used for single-phase flow, although they must accommodate sharp changes in the material properties and this can pose additional challenges. Nevertheless, most approaches to simulating multiphase flows use similar solvers for the flow field and the difference between the various methods is how the index function is advected and surface tension is included. Popular methods to advect the index function include the volume-of-fluid and the level set methods, and their many extensions, where the index function is advected directly. In our studies we use a front tracking method where the interface between the different fluids/phases is marked by connected marker points that are advected by the flow. The index function is constructed from the location of the marker points and the marker points are also used to find the surface force. This hybrid approach, using Lagrangian markers to advect the interface and a stationary Eularian grid to solve the governing equations, has been found to be both versatile and accurate. For more details, as well as discussion of the various methods for DNS of multiphase flows, see [15, 16].

Results

Figure 1 shows a part of the flow field at one time from a simulation of a few hundred clean bubbles rising in a vertical turbulent channel flow. The bubbles and the vorticity, visualized using the lambda-2 method are shown and the color indicates the direction of vorticity, with red and blue indicating the streamwise direction but rotations in the opposite directions. The results, described in more detail in [17] show that small bubbles migrate toward the walls but the larger deformable ones stay in the center of the channel. In this simulation the bubbles are not allowed

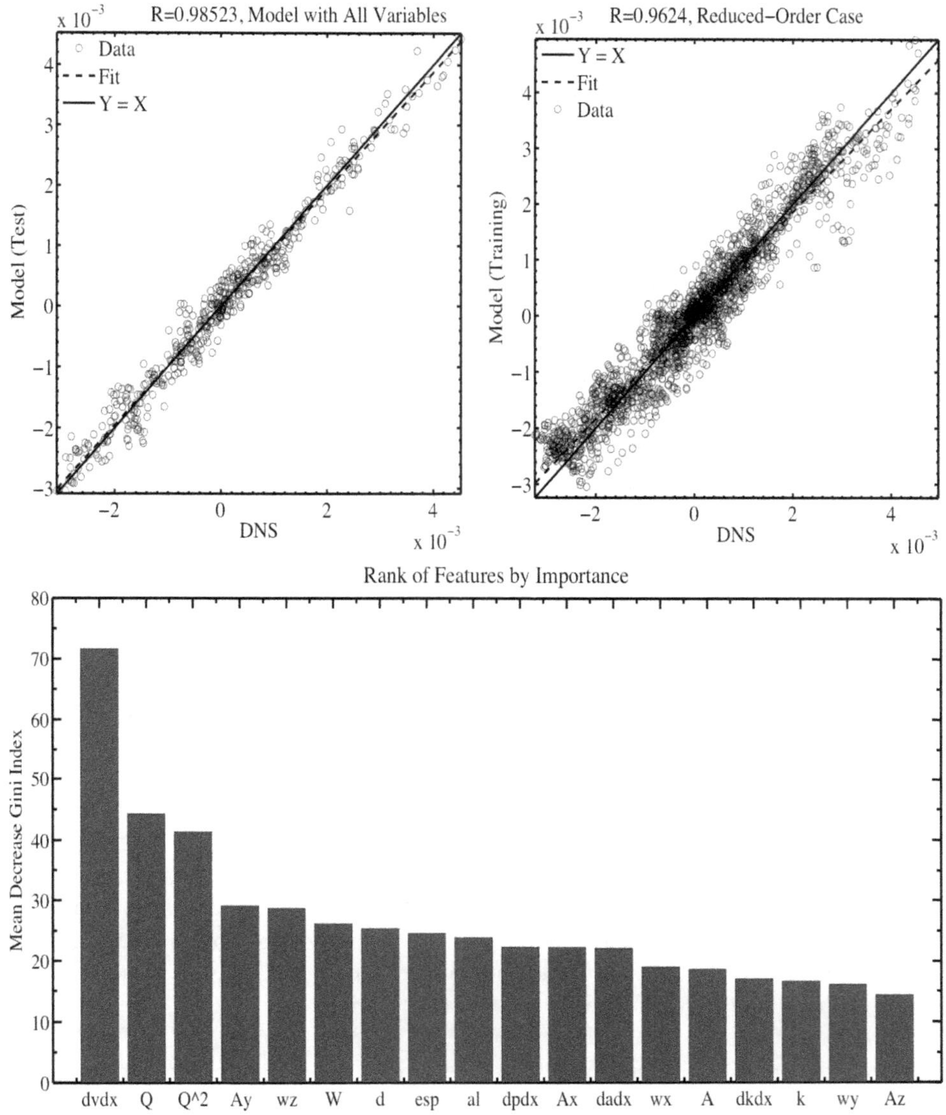

Figure 3. The correlation between the horizontal gas fluxes and various variables describing the average state of the flow is shown in the top frames. On the left the data contains 18 variables and on the right it contains 6 variables. In the bottom frame the Gini index shows what variables are most important.

to coalesce or break apart. In figure 2 we show one frame from a simulation of flows undergoing massive topology changes. Here, several deformable bubbles (high Weber number) are placed in a turbulent channel flow at a sufficiently high void fraction so that the bubbles collide and the liquid film between them becomes very thin. This film is ruptured at a predetermined thickness and the bubbles are allowed to coalesce. For low Weber numbers the bubbles continue to coalesce, eventually forming one large bubble. At high Weber numbers, on the other hand, the large bubbles break up again, sometimes undergoing repeated coalescence and breakup. For the parameters used for the simulation in figure 2, the statistically steady state consists mostly of bubbles of roughly two sizes. The evolution of the various integral quantities, such as the average flow rate, wall-shear, and interface area are monitored and compared for different governing parameters, and the microstructure at statistically steady state is quantified using low order probability functions. This simulation is described in more detail in [18] where the various ways to characterize the flow are explored. Describing the complex interface topology in statistical ways that convey the average structure and range of scales, poses considerable challenges and it is not clear at the present time what the optimal description is.

The results shown in figures 1 and 2, as well of data from many other simulations, some of which has been made available to other researchers [11], allows us to collect essentially any average and statistical quantity that we desire. This allows us to explore how closure terms, for example, depend on the various average quantities. As the flow becomes complex, finding the best correlations becomes complex and we have been exploring how the data can be analyzed in advanced ways. We have, in particular, used neural networks to find correlations between closure terms in simple models of the average flow and resolved quantities for simple laminar flows with nearly spherical bubbles [19,20]. Most recently, we have also started to use statistical learning to help determine what quantities best describe the unresolved fields in turbulent bubbly flows. Those include both variables commonly used for the modeling of turbulent flows, such as the kinetic energy and dissipation rate, as well as quantities describing the interfacial microstructure such as area concentration and various area projections. The goal is to find the variables that best correlate the closure terms and which would therefore be natural candidates for modeling. Figure 3 shows an example of such an analysis for nearly spherical bubbles in turbulent upflow in a vertical channel. The top frames show how well the horizontal gas flux correlates with the velocity gradient, the void fraction, the invariant of the velocity gradient, and various other quantities. To investigate which of these variables is most important we plot the Gini index in the frame on the bottom, which shows that the flux depends most strongly on the velocity gradient and the velocity invariance and then the dependency drops of relatively slowly. The frame in the top right corner shows the correlation of the gas fluxes with the top six variables from the frame on the bottom and while it is not as good as in the left frame, the correlation is still reasonably strong. We have examined a few other cases in the same way, but often find that the best variables depend on the particular flow regime that we are examining. It is likely that the relationships can be improved considerably by a judicious selection of nondimensional and appropriately variables, but we have not done so yet. We do, however, believe that the use of machine learning to help select the appropriate variables for modeling and closure holds considerable promise.

Conclusion

Direct numerical simulations of multiphase flows, where verified solutions of validated equations are solved for flows spanning a range of temporal and spatial scales, are now capable of providing results that can, at least in principle, be used to help generate improved models for simulations of the average flow and large eddy motion of bubbly turbulent flows. So far, relatively little has been done for cavitating flows, but those studies done so far suggest that most of the methodology carries over with only modest changes.

As simulations of bubbly turbulent flows become almost routine, new challenges are emerging and we have discussed two of those. The first is the need for more sophisticated methods to analyze the data generated by the simulations and the second is how to extend the simulations to more complex situations, such as multiphase turbulent flows where the phase boundary has a complicated structure and the topology changes due to mergers and breakups. Machine learning is emerging as one possible way to help with the former and various multiscale strategies are likely to be required for the latter. The success of DNS of bubbly flows in the past has also opened up the possibilities of exploring more complex physics, such as heat transfer and surfactants in turbulent bubbly flows [21,22].

References

[1] Yazdani, M., Radcliff, T., Soteriou, M., and Alahyari, A.A. (2016). *A high-fidelity approach towards simulation of pool boiling*. Physics of Fluids 28, 012111.

[2] Delale, C.F., Nas, S. and Tryggvason, G. (2005). *Direct Numerical Simulations of Shock Propagation in Bubbly Liquids*. Physics of Fluids. 17, 121705

[3] Delale, C.F., and Tryggvason, G. (2008). *Shock structure in bubbly liquids: Comparison of Direct Numerical Simulations and model equations*. Shock Waves 17, 433-440.

[4] Seo, J. H., Lele, S. K., and Tryggvason, G. (2010). *Investigation and modeling of bubble-bubble interactions effect in homogeneous bubbly flows*. Physics of Fluids, 22, 063302

[5] Tryggvason, G. and S. Dabiri, S. (2013). *Direct Numerical Simulation of Shock Propagation in Bubbly Liquids. Chapter 6 in Bubble Dynamics and Shock Waves.* Shock Wave Science and Technology Reference Library Vol. 8. C. F. Delale (editor). p. 117. Springer.
[6] Rossinelli, D., Hejazialhosseini, B., Hadjidoukas, P., Bekas, C., Curioni, A., Bertsch, A., Futral, S., Schmidt, S. J., Adams, N. A.,and Koumoutsakos, P. (2013). *11 PFLOP/s simulations of cloud cavita-tion collapse*. In Proceedings of the International Conference on High Performance Computing, Networking, Storage and Analysis, SC '13, pages 3:1–3:13, New York, NY, USA. ACM.
[7] Lu, J., Fernandez, A., and Tryggvason, G. (2005). *The effect of bubbles on the wall shear in a turbulent channel flow*. Physics of Fluids 17, 095102
[8] Lu, J. and Tryggvason, G. (2006). *Numerical study of turbulent bubbly downflows in a vertical channel*. Physics of Fluids 18, 103302.
[9] Lu, J. and Tryggvason, G. (2008). *Effect of Bubble Deformability in Turbulent Bubbly Upflow in a Vertical Channel*. Physics of Fluids. 20, 040701.
[10] Bois, G. (2017). *Direct numerical simulation of a turbulent bubbly flow in a vertical channel: Towards an improved second-order Reynolds stress model.* Nuclear Engineering and Design 321, 92–103
[11] Magolan, B., Baglietto, E., Brown, C., Bolotnov, I. A., Tryggvason, G., and Lu, J. (2017). *Multiphase Turbulence Mechanisms Identification from Consistent Analysis of Direct Numerical Simulation Data.* Nuclear Engineering and Technology 49, 1318-1325.
[12] Labourasse, A. T. E., Lacanette, D., Lubin, P., Vincent, S., Lebaigue, O., Caltagirone, J.-P., and Sagaut, P. (2007). *Towards large eddy simulation of isothermal two-phase flows: Governing equations and a priori tests*, International Journal of Multiphase Flow 33, 1–39.
[13] Vincent, S., Larocque, J., Lacanette, D., Toutant, A., Lubin, P., and Sagaut, P. (2008). *Numerical simulations of phase separation and a priori two-phase les filtering*, Computers and Fluids 37, 898–906.
[14] Toutant, A., Labourasse, E., Lebaigue, O., and Simonin, O. (2008). *DNS of the interaction between a deformable buoyant bubble and spatially decaying turbulence: a priori tests for les two- phase flow modelling*. Computers and Fluids 37, 877–886.
[15] Prosperetti A. and Tryggvason, G. (2007). *Computational Methods for Multiphase Flow*. Cambridge University Press.
[16] Tryggvason, G., Scardovelli, R. and S. Zaleski, S. (2011). *Direct Numerical Simulations of Gas-Liquid Multiphase Flows*. Cambridge University Press.
[17] Tryggvason, G., Ma, M. and Lu, J. (2016). *DNS–Assisted Modeling of Bubbly Flows in Vertical Channels*. Nuclear Science and Engineering, 184, 312-320.
[18] Lu, J. and Tryggvason, G. (2018). *DNS of multifluid flows in a vertical channel undergoing topology changes*. Submitted for publication.
[19] Ma, M., Lu, J. and Tryggvason, G. (2015). *Using Statistical Learning to Close Two-Fluid Multiphase Flow Equations for a Simple Bubbly System*. Physics of Fluids, 27, 092101.
[20] Ma, M., Lu, J. and Tryggvason, G. (2016). *Using statistical learning to close two-fluid multiphase flow equations for bubbly flows in vertical channels*. International Journal of Multiphase Flow. 85, 336–347.
[21] Lu, J., Muradoglu, M. and Tryggvason, G. (2017). *Effect of Insoluble Surfactant on Turbulent Bubbly Flows in Vertical Channels*. International Journal of Multiphase Flow. 95, 135-143.
[22] Dabiri, S., and Tryggvason, G. (2015). *Heat transfer in turbulent bubbly flow in vertical channels*. Chemical Engineering Science. 122, 106-113.

© 2018 ASME

Dynamics and mechanisms of intracellular calcium waves elicited by tandem bubble-induced jetting flow[a]

[1]Fenfang Li; [1]Chen Yang; [2]Fang Yuan; [1]Defei Liao, [1]Thomas Li, [3]Farshid Guilak, and [1]Pei Zhong

[1]Department of Mechanical Engineering and Materials Science, Duke University, Durham, NC 27708; [2]Huacells Corp, Natick, MA 01760; [3]Department of Orthopaedic Surgery, Washington University, St. Louis, MO 63110, and Shriners Hospitals for Children, St. Louis, MO 63110

Abstract

Ultrasound-induced cavitation microbubbles can lead to cell injury or mechanotransduction via calcium signaling processes such as intracellular calcium waves (ICWs). However, the mechanisms by which microbubbles stimulate ICWs remain unknown. Using a microfluidic platform with highly controlled bubble-cell interaction, we have identified two distinct types of ICWs – a fast response correlating with significant membrane poration, and a slow response triggered by calcium influx through stretch-activated ion channels. The fast ICWs, distinguished from those under physiological conditions, are associated with cell injuries. We further elicited ICWs without cell injury by displacing integrin-binding beads on the cell membrane under mild cavitation conditions. This study provides mechanistic insights into ICWs for guiding ultrasound therapy in tissue modification, drug delivery, and cell mechanotransduction.

Keywords: Intracellular calcium wave; cavitation bioeffects; single cell analysis; cell injury; mechanotransduction; microfluidics

Introduction

Cavitation can produce a broad and diverse range of bioeffects during ultrasound therapy. Although cavitation-induced calcium responses have been reported during sonoporation (1, 2), and ultrasonic neuromodulation (3), the mechanism whereby the calcium ion (Ca^{2+}) transient is initiated, its propagation characteristics, and relationship to downstream bioeffects such as cell injury and mechanotransduction have not been carefully examined (4), especially at the single cell level. For example, it is unclear how the Ca^{2+} transients produced during sonoporation, with or without membrane poration, differ from each other quantitatively, and whether different mechanisms are involved (2, 5). In addition, mechanotransduction applications such as sonogenetics have gained increasing attention as a non-invasive method for neuromodulation where microbubbles are required to facilitate the cellular response (3). Despite the growing interest and potential, the role of cavitation-induced Ca^{2+} transients in such mechanotransduction processes is also not well understood. Moreover, minimum injury and membrane poration are desirable in sonogenetics and other ultrasonic mechanotransduction applications. Altogether, a fundamental understanding of the mechanisms underpinning cavitation-induced Ca^{2+} response and associated bioeffects will be critical for exploiting the full potential of ultrasound in targeted molecular delivery, tissue modification, and sonogenetics through mechanosensory responses (3) that can produce the intended therapeutic outcome with minimal adverse effects (4).

Ca^{2+}-mediated signaling can be triggered when extracellular Ca^{2+} influxes into the cell through plasma membrane, or Ca^{2+} is released from intracellular stores, such as the endoplasmic reticulum (ER). This signal transduction is often accompanied by an intracellular Ca^{2+} wave (ICW). Although Ca^{2+} signaling has been well investigated in biology (6, 7) regarding the role of ion channels and intracellular release, limited work has been carried out on the Ca^{2+} response to membrane poration and cell injury, which occurs frequently in ultrasound therapy with exposure to cavitation. In particular, cavitation can generate impulsive shear flows, and high-strain-rate cell membrane deformation that may result in transient membrane poration and sublethal cell injury (8-10). Therefore, it is important to investigate cavitation-induced Ca^{2+} signaling and other cell response subjected to such high-strain rate mechanical loading. However, challenges exist for current techniques of ultrasound-generated cavitation bubbles to dissect the complex bubble(s)-cell interaction due to the randomness in bubble generation and dynamics. Furthermore, bubble-bubble interaction or bubble collapse near a boundary with cells can lead to jet formation (11, 12), which are common in therapeutic ultrasound. In this study, we therefore examined the dynamics and mechanisms of intracellular Ca^{2+} response in single HeLa cells with a microfluidic platform that offers precise control of the dynamic bubble(s)-cell interactions (9, 13), in which cells are stimulated by the directional jetting flow from laser-induced tandem bubbles (TBs) at different standoff distance S_d, corresponding to various degrees of membrane deformation.

Body

[a]A full-length version of this work has been submitted to PNAS for consideration of possible publication.

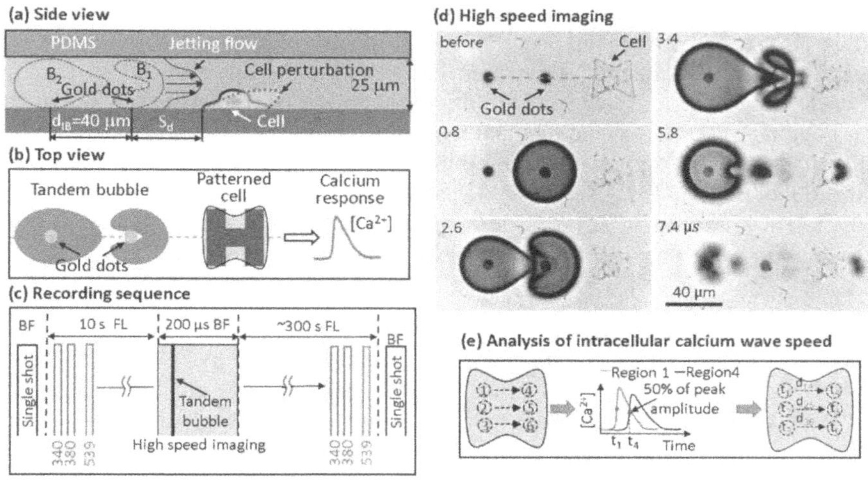

Fig. 1. Schematic diagrams of experimental design and data analysis. (a) Side view and (b) top view of TB interaction with a single cell grown on an H-shaped pattern in a microfluidic channel. (c) Recording sequences for bright field (BF) imaging of cell morphology, simultaneous monitoring of intracellular Ca^{2+} transients (340, 380 nm excitation) and membrane poration (539 nm excitation) through fluorescence (FL) imaging, and high-speed imaging of tandem bubble interaction with jet formation. (d) Example of high-speed image sequence showing the interaction of laser-generated TB from gold dots with a HeLa cell. (e) Protocol used for intracellular Ca^{2+} wave speed (C_{ICW}) analysis.

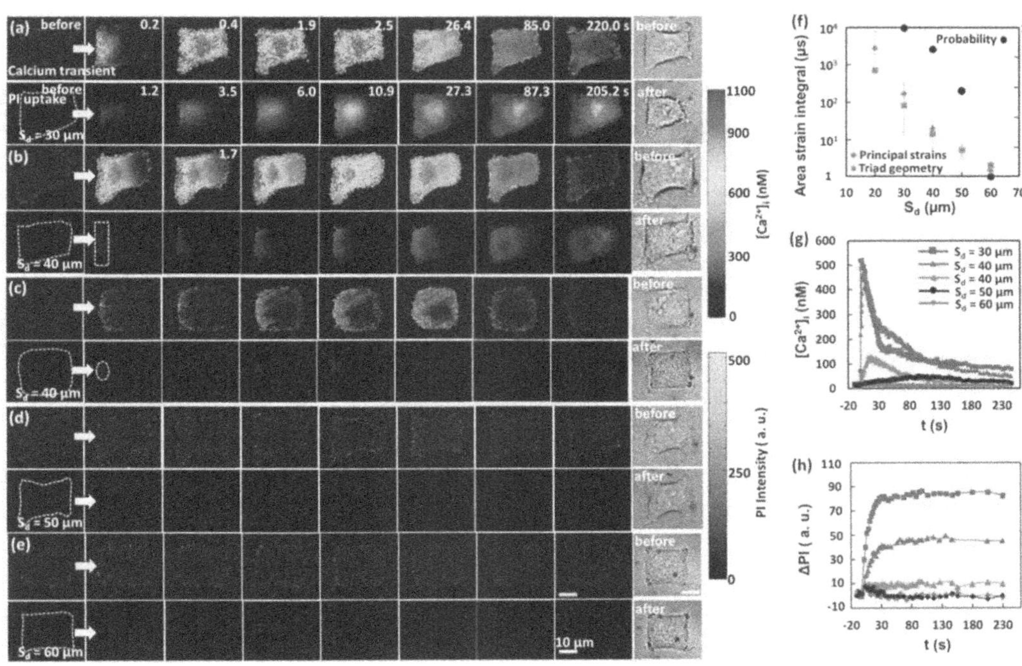

Fig. 2. Typical intracellular Ca^{2+} response and PI uptake induced by TB treatment at various S_d. (a)-(e) Image sequences showing intracellular Ca^{2+} response (top row) and PI uptake (bottom row) in five representative individual cells at S_d = 30, 40 (two cells), 50 and 60 μm, respectively. Bright field images of the cell before and after the treatment are shown on the right. The timing for the image sequences shown in (b) - (e) are the same as in (a) except as otherwise marked on the image. The white arrows indicate the jetting flow direction. The cell contour before treatment is outlined with white dashed lines in the first PI image. Cells were exposed to tandem bubbles at t = 0. (f) Ca^{2+} response probability (black circles) and area strain integral (shown in green and red symbols) at different S_d (n = 19, 21, 19, 10 for the Ca^{2+} response probability at S_d = 30, 40, 50, 60 μm, respectively). (g) and (h) are the temporal profiles of averaged intracellular Ca^{2+} concentration and PI uptake inside each individual cell shown in (a)-(e), respectively. The colors and symbols used for different S_d are the same in (g) and (h). All scale bars denote 10 μm.

Directional perturbation of single cells from TB-generated jetting flow. TBs (R_{max} = 25 ± 1 μm) were produced by focusing two pulsed Nd:YAG laser beams (λ = 532 nm, 5-ns pulse duration, 2.5 μs interpulse delay) onto a pair of gold dots (15-nm thick and 6 μm in diameter, with an interbubble distance of d_{IB} = 40 μm) patterned on the glass substrate of a microfluidic channel (H = 25 μm, Fig. 1a). An individual HeLa cell was captured and grown on a H-shaped island (32 X 32 μm) covered with fibronectin (Fig. 1 (a-b)) nearby ($\gamma = S_d/R_{max}$ = 1.2 - 2.4). Because the two bubbles are oscillating out of phase, they repel each other by the secondary Bjerknes forces (14), resulting in an outward directional jet formation between 3.0 and 5.0 μs (Fig. 1d), which imposes a highly localized shear stress and stress gradient onto the target cell. Sequential fluorescence (FL) imaging of the intracellular Ca^{2+} transients (340, 380 nm excitation of fura-2) and membrane poration (539 nm excitation of PI) was first recorded at baseline level for 10 s before high-speed imaging of the TB interaction and jet formation, followed by a sequence of FL imaging again in the target cell for 300 s (Fig. 1c). Before and after the TB treatment, changes in cell morphology were also documented by single-shot bright field (BF) images. Cell injury is assessed by PI uptake and BF cell morphology change. To quantify the propagation of intracellular Ca^{2+} transient, the Ca^{2+} response profiles from the cell leading edge (small regions 1, 2, 3) to the trailing edge (small regions 4, 5, 6) are analyzed based on the time-of-flight in 50% rise time of individual pairs (Fig. 1e).

Spatiotemporal evolution of intracellular Ca^{2+} response and PI uptake induced by TB at various S_d. Figure 2 shows the image sequences of the typical intracellular Ca^{2+} response (top row) and PI uptake (bottom row) for individual cells treated with TB at S_d =30 (a), 40 (b-c), 50 (d), 60 (e) μm, respectively. In general, the intensity of the Ca^{2+} response and PI uptake showed a clear S_d dependency, with both signals initiated at the leading edge of the cell facing the jetting flow (indicated by the while arrows) and propagated toward the trailing edge (Fig. 2 (a-c)). The probability of an evoked Ca^{2+} response was found to drop progressively with S_d (circles in Fig. 2f), varying from 100% at S_d = 30 μm to no response at S_d = 60 μm. This S_d dependency in Ca^{2+} response may be associated with the exponentially decaying shear stress experienced by the target cells at increased S_d (9, 13). The resultant area strain integral (with time) is shown by the diamonds and squares in Fig. 2f, representing their upper- and lower-bound values (9). The average $[Ca^{2+}]_i$ and PI uptake (ΔPI) within the individual cells shown in Fig. 2 (a-e) are plotted vs. time in Fig. 2g and Fig. 2h, respectively. At S_d = 30 μm (or γ = 1.2), a fast Ca^{2+} response with short rise time (t_r = 2.2 s) and large $[Ca^{2+}]_i$ change (523 nM) was produced, which was accompanied by a high PI uptake (Fig. 2 (a, g, h)). In contrast, at S_d = 50 μm (or γ = 2), a slow Ca^{2+} response with long t_r (99 s) and small $[Ca^{2+}]_i$ change (38 nM) was observed, associated with negligible PI uptake, see Fig. 2 (d, g and h). At S_d = 40 μm, a variety of responses ranging from fast to slow Ca^{2+} transients were obtained, corresponding from high to low PI uptake, see Fig.2 (b, c, g, h). Finally, at S_d = 60 μm, neither Ca^{2+} response nor PI uptake was observed (Fig. 2 (e-h)).

Quantification of calcium response and intracellular calcium wave propagation. Overall, C_{ICW} (mean ± s.d.) was found to be 57 ± 21 μm/s at S_d = 30 μm, which is significantly higher than the corresponding values of 14 ± 15 μm/s at S_d = 40 μm and 4 ± 3 μm/s at S_d = 50 μm (Fig. 3a). The corresponding probabilities of membrane poration and Ca^{2+} response at different S_d are shown in Fig. 3b. At S_d = 30 μm the probability of Ca^{2+} response and membrane poration are both 100%. In contrast, at S_d = 50 μm, 44% of the cells showed a Ca^{2+} response with negligible PI uptake. We also examined the correlations between $\Delta[Ca^{2+}]_{i(max)}$ (in filled symbols) or C_{ICW} (in open symbols) with PI uptake (Fig. 3c). The results showed two types of Ca^{2+} responses that may be delineated by thresholds at PI uptake = 10 (a. u.) and C_{ICW} = 30 μm/s, respectively (see the horizontal dashed lines in Fig. 3 (a, c)). In particular, the data points in the S_d = 50 μm group (circles) are below these thresholds (i.e., in the lower left quadrant) while most of the data in the S_d = 30 μm group (squares) are above these thresholds (i.e., in the upper right quadrant). Furthermore, the value of C_{ICW} was found to scale linearly with $(\Delta[Ca^{2+}]_{i(max)}/t_r)^{1/2}$ (Fig. 3c inset), which is consistent with previous models of ICW propagation (15, 16).

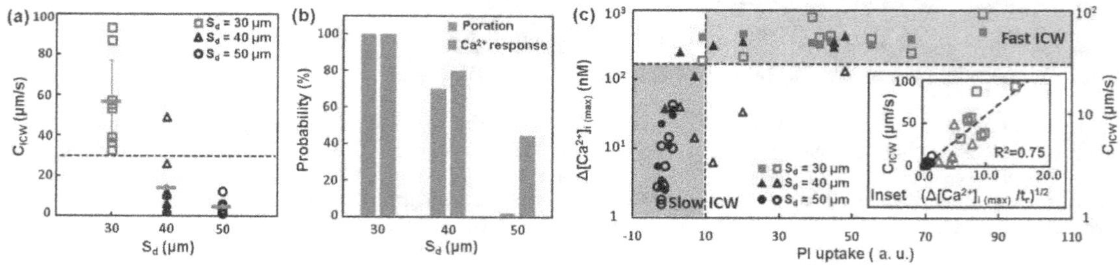

Fig. 3. Quantification of Ca^{2+} response and ICW propagation. (a) C_{ICW} vs. S_d. The error bars indicate the standard deviation. (b) Membrane poration and Ca^{2+} response probability at different S_d. (c) Correlations of the maximum intracellular Ca^{2+} concentration change $\Delta[Ca^{2+}]_{i(max)}$ (filled symbols) and C_{ICW} (open symbols) vs. PI uptake in the cytosol at S_d = 30 - 50 μm. The dashed lines indicate separation of data based on the threshold of PI uptake of 10 (a. u.), and C_{ICW} of 30 μm/s. Inset: C_{ICW} vs. the square root of concentration change rate $(\Delta[Ca^{2+}]_{i(max)}/t_r)^{1/2}$, where t_r is the rise time from baseline to peak concentration. The dashed line represents a linear fitting of the data (R^2=0.75).

Mechanistic study of intracellular Ca^{2+} response induced by TB treatment. This series of experiments was carried out at S_d = 30 and 50 μm, with three representative (high, medium, low) Ca^{2+} response curves shown in each plot of

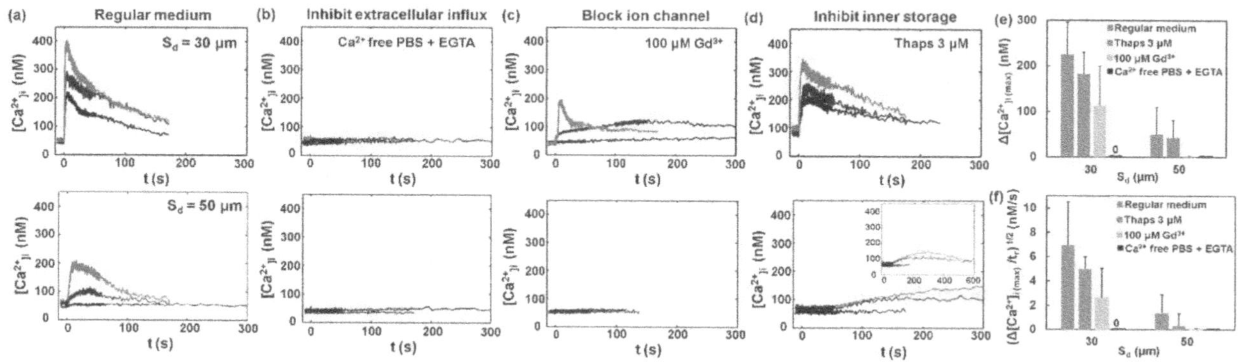

Fig. 4. Mechanistic study of intracellular Ca^{2+} response induced by TB treatment. Ca^{2+} response profiles at $S_d = 30$ and 50 μm (a) in regular culture medium; (b) in Ca^{2+} free culture medium (Ca^{2+} free 1x PBS with 0.5 mM EGTA added as Ca^{2+} chelator); (c) in culture medium with 100 μM Gd^{3+} to block non-specific mechanosensitive ion channel; (d) in culture medium with 3 μM Thaps to inhibit ER Ca^{2+}-ATPases prior to TB treatment and thus affecting Ca^{2+} release. Comparison of Ca^{2+} responses from groups of individual cells treated by TB in regular culture medium (blue), with 3 μM Thaps (red), with 100 μM Gd^{3+} (green), and in Ca^{2+} free culture medium (black) – (e) $\Delta\,[Ca^{2+}]_{i(max)}$ and (f) $(\Delta[Ca^{2+}]_{i(max)}/t_r)^{1/2}$) vs. $S_d = 30$ and 50 μm. Data are shown in mean ± standard deviation (N= 10, 3, 4, 3 at $S_d = 30$ μm, and 9, 3, 6, 5 at $S_d = 50$ μm for regular culture medium, with 3 μM Thaps, with 100 μM Gd^{3+}, and in Ca^{2+} free culture medium, respectively.)

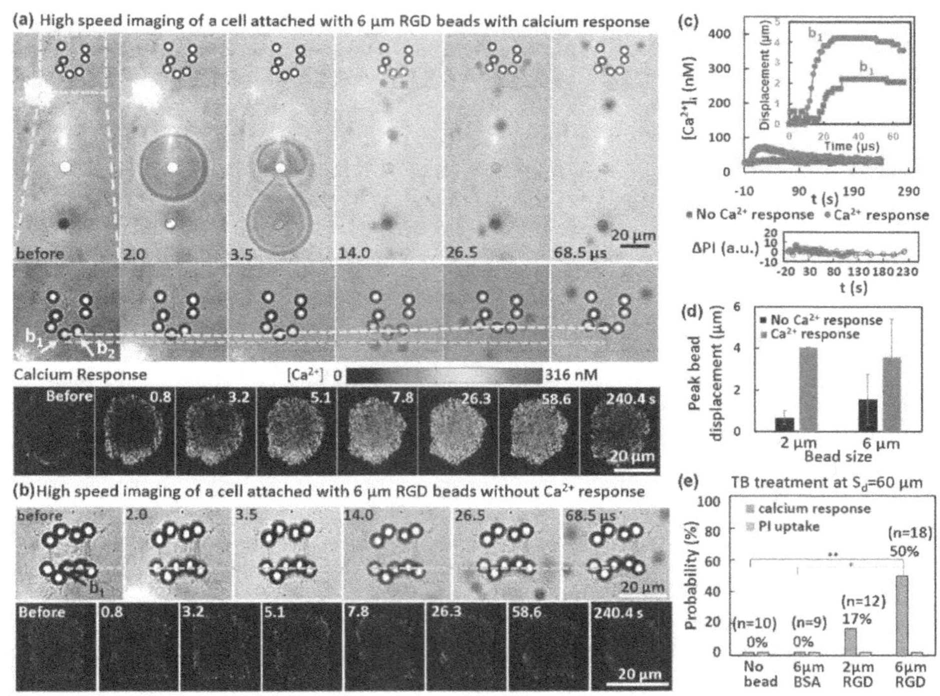

Fig. 5. Beads enhanced Ca^{2+} response at $S_d = 60$ μm. (a) Top row: TB interaction with an H-patterned cell with 6 μm RGD beads attached on the cell membrane. Middle row (enlarged region inside the dashed box in top row): the displacement of an individual bead at the cell leading edge (indicated by the yellow dashed line; the green dashed line serves as a reference). Bottom row: fluorescence image sequence showing Ca^{2+} response of the cell in (a). (b) An individual cell attached with 6 μm RGD beads were treated with TB, but with no obvious Ca^{2+} response. Top row: the displacement of an individual bead at the cell leading edge (the green dashed line serves as a reference). Bottom row: fluorescence image sequence showing no Ca^{2+} response of the cell (the same colorbar as in (a)). (c) Time evolution of $[Ca^{2+}]_i$ for the two cells shown in Fig. 5a and Fig. 5b, respectively. Inset shows the time evolution of the displacement of the beads located in the middle of the cell's leading edge (b_1). Bottom row shows there is negligible PI uptake in both cells. (d) Summary of results showing the peak displacement of the bead with the largest displacement (b_1) on the cells for each group (mean ± s.d.). (e) Probability of Ca^{2+} response and PI uptake in cells with no beads, attached with 2 and 6 μm RGD coated integrin-binding beads, or with 6 μm BSA-coated non-specific binding beads treated by TB at $S_d = 60$ μm. $^{**}p < 0.01$, $^{*}p < 0.05$ from Fisher's exact test, the p value between the Ca^{2+} response probability of no beads and 2 μm RGD beads is not significant.

Fig. 4(a-d). The control experiment with regular medium is shown in Fig. 4a. The TB-elicited Ca^{2+} response could be completely abolished at both S_d with Ca^{2+} free medium (chelated with EGTA) (Fig. 4b). At $S_d = 50$ μm where there was no detectable membrane poration, the Ca^{2+} response was completely abolished with 100 μM Gd^{3+} blocking mechanosensitive ion channels (Fig. 4c) while at $S_d = 30$ μm it was found to be either suppressed with reduced amplitude in $\Delta[Ca^{2+}]_i$, or very often, converted from fast to slow Ca^{2+} responses. With 3 μM thapsigargin (Thaps) used to inhibit intracellular Ca^{2+} release (Fig. 4d), the Ca^{2+} response was found to be suppressed in both $\Delta[Ca^{2+}]_{i(max)}$ and $(\Delta[Ca^{2+}]_{i\,(max)}/t_r)^{1/2}$ at S_d = 30 μm with a reduced C_{ICW} in the range of 4-11 μm/s, compared to the regular medium, see Fig. 4 (d-f). At S_d = 50 μm, significantly prolonged Ca^{2+} responses were observed in cells, indicating that the propagation of the Ca^{2+} transient in the cytosol was substantially delayed compared to those in the regular medium. Fig. 4e and Fig. 4f summarize the variations in $\Delta[Ca^{2+}]_{i(max)}$ and $(\Delta[Ca^{2+}]_{i(max)}/t_r)^{1/2}$ (proportional to C_{ICW}, see Fig. 3c inset) among the aforementioned culture media. Overall, Thaps, Gd^{3+}, and Ca^{2+} free media were found to affect the amplitude and propagation speed of the Ca^{2+} response. The results indicate that extracellular Ca^{2+} influx is essential for initiating both types of ICWs in the TB-treated HeLa cells: 1) through mechanosensitive ion channels (no injury) in the slow Ca^{2+} response; and 2) through both membrane poration (with injury) and mechanosensitive ion channels during the fast Ca^{2+} response. In both cases, the initiated Ca^{2+} response then propagated through the cytosol most likely via the calcium induced calcium release (CICR) mechanism from the ER (7).

Enhanced intracellular Ca^{2+} response with integrin-binding bead on the cell membrane. At S_d = 60 μm, no Ca^{2+} response in a target cell could be induced by the TB because of the diminished strength of the jetting flow (9). We hypothesized that shear flow-cell interaction and resultant membrane deformation may be enhanced by attaching micron-sized beads on the cell surface. Figure 5a shows an example of TB-cell interaction at S_d = 60 μm, in which 6 μm RGD (integrin-binding) beads were attached at the cell's leading edge (such as b_1) and found to be significantly displaced from their original locations. Consequently, a mild Ca^{2+} response was elicited inside the cytosol. Fig. 5b shows another individual cell with TB treatment (top row) that showed no Ca^{2+} response (FL images at the bottom row). The maximum bead displacement (e.g., b_1 in Fig. 5b) was found to be significantly less than its counterpart in the responsive cell in Fig. 5a (see Fig. 5c). Moreover, negligible PI uptake was detected in both cases (see ΔPI vs. t profile in Fig. 5c), indicating no cellular injury. For both 2 and 6 μm beads, the maximum bead displacement (of b_1) was found to be significantly larger in Ca^{2+} responsive cells than in non-responsive cells (Fig. 5d). This finding again suggests that large bead displacement or cell membrane stretching is critical for the enhanced Ca^{2+} response at S_d = 60 μm. We further observed that the enhancement of Ca^{2+} response was bead-size dependent, increasing from 17% with 2 μm RGD beads to 50% with 6 μm RGD beads (Fig. 5e). No Ca^{2+} response was observed in the control experiment with 6 μm BSA beads that non-specifically bind to the cell membrane ($p<0.05$ compared to RGD beads).

Discussion and conclusion

The physical characteristics of ICW are highly conserved among various species with C_{ICW} typically in the range of 3 ~ 30 μm/s at 20 °C (16, 17). Quantitatively, C_{ICW} can be scaled by $(J_{eff}D_{eff}/C_0)^{1/2}$, where J_{eff} is the effective Ca^{2+}-current density (e.g., 27 μM/s), D_{eff} is the effective diffusion constant of Ca^{2+} (e.g., 22 μm²/s), and C_0 is the $[Ca^{2+}]_i$ threshold for Ca^{2+} channel activation (e.g., 0.5 μM) (15). Assuming under physiological conditions, D_{eff} and C_0 are constants for each cell type, C_{ICW} is then determined primarily by $(J_{eff})^{1/2}$. This feature is consistent with our results in Fig. 3c inset, especially for the data in the S_d = 50 μm group and most of the S_d = 40 μm group with C_{ICW} in the range of 3 ~ 30 μm/s (at 20 °C) commonly found among eukaryotic cells (17, 18). In contrast, much higher C_{ICW} (33-93 μm/s) was found in the S_d = 30 μm group where all cells showed significant membrane poration and cell injury, which may reduce the $[Ca^{2+}]_i$ threshold for initiating an ICW (19, 20). Therefore, a high extracellular Ca^{2+} influx (i.e., greater J_{eff}) combined with a low C_0 may contribute to the significantly elevated C_{ICW} observed in the S_d = 30 μm group (Fig. 3a). Furthermore, it is worth noting that the fast ICW induced at S_d = 30 μm may reflect the injury response of the target cell, albeit sublethal, to TB-induced repairable membrane poration, which typically resealed in 30s (see Fig. 2h). The rapid Ca^{2+} influx induced under such conditions may accelerate the depolymerization of cytoskeleton filaments to facilitate internal vesicles (lysosomes) to migrate and patch the rupture site through exocytosis (21, 22), and potentially later on, even the related compensatory response of endocytosis for lesion removal (23). Although the membrane pore is resealed, such a strong Ca^{2+} response is associated with cytotoxity, manifested by the loss of F-actin stress fibers, cell shrinkage, and even apoptosis observed in this and other studies (24). These collateral injuries, while facilitating the apoptosis of cancer cells in tumor therapy, will limit the range of cavitation-mediated biomedical applications such as gene delivery and neural modulation.

To address this challenge, we have developed a novel strategy to enhance cavitation-induced Ca^{2+} response without injury by attaching micron-sized beads to the cell membrane through the RGD-integrin link (Fig. 5). This enhancement is integrin-binding specific since no Ca^{2+} response was produced with 6 μm BSA beads that were non-specifically attached to cell plasma membrane. The specificity of this response was further confirmed by additional single bubble-cell interaction experiments, which also demonstrated that the enhanced response was significantly suppressed by Gd^{3+} or Ruthenium Red –two well-known inhibitors of mechanosensitive ion channels. Further, we found that the maximum bead displacement in Ca^{2+} responsive cells is significantly larger than that in non-responsive cells attached with 6 μm integrin-binding beads (Fig. 5). Altogether, our results suggest that the amplified membrane stretch is transmitted through cell surface integrins (25) to activate mechanosensitive ion channels on the cell membrane, and is critical for the enhanced Ca^{2+} response at S_d = 60 μm.

In conclusion, we have systematically examined the dynamics and mechanisms of TB-induced ICW at the single cell level. The knowledge acquired in this work shed light on how to target different biological effects such as cell injury and apoptosis, drug delivery and mechanotransduction by tuning the magnitudes of mechanical perturbations through the high strain rate shear flows associated with inertial cavitation and dynamic bubble–bubble interactions that are prevalent in therapeutic ultrasound applications. Here we used cavitation-induced jetting flow to study the ICW in a highly-controlled manner, and the results and insights gained in this work may also apply to single inertial cavitation and other similar mechanical perturbations. Future work may focus on utilizing repeated and programmed cavitation induced shear flow to stimulate cells for long-term bioeffects such as cell proliferation, differentiation and gene expression for tissue growth and wound healing.

References

1. Leow RS, Wan JMF, & Yu ACH (2015) Membrane blebbing as a recovery manoeuvre in site-specific sonoporation mediated by targeted microbubbles. *J. Royal Soc. Interface* 12(105).
2. Fan Z, Kumon RE, Park J, & Deng CX (2010) Intracellular delivery and calcium transients generated in sonoporation facilitated by microbubbles. *J Control Release* 142(1):31-39.
3. Ibsen S, Tong A, Schutt C, Esener S, & Chalasani SH (2015) Sonogenetics is a non-invasive approach to activating neurons in Caenorhabditis elegans. *Nature Communications* 6.
4. Hassan MA, Campbell P, & Kondo T (2010) The role of Ca(2+) in ultrasound-elicited bioeffects: progress, perspectives and prospects. *Drug Discov Today* 15(21-22):892-906.
5. Zhou Y, Shi J, Cui J, & Deng CX (2008) Effects of extracellular calcium on cell membrane resealing in sonoporation. *J Control Release* 126(1):34-43.
6. Berridge MJ (2005) Unlocking the secrets of cell signaling. *Annu Rev Physiol* 67:1-21.
7. Bootman MD & Berridge MJ (1995) The elemental principles of calcium signaling. *Cell* 83(5):675-678.
8. Mohammadzadeh M, Li FF, & Ohl CD (2017) Shearing flow from transient bubble oscillations in narrow gaps. *Phys. Rev. Fluids* 2(1).
9. Yuan F, Yang C, & Zhong P (2015) Cell membrane deformation and bioeffects produced by tandem bubble-induced jetting flow. *Proc Natl Acad Sci U S A* 112(51):E7039-7047.
10. Li FF, Chan CU, & Ohl CD (2013) Yield Strength of Human Erythrocyte Membranes to Impulsive Stretching. *Biophys J* 105(4):872-879.
11. Ohl CD, *et al.* (2006) Sonoporation from jetting cavitation bubbles. *Biophys J* 91(11):4285-4295.
12. Sankin GN, Yuan F, & Zhong P (2010) Pulsating tandem microbubble for localized and directional single-cell membrane poration. *Phys Rev Lett* 105(7):078101.
13. Fenfang Li FY, Georgy Sankin, Chen Yang, Pei Zhong. (2017) A Microfluidic System with Surface Patterning for Investigating Cavitation Bubble(s)–Cell Interaction and the Resultant Bioeffects at the Single-cell Level. *JoVE* 119.
14. Yuan F, Sankin G, & Zhong P (2011) Dynamics of tandem bubble interaction in a microfluidic channel. *J Acoust Soc Am* 130(5):3339-3346.
15. Kupferman R, Mitra PP, Hohenberg PC, & Wang SSH (1997) Analytical calculation of intracellular calcium wave characteristics. *Biophys J* 72(6):2430-2444.
16. Jaffe LF (1991) The path of calcium in cytosolic calcium oscillations: a unifying hypothesis. *Proc Natl Acad Sci U S A* 88(21):9883-9887.
17. Jaffe LF (1993) Classes and mechanisms of calcium waves. *Cell Calcium* 14(10):736-745.
18. Jaffe LF (2010) Fast calcium waves. *Cell Calcium* 48(2-3):102-113.
19. Carbone E, Marcantoni A, Giancippoli A, Guido D, & Carabelli V (2006) T-type channels-secretion coupling: evidence for a fast low-threshold exocytosis. *Pflugers Arch* 453(3):373-383.
20. Weiss N, *et al.* (2012) A Ca(v)3.2/syntaxin-1A signaling complex controls T-type channel activity and low-threshold exocytosis. *J Biol Chem* 287(4):2810-2818.
21. McNeil PL & Steinhardt RA (2003) Plasma membrane disruption: repair, prevention, adaptation. *Annu Rev Cell Dev Biol* 19:697-731.
22. Reddy A, Caler E, & Andrews NW (2001) Plasma membrane repair is mediated by Ca2+-regulated exocytosis of lysosomes. *Molecular Biology of the Cell* 12:266a-266a.
23. Idone V, *et al.* (2008) Repair of injured plasma membrane by rapid Ca2+-dependent endocytosis. *Journal of Cell Biology* 180(5):905-914.
24. Heo J, Sachs F, Wang J, & Hua SZ (2012) Shear-induced volume decrease in MDCK cells. *Cell Physiol Biochem* 30(2):395-406.
25. Matthews BD, Thodeti CK, & Ingber DE (2007) Activation of mechanosensitive ion channels by forces transmitted through integrins and the cytoskeleton. *Mechanosensitive Ion Channels, Part A* 58:59-85.

© 2018 ASME

Dynamics of a Laser-induced Bubble Near a Convex Free Surface

[1]Shengji Wu; [1]Zhigang Zuo*; [1]Zibo Ren; [1]Shuhong Liu*;

[1] *State Key Laboratory of Hydro Science and Engineering, Department of Energy and Power Engineering, Tsinghua University, Beijing, China;*

Abstract

This paper experimentally documents the behavior of a laser-induced bubble near a convex paraboloidal free surface with different vertical radius of curvatures, generated in a rotating cylindrical water flask. The shape of the free surface plays a key role on bubble behaviors when bubble collapses. Due to the strong influence of the free surface, the bubble forms a microjet outwards the free surface when it collapses. And the nondimensional bubble collapse time, T_c/T_c', is smaller than 1, where T_c is the bubble collapse time and T_c' is the Rayleigh time of bubble collapse with an equivalent maximum radius. As the relative distance between the bubble center and the free surface, h/R_{max}, decreases, T_c/T_c' decreases, which is different with the bubble behavior near a flat rigid boundary, where h is the distance between the bubble center and the flat free surface or the convex free surface center, R_{max} is bubble equivalent maximum radius. Compared with the bubble dynamics near a flat free surface, the bubble near a convex free surface presents a milder collapse behavior. T_c/T_c' increases with the nondimensional radius of curvature of the free surface center, r_s/R_{max}, decreases, where r_s is the vertical radius of curvature of the free surface, which is different from the bubble behavior near a curved rigid boundary. It implies that the convex shape of free surface reduces the Kelvin impulse, and the strength of the microjet when bubble collapses decreases with a decreasing r_s/R_{max}. Meanwhile the convex free surface provides a focusing mechanism to the splash on the free surface when bubble collapses. The splash gets a higher velocity when r_s/R_{max} decreases.

Keywords: bubble dynamics; convex free surface; micro jet strength

Introduction

Cavitation phenomena is a double-edged sword for human lives. It causes cavitation erosion ship propellers [1], pipelines [2] and turbines [3]. Meanwhile, with appropriate controlling technology, it is feasible to apply the cavitation to benefit our daily life, e.g., kidney stone fragmentation [4], surface cleaning [5] and drug delivery [6]. During these cavitation process, the collapse of the cavitation bubbles, especially near a boundary, plays a key role on the cavitation erosion or the beneficial effects. Nevertheless, bubble collapse behaviors strongly depend on the features of the boundaries. Not only the physical properties but also the geometric parameters of the boundary will greatly influence on bubble dynamics.

In the past decades, extensive research been carried out on bubble dynamics near a flat boundary. Near a flat rigid boundary, the bubble presents an aspherical collapse and develops a micro jet towards the rigid boundary [7, 8], while the bubble forms a micro jet outward the free surface [9, 10] when it collapses near a flat free surface. In 1980s, Blake *et al.* [11] applied the Kelvin impulse to analyze the strength of the micro jet at bubble collapse, and Supponen *et al.* [12] developed the Kelvin impulse and use a dimensionless pressure anisotropy parameter to scale the micro jet strength with the jet feature, e.g., jet velocity, jet volume and jet impact time, etc. Most research focused on a cavitation bubble occurring near a flat boundary. However, the bubble dynamics near a curved boundary has received less attention, and the influence of the geometric parameters of the boundary remains poorly understood.

A number of studies have been dedicated to the bubble dynamics near a curved boundary. For example, Tomita *et al.* [13] generated a laser-induced bubble near a curved rigid boundary and the experiments present that the velocity of the micro jet increases while the shape of the rigid boundary changes from concave into convex. Obreschkow *et al.* [14] study the behavior of a bubble in a droplet in micro gravity. They observed two liquid jets escaping from the drop when the bubble collapses, and a shorter bubble life time in the droplet than the bubble in an infinite fluid. Also the spike and the splash on the concave free surface are not as violent as a flat free surface. In addition, Tagawa *et al.* [15] focus a laser in a capillary to generate a bubble, and the asymmetric bubble dynamics induces a thin and focused jet (velocity up to 850 m/s) on the convex free surface due to the surface tension. They stated that the convex free surface

*Corresponding Authors, Zhigang Zuo: zhigang200@mail.tsinghua.edu.cn;
Shuhong Liu: liushuhong@mail.tsinghua.edu.cn

is able to focus the shockwave and cause the supersonic jet on the surface. In this paper, we report an experimental study of bubble dynamics near a convex free surface and focus on the strength of the micro jet when bubble collapses.

Experimental setup

To simulate the dynamics of a cavitation bubble near a curved free surface, we designed a rotating experimental setup as in Fig. 1. The water flask (D = 150 mm, H = 350 mm) was filled with deionized and degassed water and all experiments took place at room temperature. When the cylindrical water flask is rotating, the flat free surface turns into a paraboloid due to the gravity and the centrifugal force. The vertical radius of curvature of the free surface $r_s = 900g/\pi^2 n^2$, where g is the gravitational acceleration, and n is the revolutions per minute of the water flask. And we used a Q-switched pulsed ruby laser (wavelength: 694.3 nm, maximum pulse energy: 1.5 J, pulse length: 20–30 ns) to generate a cavitation bubble with maximum radius of about 2-3 mm in the water flask. We used a signal generator (Rigol DG2021A) to trigger a laser and a highspeed camera (Phantom V711) simultaneously. The motions of the bubble and the free surface were recorded by the camera (at 79 069 frames/s with an exposure time of 1 μs) for subsequent analyses.

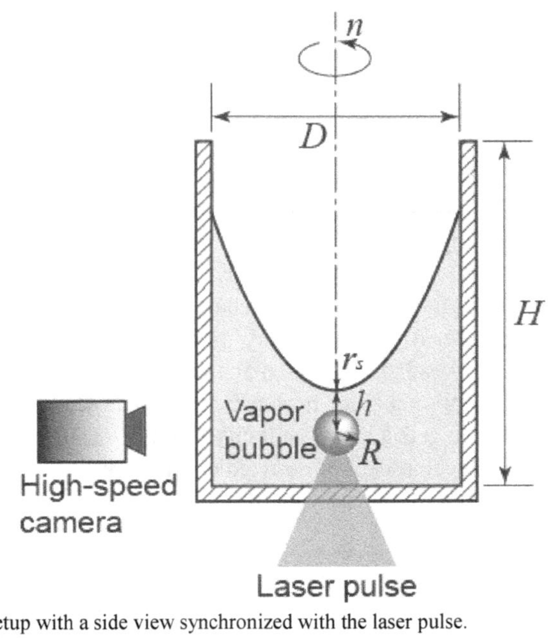

Fig. 1. Schematic of the experimental setup with a side view synchronized with the laser pulse.

Results and discussion

For the convenience of depicting the experimental results, we use three nondimensional parameters, h/R_{max}, r_s/R_{max} and T_c/T_c', where h is the distance between the bubble center and the flat free surface or the convex free surface center, R_{max} is the bubble equivalent maximum radius, T_c is the bubble collapse time and T_c' is the Rayleigh time of bubble collapse with an equivalent maximum radius. Figure 2 presents the bubble behaviors near the flat free surface with different h/R_{max}. Due to the strong influence of the free surface, the bubble collapses non-spherically and forms a micro jet outward the free surface. And the bubble collapse time, T_c, is much shorter than T_c'. T_c/T_c' increases with an increasing relative distance h/R_{max} increases, which is in contrast with the bubble behavior near a flat rigid boundary [16]. The possible explanation is that it would be much easier for bubble to push the liquid away to grow and pull the liquid back to collapse due to the deformable free surface, resulting a T_c/T_c' smaller than 1.

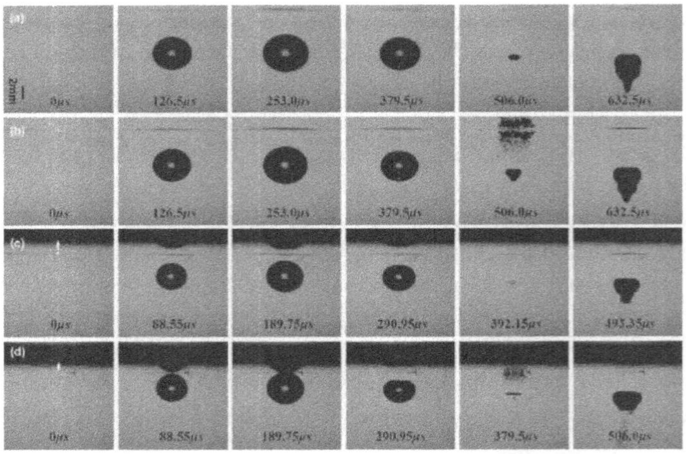

Fig. 2. Laser-induced bubble behaviors near a flat free surface with different h/R_{max}. (a) h/R_{max} =2.261, T_c/T_c'=0.916; (b) h/R_{max} =1.932, T_c/T_c'=0.904; (c) h/R_{max} =1.432, T_c/T_c'=0.862; (d) h/R_{max} =1.027, T_c/T_c'=0.846. The scale bar is only valid on the vertical direction due to the image distortion from the cylindrical water flask.

Four typical experimental results are presented when the bubble grows and collapses near a free surface with different convex shape in Fig 3. With the same h/R_{max}, T_c/T_c' increases with a decreasing r_s/R_{max}, which is different from the influence of the convex shape on the bubble behavior near a rigid boundary [13]. A more convex surface indicates more liquid to be driven into motion, especially the liquid on the bubble upper side, which would increase bubble collapse time. It implies that the Kelvin impulse and the strength of the microjet when bubble collapses near a convex free surface would be smaller than when bubble collapses near a flat free surface, because of the increasing bubble collapse time with the decreasing r_s/R_{max}, though the bubble behavior, in general, is similar with a flat free surface due to the strong influence of the free surface. It is valid to pay more attention to theoretical analysis and simulation research in the future. In the experiments, we also observed that a violent splash appears when bubble collapses, which is an important phenomenon on the bubble behavior near a free surface. The focusing of the convex free surface enhances the strength of splash. The splash gets a higher velocity when r_s/R_{max} decreases. One potential application of the high-speed and thin splash is to the development of no-needle liquid-injection systems which would be beneficial for health care worldwide.

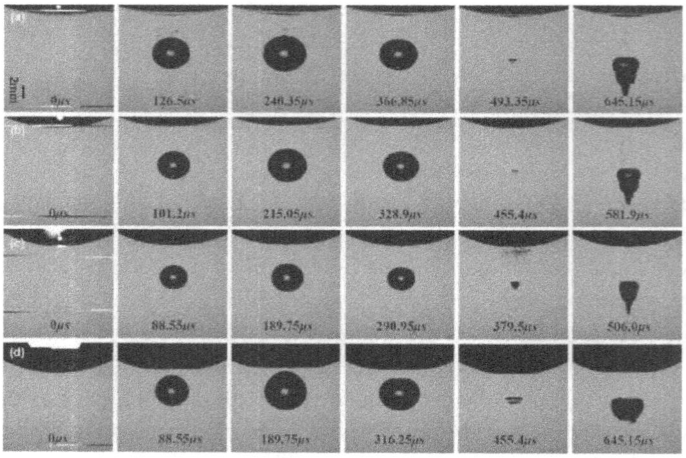

Fig. 3. Laser-induced bubble behaviors near a convex free surface. (a) h/R_{max} =1.977, r_s/R_{max} =8.292, T_c/T_c'=0.924; (b) h/R_{max} =2.125, r_s/R_{max} =5.458, T_c/T_c'=0.938; (c) h/R_{max} =2.152, r_s/R_{max} =5.710, T_c/T_c'=0.948; (d) h/R_{max} =0.977, r_s/R_{max} =4.304, T_c/T_c'=0.897. The scale bar is only valid on the vertical direction due to the image distortion from the cylindrical water flask.

Conclusions

We carried out experimental studies on the bubble dynamics near a convex free surface, which had not been studied previously. In general, the bubble presents similar collapse behavior with the bubble collapses near a flat free surface. But due to the influence of the convex shape of the free surface, some different features appear on the bubble collapse time and the jet strength. Via the bubble collapse time analysis, we found that T_c/T_c' is smaller than 1 when bubble collapses near a free surface. T_c/T_c' increases with h/R_{max} increasing and r_s/R_{max} decreasing. Compared with the bubble collapse near a flat free surface, the longer bubble collapse time near a convex free surface suggests a smaller Kelvin impulse and a milder microjet when bubble collapses. And the focusing of the convex free surface leads to a violent splash on the free surface, which would be further investigations.

Acknowledgement

The work was supported by National Natural Science Foundation of China (No. 51476083), National Basic Research Program of China (No. 613321), and State Key Laboratory of Hydroscience and Engineering (Research Fund Program, No: 2017-KY-03).

References

[1] Silberrad, D. (1912). *Propeller erosion.* Engineering, 33, 33-35.
[2] Arndt, R. E. (1981). *Cavitation in fluid machinery and hydraulic structures.* Annual Review of Fluid Mechanics, 13(1), 273-326.
[3] Duan, C. G., & Karelin, V. I. (2003). *Abrasive Erosion and Corrosion of Hydraulic Machinery*: Series on Hydraulic Machinery. World Scientific Publishing Company.
[4] Lingeman, J. E., McAteer, J. A., Gnessin, E., & Evan, A. P. (2009). *Shock wave lithotripsy: advances in technology and technique.* Nature Reviews Urology, 6(12), 660-670.
[5] Ohl, C. D., Arora, M., Dijkink, R., Janve, V., & Lohse, D. (2006). *Surface cleaning from laser-induced cavitation bubbles.* Applied physics letters, 89(7), 074102.
[6] Coussios, C. C., & Roy, R. A. (2008). *Applications of acoustics and cavitation to noninvasive therapy and drug delivery.* Annu. Rev. Fluid Mech., 40, 395-420.
[7] Lauterborn, W., & Ohl, C. D. (1997). *Cavitation bubble dynamics.* Ultrasonics sonochemistry, 4(2), 65-75.
[8] Blake, J. R., Taib, B. B., & Doherty, G. (1986). *Transient cavities near boundaries. Part 1. Rigid boundary. Journal of Fluid Mechanics*, 170, 479-497.
[9] Blake, J. R., Taib, B. B., & Doherty, G. (1987). T*ransient cavities near boundaries Part 2. Free surface. Journal of Fluid Mechanics*, 181, 197-212.
[10] Robinson, P. B., Blake, J. R., Kodama, T., Shima, A., & Tomita, Y. (2001). *Interaction of cavitation bubbles with a free surface.* Journal of Applied Physics, 89(12), 8225-8237.
[11] Blake, J. R., & Cerone, P. (1982). *A note on the impulse due to a vapour bubble near a boundary.* The ANZIAM Journal, 23(4), 383-393.
[12] Supponen, O., Obreschkow, D., Tinguely, M., Kobel, P., Dorsaz, N., & Farhat, M. (2016). *Scaling laws for jets of single cavitation bubbles.* Journal of Fluid Mechanics, 802, 263-293.
[13] Tomita, Y., Robinson, P. B., Tong, R. P., & Blake, J. R. (2002). *Growth and collapse of cavitation bubbles near a curved rigid boundary.* Journal of Fluid Mechanics, 466, 259-283.
[14] Obreschkow, D., Kobel, P., Dorsaz, N., De Bosset, A., Nicollier, C., & Farhat, M. (2006). *Cavitation bubble dynamics inside liquid drops in microgravity.* Physical review letters, 97(9), 094502.
[15] Tagawa, Y., Oudalov, N., Visser, C. W., Peters, I. R., van der Meer, D., Sun, C., Prosperetti, A & Lohse, D. (2012). *Highly focused supersonic microjets.* Physical review X, 2(3), 031002.
[16] Best, J. P., & Blake, J. R. (1994). *An estimate of the Kelvin impulse of a transient cavity.* Journal of Fluid Mechanics, 261, 75-93.

© 2018 ASME

10th International Symposium on Cavitation - CAV2018
Baltimore, Maryland, USA, May 14 – 16, 2018

Dynamics of the bubble near a triangular prism array

[1,2]Yuning Zhang*; [1,2]Shida Li; [1,2]Yongxue Zhang; [3]Yuning Zhang
[1] College of Mechanical and Transportation Engineering, China University of Petroleum-Beijing, Beijing 102249, China
[2] Beijing Key Laboratory of Process Fluid Filtration and Separation, China University of Petroleum-Beijing, Beijing 102249, China
[3] School of Energy, Power and Mechanical Engineering, North China Electric Power University, Beijing 102206, China

Abstract

The behaviors of cavitation bubbles are significantly influenced by the nearby objects. In practice, the bubble usually oscillates and collapses near complex structures. In this study, the dynamics of a laser-induced single bubble near a triangular prism array are investigated experimentally. The behaviors of the bubble, such as the collapse, the resound, and the jet formation, are revealed, which are strongly influenced by several parameters, e.g., the relative position of the bubble and the prisms, the size of the bubble and the distance between the bubble and the array.

Keywords: bubble dynamics; unconventional boundaries; triangular prism array

1. Introduction

Due to its significant mechanical effects, cavitation has been applied widely in the fields of ultrasound cleaning, sonochemistry, biomedicine, and so on. Cavitation has also contributed to the damage of the materials of various fluid machineries[1]. For understanding the underlying mechanisms of these processes, the interaction between cavitation bubbles and the nearby structures is an essential issue to study. In the literature, the bubble behaviors near flat boundary[2], curved boundary[3], cylinder[4] or sphere particles[5] have been studied intensively. However, in practice, cavitation usually occurs near more complex structures, the dynamics of which have not been fully revealed yet. In present study, the dynamics of a single cavitation bubble near a triangular prism array are investigated experimentally.

2. Experimental Methods

Figure 1(a) demonstrates the experimental system employed in this study. The triangular prism array was prepared by employing 3D printing technique with polylactic acid. As shown in figure 1(b), the array consisted of ten triangular prisms, of which the triangular faces are equilateral triangle with height of 2 mm, and the length of the prism is 25mm. The array was placed in a tank filled with deionized water. The single bubble was induced by a pulsed Nd:YAG laser through a focusing lens and was generated above the array. Cases with two kinds of the bubble position were investigated. As shown in figure 1(c), the center of the bubble was located right above the top vertex of the equilateral triangular. In the other condition, the center of the bubble was located right above the bottom vertex of the equilateral triangular, as shown in figure 1(d). The behaviors of the bubble was recorded by a high speed camera (Phantom V1212) operated at 200,000 frames per second. L represents the distance between the horizontal line crossing the bubble center and the top vertex of the triangular. The size of the bubble was controlled by adjusting the energy of the laser beam. In this study, L varied between 0.20 mm and 3.26 mm. For each value of L, the maximum bubble radius (R_{max}) varied within five grades, such as 0.89±0.05 mm, 1.04±0.05 mm, 1.23±0.05 mm, 1.38±0.05 mm, and 1.53±0.05 mm.

3. Results

3.1 The bubble above the top vertex of the triangular

The dynamics of the bubble right above the top vertex of the equilateral triangular is shown in figure 2. Figure 2(a) shows the case with R_{max}=1.48 mm and L=0.20 mm, i.e., the bubble is quite near the prism. During the collapse of the bubble, there is an indentation forming at the top of the bubble, as shown in frame 12. With continued shrinking, the bubble is separated into two parts by the prism, as shown in frame 13. Then, the two parts of the bubble begin rebounding. Meanwhile, jets are produced paralleling with the lateral faces of the prism, towards the adjacent prisms, as shown in frame 15 to 28. Figure 2(b) illustrates the case with L=1.43 mm and R_{max}=1.23 mm. Compared with figure 2(a), due to larger L, during the first collapse of the bubble, the bubble is elongated (frames 4-6) and a jet is produced vertically towards the top vertex of the prism (frames 9-12). Meanwhile, a "counterjet" formation away from the vertex of the triangle is observed, as shown in frames 8-18. This phenomenon is similar with the bubble collapse near

*Corresponding Author, Yuning Zhang: ZYNLZYN.100@163.com

a flat boundary[2]. Then in the rebound process of the bubble, it touches the edge of the prism and the lower side of the bubble is split by the prism (frames 12-21). Until the second collapse, the bubble split into two parts, and jets are generated towards adjacent prisms (frames 22-28). In figure 2(c), the bubble is located much further from the prism with L=3.26 mm and R_{max}=1.58 mm. The bubble is elongated during the first collapse (frames 5-10). When the bubble rebounds, two jets penetrate the bubble simultaneously, of which the directions parallel with the lateral faces of the prism (frames 12-15). Then in the second collapse, another jet is produced vertically downwards (frames 21-24).

Several cases with different L and R_{max} were analyzed. The patterns of bubble behaviors can be classified into three kinds by the value of the "stand-off distance", i.e., $\gamma = L/R_{max}$. When γ<0.7, the bubble will behave as figure 2(a) demonstrating. When 0.7<γ<1.6, the bubble will behave like the case shown in figure 2(b). When γ>1.6, the bubble will behave like the case shown in figure 2(c).

3.2 The bubble above the bottom vertex of the triangular

The dynamics of the bubble right above the bottom vertex of the equilateral triangular is shown in figure 3. In the case shown in figure 3(a), L=0.49 mm and R_{max}=1.53 mm. As shown in frame 1, because of the adjacent prisms, the growth of the lower side of the bubble is obstructed. During the collapse, the upper side of the bubble shrinks faster than the lower side (frames 2-14). Then two jets, which perpendicular to the lateral faces of the adjacent prisms respectively, are produced (frames 15-19). From frame 19, the bubble splits into two parts. The upper part is like a torus and the lower part consists of the jets towards the faces of the prisms. Then all the parts rebound (frames 21-28). In the case shown in figure 3(b), L=1.13 mm and R_{max}=1.38 mm. During the first collapse of the bubble, the upper side shrinks faster than the lower side (frames 4-11) and two jets perpendicular to the faces of the adjacent prisms at first (frames 12-13), which is similar with the case shown in figure 3(a). In addition, a counterjet forms, as shown in frame 13. However, the two jets towards to the faces of the adjacent prisms are covered by a vertical downward jet to the bottom vertex of the triangular very soon. This new jet develops fully, as shown in frames 15-28. In the case shown in figure 3(c), L=1.87 mm and R_{max}=1.53 mm. During the first collapse of the bubble, the bubble is elongated (frames 4-9). Then a jet is generated vertically downwards, while a counterjet is produced upwards (frames 10-23). During the rebound process, the bubble moves towards the bottom of the array, which splits into two parts (the upper one and the lower one) when it collapses again (frames 24-26).

Several cases with different L and R_{max} were analyzed. Both of these two parameters will affect the bubble behaviors. The variation of the bubble dynamics with L and R_{max} is summarized in Figure 4. The square, the circle and the star indicate the bubble behaviors similar with the cases shown in figure 3(a), 3(b) and 3(c), respectively.

4. Conclusion

In this study, the bubble dynamics near a triangular prism array are investigated experimentally. In the cases with the bubble right above the top vertex of the triangular, the bubble may be separated into two parts by the prism and generate jets in several different directions. The patterns of the bubble behaviors can be classified into three kinds by the stand-off distance. In the cases with the bubble right above the bottom vertex of the triangular, the bubble may also produce jets in several different directions. The bubble behaviors are affected by the size of the bubble and the distance between the bubble and the array.

Acknowledgements

The authors would like to acknowledge the National Natural Science Foundation of China (Project No. 51606221) and Science Foundation of China University of Petroleum, Beijing (No. 2462016YJRC003).

References

[1] Zhang, Y., Zhang, Y., Qian, Z., Ji, B., & Wu, Y. (2016). A review of microscopic interactions between cavitation bubbles and particles in silt-laden flow. Renewable and Sustainable Energy Reviews, 56, 303-318.

[2] Lauterborn, W., Bolle, H. (1975). Experimental investigations of cavitation-bubble collapse in the neighbourhood of a solid boundary. Journal of Fluid Mechanics, 72(2), 391-399.

[3] Tomita, Y., Robinson, P. B., Tong, R. P., Blake, J. R. (2002). Growth and collapse of cavitation bubbles near a curved rigid boundary. Journal of Fluid Mechanics, 466, 259-283.

[4] Zhang, A. M., Xiao, W., Wang, S. P. (2013). Experimental investigation of the interaction between a pulsating bubble and a rigid cylinder. Acta Mechanica Sinica, 29(4), 503-512.
[5] Poulain, S., Guenoun, G., Gart, S., Crowe, W., Jung, S. (2015). Particle motion induced by bubble cavitation. Physical review letters, 114(21), 214501.

Figures:

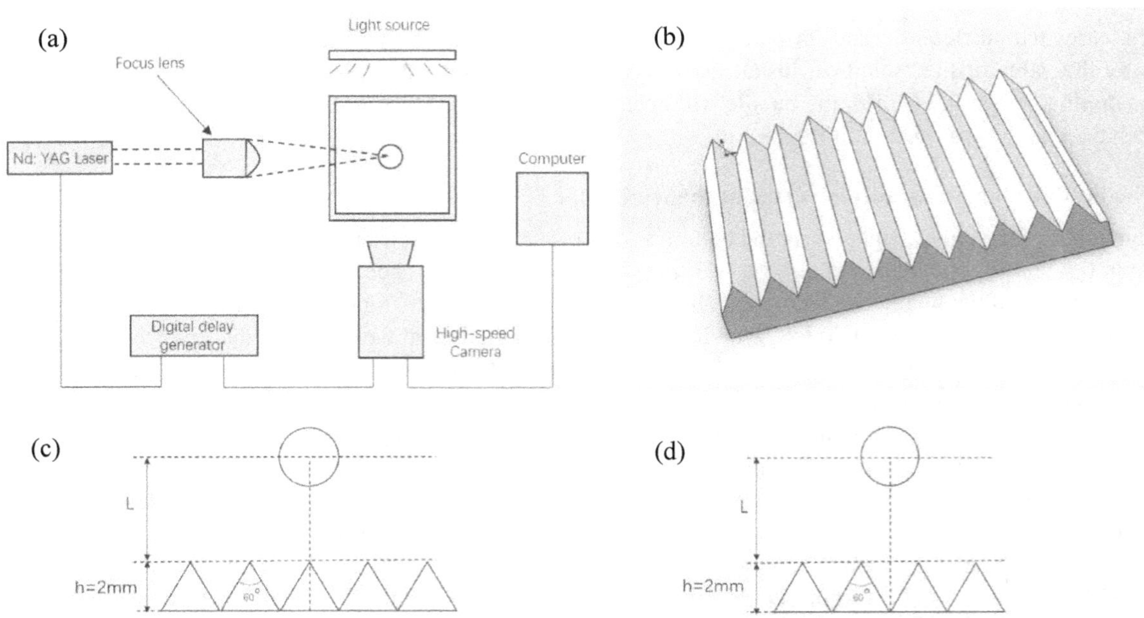

Figure 1 The experimental setup employed in this study. (a) The experimental system. (b) The triangular prism array. (c) & (d) Demonstrations of two kinds of the positions of the bubble.

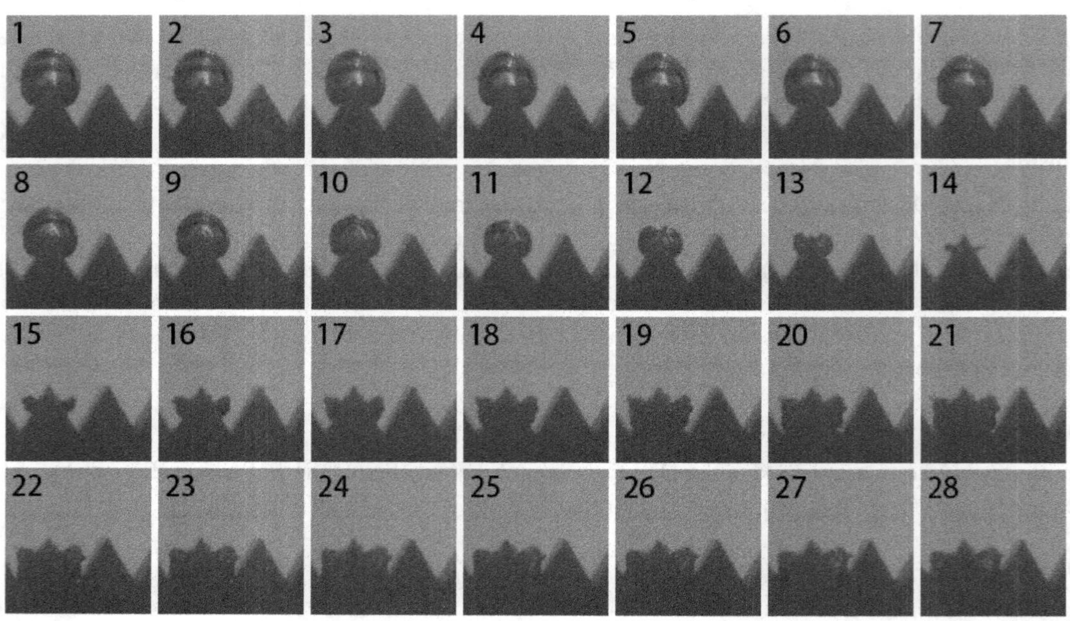

(a) L=0.20 mm and R_{max}=1.48 mm. From frames 1 to 15, the time interval between frames is 10 μs. From frames 15 to 28, the time interval between frames is 5 μs.

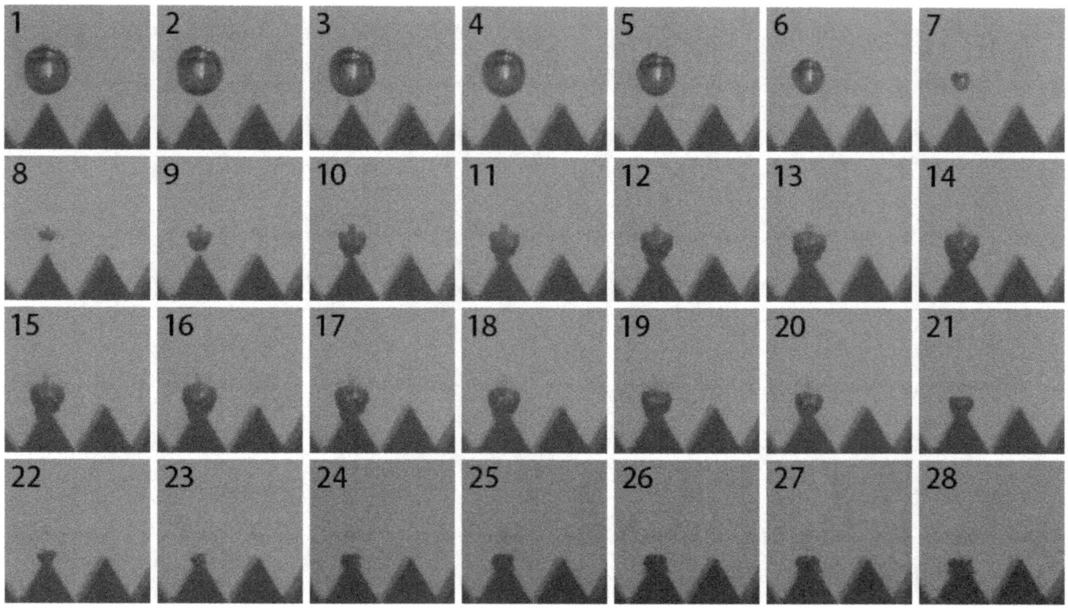

(b) L=1.43 mm and R_{max}=1.23 mm. From frames 1 to 7, the time interval between frames is 20 μs. From frames 7 to 28, the time interval between frames is 10 μs.

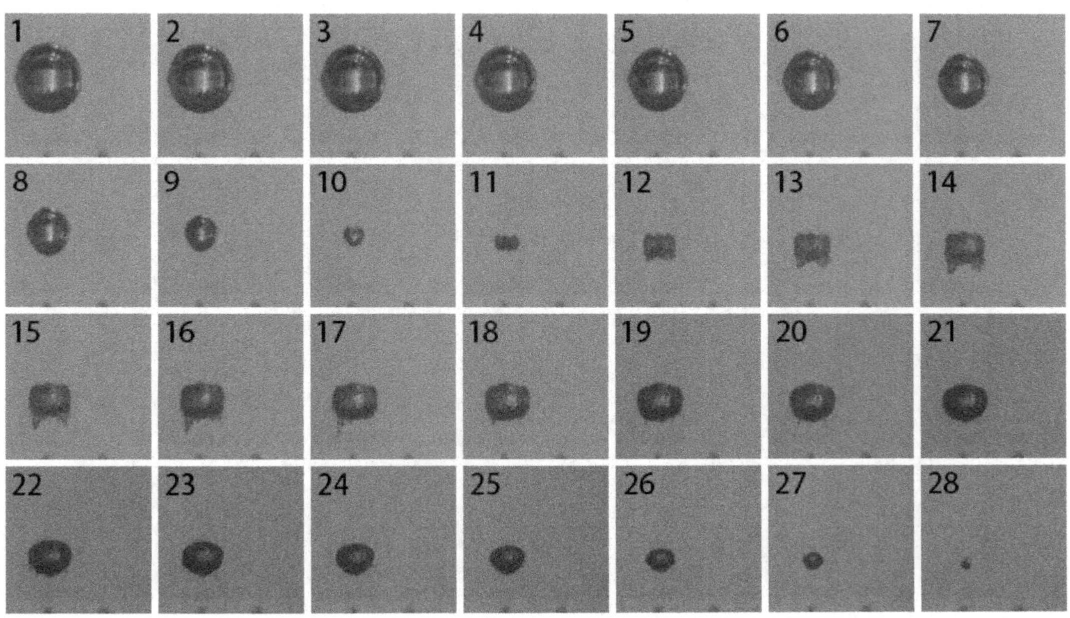

(c) L=3.26 mm and R_{max}=1.58 mm. From frames 1 to 9, the time interval between frames is 15 μs. From frames 9 to 28, the time interval between frames is 10 μs. Because of the limitation of the vision of the camera, the main body of the prisms are outside the frames, and the black points at the bottom of each frame are the top vertices of the triangulars.

Figure 2 The dynamics of the bubble right above the top vertex of the equilateral triangular (the positions of the bubble and the array are demonstrated in figure 1(c)). The frame size is 6.31mm×6.31mm.

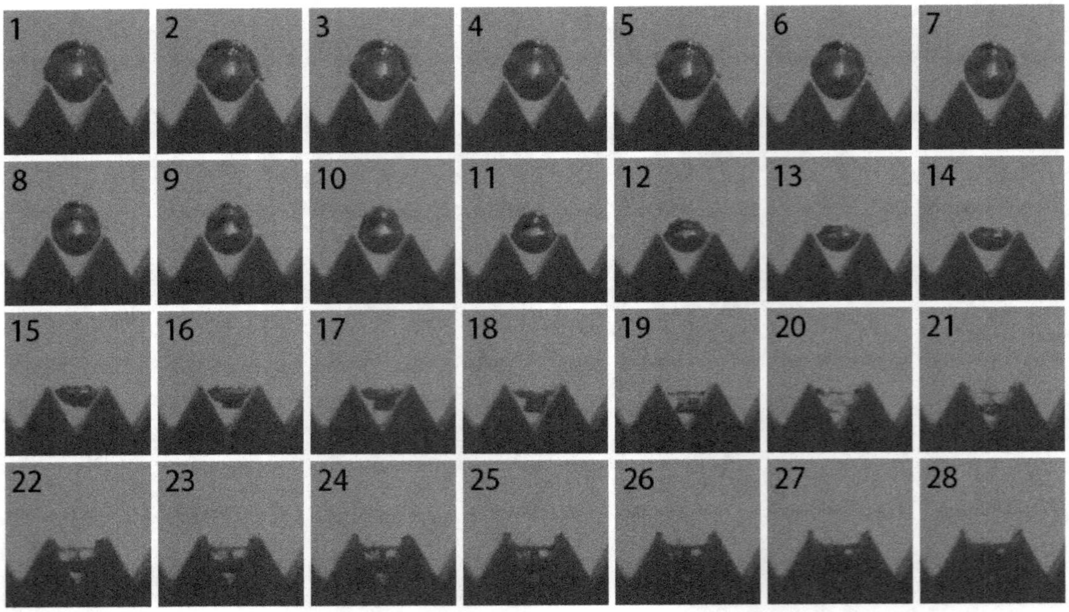

(a) L=0.49 mm and R_{max}=1.53 mm. From frames 1 to 13, the time interval between frames is 10 μs. From frames 13 to 28, the time interval between frames is 5 μs.

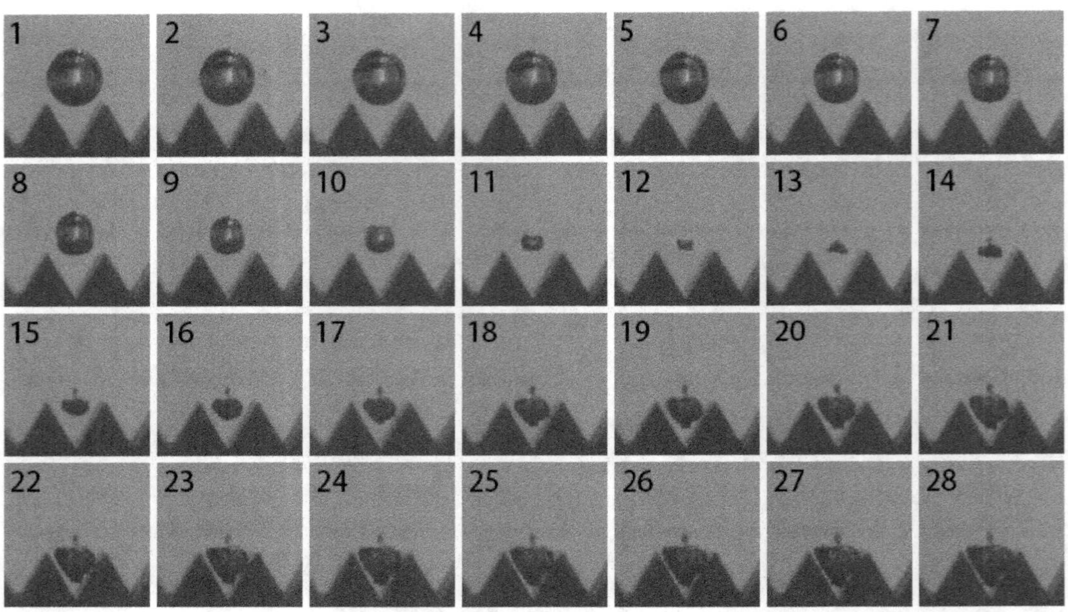

(b) L=1.13 mm and R_{max}=1.38 mm. From frames 1 to 11, the time interval between frames is 10 μs. From frames 11 to 28, the time interval between frames is 5 μs.

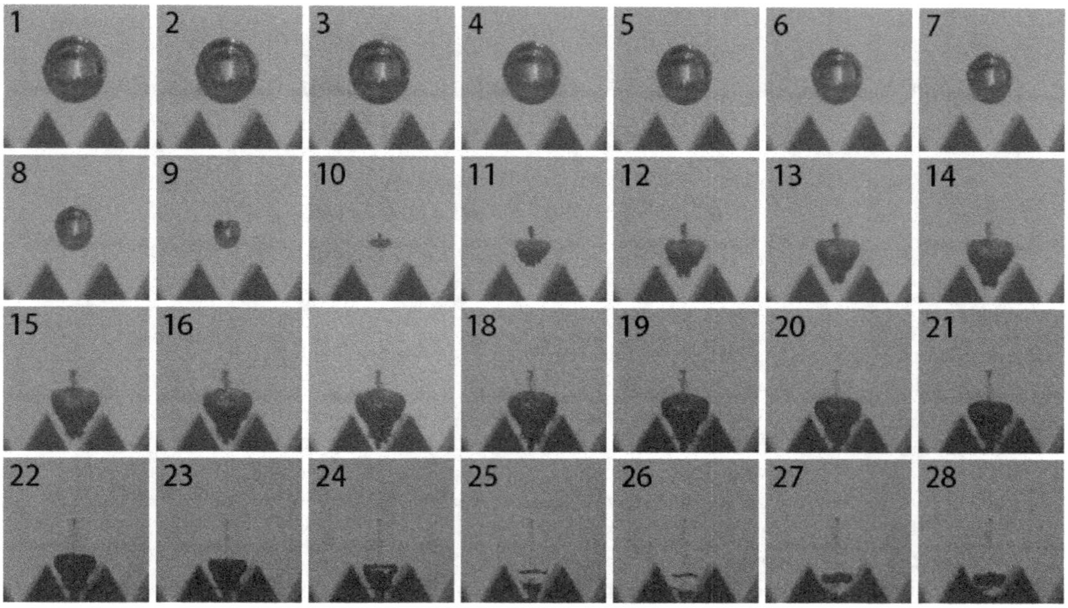

(c) L=1.87 mm and R_{max}=1.53 mm. The time interval between frames is 15 μs.

Figure 3 The dynamics of the bubble right above the bottom vertex of the equilateral triangular, (the positions of the bubble and the array are demonstrated in figure 1(c)). The frame size is 6.31mm×6.31mm.

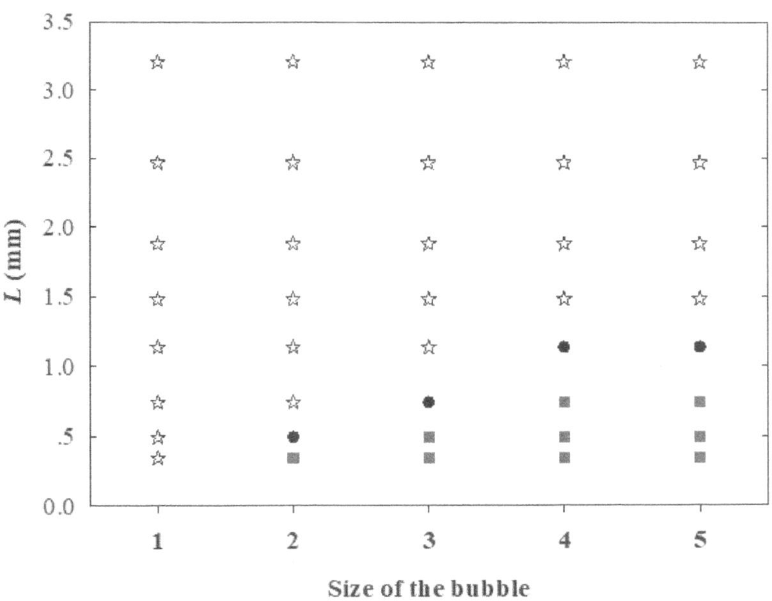

Figure 4 Variation of the dynamics of the bubble right above the bottom vertex of the equilateral triangular with the size of the bubble and the value of L. The labels of the x axis represent the grades of the maximum bubble radius: R_{max} is 0.89±0.05 mm for column 1; R_{max} is 1.04±0.05 mm for column 2; R_{max} is 1.23±0.05 mm for column 3; R_{max} is 1.38±0.05 mm for column 4; R_{max} is 1.53±0.05 mm for column 5. The square, the circle and the star indicate the bubble behaviors similar with the cases shown in figure 3(a), 3(b) and 3(c), respectively.

© 2018 ASME

Effect of Cavitation Impacts on Crack Propagation in Epoxy Resin subjected to Tensile Loading

[1]Yoshiya Kawamura; [1]Farid Triawan*; [1]Kazuaki Inaba*; [1]Kikuo Kishimoto;
[2]Hiroaki Nakamoto; [2]Keisuke Hayabusa;

[1] *Tokyo Institute of Technology, Tokyo, Japan;* [2]*Ebara Corporation, Tokyo, Japan*

Abstract

Investigation of the crack growth mechanism in epoxy resin specimen subjected to both tensile load and cavitation impact was conducted in this work. An experimental apparatus for conducting the test was self-manufactured so that a tensile load can be applied to the specimen while at the same time receiving repeated impacts loads from collapsing cavitation bubbles. The tests were done inside a tank filled with distilled water with fixed temperature. An initial crack was introduced to the specimen in order to analyze the propagation mechanism. Then, the crack growth rate was evaluated under several different loading conditions. As the results, when cavitation load was applied without a tensile load, the initiated crack did not propagate whereas erosion damages were generated on the specimen surface. When tensile load only was applied, the initiated crack propagated at a slow speed. However, when both tensile load and cavitation impacts were applied, the crack growth rate was significantly accelerated by approximately 10,000 times compared to that without cavitation load. According to the captured images, all specimens regardless of the loading condition, fractured in a similar process which began with a blunted crack, crack propagation, cracks merger, and finally specimen rupture. This finding suggests the importance of incorporating both tensile and cavitation impact loads in estimating the life-service of a component in turbomachinery application.

Keywords: cavitation impacts; tensile load; epoxy resin; crack growth

1. Introduction

In fluid machineries such as turbine, pump, and valve, components damage caused by cavitation has been a critical issue that needs a proper solution. Besides the technique for predicting the service-life of a component, the repairing technique is also important for extending the service-life and decreasing the maintenance cost. Several techniques of surface coatings are available for repairing the cavitation damage, such as by metal plating, thermal sprays, weld overlays, ceramic coatings, and polymer coatings. Among them, however, surface coatings by polymers materials have not been evaluated comprehensively.

A number of studies were done to understand the anti-cavitation performance of polymeric coating materials. Hattori et al. [1] conducted cavitation erosion tests on many kinds of polymeric materials by means of vibratory cavitation test (ASTM G32) and cavitating liquid jet test (ASTM G134). They revealed that polymeric materials show a proportionality between its fatigue strength and cavitation erosion resistance. Garcia et al. [2] evaluated the generation of damage in epoxy-based multilayer coating by vibratory cavitation test. They explained the crack growth mechanism during the incubation stage is most likely driven by plastic deformation and crazing. A recent report by Hibi et al. [3] which investigated the cavitation damage behavior of an epoxy-resin material under tensile load effect can be found. According to their experimental results, besides cavitation erosion, the specimen experienced fracture if a constant tensile load was introduced. Furthermore, they explained that the fracture mechanism was mainly driven by the combination of fatigue-like crack growth and creep deformation. However, no quantitative analysis on the crack growth rate was provided.

The present work serves as the continuation study of Hibi et al. [3]. In order to quantify the crack growth rate in epoxy resin specimen under both tensile load and cavitation impacts, experiments using the same apparatus were conducted. A new approach was implemented by introducing an initial crack-tip to the specimen. By this, we could monitor the crack growth progress more accurately by a video camera. Utilizing the captured images, observation of the crack propagation and failure mechanism could be done more easily. Finally, the crack growth rate at different loading condition was quantified and compared with others.

*Corresponding Authors, Farid Triawan: triawan.f.aa@m.titech.ac.jp, Kazuaki Inaba: inaba.k.ag@m.titech.ac.jp

2. Experimental Procedures

2.1. Test conditions

Cavitation tests based on ASTM G32 were carried out. A self-developed experimental apparatus by Hibi et al. [3] was utilized to apply several kinds of constant tensile load and cavitation impacts during the test. Figure 1 shows the experimental apparatus setup. Distilled water was used as the test liquid with a temperature of 25°C controlled by the circulation tank. A magnetostrictive vibrating device was used to generate cavitation bubbles between a stainless horn-tip (SUS304) and specimen surface. The vibration frequency was set to be 19.5 kHz with an amplitude of 50 μm based on ASTM G32 [4]. To generate different intensities of cavitation impacts, the specimen was placed at several distances (*d*) of 0.5, 1.0, 1.5, 2.0, and 3.0 mm from the horn-tip. The specimen condition during cavitation test was monitored in real-time by a video camera utilizing a mirror put at the below of the specimen. Chucks were used to apply a constant tensile load of 5, 10, and 15 kg to the specimen through a stainless steel wire using some weights. Two linear guides which can move freely were used to set the chucks. Table 1 summarizes all of the loading conditions.

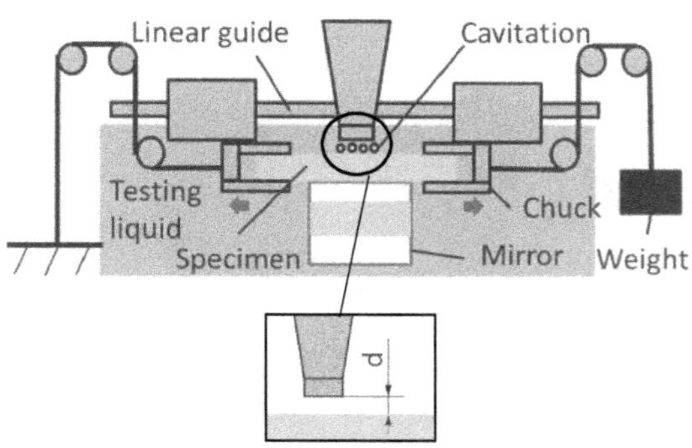

Fig. 1 Experimental apparatus of cavitation test under uniaxial constant tensile load from the weight

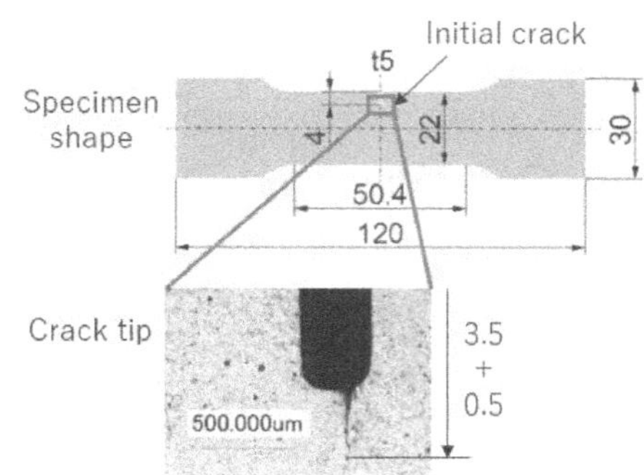

Fig. 2 Specimen shape and the initial crack-tip made by a razor. Dimension is in millimeter.

Table 1 Loading conditions applied at the experiment

		Constant Tensile Load, kg		
		5	10	15
Distance between specimen and tip d, mm	0.5	1	1	1
	1.0	1	1	1
	1.5	1	2	1
	2.0	1	1	1
	3.0	-	-	1
Creep test (tensile load only)		-	-	1

Note: the smaller the *d*, the higher the cavitation intensity

2.2. Specimen

Figure 2 shows the specimen used in the experiment which was made by epoxy-resin Bond E Set L (Konishi Co., Ltd.). A base mold of silicon rubber was manufactured for making the specimen shape. Following the manual, epoxy-resin with a ratio of base resin and curing agent of 1:1 was blended. The blend was then poured down into the mold and cured at a room temperature for 1 day until became hard enough, and subsequently put into a thermostat oven chamber (Tokyo Rikakikai Co., Ltd., WFO-450PD) at a temperature of 50°C for 72 hours to complete the curing process. A crack tip perpendicular to the direction of the tensile load was initiated at one-side of the specimen by a micro wire-cut machine (BS-300 CP-A, Meiwafosis Co., Ltd.) for 3.5 mm, and then by razor blade for 0.5 mm. The total crack length was about 4 mm as shown in Fig. 2.

3. Experimental Results and Discussion

3.1. Crack Propagation Process

Based on the monitoring results by a video camera shown in Fig. 3, all specimens tested at different tensile load and cavitation impacts acted in a similar process of crack propagation until rupture. After several minutes or hours of testing, the initiated crack became blunted while propagated in the transverse direction. As time passed by, the crack length increased, and cavitation erosion damage was generated in the specimen. Finally, the specimen fractured by the merging of the propagated crack and penetrated erosion damage. The detailed explanation of the penetrated cavitation damage was provided by Hibi et al. [3]. In addition, only the specimen tested at a tensile load of 5 kg and horn-tip distance of 2.0 mm did not fracture after 89 hours of testing. Because the location of the crack due to erosion and the time of merging of the cracks are irregular, the shape just before the rupture was different. On the other hand, in the case of a test without cavitation impact (tensile load of 15 kg), it was confirmed that the crack gradually propagated until rupture occurred. The time needed until rupture got shortened as the tensile load and the cavitation intensity was increased (3.0 mm to 0.5 mm).

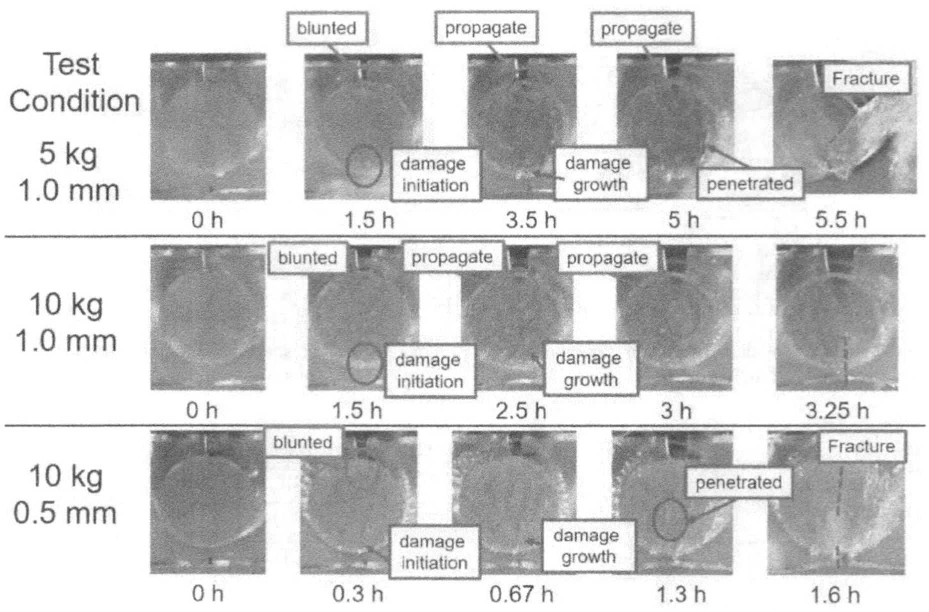

Fig. 3 Crack propagation process of specimens under tensile load and cavitation impacts

3.2. Crack Growth Rate

As the process up to rupture was qualitatively evaluated, the quantitative evaluation is thus necessary. For this, we attempted to quantify by paying attention to the crack length progression versus time. The crack length measurement was done as the following. First, images of crack growth were captured at regular time intervals from the recorded video image. Next, an image processing software, Micro Labo LT 3.6, was used to measure the crack length at every time interval. The width of the test piece was used as a reference length data. The crack growth rate was obtained by dividing the difference in crack length between two captured images by the time difference as shown in Eq. (1). The time at this capture was the average time as depicted in Eq. 2.

$$\frac{da}{dt} = \frac{(a_{n+1} - a_n)}{(t_{n+1} - t_n)} \quad (1)$$

$$t = \frac{(t_n + t_{n+1})}{2} \quad (2)$$

The stress intensity factor was calculated using Eq. (3), which is the stress intensity factor of the semi-infinite plate with edge cracks.

$$K_{\rm I} = 1.1215\sigma_{\infty}\sqrt{\pi a} \tag{3}$$

In this equation, σ_{∞} is the stress obtained from the applied tensile load divided by the cross-section, and a is the corresponding crack length. In this case, in order to obtain the crack growth rate, the crack length was taken as the average length of two captured images.

The crack lengths vs. time of all tests were summarized in Fig 4. It was found that as the tensile load got higher and the distance between the test piece and the horn-chip was smaller, the time to fracture tended to be faster. The relationship between the crack growth rate and stress intensity factor is shown in Fig. 5. It was revealed that the crack growth rate was dramatically accelerated by 10,000 times compared to that of without cavitation load. Furthermore, the growth rate was found to fluctuate and showed almost constant value regardless of the stress intensity factor with a range of 1,000 mm/s. Based on the study reported by Hibi et al. [3], it is most likely the large acceleration of crack growth rate under mixed loads of tensile and cavitation impacts was mainly caused by the increasing temperature and vibrational load effects generated by cavitation impacts. In this regard, further investigation is therefore needed.

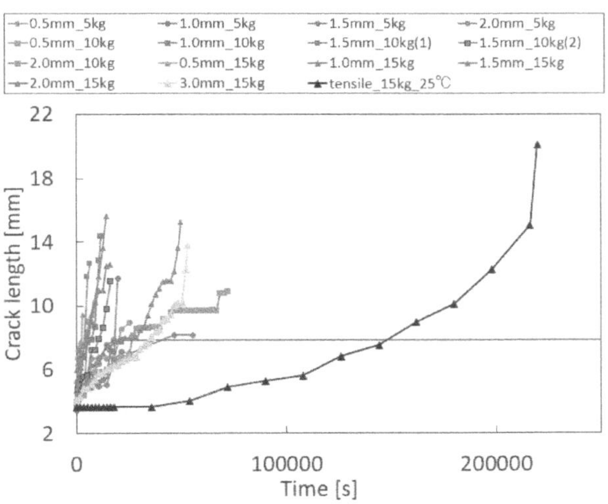

Fig. 4 Time history of crack length progression measured by image processing software based on testing results at different loading conditions

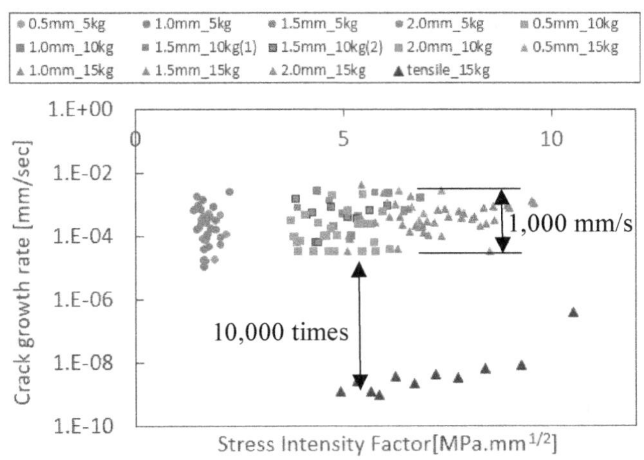

Fig. 5 Crack growth rate vs Stress intensity factor. Compared with that under a tensile load of 15 kg, the growth rate under mixed tensile load and cavitation conditions were approximately 10,000 times faster.

4. Conclusion

Investigation on the effect of cavitation impacts on the crack propagation in an epoxy-resin material under tensile load was conducted. It was revealed that under a mixed condition of tensile load and cavitation impacts, the crack propagation was significantly accelerated by 10,000 times higher than that without cavitation impacts. Further investigation to understand the mechanism which generate this phenomenon is therefore needed. This finding would be useful for improving the life-prediction technique of coating materials in turbomachinery operation.

References

[1] Hattori, S., Itoh, T. (2011). *Cavitation erosion resistance of plastics*. Wear. 271(7), pp. 1103-8.

[2] García, G.L., López-Ríos, V., Espinosa, A., Abenojar, J., Velasco, F., Toro, A. (2014) *Cavitation resistance of epoxy-based multilayer coatings: Surface damage and crack growth kinetics during the incubation stage*. Wear. 316(1), pp. 124-32.

[3] Hibi, M., Triawan, F., Inaba, K., Takahashi, K., Kishimoto, K., Hayabusa, K., Nakamoto, H. (2018). *Cavitation damage of epoxy resin subjected to uniaxial tensile loading*, Mech. Eng. J., JSME, DOI: https://doi.org/10.1299/mej.17-00151.

[4] ASTM G32, ASTM Designation, Annual Book of ASTM Standards, 2005.

© 2018 ASME

Effect of Gas Content on Tip Vortex Cavitation

[1]Ali Amini*; [2]Martino Reclari; [2]Takeshi Sano; [1]Mohamed Farhat;

[1]*École Polytechnique Fédérale de Lausanne, Lausanne, Switzerland*

[1]*Mitsubishi Heavy Industries, R&D Center, Takasago, Japan*

Abstract

Occurrence of tip vortex cavitation (TVC) is a common problem in axial hydraulic machines. The behavior of this type of cavitation is so sensitive to variations in the flow parameters and the gas content level of water. Regarding this, we have investigated TVC generated by an elliptical hydrofoil with NACA-16020 cross-section in the high-speed cavitation tunnel of EPFL. We measured the cavitation incipience and desinence thresholds for different flow conditions as well as various gas content levels of between 50 and 100 % oxygen saturation at 18 °C. We observed that the disappearance threshold of TVC increases with the gas content level, which is an indication of the fact that once the cavitation develops in the tip vortex, the dissolved gas in the surrounding supersaturated liquid diffuses into the cavity. Moreover, we found that the extent to which the desinence is delayed depends on the incidence angle and the upstream flow velocity; the parameters that determine the characteristics of the boundary layer on the hydrofoil. According to the flow visualization results, we argue that this gaseous cavitation is enhanced when a laminar separation bubble is formed on the suction side of the hydrofoil.

Keywords: *tip vortex cavitation, gas content, inception and desinence hysteresis, outgassing*

Introduction

The rotational motion of the fluid around the axis of tip vortices may result in very low static pressures in the core of these vortices, which makes them extremely vulnerable to the inception of cavitation. Occurrence of tip vortex cavitation (TVC) in axial turbines, pumps and marine propellers can lead to a severe erosion of the impeller blades and the stationary parts of the machine, with a significant increase in maintenance costs. In most of the cases, TVC is the first type of cavitation that appears in the machine [1, 2, 3, 4] and results in extensive noise emissions and structural vibrations [5]. McCormick [6] investigated the role of the boundary layer on TVC and stated that the size of the viscous core of a tip vortex scales with the boundary layer thickness on the pressure side of hydrofoils. Different authors [7, 8, 9] later proposed the following correlation for the cavitation inception number: $\boldsymbol{\sigma_i = KC_L^2Re^m}$, where C_L and Re are the lift coefficient and Reynolds number, respectively, and K and m are empirical constants. Besides the flow parameters, it is known that the gas content of water also plays a major role in the cavitation inception. Arndt and Maines [10] introduced the terms *''weak''* and *''strong''* water depending on the size and distribution of nuclei in water. They showed that in weak water with enough large nuclei, cavitation incepts when the pressure in the core of the tip vortex reaches the vapor pressure, however, strong water with fewer and/or smaller nuclei could resist significant tensions. Despite the efforts dealing with the inception of TVC, little attention was paid to its desinence. Holl et al. [11] demonstrated that hub vortex cavitation disappears at higher pressures when the gas content of water is increased. Recently, Groß et al. [12] examined the effect of gas content on the nucleation rate from a single nucleus trapped in a small hole. They demonstrated that the nucleation site does not disappear and permanently produces tiny bubbles as long as the flow is supersaturated. They have found that the nucleation rate from the wall-bounded nucleus is proportional to the velocity and the supersaturation degree of the freestream. In the present study, we address the role of the dissolved gas content in TVC for a variety of flow conditions. With this aim, the behavior of TVC is experimentally investigated in a simplified case study to determine how different flow conditions and gas content levels affect the TVC inception and desinence. The experimental methods and the corresponding results are presented in the following.

*Corresponding Author, Ali Amini, ali.amini@epfl.ch

Experimental set-up

The experimental studies were performed in the EPFL high-speed cavitation tunnel, which has a $150 \times 150 \times 750$ mm test section and a maximum inlet velocity of 50 m/s. A stainless steel hydrofoil with a NACA 16-020 cross section and an elliptical planform was used for the investigations, because it generates a well-defined and undisturbed TVC. The span and the root chord length of the profile are 9 mm and 6 mm, respectively. To monitor the gas content level, we used a Presens O_2 Dipping Probe (DP-PSt7) fitted upstream to the test section. To modify the gas content, we adopted the following procedure: A decrease of the gas content is obtained by running the tunnel with a super cavitation at 15 m/s upstream velocity, while an increase of the gas content is achieved by producing a highly turbulent free surface flow in the test section and downstream. In both procedures, the oxygen concentration is monitored in real time. Since the solubility of air in water is dependent on the temperature, the latter was kept constant at 18 °C. The procedure for the TVC inception/desinence tests was the following: For a given freestream velocity and incidence angle, the static pressure was gradually decreased from an initial value, high enough to ensure a cavitation-free regime. The cavitation inception index (σ_i) was determined upon the observation of a tiny cavitating core attached to the foil. The pressure was further reduced to allow for a well-developed TVC, and then the pressure was increased until the cavitation disappeared, which determined the cavitation desinence number (σ_d).

Results and discussion

Our observations on the inception and desinence of TVC reveal that under specific flow conditions there exists a relatively large gap between the two thresholds. Figure 1 illustrates this hysteresis for fully air-saturated water at upstream velocity of 10 m/s (Re=600,000) and 12° incidence angle. We have reported on the same figure the cavitation number and the corresponding values of the static pressure measured at the tip of the hydrofoil. One may easily observe that TVC incepts at $\sigma_i = 1.7$ and disappears at a much higher cavitation number, $\sigma_d = 4.0$, which is associated with a pressure of around 1.4 bar at the tip of the hydrofoil.

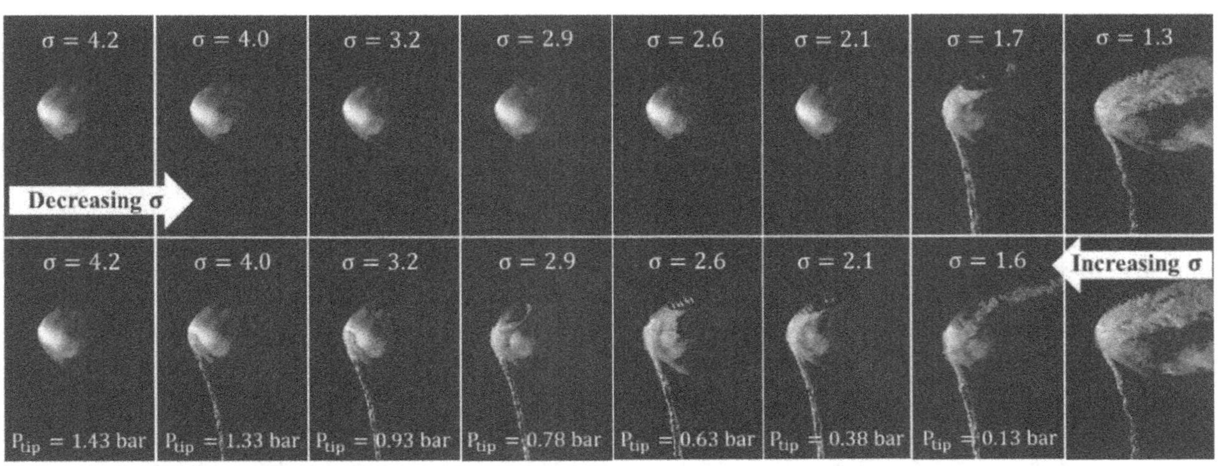

Figure 1: Hysteresis in TVC inception and desinence at $V_\infty = 10$ m/s, $\alpha = 12°$, and 100 % of O_2 saturation

Similar observations were conducted for two upstream velocities, different incidence angles and oxygen concentrations, the results of which are depicted in Figure 2. Almost for all the cases, the cavitation desinence number is significantly higher than that of the incipience. As expected, lower pressures are required for the initiation of the TVC at 50 % of oxygen saturation, because a lower gas content level is normally associated with lower amounts of active nuclei in water. We also observe that the TVC disappears at much higher pressures when water is fully saturated. In addition, the results of Figure 2 illustrate the role of the flow velocity and the angle of attack in the gap observed between the cavitation inception and desinence thresholds. The hysteresis clearly shrinks when the velocity is increased from 10 to 15 m/s. It is also found that the gap is the largest around 12° incidence angle and diminishes as we move toward 8° or 16°.

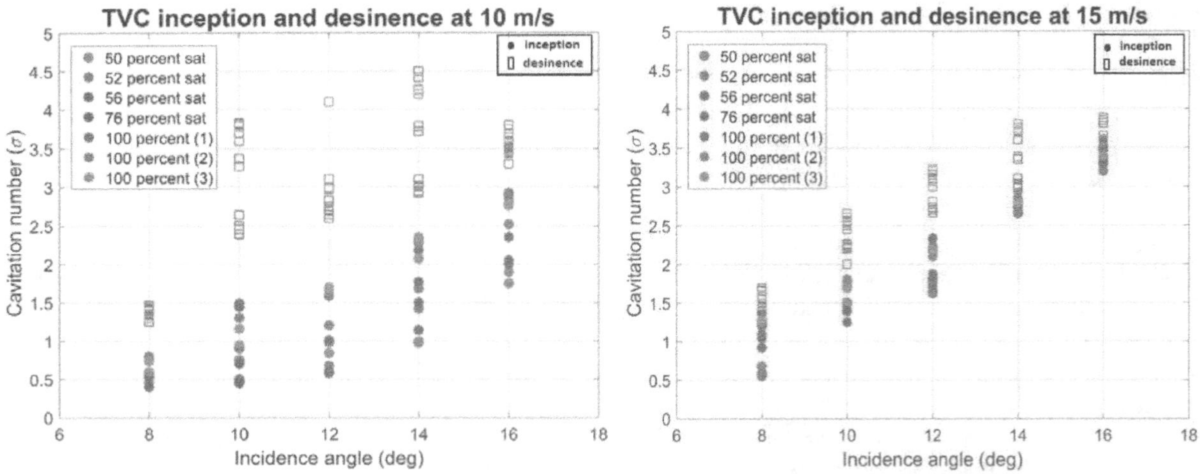

Figure 2: TVC inception/desinence for different incidence angles and gas content levels, measured at 10 and 15 m/s

Of more interest in Figure 2 is what we observe on the desinence limits of the TVC. Considering that the cavitation is only a result of the vaporization of water, the TVC must disappear as soon as the pressure inside the vortex is higher than the saturation vapor pressure. Due to the high static pressures close to the vortex core depicted in Figure 1, it is very unlikely that the evaporation process sustains the cavity. In addition, the fact that the desinence threshold varies with the gas content (Figure 2) is a clear indication that TVC resists pressure levels higher than vapor pressure because of the diffusion of the dissolved gas in water across the interface of the vortex cavity. Indeed, the local decrease in the pressure due to the rotational motion of the fluid around the vortex axis changes the local saturation level of water. For instance in the case of the 100 % of saturation, as long as the pressure within the cavitating tip vortex is below the atmospheric pressure, the water flowing around the vortex core is supersaturated.

Besides the effect of gas content, Figure 2 also shows that the delay in the desinence of the TVC depends on the flow conditions and that the hysteresis is stronger between 10° and 12° and at the lower freestream velocity of 10 m/s. We argue that the upstream velocity and the incidence angle influence the cavitation desinence through the boundary layer state on the hydrofoil suction side. We have observed that the large hysteresis is always associated with the formation of a glassy cavity at the tip of the hydrofoil, as illustrated on Figure 3. According to high-speed visualizations (not presented here), this transparent cavity could be an indication of a laminar separation of the boundary layer. Nevertheless, it is not yet clear how such a laminar separation enhances the outgassing process and delays the cavitation desinence.

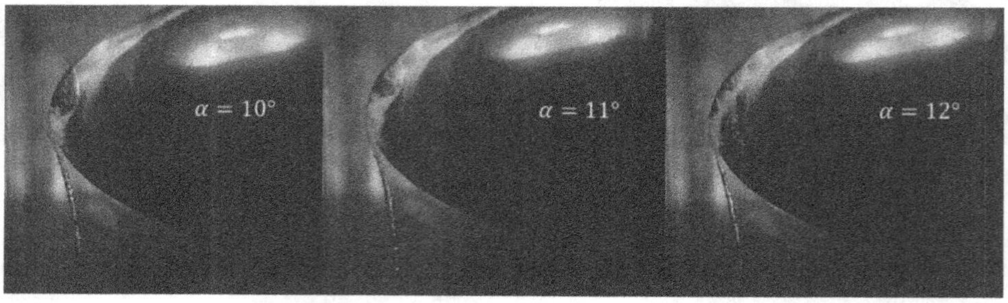

Figure 3: Evolution of the shape of the TVC with incidence angle at 10 m/s and $\sigma = 3$

The hysteresis in the TVC inception and desinence, evidenced here, poses a challenging problem for the design and the model tests of hydraulic machines. In fact, it is not clear which cavitation index (incipience/desinence) should be taken into account in the performance tests. Moreover, the actual scale-up rules always fail in predicting TVC occurrence in axial turbines. The work presented here is being pursued to address these challenges.

Conclusion

We experimentally investigated the cavitation incipience and desinence within a tip vortex, developed by an elliptical hydrofoil, in a wide range of flow parameters and gas content levels. Our observations confirm the hysteresis between cavitation incipience and desinence thresholds. Besides the gas content, we have found that this phenomenon is also dependent on flow parameters, peaking at ~12° incidence angle and 10 m/s upstream velocity. We argue that once a TVC is developed, the excessive gas within the supersaturated water diffuses into the cavity and sustains cavitation at pressure levels as high as atmospheric pressure. The combination of the flow visualizations and inception/desinence measurements indicates that a laminar separation near the hydrofoil tip likely enhances this outgassing process. Further research is underway to distinguish between vapor cavitation and gaseous cavitation, compare their consequences and ultimately define relevant criteria for cavitation tests of hydraulic machines at reduced scale.

Acknowledgement

The research leading to these results has received funding from the MSCA-ITN-ETN of the European Union's H2020 program, under REA grant agreement N°642536 and Mitsubishi Heavy Industry (Japan).

References

[1] Bailo, CDR Giuseppe M., et al. "Cavitation Committee." International Towing Tank Conference (1996).
[2] Arndt, Roger EA, et al. "Instability of partial cavitation: a numerical/experimental approach." (2000).
[3] Shen, Young T., Scott Gowing, and Stuart Jessup. "Tip vortex cavitation inception scaling for high Reynolds number applications." Journal of Fluids Engineering 131.7 (2009): 071301.
[4] Song, Mingtai, et al. "An acoustic approach to determine tip vortex cavitation inception for an elliptical hydrofoil considering nuclei-seeding." International Journal of Multiphase Flow 90 (2017): 79-87.
[5] Arndt, R., et al. "The singing vortex." Interface focus 5.5 (2015): 20150025.
[6] McCormick, B. W. "On cavitation produced by a vortex trailing from a lifting surface." Journal of Basic Engineering 84.3 (1962): 369-378.
[7] Platzer, G. P., and W. G. Souders. "Tip vortex cavitation characteristics and delay on a three-dimensional hydrofoil." 19th ATTC Conference, Madrid, Spain. 1980.
[8] Fruman, D. H., et al. "Tip Vortex Roll-Up and Cavitation," Nineteenth Symposium on Naval Hydrodynamics, Seoul, Korea, Aug. (1992).
[9] Arndt, R. E., and Christian Dugue. "Recent advances in tip vortex cavitation research." Proc. International Symposium on Propulsors Cavitation. 1992.
[10] Arndt REA, Maines BH. Nucleation and bubble dynamics in vortical flows. J. Fluids Eng. 122 (2000): 488–93.
[11] Holl JW, Arndt REA, Billet ML. Limited cavitation and the related scale effects problem. Proc. Intl. Symp. Fluid Mech. Fluidics, 2nd, (1972): 303–14. Tokyo: JSME.
[12] Groß, Tim, Gerhard Ludwig, and Peter Pelz. "Experimental and theoretical investigation of nucleation from wall-bounded nuclei in laminar flow." 16th International Symposium on Transport Phenomena and Dynamics of Rotating Machinery (2016).

© 2018 ASME

Effect of hole-to-hole spacing and row-to-row spacing as well as inclined angle of venting holes on the pressure-equalizing film along the surface of a vertical launched underwater vehicle

Guihui Ma, Fu Chen, Jianyang Yu*;

School of Energy Science and Engineering, Harbin Institute of Technology, Harbin, Heilongjiang, China

Abstract

Pressure-equalizing film is a slice of air layer attached to vehicle's exterior with nearly uniform inner pressure. Similar to ventilated cavity in composition, it is generated through exhaust process via venting holes as vehicle travels from launch tube to water surface, and could significantly weaken the pitching moment and environment disturbance that vehicle suffers. Depending on the venting structure, the forming speed of this film as well as its covering range on vehicle's exterior are the key factors that determine the improvement effect of vehicle's trajectory stability. In this article, the emerging process of vehicle through the water layer with pressure-equalizing film is simulated. Based on previous work, single-row holes with different hole-to-hole spacings and double-row holes with different row-to-row spacings as well as inclined angles are chose to investigate the influence of geometry parameters and arrangements of holes on the evolution of air film and the hydrodynamic characteristics of vehicle. Discussions about the distribution of phase and pressure, mass flow rate, and vortices motion near venting holes indicate that, as the value of pitch to diameter ratio (s/d) increases with the same total exhaust area, film length grows faster with the inner vortex structures further development. As for the double-row hole cases, reverse flow between different rows of holes is induced which has enlarged the size of film covering the exits of back holes, leading to the faster exhausting of back holes. Besides, the decreasing of row-to-row spacing inhibits the growing of film length, and simplifies the flow structures that dominate the reverse flow. While as the venting holes gradually incline to the gravity direction (inclined angle changes from 90°/90°to 90°/45°to 45°/45°), the shearing action and impeding effect of surrounding water decrease, which have weakened the reverse flow and promoted the axial extension of air film in some degree.

Keywords: underwater vehicle; pressure-equalizing exhaust; venting structure

Introduction

Ventilated cavity has been investigated for several decades for its usage in friction reduction as vehicle moves in water[1]. As a new application of bubble dynamics and could be recognized as a special thin cavity with no need for equipping cavitator, pressure-equalizing film is addressed to weaken vehicle's hydrodynamic load and ensures its trajectory stability after it is vertical ejected at launch depth. As Fig. 1 shows, vehicle's vertical ejection always carries horizontal velocity from platform with gravity center beneath its buoyancy center. Thus taking other disturbance into account (current and wave motion, transition of phase when piercing water surface et al), large rotation of vehicle even structural damage will be driven under the resultant moment stemming from the unequal load between vehicle's front and back parts. However, after adding a slice of air film upon its exterior with nearly uniform pressure, where hydrodynamic force will be unloaded, then the pitching moment and lateral load that vehicle suffering will be weaken, and the emerging trajectory of vehicle from water will be righted.

Relevant studies mentioned in literature about pressure-equalizing exhaust are limited, and its mechanism as well as how to get better performance are still inadequate. Some of the main obstacles are that it's difficult to equip facilities on an uncontrolled vehicle without disturbing its underwater motion. And multi-DOF (degree of freedom) motion simulation with multiphase flow impacting seems complicated. By contrast, a considerable amount of research about ventilated cavity, which is helpful for revealing the basic mechanism and implementing practical application, has been conducted recently. Wang[2] concludes that vortex structures that dominates the taking shape of air film mainly involve shear layer vortex, horseshoe vortex, counter rotating vortex pair, and re-entrant jet. Parameters like the ventilated rate[3,4], structure of venting holes[5], velocity of vehicle[6] all influence the formation and evolution of air film. When cavity develops, the instability forms can be concluded into the cavity oscillation, fluctuation of cavity interface and twisting of cavity interface[7]. And re-entry jet is the key factor that inducing the shedding of bubble[8,9].

*Corresponding Author, Jianyang Yu: jayoung1229@gmail.com

The fast formation and large area coverage of this slice of air film determine its timely righting effect on the deflected vehicle at an early time. This exhausting process is actually the dynamic response of pressure inside chamber to the gradually decreasing ambient pressure through exhaust structures. So with a finite amount of pre-filled air, fixed launch depth and horizontal moving velocity, the structure parameters of venting holes like the hole-to-hole space, the injection angle, the rows amount of holes as well as their arrangement play an important role in the evolution of film. In this paper, vertical launching process of an underwater vehicle with pressure-equalizing film is simulated after validation by a simplified experiment that a projectile is ejected from water with high-pressure air. Effect of hole-to-hole spacing, row-to-row spacing as well as the injection angle of venting holes on the evolution of air film and vehicle's load are studied. Flow field parameters like distribution of phase, pressure, streamlines, as well as the mass flow rate are analyzed.

Body

Simulation setup

Based on the finite volume method, Mixture multiphase model and $k-\varepsilon$ turbulence model are adopted to simulate the interaction between water and air considering the calculation amount and computational stability. The Semi-Implicit Method for Pressure-Linked Equation (SIMPLE) method is selected for the coupling of the pressure term and velocity term. The Quadratic Upwind Interpolation for Convective Kinematics (QUICK) scheme is employed for the volume fraction, which can capture a thin interface between air and water. Computational domain with only 1/8 of the columnar flow field is shown in Fig. 2 for its periodic characteristics along the circumferential direction, accordingly the lateral boundary condition is periodic. Arrangements of holes on this vehicle model with blunt head and flat bottom are also drew in Fig. 2. The initial flow field before exhausting is getting from a uniform acceleration of vehicle before reaching the fixed velocity.

Verifications of numerical methods

The gas ejection process from water of an axisymmetric body with a hemisphere head was simulated to verify the numerical method. Displacement of the projectile relative to the bottom of launch tube and trailing bubble at typical moment are given in Fig. 3, and conclusions can be found that the relative error is no more than 5%. Comparison of trailing bubble shows that the results between experiment and simulation agree well with each other both in bubble shape and length. Grid independence check is conducted among different amounts of grids involving Case1:4700000, Case 2: 8230000, Case 3: 9490000 and Case 4: 10380000 with the same first layer thickness. Pressure coefficient at typical moment in Fig. 4 shows that the relative error is no more than 1% when the mesh amount reaches nearly 9000000, so the grid refinement in Case 3 is selected among these cases in this article.

Results and discussion

Typical evolution process and of film through exhausting

The evolution of air film is presented with the distribution of phase volume fraction on vehicle's exterior as shown in Fig. 5, and typical moment when film has fully developed at $\overline{T}=2.0$ (Fig. 6). Furthermore, the pressure coefficient at the centerline of film varying from vehicle's head to bottom are also obtained to show the inside pressure of air film and its effect on vehicle. Thus some parts with changeless features can be found from these results. Three region with special compositions can be recognized, namely the air phase region (①), the mixture phase region (②), and the leakage phase region (③). The air phase region presents a hourglass shape with wide ends and thin middle, which attributes to the fast expansion at the exits of holes and stagnation when meeting the upstream motion of re-entrant jet. Mixture phase region takes up most parts of the film and is dominated by re-entry jet stemming from the film closure region. While the leakage region is separated from mixture phase region by impinging jet upon vehicle's exterior, small bubbles are then generated, and will flow downstream by the entrainment of surrounding water. The fast formation of these three parts, which could be clearly identified throughout the whole exhausting process, directly determines the performance of pressure-equalizing film.

As for the pressure which stands for the loading level of vehicle, it presents regular features within local parts along the axial direction before exhausting (when $\overline{T}=0.2$), which attributes to their special outline and shape at different axial locations. The highest pressure locating on the top of vehicle is generated by the stagnation effect, and whereafter it goes down when water bypasses vehicle's head. While the water reach the cylindrical parts, local pressure gradually rises till the bottom of vehicle since the hydrostatic pressure grows as water depth increases. When exhaust starts, nearly even pressure comes into being downstream venting holes, and it extends as vehicle moves even covering a large parts of vehicle's exterior along the axial and circumferential direction (marked ③ in Fig. 7). It is the embodiment of pressure-equalizing feature of film and mainly made up of air phase region and mixture phase region.

Exhausting process with different hole-to-hole spacings

Note that the amount of pre-filled air is always fixed for a specific vehicle, so comparison between different hole-to-hole spacings is established with the same total venting area (for the single-row hole cases) to ensure the mass flow rates are all equal at a certain moment. Thus, as hole-to-hole spacing increases, the number of holes will drop with their aperture enlarged. Fig. 8 gives the changes of film and streamlines when hole-to-hole spacing increases at three chosen moments (which could show the development of air film). Results can be figured out that when spacing between holes increases, pre-filled air is gathered and leaking from less holes, which dramatically enhance its extension in space for a single hole, resulting in the larger axial section and faster extension of axial length. While for the flow field structures recognized from these streamlines, vortices present complete similarity at these three special developing stages, only the relative larger film could bring about further development vortex structures. Note that with a certain amount of venting air at a special time, the merging process between different films are weaken or delayed when hole-to-hole spacing increases for its fast downstream extension. As for the extreme case (the value of s/d tends to 0) with a thin slot exhausting, air film covers the whole exterior along circumferential direction, which is benefit for the force balance between the front and back sides of vehicle. But its slow-growing thickness which presents weak anti-interference ability for surrounding water and its slow axial extension seem bad for air film's timely working.

Fig. 9 gives the pressure coefficient on vehicle's exterior along axial direction at three typical moment, conclusions can be gotten that when s/d increases, the pressure-equalizing region extends faster, which can also be found from the growing speed of film length in Fig. 10. Note that when s/d tends to 6, the axial extension speed of film is limited though s/d grows a lot, so choosing a reasonable value near 6 for s/d which can ensure both the fast axial extension and circumferential coalescence of air films seems to be more valuable. Based on these conclusions, the value of s/d in following research is all set as 6.65.

Exhausting process with different row-to-row spacings

The fast formation and large area coverage of air film with uniform inside pressure can meet the demands of righting vehicle's deflecting trajectory and weakening its hydrodynamic load timely. Based on the above discussions, a row of holes is in-line equipped up the basic single-row of holes with different row-to-row spacings, thus the interaction between different rows of holes and effect of row-to-row spacing on the performance of pressure-equalizing film can be investigated. Distribution of phase and streamlines near holes (including vehicle's exterior and symmetry plane) are provided in Fig. 11, while the similar results with single-row hole cases like vortex structures and pressure distribution are omitted. Apparently different from single-row hole cases, air film here is dramatically enlarged for the in-line arrangement of front row of holes, and clear reverse-flow (starting from the back holes to the front ones) can be figured out between row-to-row spacing. Furthermore, the air injection from the back holes is nearly normal to the vehicle's exterior before it impinges on water, and soon after its front part advances upstream and encounters with the downstream motion of air from front holes, forming the large-scale spanwise vortex and horseshoe vortex structures. Note that this "protective-mode" for the back holes has provide them with low-pressure and low-density environment for exhausting, as a result a large amount of air leaks from inside chamber and forms a relative larger-scale air film extending along vehicle surface compared to the single-row case (as shown in Fig. 12).

Fig. 12 gives the temporal variation of mass flow rate and film length as vehicle moves. So obvious results can be concluded that, as the row-to row space decreases the exhaust rate drops, resulting in the axial section size and axial length of air film decreasing, and the size of axial section is even close to the single-row hole cases when spacing tends to 1.5d. Besides, the relevant vortex structures like horseshoe vortex and spanwise vortex are weaken even disappear.So, a relative large distance between rows of holes is advantage for the development of air film in space.

Exhausting process with different injection angles of double-row holes

The basic physical problem of the formation of pressure-equalizing film is the air jet in water cross flow. According to some references about JICF (jet in cross flow) with single phase, the strong shearing action that air jet suffering from water crossflow could clog the fast leakage of air from inside chamber, which goes against the development of air film. So based on the above discussions on different hole-to-hole and row-to-row spacings, some results with different injection angles are provided here to see whether the weakening of stagnation between jet and crossflow could bring efficiency to the performance of pressure-equalizing air film. Fig. 13 demonstrates the film shape and flow field details near holes through distribution of phase and streamlines (both on vehicle's exterior and symmetry plane). Conclusions can be found that, as the inclining of holes to the main flow direction, reverse flow is weaken. Meanwhile horseshoe vortex demonstrated in symmetry plane is compressed and the upstream moving part of air jet from back holes (which represents the spanwise vortex) decreases, even the horseshoe vortex merges with the spanwise vortex when exhausting with two rows of inclined holes at $\overline{T} = 2.0$. At early exhaust stage the separation saddle point with two row of holes inclining has advanced to the leading edge of back holes, which should come into being at late exhaust time for the other cases. These phenomenons could bring about some advantages for the development of air film for the air jet is led to the axial direction when holes inclines, which will weaken the stagnation between air and water.

While from the temporal variation of film length demonstrated in Fig.14 we can conclude that, the axial extension velocity of air film will increase as the inclining of venting holes to the main flow direction, but the improvement seems limited compared with the inclined venting of single-row holes mentioned in reference [10]. This is probably because the exhausting process with two rows of holes has dramatically enhance the development of air film, which have weakened the effect of inclined injection of air. Though at late exhaust time cases with inclined injection bring about a relative larger film length to vehicle, but little progress will be gotten for vehicle have been near to the water surface.

Figures & Tables:

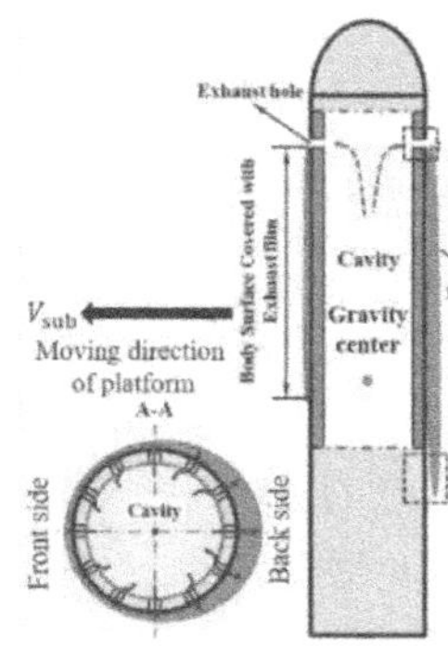

Fig. 1 Schematic of pressure-equalizing exhaust

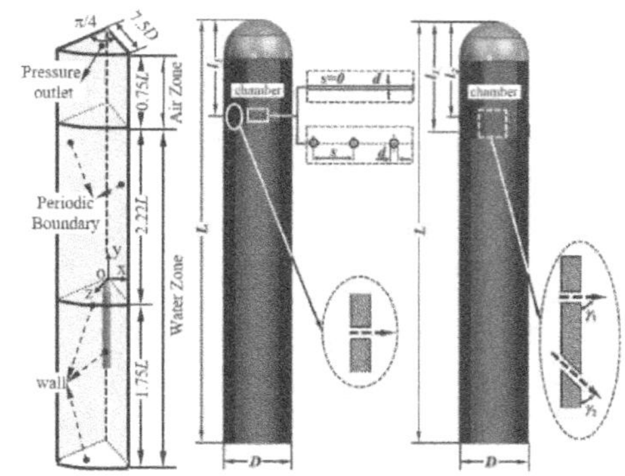

Fig. 2 Computational domain and vehicle model

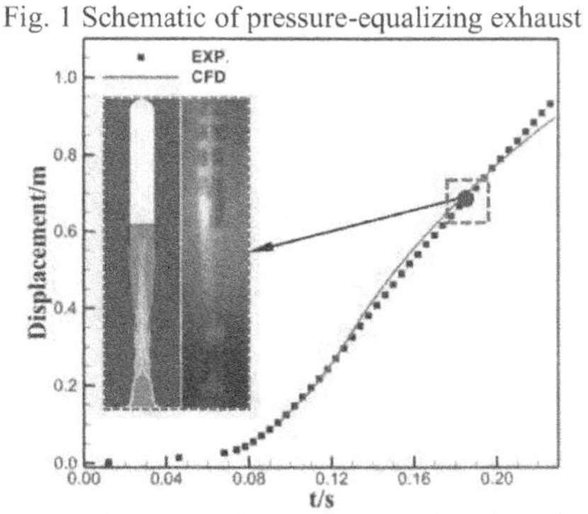

Fig. 3 Experimental and numerical results of a projectile ejected from water with high-pressure air.

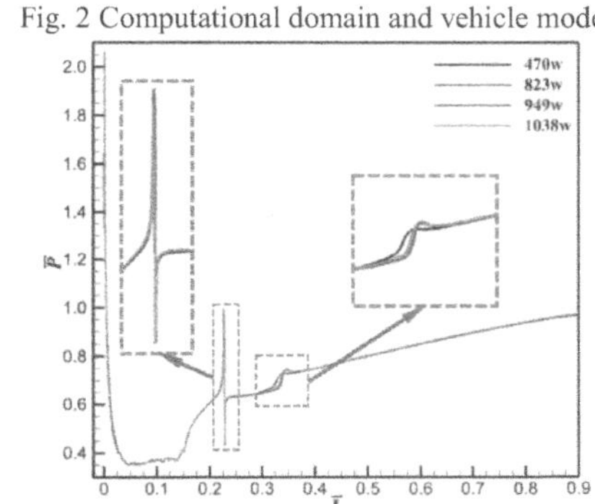

Fig. 4 Comparition of pressure coefficient $\bar{p}$ among different amounts of mesh at $\bar{T} = 1.0$

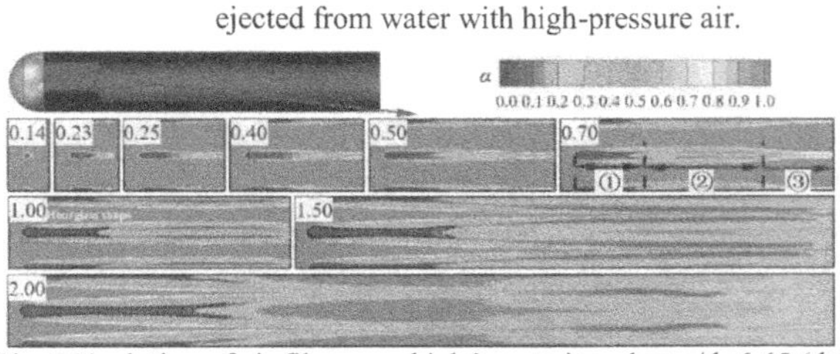

Fig. 5 Evolution of air film on vehicle's exterior when s/d=6.65 (the marked number is the dimensionless moment $\bar{T}$)

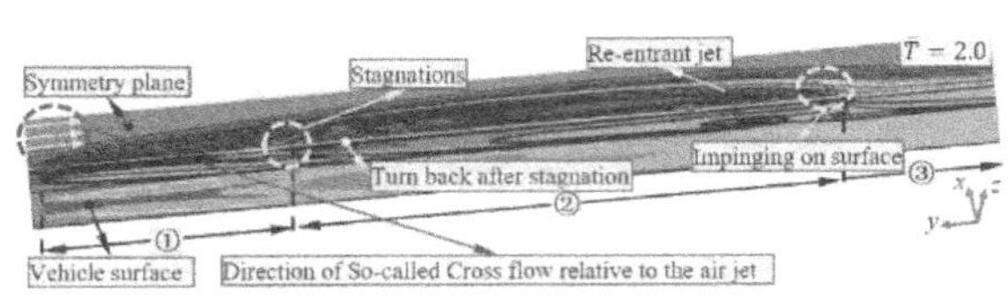

Fig. 6 Fully developed air film at $\bar{T} = 2.0$ on vehicle's exterior and symmetry plane when s/d=6.65.

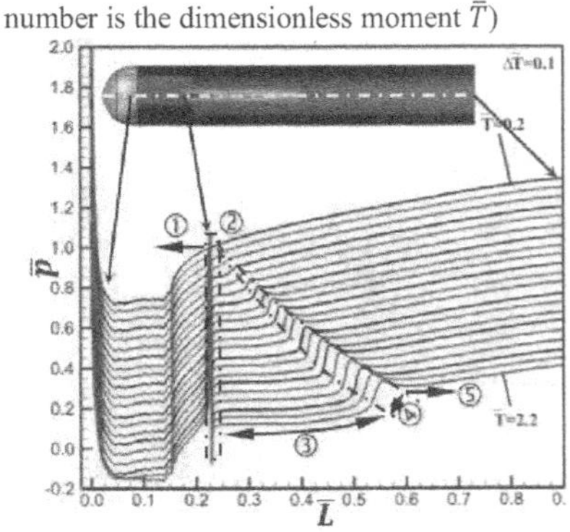

Fig. 7 Pressure coefficient from vehicle's head to bottom when s/d=6.65 ($\bar{T}$ changes from 0.2 to 2.2 represents the emerging process of vehicle from launch depth)

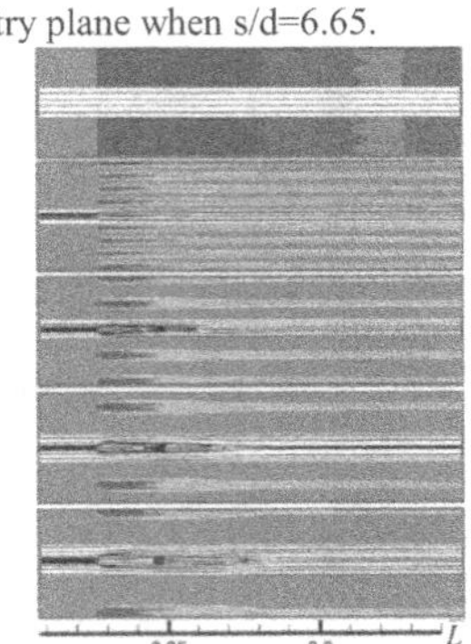

(a)$\bar{T} = 0.5$

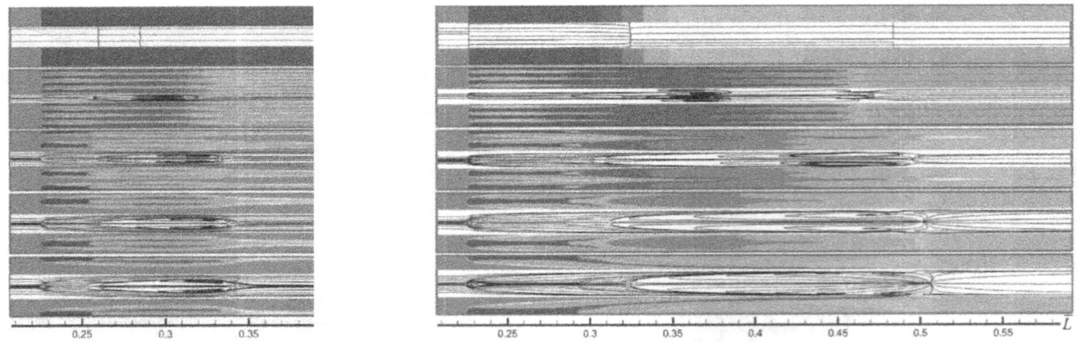

(b) $\bar{T}$ = 1.0 (c) $\bar{T}$ = 2.0

Fig. 8 Distribution of phase and streamlines on vehicle's exterior (from the top to bottom, s/d=0, 3.8. 5.38. 6.65, 7.62.)

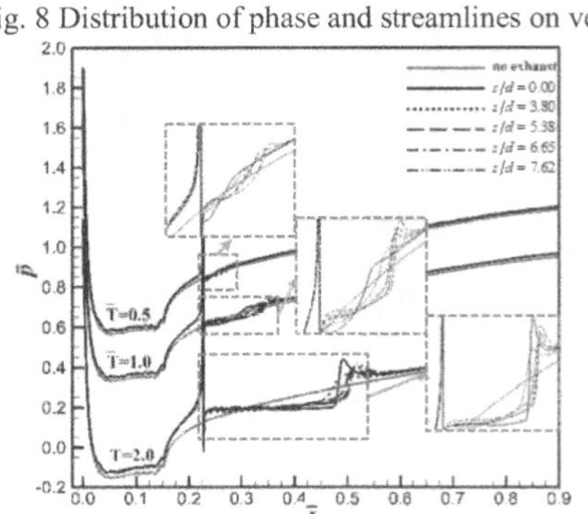

Fig. 9 Pressure coefficient when $\bar{T}$ = 0.5, 1.0, 2.0

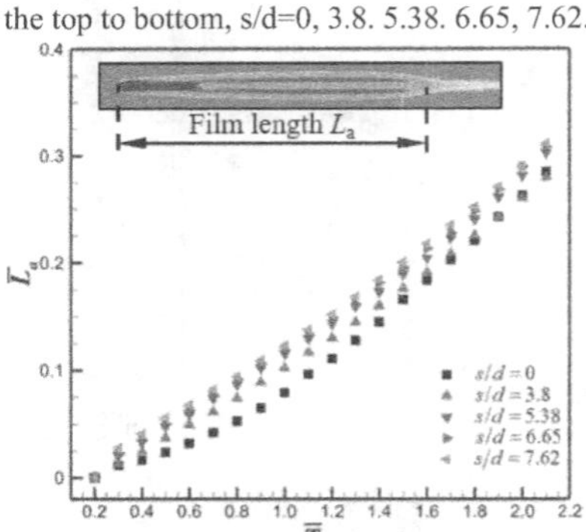

Fig. 10 Variation of film Length as vehicle moves.

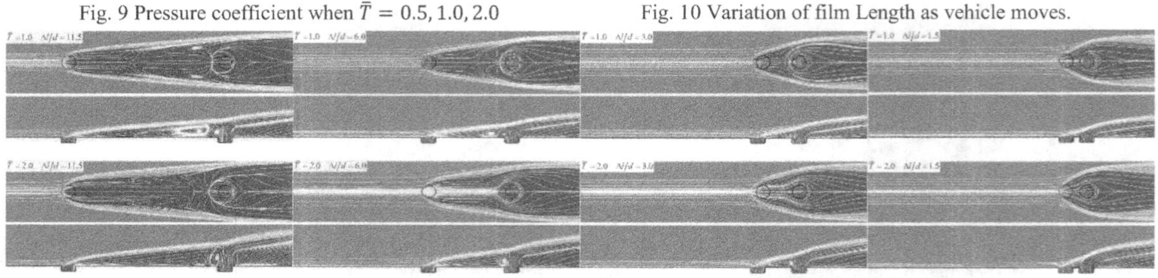

Fig. 11 Distribution of phase volume fraction near exhausting holes with different row-to-row spacings

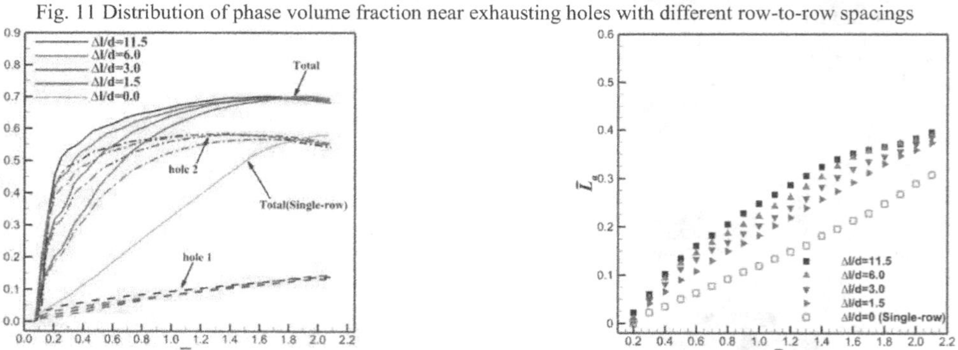

Fig. 12 Variation of mass flow rate (a) and film length (b) as vehicle moves with different row-to-row spacings.

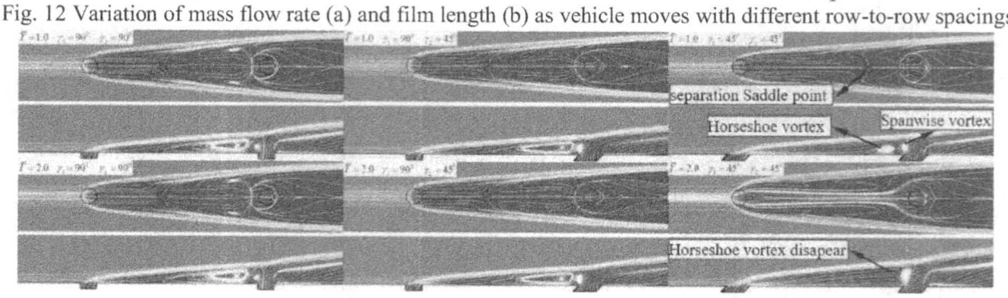

Fig. 13 Distribution of phase and streamlines with different injection angles

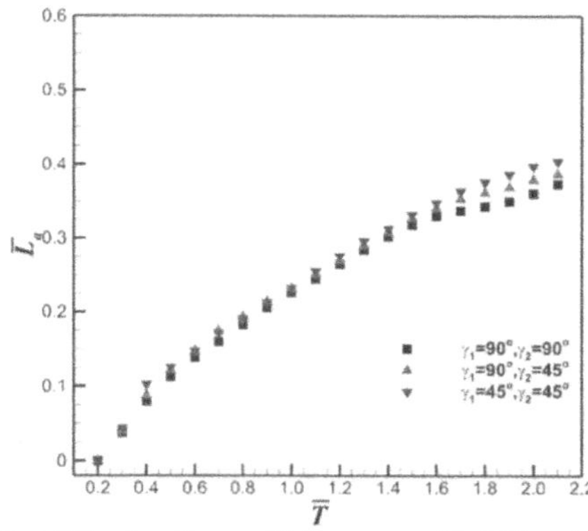

Fig. 14 Variation of film length as vehicle moves with different injection angles.

Conclusion

The vertical launching process of an underwater vehicle with different hole-to-hole spacings, row-to-row spaces as well as inclined angles of venting holes are numerical studied in this paper, the main conclusions are gathered as follow:

(1)As spacing between holes increases, pre-filled air is gathered and leaking from less holes, as a result the air film obtains a relative larger axial section and faster extension velocity of axial length, meanwhile relative further development of vortex structures come into being inside film. While the merging process between different films is weakened or delayed when hole-to-hole spacing increases for the total exhaust amount is certain at a special time.

(2)In-line equipped of another row of holes in front of the basic ones can lead to the fast leakage of air from back holes, which has significantly enhanced the fast extension of air film in space. With the different rows of holes gradually getting closer, the exhaust rate of inside chamber drops, as a result the axial length of air film decrease, and its axial section size gets smaller even close to the single-row hole cases when spacing tends to 1.5d.

(3)As the venting holes gradually incline to the gravity direction, the impeding effect and shearing action of surrounding water decrease, consequently the film grows faster and extends a longer distance along vehicle's exterior. While the improvement seems limited for the arrangement of two rows of holes has weakened the effect of inclined injection of air.

References

[1] Ceccio S L. Friction Drag Reduction of External Flows with Bubble and Gas Injection[J]. Annual Review of Fluid Mechanics, 2010, 42: 183-203.

[2] Wang G Y. Analysis on Gas-Liquid Two-Phase Flows Characteristics Around a Plane[J]. Transactions of Beijing Institute of Tecnology, 2017, 37(1): 42-45.

[3] Li G L, Shang Q, Tian J W, Yuan X J. Effect of Ventilation Cavitation Launching on Water-Exit for submarine Vehicle[J]. Physics of Gases, 2016, 1(3): 39-46.

[4] Yang W G, Zhang Y W, Deng F, Yuan X L, Fang H, Kan L. Exploring Experimentally Effect of Gas Entrainment Rate on Geometrical Shape of Supercavity[J]. Journal of Northwestern Polytechnical University, 2007, 25(3): 358-362.

[5] Zhang X S, Wang C. Research on Underwater Vehicle Based on Multiphase Flow Control. MATEC Web of Conferences. EDP Sciences, 2005; 25: 03004.

[6] Liu J. Study on Ventilated Cavitating Flow Druing Water-exit of Underwater Vehicle. M.S. thesis, Harbin Institute of Technology, Harbin, HL; .July 2012.

[7] YANG Wugang, ZHANG Yuwen, DENG Fei, et al. Exploring Experimentally Effect of Gas Entrainment Rate on Geometrical Shape of Supercavity [J]. Journal of Northwestern Poly technical University, 2007, 25(3): 358-362

[8] Yu X X, Huang C G, Du T Z, et al. Study of Characteristics of Cloud Cavity Around Axisymmetric Projectile by Large Eddy Simulation[J]. Journal of Fluids Engineering, 2014, 136: 051303.

[9] Brandner P A, Pearce B W, Graaf K L D. Cavitation about a jet in crossflow [J]. Journal of Fluid Mechanics, 2015, 768: 141-174.

[10] Chen Fu, Ma Guihui, Yu Jianyang, et al. Effect of Exhaust Angle on Evolutionary and flow Characteristics of Pressure-equalizing Film on Surface of Underwater Vehicle[J]. Journal of Ship Mechanics, 2016, 20(12):1495-1512.

© 2018 ASME

Effect of hydrodynamic induced cavitation and supercavitation on water pollutants

Andreas Schmid*

University of Applied Sciences Hof, Germany

Abstract

Hydrodynamic induced cavitation generates imploding cavitation bubbles which can lead to degradation or even mineralisation of water constituents without addition of any chemicals. This technology overcomes the restrictions of ultrasound irradiation by the local production of a cavitation cloud close to the sonotrodes. In the presented paper hydrodynamic cavitation can be stabilized and extended inside a pipe downstream the nozzle depending on the ambient pressure conditions. If the pressure is kept low, the imploding cavitation bubbles generate new cavities, analogous to a chain reaction, and elevate the radical synthesis inside the apparatus. During the pilot tests using ethan-1,2-diol, the biological degradation characteristics changes significantly depending on cavitation exposure time. A significant difference on chemical effects was observed between imploding cavitation bubbles and supercavitation. Supercavitation can mobilize chemical stable compounds for further treatment.

Keywords: industrial pollutants, hydrodynamic cavitation, supercavitation

Introduction

MTBE, ETBE and other stable chemical contaminants can be degraded under ultrasonic irradiation [1, 2, 3] by reactions with several OH•-radicals, created from imploding cavities. Using ultrasound, a cloud of cavitation bubbles is generated close to the sonotrodes, limiting the overall efficiency of the reactor [4]. To overcome these problems, hydrodynamic induced cavitation in combination with high turbulences may improve the elimination efficiency of the water contaminants. These chemicals are modified under hydrodynamic induced cavitation [5] and may even lead to complete mineralisation. Ethan-1,2-diol is commonly used as deicing-agent for airplanes and railways and should be treated to meet environmental standards [6]. This study focuses on the treatment of deicing-agent by hydrodynamic induced cavitation.

Materials and Methods

The cavitational flow regime is created by a nozzle downstream a centrifugal motor-pump and is kept up inside a pipe with reduced diameter over a length of 2 meters following the cavitation-nozzle (see figure 1). To maintain cavitation downstream the nozzle, low ambient pressure conditions are created to enable repeated cavitation bubble development, analogous to oscillating cavities [7]. When cavitation bubbles implode, the fluid close to the bubble will accelerate and as a consequence the local pressure declines, till evaporation pressure is reached and new cavities are created again. This is strongly dependent on the ambient pressure conditions and the specific energy input for cavitation induction [5].

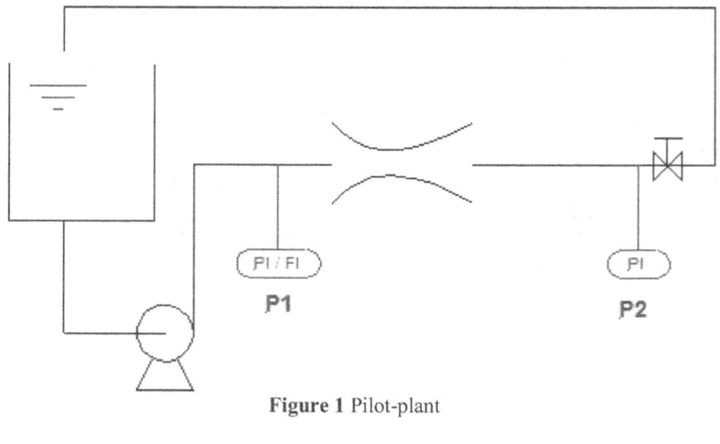

Figure 1 Pilot-plant

*Corresponding Author, Andreas Schmid: Andreas.Schmid@Hof-University.de

Results

Taking ethan-1,2-diol, the biological degradation characteristics, expressed biological oxygen demand (BOD), changes significantly depending on the exposed time of imploding cavitation bubbles / supercavitation (see figure 2&3).

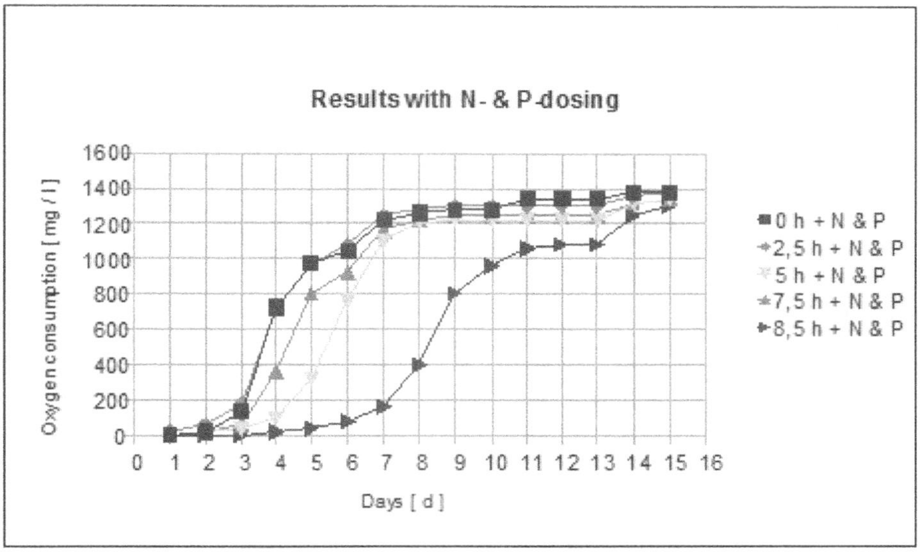

Figure 2 Changes of BOD at different durations of cavitation exposure (imploding cavities)

Especially after long cavitation exposure (8.5 h) of imploding cavities, the biological degradation characteristics deteriorates (see figure 2), despite ethan-1,2-diol is usually easily biodegradable. Unknown intermediates are generated which have a completely different behaviour of biodegradation under aerobic conditions.

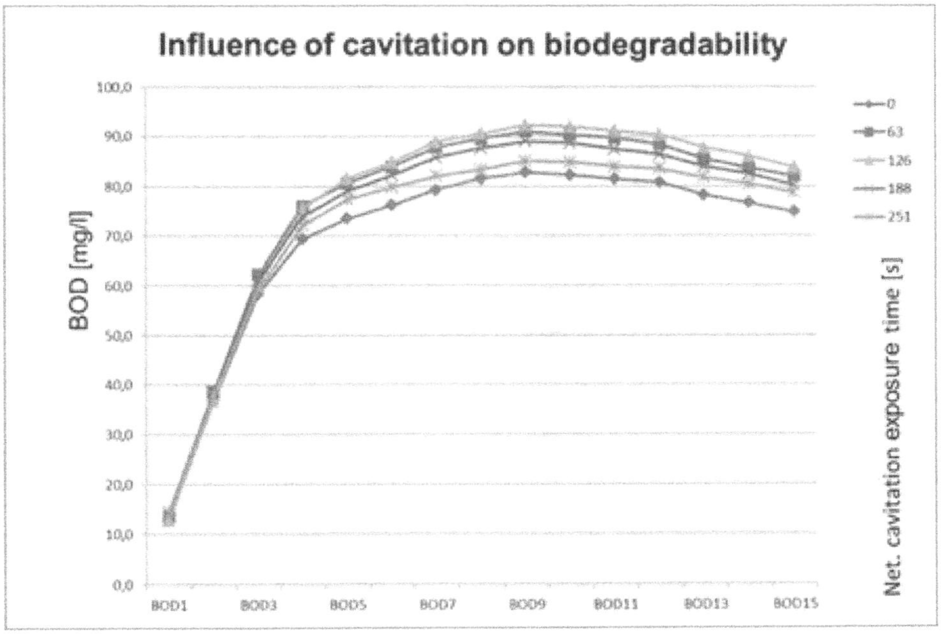

Figure 3 Changes of BOD by exposure of supercavitation

Taking the same chemical and expose supercavitation, an optimum was found, increasing the biodegradability (BOD) by app. 15% after a net supercavitation exposure time of around 2 minutes (see figure 3). Supercavitation obviously reveals completely different effects compared to imploding cavities (see figure 2).

Increasing the cavitation exposure time, using supercavitation (70 cm length), the deicing-agent apparently released new COD (chemical oxygen demand)-load (see figure 4), leading to a significant reduction of the BOD_5/COD-ratio (see figure 5), which can't be observed when imploding cavities interfere with the chemicals. Hydrodynamic cavitation, especially supercavitation, is obviously capable to mobilize unknown stable chemical compounds to improve chemical and biochemical degradation of hardly degradable substances.

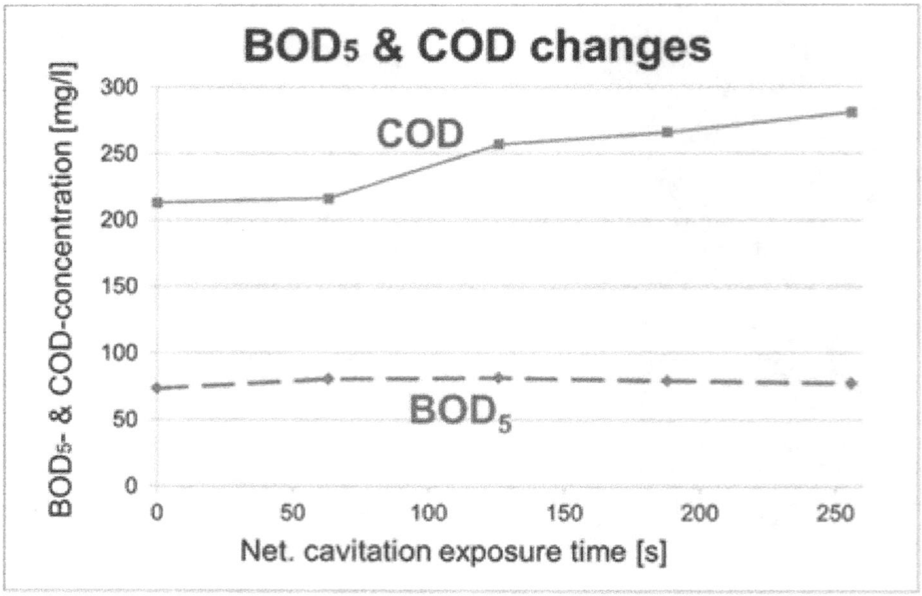

Figure 4 Changes of BOD_5 and COD under supercavitation

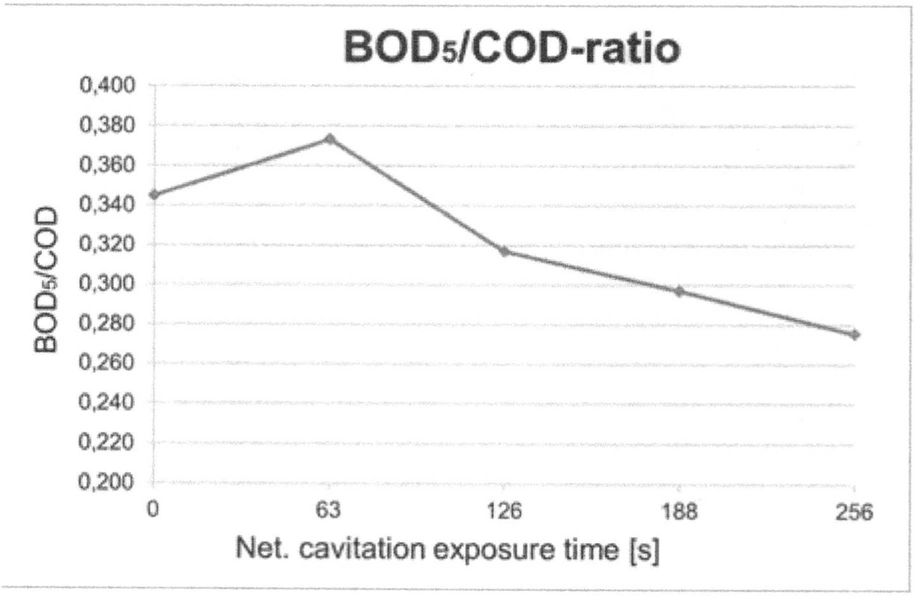

Figure 5 BOD_5/COD-ratio in dependence on supercavitation exposure

Conclusion

Hydrodynamic induced cavitation is capable to modify diverse chemicals, improve chemical as well as biochemical degradation characteristics and may even be used for complete mineralization without addition of any external chemicals. The unknown intermediates originated from ethan-1,2-diol can be degraded under aerobic conditions more easily, leading to a 15% additional oxygen requirement after supercavitation exposure of app. 2 minutes. After 1 minute COD increased, indicating that stable chemical compounds are transferred toward reactive compounds. The presented study shows an easy technology to modify water by cavitation exposure to form radicals, which are capable to increase elimination rates and change chemical structures of different trace pollutants. The system was run with a commercial centrifugal pump, generating a pressure of app. 3.5 bar at a flow rate of app. 5 l/s – these are widely used moderate operational conditions in water and wastewater systems. The chemical modifications are strongly dependent on the form of cavitation - imploding cavities or supercavitation. Supercavitation can obviously mobilize chemically stable compounds, which might be treated in a downstream biological reactor. Much more research is needed to work out optimum degradation rates in dependence on energy input and different types of cavitation.

This technology is not restricted to the investigated chemicals only; it is very likely, that this technology can be applied to other interfering water constituents, too. The focus should concentrate on changes on biodegradability for final elimination inside biological reactors to reduce energy consumption, instead of complete mineralisation.

References

[1] Kang, J., Hoffmann, M. (1998). *Kinetics and mechanism of the sonolytic destruction of methyl tert-butyl ether by ultrasonic irradiation in the presence of ozone.* Environmental Science & Technology, 32, 3194-3199.
[2] Hoffmann M, Hua I., Höchemer R. (1996). *Application of ultrasonic irradiation for the degradation of chemical contaminants in water.* Ultrasonics Sonochemistry, 3, 163–172.
[3] Nam S., Han S., Kang J., Choi H. (2003). *Kinetics and mechanisms of the sonolytic destruction of non-volatile organic compounds: Investigation of the sonochemical reaction zone using OH· monitoring techniques.* Ultrasonics Sonochemistry, 10, 139–147.
[4] Horst C., Hoffmann U. (1999). *Design, operation and characterization of ultrasound reactors. From: Tiehm A., Neis U., Ultrasound in Environmental Engineering.* TU Hamburg-Harburg Reports on Sanitary Engineering, 25, 77- 90.
[5] Schmid A. (2010). *MTBE-Degradation by hydrodynamic induced cavitation.* Water Science & Technology, 61, 10, 2591–2594.
[6] Rebsdat S., Mayer D. (2005). *Ethylene Glycol,* Ullmann's Encyclopedia of Industrial Chemistry, Weinheim, Wiley-VCH.
[7] Geisler R. (2003). *Untersuchungen zur laserinduzierten Kavitation mit Nanosekunden- und Femtosekundenlasern.* University Göttingen.

© 2018 ASME

Effect of Solid Surface Wettability on the Formation of Cavitating Wave Front with Fluid-Structure Interaction

[1]Tomohisa Kojima*; [2]Kazuaki Inaba; Yuto Takada; Farid Triawan;

[1]Meiji University, Kanagawa, Japan; [2]Tokyo Institute of Technology, Tokyo, Japan

Abstract

Fluid-structure interaction (FSI) and wave propagation occurred in engineering structure may cause severe damage and result in a leakage of hazardous material from piping in a plant. However, still limited studies could be found on the propagation of cavitating wave front along with tensile wave propagation. In addition, in those studies, the cavitating wave front is assumed to exist unconditionally. The present study aims at clarifying the relation between the interfacial condition of solid-fluid interface and the dynamic cavitation generation caused by wave propagation across the solid-fluid interface with FSI. An impact experiment was conducted with a free-falling projectile which hit the cylindrical solid buffer placed on top of the water surface within the elastic tube. Cavitation bubbles were generated at the interface of the buffer and water owing to the tensile wave propagation. Then, the surface wettability of the buffer and the surface tension of water were varied to change the intensity of cavitation. In the condition with better surface wettability, the cavitation generation from the interface was suppressed. Moreover, the cavitation generation in water was not observed. Then the cavitation generation from the solid-fluid interface was evaluated with transmitted energy from solids to water. It was revealed that the more severe the cavitation intensity on the interface with worse wettability, the more energy pressure transmittance was consumed.

Keywords: cavitation; pressure wave; interface; fluid-structure interaction

Introduction

Fluid-structure interaction (FSI) and wave propagation occurred in engineering structure may cause severe damages and result in a leakage of hazardous material from piping in a plant. In our previous study, it was observed that instantaneous cavitation occurred originating from the solid-fluid interface as a result of tensile stress wave propagation from solid medium, and it affected the wave propagation across the interface [1]. There are not many studies on the propagation of the cavitating wave front along with tensile wave propagation, and in those studies, the cavitating wave front is assumed to exist unconditionally [2-4]. However, from our previous studies, it was indicated that the interfacial condition between solid and fluid changed the cavitation occurrence and its propagation along with the tensile wave [5]. For this reason, the present study aims at clarification of the relation between the interfacial condition of solid-fluid interface and the dynamic cavitation generation caused by wave propagation across the solid-fluid interface with FSI.

Methodology

The schematic picture of experimental apparatus used in this study is presented in Fig. 1. The facility consists of a 1,000 mm long water-filled standing polycarbonate (PC) tube with its upper end closed by a buffer of 300 mm long aluminum (Al) rod. The inner diameter of the tube is 52 mm, and thickness is 4 mm. The diameter of the buffer is 50 mm. A free-falling steel projectile was used to impact the top of the buffer to generate a stress wave. The generated stress wave propagated through the solid buffer and across the interface of the solid buffer and water in the tube. Thus, the stress wave in the solid propagated to the water in the tube as a pressure wave. The deformations of the buffer and water-filled tube were measured with strain gages oriented in the longitudinal and hoop directions. The motion of the buffer and water close to their interface was observed by a high-speed video camera (VW-9000, KEYENCE) at frame rates of 230,000 fps and shutter speeds of 1/900,000.

The material properties of Al and PC are listed in Table 1. The values which were used as the material properties of water in the PC tube were also shown in Table 1. These values were used for conversion from the strain measured at the buffer and the tube wall surface to the stress and the pressure which propagating inside the buffer and water in the tube. The experiments were conducted with the drop height of the projectile of 200 mm above the top of the buffer. For each experimental condition, same experiments were conducted 10 times.

*Corresponding Author, Tomohisa Kojima: kojimat@meiji.ac.jp

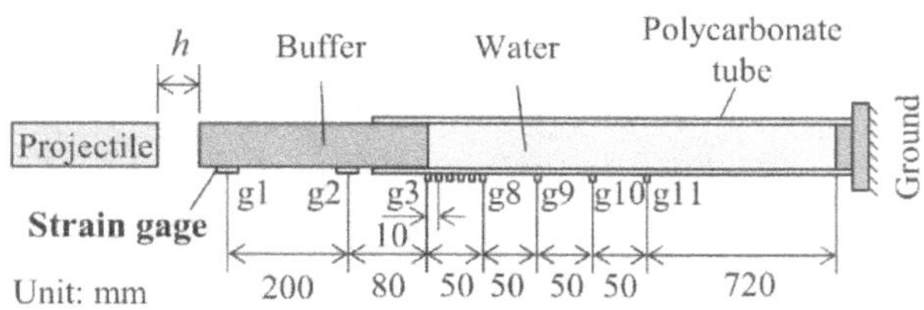

Figure 1 Experimental apparatus

Table 1 Material properties

	Aluminum (Al)	Polycarbonate (PC)	Water in PC tube
ρ (kg/m^3)	2640	1200	1000
E (GPa)	70.3	2.45	-
c (m/s)	5160	1429	429
Z (10^6 kg/m^2/s)	17.0	1.71	0.429
A (10^{-4}m^2)	4.91	4.91	5.31

As the experimental parameter, the surface wettability of the buffer at the interface with water was changed. In this study, we used a plasma treatment and surfactant to change the surface wettability for their easiness. For the plasma-treatment, the plasma-treated Al foil tape (425, 3M^{TM} treated with YHS-R, Sakigake-semiconductor) was attached the bottom of the buffer to change its surface wettability. The thickness of the aluminum foil tape was only 0.11 mm and it was plasma-treated in 180 s. As a reference, the buffer attached with the untreated Al foil tape was also prepared. For the surfactant, 0.1% solution of Nonionic surfactant (NOIGEN TDX-80D, DKS Co. Ltd.) was used instead of water. The surface tension of the solution is 28.5 mN/m which is 39% with water. With surfactant solution, untreated buffer surface is used. The surface wettability is assessed quantitatively by the contact angle with a water droplet. Figure 2 shows the contact angles measured for each surface condition. The wettability improved in the order of buffer edge, attached untreated Al tape case, attached plasma-treated Al tape case, and using surfactant solution. Thus, the contact angles decreased in the same order. The surface roughness was almost same in each condition.

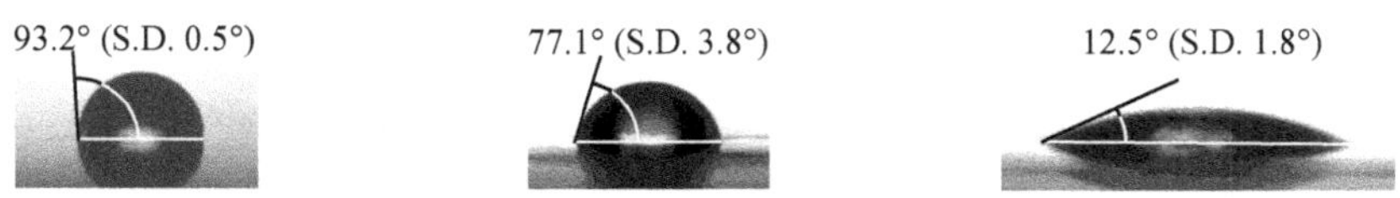

≈ 0°

No image

(a) With the buffer surface (b) With untreated Al foil tape (c) With plasma-treated Al foil tape (d) With surfactant solution

Figure 2 Contact angles of water droplets with each surface condition

Results

The images in Fig. 3 were extracted from the recorded videos with each condition of the surface wettability. Flow features can be observed at the bottom of the buffer. The images were chosen when the size of cavitation bubbles took the minimum and maximum value. The origin of the time (0 ms) is the point when the strain gage at g3 becomes greater than a threshold value. The cavitation bubbles appeared from 0 ms in cycles of 0.12 to 0.13 ms, although the water near the bottom of the buffer was compressed by the buffer continuously.

It is apparent that the cavitation bubbles appeared as a result of the transmitted tensile stress wave propagating through the buffer into the water. The tensile wave was created due to the reflection at the upper edge of the projectile or the buffer with the compression stress wave generated as a result of the projectile impact. The tensile waves are continuously reflected and propagated to the interface of the buffer and water in cycles causing the cavitation occurrence. The estimated time is 0.14 ms when the reflected stress wave on the interface of the buffer and water returns through both the buffer and the projectile, and through buffer only is 0.12 ms, which have a good accordance with the interval of the cavitation appearance.

According to the tensile wave propagation, the cavitating region spreads and propagates in water with the condition of the untreated buffer surface, forming the cavitating wave front (Fig. 3(a)). In contrast with the condition of using untreated buffer surface, with the plasma-treated Al foil tape and surfactant solution, a small number of cavitation bubbles was observed (Fig. 3(c) and (d)). Therefore, the generation of the cavitation bubbles was suppressed according to the improvement of surface wettability of the buffer edge. The cavitation generation was also slightly suppressed with the untreated Al foil tape (Fig. 3(b)). Thus, the attenuation of the transmit wave amplitude due to the tape had a strong effect on the cavitation generation. It is remarkable that in the condition with being suppressed the cavitation generation from the interface, the cavitation generation in water was not observed. It seems that the cavitating wave front would not be initiated if the cavitation generation at the starting point was suppressed by the effect of the surface wettability of the solid on the interface.

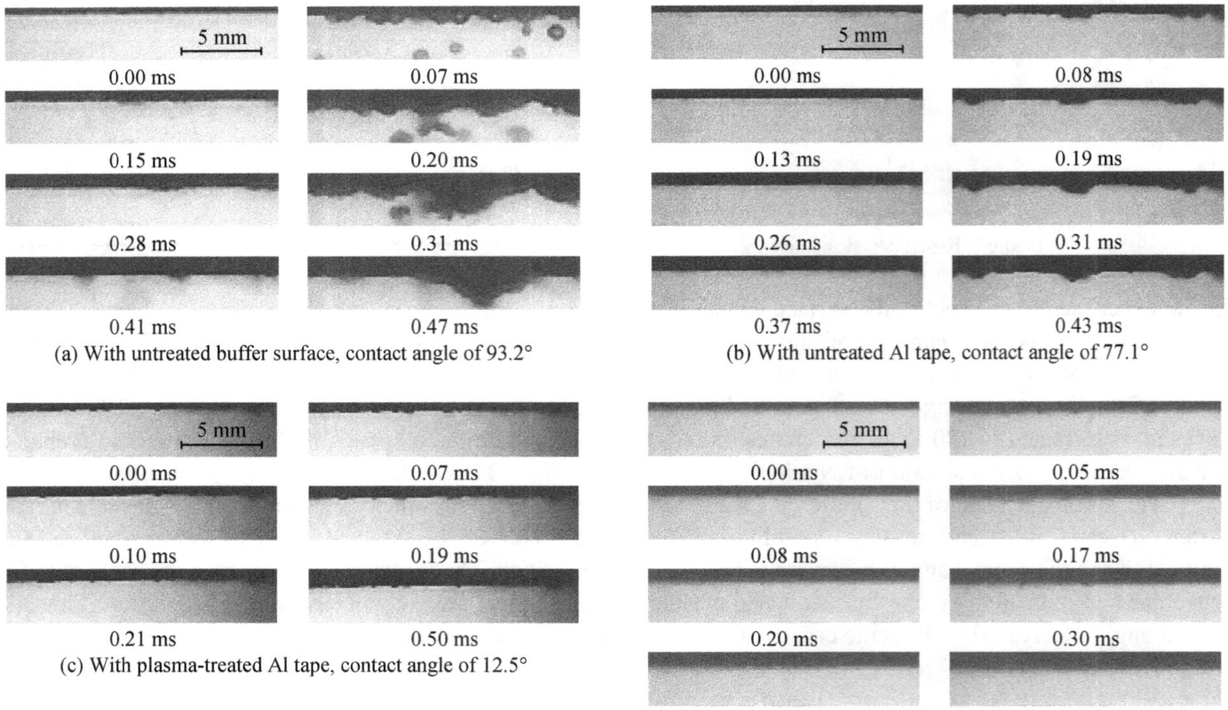

(a) With untreated buffer surface, contact angle of 93.2°

(b) With untreated Al tape, contact angle of 77.1°

(c) With plasma-treated Al tape, contact angle of 12.5°

(d) With surfactant solution, contact angle of 0°

Figure 3 Flow features around the bottom of the buffer

Discussion

In our previous study, we evaluated the effect of surface wettability on wave propagation across the solid-fluid interface by the first peak value of the transmitted pressure wave. However, the evaluation by the first peak pressure may not be appropriate since the cavitation is generated due to the tensile wave propagating after the propagation of the first wave. In this section, we evaluate the effect of the surface wettability on cavitation occurrence by variation of the transmitted energy from the buffer to water. Assuming the projectile and buffer behave as rigid bodies and they move in a body after the impact, the reduction of kinetic energy of two solids after impact can be written as follows;

$$\Delta K_s = \frac{1}{2}(M+m)\cdot(V(t))^2 \tag{1}$$

where M is the mass of the projectile, m is a mass of the buffer, and $V(t)$ is the velocity of the buffer-water interface estimated with momentum conservation. The momentum conservation equation with the two solids and water is written as follows;

$$MV_0 = (M+m)\cdot V(t) + \rho_w A_w \cdot c_w t \cdot V(t) \tag{2}$$

where V_0 is impact velocity of the projectile, ρ_w is the density of water, A_w is the cross-sectional area of water, c_w is sound speed in water, and t is time after the impact. The second term in the right side of equation (2) represents the transmitted fluid's momentum. When this term increases, the interface velocity $V(t)$ decreases over time.
In addition, the energy transmitted to water from these solids can be estimated as follows;

$$\Delta K_f = \frac{1}{2}\rho_w A_w c_w t \cdot (V(t))^2 + ptA_w V(t) \tag{3}$$

where p is a representative value of water pressure in the region where the propagating pressure wave reached. The first term in the right side of equation (3) represents the kinetic energy of the fluid, and the second term represents the work that water received from the solids. In this study, we substitute p with the estimated pressure at g4 (10 mm below the buffer-water interface), then estimate the value ΔK_f. The pressure was estimated from the hoop strain history at g4 with the conversion equation proposed by Tijsseling [6]. The equation was derived with the assumption of the quasi-static relation of pressure in water and strain of the tube. Then, we plotted ΔK_s and ΔK_f with each condition of surface wettability in Fig. 4. In Fig. 4, ΔK_s and ΔK_f show a good accordance. Until about 1.4 ms, the values of ΔK_f from measured strain exceed the theoretical value of ΔK_s. It is likely that $V(t)$ for estimating the ΔK_f is larger than the actual particle velocity in water. After about 1.4 ms, ΔK_f starts to decrease, to have a different tendency with ΔK_s. It appears that the estimated pressure at g4 would not be proper as the representative value according to the pressure wave propagating away from the interface.

To clarify the effect of surface wettability, we integrated the values in Fig. 4 in the period from 0 to 2.0 ms and compared with each other (Fig. 5). Since the estimated pressure at g4 cannot be the representative value after about 2.4 ms due to the second impact of the projectile, we decided the integration period as to 2.0 ms. In Fig. 5, the dispersion of the cumulative value is larger with smaller contact angle. Comparing Fig. 5 with the flow features shown in Fig. 3, we can observe that the more severe intensity of the cavitation on the interface with worse wettability (larger contact angle), the more energy of pressure transmittance was consumed. For the propagation of the generated cavitating wave front in water, it needs further study with larger contact angles since cavitation wave front propagated in water only in the case of the contact angle of 93.2°.

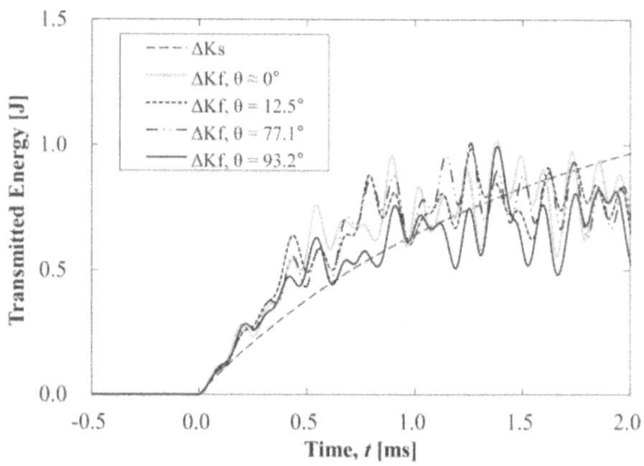

Figure 4 Kinetic energy transmitted to water

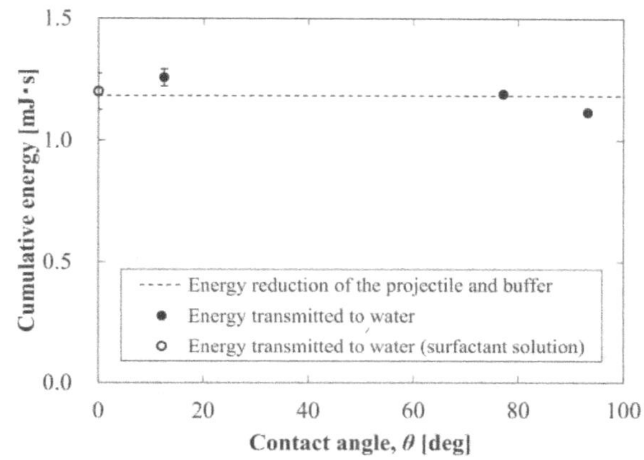

Figure 5 Cumulative energy in t = 0 to 2.0 ms

Conclusion

The effect of solid surface wettability on the formation of cavitating wave front with fluid-structure interaction was experimentally investigated and evaluated. From the result, it was observed that the generation of the cavitation bubbles from the solid-fluid interface by tensile wave propagation was suppressed according to the improvement of the surface wettability of the solid. It was notable that when the cavitation generation from the interface was being suppressed, the cavitation generation in water was not observed. It was likely that the cavitating wave front would not be initiated if the cavitation generation at the starting point was suppressed by the effect of the surface wettability of the solid on the interface. Then the cavitation generation from the solid-fluid interface was evaluated with transmitted energy from solids to water. It was revealed the more severe intensity of the cavitation on the interface with worse wettability (larger contact angle), the more energy with pressure transmittance was consumed. As for the propagation of the generated cavitating wave front in water, it will need further study with a larger contact angle.

References

[1] Kojima, T., Inaba, K., Takahashi, K., Triawan, F., Kishimoto, K. (2017). *Dynamics of Wave Propagation across Solid-fluid Movable Interface in Fluid-Structure Interaction.* Journal of Pressure Vessel Technology. 139(031308).

[2] Fan, D., Tijsseling, A. (1992). *Fluid-Structure Interaction With Cavitation in Transient Pipe Flows.* Journal of Fluids Engineering. 114(2).

[3] Inaba, K., Shepherd, J. (2010). *Dynamics of Cavitating Flow and Flexural Waves in Fluid-Filled Tubes Subject to Axial Impact.* Proceedings of the ASME 2010 Pressure Vessels and Piping Conference. 4(1).

[4] Schiffer, A., Tagarielli, V. L., Petrinic, N., Cocks, A. C. F. (2012). *The Response of Rigid Plates to Deep Water Blast: Analytical Models and Finite Element Predictions.* Journal of Applied Mechanics. 79(6).

[5] Kojima, T., Inaba, K., Takahashi, K. (2016). *Wave Propagation Across the Interface of Fluid-Structure Interaction with Various Surface Conditions of Solid Medium.* Proceedings of the ASME 2016 Pressure Vessels and Piping Conference. 4(1).

[6] Tijsseling, A. (2007). *Water hammer with fluid-structure interaction in thick-walled pipes.* Computers & Structures. 85(11-14).

© 2018 ASME

Effects of acoustic parameters and bulk fluid properties on acoustic droplet vaporization threshold of perfluoropentane droplets

[1*]Mitra Aliabouzar, 2Krishna Kumar and [1]Kausik Sarkar

[1]The George Washington University, Washington DC 20052; [2]FloDesign Sonics, Wilbraham, MA, USA

Abstract

Phase shift droplets vaporizable by acoustic stimulation are better alternatives to conventional ultrasound contrast agents in terms of higher stability and possibility of achieving smaller sizes. This study determines the threshold pressures required to induce acoustic droplet vaporization (ADV) and inertial cavitation (IC) of a suspension of droplets with a perfluoropentane (PFP) core (diameter 400-3000 nm) in a novel tubeless setup using acoustic methods. We further investigate the effects of excitation frequency, pulse lengths, temperature and dissolved gas concentration on both ADV and IC thresholds. We have found that ADV threshold decreases with temperature and increases with frequency. ADV thresholds at all the frequencies studied here occurred at lower rarefactional pressures than IC thresholds indicating that phase transition precedes inertial cavitation. The ADV and the IC thresholds did not change when the experiments were performed in degassed conditions. The scattered response from droplets above the ADV threshold was also found to qualitatively match with that of independently prepared lipid-coated microbubble suspensions in magnitude as well as trends.

Keywords: acoustic droplet vaporization; perfluoropentane droplets; scattered response; inertial cavitation; ultrasound contrast agents

Introduction

Gas-filled microbubbles encapsulated by lipids and other surfactants are highly responsive to ultrasound, which has led to their effective role as vascular agents for contrast enhanced ultrasound imaging[1]. More recently, emulsions of phase shift droplets of volatile perfluorocarbon liquids, that can be vaporized in situ into highly echogenic microbubbles by external application of ultrasound pulses, are being investigated[2, 3]. Note that phase shift droplets can be used for extravascular imaging and drug delivery mainly due to the fact that small size droplets can extravasate through leaky vasculature of cancerous tumors as a result of the enhanced permeability and retention (EPR) effect[4]. The diagnostic and therapeutic potentials of acoustic droplet vaporization (ADV) have been studied extensively [5, 6].

The phenomenon of acoustic droplet vaporization (ADV), specifically the threshold ultrasound excitation for vaporization, has been studied in the past using both acoustic and optical methods. Acoustic investigations determine ADV thresholds through the scattered signals. Optical investigations depend on direct observations of the vaporization using cameras which might suffer from limitation of resolution. Here, we have performed an acoustic investigation of the ADV threshold of lipid-coated perfluoropentane (PFP) droplets and its dependence on excitation frequency, pulse lengths, temperature and dissolved gas concentration in a tube-less setup that minimizes the effects of the containment apparatus. We also measure the acoustic signals from the lipid-coated perfluorobutane (PFB) microbubble suspension to offer a qualitative comparison with that from vaporized PFP droplets. This study also investigates the threshold pressures required to induce inertial cavitation of the vaporized droplets in the same setup.

Experimental methodology

Perfluorocarbons (PFP and PFB) and lipids (DPPC and DPPE-PEG-2000) were purchased from FluoroMed (Round Rock, Texas) and Avanti Polar lipids (Alabaster, AL), respectively. Lipid-coated PFP-droplets and PFB-microbubbles were prepared using sonication and mechanical agitation techniques, respectively. The details of the methods can be found in[7]. The size distribution and the concentration of droplets and microbubbles were determined using qNano (Izon Science™, Cambridge, MA). The droplet suspension had an average diameter of 890

*Corresponding author, Mitra Aliabouzar: mitraaali@email.gwu.edu

nm and a total concentration of 2.1×10^9 droplets/ml whereas microbubbles had an average diameter of 1.6 μm and a total concentration of 2.05×10^8.

To measure the scattered responses, droplet and microbubble suspensions were passed through a metal tube mounted right above the focal region of the two confocally aligned transducers into a chamber filled with DI water, as shown in Fig.1. For the ADV experiments at room temperature, we employed two spherically focused immersion transducers (Panametrics Transducer, Olympus NDT Corporation, Waltham, MA), with central frequencies of 2.25, 5, 10 and 15 MHz. While transmitting transducers were excited at their central frequency, the receiving transducers for measuring fundamental, sub- and second- harmonic scattering at that excitation frequency could have different central frequencies depending on the transducer bandwidth. For all the experiments performed at 37C, a broadband transducer (Sonic concepts, Woodinville, WA, USA) was placed as the receiver. The same broadband transducer was used to monitor for broadband emissions from vaporized droplets undergoing inertial cavitation.

To determine the ADV threshold, we plotted fundamental, subharmonic and second harmonic components of the scattered signal, as the ultrasound excitation amplitude was progressively increased. Liquid core droplets scatter ultrasound weakly; however, upon vaporization and turning into a gaseous bubble, the droplet diameter increases by 5-6 times resulting in a sudden jump in their scattered response. To quantitatively determine the ADV value based on this observation, we followed a curve fitting and slope-based approach similar in [8].

Results and discussion

The typical raw RF data along with the corresponding frequency spectrum for the control and droplet suspensions at the excitation frequency of 2.25 MHz and excitation amplitudes of 1 and 1.5 MPa are displayed in Figures 2 (a) and (b), respectively. It is evident from this figure that the droplet signal is significantly higher upon vaporization. Figures 3(a), (b) and (c) plot the subharmonic (1.125 MHz) and second harmonic (4.5 MHz) and the fundamental responses respectively from the droplet emulsion at the excitation frequency of 2.25 MHz at room temperature. We determined the ADV threshold by the method described earlier to be 1.22 MPa (r^2=0.99), 1.22 MPa (r^2=0.84) and 0.72 MPa (r^2=0.96) for subharmonic, second harmonic and fundamental components, respectively. The average of the three values results in the threshold value of 1.05 MPa at 2.25 MHz.

The fundamental response from droplets and microbubbles also show a qualitative match above 1 MPa. Note that, we have also investigated the effects of nonlinear propagation by studying the second harmonic response from a non-vaporizing liquid such as propylene glycol. As can be seen in Figure 3 (b), the second harmonic response from the propylene glycol stream was found to be minimal indicating minimum contribution due to nonlinear propagation through the liquid.

Similarly, we determined the ADV thresholds for higher excitation frequencies and summarized in Table I. Note that, we did not find subharmonic response at the excitation frequency of 10 MHz. We see an increasing trend for ADV threshold with frequency, as was also seen in optical investigations[9, 10]. One would expect that increasing the frequency, i.e. decreasing the time-period when the liquid is continuously under negative pressure, would increase the threshold for vaporization. The focal zone also becomes smaller at higher frequency decreasing the probability of nucleation inside the PFP droplet core.

ADV measurements performed at 37°C resulted in a substantial decrease in the vaporization threshold. Figure 4(a) displays the average ADV threshold at two temperatures as a function of frequency. Figure 4(b) shows the effect of number of cycles on the ADV threshold at 37°C at the excitation frequency of 2.25 MHz in degassed DI water. As shown in this figure, ADV threshold was independent of the pulse lengths tested here. However, the amplitude of scattered responses (fundamental, second harmonic and subharmonic) substantially decreased in case of 1 cycle excitation which is expected (data not shown).

We also studied the effects of dissolved gas concentration on ADV and IC behaviors. For ADV, we plotted the subharmonic response of the droplet suspension excited at 2.25 MHz and 8 cycles. Experiments were conducted once in DI water and once in degassed DI water at 37 °C. Figure 5(a) shows that the dissolved gas concentration does not affect the ADV threshold, however vaporized droplets oscillate more strongly in DI water due to the influx of dissolved gas and the fact that bubble dissolution decreases as the dissolved gas concentration of the bulk fluid increases[11].

It is well-documented that perfluorocarbon microbubbles are effective cavitation nuclei[12]. Vaporized PFP droplets can undergo stable oscillation or inertial cavitation depending on the excitation parameters. Since inertial cavitation is associated with generation of broadband noise, we integrated the frequency power spectrum in the range of 0.5 to 1 MHz. The integrated power for this frequency band is plotted as a function of peak negative pressure in Figure 5 (b) for an excitation frequency of 2.25 MHz. The inertial cavitation threshold was defined as the pressure at which the integrated power at this frequency band increased for at least three experiments. We found the IC thresholds of 1.7 MPa peak negative pressure (peak positive pressure of 2.3 MPa) at 2.25 MHz and 3 MPa peak negative pressure (peak positive pressure of 8.3 MPa) at 5 MHz. We did not observe any IC activity at the excitation frequency of 10 MHz up to the highest peak negative pressure of 5 MPa tested in this study.

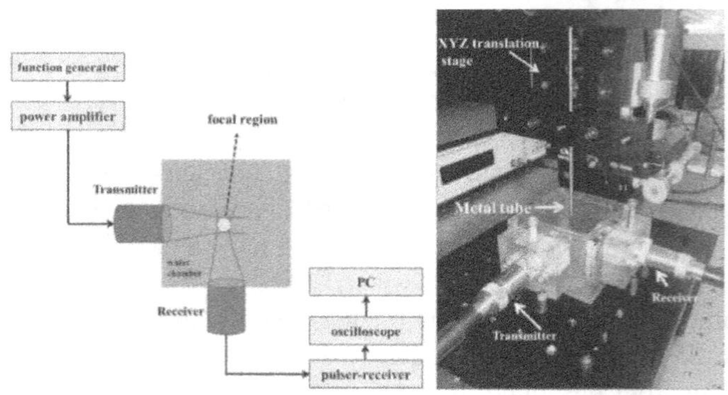

(a) (b)

Figure 1. Schematic representation (a) and a picture (b) of the experimental setup used for determining ADV threshold.

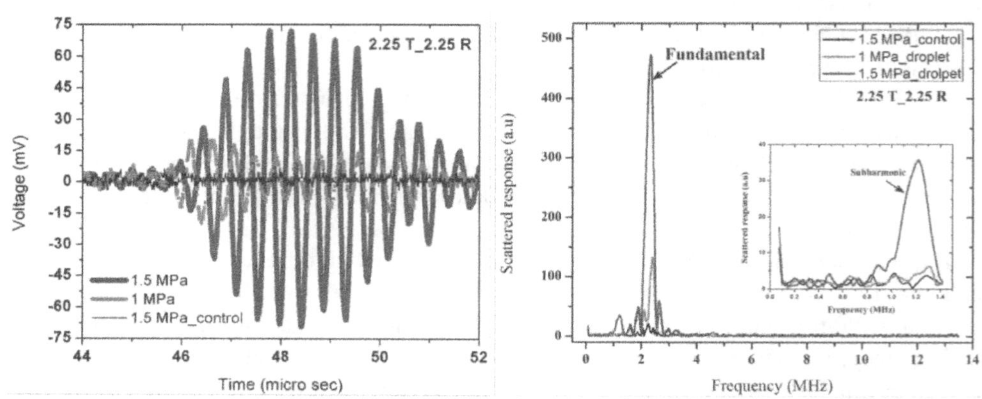

(a) (b)

Figure 2. (a) RF trace and (b) corresponding FFT from the control and droplet suspensions at 2.25 MHz.

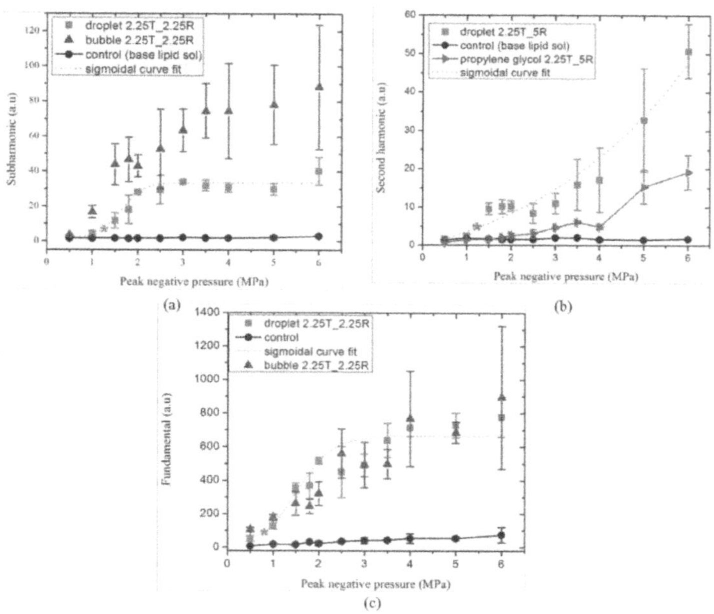

Figure 3. (a) Subharmonic, (b) second harmonic and (c) fundamental components of the scattered responses from droplets, microbubbles, propylene glycol stream and control at an excitation frequency of 2.25 MHz. Central frequencies of transmitting (T) and receiving (R) transducers are indicated in the legends. The data sets were scaled by 10^5 for easier display. The ADV threshold determined from each curve is denoted by (*).

Table I. Frequency dependence of ADV threshold.

Excitation frequency (MHz)	Threshold (fundamental) MPa	Threshold (subharmonic) MPa	Threshold (second harmonic) MPa	Threshold (average) MPa	Mechanical index
2.25	0.72	1.22	1.22	1.05	0.7
5	1.45	1.72	2.55	1.89	0.84
10	2.35	-	2.33	2.34	0.74

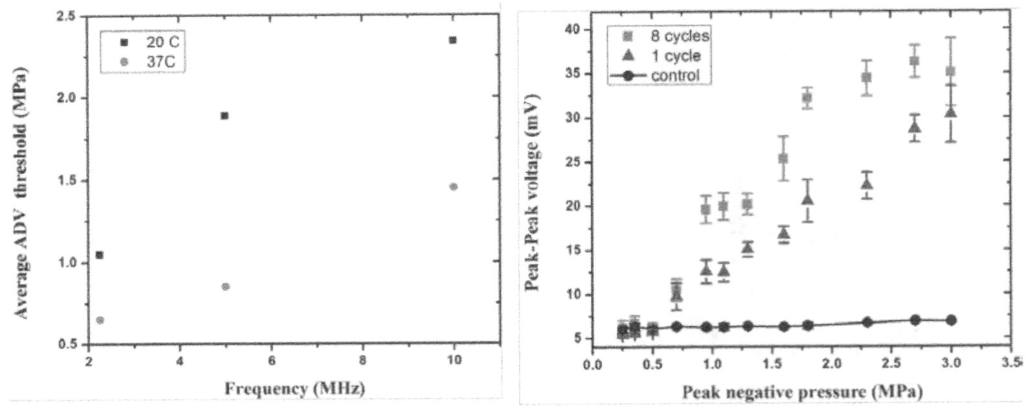

(a) (b)

Figure 4. (a) Effect of temperature on ADV threshold as a function of frequency, (b) effect of number of cycle on ADV threshold at 2.25 MHz and 37° C in degassed DI water.

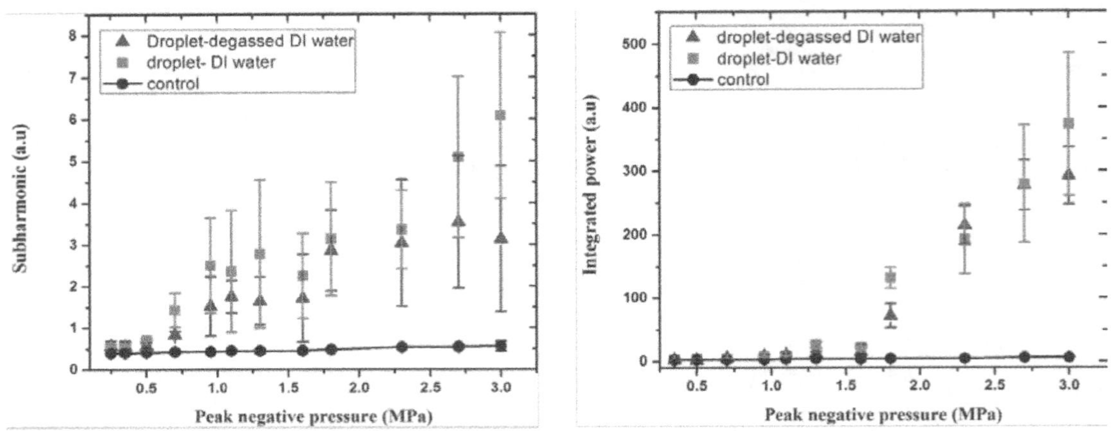

(a) (b)

Figure 5. Effect of dissolved gas concentration on (a) subharmonic response of the scattered signal from droplet suspensions at 2.25 MHz, (b) inertial cavitation threshold of the vaporized droplets excited at 2.25 MHz. All experiments were done at 37°C.

Conclusion

Lipid coated PFP droplets were excited with ultrasound pulses at various excitation frequencies and pulse lengths at two different temperatures and dissolved gas concentrations to find the threshold excitation for acoustic vaporization. Fundamental, second harmonic and subharmonic components of the scattered response of the droplet suspensions were used to determine the threshold of vaporization. The scattered response from droplets was also compared with the scattered response from a PFB microbubble suspension at the corresponding excitation pressure and frequency. The ADV thresholds were found to increase with frequency and decrease with temperature. The pulse lengths and dissolved gas concentration affected the amplitude of the scattered response from the droplets post vaporization without changing the ADV threshold. IC was measured by calculating the integrated power frequency spectrum in the range of 0.5 to 1 MHz and resulted in the peak negative pressure of 1.7 MPa and 3 MPa for 2.25 and 5 MHz, respectively. For all the frequencies studied here, ADV threshold was much lower than the IC threshold.

References

[1] A. L. Klibanov, "Microbubble contrast agents - Targeted ultrasound imaging and ultrasound-assisted drug-delivery applications," *Investigative Radiology,* vol. 41, pp. 354-362, Mar 2006.

[2] O. D. Kripfgans, J. B. Fowlkes, D. L. Miller, O. P. Eldevik, and P. L. Carson, "Acoustic droplet vaporization for therapeutic and diagnostic applications," *Ultrasound in medicine & biology,* vol. 26, pp. 1177-1189, 2000.

[3] P. S. Sheeran, V. P. Wong, S. Luois, R. J. McFarland, W. D. Ross, S. Feingold, *et al.*, "Decafluorobutane as a phase-change contrast agent for low-energy extravascular ultrasonic imaging," *Ultrasound in medicine & biology,* vol. 37, pp. 1518-1530, 2011.

[4] L. Mullin, R. Gessner, J. Kwan, M. Kaya, M. A. Borden, and P. A. Dayton, "Effect of anesthesia carrier gas on in vivo circulation times of ultrasound microbubble contrast agents in rats," *Contrast media & molecular imaging,* vol. 6, pp. 126-131, 2011.

[5] R. E. Apfel, "Activatable infusable dispersions containing drops of a superheated liquid for methods of therapy and diagnosis," ed: Google Patents, 1998.

[6] O. D. Kripfgans, J. B. Fowlkes, M. Woydt, O. P. Eldevik, and P. L. Carson, "In vivo droplet vaporization for occlusion therapy and phase aberration correction," *IEEE transactions on ultrasonics, ferroelectrics, and frequency control,* vol. 49, pp. 726-738, 2002.

[7] P. Zhang, J. A. Kopechek, and T. M. Porter, "The impact of vaporized nanoemulsions on ultrasound-mediated ablation," *Journal of therapeutic ultrasound,* vol. 1, p. 1, 2013.

[8] M. L. Fabiilli, K. J. Haworth, N. H. Fakhri, O. D. Kripfgans, P. L. Carson, and J. B. Fowlkes, "The role of inertial cavitation in acoustic droplet vaporization," *IEEE transactions on ultrasonics, ferroelectrics, and frequency control,* vol. 56, 2009.

[9] P. S. Sheeran, T. O. Matsunaga, and P. A. Dayton, "Phase-transition thresholds and vaporization phenomena for ultrasound phase-change nanoemulsions assessed via high-speed optical microscopy," *Physics in medicine and biology,* vol. 58, p. 4513, 2013.

[10] O. D. Kripfgans, M. L. Fabiilli, P. L. Carson, and J. B. Fowlkes, "On the acoustic vaporization of micrometer-sized droplets," *The Journal of the Acoustical Society of America,* vol. 116, pp. 272-281, 2004.

[11] A. Katiyar and K. Sarkar, "Stability analysis of an encapsulated microbubble against gas diffusion," *Journal of Colloid and Interface Science,* vol. 343, pp. 42-47, Mar 1 2010.

[12] W.-S. Chen, A. A. Brayman, T. J. Matula, and L. A. Crum, "Inertial cavitation dose and hemolysis produced in vitro with or without Optison®," *Ultrasound in medicine & biology,* vol. 29, pp. 725-737, 2003.

© 2018 ASME

10th International Symposium on Cavitation - CAV2018
Baltimore, Maryland, USA, May 14 – 16, 2018

CAV18-05071

Effects of non-condensable gas on cavitating flow over a cylinder

Filipe Brandao; Mrugank Bhatt; Krishnan Mahesh*

University of Minnesota, Minneapolis, MN, USA

Abstract

The effect of non-condensable gas in cavitating flow over a circular cylinder is investigated using a homogeneous mixture model. The numerical method to simulate turbulent cavitating flows developed by Gnanaskandan and Mahesh [1] for the mixture of water and vapor is extended to include the effect of non-condensable gas. The damping effects due to the presence of gas in the mixture are discussed. Finally, distribution of gas and vapor in the near wake of cylinder is investigated.

Keywords: non-condesable gas; cylinder; cavitation.

1. Introduction

Non-condensable gases (*NCG*) can behave differently from vapor in response to pressure variations as vapor can condense into liquid other than changing its volume. Presence of non-condensable gas can damp the collapse of vapor bubbles [2] and can decrease the condensation process at cavity interface leading to a higher bubble rebound [3]. The effects of injecting gas have been studied experimentally in context of sheet to cloud cavitation [4, 5]. Injection of a large amount of gas in the vapor cavity has been shown to prevent rapid collapse of vapor cloud thereby reducing the noise induced in the cloud cavitation [4]. Numerical methods based on incompressible Navier-Stokes equations have been developed to include the effect of non-condensable gas for studying ventilated supercavitation [6, 7, 8] as well as unsteady sheet cavitation [9]. In the present work, we extend the numerical method of Gnanskandan and Mahesh [1] based on a fully compressible formulation to account for non-condensable gas and study its effects on cavitating flow over a circular cylinder.

The paper is organized as follows. Section 2 discusses the extension of numerical method to account for non-condensable gas. Damping effect due to the presence of gas and distribution of gas and vapor in the near wake of the cylinder is discussed in section 3 and compared to the results obtained for the water-vapor mixture [10]. Finally, results are summarized in section 4.

2. Physical model and numerical method

Gnanaskandan and Mahesh [1] developed a numerical method to solve turbulent cavitating flows using a homogeneous mixture model for the mixture of water and vapor. Here, we extend this approach by introducing a non-condensable gas as a component of the homogeneous mixture in addition to vapor and water. This water-vapor-gas mixture is now treated as a single compressible fluid with thermal and mechanical equilibrium between phases. Compressible Navier-Stokes equations are solved for the mixture quantities along with transport equations for vapor and non-condensable gas. The system is closed by a mixture equation of state obtained from the stiffened equation of state for water and an ideal gas equation of state for both vapor and gas. The equation is given by

$$p = \rho T\left(Y_v R_v + Y_g R_g\right) + \left(1 - Y_v - Y_g\right)\rho K_l T \frac{p}{p + P_c} \quad (1)$$

Here, Y_v and Y_g are mass fractions for vapor and gas respectively. $R_v = 461.6\ \mathrm{J/(Kg\ K)}$, $R_g = 286.9\ \mathrm{J/(Kg\ K)}$, $K_l = 2684.075\ \mathrm{J/(Kg\ K)}$ and $P_c = 786.333 * 10^6$ Pa are the constants associated with equation of state for vapor, gas and liquid. The expression for speed of sound in this mixture is obtained from the mixture equation of state (1) and Gibbs equation, given by

*Corresponding Author, Krishnan Mahesh: kmahesh@umn.edu

$$a^2 = \frac{C_1 T}{C_0 - C_1/C_{pm}} \tag{2}$$
$$C_1 = (Y_v R_v + Y_g R_g)(p + P_c) + (1 - Y_v - Y_g)K_l p$$
$$C_0 = 2p + P_c - \rho T(Y_v R_v + Y_g R_g) - (1 - Y_v - Y_g)\rho K_l T$$
$$C_{pm} = Y_v C_{pv} + Y_g C_{pg} + (1 - Y_v - Y_g)C_{pl}$$

Here, C_{pv}, C_{pg} and C_{pl} are specific heats at constant pressure for vapor, gas and liquid respectively. Speed of sound obtained from equation 2 for water-vapor-air mixture is plotted as a function of vapor and gas volume fraction in Figure 1(a) and compared to the experimentally available data for water-gas mixture as shown in Figure 1(b). Note that introduction of gas leads to reduction in the sound speed of mixture.

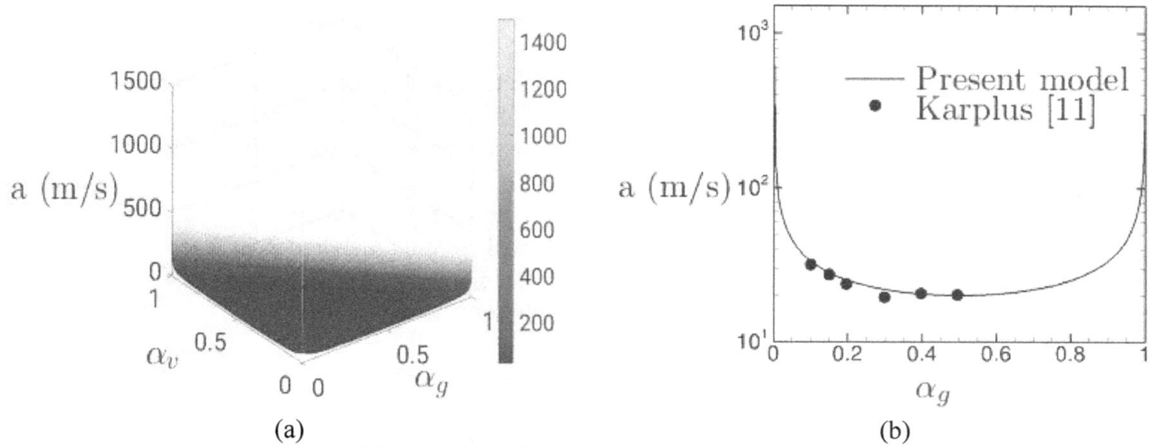

Figure 1. Speed of sound in water-vapor-gas mixture.

The numerical method is based on a predictor-corrector approach. The predictor step uses a non-dissipative scheme and corrector step applies a characteristic based filter locally in the vicinity of discontinuities. The additional equation due to the transport of non-condensable gas is coupled through Jacobians to the compressible Navier-Stokes equations, and transport equation for vapor mass fraction.

3. Results and discussion

Cavitating flow over a circular cylinder is studied at $Re = 200$ and two different cavitation numbers ($\sigma = 1.0$ and $\sigma = 0.5$). Non-condensable gas is introduced in the free stream similar to the addition of vapor nuclei in the flow. Results presented in section 3.1 correspond to $\sigma = 1.0$. Section 3.2 shows results for both $\sigma = 1.0$ and $\sigma = 0.5$.

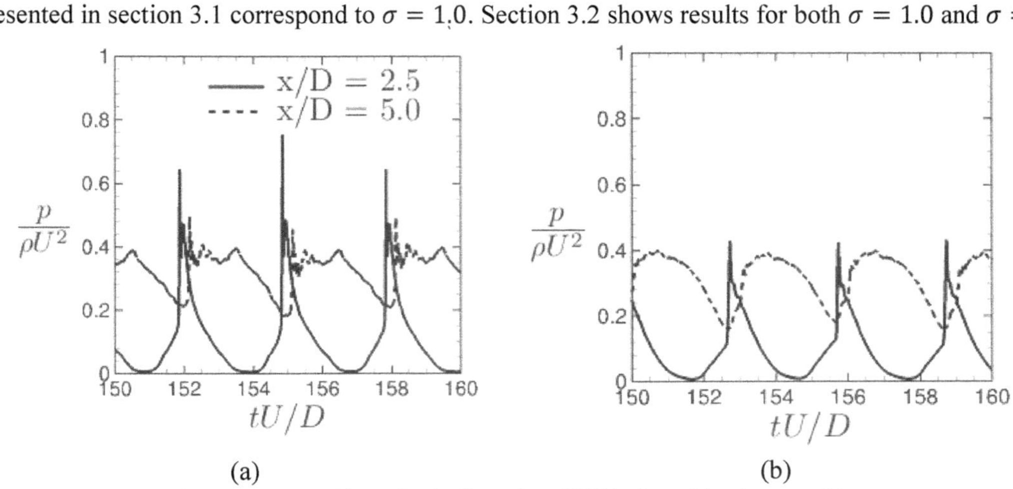

Figure 2. Pressure history for the flow where *NCG* is absent (a) and present (b).

3.1. Damping effects of gas

In order to understand the effect of gas on the pressure waves produced due to cavity collapse, we plot pressure history in the wake of cylinder at $2.5D$ and $5D$ along the wake center-line, as shown in the Figure 2 (for both flows- in presence and absence of gas). Note that at $2.5D$, the peak value of pressure is less when the gas is present. Also, the pressure wave produced due to collapse no longer has a sharp pressure peak as it reaches the station at $5D$ as seen in Figure 2(b). The distance between the peaks at $2.5D$ and $5D$ in the time domain reflects the propagation speed of the wave, which is smaller when the gas is present. This is due to the reduction in sound speed of the mixture in the presence of gas. Smaller amplitudes of pressure waves can be explained from the inviscid, isentropic relation for pressure perturbation in the medium,

$$p - \bar{p} = \bar{a}^2(\rho - \bar{\rho}) = Z\bar{a}S \tag{3}$$

where $S = (\rho - \bar{\rho})/\bar{\rho}$ is condensation ratio, $\bar{a}$ is sound speed and $Z = \bar{\rho}\bar{a}$ is acoustic impedance of the medium. Acoustic impedance is smaller for the medium when gas is present due to reduction in mean density and sound speed. Pressure perturbation scaled by $Z\bar{a}$ is plotted in Figure 3(a). Peak values of scaled pressure perturbations are smaller when the gas is present indicating smaller condensation ratio. This is also seen from the total void fraction plots in Figure 3(b). We can see that as the pressure wave passes through the medium in which the gas is present, it creates smaller drop in total gaseous phase void fraction. Reduction in sound speed, lower acoustic impedance and smaller condensation ratio entails smaller pressure perturbation in the presence of non-condensable gas as seen from equation 3. Therefore, in the presence of gas weaker pressure waves impinge on the cylinder surface. Consequently, the secondary peak in lift curve, which is due to the impingement of the pressure wave on the cylinder surface (primary peak correspond to cavity shedding) as shown in Figure 3(c), is absent in the presence of non-condensable gas.

Figure 4 shows the root mean square of density fluctuations compared for the two cases (in presence and absence of gas). It can be observed that the density fluctuations, at the cylinder surface and in the cavity closure region, are smaller in the presence of gas. This is due to the fact that the gas does not undergo any phase change (evaporation and condensation). Consequently, the flow in presence of gas is not subject to rapid changes in density compared to the flow containing vapor alone, having a stabilizing effect on the cavity being formed.

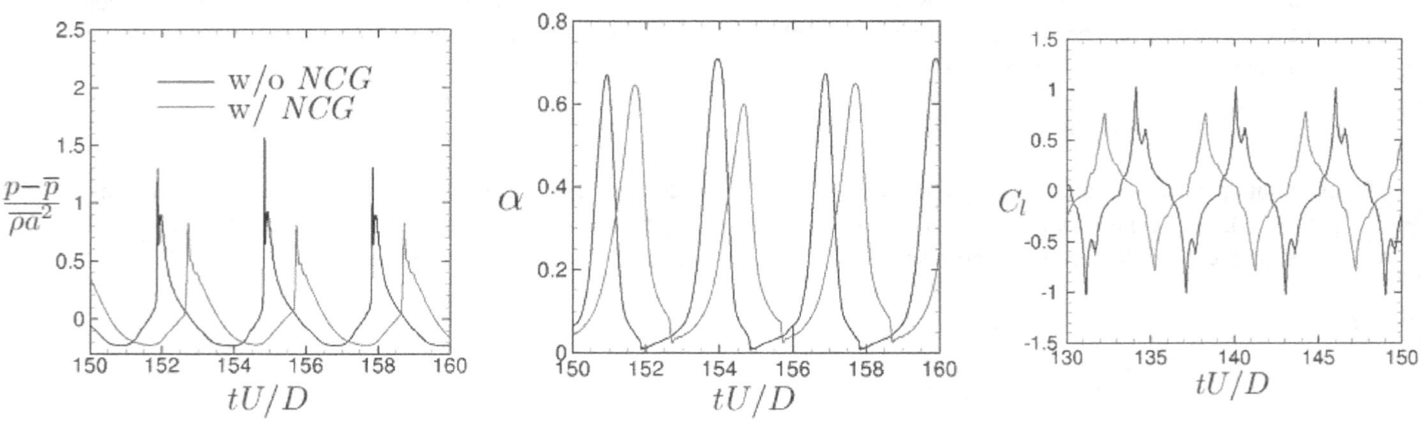

(a) (b) (c)

Figure 3. Scaled pressure pertubation (a) and total void fraction (b) along the wake centerline at $x/D = 2.5$ and lift curves (c).

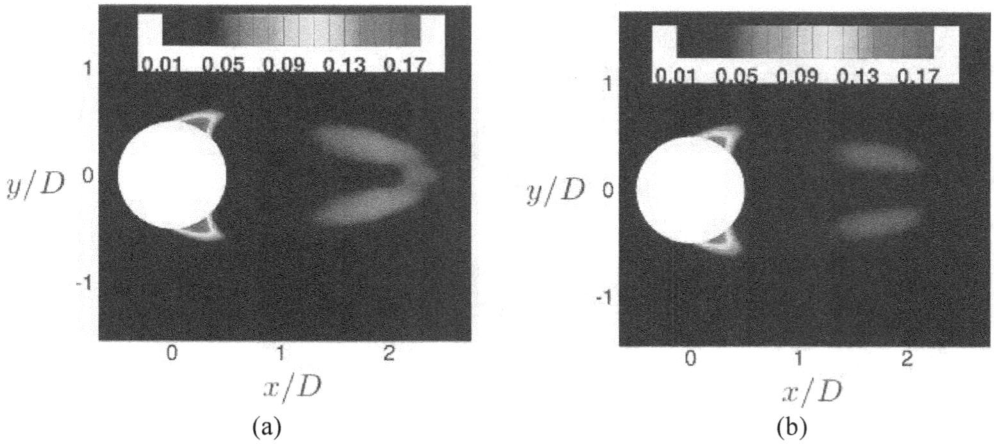

Figure 4. Density fluctuation contour for the flow where *NCG* is absent (a) and present (b).

3.2. Near wake profiles of vapor and gas

Non-condensable gas can behave differently from vapor particularly in the regions of flow which are susceptible to phase change. First, consider $\sigma = 1.0$ (higher cavitation number). Figure 5(a,b) shows evolution of gas and vapor volume fraction with time at $2.5D$ and $5D$ on wake centerline. Note that the volume fraction of vapor is higher than the volume fraction of *NCG* in the region close to the cylinder at $2.5D$, where evaporation is taking place. However, for the positions further downstream at $5D$ in the region of pressure recovery both are nearly same and close to the free stream values.

Next, we present the void fraction profiles when the cavitation number in the free stream is reduced ($\sigma = 0.5$). Figure 6(a,b) shows void fraction contours of vapor and gas respectively in the wake of cylinder at an instant when the cavity is fully attached to the body. Higher gas void fractions are observed in the shear layer and the aft portion of cavity as seen from Figure 6(b), whereas vapor is concentrated in the regions in the rear part of the cylinder as in Figure 6(a). This can be explained from the void fraction profiles taken radially on the cylinder surface as shown in Figure 6(c) at 2 different azimuthal location. For $\theta = 70°$ we see that there is no difference in gas and vapor volume fraction, also their magnitude is small since flow mostly consist of water. The distribution at $\theta = 130°$ where cavity is being formed on cylinder surface, shows that volume fraction of vapor is maximum near the surface due to the production of vapor when the pressure drops below vapor pressure. It is also noticeable that gas volume fraction is maximum in the shear layer and not at the surface of cylinder. Thus, this region of high *NCG* concentration is convected downstream when the cavity grows giving rise to the "wing" shape observed in Figure 6(b).

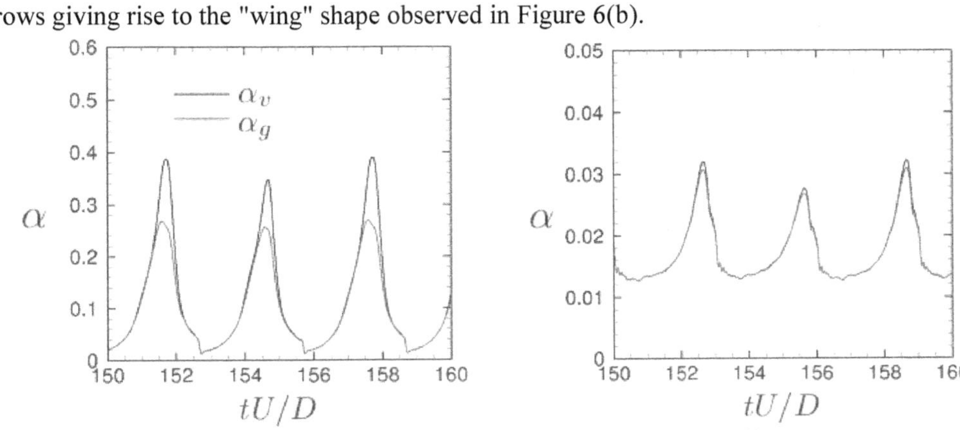

Figure 5. Time history of vapor (black) and *NCG* (red) volume fraction at two different locations along the wake centerline: (a) $x/D = 2.5$, (b) $x/D = 5.0$.

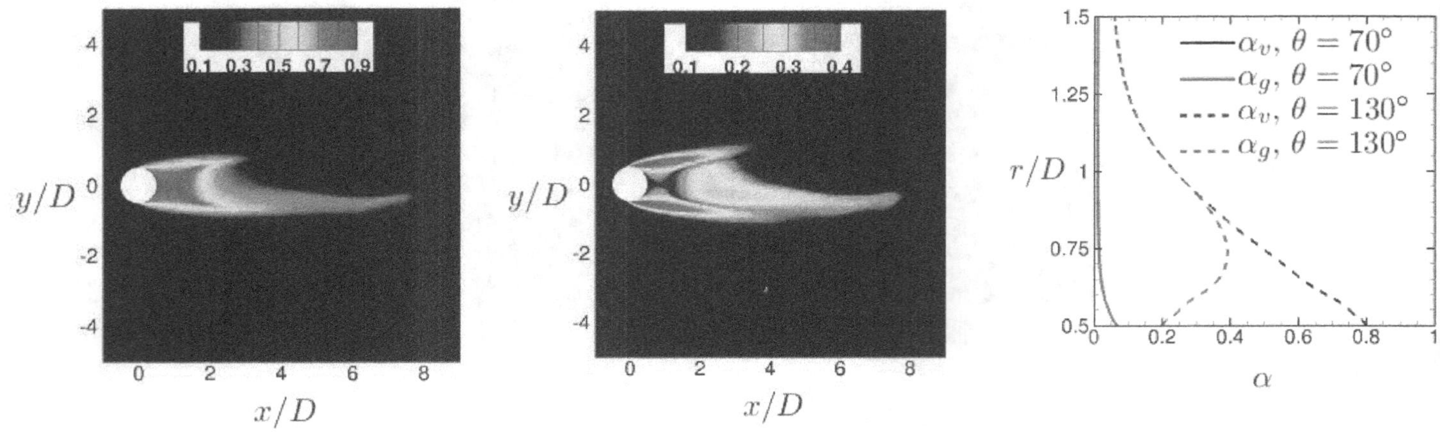

(a) (b) (c)

Figure 6. Volume fraction of vapor (a) and *NCG* (b) when the cavity is fully attached to the cylinder. Radial void fraction profiles of vapor and *NCG* at the cylinder surface (c).

4. Summary

A numerical method to simulate turbulent cavitating flows developed for the water-vapor mixture has been extended to include the effects of non-condensable gas. The speed of sound for the water-vapor-gas mixture shows good agreement with experimental data. Cavitating flow over a circular cylinder in the presence of non-condensable gas is studied at different free stream cavitation numbers. Damping of pressure perturbations due to the presence of gas in the medium is discussed. Vapor and gas uniformly introduced in the free stream, distribute themselves differently in the wake of cylinder depending upon the local flow conditions.

Acknowledgments

This work is supported by the United States Office of Naval Research under Grant ONR N00014-17-1-2676 with Dr. Ki-Han Kim as the program manager. Computing resources were provided by the Minnesota Supercomputing Institute (MSI).

References

[1] Gnanaskandan, A. and Mahesh, K. (2015). *A numerical method to simulate turbulent cavitating fows*. International Journal of Multiphase Flows 70 22-34.

[2] Franc, J-P. (2006). *Physics and Control of Cavitation.* In: Design and Analysis of High Speed Pumps. Educational Notes RTO-EN-AVT-143 2-1 - 2-36.

[3] Akhatov, I., Lindau, O., Topolnikov, A., Mettin, R., Vakhitova, N., Lauterborn, W. (2001). *Collapse and rebound of a laser-induced cavitation bubble*. Physics of Fluids 13(10).

[4] Reisman, G., Duttweiler, M., Brennen, C. (1997). *Effect of Air Injection on the Cloud Cavitation of a Hydrofoil*. ASME Fluids Engineering Division Summer Meeting.

[5] Makiharju, SA., Ganesh, H. Ceccio, SL. (2015). *Effect of non-condensable gas injection on cavitation dynamics of partial cavities*. Journal of Physics: Conference Series 656

[6] Kunz, R., Boger, F., Stinebring, D.R., Chyczewski, T.S., Lindau, J.W., Gibeling, H.J., Venkateswaran, S., Govindan, T.R. (2000). *A preconditioned Navier-Stokes method for two-phase flows with application to cavitation*. Computers and Fluids, 29(8), pp. 849-875.

[7] Singhal, A.K., Athavale M.M., Li H., Jiang Y. (2002). *Mathematical basis and validation of the full cavitation model.* Journal of Fluids Engineering, 124, pp. 617-624.

[8] Ji, B., Luo, X., Peng, X., Zhang, Y., Wu, Y.,Xu, H. (2010). *Numerical investigation of the ventilated cavitating flow around an under-water vehicle based on a three-component cavitation model.* Journal of Hydrodynamics, 22(6), pp. 753-759.

[9] Lu, N-X., Bark, G., Bensow, R. (2012). *Introducing Non-condensable Gas in Unsteady Sheet Cavitation Modelling*. 15th Numerical Towing Tank Symposium.

[10] Gnanaskandan, A. and Mahesh, K. (2016). *Numerical investigation of near-wake characteristics of cavitating flow over a circular cylinder*. Journal of Fluid Mechanics 790 453-491.
[11] Karplus, H. B. (1957). *Velocity of sound in a liquid containing gas bubbles*. The Journal of Acoustical Society of America 29 12-61.

© 2018 ASME

Elastic waves generated by laser induced bubbles in soft solids

[1]Julien Rapet*; [1,2]Claus-Dieter Ohl

[1]*Cavitation Lab, Division of Physics and Applied Physics, Nanyang Technological University*
[2]*Institut für Experimentelle Physik, Otto-von-Guericke-Universsitt*

Abstract

We experimentally investigate the generation and propagation of elastic waves in gelatin occurring during the collapse of laser induced bubbles. Sub-millimeter sized bubbles are generated with a pulsed 6ns laser within the tissue mimicking material. The stress fields and wave propagation is visualized with elastography, a method where the polarity of a polarized light is altered by the state of stress of the medium. High-speed photography captures the fast dynamics of the bubbles and is used to describe the succession of events leading to the wave emission. 4 configurations of bubble collapse are studied; spherical and non-spherical collapses in the bulk, collapse near a wall and collapse near a free surface.

Keywords: elastic waves; cavitation; laser induced bubble; soft-solids; gelatin; stress waves

Introduction

Cavitation is commonly utilized for numerous medical procedures [2]; still there is considerable lack of knowledge on the mechanisms of cavitation and the complex bubble dynamics in biological tissues. Most biological media have viscoelastic properties and as such they possess both liquid (viscous) and solid (elastic) behaviors. In experiments, it is frequent to replace the tissue with a mimicking phantom such as a gelatin-based mixture [4] which allows to cover a wide range of mechanical properties from quasi-Newtonian liquids to soft-solids by varying the gelatin/water ratio. Experiments on cavitation in soft solids have shown that longitudinal waves traveling at the speed of sound analogous to those found in liquids are produced during bubble collapse [3, 6]. But unlike liquids, soft solids elasticity allows transversal waves of much slower speed (few $m.s^{-1}$) to propagate [5].

In this paper, we experimentally study the generation and propagation of these elastic waves in gelatin created by the collapse of laser induced bubbles. The waves are made visible using elastographic methods where the polarity of a polarized light is altered by the state of stress of the medium through which they propagate. High-speed photography captures the fast dynamics of the bubbles and describes the succession of events leading to the wave emission. We describe 5 typical configurations of bubble dynamics leading to spherical and non-spherical collapse, the later are causing an emission of stress waves.

Method

Soft solids preparation

Gelatin was chosen as base to prepare the soft samples for its optical properties; i.e optical transparency and birefringence allowing high-speed recording for monitoring the bubbles dynamics and the resulting stress waves. The soft solids samples are prepared using industrial gelatin (Industrial Gelatin 250bloom, Yasin Gelatin CO.,LTD). The powdered gelatin is mixed with deionized water and is dissolved on a steering hot plate (T = 70°C, steering force = 3) until no grains are visible. Then the hot mixture is poured into a glass container and is let to cool down at lab temperature before being stored in a fridge for at least 24 hours before usage. The mass ratio of Gelatin to water so far studied can be varied between from 2% to 16%.

Bubble nucleation and recording system

The experimental setup is portrayed in a side-view in figure 1. A green Nd:YAG laser (New wave research Orion, $532nm$, $6ns$) emits a single laser pulse which is focused into the gelatin using a microscope objective (Olympus $10x$ Plan Achromat, $NA = 0.25$). At the focal point, optical breakdown occurs generating a sub-millimeter sized bubble. The dynamics of the bubble, i.e expansion, collapse and rebounds, is recorded at frame rates of up to 400,000 frames per second (fps) using a high speed camera (Photron, Fastcam, SA-X2) equipped with a macro camera lens (Nikon, 60mm f/2.8 Micro-NIKKOR AF-D).

*Corresponding Author, Julien Rapet: JULIEN001@e.ntu.edu.sg

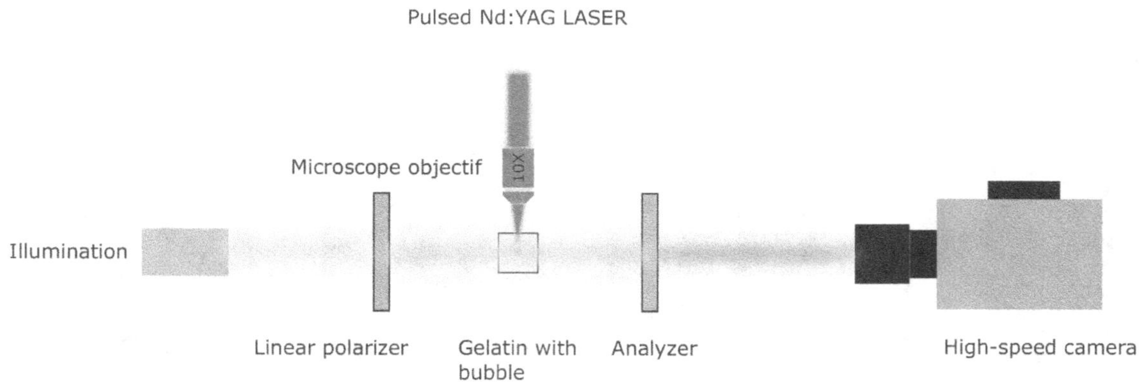

Figure 1: The experimental setup used for visualization of elastic waves generated by laser induced bubbles in soft solids

Plane polariscope

The plane polariscope is a classical device used in photoelasticity [7] to observe changes of mechanical properties of a birefringent material under mechanical deformation. It is composed of a coherent light source, a polarizer and an analyzer. The gelatin phantom is placed between the polarizer and analyzer, as shown in figure 1. Figure 2 illustrates the working principle of the plane polariscope: The light polarized by the polarizer enters and propagates through the transparent sample. Due to the birefringent properties of the gelatin, the light propagates along the principal stress directions. The light passes then through the analyzer with the optical axis oriented perpendicular to the analyzer's axis. In the absence of stress, the gelatin behave as a non-birefringent medium and the polarization is unchanged, thus the light passing through the phantom is blocked by the analyzer. If stresses are present in the gelatin, the birefringence rotates the polarization and as a result some of the light passes through the analyzer and blobs of light are recorded with the high-speed camera.

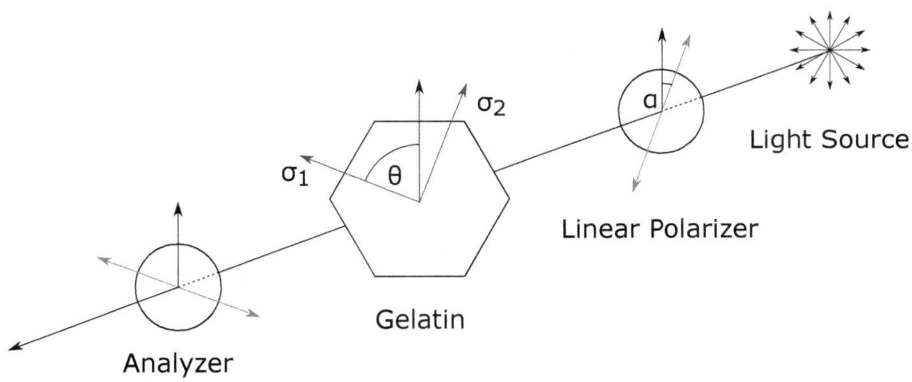

Figure 2: Optical principle of the plane polariscope; in red the optical axis of the polarizer and analyzer, in blue the principal stress direction of the gelatin

Using trigonometry, we can express the intensity of the light exiting the analyzer:

$$I = k^2\, sin^2(2\theta)\, sin^2(\frac{\delta}{2})\, cos(\alpha) \tag{1}$$

where α is the angle of the polarizer with respect to the vertical axis. Polarizer and analyser are kept at $90°$, θ is the angle between the fast axis and the vertical axis, and δ is the phase difference (retardation) between σ_1 the fast axis and σ_2 the slow axis. For white light, this equation possesses one solution for which the intensity is zero:

$$sin^2(2\theta) = 0, \quad i.e \quad \theta = 0, \frac{\pi}{2} \tag{2}$$

For this case, the principal stress directions and the axis of the polarizer and analyzer are coincident and thus the resulting intensity is zero. Those are the *isoclinics*, they indicate the stress principal directions.

Figure 3 depicts experimental and numerical examples of isoclinics. Patterns with four "stress lobes" are observed due to the presence of isoclinics along the horizontal and vertical [A][B] or at 45° [C][D].

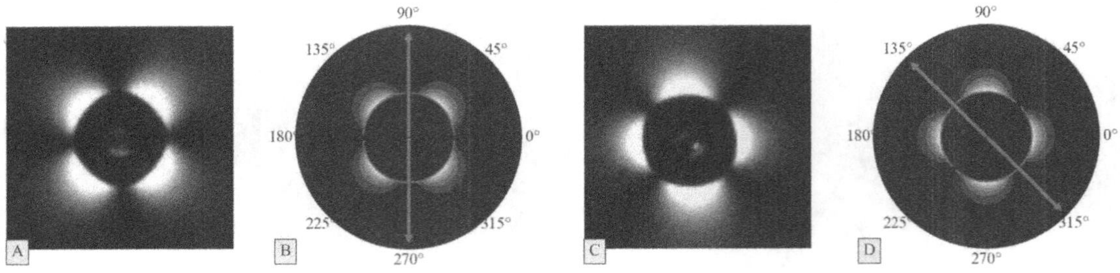

Figure 3: Stress patterns around an expanding bubble. Bright areas indicates areas of stress, red and blue lines illustrate the positions of the optical axis: $\theta = 0$ for [A] (experiments) and [B] (model), $\theta = 45$ for [C] (experiment) and [D] (model)

Results

Spherical bubble collapse

We show first the bubble dynamics in absence of polarizers before we discuss the stress patterns observed. Figure 4 shows selected frames of the dynamics of a single bubble created in the bulk, i.e far from any wall or pre-existing bubble. The time between two frames is $\Delta t = 13.3\mu s$. In the first frame, the bright dot shows the optical breakdown preceding the bubble. After $t \approx 26.67\mu s$ the bubble reaches maximum radius ($r_{max} \approx 325\mu m$), then starts to shrink before symmetrically collapsing after $t \approx 53.34\mu s$. After the first collapse the bubble rebounds and collapses a second time around $t \approx 133.3\mu s$. The bubble in figure 4 remains mostly spherical during the two first collapses and rebounds.

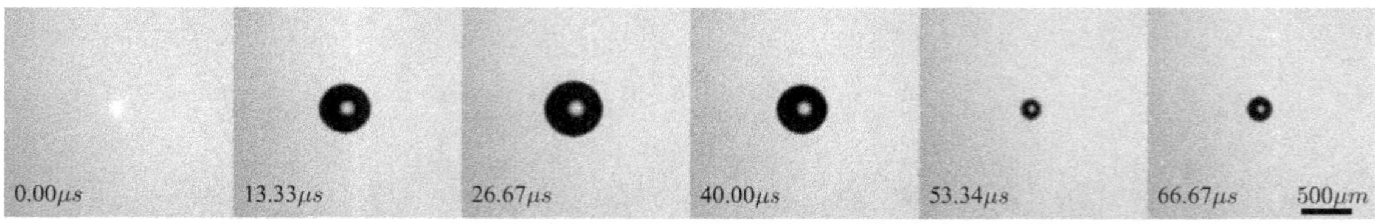

Figure 4: Dynamics of a laser-induced bubble in 4% gelatin

Figure 5, depicts a similar bubble collapsing symmetrically but now observed through the plane polariscope with $\theta = 0$. The use of the polariscope unveils the "bright lobes pattern" around the bubble. The intensity of the lobes can be related to stress intensity using equation 1. As expected the size of the lobes and thus the stress is maximum when the bubble reaches its maximum size. The experiment was repeated for different Polarizer/Analyzer orientations and demonstrates that the stress around the bubble is purely radial, i.e the gelatin near the bubble is compressed during the bubble oscillations. From recording, no apparent elastic waves are created during a symmetric collapse.

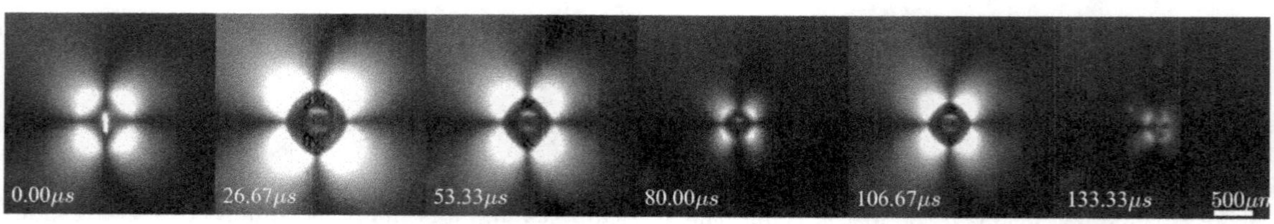

Figure 5: Spherical collapse of laser-induced bubble in 4% gelatin observed using a plane-polariscope, $\theta = 0$

Non-spherical bubble collapse

The mostly spherical collapse as observed in the previous pictures is rarely happening. Most of the times mild asymmetries in the laser breakdown, inhomogeneities of the material (e.g. from previous bubble experiments) or boundaries even far away result in non-spherical dynamics. One of these cases is depicted in figure 6. Interestingly, although the bubble reaches a spherical shape during expansion, surface modes are amplified during the collapse and it moves slightly towards the right while losing its spherical shape. During the second collapse the bubble moves back to the left.

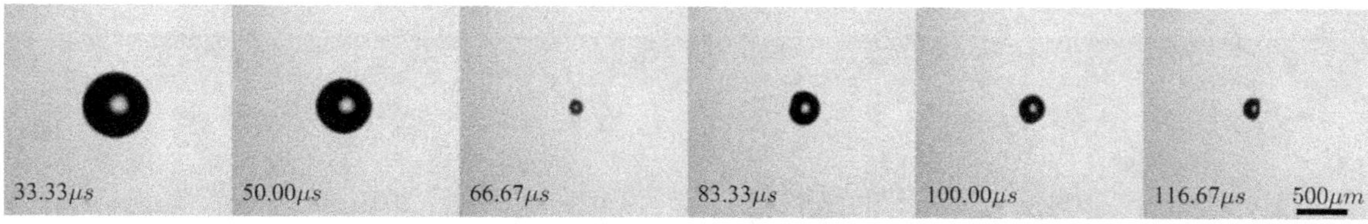

Figure 6: Non-spherical collapse of a bubble in 4% gelatin

The stress patterns during the non-spherical collapse of a different but similar oscillating bubble is depicted in figure 7. Before collapse, the pattern of four quasi-symmetric bright lobes is similar to the previous case. However, during collapse, after $t = 80\mu s$, the bubble moves toward the right by about $\approx 100\mu m$. This movement starts a faster inflow at the left bubble wall, which creates shear-stresses along the translational direction. Although we do not observe a jet piercing through the bubble, some indentation is developing on the left side of the bubble, see also in the fourth frame of figure 6. This stress propagates outwardly from its initial position with a speed of $\approx 3m.s^{-1}$, the two last frames of figure 6 show its position at later stages.

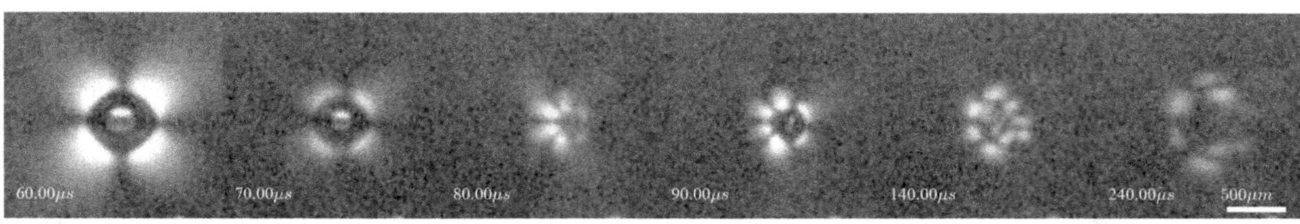

Figure 7: Non-spherical collapse of a laser-induced bubble in 4% gelatin using plane-polariscope, $\theta = 0$

Bubble collapse near a rigid wall

The experiment was repeated for a bubble created close to a solid wall. In this case, observations have shown that the bubble collapsing in gelatin resembles to some extend the bubble collapsing in water: during collapse, the bubble moves towards the wall and a deformation occurs on the wall opposing side leading to the development of a jet. In contrast to water the bubble moves back to its original position due to the elasticity of the medium.

The figure 8 shows the collapse of a bubble created at a distance $L \approx 900\mu m$ from the wall, the standoff parameter is $\gamma = \frac{L}{r_{max}} \approx 1.5$. The bright lines (indicated with arrow) propagating from the top are elastic waves created during the bubble nucleation due to the interaction of the shock-wave induced during optical breakdown and the wall. The translational motion of the bubble towards the wall launches a shear wave starting from the site opposite of the wall, which propagates outward to the left and right initially as a cylindrical wave. The amplitude of the stress waves quickly ceases with distance, likely due to symmetry and viscous dissipation in gelatin.

Figure 8: Bubble collapsing near a wall in 4%- gelatin using plane-polariscope with $\theta = 0$

Bubble collapse near a free surface

A bubble collapsing in water near a free surface will translates and jets away from the interface. In gelatin we observe similar behavior, here the bubble initially moves towards the interface during expansion but is "pulled back" to its position of generation during collapse. Figure 9 displays selected pictures of a bubble collapsing in 4%-gelatin near a free surface observed with a plane-polariscope ($\theta = 0$). Due to the presence of the curve soft boundary the gelatin is in a pre-stress condition and appears bright. Compared to the previous collapse near a wall, the situation is inverted, the bubble collapses away from the free surface and the quasi-jetting happens on the free surface side and the shear wave "tail" appears between the interface and the bubble.

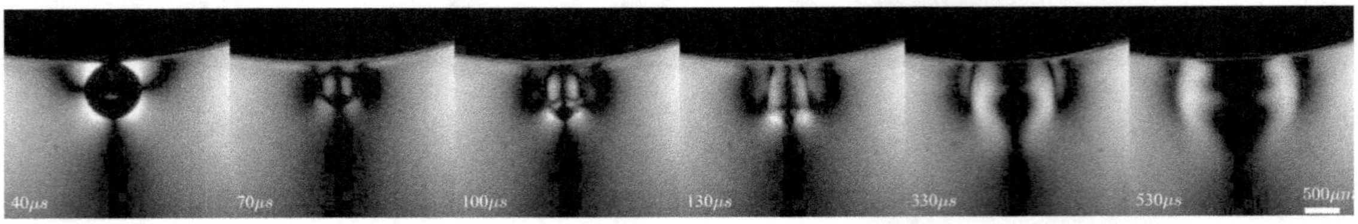

Figure 9: Bubble collapse near a free surface in 4 − % gelatin using plane-polariscope with $\theta = 0$

Discussion and conclusion

The present study is to our understanding the first report of elastic waves generated from non-spherical bubbles collapsing in a tissue-mimicking material. While spherical bubbles generate a stress field, non-spherical bubbles create stress waves due to center-of-mass translation. There we expect that the bubble moves with a speed much faster than the elastic wave velocity. The resulting wave pattern thus are Mach cones, resembling the wave generation in supersonic shearwave elastography [1]. Besides the four demonstrated cases of spherical, non-spherical, rigid and free boundary collapses, we expect also for the shock wave-gas bubble interaction the formation of stress waves. Although not shown here, the stress wave may not be generated at the location of bubble nucleation but at the location of the gas bubble impacted by the shock thus far from the origin of nucleation. This may have important consequences for medical applications of shock waves. The research presented here is only a starting point and demands for a quantitative analysis and simulations of the wave propagation. Monitoring biological cells at various distance from the bubble may allow to evaluate the importance of stress waves for cell viability or drug delivery. At last we expect that the strength of the elastic waves may be strong function of the gelatin concentration and bubble size and have some optimum at intermediate values. This again needs more experimental and numerical work.

References

[1] Jérémy Bercoff, Mickael Tanter, and Mathias Fink. Supersonic shear imaging: a new technique for soft tissue elasticity mapping. *IEEE transactions on ultrasonics, ferroelectrics, and frequency control*, 51(4):396–409, 2004.

[2] Christopher E Brennen. A review of cavitation uses and problems in medicine. In *WimRC Forum*, volume 70, 2006.

[3] Emil-Alexandru Brujan and Alfred Vogel. Stress wave emission and cavitation bubble dynamics by nanosecond optical breakdown in a tissue phantom. *Journal of Fluid Mechanics*, 558:281, July 2006.

[4] Achu G Byju and Ankur Kulkarni. Mechanics of gelatin and elastin based hydrogels as tissue engineered constructs. In *ICF13*, 2013.

[5] J.-L. Gennisson, T. Deffieux, M. Fink, and M. Tanter. Ultrasound elastography: Principles and techniques. *Diagnostic and Interventional Imaging*, 94(5):487–495, May 2013.

[6] Ryota Oguri and Keita Ando. Cloud cavitation induced by shock-bubble interaction in a viscoelastic solid. *Journal of Physics: Conference Series*, 656:012032, December 2015.

[7] K Ramesh. Digital photoelasticity, 2000.

© 2018 ASME

Erosive Aggressiveness of Collapsing Cavitating Structures

[1]Sören Schenke*; [1,2]Tom J.C. van Terwisga

[1]*Delft University of Technology, Delft, the Netherlands;* [2]*Maritime Research Institute Netherlands, Wageningen, the Netherlands*

Abstract

The erosive aggressiveness of idealized cavities collapsing on a flat surface is investigated by numerical simulation, employing the cavitation intensity approach by Leclercq et al (2017) as a measure of the local energy impact rate. We propose a more straight forward formulation of the cavitation intensity model and verify that it satisfies energy conservation requirements for the accumulated surface energy. Based on the cavitation intensity model, statistical aggressiveness indicators are proposed. The indicators account for the rapidness and frequency of the collapse events. The aggressiveness indicators are further applied to a NACA0015 hydrofoil surface.

Keywords: cavitation erosion, impact energy, aggressiveness indicator, pressure recovery

Introduction

From an energy point of view, the erosive aggressiveness of a cavity imploding close to a solid surface depends for one part on its ability to focus its impact energy on the surface. The surface distribution of impact energy accumulating throughout a collapse event partially depends on the cavity shape and its orientation relative to the surface. Starting from the potential energy approach by Vogel and Lauterborn (1988), Leclercq et al (2017) have proposed a cavitation intensity model for the instantaneous energy impact rate on a solid surface caused by collapsing cavities. The cavitation intensity model is derived from the solid angle projection of the released power on discrete triangular elements of the impacted surface. We propose a more straight forward and fully continuous form of the cavitation intensity model. This facilitates its application and allows to make analytical predictions on the amount of accumulated surface energy in certain situations, which are employed for numerical validation in this study. The erosive aggressiveness is further thought to depend on the rapidness of the collapse event and, in periodic flow, the impact frequency. Based on the cavitation intensity model, aggressiveness indicators are proposed that account for both effects. A distinct weak point of this modelling approach is that it involves the knowledge of the ambient pressure driving the cavity collapse, which is typically unknown in more complex flow situations. A solution to this problem is suggested by Arabnejad and Bensow (2017), who trace the cavity collapses individually to reconstruct the driving ambient pressure from a set of kinematic parameters employing the Rayleigh-Plesset equation. However, this approach requires any cavitating structure to be simplified to a spherical bubble of equivalent volume. Since we deliberately address the effect of cavity shape and surface orientation, we assume that the driving pressure field of a periodic cavitating flow can be approximated by the time averaged pressure field, thereby accounting for the effect of spatial pressure recovery on statistical average. The effect of pressure recovery and the feasibility of the aggressiveness indicators is demonstrated for the cavitating flow around a NACA0015 hydrofoil.

Erosion Intensity Model

Following Vogel and Lauterborn (1988), the potential energy of a cavity with volume V and liquid volume fraction $\gamma = (\rho - \rho_v)/(\rho_l - \rho_v)$ is given by $E_{pot} = (1-\gamma)(p_\infty - p_v)V$, where p_v denotes the vapor pressure and p_∞ the ambient pressure driving the cavity collapse. As the cavity collapses, the instantaneous rate of energy release is given by $\partial E_{pot}/\partial t$. Due to the linearity of these expressions, the energy release rate of the entire cavity is equal to the volume integral of the differential point source release rate $\partial e_{pot}/\partial t$. As indicated by Figure 1, this allows to reconstruct both the local surface impact rate $\partial e_S/\partial t$ as well as the overall surface integrated impact rate $\partial E_S/\partial t$ from either the local change of volume fraction γ or the volume change of the entire cavity, resulting in the notations

$$\left.\frac{\partial e_S}{\partial t}\right|_{\dot\gamma} = -\mathrm{X}\frac{\partial e_{pot}}{\partial t},\quad \left.\frac{\partial E_S}{\partial t}\right|_{\dot\gamma} = \int_{surf}\left.\frac{\partial e_S}{\partial t}\right|_{\dot\gamma} dS,\quad \left.\frac{\partial e_S}{\partial t}\right|_{\dot V} = -\int_{vol}\mathrm{X}\frac{\partial e_{pot}}{\partial t}dV \quad\text{and}\quad \left.\frac{\partial E_S}{\partial t}\right|_{\dot V} = \int_{surf}\left.\frac{\partial e_S}{\partial t}\right|_{\dot V} dS, \tag{1}$$

where X is the projection operator that converts locally released power to local surface impact power, depending on the distance from the emission source and the surface orientation relative to the source. Flageul et al (2012) determine

*Corresponding author, Sören Schenke: s.schenke@tudelft.nl

the local potential energy release $\partial e_{pot}/\partial t$ from the Lagrangian time derivative of mixture density. Alternatively, the energy release can be calculated from the Lagrangian time derivative of liquid volume fraction, which gives

$$\frac{\partial e_{pot}}{\partial t} = -\left(\frac{\partial \gamma}{\partial t}\right)^{+}_{div} (p_\infty - p_v), \quad \text{where} \quad \left(\frac{\partial \gamma}{\partial t}\right)^{+}_{div} = \max\left[-\nabla \cdot u\left(\gamma + \frac{\rho_v}{\rho_l - \rho_v}\right), 0\right]. \tag{2}$$

The '+' index indicates that only condensation is taken into account. Assuming that each point source emits its potential energy as a radial wave of infinitely large propagation speed, Leclercq et al (2017) define the potential power impact per discrete surface element ΔS from the solid angle projection of the radial source on a planar triangle as presented in the work by Van Oosterom and Strackee (1983). A continuous form of the energy impact rate on a surface location with local normal vector $\vec{n}$ is derived by employing the projection operator

$$\mathrm{X} = \frac{1}{4\pi}\left(\frac{\vec{x} \cdot \vec{n}}{|\vec{x}|^3}\right), \tag{3}$$

where $\vec{x}$ denotes the vector from the impacted surface location to the center of the emission source as shown in Figure 1. Integration of the bracket term in Equation (3) over an arbitrary convex surface gives the solid angle Ω as used in the work by Leclercq et al (2017). From the continuous form of the cavitation intensity model, it can be shown analytically that the integrated impact rate is equal to the emission rate of a volume source on any closed convex surface around the source. It is equal to half of the emission rate on a flat surface stretched to infinity, hence

$$\oint_{\substack{convex\\surface}} \left.\frac{\partial e_S}{\partial t}\right|_{\dot{V}} dS = \frac{\partial E_{pot}}{\partial t} \quad \text{and} \quad \int_{\substack{plain\\surface}}^{\infty} \left.\frac{\partial e_S}{\partial t}\right|_{\dot{V}} dS = \frac{1}{2}\frac{\partial E_{pot}}{\partial t}. \tag{4}$$

For simplicity, $\dot{e}_S$ and $\dot{E}_S$ are further on referred to as the local and the surface integrated impact rate caused by volume sources. A statistical measure for the aggressiveness distribution is given by the aggressiveness indicators

$$\langle \dot{e}_S \rangle_{e_S} = \frac{1}{e_S}\int_0^{e_S} \dot{e}_S de_S \quad \text{and} \quad \langle \dot{e}_S \rangle_f = \left(\frac{1}{T}\int_0^{e_S} \dot{e}_S de_S\right)^{1/2}, \quad \text{where} \quad e_S(t) = \int_0^t \dot{e}_S(t) dt. \tag{5}$$

The indicator $\langle \dot{e}_S \rangle_{e_S}$, being the local energy impact rate $\dot{e}_S(t)$ averaged over the accumulated energy e_S, has a tendency to amplify local extreme events, because for the same amount of accumulated energy, it is proportional to the rate at which the surface is impacted. In case of a periodic impact signal with uniform amplitude, it is independent from the impact frequency due to the normalization by the accumulated energy, whereas the indicator $\langle \dot{e}_S \rangle_f$ is also proportional to the impact frequency. Normalization by the sample time T causes it to converge as $T \to \infty$.

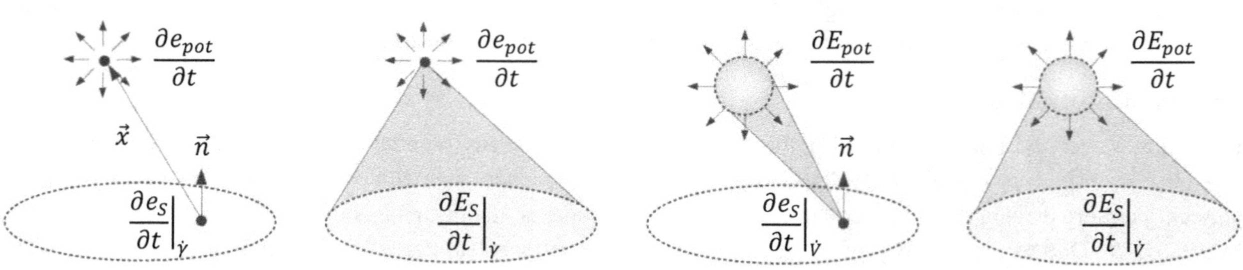

Figure 1: Terminology of energy time derivatives

Numerical Testcases

The cavitation intensity model is implemented in the open source CFD environment OpenFOAM (2017). Viscous forces and surface tension are neglected in the momentum equation of the flow model. Phase transition is modelled by the mass transfer approach, where the model by Merkle et al (1998) is employed in a slightly modified form as presented in the work by Schenke and Van Terwisga (2017). In the pure liquid and vapor limits, the flow is assumed to be incompressible and a pressure equation is solved with subsequent momentum correction. The erosive aggressiveness of three different cavity types is investigated. Those are, as shown in Figure 2, a horseshoe cavity, a parallel ring cavity and a bubble, all collapsing on a flat surface which is large enough to verify Equation (4). The initial bubble volume is equal to the horseshoe volume and the ring volume is twice the horseshoe/bubble volume. The torus diameter is R_t = 6 mm and the tube radius r = 2 mm. Both the parallel ring and the bubble are initialized at h = 0.5 mm above the surface. The inner part of the computational domain is indicated in Figure 3. It consists of a cubic block structured grid of 20 mm edge length, uniformly subdivided by 55 cells in all directions. The overall volume of the domain is 32 m^3. A time step size of Δt = 1.0e-7 s turned out to be sufficiently small to obtain a converged cavity collapse time. Additionally, a free bubble collapse is simulated and the collapse time is compared to the analytical solution of the Rayleigh-Plesset equation. As indicated in Figure 3, the solid bottom plane, treated as a slip wall, is replaced by a symmetry plane in this case. With C_c and C_v being the condensation and evaporation mass transfer coefficients of the modified Merkle model (Schenke and Van Terwisga 2017), the simulation is conducted for $C_c = C_v$ = 1000 kgs/m^5, which is sufficiently large for the collapse to be inertia driven. The pressure field is initialized with $p_v = 2340$ Pa inside the cavities and $p_\infty = 1$ bar outside the cavities. To avoid unrealistic behavior due to the initial pressure jump across the interface, the pressure equation is solved first, thereby providing a solution of the Laplace equation as the flow is at rest initially. Figure 3 exemplarily shows the pressure field close to the beginning of the horseshoe cavity collapse. It is further noted that the reconstruction of the positive Lagrangian time derivative $(\partial\gamma/\partial t)^+_{div}$ in Equation (2) involves numerical errors that may eventually violate the energy balance. To verify the cavitation intensity model itself, this effect is eliminated for the cavity collapse study by assuming that the change of vapor volume in the entire domain predicted from the divergence field differs from the actual vapor volume change predicted by the γ-time derivative by a correction factor. This factor is applied to Equation (2) at each cell individually.

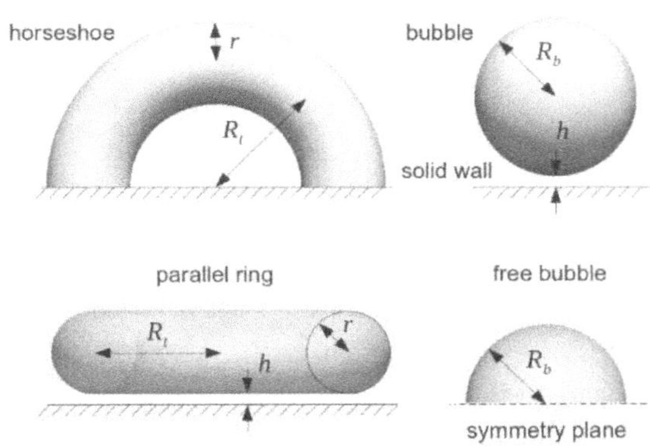

Figure 2: Idealized cavities investigated in this study

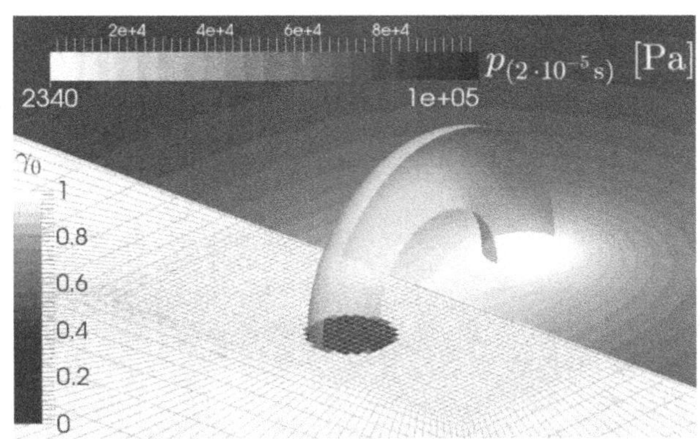

Figure 3: Initial horseshoe cavity (iso-surface for γ_0 = 0.5) and bottom plane with initial γ-field and pressure field at 2e-05 s

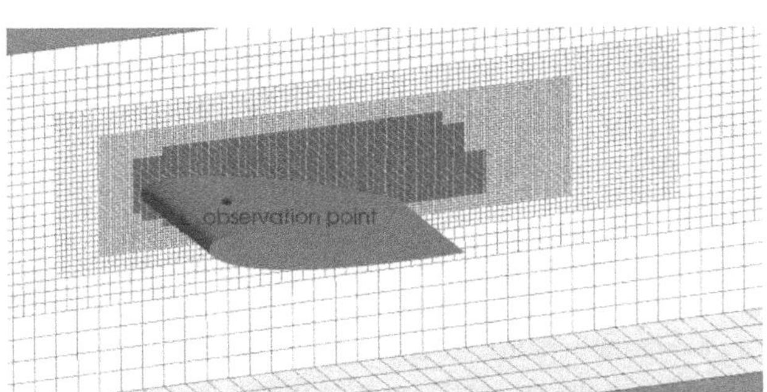

Figure 4: Grid refinement levels and observation point on the foil surface at 20% cord length

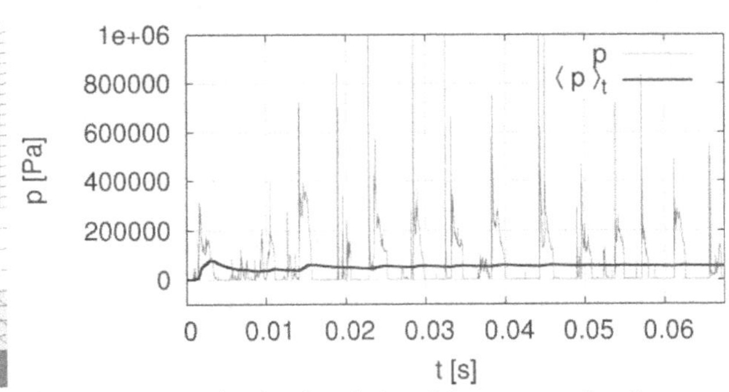

Figure 5: Pressure signal and evolution of the corresponding time averaged pressure $\langle p\rangle_t$ at the observation point in Figure 4.

The analysis is further carried out for the cavitating flow around a NACA0015 hydrofoil in a tunnel section. The flow conditions are in line with an experiment by Van Rijsbergen et al (2012), where the angle of attack is 8° at an inflow speed of 17.3 m/s and an ambient pressure of 302295 Pa. Cordlength and span are 6 cm and 4 cm, respectively. The grid, as shown in Figure 4, corresponds to what is referred to as the fine grid in previous work by Schenke and Van Terwisga (2017), where further details on grid properties and solver settings are found. The simulation is conducted for $C_c/C_v = 2$, $C_c = 5000$ kgs/m^5 and $\Delta t = 7.5\text{e-}07$ s. For these settings, a converged pressure impact frequency of 193 Hz has been found by FFT analysis, being in good agreement with the shedding frequency of 188 Hz found by Van Rijsbergen et al (2012) in their experiment. In addition to the previous work, the time averaged pressure field $\langle p \rangle_t$ under cavitating flow conditions is computed. We assume the time averaged pressure field to be the steady field driving the cavity collapses such that $\langle p \rangle_t$ locally represents the driving pressure p_∞ in Equation (2). Figure 5 exemplarily depicts the pressure signal and the evolution of the corresponding time averaged pressure at an observation on the foil surface at mid-span and 20% cord length (see Figure 4).

Results & Discussion

Figure 6 depicts the evolution of vapor volume V_v and accumulated surface energy E_S over time for the cavities shown in Figure 2. Towards the end of the collapse, the accumulated surface energy E_S converges to 50% of the initial potential energy E_{pot}, thereby verifying Equation (4). The analytical solution obtained from the Rayleigh-Plesset equation without viscous and surface tension forces is included as a reference solution for the free bubble collapse. Satisfactory agreement of the predicted collapse time is achieved. Since the presence of the wall weakens the lower half of the pressure field driving the close wall bubble collapse, the free bubble collapses faster. For the same reason the horseshoe cavity collapses faster than the parallel ring because it is not as strongly subjected to wall interaction.

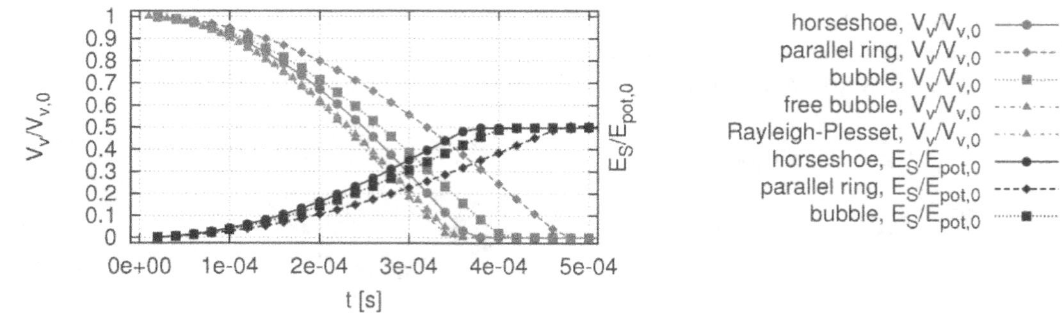

Figure 6: Evolution of vapor volume and accumulated surface energy for the cavity types depicted in Figure 2 collapsing on a flat surface

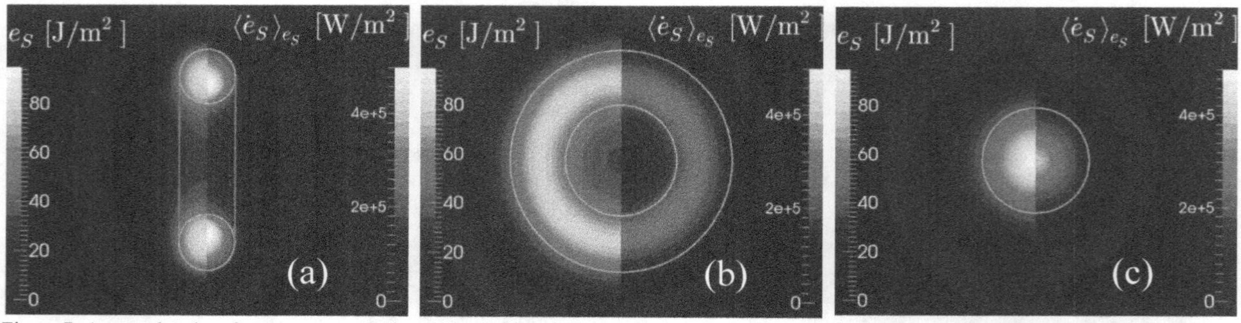

Figure 7: Accumulated surface energy e_S (left) and distribution of the aggressiveness indicator $\langle \dot{e}_S \rangle_{e_S}$ (right) for the horseshoe cavity (a), the parallel ring cavity (b) and the spherical bubble (c) with the green line indicating the outline of the corresponding initial cavity shape

To gain more insight into the energy focusing abilities of the different cavity types, Figure 7 depicts the corresponding distributions of accumulated surface energy e_S and the distributions of the aggressiveness indicator $\langle \dot{e}_S \rangle_{e_S}$. In this particular case, all three cavities exhibit similar abilities to focus their potential energy towards the surface. The different collapse times, however, clearly reflect in the distributions of the aggressiveness indicator $\langle \dot{e}_S \rangle_{e_S}$, indicating that the horseshoe cavity is most aggressive, followed by the bubble of equivalent volume. The aggressiveness indicator $\langle \dot{e}_S \rangle_{e_S}$ also predicts a more focused impact distribution than the accumulated surface energy e_S. It is further observed for the horseshoe and the ring cavity that especially the peak value of the indicator $\langle \dot{e}_S \rangle_{e_S}$ exhibits a slight eccentricity from the torus center line towards the torus center, which is motivated by the circumstance that a Laplacian pressure field exhibits a larger driving force outside from the torus center line than from the inside. This is a distinct shape effect and shows that it is not entirely correct to assume the initial potential cavity energy to be proportional to the ambient pressure p_∞. Even in the absence of wall interaction, the driving pressure may vary over the cavity surface because, depending on its shape, the cavity also interacts with itself to some extent. The same applies to cavity-cavity interaction. This makes the determination of the driving ambient pressure p_∞ a problem of distinct difficulty.

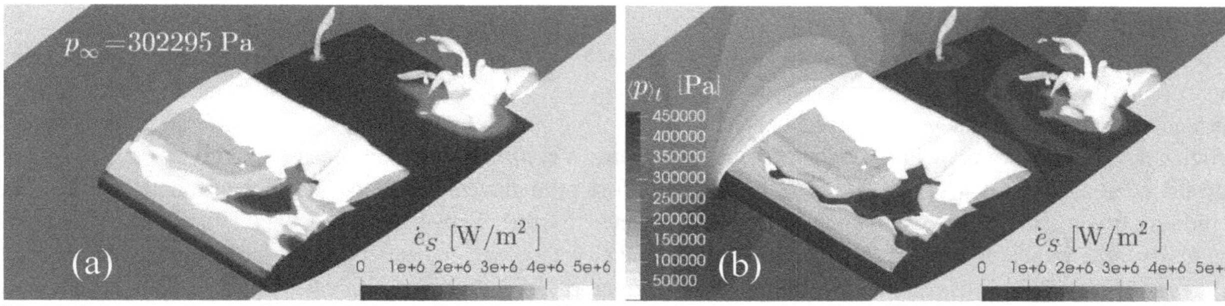

Figure 8: Instantaneous vapor structures for iso-surfaces of $\gamma = 0.5$ and energy impact rate for uniform driving pressure p_∞ (a) and variable driving pressure $\langle p \rangle_t$ (b)

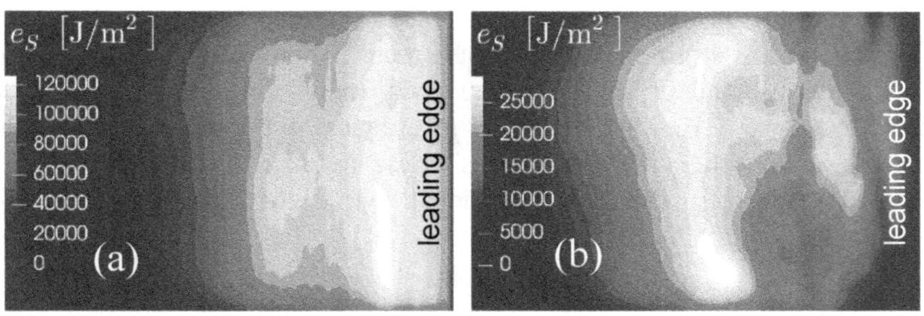

Figure 9: Accumulated surface energy on the NACA0015 hydrofoil surface for uniform p_∞ (a) and variable driving pressure $\langle p \rangle_t$ (b) after 0.27 s

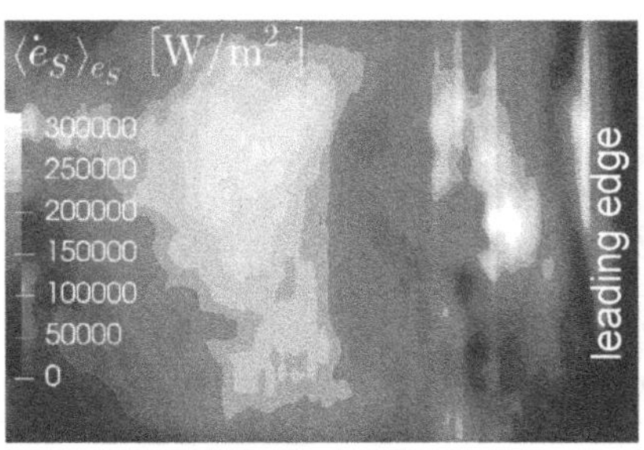

Figure 10: Distribution of the aggressiveness indicator $\langle \dot{e}_S \rangle_{e_S}$ on the NACA0015 hydrofoil surface, accounting for pressure recovery

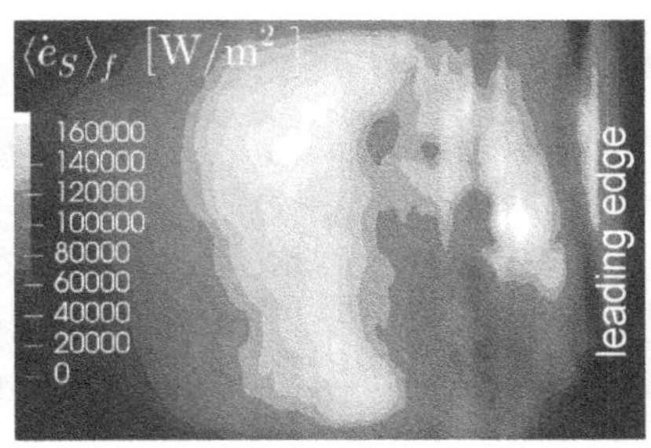

Figure 11: Distribution of the aggressiveness indicator $\langle \dot{e}_S \rangle_f$ on the NACA0015 hydrofoil surface, accounting for pressure recovery

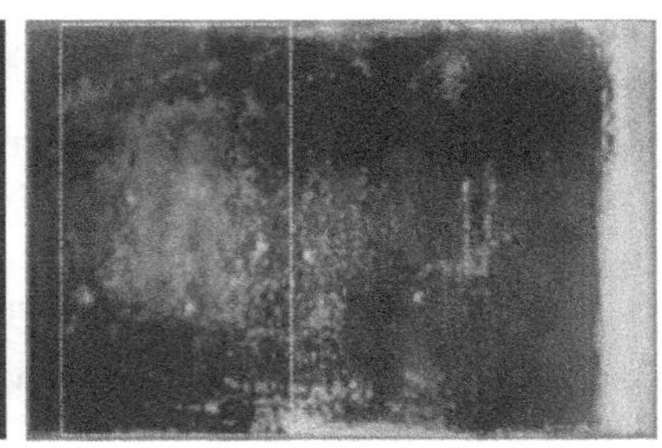

Figure 12: Damage pattern on the NACA0015 hydrofoil surface identified from paint test experiments by Van Rijsbergen et al (2012)

In case of a periodic flow, we can at least account for the effect of spatial pressure recovery on statistical average by assuming that the driving pressure field is given by the time averaged pressure field under cavitating flow conditions. Figure 8 (a) shows the distribution of the instantaneous energy impact rate $\dot{e}_S$ for uniform driving pressure p_∞. The same flow situation is depicted in Figure 8 (b), however with the driving pressure p_∞ being equal to the time averaged pressure $\langle p \rangle_t$ as depicted on the tunnel side wall. This inevitably results in a larger predicted impact rates towards the leading edge, which is also clearly reflected by the distribution of accumulated surface energy e_S, shown in Figure 9 (a) for uniform driving pressure p_∞ and in Figure 9 (b) for variable driving pressure $\langle p \rangle_t$. The rapidness of the impacts is assessed by the aggressiveness indicators $\langle \dot{e}_S \rangle_{e_S}$ and $\langle \dot{e}_S \rangle_f$ given by Equation (5). Both result in similar impact distributions, indicating regions of distinct periodic impacts caused by the re-entrant jet mechanism rather than regions of scattered impacts. The indicator $\langle \dot{e}_S \rangle_{e_S}$ further amplifies isolated events at the trailing edge because its magnitude is independent from the impact frequency. The so identified regions of high erosion risk qualitatively agree with the damage pattern obtained by Van Rijsbergen et al (2012) from experimental paint tests (Figure 12), although the main impact region identified from the simulation is somewhat closer to the leading edge and less focused towards the mid-span. The latter observation may partially be explained by the neglect of viscous forces which, if present at the tunnel wall, may cause a deflection of the re-entrant jet towards the mid-span (Dular and Petkovšek, 2015). It should further be noted that the paint test by Van Rijsbergen et al (2012) was conducted for one hour, whereas the simulation time is only 0.27 s, not yet enough to obtain a fully converged impact distribution.

Conclusion

A continuous form of the cavitation intensity model by Leclercq et al (2017) is presented in this study, providing a measure for the instantaneous impact power on a surface subjected to cavity collapses. The distribution of accumulated surface energy shows that the idealized horseshoe cavity, the parallel ring cavity and the bubble exhibit similar efficiencies in focusing their initial potential energy towards a confined space on the surface, given that the bubble and the parallel ring collapse very close to the surface. More pronounced differences have been found regarding the rapidness of the collapse event, which is strongly affected by the cavity shape and wall interaction effects. An aggressiveness indicator derived from the cavitation intensity approach indicates that the horseshoe cavity is more aggressive than a bubble of equivalent volume, which again turns out to be more aggressive than a parallel ring cavity of same torus and tube diameter as the horseshoe. Further taking into account that the impact power of the bubble and the parallel ring decreases considerably with increasing distance from the impacted surface, leads to the conclusion that the horseshoe cavity attached to the surface is most likely to result in damaging collapse events. Next to the rapidness of the collapse event, the effect of spatial pressure recovery has been identified as a second crucial effect since it strongly affects the distribution of initial potential cavity energy eventually impacting the solid surface. On time average, the flow around the NACA0015 hydrofoil investigated in this study exhibits a pronounced pressure recovery gradient, thereby increasing the flow aggressiveness towards the trailing edge. However, the exact determination of the pressure effectively driving the individual cavity collapses remains an issue of distinct difficulty and requires further research.

Acknowledgements

This research is funded by the CaFE ITN initiative and the MARIN academy. We thank both institutions for their collaboration and we highly appreciate the inspiring discussions on cavitation erosion mechanisms.

References

[1] Arabnejad, M.H. and Bensow, R. (2017). *A Methodology to Identify Erosive Collapse Events in Incompressible Simulation of Cavitating Flows.* Proceedings of the 20th Numerical Towing Tank Symposium, Wageningen, the Netherlands.

[2] Dular, M. and Petkovšek, M. (2015). *On the Mechanisms of Cavitation Erosion – Coupling High Speed Videos to Damage Patterns.* Experimental Thermal and Fluid Science. 68(2015), pp. 359-370.

[3] Flageul, C., Fortes-Patella, R. and Archer, A. (2012). *Cavitation Erosion Prediction by Numerical Simulations.* Proceedings of the 14th International Symposium on Transport Phenomena and Dynamics of Rotating Machinery, Honolulu, HI, USA.

[4] Leclercq, C., Archer, A., Fortes-Patella, R. and Cerru, F. (2017). *Numerical Cavitation Intensity on a Hydrofoil for 3D Homogeneous Unsteady Viscous Flows.* International Journal of Fluid Machinery and Systems. 10(3), pp. 254-263.

[5] Merkle, C.L., Feng, J.Z., and Buelow, P.E.O. (1998). *Computational Modelling of the Dynamics of Sheet Cavitation.* Proceedings of the 3rd International Symposium on Cavitation, Grenoble, France.

[6] OpenFOAM (2017). OpenFOAM web site, 2017.

[7] Schenke, S. and van Terwisga, T.J.C. (2017). *Numerical Prediction of Vortex Dynamics in Inviscid Sheet Cavitation.* Proceedings of the 20th Numerical Towing Tank Symposium, Wageningen, the Netherlands.

[8] Van Oosterom, A. and Strackee, J. (1983). *The Solid Angle of a Plane Triangle.* IEEE transactions on Biomedical Engineering, No. 2, pp. 125–126.

[9] Van Rijsbergen, M., Foeth, E.-J., Fitzsimmons, P. and Boorsma, A. (2012). *High-Speed Video Observations and Acoustic-Impact Measurements on a NACA 0015 Foil.* Proceedings of the 8th International Symposium on Cavitation, Singapore.

[10] Vogel, A. and Lauterborn, W. (1988). *Acoustic Transient Generation by Laser-Produced Cavitation Bubbles near Solid Boundaries.* The Journal of the Acoustical Society of America. 84(2), pp. 719-731.

© 2018 ASME

10th International Symposium on Cavitation - CAV2018
Baltimore, Maryland, USA, May 14 – 16, 2018 CAV18-05075

Experiment Research on Cavitation Control by Active Injection

[1]Shengpeng Lu; [1]Wei Wang*; [1]Tengfei Hou; [2]Mindi Zhang; [1]Jianxiong Jiao; [1]Qingdian Zhang; [1]Xiaofang Wang;

[1]Key Laboratory of Ocean Energy Utilization and Energy Conservation of Ministry of Education, Dalian University of Technology, Dalian 116024, China; [2]School of Mechanical Engineering, Beijing Institute of Technology, Haidian district, Beijing, 100081, China

Abstract

Based on the active control strategy of the flow, this paper proposed a method of suppressing cavitation by arranging water jet holes on the suction surface of the hydrofoil. The idea of cavitation control was tested in the experiments. The NACA0066 hydrofoil was placed in the test channel at the attack angle of 8°. The cavitation conditions were distinguished by cavitation number and the mass flow coefficient of the jet flow. High-speed flow field display technology was used to study the characteristic of cavitation. The influence of the jet flow mass flow on cavitation suppression was examined. The results show that the jet can suppress or weaken cavitation. With the decreasing of the cavitation number, the effect of the cavitation suppression was weakened, the mass flow of the jet needed for cavitation suppressing is increased. And the optimum mass flow coefficient of the jet flow was get under different cavitation number.

Keywords: active control; cavitation; jet hole; jet flow; re-entrant jet

Introduction

It has been proved in the results of experiments [1-3] and numerical simulations [4-6] that the re-entrant jet and the lateral jet are the main reason for sheet cavitation changing into cloud cavitation. So, it can be started with suppressing the re-entrant jets or the lateral jets to suppress the transformation of sheet cavitation to cloud cavitation. At present, there are two main methods to suppress the re-entrant jets, which are based on different flow control strategies respectively. According to the different control methods, the flow control technology can be divided into active control technology and passive control technology. The active control technology is based on the input of external energy, and the control of flow is realized by coupling appropriate disturbance model with the internal flow model, such as injecting or inhaling a liquid [7-11], gas [12] or polymer solution [13] [14] through surface of the test object. Passive control doesn't require external energy input. It only changes the energy distribution of the fluid by passive devices to control the flow. It is worth noting that, for hydraulic machinery, its working conditions are often not a single working condition. The passive control method [15] [16] (for example, the leading edge flap, vortex generator, hydrofoil surface modification, etc.) has a great influence on the hydrodynamic performance of the hydrofoil. And what's more important, once a certain passive device is applied to the surface of the test object, it is difficult to realize the interactive adjustment under different cavitation conditions. So, this paper pays more attention to the active flow control technology. Based on the active control strategy of the flow, a method of suppressing cavitation by arranging water jet holes on the surface of the hydrofoil is proposed. Through the design of the microstructures on the surface of the hydrofoil, the jet is ejected along the set channel to interfere with the re-entrants, and then the aim of suppressing cavitation is achieved. According to the degree of cavitation, by changing the flow rate of the jet, the momentum and velocity of the jet is adjusted, and thus the interactive adjustment under different cavitation conditions is realized. In the way of achieving the jet ejection by active flow control technology, the pump is used to pump the fluid into the cavity (which is set up in advance) of the hydrofoil, and then the fluid is ejected along the set flow channel. In the process, the velocity of the jet ejection is regulated by adjusting the speed of the pump. The study of this paper is based on the analysis of the visual observations of cavities on three angle of views at attack angle of 8° gathered by high-speed imaging.

Experiment Set up

All experimental study of this paper is carried out in cavitation tunnel of Beijing institute of technology. Its basic structure includes contraction, diffuser, experimental section, curved section, inlet pipe and return pipe, and the upper and lower side are equipped with three high strength transparent organic glass window to observe the flow field of the situation of water tunnel experiment section. The basic sizes of cavitation tunnel as listed in table 1. In the process of experiment, the dynamic system of tunnel is used to control and drive the water circulation in tunnel. Power system

*Corresponding Author, Wei Wang: wangw@dlut.edu.cn

mainly includes a three-phase ac motor (1480 RPM, 55 kw), axial flow pump and control frequency converter. The cavitation tunnel is also equipped with a sliding precision protractor, electromagnetic flow meter (precision level 0.5%), torque device, vacuum head meter and other facilities. The hydrofoil adopted by the experiment is NACA0066 hydrofoil, the hydrofoil is made of stainless steel, the surface machining accuracy is 3.2 (microns), thickness ratio is 11.74%, the hydrofoil chord length is 70 mm, exhibition to length is 67 mm. The hydrofoil with jet hole is shown in figure 1(a). The positions of jet holes on hydrofoil surface is 0.45C, the number of jet holes is 25, the diameter of jet holes is 1.4 mm. The photograph of the modified hydrofoil is shown in figure 1(b). High-speed flow field display technology is applied to study of cavitation phenomenon. The camera Sensor Array for the Sensor Array is 32 channels, 752 x 1128 pixels, memory is 1 GB, record speed of is up to 100000 fp/s. It can meet the requirements of cavitation flow field experiment with the characteristics of with quick speed, small power consumption.

location	contraction	experimental section	diffuser
size	1513mm	700×70×190	1043mm

Table 1 The basic sizes of cavitation tunnel

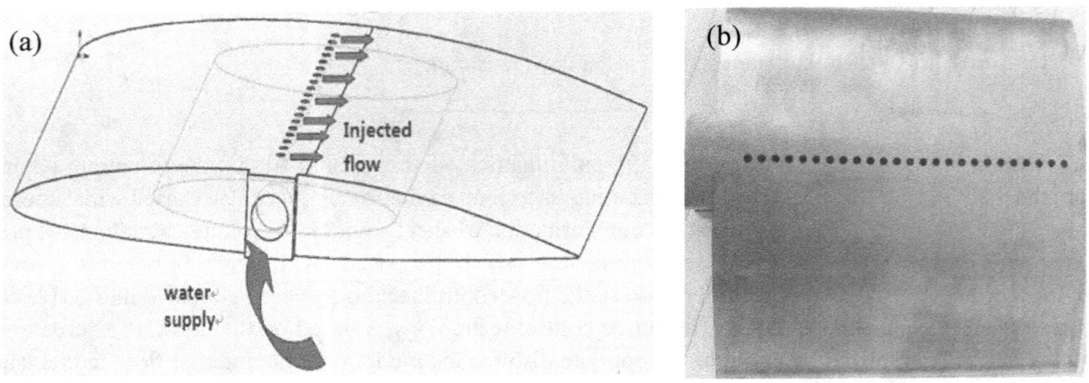

Figure 1 (a) A 3D model and (b) photograph of the modified naca0066

The measurement uncertainties

Table 2 show the absolute and the relative errors in determining the cavitation number for the maximum and the minimum flow velocity, outlet pressure, and the saturation vapor pressure (which is caused by temperature changing) at 8° angle of attack.

flow parameters		α=8°		
P_0/kpa	46.13	41.33	32.33	27.33
ΔP_0/kpa	±0.70	±0.70	±0.70	±0.70
U_0/m•s^{-1}	7.832	7.832	7.832	7.832
ΔU_0/m•s^{-1}	±0.104	±0.104	±0.104	±0.104
P_v/kpa	1.94	1.94	1.94	1.94
ΔP_v/kpa	±0.192	±0.192	±0.192	±0.192
σ	1.44	1.29	0.99	0.83
$\Delta\sigma_{max}$	0.068	0.063	0.056	0.051
$\Delta\sigma_{max}/\sigma$	0.047	0.049	0.056	0.062

Table 2 Absolute and relative errors in determining the cavitation number

$$\sigma=\frac{P_0-P_V}{\frac{1}{2}\rho U_0^{\ 2}}\Rightarrow\Delta\sigma=\frac{2}{\rho U_0^{\ 2}}(\Delta P_0-\Delta P_V-\frac{2\Delta U_0}{U_0}(P_0-P_V))\Rightarrow\Delta\sigma_{\max}=\frac{2}{\rho U_0^{\ 2}}(\left|\Delta P_0\right|+\left|-\Delta P_V\right|+\left|-\frac{2\Delta U_0}{U_0}\right|(P_0-P_V))$$

Discussions and Results

The dimensionless coefficient [9] C_Q serves for a quantitative assessment of the flowrate of the controlling flow. m_{inj} is the mass flowrate of the jet flow. Q_{inj} *(L/h)* is the jet-flow volume flow. m_0 is the equivalent mass flowrate of liquid of the main flow that would pass through the frontal (midsection) area S_0 of the hydrofoil in case of its absence, h is the height of the NACA0066 midsection, i.e. the section where the foil projection onto the y-axis depending on the attack angle α takes its maximum value (*h=Hmax* at *α=0°*). Q_0 is the mainstream volume flow. Because the density of the injection is equal to the density of the mainstream flow. Here C_Q ($C_Q=\frac{m_{inj}}{m_0}=\frac{\rho_{inj}Q_{inj}}{\rho_0 U_0 S_0}=\frac{Q_{inj}}{U_0 S_0}=\frac{Q_{inj}}{U_0 ha}$) can be regarded to be approximately equal to 0.00048 Q_{inj}/U_0. L_C^{max} is the maximum length of the cavity. Table 3 shows the summary of the regimes for modified NACA0066 model considered in this study in comparison with those for the unmodified one. It can be seen from the table, that if the flowrate of the jet flow were different, the cavitation suppression effects are different.

For sheet cavitation, the effectiveness of the cavitation suppression is more obvious. When the cavitation number is 1.44, on the occasion the non-dimension mass flowrate coefficient is C_Q =0.0245, the cavitation suppression effectiveness is the best, the cavity area is L_C^{max}=0.102C compared with the standard model, L_{C0}^{max}=0.347C decreased by70.5%; when the non-dimension mass flowrate coefficient is C_Q=0, the cavity area is L_C^{max} =0.369C, only increased by 6.5% compared with the standard model, which means that this structure has little influence on the shape of sheet cavitation. When the cavitation number is 1.29, compared with the standard model, L_{C0}^{max}=0.572C, the cavity area of without jet flow (C_Q =0) L_C^{max} =0.431C decreased by 25%; and when C_Q =0.0233, the cavitation suppression effectiveness is the best, in this case, the cavity area is L_C^{max} =0.275C, shortened by 52% compared with the standard hydrofoil.

When the cavitation number is reduced to 0.99, the cavitation forms characterize by unsteady intermittent cavitation. Without jet flow (C_Q =0), the cavity area is L_C^{max}=0.886C, compared with the standard model, decreased by 5.0%, which indicates that under this cavitation condition, the structure (without jet flow) has a little influence on cavitation. However, when the jet flow is entered, it can be seen that all the jet flow coefficient shows a certain degree of diminished effect on cavitation, and if the C_Q was different, the effectiveness is different, the Strouhal number (St) is different too. When C_Q=0.0202 and C_Q=0.0233, the weakening effectiveness of cloud cavitation are the best. But for C_Q=0.0202, the St number increased to 0.1743 indicating the frequency increased. While C_Q=0.0233, the corresponding Strouhal number is 0.1095, compared with standard model, decreased. Therefore, under this condition, C_Q=0.0233 is the optimum jet flowrate coefficient. When the cavitation number is reduced to 0.83, the flow field shows strong unsteadiness. It is different from the σ=0.99, without jet flow (C_Q =0), compared with the standard model, the cavity area length L_C^{max} decreased by 8.4%, but the Strouhal number increased to 0.1633. In another word, while the maximum length of the cavity is reduced, the shedding frequency is increased. When the non-dimension mass flowrate coefficient is changed to C_Q=0.0257, the cavity area length is the minimum, and the Strouhal number is St=0.1462 also decreased compared with the standard model. The optimum jet flowrate coefficient under this condition is C_Q=0.0257.

The jet can weaken or suppress cavitation, and there is an optimum mass flowrate coefficient under each working condition. On one hand, the optimum mass flowrate coefficient increases with cavitation number decreasing. On the other hand, under the optimum mass flowrate coefficient, with the intensity of cavitation increasing, the effectiveness of cavitation suppression is almost liner decreased. The reason can be roughly from follow aspect, that is, smaller the cavitation number is, the cavitation becomes more severe, and during the cavity developing process, more energy is carried by the fluid, so the greater the energy is needed for blocking re-entrant jet, as a result, the energy and the mass flowrate of the jet is greater. However, due to the restriction of the hydrofoil structure and the cavitation condition, the increase of momentum and energy brought by jet is smaller than that of the increase of the re-entrant jet intensity caused by the decrease of the cavitation number.

flow conditions 8° U_0=7.832m/s					
σ	/ *regime*	Q_{inj} (L/h)	C_Q	L_C^{max} / C	St
1.44	transitional regime / sheet cavity	standard		0.347	Steady regimes
		0	0	0.369	
		330	0.020	0.176	
		360	0.022	0.170	
		380	0.023	0.145	
		400	0.025	0.102	
		420	0.026	0.185	
		450	0.028	0.146	
1.29	sheet cavity	standard		0.572	Steady regimes
		0	0	0.431	
		330	0.020	0.335	
		360	0.022	0.361	
		380	0.023	0.275	
		400	0.025	0.490	
		420	0.026	0.340	
		450	0.028	0.437	
0.99	unsteady intermittent cavitation	standard		0.932	0.1373
		0	0	0.886	0.1348
		330	0.020	0.626	0.1743
		360	0.022	0.758	0.1458
		380	0.023	0.594	0.1095
		400	0.025	0.766	0.1559
		420	0.026	0.661	0.142
		450	0.028	0.68	0.1345
0.83	cloud/ unsteady intermittent cavitation	standard		1.001	0.1511
		0	0	0.916	0.1633
		330	0.020	0.963	0.1539
		360	0.022	0.955	0.1581
		380	0.023	0.825	0.1543
		400	0.025	0.870	0.1615
		420	0.026	0.780	0.1462
		450	0.028	0.936	0.1674

Table 3 Summary of the regimes for the modified NACA0066 model considered in the study in compassion with those for the unmodified one

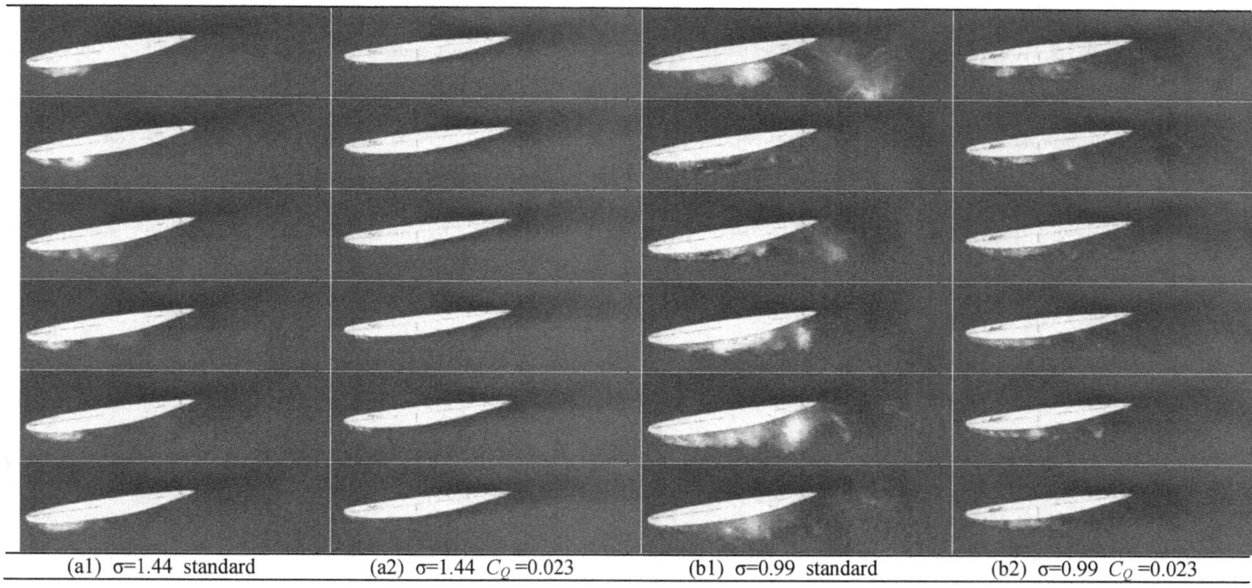

Figure 2 Instantaneous images of partial cavities (side view) of the (1) standard (without the jet holes) and modified hydrofoil model at the attack angle α = 9° for the following flow conditions: (a1) σ=1.44 standard hydrofoil sheet cavity (unmodified); (a2) σ=1.44 C_Q =0.023 (sheet cavity / subcavitating flow); (b1) σ=0.99 standard hydrofoil unsteady/cloud cavity (b2) σ=0.99 C_Q =0.023 steady cavity

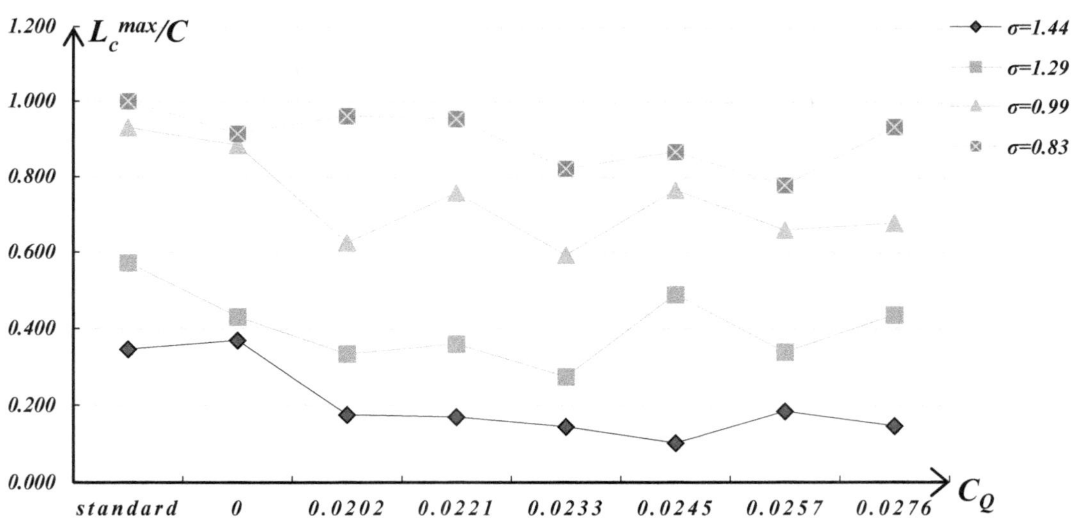

Figure 3 The dimensionless length of the cavity varies with the dimensionless mass flowrate coefficient C_Q

Conclusion

The active method for cavitation control by arranging water jet holes on the suction surface of the hydrofoil was applied to suppress cavitation. The active jet flow can completely suppress sheet cavitation, for cloud cavitation, it can weaken the development of the cavity and decrease the frequency of the shedding of the cloud cavity, but the shedding of cloud cavity cannot be avoided. The results also showed that under different cavitation conditions, the optimum mass flowrate coefficient is different. And the optimum mass flow coefficient increases with cavitation number decreasing. Regarding of these, this model show high potential of industrial application in cavitation suppression.

References

[1] Knapp, R. T. (1955). *Recent investigations of the mechanics of cavitations and cavitation damage.* Trans Asme, 77
[2] Furness, R. A., & Hutton, S. P. (1975). *Experimental and theoretical studies of two-dimensional fixed-type cavities.* Journal of Fluids Engineering, 97(4), 515-521.
[3] Kravtsova, A. Y., Markovich, D. M., Pervunin, K. S., Timoshevskiy, M. V., & Hanjalić, K. (2013). *High-speed imaging of cavitation regimes on a round-leading-edge flat plate and naca0015 hydrofoil.* Journal of Visualization, 16(3), 181-184.
[4] Yang, J., Zhou, L., & Wang, Z. (2011). *Numerical simulation of three-dimensional cavitation around a hydrofoil.* Journal of Fluids Engineering,133(8), 081301.
[5] Li, D. Q., Grekula, M., & Lindell, P. (2010). *Towards numerical prediction of unsteady sheet cavitation on hydrofoils.* Journal of Hydrodynamics Ser B, 22(5), 741-746.
[6] Asnaghi, A. (2010). *Unsteady multiphase modeling of cavitation around naca0015.* Journal of Marine Science & Technology, 18(5), 689-696.
[7] Ganesh, H., Chang, N., & Ceccio, S. (2011). *Tip vortex cavitation suppression by mass injection.* Journal of Fluids Engineering, 233-235(5), 337-340.
[8] Timoshevskiy, M. V., Zapryagaev, I. I., Pervunin, K. S., & Markovich, D. M. (2016). *Cavitation control on a 2D hydrofoil through a continuous tangential injection of liquid: Experimental study. (Vol.1770, pp.223-256). AIP Publishing LLC.*
[9] Timoshevskiy, M. V., Zapryagaev, I. I., Pervunin, K. S., Maltsev, L. I., Markovich, D. M., & Hanjalić, K. (2017). *Manipulating cavitation by a wall jet: experiments on a 2d hydrofoil.* International Journal of Multiphase Flow.
[10] WANG, W., YI, Q., WANG,Y., LU, S., & WANG,X. (2017). *Adaptability research of hydrofoil surface water injection on cavitation suppression*. Journal of drainage and irrigation machinery engineering (JDIME), 35(6), 461-466.
[11] Wang, W., Yi, Q., Lu, S., & Wang, X. (2017). *Exploration and Research of the Impact of Hydrofoil Surface Water Injection on Cavitation Suppression*. ASME Turbo Expo 2017: Turbomachinery Technical Conference and Exposition (pp.V02DT46A013).
[12] Mäkiharju, S. A., Ganesh, H., & Ceccio, S. L. (2015). *Effect of non-condensable gas injection on cavitation dynamics of partial cavities.* Reprinted from Bacteriological Proceedings May, 656(1), 012161.
[13] Chahine, G. L., Frederick, G. F., & Bateman, R. D. (1993). *Propeller tip vortex cavitation suppression using selective polymer injection.* Journal of Fluids Engineering, 115(3), 497-503.
[14] Chang, N., Yakushiji, R., Ganesh, H., & Ceccio, S. (2009). *Mechanism and scalability of tip vortex cavitation suppression by water and polymer injection.*
[15] Kawanami, Y., Kato, H., Yamaguchi, H., Tanimura, M., & Tagaya, Y. (1997). *Mechanism and control of cloud cavitation.* Journal of Fluids Engineering, 119(4), 788-794.
[16] WANG, W., LU, S., XU, R., YI, Q., WANG,Y., & WANG,X. (2017). *Numerical study of hydrofoil surface jet flow on cavitation suppression*. Journal of drainage and irrigation machinery engineering (JDIME), 35(10), 829-834.

Acknowledgments

This work is supported by the National Basic Research Program of China (2015CB057301), and special thanks to the Collaborative Innovation on Center of Major Machine Manufacturing in Liaoning for supporting the present work.

© 2018 ASME

Experimental and Numerical Investigation of the Cavitating Flows over a Modified NACA0009 Foil

[1]Mohammad Hossein Arabnejad*;[2] Ali Amini; [1]Rickard Bensow; [2]Mohamed Farhat;

[1]*Chalmers University of Technology, Gothenburg, Sweden;* [2]*Ecole Polytechnique Federale de Lausanne, Lausanne, Switzerland*

Abstract

In this study, the cavitating flow over a modified NACA0009 foil has been investigated using experimental and numerical methods. In the experimental investigation, high-speed visualization is used to study the behavior of the cavitating flow at $\sigma = 1.2, \alpha = 5^o, U_\infty = 20m/s$. In order to identify the location of erosive collapse, a preliminary soft paint test is performed. In the numerical part of this study, the flow condition subjected to the soft paint study is simulated with Large Eddy Simulation using a mixture assumption coupled with the Schnerr-Sauer mass transfer model. In order to validate the simulation, the numerical results are compared with experimental high-speed visualization of the flow at the same condition. These comparisons show that the numerical simulation is capable of reproducing the main features of the cavitating flow. The validated numerical results and the high-speed visualization are then used to explain the hydrodynamic mechanism of erosive events that have been identified in the experimental investigation.

Keywords:

Introduction

Cavitation erosion is the material loss due to the collapse of vapour structures close to the surface. In hydraulic machinery, cavitation erosion is one of the limiting factors in the design as it is usually associated with performance degradation and increase in maintenance costs. In the path towards the prediction of cavitation erosion, understanding the hydrodynamic mechanism of cavitation erosion is essential. Several studies have been devoted to identifying these hydrodynamic mechanisms using experimental observation. Bark and Bensow [1] outlined different hydrodynamic mechanisms that influence the erosiveness of a collapsing cavity. A set of analysis models for these hydrodynamic mechanisms were established using experimental observation complemented by high-fidelity numerical simulation. Dular and Petkovsek [2] studied the contribution of different cavitating structures in the cavitation erosion using synchronous observation of cavitation structure and erosion damage. They identified five different mechanisms of cavitation erosion. These mechanisms are spherical collapse of the cloud cavity, the collapse of horse-shoe vortex, the collapse of twister vortex, collapse of microbubbles at the closure of the sheet cavity, and the collapse of microcavities at the break-up region of the cloud cavity. They concluded that the collapse of the cloud cavitation structures is responsible for 80 percent of the erosion damage. Van Rijsbergen et al. [3] used high-speed visualization, paint test, and acoustic measurement to investigate the behavior of the cavitating flow over a NACA0015 foil with the aim to identify the role of different cavitating structures in the cavitation erosion. Similar to Dular and Petkovsek [2], they found that the collapse of cavitating cloudy structures is associated with high risk of cavitation erosion. More recently, Cao et al. [4] investigated the relationship between cavitation structure and erosion damage on a twisted foil using high-speed visualization and paint test. They observed a severe paint removal close to the closure line of sheet cavity where the unsteadiness of the rear part of the sheet cavity leads to the detachment and the collapse of cavity structures.

With the progress in numerical modeling of cavitating flows, high-end numerical simulations have become a useful tool to study the main features of cavitating flows. Numerical simulations provide complete access to the flow field that can be used to analyze the development of cavitating structures. These numerical analyses combined with the experimental observations promote the understanding hydrodynamic mechanisms of cavitation erosion. In this paper, numerical simulations, high-speed visualization, and paint test are employed together to study the cavitating flow over a NACA0009 foil. Based on the simulation results and the experimental images, cavitating structures responsible for the paint removal in the paint test are identified and their hydrodynamic mechanism is discussed briefly.

*Corresponding Author, Mohammad Hossein Arabnejad: mohammad.h.arabnejad@chalmers.se

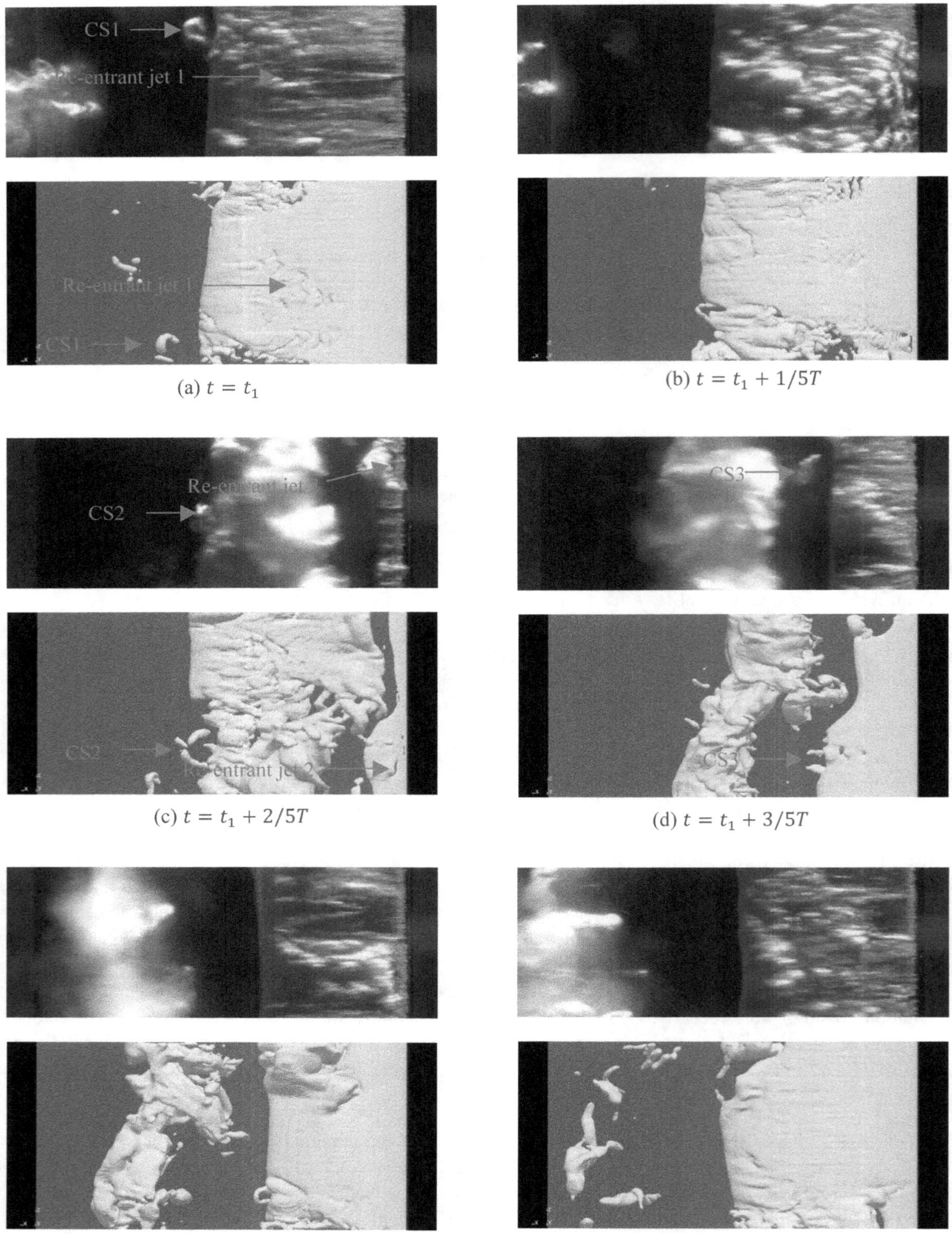

(a) $t = t_1$ (b) $t = t_1 + 1/5T$

(c) $t = t_1 + 2/5T$ (d) $t = t_1 + 3/5T$

(e) $t = t_1 + 4/5T$ (f) $t = t_1 + T$

Figure 1: Shedding of the cloud cavitation in one cycle, top) experimental high-speed visualization, bottom) numerical results

Experimental and Numerical Set-up

The experimental work presented here has been conducted in the EPFL cavitation tunnel. A modified NACA0009 foil is placed in the cavitation tunnel with 5° angle of attack. The foil has a chord length of 100 mm and span length of 150 mm. The cavitation number is 1.2, and the inlet velocity is 20 m/s. High-speed visualization and paint test are used to record the cavitation dynamics and identify areas with high risk of cavitation erosion. In the numerical part of this study, the flow has been simulated at the same conditions with Large Eddy Simulation method, using a mixture assumption coupled with the Schnerr-Sauer mass transfer model [6]. The computational grid is composed of 7.4 million hexahedral/prismatic cells. The average value of y+ around the foil is less than 1, and the maximum value of x+ on the upper surface of the foil is around 300. In the span-wise direction, a uniform mesh with z+ around 300 is used; thus the cavitation dynamics is well resolved although the surface mesh is not fully sufficient for a wall resolved LES. To decrease the computational cost, the computational domain includes only one-third of the foil span and slip boundary conditions are applied to the side planes.

Results

Figure 1 shows the shedding of the cloud cavitation during one cycle in both experimental and numerical investigations. In figure 1a, the sheet cavity has reached its maximum length, which is 55% of the chord length. At this stage, a re-entrant jet has been developed from the closure line of the sheet cavity. The front line of this re-entrant jet is marked by red arrows in figure 1a. As the re-entrant jet moves towards the leading edge, it creates disturbances on the interface of the sheet cavity. The disturbances on the surface of the sheet cavity can become larger and lead to the detachment of cavity structures. These structures are marked as CS1 in figure 1a. In figure 1b, the re-entrant jet has reached the leading edge and caused the pinch-off of the cloud cavity. The cloud cavity has a rolling motion due to the interaction between the re-entrant jet and the bulk flow. As the cloud cavity rolls up, some vapor structures of horse-shoe shape are detached from the trailing part of the cloud cavity. In figure 1b and 1c, these structures, marked as CS2, collapse as they travel downstream. In figure 1c, the new sheet cavity has appeared at the leading edge. While the sheet cavity is growing, another re-entrant jet moves upstream and reaches the leading edge. The front line of this re-entrant jet is marked by red arrows in figure 1c. This re-entrant jet does not lead to the pinch-off of the cavity, but it only creates a disturbance on the interface of the sheet cavity. This disturbance leads to the detachment and collapse of vapor structures (marked by CS3 in figure 1d) close to the closure line of the sheet cavity. In figures 1f and 1e, the sheet cavity grows to its maximum length and the cycle repeats. The comparison between the numerical simulation results and the experimental high-speed visualization shows that the numerical simulation with the current set-up and grid resolution is able to capture the main features of

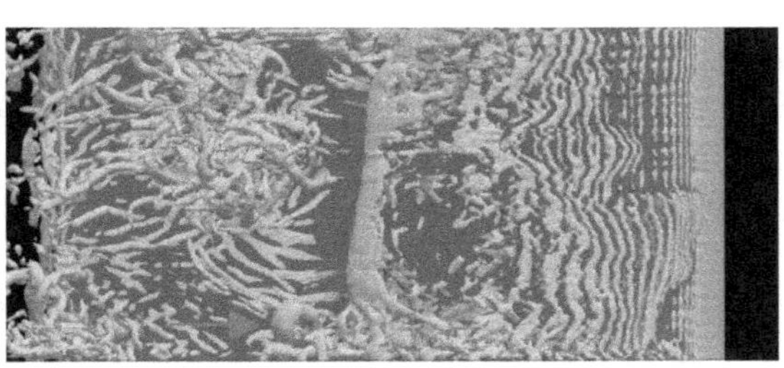

(a) $t = t_1$

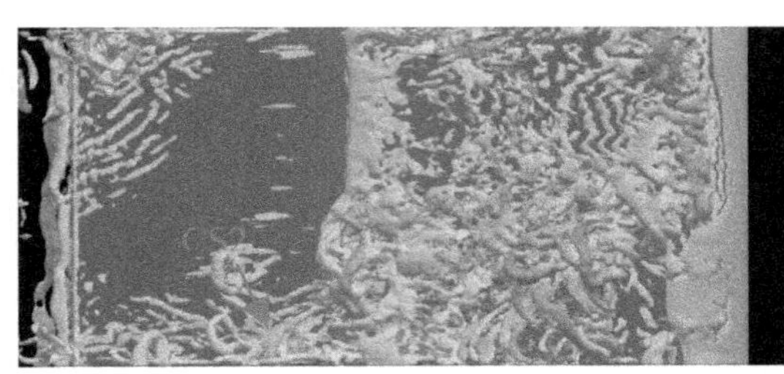

(b) $t = t_1 + 2/5T$

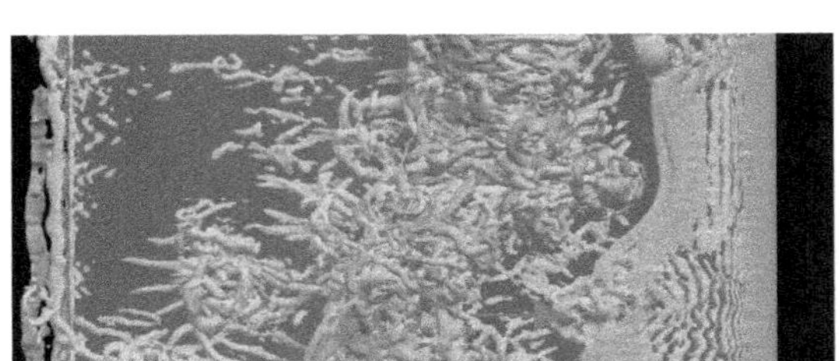

(c) $t = t_1 + 3/5T$

Figure 2: Q criteria in one shedding cycle

the cavitating flow. In the experiment results, the shed cloud turns into a U shape as it moves downstream while in the numerical results the cloud does not have any particular shape. The reason for this difference might be that the side walls in the simulation are defined as slip boundary condition, therefore the effect of the side walls are not present.

In figure 2, the vortical structures for three time instances are shown by the iso-surface of the Q criterion. These time instances corresponds to figure 1a, figure 1c, and figure 1d. As it can be seen from this figure, the structure CS1, CS2, and CS3 are horse-shoe vortical structures detached from the closure line of the sheet cavity or the downstream end of the cloud cavity.

In order to locate the areas with high risk of cavitation erosion, a stencil ink is applied to the surface of the foil. Figure 3a shows the status of the stencil ink after 10 minutes and figure 3b shows the span-wise average of the paint removal. A region with severe paint removal is observed in the stream-wise location of 0.3<X/C<0.7. This region coincides with the collapse location of cavity structures CS1, CS2, and CS3, identified in figure 1. This suggests that the cavity structures detached from the closure line of the sheet cavity and the downstream end of the cloud cavity can be responsible for paint removal seen in figure 3.

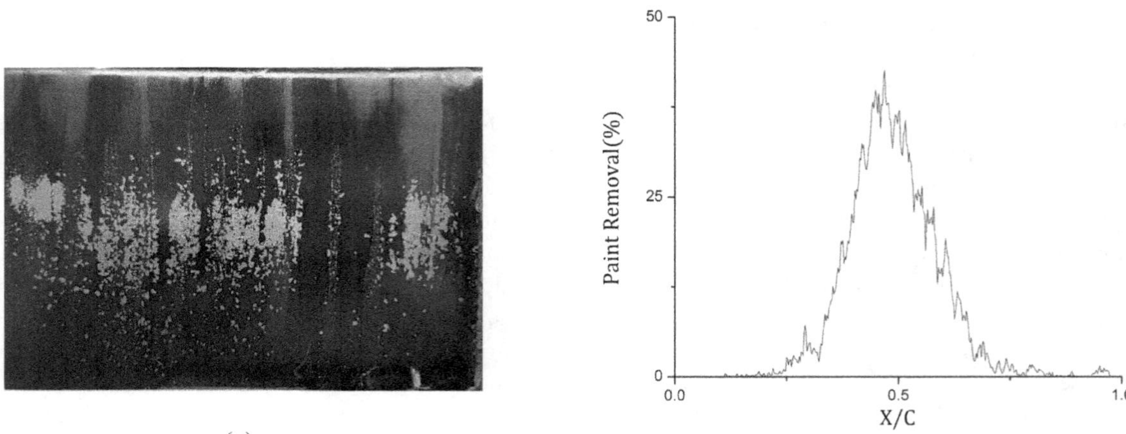

(a) (b)

Figure 2: a) Paint removal after 10 minutes. b) Span-wise average of paint removal as a function of streamwise location

Conclusion

This paper presents the numerical and experimental investigation of a cavitating flow over a modified NACA0009 foil. The behavior of the flow is studied by experimental high-speed visualization and numerical simulation results. This study shows that smaller horseshoe cavitating structures are detached from the closure line of the sheet cavity and downstream end of the cloud cavity. These structure collapse shortly after their detachment. The paint test results show that most paint removal occurs in the area close to the collapse of the identified horse-shoe structures. This suggests that the collapse of cavitating structures detached from the closure line of the sheet cavity and downstream end of the cloud cavity can be responsible for paint removal.

Acknowledgments

Financial support of this research is provided by the EU CaFE initiative.

References

[1]Bark, G., and Bensow, R. E., (2013), *Hydrodynamic Mechanisms Controlling Cavitation Erosion*,Int. Shipbuild. Prog., 60(1)

[2]Dular, M. and Petkovsek, M., (2013). *On the mechanisms of cavitation erosion – Coupling high speed videos to damage patterns*. Wear (300).

[3] van Rijsbergen, M., Foeth, E.J., Fitzsimmons, P., Boorsma, A.,(2012)*High speed video observations and acoustic impact measurements on a NACA0015 foil*, Proceedings of the 8th International Symposium on Cavitation, Singapore.

[4] Cao, Y.T., Peng, X.X., Yan, K., Xu, L.H., and Shu L.W, (2017) *A qualitative study on the relationship between cavitation structure and erosion region around a 3d twisted hydrofoil by painting method.* Fifth International Symposium on Marine Propulsors, Espoo, Finland.

[5]Bensow, R. E., and Bark, G., (2010). *Implicit LES Predictions of the Cavitating Flow on a Propeller*. Journal of Fluids Engineering, Vol. 132.

[6]Asnaghi, A., Feymark, A., and Bensow, R.E., (2017). *Improvement of cavitation mass transfer modeling based on local flow properties*. International Journal of Multiphase Flow, Vol. 93.

© 2018 ASME

Experimental and numerical investigation of vortex dynamics in ventilated cavitating flows around a bluff body

[1] Zhiying Wang; [1] Guoyu Wang *; [1] Biao Huang; [1] Mindi Zhang

[1] *Beijing Institute of Technology, Beijing, China*

Abstract

The objective of this paper is to investigate ventilated cavitating vortex shedding dynamics around a bluff body at Re=6.7×10^4 by experimental and numerical methods. The couple level-set and VOF (CLSVOF) method is applied to capture the air-water interface. The results show that the CLSVOF method can predict the air-water interface accurately. Two types of unsteady ventilated cavitating flows are investigated, namely vortex shedding structure (Q_v=0.174) and the Re-entrant jet (Q_v=0.31). The finite-time Lyapunov exponent (FTLE) and Lagrangian coherent structures (LCS) methods are applied to investigate the formation, evolution and shedding of ventilated cavitating vortices. For vortex shedding structure (Q_v=0.174), the vortices alternately shed from the cavity, forming the vortex street. For the re-entrant jet structures (Q_v=0.31), there exist the interaction between the vortex shedding and the re-entrant jet. The evolution of ventilated cavitation is divided into three stages: the vortex shedding, the development of re-entrant jet and the development of ventilated cavitation.

Keywords: ventilated cavitating flow; Couple Level-set and VOF; vortex shedding; Lagrangian coherent structures.

Introduction

Ventilated cavitation is a hot topic due to its practical importance in engineering and scientific relevance in fluid mechanics [1]. Wosnik and Arndt [2] carried out experiments in high-speed water tunnel to investigate the interaction between a ventilated supercavity and its turbulent bubbly wake. They found that the bubbly turbulent wake was resulted from the collapse of a ventilated supercavitation. Karn et al. [3] employed high-speed, high-resolution photography and pressure transducers to understand the physical mechanisms determining closure formation and transition between different closure modes. Although ventilated cavitation has been studied considerably in the literature [4], relatively fewer studies exist for bluff body [5]. For the bluff body, ventilated cavitation often results in the formation of vortex shedding before forming a stable ventilated supercavitation at relatively larger gas entrainment coefficient.

The quality of the cavity interface may have a strong effect on the amount of air entrained and the internal flow fields [6]. In order to capture the cavity interface, VOF and Level-Set methods have been extensively employed. Bonhomme et al. [7] applied a VOF method in axi-symmetric computations to investigate the dynamics of isolated air bubble crossing the horizontal interface separating two fluids. Huang et al. [8] introduced level set method into multiphase cavitating flows and pointed out that it could better resolve the cavity shape. Although the VOF and LS method have been used widely, these two methods have their advantages and disadvantages. So the Couple level-set and VOF (CLSVOF) is an inevitable trend to be developed [9].

In present work, we applied the CLSVOF method to investigate the vortex dynamics of ventilated cavitating flow around the bluff body. In addition, the FTLE is applied to study the influence of ventilated cavitation on the coherent structures and the trace of particles.

Experimental setup

Experimental studies are conducted in a closed-loop cavitation tunnel and the details of the experimental equipment are shown in Ref. [10]. The ventilated cavitation study is carried out using a bluff body. The Reynolds number is defined as, $Re=U_\infty D/v$, where U_∞ is the free stream velocity, v is the kinematic viscosity. The gas entrainment coefficient is defined as $Q_v= Q/(U_\infty D^2)$, where Q is the volumetric flow rate of the injected gas.

The cavitation patterns are observed by a high-speed digital camera (SpeedSence M310 camera) at a sampling frequency of 3260 Hz, with two light sources illuminating the flow field from different directions.

*Corresponding Author, Guoyu Wang: wangguoyu@bit.edu.cn

Numerical model

The numerical simulation is performed by FLUENT with LES Wall-Adapting Local Eddy-Viscosity (WALE) model [11] to study the unsteady cavity shedding process. And couple level-set and VOF method is used to capture the cavity interface. In this method, the VOF model offers the location of the interface, and the gradient of the level-set function determines the direction of the interface.

Results and discussion

(a)The influence of interface capture methods on the simulation of ventilated cavitating flow

To investigate the influence of interface capture methods on simulation of the ventilated cavitating flow, two flow characteristics have been considered: (i) ventilated cavitation with vortex shedding structures (Q_v=0.174) and (ii) Re-entrant jet structures (Q_v=0.31). Figure.1 displays the comparisons of instantaneous experimental photographs (left), the ventilated cavitation shape by VOF method (middle) and the Couple Level Set and VOF methods (right). Although both methods are capable of capturing ventilated cavitation flow structures, there still exist differences between two methods. When the gas entrainment coefficient Q_v is 0.174, the deformation of cavity surface is larger, and the cavity interface captured by CLSVOF method (see region I) is relatively smoother. However, when the gas entrainment is 0.31, the results from these two methods are almost similar.

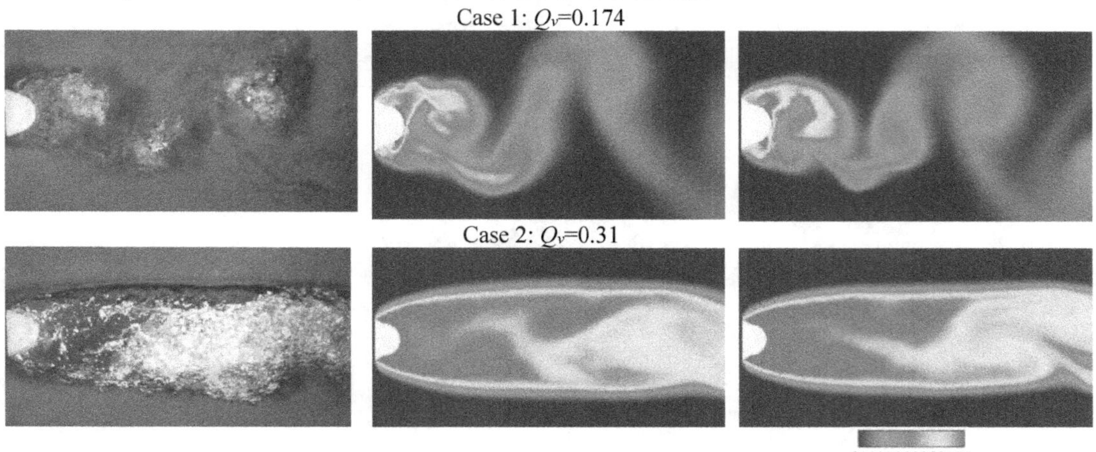

Fig.1 Visual comparisons of the predicted ventilated cavitation by VOF method (middle) and CLSVOF method (right) with observations from experiment (left)

(b)The unsteady characteristics of ventilated cavitating flow

The Lagrangian coherent structures (LCSs) are regarded as the ridges of finite-time Lyapunov exponent (FTLE) can represent the boundary of a vortex [12]. The particle trajectory is used to help elucidate the influence of ventilated cavitation on the fluid dynamics. Figure 2 and 3 display the instantaneous experimental photographs (left), the contour of air volume fraction with streamlines (middle) and the contour of finite-time Lyapunov exponent (FTLE) (right) at Q_v=0.174 and Q_v=0.31, respectively. The Strouhal numbers (St $=fD/U_\infty$) are St = 0.13 for Q_v=0.174 and St = 0.03 for Q_v=0.31.

At ventilated cavitation with vortex shedding structures (Q_v=0.174), it can be clearly seen that the wake is divided into two parts, near wake and far wake (see Fig.2a (right)). In the near wake, the vortices form, develop and shed from the attached cavity. At t_1, combing the experimental results and the LCSs in Fig.2a, it can be found that the upper circulation is split into two parts at the black point. The rear part of the upper circulation zone becomes the first cavitating vortex V_1, and the separate bubbles are entrapped into the vortex V_1. In the far wake, the vortices rotate and move downstream. The LCSs can display the rotation direction. At t_2~t_3, the lower vortex V_2 gradually rolls to grow. The upper vortex V_1 is shed from the rear part of cavity. At t_4~t_5, the process is repeated to form a cycle.

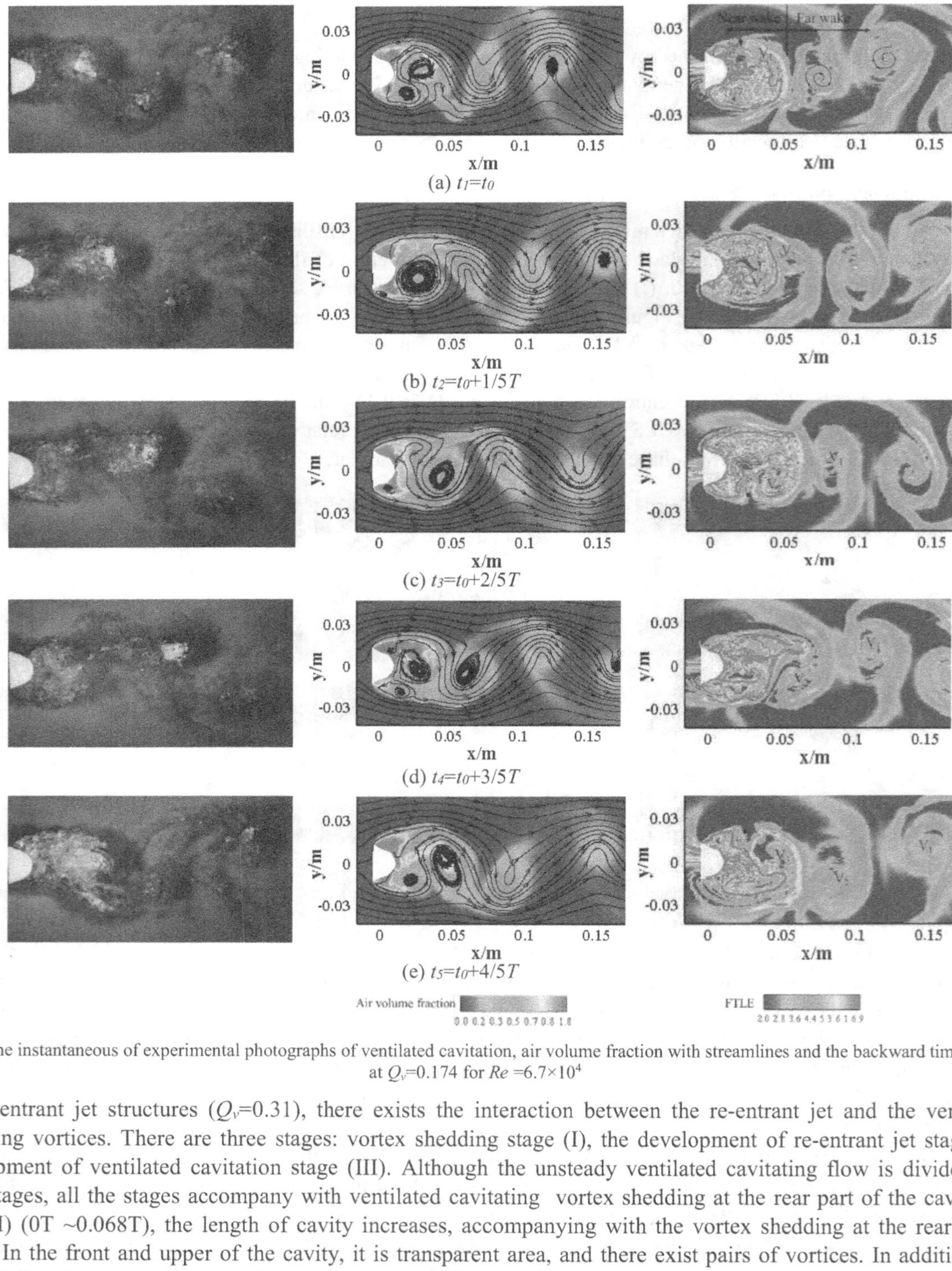

Fig.2 The instantaneous of experimental photographs of ventilated cavitation, air volume fraction with streamlines and the backward time FTLE at Q_v=0.174 for Re =6.7×10^4

At Re-entrant jet structures (Q_v=0.31), there exists the interaction between the re-entrant jet and the ventilated cavitating vortices. There are three stages: vortex shedding stage (I), the development of re-entrant jet stage (II), development of ventilated cavitation stage (III). Although the unsteady ventilated cavitating flow is divided into three stages, all the stages accompany with ventilated cavitating vortex shedding at the rear part of the cavity. At stage (I) (0T ~0.068T), the length of cavity increases, accompanying with the vortex shedding at the rear of the cavity. In the front and upper of the cavity, it is transparent area, and there exist pairs of vortices. In addition, the front of the air-water mixed area is steady, as shown by the black dotted line in Fig.3. From the contour of the FTLE, it can be found that the LCSs are also relatively stable and mainly develop along with the air-water interface. At stage (II) (0.15T ~0.63T), the re-entrant jet becomes strong and gradually moves upstream in the cavity. When the re-entrant jet moves upstream to fill the cavity, the vortex shedding also exists at the rear part of the cavity. At 0.63T, the re-entrant jet reaches and hits the bluff body. At stage (III) (0.68T ~ T), the re-entrant jet is significantly weakened and the air-water mixed area is pushed away by the injection of gas.

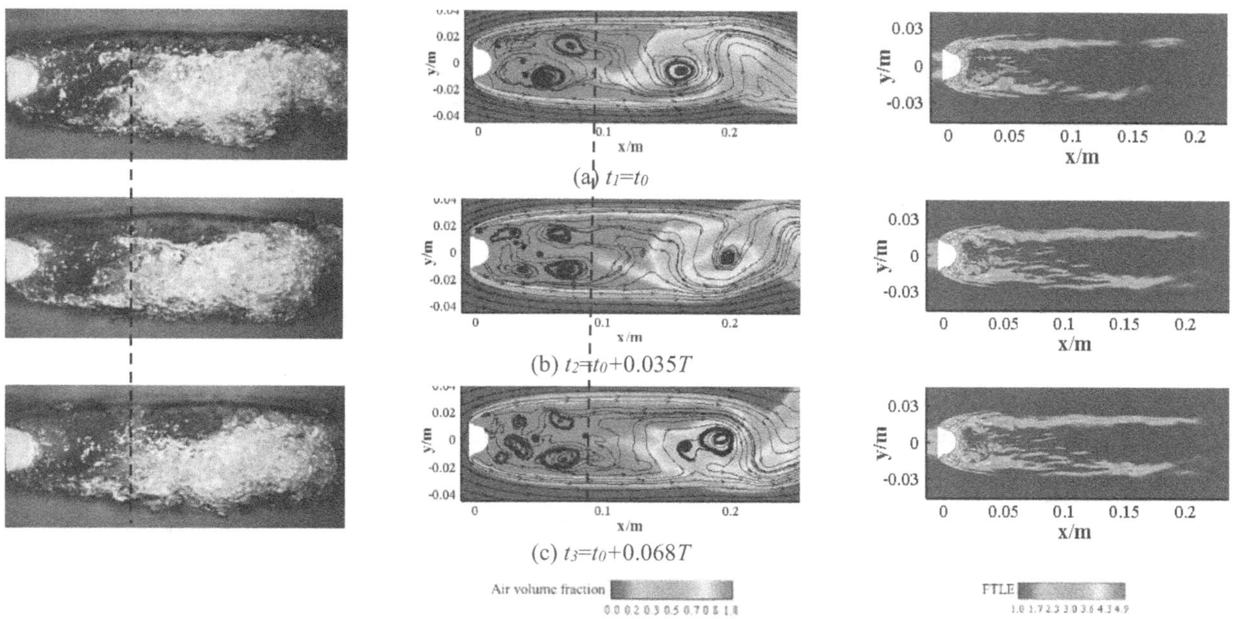

Fig.3 The instantaneous of experimental photographs of ventilated cavitation, air volume fraction with streamlines and the backward time FTLE at Q_v=0.31(stage I) for Re =6.7×10^4

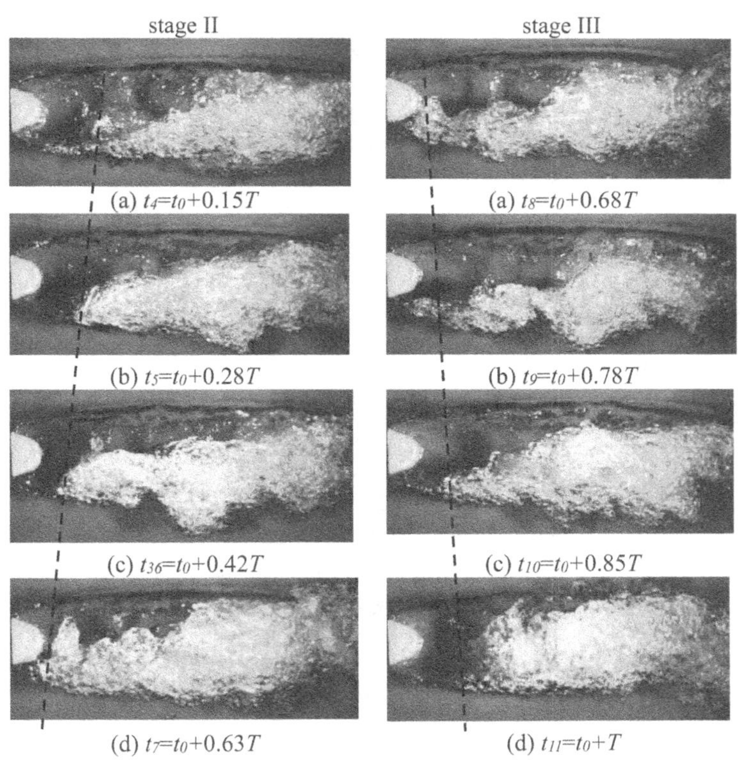

Fig.4 The instantaneous of experimental photographs of ventilated cavitation at Q_v=0.31 (stage II and stage III) for Re =6.7×10^4

In order to show the influence of ventilated cavitation on the coherent structures and the fluid motion, Fig.5 shows the contour of FTLE with particle trajectories. The initial locations of the particles are the same. At Q_v =0 and Q_v =0.174, the structures are similar and there exist the near wake and far wake. At Q_v =0, the vortices rotate and move downstream. The particle trajectory in region A can also demonstrate the rotation of vortices. Compared with non-cavitating flow, the rotation of vortices become weaken. However, at Q_v =0.31, there is not clearly boundary of the near wake and far wake, the LCSs mainly appear along the cavity interface. The particles are relatively stable and the displacement of particles are relatively small at Q_v =0.31.

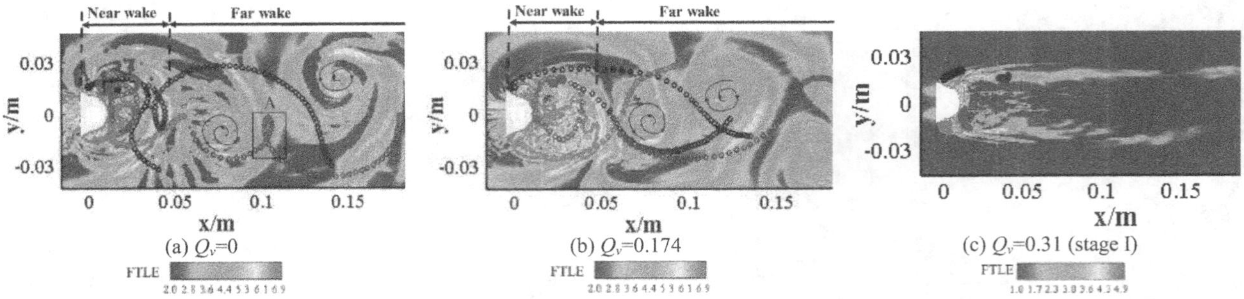

Fig.5 Lagrangian Coherent Structures and particle trajectories at different cases

Conclusion

In this study, the couple level-set and VOF (CLSVOF) method is employed to capture the air-water interface. The vortex dynamics in two types of ventilated cavitating flows are investigated, namely vortex shedding structure (Q_v=0.174) and the Re-entrant jet (Q_v=0.31).

(1) At Q_v=0.174, the vortices form, develop and alternately shed from the attached cavity, forming the vortex street. The LCSs can clearly descript the vortex shedding process in the near wake and show the rotation of vortex in the far wake.

(2) At Q_v=0.31, the re-entrant jet plays the dominant role. Although there also exist the vortex shedding, the process is different from that at Q_v=0.174. The evolution of ventilated cavitation at Q_v=0.31 can be divided into three stages, namely vortex shedding, the development of re-entrant jet and the development of ventilated cavitation. In this case, the LCSs appear along the air-water interface, and the particles are relatively stable.

Acknowledgement

The authors gratefully acknowledge support by Graduate Technological Innovation Project of Beijing Institute of Technology (2017CX10043) and China Scholarship Council (CSC) (201706030087).

References

[1] Arndt, R.E.A., Hambleton, W.T., Kawakami, E., Amromin, E.L., 2009. Creation and maintenance of cavities under horizontal surfaces in steady and gust flows. Journal of Fluids Engineering 131(11), 111301.

[2]Wosnik, M., Arndt, R.E.A., 2013, Measurements in high void-fraction bubbly wakes created by ventilated supercavitation. Journal of Fluids Engineering 135, 011304.

[3]Karn, A., Arndt, R.E.A., Hong, J.R., 2016. An experimental investigation into the physics of supercavity closure. Journal of Fluid Mechanics 789(3), 259-284.

[4] Ceccio, S.L., 2010. Friction drag reduction of external flows with bubble and gas injection. Annual Review of Fluid Mechanics 42, 183–203.

[5]Gnanaskandan, A., Mahesh, K., 2016, Numerical investigation of near-wake characteristics of cavitating flow over a circular cylinder. J. Fluid. Mech. 790, 453-491.

[6]Kinzel, M.P., Lindau, J.W., Kunz, R.F., 2009. Air entrainment mechanisms from artificial supercavities: insight based on numerical simulations. In: Proceedings of the 7th International Symposium on Cavitation, Ann Arbor, Michigan, USA, No. 136

[7] Bonhomme, R., Magnaudet, J., Duval, F., 2012, Inertial dynamics of air bubbles crossing a horizontal fluid–fluid interface. Journal of Fluid Mechanics, 707, 405-443.

[8]Huang, B., Wang, G.Y., Zhang, M.D., Chen, W.Q., 2011. Level set method for simulation of cavitating flows. J. Ship Mech. 15 (3), 207–216.

[9]Wang, Z., Yang, J., Koo, B., 2009, A coupled level set and volume-of-fluid method for sharp interface simulation of plunging breaking waves. International Journal of Multiphase Flow, 35(3): 227-246.

[10]Wang Z.Y., Huang B., Zhang M.D., Wang G.Y., Zhao X., 2017, Experimental and numerical investigation of ventilated cavitating flow structures with special emphasis on vortex shedding dynamics[J]. International Journal of Multiphase Flow, 98, 79-95.

[11]Nicoud, F., Ducros, F.,1999. Subgrid-Scale Stress Modelling Based on the Square of the Velocity Gradient Tensor. Flow Turbulence and Combustion 62, 183-200.

[12]Tang, J.N., Tseng, C.C., Wang, N.F., 2012, Lagrangian-based investigation of multiphase flows by finite-time Lyapunov exponents, Acta Mech. Sin. 28, 612–624.

© 2018 ASME

Experimental and numerical study of cavitation inside sharp-edged multi-hole orifice plate

[1]Zhi-xin Gao; [1,2]Jin-yuan Qian *; [1]Zhi-jiang Jin

[1] *Institute of Process Equipment, College of Energy Engineering, Zhejiang University, Hangzhou, 310027, PR China;* [2] *State Key Laboratory of Fluid Power and Mechatronic Systems, Hangzhou 310027, PR China*

Abstract

Cavitation can be useful in many occasions, for example, it can be adopted to improve the mass and heat transfer of the solution and it can also be used during the wastewater treatment. Among the ways to generate cavitation in hydrodynamic devices, the sharp-edged orifice is often used because it has a simple structure and the tube that has multi-holes orifice plate is more likely to generate cavitation. In this paper, experiment and computational fluid dynamic (CFD) method are carried out, the flux of the test tube is recorded during the experimental process, and the cavitation model used in the numerical study is validated by comparing the simulated results with the experimental results, then triangular, rectangular, and circular shape of holes and different kinds of distributions of holes such as circular and rectangular distributions are numerically studied. The vapor volume fraction is visualized to find out the effects of holes shape and arrangement on cavitation intensity. Besides, study focusing on the inlet pressure is achieved using CFD simulations, and the cavitation number is calculated under different cases to find out their effects on cavitation intensity. It can be concluded from the obtained results that cavitation number decreases with the increase of pressure drop, and under a certain value of pressure drop, the orifice plates with circular holes have the minimal cavitation number and maximal cavitation intensity. When the holes are circular arrangement, the orifice plates with circular and triangular holes has higher cavitation intensity, and when inlet pressure is large enough, different arrangement and shape of holes barely affect the cavitation intensity.

Keywords: hydrodynamic cavitation; multi-hole orifice; cavitation number; pressure drop

Introduction

The implosion of cavitation bubbles formed when fluid flows through a restriction induces micro-jets and shock waves, which can release high intense energy and is conducive to the formation of strong oxidation conditions and the heat transfer enhancement. Many works have been done about the utilization of hydrodynamic cavitation. Sivakumar and Pandit [1] found that multi-hole orifice used in hydrodynamic cavitation reactor could enhance the degradation of a cationic dye rhodamine B solution. Schneider et al. [2] found that the heat transfer enhancement reached 67% compared with noncavitating flow conditions. Orifice plates are often treated as hydrodynamic cavitation reactor and has been investigated a lot. Ebrahimi et al. [3] studied the cavitating flow of a thick orifice plate under high-pressure conditions. Dong et al. [4] investigated the effect of the number, the arrangement and the ratio of triangular holes on the hydraulic characteristic of the orifice plates. Rudolf et al. [5] focused on the loss coefficient and cavitation number of single and multi-hole orifices using experimental methods, and they found that there was large loss coefficient drop in multi-hole orifices. Liu et al. [6] experimentally investigated the flow and cavitation characteristics of damping orifices with different ratio of the thickness and diameter of the orifice. He et al. [7] numerically investigated the flow characteristic of a nozzle with multi-hole. He et al. [8] also experimentally and numerically investigated the cavitation inception and cloud cavitation shedding in a rectangular nozzle orifice. Ji et al. [9] studied the effect of the upstream pressure and cavitation number of multi-holes orifice using experimentally and numerically methods. Rudolf et al. [10] studied the static and dynamic characteristic of multi-hole orifices carrying out experiments. By now, the circular holes, rectangular holes, and triangular holes of orifice plates have been discussed relatively by researchers, but a comparison of different shape of holes and different arrangement of these holes have barely been done. Therefore a comprehensive study of cavitation inside different multi-hole orifice plates under different pressure drop is presented in this study, and useful information for engineering application is provided.

Experiments

The schematic of the experimental facility is shown in Fig. 1, it mainly consists of a water reservoir tank, a centrifugal pump, a gate valve, and a test section including the investigated orifice plate. The test rig is mostly made from stainless steel with a nominal diameter of 50mm. The volumetric flow rate through the pipelines before the

*Corresponding Author, Jin-yuan Qian: qianjy@zju.edu.edu

orifice is measured by a flow meter. The upstream and downstream pressure are obtained with two pressure transducers and recorded by adjusting the inlet flux. The surrounding temperature is a constant and is room temperature. All data are taken when the flow meter and pressure transducers readings are steady, and the accuracy of the flow meter and the pressure transducers is ± 0.5%.

Numerical method

Computational fluid dynamics (CFD) method is used to investigate different shape of orifice holes and different arrangements of these holes. Commercial software FLUENT is adopted and steady-state analysis is carried out. The Realizable k-ε turbulence model and Zwart-Gerber-Belamri cavitation model are used. As for the boundary conditions, a pressure inlet boundary is applied at the inlet and a pressure outlet boundary is applied at the outlet. Besides, no-slip wall boundary is applied for all walls, and standard wall function is adopted.

The working fluid is water and water vapor at 20°C, and their density and viscosity are listed in Table 1. The saturated pressure is 2,339 Pa at this temperature.

Three kinds of holes including circular, rectangular, and triangular holes and two kinds of arrangements that are circular distribution and rectangular distribution are investigated as shown in Fig. 2. The diameter of the orifice plate is 50mm, and the orifice plate thickness is 20mm. The circular, rectangular, and triangular holes have the same area, which means that the six kinds of orifice plates have the same opening ratio, and the diameter of the circular hole is 4mm.

Symmetrical model and hexahedral mesh are adopted in this study. Mesh around holes is fined and grid independence has been achieved and the number of cells is around 738,568. The orifice plate with one central circular hole is adopted in the experiment and the diameter of the circular hole is 10mm. The measured flow rate vs. pressure drop data are adopted to compare with the numerical results and the results are shown in Table 2. It can be found that the error between the experiments and the simulation is within 9.6%, so it can be inferred that the adopted numerical method is acceptable.

Results

The parameters are investigated synthetically, and the values are changed to 0.4, 0.6, 0.8, and 1MPa for the inlet pressure with different arrangements and shape of holes, as for the outlet pressure, the atmosphere that is 0.1 MPa is adopted. The water vapor volume fraction on the symmetry plane of the circular holes orifice plate with different arrangements and different inlet pressure is shown in Fig. 3. It can be found from Fig. 3 that the area of water vapor on the symmetry plane is larger under circular arrangement of holes when inlet pressure is small (0.4 MPa), while the inlet pressure in large (larger than 0.6 MPa), the water vapor distribution on the symmetry plane of the orifice plate is almost the same. Besides, it can also be seen that the water vapor distribution inside the circular holes is symmetrical about the middle plane. More water vapor appears at the outer circular holes compared with the central circular hole, and the wall of the circular holes away from the center of the orifice plate has more water vapor. In addition, the maximal water vapor volume fraction in the orifice plate with circular arrangement of circular holes is higher than the rectangular arrangement, so it can be inferred that cavitation intensity is higher when the holes of the orifice plates are circular arrangement.

The pressure distribution inside orifice plate is related to the cavitation, the static pressure along the centerline of the circular arrangement multi-hole orifice plates with different shape of holes and different inlet pressure is shown in Fig. 4. It can be seen from Fig. 4(a) that when inlet pressure is small (0.4 MPa), there is an apparent difference of the static pressure within the orifice under different shape of holes. The minimum pressure under three kinds of holes is close to each other, but after the pressure recovers within the orifice, the static pressure has the maximum when the shape of holes is triangular and has the minimum when the shape of holes is circular. At the outer of the orifice plate, the static pressure under three kinds of holes is almost the same. From Fig. 4(b), it can be found that under large inlet pressure, although there is a small fluctuation of the pressure difference between different kinds of holes, the static pressure is nearly the same.

When the water vapor volume fraction in each computational cell is higher than 0.5, the cavitation intensity is higher. Figure 5 depicts the 3D isosurfaces when the water vapor volume fraction is above 0.5 for inlet pressure 0.6 MPa. From Fig. 5, it can be found that cavitation mainly occurs near the wall of the hole but away from the center of the orifice plate, which is consistent with above analysis. When the shape of the hole is rectangular or triangular, the vapor mainly congregates around the edges, and the water vapor volume fraction on the symmetry plane is very small and can be neglected. From the water vapor distributions, it can be found that there is no apparent difference of the water vapor volume when the shape of the holes is rectangular and triangular. Besides, it can also be found that although the area of water vapor on the symmetry plane has the maximum in the central hole as shown in Fig. 3, the water vapor volume when water vapor volume fraction is above 0.5 is minimal in the central hole.

The cavitation number σ is always used to evaluate the cavitation intensity of different orifice plates and it is expressed as follows

$$\sigma = \frac{p_d - p_v}{0.5\rho v^2} \tag{1}$$

here p_d is the downstream pressure of the orifice plate, p_v is the saturated vapor pressure of water, ρ is the density of water, and v is the average velocity within the holes of an orifice plate.

Figure 5 shows the cavitation number of the six kinds of orifice plates under different inlet pressure. It can be found from Fig. 5 that the cavitation number decreases with the increase of inlet pressure for a given orifice plate, which means cavitation intensity increases with the rising of inlet pressure. Under the same inlet pressure, the orifice plate with circular holes has the minimal cavitation number and the orifice plate with triangular holes has the maximal cavitation number. For the orifice plates with circular and triangular holes, the cavitation number is larger when the holes are circular arrangement, but for the orifice plates with rectangular holes, the cavitation number is smaller when the holes are circular arrangement. Compare the six kinds of orifice plates, the orifice plate with circular arrangement circular holes has the minimal cavitation number for a given inlet pressure, and the largest cavitation number difference is 29.6% between the circular holes and the triangular holes when inlet pressure is 0.4 MPa. As the increase of inlet pressure, the cavitation number difference decreases and the cavitation number for the investigated orifice plates is almost the same when inlet pressure is 1 MPa.

Figures & Tables:

	Density (kg/m^3)	Viscosity (kg/m-s)
Water	998.16	0.0010016
Water vapor	0.0173	9.727E-06

Table 1 Water and water vapor properties used in this study

Pressure drop (Pa)	14,000	21,000	30,000	88,000
Measured flow rate (m3/h)	1.57	1.86	2.28	3.66
Simulated flow rate (m3/h)	1.42	1.74	2.13	3.46
Error	9.6%	6.5%	6.6%	5.5%

Table 2 Comparison between the simulated and the measured flow rate

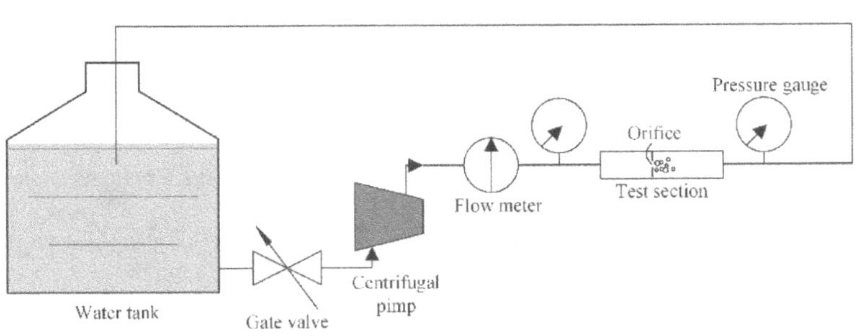

Figure 1 Schematic of the experimental facility

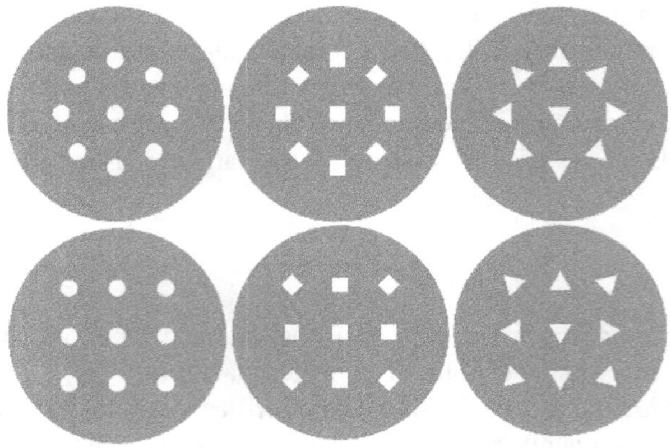

Figure 2 The shape and arrangement of the multi-holes

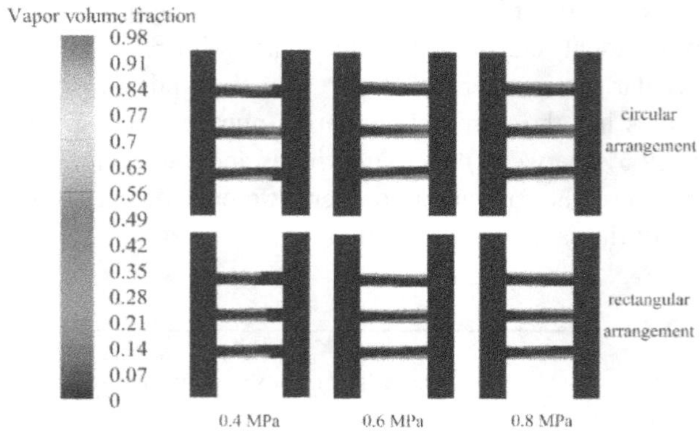

Figure 3 Vapor volume fraction on the symmetry plane with circular holes under different inlet pressure

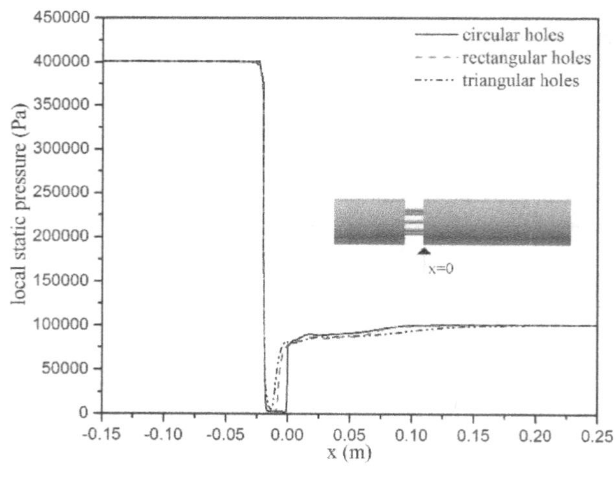

Figure 4(a) Inlet pressure is 0.4 MPa

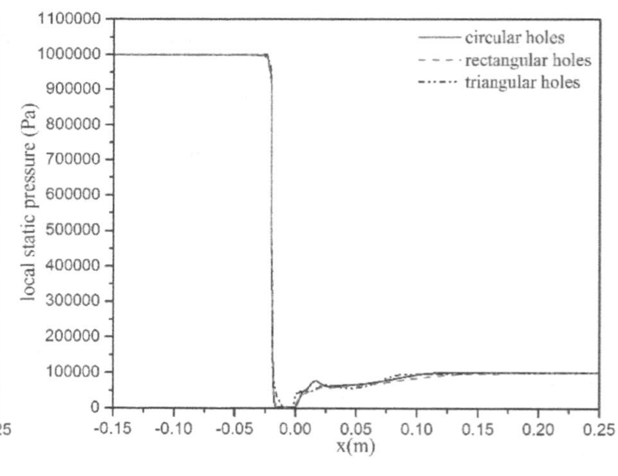

Figure 4(b) Inlet pressure is 1 MPa

Figure 4 Static pressure along the centerline of the circular arrangement multi-hole orifice plates under different inlet pressure

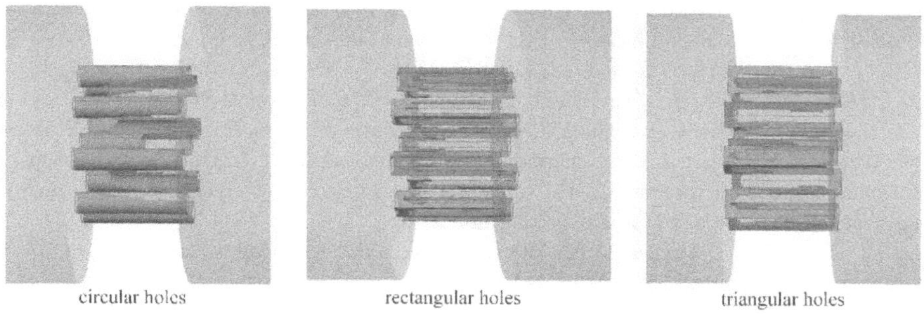

Figure 5 cavitation in circular arrangement multi-hole orifice plates

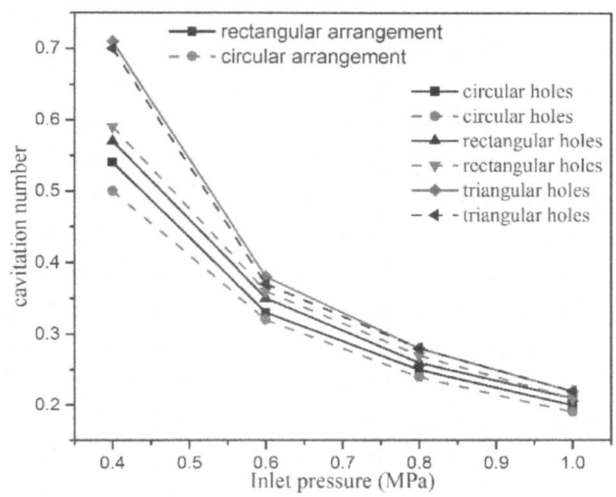

Figure 6 Cavitation number under different arrangement and shape of holes and different inlet pressure

Conclusion

A systematical numerical study is carried out with six kinds of orifice plates, and the numerical method is validated by experiment. The pressure drop, the arrangement and the shape of the holes are synthetically investigated to find out their effects on cavitation intensity. The cavitation intensity increases with the increase of pressure drop for all the studied orifice plates, and the circular arrangement orifice plate with circular holes has the maximal cavitation intensity compared with other orifice plates. For orifice plates with circular and triangular holes, the circular arrangement orifice plates have better cavitation intensity, but for orifice plates with rectangular holes, the rectangular arrangement orifice plates have better cavitation intensity. Besides, when pressure drop is small, large difference of cavitation intensity exists between different kinds of orifice plates, but under large pressure drop, the cavitation intensity of all the studied orifice plates is almost the same.

References

[1] Sivakumar, M., & Pandit, A. B. (2002). *Wastewater treatment: a novel energy efficient hydrodynamic cavitational technique*. Ultrasonics sonochemistry. 9(3), 123-131.

[2] Schneider, B., Koşar, A., Kuo, C. J., Mishra, C., Cole, G. S., Scaringe, R. P., & Peles, Y. (2006). *Cavitation enhanced heat transfer in microchannels*. Journal of Heat Transfer. 128(12), 1293-1301.

[3] Ebrahimi, B., He, G., Tang, Y., Franchek, M., Liu, D., Pickett, J., Springett, F. and Franklin, D. (2017). *Characterization of high-pressure cavitating flow through a thick orifice plate in a pipe of constant cross section*. International Journal of Thermal Sciences. 114, 229-240.

[4] Dong, Z. Y., Chen, Q. Q., Yang, Y. G., & Shi, B. (2013). *Experimental and numerical study of hydrodynamic cavitation of orifice plates with multiple triangular holes*. Applied Mechanics and Materials. 256, 2519-2522.

[5] Rudolf, P., Kubina, D., Hudec, M., Kozák, J., Maršálek, B., Maršálková, E., & Pochylý, F. (2017). *Experimental investigation of hydrodynamic cavitation through orifices of different geometries*. 143, 02098.

[6] Liu, Y., Zhu, B., Zhu, Y., & Li, Z. (2006). *Flow and cavitation characteristics of a damping orifice in water hydraulics*. Proceedings of the Institution of Mechanical Engineers, Part A: Journal of Power and Energy. 220(8), 933-942.
[7] He, Z., Zhong, W., Wang, Q., Jiang, Z., & Shao, Z. (2013). *Effect of nozzle geometrical and dynamic factors on cavitating and turbulent flow in a diesel multi-hole injector nozzle*. International Journal of Thermal Sciences. 70, 132-143.
[8] He, Z., Chen, Y., Leng, X., Wang, Q., & Guo, G. (2016). *Experimental visualization and LES investigations on cloud cavitation shedding in a rectangular nozzle orifice*. International Communications in Heat and Mass Transfer. 76, 108-116.
[9] Ji, C., He, Z., & Chen, Y. (2015). *Experimental and numerical investigation on the performance of hydrodynamic cavitation in multi-holes orifice plate*. Mechanical and Manufacturing Engineering.
[10] Rudolf, P., Kubina, D., Kozák, J., Hudec, M., & Pochylý, F. (2017). *Dynamics of the cavitating flow downstream of the orifice plate*. AIP Conference Proceedings. 1889, 020033.

© 2018 ASME

Experimental Characterization of a Cavitating Orifice

[1,2]Claudia Esposito*; [1]Jean-Baptiste Gouriet; [2]Maria Rosaria Vetrano; [2,3]Johan Steelant; [1] Jean-Marie Buchlin

[1]*von Karman Institute for Fluid Dynamics, Rhode Saint-Genèse, BELGIUM;* [2]*KUL-Leuven, BELGIUM;* [3]*ESA-ESTEC,*

Abstract

Experiments on a cavitating orifice have been performed in water. The main goal was to characterize and compare three different orifice geometries. Mass flow rates and steady pressure measurements have been used to describe the hydraulic curve of the orifice. High-speed imaging has been used to characterize the cavitation evolution. Finally, special attention has been devoted to the void fraction measurements with a three pressure probes technique.

Keywords: cavitation; orifice; void fraction;

Introduction

Cavitating flows can be faced in several industrial applications where the flow undergoes sudden pressure drops such as valves, pumps and orifices. This is why, cavitation is object of several research activities, such as (1), (2), (3), (4) and (5).

This paper presents the last experimental activities conducted at the von Karman Institute on the water facility *Becassine*. Cavitation is induced by means of an orifice. Three different orifice geometries have been considered in this study and their hydraulic performances have been compared. Synchronized time-resolved pressure measurements and flow visualizations have been performed to characterize the different cavitating stages. In addition, the void fraction has been measured. This is a key parameter since it affects the pressure drop and heat transfer in the two-phase flow. Up to now different techniques have been developed to quantify the void fraction, but many of them are either intrusive or they require big facilities, such as (6), (7). Therefore, a non-intrusive technique has been chosen within this study. It uses three pressure sensors equidistantly placed (8) and it was validated with a known void fraction content in the work by Yenigun (9).

This report develops as follows: first, the experimental set-up and the testing procedure are detailed. The second section reviews the technique to measure the void fraction. Finally, some examples of the experimental results are illustrated and conclusions are drawn.

Experimental procedure

The test campaign was performed on the water facility *Becassine*. Figure 1(a) shows the corresponding experimental set-up. The water is stored in a 750 litres tank which feeds the system in a closed loop. This tank is equipped with an electrical resistance to heat the water. The maximum operating pressure is limited to 5 bar abs in order to prevent any damage to the test section built in Plexiglas. The backpressure is adjusted either slightly by a valve located downstream of the test section, or by the vacuum pump which is connected to the top of the reservoir. A minimum backpressure of 0.2 bar abs can be achieved. A smaller reservoir is located just after the pump to calm the water flow before entering the test section.

Three orifice geometries have been tested (Figure 1(b)). The first two geometries are divergent orifices in Plexiglas: one is thick with $\beta = 0.15$ and the other is thin with $\beta = 0.30$. The last one is a cylindrical sharp-edged thick orifice with $\beta = 0.17$ and thickness $s/d = 1.2$.

The mass flow rate is measured with an electromagnetic flowmeter placed upstream the test section. The static pressure upstream of the orifice and the pressure drop across the orifice are measured by two Valydines (0.25% full-scale accuracy). The upstream pressure is measured 5D upstream the orifice and the downstream pressure is measured 16D downstream the orifice. D is the test section diameter equals to 4 cm. The test-section is also instrumented with three unsteady pressure transducers (XTL-M Kulites 0.1% full-scale accuracy). They are positioned downstream of the orifice at 4.8D, 6.4D and 7.9D, respectively. The spacing of 6 cm between consecutive transducers and the acquisition

*Corresponding Author, Claudia Esposito: claudia.esposito@vki.ac.be

frequency of 100 kHz have been set according to the previous calculations for the three-pressure probes technique validation (9).

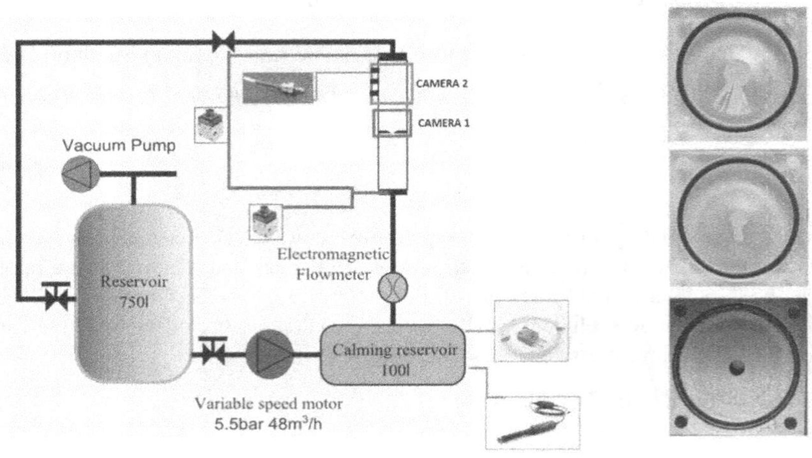

Figure 1: (a) Schematic of the facility; (b) Orifice geometry: thin divergent orifice (top), thick divergent orifice (center) and cylindrical thick orifice (bottom)

The flow is visualised at two different locations of the test section. A first camera records ~5000 images at 14 kHz just downstream the orifice. A second camera is positioned 4D downstream the orifice at the position of the three-pressure transducers. It records flow images at 1 kHz. The two cameras are synchronized between them as well as with the time-resolved pressure measurements. The water temperature is measured in the calming reservoir with a thermocouple type K. The dissolved amount of oxygen in the water is also measured in the calming reservoir with a Vernier Optical DO probe.

The water is deaerated before each test. First, the water is slightly heated at 30°C while it circulates in the closed loop at the minimum pump speed. After, the vacuum pump is activated in order to liberate the oxygen. The dissolved oxygen concentration is expressed as a percentage of the maximum amount of oxygen that water can hold at a given temperature, specifically

$$\%saturation = \left(\frac{actual\ DO\ reading\ mg/L}{saturated\ DO\ reading\ mg/L}\right) \cdot 100 \qquad \text{Eq.1}$$

Figure 2 depicts the reduction of this percentage during eight hours. The cavitation tests start when the measured dissolved oxygen content is below the 10% of the saturated dissolved oxygen at the testing temperature. During the tests, this value changes of only 2.5%. For each orifice, an average of thirty testing conditions have been recorded. Different flow rates have been tested by varying the upstream pressure. The downstream pressure was set either at ambient or at two different vacuum conditions, i.e. 0.2 bar abs and 0.5 bar abs, in order to cover the whole range from fully liquid up to the super cavitation regime.

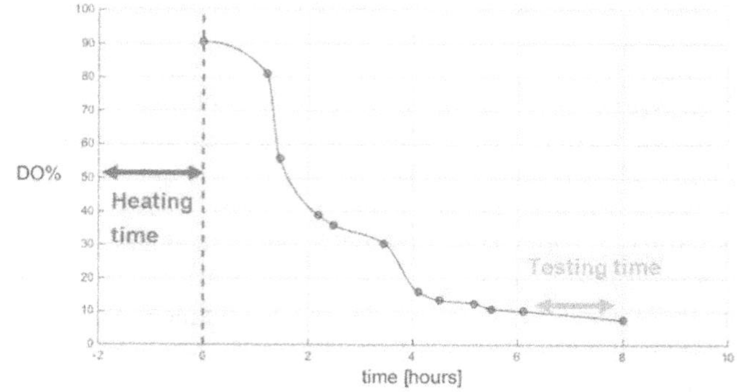

Figure 2: Evolution with time of the dissolved oxygen content

Void Fraction Measurement Technique

The void fraction has been measured during the cavitation tests by means of the three-pressure transducers technique, which was developed in 1970s by Brown and Margolis (8). This method relies on the assumption that pressure waves propagate through a pipe without changing their amplitude (10), (3). The strategy of this technique is to place three pressure sensors equidistantly. The transducer in the middle is taken as the reference one. For each location the relations reported in Eq.2 can be written

$$\begin{aligned} p_1 &= p(-L,f) = C_1(f)e^{-jkL} + C_2(f)e^{jkL} \\ p_2 &= p(0,f) = C_1(f) + C_2(f) \\ p_3 &= p(L,f) = C_1(f)e^{jkL} + C_2(f)e^{-jkL} \end{aligned} \quad \text{Eq.2}$$

A transfer function can be built from these three pressure signals in the frequency domain. This function is directly linked to the speed of sound. Thanks to the equal distance among the sensors, the transfer function reduces itself to a cosine signal and the speed of sound can be retrieved.

This velocity is the speed of sound confined to the pipe environment, (c_{conf}). This means that this value depends on the hydraulic and mechanical characteristics of the pipe. In order to retrieve the free sound velocity c_{free} , an additional equation is needed. Eq.3 takes into account the mixture density ρ_M, the pipe diameter D, the elasticity of the pipe E, its thickness e and finally the pipe Poisson's ratio ν. These terms related to the pipe and the material properties are considered to be constant. Assuming a homogeneous two-component mixture, the fluid density ρ_M is calculated as the weighted average of the vapor and liquid phase density, ρ_{gas} and ρ_{liq} respectively. In Eq.4 α stands for the void fraction.

$$c_{free} = c_{conf} \times \sqrt{1 + \frac{D}{eE}\left(1-\nu^2\right) \times c_{free}^2 \times \rho_M} \quad \text{Eq.3}$$

$$\rho_M = \alpha \times \rho_{gas} + (1-\alpha) \times \rho_{liq} \quad \text{Eq.4}$$

Finally, to close the system of equations, a relation between the speed of sound and the void fraction is needed. This can be obtained by deriving pressure over density (11), as in Eq.5.

$$\frac{1}{c_{free}^2} = \alpha \times \frac{\rho_M}{\rho_{gas}} \times \frac{1}{c_{gas}^2} + \left[\left(\frac{\rho_M}{\rho_{liq}}\right)^2 - \alpha \times \rho_M \times \frac{\rho_{gas}}{\rho_{liq}^2}\right] \times \frac{1}{c_{liq}^2} \quad \text{Eq.5}$$

Results

The first measurements concern the mass flow rate through the orifice. As an example, Figure 3 shows the hydraulic curve for the cylindrical orifice with β=0.17. Here, the mass flow rate is plotted versus the pressure drop through the orifice. A non-dimensional representation of this curve is used. The theoretical choked flow rate (Q_{chok}) and choked pressure drop (ΔP_{chok}) have been used at this purpose. In fact, these values are unique for a certain upstream pressure condition and different upstream pressures correspond to different hydraulic curves. The linear part of the theoretical curve in Figure 3 corresponds to a fully liquid behavior and it is derived by Chisholm's correlation for thick orifices (12). The choked flow is computed by modelling the hydraulic system in EcosimPro and considering an isentropic expansion (13).

Figure 4 depicts different stages of cavitation at four pressure drops $\left(\widehat{\Delta P} = \sqrt{\Delta P}/\sqrt{\Delta P_{chok}}\right)$. $\widehat{\Delta P} = 0.7$ corresponds to initial stage where cavitation is barely visisble and no significant deviation from the linear trend are noticed. $\widehat{\Delta P} = 0.9$ and $\widehat{\Delta P} = 1$ correspond to a developing cavitation. As the pressure drop increases, large vaporous structures cohexist with thin bubbles which are responsible of most of the noise. Finally, the super-cavitation stage is reached for $\widehat{\Delta P} =$ 1.2. Super-cavitation is the evolution of the choked flow. The submerged liquid jet becomes visible at the center of the test section. A thick vapor cavity surrounds it. On the walls of the test section a film of water is falling down.

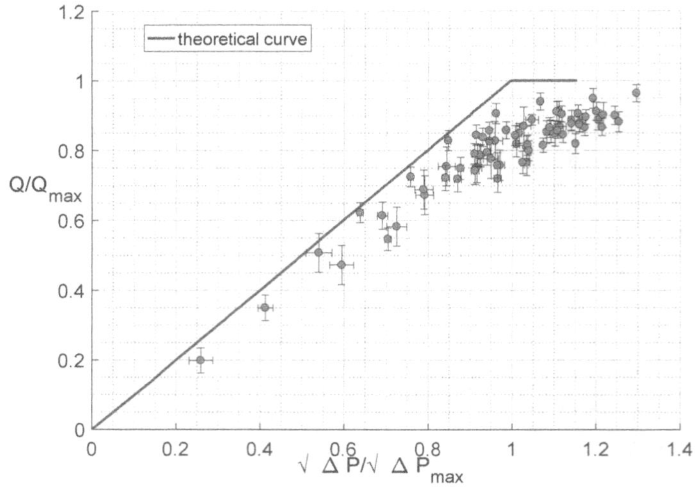

Figure 3: Hydraulic curve for the cylindrical orifice ($\beta = 0.17$)

Void fraction has been measured and these values have been related to the cavitation parameter $\left(\sigma = \left(P_{up} - P_{sat}\right)/\Delta P\right)$. An example is illustrated in Figure 5(a) for the thick cylindrical orifice. The images on Figure 5(b) are those recorded in the upper part of the test section for the cylindrical thick orifice. They show the flow topology at three different flow conditions, namely σ=1.8, σ=1.15; and σ=1. The σ values higher than 1.8 correspond to a fully liquid flow. In this range, a correct speed of sound couldn't be calculated. In fact, this technique would require an acquisition frequency higher than 100 kHz to capture such high speeds of sounds as those in a fully liquid flow (around 1500 m/s in water) (9). At σ=1.8, cavitation is barely visible at the exit of the orifice but the flow appears completely liquid in the region of the transducers (Figure 5(a)). However, a constant whistling can be heard at this condition and a free speed of sound of 1381 m/s has been computed. It corresponds to a void fraction of 3.8E-7. Figure 5(b) shows the case of σ=1.15. The vapor phase appears well dispersed in the liquid, which satisfies the homogeneous two-phase flow assumption. The void fraction is estimated at 0.0003. Finally, the highest value of void fraction (i.e. 0.0013), has been measured at σ=1. From this point, the flow starts choking and the cavitation cloud extends itself farther and farther from the orifice. In fact, the front of this cavitation cloud starts appearing in the frame as we can see on Figure 5(c). Reducing the cavitation parameter, super-cavitation appears and the technique is no longer applicable.

a) $\widehat{\Delta P} = 0.7$ b) $\widehat{\Delta P} = 0.9$ c) $\widehat{\Delta P} = 1$ d) $\widehat{\Delta P} = 1.2$

Figure 4: Flow topology downstream the orifice at different pressure drops

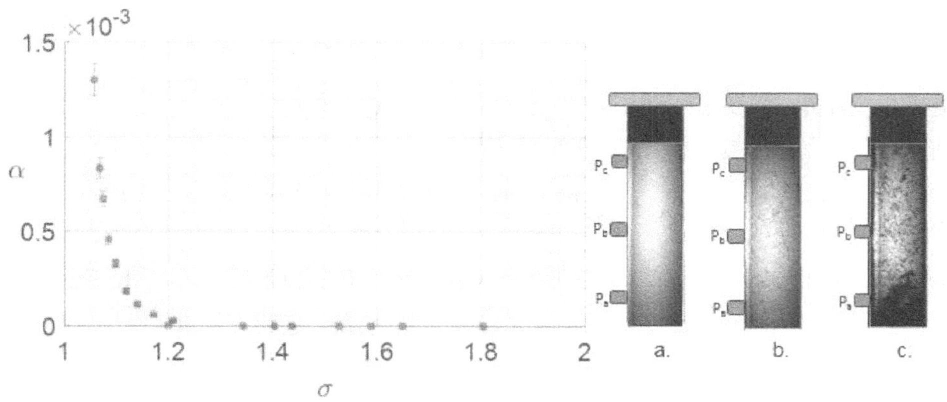

Figure 5: (a) Void fraction vs. cavitation number; (b) Flow visualization at a) σ=1.8; b) σ=1.15; c) σ=1

Conclusion

An experimental study on cavitation has been carried out. Three different orifice geometries have been analyzed, two divergent orifices, thin and thick, respectively, and a thick cylindrical orifice. The hydraulic behavior of these orifices has been observed from the fully liquid to the super-cavitation. Moreover the effect of the orifice geometry on the flow topology has been examined. Special attention has been given to the void fraction, which has been estimated by means of time-resolved pressure measurements. These results have been linked to the cavitation parameter in order to correlate the cavitation intensity to the void fraction and to question the influence of the orifice geometry on these correlations.

References

1. *Choked cavitation in micro-orifices: An experimental study.* **Cioncolini, Andrea, et al.** 2016, Vol. 74, p. 49-57.

2. *Flow regime transitions due to cavitation in the flow through an orifice.* **Yan, Y. e Thorpe, R.B.** 1990, Int. J. Multiphase Flow, Vol. 16, p. 1023-1045.

3. *Noise generated by cavitating single-hole and multi-hole orifices in a water pipe.* **Testud, P., et al.** 2007, Journal of Fluids and Structures, Vol. 23, p. 163-189.

4. *Experimental and numerical study of the flow characteristics in a cryogenic valve with liquid nitrogen and water.* **Peveroni, Laura and Pinho, Jorge and Steelant, Johan and Strengnart, Marc and Vetrano, Maria Rosaria.** 2015.

5. *Multi-Phase Investigation of the Waterhammer Phenomena using the Full Cavitation Model.* **Pinho, J., et al.** 8 January 2014, Journal of Propulsion and Power.

6. *Measurements of void fraction distribution in cavitating pipe flow using x-ray CT.* **Bauer, D., Chaves, H. e Arcoumanis, C.** s.l. : IOP Publishing, 2012, Measurement Science and Technology, Vol. 23 (5), p. 055302.

7. *Measurement of void fraction in dispersed bubbly flow containing micro-bubbles with the constant electric current method.* **Uesawa, Shin-ichiro and Kaneko, Akiko and Abe, Yutaka}.** s.l. : Elsevier, 2012, FLow Measurement and Instrumentation, Vol. 24, p. 50-62.

8. *Measurement of the propagation of long-wavelength disturbances through turbulent flow in tubes.* **Margolis, DL e Brown, FT.** s.l. : American Society of Mechanical Engineers, 1976, Journal of Fluids Engineering, Vol. 98, p. 70-78.

9. **Yenigun, O.** *Experimental characterization of two-phase flows under cavitation conditions.* 2017.

10. **Shamboran, H.** *Développement d'une méthode de mesure de la célérité du son en écoulement diphasique application aux écoulements cavitants.* 2009.

11. **Pinho, Jorge.** *Experimental investigation of cavitation in a safety relief valve using water: extension to cryogenic fluids.* s.l. : PhD Thesis, Université Libre de Bruxelles, 2015.

12. *Two-phase pressure drop in multiple thick-and thin-orifice plates* . **Kojasoy, G.** 1997, Experimental thermal and fluid science, Vol. 15.4 , p. 347-358.

13. *ESPSS Simulation Platform.* **Moral, J and Perez Vara, R and Steelant, J and De Rosa, M.** 2010, Space Propulsion.

© 2018 ASME

Experimental estimation for pressure fluctuation on ship stern induced by cavitating propeller using cavity shape measurements

[1]Koichiro Shiraishi*; [1]Yuki Sawada; [1]Daijiro Arakawa; [1]Kunihiro Hoshino;

[1] *National Maritime Research Institute, Mitaka-shi, Tokyo, Japan*

Abstract

Authors have recently developed a combination-line charge-coupled device camera-based method for measuring three-dimensional shapes that is faster and more accurate than conventional methods. Furthermore, the developed system can measure cavity shapes on a model propeller through the experiments in the large cavitation tunnel of NMRI. However, it is necessary to verify how much accuracy the system has in the cavity shape measurement. In this paper, we verify the accuracy through a comparison of the pressure fluctuation estimated by the cavity volume and measured by the pressure sensor. In the estimation, since the unsteady cavity is replaced by a spherical bubble moving with varying radius, the pressure fluctuation induced by the sphere is calculated. The pressure fluctuation is estimated by applying the proposed method to experimental data of the cavity shape. The pressure fluctuation estimated by the proposed method is in good agreement with the pressure fluctuation measured by pressure sensors. This result also showed that the cavity shape measurement system developed by the authors has sufficient accuracy.

Keywords: Cavity volume, 3D shape measurement, Combination line CCD camera method, Pressure fluctuation induced by cavitating propeller.

1. Introduction

Cavitation generated by a propeller operating in a ship's wake can cause hull vibration, thrust reduction, and erosion. To solve these problems, developing theoretical and numerical calculation methods for estimating cavitation with high accuracy at the design stage is necessary. Estimating cavitation with high precision requires measurements of cavity volume which has a strong correlation with pressure fluctuation. At present, there is a few method for measuring cavity shape with high accuracy. To fill this knowledge gap, various methods of measuring cavity shape have been proposed, including the use of stereo photography and laser beam scattering (Ukon et al 1991; Luca et al 2009; Stefan et al 2012). Recently, we have developed a three-dimensional cavity shape measurement system based on the use of a combination-line CCD camera method (Shiraishi et al 2017). However, we haven't verified the accuracy of the developed system of cavity shape measurement. In order to verify the precision of the system, we investigate the relationship of cavity volume and pressure fluctuation induced by an unsteady cavitation on propeller blades. For estimation of pressure fluctuation by cavity volume, unsteady cavitation is replaced by a spherical bubble moving with varying radius. Then the pressure fluctuation induced by the sphere is calculated by applying the potential theory.

2. Cavity Shape Measurement

2.1 Combination-line CCD camera measurement method

We used a combination of line CCD cameras to measure cavity shapes on the model propeller (Hoshino et al 2004). In this method, a laser beam is irradiated onto a measurement object and light scattered from its surface is photographed using three line CCD cameras. Based on the resulting image data, the three-dimensional surface of the object is reconstructed via triangulation. In line CCD cameras, the image-receiving elements are arranged in a row. Such cameras have the advantage of much higher scan rates and resolution than area CCD cameras. The configuration and a photo of the camera system used in this study for three-dimensional shape measuring are shown in Figures 1 and 2, respectively. We used a green laser (532 nm), a wavelength with high penetration ability in water. The specific measurement method used by the proposed system is outlined as follows. First, the laser beam is used to irradiate the position to be measured, with laser light scattered from the surface of the measurement object passed through a semi-cylindrical lens and focused onto the line CCD elements for imaging. Triangulation is then used to extract the three-dimensional coordinates of the laser light spot position from the peak coordinates of the luminance distributions of

*Corresponding Author, Shiraishi Koichiro: shiraishi@nmri.go.jp

the images captured by the cameras. Figure 3 shows the measurement principle applied by the combined-line CCD method as seen from the depth direction.

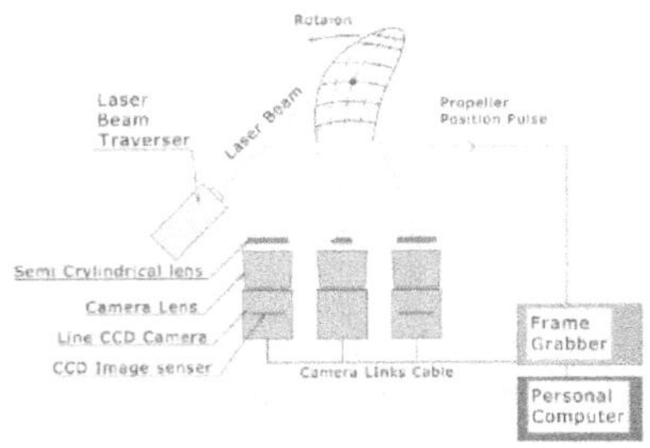

Figure 1 Schematic of measurement system

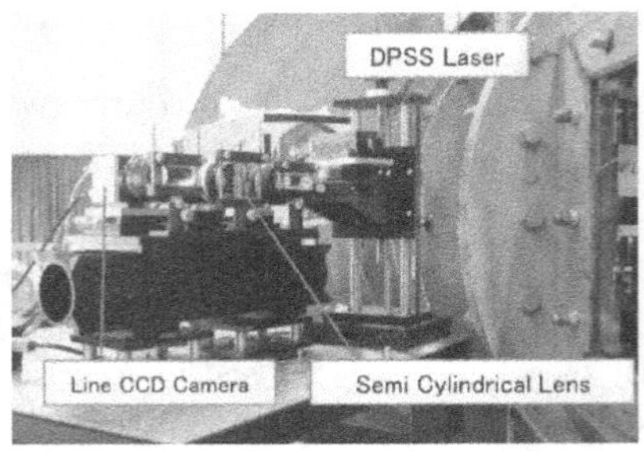

Figure 2 Photo of measurement system

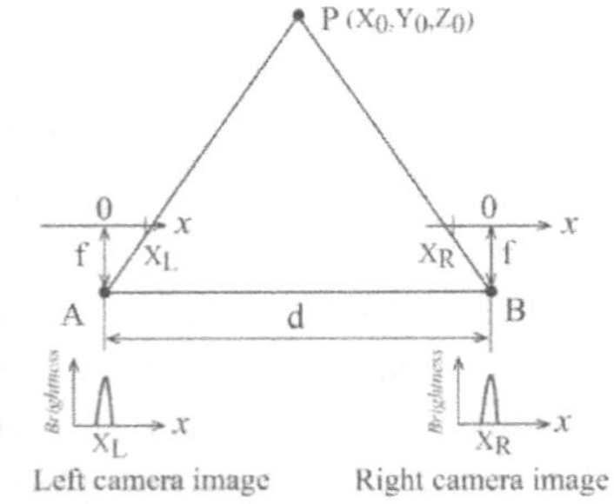

Figure 3 Illustration of the use of triangulation in the combination-line CCD camera method

2.2 Measurement results of the cavity shape

Cavity shape measurement was then performed under the cavitation condition on the Seiun-Maru-I's HSP (Shiraishi et al 2017). The detail of the calculation conditions and definitions are same to the paper. Cavity shape measurement was performed at six propeller phase angles: 15.0°, 20.0°, 30.0°, 40.0°, 50.0° and 60.0°. Figures 6–9 show the respective cavitation patterns imaged by a steel camera (left) and measured using the proposed system (right) at six propeller phase angles: 30.0°, 40.0°, 50.0° and 60.0°. It is seen from these comparisons that the proposed method can be used to measure the cavity shape over a wide range of phase angles.

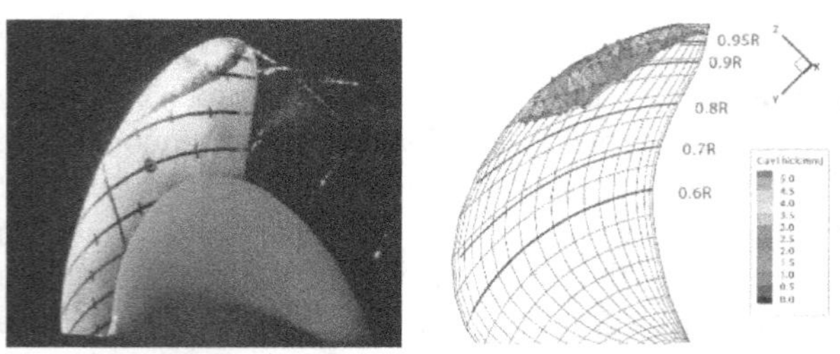

Figure 6 Image of cavitation patterns and measured cavity shapes on model propeller (Phase Angle = 30.0°)

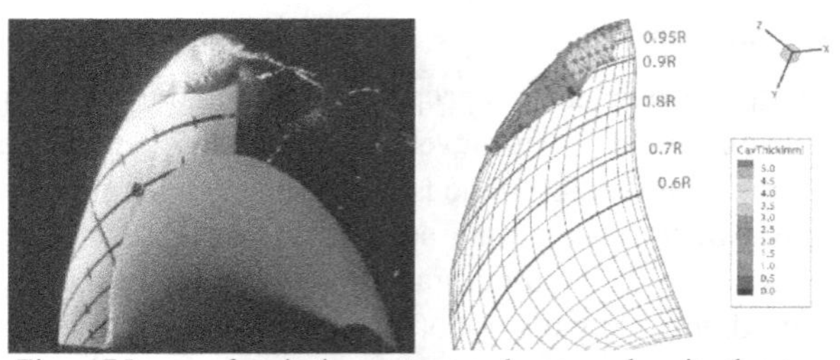

Figure 7 Image of cavitation patterns and measured cavity shapes on model propeller (Phase Angle = 40.0°)

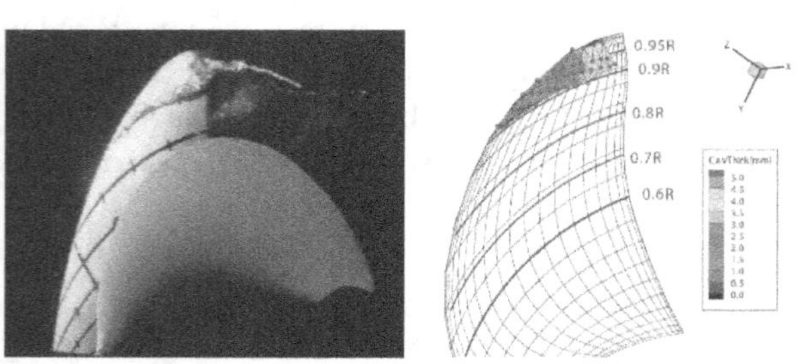

Figure 8 Image of cavitation patterns and measured cavity shapes on model propeller (Phase Angle = 50.0°)

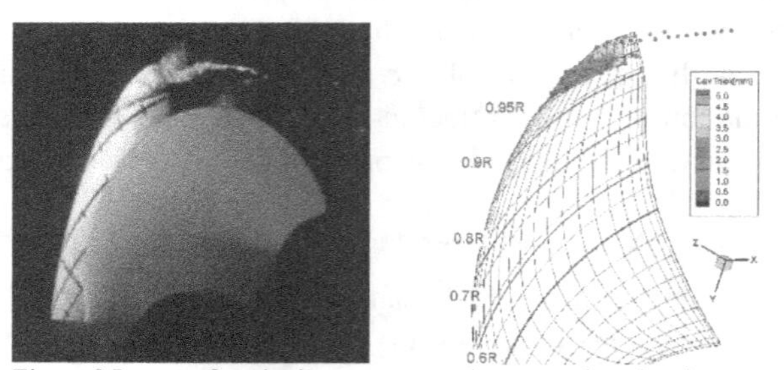

Figure 9 Image of cavitation patterns and measured cavity shapes on model propeller (Phase Angle = 60.0°)

3. Estimation for pressure fluctuations using measured cavity volume

3.1 Approximation of cavity volume with a spherical bubble

In this study, the pressure fluctuation is estimated from the cavity volume using the method of Hoshino (Hoshino, 1979). Firstly, the cavity volume is replaced with a moving spherical bubble. Then, the fluctuation of the cavity volume is considered as fluctuation of the spherical bubble. When this pressure fluctuation is calculated based on the potential theory, the fluctuating pressure is expressed by the following equation.

$$C_p = \frac{a^3}{r^3}\left(3\cos^2\theta - 1\right) + 5\frac{a^2}{r^2U}\frac{da}{dt}\cos\theta - \frac{1}{4}\frac{a^6}{r^6}\left(3\cos^2\theta + 1\right)$$
$$+ 2\frac{a}{rU}\left\{2\left(\frac{da}{dt}\right)^2 + a\frac{d^2a}{d^2t}\right\} - \frac{a^4}{r^4U^2}\left(\frac{da}{dt}\right) - 2\frac{a^5}{r^5U}\frac{da}{dt}\cos\theta \qquad (1)$$

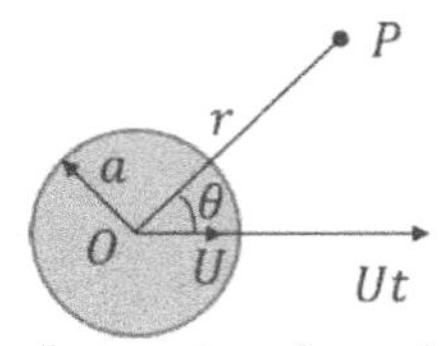

Figure The coordinate system of a moving sphere bubble

By triangulating the measurement points of the cavity shape, the surface of the cavitation is generated by a triangular panel. Then, the region surrounded by the propeller blades and the cavitation surface is assumed to be cavity volume. By approximating the obtained cavity volume as a single spherical bubble, the radius of the spherical bubble is calculated. In this study, we call this radius the equivalent spherical bubble radius. This process is executed for each propeller phase angle at which the propeller shape measurement is performed, and the data is subjected to spline interpolation to calculate time series data of the equivalent spherical bubble radius. Substituting this time series data into the equations (1), the fluctuating pressure is calculated. In this study, in order to consider the boundary effect of the flat plate, the pressure coefficient is multiplied by the boundary coefficient of 2.0.

3.2 Comparison with experiments

We apply this estimation method to the cavity measurement result of SeiunMaru I in the aquarium test and estimate the fluctuating pressure. Then, the estimated fluctuating pressure is compared with the fluctuating pressure measured by the pressure sensor, and the effectiveness of the constructed estimation method is shown. First, the results of calculating the cavity area and the cavity volume from the cavity shape measurement shown in Chapter 2 are shown in Figure 10 and Figure 11, respectively. It can be seen from these figures that the decreasing tendency of cavity area increase and the tendency of increasing cavity volume are different. It can be confirmed that the cavity volume sharply decreases as compared with the cavity area. From this, it can be seen that the cavitation characteristics cannot be sufficiently grasped only by the cavity area.

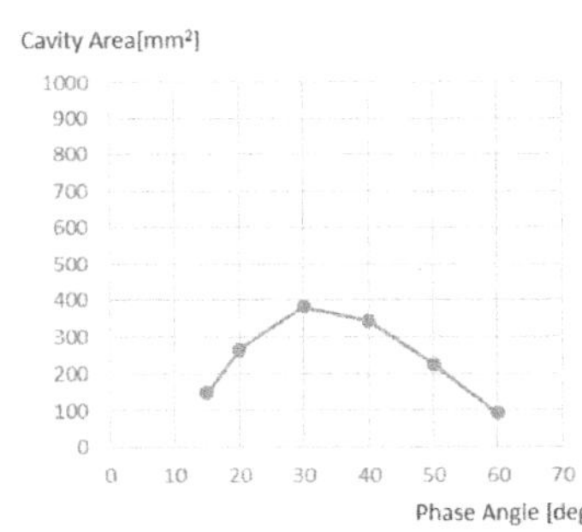

Figure 10 Measurement results of the cavity area

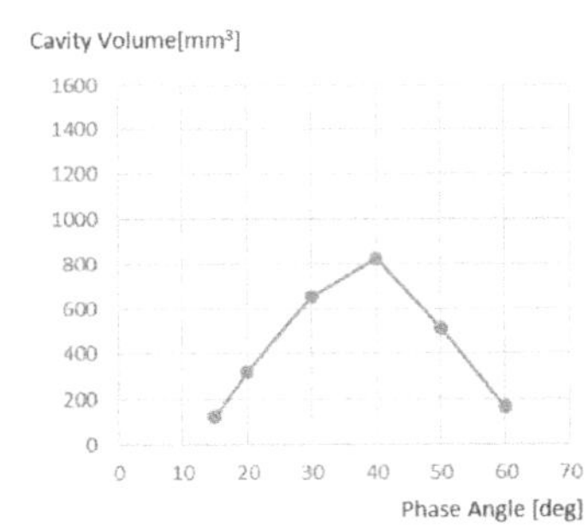

Figure 11 Measurement results of the cavity volume

Next, the equivalent spherical radius is calculated from the calculated cavity volume, and the result is interpolated with a spline. Figures 12 (a), (b), and (c) show the first derivative and the second derivative of the equivalent spherical radius. Substituting these values into the equation(1), the pressure coefficient C_p of the fluctuating pressure generated by the cavitation is calculated. The result is shown in Figure 12 (d). From these results, it was confirmed that a smooth pressure waveform is obtained without significant undulation by spline interpolation.

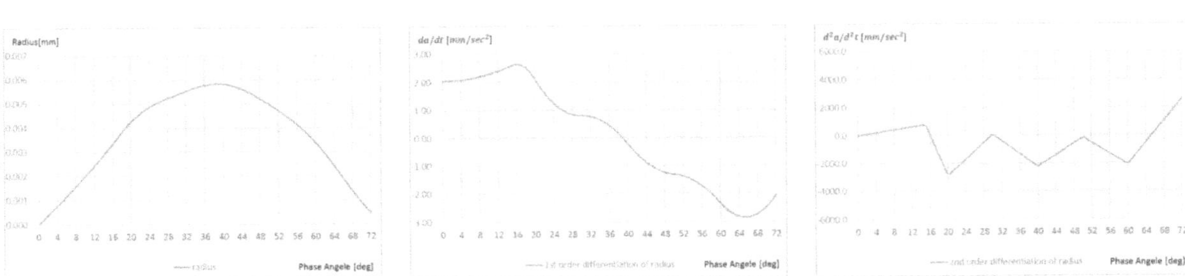

Figure 12 (a) Equivalent bubble diameter: a, (b) its first derivative value: da/dt, (c) its second derivative value: d^2a/dt^2.

Next, we compare the fluctuating pressure estimated from the cavity volume with the fluctuating pressure measured by the pressure sensor. Since the fluctuating pressure calculated above is only the component due to cavitation, the component of Non cavitation is added to this component. In this research, as a non-cavitation component, the measurement result obtained by separately measuring with a pressure sensor is used. The time series of the fluctuating pressure of both are shown in Figure 13. The blue line shows the measurement result of the pressure sensor, and the orange line shows the estimated value from the cavity volume. From this figure, it can be confirmed that the estimated value and the measured value coincide very well. In the estimated value, high-frequency components do not appear because it is not possible to consider Tip vortex cavitation. From the above, it was confirmed that the cavity shape measurement method developed by the authors has sufficient accuracy.

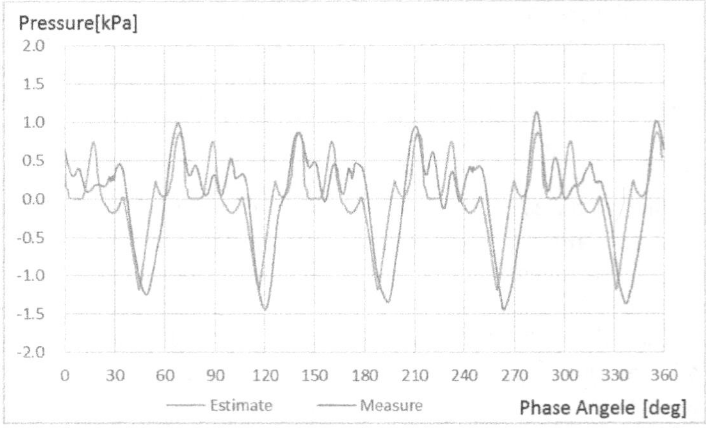

Figure 13 Time series comparison of pressure fluctuation of pressure sensor and estimation.

Conclusion

In this study, the method to estimate fluctuating pressure induced by cavitating propeller using the cavity volume measurement was proposed. The pressure fluctuation was estimated by the cavity volume measurement result of the Seiun-Maru I's HSP using the proposed method. The pressure fluctuation measured by the pressure sensors was compared with the estimated results. As a result, it was confirmed that the estimated and the measurement result coincide very well. We also showed that the cavity shape measurement method developed has sufficient accuracy.

References

[1] Luca S., Michele V., Francesco C. and Marco F. (2009a). Application of computer vision techniques to measure cavitation bubble volume and cavitating tip vortex diameter, Proceedings of the 7th International Symposium on Cavitation (CAV2009), pp. 737-748.

[2] Stefan B., Willfried K., Stephan H., Nils A. D. and Zhiliang Z. (2012), On Optical Quantification of Cavitation Properties, Proceedings of the 8th International Symposium on Cavitation (CAV 2012), pp.414-419.

[3] Ukon, Y., Kudo, T. and Kurobe, Y. (1991), "Measurement of Cavity Thickness Distribution on the Blade of Propeller Models by Laser-CCD Method", Proceedings of CAVITATION'91, ASME, FED-Vol. 116, pp.99-104.

[4] Shiraishi, K., Sawada, Y. and Hoshino, K. (2017), Cavity Shape Measurement Using Combination-Line CCD Camera Measurement Method, Proceedings of International Symposium on Marine Propulsors 2017, 5th SMP.

[5] Hoshino, K. and Tamura, K. (2004). Development of Three-Dimensional Shape Measurement Method of the Object in Water. OCEANS'04. MTTS/IEEE TECHNO-OCEAN'04, Vol.3, pp.1240-1247.

[6] Hoshino, T. (1979), Pressure Fluctuation Induced by a Spherical Bubble Moving with Varying Radius, Transactions of the West Japan Society of Naval Architects, No. 58, pp.221-234.

© 2018 ASME

Experimental Investigation of the Cavitation Effects on the Heat Generation in a Closed Loop Pumping System

[1]David Bermejo*; [1]Xavier Escaler; [2]Matevz Dular; [1]Rafael Ruíz;

[1]Universitat Politècnica de Catalunya BarcelonaTech, Barcelona, SPAIN; [2]University of Ljubljana, Ljubljana, SLOVENIA

Abstract

A series of tests were carried out in the cavitation tunnel of the Laboratory for Water and Turbine Machines of the University of Ljubljana in order to investigate the effects of cavitation in the water heating process of closed loop pumping systems. For that, the water temperature was measured at the low pressure reservoir during runs of about 10 minutes. As expected, a constant temperature increase was observed which was proportional to the pump's rotation speed. Then, the results obtained while keeping the same pump operating point and flow conditions with and without cavitation in the Venturi test section were compared. Surprisingly, it was found that the temperature increase rate was slightly higher when cavitation is present in the system. Moreover, the heat power was always higher than the hydraulic one for all the tests up to a certain cavitation level. However, this trend was inversed for higher cavitation numbers and lower hydraulic powers. Both results seem to prove that there is a clear influence of cavitation in heat generation process. Therefore, a new test rig will be built in Barcelona by the Spanish company Condorchem Envitech, S. L. with the help of Universitat Politècnica de Catalunya BarcelonaTech to continue the current research.

Keywords: hydrodynamic cavitation; heat generation; temperature increase

Introduction

The cavitation process consists on the generation, growth and collapse of a cavity fulfilled with vapour inside a liquid due to a reduction in the local pressure. As pointed out by C. E. Brennen [1], cavitation differs from boiling by the mechanism which leads the mass transfer from liquid phase to gas phase: while in boiling the leading mechanism is the rising saturation pressure up to liquid's pressure by increasing the temperature, in cavitation the change on the state is led by the pressure decrease.

Since 1873 when O. Reynolds [2] presented the phenomenon of cavitation, the studies on cavitation have been disseminated in three main fields: bubble dynamics modelling, prediction of erosion in hydraulic machinery and search for positive effects in hydraulic systems.

The most extended approach for modelling cavity dynamics is the Rayleigh-Plesset equation, introduced by Lord Rayleigh in 1917 [3]. Actually, that field of research is full of analytical and computational studies about the morphology and dynamics of bubble clusters [1].

Historically, the effort has been focused on the way to avoid the negative effects of cavitation: erosion, noise, vibrations and the performance losses [1]. The work of R. T. Knapp [4] showed that erosion appears in the downstream stagnation point of the main cavity, where the bubbles collapse near the surface due to the raising pressure, instead of the point where the bubbles nucleate.

However, the studies about positive effects of cavitation have grown up during the last decades. For instance, J. Ozonek [5], P. R. Gogate [6] or A.G. Chakinala [7] have developed acoustic and hydrodynamic cavitation devices to be used as chemical reactors. Free hydroxyl radicals are produced during bubble collapse due to the high local temperatures and pressures, and they catalyse chemical reactions in the neighbourhood of the cavities. Moreover, such efforts are also employed for waste water treatment, as R. Cirimina [8] and M. Petkovsek [9] have pointed out.

Similar steps have been done in the food processing industry, as indicated by the works of M. Martynenko [10], who tested the performance of a hydrodynamic device as homogenizer, and the invention of C. Isopo [11], which is a rotary

*Corresponding Author, David Bermejo: david.bermejo@upc.edu

device for heating and vaporizing beverages. Finally, the work of M. Petkovsek [9] also indicates that a modified centrifugal cavitation pump can enhance the disintegration of raw sludge in anaerobic digestion processes.

On the other hand, the thermodynamic or thermal effects of cavitation are well known because they control the growth and collapse of the cavity [1]. During the bubble growth, the mass transferred from liquid phase to vapour phase absorbs the liquid latent heat through the interface. As a consequence, the surrounding liquid is cooled and the saturation pressure decreases. This effect is higher in cryogenic liquids, as those used in rockets, than in water at room temperature [12]. Experimentally, the thermal effect has been visualized with a thermographic camera during the growing stage by M. Petkovsek [13].

However, there are other fields of interest related to thermodynamics and cavitation. On one hand, there are some studies, as those of B. Shneider [14] or J. Cai [15], which analyse the effects of cavitation on heat transfer through natural convection. In natural convective processes, the cavitation enhances heat transfer due the turbulence lead by the collapsing bubbles in the thermal boundary layer [14].

The prediction of very high pressures and temperatures at the end of the cavity collapse has inspired many inventions in terms of the so called hydrodynamic heat generators. Inventors and manufacturers of hydrodynamic heaters claim for the ability of cavitation to increase their efficiency. An example is the device patented by J. L. Griggs [16] which is a cylindrical drum with radial blind holes rotates at high speed surrounded by water.

Several authors have studied the performance and efficiency of such devices. The experiments carried out by K. B. Pyun [17] showed about 90% average thermal efficiency for different flow conditions. The work of S. M. Tahmid [18] described a hydrodynamic device that could rise the temperature of 113 l of water from 30° to 70° C in a closed circuit in less than half an hour with a 15 kW electric engine, while Mukesh [19] designed and tested a device with a 370 W electric engine that heated water from 25° to 60° C with a thermal efficiency of 83% at least. However, most of the companies selling such devices claim for thermal efficiencies that appear to be unrealistic.

Regarding the heat generation process associated to cavitation, the fact is that water temperature in closed pumping systems, as in cavitation tunnels [20], is known to increase during operation and, because of that, they require a cooling system to control it. However, the mechanism by which the water raises its temperature is not fully understood.

To the author's knowledge there is no clear and reliable evidence in the literature of the cavitation contribution to the heating process. In this sense, the Universitat Politècnica de Catalunya BarcelonaTech (UPC) has started a research program in collaboration with the Spanish company Condorchem Envitech, S. L. to develop an optimized heat generator for wastewater treatment technologies in Zero Liquid Discharge facilities (ZLD). As a first step, the planned objective is to put light on the opened questions about the influence of cavitation on the heat generation in pumping systems.

For that propose, it is intended to build a test rig, based on the existing cavitation tunnel at the Laboratory for Water and Turbine Machines of the University of Ljubljana [13], in order to perform the experimental tests in Barcelona. This test trig consists of a closed loop water circuit driven by a variable speed centrifugal pump. The test section where cavitation can be generated and visualized will be located between two large reservoirs. A series of instruments measuring flow rate, pressure and temperature will permit to control and monitor the system operation conditions. Since the tunnel is not ready yet, the initial tests presented here were carried out at the cavitation tunnel of the University of Ljubljana.

Description of the experiments and results

The Ljubljana cavitation tunnel was operated at fixed operating conditions and delivered hydraulic power, thus keeping the flow rate and the pump head, with and without cavitation in the test section. Several runs of about 600 s duration each were carried out at different operating conditions. During the tests, the temperature of the circulating water was monitored with a probe located at the suction side reservoir. Moreover, the pump rotation speed, head and flow rate, and the absolute pressure at the inlet of the Venturi were also controlled. A high speed video camera was used to determine the cavity morphology, size and dynamic behaviour during the tests.

Table 1 shows a couple of testing conditions and the corresponding results, where $\bar{\sigma}$ is the average Sigma value as defined by (1) and computed as the average value between the cavitation number at the start and at the end of each trial, t is the time, ω is the pump rotation frequency, Q is the flow rate, H is the pump head, H_{hyd} is the hydraulic power as defined by (2), P_{in} is the test section inlet pressure, T_{start} and T_{end} are the water temperatures at the start and at the end of each test, T_{slope} is the temperature increase rate, q is the heat power as defined by (3), ΔP is the pump pressure increase, ρ is the water density, g is the gravity, Δh is the head difference between pressure measurements, v_2 and v_1 are the mean flow velocities at the pump discharge and intake, m is the mass of water, C_P is the water specific heat, P_{sat} is the water vapour pressure and v_{throat} is the mean flow velocity at the Venturi throat.

$$\sigma = \frac{P_{in} - P_{sat}(T)}{\frac{1}{2}\rho v_{throat}^{2}} \tag{1}$$

$$N_{hyd} = Q\left[\Delta P + \rho g\left(\Delta h + \frac{v_2^{2} - v_1^{2}}{2g}\right)\right] \tag{2}$$

$$q = mC_P\frac{\Delta T}{\Delta t} \tag{3}$$

As an example, the temperature evolution with cavitation on the Venturi for a pump rotation speed of 25 Hz, a flow rate of 78 l/min, a pump pressure increase of 93.2 kPa and an inlet absolute pressure of 141.7 kPa at the test section has been plotted in Figure 1. As it can be seen, the water temperature increases continuously at a constant rate indicating that heat power is being supplied to the liquid by the system that comprises the pump, the piping, the valves, the reservoirs and the Venturi test section. For this example, the calculated slope of the linear regression indicates that the temperature increase rate is of about 0.036° C per minute.

Test	Venturi condition	$\bar{\sigma}$ [-]	t [s]	ω [Hz]	Q [l/min]	ΔP [kPa]	N_{hyd} [W]	P_{in} [kPa]	T_{start} [° C]	T_{end} [° C]	T_{slope} [° C/s]	q [W]
1	Cavitation	1.64	655	25	78.0	93.2	125.0	141.7	28.717	29.142	6.05E-04	151.8
1	No cavitation	2.79	623	25	78.0	93.9	125.9	239.0	30.699	31.056	5.69E-04	142.6
2	Cavitation	1.63	250	35	126.0	181.0	387.8	362.0	28.104	28.625	2.00E-03	500.9
2	No cavitation	1.96	433	35	128.0	181.0	394.0	448.0	28.828	29.729	1.97E-03	493.7

Table 1. Measured and computed parameters for a couple of tests with and without cavitation while keeping the same pump performance.

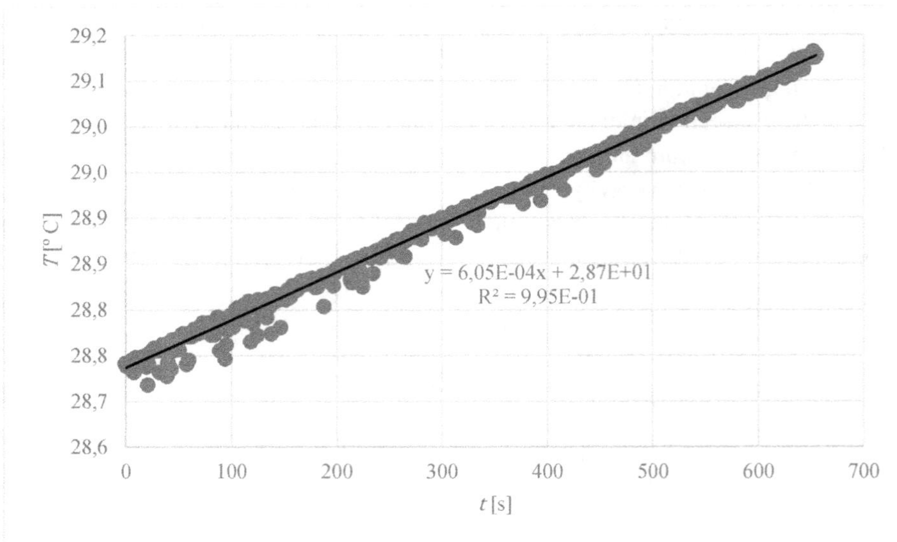

Figure 1. Water temperature inside the suction tank as a function of elapsed time with cavitation in the Venturi section, lineal trend line and equation.

Discussion

For each test, the slope of the linear regression obtained from curve fitting the temperature evolution in time was calculated. All the slope values have been plotted as function of the pump rotation speed in Figure 2. To enable a better understanding of the results, the data collected with cavitation in the test section has been represented with a red circle, meanwhile the data without cavitation has been represented with a blue triangle.

As expected, it has been confirmed that the temperature increase rate is proportional to the pump rotation speed, i.e. the energy delivered by the pump to the system. And surprisingly, the results also seem to indicate that the temperature increase rate is slightly higher when cavitation takes place in comparison with the same conditions without cavitation.

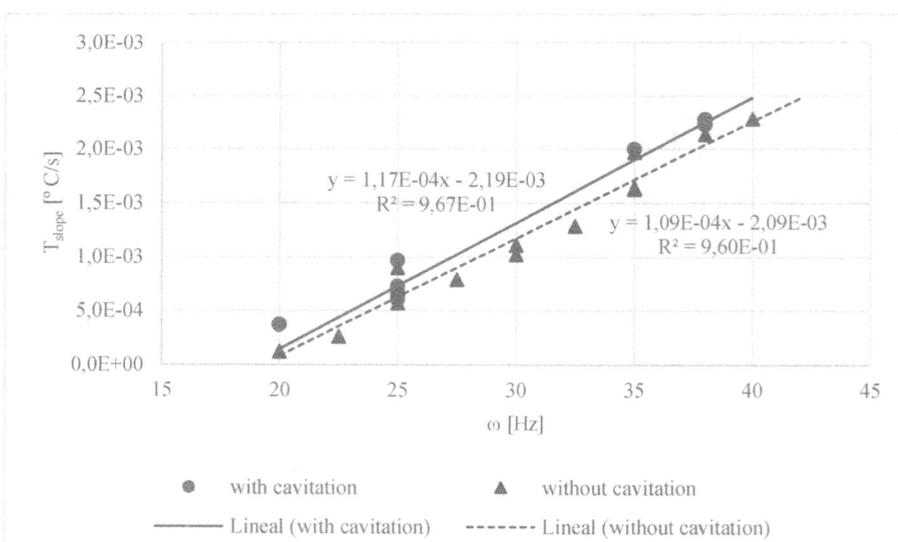

Figure 2. Temperature increase rate as a function of pump rotation frequency. Red circles indicate presence of cavitation while blue triangles indicate no cavitation in the system.

In Figure 3, the hydraulic power with cavitation (crosses) and without cavitation (pluses) and the heat power with cavitation (circles) and without cavitation (triangles) have been plotted as a function of the cavitation number. To help understanding, red colour represents again results with cavitation while blue colour represents those obtained without cavitation.

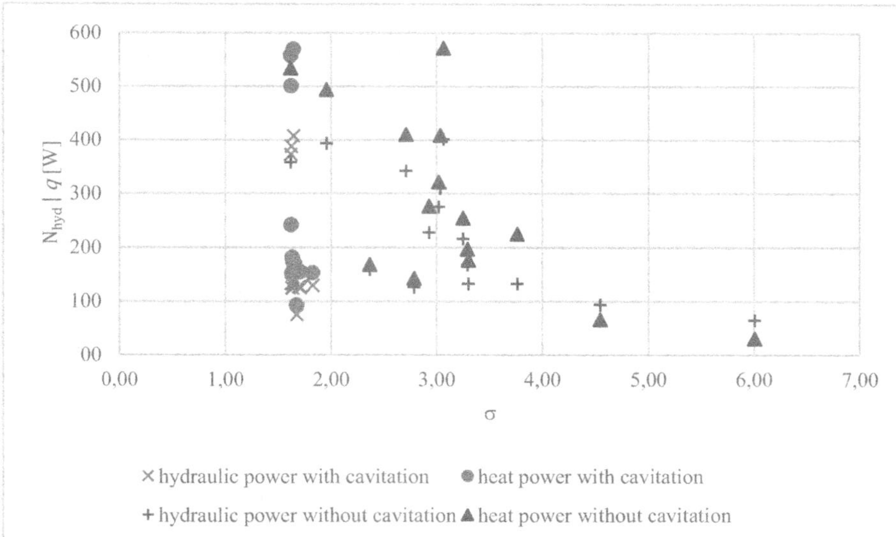

Figure .3 Hydraulic power with cavitation (crosses) and without cavitation (pluses), and heat power with cavitation (circles) and without cavitation (triangles) plotted as a function of the cavitation number. Red colour indicates cavitation while blue colour indicates no cavitation

It can be observed that the minimum cavitation number that could be achieved during the tests was around 1.62. This behaviour might be explained by a blockage of the test section. For those runs with cavitation in the Venturi, the cavitation number was always within the range from 1.62 to 1.83. Only one test without cavitation fell inside that range. Another observation is that for cavitation numbers below 4 the heat power is higher than the hydraulic power. On the contrary, for the two tests at higher cavitation numbers the tendency switches and the heat power is lower than the hydraulic power. It must be noted that in particular for these two tests, the pump hydraulic power was the lowest among all the cases and below 100 W.

Conclusion

A series of tests were carried out in the cavitation tunnel of the Laboratory for Water and Turbine Machines of the University of Ljubljana to investigate the possible relationship between flow thermodynamics and cavitation. For that, the same flow conditions were generated with and without cavitation in the test section for direct comparison. As a result, a slightly higher temperature increase rate was measured when cavitation was presented compared to the case without cavitation.

Another observation was that the heat power was higher than the hydraulic power for mean cavitation numbers below 4, but for higher values of sigma the tendency switched and the hydraulic power was higher than the heat power. It must be noted that in these cases the pump worked far from its best efficiency point.

The observed differences without and with cavitation seem to prove that cavitation has an influence in the heating generation process in closed loop pumping systems, but it does not explain the high thermal efficiency claimed by inventors and manufacturers of hydrodynamic heaters.

These experiments have provided a very valuable knowledge for designing and operating a new test rig that will be used to develop an optimized heat generator for wastewater treatment technologies in Zero Liquid Discharge facilities (ZLD). This device will be constructed in Barcelona by the Spanish company Condorchem Envitech, S. L. with the supervision of Universitat Politècnica de Catalunya BarcelonaTech.

References

[1] Brennen, C. E. (1995). *Cavitation and bubble dynamics*. Oxford University Press
[2] Reynolds, O. (1873). *The causes of the racing of the engines of screw steamers investigated theoretically and by experiment*. Transactions of the Royal Institution of Naval Architects. 14 (56 – 67).
[3] Lord Rayleigh, O. M. (1917). *VIII. On the pressure developed in a liquid during the collapse of a spherical cavity*. The London, Edinburgh and Dublin Philosophical Magazine and Journal of Science. 34 (200).
[4] Knapp, R. T. (1955). *Recent investigations of the mechanics of cavitation and cavitation damage*. Transactions of ASME. 77 (1045).
[5] Ozonek, J., Lenik, K. (2011). *Effect of Different Design Features of the Reactor on Hydrodynamic Cavitation process*. Archives of Materials Science and Engineering. 52 (2).
[6] Gogate, P. R., Pandit, A. B. (2000). *Engineering design methods for cavitation reactors II: hydrodynamic cavitation*. American Institute of Chemical Engineers Journal 46 (8).
[7] Chakinala, A. G., Gogate, P. R., Burgess, A. E., Bremner, D. H. (2006). *Intensification of hydroxyl radical production in sonochemical reactors*. Ultrasonics Sonochemistry 14 (509).
[8] Cirimina, R., Albanese, L., Meneguzzo, F., Pagliaro, M. (2016). *Wastewater remediation via controlled hydrocavitation*. Environmental Reviews 25 (2).
[9] Petkovsek, M., Mlakar, M., Levstek, M., Strazar, M., Sirok, B., Dular, M. (2015). *A novel rotation generator of hydrodynamic cavitation for waste-activated sludge disintegration*. Ultrasonics Sonochemistry 26 (408).
[10] Martynenko, A., Astatkie, T., Satanina, V. (2015). *Novel hydrodynamic food processing technology*. Journal of Food Engineering. 152 (8).
[11] Isopo, C. (2010). *Centrifugal rotary device for heating and/or vaporizing liquids*. United States Patent. US 7647896 B2.
[12] Yoshida, Y., Sasao, Y., Okita, K., Hasegawa, S., Shimagaki, M., Nakamura, N. (2006). *Influence of thermodynamic effect on synchronous rotating cavitation*. Sixth International Symposium on Cavitation CAV2006.
[13] Petkovsek, M., Dular, M. (2015). *Experimental study on the thermodynamic effect in a cavitating flow on a simple Venturi geometry*. 9th international Symposium on Cavitation CAV2015.
[14] Cai, J., Huai, X., Liang, S., Li, X. (2010). *Augmentation of natural convective heat transfer by acoustic cavitation*. Frontiers of Energy and Power Engineering in China. 4(3).
[15] Schneider, B., Kosar, A., Kuo, C. J., Mishra, C., Cole, G. S., Scaringe, R. P., Peles, Y. (2006). *Cavitation enhanced heat transfer in microchannel*. Journal of Heat Transfer. 128 (1293).
[16] Griggs, J. L. (1993). *Apparatus for heating fluid*. United States Patents. 5188090.
[17] Pyun, K. B., Kwon, W. C., Oh, K. T., Yoon, J. Y., (2011). *Investigation of the performance for heat generator using hydrodynamic cavitation*. Proceedings of the ASME-JSME-KSME AJK-Fluids2011.
[18] Tahmid, S. M., Gill, B. S., Park, K. M., Ryu, Y., Lee, S. H. (2013). *Optimization and testing of a first generation cavitation heat pump*. 120th ASEE Annual Conference & Exposition.
[19] Mukesh, Shetty, N. P. (2015). *Design and fabrication of externally driven liquid cavitation heating system*. International Research Journal of Engineering and Technology. 02 (03).
[20] Staff of Lewis research Center (1963). *Facilities and techniques employed at Lewis research center in experimental investigations of cavitation in pumps*. Symposium on Cavitation Research Facilities and Techniques.

© 2018 ASME

10th International Symposium on Cavitation - CAV2018
Baltimore, Maryland, USA, May 14 – 16, 2018

CAV18-05083

Experimental investigation of ventilated supercavitation under unsteady conditions

[1,2]Siyao Shao; [1]Yue Wu; [1,2]Joseph Haynes; [1,2]Roger Arndt; [1,2]Jiarong Hong*

[1]*Saint Anthony Falls Laboratory, University of Minnesota, Minneapolis, MN, USA;* [2]*Department of Mechanical Engineering, University of Minnesota, Minneapolis, MN, USA*

Abstract

Ventilated supercavitation is a promising technique to achieve high-speed transport underwater by generating a gas bubble enclosing a moving object through ventilation. However, implementing this technique requires clear understanding and precise control of supercavity behaviors under unsteady conditions in practical applications. In this study, we present the systematic investigation of ventilated supercavity behaviors over a broad range of unsteady conditions. The experiments are conducted in the high-speed water tunnel at Saint Anthony Falls Laboratory. The unsteady conditions are generated using a gust generator consisting of two flapping hydrofoils mounted upstream of a forward-facing cavitator. We use high-speed imaging to capture the variation of cavity dimension and flow patterns along with simultaneous pressure measurements. Measurements are conducted under fixed tunnel speed and ventilation rate with the flapping hydrofoils operating under varying angle of attack (AoA) and frequency (f_g). The visualization and pressure signals reveal five distinct states of supercavity under unsteady conditions, referred to as stable state, wavy state, pulsating state I, pulsating state II and collapsing state. The stable state occurs under low AoA and f_g, the supercavity only exhibits small amplitude of oscillation without appreciable deformation of the cavity interface. When the supercavity is at wavy state (moderate AoA and high f_g), the supercavity displays clear periodic wavelike deformation of its surface with cavity pressure varies periodically at two times of f_g. Under high AoA with low f_g, pulsating state I of the cavity shows significant fluctuations in length with intermittent shed-off of gas pockets at the rear part of the cavity. With increase of f_g, the cavity exhibits enhanced pulsating behavior and a sharp of difference between cavity pressure and test section pressure (i.e. pulsating state II). At the highest AoA (i.e. AoA 10°) and f_g above 2 Hz, the cavity collapses. The transition across different supercavity states under a broad range of unsteady conditions is summarized in the supercavity state map, showing the dependence of supercavity states on the characteristic scales of unsteadiness in comparison to longitudinal and lateral dimensions of the supercavity.

Keywords: ventilated supercavitation; unsteady flow; flow instability

Introduction

Ventilated supercavitation, i.e., a special form of cavitation in which the bubble created by injecting non-condensable gas envelopes the entire underwater object, has gained considerable research attention as a promising way to achieve high speed underwater transportation **[1]**. Conventionally, this phenomenon has been characterized using several non-dimensional parameters, including cavitation number, $\sigma = 2(P_\infty - P_c)/(\rho_w U^2)$, Froude number, $Fr = U/\sqrt{gd_c}$, and air entrainment coefficient, $C_Q = Q/Ud_c^2$, where P_∞ and P_c refer to the test-section pressure upstream of the cavitator and the cavity pressure, respectively, ρ_w corresponds to liquid density, U is the free-stream velocity, g refers to the gravitational acceleration, d_c denotes the cavitator diameter, and Q is the volumetric ventilation rate (Notes that in our study we use the volumetric air flow rate under standard condition, i.e, Q_{As}, to define air entrainment coefficient C_{Qs}) . In practical applications, the operation of a supercavitating object faces unavoidable unsteady conditions, associated with the motions of the object as well as the variation of incoming flows due to changing in depth or surface waves, which could potentially destabilize the supercavity and adversely impact the motion of the object. Therefore, it is necessary to investigate the fundamental physics dictating the behavior of ventilated supercavity in unsteady flows in order to provide guidelines for controlling supercavitating objects under practical operational conditions. However, up to now, there is only handful of investigations in this area **[2-6]**, and no systematic experimental studies have been conducted. Logvinovich first proposed supercavity independency theory for the variation of cavity geometry in unsteady flows, i.e., every lateral section of a cavity expands relative to the trajectory of the body center almost independent of the prior or the subsequent body movement, depending only on pressure and velocity at the current moment **[2]**. Based on experiments and numerical simulation, Sememenko provided some observation of supercavity behaviors (e.g., cavity pulsating, closure variation, cavity surface waves etc.) under unsteady conditions with no systematic parameterization and in-depth analysis of flow physics **[3]**. Recently, using a

*Corresponding Author, Jiarong Hong: jhong@umn.edu

gust generator in high-speed water tunnel at the Saint Antony Falls Laboratory (SAFL), Lee et al. **[4]** investigated the periodic variation of cavity geometry under a narrow range of unsteady flow conditions. With the same setup, Karn et al. investigated the closure variation for one cavitator (10 mm in diameter) under two specific gust frequencies **[5]** as well as the ventilation demand for the formation and sustenance of a supercavity under similar unsteady flows **[6]**. Following up these prior studies **[4-6]**, here we present the first systematic investigation of the behavior of ventilated supercavity over a broad range of unsteady conditions. In the following sections, we provide a summary of our experimental methodology, the results showing distinct supercavity states obtained under different unsteady conditions, and the analysis of the mechanisms governing the occurrence of each supercavity state. A cavity state map summarizing the transition across different states will also be presented and followed by a conclusion.

Methodology

The experiments are conducted in the SAFL high-speed water tunnel. The tunnel has been used for a number of supercavitation experiments in recent years [**5-9**]. A forward-facing cavitator mounted on a ventilation pipe (FFM) is employed to generate a stable ventilated supercavity with clear interface. As shown in figure 1(a), the unsteady conditions are introduced using a gust generator consisting of two parallel NACA0020 flapping hydrofoils mounted upstream of the cavitator. The hydrofoils have a chord length of 40 mm, driven by an eccentricity flywheel via pivot arms. The pivot arms apply the periodic pitching motion to the hydrofoils, which allows the hydrofoils flapping in phase to generate a periodic flow inside the water tunnel (details provided in **[10]**). We use high speed imaging to capture the variation of supercavity dimension and flow patterns along with simultaneous measurements of the test section and the cavity pressure. Specifically, high speed imaging is performed with a Photron APX-RS high speed camera with a full resolution of 1k×1k pixels up to 3000 fps. The test section pressure is recorded at one location slightly upstream of the gust generator using Validyne AP 10-50 absolute pressure transducer. The cavity pressure (P_c) is derived by monitoring the pressure difference between the test section (P_∞) and one location inside the supercavity (ΔP) using Validyne DP 15-38 differential pressure transducer. Both pressure measurements are obtained at a sampling frequency of 1 kHz with uncertainty ± 100 Pa. The ventilated supercavity is generated using a cavitator of 10 mm in diameter with air ventilation at 10 SLPM (i.e., liter per minute of gas under standard condition) at a fixed flow speed of 8.2 m/s. During the experiments, the flapping hydrofoils of the gust generator are operated under angle of attack (*AoA*) of ±2°, ±4°, ±8°, ±10° and frequency (f_g) of 1 Hz, 1.5 Hz, 2 Hz, 3 Hz, 5 Hz, 6 Hz and 10 Hz. As shown in figure 1(b), the perturbation produced from the gust generator can be characterized with its wavelength, $\lambda = U/f_g$ and wave amplitude, $\varepsilon = V_{max}/2\pi f_g$, where the V_{max} denotes the maximum vertical speed of the gust under different *AoA*s and f_g **[4, 10, 11]**. In each gust cycle, there are two maximum perturbations induced by the hydrofoils passing when they move from their neutral position upwards and downwards, respectively. It is worth noting that for a gust of frequency f_g, the frequency of test section pressure variation occurs at $2f_g$ and 90° out of phase with respect to the maximum vertical perturbation induced by the moving hydrofoils.

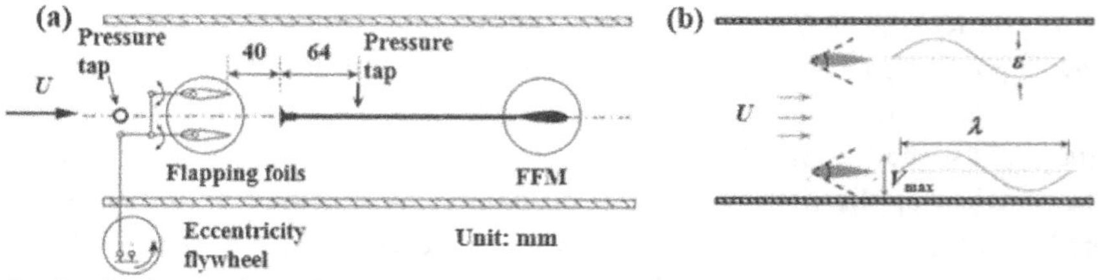

Figure 1. Schematics showing (a) the experimental setup and (b) the wavelength and amplitude of perturbation produced by the gust generator.

Results

The high-speed visualization and pressure signals reveal five distinct states of the supercavity under unsteady conditions, referred to as stable state, wavy state, pulsating state I, pulsating state II and collapsing state hereafter. The stable state cavity is observed under low *AoA* and f_g. As shown in figure 2(a), the supercavity only exhibits very small amplitude of oscillation without appreciable deformation of the cavity surface. Correspondingly, the variation of the pressure difference (ΔP) fluctuates irregularly at amplitudes equivalent to the noise level of the pressure transducer. Further, power density spectrum (PSD) of ΔP (not shown for brevity) does not show any peaks near the f_g. Overall, at stable state cavity, the perturbation introduced by flow unsteadiness is not strong enough to cause appreciable change of supercavity behavior.

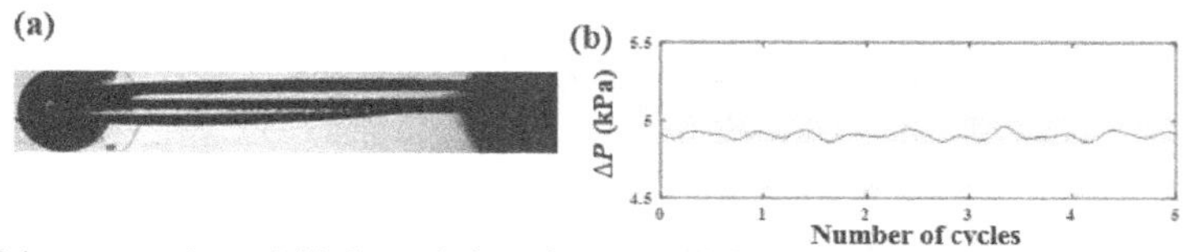

Figure 2. (a) A snapshot of the supercavity and (b) the variation of pressure difference between test section and the cavity (ΔP) at stable state.

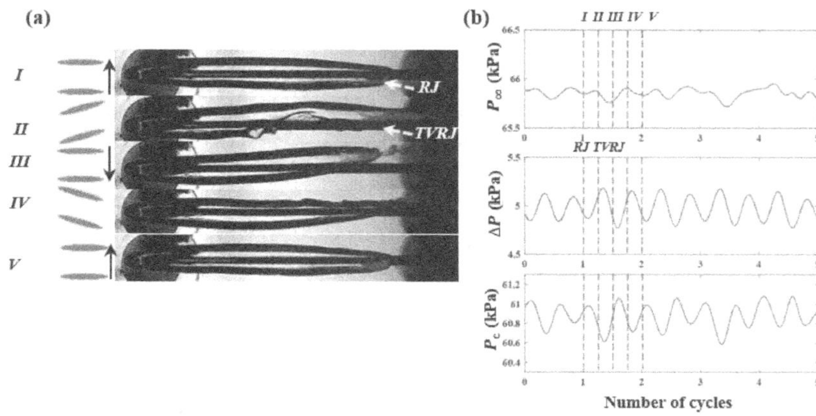

Figure 3. (a) Snapshots showing different phases of supercavity pattern corresponding to different hydrofoil positions in a gust cycle and (b) variation of pressure in test section (P_∞), pressure difference between test section and the cavity (ΔP) and cavity pressure (P_c) derived from P_∞ and ΔP at wavy state. The arrows in (a) indicate the motion of hydrofoils.

The wavy state occurs under moderate *AoA* and high f_g. The supercavity displays clear periodic wavelike undulation at its surface with a variation of closure types between re-entrant jet (RJ) and a hybrid mode of twin vortex (TV) and RJ (TVRJ) (figure 3(a)). Accordingly, as shown in figure 3(b), the ΔP and the P_c exhibit periodic variation with double peaks in each gust cycle. Particularly, the two peaks in P_c correspond to the two neural positions (i.e., *AoA*=0°, Phases ***I*** and ***III*** in the figure) of the flapping foils with a slight delay of 0.014 s matching the time required for the flow disturbance to propagate from the gust generator to the cavity pressure measurement port. Specifically, as the flapping foils start from their neutral position and move upwards (Phase ***I***), the foils have their maximum vertical velocity leading to the maximum vertical perturbation in a gust cycle. At this phase, the total loss in the water tunnel is the minimum in a gust cycle, leading to the highest tunnel velocity and local minimum of the test section pressure upstream of the gust generator. According to **[5]**, the pressure on the downstream side of the supercavity is high in the gust cycle, contributing the RJ closure mode. The maximum perturbation from the hydrofoils propagates downstream and reaches to the pressure measurement port of the cavity, causing a local maximum in the cavity pressure with 0.014 s delay. As foils moves towards their maximum upward position (Phase ***II***), the total loss in the tunnel rises, leading to a decrease of tunnel velocity and an increase of the test section pressure. Correspondingly, the perturbation from the hydrofoils causes the cavity pressure to drop from its maximum, and consequently the cavity closure transitions from RJ to TVRJ. In addition, the cavity surface deformation and the bubble shed-off from the interface due to Kevin-Helmholtz (KH) instability can be clearly observed. When the foils move downward back to their neutral location (Phase ***III***), the total loss in the tunnel decreases and the test section pressure increases accordingly. The cavity pressure begins to increase and reaches the second peak in a cycle with the same time delay as the one mentioned before. Accordingly, the closure mode shifts back to RJ from TVRJ. As the foils continue to move downwards from its neutral location (***III***), the trends of pressure variation and closure change are similar to those when the foils move from Phase ***I*** to ***II***, and the wavelike undulation of the surface occurs in an opposite phase compared to that during Phase ***I*** to ***II***.

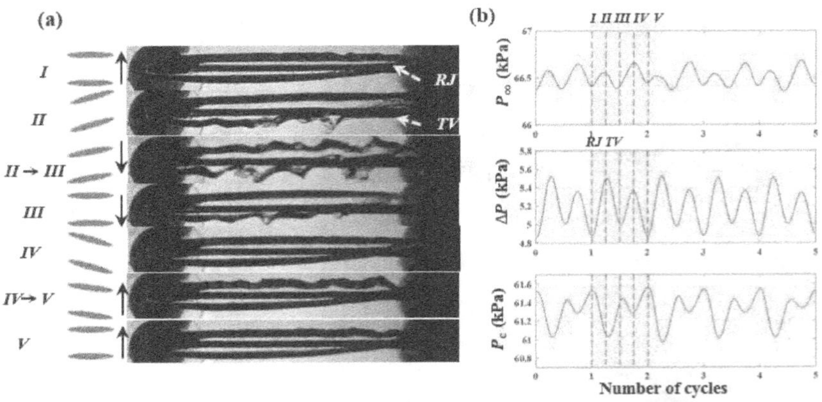

Figure 4. (a) Snapshots showing different phases of supercavity pattern corresponding to different hydrofoil positions in a gust cycle and (b) variation of P_∞, ΔP and P_c derived from P_∞ and ΔP at pulsating state I. The arrows in (a) indicate the motion of hydrofoils.

Pulsating state I is observed under high AoA and low f_g. The supercavity exhibits pulsating behavior, i.e., intermittent shed-off of gas pockets at the rear part of the cavity leading to significant fluctuation in length. As shown in figure 4(a), this process is also accompanied with wavelets generating on the cavity surfaces and the variation of closure type between TV and RJ. Due to the relatively weak vertical perturbation associated with the gust generator operating at low f_g, the supercavity at this state does not have a wavy deformation on cavity surfaces, different from that of the wavy state. Additionally, the ΔP and the corresponding P_c exhibits periodic variation with double peaks in each gust cycle, and the two peaks in ΔP (i.e., local minima of P_c) coincide with the timing of the bubble pockets shed-off when the flapping foils reach maximum AoA. In comparison to wavy state, as the hydrofoils change in a gust cycle, both the test section pressure and supercavity closure yield similar trends. However, unlike wavy state, the variation of ΔP and P_c at this stage is primarily dictated by the pressure fluctuation associated with the periodic variation of hydrofoil blockage. The propagation of such pressure fluctuation occurs at the speed of sound in water, significantly faster than the convection speed of vertical velocity perturbation induced by the hydrofoils. Therefore, little delay is observed between the phase change of hydrofoils and the signals of ΔP and P_c. Moreover, the increasing AoA of flapping foils (compared to the wave state) leads to stronger fluctuations of ΔP and P_c, which further induces significant cavity length fluctuations. Particularly, during the transition from Phases ***II*** to ***III*** and from Phases ***IV*** to ***V*** (corresponding to ***II→III*** and ***IV→V*** in figure 4(a), respectively), the cavity length shrinks drastically, resulting in enhanced KH instability and consequently the formation of clear wavelets on the cavity surface. It is also worth noting that the wavelet formation appears to be more dramatic during ***II→III*** compared to ***IV→V*** since the gravity is aligned with downward of the hydrofoils. In addition, during the pulsating state I, the closure transitions from RJ to TV, instead of TVRJ in the wavy state, due to the higher tunnel loss caused by the gust generator operated at higher AoA.

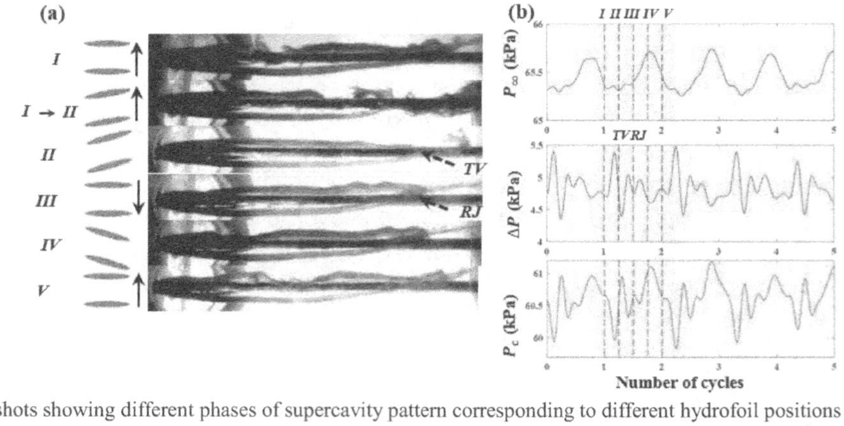

Figure 5. (a) Snapshots showing different phases of supercavity pattern corresponding to different hydrofoil positions in a gust cycle and (b) variation of P_∞, ΔP and P_c derived from P_∞ and ΔP at pulsating state II. The arrows in (a) indicate the motion of hydrofoils.

Pulsating state II is observed under *AoA* 8° with f_g of 5 and 6 Hz. Although the supercavity shows similar pulsating behavior as that in pulsating start I (figure 5(a)), ΔP displays a sharp spike followed by multiple wavelets in every gust cycle (figure 5(b)). The sharp spike results in a series of higher order spectral peaks in the corresponding PSD of the pressure signal in the frequency range above $2f_g$. Moreover, unlike pulsating state I, the spike in ΔP displays a 0.13 s delay with respect to Phase III of the hydrofoils when they are moving downward from their neural position. Similar to that in the wavy state, the signal delay matches the time required for the vertical perturbation propagating to the location where the supercavity sheds off air pockets, indicating the cavity pulsation at this state is dictated by the vertical velocity perturbation induced by the hydrofoils rather than the pressure fluctuation associated with varying blockage in pulsating state I. In addition, the variation of cavity length during pulsation (***I→II*** in figure 5(a)) is more drastic compared to that in pulsating state I (***II→III*** in figure 4(a)), which results in the abrupt drop of P_c (i.e., the spike in ΔP). Noteworthily, different from the previous state, the pulsation of the cavity in each gust cycle is always initiated from the upper surface since it is more susceptible to the KH instability due to the gravity effects.

Figure 6. A snapshot of the supercavity under collapsing state.

At the highest *AoA* (i.e., *AoA* 10°) and f_g above 2 Hz, the supercavity eventually collapses under a fixed ventilation rate and tunnel speed. At this state, the unsteadiness induced by the gust generator is so strong and rapid such that the supercavity does not have sufficient time to recover through ventilation after its breakup. Note that the reliable pressure measurements cannot be obtained due to water splashing near cavity pressure tap.

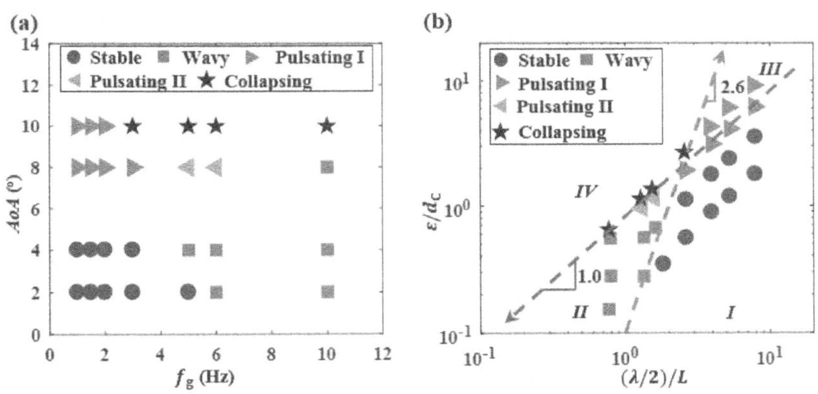

Figure 7. (a) The supercavity state map over a broad range of f_g and *AoA*, and (b) $(\lambda/2)/L$ and ε/d_C with Fr=26.5 and C_{Qs}=0.19.

The transition across different supercavity states under a broad range of unsteady conditions is summarized in the supercavity state map (figure 7). As shown in figure 7(a), with fixed f_g, the increase of *AoA* enhances the amplitude of the pressure fluctuation in a gust cycle, leading to the pulsating and eventually the collapsing of the supercavity. Similarly, under fixed *AoA*, increasing f_g amplifies the vertical velocity perturbation which results in augmenting the waviness of in the cavity geometry (i.e., stable to wavy, pulsating I to II). To further elucidate the key trends in the supercavity state transition, the state map is also presented with half wavelength ($\lambda/2$) and wave amplitude (ε), normalized by the cavity length under steady condition (L) and cavitator diameter (d_C) in logarithm scale, respectively (figure 7(b)). The non-dimensionalization of wave parameters is intended to compare the perturbation produced from the gust generator with the characteristic length scales of the supercavity. Two dashed lines, termed as critical unsteady condition lines in figure 7(b), are generated by least square fitting of the data points at the border demarcating different cavity states, which divide the map into four regions of different state. The blue dashed line with the arrow represents a dominant effect of wave amplitude (i.e, power index of 2.6) that leads to the supercavity collapsing, while moving along the red line in the direction of the arrow shows a substantial effect of decreasing wavelength that causes cavity to collapse through pulsating states. Moreover, in Region I, the supercavity is stable due to the limited effect of flow unsteadiness in terms of both the amplitude and wavelength of the perturbations. When the scale of wavelength (precisely, half wavelength) is comparable with the cavity length (i.e, $(\lambda/2)/L{\sim}1$) under limited amplitude (i.e., $\varepsilon/d_C < 1$), the cavity enters into wavy state in Region II. Comparatively, when wave amplitude increases above the characteristic lateral scale of the cavity (i.e., $\varepsilon/d_C > 1$) with limited longitudinal perturbation characterized by the wavelength, the cavity transition to pulsating state I (Region III). The supercavity collapsing (Region IV) can be achieved when both the amplitude and the wavelength of the perturbations become sufficiently strong (i.e., $(\lambda/2)/L \lesssim 1$ and $\varepsilon/d_C > 1$).

Conclusion

Five distinct states (stable state, wavy state, pulsating state I, pulsating state II and collapsing state) are observed in a systematic investigation of cavity behaviors in the unsteady flow. We attribute different cavity behaviors under varying unsteady conditions to the effects of vertical perturbation induced by gust generator motion and/or pressure fluctuation caused by varying blockage. The cavity state map suggests that the cavity states is dependent on the characteristic scale (i.e., half wavelength and wave amplitude) of the perturbation comparing to the cavity dimension. Future investigation will be focused on the influence of ventilation conditions (e.g., ventilation rate and cavitator size) on the cavity behaviors.

References

[1] Nesteruk, I. (2012). *Supercavitation.* Springer-Verlag, Berlin Heidelberg.
[2] Logvinovich, G. V. (1969) *Hydrodynamics of flows with free boundaries* Naukova Dumka, Kiev. ch.5.
[3] Semenenko, V.N. (2001) Artificial supercavitation: physics and calculation. Ukrainian academy of sciences, Kiev Institute of hydromechanics.
[4] Lee, S.J., Kawakami. E., Arndt, R.E.A. (2013). *Investigation of the behavior of ventilated supercavities in a periodical gust flow.* Journal of Fluid Engineering. 135(8).
[5] Karn, A., Arndt, R.E.A., Hong, J. (2015). *Dependence of supercavity closure upon flow unsteadiness.* Experimental Thermal and Fluid Science. 68.
[6] Karn, A., Arndt, R.E.A., Hong, J. (2016a). *Gas entrainment behavior in the formation and collapse of a ventilated supercavity.* Experimental Thermal and Fluid Science. 79.
[7] Karn, A., Arndt, R.E.A., Hong, J. (2016b). *An experimental investigation into supercavity closure mechanisms.* Journal of. Fluid Mechanics, 789.
[8] Shao, S., Karn, A, Ahn, B-K., Arndt, R.E.A., Hong, J. (2017). *A comparative study of natural and ventilated supercavitation across two closed-wall water tunnel facilities.* Experimental Thermal and Fluid Science. 88.
[9] Jiang, Y., Shao, S., Hong, J. (2018). *Experimental investigation of ventilated supercavitation with gas jet cavitator.* Physics of Fluids, 30.
[10] Korpriva, J., Arndt, R.E.A., Amromin. E.L. (2008). *Improvement of hydrofoil performance by partial ventilated cavitation in steady flow and periodic gusts.* Journal of Fluid Engineering. 130(3).
[11] Newman, J. N. (1977). *Marine hydrodynamics.* The MIT Press, Cambridge, MA.
[12] Chandrasekhar, S. (1961). *Hydrodynamic and hydromagnetic stability.* Dover Publications, NY.

© 2018 ASME

Experimental investigation on destructive potential of shock-induced bubble collapse

[1]Zhaoguang Wang *; [1]Thomas Hopfes; [1]Marcus Giglmaier; [1]Nikolaus Adams

[1]Institute of Aerodynamics and Fluid Mechanics, Technical University of Munich, Boltzmannstr. 15, D-85747 Garching, Germany

Abstract

The present study experimentally investigates shock-induced collapse of bubbles near boundaries of different plasticity and the resulting destructive erosion. Air bubbles held in gelatin are placed close to either a solid wall or a soft clay layer. High pressure impulse generated by a shock tube is employed to trigger bubble collapse. The bubble deformation is visualized by a focused shadowgraph system, and the damage caused to the boundary surface is closely examined as well. The collapse of bubbles near the soft clay layer shows differences from that near the solid wall in both qualitative description of bubble deformation and quantitative measurement of bubble dynamics. The surface of the clay is marked by the bubble collapse with a volcano-shaped imprint and some material floats into the gelatin in a mushroom shape. In this ongoing investigation, refined results under well-defined and uniform initial conditions will be provided and well suited for validation of numerical models and approaches.

Keywords: bubble collapse; cavitation erosion; shock wave

Introduction

The collapse of gas bubbles close to a boundary plays a key role in diverse research fields, such as the evaluation of cavitation damage to underwater propellers of ships [1] and the fragmentation of kidney stones in extracorporeal shock-wave lithotripsy (ESWL) [2, 3]. A deeper insight into the associated destructive mechanism helps to develop new techniques, especially in medical fields such as drug targeting and gene delivery [4]. Originating from the same physical phenomenon, topics with different emphases have been brought forward and investigated.

Bubble dynamics in the collapse near a solid boundary, which intends to analogize the surface of ship propellers or kidney stones, has been studied by a vast number of research. Benjamin & Ellis [5] are among the pioneers introducing the aspherical collapse of near-wall bubbles and demonstrating the existence of a liquid jet induced towards the wall. The liquid jet piercers through the bubble and may further impinge on the wall [6, 7]. Shima *et al.* [8] show the strength of the liquid jet is directly affected by the bubble-to-wall distance and decreases as the distance increases. Jones & Edwards [9] and Naudé & Ellis [10] first notice the emission of extra shock waves during bubble collapse. Ohl *et al.* [11] further explain such shock waves as results of jet impact upon the opposite side of the bubble and the collapse of the remaining toroidal bubble. The induced shock waves expand nearly spherically with the strength decaying and hit the wall normally [8, 12]. The appearance of counter jets directed away from the boundary at the second collapse of bubbles is initially shown by Lauterborn & Bolle [13]. Lindau & Lauterborn [14] find the counter jet disappears if the laser-produced bubble gets in contact with the boundary when expanding to the maximum radius.

Another topic which deals with bubble collapse near soft materials gains increasing interests as to evaluate the damage to soft human tissues in ESWL. Gibson & Blake [15] investigate collapse of a spark-generated bubble attached to the surface of vulcanized rubber and observe no liquid jet developed and the bubble splitting into separate parts. They conclude that the bubble behavior is notably influenced by the inertia and the stiffness of nearby boundaries. A laser-induced bubble collapsing near an elastic material which consists of a polyacrylamide mixed with water is studied by Brujan *et al.* [16]. Interesting features such as the formation of mushroom-shaped bubbles, an annular jet splitting the bubble and a pair of axial jets directed towards each other are observed. The work also shows important effects of the elastic modulus of the material and the stand-off distance on the bubble deformation. Similar results are obtained by Tomita & Kodama [17] who employ a composite surface with either rigid or soft plates glued on a foam rubber and Shaw *et al.* [18] and Turangan *et al.* [19] who replace elastic boundaries with elastic membranes.

Apart from the bubble dynamics, study of the damage pattern caused by the bubble collapse on the nearby boundaries also draws attentions. Tomita & Shima [20] and Philipp & Lauterborn [21] investigate the effect of the stand-off distance on the boundary erosion and show similar circular pit shapes. Influence of the physical properties of boundaries is studied by Kodama & Takayama [22] who conclude that even gelatin and rat livers share similar acoustic impedance, the discrepancy in their elasticity results in different damage patterns. A few researchers [23, 24] also

*Corresponding Author, Zhaoguang Wang: zg.wang@tum.de

show that another factor determining the erosion is the way to generate bubbles, such as acoustic and hydrodynamic means, electric discharge and optical breakdown.

Although extensive research has been conducted concerning bubble collapse as shown above, to the authors' best knowledge, it still remains a question to answer: what is the dominate destructive mechanism. Impingement of the liquid jet, collapse of the main bubble and collapse of the tiny bubble clouds induced by the liquid jet all contribute to the cavitation damage to certain extent, but no agreement has been achieved on the relative significance of these factors. New features in the bubble dynamics when a soft boundary appears nearby add even more complexity to this problem. Motivated to cast slightly more light on understandings of the underlying physics, the current work studies the shock-induced collapse of air bubbles in gelatin close to different boundaries, namely a solid wall and a soft clay layer. The radius change and the centroid movement of bubbles in two cases are compared. Furthermore, surface damage caused by nearby bubble collapse is analyzed.

Experimental setup

To acquire an impulsive high pressure for bubble collapse, experiments are carried out in a shock tube with an inner diameter of 290 mm. Figure 1 depicts the layout of the shock tube and the corresponding measurement system. The 3 m long driver section and the 19.5 m long driven section are separated by a metal diaphragm which is placed against a downstream X-shaped sharp blade. A 0.5 m long test section, of which the cross section is transformed by a cookie-cutter to a 190 mm×190 mm square, is attached to the end of the driven section. In the experiments, air is pumped into the driver section continuously. When the critical pressure (8.0 bar for the current setup) is reached, the diaphragm breaks and a shock wave develops rapidly propagating towards the test section.

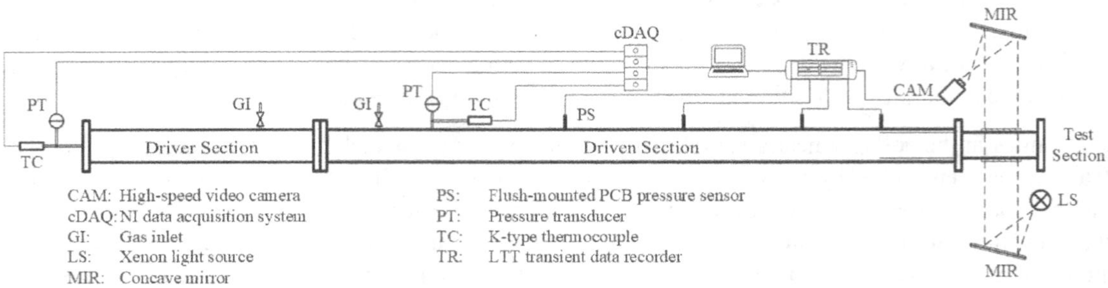

Figure 1: Experimental arrangement of the shock tube for bubble dynamics experiments

Initial experimental pressure and temperature conditions are measured by pressure transducers and K-type thermocouples, respectively. The output signals are acquired by a NI® cDAQ device at a sampling rate of 1 kHz. A few meters upstream of the test section, four PCB Piezotronics ICP® fast-response pressure sensors are flush-mounted on the shock tube, monitoring pressure changes inside the tube during the experiments. The 4 MHz transient pressure data are recorded by a LTT® data acquisition device.

A highly sensitive Z-type schlieren system is built to visualize the flow field. A 150 W xenon lamp serves as the light source and a Shimazu® Hyper Vision HPV-X ultra-high-speed camera (250×400 pixels with a frame rate of up to 5 MHz) records 128 continuous frames of the bubble deformation. For the present results, the knife edge in the optical system is removed to guarantee adequate brightness of images captured. Therefore, the original schlieren system is simplified into a focused shadowgraph.

The sketch in fig. 2 describes a new experimental test section that will be shown at the conference. However, due to delays in the manufacturing process, a workaround with an acrylic box placed inside an already existing test chamber is applied to obtain preliminary results. Deformation of the acrylic box and extra wave motion in the surrounding air of the box bring undesired effects. But main features of the bubble collapse are well captured.

In terms of the new test section, the downstream 400 mm-long part is filled with gelatin, which is made from a mixture of distilled water, Gelrite® Gellen gum and magnesium sulfate with the mass ratio of 10000 : 6 : 5. The rear wall is either exposed to serve as a solid boundary or covered with a 4 mm-thick layer of soft clay. An air bubble is produced carefully with a syringe close to the boundary surface under atmospheric conditions ($p_0 \approx 1$ bar). The solidity and the viscosity of the water-like gelatin contribute to hold the air bubble at a fix position although increasing the difficulty in producing a perfectly spherical bubble. One thing to notice is that the gelatin is a non-Newtonian fluid but behaves like water under high stresses.

The *x-t* diagram in fig. 2 illustrates the wave motion inside the test section and the temporal pressure distribution involved. When the incident shock (**S**) generated by the shock tube arrives at the air-gelatin interface, it is reflected back almost ideally (**RS**) due to the large acoustic impedance difference between air and the gelatin. Based on the pressure balance at the air-gelatin interface, a compression wave (**C**) travels through the gelatin at the sound speed of around 1500 m/s. Since the bubble-to-wall distance is small, the pressure field surrounding the bubble is raised up to a considerably high value by the upcoming compression wave (**C**) and the subsequent reflected compression wave (**RC**) in a negligibly short time. The far-field pressure for the bubble collapse remains constant until an expansion wave (**E**), which originates from the interaction of the reflected compression wave (**RC**) with the interface, arrives at the air bubble. For the example shown here, a gelatin filled test chamber with a length of ℓ = 400 mm gives a test window of Δt = 0.5 ms, which is sufficient for the primary (and secondary) collapse of a bubble with a diameter of d = 10 mm under a surrounding pressure of p = 10 bar as estimated by the Rayleigh-Plesset equation.

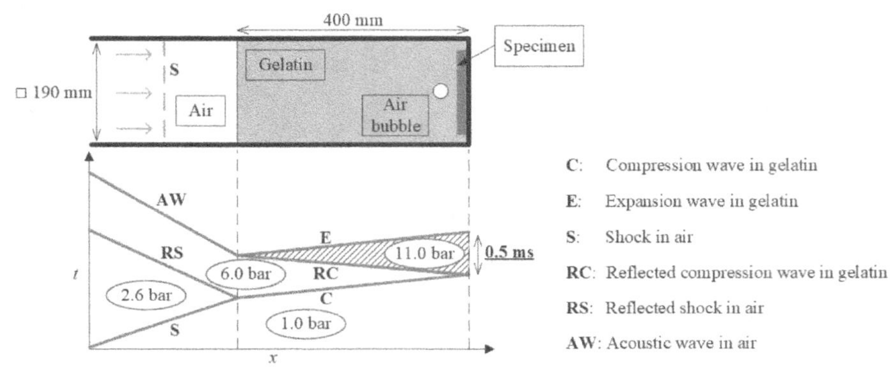

Figure 2: Sketch of the new test section with a wave motion diagram (8.0 bar in the driver section and 1.0 bar in the driven section)

Preliminary results and discussions

As aforementioned, the present work was carried out with the workaround setup which accounts for major disturbances in shadowgraph images displayed below. Also the observations to be discussed are confined to the current specific cases. The uncertainties and the repeatability will be evaluated with more upcoming experiments using the new test chamber as described in fig. 2.

Air bubble collapse close to two different types of boundaries is studied. One is a solid wall, and the other is a 4 mm-thick soft clay layer. In both experiments, the relative stand-off distance $\gamma = s/R_0$ (s: the distance from the bubble center to the boundary surface, R_0: the initial bubble radius) is similar. Therefore, it is possible to make quantitative comparisons between these cases. For all following images, considering the asymmetric deformation of bubbles, the radius of an equivalent spherical bubble consisting of the same amount of dark pixels is calculated to represent R.

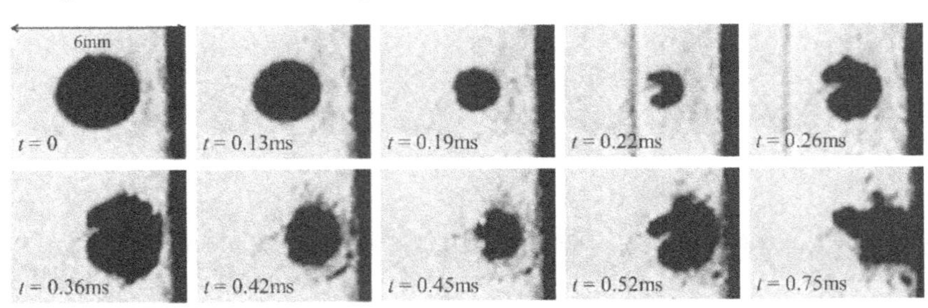

Figure 3: Bubble deformation close to the solid wall (R_0 = 2.87 mm, γ = 1.75)

Figure 3 shows the bubble deformation close to the solid wall with R_0 = 2.87 mm and γ = 1.75. Images are recorded by the focused shadowgraph system at a sampling rate of 100000 frames/s. The bubble starts to shrink after the surrounding pressure is raised instantaneously by shock waves. At time t = 0.19 ms, the off-wall side of the bubble has moved through a considerably longer distance than the near-wall side, and becomes flat or even slightly concave at middle. At t = 0.22 ms, the bubble collapses to the minimum volume. The curved off-wall surface of the bubble clearly indicates the development of a liquid jet. The straight dark lines in this and the next images are shock waves in the surrounding air of the acrylic box. At t = 0.36 ms, the re-expansion ends and the bubble recovers to nearly the initial size. The small-scale irregularities at the bubble surface facing towards the boundary suggest the liquid jet has impacted on it. Due to the aspherical bubble shape at the beginning of the second collapse, the bubble deforms into an apple shape at t = 0.45 ms. After more succeeding expansions and compressions, the bubble lands on the boundary surface at t = 0.75 ms. A crowd of smaller bubbles, which are shed off from the main bubble, are identified.

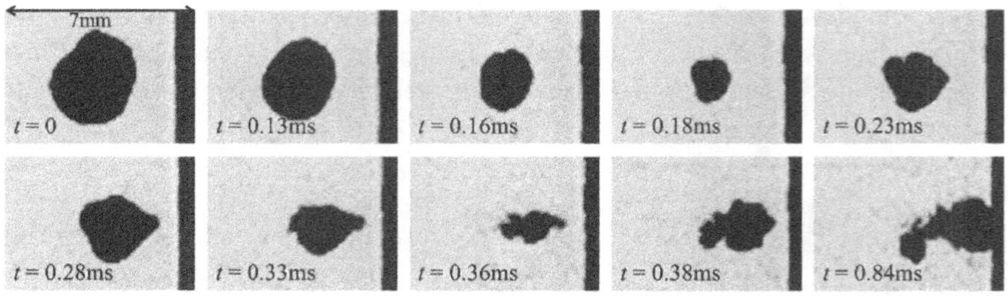

Figure 4: Bubble deformation close to the soft clay layer (R_0 = 2.95 mm, γ = 1.85)

Figure 4 shows the bubble deformation near the soft clay with R_0 = 2.95 mm and γ = 1.85. The interframe time of the full video is Δt = 0.01 ms. At t = 0.16 ms, a liquid jet starts to develop as the off-wall surface of the bubble turns flat. The minimum volume of the first collapse is reached 0.02 ms later. The bubble finishes the re-expansion at t = 0.28 ms and a "nose" extrudes out of the bubble suggesting the penetration of the liquid jet. Then the funnel-shaped bubble begins to collapse again and tends to split into two parts at t = 0.36 ms. The larger part expands further downstream and finally arrives at the boundary at t = 0.84 ms, while the center of the smaller part floats towards the opposite direction slightly.

Although liquid jets appear in both cases, bubbles in fig. 3 and fig. 4 exhibit quite differently as getting closer to the boundary in the second collapse and the following oscillations. Moreover, no shock wave emission is observed during either bubble collapse. The low trigger pressure p_∞ around the bubbles (p_∞/p_0 ~ 10) and the insufficient framing rate (100000 frames/s) are responsible for this.

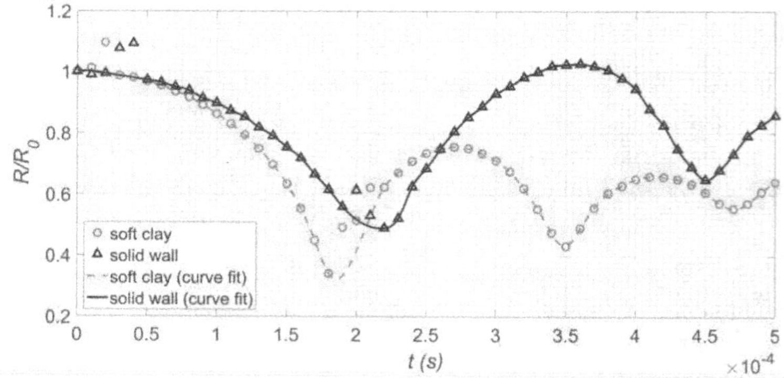

Figure 5: Comparison of the bubble radius changes between two cases (for the soft clay t_c = 0.18 ms, R_{min}/R_0 = 0.34; for the solid wall t_c = 0.22 ms, R_{min}/R_0 = 0.49)

The temporal bubble radius changes of the two cases are compared in fig. 5, where R is normalized against the initial bubble radius R_0. Both curves in the figure experience oscillations with the attenuation of amplitude. It is noticed that the bubble close to the solid wall re-expands to a radius slightly higher than the initial radius (for the dark solid line, $R/R_0 > 1$ at t = 0.35 ms). Such an unexpected behavior could result from the irregularity of the bubble shape and the lack of information about internal bubble geometries from the two-dimensional images. Nevertheless, clear differences can still be distinguished between two cases. The bubble close to the soft clay collapses faster (collapse time $t_c = 0.18$ ms) than that near the solid wall ($t_c = 0.22$ ms) (note that the former case even has a slightly larger initial bubble $R_0 = 2.95$ mm *vs* 2.87 mm), achieves a lower minimum normalized radius ($R_{min}/R_0 = 0.34$ *vs* 0.49) and has a stronger decay of the bubble oscillation. This could be explained by the previous knowledge that the pressure around the bubble is reduced more with the presence of a nearby solid boundary than a soft one.

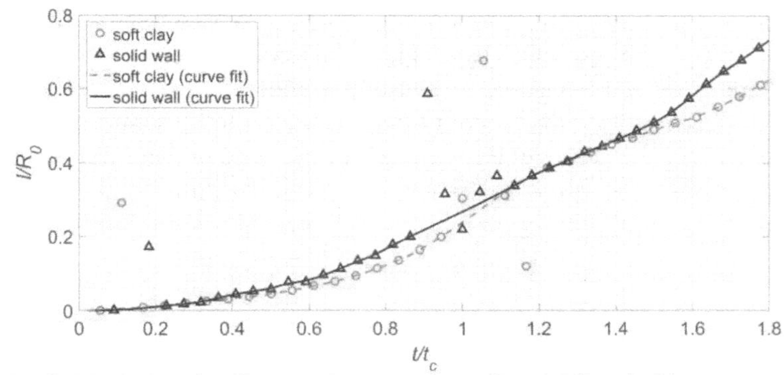

Figure 6: Comparison of the bubble centroid movement between two cases (within the first collapse $t/t_c < 1$, l/R_0 for the solid wall is higher than that for the soft clay, representing a stronger attraction for the bubble)

Comparison of bubble centroid movement is made between two cases in fig. 6, where the bubble centroid is determined as the area-weighted position of the dark pixels constituting the bubbles. The time t and the displacement l of the bubble centroids relative to the initial location are normalized by the corresponding collapse time t_c and the initial bubble radius R_0, respectively. Both bubbles travel to the boundary continuously but at different paces. During the first collapse ($t/t_c < 1$), the bubble close to the solid wall shifts more, while both bubbles share overlapped locations during the re-expansion period (roughly with t/t_c from 1 to 1.55). Overall speaking, because of the lack of deformability, the solid wall attracts nearby bubbles more.

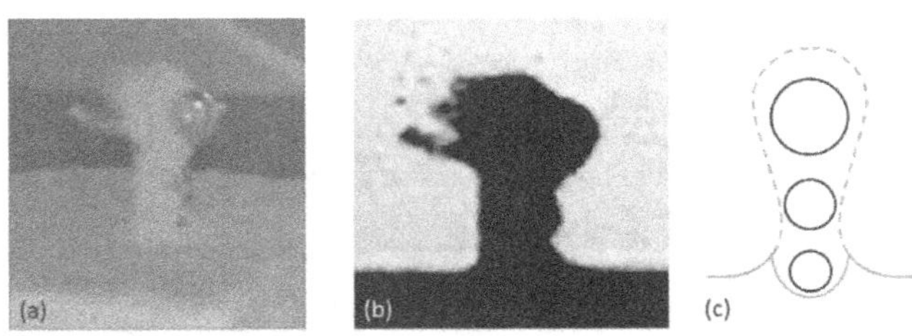

Figure 7: Surface damage of the soft clay layer caused by the bubble collapse: (a) a full-color image, (b) a shadowgraph image and (c) a simplified sketch

Apart from the bubble dynamics, the damage caused by the bubble collapse on the boundary is another concern. In terms of the solid wall, no visible imprint of damage is observed. As for the soft clay layer, however, distinct deformation is resulted. As shown in fig. 7, the bubble finally splits into three parts, with the smallest one inserted into the soft clay. Suffered from the crash of the bubbles, the surface of the boundary deforms into a volcano-shaped dimple with a mushroom-shaped clay cloud distributing around. This is a clear indication of the destructive potential of bubble collapse, which would have a significant impact on drug targeting and gene delivery. Due to the lack of physical properties of boundaries, further quantitative analysis is not given here.

Conclusion and outlooks

Air bubble collapse close to different boundaries, namely a solid wall and a soft clay layer, was studied experimentally. Major difference in the pattern of the bubble deformation was found in the second collapse, where the bubble next to the solid wall deformed into an apple shape, while in the other case the bubble split into two smaller parts. Quantitative analysis of the bubble dynamics further indicated that the bubble near the soft clay collapsed faster to a lower minimum radius and showed a stronger reduction in the amplitude of subsequent oscillations. The overall trend of the bubble centroid movement was similar in both experiments, except that the bubble moved faster towards the boundary when the solid wall was employed. In addition, the surface of the soft clay was left with a volcano-shaped dimple by the bubble collapse and some mass fell off forming a mushroom-shaped cloud surrounding the bubbles. Note that all the observations are only based on the present particular cases. More experiments will be conducted to quantify uncertainties and to estimate the repeatability before asserting any conclusion.

Since the current preliminary experimental rig was not built intentionally to investigate bubble dynamics, some important information about the experiments, like the pressure change around the bubble, is not obtainable. A new test chamber especially planned for these experiments is in preparation (as described in fig. 2), which has a stronger structure to allow for a higher operating pressure, larger glass windows with higher optical qualities to visualize the flow field better, and a more flexible arrangement to control positions of bubbles more accurately and to render a wider variety of boundaries possible. Therefore, more highly-resolved outcomes are expected in the near future and will be shown on the conference. Considering the new setup is capable to provide well-defined and uniform initial conditions for the bubble collapse, the upcoming refined results could serve to validate numerical models and approaches.

References

[1] Knapp, R. T. (1955). Recent investigations of the mechanics of cavitation and cavitation damage. *Transactions of the ASME*, *77*, 1045-1054.

[2] Crum, L. A. (1988). Cavitation microjets as a contributory mechanism for renal calculi disintegration in ESWL. *The Journal of urology*, *140*(6), 1587-1590.

[3] Delius, M., Brendel, W., & Heine, G. (1988). A mechanism of gallstone destruction by extracorporeal shock waves. *Naturwissenschaften*, 75(4), 200-201.

[4] Unger, E. C., Porter, T., Culp, W., Labell, R., Matsunaga, T., & Zutshi, R. (2004). Therapeutic applications of lipid-coated microbubbles. *Advanced drug delivery reviews*, *56*(9), 1291-1314.

[5] Benjamin, T. B., & Ellis, A. T. (1966). The collapse of cavitation bubbles and the pressures thereby produced against solid boundaries. *Philosophical Transactions of the Royal Society of London A: Mathematical, Physical and Engineering Sciences*, *260*(1110), 221-240.

[6] Brujan, E. A., Keen, G. S., Vogel, A., & Blake, J. R. (2002). The final stage of the collapse of a cavitation bubble close to a rigid boundary. *Physics of Fluids*, *14*(1), 85-92.

[7] Klaseboer, E., Hung, K. C., Wang, C., Wang, C. W., Khoo, B. C., Boyce, P., ... & Charlier, H. (2005). Experimental and numerical investigation of the dynamics of an underwater explosion bubble near a resilient/rigid structure. *Journal of Fluid Mechanics*, *537*, 387-413.

[8] Shima, A., Takayama, K., Tomita, Y., & Miura, N. (1981). An experimental study on effects of a solid wall on the motion of bubbles and shock waves in bubble collapse. *Acta Acustica united with Acustica*, *48*(5), 293-301.

[9] Jones, I. R., & Edwards, D. H. (1960). An experimental study of the forces generated by the collapse of transient cavities in water. *Journal of Fluid Mechanics*, *7*(4), 596-609.

[10] Naudé, C. F., & Ellis, A. T. (1961). On the mechanism of cavitation damage by nonhemispherical cavities collapsing in contact with a solid boundary. *Journal of Basic Engineering*, *83*(4), 648-656.

[11] Ohl, C. D., Philipp, A., & Lauterborn, W. (1995). Cavitation bubble collapse studied at 20 million frames per second. *Annalen der Physik*, *507*(1), 26-34.

[12] Klaseboer, E., Fong, S. W., Turangan, C. K., Khoo, B. C., Szeri, A. J., Calvisi, M. L., ... & Zhong, P. (2007). Interaction of lithotripter shockwaves with single inertial cavitation bubbles. *Journal of fluid mechanics*, *593*, 33-56.

[13] Lauterborn, W., & Bolle, H. (1975). Experimental investigations of cavitation-bubble collapse in the neighbourhood of a solid boundary. *Journal of Fluid Mechanics*, *72*(2), 391-399.

[14] Lindau, O., & Lauterborn, W. (2003). Cinematographic observation of the collapse and rebound of a laser-produced cavitation bubble near a wall. *Journal of Fluid Mechanics*, *479*, 327-348.

[15] Gibson, D. C., & Blake, J. R. (1982). The growth and collapse of bubbles near deformable surfaces. *Applied Scientific Research*, *38*(1), 215-224.

[16] Brujan, E. A., Nahen, K., Schmidt, P., & Vogel, A. (2001). Dynamics of laser-induced cavitation bubbles near an elastic boundary. *Journal of Fluid Mechanics*, *433*, 251-281.

[17] Tomita, Y., & Kodama, T. (2003). Interaction of laser-induced cavitation bubbles with composite surfaces. *Journal of applied physics*, *94*(5), 2809-2816.

[18] Shaw, S. J., Jin, Y. H., Gentry, T. P., & Emmony, D. C. (1999). Experimental observations of the interaction of a laser generated cavitation bubble with a flexible membrane. *Physics of Fluids*, *11*(8), 2437-2439.

[19] Turangan, C. K., Ong, G. P., Klaseboer, E., & Khoo, B. C. (2006). Experimental and numerical study of transient bubble-elastic membrane interaction. *Journal of Applied Physics*, *100*(5), 054910.

[20] Tomita, Y., & Shima, A. (1986). Mechanisms of impulsive pressure generation and damage pit formation by bubble collapse. *Journal of Fluid Mechanics*, *169*, 535-564.

[21] Philipp, A., & Lauterborn, W. (1997). Damage of solid surfaces by single laser–produced cavitation bubbles. *Acta Acustica united with Acustica*, *83*(2), 223-227.

[22] Kodama, T., & Takayama, K. (1998). Dynamic behavior of bubbles during extracorporeal shock-wave lithotripsy. *Ultrasound in medicine & biology*, *24*(5), 723-738.

[23] Dular, M., Delgosha, O. C., & Petkovšek, M. (2013). Observations of cavitation erosion pit formation. *Ultrasonics sonochemistry*, *20*(4), 1113-1120.

[24] Tzanakis, I., Eskin, D. G., Georgoulas, A., & Fytanidis, D. K. (2014). Incubation pit analysis and calculation of the hydrodynamic impact pressure from the implosion of an acoustic cavitation bubble. *Ultrasonics sonochemistry*, *21*(2), 866-878.

© 2018 ASME

Experimental Investigations of Cavitation Breakdown in an Axial Waterjet Pump

[1]Huang Chen; [1]Nick Doeller; [1]Yuanchao Li; [1]Joseph Katz*;

[1]*Johns Hopkins University, Baltimore, MD, USA*

Abstract

This paper provides detailed experimental data on the changes to the flow structure and cavitation occurring in an axial waterjet pump as the pressure is reduced causing cavitation breakdown. The database includes performance curves, high-speed images of the cavitation and velocity distributions inside the blade passage obtained using Stereo-PIV (SPIV). The corresponding pressure distribution is estimated using Bernoulli's Equation and the changes to the velocity from the inlet to the interior of the passage. Results show that cavitation breakdown is initiated by the rapid expansion of the attached cavitation on the blade suction side (SS) into the blade overlap region, blocking part of the entrance to the passage, and increasing the velocity along the pressure side (PS) of the blade. The resulting reduction in PS pressure along with the vapor pressure in the attached SS cavitation reduces the blade loading. Furthermore, entrainment of the cloud cavitation by the tip leakage vortex (TLV) also generates large scale cavitating vortical structures, which are aligned perpendicularly to the SS surface, and extend to the next blade. They reduce the throughflow in the tip region. The combined effects of reduced blade loading and tip blockage contribute to the cavitation breakdown.

Keywords: cavitation breakdown; attached cavitation; SPIV

Introduction

Cavitation breakdown refers to the rapid degradation in performance occurring when an axial pump operates below some critical low inlet pressure. Numerous past studies have attributed cavitation breakdown to different causes, such as condensation shocks in the cloud cavitation [1], flow "chocking" when the attached cavitation enters the blade overlap region [2] or reaches to the blade trailing edge [3], large scale cavitating vortical structures blocking the flow in the tip region [4], and the rapid loss of net positive pressure on the PS of the blade caused by the growing cavity [5]. In this paper, performance tests, high-speed visualizations of cavitation and SPIV measurements in the non-cavitating part of the passage are carried out to elucidate the changes to the flow field during the cavitation breakdown. The measured velocity upstream and in the passage are used for estimating the pressure distribution in the wetted part of the blade passage, including the blade PS, to elucidate the changes to the blade loading as the cavitation on the SS surface expands with decreasing cavitation index. The results elucidate the chain of events leading to a decrease in blade loading as the attached cavitation along the SS expands into the blade overlap region, and entrainment of the cloud cavitation by the TLV blocks the tip region.

Experimental setup

The experiments have been carried out in the JHU refractive index matched facility described in a series of previous papers, e.g. [4]. The AxWJ-2 water jet pump (Figs.1 a&b) designed by Michael [6] has 6 rotor blades and 8 stator blades. The rotor has a diameter of 303.8mm and tip chord length of c=274.3mm. When operating at 900rpm, the Reynolds number based on the tip velocity and chordlength is 3.6×10^6. The setup for SPIV measurements is shown in Figs. 1a-c. To measure the flow in the non-cavitating part of the rotor passage during cavitation breakdown, the laser sheet is aligned almost perpendicularly to the SS surface at the entrance to the blade overlap region, and intersects with the blade PS (Figs. 1a&b). A cylindrical coordinate system (r, θ, z) whose origin is located at the center of the pump and coincides with the blade leading edge (LE) plane, and θ=0 is aligned in the vertical direction, is used along with the corresponding velocity components (u_r, u_θ, u_z). The coordinate system of the SPIV plane is defined as (x', y', z'), with (x', y') being the in-plane components, as illustrated in Figs.1c&d. Its origin is located at θ_{PIV}=-37.5° (Figs.1 b&c), r/R = 0.995 (Figs. 1d) and z=0. The z' axis inclined by β=40° to the z direction (Figs. 1a). The resulting triangular sample area shown in Figs. 1c&d extends from r/R=1 to r/R =0.66 (R is the casing radius), covering 50% of the blade span. By fixing the laser sheet while changing the blade orientation, data are recorded in 4 planes that intersect with the blade at different distances from the blade LE. In each plane, 200 image pairs are

*Corresponding Author, Joseph Katz: katz@jhu.edu

recorded by a pair of 2048 × 2048 pixels CCD camera at several inlet pressures. All velocity components obtained in the SPIV plane (u, v, w) are subsequently transformed into the pump cylindrical coordinates, and the corresponding ensemble-averaged results are denoted as (U_r, U_θ, U_z). Further details can be found in [7]. A sample distribution of $|U_z|/U_T$ before cavitation breakdown is shown in Fig.1d. As is evident, the axial velocity is low near the blade PS tip and along the endwall casing due to the influence of the tip leakage vortex (TLV) and the endwall casing boundary layer.

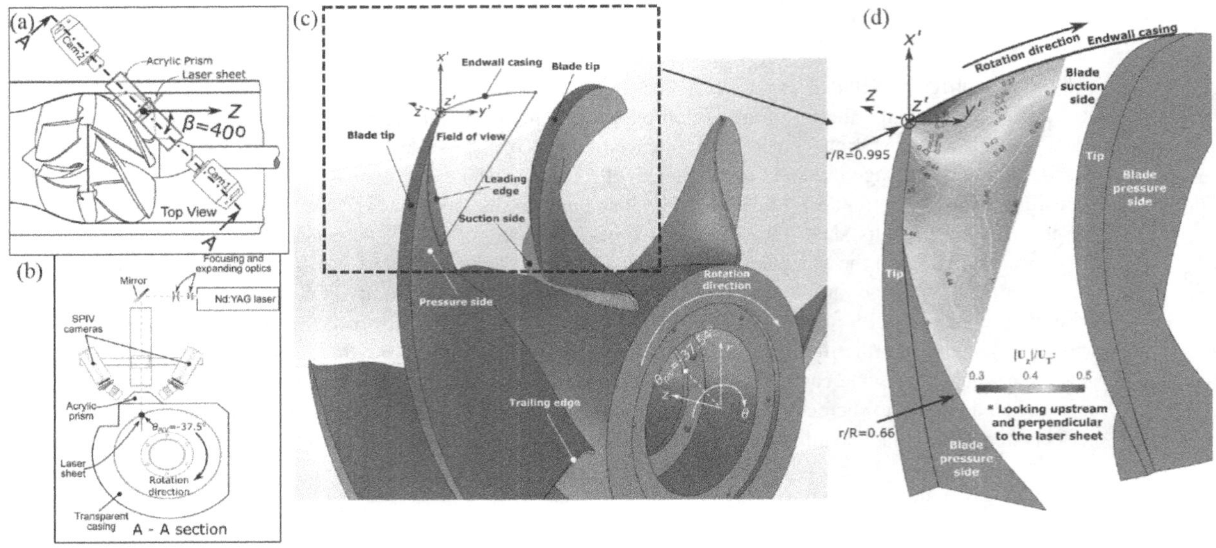

Figure 1: (a) and (b): Top and section views of the SPIV setup. (c) A perspective view of the AxWJ-2 rotor and the PIV laser sheet at LE. (d) A sample of $|U_z|/U_T$ distribution superimposed on blade drawings.

Cavitation performance

The changes in flowrate and total head rise coefficients with decreasing cavitation number are shown in Fig. 2. The cavitation number is defined as $\sigma = (p_{s,1}-p_v)/(0.5\rho U_T^2)$, where $p_{s,1}$ is the absolute pressure measured at the pump inlet, p_v is the vapor pressure of the working fluid, and U_T is the rotor tip speed. The experiments are performed at an initial flow rate of φ=0.75, very close to the peak efficiency point of this pump (φ=0.76 [8]). Here, the flowrate coefficient is defined by $\varphi=Q/(nD^3)$, in which Q is the volumetric flowrate, n is the rotor angular speed (in revolutions per second) and D is the pump inlet diameter. The total head rise coefficient is $\psi = \dfrac{\Delta p_s + 0.5\rho\left[(Q/A_2)^2 - (Q/A_1)^2\right]}{\rho n^2 D^2}$ where A_1 and A_2 are the areas of the inlet section upstream of the rotor and the outflow nozzle section just downstream of the rotor, respectively, and Δp_S is the static pressure rise across the pump. Both φ and ψ remain unchanged between $0.18<\sigma<0.78$ and decrease abruptly at $\sigma<0.17$. Just before breakdown, at σ=0.17, ψ increases slightly while φ hardly changes. In the following discussion, we refer to this point as "breakdown onset".

Appearance of cavitation

Figure 3 contains a series of images showing the progression of cavitation in the rotor passage with decreasing σ. Figures 3a&b are samples before breakdown, and Figs. 3d&e, once breakdown occurs. Figure 3c shows the cavitation at the breakdown onset point. The corresponding σ and performance parameters are indicated in each figure. In Fig.3a (σ=0.526), attached cavitation appears along the SS LE. As σ is lowered to 0.186 (Fig. 3b), just before breakdown, the sheet cavitation expands towards the blade overlap region, and tip leakage cavitation occurring in the tip gap is entrained into the cavitating TLV, creating a bubble layer. In Fig. 3c (σ=0.171), the attached cavitation already propagates into the blade overlap region. At this stage, the area covered by the SS attached cavitation oscillates in large amplitudes [7,9], from the beginning of the overlap region to about 80% of the tip chordlength. When σ is reduced further to 0.160 (Fig. 3d), representing a 2kPa decrease in the absolute inlet pressure, the head rise decreases by more than 5% (8kPa). The attached cavitation stops oscillating, and remains expanded over a broad area covering large portion of the blade SS, reaching the trailing edge near the tip, but not at mid span. Furthermore, a large scale perpendicular cavitating vortex (PCV) develops and extends from the SS of one blade to the PS of the neighboring blade. As discussed in [4], the PCV develops as the TLV entrains the cloud cavitation at the downstream end of the attached cavitation. These structures have minimal effect on the overall performance when the entrainment occurs close to the blade leading edge, but block the tip region, reducing the pump performance when they develop in the blade overlap region. The performance keeps deteriorating as the cavitation index is reduced to σ=0.155 (Fig. 3e). The attached cavitation covers large fractions of the SS, which increases with decreasing σ, and the PCVs appear to have grown into a larger scale, blocking much of the tip region.

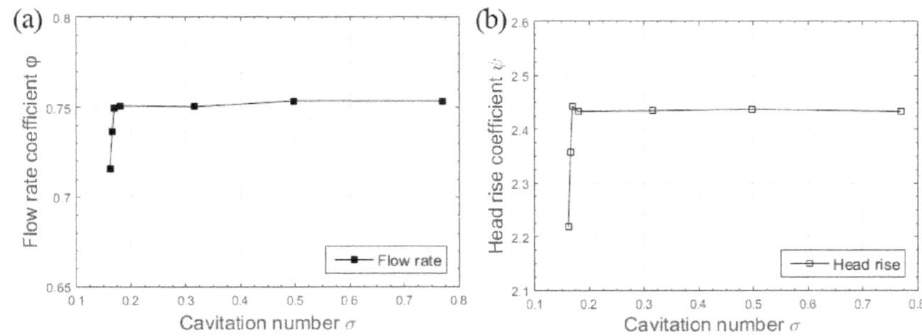

Figure 2: Cavitation breakdown curves: (a) Total head and (b) flow rate coefficient.

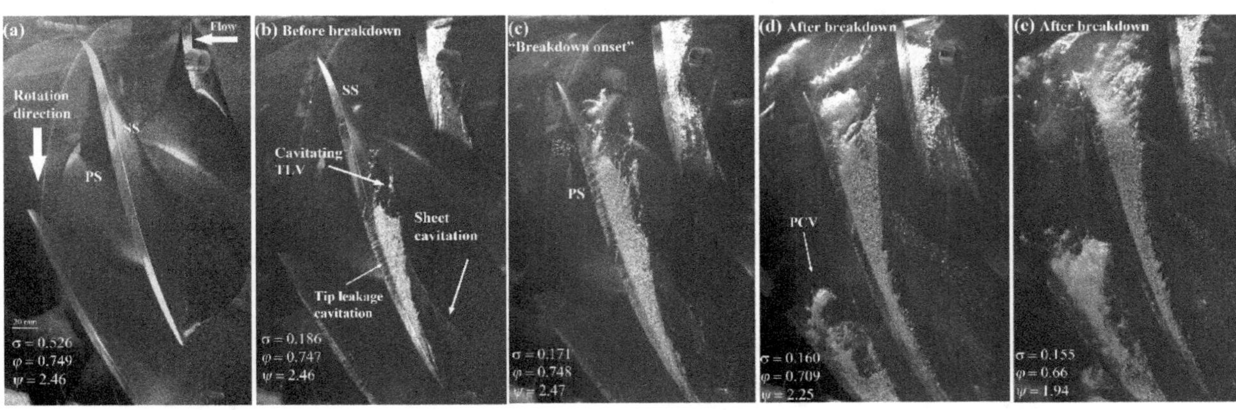

Figure 3: Evolution of cavitation phenomena in the rotor passage with decreasing cavitation number.

Velocity and pressure distributions

The evolution of axial velocity magnitude ($|U_z|/U_T$) with decreasing σ is shown in Fig. 4. The data presented here corresponds to a plane that intersects with the blade tip chord at s/c=0.065, where s is the distance from the blade LE along the tip chord. At σ=0.77, when the pressure is high enough to suppress all forms of cavitation, $|U_z|$ is low near the endwall casing due to the previously-mentioned effects of the casing boundary layer, tip leakage flows and TLV. As expected, $|U_z|$ is also low along the PS, and increases towards the SS, representing the effect of blade loading. Due to partial blockage of the optical access by cavitation, the field of view (FOV) is much smaller for the rest of Fig. 4. At σ=0.180, before breakdown, $|U_z|$ increases slightly across the FOV. At the breakdown onset point (σ=0.170), $|U_z|$ increases significantly over the entire FOV, e.g. by 6%, from $0.395U_T$ to $0.420U_T$, near the PS at x'/R=-0.35. Since the flow rate is not changed significantly under this condition, the increase in axial velocity is very likely to be associated with blockage in other regions in the rotor passage. A good candidate causing this blockage is the attached cavitation, which extends into the blade overlap region and oscillates at σ=0.170 (Fig.3c). Further decrease in the cavitation number to the early breakdown condition (σ=0.166), the flow rate and the head rise start to deteriorate. Yet, the axial velocity keeps increasing along the PS, but decreases in other regions, including the tip. Lowering the inlet pressure to deep breakdown (σ=0.161), the reduction in axial velocity expands to the entire FOV.

Using the measured three velocity components in planes located upstream of the blade and in the overlap region, it is possible to use Bernoulli's equations for estimating the pressure distributions inside the passage. Doing so involves assumptions that the passage flow is inviscid and steady (hence, having negligible Reynolds stresses) in the rotor reference frame, and that the pressure in the upstream plane is uniform. The resulting pressure distributions shown in Fig.5 are expressed as the local cavitation number, defined as $\sigma_{local} = (p(r, \theta, z) - p_v)/(0.5\rho U_T^2)$. Their values along the PS can be used as a surrogate for blade loading when the SS surface is covered with attached cavitation. The values of σ_{local} decreases substantially over the entire FOV when σ is reduced from 0.18 to 0.17 to the breakdown onset state. This reduction is caused to a small extent by the decrease in σ, and for the most part, by the increase in velocity magnitude over the entire blade overlap region. Further decrease in σ to 0.166 reduces σ_{local} along the PS, and to lesser extent, in other part of the passage. Transitioning to deep breakdown condition (σ=0.161), reduces σ_{local} slightly along the PS in the inner part of the passage.

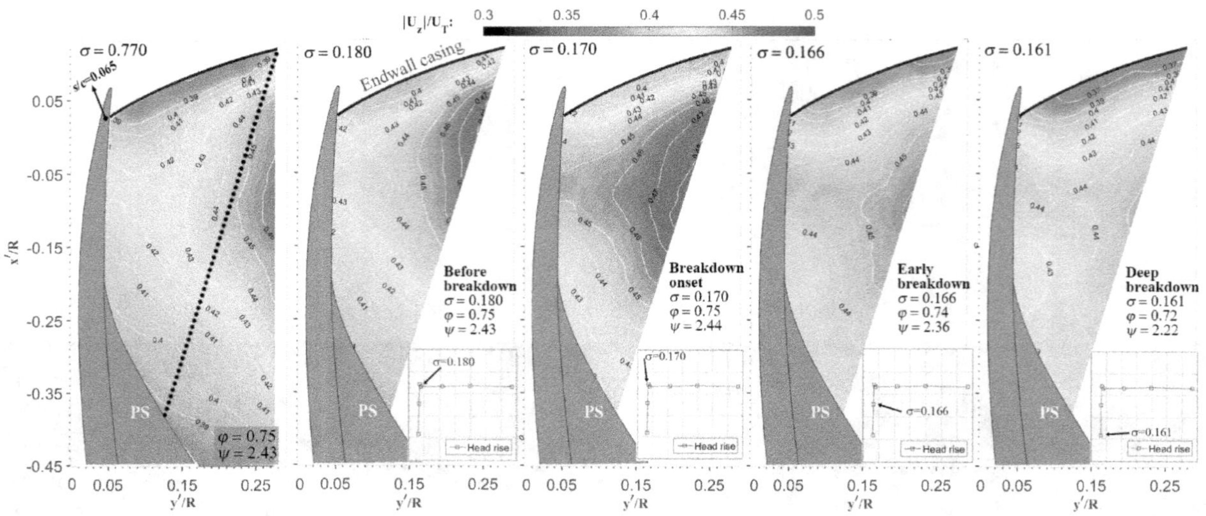

Figure 4: Evolution of $|U_z|/U_T$ with σ at s/c=0.55 inside the rotor passage: σ=0.77 (without cavitation), σ=0.180 (just before breakdown), σ=0.170 (breakdown onset), σ=0.166 (early breakdown), and σ=0.161 (deep breakdown). Dotted line in σ=0.77 indicates the boundary of the field of view in σ=0.180-0.161.

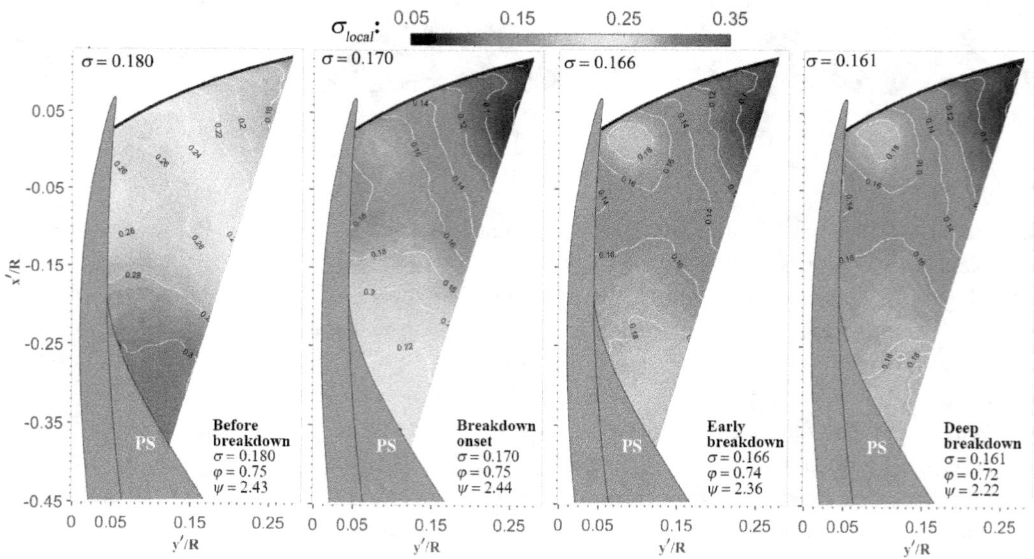

Figure 5: Evolution of σ_{local} for baseline case: σ=0.180 (just before breakdown), σ=0.170 (breakdown onset), σ=0.166 (early breakdown), and σ=0.161 (deep breakdown).

Discussion and conclusions

A plausible explanation for the mechanisms involved in cavitation breakdown can be introduced based on the experimental data. At breakdown onset, the attached cavitation expands into the blade overlap region while the flowrate is not change, the head rise increases slightly, the axial velocity along the PS decreases by 6%, and the corresponding mean pressure decreases by more than 25%. As mentioned before, the increase in axial velocity is presumably associated with partial blockage of the blade passage by the attached cavitation. Supporting evidence for this claim is obtained from previous casing endwall pressure fluctuation measurements in the same machine at two locations [4] (Fig. 6a). Transducer no.1 is located near the blade LE, and no.2, at the mid-chord. Figure 6b shows sample the ensemble-averaged pressure ($C_p = p(\theta)/(0.5\rho U_T^2)$) of Transducer no.2 for varying circumferential angles between the blade and the transducer ($\Delta\theta$). The sharp decrease in pressure at $\Delta\theta \sim 10°$ corresponds to the blade passing the transducer, with the maximum representing the PS pressure, the minimum the SS pressure, and the difference between them, ΔC_p, can be used as a surrogate for the local blade loading. Fig. 6c shows the values of ΔC_p for different σ for both transducers, and compares them to the pump performance. At $0.175<\sigma<0.25$, since the SS attached cavitation already reaches Transducer no.1, i.e. the pressure there is nearly equal to the vapor pressure, the PS pressure decreases at the same rate as inlet pressure, resulting in $\partial\Delta C_p/\partial\sigma \sim 1$. The pump performance remains nearly unchanged, indicating that the decreasing of blade loading near the LE is compensated by an increase in loading further downstream [10]. At the breakdown onset point (σ=0.171), the slight increase in head is accompanied by a rapid decrease in ΔC_p near the LE, but an increase in loading near at mid-chord (Transducer No. 2). The decrease in LE pressure is consistent with the current measurements. Conversely, the increase in ΔC_p at mid-chord is presumably caused by expansion of the SS cavitation to this region, which increases the local blade loading [10], resulting in slight, but noticeable increase in head rise.

Upon further reduction in σ, the resulting rapid decrease in performance is consistent with the decreased blade loading at the LE and mid chord (Fig. 6c). With the SS pressure remaining nearly constant at the vapor pressure, and the PS pressure decreasing, the deterioration in blade loading is inherent. This plateau LE ΔC_p under such conditions represent the combined effects of reduced flow rate and the inlet pressure. Another contributing factor discussed before is the reduction of throughflow in the tip region caused by the cavitation (PCV and cloud)-induced blockage.

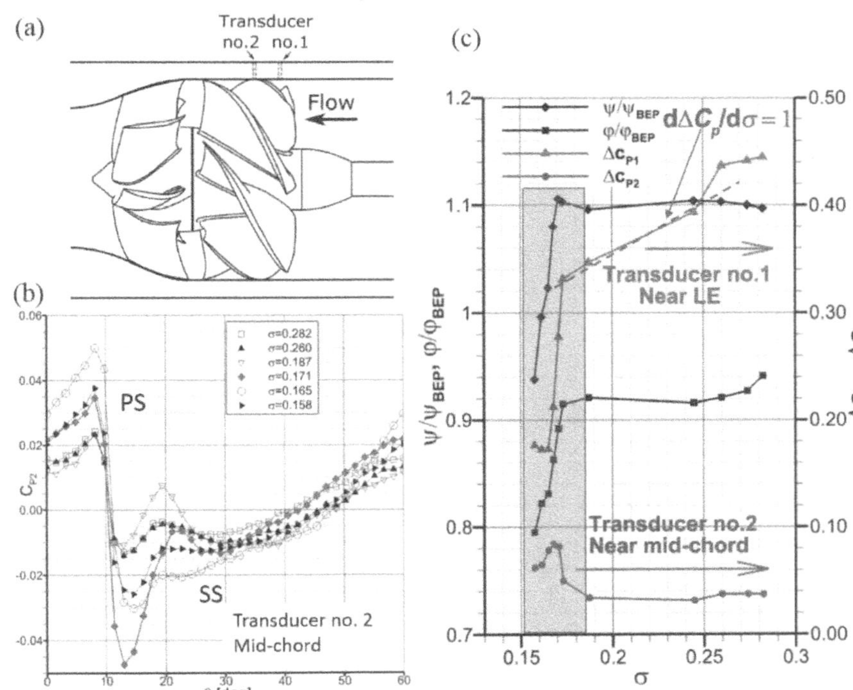

Figure 6: (a) Illustration of transducer locations. (b) Casing wall pressure variations near the mid-chord. (c) Pressure differences across the blade measured at casing wall at two different axial locations for the baseline case. (Adapted from Tan et al. 2015)

References

[1] Jakobsen, J. K., 1964, "On the Mechanism of Head Breakdown in Cavitating Inducers," J. Basic Eng., **86**(2), pp. 291–305.

[2] Pearsall, I. S., 1973, "Design of Pump Impellers for Optimum Cavitation Performance," Proc. Inst. Mech. Eng., **187**(55), pp. 667–678.

[3] Lindau, J. W., Pena, C., Baker, W. J., Dreyer, J. J., Moody, W. L., Kunz, R. F., and Paterson, E. G., 2012, "Modeling of cavitating flow through waterjet propulsors," Int. J. Rotating Mach., **2012**, p. 716392.

[4] Tan, D., Li, Y., Wilkes, I., Vagnoni, E., Miorini, R., and Katz, J., 2015, "Experimental Investigation of the Role of Large Scale Cavitating Vortical Structures in Performance Breakdown of an Axial Waterjet Pump," J. Fluids Eng., **137**(11), p. 111301.

[5] Kim, S., and Schroeder, S., 2010, "Numerical Study of Thrust-Breakdown Due to Cavitation on a Hydrofoil , a Propeller , and a Waterjet," 28th Symposium on Naval Hydrodynamics, Pasadena, California.

[6] Michael, T. J., Schroeder, S. D., and Becnel, A. J., 2008, Design of the ONR AxWJ-2 Axial Flow Water Jet Pump. No. NSWCCD-50-TR-2008/066, West Bethesda, MD.

[7] Chen, H., Doeller, N., Li, Y., and Katz, J., 2017, "Measurements of the Flow and Pressure within an Axial Waterjet Pump during Cavitation Breakdown," International Symposium on Transporat Phenomena and Dynamics of Rotating Machinery, Maui, pp. 1–12.

[8] Chesnakas, C. J., Donnelly, M. J., Pfitsch, D. W., Becnel, A. J., and Schroeder, S. D., 2009, Performance Evaluation of the ONR Axial Waterjet 2 (AxWJ-2). No. NSWCCD-50-TR-2009/089, West Bethesda, MD.

[9] Chen, H., Li, Y., Doeller, N., Koley, S. S., Keyser, B., and Katz, J., 2016, "Effects of Circumferential Grooves on the Cavitation and Performance of an Axial Waterjet Pump," 31th Symp. Nav. Hydrodyn., (September).

[10] Shen, Y., and Dimotakis, P. E., 1989, "The Influence of Surface Cavitation on Hydrodynamic Forces," American Towing Tank Conference., pp. 44–53.

© 2018 ASME

Experimental Study of Cavitation in Laminar Flow

[1]Kilian Croci *; [1]Florent Ravelet; [1]Jean-Christophe Robinet; [2]Amélie Danlos

[1]DynFluid Laboratory, Arts et Métiers PariTech, Paris, FRANCE; [2] DynFluid Laboratory, Conservatoire National des Arts et Métiers, Paris, FRANCE

Abstract

An experimental set-up has been especially developed in order to observe cavitation in laminar flows. Experiments have been carried out with a silicon oil of viscosity $\upsilon = 100\ cSt$ passing through a Venturi-type geometry with 18°/8° convergent/divergent angles respectively. The range of Reynolds numbers at the inlet section is between 350 and 1000. For these flow conditions, the Venturi only experiences air releasing gaseous cavitation for which two dynamic regimes are identified. They are characterized by two critical Reynolds numbers, induced by major hydrodynamic changes in the flow, in addition to a hysteresis effect between the inception and the disappearance of cavitation.

Keywords: laminar flows; gaseous cavitation regimes; cavitation inception; bubbles; hysteresis

1. Introduction

Several regimes can occur in turbulent cavitating flows. In studies about water flows coming across obstacles, and in particular with Venturi geometries, four main cavitation regimes can be identified:

- An **intermittent regime** which corresponds to the cavitation inception. This regime occurs for relatively high pressures and is characterized by the intermittent inception and the implosion [1] of small vapor bubbles at the Venturi throat.
- For lower pressure conditions a **"closed cavity" regime**, presenting relatively stable vapor/liquid mixture cavities, is observed. Re-entrant jets, induced by a recirculation flow, occur at the cavity rear inducing the mixing of liquid and vapor without cloud shedding.
- A **transient regime** of **partial cavitation** [2] is obtained decreasing the pressure in the test section. This highly unsteady regime is characterized by several cloud cavitation shedding dynamics induced respectively by: the re-entrant jet and by the periodical condensations and implosions of coherent cloud cavitation structures named condensation shock wave [3].
- For really low pressure conditions, cavities can reach important sizes and a **periodic regime** (also called **supercavitation**) can be established [4]. In this particular regime the condensation shock wave, which occurs periodically, is the dominant shedding dynamic in the flow.

These regimes and shedding mechanisms still remain of principal interest both experimentally and numerically. However cavitation flows have, to the authors' knowledge, been very little investigated using fluids with high-viscosity [5]. The present study focuses on the different cavitation regimes observed in laminar flows. Two gaseous cavitation regimes are highlighted and a comparison is made with the turbulent regimes previously presented. The paper is organized as follow: first the experimental set-up is described in §2. The cavitation regimes are then highlighted (§3) and finally conclusions and perspectives are proposed in §4 and §5.

2. Experimental Set-up

The experiments are conducted in a closed test loop where a pump impels a silicone oil flow with a flow rate from 21 to 60 $L.min^{-1}$. The Bluesil FLD 47 V100® silicone oil main characteristics are given in Table 1.

Viscosity υ $(m^2.s^{-1})$	10^{-4}
Volumic Mass ρ $(kg.m^{-3})$	965
Vapor Pressure P_v (Pa)	1.33
Surface Tension S $(N.m^{-1})$	20.9×10^{-3}

Table 1 - Characteristics of Bluesil FLD 47 V100® silicone oil at $25°C$ given by the constructor.

*Corresponding Author, Kilian Croci: kilian.croci@ensam.eu

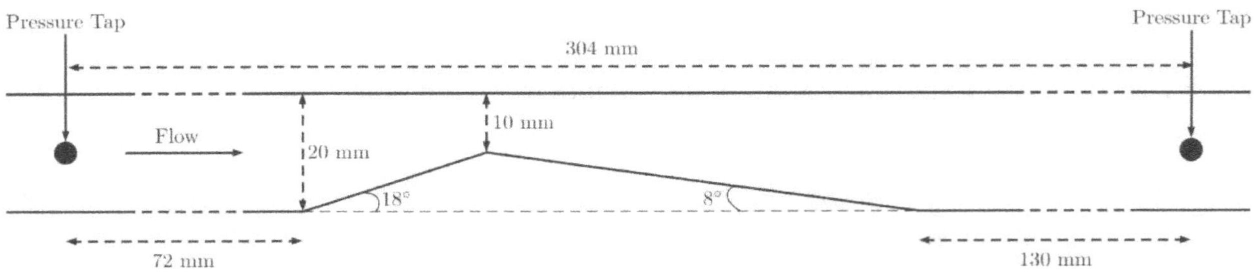

Figure 1 – Venturi geometry sketch. The width of the test section is 10 mm and the height at the inlet section h_{in} = 20 mm.

The test section is represented in figure 1. It consists of a rectangular inlet section of width 10 mm and height 20 mm and a $10 \times 10\ mm^2$ square section at the throat. The Venturi-type geometry presents 18°/8° convergent/divergent angles respectively. The velocity is measured with an ultrasonic flow-meter and studied for $1.75 < U_{in} < 5.00\ m.s^{-1}$ at the inlet test section corresponding to Reynolds numbers ($Re = \frac{U_{in} \times h_{in}}{\upsilon}$) taken from 350 to 1000. In the meantime the pressure is measured in the test section upstream (P_{in}) and downstream (P_{out}) of the throat. The cavitation number, such as the Reynolds number, is defined at the beginning of the test section as follow: $\sigma = \frac{P_{in} - P_v}{1/2 \rho U_{in}^2}$. Finally, the temperature is measured with a PT100 sensor and the kinematic viscosity υ is reevaluated accordingly based on our own calibration. In the following parts of the study a laminar flow description is introduced (figure 2) and the inception, in addition to the different cavitation regimes, will be presented and characterized as a function of the cavitation and the Reynolds numbers in the figure 3.

3. Cavitation Inception

The inception of cavitation, widely studied into water, is an intricate phenomenon which depends on several parameters. As explained in the works of Brennen [4], it might depend on the water quality regarding on the dissolved and the free gas content. For a saturated liquid experiencing a low pressure region, important outgassing and vaporization may occur in the flow generating the intermittent regime described in §1. Moreover when a nucleus goes through a region with a critical pressure drop in the surrounding liquid, its radius would be able to grow unboundedly.

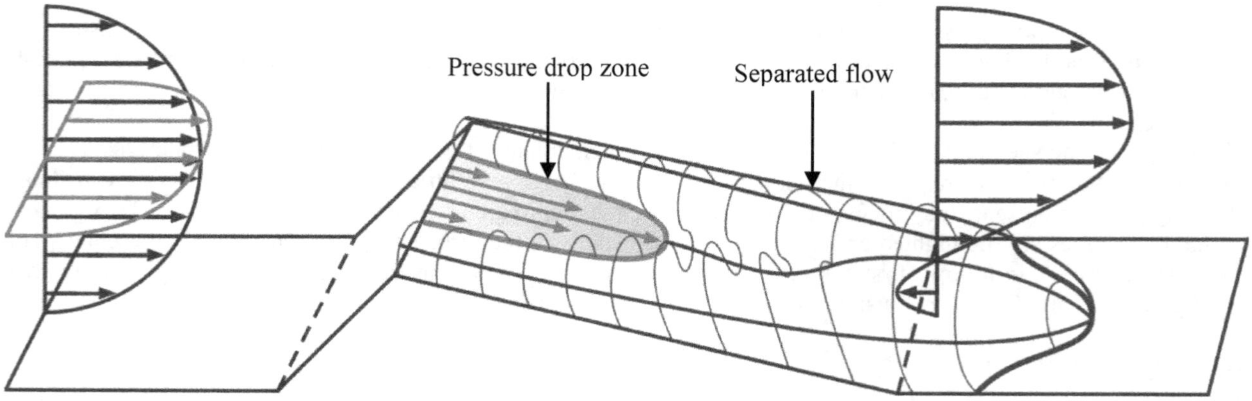

Figure 2 – Sketch of the flow at the inception. The flow separation occurs at relatively low Reynolds numbers in non-cavitating flows but when an adverse pressure gradient is generated near this region, the growth and capture of air bubbles become possible leading to attached air cavities and recirculating bubbles.

Such as in a turbulent water-flow a low pressure region can be generated in a laminar flow, inducing outgassing and nuclei explosive growths. In a Venturi geometry the velocity, as well as the pressure drop, is maximum at the center of the throat. Consequently a large air bubble might be generated in this central region which is illustrated in figure 2. After this inception, the bubble is carried away by the stream and, contrary to the water-flows, does not implode because of the high viscosity of the oil. A part of the air bubble can then be captured by the separated zone downstream of the Venturi throat which is already present in the non-cavitating flow at low Reynolds number. In summary for a critical Reynolds number Re_c^i, an adverse pressure gradient can be generated leading to an explosive bubble growth near the separated region. The recirculation region can then captures a part of the expanded bubble and make it reach back to the Venturi throat on one wall-side. After a few repetitions of this cycle, two elongated "tadpole shape" bubbles settle on both side-wall at the Venturi throat.

As a result the cavitation inception, previously defined, occurs above a critical inlet Reynolds number $Re_c^i \approx 650$ and depends on the cavitation number. The inception conditions ($Re = f(\sigma)$) are exposed in figure 3 with the blue curve. Under this curve, a dynamic regime can be defined: the **bubbly recirculation regime**, characterized by bubbles' inception and recirculation to the wall-sides of the Venturi throat. For lower Reynolds numbers and higher cavitation numbers, the recirculation regime disappears but not the two tadpole bubbles at the throat which oscillate with micro-bubble sheddings until a critical desinent Reynolds number $Re_c^d \approx 350$. It defines a second regime called the **tadpole regime**.

4. Cavitation Regimes

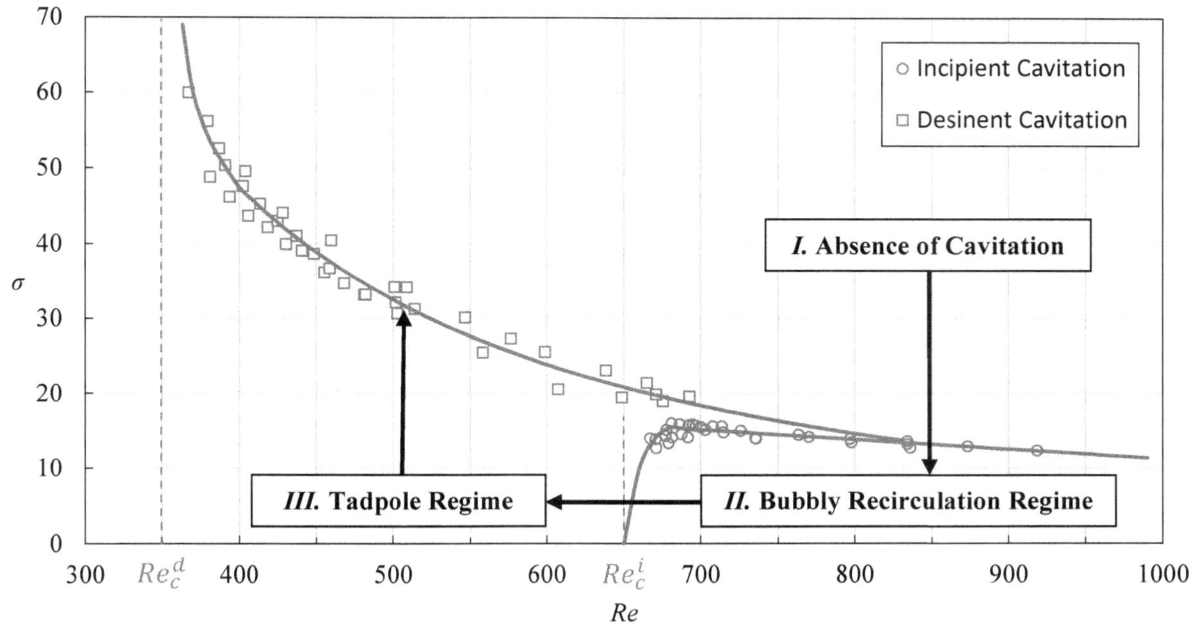

Figure 3 – Map of the dynamic regimes as a function of Reynolds and cavitation numbers. The blue and the red curves correspond repectively to the inception and to the desinence of cavitation. The blue dotted line denotes the inception critical Reynolds number Re_c^i whereas the red dotted line represents the desinent critical Reynolds number Re_c^d. The cycle of cavitation hysteresis is defined as follow. ***I*** : Absence of cavitation; (decreasing the cavitation number) ***II*** : Inception of cavitation and bubbly recirculation regime; (decreasing the Reynolds number) ***III*** : Subcritical tadpole regime and disappearance of cavitation when the cavitation number is enough increased. The uncertainties are estimated to 4 % for the Reynolds numbers and to 2% for the cavitation numbers.

The figure 3 shows the hysteresis between the inception and disappearance of cavitation with the two different cavitation regimes previously introduced. The two asymptotes, corresponding to the two critical Reynolds numbers characterized earlier, might be due to the recirculation flow characteristics. For Reynolds numbers higher than 650, the oil flow might be fast enough to generate a pressure drop zone at the throat and, in the meantime, the shear stress might be enough developed in order that a bubble generated at the throat get captured by a thick recirculation zone and go back to the Venturi throat. The desinent critical Reynolds number Re_c^d, which is around 350, might correspond to the disappearance of the recirculation zone. Near this condition, the two bubbles become less and less stable and symmetry breaks can often be observed.

Regarding to the figure 3, the two different regimes can be defined depending on the flow conditions. The **bubbly recirculation regime** is defined by the inception curve (occurring for $Re > Re_c^i$) for which the cavitation number slowly decreases with the Reynolds number. Under this curve, bubbles are generated at the throat, are carried away by the flow, get captured by the recirculation zone and its negative velocities and, finally, they reach the Venturi slope to the throat where they settle to one or the other wall-side. It results in the presence of two attached bubbles at both wall-side of the Venturi throat in addition to recirculating bubbles as illustrated in figure 4.

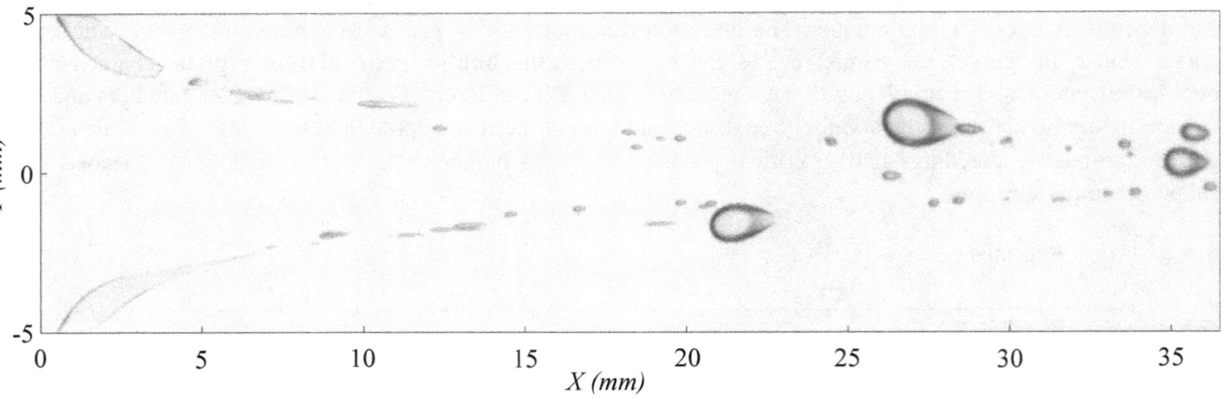

Figure 4 – Top view of the bubbly recirculation regime captured at $Re = 946$ and $\sigma = 8$. The oil stream is coming from the left to the right. The Venturi throat is positioned at $X = 0\ mm$. Dark regions are indicative for the contours of the bubbles. The two tadpole bubbles (left) are attached near the throat with micro-bubble shedding whereas recirculating bubbles (right) reach back the Venturi slope to the throat.

On the other hand the **tadpole regime** (figure 5) is the result of a hysteresis between inception and the disappearance of these two bubbles. Indeed after the settlement of the bubbles at the throat, they do not disappear immediately when the Reynolds number is decreased and the cavitation number is increased. However in these conditions the bubbly recirculation regime is stopped and only the two oscillating bubbles persist. The tadpole regime occurs for flow conditions which are below the desinent curve (red dots in figure 2) and is limited by the critical desinent Reynolds number Re_c^d. For Reynolds numbers $Re > Re_c^i$, the desinent curve seems to rejoin the inception curve meaning that in high velocity conditions the hysteresis disappears.

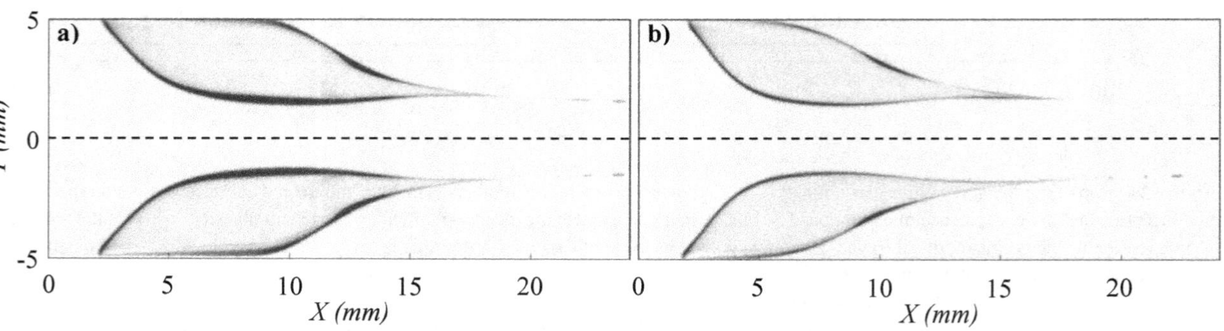

Figure 5 – Two typical "tadpole" bubbles observed with a top view for (a) $Re = 450$ and $\sigma = 42$ and for (b) $Re = 500$ and $\sigma = 34$.

In these two regimes micro-bubble sheddings can be observed, such as in figure 4 and 5. These sheddings, that occur with high frequencies, might be due to the gas diffusion [6] of the dissolved air present in the flow at interface of bubbles. This mechanism induces the growth of a bubble until a critical size leading to the shedding.

5. Conclusions

In the present paper, a first overview about the different cavitation regimes in laminar flows ($350 < Re < 1000$) has been made and the outcomes have been compared to the usual dynamical regimes observed into water-turbulent cavitating flows. The following summarizes the important conclusions.

- Two different gaseous cavitation dynamical regimes have been identified in the laminar flow, namely the **bubbly recirculation** and the **tadpole regime**.
- The inception of cavitation in the oil is similar to that one observed into water with the intermittent regime except that the bubbles generated do not condensate (and implode) in higher pressure regions, due to the high viscosity of the oil.
- The study of cavitation inception and desinence reveals a **hysteresis** which induces the tadpole regime.
- The two regimes are characterized by **two critical Reynolds numbers**, $Re_c^d \approx 350$ and $Re_c^i \approx 650$, which represent hydrodynamic transitions: namely the disappearance of the recirculation region and the emergence of an adverse pressure gradient at the throat.

6. Future Developments

In an upcoming article, the cavitation inception will be highlighted with top and side visualizations while the two regimes exposed in this paper will be characterized with the measurements of bubble sizes and shedding frequencies depending on cavitation and Reynolds numbers. In a meantime other dynamical regimes will be investigated for higher velocity inlet and thus higher Reynolds numbers. The first observations suggest the inception of a second bubbly recirculation zone at the top of the Venturi as well as a transition from gaseous to classical vaporous cavitation for higher Reynolds numbers.

Acknowledgement

The authors wish to acknowledge the great assistance of Mme Guylaine Ducouret from ESCPI Paris for the characterization of the silicone oil used in the present study.

References

[1] Brandner, P. A., Pearce, B. W., de Graaf, K. L. (2015). *Cavitation about a jet in crossflow*. J. Fluid Mech. Vol. 768, pp. 141-174.

[2] Croci, K., Tomov, P., Ravelet, F., Danlos, A., Khelladi, S., Robinet, J.-C. (2016). *Investigation of two mechanisms governing cloud cavitation shedding: experimental study and numerical highlight.* ASME Int. Mech. Eng. Congress and Exposition, *Volume 7: Fluid Engineering.*

[3] Ganesh, H., Mäkiharju, S. A., Ceccio, S. L. (2016). *Bubbly shock propagation as a mechanism for sheet-to-cloud transition of partial cavities*. J. Fluid Mech. Vol. 802, pp. 37-78.

[4] Brennen, C. E. (1995). *Cavitation and bubble dynamics*. Oxford University Press, New York.

[5] Washio, S., Takahashi, S., Yoshimori, S. (2003). *Study on cavitation starting at the point of separation on a smooth wall in hydraulic oil flow*. Proc. Inst. Mech. Eng., Part C: J. Mech. Eng. Science Vol. 217, pp. 619-630.

[6] Gross, T. F., Pelz, P. F. (2017). *Diffusion-driven nucleation from surface nuclei in hydrodynamic cavitation*. J. Fluid Mech. Vol. 830, pp. 138-174.

© 2018 ASME

Experimental Study of Disappearance Phenomenon of Unsteady Cavitation on NACA16-012

[1]Yoshiki Odaira*; [2]Wakana Tsuru; [3]Satoshi Watanabe; [4]Yuka Iga;

[1]Graduate School of Engineering, Tohoku University, Sendai, Japan; [2]Graduate School of Engineering, Kyushu University, Fukuoka, Japan; [3]Department of Mechanical Engineering, Kyushu University, Fukuoka, Japan; [4]Institute of Fluid Science, Tohoku University, Sendai, Japan

Abstract

In the previous study which was conducted by J. P. Franc et al., it was reported that on NACA16-012 hydrofoil at the angle of attack 4 degree, partial cavitation disappears although cavitation number is decreasing. We call this phenomenon "Disappearance phenomenon" and aim to clarify this mechanism. In the same previous study, it was also reported that at around 4 degree, reattachment of laminar separation leads transition to turbulence boundary layer and cavity cannot occur. In this study, in order to investigate the detail of boundary layer characteristic at the angle of attack where disappearance phenomenon occurred, we measured the pressure distribution on NACA16-012 hydrofoil and made boundary layer visible by oil flow method. In addition, fluid force was measured under the conditions from non-cavitation to super cavitation in order to estimate boundary layer characteristics during disappearance phenomenon. As the result, although the position of reattachment was different, reattachment of laminar separation described in previous study was confirmed and the lift during disappearance phenomenon was the almost same value as that of non-cavitation condition. From this result, it could be inferred that pressure distribution i.e. boundary layer characteristic during disappearance phenomenon was greatly different from that of non-cavitation condition.

Keywords: Partial cavitation, Boundary layer, Pressure distribution, NACA16-012 hydrofoil

1. Introduction

In the case of hydrofoil, various kind of cavitation occurs depending on cavitation number σ and the angle of attack α. As shown in Fig. 1, cavitation patterns on NACA16-012 is organized by J. P. Franc et al. [1]. From this figure, in the high cavitation number region (more than σ= 0.33), shear cavitation, two-phase cavitation, and partial cavitation occur according to the angle of attack, and in the low cavitation number region (less than σ= 0.33), super-cavitation and back-detached cavitation occurs in which the detachment point changes depending on the angle of attack. The change of cavitation patterns at the angle of attack around 4° is remarkable. Usually, as cavitation number decreases, cavitation occurs and develops. However, it is reported that only at about α= 4° , when cavitation number is decreased from 1.2 to 0.045, partial cavitation appears first, then it disappears and the flow is non-cavitation condition, finally super cavitation occurs. Focusing on the phenomenon of partial cavitation disappearing, we will call it "*disappearance phenomenon*". In other study, disappearance phenomenon is reported on the same NACA16-012 hydrofoil at different angle of attack [3], and also on the other hydrofoil [4]. Previous experiment by our group [5], using the same NACA16-012 hydrofoil, disappearance phenomenon has been observed at about α= 4° same as the J. P. Franc et al. result.

It is known that boundary layer characteristics greatly affect cavitation behavior [6][7], so Franc's have also investigated the boundary layer characteristics on NACA16-012 hydrofoil as shown in Fig. 2 [1]. In this figure, the location of boundary layer which develops on the suction side of hydrofoil about the angle of attack is summarized. It is reported that attached laminar boundary layer exists from leading edge and near the trailing edge at the angle of attack α from -4° to 3°. The reattachment of laminar separation leads to turbulent transition and turbulent boundary layer is consecutively present from the leading edge to the trailing edge at the angle of attack α=from 4° to 10°. From this result, it hypothesized that disappearance phenomenon occurs because cavity cannot develop due to transition of boundary layer caused by reattachment of laminar separation. However, this hypothesis merely explained that why cavitation is disappears in the middle σ region at this angle of attack. The mechanism of disappearance of partial cavitation has not been sufficiently clarified. The goal of our study is to elucidate the mechanisms involved in disappearance phenomenon.

*Corresponding Author, Yoshiki Odaira: oodaira@cfs.ifs.tohoku.ac.jp

In this study, in order to understand detail of boundary layer characteristic of NACA16-012 hydrofoil, pressure distribution on suction surface of hydrofoil is measured and visualized by oil flow method. In addition, boundary layer characteristics under the disappearance condition is estimated by fluid force measurement

.

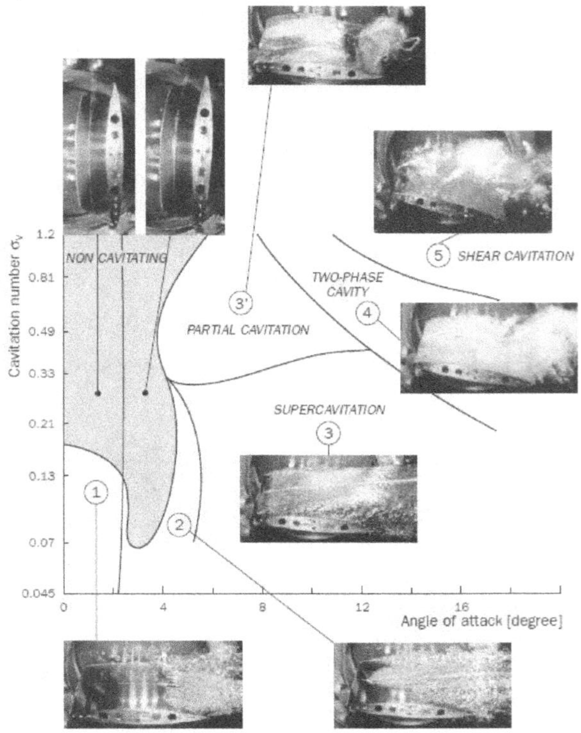

Figure 1. Cavitation patterns on NACA16-012 hydrofoil at $Re = 1.0\times10^6$ [2]

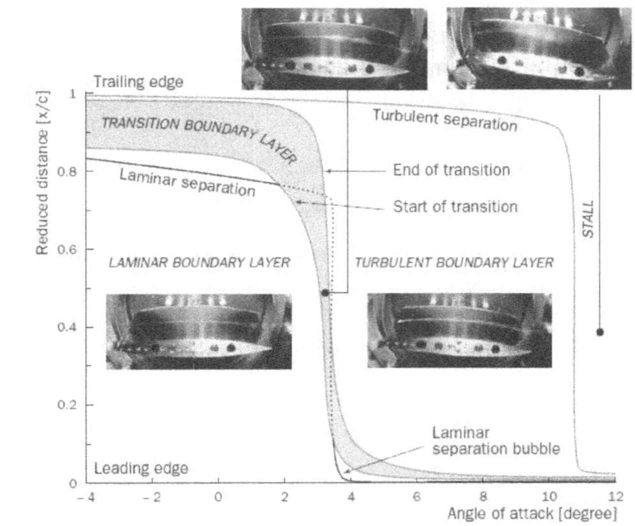

Figure 2. Location of boundary layer on NACA16-012 hydrofoil under non-cavitation condition [2]

2. Experimental apparatus and procedure

This study was conducted using the cavitation tunnel equipped at Kyushu University. The closed loop composes of tank, test section and circulating pump. The pressure of the entire tunnel is adjusted by compressor and vacuum pump connected to tank. The cross section of test part is 200 mm high and 81.5 mm wide. Two-dimensional NACA16-012 hydrofoil with the chord length C= 100 mm and the span width B= 81 mm is used as test body in this study. Shape of this hydrofoil is shown in Fig. 3. Top, side and bottom walls are made with acrylic plate for visual observations. The ratio of test section height to chord length is 100 mm/ 200 mm= 0.5 (Fig. 4). In this experiment, the blockage effect from tunnel walls is larger than that of test section at Grenoble University because the experiment conducted by J. P. Franc et al was conducted under the free-surface condition.

For pressure distribution measurement, 18 holes with a diameter of 0.5 mm were provided on hydrofoil surface and connected to pressure transducer. These holes were placed on the line which was inclined to the flow direction in order to avoid turbulence by holes. Oil flow method was used for boundary layer visualization on hydrofoil surface. In this method, the oil mixed with pigment was painted on the hydrofoil surface. After that, the hydrofoil was exposed to the flow, and from the oil film pattern the boundary layer was visualized. In this study, a mixture of vermilion pigment and pump oil was used. When oil flow method was conducted, the hydrofoil was painted black for easy observation. After the hydrofoil was exposed to the flow until oil film pattern didn't change visually, it was photographed with a camera. When we drained water from cavitation tunnel, we paid attention to prevent the oil pattern from being wiped away by passing through the free-surface.

Lift and drag forces were measured by strain gages attached on the cantilever support beam with rectangular cross section (20 mm×10 mm). 4 strain gages were attached on each sides of cross section, and two-gage method was applied for each force component in directions parallel and vertical to the hydrofoil. These forces were converted to the lift and drag forces. It was assumed that the influence of asymmetrical moment due to the tip leakage flow was sufficiently small. In this experiment, cavitation number was decreased intermittently by depressurizing inside the tunnel. Simultaneously with fluid force measurement, cavitation was observed from both the top and side of test section using high speed video camera at 8,000 fps.

These measurements were conducted with the main velocity U= 9.2 m/s (then the Reynolds number Re= 1.0×10^6). The pressure distribution measurement was conducted at α= from 2° to 7° in units of 1°. At only α= 6° and 7°, the main velocity was U= 4.7 m/s (Re= 5.0×10^5) because this experiment could not be sufficiently pressurized due to experimental equipment and cavitation occurred at these angle of attack. The fluid force measurement was conducted at α= 5° where disappearance phenomenon was observed. These measurement were performed for from 50 to 60 seconds at sampling frequency 1,000 Hz, and time-averaged value was adopted as experimental data. Cavitation number σ and the Reynolds number Re used here are defined as

$$\sigma = \frac{p_\infty - p_v}{1/2\rho U^2} \qquad \mathrm{Re} = \frac{UC}{\nu}$$

where p_∞ and p_v are the pressure at 200 mm upstream from the mid-chord and the saturated vapor pressure, ρ is the density of water, ν is kinematic viscosity of water.

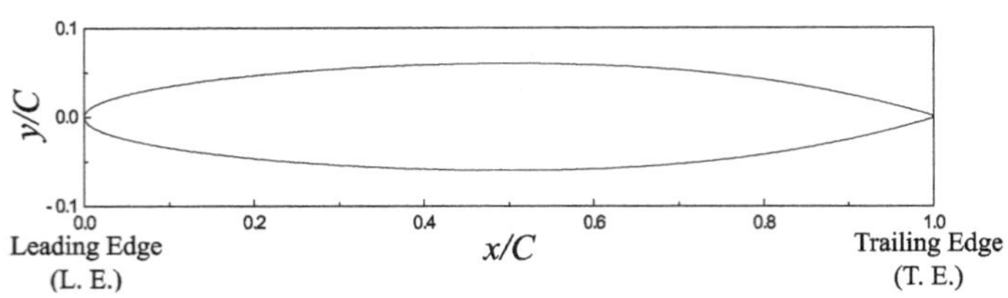

Figure 3. Shape of NCA16-012 hydrofoil

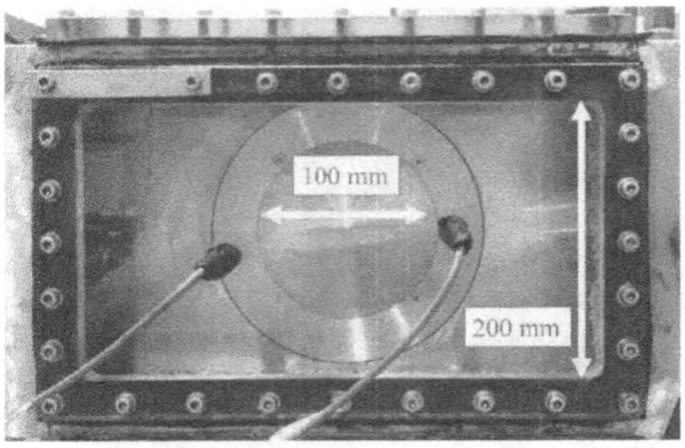

Figure 4. Test section

3. Result

In fig. 5, the time averaged pressure coefficient on the hydrofoil surface C_p= $(p-p_\infty)/(\rho U^2/2)$ are plotted against cavitation number σ for various angles of attack α and it shows the pressure distribution on NACA16-012 hydrofoil under non-cavitation condition. At α= 4°and 5°, a slight increase in pressure was seen locally at x/C= 0.5. It was considered that this increase was due to the reattachment of laminar separation. Figure 6 shows the result of oil flow visualization at α= 4.5°. The flow direction is leftward. In the vicinity of leading edge, the oil film was wiped away (seen in black color) and at immediate downstream, the oil film remains darkly (seen in vermilion color). On the further downstream, the oil film distributes in a patchy in the vicinity of trailing edge, and oil film having been wiped away was observed in an arching line near the mid-chord of hydrofoil. From this result, it was considered that laminar separation occurred at leading edge and laminar separation reattached near the mid-chord, then boundary layer turned into turbulent from the downstream of reattachment point to the trailing edge. Although the position of reattachment was different from that of previous study shown in Fig. 2, transition to turbulence due to reattachment of laminar separation could be confirmed. The difference of reattachment position was considered due to the blockage effect, main stream turbulence and surface roughness of hydrofoil.

Visualization result of cavitation using high speed video camera is shown in Fig. 7. As cavitation number decreased, cavitation inception occurs near the front part of hydrofoil at σ= 0.98 (Fig. 7 ②), and partial cavitation with periodic release of cloud cavity occurs at σ= 0.73 (Fig. 7 ④). Further cavitation number decreased, partial cavitation disappears and the flow changes to non-cavitation again in the range of σ= from 0.68 to 0.51 (Fig. 7 ⑤, ⑥). Finally, partial cavitation reappears and super cavitation with traveling bubble cavitation occurs. Figure 8 shows the lift coefficient C_L= $L/(BC\rho U^2/2)$ against cavitation number, and the number in the graph corresponds to that of in Fig. 7. The lift coefficient C_L increases from σ= 0.98 with cavitation inception to 0.73 with partial cavitation (Fig. 8 ②, ④).

When disappearance phenomenon occurs at σ= 0.68 (Fig. 8 ⑤), the lift coefficient C_L decrease and it took the same value as non-cavitation condition during disappearance phenomenon (Fig. 8 ⑤, ⑥). As cavitation number further decreases and partial cavitation reappears, the lift coefficient C_L suddenly decreased. In disappearance condition (Fig. 8 ⑤, ⑥), in order for cavitation to disappear, $-C_p$ has to be under σ value. In Fig. 5, suction peek of $-C_p$ in α= 4° in non-cavitation condition is larger than σ value in the case of ⑤, ⑥. Therefore, it could be considered that the pressure distribution on the suction side of the hydrofoil in the disappearance condition is completely changed from that in non-cavitation condition under the restriction of that the integrated values of the pressure distributions are same value in disappearance and non-cavitation conditions because the lifts take same value. For the evidence, it can be seen from Fig. 7 ⑦, ⑧ that leading edge of the reappeared cavity moves backward from that in the cavitation which appears before the disappearance.

To investigate the boundary layer characteristic under the disappearance condition, further study is going to be considered by measuring pressure distribution under the disappearance condition.

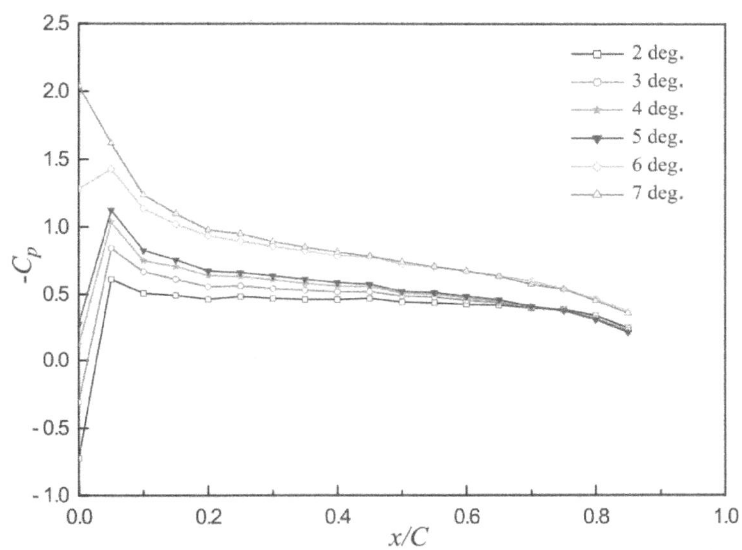

Figure 5. Pressure distribution on NACA16-012 hydrofoil under non-cavitation condition at Re= 1.0×10^6 (Re= 5.0×10^5, α= 6°and 7°)

Figure 6. Boundary layer visualization by oil flow method at Re= 1.0×10^6, α= 4.5°

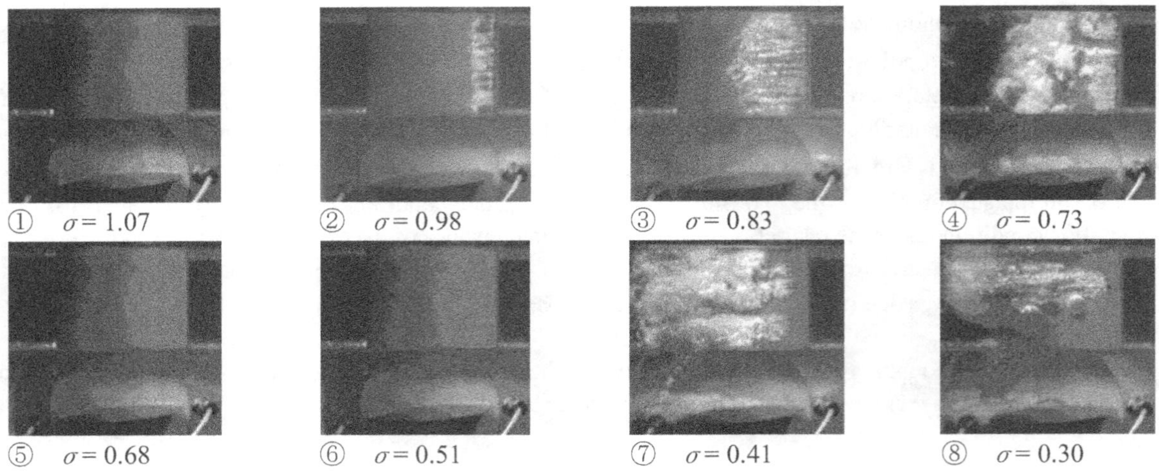

Figure 7. Aspects of cavitation observed using high speed video camera at Re= 1.0×10^6, α= 5°

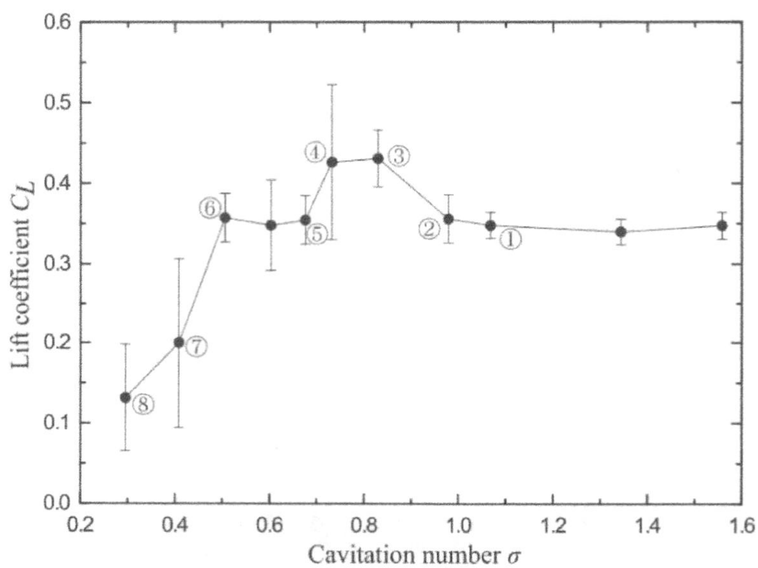

Figure 8. Lift characteristics against cavitation number at Re= 1.0×10^6, α= 5°

4. Conclusion

In this study, pressure distribution measurement on NACA16-012 hydrofoil and boundary layer visualization using oil flow method were conducted under non-cavitation condition. At the angle of attack around α=4° where disappearance phenomenon was observed, the position was different from that in previous study, and the reattachment of laminar separation and transition to turbulence could be confirmed near mid-chord. In addition, by measuring the lift coefficient for the variation of cavitation number, it was found that measured lift at both non-cavitation condition and disappearance conditions took almost the same value. From this result, it could be inferred that boundary layer characteristic and the pressure distribution in the disappearance condition was greatly different from that in non-cavitation condition.

As further study, we are planning to measure the pressure distribution under disappearance condition in order to investigate boundary characteristics under the condition and clarify why the cavity disappears in the middle condition of cavitation occurrence.

References

[1] Franc, Jean-Pierre, and Jean-Marie Michel. *Fundamentals of cavitation*. Vol. 76. Springer Science & Business Media, 2006.

[2] Franc, Jean-Pierre, and Jean-Marie Michel. "Attached cavitation and the boundary layer: experimental investigation and numerical treatment." *Journal of Fluid Mechanics* 154 (1985): 63-90.

[3] Yuka Iga, and Teppei Furusawa. "Experimental study of appearance of thermodynamic effect on cavitation in hot water." *Transactions of the JSME (in Japanese)* 83.845 (2017): 16-00377.

[4] Briançon-Marjollet, Laurence, Jean-Pierre Franc, and Jean-Marie Michel. "Transient bubbles interacting with an attached cavity and the boundary layer." *Journal of Fluid Mechanics* 218 (1990): 355-376.

[5] Yoshiki Odaira, et al. "A fundamental study on disappearance of partial cavitation on the NACA16012 hydrofoil." *Proc. Conference of Tohoku Branch 2017.52.JSME*, 2017. p. 106

[6] Oba R. et al. "Supercavitating cavity observations by means of laser velocimeter." *Trans ASME* I: *J. Fluids Eng.* 102(1980), 433-438

[7] Van Der Meulen. "Boundary layer and cavitation studies of Naca 16012 and Naca 4412 hydrofoils." In *Proc. Of the 13th symp. on Naval Hydrodynamics, Tokyo,* 1980. p. 195-219

© 2018 ASME

Experimental study of interactions between a cavitation bubble and a spherical particle near solid boundaries

[1]Yuning Zhang*; [2,3]Yongxue Zhang; [2,3]Yuning Zhang
[1] Key Laboratory of Condition Monitoring and Control for Power Plant Equipment (Ministry of Education), School of Energy, Power and Mechanical Engineering, North China Electric Power University, Beijing 102206, China
[2] College of Mechanical and Transportation Engineering, China University of Petroleum-Beijing, Beijing 102249, China
[3] Beijing Key Laboratory of Process Fluid Filtration and Separation, China University of Petroleum-Beijing, Beijing 102249, China

Abstract

In the present paper, the interactions between a laser-induced cavitation bubble and a spherical particle near solid boundaries are experimentally studied in detail with the aid of the high-speed camera. Firstly, the effects of presence of wall on the phenomenon are revealed in terms of the dynamics of cavitation bubbles. Secondly, depending on the direction of liquid jet during the first collapse stage, the behavior of bubble-particle interactions are categorized into three types. For each type, two examples are given to illustrate its main characteristics. Finally, the contributions of the present work are discussed with the possible future work.

Keywords: bubble-particle interaction; cavitation bubbles; solid boundaries

1. Introduction

Interactions between cavitation bubbles and particles are of great importance for the design and operations of hydraulic machinery. Specifically, in China, the hydroturbines built along the Yellow River suffer seriously from the damage induced by the cavitation-particle interactions. Hence, a full understanding of this phenomenon will be quite beneficial to the involved hydraulic power plants.

In the literature, the cavitation-particle interactions were usually investigated in the bulk water with certain amount of silt added. The primary interests of those research include the surface damage patterns, mass loss curve and anti-erosion materials. Those tests are quite important from the viewpoint of the engineering practice. Recently, with the rapid development of related industry, a more detailed model is further required to describe the microscopic interactions between cavitation bubbles and particles. This target is quite challenging noticing that the cavitation phenomenon itself is already very complex[1-4]. When the rapid motion of particle is further considered, a great increase of the research difficulties will be involved[5]. In our previous review, Zhang et al.[6] summarized the literature relating with the microscopic interactions between the cavitation bubbles and particles in the silt-laden flow.

In the present paper, a stationary particle is fulfilled with the aid of the sample holder and its interactions with a laser-induced cavitation bubble are investigated in detail using the high-speed camera (up to 330,000 frames per second). Both cases with and without solid boundaries are considered and compared. The influences of relative positions of the bubble, particle and wall on the phenomenon are also revealed and discussed.

2. Experimental Setups

Figure 1 shows the experimental system employed in the present work. The cavitation bubble was generated in the water chamber by a pulsed laser passing through the focus lens. In the present paper, a spherical particle with diameter 2 mm is employed. The particle was fixed on the tip of a thin needle using glue in order to make it as a stationary one. The whole dynamic event was recorded using a high-speed camera (Phantom V1212). The synchronization between the dynamics of the cavitation bubble and the high-speed camera was achieved by the digital delay generator. Figure 2 shows a detailed description of the bubble-particle-wall system with the primary parameters defined. In the present paper, our discussions mainly focus on the two cases with θ=0° and θ=90° respectively.

*Corresponding Author, Yuning Zhang: y.zhang@ncepu.edu.cn

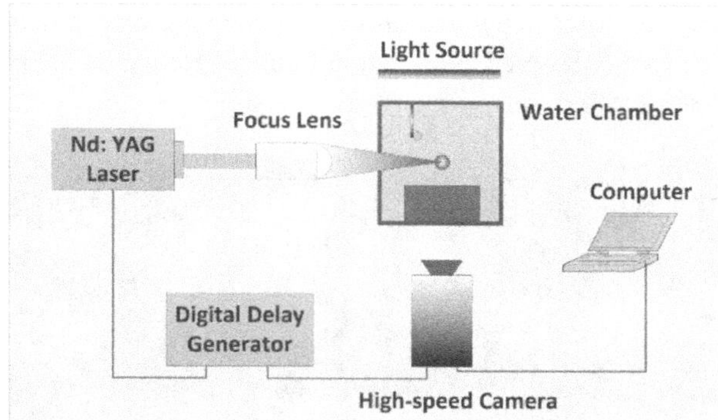

Figure 1 The experimental setup employed in the present study.

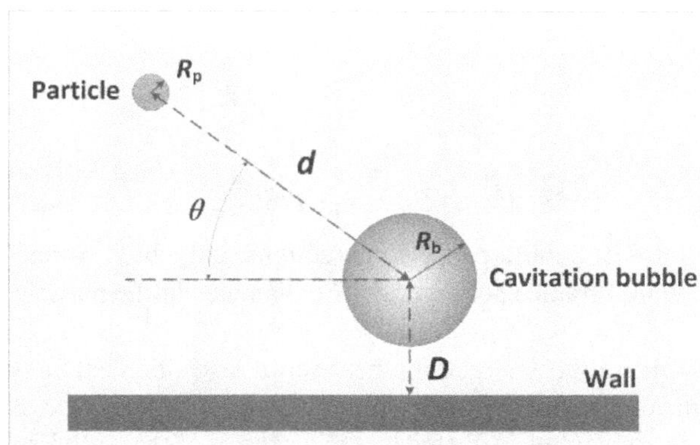

Figure 2 A detailed demonstration of the primary parameters.

3. Results and discussions

3.1 The effects of solid boundaries

Figures 3 and 4 show the interactions between a laser-induced cavitation bubble and a stationary particle without and with the presence of the solid boundaries respectively. In figures 3 and 4 (and other figures below), a selection of the frames was made in order to make the important stages (e.g. collapse and micro-jet formation) much more clear. The relative time of each frame (corresponding to the time interval between frame 1 and the present frame) are all marked on the lower right corner of each subplot.

As shown in figure 3, a typical process of cavitation bubble dynamics include:

1. Growth stage (frames 1-10).
2. Collapse stage (frames 11-15). Due to the presence of the particle, a mushroom-shape bubble will be induced during its collapse.
3. Micro-jet generation stage (frame 16). A very rapid micro jet will be generated with the directions toward the particle.
4. Rebounds and re-collapses (frames 17-24). There are usually two or more rebounds after the event of the generation of micro-jet. During the bubble redound, the original bubble could be possibly split up into two bubbles with strong interactions between them.
5. The end of the event.

During the collapse, although a strong force generated during the bubble collapse will be exerted on the particle, the movement of the particle could be totally ignored as shown in figure 3. Hence, in the present paper, the particle could be safely considered as a stationary one.

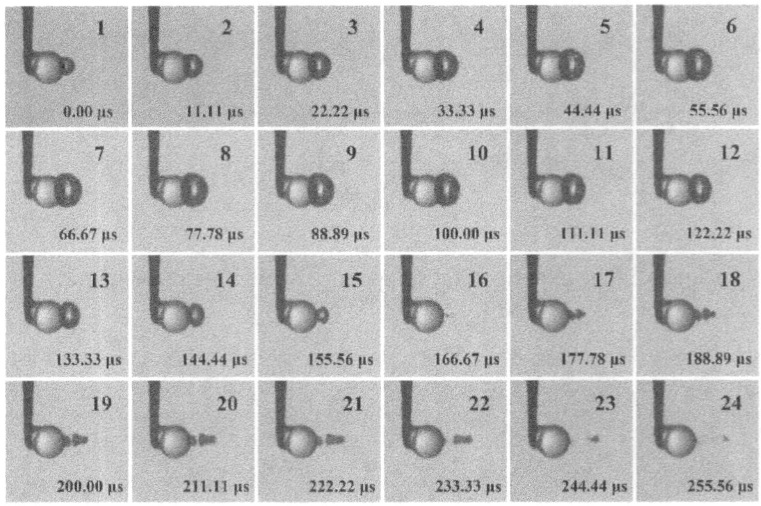

Figure 3 The experimental photos of bubble-particle interactions in the bulk water. The speed of the high-speed camera is 90,000 frames per second. The diameter of the particle is 2 mm.

Figure 4 shows the effects of wall on the interactions between a laser-induced cavitation bubble and a stationary particle. There are three main differences between the figures 3 and 4. Firstly, the shapes of the bubble during the growth and the collapse stages are both greatly affected by the presence of the wall. Especially, the shape of the bubble is non-symmetric in figure 4. Secondly, the contact line between the bubble and the particle moves quickly during the collapse. Thirdly, the direction of the micro-jet is toward the wall (as shown in frame 19 of figure 4) rather than the particle (as shown in frame 16 of figure 3). Hence, one can find that the wall could significantly alter the interactions between cavitation bubbles and particles.

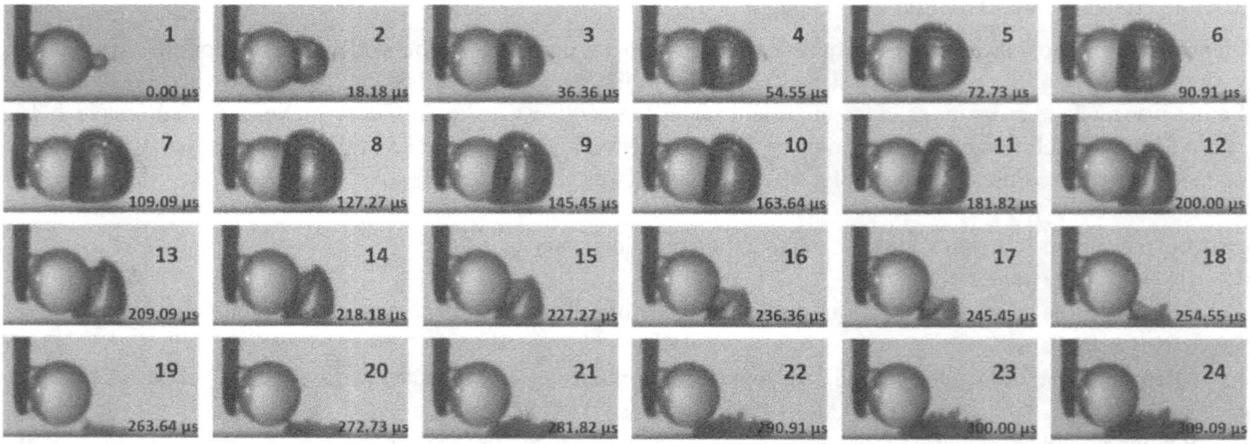

Figure 4 The experimental photos of bubble-particle interactions near the wall with θ=0° and a large bubble. The speed of the high-speed camera is 330,000 frames per second. The diameter of the particle is 2 mm.

3.2 The case with θ=0°

In this section, the interactions between a laser-induced cavitation bubble and a stationary particle near wall are studied in great detail. In this section, the cases with θ=0° are considered. As shown in figures 4-6, depending on direction of micro-jets during the bubble collapse, three types are proposed:

1. For, large bubbles, all the micro-jets generated by the bubble collapses is toward the wall (as shown in frame 19 of figure 4).
2. For medium-sized bubbles, the micro-jet during the first collapse is quite complex with several splits (as shown in frame 17 of figure 5). Sometimes, the bubble collapses strongly in the space between the particle and the bubble rather their surfaces. The micro-jet of the second collapse of the bubble is usually towards the wall surface.
3. For small bubbles, the micro-jet generated by the first and other bubble collapse is toward the particle (as shown in frame 10 of figure 6).

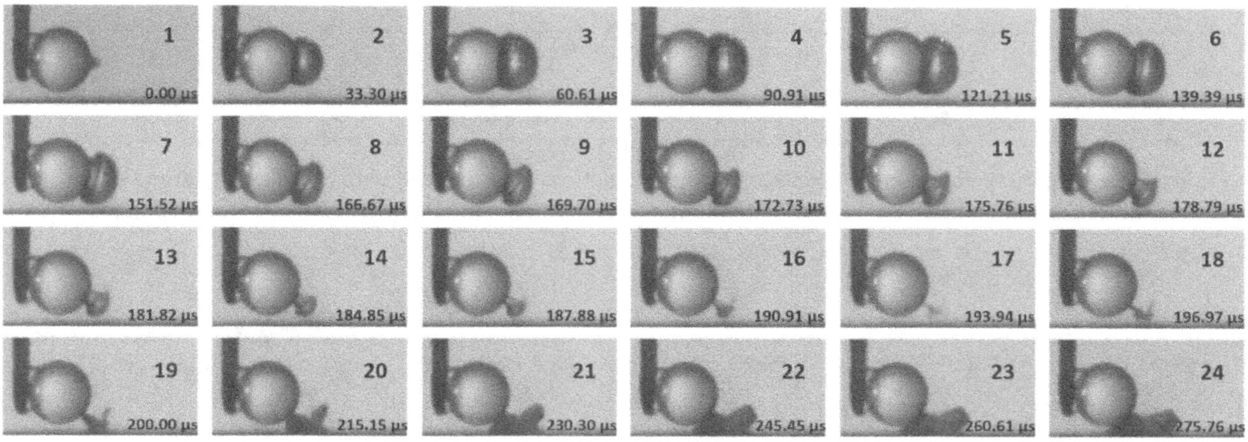

Figure 5 The experimental photos of bubble-particle interactions near the wall with θ=0° and a medium-sized bubble. The speed of the high-speed camera is 330,000 frames per second. The diameter of the particle is 2 mm.

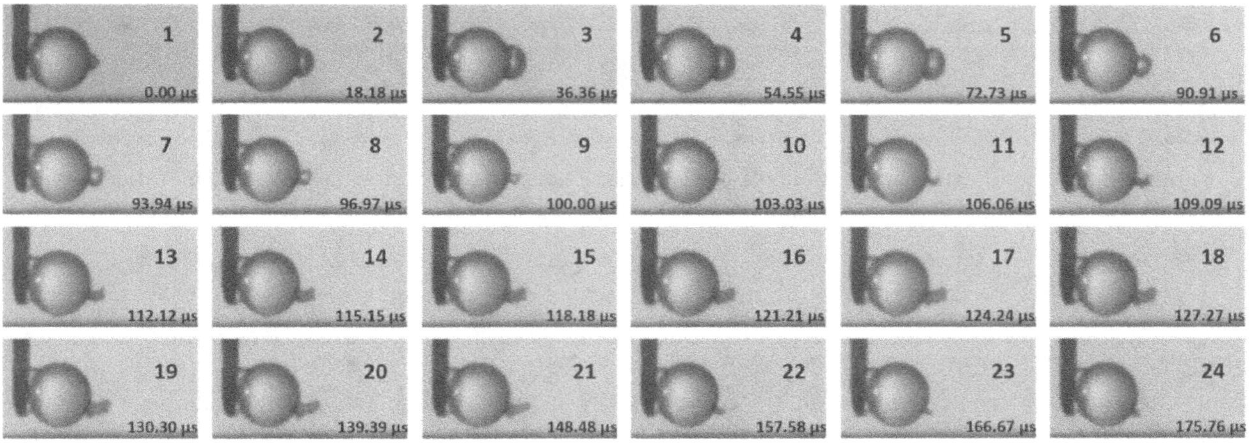

Figure 6 The experimental photos of bubble-particle interactions near the wall with θ=0° and a small bubble. The speed of the high-speed camera is 330,000 frames per second. The diameter of the particle is 2 mm.

3.3 The cases with θ=90°

To support our proposed three types of bubble-particle interaction regimes, another set of cases with θ=90° will be further introduced. Figures 7-9 correspond to types 1-3 in the section 3.2 respectively. For more general cases (e.g. with θ=varying between 0° and 90°), the research is still ongoing.

Generally speaking, cases with θ=90° could still be categorized into the aforementioned three regimes. However, it should be emphasized that the shape of bubble during collapse is quite different in this case (as shown in frames 8-20 of figure 7). In section 3.2, a mushroom shape bubble is demonstrated during the collapse. However, for this case, the bubble shows much complex and interesting shapes (cup-shape bubbles).

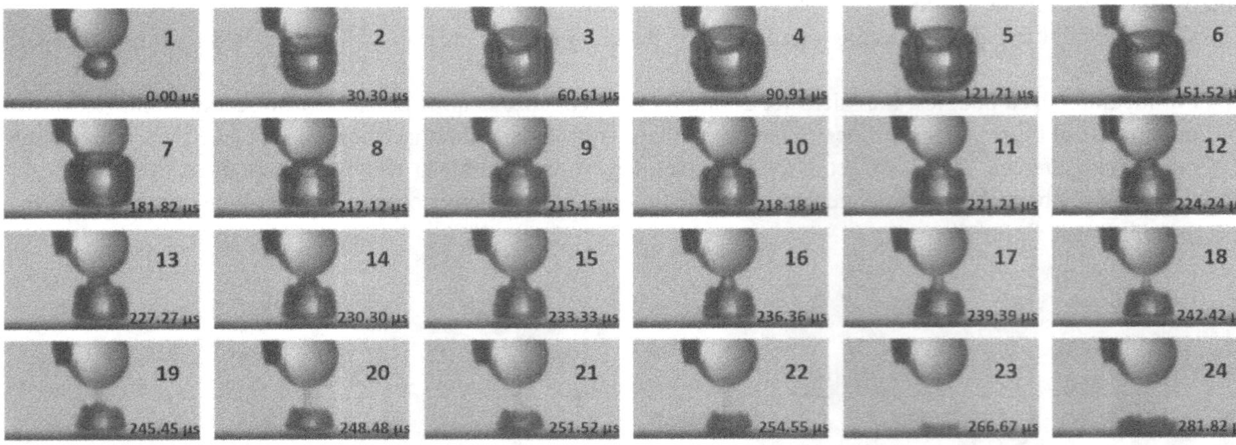

Figure 7 The experimental photos of bubble-particle interactions near the wall with θ=90° and a large bubble. The speed of the high-speed camera is 330,000 frames per second. The diameter of the particle is 2 mm.

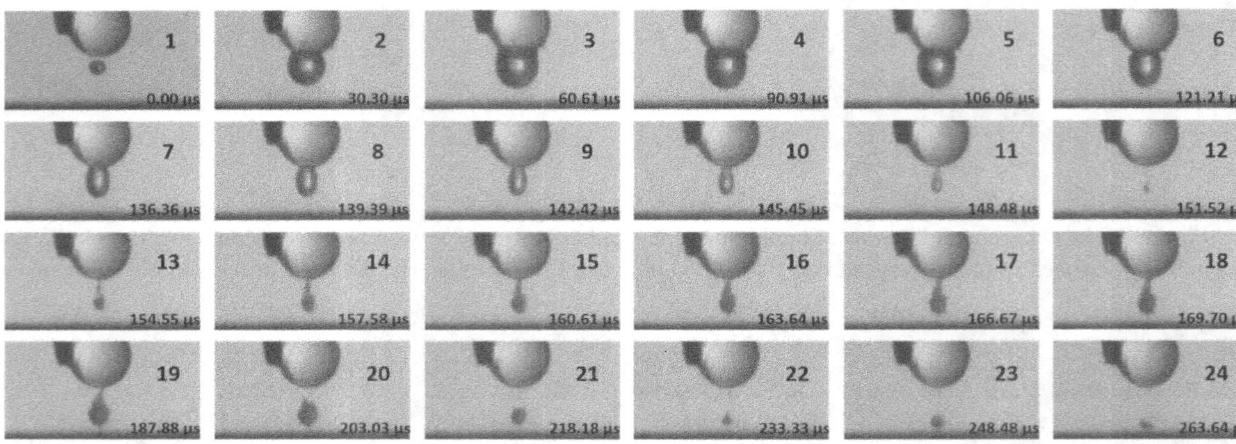

Figure 8 The experimental photos of bubble-particle interactions near the wall with θ=90° and a medium-sized bubble. The speed of the high-speed camera is 330,000 frames per second. The diameter of the particle is 2 mm.

Figure 9 The experimental photos of bubble-particle interactions near the wall with θ=90° and a small bubble. The speed of the high-speed camera is 330,000 frames per second. The diameter of the particle is 2 mm.

4. Conclusion

In the present paper, interactions between a laser-induced bubble and a stationary spherical particle have been experimentally investigated using the high-speed camera. We found that the presence of wall shows significant influences on the phenomenon. Based on the directions of micro-jet during bubble collapse, three types of the bubble-particle interactions are proposed with their characteristics introduced. In our future work, the influences of θ and the distance between bubble and solid boundary (D) on the proposed mechanisms will be further investigated in detail.

Acknowledgements

The authors would like to acknowledge the National Natural Science Foundation of China (Project Nos. 51506051 and 51606221).

References

[1] Lauterborn, W., Bolle, H. (1975). Experimental investigations of cavitation-bubble collapse in the neighbourhood of a solid boundary. Journal of Fluid Mechanics, 72(2), 391-399.
[2] Blake, J.R. and Gibson, D.C., 1987. Cavitation bubbles near boundaries. Annual review of fluid mechanics, 19(1), pp.99-123.
[3] Brujan, E.A., Nahen, K., Schmidt, P. and Vogel, A., 2001. Dynamics of laser-induced cavitation bubbles near an elastic boundary. Journal of Fluid Mechanics, 433, pp.251-281.
[4] Tomita, Y., Robinson, P. B., Tong, R. P., Blake, J. R. (2002). Growth and collapse of cavitation bubbles near a curved rigid boundary. Journal of Fluid Mechanics, 466, 259-283.
[5] Poulain, S., Guenoun, G., Gart, S., Crowe, W., Jung, S. (2015). Particle motion induced by bubble cavitation. Physical review letters, 114(21), 214501.
[6] Zhang, Y., Zhang, Y., Qian, Z., Ji, B., & Wu, Y. (2016). A review of microscopic interactions between cavitation bubbles and particles in silt-laden flow. Renewable and Sustainable Energy Reviews, 56, 303-318.

© 2018 ASME

Experimental study of planing motion of a cylinder along the nearly axisymmetric supercavity surface

[1]Vladimir Moroz; [1]Viktor Kochin; [1]Vladimir Serebryakov*; [2]John Dzielski.

[1]Institute of Hydromechanics of National Academy of Sciences, Kiev, Ukraine; [2]Stevens Institute of Technology, Davidson Laboratory, Hoboken, NJ, USA

Abstract

The aim and essence of this study is research of planing motion of an axisymmetrical cylindrical body along the cavity surface formed behind an axisymmetrical cavitator at some angle of attack, with the focus made on the study and analysis of the processes of interaction between the planing and cavity surfaces, including the development of robust methods for determining the forces and the cavity shape in the zone of this interaction and behind the planing surface

Keywords: supercavitation; planing motion; experiment

Introduction

Pioneering studies on the planing of a body within a cavity appeared in the 1950s in the known works of Kiceniuk and Greengard [1] , Waid [2] , and others. Later, these problems were studied in the well-known book [3] and in more modern publications [4] as well as many others. The base of the planning theory is the strip model assumption and impact added-mass theory. The main theoretical results of the planing theory have been developed by Paryshev [5], following the approach of Wagner [6], for the problem of a cylinder penetration into a cylindrical free surface. In case of a weak effect of planing motion on the cavity shape, this approach ensures a quite realistic estimation of the planning lift forces.. However, a number of experiments indicates a possibility of considerable interaction between the cavity and planing surface, which may significantly affect the planing forces and cavity shape. The objectives of this study are: (i) Study of cavitating flow in the zone of cavity-planing surface interaction and behind it, (ii) Estimation of the intensity of the interaction of planing and cavity formation processes.

Technique of experimental studies

Experimental studies were conducted in an open experimental basin at shallow depth $H \approx 150\text{mm}$ in the velocity range $U_\infty = 5 \div 10\text{m/s}$. As the basis, a model of planing motion of a cylinder at the angle of attack in a cavity formed behind a conic cavitator of radius $R_n = 25\text{mm}$ with a half-angle $\gamma = 20^o$ and angle-of-attack $\alpha_n = 7.5^o$ at small depth $H = 150\text{mm}$ was used.

Experimental studies , in particular, suggest that at even at shallow depths of motion the cavity shape and distortion of its axis is no longer dependent of the free boundary and is corresponding to its behavior in the infinite liquid [7]. This fact allows one to simulate the planning processes, which correspond to infinite streamlined flow, by experimental testing at shallow depth.

The planing cylinder position was designed to be in the middle part of a cavity (midsection), where the cavity surface is nearly horizontal. In addition, the drag and lift of the conic cavitator are adjusted in such way that they practically eliminate the gravity effect on the cavity axial distortion in zone of planning cylinder.

A series of tests includes planing of cylinders with two diameters D_c of 80mm and 100mm within the gap range of $\delta / D_c \sim 0.1 \div 0.3$ for three angles of attack α_0 (1.8°, 3.8°, and 6.6°) relative to the horizontal. The tests were conducted at constant flow velocity and gas volume flow rate Q_v. The latter is regulated by axial six-step fan with the maximal value $Q_v \approx 0.0066 \div 0.0070\text{m}^3/\text{s}$. During the test conduction, the value of Q_v is refined, while the difference of hydrostatic pressure and that in the cavity $P_\infty - P_c$, and value of the cylinder lift coefficient $C_{dy} = 2D_y / S_c \rho U_\infty^2$ are measured. Here $S_c = \pi D_c^2 / 4$ is cylinder cross-section area and D_y is lift.

Experimental studies were conducted under conditions close to stability loss of a ventilated cavity, which is defined via Paryshev's dependence [5]: $P_c / P_\infty > 0.54 \div 0.62$. To form quite stable supercavities with stationary streamlined flows and ensure more reliable lift measurements, a planing motion of a body along the cavity surface is realized

*Corresponding Author, Vladimir Serebryakov: serebryakov_vladimir@yahoo.com

with formation of developed hollow vortices in the cavity tail at Froude numbers by diameter d (Fr_d) being drag-equivalent to a disk with $\sigma Fr_d < 1$ outside the known inequality of Campbell and Hilborne [8] (i.e. for $\sigma Fr_d > 1$). However estimations found out here possibility of substantial influence of gas stream in a cavity on the pressure values in a cavity and accordingly on the number of cavitation especially along the front area of cavity. One of specific features of a flow of this kind is a practical essential independence of pressure in a cavity from gas injection intensity with a quite wide range of the latter variation.
Cavity deformations in the vertical plane were investigated first. Then, a mirror at an angle of 45° to the horizontal plane was placed at the distance of 300mm from the normal to the motion path in order to measure cavity deformations in the horizontal plane.

Dependences for experimental data processing and analysis

One of the basic properties of flow with free boundaries is the relative independence of expansion behavior of a cavity cross-section in the motionless liquid from the expansion behavior of its neighboring cross-sections, which was coined as "cavity expansion independence principle" [1]. The shape of the cavity front part is practically independent of the processes occurring in the planing zone. This allows to derive the cavity shape in its front part, which is not perturbed by the planing motion, and thus to assess the adjusted cavitator number σ_a, which corresponds to the infinite flow. It seems to be more accurate to apply the adjusted cavitator number σ_a, which controls the cavity shape in the infinite flow and is somewhat bigger than the cavitation number $\sigma = 2(P_\infty - P_c)/\rho U_\infty^2$, which is calculated from the measured value of pressure in the cavity P_c.
Noteworthy is that placement of a planing body into a ventilated cavity at certain cavitation number results in the cavity shape variation due to (i) planing motion, as well as to (ii) cavitation number variation induced by changes of gas loss and accordingly the pressure in cavity. If the deformation in the cavity shape and cavitation number caused by a planing hull is assessed, one can calculate the cavity shape variation induced by the planing process only. For the cavity shape diagnostics via its front part, a system of elementary equations Eq.(1a-1b) for the shape of slender axisymmetric cavity [9, 10], including solution of Eq.(1c) in the non dimensional form for cavitator radius $R_n = 1$:

$$\text{a)}\ \frac{d^2R^2}{dx^2} + \frac{\sigma(x)}{\mu_c} = 0\,,\ \text{b)}\ \left.\frac{dR^2}{dx}\right|_{x=0} = \sqrt{\frac{2(c_d - k\sigma)}{k\mu_c}}\,,\ \left.R^2\right|_{x=0} = 1\,,\ \text{c)}\ R^2 = 1 + \sqrt{\frac{2(c_d - k\sigma)}{k\mu_c}}\,x - \frac{\sigma}{2\mu_c}x^2 + \frac{\Delta\sigma(x)}{\mu_c} \quad (1)$$

For refined coefficients in Eq.(2a-2c), the known asymptotic solutions yield:

$$\text{a)}\ \mu = \ln\sqrt{\frac{\ln(10 + 2/\sigma)}{e\sigma}} \quad \text{b)}\ \mu_c = \mu\left(\frac{2}{\sqrt{1 - k\sigma/c_d} + 1}\right)^2 \quad \text{c)}\ k = 1 - \frac{2\ln(2/\sqrt{e})}{\ln(18 + 4/\sigma)} \quad \text{d)}\ \Delta\sigma(x) = 2\frac{gRn}{U_\infty^2}\int_0^x\int_0^x h_{g\alpha}dxdx \quad (2)$$

The solution (1c) is refined by adding term $\Delta\sigma(x)$ (2d), which is induced by the cavity axis distortion $h_{g\alpha}$ (3a) [1] under the gravity and cavitator's lift actions at the angle of attack α_n and Froude number $Fr = U_\infty/\sqrt{gR_n}$.

$$\text{a)}\ h_{g\alpha} = \frac{gR_n}{U_\infty^2}\int_0^x \frac{1}{R^2}\int_0^x R^2dx + \frac{c_{dy}}{2}\int_0^x \frac{R_n^2}{R^2}dx\,,\ \text{b)}\ c_d = c_{do} + \left[0.46 + 0.39\beta(\gamma) + \frac{0.89\cdot 0.017}{\beta(\gamma)^{1.36} + 0.028}\right]\sigma\,,\ \beta(\gamma) = \frac{\gamma^o}{90^o} \quad (3)$$

Here the formula for cone drag coefficient c_d (2b) $c_{do} \approx 0.206$ for $\gamma = 20^o$ is used. The cone lift coefficient formula for c_{dy} for $\gamma = 20^o$ derived by experiment is $c_{dy} = 0.835\alpha_n + 0.0158$. Given a feeble effect of some perturbations on the cavity shape under test conditions, it is expedient to use the adjusted cavitation number $\sigma = \sigma_a$, which controls the cavity shape in the infinite flow. The estimation of σ_a can be made by the solution of Eq.(4) using the experimental cavity radius R_a measured at some section $x = X_a$ in the forward part of cavity.

$$\sigma_a = \frac{2\mu_m}{(X_a - x_*)^2}\left[\sqrt{\frac{2(c_d - k\sigma R_*^2)}{k\mu_c}}(X_a - x_*) - \left(R_a^2 - R_*^2\right) + \frac{2}{\mu_c}\left(\frac{gR_n}{U_\infty^2}\int_0^{x_a}\int_0^{x_a} h_{g\alpha}dxdx\right)\right] \quad (4)$$

At the second stage of works, more exact estimates of the adjusted cavitation number in the cavity front part are made via equation (1a) at the beginning of calculation from cavity section at the distance $x = 3$ from flow separation section and correspondently corrected value of factor μ. Substantial is also an additional account of not circular deformation of the cavity sections at too small relative distance to the free border.

Experimental results: processing and analysis

Results of experiments for the cylinder of diameter $D = 80\text{mm}$ in the range of flow velocities $\approx (8.12 \div 8.14)\text{m/s}$ and the angle of rotation of the cylinder axis to the horizontal plane of 0° and 3.8^o were used (see Fig. 1).

a) b) c) d)

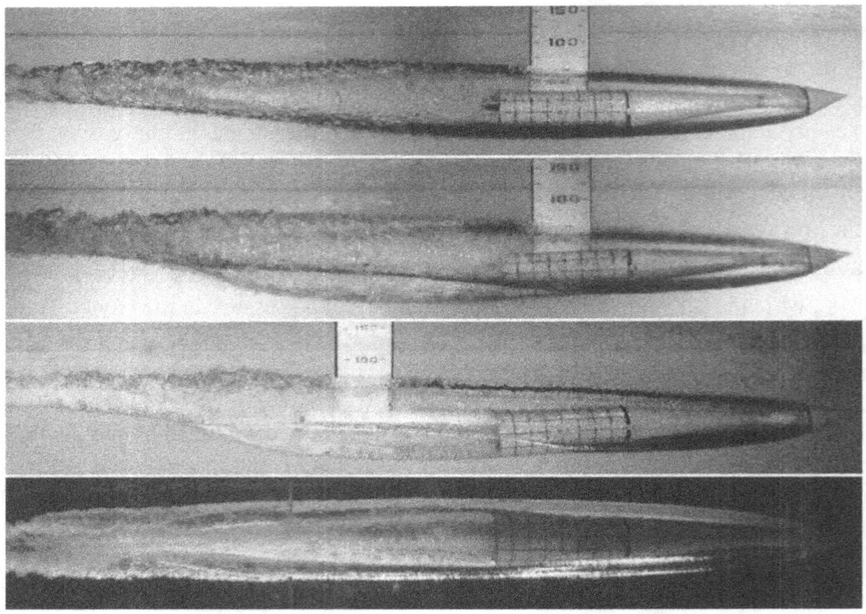

Figure 1. Results of experimental cavity photo-fixation for $D = 80\text{mm}$, $H = 150\text{mm}$ $U = 8.13 \div 8.14\text{m/s}$, $\alpha_c = 0^o$, 3.8^o

a) Experiment without mirror installation - a side view: $D = 80\text{mm}$, $U = 8.13\text{m/s}$, $\alpha_c = 0^o$, $Q_v = 0.0067\text{m}^3/\text{s}$, $\sigma = 0.034$

b) Experiment without mirror installation - a side view: $D = 80\text{mm}$, $U = 8.14\text{m/s}$, $\alpha_c = 3.8^o$, $Q_v = 0.0068\text{m}^3/\text{s}$, $\sigma = 0.036$, $c_{dy} = 0.031$

c) Experiment with mirror installation – a side view: $D = 80\text{mm}$, $U = 8.13\text{m/s}$, $\alpha_c = 3.8^o$, $Q_v = 0.0067 \div 0.0069\text{m}^3/\text{s}$

d) Experiment with mirror installation – the bottom view: $D = 80\text{mm}$, $U = 8.13\text{m/s}$, $\alpha_c = 3.8^o$, $Q_v = 0.0067 \div 0.0067\text{m}^3/\text{s}$

Results of the repeated experiment in the presence of a mirror with more accurate photo-fixation of cavity profile front parts (side and bottom views) are presented in Figure 2.

a) b)

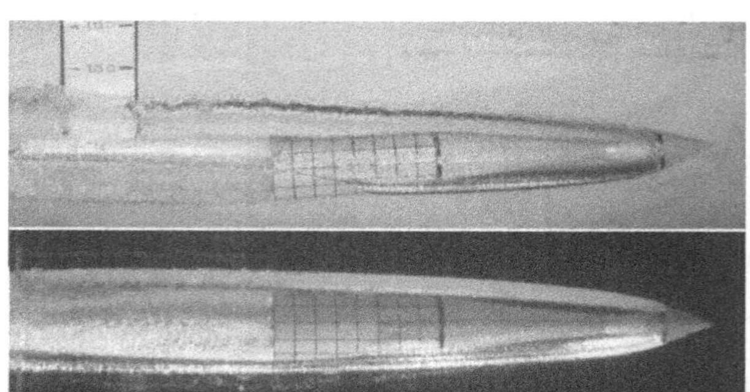

Figure 2. Experimental data obtained in the presence of a mirror with more accurate photo-fixation of cavity profiles for $D=80mm$ $U=8.13m/s$, $\alpha_c=3.8^o$, $Q_v=0.0067\div0.0069m^3/s$

a) side view: b) bottom view.

Data processing of three tests: Figures 1b, 1c, 2a made it possible to construct a composite contour of almost the total cavity in vertical plane, with the exception of portions strongly affected by the strut. The results of data processing of these tests performed with the same initial parameters are presented in the form of a cavity contour in vertical plane and compared with calculated results of cavitation number $\sigma_a\approx0.036$ defined via the profile shape identification on the basis of photo-fixation of cavity front parts (Figure 2) are illustrated by Figure 3a. Results of processing of photo-fixation of the experimental cavity profile in the horizontal plane (Figure 1d), in comparison with distribution of radii of a composite profile of a cavity in the vertical plane and cavity shape calculation for $\sigma_a\approx0.036$ are illustrated in Figure 3b.

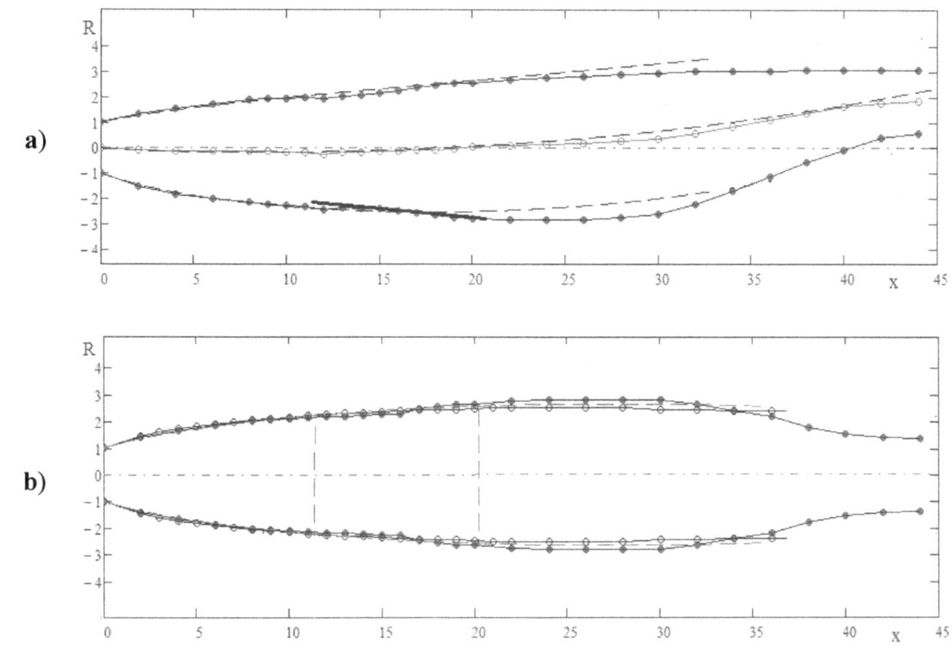

Figure 3. Comparison of contours of an experimental cavity in the vertical and horizontal planes for planing motion of a cylinder at angle of attack, in comparison with a cavity contour not perturbed by planing motion:

Cone radius $R_n=24.8mm$, cone cavitator attack angle 7.5^o, motion depth center $H=150mm$, speed of motion $U_\infty=8.13m/s$,

volumetric intensity of gas entrainment $Q_v\sim0.0067\div0.0069m^3/s$;

cylinder diameter $D_c=80mm$, angle of the cylinder axis relative to the horizontal plane $\alpha_c=3.8^o$, $c_{dy}=0.031$

a) Comparison of a profile of an experimental cavity in a vertical plane with a settlement profile of unperturbed cavity for $\sigma_a\approx0.036$

b) Comparison of profiles of an experimental cavity – in a horizontal plane with a settlement profile of unperturbed cavity for $\sigma_a\approx0.036$

—◆—◆— Profile of an experimental cavity in a vertical plane
—○—○— Profile of an experimental cavity in a horizontal plane
------ Calculation of the profile of an axisymmetric cavity via Eq (1-3), $\sigma_a\approx0.036$

Treatment of these experiments allows to present the most credible picture of flow. Comparison of data for the case of cavity form and sizes without a cylinder and with cylinders having diameters of D=80 and D=100 without of attack angles gives approximately identical results.

The change of gas pressures along cavity due to presence of models in a cavity does not excel 1%.
Maximal area sizes of cavity sections at planing and without planing on condition of identical pressure in a cavity should be approximately equal. Processing of data of experiment finds out some increase of maximal area of cavity sections at an planing as compared to a cavity without planing. It can be explained by reduction of size of circulation under the action of planing in the back part of cavity and, accordingly, change of conditions of gas loss in the back part of cavity. .Small increase of lateral sizes of cavity sections without planing explained by influence of very near free border. At the small numbers of Froude and presence of vortical cords also there is a substantial increase of cross runners of cavity after middle of cavity. Under conditional length of cavity in this case it is possible to count the section of split of surface of cavity on 2 vortical cords at the approximately permanent size of such length. Thus the known dependence included Eq. 3a for cavity axis curvature begins to give the considerably overpriced values in the conditions of these experiments beginning approximately from the sections of $x > 30$.
In reality, given the experimental data obtained, three key factors have also to be accounted for:

- Change of the relative flow velocity of cavity cross-sections of a rigid contour and cavity. This fact is proved by an essential deflection of the top contour of cavity (Figure 2a). The account of this effect can essentially reduce the lift value, in comparison with calculation via dependences [3].
- The motion of a cavity rigid section in the planing process induces the deformation of circular section of a cavity even at the invariable area of cavity section, which increases in the cross-sectional dimensions in a vertical plane, in contrast to the horizontal one. It can also essentially reduce the results of calculation, as compared to date [3].
- The account of small drag in planing motion, which leads to an increase in the cavity radius also can some reduces the lift.

Conclusion

The analysis of experimental results: Figure 2, reveals essential interaction between the cavity and planing surface. It can be used as a basis for elaboration of calculation techniques for lift and cavity shape within and behind the planing zone, which instrumental for the design and calculation of stabilizing and steering components.

Acknowledgements

The authors would like to thank Dr. Ronald Joslin, ONR and Dr. John Zimmerman, Dr. Salahuddin Ahmed, ONRG for their support and encouragement

References

[1] Kiceniuk T. and Creengard R., (1952) Measurements of Lift Coefficients for a Family of Cones planning on Water. Hydrodynamics lab Caltech, Pasadena CA, USA. Report N E-12.3, 1952, pp.15-23. (1952)
[2] Waid R. L. Forces on composite bodies in full cavity flow, ()1957, Hydrodynamics Laboratory, California Institute of Technology (1957), Report No. E-73.8.
[3] Logvinovich, G.V. (1969), Hydrodynamics of Flow with Free Boundaries, Kyiv, Naukova Dumka (in Russian)
[4] Dzielski, J., "A comparison of two planing models for a cylindrical hull on a circular free surface", 9th Int. Conference on High-Performance Marine Vehicles, Athens, Greece, 2014
[5] Paryshev E. V. (2006) "Approximate mathematical models in high-speed hydrodynamics," *J. Engineering Mathematics* (2006), No. 55, pp. 41-64.
[6] Wagner H. (1932) Uber Stoss- und Gleitvorgange an der Oberflache von Flussigkeiten. -Zeitschrift fur Angewandte
[7] Kochin V., Moroz V., Serebryakov V., and N. Nechitailo, (2015) Hydrodynamics of supercavitating bodies at an angle of attack under conditions of considerable effect of fluid weightiness and closeness of free border // J. JSOE, 2015, v. 5.
[8] Campbell I. J. and Hilborne D.V. "Entrainment behind artificially initiated cavities," Second symposium Naval Hydrodynamics, Washington (1958).
[9] Serebryakov V.V., (2009), Physical - mathematical bases of the principle of independence of cavity expansion, Proc. of the *CAV2009*, Ann Arbor, Michigan, USA, 2009, 14p.
[10] Serebryakov V.V., Arndt R. E. A., and Dzielski J. E., (2015) Supercavitation: Theory, experiment and scale effects, Proc. of Cav 2015 - J. of Physics: Conference series, v. 656, conference 1.

© 2018 ASME

Experimental Study on Interaction of Multiple Cylindrical Bubbles with Underwater Shock Wave

[1,2]Jingzhu Wang*; [3]Akihisa Abe; [1,2]Yiwei Wang

[1]Key laboratory for Mechanics in Fluid Solid Coupling System, Institute of Mechanics, Chinese Academy of Sciences, Beisihuanxi Road, Haidian District, Beijing 100190, China; [2]School of Engineering Science, University of Chinese Academy of Sciences, China;[3]Graduate School of Maritime Sciences, Kobe University, 5-1-1 Fukaeminami-machi, Higashinada-ku, Kobe 658-0022, Hyogo, Japan

Abstract

The present paper reports on an experimental investigation of the behaviors of multiple cylindrical bubbles induced by underwater shock wave. The bubbles are formed in a narrow water chamber by an injection to observe conveniently their internal flow. In this way, the number, size, and position of the bubbles are controlled. As observed by using the schlieren method, cavitation bubbles are produced behind a shear waves followed with expansion regions. The life times of these cavitation bubbles are short so that the motion of the cylindrical bubble is hardly affected. For a single bubble, the liquid jet is formed during the bubble expansion and points towards the propagation direction of shock wave. In the case of the multiple bubbles, the direction of the jet depends on the combination of the shock wave-induced action and interaction forces between the bubbles. Here, the generation of a counter jet is used to determine whether or not the interaction forces work. It is suggested that the strength of the bubble motion is a main factor related to the magnitude of the interaction force besides the inter-bubble space.

Keywords: cylindrical bubbles; underwater shock wave; liquid jet; counter jet; interaction force

Introduction

The interaction of shock waves with gas bubbles has been observed in a variety of fields such as shock wave lithotripsy, food engineering, and the treatment of the wastewater. The bubble dynamic behaviors have been studied in detail experimentally, theoretically, and numerically. To the best of the authors' knowledge, understanding of non-spherical collapse begins from the study by Kornfeld and Suvorov [1]. They suggested that cavities could collapse asymmetrically and produce a liquid jet. The jet is generated on the side of high pressure, points towards the low-pressure side, and penetrates the far surface of the bubble. After that, non-spherical collapse is characterized by the formation of a main transverse jet. In order to observe the internal processes of the cavities such as jet motion, Dear et al. [2] proposed a liquid/gel two-dimensional method to produce cylindrical bubbles. They conducted many experiments studying the collapse of a single bubble and multiple bubbles induced by shock wave. It was observed that high-speed jets were generated in the direction of the travelling shock waves. They also found the collapse took place layer by layer because of the shielding effect in the case of the clustered arrays of cavities [3]. However, the gelatin layer liquefied as soon as the shock wave went through it such that the observation was influenced by the liquefaction shadow. On the other research, Chen et al. [4] used a focused laser to generate two-dimensional bubbles in a micro gap filled with blue ink. They investigated the interaction dynamics of an existing stable microbubble and another laser-induced microbubble. They found that the strong interaction can merge the bubbles with complicated asymmetric intermediated patterns. Quinto-su and Ohl [5] recorded the dynamics of two laser-induced bubbles created at different distances in a two-dimensional geometry, and suggested that the interaction forces are inversely proportional to the inter-bubble separation.

In the present paper, an experiment is carried out to observe the behaviors of multiple cylindrical bubbles induced by a spark-induced shock wave. The cylindrical bubbles are produced by an injection in a narrow water chamber composed of two acrylic plates. Compared with the method of generating two-dimensional bubble proposed by Dear et al. [2], it is convenient to simultaneously apply various visualization methods such as the combination of an optical method with PIV or PTV technique which is not presented in the paper, for the way to use a narrow water chamber. In the experiments, the observation is carried out using the schlieren method. We discuss the jet direction and speed in the case of a single bubble and multiple bubbles after the passage of a shock wave.

Experimental Setup

Figure 1 shows a schematic of experimental setup for observing the interaction between cylindrical bubbles and shock waves. In the experiment, underwater shock waves were generated by a high-voltage pulse power supply (HPS 18K-A. Tamaoki Electronics Co., Ltd.). As shown in the figure, the narrow water chamber was composed of two acrylic

*Corresponding Author, Jingzhu Wang: wangjingzhu@imech.ac.cn

plates of 150 mm (L) ×150 mm (W) × 20 mm (D), and the width of the gap was 1-mm. In the narrow chamber, it is possible to produce the bubble of arbitrary diameters and to put them at required positions using an injection. The chamber was set horizontally to avoid the influence of the gravity on the motion of the bubbles, so that two optical mirrors were placed over and under the water chamber to introduce the observation light into the chamber. The optical observation was carried out by a high-speed camera (MEMRECAM HX-3, Nac Image Technology) with a light source (LS-M350, SUMITA optical glass Inc.). In the experiments, the electrodes were tungsten and the spark gap of the electrodes was 4 mm. The output power of the power supply was 50.0 kV.

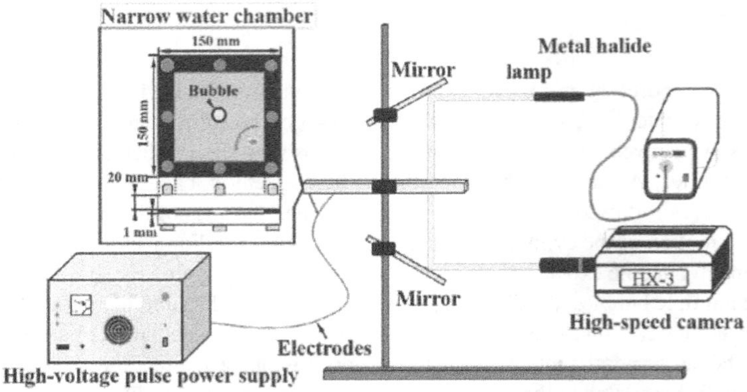

Fig. 1 Schematic of experimental setup for interaction of cylindrical bubbles with underwater shock waves

Results and Discussion

Figure 2 shows the schlieren images of the propagation of the incident shock wave and the generation of cavitation bubble in the narrow water chamber. In the following figures, the yellow arrows indicate the propagation direction of incident shock wave. The cylindrical incident shock wave (SW) passes through during the first and second frames, however there is no deformation of the quasi-cylindrical bubble. In the second frame, it is observed that another wave (EW) also passes through the cylindrical bubble. A lot of small cavitation bubbles (CB) are found behind it. The study of Wang et al. [6] suggested that the second wave is a shear wave travelling in the window material, so that expansion regions are induced in water. The shear wave is caused by the deformation of the window material as a result of the release of large energy when the electric discharge is triggered. Therefore, there are no cavitation bubbles around the cylindrical bubble.

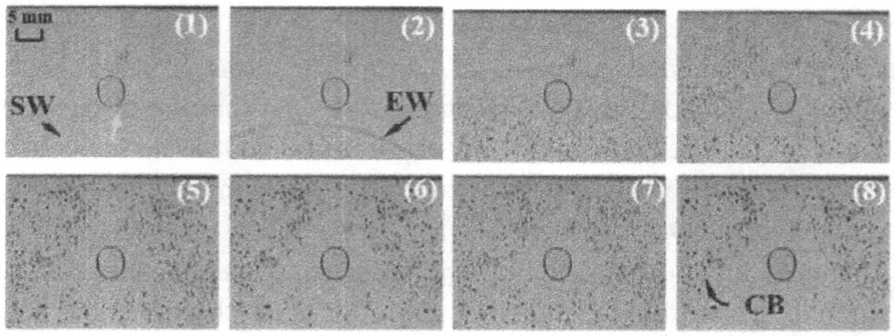

Fig. 2 Schlieren images of propagation of incident shock wave and generation of cavitation bubbles in narrow water chamber. The interval time is 10 μs. The cylindrical bubble with 5.2 mm in minor axis and 6.0 mm in major axis is placed at 50 mm from the discharge point.

Figure 3 shows the sequential images of the motion of a single bubble after the passage of the shock wave. Compared with the frame 1, change of the bubble interface is recognizable slightly after the cavitation bubbles generated by a shear wave disappear in the frame 2. Furthermore, the clear contraction starts in the frame 4. Hence, it is suggested that cavitation bubbles hardly affect the behaviors of the cylindrical bubble. Next, the liquid jet pointing towards the propagation direction of the shock wave is generated, travels inside the bubble, and impacts the far surface during the bubble contraction in the frames 5-11. Before penetrating the bubble surface, the velocity of the jet is about 10 m/s. After that, the bubble is broken up into three bubbles with its re-contraction motion. During this process, the internal flows of the bubble are not observed because the two-dimensional characteristics are not maintained for these small bubbles owing to respectively large width of the gap. On the other hand, we also investigated the correlation between the angles of the shock wave and the jets for a single bubble, as shown in Fig. 4. The results show that the liquid jet is always pointing towards the propagation direction of a shock wave when the shape of a single bubble is close to circle.

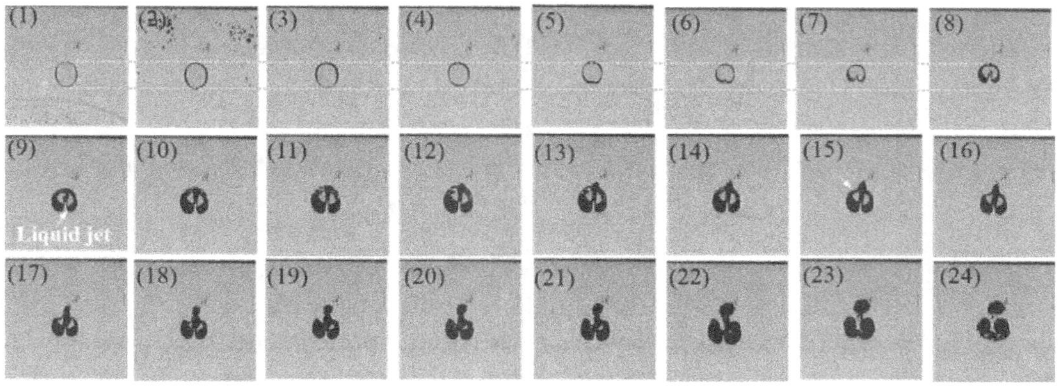

Fig. 3 Sequential images of motion of single bubble after passage of shock wave. The interval time is 100 μs. The cylindrical bubble with 5.2 mm in minor axis and 6.0 mm in major axis is placed at 50 mm from the discharge point.

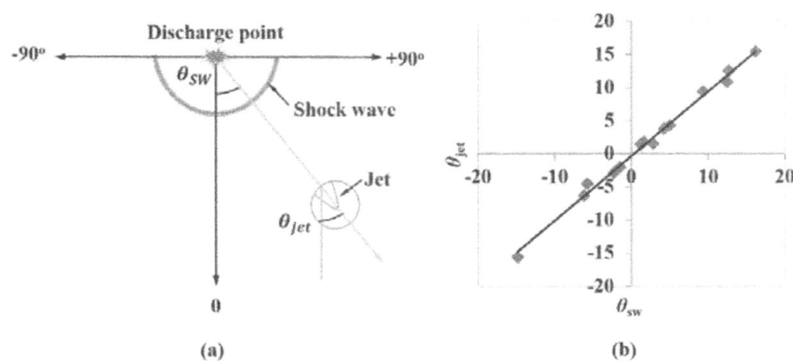

Fig. 4 Schematic of the incident angle of shock wave θ_{sw} and jet angle θ_{jet} (a), and correlation between two angles for single bubble (b).

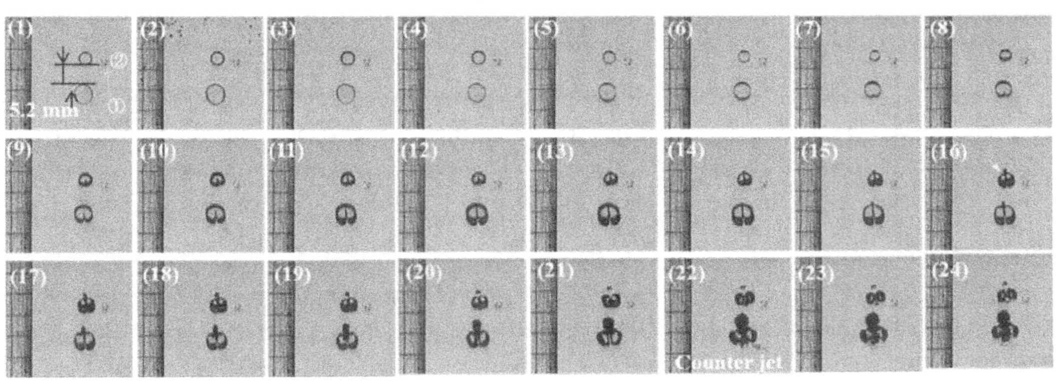

Fig. 5 Behaviors of two bubbles with 5.2-mm separation after passage of shock wave. The time interval is 100 μs. The dimensions of the bubbles are 5.2 mm in major axis and 4.6 in minor axis for ① and 3.4 mm in diameter for ②. The separation between the bubble surfaces is 5.2 mm. The distance from the discharge point to the lower surface of the bubble ① is 35 mm.

Figure 5 shows the behaviors of two bubbles after the passage of the shock wave. In the frames 1-14, it can be seen that the motions of the bubble ① and ② do not affect each other because of a large separation until the generated liquid jets penetrate the opposite bubble surfaces. During this process, the velocity of the liquid jet is 14.7 m/s for the bubble ① and 8.7 m/s for the bubble ②. Furthermore, both of them are pointing towards the propagation direction of the shock wave. After that, both of them begin to shrink. For the bubble ②, the upper surface contracts while the lower surface is stretched by the contraction of the bubble ①. Finally, the liquid jet is separated from the main bubble ② in the frame 19. For the bubble ①, the contraction motion seems not to be affected by the interaction force from the bubble ② before the fame 21. However, a counter-jet is generated in the opposite direction to the liquid jet in the frame 22, which is not captured in the case of a single bubble in Fig. 3. It suggests that the interaction force indeed works on the bubble ①. Here, a dimensionless distance γ is used to estimate the magnitude of the interaction force, i.e., the ratio of the inter-bubble separation to the bubble radius. For the bubble ①, the interaction force from the bubble ② is related to the dimensionless distance $\gamma = 3.06$. The study by Quinto-su and Ohl [5] suggests that the smaller γ is, the stronger the interaction force becomes. Besides, the generation of a counter jet can be used to judge that whether or not the interaction forces affect the bubble behaviors.

Figure 6 shows the behaviors of three bubbles after the passage of a shock wave. In the frame 1, the cavitation bubbles are captured as indicated in the red ellipse. Similarly as that in Fig. 5, the three bubbles deform successively and form the respective liquid jets. For the bubble ①, the direction of the liquid jet is the same as the propagation of the shock wave and its velocity is 13.9 m/s. It suggests that the motion of the bubble ① is not affected by the interaction forces since a counter jet is not induced, as observed in Fig. 5. However, the dimensionless distance, $\gamma = 2.83$ of the bubble ②, smaller than 3.04 in Fig. 5. Hence, the factors such as the pressure of the incident shock wave also influence the magnitude of the interaction force. In the case of the bubbles ② and ③, the jets tend to point towards each other under the combination of the interaction forces and shock wave-induced action. The velocity of the liquid jet is 14.8 m/s for the bubble ② and 14.6 m/s for the bubble ③. According to the direction angles of jets and shock wave, the magnitudes of the two forces are almost the same.

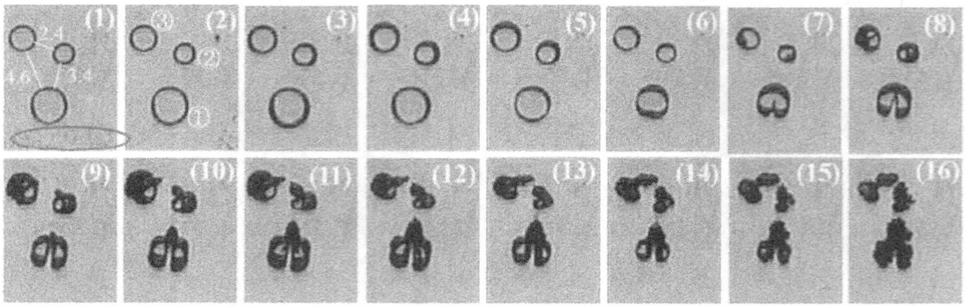

Fig. 6 Behaviors of three bubbles after passage of shock wave. The time interval is 100 μs. The dimensions of the bubbles are 4.1 mm in major axis and 4.0 in minor axis for ①, 2.4 mm in diameter for ②, and 3.4 mm in diameter for ③. The separations are 3.4 mm between ① and ②, 4.6 mm between ① and ③, and 2.4 mm between ② and ③. The distance from the discharge point to the lower surface of the bubble ① is 55 mm.

Conclusion

The paper presented an experiment to observe the behaviors of multiple cylindrical bubbles with a shock wave produced by underwater electric discharge. The cylindrical bubbles were produced by an injection in a narrow water chamber composed of two acrylic plates. In this way, the number, size, and position of the bubbles were controlled. The observation is carried out using the schlieren method. The visualization results showed the cavitation bubbles were produced due to the deformation of the window material as a result of the release of large energy at the discharge point. However, the life times of these cavitation bubbles were short so that they did not affect the subsequent motion of the cylindrical bubbles. For a single bubble, the liquid jet was formed during the bubble expansion and pointed towards the propagation direction of the shock wave. In the case of the multiple bubbles, the direction of the jet depended on the combination of the shock wave-induced action and interaction forces. Here, the generation of a counter jet was used to determine whether or not the interaction forces worked. Beside the separation between the bubbles, the strength of the bubble motion is expected to be a main factor related to the magnitude of the interaction force. A strong interaction forces requires high pressure of incident shock wave.

References

[1]Kornfeld, M., Suvorov, L. (1944) On the destructive action of cavitation. Journal of Applied Physics. 15.

[2]Dear, J. P., Field, J. E., Walton, A. J. (1988). Gas compression and jet formation in cavities collapsed by a shock wave. Nature. 332 (7).

[3] Dear, J. P., Field, J. E. (1988) A study of the collapse of arrays of cavities. Journal of Fluid Mechanics. 190.

[4] Chen, Y., Chu, H., I, L. (2006) Interaction and fragmentation of pulsed laser induced microbubbles in a narrow gap. Physical Review Letters. 96.

[5] Quinto-su, P., Ohl, C. D. (2009) Interaction between two laser-induced cavitation bubbles in a quasi-two-dimensional geometry. Journal of Fluid Mechanics. 2009.

[6] Wang, J., Abe, A., Koita, T., Sun, M. (2017) Contribution of Cavitation Generation to Shock Wave Sterilization Effects in a Narrow Water Chamber. The proceeding of 31th International Symposium on Shock Waves, in press.

© 2018 ASME

10th International Symposium on Cavitation - CAV2018
Baltimore, Maryland, USA, May 14 – 16, 2018 CAV18-05093

Experimental study on the effects of phase change during a bubble collapse

[1]Thomas Hopfes*; [1]Zhaoguang Wang; [1]Marcus Giglmaier; [1]Nikolaus Adams

[1]Institute of Aerodynamics and Fluid Mechanics, Technical University of Munich, Boltzmannstr. 15, D-85747 Garching, Germany

Abstract

A planar shock front produced in a shock tube provides an instantaneous pressure rise within a water-like gelatin mixture that contains bubbles filled with two different gases. Air represents a gas that reacts non-condensable under the experimental conditions, while refrigerant vapor R1233zd-E is a gas with a vapor pressure slightly above atmospheric pressure under the experimental conditions. We compare both types of bubbles during an aspherical collapse and highlight similarities and differences in order to analyze the effect of the gas content on collapse dynamics. The experimental set-up is presented and preliminary results are discussed. An image sequence of a recorded video and normalized radius data show the dynamic response of both investigated bubble types positioned next to each other close to a solid wall. A difference in the rebound behavior is indicated with the R1233zd-E bubble expanding more omnidirectional and to a lower relative radius.

Keywords: shock-induced bubble collapse; aspherical collapse dynamics; cavitation

Introduction

Collapsing bubbles or bubble clusters may be used in medical applications for targeted drug or gene delivery [1], for microbubble-cell interaction [2], to non-invasively dissolve blood clots [3] or to treat urinary stones in extracorporeal shock wave lithotripsy [4]. But first, the microbubbles need to be transported unharmed and in a controlled way to the desired locations in the human body and so a long residence time of the microbubbles in blood must be assured. To achieve that, bubbles are usually coated and filled with insoluble gases as for example perfluorocarbon gases like perfluorobutane or perfluoropentane. Interestingly, the vapor pressure of these gases is near atmospheric pressure [1] and hence, acoustic pressure pulses in medical treatments that are used to collapse bubbles, can also trigger condensation. It is thus possible that both effects interact and influence the bubble collapse. Consequently, the present experimental study aims at investigating this effect in the case of aspherical collapse dynamics by comparing bubbles filled with air, as a gas that reacts non-condensable under the experimental conditions, and R1233zd-E refrigerant vapor, a gas with a vapor pressure slightly above atmospheric pressure at the experimental conditions. R1233zd-E is a new generation of working fluid that is non-toxic, non- or mildly-flammable and has similar thermodynamic properties as the more common refrigerant R245fa [5]. To achieve good comparability, the collapse is initiated simultaneously for both bubbles under identical and homogeneous experimental conditions.

A wide number of high-quality experimental investigations with respect to the collapse behavior of bubbles is present in the literature, but mostly focuses on cavities produced through optical breakdown, e.g. [6, 7], or electric discharge, e.g. [8-10], where the effect of different gases cannot be investigated. Experiments investigating preexisting gas bubbles are difficult because the bubbles preferably need to be of spherical shape and must be positioned and kept at a point that can easily be observed during the collapse. Different ways to manage the problem were applied and some are mentioned here exemplarily. Tomita and Shima [9] examine bubble-shock wave interaction using a spark discharge to create a shock wave that impacts on an air bubble attached to a wall. The bubbles collapse aspherically and show the formation of a liquid jet that leaves an imprint on an impact specimen. Other groups investigate bubbles below a gelatin layer impacted by an underwater shock wave generated via a micro explosive pellet [11] or optical breakdown [12]. Rising air bubbles interacting with a high-pressure shock wave are also studied [13, 14]. Ohl and Ikink [15] analyze jetting and jet breakup of micro bubbles in water that are subject to a shock wave by an extracorporeal lithotripter, while Kodama and Takayama [16] focus on tissue-damage mechanisms during extracorporeal shock wave lithotripsy. Philipp, et al. [17] also use a lithotripter to collapse air bubbles attached to a plastic foil. Gases other than air are comparatively scarcely used, but Tomita et al. [18] mention the collapse process of a hydrogen bubble generated at the top of a needle through electrolysis.

This introduction shows that the aspherical bubble collapse is usually analyzed either for cavities or air bubbles. Other gases are rarely used and their effect on the collapse is not adequately analyzed. Nevertheless, this aspect should not

*Corresponding Author, Thomas Hopfes: thomas.hopfes@tum.de

be discarded completely as it has been shown, for example in the context of sonoluminescence, that oscillating bubbles with different gas contents show a different behavior during the collapse [19].

With this ongoing study, we will contribute further experiments to the community in order to identify the influence of different gases and the effect of phase change on bubble collapse behavior. The chosen experimental set-up offers two particular advantages: First, it allows the simultaneous investigation of diverse bubbles containing different gases and hence ensures a high comparability. Second, the homogeneous and well-defined initial conditions allow for comparison to successive numerical simulations.

Experimental equipment and procedure

A schematic diagram of the experimental set-up is shown in fig. 1. The shock tube, with an overall length of 22.5 m and an inner diameter of 290 mm, consists of three parts: the driver, the driven and the test section. A diaphragm separates the high-pressure driver section from the driven section at atmospheric pressure. After the diaphragm breaks, a shock wave forms and propagates towards the test section, where experiments are conducted by using the instantaneous pressure increase caused by the incident shock wave and subsequent shock reflection.

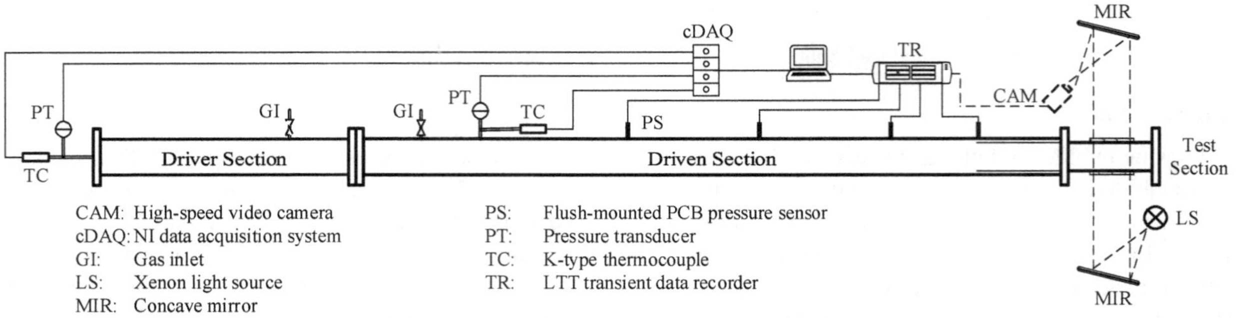

Figure 1: Experimental arrangement of the shock tube for bubble dynamics experiments.

The concept of analyzing bubble dynamics via shock tube experiments has been proofed by other groups before. Beylich and Gülhan [20], for example, investigate transient wave phenomena for glycerin-gas mixtures or shock wave-bubble interaction for single bubbles, bubble chains and bubble clouds. Even the different effect of He, N_2 and SF_6 on shock wave propagation is considered, but the influence on the collapse is not stressed. In another study, the free oscillation of bubbles after a shock tube generated pressure jump is analyzed with a needle hydrophone by monitoring acoustic waves radiated by the bubble [21]. Fujikawa and Akamatsu [22] use the expansion wave generated in a shock tube to expand gas nuclei that collapse under a following compression wave.

In the present set-up, the shock tube-generated shock wave arrives in a test section filled with gelatin. Figure 2 shows shock propagation and reflection exemplarily for a repeatedly investigated operating point (labeled as standard test case). The incident shock wave in the air arrives at the gelatin interface where it reflects almost ideally due to the high difference in acoustic impendence. However, the pressure at the interface must be equal and hence a compression wave forms and propagates into the gelatin. At the solid end of the test section, the compression wave is reflected, which further increases the pressure. Starting from that point in time, the ambient conditions around the bubbles remain constant at almost zero velocity and with a well-defined high pressure field until the compression wave reflects at the air-gelatin interface and returns as an expansion wave to the bubble position. A test time Δt_{test} of around 0.5 ms results for 400 mm gelatin and a speed of sound in gelatin similar to water. According to estimates with the Rayleigh-Plesset equation for 10 bar surrounding pressure, gas bubbles at atmospheric pressure of up to 5 mm collapse, rebound and collapse a second time within that time span, which confirms the validity of the concept for bubble dynamic experiments.

Gelrite™ Gellan Gum for microbiological applications is used in combination with magnesium sulfate as soluble salt to form a clear, agar-like gel. The Gelrite™ is dissolved in distilled water and the mixture is heated until the boiling point is reached and the magnesium sulfate is added. Subsequently, the liquid gelatin is decanted, cools down slowly and solidifies. Gelatin has already been used for bubble dynamics investigation, namely for the two-dimensional gel technique [23-25], but is now applied in a three-dimensional set-up. Bubbles are produced by carefully inserting the needle of a syringe into the partly solidified gelatin and releasing a defined amount of air or R1233zd-E.

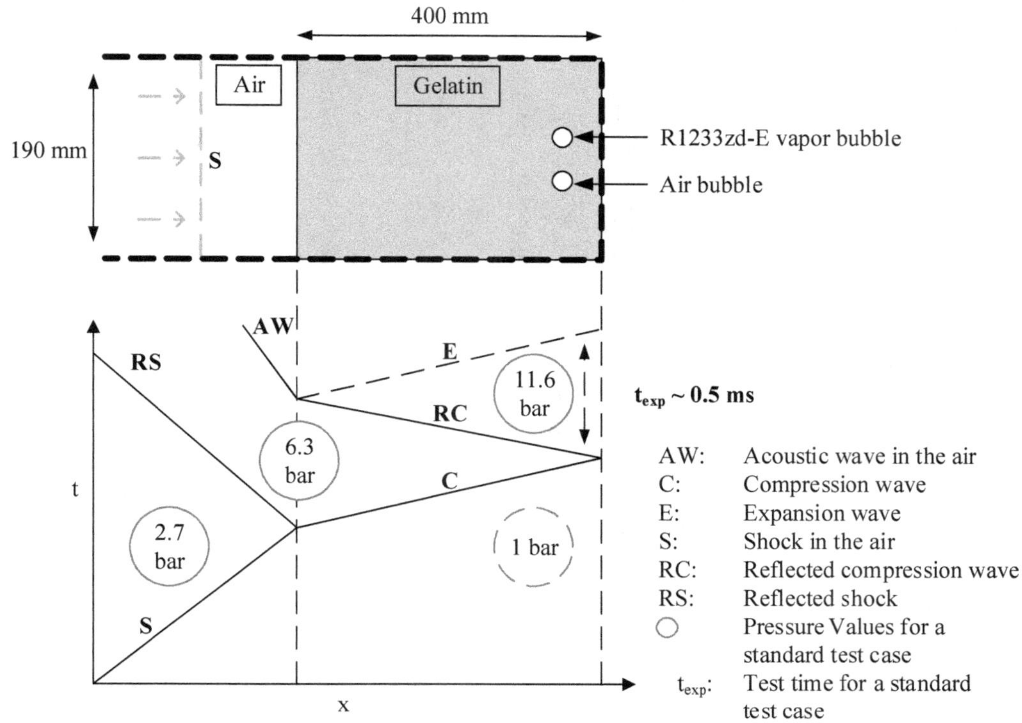

Figure 2: Sketch of the test section with a wave diagram; Pressure values and test time are given for a standard test case with 8.5 bar pressure in the driver section and atmospheric pressure in the driven section of the shock tube.

For visualization, we apply a Z-type schlieren system with a 150 W constant Xenon light source to create either schlieren images by cutting light at the second focal point with a knife edge or 'focused shadowgraph' images without the knife edge [26]. Using the knife edge is useful for shock visualization, while the direct shadowgraph provides more light and thus allows a higher frame rate or more optical zoom. The optical system projects the visualized section along the line-of-sight on a camera focal plane and thus gives a two-dimensional representation of the three-dimensional bubble. The entire collapse procedure is recorded with a Shimadzu HyperVision HPV-X ultra-high-speed camera that produces videos consisting of 128 consecutive frames with a resolution of 400x250 pixels at up to 5 million frames per second. In addition, PCB Piezotronics ICP® fast-response pressure sensors are connected to a LTT transient recorder and record the pressure during the collapse at up to 16 locations at a frequency of 4 million samples/s per channel.

A new test section was planned and is built, featuring a design that relies on experience gathered during preliminary experiments. Due to delays in deliveries and the manufacturing process, the section will be finished within January 2018 and then will allow to conduct experiments according to the simplified sketch shown in fig. 2. Despite the delay of the new test section, preliminary experiments have been conducted using acrylic boxes that are filled with gelatin and placed inside an already existing test section. This workaround did not allow for valid pressure measurement and undesired effects like deformation of the acrylic box and additional wave motion within the surrounding gas of the box occurred. The following preliminary results should thus be handled with care, but nevertheless show that the basic features of the concept work well. New results without the mentioned uncertainties are presented at the conference.

Preliminary results

Using the mentioned workaround, experiments with air bubbles have been conducted on a regular basis and capture well-known effects like the development of a liquid jet that penetrates the bubble. An image sequence of a single air bubble collapsing next to a solid wall is presented in fig. 3. The top row shows the aspherical collapse of the bubble towards the solid wall on the right. In the fourth image, a flattening of the bubble on the far side of the wall is already visible and a distinct liquid jet develops in the following images. The bottom row shows the rebound of the bubble including delicate flow features like Rayleigh-Taylor instabilities at the bubble surface causing bubble fission. Planar shock waves in the surrounding gas of the acrylic box are observed in frame 6-8, but since the new test chamber will exclude these additional waves no further attention is payed to this effect here.

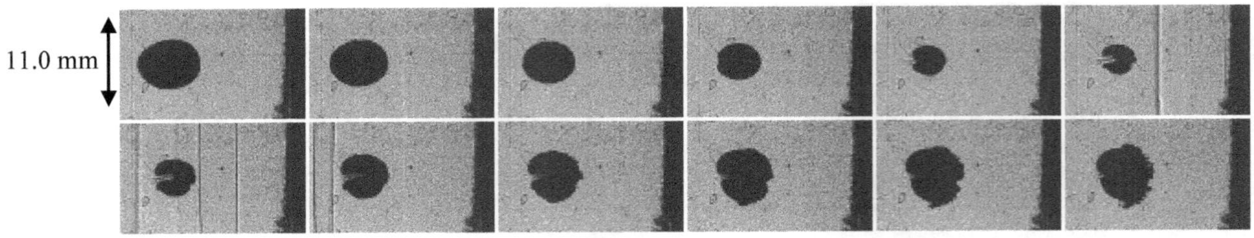

Figure 3: Image sequence of an aspherically collapsing air bubble close to a solid boundary (interframe time 30 μs).

As a next step, experiments for bubbles with different gas content were conducted and an image sequence of the recorded video of an exemplary experiment is presented in fig. 4. It shows a R1233zd-E bubble in the upper part and an air bubble in the lower part of each frame. The bubbles are positioned in the gelatin next to a solid wall in such a way that both bubbles can be observed simultaneously. The refrigerant vapor bubble is smaller and closer to the wall than the air bubble so consequently, a shorter collapse time results. However, the different bubble sizes and positions result in a similar stand-off parameter $\gamma = s/R_0$, with s as the distance between the solid boundary and the bubble center and R_0 as the bubble radius at collapse initiation. Thus, a similar collapse pattern is expected.

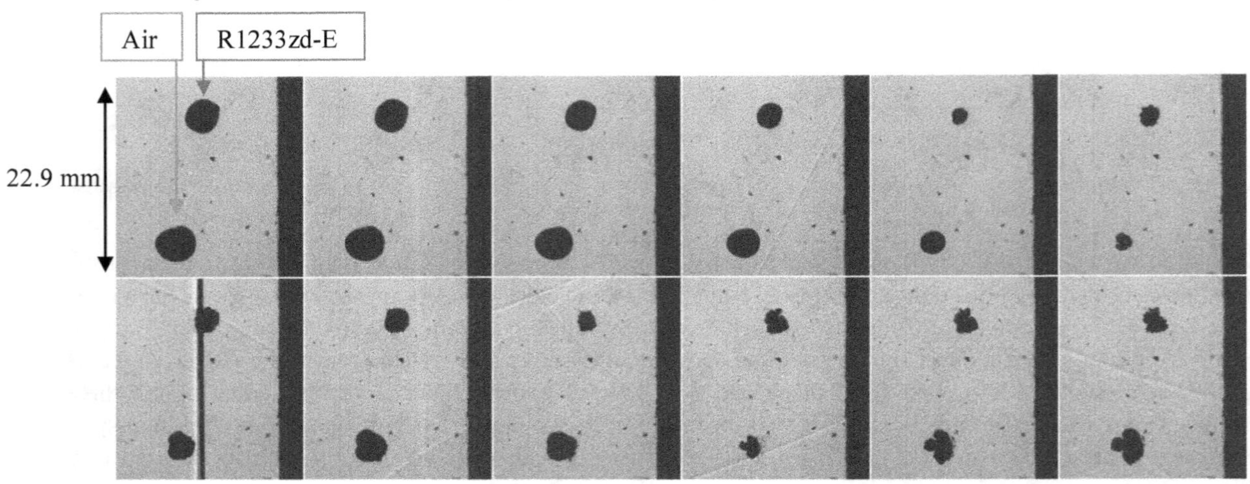

Figure 4: Image sequence of a collapsing R1233zd-E gas bubble and an air bubble close to a solid boundary (interframe time 42μs).

Comparable to the experiment with the single bubble, both bubbles collapse aspherically because of the asymmetric pressure field caused by the solid wall. The top row shows that the primary collapse occurs in a similar fashion for both bubbles, but following images indicate that there are some conspicuous differences during the rebound. The bubble with refrigerant gas creates a splash and rebounds almost uniformly in all directions while the air bubble shows a rebound clearly directed towards the solid wall. In addition, the air bubble produces a defined, mushroom-like shape after the second collapse, as can be seen in the last images, while the R1233zd-E bubble's shape remains rather arbitrary and omnidirectional.

An interesting aspect apart from the qualitative analysis is the quantitative value of the radius change. A time dependent equivalent radius $R(t)$ (determined by counting dark pixels and assuming spherical symmetry) for both bubbles of fig. 4 is extracted from the video and shown in fig. 5. The radius is normalized by the equivalent radius R_0 at collapse initiation so that the figure allows to directly compare the evolution of two bubbles with different gas content under equal experimental conditions. The R1233zd-E bubble collapses to a slightly smaller minimum normalized radius and the subsequent rebound is not as strong as for the air bubble. While the first difference is rather small and can solely depend on inaccuracies in the radius estimation, the second observation confirms the impression of the qualitative analysis that differences in the rebound behavior exist. The air bubble rebounds to a larger normalized radius than the vapor bubble and keeps a high amplitude for the oscillation during the completely observed time span. In contrast, the oscillation of the bubble filled with R1233zd-E seems to be dampened. A likely explanation is that the bubble undergoes partial condensation. Furthermore, a difference in oscillation time is observed that is caused by the different initial radii.

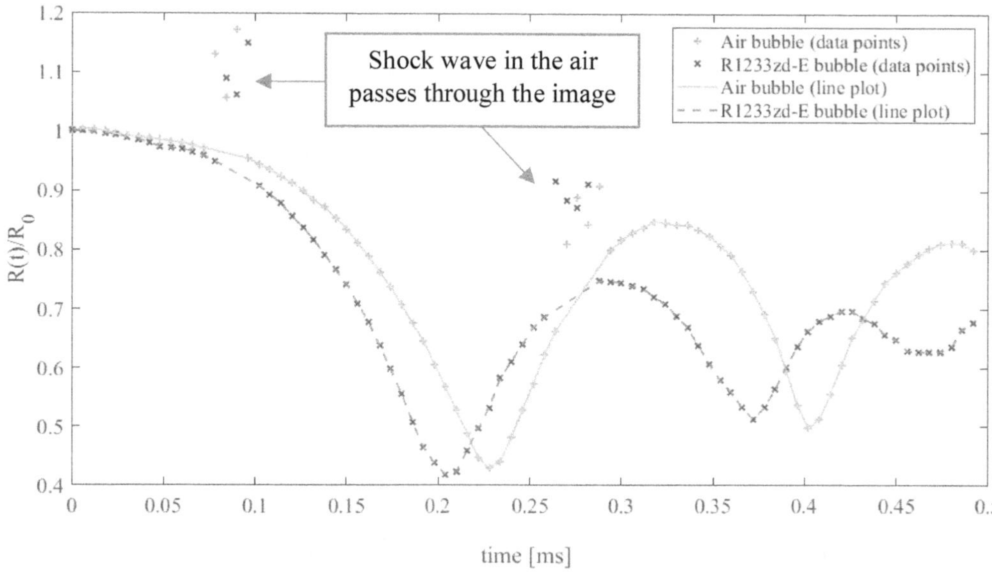

Figure 5: Temporal development of the equivalent radii of the R1233zd-E and air bubble for the experiment shown in fig. 4. Radii are normalized with the respective equivalent radii at collapse initiation.

As a final remark, the uncertainties of the preliminary workaround should be mentioned once more. The acrylic box filled with gelatin is placed in an existing non-ideal test section. The incident shock wave propagates much slower in the surrounding gas and interacts with the geometry of the box and the test section. This causes additional wave motion (visibly in fig. 4, frame 4, 7-10, 12) and a deformation of the acrylic box and both effects affect the bubble collapse. However, both bubbles undergo almost equal ambient conditions during the collapse, because the wave motion is much faster than the low-Mach number liquid flow.

Conclusion and outlook

First results prove that the concept of using a shock tube to create an instant pressure increase within gelatin is suitable to investigate bubble dynamics. Especially advantageous is the possibility that bubbles of arbitrary gas content at variable positions can be exposed to equal ambient conditions and analyzed simultaneously. For the shown experiment with an air bubble and a R1233zd-E gas bubble, a different behavior, notably concerning the bubble rebound, is indicated. A likely explanation is that partial condensation occurs whereas a complete condensation during the investigated collapse time is not observed.

To exclude the mentioned experimental uncertainties of the current workaround, further experiments will be conducted with a new test section throughout 2018 and presented at the conference. New experiments will feature a constantly high ambient pressure during the collapse and the collapse induced wave motion will be monitored by high frequency pressure measurements at distinct positions at the boundaries and near the bubble. To ensure a better comparability, these experiments will feature an equal bubble size and equal stand-off distance, allowing a more detailed analysis of the influencing parameters with a focus on the gas content. Furthermore, the non-Newtonian behavior of the gelatin and the influence on the bubble collapse will be analyzed via a separate test series with different pressures in the gelatin. The simple geometry and the well-defined initial and boundary conditions of the new configuration along with temporally highly resolved pressure distributions and schlieren pictures will provide extensive reference data to the community of bubble dynamics research. This renders a straightforward reproduction via numerical simulations possible and results can serve as numerical test cases for validation of computational models and methods.

References

[1] E. C. Unger, T. Porter, W. Culp, R. Labell, T. Matsunaga, and R. Zutshi, "Therapeutic applications of lipid-coated microbubbles," *Advanced drug delivery reviews,* vol. 56, no. 9, pp. 1291-1314, 2004.

[2] A. Delalande, S. Kotopoulis, M. Postema, P. Midoux, and C. Pichon, "Sonoporation: mechanistic insights and ongoing challenges for gene transfer," *Gene,* vol. 525, no. 2, pp. 191-199, 2013.

[3] E. Unger, T. Porter, J. Lindner, and P. Grayburn, "Cardiovascular drug delivery with ultrasound and microbubbles," *Advanced drug delivery reviews,* vol. 72, pp. 110-126, 2014.

[4] A. Coleman and J. Saunders, "A review of the physical properties and biological effects of the high amplitude acoustic fields used in extracorporeal lithotripsy," *Ultrasonics,* vol. 31, no. 2, pp. 75-89, 1993.

[5] S. Eyerer, C. Wieland, A. Vandersickel, and H. Spliethoff, "Experimental study of an ORC (Organic Rankine Cycle) and analysis of R1233zd-E as a drop-in replacement for R245fa for low temperature heat utilization," *Energy,* vol. 103, pp. 660-671, 2016.

[6] G. Sankin and P. Zhong, "Interaction between shock wave and single inertial bubbles near an elastic boundary," *Physical Review E,* vol. 74, no. 4, p. 046304, 2006.

[7] A. Philipp and W. Lauterborn, "Cavitation erosion by single laser-produced bubbles," *Journal of Fluid Mechanics,* vol. 361, pp. 75-116, 1998.

[8] C. F. Naudé and A. T. Ellis, "On the mechanism of cavitation damage by nonhemispherical cavities collapsing in contact with a solid boundary," *Journal of Basic Engineering,* vol. 83, no. 4, pp. 648-656, 1961.

[9] Y. Tomita and A. Shima, "Mechanisms of impulsive pressure generation and damage pit formation by bubble collapse," *Journal of Fluid Mechanics,* vol. 169, pp. 535-564, 1986.

[10] A. Zhang, P. Cui, and Y. Wang, "Experiments on bubble dynamics between a free surface and a rigid wall," *Experiments in fluids,* vol. 54, no. 10, p. 1602, 2013.

[11] T. Kodama, K. Takayama, and N. Nagayasu, "The dynamics of two air bubbles loaded by an underwater shock wave," *Journal of applied physics,* vol. 80, no. 10, pp. 5587-5592, 1996.

[12] T. Kodama and Y. Tomita, "Cavitation bubble behavior and bubble–shock wave interaction near a gelatin surface as a study of in vivo bubble dynamics," *Applied Physics B,* vol. 70, no. 1, pp. 139-149, 2000.

[13] Y. Tomita, K. Yamada, and K. Takayama, "Surface deformation of a gas bubble impacted by a shock wave," *Proceedings of Symposium on Shock Waves,* pp. 257-260, 1993.

[14] Y. Tomita, "Interaction of a shock wave with a single bubble," in *Shock Wave Science and Technology Reference Library*: Springer, 2007, pp. 35-66.

[15] C. Ohl and R. Ikink, "Shock-wave-induced jetting of micron-size bubbles," *Physical review letters,* vol. 90, no. 21, p. 214502, 2003.

[16] T. Kodama and K. Takayama, "Dynamic behavior of bubbles during extracorporeal shock-wave lithotripsy," *Ultrasound in medicine & biology,* vol. 24, no. 5, pp. 723-738, 1998.

[17] A. Philipp, M. Delius, C. Scheffczyk, A. Vogel, and W. Lauterborn, "Interaction of lithotripter-generated shock waves with air bubbles," *The Journal of the Acoustical Society of America,* vol. 93, no. 5, pp. 2496-2509, 1993.

[18] Y. Tomita, A. Shima, and T. Sugiu, "Mechanisms of impulsive pressure generation and damage pit formation by bubble-shock wave interaction.," *Proceedings of the Interational Symposium on Cavitation,* pp. 77–82, 1986.

[19] R. C. Srivastava, D. Leutloff, K. Takayama, and H. Grönig, *Shock Focussing Effect in Medical Science and Sonoluminescence*. Springer Science & Business Media, 2003.

[20] A. E. Beylich and A. Gülhan, "On the structure of nonlinear waves in liquids with gas bubbles," *Physics of Fluids A: Fluid Dynamics,* vol. 2, no. 8, pp. 1412-1428, 1990.

[21] K. Vokurka, A. Beylich, and H. Kleine, "Experimental study of gas bubble oscillations using a shock tube," *Acta Acustica united with Acustica,* vol. 75, no. 4, pp. 268-275, 1992.

[22] S. Fujikawa and T. Akamatsu, "Experimental investigations of cavitation bubble collapse by a water shock tube," *Bulletin of JSME,* vol. 21, no. 152, pp. 223-230, 1978.

[23] J. Dear and J. Field, "A study of the collapse of arrays of cavities," *Journal of Fluid Mechanics,* vol. 190, pp. 409-425, 1988.

[24] J. Dear, J. Field, and A. Walton, "Gas compression and jet formation in cavities collapsed by a shock wave," *Nature,* vol. 332, no. 6164, pp. 505-508, 1988.

[25] A. Swantek and J. Austin, "Collapse of void arrays under stress wave loading," *Journal of Fluid Mechanics,* vol. 649, pp. 399-427, 2010.

[26] G. S. Settles, *Schlieren and shadowgraph techniques: visualizing phenomena in transparent media*. Springer Science & Business Media, 2012.

© 2018 ASME

10th International Symposium on Cavitation - CAV2018
Baltimore, Maryland, USA, May 14 – 16, 2018 CAV18-05094

Exploration of a Possibility to Assess Erosive Cavitation by Acoustic Emission Technique

[1,2]Byoung Guk Kim*; [3]Crispin Fetherstonhough; [1]Philip A. Wilson; [1]Stephen R. Turnock;

[1]*University of Southampton, Southampton, UK;* [2]*Daewoo Shipbuilding & Marine Engineering Co. Ltd, Seoul, Rep. of Korea;*
[3]*Lloyd's Register, Southampton, UK.*

Abstract

In order to complement the conventional model scale cavitation erosion test methods, research is being undertaken to explore the use of an acoustic emission technique to provide a quantified measure of cavitation. The amplitude of AE signals within a structure, such as a propeller shaft, is assumed to change in proportion with the pressure impact loads from the collapse of vapour cavities. Analysis of the power spectrum of the signals should show some meaningful parameters in relation to the cavitation intensity and frequency, which can be used as a quantitative index of the cavitation erosion risk. The main results of a preliminary study are provided with illustration of some results of the experiment using a G-32 type vibratory cavitation apparatus combined with a CFD simulation. It is found that the AE power spectrum indicated two main peak frequencies from the acoustic excitation and the sub-harmonic oscillation of the acoustic cavities as reported in the other relevant literature. However, the acoustic driving frequency component appeared much stronger and the sub-harmonic oscillation frequency appeared as a band rather than a single peak. An investigation on the vibration characteristics of the sonotrode revealed the vibration amplitude could vary in the order of about 15 % of that desired. This might partly explain the reason of the frequency band formation around a central frequency of the sub-harmonic oscillation. Further investigation on the possible causes is underway. It has still a long way to go to establish a new model test methodology, but it appears the AE signal amplitude response has a certain relation with the magnitude of cavitation impact loadings.

Keywords: cavitation erosion; a new model test method; acoustic emission technique; sonotrode; OpenFOAM;

Introduction

For the shipbuilding industry, it is one of the most important task for the ship designers to avoid the risk of material damage from cavitation, and therefore, many efforts have been made to predict the risk of cavitation erosion in the early design stage by means of CFD and model testing. Since computer simulation to predict cavitation erosion requires too much time for practical engineering applications, the final evaluation on a design candidate is made often based on visual observation of the cavitation on a model in the cavitation tunnel and model erosion tests like paint tests ([1]). These model test methods can provide a good qualitative information such as the type of occurring cavitation events and possible location of erosion damages. However, it cannot tell the intensity of cavitation in a consistent way as desired.

Acoustic emission (AE) is the creation of transient stress waves in response to a dynamic event(s) to cause local changes in the strain inside a body ([2, 3]). It has been used for condition monitoring of many industrial structures thanks to its sensitivity to any dynamic changes inside a material ([4-7]). Boorsma and Fitzsimmons [8] reported that the AE signal could be measured inside a ship's engine room and the relevance of its rise of the peaks with the observed cavitation impact events. Since the signal amplitudes increase as the impact loadings, the AE signal amplitude appears to be a good candidate for a reliable indicator to assess severity of cavitation events and potentially the type of cavitation.

To investigate such a possibility of using AE technique for model testing to replace or complement the conventional test methods, a joint research work was started among three industrial bodies. If one can establish an AE amplitude threshold corresponding to the conventional paint tests, just one cavitation observation test while measuring AE signal shall be enough to evaluate the risk of cavitation erosion instead of waiting for another

* Corresponding author, Byoung Guk Kim: bgk1g15@soton.ac.uk

erosion tests. Therefore, it is expected to reduce the time and cost to assess the risk of cavitation erosion with more quantitatively consistent criteria by developing this new test method. As part of this wider programme, a G-32 type vibratory cavitation apparatus was used for the study. The cavitation generation by such an apparatus is regarded as reproducible, thus good for testing in the laboratory environment for this type of cavitation research in an accelerated manner ([9]). In this paper, the challenges for the exploration and some experimental results are described.

A new model test method using the AE technique

The new test scheme is in principle based on the energy approach ([10-14]). A simplifying assumption is made that the AE signal amplitude level will respond in proportion with the mechanical impact loads generated from the cavitation bubbles collapse events. Then one can analyse the aggressiveness of any cavitation events by simply analysing the AE signal amplitude and the power spectrum with the knowledge of AE signal amplitude threshold for the cavitation erosion. To do this with a G-32 type vibratory cavitation apparatus as a simpler model than the actual propeller model and the cavitation tunnel environment, it is necessary to know the difference in the power spectrums from the acoustic one and hydrodynamic one in a large cavitation tunnel. From that knowledge, one can determine an appropriate time duration of the tests equivalent to the conventional paint erosion tests in a large cavitation tunnel. For this initial stage of experiment, the equivalent time duration was set to 20 s based on literature review on the sub-harmonic oscillation frequency of the acoustic cavities and an assumption for the frequency of cavitation impact events in a cavitation tunnel that they occur once on each blade per revolution.

Reproduction of paint tests in the lab environment

A series of experiments were carried out using the G-32 type vibratory cavitation apparatus of a sonotrode according to the alternative ASTM G-32 standard method ([15]) with varying gap distances between the sonotrode horn tip and the test specimen from 30 to 15 mm. A photo of the instrumentation and a summary of the instrument specifications are shown in Figure 1 and Table 1. The objective was to find a test condition begins to remove stencil ink film from the surface of a test specimen similar to that in the conventional paint tests. Tap water was filled in the water bath until the sonotrode horn tip was submerged into water by 10 mm. A small copper alloy block specimen (dimensions: 25 x 25 x 15 mm^3) coated with a stencil ink was placed under the sonotrode horn tip. Each test was carried out with three test samples to confirm repeatability. At a test condition with 15 mm gap distance with 100 % power, the paint removal was started in 20 s and became pronounced in about 5 min. One of the test results are shown in Figure 2. The samples exhibited very good consistency in terms of location of damage and type of the damage pattern as shown in Figure 2(a), but the severity of cavitation felt from the images are fairly varying (cf. Figure 2(b).). This was regarded as to show what one could expect from the conventional paint tests. A later sonotrode vibration characteristics analysis found a possibility that some scatter in the sonotrode vibration amplitude might have contributed to such varying quality of the paint test results.

AE signal amplitude measurement

Another series of similar tests as the above were carried out to measure the AE signal amplitudes with variation of power and gap distances. A piezo-electric type AE sensor was installed just underneath a test specimen to capture the signal with a data acquisition device (National Instrument, NI USB-6251 BNC) capable up to 1.25 MS/s. Three data sets for each test condition were recorded for 5 s with 150 kHz sampling rate in a few seconds after the sonotrode began to move. This was to avoid recording any transient signal. The measured data was analysed using a FFT (Fast Fourier Transform) function implemented in MATLAB (ver. R2017a). The voltage signal from the AE sensor can be expressed in dB with reference to Hsu-Nielsen source by the following relation;

$$S_{AE} = 20\log\frac{V_{AE}}{1\times 10^{-6}}\ (dB)$$

The power spectrum of the measured AE signals are illustrated in Figure 3 and Figure 4. In the figures, one can see two peaks that are contributed from the acoustic excitation (19.96 kHz) and presumably from the sub-harmonic oscillation of the acoustic cavitation. The sub-harmonic frequency component barely changed from about 3.7 kHz as the gap distance reduced from 20 mm to 15 mm. The frequency changed to about 1.7 kHz when the gap distance reduced to 0.5 mm while the acoustic excitation frequency stayed the same in all the tested conditions. The experimental results appear to give similar results as the other references ([16-20]). However, the differences were (1) the main pressure driving component was not the sub-harmonic one and (2) the sub-harmonic frequency did not have a single value but rather a band with a central frequency. Finally, (3) the auto-correlation coefficients of the measured AE signals were more like a random process (cf. Figure 5). The last analysis result seems to explain why there appeared numerous frequency components in a wide band. Further investigation needs to be carried out on the possible causes of such departure from those previously reported.

Sonotrode vibration characteristics analysis

The sonotrode vibration characteristics was investigated by analysing the recorded data of the horn tip movement. The vibration movement was measured by a semi-conductor laser displacement sensor (Keyence, LK-G32) at a sampling rate of 50 kHz. Two sonotrode tips were available at the time of measurement. One was a slightly worn tip and used for the AE measurement reported in this paper. The other one was a brand-new one that had not yet tested before. Both tips had the same diameter of 18 mm.

The linearity of vibration amplitude response to the power control was evaluated by the least square method. The results are shown in Figure 6. Both tips did not show any significant difference in the residual squared of R^2 = 0.8531 (new tip) and 0.8799 (worn tip). However, there were significant variations of the average amplitudes between individual recording sessions outside the power band of 70 ~ 80 %. Therefore, a possibility was there to get a lower or similar amplitude with a higher power setting than a lower one. From this regard, the variance of the paint test results shown in Figure 2 might have been influenced from this variance of vibration amplitudes and it is recommended to perform the paint tests again by varying only the gap distance with a fixed power setting in 70 ~ 80 % power band. With this setup, the expected variance of the amplitude was ±13 μm assuming it follows Gaussian normal distribution. This still appears a large deviation compared with the mean amplitude of 78.6 μm at 80 % power setting. The influence of such variation needs to be evaluated using CFD simulation later.

A numerical approach to investigate the phenomenon

A study of the numerical simulation for the acoustic cavitation phenomenon has been carried out by the authors in order to understand better the relevant physics ([21, 22]). A HEM (Homogeneous Equilibrium Mixture) based compressible multiphase flow solver 'cavitatingFOAM' in an open source software, OpenFOAM (ver. 3.0.1) package was used. The description on the latest numerical model and the solver as used can be found in the reference ([22]). A test case with a 3.2 mm horn tip diameter was chosen from the experimental work of Žnidarčič, *et al.* [16]. Wallis barotropic compressibility model was used to simulate the compressibility effect and the time step size was set to 1×10^{-8} s (corresponding acoustic CFL number is 1.45). To prevent the solution from diverging at an early time stage, the minimum mixture density value had to be limited to a high value as 300 kg/m^3. The simulated time duration was up to 4.03 ms and the results are shown in Figure 7. The predicted pressure peaks appeared similar as the published experimental results, but the sub-harmonic frequency was predicted to be about twice the published results. The source of such deviation is yet to be further investigated.

Figures & Tables:

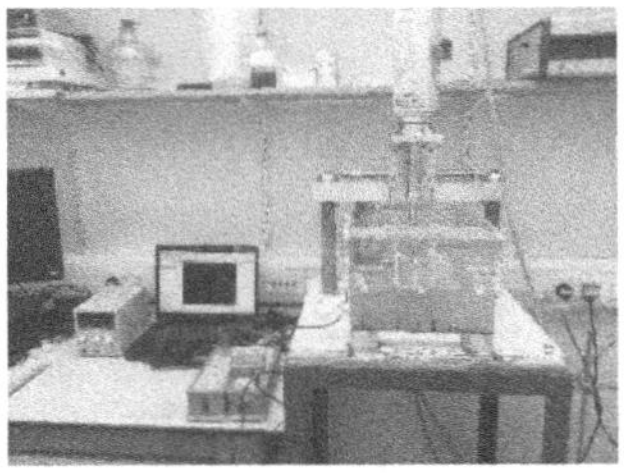

Figure 1. Instrumentation of the sonotrode to reproduce paint test results on a test specimen.

Sonotrode	Make/model: Hilscher UIP-1000HD Nominal max power output: 1000 W (adjustable range: 50 ~ 100 %) Operating frequency: 20000 ± 500 Hz Vibration amplitude (peak-to-peak): 43 (at 50 %) ~96 (at 100 %) μm Titanium horn tip diameter: 18 mm
AE sensor	Make/model: McWade, NS3303 Type: PZT. Data sampling frequency: 150 kHz. Output voltage: ± 10 V.
Water bath	Material: transparent polycarbonate acryl Dimensions: 305 x 400 x 115 mm^3

Table 1. Summary of the specifications of the instrument.

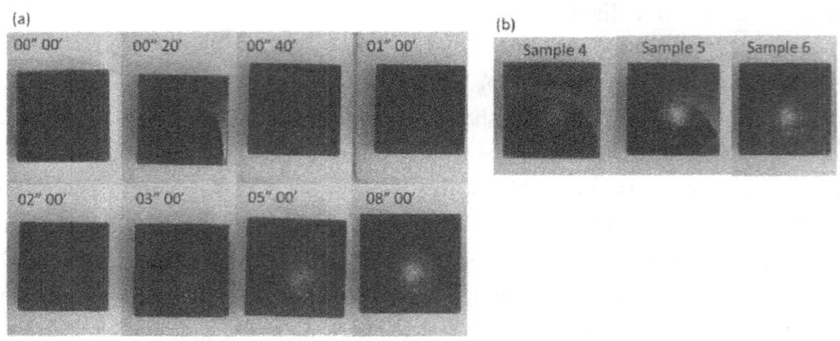

Figure 2. Damage patterns observed on the test specimens (Test condition: output power 100 %, gap distance 15 mm, exposure time: 8 min.). (a) time history of paint removal with sample 6, (b) test results showing different severity.

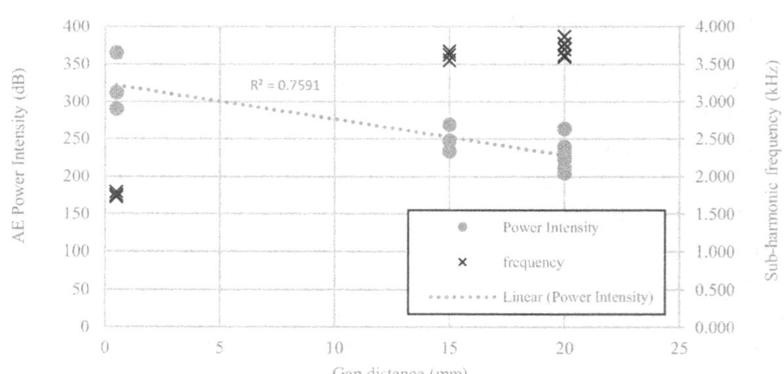

Figure 3. Measured AE power amplitude and sub-harmonic frequency vs gap distance.

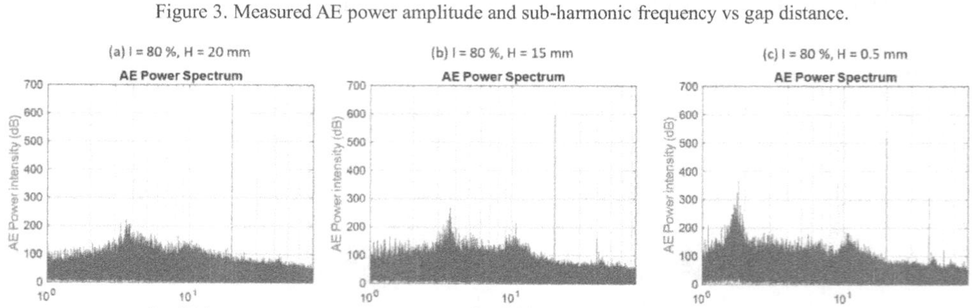

Figure 4. Power spectrum of the measured AE signals at different gap distances.

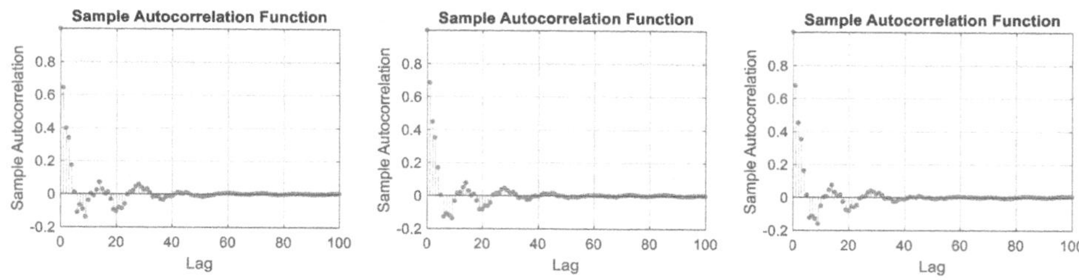

Figure 5. Auto-correlation coefficients of a set of the measured AE signals (80 % power, 15 mm gap distance).

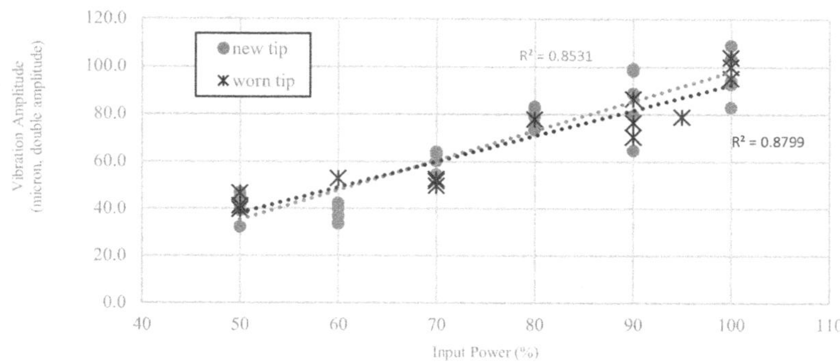

Figure 6. Mean amplitude data measured from the two sonotrod tips.

Power output setting		50 %	60 %	70 %	80 %	90 %	100 %
Worn tip	Average	42.5	-	51.4	-	77.8	99.7
	Standard deviation	3.6	-	1.5	-	8.1	4.4
New tip	Average	41.0	37.9	60.4	78.6	86.0	96.3
	Standard deviation	5.7	3.4	3.7	4.3	14.4	10.1

Table 2. Summary of the measured vibration amplitude data.

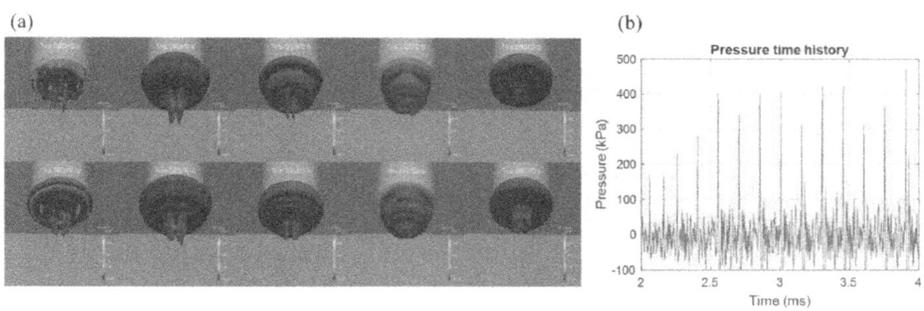

Figure 7. Predicted (a) acoustic cavity oscillation for two cycles (time interval between the frames: 30 μs, acoustic excitation period: 50 μs) and (b) time history of the pressure peaks calculated at a location laterally 7 mm apart from the horn tip end.

Conclusion

Based on the energy approach, a new model test method is being explored to utilise the acoustic emission (AE) technique to assess the risk of cavitation erosion during the early design stage of marine propellers. The key assumption is that the AE signal will respond with a certain correlation with the intensity of the cavitation events. Therefore, by measuring and analysing the power spectrum of the signals with the knowledge of the AE signal amplitude threshold for cavitation erosion, one may be able to assess the risk of cavitation erosion with more consistency without necessity of doing additional separate paint erosion tests compared with the existing model test methods.

A G-32 type vibratory cavitation apparatus was used to investigate an AE signal amplitude threshold equivalent to the conventional paint erosion tests. From the experiment, a limitation of the current paint erosion tests was seen in terms of the evaluation of severity of the involved cavitation events although a further investigation necessary to exclude the influence of variance of the vibration amplitude of the cavitation apparatus itself. Also it appears the AE signal amplitude increases as the intensity of the cavitation events become higher with a certain predictable relation. To confirm this relation, necessary several more experiments for the other gap distances between the tested points.

The power spectrum analysis results of the AE signals measured for 5 s revealed some deviation from the recent study results by other researchers. The first deviating fact was that the sub-harmonic frequency component formed a kind of band rather than a single frequency. The second fact was that the dominance of the acoustic excitation frequency component in the signal. The vibration characteristics of the cavitation apparatus was investigated to find any relevant source of such departure. It revealed a potential problem in repeatability of the acoustic excitation especially in the power range outside 70 ~ 80 % output power. Further investigation is necessary to address to main cause of such deviation in the power spectrum analysis results.

References

[1] ITTC (2005). *The specialist committee on cavitation erosion on propellers and appendages on high powered/high speed ships - proceedings of the 24th international towing tank conference.* International Towing Tank Conference. 2.
[2] Eitzen, D.G. and Wadley, H.N.G. (1984). *Acoustic-emission - establishing the fundamentals.* Journal of Research of the National Bureau of Standards. 89(1).
[3] Wadley, H.N. (1986). *Acoustic emission: Nature's ultrasound.* Review of progress in quantitative nondestructive evaluation.
[4] Rogers, L. and Carlton, J. (2010). *The acoustic emission technique: Application to marine structures and machinery.* Proceedings Lloyd's Register Technology Days.
[5] Zain, M., Jamaludin, N., Sajuri, Z., Yusof, M. and Hanafi, Z. (2010). *Acoustic emission study of fatigue crack growth in rail track material.* National Conference in Mechanical Engineering Research and Postgraduate Studies.
[6] Faisal, N., Ahmed, R. and Reuben, R. (2011). *Indentation testing and its acoustic emission response: Applications and emerging trends.* International Materials Reviews. 56(2).
[7] Ukpai, J.I., Barker, R., Hu, X. and Neville, A. (2013). *Exploring the erosive wear of x65 carbon steel by acoustic emission method.* Wear. 301(1).
[8] Boorsma, A. and Fitzsimmons, P. (2009). *Quantification of cavitation impacts with acoustic emissions techniques.* 7th International Symposium on Cavitation. (Paper No. 13).
[9] Chahine, G.L., Franc, J.-P. and Karimi, A. (2014). *Laboratory testing methods of cavitation erosion.* Advanced experimental and numerical techniques for cavitation erosion prediction. 106.
[10] Pereira, F., Avellan, F. and Dupont, P. (1998). *Prediction of cavitation erosion: An energy approach.* J Fluid Eng-T Asme. 120(4).
[11] Karimi, A. and Leo, W.R. (1987). *Phenomenological model for cavitation erosion rate computation.* Mater Sci Eng. 95.
[12] Fortes-Patella, R. and Reboud, J.L. (1998). *A new approach to evaluate the cavitation erosion power.* J Fluid Eng-T Asme. 120(2).
[13] Fortes-Patella, R., Challier, G., Reboud, J.L. and Archer, A. (2013). *Energy balance in cavitation erosion: From bubble collapse to indentation of material surface.* Journal of Fluids Engineering. 135(1).
[14] Fortes-Patella, R., Choffat, T., Reboud, J.L. and Archer, A. (2013). *Mass loss simulation in cavitation erosion: Fatigue criterion approach.* Wear. 300(1-2).
[15] International, A. (2011). *Standard test method for cavitation erosion using vibratory apparatus.*
[16] Žnidarčič, A., Mettin, R., Cairós, C. and Dular, M. (2014). *Attached cavitation at a small diameter ultrasonic horn tip.* Physics of Fluids. 26(2).
[17] Žnidarčič, A., Mettin, R. and Dular, M. (2015). *Modeling cavitation in a rapidly changing pressure field - application to a small ultrasonic horn.* Ultrason Sonochem. 22.
[18] Mottyll, S. and Skoda, R. (2016). *Numerical 3d flow simulation of ultrasonic horns with attached cavitation structures and assessment of flow aggressiveness and cavitation erosion sensitive wall zones.* Ultrasonics sonochemistry. 31.
[19] Truong, V.A. (2009). *A study of impulsive pressure distribution of cavitation generated by a high frequency vibrational probe.* J. Sci. Technol.(Technical Universities Pub. Hanoi, Vietnam). 73.
[20] Mottyll, S., Müller, S., Niederhofer, P., Hussong, J., Huth, S. and Skoda, R. (2014). *Analysis of the cavitating flow induced by an ultrasonic horn–numerical 3d simulation for the analysis of vapour structures and the assessment of erosion-sensitive areas.* EPJ Web of Conferences. 67.
[21] Kim, B.G., Wilson, P.A. and Turnock, S.R. (2016). *Numerical simulation of an ultrasonic vibratory cavitation device.* 19th Numerical Towing Tank Symposium.
[22] Kim, B.G., Wilson, P.A. and Turnock, S.R. (2017). *Simulation of acoustic stream with a sonotrode.* 20th Numerical Towing Tank Symposium.

© 2018 ASME

Exploration of cavitation-induced erosion metrics in throttle flow simulations

[1]Gina M. Magnotti*; [2]Michele Battistoni; [1,3]Kaushik Saha; [1]Sibendu Som;

[1]Argonne National Laboratory, Argonne, IL, USA; [2]Università degli Studi di Perugia, Perugia, Italy; [3]Bennett University, Greater Noida, India

Abstract

Although there have been extensive experimental and computational investigations in characterizing cavitation phenomenon in both diesel and gasoline direct injectors, much is still unknown about the mechanisms driving cavitation-induced erosion, and how this complicated fluid-structure interaction should be modeled. To explore current modeling capabilities, a numerical investigation was conducted within the CONVERGE modeling framework to assess proposed cavitation erosion metrics in the literature, and their link to the predicted cavitation cloud collapse mechanism. The multiphase flow within the Winklhofer Throttle "U" geometry was modeled using a compressible mixture model, where phase change was represented using the Homogeneous Relaxation Model (HRM) and the turbulent flow evolution was modeled using a dynamic structure approach for Large Eddy Simulations (LES). After comparing the model predictions against available experimental data, representative condensation events and potential cavitation erosion sites were identified. The cavitation cloud structures responsible for potential material damage were visualized through the evolution of the vorticity field. For the modeled throttle geometry, it was found that the horseshoe cloud implosion mechanism was predicted to occur and generate excessive impact loads at the throttle boundary.

Keywords: diesel injector; erosion index; condensation; cavitation cloud structures

Introduction

Cavitation-induced erosion has been extensively studied using experimental [1,2,3,4] and computational modeling techniques [4,5,6,7,8,9] to gain insight into the physical mechanisms driving this process for a range of applications, from pump impellers to ship rudders. Cavitation erosion has been noted in diesel injector hardware, namely within the needle seat region and along the injector needle, as well as the entrance to nozzle holes [4,6]. A common feature among these vulnerable locations is the proximity to an area contraction, where local flow acceleration and pressure reduction promote cavitation. The erosive potential of the cavitation shedding processes in these regions have been found to be highly dependent on the local geometric features and fluid properties.

Ideally, an injector hardware designer would be able to utilize a computationally efficient design tool to accurately predict these cavitation and condensation events, and to inform improved designs that mitigate the propensity and severity of cavitation-induced erosion. However, the main challenge of developing such a tool is linking the thermofluidic phenomenon of vapor generation and collapse with material damage to neighboring surfaces. In lieu of computationally expensive fluid-structure interaction modeling [9], the Eulerian mixture modeling approach has been accepted as a computationally efficient means of capturing cavitation phenomena when sufficient resolution is employed [8]. However, when employing such a framework, there remains a need to link condensation events with cavitation erosion potential. Although several cavitation erosion indicators have been proposed in the literature [5,6,7], no single metric has been identified as universally applicable across all injector geometries and injection conditions.

The overarching goal of this work is to identify cavitation cloud collapse mechanisms that are likely to occur within injector orifices, and develop parameters that adequately characterize the cavitation erosion potential of these fluid structures. In order to progress towards identifying a predictive and robust indicator, we focus our efforts on modeling the flow of pressurized fuel through a simplified throttle geometry with an inlet diameter of approximately 300 μm [10], which serves as a geometric analogue to an injector nozzle. We study the cavitation shedding process in detail, and identify cavitation cloud structures that are formed. In the initial stage of this effort, we assess the ability of the maximum predicted pressure at the throttle surface boundaries to characterize the propensity for cavitation erosion [6], and evaluate its link to the observed cavitation structures.

Model Formulation

In this work, we utilize CONVERGE [11] to model cavitation within the Winklhofer Throttle "U" geometry [10] for pressure drops ranging from 20 to 85 bar. Although no experimental data exist to characterize erosion events within

*Corresponding Author, Gina Magnotti: gmagnotti@anl.gov

this geometry, this test geometry serves as the starting point to validate the current cavitation modeling configuration, and study cavitation cloud structures that are predicted. The key parameters describing the experimental conditions explored in this work are defined in Table 1. Diesel fuel is represented using properties for tetradecane, and trace amounts of non-condensable gas, namely N_2 comprising a mass fraction of 5.0e-6, are assumed to be present in the liquid fuel. Liquid and gas phases are treated as compressible. Cavitation and condensation are represented using the homogeneous relaxation model (HRM) [12], where the rate at which the instantaneous mass fraction of vapor, x, approaches its equilibrium value, $\bar{x}$, is defined with the following relation,

$$\frac{Dx}{Dt} = \frac{\bar{x} - x}{\theta} \tag{1}$$

where θ is the relaxation time scale. In this study, cavitation and condensation are assumed to occur at equal timescales, and are calculated as $\theta = \theta_0 \alpha^{-0.54} \psi^{-1.76}$, where θ_0 is a coefficient set to 3.84e-7, α is the total void fraction, including fuel vapor and non-condensable gases, and ψ is the non-dimensional pressure ratio ($\psi = \frac{p_{sat} - p}{p_{crit} - p_{sat}}$). To adequately represent the turbulent flow structures within the throttle, the large eddy simulation (LES) dynamic structure model [13] is employed with a minimum grid size of 5.0 μm.

Liquid Fuel	Fuel Temperature [K]	Upstream Reservoir Pressure [bar]	Downstream Reservoir Pressure [bar]	Pressure Drop Across Throttle [bar]	Nitrogen Mass Fraction
Tetradecane	300	100	15 - 80	20 - 85	5.0e-6

Table 1. Modeled conditions within the Winklhofer Throttle "U" Geometry

Results and Discussion

To validate the model set-up, predictions of mass flow rate at the throttle exit are compared with the experimental data from Winklhofer et al. [10] for a range of pressure drop conditions from 20 to 85 bar, as shown in Figure 1. Overall, excellent agreement is achieved between the predicted and measured mass flow rates across the wide range of pressure drops evaluated in this study. At pressure drop conditions greater than 70 bar where choking of the flow is observed, the level of disagreement between the model predictions and experimental data increases slightly, but remains within 3% of the measured mass flow rate. This validation exercise suggests that the fluid properties and treatment of the multiphase flow in this modeling approach well represent the experimental conditions at the throttle exit.

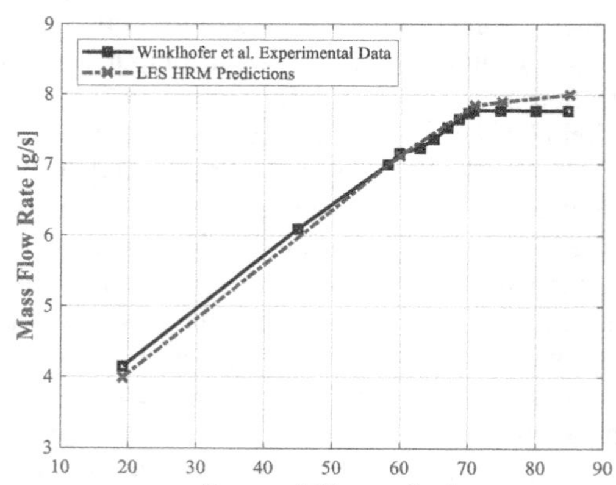

Figure 1. Comparison of measured and predicted mass flow rates for a range of pressure drop conditions across the Winklhofer Throttle "U" geometry [10].

In order to assess the ability of the modeling framework to capture the flow conditions within the throttle, model predictions of the vapor mass fraction distribution within the central plane are compared to the extent of cavitation as indicated in the experimental images of Winklhofer and co-workers [10]. A representative comparison is shown in Figure 2(a) and (b) for a pressure drop condition of 75 bar, where both the experimental and computational images have been time-averaged after steady state conditions have been reached. Cavitation inception is predicted at the inlet corners of the throttle, and extends along the wall boundary until approximately half of the length of the throttle. In comparison to the experimental results in Figure 2(a), less cavitation is predicted. Due to the sensitivity of cavitation inception and development to the inlet geometric features, it is possible that this discrepancy may be due to slight differences between the experimental and simulated geometries. Because the predicted mass flow rate at the throttle exit is within 2% of the experimental measured condition, this result also suggests that the initial flow development at the throttle entrance may not be adequately resolved to capture the flow acceleration and local pressure reduction required to initiate sufficient cavitation formation observed in the experiment. Future work will explore the influence of turbulence model choice and grid resolution on the predicted cavitation structures.

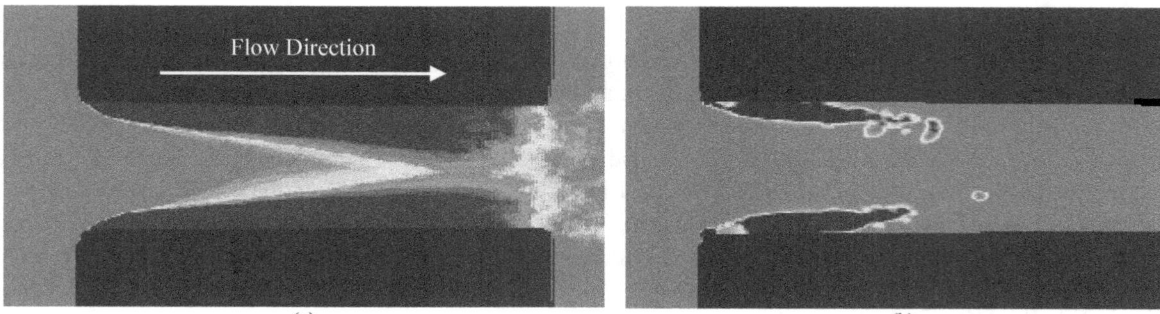

(a) (b)

Figure 2. Comparison of total void fraction between (a) experimental image from Winklhofer et al. [10], reproduced with permission, and (b) time-averaged HRM model prediction at the central slice for a pressure drop of 75 bar.

Although model predictions indicate less cavitation than was experimentally observed, the current results present an opportunity to examine the cavitation shedding process and erosive potential of predicted condensation events. In this computational exploration, the maximum predicted pressure at the throttle boundary, as proposed by Koukouvinis and co-workers [6], is implemented as an indicator of cavitation erosion. Such an indicator aids in highlighting key cavitation cloud collapse structures among the multitude of time steps. A representative cavitation cloud collapse event is shown in Figure 3, where cavitation structures are visualized using iso-contours of 1% total void fraction.

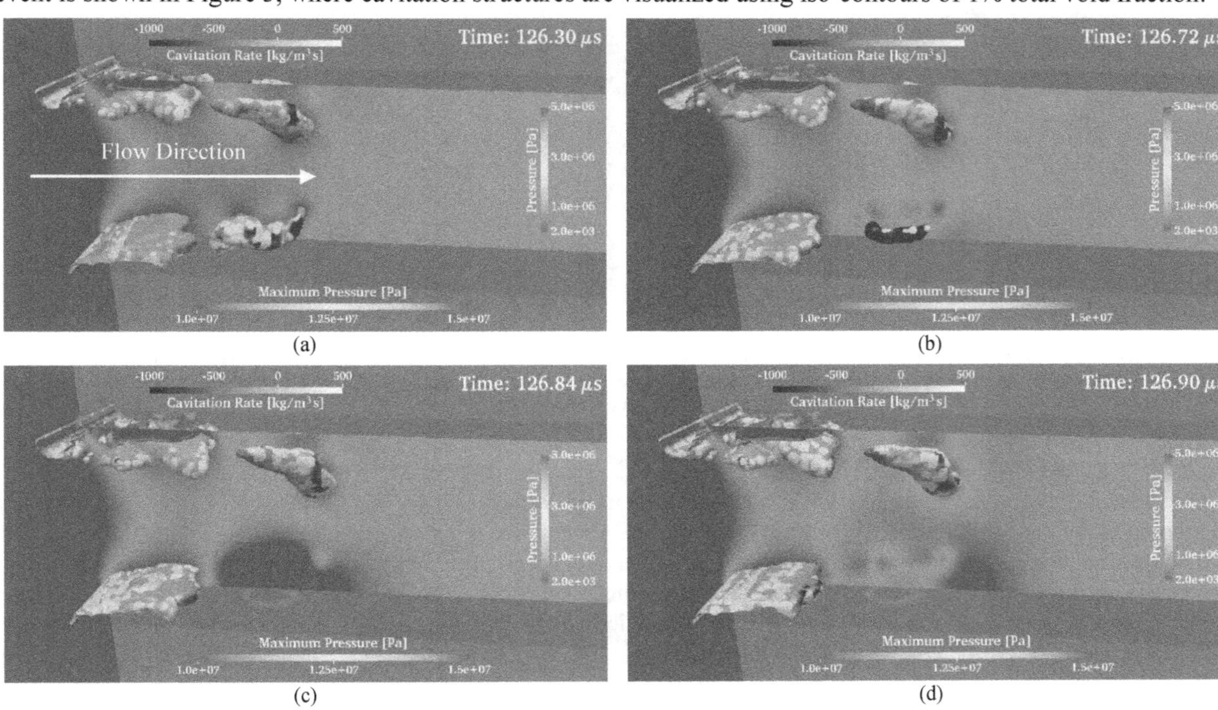

Figure 3. The temporal evolution of cavitation cloud collapse leading to substantial impact loading on the lower throttle boundary is visualized in panels (a) through (d) across a 0.60 μs time period. The boundaries are colored by maximum pressure, the central slice is colored by local pressure, and iso-contours of void fraction are colored by cavitation rate, where negative values indicate condensation.

In the time instant shown in Figure 3(a), fixed cavities form at the entrance to the throttle. The growth and breakup of this cavity lead to the formation of traveling hairpin or horseshoe-shaped cavitation structures. As these structures travel downstream, as shown in Figure 3(b), the pressure recovery region promotes condensation and vapor cloud collapse, as indicated by the dark colored void fraction iso-contours. The rapid vapor cloud collapse induces a strong pressure wave as illustrated in Figure 3(c), and an impact load in excess of 15 MPa is predicted at the neighboring throttle boundary. Although the maximum predicted pressure is below the yield strength of typical injector materials, such as steel with a yield strength greater than 200 MPa [14], it is very likely the acoustic pressure wave could generate higher erosive potential via the collapse of micro-scale cavitation structures or bubbles in the vicinity of the throttle boundary [15,16]. Following this event, a pressure wave is observed to propagate downstream, as shown in Figure 3(d).

Although the implemented index helps to mark cavitation cloud collapse events that could lead to cavitation erosion, it is not able to explain the physical mechanism governing the cavitation erosion event. In order to better understand the underlying physics, the generation and development of vortices in the cavitation shedding process was analyzed. This investigation is motivated by the work of Dular and Petkovšek [1], where the link between vortex structures and cavitation erosion potential was noted to be of critical importance for all five cavitation erosion mechanisms. The vortex structures, shown in Figure 4, are visualized using iso-contours of the magnitude of the vorticity field. Although this criterion highlights both rotational and shear structures within the flow, the development of the horseshoe vortex can be clearly visualized with appropriate selection of the vorticity magnitude threshold and filtering of the void fraction distribution. These results provide strong evidence that the predicted cavitation erosion event, indicated by the maximum pressure index, was due to the collapse of a horseshoe cloud. Although spherical cloud collapse was also observed in the simulation, the predicted maximum pressure at the wall boundaries was smaller than that from the horseshoe cloud implosion event. Future studies will implement more rigorous criteria for characterizing vortex structures in compressible flows [17] in order to identify and gather statistics regarding the occurrence of other predicted cavitation cloud collapse mechanisms. This exercise will help inform cavitation erosion metrics that are capable of linking the occurrence of cavitation cloud structures with their potential for material damage under injector relevant flow conditions.

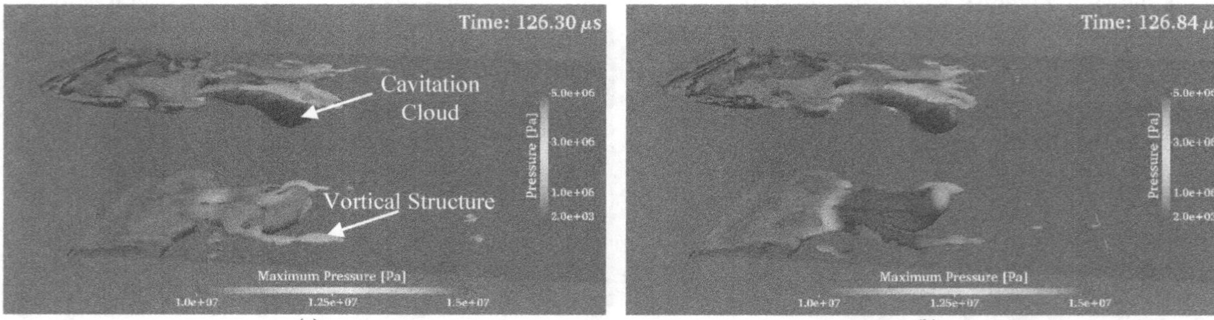

(a) (b)

Figure 4. Predicted vortical structures for the pressure drop condition of 75 bar are visualized using iso-contours of vorticity magnitude of 5.0e6 Hz, and colored by the local pressure. Cavitation cloud structures, as previously shown in Figure 3(a) and (c), are overlaid in grey to highlight the vortex-cavitation cloud interaction leading up to the predicted cavitation erosion event.

Summary

In this work, a homogeneous relaxation model was employed within a large eddy simulation to predict cavitation phenomenon in the Winklhofer Throttle "U" geometry. Potential cavitation erosion events, identified using the maximum predicted pressure at the throttle boundary, were observed to be linked to the collapse of horseshoe cloud structures. Future investigations will examine the predicted vortical structures that are formed in the cavitation shedding and condensation processes, and further assess indicators that are capable of linking the cavitation erosion potential with the observed cavitation cloud structures.

Acknowledgements

The submitted manuscript has been created by UChicago Argonne, LLC, Operator of Argonne National Laboratory (Argonne). Argonne, a U.S. Department of Energy Office (DOE) of Science laboratory, is operated under Contract No. DE-AC02-06CH11357. The U.S. Government retains for itself, and others acting on its behalf, a paid-up nonexclusive, irrevocable worldwide license in said article to reproduce, prepare derivative works, distribute copies to the public, and perform publicly and display publicly, by or on behalf of the Government.

This research was funded by DOEs Office of Vehicle Technologies, Office of Energy Efficiency and Renewable Energy under Contract No. DE-AC02-06CH11357. The authors wish to thank Gurpreet Singh, Michael Weismiller, and Kevin Stork, program managers at DOE, for their support.

The authors gratefully acknowledge the computing resources provided on Blues, a high-performance computing cluster operated by the Laboratory Computing Resource Center at Argonne National Laboratory, and Convergent Science Inc., for providing the Converge CFD software licenses.

References

1. Dular, M. and Petkovšek, M. (2015). *On the Mechanisms of Cavitation Erosion – Coupling High Speed Videos to Damage Patterns*. Experimental Thermal and Fluid Science. 68(1): 359-370.
2. Mitroglou, N., Stamboliyski, V., Karathanassis, I.K., Nikas, K.S., and Gavaises, M. (2017). *Cloud cavitation vortex shedding inside an injector nozzle*. Experimental Thermal and Fluid Science. 84(1): 179-189.
3. Arndt, R.E.A. (1981). *Cavitation in fluid machinery and hydraulic structures*. Ann. Rev. Fluid Mech. 13:273-328.
4. Gavaises, M., Papoulias, D., Andriotis, A., Giannadakis, E., Theodorakakos, A. (2007). *Link Between Cavitation Development and Erosion Damage in Diesel Injector Nozzles*. SAE Technical Paper 2007-01-0246.
5. Koukouvinis, P., Karathanassis, I.K., and Gavaises, M. (2017). *Prediction of cavitation and induced erosion inside a high-pressure fuel pump*. International Journal of Engine Research.
6. Koukouvinis, P., Gavaises, M., Li, J., and Wang, L. (2015). *Large Eddy Simulation of diesel injector opening with a two phase cavitation model*. Journal of Physics: Conference Series. 656(1).
7. Brusiani, F., Falfari, S., and Bianchi, G.M. (2015). *Definition of a CFD Multiphase Simulation Strategy to Allow a First Evaluation of the Cavitation Erosion Risk Inside High-Pressure Injector*. Energy Procedia. 81: 755-764.
8. Battistoni, M., Duke, D., Swantek, A.B., Tilocco, F.Z., Powell, C.F., Som, S. (2015). *Effects of non-condensable gas on cavitating nozzles*. Atomization and Sprays. 25(6): 453-483.
9. Chahine, G. L., & Hsiao, C.-T. (2015). *Modelling cavitation erosion using fluid–material interaction simulations*. Interface Focus. 5(5): 20150016.
10. Winklhofer, E., Kull, E., Kelz, E., Morozov, A. (2001). *Comprehensive Hydraulic and Flow Field Documentation in Model Throttle Experiments Under Cavitation Conditions*. 17th Annual Conference on Liquid Atomization & Spray Systems (ILASS-Europe), Zurich, September 2–6.
11. Richards, K.J., Senecal, P.K., and Pomraning, E. (2018). *CONVERGE 2.3 Manual*. Convergent Science, Madison, WI.
12. Bilicki, Z. and Kestin, J. (1990). *Physical aspects of the relaxation model in two-phase flow*. Proc. R. Soc. Lond. A. 428, pp. 379-397.
13. Pomraning, E. (2000). *Development of Large Eddy Simulation Turbulence Models*. Ph.D. Thesis, University of Wisconsin-Madison, Madison, WI.
14. Harvey, P. D. (1982). *Engineering properties of steel*. Metals Park, OH: American Society for Metals.
15. R. Fortes-Patella, J.L. Reboud, L. Briancon-Marjollet, A phenomenological and numerical model for scaling the flow aggressiveness in cavitation erosion. EROCAV Workshop, Val de Reuil, France, 2004.
16. G. Bark, N. Berchiche, M. Grekula. (2004). Application of principles for observation and analysis of eroding cavitation. In *The EROCAV Observation Handbook*. Edition 3.1.
17. Jiang, M., Machiragu, R., & Thompson, D. (2005). Detection and visualization of vortices. In *The Visualization Handbook* (pp. 295-309). Academic Press.

© 2018 ASME

10th International Symposium on Cavitation - CAV2018
Baltimore, Maryland, USA, May 14 – 16, 2018

Flow Enhancement and Cavitation Suppression in Nozzle Flow by Viscoelastic Additives

[1]Homa Naseri*, [1]Phoevos Koukouvinis, [1]Ioannis Karathanassis and [1]Manolis Gavaises

[1]*City, University of London, Northampton Square, UK, EC1V 0HB*

Abstract

Numerical simulations of turbulent cavitating nozzle flow of Phan-Thien-Tanner and Newtonian fluids are compared to study the effect of viscoelastic additives on cavitation dynamics. The results show that viscoelasticity can stabilize the flow-field and reduce turbulence and vorticity inside the nozzle. Subsequently small scale vortices are suppressed and larger eddies become dominant in the flowfield. Vapor formation in the core of Kelvin-Helmholtz vortices in the shear layer is delayed and overall vapor volume fraction inside the nozzle is reduced by viscoelasticity. Therefore, the density of the two-phase flow passing through the nozzle is higher and the pressure drop across the nozzle is reduced. Hence flow resistance due to turbulence and cavitation inside the nozzle is reduced and the mass flow rate is enhanced.

Keywords: viscoelastic additive; cavitation; turbulence

Introduction

Viscoelastic additives are used in pipelines or district heating/cooling systems as they can contribute to reduction of turbulent drag and cut the pumping costs, however studies about their effect on cavitation in turbulent flows are scarce in the literature. The volumetric efficiency of fuel injection systems can be improved [1] by adding viscoelastic compositions [2] of deposit control additives to fuel blends. Recent experimental evidence [3] suggest that this is due to reduction of cavitation and turbulence in the flow-field by the additive. In this study we examine numerically the effect of viscoelasticity on turbulent cavitating flows similar to the fuel flow in an injector nozzle, in an attempt to provide an explanation for the observed trends.

Viscoelastic fluids are non-Newtonian materials which can exhibit both viscous and elastic properties. They are important for a wide range of applications such as polymer and plastic industries, soft solids, tissue engineering and biological flows. Presence of microstructures such as polymers, surfactants or other particle aggregates in a liquid can often induce viscoelastic properties in fluids which can significantly change the flow behavior even in very low concentrations.

In turbulent flow conditions, polymers can stretch and absorb elastic energy from the near wall eddies, and if they release this energy beyond the buffer layer region in the streamwise direction, they can reduce the turbulence drag and increase the buffer layer thickness [4]. The extensional viscosity of polymeric solutions also inhibits the formation of vortex cavitation in propellers [5]. Moreover, viscoelasticity can alter the dynamics of collapsing bubbles, mainly near solid surfaces [6]. The re-entrant jet motion is retarded and the velocity of the jet formed by bubble collapse is reduced. Therefore the maximum pressure inside the bubble at the final stage of collapse is reduced, which means cavitation damage can be inhibited. However there is a gap in the literature regarding the effect of viscoelasticity on cavitation in turbulent flows and numerical studies can help to establish a better understanding of this effect.

Governing Equations and Simulation Setup

The test case is a step nozzle [7] with cavitation forming at the shear layer of the separated flow as water enters the nozzle with inlet pressure of 2.4 bar into atmospheric pressure and the test rig operates on a steady state condition (see Figure 1). Flow Reynolds number is 27700 based on the mean liquid velocity in the nozzle and nozzle width. Cavitation number is 1.19, defined in the experimental study as:

*Corresponding Author, Homa Naseri: homa.naseri.2@city.ac.uk

$$\sigma = \frac{p_a - p_v}{0.5\rho U^2} \qquad (I)$$

where p_a is the atmospheric pressure, p_v is the vapour pressure, ρ is the liquid density and U is the mean velocity of the liquid in the nozzle. This condition corresponds to incipient cavitation regime, therefore it is possible to track the formation, development and collapse of the cavitation structures which all happen inside the nozzle.

	Newtonian	Viscoelastic
Inlet pressure (Pa)	2380000	2380000
Outlet pressure (Pa)	101325	101325
Vapour pressure (Pa)	2340	2340
Viscosity (Pa.s)	0.00102	0.00102
Polymer Viscosity(Pa.s)	-	0.009
Relaxation time (s)	-	0.04

Table 1. Fluid properties and boundary conditions for the simulations

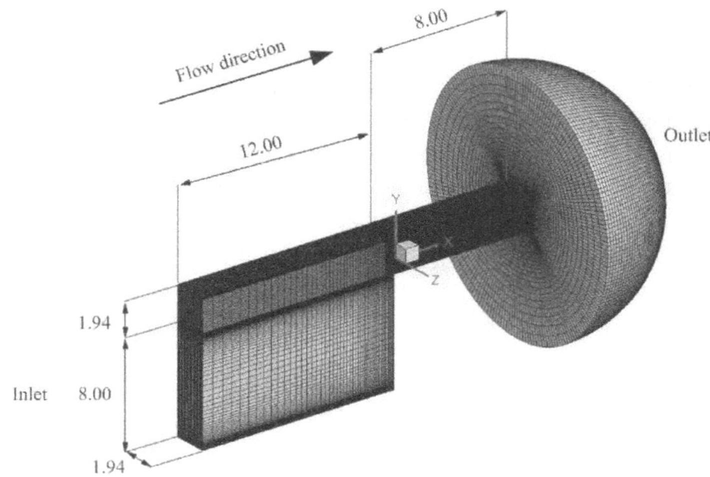

Figure 1. Simulations conditions, computational domain and geometry dimensions (in mm)

Description and details of the numerical framework employed for the Newtonian flow simulation as well as the experimental procedures can be found in [7] and [8]., therefore only a brief description of the simulation setup details are presented here. This problem concerns a two phase flow and the mixture model is used to describe the fluid. In mixture framework, the continuity and momentum equations are solved for a mixture of liquid and vapor phase, then pressure data is used by the cavitation model to obtain the volume fraction of each phase. The continuity and momentum equations are:

$$\frac{\partial \rho_m}{\partial t} + \nabla.(\rho_m \boldsymbol{u}) = 0 \qquad (II)$$

$$\frac{\partial \rho_m \boldsymbol{u}}{\partial t} + \nabla.(\rho_m \boldsymbol{u} \otimes \boldsymbol{u}) = -\nabla p + \nabla.\boldsymbol{\tau} + \nabla.\boldsymbol{\tau}_v \qquad (III)$$

where ρ_m is the mixture density, $\boldsymbol{\tau}$ is the stress tensor and $\boldsymbol{\tau}_v$ is the viscoelastic stress tensor.

The constitutive equation for the viscoelastic fluid is based on the linear Phan-Thien-Tanner model [9], this model assumes polymers form entangled networks where the junctions can be destroyed and re-created:

$$\lambda \overset{\nabla}{\boldsymbol{\tau}}_v + f(tr(\boldsymbol{\tau}_v)).\boldsymbol{\tau} = \mu_p(\nabla \boldsymbol{u} + \nabla \boldsymbol{u}^T) \qquad (IV)$$

where μ_p is the polymer viscosity and $f(tr(\boldsymbol{\tau}_v))$ is defined as:

$$f(tr(\boldsymbol{\tau}_v)) = 1 + \varepsilon \frac{\lambda}{\mu_p}(tr(\boldsymbol{\tau}_v)) \qquad (V)$$

where λ is the polymer relaxation time, ε is taken 0.02 for dilute solutions and $\overset{\nabla}{\boldsymbol{\tau}}_v$ is the Oldroyd upper convective derivative, defined as:

$$\overset{\nabla}{\boldsymbol{\tau}}_{\mathrm{v}} = \frac{D\boldsymbol{\tau}_{\mathrm{v}}}{Dt} - (\nabla \boldsymbol{u}.\boldsymbol{\tau}_{\mathrm{v}} + \boldsymbol{\tau}_{\mathrm{v}}.\nabla \boldsymbol{u}^{\mathrm{T}}) \quad (VI)$$

The polymer relaxation time is taken from properties of dilute solutions of polyisobutylene in tetradecane [10]. An artificial diffusion term is added to the viscoelastic constitutive equation for numerical stability, such that the dimensionless artificial diffusivity ($D = k/u_\tau H$, where k is the constant artificial diffusivity, u_τ is the friction velocity and H is the nozzle width) remains below 0.01.

To model cavitation, the mass transfer rate model of Schnerr and Sauer [11] is used, and turbulence is modelled using the WALE LES formulation [12]. Inside the nozzle the grid spacing is 20 μm which is below the Taylor length scale of the in-nozzle flow and near the wall the resolution is 2.5 μm giving y^+ values of 0.5. The time step is 1μs for CFL number of 0.5 and for the viscoelastic case the time step is reduced to 0.5 μs for a CFL number of 0.25. Inlet and outlet pressures and no-slip wall boundary conditions are used. Grid dependency and performance of turbulence and cavitation models was previously validated against LDV measurements of velocity and RMS of turbulence velocity [8]. The fluid properties and boundary conditions used for the simulations are summarized in table 1.

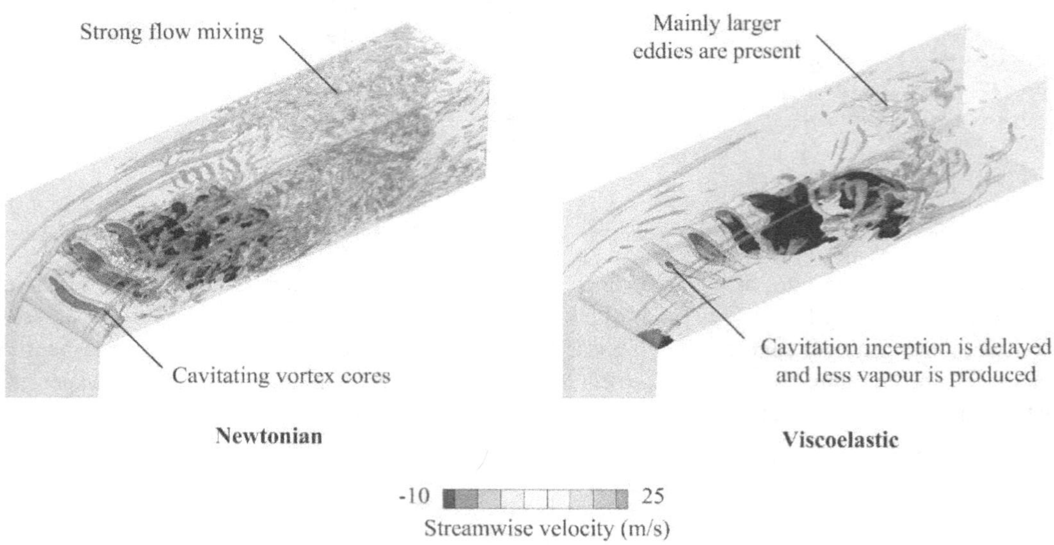

Figure 2. Isosurfaces of second invariant of velocity gradient at 1E+9 s^{-2} colored with streamwise velocity and isosufaces of 50% vapor volume fraction (black color)

The flow topology in terms of the second invariant of the velocity gradient is displayed in Figure 2 for the Newtonian and the viscoelastic fluid. It is evident that turbulent flow structures present in the Newtonian fluid are strongly suppressed by viscoelasticity. Moreover, formation of K-H vortices and development of cavitation vapor in their core is delayed in the viscoelastic fluid. Furthermore, the size of the cavitation cloud is smaller in the viscoelastic fluid. After the breakdown of the vortex sheet, a strong flow mixing happens in the Newtonian fluid, whereas the mixing is strongly suppressed in the viscoelastic fluid. Moreover, small-scale vortices are inhibited by viscoelasticity and only larger eddies last.

In Figure 3 cavitation structures in the Newtonian and the viscoelastic fluid are presented in terms of the isosurface of 50% vapor volume fraction. It is evident that the shape of the cavitation clouds is different in the two fluids; the vapor structures appear smoother in the viscoelastic fluid and the size of the vapor cloud is smaller. The more homogeneous cavitation isosurface in the viscoelastic fluid can be due to the reduced level of flow perturbations and damping of small scale turbulence in this fluid as demonstrated in Figure 2.

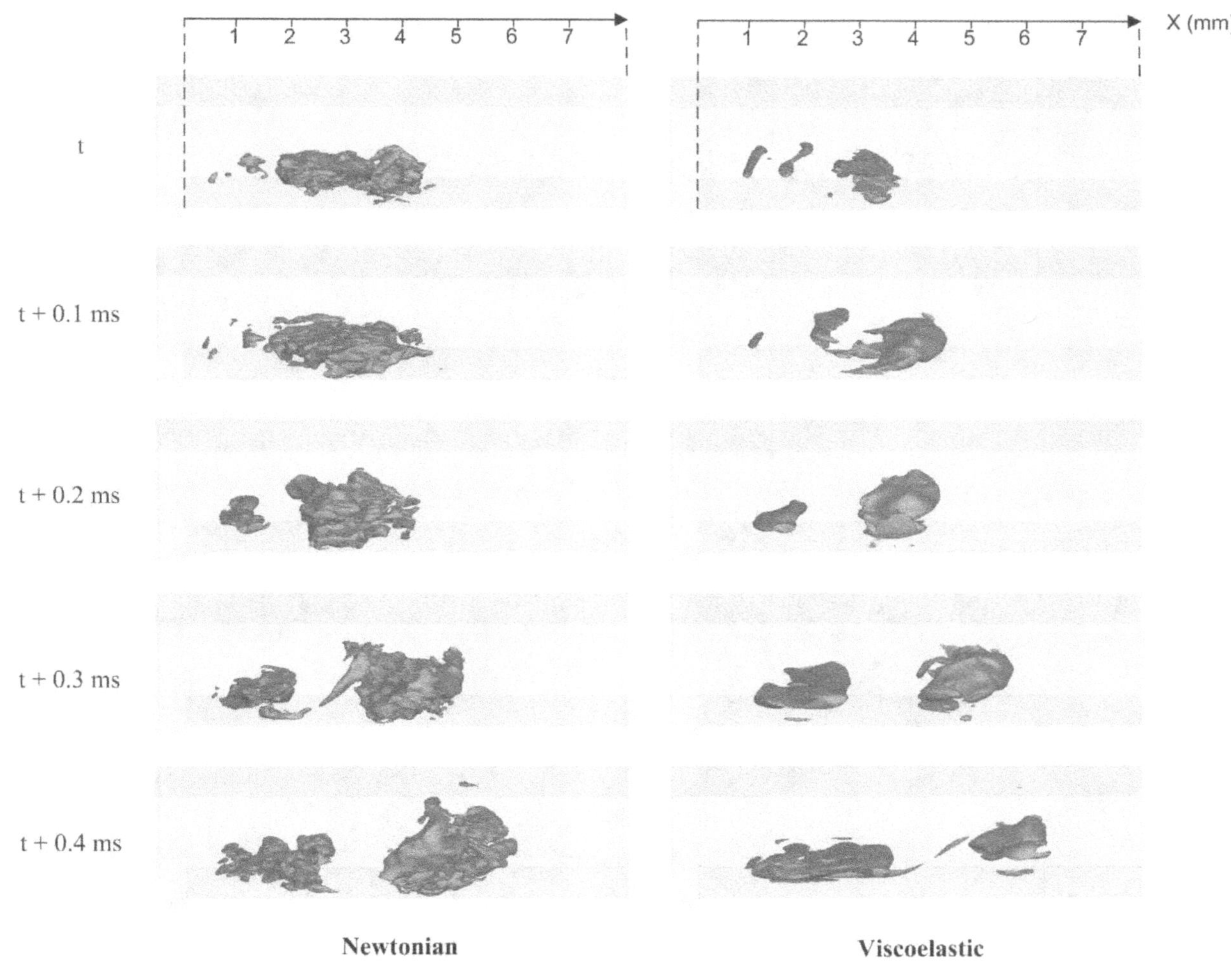

Figure 3. Evolution of the cavitation cloud topology in the Newtonian and the viscoelastic fluid, Isosurfaces represent the regions of 50% vapor volume fraction

Additionally, fluctuations of the mass flow rate at the exit of the nozzle is measured and the dominant frequencies of fluctuations are identified using FFT. For the viscoelastic fluid, a dominant frequency of 57 Hz is identified, whereas in the Newtonian fluid the frequency is increased to 168 Hz. Reduction of the frequency of mass flow rate fluctuations in the viscoelastic fluid indicates that the shedding of the cavitation cloud is also less frequent. Moreover, wall-normal velocity fluctuations are reduced along the nozzle length by ~30%, which directly shows the turbulence suppression property of the viscoelastic additive.

The time variation of the fluid density inside the nozzle is plotted in Figure 4 (a) presenting three cycles of cavitation formation and shedding. The fluid density in the Newtonian case is constantly below the viscoelastic fluid density, indicating a higher liquid fraction and hence cavitation suppression in the viscoelastic fluid. In Figure 2 it was demonstrated that the turbulent eddies observed in the Newtonian fluid are partly suppressed in the viscoelastic fluid, therefore it is expected that vorticity levels would also be lower in the viscoelastic fluid. This data is presented in Figure 4 (b) which shows the magnitude of the streamwise vorticity inside the nozzle for the two fluids. Vorticity magnitude in the Newtonian fluid is higher in all locations inside the nozzle as it can also be seen in the contour plots (Figure 4 (c)), in fact streamwise vorticity magnitude is reduced by ~88% inside the whole nozzle for the viscoelastic fluid. Moreover, regions of peak vorticity corresponding to the vortex cores in Figure 4 (c), are in agreement with observations of Figure 2 showing that smaller eddies are suppressed by viscoelasticity and large scale vortices become more dominant. It is also noted that vorticity generation downstream of the recirculation region (x>4 mm) is strongly suppressed in the viscoelastic fluid.

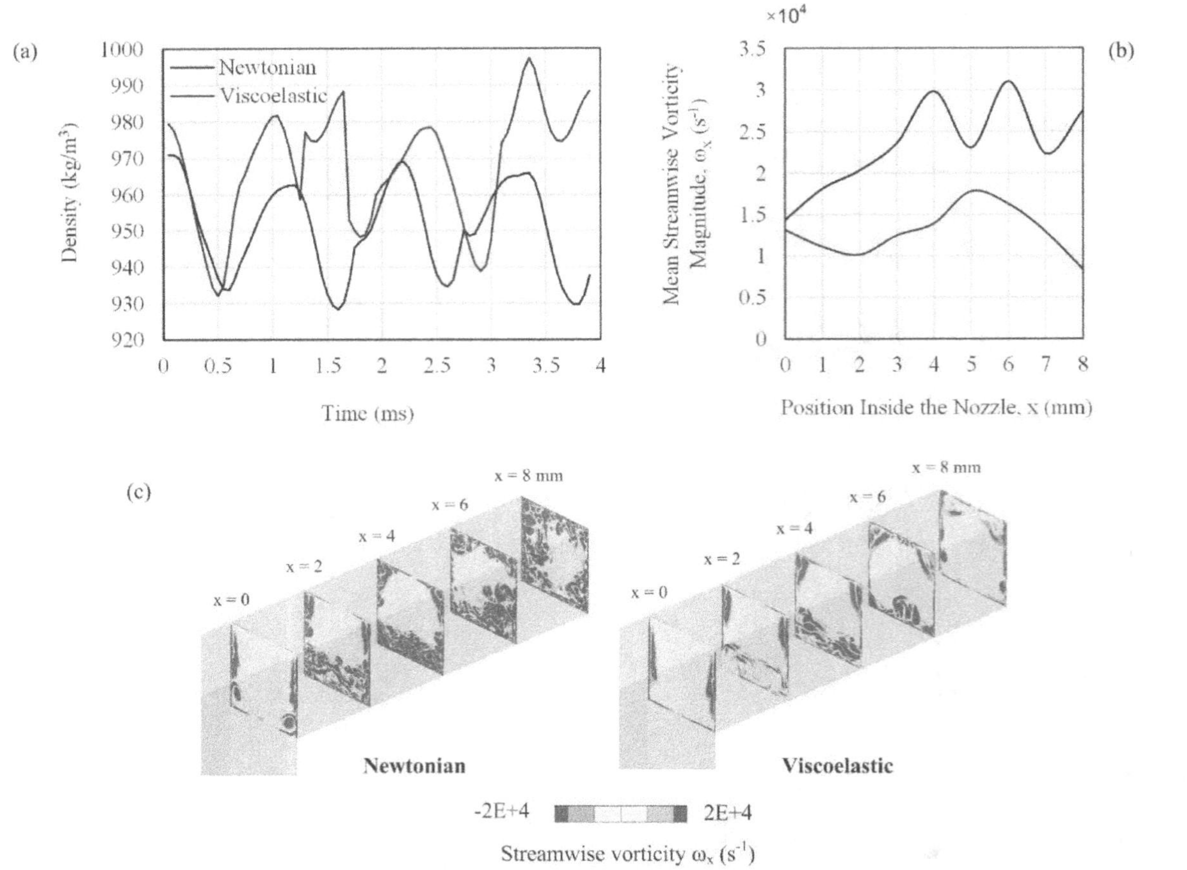

Figure 4. (a) Density variations inside the nozzle during three cavitation cycles, (b) magnitude of the streamwise vorticity, mean value is calculated in planes along the nozzle length, (c) contour plots of streamwise vorticity inside the nozzle

The time averaged pressure and vapor volume fraction data inside the nozzle are presented in Figure 5. Pressure in the cavitation cloud region (~1 mm<x<~5 mm) is increased in the viscoelastic fluid on average by ~3.5 KPa. Furthermore the pressure drop across the nozzle is also reduced in the viscoelastic fluid, which can be because of the reduction of form drag due to smaller size of the cavitation and recirculation region and also less turbulent drag. The reduction of vapor volume fraction in the viscoelastic fluid is more pronounced, where the cavitation volume fraction inside the nozzle is decreased by ~40%, indicating that more liquid is being delivered in the viscoelastic flow. Suppression of cavitation and reduction of turbulence level in the nozzle result in flow enhancement and so the mass flow rate is increased by ~2.2% through the nozzle in the viscoelastic fluid.

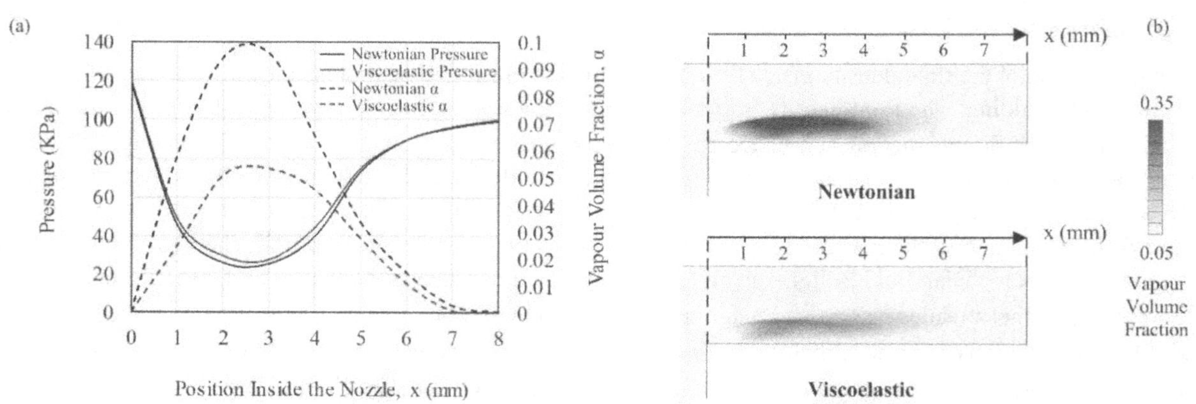

Figure 5. (a) Time-averaged data for pressure and vapor volume fraction inside the nozzle, (b) contour plots of time-averaged vapor volume fraction in the midplane of the nozzle

Conclusion

In this study we compared cavitation formation and development inside a nozzle for Newtonian and Phan-Thien-Tanner viscoelastic fluids. Turbulent flow structures visualized by the second invariant of the velocity gradient are strongly suppressed in the viscoelastic fluid. Development of vapor in the core of the shear layer K-H vortices is delayed and a smaller cavitation cloud is formed. Small scale vortices are damped and larger vortices become dominant in the flow, while streamwise vorticity is reduced. Time-averaged statistics show that vapor volume fraction is smaller in the viscoelastic fluid, hence the density of the fluid passing through the nozzle is increased.
Overall, pressure losses across the nozzle by turbulence drag and form drag due to cavitation and flow recirculation are suppressed in the viscoelastic fluid, resulting in an improved flow rate.

References

[1] R. H. Barbour, R. Quigley, and A. Panesar, "Investigations into Fuel Additive Induced Power Gain in the CEC F-98-08 DW10B Injector Fouling Engine Test," 2014.
[2] R. H. Barbour, "Method to provide power gain in an engine," 2011.
[3] H. Naseri, K. Trickett, N. Mitroglou, I. Karathanassis, P. Koukouvinis, M. Gavaises, R. Barbour, D. Diamond, S. E. Rogers, M. Santini, and J. Wang, "Turbulence and Cavitation Suppression by Quaternary Ammonium Salt Additives," 2018.
[4] Y. Dubief, C. M. White, V. E. Terrapon, E. S. Shaqfeh, P. Moin, and S. K. Lele, "On the coherent drag-reducing and turbulence-enhancing behaviour of polymers in wall flows," *J. Fluid Mech.*, vol. 514, pp. 271–280, 2004.
[5] C. T. Hsiao, Q. Zhang, X. Wu, and G. L. Chahine, "Effects of Polymer Injection on Vortex Cavitation Inception," in *28th Symposium on Naval Hydrodynamics*, 2010.
[6] E. A. Brujan, T. Ikeda, and Y. Matsumoto, "Dynamics of ultrasound - induced cavitation bubbles in non - Newtonian liquids and near a rigid boundary," *Phys. Fluids*, vol. 16, no. 7, pp. 2402–2410, 2004.
[7] A. Sou, B. Biçer, and A. Tomiyama, "Numerical simulation of incipient cavitation flow in a nozzle of fuel injector," *Comput. Fluids*, vol. 103, pp. 42–48, 2014.
[8] Koukouvinis, H. Naseri, and M. Gavaises, "Performance of turbulence and cavitation models in prediction of incipient and developed cavitation," *Int. J. Engine Res.*, vol. 18, no. 4, pp. 333–350, 2016.
[9] N. P. Thien and R. I. Tanner, "A new constitutive equation derived from network theory," *J. Nonnewton. Fluid Mech.*, vol. 2, no. 4, pp. 353–365, 1977.
[10] D. N. Sibley, "Viscoelastic Flows of PTT Fluids," University of Bath, 2010.
[11] G. H. Schneer and J. Sauer, "Physical and Numerical modelling of unsteady cavitation dynamics," *Int. Conf. Multiph. Flow*, vol. 11, no. 4, pp. 391–400, 2001.
[12] F. Nicoud and F. Ducros, "Subgrid-scale stress modelling based on the square of the velocity gradient tensor," *Flow, Turbul. Combust.*, vol. 62, no. 3, pp. 183–200, 1999.

© 2018 ASME

10th International Symposium on Cavitation - CAV2018
Baltimore, Maryland, USA, May 14 – 16, 2018 CAV18-05097

Fluid-Structure Interaction in Cavitation Erosion

[1]Yves Paquette*; [2]Marc Fivel; [1]Giovanni Ghigliotti; [3]Eric Johnsen; [1]Jean-Pierre Franc

[1]*Univ. Grenoble Alpes, CNRS, Grenoble INP, LEGI, 38000 Grenoble, France;* [2]*Univ. Grenoble Alpes, CNRS, Grenoble INP, SIMaP, 38000 Grenoble, France;* [3]*Mechanical Engineering Department, University of Michigan, Ann Arbor, MI 48109, USA*

Abstract

This numerical work focuses on the collapse of a single air bubble close to a deformable wall generated by the impact of an incident high-pressure wave. A CFD code was fully coupled to a FEM solid code in order to compute the bubble collapse and the subsequent plastic deformation in the material. The CFD code FSI4CAV was developed from the numerical model of Johnsen et al. [1]. A mobile mesh capability was added in order to account for the displacement of the fluid-structure interface. This has been done through the implementation of an ALE (Arbitrary Lagrangian Eulerian) method. The solid response to the bubble collapse was computed with the FEM software CAST3M [2] assuming an elastic-plastic constitutive law for the material. The communication between the two codes is achieved through the MPI library. The paper presents a detailed analysis of the two-dimensional bubble collapse dynamics and the corresponding solid behaviour. The damping of the impact pressure with respect to a perfectly rigid wall was computed as well as the plastic deformation developed in the material. The comparison of the simulation results with a non-coupled approach where the pressure applied to the solid is computed from CFD simulations definitely shows the importance of including the fluid-structure interaction into the simulation of cavitation erosion.

Keywords: Fluid-Structure Interaction, Cavitation, Numerical Coupling

Introduction

The prediction of cavitation erosion by numerical techniques is still a challenge for industrial applications. It faces several major difficulties including the prediction of the impact loads induced by the collapse of vapor structures (known as the erosive potential or aggressiveness of the cavitating flow) and the simulation of the response of a solid wall to these successive impact loads.

On the fluid side, recent progress has been made to predict cavitation aggressiveness in terms of impact load spectra [3] and in terms of erosion risks [4-6]. A fully compressible model is required with a high temporal resolution in order to capture the shock waves emitted during bubble collapse.

On the solid side, the response of a solid specimen to a single impact load and the prediction of the subsequent cavitation pit has been studied by several investigators such as [7] and [8]. However the dynamic response to multiple impact loads and the prediction of crack initiation and propagation and the resulting mass loss is one remaining challenge that requires developing a relevant damage model for the material under consideration.

Another challenge to be addressed is the damping effect on the amplitude of impact loads due to the wall deformation. This requires simulations of the fluid-structure interaction. Very few studies have been devoted to fluid-structure interaction in cavitation erosion. Chahine et al. investigated the case of the Rayleigh collapse of a bubble touching an elastic-plastic metallic wall [9] and confirmed the existence of a significant damping effect of the pressure peak applied to the wall. These authors predicted the fluid-solid interface displacement at each time step from the solid code but did not implement sub-iterations within the time step to correct this displacement and account for its feedback on the fluid flow during the time step.

The present work consists in developing a complete fluid-structure model which will be applied to the case of a single bubble collapsing close to an elastic-plastic specimen made of Aluminum. The bubble collapse is triggered by a plane shock wave. The bubble is assumed to be composed by air; no phase change is computed. The simulations are carried out in two dimensions assuming a plane strain state in the domain. After a short presentation of the computational strategy, the collapse of a single bubble close to a deformable aluminum medium is analyzed. Fluid-structure coupling

*Corresponding Author, Yves Paquette:Yves.Paquette@univ-grenoble-alpes.fr

is based on a two-way approach for which not only the wall deformation is affected by the flow, but the flow, and particularly the impact load, are also affected by the wall deformation, resulting in damping at the interface. This two-way approach is compared to a one-way approach, that assumes no feedback of the solid deformation on the flow, and that is thus obtained by simply applying, as a boundary condition, the pressure load obtained from a fluid-only simulation to a solid-only simulation.

Numerical approach

A CFD code for compressible fluid simulations was coupled to a FEM code for solid mechanics simulations. The CFD code is based on the code developed by Johnsen et al. [1]. This code is based on the 5-equations model: conservation of mass, momentum and energy with an additional conservation equation for the mass fraction of air and an equation of state valid for both liquid and gas. The stiffened gas equation of state was used with the same parameters as used by Beig [10]. To allow the displacement of the fluid-structure interface, the original Eulerian description in a fixed mesh was changed into an Eulerian description in a moving-deforming mesh using the Arbitrary Lagrangian Eulerian (ALE) method. The fluid mesh could therefore perfectly accommodate the deformation of the interface.

The solid deformation is computed by the FEM implicit code CAST3M developed by Commissariat à l'Energie Atomique et aux Energies Alternatives [2]. The equation of motion of the solid is solved in a dynamic way to allow wave propagation in the solid. The elastic-plastic behavior of the solid was modeled by a Ludwig-type power law connecting the Von Mises stress σ and equivalent tensile strain ε:

$$\sigma = \sigma_Y + K(\varepsilon - \varepsilon_Y)^n \qquad Y \text{ index corresponds to yield strength.} \tag{1}$$

In this first approach, the strain rate effect was not included in the model. Continuity of normal load and normal velocity is enforced at the fluid-solid interface. At each time step, each code transfers data to the other one thanks to an MPI interface. Convergence between mesh velocity at the interface in the fluid code and interface velocity calculated by the solid code is ensured at each time step by a convergence loop, implying sub-iterations at each time step.

Analysis of single bubble collapse close to an aluminum specimen

We now present the results of a typical simulation of the collapse of a 2D air bubble in water, close to a deformable wall. Air and water are initially at atmospheric pressure and a plane shock wave propagating in water towards the wall is supposed to make the bubble collapse. The amplitude of the incident shock wave is 120 MPa, the initial bubble radius is $R_0 = 50$ μm and the bubble center is initially two radii away from the solid. The solid domain is made of aluminum with a density of 2700 kg.m^{-3} and a Young modulus E of 70 GPa. Yield strength σ_Y is 500 MPa and the coefficients for the Ludwig law (Eq. 1) are K= 396 MPa and n= 0.3

We focus the following discussion on three distinct moments of the dynamic evolution:

- the impact of the incident shock wave on the fluid/solid interface.
- the impact of the shock wave due to bubble collapse on the interface.
- the generation of plasticity in the solid.

Figure 1 shows the density and pressure fields in the fluid as well as the stress component σ_{xx} and the corresponding plastic deformation ε_{xx} in the solid at various times. The reference time t_{ref} corresponds to the impact of the incident shock wave on the side of the bubble interface farther away from the wall.

(a) t_{ref} + 102ns

(b) t_{ref} + 206ns

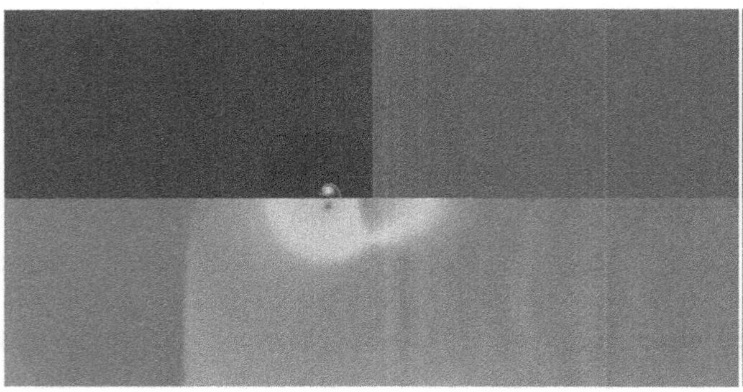

(c) t_{ref} + 260ns

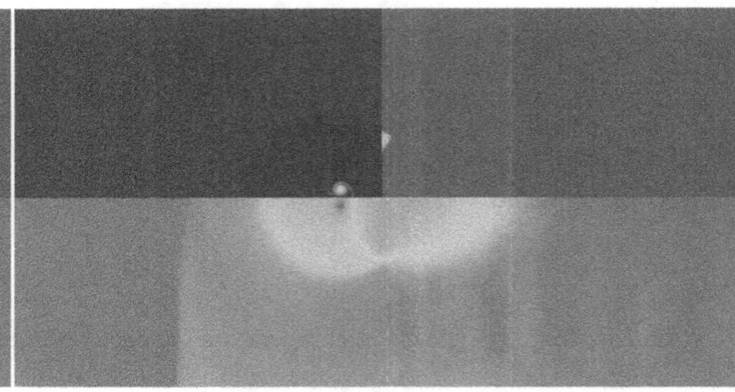

(d) t_{ref} + 273ns

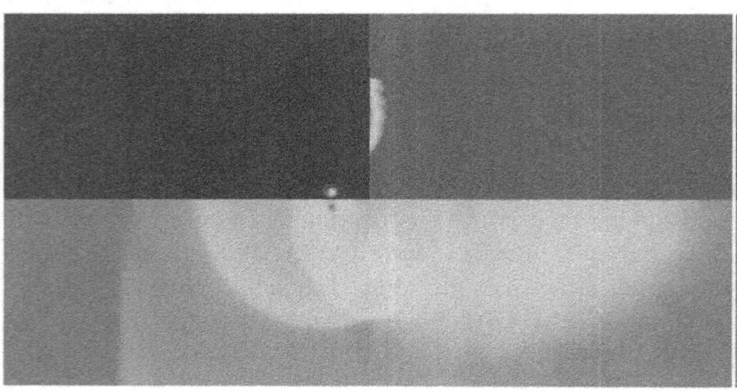

(e) t_{ref} + 315ns

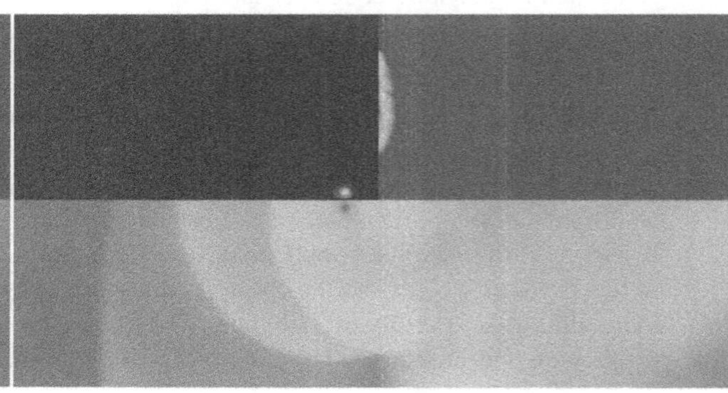

(f) t_{ref} + 348ns

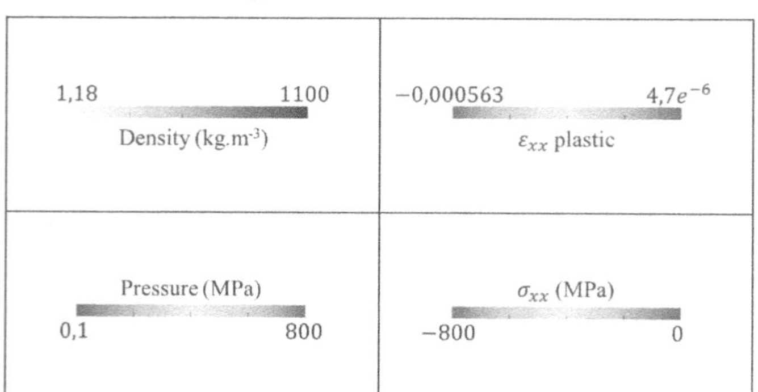

Figure 1: Dynamic evolution of a bubble collapse close to a deformable wall.

At t_{ref} + 102ns, the incident shock wave hits the wall. The wave is partially transmitted in the solid and partially reflected in the fluid. The pressure exerted on the interface is the sum of the amplitudes of the incident and reflected waves. Due to the conservation of the normal stress at the interface, the liquid pressure on the interface (bottom-left in figure 1a) is equal to the normal stress σ_{xx} in the solid at the interface (bottom-right in figure 1a). The pressure on the wall at the region facing the bubble center is smaller than elsewhere on the wall because of the screening effect of the bubble. The amplitude of the initial shock wave is too small to induce any plastic deformation in the solid (top-right in figure 1a). In the case of a rigid wall, the amplitude of the pressure peak would be twice the amplitude of the incident shock wave, i.e. 240MPa. This pressure peak is below the material yield strength of 500MPa. Note that for a compliant wall like polymers, this value could be much smaller due the damping effect of the wall displacement.

At t_{ref} + 206ns (figure 1b), water under pressure deforms the side of the bubble interface farther from the wall and initiates a microjet. When the microjet pierces the bubble (i.e. when the two-dimensional bubble is split into two parts), a radially outward-propagating shock wave is produced. When this shock wave reaches the wall, a pressure peak of 744MPa is produced. The amplitude of the shock wave generated by the bubble collapse is significantly larger than that of the incident shock wave, and is capable of inducing plastic deformation in the material.

Figure 1c shows the different fields just after the impact of the bubble-induced shock wave on the wall (t_{ref} + 260ns). Continuity between the liquid pressure and the normal stress σ_{xx} at the liquid / solid interface can be appreciated on this figure. Figure 1c also shows that the waves travel faster in the solid than in the liquid. A small, roughly circular residual air bubble is visible in figure 1c (top-left), which corresponds to pressurized air. There is still no plastic deformation in the solid (top-right in figure 1c).

After hitting the interface, the shock wave is transmitted into the solid. It will also spread along the fluid-solid interface. On the interface, a pressure peak is observed locally, which corresponds to the passage of the wave front. Regarding the solid, this pressure peak results in a high amplitude loading, applied to the surface of the material.

At t_{ref} + 315ns, the shock wave has traveled along the wall over a distance of the order of $3R_0$. Figure 1d to 1f shows that the upper limit of the plastic zone corresponds to the front of the shock wave. The plastic zone starts developing at a distance of about $1.2R_0$ from the bubble axis at time t_{ref} + 273ns (figure 1d) and progressively grows at the same speed as the wave propagates along the wall. At every instant, the plastic front in the material and the shock wave front in the fluid coincide on the wall. The growth stage of the plastic zone ends at instant t_{ref} + 348ns (figure 1f) and its maximum length is about $3R_0$. After some oscillations due to wave propagation, the deformed wall takes the form of a standard erosion pit with maximum depth on the bubble axis.

1-way/2-way comparison

In the previous section, a two-way simulation of a single bubble collapse close to a wall was presented. The two-way simulation takes into account the effect of the solid deformation on the fluid. In order to quantify this effect, the two-way simulation is compared to a one-way simulation corresponding to the solid-only simulation with the pressure load from a fluid-only simulation applied as a boundary condition. Three parameters are considered for this comparison: the amplitude of the pressure peak, the pit depth on the bubble axis and the area of the generated plastic surface. In our case, there is no damping of the wave in the solid so the wave can propagate endlessly. The interface is moving during the whole simulation without coming to rest since the incident shock wave was assumed to be a step function and not a pulse of limited width. So to define the pit depth, we consider the difference between the highest and the lowest points of the interface.

	Pressure Peak (MPa)	Pit depth (nm)	Plastic surface (mm^2)
1-way	827.4	112.2	$3.13\ 10^{-3}$
2-way	744,7	62.4	$1.52\ 10^{-3}$

Table 1: Comparison between 1-way and 2-way simulation.

Table 1 shows that the deformation of the wall damps the pressure peak by about 9.9% in comparison to a rigid wall.

In the simplified case of a plane shock wave impacting a purely elastic material, the damping coefficient depends upon the ratio α of the acoustic impedance of the solid to that of the liquid and is given by $\frac{1}{\alpha+1}$ [11]. In the present case, this equation gives a damping coefficient of 11.1% when estimating the speed of sound in the solid by $\sqrt{\frac{E}{\rho}}$. The computed value of the damping coefficient obtained by the present FSI simulation (9.9%) is relatively close to this rough estimate.

This decrease in the loading of the interface induces a drastic decrease in the generated plastic surface and in the resulting pit depth, that are roughly divided by two for the case studied. These results illustrate the importance of including the fluid-structure interaction in the simulation of cavitation erosion.

Conclusion

In this paper, fluid-structure interaction was investigated numerically in the case of a 2D air bubble collapsing close to an elastic-plastic aluminum specimen with a yield strength of 500MPa. The collapse was generated by an incident shock wave with amplitude of 120MPa traveling towards the bubble and the wall. To perform the simulation, a two-way coupling was achieved between the 2D compressible fluid code FSI4CAV originally developed by Johnsen et al. and the solid solver CAST3M. Coupling was performed by the use of MPI library.

The main steps in the process of bubble collapse and wall deformation are the following. The bubble starts collapsing when its interface is impacted by the incident wave. A microjet penetrating the bubble and moving towards the wall develops during the collapse stage. When the microjet pierces the bubble, a circular shock wave of large amplitude is generated. When this wave impacts the deformable wall, a pressure peak of amplitude 744 MPa is measured on the wall. This induces plasticity in the solid which modifies the interface profiles into a shape similar to a typical experimental pit.

This simulation based on a two-way coupling was compared to the case of a one-way coupling that assumes that the loading on the wall would be the same as on a perfectly rigid wall. It was shown that the amplitude of the impact load is damped by 9.9% due to the wall deformation. In the present case, the damping effect induces a drastic change in the extent of the plastic region together with a significant reduction in pit depth.

In conclusion, this work shows that a one-way approach that neglects the feedback of the wall deformation on the fluid dynamics may not be sufficient to accurately predict impact loads and plastic deformations. It may be critically important to take into account wall deformation and to develop a two-way approach in order to accurately simulate the processes involved in cavitation erosion.

Acknowledgements

This work has been partially supported by the LabEx Tec 21 (Investissements d'Avenir - grant agreement n°ANR-11-LABX-0030) and ONR grant N00014-12-1-0751.

References

[1] E. Johnsen, T. Colonius, Numerical simulations of non-spherical bubble collapse, 2009, *J. of Fluid Mechanics* **629**

[2] http://www-cast3m.cea.fr/

[3] M.S. Mihatsch, S.J. Schmidt, N.A. Adams, Cavitation erosion prediction based on analysis of flow dynamics and impact load spectra, 2015, *Physics of Fluids* **27**

[4] R. Fortes Patella, G. Challier, J.L. Reboud, A. Archer, Energy balance in cavitation erosion: From bubble collapse to indentation of material surface, 2012, *J. of Fluids Engineering* **135**

[5] C. Leclerq, A. Archer, R. Fortes Patella, F. Cerru, Numerical cavitation intensity on a hydrofoil for a 3D homogeneous unsteady viscous flows, 2012, *J. of Fluids Machinery and Systems* **10**

[6] A. Peters, U. Lantermann, O. el Moctar, Numerical modelling and prediction of erosion induced by hydrodynamic cavitation, 2015, *J. of Physics: Conference Series* **656**

[7] F. Pöhl, S. Mottyll, R. Skoda, S. Huth, Evaluation of cavitation-induced pressure loads applied to material surfaces by finite-element-assisted pit analysis and numerical investigation of the elasto-plastic deformation of metallic materials , 2015, *Wear* **330-331**

[8] S.C. Roy, J.P. Franc, N. Ranc, M. Fivel, Determination of cavitation load spectra-Part 2: Dynamic finite element approach, 2015, *Wear* **344-345**

[9] G.L. Chahine, C.T. Hsiao, Modelling cavitation erosion using fluid-material interactions simulations, 2015, *Interface Focus* **5**

[10] S.A. Beig, E. Johnsen, Maintaining interface equilibrium conditions in compressible multiphase flows using interface capturing, 2015, *j. of Computational Physics* **302**

[11] J.P. Franc, J.M. Michel, Fundamentals of Cavitation, 2004, *Kluwer Academic Publishers,*

© 2018 ASME

Fuel nozzle geometry effects on cavitation and spray behavior at diesel engine conditions

Brandon A. Sforzo*[1], Katarzyna E. Matusik[2], Christopher F. Powell[1], Alan L. Kastengren[2], Shane Daly[3], Scott Skeen[4], Emre Cenker[4], Lyle M. Pickett[4], Cyril Crua[5], and Julien Manin[6]

[1]*Energy Systems Division, Argonne National Laboratory, Argonne, IL 60439*
[2]*X-ray Science Division, Argonne National Laboratory, Argonne, IL 60439*
[3]*School of Mechanical, Industrial, and Manufacturing Engineering, Oregon State University, Corvallis, OR 97331*
[4]*Combustion Research Facility, Sandia National Laboratories, Livermore, CA 94550*
[5]*Advanced Engineering Centre, University of Brighton, Brighton BN2 4GJ, UK*
[6]*Artium Technologies, Inc., Sunnyvale, CA 94085*

Cavitation dynamics of diesel and gasoline injection nozzles has been a topic of ongoing research due to the effect of cavitation on the characteristics of the fuel spray, including the discharge coefficient, outlet velocity, spray angle, and atomization process. Additionally, repeated collapse of vapor cavities can damage nozzle surfaces, permanently changing the boundary conditions of the fluid flow field. Understanding the evolution and behavior of cavitation can therefore allow more precise control over its presence, as well as improved predictability of the corresponding fuel spray distribution.

Studies have shown that the inception and persistence of fuel vapor inside the spray hole is sensitive to geometric features of the injection nozzle, such as the degree of taper and inlet corner radius of curvature. For example, a hole with a cylindrical profile and sharp inlet corner more readily supports cavitation formation as compared to a monotonically converging hole with a rounded inlet. To better understand the effect of these geometric features on cavitation formation, and by extension, on the associated fuel spray, we compare the nozzle geometry and spray characteristics of two single-hole diesel injectors procured through collaboration with the Engine Combustion Network (ECN). The Spray C injector, specifically designed by the ECN for the express purpose of studying cavitating flows, features a cylindrical hole with a slight divergence near the outlet and a relatively sharp inlet corner. Its non-cavitating analog, the Spray D injector, contains a rounded inlet corner and a gently converging hole profile.

High-resolution x-ray tomography measurements coupled with optical microscopy images of both injectors provide the nozzle geometry with $\mathcal{O}(1)$ μm spatial resolution. Analysis of the measured geometries reveal that the radius of curvature of the hole inlet varies azimuthally for the modestly hydroground Spray C injector. To elucidate the effect that this asymmetric inlet condition has on cavitation formation during *operando* conditions, the fuel flow inside the nozzle hole was recorded using high-speed x-ray phase contrast imaging. These images reveal the formation of an asymmetric sheath of fuel vapor that persists throughout the injection event. Complementary x-ray and optical diagnostics of the downstream fuel spray further highlight the effect that this cavitation layer has on the spreading angle of the spray in comparison to the non-cavitating Spray D injector.

Keywords: fuel sprays; cavitation; x-ray tomography; x-ray radiography; long distance microscopy

Introduction

The study of geometry-induced cavitation in diesel fuel injection nozzles is motivated by its potentially damaging impact on the nozzle body as well as its influence on the fuel spray distribution, which in turn can affect the combustion emissions profile, performance, and fuel economy of a vehicle. The presence of cavitation has been associated with enhanced atomization, increased spreading angle, and decreased discharge of the fuel spray, to name a few effects [1–5]. Research dating back to the early 1960s has strived to understand the effect of nozzle orifice shape on the corresponding spray structure, largely focusing on simple geometries such as that of a plain-orifice nozzle [6–8]. Numerous experimental [9, 10] and numerical [11–13] works have demonstrated the importance of the inlet corner radius of curvature and hole conicity in dictating the onset of cavitation. It has also been shown that an asymmetric inlet condition, such as that of an angled nozzle, can create correspondingly asymmetric cavitation [10]. More complex geometries relevant to diesel injection have also been studied, including valve-covered orifice and sac-type, multi-hole nozzles [1, 14].

An ongoing challenge in the study of cavitation in diesel injectors is the ability to observe its formation and structure in the internal and near-nozzle region. Historically, transparent nozzles have been used as a proxy for diesel nozzles to study the flow structure optically, and have been designed out of materials such as acrylic or quartz, both of which offer an index of refraction that may be matched by the working fluid in order to reduce unwanted artifacts during visualization. Because such materials are somewhat pressure-limited, a compromise is made by creating large-scale

*Corresponding Author, Brandon Sforzo: bsforzo@anl.gov

Table 1: Summary of operating conditions for measuring the spray characteristics of ECN Spray C and Spray D.

Operating Condition	Diagnostic Technique		
	High-speed optical imaging	X-ray imaging	X-ray radiography
Ambient gas temperature (K)	440, 900	298	298
Ambient gas pressure (MPa)	2.9, 6	0.1, 0.2	2
Ambient gas density (kg m^{-3})	22.8	1.4	22.8
Ambient gas O_2 (%)	0	0	0
Fuel type	n-dodecane	n-dodecane	n-dodecane
Injection pressure (MPa)	150	150	150
Fuel temperature at nozzle (K)	440, 363	298	298
Hyd. injection duration (ms)	2.5	2.5	2.5

nozzles that can target Reynolds numbers relevant to diesel injection, albeit at lower injection pressure. However, there is evidence that scaling up the nozzle geometry while holding the Reynolds and cavitation numbers constant produces different cavitation structures [15]. To circumvent this scaling issue, nozzles manufactured from sapphire have been used due to the material's similarity in strength to steel and ability to withstand engine-relevant injection pressures [16].

Complementary to optical techniques, x-ray diagnostics can provide qualitative and quantitative measurements of in-hole cavitation as well as the near-nozzle spray structure. X-rays have the ability to penetrate through opaque material and only weakly refract from gas-liquid interfaces, making them ideal for probing optically thick regions of a spray [5]. The high flux and high energy x-ray synchrotron source of the Advanced Photon Source (APS) at Argonne National Laboratory has been used in the past to provide the requisite x-rays for measurements of the internal and near-nozzle flow structure of cavitating systems with micrometer spatial and microsecond temporal resolution using techniques such as radiography and fluorescence [17, 18]. In addition, x-ray phase contrast imaging has been previously implemented to obtain a qualitative understanding of the internal nozzle flow [19, 20]. This technique can be readily extended to actual-size diesel injectors operating at engine-relevant injection pressures.

In this paper, we leveraged both optical and x-ray diagnostic techniques to study the internal and near-nozzle flow of two single-hole diesel injectors, one of which strongly cavitates under typical diesel operating conditions. The injectors were obtained through the Engine Combustion Network (ECN) [21]. X-ray diagnostics were used to characterize the internal geometry and flow structure, as well as the near-nozzle spray distribution. Complementary optical diagnostics performed at Sandia National Laboratories extended the analysis of the spray structure to higher ambient and fuel temperatures. Previous comparison of x-ray radiography measurements and long-distance microscopy imaging of a diesel fuel spray has shown that the spray edge as defined by an optical thickness of unity corresponds to a projected fuel mass from the x-ray measurement on the order of 1 $\mu g/mm^2$ [22]. Provided with these thresholds, the spray angle as obtained through the x-ray radiography technique may be directly compared to that measured from long-distance microscopy imaging in order to explore the effect of cavitation on the jet width at variable operating conditions.

Experimental Methods

A suite of diagnostic techniques was implemented to characterize the internal geometry and spray behavior of two single-hole diesel injectors designated as Spray C #210037 and Spray D #209134 by the ECN. The target conditions used for each spray diagnostic technique are summarized in Table 1. When possible, the ECN-specified non-reacting operating conditions were utilized. Figure 1(a) illustrates the ECN coordinate system [23] convention used throughout this work. The injectors were measured from two viewing angles to explore potential asymmetries in the flow, designated as "View 1" and "View 2", corresponding to measurements taken in the X-Y plane and X-Z plane, respectively, with the origin at the nozzle tip.

Spray C and Spray D are common-rail fuel injectors fitted with two different single-hole nozzles that vary in diameter and shape. Spray D was designed to feature a relatively round hole inlet corner with a converging profile, and is intended to be the non-cavitating analog of Spray C, which, in contrast, underwent minimal hydro-erosion to help maintain a sharp inlet corner and a cylindrical hole profile.

To understand the effect of geometry on cavitation behavior, and by extension, the fuel spray distribution, high-resolution x-ray computed tomography (CT) scans of the injector geometries were obtained at the 7-BM beamline of the APS [24]. The nozzle geometry was reconstructed from a series of 2D projections using TomoPy [25]. The final spatial resolution of the geometry is 1.8 μm with a field of view (FOV) of 2.25 × 1.4 mm, capturing the full hole profile as well as part of the sac region. A more thorough discussion of the experimental setup, image processing, spatial uncertainty, and geometric analysis may be found in prior work [26].

Internal nozzle flow visualization for Spray C and Spray D was achieved with high speed x-ray phase contrast imaging conducted at the 32-ID-B beamline at APS, using similar methodology as in Ref. [19]. Visible light imaging of a scintillator was performed with a high-speed camera fitted with a 5× microscope objective, operated at 50,000 fps, gathering 250 frames per injection event repeated for 30-40 events. Spray C and Spray D imaging was performed with ambient pressures of 0.1 and 0.2 MPa, respectively.

Measurements of the time-resolved projected fuel density field were carried out at the 7-BM facility at APS using time-resolved x-ray radiography [24]. Each injector was horizontally mounted in a chamber that was pressurized to 2 MPa with N_2, which was also used to purge the environment at a flow rate of 4 standard L/min to inhibit droplet accumulation in the field of view during data acquisition. The injectors were driven by a common-rail light-duty diesel injection system, and triggered at a rate of 3 Hz for a commanded injection duration of 725 μs. The beam was focused to a 4 × 6 μm spot with the focal point aligned with the injector tip. The pressure chamber was translated over a grid to build a raster plot of measurement points, capturing the spray evolution up to 25 mm from the nozzle tip. The measured incident x-ray intensity at each spatial location was averaged over 16 injection events. X-ray photons undergo photoelectric absorption as they pass through the fuel spray, with the degree of absorption directly related to the line-of-sight density via the Beer-Lambert law [27]. The recorded beam intensity was binned over a single synchrotron orbit, providing a temporal resolution of 3.68 μs.

The fuel sprays were also injected in Sandia's optically-accessible vessel to gather high-speed microscopic images under varying fuel and ambient temperatures. High-speed images were collected at both View 1 and View 2 to ascertain the spray growth rate and transient development. At a typical View 2 condition, the camera operated at 40,000 fps, under an effective optical magnification of approximately 5.6× (digital scale factor = 5 μm/pix). Details of the experimental arrangement, and optical characterization are provided in Ref. [28].

Results

Figure 1(b) plots the radii as a function of axial distance for each of the injectors, revealing a smooth inlet corner and a monotonically converging hole profile for Spray D, both of which mitigate cavitation formation. In contrast, Spray C is characterized by a sharper inlet corner and diverging hole with a chamfer at the nozzle exit. The hole exit diameters have been verified with optical microscopy measurements performed at Sandia [29, 30].

The high spatial resolution of the geometry measurements allows the radius of curvature to be azimuthally resolved, revealing the existence of hole inlet curvature profiles that are known to influence the flowfield and encourage cavitation [10, 11]. Tomography reconstructions were processed along the hole inlet periphery in a similar manner as Ref. [31] to extract the radius of curvature as a function of the azimuthal angle, θ, as seen in figure 1(c). The more strongly hydroground Spray D injector maintains a relatively uniform radius of curvature of 112 ± 4 μm. In contrast, the hole inlet of Spray C is generally sharper and varies azimuthally, with a mean value of 41 ± 6 μm, and the sharpest edges near $\theta = 0°$.

Phase contrast images were post-processed to highlight the signal contrast between internal fluid flow and fuel vapor during an injection event. Image sequences from the 30 to 40 injection events were ensemble-median-averaged to increase the level of signal-to-noise during the event. Figure 2 depicts the time-averages of the steady portion of each injection sequence. The processing method highlights all signal variation between the no-flow period and the steady injection phase; this includes motions of the injector, which manifest themselves as intensity variations at the sac and hole edges.

Cavitation presence can be seen clearly in figure 2(a) inside the top portion of the hole, which corresponds to the $\theta = 0°$ location of the injector. This azimuthal location roughly corresponds to the hole entrance edge with the smallest radius of curvature. In contrast, the $\theta = 180°$ location on the lower side of figure 2(a) shows no cavitation, which corresponds

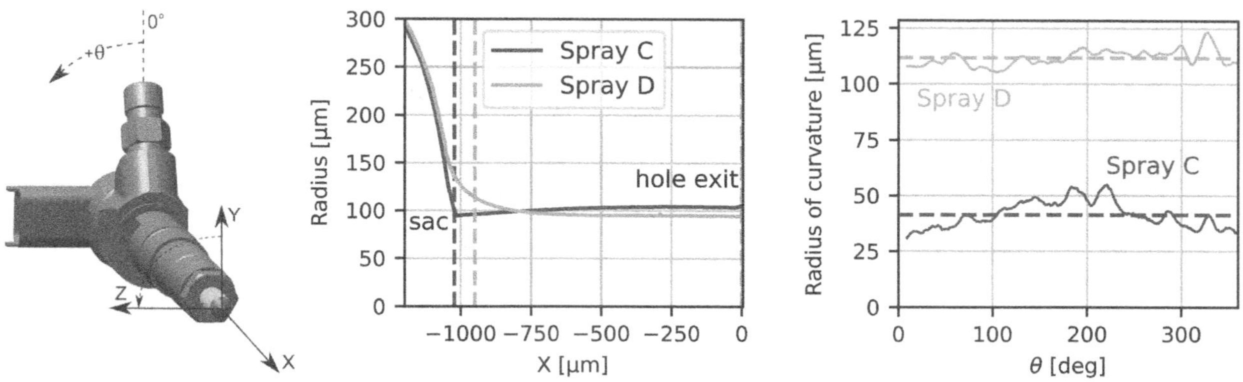

(a) Injector coordinates (b) Hole profiles (c) Inlet curvature

Figure 1: (a) ECN coordinate system for the Spray C and Spray D injectors. (b) Radial profiles of Spray C and Spray D injectors. The fuel enters from the sac and exits through the hole outlet. The vertical dashed lines indicate the start and end of each hole. (c) Hole inlet curvature as a function of azimuthal angle (θ) for Spray C and Spray D injectors. The dashed horizontal lines indicate the mean value.

to the smoothest edge of the hole entrance. The cavitation visible at the $\theta = 0°$ location grows rapidly from the entrance corner to a thickness of 47 µm at 200 µm downstream (X $= -825$ µm). This represents approximately 25% of the hole diameter. The thickness of the vapor sheath appears to asymptote by 600 µm downstream (X $= -425$ µm), to a thickness of 74 µm, corresponding to 38% of the hole diameter.

In figure 2(b), the orthogonal view depicts the presence of a thinner sheath on both the $\theta = 90°$ and 270° sides. At X $= -825$ µm, the sheath thicknesses from both sides are 27 and 23 µm, totaling to account for 26% of the hole diameter. Further downstream, the vapor regions thin to only 20% of the diameter by X $= -250$ µm. Unlike both views of the Spray C nozzle, neither view (figures 2(d) and 2(e)) of Spray D indicates any presence of cavitation. The edges of the geometry are highlighted due to vibrations caused by the needle motion. It can be deduced that at these operating conditions, and with a hole geometry that is cylindrical or divergent, the inception of cavitation is encouraged by the flow over a radius of curvature that is near or smaller than 40 µm.

Figures 2(c) and 2(f) display the steady-state projected density profiles of the fuel jet at X = 2 mm for Spray C and Spray D, respectively. Figure 2(c) clearly illustrates the asymmetric nature of the fuel distribution caused by the presence of cavitation inside the Spray C nozzle. The vapor sheath seen in figure 2(a) occludes a portion of the hole exit area, creating a unimodal spray profile that is skewed and wider in the +Y direction, while the peak projected mass shifts to the -Y direction. The projected density profile in View 2 instead reveals a bimodal distribution, where the value is lowest at the center. From this vantage, the integration line-of-sight passes normally to the thick vapor sheath interface, visible in figure 2(a), which has displaced the liquid away from the centerline. The downstream measurement, including those at axial X positions less than 2 mm, affirms that the sheath is thickest near the θ=0° location and this mass distribution propagates downstream.

Additionally, the profile from View 2, in figure 2(c), suggests that the region occluded by cavitation along the Y axis is thicker near +Z, as indicated by the lower projected mass for +Z in comparison to the peak near the -Z axis. Figure 2(f) suggests a more archetypal fuel spray distribution from Spray D, featuring a unimodal and relatively symmetric structure across both viewpoints. There is no indication in the phase-contrast images that cavitation is present in Spray D at the investigated conditions.

While the x-ray radiography profiles indicate asymmetry in distribution, this information is not typically known *a priori* for determining a threshold for optical characterization. However, previous research has shown that using a low threshold for projected mass (1 $\mu g/mm^2$) is comparable to an optical thickness of unity [22]. Therefore, a radiography full jet width was determined by setting a threshold on the steady-state projected density profile and locating the two points on the tails of the distribution. This definition of jet width actually indicates a wider Spray C at View 1 than at View 2 (0.50 mm vs. 0.47 mm, respectively) even though View 2 appears wider at higher projected mass values.

Confirming this trend, the optical microscopy measurements shown in Figure 3 also show a wider Spray C for View 1 compared to View 2, including axial distances extending to 5 mm, while Spray D is rather symmetric. Note that

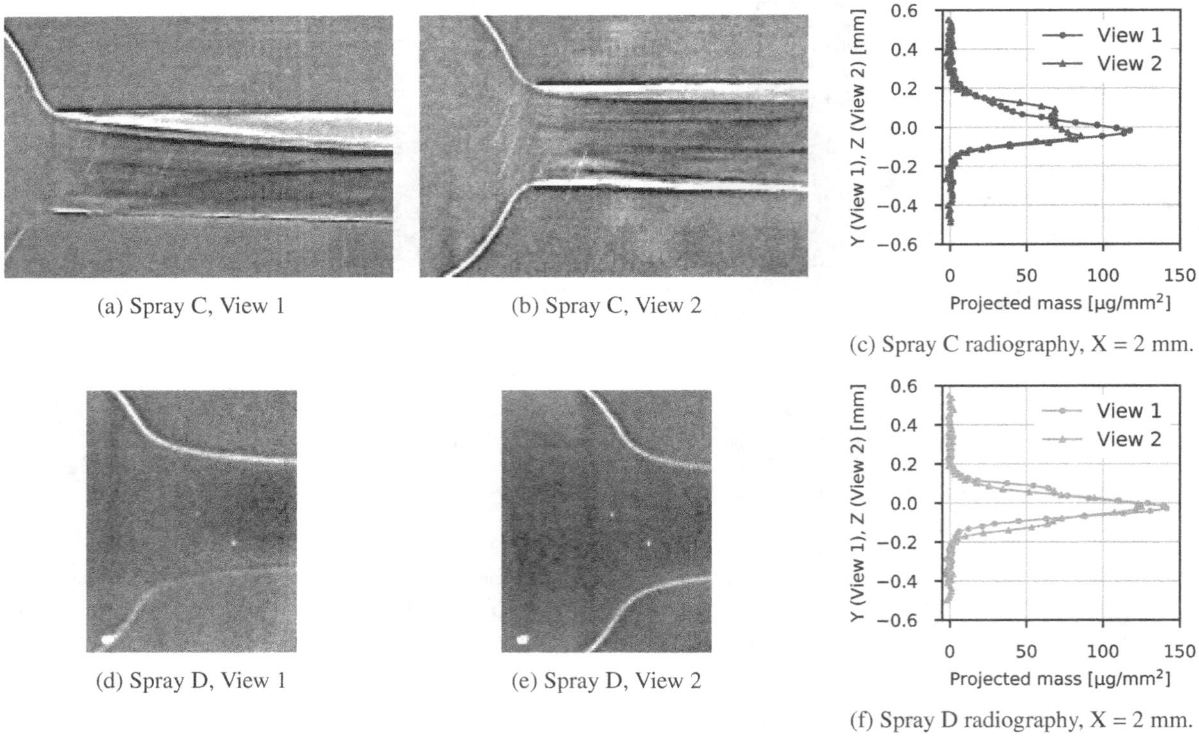

(a) Spray C, View 1 (b) Spray C, View 2 (c) Spray C radiography, X = 2 mm.

(d) Spray D, View 1 (e) Spray D, View 2 (f) Spray D radiography, X = 2 mm.

Figure 2: Phase contrast images of the steady-state internal flow for each view of the two injectors with paired radiography profiles at X = 2 mm downstream from the nozzle tips.

the optical liquid jet width depends upon the ambient and fuel temperature, becoming thinner with increased ambient temperature. Heated fuel temperature also tends to decrease spray width, though, only at increased ambient temperatures. For vaporization controlled by mixing, an increase in ambient and fuel temperature effectively increases the fuel concentration where liquid fuel exists, thereby decreasing the measured liquid jet width [32].

By comparing the measured jet width at several different ambient and fuel temperatures, the optical diagnostic exhibits a sensitivity to the distribution of fuel at low concentrations at the periphery of the spray, regions that are essentially in the noise floor of the radiography measurement. While the low-temperature (440 K) optical jet widths are larger than that of radiography at X = 2.0 mm, the high-temperature (900 K) jet widths are in closer agreement.

Unlike Spray D, the discrepancy in Spray C liquid jet width between the two viewpoints remains significant, suggesting that the impact of asymmetric cavitation existing inside the nozzle is propagated to the radial dispersion characteristics of the jet for a wide range of ambient temperatures, up to engine-relevant conditions. This suggests that as the fuel jet travels downstream, the presence of asymmetric cavitation is increasing the radial dispersion of low-density fuel in a likewise asymmetric fashion, i.e. cavitation is promoting rapid mixing and jet growth, but it occurs preferentially along the spray periphery and it is particularly aligned with the sharp inlet of the nozzle. A wider spreading angle for Spray C in comparison to Spray D at View 1 has been previously reported at ECN (900 K) conditions, including an impact on spray combustion processes [33].

Conclusions

This work investigated the influence of asymmetric injector nozzle geometry on the occurrence of cavitation and the resulting effect on fuel spray behavior at engine-like conditions. Optical and x-ray diagnostic techniques were used to investigate the internal and near-nozzle flow of two single-hole diesel injectors, Spray C and Spray D. The internal geometry of both injectors was measured with 1.8 μm spatial resolution, revealing sharp, asymmetric hole inlet contours in Spray C and a smoother and consistent curvature in Spray D. The location of the sharpest inlet feature in the Spray C geometry corresponded directly to cavitation during operation, visible through x-ray phase contrast

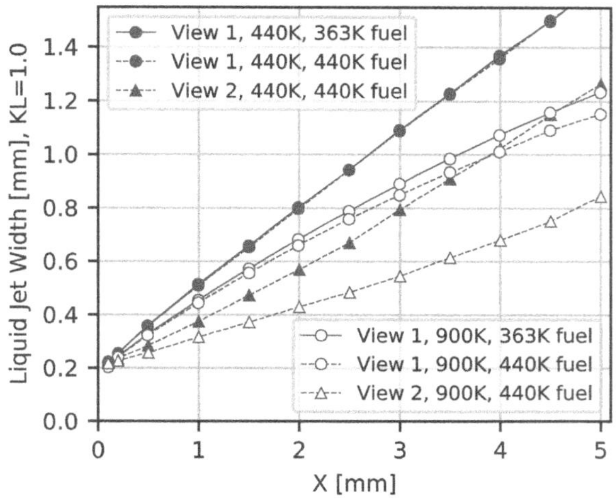

(a) Spray C

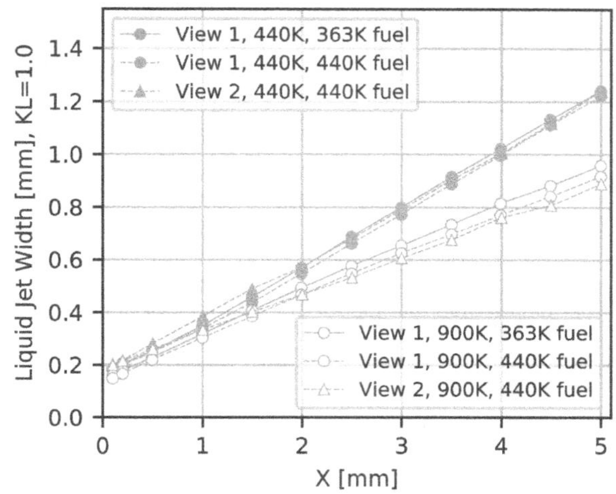

(b) Spray D

Figure 3: Liquid fuel jet width as a function of axial distance from the injector tip measured using high-speed optical imaging. Note: Spray C, View 1, 440 K ambient data overlap.

imaging, with little to no cavitation observed at locations of smoother hole entrance curvature. Cavitation was not observed in Spray D. The presence of cavitation influenced the resulting fuel spray structure, with a bias of mass away from the region of cavitation near the nozzle exit (X = 0.1 mm). Spray C profiles at X = 2 mm indicate a widening of the spray in the measurement plane with cavitation, and overall wider spray angles than Spray D, indicating increased atomization resulting in regions of low-density fuel at the periphery. Spray widening due to cavitation was also observed through optical microscopy with little influence of fuel temperature at low ambient temperatures. At higher ambient temperatures, the width decreases non-uniformly between the two perspectives, indicating a sensitivity of the spray to cavitation. This discrepancy suggests that the impact of asymmetric cavitation inside the nozzle is propagated to the radial dispersion characteristics of the jet for a wide range of ambient temperatures, up to engine-relevant conditions. These observations illustrate the influence cavitation has on fuel spray behavior and thus the significance to engine performance. Furthermore, this cavitation influence and the coupling to the asymmetric geometry impresses the importance of inclusion of these factors in predictive models for fuel spray behavior.

Acknowledgements

Research presented in this paper was performed at the 7-BM and 9-ID beamlines at the Advanced Photon Source at Argonne National Laboratory. Use of the APS is supported by the U.S. Department of Energy (DOE) under Contract No. DEAC0206CH11357. Sandia National Laboratories is a multi-mission laboratory managed and operated by National Technology and Engineering Solutions of Sandia, LLC., a wholly owned subsidiary of Honeywell International, Inc., for the U.S. Department of Energys National Nuclear Security Administration under contract DE-NA0003525. The Argonne and Sandia fuel spray research is sponsored by the DOE Vehicle Technologies Program under the direction of Gurpreet Singh and Michael Weismiller.

References

[1] Soteriou, C., Andrews, R., and Smith, M., "Direct injection diesel sprays and the effect of cavitation and hydraulic flip on atomization," *SAE Technical Paper*, 1995.

[2] Tamaki, N., Shimizu, M., Nishida, K., and Hiroyasu, H., "Effects of cavitation and internal flow on atomization of a liquid jet," *Atomization and Sprays*, Vol. 8, 1998, pp. 179–197.

[3] Sou, A., Hosokawa, S., and Tomiyama, A., "Effects of cavitation in a nozzle on liquid jet atomization," *Int. J. Heat and Mass Transfer*, Vol. 50, 2007, pp. 3575–3582.

[4] Suh, H. K. and Lee, C. S., "Effect of cavitation in nozzle orifice on the diesel fuel atomization characteristics," *Int. J. Heat and Fluid Flow*, Vol. 29, 2008, pp. 1001–1009.

[5] Sou, A., Minami, S., Prasetya, R., Pratama, R., Moon, S., Wada, Y., and Yokohata, H., "X-ray visualization of cavitation in nozzles with various sizes," *ICLASS 2015, 13th Triennial International Conference on Liquid Atomization and Spray Systems*, 2015.

[6] Bergwerk, W., "Flow pattern in diesel nozzle spray holes," *Proceedings of the Institution of Mechanical Engineers*, Vol. 173, No. 1, 1959, pp. 655–660.

[7] Hall, G. W., "Analytical determination of the discharge characteristics of cylinder-tube orifices," *Journal of Mechanical Engineering Science*, Vol. 5, No. 1, 1963.

[8] Nurick, W. H., "Orifice cavitation and its effect on spray mixing," *Journal of Fluids Engineering*, 1976, pp. 681–687.

[9] Ohrn, T., Senser, D., and Lefebvre, A., "Geometrical effects on discharge coefficients for plain-orifice atomizers," *Atomization and Sprays*, Vol. 1, No. 2, 1991, pp. 137–153.

[10] Chaves, H. and Ludwig, C., "Characterization of cavitation in transparent nozzles depending on the nozzle geometry," *Proceedings of the 20th ILASS-Europe Meeting 2005*, 2005.

[11] Schmidt, D. P., Rutland, C., Corradini, M., Roosen, P., and Genge, O., "Cavitation in two-dimensional asymmetric nozzles," *SAE Technical Paper*, 1999.

[12] Giannadakis, E., Papoulias, D., Gavaises, M., Arcoumanis, C., Soteriou, C., and Tang, W., "Evaluation of the predictive capability of diesel nozzle cavitation models," *SAE Technical Paper*, 2007.

[13] Som, S., Ramirez, A., Longman, D., and Aggarwal, S., "Effect of nozzle orifice geometry on spray, combustion, and emission characteristics under diesel engine conditions," *Fuel*, Vol. 90, 2011, pp. 1267–1276.

[14] Schmidt, D., Rutland, C., and Corradini, M., "A fully compressible, two–dimensional model of small, high-speed, cavitating nozzles," *Atomization and Sprays*, Vol. 9, 1999, pp. 255–276.

[15] Arcoumanis, C., Flora, H., Gavaises, M., and Badami, M., "Cavitation in real-size multi-hole diesel injector nozzles," *SAE Technical Paper*, 2000.

[16] Falgout, Z. and Linne, M., "Cavitation inside high-pressure optically transparent fuel injector nozzles," *Journal of Physics: Conference Series*, Vol. 656, 2015.

[17] Duke, D. J., Kastengren, A. L., Swantek, A. B., Matusik, K. E., and Powell, C. F., "X-ray fluorescence measurements of dissolved gas and cavitation," *Exp. Fluids*, Vol. 57, No. 162, 2016.

[18] Duke, D. J., Matusik, K. E., Kastengren, A. L., Swantek, A. B., Sovis, N., Payri, R., Viera, J. P., and Powell, C. F., "X-ray radiography of cavitation in a beryllium alloy nozzle," *Int. J. Eng. Research*, Vol. 18, No. 1-2, 2017.

[19] Duke, D., Swantek, A., Tilocco, Z., Kastengren, A., Fezzaa, K., Neroorkar, K., Moulai, M., Powell, C., and Schmidt, D., "X-ray Imaging of Cavitation in Diesel Injectors," *SAE Int. J. Engines*, Vol. 7, No. 2, 2014, pp. 1003–1016.

[20] Duke, D., Swantek, A., Kastengren, A., Fezzaa, K., and Powell, C., "Recent developments in x-ray diagnostics for cavitation," *SAE Int. J. Fuels Lubr.*, Vol. 8, No. 1, 2015.

[21] Pickett, L., "Engine Combustion Network," May 2014.

[22] Pickett, L. M., Manin, J., Kastengren, A., and Powell, C., "Comparison of near-field structure and growth of a diesel spray using light-based optical microscopy and x-ray radiography," *SAE Int. J. Engines*, Vol. 7, No. 2, 2014.

[23] Sandia National Laboratories, "Engine Combustion Network "Spray A" injector nozzle geometry," Accessed March 14, 2017.

[24] Kastengren, A., Powell, C., Arms, D., Dufresne, E., Gibson, H., and Wang, J., "The 7BM beamline at the APS: a facility for time-resolved fluid dynamics measurements," *J Synchrotron Rad*, Vol. 19, 2012, pp. 654–657.

[25] Gürsoy, D., Carlo, F. D., Xiao, X., and Jacobsen, C., "TomoPy: a framework for the analysis of synchrotron tomographic data," *J Synchrotron Rad*, Vol. 21, 2014, pp. 1188–1193.

[26] Matusik, K. E., Duke, D. J., Kastengren, A. L., Sovis, N., Swantek, A. B., and Powell, C. F., "High-resolution X-ray tomography of engine combustion network diesel injectors," *International Journal of Engine Research*, Vol. 0, No. 0, 2017.

[27] Als-Nielsen, J. and McMorrow, D., *Elements of modern x-ray physics*, chap. 9, John Wiley and Sons, 2011, pp. 307–313.

[28] Crua, C., Manin, J., and Pickett, L. M., "On the transcritical mixing of fuels at diesel engine conditions," *Fuel*, Vol. 208, No. Supplement C, 2017, pp. 535 – 548.

[29] Sandia National Laboratories, "Engine Combustion Network "Spray C" Injector Nozzle Geometry," 2016, Accessed July 22, 2016.

[30] Payri, R., Gimeno, J., Cuisano, J., and Arco, J., "Hydraulic characterization of diesel engine single-hole injectors," *Fuel*, Vol. 180, 2016, pp. 357 – 366.

[31] Kastengren, A., Tilocco, F., Powell, C., Manin, J., Pickett, L., Payri, R., and Bazyn, T., "Engine Combustion Network (ECN): measurements of nozzle geometry and hydraulic behavior," *Atomization and Sprays*, Vol. 22, No. 12, 2012, pp. 1011–1052.

[32] Siebers, D. L., "Scaling liquid-phase fuel penetration in diesel sprays based on mixing-limited vaporization," *SAE Technical Paper*, 1999.

[33] Westlye, F. R., Battistoni, M., Skeen, S. A., Manin, J., Pickett, L. M., and Ivarsson, A., "Penetration and combustion characterization of cavitating and non-cavitating fuel injectors under diesel engine conditions," *SAE Technical Paper*, 2016.

© 2018 ASME

Gas removal by a spark-generated bubble in a rigid tube

[1]Bo Li; [1]Chen Ji; [1]Jun Zou*; [2]Min Pan

[1]*The State Key Laboratory of Fluid Power & Mechatronic Systems, Zhejiang University, Hangzhou 310027, China*

[2]*Department of Mechanical Engineering, Center for Power Transmission and Motion Control, University of Bath, Bath BA2 7AY, United Kingdom*

Abstract

An experimental investigation of the interactions between a spark-generated bubble and a gas slug in a narrow tube is studied. The growth and collapse of the vapor bubble are recorded with the help of high-speed photography. The behavior of the trapped gas slug is affected by the dynamics of the vapor bubble in the rigid tube. In this paper, the complicated interactions of two bubbles are revealed. It is found that the direction of the post-collapse flow depends on the relative position of the vapor bubble in the tube. Likewise, the velocity of gas removal is associated with the relative positions and sizes of the bubbles. The pressure generated by the collapse of the vapor bubble plays an important role in propelling the gas out. At last the directions of the post-collapse flow are summarized. A brief "Pumping effect" in the narrow tube is discussed, which is useful for gas removal in practice. These results may be beneficial for the cavitation applications.

Keywords: spark-generated bubble; gas slug; tube; jet; post-collapse flow

Introduction

Cavitation bubbles have numerous behaviors near different boundaries such as free surface[1, 2], rigid boundary[3], elastic boundary[4, 5], and so on[6]. When a single bubble collapses near a solid boundary, it will produce a collapsing jet. The shock wave and high-speed collapse jet generated by the collapsing bubble are useful for physical and chemical reactions. Cavitation effect is useful for some applications such as surface cleaning[7, 8], extra-corporeal shock wave lithotripsy(ESWL)[9] and drug delivery[10]. With the development of ultrasound technology, the cavitation will be utilized more and more widely. Now removal of particulate is mostly associated with the cavitation implosion. Obstruction removal is another common application for cavitation bubble. Pavard investigated the particle removal effect reinforced by the bubble collapse[11]. The particle in a vertical or horizontal hole is propelled by the pressure which is associated with the expanding and collapsing of the oscillating bubble. The velocity of the solid particle removal is proportional to the stand-off distance. Xu did the similar work[12].

Nevertheless, there is a lot of gas-liquid slug flow in practical fluid systems, such as chemical equipment and hydraulic devices. The gas mixed in the liquid could induce vibration, noise and energy loss. So the gas removal is an important issue in fluid systems. In this study, the collapse jet induced by the spark-generated bubble is used to remove a trapped gas slug in a rigid tube. The correlation between the removing direction and the position of the spark-generated bubble is discussed.

Experimental Setups

The sketching of experimental setup is illustrated in Fig.1. A transparent plexiglass tank of 400mm×200mm×200mm size with an open-top, which is filled up to about 90% with tap water. The water and environment temperatures are maintained at 16℃ ±3℃, and the ambient pressure is kept at 1 atm. A quartz-glass circular tube with 3mm interior diameter and 100mm length is fixed on a rigid holder, which are all submerged in the tap water. In this work, we only consider this case that one large-size gas bubble filling up the section of the narrow tube. The gas slug attached into the tube keeps stable before the vapor bubble occurs. Just like we did before[13], the low-voltage spark method is used to generate the vapor bubble in our present work. This low-voltage circuit includes three capacitors (one of 3300μF and two of 1000μF), a 1KΩ resistor and a 45V±5V DC power. The charging and discharging of the circuit is controlled by a two-position switch. The ends of the wires are fine copper electrodes with 0.2mm diameter. Moreover, the joint point of the electrodes is adjusted along the centerline of the tube. The gas slug is trapped in the narrow tube before the electrodes are placed near it. In our experiments, we submerged half of the tube into water vertically, then rotated the tube to be horizontal and submerged the entire tube into water, so there is a section of gas bubble trapped

*Corresponding Author, Jun Zou: junzou@zju.edu.cn

in the tube. Its length and position are adjusted along the axial direction of the tube too. Specially, the gas slug should be kept away from the tube end. Besides, the vapor bubble should not be too close to the tube end and the gas edge. In our present work, the vapor bubble is kept contactless with the gas slug.

In order to record the transient process, a high-speed video camera (Phantom UHS-12 V2512) with a filming rate of 20,000 frames per second is used, which is fitted with a Nikkor 60-mm microlens. The camera is mounted in front of the tank, and a high-intensity LED lamp is placed at the back of the tank. This lamp covered with an oil paper supplies the backlight during the camera working.

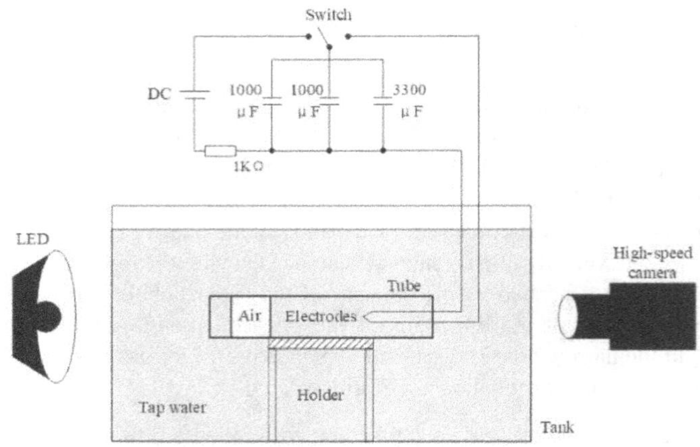

Fig.1 Experimental set-up

The typical schematic diagram of the tube is shown in Fig.2. The gas slug and vapor bubble are all located at the horizontal centerline of the narrow tube. At the moment of the vapor bubble grows to its greatest size, all the gas slug and vapor bubble occupy the cross section of the circular tube. So this case could be considered as one-dimensional motion. In order to evaluate the effect of the positions and shapes of the bubbles, the sketch of parameters is simplified. As shown in Fig.2, L and D represent the length and the interior diameter of the rigid tube respectively. In our experiments, one kind of tube has unified interior diameter of D=3mm, and length of L=100mm. The horizontal length of the gas slug and the vapor bubble are measured as L_A and L_B severally. In order to reduce the effect of the tube end, the gas slug and vapor bubble should be kept away from the tube-endings. We define the leftmost tube center as the origin point. X_A is the distance from the original point to the center of the gas slug. The length between the collapse point of the spark-generated bubble and the original point is noted as X_B. We ignored this slight shift of the collapse point away from the spark point because of the pumping effect and transformation of the gas slug. Because of the tube is horizontal and narrow, so we ignored the gravity.

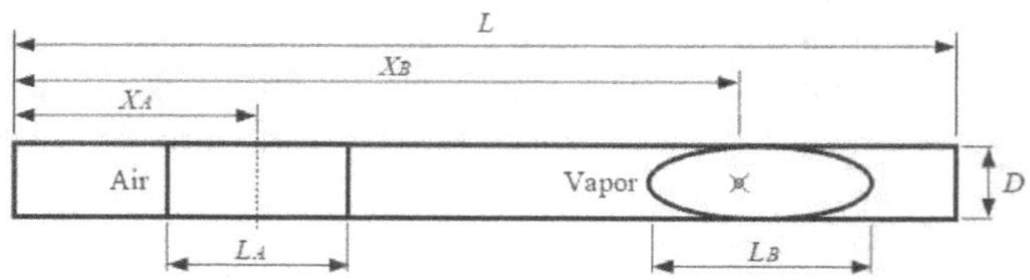

Fig.2 Sketch and parameters

Results

Obviously, the growth and collapse of the vapor bubble could induce the deformation of the gas slug. Fig.3 shows two typical sequences of the interactions between a vapor bubble and a gas slug in a rigid tube. The vapor bubble is generated at T=0ms when the capacitor voltage is applied to the electrodes. In Fig.3(a), it reaches its maximum size at 1.15ms, and then it collapses at 2.3ms. During the course of the expansion of the vapor bubble, the gas slug contracts until the vapor bubble stops expanding. The axial length of the gas slug decreases from 27.3mm to 25.7mm approximately. Conversely, the gas slug expands in the course of the contraction of the vapor bubble. The secondary cavitation occurs at 2.35ms following the collapse of the vapor bubble. The results show that the multiple oscillations of the vapor bubble induce the expansion and contraction of the gas slug repeatedly. At last the post-collapse flow moves to left from 3.6ms to 61.5ms in Fig3(a). The average velocity of the post-collapse flow is about 0.41m/s.

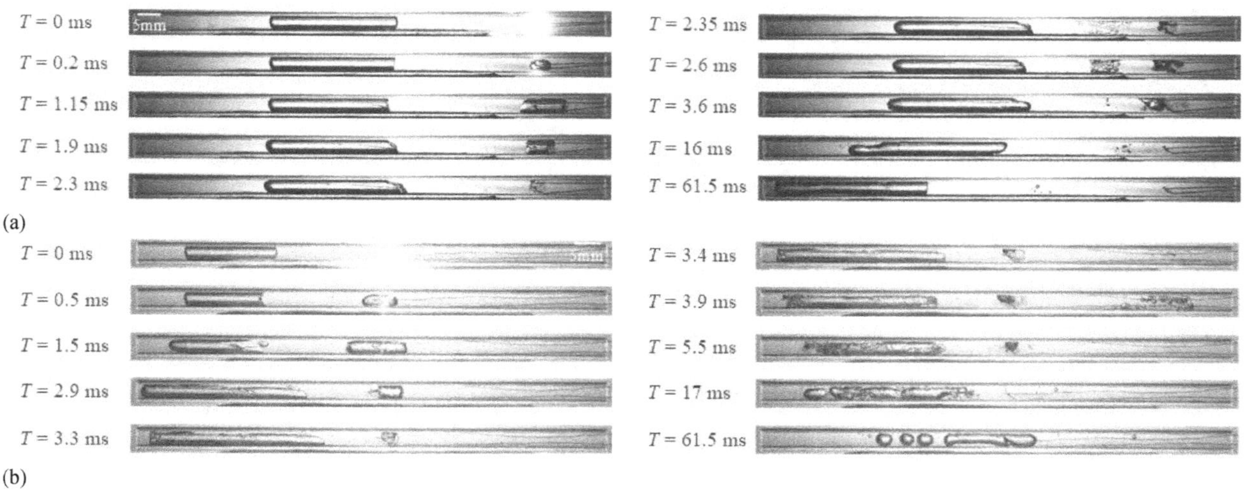

Fig.3 Gas removal by an oscillating bubble in a rigid tube. (a. The whole liquid moves towards left. b. The whole liquid moves towards right.)

Fig.3(b) shows another case where the electrodes are placed at the center of the tube. The spark-generated bubble grows to its maximum size at 1.5ms. Afterwards its left side contracts faster because of the expansion of the gas slug. Its two sides impact each other at 3.3ms. Then the gas slug expands to its maximum size at the time of the vapor bubble collapsing. Whereafter, the secondary cavitation also occurs at the same symmetric position in the tube at 3.4ms. At last the whole liquid in the tube moves rightwards after the vapor bubble collapsed. We measured that the speed of the gas removal is about 0.42m/s.

In conclusion, just like one spark-generated bubble in a rigid tube[13], secondary cavitation occurs at the same symmetric position. The gas slug edge and tube end form the new boundaries where the water-hammer shock wave is reflected. There are three types of the direction of the post-collapse flow in the tube. The three directions are moving leftwards, moving rightwards and stagnation respectively.

Discussion and Conclusion

The experimental results show that the growth and collapse of a single spark-generated bubble could induce post-collapse flow and secondary cavitation in a rigid tube. Obviously, the secondary cavitation is caused by the propagation of the pressure wave just like the results in our previous work[13, 14]. In particular, the gas slug edge and tube end form the two reflection boundaries. One-dimensional model[15, 16] is used to explain the dynamics of the cavitation bubble and gas slug. After the collapse of the spark-generated bubble, the post-collapse flow direction is based on the collapse position of the oscillating bubble. Fig.4 shows the flow directions after the oscillating bubble collapsed. The red circle represents the case of flowing towards the right, and the black square means that the flow direction is from right to left. There is a stagnated flow between moving leftwards and rightwards. The blue triangle representing the flow stagnation is between the two conditions. The blue dash line ($X_R = 0.6\,X_L$) is the fitted line. In order to distinguish the different flow directions, we define two non-dimensional parameters $X_L = \frac{X_B - L_A}{L}$ and $X_R = \frac{L - X_B}{L}$. The length of the gas column is subtracted when calculating the liquid impacting. As shown in Fig.4, the post-collapse flow moves to the right when $X_R > 0.6\,X_L$. Inversely, it will move to left in the case of $X_R < 0.6\,X_L$. Note that there is a large size of the gas slug in the case of $X_L \to 0$.

The gas removal in a rigid tube is induced by the pumping effect. Torniainen[17] studied the characteristics of the net flow induced by the vapor bubble. We will reference the similar method to study the dynamics of spark-generated bubble.

When a bubble collapses in a tube filled with a gas slug, the collapse pressure could impel the net flow. As a method of gas removal, cavitation bubble is useful and efficiently.

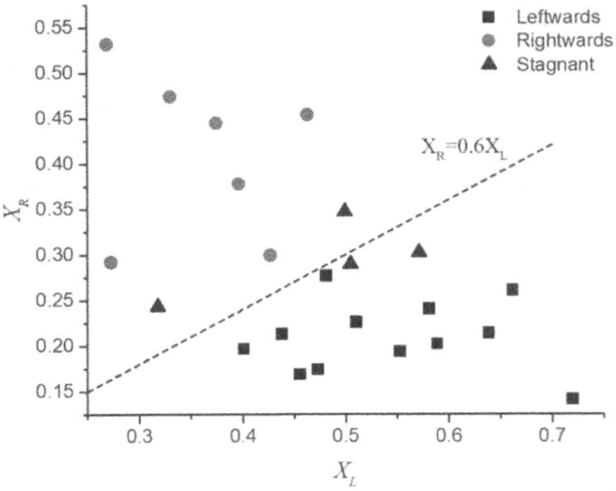

Fig.4 Flow direction after the collapse of the spark-generated bubble

Acknowledgements

This research is supported by the National Natural Science Foundation (Nos. 51475415, 51405429, and 51521064), Zhejiang Provincial Natural Science Foundation for Distinguished Young Scholars (No. LR15E050001), and Research Project of the State Key Laboratory of Mechanical System and Vibration (Grant No. MSV201706).

References

[1] Ji C, Li B, Zou J, Yang H (2017) Interaction of two spark generated bubbles beneath free surface. EXP THERM FLUID SCI 81:76-83

[2] Kedrinskii VK (1975) Dynamics of the cavitation zone during an underwater explosion near a free surface. Journal of Applied Mechanics and Technical Physics 16 (5):724-733

[3] Zhang S, Duncan JH, Chahine GL (1993) The final stage of the collapse of a cavitation bubble near a rigid wall. J FLUID MECH 257 (1):147-181

[4] Klaseboer E, Khoo BC (2004) An oscillating bubble near an elastic material. J APPL PHYS 96 (10):5808

[5] Brujan EA, Nahen K, Schmidt P, Vogel A (2001) Dynamics of laser-induced cavitation bubbles near elastic boundaries: influence of the elastic modulus. J FLUID MECH 433:283-314

[6] Zou J, Li B, Ji C (2015) Interactions between two oscillating bubbles in a rigid tube. EXP THERM FLUID SCI 61:105-112

[7] Zhu Y, Zou J, Zhao WL, Chen XB, Yang HY (2016) A study on surface topography in cavitation erosion tests of AlSi10Mg. TRIBOL INT 102:419-428

[8] Tam AC, Leung WP, Zapka W, Ziemlich W (1992) Laser - cleaning techniques for removal of surface particulates. J APPL PHYS 71 (7):3515-3523

[9] Yoshizawa S, Ikeda T, Ito A, Ota R, Takagi S, Matsumoto Y (2009) High intensity focused ultrasound lithotripsy with cavitating microbubbles. MED BIOL ENG COMPUT 47 (8):851-860

[10] Paliwal S, Mitragotri S (2006) Ultrasound-induced cavitation: applications in drug and gene delivery. Expert Opin Drug Deliv 3 (6):713-726

[11] Pavard D, Klaseboer E, Ohl SW, Khoo BC (2009) Removal of particles from holes in submerged plates with oscillating bubbles. PHYS FLUIDS 21 (8):83304

[12] Xu M, Ji C, Zou J, Ruan X, Fu X (2014) Particle removal by a single cavitation bubble. Science China Physics, Mechanics and Astronomy 57 (4):668-673

[13] Ji C, Li B, Zou J (2017) Secondary cavitation in a rigid tube. PHYS FLUIDS 29 (8):82107

[14] Chen J, Bo L, F Y L, J Z (2015) Outcomes following single bubble collapse in a rigid tube. Journal of Physics: Conference Series 656 (1):12043

[15] Yuan H, Og Uz HN, Prosperetti A (1999) Growth and collapse of a vapor bubble in a small tube. INT J HEAT MASS TRAN 42 (19):3643-3657

[16] Yuan H, Prosperetti A (1999) The pumping effect of growing and collapsing bubbles in a tube. J MICROMECH MICROENG 9 (4):402-413

[17] Torniainen ED, Govyadinov AN, Markel DP, Kornilovitch PE (2012) Bubble-driven inertial micropump. PHYS FLUIDS 24 (12200312)

© 2018 ASME

10th International Symposium on Cavitation - CAV2018
Baltimore, Maryland, USA, May 14 – 16, 2018 CAV18-05100

Global mode visualisation in cavitating flows.

[1]James A. Venning*; [1]Dean R. Giosio; [1]Bryce W. Pearce; [1]Paul A. Brandner

[1] *Cavitation Research Laboratory, University of Tasmania, Launceston, Tasmania, Australia*

Abstract

A technique for visualising global frequency modes and phasing information is presented using fast Fourier transforms applied to individual pixel intensities of high-speed photography image frames. A singular value decomposition is applied to the resulting complex data set to allow the application of windowing to the transforms. A high-speed image set of cloud cavitation about a sphere is used to demonstrate the technique and identify the dominant cavitation shedding modes. Three frequencies were found and the modal analysis show these to correspond to large-scale near symmetric shedding, intermediate-scale oblique shedding, and small-scale near symmetric shedding, respectively. Results obtained by the presented image analysis technique show close agreement to shedding frequencies obtained from pressure data.

Keywords: modal decomposition; mode visualisation; cavitation; high-speed imaging; frequency analysis

Introduction

Modal decomposition methods have been used in a variety of fluid mechanics applications to approximately describe complicated physical phenomena with few dimensions. Two commonly used techniques are the proper orthogonal decomposition (POD) and the dynamic mode decomposition (DMD). The POD ranks spatial structures according to the kinetic energy, if the data set is velocities [1]. This allows filtering based on energy so that experimental noise and turbulence can be removed. The DMD is based on growth-rates of frequencies, so that amplifying or decaying features can be extracted, but permanent features are not accentuated [e.g. 2, 3].

Here we present a technique based on the fast Fourier transform (FFT) of the pixel intensity in high-speed images. The objectives are to identify dominant frequencies in the videos, identify the regions in the spatial domain where these modes are manifested and to identify the phasing differences across the domain. This technique has been used previously in Basley et al. [4, 5] and [6], but here we extend the technique to enable the 'averaging' of multiple time instances to remove noise from the modes, as introduced by Venning [7] for velocimetry data.

In order to demonstrate the features of the decomposition, an example flow of cavitation about a sphere at a Reynolds number of $Re = U_\infty D/\nu = 1.5\times10^6$ and a cavitation number $\sigma = 2(p_\infty - p_v)/\rho U_\infty^2 = 0.8$ is presented. The sphere was sting-mounted in the centre of the tunnel test-section and the cavity was visualised with a Photron SA-5 high-speed camera acquiring megapixel images at 7000 fps. The conditions here match that of [8], and the three frequencies extracted from our video analysis correspond to those from the pressure measurements. The nature of the shedding modes of the cloud cavities are able to be distinguished using the decomposition here.

Spectral density estimation

The fast Fourier transform (FFT) is applied to the pixel intensity time series at each pixel location on the image. A Welch periodogram with Hanning windows was used to estimate the spectral density. For each data set, the window length, N, was 8192 points, representing 1.2 s or approximately 30 cycles of the dominant frequency, though this frequency was estimated as it varies with the cavitation number. The shift between windows, N_S, was 256 points. Given the time sequence is 21,000 points, there were $T = 51$ windows.

For each position (j) in the image plane, and each Welch window (t), the FFT of the pixel's intensity signal (x_n) is given by equation 1 with P and ϕ corresponding to the power and phase of each sinusoidal component of the time series, while f is the wavenumber. $\mathrm{Arg}(X_{jtf})$ returns the principal value of the angle from the real axis to a line from the origin to the complex number X_{jtf}, which in a practical sense, is $\mathrm{atan2}(\mathrm{Im}(X_{jtf}), \mathrm{Re}(X_{jtf}))$.

*Corresponding Author, James Venning: mail@jamesvenning.net

$$\begin{aligned} X_{jtf} &= \mathscr{F}(x_{jtf}) = \sum_{n=0}^{N-1} x_n \cdot e^{-2\pi i f n/N} \\ P_{jtf} &= \sqrt{\mathrm{Re}(X_{jtf})^2 + \mathrm{Im}(X_{jtf})^2} \\ \phi_{jtf} &= \mathrm{Arg}(X_{jtf}) \end{aligned} \tag{1}$$

The averaging across the T windows is trivial for the power (equation 2). For the case of the sphere, the average of the spectra across the spatial domain, $\langle \overline{P_{jf}} \rangle_j$ (figure 1) finds three key frequencies, as was reported in de Graaf et al. [8].

$$\overline{P_{jf}} = \frac{1}{T} \sum_{t=1}^{T} P_{jtf} \tag{2}$$

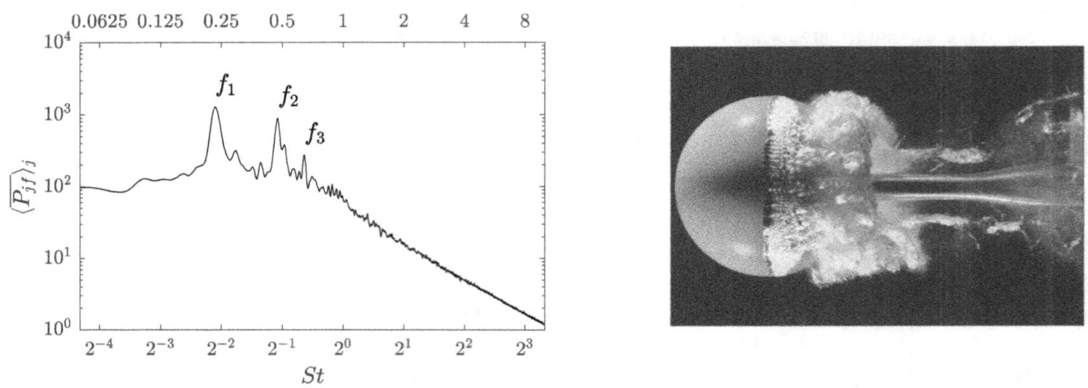

Figure 1: Spatial average (left) of all spectra across the domain identifying three peaks. Photograph (right) showing typical flow state at a cavitation number of 0.8.

It is less straightforward to produce a representative view of the phase angle, however, since it is constantly changing and thus a temporal average of ϕ_{jtf} will not be meaningful. In fact, given enough t instances, the temporal average will be zero. The phase of different windows are given in figure 2. The same general flow pattern can be seen in each, though the phase is offset by some angle between the instances. The phase offset is approximately the ratio of the time shift (N_S) to the temporal wavelength of the structure. To present an 'averaged' view of the phase information from each of the windows, the singular value decomposition (SVD) was used to decompose the Fourier coefficients and return a series of SVD-modes, the first being most representative of the spatial mode corresponding to each frequency. The matrix $\mathbf{\Phi}_f$ is constructed in which each column represents a different window (t) of the phase of one frequency – the size of this matrix is $J \times T$ (for the present study: $2^{19} \times 51$):

$$\mathbf{\Phi}_f = \overset{\text{Welch windows} \longrightarrow}{\begin{bmatrix} X_{j=1,t=1} & X_{j=1,t=2} & \cdots & X_{j=1,t=T} \\ X_{j=2,t=1} & X_{j=2,t=2} & \cdots & X_{j=2,t=T} \\ \vdots & \vdots & \ddots & \vdots \\ X_{j=J,t=1} & X_{j=J,t=2} & \cdots & X_{j=J,t=T} \end{bmatrix}} \Bigg\downarrow \text{Spatial domain} \tag{3}$$

The SVD factorizes the complex field $\mathbf{\Phi}_f$, decomposing it with:

$$\mathbf{\Phi}_f = \boldsymbol{L}_f \boldsymbol{\Sigma}_f \boldsymbol{R}_f \tag{4}$$

Which returns $\boldsymbol{L}_f$: a series of orthonormal eigenvectors of $\mathbf{\Phi}_f \mathbf{\Phi}_f^T$ which is a square matrix of size $J \times J$, $\boldsymbol{\Sigma}_f$: a rectangular matrix where the diagonal contains T values that are the square-root of the eigenvalues (these correspond to how much of the variability in $\mathbf{\Phi}_f$ is accounted for in each SVD-mode), and $\boldsymbol{R}_f$: a series of eigenvectors of $\mathbf{\Phi}_f^T \mathbf{\Phi}_f$; a square matrix of size $T \times T$.

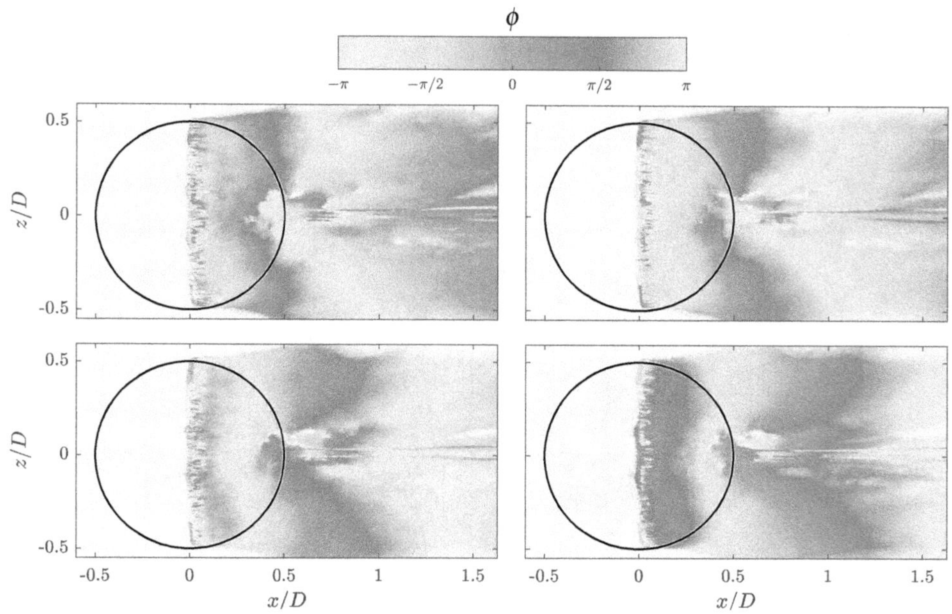

Figure 2: Phase angle of the fundamental cavitation shedding frequency about a sphere at $\sigma = 0.8$. Sequential instances of time (Welch windows) are displayed.

The T SVD-modes are the projections of $\mathbf{\Phi}_f$ onto the basis eigenvectors ($\mathbf{\Phi}_f \mathbf{R}_f$). This mode shape and the corresponding variance (from $\mathbf{\Sigma}_f$) of each SVD-mode for the second frequency (f_2 from figure 1) are given in figure 3. The first SVD-mode captures the bulk of the variance, so should be the most representative of the phase field. This is verified by comparing figure 2 with figure 3, and as such it is the only SVD-mode presented for the remainder of the paper. The most useful result of the modal analysis is the spatial phase information of each frequency across the video frame revealing the direction of the cavity flow, and more importantly the regions of the flow that are travelling in phase.

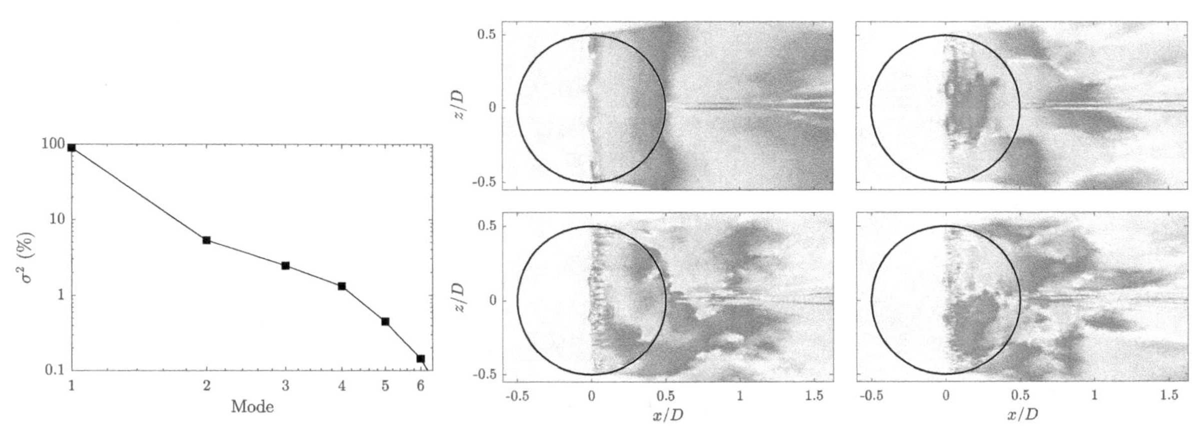

Figure 3: Variance (top) of each of the f_2 SVD-modes (bottom) as a percentage of the total variance. The first SVD-mode contributes 90.3% of the total variance, representing the most likely phase pattern in the flow, and corresponds well to the instantaneous frames in figure 2. The mode number increases across the page first.

The first SVD-mode for each of the three shedding frequencies from figure 1 are presented in figure 4. The primary large-scale shedding mode is $f_2 = 0.48$. In this mode, the cavity is shed as a series of ring-like vortices that become oblique in the wake (figure 5). Visualisation of this shedding mode suggests that it is quite powerful for most of the wake portion captured by the video frame, particularly for $0.2 < x/D < 0.7$. The mode is symmetric about the $z/D = 0$ plane, reminiscent of the ring-like vortices from a shedding sphere at $300 < Re < 420$ presented in [9]. The shed cavities condense near $x/D = 0.8$, which is why the energy content decays from then on.

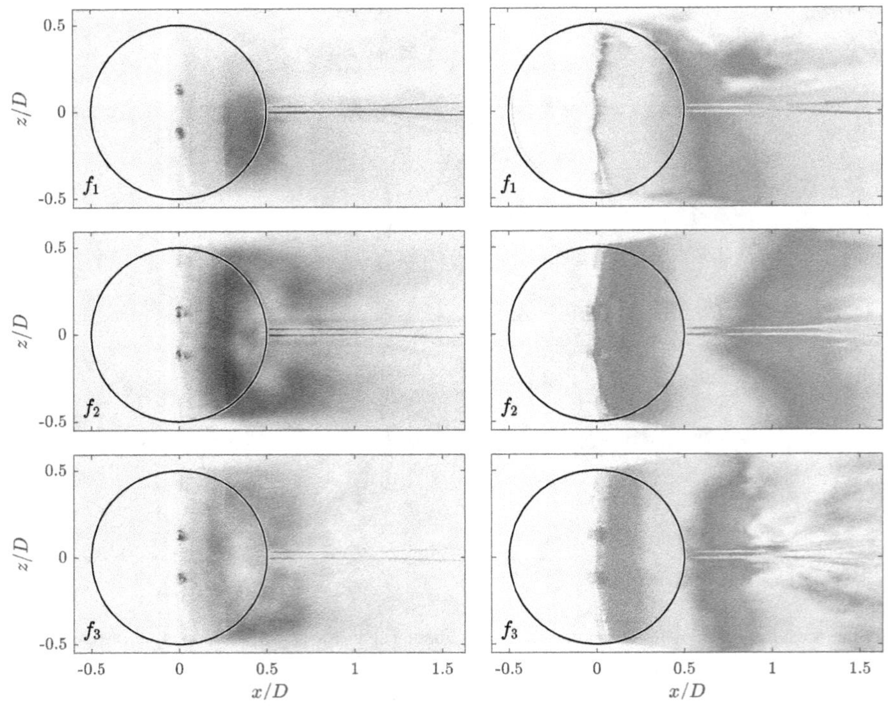

Figure 4: Frequency information for the flow around a sphere at $\sigma = 0.8$. The left hand column contains the spatial distribution of the power, and the right-hand column is the phase of those structures. The mode power is normalised to the maximum power of that frequency.

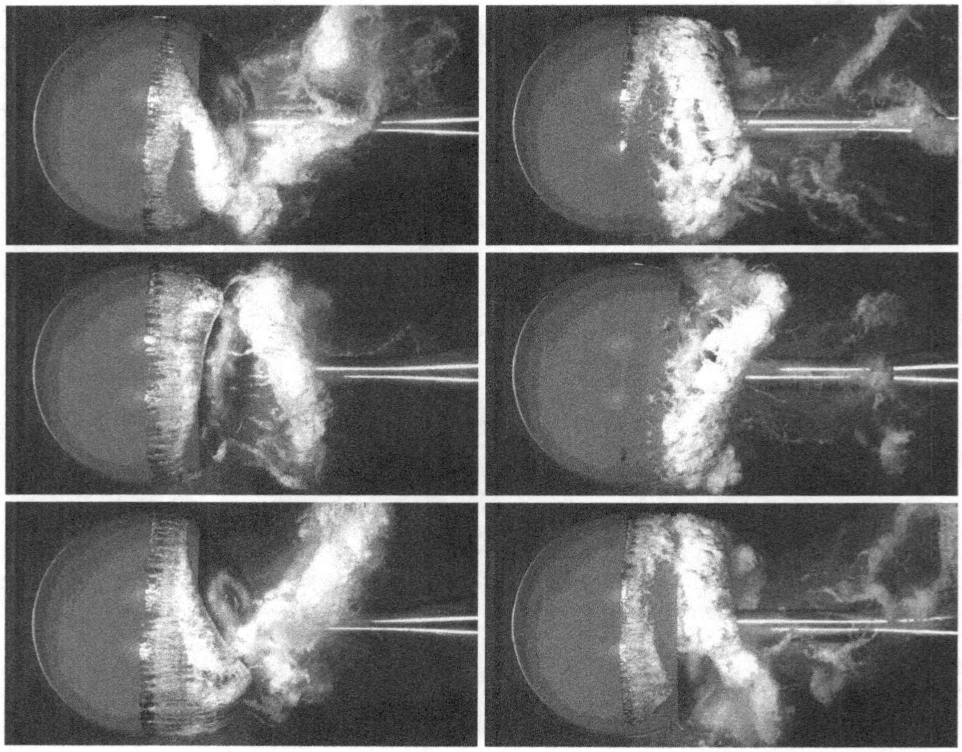

Figure 5: Six equi-spaced photographs capturing three cycles of the f_2 shedding mode, showing alternating oblique vortex shedding. The images are separated by half the period, such that each column is at a single phase.

One difficulty with this technique is that the phase information is averaged over the window of interest, so when different flow features occur at the same frequency, the results are misleading. For example, examining the high-

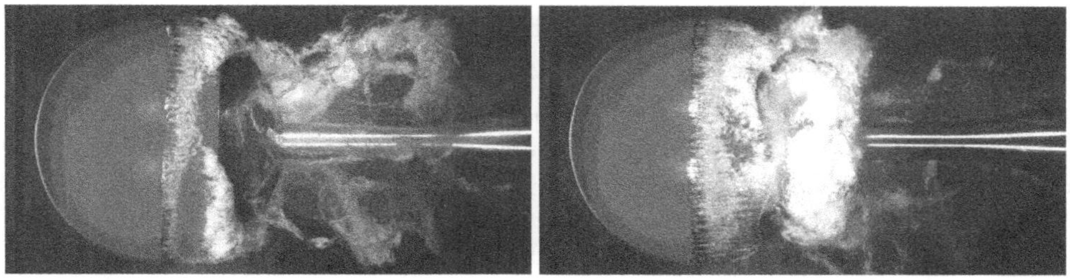

Figure 6: Photographs showing a single cycle of the f_1 shedding frequency. The images are separated by half the period.

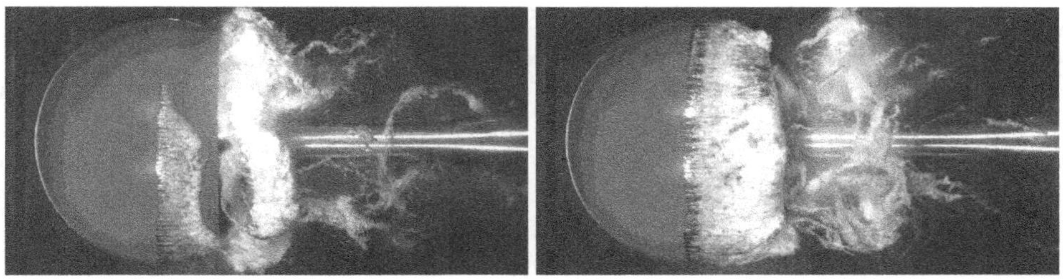

Figure 7: Photographs showing a single cycle of the f_3 shedding frequency. The images are separated by half the period.

speed footage in regions where the f_2 peak is powerful (figure 5) suggests that the shed vortices are actually oblique, alternating slope with each cycle. So, while the average of the phase is, in-fact, symmetrical, the individual structures may be oblique, thus care must be taken to investigate the original videos to understand the true nature of the modes. A clue to the true shape of this mode may be in the second of the SVD modes presented in figure 3, which shows oblique structures. Additionally, considering only one half ($z/D > 0$) of the phase diagram in figure 4 shows the angled nature of the shedding.

The first shedding mode, $f_1 = 0.23$ is concentrated in the near-wake of the sphere (figure 6). It is a sub-harmonic of the primary frequency. It is slightly more dominant in the bottom than the top, suggesting that this mode could be suppressed to some extent by the sting and/or hydrostatic differences. The shed cavities often take the appearance of hair-pin vortices (figure 6).

A third shedding mode, $f_3 = 0.64$, is also symmetric. Analysis of the video (figure 7) suggests that this frequency is associated with a very short cavity followed by a longer cavity. The initial cavity is still in the near vicinity of the body when the second cavity is growing. The energy of this higher frequency mode dissipates earlier than f_2, with this short cavity being replaced by one of the two large-scale events.

Conclusion

A technique for decomposing time-resolved visual data is presented using a Fourier transform to determine the spatial distribution of power for all frequencies of interest. The resulting reconstructed images reveal the locations within the image that experience a given frequency.

The structure of the phase can be determined by a singular value decomposition of the different phase windows. The phase information is useful to show the nature of the oscillations, but care must be taken that the spatial mode is, in fact, exemplary of the actual flow structures.

We use a high-speed video of cavitation about a sphere to show how this technique can be used to aid in the understanding of the mechanisms associated with the three shedding frequencies of interest. The main shedding mode was shown to be due to the shedding of large, oblique cavitation clouds, alternating in gradient in successive cycles. The subharmonic mode was related to symmetric, large-scale cavitation, while the harmonic mode was single-event, short-duration, cavities.

References

[1] Lawrence Sirovich. *Turbulence and the dynamics of coherent structures.* Quarterly of Applied Mathematics, 45 (3), 1987. ISSN 0033-569X. doi: 10.1090/qam/910463.

[2] Clarence W. Rowley, Igor Mezić, Shervin Bagheri, Philipp Schlatter, and Dan S. Henningson. *Spectral analysis of nonlinear flows.* Journal of Fluid Mechanics, 641, 2009. ISSN 0022-1120. doi: 10.1017/S0022112009992059.

[3] Peter J. Schmid. *Dynamic mode decomposition of numerical and experimental data.* Journal of Fluid Mechanics, 656, 2010. ISSN 0022-1120. doi: 10.1017/S0022112010001217.

[4] J. Basley, L. R. Pastur, F. Lusseyran, T. M. Faure, and N. Delprat. *Experimental investigation of global structures in an incompressible cavity flow using time-resolved PIV.* Experiments in Fluids, 50(4), 2011. ISSN 07234864. doi: 10.1007/s00348-010-0942-9.

[5] J. Basley, L. R. Pastur, N. Delprat, and F. Lusseyran. *Space-time aspects of a three-dimensional multi-modulated open cavity flow.* Physics of Fluids, 25(6), 2013. ISSN 10706631. doi: 10.1063/1.4811692.

[6] T. Mcqueen, J. Venning, and J. Sheridan. Effects of Aspect Ratio on the Wake Dynamics of the Ahmed Body. In *19th Australasian Fluid Mechanics Conference*, 2014. ISBN 9780646596952.

[7] J Venning. *Vortex structures in the wakes of two- and three-dimensional bodies.* Phd, Monash University, 2016.

[8] K. L. de Graaf, P. A. Brandner, and B. W. Pearce. *Spectral content of cloud cavitation about a sphere.* Journal of Fluid Mechanics, 812, 2017. ISSN 0022-1120. doi: 10.1017/jfm.2016.819.

[9] H Sakamoto and H Haniu. *A Study on Vortex Shedding From Spheres in a Uniform Flow.* Transactions of the ASME Journal of Fluids Engineering, 112, 1990.

© 2018 ASME

GPU-based, parallel-line, omni-directional integration of the acceleration field to obtain the 3D pressure distribution

Jin Wang[1], Cao Zhang[1] and Joseph Katz[1]

[1] *Department of Mechanical Engineering, Johns Hopkins University, Baltimore, USA*

Abstract

For recent years, PIV based pressure reconstruction from material acceleration either by solving Pressure Poisson Equation (PPE) or by direct integration has been studied. The advantage of direct integration on no need of prescribed pressure boundary encourage us to develop a 3D direct integration method. In this paper, a GPU-based, parallel-line, omni-directional integration (Omni3D) method has been developed and tested using the JHU DNS Database, comparing with three different methods. In a serious of tests, Omni3D tends out to be a robust and efficient method with only 1 minute for one realization. This method is also utilized to compute the volumetric pressure in a turbulent channel flow over a compliant surface.

Keywords: turbulence; pressure measurement;

Introduction

PIV-based pressure measurement techniques involving direct integration of material acceleration or solving the Pressure Poisson Equation (PPE) have become popular in recent years. While PPE can be readily extended to volumetric measurements [1] using e.g. time-resolved tomographic PIV data [3], extension of the direct integration method, such as the 2D Virtual-boundary Omni-directional integration (Omni2D, [2]) from 2D to 3D, has been problematic owing to the massive amount of computations involved. Yet, omni-direction integration is effective in minimizing the effect of local acceleration errors. Hence, our goal is to develop and evaluate a fast and robust 3D pressure reconstruction method.

The 3D GPU-based Parallel-line Omni-directional integration method

The new GPU-based, parallel-line, omni-directional method (Omni3D) integrates the material acceleration along a series of parallel lines aligned in all possible directions. These directions are determined by a spherical virtual grid that surrounds the sample volume. The pressure at every point is the average value obtained from all the paths crossing this point. Hence, errors associated with a certain path are minimized. Iterations (3-4) are used for correcting the initially assumed pressure distribution along the extenal surfaces of the sample volume. Hence, there is no need to prescribe a boundary condtion. Methods for identifying and circumventing regions with particularly high material acceleration errors have also been developed.

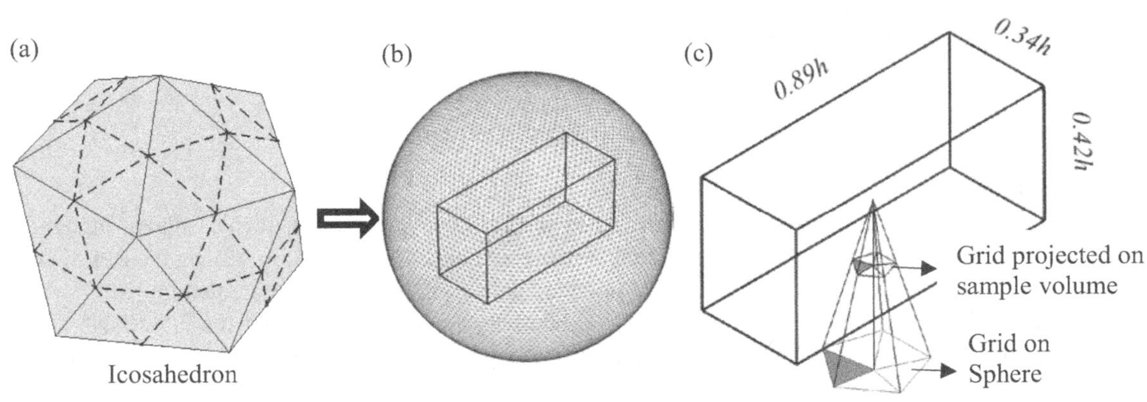

Figure 1 The grid with 20480 surfaces used for generating the parallel line integration paths.

To assure the grid homogenuity, an initially icosahedron grid is divided recursively several times, as illustrated in Figure 1. In each step each tiangular surface is divided to four parts and projected onto the surface of the sphere. Repeating this procedure 6 times generates a spherical triangular grid with 20480 faces (Figure 1(b)). As illustrated in Figure 1(c), the projected area from each spherical virtual surface should match the data resolution, yielding $4\pi(L/2)^2/N_f \sim 0.5\Delta x^2$, where L is the length of the sample volume ($4\pi/3(L/2)^3$=V, V is the volume), Δx is the data resolution, and N_f is the number of triangular faces. Hence, $N_f \sim 3\times10^4$ for a grid of 100×47×38 voxels, the grid of our recent TPIV measurements. For the present N_f =20480, the resulting total number of integration paths is in the order of 10^8, requiring more than 3 hours for integrating the acceleration in a single realization using a 3.7 GHz, 8 core Intel i7-3770k processor. This computing time is reduced to about one minute by implementing the same integration algorithm on a low cost, Tesla K40c GPU board, which has 2880 745 MHz processors.

The detailed computation steps are presented in Figure 2:

Step.1 Allocate memory needed on GPU device. Read acceleration data and transfer it to the GPU memory. The total memory needed is 3 GB.

Step.2 Use parallel-line omni-directional integration (PLODI) in parallel to obtain the pressure difference between every two different boundary nodes, denoted as PINT(nin,nout), where, nin is the integration path starting point and nout is the integration path ending point. Subsequently, parallel boundary pressure iteration is performed with the known pressure difference using

$$P_{nout}^{new} = \frac{\sum_{nin=1,nin\neq nout}^{N}(P_{nin}^{old} + PINT(nin,nout))}{N-1}$$

It can be verified that this iteration scheme is equivalent to a least square fit of the overdetermined linear system.

Step.3 In parallel, assign each thread an integration path to perform PLODI to obtain the pressure at inner nodes. The inner pressure is the averaged results of all the integration paths.

Step.4 Transfer pressure data from device to host and write it to the disk.

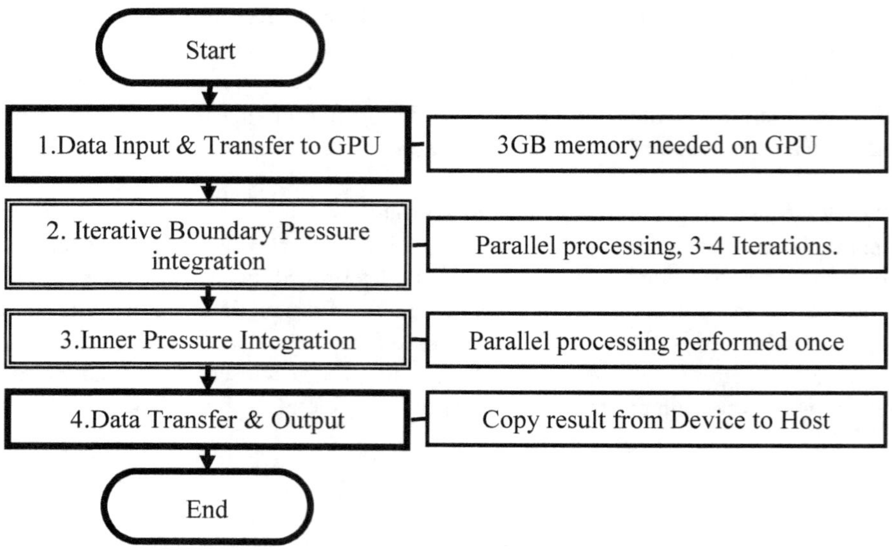

Figure 2 The flow chart of Omni3D Method

DNS Verification of Omni3D

The pressure predicted by the Omni3D method is compared to several other procedures, such as planar integration using Omni2D combined by planar integration along a perpendicular plane (Omni2D), as well as by solving the PPE with Dirichlet boundary conditions on one surface. These conditions are obtained either using the Bernoulli equation (PPE-Bernoulli [1]) or Omni2D integration along this surface (PPE-Omni2D). Validations are performed using the JHU DNS database for turbulent channel flow [4]. Randomly distributed synthetic particles are generated and displaced by the velocity field obtained from database. Subsequently, velocities are calculated via cross correlation using different interrogation volume (I.V.) ranging from 5 wall units to 20 wall units. Volume pressure is reconstructed from the material accelerations obtained by Lagrangian method [2], using four different methods.

As illustrated in Figure 3(a), for error-free acceleration (SVD Interpolation), the Omni3D and PPE-Omni2D give nearly identical results for the entire flow field, whose RMS error of p^+ is below 0.15. The PPE-Bernoulli results are erroneous in the vicinity of the Dirichlet boundary, but this error diminishes with increasing distance from this surface, becoming very small on the opposing surface, consistent with [1]. The Omni2D errors are substantially higher due to limited components of accelerations and much less number of integration paths. With larger amount of errors in acceleration (Figure 3(a), Cross

Correlation, I.V. =10 wall units), both Omni3D and PPE-Omni2D keep the pressure errors locally high near the wall. The large error near the wall is due to significant velocity gradient and large interrogation volume there. The reduction of cross correlation interrogation volume can greatly reduce those errors from 1.3 to 0.1 at y^+=5, as illustrated in Figure 3(b).

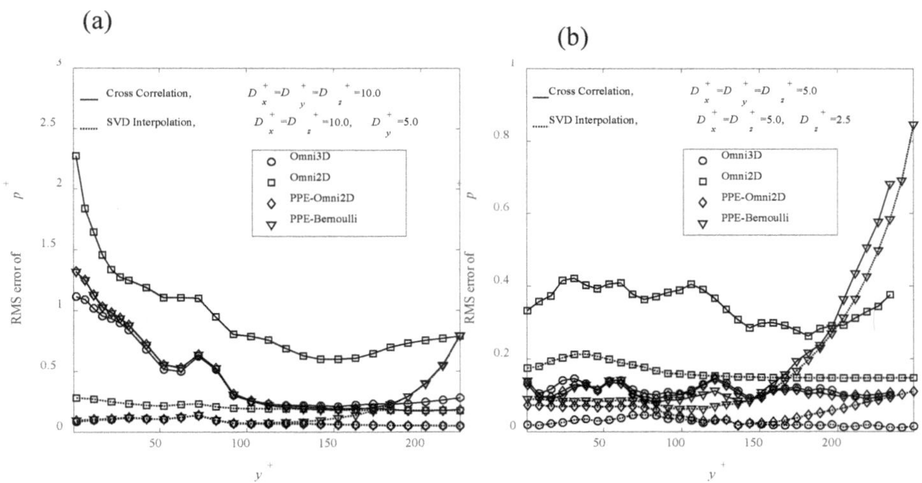

Figure 3 RMS error of pressure resulting from integration of acceleration obtained from cross correlation and SVD interpolation.

The Omni3D method can be further improved by incorporating prior knowledge about the spatial distribution of acceleration uncertainty. Firstly, 100% additive noise is added to the accelerations in the high error zone (HEZ). In the selected-path method, paths carrying error from HEZ to LEZ is circumvented while in the weighted-path method, each path is given a weight which is inversely proportional to the path dependent mean of curl of acceleration. In Figure 4(c), when there is no additive noise, the separation of HEZ and LEZ is done by setting a threshold for the curl of accelerations. In Figure 4(b), the advanced methods clearly reduced the error both in LEZ and HEZ and presented a sharp separation of pressure error between them. In Figure 4(c), error near the wall in further reduced by using the advanced methods.

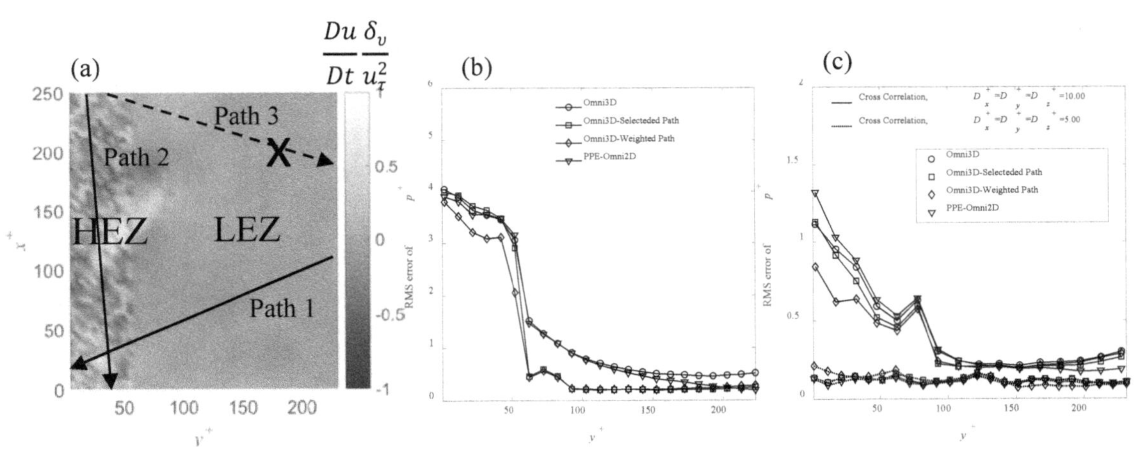

Figure 4 Error propagation in a channel flow. (a) Location of randomly distributed 100% error in acceleration, (b) effect of selected paths and $\nabla \times \frac{Du}{Dt}$ weighted paths on error propagation in (a), (c) effect of selected paths and $\nabla \times \frac{Du}{Dt}$ weighted paths on error propagation in synthetic PIV.

Application of Omni3D on Turbulent Channel Flow over a Compliant Surface

The Omni3D method has been utilized to reconstruct the measured 3D pressure field in a turbulent channel flow over a compliant surface at Re_τ=2300 [5]. The time-resolved 3D velocity distributions have been measured using TPIV, and the 2D surface deformation by the Mach-Zender Interferometry (MZI) integrated into the TPIV system. The material acceleration is calculated by a Lagrangian method [2]. Pressure field is reconstructed via four different methods. Combined plots of velocity vector, contour lines of λ_2(<-2e4 1/s^2) and pressure fluctuations are illustrated in Figure 5. We clearly see that low pressure peaks at the vortex core and high pressure peaks at the interface of sweep ejection transition, which is consistent with our previous findings [5]. Among the four methods, Omni2D has significant higher magnitude of pressure peaks and it's worse than others if we scrutinize the pressure RMS profiles. The correlations of deformation and pressure (not shown here) can tell us that Omni3D method yields the highest correlation values.

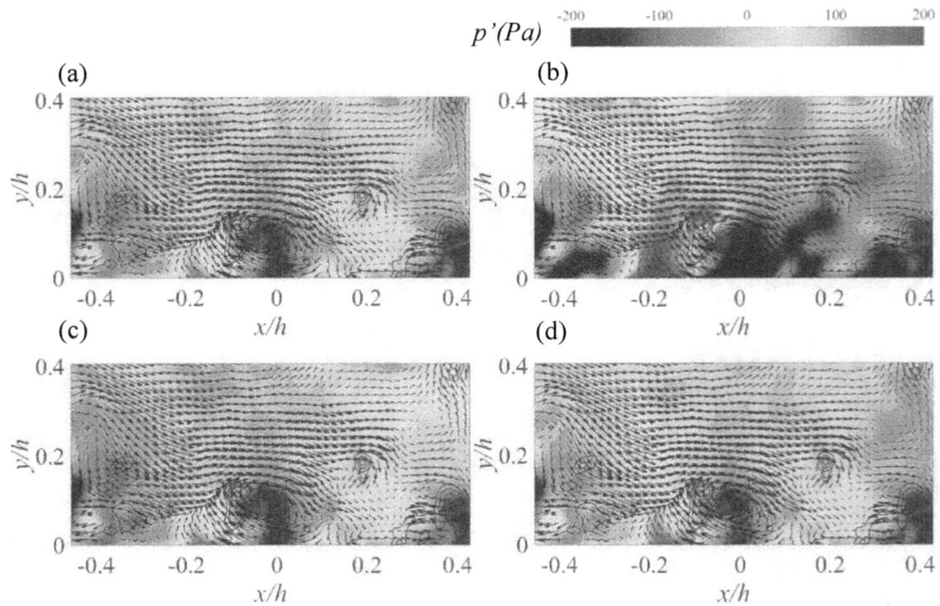

Figure 5 Instantaneous pressure field, velocity vectors and contour line of λ_2 (< -2e4 1/s^2) at *x-y* plane of four integration methods. (a) Omni3D, (b) Omni2D, (c) PPE-Omni2D, (d) PPE-Bernoulli.

Conclusion

In our work, a GPU-Based fast 3D parallel-line Omni-directional integration method is established for calculation of 3D pressure distribution. The method is compared with three other methods. The new method shows the outstanding capability of obtaining accurate pressure field based on the DNS verification and experimental applications. The advantage of Omni3D lies in no need of boundary condition, parallel computing features, as well as incorporating known acceleration uncertainty information to minimize error propagations.

References

[1] S. Ghaemi and F. Scarano, 2013. *Turbulent Structure of High-Amplitude Pressure Peaks within the Turbulent Boundary Layer*', Journal of Fluid Mechanics, 735, 381–426;

[2] Xiaofeng Liu and Joseph Katz, 2006. *Instantaneous Pressure and Material Acceleration Measurements Using a Four-Exposure PIV System*, Experiments in Fluids, 41.2 (2006), 227–40;

[3] G. E. Elsinga and others, 2005. *Tomographic Particle Image Velocimetry*, 6th International Symposium on Particle Image Velocimetry, 1–12;

[4] J. Graham and others, 2016. *A Web Services-Accessible Database of Turbulent Channel Flow and Its Use for Testing a New Integral Wall Model for LES DNS Approach and Simulation Parameters*. Journal of Turbulence, 17.2 (2016), 181–215;

[5] Cao Zhang, Jin Wang, William Blake and Joseph Katz, 2016. *Deformation of Compliant Wall in Trubulent Channel Flow*, Journal of Fluid Mechanics;

© 2018 ASME

Heating effects during bubble collapse using tabulated data

[1]Nikolaos Kyriazis*; [1]Phoevos Koukouvinis; [1]Manolis Gavaises; [2]Richard Pearson and [2]Martin Gold

[1]*City, University of London, London, UK*

[2]*BP International Ltd, UK*

Abstract

An explicit density-based solver for the compressible Navier-Stokes equations able to simulate cavitating flows has been developed and utilised for the simulation of collapsing vapour bubbles. Phase-change is considered by employing the homogeneous equilibrium model (HEM). The wide variation of Mach numbers between the liquid, vapour and mixture regimes is tackled by a Mach consistent numerical flux, suitable for subsonic up to supersonic flow conditions. Time discretisation is performed using a second order low storage Runge-Kutta scheme. Thermodynamic closure is achieved by utilising the Helmholtz energy equation of state (EoS), making feasible simulation of conditions at subcritical and supercritical regions considering the variations of liquid and vapour temperatures during bubble collapse. In order to reduce the computational cost associated with the solution of the Helmholtz EoS at each time step, a tabulated data technique has been followed. The unstructured thermodynamic table, containing the thermodynamic properties derived from the Helmholtz EoS, has been constructed for n-dodecane, which has been the considered as the working fluid. The efficiency of the method is enhanced by a static linked-list algorithm for searching among the elements of the table. In addition, a finite element bilinear interpolation is used for approximating the unknown thermodynamic properties. After validating the numerical method, parametric studies considering 2-D axisymmetric vaporous bubble collapse in the proximity of a wall have been performed at conditions realised in micro-orifice flow passages. The temperature and pressure changes on the wall are estimated as function of the surrounding liquid pressure, the initial bubble radius and the location of the wall from the center of the initial bubble, revealing the expected range of variation as function on the set parameters.

Keywords: bubble dynamics; cavitation; Helmholtz equation of state; exact Riemann solver

Introduction

Several numerical studies have been performed regarding vapour bubble dynamics, since they have numerous industrial and physical applications. Different methodologies such as potential flow solvers [1], homogeneous equilibrium mixture (HEM) models [2], interface capturing and interface tracking methods have been employed. In HEM models the phase change is predicted by the density variation and the main assumption is that the two phases are in mechanical and thermodynamic equilibrium. HEM methods have become popular due to their simplicity and because no empirical parameters are needed. On the other hand, interface tracking methods, such as front tracking methods [3], offer sharp interface and they also allow large deformations of the surface. Similar are the interface capturing methods, such as Volume of Fluid (VOF) approach [4] and Level Set Method [5]. Although the main feature of interface capturing methods is again sharp interface, in conditions close to the critical point there is not a clear distinction between the two phases. In the majority of the previous works thermal effects have been ignored, for example in barotropic models, or temperature effects have been taken into account by the use of simplified EoS, such as stiffened gas EoS or similar. The current study expands the previous work of Koukouvinis et al. [6] by examining the heating effects during the collapse of a vapour bubble. The thermodynamic closure in the HEM is achieved by the Helmholtz energy EoS from NIST Refprop databases [7], both for subcritical and for supercritical conditions. A tabulated form of the Helmholtz energy EoS has been implemented in the solver for higher numerical efficiency.

Numerical Method

The numerical method has been described in detail and validated in [8], here only the main features are discussed. The 2-D Euler equations in r-z cylindrical coordinates with a geometric source term accounting for cylindrical symmetry are explicitly solved [9]:

*Corresponding Author, Nikolaos Kyriazis: Nikolaos.Kyriazis@city.ac.uk

$$\mathbf{U}_t + \mathbf{F}(\mathbf{U})_r + \mathbf{G}(\mathbf{U})_z = \mathbf{S}(\mathbf{U}),\ in\ \Omega \quad (1)$$

Where *t, r, z* subscripts denote differentiation with respect to time, *r* direction and *z* direction respectively. **U** is the conserved variable vector, **F**(**U**) and **G**(**U**) are the fluxes at the radial (*r*) and axial (*z*) directions respectively and **S**(**U**) is the geometric source term:

$$\mathbf{U} = \begin{bmatrix} \rho \\ \rho u_r \\ \rho u_z \\ \rho E \end{bmatrix},\ \mathbf{F}(\mathbf{U}) = \begin{bmatrix} \rho u_r \\ \rho u_r^2 + p \\ \rho u_r u_z \\ (\rho E + p) u_r \end{bmatrix},\ \mathbf{G}(\mathbf{U}) = \begin{bmatrix} \rho u_z \\ \rho u_r u_z \\ \rho u_z^2 + p \\ (\rho E + p) u_z \end{bmatrix},\ \mathbf{S}(\mathbf{U}) = -\frac{s}{r} \begin{bmatrix} \rho u_r \\ \rho u_r^2 \\ \rho u_r u_z \\ u_r (\rho E + p) \end{bmatrix} \quad (2)$$

where ρ is the fluid density, u_r and u_z the radial and axial velocity components respectively, p is the pressure, E is the total internal energy, equal to $\frac{1}{2}\left(u_r^{\ 2} + u_z^{\ 2}\right) + e$, e is the internal energy of the fluid and s is the geometric source term. For cylindrical symmetry, s is equal to unity. The viscosity and the surface tension are neglected in the present study since the phenomenon is inertia driven [10].

The properties of n-Dodecane are derived from the Helmholtz energy. The Helmholtz energy is calibrated within the temperature range 263.6 K≤T≤700 K, for maximum pressure p_{max}=700 MPa and for maximum density ρ_{max}=771.62 kg/m^3 [7]. The dimensionless form of the aforementioned EoS, having as independent variables the density and the temperature is given by:

$$\frac{a(\rho,T)}{RT} = \alpha(\delta,\tau) = \alpha^0(\delta,\tau) + \alpha^r(\delta,\tau) \quad (3)$$

Where $\delta=\rho/\rho_c$, $\tau=T_c/T$. From equation (3) all the thermodynamic properties can be obtained, such as pressure, internal energy, enthalpy and speed of sound as a function of density and temperature. The saturation conditions are identified by the Maxwell criterion and the fluid properties in the saturation dome are calculated by the mixture assumption, whereas the mixture speed of sound is determined from the Wallis speed of sound formula [11].

Instead of solving the Helmholtz EoS for each time step, a tabulated technique, similar to Dumbser et al. [12] has been implemented after explicitly solving the Euler equations. An unstructured thermodynamic table of approximately 40,000 elements, refined around the saturation curve, which contains all the thermodynamic properties has been created. Then the thermodynamic table has been split into smaller data groups by static linked lists in order to search only in the group with the desired properties within its range. Since the density and the internal energy are calculated by the Euler equations, the corresponding element of the table is identified through numerical inversion from the above quantities. Any thermodynamic property φ (pressure, temperature or speed of sound) within a specific element of the table is then approximated by a Finite Element bilinear interpolation:

$$\varphi(\rho,e) = \sum_{n}^{nodes} N_n(\rho,e) b_n \quad (4)$$

Where b are the unknown coefficients of φ and N is the shape function of node n:

$$N_n(\rho,e) = 1 + (e - e_n) + (\rho - \rho_n) + (e - e_n)(\rho - \rho_n) \quad (5)$$

And the coefficients of each property are calculated by:

$$\mathbf{b} = [\mathrm{N}]^{-1} \boldsymbol{\varphi} \quad (6)$$

Where $\boldsymbol{\varphi}$ are the values of the property at the nodes of the quadrilateral element and N is given by:

$$N_{mn} = 1 + (e_m - e_n) + (\rho_m - \rho_n) + (e_m - e_n)(\rho_m - \rho_n) \quad (7)$$

A Mach number consistent numerical flux has been implemented [13] in order to handle the great variation of the Mach number between the liquid and the vapour regimes. Time discretisation has been achieved by a low storage four-stage Runge-Kutta method, 2nd order in time.

Results

In this section the results of a parametric bubble collapse are presented. The collapse of a vapour bubble in the vicinity of a wall is investigated for a different set of configurations (see Table 1 and [8] for details concerning the configuration). In Table 1 the ratio $d_w/R_{b,0}$ is the distance from the wall to the center of the bubble divided by the initial bubble radius, p_{inf} is the pressure of the liquid, T is the initial temperature for both the liquid and the vapour and τ is the Rayleigh collapse time for each case. The density of the vapour bubble for all the cases is $\rho_{vap}=33.78\ kg/m^3$. The computational domain is *20* times the initial bubble radius and *200* equally spaced cells have been used along the initial radius. This ratio has been kept for distance of $2.5R_{b,0}$ from the origin and then the mesh is coarsened with ratio *1.05* in both directions. Zero gradient boundary condition has been used for the right and the upper side, slip wall for the lower side, whereas for the y-axis of symmetry, the normal velocity component is zero.

No	$d_w/R_{b,0}$	$R_{b,0}$ (μm)	p_{inf} (bar)	T (K)	τ (s)
1	1.5	1	10	400	2.37e-08
2	1.5	10	10	400	2.37e-07
3	1.5	1	100	400	7.55e-09
4	1.5	10	100	400	7.55e-08
5	0	1	10	400	2.37e-08
6	0	10	10	400	2.37e-07
7	0	1	100	400	7.55e-09
8	0	10	100	400	7.55e-08
9	-0.5	1	10	400	2.37e-08
10	-0.5	10	10	400	2.37e-07
11	-0.5	1	100	400	7.55e-09
12	-0.5	10	100	400	7.55e-08

Table 1. *Classification of the different cases examined.*

In Figure 1 the volume of vapour (left), the maximum pressure on the wall (middle) and the maximum temperature on the wall (right) are shown with respect to time. Regarding Figure 1a, the collapse time is proportional to the initial volume of the vapour for each bubble. A rebound is noticed for cases 1-4, however it is weaker than the one noticed in [8], due to the larger distance between the wall and the center of the bubble. The early pressure peaks at *Time<0.5* for cases 1, 2 and at *Time<0.8* for cases 3, 4 (Figure 1b) are because of the collapse of a vapour region below the bubble which was originally created by a rarefaction wave (see also [8]). A pressure and temperature peak (total maximum) due to the bubble collapse is noticed in Figure 1b and Figure 1c at *Time~1* for cases 1-8 and at *Time~0.7* for cases 9-12. However, at cases 1-4 are noticed the lowest wall pressures and the lowest wall temperatures because most of the energy is transformed into kinetic and as a consequence the pressure increase is not as significant as in the rest cases; the same applies for the temperature. On the contrary, at cases 5-8 the maximum wall pressures and the maximum wall temperatures are noticed due to the maximum compression that the bubble undergoes during its collapse. After the collapse of the bubble (*Time>1.5*), additional pressure peaks are noticed for cases 1-4, because of the collapse of the regenerated vapour regions. The pressure increase due to the re-entrant jet and the shock wave after the collapse can potentially lead to erosion damage of materials. For all three figures the curves between cases with different initial radius but the same surrounding pressure and the same distance d_w coincide, as it was expected, given the fact that the time has been non-dimensionalised with τ.

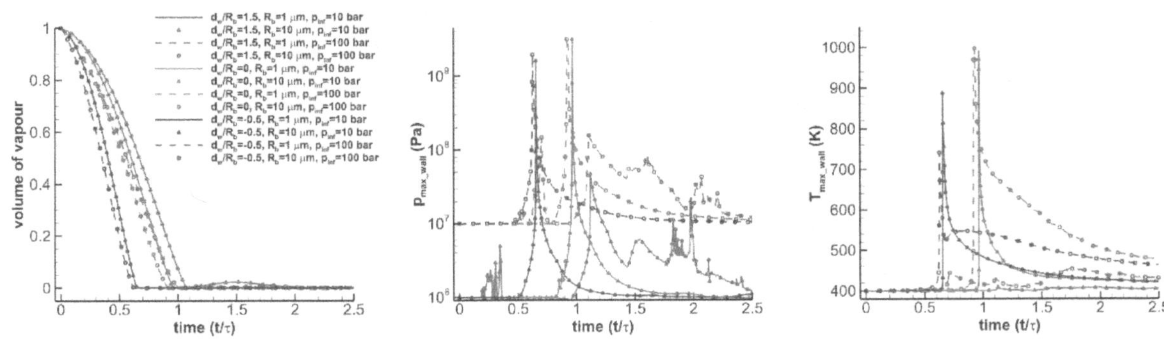

(a) (b) (c)

Figure 1. *Volume of vapour (left), maximum wall pressure (middle) and maximum wall temperature (right) with respect to time for different vapor bubble configurations. Time has been non-dimensionalized with the Rayleigh collapse time τ and the volume of vapour with the initial vapour volume of the bubble.*

In Figure 2 the pressure (left slice) and the temperature (right slice) contour fields are demonstrated for several time instances during the collapse for case No 1. At *Time=0.1* the shape of the bubble is almost spherical in comparison with [6], [8] where a heart-shape was formed. This is due to the larger distance of the bubble from the wall, therefore the effect of the boundary is weaker than the aforementioned studies. The bubble takes a non-spherical heart-like shape only at later times just before the collapse, when the micro-jet has already been formed on the top of the bubble and the pressure field has been developed (*Time=1.05* and *Time=1.07*). By that time there is no significant temperature variation. However, at the moment of the collapse (*Time=1.08*), there is a significant pressure increase up to *1600 MPa* and a temperature increase up to *550 K* due to the condensation of the bubble. Finally, at *Time=1.1* the shock wave travelling away from the axis of symmetry is shown.

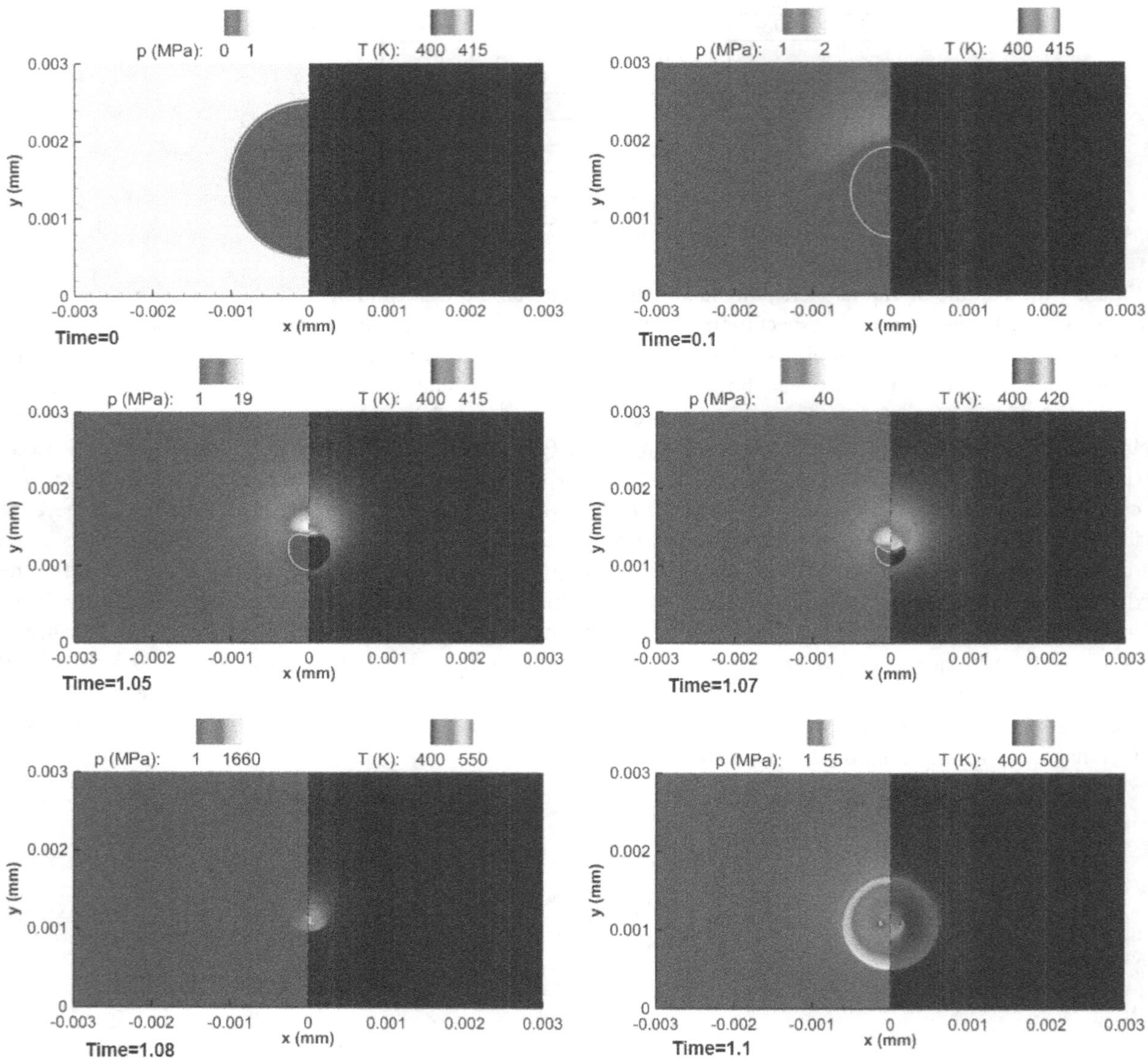

Figure 2. *Instances during the vapour bubble collapse for case No 1. Time has been non-dimensionalized with Rayleigh collapse time. Pressure field (left slice) and temperature field (right slice) are shown.*

Conclusions

In the present work, a HEM approach combined with real fuel thermodynamics has been presented and utilized for vaporous bubble collapse simulations. A parametric study of the temperature and pressure changes on the wall with respect to the surrounding liquid pressure, the initial bubble radius and the distance of the wall from the center of the initial bubble has been performed. It can be concluded that the temperature increase during bubble collapse is not negligible, for example in cases 5-8 it can reach up to *1000 K*. The maximum pressure and temperature increase is observed when the center of the initial bubble is on the wall (spherical symmetry). These heating effects during bubble collapse can possibly have implications in injector nozzles, since they can change the fuel properties and have a serious impact on the flow field.

Acknowledgements

The research leading to these results has received funding from the MSCA-ITN-ETN of the European Union's H2020 programme, under REA grant agreement n. 642536.

References

[1] Q. Wang, "Multi-oscillations of a bubble in a compressible liquid near a rigid boundary," *Journal of Fluid Mechanics*, vol. 745, pp. 509–536, 004 2014.
[2] N. A. Adams and S. J. Schmidt, "Shocks in Cavitating Flows," in *Bubble Dynamics and Shock Waves*, F. C. Delale, Ed. Berlin, Heidelberg: Springer Berlin Heidelberg, 2013, pp. 235–256.
[3] S. T. Popinet, Eacute, Phane, S. T. Zaleski, Eacute, and Phane, "Bubble collapse near a solid boundary: a numerical study of the influence of viscosity," *Journal of Fluid Mechanics*, vol. 464, pp. 137–163, 2002.
[4] B. B. Li, W. Jia, H. C. Zhang, and J. Lu, "Investigation on the collapse behavior of a cavitation bubble near a conical rigid boundary," *Shock Waves*, vol. 24, pp. 317–324, 2014.
[5] E. Lauer, X. Y. Hu, S. Hickel, and N. A. Adams, "Numerical modelling and investigation of symmetric and asymmetric cavitation bubble dynamics," *Computers & Fluids*, vol. 69, pp. 1–19, 30 2012.
[6] P. Koukouvinis, M. Gavaises, A. Georgoulas, and M. Marengo, "Compressible simulations of bubble dynamics with central-upwind schemes," *International Journal of Computational Fluid Dynamics*, pp. 1–12, 2016.
[7] E. W. Lemmon and M. L. Huber, "Thermodynamic Properties of n-Dodecane," *Energy & Fuels*, vol. 18, pp. 960–967, Jul. 2004.
[8] N. Kyriazis, P. Koukouvinis, and M. Gavaises, "Numerical investigation of bubble dynamics using tabulated data," *International Journal of Multiphase Flow*, vol. 93, no. Supplement C, pp. 158–177, Jul. 2017.
[9] E. F. Toro, *Riemann Solvers and Numerical Methods for Fluid Dynamics, A Practical Introduction*. Springer Berlin Heidelberg, 2009.
[10] S. Zhang, J. H. Duncan, and G. L. Chahine, "The final stage of the collapse of a cavitation bubble near a rigid wall," *Journal of Fluid Mechanics*, vol. 257, pp. 147–181, 1993.
[11] Brennen C. E., *Cavitation and Bubble Dynamics*. 1995.
[12] M. Dumbser, U. Iben, and C.-D. Munz, "Efficient implementation of high order unstructured WENO schemes for cavitating flows," *Computers & Fluids*, vol. 86, pp. 141–168, 5 2013.
[13] S. Schmidt, I. Sezal, G. Schnerr, and M. Talhamer, "Riemann Techniques for the Simulation of Compressible Liquid Flows with Phase-Transistion at all Mach Numbers - Shock and Wave Dynamics in Cavitating 3-D Micro and Macro Systems," in *46th AIAA Aerospace Sciences Meeting and Exhibit*, 0 vols., American Institute of Aeronautics and Astronautics, 2008.

© 2018 ASME

High speed observation of damage created by a collapse of a single cavitation bubble

[1]Matevž Dular*, [1]Žan Pirc, [1]Tomaž Požar, [1]Rok Petkovšek

[1]*Faculty of Mechanical Engineering, University of Ljubljana, Askerceva 6, 1000 Ljubljana, SI-Slovenia*

Abstract

One of the remaining open questions in the cavitation erosion research is the one on the importance of the microjet and shock wave on the formation of the pit. Up until now no successful attempt was made to study this in detail, mainly because the damage could only be detected and evaluated after several successive bubble collapses. Based on our previous work on laser induced shock waves [1] and on simultaneous observation of the dynamics of cavitation clouds and cavitation damage [2], we developed a technique which enables simultaneous observation of one single cavitation bubble collapse and the damage it creates.

A bubble with a maximum diameter of up to 5 mm was formed by a Nd:YAG laser. The damage was observed on a 9 μm thick aluminum foil which was attached to the glass substrate. Two high speed cameras were simultaneously used. One captured the dynamics of the bubble, while the other recorded the damage on the foil.

We also observed the collapse of a bubble in the presence of shear flow. Most of damage was created by the microjet mechanism. Sometimes, the collapse of the bubble rim, at rebound of the initial bubble caused pits in a well-known circular pattern.

From the recordings at the very fastest acquisition rate, we determined that the material deforms and then partially relaxes, while a significant deformation remains. The whole process is only 2-3 μs long.

Keywords: caviation; erosion; single bubble; shear flow; high speed observation

1 Introduction

Currently the most widely accepted explanation for the occurrence of cavitation erosion is that the potential energy contained in a macro cavity is transformed into the radiation of acoustic pressure waves, and further on into the erosive power contained in the micro scale cavitation structures or single bubbles that collapse in the vicinity of the material boundaries [3].

Two theories describe the last stages of life of a micro-scale cavitation structure:

- the micro jet [4],
- the spherical micro bubble collapse [5].

Figure 1 schematically shows both processes.

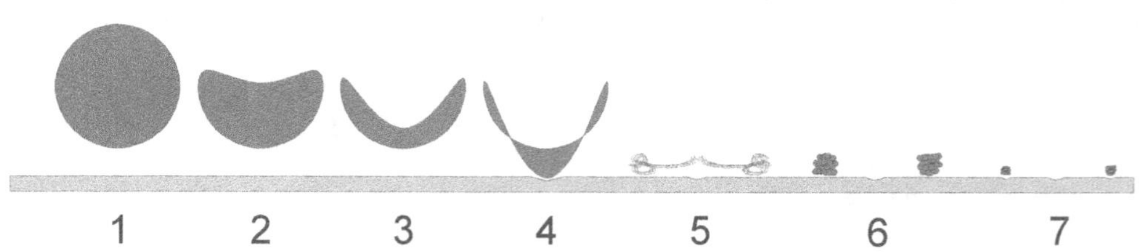

Figure 1: Two possible mechanisms of cavitation erosion.

A small bubble (mm size) reaches a higher-pressure region and becomes unstable (1). Due to the vicinity of the solid surface, its upper boundary collapses faster than the one closer to the wall (2 and 3). A microjet forms (3), which can reach a velocity of several hundred m/s. As it hits the solid surface a water hammer pressure occurs (p=ρcv), which is high enough to deform the surface (4). After the microjet induced collapse, the flow moves radially outwards (5) what can cause a secondary evaporation, or "splashing" (6). This results in a number of very small bubbles, which, due to surface tension forces, preserve their spherical shape during the collapse. This results in the formation of significant shock waves, which can again reach amplitudes, capable of deforming the material (7).

*Corresponding Author, Matevz Dular: matevz.dular@fs.uni-lj.si

Which mechanism is more pronounced was until now impossible to determine directly. With the development of a novel technique, by simultaneous observation using two high speed cameras [6], the present paper investigates this issue.

2 Experiment

Experiments were performed at the Faculty of Mechanical Engineering, University of Ljubljana, Slovenia.

2.1 Experimental set-up

Figure 2 shows the schematics of the experimental set-up.

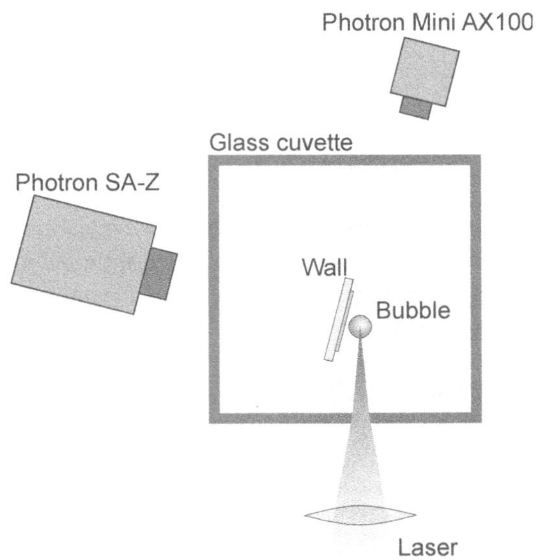

Figure 2: Experimental set-up.

A Nd:YAG laser with 1064 nm wavelenght was used to initiate the bubble growth. The energy of the 5 ns long pulse was maximally 16 mJ (10 mJ at the position of the breakdown). The focusing optics has been designed for relatively low aberrations, cone angle was 12° in water. The "standard" bubble with the diameter of 3.3 mm was generated 20 mm from the wall of the glass cuvette. The time of the bubble collapse when no wall was present was 147 μs and varied somewhat depending on the distance from the wall.

Two high speed cameras were used. Usually the slower, Photron Mini AX100, was used to observe the bubble dynamics, while the faster, Photron S-Z type 2100K-M-64GB, was used to simultaneously measure the deformation of the aluminum foil.

2.2 Aluminum foil as an erosion sensor

The idea of the experiment was to simultaneously record images of cavitation bubble and cavitation damage. The view of the upper side of the foil is obstructed by the bubble, hence one needs to look at the foil from the bottom side to see the damage. Consequently, a glass plate was used for the wall and equally important the foil had to be thin enough so that the cavitation damage which occurs on the side exposed to cavitation bubble was also visible on the other side (Fig. 3).

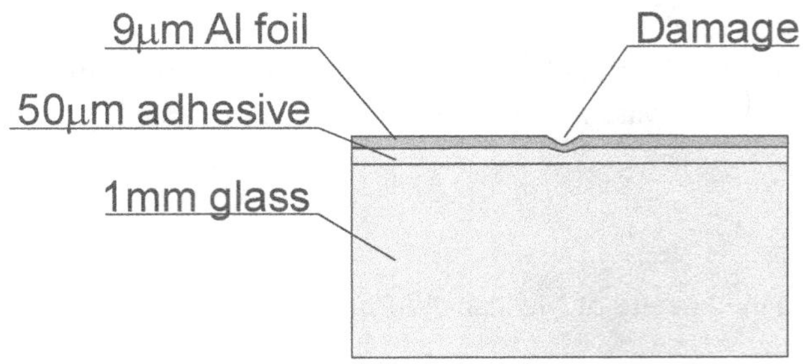

Figure 3: Construction of the "specimen".

We have chosen 9 μm thick aluminum foil and attached it to the 1mm thick glass by an optically clear two-sided adhesive tape with thickness of 50 μm (Fig. 3).

2.3 Damage detection

Since we are observing the foil with a camera, only the surface of the damaged area can be accurately measured. It is planned to estimate the geometry of the pits using stereographic and SFS (Shape From Shading) algorithms in the future [7].

For the present evaluation we used an approach that combines several evaluation procedures used before [2, 6]. We evaluated the images in pairs – the image at the time t was subtracted from the image at time t+Δt, thus eliminating most of the surface and illumination imperfections. Then we employed the pit-count method [6] which determines the pits from the darker regions in an image, while the brighter area is assumed to be undamaged surface – from each image pair we obtained the number and the area of newly appeared pits. The pit-count method gives a distribution of the number and the area of the pits and consequently, the distribution of the magnitude of cavitation erosion on the surface. We can also determine the distribution of the size of the pits. Since we were comparing pairs of two successive images we were also able to consider the possibility of pit overlapping (Fig. 4).

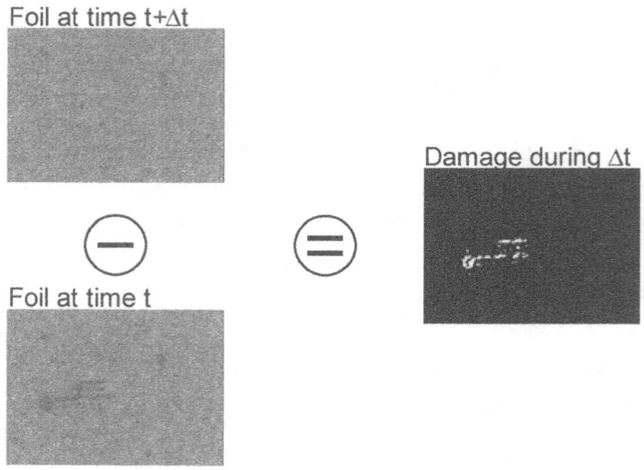

Fig. 4: Determining the damage sustained during a period of Δt.

Images of the aluminum foil were treated as matrixes A with i×j elements ($A(i,j) \in \{0,1,...255\}$) with values which can range from 0 (black) to 255 (white). Erosion was evaluated in image pairs: image matrix at time t+Δt was subtracted from image matrix at time t ($B(i,j,t)= A(i,j,t) - A(i,j,t+\Delta t)$). This way a new matrix B was obtained. When the matrix element $B(i,j)$ did not change between times t and t+Δt its value was 0 ($B(i,j,t)=0$). When change occurred the value was $B(i,j,t)>0$. Since small changes could be present due to insignificant changes in illumination, vibration etc., damage was only considered when a certain change threshold was exceeded (more than 5% of decrease in brightness – pits always appeared as dark regions). The number of the pits, their size and overall damaged area could then be easily determined.

2.4 Investigated conditions

We investigated an ensemble of conditions, where the distance of the bubble (h) from the wall was changed from $\gamma = \frac{h}{r}$ = 0.05 to 2.0. For each case 50 repetitions were made. In addition, flow was introduced in certain cases to investigate the influence of the shear forces on the specifics of bubble collapse and the induced damage.

3 Results

During the campaign we performed a vast number of experiments, which resulted in derivation of three specific cases, which are presented here in more detail. In all cases (Figs. 5 to 7) the upper sequence shows the bubble dynamics recorded by the primary camera. The bottom sequence is the result of damage evaluation (Sec. 2.3) from the images recorded by the secondary camera.

3.1 Bubble in a close vicinity of the wall (γ=0.05)

First the results of measurements in the close vicinity of the wall are presented (Fig. 5).

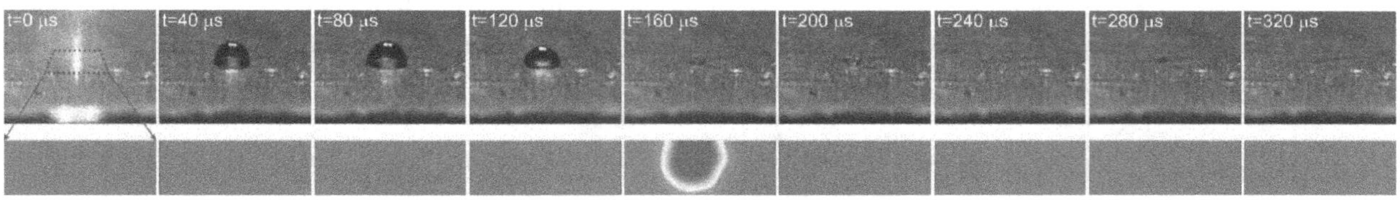

Fig. 5: Bubble and damage for collapse in the vicinity of the wall (γ=0.05)

The most pronounced damage mechanism in these cases is the micro-jet. The bubble first grows and touches the surface shortly after. The maximal size occurs about 80 μs after the laser pulse. The bubble collapses at 160 μs, and at the same instant significant damage is recorded by the second camera. The damaged area has a comparable size as the bubble shortly before the collapse, which also corresponds well to the diameter of the microjet.

3.3 Bubble in far away from the wall (γ =1.7)

Figure 6 shows the bubble and damage evolution for the case further away from the wall.

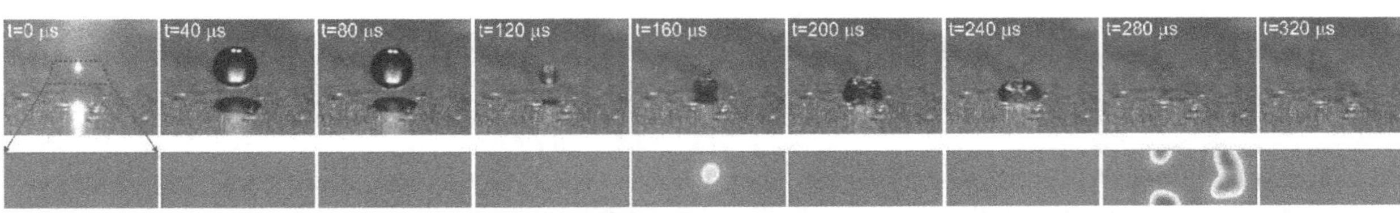

Fig. 6: Bubble and damage for collapse far away from the wall (γ =1.7)

The dynamics of the bubble is roughly the same – the same size is reached by the 80th μs but the bubble collapses a bit faster. At 160 μs the micro-jet reaches the wall and deforms it. The deformation itself is much smaller than in the case of γ =0.05 (Fig. 5). The bubble then undergoes a rebound (splashing) and a secondary collapse occurs at 280 μs. This causes significant damage in a circular pattern, which matches the diameter and the geometry of the bubble rim. One can conclude that the both mechanisms play a role in this case, but in fact the collapses of tiny bubbles inside the rim during the rebound is more aggressive.

3.4 Bubble in the presence of shear flow

Researchers argued in the past that the presence of the shear flow may significantly alter the physics of the damage occurrence. We have tested a wide range of conditions but no damage could be detected when γ exceeded 1.4 and there was no influence of the shear flow in the cases when γ was smaller than 0.5.

Fig. 7 show a representative result of this case. The flow velocity was 2.5m/s (from the right to the left) and the γ =1.0.

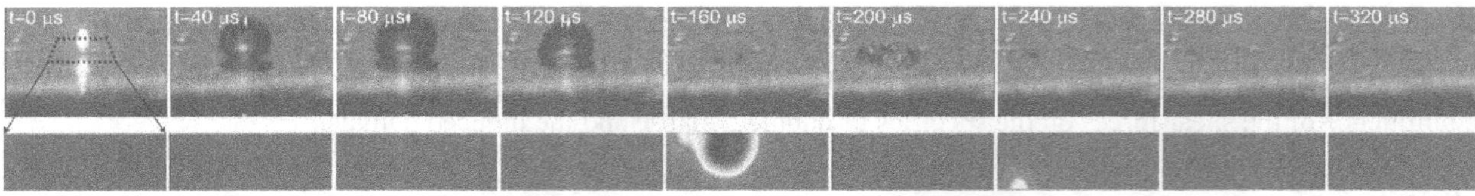

Fig. 7: Bubble and damage appearance in the presence of shear flow (γ =1.0, v=2.5 m/s)

One can see that the bubble is transported by the flow. The dynamics itself is not significantly influenced by it. The primary collapse occurs at 160 μs. The jet is deflected towards the pressure gradient by the shear forces as it was predicted by [8]. Nevertheless, its power is not diminished – it causes deformation, which is comparable to the one at stationary conditions (Fig. 5). 80 μs later, the secondary collapse occurs and causes minor damage – the position corresponds to the one of the rebounded bubble rim.

3.5 Very high-speed observation of foil deformation

To investigate further, the process of pit formation we set the camera to its fastest framerate of 2.1 Mfps. The bubble was positioned at the non-dimensional distance of γ=0.4 (without shear flow), hence we expected a single pit resulting from the micro-jet impact. Figure 8 shows the (raw) images.

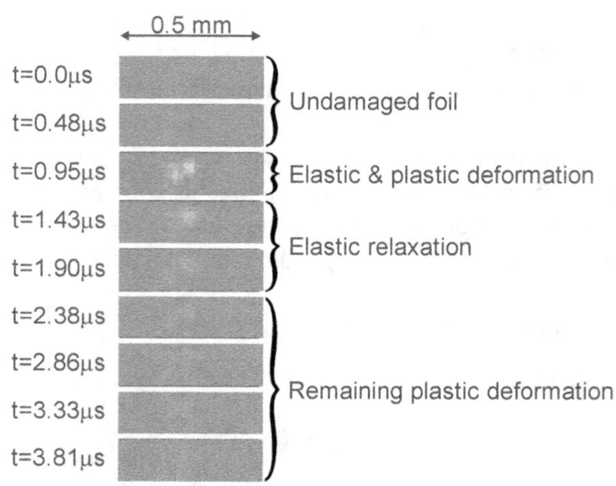

Fig. 8: Damage evolution recorded at 2.1 Mfps.

The deformation of the foil occurs in an extremely short period of time – during less than 0.5 μs. This corresponds well with the period during which the water hammer pressure is present at the jet impact. The duration of the water hammer stress is as long as the time for the impact signal to traverse the radius of the jet ($t_{wh} = \frac{r_{jet}}{c}$), which in the present case results in approximately 0.3 μs. After that time a stagnation pressure is established, which is an order of magnitude smaller than the water hammer pressure.
After the maximum deformation is reached at 0.95 μs, an elastic deformation takes place and a stationary state is established about 1.5 μs after the first contact of the micro-jet and the material.

4 Conclusions

In the study we investigated the last stage in the cascade of events that lead to the occurrence of cavitation damage – single bubble collapse and the resulting deformation of the material.
By simultaneous observation of the bubble dynamics and the deformation of a thin aluminum foil, by two high speed cameras we were able to directly observe the process.

Various distances of the bubble from the specimen were investigated. We also observed the collapse of a bubble in the presence of shear flow – this was essential, since many researchers claim that the presence of a pressure gradient diminishes the erosive potential of the micro-jet.
Based on an ensemble of measurements we can conclude that:

- when cavitation bubble implodes near the wall, the most pronounced mechanism is the impact of the micro-jet;
- when the bubble collapses further away from the wall the influence of the micro-jet diminishes and the collapse of microscopic bubbles in the rebounded cloud is more important;
- finally, in the cases where the shear flow was introduced, the bubbles near the wall behaved in the same way as without the presence of the flow (micro-jet mechanism was the driving one). In the mid-range of distances ($0.5<\gamma<1.4$) the micro-jet was still the most important mechanism, but some damage was also induced by the secondary (rim) collapse and further away from the wall the deflection of the micro-jet was too large to cause either damage or rebound of the bubble.

Further on we observed the formation of a pit at the highest framerate setting of the camera and concluded that the damage indeed results from the water hammer pressure at micro-jet impact and that one can also observe deformation and relaxation of the material by this methodology.

The work represents one of the final "pieces of the puzzle", which will lead us to a detailed understanding of the cavitation erosion process and will enable development of models for prediction of cavitation damage.

Acknowledgments

The authors acknowledge the financial support from the Slovenian Research Agency (research core funding No. P2-0401 and No. P2-0270).

References

[1] R. Petkovsek, J. Mozina, and G. Mocnik, "Optodynamic characterization of shock waves after laser-induced breakdown in water.," *Optics express*, vol. 13, no. 11, pp. 4107–12, May 2005.

[2] M. Dular and M. Petkovsek, "On the mechanisms of cavitation erosion - Coupling high speed videos to damage patterns," *Experimental Thermal and Fluid Science*, vol. 68, 2015.

[3] R. Fortes-Patella, J.L. Reboud, L. Briancon-Marjollet, A phenomenological and numerical model for scaling the flow aggressiveness in cavitation erosion, EROCAV Workshop, Val de Reuil, 2004.

[4] M.S. Plesset, R.B. Chapman, Collapse of an innitially spherical vapour cavity in the neighbourhood of a solid boundary. J. Fluid Mech. 47 (1971) 283–290.

[5] R.P. Tong, W.P. Schiffers, J.S. Shaw, J.R. Blake, D.C. Emmony, The role of splashing in the collapse of the laser-generated cavity near a rigid boundary, J. Fluid Mech. 380 (1999) 339–361.

[6] M. Petkovšek, M. Dular, Simultaneous observation of cavitation structures and cavitation erosion, Wear 300 (1/2) (2013) 55-64.

[7] M. Dular, O. Coutier-Delgosha, M. Petkovsek, Observations of cavitation erosion pit formation. Ultrasonics Sonochemistry, 2013, vol. 20, iss. 4, pp. 1113-1120

[8] P.-W. Yu, S.L. Ceccio, G. Tryggvason The collapse of a cavitation bubble in shear flows-A numerical study, Phys. Fluids 7 (ll), November 1995.

© 2018 ASME

Homogeneous Cavitation Model Used to Predict Erosion

[1]Ridha Zgolli*;[1]Ahmed Bel hadj taher; [1] Marwa Ennouri. Hatem Kanfoudi;

[1] *Laboratory of Hydraulic and Environmental Modeling, National Engineering School of Tunis, University of Tunis ELMANAR,* 1002 Tunis, Tunisia

Abstract

This study presents a numerical method using ANSYS-CFX-CFD to predict area of wall near a cavitating flow that may be eroded. It is based on the post-treatment of URANS simulations in cavitant flows. The most RANS calculations are performed with compressible consideration and the introduced cavitation model is one of the recently "homogenous model" that is based on mixture fluid with density (ρ_m) calculated as function of liquid density (ρ_{liq}), vapour or gas density (ρ_{vap}) and vapour or gas volume fraction α .

The numerical result obtained on hydrofoil geometries and compared with experimental works concerning flows characteristics (size of pocket, pressure, velocity). We present here the used cavitation model and the approach followed to evaluate the value of α fixing the shape of pocket around wall before collapsing and we limit this study to evaluate the area of wall that can be eroded by cavitation phenomenon.

Keywords: Cavitation, void fraction, micro-jet, erosion model, CFD, CLE hydrofoil.

Introduction

Several studies were carried out in recent years to have more knowledge on the various mechanisms of cavitation erosion. we present here our approach to predict the area of the solid surface where bubbles created by cavitation can be imploded. The phenomenon of micro-jet is considered the best mechanism that leads to the cavitation erosion. Several research [4-5] have made experimental studies on the phenomenon of micro-jets. They have said that the micro-jet can reach a velocity much greater than 100 m/s.

In this paper, a numerical erosion model is based on a quantitative approach based on the works of Andreas et al. [6] and Dular and Coutier-Delgosha [7] to predict the impact rate on the solid surface. A numerical investigation of unsteady flow is done by the CFX code with a coupling between CFD and an erosion model. The numerical simulations are based on a cavitation model coupling with LES turbulence calculation. This model is based on a homogeneous approach, where the liquid density is defined by a Tait equation. The impact investigation is limited to the incubation period. The implosion of the cavitation clouds causes the emission of the pressure waves and its attenuation towards the solid surface which generates the formation of the micro-jets. To validate the adopted approach, the simulation results are compared with the experimental data.

Mathematical formulation

The CFD tool applied in this study is the commercial code ANSYS-CFX with compressible calculations.

Cavitaion model

The cavitation model [8-9] adopted for these calculations has been the subject of research developments conducted in our laboratory LMHE. The transport equation with source term associated to the volume fraction α, give a closure model for the continuity and momentum equations of the mixture incompressible flow. Foer validation the numerical results are compared with experimental measurements and the recently developed cavitation models.

The density ρ of the homogeneous mixture fluid and its dynamic viscosity μ are governed by linear equations (1) and (2)

$$\rho = \rho_v \alpha + (1-\alpha)\rho_l \tag{1}$$

$$\mu = \mu_v \alpha + (1-\alpha)\mu_l \tag{2}$$

With μ_l is the liquid viscosity, μ_v is the vapor viscosity, ρ_l is the liquid density and ρ_v is the vapor density. Where α is the vapor volume fraction,

*Corresponding Author, Ridha Zgolli: ridha.zgolli@enit.utm.tn

Erosion model

The study presented is based on numerical the technique of 'volume control' resumed in the next paragraph and based on the physical criterion determining the formation of a micro-jet neighboring the solid wall caused by the collapse of the cavitation bubbles. And when the collapses of the bubbles occur at a distance close enough to a solid surface, a phenomenon of micro-jet begins to occur with velocities evaluated by equation (3).

$$V_{jet} = 8.97\gamma^2 \sqrt{\frac{P - P_v}{\rho}} \tag{3}$$

Where the dimensionless distance from the bubble center to the solid surface $\gamma = H/R_b = 1.1$ with R_b represents the bubble radius and H the distance from bubble centre to surface [6]. The corresponding water hammer pressure is given by the equation (4).

$$P_{hammer} = V_{jet}\rho_l c_l \tag{4}$$

Where c_l is the sound velocity.

The deformation time t_{def} is the time needed for the impact signal to traverse the radius of the jet r_{jet}.

$$t_{def} = \frac{r_{jet}}{c_l} \tag{5}$$

The impact of micro-jet on the solid wall is considered performed with a circular shape characterized and evaluated by d_{pit} ;

$$d_{pit} = V_{def} t_{def} \tag{6}$$

So the pit radius is calculated as follows:

$$r_{pit} = 26.7 d_{pit} \tag{7}$$

Finally, the area of the solid wall impacted by micro-jet caused by cavitation phenomenon is approched by the following expression:

$$A_{pit} = \pi . r_{pit}^2 \tag{8}$$

For average damage rates following a linear behavior during the incubation period can be extrapolated as follows

$$A_{dam} = \frac{\sum_{t=0}^{t} A_{pit}}{A_{ref}} . \frac{\tau}{t} \tag{9}$$

Where A_{ref} is a reference area.

Proposed approach

The technique of control volume used here is illustrated in Figure 1 (a). Where the face is marked in the form of a red line, the cell zone is marked as a green disk.
The calculation of the impact damage is performed at each time step. This calculation takes account of a several conditions which are presented in Figure 1 (b). However, a set of tests were instantaneously performed in each control surface (R= 1mm) verifying the next conditions;

This approach is based on three conditions
-***Vapor condition:*** This condition checks for the presence of vapor (bubble forming a micro-jet).
-***Bubble radius condition***: it is assumed that there is only one bubble in the control surface. Through the volume fraction of vapor, one can calculate the volume of the bubble then its radius.
-***Micro-jet velocity condition*****:** The micro-jet can damage the surface of the solid only if its speed is greater than a critical speed.
The different conditions adopted in the proposed approach as well as the algorithm of these processes are defined in this Figure.

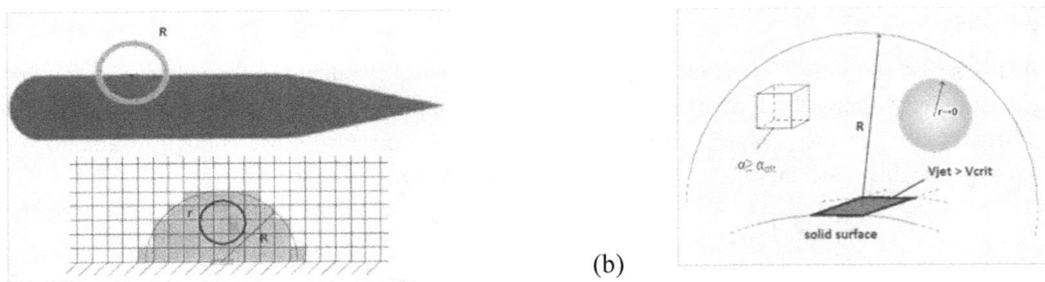

(a) (b)

Figure. 1. (a) Set of cells around a face: cells are located within the circle (green) for the face (red), (b) The different conditions applied in the control surface.

RESULT AND DISCUSSION

Simulation setup

The steady numerical solution is used to initialize the unsteady calculations. In order to be able to detect the pressure waves the time step is fixed at $\Delta t = T_{ref}/1000 = 8.92 \times 10^{-6}$ s (where $Tref = L/U_\infty$ and U_∞ is the inlet flow velocity).

The CLE hydrofoil is placed at an angle of 5 ° and a cavitation number σ=2. The C-type mesh is used because of the rounded shape of the leading edge. The computational domain is composed of 8 blocks with a structured grid, as shown in the Figure 2.

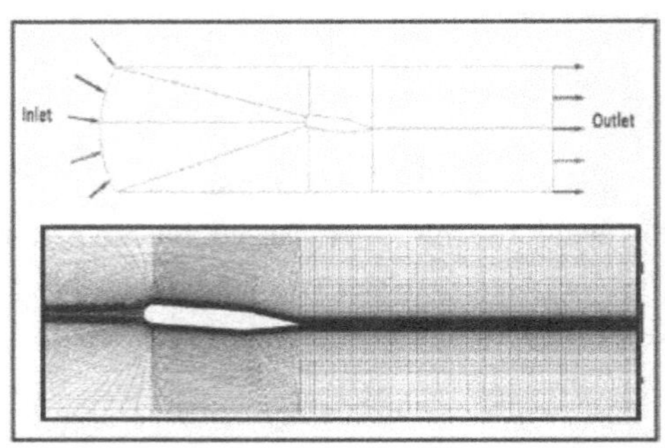

Figure 2. Dimension and domain gird of a CLE hydrofoil.

The inflow velocity is equal to $U\infty$=13 m/s and the value of the static pressure $P\infty$ is calculated as a function of the value of σ which is determined by (10);

$$\sigma = \frac{P_\infty - P_v}{\frac{1}{2}\rho_l U_\infty^2} \tag{10}$$

To capture the viscous effects of the boundary layer, a fine mesh has been chosen for the unsteady calculation with ($y^+ \approx 1$).

Grid effect

The investigation of the influence of mesh resolution was performed in steady state. This study was carried out on the CLE hydrofoil by monitoring the influence of three meshes on the values of the drag C_D and lift C_L coefficients:

$$C_D = \frac{F_D}{\frac{1}{2}\rho U_\infty^2 S} \quad C_L = \frac{F_L}{\frac{1}{2}\rho U_\infty^2 S} \tag{11}$$

Where S is the projected area of the hydrofoil, F_L and F_D are the lift and drag forces around the CLE hydrofoil.

From the results showed in Table 1, it is possible to notice that the differences were minimal between the medium and fine resolution meshes. For this reason the simulations were performed considering the medium resolution mesh.

Mesh	*Nodes*	C_D	C_L
Coarse	61 600	0.1428	0.5914
Medium	100 000	0.1449	0.7435
Fine	118 056	0.1452	0.7437

Table 1. mesh test

Validation

The numerical simulation of unsteady cavitating flow using LES is carried out around a CLE hydrofoil at an angle of incidence i=5° and the inlet flow velocity is equal $U\infty$ =13 m/s. The results of the numerical simulations are compared to results from experimental data [7]. The turbulent length scale is 0.001 m and the turbulent intensity is 1%. The bottom and top walls are modeled as Free-slip condition. The wall of the hydrofoil is modeled as nonslip.

Figure 3 present the time evolution of the shedding cavitation for ten snapshots in a cycle. The vapor pocket is located at the leading edge of the hydrofoil. The calculated vapor volume allows defining the various processes associated with its formation and collapse. At 0.1T the vapor cavities is convected by the flow of the previous cycle. A new vapor pocket appeared and began to expand at the leading edge. During this period, 0.2 T to 0.5 T, the process of cavity growth continues and the behavior of the cavity seems to be stable. The length of the cavity attains the half of the chord and she becomes thicker. It reaches its maximum length at 0.5T. As the re-entrant jet attains the region of high pressure gradient (0.6 T - T) and it makes it possible to decompose the main vapor pocket into small cavities which will be transported by the flow towards the trailing edge of the hydrofoil and a new cycle is born.

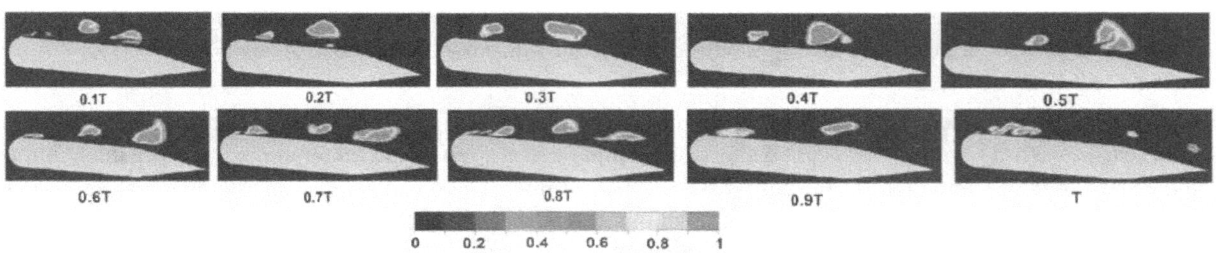

Figure 3. Time evolution of training and detachment sheet cavitation for one cycle T=0.006s.

The comparison between the experimental and time-averaged numerical velocity profiles is presented in Figure4. The average velocity is recorded over a cycle of formation and disappearance of the vapor pocket. The component u of the horizontal velocity is calculated at five positions on the surface of the CLE hydrofoil.
The numerical results are a good agreement with experimental results. The flow velocity begins to decrease as we advances towards the solid surface (the transition habitually happens roughly at the boundary of the sheet cavity). For positions x = 0mm (at leading edge) and x = 64mm (Trailing edge), no re-entrant jet is noticed. For stations x =16, 32, and 48 mm, a reverse flow is detected in numerical and experimental results. The characteristics of the re-entrant jet are properly predicted by the model.

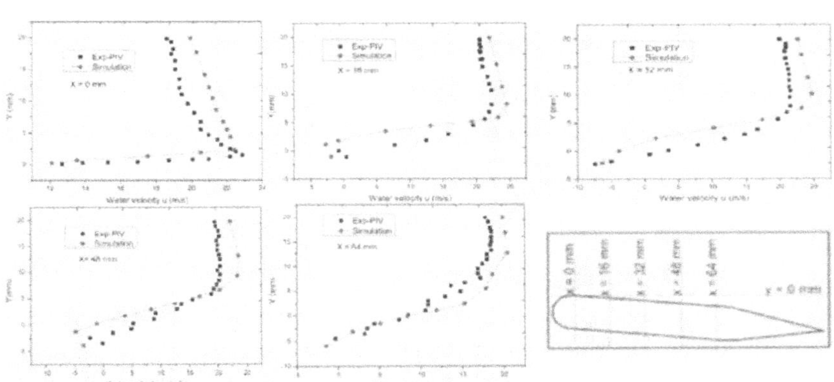

Figure 4. Comparison between numerical and experimental velocities in the x direction at the positions; x=0, 16, 32, 48 and 64 mm.

The numerical results showing the pressure oscillations obtained in four sensor positions are compared to experimental results (Figure 5).
The measures were available only for $U\infty$ = 16 m/s and σ = 2.7. These results make it possible to check the presence of the shock waves which are emitted during the collapse of the cavitation cloud will propagate in the mixture and trigger a set of events Which causes cavitation erosion. At X1=14.6 mm the pressure fluctuations attains 5 bars. At positions X2=26.6 mm and X3=36.6 mm, a significant increase of the pressure amplitude has been observed. This can be explained by collapsing of the cavitation pocket. For a very short period, the pressure peaks can reach 14 bars. An increase in pressure cannot provoke the erosion by itself but it may transport rather energy to trigger the formation of the micro-jet. At X4= 50.6 mm the magnitude of the pressure fluctuations has undergone a decrease downstream of the cavitation zone.
For this study, the sonic velocity in water is equal to c_l =1484 m/s and the densities of water and vapor at 20°C are c_l =1484 m/s and ρ_l = 998 kg/m^3.

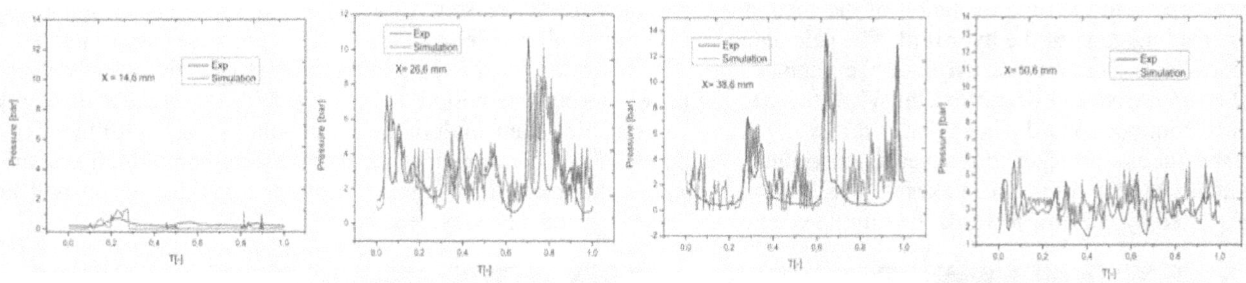

Figure 5. Time evolution of the pressure oscillations on extrados of the hydrofoil at four positions.

The calculation time is limited to 35 ms, which is enough to have statically averaged results. The simulation results are presented for σ =2.5 at $U\infty$= 13m/s. Figure 6 shows the temporal evolution of the accumulation of mechanical impacts during the short time of simulation for the case σ =2.5 at $U\infty$=13m/s. The pit accumulation rate over a short period of time is random. We are convinced that short simulations provides quite data to calculate statistical averaging and extrapolate the calculation to a longer exposure to cavitation.
A sensitivity study to evaluate the dependence for micro-jet characteristics and α_{crit} was performed. We carried out a series of tests of four values of α_{crit}. (Figure7)
We find that the variation of does not have a great influence on the evolution of A_{pit} that from α_{crit}. All this allows us to adopt the first two values of α_{crit}.
The interest of this proposed approach is to quantify the impact number of micro-jet and to calculate the intensity of these impacts. The parameter α_{crit} is considered as an essential parameter to quantify the aggressiveness of the cavitating flow and its impact on the surface of the hydrofoil.

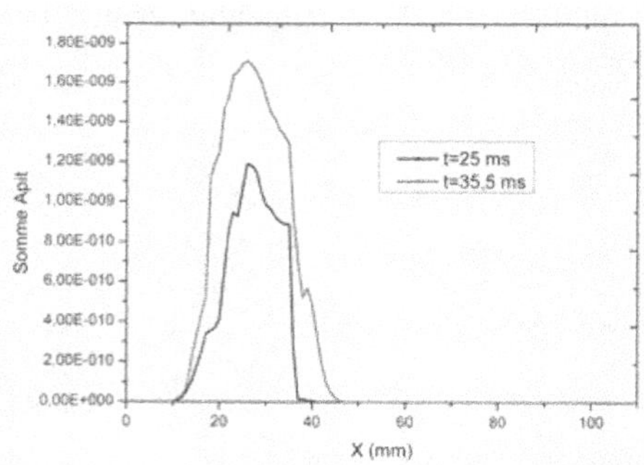

Figure 6. Accumulation of damage.

Figure 7 shows that the variation of the critical vapor volume fraction directly affects the size of the contour of the cavitation pocket and subsequently the rate of damage which increases progressively with the increase the volume of steam

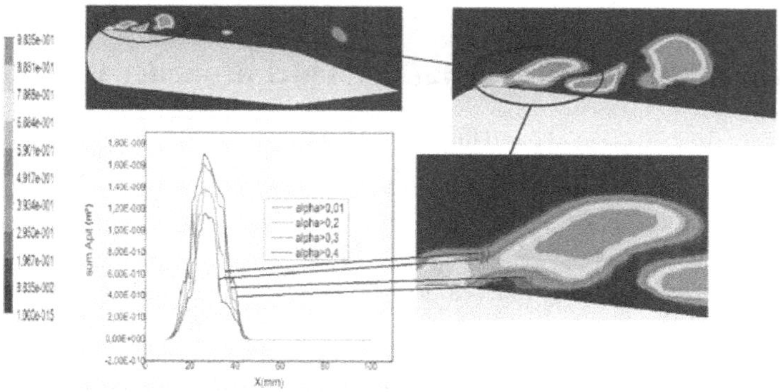

Figure 7. Evolution of the vapor pocket contour as a function of critical volume vapor fraction

The numerical distribution of Adam is compared to the measurements done by Dular and Delgosha [7]. Figure 8 shows the numerical and measured damage rate after 1 h of exposure to cavitation. We can say that the proposed model can accurately predict the magnitude and extent of damage in the forward part of the hydrofoil. An offset between the numerical and experimental result has been well noticed, especially in the rear part of the hydrofoil.

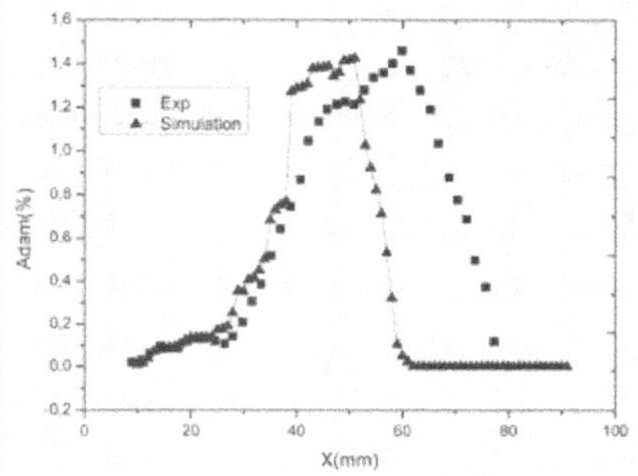

Figure 8: Predicted and measured damage on the surface. σ=2 at U_∞=13 m/s, (Exp. data from [7]).

Conclusion

The comparison with the measurements shows that the proposed approach is able to estimate the intensity of the eventual damage impact and to distinguish their location especially in the front part of the hydrofoil. On the other hand, a deviation of the numerical curve was observed with respect to the experimental curve in the rear part of the structure. However, to have an exact prediction of the impact of the damage, a correct estimate of the shock wave and its intensity is needed.

References

[1] Dupont PH. (1993), Etude de la dynamique d'une poche de cavitation partielle en vue de la prediction de l'érosion dans les turbomachines hydrauliques. - EPFL, Suisse. Thèse de doctorat.

[2] Avellan F. et Farhat M. **(1988),** Supression Générée par un Tourbillon cavitant : La Houille Blanche, pp. 527-533.

[3] Lord Rayleigh (1917), On the Pressure Developped in a Liquid During Collapse of a Spherical Cavity", Phil. Mag. 34.

[4] Benjamin TB, Ellis AT. (1966), The collapse of cavitation bubbles and the pressures thereby produced against solid boundaries. Philosophical Transactions of the Royal Society of London; 260 : 221–240.

[5] Chahine G.L. (1979), Etude Locale du Phenomene de cavitation. Analyse des Facteurs Regissant la Dynmique des Interfaces", thèse de doctorat, Université Paris.

[6] Andreas. P, Hemant S. Udo L, Ouled el Mokhtar (2015) Numerical modelling and prediction of cavitation erosion; published by Elseiver; Wear 338-339, 189–201.

[7] Dular .M; Coutier-Delgosha .O (2009) Numerical modelling of cavitation erosion; INTERNATIONAL JOURNAL FOR NUMERICAL METHODS IN FLUIDS; **61**:1388–1410.

[8] Kanfoudi .H., Zgolli.R, (2011), A numerical model to simulate the cavitating flows. International Journal of Modeling, Simulation, and Scientific Computing.

[9] Kanfoudi. H, Lamloumi H and Zgolli R, (2014). "Modelling of Unsteady Cavitating Flow Hatem Kanfoudi *, *International Journal of Renewable Energy Technology* 5(2):20–22.

© 2018 ASME

Hydro-acoustic analysis of the cavitating model propeller PPTC in oblique flow

Huiping Fu*; Jie Li

Shanghai Jiao Tong University, Shanghai, China

Abstract

Propeller cavitations and propeller induced pressure pulses in oblique flow were calculated on the propeller model of PPTC. Through the partition of tip vortex field and the refinement of the field gridding, the tip vortex cavitation was simulated based on the homogeneous mixture flow model and Zwart-Gerber-Belamri cavitation model and the computed results of the thrust coefficients and pressure pulses agreed well with the EFD data. Then, cavitation noise was calculated based on the Lighthill's acoustic analogy. Acoustic sources included monopole and dipole and partially quadrupole sources by selecting the rotating propeller blades and hub as integration surface. One of the aims is to reveal the differences between hydrodynamic pressure pulses and sound pressures. It's shown that the hydrodynamic pressure pulse and the sound pressure are both characterized with blade passage frequency (BPF); the amplitude of sound pressure is smaller than that of pressure pulse; and their phases are approximately in contrast, when pressure pulse is at the peak, the sound pressure is at the valley. The second aim is to investigate the effects of cavitation. It's shown that cavitation increases sound pressure but not hydrodynamic pressure pulse. As to the comparison among the perturbation amplitudes of three proximity points, it's shown that it's the computed SP that agrees well with the experimental PP, but not the computed PP. It may be necessary to separate and analyze distinctly the hydro-acoustic and hydrodynamic components of the recorded proximity pressure signals.

Keywords: cavitation noise; pressure pulse; sound pressure; numerical simulation

Introduction

The propeller is the main noise source of underwater vessel, especially when the propeller is cavitating. The propeller noise will produce many adverse consequences. Firstly, the noise will affect the naval environment. Secondly, the propeller noise will be transferred to the cabin and then affect the comfort of passengers. Thirdly, sonic fatigue and structural vibration induced by the propeller noise have a great impact on vessel safety. In order to mitigate and control the vessel noise it is necessary to do some research to address the two-fold task of developing effective tools to support the diagnostics and the design of ships on one hand, and to improve the understanding of the fundamental underlying mechanisms of the noise generation and radiation phenomena on the other hand.
Due to the development of computer ability, the computational science got a fast developing. There are the three means to understand the mechanism of acoustics, that is, the method of theoretical study, experiment and numerical simulation. In this paper, FW-H acoustic analogy and RANS solver embedded in CFD general code ANSYS FLUENT were utilized for the numerical investigation of propeller model of PPTC, for wetted and cavitating conditions in oblique inflow. The differences between hydrodynamic perturbation and hydro-acoustic perturbation were analyzed and the effects of cavitation were studied.

Computational Object, Mesh and Flow Simulation

The first compressible, implicit large-eddy simulation of a cavitating ship propeller was done by Budich *et al* (2016) [1]. Unlike most of the published tip vortex cavitation of PPTC are underestimated, their CFD results is an over-estimation regarding the extent of the cavity. In this paper, the homogeneous mixture flow model and Zwart-Gerber-Belamri cavitation model were used to calculate the propeller cavitations and propeller induced pressure pulses in oblique flow on the propeller model of PPTC'15 (put forward in the 4th international Symposium on Marine Propulsors, 2015). The propeller's configuration is shown in Figure 1(a) and the three locations for sensors of pressure pulses are given. Through the partition of tip vortex field and the refinement of the field gridding, as shown in Figure 1(b), the tip vortex cavitation was simulated and the computed results of the thrust coefficients and pressure pulses agreed well with the EFD data. Computed result for Case 2.2 and error analysis, including the thrust coefficients for wetted and cavitating conditions, are given in Table 1. It's shown that the relative error of CFD data to EFD data is less than 3%. Cavitation patterns are shown in Figure 2. We can see obvious tip vortex cavitation and

*Corresponding Author, Huiping Fu: fuhp@sjtu.edu.cn

sheet cavitation near the leading edge on the pressure side and sheet cavitation near the trailing edge on the suction side. The isosurface of the volume fraction α= 0.4 is more similar to the measured cavitation patterns of EFD than the isosurface of α=0.6.

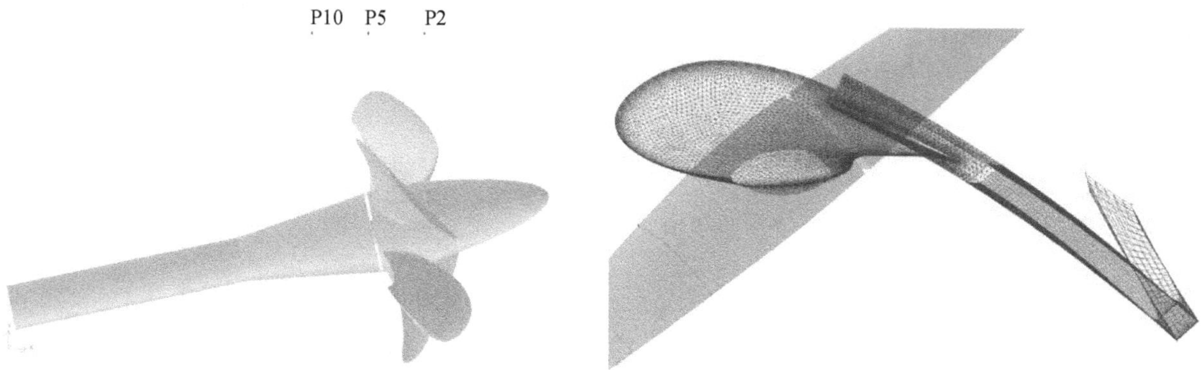

(a) Three sensors for pressure pulses (b) Mesh for tip vortex

Figure 1 PPTC'15 and Mesh

Table 1 Computed K_T and error analysis

Case 2	Wetted			Cavitating			$\triangle K_T$/%		
	EFD	CFD	*E*/%	EFD	CFD	*E*/%	EFD	CFD	*E*
Case 2.2	0.265	0.267	0.8	0.167	0.172	3.0	-37.0	-35.6	-1.4

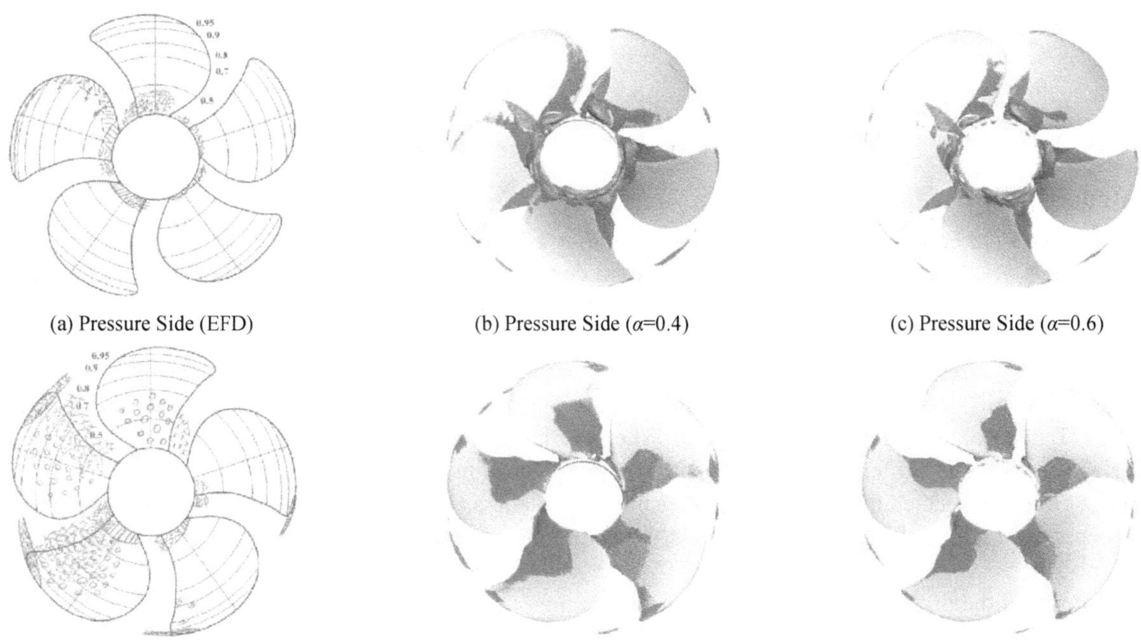

(a) Pressure Side (EFD) (b) Pressure Side (α=0.4) (c) Pressure Side (α=0.6)

(d) Suction Side (EFD) (e) Suction Side (α=0.4) (f) Suction Side (α=0.6)

Figure 2 Cavitation patterns

Table 2 Pressure pulses

	Sensor	Wetted/ kPa			Cavitating/ kPa		
		i=1	i=2	i=3	i=1	i=2	i=3
CFD	P2	0.4926	0.0135	0.0030	0.4180	0.0157	0.0022
	P5	0.3934	0.0246	0.0023	0.5415	0.0669	0.0083
	P10	0.0887	0.0094	0.0019	0.1339	0.0361	0.0056
EFD	P2	0.3277	0.0103	0.0080	0.2387	0.0206	0.0091
	P5	0.4266	0.0152	0.0108	0.4017	0.0196	0.0130
	P10	0.0711	0.0142	0.0104	0.1096	0.0227	0.0207

By means of a Fourier transformation, the harmonic frequencies and the corresponding amplitudes were evaluated. Calculated pressure pulses on 3 locations and comparisons between CFD and EFD are shown in Table 2 and Figure 3~4. For wetted condition, the greatest pulse is located in P2, otherwise experimentally in P5; for cavitating condition, the greatest pulse is located in P5, both in CFD and EFD. As to the effect of cavitation on pressure pulse, cavitation increases PP in P5 and P10 in CFD, however cavitation only increases PP in P10 in EFD.

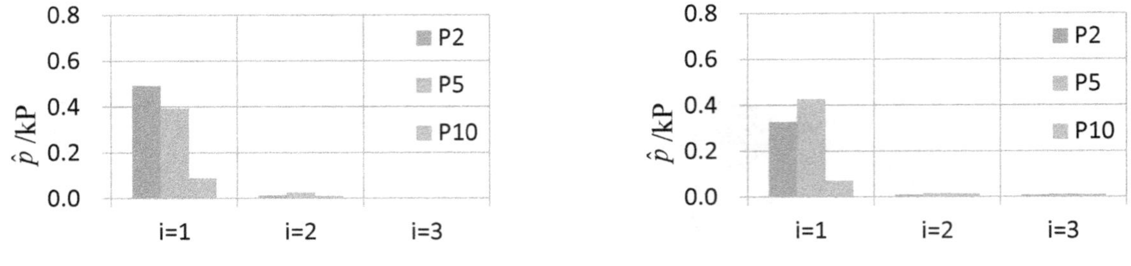

(a) CFD (b) EFD

Figure 3 Wetted pressure pulses

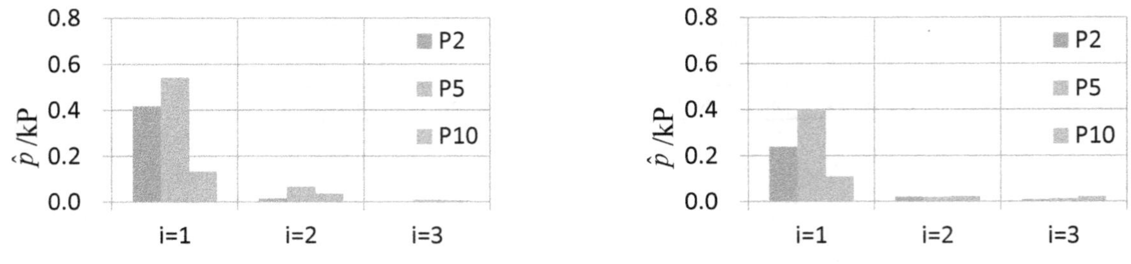

(a) CFD (b) EFD

Figure 4 Cavitating pressure pulses

Flow-Induced Noise

When the flow-field computation for wetted and cavitating conditions separately tended to be stabilized, loaded the FW-H (Ffowcs Williams and Hawkings) acoustic module to start the flow-induced noise simulation. The complete solution of FW-H equation consists of surface integrals and volume integrals. The surface integrals represent the contributions from monopole and dipole acoustic sources and partially from quadrupole sources, whereas the volume integrals represent quadrupole (volume) sources in the region outside the source surface. The contribution of the volume integrals becomes small when the flow is low subsonic and the source surface encloses the source region. In FLUENT, the volume integrals are dropped. When the integration surface coincides with an impenetrable wall, the surface integrals are often referred to as thickness and loading terms, respectively, in light of their physical meanings. Here we selected the rotating propeller blades and hub as the acoustic source surface. We made use of the post-processing module of FLUENT to process the sound signal and got the acoustic radiation results around the propeller.

In order to reveal the differences between hydrodynamic pressure pulse (PP) and sound pressure (SP), the near-field three points, P2, P5, P10, were calculated again. Firstly, take the point P2 for cavitating condition as an example. From Figure 5(a), we can see the time histories of PP and SP. The two pressures have inverse phases, that is, when PP reaches to the peak, SP reaches to the valley. Figure 5(b) shows the sound pressure level (SPL) and the pressure pulse level (PPL), the reference pressure is 1×10^{-6}Pa. We can see two peaks at blade passage frequency (BPF) for both SP and PP, and SP goes down at larger frequency, but PP keeps at some level. Figure 5(c) shows the power spectrum densities of SP and PP. We can see the peaks at the BPF are prominent all over the frequency range, and the peak of PP is higher than SP. As to the effects of cavitation on PP and SP, Table 3 shows that cavitation increases each point's SP, but not PP.

As to the comparison among the perturbation amplitudes of 3 proximity points, we can compare the peak values of SPL at the BPF from Figure 6, or just compare the average values from Table 3, and find that the computed SP shows P5>P2>P10 not only for wetted condition but also for cavitating condition. However, the computed PP shows

P5>P2>P10 for cavitating condition, not for wetted condition. It is the computed SP that agrees well with the experimental PP, not the computed PP. So it may be necessary to separate and analyze distinctly the hydro-acoustic and hydrodynamic components of the recorded near-field pressure signals. For we compute PP and SP with different modules, but EFD data all from direct pressure fluctuation measurements in the proximity of the noise source. The identification of the noise emission and propagation mechanisms is based upon direct pressure fluctuation measurements in the proximity of the noise source and/or in the acoustic far field[2]. In order to investigate the mechanisms underlying the hydro-acoustic and hydrodynamic perturbations in a rudder operating in the wake of a free running marine propeller, a novel wavelet-filtering procedure was applied to separate and analyze distinctly the acoustic and hydrodynamic components of the recorded near-field pressure signals[3].

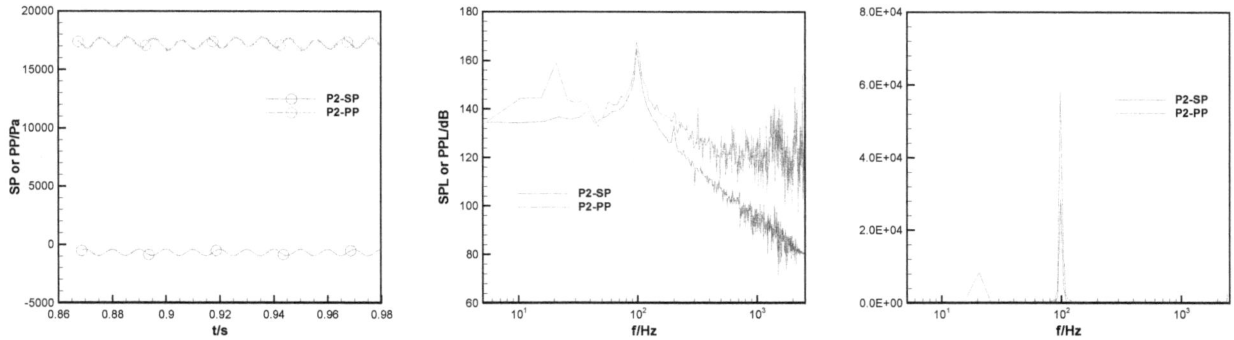

(a) Time histories (b) SPL and PPL (c) Power spectrum densities of SP and PP

Figure 5 Differences between PP and SP at point P2 for cavitating condition

Table 3 Effects of cavitation on PP and SP

Case3.2	Wetted			Cavitating			△SPL or △PPL/dB		
	P2	P5	P10	P2	P5	P10	P2	P5	P10
SPL	165.55	165.90	154.84	165.69	166.08	155.26	0.14	0.18	0.42
PPL	171.03	169.00	157.13	170.06	171.89	162.65	-0.97	2.89	5.52

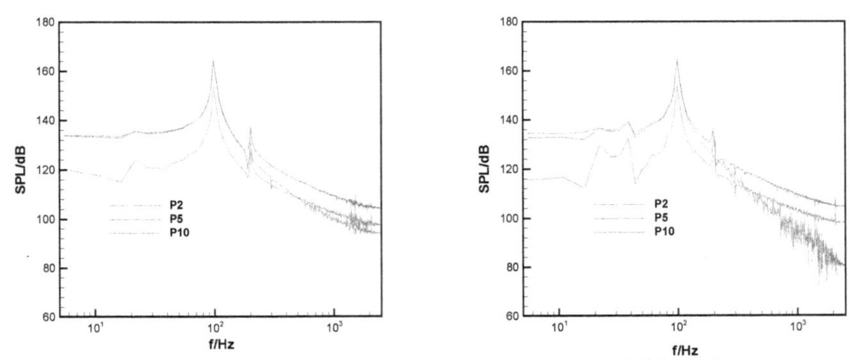

(a) Wetted (b) Cavitating

Figure 6 Comparison of SPLs at 3 points

Directivity Analysis

Extent the propeller's rotating axis to intersect with the section, located at the distance $3.106D_p$ from the point P5 along the x axis, we get a point. Centered with the point, two circles, whose diameter are respectively $1D_p$ and $2D_p$, D_p being the diameter of propeller, are selected as shown in Figure 7(a). Along each circle a point every 15 degrees being distributed, totally 48 points on the two circles are used to research directivity of PPTC's noise radiation in the wake of propeller as shown in Figure 7(b).

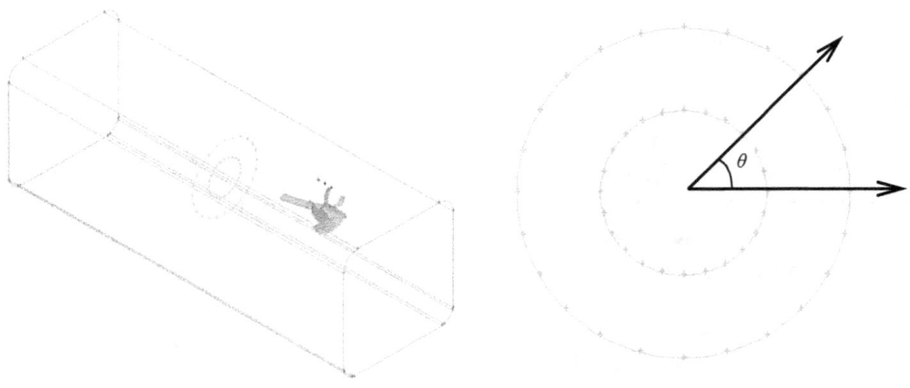

(a) Location of 48 points (b) Definition of θ (view from back to front)

Figure 7 Points used for directivity analysis

Figure 8 shows 4 curves, each is subtracted the local minimum SPL on the circle for wetted and cavitating conditions. We can see that directivity has symmetry. The symmetric axis is located at the line joining the minimum SPL and maximum SPL. For wetted condition, it's the y axis deflecting 45 degrees countering to the propeller's rotating direction; for cavitating condition, the deflecting angle is 15 degrees. All the curves distract from the circle centers and go up. The differences due to directivity on the outer circles are larger than those on the inner circles. The differences for cavitating condition are larger than those for wetted condition; the former can reach to 4.27dB, while the latter just 2.83dB.

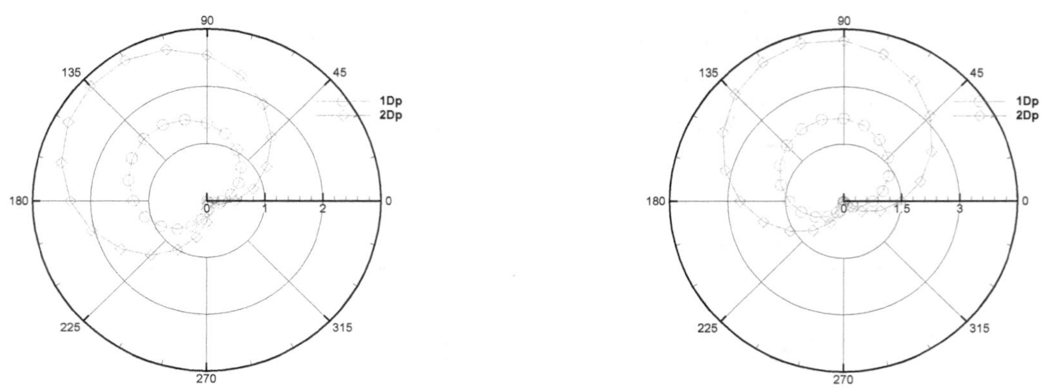

(a) Wetted (b) Cavitating

Figure 8 Directivity differences

Table 4 shows the average SPL on each circle for wetted and cavitating conditions. We can see that the mean value on the outer circle is smaller than that on the inner circle; and the mean SPL for the cavitating condition is larger than that for the wetted condition, and the difference is about 7dB. That is, cavitation increases the noise SPL in the wake of propeller more evidently than in the proximity of the propeller.

Table 4 Mean SPL on 2 circles

D	Wetted	Cavitating	Δ SPL/dB
2Dp	95.2612	102.3154	7.0542
1Dp	95.8130	102.9189	7.1059

Conclusion

It is not required that the integration surface coincide with body surfaces or walls. The formulation permits source surfaces to be permeable, and therefore can be placed in the interior of the flow. When a permeable source surface (either interior or non-conformal sliding interface) is placed at a certain distance off the body surface, the integral solutions include the contributions from the quadrupole sources within the region enclosed by the source surface.

However it's sparingly to use a non-conformal sliding interface as a permeable source surface to include the contribution from the quadrupole sources. By just using body surfaces, herein the propeller's 5 blades and hub, as the integration surface, we can get the smaller hydro-acoustic perturbation than hydrodynamic perturbation. We can get very regular directivity for both wetted and cavitating conditions. Otherwise we can't get these results when use a non-conformal sliding interface as a permeable source surface. These results can't be presented here due to the space limit of paper.

Finally, it may be necessary to separate and analyze distinctly the hydro-acoustic and hydrodynamic components of the recorded near-field pressure signals before the validation of the computation of proximity pressure pulses.

References

[1] Felli, M., Falchi, M., Dubbioso, G. (2014). *Direct and indirect experimental methodologies for the hydro-acoustic analysis of marine propellers*. Proceedings of the 30th Symposium on Naval Hydrodynamics.

[2] Felli, M., Grizzi, S., Falchi, M. (2014). *A novel approach for the isolation of the sound and pseudo-sound contributions from near-field pressure fluctuation measurements: analysis of the hydroacoustic and hydrodynamic perturbation in a propeller-rudder system*. Experiments in Fluids, 55(1): 1651-1667.

[3] Budich, B., Schmidt, S. J., Adams, N. A. (2016). *Implicit large eddy simulation of the cavitating model propeller VP1304 using a compressible homogeneous mixture model*. Proceedings of the 31th Symposium on Naval Hydrodynamics.

© 2018 ASME

Hydrodynamic Stabilization of Supercavitating Underwater Bodies

Oleksandr Mayboroda*

Kyiv State Maritime Academy, Ukraine

Abstract

In order to stabilize the motion of supercavitating underwater bodies, the damping action of the stern elastic interceptors which glide along the cavity surface is used. The results of calculation of normal forces on planing interceptors at the Mach numbers below 1.5 with a detached and attached shock wave are obtained. These results agree satisfactorily with known theoretical solutions and experimental data.

Keywords: supercavitating bodies; stabilization of moving; gliding at the Mach numbers below 1.5

Introduction

The increase the speeds of ships and other technical objects moving in the water calls new problems for hydrodynamics, in particular, the need to take into account the water compressibility, including at supersonic speeds [1]. The maximum achieved body velocity in water is about 1550 m/s and according to known data the speed range up to 2000...2200 m/s, that is corresponding to the Mach numbers below 1.5 (the Mach number is the ratio of the body speed to the sound speed in the fluid), is considered to be promising for creation of high-speed underwater bodies. Under these conditions, compressibility of water is accompanied by known gas-dynamic effects – compression shocks, shock waves etc.

The technical disadvantage of supercavitating bodies is the movement instability, since the application of resistance force in the bow stagnation point precludes the hydrodynamic damping. To stabilize the movement of such bodies, it is possible to install the stern elastic plates – planing interceptors that glide on the cavitation cavity surface and create a hydrodynamic damping. Gliding on the surface of incompressible water, that is, at the Mach numbers below 0.3, is sufficiently studied. High-speed gliding into account the gas-dynamic properties of water has not been considered yet for lack of applied interest.

Hence, it is necessary to investigate the hydrodynamic characteristics of planing plates with finite aspect ratio those glide in the range of sub-, trans- and supersonic velocities.

Body

The compressibility of water as a condensed medium, unlike a gas, has some characteristic features. The equilibrium distances of water molecules correspond to a minimum of the potential energy of their interaction. When such a medium is compressed, the internal pressure fast increases and, unlike gases, has a nonthermal, only elastic nature. This determines the main features of the water behavior during compression.

At pressures up to $3 \cdot 10^9$ Pa, corresponding to the Mach numbers below 1.5, the water state equation has a form Teta [2] with the isentropic index γ=7.15. Consequently, under the considered conditions, water is a barotropic medium, and its internal energy is the sum of two functions, one of which depends only on the specific volume, and the other – only on the entropy. Therefore, it can be shown that isentropic processes in water are isothermal. Comparison of the isentropic flow relationships of pressure and density for water and air shows that with increasing the Mach number the water exhibits under compression more greater elasticity than air.

1. Subsonic flow

According to Wagner's analogy [3], a flow under the planing plate corresponds to that under a lower surface of a thin wing. For such a surface, experimental data on the load in a wide range of the Mach numbers are known [4] and by the method of local linearization the Mach number effect on the load increase was estimated [5].

As it is shown by Kusakava [6], a pressure distribution on the lower surface of a thin wing at the Mach numbers below 1 is affinity-like. So it can be assumed that the pressure distribution on the planing plate in the subsonic flow

*Corresponding Author, Oleksandr Mayboroda: olexander.mayboroda@gmail.com

will also be affinity-like with the isentropic similarity factor k_M, which can be determined from the isentropic flow relationship as

$$k_M = \frac{2}{\gamma M^2}\left[\left(1+\frac{\gamma-1}{2}M^2\right)^{\frac{\gamma}{\gamma-1}} - 1\right]. \tag{1}$$

In Fig. 1 the Mach number effect on the relative change in the normal force coefficient on the profile lower surface according to the proposed similarity factor k_M for water and air in comparison with the well-known Prandtl-Glauert rule [2], experimental data [4] and the results of a theoretical estimate of such effect [5] is showed.

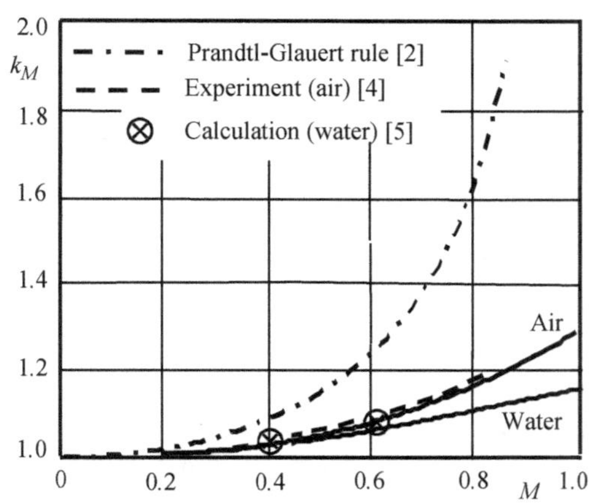

Fig. 1 Effect of the Mach number on hydrodynamic load of a profile

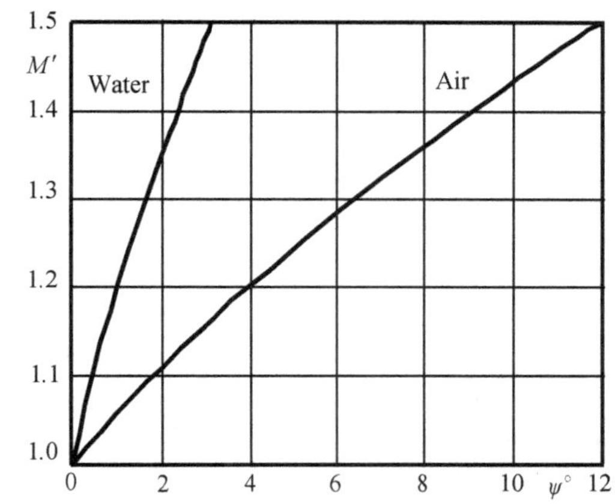

Fig. 2 Effect of a running pitch angle ψ on the Mach number M' of the shock wave attachment

Satisfactory agreement of these results allows us, for the indicated flow conditions, to recommend the similarity factor k_M (1) for taking into account the compressibility of various media at the Mach numbers below 1. In contrast to the Prandtl-Glauert rule, the similarity factor k_M takes into account the media thermodynamic properties and, as it is shown below, can be generalized to supersonic regimes with a detached shock wave.

Lack of finite-span nonlinear effects during gliding allows to generalize the well-known Jung theory [4] and to obtain the expression for the derivative of normal force C_n^{ψ} for a planing plate

$$C_n^{\psi} = \frac{\pi k_{\psi} k_M}{1+2\lambda k_{\psi} k_M}, \quad M \le 1, \ \psi \le 60^{\circ}, \ \lambda \ge 0.01, \tag{2}$$

where λ is an aspect ratio of a plate; k_{ψ} is a coefficient of effect of the running pitch angle ψ according to the nonlinear theory of Sedov [7].

Generalizing the well-known Wagner model [3] for the case of subsonic gliding, taking into account the proposed coefficient k_M (1), one can obtain an expression for a value of the relative bow backwater flow

$$\frac{l}{l_0} = \frac{1}{2}k_M\left(3+\sqrt{1+\frac{1}{\lambda_0}}\right) - 1, \quad M \le 1, \ \psi \le 60^{\circ}, \ \lambda \ge 0.01, \tag{3}$$

where l is a wetted plate length; l_0 and λ_0 are a length and initial aspect ratio of the plate relative to an undisturbed fluid surface.

The affinity-like distribution of the pressure with increasing of the Mach number suggests that the relative position of a pressure centre on the plate remains.

2. Supersonic flow

When a plate moves with supersonic speed, a shock wave moves ahead of it. The isentropic character of the water state equation makes it possible to consider a formation and propagation of this shock wave in the quasi-acoustic approximation and use the well-known dynamic compatibility conditions [8]. Such a problem has a self-similar solution only for a sufficiently weak shock [8]. Since there is no further increase of the pressure on the plate behind

the shock, we can assume that it belongs to the weak type. Taking into account the lack of a mixed flow under the plate we admit the existence of a quasi-stationary solution of the problem. An additional difficulty is a presence of a free fluid surface. At the intersection point of the shock front with the free surface, the centered Prandtl-Meyer expansion wave emerges, and the flow turns in the direction of the free surface [9].
In view of the problem complexity, first we analyze the simplified case – the gliding of the plate with infinite aspect ratio, so called the profile.
When the plate is gliding with a supersonic speed it is possible three characteristic regimes of flow around it:
$1 < M < M'$ – there is a subsonic flow behind the detached shock wave and at M' the shock wave attaches to the plate;
$M' \leq M \leq M''$ – there is a subsonic flow behind the attached shock wave, where M'' is the Mach number at which the velocity behind the shock wave becomes supersonic;
$M'' < M$ – there is a supersonic flow behind the oblique shock wave.
As it is shown in the works on supersonic gas jets [10], the limiting angles of the flow deflection in an attached shock wave are also preserved for flows with a free surface. Under these assumptions, an approximate calculation of these angles in water and air was made and is shown in Fig. 2. As it can be seen, the planing profile at running pitch angles $\psi \geq 2.6^\circ$ for the Mach numbers below 1.5 is always streamlined with an attached shock wave.
For water, within the applicability of the Teta state equation, the similarity laws of the perfect gas dynamics can be applied when the adiabatic exponent of gas is replaced by a value γ in the similarity criteria. That allows using for water the results of exact solutions and experiments obtained for similar gas flows.
In particular, an approximate evaluation of the shock wave attachment in water, shown in Fig. 2, by the value of the Karman-Cheney similarity criterion [8] agrees well with the solution of a similar problem obtained by Vincenti and Wagoner [11] by the hodograph method.

2.1 Planing profile with detached shock wave

At $1 < M < M'$ the shock wave *C* (Fig. 3) is disconnected and interacts with the fluid free surface *A*. Such a flow exists always at the running pitch angles $\psi \geq 2.6^\circ$ of the profile *L*. The shock is curved and in the convex head part, turned towards the profile movement, is straight. Supersonic flow in the main part of the shock turns into subsonic. In the remaining part, the shock intensity decreases, the flow behind it remains supersonic and the shock decomposes into a weak discontinuity. From a free surface, the shock is reflected by a centered Prandtl-Meyer expansion wave *PM* and the flow in surface vicinity remains supersonic. At the point of reflection, the pressure decreases to an unperturbed value, and the free surface deformation is determined by the distance from the shock to the profile, the so-called thickness of the shock layer.

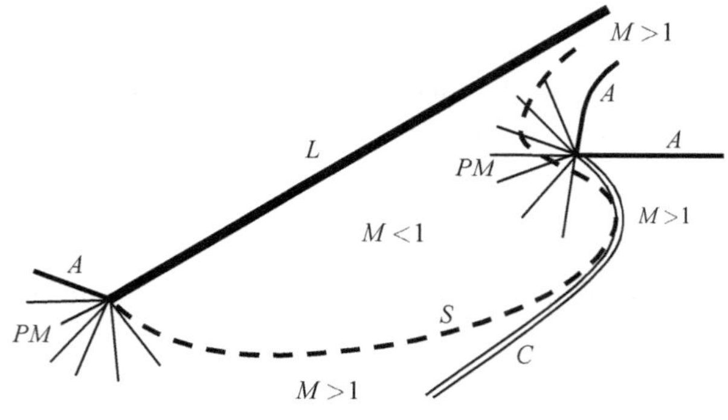

Fig. 3 The planing profile with the detached shock wave

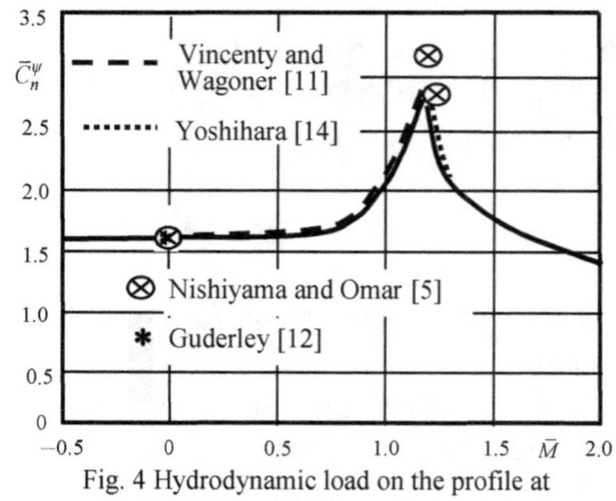

Fig. 4 Hydrodynamic load on the profile at transonic gliding with the running pitch angles below 2.6°

The profile is streamlined by the subsonic flow behind a direct shock up to the after-body Prandtl-Meyer expansion wave. From supersonic flow below and in the backwater flow on the free surface, the subsonic flow is detached by the contact discontinuity line *S*.

Using the correspondence principle [12] for the flow behind a straight shock wave we can construct in this case the following generalization of similarity factor k_M :

$$k_M = \frac{2}{nM^2}\left[\frac{\left(\frac{n+1}{2}M^2\right)^{\frac{n}{n-1}}}{\left(\frac{2nM^2}{n+1}-\frac{n-1}{n+1}\right)^{\frac{1}{n-1}}}-1\right], \quad 1<M<M', \quad \psi \geq 2{,}6^\circ. \tag{4}$$

For the supposed assumptions, it can be suggested that in the range of the Mach numbers $1<M<M'$ the change in the relative position of the pressure centre of the planing profile will be insignificant. According to Vincenti and Wagoner [11], for example, the displacement of the pressure centre to the trailing edge in this range is 4% of the profile chord.

2.2 Planing profile with attached shock wave

Such a flow exists always at the running pitch angles of the profile $\psi \leq 2.6^\circ$. As it is known, there is no stagnation point on the profile [13]. The flow behind the shock at $M=M'$ is subsonic and with a velocity increase at $M=M''$ becomes supersonic. The value of the interval $\Delta M = M'' - M'$ depends on the Mach number and for $M \leq 1.5$ in water does not exceed 0.05. By the smallness of this interval we can take $\Delta M = 0$ and as the Mach number M_S of the shock attachment consider the value $M_S = M'$.
In this case, the flow around the planing profile is similar to that of a thin symmetrical wedge and, in order to calculate the derivative of normal force C_n^ψ , we can use the corresponding results.
In Fig. 4 in the transonic similarity criteria $\bar{C}_n^\psi$ and $\bar{M}$ [8] a comparison of profile loads calculation in the range of $0.95 \leq M \leq 1.25$ according to the proposed method with known results of Nishiyama and Omar [5], Vincenti and Wagoner [11], Guderley [12] and Yoshihara [14] is showed. Here the value $\bar{M}=0$ corresponds to $M=1$.
At $M \geq M''$ the pressure centre of the planing profile is displaced to the middle of the wetted length. In the range $M' \leq M \leq M''$ the relative location of the pressure centre can be improved by interpolation on the basis of known statement on the continuity of the shock attachment with increasing flow velocity [12].

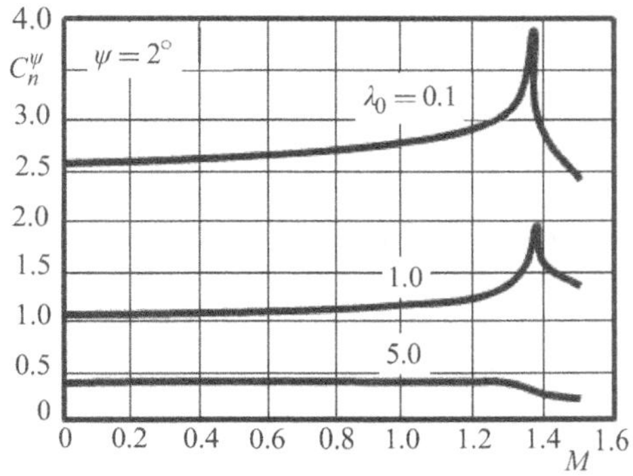

Fig. 5 Hydrodynamic load on plates at gliding with the attaching of the shock wave

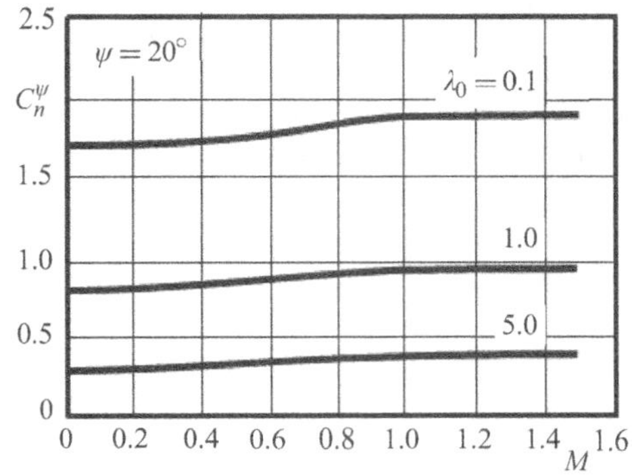

Fig. 6 Hydrodynamic load on plates at gliding with the detached shock wave

2.3 Planing plate with finite aspect ratio

The subsonic flow behind the detached shock wave in the quasi-acoustic approximation is isentropic and, therefore, potential. If to use the similarity factor k_M (4) one can apply in this case the expression (3) for the value of the relative bow backwater flow on the planing plate. Behind an attached shock wave there is no an effect of the bow backwater flow on a planing plate because the stagnation point is missing here.

Accounting of the aspect ratio influence on the hydrodynamic load of the planing plate at supersonic movement with the detached shock wave can be carried out with the generalized Young formula (2) and with an attached shock wave by the Hilton method [4].

Hydrodynamic loads on planing plates with different initial aspect ratio and running pitch angles at the Mach numbers below 1.5 are showed in Fig. 5 and 6. The values of the quantity C_n^{ψ} at $M = 0$ in Fig. 5 and 6 correspond to the known results for the gliding over the incompressible fluid surface.

Conclusion

For an isentropic equation of state of water, a quantitative estimate of the compressibility effect on the gliding of the plate at the Mach numbers below 1.5 with a detached and attached shock wave is first obtained.

A rule for conversion the pressure coefficients for the incompressible fluid flowing under the profile lower surface to their values for a given Mach number below 1.5 is proposed. Unlike the well-known Prandtl-Glauert rule, the proposed approach takes into account the thermodynamic properties of the medium.

Hydrodynamic loads on planing plates with initial aspect ratio λ more 0.01 and running pitch angles ψ below 60° at the Mach numbers below 1.5 with detached and attached shock wave are gotten.

The obtained results are in satisfactory agreement with known theoretical solutions and experimental data and can be used to calculate the damping forces on interceptors of supercavitating underwater bodies.

References

[1] Serebryakov, V. V., Kirschner, I. N., Schnarr, G. H. (2009). High speed motion in water with supercavitation for sub-, trans-, supersonic Mach numbers. Proceeding of CAV2009, Michigan, USA.
[2] Cole, R.H. (1965). *Underwater Explosions.* New York Dover Publ.
[3] Wagner H. (1932). *Über Stoss – und Gleitvorgänge an der Oberflächen von Flussigkeiten.* ZAMM. 12(4).
[4] Hilton, W.F. (1952). *High-speed Aerodynamics.* London; New York: Longmans, Green & Co.
[5] Nishiyama, T., Omar, F.K. (1981). *Compressibility Effects on Cavitation in High Speed Liquid Flow (First Report-Subsonic Liquid Flow).* Bulletin of the JSME. 24(190).
[6] Kusukawa, Ken-Ichi. (1957). *On the two-dimensional compressible flow over a thin symmetric obstacle with sharp shoulders placed in an unbounded fluid and in a choked wind tunnel.* J. Phys. Soc. Japan. 12(9).
[7] Sedov, L.I. (1965). *Two-dimensional Problems in Hydrodynamics and Aerodynamics.* New York: Wiley.
[8] Cherny, G.G. (1988). *Gas Dynamics.* Moscow: Nauka Publ. (In Russian).
[9] Grib, A.A., Ryshov, O.S., Hristianovich, S.A. (1961). *The theory of short waves.* J. Appl. Mech. Techn. Phys. № 1. (In Russian).
[10] *Supersonic gas jets.* (1983). Novosibirsk: Nauka. (In Russian).
[11] Vincenti, W.G., Wagoner, C.B. (1954). *Transonic flow past a wedge profile with detached bow wave.* NACA Rep. 1180.
[12] Guderley, K.G. (1957). *Theorie schallnaher strömungen.* Berlin.
[13] Neiland, V.J. (1975). *To the asymptotic theory of addition of a supersonic flow.* Proc. of the CAGI. 1650. (In Russian).
[14] Yoshihara, H. (1954). *On the flow over a wedge in the upper transonic region.* Proc. of the Second National Congress of Applied Mechanics. Ann Arbor.
[15] Guderley, K.G. (1954). *The flow over a flat plat with a small angle attack at Mach number 1.* J. Aeronaut. Sci. 21(4).

© 2018 ASME

Hysteresis in Cavitating Flows within Transparent Microchips

[1]Ece Ozdemir; [1]Berk Ozer; [1]Gokberk Deprem; [1]Ahmad Reza Motezakker, [2]Luis Guillermo Villanueva, [1,3]Morteza Ghorbani*; [1,3]Ali Koşar*;

[1]*Faculty of Engineering and Natural Science, Sabanci University, Orhanli, Tuzla, Istanbul 34956, Turkey*

[2]*Department of Mechanical Engineering, École Polytechnique Fedérale de Lausanne, Lausanne CH-1015, Switzerland*

[3]*Sabanci University Nanotechnology Research and Application Center, 34956 Tuzla, Istanbul, Turkey*

Abstract

Cavitation phenomenon has attracted much attention in engineering applications so the industry has provided considerable funding during recent years. Despite the simplicity and rather low price of small devices generating cavitation bubbles, the physics behind the creation and collapse of these bubble is still not well understood particularly in micro/nano scale. The assessment of size effects is vital for the design and development of new generation microfluidic devices involving phase change. Additionally, as the length scale decreases, surface nuclei dominate and dictate cavitation events. The modifications in the microchip geometry and enhancement in the micro device performance in applications involving cavitation will lead to increased cavitation bubbles number, reduced noise, improved bubbles collapse and increased energy sustainability. This study aims to investigate the creation of cavitation bubbles and classify the cavitating flow patterns in a novel roughened microchannel configuration. Cavitating flows are characterized in a transparent microchannel configuration in order to achieve a comprehensive understanding of cavitation inception and collapse in micro scale, which are crucial in the development of new-generation energy harvesting systems. In this device, a restrictive element and a big channel downstream of the restrictive element are mainly considered. The microchip consists of two main wafers, namely silicon and glass, which are anodically bonded together to withstand high pressures. The flow rate and discharge are evaluated at the outlet of the channels to characterize the chocking flow conditions in micro scale. The flow characteristics are determined to recognize differences in flow physics between smooth and roughened micro channels. Moreover, cavitation number, which is a major parameter for flow patterns, is considered in order to have valuable insights to the inception, development and collapse of the cavitation phenomenon in micro scale. Furthermore, the surface characteristics are also considered in detail in the microchip, and the effect of surface roughness on cavitating flows is investigated.

Keywords: cavitation; microchannel; roughness; pressure

Introduction

Cavitation is today's one of the hottest topic in different fields of technology. It could make it possible to acquire new power generation possibilities in order to achieve clean technology and energy savings which can be well generated in micro scale. Many studies have revealed that the collapse process of a cavitation bubble is extremely violent, and within the bubble, the gas can become extremely hot in micro/nano scale [1,2]. Hydrodynamic cavitation occurs in micro channels as a result of flow constriction. Micro channels with different structures and geometries may induce various types of cavitating flow patterns. Furthermore, thermophysical properties along with the surface topology can affect the creation of cavitation bubbles and their collapse. Upstream pressure values play a major role in the cavitation and the pressure inside the micro-channel [3,4]. The numerical studies in the micro scale hydrodynamic cavitation has shown that the intensity of the bubble generation is raised once the size of the channel is reduced [5,6]. Meanwhile, the static pressure as a significant parameter in the creation and collapse of the cavitation bubbles decreases to very low values along the micro channel particularly near the end of the channel. To study the cavitation in micro scale,

*Corresponding Author, Morteza Ghorbani: morteza@sabanciuniv.edu and Ali Koşar: kosara@sabanciuniv.edu

there is a need to shift conventional configurations to microchip format, which is called “cavitation on chip”. Therefore, it is required to develop techniques to increase the efficiency of the devices capable of generation cavitation bubbles [7]. In this study, hydrodynamic cavitation is studied in two different micro channels with different side wall roughnesses. The experiments have carried out at different upstream pressures to investigate cavitating flows. It is aimed to show the effect of different types of roughness on the generation of cavitation bubbles and utilize them in microfluidic systems for energy harvesting.

Methods and Materials

A high-pressure pure nitrogen tank (Linde Gas, Gebze, Kocaeli) supplies the required upstream pressure for the system. This tank is connected to a 1 Gallon fluid reservoir (Swagelok, Erbusco BS, Italy), which is filled with de-ionized (DI) water and serves as the working fluid. The reservoir is connected to the system with adaptor fittings. Four pressure sensors (Omega, USA) are mounted at the end of the tubing system and on the device package to measure the pressures. Two fine control valves (Swagelok) are integrated to the system to control the flow at the desired locations. A micro T-type filter (Swagelok) with a nominal pore size of 15 μm is used to filter any particles larger than 15 μm. A power LED source serves for illumination to have high quality records.

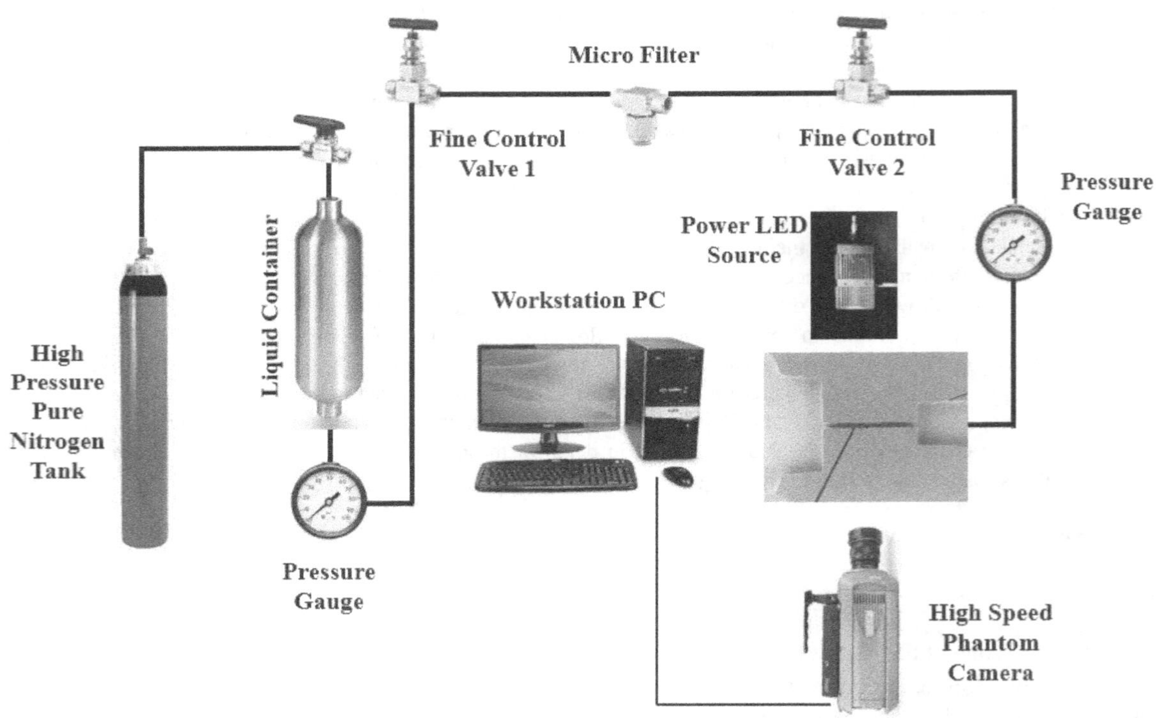

Figure 1: A schematic of the proposed system for generating cavitation bubbles and micro/mini cavitating jets.

The flow setup is shown in Figure 1. The inlet pressure varies between 5-60 bars. Although microchips have different roughness, they all have the same inlet structure which fits to the same package. Micro-chips are sandwiched to the package with screw mechanisms to allow for withstanding inlet pressures of the liquid up to around 60 bars. Liquid flow is controlled by valves at different locations. There are three pressure gauges on the setup to check for the pressure values, which are inlet, outlet, and the channel pressures. Images are recorded with a CMOS camera, which has 1280x800 resolution with very short time delay. The flow rate is measured and the effect of the discharge on the creation of the cavitation bubbles is obtained. The flow rate is measured by recording the level of the accumulated water level with respect to time.

Fabrication Process of Microchips

The device houses silicon and glass wafers, which are durable to high pressures. The channels, edges, surface roughness, inlet and outlets are formed on the silicon wafer, while the glass wafer is used as the top layer of the device for visualization purposes. The fabrication process includes three main steps, namely lithography, etching and bonding. Lithography process on the double side polished silicon wafer involves the exposure of suitable photoresist, coating of channels and development. The microfluidic device ensures a continuous access to the fluidic system including inlet, outlet and pressure ports. The fabricated devices integrated to the fabricated packages are durable to high pressures (up to 2000 Psi). The process flow for the fabrication of the microfluidic device is shown in Figure 2.

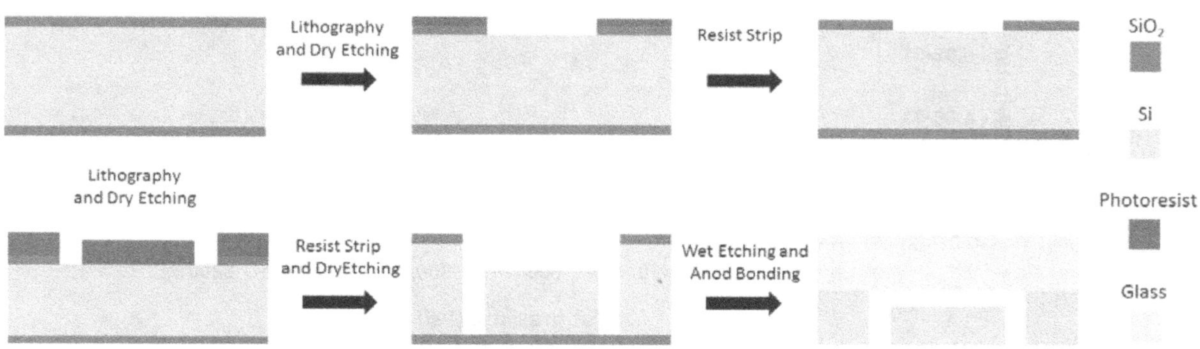

Figure 2: A schematic of the fabrication process of the microchips

Results

The intensity of cavitation along the channel is characterized by cavitation number:

$$\sigma = \frac{P_{ref} - P_V}{\frac{1}{2}\rho_l V_{ref}^2} \quad (1)$$

where ρ_l is liquid density, P_{ref} is reference pressure which is the upstream pressure in this work, P_V is vapor saturation pressure (working fluid in this work is water and vapor saturation pressure for water is 3540 Pa), and V_{ref} is reference velocity at the outlet of the microchannel, calculated with the use of mass flow rate, which varies according to the flow conditions.

Table 1 displays cavitation numbers extracted from the raw data for two different chips.

P-inlet (Psi)	Side Wall			Smooth		
	P-1 (Psi)	t(s)	σ	P-1 (Psi)	t(s)	σ
145	98	39	8.95E-01	110	51	1.20E+00
310	235	23	7.48E-01	245	32.5	1.09E+00
450	387	17	6.74E-01	380	25	9.97E-01
595	512	13	5.21E-01	525	20.6	9.36E-01
775	667	12	5.79E-01	660	18.6	9.59E-01
920	814	12	7.07E-01	790	17.8	1.05E+00
1075	944	11	6.89E-01	930	17	1.13E+00

Table 1: Cavitation number and upstream pressure for different channels

P-1 is the pressure at the entrance of the microchannel and t is the time for discharging of 20 ml liquid.

Figure 3 is extracted from Table 1 for smooth and side wall roughened micro channels. It is shown that the cavitation number has a lower value for side wall roughened micro channel in comparison to the smooth one for the identical pressures.

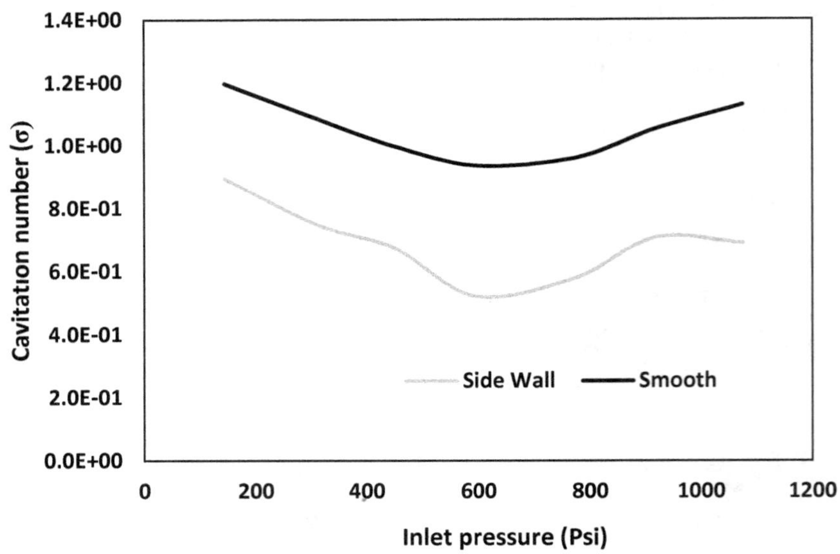

Figure 3: Cavitation numbers of two chips with respect to the input pressure

From this figure, it can be deduced that although the effect of changing the input pressure is observed as creating a "valley" that cavitation number intensifies as we move away from a minimum point, which here appears to be as around 600 Psi, the act of switching from the smooth channel to the channel with side wall roughness produces much successful results in increasing the cavitation number.

The outcomes of this study show that the cavitation phenomenon occurs as a result of the thermophysical and geometrical changes in the fabricated micro- chips. The roughness of side wall sharply facilitates the inception and development of cavitating flows. The interaction between the bubbles collapse and the solid exposed to the flow is expected to increase the thermal gradients on the surface, which could be exploited in power generation. The results in Figures 4 and 5 show that the cavitating pattern is the same for smooth microchannel at the inlet pressure of 920 Psi and the channel with side wall roughness at the inlet pressure of 450 Psi. However, the pattern for the side wall roughness channel at inlet pressure of 1070 Psi corresponds to fully developed super-cavitation, which is not seen in the similar smooth channel at this pressure. Thus, roughness leads to early transitions between flow patterns.

P-in (psi)	**450**	**595**	**775**	**920**	**1075**
Smooth	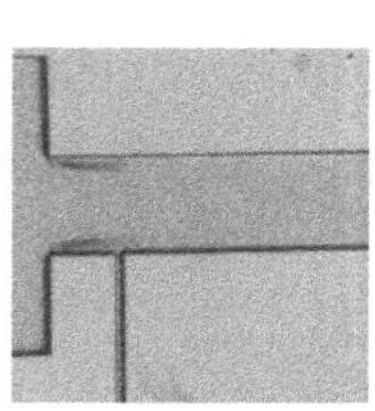	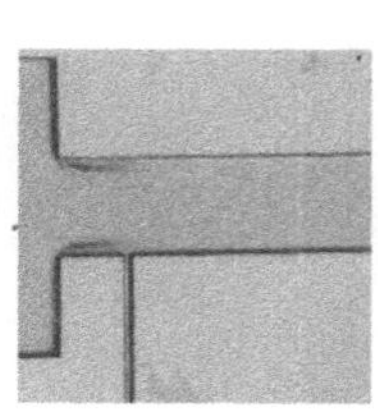	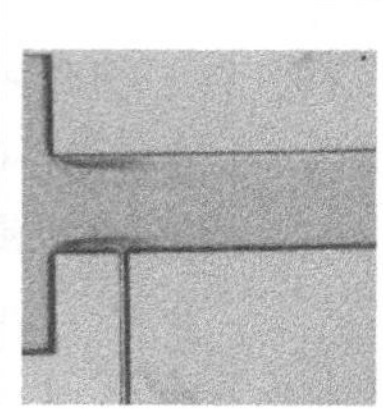	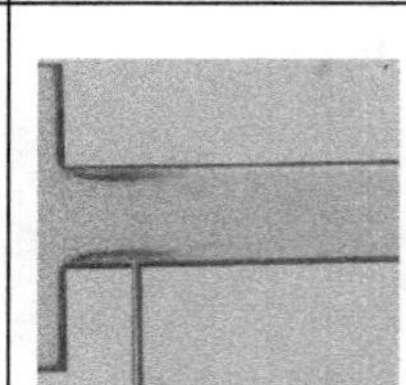	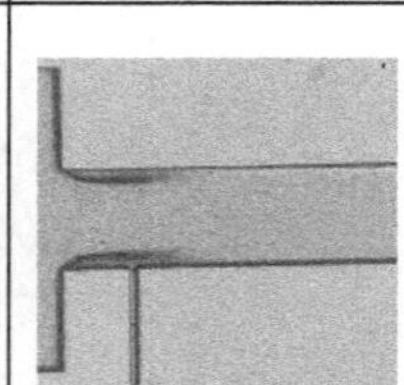

2-1 Side Wall	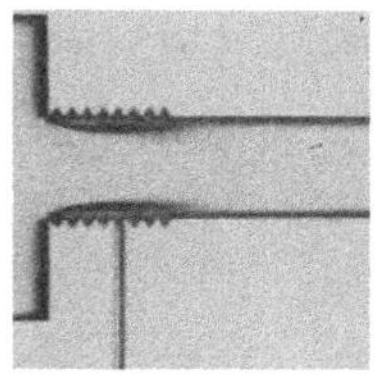	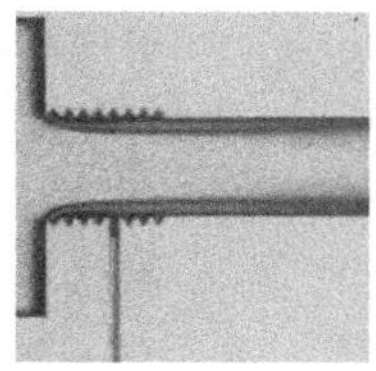	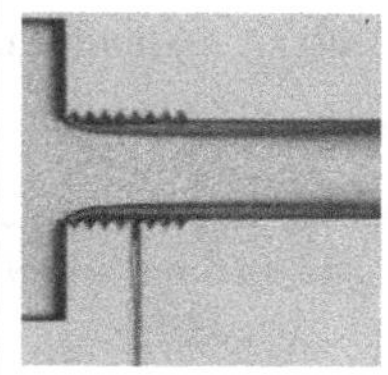	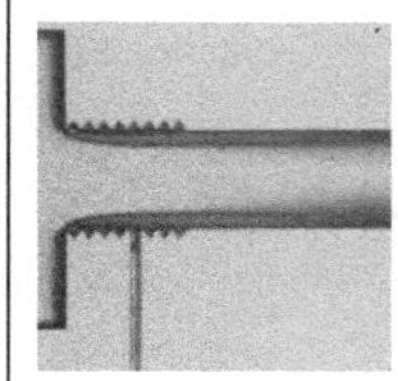	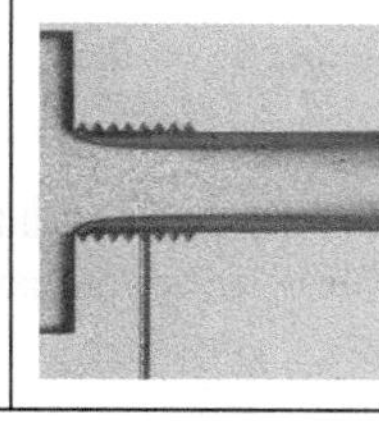

Figure 4: Cavitating flow occurrence inside the micro channel for different upstream pressures in smooth channel and channel with wall roughness

P-in (psi)	450	595	775	920	1075
Smooth				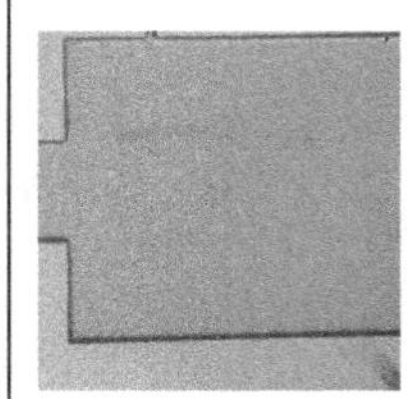	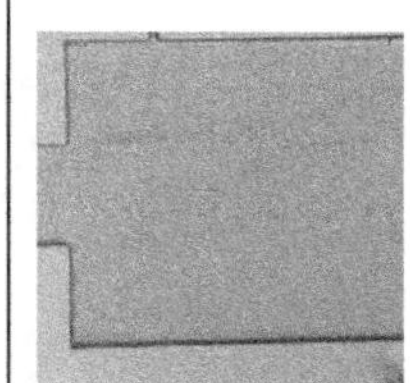
2-1 Side Wall		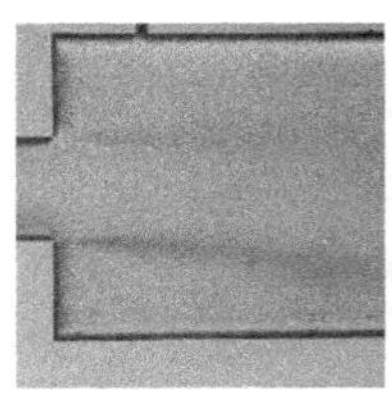	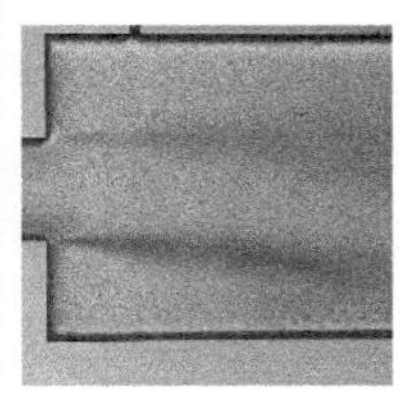	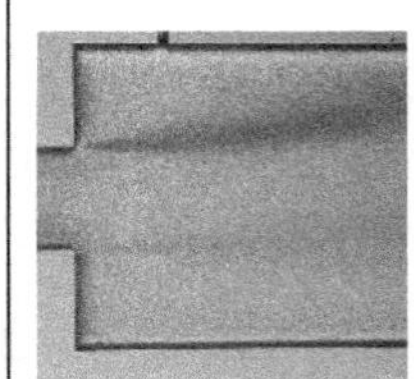	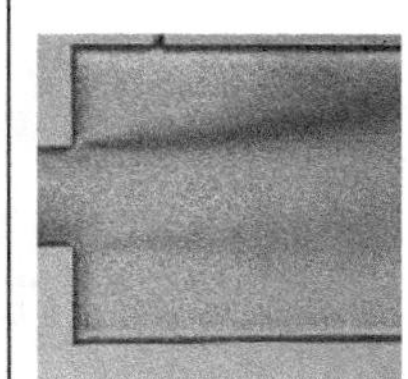

Figure 5: Cavitating flow occurrence inside the extended channel for different upstream pressures in smooth channel and channel with wall roughness

Conclusion

In this study, the cavitation phenomenon inside micro chips was studied. It is shown that the cavitation bubbles are generated inside short micro channels with any geometrical and topological properties at specific pressure drop inside the restrictive element. It is revealed that the side wall roughness has a significant effect on the generation of cavitation bubbles in a shorter period. Therefore, it is possible to reach developed cavitating flows in a side wall roughened micro channel at a lower upstream pressure. Meanwhile, it was proven that the fabricated microchips are durable to very high pressures.

References

[1] Qin, Z., & Alehossein, H. (2016). Heat transfer during cavitation bubble collapse. Applied Thermal Engineering, 105, 1067-1075. doi:10.1016/j.applthermaleng.2016.01.049.

[2] Yang, S., Jaw, S., & Yeh, K. (2009). Single cavitation bubble generation and observation of the bubble collapse flow induced by a pressure wave. Experiments in Fluids, 47(2), 343-355. doi:10.1007/s00348-009-0670-1.

[3] Ghorbani, M., Sadaghiani, A. K., Mohammadi, A., Alcan, G., Unel, M., Koşar, A., & Gozuacik, D. (2018). Characterization and pressure drop correlation for sprays under the effect of micro scale cavitation. Experimental Thermal & Fluid Science, 9189-102. doi:10.1016/j.expthermflusci.2017.10.005.

[4] Ghorbani, M., Alcan, G., Unel, M., Sabanovic, A., Uvet, H., Gozuacik, D., and Kosar, A., 'Visualization of microscale cavitating flow regimes via particle shadow sizing imaging and vision based estimation of the cone angle', Experimental Thermal and Fluid Sciences, Volume 78, 322-333, 2016.

[5] Ghorbani, M., Khalili Sadaghiani, A., Yildiz, M., and Kosar, A., 'Experimental and Numerical Investigations on the Spray Collapse under the Effect of Cavitation Phenomenon in a Microchannel', Journal of Mechanical Sciences and Technology, Volume 31, Issue 1, 235-247, 2017.

[6] Ghorbani, M., Yildiz, M., Gozuacik, D., and Kosar, A., 'Cavitating Nozzle Flow in Micro and Mini Channels under the Effect of Turbulence, Journal of Mechanical Sciences and Technology, Volume 30, Issue 6, 2565-2581, 2016.

[7] Gothsch, T., Richter, C., Beinert, S., Schilcher, C., Schilde, C., Büttgenbach, S., & Kwade, A. (2016). Effect of cavitation on dispersion and emulsification process in high-pressure microsystems (HPMS). Chemical Engineering Science, 144239-248. doi:10.1016/j.ces.2016.01.034.

© 2018 ASME

Identification of the Cavitation Flow Parameters on the Basis of Photofixing of an Experimental Supercavity

[1]Vladimir Serebryakov*

[1]*Institute of Hydromechanics of the National Academy of Sciences, Ukraine*

Abstract

The method of accelerated estimation of supercavitating flow key parameters is proposed for the experimental study of nearly axisymmetric flows. The refinement of the elemental equations controlling the cavity shape with the account of gravity and cavitator lift effects is presented.

Keywords: supercavitation; theoretical and experimental substantiation; axisymmetric flow.

Introduction

Numerous experimental study of nearly axisymmetric supercavitating flows proved a practical independence of the expansion cavity sections in immobile fluid from each other. The effect of free boundary and rigid walls at not too small depth on the cavity shape only slightly changes the initial shape observed in the infinite flow, although the flow boundaries can strongly affect the cavity size and elongation. Moreover, even in case of strong perturbations of the cavity shape within its certain portion, e.g., where a body is planing along the cavity surface, the adjoining cavity surface portion behind the planing zone exhibits very small variations. The intrinsic features of nearly axisymmetric supercavitating flows make it possible to elaborate a procedure for the estimation of the cavitating flow key parameters, including cavitation and Froude numbers, as well as the cavity shape, which would correspond to the infinite flow conditions, by matching some unperturbed part of the cavity observed under test conditions with a quite accurate equation controlling the cavity shape in the infinite flow.

Asymptotic models describing the nearly axisymmetric cavity shape and dimensions

The integro-differential equation (IDE) controlling the shape of a stationary axisymmetric cavity in subsonic flow, which has been derived via the Hydrodynamics of Slender Body approach for the small cavitator model is used as the initial basic equation in the developed technique:

$$\frac{1}{2R^2}\left(\frac{dR^2}{dx}\right)^2+\frac{d^2R^2}{dx^2}\ln\frac{\beta^2R^2}{4x(L-x)}-\int_0^L\frac{\frac{d^2R^2}{dx^2}\Big|_{x=x_1}-\frac{d^2R^2}{dx^2}}{|x_1-x|}dx_1-\frac{\frac{dR^2}{dx}\Big|_{x=0}}{x}+\frac{\frac{dR^2}{dx}\Big|_{x=L}}{L-x}=2\sigma(x), \quad (1)$$

where r, x are polar co-ordinates at $x=0$ in the flow separation section, $r=R(x)$ is the cavity shape, $\sigma=2\Delta P/\rho U_\infty^2$ is the cavitation number, $\Delta P=(P_\infty-P_c)$ is the difference of pressures in the flow P_∞ and cavity P_c, U_∞ is velocity of the incident flow, ρ is fluid mass density, parameter $\beta=\sqrt{1-M}$, where M is the Mach number. The reduction of cavitation numbers to $\sigma\to 0$ leads to the unlimited increase in the cavity dimensions and elongation $\lambda\to\infty$, making the cavitator size negligible with the order of magnitude of $O[(1/\lambda^2)\ln\lambda]$, in comparison with the cavity length. The problem analysis within the framework of MAEM (matched asymptotic expansion method) implies the structure of the singular solution for the cavity shape $R=R(x,\lambda)$ at $1/\lambda\to 0$. This structure includes: (i) an interior non-linear solution for a disk within a small area $O[(1/\lambda^2)\sqrt{\ln\lambda}]$ in the cavitator vicinity, (ii) an intermediate solution for the cavity front part having the well-known asymptotic form for flow expansions in the infinity, which is refined by adding the third term (2a), and (iii) an exterior solution in the form of ellipsoid perturbation (2b).

$$\text{a) } R^2=\frac{2\sqrt{c_{do}}\,x}{\sqrt{\ln x}}\left[1-\frac{1}{4}\frac{\ln\ln x}{\ln x}+\frac{1}{2}\frac{\ln\left(e\sqrt{c_{do}}\beta^2/2\right)}{\ln x}\right],\quad \text{b) } R^2=\frac{1}{\lambda^2}\left[(1-x^2)+\frac{x^2\ln 4-\ln(1+x)^{(1+x)}-\ln(1-x)^{(1-x)}}{\ln(\lambda/\beta)^2}\right], \quad (2)$$

The exterior solution (2a) is derived in the form of two terms of the asymptotic series at $1/\lambda\to 0$. Noteworthy is that the dimensions of zones corresponding to interior and intermediate solutions, including the cavitator size, are neglected with an order of magnitude $O(1/\lambda^2)$. In solution of (2b), $x=0$ corresponds to the cavity midsection with

*Corresponding Author, Vladimir Serebryakov: serebryakov_vladimir@yahoo.com

its normalization by the cavity semi-length L_k. The second-order dependence $\sigma = \sigma(\lambda)$ (3a) is also derived, while the inverse relationship $\lambda = \lambda(\sigma)$ is calculated via successive approximations of equations (3b-3c)

$$\text{a)}\ \sigma = \frac{\ln(\lambda/\beta)^2}{\lambda^2}\left[1 - \frac{1}{\ln(\lambda/\beta)^2}\right],\ \text{b)}\ \sigma = \frac{2\mu}{\lambda^2} \rightarrow \text{c)}\ \mu = \frac{1}{2}\ln\left(\frac{2\mu}{e\beta^2\sigma}\right);\ \text{d)}\ R_k^2 = R_n^2\frac{c_d}{\sigma}\left[1 + 2\frac{\ln 2/\sqrt{e}}{\ln(\lambda/\beta)^2}\right], \tag{3}$$

Matching of intermediate (2a) and exterior (2b) solutions yields the dependence controlling the cavity maximal radius R_k (3d), where $c_d|_{\sigma=0} = c_{do}$ is the cavitator drag coefficient. In view of a slow variation of parameters μ and k in decompositions (3a-3d), the dependences controlling the cavity basic dimensions in the form of (4a-4c) are:

$$\text{a)}\ \lambda^2 = \frac{2\mu}{\sigma};\qquad \text{b)}\ R_k^2 = R_n^2\frac{c_d}{k\sigma};\qquad \text{c)}\ L_k = \frac{R_n}{\sigma}\sqrt{\frac{2c_d\,\mu}{k}},$$

$$\text{d)}\ \mu \approx \ln\sqrt{\frac{1}{e}\left(\frac{\lambda^2}{\beta^2} + 7\right)},\ \mu \approx \ln\sqrt{\frac{1}{e\beta^2\sigma}\ln\left(\frac{2}{\beta^2\sigma} + 10\right)};\quad \text{e)}\ k \approx 1 - \frac{2\ln\left(2/\sqrt{e}\right)}{\ln\left(0.8(\lambda/\beta)^2 + 35\right)},\ k \approx 1 - \frac{2\ln\left(2/\sqrt{e}\right)}{\ln\left(4/\sigma^2 + 18\right)} \tag{4}$$

Based on the non-linear numerical simulation results [1], the dependences for μ and k were refined to the form of (4d-4e) for the case of streamlined flow past cones, including a disk, with maintenance of their asymptotic structure. This refinement improves the accuracy of equations for the cavity basic dimensions (4b-4e) within the range of cavitation numbers $\sigma = 0 \div 0.2$ to the level achieved within the framework of the second-order non-linear theory.

For small values of the inverse elongation parameter $1/\lambda_*$ of the "axisymmetric cavitator – cavity" system, IDE (1) in the limiting case of $1/\lambda_* \rightarrow 0$ is reduced to a differential equation. The latter is refined by coefficients μ and k, while the MAEM matching operation is replaced by introducing the conservation equation of energy transferred from the cavitator to the cavity during its motion in a fluid, which allows one to develop a quite effective approach for calculation of the supercavity flows [4]. Taking into account a small-scale variation of parameters μ and k, it becomes applicable for calculation of various cases of stationary and nonstationary flows. The cavity shape $R = R(x)$ for stationary streamlined flows is controlled by the set of equations (5) with the normalization of x and R by the cavitator radius R_n, which may be classified into two variants: A and B.

Variant **A**

$$\text{a)}\ \frac{d^2R^2}{dx^2} + \frac{\sigma(x)}{\mu_c} = 0,\ \text{b)}\ \frac{\mu_c}{\mu} = \left(\frac{2}{\sqrt{1 - k\sigma/c_d} + 1}\right)^2,$$

$$\text{c)}\ \left.\frac{dR^2}{dx}\right|_{x=0} = \sqrt{\frac{2(c_d - k\sigma)}{k\mu_c}},\ \left.R^2\right|_{x=0} = 1,$$

Variant **B**

$$\text{d)}\ \frac{d^2R^2}{dx^2} + \frac{\sigma(x)}{\mu_m} = 0,\ \text{e)}\ \frac{\mu_m}{\mu_c} = \left(1 - \frac{x_*}{L_m}\right)\left(\frac{1 - \dfrac{k\sigma}{c_d}}{1 - R_*^2\dfrac{k\sigma}{c_d}}\right),$$

$$\text{f)}\ \left.\frac{dR^2}{dx}\right|_{x=x_*} = \sqrt{\frac{2(c_d - k\sigma R_*^2)}{k\mu_m}},\ \left.R^2\right|_{x=x_*} = R_*^2,$$

$$\text{g)}\ x_* \geq e:\ R_*^2 = 2x_*\sqrt{c_{do}/\ln x}\,\Big|_{x=e} = 2e\sqrt{c_{do}} \tag{5}$$

$$\text{a)}\ R^2 = 1 + \sqrt{\frac{2(c_d - k\sigma)}{k\mu_c}}x - \frac{\sigma}{2\mu_m}x^2,\qquad \text{b)}\ R^2 = R_*^2 + \sqrt{\frac{2(c_d - k\sigma R_*^2)}{k\mu_m}}(x - x_*) - \frac{\sigma}{2\mu_m}(x - x_*)^2, \tag{6}$$

$$\text{a)}\ R_k = \sqrt{\frac{c_d}{k\sigma}}\ \text{b)}\ L_m = \frac{1}{\sigma}\sqrt{\frac{2\mu_c(c_d - k\sigma)}{k}},\ \text{c)}\ L_k = \frac{1}{\sigma}\sqrt{\frac{2\mu_c c_d}{k}},\ \text{d)}\ L_c = \frac{R_n}{\sigma}\sqrt{\frac{2\mu_c}{k}}\left(\sqrt{c_d - k\sigma} + \sqrt{c_d}\right) \tag{7}$$

The variant **A** with equations (5a-5c yields the solution (6a) for $\sigma = \text{const}$, which corresponds to an ellipsoidal cavity and provides a sufficient cavity shape prediction accuracy, including its maximal sizes by equations (7): radius R_k, for the cavity portion between its front part and midsection L_m and even for a limited length behind the midsection L_k/ However, it fails to predict the cavity shape in the vicinity of a disk-type cavitator. A slight refinement of parameter μ in the form of Eq. (5b) for μ_c precise cavity sizes to more realistic prediction for the cavitator vicinity zone. An essential refinement of the cavity shape prediction for its medial part (Figure 1), is reached by the application of the set of equations (5d-5g) of variant **B**. This makes it possible to predict the cavity shape starting from section $x \geq x_*$ where the first term of asymptotic dependence (5g) quite accurately controls the cavity shape. The adjustment of parameter μ_c in the form of Eq. (5e) for μ_m (5e) allows one to assess cavity dimensions via Eq. (7).

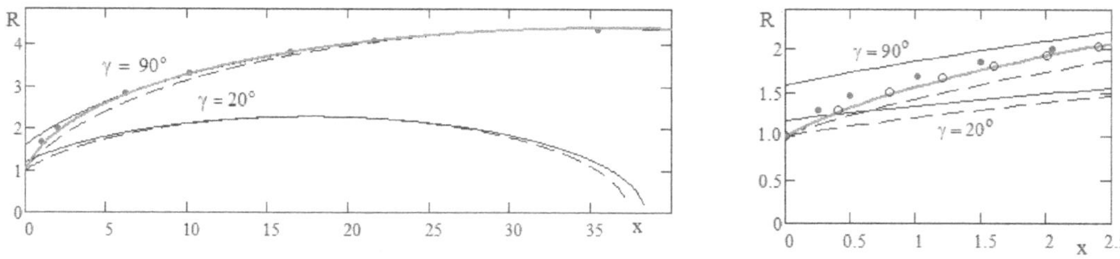

Figure 1. Shape calculation results at $\sigma = 0.05$ for a cavity behind a disk and a cone at $\gamma = 20^o$:

---- - Eq. (6a), ———— - Eq (6b), ▬▬▬ - Eq (8), • • • • - numerical simulation [1], ○ ○ ○ ○ - experiment [2]

More exact calculation of the axisymmetrical cavity shape can be performed via set of equations (9), where Eq.(9a) was derived by adding a symmetrical part to the operator of Eq.(8), which controls the asymptotic dependence (2a). Further this operator is transformed into operator (9a), converting pressure values at the paraboloid, which approximates the cavity front part, into accurate values, allowing one to refine the solution for the cavity front part shape. Finally equation (9a) approximates IDE (1) via the energy conservation equation with an approximate account of energy transfer in the axial direction with help of coefficient k.

$$\frac{1}{2R^2}\left(\frac{dR^2}{dx}\right)^2+\frac{d^2R^2}{dx^2}\ln\frac{\beta^2R^2}{4x^2}-\frac{1}{x}\frac{dR^2}{dx}=0 \tag{8}$$

$$\text{a)}\ \frac{1}{2\left(R^2+R_f^2\right)}\left(\frac{dR^2}{dx}\right)^2+\frac{d^2R^2}{dx^2}\ln\left[\frac{4(x+\Delta)^2(L+\Delta-x)^2}{\beta^2\left(R^2+R_f^2\right)L}\right]-\frac{dR^2}{dx}\left[\frac{1}{x+\Delta}-\frac{1}{L+\Delta-x}\right]=2\sigma,$$

$$\text{b)}\ \left.\frac{dR^2}{dx}\right|_{x=0}=\sqrt{\frac{2(c_d-k\sigma)}{k\ln\left[2\Delta(1+\Delta/L)/\beta\sqrt{1+R_f^2}\right]}},\quad R^2\big|_{x=0}=R_n^2;\quad \text{c)}\ R_f=\sqrt{\frac{c_d}{\ln x_*}-\frac{\kappa_r}{2x_*}},\ \text{d)}\ \Delta=\frac{1}{2}\left(\frac{1}{R_f}+R_f\right), \tag{9}$$

Here R_f is curvature radius of the paraboloid approximating the cavitator generatrix at $\kappa_r = 1$. The calculation results obtained via (9) for the medial and front part of a cavity are depicted in Fig.1.
Dependences for the cavity axial distortion induced by the effects of gravity h_g and cavitator's angle of attack α are represented by the set of equations (10a), governing dependence [2] and the solution in quadratures (10b):

$$\text{a)}\ \frac{d}{dx}\left(R^2\frac{dh}{dx}\right)-\frac{gR^2}{U_\infty^2}=0\quad \left.\frac{dh}{dx}\right|_{x=0}=h'_{xo}\quad h\big|_{x=x_o}=h_o:\ \text{b)}\ h=\frac{g}{U_\infty^2}\int_{x_o}^{x}\frac{1}{R^2}\int_{x_o}^{x}R^2dx+h'_{xo}\int_{x_o}^{x}\frac{R_o^2}{R^2}dx+h_o,\ h'_{xa}\approx\tan\alpha \tag{10}$$

$$\text{a)}\ \frac{h_g}{R_n}=\left(\frac{L_k}{R_n}\right)^2\left(\frac{gR_n}{U_\infty^2}\right)\frac{2}{3}\left\{\ln\frac{2-\bar{a}}{(2-\bar{a})-\bar{x}}+\frac{1}{4}\left[\bar{x}^2-2\bar{x}(1-\bar{a})\right]-\frac{\bar{a}^2}{4}\left(1-\frac{\bar{a}}{3}\right)\ln\frac{(2-\bar{a})(\bar{a}+\bar{x})}{\bar{a}[(2-\bar{a})-\bar{x}]}\right\},\ \bar{x}=\frac{x}{L_k},\ \text{b)}\ \bar{a}=1-\sqrt{1-\frac{k\sigma}{c_d}},$$

$$\text{c)}\ \frac{h_g}{R_n}\approx\left(\frac{L_k}{R_n}\right)^2\frac{gR_n}{U_\infty^2}\left(\frac{(\bar{x}+\bar{a})^{2.2}-\bar{a}^{2.2}}{3.2}\right),\ \text{d)}\ \frac{h_\alpha}{R_n}=\frac{c_{dy}}{2}\frac{L_k}{R_n}\frac{k\sigma}{c_d}\ln\frac{(2-\bar{a})(\bar{x}+\bar{a})}{\bar{a}[(2-\bar{a})-\bar{x}]},\ \text{e)}\ h_{g\alpha}=h_g+h_\alpha, \tag{11}$$

The dependence for the cavity shape $R = R(x)$ in the integral form of (10b) is the most accurately controlled by set of equations (9). In analytical solutions (11), for the flow separation section $x = 0$, dependence (6a) is used. The solution for h_α (11d) holds for the range of cone opening semi-angles $\gamma < 25^o \div 35^o$. For blunter cone- or disk-type cavitators, it is required to use a more accurate dependence $R = R(x)$ to in integral (10) for cavity sections $x < x_*$.

Identification of cavitation number via the experimental cavity profile photos

Given a feeble effect of some perturbations on the cavity shape under test conditions, except for the essential variation of its dimensions, it is expedient to use the adjusted cavitation number σ_a, which controls the cavity shape in the infinite flow. It may differ from that calculated via the test pressure value in cavity P_c as $\sigma = 2(P_\infty - P_c)/\rho U_\infty^2$. Estimate of σ_a can be made by the numerical solution of Eq.(12) using the experimental cavity radius R_a measured at an arbitrary section $x = X_a$. The cavity shape in Eq. (12b) is taken from Eq (6b):

$$\text{a) } \Delta\sigma(x) = -2\frac{gRn}{U_\infty^2}\int_0^x\int_0^x h_{g\alpha}dxdx\,,\ \text{b) } \sigma = \frac{2\mu_m}{(X_a - x_*)^2}\left[\sqrt{\frac{2(c_d - k\sigma R_*^2)}{k\mu_m}}(X_a - x_*) - \left(R_a^2 - R_*^2\right) + \frac{2}{\mu_m}\left(\frac{gR_n}{U_\infty^2}\int_0^{x_a}\int_0^{x_a} h_{g\alpha}dxdx\right)\right] \quad (12)$$

The actual and calculated shapes of an artificial cavity formed behind a cone of 50mm-diameter with opening semi-angle of 20° in a zero angle α_n of attack for motion at depth of 150mm are compared in Figure 2.

a)

b)

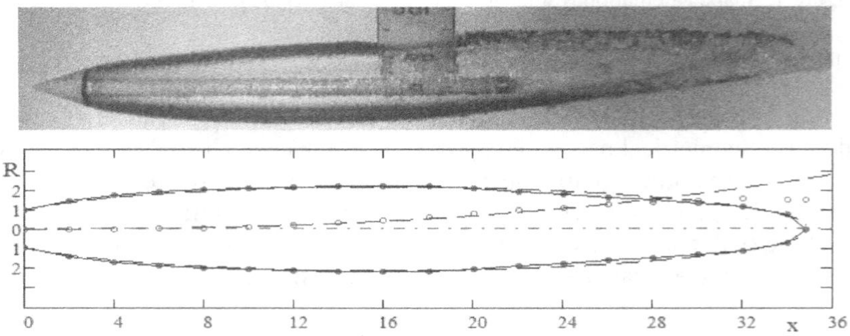

Figure 2. Photo experimental cavity behind a cone with a zero attack angle with calculated cavity contour – Eq (6a), accounting Eq. (12a) and axial distortion of it's axis – Eq.(11c)

$U_\infty = 7.12\text{m/s}$, $Q_v \approx 0.00134\text{m}^3/\text{s}$, $\alpha_n = 0$, $Fr \approx 14.38$: —+—+— Experimental data, - - - - - calculation $\sigma_a = 0.056$, Eq (12)

The adjusted cavitation number $\sigma_a \sim 0.056$ calculated via Eq. (12) used Eq. (6b) from the cavity radius measured at sections $X_a = 10$ differ from that derived in [5] with an account of free boundary effect only by ~5-6 %. However, the proposed method application requires and accurate and reliable measurement of the cavity radius, which requirement becomes more stringent with an approach of the measured section $x = X_a$ to the cavitator. For too small depth more narrow profile of the cavity forward part due to free boundary action as compared to infinite flow and small deformation of circular form of the cavity sections should be accounted additionally.

Conclusion

An essential refinement of the elementary equations of nearly axisymmetric cavities is performed. Based on this refinement, a procedure of accelerated estimate of experimental cavitation numbers and their adjustment for the respective supercavitating motion in the infinite flow is proposed and verified.

Acknowledgements

The author would like to thank Dr. Ronald Joslin, ONR for his support and encouragement.

References

[1] Guzevsky L. G. (1979). *Numerical analysis of cavitation flows.* [In Russian], Preprint No. 40-79 of CO AS USSR, Heat-Physics Institute, Novosibirsk,

[2] Logvinovich G.V. (1969). *Hydrodynamics of flows with free boundaries* [In Russian], Naukova Dumka, Kiev.

[3] Serebryakov V.V, Nechitailo N. (2015) *Hydrodynamics of supercavitating bodies at an angle of attack under conditions of considerable effect of fluid weightiness and closeness of free border*, J. JSOE, V. 5, pp. 255-265.

[4] Serebryakov V.V. (2009). *Physical-mathematical bases of the principle of independence of cavity expansion/*, Proc. CAV2009, paper No.169, Ann Arbor, Michigan, USA.

[5] Serebryakov V.V., Arndt R. E. A., Dzielski J. E. (2015). *Supercavitation: Theory, experiment and scale effects*, Proc. CAV2015, IOP: Conference series V. 656.

© 2018 ASME

Improvement of Wear Resistance of Steel by Means of a Cavitating Jet and a Submerged Pulse Laser

[1]Hitoshi Soyama*; [1]Mitsuru Sato

[1]*Tohoku University, Sendai, Miyagi pref., Japan*

Abstract

Mechanical surface treatment using cavitation impact was proposed and it is called as "cavitation peening". In order to demonstrate peening effect on improvement of wear resistance using cavitation peening, mass loss caused by wear was evaluated experimentally with and without cavitation peening. In the present paper, cavitation was produced by a cavitating jet and a submerged pulse laser. Although it was believed that impact induced by laser ablation was used for peening, impact induced bubble collapse which was generated after laser ablation was larger than that of laser ablation. It was revealed that both peening method using the cavitating jet and submerged pulse laser reduced mass loss due to wear.

Keywords: mechanical surface treatment; jet; pulse laser; wear

Introduction

Cavitation impact at bubble collapse causes severe damage in hydraulic machineries. However, the impact can be utilized for mechanical surface treatment to improve mechanical properties such as fatigue strength and stress corrosion cracking resistance. The mechanical surface modification method using cavitation impact is named as "cavitation shotless peening [1]" as shots are not required, or simply called as "cavitation peening [2]". The improvement of fatigue strength and other effects by cavitation peening have been demonstrated [2]. As one of applicable applications of cavitation peening is improvement of wear resistance, effect of cavitation peening on wear should be investigated.

It was revealed that cavitation peening could introduced compressive residual stress into titanium alloy [3], and improvement of fretting fatigue strength was demonstrated [4]. The enhancement of durability of spinal implant made of Ti-6Al-4V Extra Low Interstitial (ELI) subjected to fretting fatigue was also revealed [6]. As surface roughness by cavitation peening scarcely increased, this characteristic is suitable for improvement of fretting fatigue resistance of titanium alloy. However, effect on improvement of wear resistance of steels by cavitation peening is not clear, as oxidized particles is produced during wear of steels and they affect the wear resistance. Thus, it is necessary to investigate effect of cavitation peening on wear characteristics of steel.

In case of conventional cavitation peening, cavitation was generated by injecting a high speed water jet into water, i.e., a cavitating jet. Even though submerged laser peening, bubble is generated after laser ablation and it generates impact at bubble collapse. In the present paper, the bubble after the laser ablation was named as laser cavitation, and the effect of cavitation peening using the cavitating jet and the laser cavitation was investigated.

In the present paper, in order to make clear effect of cavitation peening on the improvement of wear resistance of steel, chrome molybdenum steel was treated by the cavitating jet and the laser cavitation, and then tested by a wear test.

Experimental Facilities and Procedures

Figure 1 illustrates schematic diagram of submerged laser peening system. The used pulse laser was Nd:YAG laser with Q-switch, whose wave length, the maximum energy, the beam diameter, the pulse width and the repetition frequency were 1,064nm, 0.35 J, 6 mm, 6 ns and 10 Hz. The laser pulse was reflected by mirrors and expanded by a concave lens then focused by convex lenses to specimen which was placed in a water filled chamber made of borosilicate glass. The focal distance of the final convex lens was 100 mm. The impact induced by laser abrasion and laser cavitation were evaluated by a handmade PVDF sensor [6],[7], a submerge shock wave sensor and a hydrophone. As peening intensity was affected by distance from the final convex lens to chamber wall in air s_a and the distance from chamber wall to the specimen surface in water s_w, the distances were optimized by measuring arc height of Almen strip changing with the distances. The specimen was treated by 4 pulse/mm^2 with black coated aluminum tape whose thickness was 80 μm.

*Corresponding Author, Hitoshi Soyama: soyama@mm.mech.tohoku.ac.jp

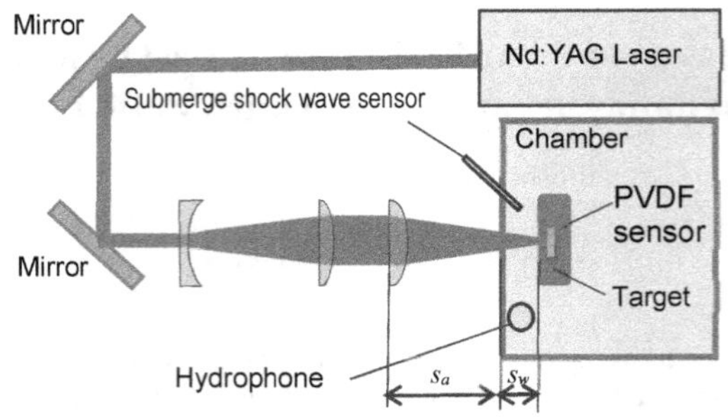

Fig. 1 Schematic diagram of submerged laser peening system

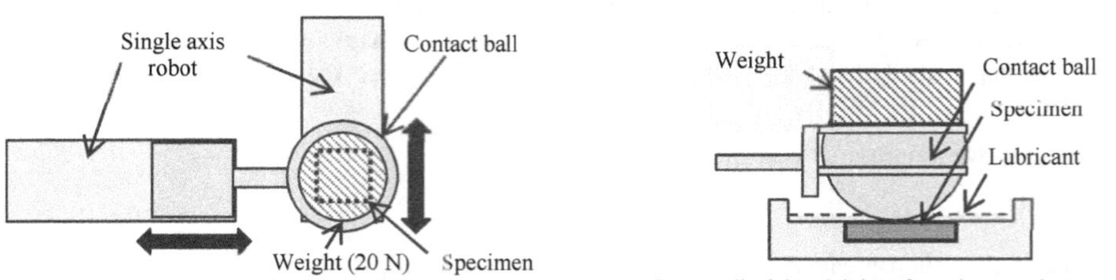

(a) Top view of wear testing machine (b) Detail of the vicinity of specimen and contact ball

Fig. 2 Schematic diagram of wear testing machine

The specimen was also treated by a cavitating jet. In the present paper, the cavitating jet in air [8] was used for cavitation peening. The injection pressure of a high speed water jet and a low speed water jet were 30 MPa and 0.05MPa, respectively. The inner diameter of the high speed water jet and the low speed water jet were 1 mm and 30 MPa, standoff distance was 45 mm. The processing time per unit length tp was defined by the flowing equation.

$$t_p = \frac{n}{v} \tag{1}$$

Here, n was number of scan and v was scanning speed.

Figure 2 shows schematic diagram of the wear testing machine. The diameter of the contact ball made of Japanese Industrial Standards JIS SUJ2 was 63.5 mm. The ball and the specimen were moved by two single axis robots in order to scan the specimen in the area of 20 mm × 20 mm. Each time of stroke was about 1.0 s and 1.3 s, respectively. Tested material was chrome molybdenum steel JIS SCM435 with vacuum oil quenching and tempering in order to get $H_{RC} = 49$. The original surface of specimen was finished by grinding.

Results

Figure 3 illustrates aspect of submerged laser peening. The laser pulse was irradiated to the target at $t = 0$ ms, and it produced the laser ablation. After the laser ablation, the bubble was generated and developed, then collapsed at $t = 1$ ms. After that, the bubble developed again and then collapsed at $t = 1.53$ ms.

Figure 4 reveals the output signals from the PVDF sensor, the hydrophone and the submerged shock wave sensor. In the case of the PVDF sensor, the maximum amplitude at $t \approx 0$ ms was 1.4 V and that at $t \approx 1$ ms was 1.96 V. And the amplitude at $t \approx 1.53$ ms had a maximum due to 2nd collapse of the laser cavitation. It means that the impact passing through the target induced by 1st collapse of the laser cavitation was 1.4 times larger than that of laser ablation. In the previous report, the cavitation aggressive intensity is increased with the square of impact force measure by the PVDF sensor [6],[7], the impact energy induced by the collapse of the laser cavitation is two time larger than that of the laser ablation. However, in the case of the pressure wave propagating in water detected by the hydrophone and the submerged shock wave sensor, the amplitude of laser ablation was larger or equivalent to that induced by laser cavitation. Note that the PVDF sensor only detected 2nd collapse. This tendency is very similar to that of the submerged laser peening using the pulse laser of 532 nm in wave length [2].

Figure 5 illustrates the relation between the mass loss Δm and the total sliding length L_t obtained from the wear test.

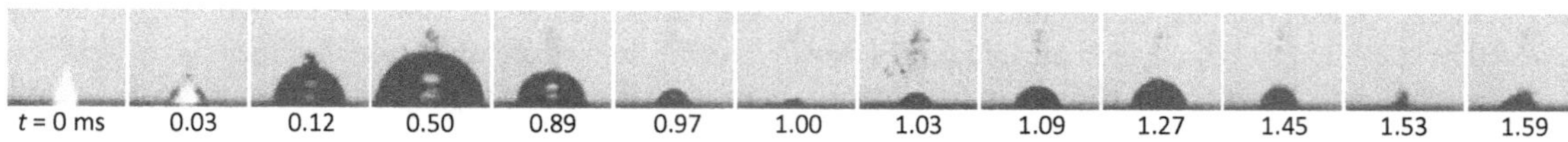

Fig. 3 Aspect of laser ablation and laser cavitation

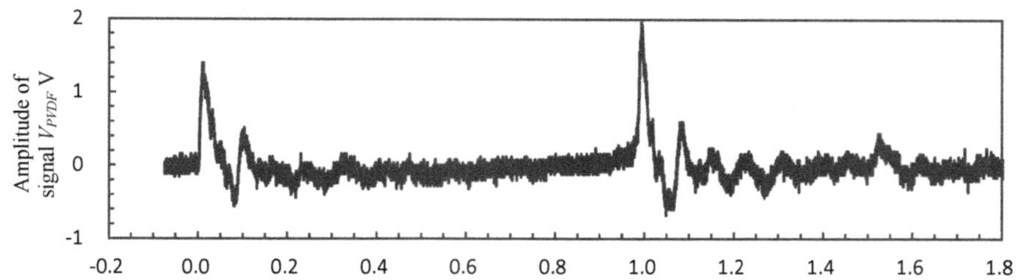

(a) PVDF sensor

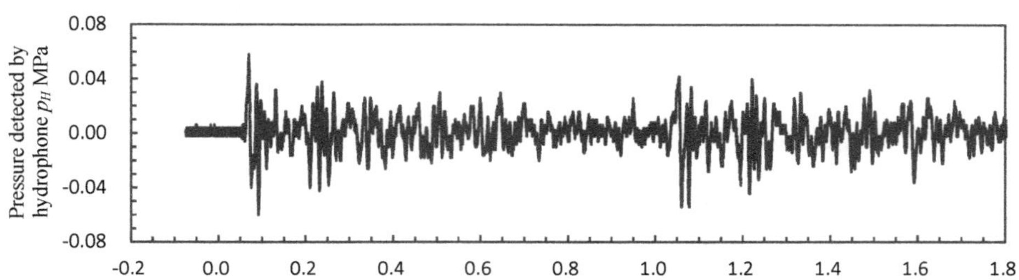

(b) Hydrophone

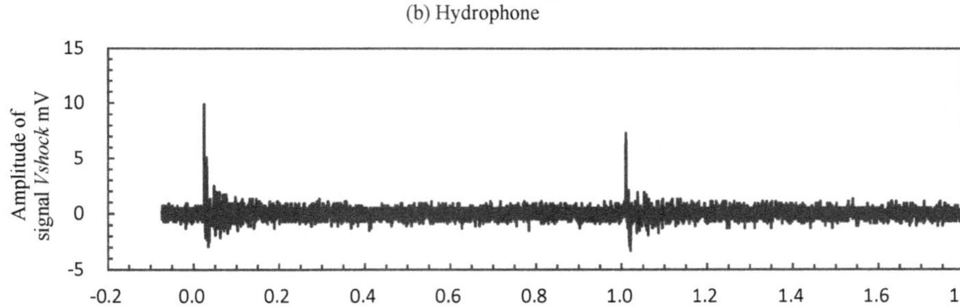

(c) Submerge shock wave sensor

Fig. 4 Output signal from sensors.

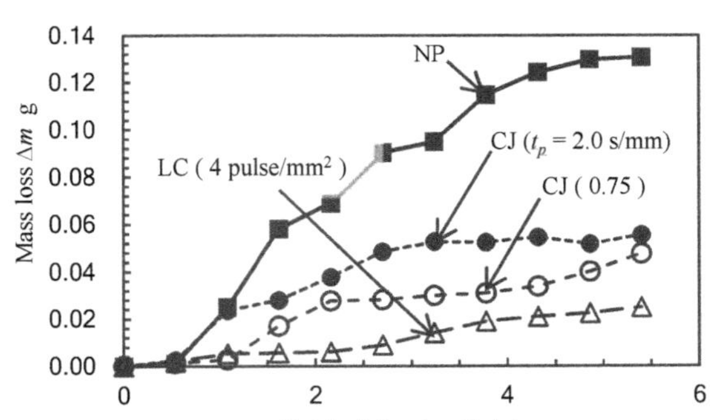

Fig. 5 Mass loss caused by wear as a function of total sliding length

In Fig. 5, "NP" mean not-peened specimen. The results of cavitation peening by using the cavitating jet in air was revealed as "CJ", and that of the laser cavitation was shown as "LC". In the case of CJ, the specimens were treated by t_p = 0.75 s/mm and 2 s/mm. For all cases, after the certain incubation period, in which the mass loss was scarcely observed, the mass loss was increased with the total sliding length. At L_t = 5.4 km, Δm of NP, CP (t_p = 0.75 s/mm), CP (2 s/mm) and LP were 0.131 g, 0.048 g, 0.055 g and 0.025 g, respectively. Both cavitation peening using the cavitating jet in air and the laser cavitation reduced Δm. Namely, both cavitation peening can improve the wear resistance of steel. In the present experiment, the mechanical surface treatment by using the laser cavitation reduced Δm about 1/5 comparing with non-peened specimen. When Δm of CP treated at t_p = 0.75 s/mm and 2 s/mm were compared, CP treated at t_p = 0.75 s/mm was better than that of 2 s/mm.

Conclusions

In order to demonstrate the improvement of wear resistance of steel by cavitation peening, chrome molybdenum steel was treated by the cavitating jet in air and the laser cavitation, and tested by the wear testing machine. It was revealed that both cavitation peening of the cavitating jet in air and laser cavitation can improve wear resistance of steel. Note that, in the case of the submerged laser peening, when the impact through the target was measured by the PVDF sensor, the impact at collapse of the laser cavitation was 1.4 times larger than that of the laser ablation.

This work was partly supported by JSPS KAKENHI Grant Number 17H03138. The author thanks Mr. M. Mikami, technician, Tohoku University for his help with the experiment.

References

[1] Soyama, H., Saito, K., Saka, M. (2002). *Improvement of fatigue strength of aluminum alloy by cavitation shotless peening*, Trans. ASME, Journal of Engineering Materials and Technology, 124 (2), pp. 135–139.
[2] Soyama, H. (2017). *Key factors and applications of cavitation peening*, International Journal of Peening Science and Technology, 1 (1), pp. 3–60.
[3] Soyama, H., Macodiyo, D.O., Mall, S. (2004). *Compressive residual stress into titanium alloy using cavitation shotless peening method*, Tribology Letters, 17 (3), pp. 501–504.
[4] Lee, H., Mall, S., Soyama, H. (2009). *Fretting fatigue behavior of cavitation shotless peened Ti-6Al-4V*, Tribology Letters, 36 (2), pp. 89–94.
[5] Takakuwa, O., Nakai, M., Narita, K., Niinomi, M., Hasegawa, K., Soyama, H. (2016). *Enhancing the durability of spinal implant fixture applications made of Ti-6Al-4V ELI by means of cavitation peening*, International Journal of Fatigue, 92 (1), pp. 360–367.
[6] Soyama, H., Lichtarowicz, A., Momma, T., Williams, E. J. (1998). *A new calibration method for dynamically loaded transducers and its application to cavitation impact measurement*, Journal of Fluids Engineering, 120 (4), pp. 712–718.
[7] Soyama, H., Sekine, Y., Saito, K. (2011). *Evaluation of the enhanced cavitation impact energy using a PVDF transducer with an acrylic resin backing*, Measurement, 44, pp. 1279–1283.
[8] Soyama, H., Kikuchi, T., Nishikawa, M., Takakuwa, O. (2011). *Introduction of compressive residual stress into stainless steel by employing a cavitating jet in air*, Surface & Coatings Technology, 205 (10), pp. 3167–3174.

© 2018 ASME

Increasing Cavitation around Dental Ultrasonic Scalers to Improve Biofilm Removal Efficiency: A High Speed Imaging and Image Analysis Study

[1]Nina Vyas, [2]Qianxi Wang, [1]A. Damien Walmsley*;

[1]*School of Dentistry, University of Birmingham, UK;* [2]*School of Mathematics, University of Birmingham, UK*

Abstract

Dental ultrasonic scalers are used to remove plaque and mineralised deposits from teeth. Cavitation occurs in the cooling water flowing over the tip and may be used to enhance removal of the biofilm. Our research shows that cavitation occurs around the bend of the scaler tip and at its free end. By increasing the amount of cavitation, the instrument can potentially be used in a novel non-contact mode. The metal tip of the scaler will not touch the tooth surface allowing the cavitation to undertake the removal. This non-touch cleaning will reduce the present damage on both teeth and dental implants, which may be damaged by contact with the metal scaler tip. The aim of this study was to modify an ultrasonic scaler tip and image it with a high speed camera to determine the increase in cavitation. An ultrasonic scaler (P5 XS Satelec, Acteon) operating at 29 kHz at the medium power setting was imaged with tip 10P using a high speed camera (Shimadzu HPV1) at 1,000,000 fps. Imaging was done with the tip immersed in distilled water preventing spray from the scaler. The outer edge of the free end of the tip was ground and polished manually (DAP-7, Struers, Ballerup, Denmark) using P500 and P4000 SiC paper until horizontal. These alterations were made to increase the pressure drag, which would create lower pressures and consequently increase the cavitation. Image segmentation was done using a semi-automatic machine learning approach to calculate the area of the cavitation occurring at the free end of the original and modified tips. The mean area of cavitation around each tip was calculated from 102 images. There was a statistically significant difference between the amount of cavitation occurring around the modified and original ultrasonic scaler tip. The area of cavitation around tip 10P was 0.27 ± 0.05 μm^2 and the area of cavitation around the modified tip was 0.35 ± 0.07 μm^2. High speed imaging also showed that there was cavitation present at the point of the free end in the modified tip but not in the original tip. This will be clinically useful as the point of the tip is the part closest to the tooth surface when in use. These results propose that cavitation from an ultrasonic scaler may be used to remove dental plaque without damaging teeth. This method also has potential to be used to clean dental implants.

Keywords: ultrasonic cleaning, biofilm disruption, cavitation, ultrasonic scalers

Introduction

Dental plaque is a biofilm, consisting of many bacterial species inside a viscous substance known as the extra cellular polymeric substance (EPS), which facilitates the adhesion of the bacteria to all surfaces of the mouth [1]. This biofilm must be removed from teeth and dental implants to prevent disease, although it typically cannot be removed effectively from such surfaces due to their micro- and nano-roughness [2]. The use of ultrasonic cavitation is being investigated as a more effective method of biofilm removal. The shock waves, high velocity micro-jets, and streaming generated by collapsing bubbles are able to disrupt the biofilm structure at the micron scale and to aid in its removal from surfaces [3].

Cavitation occurs around the metal tips of dental ultrasonic scalers, used by clinicians during cleaning of teeth [4]. Ultrasonic scaler tips vibrate at frequencies of approximately 30 kHz, and the vibrating tip is used to dislodge mineralized debris on teeth. Cooling water is passed over the tip to prevent frictional heating, and it has been speculated that cavitation occurs in this cooling water. However cavitation is not currently used as the main method of biofilm removal, and the force of the vibrating metal tip can damage the tooth surface and dental implants. There are currently no suitable methods for effectively removing biofilm from dental implants, therefore the cavitation around ultrasonic scalers is being researched as a more effective and less damaging method of removing biofilm from teeth and dental implants. We have shown using high speed imaging that cavitation occurs around different shaped ultrasonic scaler tips, and that it increases with power and tip vibration amplitude [5]. We also established that that

*Corresponding Author, Professor Damien Walmsley: a.d.walmsley@bham.ac.uk

the cavitation occurred in clusters on the tip, therefore increasing the cavitation area may be beneficial by allowing the instrument to be used in a non-touch mode. The aim of this study was to modify the shape of an ultrasonic scaler tip to increase the amount of cavitation generated around it. We use ultra-high speed imaging and image analysis to calculate the amount of cavitation occurring around the original and modified tips.

Methods

An ultrasonic scaler (P5 Newtron, Satelec, Acteon, France) was imaged with tip 10P at medium power (power setting 10) using a high speed camera operating at 1 million frames per second (fps) (HPV1, Shimadzu Corporation, Japan). The camera was attached to a zoom lens (Monozoom 7, Leica Microsystems UK Ltd) to obtain a resolution of 15.5 µm/pixel. Illumination was provided by two strobe lights which were synchronised with the camera using a flash light controller, delay generator and trigger switch. The scaler was positioned using a translation stage (PT3, Thorlabs, USA). The scaler tip was imaged in a custom-made glass container with a total volume of 10 ml (Figure 1). The container was made by cutting glass microscope slides to 2.7 x 2.7 mm and attaching 5 squares to each other using glass adhesive (Loctite, USA) to create an open cube. The scaler tip was submerged in the container in 10 ml reverse osmosis water at 20.5°C. A modified version of Tip 10P was also imaged using the same conditions. The outer edge of the free end of the tip was ground and polished manually (DAP-7, Struers, Ballerup, Denmark) using P500 and P4000 SiC paper until horizontal (Figure 2). These alterations were made to increase the pressure drag, which would create lower pressures and consequently increase the cavitation.

All image analysis was done using Fiji (distribution of the ImageJ software, US National Institutes of Health, Bethesda, Maryland, USA)[6]. Analysis was done using high speed videos containing 102 images for each tip. The high speed images were first cropped to remove any background artifacts (Figure 3a-d). The cavitation bubbles were then segmented using the Trainable Weka Segmentation plugin (Figure 3e,f)[7]. Some parts were falsely segmented, for example background noise or reflection from the scaler. Larger reflections were removed from the segmentation by manually replacing pixels with pixels of value 0 (Figure 3c,d). Any smaller background noise was removed with the Analyse Particles plugin, where objects smaller than 4 pixels were removed (Figure 3 g,h). The histogram of each image in the stack was calculated to find the area of cavitation. The mean cavitation area and standard deviation were plotted and the Mann-Whitney Rank Sum Test was used for testing for statistical significance. Data graphing and statistical analysis were performed using SigmaPlot 12.5 (Systat Software, USA).

Results

We present a high speed imaging and image analysis study to show how cavitation around dental ultrasonic scalers can be increased by changing the tip shape. We have tracked the position of the cavitation with high spatial and temporal resolution using an ultra-high speed camera and image processing of the data. This allows us to perform quantitative analysis to calculate the amount of cavitation. The high speed images show that cavitation occurred around both tips, at the front and the back of the tip. There was a statistically significant difference between the amount of cavitation occurring around the modified and original ultrasonic scaler tips ($p < 0.001$) where p is the probability value. The area of cavitation around the original tip 10P was 0.27±0.05 µm^2 and the area of cavitation around the modified tip was 0.35±0.07 µm^2 (Figure 4). High speed imaging also showed that there was cavitation present at the point of the free end in the modified tip but not in the original tip.

Figures & Tables:

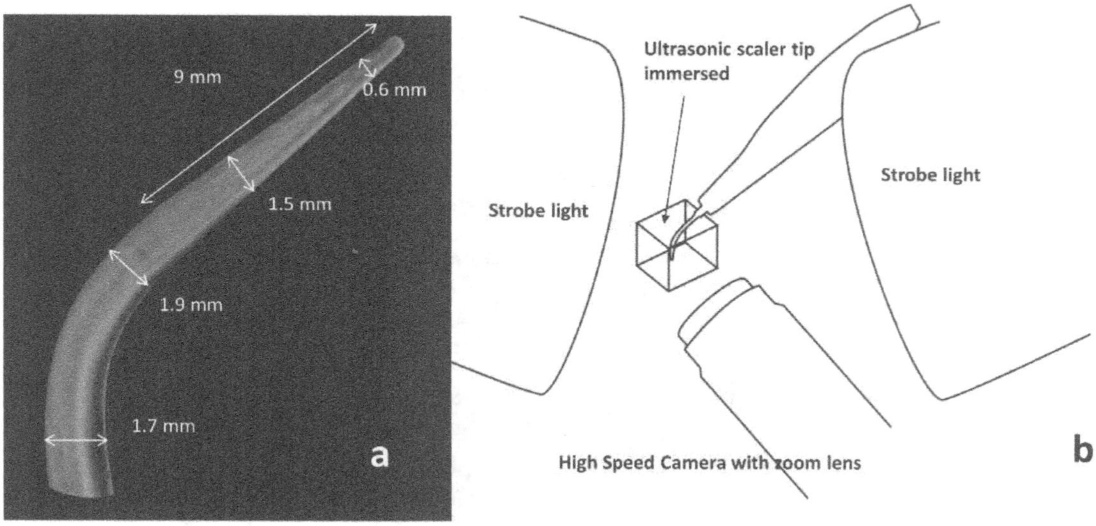

Figure 1: (a) X-ray micro-computed tomography reconstruction of the ultrasonic scaler tip used in this study, showing tip dimensions. (b) Schematic of the experiment setup for high speed imaging.

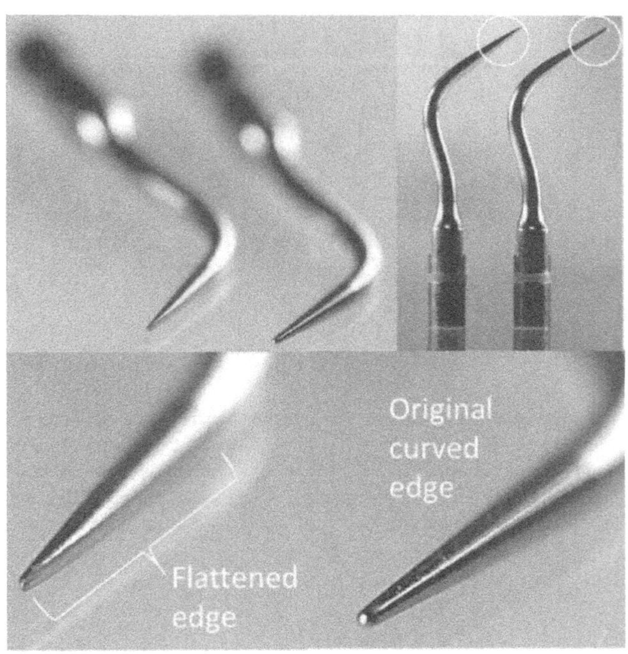

Figure 2: Photographs of the original and modified versions of the ultrasonic scaler tip used in this study.

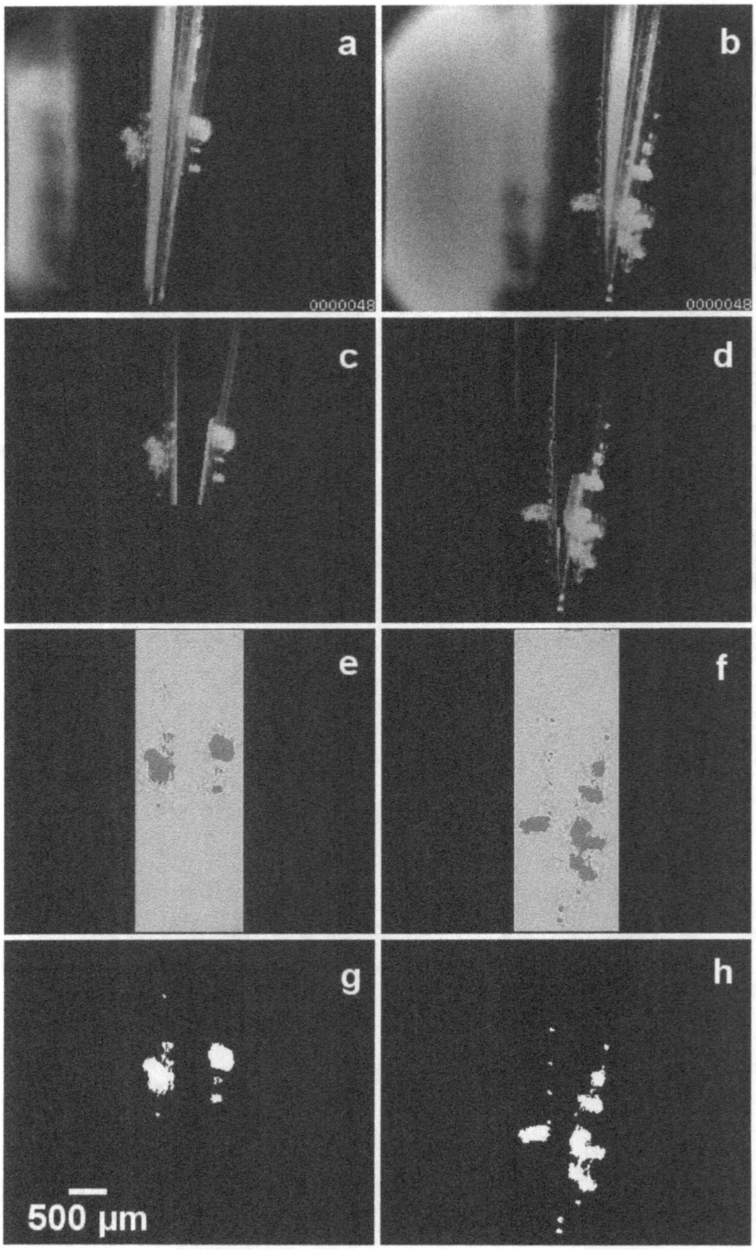

Figure 3: Series of images showing the image analysis process used to calculate the area of cavitation occurring around the tips. (a) High speed image still of tip 10P. (b) High speed image still of modified tip 10P. (c) Image still after pre-processing for segmentation – reflections from the scaler have been manually removed by setting certain pixel values to 0. (d) Corresponding pre-processed image for the modified tip. (e) Segmented image of (c) using the Trainable Weka Segmentation plugin. (f) Segmented image of (d) using the Trainable Weka Segmentation Plugin. (g, h) Segmented images after post-processing, used to calculate the cavitation area.

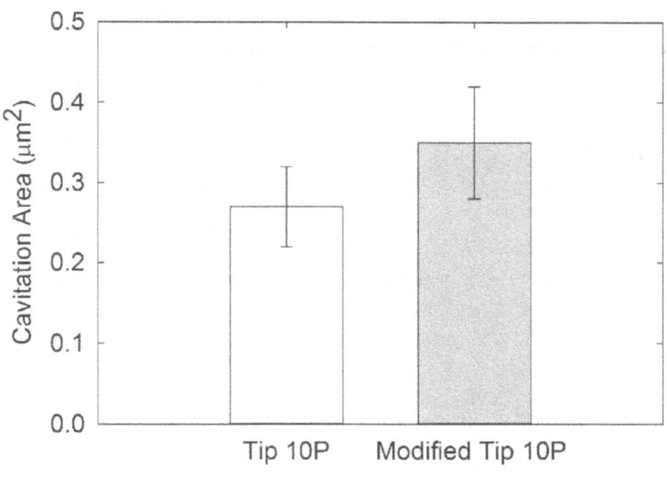

Figure 4: Graph showing the area of cavitation occurring around tip 10P and the modified version of tip 10P, calculated using image analysis from the set of high speed images taken at 1 million frames per second. There was a significant difference in the amount of cavitation between the two tips ($p < 0.001$).

Conclusions

We have used ultra-high speed imaging and quantitative image analysis to demonstrate that a novel tip modification can enhance the amount of cavitation bubbles occurring around dental ultrasonic scaler tips. We speculate that this advancement can increase the amount of biofilm removal as well as reducing the cleaning time. It can therefore help to enable ultrasonic scalers to be used in a non-contact mode, allowing teeth and dental implants to be cleaned more effectively with cavitation without causing damage to their surfaces.

Acknowledgments

The authors acknowledge support from the Engineering and Physical Sciences Research Council (EP/P015743/1).

References

1. Jacob, M., (2006). *Biofilms, a new approach to the microbiology of dental plaque.* Odontology. 94: p. 1–9.
2. Mann, M., et al., (2012). *Effect of plastic-covered ultrasonic scalers on titanium implant surfaces.* Clinical Oral Implants Research. 23(1): p. 76-82.
3. Erriu, M., et al., (2014). *Microbial biofilm modulation by ultrasound: Current concepts and controversies.* Ultrasonics Sonochemistry. 21: p. 15-22.
4. Felver, B., et al., (2009). *Cavitation occurrence around ultrasonic dental scalers.* Ultrasonics Sonochemistry. 16: p. 692-697.
5. Vyas, N., et al., (2016). *High Speed Imaging of Cavitation around Dental Ultrasonic Scaler Tips.* PLoS One. 11(3): p. e0149804.
6. Schindelin, J., et al., (2012). *Fiji: an open-source platform for biological-image analysis.* Nature Methods. 9(7): p. 676-682.
7. Arganda-Carreras, I., et al., (2017). *Trainable Weka Segmentation: a machine learning tool for microscopy pixel classification.* Bioinformatics: p. btx180.

© 2018 ASME

Influence of Collapse of Cavitation Bubble Cloud on Erosion of Solid Surface in Hydraulic Machinery

[1]Kohei Okita*; [2]Yuusuke Miyamoto; [2]Teruyuki Furukawa; [2]Shinobu Nagura; [3]Shu Takagi; [3]Hiroharu Kato;

[1]Nihon University, Chiba, Japan; [2]KOMATSU, Tochigi, Japan, [3]The University of Tokyo, Tokyo, Japan

Abstract

Collapse of oil cavitation bubble clouds near the solid surface was performed by the numerical simulation, in which fluid-structure interaction and sub-grid scale bubble dynamics were taken into account. The effect of initial void fraction, initial bubble size and initial distribution of the cloud on the stress in the solid were examined. As the result, the collapse of the attached bubble cloud on the surface with high initial void fraction and small bubble size causes high stress in the solid.

Keywords: cavitation; erosion; oil flows; numerical simulation; fluid-structure interaction;

Introduction

Construction machinery employs various types of hydraulic equipment for transmitting power by fluid. Cavitation is inevitable in hydraulic equipment where high-pressurized fluid flows through a narrow channel to low pressure region. Although cavitation itself does not disturb the work of construction machinery, cavitation noise and erosion due to the collapse of cavitation bubbles are problems to be solved [1]. Cavitation erosion in the control valve developing in long term operation of construction machinery decreases the controllability of flow in the valve. Thus, the quantitative prediction of cavitation erosion is desired in design phase by a numerical simulation.
In the present study, cavitation erosion in oil flows through a control valve was investigated as shown in Fig. 1. The high speed jet flow from the holes on the high pressure side generates cavitation bubbles. The periodic collapse of bubble cloud in the holes on the block causes the erosion of the block surface. Thus, the collapse of cavitation bubble cloud on the solid surface as shown in Fig. 2 was performed numerically with considering the fluid-structure interaction. The effect of the collapse of cavitation bubble clouds on the erosion of solid surface was discussed.

Basic Equations

To reproduce cavitation bubble behaviors, the dynamics of spherical bubble was considered by the Keller equation [2] as

$$\rho_L\left[\left(1-\frac{\dot{R}}{c_{Ls}}\right)R\ddot{R}+\frac{3}{2}\left(1-\frac{\dot{R}}{3c_{Ls}}\right)\dot{R}^2\right]=\left(1+\frac{\dot{R}}{c_{Ls}}+\frac{R}{c_{Ls}}\frac{d}{dt}\right)\left[P_L(R,t)-p(t)\right], \quad (1)$$

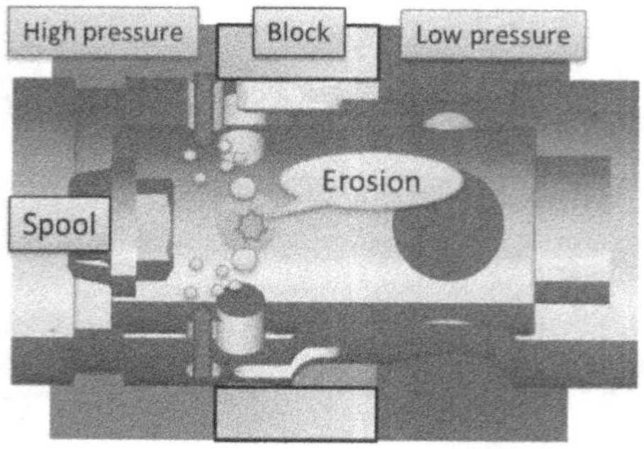

Figure 1 Oil cavitating flow in a control valve.

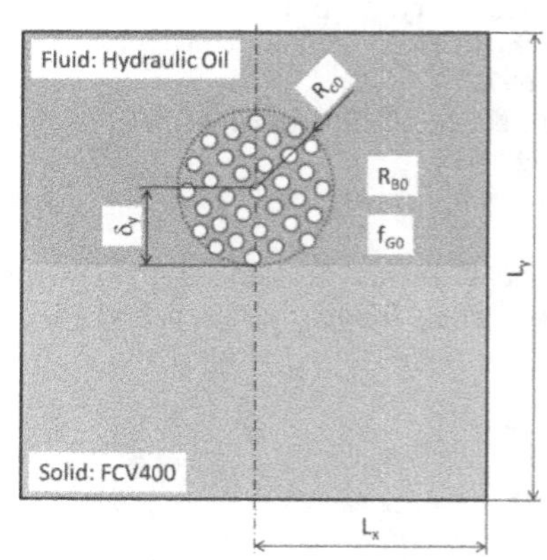

Figure 2 Numerical model for the collapse of cavitation bubble cloud on solid surface.

*Corresponding Author, Kohei Okita: okita.kohei@nihon-u.ac.jp where ρ_L and c_{Ls} are liquid density and liquid sound speed, respectively. The difference between the pressure on bubble interface P_L and the surrounding pressure p can be described by using surface tension σ and liquid viscosity μ_L as

$$P_L(R,t) - p(t) = p_G - \frac{2\sigma}{R}\left[1 - \left(\frac{R_{\min}}{R}\right)^2\right] - \frac{4\mu_L}{R}\dot{R} - p(t). \tag{2}$$

In the second term of right hand side, the reduction of surface tension was introduced to prevent the disappearance of bubble nuclei in consideration of the surfactant effect. The development of the gas pressure p_G inside bubble is described as

$$\frac{dp_G}{dt} = -3\gamma p_G \frac{1}{R}\frac{dR}{dt} + \frac{3(\gamma-1)}{R}K\left.\frac{\partial T}{\partial r}\right|_{r=R} + p_G\left(\gamma - 1 + \frac{T_i}{T}\right)\frac{1}{M_G}4\pi R^2 D_d \left.\frac{\partial C}{\partial r}\right|_{r=R}. \tag{3}$$

In the right hand side of Eq. (3), the second term is due to heat transfer through the bubble interface. The temperature gradient inside bubble is obtained by the reduced-order model [3]. The third term is due to mass transfer. Because of the high solubility of gas into oil, the mass transfer of non-condensable gas through bubble interface is important role in the growth and collapse of cavitation bubbles. The development of the non-condensable gas concentration in liquid surrounding bubble is solved to obtain the mass transfer rate [4].

Behaviour of each bubble in sub-grid scale is taken into account by a bubbly flow model, and the bubbly mixture phase is also coupled with solid phase to investigate the influences of the impulsive stress due to the violent collapse of cavitation bubble cloud on the erosion of solid [5]. The density of bubbly fluid can be assumed as $\rho_F = (1-f_G)\rho_L$ using the volume fraction of gas f_G, because the gas density is much lower than the liquid density ρ_L. Then, the mixture density of bubbly fluid and solid is described as $\rho = (1-\phi_S)\rho_F + \phi_S\rho_S$, where ϕ_S and ρ_S are volume fraction and density of solid, respectively. The translational velocity of bubble is assumed to be same as the liquid velocity. The momentum equation for the mixture can be represented by

$$\rho\frac{D\mathbf{v}}{Dt} = \nabla\cdot\boldsymbol{\sigma}, \tag{4}$$

where the stress $\boldsymbol{\sigma}$ includes elastic stress $\boldsymbol{\sigma}_E$ and viscous stress $\boldsymbol{\sigma}_V$ ($\boldsymbol{\sigma} = \boldsymbol{\sigma}_E + \boldsymbol{\sigma}_V$). Assuming the small strain, Hook's low for isotropic linear elasticity is employed for the elastic strain. The time development of elastic stress and the viscous stress can be described as

$$\frac{D\boldsymbol{\sigma}_E}{Dt} = -\frac{Dp}{Dt}\mathbf{I} + 2\mu_E\left(\mathbf{e} - \frac{1}{3}\mathrm{tr}(\mathbf{e})\mathbf{I}\right) \text{ and } \boldsymbol{\sigma}_V = \zeta_V\,\mathrm{tr}(\mathbf{e})\mathbf{I} + 2\mu_V\left[\mathbf{e} - \frac{1}{3}\mathrm{tr}(\mathbf{e})\mathbf{I}\right], \tag{5}$$

where μ_E is shear modulus, ζ_V and μ_V are bulk and shear viscosity. The total strain rate $\mathbf{e}$ is defined as $\mathbf{e} = \{\nabla\mathrm{v} + (\nabla\mathrm{v})^{\mathrm{T}}\}/2$. The total strain rate includes elastic strain rate $\mathbf{e}_{\mathrm{E}}$, thermal volumetric strain rate $\mathbf{e}_{\mathrm{T}}$ and volumetric strain rate due to bubble oscillation $\mathbf{e}_{\mathrm{B}}$ ($\mathbf{e} = \mathbf{e}_{\mathrm{E}} + \mathbf{e}_{\mathrm{T}} + \mathbf{e}_{\mathrm{B}}$). The thermal volumetric strain rate $\mathbf{e}_{\mathrm{T}}$ can be described using the thermal expansion rate at constant pressure $\beta = (1/V)(\partial V/\partial T)_p$. The volume change due to bubble oscillation $\mathbf{e}_{\mathrm{B}}$ is described using the volume fraction of bubbles f_G. Thus, the time development of pressure due to the bulk elasticity can be described by

$$\frac{Dp}{Dt} = -\zeta_E\left[\mathrm{tr}(\mathbf{e}) - \frac{\beta}{\rho C_p}\left[\nabla\cdot(k\nabla T) + Q_{Vis} + Q_{Bub}\right] - \frac{1}{1-f_G}\frac{Df_G}{Dt}\right], \tag{6}$$

where ζ_E is bulk modulus, Q_{Vis} and Q_{Bub} are heat due to viscous dissipation and the heat due to oscillation of bubbles, respectively. The temperature T is obtained by solving the energy equation as

$$\rho C_p\frac{DT}{Dt} = \beta T\frac{Dp}{Dt} + \nabla\cdot(k\nabla T) + Q_{Vis} + Q_{Bub}. \tag{7}$$

The above equations were discretized by the finite difference method and temporally developed on the basis of semi-implicit scheme. The fluid was coupled with discrete bubbles by using Euler-Lagrange method.

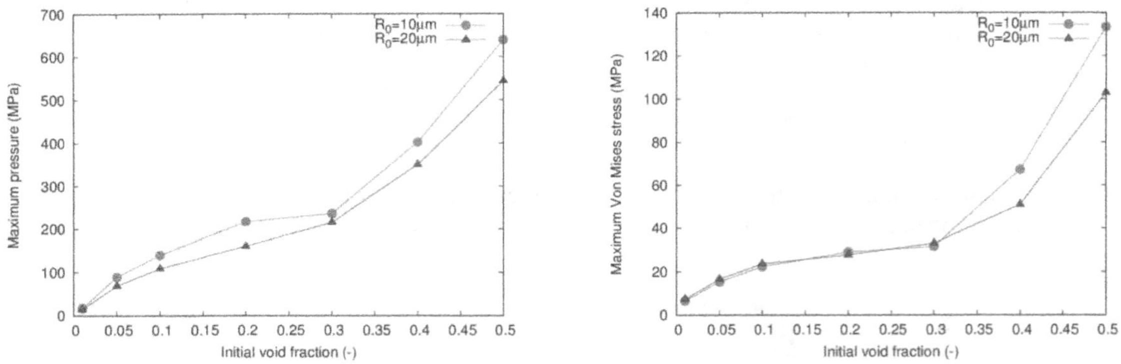

(a) Pressure in fluid (b) Von Mises stress in solid

Figure 3 Effect of initial void fraction and initial bubble radius in the bubble cloud on the maximum pressure and maximum Von Mises stress during the collapse of the bubble cloud ($\delta y = R_{c0}$= 1 mm).

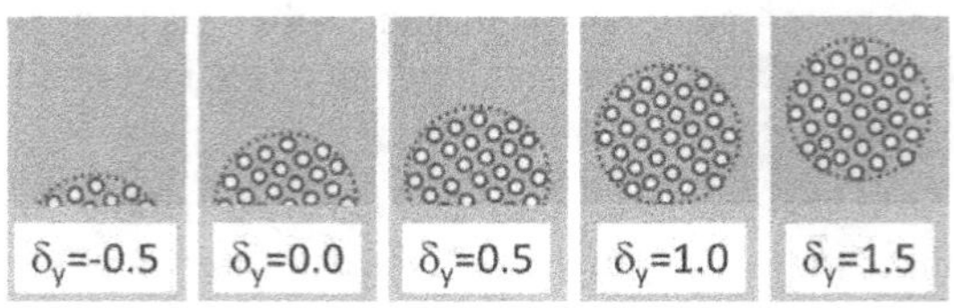

Figure 4 Initial distributions of bubble cloud for various distance from the solid surface to the center of the cloud δy (mm).

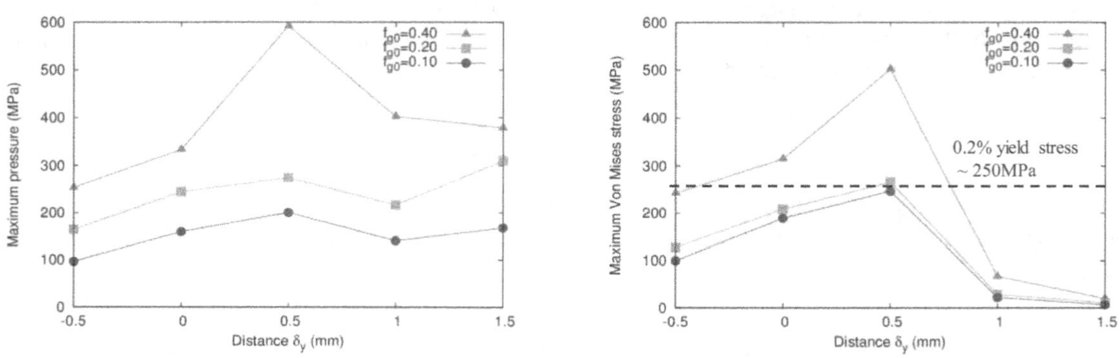

(a) Pressure in fluid (b) Von Mises stress in solid

Figure 5 Effect of initial distribution of bubble cloud on the maximum pressure and the maximum Von Mises stress during the collapse of bubble cloud for various initial void fractions (R_{B0} = 10 μm).

Results and Discussions

A collapse of spherical cavitation bubble cloud on a solid surface was considered (Fig. 2). The bubble cloud includes bubbles with initial radius R_{B0} at initial void fraction f_{G0}. The distance from the surface to the centre of the cloud is δy. The two-dimensional cylindrical coordinate was employed with assuming asymmetry. The numerical domain size is $L_x \times L_y$ =10 × 10 mm. The pressure at the upper boundary of the domain increases by 1 MPa higher than initial pressure at t = 10 μs, which causes the collapse of bubble cloud.

Figure 3 shows the effects of the initial void fraction and initial bubble radius on the maximum pressure in the fluid and the maximum Von Mises stress in the solid during the collapse of bubble cloud. As the initial void fraction increases, both pressure and Von Mises stress increase. The smaller initial bubble radius causes higher pressure and Von Mises stress. Compering Fig. 3(a) with 3(b), the pressure is much higher than the Von Mises stress for all cases.

The effect of the initial distribution of the bubble cloud was examined for various distances (Fig. 4). As shown in Fig. 5(a), the maximum pressure takes the peak at δy = 0.5 mm = 0.5R_{c0} for all initial void fractions and once decreases but increases again with increase of the distance. The pressure achieves to almost 600 MPa at f_{G0} = 0.4. Fig. 5(b) shows that the Von Mises stress is almost same order of the pressure at δy = – 0.5, 0, 0.5 mm. The Von Mises stresses at some conditions are over 250 MPa which is the 0.2 % yield stress of FCV400. The collapse of the

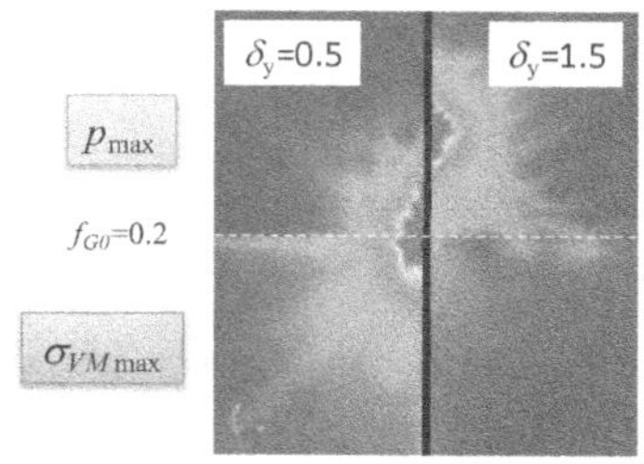

Figure 6 Distribution of the maximum pressure and the maximum Von Mises stress during the collapse of bubble cloud for δy = 0.5 (left) and 1.5 (right) mm.

bubble cloud under such conditions causes the erosion of solid. However, at δy = 1, 1.5mm, the Von Mises stress is much lower than the pressure for all initial void fraction cases.

Figure 6 shows the distribution of the maximum pressure in the fluid and the maximum Von Mises stress in the solid for δy = 0.5, 1.5 mm. The white dashed line indicates the interface between fluid and solid. It is clear that the maximum pressure and the maximum Von Mises stress cause near the solid surface in the case of δy = 0.5 mm. But the maximum pressure takes peak away from the solid surface in the case of δy = 1.5 mm, and the maximum Von Mises stress at δy = 1.5 mm is lower than that at δy = 0.5 mm in the solid region. Thus, the maximum Von Mises stress highly depends on the position of the maximum pressure caused by the collapse of bubble cloud.

Conclusion

The collapse of the oil cavitation bubble cloud near the solid surface was performed by the numerical simulation, in which the fluid-structure interaction and sub-grid scale bubble dynamics were taken into account. The higher initial void fraction of bubbles in the bubble cloud causes higher Von Mises stress in solid. The maximum Von Mises stress during the collapse of bubble cloud highly depends on the initial distribution of the bubble cloud. Especially, when the distance from the solid surface to the center of the spherical bubble cloud is half the radius of the cloud, the maximum Von Mises stress takes peak. The maximum Von Mises stress caused by the collapse of the attached cavitation cloud on surface becomes higher than the 0.2 % yield stress of the solid material depending on the condition, and causes the erosion of the solid surface.

References

[1] Okita, K., Miyamoto, Y., Kataoka, T., Takagi, S, Kato, H., (2015). *Mechanism of Noise Generation by Cavitation in Hydraulic Relief Valve*, Journal of Physics: Conference Series **656**, 012104.

[2] Prosperetti A., Lezzi A., (1986). *Bubble dynamics in a compressible liquid. part 1. first-order theory*, J. Fluid Mech., **168**, 457-478.

[3] Okita, K., Sugiyama, K., Takagi, S., Matsumoto, Y., (2013). *Microbubble behavior in an ultrasound field for high intensity focused ultrasound therapy enhancement*, J. Acoust. Soc. Am., **134**, 1576.

[4] Matsumoto, Y., Takemura, F., (1992). *Effect of Internal Phenomena on Bubble Motion (1st, Report)*, Trans. JSME Ser. B, **58**, 645-652 (in Japanese).

[5] Okita, K., Ono, K., Takagi, S., Matsumoto, Y., (2011). *Multi-scale Analysis on Cavitation Damage and its Mitigation for the Spallation Neutron Source*, E-Book Computational Methods for Coupled Problems in Science and Engineering IV, 600-610.

© 2018 ASME

Influence of dissolved gas content on Venturi cavitation at thermally sensitive conditions

[1]Haochen Zhang; [1]Zhigang Zuo*; [1]Shuhong Liu*

[1] *State Key Laboratory of Hydro Science and Engineering, Department of Energy and Power Engineering, Tsinghua University, Beijing, China*

Abstract

Influence of dissolved gas content (measured by dissolved oxygen, DO) on Venturi cavitation (10mm-scale test section) in water at different temperature, i.e., different degree of thermal sensitivity, is investigated in a blowdown-type cavitation tunnel. Investigations on the cavitation length and the unsteady cavitation behavior are presented. It is shown that the cavitation length decreases with increasing temperature (or the degree of the thermodynamic effect). In addition, the DO content has little influence on the mean cavitation length at thermally sensitive conditions.

Keywords: thermodynamic effect; dissolved gas content; cavitation

Introduction

Cavitation bubbles are formed by local pressure deficiency, and they absorb latent heat from ambient liquid as they grow. A thermal boundary is then created around the bubble. Heat transfer between the bubble and the liquid causes a local temperature decrease of the ambient liquid, which results in a slight drop of the vapor pressure [1]. Theoretically, this phenomenon suppresses the further growth of cavitation, which is known as the thermodynamic effect of cavitation [2]. This effect is required to be take into account in cavitation in cryogenic liquids, refrigerant and high temperature water, where the working temperature is close to the critical point temperature of the liquids.

In order to investigate the influence of the thermodynamic effect, cavitation experiments have been carried out using the above mentioned liquids [3-9]. In contrast to the theoretical inference, a number of experimental studies found that a stronger thermodynamic effect enhances the degree of the sheet/cloud cavitation in some cases, e.g. the cavitation length increases with increasing degree of the thermodynamic effect [4-7]. A few hypothesis have been proposed to explain this discrepancy. For example, Cervone et al. [4] proposed that it is due to the influence of the solid blockage caused by the tunnel walls. Meanwhile, Watanabe et al. [5] measured a higher pressure inside the cavitation than the vapor pressure in hydro-fluoro-ether, suggesting a possible existence of dissolved gas in the cavity. They proposed a modified cavitation number based on the pressure measured inside the cavity instead of vapor pressure. By taking into account the influence of the gas in the cavity, the cavitation length was suppressed due to the thermodynamic effect. However, in their experiments, the dissolved gas content was roughly estimated instead of controlled accurately. In addition, their proposition dose not explain the previous experimental results on unsteady cavitation characteristics and thermodynamic effect at different dissolved gas conditions. While the influences of the dissolved gas content on the inception cavitation number [10], lift/drag characteristics of a hydrofoil in the cavitating flow [11] and cavitation damage [12] have been investigated, its influence on unsteady cavitation behavior at different degree of thermal sensitivity remains unclear. In this paper, the influence of dissolved gas content (measured by dissolved oxygen, DO) on Venturi cavitation at different degree of thermally sensitive conditions in water is investigated in a blowdown-type cavitation tunnel. The cavitation length at different DO content and temperature conditions is investigated.

Experimental rig

The experiments were conducted on a blowdown-type cavitation tunnel (Fig. 1). It contains two 10-liter tanks, a transparent acrylic glass test section, connecting pipes, valves, a heater, measurement systems and a pressure regulation system, as shown in Fig. 1(a).

Both tanks are connected to a vacuum pump and an air compressor, to regulate the pressure of the tanks. In order to conduct the cavitation experiments under thermally sensitive condition, a 1 kW electrical heater mounted in the upstream tank heats and maintains water at the desired temperature, under the control of a thermostat. A convergent-divergent Venturi nozzle (divergent angle = 8°) is used for the test section (Fig. 1b), to generate cloud cavitation, which is also consistent with previous related research [13-15]. The symmetry of the flow channel is applied to achieve interchangeable flow direction.

*Corresponding Authors, Zhigang Zuo: zhigang200@mail.tsinghua.edu.cn;
Shuhong Liu: liushuhong@mail.tsinghua.edu.cn

Two Leno T21S piezoresistive pressure sensors are installed on the two tanks and coupled to ZTIC EM9118B dynamic signal acquisition device in order to measure and record the pressure. The uncertainty in the pressure measurements is ±1000Pa. Pressure signals are acquired at a sampling rate of 100 kHz for each channel. The degassing of the system is achieved by applying vacuum at the upstream tank. The dissolved oxygen (DO) content is directly measured by a DO meter (Hangzhou Puchuan Technology Co., Ltd PD-200) with an uncertainty of ±0.01 ppm, in order to quantify the dissolved gas content according to Henry's law. The temperature is measured by a Pt100 with an uncertainty of ±0.05 K. A high-speed CCD camera (Phantom v711) and a mirror system have been used to capture the spatio-temporal behavior of cavitation from two directions (front and top) simultaneously (Fig. 2). A sample rate of 100,000 fps was chosen and the images were digitally processed for further analysis of the cavitation characteristics.

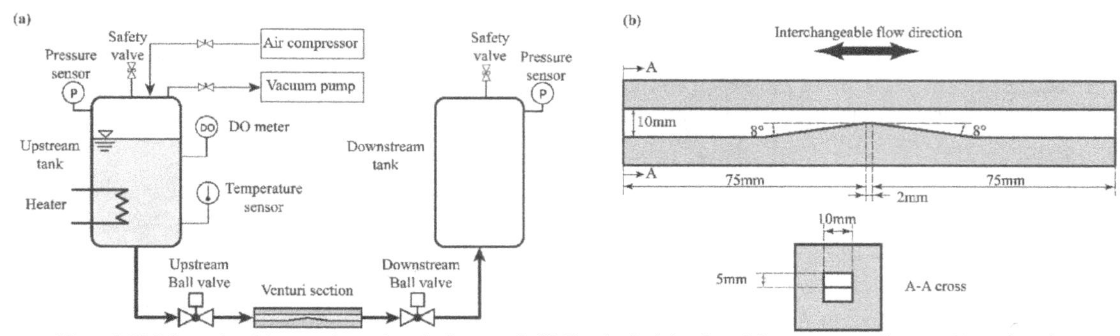

Figure 1 (a) Schematic of the experimental cavitation tunnel, (b) The detailed drawing of the convergent-divergent Venturi section

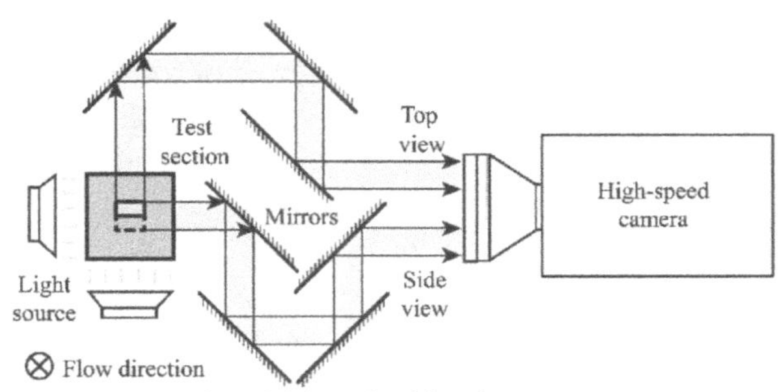

Figure 2 Schematic of the mirror system

Before the experiments, the downstream valve is closed and the upstream tank is almost filled with water. After water is degassed and heated, the upstream tank is pressurized to a desired level, while the pressure in the downstream tank remains under atmospheric condition. When the downstream ball valve is opened rapidly, water flows through the Venturi section, and cavitation occurs at the throat by the pressure difference between the two tanks. The dynamic signal acquisition device and the high-speed camera are triggered for recording the experimental data at the same time. The flow velocity changes during the 2 s long experiment due to the changing of the pressure difference between the two tanks. However, the pressure difference remains nearly constant within 0.1 s so that the experimental condition can be defined as a quasi-steady state.

Cavitation number is defined as follows with an uncertainty of 0.3%, mainly based on the pressure drop between the two tanks [15]: $\sigma = \frac{p_1 - p_v}{p_1 - p_2}$, where p_1 and p_2 are the pressures of the upstream and downstream tank, p_v is the vapour pressure. In this paper, the high (roughly 4.5-5.5ppm) and low (0.7-1.5ppm) DO conditions are conducted to obtain the qualitative tendencies of the influence of the dissolved gas content on Venturi cavitation at three different temperature conditions (298, 325, 335K).

Results

An image post-processing method of estimating the cavitation length [17] is used in this paper. Mean values and standard deviations of the grey level of the side view images of the cavitation are calculated, to evaluate the mean

cavitation length and the location of cloud separation. The closure of the sheet cavity is localized with the maximum value of the standard deviation of the grey levels calculated with the images stack. The mean cavitation length L_{cav} is then calculated from the Venturi throat to the closure line with an uncertainty of less than 2%.
Figure 3(a) and figure 4 show the mean cavitation length obtained for different temperature and for cavitation number ranging from 1.2 to 1.75 at high and low DO conditions. For each DO content and temperature condition, three experiments have been conducted and the results show good repeatability. The mean cavitation length decreases with an increasing cavitation number. In addition, when cavitation number is lower than around 1.3, the curve of cavitation length against cavitation number is concave (definition in mathematics: a line segment between any two points of the curve lies below the curve itself). When cavitation number is higher than 1.3, the curve is convex (definition in mathematics: a line segment between any two points of the curve lies above the curve itself). This difference is probably caused by different cavitation patterns occur at different range of cavitation number. Figure 3(b) and figure 3(c) show two different cavitation patterns at cavitation number of 1.24 and 1.40. At σ=1.24, the cavitation length is large and almost constant, and there is a small oscillation of the cavitation volume occurring around the closure region. At σ=1.40, there is a cavitation cloud and the periodic shedding can be seen obviously at the closure region.

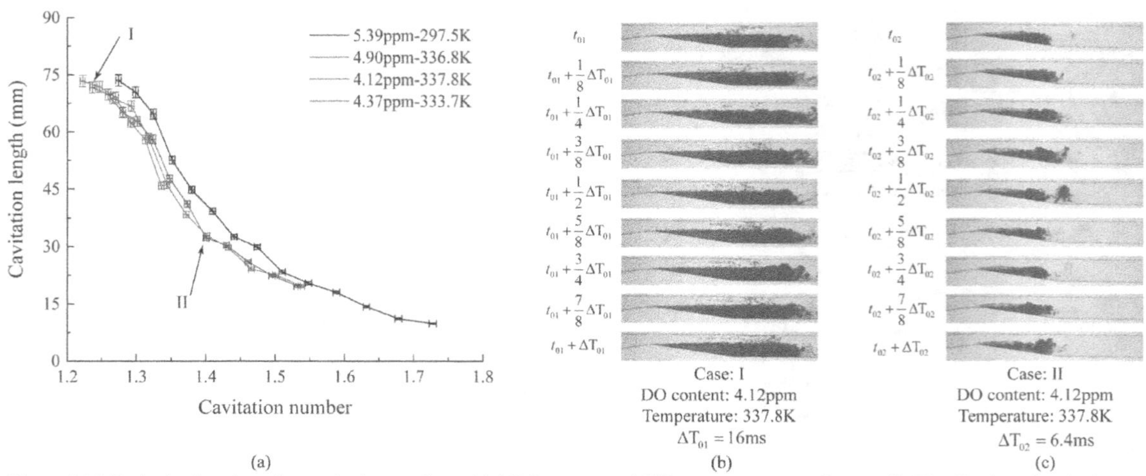

Figure 3 (a) Cavitation length against cavitation number at high DO content and different temperature of water, (b) Visualization of cavitation at σ=1.24, (c) Visualization of cavitation at σ=1.40

Besides, the cavitation length decreases with an increasing temperature (or the degree of the thermodynamic effect) at the DO content range tested, which can be explained by the suppression of the further development of cavitation due to the thermodynamic effect.
Comparing the results in the high or low DO conditions as shown in figure 5, 6 and 7, the cavitation length is almost the same at the same temperature and same cavitation number. It seems that the DO content has little influence on the mean cavitation length at thermally sensitive conditions.

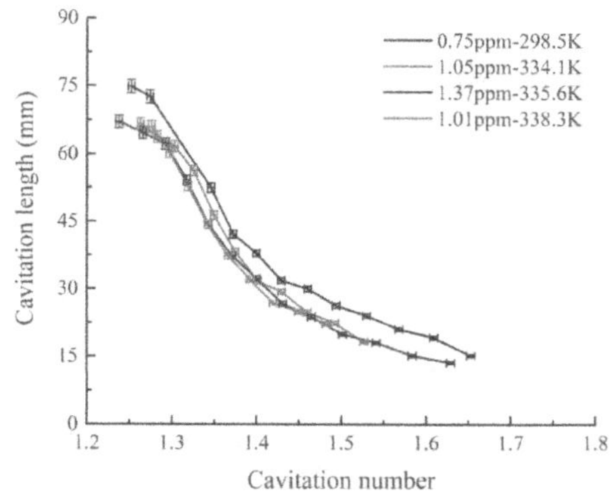

Figure 4 Cavitation length against cavitation number at low DO content and different temperature of water

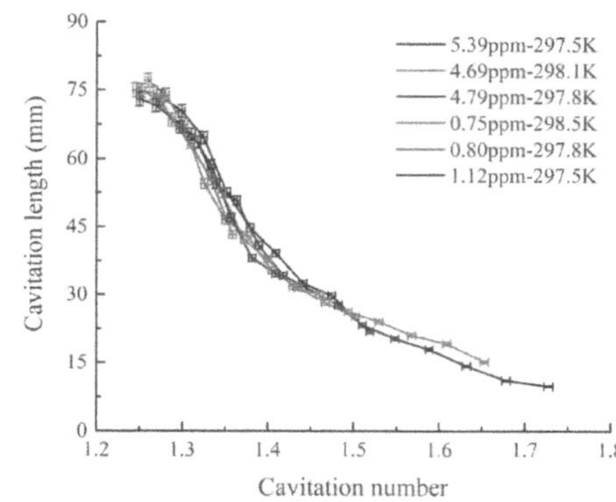

Figure 5 Cavitation length against cavitation number at the temperature of around 298K and different DO content

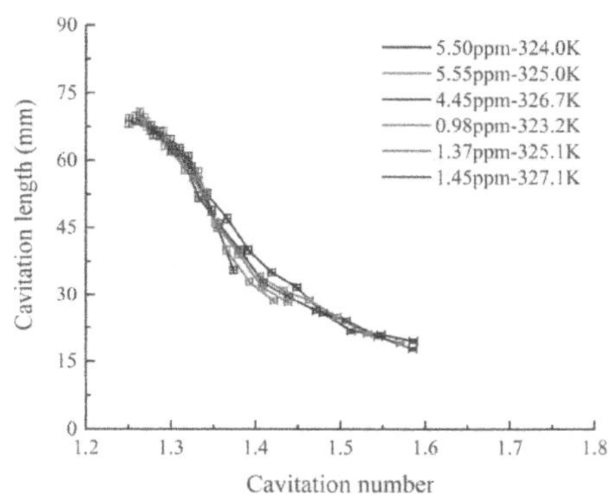

Figure 6 Cavitation length against cavitation number at the temperature of around 325K and different DO content

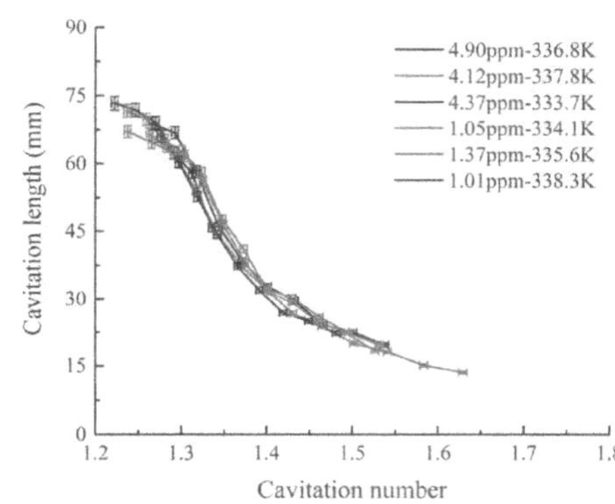

Figure 7 Cavitation length against cavitation number at the temperature of around 335K and different DO content

Conclusions

In the present study, the high-speed video observations of cavitation behavior and the mean cavitation length are discussed in high and low DO conditions to investigate the influence of the dissolved gas content on Venturi cavitation at thermally sensitive condition. Main results can be summarized as follows.

(1) The mean cavitation length decreases with an increasing cavitation number. Besides, cavitation pattern occurring at the different range of cavitation number is a possible reason for the concave or convex of the curve of cavitation length against cavitation number.

(2) The cavitation length decreases with an increasing temperature (or the degree of the thermodynamic effect) at the DO content range tested. In addition, the DO content has little influence on the mean cavitation length at thermally sensitive conditions.

Acknowledgment

The work was supported by National Natural Science Foundation of China (No. 51476083), National Basic Research Program of China (No. 613321), and State Key Laboratory of Hydro Science and Engineering (Research Fund Program, No: 2017-KY-03).

References

[1] Franc, J. P., Michel, J. M. (2006). *Fundamentals of cavitation* (Vol. 76). Springer Science & Business Media.

[2] Brennen, C. E. (2013). *Cavitation and bubble dynamics*. Cambridge University Press.

[3] Niiyama, K., Yoshida, Y., Hasegawa, S., Watanabe, M., Oike, M. (2012, August). *Experimental investigation of thermodynamic effect on cavitation in liquid nitrogen.* In Proceedings of the Eighth International Symposium on Cavitation (CAV2012), Singapore (pp. 153-157).

[4] Cervone, A., Bramanti, C., Rapposelli, E., d'Agostino, L. (2006). *Thermal cavitation experiments on a NACA 0015 hydrofoil.* Journal of fluids engineering, 128(2), 326-331.

[5] Watanabe, S., Enomoto, K., Yamamoto, Y., Hara, Y. (2014, August). *Thermal and Dissolved Gas Effects on Cavitation in a 2-D Convergent-Divergent Nozzle Flow.* In Proceedings of ASME Fluids Engineering Division Summer Meeting, Chicago, No. FEDSM2014-21902.

[6] Gustavsson, J. P., Denning, K. C., Segal, C. (2008). *Hydrofoil cavitation under strong thermodynamic effect.* Journal of Fluids Engineering, 130(9), 091303.

[7] De Giorgi, M. G., Ficarella, A., Tarantino, M. (2013). *Evaluating cavitation regimes in an internal orifice at different temperatures using frequency analysis and visualization.* International Journal of Heat and Fluid Flow, 39, 160-172.

[8] Franc, J. P., Boitel, G., Riondet, M., Janson, É., Ramina, P., Rebattet, C. (2010). *Thermodynamic effect on a cavitating inducer—Part I: geometrical similarity of leading edge cavities and cavitation instabilities.* Journal of Fluids Engineering, 132(2), 021303.

[9] Franc, J. P., Boitel, G., Riondet, M., Janson, É., Ramina, P., Rebattet, C. (2010). *Thermodynamic Effect on a Cavitating Inducer—Part II: On-Board Measurements of Temperature Depression Within Leading Edge Cavities.* Journal of Fluids Engineering, 132(2), 021304.

[10] Kodama, Y., Take, N., Tamiya, S.,Kato, H. (1981). *The effect of nuclei on the inception of bubble and sheet cavitation on axisymmetric bodies.* Journal of Fluids Engineering, 103(4), 557-563.

[11] Daido, H., Watanabe, S., Tsuda, S. I. (2015, July). *Effects of Dissolved Gas on Unsteady Cavitating Flow Around a Clark Y-11.7% Hydrofoil.* In ASME/JSME/KSME 2015 Joint Fluids Engineering Conference (pp. V02AT05A006-V02AT05A006). American Society of Mechanical Engineers.

[12] Sirotyuk, M. G. (1966). *Effect of the temperature and gas content of the liquid on cavitation processes.* Soviet Physics, Acous, 12(1), 67-71.

[13] Stutz, B., Reboud, J. L. (1997). *Experiments on unsteady cavitation.* Experiments in fluids, 22(3), 191-198.

[14] Barre, S., Rolland, J., Boitel, G., Goncalves, E., Patella, R. F. (2009). *Experiments and modeling of cavitating flows in venturi: attached sheet cavitation.* European Journal of Mechanics-B/Fluids, 28(3), 444-464.

[15] Petkovšek, M., Dular, M. (2013). *IR measurements of the thermodynamic effects in cavitating flow.* International Journal of Heat and Fluid Flow, 44, 756-763.

[16] Franc, J. P., Rebattet, C., Coulon, A. (2004). *An experimental investigation of thermal effects in a cavitating inducer.* Transactions of the ASME-I-Journal of Fluids Engineering, 126(5), 716-723.

[17] Dular, M., Bachert, B., Stoffel, B., Širok, B. (2004). *Relationship between cavitation structures and cavitation damage.* Wear, 257(11), 1176-1184.

© 2018 ASME

Influence of the Tip Clearance on the Auto-Oscillation Frequency of a Cavitating Pump

[1]Peter F. Pelz*; [1]Paul Taubert

[1]*Chair of Fluid Systems, Technische Universität Darmstadt, Darmstadt, Germany*

Abstract

This paper gives a contribution to the vortex structure, vortex dynamics and the tip clearance inertance as well as compliance of a cavitating pumps. The related vortex structure is reduced to the blade bound vortices, hub vortex and tip vortices only. At heavy part load $\varphi \to 0$, the tip vortices form a coaxial vortex ring of increasing strength. This vortex ring may break down to several wall bound vortices at severe part load. The velocity potential of the vortex ring is given by the Bessel series solution and applying a limit value analysis. The inertance of the gap flow is derived straight forward from this generic mathematical model of an axial pump at deep part load. With the compliance due to cavitation, the natural frequency of cavitation surge is discussed for the given generic mathematical model of the pump.

Keywords: tip clearance; inertance; analytics; potential flow

Introduction

The task of this paper is the determination of analytic expressions for the inertance L and compliance C of the tip clearance flow of an axial fan. The work extends the research of Brennen, 2017 [1] employing a recently developed vortex model of an axial turbomachine at part load (cf. Pelz & Taubert, 2017 [2]). This model is recapped in the following. Figure 1a shows the vortex model of the pump. The N bound vortices of strength Γ form the hub vortex of strength $N\Gamma$ and the tip vortices. At part load $\varphi := U/\Omega R \to 0$ (averaged velocity U, pipe radius R, rotational speed of the pump Ω), the tip vortices feed a coaxial vortex ring, whose radius equals the tip radius R_t. The strength Γ_t of this vortex ring is increasing in time until a critical value is reached. It is suspected that this might reason a self-excited process like the periodic cavity cloud separation (cf. Pelz & Taubert, 2017 [2]].

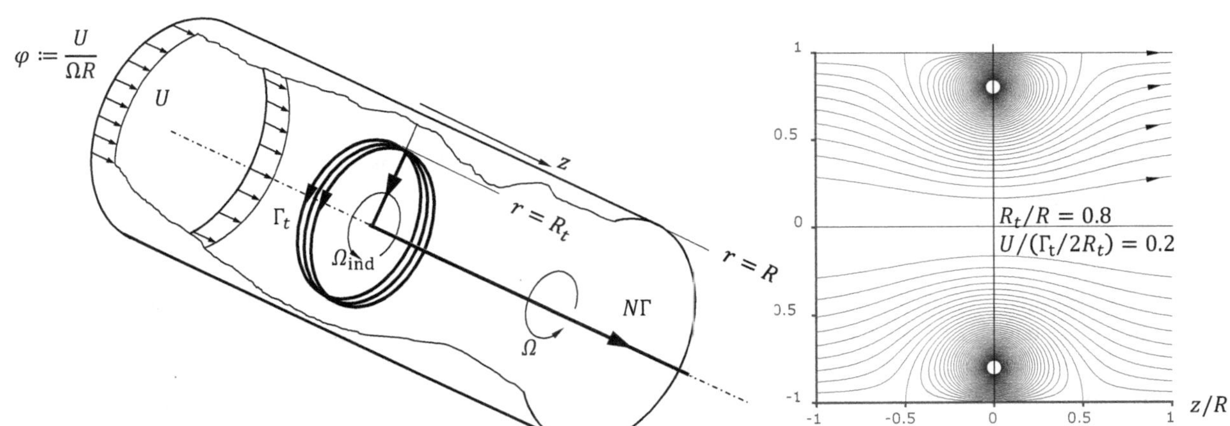

Figure 1: a) generic vortex system of an axial pump at part load given; b) streamlines in the meridian plane (cf. Pelz & Taubert, 2017 [2]).

Figure 1b shows the streamlines derived from this model in the meridian plane for the ratio $\beta := R_t/R = 0.8$ and a vortex strength of the ring vortex Γ_t of $\Gamma_t = 10\,UR_t$.
The vortex model is a coarse geometric model of the real flow which may serve gaining insights into the dynamics of a cavitating pump extending the work of Brennen.

*Corresponding Author, Peter F. Pelz: peter.pelz@fst.tu-darmstadt.de

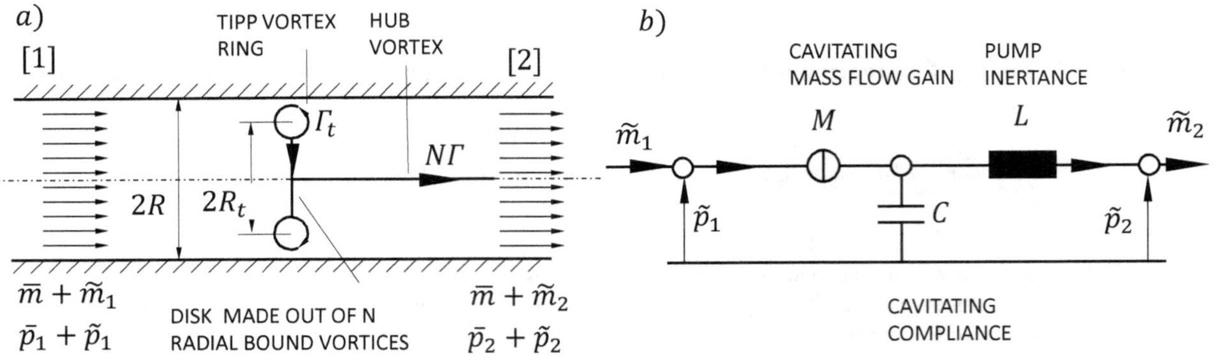

Figure 2: : a) Vortex model of an axial pump at part load [2]; b) lumped parameter model of an axial pump.

Figure 2b shows the lumped parameter model of the same pump. Brennen [1], [3] proposed a more detailed model taking a main flow and a by-pass flow into account. Nevertheless, there are indications that the tip clearance mainly influences the pressure build up and not so much the internal leakage of the flow (cf. Karstadt & Pelz, [4]). This behaviour is different to radial pumps were the internal leakage shall not be neglected. With respect to the inertance of the flow, one has to keep in mind that most of the integral kinetic energy of the flow is due to the tip vortex flow. This is mainly because the tip vortex is located at a large radius $R_t \to R$. Hence, it is sufficient in a first approach to concentrate on the inertance of the tip vortex flow. Indeed, this is the focus and novelty of this paper.

Linear perturbation expansion and dimensional analysis

Taking disturbances up to the first order into account, the usual perturbation approach is done: the pressure on the suction and pressure side are expanded as $p_1(t) = \bar{p}_1 + \tilde{p}_1(t),\ p_2 = \bar{p}_2 + \tilde{p}_2(t)$. Similar the mass flow rate on both sides of the pump are perturbed: $m_1(t) = \bar{m} + \tilde{m}_1(t), m_2(t) = \bar{m} + \tilde{m}_2(t)$. Considering only linear terms in the expansion, the transform such as $\tilde{m}_1 = \Re[\hat{m}_1 \exp(\mathrm{i}\omega t)]$ is made for all fluctuating quantities. The remaining equation for perturbations to the power zero is the pump characteristic

$$\bar{p}_2 - \bar{p}_1 = \Delta p(\bar{m})\,. \tag{1}$$

The remaining equation for the linear perturbation is the transfer characteristics given by the relation

$$\begin{pmatrix} P_{11} & P_{12} \\ P_{21} & P_{22} \end{pmatrix} \begin{pmatrix} \hat{p}_1 \\ \hat{m}_1 \end{pmatrix} = \begin{pmatrix} \hat{p}_2 \\ \hat{m}_2 \end{pmatrix}. \tag{2}$$

The momentum and continuity equation solve the elements P_{ij} of the dynamic transfer characteristics. The momentum equation in three equivalent representations yields:

$$\varrho l \pi R^2\, \dot{U}_2 = \dot{\tilde{m}}_2 l = \dot{I} = \pi R^2\,(\tilde{p}_1 - \tilde{p}_2)\,. \tag{3}$$

U_2 is the area averaged velocity at the outlet, l is the inductive length and $\dot{I}$ the rate of change of the axial momentum. With the inductive length l of the pump being determined as it will be done in the following section, the inertance L of the pump is given by $L := l/\pi R^2$. As an orientation, the inductive length of a simple aperture in an unlimited plane wall of infinite thickness is $R_t\,\pi/4 \approx R_t$ [5]. For the low pressure LOX inducer in the Space Shuttle main engine, Brennen derives the approximation $L \approx 10/R_t$ from measurements, i.e. the inductive length of this inducer is nearly thirty times the tip radius: $l = L\pi R^2 \approx 30\,R_t$. The continuity equation, ignoring cavitating mass flow gain, reads

$$\frac{\varrho V_\varepsilon}{n\,\bar{p}_1}\,\dot{\tilde{p}}_1 - \tilde{m}_1 + \tilde{m}_2 = 0, \tag{4}$$

were V_ε is the cavity volume. In the flowing, the abbreviation for the hydraulic compliance $C := \varrho V_\varepsilon/(n\,\bar{p}_1)$ is used. The polytropic exponent n equals one for a "small" cavity and equals $\gamma = 1.4$ for a "large" cavity [6]. With the abbreviations L and C and the transformation into the frequency domain, we derive

$$(1+\omega^2\,LC)\,\hat{p}_1 - \mathrm{i}\omega\,L\,\hat{m}_1 = \hat{p}_2,\\ -\mathrm{i}\omega\,C\,\hat{p}_1 + \hat{m}_1 = \hat{m}_2. \tag{5}$$

The pressure shall be non-dimensionalyzed by the dynamic pressure $\varrho\Omega^2 R_t^2$ and the mass flow rate by $\varrho\Omega R_t^3$. Dividing the first equation by $\varrho\Omega^2 R_t^2$ and the second equation by $\varrho\Omega R_t^3$ we obtain

$$(1+\omega_+^2\,L_+C_+)\,\hat{p}_{1+} - \mathrm{i}\omega_+\,L_+\hat{m}_{1+} = \hat{p}_{2+},\\ -\mathrm{i}\omega_+\,C_+\,\hat{p}_{1+} + \hat{m}_{1+} = \hat{m}_{2+}. \tag{6}$$

By inspection we gain the dimensionless products to be

$$\omega_+ := \frac{\omega}{\Omega}, \qquad L_+ := LR_t = \frac{l\beta^2}{\pi R_t}, \qquad C_+ := \frac{C\Omega^2}{R_t} = \frac{1}{n}\frac{V_\varepsilon}{R_t^3}\frac{\varrho\Omega^2 R_t^2}{\bar{p}_1} \approx \frac{2}{n}\frac{V_\varepsilon}{R_t^3}\frac{1}{\sigma} \quad \text{for } \bar{p}_1 \gg p_v\,. \tag{7}$$

Hence, the pump compliance is inverse proportional to the cavitation number σ, assuming the vapour pressure is much smaller than the suction pressure: $\bar{p}_1 \gg p_v$. This result is not new, but the dimensional analysis is indeed a nice affirmation.

We now employ the vortex model sketched in Figure 1 and Figure 2: we assume the cavity volume V_ε to be the core of the coaxial vortex ring shown in Figure 1. If the core radius is $\delta = \varepsilon R_t/2$, the volume is given by $V_\varepsilon = \pi^2/4\ R_t^3\,\varepsilon^2$. With this the dimensionless compliance yields

$$C_+ = \frac{(\pi\varepsilon)^2}{2n}\frac{1}{\sigma}. \tag{8}$$

From the mentioned performance experiments of the low pressure LOX inducer in the Space Shuttle main engine, Brennen derived the approximation $C_+ \approx 0.05\ 1/\sigma$. Comparing this datum with the result (8) we estimate $\varepsilon \approx 0.1 \ldots 0.12$ for $1 \le n \le \gamma = 1.4$. Thus, the core radius δ of the ring vortex ring is estimated to be

$$\delta \approx (0.05 \ldots 0.06)\,R_t\,. \tag{9}$$

Inertance of the pump derived from the velocity potential

In the coarse-graining limit model, the core of the vortex ring is seen as a washer-like vortex sheet located at $z = 0, R_t(1-\varepsilon) < r < R_t$. The tangential to the surface of the sheet is W at $z = 0_+$ and $-W$ at $z = 0_-$. Hence, the circulation of the vortex ring is $\Gamma_t = 2\varepsilon R_t W$. Alternatively the strength of the vortex ring is given by the dimensionless product

$$\tau := \frac{\Gamma_t}{2R_t\,U} = \varepsilon\frac{W}{U}. \tag{10}$$

From Euler's equation end the energy equation we derive the pump characteristics $N\Gamma\,\Omega/2\pi = (\Omega R_t)^2(1-\varphi/\hat{\varphi})$ and thus $N\Gamma\,\Omega/2\pi \to (\Omega R_t)^2$ for $\varphi \to 0$ [2]. In our model, the N tip vortices of strength Γ roll up in time t to form $\Gamma_t = N\Gamma\Omega t/2\pi$ [2]. With the ratio $\beta := R_t/R = 1-s$, where s is the dimensionless tip clearance, there is an upper bound for τ. The pump performance affects the upper bound to be $\Gamma_t < (\Omega R_t)^2 t$ at severe part load $\varphi \to 0$, yielding

$$\tau := \frac{\Gamma_t}{2R_t\,U} = \frac{1}{2}\frac{\Omega t}{2\pi}\frac{N\Gamma}{R_t U} < \frac{\Omega t}{2}\frac{\beta}{\varphi}. \tag{11}$$

The velocity potential ϕ_ε of the coaxial vortex ring in a parallel flow is

$$\frac{\phi_\varepsilon(r,z)}{UR} = \frac{z}{R} - 2\tau\sum_{n=1}^{\infty}\frac{1}{k_n J_0^2(k_n)}J_0\left(k_n\frac{r}{R}\right)e^{-k_n\left|\frac{z}{R}\right|}\frac{1}{\varepsilon}\int_{\beta(1-\varepsilon)}^{\beta} J_1\left(k_n\frac{r}{R}\right)\frac{r}{R}\,\mathrm{d}\left(\frac{r}{R}\right), \tag{12}$$

as derived by Pelz & Taubert, 2017 [2].The numbersk_n for $n = 1, \ldots, \infty$ are the roots of the Bessel function: $J_1(k_n) = 0$. For $\varepsilon W = \tau U = \text{const.}$ and $\varepsilon \to 0$ the integral simplifies to the expression

$$\lim_{\varepsilon\to 0}\frac{1}{\varepsilon}\int_{\beta(1-\varepsilon)}^{\beta} J_1\left(k_n\frac{r}{R}\right)\frac{r}{R}\,\mathrm{d}\left(\frac{r}{R}\right)=\beta^2 J_1(k_n\beta), \tag{13}$$

applying L'Hôpital's rule. Hence, we have the singular solution for the coaxial ring vortex in the tube with its flow potential

$$\frac{\phi_\varepsilon(r,z)}{UR}\to\frac{\phi(r,z)}{UR}=\frac{z}{R}-2\beta^2\tau\sum_{n=1}^{\infty}\frac{J_1(k_n\beta)}{k_n J_0^2(k_n)}J_0\left(k_n\frac{r}{R}\right)\,e^{-k_n\left|\frac{z}{R}\right|}. \tag{14}$$

The contour lines of the related Stokes' stream function

$$\frac{\psi(r,z)}{UR^2}=-\frac{1}{2}\left[1-\left(\frac{r}{R}\right)^2\right]+2\,\tau\beta^2\sum_{n=1}^{\infty}\frac{J_1(k_n\beta)}{k_n J_0^2(k_n)}J_1\left(k_n\frac{r}{R}\right)\,\frac{r}{R}\,e^{-k_n\left|\frac{z}{R}\right|}. \tag{15}$$

are plotted in Figure 1b. With the result (12) we are now set to derive an analytic expression for the inertance L_+ of the pump. Following Saffmann [7], the momentum used in the momentum equation (3), $\dot{I}=A(\tilde{p}_1-\tilde{p}_2)$, yields

$$\begin{aligned}I&=\varrho\int_{S_B}\phi_\varepsilon\,\vec{n}\cdot\vec{e}_z\,\mathrm{d}S_\mathrm{B}=-4\pi R^2\varrho\int_{\beta(1-\varepsilon)}^{\beta}\phi_\varepsilon\,(r,0_+)\,\frac{r}{R}\,\mathrm{d}\left(\frac{r}{R}\right)\\&\approx 8\pi R^3U\,\tau\varrho\sum_{n=1}^{\infty}\frac{J_1(k_n\beta)\beta^2}{k_n J_0^2(k_n)}\int_{\beta(1-\varepsilon)}^{\beta}J_0\left(k_n\frac{r}{R}\right)\,\frac{r}{R}\,\mathrm{d}\left(\frac{r}{R}\right).\end{aligned} \tag{16}$$

for small $\varepsilon\to 0$. The hydrodynamic impulse of the bound vorticity equals the virtual momentum since the volume of the body is zero. The integral is approximated by

$$\int_{\beta(1-\varepsilon)}^{\beta}J_0\left(k_n\frac{r}{R}\right)\,\frac{r}{R}\,\mathrm{d}\left(\frac{r}{R}\right)\approx\varepsilon\,\beta^2\,J_0(k_n\beta)\,. \tag{17}$$

Hence, the momentum yields

$$I=m\;8\,R_t\varepsilon\tau\,\beta^3\sum_{n=1}^{\infty}\frac{J_1(k_n\beta)\,J_0(k_n\beta)}{k_n J_0^2(k_n)}\;, \tag{18}$$

with the mass flux $m=\varrho U\pi R^2$. From (3) we obtain $\dot{\tilde{m}}_2 l=\dot{I}$ or with $l=L\pi R^2=L_+\pi R_t/\beta^2$ the reduced inertance of the pump

$$L_{\mathrm{red}}(\beta):=\frac{L_+}{\varepsilon\tau}=\frac{8\beta^5}{\pi}\sum_{n=1}^{\infty}\frac{J_1(k_n\beta)\,J_0(k_n\beta)}{k_n J_0^2(k_n)}\,. \tag{19}$$

The reduced inertance of the pump is plotted versus the dimensionless tip clearance $s:=1-\beta$ in Figure 3 for $0<s<0.1$.

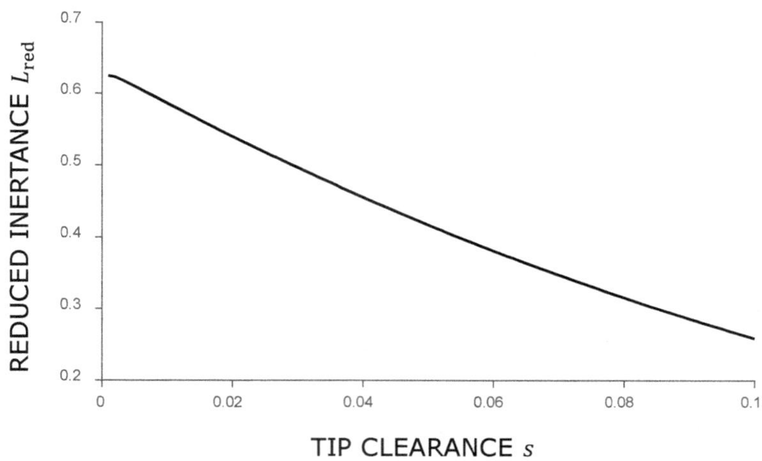

Figure 3: Reduced inertance L_{red} over tip clearance s, equation (19).

From result (19), the auto-oscillation frequency of the pump is determined to be

$$\frac{\omega_A}{\Omega} = (C_+ L_+)^{-1/2} = \left(\frac{1}{\pi^2}\frac{2n}{L_{\text{red}}(s)}\frac{\sigma}{\tau\varepsilon^3}\right)^{1/2} . \qquad (20)$$

With the results of Brennen, i.e. $\omega_A/\Omega = (2\sigma)^{1/2}$, $n \approx 1$, $\varepsilon \approx 0.1$, the typical dimensionless circulation is of the order

$$\tau(s) \approx \frac{1000}{\pi^2}\frac{1}{L_{\text{red}}(s)} \approx \frac{100}{L_{\text{red}}(s)}. \qquad (21)$$

The upper bound, equation (10), is due to the pump characteristics with the flow number $\varphi = U/R\Omega$.

References

[1] C. E. Brennen (2016). *On the Dynamics of a Cavitating Pump*. IOP Conf. Ser. Earth Environ. Sci., vol. 49, p. 052018.
[2] P. F. Pelz, P. Taubert, and F.-J. Cloos (2017). *Vortex Structure and Kinematics of Encased Axial Turbomachines*. 17th International Symposium on Transport Phenomena and Dynamics of Rotating Machinery, Maui, Hawaii, USA.
[3] C. E. Brennen (1978). *Bubbly flow model for the dynamic characteristics of cavitating pumps*. J. Fluid Mech., vol. 89, no. 2, pp. 223–240.
[4] S. Karstadt and P. F. Pelz (2012). *A Physical Model for the Tip Vortex Loss: Experimental Validation and Scaling Method*. ASME Turbo Expo 2012: Turbine Technical Conference and Exposition, Copenhagen, Denmark, pp. 73–81.
[5] J. W. S. Rayleigh (1945). *The Theory of Sound*, vol. 2. New York.
[6] P. F. Pelz and J. Buttenbender (2004). *The dynamic stiffness of an air-spring*. International Conference on Noise & Vibration Engineering ISMA, Leuven, Belgium.
[7] P. G. Saffman (1992). *Vortex Dynamics*. Cambridge University Press.

© 2018 ASME

Initial findings of an investigation on the removal of the cavitation erosion risk in a prototype control orifice inside a diesel injector

Maxwell Brunhart[1], Celia Soteriou[1], Christian Daveau[1], Manolis Gavaises[2]
Phoevos Koukouvinis[2] & Mark Winterbourn[1]

[1]Delphi Technologies, Gillingham, UK: [2]City, University of London, London, UK

Abstract

A CFD investigation is in progress to study the cavitation characteristics and potential erosion risks of a control orifice in a prototype injector. An early design of the orifice resulted in cavitation erosion after endurance testing. A design modification eliminated the erosion and subsequent prototypes were free from damage. Initial results for the two designs using different simulation methods are discussed, along with the effects of different rates of evaporating and condensing mass transfer. Preliminary findings on possible erosion risk indicators comparing the eroding with the non-eroding design are presented.

Keywords: cavitation, erosion risk, control orifice

Introduction

In some cases, the collapse of cavitation can lead to erosion of high strength metals and thereby damage system components. Being able to predict if and where such erosion is likely to occur would be beneficial to many fields, like automotive and naval. The diesel injector is a common subject in cavitation based literature as cavitation in the flow cannot be avoided, particularly so in modern systems, and so it must be understood and managed. The injector is of great importance to the performance of engines. It is an area for continued development to meet future environmental and performance standards. Hence, an understanding of the fundamental behavior of hydraulic control orifices is needed to optimize the performance and production while avoiding any potential complications due to cavitation or cavitation erosion.

The control orifice of interest in this work is a spill orifice (Fig. 1) in an automotive high-pressure fuel injector. All prototype designs are subjected to extensive testing during product development, including endurance tests. One such endurance test on a prototype component resulted in significant cavitation erosion of the orifice. Further development of the component eliminated the erosion and subsequent prototypes were free from damage.

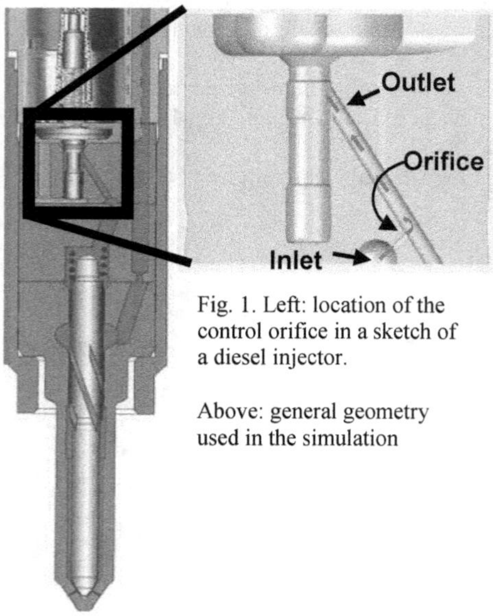

Fig. 1. Left: location of the control orifice in a sketch of a diesel injector.

Above: general geometry used in the simulation

A CFD investigation is in progress to understand the flow-field and cavitation characteristics in and around this orifice, and to explore possible indicators for cavitation erosion risk. The analyses used two design levels: the original geometry which suffered from erosion (E – eroding geometry) and a subsequent prototype design which eliminated the erosion (NE – non-eroding geometry).

The CFD study includes initial simulations using Reynolds-Averaged Navier-Stokes (RANS) turbulence modelling and the standard Zwart-Gerber-Belarmi (ZGB) cavitation model. For increased accuracy and detail, this work was then expanded upon with a hybrid LES-RANS turbulence model: Detached Eddy Simulation (DES). These types of model have shown improvement over RANS [1] [2] [3]. However, in industry the RANS method is still a useful CFD investigation tool, especially considering the heavy runtime cost of LES/DES methods. The DES simulations were first run using the standard ZGB model. Then a modified ZGB method was used, by

Corresponding Author, Maxwell Brunhart: maxwell.brunhart@delphi.com means of a UDF (user-defined function) which enabled significant changes to be specified for the rates of mass transfer [4].

The paper provides some preliminary results of this on-going project, commenting on the effects of the different simulation methods and on the two different designs, all of which is backed up with photographic evidence of damage on an early prototype component. Finally, some initial findings on indicators for cavitation erosion risk are presented.

CFD Simulation

The vapor transport equation is defined as [5]:

$$\frac{\partial(\alpha\rho_v)}{\partial t} + \nabla \cdot (\alpha\rho_v U_v) = R_e - R_c \quad \text{Eq. 1}$$

where ρ is density, α is vapor volume fraction, U is velocity, v denotes vapor and Re and Rc denote rates of evaporating and condensing mass transfer, respectively. First a standard, and then a modified form of the ZGB cavitation model was applied in the simulations. The ZGB model uses the following rates of mass transfer, first described by Zwart et al [6]:

$$R_e = F_e \frac{3\alpha_{nuc}(1-\alpha)\rho_v}{\Re_B}\sqrt{\frac{2}{3}\frac{(P_v - P)}{\rho_l}} \quad \text{Eq. 2}$$

$$R_c = F_c \frac{3\alpha\rho_v}{\Re_B}\sqrt{\frac{2}{3}\frac{(P - P_v)}{\rho_l}} \quad \text{Eq. 3}$$

where α_{nuc} is nucleation site volume faction, $\Re_B$ is bubble radius, ρ is density, P is pressure, F is a constant and v and l denote vapor and liquid respectively. In Fluent the default values for the ZGB model constants are 1e-6 m for $\Re_B$, α_{nuc} is 5e-4, F_c is 0.01 and F_e is 50. These constants can be grouped as follows:

$$F_{evap} = F_e \frac{3\alpha_{nuc}}{\Re_B} \quad \text{Eq. 4}$$

$$F_{cond} = F_c \frac{3}{\Re_B} \quad \text{Eq. 5}$$

resulting in F_{evap} equaling 75,000 m^{-1} and F_{cond} equaling 30,000 m^{-1} with the default values.

The modified cavitation method was implemented in a UDF which comes from the work of Koukouvinis et al [4]. The ZGB model is implemented with a defined mass transfer constant to create an effect closer to that of a barotropic model. A barotropic model is approached asymptotically as the mass transfer rates trend toward infinity. For this work, a large but practical value of 1e8 for both F_{evap} and F_{cond} was used.

The initial RANS CFD simulations used k-epsilon method alongside the standard ZGB cavitation model. The RANS simulations were used for the low computational costs and previously seen accuracy in predicting the Cd when implemented with due care ([7] & internal documents). Initial flow details were noted with the RANS simulations, but they were unable to display detailed transient behaviour that is likely needed in developing cavitation erosion risk assessment.

Further to that, the injector cycle experienced by the orifice was modelled in URANS (Unsteady RANS) simulations, inputting the relevant upstream and downstream pressure traces. The URANS results did show some oscillation movement in the region of cavitation, but did not add significantly to the steady state simulations. A transient injection cycle with DES would have an extremely long runtime, but may be done later when more processing power is available. For this paper though, steady state DES simulations are reviewed, which already take approximately two orders of magnitude longer then the RANS simulations.

The DES technique is an LES-RANS hybrid, implementing RANS at the boundaries and an LES model elsewhere [8]. This makes for a simulation that is less computationally intensive than full LES while still producing the higher level of detail at the areas of interest. The simulations were implemented in ANSYS Fluent v17 as IDDES (Improved Delayed Detached Eddy Simulation), as first proposed by Shur et al [8], which is a more recent version of DES that provides more flexibility and convenience for high Re flows. This setup was previously validated with experiment for a similar operating environment by Bush et al [1] and is also congruent with Koukouvinis et al [4].

The simulations were run with constant boundary conditions using an upstream pressure of 1380 bar and downstream pressure of 340 bar. These values provided the highest pressure difference across this orifice during a 2200 bar injection cycle. As the pressure drop is relatively high and the flow velocity was expected to also be high, a compressible fluid is used. The fluid is representative of Normafluid (ISO4113), a standard fluid used in automotive testing. The vapor is assumed incompressible for computational simplicity. The fluids are represented using a mixture model which assumes that the two fluids travel as a homogeneous multiphase flow.

For both designs, the entrance to the orifice is slightly rounded, the diameter is in the region of 0.22mm and there is a small divergent taper along the length of the hole [9]. Although highly significant in terms of design, the differences between the E and NE prototype geometries are not great, with the taper for NE geometry being slightly less, that is, closer to 0. To start with, CFD simulations were tested against pass-off criteria to ensure the CFD geometry of each design was an accurate replica of the hardware, along with a mesh study that found that further mesh refinement had little effect.

Experimental Tests

Newly proposed injector designs must all go through a rigorous testing process. One of these test is termed an endurance test. The test consists of exposing the prototype injector to a high typical load cycle (2200 bar cycle) for x-many cycles. This type of test is essential in determining if a prototype can withstand normal use and is ready for serial production. A failed example from an endurance tests can be seen in Fig. 2 The material used was automotive high strength steel with surface treatment.

Fig. 2- Early prototype sample showing cavitation erosion after an endurance test. Subsequent design modifications eliminated the erosion

Results

Table 1- List of the different simulations sets and results overview

Case	Description	Mass transfer constants	Negative pressure (bar)	General outcome (E vs. NE)
1	RANS with standard ZGB method and default values	Default F_{cond}=3e4, F_{evap}=7.5e4	-200	Region of recirculation and backflow in E geo.
2	DES with standard ZGB and default values	Default F_{cond}=3e4, F_{evap}=7.5e4	-200	No obvious difference
3	DES with modified mass transfer UDF	UDF F_{cond}=F_{evap}=1e8	-8	NE has significantly decreased activity in max P and max DP/Dt

Case 1- The results for the eroding geometry showed a region of recirculation and backflow at the end of the topside of the control orifice, whereas the non-eroding geometry did not (Fig. 3). This region correlated with the location of cavitation erosion seen in the failed endurance test image (Fig 2).

Case 2- Results showed no substantial differences between the E and NE geometries.

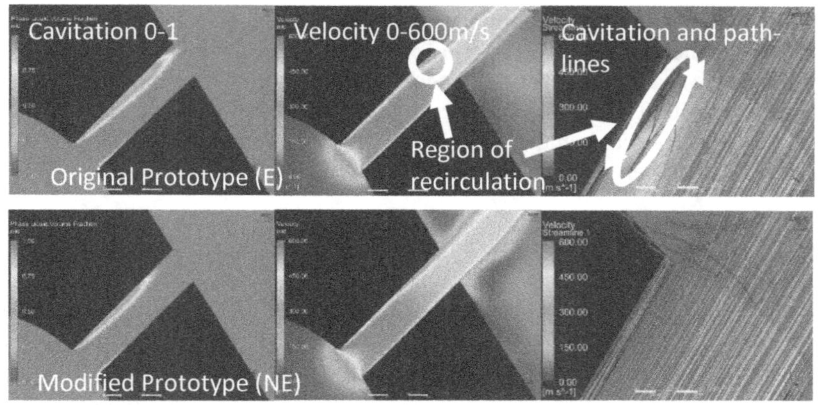

Fig. 3- Mid-plane section of the E and NE RANS simulations (Case A). Region of recirculation and backflow prominent in the E geometry.

Case 3- A key improvement using this method was the lack of negative pressure (Table 1) commonly seen in cavitating simulations which is a non-physical effect. This negative pressure is a numerical inaccuracy and is commonly capped at the vapor pressure by users during post-processing. Furthermore, the UDF enables converged solutions with the required increased mass transfer rates. However, the run time of case 3 compared to case 2 was noticeably increased.

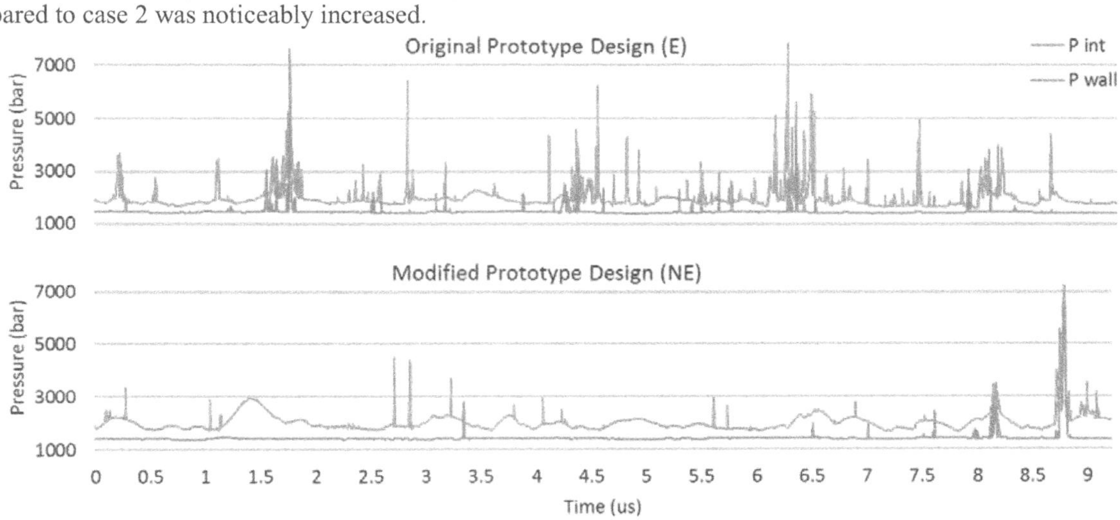

Fig. 4- Maximum pressure seen in the entire domain (P int) and on any surface (P wall) for 9 us after initialization.

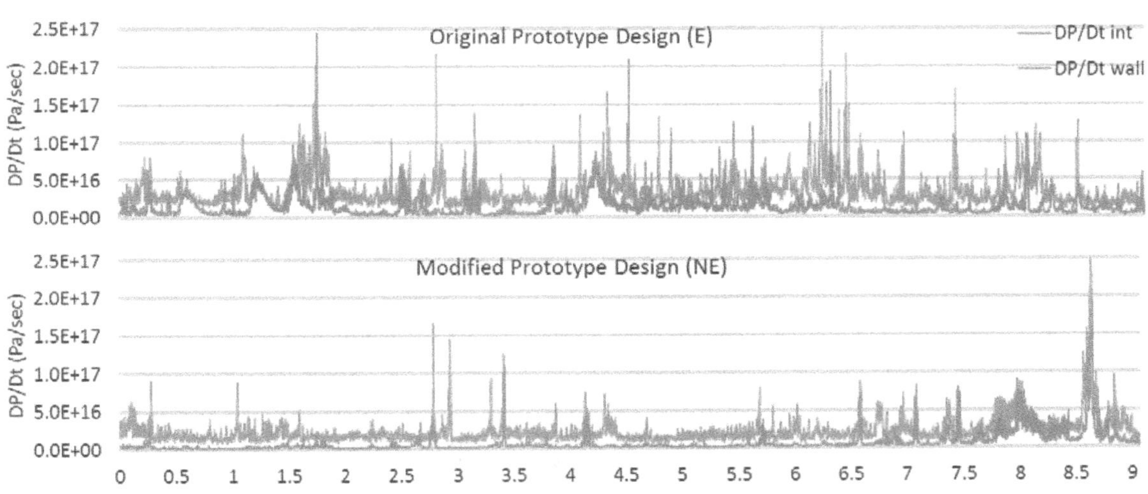

Fig. 5- Maximum total derivative of pressure seen in the entire domain (DP/Dt int) and on any surface (DP/Dt wall) for 9 us after initialization.

Fig. 4 shows the maximum pressure computed in the entire domain and on the wall for each time-step for the E and NE geometries. Both geometries experienced pressure spikes in the 7000 bar range, and further analysis (not included here) has shown that these occur in the region of erosion. However, the E geometry experienced many more spikes above 5000 bar range. The first total derivative of pressure also shows much more activity for the E geometry, both in the interior and on the wall (Fig. 5). Again, the maxima hit similar values, but the frequency of occurrence is starkly different between the two designs.

Discussion

For much of the CFD work performed in industry, the RANS technique for modelling turbulence is still the method of choice due to its acceptable time scale compared with LES/DES techniques. Hence, it is worth noting that the RANS model results (Case 1) showed a significant difference between the E and NE geometry. It is possible this difference was only case specific. However, it is interesting that the difference was in the region of erosion and it is possible the results indicate that backflow and recirculation close to collapsing vapor structures results in an increased risk of erosion.

Employing the UDF for modified mass transfer with the DES method (Case 3) enabled a more accurate and potentially useful result than any of the DES cases without the UDF. The results showed notable differences between the original prototype design (E) and the modified design (NE). As such, they are encouraging, but further work is required before concluding the outcome. This work is in progress and includes running the simulations over longer periods of time, and further data analysis and interpretation. Future work will also include stress analyses using the actual hardware geometry to understand the effect on the metal. These stress analyses will explore the impact of pressure spike levels and associated frequencies, as well as the rate of pressure change.

Acknowledgment

This work is part of the CaFE (Cavitating Flows, surface Erosion damage and material loss) Project, which is sponsored the EU Marie Sklodowska-Curie Innovative Training Network (ITN) MSCA-ITN-2014-ETN, grant agreement number 642536. Furthermore, this work would not be possible without the help of Delphi Technologies and City, University of London. In addition, the contribution of Alexandre Coutton in sourcing information is much appreciated.

References

[1] D. Bush, C. Soteriou, M. Winterbourn and C. Daveau, “Investigating hydraulic control components in high performance injectors,” in *IMechE Fuel Injection*, London, 2015.

[2] B. Befrui, P. Spiekermann, M. A. Shost and M.-C. Lai, "VoF-LES Studies of GDi Multi-Hole Nozzle Plume Primary Breakup and Comparison with Imaging Data," in *ILASS Europe Conference on Liquid Atomization and Spray Systems*, Chania, Greece, 2013.

[3] R. E. Bensow, "Simulation of the unsteady cavitation on the Delft Twist11 foil using RANS, DES and LES," in *Second International Symposium on Marine Propulsors*, Hamburg, Germany, 2011.

[4] P. Koukouvinis, N. Mitroglou, M. Gavaises, M. Lorenzi and M. Santini, "Quantitative predictions of cavitation presence and erosion-prone locations in a high-pressure cavitation test rig," *J. Fluid Mech.*, vol. 819, pp. 21-57, 2017.

[5] Ansys, Inc., "Ansys Fluent Theory Guide," SAS IP, Canonsburg, PA., 2013.

[6] P. Zwart, A. Gerber and T. Belamri, "A Two-Phase Flow Model for Predicting Cavitation Dynamics," in *5th international Conference on Multiphase Flow*, Yokohama, Japan, 2004.

[7] C. Soteriou, M. Smith and R. Andrews, "Diesel injection - laser light sheet illumination of the development of cavitation in orifices," in *IMechE C529/018/98*, 1998.

[8] M. L. Shur, P. R. Spalart, M. K. Strelets and A. K. Travin, "A hybrid RANS-LES approach with delayed-DES and wall-modelled LES capabilities," in *International Journal of Heat and Fluid Flow*, 2008.

[9] C. Soteriou, M. Lambert, S. Zuelch and D. Passerel, "The flow characteristics of high efficiency diesel nozzles with enhanced geometry holes," *Thiesel,* 2006.

© 2018 ASME

In-Nozzle Cavitation-Induced Orifice-to-Orifice Variations Using Real Injector Geometry and Gasoline-Like Fuels

[1]Roberto Torelli*; [1]Katarzyna E. Matusik; [1]Brandon A. Sforzo;
[1]Alan L. Kastengren; [1]Christopher F. Powell; [1]Sibendu Som;
[2]Yuanjiang Pei; [2]Tom Tzanetakis; [2]Michael Traver

[1]Argonne National Laboratory, Lemont, IL, USA;
[2]Aramco Services Company: Aramco Research Center – Detroit, Novi, MI, USA

Abstract

Many concurrent sources contribute to cycle-to-cycle variability (CCV) in internal combustion engines. This variability is undesirable, as it is responsible for deviations from the expected operating conditions, which in turn affect the engine efficiency and performance. Shot-to-shot variability connected to the fuel injection process is one strong candidate among CCV's many sources. Previous numerical work employing diesel fuel in a nozzle geometry measured with x-ray diagnostics has shown that manufacturing tolerances and needle radial motion are responsible for orifice-to-orifice variations in total injected mass. These variations are significant in the presence of shot-to-shot variability in the needle radial motion. The growing interest for low temperature combustion modes such as gasoline compression ignition has made the use of gasoline-like fuels in compression ignited engines desirable. However, when operating under typical diesel conditions, these fuels generally show considerable propensity for cavitation in the orifices and at the needle seat, due to their high volatility. In previous work, in-nozzle cavitation has been shown to correlate strongly with needle radial motion. This might constitute another source of CCV, as cavitation within the orifices might enhance the already present shot-to-shot variability by reducing the available cross-sectional areas and the relative orifice discharge coefficients. In addition, the systematic presence of cavitation might also result in the local erosion of the nozzle internal geometry, leading to even higher variability with time. This study focuses on the evaluation of orifice-to-orifice variability of gasoline-like fuel injection under diesel operating conditions. High-resolution x-ray geometry characterization of an eight-hole heavy-duty diesel injector has been combined with new measurements of the needle motion using a straight-run gasoline. A well-validated computational setup using the CONVERGE code was employed to perform a series of simulations, which highlighted the influence of cavitation on the internal nozzle flow. This combined effect of cavitation, needle radial motion, and orifice-to-orifice differences found is expected to have a strong influence on shot-to-shot variations.

Keywords: cavitation; gasoline compression ignition; x-ray tomography; needle motion

Introduction

During the last few decades engine manufacturers have played a major role in reducing the transportation sector share of both greenhouse emissions and criteria pollutants. The simultaneous co-optimization of engine design and existing fuels and the exploration of the chemistry of new fuels offer great opportunities of improvements. According to current projections, the partial electrification of the passenger car fleet will cause the demand for gasoline products to decrease while the demand for diesel and middle distillate products will continue to rise as a result of global economy's growth [1, 2]. Under this scenario, gasoline could become a valid alternative for commercial vehicle users if the cost associated with owning and operating such vehicles becomes competitive in comparison to traditional diesel vehicles.

Gasoline compression ignition (GCI), offers a possible solution as it can burn the fuel under low temperature combustion (LTC) conditions while increasing fuel efficiency and reducing criteria pollutant emissions [3, 4]. An interesting candidate for preliminary testing in GCI applications is Straight-run gasoline (SRG). SRG is characterized by diesel-like chemical reactivity and physical properties similar to common gasoline and therefore allows for an intermediate step toward the use of conventional gasoline in compression-ignited engines. If compared to diesel, SRG is characterized by lower viscosity and density as well as higher saturation pressure [5]. Previous work by the authors [5, 6] showed that, for internal nozzle flow applications, fuel properties such as density and viscosity are fundamental in determining the amount of fuel that the injection system is capable to deliver. If compared to diesel, SRG's viscosity enabled higher velocities and volumetric flow rates inside the injector orifices and partially compensated for the lower specific mass of SRG. On the other hand, higher saturation pressures typical of gasoline-like fuels enhance the

*Corresponding Author, Roberto Torelli: rtorelli@anl.gov

occurrence of in-nozzle cavitation. Phase change due to cavitation commonly occurs at the needle seat and at the inlet of the injector orifices where strong pressure gradients develop [7]. Cavitation's presence causes perturbations in the flow field that result in deviations of the injector behavior from the expected nominal performance. In addition to the fuel property effect, needle lateral motion is also responsible for strong perturbations of the flow field. Previous work [5-9] has shown that the lateral motion of the needle is connected to the fuel tendency to travel along preferential paths inside the nozzle resulting in orifice-to-orifice variability. Furthermore, such variability is enhanced by the presence of in-nozzle cavitation [5, 8], especially for those cases in which the amount of fuel vapor is not homogeneously distributed among the orifices due to the influence of the needle radial motion. Experiments performed by the authors have shown that the needle radial motion can be affected by considerable shot-to-shot variability [9]. As a consequence, the lateral oscillations could likely cause the location of such flow-field and cavitation non-uniformities to change for each injection events (i.e., engine cycle) possibly resulting in cycle-to-cycle variability (CCV). The authors have shown that in simulations, and for a non-cavitating fuel such as diesel, orifice-to-orifice differences as high as 10% could be found for separate injection events [9]. This might become an even more critical issue for heavy-duty GCI applications in which swirl and tumble motions are generally less intense than for light-duty engines and therefore promote lower air-fuel mixing.
The scope of this study is to point out the effect that the physical properties of a high-volatility fuel such as SRG have on in-nozzle cavitation and flow-field development. All the results obtained with SRG are compared against diesel results. The novelty of this work is in the simulation of a production eight-hole heavy-duty diesel injector whose geometry and needle motion were measured with well-established, validated x-ray techniques [9-12].

Methods

The Cummins injector geometry used for the simulations presented in this work was obtained through x-ray nozzle tomography experiments that were performed at the 7-BM beamline of the Advanced Photon Source (APS) at Argonne National Laboratory [13]. Further details on the experimental setup are reported in the authors' previous work [9].
The geometry obtained through the x-ray tomography experiments has been presented in [9] where a detailed description of the geometry preparation and characterization was provided. Table 1 summarizes the main features of the injector geometry and provides the ranges for each reported quantity. Most of the variability was found in the orifice inlet diameter as well as in the orifice included angles.

	Measured Values			Units
	Minimum	Average	Maximum	
Orifice outlet hydraulic diameter	185.43	185.83	186.46	[μm]
Orifice inlet hydraulic diameter	227.71	228.70	230.08	[μm]
K-factor	4.20	4.29	4.42	[-]
Orifice length	1.020	1.027	1.036	[mm]
L/D Ratio	5.49	5.53	5.59	[-]
Domain volume (sac + 8 orifices)	0.667 (needle completely closed)			[mm^3]
Sac diameter	1.00			[mm]
Included angle (half)	73.25	74.36	75.48	[°]

Table 1. Details of the injector geometry [9]

The injector needle motion was recorded at the 32-ID beamline of the APS [14]. A pressure chamber fitted with x-ray transparent windows was employed to host the injector at atmospheric pressure and temperature. A 5.0 standard L/min flow of gaseous N_2 was used to avoid liquid fuel accumulation within the field of view. The fuels were pressurized to the desired injection pressure using a standard common-rail diesel system. Further details about the experimental setup are reported in [9]. The tracking of the needle motion was performed in a two-step fashion, using two lines of sight that varied by 90° and recording 30 injection events per view. A cross-correlation algorithm was used on the resulting image stacks to track the in-axis and radial motions of the needle tip as a function of time. Finally, the combination of the two view's ensemble averages allowed the full three-dimensional motion to be defined. Figure 1 reports a comparison between the two fuels at the three tested injection pressures. The injector energizing times are 2.5, 2.0, and 2.5 ms at 1000, 1500, and 2500 bar injection pressure respectively. The needle lift profiles are not reported here for the sake of brevity. It is worth mentioning that the lift profiles were characterized by high repeatability and the maximum needle lifts were measured as ~450, ~470 and ~495 μm for diesel and ~425, ~453 and ~475 μm for SRG at the three injection pressures (1000, 1500, and 2500 bar). Figure 1 shows that the needle average fluctuations are characterized by some eccentricity in the needle position when the needle is at maximum lift (i.e., between 1.00 and 2.5 ms approximately). The needle vibrations are likely due to an elastic relaxation of the needle from the position it holds in sealed configuration [11]. The main difference between the two fuels is that when diesel is used, the needle tends to follow a circular (elliptical at higher pressures) trajectory around a center point whose location depends slightly on the injection pressure. In the case of SRG and for a given pressure, the needle is located at a much larger distance if compared to the corresponding diesel case. In addition, the oscillations are more limited around the average center point. An explanation to this phenomena might be linked to the different fluid-structure interactions occurring between the needle and the two fuels. In the case of diesel, the flow is characterized by a larger momentum due to the higher fuel density. This likely causes the needle to vibrate with amplitudes that are larger than those found for SRG.

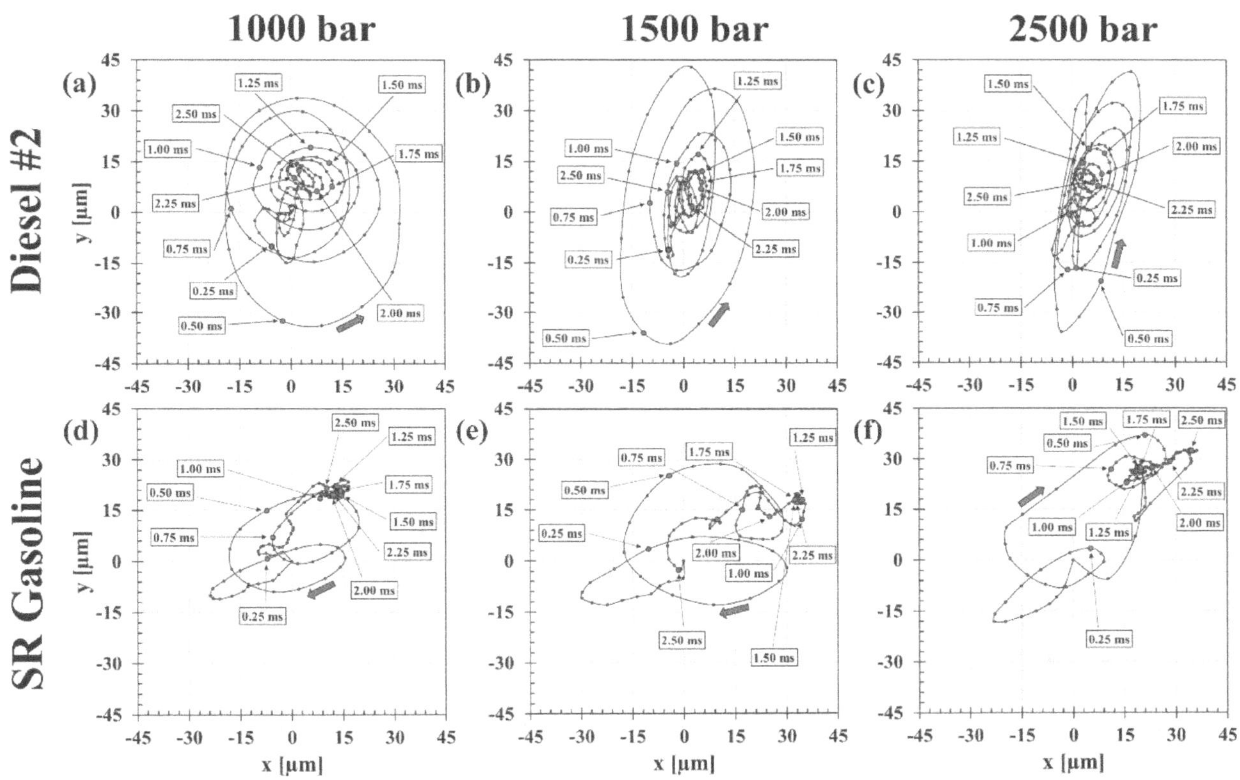

Figure 1. Average lateral motion path of the injector needle at 1000 (a, d), 1500 (b, e) and 2500 bar (c, f) injection pressures for diesel #2 (a, b, c) and SRG (d, e, f) fuels respectively. a), b), and c) are taken from [9].

All the simulations were performed with the CONVERGE CFD code [15] using a Reynolds-Averaged Navier-Stokes formulation closed by the Standard *k-ε* turbulence model. A volume of fluid (VOF) approach with the assumption of homogenous mixture was employed to describe the multiphase nature of the flow field. All phases (i.e., liquid, vapor, and non-condensable gases) were modeled as compressible. A perfect gas assumption was used for the gaseous species, while the liquid was modeled with a barotropic fluid assumption. The homogenous relaxation model (HRM) by Bilicki and Kestin [16] was adopted to account for the phase changes due to the occurrence of cavitation.

The physical properties of the two fuels, i.e., diesel #2 and SRG, were taken from the CONVERGE code database and [5] respectively. The mesh and numerical setup was consistent with the authors' previous work [9] in which the validation against experimental measurements of diesel mass flow rate at several injection pressures was reported.

Results

This section presents some results that highlight the different behavior between diesel and SRG when operating under the same nominal conditions. In all cases the fuel and ambient temperatures were set at 358 K, while the back pressure was equal to 100 bar. The results are focused on the initial stages of the injection events since cavitation was found to occur at very low lifts where the radial motion of the needle was comparable with the needle lift. Figure 2 shows a comparison of diesel and SRG's mass flow rates at the exit of the eight orifices. The mass flow rate for the two fuels is also compared by disabling the radial needle motion. It is clear how the presence of the needle motion causes high orifice-to-orifice variability to emerge. For the cases in which the needle motion was disabled (Figure 2 a and b) the orifice-to-orifice variations were mostly due to the geometric features captured through the use of the x-ray scanned geometry. Previous work on the same geometry [9] concluded that these variations were mostly due to orifice-to-orifice cross-sectional area differences and asymmetries in the needle vs. needle seat coupling.

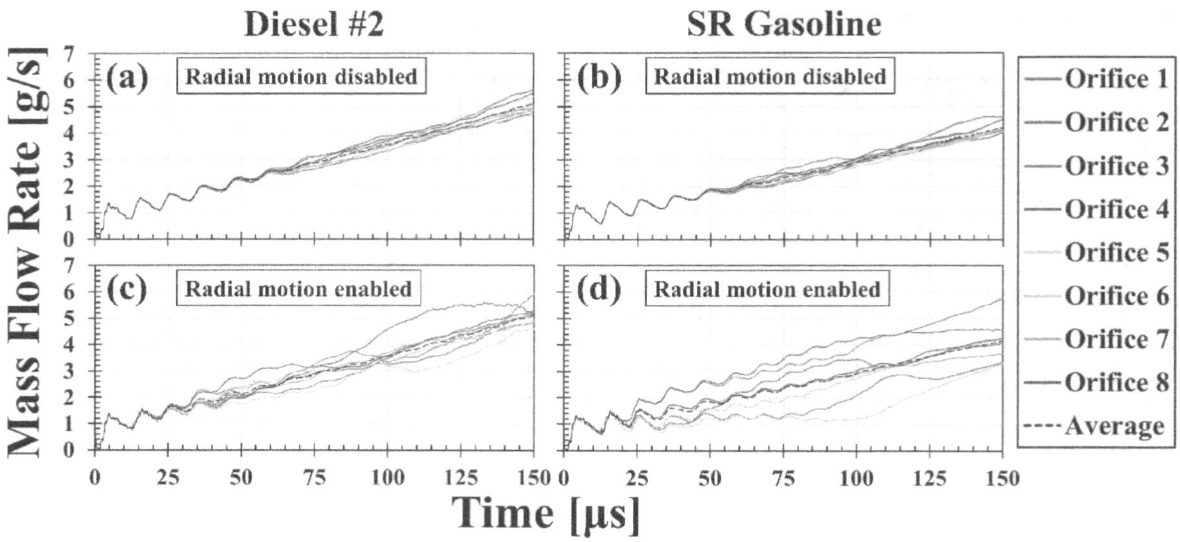

Figure 2. Orifice mass flow rates at 1000 bar injection pressure with radial motion disabled (a, b), and enabled (c, d) using diesel #2 (a, b) and SRG (c, d) fuels respectively. The results plotted in a) and c) are extracted from [9].

Figure 2 a and b also show the effects associated with the density difference. Due to SRG's lower density, mass flow rates for the gasoline-like fuel are consistently lower than diesel for all orifices, resulting in a ~20% difference at 150 μs. The effect of density becomes less distinguishable when needle radial motion is considered. Indeed, the lateral needle oscillations enhance the orifice-to-orifice variability for both fuels (cf. Figure 2 c and d). SRG was most affected by the presence of the needle lateral motion and showed a much wider spread in mass flow rate compared to diesel. This behavior could be caused by two separate sources:

- the experimentally measured needle motion is different between the two fuels, causing the flow fields to differ substantially as a direct result of different fluid-structure interactions;
- the higher saturation pressure of SRG causes the fuel to cavitate more than in the case of diesel leading to cavitation-induced perturbations of the flow field.

Both of the explanations provided above can be confirmed by the plots in Figure 3 which show the contour of fuel vapor volume fraction and velocity magnitude at 140 μs after the start of the injection. The flow structures inside the nozzle are very different between the two fuels. Indeed, in the case of diesel, the needle is located almost at the center of the sac and therefore resulting in a symmetric flow field. On the other hand, SRG's velocity contours show a much higher degree of asymmetry as the needle is located very close to the seat near orifice 4. This caused the fuel to follow a preferential path near orifice 8, which resulted in the detachment of the flow near the seat and the consequent development of a recirculation area near the top part of orifice 8. The jet coming out of the gap gained momentum in the direction of orifice 4 (diametrically opposed to 8) and entered orifice 4 with high speeds. The impact of part of the jet with the bottom of orifice 4 also resulted in a recirculation area at the inlet of the same orifice. This region was

characterized by strong velocity (and pressure) gradients, which enhanced the occurrence of cavitation as shown by Figure 3 b. A similar behavior was also observed for orifice 3 (not shown here).
Cavitation was absent for diesel at the selected time-stamp as the velocity gradients were lower than SRG's. Nevertheless, diesel was characterized by much milder cavitation due to its lower saturation pressure even at times where the needle location was similar to the one of SRG in Figure 3. Diesel's peak values of fuel vapor volume fraction never exceeded 0.1 at all times and for all tested injection pressures. On the other hand, SRG showed strong cavitation even at the lowest of the three pressures. At 1500 bar injection pressure (not shown here for the sake of brevity), cavitation started much earlier (85 μs) and persisted for a longer time than it did for the 1000 bar case. A similar and even more accentuated behavior was observed for the 2500 bar case as well (not shown here).
SRG showed some mild cavitation in all orifices and at low needle lifts even in those cases where the radial needle motion was disabled. Although not shown here, the extent of cavitation was very limited and located mainly at the needle seat and at the bottom edge of the orifice inlets (similar to what observed for orifice 8 in Figure 3 b).

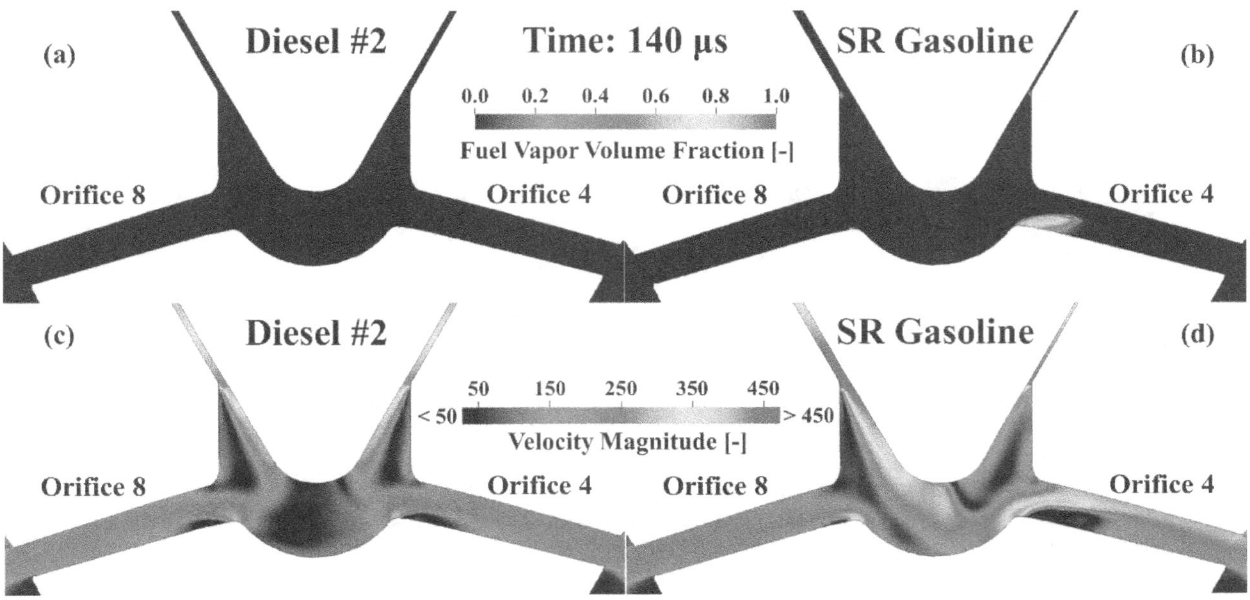

Figure 3. Contours of fuel vapor volume fraction (a, b) and velocity magnitude (c, d) at 140 μs using diesel #2 (a, c) and SRG (b, d) fuels respectively. The results are plotted on a plane that crosses through orifice 4 and 8 and includes the injector symmetry axis.

Conclusions

This study evaluated the in-nozzle flow characteristics of conventional diesel fuel and straight-run gasoline operating under typical diesel injection conditions. CFD simulations were performed to investigate geometry-induced, and orifice-to-orifice variability in a Cummins eight-hole heavy-duty injector using x-ray measured geometry and needle motion. The mass flow rate and the associated flow field were studied at three selected injection pressures and the results obtained at one of the pressures (1000 bar) were discussed. The geometry-induced orifice-to-orifice differences in mass flow rate were pointed out, confirming the unique insights that measured injector geometries can offer compared to nominal ones. The influence of the needle off-axis vibrations was evaluated by alternatively including and excluding the measured average radial motions for the two fuels. The analysis and the results confirmed that the radial displacement was effective only at low lifts. During the initial phase of the injection, jet-like structures formed in those areas where the needle-to-seat gap was larger. This provided a preferential path for the fuel to flow through the injector and reach specific orifices with very high velocities. High velocity and pressure gradients were found to be responsible for the occurrence of cavitation. SRG's different physical properties explained the differences in mass flow rate as well as cavitation magnitude compared to diesel and previous studies. Under all conditions investigated, SRG consistently showed more cavitation than diesel. This high propensity for cavitation, combined with high degree of randomness of the radial motion, is expected to contribute to the existing variability and possibly enhance the orifice-to-orifice differences among injection events. In addition, the systematic and localized occurrence of cavitation is expected to have a profound influence on cavitation erosion characteristics of heavy-duty injectors operating with light-end fuels.

Acknowledgements

The submitted manuscript has been created by UChicago Argonne, LLC, Operator of Argonne National Laboratory (Argonne). Argonne, a U.S. Department of Energy Office (DOE) of Science laboratory, is operated under Contract No. DE-AC02- 06CH11357. The U.S. Government retains for itself, and others acting on its behalf, a paid-up nonexclusive, irrevocable worldwide license in said article to reproduce, prepare derivative works, distribute copies to the public, and perform publicly and display publicly, by or on behalf of the Government.

The experimental measurements presented in this work were performed at the 7–BM and 32-ID beamlines of the APS at Argonne National Laboratory. Use of the APS is supported by the U.S. Department of Energy (DOE) under Contract No. DE–AC02–06CH11357.

The authors wish to thank:
1. Blues and Bebop High Performance LCRC cluster facilities at Argonne National Laboratory; and
2. Convergent Science Inc., for providing the CONVERGE CFD software licenses.

References

[1] US Energy Information Administration (2017). *International Energy Outlook 2017, DOE/EIA-0484*. www.eia.gov/outlooks/ieo/pdf/0484(2017).pdf, accessed 01/02/2018.

[2] ExxonMobil (2017). *2017 Outlook for Energy: A View to 2040*. www.cdn.exxonmobil.com/~/media/global/files/outlook-for-energy/2017/2017-outlook-for-energy.pdf, accessed 01/02/2018.

[3] Zhang, Y., Voice, A., Tzanetakis, T., Traver, M., Cleary, D. (2016). *An Evaluation of Combustion and Emissions Performance with Low Cetane Naphtha Fuels in a Multicylinder Heavy-Duty Diesel Engine*. Journal of Engineering for Gas Turbines and Power, Vol. 138 102805-1/10.

[4] Zhang, Y., Kumar, P., Traver, M., Cleary, D. (2016). *Conventional and Low Temperature Combustion Using Naphtha Fuels in a Multi-Cylinder Heavy-Duty Diesel Engine*. SAE Int. J. Engines, Vol. 9(2), pp. 1021–1035.

[5] Torelli, R., Som, S., Pei, Y., Zhang, Y., Voice, A., Traver, M., Cleary, D. (2017) *Comparison of In-Nozzle Flow Characteristics of Naphtha and N-Dodecane Fuels*. SAE Technical Paper 2017-01-0853.

[6] Torelli, R., Som, S., Pei, Y., Zhang, Y., Traver, M. (2017). *Influence of Fuel Properties on Internal Nozzle Flow Development in a Multi-Hole Diesel Injector*. Fuel, Vol. 204, pp. 171-184.

[7] Torelli, R., Som, S., Pei, Y., Zhang, Y., Traver, M. (2017). *Internal Nozzle Flow Simulations of Gasoline-Like Fuels under Diesel Operating Conditions*. ILASS Americas, 29th Annual Conf. on Liquid Atomization and Spray Systems, Atlanta, GA..

[8] Battistoni, M., Xue, Q., Som, S, Pomraning, E. (2014). *Effect of Off-Axis Needle Motion on Internal Nozzle and Near Exit Flow in a Multi-Hole Diesel Injector*. SAE Int. J. Fuels Lubr., Vol. 7(1), pp. 167–182.

[9] Torelli, R. Matusik, K., Nelli, K., Kastengren, A., Fezzaa, K., Powell, C., Som, S., Pei, Y., Tzanetakis, T., Zhang, Y., Traver, M., Cleary D. (2018). *Evaluation of Shot-to-Shot In-Nozzle Flow Variations in a Heavy-Duty Diesel Injector Using Real Nozzle Geometry*. SAE Technical Paper 2018-01-0303.

[10] Powell C., Kastengren, A., Liu, Z., Fezzaa, K. (2010). *The Effects of Diesel Injector Needle Motion on Spray Structure*. J. of Eng. Gas Turbines Power, Vol. 133(1), 012802.

[11] Kastengren, A., Tilocco, F., Powell, C., Manin, J., Pickett, L. M., Payri, R., Bazyn, T. (2013). *Engine Combustion Network (ECN): Measurements of Nozzle Geometry and Hydraulic Behavior*. Atomization and Sprays, Vol. 22(12), pp. 1011–1052.

[12] Swantek, A., Kastengren, A., Duke, D., Tilocco, Z., Sovis, N., Powell, C. (2015). *Quantification of Shot-to-Shot Variation in Single Hole Diesel Injectors*. SAE Int. J. Fuels Lubr. 8(1):2015.

[13] Kastengren, A., Powell, C., Arms D., Dufresne E., Gibson H., Wang J. (2012). *The 7BM Beamline at the APS: a Facility for Time–Resolved Fluid Dynamics Measurements*. Journal of Synchrotron Radiation, Vol. 19, pp. 654–657.

[14] Shen, Q., Lee, W., Fezzaa, K., Chu, Y., De Carlo, F., Jemian, P., Ilavsky, J., Erdmann, M., Long, G. (2007). *Dedicated Full-Field X-Ray Imaging Beamline at Advanced Photon Source*. Nuclear Instruments and Methods in Physics Research - Section A, Vol. 582(1), pp. 77–79.

[15] Richards, K., Senecal, P., Pomraning, E. (2016). *CONVERGE Manual (Version 2.3)*. Convergent Science Inc., Madison, WI, USA. https://convergecfd.com, accessed 01/02/2018.

[16] Bilicki, Z., Kestin, J. (1990). *Physical Aspects of the Relaxation Model in Two-Phase Flow*. Proceedings of the Royal Society London A, Vol. 428, pp. 379–397.

© 2018 ASME

10th International Symposium on Cavitation - CAV2018
Baltimore, Maryland, USA, May 14 – 16, 2018

In-Nozzle Flow Visualization of Marine Diesel Injector Nozzles with Different Inlet Radii

[1,2]Reto Balz*; [2]Andreas Schmid; [1]David Sedarsky

[1]Chalmers University of Technology, Gothenburg, Sweden; [2]Winterthur Gas & Diesel Ltd., Winterthur, Switzerland

Abstract

Injector geometries of large marine two-stroke Diesel engines differ extensively from configurations typically used in diesel engines. The multi-hole injector orifices are generally arranged asymmetrically as all the bores face a similar direction. Due to this geometric setup, the orifices are also distributed eccentrically with respect to the central bore of the injector nozzle. Experiments have shown that spray propagation from such orifices is not symmetric with respect to the nominal axis of the orifice. These spray deviations can lead to wall wetting which increases fuel consumption, emissions, and component temperatures as well as loss of lubrication film.

To further investigate the in-nozzle flow and how it affects the spray morphology for large marine two-stroke Diesel engines, experiments were performed using a transparent nozzle holder for single-hole, transparent nozzles made of PMMA. The transparent nozzles were used with Diesel fuel under real injection pressure conditions of 50 MPa. PMMA was chosen to reduce optical distortion due to refraction index differences between the fluid and transparent nozzle material. The transparent nozzles used were orthogonally arranged 0.75 mm diameter mono-hole designs with different levels of hydro-erosive grinding and therefore different inlet radii between main bore and orifice of the nozzle. These transparent nozzle designs were chosen to match those of large marine two-stroke diesel engines.

Shadowgraphy using a CCD camera, a far-field microscope and a pulsed Nd:YAG laser together with a diffusive optics were applied to visualize the cavitation in the nozzle during the complete quasi-steady-state injection process with a duration of 8 ms.

The observation of cavitation and flow behavior in the nozzles allows qualitative validation of CFD cavitation models for specific cavitation conditions relevant for marine fuel injection.

Keywords: in-nozzle flow; cavitation; fuel injection

Introduction

Approximately 90% of total world trade is moved by ships [1]. The workhorses of maritime trade are tankers, bulk carriers and container ships which are also the three vessel types that contribute the greatest carbon dioxide emissions [2]. These vessels all rely on large marine two-stroke engines for propulsion to meet the world's maritime transport needs. Their low speed allows connecting the propeller directly to the camshaft of the engine and they are very flexible regarding fuel type and fuel quality, which is of great importance given that heavy fuel oil is still the standard fuel in marine transport. Shipping has by far the highest carbon efficiency of commercial transportation, but as the industries greatest cost is fuel there is a strong economic interest to increase fuel efficiency and as a direct result further decrease carbon dioxide emissions [1]. The two-stroke design details of large marine Diesel engines, including their combustion chamber configuration, and fuel ignition properties generally require an injection strategy incorporating multiple fuel injectors. Usually, three injectors arranged by 120° are used that can inject the fuel using different strategies for different engine loads. Due to the strong swirl motion of the charged air and the circularly arranged multiple fuel injectors the injector nozzle tips also tend toward asymmetrical and eccentric designs. Experiments at Winterthur Gas & Diesel Ltd. have shown that these nozzles can lead to spray deflections on the order of 12° [4]. Further CFD simulations with an advanced cavitation model showed that the compressible, two-phase flow inside the orifices strongly influences the spray propagation [5].

In fuel injection systems, cavitation usually occurs as geometric cavitation caused by sudden reduction of static pressure in the fluid provoked through contraction, or string cavitation which forms stochastically near the center of strong vortices. Several authors ([5], [13], [15]) indicate that both cavitation phenomena are important for in-nozzle flow of Diesel fuel injectors and investigations have shown that both cavitation forms appear under realistic injection conditions for marine Diesel injectors [6].

*Corresponding Author, Reto Balz: reto.balz@wingd.com

The connection between in-nozzle flow and spray formation is currently widely using a variety of CFD models ([13], [14], [15], [17]). However, experimental data acquired under realistic conditions, especially for large marine engines, is lacking. Several authors have recently shown in-nozzle flow with transparent nozzles related to fuel injection ([8], [9], [10], [11]), but comparatively little work has been performed on large marine Diesel injectors. Hult *et. al.* [12] have shown in-nozzle flow results with orifice sizes related to marine engines, but limited fuel pressures of around 10 MPa and sharp edged nozzle designs.
The inlet radii play a dominant role for cavitation in the nozzle and have been investigated by several authors ([15], [16], [17]), however, mostly with computational methods and not under realistic conditions for large marine Diesel engines.

Experimental setup

Transparent nozzles made from polymethyl metacrylate (PMMA) were used to visualize the in-nozzle flow of large marine two-stroke Diesel engines injectors. PMMA was chosen due to its refractive index which roughly matches Diesel fuel and its superior pressure response compared with quartz glass [3]. Due to the cylindrical design of the injector nozzle bores it is necessary to use refractive index matching, since the orifice shape acts as a lens distorting the light used for optical measurements otherwise. To cope with the high fuel pressures during the quasi-steady-state injection process, the transparent nozzle is braced in a so-called transparent nozzle holder (TNH). External forces are applied from all sides to the rectangular shaped PMMA nozzle as investigations have shown that this decreases the failure probability significantly [3].

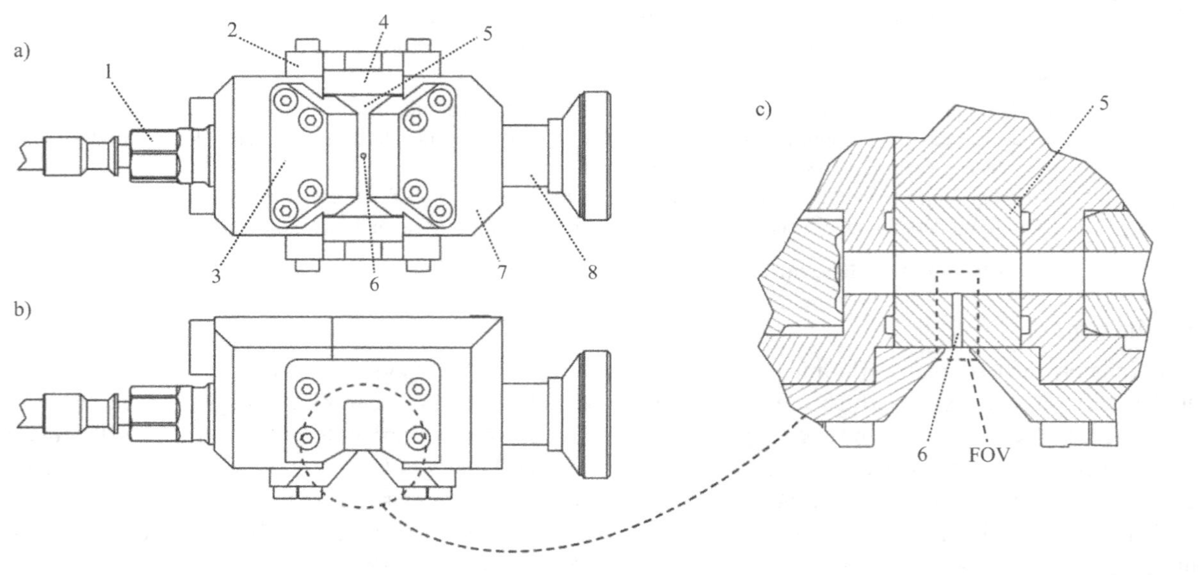

Figure 1: Schematic of transparent nozzle holder: top (a) and side (b) view and sectional view of transparent nozzle (c) with pressure sensor (1), side clamp (2), top clamp (3), sapphire brick (4), transparent nozzle (5) with orifice (6), main body (7) and injector mount (8). The sectional view (c) also depicts the approximate location of the used field of view (FOV) of the optical measurements.

The nozzles used in the work presented here are single-orifice designs to provide simple flow for comparison to CFD. Additionally, multi-hole configurations would further weaken the PMMA nozzles are challenging to work with at these pressures. The orifice diameter of the nozzles is 0.75 mm which is a representative full-size orifice for large marine Diesel engine injectors. The orifice is arranged 90° with regard to the main bore which has a diameter of 3.5 mm. Figure 1 depicts a schematic of the used TNH with a top view (a), a side view (b) and a sectional view detail showing the PMMA nozzle with the main bore and orifice (c). The whole TNH is mounted on the injector using the standard mount from a typical nozzle tip (8). The PMMA nozzle is sealed using standard O-rings with a hardness of 90 Shore-A. A piezo resistive pressure sensor (Kistler 4067C2000) is mounted as close as possible to the transparent nozzle without a tap that can decrease the frequency spectrum due to Helmholtz-resonator effects. The sensor used has a natural frequency of over 200 kHz that allows dynamic pressure data acquisition. The PMMA nozzle is pressed together by the main body (8) and the body that holds the pressure sensor. Additionally, two clamps mounted on top of the TNH (3) and two clamps on the side (2) apply a defined force to the nozzle's other surfaces. Due to optical access, the side clamps (2) apply the force via polished sapphire bricks (4). Sapphire was chosen due to its excellent optical and tensile strength properties. Due to this compression of the PMMA nozzle, it is possible to apply realistic marine two-stroke Diesel engines injection pressures. However, the PMMA construction can only withstand a limited number of injections before it fails.

The two sapphire bricks (4) together with the side clamps (2) allow the application of line-of-sight optical measurements. Here we image the flow using pulsed-light shadowgraphy. The optical setup consists of a far-field microscope (Questar QM100) with a long pass filter to block the laser source wavelength and a CCD camera (TSI PowerView 4MP-LS). A pulsed, double cavity, frequency doubled Nd:YAG laser supplies source pulses at 532 nm (Spectra Physics PIV-400) which are used to excite spontaneous emission in a dye cell diffuser (based on [7]). This cell filled with Rhodamine B was mounted before the TNH and acts to remove speckle and to guarantee a uniform background illumination. To acquire sharp images, a pulsed light source is necessary to reduce motion blur given that the velocities in the orifice are on the order of 200 m/s.

The fuel used in this work was commercially available Diesel with a density of 815.9 kg/m^3 (at 20°C) and a viscosity of 2.112 mm^2/s (at 40°C). The fuel was pressurized using a high-pressure, air driven liquid pump (Haskel DSXHF-452) together with a 5 dm^3 piston accumulator to stabilize the fuel pressure. The rail pressure was 50 MPa and the back pressure was ambient atmosphere. The injection duration was set to 8 ms but the optical arrangement only allows for two images per injection due to the laser system timing. The images shown here were taken 5 and 6 ms after start of injection.

The transparent nozzles were hydro-erosively ground using a test rig and abrasive fluid from Extrude Hone. By pushing this mixture of high-viscosity, non-Newtonian fluid and abrasive particles through the transparent nozzles, material is consistently removed forming representative inlet radii between the main bore and orifice of the nozzle. The fluid was pushed through the nozzles using a hydraulic cylinder providing a prescriptive and well-defined pressure and duration of operation for the nozzle treatment. The nozzles investigated have different strongly pronounced radii and are compared to the sharp edge design (as schematically depicted mounted in the TNH in Figure 1). The four nozzles investigated were hydro-erosively ground with a linear increasing time step of 30 s, i.e. the first nozzle was treated for 30 s, the second nozzle 60 s, the third nozzle 90 s and the fourth nozzle 120 s.

Results

All images shown in Figure 2 were taken during the quasi-steady-state condition of the fuel injection. The injector needle was fully open and the fuel pressure and volume flow are constant. The images were acquired 5 ms after start of injection.

Dark areas indicate multiphase/cavitating flow and bright areas indicate non-cavitating liquid Diesel fuel. Due to small differences in the refractive indices of the Diesel fuel used and PMMA, the main bore and the orifice wall of the nozzle can be seen in the images as dark lines.

The approximate field of view (FOV) selected is depicted schematically in Figure 1 c). The field of view focuses on the orifice, as the main bore displayed no steady-state cavitation phenomena. The fuel flow enters the main bore from the right side, enters the orifice and leaves the nozzle by flowing vertically downwards (compare with images depicted in figure 2).

The images in Figure 2 show that the level of cavitation decreases significantly with increasing radii (increasing levels of hydro-erosive grinding). Geometric cavitation is substantially reduced, this is especially apparent transitioning from the sharp-edge design (a) to the first level of hydro-erosive grinding (b). While the next step of hydro-erosive grinding (c) results in a weaker reduction, it seems there is no more supercavitation state (geometrical cavitation reaches the exit of the orifice) and no more geometrical cavitation revolving around the orifice at the inlet. The next step of hydro-erosive grinding (d) shortens the length of the geometrical cavitation pockets along the orifice wall significantly, to a length of approximately one orifice diameter. At this point there are no apparent cavitation patterns near the inlet. The last stage of hydro-erosive grinding (e) removes geometrical cavitation almost completely.

The resolution of the images is still limited to some degree by speckle and non-uniform illumination, and due to the differences in refractive index matching. The evaluation of the images with no supercavitation state is more difficult, as the walls are depicted dark as well. This can be clearly seen in Figure 2 d) and e) where the left orifice wall creates the impression of a cavitation pocket that evolves along the whole length.

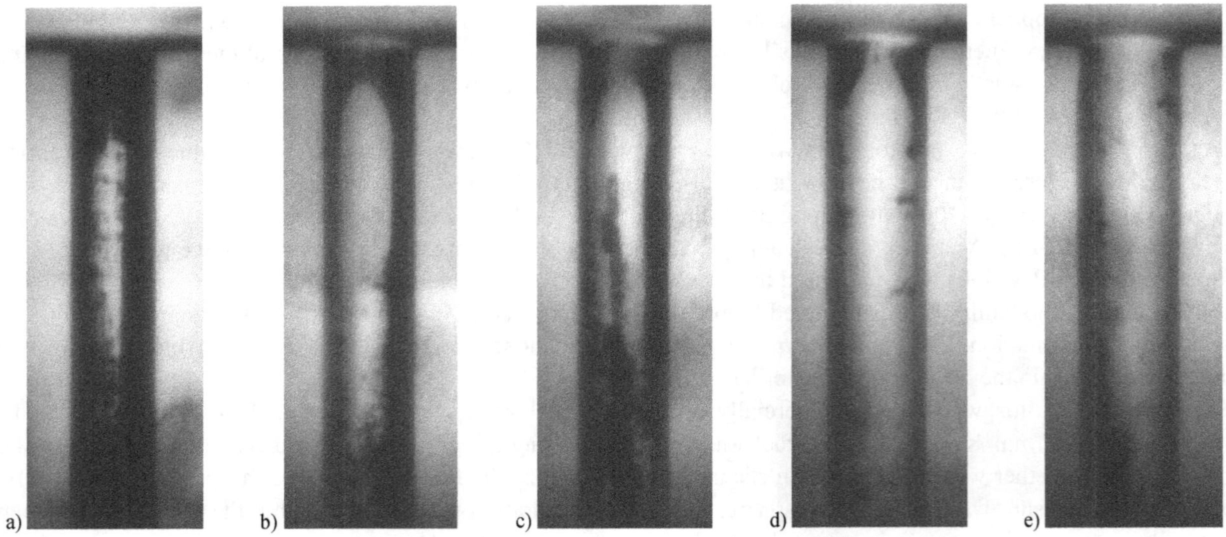

Figure 2: Acquired images of in-nozzle flow at 5 ms after start of injection, with 50 MPa rail pressure, atmospheric back-pressure and different radii between nozzle main bore and orifice. Radii increase from a) to e). Dark areas indicate cavitation.

There are transient cavitation phenomena visible such as cavitation bubbles in the center in Figure 2 c and in a smaller version closer to the right orifice wall in Figure 2 d). The increasing radii is clearly visible in the nozzle geometry, especially if one compares the orifice inlet of Figure 2 a) and e). However, the radii are comparatively small regarding their significant influence on cavitation in the orifice.

Conclusions

The influence of orifice radii on cavitation characteristics has been shown to be very important under real fuel pressure conditions for large marine Diesel injector nozzles. In this example, relatively small radii adjustments are sufficient to almost completely reduce geometrical cavitation in the orifice under ambient temperature conditions. The in-nozzle flow can be clearly visualized using the transparent PMMA nozzles although the small refractive index mismatch of the Diesel fuel used leads to darkening along the orifice and main bore walls. The hydro-erosive grinding process applied here has been shown to work well with the transparent PMMA nozzles. These in-nozzle flow results are consistent and repeatable, indicating that the abrasive grinding of the PMMA nozzle radii is a reliable approach for generating representative nozzles for testing.

Acknowledgments

The authors would like to thank the Combustion Engine Research Center (CERC), the Swedish Energy Agency and the Swiss Federal Office of Energy for funding.

References

[1] International Chamber of Shipping. (2017). Annual Review.
[2] United Nations Conference on Trade and Development. (2016). Review of Maritime Transport.
[3] Falgout, Z., Linne, M. (2016). *Novel design for transparent high-pressure fuel injector nozzles*. Review of Scientific Instruments. 87(8).
[4] Schmid, A., von Rotz, B., Schulz, R., Hermann, K., Weisser, G., Bombach, R. (2013). *Influence of nozzle hole eccentricity on spray morphology*. 25th European Conference on Liquid Atomization and Spray Systems (ILASS-Europe), Chania Greece.
[5] Schmid, A., Habchi, C., Bohbot, J., von Rotz, B., Hermann, K., Bombach, R., Weisser, G. (2014). *Influence of in-nozzle flow on spray morphology*. 26th European Conference on Liquid Atomization and Spray Systems (ILASS-Europe), Bremen, Germany.
[6] Balz, R., Schmid, A., Sedarsky, D. (2017). *Cavitation Flow Visualization in Marine Diesel Injectors*. The 9th International Conference on Modeling and Diagnostics for Advanced Engine Systems (COMODIA), Okayama, Japan.
[7] Saransh, J., Somasundaram, S., Anand, T.N.C. (2016). *A fluorescent laser-diffuser arrangement for uniform backlighting*. Measurement Science and Technology. 27(025406).

[8] Blessing, M., König, G., Krüger, C., Michels, U., Schwarz, V. (2003). *Analysis of Flow and Cavitation Phenomena in Diesel Injection Nozzles and Its Effects on Spray and Mixture Formation*. SAE Technical Paper. (2003-01-1358).
[9] Fansler, T.D., Parrish, S.E. (2015). *Spray measurement technology: a review*. Measurement Science and Technology. 26(1).
[10] Suh, H.K., Lee, C.S. (2008). *Effect on cavitation in nozzle orifice on the diesel fuel atomization characteristics*. International Journal of Heat and Fluid Flow. 29(4).
[11] Badock, C., Wirth, R., Fath, A., Leipertz, A. (1999). *Investigation of cavitation in real size diesel injection nozzles*. International Journal of Heat and Fluid Flow. 20(5).
[12] Hult, J., Simmank, S., Matlok, S., Mayer, S., Falgout, Z., Linne, M. (2016). *Interior flow and near-nozzle spray development in a marine-engine diesel fuel injector*. Experiments in Fluids. 57(4).
[13] Giannadakis, E., Gavaises, M., Arcoumanis, C. (2008). *Modelling of cavitation in diesel injector nozzles*. Journal of Fluid Mechanics. 616(2008).
[14] Yan, F., DU, Y., Wang, L., Tang, W., Zhang, J., Liu, B., Liu, C. (2017). *Effects on injection pressure on cavitation and spray in marine diesel engine*. International Journal of Spray and Combustion Dynamics. 9(3).
[15] Salvador, F.J., Carreres, M., Jaramillo, D., Martinez-Lopez, J. (2015). *Analysis of the combined effect of hydrogrinding process and inclination angle on hydraulic performance of diesel injection nozzles*. Energy Conversion and Management. 105(2017).
[16] Li, X., Chen, C., Yu, Y., Gao, G. (2015). *Numerical investigations on effects of nozzle's geometric parameters on the flow and the cavitation characteristics within injector's nozzle for a high-pressure common-rail DI diesel engine*. Energy Conversion and Management. 89(2015).
[17] Im, K.-S., Cheong, S.-K., Powell, C.F., Lai, M.-C.D., Wang, J. (2013). *Unraveling the Geometry Dependance of In-Nozzle Cavitation in High-Pressure Injectors*. Scientific Reports. 3(2013).

© 2018 ASME

10th International Symposium on Cavitation - CAV2018
Baltimore, Maryland, USA, May 14 – 16, 2018 CAV18-05117

Insight into tip vortex cavitation using velocity field measurements

Kyle M. Sinding, David R. Hanson, Michael H. Krane, and Jeremy Koncoski

Applied Research Laboratory, Penn State University, State College, PA, USA

Abstract

Statistics of tip vortex core pressure are estimated from instantaneous Stereo PIV measurements of the velocity field. The tip vortex was generated by a rectangular planform (aspect ratio 2), elliptically-loaded fin with a NACA66 foil shape. A set of 500 velocity field realizations were acquired at each of three angles of attack. Estimates of pressure for each measured velocity field were computed using a finite-element scheme that minimized error propagation from the measured velocity field. Histograms of core pressure show that the mean core pressure followed expected scaling with vortex strength, and that the width of the distribution is inversely proportional to vortex strength.

Keywords: tip vortex, cavitation, pressure estimation

Introduction

This paper presents estimates of tip vortex core pressure from instantaneous velocity fields measured in a plane intersecting a tip vortex (TV) flow over a rectangular planform. It has been proposed [1] that TV pressure coefficient fluctuations, relative to the mean, affect the mean incipient cavitation number. By non-intrusively acquiring a large number of velocity fields, and by estimating the pressure field associated with each [2], statistics of the TV pressure coefficient are derived.

Many studies have reported non-intrusive velocity measurements of a tip vortex flow over a stationary lifting surface [3-14,] but only a few include simultaneous velocity and load cell measurements [10-14]. Each study characterized vortex strength using velocity field measures, *e.g.*, circulation, core radius, or core velocity profile shape. Of these, only Stinebring, *et al.*, [9] estimated the vortex core pressure field using their velocity measurements. However, because the velocity field was measured point-by-point using Laser velocimetry, the pressure filed could only be estimated in an ensemble-averaged sense, and could only estimate an average departure from the mean using turbulent kinetic energy.

Here, a combination of load measurements and Stereo PIV measurements of velocity on a plane intersecting the TV are presented [14]. Using a finite element estimation technique [2], an instantaneous pressure coefficient, $C_{p,min}$, denoting pressure drop across the vortex core for each snapshot of the flow, was computed. From this ensemble of pressure coefficients, the probability density function and its first two moments (mean and standard deviation) are computed.

Methods

Flow measurements were conducted in the round test section of the 12inch water tunnel, located in Penn State Applied Research Laboratory's Garfield Thomas Water Tunnel facility [14, 15]. The tunnel test section flow speed ranges 0-80ft/s, and the independently controlled operating pressure ranges from 3-35psi, with free stream turbulence intensity to less than 0.5%. The tunnel test section is 18 inches long. The test section is optically accessible on three sides through flat, transparent acrylic windows. The fourth side is composed of a base plate on which the fin is mounted. A rectangular planform (aspect ratio 2) fin generated the tip vortex. The fin had a NACA 66(mod) profile and a spanwise twist distribution giving elliptical loading profile, and a rounded tip. The quarter chord was centered on the fin base/load cell axis of rotation.

Twenty-four pressure taps are located along the test section, between each window/baseplate, at 6 axial locations. A Kiel probe, located in the plenum, was used to measure total pressure. All pressures were measured using Heise DXD series, 100 PSIA digital pressure transducers. Tunnel speed was estimated using the total pressure, test section entrance pressure, and the cross section areas of the tunnel at the pressure probe locations. An AMTI Model MC1-6-250-5763

*Corresponding Author, Michael Krane: mhk5@arl.psu.edu

six-axis load cell measured the forces and moments applied to the fin during the experiment. Load cell strain gage voltages, all tunnel pressures, and water temperature data were all recorded digitally with a National Instruments DAQ system.

As illustrated in Figure 1, Stereo Particle Image Velocimetry (SPIV) measurements were performed on a plane oriented at a right angle to the tunnel axis, roughly centered on the location of the tip vortex, approximately 0.50 chord lengths downstream of the trailing edge. The plane was illuminated with a dual head Nd:YAG Solo PIV laser, operated in double pulse mode, at a frequency of 15Hz. Images were captured with two Sensicam cameras operating in doubleshutter mode. A combination of LaVision Scheimpflug lens mounts and acrylic prisms pressed onto the window exterior minimized optical distortion. The spatial calibration was performed using a 200mm^2 dual-plane, dual-side calibration target (TSI, Inc.) with markers spaced vertically and horizontally 10mm apart, and a 1mm displacement between planes. Image pairs were acquired using INSIGHT 3G software (TSI, Inc.) and processed using standard, multipass stereo-cross-correlation software (DAVIS 8.0.5, LaVision). In the final pass, a 32x32 pixel correlation window, with 50% overlap, was used, resulting in a 19x33 point vector field (627 total vectors) over the field of view.

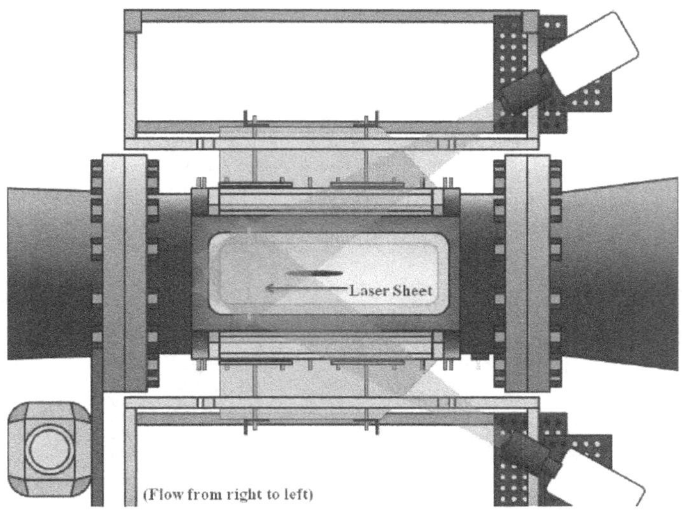

Figure 1. Schematic of SPIV setup, showing cameras, optical path (pink), prisms (gray) imaging laser sheet (green), and fin [14]. Flow is from right to left.

An instantaneous pressure field was estimated from each velocity field acquired, using a finite element formulation [2]. This method reduces the error propagation from the measured velocity into the estimated pressure field, and allows for more flexibility in applying pressure boundary conditions on the region containing the measured velocity field. The pressure and pressure gradient on the border of the measurement region was estimated using the velocity field. Finite-differencing of the velocity field was used to estimate the normal pressure gradient, at all but one point on the boundary, where the pressure was estimated using the total pressure measured in the plenum, and the measured local velocity vector, and the Bernoulli equation. This point was chosen such that it was the least likely to be located on a streamline that had passed through an upstream region containing vorticity. For each vector field, a pressure coefficient is defined in terms of the difference between minimum core pressure and ambient pressure:

$$C_{p,min} = \frac{p_{min} - p_{\infty}}{\frac{1}{2}\rho U_{\infty}{}^{2}}$$

Results

Velocity was measured in a plane located one half chord downstream of the trailing edge, nominally centered on the tip/trailing edge corner location, at U_∞ = 30 ft/s, for three angles of attack ($\alpha = 0°$, $\alpha = 2.5°$, and $\alpha = 5°$). At each angle of attack, 500 image pairs were acquired. For each resulting velocity field snapshot a corresponding pressure field was computed as described above. A typical result is shown in Figure 2. The minimum pressure coefficient $C_{p,min}$ was then computed from each field.

Figure 3 shows tip vortex pressure difference statistics as a function of tip vortex strength. Figure 3a shows probability density functions of $-C_{p,min}$ for each of the three angle of attack conditions tested. Figure 3b shows variations of the mean and standard deviations of $-C_{p,min}$ as a function of tip vortex strength $C_L^2 Re^{0.4}$ [see, *e.g.*, 16]. These results show that mean vortex core pressure coefficient increases, while variations about that mean decrease, with angle of attack (vortex strength). These results indicate that the minimum core pressure variation can be higher than 10% of the mean.

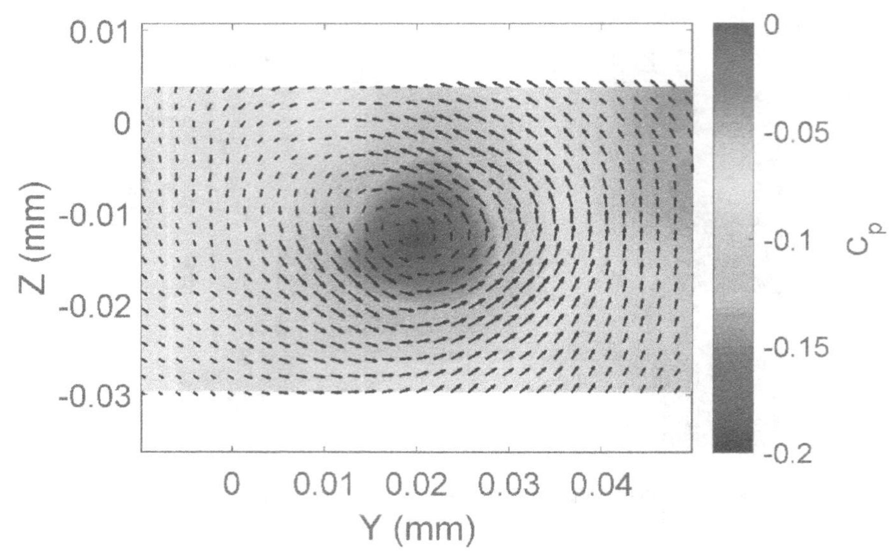

Figure 2. Instantaneous snapshot of velocity and pressure fields of tip vortex, measured one-half chord length downstream of fin trailing edge. $\alpha = 0°$, $U_\infty = 9.2$m/s. Vectors show velocity projected onto measurement plane, colormap shows pressure coefficient.

Summary

This paper presents statistics of tip vortex core pressure, estimated from instantaneous velocity measurements in a plane intersecting the vortex core. The vortex was generated using an elliptically-loaded fin. Pressure estimation was performed using a finite element scheme that minimizes propagation of velocity error into the pressure estimation. For the three fin angles of attack studied, pressure statistics were presented. These show that the mean core pressure coefficient follows the expected trend with vortex strength, and that the variations of core pressure relative to the mean have an inverse proportionality with vortex strength.

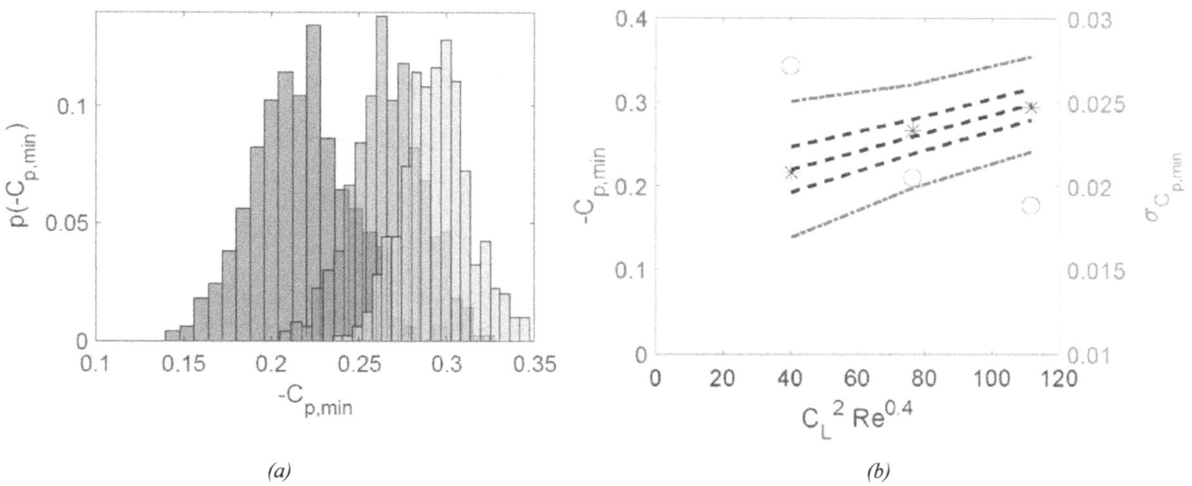

Figure 3. Tip vortex core pressure statistics as a function of tip vortex strength. *(a)* Probability density functions of $-C_{p,\min}$ for the three test conditions: *blue*: $\alpha = 0°$, *pink*: $\alpha = 2.5°$, *yellow*: $\alpha = 5°$. *(b)* variation with tip vortex strength ~ $C_L^2 Re^{0.4}$: *: mean $-C_{p,\min}$, o: standard deviation of $-C_{p,\min}$. Blue and green lines indicate one and two standard deviations from the mean, respectively. Note that as vortex strength increases, the mean core pressure difference increases, and the variation of core pressure difference decreases.

References

[1] R.E. Arndt, A.P. Keller (1992) *Water quality effects on cavitation inception in a trailing vortex*, Journal of Fluids Engineering, Transactions of the ASME 114(3), 430.
[2] Krittian, S., Lamata, P., Michler, C., Nordsletten, D., Bock, J., Bradley, C., Smith, N. (2012). *A finite-element approach to the direct computation of relative cardiovascular pressure from time-resolved MR velocity data*, Medical Image Analysis, 16, 1028.
[3] A. Shekarriz, J. Katz, H. Liu, T. Huang, T. Fu (1992) *Study of Junction and Tip Vortices Using Particle Displacement Velocimetry,* AIAA Journal 30(1), 145.
[4] A. Shekarriz, T. Fu, J. Katz, T. Huang (1993) *Near-field behavior of a tip vortex,* AIAA Journal 31(1), 112.
[5] A. Vogt, P. Baumann, J. Kompenhans, M. Gharib (1996) *Investigations of a wing tip vortex in air by means of DPIV,* in Advanced Measurement and Ground Testing Conference, p. 2254.
[6] B.S. Thurow, T. Fahringer, K. Johnson, S. Hellman (2014) *Comparison of plenoptic PIV and stereo PIV measurements in a wing tip vortex,* in The International Symposia on Applications of Laser Techniques to Fluid Mechanics, Lisbon, Portugal.
[7] Birch, D., Lee, T., Mokhtarian, F., Kafyeke, F. (2003) *Rollup and near-field behavior of a tip vortex*, Journal of Aircraft 40(3), 603.
[8] Zhang, H., Zhou, Y., Whitelaw, J. (2006) *Near-field wing-tip vortices and exponential vortex solution*, Journal of Aircraft 43(2), 445.
[9] Stinebring, D., Farrell, K., Billet, M. (1991) *The structure of a three-dimensional tip vortex at high Reynolds numbers*, Journal of Fluids Engineering 113, 496.
[10] Beresh, S., Hening, J., Spillers, R. (2009) *Planar velocimetry of a fin trailing vortex in subsonic compressible flow*, AIAA Journal 47(7), 1730.
[11] Beresh, S., Smith, J., Hening, J., Grasser, T., Spillers, R. (2009) *Interaction of a fin trailing vortex with a downstream control surface*, Journal of Spacecraft and Rockets 46(2), 318.
[12] Elsayed, O. Asrar, W., Omar, A., Kwon, K. (2011) *Evolution of NACA23012 wake vortices structure using PIV*, Journal of Aerospace Engineering 25(1), 10.
[13] Krane, M., Meyer, R., Weldon, M., Elbing, B., DeVilbiss, D. (2015) *Measurements of loading and tip vortex due to high-Reynolds number flow over a rigid lifting surface*, Journal of Fluids Engineering 137(7), 071301.
[14] Hanson, D. (2012) *Cavitation inception for non-elliptically loaded fins*. M.S. thesis, The Pennsylvania State University.
[15] Lauchle, G. Billet, M., Deutsch, S. *High-Reynolds number liquid flow measurements,* in Frontiers in Experimental Fluid Mechanics, vol. 46, ed. by M. Gad-el Hak (Springer).
[16] Arndt, R. (2002) *Cavitation in vortical flows*, Annual Review of Fluid Mechanics, 34, 143.

© 2018 ASME

Interaction between Thermodynamic Suppression Effect and Reynolds Number Promotion Effect on Cavitation in Hot Water

[1]Yuka Iga*; [1]Teppei Furusawa; [1]Hiritoshi Sasaki

[1]*Institute of Fluid Science, Tohoku University, Sendai, Miyagi, JAPAN*

Abstract In this study, in order to clarify a interaction between thermodynamic effect, which suppresses cavitation, and scale effect, which promotes cavitation, tunnel experiment was done with NACA16012 single hydrofoil in hot water. Appearance of the effects was estimated by occurrence condition of cavitation for angle of attack vs cavitation number. And the cause of thermodynamic effect is estimated by measuring the temperature depression inside the cavity. Mainstream temperature was 30 °C and 70 °C. By increasing mainstream temperature under constant Reynolds number condition, scale effect was took away to some extent and thermodynamic effect was picked out. From the result, occurrence resin of cavitation was suppressed in 70 °C hot water, where it is empirically known that thermodynamic effect does not appear in such a low temperature. The result indicates that thermodynamic effect and scale effect, which are conflicting effects, cancelled each other in water at 70 °C. Additionally, it was shown that occurrence of cavitation was drastically suppressed in a condition under specific range of angle of attack.

Keywords: Thermodynamic effect, Scale effect, Hot water, NACA16012, Thermistor

Introduction

Generally, volume of cavitation is same when a cavitation number σ is same in the flow field around same body. An effect which breaks the similarity raw is known as scale effect. The scale effect is caused by the influence of nuclei. When the Reynolds number Re in mainstream is higher, the cavity inception moves higher cavitation number condition [1,2]. Also, cavity volume is bigger when Re is higher in spite of same σ condition [3].

On the other hand, in many fluids except for room temperature water, when mainstream temperature T_∞ is higher, the cavity volume is smaller in spite of same σ condition. The effect is known as thermodynamic effect [4-8]. The mechanism of the suppression of cavity volume is that when a cavity occurs, the local temperature decreases around the cavity due to the latent heat of evaporation of the cavity. The local saturated vapor pressure decreases due to the temperature depression, and evaporation is then suppressed. The factor of the thermodynamic effect is expressed by the thermosphysical properties of each fluid in each condition, the effect is known to become obvious in cryogenic fluids such as liquid hydrogen and liquefied natural gas as well as in hot water and refrigerant. The thermodynamic effect is scale effect in a broad sense because cavity volume is different in same σ, but it is distinguished from scale effect which appears without thermal influence. Hence, the thermodynamic effect is also distinguished from scale effect in the present study.

When T_∞ increases, Re increases due to decrease of the viscosity. At that time, conflicting two effects is considered to interact with cavitation, which are suppression effect by the thermodynamic effect and promotion effect by the scale effect. Generally, nuclei are small in number in cryogenic fluid, then thermodynamic effect is obvious. On the other hand, in a tap water, nuclei are comparatively in large number and the scale effect cannot be neglected. Cervone et al. reported the opposite thermodynamic effect in a single hydrofoil experiment using hot water [9], in which the cavity volume increased according to increase T_∞ in unsteady cavitation condition. We assume it is a result from that the scale effect exceeded the thermodynamic effect in the experiment.

In the present study, experiment of NACA16012 single hydrofoil in hot water tunnel is done in order to reveal the relationship between thermodynamic effect and scale effect in hot water. Two kinds of comparison are done about occurrence range of cavitation with T_∞ change, one is in constant mainstream velocity U_∞ condition which corresponds an experiment of conventional thermodynamic effect and the other is in constant Re condition where scale effect is removed in some degree. T_∞ is increased from 30 °C to 70 °C. In the temperature range, it is empirically known thus far that the thermodynamic effect is not obvious or reversely appears. Additionally, temperature depression inside the cavity which is a factor of the thermodynamic effect.

*Corresponding Author, Yuka Iga: iga@ifs.tohoku.ac.jp

Experimental Condition

The present experiment is done in high-temperature water cavitation tunnel in Institute of Fluid Science, Tohoku University [10]. The mainstream temperature T_∞ is increased from 30 °C to 70 °C in two conditions of constant mainstream velocity U_∞ and constant Reynolds number Re. Although in the temperature range, it is empirically known that the thermodynamic effect is not obvious, it is widest range in constant Re condition in the present tunnel. Mainstream velocity is U_∞ = 6.9 m/s and U_∞ = 13.4 m/s. The test body is NACA16012 single hydrofoil of chord length 30 mm. The angle of attack is incremented by 1 degree from 0 to 9 degree. Definitions of Cavitation number s, Reynolds number Re, thermodynamic parameter Σ [5] and non-dimensional thermodynamic parameter Σ^* [11] are follows.

$$\text{Cavitation Number :} \quad \sigma = \frac{p_\infty - p_v}{1/2\rho_\infty U_\infty{}^2} \tag{1}$$

$$\text{Reynolds Number :} \quad \mathrm{Re} = \frac{U_\infty C}{\nu_\infty} \tag{2}$$

$$\text{Thermodynamic Parameter :} \quad \Sigma = \frac{\rho_v{}^2 L^2}{\rho_l{}^2 C_{pl} T_\infty \sqrt{a_l}} \tag{3}$$

$$\text{Non-dimensional Thermodynamic Parameter :} \quad \Sigma^* = \frac{\rho_v{}^2 L^2}{\rho_l{}^2 C_{pl} T_\infty}\sqrt{\frac{C}{a_l U_\infty^3}} = \Sigma\sqrt{\frac{C}{U_\infty^3}} \tag{4}$$

In above equations, p_∞ [Pa], ρ_∞ [kg/m^3], ν_∞ [Pa s] and p_v [Pa] are static pressure, density, viscosity and saturated vapor pressure in mainstream, respectively. ρ_v [kg/m^3] is vapor density at mainstream pressure and temperature. L [J/kg] is latent heat. C_{pl} [J/(kg・K)] and a_l [m^2/s] are isobaric specific heat and thermal diffusivity of water. C [m] is chord length of hydrofoil. The bigger Σ become, the larger thermodynamic effect become, then cavity volume suppressed. Σ^* takes into account time interval for heat transfer. Then, the lower mainstream velocity is, the bigger thermodynamic effect becomes.

The fluid is degassed tap water, degree of oxygen saturation is 30%(±3.0%以内) at 30 °C. In addition, the oxygen saturation cannot be measured after rising T_∞. And number of nuclei cannot be controlled. Therefore, the scale effect cannot be completely removed away although in constant Re condition.

Results

A. Influence of Mainstream Temperature under Constant Velocity Condition

Occurrence range of cavitation and the pattern is determined from visualization of the aspect and frequency analysis of pressure fluctuation downstream of the test section. The resulting cavitation pattern map of NACA16012 hydrofoil is shown in Fig. 1 (a) about angle of attack α versus cavitation number σ in a condition of mainstream temperature T_∞ = 30 °C and mainstream velocity U_∞ = 13.4 m/s. In the flow condition, five kinds of cavitation patterns are observed, which are attached sheet cavitation, sheet-cloud cavitation, supercavitation, bubble cavitation and back detachment cavitation. In addition, a case in which cavitation pattern changes from one to another and the pattern is hard to distinguish is expressed by transition in the map.

In Fig 1 (a) where T_∞ = 30 °C and U_∞ = 13.4 m/s, according to increase of α, occurrence range extends to higher σ condition. At higher α condition from 6 deg. to 9 deg., attached sheet cavitation, sheet-cloud cavitation and supercavitation occurs in order according to decrease of σ. At α = 4 and 5 deg., bubble cavitation occurs in lower σ condition. At lower α condition from 0 deg. to 3 deg., back detachment cavitation occurs. The cavitation pattern map of NACA16012 was also made by Franc in room temperature water [12]. Although the cavitation pattern was no so far from the present result, but the occurrence condition of each pattern was different in some degree. It is because the tunnel scale and water quality were different from the present tunnel and the experiment was done with free surface condition.

Usually, thermodynamic effect is discussed by increasing T_∞ in a condition of constant U_∞. Then, the case where T_∞ is increased to 70 °C in the constant U_∞ condition at U_∞ = 13.4 m/s is shown in Fig 1 (b) and compared. In the condition, although the occurrence range slightly decreases at α = 4 deg. and over 7 deg., but there is not so large suppression occurrence range in Fig 1 (b) compared with Fig 1 (a). Additionally, although back detachment cavitation changes to bubble cavitation at α =2 deg. and 3 deg., but there is not so large difference in the cavity patterns except for that. The result agrees with common knowledge in which thermodynamic suppression effect does not appear in a hot water under T_∞ = 70 °C.

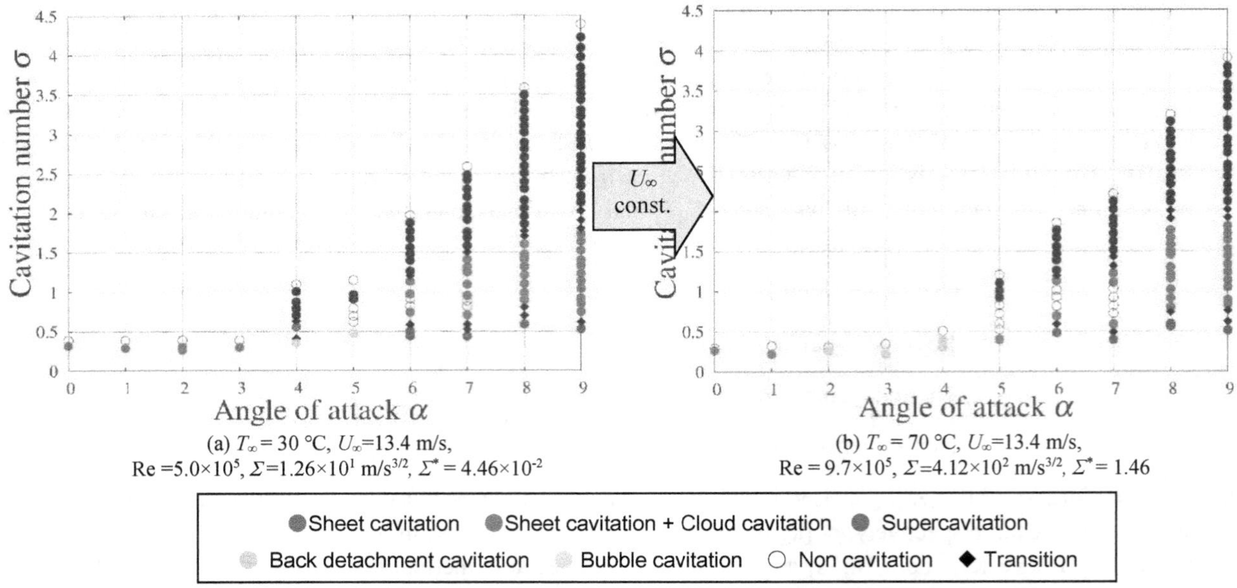

(a) T_∞ = 30 °C, U_∞=13.4 m/s, Re =5.0×10^5, Σ=1.26×10^1 m/s$^{3/2}$, Σ^* = 4.46×10^{-2}

(b) T_∞ = 70 °C, U_∞=13.4 m/s, Re = 9.7×10^5, Σ=4.12×10^2 m/s$^{3/2}$, Σ^* = 1.46

Fig. 1 Influence of increase of mainstream temperature on cavitation pattern map in a condition of constant mainstream velocity U_∞.

B. Influence of Mainstream Temperature under Constant Reynolds Number Condition

In the comparison in Fig 1, Reynolds number Re also increases at T_∞ = 70 °C in Fig 1 (b). Then, scale effect which is promotion effect of cavitation is considered to also affect on the cavitation with thermodynamic suppression effect in Fig 1 (b). Therefore, in order to reduce the scale effect, T_∞ is increased from 30 °C to 70 °C under constant Re condition at Re =5.0×10^5 by decreasing of U_∞. The comparison of cavitation pattern map under constant Re condition is shown in Fig 2, where Fig 2 (a) is same figure in Fig 1 (a).

In the condition at T_∞ = 70 °C in Fig 2 (b), occurrence range of cavitation is drastically suppressed as indicated by the region enclosed by the red line. The result indicates that thermodynamic effect is hiding in 70 °C water which is relatively low temperature and it appears when scale effect is reduced. In addition, the result supports our assumption of the reason of opposite thermodynamic effect in the hot water experiment by Cervone [9], in which the scale effect exceeded the thermodynamic effect due to the tunnel condition.

The extension of occurrence range of cavitation from T_∞ = 70 °C, U_∞=6.9 m/s in Fig 2 (b) to T_∞ = 70°C, U_∞ = 13.4 m/s in Fig 1 (b) corresponds to the amount of promotion by scale effect. At the same time, it is also explained by decrease of non-dimensional thermodynamic parameter Σ^* in which time interval for heat transfer is reduced by increasing the mainstream velocity. But, the value of decreasing of Σ^* from Fig 2 (b) to Fig 1 (b) is 3.94 to 1.46, it is sufficiently smaller than that from Fig 2 (b) to Fig 2 (a) which is 3.94 to 4.46×10^{-2}. Then, the contribution of reduction of thermal suppression effect is considered to be small in the extension of occurrence rage from Fig 2 (b) to Fig 1 (b). It is considered to be caused mainly by increase of scale effect.

The red line in Fig 2 (b) shows that the occurrence range of cavitation is drastically suppressed at the condition of angle of attacked between α = 4 deg. to 7 deg. It is important evidence that the degree of appearance of thermodynamic effect changes according to the change of local situation around a body although the mainstream condition is same. Franc et al investigated the relationship between the condition of boundary layer of suction side of a hydrofoil and cavitation pattern in same NACA16012 hydrofoil [12]. Although the Franc's tunnel condition was different from that in the present experiment as mentioned before, it was reported that the boundary layer changes from attacked laminar boundary layer with separation in around trailing edge, attached turbulent boundary layer with separation in around trailing edge to turbulent boundary layer separated from leading edge according to increase of angle of attack of the hydrofoil, and back detachment cavitation occurred in a condition of laminar boundary layer. Then, we are considering that thermodynamic effect is easy to appear in a boundary layer condition of attached turbulent boundary layer and now investigating. If the relationship between boundary layer condition and appearance of thermodynamic suppression effect will be clarified, it will be helpful in a design and development in fluid machinery.

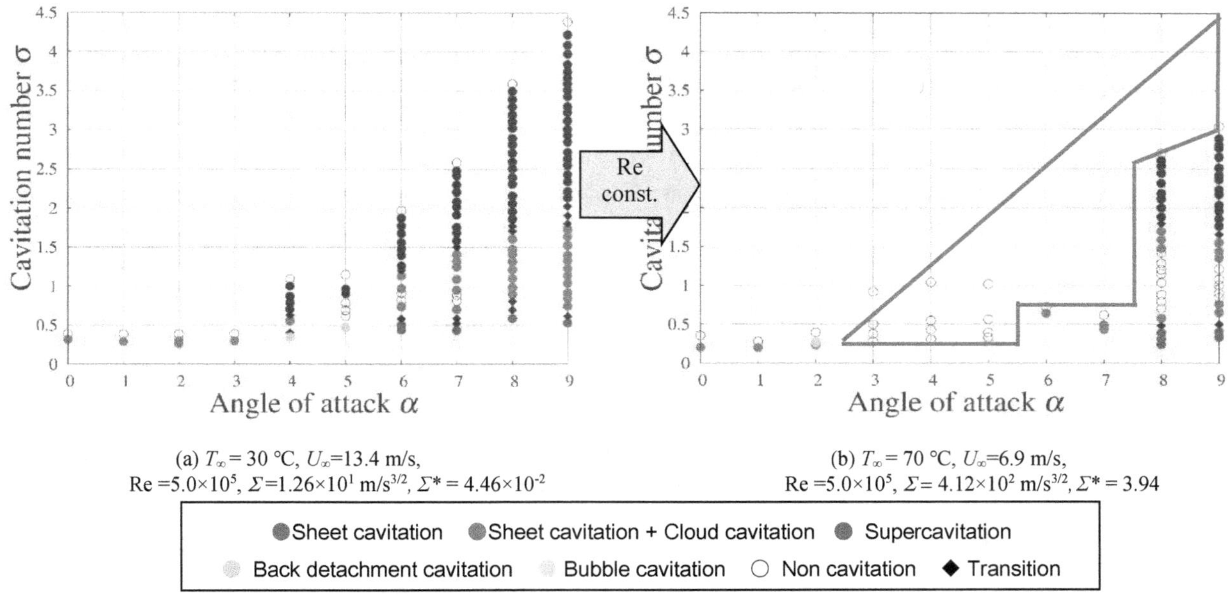

Fig. 2 Influence of increase of mainstream temperature on cavitation pattern map in a condition of constant Reynolds number.

C. Temperature inside a Cavity

Temperature depression inside a cavity is measured, which is a cause of thermodynamic effect and the evidence of latent thermodynamic effect. The temperature is measured by thermistor probe which is inserted from the side wall of the cavitation tunnel to inside the cavity directory [10]. The temperature depression ΔT is estimated using the temperature inside a cavity subtracted by mainstream temperature which is also measured by thermistor probe inserted from the side wall to the mainstream without cavity in the test section. The result of ΔT is shown in Fig 3, where temperature depression is measured in the three cases in which U_∞=13.4 m/s at T_∞ = 30 °C, U_∞=6.9 m/s and 13.4 m/s at T_∞ = 70 °C at angle of attack α = 9 deg. in all cases. And the error bar expresses expanded uncertainty of the temperature probe, which includes calibration and the repeated measurement. Because the time constant of the temperature probe is large, unsteady temperature cannot be measured and the measured temperature in unsteady

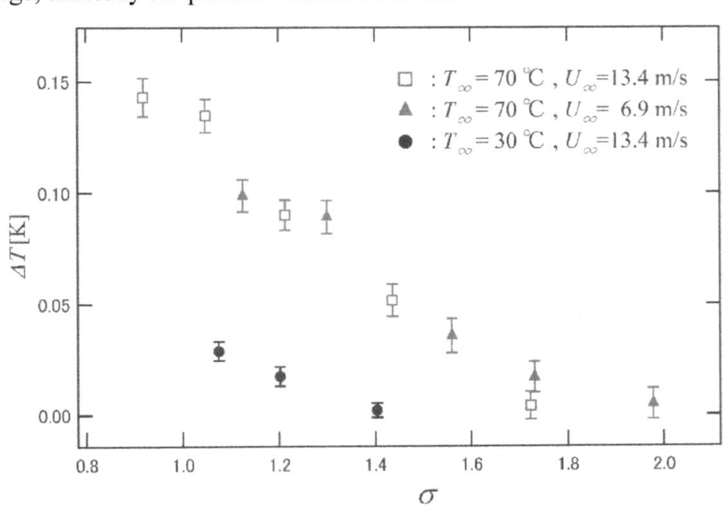

Fig.3 Temperature depression inside a cavity at angle of attack α = 9 deg.

cavitation condition is time averaged value in some degree, but it is not strict time-average.

In Fig 3, the ΔT is larger in $T_\infty = 70$ °C than that in $T_\infty = 30$ °C, it means thermodynamic effect become larger in $T_\infty = 70$ °C than that in $T_\infty = 30$ °C. Additionally, in the two cases in $T_\infty = 70$ °C, the ΔT are almost same. Then, it is confirmed that thermodynamic effect is also hiding in the case of $T_\infty = 70$ °C and U_∞=13.4 in Fig 1 (b) in same degree in Fig 2 (b). Furthermore, the same ΔT in the two cases in $T_\infty = 70$ °C also confirms that the extension of the occurrence region of cavitation from Fig 2 (b) to Fig 1 (b) is not by the decrease of Σ^* but by the scale effect. Therefore, it is confirmed that two conflicting suppression and promotion effects, which are thermodynamic effect and scale effect, are superimposed and cancelled each other in the case in Fig 1 (b).

Conclusion

In the present study, through visualization of the aspect of cavitation and measurement of temperature inside the cavitation around NACA16012 single hydrofoil in hot water, relationship between thermodynamic effect and scale effect is investigated. The obtained results are summarized as follows;

1. By increasing mainstream temperature under a constant Reynolds number condition, because the scale effect is reduced in some degree, the thermodynamic suppression effect drastically appears. Then, it can be shown that thermodynamic effect is hiding also in 70 °C water in which it is empirically known that the thermodynamic effect is not obvious or reversely appears.

2. When mainstream temperature is increased under a constant mainstream velocity condition, the occurrence range of cavitation is not so suppressed. The reason is confirmed by temperature measurement inside the cavity that two conflicting suppression and promotion effects, which are thermodynamic effect and scale effect, are superimposed and cancelled each other in the case.

3. It is shown that occurrence of cavitation is drastically suppressed in a condition under specific range of angle of attack. It is important evidence that the degree of appearance of thermodynamic effect changes according to the change of local situation around a body although the mainstream condition is same.

References

[1] Holl, J. W. and Wislicenus, G. F. (1961), *Scale effects on cavitation*, Journal of Basic Engineering, Vol. 83, pp.385-395.
[2] Arakeri, V. H. (1979), *Cavitation inception*, Proceedings of Indian Academy of Science, Vol. C2, Pt. 2, pp. 149-177.
[3] Keller, A. P. (2001), *Cavitation scale effects empirically founded relation and the correlation of cavitation number and hydrodynamic Coefficients*, Proceedings of CAV2001, Pasadena, No. lecture001, pp.1-18.
[4] Stepanoff, A. J. (1964), *Cavitation properties of liquids*, Journal of Engineering for Power, Vol. 86, pp.195-200.
[5] Brennen, C. E. (1973), *The dynamic behavior and compliance of a stream of cavitating babbles*, Journal of Fluids Engineering, Transaction ASME, Vol. 95, pp. 533-541.
[6] Kato, H. (1984), *Thermodynamic effect on incipient and developed sheet cavitation*, Proceedings of International Symposium on Cavitation Inception, FED-Vol. 16, pp. 127-136.
[7] Fruman, D. H., Reboud, J-L. and Stutz, B. (1999), *Estimation of thermal effects in cavitation of thermosensible liquids*, International Journal of Heat and Mass Transfer, Vol. 42, pp. 3195-3204.
[8] Franc, J. P., Rebattet, C. and Coulon (2004), A., *An experimental investigation of thermal effects in a cavitating inducer*, Journal of Fluids Engineering, Transaction ASME, Vol. 126, pp. 716-723.
[9] Cervone, A., Bramanti, C., Rapposelli, E. and d'Agostino, L. (2006), *Thermal cavitation experiments on a NACA 0015 hydrofoil*, Journal of Fluids Engineering, Transaction of ASME, Vol. 128, pp. 326-331.
[10] Iga, Y. and Yamaguchi, Y. (2015), *Temperature measurement inside a cavity in high temperature and high pressure water tunnel*, Turbomachinery, Vol. 43, No. 3, pp. 177-184 (in Japanese).
[11] Brennen, C. E. (1974), *Hydrodynamics of pumps*, Concepts ETI Inc. and Oxford Science Publications.
[12] Franc, J.P. and Michel, J. M. (1985), *Attached cavitation and the boundary layer: Experimental investigation and numerical treatment*, Journal of Fluid Mechanics, Vol. 154, pp.63-90.
[13] Iga, Y. and Yamaguchi, Y. (2016), *Temperature inside a cavitation on a hydrofoil in hot water*, Transactions of the JSME, Vol.82, No.837, DOI: 10.1299/transjsme 15-00548 (in Japanese).

© 2018 ASME

Interlaboratory Study on Standard Test Method for Erosion of Solid Materials by a Cavitating Jet

[1]Hitoshi Soyama*; [2]Seiji Shimizu; [3]Shuji Hattori; [4]Georges Chahine

[1]Tohoku University, Sendai, Miyagi pref., Japan
[2]Nihon University, Koriyama, Fukushima pref., Japan
[3]Fukui University, Fukui, Fukui pref., Japan
[4]Dynaflow, Inc., Jessup, MD, USA

Abstract

A test method for erosion of solid materials by using a cavitating jet was standardized by ASTM G02 committee in 1995 as ASTM G134. However, an interlaboratory study ILS was not carried out. In order to revise ASTM G134, ILS was carried out in collaboration with four institutions. Five materials, i.e., pure aluminum A1070B, bronze CAC402, popular nickel NW2201, nickel 200 NW2200 and stainless steel SUS316L were tested in the ILS at each institution by using cavitating jet apparatuses, which were compliant with ASTM G134. It was revealed that the variation of the repeatability coefficient was from about 2 % to 7 % and the variation of the reproducibility coefficient was from 6 % to 22 %. Note that the variation of the coefficient in the same institute was about 1 % - 10 %.

Keywords: erosion; cavitating jet; material test

Introduction

As cavitation erosion causes severe problems on hydraulic machineries such as pumps and valves, experimental evaluation of material resistance to cavitation erosion is a big issue. A test method of cavitation erosion by using a cavitating jet is useful, as the effects of hydraulic parameters such as velocity and cavitation number on cavitation erosion can be examined.

The standard test method for cavitation erosion by using a cavitating jet was standardized in American Standard Testing of Materials ASTM as G134-95 [1], by the ASTM subcommittee G02.10 "Erosion by Solid and Liquids", based on Lichtarowicz's research [2, 3]. However, there were no results of interlaboratory study (ILS) presented in ASTM G134. In order to keep ASTM G134, an ILS was required, as it is necessary to investigate repeatability and reproducibility of ASTM G134. In order to maintain ASTM G134, an ILS was carried out by collaboration with four different institutions, i.e., Tohoku University, Nihon University, Fukui University and DYNAFLOW INC. under the task group chaired by Soyama, Tohoku University, who has worked with Lichtarowicz from 1994 to 1996 at the University of Nottingham. The revised ASTM G134 was proposed and voted in 2017, it is now available as ASTM G134-17 [4]. The present paper presents the ILS of ASTM G134-17, and the results of erosion tests on five materials using cavitating jet apparatuses, which were compliant with ASTM G134 at four different institutions are shown.

Experimental Facilities and Procedures

Figure 1 shows a schematic diagram of a test chamber of a cavitating jet apparatus which is compliant with ASTM G134. The pressurized water is injected into the chamber through a nozzle. The size of the test chamber in each institution is shown in Table 1. Table 2 shows test conditions of ILS. Figure 2 shows the geometry of the ASTM G134 nozzle. The aggressive intensity of a cavitating jet depends on the outlet geometry of the nozzle, and the aggressive intensity of the jet issuing from the ASTM G134 nozzle is larger than that of nozzles for water jet [5]. Thus, the nozzle geometry of ASTM G134-95 [1] was used in this ILS and the geometry of nozzle was not changed in ASTM G134-17 [4].

Table 1 Size of test chamber

	Diameter mm	Length mm
Tohoku University	50	50
Nihon University	50	50
Fukui University	40	44
Dynaflow Inc.	44.5	50

Table 2 Test conditions

Diameter of nozzle d	0.4 mm
Upstream pressure p_u	20 MPa
Static pressure in the downstream chamber p_d	0.28 MPa
Cavitation number σ	0.014
Temperature	297 ± 2 K

*Corresponding Author, Hitoshi Soyama: soyama@mm.mech.tohoku.ac.jp

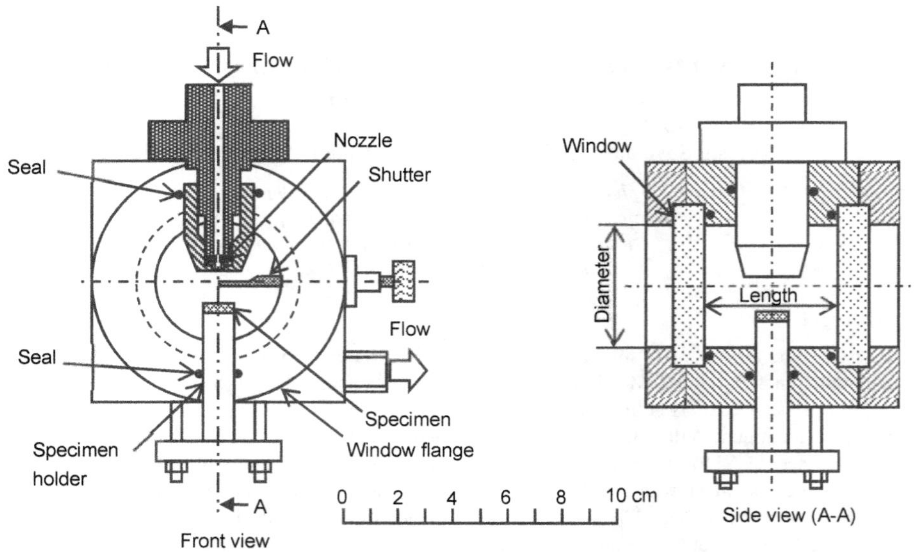

Figure 1 Schematic diagram of test chamber

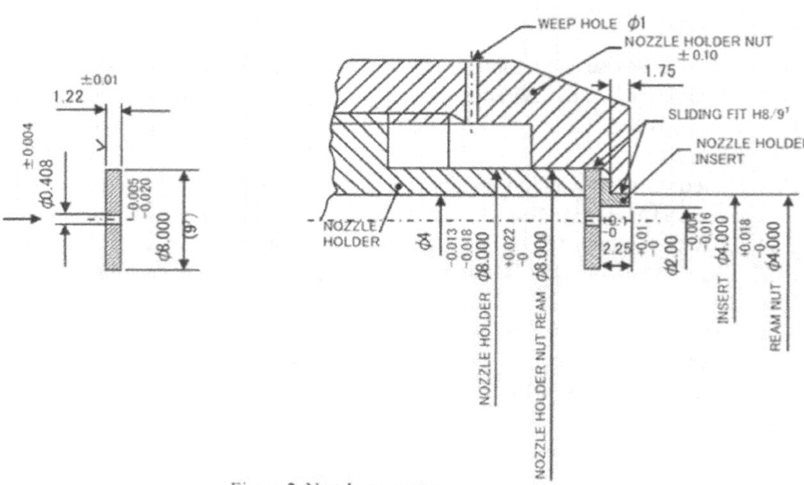

Figure 2 Nozzle geometry

The tested materials in this ILS were pure aluminum Japanese Industrial Standards JIS A1070B, bronze JIS CAC402, popular nickel JIS NW2201, nickel 200 JIS NW2200, and stainless steel JIS SUS316L. The erosion tests were carried out three times using the same materials in each institution.

Results

Figure 3 reveals the mass loss as a function of erosion time for five materials at each institution. As mentioned above, erosion test was carried out three times using each material, and the averaged value of the three results is shown in Fig. 3. For all cases, the erosion curve reveals an incubation period, an acceleration period, a steady state period, and an attenuation period. In ASTM G134-95, the maximum cumulative erosion rate is strongly recommended as the parameter to reveal a material resistance to cavitation. In order to investigate repeatability and reproducibility, the maximum cumulative erosion rate was examined. In Fig. 3, in the case of soft material such as pure aluminum A1070B, the difference between institutions is rather small. On the other hand, in the case of hard material such as NW2200 and SUS316L, the differences are relatively large.

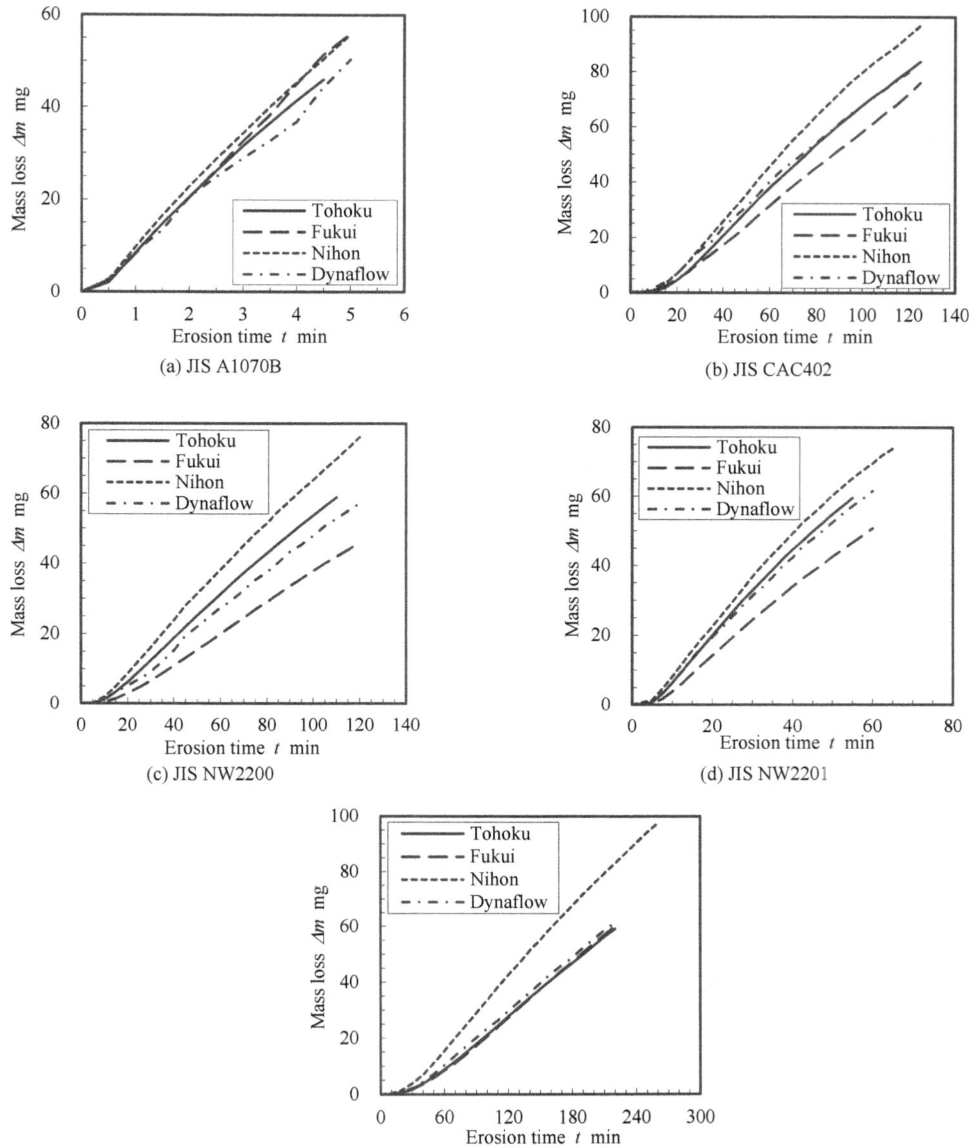

(a) JIS A1070B

(b) JIS CAC402

(c) JIS NW2200

(d) JIS NW2201

(e) JIS SUS316L

Figure 3 Mass loss changing with erosion time

Table 3 shows the maximum cumulative erosion rate at each institution, and repeatability and reproducibility of the maximum cumulative erosion rate. The coefficient of variation of three times in the same laboratory depends on the materials and it was about 1 % - 10 %. The repeatability and reproducibility were obtained as follows. For example, in the case of pure aluminum A1070B, the average values of each laboratory were 10.54, 11.45, 11.69 and 10.32 mg/min, respectively. Thus, the average of all laboratories averages was 11.00 mg/min. The standard deviation of each laboratory were 0.53, 0.81, 0.70 and 0.95 mg/min, respectively, and the average of laboratory variations was 0.75 mg/min. Thus, the repeatability coefficient of the variation was obtained by dividing 0.75 by 11.00, it was 0.068 = 6.8 %. Reproducibility was obtained from the standard deviation of 10.54, 11.45, 11.69 and 10.32 mg/min, and it was 0.67 mg/min. Then, the reproducibility of the variations was obtained by dividing 0.67 by 11.00, it was 0.061 = 6.1 %.

Table 3 Maximum Cumulative Erosion Rate (mg/min)

	A1070B	CAC402	NW2200	NW2201	SUS316L
Tohoku University	10.54±0.53	0.685±0.028	0.538±0.017	1.119±0.013	0.271±0.003
coefficient of variation %:	5.1	4.0	3.2	1.2	0.9
Fukui University	11.45±0.81	0.608±0.007	0.383±0.035	0.858±0.024	0.274±0.019
coefficient of variation %:	7.1	1.2	9.2	2.8	6.9
Nihon University	11.69±0.70	0.812±0.055	0.648±0.021	1.244±0.032	0.379±0.005
coefficient of variation %:	6.0	6.7	3.2	2.6	1.4
Dynaflow Inc.	10.32±0.95	0.691±0.072	0.484±0.023	1.077±0.034	0.280±0.015
coefficient of variation %:	9.2	10.4	4.8	3.2	5.3
Average of laboratory averages:	11.00	0.699	0.513	1.074	0.301
			Pooled Variabilities—Absolute Values		
"Repeatability" standard deviation	0.75	0.040	0.024	0.026	0.010
"Reproducibility" standard deviation	0.67	0.084	0.111	0.161	0.053
			Pooled Variabilities—Normalized Values		
"Repeatability" coefficient of variation %	6.8	5.8	4.7	2.4	3.5
"Reproducibility" coefficient of variation %	6.1	12.1	21.6	15.0	17.5

The repeatability and the reproducibility depend on the materials and they varied as shown in Table 3 from 2 % - 7 % and 6 % to 22 %.

In the case of ASTM G32, repeatability and reproducibility coefficient of variation using Nickel 200 were 4.2 % and 10.3 %, respectively [6].

Conclusion

In order to keep ASTM G134, an interlaboratory study ILS was carried out in collaboration with four institutions using five materials, and the maximum cumulative erosion rates obtained were presented. It was revealed that the "Repeatability" coefficient of variation depends on the materials, and it varied from 2 % to 7 %. The "Reproducibility" coefficient of variation also changes from material to material from 6 % to 22 %. Here, "Repeatability" means the average of the standard deviation of each institution. "Reproducibility" was obtained from the standard deviation of the average of each institution. Note that the coefficient of variation in the same institute was about 1 % - 10 %.

References

[1] ASTM G134-95 (1995). *Standard test method for erosion of solid materials by cavitating liquid jet*, ASTM, pp. 1–14.

[2] Lichtarowicz, A. (1972). *Use of a simple cavitating nozzle for cavitation erosion testing and cutting*, Nature: Physical Science, 239(91), pp. 63–64.

[3] Lichtarowicz, A. (1979). *Cavitating jet apparatus for cavitation erosion testing*, Erosion: Prevention and Useful Applications, ASTM STP 664, pp. 530–549.

[4] ASTM G134-17 (2017). *Standard test method for erosion of solid materials by cavitating liquid jet*, ASTM, pp. 1–17.

[5] Soyama, H. (2013). *Effect of nozzle geometry on a standard cavitation erosion test using a cavitating jet*, Wear, 297(1-2), pp. 895–902.

[6] ASTM G32-10 (2010). *Standard test method for cavitation erosion using vibratory apparatus*, ASTM, pp. 1–16.

© 2018 ASME

10th International Symposium on Cavitation - CAV2018
Baltimore, Maryland, USA, May 14 – 16, 2018

CAV18-05120

Internal and near nozzle flow simulations of gasoline multi-hole injector (ECN Spray G) with transient needle motion

[1]Balaji Mohan*; [1]Mohammed Jaasim; [1]Francisco Hernandez Perez; [2]Jaeheon Sim; [1]William Roberts; [1]Hong Im

[1]King Abdullah University of Science and Technology, Thuwal, KSA; [2]Saudi Aramco, Dhahran, KSA

Abstract

In this work, the effect of transient needle motion on gasoline multi-hole injector's internal and near nozzle flow was studied. Spray G nozzle, an eight hole counter-bore injector from the Engine Combustion Network (ECN), was considered as the fluid domain. Simulations considered the effect of turbulence, cavitation, flash-boiling, compressibility and non-condensable gases under transient needle lift. To model the two phase flow inside and outside the nozzle, the Homogeneous Relaxation Model (HRM) coupled with the Volume of Fluid (VOF) approach was used. HRM model is used in this study because it uses an empirical timescale to reproduce a range of vaporizations mechanisms i.e. both cavitation and flash boiling mechanisms. To model turbulence, the RNG $k - \varepsilon$ model was used. Simulations were performed with two different boundary conditions for the outlet domain to investigate non-flashing and evaporative (Spray-G) and flashing (Spray-G2) conditions. The simulation results were qualitatively validated against the experimental images and quantitatively against the experimental rate of injection (ROI) profile. The liquid plume angle showed good agreement with that of experimental measurement. The results show that the simulation is capable of capturing the ROI accurately with upstream pressure boundary condition and transient needle lift profile. Furthermore, the results show that the hole-to-hole variation in the total injected mass was not very significant. The simulation was also able to capture the cavitation phenomena inside the nozzles and flash boiling in the near nozzle region.

Keywords: Spray G injector, flashing, non-flashing, rate of injection (ROI)

Introduction

Gasoline compression ignition (GCI) engines are becoming popular to exploit the advantage of higher fuel volatility and potentially lower aromatic content [1]. To improve the fidelity of simulation-based GCI combustion system development, it is mandatory to enhance the prediction of injection characteristics. Fuel injection always involves phase-change phenomena. The two most common phase-change phenomena are cavitation and flash boiling. The phase-change process is highly complex in nature. Cavitation takes place in high pressure diesel injectors and is a pressure-driven vaporization taking place at low temperatures. At high fuel temperatures, more energy is required for phase-transition per unit volume of vapor due to high saturation vapor density compared to that of low temepraturefuel. Thus, flash boiling is a thermal non-equilibirium process unlike cavitation. The non-dimensional Jacob number ($Ja = \frac{\rho_l C_P \Delta T}{\rho_v h_{l_v}}$) which is the ratio of sensible heat energy available to the energy required for vaporization can be used to explain this phenomenon [2]. The high Jacob number indicates that there is abundance of energy available in the liquid to generate vapor. This means that the process is close to equilibirium since heat transfer time-scale is much lower than flow time-scale, and hence cavitating. Alternatively, when the Jacob number is low, the process is non-equilibirium, i.e. flash boiling and the heat transfer time scales will be of the same order as that of the flow time scales [3].

There are several numerical studies on flash boiling for GDI injectors available in the literature [3,4,5,6,7]. Moulai et al. [6] used the OpenFOAM® CFD package to simulate the flash boiling condition by using a velocity boundary condition calculated from the rate of injection (ROI) data from experiments. Recently, Baldwin et al. [7] used the OpenFOAM® package to simulate the flash boiling under transient needle motion. Other simulation works have used

*Corresponding Author, Balaji Mohan: balaji.mohan@kaust.edu.sa

the CONVERGE CFD code and successfully demonstrated its potential to predict flash boiling conditions either at fully opened condition or at different static needle lift positions [3,4,5]. In this work, the CONVERGE CFD package is employed to study the flash boiling phenomenon with transient needle motion. In what follows, the numerical model is described and results are presented and discussed.

Numerical methodology

The simulations were performed using the CONVERGE v2.4 CFD package. CONVERGE solves the conservation equations for mass, momentum, and energy, with the addition of a turbulence closure model. To model turbulence, the Reynolds Averaged Navier-Stokes (RANS) approach was employed and, in particular, the $k-\varepsilon$ model was used.

The Volume of Fluid (VOF) approach is used to model the two-phase flow. In the VOF method, a function α is used to represent the void function which is defined as follows

- $\alpha = 0$: the cell contains only liquid
- $0 < \alpha < 1$: the cell contains both liquid and gas
- $\alpha = 1$: the cell contains only gas

The density in the cell is computed by the following equation

$$\rho = \alpha\rho_g + (1-\alpha)\rho_l \tag{1}$$

where, ρ_g is the gas density and ρ_l is the liquid density.

The void fraction is solved by the following conservation equation

$$\frac{\partial \alpha}{\partial t} + u_i \frac{\partial \alpha}{\partial x_i} = S \tag{2}$$

However, the void fraction is not directly transported as in Eq. (2), instead the species are first solved using the species transport equation

$$\frac{\partial \rho_m}{\partial t} + \frac{\partial \rho_m u_j}{\partial x_j} = \frac{\partial}{\partial x_j}\left(\rho D \frac{\partial Y_m}{\partial x_j} + S_m\right) \tag{3}$$

where, $\rho_m = Y_m \rho$

The void fraction is then computed based on the liquid and gas mass fractions as follows

$$\alpha = \frac{m_g/\rho_g}{m_g/\rho_g + m_l/\rho_l} \tag{4}$$

where, m_g is the gas mass fraction (i.e. fuel vapor and dissolved gas) and m_l is the liquid mass fraction.

The Homogenous Relaxation Model (HRM) [8] approach is used to evaluate the source term S_m in the species conservation Eq. (3). This approach describes the rate at which the instantaneous mass fraction of vapor in a two-phase mixture will approach its equilibrium. A simple linearized form for this rate was proposed by Bilicki and Kestin [9] as

$$\frac{Dx}{Dt} = \frac{\bar{x} - x}{\theta} \tag{5}$$

where, $\frac{Dx}{Dt}$ is the rate of change of local vapor quality which gives the estimate for S_m, x is instantaneous mass, $\bar{x}$ is equilibrium mass and θ is the time scale over which x reaches to $\bar{x}$. For evaporation, the time scale θ is calculated using the following equation

$$\theta = \theta_0 \alpha^a \varphi^b \tag{6}$$

where, θ_0 is $3.84 * 10^{-7}[s]$, α is the fuel void fraction, $a = -0.54$, $b = -1.76$, and φ is a dimensionless pressure defined as

$$\varphi = \left|\frac{P_{sat} - P}{P_{cr} - P_{sat}}\right| \tag{7}$$

where, P_{sat} is the pressure at saturation temperature and P_{cr} is the critical pressure

Computational domain

The computational domain used in this current study is the internal nozzle geometry of spray G injector from Engine Combustion Network (ECN) [10]. The nozzle has 8 counter-bored holes with 5 dimples as shown in Fig. 1. In order to capture the near nozzle flow, a hemispherical plenum is necessary to extend the outlet domain. In this study, a hemispherical plenum of 9 mm diameter is used based on the studies done by Saha et al. [3,4]. The vertical cut-plane showing the mesh with the 9 mm outlet domain is shown in Fig. 2. A base cellsize of 150 μm is used in this study. Fixed embedding has been used near the walls, inside the holes and inside the hemispherical outlet domain near the nozzles to capture the sharp gradients in velocity, temperature, species, etc. Three levels of fixed embedding have been used. Therefore, the smallest cell size is 8 times smaller than the base grid, i.e., smallest cell size is $150\ \mu m * 2^{-3} = 18.75\ \mu m$. According to Moulai et al. [5], reasonable predictions were feasible with even a 22.5 μm of minimum grid size.

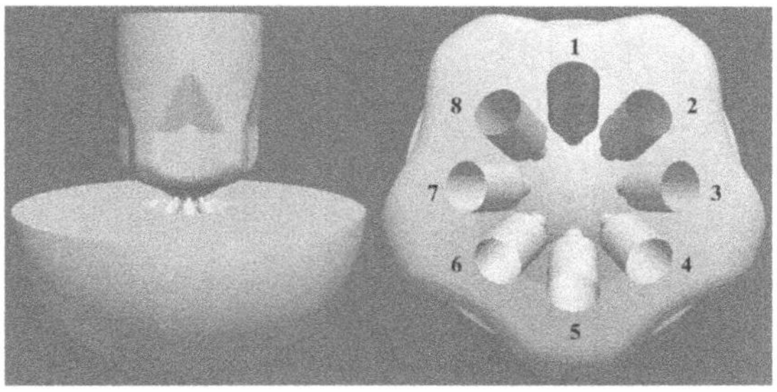

Figure 1: Spray G nozzle geometry with 8 nozzle holes and 5 dimples obtained from ECN [10].

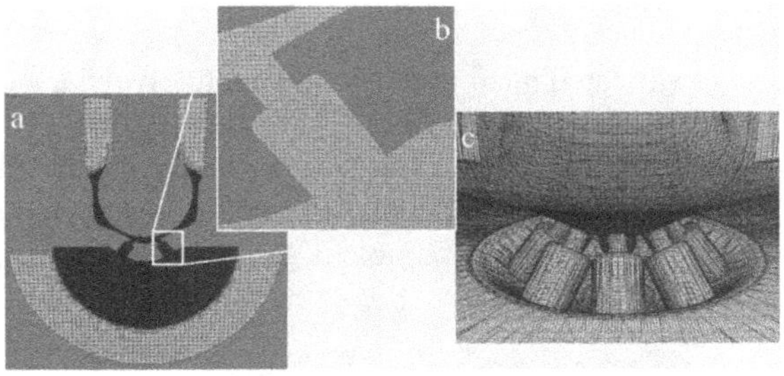

Figure 2: (a) Vertical cut-plane showing the mesh with 9 mm hemispherical outlet domain at fully opened conditions, (b) Zoomed in view of mesh in the nozzle region and (c) 3D mesh in the nozzle region.

Initial and boundary conditions

At the inlet and outlet, a pressure boundary condition was used. For walls and the interface between the liquid and vapor phase, a no-slip boundary condition was enabled. Turbulent kinetic energy and dissipation rate were given as the turbulence boundary condition at both the inlet and outlet of the domain. As liquid fuel cannot be 100% pure, a very small amount of non-condensable gas was considered to be present in the liquid fuel. This non-condensable gas also acts as nucleation sites for cavitation inception. A more realistic needle lift profile obtained from experiments was used in the simulation. The needle motion was changed to make it monotonic for a finite positive initial needle lift. The needle position was started from 2 μm of lift and remained stationary until 2.5 μs into the simulation. The time of 2.5 μs was chosen because this was the amount of time the experimental data indicated it would take to reach 2 μm of lift. After this point, the needle position followed the experimental lift generated by the experiment [10]. Fig. 3 shows the experimental ensembled-averaged needle lift and lift profile used in the simulations. The details of the cases studied in this work are summarized in Table 1.

Table 1: Cases simulated.

	Spray-G	Spray-G2
Fuel	*iso*-octane	*iso*-octane
Fuel temperature	363.15 K	363.15 K
Injection pressure	20 MPa	20 MPa
Back pressure	600 kPa	53 kPa
Ambient temperature	573.15 K	333.15 K

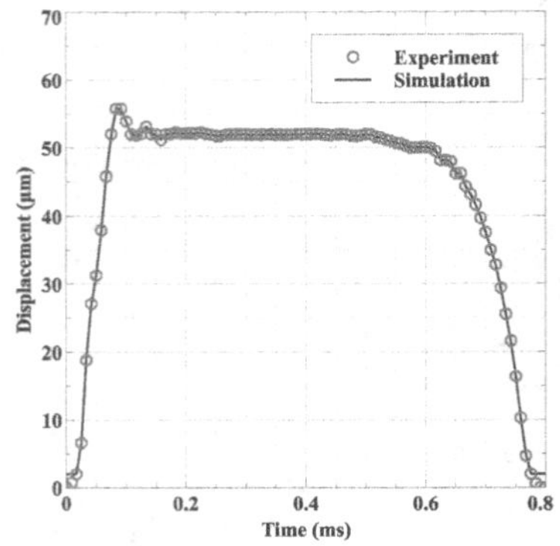

Figure 3: Experimental [10] ensembled-averaged needle lift and needle lift profile used in simulation which starts at 2 μm.

Results and discussion

The numerical simulation results were validated quantitatively and qualitatively against the experimental results available from literature. Fig. 4 shows the qualitative comparison of both flashing and non-flashing conditions. The volume rendered fuel mass fraction of *iso*-octane is compared with the spray images obtained from the experiments. It can be observed that there is a uniform contact of fuel with counter-bore under flashing conditions, while non-flashing case shows gaps between the plumes allowing ingestion of ambient gas. This phenomenon is very well captured by the simulations.

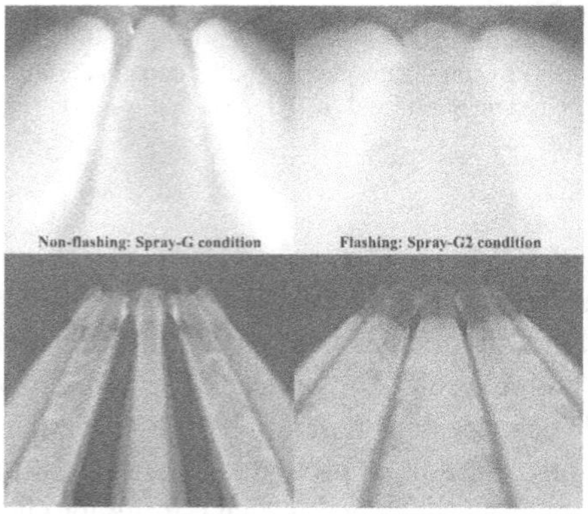

Figure 4: Experimental images (top) [6] of non-flashing (left) and flashing (right) and volume rendered fuel mass fraction from the simulations (bottom).

Fig. 5a shows the raw and smoothed rate of injection profiles obtained from the simulation. The raw data exhibited a high degree of transient variability as an inherent nature of two-phase flow. The raw data have been filtered using a Fast Fourier Transform (FFT) method to de-noise the simulation result. Fig. 5b, on the other hand, shows the comparison between experimental and smoothed simulated data. It can be found that there is a good agreement between the experimental and simulated data. This demonstrates the capability of the methodology to predict the rate of injection profile under flash boiling conditions incorporating transient needle motion.

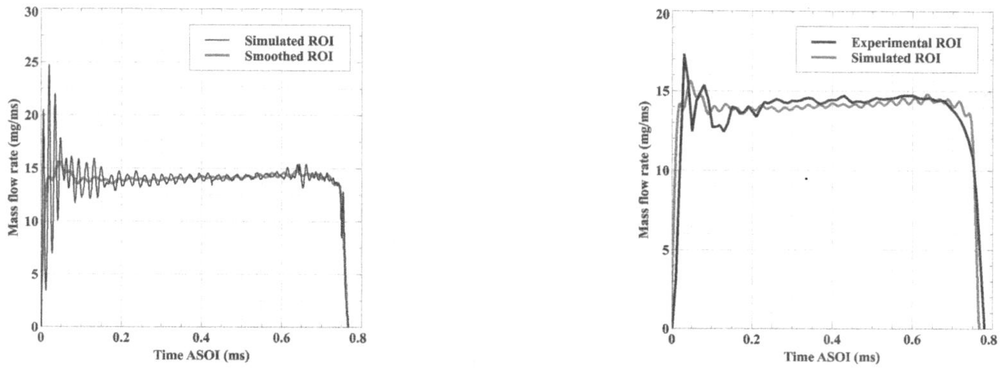

(a) Raw predicted rate of injection and smoothed rate of injection using FFT filter.

(b) Comparison of experimental and simulated rate of injection.

Figure 5: Rate of injection (ROI).

The total injected mass was calculated by integrating the mass flow rate across the outlet of each hole throughout the injection duration. The total mass agreed well with experimental data of 10 mg as specified by ECN. The hole-to-hole variation is shown in Fig. 6. It was found that the variation was very small in the order of 2-3%. No apparent dependance was found on the asymmetry of the nozzle, as expected in previous studies [4,5]. This indicates that the nozzle asymmetry has negligible importance on hole-to-hole variations and if any were found experimentally, it may be due to the manufacturing imperfections in the nozzle.

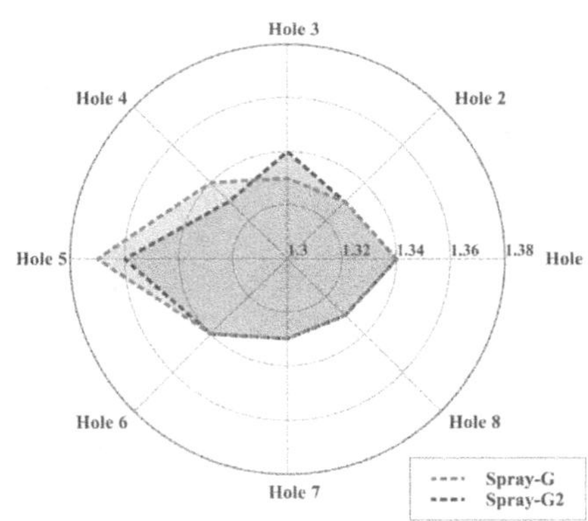

Figure 6: Hole-to-hole variation in mass flow rate.

Fig. 7 shows the comparison of liquid spray half-cone angle between simulation and experiment. As the phase change occurs inside the nozzle holes, it forces liquid to flow out at a different angle and it displayed a good agreement with the experimental result.

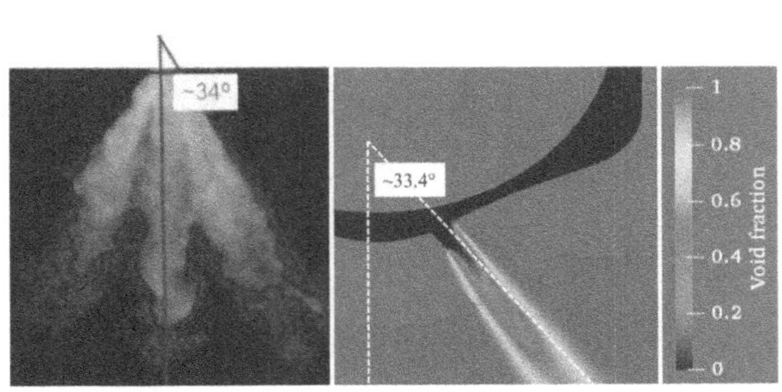

Figure 7: Comparison of liquid spray half cone angle between simulation and experiment [11].

Fig. 8 shows the comparison of cavitation inception between spray-G and spray-G2 conditions. It can be found that mild cavitation like phenomenon takes place inside the injector nozzle holes. The higher differential pressure in the spray-G2 case under same fuel temperature leads to more vapor formation inside the holes as compared to the spray-G case.

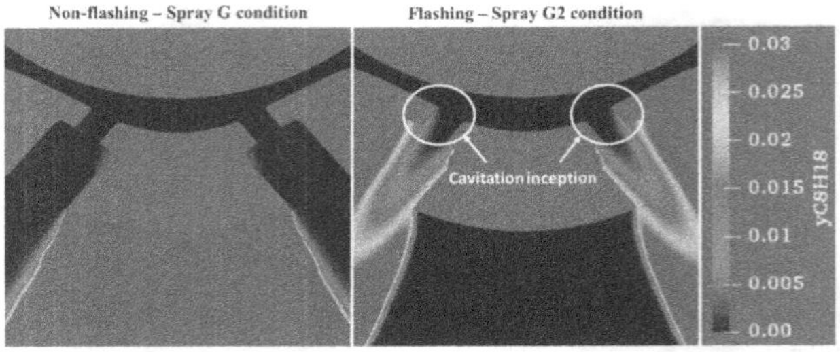

Figure 8: Comparison of cavitation inception between Spray-G and Spray-G2 conditions.

Conclusions

Numerical simulations of internal and near nozzle flow under flashing and non-flashing conditions with transient needle motion were succesfully done with the CONVERGE v2.4 CFD code. The simulations predicted the plume-to-plume interactions under flashing conditions and rate of injection in good agreement with the experiments. It was found that the hole-to-hole variation due to nozzle asymmetry was not significant. Furthermore, the simulations captured mild cavitation like phenomenon inside the nozzle holes due to higher differential pressure.

References

[1] Zhang, Y., Kumar, P., Traver, M., Cleary, D. (2016) *Conventional and Low Temperature Combustion Using Naphtha Fuels in a Multi-Cylinder Heavy-Duty Diesel Engine.* SAE International Journal of Engines. 9.

[2] Neroorkar, K. (2011). *Modeling of flash boiling flows in injectors with gasoline-ethanol fuel blends.* University of Massachusetts Amherst.

[3] Saha, K., Som, S., Battistoni, M., Li, Y., Quan, S. and Senecal, P.K. (2015). *Numerical simulation of internal and near-nozzle flow of a gasoline direct injection fuel injector.* In Journal of Physics: Conference Series. 656(1).

[4] Saha, K., Som, S., Battistoni, M., Li, Y., Quan, S. and Senecal, P.K. (2016). *Modeling of internal and near-nozzle flow for a gasoline direct injection fuel injector.* Journal of Energy Resources Technology. 138(5).

[5] Saha, K., Som, S., Battistoni, M., Li, Y., Pomraning, E. and Senecal, P.K. (2016). *Numerical investigation of two-phase flow evolution of in-and near-nozzle regions of a gasoline direct injection engine during needle transients.* SAE International Journal of Engines. 9(2016-01-0870).

[6] Moulai, M., Grover, R., Parrish, S. and Schmidt, D. (2015). *Internal and near-nozzle flow in a multi-hole gasoline injector under flashing and non-flashing conditions* SAE Technical Paper. (No. 2015-01-0944).

[7] Baldwin, E.T., Grover, R.O., Parrish, S.E., Duke, D.J., Matusik, K.E., Powell, C.F., Kastengren, A.L. and Schmidt, D.P. (2016). *String flash-boiling in gasoline direct injection simulations with transient needle motion.* International Journal of Multiphase Flow. 87.

[8] Schmidt, D., Rakshit, S., and Neroorkar, K. (2009). *Thermal and Inertial Equilibrium in Small, High-Speed, Cavitating Nozzle Simulations.* 11th International Conference on Liquid Atomization and Spray Systems (ICLASS-2009).

[9] Bilicki, Z. and Kestin, J. (1990). *Physical aspects of the relaxation model in two-phase flow.* Proceedings of the Royal Society of London A: Mathematical, Physical and Engineering Sciences. 428.

[10] Engine Combustion Network. (accesed on 29 October, 2017). *https://ecn.sandia.gov/gasoline-spray-combustion/target-condition/spray-g-operating-condition/.*

[11] Manin, J., Jung, Y., Skeen, S., Pickett, L. et al. (2015). *Experimental Characterization of DI Gasoline Injection Processes.* SAE Technical Paper. (No. 2015-01-1894).

Acknowledgements

This work was sponsored by King Abdullah University of Science and Technology. The computational simulations utilized the clusters at KAUST Supercomputing Laboratory. The authors thank Convergent Science Inc. for providing the CONVERGE license.

© 2018 ASME

Investigation of acoustic streaming and cavitation intensity in water as an analogue for liquid metal

[1,5] Iakovos Tzanakis*; [2]G. S. Bruno Lebon; [2,3]Dmitry Eskin; [4]Martin Hyde, [5]Patrick S. Grant

[1]Oxford Brookes University, Oxford, UK; [2]Brunel University London, Uxbridge, UK; [3]Tomsk State University, Tomsk, Russia; [4]TSI Instruments, UK; [5]University of Oxford, Oxford, UK

Abstract

This paper presents an investigation of the evolution of flow structures and cavitation intensity in water as an analogue for a liquid metal under ultrasonic excitation. Results are presented for 20 kHz high-power ultrasound. The input power ranged from 50% (8.5 μm p-p) to 100% (17 μm p-p). To identify the streaming structures and understand the recirculation flows for different vibrational amplitudes of the sonotrode, particle image velocimetry (PIV) measured the velocity field. Simultaneously, a calibrated cavitometer probe measured acoustic intensity in the fluid. The cavitation intensity away from the acoustic source decreased with increasing input acoustic power, but was relatively constant inside the cavitation zone (irrespective of the input power). PIV measurements showed that the direction of the flow pattern was strongly related to the vibrational amplitude of the sonotrode. These results are compared with the predictions of an acoustic cavitation model. The outcome of the present work will help to determine the efficient optimization of ultrasonic processing of liquid metals that is of increasing technological importance.

Keywords: Particle Image Velocimetry (PIV); acoustic streaming; cavitation intensity, ultrasonic melt processing, aluminium

Introduction

Ultrasonic melt treatment (UST) is an environment friendly and economical alternative to many conventional melt processes used to control liquid metal quality, such as fluxing, modification, and gas lancing. UST involves introducing high-intensity ultrasonic waves into liquid metal to induce acoustic cavitation [1]. Cavitation bubbles expand non-linearly, become unstable upon reaching a critical size, and collapse catastrophically, thereby generating high-speed liquid jets, shockwaves, and local hot spots [2–4]. Laboratory tests showed that UST offers beneficial effects, such as accelerated diffusion, activation of inclusions, improved wetting, dissolution, deagglomeration, and dispersion of particles leading to degassing, refined, equiaxed solidification microstructure, and uniform distribution of constituent phases [5–9]. However, an insufficient understanding of the phenomena governing UST (e.g. cavitation intensity distribution and acoustic streaming) hinders more widespread implementation of UST technology in industry [10].

The understanding and quantification of recirculation patterns within the acoustic cavitation zone and the mass exchange between the cavitation zone and the surrounding fluid are important for optimizing UST e.g. for the grain refinement of the as-cast structure. Recently, Wang *et al.* [11] showed that a continuous recirculation pattern created a low temperature gradient throughout the melt: this favours an equiaxed grain structure, which is usually preferred by industry. In direct-chill casting, induced-acoustic streaming promotes forced convection that is opposed and much stronger to natural convection in the melt, facilitating further activation (wetting) of extrinsic particles, solid fragmentation and self-grain refining, mixing and elemental homogenisation within the bulk melt [5].

So far, no accurate comprehensive theory of acoustic streaming with cavitation has been developed. There were some successful attempts to model acoustic streaming using a sonotrode as a flow/pressure source and getting average distribution of velocities [12,13] as well as performing PIV measurements of acoustic flows generated by a sonotrode[13,14]. The calculated range of velocities was between 0.2 and 2 m/s [12] while PIV measurements gave the range up to 0.1 m/s [13] for similar ultrasonic conditions. Linking the intensity and details of cavitation to acoustic streaming in modelling approaches remains a major challenge. Barthès *et al.* [14] used PIV techniques to demonstrate the interaction between the acoustic streaming below a sonotrode with a transverse horizontal flow in a rectangular duct, showing that cavitation bubbles can be transported significant distances by the flow and remain active, which may be important for practical applications [15]. To get more insight into the inter-related effects of cavitation and acoustic streaming, a combination of visualisation techniques and simultaneous acoustic measurements, and more sophisticated modelling, is required.

* Corresponding Author, Iakovos Tzanakis: itzanakis@brookes.ac.uk

In this study, we use a calibrated cavitometer [16] to characterise cavitation development in water and PIV to study acoustic streaming inside and near the cavitation zone. Experiments were performed in water because it is both a convenient and suitable analogue (among the range of possible transparent liquids) to molten aluminium in studies of acoustic cavitation [17].

Methodology

A Hielscher 20-kHz ultrasonic transducer introduced continuous acoustic waves in a rectangular, glass tank of base area 300 mm × 200 mm filled with distilled water (Figure 1). The liquid height was 150 mm and the radiating surface (radius 20 mm) of the vertically mounted titanium sonotrode was immersed 20 mm below the liquid surface. Acoustic power was initially at 50% and was gradually increased to 100%. The amplitude of the vibrating surface at 100% power was 17 μm peak-to-peak. A wattmeter was integrated to the transducer device to record power. The acoustic power of the sonotrode was obtained by subtracting the input electric power of the transducer when the sonotrode operated in the studied liquid from that measured in air, the reference medium. The acoustic power of the sonotrode was 90±3 Watt when operating at 100% power. The temperature was continuously monitored during the experiments and was maintained at 21±1 °C.

Cavitation intensity measurements were taken with a cavitometer probe in different positions, along the z-axis (Figure 1). Measurements between the sonotrode and the sidewall of the tank were taken at 6 different points of interest: under the sonotrode and at 5, 7, 9, 11 and 13 cm from the sonotrode centre-point. The cavitometer probe had a spatial resolution of about ~50 mm, allowing measurements of acoustic intensities of frequency components up to 10 MHz. The 4 mm waveguide of the cavitometer probe was submerged to a depth of 80 ± 2 mm. A full account of the cavitometer design and performance can be found elsewhere [16].

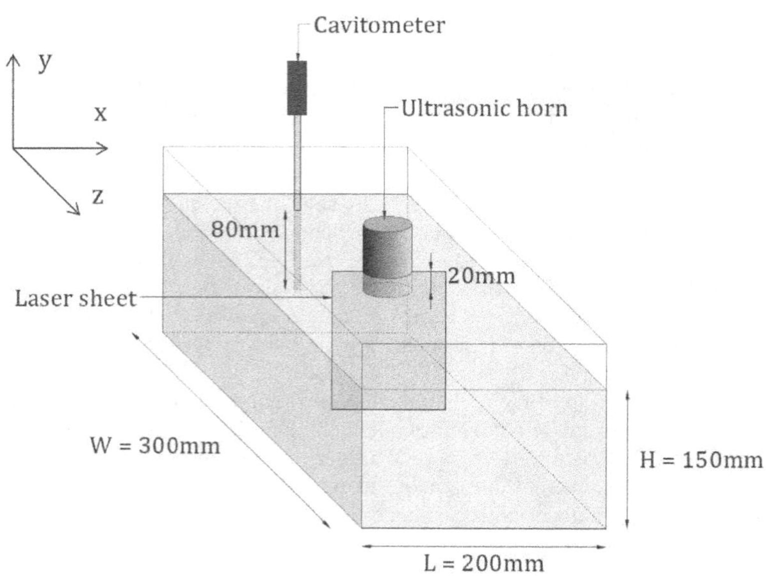

Figure 1: Graphical representation of the experimental test rig for measuring cavitation intensity and acoustic streaming

A TSI particle image velocimetry (PIV) system measured acoustic streaming velocities along the cavitation zone and in the bulk volume. A dual cavity Nd YAG laser source – wavelength 532 nm – generated a ~1 mm thick 2D pulsed-laser sheet. A short time interval Δt separated the two pulses; Δt was set at 250 μs or 500 μs, depending on the operating conditions. For each operating condition, a set of 100 pairs of images was captured using a TSI PowerView 4MP-LS digital CCD camera with an array of 2352 by 1768 pixels and a 50mm F1.8 Nikon lens operated at a maximum frame rate of up 16 frames per second and positioned normal to the illuminated plane. The PowerView 4MP-LS camera utilised frame straddling to enable short inter-frame times. Glass hollow sphere seeded particles (~10 μm in diameter) with density close to water were added at a concentration such that 5–10 particles were recorded in each frame and typically 5 pair of images were recorded per second. Acoustic streaming measurements were initiated by positioning the acoustic propagation axis within the laser sheet with measurements taken in the plane $z = 0$ (Figure 1) with the origin at the radiating surface centre. With the particles in the illuminated plane in focus, the resultant x-y field of view was 85 mm × 95 mm. Using TSI's Insights software, an interrogation area of 64 x 64 pixels with 50% overlap was set, and the same software used for PIV analysis and the production of 2D time-averaged vector flow fields.

Results and Analysis

Cavitation Intensity Measurements. To characterise cavitation development, a series of acoustic measurements using a calibrated cavitometer were performed across the experimental tank. As expected, the cavitation spectrum consisted of acoustic emissions from the fundamental frequency and emissions from the collapsing bubbles that were strongly dependent on the acoustic power and the distance from the source.

Figure 2 shows a range of measured cavitation intensity values: at distances of 5 cm, 7 cm and 9 cm from the sonotrode axis. The cavitation intensity gradually decreased with acoustic power up to a point (around 80–90% of the maximum) where it rebounded in magnitude. At longer distances, the intensity continuously reduced across the acoustic power range because of the shielding and scattering effects commonly generated in sonicated environments. As the sonotrode power increased, the number of the bubbles below the horn also increased, hindering sound wave propagation into the bulk liquid. Concomitantly, neighbouring bubbles shielded bubbles inside the bubbly cloud to some extent, inhibiting their growth. Upon collapse, these interior bubbles produced less powerful impact pressures in contrast to the fully developed (based on the fundamental frequency [17]) cavitation bubbles. Consequently, weaker acoustic waves reached the cavitometer probe [18]. At a higher power of 100%, acoustic streaming pushed the bubbly clouds away and a cone-like bubbly structure was formed, facilitating the propagation of the cavitation and acoustic waves to the liquid. Hence, the overall cavitation intensity increased. Nevertheless, the intensity remained approximately 20% lower than that measured at 50% power.

Under the sonotrode, the largest intensity was measured at 100%, with an approximately linear relationship between cavitation intensity and acoustic power (see plot "under" in Figure 2). However, at 80% power there was a drop in intensity and deviation from this trend. This is likely attributed to the shielding effects as explained in [16]. A further increase followed as with more power acoustic streaming pushed the accumulated bubbly structures further apart and released the pathway for sound wave propagation. This highlighted the inter-related role of the cavitation zone and bubbly structures in regulating acoustic pressure propagation.

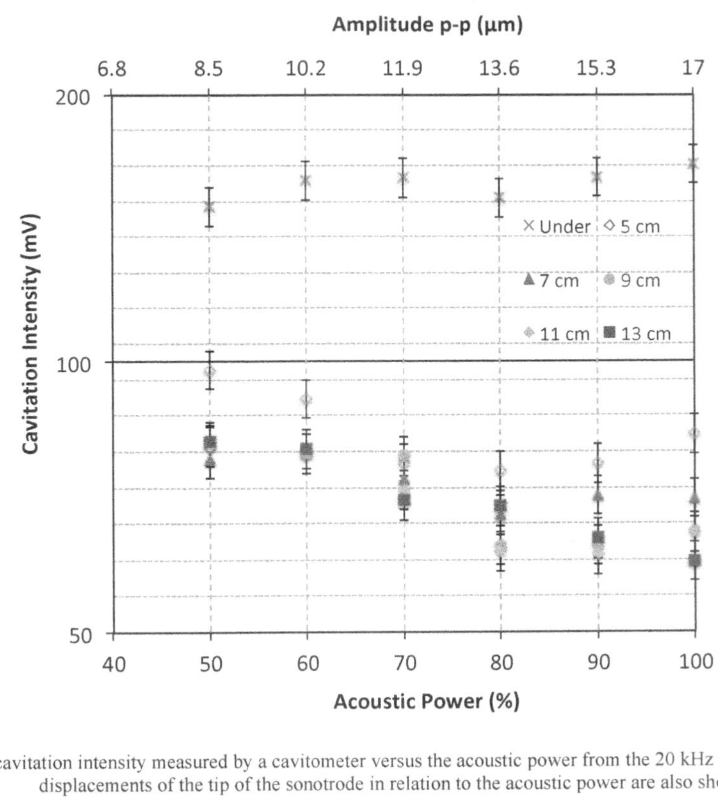

Figure 2: The cavitation intensity measured by a cavitometer versus the acoustic power from the 20 kHz transducer. Peak to peak displacements of the tip of the sonotrode in relation to the acoustic power are also shown.

Interestingly, for all other cases away from the cavitation zone, the 50% power produced the most efficient cavitation activity. At 50%, fewest bubbles were generated resulting in bubble displacement without excessive disruptions from other neighbouring bubbles. Bubble growth was comparatively smooth and not severely affected by the shielding effect. This effect was advantageous because the generation of the maximum cavitation intensity occurred below 100% sonotrode power resulting in potential environmental and economic benefits for ultrasonic aided technologies.

Variation of the maximum intensity with the distance from the source is also depicted in Figure 2. The highest cavitation intensity measured by the cavitometer probe was near the sonotrode. In general, there was a reduction of about 25% as the measuring point moved from 5 cm to 13 cm and results are in a good agreement with [18,19] where a decay in intensity with distance from the source was also reported.

PIV Measurements. Figure 3 shows typical PIV measurements for the two extremes of the acoustic power range of 50% (left) and 100% (right). Each vector represents the time-averaged distribution of velocities, from 100 pair of images.

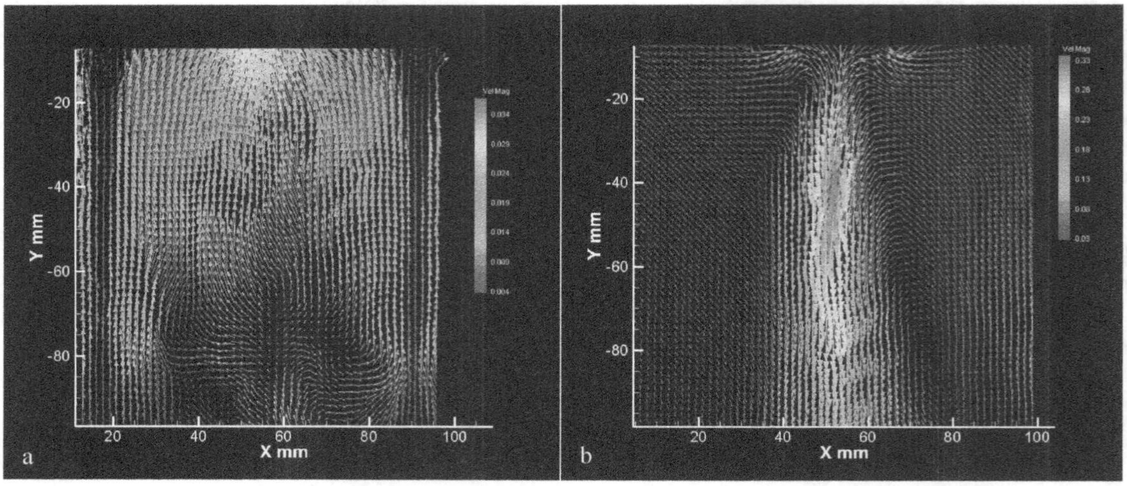

Figure 3: Typical images of PIV showing spatial distribution of acoustic flow structures under different input power at 50% (a, left) and 100% (b, right) acoustic power from the transducer. Legend: velocity magnitude in m/s (0.004–0.034 left, 0.03–0.33 right).

Acoustic streams at 50% power moved towards the horn and then recirculated, following the same pattern on either of the side of the sonotrode, as previously shown in [20]. In contrast, at 100 %, the flow had a distinct downward movement, with almost an order of magnitude increase in peak velocity, with comparatively weak recirculating patterns formed on both sides along the symmetry line of the acoustic stream, which is a typical streaming pattern [13]. Overall, the recirculation patterns and velocity magnitudes and directions were markedly different at the two power settings. From a technological viewpoint, the less intuitive – perhaps "unusual" – flow pattern (at 50%) may facilitate the application of ultrasound to liquid metals in the cavitation zone because the melt residence time in the cavitation zone is increased due to slower streams, while the "back flow" towards the horn continuously re-cycles a stream of melt into the cavitation zone. At the same time, the acoustic intensity is still high enough for effective treatment of the liquid volume (Figure 2). This flow behaviour was monitored across the tip of the sonotrode and repeated using differently positioned optical 2D slices to ensure the findings in Figure 3 were reproducible and representative.

Careful examination of the acoustic stream development on increasing the acoustic power from 50% to 100% suggested that the switch from upward to downward flow occurred at approximately 55% acoustic power. For a firmer conclusion of this transition, a higher temporal resolution in the PIV image pair acquisition would be helpful.

A possible interpretation for the predominantly "upward flow" at 50% power in Figure 3 is that the sound wave may penetrate the bubbly zone and establish a standing wave, creating bubble streams that may induce net upward flow, even if the sound radiation is propagating in the opposite direction. Thus this regime might be postulated as a "standing wave/stream filament" regime, as suggested in [21]. This may be similar to the behaviour of larger bubbles or bubbly clouds that can move against the flow towards the sonotrode tip, even when bulk flow moves in the opposite direction: Yasui *et al.* [22] used a 10 mm sonotrode with the same peak-to-peak amplitude used here (set at 8.5 μm) to show that some cavitation bubbles at distances of 1 cm or more from the horn tip moved toward the horn tip due to the secondary Bjerknes force acting on the bubbles near the horn tip. It is speculated that here a similar mechanism is dragging acoustic streams (flow and bubbles) towards the horn tip. At higher transducer powers, the shielding effect sets in and changes the wave from "standing" to "traveling" via more efficient dissipation at the higher induced bubble density, in the immediate vicinity of the end of the horn, and net downwards acoustic streaming at a higher velocity dominates the flow.

At approximately 8 cm below from free surface at 50% power, acoustic streams bifurcated (Figure 3a) and there was both some upward flow towards the transducer but also more isotropic flow (at least in 2D). From a UST point of view, this reproducible phenomenon i.e. that within the same flow there are distinctive secondary multi-directional flows with dynamic recirculating flow patterns, together with strong eddies near the transducer and free surface under the same experimental conditions [20], may provide for a more efficient treatment of the liquid, even though velocity magnitudes are lower than at 100% power. This arises because flow paths are more varied and excessive recirculation (vortex) regions are avoided, helping to ensure macro-scale homogenisation of the liquid, noting that even at 50% power cavitation remains vigorous, as described in [23].

Numerical Modelling. In order to support the experimental insights, numerical modelling was investigated. The numerical model was based on the homogeneous cavitation model of Schnerr and Sauer [24] as implemented in the OpenFOAM solver *interPhaseChangeDyMFoam* [25], with turbulence handled using the kEqn LES model [26]. Numerical results were in broad qualitative agreement with the flow patterns at 100% acoustic power, but could not reproduce the reverse flow at 50% power, with reverse flow shown only very close to the sonotrode (Figure 4a). The pre-dominant downward flow (Figure 4b) at 100% power is well-explained by the net difference between the high pressure near the sonotrode and the lower pressures in regions away from the sonotrode, with the same overall trend pressure distributions at 50 % power (Figure 4a). The calculated flow velocity was higher than experiment at both acoustic powers, and was expected because a homogeneous cavitation model of the type used here for convenience cannot predict pressures (and therefore velocities) accurately at these high powers typical or technological applications. In future work, acoustic streaming will be integrated into a high-order acoustic model [19], and the present work underlines how some of the more convenient current modelling approaches cannot reproduce robustly experimental results at powers relevant to UST.

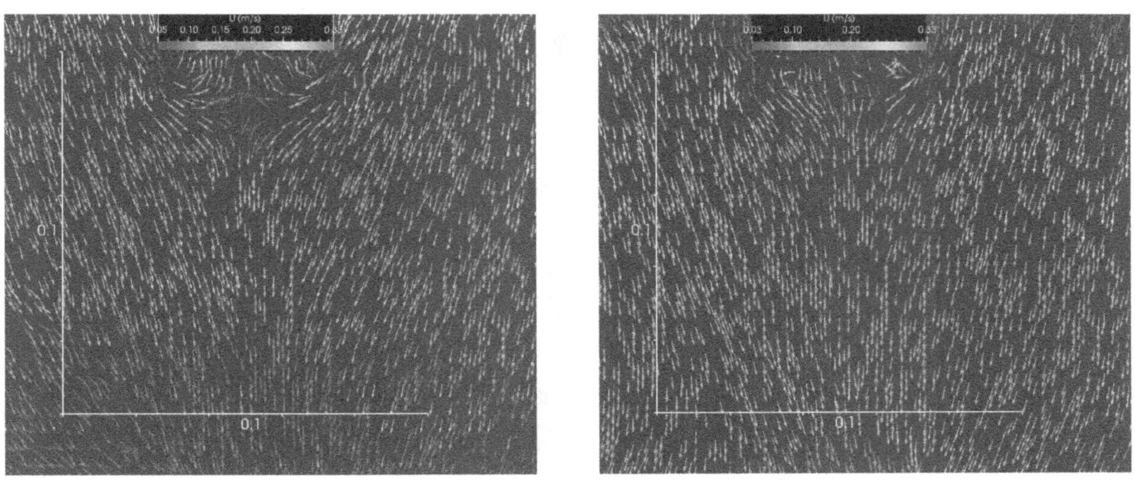

(a) (b)

Figure 4: Numerical prediction of flow pattern with (a) 50 % input power and (b) 100 % input power. The colour scale for velocity magnitude corresponds to scale in Figure 3 (b). The white axes delimit the corresponding field of view in the PIV measurements.

Conclusions

Experimental measurements showed that cavitation intensity in water used as a liquid metal analogue significantly reduced with increasing acoustic power away from the immediate vicinity of the sonotrode. PIV macro-scale measurements showed that flow velocity and direction were dependent on the operating power of the sonotrode transducer in a manner not captured by numerical modelling. Future numerical models need to better integrate microscopic cavitation, meso-scale bubble cloud dynamics and macro-scale acoustic streaming to reproduce better experimental behaviour. The approach described here of simultaneously investigating cavitation intensity distribution and acoustic streaming could be used to optimize the geometrical features of a casting arrangement and the liquid treatment efficiency in ultrasonic melt treatments. Looking further to the future, combining several transducers along the melt volume with variable acoustic power outputs, guided by improved modelling capability, could be used to flexibly generate a range of novel cavitation and flow regime combinations to optimize ultrasonic melt treatment.

Acknowledgements

Financial support from EPSRC (UK) under projects LiME Hub (EP/N007638/1), UltraMelt (EP/K00588X/1 and EP/K005804/1) and UltraMelt2 (EP/R011001/1, EP/R011044/1, EP/R011095/1) is gratefully acknowledged.

References

[1] Tzanakis, I., Lebon, G.S.B., Eskin, D.G., and Pericleous, K.A. (2016) *Characterisation of the ultrasonic acoustic spectrum and pressure field in aluminium melt with an advanced cavitometer*. J. Mater. Process. Technol., 229, 582–586.

[2] Xu, W.W., Tzanakis, I., Srirangam, P., Mirihanage, W.U., Eskin, D.G., Bodey, A.J., and Lee, P.D. (2016) *Synchrotron quantification of ultrasound cavitation and bubble dynamics in Al–10Cu melts*. Ultrason. Sonochem., 31, 355–361.

[3] Flannigan, D.J., and Suslick, K.S. (2005) *Plasma formation and temperature measurement during single-bubble cavitation*. Nature, 434 (7029), 52–55.

[4] Tzanakis, I., Eskin, D.G., Georgoulas, A., and Fytanidis, D.K. (2014) *Incubation pit analysis and calculation of the hydrodynamic impact pressure from the implosion of an acoustic cavitation bubble*. Ultrason. Sonochem., 21 (2), 866–878.

[5] Eskin, G.I., and Eskin, D.G. (2015) *Ultrasonic treatment of light alloy melts*, Taylor & Francis, CRC Press, Boca Raton.

[6] Tzanakis, I., Xu, W.W., Eskin, D.G., Lee, P.D., and Kotsovinos, N. (2015) *In situ observation and analysis of ultrasonic capillary effect in molten aluminium*. Ultrason. Sonochem., 27, 72–80.

[7] Atamanenko, T.V., Eskin, D.G., Zhang, L., and Katgerman, L. (2010) *Criteria of Grain Refinement Induced by Ultrasonic Melt Treatment of Aluminum Alloys Containing Zr and Ti*. Metall. Mater. Trans. A, 41 (8), 2056–2066.

[8] Eskin, D.G., Al-Helal, K., and Tzanakis, I. (2015) *Application of a plate sonotrode to ultrasonic degassing of aluminum melt: Acoustic measurements and feasibility study*. J. Mater. Process. Technol., 222, 148–154.

[9] Shu, D., Sun, B., Mi, J., and Grant, P.S. (2012) *A High-Speed Imaging and Modeling Study of Dendrite Fragmentation Caused by Ultrasonic Cavitation*. Metall. Mater. Trans. A, 43 (10), 3755–3766.

[10] Lebon, G.S.B., Tzanakis, I., Djambazov, G., Pericleous, K., and Eskin, D.G. (2017) *Numerical modelling of ultrasonic waves in a bubbly Newtonian liquid using a high-order acoustic cavitation model*. Ultrason. Sonochem., 37, 660–668.

[11] Wang, G., Croaker, P., Dargusch, M., McGuckin, D., and StJohn, D. (2017) *Simulation of convective flow and thermal conditions during ultrasonic treatment of an Al-2Cu alloy*. Comput. Mater. Sci., 134, 116–125.

[12] Trujillo, F.J., and Knoerzer, K. (2011) *A computational modeling approach of the jet-like acoustic streaming and heat generation induced by low frequency high power ultrasonic horn reactors*. Ultrason. Sonochem., 18 (6), 1263–1273.

[13] Schenker, M.C., Pourquié, M.J.B.M., Eskin, D.G., and Boersma, B.J. (2013) *PIV quantification of the flow induced by an ultrasonic horn and numerical modeling of the flow and related processing times*. Ultrason. Sonochem., 20 (1), 502–509.

[14] Barthès, M., Mazue, G., Bonnet, D., Viennet, R., Hihn, J.-Y., and Bailly, Y. (2015) *Characterization of the activity of ultrasound emitted in a perpendicular liquid flow using Particle Image Velocimetry (PIV) and electrochemical mass transfer measurements*. Ultrasonics, 59, 72–78.

[15] Tzanakis, I., Lebon, G.S.B., Eskin, D.G., and Pericleous, K. (2016) Optimization of the Ultrasonic Processing in a Melt Flow, in *Light Metals 2016* (eds.Williams, E.), Springer International Publishing, Cham, pp. 833–836.

[16] Tzanakis, I., Hodnett, M., Lebon, G.S.B., Dezhkunov, N., and Eskin, D.G. (2016) *Calibration and performance assessment of an innovative high-temperature cavitometer*. Sens. Actuators Phys., 240, 57–69.

[17] Minnaert, M. (1933) *XVI.* On musical air-bubbles and the sounds of running water. Lond. Edinb. Dublin Philos. Mag. J. Sci., 16 (104), 235–248.

[18] Tzanakis, I., Lebon, G.S.B., Eskin, D.G., and Pericleous, K. (2016) *Investigation of the factors influencing cavitation intensity during the ultrasonic treatment of molten aluminium*. Mater. Des., 90, 979–983.

[19] Lebon, G.S.B., Tzanakis, I., Pericleous, K., and Eskin, D. (2018) *Experimental and numerical investigation of acoustic pressures in different liquids*. Ultrason. Sonochem., 42, 411–421.

[20] Eskin, D.G., Tzanakis, I., Wang, F., Lebon, G.S.B., Pericleous, K., Lee, P.D., Connolley, T., and Mi, J. (2017) *Fundamental studies of ultrasonic melt processing*. Proc. 6th Decenn. Int. Conf. Solidif. Process., 546–549.

[21] Nowak, T., Cairós, C., Batyrshin, E., and Mettin, R. (2015) *Acoustic streaming and bubble translation at a cavitating ultrasonic horn*. 020002.

[22] Yasui, K., Iida, Y., Tuziuti, T., Kozuka, T., and Towata, A. (2008) *Strongly interacting bubbles under an ultrasonic horn*. Phys. Rev. E, 77 (1).

[23] Tzanakis, I., Lebon, G.S.B., Eskin, D.G., and Pericleous, K.A. (2017) *Characterizing the cavitation development and acoustic spectrum in various liquids*. Ultrason. Sonochem., 34, 651–662.

[24] Schnerr, G.H., and Sauer, J. (2001) *Physical and Numerical Modeling of Unsteady Cavitation Dynamics*. Proc. 4th Int. Conf. Multiph. Flow, 1.

[25] Weller, H.G. (2017) *OpenFOAM*, OpenCFD Ltd (ESI Group).

[26] Yoshizawa, A. (1986) *Statistical theory for compressible turbulent shear flows, with the application to subgrid modeling*. Phys. Fluids, 29 (7), 2152.

© 2018 ASME

Investigation of an implicit solver for the simulation of bubble oscillations using Basilisk

D. Fuster[1,*]and S. Popinet[1]

[1] Sorbonne Universit´es, UPMC Univ Paris 06, CNRS, UMR 7190
Institut Jean Le Rond d'Alembert, F-75005 Paris, France

Abstract

In this work we investigate an extension of an implicit all-Mach formulation using the Volume of Fluid (VOF) method. The numerical scheme is implemented in Basilisk [1], an open source program widely used and extensively validated for various problems involving multiple phases in the incompressible limit that allows for Adaptive Mesh Refinement using octree grids. The numerical method is adapted to ensure that the numerical scheme exactly recovers the classical fractional step method used for incompressible fluids. The new formulation is shown to capture well the dynamic response of bubbles in a liquid by validating it in two different scenarios: (a) The linear oscillation of a bubble in a slightly compressible liquid and (b) the Rayleigh collapse problem.

Keywords: CFD, Numerical methods, multiphase flows, bubble dynamics.

1 Introduction

The development of robust and accurate compressible codes for multiphase flows is important for a better understanding of processes related to wave propagation and cavitation. All-Mach formulations have been proposed by various authors (see for example [2]). The main advantage of these solvers is that it is possible to solve problems in which one of the phase (e.g. liquid) is considered as an incompressible substance circumventing the classical timestep restriction of solvers based on the solution of Riemann problem. Another advantage of these solvers is that it naturally recover the method developed to solve the Navier-Stokes equations for multiphase flows in the incompressible limit. In this work we present preliminary results on the performance of a variant of this method implemented in Basilisk, an open free-software for the solution of the fluid equations that allows for mesh adaptation. Two classical problems related to the dynamic response of bubbles in liquids are used: The linear oscillation of a bubble in a slightly compressible liquid and The Rayleigh collapse problem. The robustness of the method is shown for the collapse of a gas bubble in water induced by a shock wave.

2 Modelling equations

In this work we solve for the Navier-Stokes equations neglecting surface tension effects:

$$\frac{\partial \rho}{\partial t} + \nabla \cdot (\rho u) = 0 \tag{1}$$

$$\frac{\partial \rho \boldsymbol{u}}{\partial t} + \nabla \cdot (\rho \boldsymbol{u}\boldsymbol{u}) = -\nabla p + \nabla \cdot \tau' \tag{2}$$

where $\tau' = \mu \left(\nabla \boldsymbol{u} + (\nabla \boldsymbol{u})^T\right) - \frac{2}{3}\mu \nabla \cdot \boldsymbol{u}\delta$. In the compressible framework we need to add any form of the energy equation. For instance, in absence of thermal diffusion, the total energy equation is,

$$\frac{\partial \rho e + 1/2\rho \boldsymbol{u}^2}{\partial t} + \nabla \cdot (\rho e \boldsymbol{u} + 1/2\rho \boldsymbol{u}^2) = -\nabla \cdot (\boldsymbol{u}p) + \nabla \cdot (\tau' \boldsymbol{u}) \,. \tag{3}$$

*Corresponding author, Daniel Fuster: fuster@dalembert.upmc.fr

The system is closed by adding an state equation that can establishes the relation between the various thermodynamic variables. From the state equation we can determine various fluid properties such as the speed of sound $c = \left(\frac{\partial p}{\partial \rho}\right)_{S=cte}$ and the thermal dilatation coefficient $\beta_T = \frac{-1}{\rho}\left(\frac{\partial \rho}{\partial T}\right)_p$.

In the all-Mach formulation we rewritte the continuity equation to express density derivatives as a function of pressure and temperature derivatives. The density variations between two thermodynamic states can be obtained as a sum of the changes of density in an isothermal process plus the density change in an isobaric process, $d\rho = \left(\frac{\partial \rho}{\partial p}\right)_T dp + \left(\frac{\partial \rho}{\partial T}\right)_p dT$. Using the definitions of the speed of sound, the polytropic coefficient, γ and the thermal dilatation coefficient we find [3]

$$d\rho = \frac{\gamma}{c^2}dp - \rho\beta dT. \tag{4}$$

The temperature changes can be obtained from the internal energy equation, which in absence of thermal diffusion effects is written as

$$\rho c_p \frac{DT}{Dt} = \beta_T T \frac{Dp}{Dt}. \tag{5}$$

Using Eqs. 4-5 we express the density changes in the continuity equation as a function of pressure changes

$$\left[\frac{\gamma}{\rho c^2} - \frac{\beta_T^2 T}{\rho c_p}\right]\frac{Dp}{Dt} = -\nabla\cdot\boldsymbol{u}. \tag{6}$$

Note that in the limiting situation of an ideal gas, where $c = \sqrt{\gamma RT}$ and $\beta_T = 1/T$, and water $c \approx 1500$ m/s and $\beta_T \approx 0$, it is easily verified that $\frac{1}{\rho c^2} = \left[\frac{\gamma}{\rho c^2} - \frac{\beta_T^2 T}{\rho c_p}\right]$.

In the case of two phase mixtures, we need to add an advection equation for the evolution of the volume fraction for each phase,

$$\frac{\partial f}{\partial t} + \boldsymbol{u}\cdot\nabla f = 0. \tag{7}$$

3 Numerical method

In this work we are interested in investigating the time-split scheme that solves the set of equations above in two steps: The advection and the projection step.

In the *advection step* we solve for the advection of the color function f rewritting it as

$$\frac{f^{n+1} - f^n}{\Delta t} + \nabla\cdot(\boldsymbol{u}c) = c\nabla\cdot\boldsymbol{u}. \tag{8}$$

This equation is solved using the Volume of Fluid (VOF) method proposed by Weymouth & Yue [4], where we obtain the updated value of f from the flux, uf, obtained geometrically using the onto-cell implicit linear mapping of the interface. The advection step for the rest of the conservative variables is

$$\frac{\rho^{n+1} - \rho^n}{\Delta t} + \nabla\cdot(\rho u) = 0 \tag{9}$$

$$\frac{(\rho\boldsymbol{u})^* - (\rho\boldsymbol{u})^n}{\partial t} + \nabla\cdot(\rho\boldsymbol{u}\boldsymbol{u}) = 0 \tag{10}$$

$$\frac{(\rho e + 1/2\rho\boldsymbol{u}^2)^* - (\rho e + 1/2\rho\boldsymbol{u}^2)^n}{\Delta t} + \nabla\cdot(\rho e\boldsymbol{u} + 1/2\rho\boldsymbol{u}^2) = 0. \tag{11}$$

This step is solved with a Bell-Collela-Glaz advection scheme, where the flux of each conservative quantity is computed separately for each phase using the flux of f. This ensures the consistency on the advection of the various conservative quantities and the transport of the color function f avoiding any diffusion of the conservative quantities during the advection step.

The solution of the equation for pressure (Eq. 6) is also split in time, therefore, a priori one needs to solve for $\frac{Dp}{Dt} = 0$ to obtain a provisional pressure p^* in the advection step. However, this pressure is typically obtained from the provisional value of the total energy obtained after advection using the equation of state.

In the *projection step* we take the divergence of the remaining terms in the momentum equation and we replace the expression of the divergence at the end of the time-step as a function of the pressure temporal derivative given by Eq. 6. Thus, we obtain a Helmholtz-Poisson equation for pressure that is solved with the multigrid solver available in Basilisk

$$\left[\frac{\gamma}{\rho c^2} - \frac{\beta_T^2 T}{\rho c_p}\right] \frac{p^{n+1} - p^*}{\Delta t} + \nabla \cdot \boldsymbol{u}^* = \nabla \cdot \left(\frac{\Delta t}{\rho} \nabla p\right). \tag{12}$$

Once the pressure is obtained we update the value of velocity and total energy as

$$\boldsymbol{u}^{n+1} = \boldsymbol{u}^* - \frac{\Delta t}{\rho} \nabla p, \tag{13}$$

$$(\rho e + 1/2 \rho \boldsymbol{u}^2)^{n+1} = (\rho e + 1/2 \rho \boldsymbol{u}^2)^* - \nabla \cdot (\boldsymbol{u} p) \Delta t + \nabla \cdot (\tau' \boldsymbol{u}) \Delta t. \tag{14}$$

4 Results

4.1 Linear oscillations

We start investigating the response of a bubble to a low amplitude sinusoidal wave. We take the bubble radius R_0, the bubble natural resonance frequency $\omega_N = \frac{1}{R_0}\sqrt{\frac{3\gamma p_0}{\rho_l}}$ and the liquid's density as characteristic values of the problem. The bubble is placed at the center of a cubic domain of length $L/R_0 = 2560$. The nondimensional reference pressure is $p_0^* = 1$ and the initial gas density is $\frac{\rho_g}{\rho_l} = 10^{-3}$. We initialize a traveling planar sinuisoidal pressure wave of wavelength $\lambda^* = \lambda/R_0 = 640$ in the liquid. The amplitude of the pressure wave is initialized to a very small value to guarantee that the response of the bubble is dominated by linear terms. The nondimensional speed of sound is $c_l^* = 146.5$, which implies that $\frac{\omega}{\omega_N} = c_l^*/\lambda^* = 0.25$. The maximum resolution is applied close to the bubble interface where the grid size is $\Delta x^* = 0.156$, which gives around 12 points per diameter. The grid is coarsened far from the bubble.

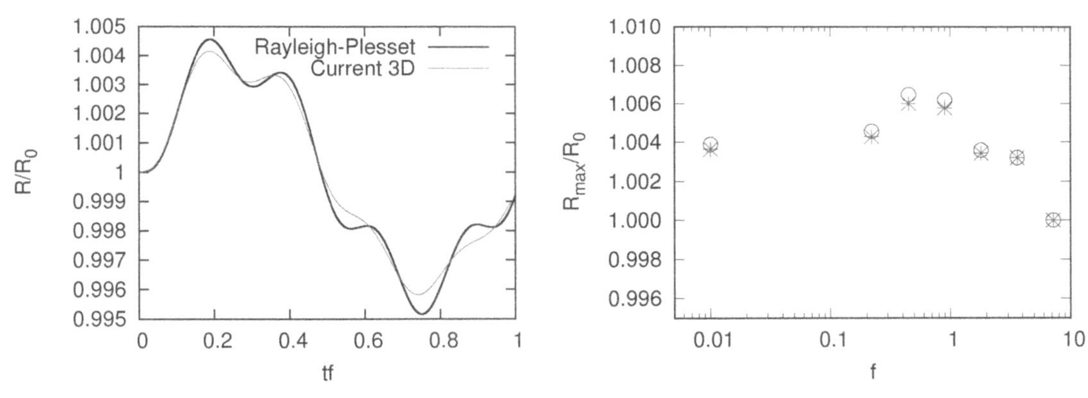

Figure 1: Left: Temporal evolution of the averaged bubble radius compared with the Rayleigh-Plesset model for the low frequency limit case $\frac{\omega}{\omega_N} = 0.25$. Right: Amplitude of the bubble oscillation as a function of the excitation frequency with the current method and the Rayleigh-Plesset model.

Figure 1 compares the evolution of the averaged radius defined as $R_{avg} = \left(\frac{V_{bubble}}{4/3\pi}\right)^{1/3}$ with the solution obtained by the Rayleigh-Plesset equation. We obtain good agreement even for the relatively low resolution used. In the same figure we also compare the maximum radius reached during the first cycle of the bubble oscillation as a function of frequency for both, the current 3D numerical simulations and the solution obtained using a Rayleigh-Plesset model. The bubble remains frozen at large frequencies, it reaches an asymptotic limit for low frequency oscillations and the amplitude of the oscillation is maximum at resonance conditions. Good agreement is found between the two models irrespective of the excitation frequency used.

4.2 Rayleigh collapse problem

The Rayleigh collapse problem is a model problem that considers the collapse of an spherical bubble at initial pressure $p_{g,0}$ in a liquid where the initial pressure is set to $p_{l,0}$. In this case we use the initial radius R_{max}, the Rayleigh collapse time $t_c = 0.915 R_{max}\sqrt{\frac{\rho_l}{\Delta p}}$ and the liquid's density as characteristic values of the problem. The initial gas density is $\rho_g^* = 10^{-3}$ in all cases and the nondimensional liquid's speed of sound is $c_l^* = 50$. A large domain, $L/R_0 = 200$ is used to avoid the influence of boundaries on the dynamic response of the bubble. Fixing these parameters, the solution of the problem depends on the ratio between the initial liquid pressure and gas pressure and the Reynolds number $\mathrm{Re} = \frac{\sqrt{\Delta p \rho_l} \mathrm{R_{max}}}{\mu_l}$, defined using the characteristic velocity $U_c = 0.915 R_{max}/t_c$. In this case the minimum grid size is set to $\Delta x^* = 0.012$ (around 160 grid points per diameter).

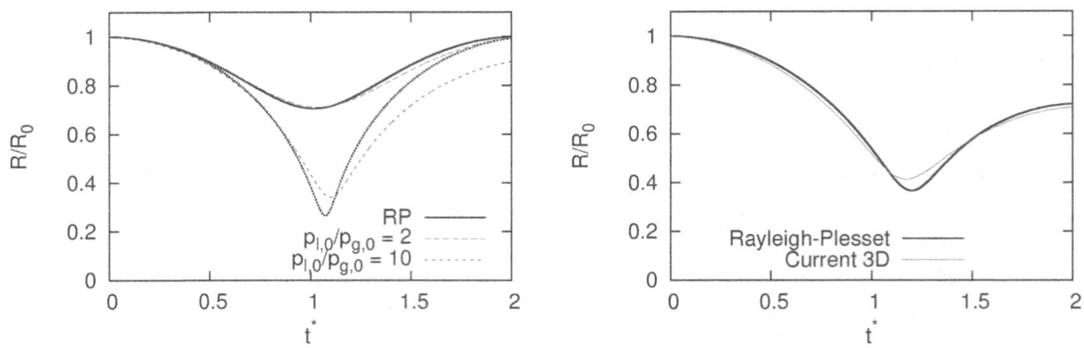

(a) Re = ∞ (b) Re = 100

Figure 2: Bubble radius evolution for two different collapse intensities, $p_{l,0}/p_{g,0} = 2$ and $p_{l,0}/p_{g,0} = 10$. Color lines represent the solution obtained with current 3D simulations, black lines represent the solution obtained with a Rayleigh Plesset model.

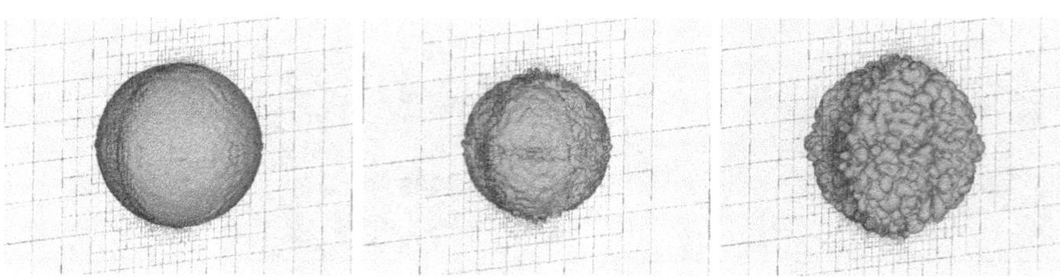

Figure 3: Evolution of the bubble interface for $p_{l,0}/p_{g,0} = 2$ and Re $= \infty$ at $t^* = \frac{2}{3}, 1, \frac{4}{3}$.

Figure 2a shows good agreement between the numerical results and the results obtained from the Rayleigh-Plesset equation for the case of a low intensity collapse ($p_{l,0}/p_{g,0} = 2$) in the inviscid case.

The three dimensional numerical simulation shows the development of a Rayleigh-Taylor instability that become especially visible right after the instant of minimum radius, when the light gas phase pushes the heavy liquid surrounding it (Fig 3). In absence of viscous and surface tension effects the bubble is always unstable during these instants. The role of the Rayleigh-Taylor instability on the bubble collapse becomes more relevant as the intensity of the collapse is increased, which may be part of the reason why the full 3D simulation and the solution obtained with a perfectly spherical bubble (Rayleigh-Plesset model) do not fit well for $p_{l,0}/p_{g,0} = 10$ after the gas start exerting resistance to the bubble collapse. The possibility to dynamically adapt the grid is critical to accurately capture the development of interfacial instabilities during the collapse process.

It is possible to suppress the development of the Rayleigh-Taylor instability by adding viscosity. Simulations at $p_{l,0}/p_{g,0} = 10$ and Re=100 reveal that the bubble collapse is almost perfectly spherically symmetric (not shown here). In this case the evolution of the bubble radius predicted by the Rayleigh-Plesset model is in agreement with the result obtained from the full 3D simulations (Figure 2b).

4.3 2D Shock wave/bubble interaction problem

Following previous works (see for example [5]), the roboustness of the solver for strong collapses is tested using the 2D interaction of a shock wave with a gas bubble in water. The strength of the incoming wave moving from left to right is set to $p_{shock}/p_0 = 1000$. The rest of variables are initialized according to the Rankine-Hugoniot conditions. Figure 4 shows the violent process of the bubble collapse. A characteristic jet previously observed in this type of simulations generate small bubbles as a consequence of the breakup of the gas sheet when the jet traverses the bubble.

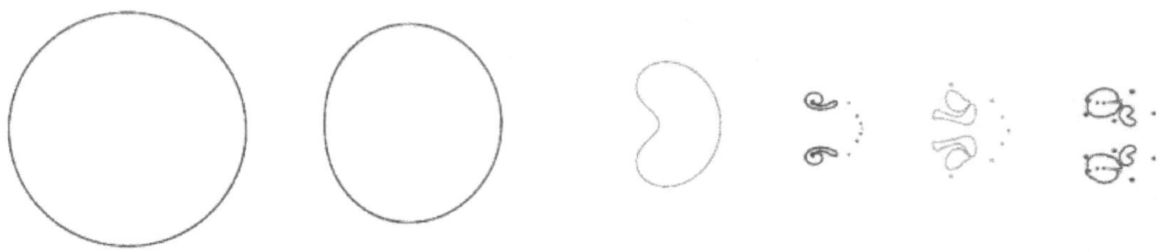

Figure 4: Interface contours for the 2D Shock wave-bubble interaction problem at different instants. $p_{shock}/p_0 = 1000$, $\rho_{g,0}/\rho_{l,0} = 10^{-3}$.

5 Conclusions

In this work we present an extension of an implicit all-Mach formulation using the Volume of Fluid (VOF) method. The discretized numerical scheme is implemented in Basilisk, an open source software widely used and extensively validated for various problems involving multiple phases in the incompressible limit that is parallel and allows for Adaptive Mesh Refinement (AMR) using octree grids. The current numerical scheme accounts for compressible effects ensuring that it exactly recovers the classical fractional step method used for incompressible fluids.

The new formulation is shown to perform well in cases involving multiple phases related to the dynamic response of bubbles in a liquid. The code is validated in two different scenarios: The linear oscillation of a bubble in a slightly compressible liquid and the Rayleigh collapse problem. Finally, the code is shown to be robust in the case of the shock induced collapse of a gas bubble in water.

References

[1] *www.basilisk.fr*

[2] S.Y. Yoon, T. Yabe *The unified simulation for incompressible and compressible flow by the predictor-corrector scheme based on the CIP method*, J. Comp. Phys, 119, 149-158, 1999.

[3] D. Fuster. *Modelling and numerical simulation of the dynamics of liquid-bubble cavitating systems with chemical reaction processes*, PhD dissertation, University of Zaragoza, 2007.

[4] G.D. Weymouth, D.K.P. Yue. *Conservative volume-of-fluid method for free-surface simulations on Cartesian-grids*, J. Comp. Phys, 229, 8, 2010.

[5] E. Johnsen, T. Colonius. *Numerical simulations of non-spherical bubble collapse*, J. Fluid Mech, 629, 231-262, 2009.

© 2018 ASME

10th International Symposium on Cavitation - CAV2018
Baltimore, Maryland, USA, May 14 – 16, 2018

CAV18-05123

Investigation of Cavitation Bubble Cloud with discrete Lagrangian Tracking

Xiuxiu Lyu, Xiangyu Hu, Nikolaus A. Adams

Chair of Aerodynamics and Fluid Mechanics, Department of Mechanical Engineering, Technical University of Munich
85748 Garching, Germany

Abstract

Here, to study more about the mechanism of the cavitating bubble cloud, we simulate numerically microbubble cloud inside bulk fluid with Lagrangian tracking. The bubble-flow mixture is modelled in a two-way coupled Eulerian-Lagrangian scheme, continuum description of the bulk fluid with discrete Lagrangian tracking of the gas bubbles. The Euler equations are solved based on the multi-resolution conservative compressible fluid method. Dynamics of the dispersed cavitating gas bubbles is modelled in a Lagrangian framework, describing the bubble location and bubble size variation. The sub-grid interface is allowed to diffuse on the computational cells around it within the kernel width. To close the system, the equations of state (EOS) of liquid-gas mixture are given in a stiff gas equation of state, along with one isobaric closure of the pressure. Stability and accuracy of the interface are studied by some classical cases. Pressure perturbation induced by the bubble cloud inside flow is simulated based on two-way coupling framework.

Keywords: Bubble cloud, Euler-Lagrange method, Bubble dynamic, Cavitation

1. Introduction

Bubbly flow has been investigated as applications in efficient diagnostic ultrasound treatment and chemistry process, where the dynamics of bubble cloud is of fundamental importance [1, 2, 3]. Our motivation comes from simulating numerically the bubbly cloud and at the same time tracking accurately the under-resolved or subgrid interface between micro-bubbles and bulk fluid.

In literature, dynamics of interface flows has been normally cataloged into three major models: continuum model, level set method, and Lagrangian dispersed model. Continuum model solves the Euler equation coupled with transport equations [4, 5, 6]. The effect of bubble cloud on the computational cells is represented by volume fraction. The actual position of the interface is recovered at the end of each computational step by an interface reconstruction process. In level set method introduced by Osher and Sethian [7], the interface is described as signed distance function, and it is popularly applied in computational sharp interface multi-phase fluid [8, 9]. This method is mostly applied for the numerical simulation of single bubble and suffers from drawback when the complexity increases to 3D. In the Lagrangian dispersed model [10, 11], the interface is discetized into particles which move dynamically at each computational time step. It is generally accurate in predicting the bubble dynamics and transportation when all bubbles are tracked with a much higher spatial resolution. Limitation of Lagrangian dispersed method is that equation of motion for each bubble has to be solved and larger number of bubbles can lead to higher computational cost especially when bubble-bubble interaction is taken into account. However, due to the simplicity in operation, low computational cost, Lagrangian dispersed model is quite suitable for the micro-bubble cloud simulation and meanwhile can be easily developed into 3D. Assuming that in each computational cell for the bubble-fluid mixture there's only one single velocity and pressure, the complete system for Eulerian-Lagrangian model includes the basic conservation law of mass, momentum and energy, one transport equation of the void fraction (bubble dynamics equation of each particle) and one effective Eulerian-Lagrangian coupling model.

Email addresses: `xiuxiu.lyu@tum.de` (Xiuxiu Lyu), `xiangyu.hu@tum.de` (Xiangyu Hu), `nikolaus.adams@tum.de` (Nikolaus A. Adams)

In [11], Darmana et al studied a new paralleled algorithm in application of to Euler-Lagrange in specific for dispersed gas-liquid flow. In Ref. [12], Fuster and Colonius proposed a new model for bubbly cavitating flow, in which they solve the volume-averaged Eulerian equations for the bulk fluid coupled with Lagrangian equations for each gas bubble. Their results show that their method can capture the pressure disturbance induced in the liquid. However, the closure law in those models is quite easy, which is only one expression of the liquid pressure, not accurately the pressure of the mixture. In [13], the mixture density is not an explicit function of the pressure of EOS. Here, we do the numerical simulation to study the dynamics of cloud of microbubbles to the pressure wave, in order to understand the detail mechanism of the bubble cloud under pressure wave and also the resulting stable bubble distributions afterwards. The background cavitating flow is treated as Eulerian flow with initially dispersed cavitating gas bubbles with radius of several micrometers.

2. Physical Models and numerical methods

2.1. Single bubble dynamics

The dynamics of the dispersed phase is treated in a Lagrangian framework, describing the volume and the motion of the gas bubbles. The bubbles are assumed to be spherical, as they are extremely small in the scale of nanometres or micrometers that the surface tension is large enough to maintain the spherical shape. Bubble-bubble interaction, bubble coalescence and collision are not considered.

The dynamics and motions of bubbles is controlled by the bulk flow. The Rayleigh-Plesset equation proposed by Rayleigh [14] and Plesset [15] has been widely used for the bubble dynamics modelling in hydrodynamic cavitations, multi-phase bubbly flows and underwater explosion bubbles. Based on Rayleigh-Plesset equation, dynamics of a spherical bubble surrounded by a weekly compressible liquid in infinite can be expressed as:

$$\rho_m\left[R\frac{d^2R}{dt^2}+\frac{3}{2}\left(\frac{dR}{dt}\right)^2\right]=p_B-p_\infty-\frac{2S}{R}-\frac{4\mu}{R}\frac{dR}{dt}, \tag{1}$$

n where ρ is the density of the liquid-gas mixture, R is the bubble radius, t is the time, S is surface tension, μ is the viscosity, p_B is the pressure inside the bubble, p_∞ is the far-field pressure of the surrounding field. In general, the bubble inside the water contains vapour whose pressure is labelled as p_v and some quantity of contaminant gas pressure p_{g0} at a reference bubble size R_0. And p_v, is considered constant when the temperature T_b is constant. We assume the mass of the gas m_g in the bubble and its temperature T_b remain the same, the pressure inside a bubble has the expression as flows:

$$p_B=p_v+p_{g0}\left(\frac{R_0}{R}\right)^{3\gamma_g}, \tag{2}$$

with $\gamma_g=1.4$.

2.2. Liquid equations

We assume the bulk flow represents the liquid-gas mixture, which means the velocity of the mixture $\boldsymbol{u}$ equals to the velocity of the fluid $\boldsymbol{u}_l$. So that each computational cell is homogeneous and has only one velocity vector $\boldsymbol{u}$ and one pressure P. Also we assume the bubble-fluid mixture is inviscid and compressible. The conservation law of the mixture can be written as

$$\frac{\partial\mathbf{U}}{\partial t}+\nabla\cdot\mathbf{F}(\mathbf{U})=0, \tag{3}$$

where $\mathbf{U}=(\rho_m,\rho_m\boldsymbol{u},E)^T$ is the average local mixture density of the mass, momentum and energy, here $E=\rho_m e_m+\frac{1}{2}\rho_m\boldsymbol{u}^2$, $\mathbf{F}$ represents the corresponding flux functions.

To express the amount of the gas per unit volume, we define the void fraction of gas α_g and the void fraction of the liquid α_l,

$$1=\alpha_l+\alpha_g. \tag{4}$$

Also, according this definition, the averaged density and inner energy of the mixture can be written as

$$\rho_m=\alpha_l\rho_l+\alpha_g\rho_g, \tag{5}$$

$$\rho_m e_m = \alpha_l \rho_l e_l + \alpha_g \rho_g e_g, \tag{6}$$

where the subscripts l and g denote respectively the liquid and the gas.

2.3. The isobaric closure law

For the computation of gas-liquid mixture, we apply the expressions of pressure in multi-component five-equation model in consideration of void fraction. The stiffed-gas EOS for each fluid is written as

$$P_i = (\gamma_i - 1)\rho_i e_i - \gamma_i B_i \tag{7}$$

involving only constants γ and reference pressure B for each fluid. In our simulation, for water $\gamma_l = 7.1, B_l = 300$MPa, for gas $\gamma_g = 1.4, B_g = 0.1$MPa.

As mentioned before, the mixture pressure must apply the property: mixing of two fluids with the same pressure should result in a mixture equilibrium pressure with the same value. Failing to make this equilibrium will generate numerical vibration near the location of interface. The EOSs for liquid-gas mixture must follow

$$P_l = P_g = P_m. \tag{8}$$

Introducing $\xi_g = \dfrac{1}{\gamma_g - 1}$, $\xi_l = \dfrac{1}{\gamma_l - 1}$ and $\xi_m = \alpha_l \xi_l + \alpha_g \xi_g$, an implicit expression for P is given as

$$\rho_m e_m = P_m \xi_m + \alpha_l \gamma_l B_l \xi_l + \alpha_g \gamma_g B_g \xi_g \tag{9}$$

which yields the expression

$$P_m = \frac{\rho_m e_m}{\xi_m} - \frac{1}{\xi_m}\left(\alpha_l \gamma_l B_l \xi_l + \alpha_g \gamma_g B_g \xi_g\right). \tag{10}$$

2.4. Euler-Lagrange coupling

The void fraction distribution is derived from the instantaneous bubble sizes and locations. The concept of kernel function has been applied widely in the smooth particle hydrodynamics (SPH) [16]. And to smooth the discontinuities and improve stability of the vapor void faction in the computational grids, we apply a Gaussian kernel volume distribution scheme. The Gaussian interpolation function is given by

$$\zeta_\sigma(\mathbf{x},\mathbf{x}_b) = \frac{1}{\left(\sigma\sqrt{2\pi}\right)^3}\mathbf{exp}\left(\left[-\frac{\sum_{k=1}^{3}(x_k - x_{b,k})^2}{2\sigma^2}\right]\right) \tag{11}$$

where σ is the kernel width, x_k and $x_{b,k}$ are the grid cell center location and bubble location respectively. In order to enforce mass conservation, the kernel function is normalized over the volume of integration by

$$\int_{V_{cv}} \zeta_\sigma(\mathbf{x}_{cv},\mathbf{x}_b)\, dV = 1. \tag{12}$$

Using the above kernel function, volume fraction of the gas bubbles in every controlled volume can be calculated as:

$$\alpha_g = \sum_{i=1}^{N} \frac{4}{3}\pi R_i{}^3 \zeta_\sigma. \tag{13}$$

In this formulation, the volume occupied by the gas bubble in a fluid control volume is accounted by α_g and fluid void fraction α_l ($\alpha_g + \alpha_l = 1$).

3. Results and Discussion

3.1. Simulation set-up

In our simulation, we solve the Euler equations on a Cartesian grid, with the details referred in Ref. [8, 17]. The equations are discretized with the fifth-order WENO method [18] and the second-order TVD Runge-Kutta scheme [19]. All the computations are carried out with the CFL number of 0.3.

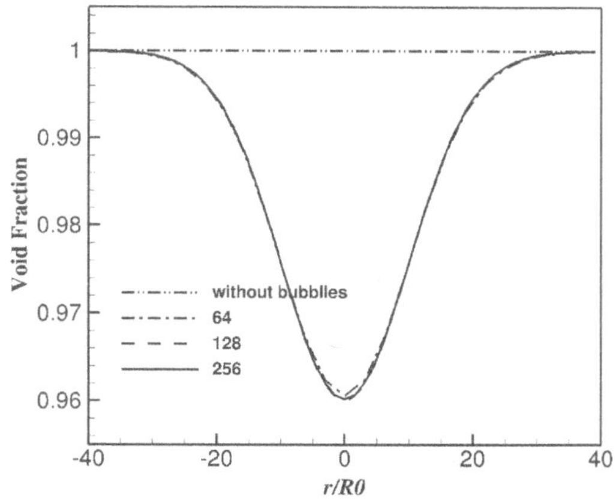

Figure 1: Void fraction α_l along the bubble radius under different resolutions.

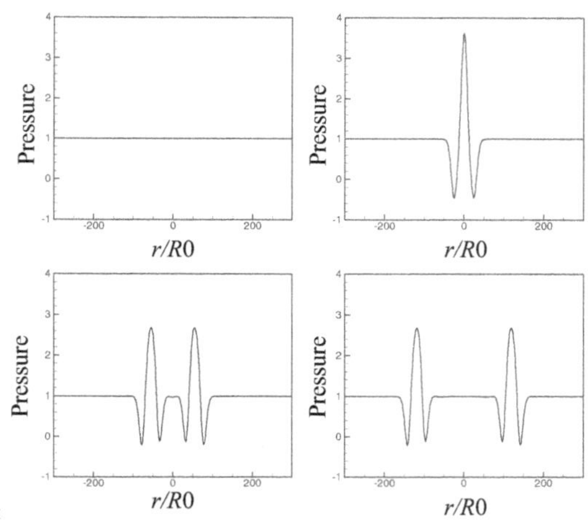

Figure 2: Pressure wave propagation induced from an oscillating gas bubble in the bulk fluid .

3.2. One bubble oscillating

With the variable time-step numerical algorithm in Ref. [20], we can solve the bubble dynamics even when the bubble size has extreme size variation ratio at one time step. Here, we consider a single gas bubble placed inside water in a 1D domain. The bubble radius change dynamically in time as $R = R_0(1 - \varepsilon \sin\omega t)$, where ε is the perturbation magnitude and ω is the frequency.

We set $R_0 = 100\mu m$, $\varepsilon = 0.1$ and $\omega = 10$kHz. The 1D dimension size is $2L$, where $L = 300R_0$. The width of the support of the kernel $\delta = 10.0R_0$. Figure.1 indicates the void fraction of water along the radius away from the bubble center under different resolutions. Here, the bubble oscillate actively only in the first round, afterwards, the volume of the bubble goes back to the equilibrium state. The results show the convergence of our method. Figure. 2 shows the pressure wave induced by the oscillating bubble at $t^* = tf = 0.0, 1.0, 3.0$ and 5.0. The pressure wave is well resolved and propagates to left and right sides. It indicates that the local variables are well predicted based on our method for a single sub-grid bubble.

3.3. Multi-bubbles under pressure pulse

Here in this section, we simulate the dynamics of a 2D cylindrical bubble interacting with a sinusoidal pulse in the bulk flow. We investigate a 2D bubble cloud with 200 bubbles randomly radius distribution between 1μm and 5μm , the radius of the cluster being R_c = 2mm. The sinusoidal pressure wave moving from left to right is set to be $p = p_0 + \Delta p \sin\left(2\pi\frac{x}{\lambda}\right)$, whose wave length is $\lambda = R_c$ and wave amplitude is $\Delta_p = 3.0p_0$ with pressure at infinity p_0 = 1.0atm. The initial pressure wave center locates 0.4cm right side of the bubble cloud center. The negative pressure will reach the bubble cloud firstly, which makes the bubbles expand firstly and collapse secondly. The kernel width for each gas bubble is $\sigma = 2.0R_i$, R_i is the radius of the bubble i. The fluid void fraction at initial ranges from 0.557 to 1.0, with a gray scale contour in Figure. 3a. The computational domain size is set to be (2.0cm ,1.0cm) with the finest resolution (1024, 512).

As shown in Figure. 3, the v-velocity in the mixture is plotted at 1.2μs,2.0μs,3.0μs, and 4.0μs after the sinusoidal pulse reaches the cloud. Simultaneously, pressure field at 2.0μs,3.0μs, and 4.0μs are also shown respectively in Figure. 4. As shown in this figure, the pressure pulse inside the bulk flow induces the bubbles vibration, in reverse, the vibration of the gas bubbles induces the pressure noise in the bulk flow.

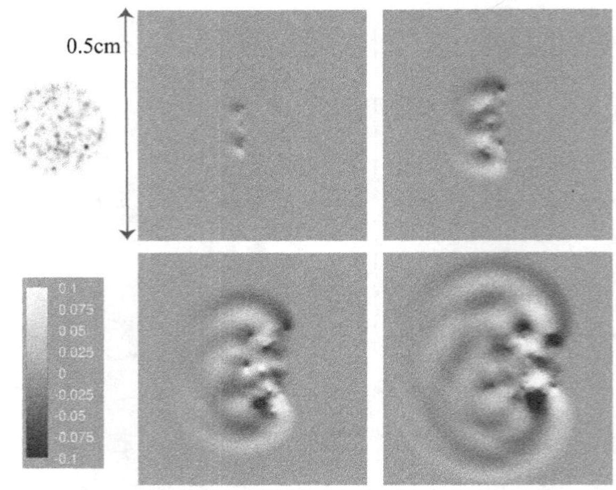

Figure 3: Bubble cluster interacting with a sinusoidal pulse, the v velocity field induced by the bubble cluster.

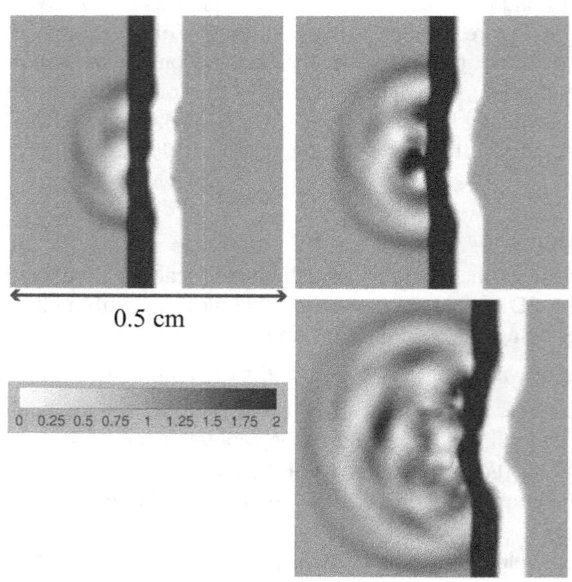

Figure 4: The pressure field induced by the bubble cluster when bubble cluster interacting with a sinusoidal pulse.

4. Conclusion

In this paper, a Lagrangian-Eulerian coupling scheme is implemented based on one sharp interface multi-resolution conservative Euler solver for bubbly flow. The main conclusion of this paper can be drawn as follows: the current method convergences; the pressure pulse inside bulk flow induces the bubbles vibration, in reverse, the vibration of the gas bubbles induces the pressure noise in the bulk flow; the pressure wave induced by the oscillating bubble can be predicted by our method.

5. References

[1] Y. Matsumoto, J. S. Allen, S. Yoshizawa, T. Ikeda, Y. Kaneko, Medical ultrasound with microbubbles, Experimental thermal and fluid science 29 (3) (2005) 255–265.
[2] C. C. Coussios, R. A. Roy, Applications of acoustics and cavitation to noninvasive therapy and drug delivery, Annu. Rev. Fluid Mech. 40 (2008) 395–420.
[3] E. Vlaisavljevich, K. W. Lin, M. T. Warnez, R. Singh, L. Mancia, A. J. Putnam, E. Johnsen, C. Cain, Z. Xu, Effects of tissue stiffness, ultrasound frequency, and pressure on histotripsy-induced cavitation bubble behavior, Physics in medicine and biology 60 (6) (2015) 2271.
[4] C. W. Hirt, B. D. Nichols, Volume of fluid (vof) method for the dynamics of free boundaries, Journal of computational physics 39 (1) (1981) 201–225.
[5] G. Allaire, S. Clerc, S. Kokh, A five-equation model for the simulation of interfaces between compressible fluids, Journal of Computational Physics 181 (2) (2002) 577–616.
[6] F. Örley, T. Trummler, S. Hickel, M. Mihatsch, S. Schmidt, N. Adams, Large-eddy simulation of cavitating nozzle flow and primary jet break-up, Physics of Fluids 27 (8) (2015) 086101.
[7] S. Osher, J. A. Sethian, Fronts propagating with curvature-dependent speed: algorithms based on hamilton-jacobi formulations, Journal of computational physics 79 (1) (1988) 12–49.
[8] X. Y. Hu, B. C. Khoo, N. A. Adams, F. L. Huang, A conservative interface method for compressible flows, Journal of Computational Physics 219 (2) (2006) 553–578.
[9] E. Lauer, X. Y. Hu, S. Hickel, N. A. Adams, Numerical modelling and investigation of symmetric and asymmetric cavitation bubble dynamics, Computers & Fluids 69 (2012) 1–19.
[10] E. Delnoij, F. Lammers, J. Kuipers, W. Van Swaaij, Dynamic simulation of dispersed gas-liquid two-phase flow using a discrete bubble model, Chemical Engineering Science 52 (9) (1997) 1429–1458.
[11] D. Darmana, N. G. Deen, J. Kuipers, Parallelization of an euler–lagrange model using mixed domain decomposition and a mirror domain technique: Application to dispersed gas–liquid two-phase flow, Journal of Computational Physics 220 (1) (2006) 216–248.
[12] D. Fuster, T. Colonius, Modelling bubble clusters in compressible liquids, Journal of Fluid Mechanics 688 (2011) 352–389.
[13] J. Ma, C.-T. Hsiao, G. L. Chahine, Numerical study of acoustically driven bubble cloud dynamics near a rigid wall, Ultrasonics sonochemistry 40 (2018) 944–954.
[14] L. Rayleigh, Viii. on the pressure developed in a liquid during the collapse of a spherical cavity, The London, Edinburgh, and Dublin Philosophical Magazine and Journal of Science 34 (200) (1917) 94–98.
[15] M. S. Plesset, The dynamics of cavitation bubbles, Journal of applied mechanics 16 (1949) 277–282.
[16] J. J. Monaghan, Why particle methods work, SIAM Journal on Scientific and Statistical Computing 3 (4) (1982) 422–433.
[17] L. H. Han, X. Y. Hu, N. A. Adams, Adaptive multi-resolution method for compressible multi-phase flows with sharp interface model and pyramid data structure, Journal of Computational Physics 262 (2014) 131–152.
[18] G. S. Jiang, C. W. Shu, Efficient implementation of weighted eno schemes., Tech. rep., DTIC Document (1995).
[19] C. W. Shu, S. Osher, Efficient implementation of essentially non-oscillatory shock-capturing schemes, Journal of Computational Physics 77 (2) (1988) 439–471.
[20] H. Alehossein, Z. Qin, Numerical analysis of rayleigh-plesset equation for cavitating water jets, International Journal for Numerical Methods in Engineering 72 (7) (2007) 780–807.

© 2018 ASME

Investigation of cloud cavitation passive control method using Cavitating-bubble Generators (CGs)

Ebrahim Kadivar*; Ould el Moctar

University of Duisburg-Essen, Duisburg, Germany

Abstract

In this article, we propose a passive method to control unsteady cloud cavitation on hydrofoils using cavitation-bubble generator (CGs). This method may be used in many engineering applications, in particular in marine and turbomachinery. First, we used a Partially-averaged Navier Stokes (PANS) model for turbulence to simulate the unsteady cavitating flow and validated it based on experimental data. This model was coupled with a mass transfer model and implemented to the open source software package OpenFOAM. Second, the effect of a proper design of CGs on qualitative parameters such as cavitation structure and the shape of cavity were studied. The effect of CGs on the destructive effects of cavitation such as vibration, turbulent velocity fluctuations and high-pressure amplitude were analyzed. Our results showed that a proper design of CGs may reduce the amplitude of the force fluctuations on the hydrofoil substantially. Further on, the local boundary layer around the hydrofoil surface was altered and the turbulent velocity fluctuation was reduced significantly using this technique.

Keywords: Passive flow control; unsteady cloud cavitation; vortex generators; PANS

Introduction

Unsteady cloud cavitation is a flow phenomenon that may have destructive effects such as pressure pulsation, noise and erosion [1] - [2]. The control of the unsteady cavitation is expected to improve the performance of blades in turbines, propellers and rudders specially in marine and turbomachinery applications. In the past decades, most of the researchers were focused on resolving cavitation phenomenon numerically and experimentally but how the destructive effects of cavitation can be controlled has been studied scare. Few investigations were conducted to control the cavitation and reduce its undesirable behavior in the supercavitation and cloud cavitation regimes. Crimi et al. [3] investigated the effect of introducing a sweep angle to a hydrofoil. They carried out tests on semi-span hydrofoils with sweep angles of 0, 15, 30 and 45 degrees. They showed that the sweep angle may alleviate the problem of erosion due to cavitation. Kuiper [4] improved the inception behavior of tip vortex cavitation on propellers. He added forward skew to the propeller blade and compared the result to those obtained from cases with no skew. Ausoni et al. [5] investigated the hydrofoil roughness effects on von karman vortex shedding. They showed that with the help of a distributed roughness, the transition to turbulence is triggered at the leading edge, which reduces the span-wise non uniformities in the boundary layer transition process. Xiang et al. [6] investigated the effect of the ventilated partial cavity on the drag reduction. Their results showed that a remarkable drag reduction may be achieved for the lower and higher cavitation cases. Kadivar et al. [7] investigated supercavitation flow over different 30, 45 and 60 degree wedge cavitators. They showed that the drag coefficient decreases by the reduction of the wedge angle of cavitator. They illustrated that the wedge angle of the cavitator affects significantly the shape and type of supercavitation. Coutier-Delgosha et al. [8] studied the effect of the surface roughness on the dynamics of sheet cavitation on a two-dimensional foil section. They presented that the roughness in the downstream end of the sheet cavity plays a major role in the arrangement of the cavitation cycle. According to the previous studies it is well known that the control of cavitation on the immersed bodies has a significant influence on the reduction of the destructive effects of cavitation such as loss of efficiency and vortex-induced vibration. We introduce in this work a method to control the unsteady cloud cavitation. The idea was adapted from vortex generators that are common in boundary layer control around airfoils in aerospace engineering applications. This analogy is used to control boundary layer and as a consequence the upper surface pressure distribution and to reduce the destructive effects of cavitation.

*Corresponding Author, Ebrahim Kadivar: ebrahim.kadivar@uni-due.de

Numerical Modelling

In this paper PANS model was used for the simulation of unsteady cavitating flow. The PANS model is a hybrid method of Reynolds-averaged Navier Stokes (RANS) and Direct Numerical Simulation (DNS), which was first proposed by Girimaji [9]. The results of other publications using this model showed the improvement of the accuracy of numerical simulation, e.g. Song and Park, [10] and Ji et al. [11]. This approach is coupled with a mass transfer model which was implemented to the open source software package OpenFOAM. The interface between the liquid and vapor phases is captured by using the volume of fluid (VOF) method and the Schnerr and Sauer [12] cavitation model is chosen for numerical simulation of cavitation.

Passive cavitation control

Cavitation bubbles are generated artificially to influence the whole processes of vaporization, bubble generation and bubble implosion, which occur at normal condition without any control. This is done by inserting a type of micro vortex generators called cavitating-bubble generators (CGs) on the upper surface of the hydrofoil where it is expected that the cavitation is produced naturally. The 2D schematic view of a hydrofoil with CGs on the suction side of the hydrofoil and the 3D view of CGs located on the hydrofoil are shown in Figure 1. "L" and "H" are the dimensionless length and height of the CGs with regard to the chord length of the hydrofoil. Our investigations on the size of the CGs show that it should be small enough so that it does not have a significant effect on the hydrodynamics performance of the hydrofoil. Because of unpleasant side effects that may occur, the shape, the size and the location of the CGs are curtail.

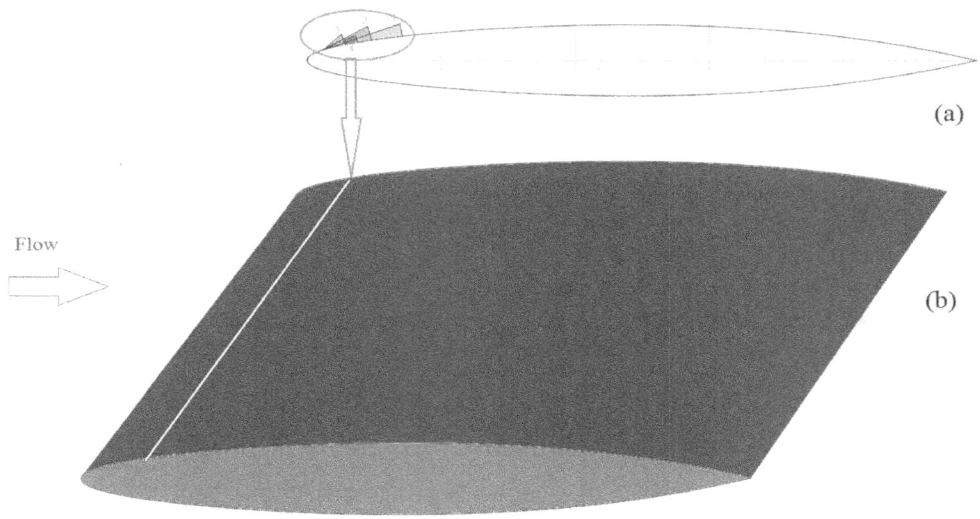

Figure 1. a) 2D view of some example CGs located on the hydrofoil with different sizes and angles, b) 3D view of a spanwise CG located on the upstream of the hydrofoil suction side

Results and discussion

Results of unsteady cavitating flow over the CAV2003 benchmark hydrofoil with and without of the cavitation control are presented here. The geometry of this hydrofoil was perfected at the CAV2003 workshop, [13]. To confirm the results independency on the mesh size and on the time step value these two parameters are tested in this work. The effect of using three different mesh sizes on time-averaged lift and drag coefficients and Strouhal number based on the chord $St_c = f \times l_{ref}/V_{ref}$ are shown in Table. 1, which f and V_{ref} are the cavity self-oscillation frequency and reference velocity respectively. The Grids A to C have 1,150,000, 3,450,000 and 7,257,600 cells respectively. At first the time step for the unsteady computation was set to 4e-6. The results show that the changes between the last two grids are small. Therefore, we performed our simulations using grid B in the entire computational domain.

	Time-averaged C_l	Time-averaged C_d	St_c
Grid A	0.48	0.075	0.107
Grid B	0.44	0.073	0.11
Grid C	0.43	0.072	0.11
Delgosha-Simulation	0.45	0.07	0.108
Delgosha-Experiment	-	-	0.15

Table 1. Comparison of the time-averaged lift and drag coefficients and Strouhal number based on the chord at cavitation number $\sigma = 0.8$

After that for grid B we have simulated with two different time steps 4e-6 s and 1e-5 s. The results showed that the difference between these two time steps is not significant. Time history of lift C_l and drag coefficients C_d from our numerical simulation were compared with the work by Delgosha et al. as shown in Fig. 2.

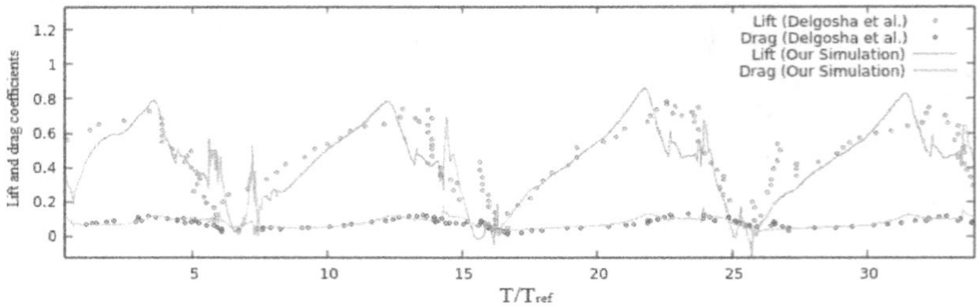

Figure 2. Time history of lift and drag coefficients, present work and Delgosha et al. (2007)

Figure 3 shows that the amplitude of the force fluctuations using proper CGs was reduced significantly and the cavitating flow reached a quasi stable situation. This figure shows that using proper size and position of CGs on the hydrofoil surface the advantage of reduction of the pressure drag force around the hydrofoil will be obtained. The pressure drag was reduced more in comparison with the lift reduction. That means with the proper design of passive controller the hydrodynamic efficiency was increased and the unsteady behavior of the cavitation was suppressed. Figure 4 and 5 present the pressure distributions around the hydrofoil with and without using passive controller in one typical cycle.

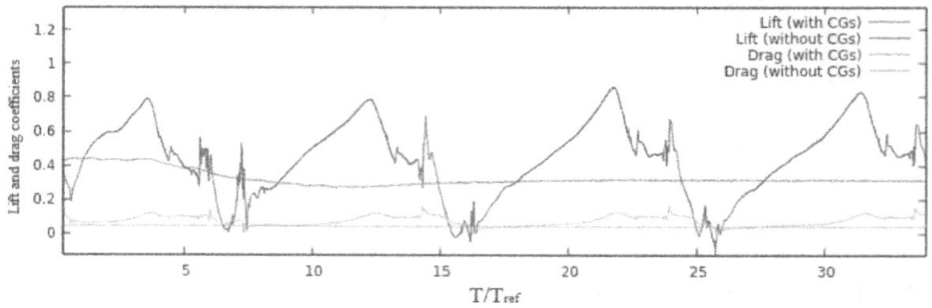

Figure 3. Time history of lift and drag coefficients with a proper CG on the hydrofoil and the hydrofoil without using CGs.

When the large number of vapor structures as bubble clusters and small-scale vortices reach the high pressure region on the surface of hydrofoil, they collapse. This process may induce different destructive effects such as high pressure peaks on the solid surface of hydrofoil and wall-pressure fluctuations. As be seen in Figure 4 for the hydrofoil without CGs high pressure peaks are captured during the cavitation process. It can be seen from the instantaneous figures that these pressure peaks were occurred near the rear of attached cavity at the cavity collapse region and at the trailing edge of the hydrofoil. Figure 5 shows that for the hydrofoil with proper CGs no pressure peaks are observed. The high peak values of the pressure using proper design of CGs were reduced significantly and the cavitating flow reaches a quasi-steady state situation which shows no periodical large cloud cavitation.

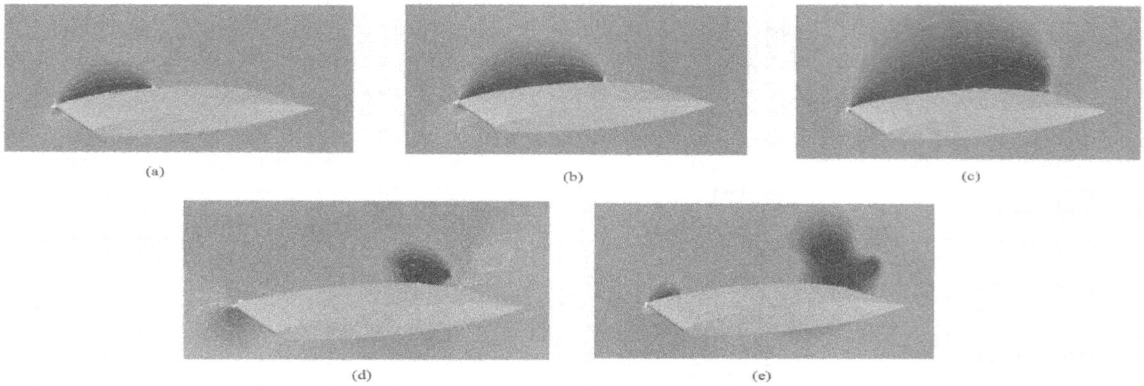

Figure 4. Pressure distributions at the middle of spanwise of the hydrofoil without CGs in one typical cycle.

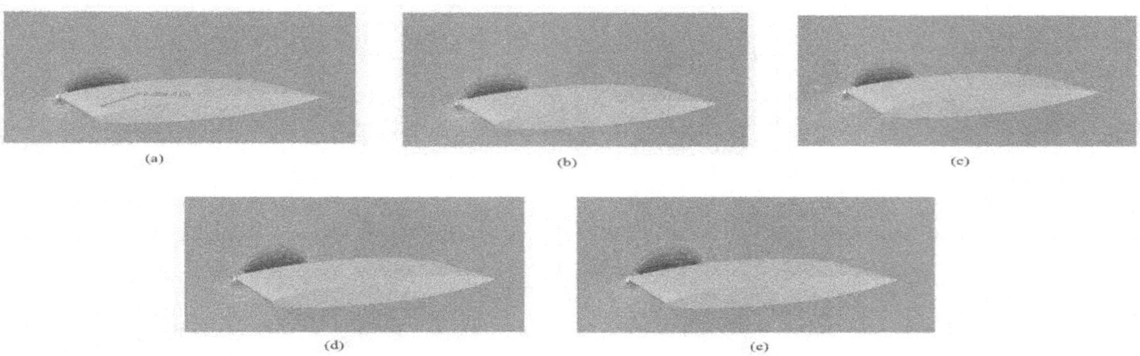

Figure 5. Pressure distributions at the middle of spanwise of the hydrofoil with using a proper CGs in one typical cycle.

For the case without CGs four different peak frequencies f_1 = 6.6 Hz, f_2 = 12.4 Hz, f_3 = 19.13 Hz and f_4 = 25.8 Hz and for the case with CGs two different peak frequencies f_1 = 10.23 Hz and f_2 = 16.8 Hz were observed. These values show that the dominant frequency for the case of using CGs is increased insignificantly in comparison with the simple one. However, the amplitude of the dominant frequency using proper CGs was reduced remarkably. The dominant frequency with the highest amplitude corresponding to the cavitation shedding events is considered to find the Strouhal number. For the case of ith CGs the dominant frequency corresponds to Strouhal number of St_c = 0.17 and St_l = 0.093, while for the case without CGs the Strouhal number values are about St_c = 0.11 and St_l = 0.077. Since the unsteady cloud cavitation and the shedding of sheet cavity this are complex phenomena, Q-criterion reflected the structure of the cloud cavitation and vorticity distribution properly. This criterion defined as the positive second invariant of the velocity gradient tensor, [14]. Figure 6 and 7 show the dynamics of the vortex structure evolution as visualized by the Q-criterion during one oscillation cycle with and without using a proper manner of passive controller. The vortex structures and the sudden changes of cavity cause to the significant changes at the downstream of the sheet cavitation, which induce a strong variation of lift and drag forces on the hydrofoil surface. The figures show that the sheet cavity shedding and the vapor cloud cavity have a substantial influence on the vortex structures. The main mechanism for the generation of the cavitating horse-shoe vortex could be the interaction between the circulating flow and the shedding of vapor cloud. It can be seen from Figure 7 (a-h) that the local flow structures and the local boundary layer don't change significantly. They show a quasi-constant turbulent velocity process in the cavitation regimes around the hydrofoil. According to instantaneous images, it is clear that the local boundary layer and wake flow changes insignificantly under the effect of CGs on the hydrofoil surface which means that the vortex structure of the cavity in the closure and wake regions changed remarkably in comparison with the hydrofoil without CGs.

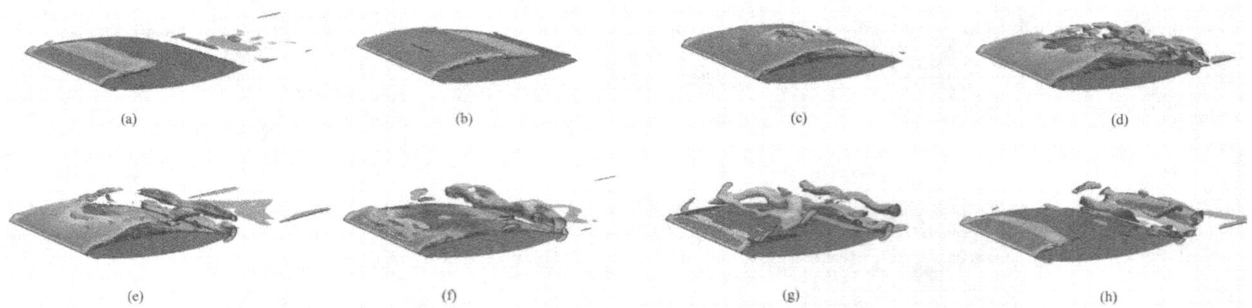

Figure 6. Instantanous Iso-surface of the Q-criterion with 100,000 [s^{-2}]) colored by streamwise velocity without using CGs.

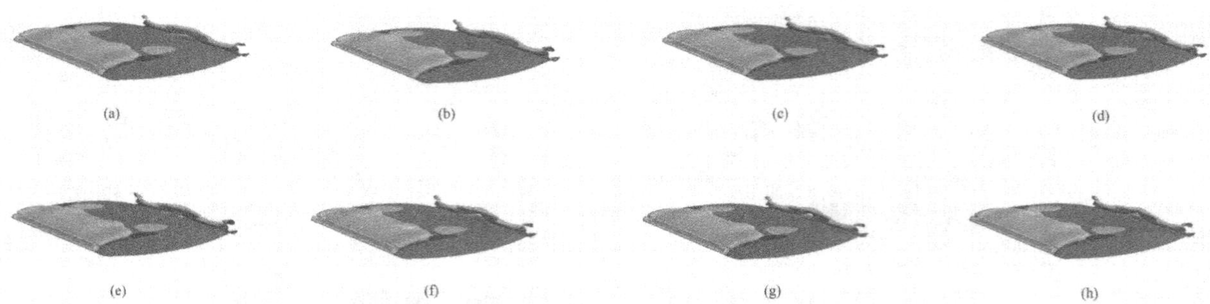

Figure 7. Instantanous Iso-surface of the Q-criterion with 100,000 [s^{-2}]) colored by streamwise velocity with using CGs.

Conclusions

In the present study, a method to control unsteady cloud cavitation around the CAV2003 benchmark hydrofoil using passive cavitation controllers called cavitation-bubble generator (CGs) was studied. The different sizes and locations of the CGs were investigated to find a proper design of CGs to control the cloud cavitation. First, the unsteady cloud cavitation around the hydrofoil without CGs has been simulated using a Partially-averaged Navier Stokes (PANS) method to evaluate the numerical simulations based on experimental data. Second, the effect of a proper CGs as cavitation controller on the qualitative parameters such as cavitation structure and the shape of cavity was presented. The effect of passive controller on the different destructive effects of cavitation such as unsteadiness of the cloud cavitation, turbulent velocity fluctuations, high wall-pressure peaks and reduction of hydrodynamic performance has been analyzed. Our results showed that the appropriate design of the CGs on the hydrofoil surface led to the reduction in high-pressure amplitude on the wall surface of the hydrofoil. The results showed that the cyclic behavior of the cloud unsteady cavitation was suppressed partially and the hydrodynamic efficiency of the hydrofoil increased. The high wall-pressure peaks was reduced significantly. In conclusion, using this method of passive control on the surface of immersed bodies such as hydrofoil, propeller and water turbine blades in cloud cavitation regime a significant reduction in noise generation, unpleasant unsteady side force effects, high-pressure peaks on the surface, flow-induced vibration and surface erosion can be expected. The investigation of the cavitation control using passive control methods with LES simulation and experimental studies will be studied by authors in future work.

Acknowledgement

The authors would like acknowledge Dr. Javadi from sharif university for helpful discussions. The authors would like to thank the support of computer resources by MagnitUDE supercomputers of the Center for Computational Sciences and Simulation (CCSS) at University of Duisburg-Essen.

References

[1] Reisman, G., Wang, Y., Brennen, C. (1998). *Observations of shock waves in cloud cavitation*. Journal of Fluid Mechanics . 355 255-283.

[2] Dular, M., Bachert, B., Stoffel, B., Sirok, B. (2004). *Relationship between cavitation structures and cavitation damage*. Wear. 257 (11) 1176-1184.
[3] Crimi, P. (1970). *Experimental study of the effects of sweep on hydrofoil loading and cavitation (Sweep angle relationship to cavitation inception on hydrofoils and to hydrofoil performance deterioration due to cavitation)*. Journal of Hydronautics. 4 (1) 3-9.
[4] Kuiper, G. (2001). *New Developments around Sheet and Tip Vortex Cavitation on Ships Propellers*. In: CAV 2001, Fourth International Symposium on Cavitation.
[5] Ausoni, P., Farhat, M., Avellan, F. (2007). *Hydrofoil roughness effects on von karman vortex shedding*. Proceedings of the 2nd IAHR International Meeting of the Workgroup on Cavitation and Dynamic Problems in Hydraulic Machinery and Systems, Timisoara, Romania. October 24-26.
[6] Xiang, M., Cheung, S.C.P., Tu, J.Y., Zhang, W.H. (2011). *Numerical research on drag reduction by ventilated partial cavity based on two-fluid model*. Ocean Eng. 38 2023-2032.
[7] Kadivar, E., Kadivar, E., Javadi, Kh., Javadpour, S.M. (2017). *The Investigation of Natural Super-Cavitation Flow behind Three-Dimensional Cavitators: Full Cavitation Model*. Appl. Math. Model. 45 165-178.
[8] Coutier-Delgosha, O., Devillers, J., Leriche, M., Pichon, T. (2005). *Effect of Wall Roughness on the Dynamics of Unsteady Cavitation*. ASME. J. Fluids Eng. 127(4) 726-733.
[9] Girimaji, S.S. (2006). *Partially-Averaged Navier Stokes model for turbulence. A Reynolds-Averaged Navier Stokes to direct numerical simulation bridging method* . Appl. Mech.-T. ASME. 73 (3) 413-421.
[10] Song, C.S., Park, S.O. (2009). *Numerical simulation of flow past a square cylinder using partially-averaged Navier-Stokes model*. J. Wind Eng. Ind. Aerod. 97, 37-47.
[11] Ji, B., Luo, X., Wu, Y., Peng, X., Duan. Y. (2013). *Numerical analysis of unsteady cavitating turbulent flow and shedding horse-shoe vortex structure around a twisted hydrofoil*. International Journal of Multiphase Flow . 51 33-43.
[12] Schnerr, G., Sauer, J. (2001). *Physical and numerical modeling of unsteady cavitation dynamics*. In: 4th Int. conference on multiphase flows, May 27June 1, New Orleans, USA.
[13] Franc, J.P., Schnerr, G.H. (2003). *Workshop on Physical Models and CFD Tools for Computation of Cavitating Flows*. Call for participation in present workshop, CAV2003, Osaka, Japan.
[14] Hunt, J.C.R., Wray, A.A., Moin, P. (1988). *Eddies, streams, and convergence zones in turbulent flows*. Center for Turbulence Research Report CTR-S88, 193-208.
[15] Coutier-Delgosha, O., Deniset, F., Astolfi, J., Leroux, J. (2007). *Numerical Prediction of Cavitating Flow on a Two-Dimensional Symmetrical Hydrofoil and Comparison to Experiments*. ASME. J. Fluids Eng. 129 279-292.

© 2018 ASME

Investigation of Reynolds Number Scale Effects on Propeller Tip Vortex Cavitation and Propeller-Induced Hull Pressure Fluctuations

Roland Gosda*, Stephan Berger and Moustafa Abdel-Maksoud

Institute for Fluid Dynamics and Ship Theory, Hamburg University of Technology, Hamburg, Germany

Abstract

A two-dimensional viscous vortex model with a cylindrical cavity at its center is applied to investigate Reynolds number scale effects on propeller tip vortex cavitation. First, the behaviour of small isolated cavities inside the viscous core are studied. In order to study Reynolds number scale effects, the viscous core radius of the vortex is varied. Thereafter, the investigation is extended to the cavitating tip vortex of a propeller as a whole. A generic scenario is set up which is representative for real propeller flows. It can be shown that apart from the viscous core radius, the blade load and the presence of sheet cavitation influence hull pressure fluctuations due to tip vortex cavitation in a significant manner.

Keywords: Tip Vortex Cavitation, Marine Propellers, Cavitation-Induced Noise and Vibration

1 Introduction

Propeller tip vortex cavitation is known to contribute to underwater noise and extended hull pressure fluctuations. In practice, the assessment of the propeller's cavitation behaviour is usually carried out by means of model tests, i.e. a geometrically similar model of propeller and ship with a scale factor of, say, $1/40 \ldots 1/25$ is investigated. Due to Reynolds number scale effects, extrapolating the results to the full-scale propeller is then afflicted with some uncertainties. In the present work, some light shall be shed on these uncertainties. Two aspects are considered to have a great impact:

(1) There is a correlation between the viscous core radius of the trailing tip vortex and the boundary layer thickness of the lifting surface [1], i.e. the propeller blade. Assuming typical operation conditions and length dimensions for model tests, the viscous core radius of the tip vortex of the model propeller can be up to 60 % relatively larger than the full-scale pendant [3]. This certainly will have an effect on the dynamic behaviour of the cavities inside the vortex core.

(2) The propeller operates in the ship's wake field, which in turn is highly affected by the different Reynolds numbers of model and ship. Since the unsteady propeller load depends on the inhomogeneous inflow, the circulation of the tip vortex and – if present – the extent of sheet cavitation at the propeller blade tip will also be influenced accordingly. These factors are assumed to have a considerable impact on the dynamical behaviour of tip vortex cavitation.

In the present paper, both effects are discussed and separated from each other. The numerical simulations carried out for this purpose are based on a two-dimensional vortex model, where the behaviour of a cylindrical cavity in the center of a viscous line vortex is observed. In order to capture the interaction between cavity and vortical flow, the method solves both the circumferential as well as the radial momentum equation in a coupled manner.

The paper is organised as follows: After introducing the governing equations of the model and making some notes on the numerical implementation in Sections 2 and 3, the focus in Section 4 is put on point (1) mentioned above by carrying out an extensive numerical study for cylindrical bubbles in the viscous core of a line vortex at various Reynolds numbers and initial conditions. Finally, in Section 5, a simplified generic test scenario is set up in order to elucidate those scale effects subsumed under point (2).

2 Governing Equations and Boundary Conditions

Modelling a viscous line vortex with a cylindrical cavity at its center consists of an initial boundary value problem of a set of coupled partial differential equations. As derived in [6] or [8] for example, the equations are the Rayleigh-Plesset equation (integral form of the radial momentum equation)

$$(\ddot{r}_c r_c + \dot{r}_c^2)\ln\left(\frac{r_D}{r_c}\right) + \frac{\dot{r}_c^2 r_c^2}{2}\left(\frac{1}{r_D^2} - \frac{1}{r_c^2}\right) = \frac{1}{\rho}\left(p_v + p_{g,\mathrm{ref}}\left(\frac{r_{c,\mathrm{ref}}}{r_c}\right)^2 - 2\mu\frac{\dot{r}_c}{r_c} - \frac{S}{r_c} - p_D\right) + \int_{r_c}^{r_D}\frac{v_\varphi^2}{r}\partial r \quad (1)$$

*Corresponding author, contact: roland.gosda@tuhh.de

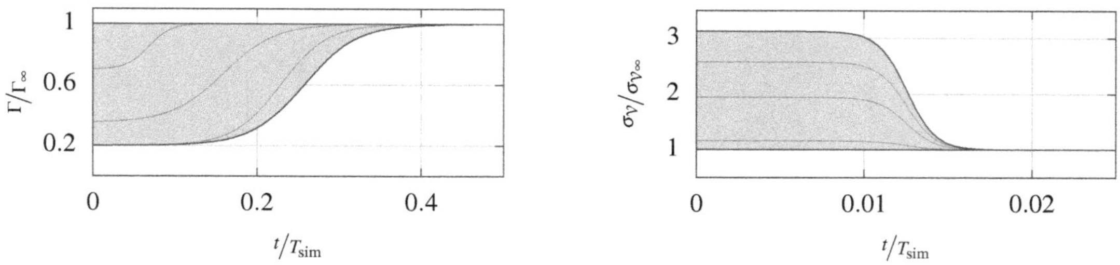

(a) Ciculation Γ.

(b) Cavitation number σ_V.

Figure 1: Ranges and evolution of the circulation Γ and cavitation number σ_V over time t with respect to the duration of the simulation T_{sim}. Γ_∞ and $\sigma_{V\infty}$ denote the final circulation and cavitation number, respectively. While the thick lines show the limits for possible slopes, the thin lines hint randomly varying slopes.

and the circumferential equation of momentum

$$\rho\left(\frac{\partial v_\varphi}{\partial t}+\frac{\dot r_c r_c}{r}\frac{\partial v_\varphi}{\partial r}+\frac{\dot r_c r_c}{r^2}v_\varphi\right)=\mu\left(\frac{\partial^2 v_\varphi}{\partial r^2}+\frac{1}{r}\frac{\partial v_\varphi}{\partial r}+\frac{1}{r^2}v_\varphi\right)+f_\varphi, \tag{2}$$

where the kinematic relation $v_r r = \dot r_c r_c$ has been used. Here, v_φ denotes the circumferential velocity, v_r the radial velocity, and r_c is the cavitation radius with its respective derivatives in time $\dot r_c$ and $\ddot r_c$. A large arbitrary outer domain radius r_D is required for modelling the radial equation of momentum. Further constants and variables are: the density ρ, the partial vapour saturation pressure p_v, the tuple $\{p_{g,\text{ref}}, r_{c,\text{ref}}\}$ regarding the non-condensable gaseous content inside the cavity, the pressure p, the interface tension coefficient S, the viscosity μ, the radial coordinate r and the time t. The use of the momentum source term f_φ will be explained later. The spatial domain is bounded by $r \in [r_c, r_D]$. The boundary conditions for (2) are:

$$\begin{aligned}\frac{\partial v_\varphi}{\partial r}-\frac{v_\varphi}{r}&=0,\ \text{at } r=r_c>0,\\ v_\varphi(r_c)&=0,\ \text{if } r_c=0,\end{aligned} \tag{3}$$

stating that no shear forces act across the liquid-vapour interface. Furthermore, $v_\varphi(r \geq r_D) = \frac{r_D}{r} v_\varphi(r_D)$ is used as an approximation for the velocity field outside the domain.

3 Numerical Implementation

The set of governing equations (1) and (2) is solved using finite differences and a Crank-Nicolson-scheme as suggested by [5], while the spatial domain is transformed to a uniform grid as described in [8]. For previous results and validation, please refer to [2, 3].

4 Reynolds Number Effects on an Isolated Vortex Cavity

In the following, the behaviour of small cavitation bubbles in the viscous core of a propeller tip vortex is studied. There are two significant simplifications: (1) the three-dimensional character of the flow is neglected, i.e. cylindrical 'bubbles' are considered, and (2) these bubbles are located in the center of the vortex, whereas the passage to the vortex center is not taken into account. Despite these simplifications, some important conclusions can be drawn from this scenario.

Predominant parameters governing the tip vortex flow are the circulation Γ of the tip vortex, which increases due to the roll-up process, the ambient pressure p_D, which may also undergo changes, as well as the viscous core radius r_a, in which Reynold number scale effects are assumed to be reflected.

To study the influence of the different parameters, 10000 simulations of single bubbles have been carried out. A characteristic quantity of the problem is the vortex cavitation number σ_V, which is defined here as

$$\sigma_V=\frac{p_D-p_v}{1/2\rho\mathcal{V}_\Gamma^2}\quad\text{with}\quad \mathcal{V}_\Gamma=\frac{\Gamma_\infty}{2\pi r_D} \tag{4}$$

and r_D being the outer domain radius introduced in Section 2. The initial data has been chosen s.t. Reynolds number effects of the core radius are included by selecting discrete values $r_{a0}/r_D \in \{0.02, 0.04, \ldots, 0.2\}$. Furthermore, the increase of circulation and changes in the ambient pressure p_D as the bubble travels downstream along

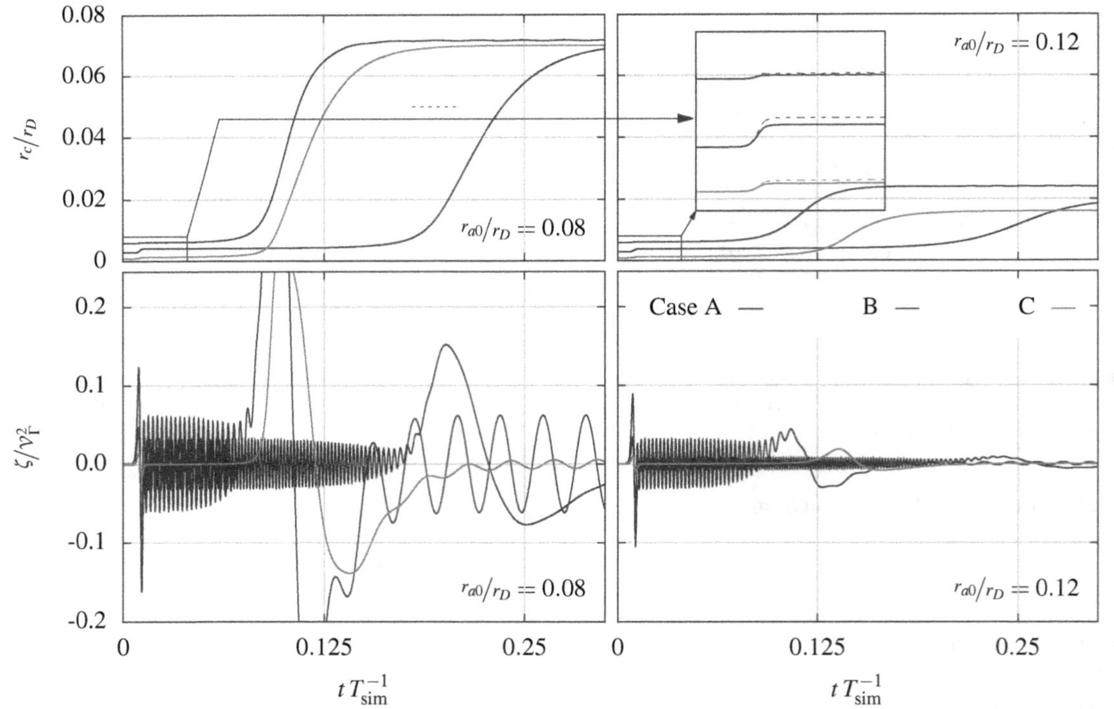

Figure 2: Three exemplary bubble evolutions for the cases listed in the table below for $r_{a0}/r_D = 0.08$ (left) and $r_{a0}/r_D = 0.12$ (right) resembling full-scale and model conditions, respectively. Top: evolution of the cavity radius $r_c(t)$; bottom: pseudo radiated pressure ζ (see text for definition).

Case	Colour	r_{c0}/r_D	r_{a0}/r_D	t/T_{sim} when $\Gamma \geq$		$\sigma_{V0}/\sigma_{V\infty}$
				$105\,\%\,\Gamma_0$	$99\,\%\,\Gamma_\infty$	
A	• blue	$5.870 \cdot 10^{-3}$	0.08, 0.12	0.080	0.226	1.048
B	• black	$2.852 \cdot 10^{-3}$	0.08, 0.12	0.182	0.478	1.641
C	• red	$8.658 \cdot 10^{-4}$	0.08, 0.12	0.058	0.278	1.814

The magnifying glass in the upper right diagram helps to observe the impact of the change of the cavitation number σ_V from σ_{V0} to $\sigma_{V\infty}$. Within the magnifying glass, the solid lines correspond to $r_{a0}/r_D = 0.12$, whereas the dashed lines display the results for $r_{a0}/r_D = 0.08$. The influence of the viscous core radius is relatively small at this stage.

the vortex axis are varied randomly in intervals $\Gamma_0/\Gamma_\infty \in [0.2, 1.0]$ and $\sigma_{V0}/\sigma_{V\infty} \in [1.0, 3.0]$, respectively, using an uniform distribution, while the initial cavitation radius is chosen to be in the range of $r_{c0}/r_D \in [2 \cdot 10^{-5}, r_{a0}/r_D]$. Here, subscripts with $(\cdot)_0$ denote respective parameters at t_0, i.e. at the initialisation of the simulation, whereas $(\cdot)_\infty$ refers to the final state of a simulation.

Both the increase of circulation and the drop of pressure have been prescribed using a hyperbolical tangent (see Figure 1). The initial circumferential velocity $v_\varphi(r)$ is modelled using the analytical model proposed by [4] – basically a modified Burgers vortex-like radial distribution of circumferential velocity. Thus, r_{a0} refers to the core radius of the non-cavitating vortex. In order to mimic the increase of circulation, the momentum source term f_φ of the circumferential momentum equation (2) is manipulated according to the prescribed increase of circulation. All simulations have been initialised with a corresponding partial pressure of the non-condensable gaseous content $p_{g,\text{ref}}$ s.t. the cavity is in equilibrium at t_0.

Exemplary simulation results are shown in Figure 2 for the three cases A, B and C and two viscous core radii. In the figure, the evolution of the cavity radius $r_c(t)$ is shown in the upper diagrams, and the pseudo radiated pressure resulting from the oscillation of the cavity radius in terms of

$$\zeta \equiv \frac{\partial^2 \left(r_c^2\right)}{\partial t^2} \tag{5}$$

is plotted in the lower diagrams. These three cases demonstrate the strong dependence of the evolution of the bubble radius on the initial conditions and the viscous core radius. It can be observed that in the present scenario, a bubble undergoes three phases. The first phase starts directly after the pressure drops from p_{D0} to $p_{D\infty}$, and the third phase is characterised by the steady state oscillations arising when the vortex circulation Γ reaches its final

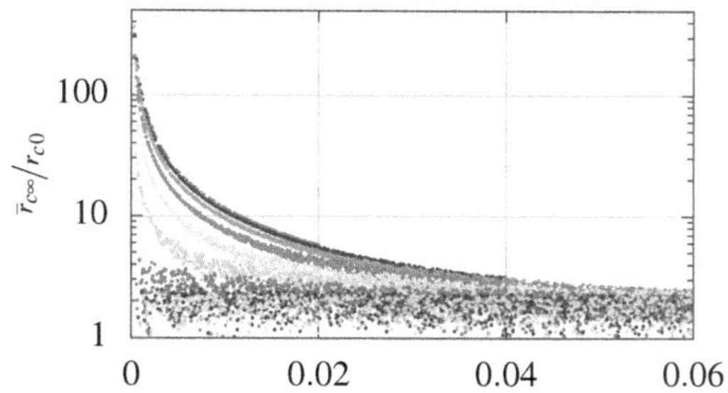

(a) Mean cavitation radius $\bar{r}_{c\infty}$.

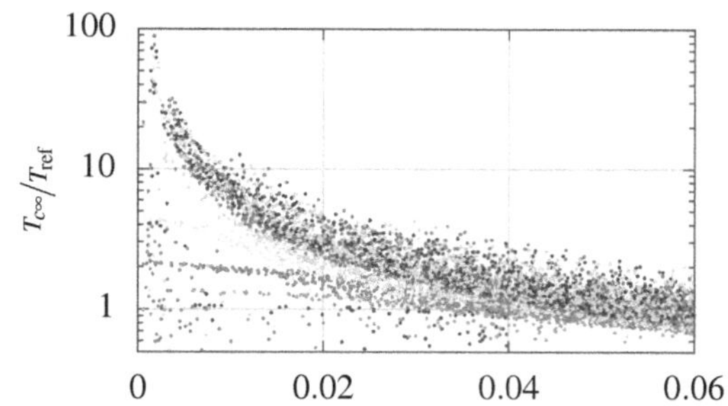

(b) Oscillation period $T_{c\infty}$.

Figure 3: Exemplary simulation results with respect to oscillation period $T_{c\infty}$ and mean cavitation radius $\bar{r}_{c\infty}$ at the end of the simulation. Each dot indicates one simulation result. Colour code: $r_{a0}/r_D = \{\bullet\, 0.02, \bullet\, 0.04, \bullet\, 0.06, \bullet\, 0.08, \bullet\, 0.1, \bullet\, 0.12, \bullet\, 0.14, \bullet\, 0.16, \bullet\, 0.18, \bullet\, 0.2\}$.

value. The second phase refers to the transition from the first to the third phase. In the following, the steady-state results of the third phase will be analysed.

Figure 3(a) shows the average cavitation radius $\bar{r}_{c\infty}$ of the steady-state oscillations. Obviously, a larger viscous core radius r_a dampens the growth of the cavitation radius $\bar{r}_{c\infty}/r_{c0}$. For large cavitation radii, the results for cases with different core radii merge, which is due to the increasing similarity to the inviscid solution that starts to dominate the viscous effects.

The results presented in Figure 3(b) show the effect of r_{c0} on the oscillation period $T_{c\infty}$ in the steady state with respect to the linearised inviscid oscillation period [7]:

$$T_{\mathrm{ref}} = 4\pi^2 \bar{r}_{c\infty}^2 \Gamma_\infty^{-1} \sqrt{\ln\left(r_D \bar{r}_{c\infty}^{-1}\right)}. \tag{6}$$

The deviance of the oscillation period $T_{c\infty}$ to the reference period T_{ref} is more dominant if r_{a0} is larger and if the relative difference of the initial cavitation radius r_{c0} to r_{a0} is large. This emphasises the effect that the viscous core has: A larger viscous core radius effectively slows down bubble oscillations.

In the following, the implications on hull pressure fluctuations due to a cavitating propeller tip vortex are investigated.

5 Reynolds Number Effects on a Generic Propeller Tip Vortex

So far, the behaviour of isolated vortex cavities ('bubbles') has been studied. In this section, Reynolds number scale effects on the behaviour of a cavitating propeller tip vortex as a whole are considered. The applied model originates from [9, 10] and is described in detail in [2, 3]. It considers the propeller tip vortex to be a large sequence of vortex segments. Each of them is treated separately by the method used so far in this paper. The scheme is shown in Figure 4. A segment k originates at the propeller tip at $t = t_{0,k}$. According to the selected time step size $\mathrm{d}t$, each element has the length $\mathrm{d}l_k = \mathrm{d}l = \mathrm{d}t\left((\pi n D)^2 + V_s^2\right)^{0.5}$, where n is the number of revolutions, D is the propeller diameter and V_s is the ship speed. The segment k is swept away by the flow making room for the new segment $k+1$ originating at the propeller tip at $t = t_{0,k+1}$. In order to approximate the tip vortex roll-up, the circulation $\Gamma_k^\star\left(t_{0,k}, t^\star\right)$ assigned to the segment grows from an initial value $\Gamma_k^\star\left(t_{0,k}, 0\right) = \gamma_{\mathrm{ini}}\Gamma_b\left(t_{0,k}\right)$ to $\Gamma_k^\star\left(t_{0,k}, t^\star\right) = \Gamma_b\left(t_{0,k}\right)$ as the segment travels downstream and the segment's age $t^\star$ exceeds a certain value. Apart from the circulation, the initial viscous core radius r_{a0} of the tip vortex and the initial cavitation radius $r_{c0,k}$ of a segment k at $t = t_{0,k}$ are governing influence parameters of the problem.

Once the unsteady cavitation radius r_c is known for every segment, the unsteady pressure p'_{TV} due to the cavitation volume fluctuation of the tip vortex observed in a point O sufficiently far away from the propeller can be approximated by the linearised Bernoulli equation:

$$p'_{\mathrm{TV}} = -\rho \frac{\partial \Phi_{\mathrm{TV}}}{\partial t}, \tag{7}$$

where $\Phi_{\mathrm{TV}} = \Phi_{\mathrm{TV}}(t) = \sum_k \phi_k^\star$ is the accumulated potential of all segments k with

$$\phi_k^\star = -\frac{\sigma_{c,k}^\star}{4\pi d_k} \quad \text{and} \quad \sigma_{c,k}^\star = \pi \mathrm{d}l \frac{\partial\left(r_{c,k}^2\right)}{\partial t}. \tag{8}$$

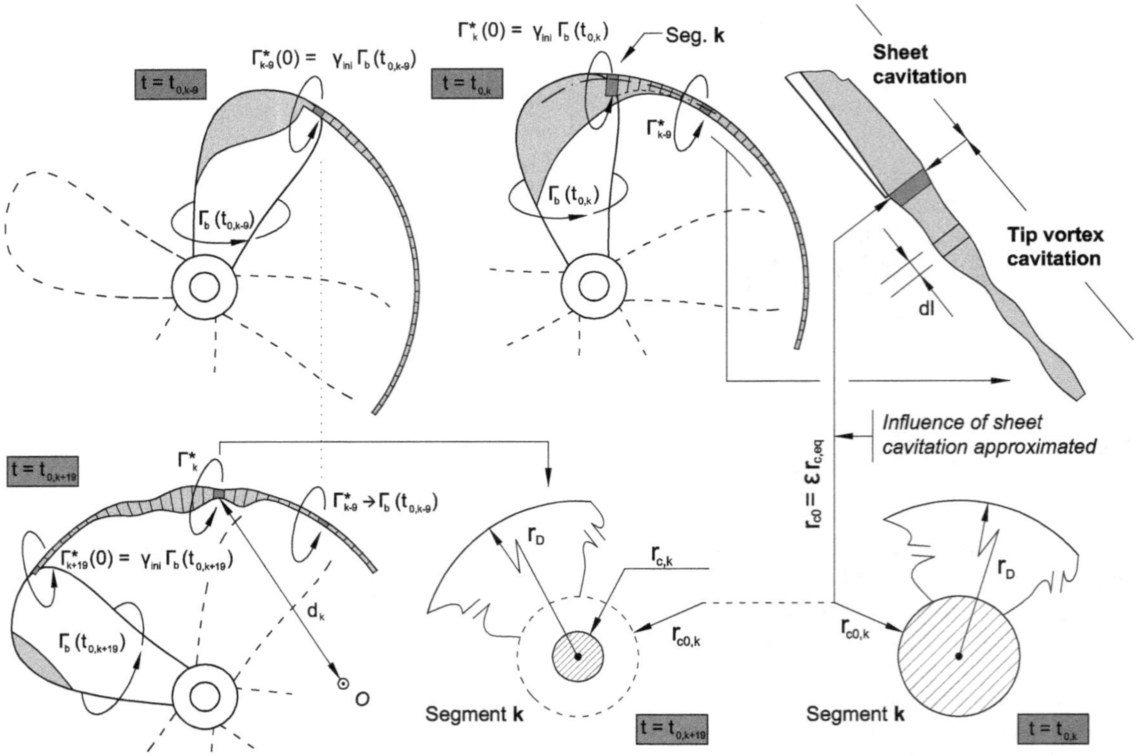

Figure 4: Segmentation scheme of the propeller tip vortex cavitation model. Also showing the influence of sheet cavitation on the dynamic behaviour of tip vortex cavitation.

In order to study Reynolds number effects on pressure fluctuations on the ship hull caused by the cavitating tip vortex, a generic scenario has been developed that reproduces the flow conditions observed at real propeller–hull configurations and covers both effects (point 1 and 2) mentioned in the introduction in Section 1. First, consider the circulation Γ_b bound to the blade as shown for one revolution in the upper diagram of Figure 5a. Here, $\vartheta = 0°$ and $\vartheta = 360°$, respectively, refer to the 12 o'clock-position in the wake field where the velocity deficit due to the ship hull's boundary layer is maximum. The black curve refers to full-scale conditions. In order to arrive at the conditions for the model (represented by the dashed red curve), the plateau of high blade load has been broadened according to the more pronounced wake peak one would expect for model conditions, and the maximum at 0° has been increased slightly for the same reason. In both cases, the average propeller thrust is the same. For the sake of simplicity, it is assumed that the initial circulation of the tip vortex directly behind the trailing edge is half of the circulation bound to the blade, i.e. $\gamma_{\mathrm{ini}} = 0.5$.

Sheet cavitation in the propeller tip region is assumed to excite pressure fluctuations induced by the tip vortex by widening the vortex cavity beyond the equilibrium cavity radius, such that oscillations of the cavitation radius are stimulated. This phenomenon will usually be observed in the wake peak region and is discussed in [2, 3]. To approximate this effect without modelling sheet cavitation, the initial cavitation radius r_{c0} of a new tip vortex segment is chosen to be the equilibrium cavitation radius $r_{c,\mathrm{eq}}$ for given flow conditions which is then exaggerated by a factor $\varepsilon \in [1.0, \varepsilon_{\max}]$ in the wake peak region. In the generic scenario investigated here, the following procedure is used:

$$r_{c0} = \begin{cases} \varepsilon r_{c,\mathrm{eq}} & \text{in the wake peak region} \\ r_{c,\mathrm{eq}} & \text{elsewhere} \end{cases} \tag{9}$$

with $r_{c,\mathrm{eq}}$ at $t = t_0$ for every segment k and given tip vortex flow conditions (see lower diagram of Figure 5a).

In Figure 5b, exemplary pressure signals p'_{TV} are shown in the frequency domain for two extreme cases in a dimensionless manner in terms of $k_{\hat{p}} = \hat{p}_{\mathrm{TV}}/(\rho n^2 D^2)$ and f/f_b with the blade frequency $f_b = n n_b$. Obviously, pressure fluctuations are predicted to be less intense under model conditions than under full-scale conditions. To clarify this further, the influence of the viscous core radius and the exaggeration factor $\varepsilon_{\max}$ on pressure fluctuations $k_{\hat{p}[3]}$ and $k_{\hat{p}[4]}$ of third and fourth order, i.e. occurring with $3f_b$ and $4f_b$, has been studied. The results are shown in Figure 6.

Obviously, the amplitudes of pressure fluctuations due to tip vortex cavitation generally grow with the exag-

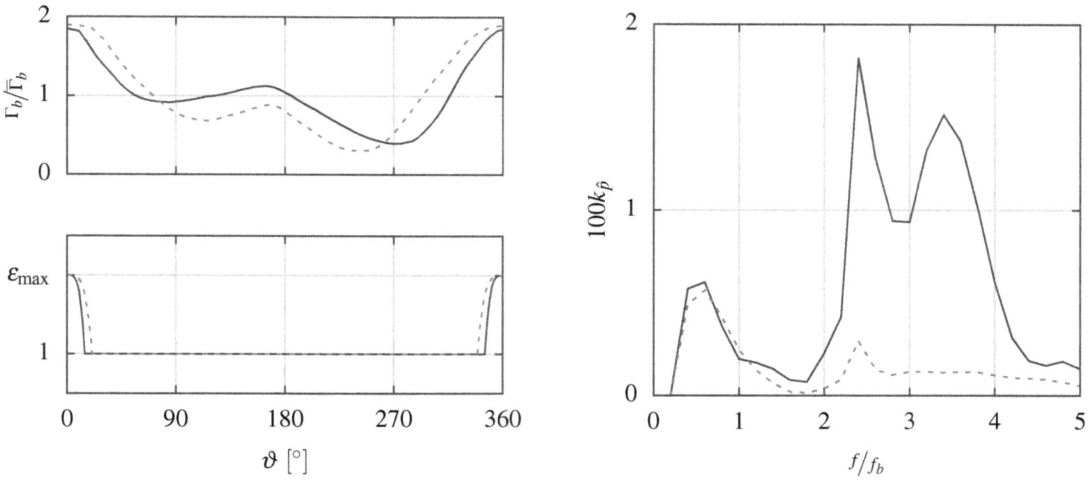

(a) Γ_b and ε over ϑ for full-scale and model configuration.

(b) Exemplary pressure amplitude spectra.

Figure 5: Circulation distributions $\Gamma_b(\vartheta)$ and exaggeration factors $\varepsilon(\vartheta)$ (a) and exemplary pressure amplitude spectra due to tip vortex cavitation (b) for representative full-scale (solid line) and model configurations (dashed line) with $\varepsilon_{max} = 1.225$ and $2 \cdot 100 r_{a0}/D = \{\bullet\ 1.067, \bullet\ 1.867\}$, respectively. These extreme cases are also included in the diagrams displayed in Figure 6 using filled circles with the same colour code.

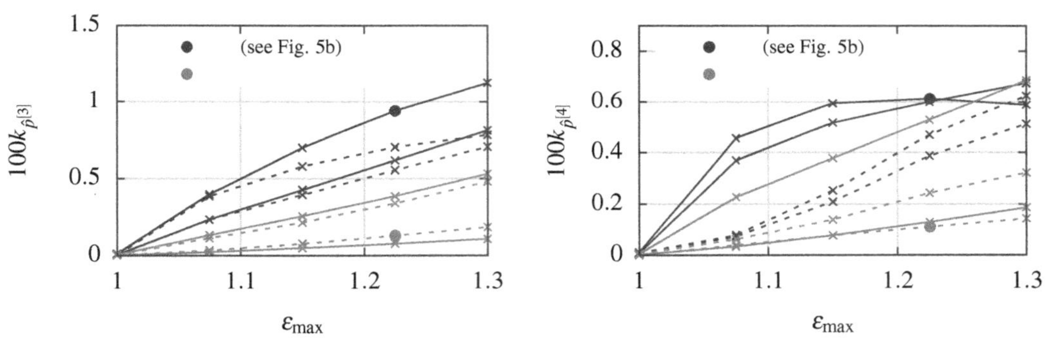

(a) Third order $f = 3f_b$.

(b) Fourth order $f = 4f_b$.

Figure 6: Influence of the viscous core radius r_{a0} and the exaggeration factor ε_{max} for the full-scale configuration (solid lines) and the model configuration (dashed lines). Colour code: $2 \cdot 100 r_{a0}/D = \{\bullet\ 1.067, \bullet\ 1.333, \bullet\ 1.6, \bullet\ 1.867\}$. The filled circles correspond to the exemplary cases in Figure 5b.

geration factor ε_{max} and – similar to the observation made in the previous section – a larger viscous core radius dampens pressure fluctuations. No significant pressure fluctuations are induced if sheet cavitation is not present, i.e. $\varepsilon_{max} = 1.0$. However, it is important to note that, apart from this, the angular distribution of blade load in terms of $\Gamma_b(\vartheta)$ can have a significant impact on the results.

6 Discussion, Conclusions and Future Outlook

In the present paper, Reynolds number scale effects on tip vortex cavitation have been investigated by means of a method which is based on the model of a cavitating axisymmetrical viscous line vortex. The important interaction between the distribution of circumferential velocity $v_\varphi(r)$ and the cavity inside the vortex is captured by the present method. With respect to isolated vortex cavities (cylindrical 'bubbles') inside the viscous core of the vortex, it has been shown in Section 4 that a larger viscous core radius effectively dampens both the growth of the cavity radius and the frequency of its oscillations.

The problem gets more complex when pressure fluctuations radiated by the propeller tip vortex as a whole are addressed. Apart from the viscous core radius, which tends to dampen oscillations as it is increased, the blade load in the wake peak region around the 12 o'clock-position and sheet cavitation at the blade tip also influence pressure fluctuations in a significant manner. Within the simulations carried out, the latter has been taken into account by a simplified model. More knowledge is required about the complex and relevant cavitation pattern in the blade tip region, which is characterised by the transition from sheet cavitation to tip vortex cavitation.

With respect to hull pressure fluctuations due to cavitating propellers, the results obtained can be valuable to better extrapolate model test results to the full-scale ship. Further investigations are required to obtain a reliable general rule for the extrapolation.

Acknowledgements

The authors would like to thank the German Federal Ministry for Economic Affairs and Energy for the financial support within the research projects NoiseLES-eNoiSim and HiOcav-HiOsim.

References

[1] R. E. A. Arndt: *Vortex Cavitation*. Fluid Vortices. Springer, Dordrecht, 1995. 731–782.

[2] S. Berger, R. Gosda, M. Scharf, R. Klose, L. Greitsch, M. Abdel-Maksoud: *Efficient Numerical Investigation of Propeller Cavitation Phenomena Causing Higher-Order Hull Pressure Fluctuations*. 31st Symposium on Naval Hydrodynamics, Monterey, California, 2016.

[3] S. Berger: *Numerical Analysis of Propeller-Induced Higher-Order Pressure Fluctuations on the Ship Hull*. PhD Thesis, Hamburg, 2017 (`https://doi.org/10.15480/882.1605`).

[4] J. Bosschers: *On the influence of viscous effects on 2-D cavitating vortices*. Journal of Hydrodynamics, Ser. B 22.5, 2010.

[5] G.L. Chahine: *Bubble interactions with vortices*. Fluid Vortices. Springer, Dordrecht, 1995. 783-828.

[6] J. Choi, C. Hsiao, G. Chahine, S. Ceccio: *Growth, oscillation and collapse of vortex cavitation bubbles*. Journal of Fluid Mechanics, Cambridge University Press, 2009.

[7] J.-P. Franc, J.-M. Michel: *Vortex Cavitation*. Fundamentals of Cavitation. Kluwer Academic Publishers, 2004. 223–246.

[8] R. Gosda: *Numerical Investigation of the Interaction between a Viscous Line Vortex and a Cylindrical Cavitation Bubble in the Viscous Core Region*. Bachelor thesis, Hamburg University of Technology, 2015; (`www.panmare.de` → Publications → 2016).

[9] T. Kanemaru, J. Ando: *Numerical Analysis of Tip Vortex Cavitation on Marine Propeller Using a Simple Surface Panel Method SQCM*. Fourth International Symposium on Marine Propulsors, Austin, Texas, 2015.

[10] J. Szantyr: *A Computational Model of Propeller Cavitating Tip Vortex Interacting with the Rudder*. CAV2006, Wageningen, Netherlands, 2006.

© 2018 ASME

10th International Symposium on Cavitation - CAV2018
Baltimore, Maryland, USA, May 14 – 16, 2018

Investigation of sheet to cloud transition due to the propagation of condensation fronts over a sharp wedge using large eddy simulations

Mrugank Bhatt; Krishnan Mahesh*
University of Minnesota, Minneapolis, MN, USA

Abstract

Sheet to cloud cavitation over a sharp wedge is studied at Re = 203, 000 using large eddy simulations. Transition of sheet cavity into the cloud due to the propagation of condensation fronts as observed in the experiments by Ganesh et. al. [1], is investigated using a finite rate homogeneous mixture model. Presence of condensation front in the simulations is illustrated using the instantaneous flow field. Mean void fractions obtained using LES as well as from simulations performed without considering the spanwise variation (i.e. in 2D) are compared to the X-ray densitometry showing good agreement. Strouhal frequency of periodic shedding of cloud and propagation speed of the front is computed. The transport of vorticity is investigated using the vorticity transport equation.

Keywords: cavitation; sheet/cloud transition; bubbly shock waves

Introduction

Understanding cavitation inception on marine lifting surfaces, the unsteady behavior of sheet cavities and their transition to cloud (referred to as "sheet to cloud") is important, as the collapse of these large regions of vapor cloud often cause erosion and noise. The sheet to cloud transition is often related to the formation of re-entrant jet in the aft portion of the sheet cavity as the liquid flow reattaches. Several studies have investigated the role of re-entrant jet in the shedding process, both experimentally and computationally. [2, 3, 4]. Recently, experiments performed over a sharp wedge by Ganesh et. al. [1] pointed out bubbly shock propagation as a mechanism for sheet to cloud transition. In the experiments, they observed sheet to cloud transition due to both re-entrant jet and by the propagating bubbly shock. Subsequent computational studies by Budich et. al. [5] and Schenke and Terwisga [6] of the experimental configuration also showed the presence of condensation front in their simulations. Gnanaskandan and Mahesh studied the experimental configuration in the transitory regime where they investigated the sheet to cloud transition due to the formation of re-entrant jet [3]. In the present work, we investigate the performance of the same finite rate homogeneous mixture model used by Gnanaskandan and Mahesh [3] in predicting periodic shedding of the cloud due to the propagation of bubbly shock wave.

The paper is organized as follows. Section 2 describes the cavitation model and numerical method used for the simulations. The computational setup and procedure to match the experimental conditions, instantaneous solution obtained from LES, comparison with experimental data and finally, the transport of vorticity in the flow is discussed in section 3. Conclusions are summarized in section 4.

Physical Model and Numerical Method

The simulations use the homogeneous mixture model, where the mixture of water and vapor is treated as a single compressible fluid. Mechanical and thermodynamic equilibrium is assumed between the constituent phases. Compressible Navier Stokes equations are solved for mixture quantities along with the transport equation for vapor mass fraction employing finite rate mass transfer between the two phases. The cavitation model of Saito et al. [7] is used to model evaporation and condensation processes. The system is closed by a non-barotropic mixture equation of state obtained from stiffened equation of state in water and ideal gas equation state in vapor. The sound speed of the mixture is derived from the mixture equation of state and the Gibbs equation.

The numerical method is based on a predictor-corrector approach. Predictor step uses non-dissipative finite volume scheme and the corrector step applies a characteristic based filter locally in the vicinity of discontinuities. Time integration is performed using the explicit Adams-Bashforth scheme. The governing equations are spatially Favre filtered for LES and subgrid terms arising due to the filtering are modeled using the Dynamic Smagorinsky Model. The details of the method are given by Gnanaskandan and Mahesh [8]

*Corresponding Author, Krishnan Mahesh: kmahesh@umn.edu

Results

1. Problem setup

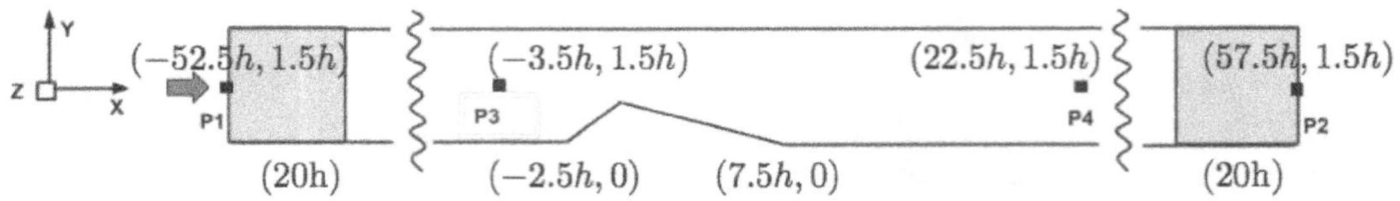

Figure 1. Schematic of the x − y plane of the computational setup (figure not to scale)

Figure 1 shows the schematic of the computational setup. The apex of the wedge is located at (0, h) in the x–y plane (symmetry plane, z = 0), where h (1 inch) is the height of the wedge. The mean flow direction is along the x-axis indicated by the red arrow. The flow enters the domain through 3h×3h cross section (y–z plane) which corresponds to the reduced test section in the experiments [1]. The domain is extended 52.5h upstream and 57.5h downstream from wedge apex to minimize the reflection of acoustic waves through the boundaries. In addition, acoustically absorbing sponge layers [9] as indicated by blue regions in the figure 1, are applied at both inlet and outlet boundaries. Homogeneous inflow velocity is prescribed at location P1 and pressure is prescribed at the outflow location P2 of the computational domain. Inflow-outflow conditions are changed iteratively in order to match the experimental values in the reduced test section at P3 and P4. No-slip boundary conditions are imposed on the top and bottom walls. Periodic boundary conditions are enforced in the spanwise direction.

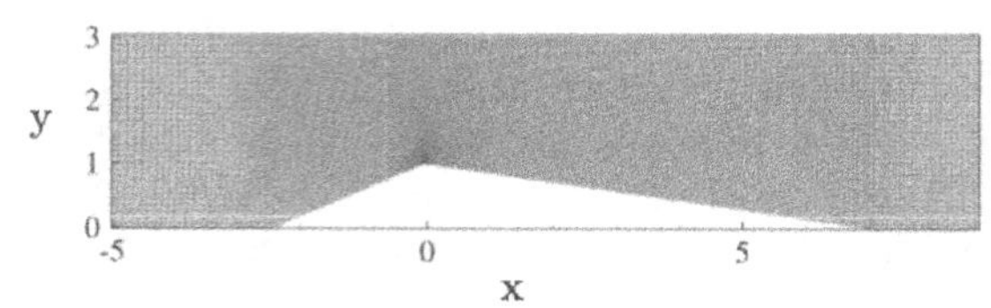

Figure 2. 2D-slice of the computational mesh close to the wedge (dimensions are in inches)

The computational mesh consists of approximately 3 million hexahedral cells. The mesh is made very fine in the convergent-divergent section of the wedge as shown in the figure 2 with minimum spacing of 0.005h in both streamwise and spanwise direction near the wedge apex. In order to accelerate the process of iteratively prescribing inflow-outflow, solution is first obtained on a two dimensional grid and interpolated along the span, to provide an initial condition to the LES simulations. The mesh used for LES has 80 cells along the spanwise direction. Inflow velocity obtained at station P3 is 7.98 m/s and back pressure obtained at station P4 is 47.05 kpa in simulations, as compared to the velocity of 8 m/s at inflow and back pressure of 52 kpa in the experiments. Non-dimensional time step used in the simulation is 5×10^{-5}. Reynold's number based on wedge height (h = 1 inch) and inflow velocity (7.98 m/s) is 2.03×10^{5}.

2. Instantaneous flow field

The instantaneous flow field is illustrated by iso-contours of vapor volume fraction (α) and pressure contours in figure 3. The incoming flow accelerates in the convergent section of the wedge resulting in pressure drop as evident from the pressure contours. As the pressure drops below vapor pressure at the apex of wedge, flow cavitates leading to the formation of vapor sheet cavity on the wedge. Figure 3 shows an instant when this sheet cavity transitions into the cloud by propagation of condensation front. This front is evident from the figures 4.a) and b), which represent the volume fraction contours in the symmetry plane and averaged along the spanwise direction ($< \alpha >_z$) respectively, at the same instance of time as figure 3. It is observed that this front spans across the entire thickness of the sheet cavity, unlike the layer of re-entrant jet which is often restricted to the regions close to the wall. Also, the vapor volume fractions obtained from the spanwise averaged data inside the sheet cavity are much higher (0.85) in the simulations compared to the case when re-entrant jet is formed [3]. This was also observed in the experiments. [1].

3. Comparison to experiment

Figure 5(a) shows preliminary comparison of mean void fraction data obtained for 3 shedding cycles of simulations with that of experiment. Void fraction profiles are taken along the direction normal to the surface at different streamwise locations along the wedge. The distance along the wedge is denoted by "s" and the distance normal to the wedge is denoted by "n" , which is in turn normalized with respect to the length of the wedge (L_w).

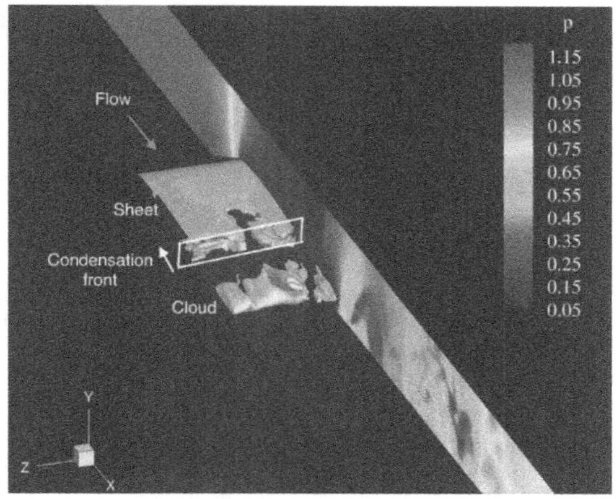

Figure 3. Iso-contours of volume fraction with pressure contours on vertical wall.

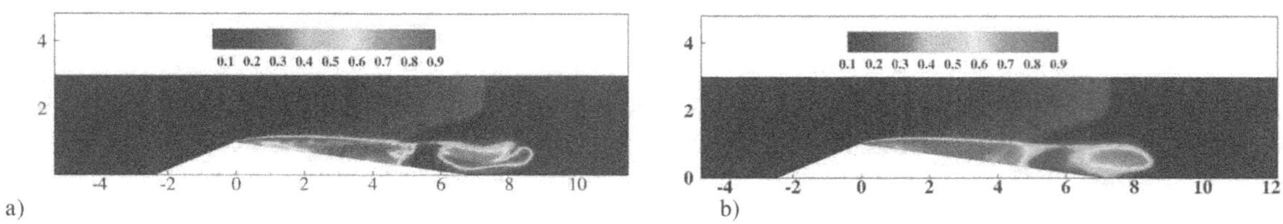

Figure 4. Instantaneous volume fraction contours in the a) symmetry plan (z=0) and b) span-wise averaged.

Results show good agreement for void fraction profiles at streamwise stations close to the wedge apex (s/L_w = 0.02 and 0.014). As one moves further downstream, the overall cavity size obtained from the computation is larger as compared to the experiments. This is expected due to lower back pressure in the simulations (47kpa) as compared to experiments (52kpa). Also, since each shedding cycle is not identical (leading to cycle to cycle variation), a better comparison would require averaging over more number of cycles. Figure 5(b) shows void fraction comparison obtained from 2D simulations averaged over 18 shedding cycles with the back pressure of 52kpa showing much better comparison as expected. PSD of pressure and void fraction spectra taken inside the cavity for the same simulations is shown in the Figure 6(a). Spectra due to both pressure and void fraction show identical values of peak frequency (18.2 Hz) indicating the fact that pressure wave propagation is responsible for cloud shedding mechanism. The Strouhal number of the shedding process obtained based on cavity length (0.106m) and free stream velocity (8m/s) is 24.12, in close agreements with experiments (26 ± 2). Figure 6(b) represents s-t diagram of vapor volume fraction taken parallel to the wedge surface (at 0.1h distance normal to the surface). The speed of condensation front is obtained from the slope of the discontinuity in the volume fraction, similarly as done in the experiments. Speed of the front thus obtained is −5.6m/s of the order of flow speed in the reduced test section, as seen in the experiments.

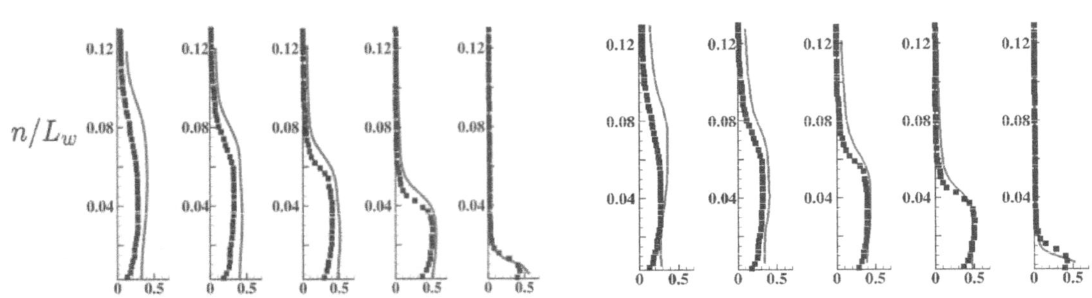

Figure 5. Surface normal profiles of the temporal average of spanwise averaged (for 3D) volume fraction data ($< \alpha >_z$) at different stations along the wedge surface, s/L_w = 0.49, 0.37, 0.26, 0.14, 0.02 respectively from left to right Ganesh et al symbols [1], (-) LES.

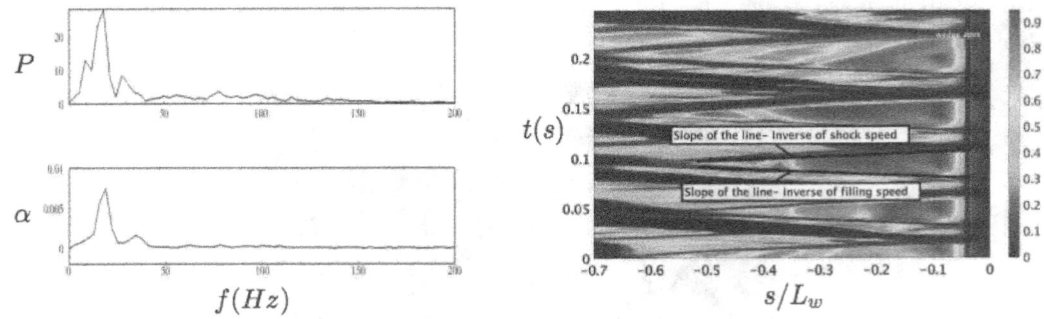

a) b)

Figure 6. a) PSD of pressure and volume fraction taken inside cavity at 1.5h downstrem of wedge apex. b) Contours of vapor volume fraction taken parallel to the surface of wedge for 0.25s (roughly 4-5 shedding cycles)

4. *Vorticity transport*

In order to investigate the interaction of cavitation mechanism with the vorticity in the flow, we consider the vorticity transport equation,

$$\frac{D\omega}{Dt} = \frac{1}{\rho^2}\left(\nabla\rho \times \nabla p\right) - \omega(\nabla \cdot u) + (\omega \cdot \nabla)u + \nabla \times \frac{\nabla \cdot \tau}{\rho}$$

Lagrangian time rate of change of vorticity in a given material volume is due to the production of vorticity due to the misaligned pressure and density gradients (baroclinic torque), modulation of vorticity because of the compressibility of the medium, stretching/tilting of the vortex filament in the presence of velocity gradients and viscous stress. Individual contributions are given by the terms on the RHS of the equation 1 and plotted in the figure 7. Contours are plotted at an instant when the condensation front travels upstream detaching the sheet cavity into the cloud. The cavity interface is indicated by the dashed lines. As seen in figure 7(a) vorticity is produced at the material interface as the condensation front travel upstream, where pressure gradients in the front are misaligned (nearly perpendicular) to the density gradients at the material interface. Although, its contribution is restricted to the regions close to material interface in the cavity closure region and is small in the rest of the cavity and interface. Figure 7(b) shows vorticity dilatation spanned across the entire material interface extending to the cloud downstream with a significant magnitude. This suggests that compressibility of the medium plays major role in the transport of vorticity, particularly at the material interface which undergoes volumetric changes in density due to phase change. Vortex stretching/tilting is observed in the region downstream of the cavity and its magnitude is small in the sheet cavity as seen from the figure 7(c). This can be explained by the breaking up of the spanwise vortices downstream of the condensation front which is evident from the iso-contours of Q-criterion plotted in the figure 8. Also, the contribution due to vortex stretching/tilting is an order of magnitude smaller than the vorticity dilatation.

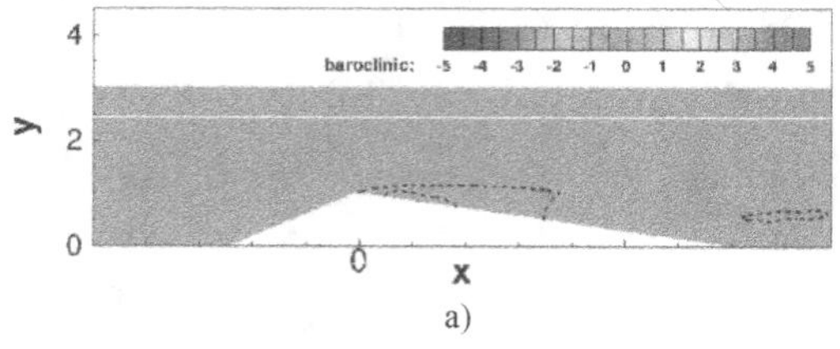

a)

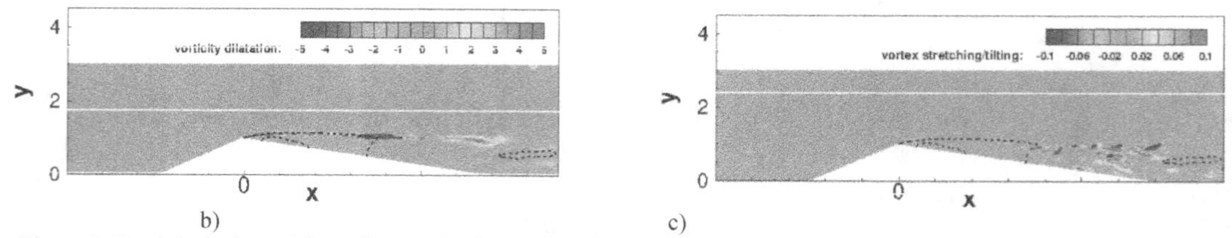

b) c)

Figure 7. Vorticity budget: a) baroclinc production, b) vorticity dilatation and c) vortex stretching/tilting averaged along the span

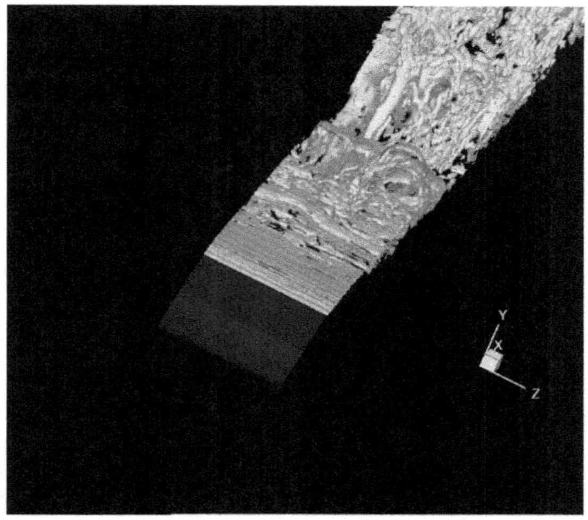

Figure 8. Iso-contours of Q Criterion colored by stream-wise velocity representing vortical structures.

Conclusion

Large Eddy Simulation of sheet to cloud cavitation is performed at Re = 203, 000 over a sharp wedge. Sheet to cloud transition due to the propagation of condensation front is observed in LES simulations, as seen in the experiments. The overall flow-field shows reasonable agreement with the experiments suggesting the suitability of the current finite rate homogeneous mixture model in predicting the periodic shedding of the cloud. Interaction of cavitation with vorticity in the flow is discussed. Vorticity production due to the baroclinic torque is generally smaller as compared to the vorticity dilatation and was restricted to the closure regions of cavity at material interface. Vortex stretching/tilting is observed in the regions downstream of cavity due to breaking up of the span-wise vortices due to the propagation of condensation front.

Acknowledgments

This work is supported by the United States Office of Naval Research under Grant ONR N00014-14-1-0290 and ONR N00014-17-1-2676 with Dr. Ki-Han Kim as the program manager. Computing resources were provided by the Minnesota Supercomputing Institute (MSI) and US Army Engineer Research and Development Center (ERDC).

References

[1] Ganesh, H., Makiharju, S.A., Ceccio, S.L. (2016). *Bubbly shock propagation as a mechanism for sheet-to-cloud transition of partial cavities.* J. Fluid Mech. 802, 37-78.
[2] Arndt, R.E.A., Song, C.C.S., Kjeldsen, M., He, J., Keller, A. (2000). *Instability of partial cavitation: a numerical/experimental approach.* In: Proceedings of the 23rd Symposium on Naval Hydrodynamics. National Academies Press, pp. 519-615.
[3] Gnanaskandan, A., Mahesh, K.(2016). *Large Eddy Simulation of the transition from sheet to cloud cavitation over a wedge.* International Journal of Multiphase Flow. 83, 86102
[4] Laberteaux, K.R., Ceccio, S.L. (2001). *Partial cavity flows. Part 1. Cavities forming on models without spanwise variation.* J. Fluid Mech. 431, 1-41.
[5] Budich, B., Neuner, S., Schmidt, S.J., Adams, N.A. (2015). *Numerical investigation of shedding partial cavities over a sharp wedge.* Journal of Physics: Conference Series 656 012122
[6] Schenke, S., Terwisga, T.JC. (2015). *Simulating Compressibility in Cavitating Flows with an Incompressible Mass Transfer Flow Solver.* Fifth International Symposium on Marine Propulsors.
[7] Saito, Y., Takami, R., Nakamori, I. Ikohagi, T. (2007). *Numerical analysis of unsteady behavior of cloud cavitation around a NACA0015 foil.* Compt. Mech. 40, 85-96.
[8] Gnanaskandan, A., Mahesh, K. (2015). *A numerical method to simulate turbulent cavitating flows.* Int. J. Multiphase Flow 70, 22-34.
[9] Colonius, T. (2004). *Modeling artificial boundary conditions for compressible flow.* Ann. Rev. Fluid Mech. 36, 315345

© 2018 ASME

10th International Symposium on Cavitation - CAV2018
Baltimore, Maryland, USA, May 14 – 16, 2018

Investigation of the Energy Shielding of Kidney Stones by Cavitation Bubble Clouds during Burst Wave Lithotripsy

[1]Kazuki Maeda*; [2,3]Adam D. Maxwell; [2]Wayne Kreider; [1]Tim Colonius; [2,3]Michael R. Bailey

[1]*Division of Engineering and Applied Science, California Institute of Technology*

1200 E California Blvd, Pasadena, CA 91125, USA

[2]*Center for Industrial and Medical Ultrasound, Applied Physics Laboratory, University of Washington*

1013 NE 40th St, Seattle, WA 98105, USA

[3]*Department of Urology, University of Washington School of Medicine, 1959 NE Pacific St, Seattle, WA 98195, USA*

Abstract

We conduct experiments and numerical simulations of the dynamics of bubble clouds nucleated on the surface of an epoxy cylindrical stone model during burst wave lithotripsy (BWL). In the experiment, the bubble clouds are visualized and bubble-scattered acoustics are measured. In the numerical simulation, we combine methods for modeling compressible multicomponent flows to capture complex interactions among cavitation bubbles, the stone, and the burst wave. Quantitative agreement is confirmed between results of the experiment and the simulation. We observe and quantify a significant shielding of incident wave energy by the bubble clouds. The magnitude of shielding reaches up to 80% of the total acoustic energy of the incoming burst wave, suggesting a potential loss of efficacy of stone comminution. We further discovered a strong linear correlation between the magnitude of the energy shielding and the amplitude of the bubble-scattered acoustics, independent of the initial size and the void fraction of bubble cloud within a range addressed in the simulation. This correlation could provide for real-time monitoring of cavitation activity in BWL.

Keywords: cloud cavitation; lithotripsy; numerical simulation; high-speed imaging; acoustic imaging

Introduction

Cavitation bubble clouds nucleated in the human body during the passage of tensile components of ultrasound pulses are of critical importance for the safety and efficacy of the treatment of lithotripsy [1,2]. Shock wave lithotripsy (SWL), uses a shock wave form of ultrasound pulses with a peak amplitude of O(10-100) MPa. The large amplitude of the pulses followed by a long tensile tail tends to results in formation of a large bubble cluster that can violently collapse to cause tissue injury, which has been seen as a major disadvantage of SWL [3,4]. Burst wave lithotripsy (BWL) is a newly proposed alternative to SWL that uses pulses of a continuous form of pressure wave (burst wave) with an amplitude of O(1-10) MPa and a frequency of O(100) kHz for stone comminution [5]. Due to the low peak amplitude of the wave, resulting cloud cavitation in BWL is expected to involve smaller bubbles and less violent bubble dynamics. Recent *in vitro* experiments show that cavitation bubble clouds with a size of O(1) mm are formed during a passage of the burst wave [6]. Such bubble clouds can scatter and absorb the incoming burst wave to cause energy shielding of a nearby kidney stone. The magnitude of this shielding has not been quantitatively analyzed, despite its importance for the efficacy of stone comminution.

In the present study, we quantify the energy shielding by bubble clouds nucleated on the surface of an epoxy stone model during the passage of a burst wave with an amplitude of 7 MPa and a frequency of 340 kHz generated by a focused ultrasound transducer in water through experimental measurements and numerical simulation. In the experiment, we visualize the evolution of bubble clouds using a high-speed camera and measure the back-scattered acoustics from the bubbles with transducer array elements. In the simulation, we combine novel numerical methods for modeling compressible multiphase flows to capture complex interactions among bubbles, the stone, and the burst wave. Simulated evolution of the bubble cloud and bubble scattered acoustics quantitatively agree with the results of the experiment. We vary the initial void fraction and size of bubble cloud to assess the magnitude of the shielding as well as its correlation with the bubble-scattered acoustics. Results of the study indicate that the magnitude of the shielding reaches up to 80% of the total acoustic energy of the incident burst wave. We further discovered a strong

*Corresponding Author, Kazuki Maeda: maeda@caltech.edu

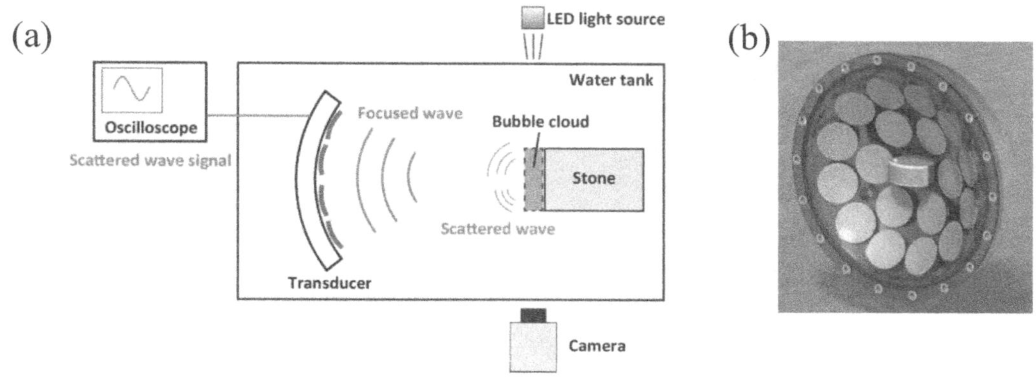

Figure 1. (a) Schematic of the experimental setup; (b) Multi-element array medical transducer.

linear correlation between the magnitude of the shielding and the amplitude of the back-scattered acoustics, independent of the initial condition of the cloud. The results of the study can be used to further identify appropriate parameters and improve the efficacy of stone comminution in BWL.

Experimental and numerical setups

Fig. 1a shows a schematic of the experimental setup. We generate pulses of 10 cycles of a sinusoidal pressure wave with a frequency of 340 kHz and a peak focal pressure of 7 MPa from a multi-element array focused transducer with an aperture of 180 mm and focal length of 150 mm (fig. 1b) toward a cylindrical shape of epoxy stone model with its axis aligned on the acoustic axis of the transducer. The pulse repetition frequency is 100 Hz. The focal point of the transducer is located at the center of the stone. The height and radius of the stone are 10 and 6.25 mm, respectively. A high-speed camera captures a thin layer of bubbles forming a cloud nucleated on the proximal base of the stone. We concurrently sample the back-scattered acoustics from the bubble cloud and the stone by using the transducer array elements.

In simulations, we formulate the dynamics of the multi-component mixture using the compressible, multi-component, Navier-Stoke equation. We model the stone as an elastic solid with zero shear modulus with a density of 1200 kgm^{-3} and a longitudinal sound speed of 2500 ms^{-1}. The coupled dynamics of the stone and the surrounding water are modeled using an interface capturing method [7]. We model the transducer as a portion of a spherical surface with an aperture of 60 mm and a radius of 50 mm with its center located at the origin. We generate burst waves at the modeled surface by using a source term approach [8]. The governing equations are discretized on an axisymmetric grid and integrated using a fifth-order, finite volume WENO scheme. The grid size is uniform with a radial and axial width of 100 um in the stone-wave interaction region. For modeling bubbles, we utilize an Eulerian-Lagrangian method [9]. We initially distribute spherical bubble nuclei with a radius of 10 um in the cylindrical region that faces the proximal base of the stone, then track the radial evolution of the bubbles at the sub-grid scale and resolve the bubble-scattered acoustics on the grid. For the parametric study, we run simulations varying the thickness of the bubbly layer, h, and the initial void fraction, β_0, of the cylindrical cloud, within ranges of $h \in [0.25, 1.5]$ mm and $\beta_0 \in [1.0, 8.0]\times10^{-5}$, respectively. We capture the evolution of the bubbles and sample the back-scattered acoustics at points on the acoustic source surface corresponding to the locations of the center of the transducer array elements.

Results and discussion

In fig. 2a we compare the evolution of the focal pressure during the passage of the burst wave obtained from the experiment and a simulation without the stone. The head of the wave arrives at the focal point at $t = 0$. The simulated waveform and the peak pressure agree well with those of the measurement, except for the small oscillations due to ring down after $t = 30$ µs. In fig. 2b we compare the evolution of the area occupied by the bubbles projected on a 2D plane, A mm^2, obtained in the experimental visualization and the simulation with $h = 1$ mm and $\beta_0 = 1.0\times10^{-5}$. Each data point of the measurement is obtained by averaging 440 independent realizations of the bubble clouds. The shaded area corresponds to the standard deviation resulting from the randomness in the clouds. The result of the simulation shows that the cloud experiences a transient growth with rapid oscillations during the passage of the wave, up to about t =30 µs, followed by a smooth decay after the passage of the wave. The result of the measurement cannot capture the

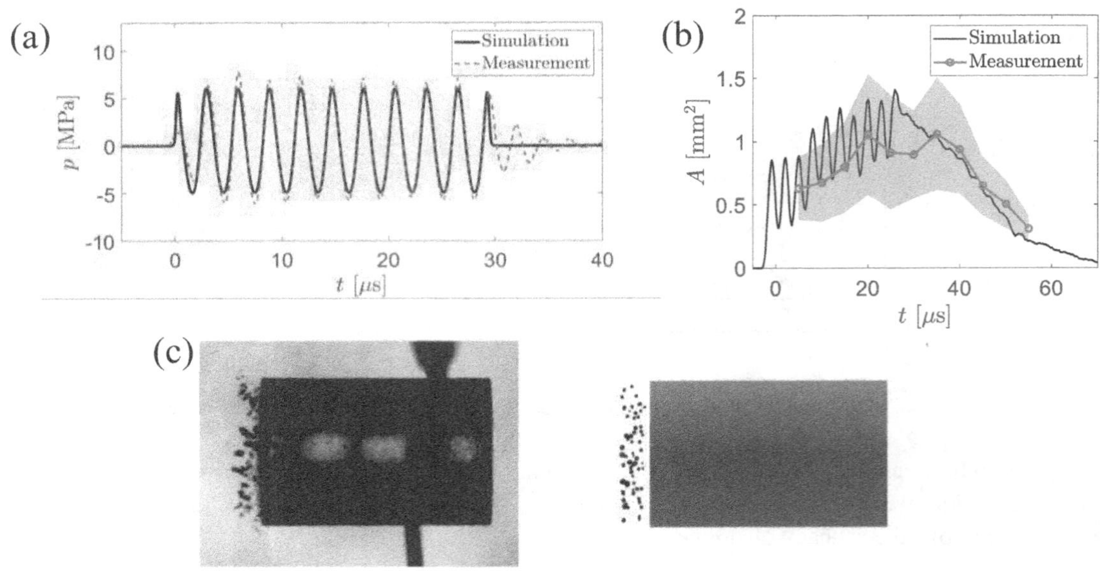

Figure2. (a) Evolution of the focal pressure. (b) Evolution of the projected area of bubble cloud. Result of the simulation with h = 1 mm and $\beta_0 = 1.0\times10^{-5}$ is shown. (c) Images of representative bubble clouds obtained at t = 20 us from the experiment and the simulation.

oscillations, even though they may be present in the experiments, as the present sampling frequency was lower than that of the oscillation, while its growing trend until around at t = 30 μs and decay after the passage of the wave agree well with the result of the simulation. Fig. 2c shows representative images of bubble cloud obtained from the experimental visualization and the simulation at t = 20 μs. A layer of dispersed bubbles is present on the proximal base of the stone in both images.

Fig. 3 shows pressure contours at different instants in time during the simulation. A burst wave is generated and focused toward the stone (fig. 3a). A bubble cloud is excited during the passage of the wave (fig. 3b). A portion of the wave is scattered back from the bubbles and the stone, and the rest of the wave transmits through the stone and is scattered forward (fig. 3b). After the passage of the incident wave, weak reverberation of the pressure wave is observed in the stone (fig. 3d). Strong collapse of bubble cloud is not observed. This may be due to the small void fraction and size of the cloud.

In order to quantify the amplitude of the back-scattered acoustics, we compute an imaging functional using the acoustic signals sampled in both the experiment and the simulation. The imaging functional at coordinate z is defined as:

$$I^B(\boldsymbol{z}) = \sum_{j,l=1}^{N_{sensor}} C\big(\Delta t(\boldsymbol{z}, \boldsymbol{x}_j, \boldsymbol{x}_l), \boldsymbol{x}_j, \boldsymbol{x}_l\big),$$

where C is the cross correlation of the signals obtained at coordinates $\boldsymbol{x}_j$ and $\boldsymbol{x}_l$. Δt is the difference between the acoustic travel time from $\boldsymbol{z}$ to $\boldsymbol{x}_l$ and that from $\boldsymbol{z}$ to $\boldsymbol{x}_j$. Details of the algorithms to obtain the functional can be found in [10]. Fig. 5a and 5b show normalized contours of the imaging functional obtained in the experiment and the simulation. Coordinates with a large value of the imaging functional correspond to estimated locations of the acoustic source. In both plots, a region of a large value of the imaging functional is localized within the area of the stone. This result suggests that the acoustics scattered from the stone (and the bubbles) are captured in both the experiment and the simulation, and their magnitude can be quantified by the imaging functional. Fig. 5c shows the evolution of the acoustic energy in the stone during three distinct simulation cases: without bubbles; with a thin and dilute cloud (h = 0.25 mm and $\beta_0 = 1.0\times10^{-5}$); with a thick and dense cloud (h = 1.5 mm and $\beta_0 = 8.0\times10^{-5}$). In all the cases, the energy steadily grows after arrival of the wave until around at t = 10 μs, then oscillates around a constant value during the passage of the wave until around at t = 30 μs. After the passage of the wave, the energy steadily decays and reaches zero around at t = 40 μs. The energy is highest for all t when there are no bubbles. The energy level is slightly decreased with the thin and dilute cloud, while it is drastically reduced with the thick and dense cloud. This result indicates that the energy shielding by bubbles is enhanced by increasing the thickness and/or the void fraction of the cloud. To

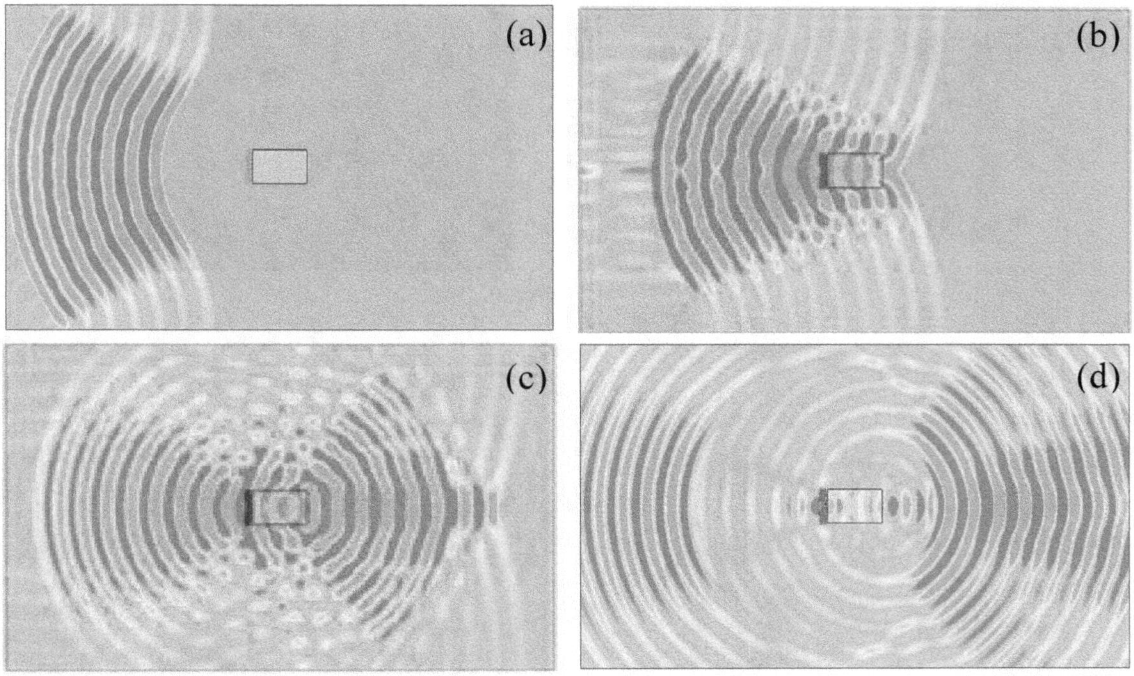

Figure3. Snapshots of the pressure contour on the axial-plane during the simulation with h = 1 mm and $\beta_0 = 1.0\times10^{-5}$ at (a) t = 10, (b) 20, (c) 30, and (d) 40 us. Stone and the void fraction of bubbles are indicated with black line and shading, respectively

quantify the correlation between the shielding and the scattered acoustics, in fig 5d we plot the shielding factor S as a function of the maximum value of the imaging functional, I^B_{Max}. The shielding factor is defined as

$$S = 1 - \frac{P}{P_{ref}},$$

where P is the total work done by the acoustic energy to the stone during each simulation: $P = \int E_{stone}\,dt$. P_{ref} is the reference value of P obtained in the case without bubbles. Note that $S = 0$ and $S = 1$ indicates no shielding and perfect shielding (no energy transmission into the stone), respectively. The maximum value of the shielding factor in the plot is approximately 0.8, indicating that up to 80% of the total energy of the incident burst wave can be absorbed/scattered by bubbles that otherwise transmits into the stone. Interestingly, the plot also indicates a strong linear correlation between the shielding factor and the maximum imaging functional within the global range of the shielding factor, independent of the initial condition of bubbles. This correlation suggests that the scattered acoustics can be directly used to estimate the magnitude of the shielding regardless of the bubble dynamics, at least within the range of parameters of bubble cloud addressed in the present study.

Conclusion

We quantified the energy shielding of kidney stones by a bubble cloud nucleated on the stone surface during the passage of a burst wave through a combined experimental and numerical approach. Simulated evolution of the bubble cloud and the bubble-scattered acoustics using a compressible multicomponent flow solver showed quantitative agreement with the results of high-speed photography and acoustic measurements. Results of the simulation revealed that the magnitude of the energy shielding by a thin layer of bubble cloud can reach up to 80% of the total energy of the burst wave, indicating a large potential loss of efficacy in the treatment of BWL due to cloud cavitation. Furthermore, we discovered a strong linear correlation between the magnitude of the shielding and the amplitude of the back-scattered acoustics. This correlation could be used, for example, to monitor the magnitude of the shielding in a human body by using ultrasound imaging in real time during the treatment of BWL, and adjust parameters to minimize shielding during a procedure. Future work can include further detailed analysis of the bubble-scattered acoustics as well as assessment of the effect of the amplitude and frequency of the burst wave on the energy shielding.

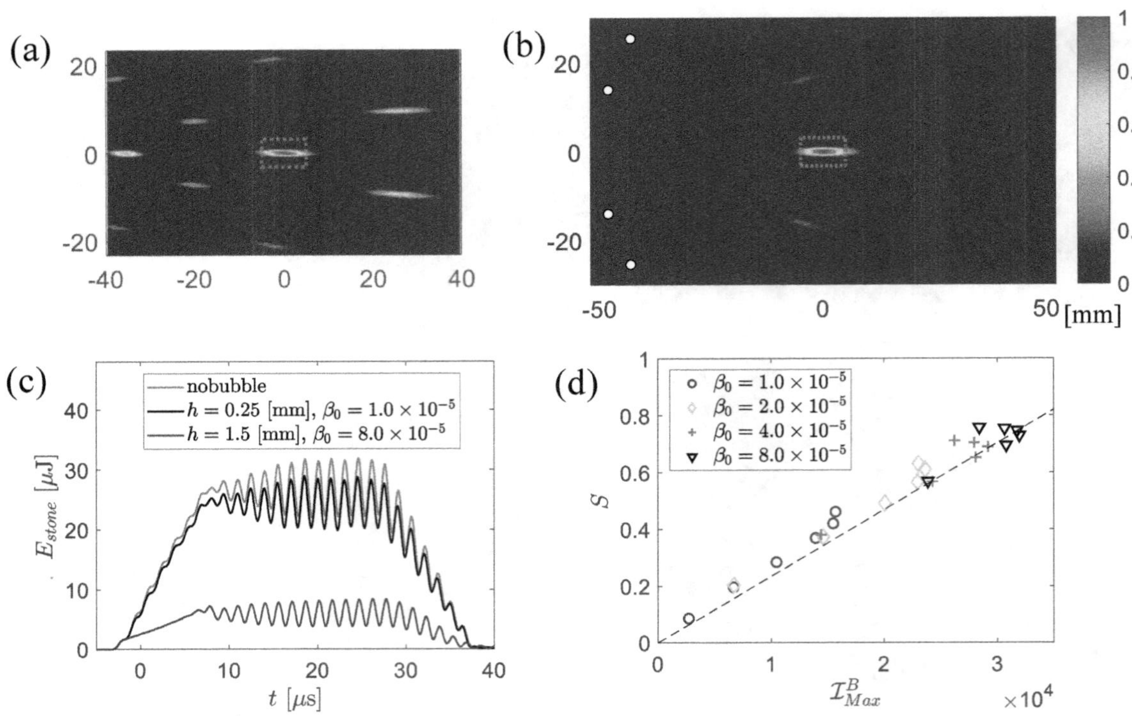

Figure 4. (a) Normalized contour of the imaging functional obtained from the experiment. Stone is indicated by a dotted line. (b) That obtained from the simulation. Sensor locations are indicated by white dots. (c) Evolution of the acoustic energy in the stone. (d) Correlation between the shielding factor and the maximum value of the imaging functional.

Acknowledgement

K.M would like to acknowledge the Funai Foundation for Information Technology, for the Overseas Scholarship. This work was supported by the National Institutes of Health under grants P01-DK043881 and K01-DK104854. The simulations presented here utilized the Extreme Science and Engineering Discovery Environment, which is supported by the National Science Foundation grant number CTS120005.

References

[1] Pishchalnikov, Y.A., Sapozhnikov O.A, Bailey, M.R., Williams, J.C., Cleveland, R.O., Colonius, T., Crum, L.A., Evan, A.P., McAteer, J.A. (2004). *Cavitation Bubble Cluster Activity in the Breakage of Kidney Stones by Lithotripter Shockwaves.* Journal of Endourology. 17(7).

[2] Ikeda, I., Yoshizawa, S., Tosaki, M., Allen, J.S., Takagi, S., Ohta, N., Kitamura, T., Matsumoto, Y. (2006). *Cloud cavitation control for lithotripsy using high intensity focused ultrasound.* Ultrasound in Medicine & Biology. 32(9).

[3] McAteer, J.A, Williams, Jr, J.C., Bailey, M.R., Cleveland, R.O., and Evan, A.P. (2005). *Strategies for improved shock wave lithotripsy.* Minerva Urologica E Neffrologica. 57(271).

[4] Bailey, M.R., McAteer, J.A., Pishchalnikov, Y.A., Hamilton, M.F., and Colonius, T. (2006). *Progress in Lithotoripsy Research.* Acoustics Today. 3, 18-29.

[5] Maxwell, AD., Cunitz, BW., Kreider, W., Sapozhnikov, OA., Hsi, RS., Harper, JD., Bailey, MR., Sorensen, MD. (2015). *Fragmentation of urinary calculi in vitro by burst wave lithotripsy.* Journal of Urology. 193(1).

[6] Maeda, K., Kreider, W., Maxwell, D., Cunitz, B., Colonius, T., Bailey, M. (2015). *Modeling and experimental analysis of acoustic cavitation bubbles for Burst Wave Lithotripsy.* Journal of Physics: Conference Series. 656(1).

[7] Coralic, V., Colonius, T. (2014). *Finite-volume WENO scheme for viscous compressible multicomponent flows.* Journal of Computational Physics. 274, 95-121.

[8] Maeda, K., Colonius, T. (2017). *A source term approach for generation of one-way acoustic waves in the Euler and Navier-Stokes equations.* Wave Motion. 75, 36-49.

[9] Maeda, K., Colonius, T. (2017). *Eulerian-Lagrangian method for simulation of cloud cavitation.* arXiv preprint arXiv:1712.00670.

[10] Garnier, J., Papanicolaou, G. (2009). *Passive sensor imaging using cross correlations of noisy signals in a scattering medium.* SIAM Journal on Imaging Sciences. 2(2).

© 2018 ASME

Investigation of The Improvements in Cavitation And Pressure Pulse Due to The Endplate Propeller

[1] Young-Zehr Kehr; [1] Huan-Jia Xu; [1] Jui-Hsiang Kao*; [1] Yan-Jhang Lin;

[1] *Department of Systems Engineering and Naval Architecture National Taiwan Ocean University Keelung, Taiwan, R.O.C*

Abstract

The aim of this paper is to investigate cavitation and pressure pulse on the stern due to the operating propellers of a container ship. Three optimally designed propellers are tested in this paper. Two propellers are conventional type. The third one is unconventional, and formed by a tip-end plate. These three propellers are design optimally at the same design point, and compared according to the experimental results. The tests are carried out in the large cavitation tunnel of National Taiwan Ocean University (NTOU), and evidence is offered regarding the remarkable improvements in the cavitation extension and the exciting force on the stern achieved by the unconventional propeller.

Keywords: cavitation; pressure pulse; tip-end plate; cavitation tunnel

1. Introduction

Generally, some skills, such as high skew, increasing area ratio, and tip unloading are often applied in conventional propellers to reduce the cavitation. The cavitation of the marine propellers can be classified into several types, and the tip vortex cavitation always appears first. As the tip vortex cavitation is controlled well, the inceptions of other types of cavitation can be delayed.

In Gómez (1976), the ideal of tip loaded propeller was first proposed to restrain the tip vortex cavitation. The tip loaded propeller was shaped by a bent tip, and named as TVF (tip vortex free). Since 1980, Gómez's company, SISTEMAR, has remained devoted to the development of the CLT (Contracted and Loaded Tip) propeller. Gómez and Adalid (1995) derived the New Momentum Theory, which mathematically proves that the induced velocity at the propeller disk of a CLT propeller is much lower than that of a conventional propeller. The ideal efficiency of the CLT propeller is greater than that of the conventional propeller due to the increased contribution of the over-pressure on the pressure side. Some important features of the flow, like the leading edge vortex of skewed blades and the vortex at the outer region of the endplate on the suction side were identified by Antonio et al. (2006). Adalid and Gennaro (2011) proved that CLT propellers grant several significant advantages over conventional propellers, the most important being: (1) 5 to 8% higher efficiency over the entire operation; (2) lower induced noise and vibrations; and (3) Improved ship maneuverability characteristics. Several types of modifications including variations in plate contraction angle, and in plate swept and flap angle are studied in Antonio et al. (2012). Stefano Gaggero and Stefano Brizzolara (2012) conducted numerical computations to predict the development of a smaller sheet cavity bubble (from the leading edge) on the outer side of the tip plate side; the numerical result is qualitatively in agreement with the experimental observations. It is posited by Michael et al. (2014) that the CLT propeller with reduced angle-of-attack loading would have higher efficiency. Stefano et al. (2016) applied tip unloading in the CLT propeller to reduce the tip vortex cavitation; they found that the optimization of the tip plate improves cavity reduction by 10–20%. In this paper, a container ship equipped with three propellers separately is tested in the cavitation tunnel. The improvements in cavitation and pressure pulse on the stern due to the endplate propeller are investigated.

2. Simulated Conditions and Models

Three model propellers (P1, P2, and P3) are tested in this paper. The simulated conditions are as shown in Table 1.

*Corresponding Author, Jui-Hsiang Kao: JHKAO@mail.ntou.edu.tw

Main Engine	M/E Type : MAN B&W 10K98MC DMCR : 41,200 KW x 84.2 RPM
Service Speed	NOR with 15% SM on design draft (Based on BFO, Transmission Efficiency 0.99)
Shaft Immersion (m)	7.6
Ship Speed (kn)	23.35 (P1 propeller)/ 23.38 (P1 propeller)/ 23.39 (P3 propeller)
Shaft Speed (RPS)	1.43 (P1 propeller)/ 1.43 (P1 propeller)/ 1.45 (P3 propeller)
Max. Propeller Diameter (m)	8.7

Table 1. Simulated conditions for the ship

The tested model propellers are list in Table2, and the outlines of these three model propellers are shown in Figure 1. P1 and P2 are conventional type. The area ratio of P2 is increased for reducing the cavitation and exciting force on the stern. P3 is the unconventional type, and with an endplate in order to restrain the tip vortex cavitation.

Model Propeller No	P1	P2	P3
Diameter	269mm	269mm	266mm
Hub Diameter	48mm	48mm	48mm
Pitch-Dia.-Ratio at 0.7R	1.086	1.07	1.073
Blade Area Ratio	0.56	0.663	0.53
No. of Blades	5	5	5

Table 2. The main dimensions for the model propellers

Figure 1. The model propellers

3. Test Arrangements

The tests are carried out in the large cavitation tunnel of NTOU, Figure 2. Its test section is 10m long, 2.6m wide and 1.5m high and allows the installation of the whole ship model. The free surface is substituted by aluminum plates, suppressing the wave system of the model. In order to compensate for the deceleration of the flow due to the boundary layer below these wave suppressing plates, the model draught in the tunnel is increased by 50mm.

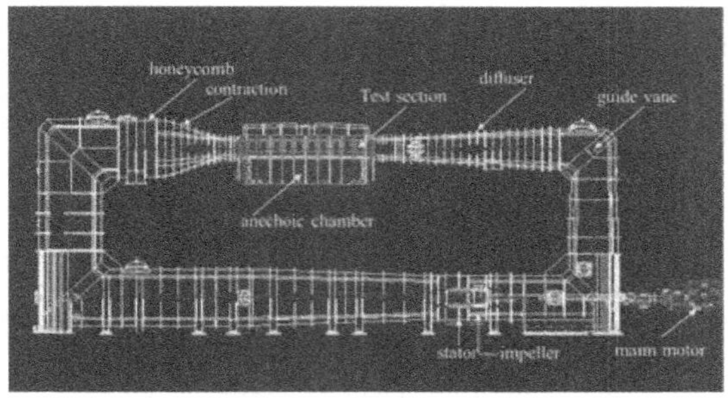

Figure 2 The NTOU large cavitation tunnel

The model scale is 1/32.34. The ship model is equipped with rudder, propeller and shaft line. The stern arrangement is shown in Figure 3.

Figure 3 The hull model with model propeller

In the ship model, the R51 dynamometer made by Cusson was equipped to measure the thrust, torque and rotating speed. Eight pressure pick-ups were installed inside the model above the propeller as shown in Figure 4.

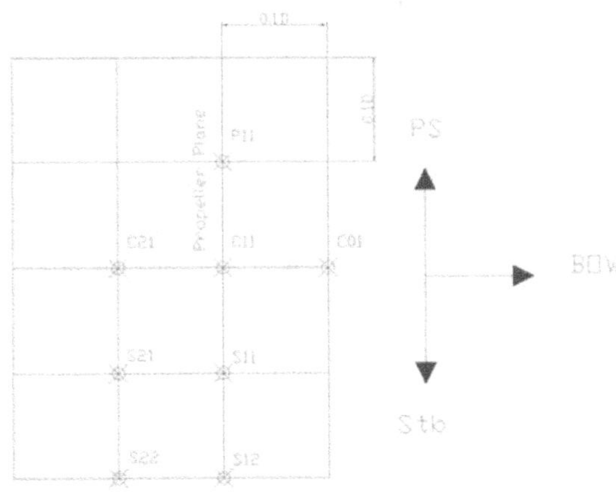

Figure 4 The arranged pressure pick-ups

4. Result of The Propeller Cavitation Observation

The definition for the angular positions of the propeller is given in Figure 5. The cavitation extensions are given in Figures 6 to 8.

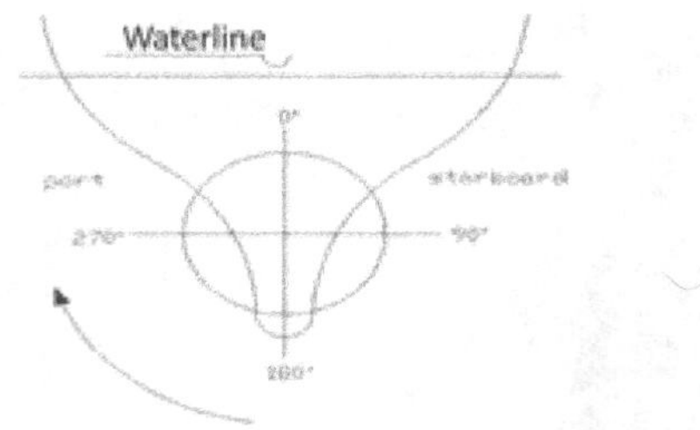

Figure 5 The definition for the angular positions

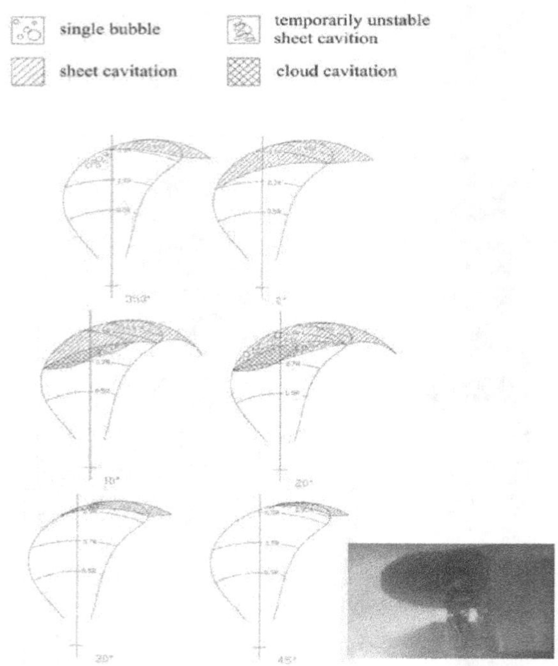

Figure 6 Cavitation phenomena sketches for the P1 propeller

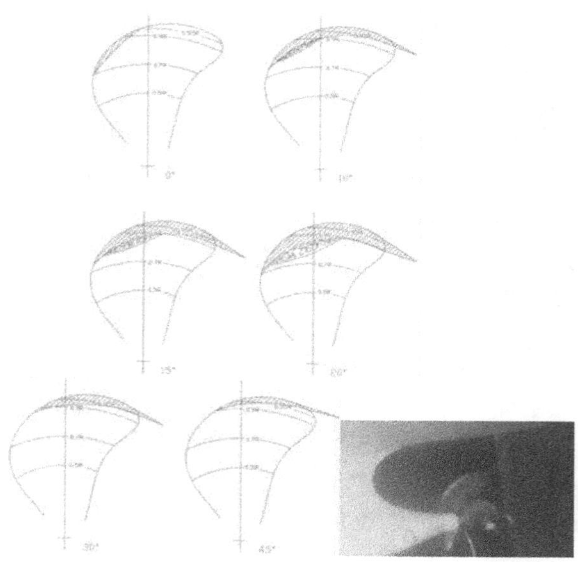

Figure 7 Cavitation phenomena sketches for the P2 propeller

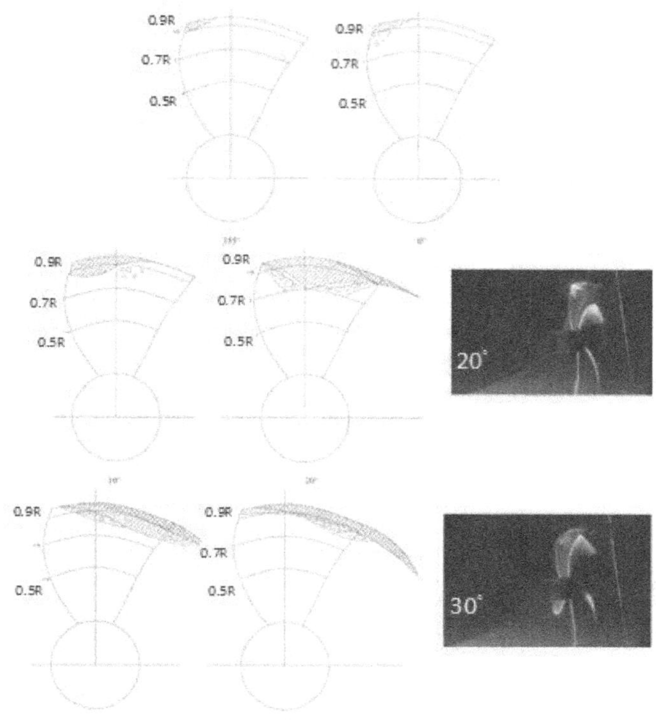

Figure 8 Cavitation phenomena sketches for the P3 propeller

The cavitation extension of P1 appears from 350 to 45 deg. From 0 to 20 deg., sheet cavitation extends from 0.7R, and the thickness of the sheet cavitation is very thick. Because the volume of the sheet cavitation varies very rapidly and complexly; thus, cloud cavitation and unstable sheet cavitation are also found. The cavitation extensions at 10 and 20 deg. nearly cover the whole blade back side from 0.8R. Besides, serious tip vortex cavitation bursting can be found from 10 and 20 deg. The cavitation extension of P2 appears from 0 to 45 deg. As compared with P1, the cavitation extension caused by P2 is smaller and thinner. From 0 to 20 deg., sheet cavitation extends from 0.7R, and some unstable sheet cavitation also appears. Though, the cavitation extensions of P3 are similar with those of P2, the volume of sheet cavitation near the tip of P3 is smaller due to the endplate which effectively reduces the strength of the tip vortex.

5. Result of The Propeller Pulse Measurement

The measured pressure pulses are converted to full-scale ones by using the non-dimensional factor Kp. The conversion is shown in Equation 1.

$$\Delta P_{ship} = \left(K_p\right)_{model} \times \rho n_{ship}{}^2 D_{ship}{}^2 \times \sqrt{2} = \left[\frac{(\Delta P_{model})}{\rho n_{model}{}^2 D_{model}{}^2}\right] \times \rho n_{ship}{}^2 D_{ship}{}^2 \times \sqrt{2} \qquad (1)$$

where ρ is the fluid density, n the shaft speed and D the propeller diameter. In Equation (1) ΔP_{model} is the root-mean-square pressure pulse; thus √2 is required to transfer the root-mean-square value to the maximum one, ΔP_{ship}.
The full-scale results are given in Figures 9. Figure 9 shows the P1 propeller causes the highest harmonic amplitude, 9 KPa. The pressure pulse due to P3 propeller is the smallest, 5.5 KPa.

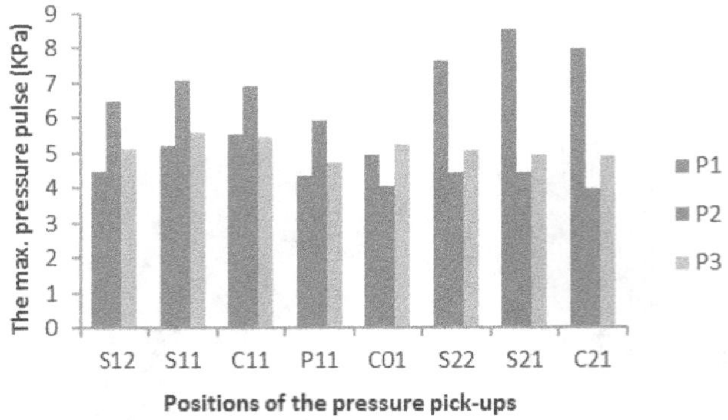

Figure 9 The maximum full-scale amplitude of the pressure pulses for the model propellers

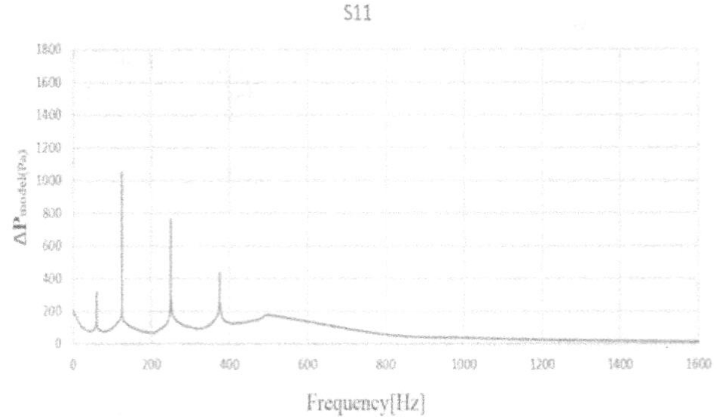

Figure 10 Frequency spectra at pick-up S11 of the P3 model propeller

Figure 10 shows the frequency spectra of P3 model propeller at pick-up S11, having the highest full-scale first harmonic amplitude, 5.5 KPa. It is indicated in Figure 10 that the amplitudes in high frequency decay rapidly, and the blade-rate amplitudes caused by sheet cavitation in low frequency dominate. In other words, the tip vortex cavitation is controlled well; thus, the high-frequency amplitudes disappear.

6. Conclusion and Recommendations

From above results, it is known that the maximum pressure pulse and extended sheet cavitation volume of P3 propeller are less than those of P1 and P2 propellers. Besides, the pressure pulses of P3 propeller at high frequency disappear. The reason for these is due to the endplate which successfully prevents the tip vortex form growing, and reduces the strength of the tip vortex.

It is recommended that the chord lengths near the blade tip should be increased for the P3 propeller in order to restrain the growing of the cavitation volume more.

References

[1] Gómez, G.P. 1976. *Una Innovación en el Proyecto de Hélices*. Ingeniería Naval.

[2] Gómez, G.P. and Adalid, J.G. 1995. *Tip Loaded Propellers (CLT) Justification of Their Advantages Over Conventional Propellers Using the New Momentum Theory*. Report no.429, pp.5-60, SISTEMAR, S.A.Spain.

[3] Antonio Sánchez-Caja, Sipilä, T. P., and Pylkkänen, J.V. 2006. *Simulation of the Incompressible Viscous Flow Around an Endplate Propeller Using a RANSE Solver*. 26th Symposium on Naval Hydrodynamics Rome, Italy, 17-22.

[4] Adalid, J.G., and Gennaro, G. 2011. *Latest experiences with Contracted and Loaded Tip (CLT) propellers. Sustainable Maritime Transportation and Exploitation of Sea Resources*, 1st ed., Vol. 1, Taylor & Francis Group, London, 47-53.

[5] Antonio Sánchez-Caja, J. González-Adalid, M. Pérez-Sobrino, and I. Saisto, 2012. *Study of End-Plate Shape Variations for Tip Loaded Propellers Using a RANSE Solver*. 29th Symposium on Naval Hydrodynamics Gothenburg, Sweden, 26-31.

[6] Stefano Gaggero, and Stefano Brizzolara. 2012. *Endplate effect propellers: A numerical overview. Sustainable Maritime Transportation and Exploitation of Sea Resources* – Rizzuto & Guedes Soares (eds) Taylor & Francis Group, London, ISBN 978-0-415-62081-9.

[7] Michael Brown, Antonio Sánchez-Caja, Juan G. Adalid., Scott Black, Mariano Pérez Sobrino, Phillip Duerr, Seth Schroeder, Ilkka Saisto, and Saisto, I. 2014. *Improving Propeller Efficiency Through Tip Loading*. 30th Symposium on Naval Hydrodynamics Hobart, Tasmania, Australia, 2-7.

[8] Stefano Gaggeroa, Juan Gonzalez-Adalidb, and Mariano Perez SobrinoI. 2016. *Design of Contracted And Tip Loaded Propellers by Using Boundary Element Methods And Optimization Algorithms*. Applied Ocean Research ,55, 102–129.

© 2018 ASME

Investigation of the Primary Mechanisms of Cavitation-Induced Damages

[1]Ben Zhao; [1]Shunxiang Cao; [1]Kevin Wang; [1, 2]Olivier Coutier-Delgosha;

[1]*Virginia Tech, Kevin T. Crofton Aerospace and Ocean Engineering Dept., Blacksburg, VA, USA*
[2]*Arts et Metiers ParisTech / LMFL, Lille, France*

Abstract

Erosion of solid surfaces due to cavitation has been studied for decades. However, it has been a long debate that which mechanism, namely shockwaves, microjets towards the surface, or both, during the cavitation bubble collapse is the primary factor responsible for that erosion. In this project we investigate the small-scale mechanisms of material erosion induced by the collapse of a single cavitation bubble close to a wall. More specifically, our experimental setup includes modification of the initial nucleus size, the maximum bubble radius, the stand-off distance to the wall, the material softness, and the initial flow temperature. We record the evolution of the bubble using high speed cameras as well as the local impacts on the materials. With the help of specifically designed cold-wires, we also measure the temperature in the liquid and in the bubble. Two different methods are used to generate the bubble: (i) an acoustic shockwave of variable intensity, (ii) a YAG laser, which may introduce a high temperature at the start. We also combine the two methods in which the laser initially creates a nucleus, then the shockwave triggers the expansion of the bubble. The objectives of the project are included in this paper, while some first results will be presented at the CAV2018 conference.

Keywords: cavitation; shockwaves; microjets; material erosion; small-scale mechanisms;

1. Introduction

Cavitation occurs in case of a rapid pressure decrease in a liquid. It consists of formation, growth, collapse, and possible rebound of bubbles. The collapse of cavitation bubbles will release both shockwaves (up to 200 MPa) and microjets (up to 280 m/s) in addition to a temperature increase (up to 5000 Kelvin): it is a consensus in the scientific community that these phenomena are generally involved in the erosion of nearby solid surfaces. There are many experiments and computational works that have investigated the fluid dynamics of cavitation bubble collapse near a rigid wall [1-10]. Material deformation caused by cavitation has been also experimentally measured for materials with different hardness [11-13].

However, the transient, two-way interaction between the collapse of the cavitation bubble and the material still remains unclear. On one hand, almost nothing is known about the dynamic response of the material to cavitation-induced loads, such as the magnitude, profile, and propagation of the surface and body (P- and S-) elastic waves, and the effects of these waves on material failure. On the other hand, although a few experimental studies have revealed — using high-speed photography and acoustic measurements — significant impact of the material's Young's modulus on bubble dynamics [14, 15], the detailed reciprocal effects of the acoustic, elastic, and viscoelastic properties of the material on the two-phase bubble/liquid flow field is still unknown. Moreover, the issue of whether cavitation damage is caused by liquid jets, shock waves, or both has been debated for many years. For soft materials, much of the literature supports that the liquid jets play a dominating role. However, for harder materials, there has not been a consensus, or a systematic study that yields a complete theory (e.g. [1, 2, 16, 17]). Table 1 summarizes a number of representative studies which suggest different dominating damage mechanisms.

The analysis of the mechanisms of cavitation erosion has been made essentially by measurements of the pressure wave intensity generated by the bubble collapse, and visualization of the shock wave propagation [1, 18-20]. The impact force on the material has been measured in some cases by the use of PVDF (Polyvinylidene Fluoride) transducer [21]. Abouel-Kasem et al. [17] identified two different types of pits, small pits (of the order of micrometer) that do not change with time, and large ondulations (tens of micrometers) whose size and depth increase with time. The authors claim that the first type is due to microjets, while the second type is the effect of the pressure wave. They also conclude that the large pits are mostly responsible for the erosion process, which is consistent with a fatigue process, according to their observations. Although these explanations make sense, there is no real evidence of their validity, yet, since they are not supported by any direct observation of the respective effects of the microjet and the pressure waves emitted during bubble collapse. Philipp and Lauterbaun [1] show that the damage due to a single bubble collapse depends much on the distance to the wall at the instant of the collapse. According to their observations, only bubbles in contact with the wall generate high speed microjets that can contribute significantly to the wall pitting. They also mention the occurrence of multiple pits in case the bubble adopts a toroidal shape during the collapse. Roy [22] confirms that for bubbles close to the solid surface, both the microjet and the pressure wave impact the wall, with respective intensities that depend on the bubble size, its initial distance to the wall, and the pressure gradient that drives the collapse.

More specifically, the limitations of the current state of the art can be summarized as follows:

- The re-entrant jet velocity has not been characterized yet.
- In most of the previous studies, the bubbles have been generated with a laser. It results in bubbles filled with vapor, which may be significantly different from "real" bubbles, usually partially filled with air.
- Local efforts generated by the bubble collapse on the wall have never been measured.
- The temperature variations induced by the bubble collapse have never been measured.

Authors	Cause of Bubble Collapse	Solid Material	H/R	Dominating Mechanism
Tomita and Shima (1986)	Inertial	Epoxy	0.8 – 1.2	Liquid jet
Philipp and Lauterborn (1998)	Inertial	Aluminum	0 – 2	Shock
Shaw et al. (2000)	Inertial	Polymethylmethacrylate (perspex) block	0.65 & 0.54	Shock (0.65) / Jet (0.54)
Sankin and Zhong (2006)	Shock-induced	Rubber	0 – 1	Liquid jet
Tzanakis et al. (2014)	Inertial	Steel		Liquid jet
Abouel-Kasem et al. (2009)	Ultrasound	Aluminum		Shock

Table 1: Representative studies on the dominating mechanism of cavitation damage on solid materials. H: stand-off distance, i.e. distance between bubble and solid material. R: bubble radius before collapse.

2. Objective

The bubble-material interaction problem described above is a challenging multiphysics and multiscale problem involving a strong coupling between the fluid dynamics and the wall deformation. The dynamic process is highly nonlinear, featuring shock waves, high speed flows, large deformation and topological change of liquid-gas interface, and shock-induced fracture. The effects of the bubble size, distance to the wall and characteristic time of the collapse on the effects on the wall are currently an open question. More specifically, the respective impacts of the microjet and the shock waves, according to these different parameters, in terms of local efforts, elastic or plastic deformation, and potential mass loss has to be clarified. Both the effects of a single bubble collapse and the cumulative effects of the collapses are of interest, to eventually determine the primary mechanisms that are responsible for the damages. The final objective is the clarification of the various couplings between the flow dynamics and the material response. The analysis will focus on the characterization of small scale mechanisms induced by the collapse, i.e. the bubble evolution, the microjet propagation, the temperature variations, and the local stresses and deformations in the material.

3. Experimental Setups

The general objective of the experiments is to investigate the microscale mechanisms that govern the collapse of a cavitation bubble and its interaction with a solid surface located close to the bubble. The study is focused on quite large bubbles (the size will be typically varied between 1 and 10 mm), which are characterized by a life time of a few milliseconds.

We will use three techniques to create the bubbles. The first one is laser-induced cavitation. In this setup we use Q-Switched Nd: YAG laser (532 nm), which is commonly used in such tasks, to introduce cavitation bubbles. The laser is expanded by a beam expander, then converged using a focal lens. In the second setup we will use an acoustic method to trigger a single nucleus expansion in the liquid, specifically with a shaker moving up and down to create shockwave. The pressure at a specific height is given by

$$p_h = p_0 + \rho h(g + \frac{dz(t)^2}{d^2t})$$

where p_h is the pressure at height h of bubble position, p_0 is the pressure above liquid, h is the distance to free surface, ρ is the liquid density, g is gravitational acceleration, and z is the vertical displacement. This setup can generate more realistic cavitation bubbles, partially filled with air, but harder to control. The third option will be to combine the two techniques above to create a single bubble, as an intermediate way to balance the merits of both previous techniques We also want to control p_0 so we will connect a pump to the water tank. We use a high-speed camera (Phantom VEO 410) to record the evolution of cavitation bubbles at a frequency of about 50,000 Hz. A continuous lighting is applied to illuminate the bubbles and appropriate filters are used to cutoff the light induced by the laser. With these experimental setups, we will investigate various sizes, different stand-off distances to the wall, and evaluate the differences between the three techniques of bubble inception.

The experimental setup will provide measurements of the following quantities:

(1) The bubble size and shape at all instants.

(2) The velocity fields in the liquid and vapor phases, based on Particle Image Velocimetry (PIV) with optical means [23].

(3) The temperature inside the bubble and in the liquid will be measured with cold wires. This technique is currently in development in a similar configuration of growth and collapse of a cavitation bubble [24]. Cold wires are homemade sensors composed of an internal fiber that provides the mechanical resistance, a first coating with a high temperature coefficient, and a second external layer of low resistivity, which is applied everywhere but at the sensor location (about 1 mm long). A small current intensity is used to measure the variations of the wire resistance according to the flow temperature.

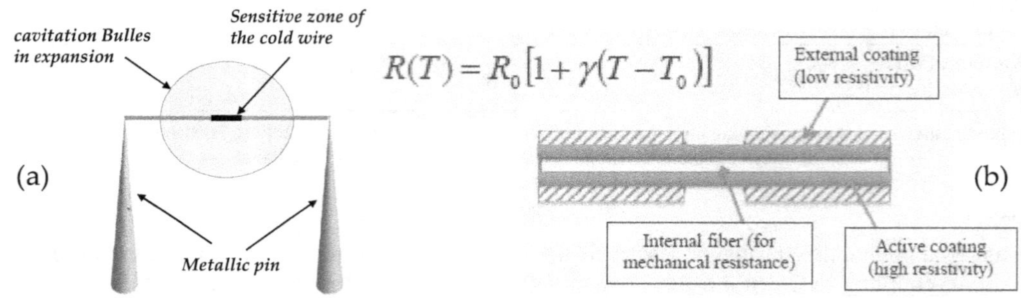

Figure 1: Cold wires for temperature measurements [24]. They are composed of an internal glass fiber and two successive coatings (see Figure 1b). The small diameter of the wires (about 15μm) enables to achieve very short time response (of the order of the microsecond or less) in order to capture the rapid temperature peak expected at the end of the bubble collapse. Using several wires will enable to obtain the temperature distribution inside the bubble and in the surrounding liquid. This device will be used in the present project to characterize the temperature variations during all investigated bubble collapses.

(4) Measurements of local efforts will be performed on the surface of the material, using PVDF (polyvinylidene fluoride) coating on the wall. Previous experiments have used such sensors to monitor at high frequency the global effort variations, while it is planned in this work to manufacture a specific array sensor that will provide the local efforts due to the various impacts related to the bubble collapse. For that purpose, a distributed sensor composed of a segmented electrode etched on a continuous PVDF layer using classical lithography techniques will be manufactured [25].

4. Current Work

We have already prepared all the required components to generate one single bubble using a laser. Currently we have designed most of the experimental setup, such as the tank shape and different optical sets (e.g. beam expander vs concave lens, parabolic lens vs convex lens). We will be evaluating a variety of equipment combinations until we reach the desired performance. In the next months we will start some preliminary tests. By the time of conference, we should be able to present some first results of bubble collapse and related erosion on the wall, for several stand-off distances of the bubble.

References

[1] Philipp, A. and W. Lauterborn, *Cavitation erosion by single laser-produced bubbles*. Journal of Fluid Mechanics, 1998. 361: p. 75-116.

[2] Tzanakis, I., et al., *Incubation pit analysis and calculation of the hydrodynamic impact pressure from the implosion of an acoustic cavitation bubble*. Ultrasonics sonochemistry, 2014. 21(2): p. 866-878.

[3] Brujan, E.A., et al., *The final stage of the collapse of a cavitation bubble close to a rigid boundary*. Physics of Fluids, 2002. 14(1): p. 85-92.

[4] Brujan, E.-A. and Y. Matsumoto, *Collapse of micrometer-sized cavitation bubbles near a rigid boundary*. Microfluidics and nanofluidics, 2012. 13(6): p. 957-966.

[5] Zhang, S., J.H. Duncan, and G.L. Chahine, *The final stage of the collapse of a cavitation bubble near a rigid wall*. Journal of Fluid Mechanics, 1993. 257: p. 147-181.

[6] Tong, R., et al., *The role of 'splashing' in the collapse of a laser-generated cavity near a rigid boundary. Journal of Fluid Mechanics*, 1999. 380: p. 339-361.

[7] Kobayashi, K., T. Kodama, and H. Takahira, *Shock wave–bubble interaction near soft and rigid boundaries during lithotripsy: numerical analysis by the improved ghost fluid method*. Physics in medicine and biology, 2011. 56(19): p. 6421.

[8] Zhang, Z.-y. and H.-s. Zhang, *Surface tension effects on the behavior of a cavity growing, collapsing, and rebounding near a rigid wall*. Physical Review E, 2004. 70(5): p. 056310.

[9] Wang, Q., *Multi-oscillations of a bubble in a compressible liquid near a rigid boundary*. Journal of Fluid Mechanics, 2014. 745: p. 509-536.

[10] Johnsen, E. and T. Colonius, *Numerical simulations of non-spherical bubble collapse*. Journal of fluid mechanics, 2009. 629: p. 231-262.

[11] Ohl, C.-D., et al., *Sonoporation from jetting cavitation bubbles*. Biophysical journal, 2006. 91(11): p. 4285-4295.

[12] Dular, M., Coutier-Delgosha, O., and Petkovšek, M., *Observations of cavitation erosion pit formation*. Ultrasonics sonochemistry, 2013. 20(4): p. 1113-1120.

[13] Chahine, G.L. and C.-T. Hsiao, *Modelling cavitation erosion using fluid–material interaction simulations*. Interface focus, 2015. 5(5): p. 20150016.

[14] Brujan, E-A., et al., *Dynamics of laser-induced cavitation bubbles near an elastic boundary*, Journal of Fluid Mechanics, 2001. 433: p. 251-281.

[15] Sankin, G. and P. Zhong, *Interaction between shock wave and single inertial bubbles near an elastic boundary*. Physical Review E, 2006. 74(4): p. 046304.

[16] Tomita, Y. and A. Shima, *Mechanisms of impulsive pressure generation and damage pit formation by bubble collapse*. Journal of Fluid Mechanics, 1986. 169: p. 535-564.

[17] Abouel-Kasem A., Ezz El-Deen A., Emara K. M. and Ahmed S. M., *Investigation into Cavitation Erosion Pits*, J. of Tribology 131: p. 031605-1.

[18] Bidin N., *Shock Wave Emission during Cavitation Bubble Collapse in Free Liquid*, 1995, PertanikaJ. Sci. & Techno. 3(1):51-55.

[19] Vogel A., Busch S., Parlitz U., *Shock wave emission and cavitation bubble generation by picosecond and nanosecond optical breakdown in water*, 1996, J. Acoustic Soc. Am. 100(1): 148-165.

[20] Sugimoto Y., Sato K., Oojimi S., *Visualization of pressure wave generated by collapse of cavitation cloud using frame difference method*, 2008, ISFV13 - 13th International Symposium on Flow Visualization.

[21] Hattori S., Hirose T., Sugiyama K., *Prediction of cavitation erosion based on the measurement of bubble collapse impact loads*, 2009, Proc. of the 7th Int. Symposium on Cavitation CAV2009, Ann Arbor, Michigan, USA.

[22] Roy, Samir & Franc, Jean-Pierre & Fivel, Marc. (2015). *Cavitation erosion: Using the target material as a pressure sensor*. Journal of Applied Physics. 118. 1-11. 10.1063/1.4934747.

[23] Fuzier S., Coudert S., Coutier-Delgosha O., *Two phase velocity measurements using LIF-PIV inside the cavitation sheet generated in a venturi*, Proceedings of the Int. Conf. on Multiphase Flow, Jeju, Korea, May 26-31, 2013.

[24] Hamdi M., Coutier-Delgosha O., Baudoin M., *Measurements of the temperature variations during the growth and collapse of cavitation bubbles*, CAV 2018 Conference.

[25] Preumont A., Francois A., De Man P., Piefort V., *Spatial filters in structural control*, J. of Sound and Vibrations, 2003, 265 (2003) 61–79.

© 2018 ASME

Investigation on cavitating flow of a novel high pressure nozzle with grooved needle

[1]Li-yun Fan; [1]Yi-mo Zhang; [1]Peng Liu; [2]Manolis Gavaises

[1]Harbin Engineering University, Harbin, Heilongjiang Province, China; [2]City University London, London, UK

Abstract

In this paper, a computational study was carried out to investigate the influence of the use of grooved needles on the cavitation flow with needle movement inside a injector nozzle. With this aim, high pressure nozzles with needle movement have been simulated and analysed for 2 different nozzles: one with grooved needle valve and one ordinary nozzle. The internal cavitating flow during the full opening period has been modeled with a multi-hole nozzle model that considers the cavitating flow in thermodynamic and homogeneous equilibrium state. The comparison between two type nozzles has been made in terms of velocity, vorticity, vapor volume fraction and mass flow rate over the needle opening period. The results of the investigation have shown that the groove leads to an increment of the cavitation bubbles of the injector orifice entrance.

Keywords: cavitation; injector nozzle; needle grooves

Introduction

High pressure fuel injection tech has been widely used in modern common rail injection systems in order to meet the demanding emission standards with some injection pressures of even more than 2000 bar. However, during the process of fuel injection, the high pressure fuel flows through areas with sudden cross-sectional change inside injector. Especially in injector nozzles, the pressure drops and speed mutates sharply due to small diameter of nozzle hole. Consequently, when the decreased local pressure is lower than saturation vapor pressure, cavitation occurs in nozzles. Cavitation phenomenon of fuel can cause fluctuation of pressure and turbulence intensity, which plays an essential role in the primary spray process at the combustion chamber, and further affects the combustion and emission performance of diesel engine[1]. Apart from that, cavitating flow may lead to choked conditions and impact on the control of the fuel injection quantity. Several studies have shown that the internal nozzle flow characteristics are strongly influenced by geometry structure of injector nozzle, including the design of pressure chamber[2, 3] and the shape of nozzle holes[4, 5], which indicates that the geometry of the nozzle changes the cavitating flow of the injector orifices.

In the present work, we proposed a novel nozzle with grooved needle under high pressure and investigated the cavitating flow in it by comparing with an ordinary nozzle. A mixed two-phase three dimensional transient cavitation model was applied to simulate the complicated nozzle flow process with needle movement.

Numerical Method

Computation models used in this work are based on the solution of Reynolds Average Navier-Stokes. There are two main approaches for modelling cavitating flows inside the injector nozzle: mixed two-phase flow models and homogeneous equilibrium models. For the cavitation model, the Zwart-Gerber-Belamri cavitation model developed by Rayleigh-Plesset equations was used to simulate the growth and collapse of the cavitation bubbles[6]. In the two-phase flow model, the liquid of the fuel was set as main phase while the vapor phase was set as secondary phase. The compressibility of the gas is neglected, and then the nozzle mixture flow is modeled as incompressible. In addition, the gas-liquid flow is supposed in thermodynamic equilibrium state and those two phases possess the same mixture dynamic viscosity[7]. Thus, the slip velocity between the gas phase and the liquid phase is ignored, and the gas phase and the liquid phase are described with the same velocity field.

Computational grids and setup

In order to analyse the internal cavitating flow characteristics of the nozzles, a three-dimensional geometric model of a special nozzle was built. This type of nozzle was developed on the basis of a ordinary diesel injection nozzle, which is a symmetrical nozzle with 6 nozzle orifices. In this study, the novel nozzle is characterized with grooves at the tip of needle valve. The groove is semi-cylindrical in shape and is distributed in the surface of the lower part of

*Corresponding Author, Li-yun Fan: fanly_01@163.com

needle valve, which begins from the inlet of the sealed conical surface and ends at the SAC volume chamber, as shown in Figure 1. To save the computational resources, the 1/6 geometric nozzle model was established for calculation. Figure 2 shows the mesh used around the nozzle hole. Local meshes were refined to ensure the accuracy of computing, including the area of the sealing cone between the needle and needle seat, the groove and the orifice.

The boundary conditions are shown in Figure 1. For the needle seat, the periodic cell zone conditions were used in the wall. As for the inlet and outlet sections, the inlet pressure was kept at 2500bar, and the outlet pressure was set to 10bar. Applied the user defined function in the calculation model, the needle velocity was loaded into the solver. The data of needle lift motion were obtained from the simulation results of the fuel injection system[8]. According to the needle motion curve, the needle lifts from 0.474ms, and remains at full needle lift until 1.204ms. For the turbulence model, the RNG $k-\varepsilon$ turbulence closure was adopted, with scalable wall functions for near wall surface treatment[9]. The spatial discretization of momentum and turbulent kinetic energy were based on the two order upwind scheme, and the SIMPLEC algorithm was used.

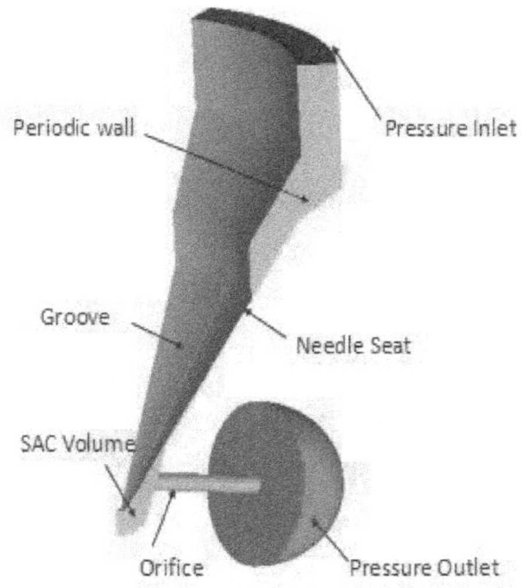

Figure 1 Nozzle geometric model for simulation

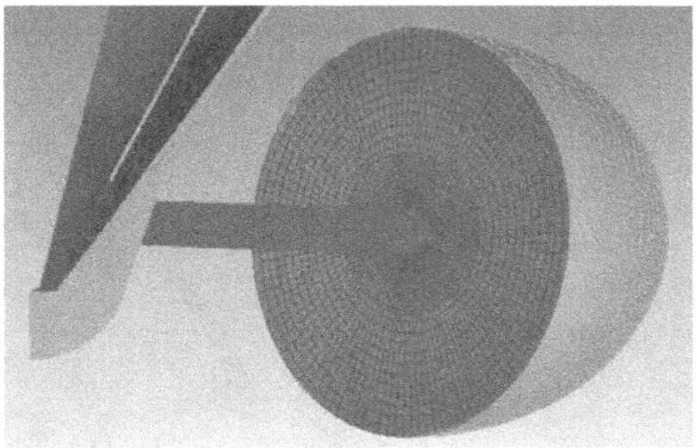

Figure 2 Details of local nozzle mesh

Results and Discussion

In Figure 3, the simulated velocity profiles and vapor distribution between the ordinary and grooved nozzle are presented. In both nozzles, cavitation is formed near the upper wall of the orifice entrance and develops downstream to the outlet of the orifice. At the early time of the opening period, the fuel flow inside injection nozzle is unsteady at 0.5ms, while at the 0.7ms to 1.1ms, the cavitation state is gradually stable as flow moves, in Figure 3(b)-(d). With the fuel flowing inside the orifice, string cavitation occurs at the 1/3 length of the injector orifice. From sectional contour of the vapor volume fraction, as shown in Figure 3(right), the large vapor volume fraction cavitation bubbles mainly locate at the center of the cavitation area. Also, the region of string cavitation is expanding downstream and even mixes with the upper cloud cavitation near the outlet section of the orifice. Particularly in the grooved nozzle, the mixing range is wider.

Cavitation occurrence in grooved nozzle is basically similar to the typical nozzle, however there are some significant differences, shown in Figure 3 as well. First of all, the numerical value of the vapor volume fraction in the cavitation region, at the injection orifice entrance, is larger than the equivalent value of the ordinary one, (see Fig. 4). Moreover, as show in the vapor volume fraction contour, the groove also has positive influence on the string cavitation occurrence at lower part of the orifice. The noticeable increment of vapor volume fraction can be explained by the velocity and vorticity contour in the left of the Figure 3. The fuel flows out of the groove with a higher vorticity at the flow passage between the needle and needle seat, which is just in front of the orifice entrance. Consequently, the velocity of the grooved nozzle in the pressure chamber is obviously less than that of the ordinary nozzle. The groove introduce a channel which lead to the high vorticity fuel flowing into the orifice rather than the pressure chamber and, as a result, the cavitation appearance is promoted.

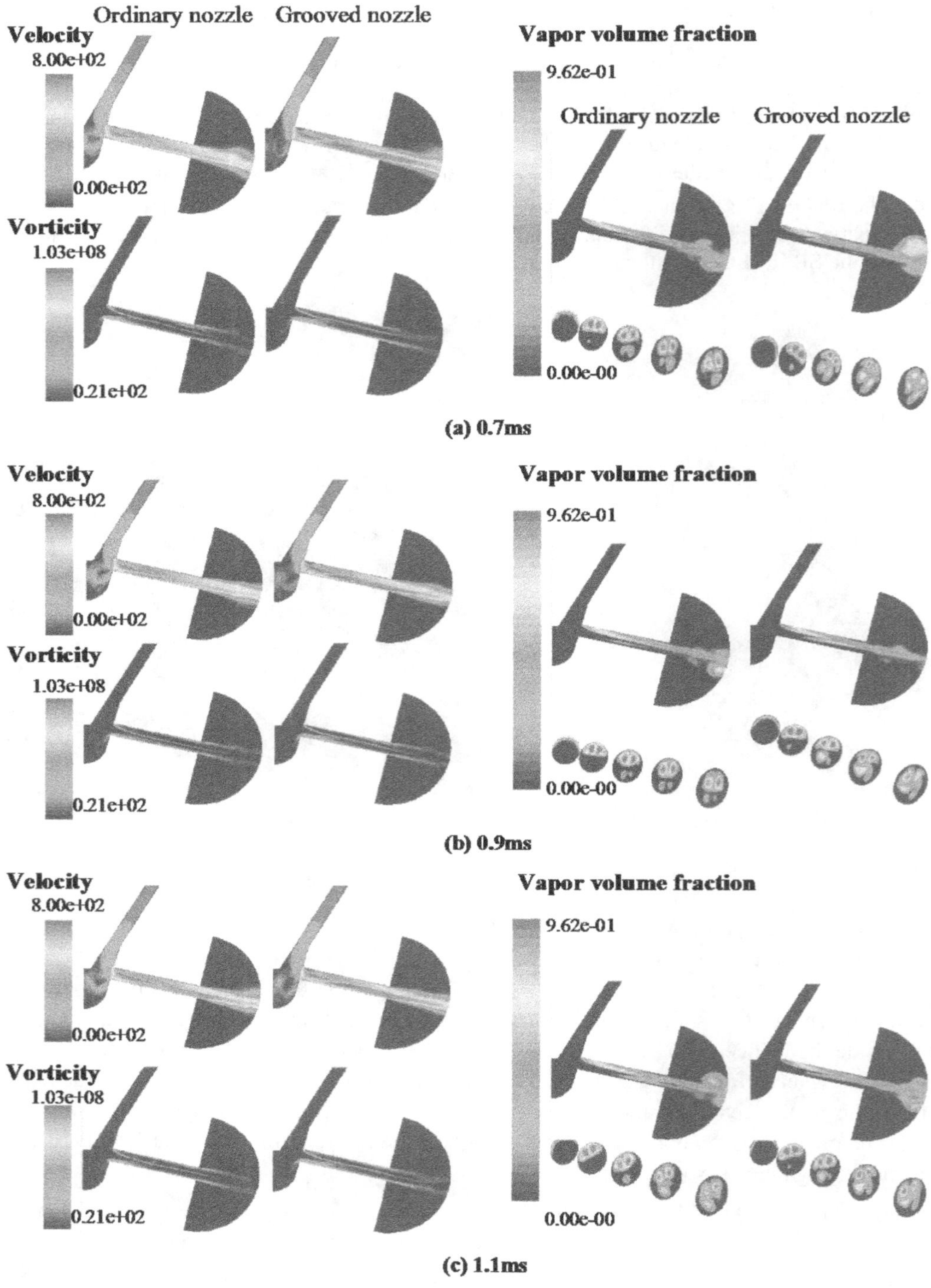

Figure 3 Instantaneous velocity (left, top), vorticity (left, bottom) and vapor volume fraction(right) in mid-plane of injector

The average vapor volume fraction and mass flow rate at the inlet of the nozzle orifice with the needle movement are shown in Figure 4. As can be seen from the graph, the vapor volume fraction and mass flow rate increase with the needle rising. After the needle reaches the full lift at 0.474ms, the vapor volume fraction and mass flow rate no longer increase and remain relatively stable during full opening period of the needle movement. In the average vapor volume fraction graph, Figure 4(a), at the very beginning of needle going up, the vapor volume fraction starts falling until 0% due to the initial value setting. Once the effect of the initial cavitating flow disappears, the cavitation bubbles increase with the needle lifting. Then the steady cavitation can be observed at full needle lift condition, as discussed earlier. It is clear that the vapor volume fraction increases in grooved nozzle, with the value of around 12%, compared with 10% in typical nozzle during full opening period. As before, this is related to the imposed groove which is contributed to the cavitation appearance in the entry of the orifice. In Figure 4(b), the mass flow rate stably keeps at approximately 7.0g/s over the opening period. At the sme time, the groove almost has no impact on the mass flow rate owing to its small size.

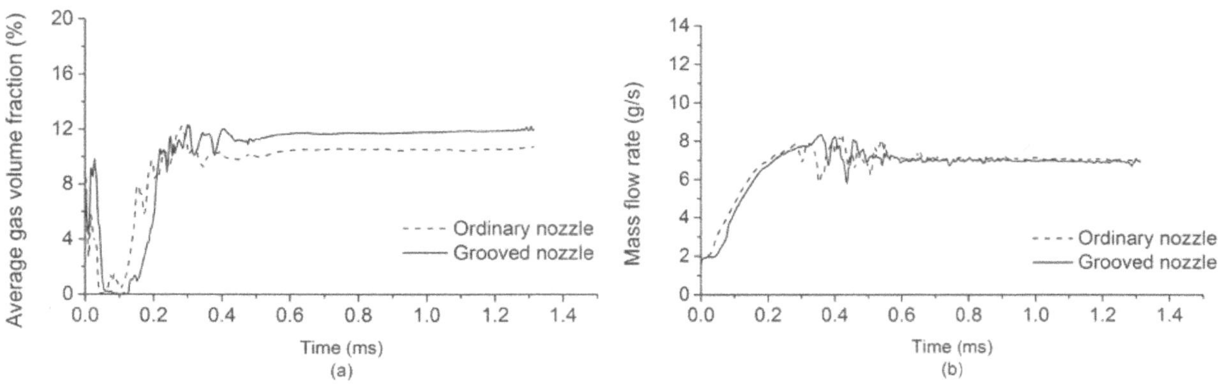

Figure 4 Average vapor volume fraction and mass flow rate through the orifice entrance of the examined injector

Conclusion

This work intends to investigate the influence of the grooves at the novel nozzle needle on the cavitating flow characteristics. The results indicate clear differences in cavitation appearance and generation between two injector nozzles, ordinary nozzle and grooved one. The grooves contribute a higher vorticity area upstream the orifices, which increases the vapor volume fraction at the entrance of the orifice and provides a negligible fluctuation on mass flow rate during the needle opening period.

Acknowledgments

This paper is funded by the International Exchange Program of Harbin Engineering University for Innovation-oriented Talents Cultivation.

References

[1] Desantes JM, Payri R, Salvador FJ, De la Morena J. Influence of cavitation phenomenon on primary break-up and spray behavior at stationary conditions. Fuel 2010;89:3033-41.

[2] Lee W G, Reitz R D. A numerical investigation of transient flow and cavitation within minisac and VCO diesel injector nozzles. Proceedings of ASME Internal Combustion Engine Division 2009 Spring Technical Conference, May 3-6, 2009, Wisconsin, Milwaukee, USA.

[3] Phoevos Koukouvinis, Manolis Gavaises, Jason Li. Large eddy simulation of diesel injector including cavitation effects and correlation to erosion damage. Fuel 2015; 175(2016): 26-39.

[4] Som S, Ramirez AI, Longman DE, Aggarwal SK. Effect of nozzle orifice geometry on spray, combustion, and emission characteristics under diesel engine conditions. Fuel 2011; 90(3): 1267-76.

[5] Salvador F J, Carreres M, Jaramillo D, Martínez-López J. Analysis of the combined effect of hydrogrinding process and inclination angle on hydraulic performance of diesel injection nozzles. Energy Conversion & Management, 2015, 105:1352–65.

[6] P. J. Zwart, A. G. Gerber, and T. Belamri. A Two-Phase Flow Model for Predicting Cavitation Dynamics. In Fifth International Conference on Multiphase Flow, Yokohama, Japan, 2004.

[7] Som S, Aggarwal S, El-Hannnouny R, DE Longman. Investigation of nozzle flow and caviation characteristics in a diesel injector. Journal of Engineering for Gas Turbines and Power, 2010; 132: 042802.

[8] Bai Y, Fan LY, Ma XZ, et al. Effect of Injector Parameters on the Injection Quantity of Common Rail Injection System for Diesel Engine. International Journal of Automotive Technology, 2016, 17: 567-579.
[9] Salvador F J, Mratine-Lppez J, Caballer M. Study of the influence of the needle lift on the internal flow and cavitation phenomenon in diesel injector nozzles by CFD using RANS methods. Energy Conversion and Management, 2011; 66(2013): 246-256.

© 2018 ASME

Investigation on Numerical Prediction of Propeller Induced Hull Pressure Pulses

Muye Ge[1]*; Rickard E. Bensow[1]; Urban Svennberg[2];
[1]*Chalmers University of Technology, Sweden;* [2]*Rolls Royce Hydrodynamic Research Center, Sweden*

Abstract

Simulation of a cavitating propeller in behind conditions and analysis of induced hull pressure fluctuations are presented. All the simulations were performed using RANS method in the commercial package Star-CCM+. Cavitation patterns show good agreement with experimental measurements, especially the blade tip refined meshes which captured the dynamic behaviour of tip vortex cavitation. The predicted pressure pulse amplitudes agree reasonably well with experimental measurements up to 3rd to 4th order of blade passing frequency.

Keywords: propeller; cavitation; hull pressure fluctuation; tip vortex cavitation; RANS

Introduction

The demand for the prediction of pressure pulses from operating marine propellers is increasing, due to concern for environmental impact and the comfort of onboard passengers and crew. The quest for higher efficiency also calls for more accurate estimations of the pressure pulses, as there is often a trade-off between the two. Numerical prediction of pressure pulses is a challenging task because of the complexity of involved physical phenomenon, including the interaction between the ship wake and propeller, different kinds of cavitation occurring on the propeller and in the tip vortex. The Blade Passing Frequency (BPF) is usually used to analyse the propeller induced pressure pulses, and it is believed that the 1st and 2nd order BPF fluctuations are caused by the blade load and growth and shrinkage of propeller sheet cavitation while tip vortex cavitation is the source of higher order pressure fluctuations, even though the mechanisms are still unclear. In the present study, a model scale container vessel with a five bladed propeller was simulated and results are compared with experiments performed by HSVA in the VIRTUE and SONIC EU projects.

Successful predictions of the sheet cavitation extent and related 1st order pressure fluctuations have been reported [1] [2]. Regarding prediction of the tip vortex, its strength, at least the minimum pressure, could be predicted well close to the propeller tip [3] [4]. Using scale-resolved RANS models with tip mesh refinement, also tip vortex cavitation has been predicted [5] [6]. In the present study, the commercial package Star-CCM+ was used to solve the incompressible flow with the $k-\omega\ SST$ turbulence model and the Schnerr-Sauer mass transfer model. Different meshes have been tested where the finest mesh consists of about 30 million cells, including a propeller blade tip refinement region used to capture the tip vortex and its interaction with the sheet cavitation.

Simulation set up

The studied container vessel was developed in 2002 with bulbous bow and single screw five bladed fixed pitch propeller of diameter D=7.9m with tip clearance of $0.277D$. The model scale ship (scale ratio 1:29.1) was tested in the cavitation tunnel HYKAT at HSVA. It was installed according to full scale draft of 11.3m while the free surface was substituted by flat plates. 13 probes were placed on the hull body to measure the propeller induced hull pressure fluctuations. The arrangement of probe locations is shown in figure 1. Two operating model conditions have been considered in the present study, with same thrust coefficient $K_T = \frac{T}{\rho n^2 D^4}$ and cavitation number $\sigma_n = \frac{p-p_v}{0.5\rho(\pi n D)^2}$ but based on different approaches of scaling correspond to towing tank condition and cavitation tunnel condition, as shown in table 1. Both conditions correspond to full scale ship cruising speed of 23.76 knots.

Case	Inlet velocity	Rotation rate	Thrust coefficient	Cavitation number
Condition 1	7.268 m/s	28 rps	0.2234	0.2354
Condition 2	2.216 m/s	8.5 rps	0.2234	0.2354

Table 1: Simulation conditions

*Muye Ge, muye.ge@chalmers.se

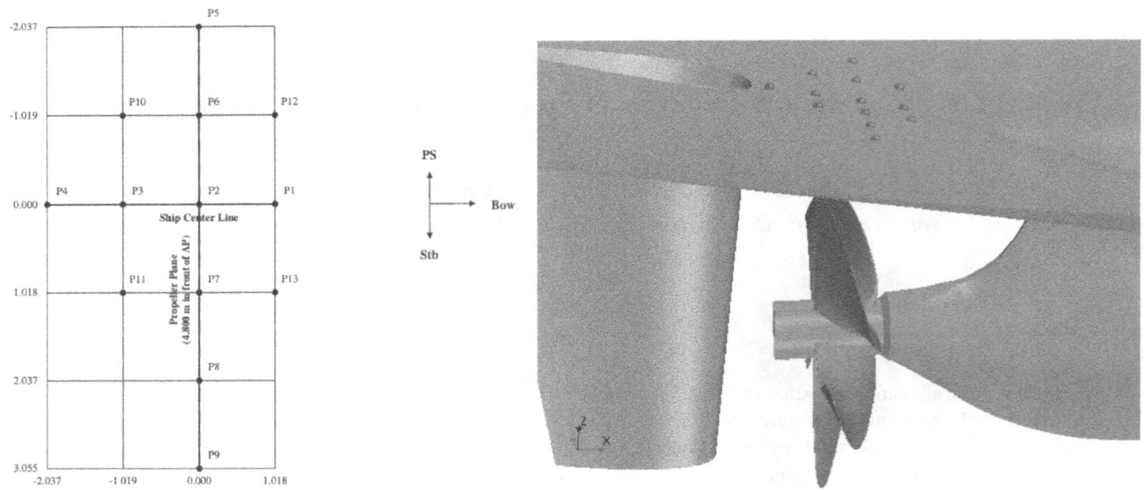

Figure 1: Pressure sensors configuration during model test

Mesh

The simulation domain corrsponds with the model test tunnel section, as shown in figure 2. The meshes were generated using Pointwise. Structured mesh was applied on the blades (31 layers) and hull body surfaces (34 layers) and hybrid extrusion method was used to create the boundary layer prisms. Tetrahedral cells were used to fill the domain volume. A cylinder region was created for the propeller rotation and connected with the outer region by a sliding interface. The propeller blades tip regions were further refined by structured meshed helix-like regions (core size 0.086 mm) to have a better capture of the tip vortex behaviour. The target y^+ value is 1 on both propeller blades and hull body.

#cells/10^6	Tets	Pyramids	Prisms	Hexes	Total cells	Inner region	Outer region
Condition 1 with refinement	19.46	0.47	0.11	12.66	32.69	19.77	12.92
Condition 1 without refinement	16.05	0.38	0.11	11.03	27.57	14.65	12.92
Condition 2 with refinement	18.29	0.46	0.11	13.87	32.73	21.57	11.16
Condition 2 without refinement	14.66	0.38	0.11	12.24	27.38	16.22	11.16

Table 2: Summary of mesh type counts

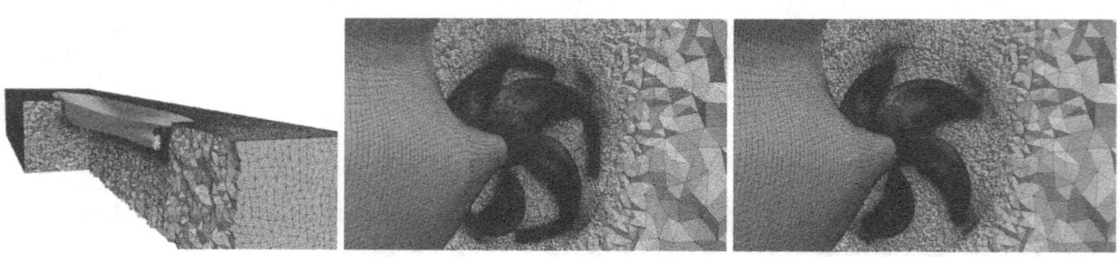

Figure 2: Simulation domain and mesh close to the propeller region

Results

The major difference between the two studied conditions is the Reynolds number, which would influence the boundary layer and thus the wake where the propeller is operating in. The inlet flow velocity was adjusted to produce same thrust coefficients with the experiment by performing transient non-cavitating simulations. Predicted wakes are shown in figure 3a and figure 3b. As shown in the figures, condition 2 has a more compressed wake field than condition 1, but qualitatively similar. The predicted thrust coefficients K_T are shown in figure 3c.

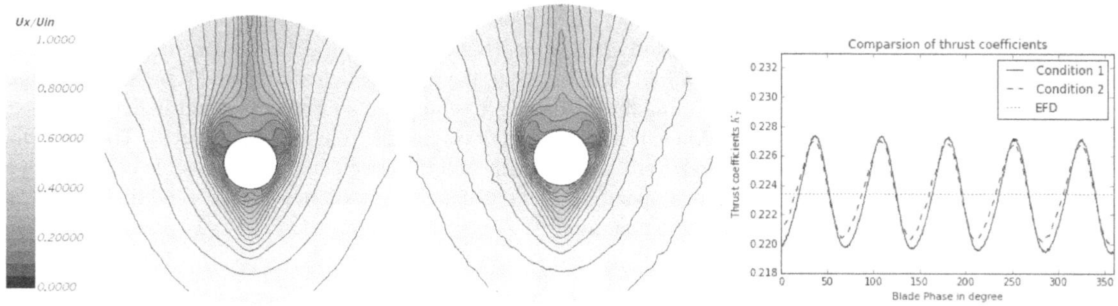

(a) Nominal wake ,condition 1 (b) Nominal wake ,condition 2 (c) Predicted K_T in 1 revolution

Figure 3: Predicted wake and thrust coefficients for non-cavitating cases

For the cavitating conditions, the time step was set to equivalent to 2048 steps per revolution, which is about 5.69 steps per 1 degree revolution. The predicted cavity extent from the tip-refined mesh in condition 1 is shown in figure 4. During the experiment, the cavitation pattern did not repeat totally from one revolution to the next, and the variation was quite significant. The sketches from experiment demonstrate the stable cavitations in cross lines, which is largely consistent with numerical predictions.

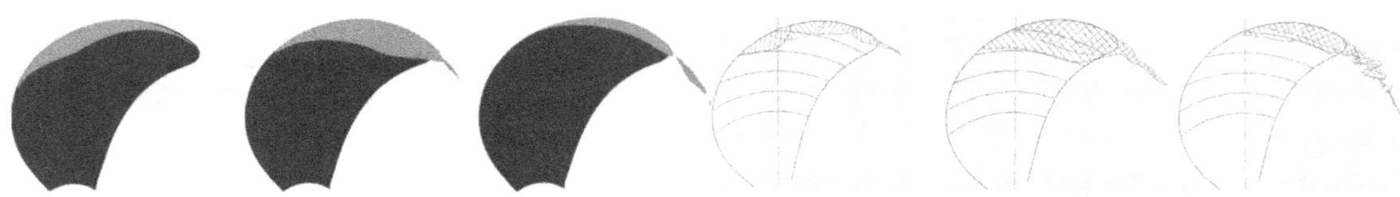

$\phi = 180°$ $\phi = 210°$ $\phi = 240°$ $\phi = 180°$ $\phi = 210°$ $\phi = 240°$

Figure 4: Predicted iso-surfaces of $\alpha = 0.5$ in condition 1 and cavity sketches from experiment

The predicted tip vortex cavitation show a very unstable behaviour. In figure 5 the cavity is shown close to the blade tip region. The major tip vortex cavitation show a rolling-up structure in a short distance after the blade tip. Unlike the non-cavitating case in which only one major tip vortex is predicted close to the blade tip, some secondary vortex structures are formed and rolling with the major tip vortex cavitation, and these secondary vortex would be cavitating in certain conditions.

Figure 5: $\phi = 232°$, tip vortex cavitation

The pressure at each probe and total vapor volume $V_{t,vapor} = \Sigma\alpha_{t,i}V_{t,i}$was recorded during the simulation, as shown in figure 6. Compared to non-cavitating conditions, the pressure fluctuations increased significantly due to the dynamics of vapor structures. The recorded pressure fluctuations follow the same trend as $p \sim d^2V_{vapor}/dt^2$. The prediction of tip vortex cavitation increase the total vapor volume and induce more pressure fluctuations.

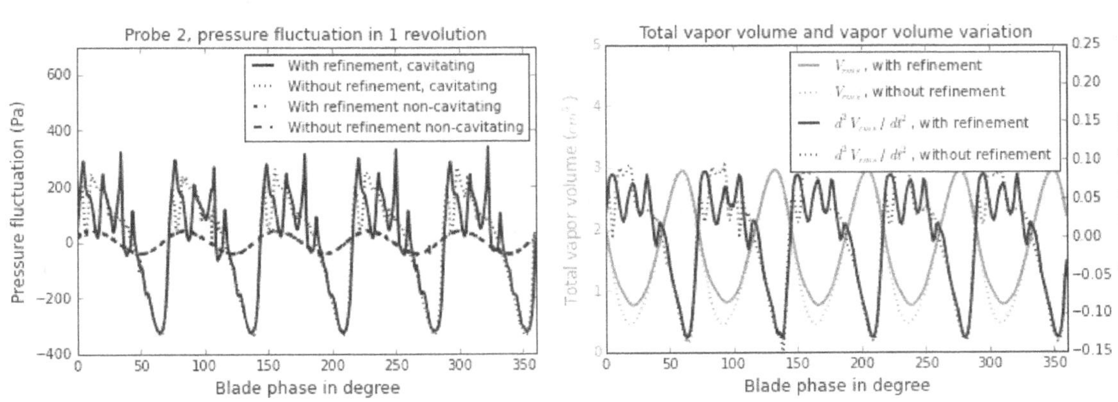

Figure 6: Recorded pressure fluctuations and rms values of total vapor volume in condition 1

The predicted hull pressure amplitudes equivalent to full scale are shown in figure 7. The results from tip refinement meshes predicted lower amplitudes than meshes without refinement for 1st order pressure pulses, slightly higher amplitudes for 2nd order pressure pulses and significantly higher and better results for 3rd order pressure pulses. This is more or less expected, since part of sheet cavity would be rolled up to the tip vortex cavitation and results in a stronger higher order blade passing frequency pulses. Regarding the 4th order pressure pulse amplitudes, rather noticeable values were predicted, but further study is needed to determine if it is physical or originates from numerical errors. The predicted values from condition 2 are overall lower than condition 1 and experimental measurements. One possible explanation is that for condition 2, the residence time of each blade operating in the wake is much shorter than in condition 1, and the recorded total vapor volume is about 70% compared to condition 1.

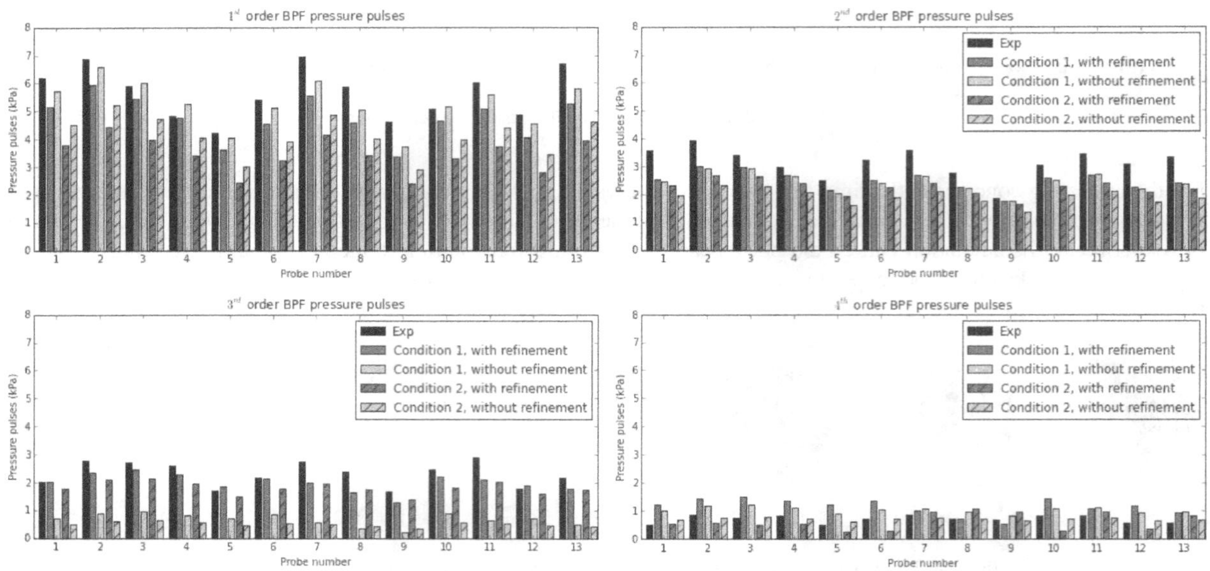

Figure 7: Predicted and measured pressure pulses amplitudes

Conclusion

Numerical studies for the prediction of pressure pulses induced by an operating marine propeller in behind conditions were presented. Regions of tip refinement were applied close to the propeller tip regions which give better predictions of the dynamic behaviour of tip vortex cavitation. The predicted pressure pulse amplitudes agreed well with experimental data for the simulation condition 1, while about 70% of the amplitudes were predicted for the simulation condition 2. The simulation results also demonstrate the higher order BPF pressure pulses are highly related to the tip vortex cavitation behaviour, and for this case it is possible to be captured by using RANS method.

Acknowledgements

Financial support for this work has been provided by Rolls-Royce Marine through the University Technology Centre in Computational Hydrodynamics hosted at the Department of Mechanics and Maritime Sciences at Chalmers University of Technology. The computations were performed on resources at Chalmers Centre for Computational Science and Engineering (C3SE) provided by the Swedish National Infrastructure for Computing (SNIC).

References

[1] K.-J. Paik, H.-G. Park, and J. Seo, "Rans simulation of cavitation and hull pressure fluctuation for marine propeller operating behind-hull condition," *International Journal of Naval Architecture and Ocean Engineering*, vol. 5, no. 4, pp. 502–512, 2013.

[2] P. Perali, T. Lloyd, and G. Vaz, "Comparison of urans and bem-bem for propeller pressure pulse prediction: E779a propeller in a cavitation tunnel," in *Proceedings of 19th Numerical Towing Tank Symposium, 3rd-4th October, St. Pierre dOléron, France*, 2016.

[3] C. Hsiao and G. Chahine, "Scaling of tip vortex cavitation inception for a marine open propeller," in *27th Symposium on Naval Hydrodynamics, Seoul, Korea*, 2008, pp. 5–10.

[4] A. Asnaghi, R. E. Bensow, and U. Svennberg, "Comparative analysis of tip vortex flow using rans and les," 2017.

[5] K. Fujiyama, "Numerical simulation of ship hull pressure fluctuation induced by cavitation on propeller with capturing the tip vortex," in *Fourth International Symposium on Marine Propulsors*, 2015.

[6] S. Gaggero, G. Tani, M. Viviani, and F. Conti, "A study on the numerical prediction of propellers cavitating tip vortex," *Ocean Engineering*, vol. 92, pp. 137–161, 2014.

© 2018 ASME

10th International Symposium on Cavitation - CAV2018
Baltimore, Maryland, USA, May 14 – 16, 2018

Jetting of Viscous Single Droplets from Cavitation Induced Rayleigh-Taylor Instability

[1]Qingyun Zeng; [1]Silvestre Roberto Gonzalez-Avila; [1]Sophie Ten Voorde; [1,2]Claus-Dieter Ohl*

[1]*Nanyang Technological University, Singapore;* [2]*Otto-von-Guericke University of Magdeburg, Germany.*

Abstract

Cavitation bubbles oscillating inside liquid droplets can produce high speed liquid jets from the surface of the droplets. This type of jetting is induced by the spherical Rayleigh-Taylor instability, in which the radial acceleration is due to the oscillation of an internal cavitation bubble. The experiment is accomplished by levitating the droplet with an acoustic trap and creating a vapor bubble within the droplet with a pulsed laser. Using high speed photography, we observed jets emerging from the droplet surface which pinch off into finer secondary droplets. The phenomenon is then reproduced with a computational fluid dynamics (CFD) simulation model to study the jetting mechanism. Additionally, we developed an analytic model to calculate the droplet surface perturbations growth based on the spherical Rayleigh-Taylor instability, which allows to evaluate the surface stability over a large parameter space. The analytic model predicts correctly the onset of jetting as a function of Reynolds number and normalized internal bubble energy.

Keywords: Jetting, Droplet, Cavitation, Bubble Oscillation, Spherical Rayleigh-Taylor Instability

Introduction

The formation of liquid jet is a very interesting and important phenomenon in fluid mechanics related to many industrial and medical applications [1,2]. Recently Gonzalez-Avila and Ohl [3] reported that cavitation bubble can be used to produce the jetting from single droplets, in which liquid jets were excited from an oscillating spherical liquid surface driven by an internal cavitation bubble. The origin of the jets was related to a Rayleigh-Taylor unstable [4] air-water interface caused by the spherical geometry undergoing rapid radial acceleration; this radial acceleration is caused by the bubble oscillation within the droplet. The surface disturbances grow due to the amplification of Rayleigh-Taylor instability and later form the jets under the strong interaction between the corrugated surface and the pressure wave created by the bubble collapse. However, viscosity inhibits the Rayleigh-Taylor instability thus suppresses the jetting. In this study we focus on the mechanism leading to this type of jetting and the effect of viscosity.

Droplet & Bubble Dynamics

Figure 1(a) shows the experimental setup. The transducer at the top and the spherical reflector create a standing wave at a frequency of 27.4±0.1 kHz to levitate the droplet. A pulsed laser beam is focused inside the droplet and nucleates a cavitation bubble. A high-speed camera records the experiment while the illumination is provided with intense back illumination. The camera and laser are triggered and synchronized with a signal delay generator. Experiment details are described in [3].

The droplet and bubble dynamics are also reproduced with CFD simulations, which are done based on the OpenFOAM framework [5] by solving the compressible Navier-Stokes equations with the Finite Volume Method (FVM) and using the Volume of Fluid (VOF) based method for interface capturing. The complex physics of the laser induced cavitation bubble nucleation is avoided by starting with a spherical non-condensable gas bubble with a small initial radius centered in the droplet. The two-phases gas and liquid are treated as a compressible and immiscible Newtonian fluids while neglecting heat and mass transfer. For more details, see [6].

Figure 1(b) compares the experiment with simulations done in 2D axisymmetric symmetry and in 3D for a water droplet. In the experiments, the droplet is initially an ellipsoid with semi-principal axes of 1.58 mm, 1.17 mm, and 1.58 mm. The cavitation bubble is nucleated by a pulse laser with an energy of 2.2 mJ. 7.1±4.0% of this laser energy is converted into the mechanical energy of the bubble [3]. In simulation we start the gas bubble with an initial radius of 100 μm and pressure of 400 bar, the initial radius and pressure together reflects the bubble energy and we get similar maximum size of the bubbles in simulations as in the experiment. The result of 2D-simulation is shown as the slice through the center of the droplet, in which the bubble dynamics are clearly seen. Due to the axisymmetry,

*Corresponding Author, Claus-Dieter Ohl: claus-dieter.ohl@ovgu.de

the column-like jets in 2D-simulation are not the real jets but is an annular surface while in 3D-simulation the simulated ligaments are thin jets as observed in the experiments. Comparing the results, we find good agreement between the simulations and experiments of the bubble dynamics, and the onset of disturbances and their velocity.

Since all the results depict overall very similar scenario, we now focus on the 2D-simulation which present the clearest shape of bubble and droplet to analyze the jetting dynamics. The cavitation bubble expands and reaches a maximum volume at $t \approx 50$ μs and collapses around 117 μs. During the collapse of the bubble, two liquid jets from the top and bottom of the droplet accelerate to the droplet center. The impact of these two jets makes the bubble collapse into an approximate toroidal shape. At the late stage of the bubble expansion (t = 50 μs), some roughness appears on the upper and lower poles of the droplet surface, in both experiment and simulations, and the height of these ripples increase during the collapse. During the second expansion (t = 117 μs to 175 μs) the troughs on the surface convert into crests and develop into thin ligaments. Dimples now are seen all over the droplet and also develop into ligaments during the third expansion (t = 225 μs to 300 μs), undergoing a same process as the previous ripples. It is noted that the jets are fastest at both poles, at places closest to the bubble which are also the places of highest acceleration.

Overall, the jets are formed when the troughs are accelerated by the re-expansion of bubble, which is similar to the formation of jets under impulsive acceleration of a curved free surface [7]. The obvious difference in our present study is that the initial interface is convex, thus the jetting dynamics undergo two steps: first the droplet grows a corrugated surface due to the Rayleigh-Taylor instability; then the pressure impulse created by the collapse of the bubble and the re-expansion of the bubble accelerate on the troughs of corrugated surface, producing the fine ligaments.

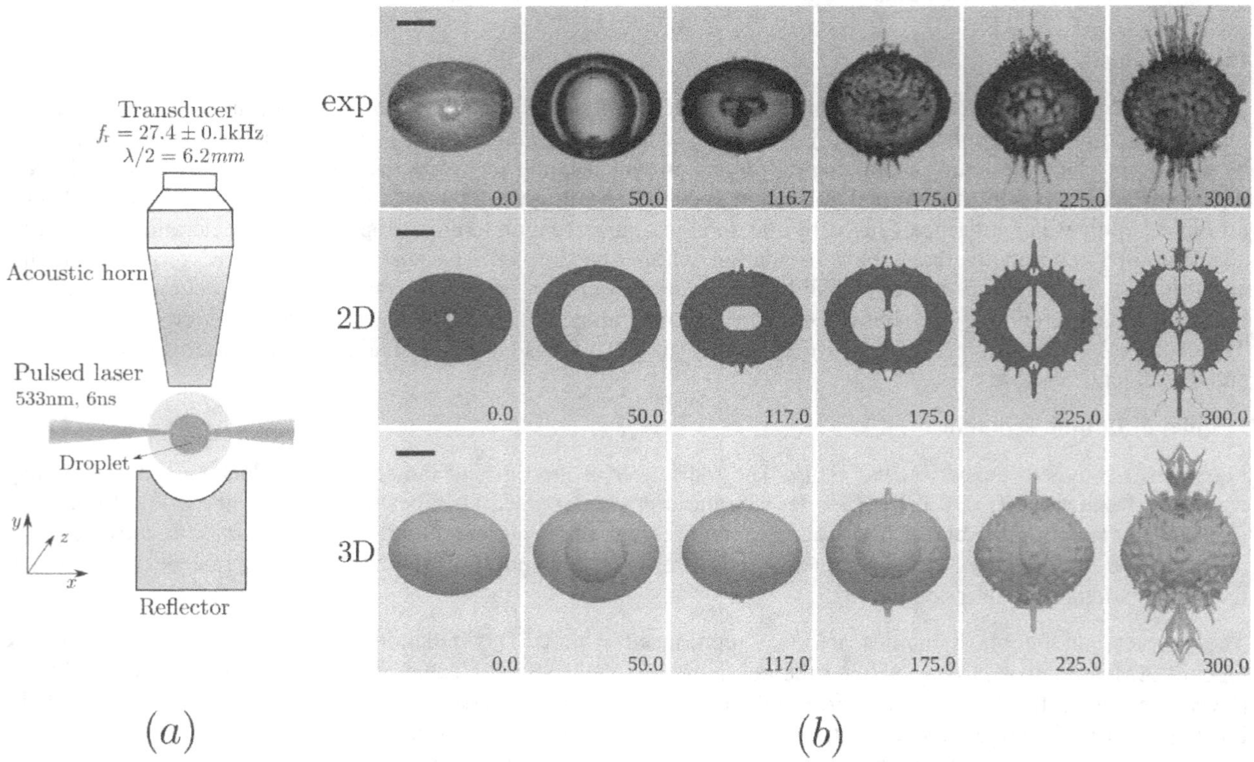

Figure 1 (a) Experimental setup; (b) Comparison of experiment (row 1), 2D axisymmetric (row 2), and 3D simulation (row 3) results of water droplet. Time in μs; the scale bar represents 1 mm.

Viscous effect

Figure 2 summarizes the results of 2d axisymmetric simulations with increasing droplet viscosity. The initial droplet is with semi-principal axes of 1.20 mm, 0.74 mm, and 1.20 mm while the bubble is with an initial radius of 100 μm and pressure of 250 bar. The simulations indicate that the bubble dynamics are almost the same but the jetting dynamics is very different: fast jets occur in 10cst and 20cst oil droplet while no jetting is seen in 50cst oil droplet. If we look carefully the surface become smoother as viscosity increases, which means viscosity has a pronounced effect on the stability on this spherical R.T. instability. Next we will study more details on this viscous effect with an analytic model.

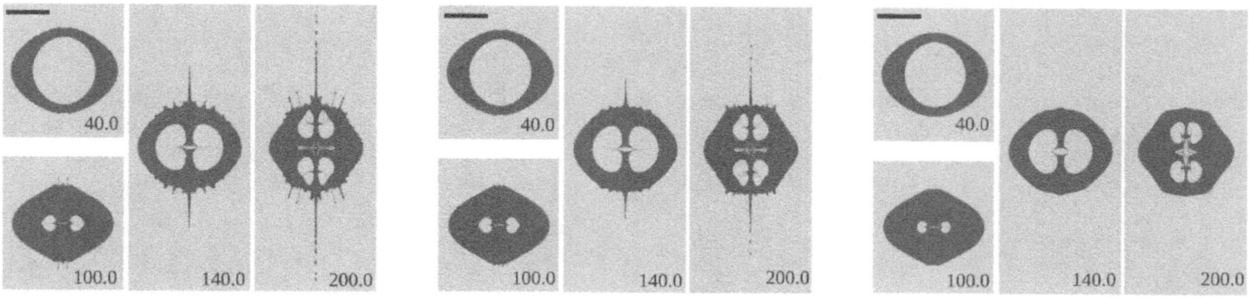

(a) 10cst (b) 20cst (c) 50cst

Figure 2 Comparison of 2D axisymmetric simulation results under different viscosity and same initial bubble energy: (a) 10cst oil droplet; (b) 20cst oil droplet; (c) 50cst oil droplet; Time in μs; the scale bar represents 1 mm.

Analytic Model

Firstly, we sketch the problem as follow: the droplet is a sphere with a radius of R, and density ρ and dynamic viscosity μ; the outer interface is surrounded by a lighter fluid, i.e. the surrounding gas, whose density and viscosity are neglected; a bubble of initial radius R_{b0} and uniform high pressure P_0 is located at the center of the droplet. The droplet and the surrounding gas are treated as immiscible and incompressible fluids. The bubble R_b and droplet R(t) radii are related through conservation of volume: $R(t)^3 - R_b(t)^3 = R_0^3 - R_{b0}^3$. Thus, the spherical bubble dynamics within a droplet can be obtained from the unsteady Bernoulli equation similar to [8]:

$$\frac{P_b - P_\infty}{\rho} = \frac{3}{2}\dot{R_b}^2 + R_b\ddot{R}_b + \frac{4\mu}{R_b\rho}\dot{R}_b + \frac{2\sigma}{\rho R_b} + \frac{2\sigma}{\rho R} - 2\dot{R_b}^2\lambda - R_b\ddot{R}_b\lambda + \frac{1}{2}\dot{R_b}^2\lambda^4$$

where $\lambda = R_b(t)/R(t)$, and σ is the surface tension coefficient.

To evaluate the stability of the droplet's spherical surface, i.e. the stability of the surface from which the jets are ejected from, we assume the droplet is with a surface perturbation, which could be modeled as a spherical harmonic Y_n with mode n, i.e. $r_s(t) = R(t) + a(t)Y_n$, a(t) is the amplitude of the perturbations, starting from some small initial value $a(t = 0) = 0.001R(t = 0) \ll R(t = 0)$. The evolution equation of this perturbation amplitude to the first order has been derived by Prosperetti [9] for incompressible and viscous spherical flows, and we simplify it as follow:

$$\ddot{a_n} + B_n(t)\dot{a_n} - A_n(t)a_n = 0$$

with

$$A_n(t) = -(n+2)\frac{\ddot{R}}{R} - (n-1)n(n+2)\frac{\sigma}{\rho R^3} - (n-1)(n+2)\frac{2\mu\dot{R}}{\rho R^3}$$

and

$$B_n(t) = \frac{3\dot{R}}{R} + \frac{2(n-1)(2n+1)\mu}{\rho R^2}$$

The parameters affecting jetting are bubble energy, droplet size, liquid viscosity, surface tension, and the ambient pressure. We can formulate three dimensionless quantities: normalized energy E, Reynolds number Re, and Weber number We: $E = \frac{E_{bo}}{V_{do}P_\infty} = \frac{V_{bo}P_{bo}}{V_{do}P_\infty} = \frac{R_{bo}^3 P_{bo}}{R_{do}^3 P_\infty}$, $Re = \frac{\rho U_{ref} R_{do}}{\mu}$, $We = \frac{\rho U_{ref}^2 R_{do}}{\sigma}$.

The stability of the droplet surface is quantified by calculating the growth of the small initial surface perturbation with mode n of 16. Figure 3(a) depicts in the upper graphs the radius-time curve of the bubble and the droplet surfaces and in the lower graph the non-dimensional disturbance. In figure 3(a) we show the perturbation growth varying the viscosity and keep the energy constant. The analytic results correspond to the results in figure 2 while we are using a spherical droplet with a radius of 1mm, which has the same volume as the droplet in figure 2.

For the 10cst oil droplet in figure 3(a) solid line, the normalized perturbation drops from 0.1% (see inset) during the early growth of the bubble, and then grows to 1.2% during the late expansion and the early collapse. Later on the perturbation decrease to 1.1% at the moment when bubble reaches its minimum volume. However, this incompressible model is only appropriate for the first bubble oscillation cycle since it does not account for acoustic wave emitted at the bubble collapse and their interaction with the surface. From the CFD simulation we know that it is the acoustic wave and the re-expansion of the bubble accelerate the troughs of the droplet surface into thin jets. Thus the relative amplitude $|a(t = t_c)|/R_d$ grown by the R.T. instability at time of bubble collapse, $t = t_c$ become the key quantity to evaluate whether the disturbances of the corrugated droplet surface is large enough to eject the jets. The droplet and bubble dynamics are very similar for all three cases, however the amplitude for 20cst oil has markedly decreased to 0.8% and the instability of 50cst oil is nearly fully prevented, which coincides the simulations.

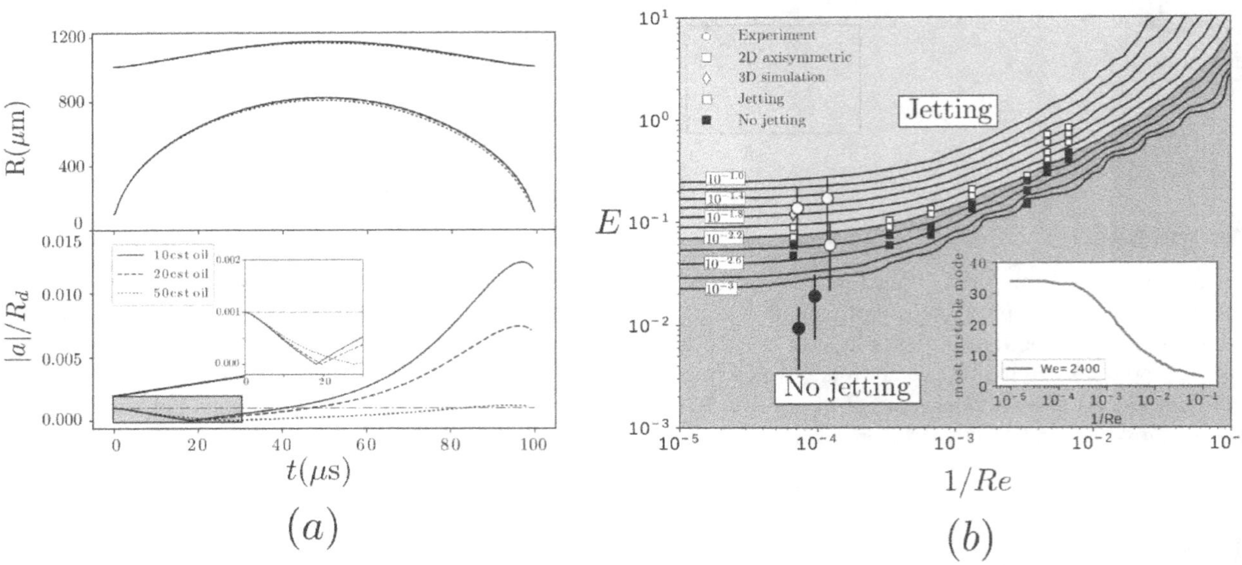

Figure 3 (a) The radius and normalized perturbation amplitude. Radius time traces (top) and non-dimensional perturbation amplitude (bottom) of droplets with varying viscosity and constant normalized energy E=0.4: for water Re=10000, for10cSt oil Re=1000, for 100cSt oil Re = 100. The mode is 16 for all the cases in this figure. The dash-dot line indicates the origin amplitude of the perturbation. (b) Jetting regimes. The parameter map consists of 1/Re in horizontal axis and the normalized internal energy E in vertical axis. The contour represents the normalized amplitude of surface perturbations $|a|/R_d$ at the bubble collapse in logarithm scale. The circle symbols are experiment data; square symbols are 2D axisymmetric simulation results; diamond symbol are 3D simulation results. The symbols filled are the cases of jetting while the unfilled are the cases of no jetting. Inset is the most unstable mode vs. 1/Re under We=2400.

Before we use the model to predict the jetting regime, we assume that jetting occurs once the surface perturbations are larger than some threshold needed for pressure impulse driven jetting. The parameter map is shown in figure 3(b) for the case We = 2400, which consists of inverse Re number in the horizontal direction and the normalized energy E in the vertical axis. Experimental results, simulation results, and the analytical results are included in the maps. Moreover, the most unstable mode is a function of We number (for more see [6]) only is shown in the inset in figure 3(b) for the same inverse Re number regime as in the main plot. Normalized amplitude of the surface perturbation with the most unstable mode at minimum bubble volume are plotted as contour with 11 lines between 0.001 and 0.1. Comparing all the results, the threshold of $|a|/R_d = 10^{-2.2} \approx 0.63\%$ seems to provide a good criterion between jetting and no-jetting. The required energy of the transition level is increasing from small 1/Re to large values, as expected that less energy is needed for low viscosity droplets to induce jetting. Interestingly, for small 1/Re the energy required for jetting remains approximately constant, indicating that surface tension and not viscosity is the limiting parameter stabilizing the droplet. Yet a further increase, i.e. from $1/\mathrm{Re} \approx 10^{-4}$ the energy grows approximately linearly in the double logarithmic plot in figure 3(b).

Conclusion

The oscillation of cavitation bubble inside a droplet could induce jetting from droplet surface. The droplet surface becomes corrugated under the cavitation induced Rayleigh-Taylor instability. The pressure impulse created by the collapse of the bubble then focuses the troughs of the perturbed surface into thin jets. 2D axisymmetric and 3D CFD simulations were performed, good agreement of the bubble and jetting dynamics are found between experiments and simulations. The analytic solution including viscous effect re-derived from Prosperetti's work [9] well describes the instability dynamics and predicts the onset of jetting. Surface tension and not viscosity is found as the limited parameter stabilizing the droplet. The results may provide a pathway to explore flashing sprays, e.g. sprays consisting initially of large droplets which are governed initially by inertia and through some physical or chemical mechanism become surface unstable and eject fine droplets for a good surface coverage.

References

[1] Eggers, J. (1997). *Nonlinear dynamics and breakup of free-surface flows*. Reviews of modern physics, 69(3)

[2] Eggers, J., Villermaux, E. (2008). *Physics of liquid jets*. Reports on progress in physics, 71(3)

[3] Gonzalez-Avila, S.R., Ohl, C.D. (2016). *Fragmentation of acoustically levitating droplets by laser-induced cavitation bubbles*. Journal of Fluid Mechanics, 805

[4] Taylor, G. (1950). *The instability of liquid surfaces when accelerated in a direction perpendicular to their planes. I.* In Proceedings of the Royal Society of London A (Vol. 201, No. 1065). The Royal Society.

[5] Weller, H.G., Tabor, G., Jasak, H., Fureby, C. (1998). *A tensorial approach to computational continuum mechanics using object-oriented techniques*. Computers in physics, 12(6).

[6] Zeng, Q., Gonzalez-Avila, S.R., Voorde, S.T., Ohl, C.D., *Jetting of Viscous Droplets from Cavitation Induced Rayleigh-Taylor Instabilities*. Submitted to Journal of Fluid Mechanics.

[7] Antkowiak, A., Bremond, N., Le Dizes, S., Villermaux, E. (2007). *Short-term dynamics of a density interface following an impact*. Journal of Fluid Mechanics, 577.

[8] Obreschkow, D., Kobel, P., Dorsaz, N., De Bosset, A., Nicollier, C., Farhat, M. (2006). *Cavitation bubble dynamics inside liquid drops in microgravity*. Physical review letters, 97(9).

[9] Prosperetti, A. (1977). *Viscous effects on perturbed spherical flows*. Quarterly of Applied Mathematics, 34(4).

© 2018 ASME

10th International Symposium on Cavitation - CAV2018
Baltimore, Maryland, USA, May 14 – 16, 2018 CAV18-05133

Large eddy simulation of a collapsing vapor bubble containing non-condensable gas

Theresa Trummler*; Lukas Freytag; Steffen J. Schmidt and Nikolaus A. Adams

Chair of Aerodynamics and Fluid mechanics, Department of Mechanical Engineering, Technical University of Munich

Abstract

We investigate numerically the effect of non-condensable gas inside a vapor bubble on the bubble dynamics and the collapse pressure. Free gas in the vapor bubble has a cushioning effect that can weaken the pressure wave and enhance the bubble rebound. In order to access this effect numerically, simulations of collapsing vapor bubbles containing non-condensable gas are performed. In the simulations, the effects of the gas on the rebound and the shockwave energy are monitored for different operating points with varying initial gas contents $p_{g,0}$ inside the bubble and driving pressures Δp. For the cavitating liquid and the non-condensable gas we employ a homogeneous mixture model with a coupled equation of state for all components and the cavitation model is a barotropic thermodynamic equilibrium model, which is embedded in a higher order implicit large eddy approach for narrow stencils. Compressibility of all phases is considered, to capture the shockwave of the bubble collapse. The effect of the free gas on the rebound and the dampening of the emitted shockwave by the gas contained in the bubble are well captured by our simulations.

Keywords: bubble collapse; vapor bubble containing gas; energy partitioning; rebound; shock wave energy

Introduction

Liquids may contain dissolved gases that can be set free by pressure reduction or cavitation, leading to the presence of gas in vapor cavities. In experiments of bubble collapses the gas content in bubbles was shown to depend on whether they are generated using a laser or a spark and bubble dynamics were found to be influenced by this [1]. Gas inside the vapor bubble has a cushioning effect that can weaken the pressure wave and enhance bubble rebound. For spherical bubble collapses, this has already been investigated analytically and experimentally. However, for more complex configurations, the effect of the gas is not yet clarified and could be investigated with three-dimensional time resolved numerical simulations. Thus, the presented work serves as a validation of our modeling approach and a basis for later investigations.

In a vapor bubble containing free gas the pressure is $p = p_{\text{vap}} + p_g$, where p_{vap} denotes the vapor pressure and p_g the partial pressure of the gas. p_g depends on the compression of the bubble and assuming an adiabatic transformation of the gas it can be described with $p_g(t) = p_{g,0} \cdot (R/R_0)^{1/\gamma}$. The index 0 indicates initial values and γ is the adiabatic index.

Tinguely et al. [2] experimentally and theoretically investigated the effects of the driving pressure difference $\Delta p = p_\infty - p_{\text{vap}}$ and the gas content $p_{g,0}$ on bubble dynamics and shockwave emission. For their theoretical model they used an inviscid Keller Miksis model [3] taking advantage of the fact that it can be treated to first order [4]:

$$\rho((1-v)\ddot{R}R + 3/2\dot{R}^2(1-v)) = (1-v)(p_g - \Delta p) + R\dot{p}_g/c \tag{1}$$

with $v = \dot{R}/c$. Based on this model and their experimental measurements, they could show that the initial energy of a bubble $E_0 = \frac{4\pi}{3}R_0^3\Delta p$ [5] partitions mainly into rebound energy $E_{\text{reb}} = \frac{4\pi}{3}R_{\text{reb}}^3\Delta p$ and shockwave energy $E_{\text{SW}} = \frac{4\pi d^2}{\rho c}\int p(t)^2 dt$ [6]. In addition to this, they postulated that the energy partitioning between rebound ε_{reb} and shock wave energy ε_{sw} depends on a single parameter

$$\xi = \frac{\Delta p \gamma^6}{{p_{g,0}}^{1/\gamma}(\rho c^2)^{1-1/\gamma}}, \tag{2}$$

where ρ, c are density and speed of sound in the liquid. The shockwave energy increases with ξ and thus with the driving pressure difference and decreases with the partial pressure of free gas. On the other hand, rebound is enhanced for a lower driving pressure difference and a higher gas content.

The aim of the presented work is to numerically investigate and reproduce the findings of Tinguely et al. [2] and to thereby validate our gas modeling approach.

*Corresponding Author, Theresa Trummler: *theresa.trummler@tum.de*

Thermodynamic model

We employ a multi-component homogeneous mixture model [7]. The cavitating liquid (lv) and the non-condensable gas (g) are described with one mixture fluid, which is defined by the volume averaged density inside a computational cell $\rho = \sum \beta_\phi \rho_\phi$. β_ϕ denotes the volume fraction and ρ_ϕ the density of each component $\phi = \{\text{lv}, \text{g}\}$. This single fluid approach implies that within a computational cell all phases have the same velocity, temperature and pressure. Since the pressure in the bubble is $p = p_{\text{vap}} + p_{\text{g}}$, the pressure acting on the vapor is $p_{\text{vap}} = p - p_{\text{g}} = (1 - \beta_{\text{g}})\,p$. A coupled equation of state (EOS) gives the pressure as a function of the mean density $p = p(\rho)$ and is obtained by expressing the densities ρ_ϕ with the corresponding thermodynamic relations.

The cavitating water is described with a barotropic EOS, derived by integration of the isentropic speed of sound $\rho_{\text{lv}} = \rho_{\text{sat,liq}} + (p_{\text{lv}} - p_{\text{sat}})/c^2$, where $\rho_{\text{sat,liq}}$ is the saturation density for liquid water and p_{sat} the saturation pressure. Phase change is modeled assuming local thermodynamic equilibrium. For $p > p_{\text{sat}}$, there is purely liquid water and $c = 1482.35\,\text{m/s}$. For $p < p_{\text{sat}}$, there is a liquid vapor mixture with $c = 0.1\,\text{m/s}$ as a typical value for an equilibrium isentrop.

The non-condensible gas phase is described with $\rho_{\text{g}} = \rho_{\text{g,ref}}(p/p_{\text{ref}})^{1/\gamma}$, where $\rho_{\text{g,ref}}$ is the reference density at the reference pressure p_{ref}. In the presented results the gas is modeled isothermal with $\gamma = 1$.

Viscous effects are considered in our simulations using a linear blending with volume fractions for the mixture viscosity.

Numerical Method

The thermodynamic model is embedded in a density-based fully compressible flow solver. For the numerical model a higher order implicit large eddy approach for narrow stencils [8] is utilized. In implicit SGS the truncation error of the discretization serves as a subgrid-scale model for turbulence. Shockwaves and pseudo phase boundaries are detected with a sensor and in such regions an upwind based second order scheme is used. In smooth regions, on the other hand, a high order central discretization scheme with regularization terms is applied. Time integration is done using a 4-stage Runge-Kutta method with the Courant Friedrichs Lewy (CFL) number set to 1.4.

In [8], simulations of a spherical single vapor bubble collapse have been performed, comparing a more dissipative scheme and the applied one. The higher order discretization scheme was chosen for this work, since it resolves the pressure peaks more accurately.

Setup

A vapor bubble with an initial radius R_0 of $4 \cdot 10^{-4}$ m is placed in the center of a box with dimension $2\,L$ in each direction, where $L = 0.1$ m. Taking advantage of the symmetry, only an eighth of a vapor bubble is simulated. The domain is discretized with an equidistant grid within a cubic sub-domain with an edge length of $1.25\,R_0$ and for the outer part a grid stretching is applied. Simulations are performed on two different grids with 20 (coarse) and 80 (fine) computational cells over the initial bubble radius $1.25\,R_0$.

For the investigation, the initial gas content in the bubble $p_{\text{g},0}$ and the driving pressure difference $\Delta p = p_\infty - p_{\text{sat}}$ are varied covering different combinations of $\Delta p = \{10\,\text{kPa}, 30\,\text{kPa}, 80\,\text{kPa}\}$ and $p_{\text{g},0} = \{0\,\text{Pa}, 100\,\text{Pa}, 1000\,\text{Pa}\}$. During the simulations, the pressure signal is recorded by pressure probes placed in radial direction between $0.25\,R_0$ and $1.5\,R_0$ with a distance of $0.25\,R_0$.

Results

We mainly focus on the operating points $\Delta p = 10$ kPa with $p_{\text{g}} = 0$ Pa and $p_{\text{g}} = 100$ Pa and its ξ-equivalent $\Delta p = 100$ kPa; $p_{\text{g}} = 1000$ Pa. Fig. 1 (a) compares the temporal evolution of the normalized bubble radius R/R_0 for different gas contents and grid resolutions. The time is normalized to the Rayleigh collapse time $\tau_C = 0.915 \cdot R_0 \sqrt{\rho/\Delta p}$ [9]. Without gas, there is a very small rebound after the shockwave has propagated outwards. In configurations with gas the bubbles rebound significantly, and rebound increases with grid resolution, nearly matching the theoretical inviscid prediction of 42.7 % in maximum relative rebound radius at 41% for the fine grid. Moreover, the results of the ξ-equivalent set-up on the coarse grid are shown and verify that there is a ξ - similarity for the rebound.

Besides the rebound, the non-condensable gas in the vapor bubble also affects the intensity of the emitted pressure wave. Fig. 1(b) shows the monitored pressure for the probes at different positions for different gas contents and grid resolutions. For all configurations, the radial decay of the maximum pressure is clearly visible and the presence of gas reduces the maximum pressure. The damping effect of the gas is, in general, more distinct for probes closer to the bubble center and stronger on the finer grid resolution. Additionally, the pressure signals reveal that the collapse time is perfectly matched. On the coarse grid the collapse time is slightly smaller, which is in agreement with the observations of [8]. The bubbles containing gas collapse later since the effective pressure difference is reduced by the gas.

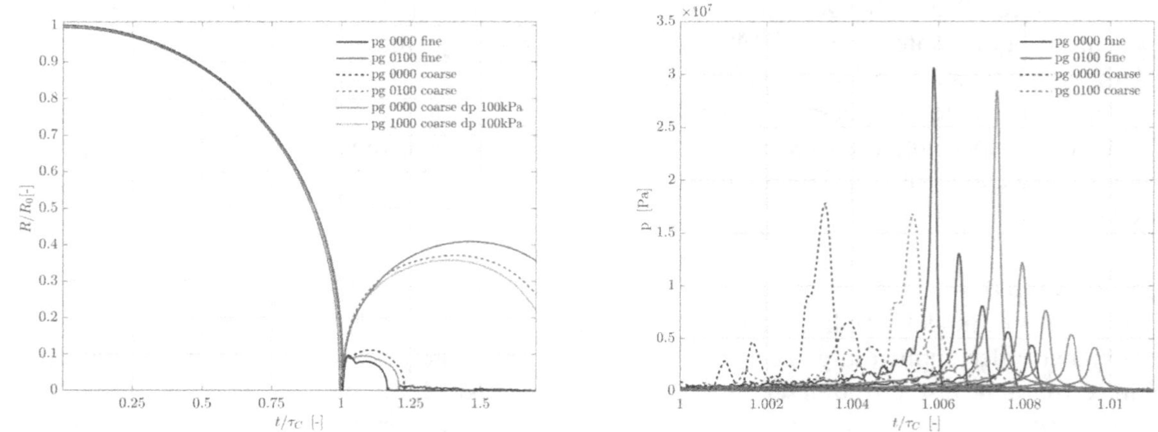

(a) Collapse dynamics of different configurations. (b) Pressure signals from the probes $0.5\,R_0$-$1.5\,R_0$

Figure 1. Simulation results

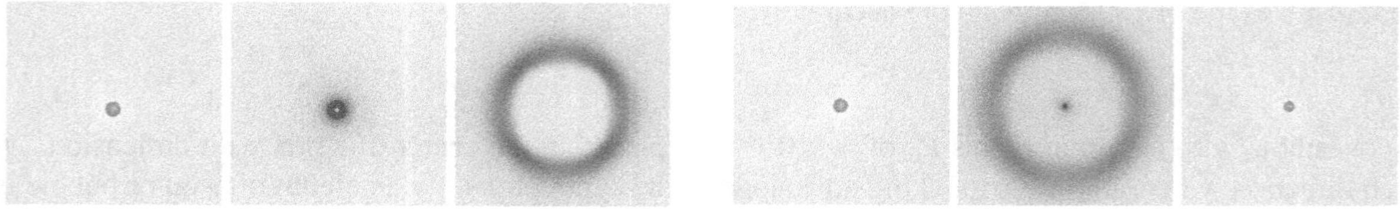

(a) $t = 0\ \tau_c$ $p_{g,0} = 0\,\text{Pa}$ (b) $t = 0.99\ \tau_c$ (c) $t = 1.02\ \tau_c$ (d) $t = 0\ \tau_c$ $p_{g,0} = 1000\,\text{Pa}$ (e) $t = 1.05\ \tau_c$ (f) $t = 1.44\ \tau_c$

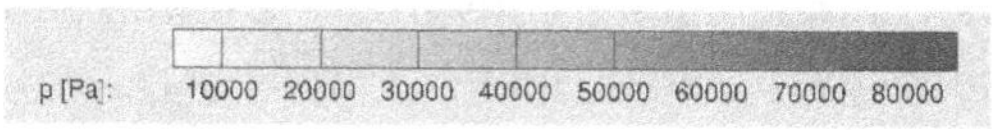

Figure 2. Time series for $\Delta p = 10\text{kPa}$ with $p_{g,0} = 0$ and 1000Pa.

Fig. 2 depicts the bubble collapse and the rebound at different time steps for $\Delta p = 10\text{kPa}$ with $p_{g,0} = \{0\ \text{Pa}, 1000\ \text{Pa},\}$. The left time series presents the bubble collapse without gas, showing the initial bubble, the situation shortly before the collapse and the emitted shockwave after collapse. Analogously, the dynamics of a bubble with a high gas content is visualized in the right time series. In this case the rebound is clearly visible at $t = 1.44\ \tau_c$

In order to evaluate the energy partitioning, the normalized rebound energy ε_{reb} is obtained from the maximal radius of the bubble in the first rebound $\varepsilon_{\text{reb}} = (R_{\text{reb}}/R_0)^3$. For the normalized shockwave energy ε_{sw}, the pressure signals are numerically integrated and set in relation to the respective values without gas.

Fig. 3 compares the simulation results with the theoretical energy partitioning based on the solution of the adapted Keller-Miksis equation Eq. 1 over ξ. Our results are in good agreement with the theoretical ones and grid refinement leads to a better approximation to the theoretical model. Our simulation results confirm a ξ equivalence. The rebound for the ξ-equivalent configurations is the same, the normalized shockwave energy is higher for the higher pressure difference and better matches the theoretical model.

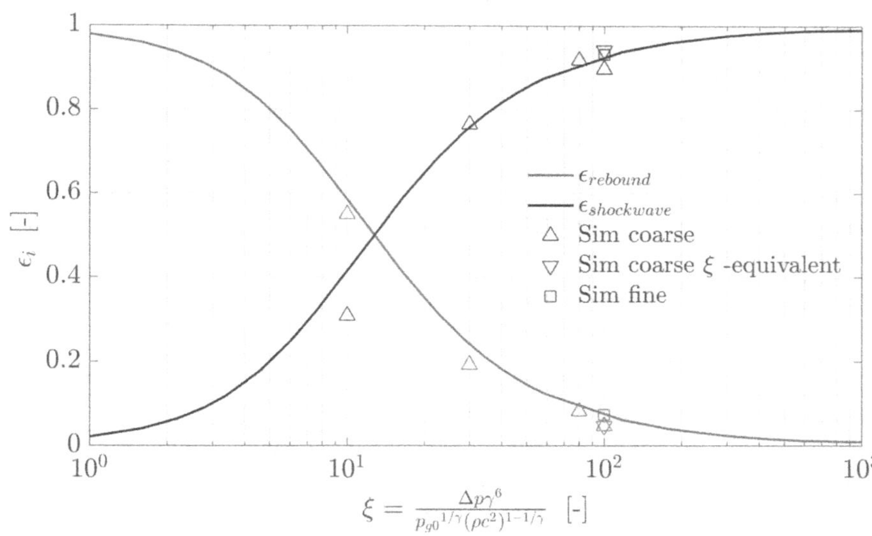

Figure 3. Simulation results in comparison with the theoretical energy partitioning proposed by Tinguely et al. [2].

Conclusion

With our simulations, we could reproduce the physical effects of gas inside a vapor bubble. Free gas in a vapor bubble leads to a stronger rebound and dampens the emitted shockwave. This effect is already visible on coarse grid resolutions, but gets more pronounced for higher grid resolutions. Additionally, we were able to reproduce the partitioning into rebound and shockwave energy proposed by Tinguely et al. [2] and could confirm a $\xi = (\Delta p \gamma^6)/(p_{g,0}{}^{1/\gamma}(\rho c^2)^{1-1/\gamma})$ equivalence, i.e. same ξ-values result in the same rebound behavior and reduction of the shockwave energy. This validation allows us to investigate the effect of free gas on more complex configurations such as a bubble collapse at a wall.

References

[1] Sato T, Tinguely M, Oizumi M and Farhat M 2013 *Applied Physics Letters* **102** 074105–5
[2] Tinguely M, Obreschkow D, Kobel P, Dorsaz N, de Bosset A and Farhat M 2012 *Physical Review E* 1–7
[3] Keller J B and Miksis M 1980 *The Journal of the Acoustical Society of America* **68** 628–633
[4] Prosperetti A 1987 *Physics of Fluids* **30** 3626–4
[5] Obreschkow D, Kobel P, Dorsaz N, De Bosset A, Nicollier C and Farhat M 2006 *Physical review letters* **97** 094502
[6] Vogel A and Busch S 1996 *The Journal of the Acoustical Society of America* 1–18
[7] Örley F, Trummler T, Hickel S, Mihatsch M S, Schmidt S J and Adams N A 2015 *Physics of Fluids* **27** 086101–28
[8] Egerer C P, Schmidt S J, Hickel S and Adams N A 2016 *Journal of Computational Physics* **316** 453–469
[9] Rayleigh L 1917 *The London, Edinburgh, and Dublin Philosophical Magazine and Journal of Science* **34** 94–98

© 2018 ASME

10th International Symposium on Cavitation - CAV2018
Baltimore, Maryland, USA, May 14 – 16, 2018 CAV18-05134

Large Eddy simulation of cavitating nozzle flows and primary jet break-up with gas-entrainment into the nozzle

Theresa Trummler*; Daniel Rahn; Steffen J. Schmidt and Nikolaus A. Adams

Chair of Aerodynamics and Fluid mechanics, Department of Mechanical Engineering, Technical University of Munich

Abstract

We present results of a Large Eddy simulation of a cavitating nozzle flow and its injection into gas. The focus of the investigation is set on gas-entrainment into the nozzle, which occurs when vapor reaches the nozzle outlet and gas enters the nozzle from the outflow region driven by the pressure gradient. The numerical setup is based on a reference experiment, where liquid water is discharged into ambient air through a step nozzle. For the cavitating liquid a barotropic equilibrium cavitation model is utilized, which is embedded in a homogeneous mixture model for the liquid-vapor mixture and the non-condensable gas phase. Full compressibility of all components is taken into account by the high order implicit Large Eddy approach and thus enables us to capture the effects of collapsing vapor structures. Four different operating points are investigated. Special emphasis is put on the operating points with gas-entrainment and its effect on the jet characteristics. Additionally, mass flux and erosion are analyzed.

Keywords: multi-phase-flow; cavitation; jet; primary jet break-up; gas-entrainment; mass flux; erosion

Introduction

Cavitation takes place in the nozzle of pressure atomizers, such as fuel injectors, and can have various effects. The vigorous collapse of vapor structures can lead to damage of the injector. Enhancement of primary jet breakup, on the other hand, can be an advantageous and desired effect of cavitation in nozzles. This interaction between cavitation in the nozzle and the jet characteristics has already been investigated experimentally by e.g. [1] and numerically by [2, 3]. Another phenomenon which can occur in cavitating injection systems is gas-entrainment in the nozzle. When vapor reaches the nozzle outlet, there is a pressure gradient from the outflow region to the vapor region and gas is sucked into the nozzle. Methods applied in experiments generally distinguish between liquid and gaseous phase and thus provide no information on gas in the nozzle. Using a fully compressible implicit Large Eddy Simulation (LES) code, we numerically investigate effects of gas-entrainment in the nozzle on the jet characteristics, the mass flux and the cavitation erosion. All three aspects are important for example in fuel injectors as spray break up and mass flux are crucial parameters for combustion and erosion affects the durability of the components.

Method

In order to access these mutual interactions of cavitation and gas-entrainment with the jet, we employ a multi-component cavitation model as introduced by Örley et al. [2]. Cavitating liquid and non-condensable gas are represented with a homogeneous mixture approach. For the thermodynamic closure relation, a barotropic equation of state for the cavitating liquid is coupled with the ideal isothermal gas law for the gas component over the averaged density. The cavitation model is based on the assumption of local thermodynamic equilibrium and has been successfully validated for wall bounded LES simulations [4].

The thermodynamic model is embedded in a density-based fully compressible flow solver, which uses a higher order implicit LES approach for narrow stencils [5]. In implicit LES the truncation error of the discretization serves as a subgrid-scale model for turbulence. Shockwaves and pseudo phase boundaries are detected with a sensor and in this region an upwind based second order scheme is used. Whereas in smooth regions, a high order central discretization scheme with regularization terms is applied. For the investigated multi-component flow also the accuracy of the gas transport has been increased with a thermodynamic coupling of the numerical flux function. Which means that the transported gas is consistent with the reconstructed density and the pressure at the cell face.

*Corresponding Author, Theresa Trummler: *theresa.trummler@tum.de*

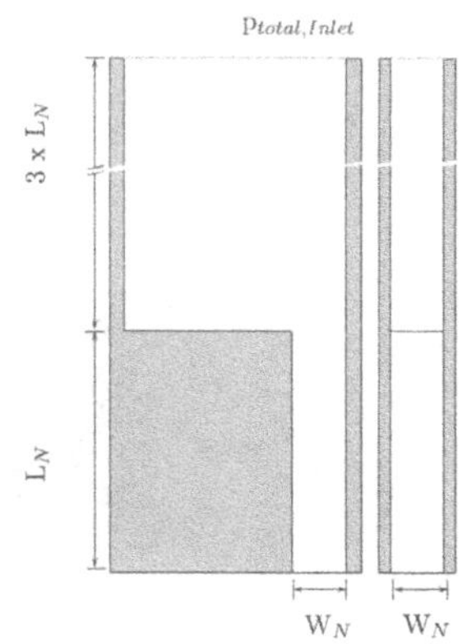

(a) Setup

Figure 1. Sketch of the numerical setup.

Setup

The numerical setup is based on the experiments performed by Sou et al. [6] and Bicer and Sou [7], where water is discharged trough a large-scale step nozzle into ambient air. A sketch of the numerical setup is depicted in Fig. 1. The nozzle cross-section is quadratic with $W_N = 1.94$ mm and the length is $L_N = 8.0$ mm. Applying the same approach as Koukouvinis et al. [8] the upstream region has been elongated and at the inlet a total pressure determined in preliminary studies is prescribed. The outflow region was initialized with gas and $p_{\text{outlet}} = 1 \cdot 10^5$ Pa. Four operating points were simulated covering different cavitation and jet regimes. The operating points are defined with the cavitation number $\sigma = (p_{\text{outlet}} - p_{\text{sat}})/(0.5\,\rho\,\bar{u}^2)$ and determined by the mean velocity $\bar{u}$, which is tuned with the total inlet pressure.

The domain is discretized on a block-based, structured mesh. To reduce the computational costs a grid sequencing strategy was applied using four different grid levels, where the finest grid has 51.5 million cells and the smallest cell in wall-normal direction is 2.5 μm . The grid is also refined around the sharp edge at the inlet and the outlet area as well as in the region of the shear layer and in the outlet region of the nozzle where primary jet break-up occurs. For the simulations the Courant Friedrichs Lewy (CFL) number is set to 1.4, resulting in a timestep of approximately $0.7 \cdot 10^{-9}$ s on the fine grid. The statistical averaging on the fine grid was performed over 4 ms, which corresponds to roughly 6 flow through times. During the simulation the mass flux and the average velocity at the nozzle inlet and outlet was monitored, as well as the vapor content in the nozzle area and in the whole domain.

Results

Fig. 2 compares time steps from the simulation with the experimental data [6]. For the operating points with the lower cavitation numbers $\sigma = \{0.84, 0.73\}$ the applied approach with the total pressure inlet boundary condition led to a little too low average velocity in the nozzle. Thus, the cavitation number from the experiment could not be matched completely. In general, there is a good agreement of the cavitation pattern and the jet characteristics. For the higher cavitation numbers $\sigma = \{1.19, 0.94\}$ the experiment predicted *developing cavitation*. In the simulations, this cavitation pattern can be well reproduced as well as the jet characteristics, albeit the cavitation is slightly over-predicted for both cases compared to the experimental data. At $\sigma = 0.84$ the vapor regions collapse close to the nozzle outlet and a more disturbed jet surface and significant increase of the jet angle are visible. This is also observed in the corresponding experimental operating point. In the case of $\sigma = 0.73$ the simulation is in the transition to the hydraulic flip, whereas in the experiment the cavitation number was lower and the flip occurred.

In order to analyze the effects of the gas-entrainment in detail, time series for $\sigma = 0.84$ and $\sigma = 0.73$ are shown in Fig. 3. In the top the cavitation in the nozzle, the entrained gas and the jet are visualized for

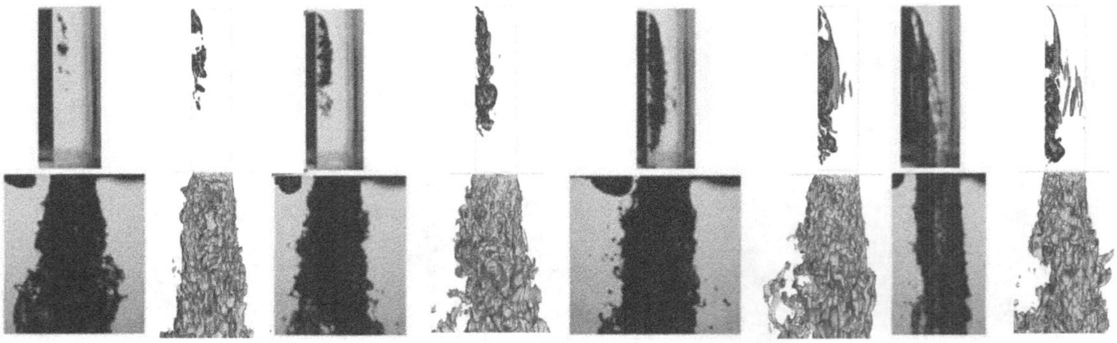

(a) $\sigma = 1.19$ (b) $\sigma = 0.94$ (c) $\sigma = 0.84^{*}$ (d) $\sigma = 0.73^{**}$

Figure 2. Comparison of the experimental data (reprinted from Sou et al. [6], with permission from Elsevier) (left) and the simulation results adopted from [9] (right) (Isosurfaces: $\alpha = 0.1$ and jet surface: $\beta_G = 0.75$) $* \ \sigma_{Experiment} = 0.83$ $** \ \sigma_{Experiment} = 0.71$

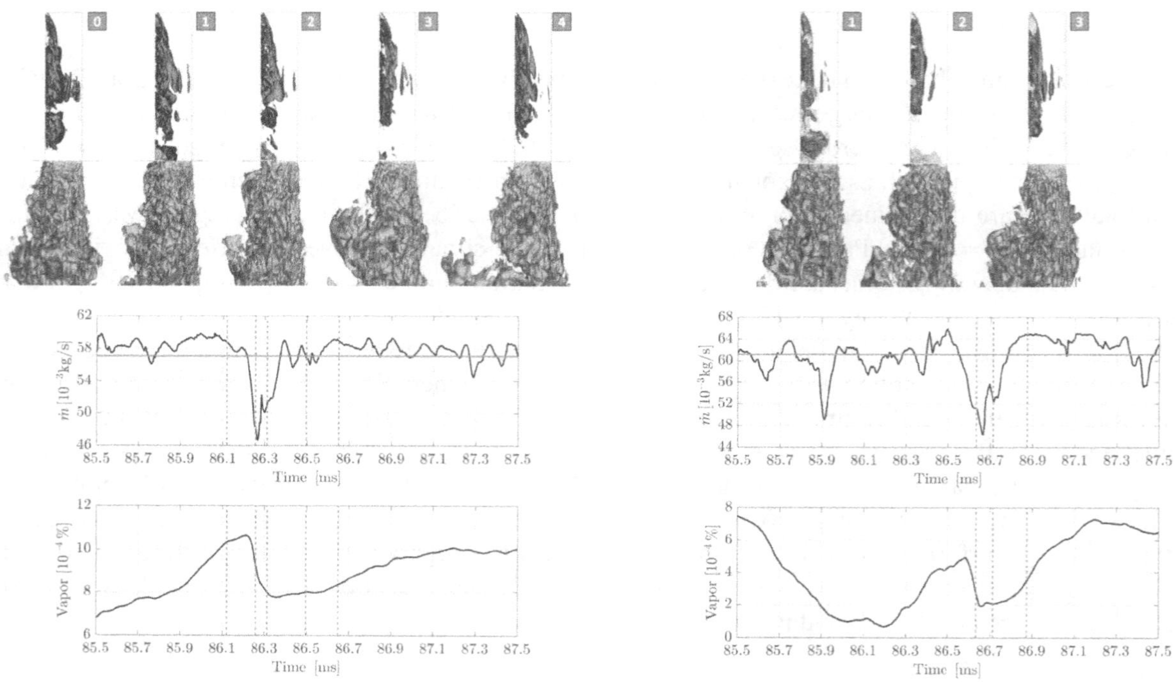

(a) $\sigma = 0.84$ (b) $\sigma = 0.73$

Figure 3. Time series for $\sigma = 0.84$ and $\sigma = 0.73$. Top: Time series showing the vapor in the nozzle (grey,10%), the jet and the gas sucked into the nozzle (green, 25%), adopted from [9]. Bottom: Mass fluxes at the nozzle outlet and the averaged mass flux and the vapor content in the nozzle. Depicted timesteps are marked with vertical dashed lines.

selected timesteps. In the bottom the monitored mass flux at the nozzle outlet and the vapor content in the nozzle are shown. The timesteps are marked with vertical lines. In the time series of $\sigma = 0.84$ the following can be observed: In the timestep **1** a vapor structure reaches the outlet and gas is sucked into the nozzle due to the pressure gradient; **2** the jet surface bulges where the gas is sucked in. This means that the spray angle near the outlet is suddenly strongly increased. **3** The jet bulge is transported downstream and bursts away. **4** The jet bulge detaches more. This effect was also described and observed in previous simulations [2, 3].

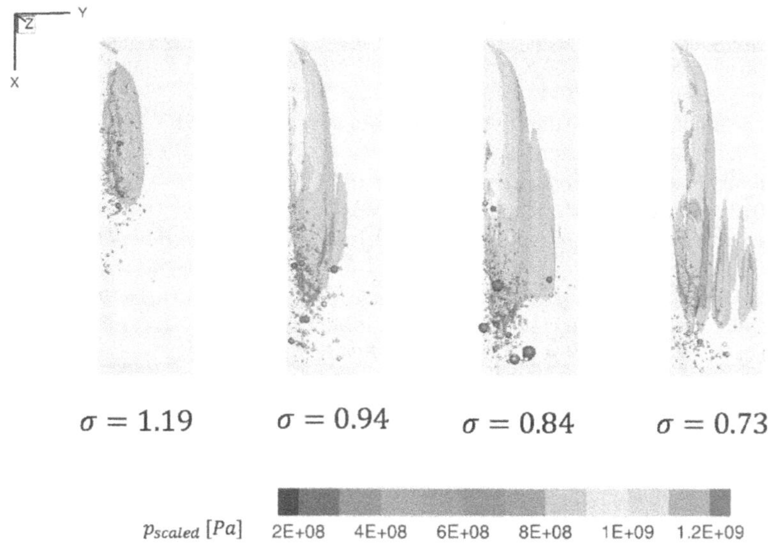

Figure 4. Detected collapses of vapor structures and averaged vapor content (10%) for all investigated operating points over 4 ms. Color and size of the collapses both correspond to the scaled collapse pressure.

For the operating point $\sigma = 0.73$, which cavitates more strongly, the same can be observed. However, in this case there is already a permanent gas content in the nozzle mixed with the vapor. At $\sigma = 0.73$ the effect on the jet is not as distinct as for the higher cavitation number $\sigma = 0.84$. Besides the jet characteristics, the gas-entrainment also affects the mass flux. Observing the mass fluxes reveals that gas-entrainment reduces the mass flux at the nozzle outlet with respect to its average by 18 % for $\sigma = 0.84$ and by nearly 25 % for $\sigma = 0.73$. Moreover, the vapor content in the nozzle at $\sigma = 0.73$ is a lot smaller than at $\sigma = 0.84$, which shows that apparently the permanent gas content suppresses and replaces the vapor.

Based on our observations we propose to distinguish between different forms of gas-entrainment depending on the strength of the cavitation: The *partial gas-entrainment*, where from time to time detached vapor pockets reach the outlet and a small amount of gas enters the region close to the outlet, leading to a bulging of the jet and *permanent gas-entrainment*, occurring when the cavitation region is constantly very close to the outlet, gas mixes with the vapor and forms a stable gas-vapor sheet. A further decrease of the cavitation number then leads to the so-called *hydraulic flip*, where the flow is completely detached from the side wall [1, 6, 7].

As a last aspect we focus on the erosion prediction for all simulated operating points. Fig. 4 shows the detected collapses over the analyzing time of 4 ms together with the averaged vapor content. The collapse detection algorithm and the scaling law to avoid grid resolution effects by Mihatsch et al. [10] was utilized. The collapses mostly occur at the end of the vapor sheet, where vapor structures detach and collapse.

Conclusion

Besides the effects of the gas-entrainment on the jet, our simulations revealed a significant drop of the mass flux at the nozzle outlet due to the displacement effect of the back-flowing gas. Moreover, we find that the cavitation regime is very sensitive to the cavitation number. And therefore it can be challenging to exactly match one regime.

Acknowledges

The authors are grateful to acknowledge the Gauss Centre for Supercomputing e.V. for providing computing time on the GCS Supercomputers SuperMUC at Leibniz Supercomputing Centre (LRZ, www.lrz.de).

References

[1] Sou A, Hosokawa S and Tomiyama A 2007 *International Journal of Heat and Mass Transfer* **50** 3575–3582
[2] Örley F, Trummler T, Hickel S, Mihatsch M S, Schmidt S J and Adams N A 2015 *Physics of Fluids* **27** 086101–28
[3] Edelbauer W 2017 *Computers & Fluids* **144** 19–33
[4] Egerer C P, Hickel S, Schmidt S J and Adams N A 2014 *Physics of Fluids* **26** 085102
[5] Egerer C P, Schmidt S J, Hickel S and Adams N A 2016 *Journal of Computational Physics* **316** 453–469
[6] Sou A, Biçer B and Tomiyama A 2014 *Computers and Fluids* **103** 42–48
[7] Biçer B and Sou A 2015 *Atomization and Sprays* **25** 1063–1080
[8] Koukouvinis P, Gavaises M, Georgoulas A and Marengo M 2016 *International Journal of Computational Fluid Dynamics* **30** 129–140
[9] Rahn D 2017 *Large-eddy simulation of a cavitating nozzle flow and primary jet breakup* Master's thesis Chair of Aerodynamics and Fluid mechanics, Technical University of Munich
[10] Mihatsch M S, Schmidt S J and Adams N A 2015 *Physics of Fluids* **27** 103302–22

© 2018 ASME

10th International Symposium on Cavitation - CAV2018
Baltimore, Maryland, USA, May 14 – 16, 2018

Large Eddy Simulation of the internal injector flow during pilot injection

[1,2]Marco Cristofaro*; [1]Wilfried Edelbauer; [2]Phoevos Koukouvinis; [2]Manolis Gavaises

[1]AVL List GmbH, Hans List Platz 1, 8020 Graz, AT; [2]City University London, Northampton Square EC1V 0HB, UK

Abstract

The aim of this work is to simulate the internal flow of a Diesel injector during an entire pilot injection event. In common rail systems a small quantity of fuel can be injected before the main injection is started. This increases the temperature in the combustion chamber and improves the combustion, leading to higher engine efficiency and reduced emissions. The internal nozzle flow during this short event is highly dynamic and vapor cavities may appear at the end of the injection. In order to study the flow characteristics, a numerical methodology based on the Eulerian multi-fluid approach is adopted. The filtered Navier-Stokes equations are discretized with the finite volume method and then solved with an implicit pressure-based solver. The flow field is modelled considering single pressure and velocity fields. The Coherent Structure Model is used to derive the eddy viscosity applied to the Large Eddy Simulation approach. The liquid evaporation rate is evaluated with a cavitation model based on the Rayleigh-Plesset equation for a single bubble. Even though thermodynamic equilibrium is not satisfied a priori, the main parameter is adjusted in order to limit the thermodynamic states to be in a range close to the equilibrium conditions. The liquid compressibility is modelled with a linear correlation between pressure and density variations. The needle longitudinal movement obtained from the experiments is applied to the simulation. The adopted geometry is the Spray A case defined by the Engine Combustion Network. It is an asymmetric single hole Diesel injector that has been extensively studied in the past both experimentally and numerically. The injection pressure is 1,500 [bar] and the ambient pressure is 60 [bar] with a fuel temperature of 363 K inside the injector. Pure n-dodecane is used as fluid in order to have a precise specification of the physical properties. Although both experiments and simulations showed no cavitation for completely open needle at fixed position, recent studies demonstrated that phase-change of the liquid can appear during the needle closing phase. Cavitation erosion prone locations are then evaluated by recording the maximum intensity of pressure on the surface.

Keywords: LES nozzle flow; pilot injection; needle closing cavitation

Introduction

Gradually stricter emissions regulations challenged the design of Diesel engines in the last decades. In the late 90s the common rail injection system has been introduced in the automotive field. The usage of a common high pressure rail and electronically controlled injectors allows to decouple the injection time and the injection pressure leading to a high flexibility in controlling the fuel injection. A pre-injection (also known as pilot) can be adopted before the main injection or multiple shorter injections can substitute the single injection. The usage of multiple injections is today a well consolidated approach to reduce the generation of NOx and soot from the combustion, as well as engine noise and vibrations. The adoption of electronic control units allows to map different injection strategies depending on the operation points (e.g. temperature, throttle and speed) of the engines. An optimization process can then be run on each operation point to obtain the best injection strategy. This process can be automated with validated numerical models of the injection, spray and combustion processes. Even though computationally more expensive compared to one-dimensional models, numerical approaches based on multi-phase 3-dimensional Computational Fluid Dynamics (CFD) are able to provide useful information about the underlying physics of the phenomena. The aim of this study is to obtain a Large Eddy Simulation (LES) of the injector internal flow during an entire pilot injection event. The understanding of the fluid dynamics during a pilot injection can be very useful for the understanding of any short time injection event. The "Spray A" case from the Engine Combustion Network (ECN) described in [1], [2] is used as test case for the presented study thanks to the availability of detailed description of the geometry, fluid properties and operation conditions as well as extensive experimental data. In [3] the LES methodology was used to present the main flow dynamics aspects during the start and end of injection of "Spray A". The simulations were obtained with a pressure based solver with a cavitation model based on thermodynamic equilibrium combined with a relaxation time. The needle lift profile was however approximated as linear in both the start and end of injection. In [4] a density-based solver with a cavitation model based on thermodynamic equilibrium and LES assumptions was applied to the main injection of a 9-hole injector with a realistic needle-lift profile. Two different designs of an injector showing erosive cavitation were instead simulated in [5].

*Corresponding Author, Marco Cristofaro: Marco.Cristofaro@avl.com

A tunable cavitation model based on single bubble dynamic was applied on a pressure-based solver in order to reduce the presence of negative pressure.
The present work exploits the pressure-based commercial solver FIRE™ to numerically resolve a three-phase flow [6]. Implicit time integration is adopted, thus avoiding stability constraints on the time step. Liquid phase compressibility is considered with a linearized equation of state and slip velocities between the phases are neglected.

Numerical model

The current numerical approach models a system with three-phases: liquid, vapor and air. The volume fraction of each phase is transported by a corresponding volume fraction equation but single pressure and velocity fields for the mixture are computed (homogeneous mixture hypothesis). The system is supposed to be isothermal at the temperature of 90° [C]. The effect of high temperature outside the nozzle is neglected in the current study, and the spray simulation is beyond the aim of this work. The SIMPLE algorithm is used to solve the pressure-velocity coupling. In the current study the Coherent Structure Model is applied as closure for solving LES as proposed in [7]. A second order differencing scheme is used for the convection terms of the momentum equation. A blending factor of 0.95 with the upwind scheme avoids numerical oscillations. Time derivatives are also obtained with a second order accurate three time level scheme. For further details about the simulation methodology the reader is referred to [6], [8].
In the current approach the cavitation phenomenon is modeled by introducing a mass source term in the liquid and vapor continuity equations. The current model is based on the hypothesis of monodispersed vapor bubbles distribution and on single bubble dynamics considerations. The mass transfer between liquid and vapor is supposed to be the sum of the contribution of a certain number of bubbles behaving identically. The bubbles growth or collapse is computed for a representative bubble size with a simplified version of Rayleigh-Plesset equation, neglecting the acceleration term and the effects of viscosity, temperature and non-condensable gas [9], [10]. The mass transfer term is modeled as:

$$\Gamma = -\rho_v N\, 4\pi R^2\, \dot{R} = \pm \rho_v N\, 4\pi R^2 \sqrt{\frac{|p - p_v|}{\rho_l}}$$

In the presented Equation Γ indicates the mass source applied to the liquid continuity equation, ρ_v and ρ_l are vapor and liquid densities respectively, p is the local pressure and p_v is the vapor pressure. The vapor bubbles number density is expressed with N, while the characteristic bubble dimension and dynamic are represented by an average radius, R, and its time derivate, $\dot{R}$. Local pressure values below the vapor pressure leads the evaporation of liquid, thus the negative sign is taken. Oppositely, pressure higher than vapor pressure causes the vapor to condensate back into liquid, so positive mass source. The same mass source is added to the vapor mass conservation equation, but with opposite sign. The presented single bubble dynamics models do not satisfy *a priori* the thermodynamic equilibrium, and a finite relaxation time is required for the phase-change to happen depending on the mass transfer and time step. In the current model the mass transfer is tuned with a single parameter, similarly to [5], in order to obtain physically acceptable thermodynamic states. The bubble number density, N, is then artificially increased to higher values than experimental measurements. The usage of these higher values aims to model the cavitation process to lay between the thermodynamic equilibrium state and the Rayleigh-Plesset equation solution. Furthermore the usage of an artificially increased number density accelerates the condensation process, leading to the formation of pressure waves that can be linked to a cavitation erosion mechanism. In this work a bubble number density value of 10^{18} [m^{-3}] is adopted.
The used liquid is n-dodecane and the fluid properties are extracted from [11]. The liquid is treated as weakly compressible with a constant speed of sound of 1336 [m/s] and a reference density at 1 [bar] of 704.2 [kg/m^3], leading to a liquid density variation in the domain in the order of 12%. Since the vapor should exist only close to the saturation condition, the vapor phase is considered incompressible and the properties are computed at the saturation pressure of 1,276 [Pa]. A constant a value of 57.6 [kg/m^3] is instead used for the air density to reduce the problem complexity. Although air compressibility may play a role in the spray formation, it should not in the internal nozzle flow. The n-dodecane viscosity is obtained from a linear fit of the experimental data presented in [11]. The derived formulation is: $\mu(p) = 0.493 + 0.009\, p$, being p the pressure expressed in [MPa] and μ the molecular dynamic viscosity expressed in [mPa s]. The viscosity of vapor and air are taken constants with values of 5.23 and 18.24 [μPa s] respectively.
Simulations are initialized with air at low pressure in the nozzle, sac and discharge volume, while with high-pressurized liquid in the remaining upstream region. Constant pressures of 1,500 [bar] at the inlet and 60 [bar] at the outlet of the discharge volume are imposed over the entire injection period. Considering a pure liquid flow, the Bernoulli velocity of 640 [m/s] corresponds to a Reynolds number of 80,000 and a Mach number of 0.48 for the liquid. The needle lift is obtained from the experiments presented in [2], however for the pilot injection with a duration of 300 [μs], the ascending and descending profiles are joined at maximum needle lift of 120 [μm].

The geometry is obtained from X-ray tomography as it refers to "Spray A" of the ECN database with serial number 210677. Two meshes with different resolutions were created, depending on the simulation target: a coarse mesh of 330,000 cells, with characteristic length inside the nozzle of 10 [μm] and a fine mesh of 15 million cells, with characteristic length inside the nozzle of 2.5 [μm] and first cell height close to the wall of 0.11 [μm]. For the coarse mesh a time step of 20 [μs] (CFL=1,000) is adopted while for the fine mesh a value of 0.004 [μs] (CFL=0.8) is used. Needle lift is modeled by deforming the meshes with the cell stretching technique.

Results

Main and pilot injections are compared in Figure 1. The mass flow through the nozzle outlet obtained from the simulations is compared with the experimental measurements for the main injection together with the needle lift profile (Figure 1(a)). For the pilot injection event only the simulation mass flow is presented with the corresponding used needle lift (Figure 1(b)). Due to the available computational budget, the coarse mesh is firstly used for the main injection to validate the simulation methodology while the fine mesh is used for the pilot injection to obtain a more detailed flow description. It has to be noticed that the flow for high values of needle lift is generally quasi-steady, while the most of unsteady phenomena happens at the start and end of the injection. While experimental values are available for the needle lift of the main injection, data do not exist for a pilot injection. Results show a small delay time of 20 [µs] from the needle opening to the actual start of the injection (a reference mass flow of 2 [g/s] is considered) in both pilot and main injection events due to the time required for the fuel to fill the sac and the nozzle. Due to the "Spray A" single-hole geometrical configuration, both experiments and simulation showed that the maximum mass flow is reached already at a very small needle lift. Oscillations in the mass flow are clearly visible in the experimental results for the main injection. In the simulation one fluctuation is present at the beginning, however this is damped immediately and the mass flow remains constant over the rest of the injection probably due to the very large time step. The error for the average mass flows is in the order of 5%. In the pilot injection, higher frequency fluctuations are visible with a period in the order of 20 [µs] that corresponds with the lift profile time discretization. This shows that the fine simulation is strongly affected by changes in the needle velocity. A relation between needle velocity oscillations and mass flow is also visible in the experiments presented in [2].

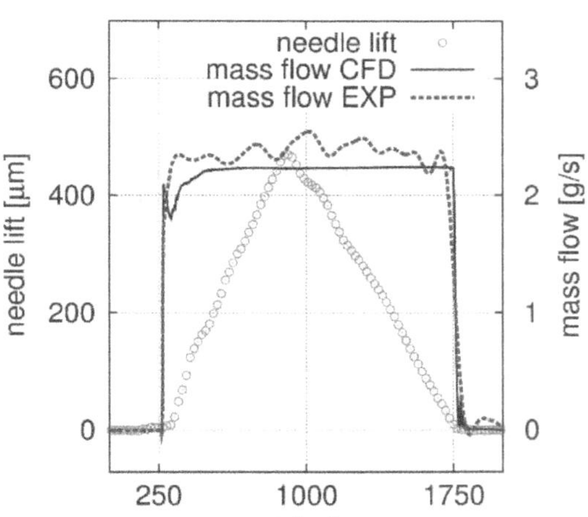

Figure 1(a)

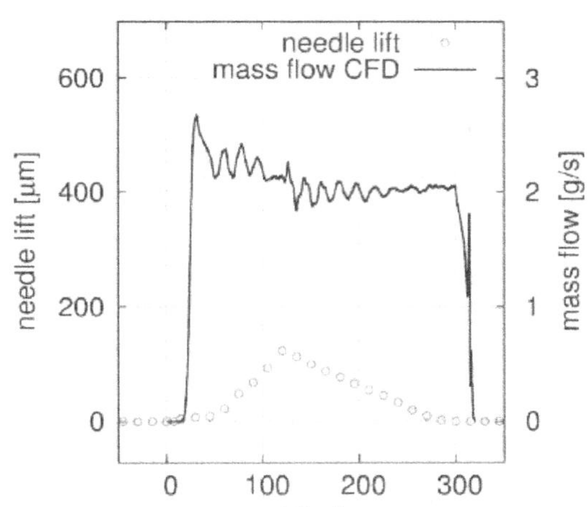

Figure 1(b)

Figure 1 – Main and pilot injection events: needle lift profiles (red circles) and simulations mass flow (continuous black line) for both pilot and main injections and experimental mass flow (blue dashed line) for the main injection only.

The injector flow does not show cavitation during the start and central injection periods, however the highly dynamic flow during the closing phase can lead to cavitation. At the end of injection, the needle touches the seat leading to a sudden flow blockage. During the simulations this phenomenon was modeled by decoupling the high pressure region (above the needle touching point) with the rest of the domain. The main flow characteristics modeled with the fine mesh are shown in Figure 2. The high momentum of the liquid causes a pressure decrease as soon as the upstream flow stops. Vapor is first generated at the nozzle inlet, where the flow velocity is the highest due to the presence of a recirculation zone, and then convected downstream (Figure 2(a)). The long cavity is then disintegrated into smaller cavities that interact within each other and with the surrounding flow, significantly increasing the turbulent structures in the flow (Figure 2(b)). A pressure wave causing sudden condensation close to the nozzle inlet and moving upstream is also visible in Figure 2(b), dividing the laminar vapor cavity from the collapsing turbulent one. The collapses close to the nozzle inlet lead to pressure peaks on the walls and more vapor cavities appear then inside the sac (Figure 2(c)). The big vapor cavities in the sac collapse as soon as the pressure is recovered, generating values of pressure on the surrounding walls up to 5,000 [bar] (Figure 2(d)).

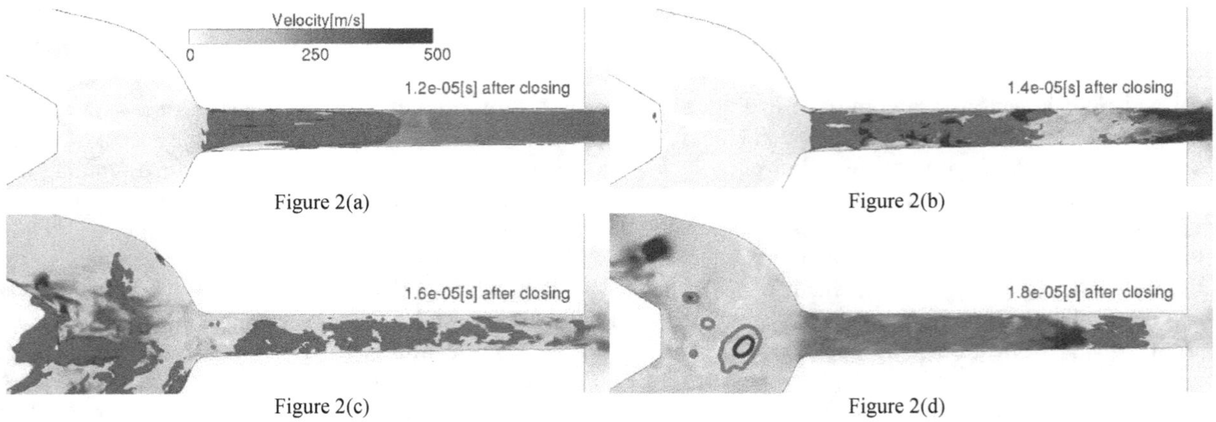

Figure 2 – Flow fields at different instants after the needle touches the wall. Cut along the injector nozzle axis of the velocity field, blue iso-surface at 0.9 of the vapor volume fraction and iso-lines at 3,500 [bar] (red) and 5,000 [bar] (blue) of the recorded pressure peaks on the walls.

Conclusion

The results of a Large Eddy Simulation during an entire pilot injection of the "Spray A" is presented. The needle lift profile is extracted from the experiments and mass flow results are shown. The model validation was obtained from a main injection event simulated on a coarser mesh in order to reduce the computational cost. Even if no cavitation appears during the steady injection phase, vapor formation appears after the needle touches the seat. Using the fine mesh, very high values of pressure peaks are recorded on the sac and nozzle walls due to the violent collapse of vapor cavities.

Acknowledgments

Financial support from the MSCA-ITN-ETN of the European Union's H2020 programme, under REA grant agreement n. 642536 is acknowledged. The authors acknowledge the support of Zoran Pavlovic and Mandumpala Devassy Bejoy for the technical support for the creation of the meshes and the needle movement.

References

[1] L. M. Pickett *et al.*, "Comparison of Diesel Spray Combustion in Different High-Temperature, High-Pressure Facilities," *SAE Int. J. Engines*, vol. 3, Oct. 2010.
[2] A. L. Kastengren *et al.*, "Engine Combustion Network (ECN): Measurements Of Nozzle Geometry And Hydraulic Behavior," *At. Sprays*, vol. 22, Jan. 2012.
[3] M. Battistoni, Q. Xue, and S. Som, "Large–Eddy Simulation (LES) of Spray Transients: Start and End of Injection Phenomena," *Oil Gas Sci. Technol.*, 2016.
[4] F. Örley, S. Hickel, S. J. Schmidt, and N. A. Adams, "Large-Eddy Simulation of turbulent, cavitating fuel flow inside a 9-hole Diesel injector including needle movement," *Int. J. Engine Res.*, vol. 18, Apr. 2016.
[5] P. Koukouvinis, M. Gavaises, J. Li, and L. Wang, "Large Eddy Simulation of Diesel injector including cavitation effects and correlation to erosion damage," *Fuel*, 2016.
[6] AVL LIST GmbH, "FIRE CFD Solver – User Guide." Hans–List–Platz 1, 8020 Graz, Austria.
[7] H. Kobayashi, "The subgrid-scale models based on coherent structures for rotating homogeneous turbulence and turbulent channel flow," *Phys. Fluids*, vol. 17, pp. 45107–2225, 2005.
[8] M. Cristofaro, W. Edelbauer, M. Gavaises, and P. Koukouvinis, "Numerical simulation of compressible cavitating two-phase flows with a pressure-based solver," in *ILASS 2017 - 28th European Conference on Liquid Atomization and Spray Systems.*
[9] J. W. S. Lord Rayleigh, "On the pressure developed in a liquid during the collapse of a spherical cavity," *Philos. Mag. Ser. 6*, vol. 34, pp. 94–98, 1917.
[10] C. E. Brennen, *Cavitation and Bubble Dynamics*. Oxford University Press, 1995.
[11] D. R. Caudwell, J. P. M. Trusler, V. Vesovic, and W. A. Wakeham, "The Viscosity and Density of n-Dodecane and n-Octadecane at Pressures up to 200 MPa and Temperatures up to 473 K," *Int. J. Thermophys.*, vol. 25, 2004.

© 2018 ASME

Large Eddy Simulations of a confined tip-leakage cavitating flow with special emphasis on vortex dynamics

[1,2]Huaiyu Cheng; [1,2]Bin Ji*; [1,2]Xinping Long; [3]Xiaoxing Peng;

[1] Hubei Key Laboratory of Waterjet Theory and New Technology, Wuhan University, Wuhan 430072 China

[2] School of Power and Mechanical Engineering, Wuhan University, Wuhan 430072 China

[3] National Key Lab on Ship Vibration and Noise, China Ship Scientific Research Center, Wuxi 214082, China

Abstract

In current paper, Large Eddy Simulations are conducted to investigate the tip-leakage cavitating flow around a NACA0009 hydrofoil with a gap. The numerical result is compared with experimental data and a reasonable agreement is obtained. The vortex structures in the tip-leakage cavitating flow are well reproduced. Three types of vortex structures in tip-leakage cavitating flow are observed near the gap, the tip-leakage vortex, tip-separation vortex and induced vortexes. Compared with the tip-separation and induced vortexes, the tip-leakage vortex is much stronger and impacts significantly the development of the formers. The simplified vorticity transport equation suggests that the dilation term dominates in the whole process of vortex development, while the stretching term mainly concentrates near the end-wall of foil. The baroclinic torque is the major source of the vorticity production during the collapse of cavitation. Our work is helpful to understand the interactions between cavitation and vortexes in a tip-leakage cavitating flow.

Keywords: gap flow; tip leakage vortex; cavitation; CFD

1 Introduction

Tip-leakage cavitating flow, as one of primary cavitation types, has attracted much attention in recent decades[1, 2]. A number of experimental and numerical investigations have been conducted to obtain a better knowledge on this flow. Inoue et al.[3] measured the flow field in an axial compressor rotor with various tip clearances in detail, which suggested that the increase of gap leaded to a stronger tip-leakage vortex. You et al.[4] numerically studied the tip-clearance flow in an axial turbomachines with LES. The velocity and pressure fields were systematically analyzed to reveal the underlying mechanisms for cavitation-inducing low-pressure fluctuations. Dreyer et al.[5] conducted a series of stereo-PIV measurements and flow visualizations for the tip-leakage cavitating flow with special emphasis on the influence of the clearance size on the tip vortex structure in a simplified case study. However, the knowledge about the vortex structures in a tip-leakage cavitating flow is still inadequate.

In current paper, with the measured data[5], the tip-leakage cavitating flow around a NACA0009 hydrofoil with a gap is simulated with LES, the complex vortex structures near the gap is discussed and the interaction between vortex and cavitation is analyzed with a vorticity transport equation.

Simulation setup

The computational domain and boundary conditions are shown in Figure. 1. A truncated NACA0009 hydrofoil with a chord length of C=100mm and attack angle of α=10° is located in a channel to generate a tip-leakage cavitating flow. The gap size is set as 0.1C and the velocity of main flow is 10m/s in the present paper. A more detailed configuration for the computational domain can be referenced in Refs [5, 6]. The ZGB cavitation model combined with LES is utilized to reproduce the unsteady cavitating flow. Three meshes with different resolutions are tested, and the middle resolution mesh is used as the final mesh, which provide a reasonable agreement between the numerical and experimental results, as shown in Figure. 2.

*Corresponding Author, Bin Ji: jibin@whu.edu.cn

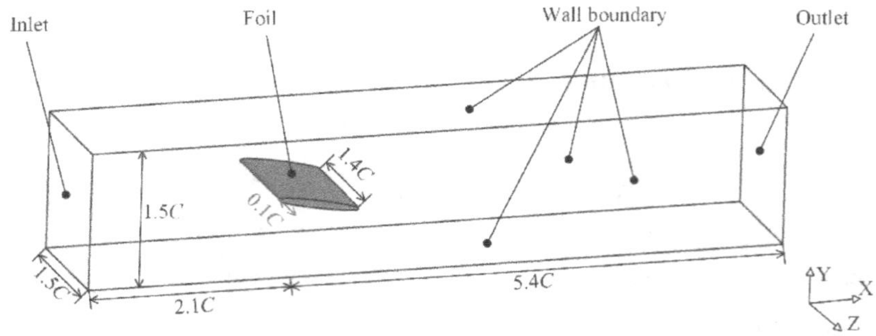

Figure. 1 Computational domain and boundary conditions

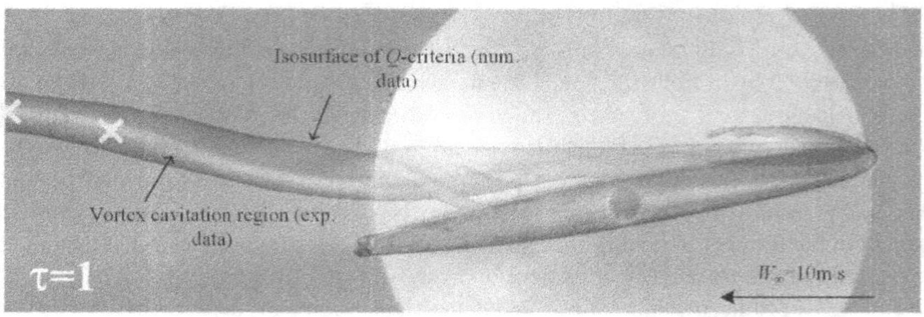

Figure. 2 Comparison of the time-averaged experimental picture [5] and numerical data

Results and discussion

Figure. 3 shows the time-averaged cavities displayed with the iso-surface of vapor volume fraction α_v=0.1. The general features of the attached cavity, tip-leakage cavity and tip-separation cavity are captured though the tip-leakage cavity is underestimated to a certain extent, which is mainly caused by the imperfection of the ZGB cavitation model. The time-averaged streamlines near the gap are also shown in Figure. 3 to give an intuitive description for the tip-leakage cavitating flow. It indicates that the streamlines starting near the foil tip, moves downstream spirally and wrap with each other, which are significantly influenced by vortex structures. To show the vortex structures more clearly, Figure. 4 shows the iso-surface of λ_2 based on the time-averaged velocity. It suggests that though the cavitation is underestimated, the vortex structures in tip-leakage cavitating flow are reproduced well. The tip-leakage and tip-separation vortexes are induced near the leading edge and then the tip-separation vortex is entrained to the suction side, and then merges with the tip-leakage vortex. Two group induced vortexes can be observed near the gap wall, which are much weaker when compared with the tip-leakage vortex.

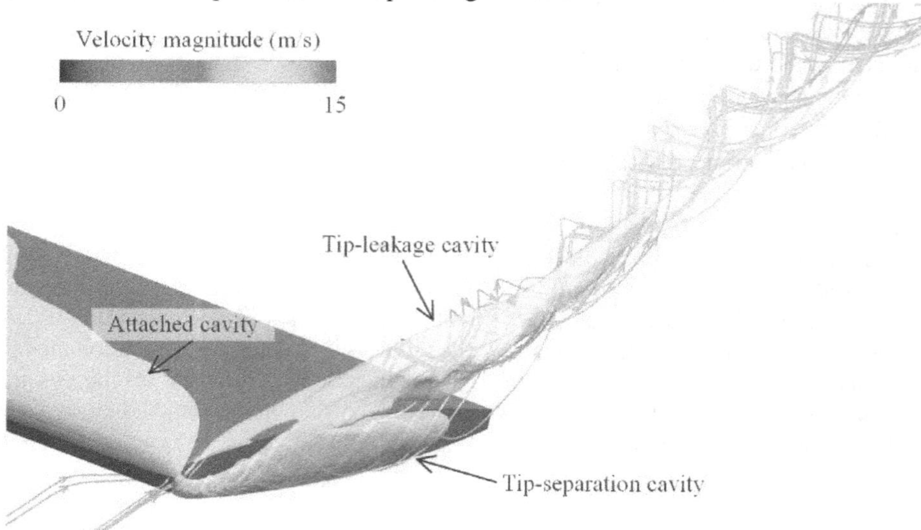

Figure. 3 Time-averaged cavities (iso-surface of vapor volume fraction α_v=0.1) and streamlines near the gap

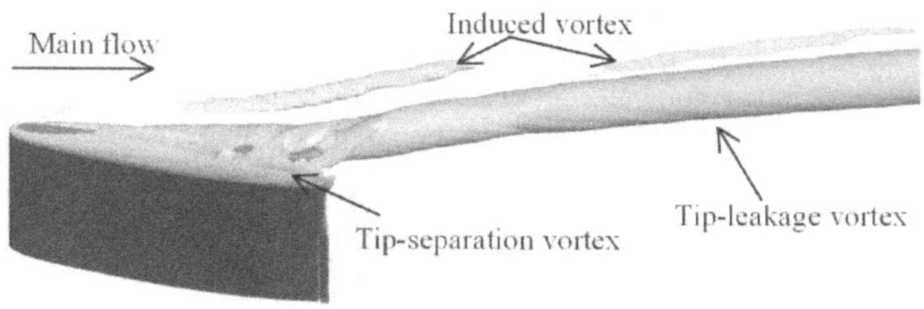

Figure. 4 Vortex structures displayed with the iso-surface of λ_2 based on the time-averaged velocity

The time-averaged streamlines on several typical sections are shown in Figure. 5 to provide a detailed view on the vortex structures near in the tip-leakage cavitating flow, which describes the mergence of tip-leakage and tip-separation vortexes. It indicates that the induced vortexes can be observed at x/C=0.9 and 1.1, near the trailing edge of hydrofoil. Moreover, an induced vortex is also found at x/C=0.1 in the scope of the attached cavity, but vanishes quickly, which should impact the development local cavitation.

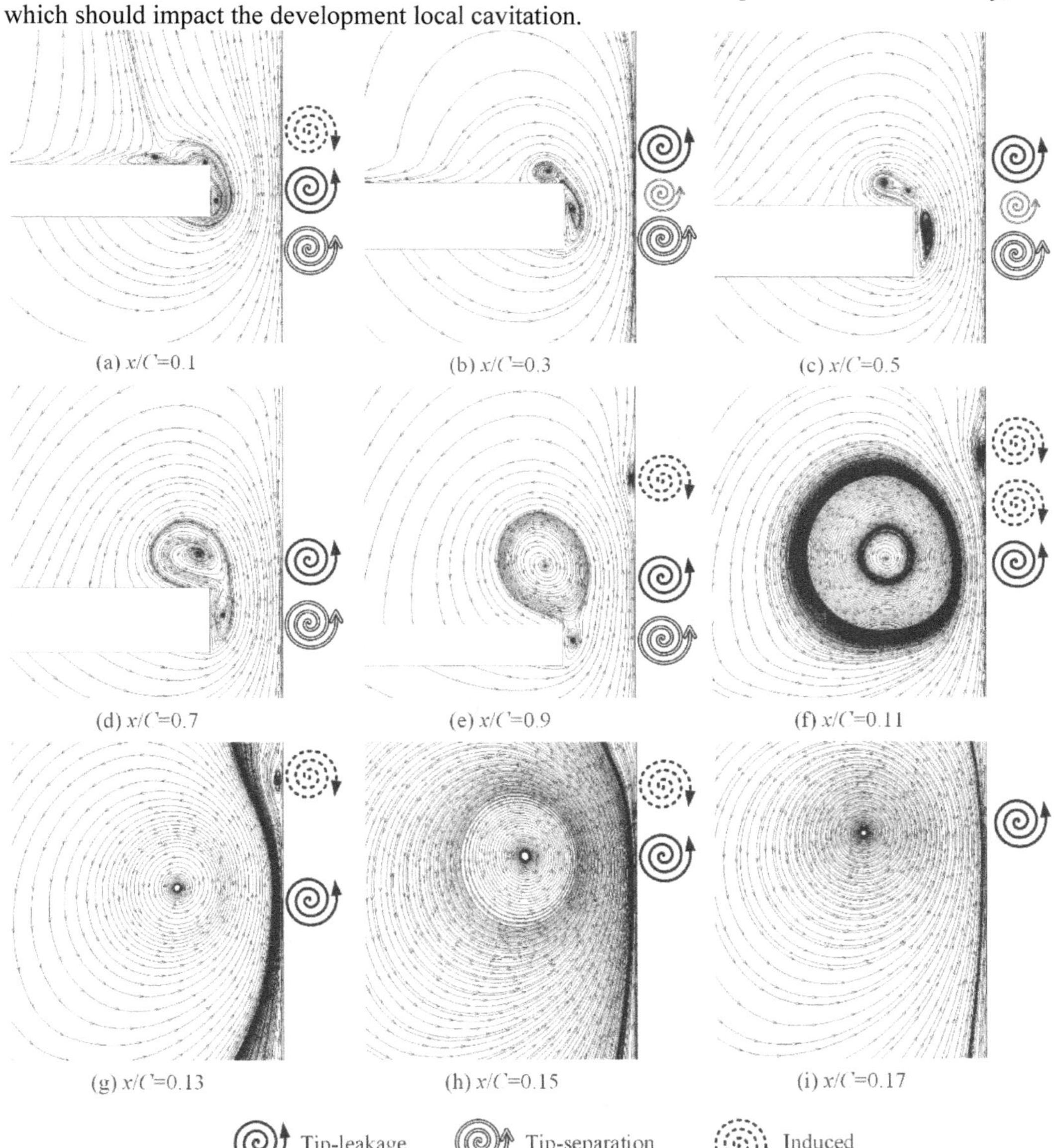

Figure. 5 Time-averaged streamlines on typical sections

To further study the interactions between cavitation and vortex in the tip-leakage cavitating flow, a simplified vorticity transport equation is used. Figure. 6 shows the distributions of stretching, dilatation and baroclinic torque terms based on the time-averaged velocity field. It demonstrates that the dilatation of vortex, not the stretching, dominates the whole process of vorticity transport in tip-leakage cavitating flow, which is different from the cavitating flow around a hydrofoil without a gap. Baroclinic torque term concentrates along the liquid-vapor interface with the large mass transfer rate, for example in the cavity collapse regions as shown in Figure. 6.

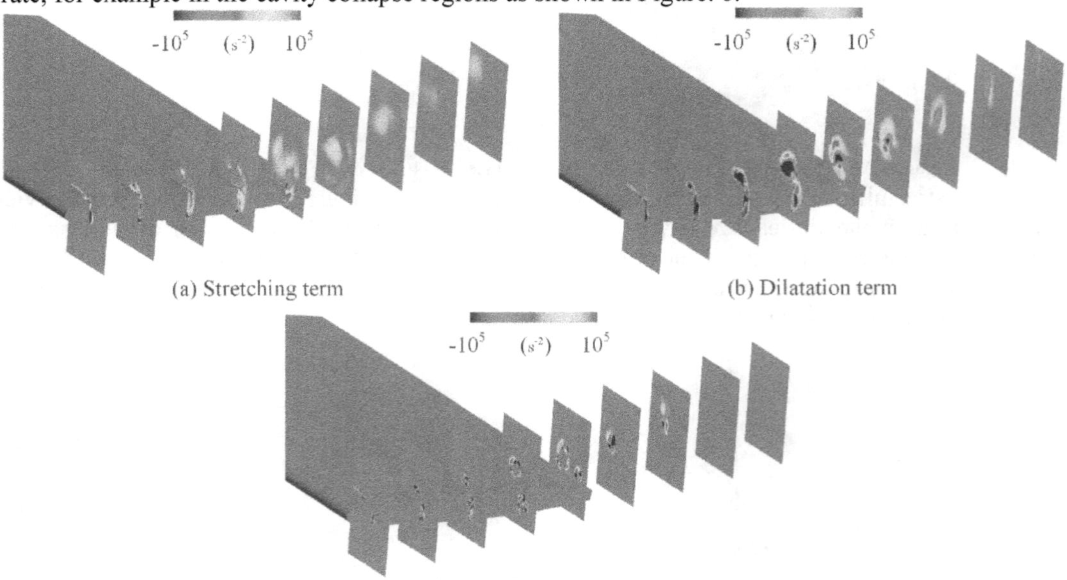

(a) Stretching term (b) Dilatation term

(c) Baroclinic torque term

Figure. 6 Distributions of stretching, dilatation and baroclinic torque terms based on the time-averaged velocity

Conclusion

In current paper, the tip-leakage cavitating flow around a NACA0009 hydrofoil with a gap is simulated with LES and the numerical result shows a reasonable agreement with experimental data[5]. The main conclusions can be draw as follow,

1. The Large Eddy Simulations of tip-leakage cavitating flow combined with the ZGB cavitation model conducted in current paper provide a reasonable result when compared with experimental data. The vortex structures in the tip-leakage cavitating flow are well reproduced.
2. Three types of vortex structures in tip-leakage cavitating flow are observed near the gap, the tip-leakage vortex, tip-separation vortex and induced vortexes. Compared with the tip-separation and induced vortexes, the tip-leakage vortex is much stronger and impacts significantly the development of the formers.
3. The simplified vorticity transport equation suggests that in the tip-leakage cavitating flow, the dilation term dominates in the whole process of vortex development, while the stretching term mainly concentrates near the end-wall of foil. The influence of baroclinic torque is also not negligible. It is the major source of the vorticity production during the collapse of the cavity.

References

[1]. Zierke, W. C., Farrell, K. J., & Straka, W. A. (1995). Measurements of the Tip Clearance Flow for a High-Reynolds-Number Axial-Flow Rotor. *Journal of Turbomachinery, 117*(4), 522-532.

[2]. Arndt, R. E. A. (1981). Cavitation in fluid machinery and hydraulic structures. *Annual Review of Fluid Mechanics, 13*, 273-328.

[3]. Inoue, M., Kuroumaru, M., & Fukuhara, M. (1986). Behavior of Tip Leakage Flow Behind an Axial Compressor Rotor. *Journal of Engineering for Gas Turbines and Power, 108*(1), 7-14.

[4]. You, D., Wang, M., Moin, P., & Mittal, R. (2006). Effects of tip-gap size on the tip-leakage flow in a turbomachinery cascade. *Physics of Fluids, 18*(10), 105102.

[5]. Dreyer, M., Decaix, J., Münch-Alligné, C., & Farhat, M. (2014). Mind the gap: a new insight into the tip leakage vortex using stereo-PIV. *Experiments in Fluids, 55*(11), 1849.

[6]. Decaix, J., Balarac, G., & Münch, C. (2016). Rans Computations of a Cavitating Tip Vortex. In P. Gourbesville, J. A. Cunge & G. Caignaert (Eds.), *Advances in Hydroinformatics: SIMHYDRO 2014* (pp. 519-529). Singapore: Springer Singapore.

© 2018 ASME

Laser Induced Breakdown and Bubble Cavitation

[1]Giorgia Sinibaldi*; [2]Agostino Occhicone; [3]Francisco Alves Pereira; [1]Davide Caprini; [1]Luca Marino; [2]Francesco Michelotti; [1,4]Carlo Massimo Casciola

[1]Department of Mechanical and Aerospace Engineering, La Sapienza, Rome, Italy;
[2]Department of Basic and Applied Sciences for Engineering, La Sapienza, Rome, Italy;
[3]Marine Technology Research Institute, CNR-INSEAN, Rome, Italy;
[4]CLNS@Sapienza, Istituto Italiano di Tecnologia, Rome, Italy

Abstract

This work presents the results of an experimental investigation of laser induced cavitation. We find that the breakdown plasma can be randomly split in different branches depending on the laser energy and on the focusing angle. This behavior strongly affects the successive dynamics of the bubble and limits the reproducibility of the process. Using a custom fiber optic hydrophone, we correlate the elongated shape of the plasma with the number of shock waves detected at breakdown. The conditions for single breakdown shockwave emission and spherical expansion are thus identified, thus improving the reproducibly for laser-induced cavitation bubble.

Keywords: cavitation; laser induced bubble; fiber optic probe hydrophone

Introduction

Bubble cavitation is one of the most discussed topics in fluid dynamics. Generally, it is associated with erosion damage and high-frequency noise [1], but it is currently reconsidered for a wide range of modern applications within medicine [2], microfluidics [3] and other fields. The life of a cavitation bubble can be divided into four different phases: nucleation, growth, collapse and rebound. In this study, we focus on the bubble dynamics in pure water. In the context of the classical nucleation theory (CNT) [4], homogeneous nucleation is the condition where a single spherical bubble is formed in a volume of water due to energy fluctuations of the system. Laser-induced cavitation has been extensively used to create this condition experimentally as it allows the control of the breakdown position and a high reproducibility of the bubble dynamics, see e.g. Lauterborn (1972) [5], Vogel et al. (1996) [6], Philipp & Lauterborn (1998) [7], Vogel et al. (1999) [8], Ohl et al. (1999) [9], Vogel et al. (2005) [10], Lim et al (2010) [11], Obreschkow et al. (2013) [12], Tagawa et al. (2016) [13].

However, the characteristics of the setup such as focusing optics, laser energy and water quality can introduce deviations from the ideal single bubble condition. Here we report on the observation of multiple plasmas formation and analyze the effect of both the focusing angle and laser energy on the breakdown event and on the bubble growth. The pulsed laser beam is expanded, collimated and then focused by a parabolic mirror immersed in water to minimize spherical aberrations [12]. The characterisation of the phenomenon is done through the use of high-speed imaging and pressure measurements. In particular, a fiber optic hydrophone has been purposely developed to quantify the pressure peak of the shock waves emitted upon optical breakdown of the water medium and bubble collapse.

The first part of the study addresses the impact of the focusing optics on the bubble dynamics, specifically the effect of the converging angle and of the laser energy on the size and potential energy of the bubble. In a second part, we show how these two factors can affect the number and shape of plasmas, and consequently the pressure field.

Experimental setup

The setup consists of a stainless-steel chamber (120x120x120 mm^3) equipped with quartz windows for optical access (cavitation box in Fig. 1). Bubble generation is obtained by focusing a Q-switched Nd:YAG laser (Litron Nano S 35-15), which can deliver light pulses at the wavelength $\lambda = 532$ *nm*, with a duration of 8 *ns*, a maximum repetition rate of 15 *Hz* and a pulse energy up to 30 *mJ*. At the laser exit, 4% of the energy is extracted by a beam splitter and sent to a pyroelectric energy sensor in order to accurately measure the laser energy to be focused, see

*Corresponding Author, Giorgia Sinibaldi: giorgia.sinibaldi@uniroma1.it

Fig. 1. The beam, with energy E_L, then passes through a custom Galilean beam expander and is collimated to reach an off-axis parabolic aluminum mirror placed inside the cavitation box filled with pure water. The parabolic mirror was chosen to avoid spherical aberrations [12] and is off-axis to prevent laser beam back reflection into the laser source. Different beam expander magnifications (×8, ×6 and ×3) were tested to evaluate the effect of the focusing angle γ (53°, 33° and 17°, respectively).

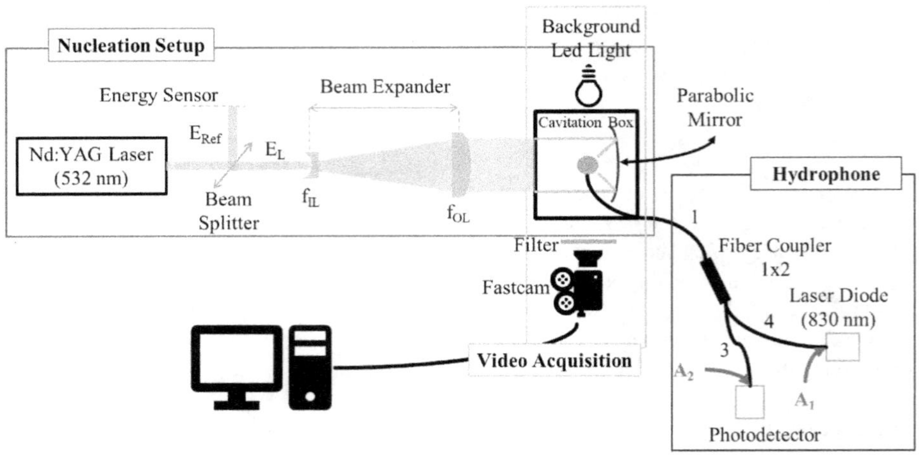

Fig. 1 Schematic of the experimental setup. System for bubble generation (red box), system for high speed acquisition (yellow box), custom fiber optic hydrophone setup (blue box).

High speed imaging is performed with a Photron Mini UX100 fast camera and a LED light for background illumination (see the yellow box in Fig. 1). The recordings are then analysed to investigate the bubble dynamics and the plasma characteristics.

A fiber optic probe hydrophone (FOPH) has been developed for local pressure measurements of the breakdown shock wave. The fiber optic allows to measure the refractive index change related to the density variation at shock wave passage. The FOPH setup is schematized in the blue box of Fig. 1. The light from a 1 *W* laser diode emitting at 830 *nm* is injected into a 2×1 fiber coupler (arm 4), and feeds an exit fiber that is immersed in water in the cavitation box and serves as the pressure probe (arm 1). This light reflects at the fiber-water interface back into the fiber coupler and is finally collected by an avalanche photodetector (arm 3). The output signal is acquired by a system with a bandwidth of 200 *MHz* and is recorded and analysed by a custom LabView software.

Results

In Fig. 2a, the values of the maximum bubble radius (R_{max}) measured with the fast camera are plotted as a function of the laser energy (E_L) for different focusing angles, 53° (black curve), 33° (red curve) and 17° (blue curve). The R_{max} values are the average over 64 consecutive events acquired at constant energy, and are plotted with their standard deviation. We used R_{max} to estimate the bubble potential energy as $E_B=\frac{4}{3}\pi(p_{amb}\text{-}p_v)R_{max}^3$, where p_{amb} is the ambient pressure, p_v is the vapor pressure (2.33 *kPa* in water at 20 °*C*). The E_B values are plotted in Fig. 2b together with the case of Vogel et al (1994) [14] for reference. The ratio between E_B and E_L represents the amount of laser energy converted into mechanical energy, ultimately responsible for shock wave emission and cavitation. Increasing the focusing angle, the conversion efficiency of our setup spans between 9% and 15%, while in the experiment of Vogel et al (1994) [14] a conversion efficiency of the 25% was obtained for ns-laser pulses. We note that the optical setups differ between the two experiments.

In Fig. 3 the experimental measurement of the temporal evolution of the bubble radius is shown. The effect of the focusing angle is shown in Fig. 3a for a fixed laser energy E_L = 26 *mJ*. Here, single bubble acquisitions at 160.000 *fps* are shown for each case. For greater focusing angles the first two rebounds are also shown. For the case at γ~17° no clear rebound can be observed and data are therefore omitted in the plot. In Fig. 3b, the experimental temporal evolution of the bubble radius is shown for high and low laser energy at a constant focusing angle γ~53°. A good agreement is found between the experimental data and the theoretical dependencies (solid line on the plot). The latter was obtained from the integration of the Keller-Miksis model [4, 15], a modified form of the Rayleigh-Plesset model.

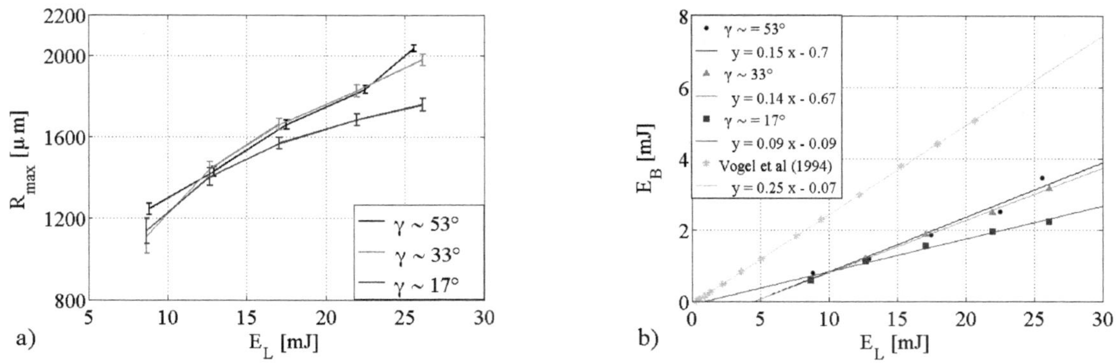

Fig. 2 Maximum bubble radius (a) and bubble potential energy (b) as a function of laser energy (bubble was imaged at 64.000 *fps*).

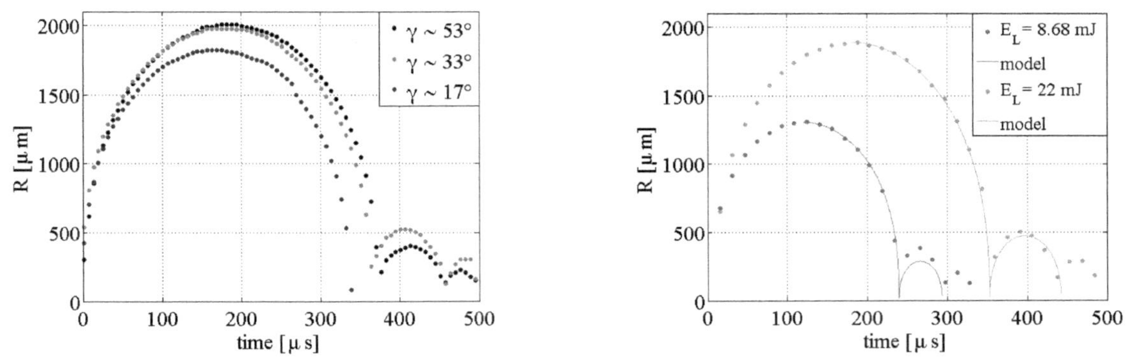

Fig. 3 Bubble radius evolution. (a) Effect of the focusing angle at E_L= 26 *mJ* (160000 *fps*). (b) Effect of the laser energy and comparison with the Keller-Miksis model at γ ~ 53° (64000 *fps*).

The dependency of the plasma shape on the main properties of the experimental set up has been already investigated, e.g. in [6, 13, 16, 17]. The main evidences are that the formation of multiple plasma depends on spherical aberrations, numerical aperture/focusing angle, liquid impurities and input laser energy. A small focusing angle favours the occurrence of multiple point-like plasmas. Moreover, multiple plasmas may be present even with “perfect” focusing, i.e. with no spherical aberrations. This is because the plasma occurs where the breakdown threshold is reached, which is not necessarily at the focusing point [13].

In Fig. 4a, we show the superposition of the images of the plasma (yellow) and of the bubble at its maximum expansion (red disk). The direction of the laser beam is marked by the green arrow. Such images were captured at 64000 *fps* for the different focusing angles and at a laser energy of 22 *mJ*. The change of the plasma length with the laser energy is shown in Fig. 4b for all focusing angles. From such figures it is evident that, at larger focusing angles, the plasma has a compact appearance and its length (see black curve) is almost constant with laser energy. Moreover, it is possible to appreciate from Fig. 4a that the plasma stretches with decreasing focusing angle, and is accompanied by a smaller bubble.

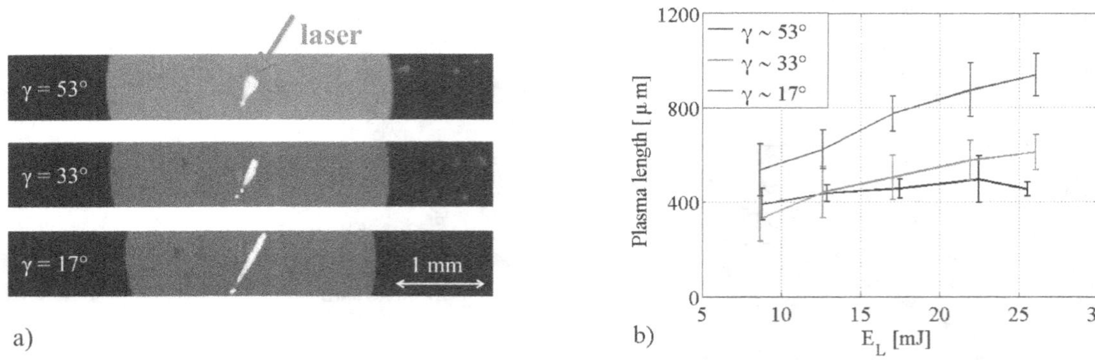

Fig. 4 Plasma morphology acquired at 64000 *fps*: a) superimposition of plasma (yellow) and bubble at maximum radius (red disk) at different focusing angles $E_L = 22$ *mJ*; b) plasma length as a function of the laser energy.

In Fig. 5 the grayscale intensity of the plasma (top frames) and the related intensity profile along the plasma axis (bottom) are shown. Each case is averaged over 64 bubble nucleation events. The averaging procedure permits to reject the random effect of the impurities and to obtain the deterministic influence of energy and focusing angle on the plasma shape. Two different focusing angles are reported, γ~17° (Fig. 5a) and γ~53° (Fig. 5b). As expected, by increasing the laser energy, the tip of the plasma remains fixed and the tail shifts backward in the direction of the light source [14, 16, 17]. Moreover, the region of highest intensity extends and the maximum intensity increases.

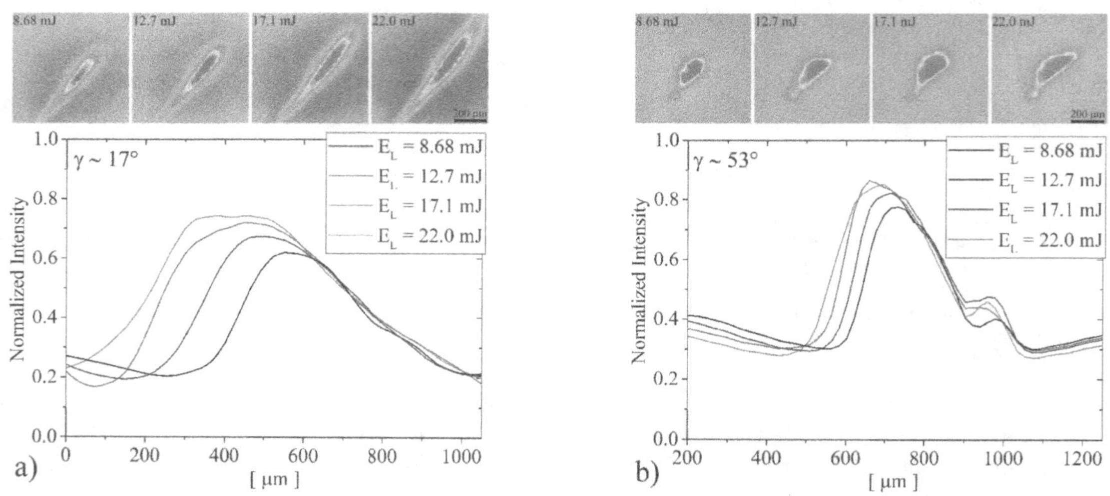

Fig. 5 Contour plot of the gray scale intensity of the plasma (top) and intensity profile along the plasma axis (bottom). a) γ ~ 17°, b) γ ~ 53°.

The main outcome of this analysis is the elongated shape of the plasma along the propagation direction of the laser beam. Considering that spherical aberrations have been substantially avoided by using the parabolic mirror, the plasma shape can depend only on the magnification parameters, i.e. the focusing angle, on the laser pulse energy and the random effects of the impurities in the liquid. Such elongation is the source of asymmetry of the breakdown shock wave, which can be detected from the pressure measurement obtained by placing the tip of the FOPH at different locations, defined with the angle θ of Fig. 6a [13]. We expect to measure a number of peaks equal to the number of shockwave fronts reaching the fiber optic. From the three graphs shown in Figs 6b, 6c and 6d, the anisotropy of the shockwave system is clear. In the case θ = 90° (Fig. 6b) a single peak for each focusing angle is present, thus indicating that only one shock wave is detected. In the cases θ = 45° (Fig. 6c) and θ = 0° (Fig. 6d) more than a peak is found, evidencing the existence of multiple shockwave fronts. It can be observed that, moving from 90° to 0°, the peak widen in time. The effect is more pronounced for a reduced focusing angle, in good agreement with the plasma shape observed in the video frames shown above (Fig. 4).

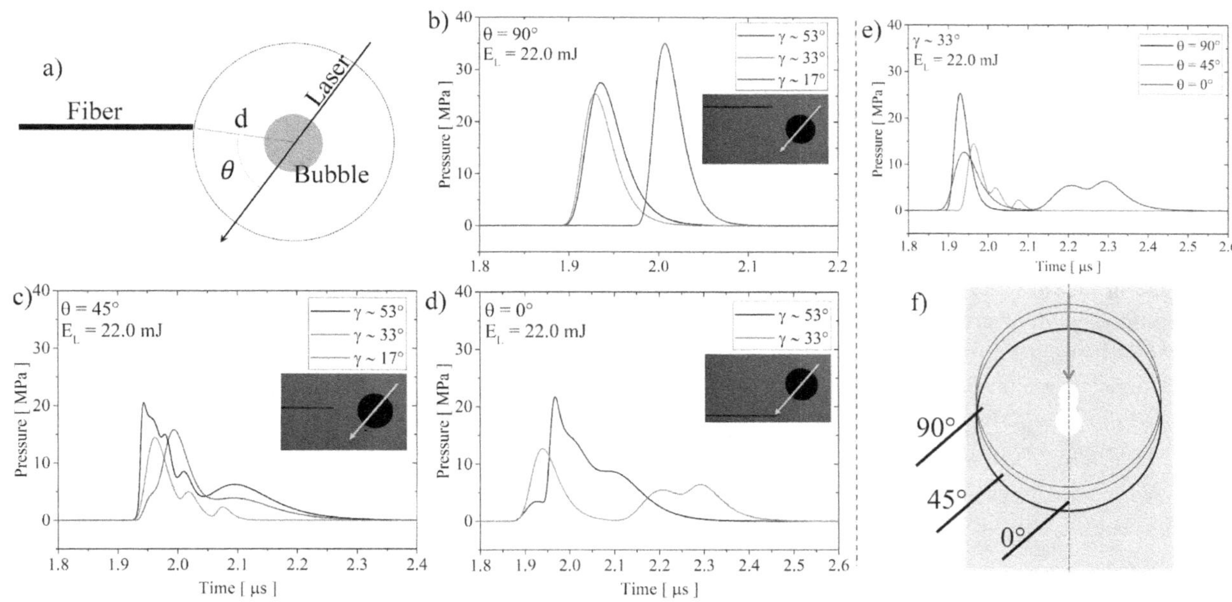

Fig. 6 Pressure signal at different probe locations as a function of focusing angle for a laser energy of 22 mJ: a) Sketch of the fiber position with respect to the laser direction; b) pressure signals at θ=90°; c) pressure signals at θ=45°; d) pressure signals at θ=0°; e) pressure signals for a focusing angle γ~33°; f) reconstruction of the wavefront system and the bubble initial configuration.

From the information collected with the hydrophone we can schematically reconstruct the multiple structure of the initial shock wave and the number of bubbles present in the first few nanoseconds. We assume that each shock wave is the superposition of different spherical components originating from different portions inside the plasma region [13]. An example of this reconstruction is shown in Figs 6e and 6f, related to the case which corresponds to $\gamma \sim 33°$ and $E_L = 22$ *mJ*. In Fig. 6e the number of shock waves reduces from three to one as the probe moves from $\theta = 0°$ to $\theta = 90°$, where the three pressure peaks reach the FOPH with the same delay and the overall shock wave appears as a single pressure pulse with the maximum amplitude, about 25 *MPa*. The schematic reconstruction gathered from this analysis is shown in Fig. 6f, with the three shockwaves and the subsequent bubble initial configuration.

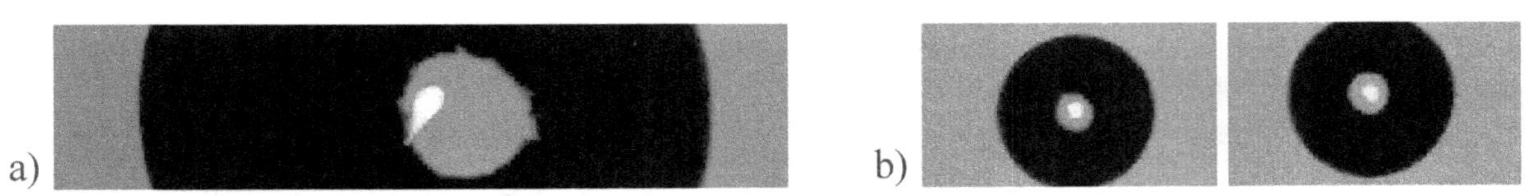

Fig. 7 Superimposition of plasma (yellow), bubble at maximum radius (black disk) and maximum radius of the first rebound (small red disk). (a) $E_L = 26$ *mJ*, $\gamma \sim 53°$ and R_{max}=1970 *μm*. (b) $E_L = 9.5$ *mJ* and $\gamma \sim 12°$ and $R_{max} \approx 550$ *μm*.

We found that almost all the investigated cases present an elongated plasma along the beam axis, a condition that leads to multiple shock waves emission, which in turn leads to multiple collapsing events with non-spherical bubbles at rebound, hence very poor reproducibility. However, we have identified two configurations that allow single spherical bubble nucleation with high repeatability as well as single rebound bubble after the first collapse: i) high energy and wide focusing angle; ii) low energy and narrow focusing angle. These two configurations are illustrated in Fig. 7a and 7b, respectively. Here, the superimposition of three frames for each case is reported: plasma (yellow), bubble maximum expansion (black disk) and the maximum expansion of the first rebound (small red disk). We note that in the first situation, the plasma has a conical form and remains located at the focusing point. In the second configuration, the plasma is instead point-like, however its location changes within the light cone, see the shifted position of the two bubbles in Fig. 7b.

Conclusion

In order to investigate laser-induced bubble cavitation in water, we assembled a laser nucleation set-up and combined high-speed visualizations with a fiber optic probe hydrophone (FOPH) to determine the pressure field induced during the process. A parabolic mirror immersed in water was used to minimize spherical aberrations [12]. However, multiple plasma sites were observed depending on the focusing angle and laser energy. To understand these effects, the characteristics of the laser-induced breakdown and bubble formation were first investigated for different laser beam convergence angles and different laser energies. The plasma length was found to increase with the laser energy, extending along the focusing cone towards the laser source. Shock pressure waves were measured with the FOPH in different positions with respect to the focusing location. We found that the laser-induced shock waves are in fact a collection of spherical shock waves emitted from the different plasma spots, demonstrating that multiple shock waves are directly related to the plasma shape. We found that single spherical bubbles could be produced with high reproducibility, at high energy and large focusing angle or at low energy and small focusing angle. However, we observed in this latter configuration that the bubble position in space could vary significantly because of the small energy density gradient along the focusing axis. In all other configurations, multiple breakdown sites and elongated plasmas are observed.

Future work will focus on the effect of ambient pressure on the bubble dynamics and on the interaction of the cavitation bubble with solid/membrane boundaries.

Acknowledgement

The research leading to these results has received funding from the European Research Council under the European Union's Seventh Framework Programme (FP7/2007-2013)/ERC Grant Agreement No. [339446].

References

[1] Plesset, M.S., Devine, R.E. (1966), *Effect of Exposure Time on Cavitation Damage*, Journal of Basic Engineering, 88(4).
[2] Brennen, C.E. (2015), *Cavitation in medicine*, Interface Focus, 5(5).
[3] Zwaan, E., Le Gac, S., Tsuji, K., Ohl, C.D. (2007), *Controlled Cavitation in Microfluidic Systems*, PRL, 98.
[4] F. Caupin and E. Herbert, "Cavitation in water," *C.R. Physique,* vol. 7, pp. 1000-1017, 2006.
[5] Lauterborn, W. (1972), *High-speed photography of laser-induced breakdown in liquids*, Applied Physics Letters, 21(1).
[6] Vogel, A., Busch, S., Parlitz, U. (1996), *Shock wave emission and cavitation bubble generation by picosecond and nanosecond optical breakdown in water*, The Journal of the Acoustical Society of America, 100(1).
[7] Philipp, A., Lauterborn, W. (1998), *Cavitation erosion by single laser-produced bubbles*, Journal of Fluid Mechanics, 361.
[8] Vogel, A., Noack, J., Nahen, K., Theisen, D., Busch, S., Parlitz, U., Hammer, D.X., Noojin, G.D., Rockwell B.A., Birngruber, R. (1999), *Energy balance of optical breakdown in water at nanosecond to femtosecond time scales*, Applied Physics B: Lasers and Optics, 68(2).
[9] Ohl, C.D., Kurz, T., Geisler, R., Lindau O., Lauterborn, W. (1999), *Bubble dynamics, shock waves and sonoluminescence*, Philosophical Transactions of the Royal Society of London A: Mathematical, 357(1751).
[10] Vogel, A., Noack, J., Huttman, G., and Paltauf, G. (2005), *Mechanisms of femtosecond laser nanosurgery of cells and tissues*, Applied Physics B, 81(8).
[11] Lim, K.Y., Quinto-Su, P.A., Klaseboer, E., Khoo, B.C., Venugopalan, V., Ohl, C.D. (2010), *Nonspherical laser-induced cavitation bubbles*, Physical Review E, 81(1).
[12] Obreschkow, D., Tinguely, M., Dorsaz, N., Kobel, P., de Bosset A., Farhat, M. (2013), *The quest for the most spherical bubble: experimental setup and data overview*, Experiments in Fluids, 54(5).
[13] Y. Tagawa, S. Yamamoto, K. Hayasaka and M. Kameda, "On pressure impulse of a laser-induced underwater shock wave," *J. Fluid Mech.,* vol. 8080, pp. 5-18, 2016.
[14] Vogel, A., Busch, S., Jungnickel, K., Birngruber, R. (1994), *Mechanisms of intraocular photodisruption with picosecond and nanosecond laser pulses*, Lasers in surgery and medicine, 15(1).
[15] J.B. Keller and M. Miksis, «Bubble oscillations of large amplitude,» *J. Acoust. Soc. Am.,* vol. 68, n. 2, pp. 628-633, 1980.
[16] Vogel, A., Nahen, K., Theisen, D., Noack, J. (1996), *Plasma formation in water by picosecond and nanosecond Nd: YAG laser pulses. I. Optical breakdown at threshold and superthreshold irradiance*, IEEE Journal of Selected Topics in Quantum Electronics, 2(4).
[17] Nahen, K., Vogel, A. (1996), *Plasma formation in water by picosecond and nanosecond Nd: YAG laser pulses. II. Transmission, scattering, and reflection*, IEEE Journal of Selected Topics in Quantum Electronics, 2(4).

© 2018 ASME

LES Investigation with an Eulerian Stochastic Field Cavitation Model

Boxiong Chen*, Michael Oevermann

Chalmers University of Technology, Gothenburg, Sweden

Abstract

Due to an ongoing trend of high injection pressures in the realm of internal combustion engines, the role of cavitation that typically happens inside the injector nozzle has become increasingly important. In this work, a large eddy simulation (LES) with cavitation modeled on the basis of an Eulerian Stochastic Field (ESF) method and a homogeneous mixture model is performed to investigate the role of cavitation on academic injector. The ESF cavitation model is coupled to a pressure based solver for the flow, which lowers the computational cost, thereby making the methodology highly applicable to realistic injector geometries. Moreover, the nature of the ESF method makes it more convenient to achieve a high scalability when applied to parallel cases, which gives the method an edge over cavitation models that are based on Lagrangian tracking. The result of the ESF simulation is compared against that from a typical single volume fraction solver for validation. The capability of the ESF model to capture the probability density function (PDF) of vapor bubble radii is also demonstrated.

Keywords: multiphase flow; volume of fluid; Eulerian Stochastic Field; cavitation

Introduction

In the realm of internal combustion engines, extensive effort has been made to improve fuel efficiency. The formation and breakup of fuel sprays play an important role in lean combustion engines such as diesel engines and gasoline direct injection (GDI) applications. Along with the trend of high injection pressures nowadays, cavitation inside the injector nozzle has been attracting an increasing amount of interest from both industry and academia. The impact of cavitation on internal combustion engines is twofold: First, it has a tendency to enhance the breakup of the fuel spray outside the injector nozzle as observed in previous studies, e.g. [1]. Second, the pressure wave induced by this intense phenomenon could cause brittling and erosion to the injector wall, thereby reducing the life span of the injector. Spray studies on cavitation inside injector nozzles have been carried out extensively through both experimental and simulation efforts. However, experimental studies of cavitation on realistic nozzle geometries and injection conditions remain extremely difficult, because the geometric size and the residence time of the bubbles require restrictively high resolutions of both time and space. Consequently, the majority of experimental studies with optical access up to now are set up on up-scaled nozzles and relatively low pressure conditions, e. g. [2, 3].

Computational cavitation studies with various modelling methodologies have been performed in numerous previous investigations [4-6]. Lagrangian description of individual bubbles or bubble clusters have been used to represent cavitation [4]. However, the complicated implementation and the well-known difficulties involved in the scalability of Lagrangian based methods on parallel computations make models of this type less feasible in industrial applications. In the realm of Eulerian modelling, interface diffuse models [6] regard bubbly flow as a continuous dispersed phase. This approach spends no effort tracking the fronts of each individual bubbles, but regard the bubbly flow as a mixture instead. The phases are represented by volume fractions and the interface is treated as a zone where the two phases coexists. Since physically bubbles exist on the sub-grid level, the volume fraction field does not define a sharp interface between the phases but provides the volumetric fraction of liquid and vapor within a computational cell. Yuan et al. in [7, 8] and Chen et al. in [9] adopted the volume of fluids method, combined with various turbulence models to describe the flow field. A simplified Rayleigh-Plesset relation [8, 10], which reflects spherical bubble dynamics, was used to model the phase change. Along the same lines, Mireia and Fuchs [11] investigated the interaction between cavitation and turbulence.

As it is shown in [12], the volume fraction field of the vapor phase is associated with only one bubble radius (for each computational cell). It is therefore natural to think of a representation through which a distribution of bubble radii can be rendered. Models that use probability density functions (PDF) have attained a certain level of popularity first in the simulations of combustion applications. A full Eulerian PDF solving framework has been developed in [13] and applied successfully on cavitation in [14]. Applying certain stochastic procedures, this approach uses an ensemble of Eulerian fields to represent a distribution of bubble radii. In this work, the Eulerian Stochastic Field model similar to [14] is applied to represent the liquid volume fraction, and an incompressible solver that is based on pressure-velocity coupling is used as a flow solver in order to avoid the high computational cost typically involved in a density based solution. A solver is developed here within the open source C++ toolbox OpenFOAM extended 3.1. A numerical study on the flow inside a simplified geometry resembling features of a diesel injector, which corresponds to an on-going experimental campaign based on the same geometry, is performed to demonstrate the capability and behavior of the stochastic model.

Governing equations for mixture

In the current work, the two phase flow is regarded as a homogeneous mixture of liquid and vapor, therefore only one set of mass and momentum equation is involved for the mixture:

$$\frac{\partial \rho}{\partial t} + \nabla \cdot (\rho \vec{u}) = 0, \tag{1}$$

$$\frac{\partial}{\partial t}(\rho \vec{u}) + \nabla \cdot (\rho \vec{u}\vec{u}) = \nabla \cdot (\mu \nabla \vec{u}) - \nabla p, \tag{2}$$

where $\vec{u}$ denotes the velocity and p the pressure. Here we adopt the isothermal mixture assumption, thus no energy equation is involved in solving the system, and pressure p will be determined by pressure-velocity coupling. The density ρ and the dynamic viscosity μ are closed by the following relations in line with the homogeneous mixture assumption:

$$\rho = (1 - \alpha_l)\rho_v + \alpha_l \rho_l, \tag{3}$$

$$\mu = (1 - \alpha_l)\mu_v + \alpha_l \mu_l, \tag{4}$$

where indices l and v correspond to liquid and vapor phases, respectively. α_l denotes the volume fraction of the liquid phase, the definition of which will be given later. Turbulence has been found to be closely relevant to cavitation studies in previous literatures [11, 15], therefore a large eddy simulation (LES) framework is applied in this work and the behavior of subgrid scale turbulence is modeled by a standard Smagorinsky model [16]. Thus, μ_{eff} is used in the context of LES in place of the molecular viscosity μ in equation (4):

$$\mu_{eff} = \mu + \mu_t, \tag{5}$$

where μ_t denotes the turbulent viscosity which is provided by the turbulence model.

Single field cavitation model

It can be observed in equation (3) and (4) that both density ρ and viscosity μ depend on α_l. Therefore, it takes an additional equation for α_l to close the Navier-Stokes equation for the mixture. In the typical single volume fraction model, the vaporized gas phase is assumed to consist of monodispersed spherical bubbles in each control volume. Therefore, the liquid volume fraction can be written as

$$\alpha_l = \frac{1}{1 + n \times 4\pi R^3/3}, \tag{6}$$

where R denotes the bubble radius and n is the nuclei number density, which is defined as the number of nuclei per unit volume of liquid and is assumed to be constant, which will be discussed later. Taking the material derivative of α_l, the transport equation for the (single) liquid volume fraction can be obtained:

$$\dot{\alpha}_l + \vec{u} \cdot \nabla \alpha_l = -3\alpha_l(1 - \alpha_l)\frac{\dot{R}}{R}. \tag{7}$$

A model for the bubble growth rate $\dot{R}$ is needed to close equation (7). A simple, yet effective description by Rayleigh [17] is widely applied in cavitation modelling. Neglecting thermal effects and surface tension forces, we have

$$\dot{R} = \sqrt{\frac{2}{3}\frac{p(R) - p_\infty}{\rho_l}}, \tag{8}$$

where $p(R)$ is the pressure in the liquid at the liquid-bubble interface, which in the current model is assumed to be equivalent to the saturation pressure, and p_∞ is the pressure at a large distance from the bubble which is in practice assumed to be equivalent to the local pressure in the control volume in which the bubble is situated.

PDF equation and stochastic model

The standard single volume fraction multiphase flow model described above solves the transport equation of one single liquid volume fraction (7). Such an approach takes no consideration of local fluctuations and bubble size distributions. As pointed out in [14], cavitating flows are inherently stochastic because of the uncertainty involved in water quality (nuclei size and number density variations) and turbulence-cavitation interaction. Given the restrictions on computational and experimental capacities, solving for the instantaneous number and size of bubbles at specific locations would be neither economically feasible nor necessary for most practical flow applications. Therefore, the purpose of the current study is to develop and refine a model that is capable of reproducing statistical properties of the flow, here the size distribution of vapor cavitation bubbles. Numerical models based on probability density functions (PDF) have gained certain popularity to describe stochastic phenomena, especially in the field of turbulent combustion, where PDF methods have demonstrated their capability of dealing with highly nonlinear problems in turbulent flow. In such methods, the transport equation of the joint PDF instead of an ensemble of single realizations is solved. Here we use the PDF approach to model stochastic variations of the liquid volume fraction α_l, which implies a vapor bubble size distribution according to equation (6). In our numerical solution of the liquid volume fraction α_l, similar to that of the dynamically passive scalar in [13] but without the molecular diffusion term which is not present in the transport equation for the liquid volume fraction (7), we have the transport equation of the liquid volume fraction PDF, P_{α_l}, as follows:

$$\frac{\partial P_{\alpha_l}}{\partial t} + u_i \frac{\partial P_{\alpha_l}}{\partial x_i} + \frac{\partial}{\partial \psi}\left[S(\alpha_l) P_{\alpha_l}\right] = \frac{\partial}{\partial x_i}\left(D' \frac{\partial P_{\alpha_l}}{\partial x_i}\right), \tag{9}$$

where $P_{\alpha_l}(\psi)$ is the probability of $\psi < \alpha_l < \psi + \mathrm{d}\psi$. $\boldsymbol{u} = u_i$ is the filtered velocity in the present LES framework, D' the turbulent diffusivity that accounts for subgrid scale turbulent transport which is also calculated from the Smagorinsky model [16]. $S(\alpha_l)$ is the volumetric source term given by the RHS of equation (7).

Several previous studies of combustion problems, e.g., [18], [19] and [20], have applied Monte Carlo methods on the numerical solution of the PDF. Monte Carlo methods are a broad class of computational algorithms that are used on a wide range of mathematical and physical problems. The most common approach at present to solve PDF transport equations with Monte-Carlo methods is the representation of the PDF via Lagrangian particles, which is grid free as particles carry their own velocity, position, and scalar state. In [13] a novel PDF representation is developed, in which the description of the PDF does not rely on Lagrangian particles [18], but is based on Eulerian stochastic fields (ESF) instead. In the current study, through the random sampling process, a series of stochastic fields are taken as representatives that share the same expectation with the original volume fraction. The essence of the method is to approximate the PDF with a finite number of stochastic fields $\alpha_l(x,t)$:

$$P_{\alpha_l}(\psi; x, t) = \frac{1}{N}\sum_{n=1}^{N} \delta[\psi - \alpha_l^n(x,t)] := \langle \delta[\psi - \alpha_l(x,t)] \rangle, \tag{10}$$

where δ represents the Dirac delta function and N is the number of stochastic fields. It is important to note that the stochastic fields $\alpha_l^n(x,t)$ are continuous and twice differentiable in space. As Valino [13] stated, the N representative fields in the above equation are not any particular realization of the real field, but a stochastic system allegedly equivalent, which provides an unbiased estimation of possible scalar values on each value of x and t. It is straightforward to obtain an estimation of moments for $\alpha_l(x,t)$ from $\alpha_l^n(x,t)$, i.e. the mean value of the liquid volume fraction can be evaluated as

$$\langle \alpha_l \rangle = \frac{1}{N} \sum_{i=1}^{N} \alpha_l^i. \tag{11}$$

According to the law of large numbers, the approximation of the PDF is exact for an infinite number of stochastic fields. However, in practical simulations, too large a number of stochastic fields would entail an unaffordable computational cost. According to [13, 14], 8 fields represent a good compromise between stability and efficiency. However, in the current study we take 50 stochastic fields in order to fully capture the transient behavior of the PDF. Such choice inevitably involves formidable overhead on computational time and even memory requirement, however, it is justified by the relatively short simulation time using the stochastic solver here. And similar to [14], since LES simulation is performed to capture unsteady behavior of both the flow and stochastic fields, the joint evolution of velocity and stochastic fields is captured.

In [21] it is proved that a stochastic process described by equation (9) is equivalent to the corresponding Ito stochastic differential equation (SDE). Therefore, transport equations of stochastic fields can be written as:

$$\mathrm{d}\alpha_l^n = \left[-u_i \frac{\partial \alpha_l^n}{\partial x_i} + \frac{\partial}{\partial x_i} \left(D' \frac{\partial \alpha_l^n}{\partial x_i} \right) + S(\alpha_l^n) \right] \mathrm{d}t + \sqrt{2D'} \frac{\partial \alpha_l^n}{\partial x_i} \mathrm{d}W, \tag{12}$$

where W is known as a Wiener process, which is a continuous but non-differentiable Gaussian stochastic process with

$$\mathrm{d}W = \epsilon \sqrt{\mathrm{d}t}, \tag{13}$$

where $\epsilon \sim N(0,1)$, with N representing a normal distribution with zero mean and variance one. From the definition of α_l in equation (6), each stochastic field obviously corresponds to a field of bubble radii through the following relation:

$$R^n = \sqrt[3]{\frac{3(1-\alpha_l^n)}{4\pi n \alpha_l^n}}. \tag{14}$$

It is obvious that a numerical scheme that is devised specifically to integrate both the deterministic term over time and the contribution from the stochastic term is required to perform the stochastic integration on equation (12) in a theoretically consistent and numerically accurate way. In this work, a second order stochastic Runge-Kutta scheme that was proposed in [22, 23] is adopted.

Result and discussions

The geometry involved in the current study is shown in Figure 1(a). Diesel fuel is enters at a pressure value of 390 bar from the inlet (the surface on the top) and leaves the injector at the outlet at a back pressure of 1 bar. The physical properties of the diesel used in the simulation are similar to those in [11]. A mesh with 7.5 million grid nodes is generated using ICEM-CFD, with the nozzle region (smaller cylinder) much finer than the sac volume. Lines L0, L1, and L2, and the corresponding points P0, P1, and P2 are shown in Figure 1(b) in a zoomed-in view of the mid-section of nozzle area as reference locations for data and results analysis.

In order to observe the fluid dynamic characteristics of the case of interest and validate the stochastic model, time averaged velocity and pressure profile along L1 that are calculated by the stochastic model and a typical single volume fraction solver, namely interPhaseChangeFoam, are compared in Figure 2. The velocity profiles achieved a very close agreement. The pressure profiles agree very well in their trend along the spatial location, with the absolute value predicted by the stochastic model slightly higher than in the single volume fraction solver result. The difference is relatively small considering the pressure difference close to 400 bar involved in the current case of study. Such difference is likely caused by the perturbation in volume fraction. It is obvious that the pressure field solution is more sensitive to the change in cavitation model as compared to velocity field in the case of interest. Similar agreements have been observed in profiles along P0, and P2 (not included).

A zoomed-in view on the mid-section is applied in Figure 3 to demonstrate the comparison between volume fraction predicted by the stochastic model and interPhaseChangeFoam. With the large pressure difference involved in the case, the in-nozzle flow is highly turbulent and three dimensional as it can be expected. In Figure 3, the turbulent flow structure is captured well by both the stochastic model and interPhaseChangeFoam.

As discussed earlier, with the proposed stochastic model, a distribution of vapor radii is solved, based on which the PDF at different spatial location P0, P1, and P2 at time 0.0003s is shown in Figure 4. Since the main spot of cavitation inception is expected to be close to the sharp contraction at nozzle inlet, upstream location P0 has a stronger cavitation than downstream location P1 and P2, with P1 slightly stronger than P2. The trend is correctly captured by the PDF. PDF evolution at P1 over time has also been captured (not included).

Figures & Tables:

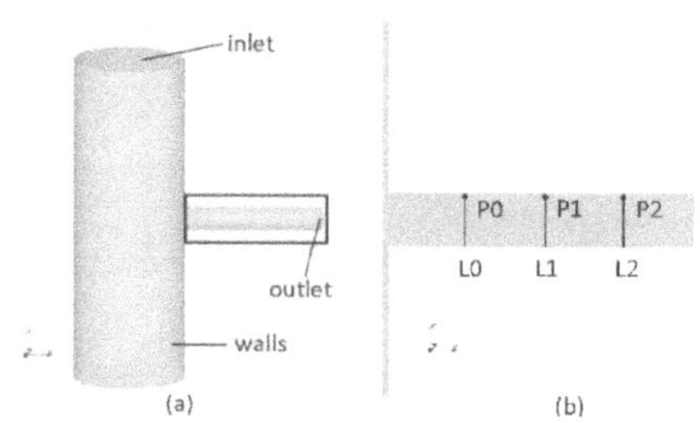

Figure 1. The academic diesel injector geometry and the probe locations for results analysis.

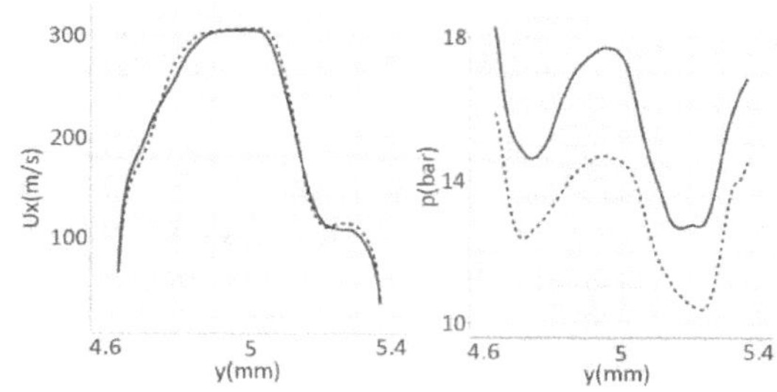

Figure 2. Time averaged filtered velocity profile (left) and pressure profile (right) along L1. Solid lines represent velocity and pressure computed by stochastic model, dash lines by interPhaseChangeFoam.

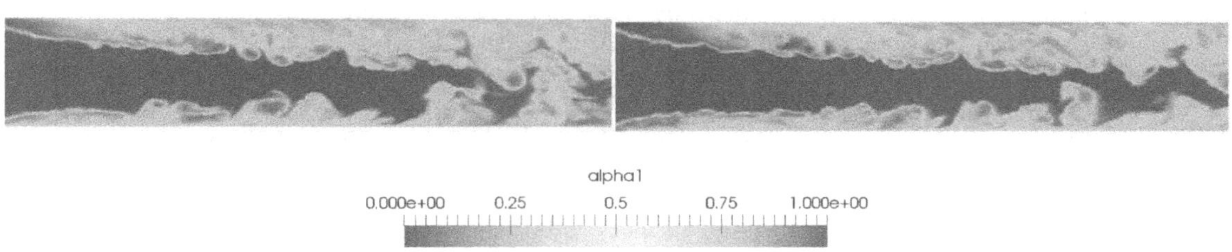

Figure 3. Instantaneous mean volume fraction field from the stochastic model (left) and volume fraction field from interPhaseChangeFoam (right).

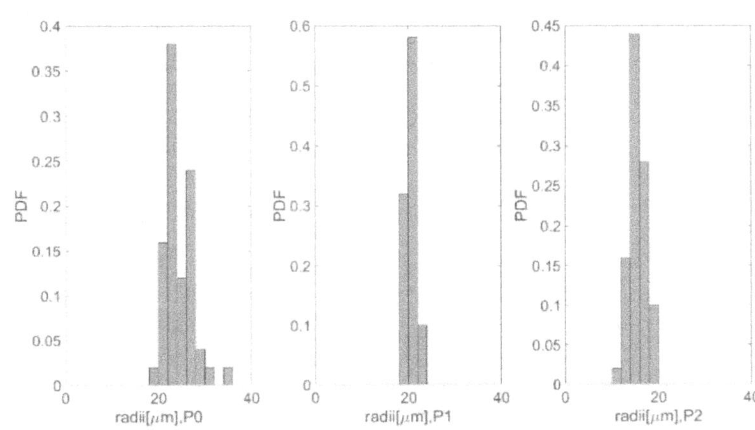

Figure 4. PDF of bubble radii at probe locations P0, P1, and P2.

Conclusion

A three dimensional simulation of diesel flowing through an academic injector nozzle featuring an Eulerian stochastic field method combined with a Large Eddy Simulation approach has been performed. The cavitating flow is modelled as a homogenous mixture in a volume of fluids framework. A simplified Rayleigh-Plesset model has been adopted to model the mass transfer between the liquid and the vapor phase. A fully Eulerian fields based Monte-Carlo method has been implemented in the open source C++ toolbox Foam extended 3.1. Results from the stochastic model have been compared with those from a typical single volume fraction solver, namely interPhaseChangeFoam. Good agreement has been achieved with respect to the prediction of flow behavior and mean liquid volume fraction. The study of the current case has clearly demonstrated the capability of the stochastic model to provide access to both steady and transient behavior of bubble radii PDF.

References

1. Sou, A., H. S., and A. Tomiyama, *Effects of cavitation in a nozzle on liquid jet atomization.* International Journal of Heat and Mass Transfer, 2007. **50**: p. 3575-3582.
2. Roosen, P., *Investigation of the Transient Behaviour of Cavitation Effects in Liquid Injection Nozzles*, in *Transient phenomena in multiphase and multicomponent systems*. 2007, Wiley-VCH Verlag GmbH. p. 72-83.
3. Winklhofer, E., et al. *Comprehensive hydraulic and flow field documentation in model throttle experiments under cavitation conditions*. in *ILASS-Europe*. 2001. Zurich.
4. Ginnadakis, E., M. Gavaises, and C. Arcoumanis, *Modeling of cavitation in diesel injector nozzles.* Journal of fluid mechanics, 2008. **616**: p. 153-193.
5. Dabiri, S., W.A. Sirignano, and D.D. Joseph, *A numerical study on the effects of cavitation on oriface flow.* Physics of fluids, 2010. **22**(042102): p. 1-13.
6. Rodio, M.G. and R. Abgrall, *An innovative phase transition modelling for reproducing cavitation through a five-equation model and theoretical generalization to six and seven-equation modelss.* International journal of heat and mass transfer, 2015. **89**: p. 1386-1401.
7. Yuan, W.X., J. Sauer, and G.H. Schnerr, *Modeling and computation of unsteady cavitation flows in injection nozzles.* Mecanique& Industries, 2001. **2**(5): p. 383-394.
8. Yuan, W. and G.H. Schnerr, *Numerical simulation of two-phase flow in injection nozzles: interaction of cavitation and external jet formation.* Journal of fluids engineering, 2003. **125**: p. 963-969.
9. Chen, B. and M. Oevermann. *LES modelling of cavitation flow in a diesel injector nozzle* in *International Conference on Multiphase Flow*. 2016. Florence.
10. Rayleigh, L., *On the pressure developed during the collapse of a spherical cavity.* Philosophical Magazine Series 6, 1917. **34**(200): p. 94-98.
11. Altimira, M. and L. Fuchs, *Numerical investigation of throttle flow under cavitating conditions.* International Journal of Multiphase Flow, 2015. **75**: p. 124-136.
12. Yuan, W. and G.H. Schnerr, *Numerical simulation of two-phase flow in injection nozzles: Interaction of cavitation and external jet formation.* Journal of Fluids Engineering, 2003. **125**(6): p. 963-969.
13. Valiño, L., *A field Monte Carlo formulation for calculating the probability density function of a single scalar in a turbulent flow.* Flow, turbulence and combustion, 1998. **60**(2): p. 157-172.
14. Dumond, J., F. Magagnato, and A. Class, *Stochastic-field cavitation model.* Physics of Fluids, 2013. **25**(7): p. 073302.
15. Sou, A., S. Hosokawa, and A. Tomiyama, *Effects of cavitation in a nozzle on liquid jet atomization.* International Journal of Heat and Mass Transfer, 2007. **50**(17): p. 3575-3582.
16. Smagorinsky, J., *General circulation experiments with the primitive equations: I. The basic experiment.* Monthly weather review, 1963. **91**(3): p. 99-164.
17. Rayleigh, L., *On the pressure developed in a liquid during the collapse of a spherical cavity, VIII.* The London, Edinburgh, and Dublin Philosophical Magazine and Journal of Science, 1917. **34**(200): p. 94-98.
18. Pope, S.B., *PDF methods for turbulent reactive flows.* Progress in Energy and Combustion Science, 1985. **11**(2): p. 119-192.
19. Delarue, B. and S. Pope, *Calculations of subsonic and supersonic turbulent reacting mixing layers using probability density function methods.* Physics of Fluids, 1998. **10**(2): p. 487-498.
20. Pope, S.B., *A Monte Carlo method for the PDF equations of turbulent reactive flow.* Combustion Science and Technology, 1981. **25**(5-6): p. 159-174.
21. Gardiner, C.W., *Handbook of stochastic methods*. Vol. 3. 1985: Springer Berlin.
22. Rößler, A., *Runge-Kutta methods for Itô stochastic differential equations with scalar noise.* BIT Numerical Mathematics, 2006. **46**(1): p. 97-110.
23. Zahri, M., *ON NUMERICAL SCHEMES FOR SOLVING A STOCHASTIC ADVECTION-DIFFUSION.* International Journal of Pure and Applied Mathematics, 2012. **77**(5): p. 681-694.

© 2018 ASME

Luminescence based measurements in micro cavitating flow

D. Podbevšek[1], D. Colombet[2], F. Ayela[2], H. T. T. Lai[1], M. Martini[1], O. Tillement[1], G. Ledoux[1],*

[1]*Institut Lumière Matière, Université Claude Bernard Lyon 1, 69622 Villeurbanne, France.*

[2]*Laboratoire des Écoulements Géophysiques et Industriels, Domaine Universitaire, 38400 Saint Martin d'Hères, France.*

Abstract

Microchannels are often used to study fluid dynamics. For steady state problems like cavitating flow, fluorescent microscopy, with the addition of temperature sensitive nanoprobes into the observed fluid, can be used to determine the temperature at a chosen point, averaged over the integration time. Coupled with a confocal microscope setup, we are able to produce two and three dimensional temperature maps of the flow in the microchannel by the use of ratiometric intensity measurements. The nanometric scale of the probes assures fast thermalization of the probes and below certain concentrations does not modify the properties of the studied liquid. Since the probes are not present in the vapor phase, the relative intensity map also corresponds to the average void fraction in the flow. These nanoprobes are composed of a gold core and a polysiloxane shell containing fluorescent dyes (FITC, RBITC). Organic dyes were chosen due to their compatibility with the shell and primarily for the fast luminescence lifetime, which is essential due to the rapid flow in the microchannel and the consequent short dwell time of an individual nanoprobe in the excitation volume. The temperature information in each measured point is obtained from the temperature sensitive spectrum of the dye. The shell protects the dye from the environment and allows for the functionalization of the surface to prevent agglomeration, while the gold core mitigates photo bleaching. The technic allowed us to observe temperature gradients in microfluidic two-phase flow and observe the thermal effect associated with phase transition. Typically, a region of decreased temperature is observed downstream the orifice in the liquid-vapor stream, attributed to the cooling of the liquid due to the latent heat of the phase change. However, small changes in the diaphragm geometry can induce recirculating vortices, where the vapor bubbles condensate and induce a region of increased temperature.

Keywords: Cavitation, microchannel, nanoprobes, temperature, thermal effect.

Introduction

Microfluidic devices are frequently used to study fluid dynamics, primarily when specialized liquids are used, which are often hard to synthesize in volume (nanofluids). A narrow opening in a microfluidic channel will accelerate the liquid and, in accordance with Bernoulli's principle, cause the pressure locally to drop. When the pressure decreases below vapor pressure, the liquid will become metastable and eventually vapor bubbles appear. These bubbles grow in the low pressure region and collapse violently upon exiting, which leads to a steady state two-phase flow downstream of the constriction (figure 1a). Due to the latent heat linked to the phase transition, temperature gradients are expected in the region of the growth and collapse of the bubbles [1–4]. Radical production has also been observed with cavitation, linked to the violent bubble collapse [5]. The addition of luminescent species into the liquid allows the use of luminescent based technics to yield the information about the local environment [6].

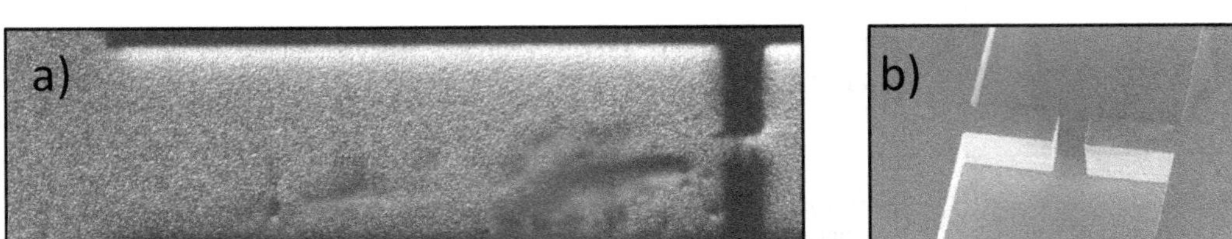

Figure 1: a) Example of cavitating flow in microfluidic channel (8bar driving pressure); flow from right to left. b) SEM image of etched silicon, before sealing the channel with a Pyrex cap. Opening of the constriction in both cases is about 80μm.

The temperature information in each measured point is obtained from the emission spectrum of the dye. On the other hand, with the use of an aqueous solution of luminol as the working fluid, we can probe the radical yield following the violent bubble collapse, via the chemiluminescent pathway. Our aim is to demonstrate the use of luminescent based techniques for sensing in microcavitating flow and show recent developments in the field.

* Corresponding author: gilles.ledoux@univ-lyon1.fr

Body

The use of temperature sensitive luminescent nanoprobes in the observed liquid (water) and the confocal microscope setup for the excitation and detection, makes 2 and 3D characterization of the flow downstream the constriction possible (figure 2). The thermometric method used is called the ratiometric intensity measurement and is the least sensitive to variation in concentration and perturbations due to vapor bubbles in the two-phase flow [1]. It uses the temperature sensitive emission spectra of the fluorophores in the nanoprobe's shell, where the peak shape is altered when the temperature varies. As shown in figure 3b, the ratio of two sections from the normalized emission spectrum corresponds to a given temperature.

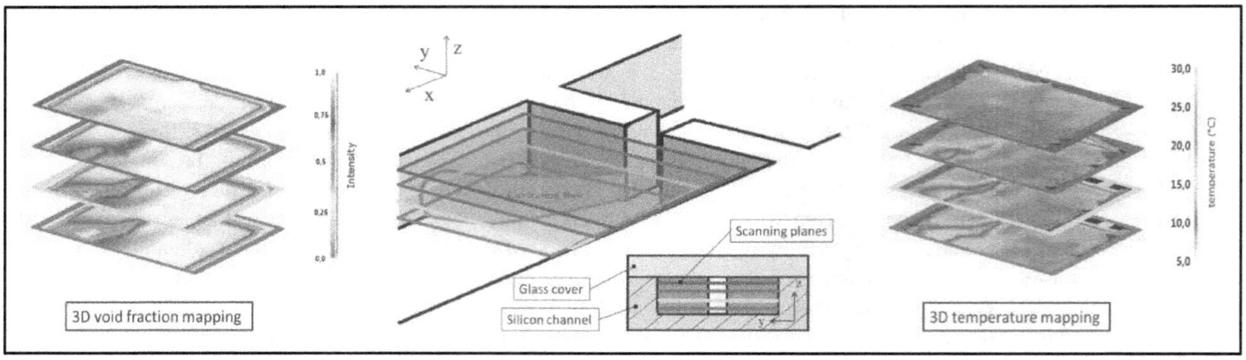

Figure 2: Raster scanning 2d planes at different heights makes 3d characterization of the channels possible. The recorded planes are recorded 20, 50, 70 and 90μm from the transparent wall (cover glass) of the microchannel. The observed two-phase flow and the associated cooling is predominantly located towards the silicon channel wall

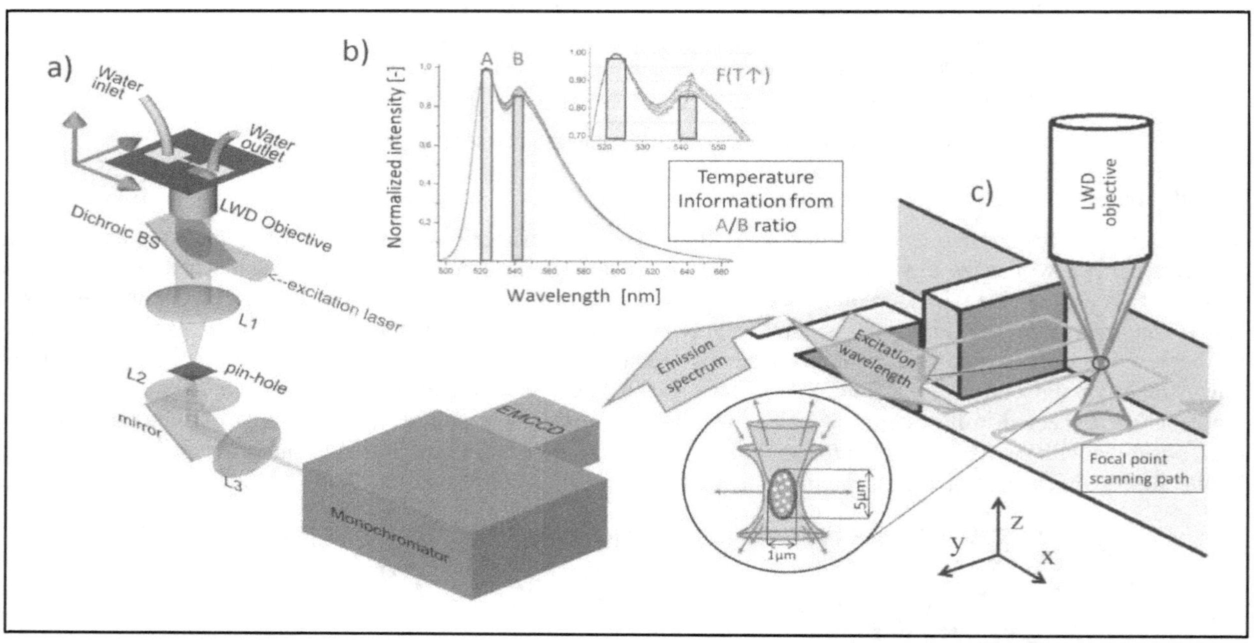

Figure 3: a) confocal microscope detection system, b) emission spectra varying with temperature, ratio of the two peaks yields a value corresponding to a certain temperature, c) focal point raster scanning and emission voxel with excited fluorophores in the nanoprobes producing the emission spectrum.

The encapsulation of the fluorophores into the nanoprobe shell protects them from the hostile environment produced by the violent bubble collapses. Organic dyes were chosen due to their compatibility with the shell and primarily for the fast luminescence lifetime (ns range), which is essential due to the rapid flow in the microchannel (20 – 30 m/s) and the consequent short dwell time of an individual nanoprobe in the excitation volume. The polysiloxane shell protects the dye and also allows for the functionalization of the surface to prevent agglomeration, while the gold core mitigates quenching [7]. The nanometric scale (diameter of 60nm) of the probes assures fast thermalization of the probes and below certain concentrations does not modify the properties of the studied liquid [8]. The channels used are made by DRIE process on a silicon wafer and anodic bonding of a Pyrex top, described in detail in [8]. These kind of channels have been used extensively by our team [8–10] and others [11–14]. Channel geometries most used are micro diaphragms (figure 1b) and micro Venturis.

The pinhole in the detections system of the confocal microscope is acting as a spatial filter which makes our emission voxel (volume where the signal is collected from) confined in all three directions (Figure 3a,c). By raster scanning an area of interest with this voxel and recording the spectrum in each point, we are able to produce two and three dimensional temperature maps of the flow (figure 2). Furthermore, since the probes are not present in the vapor phase, the relative intensity map also corresponds to the average void fraction in the flow (averaged over the integration time of a single acquisition point).

An example of the recorded temperature gradients and the corresponding intensity map (indication of void fraction) downstream the constriction are show in figure 4 (flow pattern similar to figure 1a). The low temperature region corresponds neatly with the two-phase flow, indicated by lower intensity on the relative intensity map. This indicates that the cooling is most likely linked to the phase transition (latent heat). Using the same technique, an area of heating was identified in a recirculating vortex [1], caused by the bubble collapse (condensation). Although using the same channel design, even slight variations of the channel geometry can cause differences in the resulting flow pattern and produce a stabile recirculating vortex, where bubble collapses are localized.

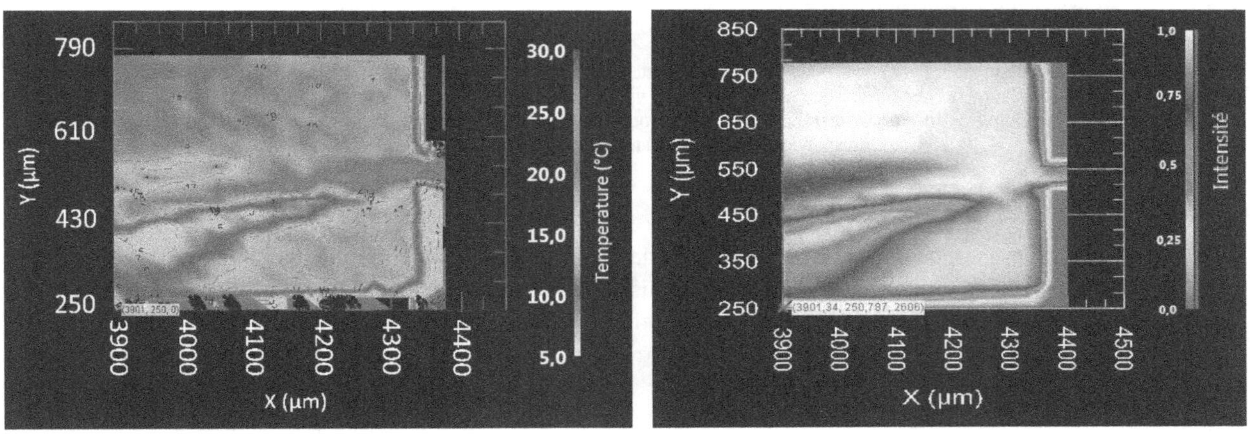

Figure 4: Preliminary temperature and intensity map of flow conditions corresponding to figure 1. Intensity map can be considered the average void fraction mapping, since the dwell time of the vapor phase in the excitation voxel during recording, will yield a proportional drop in intensity. The cooling region and the lower intensity correspond to the two-phase flow region, seen in figure 1. The pressure drop over the microdiaphragm constriction is 10bar, corresponding to flow conditions of about 320µl/s.

A different technique recently developed, allows us to quantify OH radical production in hydrodynamic cavitation. Radical formation has been frequently observed with cavitation and is believed to be linked with the extreme conditions in the bubble implosion. The well-known chemiluminescent reaction of luminol with radical species has been used with ultrasonic cavitation [15–18], while for hydrodynamic cavitation only one previous application was found by the authors [19]. Due to the different bubble dynamics between ultrasonic and hydrodynamic cavitation [20], there were concerns whether there will be significant radical production also for hydrodynamic cavitation [14]. Using an aqueous solution of luminol and the same microfluidic devices as before, the chemiluminescent pathway was used to quantify OH radical production, by a photon counting technique [21]. By placing the photomultiplier tube (PMT) on top of the microfluidic channel, as close as possible to the cavitation active area and using a simple microphone to coordinate when cavitation was occurring, we could observe the well correlated relationship between the cavitation noise and the chemiluminescent signal from the OH/luminol reaction (figure 5). The PMT detects individual photons being produced and by logging the arrival time, the photon yield could be obtained. As the solid detection angle was known and the efficiencies (PMT, quantum yield) estimates in place, we could therefore estimate the actual radical production rate at specific flow conditions. A linear relationship was observed between the flow rate and photon production rate for several microfluidic devices. Considering the relative simplicity of the technique and the ability to quantify radical production, it is a viable option for optimizing radical yield in hydrodynamic flows (biological or chemical wastewater treatment).

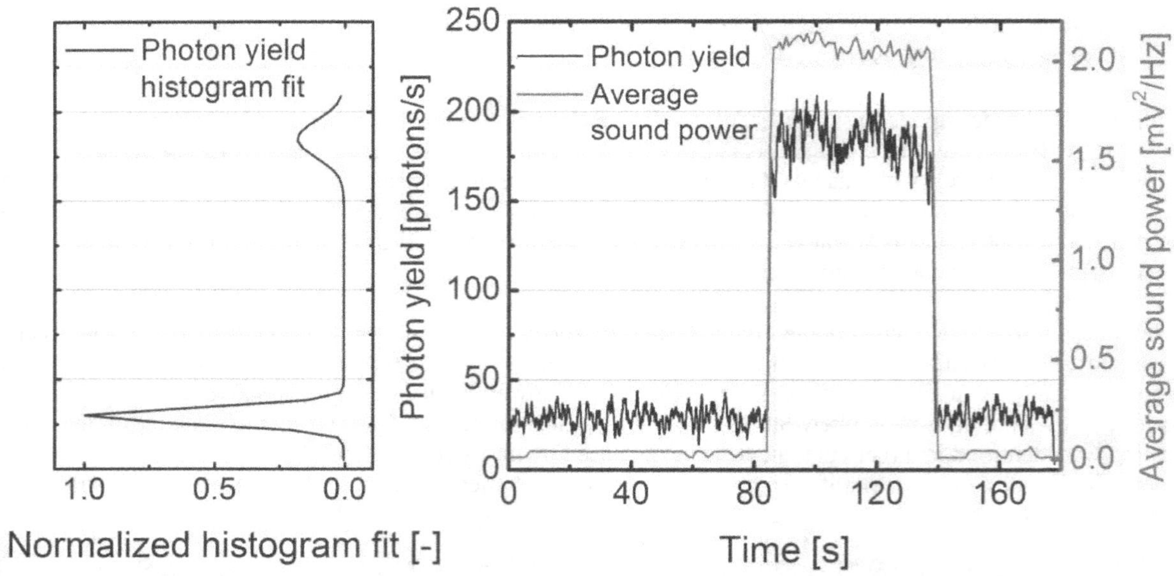

Figure 5: Photon production rate (black) and sound power (red) plotted with time (right) and the fit of the histogram data from the PMT (left). Clearly visible is the correlation between cavitation noise and photon yield. The background signal from the PMT is around 25photons/s which is considered the chemiluminescence off state (no photons being produced). The pressure drop over the microdiaphragm was 10 bar which corresponds to 377μl/s and an average jet velocity of 35.8m/s.

Conclusion

Two luminescence based optical techniques have been demonstrated as a way to probe into the local environment of a microcavitating flow. The temperature dependent emission spectrum of the nanoprobes dispersed in the working liquid, allowed us to observe temperature gradients in microfluidic two-phase flow and the thermal effect associated with phase transition. The preliminary results show a region of decreased temperature observed downstream the orifice in the liquid-vapor stream, as might be expected due to the cooling of the liquid by the phase change transition. The second technique used an aqueous solution of luminol as the working fluid, which allowed the quantification of OH radicals produced by the hydrodynamic cavitation via the chemiluminescent pathway. By using a PMT with photon counting capabilities we could show the correlation of chemiluminescent photon yield to the cavitation noise. As this is a direct and quantitative evidence of OH radical formation in hydrodynamic cavitation, this simple technique could be used for radical production optimization in cavitating flows.

References

[1] Ayela, F., Medrano-Muñoz, M., Amans, D., Dujardin, C., Brichart, T., Martini, M., Tillement, O., and Ledoux, G., 2013, "Experimental Evidence of Temperature Gradients in Cavitating Microflows Seeded with Thermosensitive Nanoprobes," Phys. Rev. E Stat. Nonlin. Soft Matter Phys., **88**(4), p. 043016.

[2] Petkovšek, M., and Dular, M., 2013, "IR Measurements of the Thermodynamic Effects in Cavitating Flow," Int. J. Heat Fluid Flow, **44**(Supplement C), pp. 756–763.

[3] Petkovšek, M., and Dular, M., 2017, "Observing the Thermodynamic Effects in Cavitating Flow by IR Thermography," Exp. Therm. Fluid Sci., **88**(Supplement C), pp. 450–460.

[4] Rimbert, N., Castanet, G., and Funfschilling, D., 2013, "Measurement of Thermal Effects in a Cavitating Channel Flow by 2cLIF."

[5] Arrojo, S., Nerín, C., and Benito, Y., 2007, "Application of Salicylic Acid Dosimetry to Evaluate Hydrodynamic Cavitation as an Advanced Oxidation Process," Ultrason. Sonochem., **14**(3), pp. 343–349.

[6] Valeur, P. D. B., and Berberan-Santos, P. M. N., *Molecular Fluorescence: Principles and Applications, Second Edition.*

[7] Martini, M., Perriat, P., Montagna, M., Pansu, R., Julien, C., Tillement, O., and Roux, S., 2009, "How Gold Particles Suppress Concentration Quenching of Fluorophores Encapsulated in Silica Beads," J. Phys. Chem. C, **113**(41), pp. 17669–17677.

[8] Medrano, M., Zermatten, P. J., Pellone, C., Franc, J. P., and Ayela, F., 2011, "Hydrodynamic Cavitation in Microsystems. I. Experiments with Deionized Water and Nanofluids," Phys. Fluids, **23**(12), p. 127103.

[9] Medrano, M., Pellone, C., Zermatten, P. J., and Ayela, F., 2012, "Hydrodynamic Cavitation in Microsystems. II. Simulations and Optical Observations," Phys. Fluids, **24**(4), p. 047101.
[10] Mossaz, S., Colombet, D., and Ayela, F., 2017, "Hydrodynamic Cavitation of Binary Liquid Mixtures in Laminar and Turbulent Flow Regimes," Exp. Therm. Fluid Sci., **80**(Supplement C), pp. 337–347.
[11] Mishra, C., and Peles, Y., 2005, "Flow Visualization of Cavitating Flows through a Rectangular Slot Micro-Orifice Ingrained in a Microchannel," Phys. Fluids, **17**(11), p. 113602.
[12] Mishra, C., and Peles, Y., 2004, "Cavitation in Flow through a Micro-Orifice inside a Silicon Microchannel," Phys. Fluids, **17**(1), p. 013601.
[13] Mishra, C., and Peles, Y., 2006, "An Experimental Investigation of Hydrodynamic Cavitation in Micro-Venturis," Phys. Fluids, **18**(10), p. 103603.
[14] Rooze, J., André, M., Gulik, G.-J. S. van der, Fernández-Rivas, D., Gardeniers, J. G. E., Rebrov, E. V., Schouten, J. C., and Keurentjes, J. T. F., 2012, "Hydrodynamic Cavitation in Micro Channels with Channel Sizes of 100 and 750 Micrometers," Microfluid. Nanofluidics, **12**(1–4), pp. 499–508.
[15] McMurray, H. N., and Wilson, B. P., 1999, "Mechanistic and Spatial Study of Ultrasonically Induced Luminol Chemiluminescence," J. Phys. Chem. A, **103**(20), pp. 3955–3962.
[16] Hatanaka, S.-I., Mitome, H., Yasui, K., and Hayashi, S., 2002, "Single-Bubble Sonochemiluminescence in Aqueous Luminol Solutions," J. Am. Chem. Soc., **124**(35), pp. 10250–10251.
[17] Price, G. J., Harris, N. K., and Stewart, A. J., 2010, "Direct Observation of Cavitation Fields at 23 and 515kHz," Ultrason. Sonochem., **17**(1), pp. 30–33.
[18] Cao, H., Wan, M., Qiao, Y., Zhang, S., and Li, R., 2012, "Spatial Distribution of Sonoluminescence and Sonochemiluminescence Generated by Cavitation Bubbles in 1.2MHz Focused Ultrasound Field," Ultrason. Sonochem., **19**(2), pp. 257–263.
[19] Schlender, M., Minke, K., and Schuchmann, H. P., 2016, "Sono-Chemiluminescence (SCL) in a High-Pressure Double Stage Homogenization Processes," Chem. Eng. Sci., **142**(Supplement C), pp. 1–11.
[20] Ashokkumar, M., 2011, "The Characterization of Acoustic Cavitation Bubbles – An Overview," Ultrason. Sonochem., **18**(4), pp. 864–872.
[21] Podbevsek, D., Colombet, D., Ledoux, G., and Ayela, F., "Observation of Chemiluminescence Induced by Hydrodynamic Cavitation in Microchannels.," Ultrason. Sonochem. (In Press, Accepted Manuscript - doi:10.1016/j.ultsonch.2018.01.004).

© 2018 ASME

Material Bend-Twist Coupling Effects on Cavitating Response of Composite Hydrofoils

Yin L. Young[1,*], Nitin Garg[1], Paul A. Brandner[2], Bryce W. Pearce[2], Daniel Butler[3], David Clarke[3], Andrew W. Phillips[3]

[1] Department of Naval Architecture and Marine Engineering, University of Michigan, MI, USA
[2] Australian Maritime College, University of Tasmania, Launceston, TAS, Australia
[3]Maritime Division, Defence Science and Technology Group, Fishermans Bend, VIC, Australia

Abstract

The objective of this work is to investigate the effects of material bend-twist coupling on the cavitating response of adaptive composite hydrofoils. Experimental results are shown for two composite hydrofoils and one stainless steel (SS) hydrofoil. All three hydrofoils have identical unloaded geometry and are tested in the same cantilevered configuration at the Cavitation Research Laboratory variable pressure tunnel at the University of Tasmania. The results show that material bend-twist coupling that leads to nose-up twist (N30 hydrofoil) will increase the mean hydrodynamic load, accelerate cavitation inception, increase the maximum cavity length, and lower the Type II cavity shedding frequency compared to the SS hydrofoil. The opposites are true for material bend-twist coupling that leads to nose-down twist (P30 hydrofoil). For all three hydrofoils, Type I shock-wave driven cavity shedding is observed when the maximum cavity length normalized by the chord is between 0.75 and 1.7, while Type II re-entrant jet driven cavity shedding is observed for the full range of cavitating flow. In addition, significant load amplification is observed when the Type II cavity shedding frequency is near the first wetted natural frequency of the P30 hydrofoil. To complement the experimental studies, semi-empirical relations have been developed to predict the hydroelastic response of the hydrofoils in cavitating flow, and good agreement is observed between predictions and measurements. The results show that by accounting for the flow-induced twist in the effective angle of attack (α_{eff}), and using $\psi = \sigma/2\alpha_{eff}$ as the effective cavitation parameter (with σ as the cavitation number), the hydroelastic response for all three hydrofoils can be collapsed onto the same trend lines.

Keywords: bend-twist coupling, composite, flexible, hydrofoil, cavitation

Introduction

Advanced materials such as carbon fiber reinforced polymers (CFRP) and glass fiber reinforced polymers (GFRP) are attractive for maritime applications because they have high specific strength and specific stiffness, as well as better resistance to seawater corrosion than typical metallic alloys. Moreover, the intrinsic bend-twist coupled behavior of anisotropic composites can be exploited to passively tailor the load-dependent deformations of lifting bodies to improve performance. In particular, recent works have explored the use of material bend-twist coupling to enable passive pitch adaptation to improve the performance of propellers, turbines, hydrofoils and other control surfaces [12]. Nevertheless, most of the work thus far on composite hydrodynamic lifting devices are limited to fully wetted conditions. Since such devices are susceptible to cavitation, the objective of this work is to investigate the effects of material bend-twist coupling on the cavitating response of adaptive composite hydrofoils.

Results

Experimental results are shown for two composite hydrofoils and one stainless steel (SS) hydrofoil, where the SS hydrofoil is used to serve as the rigid reference. All three hydrofoils have identical unloaded geometry and are tested in the same cantilevered configuration at the Cavitation Research Laboratory (CRL) variable pressure tunnel at the University of Tasmania. Details of the experimental setup and procedure can be found in [9,11]. All the hydrofoils have modified NACA 0009 cross sections with a loaded span of $s = 0.3$ m and a linearly tapered planform with a root chord of 0.12 m, a tip chord of 0.06m, and a mean chord of $c = 0.09$ m. The two composite hydrofoils are both made of epoxy resin reinforced with the same nominal layup of CFRP and GFRP. The only difference between the composite hydrofoils is the orientation of the structural CFRP layers. For simplicity, the two composite hydrofoils

*Corresponding author, E-mail: ylyoung@umich.edu, Tel.: +001-734-647-0249.

are called P30 and N30 from herein to indicate the +30° and -30° orientation, respectively, of the structural CFRP layers relative to the spanwise axis, with the positive direction towards the foil leading edge.

The influence of material bend-twist coupling on the fully wetted response of adaptive composite hydrofoils for a wide range of flow conditions have been investigated in [11]. They presented experimental results for the SS, P30 and N30 hydrofoils mentioned here along with another composite hydrofoil with identical geometry and nominal material composition but with the structural CFRP layers oriented parallel to the spanwise axis (P00). They also presented a simplified two-degrees-of-freedom model to predict the hydroelastic response of adaptive composite hydrofoils in steady-state fully wetted flows. They explained that material bend-twist coupling results when the structural CFRP layers are oriented away from the spanwise axis of the hydrofoil. At positive angles of attack, the hydrodynamic lift induces spanwise bending towards the suction side for all the hydrofoils. For fully attached flow, the center of pressure is upstream of the elastic axis, which leads to a slight nose-up twist for isotropic material response (e.g. SS hydrofoil). However, the effect of flow-induced bend-twist coupling on the hydroelastic response is small in comparison to material-induced bend-twist coupling. The results show that the P30 hydrofoil underwent increasing nose-down twist, while the N30 hydrofoil underwent increasing nose-up twist, with increasing hydrodynamic load caused by higher flow speed (or Reynolds number, Re) and/or higher geometric angle of attack (α_o). Hence, while the SS and P00 hydrofoils showed negligible dependence on Re until near stall, the P30 hydrofoil resulted in decreasing lift and moment coefficients with increasing Re, while the N30 hydrodoil resulted in increasing lift and moment coefficients with increasing Re. Moreover, compared to the SS hydrofoil, material bend-twist coupling lead to delayed stall for the P30 hydrofoil and accelerated stall for the N30 hydrofoil.

The focus of this work is on the cavitating response of the SS, P30, and N30 hydrofoils. Experimental studies were conducted in the CRL for all three hydrofoils at $\mathrm{Re} = \frac{\mathrm{Uc}}{\nu_\mathrm{f}} = 0.8 \times 10^6$ and $\alpha_o = 6^o$ across a range of cavitation number, $\sigma = (P_\infty - P_v)/(0.5\rho_f U^2)$. U is the inflow speed, ρ_f and ν_f are respectively the density and kinematic viscosity of the demineralized water. P_∞ is the absolute static pressure at the test section centerline, and P_v is the vapor pressure. The measured variations of the instantaneous lift coefficients, $C_L = L/(0.5\rho_f U^2 sc)$, with σ for all three hydrofoils are shown in Figure 1(a). The symbols represent the mean values, and the error bars represent the amplitude of the instantaneous load fluctuations caused by unsteady cavity shedding. At $\mathrm{Re} = \frac{Uc}{\nu_f} = 0.8 \times 10^6$ and $\alpha_o = 6^o$, material bend-twist coupling lead to nose-up twist for the N30 hydrofoil and nose-down twist for the P30 hydrofoil. Hence, the mean C_L for the N30 and P30 hydrofoils are respectively higher and lower than the SS hydrofoil in fully wetted flow, i.e. when $\sigma > 3$. To remove differences in C_L and to account for the change in effective angle of attack caused by twist, the instantaneous lift coefficients are normalized by their respective wetted values, $C_{L,wet}$, and are plotted against the effective cavitation parameter, $\psi = \sigma/(2\alpha_{eff})$, in Figure 1(b). $\alpha_{eff} = \alpha_o + \frac{2\theta_{tip}}{\pi}$ is the effective angle of attack where α_o is the geometric angle of attack and θ_{tip} as the tip twist angle in fully wetted flow. The $2/\pi$ factor for θ_{tip} is used to account for the spanwise variation in twist. The first twist mode of the cantilevered hydrofoils generally follows a half sine curve that goes from zero at the root to maximum value at the tip, and integral of the first twist mode shape function along the span yields $2/\pi$. It should be noted that θ_{tip} was not measured in separated or cavitating flows because of challenges bought by flow-induced vibrations. Hence, the fully wetted θ_{tip} is used for simplicity, and it is equal to 0°, -1.2°, and +3.36° for the SS, P30, and N30 hydrofoils, respectively, at $Re = 0.8x10^6, \alpha_o = 6^o$. Also shown on Figure 1(b) are the measured maximum cavity lengths normalized by the mean chord ($L_{c,max}/c$) at mid-span afor all 3 hydrofoils. The results show that both the normalized mean lift coefficients and the normalized maximum cavity lengths for all three hydrofoils can be collapsed on the same trend lines when plotted against ψ instead of σ.

Cavitation inception occurred in the form of bubble cavitation. As σ lowers, bubble cavitation transitions to leading edge partial sheet cavitation, then to sheet-cloud cavitation, and finally to supercavitation. The mean C_L increases slightly with the growth of the leading edge partial cavitation because of the increase in effective camber, while the amplitude of the load fluctuations increase significantly due to periodic growth and collapse of the cavities. As σ further lowers, supercavitation develops when the maximum cavity extent reaches to beyond the foil trailing edge, which leads to rapid reduction in the mean C_L, while the amplitude of the C_L fluctuations continue to grow until the maximum cavity extent reaches to near 1.7 times the local chord length. Stable supercavitation develops when the cavity length reaches beyond approximately 2 times the local chord.

To complement the experimental measurements, semi-empirical relations (Equations (1)-(5)) have been developed to predict the variations in the lift coefficients, maximum cavity lengths, and cavity shedding frequencies.

3-D lift coefficient $$C_L = a_o^{3D}\alpha_{eff};\ a_o^{3D} = \frac{a_o^{2D}}{1+\frac{a_o^{2D}}{\pi A_R}};\ \alpha_{eff} = \alpha_o + \frac{2\theta_{tip}}{\pi} \quad (1)$$

2-D lift slope $$a_o^{2D} = \pi * \max(0.5, 2 - 5.19e^{-1.05\psi});\ \psi = \frac{\sigma}{2\alpha_{eff}} \quad (2)$$

Maximum cavity length $$\frac{L_{c,max}}{c} = 2.326e^{-0.316\psi} \quad (3)$$

Type I cavity Strouhal No: $$St_1 = \frac{f_1 c}{U} = 0.12 \text{ for } 0.75 \le \frac{L_{c,max}}{c} \le 1.7 \quad (4)$$

Type II cavity Strouhal No: $$St_2 = \frac{f_2 c}{U} = 0.0052\psi^3 + St_1 \quad (5)$$

Equation 1 assumes elliptic lift distribution along the span for both fully wetted and cavitating flows. Numerical computations using a Reynolds Averaged Navier Stokes (RANS) solver for the rigid SS NACA 0009 hydrofoil in fully wetted flow confirmed that the sectional (2D) lift distribution is very close to an elliptic distribution [7,8]. Since the NACA 0009 cross-section is relatively thin, the fully wetted (i.e. $\psi = \infty$) 2D lift slope (a_o^{2D} in Equation 2) is assumed to be 2π. The effective aspect ratio of the hydrofoil is $A_R = 2s/c$=6.7 because of the solid wall at the cantilevered end of the hydrofoil. Following the elliptic load assumption, Equation (1) yields a 3D lift slope (a_o^{3D}) of 1.54π in fully wetted flow, which agrees very closely with the experimental fitted slope of 1.53π given in [11]. They showed that the measured C_L for all the SS and the composite hydrofoils in the pre-stall regime, regardless of α_o and Re, can be collapse onto the same linear line when plotted against α_{eff}.

The 2D lift slope in cavitating flow, Equation 2, is obtained by fitting the experimental data of the SS hydrofoil and by recognizing the theoretical limits given in [1]: $a_o^{2D} \to 2\pi$ as $\psi \to \infty$ and $a_o^{2D} \to \pi/2$ as $\psi \to 0$. Note that the small rise in the mean lift coefficient due to development of partial cavitation is ignored for the sake of a simple semi-empirical equation. The predicted variation of the mean lift coefficient obtained using Equations (1)-(2) are shown as solid lines in Figures 1(a) and 1(b), which are in good agreement with the measured values for all three hydrofoils. Moreover, Figure 1(b) showed that the cavitating C_L values of all three hydrofoils can be collapsed onto the same curve when plotted against ψ, as it takes into account the twist (θ_{tip}) in addition to α_o and σ.

Equation 3 gives the variation of the maximum cavity length ($L_{c,max}$) with ψ. It is obtained by fitting the experimental data of the measured maximum cavity length at mid-span of the SS hydrofoil and by recognizing that $\frac{L_{c,max}}{c} \to 0$ as $\psi \to \infty$. As shown in Figure 1(b), good agreement is observed between the measured and predicted values for all three hydrofoils, and they all collapse onto the same curve given by Equation 3.

As observed in previous experimental studies by [1,2,4,5,6,10], there exists two primary cavity shedding mechanisms. Type I shedding is driven primarily by shock-wave propagation when the local flow speed exceeds the sound speed of the liquid-vapor mixture, which can be reduced to near 3.5-10 m/s when the void fraction is near 50% for $0.3 < \sigma < 1$ based on the equations presented in [6]. Type I shock-driven cavity shedding is called transitional cavity instability in past literature, and is typically found when $\frac{L_{c,max}}{c} > 0.75$ with a relatively constant shedding frequency that yields a chord-based Strouhal number, St_1, of 0.1-0.2 [2,5,10]. Type I cavity shedding frequency calculated using Equation 4 is 10.8 Hz for all three hydrofoils, and is shown as the thick black dotted lines along with the frequency spectra of the measured instantaneous lift coefficients in Figures 2(a)-2(c). Note that Type I cavity shedding is assumed to take place only when $0.75 \le \frac{L_{c,max}}{c} \le 1.7$. Cavities with $\frac{L_{c,max}}{c} < 0.75$ will tend to have lower void fraction and will be driven primarily by re-entrant jet flow. Cavities with $\frac{L_{c,max}}{c} > 1.7$ will have a high enough void fraction such the sound speed of fluid mixture is higher than the local flow speed, preventing shock wave development, and hence the supercavity stabilizes with relatively low load fluctuation. Since $L_{c,max}$ depends on α_{eff} and thus θ_{tip}, the σ range where Type I shedding is expected is different for the SS, P30, and N30 hydrofoils, as indicated in the thick black dotted lines in Figures 2(a)-(c). Type II cavity shedding is driven primarily by re-entrant jet flow that develops at the cavity trailing edge, and is commonly called partial cavity instability or cloud cavity instability in previous literature [1,2,3,4,5,6,10]. The mean-chord based Strouhal number for Type II re-entrant jet driven cavity shedding, St_2, is dependent on the maximum cavity length, and hence the effective cavitation parameter. Type II cavity shedding frequency calculated using Equation 5 is shown as the thick red dashed lines along with the frequency spectra of the measured instantaneous lift coefficients in Figures 2(a)-2(c). Note that Equations 4 and 5 are equivalent to the experimentally derived equations given in [9] for the SS and quasi-isotropic composite hydrofoils. For Type II cavity shedding, St_2 is expressed in terms of ψ (instead of just σ in [9]) to account for changes in the effective angle of attack by twist. As shown in Figures 2(a)-(c), Type II cavity shedding frequencies are higher for the P30 hydrofoil than the SS hydrofoil because of the shorter maximum cavity lengths caused by the reduced effective angles of attack with the nose-down twist. Following a similar logic, Type II cavity shedding frequencies are lower for the N30 hydrofoil because of the longer maximum cavity lengths induced by the nose-up twist. In general, the peaks of the lift frequency spectra falls along the Type I and Type II cavity frequency curves given by Equations (4) and (5). In Figure 2(b), two additional high intensity peaks near 30 Hz can also be observed for the P30 hydrofoil, which corresponds to when the Type II cavity shedding frequency is near the foil's first wetted natural frequency. Although the first wetted natural frequency of the N30 hydrofoil should also be around 30 Hz, only minor peaks can be observed along the 30 Hz line because of the much lower cavity shedding frequencies of the N30 hydrofoil, as shown in Figure 2(c). The first wetted natural frequency of the SS hydrofoil is 54 Hz, and hence is not shown in Figure 2(a).

It should be noted that the cavity length and cavity shedding frequency estimations shown in Figures 1(b), 2(a)-(c) are based on the mean chord length, but the foils have a tapered planform so the cavity lengths and cavity shedding frequencies vary along the span due to 3D effects. Comparisons of the observed and predicted cavitation patterns for the SS, P30, and N30 hydrofoils at $\sigma = 0.6, Re = 0.8x10^6, \alpha_o = 6^o$ are shown in Figures 2(d) and 2(e), respectively. The corresponding predicted spanwise variation of the Type I and Type II cavity shedding frequencies are shown in Figure 2(f). The predictions are obtained by assuming elliptic load distribution using Equations (1)-(5). Although the maximum cavity extent is slightly under-predicted for all three hydrofoils, good agreement is observed on the cavity shapes. The results in Figure 2(f) show that while the SS and P30 hydrofoils exhibit only Type II cavity shedding, the N30 hydrofoil exhibit both Type I and Type II cavity shedding along majority of the span, which explains the high intensity peaks along the 10.8 Hz line in Figure 2(c).

Conclusion

In this paper, experimental measurements and semi-empirical relations are presented to study the effects of material bend-twist coupling on the cavitating response of adaptive composite hydrofoils. The results show that material bend-twist coupling that leads to nose-up twist will increase the hydrodynamic load coefficients, accelerate cavitation inception, increase the maximum cavity length, and lower the Type II cavity shedding frequency compared to the SS hydrofoil. The opposites are true for material bend-twist coupling that leads to nose-down twist. Both Type I shock-driven and Type II re-entrant jet driven cavity shedding mechanisms are observed for all three hydrofoils, but the Type I shedding mechanism is only observed when the maximum cavity length normalized by the chord is between 0.75 to 1.7. Significant load amplification is also observed when the Type II cavity shedding frequency is near the first wetted natural frequency of the P30 hydrofoil. Good agreement is observed between the semi-empirical predictions and experimental measurements. The results show that by accounting for the flow-induced twist in the effective angle of attack (α_{eff}), and using $\psi=\sigma/2\alpha_{eff}$ as the effective cavitation parameter, the hydroelastic response for all three hydrofoils can be collapsed onto the same trend lines.

Acknowledgments

This project was supported by the US Office of Naval Research (ONR Grant No. N00014-16-1-2972 managed by Ms. Kelly Cooper and Grant No. N00014-16-1-2433 managed by Dr. Ki-Han Kim), ONR Global (NICOP S&T Grant No. N62909-11-1-7013 managed by Dr. Woei- Min Lin), as well as the Defense Science and Technology Group (Mr. Brendon Anderson) and the University of Tasmania. The authors wish to thank Mr. Robert Wrigley for his valuable help setting up the experiments, as well as Dr. Nigel St. John and Mr. Russell Cairns for their help in manufacturing the composite hydrofoils.

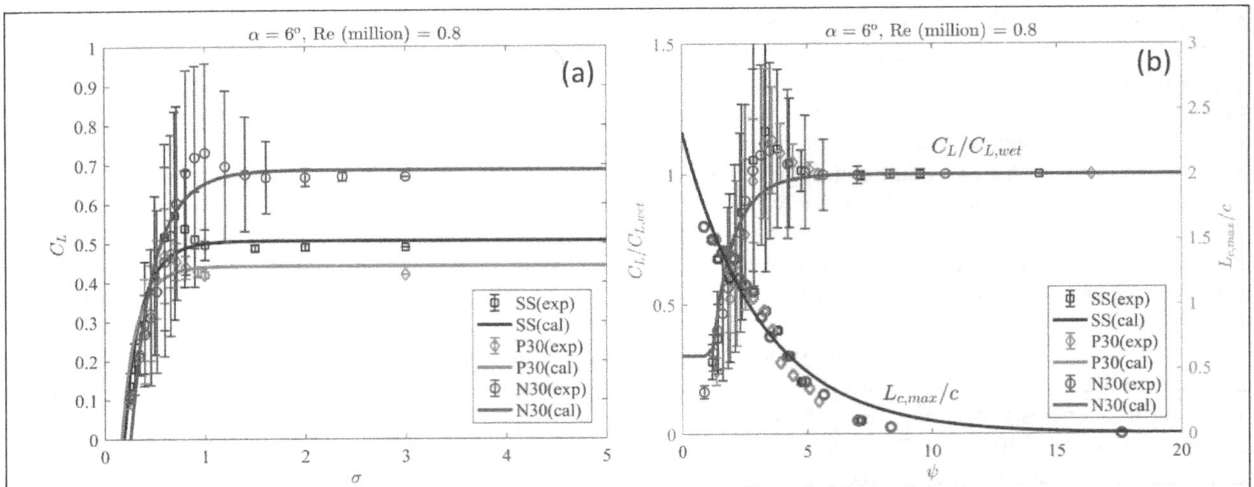

Figure 1: (a) Variation of the measured lift coefficient (C_L) against the cavitation number (σ) for all 3 hydrofoils. (b) Variation of the measured lift coefficients normalized by their respective fully wetted values ($C_{L,wet}$), and the measured maximum cavity length normalized by the chord at midspan ($L_{c,max}/c$), against the effective cavitation parameters ($\psi = \sigma/2\alpha_{eff}$) for all 3 hydrofoils, where $\alpha_{eff} = \alpha_o + 2\theta_{tip}/\pi$ with θ_{tip} as the tip twist angle in fully wetted flow. Also shown in solid lines in both plots are the predicted mean lift coefficients using Equations 1-2, and the predicted normalized maximum cavity lengths using Equation 3. θ_{tip} is equal to 0°, -1.2°, and +3.36° for the SS, P30, and N30 hydrofoils, respectively, at $Re = 0.8x10^6, \alpha_o = 6^o$. The $2/\pi$ factor is used to account for the spanwise variation in twist.

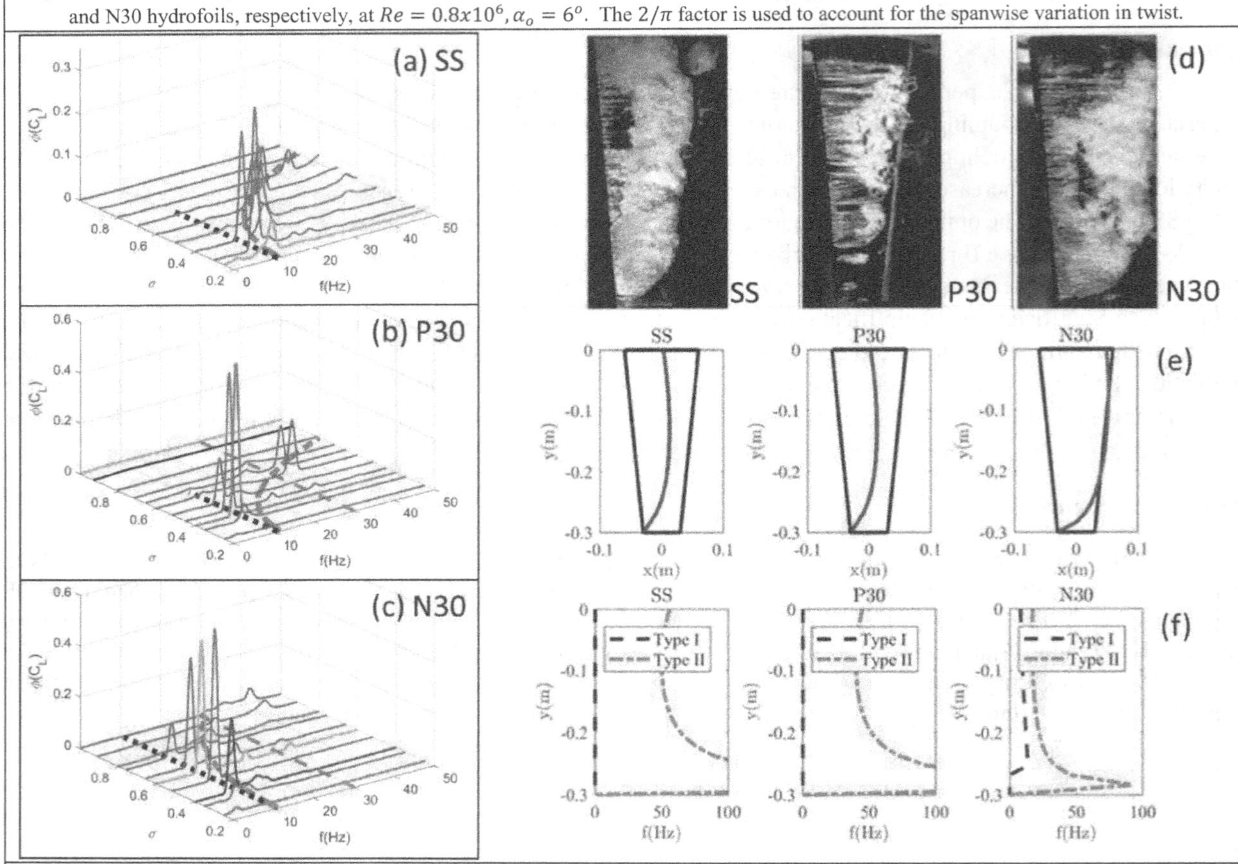

Figure 2: Measured lift coefficient spectra for the (a) SS, (b) P30, and (c) N30 hydrofoils for $Re = 0.8x10^6, \alpha_o = 6^o$. The thick black dotted line and the red dashed lines on the bottom of each plot respectively indicate the Type I shock-driven and Type II re-entrant jet driven cavity shedding frequencies calculated using Equations 4 and 5. The thinner magenta dashed lines in (b) and (c) indicate the first wetted natural frequency of 30 Hz for the P30 and N30 hydrofoils, respectively. The right plots compare the (d) observed and (e) predicted cavitation patterns for the SS, P30, and N30 hydrofoils at $Re = 0.8x10^6, \alpha_o = 6^o$, and $\sigma = 0.6$. The bottom plots in (f) show the corresponding predicted Type I and Type II cavity shedding frequencies according to Equations 4 and 5.

References

[1] Brennen, C.E. (1995). *Cavitation and bubble dynamics.* Oxford University Press. New York, NY.

[2] Coutier-Delgosha, O., Devillers, J.F. Pichon, T., Vabre, A., Woo, R. and Legoupil, S. (2006). *Internal structure and dynamics of sheet cavitation.* Physics of Fluids. 18: 017103.

[3] Duttwiler, M. E. and Brennen, C.E. (2002). Surge instability on a cavitating propeller. Journal of Fluid Mechanics. 458: 133-152.

[4] Franc, J.P. (2001). *Partial cavity instabilities and re-entrant jet.* In: CAV 2001: Fourth International Symposium on Cavitation, June 20-23, 2001, California Institute of Technology, Pasadena, CA USA.

[5] Fujii A., Kawakami, D.T., Tsujimoto, Y. and Arndt, R.E.A. (2007). *Effect of hydrofoil shapes on partial and transitional cavity oscillations.* Journal of Fluids Engineering. 129: 669-673.

[6] Ganesh, H., Makiharju, S.A., and Ceccio, S.L. (2016). *Bubbly shock propagation as a mechanism for sheet-to-cloud transition of partial cavities.* Journal of Fluid Mechanics. 802: 37-78.

[7] Garg, N., Kenway, G. K. W., Martins, J. R. R. A., Young, Y. L. (2015). *High-fidelity hydrodynamic shape optimization of a 3-D hydrofoil.* Journal of Ship Research. 59(4): 209-226.

[8] Garg, N., Kenway, G. K. W., Martins, J. R. R. A., Young, Y. L. (2017). *High-fidelity multipoint hydrostructural shape optimization of a 3-D hydrofoil.* Journal of Fluids and Structures. 71: 15-39.

[9] Smith, S.M., Venning, J.A., Brandner, P.A., Pearce, B.W., Giosio, D.R., and Young, Y.L. (2018). *The influence of fluid-structure interaction on cloud cavitation about a hydrofoil.* 10th International Symposium on Cavitation (CAV2018). May 14 – 16, 2018. Baltimore, Maryland, USA.

[10] Tsujimoto, Y., Watanabe, S. and Horiguchi, H. (2008). *Cavitation instabilities of hydrofoils and cascades.* International Journal of Fluid Machinery and Systems. 1(1): 38-46.

[11] Young, Y.L., Garg, N., Brandner, P.A., Pearce, B.W., Clarke, D. and Philips, A.W. (2017). *Load-dependent bend-twist coupling effects on the steady-state hydroelastic response of composite hydrofoils.* Composite Structures. https://doi.org/10.1016/j.compstruct.2017.09.112, 2017.

[12] Young, Y.L., Motley, M.R., Barber, R., Chae, E.J., Garg, N. (2016). *Adaptive composite marine propulsors and turbines: progress and challenges.* Applied Mechanics Review. 68(6): 060803.

© 2018 ASME

10th International Symposium on Cavitation - CAV2018
Baltimore, Maryland, USA, May 14 – 16, 2018

CAV18-05141

Measurements of Cavitation Compliance in the Draft Tube Cone of a Reduced Scale Francis Turbine Operating at Part Load

[1]João Gomes P. Jr.*; [1]Arthur Favrel; [2]Christian Landry, [2]Christophe Nicolet; [1]François Avellan;

[1]*École Polytechnique Fédérale de Lausanne, Lausanne, VD, Switzerland;* [2]*Power Vision Engineering Sàrl, Ecublens, VD, Switzerland*

Abstract

Francis turbines operating at part load conditions typically experience a cavitation vortex rope immediately at the runner outlet. As the compliance of this cavitation flow is much higher than the one in cavitation-free conditions, a decrease in the value of the eigen frequencies of the hydraulic circuit is observed. One of the eigen frequencies can eventually match the frequency of the vortex precession, which acts as an excitation source for the hydraulic system, inducing strong pressure pulsations and output power swings. To predict such undesirable resonance conditions in hydropower plants, the cavitation compliance in the draft tube cone can be identified beforehand at the model scale. This paper presents the identification of the cavitation compliance on a reduced scale model of a Francis turbine over a wide range of part load operating conditions, for different Thoma and Froude numbers. The first eigen frequency of the test rig is firstly identified by modal analysis. The cavitation compliance is then defined by adjusting a 1-D numerical model of the test rig to match this first eigen frequency. These compliance results could then be transposed to the prototype scale, enabling the prediction of the first eigen frequency of the hydropower plant in any part load condition.

Keywords: Cavitation vortex rope; resonance; Francis turbine; hydro-acoustic model; cavitation compliance; pressure pulsations

Introduction

At off-design conditions, Francis turbines experience the development of a swirling flow in the draft tube, which intensity depends mainly on the value of the flow discharge. At part load conditions, with a discharge ranging usually from about 50 % to 90 % of the rated value, a vortex breakdown arises immediately at the runner outlet and produces a flow recirculation zone in the draft tube, accompanied by a helical cavitation vortex rope.

The precession of this vortex acts as a pressure excitation source for the system, leading to the propagation of synchronous pressure pulsations in all the hydraulic circuit [1]. Further consequence of the cavitation vortex development is the strong impact on the eigen frequencies of the hydraulic circuit, as it increases the flow compliance inside the draft tube [2]. One of these eigen frequency values may match the frequency of the synchronous pressure pulsations, causing resonance phenomena and potentially electrical output power swings of the generating unit [3]. To predict these undesirable resonance conditions at the prototype scale, the hydro-acoustic parameters describing the draft tube cavitation flow in transient 1-D numerical simulations must be known. They can be determined by performing hydro-acoustic modal analysis of a test rig facility containing the reduced scale physical model of the Francis turbine prototype operating in similar conditions. Once the experimental eigen frequencies of the test rig are known, the parameters of the 1-D hydro-acoustic numerical model of the test rig, including the draft tube cavitation flow, can be adjusted to match the measurements [4].

In this paper, the hydro-acoustic parameters of the hydraulic circuit are initially adjusted by the experimental results obtained in cavitation-free conditions. Measurements are then performed at part load conditions, for which the cavitation vortex rope is experienced. The compliance of the draft tube cavitation flow is then modified in the 1-D numerical model to match the experiments, while the hydro-acoustic parameters of the other parts of the test rig remains unchanged. Finally, the obtained values of cavitation compliance are presented and comparisons are made between different IEC speed and discharge factors, n_{ED} and Q_{ED} respectively, and Thoma and Froude numbers, σ and Fr respectively (definition of these factors and numbers can be found in [5]). The possibility of using the swirl number S to reduce the number of measurements at different n_{ED} values is then analyzed and discussed.

*Corresponding Author, João Gomes P. Jr.: joao.gomes@epfl.ch

Resonance in part load conditions

For a given longitudinal position x along the draft tube cone center-line, the local plane wave speed a can be calculated as in Equation (1). The variable C_{eq} represents the equivalent compliance of the draft tube, including the cavitation vortex, the water and the pipe, as presented in Equation (2). The cavitation compliance C_c is defined in Equation (3), [6].

$$a(x)=\sqrt{\frac{gA_p(x)dx}{C_{eq}(x)}} \tag{1}$$

$$C_{eq}=C_c+C_{water}+C_{pipe} \tag{2}$$

$$C_c(x)=-\frac{\partial V_c(x)}{\partial h}=-dx\frac{\partial A_c(x)}{\partial h} \tag{3}$$

where g is the local gravity, A_p is the pipe cross-section, A_c is the cavitation vortex cross section and h is the piezometric head.

Equation (1) implies that higher values of C_{eq} lead to lower wave speeds. As the wave speed in cavitation conditions is much smaller than the one in cavitation-free conditions, the approximation $C_{eq}\approx C_c$ can be made. Consequently, the cavitation compliance becomes the most important factor determining the wave speed that, in its turn, will have a major impact on the value of test rig eigen frequencies.

In Figure 1(a), the influence of the cavitation Thoma number on both the first eigen frequency of the test rig and the vortex rope frequency is presented. The influence of the discharge factor on both frequency values is given in Figure 1(b). For the conditions presented in this figure, the vortex rope frequency remained approximately constant for any σ and presented a small decrease with the increase in Q_{ED}. Furthermore the first eigen frequency of the test rig presents an important decrease for both lower σ and Q_{ED} values. In both cases, resonance occurs when the first eigen frequency of the test rig and the precession frequency of the vortex match, notably for $\sigma=0.112$ and $Q_{ED}=0.303$ in the shown test cases. It is reminded, however, that this kind of resonance at part load is commonly encountered during reduced scale model tests at low σ and does not necessarily occur on the prototype, as the plant and the turbine can be properly designed to avoid this type of risk.

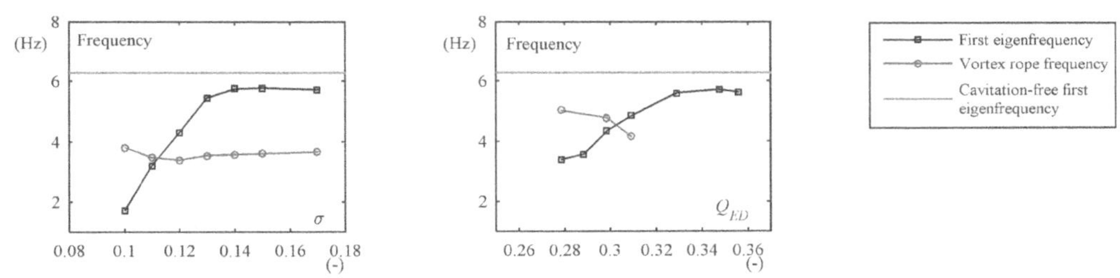

(a) $Q_{ED}=0.282$, $n_{ED}=0.478$, Fr=5.6 (b) $\sigma=0.15$, $n_{ED}=0.532$, Fr = 5.6

Figure 1 – Influence of the Thoma number (a) and discharge factor (b) on both the first eigen frequency of the test rig and the vortex rope frequency.

Experimental setup and cavitation compliance determination

The methodology to determine the first eigen frequency and the cavitation compliance C_c is similar to the one described by Landry et al. in [4]. For the present study, the two 400 kW centrifugal pumps are installed in parallel and the test case is a reduced scale model of a Francis turbine with specific speed $N_{QE}=0.28$ and a reference diameter $D=0.35$m.

An external excitation system is used to identify the first eigen frequency of the test rig. It consists of a rotating valve that is installed connecting one pipe at the turbine upstream pipe to a high-pressure reservoir, allowing the injection of periodical fluctuations at a given frequency into the test rig. By using pressure sensors installed along the hydraulic circuit, the first eigen frequency is identified as being the frequency leading to the maximum response.

a)

b)

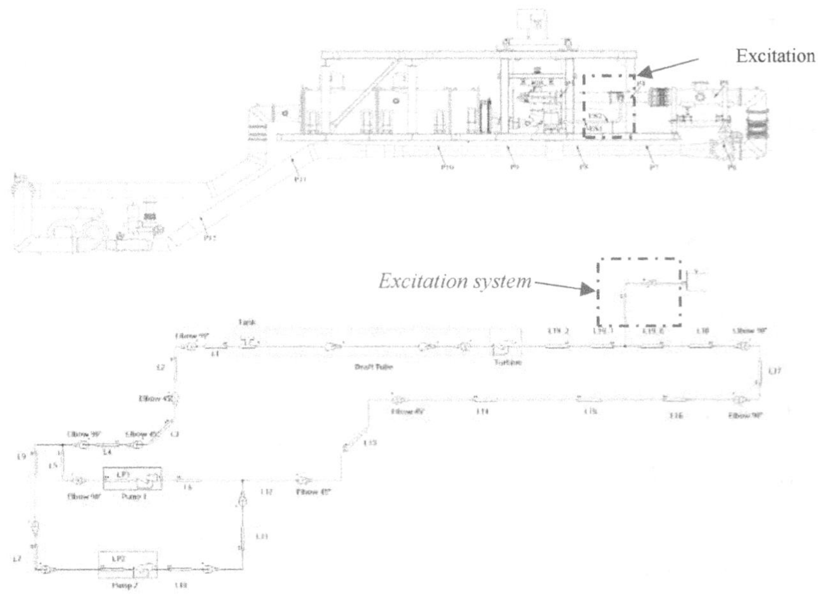

Figure 2 – Complete test rig and excitation system experimental setup (a) and the 1-D numerical model (b).

The compliance and bulk viscosity values of the test rig pipes are first identified by adjusting their values in the 1-D numerical model to match the first eigen frequency and the system response of the test rig obtained experimentally in a cavitation-free condition. The value of first eigen frequency in cavitation-free conditions is 6.28 Hz and is indicated by the grey line in Figure 1.

In cavitation conditions, a lower eigen frequency is made apparent and the cavitation compliance values in the first part of the draft tube of the numerical model is adjusted to obtain the same value of eigen frequency. The cavitation compliance is applied to the first one meter long ($2.86 \times D$) upper part of the draft tube, starting from the runner outlet until the beginning of the draft tube elbow. The remaining parts of the draft tube (roughly half of the elbow and the diffuser), and the rest of the test rig keep the same compliance as in cavitation-free condition. The bulk viscosity remains unchanged and equal to 0.4818 MPa·s in the full circuit. More details on the 1-D model applied in this paper can be found in [6] and [7]. General aspects and discussions on modeling the cavitation compliance can be found in [8].

The investigated operating conditions are summarized in Table 1. Firstly, it presents the cavitation-free operating condition used to calibrate the model. Secondly, it presents the tests where the influence on the discharge factor variation is focused. These tests are made for a combination of two values of Thoma number, σ_{min} and σ_{max}, two Froude numbers, and five different speed factors. The two Thoma numbers, σ_{min} and σ_{max}, are defined using a similar approach to the one adopted in [9]. The last line of Table 1 presents additional measurements focused on the effects of the Thoma number variation on the cavitation compliance.

Swirl number

The swirl number S is defined as the ratio between the axial flux of angular momentum and the axial flux of axial momentum and was originally proposed by Gupta et al. [10]. As the swirl intensity and the cavitation compliance in a vortex rope are strongly related, Favrel et al. proposed Equation (4) for the calculation of the swirl number at the Francis turbines outlet [9]. As presented in [9], points with similar σ, same Froude number and varying n_{ED} presented resonance conditions when operating at the same swirl number.

$$S = n_{\text{ED}} \frac{\pi^2}{8} \left(\frac{1}{Q_{\text{ED}}} - \frac{1}{Q_{\text{ED}}^{S=0}} \right) \tag{4}$$

Table 1 – Tested operating conditions

	Fr (-)	n_{ED} (-)	Q_{ED} (-)	σ(-)
Cavitation-free test	5.6	0.478	0.282	0.917
Q_{ED} variation tests	5.6 and 7.10	0.457	0.183 → 0.287	$\sigma_{min} = 0.110$ and $\sigma_{max} = 0.128$
		0.478	0.204 → 0.322	$\sigma_{min} = 0.120$ and $\sigma_{max} = 0.140$
		0.506	0.229 → 0.314	$\sigma_{min} = 0.134$ and $\sigma_{max} = 0.157$
		0.532	0.245 → 0.356	$\sigma_{min} = 0.149$ and $\sigma_{max} = 0.174$
σ variation tests	5.6	0.478	0.220 and 0.282	0.10 → 0.17

Measurements results and discussion

The first eigen frequency measured on the test rig and the average wave speed in the draft tube cone as a function of the cavitation compliance are presented in Figure 3. The relation between C_c and the first eigen frequency is provided by the 1-D numerical model and the local wave speed is calculated as in Equation (1). The graph contains all the operating conditions described in the Table 1. With the exception of the result in cavitation-free condition, all the eigen frequency data collapses into one single curve, regardless of variations in discharge, rotation speed, Froude and Thoma numbers. The presented average wave speed values are below 50 m/s, far lower than the value in cavitation-free conditions, for which a wave speed above 750 m/s is expected.

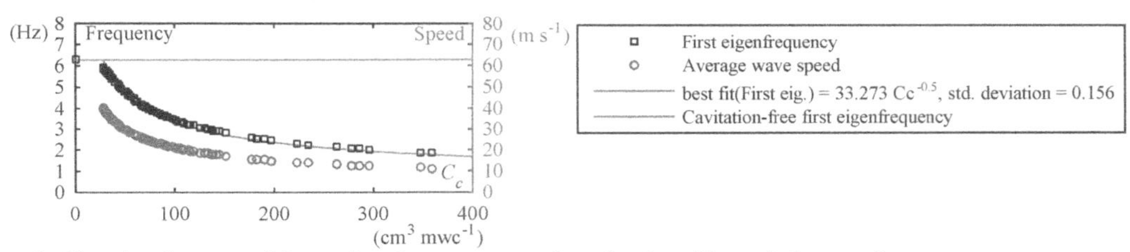

Figure 3 – First eigen frequency of the test rig and mean wave speed as a function of the cavitation compliance.

In Figure 4, the influence of the discharge factor (Figure 4(a)) and the swirl number (Figure 4(b)) on the value of C_c is presented for four different n_{ED} values with $\mathrm{Fr} = 5.6$ and $\sigma = \sigma_{min}$. While using the swirl number instead of Q_{ED}, with the exception for $n_{ED} = 0.457$, the measurements are slightly better collapsed into a single curve. This deviation for $n_{ED} = 0.457$ may be related to the presence of inter-blade vortices, expected to be more intense for this speed factor. Even though this type of cavitation is not developing in the draft tube cone, it affects the value of the first eigen frequency and is then assimilated by the numerical model into the value of cavitation compliance in the draft tube.

The value of C_c tends to grow exponentially with an increase in the value of the swirl number. For each tested n_{ED} value, measurements are performed until a Q_{ED} value for which the cavitation vortex dynamics lose its coherence and the first eigen frequency cannot be determined accurately. This phenomenon is discussed in more details in [9].

Figure 5 presents the cavitation compliance as a function of the swirl number for different values of σ. The speed factor and the Froude number are kept constant ($n_{ED} = 0.478$, $\mathrm{Fr} = 5.6$). As expected, smaller values of σ are characterized by higher values of cavitation compliance. The cavitation compliance increases exponentially with the decrease in σ, explaining the sharp drop in the first eigen frequency shown in Figure 1(a) for low σ values. It can also be noticed that σ has a greater influence on C_c as the swirl increases.

The effect of the Froude number is presented in Figure 6 and Figure 7. The influence of the Froude number on the cavitation compliance for σ_{min} and σ_{max} at $n_{ED} = 0.506$ is given in Figure 6. It can be noticed that for higher swirl numbers, an increase in the value of Froude number tends to decrease the cavitation compliance. The same observation is made for all n_{ED} values in Figure 7 for σ_{min} and σ_{max}.

The tested σ values were rather low when compared to typical rated values of turbines with a similar specific speed. For this reason, inter-blade cavitation was observed even at swirl-free conditions, keeping all the cavitation compliance curves slightly above zero.

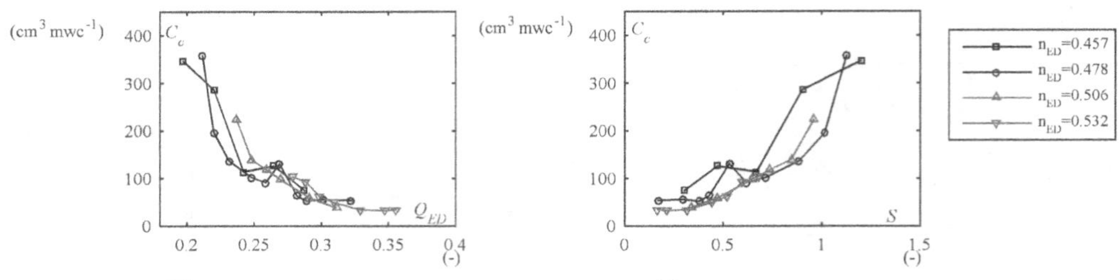

Figure 4 – Cavitation compliance for four different values of n_{ED} as a function of (a) Q_{ED} and (b) swirl number S for Fr = 5.6 and $\sigma = \sigma_{min}$.

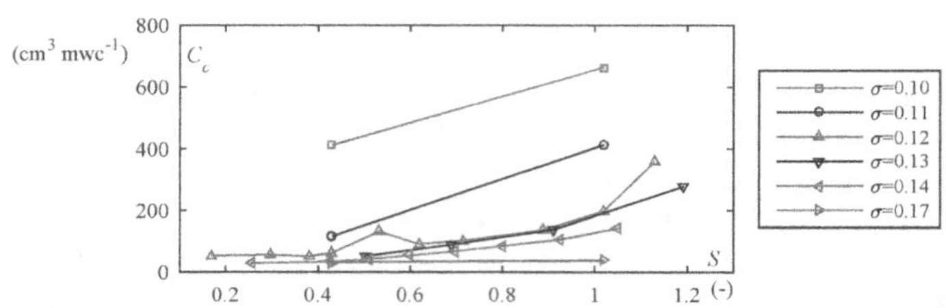

Figure 5 – Influence of σ on the cavitation compliance for a constant Froude number and n_{ED} value.

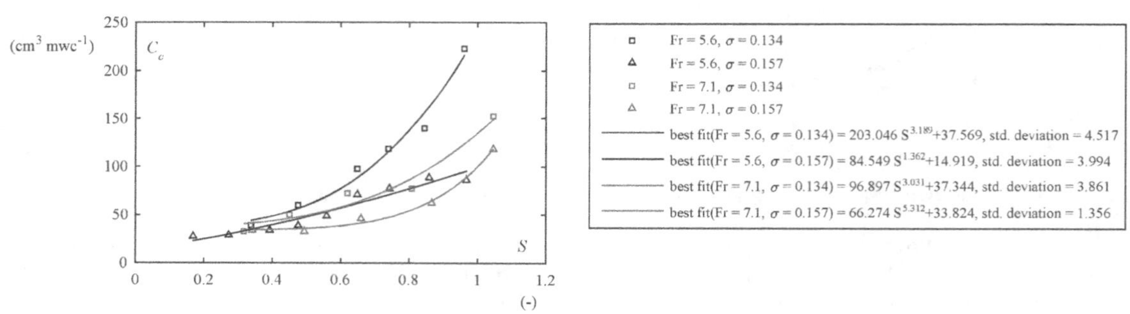

Figure 6 – Influence of Froude number and σ on the cavitation compliance for a constant n_{ED} value (n_{ED} = 0.506).

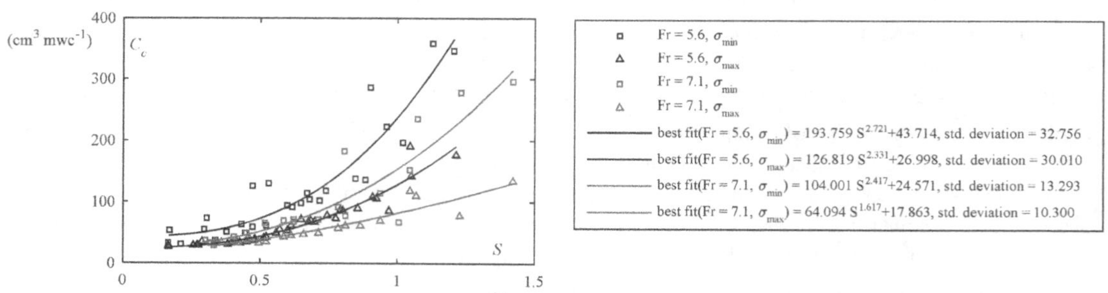

Figure 7 – Influence of the Froude number and σ on the cavitation compliance for all the tested n_{ED} values.

Conclusion

This paper presents the results of measurements of the cavitation compliance in the draft tube cone of a Francis turbine model operating in part load conditions. The results are intended to improve the understanding of the influence of the operating parameters on the compliance, so that resonance conditions can be predicted in the future.

A simple relation between the eigen frequency and the cavitation compliance is first presented. A similar behavior is expected in a prototype hydraulic circuit. By performing measurements such as the one presented in this paper, combined with a proper transposition from model to prototype of the cavitation compliance, the first eigen frequency of the power plant can then be predicted for any part load operating condition.

Furthermore, the swirl number can be used to obtain a good overall tendency of the cavitation compliance behavior at different n_{ED} values. However, additional phenomena that do not depend directly on the swirl number, such as inter-blade cavitation vortices development, affect the calculated cavitation compliance, making the only use of the swirl number an approximate approach in certain cases.

Assuming a constant cavitation compliance located only in the upper part of the draft tube does not reflect the real size and evolution of the cavitation volume along the draft tube. In addition, other types of cavitation can develop in the runner. However, the proposed approach largely simplifies the comparison of the cavitation compliance between different operating points and, eventually, different turbine designs. The results presented in this paper can be quickly compared with measurements performed with other test cases. This methodology can in the future become a standard procedure during model testing for the assessment of risk of resonance in any hydropower plant.

Acknowledgements

The authors would like to kindly thanks Nippon Koei Co., Ltd. for making available the reduced scale model, in particular to Mr. Kazuhiko Yamaishi. The authors also acknowledge the commitment of the LMH technical staff Loïc Andolfatto, Buzzi David Ayn, Alain Renaud, Phillipe Faucherre and Vincent Berruex.

Bibliography

[1] J. Arpe, C. Nicolet and F. Avellan, "Experimental Evidence of Hydroacoustic Pressure Waves in a Francis Turbine Elbow Draft Tube for Low Discharge Conditions," *Journal of Fluids Engineering,* vol. 131, no. 8, pp. 081102-081102-9, 2009.

[2] C. Brennen and A. J. Acosta, "The Dynamic Transfer Function for a Cavitating Inducer," *Journal of Fluids Engineering,* vol. 98, no. 2, pp. 182-191, 1976.

[3] W. J. Rheingans, "Power Swings in Hydroelectric Power Plants," *Trans. ASME,* vol. 62, pp. 171-184, 1940.

[4] C. Landry, A. Favrel, A. Müller, c. Nicolet and F. Avellan, "Local wave speed and bulk flow viscosity in Francis turbines at part load operation," *Journal of Hydraulic Research,* vol. 54, no. 2, pp. 185-196, 2016.

[5] International Electrotechnical Commission, IEC 60193 - Hydraulic turbines, storage pumps and pump-turbines - Model acceptance tests, 2 ed., Geneva, 1999.

[6] C. Landry, Hydroacoustic modeling of a cavitation vortex rope for a Francis turbine, Thèse n. 6547 ed., Lausanne: EPFL, 2015.

[7] C. Nicolet, "Hydroacoustic modeling and numerical simulation of unsteady operation of hydroelectric systems," PhD Thesis number 3751, EPFL, Lausanne, Switzerland, 2007.

[8] P. K. Dörfler, "Cavitation compliance in 1D part-load vortex models," *International Journal of Fluid Machinery and Systems,* vol. 10, no. 3, 2017.

[9] A. Favrel, J. Gomes, C. Ladry, C. Nicolet, F. Avellan and A. Müller, "New insight in Francis turbine cavitation vortex rope: role of the runner outlet flow swirl number," *Journal of Hydraulic Research,* 2017.

[10] A. K. Gupta, D. G. Lilley and N. Syred, Swirl flows, Trundbridge Wells, Kent: Abacus Press, 1984.

© 2018 ASME

Measurements of the temperature variations during the growth and collapse of cavitation bubbles

Merouan HAMDI[1], Olivier COUTIER-DELGOSHA[2, 1]Michael BAUDOIN[3]

[1]*Arts et Metiers ParisTech, Laboratoire de Mécanique des Fluides de Lille, 8 bld Louis XIV, Lille, France*

[2]*Virginia Tech, Kevin T. Crofton dept. of Aerospace and Ocean Engineering, 460 Old Turner St, Blacksburg, VA 24061, USA*

[3]*Université Lille, CNRS, EC Lille, ISEN, Université Valenciennes, UMR 8520IEMN, F-59000 Lille, France*

Abstract

The present work focuses on the analysis of the extreme conditions encountered during the process of collapse of cavitation bubbles. The objective is to characterize the temperature variations inside the vapor/gas bubble, and also in the surrounding liquid. This information is especially relevant to address the issues of cavitation erosion and develop appropriate models that take into account the thermal effects in cavitation. The work is based on an experimental approach where temperature measurements are performed with cold wires technique. This technique is characterized by a very small time response, and can be applied to both phases during the growth and collapse of the bubbles.

Keywords: cavitation bubble; temperature measurements; cold wire

1. Introduction

In rotating machinery, cavitation occurs preferentially on the suction side of the blades and at the tip region, because of i) the pressure drop due to the gap between the rotor and the casing for turbomachinery and ducted propulsors and ii) the local high velocities due to the rotation speed. It leads to performance decrease and/or to perturbations due to the development of instabilities. In ships and submarines, cavitation causes undesirable effects such as noise, hull vibration and material erosion. In addition to these effects, erosion due to cavitation is also a major limitation in the lifetime of hydraulic devices and rotating machinery operating in cavitating conditions. Erosion is due to the extreme conditions that occur during the collapse of the vapor bubbles. Very brief intense local heating and high pressure is generated at the location of bubble implosions, because of the fast compression of the bubbles at the end of the implosion process. During a very short time at the end of the implosions, the local heating is more rapid than the thermal exchanges, which generates temperatures probably up to 5000 K in configurations of cloud cavitation [2]. Such conditions may explain why melting spots have been observed on the special coating used in naval environments to protect the surfaces submitted to cavitation [3]. Such results suggest that small scale mechanism of heat transfer during the collapse may play a major role in the damage of the surface.

In case of single bubble configuration, a higher temperature of about 15000°K is expected, because the bubble collapse is spherical, which results in a greater compression of the gas. It has been found that such bubble contracts by more than a factor 100 within the few ms of implosion. Some authors even expect higher core temperature, because plasma formation at such extreme conditions makes an optically opaque area at the center of the bubble, so the estimated temperature may be the one at the limit of this area, not the highest one inside the core. These intense and brief temperature peaks expected in collapsing bubbles have not been confirmed yet, by any direct measurement. More specifically, no quantification of the temperature condition as a function of space location inside the bubbles has been reported. Indeed, specific instrumentation is required to detect with enough precision the high-frequency temperature variations in the flow.

Recent attempts have focused on the temperature measurement in the liquid surrounding the bubbles [1], [4] with measurements based on infrared imaging or fluorescence thermometry. These techniques are currently characterized by a limited acquisition frequency and accuracy, so the modeling of energy release and heat transfers during this process is still an open question, which dramatically hampers the prediction of the effects of cavitation erosion.

The present work focuses on the analysis of the extreme conditions encountered during the process of collapse of cavitation bubbles. The objective is to characterize the temperature variations inside the vapor/gas bubble, and also in the surrounding liquid. This information is especially relevant to address the issues of cavitation erosion and develop appropriate models that take into account the thermal effects in cavitation. The work is based on an experimental approach where temperature measurements are performed with cold wires. This technique is characterized by a very small time response, and can be applied to both phases during the growth and collapse of the bubbles.

2. Methods

The experimental setup is based on the so-called tube arrest method. It consists of a vertical tube containing liquid which is pulled downwards against (tensioned) supports and released. After the tube's arrest the liquid continues its upward motion generating a tension pulse. The acrylic (1) glass tube has 5cm diameter and 100cm length, with an open upper end. It is put on slides (2) so it can move freely in the vertical direction. The tube is filled with water up a level varying between 50 and 80cm. Distilled and degassed water was used in order to remove as many cavitation nuclei as possible before the experiments. The tube is fastened indirectly to a spring (3) pinned at one end. For the generation of a cavitation bubble, the tube is pulled downward for some length, the spring thus being stretched. The former is then released, and is quickly pulled upward by the spring, and then suddenly arrested by a barrier (4) after shooting up a short distance (typically a few cm). A thin rod (5) with a diameter of 2 mm was inserted vertically into the water. At the bottom part it had a small dent so that an air bubble nucleus with a diameter of 0.5 mm could be formed in there. The rod was suspended along the centerline of the tube. Due to the sudden arrest, a tension wave is produced in the water column, starting from the tube bottom and traveling upward, which results in the growth of the bubble. When the velocity of the tube before the arrest is in the range 1 - 2 m/s, the resulting bubble has a maximum diameter varying between 10 and 20 mm, depending also on the height of water in the tube. The resulting lifetime of the bubble is of the order of 5 to 10 ms.

For observation of the bubble, one high speed camera (Motion Blitz EoSens mini capturing images from side view is used. It enables capturing at 506 fps at 1Mpixel resolution. For the present experiment, the camera is recording at several thousands fps at a reduced resolution. This setup has enabled to obtain single bubbles of the appropriate size and characteristic lifetime, so it is currently used to test the sensors (cold wires and micro-thermocouples).

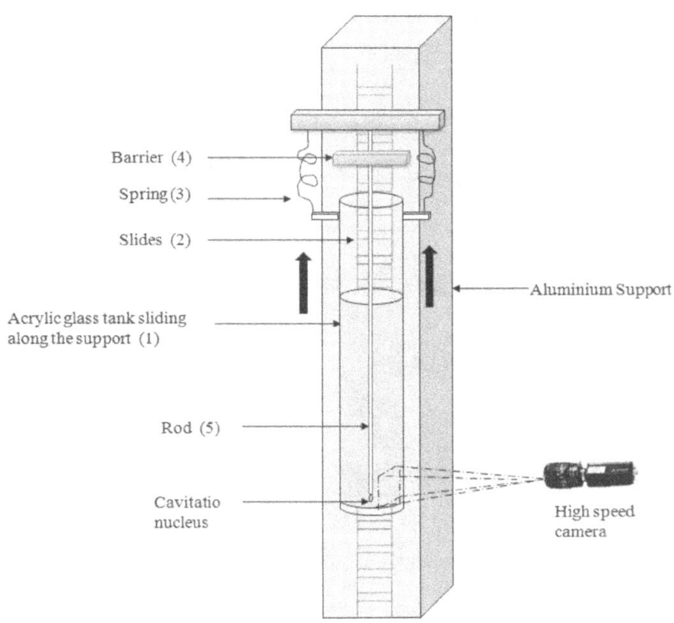

Figure 1. Experimental setup for single bubble growth and collapse

2.2. Temperature measurements

The sensor developed for high frequency temperature measurements is a cold wirethermometerwith15μm wire diameter. It is intended to measure the temperature with a very small time response. For that purpose, specific thin cold wires obtained by Nickel metallic coating, whose resistance varies according to the local temperature, have been developed. A low intensity current (lower than 0.5 mA) is applied to avoid any heating by Joule effect, which means that the wire resistance depends on the local flow temperature only. The use of the cold wire is based on the configuration shown in figure 2: the wire should be located very close to the cavitation nucleus, so that the sensitive part (2 mm) of the sensor will be included inside the bubble shortly after the start of the expansion. Each cold wire is calibrated before being mounted in the experimental setup. A static calibration is performed between 20°C and 100°C.

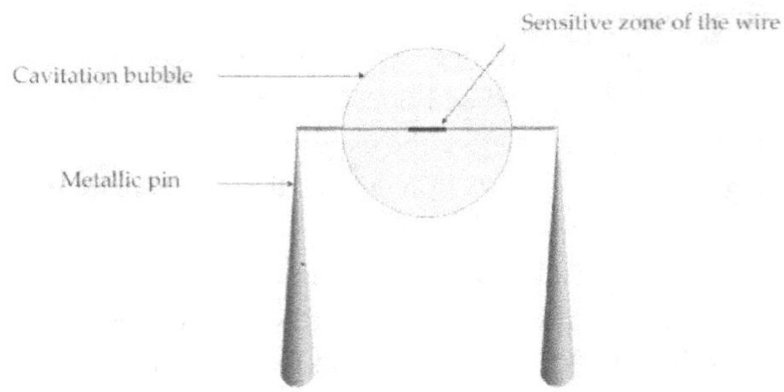

Figure 2. Temperature measurement with a cold wire - Scheme of the experimental setup

The temperature measurements provided by the cold wire are recorded at 250 KHz and synchronized with the high speed camera, using an accelerometer mounted on the barrier (see figure 1), which detects the exact time of the tube arrest and triggers the start of the acquisitions.

3. Results and discussions

In the tests, the cavitation nucleus is located very close to the wire, so the sensor is located inside the cavitation bubble during most of the grow thand collapse phases (see figure 2). In the case presented here, the nucleus is located 10 mm above the bottom wall, which results in an elliptical bubble. The duration of bubble expansion is about 7ms in this case (images 1 to 6 in figure 4),and it is followed by a compression phase of similar duration (images 7 to 10), and eventually the complete collapse of the bubble (images 11 and 12). The signal of the cold wire is given in figure 5.

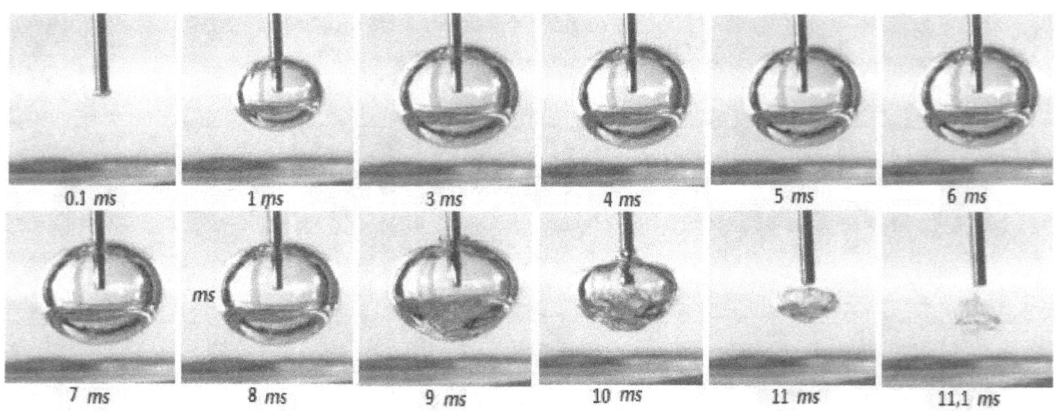

Figure 4. High speed visualization of the cavitation bubble

Based on the synchronization between the camera and the sensor, the expansion and compression steps are roughly indicated in figure 5: during this period, the temperature inside the bubble is fluctuating between 20 to 65 °C. The peak observed on the signal is obtained in the final stage of the bubble collapse, as expected.

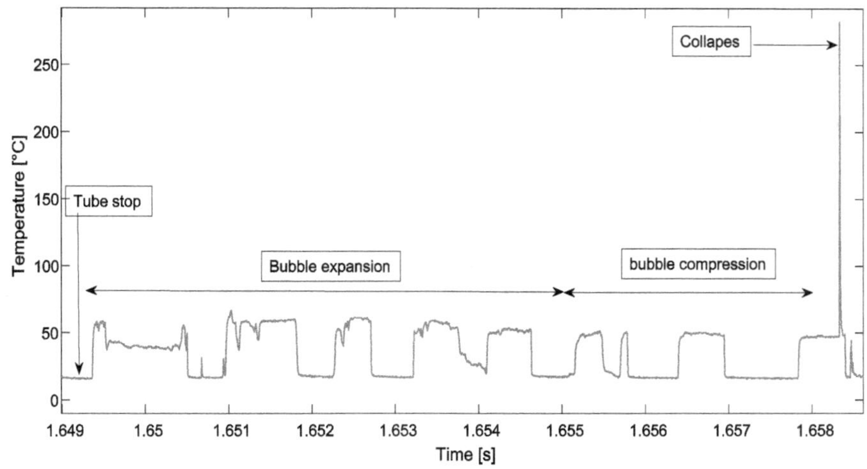

Figure 5. The temperature signal during bubble lifetime

Based on the calibration data previously performed, the maximum temperature obtained in this test is about 280°C. In most of other tests, the temperature peak magnitude is varying between 300 and 600°C. It strongly depends on the shape, the diameter ant the distance of the bubble to the wall, and even more on the wire position during the bubble collapse. Note that the temperature peak is very brief, about 10 to 30 μs.

4. Conclusion

Temperature measurements have been performed during the growth and collapse of cavitation bubbles, using cold wire sensors. In the particular case presented here, the temperature at the end of the collapse is obtained for a non-spherical bubble, so it is expected that much higher values may be obtained in case of spherical bubbles, at a higher distance from the bottom wall. Future tests will be performed with a better control of the wire location and at various distances to the bottom wall, which will enable a more in-depth analysis.

To have a better evaluation of the time response of the cold wires, several wires with different values of the Nickel coating thickness will be made in the future, and tested with the same bubble position and dynamic. It will be checked if the magnitude of the temperature peak variesas the layer thickness is reduced, or not.

Références

[1] Dular M., Coutier-Delgosha O. (2013), "*Thermodynamic effects during Expansion and Collapse of a Single Cavitation Bubble*", J. of Fluid Mechanics, 736, pp. 44-66.

[2] Suslick K.S., McNamara III W.B., Didenko Y. (1999), *"Hot spot conditions during multibubble cavitation, in Sonochemistry and Sonoluminescence*, Crum, L. A.; Mason, T. J.; Reisse, J.; Suslick, K.S., eds., Kluwer Publishers: Dordrecht, Netherlands, 1999, pp. 191-204.

[3] Paik B.G., Kim K.S., Kim K.Y., Ahn J.W., Kim T.G., Kim K.R., Jang Y.H., Lee S.U. (2011), *"Test method of cavitation erosion for marine coatings with low hardness", OceanEngineering* **38**(13), pp. 1495-1502.

[4] Quinto-Su P. A., Suzuki M. & Ohl C-D. (2014), *"Fast temperature measurement following single laser-induced cavitation inside a microfluidic gap", Scientific Reports* **4**, Article number: 5445

© 2018 ASME

10th International Symposium on Cavitation - CAV2018
Baltimore, Maryland, USA, May 14 – 16, 2018

Mechanism of Lift Increase of Cavitating Clark Y-11.7% Hydrofoil

[1]Wakana Tsuru*; [1]Yushin Ehara; [1]Soichiro Kitamura; [2]Satoshi Watanabe; [2]Shin-ichi Tsuda;

[1]Graduated School of Kyushu University, Fukuoka, Japan; [2]Kyushu University, Fukuoka, Japan

Abstract

It is known that the lift force produced by hydrofoil slightly increases with some amount of cavitation. In the present study, cavitating flow around a Clark Y-11.7% hydrofoil in a water tunnel is experimentally studied in order to understand the detailed mechanism of lift increase. The experiments are carried out in three angles of attack conditions; 2.0, 8.0 and 20.0 degrees and in various gas content conditions. Although, at 2.0 degrees, the cavitation appearance and time-averaged lift and drag force differ due to differences of gas content, only the cavitation appearance differs at 8.0 degrees. In addition, at 20.0 degrees, the lift and drag force increase just after the inception of cavitation.

Keywords: cavitation; Clark Y hydrofoil; lift increase

Introduction

It is known that the lift force produced by a single hydrofoil slightly increases with some amount of cavitation before the sudden drop of lift (lift breakdown) at some angles of attack. This has been qualitatively explained by the additional camber effect, since the streamline of working liquid changes where the sheet cavity forms [1]. However, such sheet cavitation is often unsteady and in some cases the periodic large-scale vertical cloud cavities are generated. In such cases, it seems to be difficult to explain the mechanism of lift increase due to cavitation by stationary "camber effect". In addition, it is important to note that an increase of lift force is difficult to be predicted by even well-developed cavitation model in CFD [2]. Previous studies in the past have shown that the lift increase is related to the cavitation pattern. Various factors are involved in the cavitation pattern, and in experiments of single hydrofoil, cavitation shows different appearance at various angles of attack [3]. In addition, the gas content in water also affects cavitation, and it is shown that there are differences in the lift characteristics and the aspect of cavitation by amount of dissolved air [4]. Therefore, it is necessary that the mechanism of lift increase is understood by conducting more-detailed experimental investigation considering the flow features such as boundary layer types as well as water quality.

The measurement of lift and drag forces and the observation of cavitation appearance have been carried out in various angles of attack and dissolved air conditions in order to investigate the mechanism of lift increase. In the present study, the appearance and time-averaged lift and drag force of three angles of attack is investigated, and at the angle of attack of 20 degrees, the mechanism of lift increase is investigated.

Experimental apparatus and method

Experiments are carried out in a closed-loop cavitation tunnel consisting of upstream tanks, a rectangular test section with the width of 200mm and the span of 81.5mm and a circulating pump. The transparent acrylic windows are set at one-side, top and bottom walls for easy access for visual observations. In the present study, a two-dimensional Clark Y-11.7% hydrofoil with the chord length of C=100 mm and the span of S=81.0 mm as shown in Fig. 1 (a) is used.

The lift and drag forces (F_L, F_D) are measured by the strain gauges attached to the cantilever supporting the test hydrofoil under the assumption in which the flow is symmetrical with respect to the span center. The sampling frequency is 1,000 Hz and the sampling time is 40 s, whereas the resonance frequencies of cantilever system are 208 Hz and 74 Hz in bending modes normal and tangential to the hydrofoil chord respectively. The time-averaged surface pressure distribution is measured using the hydrofoil with pressure holes as shown in Fig. 1(b). In the high-speed observation of cavitation appearance, only one camera is used with mirrors to record two views from two directions, the top and the side, simultaneously. The frame rate is 8,000 fps and total time of recording is 1s. The measurements of lift and drag forces are synchronously carried out with high-speed recording.

Figure 2 shows the dependence of boundary layer developing along the suction surface of test hydrofoil on the angle of attack α in non-cavitating conditions, which has been specified from the measured surface pressures and the

*Corresponding Author, Wakana Tsuru: tsuru.w.123@kyudai.jp

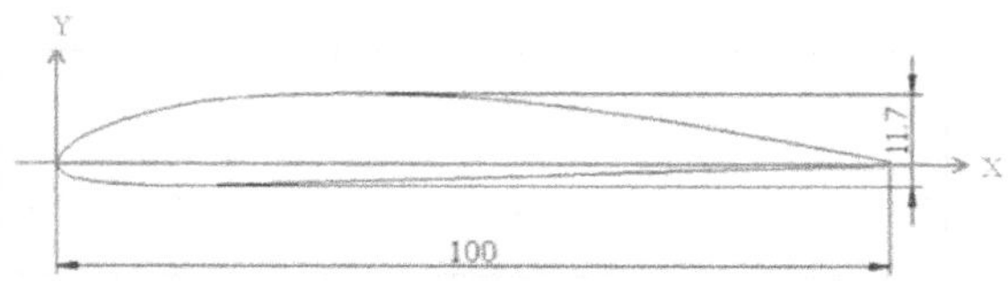

(a) Normal hydrofoil

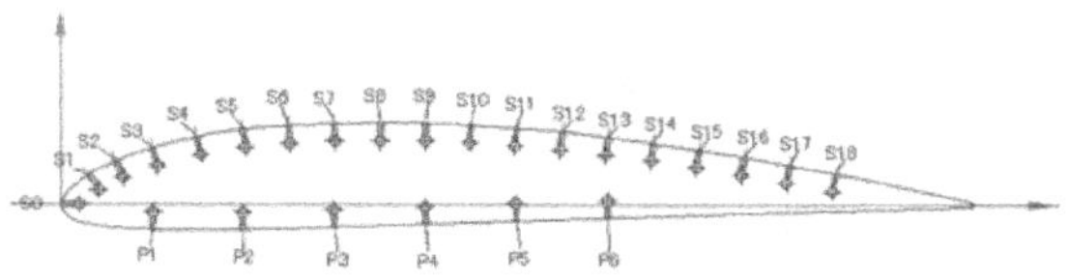

(b) Hydrofoil with pressure taps

Figure 1 Clark Y-11.7% hydrofoil

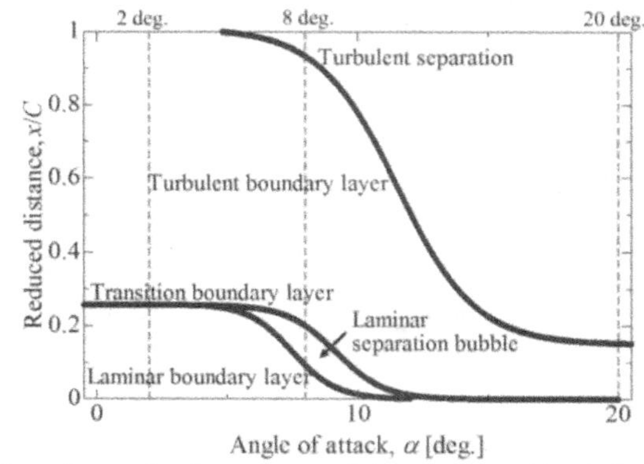

Figure 2 Boundary layer characteristics of suction side of Clark Y hydrofoil (Non-cavitation, R_e=4.5x10^5-1.0x10^6).

oil film visualizations. The surface limiting streamlines are observed from oil film composed of pigment, oleic acid, lubricant and grease after the operation with the main flow velocity of U=5.8m/s during 150s to 300s. From this chart, we have selected α=2.0, 8.0 and 20.0 degrees as the test conditions; at α =2.0 degrees, the laminar boundary layer smoothly transit to the turbulent one around the minimum pressure location. At α=8.0 degrees, the laminar separation bubble which promotes the turbulence transition forms near the leading edge of hydrofoil. At α =20.0, the turbulent flow separation occurs near the leading edge. The flow velocity is set to approximately U=8.1m/s at α =2.0 and 8.0 degrees, and U=5.8m/s at α=20.0 degrees. The gas content in water is evaluated by the amount of dissolved oxygen (DO) which is measured before and after every experiment. On the basis of measured DO value, we categorize the condition of gas content into three; high DO (70% or more of saturated value), medium DO (around 50%) and low DO (less than 30%) conditions.

The results are summarized in the non-dimensional forms using the cavitation number σ, the lift and drag coefficients C_L and C_D and blade surface pressure coefficient C_p, which are defined by $\sigma=(p_t-p_v)/0.5\rho U^2$, $C_L=F_L/0.5CS\rho U^2$, $C_D=F_D/0.5CS\rho U^2$ and $C_p=(p_x-p_t)/0.5\rho U^2$, respectively, where p_t is the test section pressure, p_v is the saturated vapor pressure, ρ is the water density, and F_L and F_D are lift and drag forces. Also, x is the distance in the chord length direction from the hydrofoil leading edge, and p_x is the blade surface pressure at x.

Results and discussion

Figure 3 shows the time-averaged lift and drag coefficients for α= 2.0, 8.0 and 20.0 degrees, respectively. The plots of red, green and blue show high DO condition, medium DO condition and low DO condition, respectively. In these figures, the inception cavitation number detected by naked eye observation is roughly indicated by the broken lines. The inception cavitation number is defined as the cavitation number when the cavitation occurs on the blade surface and cavitation growing begins.

At α=2.0 degrees, the drag force increases just after the inception of cavitation in the high and medium DO conditions, and at the same instance the lift force starts to decrease. On the other hand, in the low DO condition, the increase of the drag force is not seen and the lift force decrease is delayed to lower cavitation number. In addition, as shown in Fig. 4, in the high and medium DO conditions, many traveling bubble cavitations are observed, whereas in

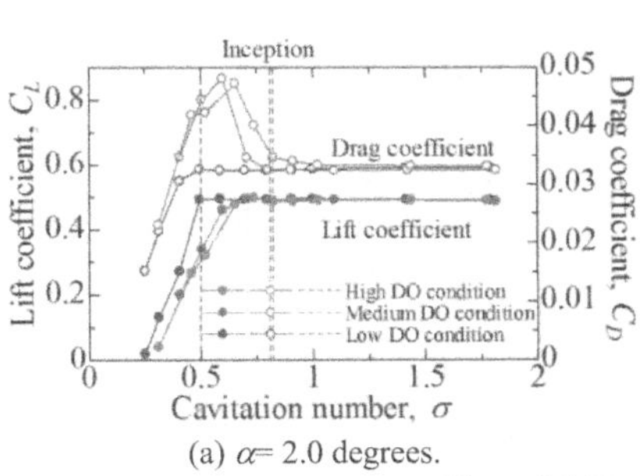

(a) α= 2.0 degrees.

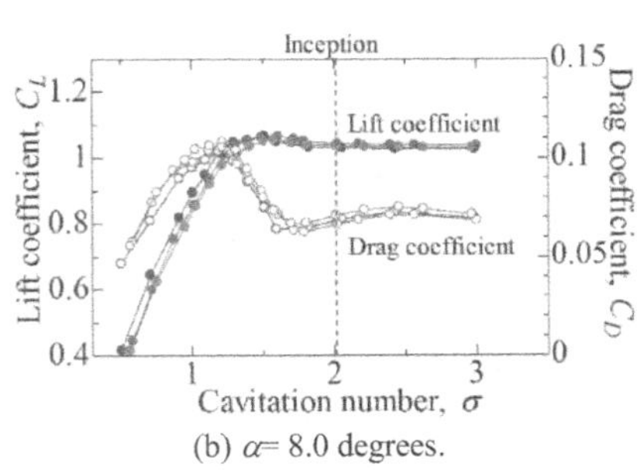

(b) α= 8.0 degrees.

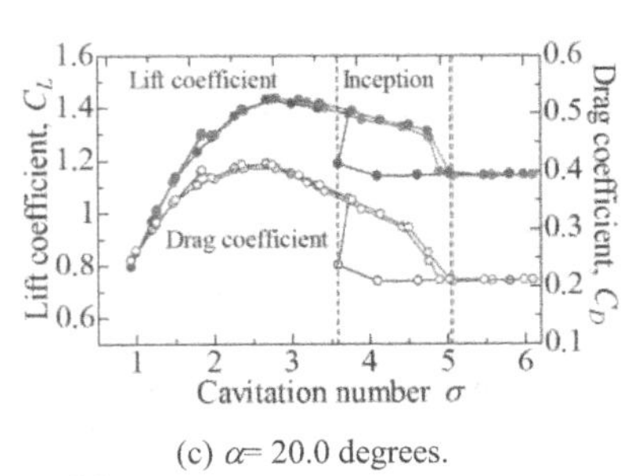

(c) α= 20.0 degrees.

Figure 3 Lift and drag coefficient curves for different gas content conditions.

the low DO condition sheet cavitation forms immediately after the inception. The increase of drag coefficient at high and medium DO conditions are supposed to be caused by the increase of momentum defect due to such bubble cavitations.

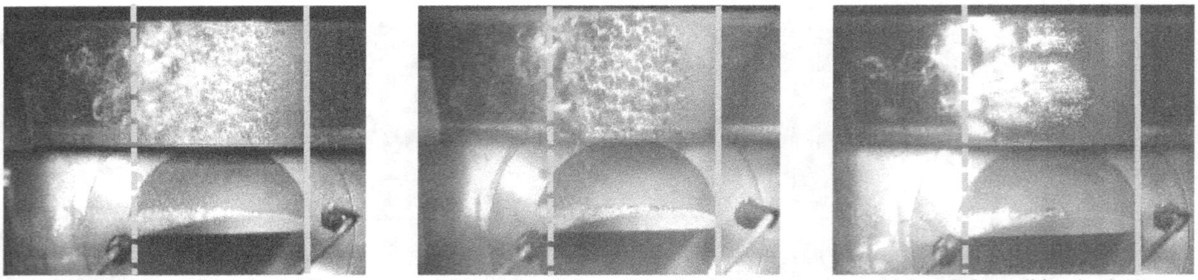

(a) High DO condition (σ=0.46) (b) Medium DO condition (σ=0.41) (c) Low DO condition (σ=0.40)

Figure 4 Cavity pattern in various DO conditions at α=2.0 degrees.

At α=8.0 degrees, in all DO conditions, the lift coefficient increases slightly after the inception of cavitation and decreases suddenly from around σ=1.3 at which the sheet cavity shows the strong unsteadiness due to the large-scale cloud cavity shedding. However, the appearance of cavitation is very different with the amount of gas content as shown in Fig. 5. Sheet cavitation consists of many travelling bubbles in the high and medium DO conditions, while the streak or film-like sheet cavitation is observed in the low DO condition. It is surprising that the influences of gas content on the time-averaged lift and drag coefficients as well as on the cavitation inception are small despite the different appearance of cavitation.

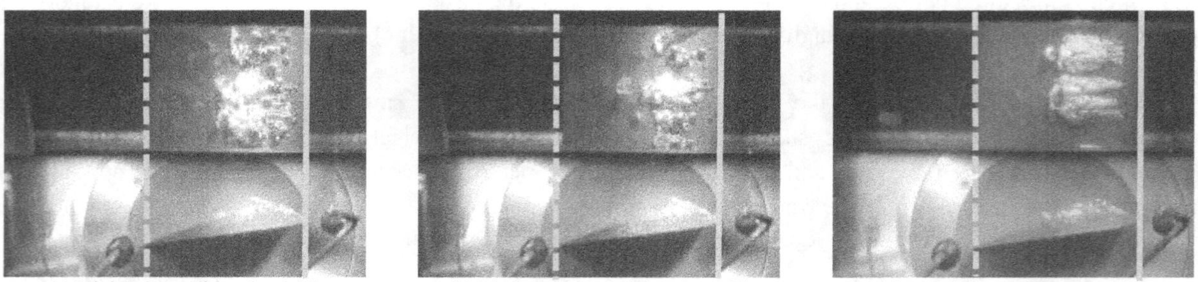

(a) High DO condition (σ=1.50) (b) Medium DO condition (σ=1.57) (c) Low DO condition (σ=1.51)

Figure 5 Cavity pattern in various DO conditions at α=8.0 degrees.

At α=20.0 degrees, as shown in Fig. 3(c), sudden increases of the lift and drag coefficients are observed immediately after the cavitation inception, and they reach the maximum value around σ=2.7. In addition, in the low DO condition, both the inception cavitation numbers and the cavitation number at the sudden decreases of lift and drag forces are small compared in the high and medium DO conditions. Figure 6 shows the blade surface pressure distributions in non-cavitating and cavitating conditions with large gas content. It can be seen that the pressure distribution changes all over the suction surface just after the inception of cavitation. In particular, in cavitation conditions, the minimum pressure is much larger than the vapor pressure as can be understood from $-C_{p,min}<\sigma$ ($C_{p,min}$ is the pressure coefficient at the suction peak). This indicates that the flow is highly unsteady, otherwise, we must not observe any cavitation bubbles in such a high pressure condition. Actually, Fig. 7 shows the cavity patterns and the time histories of lift and drag coefficients at σ=2.7 in the low DO condition, from which we can see the clear and significant periodic phenomenon; the cavity occurs from the leading edge of hydrofoil disappears after it has grown to a certain size. The lift and drag coefficients significantly fluctuate at the frequency corresponding to the cavity oscillations. In addition, as shown in Fig. 7, the lift and drag coefficients when there is no cavitation on the blade surface is approximately equal to those in non-cavitating condition. Although the results are not shown due to the limited space, it is confirmed that, as the cavitation number is decreased, the volume of cavity increases, and the fluctuation amplitudes and frequency of lift and drag coefficients also increase. The difference of cavity appearance and time histories of lift and drag coefficients due to the difference of DO are hardly observed once the cavitation starts to occur.

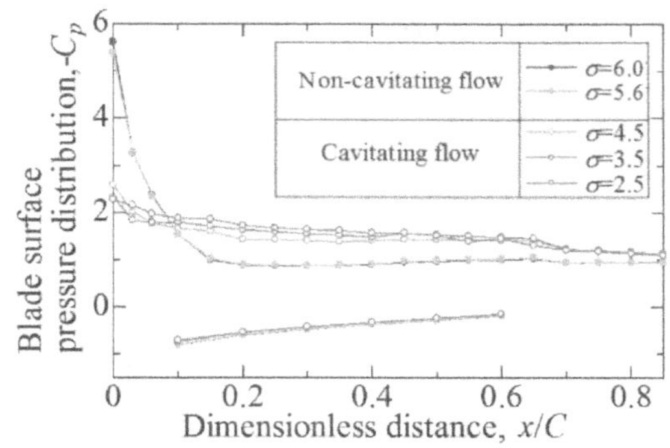

Figure 6 Surface pressure distributions at α=20.0 degrees.

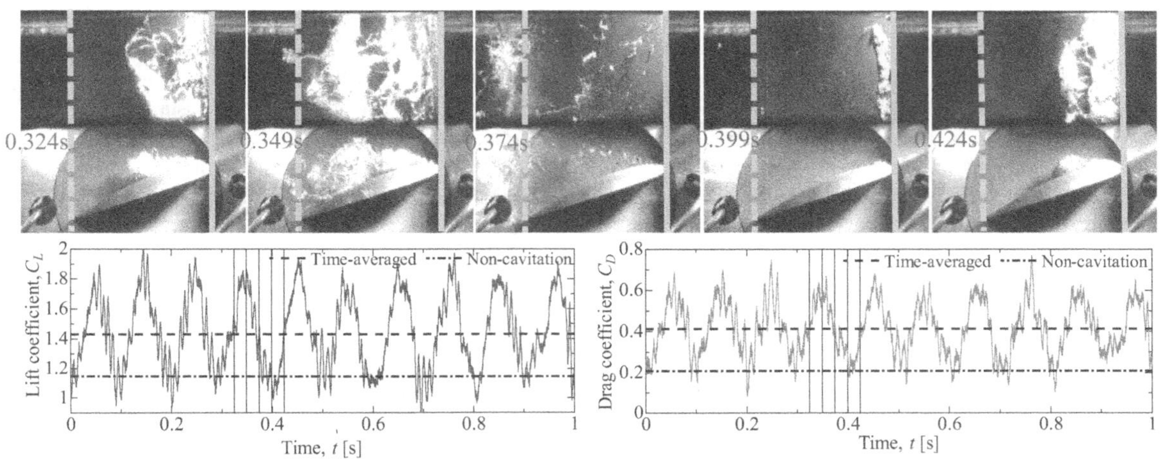

Figure 7 Cavity patterns and the corresponding lift and drag coefficients fluctuations at α=20.0 degrees (σ=2.7).

Conclusion

The mechanism of lift increase of a single hydrofoil with some occurrence of cavitation is investigated under the angles of attack of 2.0, 8.0 and 20.0 degrees where the boundary layer characteristics is different. The effect of gas content is also examined. Results are summarized as follows:

1. At the angle of attack of 2.0 degrees at which laminar to turbulent boundary layer transition occurs around the minimum pressure location, the influence of dissolved air is observed in both the appearance of cavitation and lift and drag characteristics.
2. At the angle of attack of 8.0 degrees at which the laminar separation bubble forms near the leading edge of hydrofoil, the difference due to the gas content is hardly observed in the time averaged lift coefficients although the cavitation appearance is so different.
3. At the angle of attack of 20.0 degrees at which the turbulent flow separation occurs near the leading edge, the blade surface pressure distribution and the time-averaged lift and drag forces significantly change once the cavitation occurs. This is seen regardless of the gas content, although the inception cavitation number is different. In addition, the lift and drag forces fluctuate greatly with the growth and extinction of the cavitation on the blade surface, resulting in the increases of time-averaged lift and drag coefficients.

References

[1] Knapp, R. T., Daily, J. W., Hammitt, F. G. (1970). *Cavitation*. McGraw-Hill.

[2] Kato, C., (2011). *Industry-university collaborative project on numerical predictions of cavitating flows in hydraulic machinery*. ASME-JSME-KSME Joint Fluids Engineering Conference 2011. AJK2011-06084. pp. 445-453

[3] Franc, J.-P., Michel J.-M. (2004). *Fundamentals of cavitation*. Kluwer Academic Publishers.

[4] Daido, H., Watanabe, S., Tsuda, S. (2015), *Effects of dissolved gas on unsteady cavitating flow around a Clark Y-11.7% hydrofoil*. ASME-JSME-KSME Joint Fluids Engineering Conference 2015. AJK2015-05488. V02AT05A006

© 2018 ASME

10th International Symposium on Cavitation - CAV2018
Baltimore, Maryland, USA, May 14 – 16, 2018 CAV18-05144

Microbubbles and Cavitation: Microscales to Macroscales

Paul A Brandner

Australian Maritime College, University of Tasmania, Launceston, Australia

Abstract

Microbubbles are systemic to cavitation physics. They are present in all practical liquid flows, they provide nuclei for inception, their dynamics play a role in many of the phenomena and processes involved in cavitation and ultimately they are generated by cavitation collapse and condensation. Various techniques for the generation, measurement and dispersion of microbubbles for experimental modelling of cavitation nucleation and dynamics in hydrodynamics test facilities are presented. Controlled generation of microbubbles are also required for development and calibration of optical measurement techniques and for flow diagnostics. Several techniques for generation of mono-, and poly-disperse microbubble populations in the size range 1 to 100 microns are discussed. Microbubble concentrations generated by devices or from cavitation itself vary from 1000 bubbles per cubic cm down to 0.01, in addition to the size range, necessitating the use of several methods for measurement of both size range and concentration. Experimental results demonstrating the influence of nuclei populations on stable and unstable cavitation in canonical flows about spheres and hydrofoils are presented. Nuclei populations affect shedding mechanisms and modes, spectral content, amplitudes of forces and microbubble content in downstream wakes.

Keywords: cavitation; microbubbles; nucleation

Introduction

Hydrodynamic cavitation in practical flows invariably involves heterogeneous nucleation and many flows generally in naval hydrodynamics involve microbubble disperse flows. Oceans and waterways contain naturally occurring populations of bubbles that vary with sea state and depth. These populations are generated by various mechanisms [1,2] and whilst buoyancy limits larger bubbles to near surface depths microbubbles persist at sufficient depths [3] to be a consideration for submarines as well as ships. In addition to natural populations, ships are prolific sources of poly-disperse bubble populations due to significant surface disturbances and turbulence generation. Entrained air is broken up to micro scales via various mechanisms and globally dispersed about the hull to ultimately affect, and be affected, by highly loaded lifting surfaces of propulsion and control equipment [4,5]. Cavitation occurrence about such lifting surfaces are modulated by both natural and ship generated microbubble sources. New populations of microbubbles are then generated through a complex interaction of bubble dynamics, condensation, shockwave, diffusion and turbulence phenomena [6,7]. Sub-cavitating or incipient flow about lifting surfaces may also modify microbubble populations via such processes.

These flows present many challenges due to the large range of length/time scales and physical phenomena involved. In developing test flows these include the generation, injection, dispersion, measurement and separation of microbubbles. An overview of the development of experimental capabilities for modelling microbubble disperse flows in the Cavitation Research Laboratory (CRL) at the Australian Maritime College (AMC) is presented. Mono-, and poly-disperse microbubble populations are desirable for modelling nucleation and cavitation dynamics but also for diagnostics and calibration of optical measurement techniques. Microbubble generators have been developed and assessed in small-scale bench setups and later implemented in the CRL large-scale variable pressure tunnel (or cavitation tunnel). The implementation of these devices must also involve means to vary production rates to achieve concentrations varying over several orders of magnitude. Microbubble bubble populations generated by cavitation itself are of much higher concentrations than those injected for nucleation. Different optical measurement techniques have therefore been developed to measure these populations in the tunnel. Cavitation tunnels also contain a naturally occurring background nuclei population too small and sparse for optical measurement necessitating an alternative technique for their measurement. Finally recent results from experiments implementing these techniques to investigate the influence of freestream microbubble populations on various cavitating flows are presented.

*Corresponding Author, Paul; Brandner: P.Brandner@utas.edu.au

Experimental modelling of microbubble disperse flows

The AMC CRL has several experimental facilities for investigating cavitation over a range of physical scales, microbubble generation and microbubble disperse flows generally. The principal facility is a variable pressure water tunnel for the modelling of larger scale flows [8]. A schematic of the circuit with relevant ancillaries indicated is shown in figure 1. The tunnel test section is 0.6 m square by 2.6 m long, the nominal operating velocity and pressure ranges are 2 to 12 m/s and 4 to 400 kPa absolute respectively and the tunnel volume is 365 m^3 with demineralised water the working fluid. The circuit architecture has been developed for continuous elimination of large volumes of injected incondensable gas (up to 200 ℓ/s) and microbubbles be they injected or created by cavitation. This is achieved through a combination of coalescence/gravity separation in a downstream tank and dissolution via extended residence in a resorber. Ancillaries include a rapid degasser capable of 20 % of atmospheric saturation in 2 hours and test section ceiling boundary manipulator capable of total thicknesses between 10 and 130 mm. The circuit has low background noise and vibration levels through low circuit velocities and isolation of the circuit from the surrounding building and all noise generating machinery. Two separate arrays of microbubble injectors are located upstream of the honeycomb for nuclei and diagnostics. The nuclei array is a triangular grid of 80 mm spacing configurable for targeted seeding in rectangular or circular images and the diagnostic array for line images. Close spacing of nuclei array is required to ensure homogeneous seeding. The microbubble generators used for these arrays are described in detail below. For the purposes of nuclei seeding devices used generate microbubbles of the order of diameters 10 to 100 μm at rates of 10^6 s^{-1} to achieve seeding concentrations ranging from 0.01 to 10 cm^{-3} via direct or dilute injection. For diagnostics diameters of the order 10 to 20 μm at concentrations of 10 mm^{-3} are required in confined volumes.

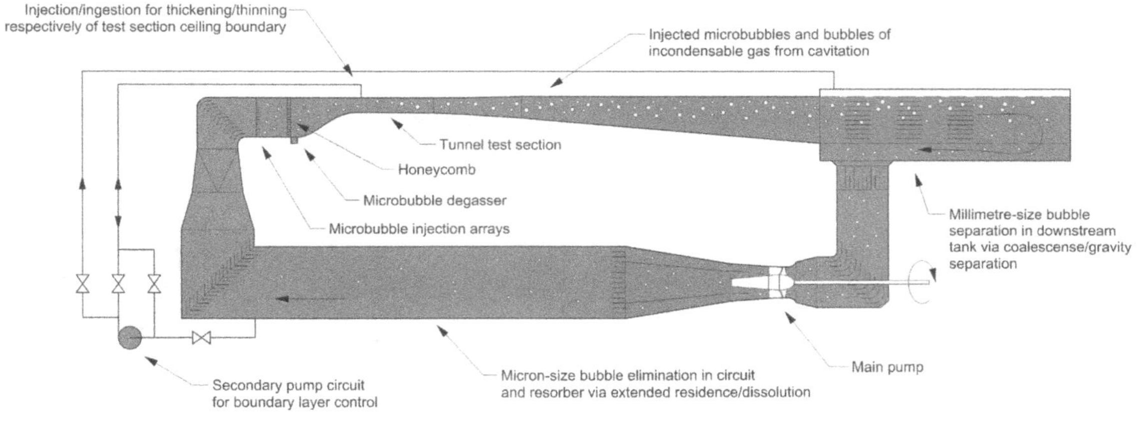

Figure 1 Schematic of AMC, CRL variable pressure water tunnel showing circuit architecture for continuous separation of large volumes of incondensable gas or microbubbles be they either injected or created by cavitation. Ancillaries for microbubble injection, degassing and manipulation of the test section ceiling boundary layer are also shown.

Techniques for generation of mono- and poly-disperse microbubbles

The microbubble generation capabilities employed in the CRL for the tunnel and diagnostic developments are subject to ongoing refinement for which currently three techniques are used all involving micro-scale flows. For nuclei seeding so called 'mini-tube' microbubble generators [9] are used. These have been investigated extensively in the CRL [10] with further experiments planned. Supersaturated water at high pressure is released through a 0.5 mm sharp edged orifice discharging into a 200 mm long 1 mm diameter hypodermic. Microbubbles are generated in the small scale turbulent shear layers of the confined turbulent jet through cavitation and gaseous diffusion. Results from a recent CRL investigation using dilution and Hele-Shaw shadowgraphy are shown in figure 2 [10]. These indicate poly-disperse populations generated at rates of about 10^6 s^{-1} with dominant size of 10 to 15 μm. The microbubble size distribution and production rate are a function of the cavitation number and Reynolds numbers which is the subject of a current test campaigns to fully map the operating envelope relevant to CRL tunnel operation. Devices for generating microbubbles suitable for diagnostic use, namely particle imaging velocimetry, have been under development for some time. The use of microbubbles avoids the requirement to seed the large volume of the CRL tunnel, avoids circuit and ancillary contamination and any ambiguity as to active nuclei. Devices developed for this purpose also use supersaturated water in a confined flow to create cavitation and diffusion as with the mini-tube devices. These employ a confined radial jet [11] similar to that which has been used for erosion studies [12]. Such devices are also used in the degassing system in the CRL tunnel [13, 14]. Experiments have shown that if the discharge is directed to maintain high velocity coalescence is avoided [15] and similar devices have been reported on previously [16]. Typical geometries used involve a 0.5 mm orifice to create the radial cavity in a 50 μm deep radial passage 9 mm diameter discharging through one side, as shown in figure 2. Measurements have been made near range microscopic shadowgraph giving populations generated at rates of the order 10^7 s^{-1} with at dimeters less than 20 to 30 μm, as shown in figure 2.

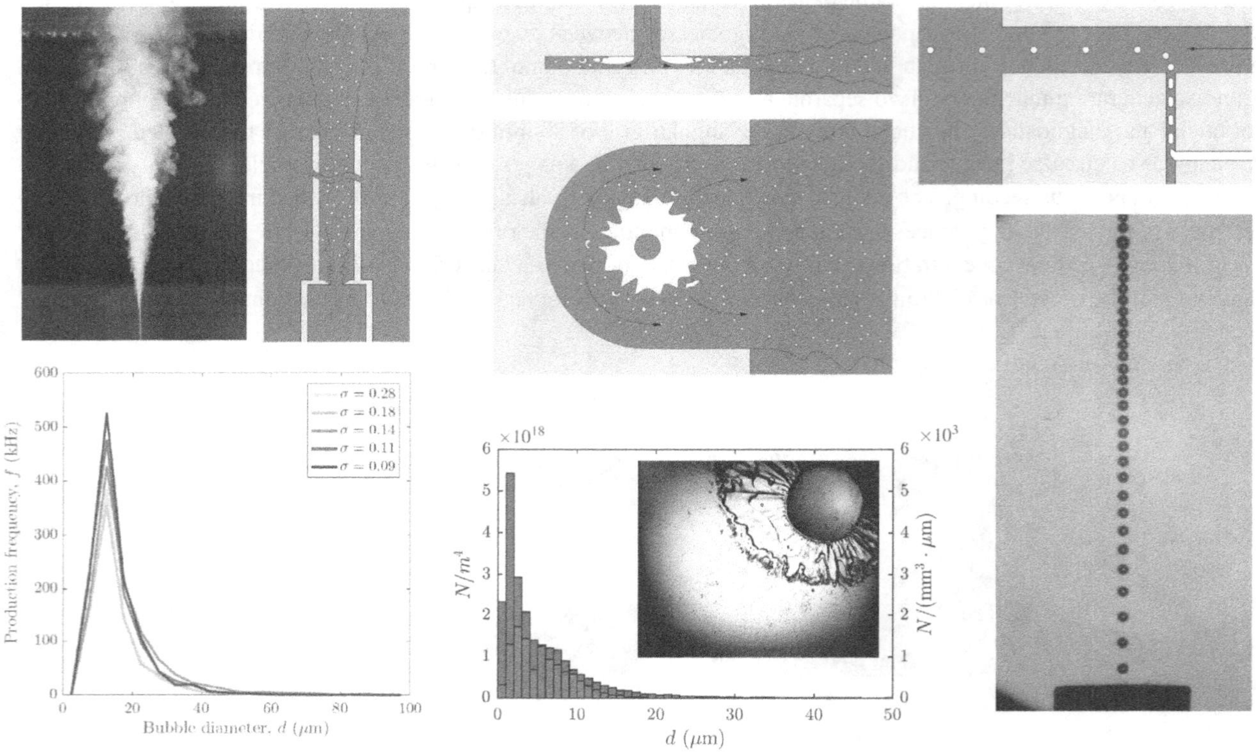

Figure 2 Techniques for generation of mono- and poly-disperse microbubbles in the CRL involving micro-scale flows. Cavitation and gaseous diffusion about a confined round turbulent jet discharging into a hypodermic used for nuclei seeding (left), Cavitation and gaseous diffusion about a directed confined radial jet used for diagnostics (middle), and microfluidic 'T' junction combined with dispersing shear flow for nuclei and for calibration of optical techniques (right).

Devices for generation of mono-disperse bubbles have also been investigated for development and calibration of measurement techniques in bench setups and ultimately for potential use in the CRL tunnel as nuclei. Microfluidic 'T' junctions developed by YLec consultants, France, that do not require special materials or surfactants have been used for microbubble generation in the size range 50 to 150 μm at production rates or the order 10^3 s^{-1} [17]. Efforts thus far have been for calibration of interferometric Mie techniques for nuclei measurement similar to [17]. For this purpose setups have been developed where the bubble train is injected into a co/shear flow to manage inter-bubble spacing and coalescence, as shown in figure 2.

Microbubble measurement techniques

For small-scale bench tests microscopic shadowgraphy with diffused laser illumination has shown to be the most versatile. However, for measurements in the CRL tunnel several techniques have been developed to measure the large range of size/critical pressures and concentrations involved in general microbubble disperse flows. A summary of recent CRL tunnel measurements from several flows is given in figure 3. These techniques include the Cavitation

Susceptibility Meter (CSM), Interferometric Mie Imaging (IMI) and Long-range Microscopic Shadowgraphy (LMS). The CRL tunnel natural or background nuclei population has been measured using CSM [18] since these are too small and sparse to be measured optically. This population has been found to obey a power law relationship between cumulative concentration and critical pressure and to be invariant of normal operating conditions [19]. The results shown have been converted to equivalent bubble size from bubble equilibrium theory. This population has been found to not be active for typical higher cavitation number flows such as sheet and cloud but to be important for lower cavitation number flows such as tip vortex. Recent comparative studies show similarities of background populations between different tunnels suggesting perhaps some universality although more data are required [20]. IMI is particularly suited for typical seeded nuclei sizes and concentrations and has appropriate working distance and field of view. It is a technique subject ongoing development in the CRL as in other facilities internationally [17]. Current experiments are planned to obtain more accurate calibrations for size and measurement volume. Microbubble populations generated by cavitation are of the order of 10^3 cm^{-3} which is too high for IMI and must be measured using LMS. Although the magnifications and working distances required reduce the measurement to effectively a point such that multiple magnifications are required to obtain all bubbles sizes [6, 7].

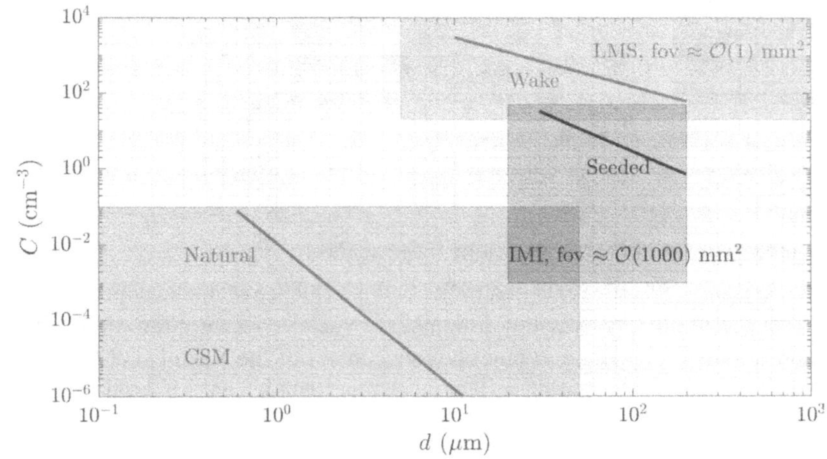

Figure 3 Nuclei/microbubble cumulative histograms from various experiments in the CRL cavitation tunnel measured using three different methods depending on the population size/critical pressure and concentration range. The shaded regions show the application range of each technique. The CSM has been used for measurement of the tunnel natural or background nuclei population, IMI for seeded nuclei populations and LMS for microbubble populations generated from collapse and condensation of large-scale sheet and cloud cavitation. These results demonstrate the large range of sizes/critical pressures and concentrations in microbubble disperse flows and the need for several techniques for their measurement.

Nucleation effects on cavitation physics

The influence of nucleation on cloud cavitation about spheres and hydrofoils has been investigated in the CRL tunnel. Sample results for nuclei deplete (background) and abundant (high concentration) cases for a sphere and a hydrofoil are shown in figure 4. Cavitation about a sphere at intermediate cavitation numbers in a nuclei deplete flow involves large-scale predominantly bimodal shedding driven by coupled re-entrant jet formation and shock wave propagation. Although a third frequency is also evident in spectra from surface pressure and high-speed imaging that only occurs at a cavitation number of about 0.8. The two dominant frequencies are attributed to axisymmetric and asymmetric global modes and the third to due local flow effects [21, 22]. Recently similar experiments have been made with a nuclei abundant flow. Due to the constant supply of activated nuclei the cavity becomes discrete or discontinuous. Similar global modes develop but with increased frequencies and reduced amplitudes, although the local mode occurs at the same frequency with increased amplitude. For unsteady cavitation about a hydrofoil there are more significant differences involving shedding mechanisms, the number of modes and frequencies [23]. The deplete case has been found to have three modes, all driven by shockwave propagation, the second of which is the fundamental involving large-scale shedding, the subharmonic associated with local shedding near the tip and the first harmonic with both tip and root activity. For the abundant case continuous supply of activated nuclei alter shedding physics to bimodal driven by lower frequency shockwave propagation, as shown in the lift spectra in figure 4 (top). Two alternate shockwaves of differing strength drive the shedding the stronger of which is associated with the fundamental causing large-scale shedding and the weaker the harmonic causing small scale shedding.

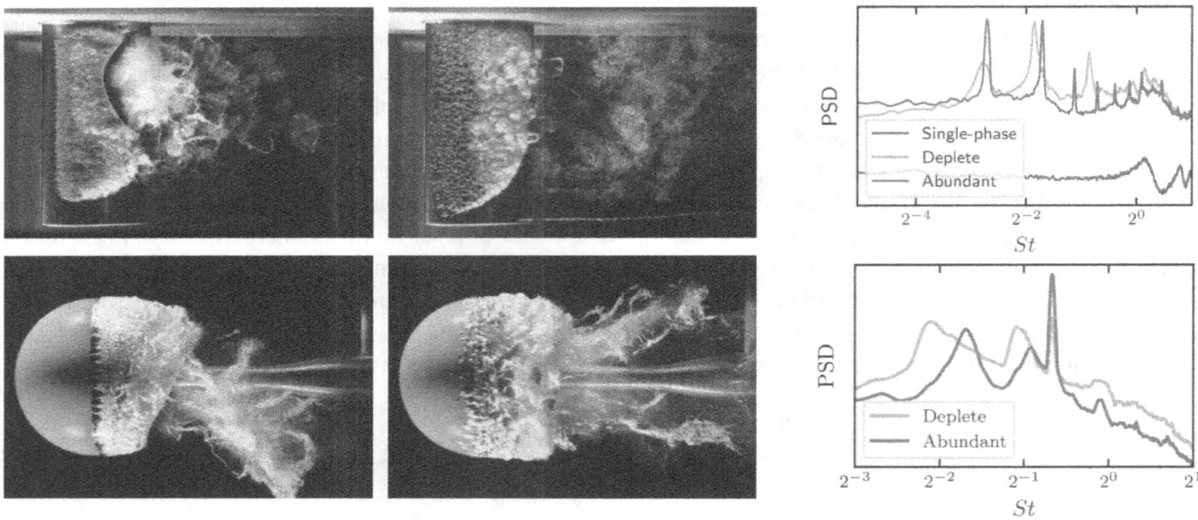

Figure 4 Photographs and Spectral data demonstrating the influence of nuclei concentration on shedding frequencies of unsteady cavitation about a hydrofoil (top) and sphere (bottom) in nuclei deplete and abundant flows. The Power Spectral Density (PSD) plots for the hydrofoil is of lift force and for the sphere of surface pressure.

Microbubble populations generated by cavitation at large scales

The set of images shown in figure 5 demonstrate qualitatively the influence of nuclei concentration on stable cavitation about a hydrofoil for fixed Reynolds and cavitation numbers. Nuclei concentration affects the physics of the bulk cavity, condensation and turbulent breakup and microbubble content of the wake [7]. The supply of activated nuclei make the cavity discontinuous and concentrations affect the extent to which each bubble can grow, the surface area to volume ratio and hence levels of diffusion and turbulent breakup which ultimately microbubble generation. Wake microbubble measurements in the size range of 5 to 50 μm show that bubble concentrations are highest for the lower concentrations demonstrating that bubble growth is a significant factor in the diffusion and possibly breakup processes.

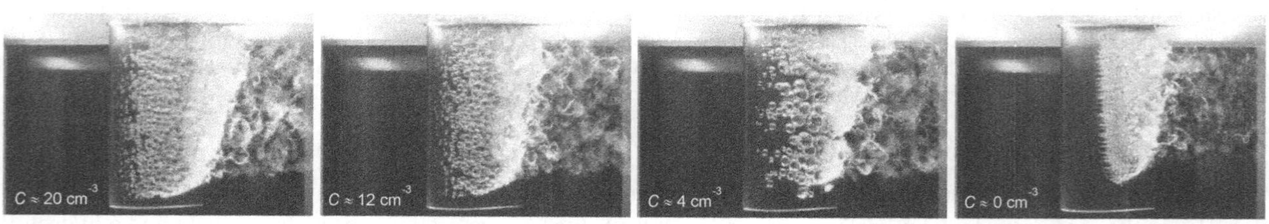

Figure 5 The influence of nuclei concentration on globally stable hydrofoil cavitation and microbubble disperse wake generated from cavity condensation and turbulent breakup. Nuclei populations for the images shown in the three left images where generated from an array of mini-tube microbubble generators operated at different cavitation numbers to produce the estimated concentrations shown.

Conclusions

Techniques for microbubble generation and homogenous dispersion in test flows have been developed in the Cavitation Research Laboratory at the Australian Maritime College. More is needed on the flow physics of microbubble generators even though latest work has produced several controlled concentration levels. Microbubble measurements in both small- and large-scale domains remain technically intensive. Several measurement techniques are required to cover the large size and concentration ranges. Interferometric Mie appears the promising technique for final development of turnkey applications for medium concentrations. The understanding of microbubble generation by macroscale or high Reynolds number cavitation remains at an early stage and more research is required. There is a nuclei permanent population in all practical test facilities important for tip vortex cavitation or other low cavitation number flows but is often not active for many higher cavitation number flows such as sheet cloud cavitation. Beyond inception microbubble populations have a dramatic effect of cavitation dynamics and physics necessitating their inclusion in all modelling. Recent work suggests future possibilities for laboratory and full-scale real-time measurement of microbubbles and active control or modulation of cavitation using free bubbles/nucleation as well as surface treatments/nucleation. More work is also required on mechanisms for microbubble generation from bubble breakup in turbulent flows. In summary the very small must be studied to understand the large.

Acknowledgements

The author wishes to acknowledge the contributions of Bryce Pearce, Gregory Walker, Yves Lecoffre, David Clarke, James Venning, Dean Giosio, Patrick Russell, Matthew Khoo, Steven Ceccio, Luka Barbaca, Katrina de Graaf, Robert Wrigley and Steven Kent and the support of the Australian Commonwealth and Defence Science and Technology Group, the United States Office of Naval Research, the Australian Research Council and the University of Tasmania.

References

[1] Deane, G.B. Stokes M.D. (2002). *Scale dependence of bubble creation mechanisms in breaking waves*. Nature. 418: 839-844.
[2] Terrill, E. J., Melville, W.K. and Stramski, D. (2001), *Bubble entrainment by breaking waves and their influence on optical scattering in the upper ocean*, Journal of Geophysical Research, 106: 16815-16823.
[3] Gindroz.,B., *Propeller Cavitation Characteristics The Practical Interest of Nuclei Measurements in Test Facilities and at Sea*, FED Vol. 226 Cavitation and Gas-liquid Flow in Fluid Machinery Devices ASME 1995.
[4] Li, J., Castro, A. M., Carrica. P. M. (2016). *Progress on Prediction of Bubbly Flows around Ships*. ASME 2016 FEDSM.
[5] Hsiao, C.-T., Chahine, G.L. (2012), *Effect of a propeller and gas diffusion on bubble nuclei distribution in a liquid*, Journal of Hydrodynamics, Series B 24(6): 809-822.
[6] Russell, P.S., Giosio, D.R., Venning, J.A., Pearce, B.W., Brandner, P.A., Ceccio, S.L. (2016). *Microbubble generation from condensation and turbulent breakup of sheet cavitation*. 31th Symposium on Naval Hydrodynamics. Monterey. USA.
[7] Russell, P.S., Venning, J.A., Brandner, P.A., Pearce, B.W., Giosio, D.R., Ceccio, S.L. (2018). *Microbubble disperse flow about a lifting surface*. 32nd Symposium on Naval Hydrodynamics. Hamburg. Germany.
[8] Brandner, P.A. Lecoffre, Y. and Walker, G.J. (2007). *Design considerations in the development of a modern cavitation tunnel*. 16th Australasian Fluid Mechanics Conference. Gold Coast.
[9] Oldenziel, D.M, Jansen, R.H.J, Keller, A.P., Lecoffre, Y. van Renesse, R.L. (1982) *Comparison of instruments for detection of particles and bubbles in water during cavitation studies*. Proceedings on Operating Problems of Pump Stations and Power Plants IHAR Amsterdam. Netherlands.
[10] Giosio, D., Pearce, B. W. Brandner, P. A. (2016). *Influence of Pressure on Microbubble Production Rate in a Confined Turbulent Jet*. 20th Australasian Fluid Mechanics Conference.
[11] Brandner, P. A., de Graaf, K. L., Pearce, B. W. and Burgess, J. S. (2014) *Microbubble Generation in a Confined Radial Jet*, 19th Australasian Fluid Mechanics Conference, Melbourne. Australia.
[12] Chahine, G., Franc, J.-P., Karimi, A. (2014). *Laboratory Testing Methods of Cavitation Erosion*. Advanced Experimental and Numerical Techniques for Cavitation Erosion Prediction. Springer.
[13] Lecoffre, Y. (1999) *Cavitation: Bubble Trackers*, A.A. Balkema.
[14] Lecoffre, Y. (1986) *Méthodes practiques de control de la teneur en air libre et dissous dans les moyens d'esssais*. 12th Symposium IAHRAIRH. Montreal. Canada.
[15] Venning, J. A., De Vincentis, S., Pearce B. W., Brandner P. A. *Microbubble generation for PIV seeding*. 20th Australasian Fluid Mechanics Conference, Perth. Australia.
[16] Franc, J.P., Avellan F., Belhadji B., Billard J.Y., Briançon-Marjollet, L., Fréchou, D., Fruman, D.H., Karimi, A., Kueny J.L., Michel J.M. (1995). *La Cavitation : Mécanismes Physiques et Aspects Industriels*. Grenoble, Presse Universitaires de Grenoble.
[17] Boucheron, R. Aumelas, V. Donnet, M. Fréchou, D., Poidatz, A. (2018) *Comparative study of optical experimental methods for micro-bubble sizing*. 19th International Symposium ALITFM. Lisbon, Portugal.
[18] Khoo, M., Venning, J., Pearce, B., Brandner, P. and Lecoffre, Y. (2016). *Development of a cavitation susceptibility meter for nuclei size distribution measurements*. 20th Australasian Fluid Mechanics Conference, Perth.
[19] Venning J.A. Khoo M.T. Pearce B.W. Brandner P.A. (2018) *Background nuclei measurements and implications for cavitation inception in hydrodynamic test facilities*. Exp. Fluids **59**, 71.
[20] Khoo, M. T. Venning, A. J. Takahashi, K. Arai, J. Mori, T. Pearce, B.W. Brandner, P.A. Ranmuthugala, D. (2017). *Joint Research between Australia and Japan on the Cavitation Inception of Marine Propellers and Control Surfaces*, MAST Asia. Japan.
[21] de Graaf, K. L. Brandner P. A. Pearce B.W. (2017) *Spectral content of cloud cavitation about a sphere*. J. Fluid Mech., 812, R1.
[22] Brandner P.A. Venning, J.A. Pearce B.W. (2018) *Wavelet analysis techniques in cavitating flows*. Phil. Trans. R. Soc. A **376**: 20170242.
[23] Venning, J.A., Giosio, D.R., Smith, S. Pearce B.W., Brandner P. A. (2018). *The influence of nucleation on the spectral content of cloud cavitation about a hydrofoil*, CAV 2018. Baltimore. USA.

© 2018 ASME

10th International Symposium on Cavitation - CAV2018
Baltimore, Maryland, USA, May 14 – 16, 2018

Modeling Fluid-Structure Interaction in Cavitation Erosion using Smoothed Particle Hydrodynamics

Shrey Joshi[1,2], Jean Pierre Franc[2], Giovanni Ghigliotti[2], Marc Fivel[1]

[1] *Univ Grenoble Alpes, CNRS, Grenoble INP, SIMaP, 38000 Grenoble, France*

[2] *Univ. Grenoble Alpes, CNRS, Grenoble INP, LEGI, 38000 Grenoble, France*

Abstract

In the present study a meshless Smoothed Particle Hydrodynamics cavitation solver is developed. The fluid bubble collapse solver is validated against analytical Rayleigh-Plesset equation and shows good agreement. The solid solver capable of solving elastic-plastic deformation and material damage is developed and is validated against FEM results. A fluid structure interaction solver capable of solving cavitation erosion is presented. A single bubble collapse is demonstrated in the paper.

Keywords: Smoothed Particle Hydrodynamics, fluid-structure interaction, cavitation, material damage

Introduction

Cavitation erosion is a major issue amongst a wide range of equipment such as hydraulic devices, diesel injectors, artificial heart valves etc. Cavitation is defined as the appearance of vapour cavities inside a continuous and homogeneous liquid medium. The generation of vapour cavities could happen due to various reasons, but the cause is mostly associated to the drop of local pressure below the vapour pressure. During cavitation the continuum liquid medium breaks down to form vapour cavities. The vapour formation occurs in cavitation when the pressure drops below the vapour pressure as the new pressure and temperature lies in the vapour phase zone.

These vapour bubbles can collapse when the ambient pressure increases above the vapour pressure. The collapse of these bubbles can be very different depending on the position of the bubbles wrt to a solid surface. A collapse of bubble near the surface can lead to formation of a high velocity micro jet and subsequent shock wave. The asymmetric collapse in presence of the wall near the bubble is due to the lack of fluid flow from the side of the wall. This micro-jet then hits the other side of the bubble producing an intense shock wave. The solid surface experiences a high pressure due the shock wave and the high velocity micro jet hitting the surface. The magnitude of the pressure acting on the surface could be high enough to cause plastic damage in the material. Although the experimental measurement of these pressures acting on the surface still remains a challenge, various studies have estimated it to be around a few GPa [1-5]. Moreover, experimental investigation offers only limited information about bubble collapse and the subsequent formation of micro jet and pressure wave since any intrusive measurement can cause deviations. Such difficulties in experimental investigation has led to the use of numerical studies to understand and analyze cavitation.

A thorough cavitation study requires a two-way fluid-structure interaction coupling to get realistic results for cavitation erosion. A comprehensive cavitation numerical model should include the following elements in the solver:

- A fluid model capable of capturing the dynamics of a collapsing bubble including the details about the micro-jet and the shock wave.
- A solid solver to solve for the material response due to the pressure acting on the material surface as a result of the micro jet and the shock wave. This solver must be capable of solving an elastic-plastic behavior as well as damage in the material.
- A fluid structure interaction solver: During a cavitation bubble collapse, high intensity shock waves are produced along with micro-jet. The shock wave travels through the fluid to the solid, a part of the wave is reflected back into the liquid and the rest gets transmitted to the solid [6]. In the case of two elastic media, transfer of the energy and the solid-liquid interface velocity depends on the ratio of acoustic impedances of

*Corresponding Author, Shrey Joshi: shrey.joshi@simap.grenoble-inp.fr the liquid and solid. Acoustic impedance could be estimated simply as the product of density by the speed of sound through the medium. A two-way coupling is required to model this behavior properly.

Conventionally, ALE (Arbitrary Lagrangian Eulerian) methods have been used to simulate such a problem. The problem is quite complex since the fluid solver (Finite Volume Method (FVM) code) and a solid solver (Finite Element Method (FEM) code) are generally two different codes which then need to be coupled in order to communicate data across the two solvers. To overcome the above complexity, in the present study a first attempt has been made to solve cavitation using a meshless method Smoothed Particle Hydrodynamics (SPH). The method offers the following advantages:

- Both fluid and solid response can be captured using the same solver and the same numerical method i.e. SPH, hence eliminates the use of two different solvers while numerically solving for cavitation.
- Coupling an FVM solver to an FEM solver can be quite complicated and requires a coupling for data transfer across the two codes. The problem is eliminated as the solver uses SPH for both solid and fluid and no such coupling is required.
- The material can exhibit plastic deformation during cavitation, while using mesh methods would require a mesh re-construction algorithm to account for the deformed material. The reconstruction of mesh for fluid is required which is complex and can slow the code significantly. However, this problem is eliminated while using meshless methods like SPH.

Axisymmetric SPH solver

In the present work a 2D axisymmetric Smoothed Particle Hydrodynamics solver is developed. An existing fluid 2D SPH solver SPHYSICS [7] is used as the basis to develop the solver further. For a 2D axisymmetric fluid solver the cylindrical form of SPH equations are used [8] and the original solver is modified accordingly. A cavitation bubble collapse simulation was carried out. Figure 1 shows the domain used for a bubble collapse simulation in an infinitely large medium. The wave maker shown in the figure is a set of repulsive particles in SPH which apply force on the nearby particles hence acting as a pressure wave generator. Figure 2 shows the comparison for a bubble collapse radius wrt time against 3D Rayleigh-Plesset equation which shows good agreement.

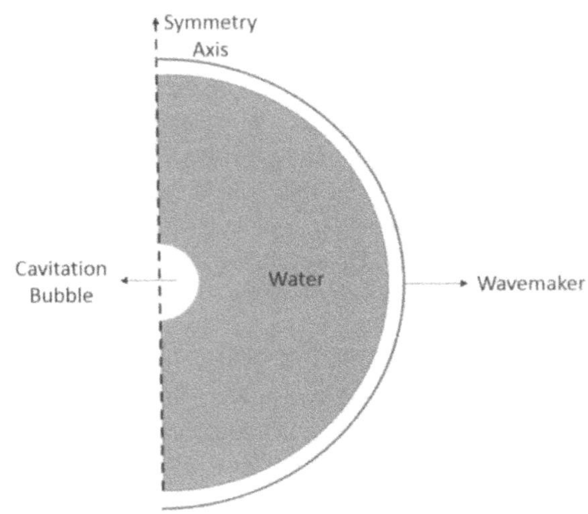

Figure 1. Simulation domain for bubble collapse simulation in 2D axisymmetric, a vacuum bubble of radius 0.095 mm is placed in a spherical domain which is 7 times the bubble radius. A pressure wave of 60 MPa is generated by the wavemaker whereas the initial pressure inside the bubble is 0 Pa, all the domain walls are given a non-reflective boundary condition to avoid any wave reflections hitting the bubble. The flow is treated as non-viscous and no surface tension forces are considered in the model.

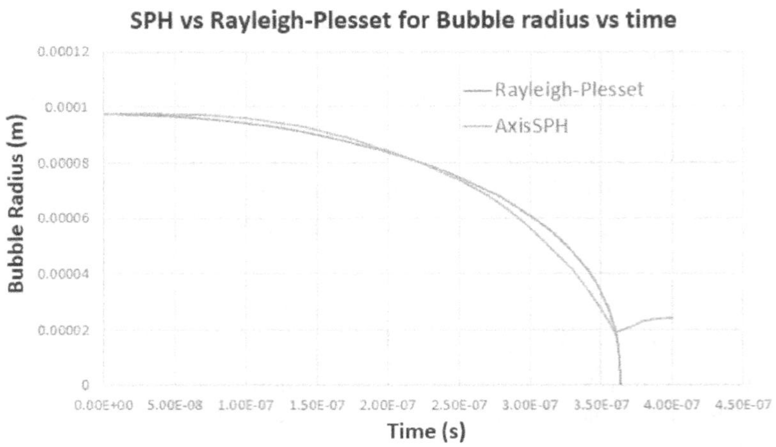

Figure 2. Comparison of bubble radius vs. time, SPH against Rayleigh-Plesset solution for a bubble collapse simulation (domain shown in figure 1).

For the solid SPH solver, similar to the fluid solver the momentum equations are modified to cylindrical SPH equations. However, the additional terms due to density correction near axis have only been derived for a fluid solver [8]. For a similar density correction, a novel scheme is implemented with new momentum equation for a solid with additional terms occurring due to density corrections. Plasticity behavior is modelled using Yield criteria from Johnson-Cook model [9]. The elastic-plastic solver is then validated against FEM (obtained from FEM solver Castem [11]) results for an indentation simulation. Figure 3 shows the domain used for indentation simulations, the boundary particles are given a downward velocity to produce an indent in the solid specimen. Figure 4 shows comparison between SPH and FEM solvers which shows an excellent agreement. The solid SPH solver is thus validated.

To account for mass loss a damage criterion is to be introduced into the model. A simple damage criterion based on the plastic strain is used in the present simulations. A critical value of fracture strain is used to predict whether the SPH particle is damaged or not. If the plastic strain is above the fracture strain the material can sustain, the particle is assumed damaged and removed from the simulation. The material used for simulation is stainless steel A-2205 with fracture strain (ε_f) 0.03. Figure 5 shows a sequence of images showing the evolution of damage for one single indentation. Further simulations were carried out to understand the effect of indenter size (that might to some extent correspond to bubble size) while keeping the indentation depth unchanged and the results are shown in figure 6. The graph on figure 6 shows that the larger radius of indenter can produce high mass loss rate but the incubation time appears to be smaller for smaller indentor which is counter-intuitive. The reason is that the plastic deformation has a direct dependence on von Mises stress and the smaller radius indenter can produce significantly higher shear stress compared to the larger radius indenter for the same depth of indent, the von Mises stress is thus higher in case of smaller indenter leading to lower incubation time. However, the larger indenter can plastify a larger volume of the specimen and hence could produce higher mass loss rates over a longer period of time.

The two solvers i.e. the fluid SPH and the solid SPH solvers have been separately validated and used to produce mass loss curves. Eventually we then put together the two solvers for a fluid-structure interaction simulation and a wave propagation test is carried out across the interface to determine whether the transmission of pressure wave across the interface follows the analytical solution. Once the wave propagation is validated against analytical results a single bubble collapse simulation was carried out. Figure 7 demonstrates a single bubble collapse simulation carried out by the SPH FSI cavitation solver. Figure 7 (a) shows the domain of the simulation where a bubble of radius 0.15 mm with its center placed at a distance of 0.2 mm from the interface is collapsed using a pressure wave of 50 MPa. Plastic strain developed in the material is shown in figure 7(c), the magnitude of plastic strain is however low (0.0012) and couldn't result in damage. So to produce enough plasticity to cause damage, multiple bubble collapse at the same point on the material is required, a single bubble collapse in the present case was not enough to cause damage.

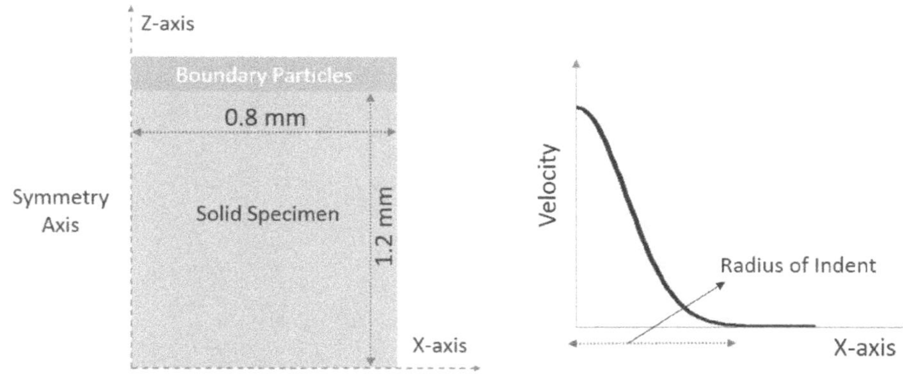

Figure 3. The figure on the left show the computational domain for solid simulation, the boundary particles marked in blue are given a downward velocity plotted on the right. (a Gaussian distribution is provided as the indent velocity as shown on the right, the radius of indent is defined as the distance from the center where the velocity is 1% of the peak value, a constant velocity is given to the indenter wrt time)

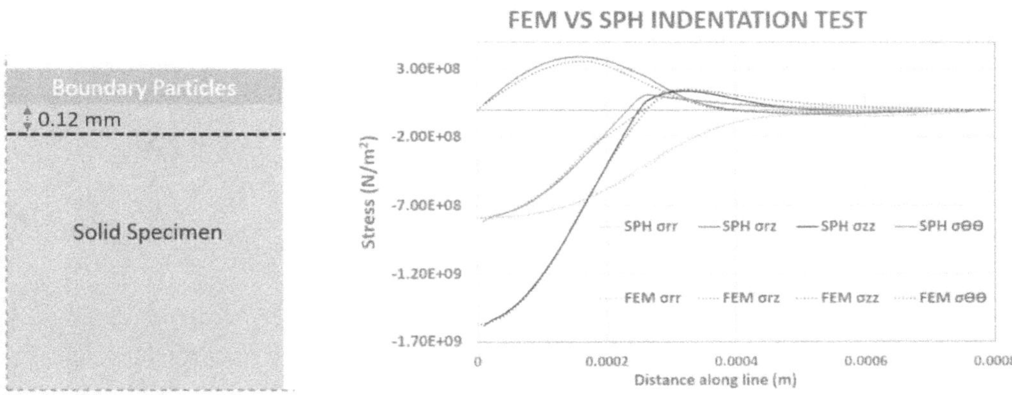

Figure 4. FEM results are compared against SPH for indentation of 6 microns in depth and 0.4 mm in radius for a stainless steel A-2205 specimen, the plots on right are plotted on the horizontal black dotted line (0.12 mm below the top surface) shown on the domain on the left.

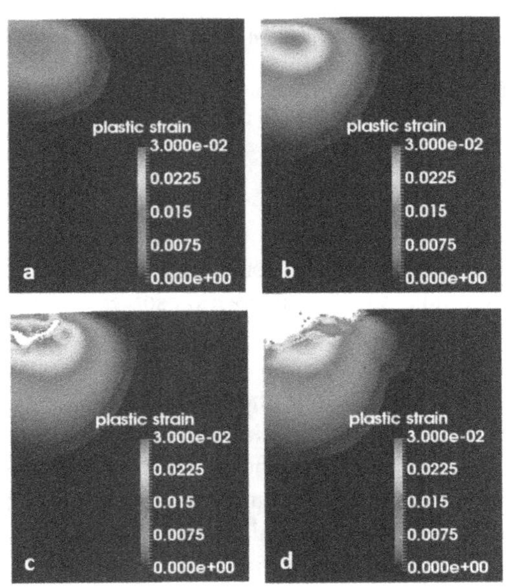

Figure 5. Sequence of images from damage simulation of a stainless steel A-2205 specimen, (a) Shows the plastic strain accumulation in material without damage, (b) high plastic strain zone just beneath the top surface, (c) damage initiation from the point of highest plastic strain, (d) material loss due to a single indent (domain size shown in the figure is same as in figure 3)

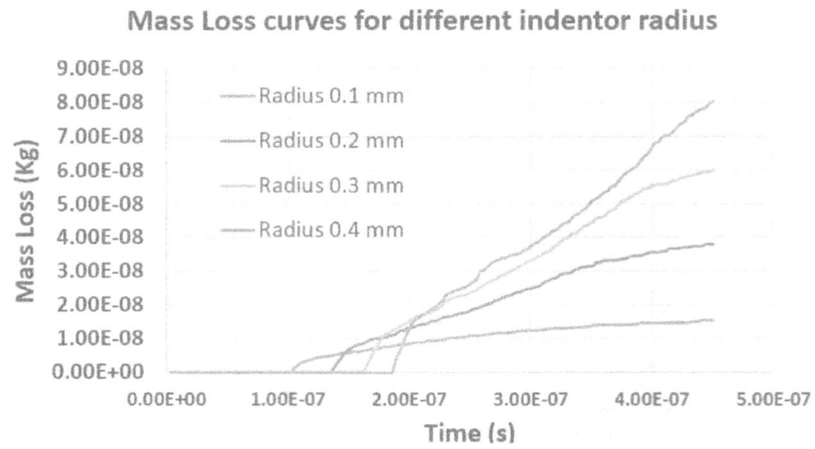

Figure 6. Mass loss curves for different indenter radius on a stainless steel A-2205 specimen obtained from SPH simulation, indentation depth is kept constant at 6 microns for all cases while the indenter radius is varied from 0.1 mm to 0.4 mm.

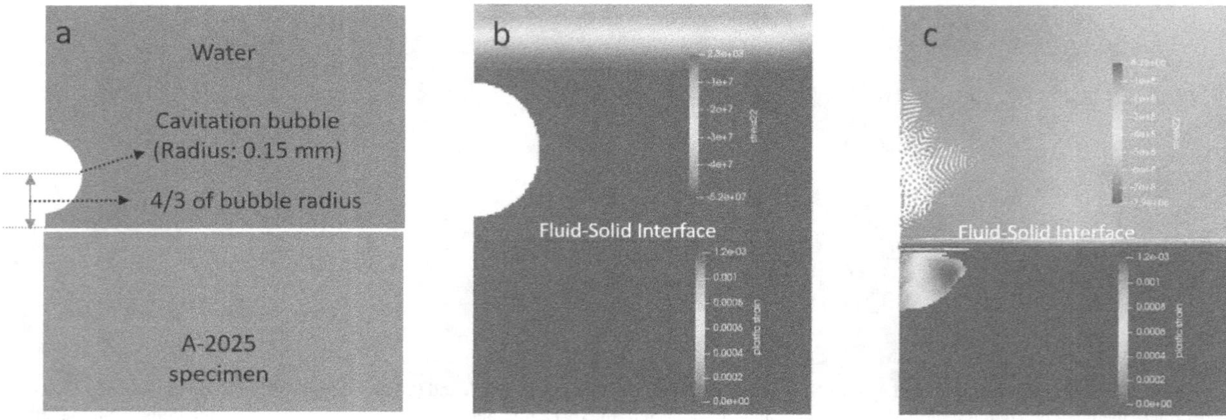

Figure 7. Fluid-structure interaction simulation of a bubble collapsing over a solid specimen: (a) Shows the computational domain for the simulation of a bubble of radius 0.15 mm with the center of the bubble at a distance of 4/3 times the radius from the interface, (b) SPH simulation where a pressure wave is generated from the top of the domain using repulsive particles producing a 50 MPa pressure wave approaching the bubble, (c) plastic strain developed in the material after the bubble collapse (contours in figure 7 (b) & (c) are for pressure in the fluid and plastic strain in the solid)

Conclusion

A 2D axisymmetric fluid-structure interaction cavitation solver is developed. The fluid solver is used to simulate bubble collapse and a solid axisymmetric solver is used for solving elastic-plastic damage and material loss. The two solvers were put together to produce a fluid-structure interaction solver capable of solving a single bubble collapse near a solid surface.

Damage simulations were carried out to demonstrate the capability of the solver to solve for material damage. However, the damage criterion based on only plastic strain is simplistic. In reality the damage parameter should depend on three aspects [10]: Plastic strain, Triaxiality (the ratio of mean stress to von Mises stress which represents the pressure state of the material i.e. tension or compression) and deviatoric stress. Various predictive models have been proposed to determine fracture strains considering all the above parameters. However, using the above criteria leads to no damage in a single bubble collapse as shown in figure 7 since plastic strain is relatively low. To produce high plastic strain and therefore create damage needs multiple bubble collapse. The solver is presently being modified to accommodate multiple bubble collapse cases.

References

[1] A. Philipp, W. Lauterborn, Cavitation erosion by single laser-produced bubbles, J. Fluid Mech. 361, 1998, pp 75–116. doi:10.1017/S0022112098008738.J.
[2] A. Vogel, W. Lauterborn, R. Timm, Optical and acoustic investigations of the dynamics of laser-produced cavitation bubbles near a solid boundary, J. Fluid Mech. 206 (1989) 299–338..
[3] S.M. Ahmed, K. Hokkirigawa, R. Oba, Fatigue failure of SUS 304 caused by vibratory cavitation erosion, Wear. 177 (1994) 129–137. doi:10.1016/0043-1648(94)90238-0..
[4] I.R. Jones, D.H. Edwards, An experimental study of the forces generated by the collapse of transient cavities in water, J. Fluid Mech. 7 (1960) 596–690. doi:10.1017/S0022112060000311..
[5] T. Momma, A. Lichtarowicz, A study of pressures and erosion produced by collapsing cavitation, Wear. 186-187 (1995) 425–436. doi:10.1016/0043-1648(95)07144-X..
[6] Par Francois Axisa, Jose Antunes, "Modelling of Mechanical Systems: Fluid-Structure Interaction," 2007, Elsevier, ISBN 10: 0-7506-6847-4.
[7] Gómez-Gesteira, M., Rogers, B.D., Dalrymple, R.A., Crespo, A.J.C. and Narayanaswamy, M., 2010, User Guide for the SPHysics Code v2.0. http://wiki.manchester.ac.uk/sphysics
[8] D. García-Senz, A. Relaño, R. M. Cabezón, E. Bravo, Axisymmetric smoothed particle hydrodynamics with self-gravity, Monthly Notices of the Royal Astronomical Society, Volume 392, Issue 1, 1 January 2009, Pages 346–360, 2008, https://doi.org/10.1111/j.1365-2966.2008.14058.x
[9] G.R. Johnson, W.H. Cook, Fracture characteristics of three metals subjected to various strains, strain rates, temperature and pressure, Engineering Fracture Mechanics, Vol. 21, No. 1, pp. 31-48, 1985.
[10] André Pineau, Amine A Benzerga, T Pardoen, Failure of metals I: Brittle and ductile fracture, Acta Materialia, Vol. 107, pp. 424-483, 2016.
[11] http://www-cast3m.cea.fr

© 2018 ASME

Multiscale Modeling of Cavitation using a Level Set Method with Cavity Detection

[1]Jingsen Ma*; [1]Chao-Tsung Hsiao; [1]Georges L. Chahine

[1]*DYNAFLOW, INC., 10621-J Iron Bridge Road, Jessup, MD 20794, USA*

Abstract

An Eulerian-Lagrangian multi-scale two-phase flow model is presented, which includes a micro-scale model for tracking the microbubbles, a macro-scale model for describing the overall flow field and large cavity dynamics, and a transition model to bridge the micro and macro scales. The method also includes a cavity detection scheme which marks each cavity with an ID index so that each can be tracked and treated individually. This approach was implemented in our CFD code 3DYNAFS© and used to simulate, for illustration, multiple bubble dynamics simultaneously released at different locations in a Rankine line vortex. The method captures their growth, entrainment into the vortex core, elongation along the vortex centerline, followed by potential splitting and breakup.

Keywords: Multiscale Model; Level Set Method; Cavity Detection

Introduction

Modelling of cavitation is challenging as it involves a wide range of characteristic lengths and phenomena including nucleation, large deforming free surfaces, bubble coalescence, bubble breakup into very fine microbubbles, bubble clouds collective effects, … etc. [1]. Motivated by the need for a physics-based modeling of the dynamics, we are developing a multiscale approach which tracks dispersed microbubbles in a Lagrangian fashion while solving the overall two-phase viscous flow in an Eulerian fashion. The microbubbles are modelled as singularities using a Discrete Singularity Model (DSM) until they grow to no longer be sub-grid or merge into macro cavities or pockets. They are then discretized and fully resolved in an Eulerian grid using a Level Set Method (LSM) [2, 3]. Afterwards, in order to continuously track each individual cavity/pocket, a cavity ID detection scheme [4] is used. This cavity tracking allows to follow each cavity and to assess the properties of its content individually. This capability is particularly highlighted in this presentation, for the simulation of multiple bubble entrainment, as well as bubble breakup in vortex flows.

Multiscale Two-phase Model

Eulerian Continuum Model for Viscous Mixture Flow

The two-phase medium is treated as a continuum, which satisfies the following continuity and momentum equations:

$$\frac{\partial \rho_m}{\partial t} + \nabla \cdot \left(\rho_m \mathbf{u}\right) = 0, \qquad \rho_m \frac{D\mathbf{u}}{Dt} = -\nabla p + \mu_m \nabla^2 \mathbf{u} + \rho_m \mathbf{g}, \tag{1}$$

where $\mathbf{u}$ is the mixture velocity, p is the pressure and $\mathbf{g}$ is the acceleration due to gravity. The mixture density, ρ_m, and the mixture viscosity, μ_m, are related to the void fraction, α, and to the liquid and gas properties through

$$\rho_m = \left(1-\alpha\right)\rho_l + \alpha \rho_g, \qquad \mu_m = \left(1-\alpha\right)\mu_l + \alpha \mu_g. \tag{2}$$

α is deduced from the location and size information of the bubbles using a Gaussian distribution scheme which smoothly "spreads" each bubble's volume over its surroundings while conserving the total bubble volumes [2].

Lagrangian Discrete Bubble Model for Microbubbles

The bubble model uses a Surface Average Pressure approach [1] to average fluid quantities along the bubble surface and model its dynamics using a Rayleigh-Plesset-Keller-Herring equation:

$$\left(1-\frac{\dot{R}}{c_m}\right)R\ddot{R} + \frac{3}{2}\left(1-\frac{\dot{R}}{3c_m}\right)\dot{R} = \frac{\mathbf{u}_s^2}{4} + \frac{1}{\rho_m}\left(1+\frac{\dot{R}}{c_m}+\frac{R}{c_m}\frac{d}{dt}\right)\left[p_v + p_g - p_{enc} - \frac{2\gamma}{R} - 4\mu_m\frac{\dot{R}}{R}\right], \tag{3}$$

where R is the bubble radius, p_v is the liquid vapor pressure, p_g is the bubble gas pressure, γ is the surface tension. $\mathbf{u}_s = \mathbf{u}_{enc} - \mathbf{u}_b$, where $\mathbf{u}_b$ is the bubble travel velocity, while $\mathbf{u}_{enc}$ and p_{enc} are respectively the liquid velocity and the

*Corresponding Author, Jingsen Ma: jingsen@dynaflow-inc.com

ambient pressure "seen" by the bubble during its travel, and are computed using surface averaged values of the liquid quantities at the bubble interface. c_m is the local sound velocity in the two-phase medium. The bubble trajectory is obtained from the following bubble motion equation [2]:

$$\frac{d\mathbf{u}_b}{dt} = \left(\frac{\rho_l}{\rho_b}\right)\left[\frac{3}{8R}C_D\,|\,\mathbf{u}_s\,|\,\mathbf{u}_s + \frac{1}{2}\left(\frac{d\mathbf{u}_{enc}}{dt} - \frac{d\mathbf{u}_b}{dt}\right) + \frac{3\dot{R}}{2R}\mathbf{u}_s - \frac{\nabla p}{\rho_l} + \frac{(\rho_b - \rho_l)}{\rho_l}g + \frac{3C_L}{4\pi}\frac{\sqrt{\nu}}{R}\frac{\mathbf{u}_s \times \mathbf{\Omega}}{\sqrt{|\mathbf{\Omega}|}}\right], \quad (4)$$

where ρ_b is the bubble contents density, C_D, C_L are the drag and lift coefficients, and Ω is the deformation tensor.

Level-Set Method for Macro Cavities

Once the microbubbles treated as singularities increase size to be able to be resolved in the Eulerian frame, it becomes important to describe their fine details, interface dynamics and deformation. A criterion based on bubble size is set to "activate" the bubbles and define their interface using the level-set approach. Afterwards, based on the initialization distance function field from "activated" bubbles, a Level-Set method is used to simulate large free surface deformations [3]. A smooth function, $\varphi(x,y,z,t)$, , whose zero level coincides with all liquid/gas interfaces when the level set is introduced, is defined in the whole physical domain (i.e. in both liquid and gas phases) as the signed distance $d(x,y,z)$ from the interface $\varphi(x,y,z,0) = d(x,y,z)$. This function is enforced at each time step to be a material surface evolving with the velocity field $\mathbf{u}$:

$$\frac{d\varphi}{dt} = \frac{\partial\varphi}{\partial t} + \mathbf{u}\nabla\varphi = 0. \quad (5)$$

In order to continuously track all individual cavities, a cavity ID detection scheme is implemented. This is used to detect, track, and distinguish the properties of the contents of each individual cavity. Each cavity and all of its enclosed fluid cells are assigned a unique ID number, and an ideal gas law is applied independently to each separate cavity.

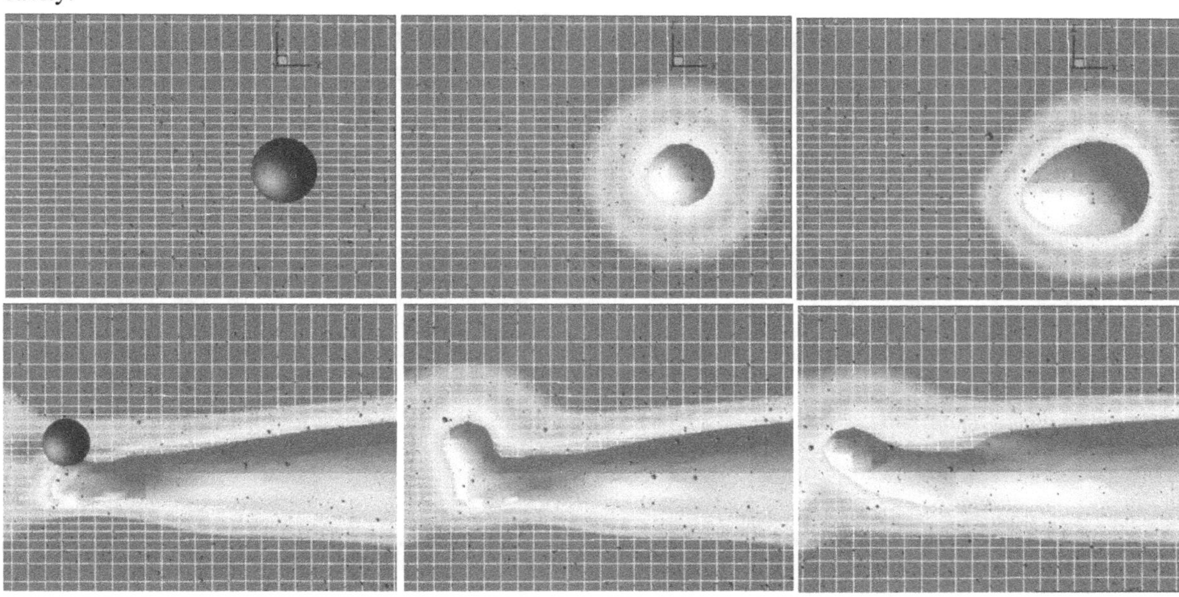

Figure 1: Sequences of multiscale method modeling [2] of the cavitation and bubble capture in a line vortex. Rankine vortex of viscous core radius, a_c=5mm, and circulation, Γ=1.255m^2/s. (Top) Bubble deformation; (Bottom) Bubble merging into the developing vapor core. The dark blue spheres are DSM bubbles while the grey iso-surface is the liquid-vapor interface represented by level set method. The grey mesh is the computation grid used by the Eulerian solver and the background color denotes the value of level set at each cell.

Simulation of Bubble Cavitation in a Line Vortex

A Rankine vortex is defined by the angular velocity, u_θ, and pressure, p_ω, of the vortical flow given by

$$u_\theta(r) = \left\{\frac{\Gamma r}{2\pi a_c^2},\ r \le a_c;\quad \frac{\Gamma}{2\pi r},\ r > a_c\right\}, \quad (6)$$

$$p_{\omega}(r)=\begin{Bmatrix} p_{\infty}-\dfrac{\rho\Gamma^2}{4\pi^2 a_c^{\ 2}}+\dfrac{\rho\Gamma^2 r^2}{8\pi^2 a_c^{\ 4}}, & r\le a_c \\ p_{\infty}-\dfrac{\rho\Gamma^2}{8\pi^2 r^2}, & r>a_c \end{Bmatrix}, \quad (7)$$

where, r is the radial coordinate, Γ is the circulation strength and a_c is the viscous core radius. Figure 1 illustrates how the present multiscale method tracks a DSM bubble, which grows large enough to be switched to be represented by the level set free surface. From there on, fine details of the bubbles can be captured. These include its deformation, elongation along the axis of vortex, as well as its merging into the cavity… etc.

Single Bubble Traveling and Entrainment

The following simulations show the resolution of bubble shape details obtained by the level set modeling. An initially spherical bubble of a radius of 1mm is released in a Rankine vortex flow with core radius a_c=5mm and a circulation Γ=0.097m^2/s. The bubble grows and deforms while being drawn into the core center. Figure 2 displays the history of bubble trajectory and shape during this process. It is seen that the present model successfully captures the process of bubble deformation, entrainment and elongation in the vortex, as well as its interaction with the surrounding flow. As seen in the right panel of Figure 2, similar phenomena are also observed in a high speed video taken in the DYNAFLOW lab [5].

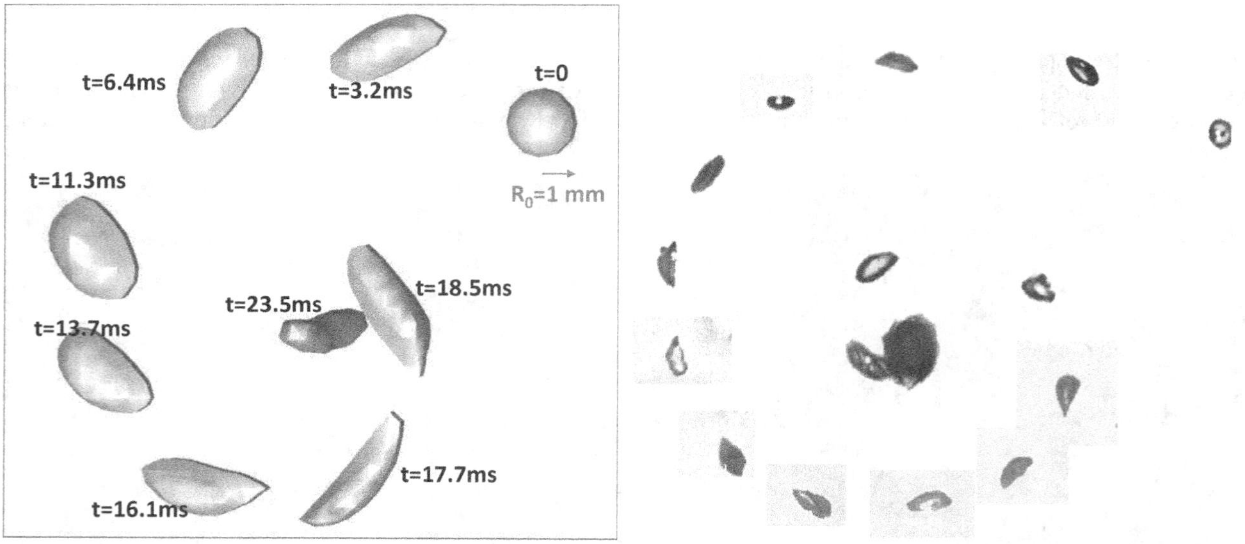

Figure 2: (Left) 3DYNAFS© simulation of the history of a 1mm bubble trajectory and shape while being captured in a Rankine vortex of a core radius a_c=5mm and a circulation Γ=0.097m^2/s; (Right) Images extracted from high speed video taken in the DYNAFLOW lab [5].

Multiple Bubble Tracking

In this section, the model is used to simulate two bubbles released simultaneously from two field locations in the Rankine line vortex: one on the vortex axis and the other outside of the core. Though the bubbles have the same initial size and gas pressure, due to the different pressures they encounter along their trajectories, they show very different bubble volume oscillations and gas pressure history. The bubble, initially located on the axis, grows to a much larger size immediately after the release. On the contrary, the outside bubble, which experiences much higher pressures, displays strong oscillations of both its volume and gas pressure, even though overall it grows as it moves towards the vortex axis.

It is also interesting to see the dissymmetry in the pressure field in the contour plots in Figure 3, Left due to the interaction between the two bubbles.

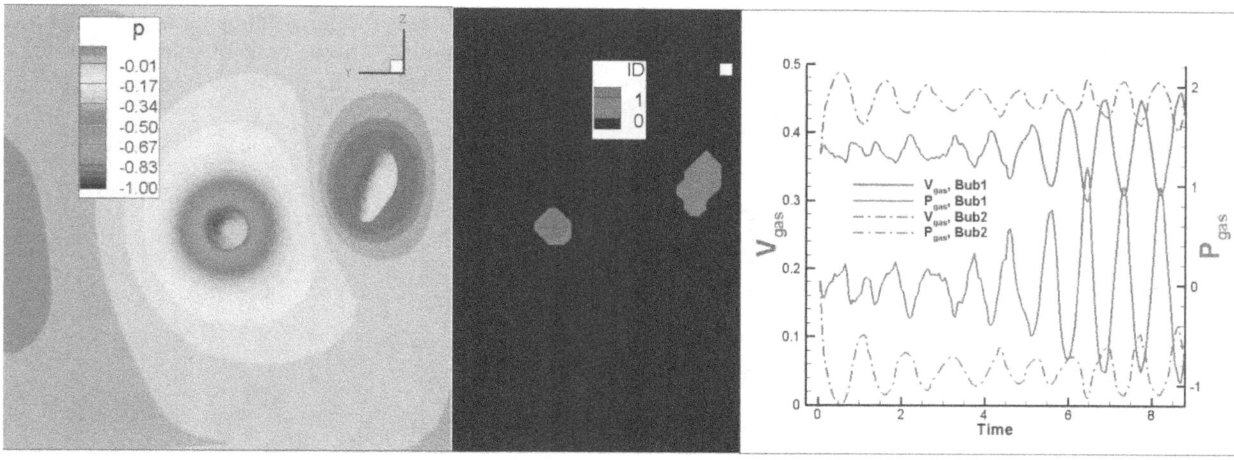

Figure 3: 3DYNAFS© simulation of two bubbles released at different initial locations in a Rankine line vortex. (Left) Bubble Iso-surface (in silver) and pressure contour at the central plane crossing two bubbles. (Middle) ID field at the central plane crossing two bubbles. (Right) Volume and gas pressure history of the two bubbles while they are captured in the vortex core.

Bubble Breakup

In this section we look at bubble breakup modeling. Figure 4 demonstrates bubble elongation along the axis of vortex. As this continues, the bubble breaks up into two separate bubbles with the new daughter bubble given a new ID index. This illustrates the process of bubble tracking and bubble ID assignment.

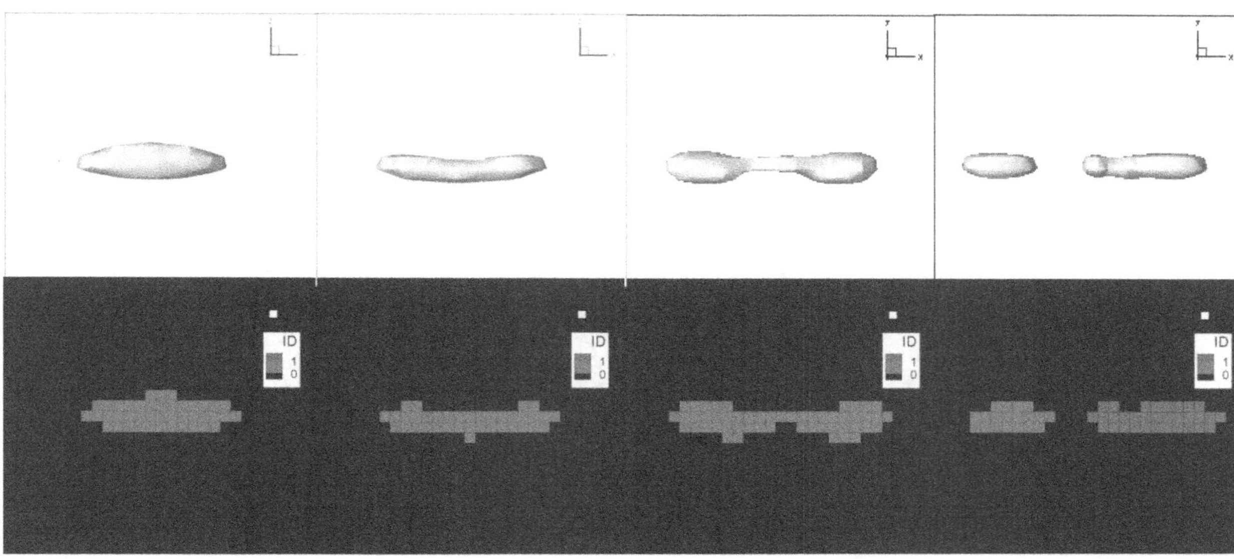

Figure 4: Simulation of a bubble breakup event in a Rankine vortex using 3DYNAFS©. (Top) Bubble interface showna as an iso-surface. (Bottom) Color contour of bubble ID field at the central plane crossing bubble(s), with computational grids in background

This capability to track generated daughter bubbles is of particular interest to the study of bubble dynamics in generalized and more complex configurations such as on lifting surfaces, where the daughter bubble number and size distributions could play a critical role. For example, in [6] a 3D simulation of a traveling cavitation bubble close to a NACA0015 foil surface, its 3D deformation, and its breakup into multiple smaller bubbles is presented.

This is illustrated in Figure 5, which shows a simulation starting with a small bubble released near the leading edge. The small bubble grows large as it travels through the low pressure on the suction side of the lifting surface. It then elongates stream wise while flattening in both the direction perpendicular to the foil surface and that along the span direction. This is particularly highlighted in the lowest image sequence in Figure 5, which displays a zoomed front view of the bubble interacting with the liquid flow around it. This indicates that the bubble elongates in the stream wise direction due to the foil viscous shear flow. At the same time a reverse flow is seen under the bubble, which creates local vorticity and in turn shears off the downstream bubble part, introducing a strong secondary flow underneath. As it travels further downstream and encounters higher pressures, the large elongation and flattening lead the bubble during its collapse to break up into multiple daughter bubbles. Those small bubbles then become new seed nuclei in the viscous wake, which feed the long lived bubbly wake behind the lifting surface [6].

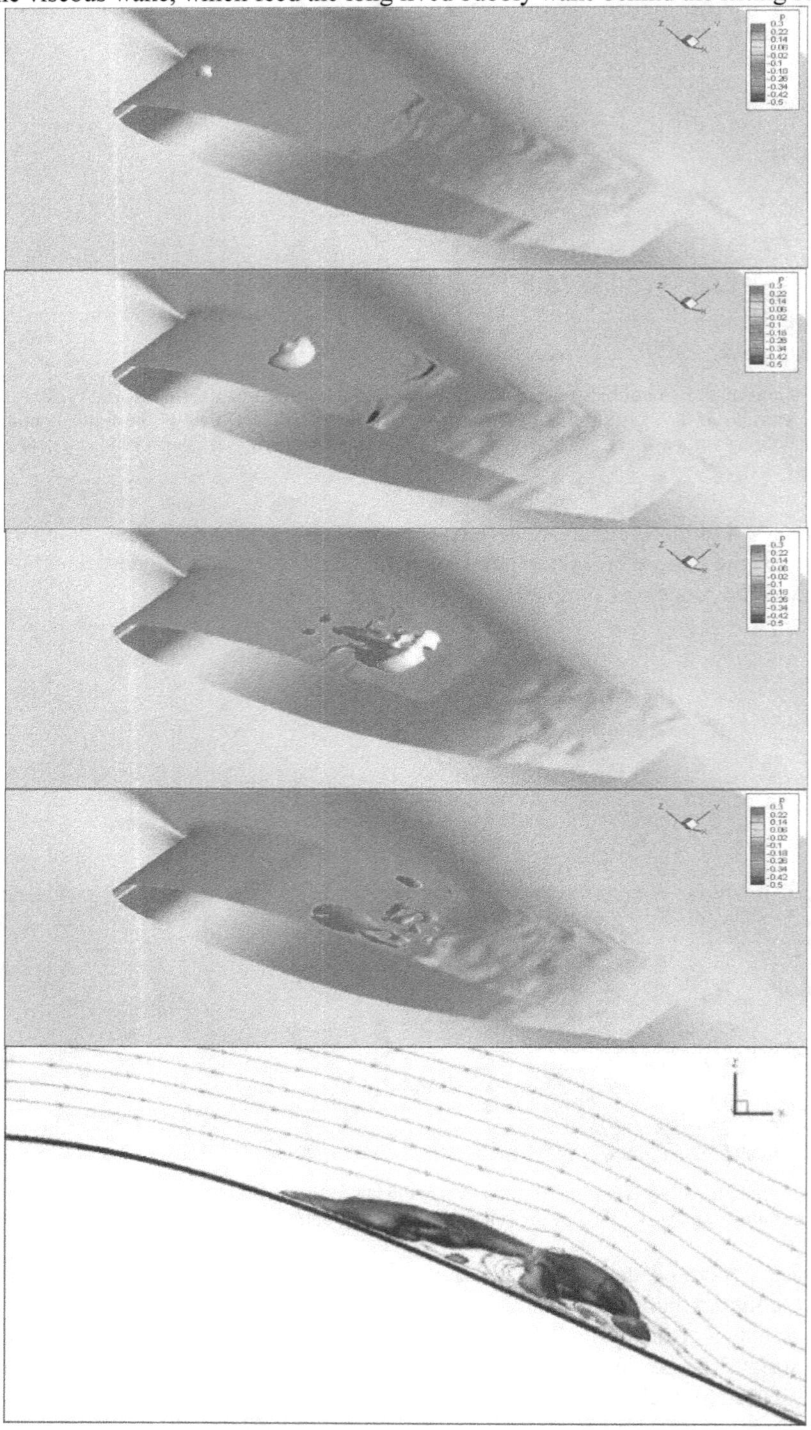

Figure 5: 3D simulation of a non-spherical traveling cavitation bubble on a NACA0015 foil using 3DYNAFS© [6], where the silver iso-surface represents the bubble gas/liquid interface and the background flow is contoured by flow pressure. The bottom row is a front view zoom on the bubble interacting with the liquid flow around it, where the blue iso-surface represents the 3D bubble surface and the purple lines are streamlines in the mid plane.

In on-going studies, the simulations are being refined to individually identify and quantify the local flow parameters (e.g., shear rate, vorticity etc.) around each daughter bubbles. The ultimate goal is to uncover breakup criteria to correlate the breakup parameters with local flow conditions such as vorticity and turbulence, etc.

Summary

A multi-scale framework is being developed for the simulation of cavitation problem. An example of bubble cavitation in a line vortex is presented to show how the model successfully captures bubble growth, entrainment into the core center to form a gaseous/vaporous core, as well as bubble elongation and breakup in the vortex core. The same model is also applied to simulations of sheet cavity formation and breakup on a foil.

Acknowledgements

This work was funded by NASA under award No. NNX16AB85A monitored by Mr. John McQuillen and Dr. Enrique Rame, and ONR under Contract No. N00014-15-C-5029 monitored by Dr. Ki-Han Kim.

References

[1] Hsiao, C-T, Chahine, G. L. and Liu, H-L (2003) *Scaling effect on prediction of cavitation inception in a line vortex flow*, ASME J. Fluids Eng. 125 53–60

[2] Ma, J., Hsiao, C-T, and Chahine, G. L., (2017) *A Physics Based Multiscale Modeling of Cavitating Flows*, Computers and Fluids, 145, pp: 68-84.

[3] Hsiao C-T, Wu X, Ma J and Chahine G L, (2013) *Numerical and Experimental Study of Bubble Entrainment Due to a Horizontal Plunging Jet*, Int. Shipbuild. Prog. 60 435–69

[4] Fang, J., Rasquin, M., Bolotnov, I.A., (2017) *Interface tracking simulations of bubbly flows in PWR relevant geometries*, Nuclear Engineering and Design, vol. 312, pp. 205-13

[5] Wu, X., Loraine, G., Mansouri, A. Chahine, G., (2017) *Ground Studies of a DYNASWIRL® Phase Separator for Space Applications*, ASGSR 2017, Seattle, Washington, October 25-28.

[6] Hsiao, C-T., Ma, J., Chahine, G.L., (2017) *Numerical Study of the Effects of Gas Diffusion Bubble Distribution about a Lifting Surface*, 2017 Symposium of Marine Propulsion, Espoo , Finland , June 12-15.

Noise from Ventilated Supercavities and its Utility for Inferring Cavity Dynamics

[1]Samuel Hansford, [1]Timothy Brungart*, [1]Jules Lindau, [1]Michael Moeny

[1]*Penn State Applied Research Lab, PO Box 30, State College, PA, USA,*

Abstract

Ventilated supercavities are shown to be monopole noise sources over a wide range of frequencies. As such, the radiated noise may have utility for providing insight into cavity characteristics, such as the entrainment rate of gas, cavity closure regime, and cavity stability. Lower frequency waves on the cavity interface define its shape and closure regime. When high amplitude, monotonic waves are present on the interface, corresponding high amplitude tones are present in the radiated noise. High frequency interface waves, associated with interface roughness, give rise to higher frequency broadband noise. Increased roughness increases the shear entrainment rate and ventilation requirements.

Keywords: Supercavitation, Supercavitation Acoustics

Introduction

It is widely accepted that the interface motion of a ventilated supercavity is related to its radiated noise. As such, this relationship presents an attractive means of inferring ventilated supercavity dynamics through its radiated noise. Previous work has studied ventilated supercavity noise either analytically or numerically and has shown that low frequency noise is expected to radiate as an acoustic monopole [1,2]. While these studies considered mainly the impingement of a gas jet on the cavity interface, Skidmore *et al.* [3] demonstrated that the noise from pulsating cavities is related to the dominant wave on the cavity interface and radiates as a monopole source at the wave frequency.

In addition to gross cavity motion, such as that occurring during pulsation, it is possible for several waves to be superimposed on the cavity interface with varying amplitudes that are controlled by the Kelvin-Helmholtz (KH) and Rayleigh-Taylor (RT) instabilities. The density and vortex sheets are colocated at the cavity interface, hence the existence of waves on the cavity interface. Perturbations that are amplified by these two instabilities are either generated by fluidic or mechanical sources [4]. The disturbances that are amplified by the KH and RT instabilities cause pressure fluctuations in the cavity gas. The cavity is unable to sustain pressure oscillations and the cavity deforms, resulting in volume oscillations. These volume oscillations not only act as monopole noise sources, thus cavities are efficient acoustic sources, but can also lead to cavity stability issues and alter the shear entrainment rate, depending on the disturbance scale.

Methods

Experiments were conducted in the 0.305 m diameter water tunnel at the Penn State University Applied Research Laboratory. The test setup is shown in Figure 1a. Gas from 6 gas bottles entered through the strut and sting, and then into the cavity through the vent ports. The cavitator disk diameter was 34.29 mm, the freestream speed was set to 1.7 m/s (Fr=3.0), and the tunnel pressure was 103 kPa. To facilitate pulsation generation, the test section was filled just until there was no free surface visible. A Measurement Specialties XPM5 pressure sensor measured the interior cavity pressure while 2 Benthowave BII-7071 hydrophones measured the radiated sound pressures. Only the downstream hydrophone measurements are presented. The resonance frequencies of the sensors were well above the frequency range of interest. The sampling rate and sample time was 25 kHz and 10 seconds, respectively. An in-situ calibration was performed to remove the effects of the tunnel walls and transfer the signal to the equivalent free-field condition. Ventilation rates were measured by a Sierra Instruments FlatTrak 780S thermal gas flow meter and manufacturer supplied conversion factors were used to compensate for different ventilation gases.

*Corresponding Author, Timothy Brungart: tab7@arl.psu.edu

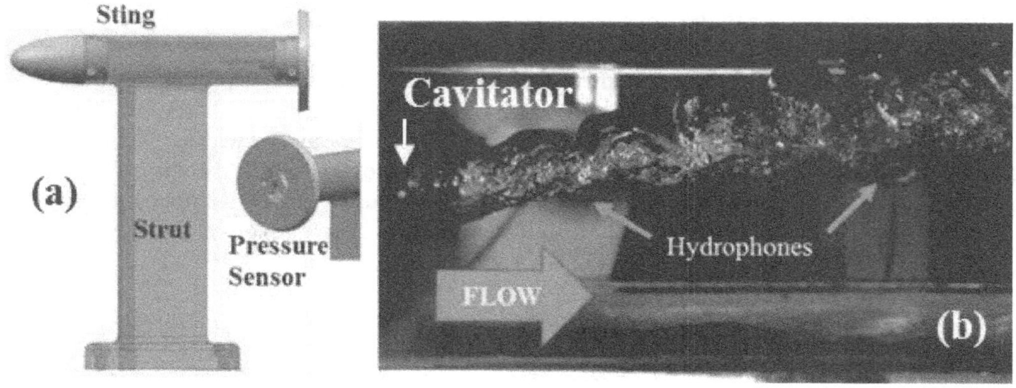

Figure 1: (a) Experimental setup in the 0.305 m diameter water tunnel tests. (b) High speed image of a 2nd order pulsating air cavity.

Results

Experimentally measured 2nd order pulsating air cavities have a frequency of 33 Hz and the surface area of the cavity can be estimated using the Song model [5]. A high speed image of a pulsating cavity is shown in Figure 1b. The interior pressure spectrum for the pulsating cavity is represented by the black line in Figure 2a. In addition to a pulsating cavity, there is a tone in an air twin vortex cavity that occurs at 60 Hz (represented by the red line in Figure 2a); a high speed image of this cavity is shown in Figure 2b. There is a tone around 30 Hz in the red pressure spectrum but the feedback tone is dominant. The twin vortex tone appears to be a due to a complex feedback loop that depends on the sound speed of the gas, cavity length, and delay time between an acoustic perturbation and the resulting disturbance rollup. For the cavities studied, there were no tones measured for re-entrant jet cavities and the broadband pressure spectral levels were lower than the broadband pressure spectral levels for twin vortex or pulsating cavities.

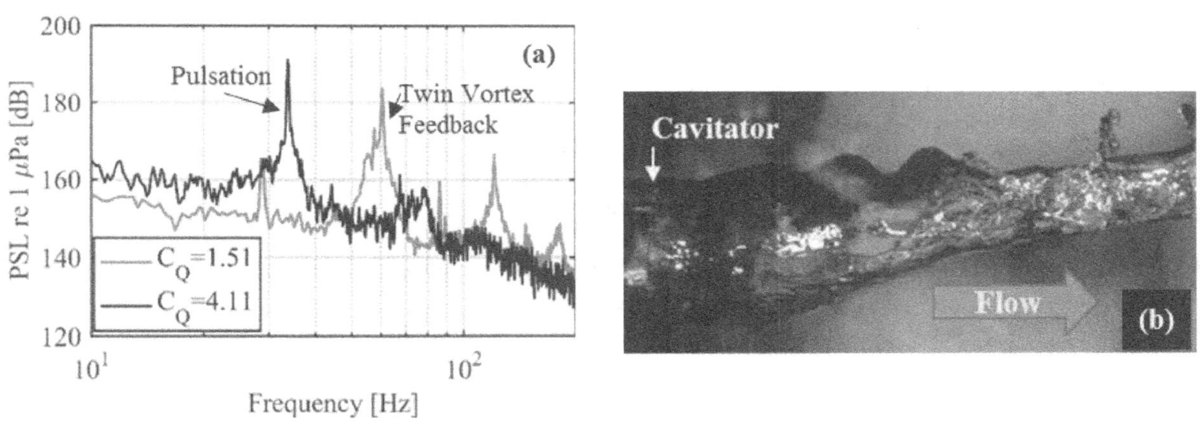

Figure 2: (a) Cavity interior pressure spectra for a pulsating (black) and twin vortex (red) cavity. (b) An air twin vortex cavity.

While the pulsation and feedback tones are tonal monopole noise sources, ventilated supercavities also act as broadband monopole noise sources. Given in Figure 3a, the cavity interior and the +25 dB shifted hydrophone pressure spectra show a range of frequencies where the pressure spectral levels are the same. The +25 dB shift compensates for the reduction in pressure spectrum level due to spherical spreading from the cavity interface to the hydrophone. To verify cavities are broadband monopole noise sources, the coherence between the cavity sensor and hydrophone for the Figure 3a cavity was found. The coherence, given in Figure 3b, is high between the two sensors and ensures a high degree of certainty that the measured pressure fluctuations are generated by the cavity. Using the coherence as a condition for a good signal, along with the signal-to-noise ratio, the phase of the cross spectrum between the sensors was found, given in Figure 3b. The frequencies are strongly correlated and the phase difference is around 0 degrees for a range of frequencies, indicating there is acoustic wave propagation between the sensors.

The effects of interface waves on cavity dynamics depends on their wavelength. Low frequency noise is related to deformations of the cavity interface while high frequency noise is related to roughness. The demarcation between these scales is currently unknown. Deformations mean the cavity interface is unstable. Pulsating cavities, such as those shown in Figure 1b, have wave motion on both the upper and lower interface, a hallmark of pulsating cavities. Conversely, a twin vortex cavity has a coherent wave only on the upper cavity interface. Both tones in the cavity interior pressure spectra for the pulsating and twin vortex cavities are over 180 dB, meaning both tones have strong volume velocities, $O(10^{-3}\text{-}10^{-4})$ m^3/s; therefore, these tones can cause cavity stability issues. It appears that deformations will cause a contraction in the cavity length at a given mass ventilation rate.

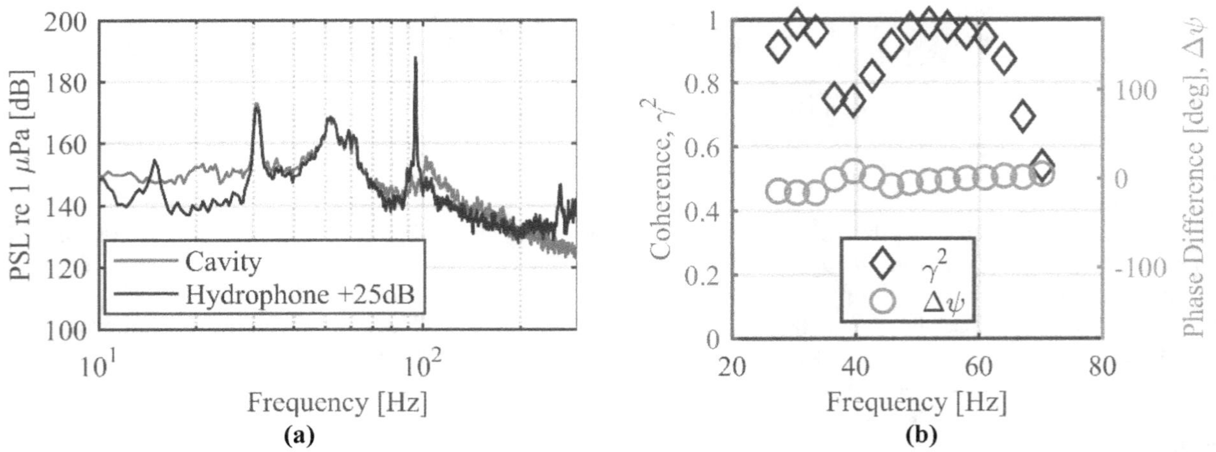

Figure 3: (a) Cavity interior and shifted radiated pressure spectra with C_Q=1.72. (b) Coherence and phase difference between cavity interior pressure sensor and downstream hydrophone for C_Q=1.72.

Gas jets provide broadband forcing to the cavity and, as a result, are responsible for radiated noise over a range of frequencies. Thus, higher frequency perturbations are present. The higher frequency or shorter wavelength cavity disturbances are expected to increase the ventilation rate by increasing the shear entrainment rate caused by the rougher cavity interface. Assuming an average wavelength of $O(10^{-3})$ m for surface roughness, the frequency of the waves that are radiated by those surface roughness disturbances are $O(10^3)$ Hz for the given freestream speed. Unfortunately, external noise sources, such as vibration induced noise and flow noise, dominated the cavity interior pressures at frequencies near 1 kHz but high speed images of ventilated cavities can be used to ascertain roughness.

Impingement of the ventilation gas jet and the acoustic disturbances generated by the turbulence of the gas jet which perturb the interface at the separation point are two sources of disturbances on the cavity interface. To reduce the ventilation speed, a denser gas was used to generate an equivalent cavity at an equivalent mass flow rate. Lower velocity gases are expected to result in lower broadband pressure spectral levels for this cavity. In other cavities tested, however, it was observed that helium ventilated cavities possessed smoother cavity interfaces than air ventilated cavities; therefore, there is a complex dependence on the ventilation gas speed and density of the vent gas on cavity interface disturbances. Given in Figure 4 are the cavity interior pressure spectra for air and SF_6 ventilated cavities at the same mass flow rate. The ventilation speed decreases by a factor of 5 with SF_6. For this operating condition and cavitator, the reduction in vent speed leads to a lower cavity interior pressure spectrum level, which means a lower volume velocity. Disturbances on the cavity interface drive pressure oscillations in the gas that cause volume oscillations of the cavity. At these lower frequencies (<100 Hz), cavities behave as broadband monopole noise sources so the cavity is quieter because of the reduced volume velocity and the cavity experiences less intense radial size fluctuations. The lower amplitude deformations should lead to less contraction of the cavity length. In addition to the lower kinematic viscosity, the roughness of the cavity interface is less due to the lower vent rate so the shear entrainment should decrease on a volumetric basis.

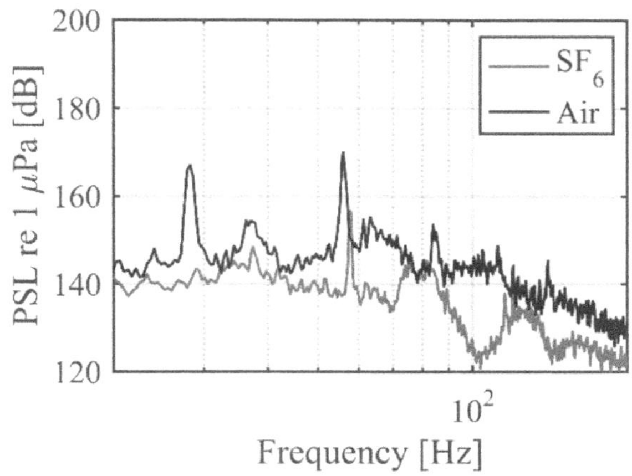

Figure 4: The cavity interior pressure spectra for air and SF_6 ventilation gases at similar mass flow rates.

Conclusions

This work has shown that ventilated supercavities are broadband as well as tonal monopole noise sources. High amplitude, monotonic waves on the cavity interface give rise to corresponding tones in the cavity interior pressure and radiated noise spectra, whereas lower amplitude waves give rise to broadband cavity interior pressures and radiated noise. The high amplitude tones are present in pulsating and twin vortex cavities. Cavity dynamics are affected by both the amplitude and the length of the interface waves. Long length interface waves appear to lead to a significant contraction of the mean cavity length to maintain the cavity surface area. Small length interface waves give rise to cavity roughness that results in high frequency broadband cavity interior pressures and radiated noise. Cavity roughness affects the shear entrainment rate and ventilation rate requirements. Rougher cavity interfaces entrain gas at a faster rate than smooth cavity interfaces for a given Reynolds number for the gaseous boundary layer and require increased ventilation rates. For the experimental setup reported herein, changing the ventilation gas while maintaining the same mass flow rate, has shown the broadband noise levels are reduced and the cavity motion and roughness are reduced using a denser gas.

Acknowledgments

This work was sponsored by the Office of Naval Research under Grant N00014-11-10322 and monitored by Dr. David Drumheller.

References

[1] Foley A W, Howe M S and Brungart T A 2010 Spectrum of the sound produced by a jet impinging on the gas-water interface of a supercavity *J. Sound Vib.* **329** 415–24

[2] Ramesh S S, Lim K M, Zheng J and Khoo B C 2014 Numerical analysis of flow induced noise propagation in supercavitating vehicles at subsonic speeds. *J. Acoust. Soc. Am.* **135** 1752–63

[3] Skidmore G M, Brungart T A, Lindau J W and Moeny M J 2015 Noise generated by ventilated supercavities *Noise Control Eng. J.* **63** 94–101

[4] May A 1975 Water Entry and the Cavity-Running Behavior of Missiles *DTIC ADA020429*

[5] Song C S 1962 Pulsation of Ventilated Cavities *J. Sh. Res.* **5**

[6] Skidmore G 2016 *The mitigation of pulsation in ventilated supercavities* (Pennsylvania State University)

© 2018 ASME

Non-spherical Dynamics of Gas Bubbles in Soft Matter

[1]Kazuya Murakami*; [2]Renaud Gaudron; [1]Eric Johnsen

[1]*University of Michigan, Ann Arbor, MI, USA;*
[2]*Université Paris Saclay, Grande voie des vignes, Châtenay-Malabry Cedex 92295, France*

Abstract

The non-spherical dynamics of gas bubbles in soft matter are analytically investigated. We develop a model using Cauchy's equation of motion in conjunction with a Kelvin-Voigt viscoelastic model, where it is assumed that the soft matter behaves in an irrotational fashion. The model is solved numerically and validated against experiments of non-spherical bubble oscillations in gelatin. The natural frequency of each non-spherical mode is obtained from the model, which is used to predict the unstable mode due to parametric resonance. Furthermore, we investigate the non-spherical bubble dynamics under ultrasound irradiation and examine viscoelastic effects on the time evolution of non-spherical mode amplitudes. Viscosity stabilizes non-spherical oscillations, while elasticity increases the natural frequency, which also changes the unstable mode.

Keywords: bubble dynamics, viscoelastic medium, parametric resonance

Introduction

Bubble dynamics play an important role in a variety of therapeutic ultrasound applications [1]. The tension produced by the passage of ultrasound causes cavitation nucleation and possibly large, inertially dominated oscillations of the bubbles. When bubbles collapse inertially, the high pressure and temperature generated inside the bubbles can have therapeutic effects such as ablating a tumor or breaking up a blood clot. In practice, the bubble often deviates from a spherical shape, which results in less violent collapse [2]. Moreover, soft tissue has high viscoelasticity which significantly influences the bubble dynamics. Therefore, it is important to understand the non-spherical dynamics of gas bubbles in soft matter to predict tissue damage.

The study of non-spherical bubble dynamics was pioneered by Plesset [3] who applied spherical harmonics to express the non-spherical bubble surface, and reduced the Bernoulli equation to two ordinary differential equations (ODEs): the Rayleigh-Plesset equation for the mean bubble radius and an equation for the leading-order non-spherical mode amplitude. Prosperetti [4] extended this inviscid model to incompressible and rotational flow. Based on this model, the non-spherical oscillations of a bubble [5] and an encapsulated microbubble [6] have been analytically investigated. Some experimental studies showed the direct observations of non-spherical oscillations of a bubble in water [7] and in gelatin [8]. However, these modeling studies focused on a bubble in water: the analysis has yet to be extended to non-spherical bubbles in soft matter.

The goal of this study is to extend Plesset's procedure [3] to bubbles in soft matter and investigate the non-spherical bubble dynamics. We derive our model from Cauchy's equation of motion in conjunction with a Kelvin-Voigt viscoelastic model by assuming that the soft matter is incompressible and irrotational. The model is solved numerically and validated against the experiments of Hamaguchi & Ando [8], who studied non-spherical bubble oscillations in gelatin. Furthermore, we investigate non-spherical bubble dynamics under pulse-wave ultrasound irradiation and examine viscoelastic effects on the time evolution of non-spherical mode amplitudes.

Physical model

We consider an isolated free gas bubble in soft matter. Assuming axisymmetry, the non-spherical bubble surface can be expressed by the superposition of spherical harmonics [4],

$$F(r,\theta,t) = r \quad R(t) \quad \underset{n}{} a_n(t)Y_n(\theta) = 0, \tag{1}$$

where R is the mean bubble radius and a_n is the amplitude of the n-th mode of a non-spherical perturbation, with $\nabla a_n \nabla O$ R and where Y_n is the n-th spherical harmonic.

The dynamics of soft matter are described by Cauchy's equation of motion,

$$\frac{\partial \mathbf{u}}{\partial t} + (\mathbf{u} \ll)\mathbf{u} = \frac{1}{\rho}\mathrm{div}\mathbf{T}, \tag{2}$$

*Corresponding Author, Kazuya Murakami: mkazuya@umich.edu

where $\mathbf{u}$ is the velocity vector, ρ is the density and $\mathbf{T}$ is the Cauchy stress tensor. Since soft tissue generally deforms relative to an original configuration, we use the Kelvin-Voigt viscoelastic model, $\mathbf{T} = \mathbf{T}_v + \mathbf{T}_e$, where the viscous stress is Newtonian and the elastic stress is neo-Hookean [9]. This model has been validated for spherical bubble dynamics in polyacrylamide gels [10].

For a bubble in soft matter, the boundary layer is confined near the bubble because the diffusive time scales are far slower than the relevant oscillation time scales. Thus, we assume the surroundings to behave in an irrotational fashion. In addition, compressible effects on the non-spherical mode are negligible under small perturbations. Therefore, we also assume the dynamics of the surroundings to be incompressible.

In general, there are four boundary conditions at the bubble surface [4]. Two are in the direction normal to the bubble surface (no-penetration and normal stress continuity):

$$\frac{\partial F}{\partial t} + (\mathbf{u}\cdot\nabla)F = 0, \tag{3}$$

$$\mathbf{n}\cdot\mathbf{T}_2\mathbf{n} - \mathbf{n}\cdot\mathbf{T}_1\mathbf{n} = S\,\nabla\cdot\mathbf{n}, \tag{4}$$

where the subscripts 1 and 2 represent the gas and soft matter, respectively, $\mathbf{n} = \nabla F/|\nabla F|$ is the normal unit vector and S is the surface tension. The other two are in the tangential direction (no-slip and shear stress continuity), which are not applied under the irrotational assumption of soft matter.

Plesset [3] derived the velocity potential φ by satisfying no-penetration (3),

$$\varphi = -\frac{R^2\dot R}{r} - \frac{R^{n+2}}{(n+1)r^{n+1}}\left(\dot a_n + \frac{2\dot R}{R}a_n\right)Y_n. \tag{5}$$

Then, the velocity is obtained from $\mathbf{u} = \nabla\varphi$, and the pseudo-pressure [9] is determined by integrating Cauchy's equation of motion (2) from the bubble surface to infinity,

$$\begin{aligned}\wp &= p_\infty + \rho\left[R\ddot R + \frac{3}{2}\dot R^2 + \frac{3\dot R\dot a_n + R\ddot a_n - (n-1)\ddot R a_n}{n+1}Y_n\right] \\ &\quad + \frac{\eta}{2}\left[5 - 4\left(\frac{R_0}{R}\right) - \left(\frac{R_0}{R}\right)^4\right] - \frac{\eta}{R}\left[2\frac{R_0}{R} - 2\left(\frac{R_0}{R}\right)^4 + \frac{n(n+1)}{1+\frac{R_0}{R}+\left(\frac{R_0}{R}\right)^2}\right]a_nY_n,\end{aligned} \tag{6}$$

where p_∞ is the far-field pressure, η is the elasticity and R_0 is the equilibrium bubble radius.

Normal stress continuity (4) reduces to

$$2\mu\frac{\partial u_r}{\partial r} + \eta\left(\frac{r_0}{r}\right)^4 - \wp + p_b = \frac{2S}{R} + \frac{(n-1)(n+2)}{R^2}Sa_nY_n, \tag{7}$$

where μ is the viscosity, u_r is the r-component of the velocity vector and r_0 is the original configuration of soft matter. It is important to note that, even if the soft matter is incompressible and irrotational, there is viscous contribution from this dynamic boundary condition. The bubble pressure p_b is described by an effective polytropic index [11]. Substituting the velocity u_r and pseudo-pressure $\wp$ into equation (7) at the bubble surface gives, within order $\mathcal{O}(1)$, the nonlinear ODE for mean bubble radius

$$R\ddot R + \frac{3}{2}\dot R^2 = \frac{p_b - p_\infty}{\rho} - \frac{2S}{\rho R} - \frac{4\mu\dot R}{\rho R} - \frac{\eta}{2\rho}\left[5 - 4\frac{R_0}{R} - \left(\frac{R_0}{R}\right)^4\right], \tag{8}$$

which is the Rayleigh-Plesset equation accounting for (nonlinear) elasticity of the surroundings [9]. To order $\mathcal{O}(a_n)$, the linear ODE for mode n is obtained as

$$\ddot a_n + A_n\dot a_n + B_na_n = 0, \tag{9}$$

where

$$A_n = 3\frac{\dot R}{R} + 2(n+1)(n+2)\frac{\mu}{\rho R^2}, \tag{10}$$

$$\begin{aligned}B_n &= -(n-1)\frac{\ddot R}{R} + 4(n-1)(n+1)\frac{\mu\dot R}{\rho R^3} + (n-1)(n+1)(n+2)\frac{S}{\rho R^3} \\ &\quad + (n+1)\frac{\eta}{\rho R^2}\left[2\frac{R_0}{R} + 2\left(\frac{R_0}{R}\right)^4 + \frac{n(n+1)}{1+\frac{R_0}{R}+\left(\frac{R_0}{R}\right)^2}\right].\end{aligned} \tag{11}$$

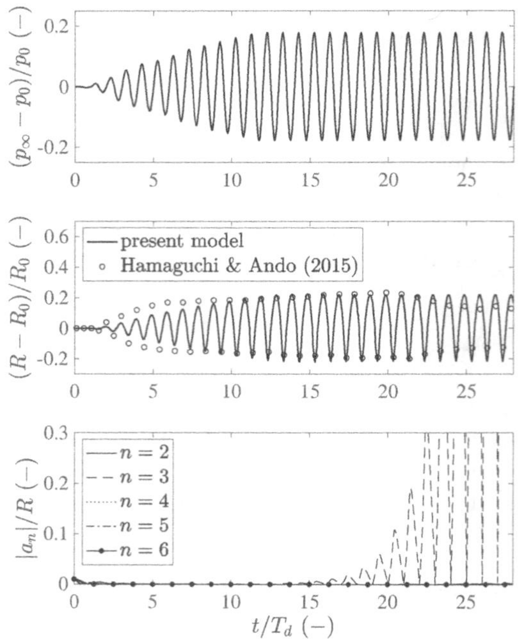

Figure 1: The solution of the present model compared to the experiments of Hamaguchi & Ando [8]: far-field pressure (top), mean bubble radius (middle) and mode amplitudes (bottom). The third mode becomes large after several oscillations due to parametric resonance.

Since order $\mathcal{P}(a_n^2)$ and higher are neglected, the equations for the different modes are not coupled. Note that, when $\mu = 0$ and $\eta = 0$, our model reduces to Plesset's inviscid (non-elastic) model [3]. In summary, the non-spherical bubble surface in soft matter is determined by solving two ODEs (8) and (9) simultaneously.

Results and discussion

For validation purposes, we compare our solutions to the experiments of Hamaguchi & Ando, who examined non-spherical bubble oscillations in gelatin [8]. The time evolutions of the mean bubble radius and non-spherical mode amplitudes are obtained by numerically solving the ODEs (8) and (9) with the Cash-Karp Runge-Kutta method. Following their experiments, we consider a bubble with $R_0 = 100$ μm in a viscoelastic medium of density $\rho = 998$ kg/m^3, surface tension $S = 0.04$ N/m, viscosity $\mu = 14.1$ mPa ≪ and elasticity $\eta = 1.7$ kPa, under atmospheric pressure. We prescribe an ultrasound pressure wave with driving frequency $f = 28$ kHz and pressure amplitude $p_a = 18$ kPa. The initial mode amplitudes are set to $a_n = 1$ μm. We use an effective viscosity for equation (8) to represent thermal and acoustic damping effects [12].

Figure 1 shows the solutions compared to the experiments of Hamaguchi & Ando [8], including the time history of far-field pressure, mean bubble radius and the second to sixth mode amplitudes normalized by the mean bubble radius. Overall, the results agree well with the experiments. We can see that the bubble oscillations remain spherical initially. However, the third mode becomes large after several oscillations, while other modes remain stable. This instability is caused by the parametric resonance between the mode amplitude and mean bubble radius under simple harmonic forcing with an ultrasound wave, which occurs when $\omega = 2\omega_n$ where $\omega = 2\pi f$ is the driving frequency and ω_n is the natural frequency of each non-spherical mode [7]. Since equation (9) is a second-order linear ODE, the natural frequency of the n-th mode is given by

$$\omega_n^2 = (n-1)(n+1)(n+2)\frac{S}{\rho R_0^3} + (n+1)\frac{\eta}{\rho R_0^2}\left[4 + \frac{n(n+1)}{3}\right]. \tag{12}$$

Note that, when $\eta = 0$, equation (12) reduces to the natural frequency for a bubble in water [13]. For the given parameters, equation (12) yields the most unstable mode number $n = 3.1$, which is consistent with our observation that the third mode is unstable.

The experiments of Hamaguchi & Ando [8] showed that the third mode appeared after several oscillations, which our

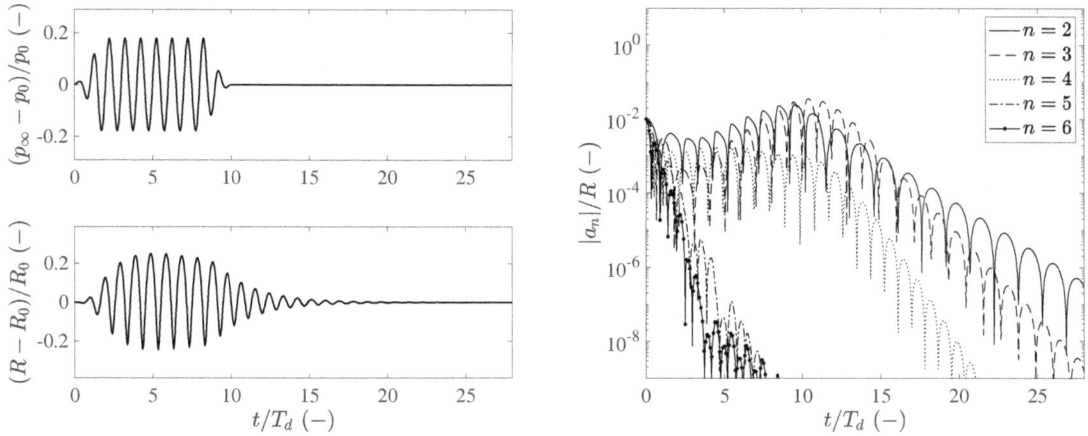

Figure 2: The solution of the present model under pulse-wave ultrasound irradiation: far-field pressure (left top), mean bubble radius (left bottom) and mode amplitudes (right).

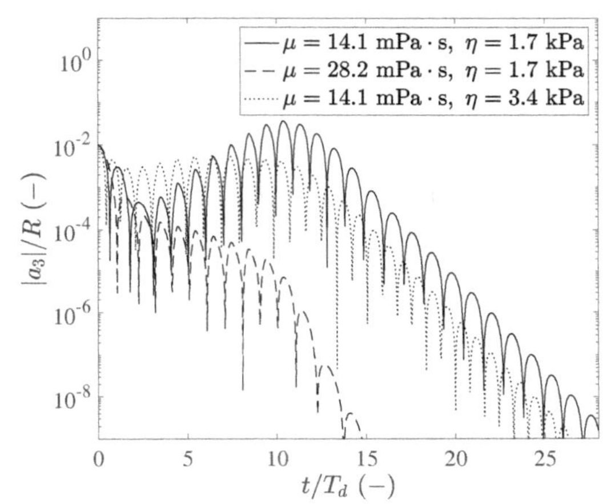

Figure 3: The time history of the third mode in three different cases: the same as in figure 2, the same but with double the viscosity, and the same but with double the elasticity. Viscosity stabilizes the dynamics, while elasticity changes the unstable mode, thus stabilizing the third mode.

model also demonstrates. Note that the mean bubble radius is damped after the third mode appears in the experiments, as shown in Figure 1. This behavior occurs in the experiments because the kinetic energy of the radial mode is transferred to the non-spherical mode and dissipated [14]. However, our linear analysis cannot capture this kind of mode interaction.

Next, we investigate non-spherical bubble dynamics under pulse-wave ultrasound irradiation. We use the same parameters as those listed above. Figure 2 shows the time history of the far-field pressure, mean bubble radius and the second to sixth mode amplitudes. We can see that the second, third and fourth modes grow, while the fifth and sixth modes decay exponentially. This behavior occurs because the third mode is excited the most due to parametric resonance. Since the fifth and sixth modes are far from the most unstable mode number $n = 3.1$ obtained from equation (12), these modes are hardly excited by the pulse-wave. Every mode starts decaying after the forcing ends, and the high mode decays faster than the low mode.

In order to examine the viscoelastic effects on non-spherical mode amplitudes, we compare the third mode presented in figure 2 with solutions where the viscosity is doubled (but the elasticity is kept the same) and another where the elasticity is doubled (but the viscosity is kept the same), as shown in figure 3. Although the mean bubble radius is slightly different in each case, it has only a small impact on the mode amplitudes. Viscosity makes the mode amplitudes more stable and increases the decay rate. Elasticity increases the natural frequency of the n-th mode (12), which yields the most unstable mode number $n = 2.2$ for $\eta = 3.4$ kPa. As a result, it makes the third mode more stable, though it does not change the decay rate.

Conclusion

We developed a model to investigate the non-spherical dynamics of gas bubbles in soft matter. Although the experimental data is limited, our study shows agreement between our model and available experiments. We found the natural frequency of each non-spherical mode and predicted the unstable mode due to parametric resonance. Furthermore, we investigated the viscoelastic effects on the non-spherical mode amplitudes. Viscosity has a stabilizing effect and increases the decay rate. Elasticity increases the natural frequency, which changes the unstable mode, though it does not affect the decay rate.

The irrotational assumption of soft matter is one of the limitations of our model. The rotational field is important for viscous damping [4] and likely for elastic effects as well. In this study, we analyzed the shape instability during linear oscillations, but our model would be useful to study the shape instability during violent collapse as well, which is more interesting for tissue damage in therapeutic ultrasound applications. In the case of investigating the nonlinear effects (large non-spherical deformation, mode interaction), different approaches like direct simulations of the full equations of motion are required.

Acknowledgement

This work was supported in part by ONR grant N00014-17-1-2058 (under Dr. T. Bentley) and NSF grant CBET 1253157. KM acknowledges the Overseas Scholarship from the Funai Foundation for Information Technology.

References

[1] Mitragotri, S. (2005). *Healing sound: the use of ultrasound in drug delivery and other therapeutic applications.* Nature Rev. Drug Discov. **4**. 255-260.

[2] Brennen, C. E. (2015). *Cavitation in medicine.* Interface Focus. **5**. 20150002.

[3] Plesset, M. S. (1954). *On the stability of fluid flows with spherical symmetry.* J. Appl. Phys. **25**. 96-98.

[4] Prosperetti, A. (1977). *Viscous effects on perturbed spherical flows.* Quart. Appl. Math. **34**. 339-352.

[5] Brenner, M. P., Hilgenfeldt, S., Lohse, D. (2002). *Single-bubble sonoluminescence.* Rev. Mod. Phys. **74**. 425-484.

[6] Liu, Y., Sugiyama, K., Takagi, S., Matsumoto, Y. (2012). *Surface instability of an encapsulated bubble induced by an ultrasonic pressure wave.* J. Fluid Mech. **691**. 315-340.

[7] Versluis, M., Goertz, D. E., Palanchon, P., Heitman, I. L., van der Meer, S. M. (2010). *Microbubble shape oscillations excited through ultrasonic parametric driving.* Phys. Rev. E. **82**. 026321.

[8] Hamaguchi, F., Ando, K. (2015). *Linear oscillation of gas bubbles in a viscoelastic material under ultrasound irradiation.* Phys. Fluids **27**. 113103.

[9] Gaudron, R., Warnez, M. T., Johnsen, E. (2015). *Bubble dynamics in a viscoelastic medium with nonlinear elasticity.* J. Fluid Mech. **766**. 54-75.

[10] Estrada, J. B., Barajas, C., Henann, D. L., Johnsen, E., Franck, C. (2018). *High strain-rate soft material characterization via inertial cavitation.* J. Mech. Phys. Solids **112**. 291-317.

[11] Prosperetti, A., Crum, L. A., Commander, K. W. (1988). *Nonlinear bubble dynamics.* J. Acoust. Soc. Am. **83**. 502-514.

[12] Chapman, R. B., Plesset, M. S. (1971). *Thermal effects in the free oscillation of gas bubbles.* J. Basic Eng. **33**. 373-376.

[13] Lamb, H. (1932). *Hydrodynamics.* Cambridge University Press.

[14] Liu, Y., Sugiyama, K., Takagi, S., Matsumoto, Y. (2011). *Numerical study on the shape oscillation of an encapsulated microbubble in ultrasound field.* Phys. Fluids **23**. 041904.

© 2018 ASME

Nucleation of bubbles in perfluoropentane droplets under ultrasonic excitation

[1]Krishna N Kumar*; Mitra Aliabouzar; Kausik Sarkar;
George Washington University, Washington, DC, USA
([1]Currently *in FloDesign Sonics, Wilbraham, MA, USA*)

Abstract

Lipid coated perfluorocarbon (PFC) nanodroplets can be vaporized by an external ultrasound pulse to generate bubbles. These bubbles produced in situ have potential applications in tumor imaging and drug delivery. Here we employ the classical nucleation theory (CNT) to investigate the peak negative pressure required to induce acoustic droplet vaporization (ADV). The theoretical analysis predicts that the ADV threshold increases with increasing surface tension of the droplet core and increasing excitation frequency. It decreases with increasing temperature and droplet size. The predictions are in qualitative agreement with experimental observations. We also estimate and discuss the energy required to form critical cluster to argue that nucleation occurs inside the droplet, as was also observed by high-speed cameras. To find the stability of these PFC droplets to temperature fluctuations, their limit of superheat was also calculated. These CNT based theoretical predictions prove to be a promising tool to design phase-shift nano emulsions with low threshold pressures.

Keywords: Acoustic droplet vaporization; classical nucleation theory; perfluoropentane

Introduction

Nanodroplets with perfluoropentane core and lipid encapsulation have potential applications in drug delivery, tumor imaging and gas embolotherapy. For all these applications, the nanodroplets need to be vaporized in situ. Droplet vaporization can be obtained either through heating (boiling) locally using a laser, or a sudden reduction in the pressure (cavitation) by using an acoustic wave. Here, ultrasound, which has been widely used as a non-invasive imaging modality, has been used to vaporize the droplets. Experimental studies have shown that the ADV threshold depends on the material properties of the droplet core as well as the acoustic parameters of excitation. Here we employ classical nucleation theory (CNT) to model the nucleation of a gas bubble into a droplet. Using CNT, we predict the threshold of vaporization and its dependence on size of droplets, ambient conditions (temperature, dissolved gas concentration), material properties (surface tension of the core liquid, interfacial tension at the droplet interface), and acoustic parameters (frequency of excitation). In addition, we predict whether the nucleation happens inside the droplet core (homogenous nucleation) or at the droplet interface (heterogeneous nucleation). To find the stability of PFC droplets to temperature fluctuations, the limit of superheat of these droplets were also calculated. The effects of mixing of two PFC on the threshold were also explored. Our theoretical predictions about the ADV threshold and its dependence on different parameters are in good qualitative agreement with the experimental results. Our prediction that the nucleation will occur inside the droplet and not at the interface also matches with high-speed camera observations. Detailed derivations of the expressions presented in this paper can be found in Kumar (2018). The expressions for different phenomena pertaining to droplet vaporization are provided. In the comparison section, the CNT predictions are compared with available experimental data. Finally, we conclude that the classical nucleation theory proves to be a tool in the design of phase-shift nano-droplets.

Background classical nucleation theory

Classical nucleation theory has been used in the past to model various nucleation phenomena such as crystal formation (Markov, 2016) , drop formation (Abraham, 2012) and bubble formation (Skripov, 1974). It is successful in capturing the qualitative features of the nucleation process. The quantitative agreement between the experimental measurements and CNT predictions is achieved only in few cases. In others, the agreement is at times reasonable. For the case of droplet vaporization, when a droplet is exposed to an increasing negative pressure, the liquid faces tensile stress and attains initially a metastable equilibrium. At sufficiently high degree of metastability, nucleation occurs. Classical nucleation theory has been an active area of research since early 1930's and in this work, the buildup framework of CNT was used to discuss the nucleation of a bubble in a drop under influence of acoustic waves. Readers can refer to Blander and Katz (1975) for an excellent review on bubble nucleation. It is to be noted that the classical nucleation theory was developed for the bulk of a material. Here, we applied it to a drop. Another recent application of CNT in droplet vaporization was done by Miles *et al.* (2016). The CNT formulation depends on the critical cluster and the

*Corresponding Author, Krishna N Kumar: k.kumar@fdsonics.com

size of a typical critical cluster is from 0.5 nm to 2 nm, which is small compared to our smallest droplets (~100 nm). So, the CNT could be applied to nucleation of a bubble in a drop.

Nucleation of bubble inside the droplet using acoustic waves

According to CNT, the rate of nucleation events/volume J is proportional to the probability of formation of a critical cluster, which in turn is determined by the free energy of the critical cluster formation W_c (Kelton and Greer, 2010)

$$J = J_0 \exp\left(\frac{-W_c}{k_B T}\right), \qquad W_c = -\frac{16\pi\sigma^3}{3\left(P_{sat} - P'\right)^2}, \qquad J_0 = n\sqrt{\frac{2\sigma}{\pi m}}, \tag{1}$$

where r_c is the radius of the critical cluster, k_B is the Boltzmann constant, T is temperature, P_{sat} is the saturation vapor pressure, σ, the surface tension of the liquid-vapor interface, n is the number density of the liquid molecules, and m is the mass of a liquid molecule. The value of J_0, is almost constant for all the applications as noted by Blander and Katz (1975). Our aim here is to find pressure P' at which nucleation starts and for this, one needs to find J at threshold, discussed in the next section.

Threshold pressure criterion

Formation of nuclei at a given supersaturation can be treated as a sequence of random events. The probability of finding certain number of nucleation events within the time interval τ can be calculated by the Poisson distribution (Skripov, 1974). In this study, the threshold pressure is the pressure at which at least 1 nucleation event occurs in volume V and time τ and its probability is 50%. This 50% threshold criterion has been used in the past (Herbert *et al.*, 2006; Maris, 2006) for nucleation in liquid helium leading to

$$\ln 2 = JV\tau \tag{2}$$

To find the threshold pressure P', the effective volume of nucleation (V) and the time scale of nucleation τ have to be estimated.

Estimation of effective volume of nucleation

Let V be the volume of the system where nucleation takes place. In this study, the nucleation results from the negative pressure of the acoustic wave. This high intensity negative pressure zone is only present at the focus of the acoustic transducer. The size of the focal volume of the acoustic transducer is directly proportional to λ^3; where λ is the wavelength of acoustic wave. In this study, we have used frequencies (2.25 MHz, 5 MHz and 10 MHz) relevant to diagnostic ultrasound. However, the size of the droplets (200nm- 2μm) is very small compared to the wavelength of these frequencies (177 μm for 2.25 MHz). Since the focal volume is larger than the droplet volume, it contains both water and PFP droplet. It is well known that water vaporizes at very high peak negative pressures (Herbert *et al.*, 2006) and here the range of peak negative pressure is relatively low, so actually only the droplets contributes to nucleation. Therefore, the effective volume for droplet nucleation is the volume of the drop:

$$V = \pi d^3 / 6, \quad \text{where } d \text{ is diameter of droplet.} \tag{3}$$

Estimation of time scale of nucleation

Because it is the negative pressure which is responsible for phase transition, the half period during which the pressure is negative is taken as the time scale. Note an underlying simplifying assumption of square wave excitation (pressure remains same the entire half period), whereas experiments used a sine wave for the excitation.

$$\tau \sim 1/2f \tag{4}$$

Acoustic droplet vaporization threshold pressure

Using equations , (2), (3) and (4) we obtain:

$$P_{threshold} = P_{sat} - \sqrt{\left(\frac{16\pi\sigma^3}{3k_BT}\frac{1}{\ln(\pi J_0 d^3/12 f \ln 2)}\right)}. \tag{5}$$

Equation (5) shows that the ADV threshold strongly depends on σ and T, whereas it weakly depends on d and f (inside the logarithmic term). Based on equation (5) it can be predicted that the threshold is directly proportional to surface tension of liquid-vapor interface and the frequency of excitation, whereas it is inversely proportional to temperature and size of droplets. It should be noted that the above expression of free energy of formation of a critical cluster depends on the fact that nucleation occurs inside the droplet. In the next section, we explore the situation where the nucleation of bubble occurs at the interface of a droplet.

Nucleation at a liquid-liquid interface

For the sake of simplicity, we assume the liquid-liquid interface to be planar and only one of the liquid vaporizes. In our experimental investigation (Kumar, 2018) , the two liquids used are perfluorocarbon and water, where only the former is volatile and water requires very high negative pressure (Herbert *et al.*, 2006) to vaporize. The free energy of formation of a bubble nucleation at the liquid-liquid interface involves displacement work, change in chemical potential, interfacial tension work and surface tension work. In the past, three independent investigations of nucleation at liquid-liquid interface (Apfel, 1971; Jarvis *et al.*, 1975; Avedisian and Andres, 1978) for boiling nucleation has been investigated.

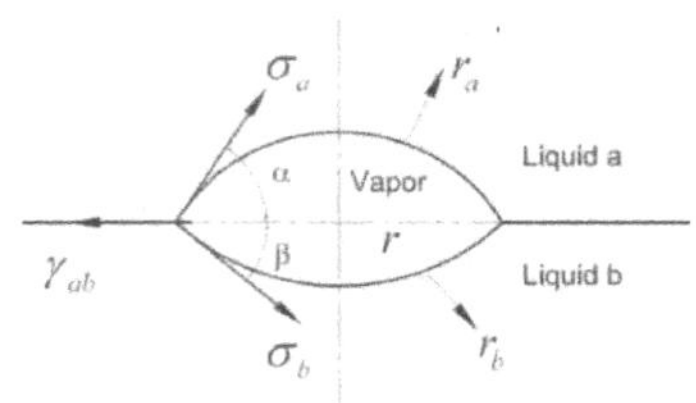

Figure 1 Lenticular bubble

Below we use the subscript 'b' for the volatile liquid, and 'a' for the stable one. The vapor formed inside the lenticular bubble in Figure 1 will be from liquid 'b'. At critical condition the critical radius r_c and W_c can be derived as:

$$r_c = \frac{2\sigma_a}{\left(P_{vap}-P_L\right)}\sqrt{1-\left(\frac{\gamma_{ab}^2+\sigma_a^2-\sigma_b^2}{2\gamma_{ab}\sigma_a}\right)^2}$$

$$W_c = \frac{16\pi}{3\left(P_{vap}-P_L\right)^2}\left[\frac{\sigma_a^3+\sigma_b^3}{2}+\frac{\gamma_{ab}^3}{16}-\frac{3}{16\gamma_{ab}}\left(\sigma_a^2-\sigma_b^2\right)^2-\frac{3\gamma_{ab}}{8}\left(\sigma_a^2+\sigma_b^2\right)\right] \tag{6}$$

We apply the theory to the case of a perfluoropentane droplet by computing the radius of a critical cluster. Using the the values of surface tension and interfacial tension for perfluoropentane and water σ_a =73.2, σ_b =10.2 and γ_{ab}=51 (mN/m), we find the critical cluster radius for vaporization at the interface and inside PFP droplet. For the vaporization at the interface, the predicted cluster radius was imaginary, while for inside the droplet it is a positive number. We infer that that nucleation occurs inside the droplet. This theoretical prediction matches with the high-speed camera observations (Kripfgans *et al.*, 2004; Li *et al.*, 2014; Shpak *et al.*, 2014).

Nucleation in a binary mixture of perfluorocarbons

The objective here is to derive an expression for the threshold pressure for a binary mixture of two PFC's, one of them volatile and the other relatively stable at room temperature. Let the volume fraction of the more volatile PFC be v_f and the corresponding mole fraction in the liquid mixture be x_f, *M* is molar molecular weight. Following Poling *et al.* (2001), one can compute the surface tension and saturation pressure of a binary mixture at a temperature lower than critical temperature of the pure components:

$$x_f = \left(\frac{\rho_1 v_f}{M_1}\right) / \left(\frac{\rho_1 v_f}{M_1} + \frac{\rho_2 (1 - v_f)}{M_2}\right),$$

$$\sigma_{mix} = \sigma_1 x_f + \sigma_2 (1 - x_f), \tag{7}$$

$$P_{sat-mix} = P_{sat1} x_f + P_{sat2} (1 - x_f),$$

And then one obtains:

$$J = N \sqrt{\frac{2\sigma_{mix}}{\pi}} \left(\frac{x_f}{\sqrt{m_1}} + \frac{1 - x_f}{\sqrt{m_2}}\right) \exp\left(\frac{16\pi}{3k_B T} \frac{\sigma_{mix}^3}{(P_{sat-mix} - P')^2}\right) \tag{8}$$

Using $\ln 2 = JV\tau$ in (8), one can determine P'.

Kinetic limit of superheat

The maximum temperature a liquid can sustain before it vaporizes at a given pressure is referred to as the limit of superheat. Since the derivation of the problem considers kinetic considerations such as collision at the interface, evaporation and condensation, it is called the kinetic limit of superheat. This quantity is important for volatile PFC droplets to be used for creating a contrast in tumor tissues; the droplets have to be stable enough not to vaporize due to temperature fluctuations before they reach the tumor tissues. The kinetic limit of superheat gives us a range of temperature where the droplets could be used. One obtains kinetic limit of superheat to be used below:

$$T = \frac{-16\pi\sigma^3}{3k(P_{sat} - P')^2 \ln \frac{J}{n} \sqrt{\frac{\pi m}{3\sigma}}} \tag{9}$$

σ and P_{sat} are function of temperature (T). To find T we use the experimentally observed value of J=10^{12} bubbles/m^3/s (Skripov, 1974), which is a standard practice in finding the limit of superheat of many liquids. The variation of P_{sat} and σ with temperature for all the PFC were obtained from NIST database. The value of J_0 for superheating is found from Blander and Katz (1975).

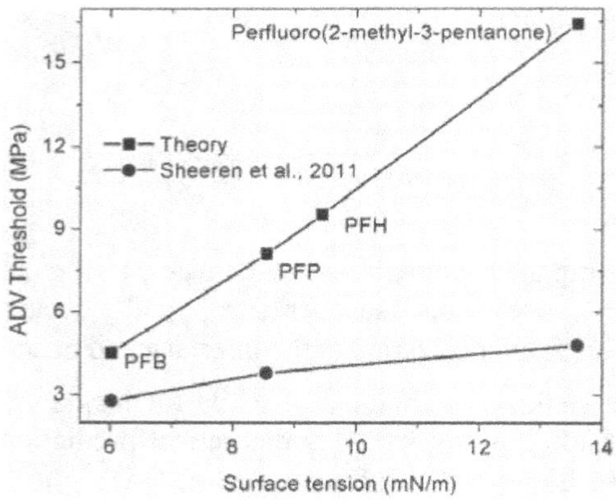

Figure 2 ADV threshold of different materials

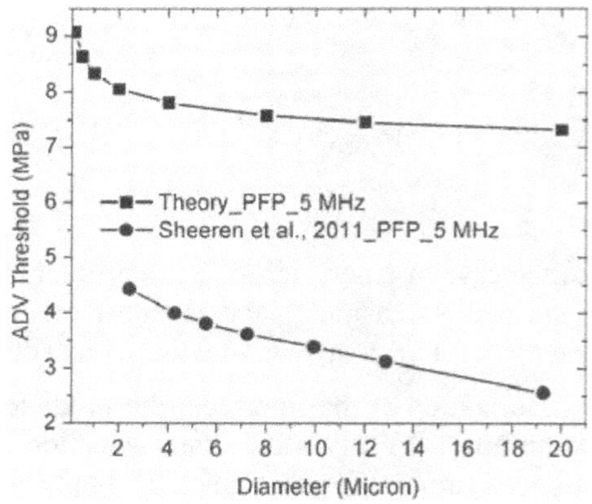

Figure 3 ADV threshold of PFP droplets of different sizes.

ADV threshold and limit of superheat: comparison of theoretical predictions with experiments

Here we compare the CNT predictions with experimental results. In particular the effects of the surface tension of liquid, size of the drop, temperature of the drop, frequency of excitation on ADV threshold are presented. The kinetic limit of superheat was also computed for different PFCs. Figure 2 shows an increasing trend in the ADV threshold with surface tension of the droplet core. The results are in qualitative agreement with our experimental results---threshold for PFP is 1.5 MPa and for perfluoro hexane is above 5 MPa at 2.25 MHz. As low surface tension material has a low threshold, it is recommended that one uses low surface tension liquids as the droplet core. Figure 3 shows that the ADV threshold according to CNT predictions decreases with an increase in drop size. again in qualitative agreement with high speed camera observations (Sheeran *et al.*, 2011) of the vaporization of individual droplets.

Figure 4 shows that the ADV threshold decreases with increase in temperature both by theory and experimental data. Figure 5 shows the theory and our experimental results indicate that the ADV threshold for perfluoropentane increases with the excitation frequency.

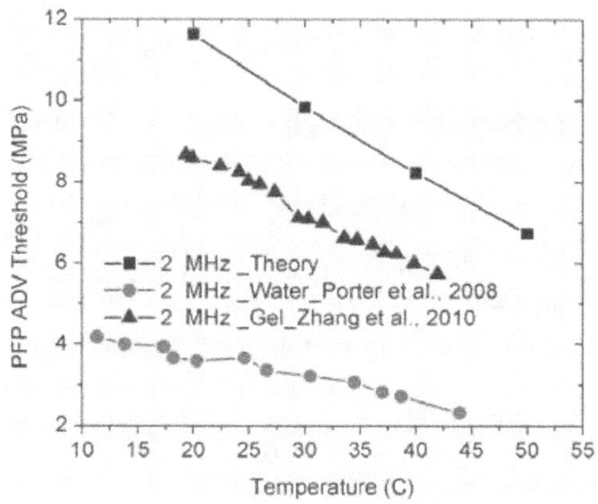

Figure 4 ADV threshold at different temperatures.

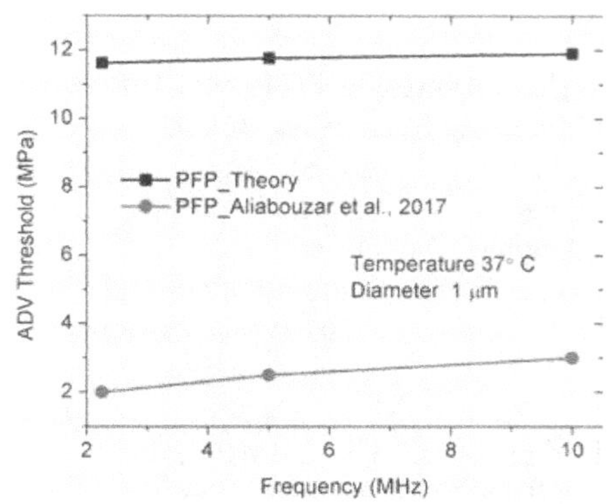

Figure 5 ADV threshold at different frequencies

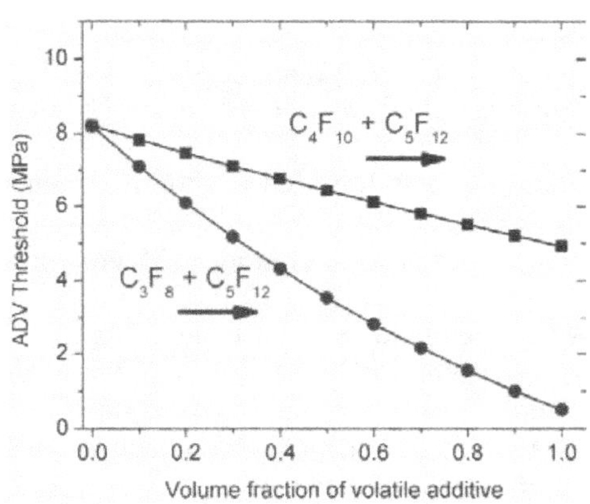

Figure 6 ADV threshold of binary mixtures

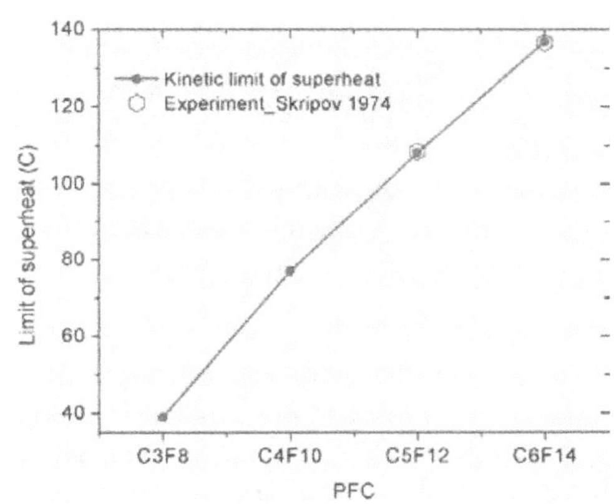

Figure 7 Limit of superheat of different PFC's

Although materials with low surface tension require a lower peak negative pressure for vaporization, they suffer from stability issues because of their volatility. The motivation of mixing two PFCs is to obtain a liquid which has low threshold and at the same time is easy to handle at room temperature. Perfluoro-propane (-32°C b.p) and Perfluoro-butane (-2°C b.p) are highly volatile PFCs having low threshold pressures. These volatile PFCs are mixed with Perfluoropentane (stable 29°C b.p) in different volume fractions. Figure 6 demonstrates that the threshold decreases with increasing volume fraction of the volatile PFC. Figure 7 shows an excellent match between the CNT predictions and experimental observation of kinetic limit of superheat (Figure 7). The experimental results are available only for perfluoropentane and perfluorohexane. One point to be noted is that the perfluoropropane droplets may be sensitive to change in temperature in the body as their limit of superheat is around 39°C, which is close to the normal body temperature.

Conclusion

Here classical nucleation theory was applied to study the vaporization of a PFC droplet under acoustic excitation. A theoretical expression for the threshold pressure was obtained, which demonstrates that ADV threshold is a strong function of the surface tension of the liquid vapor interface and temperature, and a weak function of the droplet diameter and the excitation frequency. We also investigate the possibility of bubble nucleation at the interface of PFC water to find that in case of perfluoropentane droplets nucleation occurs in the droplet core, and not at the PFP-water interface. To check the stability of PFC droplets to temperature fluctuations their kinetic limit of superheat (KLS) was computed. The effects of mixing two PFCs on the ADV the threshold was discussed theoretically. The theoretical analysis predicts that the ADV threshold increases with increasing surface tension of the droplet core and the frequency of excitation, while it decreases with increasing temperature and droplet size. The predictions are in qualitative agreement with experimental observations.

References

Abraham, F. (**2012**). *Homogeneous nucleation theory: the pretransition theory of vapor condensation* (Elsevier).

Apfel, R. E. (**1971**). "Vapor nucleation at a liquid–liquid interface," The Journal of Chemical Physics **54**, 62-63.

Avedisian, C. T., and Andres, R. P. (**1978**). "Bubble nucleation in superheated liquid-liquid emulsions," Journal of Colloid And Interface Science **64**, 438-453.

Blander, M., and Katz, J. L. (**1975**). "Bubble nucleation in liquids," AIChE Journal **21**, 833-848.

Herbert, E., Balibar, S., and Caupin, F. (**2006**). "Cavitation pressure in water," Physical Review E **74**, 041603.

Jarvis, T. J., Donohue, M. D., and Katz, J. L. (**1975**). "Bubble nucleation mechanisms of liquid droplets superheated in other liquids," Journal of Colloid And Interface Science **50**, 359-368.

Kelton, K., and Greer, A. L. (**2010**). *Nucleation in condensed matter: applications in materials and biology* (Elsevier).

Kripfgans, O. D., Fabiilli, M. L., Carson, P. L., and Fowlkes, J. B. (**2004**). "On the acoustic vaporization of micrometer-sized droplets," The Journal of the Acoustical Society of America **116**, 272-281.

Kumar, K. N. (**2018**). "Acoustic Studies on Nanodroplets, Microbubbles and Liposomes," (The George Washington University).

Li, D. S., Kripfgans, O. D., Fabiilli, M. L., Brian Fowlkes, J., and Bull, J. L. (**2014**). "Formation of toroidal bubbles from acoustic droplet vaporization," Applied physics letters **104**.

Maris, H. J. (**2006**). "Introduction to the physics of nucleation," Comptes Rendus Physique **7**, 946-958.

Markov, I. V. (**2016**). *Crystal growth for beginners: fundamentals of nucleation, crystal growth and epitaxy* (World scientific).

Miles, C. J., Doering, C. R., and Kripfgans, O. D. (**2016**). "Nucleation pressure threshold in acoustic droplet vaporization," Journal of Applied Physics **120**, 034903.

Poling, B. E., Prausnitz, J. M., and O'connell, J. P. (**2001**). *The properties of gases and liquids* (Mcgraw-hill New York).

Sheeran, P. S., Wong, V. P., Luois, S., McFarland, R. J., Ross, W. D., Feingold, S., Matsunaga, T. O., and Dayton, P. A. (**2011**). "Decafluorobutane as a phase-change contrast agent for low-energy extravascular ultrasonic imaging," Ultrasound in medicine & biology **37**, 1518-1530.

Shpak, O., Verweij, M., Vos, H. J., de Jong, N., Lohse, D., and Versluis, M. (**2014**). "Acoustic droplet vaporization is initiated by superharmonic focusing," Proceedings of the National Academy of Sciences **111**, 1697-1702.

Skripov, V. P. (**1974**). "Metastable Liquids," Metastable Liquids.

© 2018 ASME

Numerical 3D simulation of the NPSH characteristics of centrifugal pumps with local and integral analysis of void structures

[1]Phillip Limbach*; [1]Romuald Skoda;

[2]Tim F. Groß; [2]Gerhard Ludwig; [2]Peter F. Pelz

[1]*Chair of Hydraulic Fluid Machinery, Ruhr-Universität Bochum, Universitätsstr. 150, 44801 Bochum, Germany;*

[2]*Chair of Fluid Systems, Technische Universität Darmstadt, Otto-Berndt Str. 2, 64287 Darmstadt, Germany*

Abstract

3D numerical simulations with mass transfer cavitation model are performed in a centrifugal pump (n_q=26 min^{-1}). Simulation results are compared to measurements of $NPSH$ characteristics and optical measurements of void structures. Transient single bubble cavitation is observed in the experiments with hydraulically smooth blades, whereas in the simulations, the cavitation region appears as stationary sheet cavitation for the entire operation range. Local cavitation structures in the impeller are spatially integrated for measurements and simulation results. The resulting cavitation intensity provides additional information on the extent of cavitation zones at $NPSH$ levels well above $NPSH_{3\%}$, where the head is yet unaffected by cavitation. The $NPSH_{3\%}$ curve as well as cavitation intensity are well predicted, although local void structures show a mismatch to measurements.

Keywords: centrifugal pump, mass transfer cavitation model, NPSH, void structures, cavitation intensity

1. Introduction

Different cavitation regimes may develop in centrifugal pumps varying in their transient occurrence and erosive impact. An accurate prediction of the cavitation dynamics is therefore of particular importance for reliable pump design and performance. For the prediction of cavitation in centrifugal pumps, URANS methods in combination with mass transfer models that are based on simplified bubble dynamics, i.e. Rayleigh equation [1], are the most commonly used model class, because of the moderate computational effort. Due to crucial simplifications in this approach, modifications of the cavitation model parameters and mixture eddy viscosity may be necessary to capture the relevant flow characteristics, e.g. cloud cavitation [2–4]. Cavitation model parameter variations for centrifugal pump flow usually aim at matching the measured $NPSH_{3\%}$ [5, 6] and do not consider local cavitation structures. Therefore, in the present study, an evaluation of void structures is performed in addition to $NPSH$ characteristics.

As found in previous investigations of the present authors [7] the reduction of the mixture eddy viscosity [2, 3] is indispensable for the prediction of cloud cavitation at external (i.e. hydrofoil) as well as internal flow (i.e. orifice flow). In order to capture cloud cavitation over a wide range of operation points and at the same time accurately predict the mass flow rate, a modification of the pre-factors for evaporation and condensation in the cavitation model is necessary, leading to different parameter sets for hydrofoil and orifice flow. However, the mass flow rate prediction is predominantly influenced by the turbulent wall treatment, rather than by the cavitation model modifications, i.e. a low-Re approach is indispensable if viscous flow separation is present. The mass flow rate at choking conditions in the orifice is well predicted even with original cavitation model parameters and un-modified mixture eddy viscosity. Transferring these findings to cavitating low specific speed pump flow with n_q=12 min^{-1} [8], we confirmed that a low-Re approach is in fact indispensable for an accurate prediction of choking and $NPSH_{3\%}$ at overload operation, because cavitation zones switch from the impeller to the volute tongue, where significant incidence and viscous separation occur. On the other hand, wall functions are appropriate for moderate incidence flow in the impeller as far as no viscous separation occurs.

To evaluate these conclusions of the previous studies, in the present study a centrifugal pump with n_q=26 min^{-1} is investigated, where no significant viscous flow separation is present at the considered operation points. Local

*Corresponding Author, Phillip Limbach: phillip.limbach@rub.de

visualizations of the cavitation zones are used to validate and assess the simulation results concerning the head drop mechanisms.

2. Experimental setup

The centrifugal pump with specific speed of n_q=26 min^{-1} is part of the closed-loop test rig at the chair of Fluid Systems of the Technical University of Darmstadt. A wide range of rotational speeds, flow rates and gap sizes between the impeller shroud and casing at the suction side have been considered. A detailed description of the test rig, experimental setups as well as a comprehensive discussion of the results is provided by e.g. [9] and briefly summarized in the following. The pump is equipped with a six blade impeller. The impeller outlet diameter and outlet width are D_2=260 mm and b_2=23.5 mm, respectively. A sketch of the pump and an image of the impeller with de-mounted shroud is presented in Fig. 1a. A rotation-symmetrical annulus chamber instead of a volute yields essentially uniform flow conditions in circumferential direction downstream of the impeller. The casing is connected to the pressure pipe via twelve outflow tubes, which are distributed symmetrically over the casing. The static outlet pressure p_{out} is measured at D_{out}=500 mm in the annulus chamber and the static pressure on the suction side p_s is measured in the suction pipe ~550 mm upstream of the impeller inlet. The nominal load operation points at rotational speeds of n=1750 min^{-1} and n=2000 min^{-1} are specified as Q_{opt}=125 m^3/h and Q_{opt}=143 m^3/h, respectively. A transparent suction pipe, front casing and shroud are utilized, and cavitation voids are visualized using stroboscopic lighting and a 90° borescope that is positioned in front of the impeller. A high-speed camera system enables image recordings with 10,600 frames per second and a resolution of 1920x1080 pixels. The experiments in the present study are primarily performed with hydraulically smooth blades. The impeller suction side roughness has been varied exemplarily by applying thin layers of glue and small particles of silicon carbide with a particle size of 125 μm that are distributed over a distance of 3 mm from the leading edge.

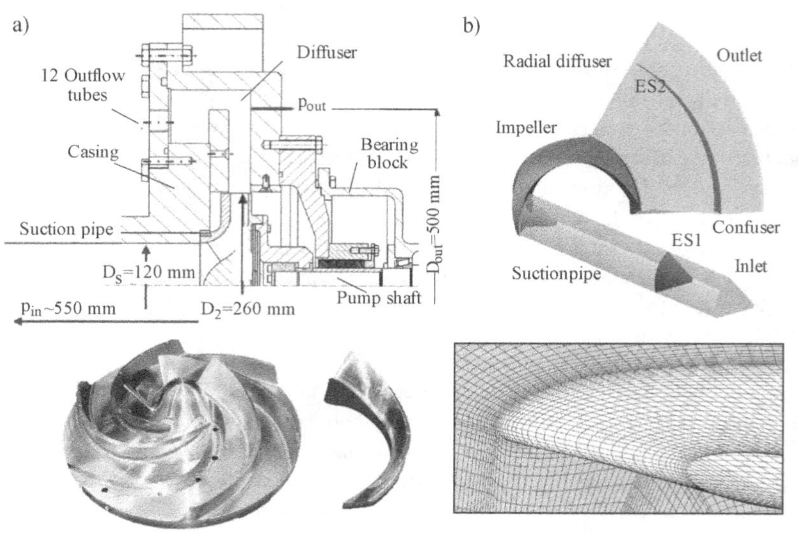

Fig. 1: a) Section view of the centrifugal pump and image of the impeller with de-mounted shroud; b) Computational domain and grid details of grid G02 in the vicinity of the leading edge.

3. Numerical setup

Due to the rotation-symmetrical annulus chamber, periodic flow conditions are assumed and only one impeller channel (60° segment) is considered in the numerical model applying periodic boundary conditions in circumferential direction, see Fig. 1b. The suction pipe and the discharge of the impeller channel are elongated to reduce the influence of the boundary conditions on the impeller flow. A confuser is added at the radial outlet section to avoid positive pressure gradients in that region. A block structured grid with a resolution of approximately 55,000 (G01) and 480,000

(G02) nodes is used, see Fig. 1b. The grid G02 is generated by halving the length of the cell edges of grid G01 in all directions. No significant grid dependence could be observed for the subsequently discussed results, so that only G02 results are presented in section 4. According to the experiments, different gap sizes, s, are considered in the numerical model (s=0; 0.15; 0.2 mm). Simulations are performed with the commercial flow solver ANSYS CFX 16.1 and cavitation model of [10]. Since previous studies [7, 8] revealed that choking mass flow rate is well predicted by default settings, the default pre-factors for evaporation (F_{vap}=50) and condensation (F_{cond}=0.01) are maintained, and no modification of the mixture eddy-viscosity is done. Normal velocity is specified at the inlet and static pressure is defined at the outlet, which is successively reduced in several simulation runs for the calculation of the head drop and $NPSH_{3\%}$. The k-ω SST turbulence model is applied with an automatic wall treatment [11, 12]. Although all simulations are performed in the relative frame of reference, transient simulation runs are performed because for several stationary simulations, the residuums do not drop by more than two orders of magnitude. Second order schemes are used for spatial and temporal discretization. The convergence criterion is set to a drop of maximum residuum R_{Max} by four orders of magnitude and imbalances < 1%. The head H, is evaluated in accordance to the experiments by the static pressure difference between pressure and suction side $p_{s,ES1,2}$. The $NPSH$ is determined by the difference between total pressure on the suction side, $p_{t,ES1}$, and the vapor pressure, p_v.

4. Results

For experiments with hydraulically smooth surface, single bubble cavitation is primarily observed, see Fig. 2a and for details [9], and no cloud cavitation is observed. These bubbles appear to be inherently unsteady, even after transformation of the images into the relative frame of reference, i.e. rotating every picture against the direction of rotation to a certain amount to freeze the impeller.

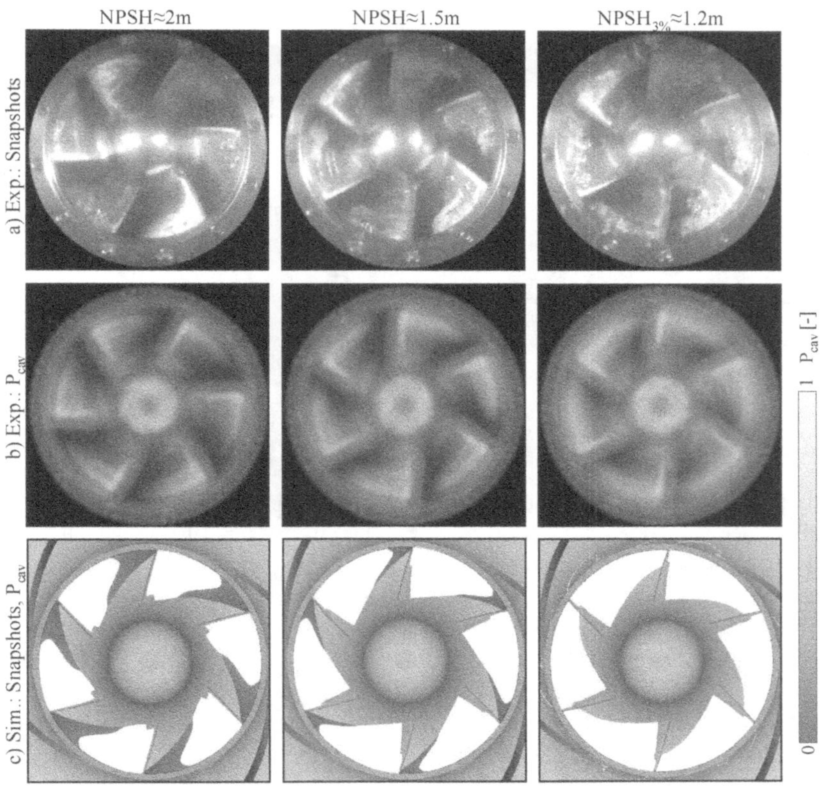

Fig. 2: a) Snapshots of the cavitating flow at arbitrary time instances and b,c) cavitation probability, P_{cav}, for varying NPSH. The single-channel simulation results are copied in circumferential direction according to the number of blades for a full-impeller illustration and comparison with measurements.

The simulations a priori cannot capture this cavitation regime with distinct single bubbles, due to the assumption of a homogenous mixture, e.g. a particular amount of individual bubbles and nuclei within each computational cell [10]. Hence, the simulation results show essentially stationary cavitation sheets, which are attached at the blades and illustrated in Fig. 2c by means of iso-surfaces of 10% void volume fraction. Note that the absence of transient cavitation patterns in the impeller simulation is not necessarily only a consequence of model simplifications, since results of preliminary simulations with cavitation model settings adopted from [7, 8], i.e. parameters optimized for cloud cavitation simulations and a locally reduced mixture eddy-viscosity, show essentially stationary cavitation sheets as well.

The $NPSH_{3\%}$ is predicted accurately over a wide range of normalized operation points $q = Q/Q_{opt}$ and gap widths, as shown in Fig. 3 and discussed in the following. The influence of the grid resolution on $NPSH_{3\%}$ is below $\Delta NPSH_{3\%}$<1% and therefore minor (not shown). Although the cavitation regime is predicted to be stationary in contrast to the measurements, the time-averaged spatial extent of the measured cavitation zones, represented by the cavitation probability, P_{cav}, is qualitatively well predicted, see Fig. 2b and c. The cavitation probability is evaluated equally in measurement and simulation: by averaging the grayscale of a pixel over a distinct number of single-shot images, it provides the qualitative distribution of the probability of the existence of voids at a certain location. Due to essentially stationary voids in the simulation results, P_{cav} in Fig. 2c is essentially identical to the instantaneous images of void fraction iso-surfaces. Note that the background grayscale is not subtracted from the cavitation probability and the impeller also appears gray in Fig. 2b and c. At ~$NPSH_{3\%}$, the cavitation structures on the suction side extend to the impeller throat area, leading to choking and the head drop. The decrease of $NPSH_{3\%}$ due to closing the gap is captured qualitatively by the simulations. Interestingly, the application of modified cavitation model parameter sets for the prediction of cloud cavitation [7, 8] does not lead to significant differences of both, $NPSH_{3\%}$ and the local cavitation structures.

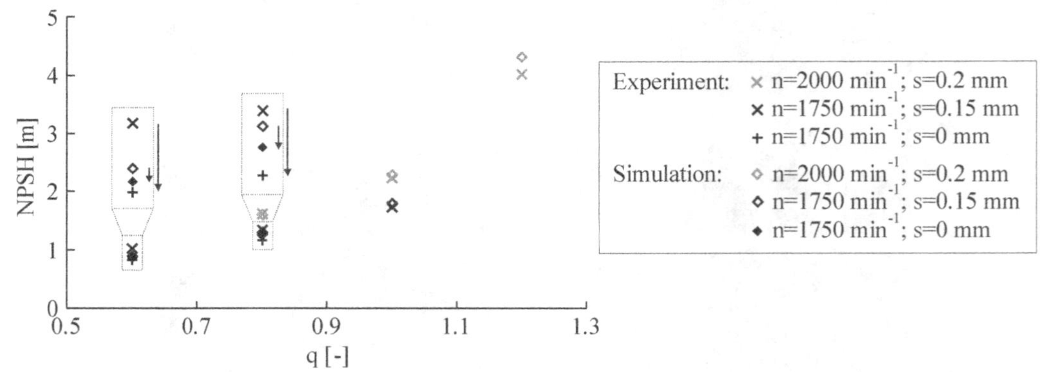

Fig. 3: $NPSH_{3\%}$ curve for different rotational speeds (n=1750; 2000 min^{-1}) and gap widths (s = 0, 0.15 and 0.2 mm)

For a quantitative analysis at q=0.6 and 0.8, Fig. 4 shows the cavitation intensity, I_{cav}, that is evaluated by integrating the grayscale values over all pixels of P_{cav} in the experiments. In the simulations, the vapor volume fraction is integrated over the computational domain and subsequently time-averaged to evaluate I_{cav}. For a comparison of simulation and measurement results, I_{cav} is normalized. For experimental data, the background intensity, I_{ref}, i.e. the cavitation intensity at a high pressure level and single-phase flow conditions, is substracted from the respective I_{cav} values before normalization. Normalization to I_{cav} e.g. at 3% head drop ($NPSH_{3\%}$) is not appropriate, because cavitation zones on the pressure side of the blades as well as in the impeller throat area are not visible in the experiments and therefore do not contribute to the evaluation of I_{cav}. On the other hand, these hidden voids do contribute to the evaluation of I_{cav} in the simulations. For $NPSH$=1.5 m, the cavitation zones do not extend to the throat area of the impeller. According to the simulation results, no cavitation is observed on the pressure side of the impeller blades. Therefore, I_{cav} is normalized by $I_{cav,1.5}$ at $NPSH$=1.5 m, where no head drop is discernible yet.

$I_{cav}/I_{cav,1.5}$ thus per definition equals one at $NPSH$=1.5 m, for measurement and simulation. A good agreement between simulations and experiments is obtained for both flow rates, q=0.6 and 0.8, at high and moderate $NPSH$, see Fig. 4. The cavitation intensity is slightly overestimated at low $NPSH$ values. We attribute this overestimation to the limited visual access in the experiments, as described above.

In Fig. 4, the head drop curves are also included and show essentially the same drop of H for measurement and simulations, resulting in good agreement of $NPSH_{3\%}$ in Fig. 3. A significant head drop occurs rather abruptly only at low $NPSH$ levels where pronounced void structures are present. Thus, the cavitation intensity provides additional information on the extent of cavitation at $NPSH$ levels, where smaller cavitation zones are present that do not influence the pump performance yet.

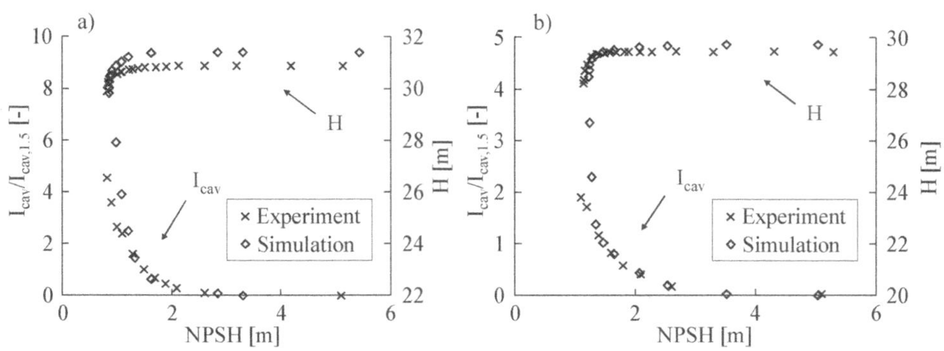

Fig. 4: Cavitation intensity, I_{cav}, for n=1750 min^{-1} and s=0 mm, at a) q=0.6 and b) q=0.8.

Interestingly, the inclusion of roughness elements at the leading edge of the impeller blades leads to the occurrence of sheet cavitation even in the measurements (details are available in [9]) and a good agreement to the cavitation zones that are observed in the simulations, as exemplarily shown in Fig. 5. We attribute this to a more homogeneous distribution of nuclei at the roughness elements and wall crevices along the leading edge and therefore a better fulfilment of one main assumption of the cavitation model, i.e. a homogenous nuclei distribution within each computational cell. A general analysis of the role of surface nuclei for the inception of hydrodynamic cavitation has recently been studied by some of the authors of this paper [13].

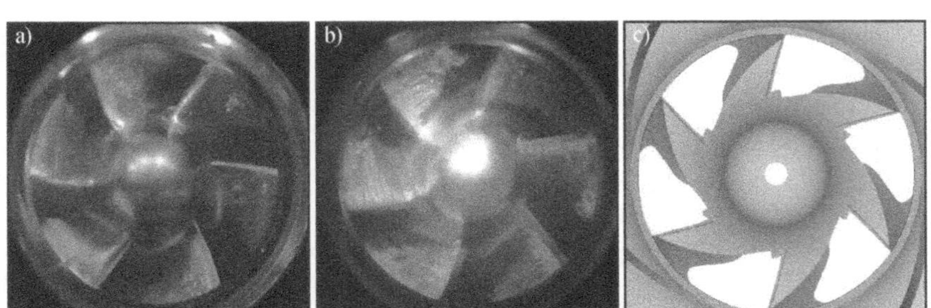

Fig. 5: Exemplary snapshots of the local cavitation structures on the suction side of the impeller blades at n=1750 min^{-1}, q=0.8, s = 0 mm and NPSH=2 m; a) smooth blades; b) roughened leading edge on the suction side; c) simulation results.

5. Conclusion

In addition to head drop and $NPSH_{3\%}$ characteristics, the cavitation probability and intensity provide complementary information on the extent of cavitation zones at moderate $NPSH$ levels, where the head is not yet affected by cavitation. Although the highly unsteady appearance of the cavitation zones for smooth surface is not reflected by the simulation results, the head drop and $NPSH_{3\%}$ characteristics are well predicted. In former investigations this holds even for cloud cavitation [7] where the low-Reynolds resolution of viscous wall boundary layer separation has been shown to be more critical for $NPSH_{3\%}$ prediction than the resolution of unsteady shedding of voids. Since for the operation conditions considered in the present study viscous separation is not significant, the utilization of turbulent wall functions seems feasible. It may be concluded that for prediction of $NPSH_{3\%}$ and even for prediction of integral void structures it may be sufficient to capture the time-averaged spatial extent of the cavitation zones, without capturing the unsteady cavitation regime. This assumption need to be verified for a wider operation range and range of specific speeds in further studies.

6. References

[1] Rayleigh (1917). *On the pressure developed in a liquid during the collapse of a spherical cavity.* Philosophical Magazine Series 6; 34, 94–8.

[2] Reboud, J. L., Stutz, B., Coutier, O. (1998). *Two-phase flow structure of cavitation: Experiment and modeling of unsteady effects.* In 3rd International Symposium on Cavitation, Grenoble, France.

[3] Coutier-Delgosha, O., Fortes-Patella, R., Reboud, J. L. (2003). *Evaluation of the Turbulence Model Influence on the Numerical Simulations of Unsteady Cavitation.* Journal of Fluids Engineering; 125, 38–45.

[4] Frikha, S., Coutier-Delgosha, O., Astolfi, J. A. (2008). *Influence of the Cavitation Model on the Simulation of Cloud Cavitation on 2D Foil Section.* International Journal of Rotating Machinery; 2008, 1–12.

[5] Chatagny, L., Berten, S. (2016). *Challenges and open questions in cavitation simulations for centrifugal pump applications.* In 3rd International Rotating Equipment Conference (IREC), Pumps, Compressors and Vacuum Technology, 765–75, Düsseldorf, Germany.

[6] Salvadori, S., Cappelletti, A., Martelli, F., Nicchio, A., Carbonino, L., Piva, A. (2012). *Numerical prediction of cavitation in pumps.* In 15th International Conference on Fluid Flow Technologies, Budapest, Hungary.

[7] Limbach, P., Skoda, R. (2017). *Numerical simulation of cloud cavitation in hydrofoil and orifice flows with analysis of viscous and non-viscous separation.* Journal of Fluids Engineering; submitted.

[8] Limbach, P., Skoda, R. (2017). *Numerical and experimental analysis of cavitating flow in a low specific speed centrifugal pump with different surface roughness.* Journal of Fluids Engineering; 139, 101201.

[9] Groß, T., Ludwig, G., Pelz, P. (2016). *Influence of surface roughness on cavitation in a centrifugal pump.* In 3rd International Rotating Equipment Conference (IREC), Pumps, Compressors and Vacuum Technology, 801–11, Düsseldorf, Germany.

[10] Zwart, P. J., Gerber, A. G., Belamri, T. (2004). *A two-phase flow model for predicting cavitation dynamics.* In 5th International Conference on Multiphase Flow, Yokohama, Japan.

[11] Menter, F. R. (1994). *Two-equation eddy-viscosity turbulence models for engineering applications.* AIAA Journal; 32, 1598–605.

[12] Menter, F. R., Kuntz, M., Langtry, R. (2003). *Ten years of industrial experience with the SST turbulence model.* Turbulence, heat and mass transfer; 4, 625–32.

[13] Groß, T. F., Pelz, P. F. (2017). *Diffusion-driven nucleation from surface nuclei in hydrodynamic cavitation.* Journal of Fluid Mechanics; 830, 138–64.

© 2018 ASME

Numerical analysis for the prediction of hull pressure fluctuation and underwater radiated noise induced by marine propeller cavitation

[1]Jeong-yong Park*; [2]Milovan Peric; [2]Mark Farrall; [3]Cheolsoo Park;

[1]*Maritime Research Institute, Hyundai Heavy Industries Co., Ltd., Ulsan, Republic of Korea;*
[2]*Siemmens Industry Software Limited, London, United Kingdom;*
[3]*Korea Research Institute of Ships & Ocean engineering, Daejeon, Republic of Korea;*

Abstract

In this study, viscous flow analysis by computational fluid dynamics (CFD) was performed on a five bladed conventional marine propeller in behind condition on the purpose of predicting hull pressure fluctuations and underwater radiated noise (URN) induced by the propeller. The computation was conducted in model and full scale using the total underwater geometrical model of the propeller, hull and rudder for the direct comparison with the experiment and full scale measurement, respectively. The cavitation pattern and subsequent fluctuating pressure were investigated in model scale. A good agreement in cavitation pattern was found between the numerical analysis and experiment, yet the resolution of tip vortex cavitation needs to be improved. The tendency and amplitude of pressure fluctuations were reasonably predicted, especially for the 1st blade passing frequency (BPF). For the prediction of sound pressure level (SPL) in full scale, a hybrid approach based on CFD and Ffowcs-Williams and Hawkings (FW-H) method is applied. In case of URN for full scale, SPL from the numerical computation was compared with the result from full scale measurement. The numerical analysis generally underestimated SPL in comparison with the result of full scale measurement. For the validation of numerical analysis, only the computational result of propeller and rudder radiated noise is compared with the result from the full scale measurement.

Keywords: Marine propeller, Cavitation, Hull pressure fluctuation, URN, RANS, CFD, FW-H, SPL

Introduction

A rotating marine propeller and unsteady cavitation induce pressure fluctuations on a ship. They can cause the deterioration of comfort on board and the fatigue damage on the ship structure. In addition, ship radiated noise has been of a significant interest to international maritime community in terms of passenger's comfort as well as environmental protections. The propeller cavitation is also the main source of ship radiated noise. In this point of view, International Maritime Organization (IMO) released guidelines for the reduction of underwater noise in 2014 even if it is non-mandatory [1]. Also, considerable studies have been recently conducted in the European research project like AQUO and SONIC to improve both experimental and numerical prediction method of cavitation noise. For these reasons, the prediction of unsteady cavitation, hull pressure fluctuation and URN by propeller is very essential in the propeller design phase. Regarding the prediction of hull pressure fluctuation, Paik [2] showed that Reynolds Averaged Navier-Stokes (RANS) equation provides a good agreement with experimental results in the 1st BPF of hull pressure fluctuations. Another recent study presents numerical simulations for the prediction of hull pressure fluctuation induced by cavitation on propeller with capturing the tip vortex [3]. Concerning the numerical analysis for URN by a propeller, it can be calculated by complementary use of CFD and FW-H equation [4].

In this study, numerical analyses are conducted for the prediction of hull pressure fluctuations and underwater radiated noise induced by marine propeller cavitation. For the validation, numerical results are compared with the hull pressure fluctuation by model test and URN from the full scale measurement on a modern ultra large container carrier Especially, the URN is directly predicted by flow analysis and FW-H in full scale while most of previously conducted researches focused on the analysis in model scale or its scale up results. And the current work presented in this paper is an exploration study of the feasibility to predict the URN by a ship propeller from a practical perspective for the propeller design, using a commercial CFD tool STAR-CCM+ which provides flow simulation and acoustic analogy.

*Corresponding Author, Jeong-yong Park: pjy6262@hhi.co.kr

Prediction of cavitation and hull pressure fluctuation in model scale

For the numerical analysis, RANS is applied for the prediction of cavitation and hull pressure fluctuation in model scale. The cavitation is simulated by using the Schnerr-Sauer model which is based on a simplified form derived from the general Rayleigh-Plesset model by excluding higher order terms, viscous and surface tension effects. For the propeller rotation, overlapping grids are adopted. The background grid includes hull and rudder and is meshed as if the propeller was not present, while a separate grid region is created around the propeller and rotates with it. Table 1 describes the detailed numerical setup and analysis condition. In this table, T, ρ, n and D represent thrust, density, rps and propeller diameter, respectively. $P_{0.7R}$ is a static pressure at the 70 % of propeller radius above the propeller shaft center and P_v is a vapor pressure.

Software	STAR-CCM+ (v12.02)
Turbulence Model	RANS (k-ε model)
Cavitation model	Schnerr-Sauer
Wall treatment	All y+ wall treatment
Propeller rotation rate (rps)	40 rps
Time step	2.5E-5s (0.36 deg/s)
Inflow velocity	7.5 m/s
Thrust oe i ient $\left(K_T = \frac{T}{0.5\rho n^2 D^4}\right)$	0.1850
Cavitation number $\left(\sigma_{0.7R} = \frac{P_{0.7R} - P_v}{0.5\rho n^2 D^2}\right)$	1.45
Scale ratio	45.7613
Draft	Ballast draft
Power	85% MCR
Free surface	Neglected
No. of cells	Abt. 12.5 millions

Table 1 Numerical setup and analysis condition

For the validation of numerical analysis, the predicted cavitation pattern is compared with the result from the model test performed at Large Cavitation Tunnel (LCT) in KRISO. Figure 1 depicts the cavitation pattern. A good agreement was found between the numerical analysis and experiment, yet the resolution of tip vortex cavitation needs to be improved. Also, the predicted pressure fluctuation by computation is also compared with the model test results on several sensor locations as shown in Figure 2. Figure 3 shows that the tendency and amplitude of hull pressure fluctuations are reasonably predicted, especially for the 1st BPF. In this figure, the hull pressure fluctuation is expressed as pressure coefficient, $K_P \left(= \frac{\Delta p}{\rho n^2 D^2}\right)$ where Δp is the pressure lu t uation amplitude rom the mean pressure.

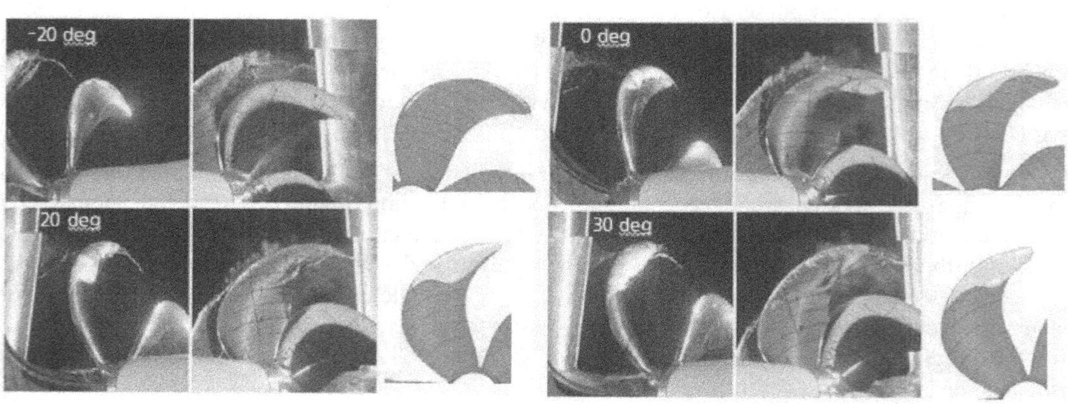

Figure 1 Comparison of cavitation between experiment and numerical analysis
[left: experiment (starboard side view), center: experiment (portside view), right: numerical analysis]

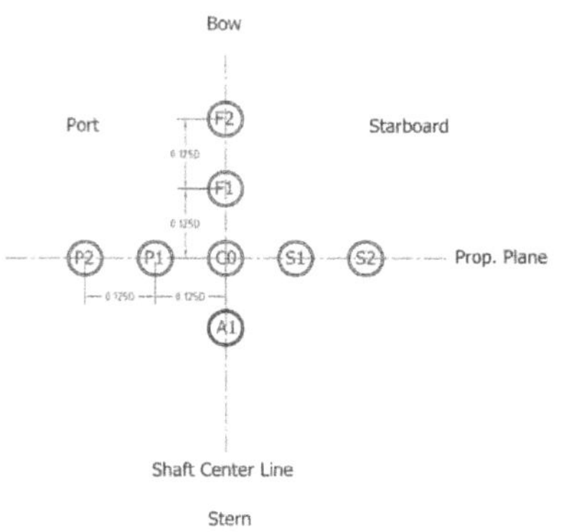

Figure 2 Sensor locations to measure hull pressure fluctuation

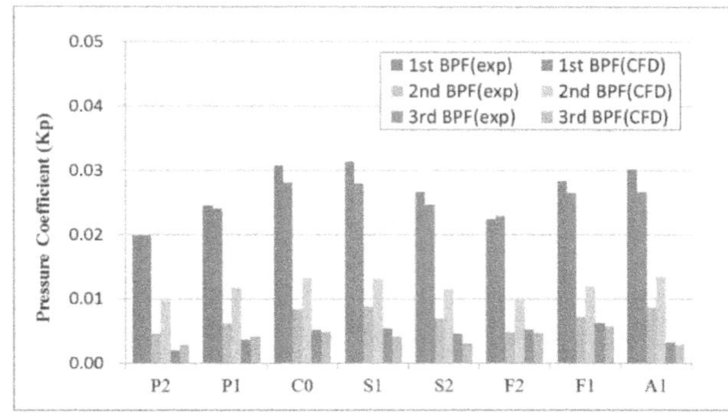

Figure 3 Comparison of Pressure Coefficient

Prediction of cavitation and SPL in full scale

Numerical predictions on cavitation and SPL were conducted in full scale using a grid with about 19.5 million cells. Unlike the computation in model scale, free surface was considered as shown in Figure 4. The ship speed and propeller rotation rate was about 25 knots and 79 rpm, respectively. The simulation was first carried out with larger time steps until free surface waves were developed and became nearly steady; the time step was then gradually reduced to values suitable to resolve propeller rotation and unsteady cavitation (2.5E-4s). Figure 5 indicates the computed cavitation pattern in full scale. The transient data obtained by CFD is transferred and used in FW-H formulation as the acoustic solver. In this study, FW-H embedded in STAR-CCM+ was also used. Monopole and dipole are included while quadrupole is excluded in the numerical simulation. The impermeable surfaces were selected as shown in Figure 6.

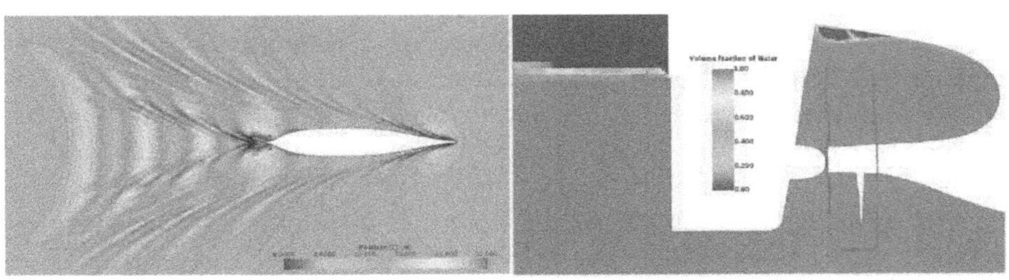

Figure 4 Predicted free surface waves

Figure 5 Cavitation pattern

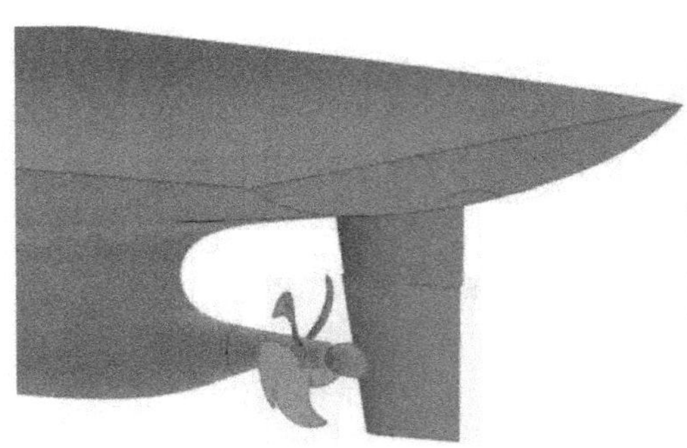

Figure 6 Defined impermeable surfaces

For the validation of numerical analysis, only the computational result of propeller radiated noise is compared with the result from the full scale measurement because cavitation observation and measurement of hull pressure fluctuations were not carried out. The obtained data from computation is corrected to an equivalent 1 Hz bandwidth and 1 m source spectrum level. ITTC suggests that the sound pressure level be corrected to a standard measuring distance of 1m using the following relationship: $SPL = SPL_1 + 20\log(r)$, where SPL and SPL_1 are the sound pressure level in 1 Hz band in dB relative to 1 μPa at 1 m and 1 Hz band in dB relative to 1 μPa, respectively; r is the distance of the location of the receiver distance from the propeller center. This correction is adopted in this study.

Figure 7 represents the comparison of source spectrum level between full scale measurement and numerical analysis plotted in narrowband and 1/3 octave band. BPF is clearly observed in the computational result from the narrowband plot. The SPL from numerical analysis is generally matched with that from full scale measurement, however, there are some discrepancies including 1st BPF which is over estimated about 11dB. Additional peaks excluding BPF shown in full scale measurement results are regarded to be due to the excitation force from engine, diesel generator and etc. In 1/3 octave band plot, SPL from the numerical computation is generally in a good agreement with the result of full scale measurement but slightly underestimated. This underestimation can be related with the lack of machinery noise, the lack of predicted tip vortex cavitation, the exclusion of quadrupole source and etc.

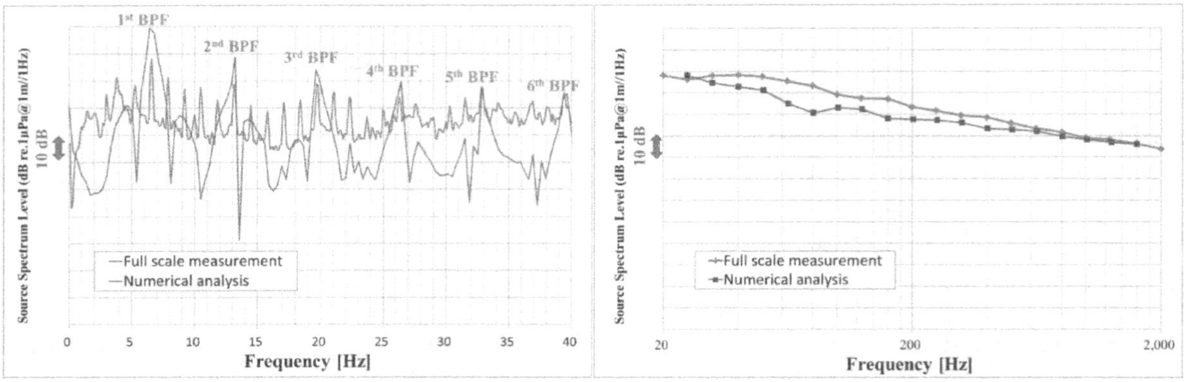

Figure 7 Source Spectrum Level (left: Narrowband, right: 1/3 octave band)

Conclusions

In spite of some limitations in capturing the growth and collapse of cavitation considering that Schnerr-Sauer model is used in this study, reasonable sheet cavitation and hull pressure fluctuation are obtained from the numerical analysis in model scale. It can provide useful information at propeller design stage. For the prediction of SPL in full scale, a hybrid approach based on CFD and FW-H is adopted. URN is predicted directly by flow analysis and FW-H in full scale. Although SPL from numerical analysis is somewhat underestimated in comparison with the result of full scale measurement, it may be the result of having the propeller and rudder as the only two sources of noise in the simulation while in reality there will be contribution from some other sources as well. We could find a possibility to obtain a reasonable SPL in full scale within practical use of commercial CFD software. However, it is necessary to perform additional works on the broadband noise source and other case studies in the future for making the current approach on prediction of URN more reliable. This approach is regarded as almost unprecedented in respect that URN is predicted directly by flow analysis and FW-H in full scale. It is also considered to be a challenging task in consideration of the fact we apply it to very large commercial ship and compare the computation results with those from full scale measurement.

Acknowledgement

This work was supported by the Ministry of Trade, Industry and Energy (MOTIE) (project code: 10045337, "Development of fundamental technology for ship propeller noise and key te hnology or noise redu tio n design").

References

[1] International Maritime Organization, 201, MEPC.1/Circ.833

[2] Paik, K.J., Park, H.,G., Seo, J. (2013). *RANS simulation of cavitation and hull pressure fluctuation for marine propeller operating behind-hull condition*. Int. J. Nav. Archit. Ocean Eng. 5:502~512.

[3] Keita Fujiyama, (2015). *Numerical simulation of ship hull pressure fluctuation induced by cavitation on propeller with capturing the tip vortex*. Proceeding of 4th International Symposium on Marine Propulsors. 649~655.

[4] Artur K. Lidtke. et al., (2015). *Use of Acoustic Analogy for marine propeller noise characterization,* Proceeding of 4th International Symposium on Marine Propulsors. 231~239.

© 2018 ASME

Numerical investigation of cavitation scale effects by bubble acoustic

[1]Patrick Schiller,* [1]Moustafa Abdel-Maksoud

[1]*Institute for Fluid Dynamics and Ship Theory, Hamburg University of Technology (TUHH), Hamburg, Germany*

Abstract

This paper focuses on a number of issues related to scale effects based on acoustic cavitation inception. The acoustic pressure emitted by flow bubbles is used to numerically detect the acoustic cavitation inception. The acoustic pressure, the motion as well as the bubble size are calculated by a Lagrange method, which is implemented in the in-house RANS code FreSCo$^+$. In a numerical study on a 2D hydrofoil, various aspects related to water quality, such as e.g. bubble size and bubble concentration, are systematically investigated and analyzed. The inflow velocity and dimension of the investigated body are varied. For a precise definition of acoustic cavitation inception two different criteria are used. They are based on an event rate and an acoustic pressure, which have to be exceeded. The results of the numerical study show that the known cavitation scale effects have a significant influence on the detected cavitation inception number. Based on the analysis of the simulation results for the velocity and body size variation, an extrapolation method is proposed and validated, which is also presented in the paper.

Keywords: acoustic cavitation inception, bubble acoustic, Lagrange

Introduction

It is well known from systematic investigations as well as from practice that scale effects on cavitation phenomena deviate from classical similarity laws, such as the cavitation number. This means the extrapolation from model scale to full scale according to the Eulerian law of similarity will not always lead to reasonable results. For example, developed cavitation might be observed on parts of the prototype, whereas no or incipient cavitation is detected in model scale at the same cavitation number. Further, noticeable variations of the cavitation pattern on geometrically similar bodies can be observed if they are tested in model scale at the same cavitation number but at different inflow velocities. The viscosity and the intensity of turbulence of the fluid can have considerable influences on the cavitation inception and extent. These effects are often categorized as 'scale effects'.

Scale effects are often sensitive to variations in water quality, e.g. concentration and size of nuclei in the water, tensile strength, and Reynolds number (model scale, flow velocity, viscosity, and intensity of turbulence) [1, 2]. Quantifying these scale effects is important for predicting the prototype cavitation behavior based on model tests, such as water pumps and ship propellers.

It has been shown that the in-house RANS code FreSCo$^+$ implemented Euler-Lagrange cavitation model is able to predict scale effects regarding water quality, velocity and model dimensions in general (developed cavitation) [3, 4]. However, cavitation inception is also an important aspect. Two methods, visually or acoustically, can be used to determine cavitation inception. The visual criterion, mostly used in model tests, is subjective because it based on different perceptions of different observers. Further, in CFD, it is difficult to decide which vapor volume fraction or vapor volume should be used for identifying the inception point. By defining a certain cavitation event rate (events per time unit), a more precise criterion based on the acoustic behavior can be applied. This acoustic criterion can be based on the acoustic pressure emitted by a spherical bubble that arises due to changing bubble volume. The acoustic pressure emitted by a spherical bubble can be calculated via a formula according to Fitzpatrick and Strasberg, which is additionally implemented into the Lagrange code. With the help of two different event rates and peak limits that must be exceeded for the acoustic pressure, an acoustic cavitation inception criterion is defined and different flow scenarios for an 2D NACA66(2)-415 a= 0.8 hydrofoil are numerically investigated for the acoustic cavitation inception point.

In addition to determining a more precise definition of the cavitation inception criterion by bubble acoustics, this method has the advantage of less computational effort because the negligible amount of vapor the bubbles contain means there is no need to carry out a two-way coupling computation between the Euler and Lagrange phase. Thus, a frozen Eulerian flow field can be used for the Lagrangian part, where only the ambient pressure is changed to ensure the correct cavitation number.

* Corresponding Author, Patrick Schiller: patrick.schiller@tu-harburg.de, Web-page: www.tu-harburg.de/fds

Numerical Method

The results of the numerical simulations presented here have been performed with the in-house RANS code FreSCo$^+$ [5], which is jointly developed by the Hamburg Ship Model Basin (HSVA) and the Institute for Fluid Dynamics and Ship Theory (FDS) at Hamburg University of Technology (TUHH). Two-phase flows can be addressed by interface-capturing methods based on the Level-Set or Volume-of-Fluid (VOF) technique. Cavitation can be modelled using either mass-transfer models – i.e. Sauer [6], Zwart et al. [7] - or two-way coupled Euler-Lagrange approaches. The method is parallelized using a domain-decomposition technique based on a Single Program Multiple Data (SPMD) message-passing model. For the Euler-Lagrange simulation, a hybrid MPI/OpenMP parallelization strategy is applied. The dynamic decomposition of the particles follows the Eulerian background phase.

The implemented Euler-Lagrange cavitation model is used to calculate the bubbles. In the Euler-Lagrange approach, separate equations for calculating bubble size and motion are solved for each of the bubbles. More details have already been presented in [3, 4, 8].

The acoustic pressure and the radiated noise of a bubble is the consequence of the temporal variation of the bubble volume $V_b(t)$ due to vaporization and condensation of the fluid. These pressure fluctuations in the surrounding liquid are the reason for the cavitation noise. Most of the sound emitted by cavitating bubbles occurs during the collapse phase, which is characterized by a smaller time-scale than the growing phase (see e.g. Ross [9]). Instantaneously emitted acoustic pressure p_{ac} is related to the volume acceleration of a vapor–gas bubble:

$$p_{ac} = \frac{\rho \cdot d^2 V_b}{4\pi L\, dt^2} \quad with \quad V_b = \frac{4}{3}\pi R^3 \,. \tag{1}$$

Length L is the distance between the bubble and the observer position. When $L \gg R$, this equation becomes the expression for the acoustic pressure $p_{ac}(t)$ of Fitzpatrick and Strasberg [10] after introduction of the delayed time t' due to the finite speed of sound a, which is

$$t' = t - L/a \,. \tag{2}$$

The time t' is the time when the bubble event takes place. The information is transported to an observer with the speed a over the distance L between the observer and the instantaneous position of the bubble. The acoustic pressure reads then

$$p_{ac}(t) = \frac{R(t')}{L}\rho\left(R(t')\cdot\ddot{R}(t') + 2\dot{R}^2(t')\right). \tag{3}$$

One approach for calculating the volume acceleration is the classical theory of Rayleigh, expressed by the Rayleigh–Plesset equation. This information is already available in Lagrange method for the bubbles. So, the acoustic pressure can be directly calculated.

To evaluate the acoustic pressure for an arbitrary recording position, the acoustic pressure is calculated for each bubble and the values are saved in a timeline, where the delay time t' is considered for each individual bubble. For the traveling speed a, the speed of sound in water, a value of 1500 m/s is used. A damping of the acoustic pressure signal (aside from the loss by the divergence) on its way to the observing point is not considered because the distances between the bubbles and the observing points are too small in the cases considered in this work. Therefore, the damping is negligibly small.

Investigated Case and Numerical Setup

The numerical investigations are conducted for a 2D hydrofoil. The section shape is a NACA66(2)-415 a=0.8 hydrofoil with constant chord length of c_0=0.2025m. It was rigorously investigated in the cavitation tunnel of SVA Potsdam during the KonKav I joint project under various flow conditions and already used for numerical investigations in terms of the Euler-Lagrange cavitation model (see [1, 2]). The 2D hydrofoil was selected for this study because it requires less computational effort in comparison to a more complicated geometry such as a ship propeller, and in turn, a wider range of parameters can be varied and their influences can be analyzed.

The simulation grid used is a quasi 2D-structured grid with only one cell in span wise direction and generated with HEXPRESS™ (see Figure 1(a)). The grid is scaled to achieve different foil sizes and the height of the first boundary cell for the foil surface is selected in a way that it has a Y+-value ≤ 1 in the whole investigated parameter range for flow velocity and foil size. The top and the bottom of the tunnel domain are treated as slip walls and at the sides of the domain symmetry boundary conditions are used. The inflow velocity is given at the inlet, and a pressure boundary condition is applied at the outlet. A reference or a baseline case is defined for the numerical study. The baseline case has an inflow velocity of v_0=5.5m/s and a chord length of c_0=0.2025m at an angle of attack of AoA=7.0°.

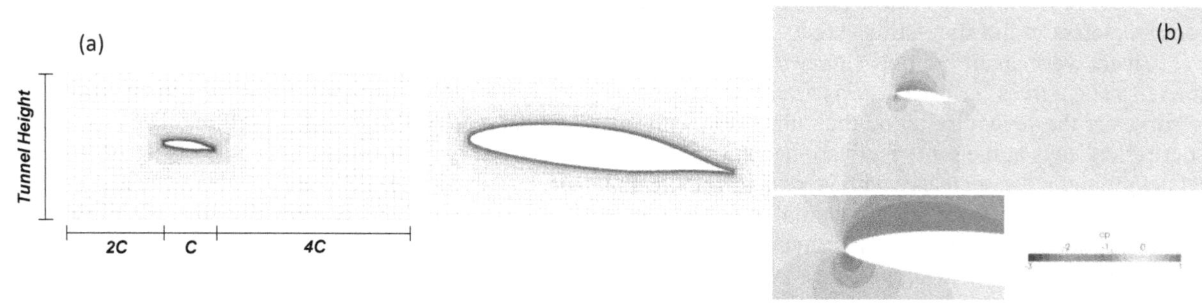

Figure 1: Details of the employed computational mesh (a) and normalized pressure distribution (b) around the hydrofoil at AoA=7°.

For this investigation of acoustic cavitation inception, a frozen flow field is used and the two-way coupling is disabled because there is no need for a vapor-volume transfer from the Lagrange to the Eulerian flow field for conditions with almost no vapor-volume. This saves a lot of computational effort.

The bubbles are released in front of the hydrofoil, as illustrated in Figure 2. The release area is located 0.5 chord length in front of the foil and has a height so that the suction as well as the pressure side is adequately covered with bubbles. To reduce the computational effort, the calculation domain is not completely filled with bubbles and the bubbles are tracked only until 75% of the chord length (see Figure 2). The bubble concentration is kept constant for all simulations by means of a constant number of bubbles per fluid volume. The concentration is n_0=6.22/cm² in this study and the bubble diameter for the baseline case is d_{b0}=90µm. The bubble release area and the number of bubbles released in each timestep is adjusted if necessary when the chord length or inflow velocity is changed to keep the concentration constant. The bubbles are released with a delta t of 10^{-4}s, which corresponds to the Euler timestep for a two-way simulation, but tracked with the standard Lagrange timestep of 10^{-7}s.

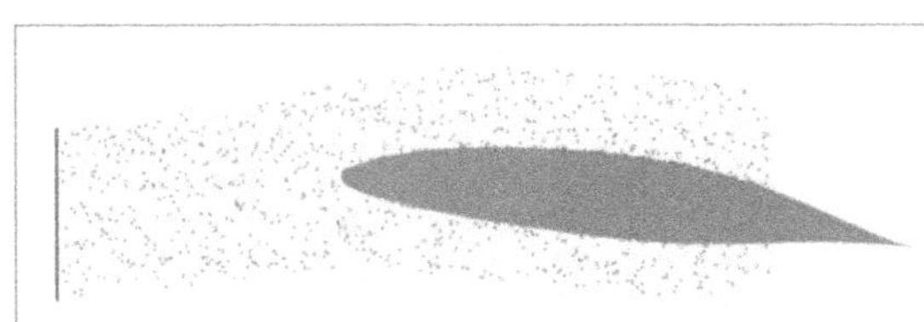

Figure 2: Bubble release position (red line) relative to the hydrofoil and area filed with bubbles (green points).

For detecting the acoustic cavitation inception point, two criteria are used with the following event rates and peak limits:

1. Criterion: 10peaks/s over 10Pa
2. Criterion: 50peaks/s over 40Pa.

They are chosen from a systematic variation of peak/s (from 10 to 50 in steps of 10) and different limits (from 10Pa to 50Pa in steps of 10Pa) and found to be representative for a more or less stringent criterion.

A total simulation time of 10s is applied in order to ensure a sufficient statistical scope for the acoustic timeline. This means that 100 and 500 Peaks above the corresponding peak limit must be counted in this time, respectively, in order to fulfill the above criteria. As a recording position of the acoustic signal, a position directly above the leading edge of the hydrofoil at a distance of 1m is chosen. This position is kept fixed for all investigated cases. The definition of the cavitation number sigma used in this work is:

$$\sigma = (p_\infty - p_v)/(0.5 \cdot \rho \cdot v^2) \tag{4}$$

Results

Figure 1(b) shows the normalized pressure distribution for the baseline case. The comparison of the simulation result with the SVA measurements show a good agreement for the forces acting on the foil as well as for velocity profiles at 6 different positions before, along and behind the hydrofoil. These results are not presented here in detail but indicate that the accuracy of the calculated flow field is sufficient for the acoustic cavitation inception study. In the following, the results of the systematic investigation with respect to water quality (bubble size and number), flow velocity and foil size (chord length) are presented.

In a first step, the influence of different bubble diameters (d_b) on the acoustic cavitation inception is studied because the bubble diameter is an important parameter with respect to the water quality. The investigated bubble diameters range from 10 to 360μm and are made dimensionless by the bubble dimeter of the baseline case (d_{b0}). The other parameters in the simulations are kept constant and correspond to the baseline case. The results in Figure 3(a) show that for very small bubbles sigma increases drastically and then turns into a constant slope for larger bubble diameters. The curves for both criteria are close together and show the same trend until a bubble diameter of 120μm. The curve for the second criterion then turns into a constant level at a bubble diameter of 360μm. The curve for the first criterion stays at the same inclination and a strong increasing trend is visible from 240μm. This is due to the fact that background noise increases with increasing bubble diameter. Larger bubbles generate a higher acoustical pressure amplitude even when there is only a small and slow change in the bubble diameter. But the sum of the acoustical signal of all tracked bubbles can then reach the specified criterion, which is here the case for first one. A higher-pressure level is required to suppress this phenomenon, which then automatically leads to higher cavitation inception numbers, as one can see here.

An additional property with respect to water quality is bubble concentration. It is found that the number of peak counts that exceeds a certain limit either increases or decreases linearly with the change of the concentration n (Figure 3(b)), which is a quite plausible phenomenon. Due to the fact that the development of the peak count number is not linear over the sigma value, the change of the acoustic cavitation inception point is also nonlinear. In the simulation, the ambient pressure must be adjusted until the first or the second defined criterion is fulfilled. Changing the pressure leads to a different cavitation number. The change depends on the characteristic curve of the peak counts over the cavitation number and when it reaches the defined criterion. The increase of this peak count number is low for higher sigma values und starts to increase drastically at a certain point when sigma is reduced. This point is close to the point of cavitation inception found by the second criterion. Particularly at low concentrations there can be a remarkable difference when the intersection of the characteristic curve with the inception criterion lies in the flat rising part of the curve.

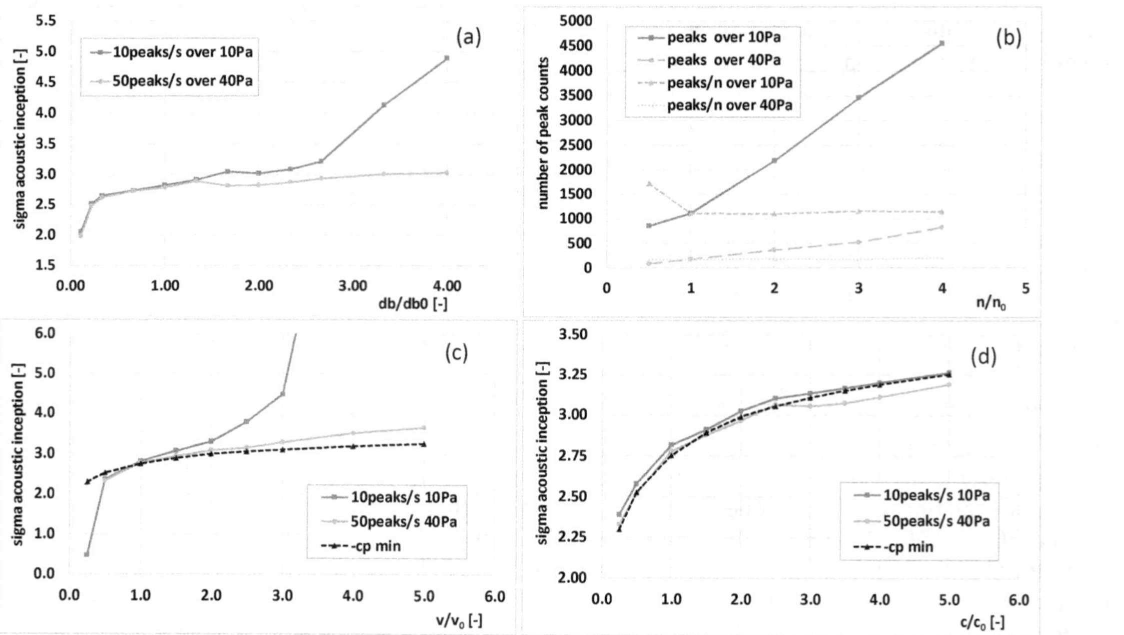

Figure 3: Influence of bubble concentration on number of peak counts (b) and development of the acoustic cavitation inception number for bubble diameter variation (a), inflow velocity variation (c) and chord length variation (d).

To investigate the known velocity scale effect, the inflow velocity is varied in a range from 1.375 to 27m/s and the other parameters are kept in line to the baseline case. The result is given in Figure 3(c), where the development of the acoustic cavitation inception point is plotted over the velocity ratio with respect to the velocity of the baseline case. The expected trend, lower acoustic cavitation inception numbers at low flow velocities and vice versa, is represented correctly. But the development of the curve for the two criteria is different. Due to the increasing background noise, the curve for the first criterion increases rapidly for velocities above 11m/s and 16.5m/s, respectively. This shows that the chosen value for this criterion is not convenient for high velocities. But it works for low velocities where the second criterion cannot be reached. The development of the curve for the second criterion shows a constant increasing trend with decreasing gradient. It seems that there might be a certain limit for the cavitation inception number at very high velocities.

The model scale of the cavitating object is another important influence factor for scale effects. The results of the detected acoustic cavitation inception numbers for different foil chord lengths (variation between 0.0506 to 1.012m) are depicted. Figure 3(d) shows the dependency of the calculated acoustic cavitation inception number on the chord length. Again, the chord length is made dimensionless by the corresponding value of the baseline case. In general, cavitation inception numbers detected by both criteria increase as the inflow velocity increases. This observed dependency is also in line with the trend reported by Keller [2]. In fact, the acoustic cavitation inception number increases with increasing the chord length. However, the increase of the acoustic cavitation inception number is getting smaller with rising foil size. The curves of cavitation inception numbers detected by both criteria show a strong similarity to the calculated minimum pressure (cp_{min}) coefficient on the foil. However, a nearly constant difference between the curves of both criteria is visible (see Figure 3(d)). The difference becomes larger for ratios above 2.5. Further investigations are needed to clarify the reasons for this small decrease in the cavitation inception number estimated by the second criterion. It could be an effect of the statistical reliability of the simulations. The characteristic increase of the cavitation inception number when increasing the foil size is reasonable due to the fact that the area captured by low pressure regime becomes greater with larger object size. A doubling of the chord length results in four times greater low-pressure regime and thus more bubbles are captured in this area. It is remarkable that there seems to be a constant increase of the sigma by a value of 0.2 - 0.25 when the chord length is doubled.

Extrapolation Method

The aim of the developed extrapolation method is to predict the acoustic cavitation inception point for other flow velocities and object sizes or any combinations of it, when it is known only for model-scale configurations. It is based on the curves of the second criterion and the following assumption:

$$\sigma_{in} = -cp_{min} + \Delta\sigma_c + \Delta\sigma_v, \tag{5}$$

where $\Delta\sigma_c$ and $\Delta\sigma_v$ are the deviations from $-cp_{min}$ with respect to the body size and the flow velocity, respectively. The minimal pressure coefficient is used because cavitation mainly depends on the minimal pressure in a flow.

The minimal pressure along a foil depends on the Reynolds number and therefore on the thickness of the boundary layer. With increasing the Reynolds number, the impact of the frictional part on the total resistance decreases and the velocity in the boundary layer increases and thus the minimal pressure. It is found on the basis of the resulting friction resistance coefficients cD_f based on the simulation results, that in the present case the following relationship between cD_f and cp_{min} exists (compare Figure 4(a)):

$$\frac{cp_{min}}{cp_{min\,0}} \cong 1 + \frac{cD_{f\,0} - cD_f}{cD_{f\,0}} = 1 + \frac{cD_{f\,0}}{cD_{f\,0}} - \frac{cD_f}{cD_{f\,0}} = 2 - \frac{cD_f}{cD_{f\,0}}. \tag{6}$$

Values with the index 0 refer to a known starting simulation condition and without the index 0 to the new condition. However, this approach is only applicable if the friction part of the drag is less than ≈ 50% of the total resistance. The friction resistance coefficient cD_f can be determined by known friction lines as a function of the Re number. In this case, the cf value from ITTC 1978 friction line fits quite well, as can be seen in Figure 4(b). With this approach, cp_{min} of a known configuration can be converted to another one with the help of the Reynolds number and the ITTC 1978 friction line.

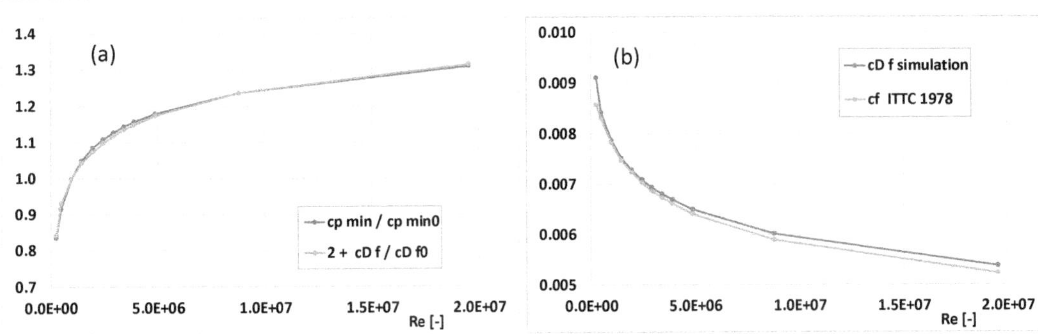

Figure 4: Development of $cp_{min}/cp_{min\,0}$ and $2 - cD_f/cD_{f\,0}$ (a) and frictional resistance coefficient (b) over Reynold number.

As it can be seen from Figure 3(d), sigma follows the minimal pressure coefficient $-cp_{min}$ quite well when the chord length is changed. Thus, $\Delta\sigma_c$ is a constant value that also depends on bubble diameter. The presented results in Figure 3(d) are obtained for a bubble diameter of 90µm.

For considering the influence of the variation of the inflow velocity, an additional increasing difference $\Delta\sigma_v$ to $-cp_{min}$ must be taken into account (see Figure 3(c)). Figure 5 shows that there is a linear relationship between $\Delta\sigma_v = \sigma_{in} - (-cp_{min})$ and the pressure difference of the bubble internal pressure p_{g0} and the minimal pressure p_{min} in a certain range, which corresponds to a velocity ratio v/v_0 from 1 to 4 and covers the common velocity ratios between model and full scale. The linear relationship can be expressed by equation (7) and the term $p_{g0} - p_{min}$ by equation (8). Herein S is the surface tension and R_0 the initial bubble radius. The variables a and b can be determined by evaluating the results of from model tests or CFD calculations at two different inflow velocities, where all other parameters are known. Then $\Delta\sigma_v$ can be calculated for a different velocity.

$$\Delta\sigma_v = a \cdot \frac{(p_{g0} - p_{min})}{p_{min0}} + b \quad with\ a = 1.07 \cdot 10^{-3}\ and\ b = 3.09 \cdot 10^{-3} \tag{7}$$

$$p_{g0} - p_{min} = -cp_{min} \cdot \frac{\rho}{2} \cdot v^2 - p_v + \frac{2S}{R_0} \tag{8}$$

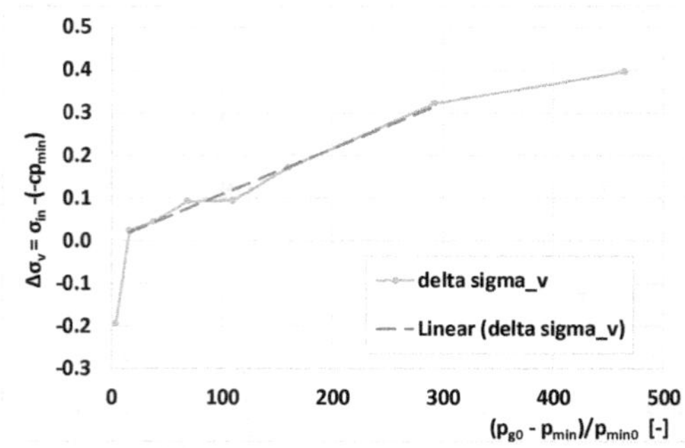

Figure 5: Development of $\Delta\sigma_v$.

In order to validate the extrapolation method, different combinations of inflow velocity and chord length are simulated and the detected acoustic cavitation inception points are compared with the one predicted by the extrapolation method. This is done for an angle of attack of 7° as before and additionally for 12°. The results are given in Table 1 and show a good agreement for both cases with errors less than ±3%. Exceptions are the combinations with a velocity ratio of 0.5 at 7°. But this is not surprising because the method isn't valid for this range.

Case	velocity ratio [-]	size ratio [-]	velocity [m/s]	chord length [m]	cp min [-]		sigma acoustic cavitation inception for		Error [%]
					estimated	from simulation	prediction	50peaks/s over 40Pa	
7deg - 1	0.5	0.5	2.75	0.10125	-2.321	-2.300	2.328	2.098	10.946
7deg - 2	2	0.5	11	0.10125	-2.757	-2.757	2.828	2.829	-0.041
7deg - 3	0.5	2	2.75	0.405	-2.757	-2.757	2.765	2.512	10.058
7deg - 4	2	2	11	0.405	-3.169	-3.191	3.250	3.252	-0.049
7deg - 5	3	3	16.5	0.6075	-3.409	-3.411	3.600	3.518	2.323
7deg - 6	4	5	22	1.0125	-3.630	-3.614	3.988	3.905	2.132
12deg - 1	2	3	11	0.6075	-7.335	-7.179	7.517	7.282	3.233
12deg - 2	4	5	22	1.0125	-8.058	-7.775	8.849	8.691	1.817
12deg - 3	3	1	16.5	0.2025	-6.900	-6.842	7.283	7.409	-1.707
12deg - 4	1	2	5.5	0.405	-6.654	-6.641	6.698	6.684	0.212

Table 1: Comparison of predicted and simulated acoustic cavitation inception point for different combinations of flow velocity and chord length at AoA=7° and 12°.

Conclusion

The study and results presented in this paper show that with the Lagrange model and its ability to calculate the acoustic pressure emitted by a bubble, cavitation scale effects can be calculated with the help of the acoustic cavitation inception point. Two inception criteria are proposed that are based on an event rate and peak limits.

The results obtained with the chosen acoustic criteria for cavitation inception basically show a similar trend. Many aspects that can have a significant influence on the cavitation inception are investigated. The results show a linear relationship between the event rate and bubble concentration. By increasing the bubble diameter, the inception cavitation number increases rapidly at very small bubble diameters and the increase continuously declines as bubble diameters get larger. This trend can also be observed when the chord length and/or the inflow velocity is increased. In general, the results of the numerical investigations show the known trends regarding scale effects, e.g. lower inception cavitation numbers for lower flow velocities or object sizes.

Further, it becomes clear that different inception criteria can lead to different results, especially if a high background noise level is present. Based on the results for velocity and size scale effects, an extrapolation method is proposed to predict acoustic cavitation inception for any combinations of these two parameters. The validation on two cases reveals a good agreement between the extrapolated results and the simulation. But it should be mentioned that the described extrapolation method is only applied for a 2D case and the applicability to other flow objects and water qualities is still to be investigated.

References

[1] Keller, A. P., Rott, H. K., Stoffel, B., Striedinger, R. (1999). *Maßstabseffekte bei der Strömungskavitation*. Forschung im Ingenieurwesen 65. Springer-Verlag.

[2] Keller, A. P. (2001). *Cavitation scale effects empirically found relations and the correlation of cavitation number and hydrodynamic coefficients.* CAV2001.

[3] Maquil, T., Yakubov, S., Cankurt, B., Schiller, P., Abdel-Maksoud, M., Rung, T. (2011) *Simulation of Water-Quality Effects for Cavitating Engineering Flows*. WIMRC 3rd International Cavitation Forum. University of Warwick. UK.

[4] Schiller, P., Abdel-Maksoud, M., Rung, T., Cankurt, B., Maquil, T., Yakubov, S. (2012) *Investigation on Cavitation Scales Effects with different Cavitation Models.* 8th International Symposium on Cavitation. Singapore.

[5] Rung, T., Wöckner, K., Manzke, M., Stück, A., Brunswig, J., Ulrich, C. (2009). *Challenges and Perspectives for Maritime CFD Applications.* Jahrbuch der Schiffbautechnischen Gesellschaft (Vol. 103).

[6] Sauer, J. (2000). *Instationär kavitierende Strömungen - Ein neues Modell, basierend auf Front Capturing (VoF) und Blasendynamik*. PhD thesis, Universität Karlsruhe.

[7] Zwart, P.J., Gerber, A.G., Belamri, T. (2004). *A two-phase flow model for predicting cavitation dynamics*. In ICMF 2004 International Conference on Multiphase Flow. Number 152. Yokohama. Japan.

[8] Yakubov, S., Cankurt, B., Abdel-Maksoud, M., Rung, T. (2012). *Hybrid MPI/OpenMP parallelization of an Euler-Lagrange approach to cavitation modelling*. Computers and Fluids. Doi:10.1016/j.compfluid.2012.01.020.

[9] Ross, D. (1976). *Mechanics of Underwater Noise.* Pergamon Press. New York, Oxford, Frankfurt, Paris.

[10] Fitzpatrick, N., Strasberg, M. (1958). *Hydrodynamic sources of sound.* In: 2ndSymposium on Naval Hydrodynamics. pp. 201–205.

© 2018 ASME

Numerical investigation of cavitation-regimes in a converging-diverging nozzle

[1]Polina Gorkh, [1]Steffen J. Schmidt, and [1]Nikolaus A. Adams

[1]*Institute of Aerodynamics and Fluid Mechanics, Department of Mechanical Engineering, Technical University of Munich*

Abstract

In the present study the authors analyze cavitation dynamics, instabilities and detachment mechanisms in an axisymmetric converging-diverging nozzle. Numerical simulation is used to investigate a recent experiment. Following the experimental setup, three operating points are simulated to identify two different cloud detachment mechanisms: re-entrant jet and condensation shock. A homogeneous mixture model is applied to model cavitating two-phase flows. Pure liquid and liquid-vapor mixtures are both treated as fully compressible substances to enable the computation of propagating pressure waves into the liquid bulk. Because the study focuses on inertia-dominated mechanisms, the flow is modeled as inviscid.

Our numerical simulation predicts a slightly different cavitation behavior than the one published for the reference experiment. We observe a high-frequency shedding that generates vapor pockets inside the channel. These vapor pockets partially collapse but frequently they integrate into a large coherent structure. Once a coherent vapor cloud reaches a sufficient length, it detaches and is advected downstream. As a result, the total vapor volume fraction shows a low frequency oscillation which is about one order slower than the shedding frequency. Two detachment mechanisms are present during one cycle: the re-entrant jet for shorter cavities and the condensation shock for longer cavities. The maximum cavity length is not constant but features significant cycle-to-cycle variation. Even a slight difference in the cavitation number leads to a difference in the detachment mechanisms: the lower cavitation number leads to a bubbly shock as the dominant detachment mechanism, while for a higher cavitation number shock-driven detachment occurs only once per cycle.

Keywords: cavitation, multi-phase flow, homogeneous mixture, nozzle flow

Introduction

In a broad range of fluidic applications, such as hydraulic machinery or diesel injection systems, understanding and assessment of transient cavity dynamics play a key role. In liquids, cavitation occurs when the pressure drops beneath a certain level. It can decrease machinery performance, cause pressure oscillations and, moreover, may lead to material erosion during the re-condensation process. Thus, prediction and investigation of various cavitation regimes are important.

Two main mechanisms cause the periodic cloud-detachment process: the classical re-entrant jet as described i.e. in [1], and condensation shock waves, propagating through a two-phase medium, as recently investigated in [2]. The latter can appear because the speed of sound within a two-phase mixture is significantly lower than within any of the pure states. Recent experimental observations of condensation shocks resulting in periodic shedding have been published for the case of a sharp-edged wedge [3]. In [4], the authors performed a simulation, proving that, for some boundary conditions, cloud detachment is dominated by the periodic occurrence of a bubbly shock wave.

In another recent study [5], the authors conducted flow experiments in a converging-diverging axisymmetric nozzle. Their findings are used in this work as the experimental reference. In [5], the authors applied high-speed imaging techniques to investigate three test cases with decreasing cavitation numbers. Depending on the cavitation number, they observe three different processes: periodic shedding driven by the re-entrant jet, transitional aperiodic behavior, and periodic shedding driven by the condensation shock wave.

The present study provides a numerical simulation of the convergent-divergent nozzle flow, accompanied by a further analysis of transient cavity dynamics and cavitation patterns. The numerical simulation reproduces the conditions of the reference experiment at all three operating points in order to assess different shedding mechanisms. Quantitative measurements of vapor volume frequencies, shedding frequencies, and the cavity length are compared to experimental data. In comparison with the wedge experiment, where the flow can be captured from the top or from the side, high-speed imaging of the axisymmetric flow lacks this opportunity. In contrast, simulation results allow to analyze complex 3-D flow features inside the axisymmetric nozzle.

Numerical Method

A density-based Finite Volume Method (FVM) is used in the simulations. A homogeneous mixture model [6] is applied to model the mixture region between liquid water and water vapor. All phases are modeled as being compressible to enable propagation of the pressure waves into the liquid bulk. Thermodynamic equilibrium is assumed; water is modeled as a barotropic fluid. Phase change from pure liquid to saturated mixture is assumed to follow an equilibrium isentrope. The study is focused on inertia-dominated mechanisms, so the flow is modeled as inviscid. Solved and non-condensable gas is neglected.

The governing equations are the compressible Euler equations in conservative form. These are discretized on a 4-cell stencil for computational efficiency on structured body-fitting grids. A 4-stage 2-nd order Runge-Kutta method is used. Pressure dependency of density is described by modified Tait equation for pure liquid state [7] and is obtained by integrating the equilibrium speed of sound $c^2 = \left.\frac{\partial p}{\partial \rho}\right|_s$. The vapor volume fraction α is computed by $\alpha = \left(\rho - \rho_{l,sat}(T)\right)/\left(\rho - \rho_{v,sat}(T)\right)$, where $\rho_{l,sat}$ and $\rho_{v,sat}$ are saturation densities of liquid and vapor phases, respectively.

Numerical Setup

The baseline experimental facility is described in [5] in detail. A schematic of the test section is shown in Figure 1 - all dimensions are given in mm. Flow direction is from left to right. The convergence and divergence angles are 36° and 16°, respectively. The diameter of the pipe section is 50 mm. The throttle diameter is 16,67 mm and the diameter of the exit is 50 mm. This leads to an area ratio of 1:9. The authors who performed the experiments kindly provided mass flux data at the inlet, from which uniform inflow velocities where computed and used as boundary conditions for the simulations. The inflow velocities, as well as the outlet pressure and the temperature in the domain (all measured in the experiments and used for our simulation) are summarized in Table 1 for three cases: re-entrant jet cavities detachment mechanism (Case 1), transition (Case 2), and condensation shock-induced cavity detachment mechanism (Case 3) [5].

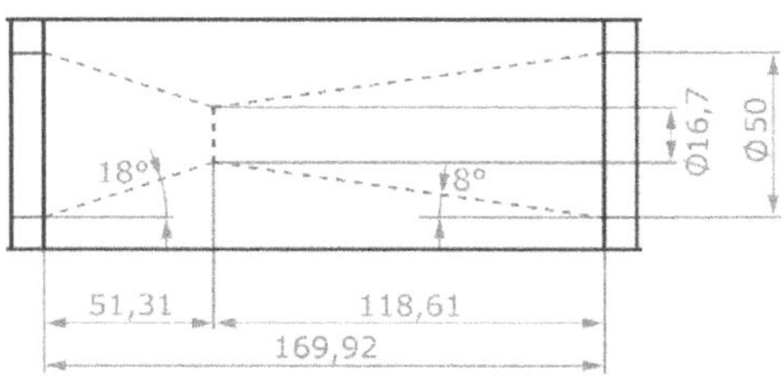

Figure 1: Test section geometry.

	$u_{in}, {}^{m}/_{s}$	P_{out}, Pa	T, K
(1) re-entrant jet	1,479	80580	287,66
(2) transition	1,578	80050	288,97
(3) condensation shock	1,525	25701	292,28

Table 1: Boundary and initial conditions.

In the experiments, pipe sections are added upstream and downstream of the test section. For the simulation, a pipe section ($l = 0.50\ m$) is added in front of the test section and a second one ($l = 0.83\ m$ for Cases 1 and 2 and $l = 1.08\ m$ for Case 3) is added to the end of the test section (see Figure 2(a)). The computational domain for the latter case is longer, because the predicted cavitation length is larger than for the two other cases. For the same reason, the grid in the diverging region for the latter case is finer – see Figures 2(b) and 2(c) for a comparison of the meshes. Grid cells decrease in height toward the walls as water vapor is experimentally predicted to appear mainly close to the walls. The resulting cross-section topology along the whole domain can be seen in Figure 2(d).

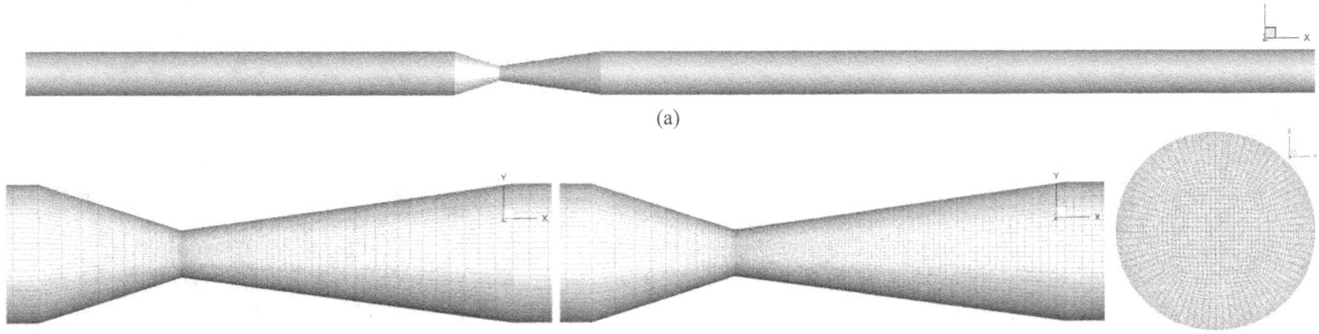

(a) (b) (c) (d)

Figure 2: (a): Overview of the entire computational domain for re-entrant jet and transition cases; (b), (c): numerical grid in the vicinity of the throat, (b): for the re-entrant jet and transition cases, (c): for the condensation shock case; (d): grid cross-section.

To speed up the computations, grid sequencing was used: results from a coarser grid level are used as restart data for the finer grid level. In order to ensure suitable initial conditions, conservative interpolation techniques are applied. Mesh sizes for the final grid level are 280k cells for Case 1 and 2, and 400k cells for Case 3. Simulations on coarse grids were performed to establish statistically converged pressure and velocity distributions.

Results and Discussion

Several definitions of the cavitation number can be found in the literature. The definition $\sigma = (p - p_{sat})/\left(\frac{1}{2}\rho u^2\right)$ is

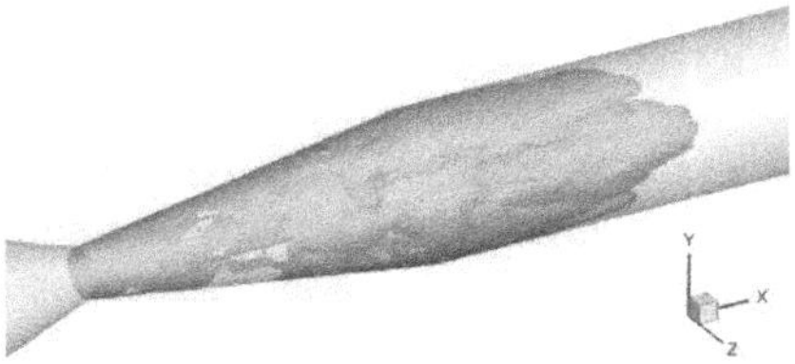

Figure 3: Instantaneous vapor volume fraction for case 3: vapor propagates into the pipe.

case	$P_{in,exp}, Pa$	$P_{in,sim}, Pa$
1	100277.6	130259
2	120469.9	145924
3	110383.6	124379

Table 2: Pressure at the inlet, measured in the reference experiment and in the simulation.

f_{exp}, Hz	L_{exp}, mm	f_{sim}, Hz	L_{sim}, mm
298.6586	15.9	251.1	13 - 37
101.1151	23.1	198.0	16 - 45

Table 3: Comparison of shedding frequencies and maximum cavity length in reference experiments and present simulations.

commonly used [8]. Here, p and u are pressure and velocity at the reference point, respectively (usually upstream of the cavitating region). However, for venturi flows, $p = p_{out}$ is typically chosen [5]. In this case, density and (uniform) velocity at the inlet, as well as the static pressure at the outlet, affect the cavitation number. The static pressure at the inlet is not taken into account into the cavitation number, although it is one of the most important quantities.

Our simulations predict higher pressures at the inlet than in the reference experiments (see Table 2). This raises the pressure difference between inflow and outflow and increases the driving force, which results in longer cavities and lower shedding frequencies. For Case 3, vapor structures are significantly elongated as compared to the reference experiment. The coherent vapor structures even exceed the divergence region and are propagating into the pipe region (see Figure 3). Up to now, we do not have a profound explanation for this behavior and thus, we excluded Case 3 from this report. In the following, our focus is put on Cases 1 and 2.

Figures 4(a) and 4(d) show the temporal development of the vapor volume fraction for Cases 1 and 2, respectively. An FFT of this data (Figures 4(b) and 4(e)) yields frequencies of 27.9 Hz and 22.0 Hz, which are significantly lower than in the reference experiments. As expected, the frequency is lower for the lower cavitation number (Case 2).

For cases 1 and 2, with an equal number of time steps, Figures 4(c) and 4(f) show an x-t-diagram of axially integrated vapor volume fraction α, computed at the divergent part of the nozzle. As the computational time steps are identical for Cases 1 and 2 (28.3 ns), the durations of the corresponding x-t-diagrams are comparable as well.

Figure 4(a) indicates a single frequency in the "vapor-volume versus time" graph ($f = 27.9\ Hz$, peak frequency shown in Figure 4(b)), which is much lower than the experimental shedding frequency. However, in Figure 4(c) an x-t diagram of the shedding is plotted (in a vertical orientation) covering just about 1.3 periods. Multiple shedding processes are visible – although the integrated vapor volume fraction undergoes just one single oscillation! Figures 4(a) and 4(b), do not show the

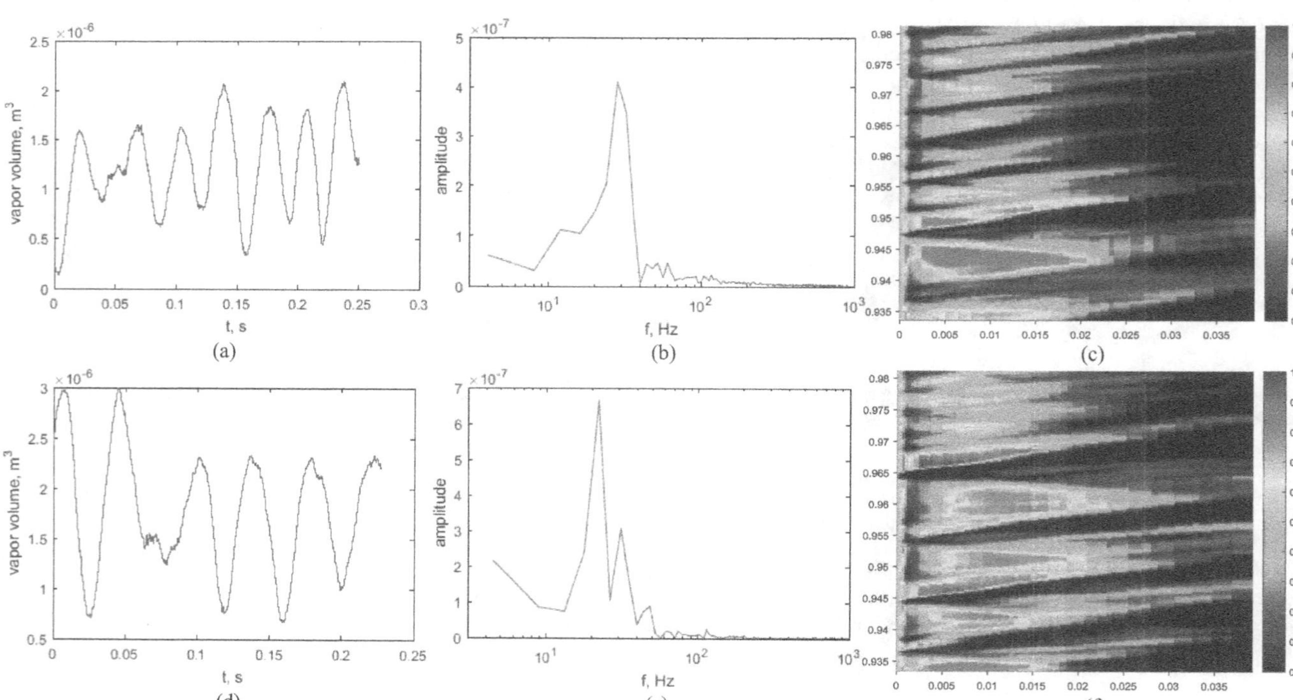

Figure 4: (a), (d): Vapor volume fraction development for Cases 1 and 2, respectively; (b), (e): FFT of vapor volume fraction for Cases 1 and 2, respectively; (c), (f): x-t-diagram of axially integrated vapor volume fraction for Cases 1 and 2, respectively.

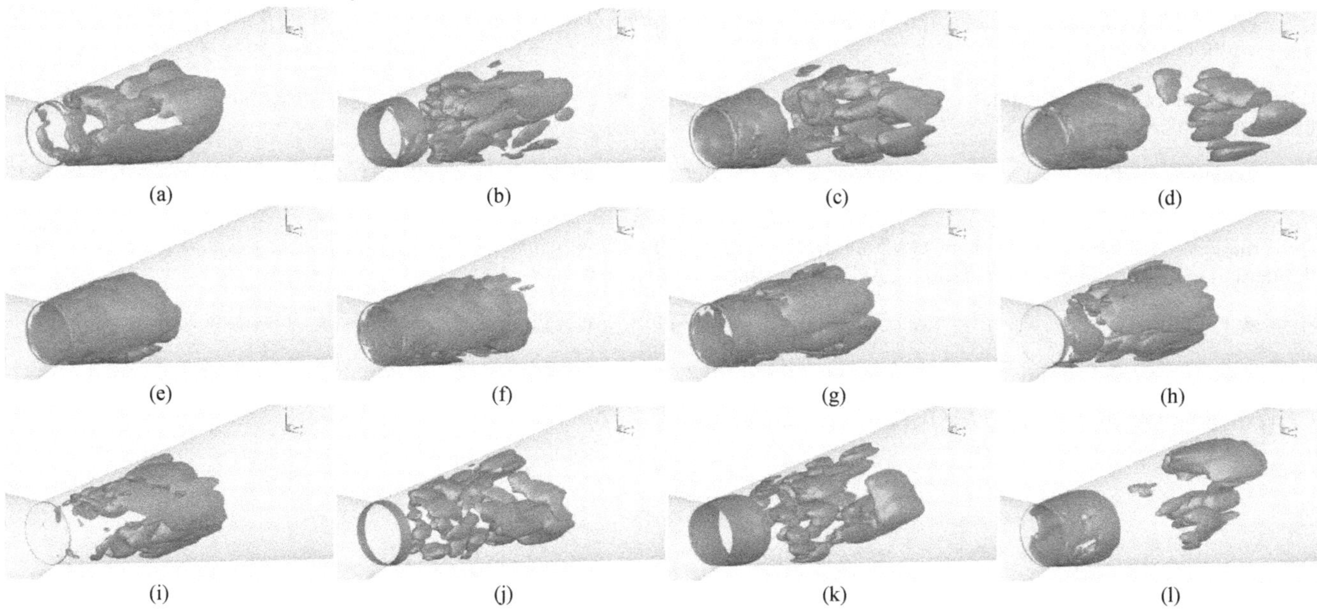

Figure 5: Development and shedding of two vapor clouds for Case 2 with different detachment mechanisms.

shedding frequency but an accumulated low frequency behavior. It entirely hides the shedding process which is visible on the x-t diagram of the cavity length (Figure 4 (c)). The process "hidden" in Figures 4(a) and 4(b) has a much higher frequency and agrees much better to the experimental observations of the shedding frequency.

Our interpretation is as follows: The high frequency shedding generates vapor pockets inside the channel. These pockets sometimes collapse, but mostly they integrate into larger vapor clouds. Once such a vapor cloud reaches a sufficient length, it detaches and is advected downstream. This causes the low frequent oscillation of the integrated vapor volume fraction. For both Cases, 1 and 2, two detachment mechanisms are present during the same cycle: re-entrant jet for shorter cavities (i.e. at $t = 965\ ms$ and $t = 975\ ms$ for Case 1, Figure 4(c)) and condensation shock for longer cavities (i.e. at $t = 943\ ms$ for Case 1, Figure 4(c)). However, the maximum cavity length is not constant but features a strong cycle-to-cycle variation.

In contrast to the reference experiment, both mechanisms, the re-entrant jet and the condensation shock, are simultaneously present in both simulations. Case 1 features just one instance of a condensation shock, while Case 2 shows 3-4 such instances. Using x-t-diagrams, one can count the approximate number of shedding processes and then compute the shedding frequencies. We obtain $251\ Hz$ for Case 1 and $198\ Hz$ for Case 2. These frequencies and cavity lengths are compared to the reference data in Table 3.

Figure 5 demonstrates the formation of cavitation patterns for Case 2 for a time span ranging from $954.5\ ms$ to $968.5\ ms$. This time period includes two shedding processes (see Figure 4(f)): one shock-induced detachment (Figures 5(a)-5(i)), and afterwards one re-entrant jet detachment (Figures 5(i)-5(l)). Figure 5(a) at $t = 954.47\ ms$ shows a previous cavity that has just detached and a new one is starting to form. Its development is shown in Figures 5(b), 5(c), and 5(d), until it reaches a length of $L = 33\ mm$ at $t = 960{,}90\ ms$, see Figure 5(e). After that, a bubbly shock starts to propagate upstream through the vapor cloud (Figures 5(f), 5(g), 5(h)). Detachment of the cloud appears at $964.48\ ms$, reaching a maximum length $L_{max} \approx 45\ mm$, captured in Figure 5(i). Next, the new cavity starts to grow. This process is shown in Figures 5(j) and 5(k). A second shedding process, in this case induced by a re-entrant jet, happens at $t = 958.62\ ms$. It is shown in Figure 5(l).

Acknowledgments

This study is part of a project that has received funding from the European Union's Horizon 2020 research and innovation programme under grant agreement № 642536. The authors gratefully acknowledge the Gauss Centre for Supercomputing e.V. (www.gauss-centre.eu) for funding this project by providing computing time on the GCS Supercomputer SuperMUC at Leibniz Supercomputing Centre (www.lrz.de). The authors further greatly acknowledge Mr. Saad Jahangir and Mr. Willian Hogendoorn from Delft University of Technology for sharing experimental data and for fruitful discussions on the subject.

References

[1] Franc, J.-P., Michel, J.-M. (2005). *Fundamentals of Cavitation.* Springer Science + Business Media.

[2] Ganesh, H. (2015). *Bubbly shock propagation as a cause of sheet to cloud transition of partial cavitation and stationary cavitation bubbles forming on a delta wing vortex.* Ph.D. thesis, University of Michigan.

[3] Ganesh, H., Mäkiharju, S., Ceccio, S. (2016). *Bubbly shock propagation as a mechanism for sheet-to-cloud transition of partial cavities*. Journal of Fluid Mechanics. 802.

[4] Budich, B., Neuner, S., Schmidt, S., Adams, N. (2015). *Numerical investigation of shedding partial cavities over a sharp wedge.* Journal of Physics: Conference Series. 656.

[5] Hogendoorn, W. (2017) *Experimental investigation of cavitation regimes in a converging-diverging nozzle.* Master's Thesis, Delft University of Technology.

[6] Schmidt, S., Thalhamer, M., Schnerr, G. (2009). *Inertia controlled instability and small scale structures of sheet and cloud cavitation.* 7th International Symposium on Cavitation, CAV2009, Michigan, USA.

[7] Sezal, I. (2009). *Compressible dynamics of cavitating 3-D multi-phase flows.* Ph.D Thesis, Technische Universität München.

[8] Brennen, C. (2013). *Cavitation and bubble dynamics*. Cambridge University Press.

*Corresponding Author, Polina Gorkh: polina.gorkh@tum.de

© 2018 ASME

Numerical investigation of the dynamics of pressure loading on a solid boundary from a collapsing cavitation bubble

[1]Prasanta Sarkar*; [1]Giovanni Ghigliotti; [2]Marc Fivel; [1]Jean-Pierre Franc

[1]*Univ. Grenoble Alpes, CNRS, Grenoble INP**, LEGI, 38000 Grenoble, France;* [2]*Univ. Grenoble Alpes, CNRS, Grenoble INP**, SIMaP, 38000 Grenoble, France*

Abstract

The motivation behind this research lies in understanding the physical mechanism of cavitation erosion in compressible liquid flows, with applications in the field of aerospace, hydrodynamics, diesel injectors etc. As a consequence of collapsing vapor cavities in cavitating flow near solid boundaries, high pressure impact loads are generated. These pressure loads are believed to be responsible for the erosive damages on solid surface observed in most applications. For our investigation, the initial geometry is a single vapor bubble near a solid boundary collapsing due to the pressure difference between the bubble and surrounding liquid. The numerical approach employs a simplified homogenous mixture or '*single fluid*' model with barotropic assumption in a fully compressible finite-volume fluid solver. The numerical method is validated against the well-known Rayleigh collapse of a pure *3D* vapour bubble. It is then used for the simulation of a *2D* vapour bubble collapsing in the proximity of a solid boundary placed at a specified distance from the centre of the bubble. The pressure loads are computed from the evolving dynamics of collapsing bubble near a solid boundary which can be used to determine the resulting surface deformation. The developed compressible cavitation solver in the *CFD* code *YALES2* can efficiently model small and large scale cavitating structures in a fully resolved three dimensional flow.

Keywords: cavitation; cavitation erosion; bubble dynamics; fluid-structure interaction

Introduction

The collapse of cavitation structures formed due to the breaking down of the continuous liquid medium under very low pressures, near a solid surface leads to surface damage known as cavitation erosion. The driving mechanism for the cavitating structures is the local decrease in liquid pressure below typically vapor pressure at a given temperature [1]. The cavitating regions consist number of cavitation bubbles due to the rapid growth of initially present air nuclei in the liquid flow. These cavitation bubbles move along the liquid flow until they collapse in the areas of high pressure and disappear. The presence of air nuclei, amount of dissolved gas in the flow are determining factors for the amount and size of cavitation bubbles. The motivation of this research is to numerically investigate the unsteady compressible cavitation bubble behaviour in a continuum liquid flow, with associated compressibility effects inside the flow domain. It will be used to predict accurately the material surface response to these small time-scale, highly unsteady flow characteristics. The vapor bubble collapse is usually characterized by the shrinking of bubble surface, acceleration of liquid flow towards the centre of bubble and shock propagation during the collapse. During the final stages of the collapse, due to the collision of bubble surface and liquid jet at the center, a high impulse shock wave is produced and propagated into the surrounding liquid. The collapse of a cavitation bubble close to a solid boundary is significantly affected by the non-dimensional *stand-off parameter* (γ), defined as the ratio of distance of the bubble center from the solid boundary to the maximum bubble radius. The dynamics of the bubble collapse and its interaction with the solid boundary such as the liquid micro-jet formation, splashing and surface erosion is strongly dependent on the stand-off parameter. Comprehensive description of the effect of the stand-off parameter may be found in the paper of *Philipp* and *Lauterborn* [2]. For an isolated bubble, the collapse is basically symmetric and a shock wave forms near the center of the bubble which propagates into the surrounding liquid. For an asymmetric bubble collapse near a solid boundary, a liquid re-entrant micro-jet pierces through the bubble surface opposite to the solid boundary accelerating one side of the bubble towards the opposite surface. The liquid jet is directed towards the solid boundary and the estimated speed of the micro-jet is usually very high, several hundred meters per second. Multiple shock waves are emitted due to the impact of the liquid jet with the opposite bubble surface and eventually, the solid boundary. The high amplitude, localized pressure impulse at the solid boundary due to the emitted shock and micro-jet results in material surface damage or erosion. The collapse of a pure vapor bubble surrounded by a liquid can be represented as

*Corresponding Author, Prasanta Sarkar: prasanta.sarkar@univ-grenoble-alpes.fr
**Institute of Engineering Univ. Grenoble Alpes

as figure 1 where P_{bubble} is the pressure inside the bubble and P_{liquid} is the surrounding liquid pressure such that $P_{liquid} \gg P_{bubble}$.

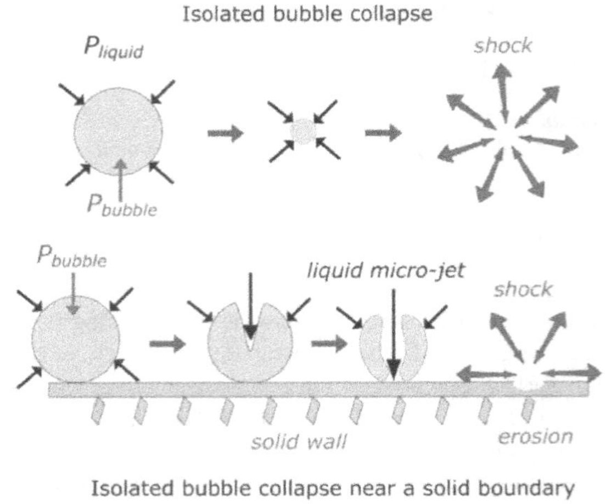

Figure 1: Vapor bubble collapse due to imposed pressure difference P_{liquid} >> P_{bubble}; collapse of an isolated bubble - symmetric collapse (top); bubble collapse near a solid boundary - asymmetric collapse (bottom)

Since one of the primary consequence of cavitation is its effect on nearby solid boundaries, a detailed investigation is needed on the final stage of bubble collapse with high temporal and spatial resolution. A feedback of solid material response can then be introduced into the fluid domain using a two-way coupled fluid-structure interaction model. The numerical investigation of collapsing cavitation bubble is challenging due to the need of high spatial and temporal resolution to capture the instantaneous high amplitude pressure loads.

CFD Solver *YALES*2

A pressure-based semi-implicit algorithm is used in the multi-physics solver YALES2 [3] for compressible flows with a fractional-step method [4] which is based on a characteristic splitting of the Navier-Stokes equation. The method consists of an advection or predictor step and a pressure-correction step separating the acoustics from the advection. In the pressure-correction step, a Helmholtz equation for pressure is solved implicitly to remove the acoustic *CFL* limitation. The compressible formulation of the governing flow equations allows its hyperbolic treatment to include the time dependent flow characteristics in the solution. Liquid flows with cavitation undergo strong negative pressure gradients for liquid break up to initiate cavitation. The speed of sound may be quite small in the two-phase region of a cavitating flow. For cavitation phase transition in a compressible liquid flow numerical methodology should resolve a wide range of Mach numbers (M), defined as the ratio of flow velocity to sound speed in the flow field. In the two-phase liquid-vapor mixture, local Mach number can be extremely high $M \gg 1$ whereas pure liquid phase stays close to low Mach number regime $M \rightarrow 0$. In the low Mach limit, the compressible *YALES*2 solver tends towards an incompressible solver inheriting the same stability and efficiency. In the solver, the wall boundary condition is treated implicitly by imposing a no-slip velocity in the advection or predictor step. Outlet boundary conditions are treated explicitly with classical Navier–Stokes Characteristic Boundary Conditions $(NSCBC)$ [5] to enforce acoustically non-reflecting boundaries. A consistent equation of state is used for the pure liquid phase using the Tait's equation of state and for the two phase liquid-vapor mixture following an isentropic phase transition model proposed in *Egerer et al.* [6]. The two phase modelling using a homogenous mixture or '*single fluid*' is widely accepted and implemented with the barotropic fluid assumption which specifies that the pressure is only a function of the fluid density. The model uses a single set of governing equations for all phases with consistent equations of state and fluid viscosities. Since the cavitation model is assumed to be barotropic, the energy equation is decoupled from the system of governing equations in the solver. The speed of sound in the pure liquid phase is computed as $c = \sqrt{\partial P/\partial \rho}$ whereas in the liquid-vapor phase it is constant equal to the speed of sound at ρ_L^{sat}. This is implemented to eliminate numerical instability of the pressure correction equation due to steep change in the speed of sound at the interface of pure liquid and liquid vapor mixture phase.

Results

The two phase model for compressible liquid is validated for an isolated *3D* single bubble collapse case with the analytical model of Rayleigh-Plesset described in [1]. The analytical model considered assumes liquid incompressibility and no liquid viscosity, gravity and surface tension. The analytical model considers a bubble saturated with pure vapor whose pressure is equal to vapor pressure $P_{vap} = 2194\ Pa$ at $292.15\ K$ and surrounding liquid pressure $P_\infty = 10\ MPa$. The numerical simulation is carried out with a bubble initially filled with two phase mixture of 99% saturated vapor content inside or simply, a vapor void fraction $\alpha = 0.99$ corresponding to a pressure of $2194\ Pa$ at temperature $292.15\ K$ and liquid-vapor mixture density of $10\ kg/m^3$. The isentropic cavitation model implemented assumes that phase change from pure liquid to liquid-vapor mixture starts at $P_{vap} = 2340\ Pa$ which corresponds to the vapor pressure at temperature $293.15\ K$. The pressure decreases along the isentropic path with increasing vapor void fraction i.e. with decreasing liquid-vapor mixture density. The temperature difference between $293.15\ K$ and $292.15\ K$ corresponds to the increase in temperature along the isentropic path assuming that all the initial vapor content of the bubble would turn into liquid. The *3D* spherical bubble is initialised inside the flow domain at time $t = 0\ s$ with initial bubble radius $R_0 = 500\ \mu m$. The surrounding liquid is initialized with a $P_\infty = 10\ MPa$ and the bubble starts to collapse at $t > 0\ s$ under the influence of the surrounding liquid pressure. The computational domain for the *3D* simulation is cubical in shape with numerical boundaries located at $25\ R_0$ away from the bubble center in all dimensions. The bubble is spatially resolved with *100* cells in the initial bubble radius R_0. The bubble interface is defined by the variation of vapor void fraction $\alpha = 0$ in the pure liquid to $\alpha = 0.99$ inside the bubble.

The time step for the resolved problem is $\Delta t_{CFD} = 5{\cdot}10^{-10}\ s$ based on CFL and acoustic CFL condition. The total simulated collapse is completed within the computed Rayleigh time $t_{rayleigh}$ of $4.6{\cdot}10^{-6}$ s. The comparison of the simulated *3D* bubble collapse by the developed $YALES2$ compressible cavitation solver and theoretical Rayleigh-Plesset collapse is shown below in figure 2. The plot shows the evolution of bubble radius R with respect to time t, both normalized by initial bubble radius R_0 and Rayleigh time $t_{rayleigh}$ respectively. The solid line is the solution obtained from Rayleigh-Plesset analytical equation and the dots represent the equivalent radius from the bubble vapor volume content obtained from the simulation. A decent agreement between the analytical and numerical result is obtained during the collapse, demonstrating the ability of the developed compressible cavitation solver to predict the dynamics of bubble collapse effectively.

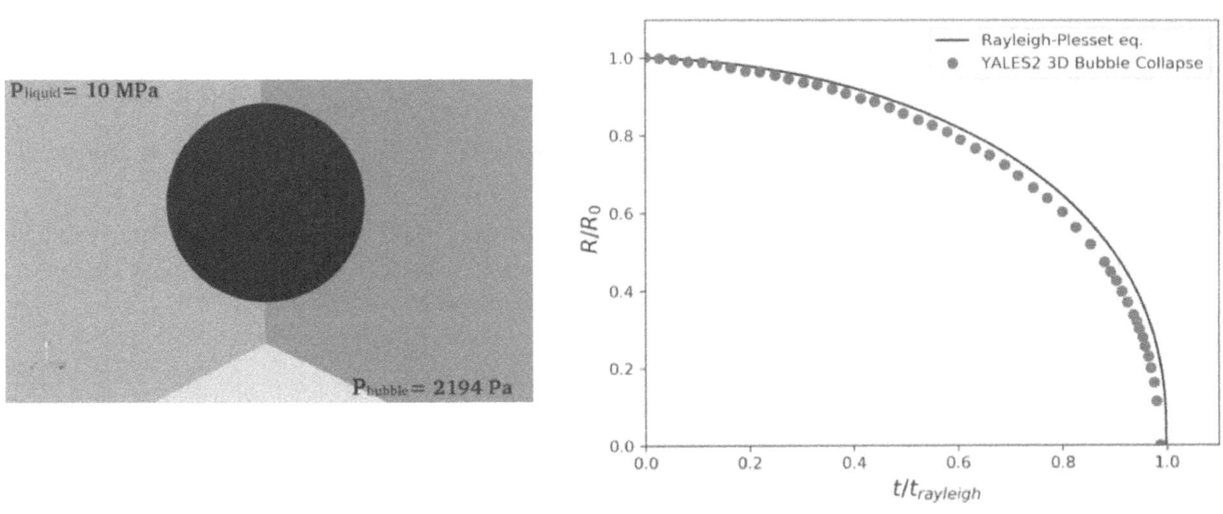

Figure 2 :Simulation of a single *3D* spherical bubble collapse with $YALES2$ and comparison with Rayleigh Plesset collapse model for $R_{max}= 500\ \mu m$, $P_\infty=10\ MPa$. Numerical bubble collapse setup (left); evolution of non-dimensional radius with respect to non-dimensional time (right)

Further, the collapse of an isolated vapor bubble attached to a solid boundary is investigated in a *2D* setup. The initial radius of the bubble is $R_0 = 500\ \mu m$ and the bubble is placed near the boundary as such to obtain a stand-off parameter of 0.8. The bubble is initialised with a vapor void fraction $\alpha = 0.99$ saturated inside at vapor pressure $P_{vap} = 2194\ Pa$ and surrounded by liquid pressure of $P_\infty = 100\ MPa$. The bubble is placed on the bottom surface of a rectangular domain which is numerically treated as a rigid boundary. An acoustically non-reflecting outlet boundary condition is imposed at other faraway boundaries of the rectangular domain situated $25R_0$ distance away from bubble center to avoid the interaction of reflected shock waves with the final stages of the collapse. The bubble is spatially well resolved with *100* cells along the initial bubble radius R_0. The problem was simulated with a temporal resolution of $\Delta t_{CFD} = 5.10^{-10}\ s$, chosen from the minimum time step computed from *CFL* and acoustic *CFL* condition during the moment of collapse and kept constant throughout the computation. Figure 3a shows the initial set up of the *2D* bubble geometry at $t = 0\ s$. The bubble interface remains well defined during the complex collapse phase and a liquid micro-jet is formed during the final stages of the collapse from the opposite side of the rigid boundary due to asymmetry of the problem. The pressure evolution at a point *K* on the bottom boundary surface is presented in figure 3b, extracted at every $0.1\ ns$ during the simulated collapse. The considered point *K* is located right under the bubble to capture the pressure evolution due to impacting liquid jet and subsequent shock waves. The initial pressure on the point is the saturation vapor pressure $P_{vap} = 2194\ Pa$ due to the attached bubble on the boundary. It is assumed that there is no liquid film layer between the attached bubble and solid boundary. Until $t = 3.0326\ \mu s$, the pressure at the point remains at P_{vap} as the bubble shrinks and formation of micro-jet starts. Two distinct pressure peaks can be seen in the pressure plot at $t = 3.0715\ \mu s$ and $t = 3.2079\ \mu s$ of $1986\ MPa$ and $3920\ MPa$ respectively. The first observed pressure peak is due to liquid jet impact and second peak due to the collapse of the bubble ring or torus, similar to the observations made in *Chahine et al.* [7].

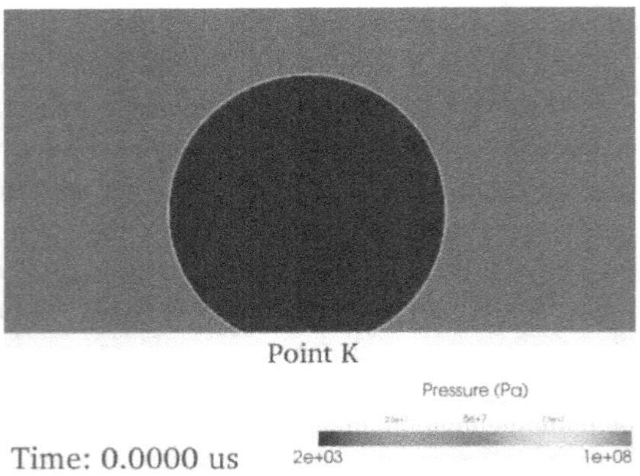

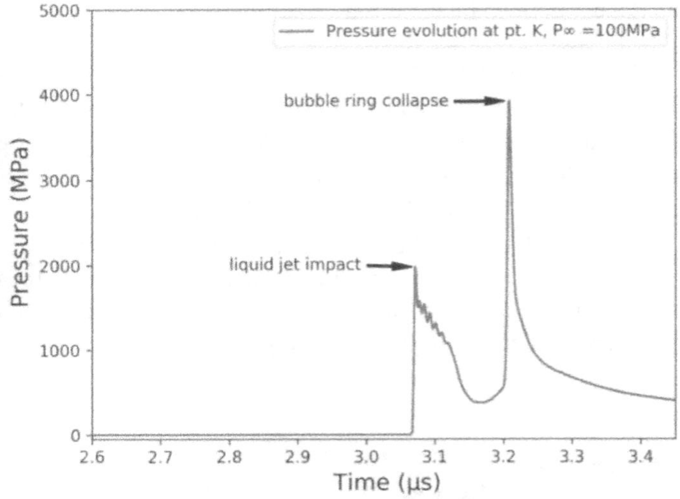

Figure 3a (left) : : Initial bubble setup with represented point *K* in the bottom solid boundary where pressure evolution is plotted; Figure 3b(right): Pressure evolution at point *K* in the bottom solid boundary, $R_{max} = 500\ \mu m,\ \gamma = 0.8,\ P_\infty = 100\ MPa$

The corresponding pressure and density contours for the observed pressure peaks are shown next in figure 4 and figure 5. The first pressure peak of $1986\ MPa$ is the liquid jet with an estimated velocity of $880\ m/s$ impacting the solid boundary and producing an instantaneous water hammer like pressure load at $t = 3.0715\ \mu s$.

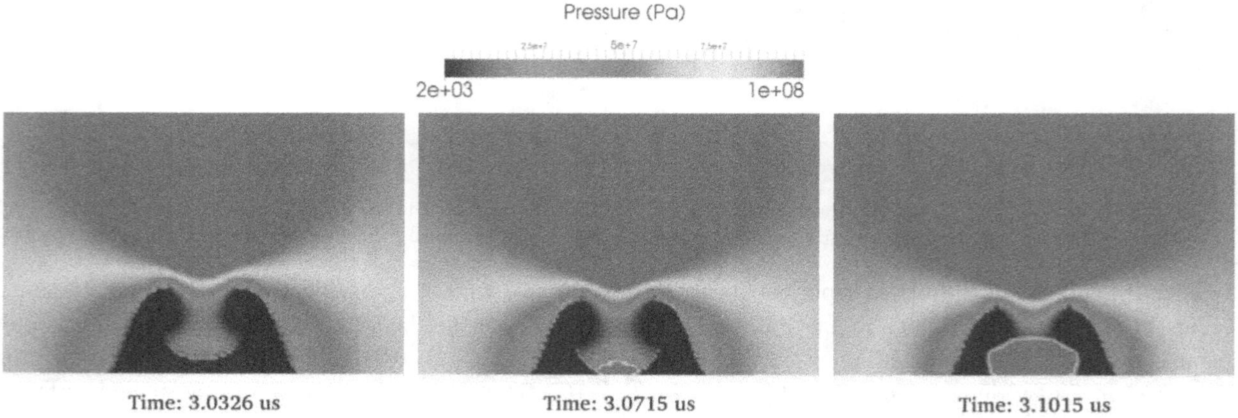

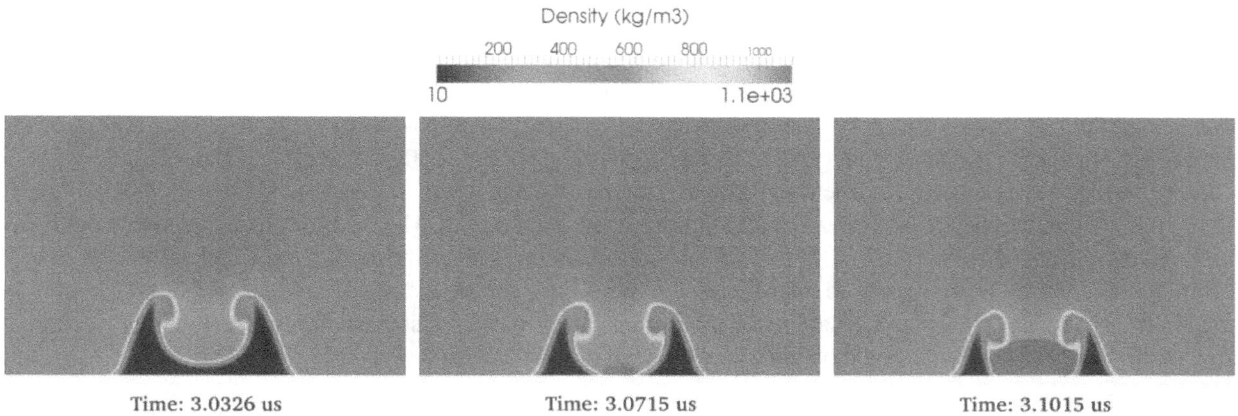

Figure 4: Pressure (top) and density (bottom) contour for a *2D* bubble collapse showing the liquid micro-jet impacting the solid boundary, corresponding to the first pressure peak of *1986 MPa* seen at time $t = 3.0715\ \mu s$

The remaining bubble ring after the liquid micro-jet impact contracts due to the surrounding liquid pressure towards the center of the bubble. The impacting liquid jet rebounds from the solid boundary and contributes to the shrinking of remaining vapor bubble ring. The remaining bubble ring is thus, impacted by an inward moving flow surrounding the bubble and an outward moving flow from the liquid jet rebound. As the bubble ring collapses, two shock waves are emitted that travel through the domain with liquid speed of sound. The second pressure peak of 3920 MPa is created at $t = 3.2079\ \mu s$ due to the interaction of the travelling shock waves at the bubble center as seen in figure 5.

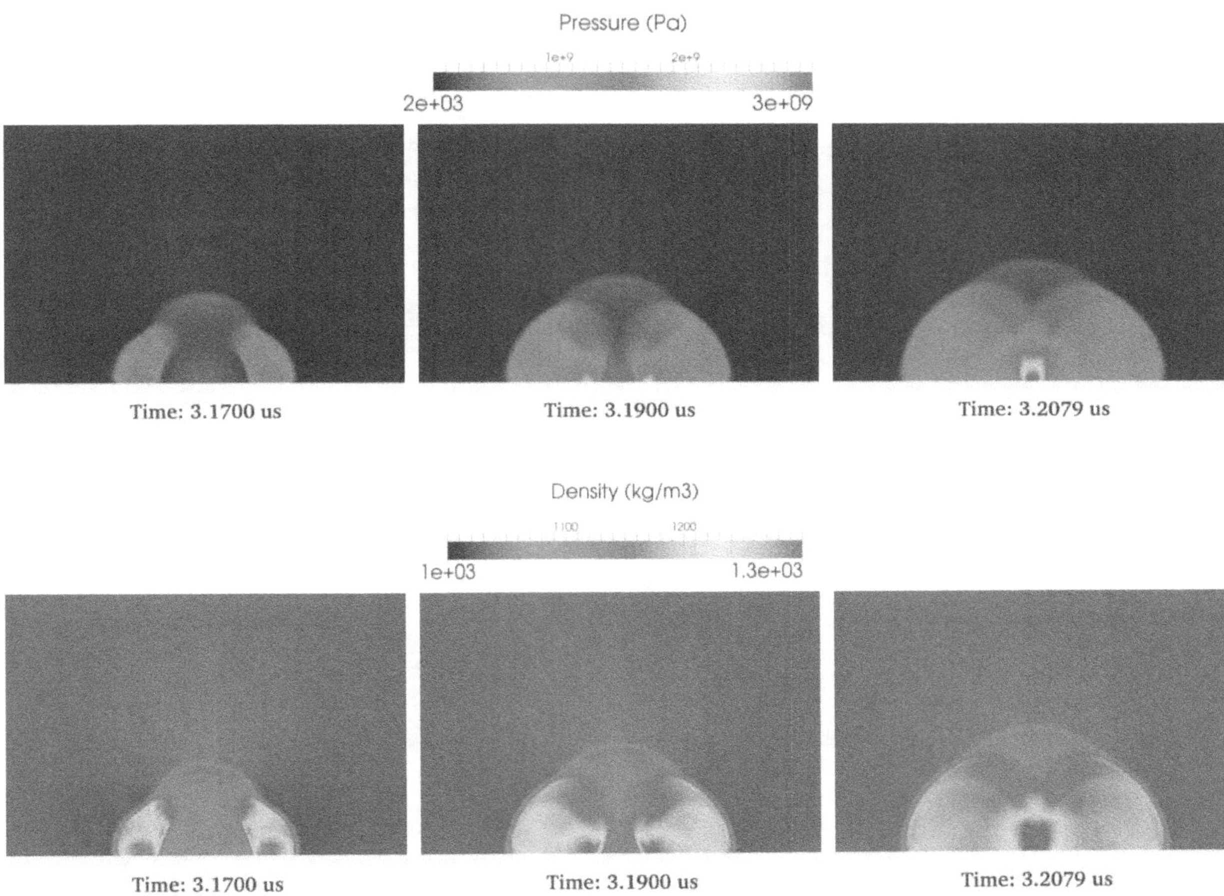

Figure 5: Pressure (top) and density (bottom) contour for a *2D* bubble collapse showing the interaction of shock waves after the collapse of the bubble torus, corresponding to the second pressure peak of *3920 MPa* at $t = 3.2079\ \mu s$

Conclusion

This work has been devoted to the investigation of the dynamics of a single vapor bubble collapse. The fluid is assumed to be compressible and viscous for the problem. Using the described numerical method, a *3D* bubble collapse in an unbounded fluid has been realised and compared with the analytical Rayleigh-Plesset result. There is good agreement between the numerical simulation and analytical result for the *3D* bubble collapse case. The compressible cavitation solver developed is used to investigate the collapse dynamics of a vapor bubble near a solid boundary in a *2D* set up, for an initial bubble radius $R_0 = 500\mu m$ and stand-off parameter $\gamma = 0.8$. The bubble collapses under the influence of surrounding liquid pressure $P_\infty = 100\ MPa$ and two distinct pressure peaks are observed. The first peak of $1986\ MPa$ is due to impacting liquid micro-jet and a much stronger second pressure peak of $3920\ MPa$ from the interacting shock waves after the collapse of bubble ring. This predicted instantaneous pressure will be used in the *FSI* methodology to predict the solid boundary response.

References

[1] Franc, J.-P., Michel, J.-M. (2004). *Fundamentals of Cavitation*. Kluwer Academic Publishers, Dordrecht.

[2] Philipp, A., Lauterborn, W. (1998). *Cavitation erosion by single laser-produced bubbles*. J. Fluid Mech., vol. 361.

[3] Moureau, V., Berat, C., Pitsch, H. (2007). *An efficient semi-implicit compressible solver for large-eddy simulations*. J. Comput. Phys. 226.

[4] Kim, J., Moin, P. (1992). *Application of a fractional step method to incompressible Navier Stokes equations*. J. Comput. Phys. 101.

[5] Poinsot, T., Lele, S. (1992). *Boundary Conditions for Direct Simulations of Compressible Viscous Flows*. J. Comput. Phys. 101.

[6] Egerer, C. et al. (2013). *LES of turbulent cavitating shear layers*. in Nagel, W.E. et al. (eds.), High Performance Computing in Science and Engineering, Springer.

[7] Chahine, G.L., Hsiao, C.-T. (2015) *Modelling cavitation erosion using fluid-material interaction simulations*. Interface Focus 5: 20150016.

© 2018 ASME

Numerical Investigation of the Effect of Dissolved Non-Condensable Gases on Hydraulic Flip in Cavitating Nozzles

Mathis Bode[a,*], Florian vom Lehn[a], Heinz Pitsch[a]

[a] *Institute for Combustion Technology, RWTH Aachen University, 52056 Aachen, Germany*

Abstract

This work focuses on hydraulic flip in cavitating nozzles and systematically investigates the effect of dissolved non-condensable gases on the concurrent phase change processes. First, different sets of equations known as 6-, 5-, and 4-equation models are evaluated. All models use a bisection-based Gibbs free energy relaxation algorithm for phase change after consecutively imposing velocity equilibrium, pressure equilibrium, and temperature equilibrium either in a hyperbolic solver or in subsequent relaxation steps. The well-established stiffened-gas equation-of-state is employed for thermodynamic closure. The impact of the different equilibrium assumptions on the resulting mixture speed of sound is discussed and the necessity of high-order numerics is investigated. Second, simulations of injector nozzles at gasoline direct injection (GDI) conditions were performed, resulting in flow regimes dominated by either hydraulic flip or cavitation. The underlying mechanism of hydraulic flip is analyzed and the competitive relationship between mechanical gas expansion and vapor formation in the cavitating flow regime is emphasized. It is shown that the amount of dissolved non-condensable gases has a strong, non-linear impact on the effective fuel cross section in the nozzle. Finally, the effect of viscosity on this non-linear behavior is studied.

Keywords: Cavitation; Hydraulic Flip; Non-Condensable Gas; Phase Change; N-Equation Models; Injector Flows

Introduction

Non-condensable gases are typically present in liquid fuel injection systems. They can originate from dissolved gases in the liquid fuel, which diffuse at small scale, or penetrate from the low-pressure chamber into the injector nozzle on a large scale. The large-scale phenomenon is known as hydraulic flip. While non-condensable gases including hydraulic flip are still often neglected in injector simulations, it is generally assumed that their presence influences the actual flow patterns [1]. This is supported by recent studies, which report a strong impact of non-condensable gases on important injector parameters such as the amount of vapor formed in the nozzle, the mass flow rate, the effective cross section, and the droplet size distribution resulting from the subsequent breakup processes [2, 3]. However, the underlying mechanisms of hydraulic flip and its interaction with in-nozzle cavitation are not well understood yet and therefore investigated in this work.

For multiphase flow systems, N-equation models have been developed over recent years [4, 5, 6, 7, 8, 9]. Originally developed for 2-phase systems considering only liquid and vapor, which results in seven governing equations, three conservation equations for each phase and an additional volume fraction equation, the so-called 7-equation model (7EM) is the most general approach. Applying kinematic, mechanical, and thermal equilibrium, it can be reduced to the 6-equation model (6EM), 5-equation model (5EM), and 4-equation model (4EM), respectively. Extensions towards an arbitrary number of additional non-condensable phases have been already shown. A corresponding 7EM has been applied to simulate injector flows with non-condensable gases in a simple geometry [9]. Later, Habchi [10] used a similar model and identified two cavitation regimes due to mechanical expansion and phase change, denoting them gaseous and vaporous cavitation regimes. However, hydraulic flip was not discussed.

*Corresponding Author, Mathis Bode: m.bode@itv.rwth-aachen.de

For a systematic discussion of the interplay of cavitation and hydraulic flip at engine-relevant conditions, 6EM, 5EM, and 4EM are applied in this work to simulate the flow under typical gasoline direct injection (GDI) conditions in a simplified 2D geometry. The amount of non-condensable gases dissolved in the liquid fuel and the presence of phase change are evaluated in terms of their effect on the resulting flow patterns. Furthermore, the different reduced models are compared in order to identify relevant differences in simulation results as well as their causes. This involves a discussion of the corresponding mixture speeds of sound, which are known to be a crucial difference between the models [11]. Finally, the effect of viscosity on the flow patterns of the GDI case is evaluated in order to understand the accuracy of inviscid simulations, which are still the most common approach for N-equation models in literature.

Numerical Models

The 3-phase system comprising of liquid fuel, fuel vapor, and non-condensable gases is simulated by extended 6EM, 5EM, and 4EM considering phase change and viscosity in this work. The full equilibrium assumption is used to close the appearing source terms with respect to volume, heat, and mass transfer and enforced by model-consistent relaxation steps such as a bisection-based Gibbs free energy relaxation algorithm for phase change. Vapor and non-condensable gases are modeled as ideal mixture, which is also denoted as gas phase. Liquid phase and gas phase are assumed to be immiscible. Separate stiffened-gas equations-of-state (SG-EOSs) are used for each phase and Stokes' hypothesis of a zero bulk viscosity is used for the viscous stress tensor. The dynamic viscosity of the gaseous components is modeled by Sutherland's law and the liquid viscosity is approximated by an exponential law [12]. The gas mixture viscosity is calculated out of the pure component viscosities by means of the Herning and Zipperer equation [13]. As flow solver, a finite volume method is used, where the intercell numerical fluxes are evaluated by an HLLC approximate Riemann solver [14], which is modified to account for non-conservative terms.

Results and Discussion

Model Comparison

The implementation of the N-equation models has been validated and studied by means of well-known and well-characterized 1D test cases [4, 6, 7]. Namely, a rarefaction tube and a shock tube test case were simulated and compared with data from literature. It has been found that the results of the 5EM and the 4EM are identical while the 6EM results differ. Thus, only simulation results from the 4EM, which respresent also the 5EM results, will be shown for the remainder of this section. The term '4/5EM' will be used in the discussion if both 4EM and 5EM are meant. The differences between the 4/5EM and the 6EM are shown in Figure 1a, which presents shock tube results for the 4/5EM, the 6EM, and 6EM data published by Pelanti and Shyue [7]. The most important observation is made at the shock wave: The shock propagation is predicted to be faster by the 4/5EM than by both 6EM (the differences between both 6EMs arises from the different numerical solvers).

In all models, instantaneous pressure, temperature, and Gibbs free energy equilibrium is assumed, but the models differ with respect to the numerical method: For the 6EM, the phase total energy equations are first solved separately and the phase pressures are then relaxed towards equilibrium by means of a linearized pressure relaxation. In contrast, the mixture energy equation is solved for the 4/5EM and the equilibrium pressure is directly computed by the respective mixture SG-EOS. The relaxation steps for temperature and Gibbs free energy are similar for the 5EM and the 6EM, which gives rise to the assumption that the differences must originate from the hyperbolic step in conjunction with pressure relaxation. In terms of wave propagation in the hyperbolic step, the 6-equation model differs strongly from the 4/5EM: While the latter exhibit wave propagation with the low Wood speed of sound or the even lower equilibrium speed of sound of the 4EM in the two-phase region, the mixture speed of sound of the 6EM is monotonic with respect to the volume fraction and has much higher values in the two-phase region as shown in Figure 1b. Due to the mass fraction-weighted averaging of the squared single-phase speeds of sound resulting as multiphase speed of sound from the hyperbolic part of the 6EM, the multiphase speed of sound is unphysically dominated by the heavier phase. For the resulting wave propagation at pressure equilibrium of the 6EM to be similar to that of the 5EM, the combined 6EM hyperbolic step and the pressure relaxation step would have to yield the same results as the hyperbolic step with the equilibrium volume fraction equation of the 5EM.

To further investigate the differences between the 4EM and the 6EM, a simplified GDI nozzle is studied, which corresponds to the GDI case discussed by Bode et al. [2]. In agreement with Habchi [10], an initial air mass fraction of $Y_{\text{o,L}} = 10^{-8}$ is assumed as amount of dissolved non-condensable gas in the liquid. Details about the development of hydraulic flip in this nozzle will be given in the next subsection. The averaged gas volume fractions $\hat{\alpha}$ in the nozzle are shown for the inviscid 4EM and the inviscid 6EM in Figure 1c. The 6EM predicts hydraulic flip to occur more than 2 µs later. However, this is not the only difference observed here: The 6EM also predicts a lower gas volume fraction and a small non-negligible amount of remaining vapor in the nozzle at steady state. Finally, the period of air penetrating into the nozzle is predicted to last longer by the 6EM.
It was found that one way to overcome the issue of different wave propagation speeds is to interpret the given speed of sound of the 6EM as a numerical speed of sound, which must be used, for example, for stability considerations with respect to the flow solver, and use the 5EM speed of sound as physical speed of sound for wave advection. By doing so, the results of the 4/5EM and 6EM matched for the studied 1D and 2D cases.

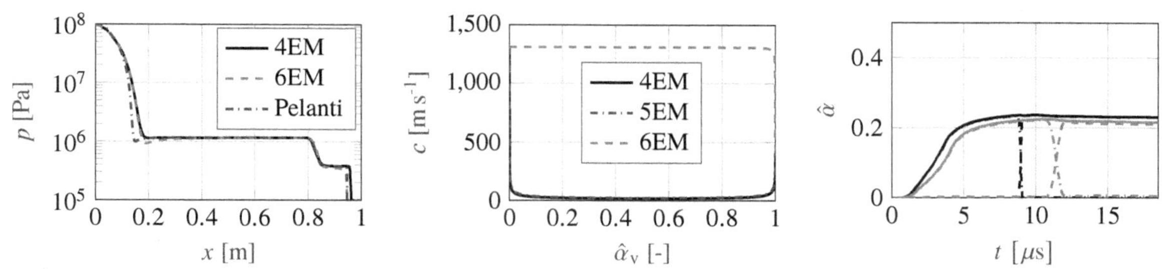

(a) Pressure (b) Speed of sound (c) Volume fractions

Figure 1: Comparison of dodecane liquid-vapor shock tube results at $t = 473$ µs computed with the 4EM and the 6EM as well as data from Pelanti and Shyue [7] (left plot). Mixture speed of sound of the 4EM, the 5EM, and the 6EM (middle plot). Comparison of nozzle-averaged volume fractions computed with the 4EM (black) and the 6EM (red): solid line - gas volume fraction, dashed line - air volume fraction, dash-dotted line - vapor volume fraction (right plot).

Cavitation and Hydraulic Flip

For understanding the interaction of cavitation and hydraulic flip during the injection process, inviscid 4EM simulation results of the GDI case are presented in Figure 2. Shortly after start of injection, the shock wave of the initial pressure discontinuity propagates from the high pressure chamber through the nozzle into the low pressure chamber. The liquid-gas interface starts moving from the time when it is hit by the shock wave and it is well visible how the bulk air has been completely pushed out of the nozzle by the liquid at 3 µs. Simultaneously, vapor sheets are formed due to cavitation at the inlet corners and are transported downstream. They eventually reach the nozzle orifice at about 6 µs where the vapor quickly condensates indicating a higher pressure in the right chamber than in the near-wall regions of the nozzle over its entire length. Once the vapor sheets have reached the nozzle orifice, air from the low-pressure chamber is drawn back into the nozzle and quickly replaces the vapor at the nozzle wall. At steady-state conditions ($t = 18$ µs), no vapor is visible at all and the air pockets at the nozzle walls almost stretch to the inlet corners. Thus, cavitation has been completely replaced by hydraulic flip. These observations may allow the assumption that the vapor production due to cavitation is the main responsible mechanism for inception of hydraulic flip.
In order to evaluate this assumption, the air mass fraction in the liquid fuel of the GDI case has been varied while keeping all other initial conditions unchanged. Figures 3a and 3b show the development of nozzle-averaged vapor and air mass fractions for those cases. Multiple striking differences can be observed: First, an increased air mass fraction in the liquid fuel results in lower vapor production by cavitation and thus lower vapor volume fractions. These lower vapor volume fractions are on the other hand balanced by higher air volume fractions due to mechanical expansion of the dissolved air. With higher dissolved air mass fraction in the liquid, vapor volume is apparently replaced by air volume while the total gas volume hardly changes. Second, it is also visible that with higher dissolved air mass fraction in the liquid, the inception of hydraulic flip is slightly retarded. This applies especially to the case with $Y_{\text{o,L}} = 10^{-5}$. Moreover, a kink is visible for this case in the air volume fraction plot as well as in the total gas volume fraction plot.

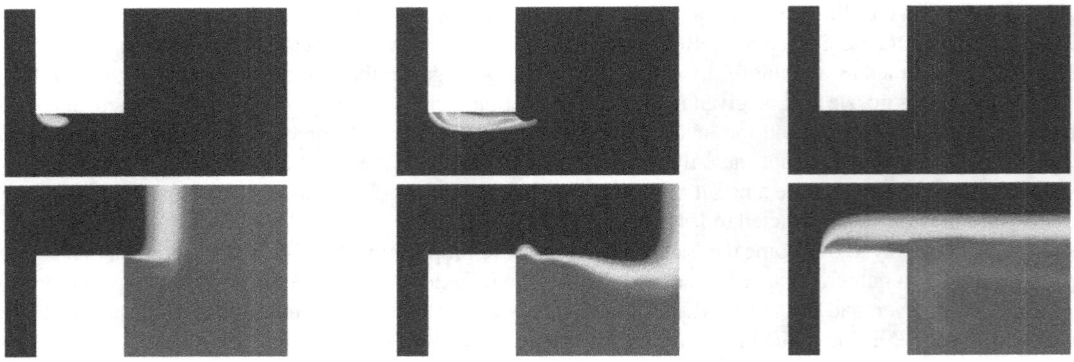

Figure 2: Temporal evolution of the vapor mass fraction (top) and air mass fraction (bottom) at 3 µs, 6 µs, and 18 µs (left to right). Logarithmic colormap from 10^{-8} (blue) to 1 (red).

This temporary decrease in gas volume can be explained as follows: At inception of hydraulic flip, the higher pressure of the right chamber quickly penetrates into the gas pockets, starting at the nozzle outlet. Comparing the development of pressure and air mass fraction in the gas pockets at hydraulic flip inception, one observes that the pressure level in the gas pockets increases slightly faster than air mass fraction. Hence, in a small region pressure has already risen while air mass fraction is still at low level, resulting in lower air volume fraction in that region which shows up as a kink in the respective plots. Since no kink is visible in the other cases with lower dissolved air mass fractions in the liquid fuel, the higher air mass fraction in the gas pockets before inception of hydraulic flip is probably co-responsible for this behavior. Possibly, this higher amount of air in the gas pockets generates a certain back-pressure against the penetrating air from the low-pressure chamber that might also be responsible for the retarded inception of hydraulic flip in this case.

The result that the amount of produced vapor decreases with increasing air mass fraction in the liquid fuel may reveal that cavitation is only one possible mechanism associated with the massive pressure drop at the inlet corners leading to gas pocket formation and eventually to hydraulic flip, but mechanical expansion of dissolved gas is another possible mechanism with almost the same effect on hydraulic flip. For a quantitative evaluation of this, the discussed 4EM GDI case simulation with $Y_{\mathrm{o,L}} = 10^{-5}$ is compared to a 4EM simulation with phase change turned off in Figure 3c. As expected, no vapor volume fraction is present in the nozzle if phase change is deactivated. Higher air volume fraction is visible instead due to stronger mechanical expansion of the air, even though the initial air mass fraction was the same in both cases. The development of gas volume fractions is identical in both cases. This result is surprising since it indicates that phase change does not affect the total gas volume in the nozzle at all. Instead, if phase change is activated, the vapor only partly replaces air in the gas mixture, yet the mixture gas volume stays constant. At steady state, even the gas mixture's composition is almost similar since cavitation is suppressed in both cases by hydraulic flip. This observation confirms the assumption that cavitation is only a result of the pressure drop, but in the presence of dissolved gas in the liquid, phase change is not required for the inception of hydraulic flip. Instead, hydraulic flip is caused by low-pressure gas pockets reaching the nozzle outlet whose composition is not important. Since fuels always contain a certain amount of dissolved gas, these effects should also apply to injector flows through sharp-edged nozzles.

Viscosity

Finally, the impact of viscosity on the discussed results is quantitatively studied in this subsection. Adding viscosity to the initially hyperbolic equations, parabolic equations are obtained allowing the prescription of no-slip wall boundary conditions. Since the viscosity is proportional to the shear, the flow is mostly affected in the highly sheared near wall region within the gas pockets. The velocity u is plotted at four equidistantly spaced locations along the nozzle for an inviscid and a viscous 4EM simulation in Figure 4. The nozzle diameter D and the inviscid centerline velocity are used for non-dimensionalization. As expected, the tangential velocity at the wall is finite in the inviscid simulation.

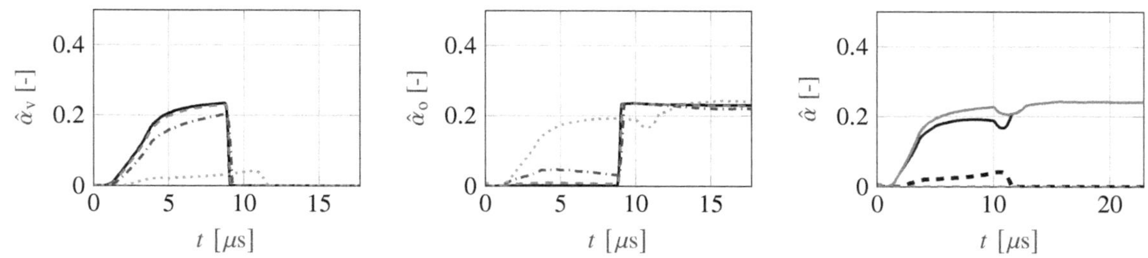

(a) Vapor volume fractions (b) Air volume fractions (c) Volume fractions

Figure 3: Comparison of the temporal evolution of nozzle-averaged vapor and air volume fractions for different initial air mass fractions $Y_{o,L}$: black line - $Y_{o,L} = 10^{-8}$, red line - $Y_{o,L} = 10^{-7}$, blue line - $Y_{o,L} = 10^{-6}$, cyan - $Y_{o,L} = 10^{-5}$ (two left plots). Comparison of nozzle-averaged volume fraction results of the 4EM with phase change (black) and without phase change (red): solid line - air volume fraction, dashed line - vapor volume fraction (right plot).

However, due to the enclosed liquid near the wall with its high density, it acts like an obstacle, slowing down the surrounding backwards flowing gas. In the viscous case, a smoother velocity decay with much lower shear rates, not only inside the nozzle, but also farther downstream of the nozzle outlet, can be seen. However, it was found that the impact of viscosity on the total discharge coefficient for the case considered here is negligible. The most notable difference concerns the convergence towards steady-state. While the viscous flow seems to have reached steady conditions at already $t = 20\,\mu s$ with a nearly unchanged velocity field and profile, the gas phase velocity in the inviscid case still undergoes a drastic rise in the maximum magnitude, indicating a higher intensity of hydraulic flip and recirculation within the gas pockets. Note that the velocity at which the gas is sucked into the nozzle is almost doubled between $t = 20\,\mu s$ and $t = 30\,\mu s$ and even for much larger times the solution keeps changing. This phenomenon might partially also originate from the characteristic boundary conditions applied at the outlet. During the simulation, parts of the outlet become effectively inflows, which technically try to avoid a too large backwardly directed flow and are treated in a fully non-reflecting manner. For the viscous flow, this problem is less serious because the flow has a natural resistance which will act as a sink for momentum and overall mechanical energy.

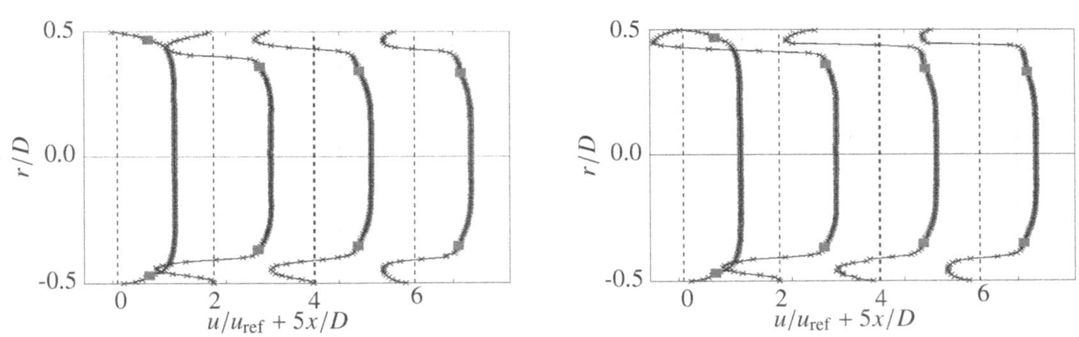

Figure 4: Streamwise velocity profiles at two time instances $t = 20\,\mu s$ and $t = 30\,\mu s$ of an inviscid (upper half) and a viscid (lower half) 4EM simulation. The velocity is non-dimensionalized by the inviscid centerline velocity u_{ref} and the spatial coordinates by the nozzle diameter D. A stretching factor (5) is used to enable better plotting. Red markers indicate the gas height.

Conclusions

N-equation models incorporating phase change and viscosity have been used to study hydraulic flip in injector flows in this work. It was found that 4EM and 5EM give similar results while 6EM results differ. The mixture speed of sound was identified as main reason and causes were discussed. Furthermore, a non-linear effect of dissolved non-condensable gases on hydraulic flip was shown and it was explained that cavitation is no requirement for the occurrence of hydraulic flip. Although viscosity strongly affects the near-wall gas phase velocity in the nozzle, its effect on the total discharge coefficient for the considered GDI case was found to be negligible.

Acknowledgements

The authors gratefully acknowledge funding by the Cluster of Excellence "Tailor-Made Fuels from Biomass" and Honda R&D. Also, the fruitful discussions with Kazuki Maeda and Tim Colonius from the California Institute of Technology are kindly acknowledged. Additionally, the computing time and support for simulations performed on the national supercomputer Cray XC40 at the High Performance Computing Center Stuttgart (HLRS) under the grant number GCS-MRES is acknowledged.

References

[1] M. Battistoni, D. Duke, A. B. Swantek, F. Z. Tilocco, C. F. Powell, S. Som, Effects of noncondensable gas on cavitating nozzles, Atomization and Sprays 25 (6) (2015) 453–483.
[2] M. Bode, T. Falkenstein, M. Davidovic, H. Pitsch, H. Taniguchi, K. Murayama, T. Arima, S. Moon, J. Wang, A. Arioka, Effects of cavitation and hydraulic flip in 3-hole gdi injectors, SAE International Journal of Fuels and Lubricants 10 (2) (2017) 380–393.
[3] M. Bode, S. Satcunanathan, K. Maeda, T. Colonius, H. Pitsch, An equation-of-state tabulation approach for injectors with non-condensable gases: Development and analysis, in: 10th International Cavitation Symposium, Baltimore, USA, 2018.
[4] R. Saurel, F. Petitpas, R. Abgrall, Modelling phase transition in metastable liquids: application to cavitating and flashing flows, Journal of Fluid Mechanics 607 (2008) 313–350.
[5] R. Saurel, F. Petitpas, R. A. Berry, Simple and efficient relaxation methods for interfaces separating compressible fluids, cavitating flows and shocks in multiphase mixtures, Journal of Computational Physics 228 (2009) 1678–1712.
[6] A. Zein, M. Hantke, G. Warnecke, Modeling phase transition for compressible two-phase flows applied to metastable liquids, Journal of Computational Physics 229 (8) (2010) 2964–2998.
[7] M. Pelanti, K.-M. Shyue, A mixture-energy-consistent six-equation two-phase numerical model for fluids with interfaces, cavitation and evaporation waves, Journal of Computational Physics 259 (2014) 331–357.
[8] R. Saurel, P. Boivin, O. Le Métayer, A general formulation for cavitating, boiling and evaporating flows, Computers and Fluids 128 (2016) 53–64.
[9] Y. Wang, L. Qiu, R. D. Reitz, R. Diwakar, Simulating cavitating liquid jets using a compressible and equilibrium two-phase flow solver, International Journal of Multiphase Flow 63 (2014) 52–67.
[10] C. Habchi, A gibbs energy relaxation (germ) model for cavitation simulation, Atomization and Sprays 25 (4) (2015) 317–334.
[11] T. Flatten, H. Lund, Relaxation two-phase flow models and the subcharacteristic condition, Mathematical Models and Methods in Applied Sciences 21 (2379) (2011) 2379–2407.
[12] E. Goncalves, R. F. Patella, Numerical simulation of cavitating flows with homogeneous models, Computers & Fluids 38 (9) (2009) 1682–1696.
[13] B. E. Poling, J. M. Prausnitz, J. P. O'connell, et al., The properties of gases and liquids, Vol. 5, Mcgraw-hill New York, 2001.
[14] E. F. Toro, M. Spruce, W. Speares, Restoration of the contact surface in the hll-riemann solver, Shock Waves 4 (1994) 25–34.

© 2018 ASME

Numerical prediction of cavitation performance for rim driven thruster

[1]Li-wei Zhang; [1]Zi-ru Li; [1,*]Wei He; [1]Ling-yu Zhu

[1]*Wuhan University of Technology, Wuhan, Hubei, China*

Abstract

The open-water characteristics and cavitation behavior of a rim-driven thruster (RDT) are examined by using the Reynolds-averaged Navier-Stokes (RANS) method. The total thrust coefficients and torque coefficients calculated by RANS solver with the SST k-ω turbulence model show relatively better agreement with the experimental data than other two-equation turbulence models. The cavitation behavior of the RDT with a cavitation number of σ_n=3.0 in uniform and non-uniform flow are then separately explored by using the tow-phase RANS method with the SST k-ω turbulence model and the Schnerr-sauer cavitation model, as well as the sliding mesh technique. It can be observed that a sheet cavitation occurs on the suction side of the blades, occupying almost 35% blade area, and kind of cloudy structures are captured to be shed from the sheet cavity for the RDT in uniform flow. Whereas, the sheet cavitation is observed to take up more than 50% blade area for the RDT in a 'Triple Peak' wake field.

Keywords: rim-driven thruster; Cavitation performance; Non-uniform flow; Schnerr-sauer model

Introduction

The rim-driven thruster (RDT) is a relatively new marine propulsion device that uses a motor in its casing to drive a propeller by its rim[1]. The well-known advantages of the rim driven thruster are its low noise emission, vibration generation and space requirement. In addition, cavitation, especially for the annoying tip vortex cavitation, that often occurs when the propeller rotates at a high speed will be alleviated and even eliminated.

Recent years, RDT has been investigated by many researchers. Song et al. (2015) applied Reynolds-averaged Navier-Stokes (RANS) method with moving reference frame (MRF) model to investigate the open-water performance of the RDT by using the steady solver without experimental validation[2]. Huyer et al. (2010) used the commercial software FLUENT to calculate and discuss the influence of the duct geometry, front stator and rotor on the open-water performance of the RDT[3]. Dubas et al. (2015) calculated the influence of the gap between the stator and the blade on the torque in OpenFoam by using the transient solver with experiment validation[4]. It was found that the SST k-ω turbulence model is more suitable to capture the flow separation at low advanced velocities while the RNG k-ε turbulence model is better at high advanced velocities. Yang et al. (2017) analyzed the cavitation performance of the RDT in an uniform flow with a cavitation number of σ_n=3.0 and found that the tip-vortex cavitation that often occurs at the tip of the traditional shaft driven thrusters was not observed around the tip of the blades due to the special configuration of the RDT[5]. However, the propulsion system often operates behind the ship with a non-uniform wake in practical engineering. This would then rationalize further investigation of the cavitation behavior of the RDT in non-uniform flow, which would benefit the subsequent research of the vibration and noise characteristics by analysis of the hydro-acoustic noise field.

The purposes of the current study are to predict the open water performance and study the cavitation behavior of a RDT in uniform and non-uniform flow using computational fluid dynamics method. Firstly, the open-water characteristics of a RDT are calculated by using the RANS method and MRF model. Secondly, the cavitation performance of the RDT is predicted by using the unsteady two-phase RANS solver and the Schnerr-sauer cavitation model together with the sliding mesh technique. Then, a 'Triple Peak' wake field is applied at the inlet boundary for the simulation of the cavitation behavior of the RDT that operates in a non-uniform flow. Finally, the characteristics of the wake field are explored by analysis of the pressure fluctuations at some distance behind the RDT.

Case Description

Rim-driven Thruster Geometry

The rim-driven thruster that is investigated in the current study is composed by a duct and an embedded rotor with seven blades. Part of the geometry data of the RDT is listed in Table 1. The open-water experiment of the RDT was conducted at the towing tank of Wuhan University of Technology. A hub and shaft were fitted to reduce the processing

*Corresponding Author, Wei He: hwcuduca@163.com

difficulties and test unsteadiness. Thus, a hub is added to the three-dimensional RDT model in the numerical simulation of the open-water performance for proper validation against the experimental observations.

Items	Nomenclature	Unit	Value
Diameter	D	[m]	0.2
Design pitch ratio/R=0.7	$P_{0.7}/D$	[-]	-
Area ratio	A_E/A_0	[-]	0.7
Number of blades	Z	[-]	7
Direction of rotation		[-]	right

Table 1. Geometry data of RIM-driven thruster model

Computational Settings

A cylindrical domain with the RDT model along its central axis has been established for the numerical simulation in non-cavitating and cavitating conditions. According to our previous studies[5,7], the radius of the overall cylindrical domain is set to be 4D with an extension of 8D ahead of the thruster's mid plane and 10D behind the thruster's mid plane by considering both of the computation time and resource. As two coordinate systems are involved in the computation of the hydrodynamic performance of the RDT, the whole domain are divided into a rotating region and a static region. The information exchange in the fluid filed is achieved by the interface between the two regions.
For the investigation of the open-water performance of the RDT, the number of revolution of the thruster is fixed to n=1000rpm with advance coefficients varying from 0.1 to 0.5. The density and dynamic viscosity of the water is set to be the water properties at 20 degree. The inlet boundary is set as velocity inlet condition with the inlet velocities corresponding to various advance coefficients. No-slip wall is applied to the overall configuration of the RDT and the outer boundary of the cylindrical domain. The outlet boundary is set as pressure outlet with no assigned pressure. Whereas for the numerical simulation of the cavitation behavior of the RDT, the advance coefficient is set as the design point with a value of J=0.3. And the pressure at the outlet should be specified according to the cavitation number when the cavitation performance of the RDT is examined.

Meshing

Mesh generation is performed using the trimmed mesher, which provides a robust and efficient method of producing high-quality mesh for the blade surface where large curvature variation occurs. Comparison of results calculated by three grids that have been refined substantially with similar topology is conducted to study the grid sensitivity for the calculation of the open-water performance. The detailed information of these three grids are shown in Table 2. The mesh around the rim part and the seven blades is shown in the figure 1 (a). The leading and trailing edge of the blades are refined due to the characteristics of high curvature and small thickness, as shown in figure 1 (b). Owing to the tiny gap between the rim part and the duct part with around 1mm thickness in the model scale, eight prism layers are applied to improve the flow solution near the duct.

Grid	Total cells	Level	$Y^{+}max$	Prism layers
G1	2.4 million	Coarse	101.2	5
G2	4.5 million	Medium	76.3	8
G3	8.9 million	Fine	38.6	8

Table 2. Grid features of the rim-driven thruster

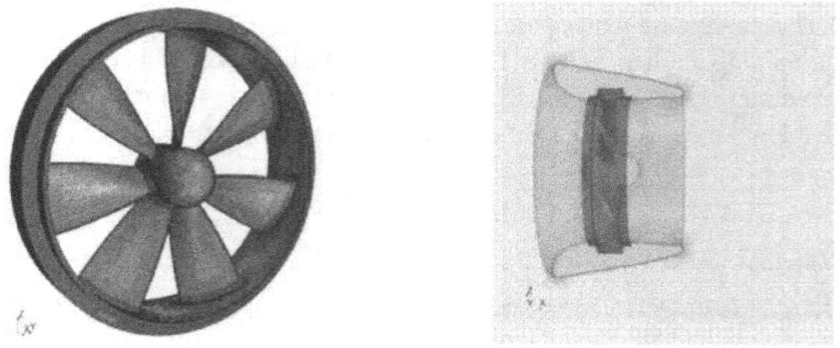

(a) Mesh of the RIM and blades (b) The refined mesh region
Figure 1. Details of the medium grid G2 of the RDT

Results and Discussions

Open-water Characteristics

Validation of the numerical open-water characteristics of the RDT against the experimental data from Wuhan University of Technology has been performed by using the RANS method with the two-equation turbulence models and a moving reference frame. Firstly, the sensitivity of the grid densities is examined on three geometrically similar girds with substantially refined levels (as shown in Table 2). In Figure 2. , the thrust and torque coefficients that are predicted by the three grids are compared with the experimental measurement. It can be observed that the benefit from the refinement of grids in the current study is very small for the grids G2 and G3. Thus, comparison of results that are calculated by different two-equation turbulence models has been made by using the grid in medium level G2 with consideration of accuracy and computational resource.

The total thrust coefficients and torque coefficients calculated by the SST k-ω turbulence model with the advance velocities varying from 0.1 to 0.5 show relatively better agreement with the experimental data than other two-equation turbulence models. Figure 3 shows the open-water characteristics obtained by the medium grid G2 with the SST k-ω turbulence model. The calculated torque coefficients K_Q match well with the experimental measurement with errors less than 2.5%, whereas the thrust coefficients K_T at low advance ratios are underestimated with errors within 6.7%, which are acceptable for the engineering application. By comparison with our previous work[5], the differences between the predicted thrust coefficients and the experimental data decrease around 1% by the increment of the number of the prism layers from five to eight, but with a reduction of total grid number from 4.8 million to 4.5 million. It is suggested that the sufficient resolution in tiny gap between the duct and the rim has more important influence than the total grid number on the prediction of the hydrodynamic performance of the RDT.

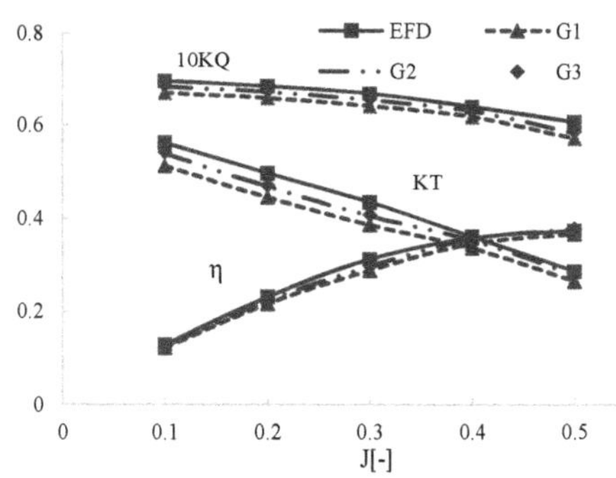

Figure 2. The open-water characteristics of the RDT with different grids

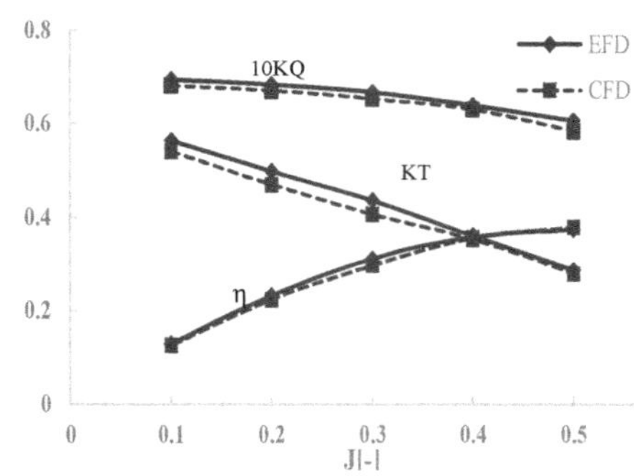

Figure 3. The open-water characteristics of the RDT with the SST k-ω turbulence model by employing grid G2

Cavitation Behavior of the RDT in Uniform Flow

As the numerical simulation of the cavitating flow demands higher grid resolution, the fine grid G3 has been selected to better simulate the cavitation behavior of the RDT in uniform flow and non-uniform flow. In order to simulate the real shaft-less situation as much as possible, the hub is ignored in the cavitation simulation. The cavitation behavior of the RDT at a specific cavitation regime with a cavitation number of σ_n=3.0 in uniform flow is explored by using the tow-phase RANS method with SST k-ω turbulence model and the Schnerr-sauer cavitation model. The operation condition for the current study is at an advance coefficient of J=0.3, which is the design point of the RDT. Transient computations are performed with a time step size of 1.7e-4s corresponding to 1° rotation of the RDT. The propeller rotation is simulated by a rigid body motion with a sliding mesh technique.

Figure 4 shows the iso-surface plots of instantaneous vapor volume fraction of α =0.2 observed in the current study with finer grid and larger number of the prism layer and those observed in our previous work [5]. In both cases, a sheet cavitation has been captured on the suction side of the blades, and none of the tip-vortex cavitation occurs at the blade tip due to the high integration between the rim and the blades. However, a larger extent of the sheet cavitation is observed at the leading edge but covers less area of around 35% of the blades' area. Moreover, kind of cloudy structures are captured to be shed from the sheet cavitation in the current study. When comparing the hydrodynamic performance of the RDT under the cavitation condition and the non-cavitation condition in uniform flow, it is found that the cavitating behavior causes the thrust fall down about 9.12%. Whereas a loss of 12.665% of the thrust was predicted by using the same numerical method but with the coarser grid and less number of prism layers[5]. It is supposed that the grid resolution at the gap between the rim part and the blades also has an impact on the simulation of the cavitation behavior of the RDT. However, the influence of the hub on the cavitation performance of the RDT demands further investigation.

(a) current study (no hub) (b) previous work by Tao Yang (2016)

Figure 4. Cavitation pattern of the rim-driven thruster in uniform flow at J=0.3, σn=3.0

Cavitation Behavior of the RDT in Non-uniform Flow

To investigate the cavitation behavior of the RDT in non-uniform, a 'Trible Triple Peak' wake field obtained from Boswell'experiment[6] has been applied in the current study, as shown in figure 5. And the inlet boundary changes to 1/3 D to avoid numerical dissipation while other numerical methods remain the same as those in the uniform flow condition. As we can see from figure 6, a sheet cavitation starts from the leading edge and extends to the trailing edge with kind of cloudy structures tending to shed form the whole sheet. The occupied area and the cavitation pattern on each blade are qualitatively similar as each other but with quantitative differences. It can be observed that the averaged cavitation area covers almost 50% of the blades which may result in larger loss of the thrust and torque by comparison with the hydrodynamic performance predicted when the cavitating RDT operates in uniform flow.

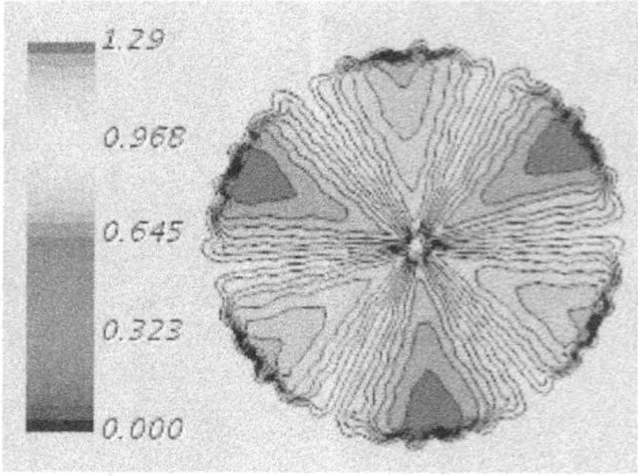

Figure 5. The velocity magnitude of the 'Triple Peak' wake field

Figure 6. Cavitation pattern in non-uniform flow

Pressure Fluctuations in Non-uniform Flow

Even though in the non-cavitation condition, noise could be generated due to the water displacement by blades and the pressure difference between the suction and pressure sides of the blades of the RDT. When it comes to the cavitation condition, the dynamic behavior of large volume of cavitation and the sudden collapses of complex and various cavitation pattern generate strong pressure fluctuations, which would intensify the vibration and noise and has a close correlation with the erosion risk. In the present study, the pressure fluctuations generated by the RDT in the 'Triple Peak' wake field are examined by monitoring the pressure of four points that are located closely behind the RDT. The specific location of the four points are P1(0.015m,0m,0.1m), P2(0.015m,0m,-0.1m), P3(0.015m,-0.1m,0m), P4(0.015m,0.1m,0m).

For convenience of the analysis, a dimensionless physical quantity K_P is used to describe the pressure signal:

$$K_P = \frac{P_k}{\rho n^2 D^2} \tag{1}$$

where : ρ is the density of the water, n is the rotation speed, D is the propeller diameter, P_k is the static pressure. The frequencies of the pressure fluctuations are obtained by transforming the pressure signals in the time domain into the frequency domain by Fast Fourier Transformation (FFT), as shown in figure 7. It can be found that strong pressure fluctuations are generated at blade passing frequency and its multiples, which can be related to the dynamic behavior of large volume of the observed sheet cavitation on the suction side of the RDT. The typical pressure fluctuating frequencies vary from the first order frequency of 116.67HZ, to the second order fluctuating frequency around 233HZ with an amplitude decrease of 60%.

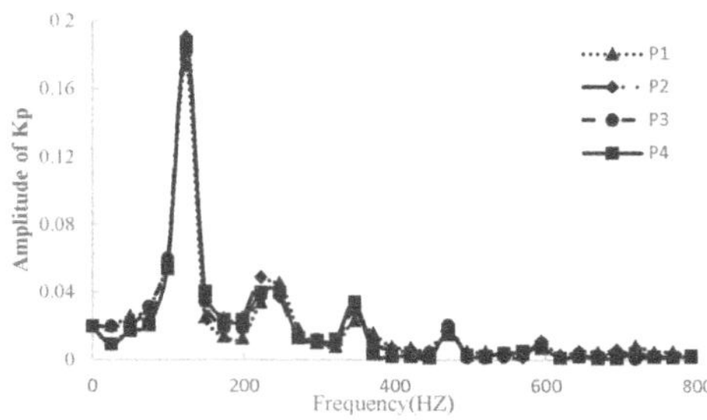

Figure 7. The pressure fluctuations observed at four specific locations in frequency domain

Conclusion

The following conclusions are drawn from the present study:

- The open-water performance of the RDT with the SST k-ω turbulence model by employing the medium grid G2 fairly agrees with the experimental data to within 6.7%. It is suggested that the sufficient resolution in the tiny gap between the duct and the rim has more important influence than the total grid number on the prediction of the hydrodynamic performance of the RDT.
- A sheet cavitation occurs on the suction side of the RDT in the uniform flow with a cavitation number of σ_n=3.0, and kind of cloudy structures are observed to be shed from the sheet cavity. None of the tip-vortex cavitation appears due to the high integration between the rim and the blades.
- The sheet cavitation that captured for the RDT in a 'Triple Peak' weak field occupies almost 50% of the blades, resulting in larger loss of the thrust and torque than the cavitating RDT operates in the uniform flow.
- A preliminary study of the pressure fluctuations is performed by examining four points closely behind the RDT. Strong pressure fluctuations that can be related to the dynamic behavior of large sheet cavitation are found to be generated at blade passing frequency and its multiples. Further investigations are recommended for the cavitation behavior of the RDT in complicated practical wake field.

ACKNOWLEDGEMENT

This work is financially supported by the National Natural Science Foundation of China (Project No.51609189, No.51309184), and the work is also supported by the Fundamental Research Funds for the Central University (WUT: 2016IVA010 and 2017IVB008).

References

[1] Tuohy, P., M., Smith, A. C., Husband, M. (2010). *Induction rim-drive for a marine propulsor.* International Conference on Power Electronics, Machines and Drives. (pp.1-6). IET.

[2] Song, B., W., Wang, Y, J., Tian, W, L. (2015). *Open-water performance comparison between hub-type and hubless rim driven thrusters based on CFD method.* Ocean Engineering. 103:55-63.

[3] Huyer, S, A., Dropkin, A. (2010). *Integrated motor/propulsor duct optimization for increased vehicle and propulsor performa nce.* Journal of Fluids Engineering, 133(4), 797-807.

[4] Dubas, A, J., Bressloff, N, W., Sharkh, S, M. (2015). *Numerical modelling of rotor–stator interaction in rim driven thrusters.* Ocean Engineering. 106:281-288.

[5] Yang T., Li, Z., Chen, K., He, W. (2017). *Prediction of Hydrodynamic Performance for Rim-Driven Thruster*. International Society of Offshore and Polar Engineers. 17-409.
[6] Ye, J., Xiong, Y., Gao , X., Liu, Z. (2013). *Prediction method of low-order blade frequency noise of non-cavitation propeller in time domain.* Journal of Harbin Engineering University. 34(1), 1-6.
[7] Zhang, G, M., Li, Z, R., He, W. (2016). *Numerical simulation of open-water characteristics and cavitation performance for a highly skewed propeller*," 26th Int Offshore and Polar Eng Conf, Greece, ISOPE.

© 2018 ASME

Numerical simulation of collapsing vapor bubble clusters close to a rigid wall

[1]Daria Ogloblina*; [1] Steffen J. Schmidt; [1] Nikolaus A. Adams;

[1]*Institute of Aerodynamics and Fluid Mechanics, Technical University of Munich, Germany;*

Abstract

Numerical simulation is a promising way for improvement of physical insight into collapse dynamics for clouds with strong bubble-bubble interaction. The current study adopts the theory proposed by Brennen et al [6] based on "cloud interaction parameter", although the theory is a simplification of realistic cloud collapses, it shows main features, such as shock focusing and enhancement of the resulting shock strength. We offer high quality CFD Simulations of collapsing vapor-bubble clouds close to a rigid wall. The simulations contain physical details such as wave dynamics, bubble deformation and vapor bubble rebound. The simulation time captures several sequential collapses. Our results show different collapse structures as well as different interaction mechanisms of bubbles. We present focusing features as well as unfocused collapses and discuss their effects on material loads. . Increasing the interaction parameter leads to an enhancement of the collapse pressure of individual bubbles, enhanced vapor production during rebound and enhanced load at the wall. An increase in stand-off distance changes the collapse dynamics significantly. Depending on the interaction, rebounds may occur close to the wall, leading to intense secondary collapses.

Keywords: cloud cavitation; cavitation erosion; bubble cluster collapse; bubble interaction parameter;

Introduction

Collapsing single bubbles have been extensively investigated in the past and results obtained by experiments, theory and simulations are broadly represented in the literature [1], [2]. Significantly fewer publications can be found where larger bubble clusters are considered [3-5]. Although there is a simple theory provided by Brennen et al [6], clusters of bubbles are difficult to investigate experimentally. Therefore, numerical simulation of collapsing bubble clusters is a suitable way to enhance physical insight into collapse processes.

To estimate the interaction among bubbles a cloud interaction parameter β was introduced [4]. This parameter is defined as $\alpha_0 \cdot (1-\alpha_0) \cdot A_0^2/R_0^2$, where α_0 is initial cloud void fraction and A_0 and R_0 are the cloud and the bubble radius, respectively. It was shown in previous studies [7], [8] that cloud collapse mechanisms are significantly depending on the cloud interaction parameter. In this investigation the interaction parameter is chosen to be 4, 8 and 11, indicating that bubble interaction is relevant.

In the present work a single-fluid approach coupled with a thermodynamic equilibrium model is adopted. When the grid resolution is sufficiently fine, the thermodynamic equilibrium model implicitly resolves the interface and becomes comparable to a sharp interface technique without surface tension [3]. It enables the prediction of re-evaporation processes, such as bubble rebound.

Two major physical aspects, namely the cloud interaction parameter and the standoff-distance, are investigated. First, a series of three bubble clusters at a fixed stand-off distance from a wall are generated. The radii of the approximately spherical clusters are kept constant while the common radii of the involved bubbles are varied. This leads to three different cloud interaction parameters. In a second series of computations, the cloud interaction parameter is kept constant while four different stand-off distances are investigated.

For all configurations, the dominant collapse mechanism, rebound structures and maximum impact loads are analyzed. A major finding is that the secondary collapse after the first rebound may lead to the highest loads at the solid wall. This indicates that the secondary collapse can considerably contribute to cavitation induced material erosion.

Methodology

The main focus of this investigation is on inertia-driven effects and wave dynamics arising during the collapse of a bubble cloud. Surface tension and viscous effects are neglected and the governing equations are the compressible Euler equations. For this investigation the flow simulation code CaTUM is used [9]. All numerical procedures have been extensively validated and the results have been published in the literature [10-14]. The compressibility of the fluid is taken into account in order to capture shock formation and wave propagation.

Phase change is modelled with equilibrium phase transition, i.e. both phases are in thermal, mechanical and phase equilibrium. In this study, the thermodynamic properties of water (in its liquid and vapor phases) are described by efficient barotropic equations of state. Pure liquid can be modelled by a Tait equation while saturated mixtures are

*Corresponding Author, Daria Ogloblina: daria.ogloblina@tum.de

modelled by a barotropic relation $p = p(\rho)$, which is obtained by integration of the equilibrium speed of sound along an isentrope. The vapor volume fraction of a saturated mixture of liquid and vapor can be derived from the saturation densities of both phases.

As there is insufficient experimental data of "reference clouds" available, a random procedure proposed by *Schmidt et al* [3] is utilized to generate bubble clusters. The following criteria are assumed:

1. The bubble cluster covers a spherical domain with a diameter of 30 mm, 150 spherical bubbles of equal radii are randomly generated within it, and the minimum distance between bubbles is one bubble radius. The bubble number density has its maximum at the center of the cloud and follows a *K/r* rule. Here, *K* is a positive constant dependent on the selected bubble radius and *r* is the radial coordinate from the center to the outside of the bubble cluster; **2.**The generated cloud is embedded within a small computational domain of 35x35x40 mm³ which itself is embedded in a larger domain of 4x4x2 m³. In the small domain a fine Cartesian grid is applied in order to minimize numerical errors. The large outer domain contains a hexahedral grid where grid cells are stretched towards the boundary, which are inviscid adiabatic walls; **3.** All bubbles are filled with water vapor while the surrounding domain contains liquid water at an initial pressure of p_∞ = 100 bar. The initial pressure inside the bubbles is equal to the vapor pressure p_{sat} = 2340 Pa at T = 293K. The velocity field is initially at rest. **4.** The effects of surface tension, viscosity, gravity and non-condensable gas content are neglected.

Problem setup

The aim of this investigation is to examine the influence of two parameters (cloud interaction parameter β and stand-off distance d) with respect to the collapse process and on maximum loads at the wall. Therefore, two setups are generated and simulated. For both setups the radius of the generated cloud is A_0 = 15mm and the number of bubbles equals 150 for all cases. Setup details are shown in the **Table 1.**

The first setup contains three random clouds with different initial bubble radii of R_0 = 0.5 mm, 1.0 mm and 1.5 mm, and the interaction parameter β equals to β = 4, 8 and 11 respectively. The positions of individual bubbles within this setup are case dependent. The stand-off distance d = 3.2 mm of the cloud from the wall is kept constant.

The second setup is designed to investigate the effects of different stand-off distances to the wall. The clouds are identical in terms of bubble distribution and initial bubble radius, which is R_0 = 1 mm. The stand-off distance is defined as the minimum distance between the virtual surface of the cloud and the bottom wall and equals to d = 1.2 mm, 3.2 mm, 6.0 mm and 10.0 mm.

Parameters	Setup I	Setup II
Bubble radius, R_0 [mm]	R_0 = 0.5, 1.0, 1.5	R_0 = 1.0
Vapor ratio, α_0 % [-]	α_0 = 0.4%, 3%, 8%	α_0 = 3%
Cloud interaction parameter, β [-]	β = 4, 8, 11	β = 8
Standoff distance, d [mm]	d = 3.2	d = 1.2, 3.2, 6.0, 10.0

Table 1: Properties of the designed setups.

Results

Setup I: Variation of interaction parameter β. The variation of this quantity will affect the cloud collapse behaviour, e.g. the strength of bubble-bubble interaction, intensities of collapses of individual bubbles and of the focused collapse at the center of the bubble cluster.

Figure 1 shows the dimensionless temporal evolution of the equivalent radius R_{equiv} as well as of the output from the numerical pressure transducer at the wall for all three interaction parameters. It can be seen that the relative duration of the cloud collapse is about 60% of the case-specific Rayleigh time [15] for all three configurations. However, the subsequent re-evaporation is quite different between the three cases. The smaller the interaction parameter, the more pronounced is the delay between collapse and rebound. This tendency seems to hold for subsequent collapses of re-evaporated structures as well.

With respect to the evolution of the pressure recorded at the wall, the following trends are found. Weak interaction leads to a low frequency increase of the pressure. Higher interaction causes high frequency pressure oscillations and larger amplitudes (see **Fig. 1**). However, a comparison of the maximum pressure recorded by the numerical transducers subsequent to the collapse phases show an unexpected behaviour, as depicted in **Figure 2**. While the load at the wall decays more or less monotonically in case of strong interaction, it stays approximately constant in case of very weak interaction. In case of intermediate interaction (here for $\beta = 8$) a nonlinear trend is predicted. In this case, the most intense load at the wall occurs after the secondary collapse.

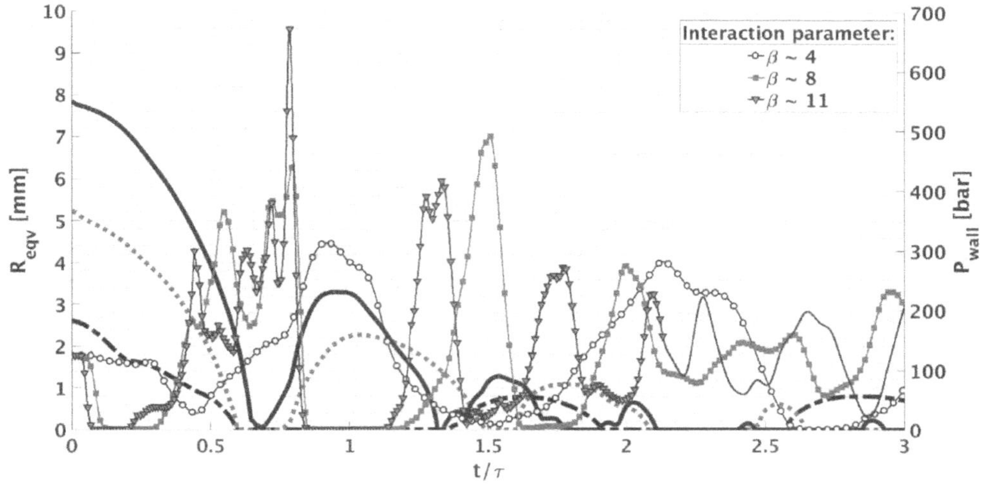

Fig. 1: Setup I: Thick line – equivalent bubble radius evolution; thin line - pressure recorded by the wall transducer; both against dimensionless time scaled with Rayleigh time for equivalent bubble

Setup II: Variation of stand-off distance d. Figure 3 shows the dimensionless temporal evolution of the equivalent radius R_{equiv} as well as of the output from the pressure transducer at the wall for all four stand-off distances. Again, the duration of the collapse of the primary cloud is approximately 60% of the Rayleigh time of an equivalent bubble for all cases. It can be seen that the duration of the primary collapse is slightly shorter when the stand-off distance is increased. The wall partially confines the flow field for small stand-off distances, while it has minor effects when the cloud is located sufficiently far away from it. As $\beta = 8$, moderate interaction intensity and collapse focussing occurs for all cases.

An unexpected behaviour is seen for the second and the third rebound of vapour structures. The maximum amount of vapour is inversely proportional to the initial stand-off distance during the first rebound. This is, however, not the case for the second rebound, where the tendency seems to be reversed. The output of the numerical pressure transducer at the wall shows that the pressure increases during the end of a collapse phase and features several maxima during collapse and rebound (**Fig. 3**). This observation holds for all four stand-off distances and all computed collapse-rebound-cycles.

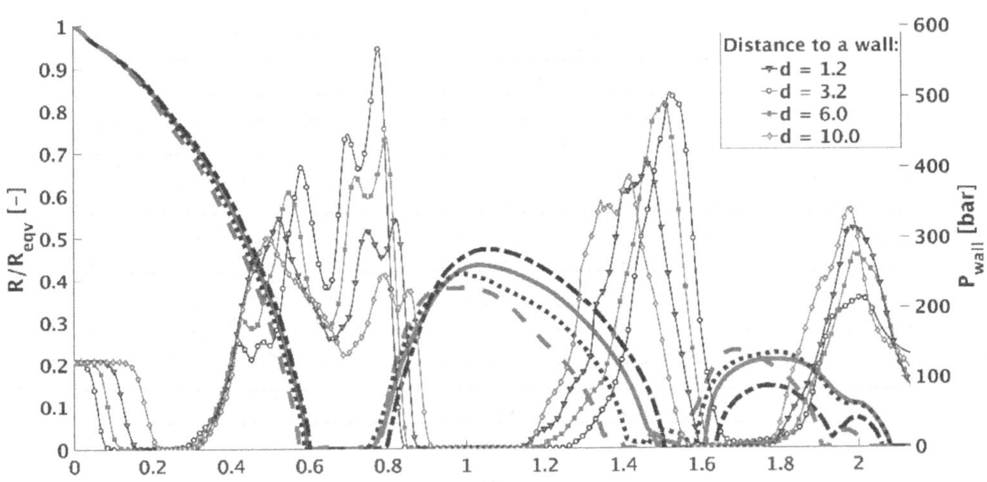

Fig. 3: Setup II: Thick line – equivalent bubble radius evolution; thin line - pressure recorded by the wall transducer; both against dimensionless time scaled with Rayleigh time for equivalent bubble

Figure 4 shows maximum pressures recorded at the wall after three collapse phases for all investigated stand-off distances. As expected, the intensity of the primary collapse is inversely proportional to the initial stand-off distance. This trend can be explained by the location of the focal point of the collapse, which locates more closely to the wall if the stand-off distance is smaller. Interestingly, the pressures recorded after the secondary collapses and especially subsequent to the third collapses show a nonlinear behaviour: for a stand-off distance of d = 1.2 mm a decrease in pressure is predicted for subsequent collapses. In case of d = 3.2 mm to d = 10 mm the pressures recorded after the secondary collapses exceed those recorded after the primary collapses. Although the collapse intensities decrease when comparing collapses 2 and 3, an unexpected trend is found: the pressures recorded after the third collapse are now proportional to the initial stand-off distance.

In order to understand this behavior, the maximum vapor volume during rebound and the position of the focal point of the subsequent collapse are investigated. The maximum vapour volume (or the radius of the equivalent bubble) is reversed in order when comparing rebound 1 to rebound 2. Additionally, the focal points are shifted towards the wall in all cases. Both contributions affect the pressure recorded at the wall. However, the maximum pressure at the focal point of a (focussed) cloud collapse is not necessarily related to the pressure recorded at the wall. For example, in case of d = 10 mm, the pressure detected at the focal point is much higher during the primary collapse compared to the subsequent collapses.

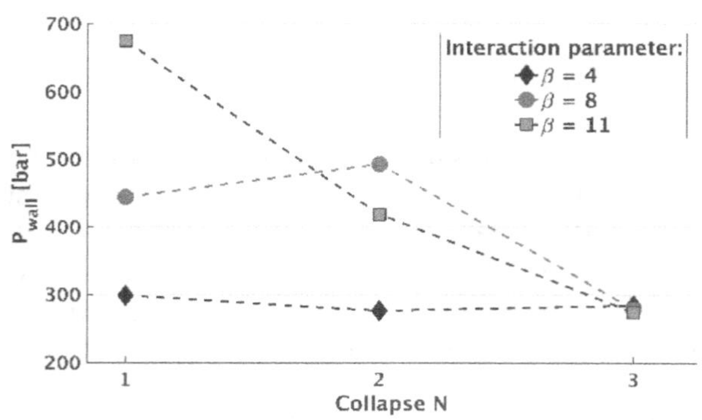

Fig. 2: Maximum pressure peaks recorded by the numerical wall pressure transducer subsequent to consecutive collapses

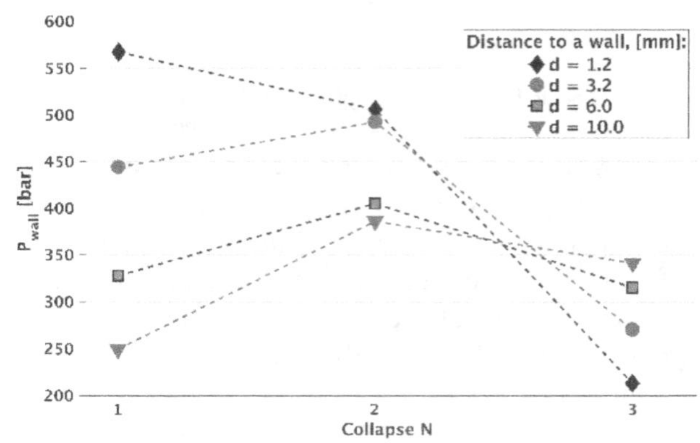

Fig. 4: Pressure peaks caused by primary collapse and two rebounds. Pressure of primary collapse is proportional to cloud distance to the wall. Intensity of second rebound collapse reverses intensity of primary collapse

Conclusion

This study shows the effects of the cloud interaction parameter and of the stand-off distance on the collapse of a cloud of vapor bubbles in a liquid ambient. Maximum loads at the surface are recorded by application of a numerical force/pressure transducer. Two series of numerical simulations with in total seven cases were performed and analyzed. The major findings are:

1. The cloud interaction parameter proposed by Brennen [4] is a suitable measure for bubble interactions and shock focusing. From this investigation it can be concluded that an interaction parameter of O(1) or smaller results in weak interaction without shock/collapse focusing. Once the interaction parameter is O(10) or higher, strong interaction and pronounced focusing are to be expected. Increasing the interaction parameter β leads to an enhancement of geometrical focusing and of the collapse pressure of individual bubbles. Additionally, the intensification of vapor production close to the wall during rebound is observed and, consequently, an increase of maximum loads at the wall is detected.

2. The stand-off distance is well suited to quantify the intensity of the primary cloud collapse with respect to its erosive potential at an adjacent material surface. However, this investigation shows that secondary and even third collapses subsequent to rebound processes can cause intense loads at material surfaces, too. Therefore, primary collapses and collapses of rebounded vapor structures are both potentially relevant for erosion. The primary collapse induces a velocity field which targets towards the wall and rebounding vapor patters are advected with this velocity field. Secondary and tertiary collapses occur closer and closer to the wall and the resulting load at the wall increases as well. Further studies of this work will include non-homogeneous initial bubble radii in order to assess effects of micro-scale details aside from macro-scale quantities such as the interaction parameter. Further numerical developments will be the inclusions of viscous effects and non-condensable gas content.

a = 0.4%, β = 4 *a = 3%, β = 8* *a = 8%, β = 11*

a – c: Pressure field and vapor iso-surface after the prior collapse before the second. In the first case no pronounced cloud appears during rebound. In the other two cases vapor rebound occurs right at the wall, a secondary cloud in the center of the domain. Clouds collapse from top to bottom in both latter cases.

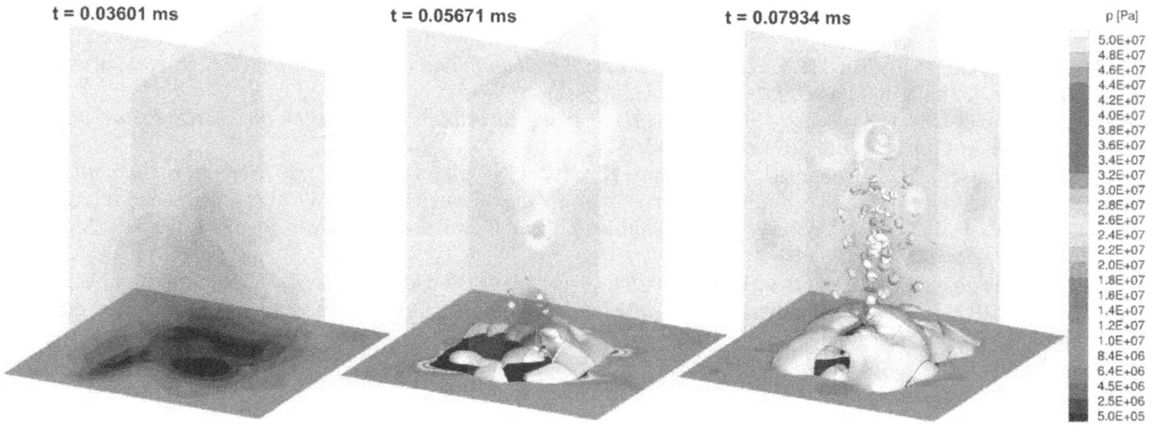

d – f: Pressure field and vapor iso-surface at the second collapse. In the first case relatively small collapse of $1.1 \cdot 10^4$ bar appears above the wall. In the middle case vapor at the wall collapses producing pressure peak of $1.78 \cdot 10^4$ bar. The last case shows the most violent collapse at the wall of $2.26 \cdot 10^4$ bar. This type of collapse is supposed to be highly relevant for cavitation erosion.

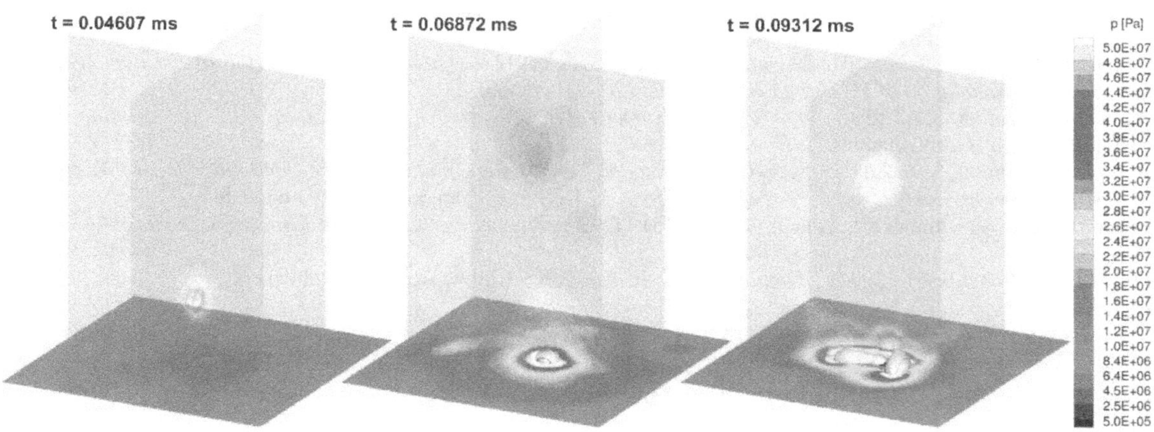

g – i: Pressure field and vapor iso-surface at the third collapse. In the first case the position of a collapse shifts further towards the wall and it occurs right at the wall with pressure magnitude $0.4 \cdot 10^4$ bar. In the middle case the third collapse of $0.8 \cdot 10^4$ bar is also at the wall. In opposite, the last case shows no significant impact on the wall since vapor structures are driven far above the wall and then collapse with $0.07 \cdot 10^4$ bar. Two latter clouds collapse from bottom to top.

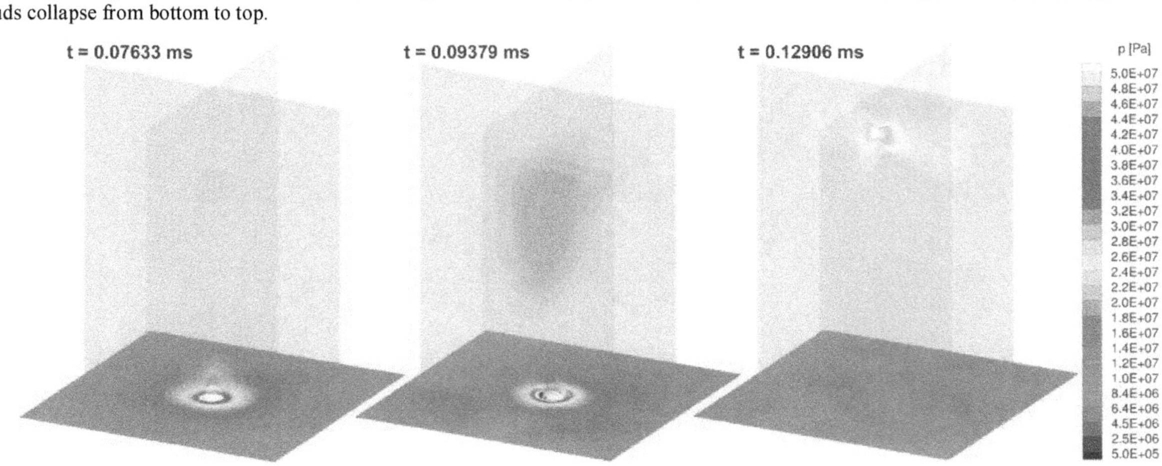

Fig. 5. Setup I: Pressure field depicted on two intersecting slices of the inner domain. White structures are iso-surfaces of vapor ratio equal to 5%. The initial vapor parameter and therefore interaction parameter increase from left to right.

Acknowledgment

This study is part of a project that has received funding from the European Union's Horizon 2020 research and innovation programme under grant agreement № 642536.

References

[1] Lauer, E. Hu, X. Y. Hickel, S. Adams, N. A. (2012). *"Numerical modelling and investigation of symmetric and asymmetric cavitation bubble dynamics".* Comput. Fluids 69, 1–19.
[2] Fujikawa S. and Akamatsu T.(1980). *"Effect of the non-equilibrium condensation of vapour on the pressure wave produced by the collapse of a bubble in a liquid."* J. Fluid Mech. 97, 481–512.
[3] Schmidt, S.J., Mihatsch, M., Thalhamer, M., Adams, N.A. (2011). *"Assessment of the Prediction Capability of a Thermodynamic Cavitation Model for the Collapse Characteristics of a Vapor-Bubble Cloud."* WIMRC Cavitation Forum..
[4] Wang, Y.-C., Brennen, C. E. (1999). "*Numerical Computation of Shock Waves in a Spherical Cloud of Cavitation Bubbles,"* J. Fluids Eng 121(4).
[5] Rossinelli, D., Hejazialhosseini, B., Hadjidoukas, P., Bekas, C., Curioni, A., Bertsch, A., Futral, S., Schmidt, S. J., Adams, N. A., Koumoutsakos, P. (2013). *"11 PFLOP/s simulations of cloud cavitation collapse"* International Conference on High Performance Computing, Networking, Storage and Analysis.
[6] Brennen C. E. (1995) "*Cavitation and Bubble Dynamics"*, Cambridge University Press.
[7] d'Agostino, L., Brennen, C. E., (1983). *"On the acoustical dynamics of bubble clouds".* AESME Cavitation and Multiphase Flow Forum, 72-75.
[8] d'Agostino, L., Brennen, C. E. (1999). *"Acoustical absorption and scattering cross sections of spherical bubble clouds,"* J. Acous. Soc. Am. 84, 2126-2134.
[9] Sezal I. H. (2009). "*Compressible Dynamics of Cavitating 3-D Multi-Phase Flows"*, PhD Thesis, TUM.
[10] Egerer, C. P., Schmidt, S. J., Hickel, S., Adams, N. A. (2016). *"Efficient implicit LES method for the simulation of turbulent cavitating flows."* J. Comp Phys.
[11] Mihatsch, M., Schmidt, S.J., Thalhamer, M., Adams, N.A. (2012). *„Quantitative Prediction of Erosion Aggressiveness through Numerical Simulation of 3D Unsteady Cavitating Flows."* CAV2012.
[12] Budich, B., Schmidt, S. J., Adams, N. A. (2016). *"Numerical Investigation of Condensation Shocks in Cavitating Flow."* 31st Symposium on Naval Hydrodynamics.
[13] Budich, B., Schmidt, S. J., Adams, N. A. (2016) *"Implicit Large Eddy Simulation of the Cavitating Model Propeller VP1304 using a Compressible Homogeneous Mixture Mode."* 31st Symposium on Naval Hydrodynamics.
[14] Beban B., Legat S., Schmidt S.J., Adams N. A. (2015). "*On instationary mechanisms in cavitating micro throttles."* CAV2015, J. Phys 656.
[15] Franc, J.-P., Michel, J.-M. (2004). "*Fundamentals of Cavitation."* Kluwer Academic Publishers.

© 2018 ASME

Numerical simulation of High Intensity Focused Ultrasound (HIFU) using a fully compressible multiscale model

Aswin Gnanaskandan*; Chao-Tsung Hsiao; Georges Chahine

DYNAFLOW, INC., MD, USA

Abstract

A fully compressible multiscale model for simulation of bubble enhanced High Intensity Focused Ultrasound (HIFU) is presented. The non-linear ultrasound field is modeled using compressible Navier-Stokes equations on a fixed grid, while the microbubbles are tracked as discrete singularities in a Lagrangian fashion. These two models are two-way coupled to each other such that both the acoustic field and the bubbles influence each other. The energy absorbed by the medium locally due to the focused ultrasound and bubble dynamics is then used to compute the temperature rise in the focal and surrounding regions by solving a heat transfer equation over the insonation time period. We first simulate the HIFU field without microbubbles and characterize the pressure and temperature fields and compare against available experiments. Experimental validation in the presence of microbubbles is then carried out to demonstrate the accuracy of the model. We then study the effect of the microbubbles on altering the temperature rise obtained in the focal region.

Keywords: HIFU; Euler-Lagrange method; microbubbles

Introduction

High Intensity Focused Ultrasound (HIFU) uses the focused energy of sound waves to elevate temperature locally, causing thermal ablation of tissues. For deep seated cancers, higher intensity sound waves are required to reach the target area [1]. However, this may cause unwanted tissue damage outside of the targeted region. A major impediment with the current use of HIFU technique to efficiently treat deep seated cancer is the long treatment time whilst using higher intensity for deeper penetration. In order to reduce undesirable damage of surrounding tissues, the resting time for cooling in the pre-focal regions has to be increased. It is therefore desirable to obtain higher temperature elevations at the focal region while still using moderate intensity levels (100 – 1000 W/cm^2) [2]. Introducing microbubbles in the form of contrast agents has been shown to increase temperature levels in both *in vitro* [3], [4] and *in vivo* experiments [5], [6]. However the exact mechanism of how the microbubbles contribute to the heat increase has only been speculative due to the difficulty in both experimental measurements and computational modeling.

The physical mechanism of heat enhancement through bubbles is thought to be primarily from three mechanism viz. acoustic emission, viscous dissipation, and thermal exchange [7]. Acoustic damping of the bubble oscillations arises because the bubble radiates acoustic energy and becomes important primarily when the bubble undergoes inertial oscillations. Viscous damping arises primarily from viscous dissipation in the relatively thin layer of host medium surrounding the bubble that moves during the bubble oscillations. The dissipation is directly proportional to the bubble interfacial velocity and the shear viscosity of the host medium. Thermal damping arises from the heat flow into the bubble during its expansion phase and out of the bubble during its collapse phase. The net amount of such heat flow for a particular bubble oscillation depends on the oscillation amplitude, which, in turn, is a strong function of the insonation pressure, frequency, and equilibrium bubble size. The relative contribution of these terms, as explained above, depends on a variety of factors and numerical simulations can shed light on the importance of these terms, if properly modeled. The main objective of this work is to develop a numerical model to simulate HIFU in the presence of microbubbles, verify its validity and apply it to understand the effect of bubbles on heat enhancement.

Physical model and numerical method

A fully compressible Euler-Lagrange model is employed in this study. The background acoustic and flow field is solved using the compressible Navier-Stokes equations on a fixed grid, while the presence of microbubbles are modeled by solving Keller-Herring equation given by,

*Corresponding Author, Aswin Gnanaskandan: aswin@dynaflow-inc.com

$$(1-\frac{\dot{R}}{c})R\ddot{R}+\frac{3}{2}(1-\frac{\dot{R}}{3c})\dot{R}^2=\frac{1}{\rho}(1+\frac{\dot{R}}{c}+\frac{R}{c}\frac{d}{dt})\left[p_v+p_g-p_l-\frac{2\gamma}{R}-4\mu\frac{\dot{R}}{R}-\frac{4}{3}G\left[1-\left(\frac{R_0}{R}\right)^3\right]\right]+\frac{(\mathbf{u}_l-\mathbf{u}_b)^2}{4}. \quad (1)$$

Here, R is the radius of the bubble, $\dot{R}$ is the bubble interface velocity, ρ, c and μ are the density, sound speed and viscosity of the surrounding medium. p_v, p_g are the vapor and gas pressure inside the bubble and p_l is the liquid pressure driving the bubble dynamics. γ is the surface tension and G is the shear elasticity of the medium. We account for the motion of the bubbles by solving a motion equation, which accounts for effect of the slip velocity between the bubble of velocity $\mathbf{u}_b$ and the liquid of velocity $\mathbf{u}_l$.

$$\frac{d\mathbf{u}_b}{dt}=-\frac{3}{\rho_l}\nabla p+\frac{3}{4}\frac{C_D}{R}(\mathbf{u}_l-\mathbf{u}_b)\mathbf{u}_l-\mathbf{u}_b+\frac{3\dot{R}}{R}(\mathbf{u}_l-\mathbf{u}_b), \quad (2)$$

where C_D is a drag coefficient and the last term is a Bjerknes force term due to coupling between bubble volume oscillations and bubble motion.

The ultrasound source is modelled as an imposed pressure distribution with phasing on the inlet boundary (z=0) [8],

$$p(z=0,r,t)=p_0\sin\left[2\pi f_0\left(t+\frac{r^2}{2cF}\right)\right], \quad (3)$$

where p_0 and f_0 are the amplitude and frequency of the ultrasound and r and F are the radius and focal length of the transducer. Since the actual insonation time is of the order of 1 s, orders of magnitude higher than the acoustic field CFD computation, we used a decoupled approach [2], [9], [10], where after solving the dynamic equations, we separately solve a heat transfer equation,

$$\rho C_p\frac{\partial T}{\partial \tau}=K\nabla^2 T+q_{US,AC}+q_{VIS}+q_{THER}, \quad (4)$$

where ρ, C_p and K are the density, specific heat and thermal conductivity of the medium, T is the temperature and τ is time. The heat sources, q_*, in this equation are due to the viscous dissipation of ultrasound and acoustics, and due to viscous and thermal damping of the bubbles. The heat source due to ultrasound absorption and acoustic damping of the bubble is given by

$$q_{US,AC}=\mu_b\varepsilon_{kk}^2+2\mu\left(\varepsilon_{ij}-\frac{1}{3}\varepsilon_{kk}\delta_{ij}\right)^2, \quad (5)$$

where μ and μ_b are the shear and bulk viscosity of the medium and $\boldsymbol{\varepsilon}$ is the strain rate tensor. The heat addition due to the viscous damping of the bubble in a given control volume is given by

$$q_{VIS}=\sum_{j=1}^{N_i}\frac{q_{vis}^b f_{i,j}}{\sum_{k=1}^{Ncells}V_k^{cell}f_{k,j}}, \qquad q_{vis}^b=\left(4\pi R_b^2\right).\left(4\mu\frac{\dot{R}_b^2}{R_b}\right), \quad (6)$$

where $f_{i,j}$ is the smoothing kernel [11], V_k is the volume of the control volume. The heat addition due to thermal damping of bubbles is not modelled in this study and will be considered in future studies.

For closing the Navier Stokes equations, a stiffened gas equation of state is used. For details regarding the numerical method for the Eulerian and Lagrangian approached refer to [12] and [11]. All the results in this study have been obtained using axisymmetric simulations for the Eulerian phase, while bubble distributions were fully 3D.

Results and discussion

The acoustic calculation is first verified by computing a focused ultrasound field in water in both linear and non-linear regimes. For the linear regime, we compare with the experiments in [13], where a 1.1 MHz transducer of radius 35 mm and focal length 60 mm is used with a source pressure amplitude P_a = 0.01 MPa. We compare the focal scan data obtained along the axis with the numerical results in Figure 1(a). The focal scan showing the magnitude of pressure peaks along the axis agrees well with the experiment indicating that the area in which the ultrasound is focused is predicted properly. In Figure 1(b), we compare the pressure history obtained at the focal point for a 2.2 MHz transducer of 20 mm radius and 44 mm focal length at a source pressure of 0.29 MPa corresponding to the experiments in [8]. At this source pressure, non-linear steepening of the acoustic waves results in a shock wave at the focus, which is captured well by the numerical simulations.

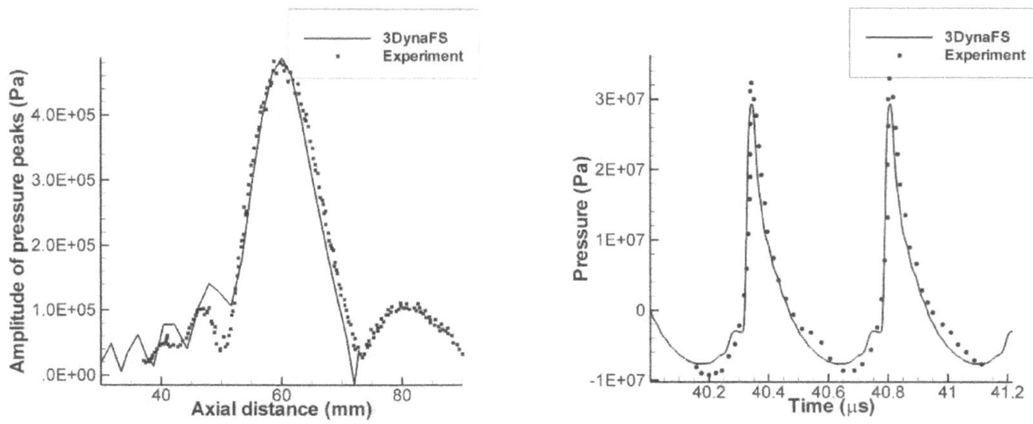

Figure 1: Characterization of the acoustic field in linear and non-linear regime in water. (a) Comparison of focal scan data along the axis with the experiments in [13] for 1 MHz transducer with source pressure Pa = 0.01 MPa, linear regime. (b) Comparison of time history of pressure at the focus with the experiments in [8] for 2.2 MHz transducer with source Pa = 0.28 MPa, non-linear regime. Profile shows non-linear steepening of acoustic waves.

Next, we validate the numerical method for a HIFU procedure in a phantom tissue without introducing any bubbles. The experiments are from [10], where a phantom tissue made of Agar is subjected to focused ultrasound from a 1.1 MHz transducer of radius 35 mm and focal length 60mm. The peak rarefaction at the focus is 1.1 MPa. The considered properties of the Agar are ρ=1,044 Kg/m^3, μ= 0.3 Pa.s and shear elasticity is neglected. The focused ultrasound field is depicted in Figure 2(a), where the acoustic waves are seen to focus at the geometric focus. The heat absorbed by the phantom tissue is obtained as a time averaged quantity over 0.01 ms and is further used as source term in the heat equation. The time averaged contours of heat release, depicted in Figure 2(b), shows heat concentration around the focal region. Note that the contours in Figure 2(b) are in logarithmic scale for clarity and that the highest value around the focus is at least an order of magnitude higher than that in the pre and post focal regions.

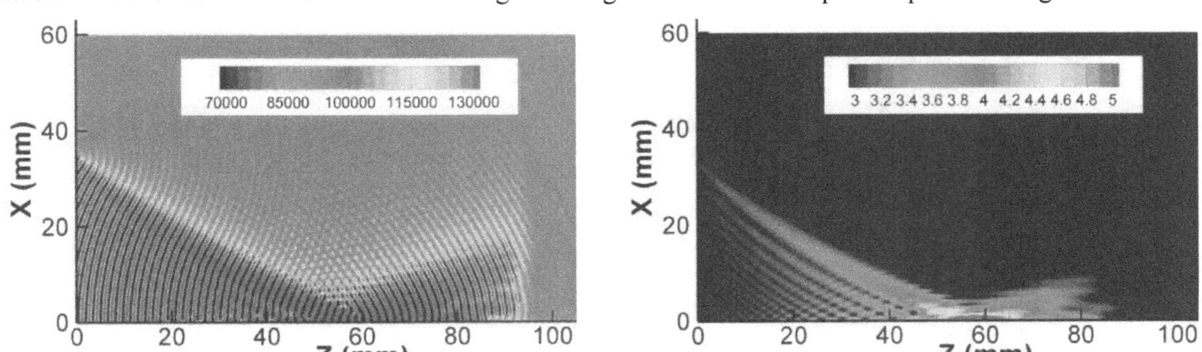

Figure 2: (a) Instantaneous contours of pressure showing focusing of the wave at z = 60 mm. (b) Contours of the time averaged (for 0.01ms) heat release computed from the hydrodynamic field. Logarithm of heat release is plotted for clarity. The simulations are based on the experiments in [10] with a 1 MHz transducer.

Comparison of the temperature history between simulation and experiments at the focus and at one off-focus location is depicted in Figure 3. The ultrasound source is on for 1s followed by a cooling time of 4 s. The simulations results show good agreement with the experiments indicating the accuracy of the physical model and the numerical method in simulating HIFU flow field.

Finally, the model is applied to study HIFU in the presence of microbubbles. The simulations correspond to the *invitro* experiments in [3], where a 2.2 MHz transducer of radius 20 mm and focal length of 40 mm is used to insonate a phantom tissue made of Polyacrylamide gel. The density of the gel is 1,060 Kg/m^3, the viscosity is 0.01 Pa.s, and the shear elasticity is 0.1 MPa. Bubbles of radius 1.3 μm are distributed uniformly in a cylindrical volume of length 10 mm and diameter 10 mm centered at the geometric focus (z = 40 mm). Acoustic calculations are carried out for 0.1 ms and the time averaged source terms for the heat equation are computed over 0.01 ms. The heat equation is then solved for 60 s with the source terms. Figure 4(a) shows the instantaneous pressure contours in the presence of bubbles. Two different void fractions $\alpha = 1x10^{-4}$ and $1x10^{-5}$ (shown in Figure 4 (a)) are used. The scattering of pressure waves due to the bubbles is evident; the bubbles also attenuate the ultrasound waves before they reach the focus. The temperature history obtained at several locations near the geometric focus are compared to the experimental data [3] in Figure 4 (b). Since the location of the temperature measurement in the experiment is not known, we only compare qualitatively the temperature profile at a few locations around the geometric focus. Although the quantitative agreement is not satisfactory, the simulations reproduce the qualitative trend in the experiment.

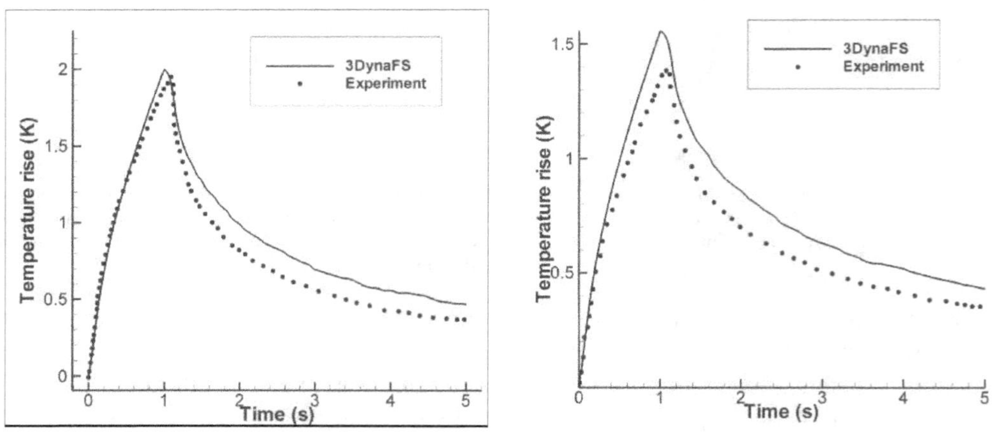

Figure 3: Comparison of the temperature history obtained from HIFU simulations in phantom tissue without bubbles with experiments in [10] for 1s of insonation followed by cooling for 4s. (a) At the focal point, z = 60mm. (b) At a point away from the focus, at x = 0.1 mm and z = 60 mm.

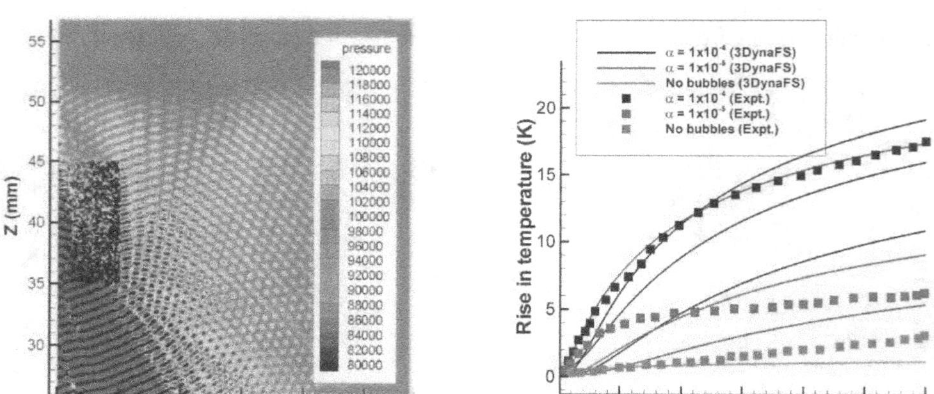

Figure 4: (a) Instantaneous pressure contours from HIFU simulations in phantom tissue with bubbles. The bubbles are distributed uniformly inside a cylinder of length 10mm and diameter 10 mm centered at the geometric focus as in the experiments in [3]. The bubble sizes have been magnified by 125 times to enhance visibility. (b) Comparison of the temperature history with the experimental data at a few locations in the focal region.

It is evident that the presence of bubbles leads to an increase in temperature. However, it is known that the presence of bubbles can attenuate the ultrasound propagation and hence it is necessary to ascertain whether the increased temperature is obtained in the desired focal region or not. Figure 5(a) shows the distribution of temperature along the axis for different void fractions. It is evident that the presence of bubbles leads to an increased temperature; however, the location of the peak temperature shifts to the geometric pre-focal region as the void fraction is increased. The attenuation is larger for larger void fractions, which also results in the shifting of the focal region towards the transducer. Figure 5(b) shows the effect of the localization of bubbles for $\alpha = 1\text{x}10^{-5}$. The results with the bubbles distributed in a fictitious cylinder of length 5 mm and diameter 5 mm, are compared with the experimental configuration of 10 mm x 10 mm cylinder. With the 5 mm cylinder, the bubbles are present only near the focus, thereby the attenuation of the ultrasound by the bubbles is reduced leading to a higher temperature increase.

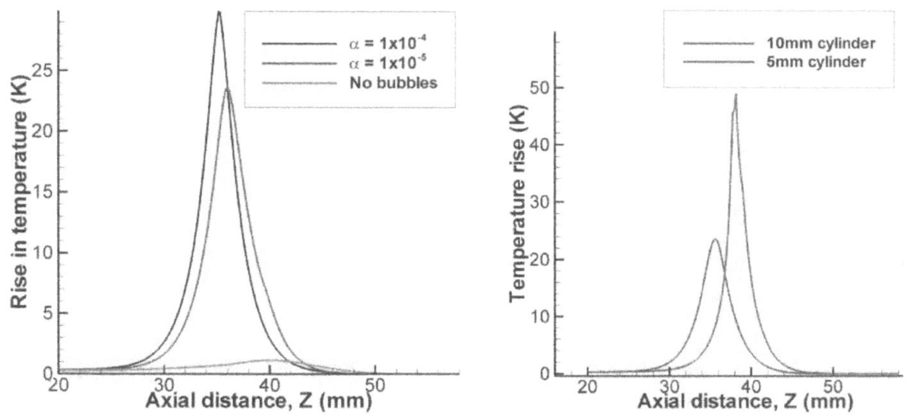

Figure 5: (a) Distribution of the temperature along the Z-axis obtained from the 3DYNAFS-COMP© / 3DYNAFS-DSM© coupled simulations for different void fractions. (b) Effect of bubble localization around the focus (z = 40 mm) on the temperature distribution along the Z-axis.

The individual contributions of the heat source due to ultrasound absorption, acoustic damping and viscous damping of the bubbles are plotted Figure 6(a) and (b). The heat due to ultrasound absorption and acoustic damping are combined as a single term. The viscous damping contribution is only at the location of the bubbles and its magnitude is approximately 2 orders of magnitude higher than the contribution of the acoustic dissipation to heat release. Both contours are represented using their respective logarithmic values for clarity. This demonstrates that for the current simulations, viscous damping of the bubbles plays the main role in adding heat locally in the vicinity of the bubbles. However, due to the long insonation time (60s), the heat diffuses away from the bubbles.

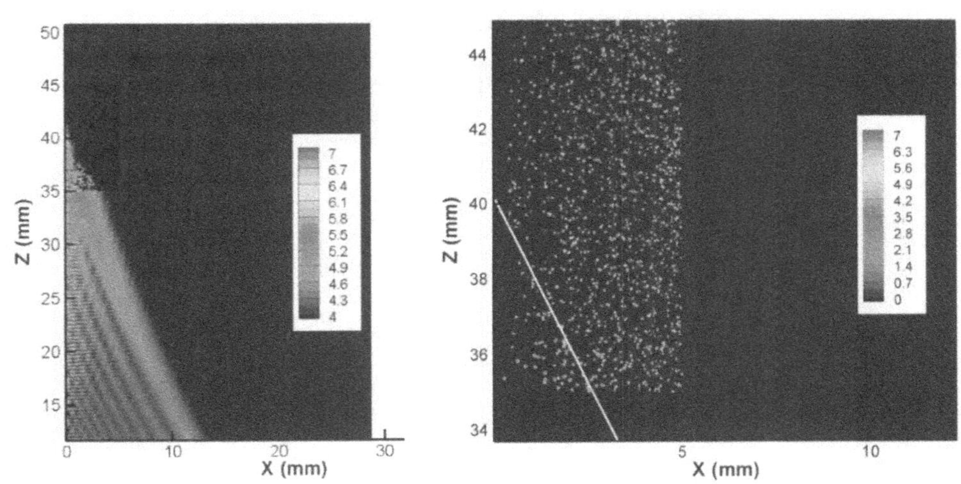

Figure 6: (a) Time averaged contours of heat source due to the absorption of ultrasound by the medium and acoustic damping of bubbles. (b) Time averaged contours of heat released due to the viscous damping of bubble oscillations by the surrounding medium.

Conclusion

A fully compressible multiscale numerical model is presented for the simulation of a bubble enhanced HIFU field. The non-linear acoustic field modeled in an Eulerian framework is coupled to a bubble dynamics solver, where the individual bubbles are tracked in a Lagrangian fashion. The method is first validated using *in vitro* experiments in the absence of microbubbles. Good agreement for temperature profile is obtained in and around the focal region. The method is then applied to HIFU simulation in the presence of microbubbles. Relatively good agreement of the temperature profiles around the focal region is obtained for two different void fractions. The effect of localization of bubbles is then demonstrated by concentrating the bubbles close to the focal location, which results in higher temperature increase due to reduced attenuation of the ultrasound.

Acknowledgements

We gratefully acknowledge the support from NIH under SBIR Grant No. 1R43CA213866-01A1

References

[1] J. E. Kennedy *et al.*, "High-intensity focused ultrasound for the treatment of liver tumours.," *Ultrasonics*, vol. 42, no. 1–9, pp. 931–5, Apr. 2004.

[2] P. Hariharan, M. R. Myers, and R. K. Banerjee, "HIFU procedures at moderate intensities—effect of large blood vessels," *Phys. Med. Biol.*, vol. 52, no. 12, pp. 3493–3513, Jul. 2007.

[3] K. Kajiyama, K. Yoshinaka, S. Takagi, and Y. Matsumoto, "Micro-bubble enhanced HIFU," *Phys. Procedia*, vol. 3, no. 1, pp. 305–314, Jan. 2010.

[4] D. Razansky, P. D. Einziger, and D. R. Adam, "Enhanced heat deposition using ultrasound contrast agent - modeling and experimental observations," *IEEE Trans. Ultrason. Ferroelectr. Freq. Control*, vol. 53, no. 1, pp. 137–147, Jan. 2006.

[5] Y. Kaneko *et al.*, "Use of a microbubble agent to increase the effects of high intensity focused ultrasound on liver tissue," *Eur. Radiol.*, vol. 15, no. 7, pp. 1415–1420, Jul. 2005.

[6] D. J. Chung, S. H. Cho, J. M. Lee, and S.-T. Hahn, "Effect of microbubble contrast agent during high intensity focused ultrasound ablation on rabbit liver in vivo," *Eur. J. Radiol.*, vol. 81, no. 4, pp. e519–e523, Apr. 2012.

[7] R. G. Holt and R. A. Roy, "Measurements of bubble-enhanced heating from focused, MHz-frequency ultrasound in a tissue-mimicking material," *Ultrasound Med. Biol.*, vol. 27, pp. 1399–1412, 2001.

[8] M. S. Canney, M. R. Bailey, L. A. Crum, V. A. Khokhlova, and O. A. Sapozhnikov, "Acoustic characterization of high intensity focused ultrasound fields: a combined measurement and modeling approach.," *J. Acoust. Soc. Am.*, vol. 124, no. 4, pp. 2406–20, Oct. 2008.

[9] K. Okita, K. Sugiyama, S. Takagi, and Y. Matsumto, "Microbubble behavior in an ultrasound field for high intensity focused ultrasound therapy enhancement," *J. Acoust. Soc. Am.*, vol. 134, no. 2, pp. 1576–1585, Aug. 2013.

[10] J. Huang, R. G. Holt, R. O. Cleveland, and R. A. Roy, "Experimental validation of a tractable numerical model for focused ultrasound heating in flow-through tissue phantoms," *J. Acoust. Soc. Am.*, vol. 116, no. 4, pp. 2451–2458, Oct. 2004.

[11] J. Ma, G. L. Chahine, and C.-T. Hsiao, "Spherical bubble dynamics in a bubbly medium using an Euler–Lagrange model," *Chem. Eng. Sci.*, vol. 128, pp. 64–81, May 2015.

[12] A. Kapahi, C. Hsiao, and G. Chahine, "A multi-material flow solver for high speed compressible flows," *Comput. Fluids*, 2015.

[13] J. Huang, "Heating in Vascular Tissue and Flow-through Tissue Phantoms Induced by Focused Ultrasound," *PhD Thesis*, 2002.

© 2018 ASME

Numerical Simulation of the Bubble Cloud Dynamics in an Ultrasound Field

Kazuki Maeda* and Tim Colonius

Division of Engineering and Applied Science, California Institute of Technology

1200 E California Blvd, Pasadena, CA 91125, USA

Abstract

We use a coupled Eulerian-Lagrangian method to simulate the dynamics of a spherical bubble cloud with various void fractions excited by high-amplitude ultrasound pulses. We consider two cases: a single cycle of a sinusoidal waveform whose wavelength is large compared to the cloud diameter, and multiple cycles with a short wavelength. For the long wavelength, bubble cloud dynamics are nearly spherically symmetric. Bubbles near the periphery grow more than the those close to the center, and the collapse of bubbles propagates inward from the periphery of the cloud. The structure and the dynamics of the cloud are scaled with the cloud interaction parameter introduce by d'Agostino and Brennen. It is shown that polydispersity does not significantly alter the cloud dynamics. In the short wavelength case, the clouds develop an anisotropic structure in the direction of the incident wave propagation. Over a wide range of the void fraction, the distal side of the cloud is shielded from the incident wave and bubbles grow less. As characterized by the center of volume of the cloud, the anisotropy is similar over the range of volume fractions considered. The results of the study can be used to characterize the acoustic cavitation in ultrasound therapies.

Keywords: cloud cavitation; numerical simulation; Eulerian-Lagrangian method

Introduction

The dynamics of bubble clouds nucleated by ultrasound are important for successful outcomes of ultrasound therapy, including high-intensity focused ultrasound (HIFU) and lithotripsy. Bubbles can collapse violently to cause tissue injuries as well as to scatter and absorb incident ultrasound waves and reduce the energy focused on the target [1-3]. Due to the fast, unsteady nature of the bubble cloud dynamics of the bubble clouds, accurate, quantitative experimental measurements are limited, and therefore modeling and numerical simulations have been essential tools for understanding the physical mechanisms at play. Many previous studies have focused on the dynamics of monodisperse, spherical bubble clouds in an otherwise incompressible liquid so that the wavelength of the pressure excitation is much larger than the size of the cloud. D'Agostino and Brennen [4] used a volume averaging approach to analyzed the linearized dynamics of spherical bubble clouds under weak pressure excitation and found that the cloud interaction parameter, $B = \beta R_C^2/R_b^2$, dictates the effect of inter-bubble interaction on the resulting dynamics of the cloud, where β is the void fraction, R_C and R_b are the radii of the cloud and the bubbles, respectively. Wang and Brennen [5] extended the study to the nonlinear regime, observing strong collapses involving bubbly shock waves that propagate inward from the periphery of the cloud. Matsumoto and Yoshizawa used a similar approach to quantify amplifications of the collapse when excited by HIFU waves with a resonant frequency of the cloud, and discussed applications of the collapse energy to ablate a target in medical applications [6]. More recently, Ma *et al* used an Eulerian-Lagrangian approach to investigate the dynamics of bubble clouds near a rigid wall excited by pressure waves [7].

In practical applications of HIFU and lithotripsy, the scale separation invoked above does not necessary hold; bubble clouds with a size of $O(1)$ mm have been observed *in vitro* during the passage of a strong ultrasound wave with a wavelength of $O(1)$ mm [8]. In order to gain insights into such clouds, we simulate the dynamics of spherical bubble clouds with a radius of 2.5 mm containing bubbles with a mean initial radius of 10 um excited by an ultrasound wave with an amplitude of 1 MPa. The simulation uses a coupled Eulerian-Lagrangian method for bubbly-liquid that can capture the dynamics of individual spherical bubbles and the fine structure of the pressure field. The numerical method and its validation have been described elsewhere [9]. We use the following two kinds of ultrasound wave: 1 cycle of a sinusoidal wave with a wavelength of 30 mm, and 10 cycles of a sinusoidal wave with a wavelength of 5 mm. These cases are chosen to explore the differing bubble cloud dynamics when the wavelength is either long or short compared to the cloud size. For each wave, we conduct a parametric study varying the volume fraction of bubbles, and thus cloud interaction parameter. In the long wavelength case, we qualitatively reproduce the results of the aforementioned studies invoking the scale separation. We show that the dynamics of the clouds are not significantly altered by the

*Corresponding Author, Kazuki Maeda: maeda@caltech.edu

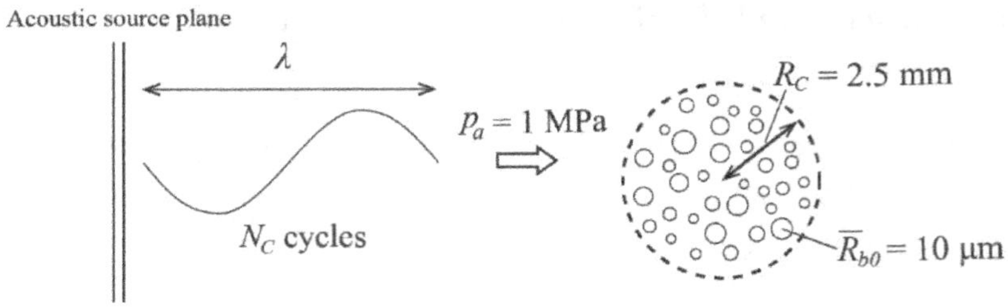

Figure 1. Schematic of the problem.

	σ	N_C	λ / R_C
Case 1	0	1	12
Case 2	0.7	1	12
Case 3	0	10	2

Table 1. Parameters used in the simulation cases

polydispersity of the bubbles. For the short wavelength case, we observe an anisotropic structure in the cloud during the passage of the wave, in that the proximal bubbles grow to larger radius than those in the distal side, within a wide range of the parameters of the cloud. This anisotropy is the result of a shielding of the distal bubbles from the incident wave. We discovered that the anisotropy is well characterized by the center of volume in the clouds.

Problem statement

Fig. 1 shows a schematic of the problem. We send N_C cycles of a planer, sinusoidal pressure wave with an amplitude of 1 MPa and a wavelength of λ mm from an acoustic source plane to a spherical bubble cloud with a radius of 2.5 mm. Bubbles are randomly seeded in the cloud in the initial condition. The initial radii of the bubbles, R_{b0}, are selected from a log normal distribution that follows $\ln(R_{b0}/\bar{R}_{b0}) \sim N(0, \sigma^2)$, where $\bar{R}_{b0}$ is the reference value of the initial radius: $\bar{R}_{b0}$ = 10 um. We characterize the bubble cloud by using the cloud interaction parameter defined as $B = \beta R_C^2/R_b^2 \sim N_b \bar{R}_b/R_C$, where N_b is the total number of the bubbles in the cloud. This expression of B is useful since it is independent of the polydispersity. The combinations of N, λ and σ addressed in the simulation are summarized in table 1. In all the cases, we vary N_b to realize the four distinct values of B: B = [0.625, 1.25, 2.5, 5]. Note that in the monodisperse case, B is a linear function of the void fraction. For each set of parameters, we run five independent simulations with randomized bubble positions in order to assess the variability with respect to the distribution of bubbles.

Numerical setup

We simulate the dynamics of bubbly-mixture using an Eulerian-Lagrangian method [9]. We discretize the governing equations on an axi-symmetric grid and integrate it using a finite volume, fifth-order WENO scheme. The incident pressure wave propagates in the positive axial (+x) direction. The grid size is uniform with a radial and axial grid spacing of 100 μm in the wave-cloud interaction region. The bubbles are modeled as sub-grid-scale spherical cavities that experience volumetric oscillations excited by pressure fluctuations in the surrounding liquid. The radial evolutions of each bubble is found by solving the Keller-Miksis equation. Heat and mass transfer across the bubble-water surface are accounted for using a reduced order model. Bubble-scattered pressure waves are resolved on the grid. The incident pressure wave is generated with a source-term approach [10].

Results and discussion: long wavelength cases

In this section we discuss the results of the long wavelength cases (case 1 and 2). Fig. 1a and b show the evolutions of the void fraction for several different values of B for the monodisperse (case 1) and polydisperse (case 2) clouds, respectively. In all clouds, after the passage of the incident wave, the void fraction steadily grows to reach its maximum

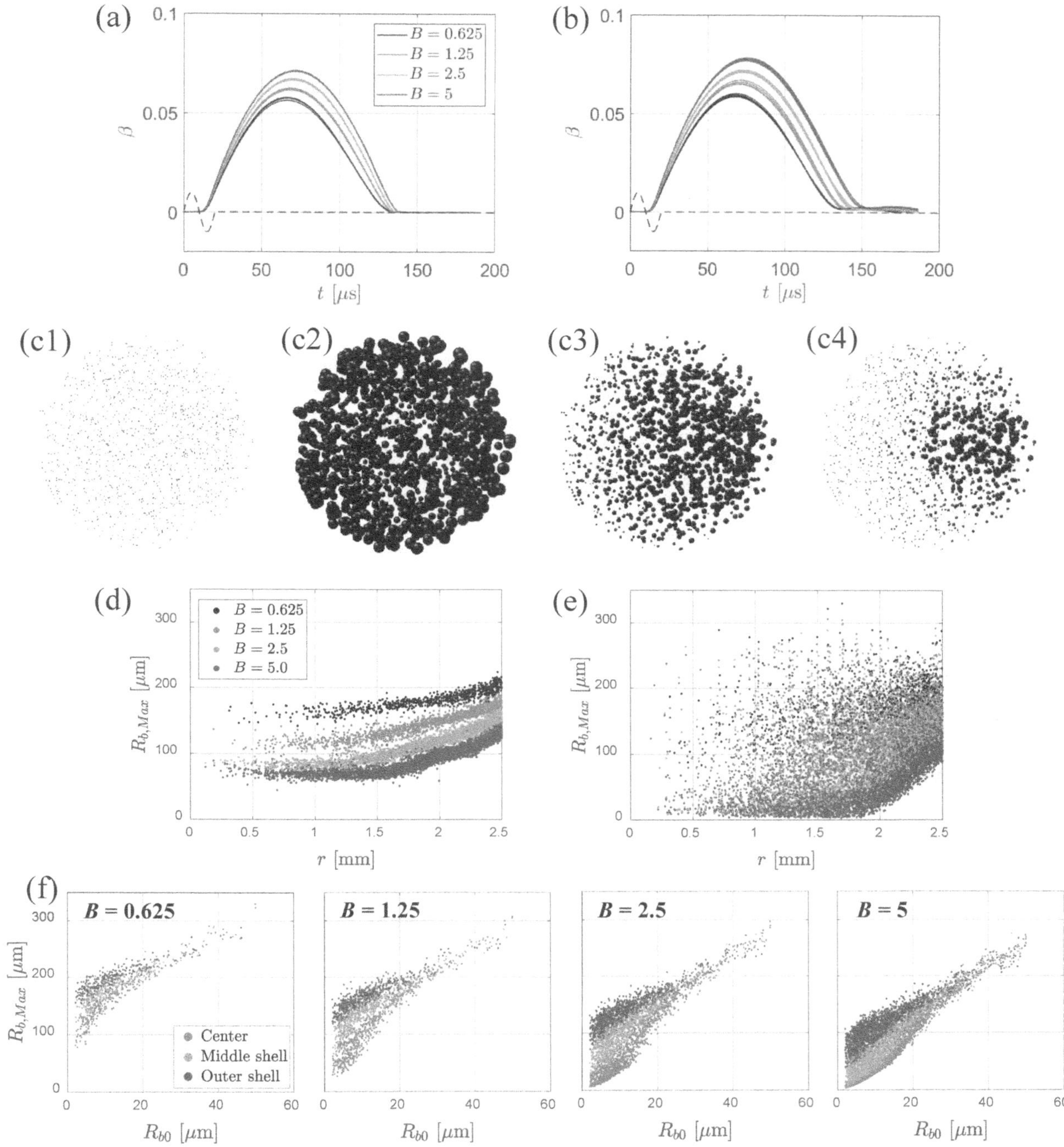

Figure2. (a,b) Evolution of the void fraction of the clouds. The dotted lines indicate the (scaled) incident wave pressure at the cloud center without bubble. For each value of B, all 5 runs with different randomized bubble positions are shown. (c1-4) Images of the initially monodisperse bubble cloud with $B = 5$ at t = 0, 76, 138, and 145 us. (d,e) Scatter plot of the maximum radii of each bubble over time and their radial coordinates. (f) Scatter plot of the maximum radii of the bubbles and their initial radii in the clouds of case2.

value around 0.05 – 0.08 at around t = 70 – 80 us, then collapses to zero at around t = 140 us. Slight rebounds are observed in the clouds in case2 after the collapse. The spread in the curves with the same value of B (same colors), resulting from the randomness of the coordinates of bubbles, are small in both cases. Interestingly, the results of case 1 and 2 do not show significant differences, suggesting that the effect of polydispersity is small. Fig. 1 c1-c4 show images of a representative cloud in case1 with B = 5 at the initial condition (c1), at its maximum void fraction (c2), and during the collapse (c3 and 4). The structures of the clouds possess nearly spherical symmetry, as bubbles

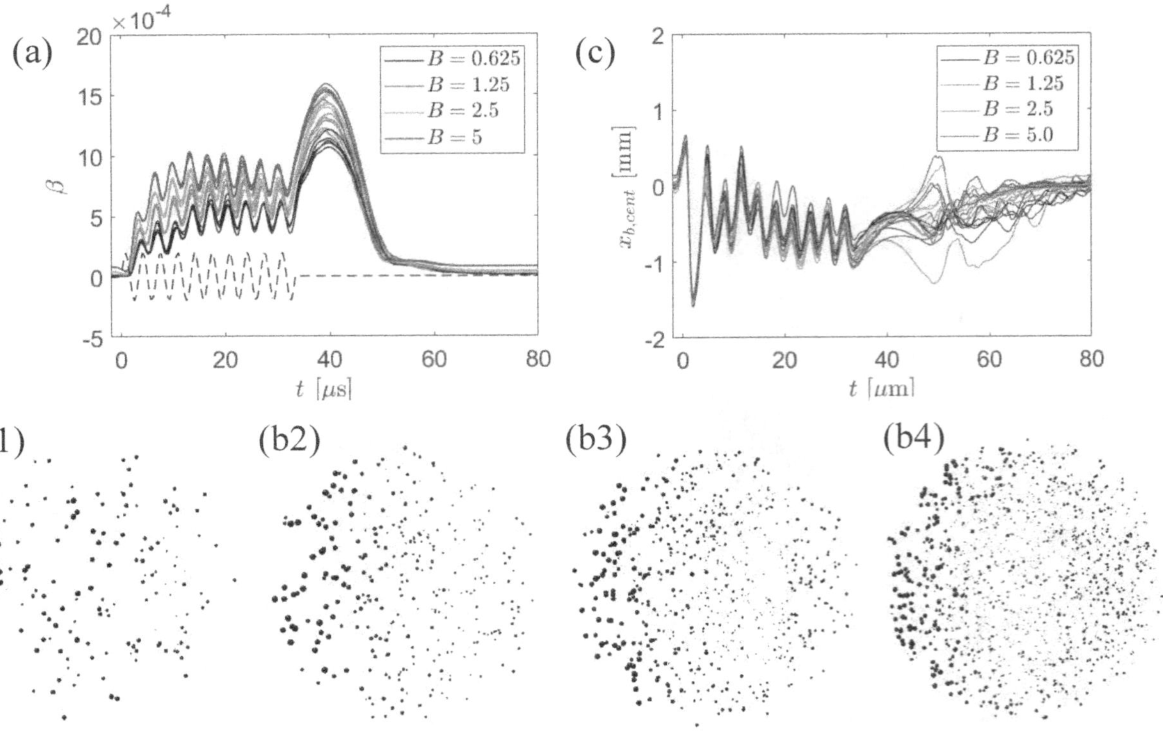

Figure 4. (a) Evolutions of the void fraction of the clouds. (b1-4). Images of bubble clouds with distinct values of B: B = 0.625, 1.25, 2.5 and 5, at t = 33 us. (c) Evolution of the volumetric center of the clouds.

experience uniform back-ground pressure during the passage of the incident wave due to its long wavelength. At its maximum void fraction, bubbles near the periphery of the cloud are larger than those near the center (c2). This can be explained by a shielding of the inner bubbles by the outermost layer; the outer bubbles scatter/absorb the acoustic energy of the incident wave to mitigate its penetration into the center. During the collapse, it is observed that the bubbles near the periphery collapse earlier, followed by the inner bubbles. This inward propagation of the collapse corresponds to the bubble cloud collapse involving bubbly-shockwave observed in the previous simulation [9] and experimental study [11]. The geometric center of the collapse is slightly offset to the right from the cloud center of the cloud in the present result. This offset can be explained by the slight delay in the arrival of the incident wave at the right side of the cloud due to the finite wavelength and the sound speed that were not considered in the previous studies.

In order to quantify the anisotropy in the growth of bubbles in the cloud, in fig. 2d and e we plot the maximum radius of the bubbles as a function of their radial coordinates (distance from the origin) for each value of B, for the monodisperse and polydisperse cases, respectively. In the monodisperse case (fig. 2d), the maximum radius has a clear monotonic, positive correlation with the radial coordinate. The slope of the correlation is increased with B, though the values of the radius tend to decrease with B. This results suggest that with increasing the inter-bubble interaction, the bubble growth is suppressed and anisotropy in the bubble growth is enhanced, as discussed in the aforementioned studies [4,5]. In the polydisperse case (fig. 2e), we do not observe clear correlations except for $B = 5$ indicated by a blue scattered points, due to higher dispersions in the bubble radius. For a better analysis of the polydisperse clouds, in fig. 2f we scatter the data points of the maximum radius of each bubble as a function of its initial radius separately for distinct values of B, with distinct colors based on bubble's radial coordinate: $r \in [0, 1.74]$ (center); [1.74, 2.18] (middle shell); [2.18, 2.5] mm (outer shell). Note that these three regions have approximately the same volume. The results show that for each value of B, the bubbles attain the largest radii in the outer shell, followed by the middle shell. The maximum radius tends to decrease by increasing B. Thus, it is indicated that the correlations observed in the monodisperse clouds hold true in the polydisperse case.

Results and discussion: short wavelength case

In this section we discuss results of the short wavelength case (case 3). We note that a single simulation of a similar cloud $B = 2.5$ and a weak polydispersity was reported in our previous study [9]. Fig. 4a shows the evolution of the void fraction of the clouds. For all the clouds, during the passage of the wave around until $t = 35$ us, the void fraction oscillates around at 0.5 - 1.0×10^{-3}. After the passage of the wave, it grows to reach its maximum value of 1.0 - 1.6×10^{-3} around at $t = 40$ us, then decays to reach zero around at $t = 60$ us. Variability due to the random positions of the bubbles is small. Fig. 4b1-4 show images taken during the passage of the incident wave with distinct values of B at $t = 33$ us. In all cases, we observe anisotropic structures in the cloud, whereby that the proximal side of bubbles grow more than those in the distal side. In the previous study, it was discussed that the anisotropy was the result of a shielding of the distal bubbles by the proximal ones. In order to quantify the anisotropy, we define the center of volume of the cloud:

$$x_{b,cent} = \frac{\sum_{cloud} \frac{4\pi}{3} R_b^3 x_b}{\sum_{cloud} \frac{4\pi}{3} R_b^3},$$

where x_b is the axial coordinate of each bubble. Fig. 4c plots the result. Interestingly, during the passage of the wave, until around $t = 35$ us, the evolution of $x_{b,cent}$ is similar for all simulations and oscillates between 0 and -1 mm. After the passage of the wave, $x_{b,cent}$ decays to zero with high variability. This result suggests that the structural anisotropy is always excited in the cloud during the passage of the wave and it is characterized by $x_{b,cent}$, regardless of the initial condition of the cloud within the range of parameters addressed in the present study.

Conclusion

We simulated the dynamics of bubble clouds with various cloud interaction parameters. With a single wavelength much larger than the cloud, the resulting cloud dynamics are nearly spherically symmetric, involve shielding by outer bubbles, and coherent cloud collapse propagating from the periphery toward the center. The results were scaled with the cloud interaction parameter, as formulated in the long wavelength limit [5]. We showed that the initial polydispersity of bubbles does not significantly alter the resulting dynamics of the clouds. With 10 cycles of a wave with a wavelength shorter than the cloud, an anisotropic structure is observed. The proximal bubbles in the cloud shield the distal bubbles. As characterized by the center of volume of the cloud, the anisotropy was similar for all cases simulated. The results of the study could be used to further understand and characterize the dynamics of cavitation bubble clouds formed during HIFU therapy and lithotripsy. Future work includes assessment of the far-field acoustics and the effect of polydispersity on the cloud dynamics in the short wavelength case as well as experimental validation.

Acknowledgement

K.M would like to acknowledge the Funai Foundation for Information Technology, for the Overseas Scholarship. This work was supported by the National Institutes of Health under grant P01-DK043881 and in part by the Office of Naval Research under grant N00014-17-1-2676. The simulations presented here utilized the Extreme Science and Engineering Discovery Environment, which is supported by the National Science Foundation grant number CTS120005.

References

[1] Pishchalnikov, Y.A., Sapozhnikov O.A, Bailey, M.R., Williams, J.C., Cleveland, R.O., Colonius, T., Crum, L.A., Evan, A.P., McAteer, J.A. (2004). *Cavitation Bubble Cluster Activity in the Breakage of Kidney Stones by Lithotripter Shockwaves.* Journal of Endourology. 17(7).

[2] Ikeda, I., Yoshizawa, S., Tosaki, M., Allen, J.S., Takagi, S., Ohta, N., Kitamura, T., Matsumoto, Y., (2006). *Cloud cavitation control for lithotripsy using high intensity focused ultrasound.* Ultrasound in medicine & biology. 32(9).

[3] Bailey, M.R., McAteer, J.A., Pishchalnikov, Y.A., Hamilton, M.F., and Colonius, T. (2006). *Progress in Lithotoripsy Research.* Acoustics Today. 3, 18-29.

[4] D'Agostino, L. and Brennen, C.E. (1989). *Linearized dynamics of spherical bubble clouds.* Journal of Fluid Mechanics. 199, 155-176.

[5] Wang, Y-C. and Brennen, C.E. (1999). *Numerical Computation of Shock Waves in a Spherical Cloud of Cavitation Bubbles.* Journal of Fluids Engineering. ASME Journal of Fluid Engineering. 121(4).

[6] Matsumoto, Y. and Yoshizawa, S. (2005). *Behaviour of a bubble cluster in an ultrasound field.* International Journal for Numerical Methods in Fluids. 47, 591-601.
[7] Ma, J., Hsial, C-T., Chahine, G.L. (2018). *Numerical study of acoustically driven bubble cloud dynamics near a rigid wall.* Ultrasonics Sonochemistry. 40, 944-954.
[8] Maeda, K., Kreider, W., Maxwell, D., Cunitz, B., Colonius, T., Bailey, M. (2015). *Modeling and experimental analysis of acoustic cavitation bubbles for Burst Wave Lithotripsy.* Journal of Physics: Conference Series. 656(1).
[9] Maeda, K. and Colonius, T. (2017). *Eulerian-Lagrangian method for simulation of cloud cavitation.* arXiv preprint arXiv:1712.00670.
[10] Maeda, K. and Colonius, T. (2017). *A source term approach for generation of one-way acoustic waves in the Euler and Navier-Stokes equations.* Wave Motion. 75, 36-49.
[11] Arora, M., Ohl, C.D., Lohse, D. (2007). *Effect of nuclei concentration on cavitation cluster dynamics.* The Journal of the Acoustical Society of America. 121, 3242.

© 2018 ASME

Numerical Simulation of the Collapse of Bubbles Near a Rigid Wall

Jing Zhang, Wei Lv, Xueming Shao, Lingxin Zhang

State Key Laboratory of Fluid Power Transmission and Control,

Department of Mechanics, Zhejiang University, Hangzhou, 310027, China

Email: zhanglingxin@zju.edu.cn

Abstract

A DNS study which employs VOF method is presented to investigate the collapse of cavitation bubbles near a rigid wall. The motion of bubbles is driven by the pressure difference between the interfaces of bubbles. We analyze the evolution of bubble deformation, pressure distribution and the impact received on the wall. The results shows that the bubbles far from the rigid wall collapse first, while other bubbles appear to collapse toward the wall subsequently. When the last bubble disappears, the impact received on the wall reaches a maximum value. Compared to the single bubble, multiple bubbles can generate greater degree of impact on the wall.

Keyword: cavitation bubbles, VOF, bubble collapse, multiple bubbles, rigid boundary.

1. Introduction

Cavitating bubbly flow can be observed in hydraulic system [1-3]. Numerous researchers have investigated the feature of motion of single bubble which collapses near a wall [4-8]. In the final stage of the collapse, a high-speed liquid jet and a high-pressure impulse are formed and impact on the wall, consequently lead to cavitation erosion. Rattray[5] studied the non-spherical collapse of single bubble near the wall and derived a formula for the prediction of the prolongation factor κ . Kim et al.[6] simulated the collapse of a gas bubble and found that the degree of pressure impulse is independent of the bubble size, but it is a function of the dimensionless distance between the bubble center and the wall. However, there is few work related to the motion of multiple bubbles. The motion of bubbles is more complicated, which involves the interaction between bubbles and the effect of the wall. In present study, we present the results of simulation of bubbles which collapse near a rigid wall. In first part, numerical methods are introduced briefly, then in second part, results and discussion is given.

2. Numerical method

The bubbles are set as vapour bubbles without incompressible gas. Phase transition and mass exchange around the interface is not considered, but the effects of surface tension, viscosity and compressibility are computed in simulation. The governing equations of this two-phase flow are shown as:

$$\nabla \cdot \vec{V} + \frac{1}{P}\frac{dP}{dt} = 0 \tag{1}$$

$$\frac{\partial(\rho\vec{V})}{\partial t} + \nabla \cdot \left(\rho\vec{V}\vec{V}\right) = -\nabla P + 2\nabla \cdot \left(\mu\vec{E}\right) + \sigma\, k\vec{N} \tag{2}$$

$$\frac{d\alpha}{dt} = \frac{\partial\alpha}{\partial t} + \left(\vec{V} \cdot \nabla\right)\alpha = 0 \tag{3}$$

where, ρ is the mixed density, k is the surface curvature, $\vec{N}$ is the unit normal vector of the interface, α is the volume fraction of liquid phase. A pressure equation can be derived from the continuity equation, and the pressure-velocity coupling is employed to solve the governing equations. Here, the compressibility of liquid phase has been considered in the modelling.

3. Computational configuration

Figure.1 shows the computational domain, where 27 cavitation bubbles are initially set near the bottom. The bubbles are evenly placed and each layer has nine bubbles. In order to reduce calculation parameters, all bubbles are initially set as spherical bubbles with same radius R_{max}. As shown in Figure.2, D is the distance between two bubbles, and L is the distance between the centre of the lowest bubble and the wall.

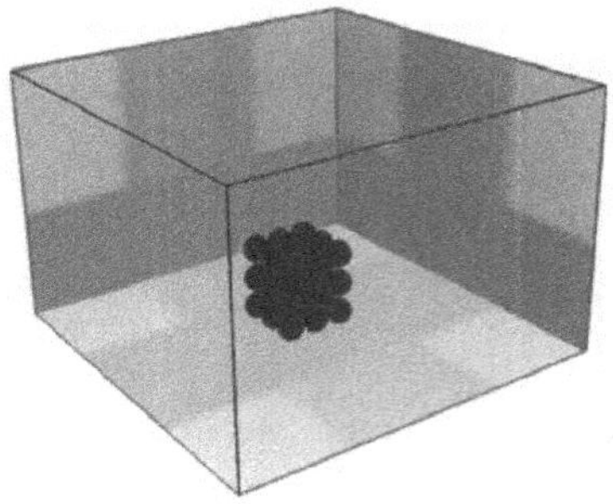

Figure 1. Computational domain

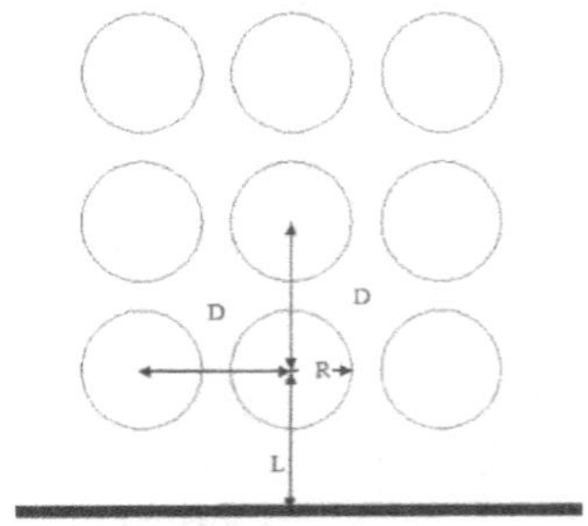

Figure 2. Geometry setup

For simplicity, we define the following dimensionless parameters:

$$\gamma = \frac{L}{R_{max}} \qquad P^* = \frac{P}{P_\infty - P_0} \qquad t^* = \frac{t}{t_c}$$

Where γ is dimensionless distance between lowest bubbles and rigid wall, P^* is dimensionless pressure, P_∞ is the pressure in liquid, P_0 is the pressure in the bubbles, t_c is the collapse time of spherical bubble in open water condition. The initial flow field is static, and $\gamma = 1.5$, $R_{max} = 1mm$, $D = 2.4mm$, $P_\infty = 101325Pa$, $P_0 = 3169Pa$.

4. Results and discussion

Figure.3 presents the results of evolution of bubbles deformation, with pressure contours at different times. It can be observed that the bubbles of top layer collapse first, with the delivery of high pressure into lower bubbles. This phenomenon produces a higher pressure environment for the rest of the bubbles which have not collapse. Due to the effect of the wall, the bubbles near the wall collapse slower and later than the upper layer of bubbles. These two factors, i.e., the interaction of bubbles and the effect of the wall, make the bubbles collapse from top to wall and layer by layer. In final stage of simulation, when last bubble collapses, it generates a strong pressure wave propagating in the flow field. The wall receives this impact of pressure wave very soon.

Figure.4 compares the volume change between single bubble and 27 bubbles during the collapse. The curve illustrates that with same dimensionless distance, driving pressure and distance D, the collapse time of 27 bubbles is much longer than the single bubble's. In this case, the prolongation factor of 27 bubbles is complicated, not just a function of the dimensionless distance. Figure.5 shows the time histories of the impulsive pressure of single bubble and 27 bubbles, respectively. At the final stage of collapse, the impact pressure received by the wall increases rapidly. The maximum value of the pressure is about 20 times the driving pressure for single bubble, while 90 times for 27 bubbles. This result indicates that multi-bubbles can produce larger impact than single bubble. In fact, the pressure generated in the flow field is much higher than the pressure received on the wall. The process of pressure wave propagation needs further observation.

5. Conclusions

The collapse process of 27 bubbles near a rigid wall has been investigated by a DNS approach. It is observed that the bubbles collapse layer by layer from top to the wall due to the interaction of bubbles and effect of the wall. The collapse time of 27 bubbles is longer than that of single bubble. Meanwhile, the maximum pressure received on the wall is much larger than that of the single bubble. The interaction of bubbles is responsible for the delay of the volume change and the higher impulsive pressure.

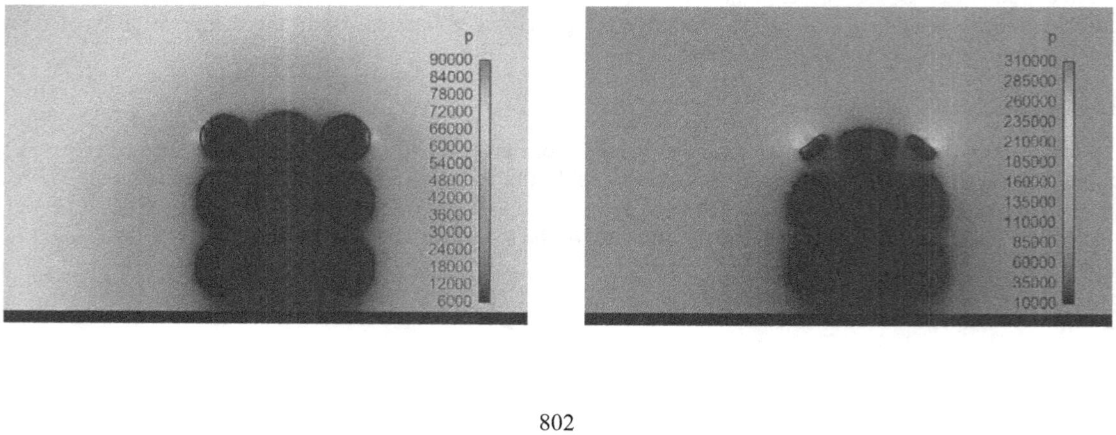

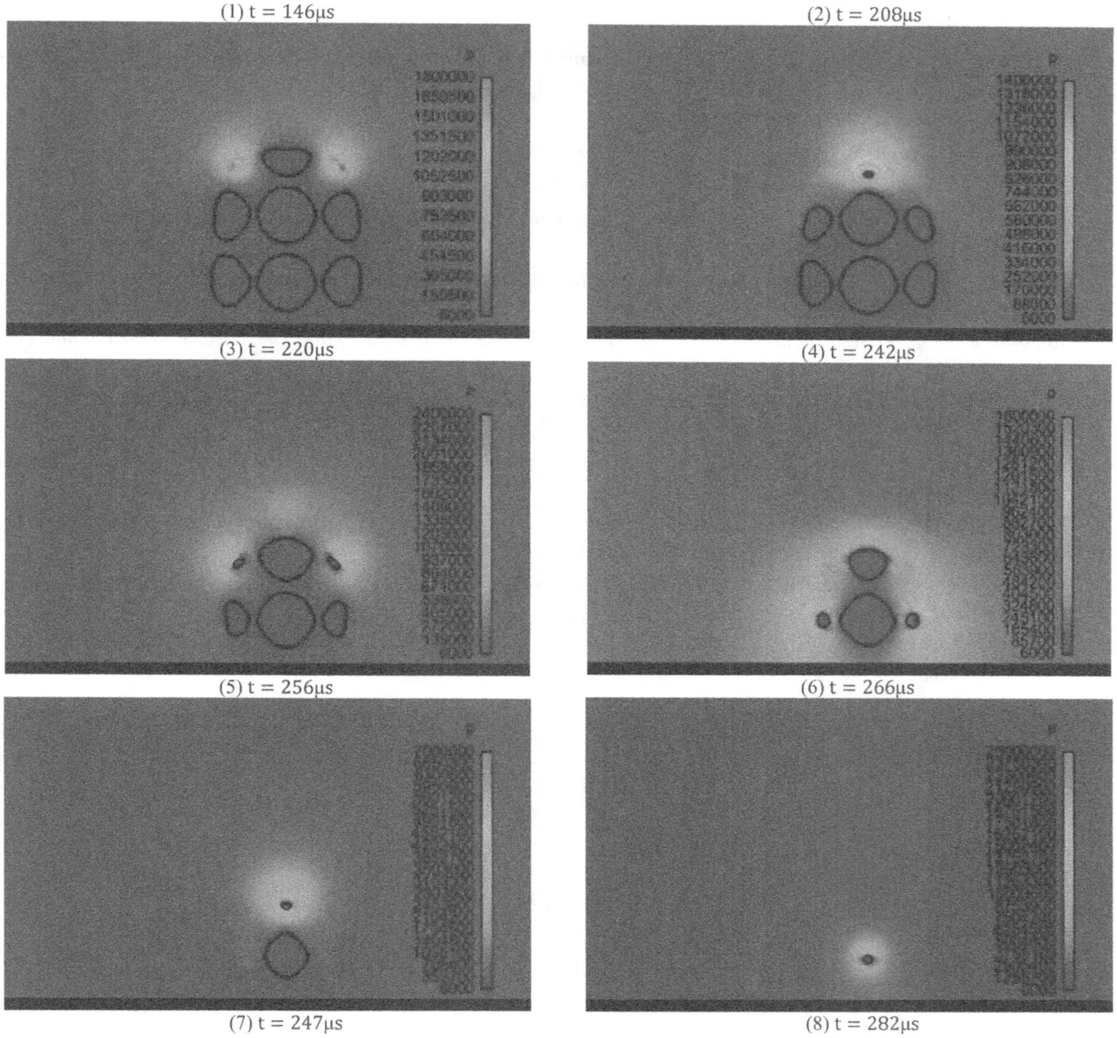

Figure 3. Bubble shapes and pressure contours on the centre section

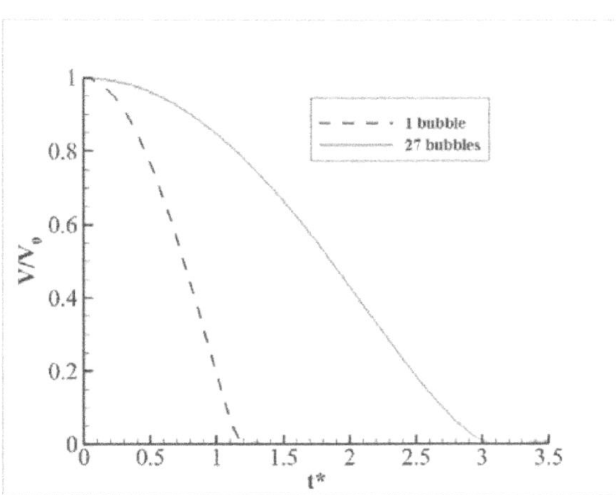

Figure 4. Volume evolution of bubble(s)

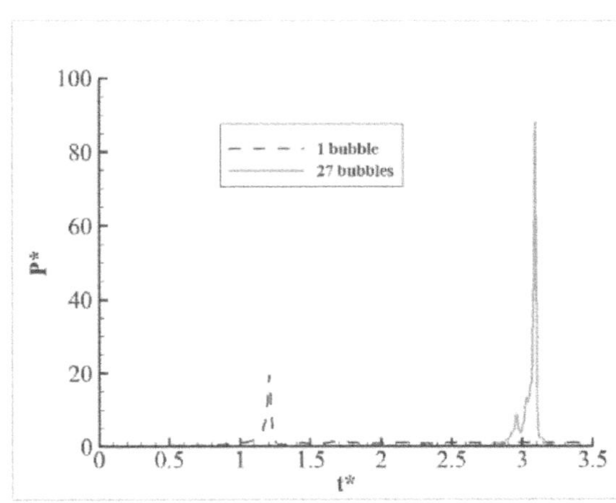

Figure 5. Time distribution of pressure at the middle point of the wall

Acknowledgement

This project was supported by the National Natural Science Foundation of China (NO. 11772298, NO. 11272284) and the State Key Program of National Natural Science of China (NO. 11332009).

References

[1] Li S 2000 *Cavitation of hydraulic machinery* World Scientific

[2] Arndt R E 1981 *Cavitation in fluid machinery and hydraulic structures* Annual Review of Fluid Mechanics 13 pp 273-326

[3] Escaler X, Egusquiza E, Farhat M et al. 2006 *Detection of cavitation in hydraulic turbines* Mechanical systems and signal processing 20 pp 983-1007

[4] Brujan E, Keen G, Vogel A et al. 2002 *The final stage of the collapse of a cavitation bubble close to a rigid boundary* Physics of Fluids 14 pp 85-92

[5] Plesset M S, Chapman R B 1971 *Collapse of an initially spherical vapour cavity in the neighbourhood of a solid boundary* Journal of Fluid Mechanics 47 pp 283-290

[6] Kim K H, Chahine G, Franc J P et al. 2014 *Advanced Experimental and Numerical Techniques for Cavitation Erosion Prediction* Springer

[7] Shaw S, Schiffers W, Gentry T et al. 2000 *The interaction of a laser-generated cavity with a solid boundary* The Journal of the Acoustical Society of America 107 pp 3065-3072

[8] Tomita Y, Robinson P, Tong R et al. 2002 *Growth and collapse of cavitation bubbles near a curved rigid boundary* Journal of Fluid Mechanics 466 pp 259-283

© 2018 ASME

Numerical simulation of the micro-jet velocity and cavitation-erosion on an axisymmetric nozzle

[1]Víctor Hidalgo*; [2]Xianwu Luo; [3]Xavier Escaler; [1]Esteban Valencia; [1]Albaro Aguinaga

[1]Escuela Politécnica Nacional, Quito, Ecuador; [2]Tsinghua University, Beijing, China; [3]UPC, Barcelona, Spain

Abstract

The present study deals with the numerical study of the micro-jet velocity and the cavitation- erosion on an axisymmetric nozzle, which presents high flow unsteadiness and aggressiveness. The proposed model is based on energy conservation for the micro-jet velocity prediction, and assumes the potential and kinetic energies, at the start and at the end of the cavity collapse, respectively. The mass transport between liquid and vapor is closely related to Rayleigh Plesset's equation. Results have been obtained with a numerical scheme assuming homogeneous mixture ow, implicit LES and Zwart-Gerber-Belamri cavitation model. The 3D unsteady flow simulation has been solved using OpenFOAM, whilst for the micro-jet estimation, Python language coupled with OpenFOAM's calculator have been used. Results clearly show that the implemented model captures adequately the phenomenon and enables to identify attacked areas. Outcomes agree reasonably with the experimental results obtained by Franc in the Laboratoire de Écoulements Géophysiques et Industriels. Furthermore, results of the implemented Zwart-Gerber-Belamri cavitation model with adaptive time step, showed more details related to the cavitation shedding and detachment, which are important factors to predict erosion.

Keywords: cavitation-erosion; micro-jet velocity; OpenFOAM; axisymmetric nozzle

Introduction

To study cavitation-erosion is important not only to prevent the damage on hydraulic machinery components as blades and guide vanes, but also for optimized designs [1, 2].The basic ideas of the aforementioned topic are based on the work carried out by Hammitt [3, 4], who proposed that the energy contained in a cavity is released as a water shock, which can be higher than the material threshold strength. Kubota, Kato and Yamaguchi [5] proposed that the cavity collapse must be in a trajectory close to the walls. Bark et al. [6] and Patella et al. [7] proposed that potential energy stored in cavities is released during their collapse in cascade with the consideration of an effective trajectory. Dular et al. [8, 9] proposed a numerical modeling of cavitation erosion using the concepts of potential energy liberated in cascade and the high speed micro-jet [10]. The model explains that the bubble collapse located in a specific layer releases energy as a wave, which can re-energize bubbles close to hydrofoil walls. The collapse of the bubbles cluster generates a high speed micro-jet and an impact load on the material similar to a water hammer.

Bearing in mind these aspects, Hidalgo [11] proposed a cavitation-erosion model using OpenFOAM and python language, which joined the potential energy in cavities with the kinetic energy in the micro-jet. In this context, the proposed model has been applied to study cavitation- erosion on a nozzle due to the highest aggressiveness of the phenomena, which was observed on Franc's studies [12].

Model Description

Homogeneous mixture flow assumption, implicit large eddy turbulence model (ILES) and the Zwart-Gerber-Belamri cavitation model (ZGB) have been used for the numerical simulation of unsteady cavitating flow, which is based on previous research [13].

*Corresponding Author, Víctor Hidalgo: victor.hidalgo@epn.edu.ec

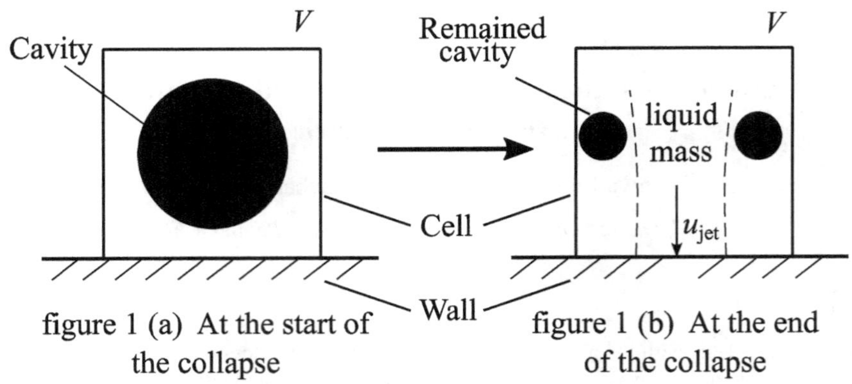

Figure 1: Collapse of a vapor cavity structure for one cell located over the solid wall

The model considers the collapse of a vapor structure in one cell of volume V as indicated in figure 1. The dimensionless micro-jet velocity, c_{flow}, is expressed in (1) as a function of the micro-jet velocity, u_{jet}, and the undisturbed flow velocity, U_∞ (31 m/s).

$$c_{\text{flow}} = \begin{cases} 8.97\gamma^2\sqrt{\left(\frac{p-p_{\text{v}}}{\rho_{\text{l}}U_\infty^2}\right)\left(\frac{\alpha_{\text{o}}-\alpha_{\text{f}}}{\alpha_{\text{o}}}\right)} & , \text{ if } (\alpha_{\text{o}}-\alpha_{\text{f}}) > 0 \text{ and } (p-p_{\text{v}}) > 0 \\ 0 & , \text{ if } (\alpha_{\text{o}}-\alpha_{\text{f}}) < 0 \text{ and } (p-p_{\text{v}}) < 0 \end{cases}, \tag{1}$$

where γ is a dimensionless distance from the bubble centre to the wall [8], p is the static pressure, p_v is the water saturation pressure, ρ_l is the density of the liquid water, α is the vapor volume fraction, and $_o$ and $_f$ are subscripts for the initial and final instant of the cavity collapse.

Furthermore, a qualitative value of flow aggressiveness, $\bar{c}_{flow}$, can be estimated using (2) to predict the affected region in relation with its maximum value after one characteristic cavitation cycle. Where $_i$ is the subscript for each affected cell, and t_o and t_f are the initial and final time of the characteristic cavitation cycle respectively.

$$\bar{c}_{\text{flow}} = \frac{\left[\sum_{t_o}^{t_f} c_{\text{flow}}\right]_i}{\max\left(\left[\sum_{t_o}^{t_f} c_{\text{flow}}\right]_i\right)}. \tag{2}$$

Computational domain and boundary conditions

The computational domain and boundary conditions indicated in figure 2 were based on the experimental studies carried out by Franc [12], and the numerical studies carried out by Koukouvinis et al. [14]. Structured mesh with a mean y^+ equal to 1.3 and 4.7 x 10^5 cells was used for the numerical simulation, which is in accordance with previous research [11, 15]. Furthermore, the mesh was refined close to the bending part, walls and the possible erosion target.

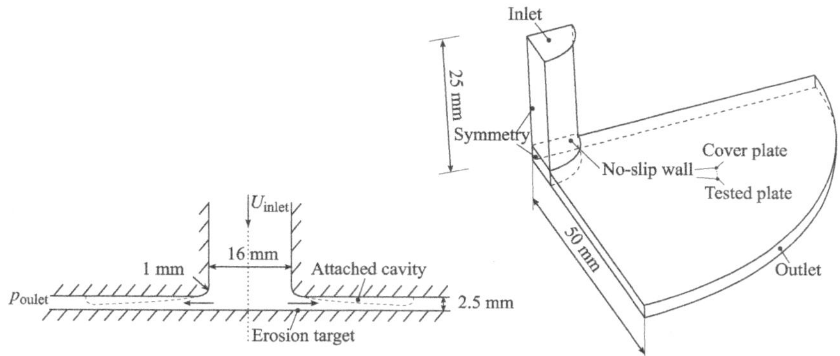

figure 2 (a) Axial view of the tested nozzle figure 2 (b) Computational domain

Figure 2: View with dimensions of the quarter part of the simulated nozzle

Results and discussion

Figure 3 (a) shows the comparison of erosion results between Franc's studies [12] and the proposed numerical simulation for the axisymmetric nozzle. The numerical result matched fairly well the affected region of the experimental result, which is between radius of 19 mm and 32 mm with a maximum $\bar{c}_{flow}$ between these two radius. A small region was selected from the numerical part for a better understanding of the phenomenon and it is indicated in figure 3 (b). From that region, a point called 1 was selected to study the u_{jet} and to compare with the critical velocity that the material can absorb, $u_{cri.}$, which is based on Lush's observations [16].

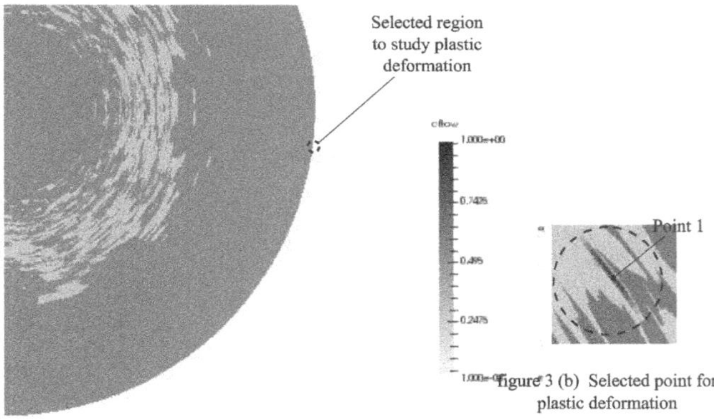

figure 3 (b) Selected point for plastic deformation

Figure 3: Comparison of erosion results between experimental studies and numerical simulations

Moreover, the pressure coefficient, C_p, and the dimensionless time, ξ, were estimated using equations (3) and (4) respectively.

$$C_p = \frac{p - p_\infty}{0.5\rho U_\infty^2} = -\sigma + \frac{p - p_v}{0.5\rho U_\infty^2}, \tag{3}$$

$$\xi = \frac{t - t_o}{t_f - t_o}, \tag{4}$$

where σ is the estimated cavitation number, which is equal to 0.9 [11, 12] and p_∞ is the undisturbed pressure [13]. Bearing in mind these aspects, c_{flow}, α and C_p have been plotted in function of ξ as indicated in function as indicated in figure 4.

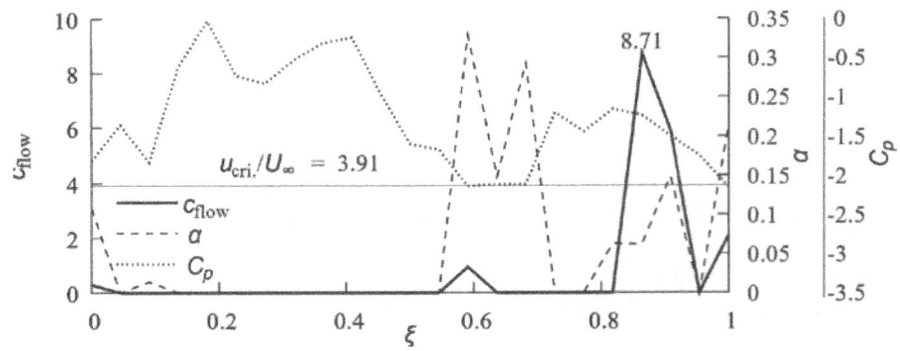

Figure 4: c_{flow}, α and C_p in function of ξ for point 1

The aforementioned figure shows that C_p fluctuates between -2 and 0 up to ξ = 0.6 mainly due to the nozzle configuration and the turbulence [13]. In this range α is small and hence it is not significant for cavity collapses. The maximum c_{flow} is given after ξ = 0.8, which is mainly relates to the fall of C_p with a value close to $-\sigma$ and the presence of α. Furthermore, the maximum c_{flow} is two times higher than the $u_{cri.}/U_\infty$ for steel, which was estimated based on Lush [16]. In this context, the tested plate of figure 3 can be eroded in a short period of time, which agrees with the experimental data [12].

Conclusions

Results of the present research show that the proposed model can reproduce the cavitation- erosion problem on an axisymmetric nozzle with enough accuracy compared with experimental data. The model can be applied to study the induced erosion by cavitation on industrial applications. Moreover, the maximum c_{flow} of 8.71 found in the assessment is equivalent to u_{jet} = 270 m/s, and it highlights the necessity of compressible correlations for further analysis.

Acknowledgments

The authors gratefully acknowledge the financial support provided by Escuela Politécnica Nacional for the development of the project PIMI 14-03.

References

[1] Escaler X, Farhat M, Avellan F and Egusquiza E 2003 *Wear* 254 441–449
[2] Escaler X, Farhat M, Egusquiza E and Avellan F 2007 *J. of Fluids Eng. of The ASME* 129 886–893
[3] Hammitt F, 1968. *Proceedings of the Institution of Mechanical Engineers* 183 31–50 ISSN 0020-3483
[4] Hammitt F, 1975. *Journal of Hydraulic Research* 13 1–17 ISSN 0022-1686
[5] Kubota A, Kato H and Yamaguchi H, 1992. *Journal of fluid Mechanics* 240 59–96
[6] Bark G, and et al 2004 *Appl. of princ. for obs. and analysis of eroding cavitation* vol 3.1 (EROCAV)
[7] Reboud J, Coutier-Delgosha O and et al 2003. *Fifth Int. Symposium on Cavitation (CAV2003)* pp 1–10
[8] Dular M, Bachert R, Stoffel B and Širok B 2005. *European Journal of Mechanics-B/Fluids* 24 522–538
[9] Dular M and Coutier-Delgosha O 2009. *Int. Journal for Num. Methods in Fluids* 61 1388–1410
[10] Plesset M S and Chapman R B 1971. *Journal of Fluid Mechanics* 47 283 ISSN 0022-1120, 1469-7645
[11] Hidalgo V 2016 *Num. study on unst. cav. flow and er. based on hom. mixt. assump.* PhD thesis Tsinghua
[12] Franc J P and Riondet M 2012 *Wear* 274 248–259 00019
[13] Hidalgo V, Luo X, Escaler X and et al 2015. *Journal of Hydrodynamics, Ser. B* 27 815–823 ISSN 1001-6058
[14] Koukouvinis P, Bergeles G and Gavaises M 2015. *J. of Hydrodynamics, Ser. B* 27 579–586 ISSN 1001-6058
[15] Peters A, Sagar H, Lantermann U and el Moctar O 2015 *Wear* 338339 189–201
[16] Lush P A 1983. *Journal of Fluid Mechanics* 135 373–387 ISSN 1469-7645

© 2018 ASME

Numerical simulation of tip leakage flow and cavitation observation on two-dimensional hydrofoil*

[1,2] Yuwen Liu*; [1,2] Xiaoxing Peng; [1,2] Lianghao Xu; [1,2] Yantao Cao

[1]*National Key Laboratory on ship Vibration & Noise, China Ship and Scientific Research Center, Wuxi, China*

[2]*Jiangsu Key Laboratory of Green Ship, China Ship and Scientific Research Center, Wuxi, China*

Abstract

Two-dimensional hydrofoil model of NACA0012 is chosen as the object for researching tip clearance flow in different gap width between the foil tip and the endwall by numerical simulations. The commercial computational software is applied to obtain the tip leakage flow field and analyze its variation with the different tip gap sizes. In the same time, the tip clearance cavitation observation experiments of NACA0012 hydrofoil models were carried out in cavitation mechanism tunnel of CSSRC. The cavitation inception number in different operating condition is studied in present paper.

Keywords: tip leakage flow; tip leakage vortex; tip clearance cavitation

Introduction

Tip leakage flow exists in many industrial applications, such as axial turbines and compressors, which require a small gap between the tip of the blades and the casing in order to ensure a rotor movement without rubbing. In an objective to minimize the energy losses, the tip clearance is usually kept as small as possible. Due to the pressure difference between the pressure side and the suction side, there is a tip leakage flow through the clearance region driven by the pressure gradient. When the leakage flow pass through the tip edge of the pressure and suction sides, a relatively steady eddy will come forth, because of the flow separation or the vorticity shed by leakage flow, which is named 'tip leakage vortex (TLV)' equivalent to a tip vortex developing in the vicinity of a wall. The TLV often makes some adverse effects include efficiency losses, rotating instabilities, cavitation in pumps and inducers.

The tendency of continuously increasing the output power of axial hydraulic machines, such as Kaplan, bulb and propeller turbines, is bringing the cavitation erosion issue to the front. Cavitation is likely to develop in the low pressure region that prevailsin the core of TLVs as well as in the clearance region. Up to now there are some progress achieved in the understanding of tip vortex cavitation to predict the onset and development of TLV cavitation for marine propellers and axial turbine. However the accurate numerical simulations and scale model tests are still great challenge.

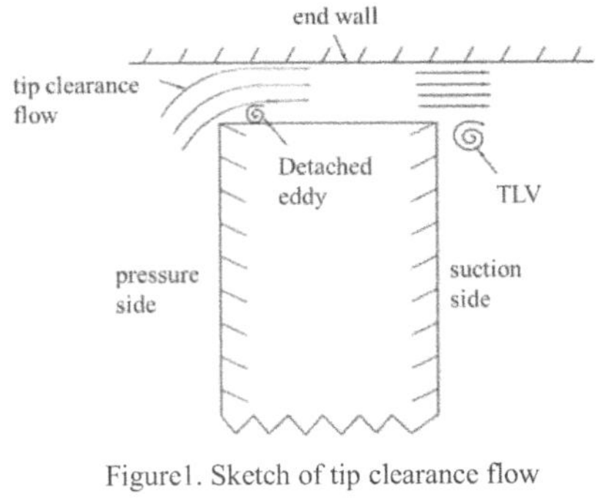

Figure1. Sketch of tip clearance flow

1. Numerical prediction

*Corresponding Author, Yuwen Liu: wsliuyuwen@163.com

This work was supported by the National Natural Science Foundation of China (Project No. 11332009 & 11072223).

Numerical simulation by commercial software to predict the flow field is widely used by many scholars for scientific research and engineering applications. And here the commercial computational software with unstructured grids is adapted for simulating the tip leakage flow. The two-dimensional hydrofoil model of NACA0012 with invariable chord length c =130mm along the span is chosen as object of this research. As shown in the figure2, the cuboid computational domain is setup with the section size 225mm×225mm which is same with the section size of high speed cavitation tunnel, and 1040mm length which is 8 times chord length of the model. The origin of coordinate system is located on the side wall with the 112.5mm distance to the bottom.

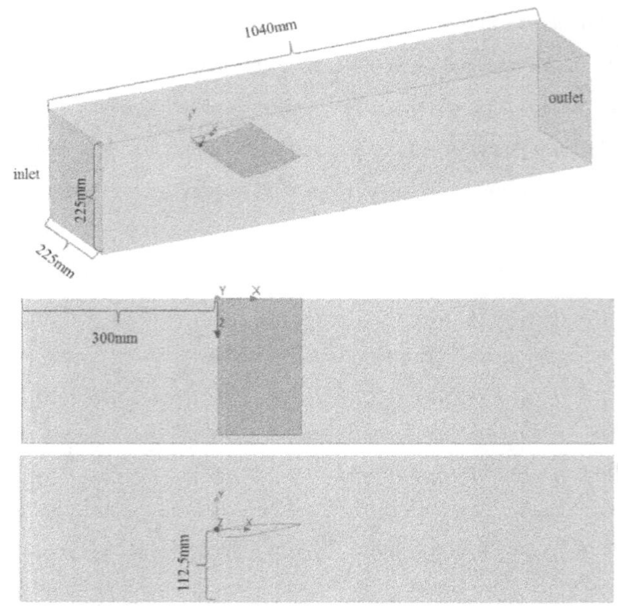

Figure2. coordinate system of computational region

The total number of unstructured cell in the computational domain is about 25million with some of local mesh refinement in special region, as shown in figure3. The incidence angle of the hydrofoil is set as α=7°. The clearance width δ between the foil tip and wall is nondimensionalized by the maximum thickness t of the model, $\tau=\delta/t$, and t=15.6mm. The inlet velocity V_0 is set as 8m/s.

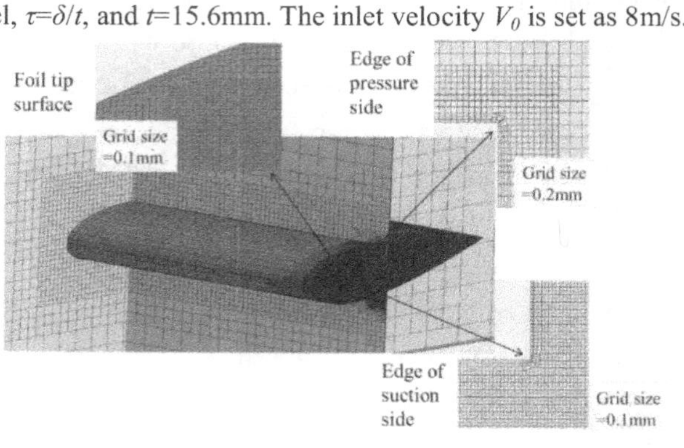

Figure3. grid setup and refinement of computational region

2. Experiment setup and method

The cavitation observation experiment is mainly aimed to study the influence of gap width variation on the tip leakage flow through the measurement of cavitation inception number σ, which is defined as follow:

$$\sigma = \frac{p_0 - p_v}{0.5\rho V_0^2}$$

where V_0 and P_0 are the incoming velocity and the static pressure, P_v saturation vapor pressure in the test water temperature and ρ water density. As shown in the figure4, a high-speed video camera of type of Photron APX with resolution 2048×2048 pixels and 4000fps was set on the tip side of test section to visualize the inception and development of tip leakage vortex cavitation. The gap width and attack angle of hydrofoil is adjusted by the special gear case.

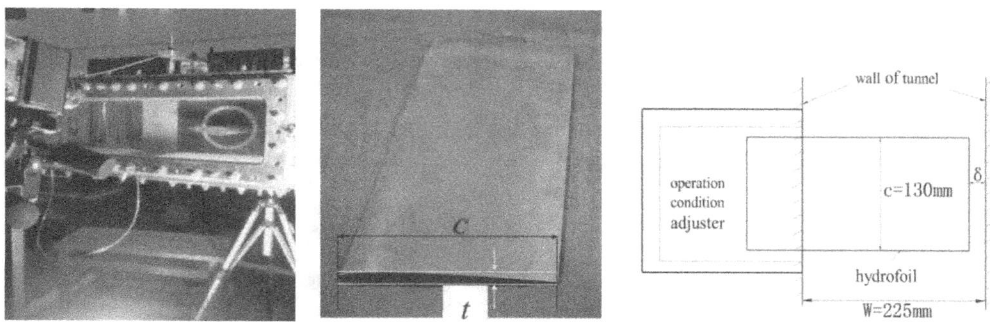

(a) (b) (c)

Figure4. experimental model and sketch of operation condition

3. Result analysis

As shown in the figure5-6, the result of numerical simulation shows that the tip leakage flow in different gap width from pressure side to suction side could provoke the formation of vortex system, and the most stabilized one beneath the suction side is so called 'tip leakage vortex', shorted as 'TLV'. Especially for the small gap width, TLV is the only distinct one, nevertheless, a quite obvious small vortex structure adjacent to the TLV, which moves downstream with the increase of gap width, as shown in the figure5(a). This phenomenon could be explained by the induced velocity produced by TLV. When the tip surface draw close to the wall, the tip leakage vortex will affect the velocity distribution in the boundary layer, which could generate the reverse vorticity, as shown in the vorticity distribution of figure6(a) for small gap width, such as τ=0.08, 0.17, 0.25.

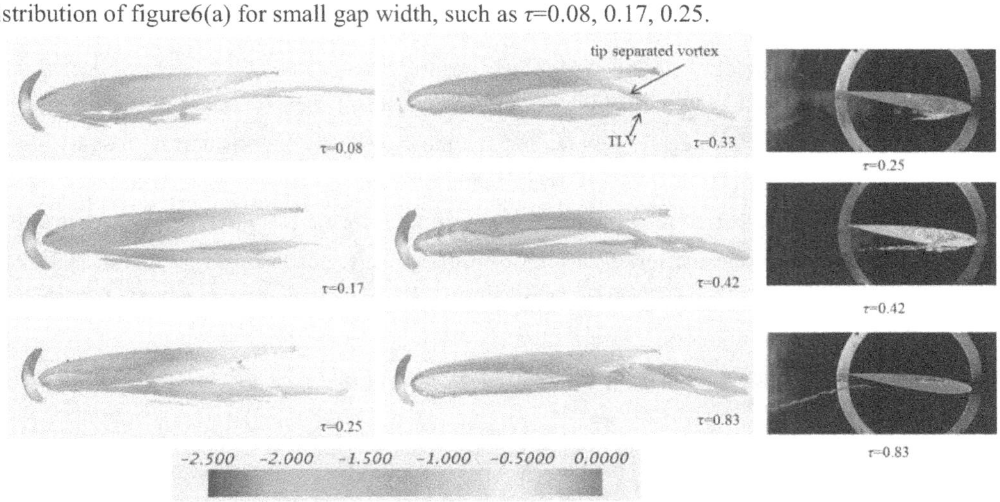

(a) (b)

Figure5. a: pressure coefficient distribution in main flow direction(x axial) under Q criterion (isovalue=500000)

b: tip clearance cavitation captured by high speed camera, V_0=8m/s, σ=1.6

As the tip clearance increase, another vortex structure located on the tip edge of pressure side can be seen in width τ=0.08, 0.17, 0.25 of the isosurface map in Figure 5, which is formed by leakage flow separation named as tip separated vortex, twines together with TLV developing towards downstream. Correspondingly, the vorticity distribution map of figure 6(a) shows that there are two obvious vorticity regions. Similarly, the cavitation picture captured by high speed camera could also display the vortex development intuitively in some extent. For the small gap width, the velocity of leakage flow will increase which resulted in the lower pressure area on the tip endwall, as shown in the pressure coefficient isosurface map, and then the tip clearance cavitation, tending to be in the form of sheet cavity, becomes relatively more intense in this case, such as for τ=0.25, 0.42 in the figure 5(b). While in the case of τ=0.83, tip clearance flow get rid of the low pressure influence, the tip clearance cavitation just appears in the vortex region which produces the lowest pressure area around of the vortex center.

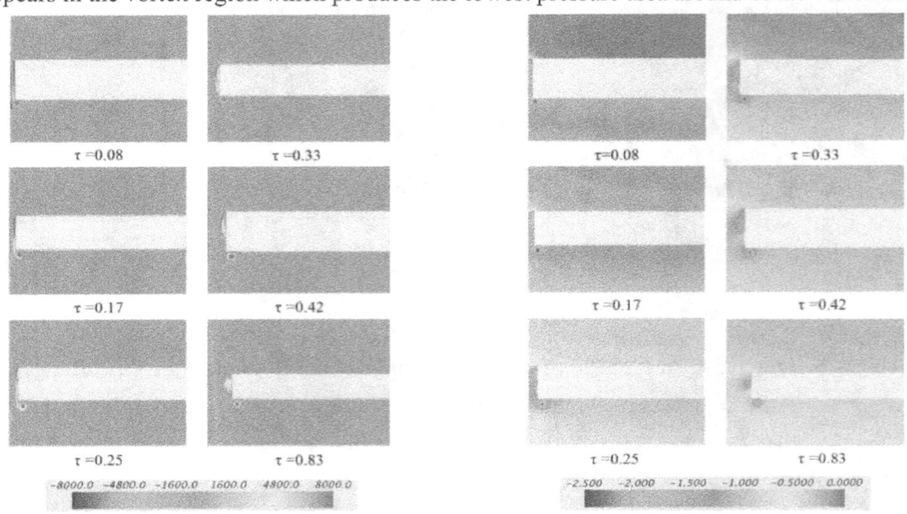

Figure6. conture cloud map of vorticity and pressure coefficient distribution for different gap size

Conclusion

Tip leakage flow is a fairly complicated clearance flow which comes with several different sizes of vortex structures due to flow separation on the tip endwall. This research gives numerical simulation of the flow field for different gap width of NACA0012 hydrofoil, and analysis of the tip leakage vortex and its cavitation form changed with the clearance size. The result shows that tip clearance size have great influence on the tip leakage flow field, and the most obvious phenomenon is the variation of vortex structure topology. In the case of small gap width, tip leakage vortex is relative strong shown by the vorticity distributions. But for the tip clearance cavitation, there are two kind of cavity for this case, one is sheet cavitation due to the low pressure of tip clearance region induced by increased leakage flow velocity and the other is vortex cavitation due to the low pressure in the center of TLV. Therefore, to predict the cavitation of tip leakage flow accurately just by understanding the intensity of vorticity of TLV is not enough, the pressure in other positions beside the TVC in tip clearance should be considered

References

[1] Matthieu Dreyer, Jean Decaix, Cecile Munch-Alligne, Mind the gap: a new insight into the tip leakage vortex using stereo-PIV [J]. Exp Fluids, 2014, 55:1849.

[2] J.Decaix, G.Balarac et al. RANS and LES Computations of the Tip-Leakage Vortex for Different Gap Widths [J]. Journal of Turbulence, 2015, 16(4):309-341.

[3] A. Cahuzac, J. Boudety and M.C. Jacob, Large-Eddy Simulation of a Rotor Tip-Clearance Flow. 17th

AIAA/CEAS Aeroacoustics Conference, 2011.

© 2018 ASME

Numerical Simulation of Unsteady Cavitation in a Submerged Water Jet by Compressible Bubbly Mixture Flow Method

[1]Guoyi Peng *; [1]Yasuyuki Oguma

[1]*College of Engineering, Nihon University, Koriyama, Fukushima, Japan*

Abstract

Focused on the unsteady behavior of high-speed water jets with intensive cavitation a numerical analysis is performed by applying a compressible mixture-flow bubble-cavitation model. The mean flow of two-phase mixture is calculated by URANS for compressible flow and the intensity of cavitation in a local field is evaluated by the volume fraction of gas phase varying with the expansion and contraction of bubbles. High-speed submerged water jet issuing from a sheathed nozzle is treated when σ=0.1 and $\mathrm{Re} = 4.5\times10^5$. The distribution of cavitation clouds predicted approximately agrees with experiment data of visualization observation and the unsteady behavior of cavitation cloud shedding is captured acceptably. Computational results reveal that cavitation occurs at the entrance of nozzle throat and bubble clouds develop and flow downstream along the boundary layer. Developed bubble clouds split into small blocks and shed downstream periodically near the exit of sheath.

Keywords: Water jet, cavitation, multi-phase flow, bubble dynamics

1. Introduction

High-speed water jet injected into still water, which is called submerged water jet, has received much attention for its capacity of generating very high cavitation impact pressure in the collapse of cavitation bubbles, and widely applied to such as peening of metal materials, decomposition of toxic substances and purification of sewages. Until now, many experimental studies have been made concerning jet driven pressure, shape and size of a nozzle, cavitation number etc. However, the flow structure of cavitating jet and the behavior of unsteady cavitation clouds are still unclear for the difficulty to observe the interior of cavitating flow [1] [2]. For the purpose of performance prediction and optimum design of water jet devices numerical simulation of high-speed water jets with intensive cavitation becomes a task full of challenges [3] [4].

In the present work, unsteady turbulent cavitating flows caused by high-speed submerged water jet are calculated numerically by applying a simplified compressible mixture-flow bubble-cavitation model. The fluid media of cavitating flows are treated as a two-phase mixture of liquid and gas bubbles dispersing uniformly. The mean flow of two-phase mixture is calculated by the set of Unsteady Reynolds Averaged Navier-Stocks (URANS) equations for compressible flow and the intensity of cavitation is evaluated by the volume fraction of gas bubbles whose radius is estimated with a simplified Rayleigh-Plesset equation. High-speed submerged water jet issuing from a sheathed nozzle is treated and the temporal variation of cavitating flow is investigated. The distribution of cavitation clouds predicted approximately agrees to experiment data of visualization observation and the unsteady behavior of periodical shedding of cavitation clouds is captured acceptably.

2. Compressible Mixture Flow Bubble-Cavitation Model

The fluid media of cavitating flow are taken as a two-phase mixture of working liquid and cavitation bubbles. The gas bubbles are supposed to disperse in the liquid phase uniformly and its volume faction is denoted as α_G. The liquid volume fraction is written to be α_L. Then, the density of two-phase mixture can be defined as follows by volume averaging.

$$\rho_M = \rho_L \alpha_L + \rho_G \alpha_G \tag{1}$$

where ρ denotes fluid density and the subscripts L, G, and M respectively do the liquid phase, the gas phase and the two-phase mixture. Then variation of mixture density can be obtained by taking the differential of above equation and it is then arranged to the following form.

*Corresponding Author, Guoyi Peng: peng@mech.ce.nihon-u.ac.jp

$$\frac{1}{\rho_M}\frac{d\rho_M}{dt}=\frac{1}{\rho_M c_M^2}\frac{d\,p_L}{d\,t} \tag{2}$$

where

$$\frac{1}{\rho_M c_M^2}=\frac{\alpha_L}{\rho_L c_L^2}+\frac{\alpha_G}{\rho_G c_G^2}\frac{\partial p_G}{\partial p_L} \tag{3}$$

in which c denotes the mean sonic speed and p the static pressure. The liquid phase is supposed to be continuous one and the gas phase disperses in the form of small bubble. Thus, their pressure variations convey each other.

According to the above equation, we understand that the compressibility of the bubbly mixture mainly depends upon the volume fraction of gas phase as well as the variation of gas pressure, which is determined by bubble size. To determine the compressibility of mixture media both the liquid and the phase are supposed to work exponentially and the following state equations are applied.

$$(p+B)/(p_0+B)=(\rho/\rho_0)^n \tag{4}$$

where the subscript 0 does a reference state which is taken to be the atmospheric one. The superscript n denotes the specific heat ratio of the working liquid or gas phase including in bubbles. B is a constant given to be 3.049×10^8 Pa for the liquid phase and zero for the gas phase respectively. Then the sonic speeds in liquid and gas phases are given as follows.

$$\rho\, c^2 = n\,(p+B) \tag{5}$$

The gas phase included in a bubble is assumed to consist of non-condensation gas and vapor which are denoted by subscripted g and v, respectively. The pressure of non-condensation gas in a bubble is related to the bubble radius R_b, that is to say,

$$p_g = p_0\left(R_{b0}/R_b\right)^{3n_g} \tag{6}$$

where, R_{b0} denotes the initial bubble radius at the reference state. Bubble radius varies rapidly when cavitation occurs corresponding to a sharp decrease of surrounding liquid pressure. It may be estimated by applying the Keller equation or similar ones [5]. However, the calculation of these equations is time consuming for the high frequency of bubble oscillation. For simplifying the numerical calculation, the cavitation bubbles are treated as quasi-still one in this work and an estimation of bubble radius is estimated by solving a simplified Rayleigh-Plesset equation [6].

$$\frac{dR_b}{dt}=Sign\,(\Delta P)\sqrt{\frac{2}{3}\frac{|\Delta P|}{\rho_{L0}}} \tag{7}$$

where $\Delta P = p_g + p_v - p_L$ and $R_b = R_b(\boldsymbol{x}, t)$, which denotes the mean radius of bubbles in a local area $\boldsymbol{x}$ at an arbitrary instant t. Then, the bubble radius may be estimated according the liquid pressure and the compressibility of bubbly mixture may be evaluated by equation (4). The variation of gas volume fraction is then determined by mass conversation of the gas phase under the effect of bubble radius oscillation.

$$\frac{\partial \alpha_G}{\partial t}+\boldsymbol{u}\cdot\nabla\alpha_G=-\alpha_G\left(\nabla\cdot\boldsymbol{u}-\frac{3}{R_b}\frac{dR_b}{dt}\right) \tag{8}$$

If the effect of bubble surface is negligible then the gas phase within a bubble expands and/or contracts freely with the surrounding pressure. Then, we goes to $p_L \cong p_G$. The above bubble cavitation model will decline to homogenous one, and the compressibility of cavitating mixture can be estimated by Eq. (3) simply.

Based on the above simplified bubble cavitation model, the flow of bubbly fluid consists of bubbles and surrounding is calculated by using the set of URANS equations for compressible fluid. The variation of temperature caused by cavitation is thought to be very small in the whole flow field and the conservation equation of energy is omitted. Conservation equations of mass and momentum are then closed by introducing the transportation equation of pressure as well as the transport equation of gas volume fraction. The turbulence effect is modelled by adopting the RNG k-ε model for high Reynolds number flow with a modification on the eddy viscosity considering the influence of cavitation. In order to deal with the great variation of compressibility caused by cavitation the set of flow governing equations are solved by using the CIP-CUP (Cubic-Interpolated Propagation/Combined and Unified Procedure) method based on the time splitting technique [7].

Computation Results and Discussions

Figure 1 shows the schematic arrangement of submerged water jet to be concerned, where a sheathed orifice nozzle is set up at the bottom of a closed cylindrical chamber full of tap water. Pressured tap water supplied by a plunger pump is injected into the water chamber and the depth from the water surface to the nozzle exit is keep to a constant. The throat diameter of the nozzle d = 1.0 mm and the length of nozzle throat is 5.0d. The inner diameter of the pipe-like sheath is 3.0d and its length is 10.0d. Focused on the upstream axisymmetric structure the assumption of axisymmetric flow is adopted. The computation domain is taken from 15.0d upstream of the nozzle inlet to 70.0d downstream of the nozzle exit. The width of computational domain is 25.0d in the radial direction. The computation domain is then discretized with quadrilateral structure grid. As for the boundary conditions, a pressure condition is imposed at the inlet according to the given total pressure and the intensity of turbulence was given to be 1.0% of the inflow velocity. At the outlet, a given static pressure is imposed and the Neumann condition was applied to other flow variables such as velocity etc. All the wall boundaries such as the nozzle and the sheath geometries are treated as no-slip walls by applying a universal wall function. As an index for the similarity of cavitation dynamics, the cavitation number σ for the present water jet device is defined as follows.

$$\sigma = \frac{p_o - p_v(T_\infty)}{P_i - p_o} \tag{9}$$

where P_i denotes the injection pressure, p_o the static surrounding pressure at the nozzle exit and p_v the saturated vapor pressure at the reference temperature T_∞.

Figure 2 shows, as an example, the pattern of cavitation cloud shedding in submerged water jet when the injection pressure $\overline{P_i} = 1.1$ MPa and the discharge pressure p_o = 0.1MPa. The cavitation number σ is calculated to be 0.1 and the Reynolds number Re is about 5.0×10^4 for the concerned case. Figure 2 (a) shows the computational results of developed cavitating jet, where instantaneous distributions of gas volume fraction are demonstrated by contour maps in a time sequence. The areas of high α_G value denoted by red and green colours present axial sections of cavitation cloud and the blue colour indicates the liquid phase. The dark vectors denote the mean velocity in local flow field. It is demonstrated that cavitation clouds expand within the nozzle and the sheath along the shear layer around the core region, and then split into small blocks at the downstream the sheath exit. As shown by the dashed line cavitation clouds denoted by green-like colours stretch out of the sheath and shed downstream periodically. As a comparison, figure 2 (b) shows the experimental results of high-speed video camera visualization observation of developed cavitating jet under the same working condition, where the appearance of cavitation jet is demonstrated by a sequence of pictures taken at imaging speed of 27,000 fps. These pictures were taken under penetrating light condition, and the area full of clear water is observed to be bright but the area of cavitation cloud including many bubbles is observed to be dark one since the cavitation cloud is almost impermeable to light for the light reflection at bubble surfaces. For the difficulty of observing the interior of the narrow nozzle throat flow visualization observation was focused on the interior of pipe-like sheath and the shooting area was taken from the nozzle exit (left) to the sheath exit (right). As shown in the figure, cavitation clouds demonstrated by the first picture start to stretch from its minimum length, and gradually expand to the out of the pipe-like sheath along the shear layer (the 2nd & 3rd pictures) and finally shed downstream (the 4th & 5th pictures). The dashed line denotes periodical oscillations of cavitation clouds within the sheath.

Comparing figure 2 (a) and (b) we understand that the temporal distributions of gas volume fraction α_G predicted by computation approximately agree with the experimental results of high-speed camera observation and the periodically shedding of cavitation clouds in water jet is reliably reproduced.

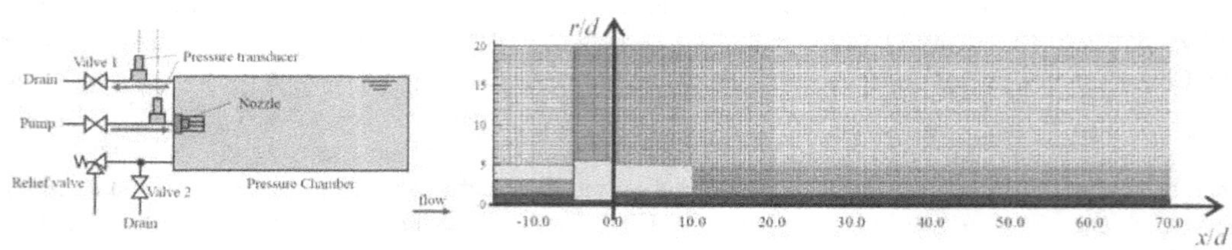

Figure 1. Schematic diagraph of water jet system (left) and computation domain of numerical simulation (right)

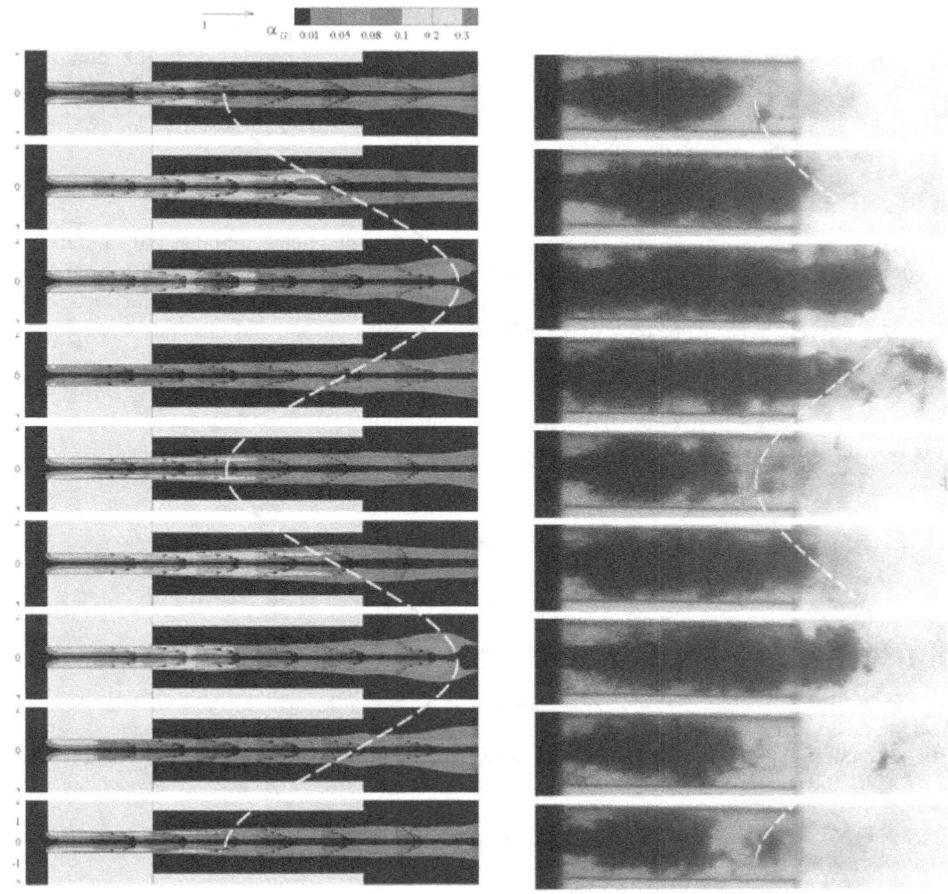

Figure 2. Images of cavitation cloud shedding in time sequence: (a) Computational result of GVF distributions in time sequence (left) and (b) Sequential images of cavitation clouds observation by high-speed video camera (right).

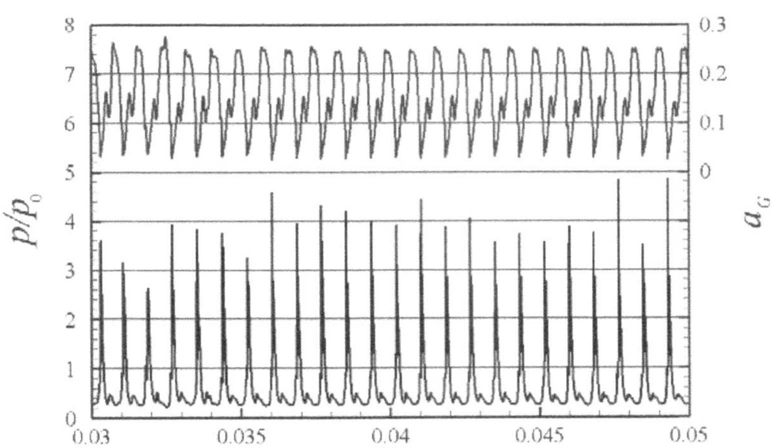

Figure 3. Oscillations of pressure and GVF at the reference position (x/d = −4.5, r/d = 0.45)

Figure 3 shows, as an example, the oscillation of pressure p (absolute) normalized by atmospheric pressure p_0 and GVF α_G at the given monitoring position of (x/d =−4.5 , r/d =0.45), which is locates at the entrance of nozzle throat near the wall. The black solid line denotes the temporal variation of p and the blue dashed line with quadrilaterals does the variation of α_G. The figure demonstrates that the pressure and the GVF vary periodically almost at the same frequency. When GVF decreases to a minimum with collapsing of bubble cloud a high impact pressure released impulsively is demonstrated. FFT (Fast Fourier Transform) analysis reveals that the waveforms fluctuate at two dominant frequencies: f_1=120 Hz and f_2=240 Hz. It may be related to the bubble clouds expanding and collapsing corresponding to the variation of local flow field [8].

Conclusion

Focused on the unsteady behavior of intensively cavitating jets, a practical compressible mixture-flow bubble-cavitation model is developed for the simulation of turbulent cavitating flows. The model has been applied to treat the unsteady cavitating flow caused by submerged water jets. The distribution of cavitation clouds predicted approximately agrees with experiment data of visualization observation and the unsteady behavior of cavitation cloud shedding is captured acceptably.

Computational results demonstrate that: (1) Cavitation occurs at the entrance of nozzle throat and cavitation bubble clouds expand and develop while flowing downstream along the boundary layer. (2) Developed bubble clouds split into small blocks and sheds downstream periodically near the exit of sheath.

References

[1] Kalumuck, K.M., Chahine, G.L., (2000). *The use of cavitating jets to oxidize organic compounds in water.* J. Fluids Eng., 122, 465-470.

[2] Nishimura, S. et al., (2012). *Similarity law on shedding frequency of cavitation cloud induced by a cavitating jet.* J. Fluid Sci. & Tech., 7(3), 405-420.

[3] Qin, Z., Bremhorst, K., Alehossein, H., (2007). *Simulation of cavitation bubbles in a convergent–divergent nozzle water jet.* J. Fluid Mech., 573, 1–25.

[4] Iga, Y., Konno, T., (2012). *Numerical analysis of the influence of acceleration on cavitation instabilities that arise in cascade.* Int. J. Fluid Machinery and Systems, 5, 1-9.

[5] Keller, J.B., Kolodner, I.I., (1956). *Damping of underwater explosion bubble oscillations.* J. Appl. Phys., 27, 1152-1161.

[6] Peng, G., Yang, C., Oguma, Y., Shimizu, S., (2016). *Numerical analysis of cavitation cloud shedding in a submerged water.* J. Hydrodynamics (B), 28(6), 986-993.

[7] Peng, G., Ishizuka, M., Hayama, S., (2001). *An improved CIP-CUP method for submerged water jet flow simulation.* JSME Int. J. (B): Fluids and Thermal Eng., 44(4), 497-504.

[8] Zhou, Y. et al. 2016, *Fluid-Structure-Sound Interactions and Control*, Springer-Verlag, Berlin Heidelberg, 229-234.

© 2018 ASME

10th International Symposium on Cavitation - CAV2018
Baltimore, Maryland, USA, May 14 – 16, 2018

Numerical simulations of the shockwave induced collapse of bubbles near surfaces

[1]Nikolaos Bempedelis*; [1]Yiannis Ventikos

[1]*Department of Mechanical Engineering, University College London, WC1E 7JE, United Kingdom*

Abstract

In the study of collapsing bubbles and their relation to surface erosion, two mechanisms are identified as of major significance: the high-speed liquid jet and the water-hammer shock wave that is subsequently formed. In the current work, we suggest that secondary mechanisms such as the emission of shockwaves following the collapse of the bubble remains and the late-time wave interactions may also be of importance when it comes to surface erosion. We examine our hypothesis by considering the collapse of a bubble by a shock wave that runs in a direction that is parallel to a solid surface.

Keywords: cavitation; shock waves; bubble collapse; front tracking;

Introduction

Cavitation is a phenomenon that occurs in various applications and domains such as marine engineering, energy production and clinical medicine. The structural damage caused by bubbles collapsing near a surface has been a major drive in studies of cavitation. Two mechanisms have been suggested as primarily responsible for erosion: (i) the formation of a high-speed liquid jet and (ii) the water-hammer shock wave that is generated upon its impact on the bubble walls.

In the majority of applications, bubbles are present in clouds. Collapse dynamics are therefore dominated by inter-bubble effects: the presence of nearby bubbles disturbs the pressure field, whereas a bubble may collapse due to it being impacted by waves emitted from a neighboring bubble.

The shock-induced collapse of bubbles near surfaces has been mainly considered in the context of shockwave lithotripsy, where waves run in a direction that is normal to the surface [1,2]. The potential for damage in such configurations is great, as the main jet is directed towards the surface and the impact occurs at the side of the bubble that is closer to the wall. The emitted shockwave originates close to the surface and induces large pressures on it before decaying in strength. However, it has been reported that in the shock induced collapse of bubbles in a free field, peaks in pressure are found not only in the downstream propagating water-hammer wave [3]. Locally high pressure regions also develop upstream of the bubble, following the collapse of the bubble remains and the late-time wave interactions. With respect to erosion, these mechanisms are not of importance when the incident wave propagates in a direction that is normal to the surface. However, in the case where the shockwave impacts the bubble from a different direction, the aforementioned mechanisms could be inducing loads of non-negligible magnitude on the nearby surface.

Results

A front tracking method is employed for the localization of the interface [4]. The flow is modeled by the compressible Euler equations. A fifth-order WENO scheme is used for the convective terms, while integration is carried out by a second-order Runge-Kutta method. The fluids are modeled by the stiffened polytropic equation of state, which reduces to the ideal gas law in the case of air.

A cylindrical air bubble of diameter d=0.001m is placed in a volume of water next to a solid surface. The distance of the bubble center from the wall is given in terms of the non-dimensional parameter $\gamma=\delta/R$, where δ is the distance from the wall and R the bubble radius. The computational domain has a size of $x\in[0m, 0.002m]$, $y\in[0m, 0.0025m]$ and is discretized with a resolution that yields a number of 200 points per initial bubble radius. The bubble center is placed along the y=0.00075m line. Both fluids are initially at rest at atmospheric pressure. A shockwave of strength p=1GPa ($M\approx1.42$) propagates from the lower boundary (y=0m) of the domain. The wall at x=0m is modelled by appropriate reflecting conditions, while non-reflecting conditions are set on the remaining boundaries.

We consider the case where $\gamma=1.25$. As the incident shock impacts the bubble interface a rarefaction wave is reflected back in the liquid. This wave gets reflected at the wall and expands the liquid to the point that it starts to cavitate. The density gradient fluctuations in the vicinity of the solid surface are an artifact of the employed pressure cutoff model, which in general has little to no effect on the overall flow dynamics. The delayed compression of the wall-adjacent bubble half can be observed at the moment just before the impact of the main jet (figures 1(a), 2(a)). In the same frame, we may observe how the shockwave in the vicinity of the wall runs slower and is weaker than its free-field counterpart, following its interaction with the reflected rarefaction wave. In the following frame (figures 1(b), 2(b)), we observe the water-hammer shockwave and the delayed collapse of the lobe that is close to the wall. The microjetting

*Corresponding Author, Nikolaos Bempedelis: nikolaos.bempedelis.16@ucl.ac.uk

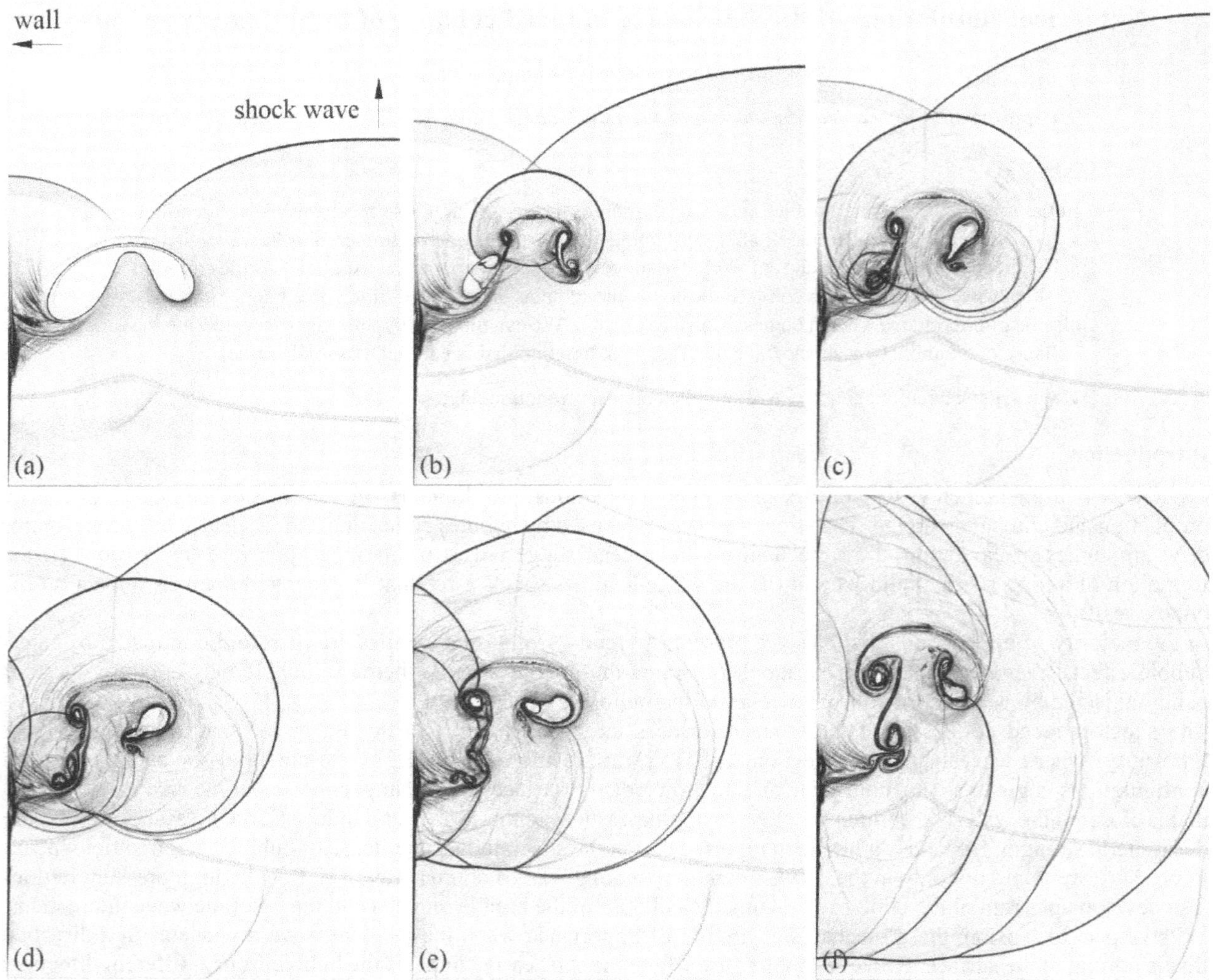

Figure 1:Magnitude of density gradient contours at various time instants after shock impact, γ=1.25.
(a) t=660ns, (b) t=820ns, (c) t=940ns, (d) t=1010ns, (e) t=1120ns, (f) t=1340ns.

that is formed following the impact of the main jet at the bubble walls is observed on the wall-adjacent lobe only, as it has already impacted the sides of the wall-distal lobe, which is subsequently getting collapsed by the water-hammer shockwave. In the third frame (figures 1(c), 2(c)), the flow field is depicted right before the water-hammer wave impacts the wall. The waves that were formed upon the collapse of the wall-adjacent lobe follow shortly after, and are shown to interact with the reflected water-hammer shock and the wall at the fourth frame (figures 1(d), 2(d)). The gas in the wall-distal lobe has started expanding, whereas its motion is strongly vortical. The region of cavitating liquid is reduced as the water-hammer shockwave traverses it and raises the pressure. The bubble remains get compressed once again by the reflected shocks. In the final frame (figures 1(f), 2(f)), we observe the waves running along the solid surface while decaying in strength. Strong interactions and reflections have ceased. Due to the multiple interactions and shock passages, the gas in the wall-adjacent lobe does not expand as its right counterpart.

The interface of the bubble is depicted in figure 3 at four different instants after shock impact (starting from the moment of jet impact on the bubble wall and following increments of 100ns). In these figures, results from simulations with different stand-off parameters have been superposed on those from a bubble collapsing in a free field. We observe that the dynamics of the first phase (i.e. until the jet impact) are not severely altered. Even though the wall-adjacent lobe is compressed with delay, there is no significant effect on the time that the jet takes to impact the far bubble wall. However, in later stages, the delay gets pronounced for the dynamics of both lobes. As the stand-off parameter decreases, the generated vorticity is reduced. In the case where γ=1.02 the wall-adjacent upper half of the bubble is practically stagnant. As a result, cavitation is observed only upstream of the bubble. In general, the patterns of the collapse are similar for all stand-off parameters; they just manifest in different intensity and dynamics.

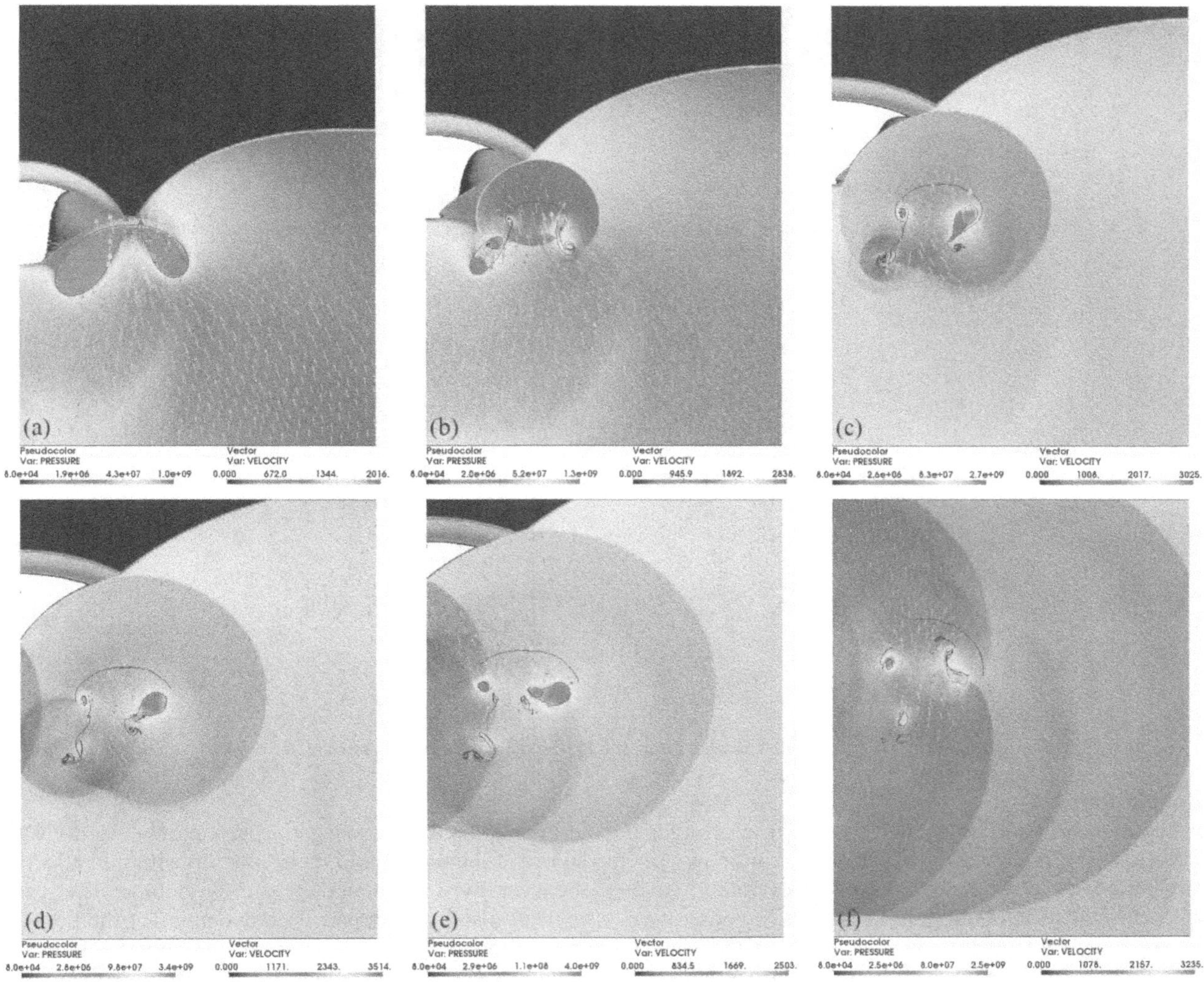

Figure 2: Pressure contours and velocity vectors at various time instants after shock impact, γ=1.25. Regions of cavitation are shown in white. (a) t=660ns, (b) t=820ns, (c) t=940ns, (d) t=1010ns, (e) t=1120ns, (f) t=1340ns.

With respect to erosion, the pressure induced on the wall along time is presented in figure 4. Although the plotted results are from the γ=1.25 case, the same mechanisms may be identified in all other simulations. However, their relative importance, magnitude and dynamics vary with the stand-off parameter γ. A first load is induced on the surface following the impact of the water hammer shockwave (at t=960ns after shock impact). However, this is not a direct impact as in the normal-to-the-wall collapse, and the resulting pressures are weak. The wall experiences a first significant pressure load when impacted by the waves formed during the collapse of the wall-adjacent bubble half (around t=1000ns after shock impact). The intensity of these waves at impact is increasing as the bubble gets closer to the wall. This is expected as the waves originate closer to the wall and have therefore not significantly decayed in strength. Prior to impacting the wall, these waves might experience an increase in strength, in case of interaction with the reflection of the water-hammer shockwave. The peak in the load around t=1120ns occurs after this type of interaction. At later times, the interactions continue as the waves run along the wall. However, all these late-time interactions and reflections are of lower intensity and the waves propagate away from the wall while decaying in strength.

The maximum wall pressures recorded for all simulated stand-off parameters are presented in figure 4. The results are compared against the respective normal-to-the-wall collapse configurations. The induced pressures are clearly not negligible in magnitude but are smaller than a direct water-hammer shock impact by a percentage of about $\approx$50-60%.

Figure 3: Bubble interface at four different time instants after shock impact for various stand-off parameters. (a) t=680ns, (b) t=780ns, (c) t=880ns, (d) t=980ns.

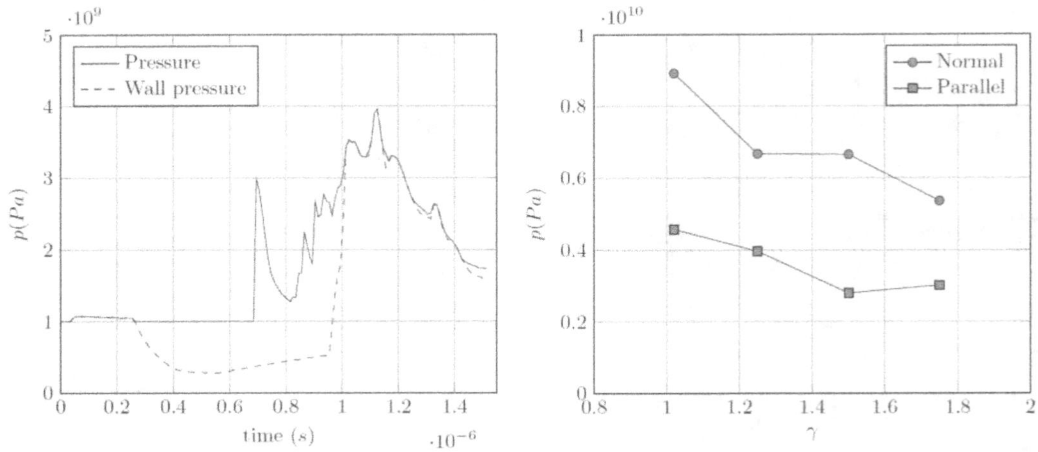

Figure 4: (left) Maximum pressure along time after shock impact, γ=1.25, (right) maximum wall pressures for various stand-off parameters.

Conclusions

In the present work we studied the collapse of a gas bubble by a shockwave propagating in a parallel-to-the-wall direction. The shockwave weakens and slows down in the vicinity of the wall due to its interaction with the reflected rarefaction wave, whereas a region of cavitation develops between the solid boundary and the bubble. The wall-adjacent half of the bubble does not get as compressed as its wall-distal counterpart. The dynamics of the post-jet impact stage are asynchronous; the wall-distal lobe collapses first followed by its wall-adjacent counterpart. Following the reflection of the waves at the solid surface the bubble remnants get collapsed again. With respect to erosion, we demonstrated that the peak wall pressures are generated at the impact of the shocks that are formed during the collapse of the wall-adjacent bubble remains. The loading was shown to be comparable in magnitude to cases of normal-to-the-wall shock induced collapse. In the future, the collapse of cavity arrays will be considered, since pressure is amplified in such configurations. Moreover, the investigation will be extended to the case of spherical bubbles, where the focusing (and hence the pressure) is also more intense.

References

[1] Johnsen, E., Colonius, T. (2008). *Shock-induced collapse of a gas bubble in shockwave lithotripsy.* J. Acoust. Soc. Am, 124(4).

[2] Jamaluddin, A.R., Ball, G.J., Turangan, C.K., Leighton, T.G. (2011). *The collapse of single bubbles and approximation of the far-field acoustic emissions for cavitation induced by shock wave lithotripsy.* J. Fluid Mech, vol. 677, pp. 305-341.

[3] Hawker, N.A., Ventikos, Y. (2012). *Interaction of a strong shockwave with a gas bubble in a liquid medium: a numerical study.* J. Fluid Mech, vol. 701, pp. 59-97.

[4] Bo, W., Liu, X., Glimm, J., Li, X. (2011). *A robust front tracking method: verification and application to simulation of the primary breakup of a liquid jet.* SIAM J. Sci. Comput., 33(4).

© 2018 ASME

Numerical simulations of tip leakage vortex cavitation flows around a NACA0009 hydrofoil

[1,2]Benlong Wang*; [1]Zhihui Liu;

[1]*Shanghai Jiao Tong University, Shanghai, China;* [2]*MOE Key Laboratory of Hydrodynamics (SJTU), Shanghai, China*

Abstract

Numerical simulations of full wet and cavitating tip leakage vortex flows are studied. By comparing vortex structures, strain rates and Reynolds shear stresses distributions in the vortex region for full wet flows between numerical and experimental results, it is found that the numerical results with the nonlinear k-ε turbulence model is in good agreement with the experimental results. Both full wet and cavitating flows around a NACA0009 blade with various gaps between the tip and end-wall are studied to investigate the influence of gap width on the development of tip vortex and tip vortex cavitation. Good agreements are obtained on the trajectories of tip leakage vortex by comparing the numerical results and experimental data. When the dimensionless gap size $\tau < 0.7$, numerical results show that there is an apparent difference on the vortex trajectories behind the hydrofoil between full wet and cavitating flows.

Keywords: tip-leakage vortex; cavitation; turbulence model

Introduction

Cavitation is a phenomenon that plays an important role in surface sea-going vessels, as well as in hydraulic equipments including hydrofoils, propellers, turbines and pumps. The tip-leakage vortex cavitation, which can lead to thrust breakdown and blade vibration, can be observed frequently in turbo-machine with a radial clearance between the rotor-blade tip and the casing wall. Complex vortex structures, including the induced vortex, tip-leakage vortex and tip-separation vortex, can be observed in the gap region (You et al, 2007). Due to the difficulties of numerical simulation of vortex flows, more advanced numerical models are desired to capture the main characteristic of the flow structures. Using both LES and RANS, Decaix et al. (2015) simulated the wet flow around the NACA0009 hydrofoil, and proposed an empirical formulation for the vortex trajectory from the leading edge to the mid-chord. Although DES and LES have been successfully applied in many cavitating flows, the Reynolds-averaged Navier-Stokes (RANS) model remains the primary CFD solver used in practical applications for its low computational costs. Most of the RANS models are based on the Boussinesq turbulent-viscosity hypothesis, which has been used successfully in many turbulence flows dominated by shear stress. In this kind of flow, the orientations of the strain rate tensor and Reynolds stress tensor coincide. However, this type of linear model fails when the flow stream is strongly curved or rotated. Experimental observations show that the strain rate and Reynolds stress are not aligned in their tip vortex region (Chow et al. 1997). Churchfield & Blaisdell (2009) explored several turbulence models and corrections for rotating flows with curvature streamlines, including one-equation model, two-equation models and shear stress transport model. It was found that none of these turbulence models can accurately capture the spatial lag between the Reynolds stress components and the corresponding strain rate components within the vortex. Thus, a nonlinear k-ϵ model is employed in the present study which has been successfully applied for tip vortex simulations in previous studies (Liu et al 2016). Numerical simulations for various gap size τ are conducted for both full wetting and cavitating flows.

Problem Descriptions

The numerical models are solved by using a flow solver OpenFOAM with finite volume method. The PISO scheme is used to solve the velocity/pressure coupling between the momentum and continuity equations. A second-order upwind scheme is used for the spatial discretization in the momentum equations and transport equations for k and ϵ. The QUICK scheme is used in the advection of the volume fraction function. The viscous terms are solved using a second-order central difference scheme.

In present work, turbulence models for RANS are investigated. The nonlinear k-ϵ model proposed by Zhu & Shih(1994) is used. In this turbulence model, the rotation rate is also included in the constitute relationship. The nonlinear k-ϵ model is a high-Reynolds-number model with wall functions, denoted as NonlinearKEShih in OpenFOAM, which is also known as the realizable Reynolds stress algebraic equation model. Although the theoretical

*Corresponding Author, Benlong Wang: benlongwang@sjtu.edu.cn

foundation of the nonlinear k-ϵ model is significantly different from that of the standard k-ϵ model, the nonlinear k-ϵ model using the same transport equations for the turbulent kinetic energy k and dissipation rate ϵ as the standard k-ϵ model (Launder & Spalding, 1974).

To simulate the phase transition between the liquid and vapor phases, a cavitation model is needed. The transport equation model has been widely used in recent numerical studies of cavitating flows. These types of cavitation models have been successfully used in simulations of various unsteady cavitating flows. The transport equation of phase function proposed by Schnerr & Sauer (2011) is used in the present work.

Numerical Results and Discussions

When the tip of a hydrofoil is located close to a solid wall, the tip vortex is confined in a narrow gap region, which is termed as tip leakage or tip clearance flows. The tip leakage flows are important topics in rotating machineries such as axial pump impellers and ducted propellers, which have been extensively studied in different fields. Due to the sharp difference between the chord-wise pressure gradients and the normal pressure gradients in the clearance region, the tip clearance flows can be approximately decomposed into independent through-flow and cross-flow(Chen et al, 1991), which provide a theoretical basement to simplify the complex flows in rotating machineries. Neglecting the chord-wise rotating flows inside a casing, flow around a hydrofoil near a plate with a narrow gap is a simple model for both experimental and numerical studies, which provides the most important characteristics of tip leakage flows.

There are two different types of flows: strong tip vortex cavitating flow relating to cavitation number σ; separation flows at the edge of the hydrofoil's tip relating to Reynold number Re. To quantitatively describe the tip leakage flow, both of these two kind flows should be properly simulated. Besides the general requirements for the tip vortex cavitating flows, the interactions between the separation bubble and the wall boundary layer, which depends heavily on the Reynold number, is another crucial issue for the numerical simulations.

The unsteady, secondary vortical structures were expected to be the trigger mechanism for tip-leakage vortex cavitating inception. Therefore, advanced numerical models, especially the turbulence models, are one of the critical factors to predict the inception. Indeed, it is far away from predicting inception numerically. Herein, the flows around a truncated NACA0009 hydrofoil have been simulated with RANS equations and the nonlinear k -ϵ turbulence model (Liu et al 2015). The setup follows the water tunnel experiments of Dreyer et al. (2015). The hydrofoil is mounted on a vertical plate with an angle of attack $= 11^o$. There is a gap between the free end and the vertical plate. The incoming velocity is 10m/s and the corresponding Reynolds number is $Re = 1 \times 10^6$, and the cavitation number is σ= 1.94. Mesh convergence tests have been conducted. Around 6 to 10 million mesh are used in following numerical simulations.

The gap size τ, usually normalized by the maximum thickness of the hydrofoil, has a significant influence on the tip-leakage vortex cavitating flows. Numerical simulations with various dimensionless gap size τ are conducted. One example of numerical results is shown in Fig.1. Both cases of gap size smaller and larger than the plate boundary thickness are investigated. According to the measurements of the cavitation flows around a 3D NACA0009 hydrofoil by using stereo-PIV, it was found that there exist a specific tip clearance for which the vortex strength is maximum and most prone to generating cavitation (Dreyer et al., 2014). It was found that the most strongest tip-leakage cavitation occurs near τ= 0.2. From the numerical study we found that the boundary layer on the side plate near the free end of the hydrofoil plays an important role in the upward tendency of the tip-leakage vortex. When the free end is inside the boundary layer, there is an obvious upward tendency of the vortex structure. It was observed that the tip leakage vortex trajectory moves closer to the hydrofoils as the clearance is increased in early experimental study by Gopalan et al. (2002). The meandering of the vortex core is substantial in the experiments, but have not been predicted in present numerical simulations.

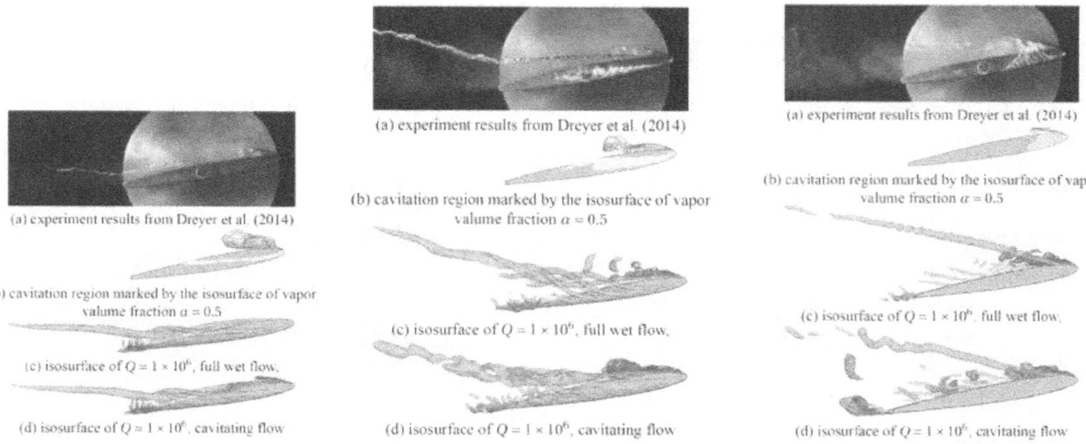

Figure.1 Comparison on the numerical results and instantaneous experimental high-speed photography, $\tau = 2.0, 0.7$ and 0.1.

An empirical law that matches the vortex trajectory from the leading edge to the mid-chord has been proposed based on the numerical results on a non-cavitating tip-leakage flow around NACA0009 hydrofoil (Decaix et al. 2015). From the present numerical results, there is a difference in the trajectories of cavitating and non-cavitating tip-leakage vortex flows, as shown in Fig.2, when the dimensionless gap size is smaller than 1.0. This kind of difference deserves to be investigated in future studies, which will help to establish an empirical law for the vortex trajectories in the cavitating tip-leakage flows.

Several early papers have dealt with the cavitation in tip leakage or tip clearance flows. The location of the cavitating vortex origin moved along the streamwise direction, depending on the relative gap size τ. The influence of Reynolds number among experiments is smaller than the general propellers and hydrofoils in open water. The measured dependence on inception number for the propulsor is 0.21 (Chesnakas & Jessup, 2003), which is significantly lower than the empirical values 0.3 (Farrell & Biller, 1994), or 0.4-0.5 (Arndt, 2002). The visual inception and the sub-visual cavitation events occur approximately at the merge location of tip-leakage vortex and trailing edge vortex. Careful investigations on the merging process may help to resolve the mismatch between the lowest pressure location and inception location.

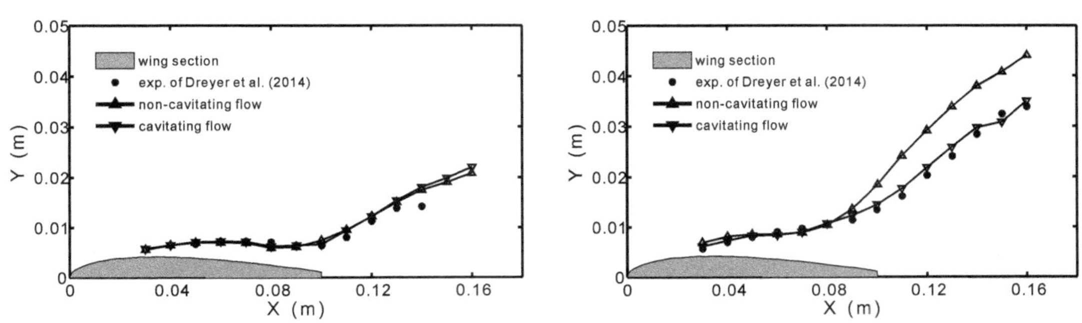

(a) tip-leakage vortex trajectory of τ =2 (b) tip-leakage vortex trajectory of τ =0.7

Figure.2 Comparison on the trajectories of cavitating and non-cavitating tip-leakage vortex flows .

When there is no relative motion between the tip and the end wall, experiments do not show a minimum in the cavitation inception index (Gopalan et al, 2002), which is different from the experience obtained from the experiments for a rotating turbomachine (Farrell & Biller, 1994). It should be emphasized that simulations of 3D cavitating tip-leakage flow field around rotators are important to understand the realistic flow field in turbomachinery. A few 3D

RANS simulations, using SST k -ω and other two-equation turbulence models, have been conducted for the full rotation machinery recently, in which flow rate and TLV trajectory can be quantitatively predicted. Using LDV and PIV, detailed tip-leakage flow fields inside ducted propulsors at Reynolds number 10^6 have been measured (Oweis et al, 2006a,b). Comparisons on the turbulence flow field with these experimental results are important and challenging tasks, which require high computational efforts as well as accurate cavitation models and turbulence models.

Conclusion

We perform the numerical simulations for the tip leakage vortex for both full wet and cavitating flows by solving the RANS equations with the nonlinear k-ϵ turbulence model. From the comparisons between the numerical results and experimental data, it is found the proposed approach can capture the main characteristics of the tip leakage vortex flows at a moderate computational cost. More detail analysis on the flow field will be reported at the conference.

References

[1] Arndt, R.E.A. (2002) *Cavitation in vortical flows*, Annual Review of Fluid Mechanics, 34:1561.

[2] Chen G.T., Gretzer E.M., Tan C.S., Marble F.E. (1991) *Similarity analysis of compressor tip clearance flow structures*. ASME Journal of Turbomachinery, 113(2):260.

[3] Chesnakas, C.J., Jessup, S.D. (2003) *Tip-vortex induced cavitation on a ducked propulsor*. Proceedings of ASME FEDSM'03, 4th ASME_JSME Joint Fluids Engineering Conference, Honululu, Hawaii.

[4] Chow, J.S., Zilliac, G.G., Bradshw, P. (1997) *Mean and turbulence measurements in the near field of a wingtip vortex*. AIAA Journal. 35(10):1561.

[5] Churchfield, M.J., Blaisdell, G.A. (2009) *Numerical simulations of a wingtip vortex in the near field*. Journal of Aircraft, 46(1):230.

[6] Decaix, J., Balarac, G., Dreyer, M., Farhat, M., Munch, C. (2015) *RANS and LES computations of the tip-leakage vortex for different gap widths*. Journal of Turbulence, 16(4):309.

[7] Dreyer, M., Decaix, J., Munch-Alligne, C., Farhat, M. (2014) *Mind the gap: a new insight into the tip leakage vortex using stereo-PIV*. Experiments in Fluids, 55(11):1849.

[8] Farrell, K.J., Biller, M.L. (1994) *A correlation of leakage vortex cavitation in axial-flow pumps*, ASME Journal of Fluids Engineering, 116:551.

[9] Gopalan, S., Katz, J., Liu, H.L. (2002) *Effects of gap size on tip leakage cavitation inception, associated noise and flow structure*. ASME Journal of Fluids Engineering, 124:994.

[10] Launder, B.E., Spalding, D.B. (1974) *The numerical computation of turbulent flows*. Computer Methods in Applied Mechanics and Engineering, 3(2):269

[11] Liu, Z.H., Wang, B.L., Peng, X.X., Liu, D.C. (2016) *Calculation of tip vortex cavitation flows around three dimensional hydrofoils and propellers using a nonlinear k-ϵ turbulence model*. Journal of Hydrodynamics, 28(2):227.

[12] Oweis, G.F., Fry, D., Chesnakas, C.J., Jessup, S.D., Ceccio, S.L. (2006) *Development of a tip-leakage flow-Part 1: the flow over a range of Reynolds number*, ASME Journal of Fluid Engineering, 128:751

[13] Oweis, G.F., Fry, D., Chesnakas, C.J., Jessup, S.D., Ceccio, S.L. (2006) *Development of a tip-leakage flow-Part 2: comparison between the ducted and un-ducted rotor*, ASME Journal of Fluid Engineering, 128:751

[14] Schnerr, G.H., Sauer, J. (2001) *Physical and numerical modeling of unsteady cavitation dynamics*. ICMF-2001, 4th International Conference on Multiphase Flow. New Orleans, USA.

[15] You, D., Wang, M., Moin, P., Mittal, R. (2007) *Large eddy simulation analysis of mechanisms for viscous losses in a turbomachinery tip-clearance flow*. Journal of Fluid Mechanics, 586:177.

[16] Zhu, J., Shih, T.H. (1994) *Computation of confined coflow jets with three turbulence models*. International Journal for Numerical Methods in Fluids, 19(10):939.

© 2018 ASME

Numerical study of cavitating structure near wake of a circular cylinder

[1,2] Pankaj Kumar*; [1]Dhiman Chatterjee; [1]Shamit Bakshi

[1]*Department of Mechanical Engineering, IIT Madras, Chennai 600 036*

[2]*Department of Mechanical Engineering, SRM University, Kattankulathur, Chennai 603203*

Abstract

A study of the existing literature on flow past circular cylinder reveals that less work has been reported in the upper sub-critical flow regime which corresponds to Reynolds number (Re_D) in the range $2x10^4 < Re_D < 2x10^5$. Hence, this work brings out the flow features in the near wake of a cylinder for a sub-critical flow regime under non-cavitating as well as under different cavitating conditions. A three-dimensional numerical study of flow around a circular cylinder was carried out at $Re_D=6.4x10^4$ for a wide range of cavitation number ($1.06 \geq \sigma/\sigma_i \geq 0.5$). Unsteady Reynolds averaged Navier-Stokes (URANS) approach coupled with Schnerr-Sauer cavitation model available in ANSYS-Fluent was used for this purpose. With the help of fine mesh around the cylinder and choice of suitable time step size, this study identifies factors influencing the occurrence of cavitation and portrays dynamics of cavities downstream of the cylinder. As a part of the validation of the numerical approach, variations of mean pressure coefficient around the cylinder as well as mean and fluctuating coefficients of base pressure have been compared with experimental data available in literature and a good agreement between the two have been observed. With a reduction in cavitation number, cavitation activity extends downstream. Based on the fast Fourier transformation (FFT) of pressure and vapor fraction data, it can be concluded that vortex shedding is suppressed due to the formation of extended cavities in the wake. Another interesting observation is the interplay between flow turbulence and cavitation structures. Simulations indicate that with increasing downstream distance from $x/d=0.3$ (near the cylinder) to $x/d=3$ (away from the cylinder), pressure recovers and turbulent kinetic energy decreases in case of non-cavitating flow. However, in case of cavitating flow, turbulent kinetic energy is higher at the region of collapse of vapor pockets than that at near the cylinder which is lower than that corresponding to the non-cavitating case at the same location.

Keywords: Circular cylinder, sub-critical flow regime, turbulent kinetic energy.

Introduction

Flow around a circular cylinder has been studied extensively in the past as it exhibits rich flow physics like wake structures, shear layer interaction and boundary layer. Roshko [1] categorized flow based on Reynolds number (Re) as subcritical ($300 < Re < 3 x 10^5$), critical ($3 x 10^5 < Re < 3.5 x10^5$) and supercritical ($Re > 3.5 x 10^5$). This categorization is now widely used in the literature [2-3]. The difference in flow separation occurring on the cylinder surface is the main reason for the distinctive features observed in each of these flow regimes. Different aspects and properties of this flow, like lift and drag forces, phenomenon of vortex shedding and effect of aspect ratio of cylinder was reported past [4-6] but most of the work falls in critical and super critical flow regime.

In comparison with non-cavitating flow past a circular cylinder, there has been relatively less research on cavitating flow past a circular cylinder. Furthermore, investigations have mostly examined the critical and supercritical flow regimes. It has been discussed, in the context of non-cavitating flow that for flow past a bluff body, instantaneous lowest pressure is not necessarily located on the surface but may also occur in the centre of a fully formed eddy in the wake. Thus, cavitation is most likely to originate in these places. Specifically, in case of circular cylinder, cavitation may initiate on the surface of the cylinder, in the shear layer or in the wake depending on the Reynolds number. Wykes [7] had observed the formation of cavity for a circular cylinder, for different flow regimes. He found that, for subcritical flow regime, the cavity inception took place in shear layer or wake whereas for critical flow regime, cavitation inception took place on the cylinder surface. For supercritical flow regime, the cavity formed a wedge-like structure. Ramamurthy and Bhaskaran [8] studied cavitating flow past a circular cylinder for varying blockage factors. They found that the wall interference effect was less at high cavitation number but at low cavitation number this effect was high and drag coefficient dropped rapidly with a reduction in cavitation number.

*Corresponding Author, Pankaj Kumar: pankaj20221@gmail.com

Ihara and Murai [9] conducted an experiment to measure mean and fluctuating pressure distributions around a circular cylinder for Re between 3.2 x 10^5 and 3.6 x 10^5 in critical flow regime. Matsudaira et al.[10] measured the cavity collapse pressure and intensity distribution in wake with help of pressure transducers for the range of 4.5 x 10^5 < Re < 6 x 10^5. They found that due to cavity collapse, high imperative forces appeared in a regular fashion in wake and it was closely related with Karman vortex shedding frequency. In the subcritical range of flow, cavitation damage test was done by Sato and Saito [11] on Shalnev-type venturi in which straight cylindrical pin is mounted at the throat of venturi. They found that with a decrease in cavitation number, cavity length increases and was accompanied by an increase in surface damage.

Cavitation being a two-phase phenomenon with varying phase change, it is very difficult to model and several attempts have been made in the past to address this issue. The bottom line is that cavitating flows are usually high Reynolds number flows. Accurate representation of turbulent cavitating flow is quite challenging and yet progressive. The homogeneous flow model [12-14] assumes the mixture of constituent phases to be a single fluid and the mixture flow is governed by only one set of equations. Here both the phases are assumed to be in thermal and mechanical equilibrium and the governing equations are conservation laws written for the mixture. The modeling process is simplified with the help of some basic assumptions like, both the phases are strongly coupled and moves with same velocity. As far as the numerical simulation of cavitating flow past a circular cylinder is concerned, there are only a few studies available in literature. Seo et al. [15] used density-based homogeneous equilibrium model solver in DNS to compute sound produced by cavitating flow over a 2D cylinder at a Reynolds number of *200*. They used mixture density as a function of void fraction, mass fraction, pressure and temperature of liquid and vapour phase. This model captures the formation, detachment, convection, and collapse of the cavitation bubble very well, besides the generation and propagation of the linear and non-linear waves in the entire field. They concluded that the main source of noise found in spectrum was the collapse of vapour cavities. Gnanaskandan and Mahesh [16] conducted LES simulation on cylindrical geometry at Re=*200 & 3900* for different cavitation stages. They used homogeneous mixture model and characteristic based filtering to solve compressible Navier Stokes equation. They found that cavitation phenomenon suppressed turbulence and three-dimensionality.

So above mentioned literature survey reveals that there is a lack of study for upper subcritical range ($2x10^4 < Re < 2x10^5$), especially in the near wake ($x/d = 10$). This lack is more pronounced in case of cavitating flow. The present work aims to address this gap with emphasis of turbulent energy interaction with cavities.

2. Computational Procedure

The described case was simulated by using CFD software ANSYS Fluent 14. The formation of cavity behind cylinder is governed by the combination of turbulence and cavitation. Thus, in this study, a combination of *k-ω-SST* turbulent model and Schnerr–Saur cavitation model was used to simulate the transient behavior of wake cavitation.

2.1. Governing Equations

The governing equations for unsteady, incompressible flow with constant fluid properties are:

$$\frac{\partial u_i}{\partial x_i} = 0 \tag{1}$$

$$\frac{\partial u_i}{\partial t} + u_j \frac{\partial u_i}{\partial x_j} = -\frac{\partial p}{\rho \partial x_i} + \frac{\partial}{\partial x_j}[\vartheta \frac{\partial u_i}{\partial x_j}] \tag{2}$$

For turbulent flow, the 'closure' is achieved with the help of a suitable turbulent model. Works of several researchers (e.g. Unal et al.[17]) show that *k-ω-SST* model of Menter [18] performs better than other turbulent models for flow past a circular cylinder. Hence in this work, *k-ω-SST* model is adopted.

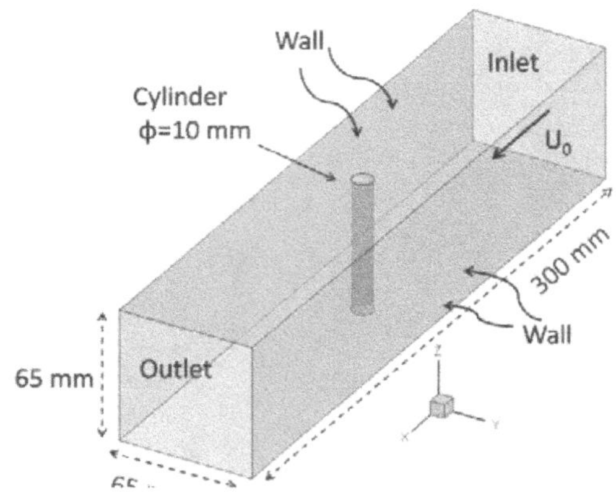

Figure 1: Computational domain showing the domain size, inlet, outlet and cylinder.

2.2. Computational Domains and Boundary Conditions

The choice of computational domain was mainly dictated by the experimental set up which is *65 mm x 65 mm* in cross-section and *300* mm long. A circular cylinder of length *65* mm was placed at the center of the computational domain shown in figure 1. A 3-D structured mesh was built in ICEM-CFD™ with multi-blocks using O-grids at the central block near the cylinder to ensure high density near the cylinder and wall. Proper attention was given to adjust the maximum and minimum acceptable angles (*126* and *54*), control skewness (*0.7 - 1*) and achieve an orthogonal quality of *0.9*. Total number of mesh elements of 4.43×10^5 is used in all the simulations obtained from grid independent. Unsteady simulation was carried out and time step sensitivity study was taken up to ascertain the effect of time step size. Based on the time-independent study, time step size of *100* μs was chosen and kept constant in all simulations. Uniform velocity was applied at the inlet and pressure condition was imposed at the exit of the domain. Side walls and cylinder surface were given a no-slip wall boundary condition. SIMPLE (Semi-Implicit Method for Pressure-Linked Equation) algorithm was used for pressure-velocity coupling. In spatial discretization, second order upwind scheme was used for momentum equation. Standard scheme was used for pressure terms and the quadratic upstream interpolation for convective kinetics (QUICK) scheme was used for vapor mass fraction transport equation. At each time step, the residue for all the equations solved was brought down to a convergence criterion of 10^{-5}. For each simulation, once the initial transience has decayed, *500* data files were stored for further post processing. MATLAB® was used to obtain mean and fluctuating contours of pressure, velocity, void fraction and turbulent kinetic energy.

3. Results and Discussion

Mean pressure coefficient ($\overline{C_P} = \frac{P_\theta - P_T}{\frac{1}{2}\rho v^2}$) was estimated for non-cavitating case. Comparison of numerical results with non-cavitating data is presented in figure 2(a). This serves as a validation of numerical simulation used for the present study. Mean pressure coefficient ($\overline{C_P}$) variation for different cavitation number is shown in figure 2(b) and shows that the surface pressure is influenced by cavitation. Here after angle *90°* ($\overline{C_P}$) is constant which confirms that separation angle is at *90°* for subcritical flow range [4]. The frequency obtained from pressure fluctuation ($C_P' = C_P - \bar{C}_P$) data at base point (at *180°* from front stagnation point) is *146* Hz. Strouhal number ($\frac{fD}{U}$) is obtained from this frequency is close to *0.2* which is also well validated for this range of Reynolds number [1, 19]. There is a very good agreement between the numerical prediction and experimental determination of the variation of Strouhal number with different cavitation numbers (figure 3(a)). The decrease in vortex shedding frequency is closely related with the elongation of cavity structure as reported by Young and Hall [20] and Ramamurthy and Bhaskaran [8]. Here the figure 3(b) brings out a comparison of instantaneous cavity and vortex structures and establishes the fact that wake cavitation is intricately related with vorticity dynamics.

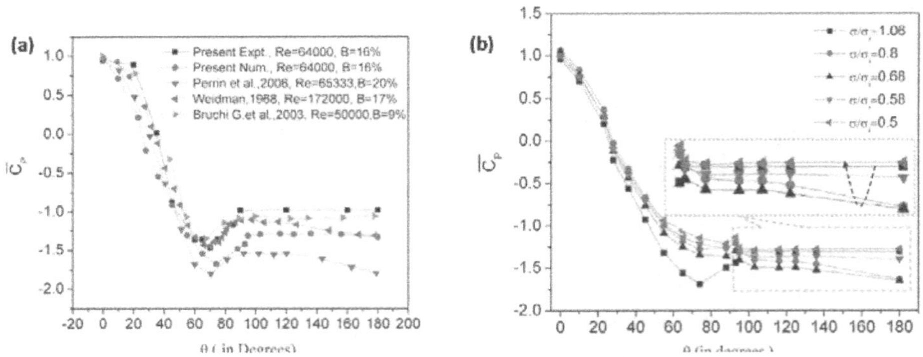

Figure 2: (a) Non-cavitating mean pressure coefficient (C_P) for σ/σ_i=*1.06* (b) Mean pressure is plotted for different cavitation numbers. Present Expt. is referred to experiments conducted [19].

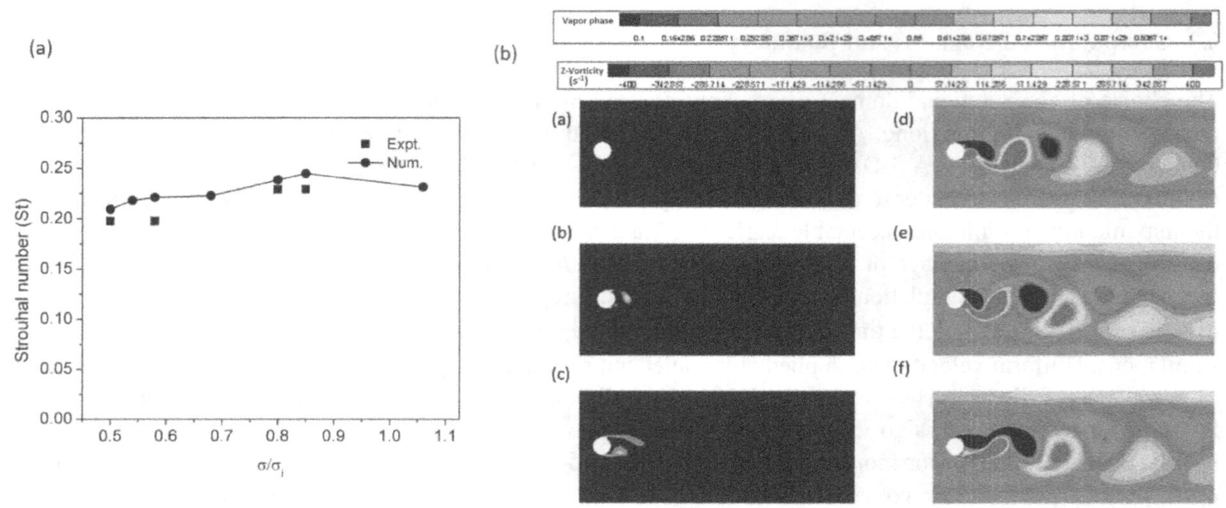

Figure 3: (a) Variation of Strouhal number obtained at base point (*180°*) for different cavitation numbers. "Expt." is pressure fluctuation at base obtained experimentally [21] whereas "Num." is pressure fluctuation at base obtained numerically. (b) Variation of vapour fraction and Z-vorticity obtained numerically for cavitation number (σ/σ_i) *1.06* (a & d), *0.58* (b & e) and *0.5* (c & f). Here (a),(b),(c) are instantaneous vapour fraction and (d),(e),(f) are corresponding vorticity contours.

The effect of cavity formation and collapse in the wake is studied in terms of turbulent kinetic energy of the cavitating flow. The mean vapour fraction together with mean turbulent kinetic energy contour is shown in figure 4 for different cavitation numbers (σ/σ_i) *0.85, 0.54* and *0.5*. Dashed line in this figure indicates the location of cavity collapse which is also the position where turbulent kinetic energy is high. The collapsing bubble produces pressure oscillations and increases turbulent kinetic energy locally. This phenomena is more clearly explained by time variation plots of pressure, turbulent kinetic energy and vapour fraction for non-cavitating flow (σ/σ_i = *1.06*) in figure 5. It is clear from figure 5(a) that moving from x/d *=0.6* to x/d*= 3,* pressure increases and turbulent kinetic energy decreases for non-cavitating flow. It also shows that near cylinder at x/d*=0.6*, temporal variations of pressure and turbulent kinetic energy are in phase but away from cylinder at x/d *=3*, they become out of phase. Figure 6 shows the time variation plots of pressure, turbulent kinetic energy and vapour fraction for cavitating flow (σ/σ_i *=0.5*). A closer look at figure 6(a) shows that a reduction in pressure is accompanied by an increase in volume fraction and conversely, a collapse in vapour volume accompanies a pressure rise. What is interesting is the fact that a rise in vapour volume also times well with low turbulent kinetic energy while a collapse of the cavity is followed by an increase in turbulent kinetic energy. In the non-cavitating case, (figure 5), an increase in downstream distance

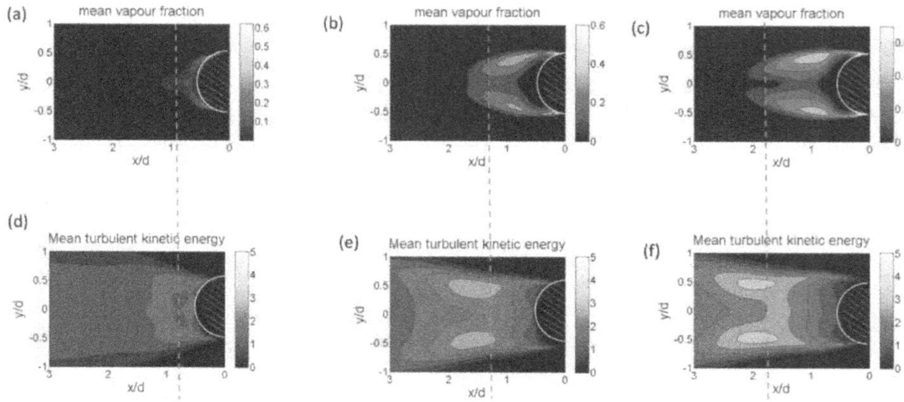

Figure 4: Mean contour of turbulent kinetic energy and vapour fraction shown for different cavitation number (σ/σ_i) : *(a, d) 0:85, (b, e) 0:54, (c, f) 0:5.* Dashed vertical lines shows the position of cavity collapse as well peak of turbulent kinetic energy.

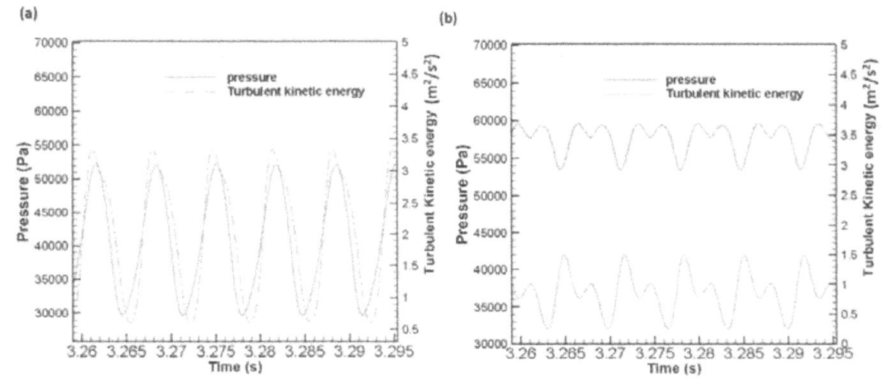

Figure 5: Variation of pressure and turbulent kinetic energy for cavitation number *(σ/σ_i=1.06)(a) at x/d=0.6 (b) at x/d=3.*

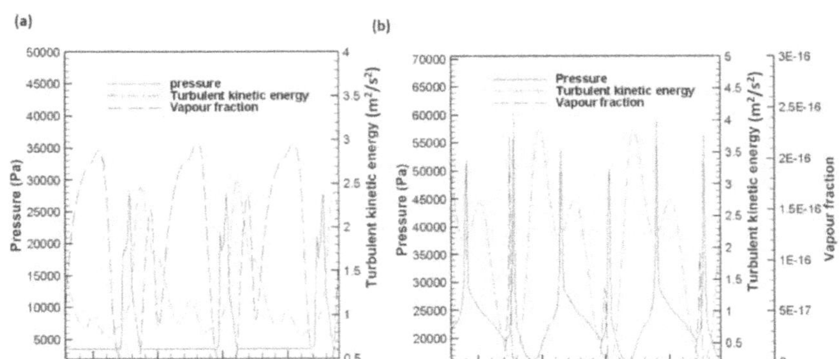

Figure 6: Variation of pressure, turbulent kinetic energy and vapour fraction for cavitation number *(σ/σ_i=0.5))(a) at x/d=0.6 (b) at x/d=3.*

leads to a significant reduction in turbulent kinetic energy. However, in case of cavitation (figure 6(b)), turbulent kinetic energy peak value actually increases when compared with that at the near cylinder (figure 6(a)). Also similar to near cylinder case presented in figure 6 (a), there is an increase in turbulent kinetic energy and pressure along with a reduction in vapour content. Thus figure 4 and figure 6 together present a picture of an energy exchange —

turbulent kinetic energy is used for the growth of a cavity while a collapsing cavity leads to an increase in turbulent kinetic energy.

4. Conclusion

Numerical results agreed with the experiment in terms of mean pressure, base pressure fluctuations and corresponding frequency. The reduction of shedding frequency with cavitation number is explained in terms of vorticity. The variation of mean pressure, vapour fraction and turbulent kinetic energy contour and plots for different cavitation number suggest the presence of an energy interaction in which local turbulence promotes bubble formation and growth and there is an increase in turbulent intensity in regions of cavity collapse due to recovery of mean pressure.

References

[1] Roshko, A. (1961). *Experiments on the flow past a circular cylinder at very high Reynolds number.* J. Fluid Mech., 10, 345–356.

[2] Schewe, G. (1983). *On the force fluctuations acting on a circular cylinder in cross-flow from subcritical up to transcritical Reynolds numbers.* J Fluid Mech., 133, 265–285.

[3] Sumer, B. M. and Fredsoe J.(2006). *Hydrodynamics around cylindrical structures.* World Scientific Publishing Ltd., Singapore.

[4] Achenbach, E. (1968). *Distribution of local pressure and skin friction around a circular cylinder in cross-flow upto Re = 5 x 10^5.* J. Fluid Mech., 34(4), 625–639.

[5] Bearman, P. W. (1969). *On vortex shedding from a circular cylinder in the critical Reynolds number regime.* J. Fluids. Mech., 37, 577–585.

[6] Norberg, C. (1994). *An experimental investigation of the flow around a circular cylinder: influence of aspect ratio.* J. Fluid Mech., 258, 287–316.

[7] Wykes, M. E. P. (1978). *Experimental studies of viscous effects on cavitation.* Ph.D. thesis, St. John's College, University of Oxford, United Kingdom.

[8] Ramamurthy, A. and Bhaskaran P. (1977). *Constrained flow past cavitating bluff bodies.* ASME J. Fluids Eng., 99, 717–726.

[9] Ihara, A. and Murai H.(1986). *Cavitation inception on a circular cylinder at critical and* supercritical flow range. ASME, J. Fluids Eng., 108, 421–427.

[10] Matsudaira, Y., Gomi Y., and Oba R.(1992). *Characteristics of bubble-collapse pressures in a karman-vortex cavity.* JSME INT. J., Ser. II, 35, 179–185.

[11] Saito, Y. A. and Sato K.,(2003). Cavitation bubble collapse and impact in the wake of circular cylinder. 5th Int. Symposium on Cavitation (CAV2003), Osaka, Japan.

[12] Kunz, R. F., Boger D.A., Stinebring D.R., Chyczewski T.S., Lindau J.W., Gibeling H.J., Venkateswaran S. and Govindan T. (2000). *A preconditioned navier stokes method for two-phase flows with application to cavitation prediction.* Comput. Fluids, 29, 849875.

[13] Schnerr, G. H. and Sauer J. (2001). *Physical and numerical modeling of unsteady cavitation dynamics.* 4th int. conference on multiphase flow, Orleans, USA.

[14] Singhal, A. K., Athavale M., Li H., and Jiang Y.(2002). *Mathematical basis and validation of the full cavitation model.* J. Fluids Eng., 124, 617–624.

[15] Seo, J. H., Moon Y.J., and Shin B. R. (2008). *Prediction of cavitating flow noise by direct numerical simulation.* J. Comput. Phys., 13, 65116531.

[16] Gnanaskandan, A. and Mahesh K.(2016). *Numerical investigation of near-wake characteristics of cavitating flow over a circular cylinder.* J. Fluid Mech, 790, 453–491.

[17] Unal, M. F. and Rockwell D. (1988). *On the vortex formation from a cylinder part 1.the initial instability.* J. Fluid Mech., 190, 91–512.

[18] Menter, F. (1994). *Two equation eddy viscosity turbulence models for engineering applications.* AIAA J., 32, 1598–1605.

[19] Kumar P., Bakshi S. and Chatterjee D.(2017). *Experimental investigation of cavitation behind the circular cylinder* J. Thermal Sci. Eng. App., ASME, 9(3), 031004-1-6.

[20] Young, J. O. and Hall W. J.(1966). *Effects of cavitation on periodic wakes behind symmetric wedges.* ASME, J. Basic Eng., 88, 163–176.

© 2018 ASME

Numerical study on the effect of turbulence and cavitation model for propeller induced hull pressure fluctuation

[1]Jaewook Hur*, [1]Hyungjun Kim, [1] Hyoungsuk Lee

[1]*Maritime Research Institute, Hyundai Heavy Industries Co., Ltd., Ulsan, Korea*

Abstract

In this paper, the effect of turbulence and cavitation model is investigated to predict the propeller induced hull pressure fluctuation accurately. Experimental cavitation test results are also presented in order to assess the suitability of turbulence and cavitation models. Reynolds averaged Navier-Stokes (RANS) and detached-eddy simulation (DES) approaches are adopted as a turbulence model, and Schnerr-Sauer and classical Rayleigh-Plesset model are considered as a cavitation model. Based on various numerical tests, it is demonstrated that the hull pressure fluctuations for marine propellers are less influenced by turbulence and cavitation models.

Keywords: Marine propeller, Cavitation, RANS, DES, Hull pressure fluctuation, Classical Rayleigh–Plesset cavitation model

Introduction

Hull pressure fluctuation is one of the most important factors to assess the cavitation performance of a marine propeller. When a propeller blade is undergoing sheet or tip vortex cavitation, pressure field around the blade is fluctuated due to the periodic development and collapse of bubbles. This pressure fluctuation occasionally plays a role in vibration problems at ship structures. Within this context, the prediction of hull pressure due to cavitation is essential for a marine propeller design. Recently, many researchers have tried to analyze the cavitation phenomenon using the computational fluid dynamics (CFD) approach. Kim [3] and Bensow [1] studied the effect of turbulence models on the cavitation analysis and showed that the turbulence model plays an important role in cavitation predictions. Thus, the detached-eddy simulation (DES) and the large-eddy simulation (LES) are more suitable for the prediction of highly unsteady flow than Reynolds averaged Navier-Stokes (RANS) simulation. Regarding the cavitation analysis of marine propellers, Hasuike [2] studied the erosion index through cavitation analyses. Paik [6] showed that RANS gave a good agreement with experimental results in the 1st blade passing frequency (BPF) components of hull pressure fluctuations. Sauer and Schnerr [5] suggested a simplified Rayleigh-Plesset equation which has limited accuracy in predicting flows with highly fluctuating cavitation because the higher order terms are neglected. Recently, Muzaferija [4] studied the two cavitation models such as classical Rayleigh-Plesset model and Schnerr-Sauer model for the vortex shedding flows around a hydrofoil. This paper presents the effect of turbulence models on cavitation models for propeller induced pressure fluctuations. To validate the suitability of the numerical method, the CFD results are compared with the experimental data. Numerical studies involve various cases with a combination of RANS/DES and classical Rayleigh-Plesset/Schnerr-Sauer cavitation model to investigate the characteristics of propeller induced pressure fluctuations.

Hull pressure fluctuation and Cavitation observation test

Cavitation erosion and pressure fluctuation characteristics are important factors in evaluating propeller performance. For this reason, the prediction of pressure fluctuation is important for propeller design. In this study, a cavitation test for a large crude oil carrier was performed on two propellers at KRISO LCT. The model test was carried out with the model ship installed and pressure on the hull surface is measured by pressure sensors installed at the hull surface above the top of the propeller. The principal dimensions and test conditions are as shown in Table 1.

*Corresponding Author, Jaewook Hur: jwook1009@hhi.co.kr

	Prop. A	Prop. B
Diameter (mm)	240.00	240.00
P/D_{mean}	0.7423	0.7297
EAR	0.4700	0.5300
K_T	0.1707	0.1671
σ	1.77	1.75

Table 1 Principal dimensions and test conditions for Propeller A, B

Numerical method

For the prediction of propeller induced pressure fluctuations, the effect of turbulence and cavitation model is studied. RANS and DES are used for turbulence model using STAR-CCM+. Cavitation model is developed for the growth and collapse of cavitation as pressure changes. Schnerr-Sauer (S-S) model is a simplified Rayleigh-Plesset (R-P) equation neglecting the influence of the higher order terms. Muzaferija [4] proposed classical R-P model considering bubble's inertial effects. In this study, S-S model and classical R-P cavitation model were compared for prediction of propeller induced pressure fluctuations. Detailed information for numerical simulation is shown in Table 2.

Turbulence Model	k-ω SST (KW), IDDES (DES)
Cavitation model	Schnerr-Sauer (S-S) classical Rayleigh-Plesset (R-P)
Temporal discretization	2nd order implicit
Time step	time to pass 0.2 deg

Table 2 Numerical simulation setting

Result of numerical simulation

To understand the effect of turbulence and cavitation model, the time history and FFT result of pressure from both model tests and numerical simulations are compared. In the model test, sheet cavitation and tip vortex cavitation are main source of propeller induced pressure fluctuations. In the previous research[3], it was found that cloud cavitation on hydrofoil is well resolved with DES and large eddy simulation (LES), meanwhile RANS computation give poor prediction on lift and drag fluctuation . Cavitation, which causes fluctuating pressure or erosion, is known to be influenced by turbulence and cavitation model [3]. And, for highly unsteady flow, cavitation and turbulence model is important to resolve the pressure peak. However, the main cavitation patterns of propellers applied in this study are sheet and tip vortex cavitation not including the cloud or bubble cavitation. Therefore, the accuracy of the propeller induced pressure fluctuations for propeller A and B was similarly predicted regardless of the changes in the turbulence model and the cavitation model for propeller A and B as shown in Figures 1 and 2. The pressure is measured and compared on top of propeller generating line position. In this figure, hull pressure fluctuation is expressed as pressure coefficient, $K_P \left(= \frac{\Delta p}{\rho n^2 D^2}\right)$ where Δp is the pressure fluctuation amplitude from the mean pressure, where n and D represent rps and propeller diameter, respectively.

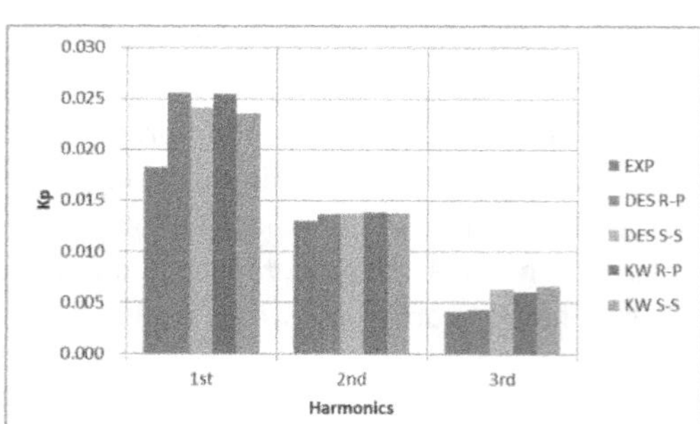

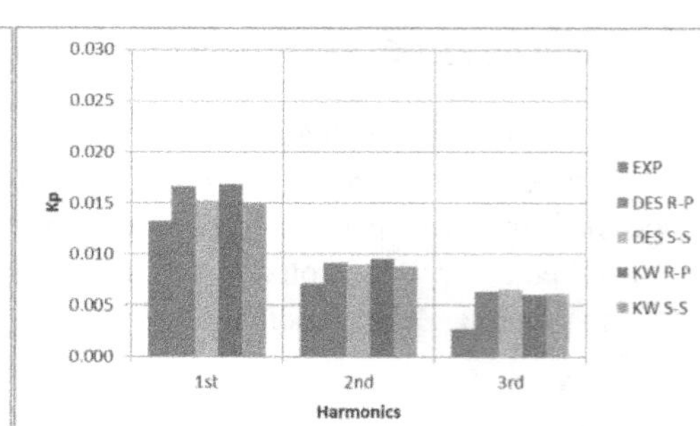

Figure 1 Amplitude of hull pressure fluctuation (Kp) - left: Prop. A, right: Prop. B

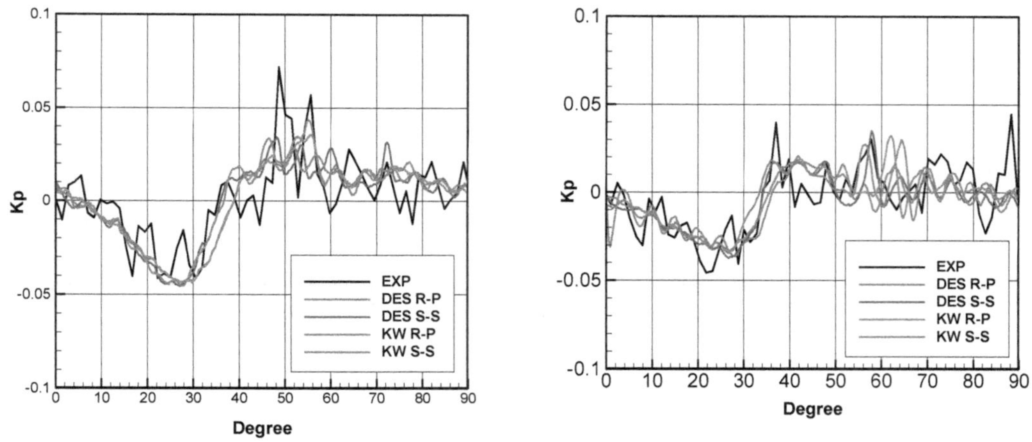

Figure 2 Time history of hull pressure fluctuation (Kp) - left: Prop. A, right: Prop. B

The cavitation pattern was compared with the model test for DES and R-P model with propeller A in Figures 3 and 4. Overall, simulations for the variation of cavitation extent are similar to the experimental result. The result of numerical analysis shows the difference of cavitation due to the propeller design such as expanded area ratio and pitch distribution. The characteristics of pressure measured on the hull were compared in Figure 5. In the propeller A, sheet and tip vortex cavitation is generated and collapses, which affects propeller induced pressure fluctuations. In step 1, the pressure on the hull is reduced due to the pressure drop caused by sheet cavitation. At step 2, the pressure recovers due to the collapse of the sheet cavitation. When cavitation does not occur, the pressure fluctuation amplitude is low than cavitating condition and there is no high order harmonic component. As a whole, a large pressure fluctuation occurs due to the generation and collapse of sheet cavitation, and tip vortex cavitation in steps 3~5. To investigate the cavitation behavior, p' (p'= p $-p_{average}$, $p_{average}$ is mean value of pressure during 1 revolution) is shown in Figure 6 (middle). When the sheet cavitation is at growth stage, p' decreases according to cavitation extension. And, in the stage of collapse, p' increases and this make the propeller induced pressure fluctuations around propeller as shown in Figure 6 (bottom). Moreover, considering the period of the pressure change by the generation and collapse of sheet cavitation (step 1 ~ 2), tip vortex cavitation (step 3~5), the 1st BPF was attributed to the sheet cavitation, and the higher order component was due to the change in the tip vortex.

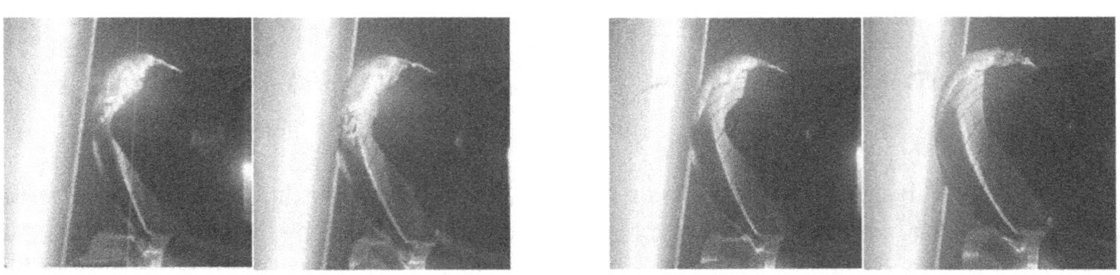

Figure 3 Cavitation pattern of EXP - left : Prop. A, right : Prop. B

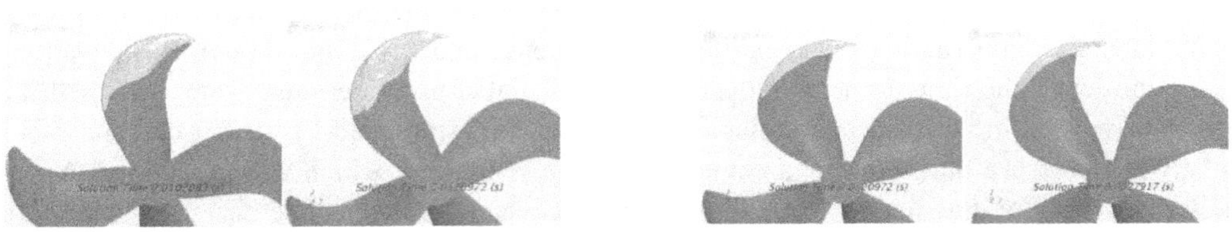

Figure 4 Cavitation pattern of CFD - left : Prop. A, right : Prop. B

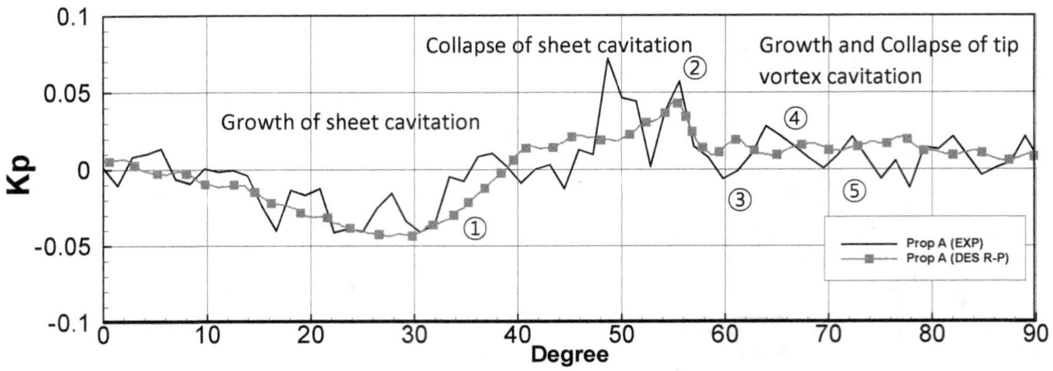

Figure 5 Time history of hull pressure fluctuation (Kp) for Prop. A

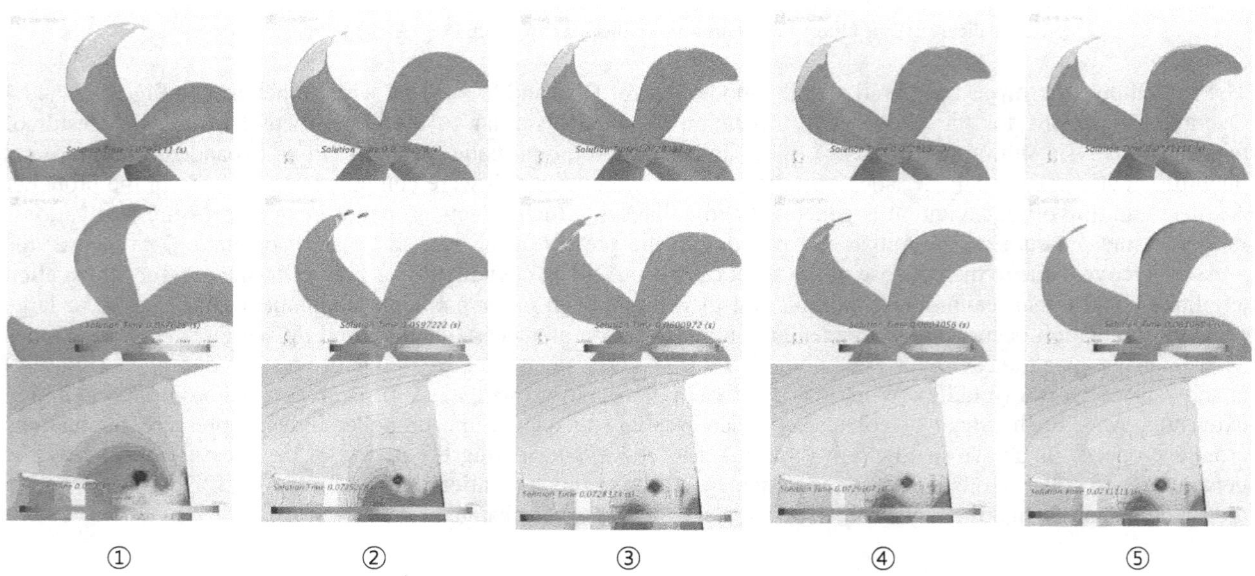

Figure 6 Cavitation pattern (top) and p' (middle, p'= p $-p_{average}$) and pressure around propeller (bottom), Prop. A

Conclusion

The computational fluid analysis of the cavitation on the propeller in the behind-hull condition was performed. From the calculation result, unsteady cavitation and the pressure fluctuation induced by the cavitation were evaluated. The CFD simulation achieves good agreement in prediction of the sheet cavitation which occurs on the blade surface. The harmonic component levels of pressure fluctuation on a model ship hull induced by the sheet cavitation because both the Schnerr-Sauer model and the Rayleigh-Plesset model with RANS/DES well matched unsteady sheet and tip vortex cavitation with model test. From this result of numerical simulation, the prediction of propeller induced pressure fluctuations give good agreement regardless of turbulence and cavitation model. Besides the prediction of pressure fluctuation level, the time domain analysis of the calculated pressure pulse from DES with the Rayleigh-Plesset model showed that the growth and collapse procedures well simulated. By using this analysis method, the relationship between pressure fluctuation and cavitation can be visualized, which will be applied to more advanced initial propeller design. For the future work, it will be necessary to investigate the effects of turbulence and cavitation model for various cavitation patterns on marine propeller.

References

[1] Bensow, R.,E. , 2011, *Simulation of the unsteady cavitation on the the Delft Twist11 foil using RANS, DES and LES,* Second International Symposium on Marine Propulsors, Hamburg, Germany, June

[2] Hasuike, N., Yamasaki, S., Ando, J. and Okazaki, A., 2010. *Numerical study on cavitation erosion risk of marine propellers operating in wake flow,* International Propulsion Symposium, Okayama, Japan, April

[3] Kim, S.E., Brewton, S., 2008, *A Multiphase Approach to Turbulent Cavitating Flows*, 27th Symposium on Naval Hydrodynamics, Seoul, Korea, October

[4] Muzaferija, S., Papoulias, D., Peric, M., 2017 *VOF Simulations Of Hydrodynamic Cavitation Using The Asymptotic And Classical Rayleigh-Plesset Models* Fifth International Symposium on Marine Propulsion, Espoo, Finland, June

[5] Schnerr, G. H. and Sauer, J., 2001. *Physical and numerical modeling of unsteady cavitation dynamics.* 4th International Conference on Multiphase Flow, New Orleans, United States, May [5]

[6] Paik, K.J., Park, H.,G., Seo, J. , 2013, *RANS simulation of cavitation and hull pressure fluctuation for marine propeller operating behind-hull condition*, Int. J. Nav. Archit. Ocean Eng. (2013) 5:502~512

© 2018 ASME

On Cavitation Aggressiveness and Cavitation Erosion on Marine Propellers using a URANS Method

[1,2]Themistoklis Melissaris*; [1]Norbert Bulten; [2,3]Tom van Terwisga;

[1]Wärtsilä Netherlands B.V., Drunen, The Netherlands; [2]Delft University of Technology, Delft, The Netherlands; [3]Maritime Research Institute (MARIN), Wageningen, The Netherlands

Abstract

Cavitation occurrence on the propeller blades may have severe effects, such as cavitation noise, vibrations in the aft ship, and material damage on propellers or rudder. While noise and vibrations affect comfort and stability, cavitation erosion might lead to a significant reduction of the propeller efficiency and structural integrity. Cavitation erosion prediction is therefore extremely important. However, it still remains a challenging issue, especially on marine propeller applications where computational time limit in a practical propeller design environment is imperative. This paper focuses on cavitation aggressiveness on the propeller blades of the KCD-193 model propeller. Cavitation intensity is assessed by the potential power density. Following the idea that the potential power is the basis of the cavitation aggressiveness, we found that by computing the potential power of the first layer of cells, in contact with the blade surface, it gives a useful first estimate of the regions that are exposed to a high erosion risk. The use of a mean field pressure, which is the time average of the local pressure, as the driving pressure behind the cavity collapse, was necessary to get a good match with experimental observations.

Keywords: Cavitation aggressiveness; model propeller; potential power; driving pressure.

Introduction

Cavitation erosion strongly impacts the efficiency and structural integrity of marine propellers. Although much is known about bubble dynamics and the material response, the problem of predicting cavitation damage remains unresolved. Nevertheless, the development of numerical flow simulations seems to offer a good prospect for predicting cavitation and cavitation erosion using computational methods. As propeller design applications require reliable results in acceptable timespan, any numerical method for turbulence modelling other than unsteady RANS is currently considered as too time consuming.

This study is based on the assumption that the collapse of vapor cavities is characterized by potential energy E_{pot}, related to the pressure gradient and the cavity volume [1-4]. During the collapse, this potential energy converts mainly into acoustic energy, which is transmitted to the surface, causing an impact on the wall. The magnitude of the impact depends on the distance between the center of the collapse and the surface, as well as on the direction of the acoustic wave towards the surface wall [9]. Moreover, the water quality and the gas content within the vapor cloud before the collapse, also influence the collapse dynamics and the propagation of the acoustic energy [10].

The process to calculate the contribution of each collapse to the surface, including all the aforementioned parameters, can be very time consuming, resulting in a large computational time. To this end, as a first step, only the contribution of the first layer of cells in contact with the blade surface is taken into account. As driving pressure, the time average of the local pressure in every cell is computed and used in the definition of the potential energy. It was found that using this mean field pressure, instead of the direct local pressure, gives results that agree well with the experiments.

Numerical Modeling

The equations solved are the Reynolds Averaged Navier-Stokes (RANS) equations. An incompressible segregated flow model is selected solving the integral conservation equations of mass and momentum in a sequential manner combined with the SIMPLE pressure-velocity coupling algorithm. A Eulerian multiphase model is used, treating the fluid as a single continuum, assuming a no-slip condition between liquid and vapor phase, with varying properties in space according to its composition. The volume fraction of the components is determined from the condition $a_v + a_l = 1$ while density and viscosity are defined as $\rho = a_l\rho_l + \alpha_v\rho_v$ and $\mu = a_l\mu_l + a_v\mu_v$ respectively.

The turbulence model used in this study is the k-ω SST turbulence model developed by Menter [5]. An empirical reduction of turbulence dissipative terms in two-phase regions has been applied, by modifying turbulent viscosity:

*Corresponding Author, Themistoklis Melissaris: themis.melissaris@wartsila.com

$$\mu_t = f(\rho)C_\omega \frac{k}{\omega}; \quad f(\rho) = \rho_v + \frac{(\rho_m - \rho_v)^n}{(\rho_l - \rho_v)^{n-1}}; \quad n \gg 1 \tag{1}$$

where ρ_v is the vapor density, ρ_l the liquid density and ρ_m the mixture density. For the constant n, a recommended value $n = 10$ has been used. This empirical modification improves the artificial increase of dissipation, generated by the no-slip condition between the two phases. The arbitrary reduction of the turbulent viscosity in the two face regions allows stronger shear flows, while dissipation in the pure liquid remains the same [6].

The cavitation model used is the model proposed by Schnerr-Sauer [7] based on a simplified Rayleigh-Plesset equation, which neglects the influence of bubble growth acceleration, as well as viscous and surface tension effects.

Erosion Modeling

Cavitation intensity is addressed by monitoring the initial potential power of the vapor structures. According to the approach proposed by [2], a vapor cloud contains potential energy which can be defined as follows:

$$E_{pot} = \Delta p \cdot V_{vap} \tag{2}$$

where $\Delta p = (p_\infty - p_{vap})$ is the difference between the surrounding pressure P_∞ and the vapor pressure P_{vap} and the vapor volume V_{vap}. Then, the potential power within the structure can be calculated as the time derivative of the potential energy

$$P_{pot} = \Delta p \cdot \frac{dV_{vap}}{dt} + \frac{dp}{dt} \cdot V_{vap} \tag{3}$$

and a potential power density can be estimated from:

$$\frac{P_{pot}}{V_{cell}} = \Delta p \cdot \frac{da}{dt} + \frac{dp}{dt} \cdot a \tag{4}$$

where $a = V_{vap}/V_{cell}$. To evaluate the cavitation aggressiveness on the surface, the contribution of every vapor structure should be taken into account. The transmitted energy on the surface, due to the collapse of a vapor cloud, depends on several factors, such as the distance of the cloud from the wall and the angle with which the emitted pressure wave hits the surface, the amount of dissolved gas within the initial bubbly cloud, and the water quality [9, 10]. The final erosion damage depends also on the material properties and on the material response to the pressure wave impact. In this study, we focus first on the evaluation of the cavitation intensity, considering only the vapor content of the cells in contact with the blade surface, and we neglect any collapse event away from the surface. The contributions of both time derivatives to the potential power density are computed separately, and compared with each other.

Another important element for the assessment of the cavitation intensity is the driving pressure P_∞ occurring in the Δp term of eq. 3. The driving pressure should describe the flow characteristics while being independent of the cavitation dynamics. The local instantaneous pressure is strongly influenced by the cavitation dynamics. Therefore, the time average of this local pressure is considered as the driving pressure. In this way, we neglect the high pressure peaks due to bubble collapses.

Case Description

The case used in this study is the King's College-D (KCD)-193 model propeller. This propeller has been tested at the Emerson Cavitation Tunnel (ECT) of Newcastle University [8]. Erosion tests and performance measurements in cavitating flow have been conducted. A 2D wake screen was used, located 0.4572 m from the propeller center, to simulate the wake field of the propeller. Figure 1 shows the propeller geometry within the computational domain, and the velocity distribution right before the propeller. The tunnel and propeller geometry, and the inlet boundary condition were provided by University of Strathclyde, Glasgow. Table 1 gives the propeller characteristics. The propeller was tested in two cavitating conditions, one in atmospheric and one in vacuum condition (see Table 2). The cavitation number is defined as

$$\sigma = \frac{P_{ref} - P_v}{0.5\rho\ (nD)^2} \tag{2}$$

where P_{ref} is the reference pressure, P_v the vapor pressure, ρ the liquid density, n the rotational speed and D the propeller diameter.

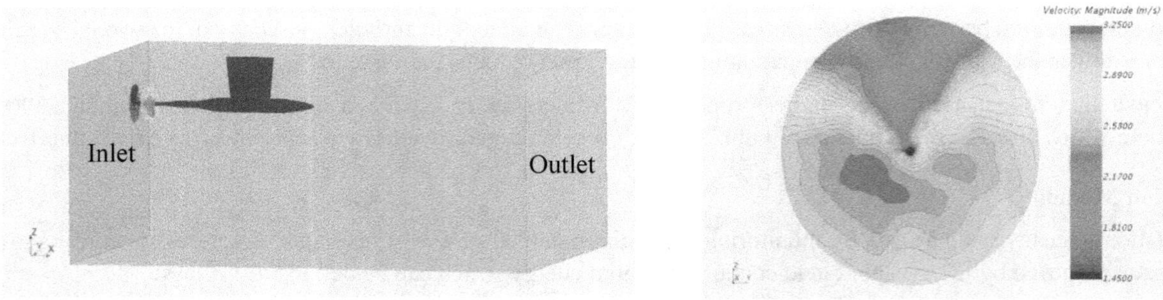

Figure 1. Computational domain (left) and the velocity distribution right before the propeller geometry (right).

Propeller Diameter	0.3048 m
P/D	1
Blade Area Ratio	0.65
Number of Blades	4

Table 1. Propeller Characteristics

Condition	P_{ref} (kPa)	n (RPM)	V_A (m/s)	J	σ
Atmospheric	116.72	1500	3	0.393	3.88
Vacuum	77.016	1500	3	0.392	2.52

Table 2. Conditions at which measurements were taken during the experiments.

For the grid generation, trimmed hexahedral cells were used with local refinements and prism layers along the wall, with such first cell distance so the average y+ value was well below 1 to resolve the viscous sublayer. An extra refinement was applied around the tip area to capture the tip vortex cavitation (Figure 2). After a sensitivity study, a grid with around 8.5 million cells was used, with 6 million cells at the rotating region. The rotation rate was 0.1 degrees per time step.

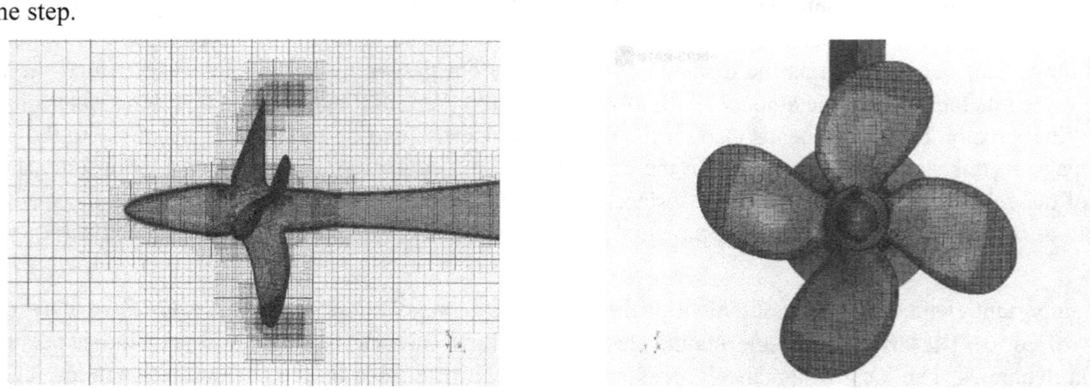

Figure 2. Computational grid. A refinement was used around the tip area to capture cavitation dynamics in detail.

Results

The propeller thrust and torque were calculated for both conditions (atmospheric and vacuum) and they were compared with experimental data. For the thrust, two values were computed, one including only the thrust of the propeller blades, and one including also the thrust and drag of the hub and the hubcap respectively. This was because the way the thrust coefficient was measured during the experiment is not known. The results are shown in Table 3. The uncertainties of the experiment are also unknown, and the presence of the wake screen, introduces an extra uncertainty in the numerical modelling. No uncertainty assessment was conducted for the computations in the current study, but we ensured compliance with existing Best Practice Guidlines. Following these guidelines and using previous studies on uncertainties we consider the discrepancy between the measurements and the CFD results as acceptable. The computed cavities showed a good qualitative agreement with the cavitation observations for both conditions. However, no quantitative comparison was possible as there is no experimental information regarding the total vapor volume. The length of the tip vortex was slightly underestimated, but it is not of high importance in this case. Figure 3 and Figure 4 show the qualitative comparison.

		$K_{T_{Blades}}$	$K_{T_{Propeller}}$	$10K_Q$	Vapor Volume (m^3)
Atmospheric Condition	**Experiment**	0.374	0.374	0.541	-
	CFD	0.389	0.355	0.580	2.096E-05
	Difference %	4.14%	-5.18%	7.29%	-
Vacuum Condition	**Experiment**	0.371	0.371	0.554	-
	CFD	0.383	0.360	0.595	6.974E-05
	Difference %	3.14%	-2.95%	7.5%	-

Table 2. Thrust and torque coefficients and total vapor volume as computed with CFD for both atmospheric and vacuum conditions. K_T and K_Q are compared with the experimental measurements.

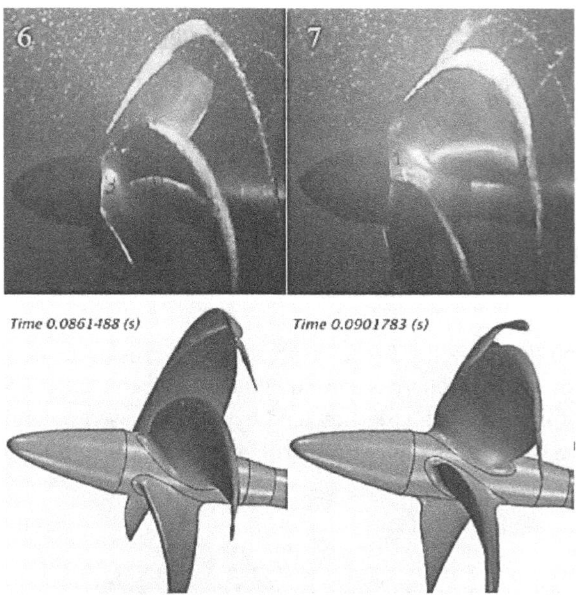

Figure 3. Qualitative comparison of the vapor volume between the experimental observations and the computations with iso-value of $a = 0.1$ in atmospheric condition [8].

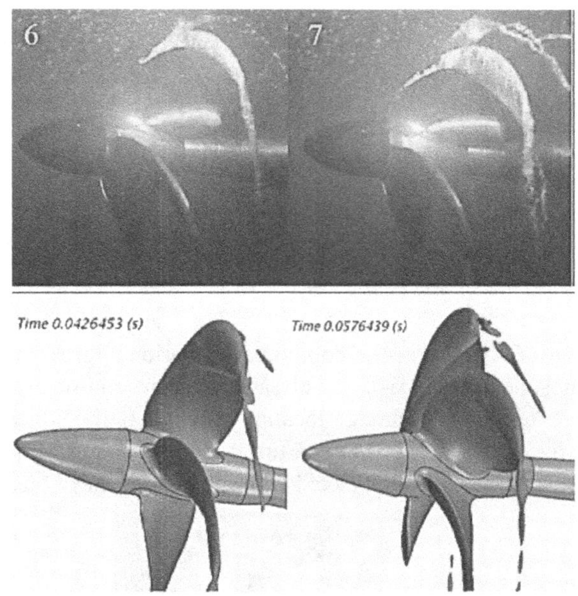

Figure 4. Qualitative comparison of the vapor volume between the experimental observations and the computations with iso-value of $a = 0.1$ for the vacuum condition [8]

The cavitation intensity on the propeller blades was estimated by calculating the potential power density during two propeller rotations. The statistical uncertainty of the total vapor volume was estimated to be around 0.5%. The two contributions to this power density (the relative variation of the vapor volume and the pressure) were computed. Two different interpretations for the driving pressure were used; the time average of the local pressure in every cell, in addition to the local instantaneous pressure. Not surprisingly, the results show that the time derivative of the void fraction (Figure 5a) contributes much more than the averaged pressure time derivative term (Figure 5b). This result is in line with Leclercq et. al. [8]. On the other hand, when the local instantaneous pressure would have been used for the computation of the two terms, either the cavitation intensity would be much lower (Figure 5c) or a high intensity would be located only at the leading edge (Figure 5d), which was not observed in the experiment [8].

Li et. al. [4] found that the variation of the local pressure in time showed higher contribution to the cavitation aggressiveness than the variation of the volume fraction, for two NACA hydrofoils. Their outcome agrees with our results with the local instantaneous pressure as the driving pressure behind the collapse (figure 5c and 5d). However the high intensity areas in our case do not agree with the experiments. On the other hand, using the mean local pressure as the driving pressure, shows a higher contribution of the vapor volume variation to the cavitation aggressiveness. In addition to that, the results agree well with the paint test (Figure 6), showing the importance of a correct definition of the driving pressure when assessing the erosion intensity.

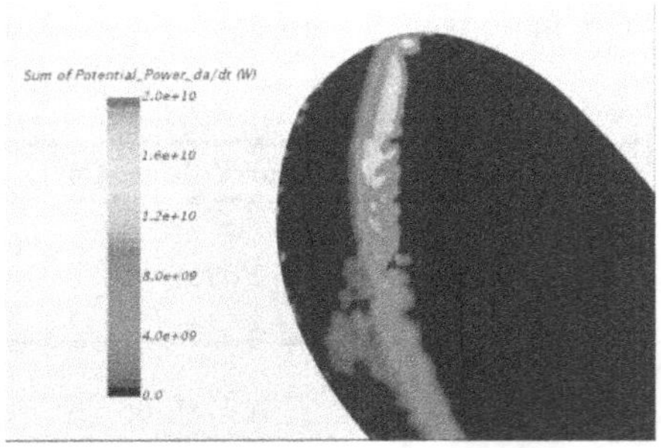

Figure 5a. The sum of $\Delta p \cdot \frac{da}{dt}$ on the blade surface after two rotations, with the time averaged local pressure for p_∞ in Δp.

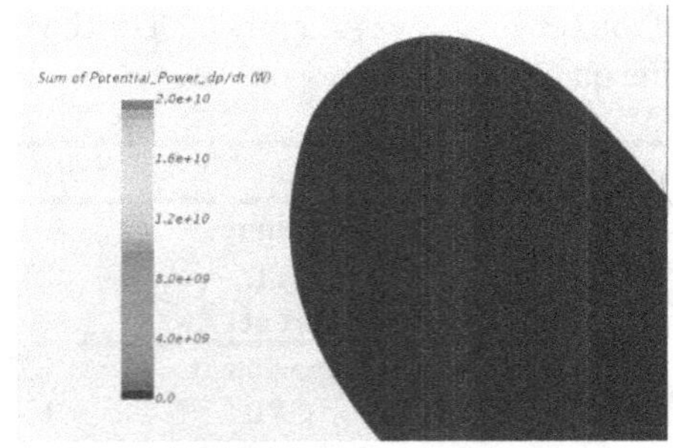

Figure 5b. The sum of $\frac{dp}{dt} \cdot a$ on the blade surface after two rotations, with the time averaged local pressure for p_∞ in Δp.

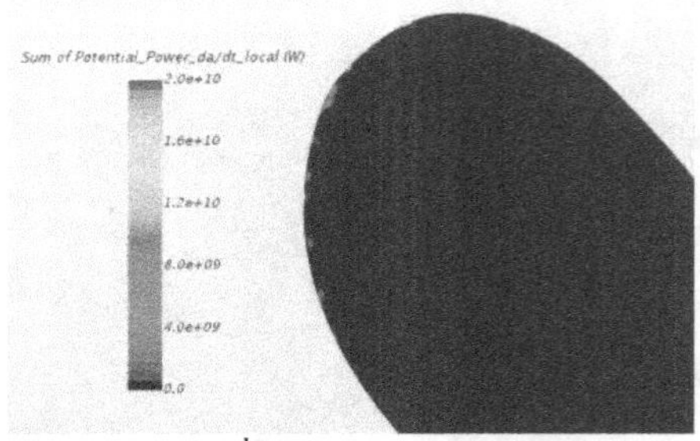

Figure 5c. The sum of $\Delta p \cdot \frac{da}{dt}$ on the blade surface after two rotations, with the local instantaneous pressure for p_∞ in Δp.

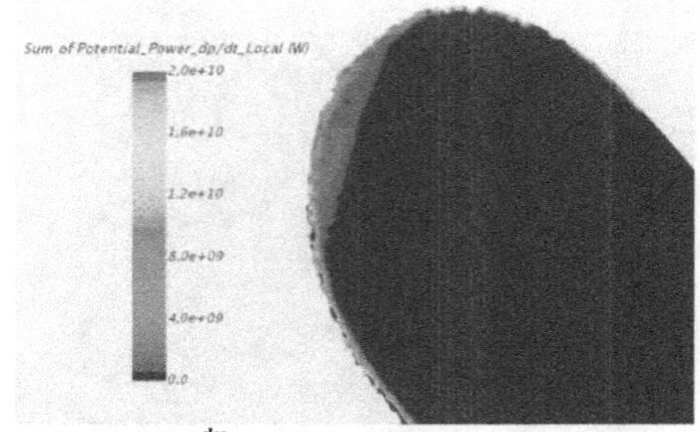

Figure 5d. The sum of $\frac{dp}{dt} \cdot a$ on the blade surface after two rotations, with the local instantaneous pressure for p_∞ in Δp.

Figure 6, compares the computed cavitation aggressiveness on the propeller blades with the paint erosion tests, for atmospheric (Figure 6a and 6b) and vacuum condition (Figure 6c and 6d). There is a fair qualitative agreement between the experimental damage location and the computed region of high erosion risk indicated by the increased potential power density. Nevertheless, this can only be a first indication of the areas with high erosion risk, because only the first layer of cells, connected with the surface, have been taken into account.

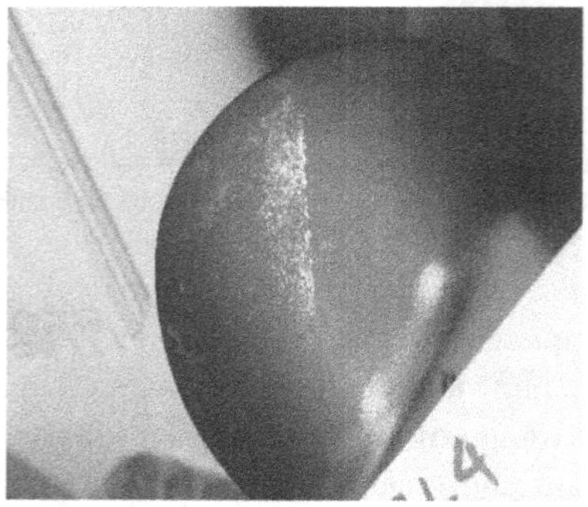

Figure 6a. Experimental erosion observation in atmospheric condition [8].

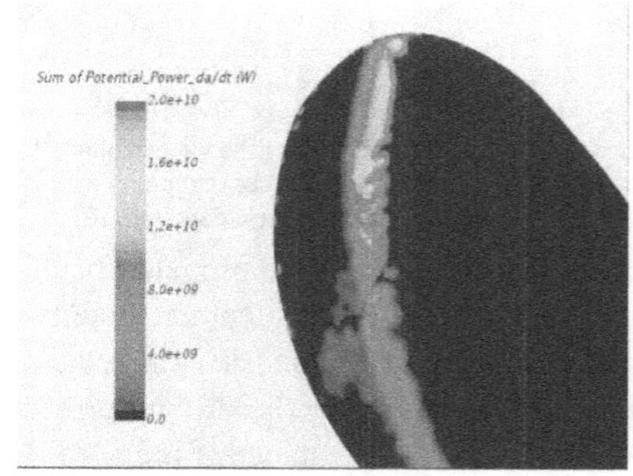

Figure 6b. Computed potential power after two propeller rotations in atmospheric condition.

Figure 6c. Experimental erosion observation in vacuum condition [8].

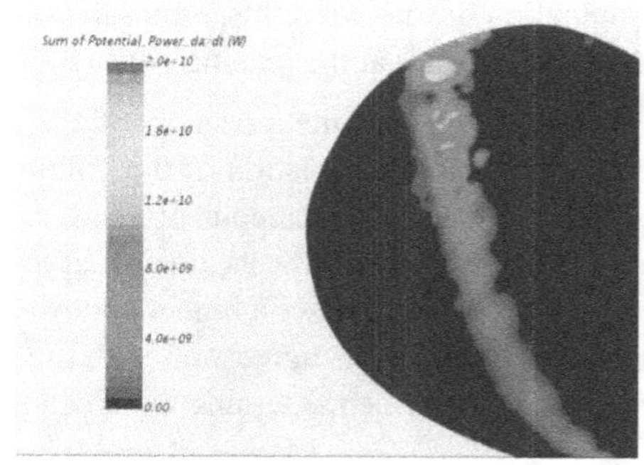

Figure 6d. Computed potential power after two propeller rotations in vacuum condition.

Additional contribution should be accounted for all the structures close to the surface, by considering the distance and the direction of the collapse towards the surface. Moreover, the impact of the water quality and the bubble gas content has also been neglected. Finally, the material response should be taken into account as well, to be able to calculate the actual damage and material loss on the propeller blades.

Conclusion

Cavitation erosion can cause severe damage to the propeller blades, and gives opposite design trends to fuel efficiency, rendering cavitation erosion risk assessment imperative, already in the design stage. Making an early stage propeller erosion assessment requires the right selection of an erosion model (physical modelling), in addition to the numerical modelling. In this study, the energy balance approach has been adopted [2, 10], to estimate cavitation intensity. The contribution of both the vapor volume and pressure derivative terms to the potential power density has been evaluated. In every cell, the time average of the local pressure was used as the driving pressure p_∞, and we compared the results with the results based on the local instantaneous pressure. As a first step, only the first layer of cells, connected with the surface were considered, to facilitate the erosion intensity assessment. We found that the areas with high potential power density agree well with the erosion observations for both atmospheric and vacuum condition, when the mean field pressure is used. However, the contribution of all the vapor structures in the domain should also be considered, to get a representative estimate of the cavitation erosion risks. If, furthermore, an estimate of the damage would be needed, the material response would have to be invoked, in order to compute the material loss on the propeller surface. This will be further investigated in the future. We conclude that this method can be used for an initial and quick evaluation of the high erosion risk areas on the propeller blades, useful for propeller design optimization. Further studies are planned to evaluate applicability of this erosion criterion for propeller optimization in behind conditions.

Acknowledgements

The kind cooperation of Strathclyde University by making their erosion propeller data available to our computational study is highly appreciated and acknowledged.

References

[1] Hammit, F. G., (1963). *Obserbations on Cavitation Damage in a Flowing System*. ASME J. Basic Engineering, **85** (3), p. 347-359.

[2] Fortes-Patella, R., and Reboud, J. L., (1998). *A New Aproach to Evaluate the Cavitation Erosion Power*. ASME Journal of Fluids Engineering, vol 120, p. 335-344.

[3] Bark, G., Berchiche, N., and Grekula, M., (2004), *Application of Principles for Observation and Analysis of Eroding Cavitation-The EROCAV Observation Handbook*, 3.1 ed., Chalmers University of Technology, Goteborg, Sweden.

[4] Li, Z., and Terwisga, T. J. C. (2014), *Assessment of Cavitation Erosion with a URANS Method*. ASME J. Fluids Eng., 136, p. 041101-1.

[5] Menter F., (1994). *Two-equation eddy-viscosity turbulence modeling for engineering applications*. AIAA Journal, vol. 32, p. 1598-1605.

[6] Reboud J.-L, Stutz, B., and Coutier, O., (1998). *Two-phase flow structure of cavitation: experiment and modelling of unsteady effects*, in Third International Symposium on Cavitation, Grenoble, France.

[7] Sauer J., (2000). "*Instationär kavitierende strömungen - Ein neues modell, basierend auf fron capturing (VoF) und blasendynamik*" Ph.D. Thesis, Karlsruhe University, Karlsruhe, Germany.

[8] Mantzaris, A., Aktas, B., Fitzsimmons, P., .and Atlar, M., (2015). "*Establishment and verification of reproducible method for coating propeller blades for erosive cavitation detection*", Forth International Conference on Advance Model Measurement Technologies for the Maritime Industry, Istanbul, Turkey

[9] Leclercq, C., Archer, A., and Fortes-Patella, R., (2016). *Numerical investigations on cavitation intensity for 3D homogeneous unsteady viscous flows,* IOP Conference Series: Earth and Environmental Science, 49(9), p. 092007, IOP Publishing

[10] Fortes-Patella, R., Challier, G., Reboud, J. L., .and Archer, A., (2013), "*Energy Balance in Cavitation Erosion: From Bubble Collapse to Indentation of Material Surface*", ASME J. Fluids Eng., 135, p. 011303-1.

[11] Usta, O., Aktas, B., Maasch, M., Turan, O., Atlar, M., and Korkut, E., (2017), "*A study on the numerical prediction of cavitation erosion for propellers*", Fifth International Symposium on Marine Propulsion, Espoo, Finland.

© 2018 ASME

 CAV18-05169

On Higher Order Blade Harmonics of Propeller-Excited Hull Pressures Due to Cavitation - A Review and Discussion

Ernst-A. Weitendorf
Consultant, Hamburg, Germany

Abstract

Cavitating tip vortices, higher harmonic order hull pressure fluctuations, and PD- as well HDNC-technique for cavitation nuclei measurements are treated in this paper. - In section 2 earlier measurements of propeller excited pressure fluctuations with the detection of the 2nd harmonic blade order of a three-bladed propeller are described. This 2nd order is caused by the displacement effect of the tip vortex with nodes, that in section 3 is theoretically treated by a velocity potential for a hollow vortex with nodes based on a theory by Ackeret [1]. Calculated results and experiments agree reasonably well. Section 4 contains reflections and discussions on further flow effects of cavitating tip vortices, e. g. vortex bursting. In section 5 higher order hull pressure fluctuations from two water tunnels and full scale are compared, and the occurred scale effects between model and full scale results are discussed. In section 6 a comparison of nuclei concentrations from the PD- and HDNC-techniques is shown; also the question, whether solid particles serve as cavitation nuclei, is addressed. Finally in section 7 a conclusion is drawn.

Keywords: Propeller cavitation; higher blade harmonics of hull pressures; cavitation nuclei measurements

Motto: "… with the large recapitulating formulations of laws of nature, as they were possible for the first time in Newton's mechanics, it concerns idealizations of the reality, not the reality itself …"
W. Heisenberg (Physics Nobel-Price Winner 1932), Essay, Süddeutsche Zeitung, Oct. 6th, 1970
(Translated by E.-A. Weitendorf)

1 Introduction

In a recent paper by Berger al. [4] higher order hull pressure fluctuations are treated. They can also be caused by interactions between tip vortices and sheet cavitation of skewed propellers. A further work on this subject is by Gosda et al. [10]. These papers, as well as those by Bosschers [6] and Wijngaarden et al. [27] suggest a review backwards on this field including the cavitating tip vortex with nodes.

Whereas an earlier paper by Heinke et al. [11] contains results of nuclei concentrations and cavitation extents in two water tunnels, the present paper treats, added to nuclei concentrations, also hull pressure fluctuations. Surprisingly, as shown in [11], the propeller cavitation extents in the two tunnels agreed almost, in spite of different air contents and nuclei concentrations. The more it is interesting, how the comparison of the pressure fluctuations from the two tunnels would look like in comparison to the full scale values.

In the present paper also the influence of the free air content (i.e. nuclei content) of the tunnel water on the hull pressure fluctuations is touched. Here the techniques of measuring the free air content become relevant. Whereas up to now the free air content is often (see [14]) measured by PD- (Phase Doppler Anemometry), recently the HDNC- (Hydrodynamic Nuclei Concentration) technique is applied. Such results will be shown here. While the PDA-technique measures mainly only micro bubbles (i.e. nuclei), the HDNC-technique provides also solid particles; yet this leads to the critical and ever lasting question, whether and how the latter participate in the cavitation process. This will be discussed in some detail within the present paper, yet not only for model tests and laboratory investigations, but also for full scale measurements in the free sea.

It may be admitted, topics, as tip vortex, higher hull pressure fluctuations, as well as PD- and HDNC-technique, seem to cover a far field, yet the cavitation process is complex and these topics are interdependent; therefore, it may be allowed to treat and discuss them here.

2 First Knowledge on Higher Blade Harmonic of Hull Pressures - Experimental Investigations

In the early 1970-ties an understanding of ship hull vibrations by the propeller was urgently needed. Therefore, a measuring campaign of propeller-excited pressure fluctuations was performed in the medium sized cavitation tunnel K22 of HSVA (Hamburgische Schiffbau Versuchsanstalt). The tunnel has a test section of 570mm x 570mm and the pressure fluctuations caused by a three-bladed frigate propeller with optimal circulation on the blades were measured by pressure pick-ups installed on a flat plate as shown in Figure 1.

The measuring signals were digitized and harmonically analyzed by a fast Fourier transformation on a computer.

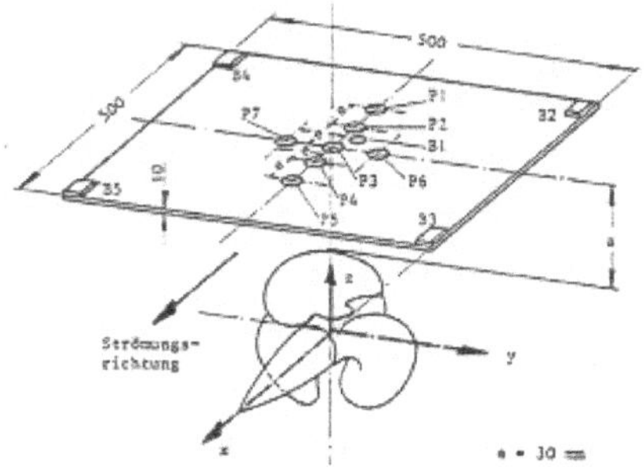

Figure 1: Measuring Positions with Coordinates

Further details may be found in paper [22]. The locations of the pick-us P1 to P5 in longitudinal direction are those where the relevant results will be shown in the present paper.

The pressure fluctuation amplitudes are documented (e.g. in in Figure 2) as non-dimensional values

$$K_{pi} = p / (\rho\, n^2\, D^2) \qquad (1)$$

with p the pressure amplitude of the harmonic order i ,
ρ the mass density of the fluid, n the rate of revolutions,
D the propeller diameter, and N the blade number.

(Here should be mentioned that the harmonic order i is related to the rate of revolutions, i.e. for this three-bladed propeller i = N = 3 means the blade frequency.)

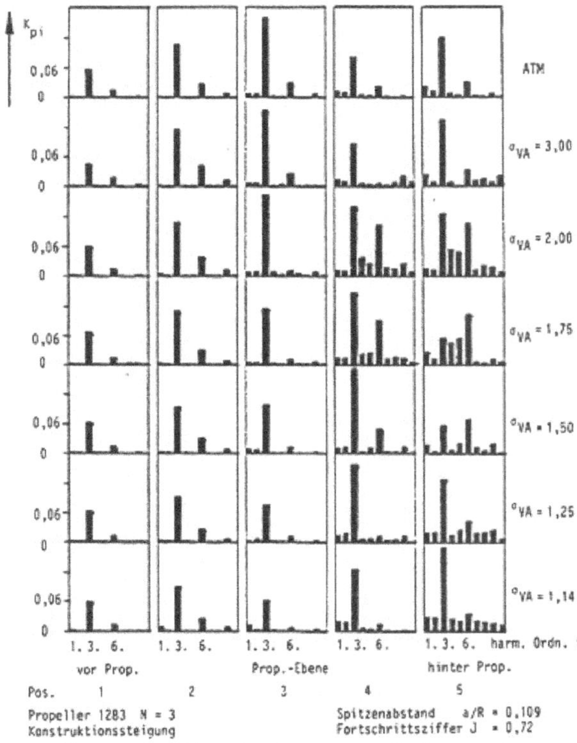

Figure 2: Non-dimensional pressure amplitudes up to the 10th harmonic order i in homogeneous flow

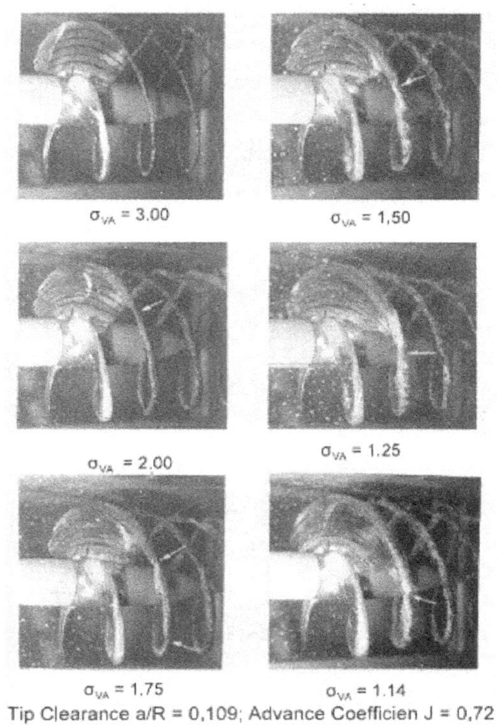

Figure 3: Cavitating tip vertices with nodes in homogeneous inflow

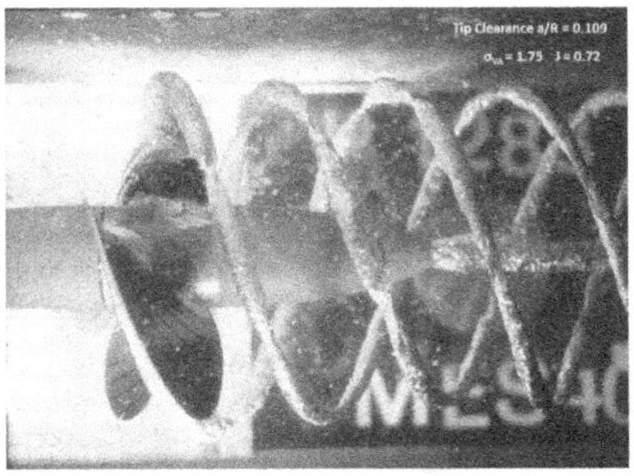

Figure 4: Cavitating Tip Vortex with Nodes in Homogeneous Inflow

igure Figure 2 shows pressure fluctuation amplitudes K_{pi} in longitudinal direction at the positions Pos. 1 to 5 and six varied cavitation numbers

$$\sigma_{VA} = (p - p_V) / 0.5\, \rho\, V_A^2 \quad (2) \quad ,$$

which are related to the speed of advance V_A. Further relevant values are given in the figure.

Interesting for Figure 2 is that at the position Pos. 5 for $\sigma_{VA} = 1.75$ the amplitude of the sixth harmonic order attains almost two times the value of the blade frequency, what is of course also found at the registered measuring signals at Pos. 5. The reason for this fact can be found by a detailed view on the photographs in Figure 3. Here, i.e. on each of five photographs, small little white arrows mark the position of a node on the surface of the propeller tip vortex. The position of the node depends on the respective cavitation number. Clearly visible on the photo for $\sigma_{VA} = 1.75$ are two arrows. An estimation for the distance between those respective two nodes results in a wave length of $\chi_0 = 60°$, what is two times the division between two blades of this tested three-bladed propeller. This fact explains the occurrence of an increased 6th harmonic (related to the rate of revolutions) by the displacement effect of the tip vortex with nodes. The theoretical investigation of the hydrodynamic displacement effect of such vortices with different wave length χ_0 between the nodes will be treated in the next section. Yet, before ending this section on experimental investigations of the tip vortex a single photograph is presented in Figure 4, showing impressively the occurrence of those nodes on the vortex for the cavitation condition $\sigma_{VA} = 1.75$.

3 Theoretical Investigation of the Tip Vortex with Nodes

One of the probably first theoretical researches on a cavitating stationary hollow vortex with nodes on its surface was presented by Ackeret [1]. His model was combined in the here relevant investigation with a velocity potential source-sink distribution for the cavitating propeller tip vortex with varying cross section (Isay [13]) in order to calculate the hull pressures.

The shape of the tip vortex with nodes was treated in [23] with potential flow expressions (source-sink distributions) in combination of an elliptically-shaped hollow vortex (Sketch 1) with a wavy hollow vortex (Sketch 2). The resulting shape for the simulation of a tip vortex with nodes is shown in Sketch 3. The shape of a tip vortex creates a non-dimensional pressure fluctuation K_p by displacement which is proportional to the change of the cross section area $F(\chi)$ of the cavitating tip vortex

$$K_p \sim \Delta p \sim \frac{dF(\chi)}{d\chi} \qquad (3)$$

The mean vortex radius R_K depends on the thrust coefficient K_T (or the circulation Γ around the vortex) and the cavitation number $\sigma_n = (p - p_V) / 0.5\rho(\pi n D)^2$, i.e.

$$R_K \sim K_T \sqrt{\frac{1}{\sigma_n}} \sim \Gamma \sqrt{\frac{1}{\sigma_n}} \qquad (4)$$

The wave length L, i.e. the distance between two nodes (also designated by χ_0), follows the theory by Ackeret [1]:

$$L \sim \left(\frac{U}{c_t}\right)^2 R_K \cdot (Hank) \qquad (5)$$

where U is the velocity in vortex direction and c_t in tangential direction, and where Hank are the Hankel-functions H_0 and H_1.

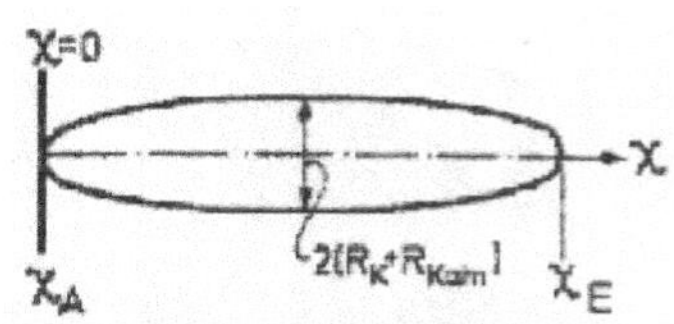

Sketch 1. Ellipsoidally hollow vortex

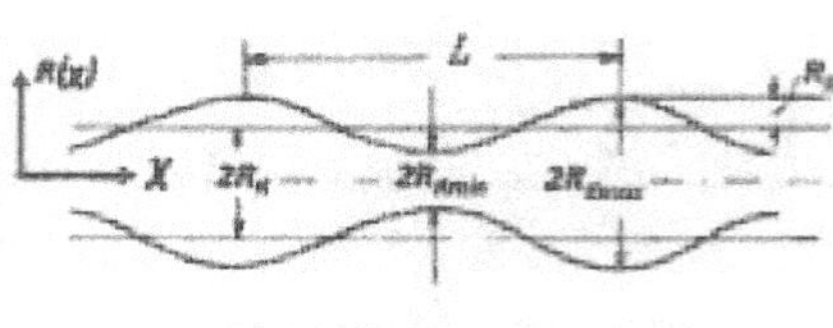

Sketch 2. Wavy hollow vortex

Sketch 3. Example of a tip vortex with nodes

Further, it should be mentioned that the amplitude R_{kam} cannot be determined by Ackeret's theory; i.e. when applying potential flow theory the amplitude could not be derived by closed mathematical solutions, yet only from the photographs of the experiments, as made in [23].

The calculations containing the displacement effect of the cavitating vortex showed that the amplitudes enlargement of the blade frequency hull pressures or its higher harmonics depends additionally on the wave length χ_0 between the nodes. Results are given in Figure 5. The figure shows that for $\chi_0 = 60°$ (i.e. half the division between two blades of the 3-bladed propeller) the non-dimensional hull pressure amplitude K_{pi} of the second blade harmonic attains a value at least two-times larger than the first one. Moreover, it has to be remarked that the above mathematical relations for the vortex with nodes are formulated for a cavitating optimal propeller in homogeneous inflow. The behavior of tip vortices in a ship wake means another complex task (Berger [5]). Nevertheless, the above expressions yield some insight into the excitation of higher blade order hull pressure fluctuations by cavitating tip vortices with nodes.

A comparison between the calculated results with measured ones both in homogeneous inflow for the 3rd harmonic

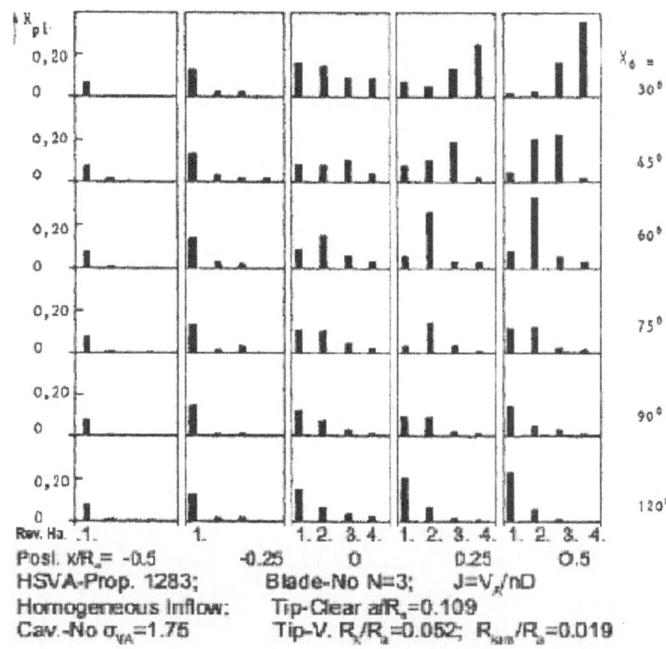

Figure 5: Calculated non-dimensional hull pressure amplitude with the influence of the wave length χ_0

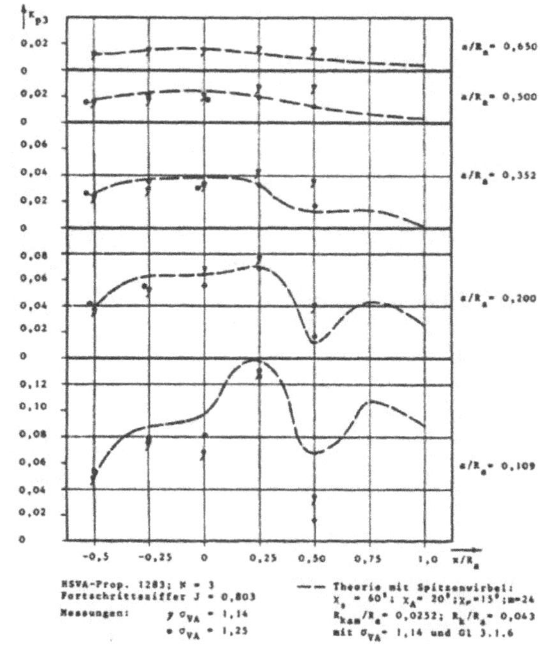

Figure 6: Calculated and measured non-dimensional hull pressure amplitudes K_{p3} with variation of the propeller tip clearance a/R_a

order (i.e. the blade frequency) is given in Figure 6. This figure displays results in longitudinal flow direction x/R_a (where R_a is the outer radius of the propeller) for the cavitation numbers $\sigma_{VA} = 1.25$ and $\sigma_{VA} = 1.14$ as well as for different tip clearances a/R_a (a is the clearance, R_a the propeller radius; for further numbers in Figure 6, see [23]). The agreement between measurements and calculations in Figure 6 is quite reasonable, especially at the longitudinal position $x/R_a = 0.25$.

4 Reflections and Discussion on Cavitating Tip Vortices

In the following some discussions on different flow phenomena will be carried through, that are the source for different excitations of the blade frequency and higher harmonic orders.

Regarding higher harmonic orders of hull pressure fluctuations induced by cavitating propellers recently a work was carried out by Berger et al. [5]. In this work coherent structures of sheet cavitation on the propeller blades and tip vortex cavitation have been considered. It could be shown that in particular the interaction between sheet cavitation and developed tip vortex cavitation is responsible for notably higher-order pressure fluctuations. The results have been found with a novel simulation technique and the panel code panMARE as well as the RANSE solver ANSYS CFX. The vortex cavity has been divided into a large number of cylindrical segments with ship wake adapted circulations that were treated separately in a 2D- instead of a 3D-manner. The numerical results have been compared with experimental ones for three vessels with skew propellers. This comparison was regarded as very promising. In continuation of the latter work, results of an extended numerical study on Reynolds number scale effects of pressure fluctuations due to tip vortex cavitation are presented in [10]. In this study, the influence of the viscous core radius of the vortex as well as the influence of different propeller loads with respect to model and full-scale conditions have been investigated. The effect on the dynamical behavior of the cavities inside the vortex core is treated. Also, the scale effect of the ship's wake field on the cavitating vortex has been investigated. The results up to now of both influences, i.e. the dynamical cavitation behavior inside the vortex core and the varying circulation in the ship's wake field, have been considered.

Another relevant flow phenomenon of the cavitating tip vortex in the ship's wake showed a substantial influence on the 2nd and higher blade pressure harmonics, namely the bursting of the propeller tip vortex. Those experimental results with a ship dummy model in the former water tunnel no 2 at NMI were presented by English [9]; he drew the attention on this phenomenon for the first time. The vortex bursting attained English's attention at these model tests and he reported also about respective full scale observations. Further, he derived a remedy at the model tests for suppressing higher blade frequency amplitudes by air injection into the bursting vortex.

Luecke [16] performed a rather detailed investigation on the bursting of the propeller tip vortex. Model tests in the medium sized HSVA cavitation tunnel (Test section 570 x 570mm²) and highspeed video records were performed with a frame rate of 4500 f/s. The tests revealed the temporal vortex behavior during bursting processes, - processes, because namely it turned out that there are two types of bursting cavitating vortices. Firstly, there is the vortex bursting due to hydrodynamic instability which is found on marine propellers as well as in aerodynamics. The second type is also called bursting of a propeller tip vortex, yet its calm structure is "destroyed" by the rolled-up sheet cavity. The reason for the bursting of the first type is described by different stability criteria, one of them is contained in Ludwig's stability map. Here the gradients of the axial and tangential velocities inside the vortex are from crucial influence concerning the hydrodynamic instability. Sometimes the pressure side of a ship rudder has a crucial influence on the bursting of the propeller tip vortex, especially when the axial speed of the vortex is decreased. Finally, the work by Luecke contains CFD calculations with a RANSE code; those have been compared with the tests results. The gained competence led to guidelines for propeller designs.

Higher order blade frequencies of the hull pressures are also treated in a paper by Bosschers [6]. From measured model hull pressure fluctuations, generated by the cavitating vortex on a two-bladed propeller, was found that, beside the braodband energy content of the spectrum, of course also harmonics of the blade frequency will be excited. The latter are caused by waves on the cavitating vortex. Bosschers characterized three possible modes (see Figure 8 in [6]) for spatial deformations of cavitating vortices according to a theory by Thomson [21]. These modes are:

$n = 0$ characterizes a breathing vortex with volume variation

$n = 1$ characterizes a serpentine vortex with constant cross section and waving center line

$n = 2$ characterizes a double helix with an elliptical shape of the vortex core

What Bosschers called "breathing vortex with volume variation" (corresponding to $n = 0$) is obviously the same that is denoted as "tip vortex with nodes" in sections 2 and 3, as well as in sketches 2 and 3 of this paper. Finally it should be remarked regarding paper [6] that it additionally contains valuable investigations on the collapse and rebounds of the cavitating vortex in combination with a kind of sheet cavitation. It seems that not only this combination needs further research, yet also the bubble-dynamic rebound. Also, the near physical connection of the vortex with nodes with the singing vortex investigated by Maines et al. [17] and Arndt [3] is to be mentioned here.

Finally, a paper will be considered that was presented by van Wijngaarden et al. [27]. Regarding the cavitating tip vortex with nodes, as here in section 2 and 3, was criticized in [27] that the nodules (designation in [27], p. 4) „ have been considered as cause for hull excitation" and that "... the cavitating core is no longer cylindrical, but has a ribbon type shape ..., [and] in reality the cavity volume is not changing". Yet, the latter remark agrees with the designation for a cavitating tip vortex with the mode $n = 2$ as listed by Bosschers [6]. On the other hand, the clear demonstration of the excitation of higher order blade frequencies in section 2 and 3, supported by the existence of a volume breathing cavitating tip vortex with the mode $n = 0$ according to Bosschers, proofs that the cavitating tip vortex with nodes is very well a contributing source to the hull pressure excitation. - Very instructive are the full scale photographs in [27] showing in its Figure 10 bursting vortices, and concerning the understanding of those vortices it may be recommended to look into the paper by Luecke [16]. - And still because for the sake of completeness with vortices: Huse mentioned in his dissertation ([12], see his Fig. 35) a **P**ropeller **H**ull **V**ortex (PHV) that at very slow propeller inflow can occur and touches the ship hull, causing "extremely high local pressures on the hull" of irregular frequencies.

5 Higher Blade Harmonics Results from a Recent Research Project

In the German R&D-Project *KonKav ...on correlation of cavitation effects...*[26] hull pressure fluctuations due to propeller cavitation were measured in the K15A-Tunnel of SVA, Potsdam and in the HYKAT of HSVA, Hamburg. The tests in the K15A-Tunnel (Test Section 850 x 850 mm²) were carried out with a dummy ship model with simulated full scale wake, in the HYKAT (Test Section 2.8 x 1.6 m²) with a complete ship model. The full scale tests were carried out on the single screw Con-Ro vessel "Amandine". Details on the model propeller and the full scale ship are given in [11].

At this project on hull pressure fluctuations also PD (Phase-Doppler technique) nuclei concentrations at different O_2-saturations of the water were determined in both tunnels. For the propeller loading at the model tests the thrust identity was applied, i.e. $K_T = T/(\rho n^2 D^4)$, with T = thrust, n = rate of revolution, and D = propeller diameter; the cavitation occurrence was characterized by the cavitation number $\sigma_n = (p_A - p_V)/0.5\rho(nD)^2$, with p_A the tunnel pressure, and p_V the vapor pressure. - The measured amplitudes were converted to non-dimensional values $K_p = p/(\rho\, n^2 D^2)$, with p the pressure amplitude of the respective blade order.

For the cavitation behavior during the pressure fluctuation tests two different measuring procedures were used: Firstly,

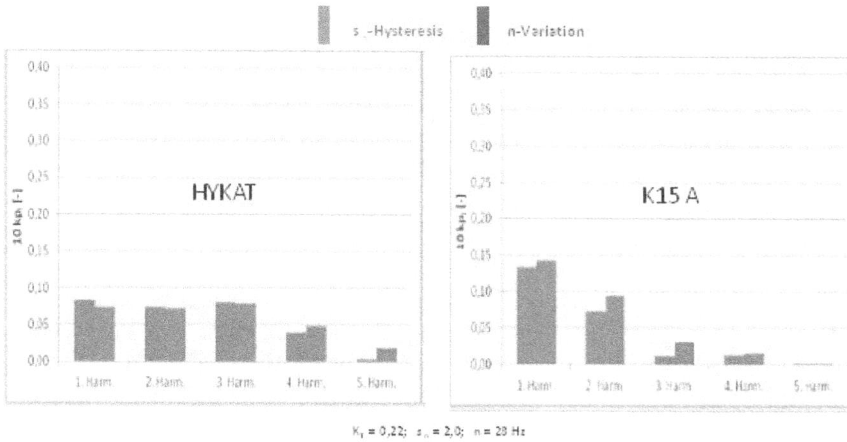

Figure 7: Pressure fluctuations of harmonic blade orders at identical conditions, i.e. for σ_n-hysteresis and n-variation, for HYKAT at 80% and K15A-tunnel at 60% O_2-content

the cavitation number was varied by setting the tunnel pressure according to the selected number of revolutions n = 28 Hz, this procedure was called Sigma-Hysteresis. Secondly, the cavitation number was varied by changing the numbers of revolution that resulted in different tunnel pressures, this procedure was called n-Variation. Both cases were carried out in compliance with the to be tested propeller thrust and the necessary tunnel water speed. Then, for identical conditions of both series the pressure fluctuations have been compared. Figure 7 with $10xK_p$-values is selected out of all possible comparisons that have been shown in paper [26] (there Figs. 9 and 10). Regarding the comparison of the different harmonic blade orders from both tunnels small differences are observable in the range of measuring uncertainties. For the blade harmonic it has to be mentioned that the propeller at design condition showed almost no sheet cavitation that also resulted in rather small hull pressure fluctuations which are well below the normally recommended values from 4 to 6 kPa.

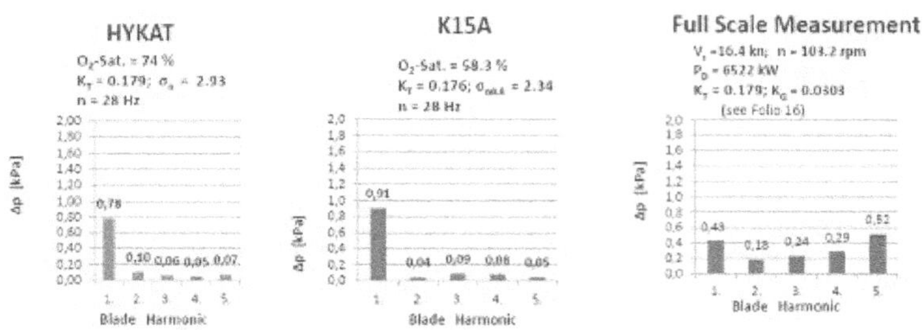

Figure 8: Pressure fluctuation amplitudes from HYKAT and K15A-Tunnel compared with those from full scale

As said before, at trial condition of the full scale tests almost no sheet but only tip vortex cavitation occurred. The model to full scale comparison in Figure 8 shows, the first blade harmonic amplitudes of the tunnel tests are larger than those from the ship trial, and for the higher harmonics opposite results occur. Concerning these differences one could think on ship wake scale effects in the tunnels, especially affecting the blade harmonic. Further details on these effects may be found in [26]. Yet, two short remarks may be given here: The scale effects of the blade harmonic and the higher ones are probably caused by different flow phenomena; those of the higher harmonics could be excited by cavitating tip vortices with nodes as explained in section 2 and 3 of this paper; and, different thicknesses of the cavitating vortices in model and full scale may come from a scale effect due to different absolute cavitation pressures, which are in full scale very probably more below vapor pressure than in model scale. This bubble-dynamic scale effect was explained for sheet cavitation in [25] (there p. 535); it is very probably also valid for vortex cavitation.

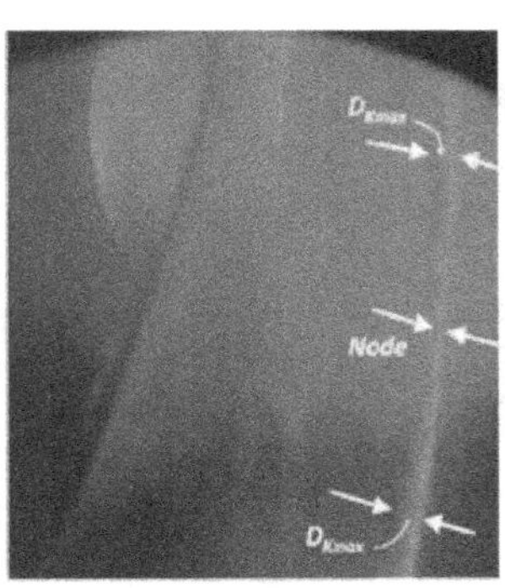

Figure 9: Full scale view onto pressure side on board MS „Amandine";
left turning propeller at n = 103 rpm

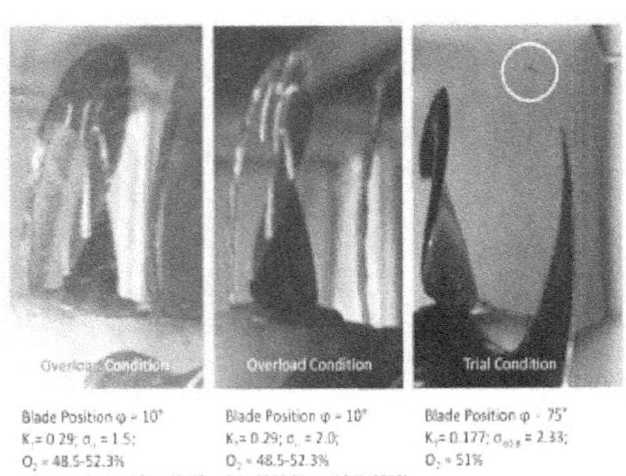

Figure 10: Tip vortices with nodes at model tests

The different thicknesses of the cavitating tip vortices are shown here in two photographs: Figures 9 concerns the propeller viewing aboard the Con-Ro vessel "Amandine", and Figure 10 displays propeller tip vortices at model tests. The right hand photo in Figure 10 with almost no vortex (in the white circle) has to be compared with Figure 9, resulting in the just mentioned pressure issue of thicker vortices in full scale than in model. Further, in full scale (Figure 9) a tip vortex is seen whose diameter changes between D_{Kam} and the diameter of the node. An estimate of the wave length between two nodes, i.e. the angle (measured along the vortex and down to the propeller center) results in abt. 18° to 30°; the value 18° would correspond to 4 nodes within 90° (which is the angle between two blades of a 4-bladed propeller) and that would create the fifth harmonic blade order of the pressure fluctuations by displacement effects; and actually, the fifth harmonic amplitude from full scale is the largest one in Figure 8.

The right hand photograph in Figure 10 from model tests corresponds to the full scale trial condition (K_T = 0.177; $\sigma_{no.8}$ = 2.33 with $\sigma_{no.8}$ referenced to the 0.8-radius); it shows no cavitation on the blade but a rather tiny piece of an irregularly flushing tip vortex (see the white circle); obviously the cavitating vortex is not yet fully generated.

Finally, one may state that the hydrodynamic process of propeller excited pressure fluctuations is very complex. From this point of view the results of the here treated model project may be regarded as almost satisfying, when predicting full scale pressure fluctuations by model tests. The predicted values of Figure 8 are clearly below 1 kPa and this is well below the upper border of about 4 to 6 kPa for avoidance of unacceptable ship vibrations. And actually, the Con-Ro vessel "Amandine" showed practically no ship structure vibrations as the author felt on a voyage from Rotterdam to Dublin.

6 Recent Nuclei Measurements (PD and HDNC) - Consequences and Constant Challenges

Since model hull pressure fluctuations are obviously influenced by the free air content of the tunnel water (i.e. by its nuclei) as shown e.g. in [26], techniques for measuring the latter are relevant. PD- (Phase Doppler-) and HDNC- (Hydrodynamic Nuclei Concentration-) technique results from the project *KonKav II* are shown in Figure 11 (Ebert et al., [8]). From this figure can be concluded "that the HDNC-technique reaches similar capabilities in number concentration estimation as the Phase Doppler (PD) technique" [8]. Further, it may be pointed out, when comparing the two techniques the HDNC-technique contains the advantages of a reduced optical setup and higher data rate.

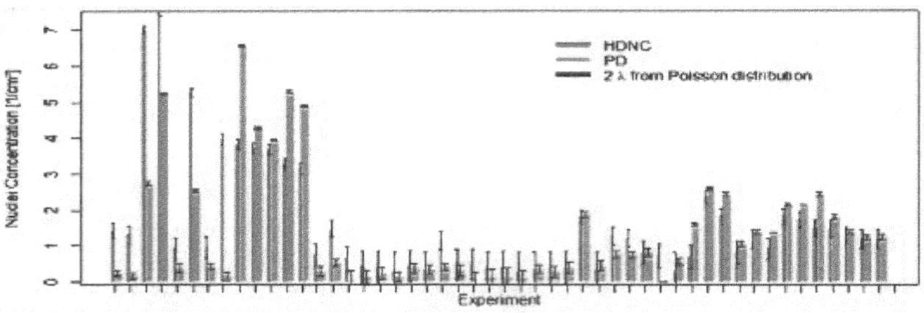

Figure 11: Comparison of HDNC and PD nuclei concentrations in SVA-Tunnel K15A from [8]

While the PD-technique determines mainly only micro bubbles (nuclei) by applying the linear relation between the phase of a Doppler burst from two photomultipliers and the bubble diameter, the HDNC-technique provides additionally solid particles by the analysis of scattered light from particles. This is displayed in Figure 12 where optical interference patterns of the HDNC-technique are given in the upper row, as e.g. for bubbles or solid particles. Also, in the figure a diagram is displayed for a particle classification resulting in a relation of about 10 solid particles to about 1 bubble. This result was found on board the Con-Ro ship "Amandine" at the exit of she North Sea to the Irish Sea, where the sea water is not so clear compared to open oceans.

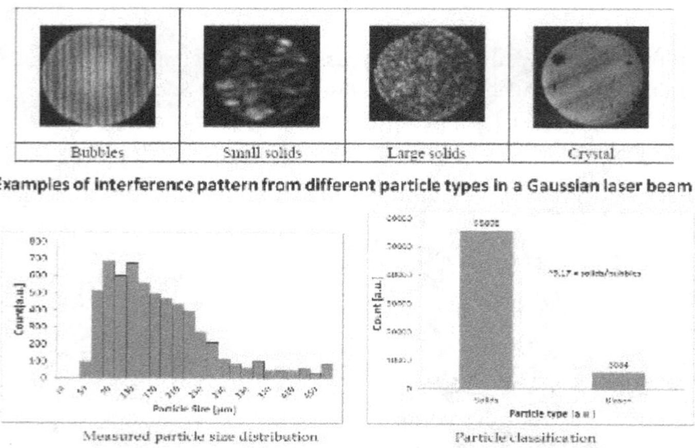

Figure 12: Interference patterns of measured particle size distribution and classification accordimg to Ebert et al. [7]

In oceanography open ocean waters are indicated as case 1 water, and coastal waters as case 2 ([18] and [20]). Interesting in Figure 1 of the latter, rather fundamental paper are the acoustically measured bubble size distributions at four different depths (0.7 to 4.1m) below the ocean surface with breaking waves during 30 seconds. The determined average bubble diameters range from about 60 to 960 μm. Those bubbles could well serve as cavitation nuclei.

Acosta et al. [2] touched the problem which oceanic particulates down to abt. 30 m depth serve as cavitation nuclei. For these particulates the authors indicate (see [2] Table I) that 10 % of them are microbubbles which are of course nuclei; the remainder appear as organic and inorganic particulates. This relation 10 % bubbles to 90 % particulates by Acosta et al. [2]) agrees rather well with the result by Ebert et al. [7]. Acosta et al. care about the problem which oceanic particulates serve as cavitation nuclei, yet this issue is still open with Ebert et al. Already Weitendorf and Keller [24] tried to measure the free air content by Keller's scattered light method in front of the full scale propeller of the "Sydney Express". In the discussion of paper [24] was stated: "The propeller will always find in the flow a sufficient number of small nuclei leading to cavitation." This sentence was formulated after some detailed studies of papers on hydrobiology and personal contacts to respective experts (see References in [24]).

In Figure 12 of this paper (from Ebert et al. [7], upper row) different kinds of solid particles are shown; but it is not clear whether they participate in the cavitation process as nuclei or how they do it. One proceeding work on this issue is a theoretical paper on the nucleation of hydrophobic nonspherical solid particles at acoustic cavitation by Mørch [19]. The work describes numerical calculations for different parameters until cavitation cluster formations are theoretically reached which then were compared with experiments. The differences were discussed. This work is mentioned here as an attempt for widening the understanding of the nucleation of solid particles. Similar approaches for this problem could be helpful.

Finally can be stated that it is rather favorable to have the capability of new techniques for the determination of particles in sea water like the HDNC, yet, further experiences with the latter technique are necessary and maybe helpful solving the problem what are cavitation nuclei (as mentioned by e.g. Acosta et al. [2]). Finally, when further research on nucleation will be pursued, one should not forget the extensive efforts and compilations since years on nuclei by the Cavitation Committees of the ITTC (Intern. Towing Tank Conference); in 2002 the ITTC even published a report of its *Specialist Committee On Water Quality and Cavitation* [14].

7 Conclusions

The sections no 2 to 4 of this paper deal with cavitating vortices exciting higher propeller blade harmonics. In section 2 the experimental detection of the cavitating tip vortex with nodes as source of higher blade harmonics is shown. This phenomenon has been reasonably and clearly explained by theoretical simulations with a velocity potential for the vortex with nodes (section 3). The wave length between the nodes (sketch 2 and 3) has been determined according to a theory by Ackeret [1]. For the calculations, the amplitudes of the vortex diameter have been taken from the experiments. Yet, those amplitudes - as shown in section 4 - could have been taken from a theory by Thomson [21], that describes three modes, i.e. n = 0 to n = 2, for the shape of the cavitating vortex. Bosschers [6] also showed that between the passages of propeller blades waves on cavitating vortices occurred; this is finally the same that is treated in sections 2 and 3 of the present paper as cavitating tip vortex with nodes.

Further, in section 4 the bursting of cavitating vortices, that can contribute to higher order blade frequencies, is treated with the first mention of this phenomenon by English [9] and its rather detailed treatment by Luecke [16]. The latter found two types of bursting cavitating vortices. The first one may be called bursting vortex by hydrodynamic instability, and the second one interacting bursting vortex, i.e. interacting with the sheet cavitation.

In section 5 (Figure 7) comparisons of blade frequency and higher harmonics from two water tunnels, (K15A-Tunnel, SVA, Potsdam and HYKAT, HSVA, Hamburg) for the Con-RoFerry "Amandine" are shown. They reveal scale effects, which for the model blade frequency are ascribed to the relatively wider ship model wake field and for the higher blade harmonics to the thicker full scale propeller tip vortex. This thicker vortex is argued to come from a bubble dynamic scale effect due to the larger absolute negative pressures in full scale. These pressures let a larger number of sea water nuclei grow than this may happen in the water tunnels. Of course, these deliberations are somewhat hypothetical, but rather probable. It would be worthwhile to concentrate a great deal of research on this issue.

The sea water nuclei contents (treated in section 6) measured by PD- and HDNC-techniques show the relation of 10 to 1 for solid particles to microbubbles. Here the problem emerges which sea water particulates serve as cavitation nuclei. Further challenging research on the nuclei participation problem of solid particles (whatever they are, crystals, dust, sand, maybe algae) is necessary. Methods from other research fields, e.g. acoustical measuring [20] or digital holography [15] and numerical investigations, as e.g. [19], may be stimulating. These endeavors would also support the reliability of cavitation model tests and the understanding of various areas within the cavitation field.

8 References

[1] Ackeret, J. (1930). *Über stationäre Hohlwirbel.* Ing.-Archiv, p. 399-402

[2] Acosta, A., O'Hern, T., Katz, J. (1986). *Some Recent Trends in Cavitation Research*, Intern. Symp. on Cavitation. Vol. 1, Sendai, Japan, p 1

[3] Arndt, REA. (2006). *From Wageningen to Minnesota and Back: Perspectives On Cavitation Research.* 6th Intern. Symp. on Cavitation, Wageningen

[4] Berger, S., Gosda, R., Scharf, M., Klose, R., Greitsch, L., Abdel-Maksoud, M. (2016). *Efficient numerical investigation of propeller cavitation phenomena causing higher order hull pressure fluctuations.* 31st Symposium on Naval Hydrodynamics Monterey, California,

[5] Berger, S. (2018). *Numerical Analysis of Propeller-Induced Higher-Order Pressure Fluctuations on the Ship Hull.* Technical University Hamburg, Schriftenreihe Schiffbau, Rep. 705 / 2018, http://www.tuhh.de/vss

[6] Bosschers, J. (2009). *Investigation of hull pressure fluctuations generated by cavitating vortices.* First International Symposium on Marine Propulsors, smp'09, Trondheim, Norway

[7] Ebert, E., Kleinwächter, A., Kostbade, R., Damaschke, N. (2014). *Interferometric Particle Imaging for cavitation nuclei characterization in cavitation tunnels and in the wake* flow. 17th Intern. Symp. on Applications of Laser Techniques to Fluid Mechanics, Lisbon, Portugal

[8] Ebert, E., Kröger, W., Damaschke, N. (2015). *Hydrodynamic Nuclei Concentration Technique in Cavitation Research and Comparison to Phase-Doppler Measurements.* 9th Intern. Symp. on Cavitation, Lausanne

[9] English, J. W. (1979). *Cavitation Induced Hull Surface Pressures. Measurements in Water Tunnels.* RINA Symposium on Propeller-Induced Ship Vibration, London

[10] Gosda, R., Berger, S., Abdel-Maksoud, M. (2018). *Investigation of Reynolds Number Scale Effects on Propeller Tip Vortex Cavitation and Propeller-Induced Hull Pressure Fluctuations.* 10th Intern. Symp. on Cavitation, Baltimore

[11] Heinke, H.-J., Johannsen, C., Kröger, W., Schiller, P., Weitendorf, E.-A. (2012). *On Cavitation Nuclei in Water Tunnels.* 8th Intern. Symp. on Cavitation, Singapore

[12] Huse, E. (1972). *Pressure fluctuations on the hull induced by cavitating propellers.* Norwegian Ship Model Experiment Tank Publication No. 111

[13] Isay, W.-H. (1970). *Moderne Probleme der Propellertheorie.* Springer-Verlag

[14] ITTC (2002). *Specialist Committee on Water Quality and Cavitation.* 23rd International Towing Tank Conference, Venice (https://ittc.info/media/1465/water-quality.pdf)

[15] Katz, J.. Sheng, J. (2010). *Applications of Holography in Fluid Mechanics and Particle Dynamics.* 42:531-55. Annu. Rev. Fluid Mech.

[16] Luecke, T. (2006). *Investigations of Propeller Tip Vortex Bursting.* NAV International Conference on Ship & Maritime Research, Genova

[17] Maines, B. H., and Arndt, REA. (1997). *The Case of the Singing Vortex.* J. Fluids Eng. 119: 271-276

[18] Morel, A. and Prieur, L. (1977). *Analysis of variations in ocean color.* Limnological Oceanogr., 22, 709–722

[19] Mørch, K. A. (1986). *Cavitation Nucleation by Solid Particles in a Focussed Acoustic Field.* Intern. Symp. on Cavitation. Vol. 1, Sendai, Japan, p 43

[20] Terrill, E. J., Kendall M. W., Stramski, D. (2001). *Bubble entrainment by breaking waves and their influence on optical scattering in the upper ocean.* Journal of Geophysical Research. Vol. 106, No. C8,

[21] Thomson, W. (Lord Kelvin) (1880). *Vibrations of a columnar vortex*, Proceedings of the Royal Society of Edinburgh, Scotland, March 1, Phil. Mag. X., pp. 152-165

[22] Weitendorf, E.-A. (1973). *Experimentelle Untersuchungen der durch kavitierende Propeller erzeugten Druckschwankungen.* Schiff und Hafen, No 11 11, Vol 25, pp 1040-1060 (Engl. version from Brit. Ship Research Ass. BSRA No 29,215)

[23] Weitendorf, E.-A. (1976). *Kavitationseinflüsse auf die vom Propeller induzierten Druckschwankungen.* IfS-Ber. 338, Technical University Hamburg

[24] Weitendorf, E.-A. and Keller, A. P. (1978). *A determination of the free air content and velocity in front of the "Sydney-Express"-Propeller in connection with pressure fluctuation measurements.* 12th Symp. Navel Hydrodynamics, Washington DC

[25] Weitendorf, E.-A. (1981). *Cavitation phenomena, propeller excited hull pressure amplitudes and cavitation scale effect. Ocean Engineering,* Vol. 8, No. 5, pp. 517-539

[26] Weitendorf, E.-A., Lücke, T., Richter, Heike, Schiller, P. (2013). *Nuclei and cavitation induced pressure fluctuations - Investigations in two cavitation tunnels and full scale.* Jahrbuch Schiffbautechnische Gesellschaft, pp 83-96

[27] Wijngaarden, E. van, Bosschers, F., Kuiper, G. (2005). *Aspects of the cavitating propeller tip vortex as a source of inboard noise and vibration.* ASME Fluids Engineering Division, Summer Meeting, Houston TX, USA

9 Acknowledgment

The author would like to acknowledge the open support of the colleagues from the Institute for Fluid Dynamics and Ship Theory (Technical Univ. Hamburg), Institute General Electrical Engineering (Univ. Rostock) as well as Hamburg Ship Model Basin (HSVA) and Ship Model Basin (SVA) Potsdam. They provided not only their cited papers but also contributed by discussions to the improvement of this paper.

© 2018 ASME

On the Dispersion Relation of a Vortex Cavity

Johan Bosschers[1]

Maritime Research Institute Netherlands (MARIN), Wageningen, The Netherlands

Abstract

A dispersion relation of a vortex cavity describes the relation between spatial wave numbers and frequencies of perturbations of the cavity radius. The paper presents an update of the theory and the interpretation of experimental data for the dispersion relation of a vortex cavity as presented by Pennings *et al.* [JFM 778:288-313 (2015)]. Especially, the standing wave pattern generated by the attachment of the cavity to the foil is now identified in the theoretical and experimental wave number-frequency diagram.

Keywords: cavitation, vortex, dispersion relation

Introduction

The frequencies of perturbations of a vortex cavity depend on the wave lengths of the perturbation through a dispersion relation. The dispersion relation for a vortex cavity of infinite length has been derived by Lord Kelvin (Thomson, 1880) for incompressible flow and has been extended to compressible flow by, among others, Morozov (1974). The relation has been used to derive criteria for standing waves on cavitating vortices observed in model scale facilities (Ackeret, 1930; Keller and Escudier, 1980) and to derive criteria for the frequency of tonals in the radiated noise by cavitating wing-tip vortices (Briançon-Marjollet and Merle, 1996; Maines and Arndt, 1997). The dispersion relation has also been used to derive criteria for the resonance frequency in the collapse of the cavitating tip vortex of a ship propeller as observed in the hull pressure fluctuations and underwater radiated noise (Bosschers, 2009a, 2017).

Experimental evidence of the dispersion relation of a vortex cavity by a wave number-frequency diagram has been presented by Pennings *et al.* (2015a) in which two high-speed video camera's were used to observe the perturbations of a cavitating tip vortex of a wing in stationary conditions in a cavitation tunnel. The perturbations were transformed to the wave number-frequency domain by a 2-D FFT. The experimental results were compared with theory making use of the formulation derived by Morozov (1974), but extended with a convection term and an ad-hoc correction for viscous effects (Bosschers, 2008, 2009b). A reasonable agreement between theory and experiment was obtained but not all aspects of the experimental dispersion curves were understood. Also, the values for the fitted parameters used in the theoretical model showed relative large differences to what was expected from flow field measurement data and theoretical considerations.

The present paper provides some further information on both the theoretical formulation and the experimental wave number-frequency diagram as presented by Pennings *et al.* (2015a). This leads to a better understanding of the dispersion relation for a vortex cavity.

Theory

The formulation for the dispersion relation can be derived from the convected Helmholtz equation for a disturbance potential due to a potential flow vortex in a cylindrical coordinate system (r, θ, z). The distortion of the cavitating vortex with average radius r_c in time t is described by a number of modes characterized by the axial wave number k_z, azimuthal wave number n, angular frequency ω and amplitude $\hat{r}$. For small amplitudes, the local cavity radius η is given by

$$\eta(r,\theta,z,t) = r_c + \hat{r}\exp[i(k_z z + n\theta - \omega t)]. \quad (1)$$

The mode $n = 0$ corresponds to a breathing mode and involves volume variations. Mode $n = 1$ corresponds to a serpentine mode, also called bending mode, helical mode or displacement mode, as it is a mode that leads to a displacement of the vortex centre line only. The mode $n = 2$ is the bell mode or double helix or fluted mode, and leads to an elliptical shape of the vortex cavity. The distortions are transversely propagating inertial waves and are sometimes referred to as Kelvin waves (Saffman, 1992).

[1]j.bosschers@marin.nl

For the kinematic and dynamic boundary condition at the cavity interface, an ad-hoc correction is used to account for viscous effects by making use of the vortex model of Rosenhead (1931) for which a simple relation between azimuthal velocity and pressure exists. The resulting dispersion relation involves Hankel functions. However, at low frequencies for which the phase velocity of the axial deformations is much smaller than the speed of sound, the Hankel function reduces to a modified Bessel function of the second kind K_n. The non-dimensional dispersion relation is then given by

$$\overline{\omega}^{\pm}(\kappa,n) = \frac{\omega^{\pm} r_c}{W_\infty} = \overline{W}_c \kappa + \overline{V}_c n \pm \sqrt{K_\sigma} \sqrt{\frac{-|\kappa| K_n'(|\kappa|)}{K_n(|\kappa|)}}, \quad (2)$$

with non-dimensional wave number κ defined as $\kappa = k_z r_c$, and $\overline{W}_c$ and $\overline{V}_c$ denote the non-dimensional axial and azimuthal velocity at the cavity interface, respectively. Both $\overline{W}_c$ and $\overline{V}_c$ have been made non-dimensional with the free-stream velocity W_∞. Parameter K_σ is a 'stiffness' coefficient, that is defined as

$$K_\sigma(\sigma_W, r_c, r_v) = \sigma_W \frac{r_c^2}{r_v^2 + r_c^2}, \quad (3)$$

where σ_W is the cavitation number and r_c and r_v denote the vortex cavity and viscous core size, respectively. The formulation of K_σ is obtained for the Rosenhead vortex. Each mode n has two branches corresponding to the plus and minus sign in the dispersion relation. The absolute value of κ is used in K_n, its derivative K_n' and in the square root term to allow for negative axial wave numbers. Equation (2) has a different formulation than presented in Pennings *et al.* (2015a) and somewhat better describes the underlying physics. The formulations become identical for a potential flow vortex.

Comparison of theory with experiments

The theoretical formulation is compared to the experimental data of Pennings *et al.* (2015a), see also Pennings (2016). The test set-up consists of a half-wing of elliptical planform with an aspect ratio AR= 3 positioned at various angles of attack in the Delft cavitation tunnel. The wing has a chord length of the root of $c_0 = 0.1256$ m and half span of 0.150 m with the tip positioned in the centre of the test section. The cavitating tip vortex is observed by two high speed video camera's located above and aside the test-section. The frame rate was 5×10^3 frames per second and total recording time was five s econds. The top view camera observes the cavitating vortex between $x/c_0 = 0.19$ to $x/c_0 = 2.48$ with an equivalent pixel size of 0.16 mm. The side view camera observes the cavitating vortex between $x/c_0 = 1.12$ to $x/c_0 = 2.60$ with an equivalent pixel size of 0.10 mm. The edge of the cavity was detected using the gradient in light intensity from which the cavity meanline and diameter was determined by Pennings.

For the test-case considered, the angle of attack was 7° and the cavitation number equals $\sigma_W = 0.87$. The vortex cavity is attached to the blade and has an elliptical shape that, overall, remains stationary, see Figure 1. The mean radius of the vortex cavity is r_c= 2.8 mm. The size of the viscous core, r_v= 1.2 mm, has been obtained from Stereoscopic Particle Image Velocimetry (SPIV) measurements published in Pennings *et al.* (2015b). Here, only the variations of the cavity radius are analysed. The variation of the cavity mean line, described by mode $n = 1$, are not considered. The wave number-frequency diagrams, presented in Figure 2, show the coherent phase difference and the cross-power spectral density (CPSD) of the top and side view. The theoretical dispersion lines are computed for positive wave numbers for which the frequency may have positive and negative values. Because the 2-D FFT is symmetric with respect to the origin, the quadrant with positive wave numbers and negative frequencies is identical to the quadrant with negative wave numbers and positive frequencies. Because it was decided to present the quadrant with negative wave numbers, the theoretical results with negative frequencies were mirrored to the quadrant with negative wave numbers.

The phase difference diagram of the experimental data is dominated by two regions with a phase difference of 180 degrees. These regions correspond to the modes $n = 2^-$ and $n = -2^+$. The introduction of negative values for the azimuthal wave number is a new finding with respect to Pennings *et al.* (2015a). The standing elliptical wave pattern as observed by the video camera's can be constructed from these two modes for $\overline{\omega} = 0$. Also for $n = 0$ a standing wave is present of which the wave number is almost twice the value as for the elliptical standing wave. The attachment of the vortex cavity to the blade causes the elliptical cross-section. As the cavity on the blade can not sustain any perturbation in azimuthal direction, a standing wave is formed. The dispersion line for mode $n = 0^-$ can also clearly be identified. Other regions with zero degrees phase difference seem to correspond to harmonics of presented modes and are therefore not represented by a theoretical line. The modes $n = 2^+$ and $n = -2^-$ can also be identified in the phase diagram but the amplitudes are very small. The theoretical lines have been fit to the experimental data using $\overline{W}_c = 0.95$, $\overline{V}_c = 0.58$ and $K_\sigma = 0.36$. The value for $\overline{W}_c$ is in very good agreement with the SPIV measurements while the value for $\overline{V}_c$ is approximately 30% larger than measured. The value for K_σ is 60% of the value for a potential flow vortex and 70% of the value for the Rosenhead viscous flow vortex.

Discussion

Overall, it is seen that the theoretical dispersion line is in good agreement with the measurements although the values for the fitted parameters show differences with respect to measured values and values obtained from theory. Limitations of the theory are that it is a linearised theory that is derived for potential flow. A very simple viscous flow vortex model has been used to introduce the influence of viscosity and it is known that this vortex model does not accurately describe measured azimuthal velocity distributions. Also, the cavity size in the measurements can be influenced by diffusion of non-condensable gas and by the attachment of the vortex cavity to the blade. These aspects are not included in the theory. Furthermore, the influence of the tunnel walls on the dispersion relation has not been taken into account.

One of the motivations to investigate the dispersion relation is to derive a criterion for the resonance frequency of a cavitating tip vortex of ship propellers observed in the hull pressure fluctuations and underwater radiated noise (Bosschers, 2009a, 2017). The derived formulation for the dispersion relation shows only neutrally stable modes and does therefore not directly give an indication of a resonance frequency. There are however two aspects in the dispersion diagram that are interesting from this point of view. First, it is observed that the mode $n = 0$ crosses the mode $n = 2$ at two locations opposite and very close to $\kappa = 0$. As such a location, energy between modes can be exchanged. The second aspect is the presence of a local maximum of the mode $n = 0^-$ at which the group velocity becomes zero. A zero group velocity indicates that energy does not propagate which may lead to local resonance. The measurements do not indicate at which location resonance may occur. Analysis of the theoretical formulation shows that both locations coincide for very specific combinations of the viscous core and the cavitation number. These aspects need to be further investigated.

Conclusions

An update has been given on the comparison between the experimental and theoretical dispersion relation for cavitating vortices as presented by Pennings *et al.* (2015a). The theoretical dispersion relation is presented in a slightly different formulation that better shows the different parameters involved. An important new finding is that negative azimuthal wave numbers are present through which standing waves are modelled. The experimental dispersion lines are well described by theory if parameters are fitted.

References

Ackeret, J. (1930). Uber stationare hohlwirbel.pdf. *Ing.-Archiv, I Band*, pages 399–402.

Bosschers, J. (2008). Analysis of inertial waves on inviscid cavitating vortices in relation to low-frequency radiated noise. In *WIMRC Cavitation Forum*, Warwick University, UK.

Bosschers, J. (2009a). Investigation of hull pressure fluctuations generated by cavitating vortices. In *First International Symposium on Marine Propulsors, smp'09*, number June, Trondheim, Norway.

Bosschers, J. (2009b). Investigation of the resonance frequency of a cavitating vortex. In *NAG/DAGA 2009, Rotterdam.*

Bosschers, J. (2017). A Semi-empirical method to predict broadband hull pressure fluctuations and underwater radiated noise by cavitating tip vortices. In *Proceedings of the Fifth International Symposium on Marine Propulsors smp'17*, number June, Espoo, Finland.

Briançon-Marjollet, L. and Merle, L. (1996). Inception, development, and noise of a tip vortex cavitation. In *21st Symposium on Naval Hydrodynamics, Trondheid, Norway.*

Keller, J. J. and Escudier, M. P. (1980). Theory and observations of waves on hollow-core vortices. *Journal of Fluid Mechanics*, 99(3):495–511.

Maines, B. and Arndt, R. E. A. (1997). The case of the singing vortex. *Journal of Fluids Engineering*, 119.

Morozov, V. P. (1974). Theoretical analysis of the acoustic emission from cavitation line vortices. *Sov. Phys. Acoust.*, 19(5).

Pennings, P. C. (2016). *Dynamics of vortex cavitation.* PhD thesis, Delft University of Technology.

Pennings, P. C., Bosschers, J., Westerweel, J., and van Terwisga, T. J. C. (2015a). Dynamics of isolated vortex cavitation. *J. Fluid Mech.*, 778:288–313.

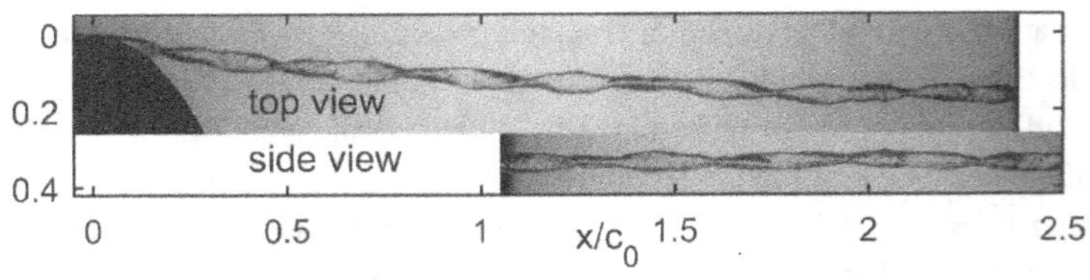

Figure 1: Top view and side view of the cavitating vortex trailing from the tip of a wing of elliptical planform.

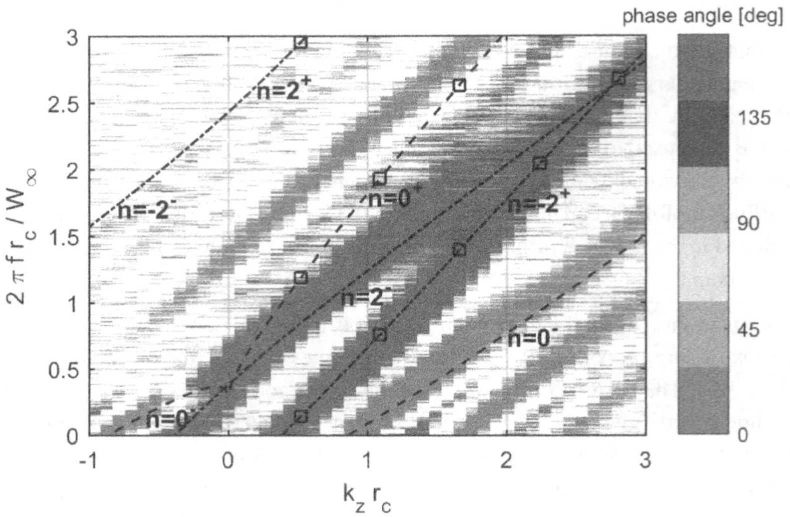

a) Phase angles for coherence larger than 0.4

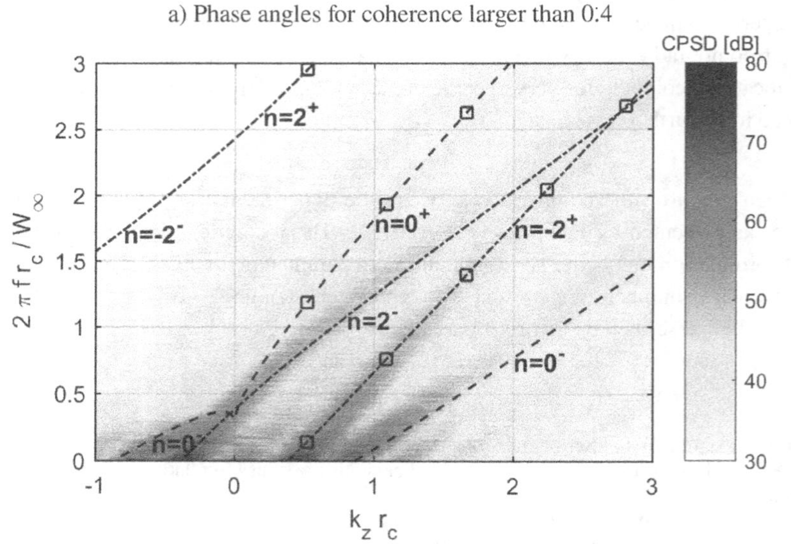

b) Cross-power spectral density values in dB

Figure 2: Wave number-frequency diagrams of the measured spatial and temporal variations of the vortex cavity radius.

Pennings, P. C., Westerweel, J., and van Terwisga, T. J. C. (2015b). Flow field measurement around vortex cavitation. *Experiments in Fluids*, 56(11):1–13.

Rosenhead, L. (1931). The formation of vortices from a surface of discontinuity. *Proc. R. Soc. Lond. Ser. A*, 134.

Saffman, P. G. (1992). *Vortex dynamics*. Cambridge University Press.

Thomson, W. L. K. (1880). Vibrations of a columnar vortex. *Philosophical Magazine*, X.

© 2018 ASME

10th International Symposium on Cavitation - CAV2018
Baltimore, Maryland, USA, May 14 – 16, 2018

On the Dynamics of a Torus Cloud Employing the Van Wijngaarden Ansatz and the Gilmore Bubble Dynamics Equation

[1]Paul Taubert*; [1]Peter F. Pelz; [1]Johannes Buttenbender

[1]*Chair of Fluid Systems, Technische Universität Darmstadt, Darmstadt, Germany*

Abstract

The picture of the cloud cavitation consists of three parts, which are at the same time the three phases of the bubble: (i) birth, i.e. bubble nucleation (cf. Groß and Pelz, 2017 [1]), (ii) midlife, i.e. transition from sheet to cloud cavitation (cf. Pelz et al., 2017 [2]), and (iii) death, i.e. cloud collapse (cf. Buttenbender and Pelz, 2012 [3]). To model cavitating flows, these three phases have to be analysed separately. The present work analyses the dynamics of cloud cavitation in a compressible flow, focusing on part (iii) mentioned above. Recent experiments highlight, that a common cloud geometry for cavitation clouds is a horseshoe. With the Helmholtz vortex theorem in mind, the horseshoe is artificially completed to be a generic torus shaped cloud. Following van Wijngaarden, the mixture of cavitation bubbles and liquid inside the cloud is treated as a continuous medium, i.e. a homogenous model with the share of vapour α and the dimensionless single bubble radius R. Hence, the bubble radii are a function of the radial position inside the cloud only, $R = R(r,t)$. The flow outside the cloud is modeled by a potential flow. The excitation of the cloud is carried out dynamically by applying a pressure history at infinity. With the Gilmore equation, the resulting system of partial differential equations is a parabolic system. This is due to the compressibility of the flow, which is taken into account. Previous investigations (cf. Buttenbender and Pelz, 2012 [3]) considered an incompressible flow, resulting in a hyperbolic system. The resulting pressure coefficient C_p and the bubble radii R inside the cloud are highlighted and analysed. They vary in time t and position r and depend on the pressure excitation of the cloud $C_{p\infty}$, the Mach number Ma, the Reynolds number Re and the Weber number We.

Keywords: cloud cavitation; analytics; Gilmore equation; van Wijngaarden

Introduction

Depending on the Reynolds and the cavitation number, the cavitation regime transitions from sheet to cloud cavitation, as Pelz et al. [2] describe. This effect was observed in our experiments in a convergent-divergent nozzle [4]. With the Helmholtz vortex theorem in mind, a torus is considered to be a common cloud geometry inside turbomachines (cf. Kawanami et al., 1998 [5]; Keil et al., 2012 [6]; Pelz et al., 2014 [7]).

The present work analyses the dynamics of a torus cloud employing the van Wijngaarden ansatz [8]. This method combines the continuity equation, the Euler equation and the Gilmore bubble equation to obtain a solution for a bubbly cloud. The flow inside the cloud is treated quasi one-dimensional leading to a single phase model introducing the share of vapour α. A potential flow models the flow outside the cloud. The excitation of the cloud is carried out dynamically applying a given pressure distribution at infinity. The cloud is composed of individual bubbles whose radii are a continuous function of space inside the cloud and time. Furthermore, the pressure distribution inside the cloud is also a continuous function of space and time. Applying initial and boundary conditions to the flow, the complete cloud characteristics are analyzed. Hence this study generates a deeper understanding of the process of cloud cavitation and is applicable to numerical methods solving cavitating flow problems.

Other than previous investigations [3], this analysis takes the compressibility of the liquid into account. This results in a change of the type of the resulting system of differential equation to a parabolic system.

Torus Cloud Model

The torus shown in Figure 1 is an idealized cloud geometry, which permits to investigate the dynamic behavior of a torus cloud. Two geometrical quantities describe the torus. These are the major radius $\tilde{R}_T$, and the minor radius $\tilde{R}_{\mathrm{Cl}}$. In the following analysis, quantities with a tilde are dimensional quantities and without tilde are dimensionless. As a generic model we consider the cloud in an infinite ambiance. Hence each cross section of the cloud behaves equally which leads to a quasi one dimensional problem.

*Corresponding Author, Paul Taubert: paul.taubert@fst.tu-darmstadt.de

It is important to mention: In contrast to many other approaches the entire flow (liquid and gas-phase) is treated in the framework of continuum mechanics. The flow field is divided into a field outside the cloud, in which cavitation is excluded and the compressible cloud interior.

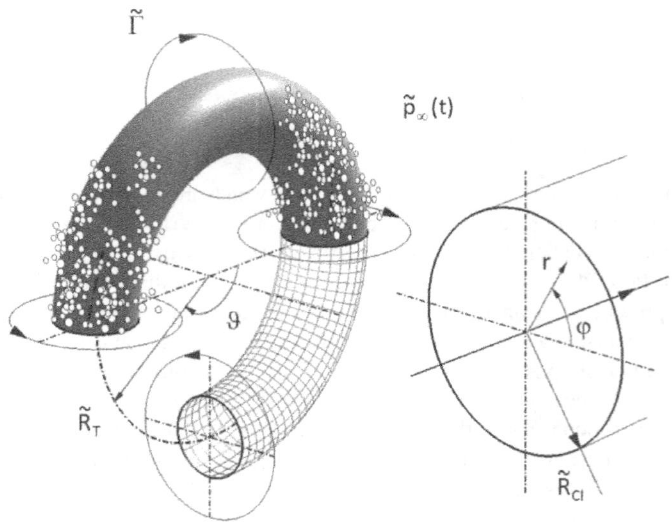

Figure 1: Idealized torus geometry and the one-dimensional spatial coordinate for the behavior interior of the cloud.

In the interior of the cloud a mixture of vapor bubbles of radius $\tilde{R}$ and inviscid fluid of constant density is assumed. The flow of this dispersion is modeled as a homogeneous medium neglecting the relative motion between the phases. This assumption is justified provided (i) the bubble size $\tilde{R}$ is small compared to the cloud size $\tilde{R}_{Cl}$, and (ii) the bubble density is small. The density of the mixture is then given by the void fraction α, the densities of the vapor within the bubbles $\tilde{\varrho}_V$ and the liquid density $\tilde{\varrho}_L$:

$$\tilde{\varrho} = \alpha\tilde{\varrho}_V + (1-\alpha)\tilde{\varrho}_L \approx (1-\alpha)\tilde{\varrho}_L. \tag{1}$$

The non-equilibrium bubble dynamics, which is essential to study the processes of cloud collapse, requires a description of the density on the geometry of the individual bubbles. This is achieved by a coupling the Gilmore bubble equation to mass and momentum balance equations for the mixture (cf. van Wijngaarden, 1964 [8]).
With the bubble population $\tilde{\eta}$, that is the number of bubbles per volume of liquid, the void fraction can be expressed by the bubble volume $\tilde{V}$, which is a continuous spatial function:

$$1-\alpha = \frac{1}{1+\tilde{\eta}\tilde{V}} = \frac{\tilde{\varrho}}{\tilde{\varrho}_L} = \varrho. \tag{2}$$

One assumes that no bubbles will be created or destroyed, which is equivalent to a vanishing change of bubble population along the particle path. In the following, the quantities are non-dimensionalized using the liquid density $\tilde{\varrho}_L$, the equilibrium bubble radius $\tilde{R}_E$ as length, the mean channel velocity $\tilde{U}$ and therefore $\tilde{R}_E/\tilde{U}$ as time. The flow within the cloud interior is not divergence free and not irrotational. The mixture is assumed to be inviscid but not the surrounding of the single bubble.
The system of equations is closed by van Wijngaarden's continuums-mechanical form of the Gilmore equation. This form follows from an application of the Gilmore equation along the particle path and couples the pressure resulting from the bubble motion to the momentum equation and the bubble size to the mixture density:

$$R\frac{\mathrm{D}^2R}{\mathrm{D}t^2}\left(1-M\frac{\mathrm{D}R}{\mathrm{D}t}\right)+\frac{3}{2}\left(\frac{\mathrm{D}R}{\mathrm{D}t}\right)^2\left(1-\frac{M}{3}\frac{\mathrm{D}R}{\mathrm{D}t}\right) = H\left(1+M\frac{\mathrm{D}R}{\mathrm{D}t}\right)+MR\left(1-M\frac{\mathrm{D}R}{\mathrm{D}t}\right)\frac{\mathrm{D}H}{\mathrm{D}t}, \tag{3}$$

with the Mach number M and the pressure coefficient $H := 1/\tilde{U}^2 \int_\infty^R \mathrm{d}\tilde{p}/\tilde{\varrho} = H(R, C_p, C_{pR})$. The coefficient H is a function of the bubble radii R, the dimensionless pressure inside the bubble $C_p := 2(\tilde{p}-\tilde{p}_0)/\tilde{\varrho}_L\tilde{U}^2$ and the dimensionless pressure at the bubble surface $C_{pR} := 2\left(\tilde{p}(\tilde{R})-\tilde{p}_0\right)/\tilde{\varrho}_L\tilde{U}^2$ and is solved applying Taits law (Gilmore 1952, [9]).

Equation (3) results in an equation explicit in the highest time derivative of pressure C_p and bubble radius R:

$$\frac{\mathrm{D}C_p}{\mathrm{Dt}} = \frac{1}{F_{Cp}}\left[(R\beta - F_{\ddot{R}})\frac{\mathrm{D}^2 R}{\mathrm{D}t^2} + \frac{3}{2}\left(\frac{\mathrm{D}R}{\mathrm{Dt}}\right)^2 \beta^* - F_R\right]. \tag{4}$$

Hereby, the terms $F_{Cp}, F_{\ddot{R}}, F_R, \beta, \beta^*$ are functions of $M, R, \mathrm{D}R/\mathrm{Dt}, C_p$, as well as of the polytropic exponent k, cavitation number σ, Reynolds number Re and Weber number We:

$$\sigma := \frac{2(\tilde{p} - \tilde{p}_v)}{\tilde{\varrho}_L \tilde{U}^2}, \qquad Re := \frac{\tilde{R}_E \tilde{U} \tilde{\varrho}_L}{\tilde{\mu}_L}, \qquad We := \frac{\tilde{\varrho}_L \tilde{U}^2 \tilde{R}_E}{\tilde{S}}, \tag{5}$$

including the vapor pressure $\tilde{p}_v$, dynamic viscosity of the fluid $\tilde{\mu}_L$ and the surface tension $\tilde{S}$.
To describe the cloud interior, i.e. its pressure C_p and individual bubble radii R, two additional equations are necessary to close the system of equations. The missing equations are the Euler equation

$$\frac{\mathrm{D}\vec{u}}{\mathrm{Dt}} = \frac{\mathrm{D}^2\vec{r}}{\mathrm{D}t^2} = -\frac{\nabla C_p}{2(1-\alpha)} \tag{6}$$

and the continuity equation

$$\frac{\mathrm{D}\alpha}{\mathrm{Dt}} = (1-\alpha)\,\mathrm{div}(\vec{u}). \tag{7}$$

It is important to note, that the system of equations (4),(6),(7) represents an ordinary system of equations for the variables R, C_p. Both, the bubble radius R and the pressure C_p are a continuous function of the radial position r and time t. A variable-step, variable-order solver based on the numerical differentiation formulas of order 1-5 solves the system of equations (4),(6),(7).

The dynamic excitation (excitation time t_E) is carried out by the dimensionless pressure

$$\begin{aligned} C_{p\infty} &= \frac{C_{p,\min}}{2}\left(1 - \cos\left(2\pi\frac{t}{t_E}\right)\right) && \text{for } \frac{t}{t_E} \leq 1 \\ C_{p\infty} &= 0 && \text{for } \frac{t}{t_E} > 1 \end{aligned} \tag{8}$$

at infinity. A minimum excitation pressure of $C_{p,\min} = -0.5$ is used. The excitation function is harmonic and remains above the dimensionless vapor pressure $-\sigma$. A variable-step, variable-order solver based on the numerical differention formulas of orders 1 to 5 solves the system of equations (4) - (7).

The problem contains three time scales. The first time scale is the excitation time $\tilde{t}_E$. The second time scale is the typical time of an individual bubble, given by its natural frequency (Plesset & Prosperetti, 1977 [10]):

$$t_B = \frac{2\pi}{\omega_B} = 2\pi\left(3k\frac{\sigma}{2} + (3k-1)\frac{2}{We}\right)^{-1/2}. \tag{9}$$

The third time scale of the given problem is the typical time of the cloud, given by the phase velocity of the dispersion:

$$t_{\mathrm{Cl}} = \frac{R_{\mathrm{Cl}}}{\omega_B}\sqrt{3\alpha_0(1-\alpha_0)}, \tag{10}$$

with the initial void fraction α_0. The interaction parameter

$$\beta_0 := R_{\mathrm{Cl}0}\alpha_0(1-\alpha_0) \sim \left(\frac{t_{Cl}}{t_B}\right)^2 \tag{11}$$

is used, cf. Brennen 2005 [11]. This parameter influences and specifies the initial cloud radius $R_{\mathrm{Cl}0}$.

Results

The growth of bubbles in a cloud depends on the bubble parameters k, σ, Re, We. In this study we concentrate on the set of parameters $k = 1.4,\ Re = 20,\ We = 200,\ \sigma = 1.0$ which are typical for cavitation.
The interaction parameter β_0 has a high influence on the behavior of the cavitation cloud, i.e. the bubble radius and the pressure. Hence only this dependency is shown in the framework of this investigation for simplification. Figure 2 and Figure 3 show the individual bubble raddi R and pressure C_p over time t/t_E and radial position $r = \tilde{r}/\tilde{R}_{Cl0}$ for an interaction parameter $\beta_0 = 1$ and 10 respectively. The position $r = 0$ represents the center of the torus minor radius, $r = 1$ is the surface of the torus, see Figure 1. The dynamic excitation ends at $t/t_E = 1$ as equation (8) describes.

$$\beta_0 = 1, \quad k = 1.4, \quad Re = 20, \quad We = 200, \quad \sigma = 1.0$$

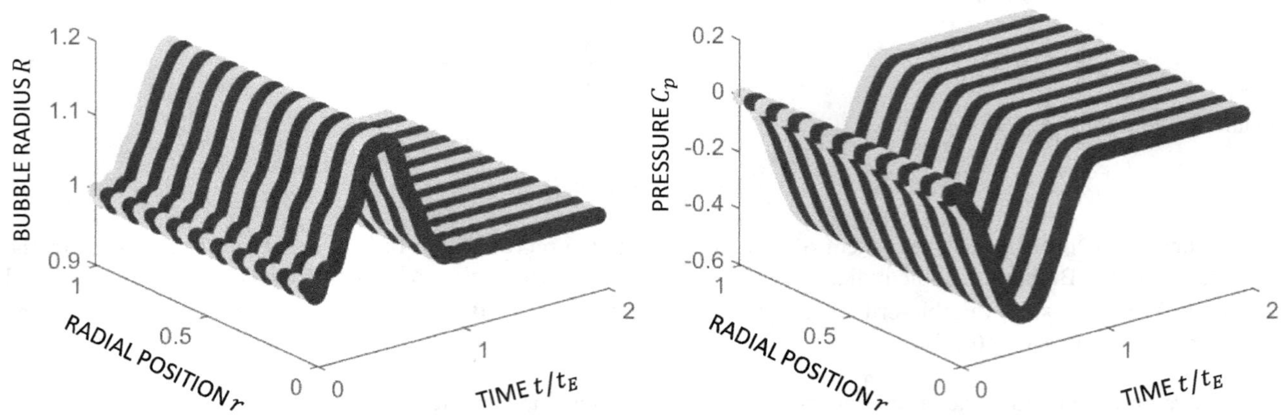

Figure 2: Bubble radius and pressure over radial position and time for $\beta_0 = 1$.

$$\beta_0 = 10, \quad k = 1.4, \quad Re = 20, \quad We = 200, \quad \sigma = 1.0$$

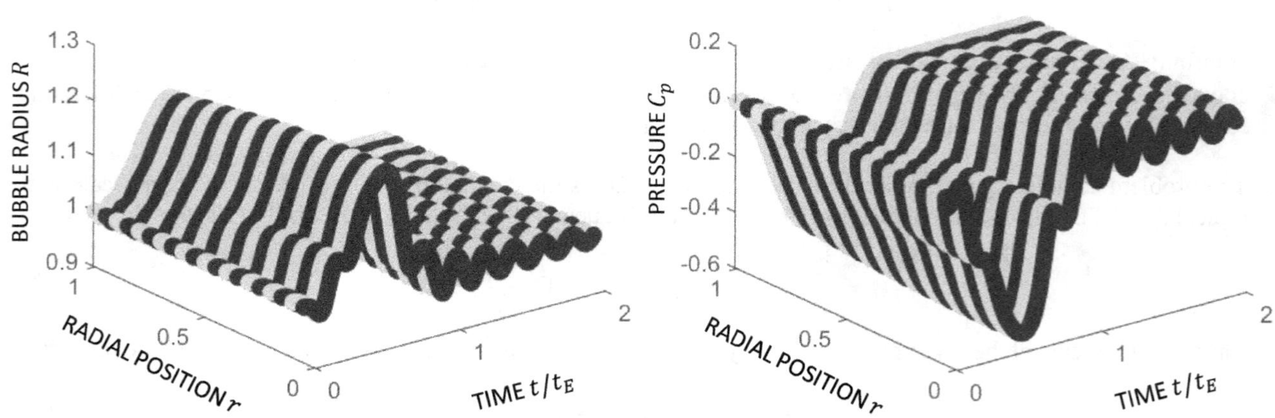

Figure 3: Bubble radius and pressure over radial position and time for $\beta_0 = 10$.

For smaller interaction parameter $\beta_0 = 1$, i.e. a relatively small cloud, the excitation is transmitted into the center of the cloud. For this case the growth of the bubble raddi R does not depend on r. The individual bubbles inside the cloud show synchronous dynamics, i.e. the radius R is a function of time only. This behavior changes for higher $\beta_0 = 10$, see Figure 3. For this case, the bubble radius R and the pressure C_p oscillate as the positions tends to the cloud center $r \to 0$. At $r = 1$, both R and C_p behave in the same way for $\beta_0 = 1$ and $\beta_0 = 10$ due to the same dynamic excitation at infinity.

Summary and Conclusion

This study analyses the dynamics of a toroidal cavitation cloud under the influence of an external dynamic excitation. The cloud is composed of smaller bubbles and treated as a continuous medium. Hence, the bubble radii R are a function of space and time inside the cloud. Van Wijngaarden's continuums-mechanical form of the Gilmore equation including the Euler and the continuity equation solve this problem. Both, bubble radius R and pressure C_p depend on the interaction parameter β_0. For higher β_0, i.e. a bigger torus cloud, the time history of bubble radius and pressure show an oscillation becoming stronger in the center of the cloud. For smaller β_0 no oscillation appears, the bubbles behave rather synchronous without depending on the position but on time only. At the cloud surface, the behaviour does not depend on β_0 but on the external dynamic excitation only.
Further investigations on this field such as analyzing dominating frequencies, influences of the share of vapor, the excitation characteristics or other parameters will provide additional data leading to a better understanding of cloud cavitation.

References

[1] T. F. Groß and P. F. Pelz (2017). *Diffusion-driven nucleation from surface nuclei in hydrodynamic cavitation.* Submitted to J. Fluid Mech.
[2] P. F. Pelz, T. Keil, and T. F. Groß (2017). *The transition from sheet to cloud cavitation.* J. Fluid Mech., vol. 817, pp. 439–454.
[3] J. Buttenbender and P. F. Pelz (2012). *The Influence of Imposed Strain Rate and Circulation on Bubble and Cloud Dynamics.* Proceedings of the 8th International Symposium on Cavitation, CAV2012, Singapore.
[4] T. Keil, P. Pelz, U. Cordes, and G. Ludwig (2011). *Cloud cavitation and cavitation erosion in convergent divergent nozzle.* WIMRC 3rd International Cavitation Forum, University of Warwick, UK.
[5] Y. Kawanami, H. Kato, and H. Yamaguchi (1998). *Three-dimensional characteristics of the cavities formed on a two-dimensional hydrofoil.* Proceedings of the 3rd International Symposium on Cavitation, CAV1998, Grenoble, France, vol. 1, pp. 191–196.
[6] T. Keil, P. F. Pelz, and J. Buttenbender (2012). *On the Transition from Sheet to Cloud Cavitation.* Proceedings of the 8th International Symposium on Cavitation, CAV2012, Singapore.
[7] P. F. Pelz, T. Keil, and G. Ludwig (2014). *On the Kinematics of Sheet and Cloud Cavitation and Related Erosion.* Advanced Experimental and Numerical Techniques for Cavitation Erosion Prediction, vol. 106, K.-H. Kim, G. Chahine, J.-P. Franc, and A. Karimi, Eds. Dordrecht: Springer, pp. 221–237.
[8] L. van Wijngaarden (1964). *On the Collective Collapse of a Large Number of Gas Bubbles in Water.* Proceedings of the 11th International Conference of Applied Mechanics, pp. 854–861.
[9] F. R. Gilmore (1952). *The growth or collapse of a spherical bubble in a viscous compressible liquid.* Hydrodyn. Lab. Calif. Inst. Technol. Pasadena CA Rep. No 26-4.
[10] M. S. Plesset and A. Prosperetti (1977). *Bubble dynamics and cavitation.* Annu. Rev. Fluid Mech., vol. 9, no. 1, pp. 145–185.
[11] C. E. Brennen (2005). *Fundamentals of multiphase flow.* Cambridge [England] ; New York: Cambridge University Press.

© 2018 ASME

On the Influence of Eddy Viscosity in the Numerical Modelling of Air Cavities

[1]Gem Rotte*; [2]Maarten Kerkvliet; [1,2]Tom van Terwisga

[1]Delft University of Technology, Delft, NL; [2]Maritime Research Institute Netherlands, Wageningen, NL

Abstract

This article aims to link the physical modelling of air cavities to their numerical modelling with a RaNS solver. The largest challenge in predicting the cavity characteristics numerically lies in correctly modelling their closure region. According to Reboud et al. [1], the commonly over-predicted eddy-viscosity at the water-gas interface reduces the development of the re-entrant jet and can thus prevent the occurrence of shedding. Computations were carried out using two different eddy-viscosity correction functions: a power function as described by Reboud et al. [1] and Coutier-Delgosha et al. [2] and a Gaussian function proposed by Rotte et al. [3]. The Gaussian correction function does not contain large gradients when the local density approaches pure gas or liquid. It also has no bias towards one of the fluids and is expected to improve the stability of the simulation and the physical modelling of the problem.

Keywords: RaNS turbulence modelling; air lubrication, re-entrant jet

Introduction

The use of air lubrication techniques for the reduction of frictional resistance has the potential to significantly reduce a ship's fuel consumption, benefiting sustainable employability and cost reduction. Several techniques are available of which an overview is given by Rotte et al. [4]. This article focuses on the air cavity techniques. An external cavity (figure 1(a)) is created beneath the ship's hull with the help of a cavitator, which is located directly upstream of an air injection point. It is used to separate the mean water flow from the hull, providing a stable air layer with a length prescribed by half the wavelength of the gravity-dominated surface wave [5]. Next to the external air cavity technique, hull integrated or internal cavities (figure 1(b)) are characterized by the presence of a recess underneath the ship's hull. In contrast to the external cavity, the free surface is not limited by the wavelength of the gravity-dominated surface wave when the recess is of sufficient depth. Air cavity applications are claimed to lead to propulsive power reduction of 10-20% ([5], [6]). However, a complete understanding of the relevant two-phase flow physics, particularly in the closure region, is still lacking. The largest challenge in predicting the cavity characteristics numerically lies in correctly modelling this closure region ([7], [8]). The air cavity characteristics are strongly related to this region which predicts the power reduction, operability and safety of the system. This lack of information is holding back the application of the application of air cavity techniques in the shipping industry.

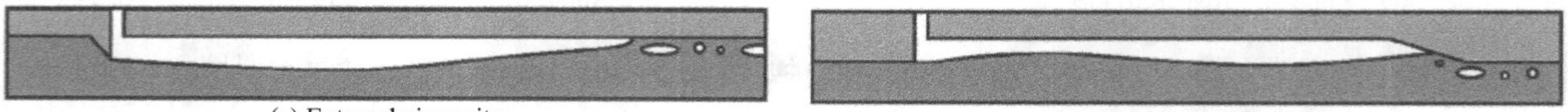

(a) External air cavity (b) Internal air cavity

Figure 1: Air cavity techniques (flow from left to right)

Both the re-entrant jet mechanism and the wave pinch-off mechanism are cited as possible mechanisms for air discharge from the cavity in the closure region (e.g. [5], [7], [9]). The re-entrant jet mechanism is provisionally assumed to be similar to the re-entrant jet mechanism that is responsible for the break-up of vapor sheet cavities. The wave pinch-off mechanism is governed by waves on the air-water interface that are hypothesized to be induced by turbulent eddies. If the wave amplitudes are of the same magnitude as the cavity thickness, the free surface interface will hit the ship's bottom and pockets of air will be shed from the cavity. This article aims to link the physical modelling of the relevant phenomena to their numerical modelling, with an emphasis on the modelling of the re-entrant jet mechanism for air filled internal cavities with a RaNS turbulence model. A comparison is made between the internal and external cavity cases, of which the latter case is reported in Rotte et al. [3].

The numerical solver used for all simulations is ReFRESCO [10]. It is a multiphase (unsteady) incompressible viscous flow solver using the RANS equations, complemented with turbulence and cavitation models. Time integration is performed implicitly with first or second-order backward schemes. For turbulence modelling, both RANS/ URANS and Scale-Resolving Simulations (SRS) models such as SAS, DDES/IDDES, XLES, PANS and LES approaches can be used.

*Corresponding Author, Gem Rotte: G.M.Rotte@tudelft.nl

Numerical settings

The (2D) simulations for the internal cavity are based on the geometry as reported by Shiri et al. [7]. The geometry as applied in the current simulations is scaled down with a factor 4.5, to enable validation experiments in the Delft Cavitation Tunnel. The cavitation tunnel has a test section length of 2m and a 0.3x0.3m cross-sectional area. A schematic of the domain including the applied boundary conditions and close-ups of the step initiating the internal cavity and the beach area is shown in figure 2. Close to the wall, grid clustering was applied to ensure y+ values well below 1 and refinement zones were applied around the step and closure area. The total number of grid cells is around 132000. Unsteady simulations were carried out with a water tunnel inlet velocity of 0.943 m/s. For the external cavity, the length of the computational domain upstream of the cavity is chosen such that the boundary layer profile in de simulations approaches the one from the measurements as performed by Zverkhovskyi [5]. The computational grid includes refinement zones in the region around the cavitator, the cavity closure and the expected cavity interface. Close to the wall, y+ values well below 1 are ensured. The total number of grid cells is around 185000. Unsteady simulations were carried out with an inlet velocity of 1 m/s.

Zverkhovskyi et al. [8] found that prescribing the air flux at the air inflow did not yet give physical solutions. Instead, it is chosen to prescribe the pressure at the air inlet for both the internal and external cavity simulations. All computations were carried out using the $k - \sqrt{k}L$ (KSKL) turbulence model as described by Menter et al. [11]. A second order backward scheme is used for the time discretization and the time step used in the calculations was 0.5 ms in case of the internal cavity and 0.04 ms in case of the external cavity to ensure Courant numbers smaller than 0.8 and 0.2 in the domains. This corresponds to dimensionless time steps $\frac{U_\infty dt}{\lambda_{lin}}$ of $8{,}3 * 10^{-4}$ and $6{,}25 * 10^{-5}$ respectively, where λ_{lin} is the wave length of the gravity dominated free-surface wave following from linear theory. The discretization of both the momentum and turbulence equations is treated by a QUICK scheme and the transport equation for the volume fraction is handled by a SUPERBEE scheme, which is flux limited, compressive and TVD. Regarding the iterative convergence of the simulations presented for most cases, the L_∞ norm is in the order of 10^{-5} for all residuals, except the residual of the momentum equation in x and y-direction which are in the order of 10^{-3}. It is noted that this is a relatively poor convergence level. However, the regions where these levels are found are very local. It is condensed to regions inside the cavity and mostly does not translate to the water phase. Away from the interface in the water phase all residuals drop below 10^{-7} for the L_∞ norm.

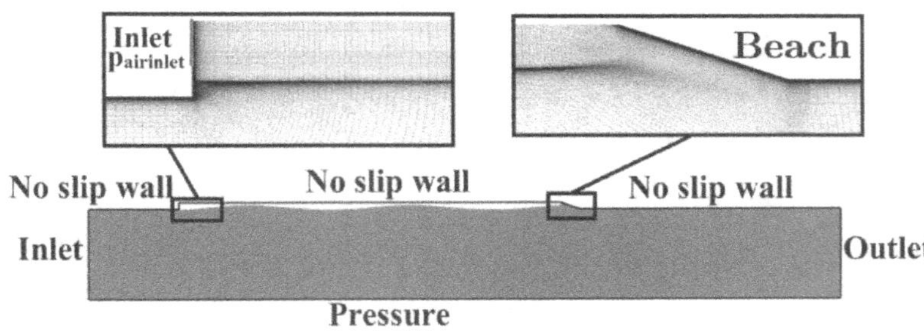

Figure 2: Computational domain including boundary conditions for the internal cavity simulations

Cavity profiles

Figures 3 and 4 show the internal and external cavity profiles for simulations with different air inlet pressures respectively. The profiles are plotted using the contour where the air volume fraction equals 0.5. For the internal cavity (Figure 3), the wavelength of the wave stays constant, $\lambda = 0.562 \pm 0.013$. For increasing pressure difference, the amplitude of the wave grows and the stationary wave slowly shifts upstream, which is also observed for transom flows in the dry transom regime. The transom flow must be such that the pressure difference between the cavity and the pressure at the step is cancelled. The flow must curve upward when the pressure inside the cavity is smaller, the convexity of the streamlines then produce the required pressure difference. When the transom immersion is increased (essentially: when the pressure difference is increased), the upward curvature increases, the wave height grows and the stationary wave system shifts upstream [12]. The flow behavior at the closure gets more chaotic with increasing pressure difference. The area of mixing water and air increases in the closure region and it is questionable whether the size of this region is physically correct. Some mixing is expected, due to the breaking wave and/or re-entrant jet in the closure area. It must be noted that the residuals for the momentum equation in x and z-directions stagnated for the cases with pressure differences higher than around 40Pa.

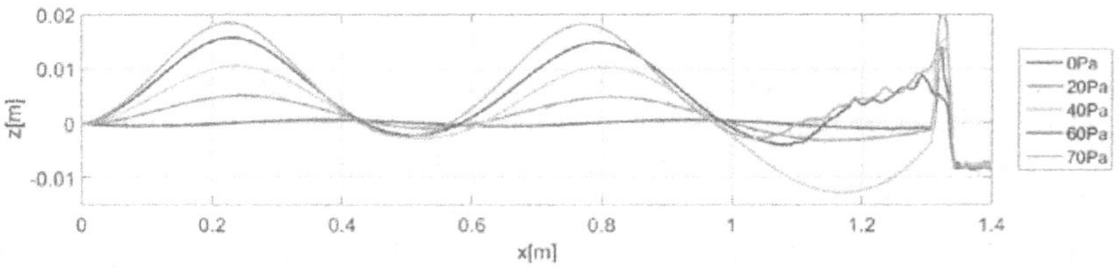

Figure 3: Internal cavity mean profiles as function of air inlet pressure, flow from left to right

For the external cavity, the length as well as the thickness of the cavity increase with increasing pressure difference (Figure 4). Based on linear wave theory and the experiments from Zverkhovskyi [5], a cavity length 0.32 m and a cavity thickness of around 0.01 m is expected, and the length of the cavity should not be dependent on the pressure inside the cavity. The external cavity simulations did not show correct cavity closure behavior when compared to the experiment [5]. From the experiment at this flow velocity it is observed that the cavity remained very stable, and only sometimes small pockets of air were lost. In the simulations, a thin layer of air is constantly present at the closure and downstream of the cavity, influencing the computed wall shear stress. However, the ensemble averaged velocity profiles in the boundary layer upstream, in the middle and downstream of the air cavity compare reasonably well with the measured time-averaged profiles [3]. Unfortunately the results of the experimental measurements close to the wall are of insufficient resolution to quantitatively compare the viscous sub-layer of the boundary layer, which determines the wall shear stress.

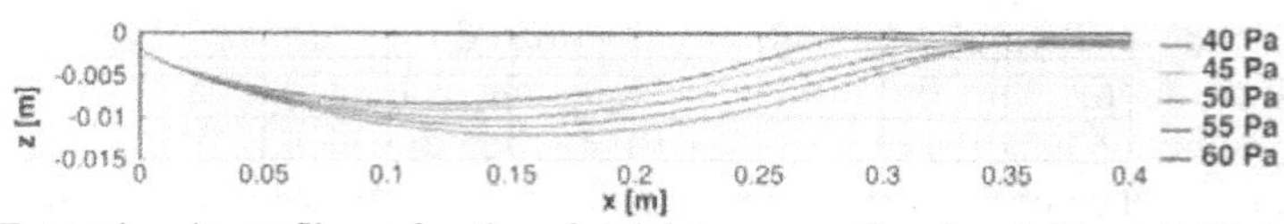

Figure 4: External cavity profiles as function of air inlet pressure, flow from left to right (Rotte et al. [3])

In order to study the cavity profiles more quantitatively, both the normalized cavity wavelength (Figure 5 (a)) and steepness (Figure 5 (b)) of the cavity profile is compared with the profile following from linear wave theory and plotted as a function of the cavity cavitation number σ. The cavitation number for the internal and external cavities is here defined as the pressure difference between inside and outside the cavity divided by ½ ρU_∞^2. The wavelength for undisturbed deep water waves $\lambda_{lin} = (2\pi U_\infty^2)/g$.

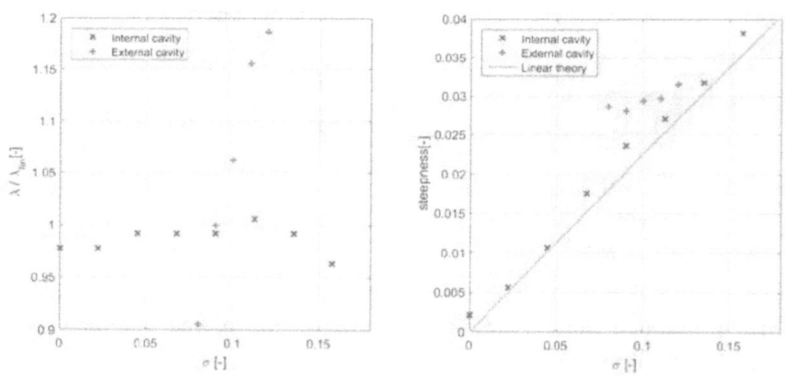

(a) (b)

Figure 5: Normalized wavelength as function of cavitation number (a) and steepness of cavity profile as function of cavitation number (b)

For the cavity wavelengths (Figure 5(a)), all simulation results for the internal cavity show a small under-prediction (within 5%) of the wavelength as predicted by linear theory and do not show a dependency of the pressure inside the cavity. For the external cavity however, this is not the case. Based on the experimental observations of Zverkhovskyi [5] there should be no dependency of the cavity length on the pressure, since this should only be a function of the free stream velocity. The cavity steepness as plotted as a function of the cavitation number in figure 5(b) is determined by dividing the wave height $2A$ by the wave length λ. The wave amplitude A as determined by linear wave theory is a function of the pressure difference dp and can be derived by setting the power input into the flow equal to the energy flux of a sinusoidal wave in a control volume of one wavelength and is given by:

$A = dp/\rho g$. For the external cavity, the wave height is determined by multiplying the cavity thickness by two, where the cavity thickness is defined from the flat plate to the thickest point of the cavity. Both for the internal and external cavity case, an over-prediction of the wave steepness is found, but especially the results for the internal cavity compare reasonably well to linear theory.

Concluding this part, the cavity profile until the closure region for the internal cavity is well predicted since the wavelength remains practically constant and the amplitude of the wave follows the same trend as predicted using linear theory. However, the exact mechanism for air discharge at the cavity closure is still not clear. For the lower cavitation numbers the behavior at the closure is stable (Figure 6(a)) and a growing region of higher pressure can be observed, indicating a stagnation point and possibly the onset of a re-entrant jet. A thin stream of water is flowing parallel to the wall in the upstream direction inside the cavity. At some point, this thin stream is mixing with the air phase, but no pocket of air is shed from the cavity. It is hypothesized that the momentum of the jet is lost as a result of the decrease in local density due to the mixing. For the higher cavitation numbers however, water and air already mix upstream of the beach in the simulations (Figure 6(b)), which is most probably due to numerical problems occurring due to lack of local convergence of the momentum equations. Another possibility is that this particular turbulence and/or free surface model is unable to predict the closure behavior. There is no solution to this problem yet.

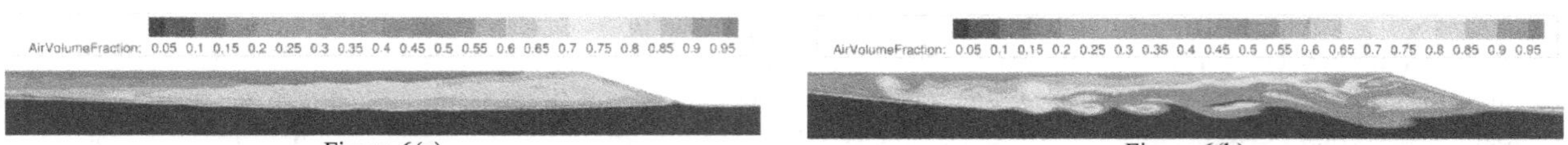

Figure 6(a) Figure 6(b)

Figure 6: Air volume fraction for internal air cavity closure region for σ=0.045 (a) and σ=0.16 (b), flow from left to right

For the external cavity case on the other hand, the computed wavelength does not stay constant when the air inlet pressure is varied. This is hypothesized to follow from the incorrect turbulence and/or free surface modelling in the closure region. In the simulations, one can indeed find a region of higher pressure just behind the cavity, but it is questionable whether this pressure difference is high enough to force a thin stream of liquid inside and thus induce a re-entrant jet shedding behavior.

Eddy viscosity correction

For the numerical modelling of the re-entrant jet in the air cavity cases, it is questionable whether RaNS methods can directly be used. The commonly used two-equation turbulence models, such as the $k - \varepsilon$ and $k - \omega$ models, were originally developed for RaNS simulations of single-phase flows. A frequently encountered problem with these turbulence models that are based on the Boussinesq assumption is the over-prediction of the eddy-viscosity around liquid-gas interfaces and thus in and downstream of the cavity closure region. According to Reboud et al. [1], this over-predicted eddy-viscosity reduces the development of the re-entrant jet and can thus prevent the occurrence of shedding. Computations were carried out using two different eddy viscosity correction functions: a power function as used by Reboud et al.[1] and described in Coutier-Delgosha et al. [2] and a Gaussian function which is proposed by Rotte et al [3]. In these cases, the originally computed turbulence viscosity μ_t is artificially lowered by multiplying it by a correction factor based on either the power function or Gaussian function of the local density $f(\rho)$. More information on these eddy-viscosity correction models can be found in [1], [2] and [3]. Figure 7 shows the different correction functions for different values of power exponent for cases of the power correction function and different values of amplitude A, width W and power p for Gaussian correction cases. The power function is asymmetrical and has a bias towards the lower density fluid. It also has large gradients when the mixture approaches pure air or pure water, which can possibly introduce numerical difficulties. For the Gaussian functions the correction only is active when the cell is marked as an interface cell and the function is symmetric.

For the internal cavity case, only the Gaussian correction function was applied with A=1 and W=1 and p=1, 10 and 100 respectively, for $\sigma = 0.135$. The cavity profiles for the simulations including correction are compared to the case without eddy viscosity correction. The mixing of water and air already takes place upstream of the beach in the simulations without correction and therefore the eddy-viscosity correction only increases the amount of mixing and no re-entrant jet is observed. The correction function makes the interface even more diffuse and less stiff. The computed mean cavity profiles remain constant and compare well to the simulation without correction, but the cavity closure region gets more chaotic with increasing eddy viscosity correction power. The required air flux for all cases still needs to be computed, so the influence of the correction function on this parameter is not yet known.

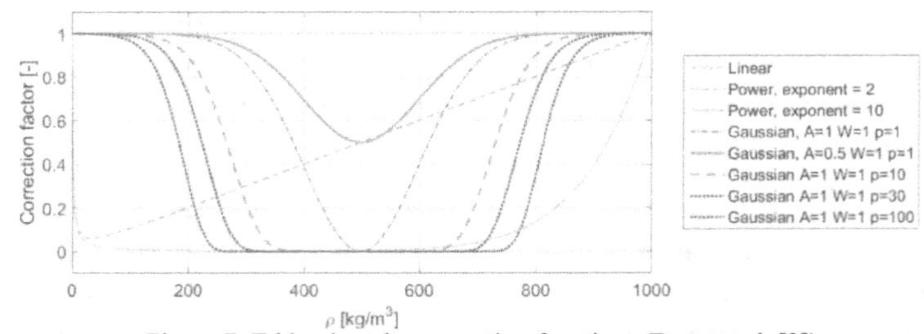

Figure 7: Eddy viscosity correction functions (Rotte et al. [3])

For the external cavity cases, a different behavior is observed. These cases are more extensively reported in [3]. Figure 8 shows the computed cavity profiles with σ = 0.135 for three representative cases: a base case without correction, one case with power function correction and one case with Gaussian correction. All simulations performed with this correction function active produced the same stable cavity profile as found for the base case, with a steady layer of air escaping from the closure. The Gaussian function was applied with A=1, W=1 and p=1, 30 and 100 respectively.

However, all the external cavity profiles as computed including the power function correction are unstable. In all these cases, the cavity grows to its stable length which is similar to the profile of the case with no eddy-viscosity correction and then becomes unstable. Mixing occurs and due to the bias of the power function towards the low density fluid, all eddy-viscosity μ_t inside the cavity is reduced such that there is no eddy-viscosity at all. The total viscosity difference between inside and outside the cavity is approximately ten times lower for the base case when compared to the power function cases. The total eddy-viscosity just outside the cavity is rather high, due to the developed boundary layer upstream of the cavity ($Re_x > 10^6$). Due to the recirculating flow inside the cavity and the lack of eddy-viscosity in the air layer, the velocity differences as well as the vorticity are larger in the power function case and instabilities are more likely to occur.

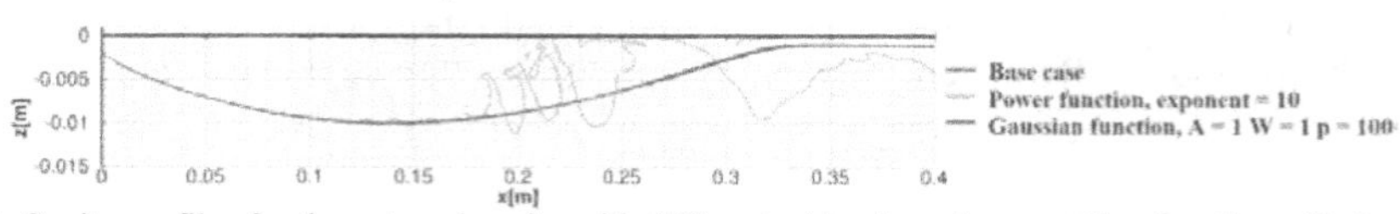

Figure 8: Cavity profiles for the external cavity with different eddy viscosity correction functions (Rotte et al. [3])

These findings lead to the conclusion that the reduction of the eddy-viscosity on the air-water interface is not the only requirement for the re-entrant jet to form in RaNS simulations of air cavities. However, the computed eddy-viscosity at the cavity interface as well as inside the cavity have a large influence on the interface stability for both internal and external cavities. For the internal cavity case, the Gaussian correction function has more influence than for the external cavity case because the interface is already destabilized. The source of this instability is still unknown, but it is likely a numerical modelling issue.

Regarding the re-entrant jet air loss mechanism for both types of cavities, the previously made assumption that the closure mechanisms are similar may not be correct. The method for generating the cavities as well as the geometry of the ship bottom in the closure region differ significantly for both types of cavities. The angle at which the cavity profile enters the ship hull can be larger in the case of the internal cavity, due to the beach geometry and steepness of the wave profile. Depending on the pressure inside the cavity as well as the flow speed, the reattachment is either smooth or steep. In case of a steep reattachment for the internal cavity case, it is hypothesized that the pressure directly behind the reattachment can get higher than in the external cavity case, possibly leading to another shedding mechanism or wave breaking. This may also indicate that a different numerical model is required for both cases.

For the modelling of the re-entrant jet mechanism, one of the differences with respect to natural sheet cavitation modelling is the lack of a condensation term. There is no condensation of gas at the cavity closure in the case of air filled cavities. Additionally, the pressure difference between inside and outside the cavity is around one order of magnitude lower for air cavities when compared to vapor cavities. The question then rises whether this pressure difference is high enough to create a re-entrant jet. Also, the RaNS modelling of natural cavitation problems usually asks for a low-order and thus diffuse discretization scheme for the volume fraction equation, due to convergence problems that occur when higher order schemes are used. Here a higher order scheme is used, since in the current approach the stable cavity interface till the closure should be solved as well, especially when one is also interested in the wave pinch-off mechanism and the total air flux leaving the cavity.

Conclusions and further work

A comparison is made between the numerical simulations of internal and external air cavity characteristics. For the internal air cavity, the free surface profiles up until the closure area agree with linear theory for both the wave length and steepness of the waves. But, for increasing cavitation number, the interface in the closure area gets more diffuse and a more chaotic closure behavior is observed.

For the external cavity, the wavelength of the free surface wave and thus the cavity length do not agree with linear theory and the experimental findings from Zverkhovskyi [5]. The cavity length depends on the pressure prescribed in the cavity while it should only be a function of free stream velocity and the cavity interface remains stiff and thin for all computed cavitation numbers.

The previously made assumption that the closure mechanisms for internal and external cavities are similar may not be correct. The method for generating the cavities as well as the geometry of the ship bottom in the closure region differ significantly for both types of cavities. In case of a steep reattachment for the internal cavity case, it is hypothesized that the pressure directly behind the reattachment can get higher than in the external cavity case, possibly leading to another shedding mechanism or wave breaking. This may also indicate that a different numerical model is required for both cases.

It has also been found that RaNS two-equation turbulence models like KSKL model show serious deficiencies for the modelling of the shedding mechanisms for air cavities. Also after including an ad-hoc eddy-viscosity correction model as proposed by Reboud et al. [1] and Rotte et al. [3]. Next to the modelling of the re-entrant jet, the simulation of the wave pinch-off mechanism also needs to be included to be able to capture the cavity closure dynamics. This mechanism is hypothesized to be governed by waves formed by turbulence structures disturbing the interface. A scale resolving or hybrid model such as LES, RaNS-LES or PANS is then required. These models will first be used to identify the physical conditions in the closure region and to study and identify scaling influences. Thereafter, a practical engineering correction which can be used with RaNS models for the numerical modelling of air cavities could be developed. The dispersion of the gas phase can then e.g. be modelled by an extra convection-diffusion term to break up the air layer.

Acknowledgements

We thank the Netherlands Organization for Scientific Research NWO, by whom this work is financially supported.

References

[1] Reboud, J., Stutz, B. and Coutier, O. (1998). *Two-phase Flow Structure of Cavitation: Experiment and modelling of Unsteady Effects*. Third International Symposium on Cavitation, Grenoble, France

[2] Coutier-Delgosha, O., Fortess-Patella, R. and Reboud, J.L. (2003). *Evaluation of the turbulence model influence on the numerical simulations of unsteady cavitation*. J. of Fluids Eng. 125(1), 38-45

[3] Rotte, G., Kerkvliet, M. and van Terwisga, T. (2017). *On the Turbulence Modelling for an Air Cavity Interface*. Numerical Towing Tank Symposium, Wageningen, The Netherlands

[4] Rotte, G., Zverkhovskyi, O., Kerkvliet, M. and van Terwisga, T. (2016). *On the physical mechanisms for the numerical modelling of flows around air lubricated ships.* International Conference on Hydrodynamics, Wijk aan Zee, The Netherlands

[5] Zverkhovskyi, O. (2014). *Ship Drag Reduction by Air Cavities*. Ph.D. Thesis, TU Delft

[6] Gorbachev, Y. and Amromin, E. (2012). *Ship Drag Reduction by Hull Ventilation from Laval to Near Future: Challenges and Successes.* ATMA 2012

[7] Shiri, A., Leer-Andersen, M.,Bensow, R.E. and Norrby, J. (2012). *Hydrodynamics of a Displacement Air Cavity Ship*. 29th Symposium on Naval Hydrodynamics, Gothenburg, Sweden

[8] Zverkhovskyi, O., Kerkvliet, M., Vaz, G. and van Terwisga, T. (2015). *Numerical Study on Air Cavity Flows*. Proceedings of NuTTS 2015, Cortona, Italy

[9] Mäkiharju, S.A. (2012). *The Dynamics of Ventilated Partial Cavities over a Wide Range of Reynolds Numbers and Quantitative 2D X-Ray Densitometry for Multiphase Flow*. Ph.D. Thesis, The University of Michigan

[10] ReFRESCO (2017) *http://www.refresco.org*, Online; accessed 23 March 2017

[11] Menter, F.R., Egorov, Y. and Rusch, D. (2006). *Steady and unsteady flow modelling using the* $k - \sqrt{k}L$ *model.* Ichmt Digital Library Online, Begel House Inc.

[12] Starke, B., Raven, H. and Van der Ploeg, A. (2007). *Computation of transom-stern flows using a steady free-surface fitting RANS method,* 9th International Conference on Numerical Ship Hydrodynamics, Ann Arbor, Michigan, USA

© 2018 ASME

10th International Symposium on Cavitation - CAV2018
Baltimore, Maryland, USA, May 14 – 16, 2018

CAV18-05173

On the rebounds of spherical and deformed cavitation bubbles

[1]Outi Supponen, [2]Danail Obreschkow and [1]Mohamed Farhat*

[1]*Laboratory for Hydraulic Machines, École Polytechnique Fédérale de Lausanne, Switzerland*
[2]*International Centre for Radio Astronomy Research, University of Western Australia, Australia*

Abstract

As a single cavitation bubble collapses, its energy is distributed into several powerful phenomena that contribute to its erosive potential. These events are the emission of shock waves that can reach GPa-level pressures, micro-jets reaching speeds of hundreds of ms^{-1}, and the extreme heating up to temperatures in the order of 10,000 K, which may yield light emission called luminescence. Much of what is left of the bubble's energy from its redistribution to these violent collapse effects goes to the formation of a rebound bubble, which will undergo its own collapse and can produce further damage. Here, we experimentally investigate the effect of a bubble's deformation on its rebound using single laser-induced bubbles deformed by gravity. While the driving pressure also affects the rebound dynamics, the bubble deformation dominates over its effect on the rebound beyond a certain level of pressure field anisotropy. Thanks to micro-gravity conditions, we are able to decouple the effects of the driving pressure and the gravity on the rebound.

Keywords: cavitation, bubble, rebound

Introduction

As a cavitation bubble's interface bounces off its violently compressed gaseous contents, which act as a spring during its final collapse stage, a rebound bubble is formed. The rebound bubble is not a damaging 'event' *per se*, yet its characteristics (e.g., geometry, distance to a surface) may differ from those of its mother bubble, and its subsequent collapse can lead to comparably - if not more - damaging effects. As has previously been established in numerous studies, the main micro-jet characteristics such as the jet speed, impact timing, bubble displacement, jet volume [1,2], as well as the shock wave strength (peak pressure and energy) [3,4] and the luminescence energy [5,6], vary as a function of the bubble's level of deformation. We quantify this with an anisotropy parameter ζ, which represents a dimensionless equivalent of the linear momentum of the liquid at the bubble's collapse (Kelvin impulse) [7] and which is defined for different sources of bubble deformation, such as near surfaces and a uniform pressure gradient [1,2]. With an increasing bubble deformation, i.e., with increasing ζ, the jet speed decreases and so does the resulting water hammer pressure and shock energy (up to $\zeta \sim 0.1$) [4]. We therefore expect also the rebound energies to vary with ζ. Previous studies have suggested larger rebounds for more deformed bubbles as less energy is lost to the shock wave emission [8,9]. Here, we experimentally investigate the direct role of a bubble's deformation on the formation of its rebound bubble.

For bubbles collapsing in a pressure gradient, such as the one induced by gravity, the deformation-quantifying anisotropy parameter reads $\zeta = \nabla p R_0 \Delta p^{-1}$, where ∇p is the pressure gradient ($= \rho g$ for gravity, where ρ is the liquid density and g the gravitational acceleration), R_0 is the maximum radius of the initial bubble and Δp is the driving pressure ($= p_0 - p_v$, where p_0 is the static pressure of the liquid and p_v is the vapour pressure). For comparison with bubbles collapsing near a flat free or rigid surface, the anisotropy parameter can be expressed in terms of the dimensionless stand-off distance of the bubble to the surface $\gamma = h/R_0$ (where h is the smallest distance between the initial bubble centre and the surface) as $\zeta = 0.195\gamma^{-2}$ [1]. In the following, we investigate the radial evolution of spherical bubbles and bubbles deformed by gravity.

Experimental setup

The central elements of our experimental setup are illustrated in the schematic of Figure 1(a). The cavitation bubble is generated by means of a pulsed laser (Nd:YAG, 532 nm, 8 ns), focused by an immersed parabolic mirror in the middle of a cubic test chamber (18 cm×18 cm×18 cm) filled with demineralised and partially degassed water. The initial plasma is point-like and thereby generates initially so spherical bubbles that, when they collapse without any external sources of deformation, we may easily observe luminescence from a single collapse in images and in measured spectra [6]. The timing of the bubble generation and the first and second bubble oscillations are recorded by a pre-amplified piezoelectric needle hydrophone (Precision Acoustics, 75 μm sensor), which detects the passage of the corresponding shock waves and is connected to an oscilloscope. The needle is sufficiently thin

*Corresponding author, Mohamed Farhat: mohamed.farhat@epfl.ch

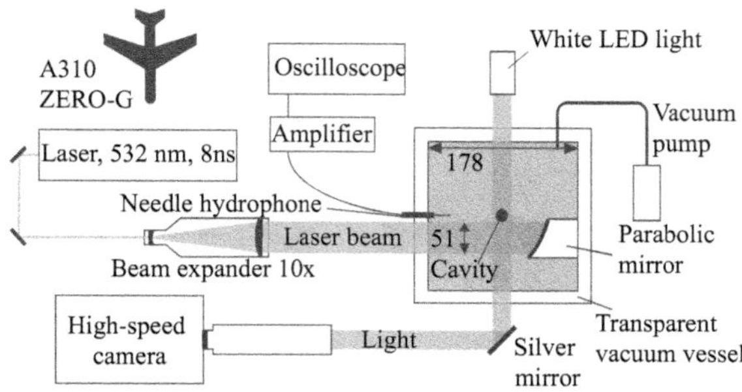

Figure 1: Schematic of the experimental setup and direction with respect to aircraft.

(0.3 mm diameter) and far away (33 mm from bubble centre) to justify the assumption of its negligible effect on the bubble dynamics. We have previously used this hydrophone to quantify the energy and the peak pressure of the various shock waves from collapsing bubbles to study the effect of bubble deformation on them [4]. We observe the bubble dynamics through high-speed imaging (Shimadzu HPV-X2, up to 10 million f/s), and obtain the radii of the various bubble oscillations by fitting a circle on the obtained images using automated image processing. In order to verify the accuracy of the measured bubble radii, as obtaining an accurate measure from images of the rebounds of strongly jetting bubbles with complex geometries becomes troublesome, we compare the image-obtained radii with the equivalent radii computed with the shock timings recorded by the hydrophone through the Rayleigh collapse time $T_{\rm C} = 0.915 R_{\rm B}(\rho/\Delta p)^{1/2}$, where $R_{\rm B}$ is the maximum radius of the initial or the rebound bubble. The differences in the rebound radii from images and hydrophone remain below 5% for the cases covered in this study ($\zeta < 10^{-2}$).

By adjusting the laser energy and the static pressure inside the test chamber using a vacuum pump, we can create bubbles of maximum initial radii within the range $R_0 = 0.5$–10 mm collapsed by a driving pressure within the range $\Delta p = 8$–100 kPa. The experiments were conducted during the 62$^{\rm nd}$ and 67$^{\rm th}$ European Space Agency (ESA) parabolic flight campaigns, where we could vary the perceived gravitational acceleration within the range 0–2×9.81 ms^{-2}. The interest in deforming the bubbles by the gravity-induced uniform pressure gradient is that the rebound bubble feels the same pressure gradient than the initial oscillation, which is not the case for bubbles near boundaries. It is also a convenient method to carefully probe between spherical and weakly deformed bubbles in a highly controlled way. Further details on the experimental setup may be found in ref. [10].

Results

Figure 2 displays, in nondimensional coordinates, the radial evolution of various bubbles with different driving pressures and maximum radii oscillating spherically in micro-gravity. It is clear from this figure that the driving pressure Δp plays an important role in the rebound formation, which has already been reported in ref. [11]. However, the highest energy fraction of the rebound with respect to the initial bubble energy (the cyan and blue curves) in this figure is only 7% (the bubble energy being computed as $E_{\rm B} = (4\pi/3)R_{\rm B}^3\Delta p$). Another interesting observation from this figure is that the blue ($R_0 = 3.90$ mm) and the cyan ($R_0 = 7.44$ mm) curves have been obtained at the same driving pressure Δp but have very different sizes, yet the multiple rebounds up to the fourth bubble oscillation are almost identical, indicated by these curves being superimposed. Therefore, the absolute bubble size does not seem to affect the rebound formation the range covered here.

Figure 3 shows, in normalised coordinates, the radial evolution of three bubbles at different perceived levels of gravity, with the subfigures 3(a), 3(b) and 3(c) each having a different Δp and input laser energy. Figure 4 displays a sequence of images of bubbles in the same conditions except for gravity. The maximum radii are given as approximations since they experience a slight variation depending on the perceived gravity level, in particular for Figure 3(c) where the maximum radii are $R_0 = 3.90$ mm for micro-gravity, $R_0 = 3.16$ mm for normal gravity and $R_0 = 2.55$ mm for hyper-gravity despite the same input laser energy. This is due to the bubble growth also being affected by the pressure gradient. However, the absolute bubble size should not affect the relative rebound size, as concluded from Figure 2. At lower Δp, the bubble's dynamics is more subject to the pressure gradient induced by gravity, which is directly implied by the anisotropy parameter ζ which comprises Δp. As Figure 3 shows, there is a clear effect of gravity on the rebound formation, as the bubble collapsing in normal and hyper-gravity forms a relatively larger rebound compared to the bubble collapsing in micro-gravity, despite all other experimental parameters being kept the same (also visible in the images in Figure 4). This difference becomes more important at lower Δp, as the anisotropy in the pressure field increases, i.e., when ζ increases. The

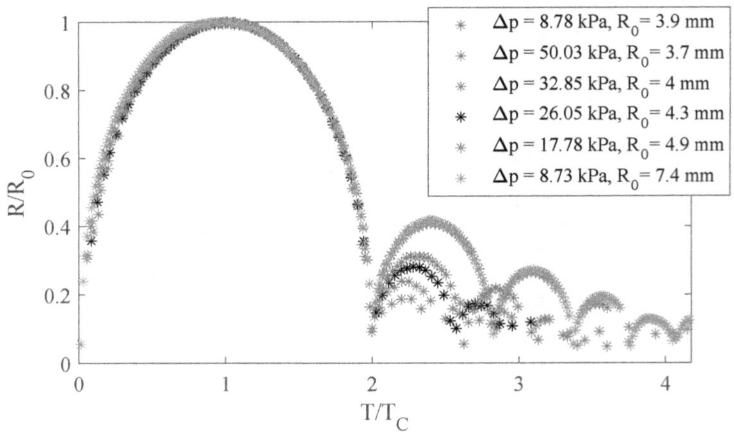

Figure 2: Observed bubble radius evolution in normalised coordinates (radius R normalised to maximum bubble radius R_0 and time T normalised to collapse time $T_C = 0.915R_0(\rho/\Delta p)^{1/2}$) for bubbles with various maximum radii R_0 and driving pressures Δp collapsing at micro-gravity conditions ($\zeta < 1.5 \times 10^{-3}$).

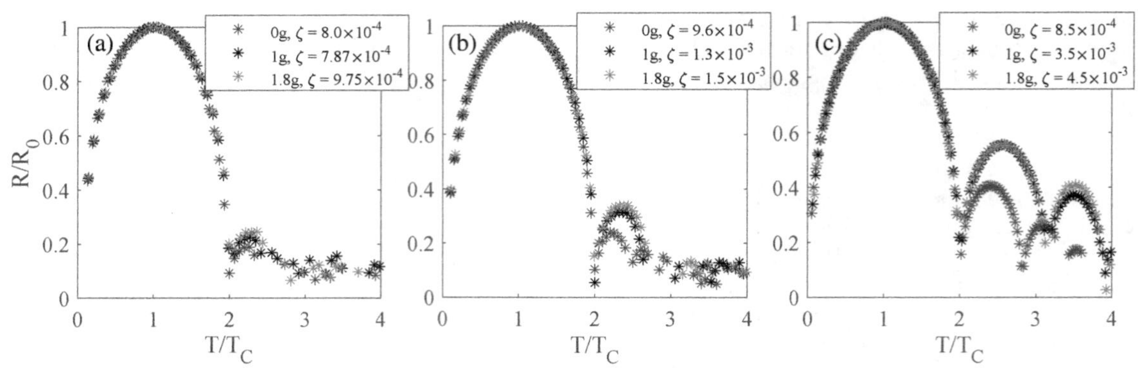

Figure 3: Observed bubble radius evolution in normalised coordinates (radius R normalised to maximum bubble radius R_0 and time T normalised to collapse time $T_C = 0.915R_0(\rho/\Delta p)^{1/2}$) for bubbles collapsing at three different gravity levels: 0 g, 1 g, and 1.8 g, where $g = 9.81$ ms^{-2}. (a) $R_0 \approx 3.5$ mm and $\Delta p = 50$ kPa, (b) $R_0 \approx 3.8$ mm and $\Delta p = 33$ kPa, and (c) $R_0 \approx 3$ mm and $\Delta p = 9$ kPa.

non-zero ζ for the bubble in micro-gravity is explained by the nearest surface (parabolic mirror at 55 mm from bubble) being taken into account when determining ζ. In Figure 3(c), the rebounds of the bubbles in normal and in hyper-gravity both take 17% of the bubble's initial energy, while in micro-gravity the energy fraction is as low as 7%. Here, there is not an observable difference between the first rebounds in normal and hyper-gravity, but the difference in the initial deformation is seen in the second rebound, which is larger in hyper-gravity compared to normal gravity. Overall, the highest energy fractions of the rebounds obtained from bubbles in hyper-gravity at very low pressures and high deformations ($\zeta \sim 10^{-2}$) were as high as 30%. Therefore, the bubble deformation clearly has a substantial effect on the rebound formation.

Discussion

The formation of the rebound bubble naturally also depends on the gas inside the collapsing bubble, which we have not varied in a controlled way in this work. Varying Δp likely also varies this gas content, which varies the energy partition into the rebound for spherical collapses [11], and could be the reason for lower Δp yielding relatively bigger rebounds in Figure 2. Indeed, the gas inside the bubble, in our experiment, likely comprises i) laser-generated gas, as has been reported in the past [12] - we assume this gas pressure to be proportional to the energy deposited by the laser to generate the bubble; ii) non-condensible gas, for which we assume the partial pressure to be proportional to the bubble volume; iii) diffused gas, for which we assume the partial pressure to be proportional to the total exposed bubble surface being covered during its lifetime; and iv) vapour, the partial pressure of which is assumed to be the vapour pressure of the liquid, p_v, during most of the bubble's lifetime, but which may act as a non-condensible gas in the final collapse stages due to its condensation rate not being

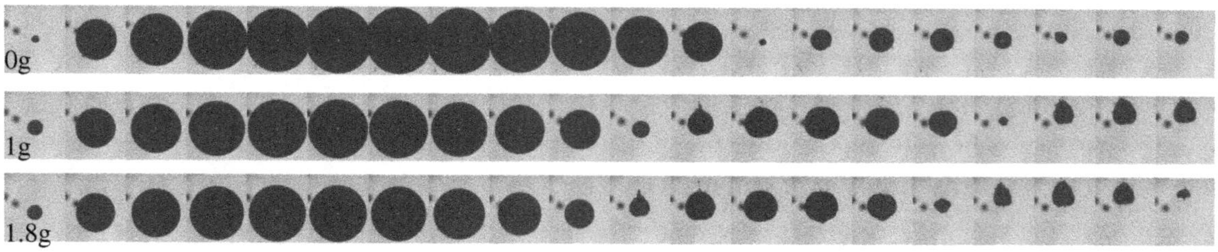

Figure 4: Visualisation of the first three oscillations of a bubble at micro-gravity (top), normal gravity (middle) and hyper-gravity (bottom) at $\Delta p = 11$ kPa. The frame rate is 250 μs, exposure time 100 ns, and the maximum bubble radius of the bubble in 1.8 g is 4.4 mm.

able to keep up with the bubble's rapid reduction in volume [13]. Much more thorough modeling is needed to obtain proper predictions of the gas effects on the rebound dynamics. However, as $\zeta > 10^{-3}$ (equivalently, when $\gamma < 14$) - which we have previously defined as being the limit between the *weak* and *intermediate jet regimes* [1], i.e., pressure field anisotropy limit below which the bubbles appear mostly spherical and above which jets can be observed - the effect of deformation starts to dominate the rebound dynamics over the effect of the different Δp, as demonstrated in Figure 3. This result implies that care should be taken with theoretical models aiming to accurately model the rebound dynamics of bubbles that are not perfectly spherical, which is the case with most bubbles in reality. Furthermore, if the rebound bubble is big enough and is able to undergo its own violent collapse, it might be the rebound that is the most damaging to, for example, a nearby boundary. It has been shown that for bubbles collapsing at a certain range of distances to a solid boundary ($0.5 < \gamma < 2$), the shocks from the rebound collapse may carry a comparable amount of energy to the fist collapse. In fact, the rebound collapse shocks were even found to be higher than the first collapse between $0.75 < \gamma < 1.4$ [3]. In this range, the rebound bubbles are large enough thanks to the sufficient deformation of the initial bubble, and spherical enough to be sufficiently compressed to emit relatively strong shocks upon their collapse. In addition, if the rebound bubbles find themselves closer to a rigid boundary due to the bubble migration towards it, the damaging effects may be even stronger. Therefore, the prediction of the rebound bubble's behaviour is invaluable in determining the harm or the benefit that collapsing bubbles can offer. It would be interesting, in the future, to understand the rebound's dependence on the origin of the bubble's deformation and to define and evaluate a parameter describing the anisotropy felt at the second bubble collapse.

Conclusion

The present study presents experimental observations to uncover the direct effect of a collapsing bubble's deformation on the formation of a rebound bubble. The main outcome is that both lower driving pressures and the bubble deformation yield relatively bigger rebounds.

Acknowledgements

We gratefully acknowledge the support of the Swiss National Science Foundation (Grant no. 200020-144137), the University of Western Australia Research Collaboration Award obtained by D.O. and M.F., and the European Space Agency.

References

[1] Supponen O., Obreschkow D., Kobel P., Tinguely M., Dorsaz N., Farhat M. (2016). *Scaling laws for jets of single cavitation bubbles.* Journal of Fluid Mechanics. **802** 263.

[2] Obreschkow D., Tinguely M., Dorsaz N., Kobel P., De Bosset A., Farhat M. (2011). *Universal Scaling law for Jets of Collapsing Bubbles.* Physical Review Letters. **107**(20) 204501.

[3] Vogel A., Lauterborn W. (1988). *Acoustic transient generation by laser-produced cavitation bubbles near solid boundaries.* Journal of the Acoustical Society of America. **84**(2) 719.

[4] Supponen O., Obreschkow D., Kobel P., Tinguely M., Dorsaz N., Farhat M. (2017). *Shock waves from nonspherical cavitation bubbles.* Physical Review Fluids. **2**(9) 093601.

[5] Ohl C.-D. (2002). *Probing luminescence from nonspherical bubble collapse.* Physics of Fluids. **14**(8) 2700.

[6] Supponen O., Obreschkow D., Kobel P., Farhat M. (2017). *Luminescence from cavitation bubbles deformed in uniform pressure gradients.* Physical Review E. **96**(3) 033114.

[7] Blake J.R. (1988). *The Kelvin impulse: Application to cavitation bubble dynamics.* Journal of Australian Mathematical Society Series B. **30**(2) 127.

[8] Zhang A. M., Cui P., Cui J., Wang Q. X. (2015). *Experimental study on bubble dynamics subject to buoyancy.* Journal of Fluid Mechanics. **776** 137.

[9] Wang Q. (2016). *Local energy of a bubble system and its loss due to acoustic radiation.* Journal of Fluid Mechanics. **797** 201.

[10] Obreschkow D., Tinguely M., Dorsaz N., Kobel P., De Bosset A., Farhat M. (2013). *The quest for the most spherical bubble: Experimental setup and data overview.* Experiments in Fluids. **54**(4) 1503.

[11] Tinguely M, Obreschkow D., Kobel P., Dorsaz N., De Bosset A., Farhat M. (2012). *Energy partition at the collapse of spherical cavitation bubbles.* Physical Review E. **86**(4) 046315.

[12] Sato T., Tinguely M., Oizumi M., Farhat M. (2013). *Evidence for hydrogen generation in laser- or spark-induced cavitation bubbles.* Applied Physics Letters. **102**(7) 074105.

[13] Fujikawa S., Akamatsu T. (1980). *Effects of the non-equilibrium condensation of vapor on the pressure wave produced by the collapse of a bubble in a liquid.* Journal of Fluid Mechanics. **97** 481.

© 2018 ASME

Phase Field/Fluctuating Hydrodynamics approach for bubble nucleation

[1]Mirko Gallo; [1]Francesco Magaletti; [1] Carlo Massimo Casciola*

[1]*Dept. of Mechanical and Aerospace Engineering, Sapienza University of Rome, Roma, Italy*

Abstract

Vapour bubbles form in liquids by two main mechanisms: boiling, by increasing the temperature over the boiling threshold, and cavitation, by reducing the pressure below the vapour pressure threshold. The liquid can be held in these metastable states (overheating and tensile conditions, respectively) for a long time without forming bubbles. Bubble nucleation is indeed an activated process, requiring a significant amount of energy to overcome the free energy barrier and bring the liquid from the metastable conditions to the thermodynamically stable state where vapour is observed. Nowadays molecular dynamics is the unique tool to investigate such thermally activated processes. However, its computational cost limits its application to small systems (less than few tenth of nanometers) and to very short times, preventing the study of hydrodynamic interactions. In this work a continuum diffuse interface model of the two-phase fluid has been embedded with thermal fluctuations in the context of the so-called Fluctuating Hydrodynamics (FH), enabling the description of the liquid-vapour transition in extended systems and the evaluation of bubble nucleation rates in different metastable conditions by means of numerical simulations. Such an approach is expected to have a huge impact on the understanding of the nucleation dynamics since, by reducing the computational cost by orders of magnitude, it allows the unique possibility of investigating systems of realistic dimensions on macroscopic time scales.

Keywords: boiling; cavitation; diffuse interface; fluctuating hydrodynamics; thermal fluctuations

Introduction

Thermal fluctuations play a dominant role in mesoscopic phenomena occurring in fluids. Since the pioneering work of Landau and Lifshitz in the late fifties [1], several models have been developed [2] to properly include stochastic terms in the governing equations contributing to the growing field of "Fluctuating Hydrodynamics" (FH). Also motivated by applications, these efforts were followed by the exponential increase of specialised numerical methods for stochastic differential equations [3-5]. Fluctuations are crucial in a number of contexts, e.g., among others, flow-driven micro-devices, biological and lipid membranes [6], systems exploiting ratchet motion [7] like Brownian engines and artificial molecular motors [8, 9]. Presumably, nucleation - the precursor of the phase change in metastable systems - is however the most common fluctuation-dominated phenomenology found in applications. Examples are cavitation [10], freezing rain [11], or the intrusion of liquids in microporous materials [12], to cite a few. Since the energy barriers for phase transition is overcome by thermal fluctuations [13-17], the nucleation time may be long and the phenomenon is labeled as a "rare event". Such problems are typically addressed by molecular dynamics (MD) simulations [18] which, for their computational cost, are limited to systems too small to rise a direct technological interest.

Here homogeneous vapour bubble nucleation in a metastable liquid is addressed through a novel approach based on a diffuse interface description [19] of the two-phase system embedded with thermal fluctuations [20] that can describe both the thermodynamic and the fluid dynamics fields [21, 22], see [23] for applications to the spinodal decomposition in a liquid--vapour system.

Mathematical model

The model is based on the Van der Waals's square gradient approximation [19] of the Helmholtz free energy

$$F\left[\rho,\theta\right] = \int_V dV \left(f_0\left(\rho,\theta\right) + \frac{1}{2}\lambda \nabla\rho \cdot \nabla\rho \right), \qquad (1)$$

*Corresponding Author, Carlo Massimo Casciola: carlomassimo.casciola@uniroma1.it where f_0 is the classical bulk term, function of density ρ and temperature θ, and λ is related to the (equilibrium) surface tension γ and interface thickness [1,2].

The (fluctuating) hydrodynamic fields are governed by mass, momentum and energy conservation, with the addition of stochastic contributions:

$$\frac{\partial \rho}{\partial t} + \nabla \cdot \left(\rho \mathbf{u}\right) = 0\,,$$

$$\frac{\partial \rho\,\mathbf{u}}{\partial t} + \nabla \cdot \left(\rho\,\mathbf{u} \otimes \mathbf{u}\right) = -\,\nabla p + \nabla \cdot \boldsymbol{\Sigma} + \nabla \cdot \boldsymbol{\delta \Sigma}\,,$$

$$\frac{\partial E}{\partial t} + \nabla \cdot (\mathbf{u}E) = \nabla \cdot \left(-p\mathbf{u} + \mathbf{u} \cdot \boldsymbol{\Sigma} - \mathbf{q}\right) + + \nabla \cdot \left(\mathbf{u} \cdot \boldsymbol{\delta \Sigma} - \boldsymbol{\delta q}\right)\,,$$

where $\mathbf{u}$ is the fluid velocity, $p = \rho^2 \partial(f_0/\rho)/\partial\rho$ the pressure and $E = \mathcal{U} + 1/2\rho\,|\,\mathbf{u}\,|^2 + 1/2\lambda\,|\,\nabla\rho\,|^2$ the energy density, with $\mathcal{U}$ the internal energy per unit volume. $\boldsymbol{\Sigma}$ and $\mathbf{q}$ are the deterministic stress tensor and energy flux, respectively. The prefix δ denotes the stochastic contributions, which statistical properties follow by enforcing the *fluctuation-dissipation balance* (FDB):

$$\left\langle \boldsymbol{\delta \Sigma}(\hat{x},\hat{t}\,) \otimes \boldsymbol{\delta \Sigma}^{\dagger}(\tilde{x},\tilde{t}\,)\right\rangle = 2\mathrm{k}_B\theta\mu\left(\delta_{\alpha\nu}\delta_{\beta\eta} + \delta_{\alpha\eta}\delta_{\beta\nu} - 2/3\delta_{\alpha\beta}\delta_{\nu\eta}\right)\delta(\hat{x} - \tilde{x})\delta(\hat{t} - \tilde{t}\,),$$

$$\left\langle \boldsymbol{\delta q}(\hat{x},\hat{t}\,) \otimes \boldsymbol{\delta q}^{\dagger}(\tilde{x},\tilde{t}\,)\right\rangle = 2\mathrm{k}_B\theta^2 k\,\delta_{\alpha\beta}\delta(\hat{x} - \tilde{x})\delta(\hat{t} - \tilde{t}\,),$$

with μ and k the dynamic viscosity and the thermal conductivity, respectively. The cross-correlation between tensors of different rank vanishes due to the Curie-Prigogine principle, $\left\langle \delta\mathbf{q}^{\dagger}(\tilde{x},\tilde{t}\,) \otimes \delta\boldsymbol{\Sigma}(\hat{x},\hat{t}\,)\right\rangle = 0$. $\boldsymbol{\Sigma}$ and $\boldsymbol{q}$ follow from standard non-equilibrium thermodynamics considerations [1,2] as

$$\boldsymbol{\Sigma} = \left(\frac{\lambda}{2}|\nabla\rho|^2 + \rho\nabla\cdot(\lambda\nabla\rho)\right)\boldsymbol{I} - \lambda\nabla\rho\otimes\nabla\rho + +\mu\left[\left(\nabla\mathbf{u} + \nabla\mathbf{u}^T\right) - \frac{2}{3}\nabla\cdot\mathbf{u}\boldsymbol{I}\right],$$
$$\boldsymbol{q} = \lambda\rho\nabla\rho\nabla\cdot\mathbf{u} - k\,\nabla\theta\,.$$

Bubble-Nucleation Simulations

Homogeneous vapour bubble nucleation in a metastable liquid enclosed in a cubic box with periodic boundary conditions is studied by numerically solving the equations presented in the previous section. The equation of state corresponds to a Lennard-Jones fluid [24] to allow comparison with available MD results. The system volume $V* = (600)^3$ is discretised on a uniform grid with 50 cells per direction and several metastable conditions are investigated. Here we report the *long-time dynamics* evolving from the metastable thermodynamic condition $\theta = 1.25\,\rho_L = 0.46$ where strong hydrodynamic interactions, e.g., bubble collapse, coalescence and expansion, take place during nucleation. A few snapshots of the system evolution in the different metastable conditions are shown in the left panels of Fig. 1. Starting from the homogeneous liquid, fluctuations lead the system to nucleate the vapour phase. The initial shape of the nuclei is far from spherical, as observed in MD [18]. After reaching the critical size, the bubbles start expanding and the system stabilises in a state characterised by several vapour bubbles in equilibrium with the surrounding liquid. Bubble number and dimensions is dictated by the initial metastable condition. FH provides access the bubbles long-time dynamics in a very large system, taking in full account their hydrodynamic interactions. Before addressing the nucleation dynamics in detail, it is worth introducing the notion of critical bubble, in order to be able to identify the genuinely nucleated bubbles and their long term evolution.

The critical bubble

It is known that bubbles smaller than critical shrink and disappear, while supercritical one may grow to macroscopic size. The demarcation line is the critical bubble which may be determined from the Helmholtz free energy, eq. (1). Minimising F identifies the critical state where the generalised chemical potential $\mu_c = \mu_c^b(\rho) - \lambda \nabla^2 \rho$ equals the external chemical potential μ_{eq}. The critical density profile under spherical symmetry can be found by the string method proposed in [25], see also [26], which identifies both the critical nucleus and the transition path joining the metastable liquid condition to the cavitated state. The method consists in initialising a set of N_s density fields $\{\rho^k(r)\}$, $k = 1,\ldots, N_s$, the so-called *string,* joining the metastable liquid state and the critical state. The head of the string is the uniform density of the liquid $\rho^1(r) = \rho_L$; the tail, $\rho^{N_s}(r)$, is initialised as a tanh profile approximating the critical nucleus which is obtained from classical nucleation theory (CNT) [27]; the intermediate fields are initialised by interpolation of these two density profiles, $\rho^k(r) = (1 - k/N_s)\rho^1(r) + k/N_s \rho^{N_s}(r)$. The string algorithm involves two steps. 1) All the images $\rho^k(r)$ are evolved over one pseudo-time step $\Delta\tau$ following the steepest-descent algorithm consisting in evolving the following relaxation dynamics in the pseudo-time

$$\frac{\partial \rho}{\partial \tau} = \mu_{eq} - \left[\mu_c^b(\rho) - \frac{\lambda}{r^2}\frac{\partial}{\partial r}\left(r^2 \frac{\partial \rho}{\partial r}\right)\right] .$$

Special attention is dedicated to the tail of the string that needs to climb the energy landscape up to the saddle point. An extra driving force, directed in the direction tangential to the string at the tail, is applied to allow the climbing motion. 2) The images are redistributed along the string following a re-parameterisation procedure by equal arc length. The algorithm is arrested when the string converges within a prescribed tolerance thereby identifying the transition path as the sequence of density fields along the string. Critical radius and free energy barrier $R_c = \int_0^\infty r(\partial\rho_c/\partial r)^2 r^2\,\mathrm{d}r / \int_0^\infty (\partial\rho_c/\partial r) r^2\,\mathrm{d}r$ [28] and $\widetilde{\Delta\Omega} = \int_0^\infty \left[w(\rho_c(r)) - w(\rho_L)\right] 4\pi r^2 \mathrm{d}r$, respectively, follow from the critical density profile $\rho_c(r)$. In the following sections the critical bubble, and its size in particular, will be used to discriminate whether the bubbles found in the FH simulations to be discussed are sub- or supercritical, thus allowing to evaluate the nucleation rate as the number of supercritical bubbles formed per unit volume and time.

Bubble Kinematics

In order to extract the bubble kinematics and track bubble motions, the evolution of the density iso-surface $\rho[(\mathbf{x}(\eta,t),t)] = \rho_0$, representing the nominal interface represented as $\mathbf{x} = \mathbf{x}(\eta,t)$, is considered. Its time derivative $d\rho[(\mathbf{x}(\eta,t),t)]/dt = \partial\rho/\partial t + \dot{\mathbf{x}}\cdot\nabla\rho = 0$ defines the interface velocity $\mathbf{w}(\mathbf{x},t) := \dot{\mathbf{x}} = -\,\partial\rho/\partial t\,\nabla\rho/|\nabla\rho|^2$. Hence the evolution of a generic physical quantity $\phi(\mathbf{x},t)$ related the bubble B_i is given by

$$\dot{\Phi} = \frac{d}{dt}\int_{\mathbf{B}_i(t)} \phi(\mathbf{x},t)\, dV = \frac{d}{dt}\int_{\mathbf{B}_i^\eta} \phi\left(\mathbf{x}(\eta,t),t\right) \mathbf{J}^\eta\, dV^\eta = \int_{\mathbf{B}_i(t)} \frac{D^{(w)}}{Dt}\phi + \phi\nabla\cdot\mathbf{w}\, dV, \qquad (2)$$

where $D^{(w)}/Dt = \partial/\partial t + \mathbf{w}\cdot\nabla$ is the total derivative following the $\mathbf{w}$ velocity field and $\mathbf{J}^\eta = |\partial\mathbf{x}/\partial\eta|$ is the Jacobian determinant, with derivative $\dot{\mathbf{J}}^\eta = \mathbf{J}^\eta\nabla\cdot\mathbf{w}$. Defined the peculiar velocity $\mathbf{w}_i'(\mathbf{x},t) := \mathbf{w}_i(\mathbf{x},t) - \mathbf{u}(\mathbf{x},t)$ and making use of eq. (2) with $\phi(\mathbf{x},t) = \rho(\mathbf{x},t)$ combined with mass conservation $\partial\rho/\partial t + \nabla\cdot(\rho\mathbf{u}) = 0$, the bubble mass is found to obey

$$\dot{M}_{B_i} = \frac{d}{dt}\int_{\mathbf{B}_i(t)} \rho(\mathbf{x},t)\, dV = \oint_{\partial\mathbf{B}_i(t)} \rho(\mathbf{x},t)\, \mathbf{w}_i'(\mathbf{x},t)\cdot\mathbf{n}\, dS,$$

with $\mathbf{B_i(t)}$ the bubble domain. The bubble centre of mass velocity is then

$$\dot{\mathbf{X}}_{g_i} = \frac{1}{M_{B_i}}\frac{d}{dt}\int_{\mathbf{B_i(t)}} \mathbf{x}\rho(\mathbf{x},t)\, dV - \frac{\dot{M}_{B_i}}{M_{B_i}}\mathbf{X}_{g_i} = \frac{1}{M_{B_i}}\int_{\mathbf{B_i(t)}} \rho(\mathbf{x},t)\mathbf{u}(\mathbf{x},t) - \nabla\cdot\mathbf{G}_i\left(\mathbf{x},t,\mathbf{X}_{g_i}\right) dV,$$

where $\mathbf{G}_i\left(\mathbf{x},t,\mathbf{X}_{g_i}\right) = \rho\left(\mathbf{x},t\right)\left(\mathbf{x}-\mathbf{X}_{g_i}\right) \otimes \left(\partial\rho/\partial t \, \nabla\rho/|\nabla|^2 - \mathbf{u}\left(\mathbf{x},t\right)\right)$. Similarly the time derivative of the bubble volumes reads $\dot{V}_{B_i} = \oint_{\partial \mathbf{B_i(t)}} \mathbf{w}_i^{'}(\mathbf{x},t)\cdot\mathbf{n}\, dS$.

On the basis of the above analysis, an algorithm can be implemented to identify and track the distinct bubbles and their properties. First, a *cell-flagging* procedure based on the local density is used to identify the grid cells considered as non-liquid (cut-off density $\rho_{cut} = \rho_c(R_c)$, corresponding to the critical density at the nominal radius of the critical bubble). A clustering technique based on a region growing procedure is then applied to the non-liquid cells to determine when a neighbouring cell should be added to the current cluster. The process is iterated until all the detected vapour (non-liquid) cells has been clustered into distinct bubbles. The volume of the *k*-th bubble is used as the order parameter to identify the supercritical bubbles defined as those which volume exceeds the critical volume. For each bubble present in the field, this procedure allows to obtain mass centre trajectory and volume evolution and to identify the coalescence events, see the left panel of Fig. 1.

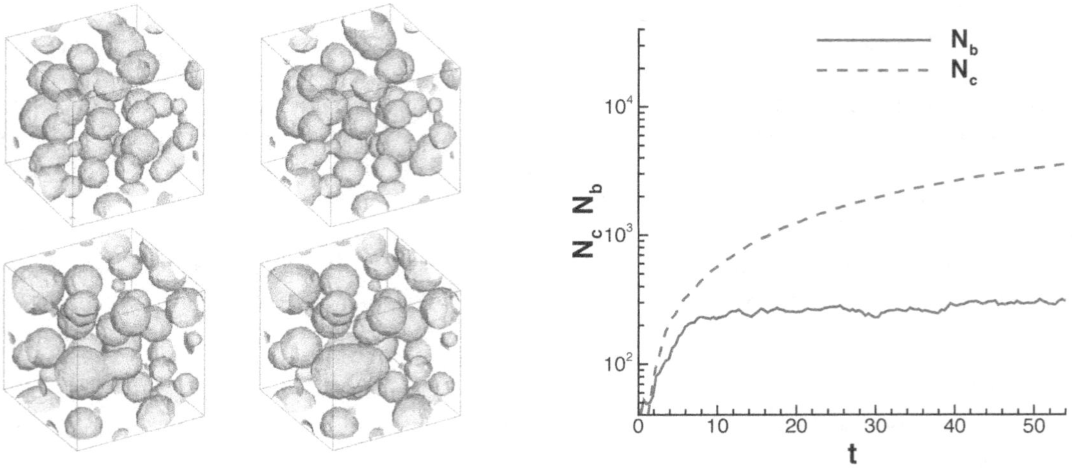

Figure 1. Left panel: snapshots of bubble configurations during the nucleation process to highlight the coalescence process. Right panel: Time evolution of the number of supercritical bubbles and the total collapsed supercritical bubble up to time instant t.

The central observable is the nucleation rate J, defined as the number of supercritical bubbles formed per unit volume and time. CNT [27] provides an estimate of the energy barrier $\widetilde{\Delta\Omega}_{CNT}$ and consequently of the nucleation rate as $J_{CNT} = n_L\sqrt{2\gamma/m\,\pi}\,\exp(-\,\widetilde{\Delta\Omega}_{CNT}/k_B\theta)$, where n_L is the liquid number density. CNT can be compared with the nucleation rate J_{FH} from FH simulations, estimated as the slope of the linear initial part of the curve giving the number of supercritical bubbles N_b vs time in the right panel of Fig. 1.

Discussion of the results

As known from the literature, CNT over-estimates the energy barrier near the spinodal condition and tends to behave better at lower metastabilities. Consistently the barrier obtained from the string method are smaller than CNT ones, Table 1. Notwithstanding this point the rates, J, evaluated with CNT are puzzlingly larger than those from FH. In order to correctly understand this point, it is crucial to take into account the confinement affecting a constant mass/ constant volume system. In these conditions, the larger the number of nucleated bubbles the more the liquid is compressed. As a result the nucleation process is discouraged and consequently the nucleation rate is reduced, thus explaining the lower nucleation rate of FH with respect to CNT. In fact, by tracking all the vapour clusters in the system during the initial nucleation stage (characterised by constant rate) we observed that the number of collapsed bubbles after crossing the critical size is of the same order of (and often greater than) the number of the "survived" ones. In fact, if all the bubbles survived we would have obtained a rate larger than predicted by CNT, consistently with the reduced free energy barrier. This point is illustrated in the right part of Fig. 1 where the evolution of the number of supercritical bubbles N_b and of the number of supercritical bubbles N_c collapsed up to current time is

$\widetilde{\Delta\Omega}/\theta_0$	$\widetilde{\Delta\Omega}_{CNT}/\theta_0$	J_{FH}	J_{CNT}	J_{FH}^{tot}
5.90	14.05	$1.3 \cdot 10^{-8}$	$7.0 \cdot 10^{-8}$	$3.2 \cdot 10^{-7}$

Table 1. Comparison between CNT and FH rate, the free energy barrier for bubble nucleation from the liquid is evaluated with the string method.

shown. In absence of collapses, the total number of bubbles in the system would have been $N_{tot} = N_b + N_c$ and the rate would have been larger by roughly a factor 10^2, consistently with the lower free energy barriers reported in Table 1.

Conclusion

In conclusion, we have shown that a Van der Walls/Navier-Stokes/Landau-Lifshitz model can properly capture the vapour bubble nucleation process. The model works equally well in the context of cavitation, where nucleation is induced by pressure decrease, and boiling, where the phase transition is due to increased temperature. Both processes involve significant metastabilities. The continuum description of the two phase system endowed with proper thermal fluctuations satisfying the appropriate form of fluctuation dissipation balance is able to describe the thermally activated transition process and predict the nucleation rate. Unique feature of the model is the ability to deal with extended systems, under non-equilibrium and unsteady conditions like those most often encountered in applications involving liquid-vapour phase transition. Particularly significant is the interaction of the microscopic process of nucleation with the hydrodynamic scales that couples in both directions the phase transition to the macroscopic flow.

Acknowledgments

The research leading to these results has received funding from the European Research Council under the European Union's Seventh Framework Programme (FP7/ 2007-2013)/ERC Grant agreement no. [339446].

References

1.Landau, L. and E. Lifshitz, *Statistical physics, vol. 5.* Course of theoretical physics, 1980. **30**.

2.Fox, R.F. and G.E. Uhlenbeck, Contributions to Non-Equilibrium Thermodynamics. I. Theory of Hydrodynamical Fluctuations. The Physics of Fluids, 1970. **13**(8): p. 1893-1902.

3.Donev, A., et al., *On the accuracy of finite-volume schemes for fluctuating hydrodynamics.* Communications in Applied Mathematics and Computational Science, 2010. **5**(2): p. 149-197.

4.Balboa, F., et al., Staggered schemes for fluctuating hydrodynamics. Multiscale Modeling & Simulation, 2012. **10**(4): p. 1369-1408.

5.Donev, A., et al., *Low Mach number fluctuating hydrodynamics of diffusively mixing fluids.* Communications in Applied Mathematics and Computational Science, 2014. **9**(1): p. 47-105.

6.Naji, A., P.J. Atzberger, and F.L. Brown, Hybrid elastic and discrete-particle approach to biomembrane dynamics with application to the mobility of curved integral membrane proteins. Physical review letters, 2009. **102**(13): p. 138102.

7.Chinappi, M., et al., Molecular dynamics simulation of ratchet motion in an asymmetric nanochannel. Physical review letters, 2006. **97**(14): p. 144509.

8.Peskin, C.S., G.M. Odell, and G.F. Oster, Cellular motions and thermal fluctuations: the Brownian ratchet. Biophysical journal, 1993. **65**(1): p. 316-324.

9.Chow, C.C. and M.A. Buice, Path integral methods for stochastic differential equations. The Journal of Mathematical Neuroscience (JMN), 2015. **5**(1): p. 8.

10.Brennen, C.E., *Cavitation and bubble dynamics*. 2013: Cambridge University Press.

11.Cao, L., et al., Anti-icing superhydrophobic coatings. Langmuir, 2009. **25**(21): p. 12444-12448.

12.Tinti, A., et al., *Intrusion and extrusion of water in hydrophobic nanopores.* Proceedings of the National Academy of Sciences, 2017: p. 201714796.
13.Jones, S., G. Evans, and K. Galvin, *Bubble nucleation from gas cavities—a review.* Advances in colloid and interface science, 1999. **80**(1): p. 27-50.
14.Kashchiev, D. and G. Van Rosmalen, *Nucleation in solutions revisited.* Crystal Research and Technology, 2003. **38**(7‐8): p. 555-574.
15.Giacomello, A., et al., Geometry as a catalyst: how vapor cavities nucleate from defects. Langmuir, 2013. **29**(48): p. 14873-14884.
16.Amabili, M., et al., Wetting and cavitation pathways on nanodecorated surfaces. Soft matter, 2016. **12**(12): p. 3046-3055.
17.Lohse, D. and A. Prosperetti, Homogeneous nucleation: Patching the way from the macroscopic to the nanoscopic description. Proceedings of the National Academy of Sciences, 2016. **113**(48): p. 13549-13550.
18.Diemand, J., et al., Direct simulations of homogeneous bubble nucleation: Agreement with classical nucleation theory and no local hot spots. Physical review E, 2014. **90**(5): p. 052407.
19.Lutsko, J.F., Density functional theory of inhomogeneous liquids. IV. Squared-gradient approximation and classical nucleation theory. The Journal of chemical physics, 2011. **134**(16): p. 164501.
20.Gallo, M., F. Magaletti, and C.M. Casciola, *Fluctuating Hydrodynamics As A Tool To Investigate Nucleation Of Cavitation Bubbles.* International Journal of Computational Methods and Experimental Measurements, 2017. **6**(2): p. 345-357.
21.Magaletti, F., L. Marino, and C.M. Casciola, Shock Wave Formation in the Collapse of a Vapor Nanobubble. Physical Review Letters, 2015. **114**(6): p. 064501.
22.Magaletti, F., et al., Shock-induced collapse of a vapor nanobubble near solid boundaries. International Journal of Multiphase Flow, 2016. **84**: p. 34-45.
23.Chaudhri, A., et al., Modeling multiphase flow using fluctuating hydrodynamics. Physical Review E, 2014. **90**(3): p. 033014.
24.Johnson, J.K., J.A. Zollweg, and K.E. Gubbins, *The Lennard-Jones equation of state revisited.* Molecular Physics, 1993. **78**(3): p. 591-618.
25.Weinan, E., W. Ren, and E. Vanden-Eijnden, Simplified and improved string method for computing the minimum energy paths in barrier-crossing events. The Journal of chemical physics, 2007.
26.Giacomello, A., et al., Mechanism of the Cassie-Wenzel transition via the atomistic and continuum string methods. The Journal of chemical physics, 2015. **142**(10): p. 104701.
27.Blander, M. and J.L. Katz, *Bubble nucleation in liquids.* AIChE Journal, 1975. **21**(5): p. 833-848.
28.Dell'Isola, F., H. Gouin, and G. Rotoli, Nucleation of spherical shell-like interfaces by second gradient theory: numerical simulations. arXiv preprint arXiv:0906.1897, 2009.
29.Magaletti, F., et al., The sharp-interface limit of the Cahn–Hilliard/Navier–Stokes model for binary fluids. Journal of Fluid Mechanics, 2013. **714**: p. 95-126.

© 2018 ASME

10th International Symposium on Cavitation - CAV2018
Baltimore, Maryland, USA, May 14 – 16, 2018

Prediction of the Propeller-induced Hull Pressure Fluctuation via a Potential-based Method: Study of the Influence of Cavitation and Different Wake Alignment Schemes

[1]Seungnam Kim*; [1]Yiran Su; [1]Spyros A. Kinnas;

[1]*Ocean Engineering Group, The University of Texas at Austin, Austin, TX, USA*

Abstract

The propeller-induced hull pressure fluctuation is one of the main sources of the noise and vibration on ships. It has two main contributors which are equally important: the pressure induced by the displacement of mass, including the blade thickness and the cavitation bubble, and the pressure induced by the blade loading. The work in this paper is a continuation of previous work which introduced a sequence of numerical methods, which, in combination, can efficiently predict the propeller-induced hull pressures given the hull/rudder/propeller geometries, the ship speed, and the propeller RPM. A hybrid BEM/RANS method is first used to evaluate the effective wake on the propeller plane. Then, a BEM solver is used to calculate the propeller-induced hull pressures. The importance of the proper wake alignment model on the accuracy of the predicted pressure field in non-cavitating situations, was addressed before. In this paper, the method is extended to cavitating propellers. First, the wake alignment model is adjusted and improved for cavitating propeller conditions. Then, the unsteady wake alignment model is used to predict the propeller-induced hull pressures. The results are compared to experimental data, and the influence of different wake alignment models on the results is investigated.

Keywords: cavitating propeller, unsteady wake alignment model, hull pressure fluctuations, BEM, hybrid BEM/RANS method.

Introduction

In recent years, the onboard comfort of crew and passengers has become more and more crucial in the maritime industry. One key factor in evaluating the comfort level is the onboard noise and vibration level at the ship stern. The propeller-induced hull pressure fluctuation is one of the main sources of the noise and vibration on ships. Therefore, it is crucial to effectively predict the pressure fluctuation frequency magnitude at the design stage. BEM can be a proper tool to predict the hull pressure fluctuation at the design stage since BEM significantly reduces computing time compared to RANS solvers. This paper will introduce the unsteady wake alignment model to account for the non-axisymmetric variation of the incoming flow to the propeller. The incoming flow is evaluated by a hybrid BEM/RANS method in the form of the 3-D effective wake (Su et al, 2017b). Then, an acoustic BEM solver is used to calculate the propeller-induced hull pressure. When evaluating the hull pressures due to the propeller and its wake, it was found before that the full unsteady wake alignment model improved the accuracy of the predicted pressure field in non-cavitating conditions. In this paper, two different wake alignment models, i.e., PSF-2 of (Greeley & Kerwin, 1982) and unsteady Full Wake Alignment model (Kim, 2017), are applied to cavitating conditions. The results from the two different alignment schemes are compared to experimental data of (Li et al 2015).

Unsteady Wake Alignment Model

An unsteady wake alignment model was introduced by (Lee 2002), (Lee & Kinnas, 2004). This model was recently improved by (Kim 2017) and coupled with the Full Wake Alignment model of (Tian & Kinnas, 2012), which models the trailing vortices as material lines, following the local flow. As a result, both the numerical stability and accuracy have been improved such that the various input conditions, i.e., effective wake and propeller geometry, can be considered in the unsteady state. Steady wake alignment model is the appropriate option for the case of axisymmetric inflow due to its fast calculation. However, in the case of unsteady flow, such as in the case of a propeller working in the wake of a ship, an unsteady wake alignment scheme is required to account for the variation of incoming flow with space and time. In case of non-uniform inflow, the hub effect on trailing wake panels should also be considered. However, when, due to local numerical error, the trailing wake penetrates the hub, the radial distance of the wake is set equal to the hub radius. Considering that the hull pressure due to the rotating propeller is mainly affected by the wake near the hull, the radial modification of the wake panels around hub surface should not affect the predicted hull pressures considerably.

*Corresponding Author, Seungnam Kim: naoestar@utexas.edu

The basic algorithm of potential flow assumptions in inviscid and irrotational flow is used to derive the induced velocity on the wake surface. Then, the numerical method, which is implemented to compute the aligned wake geometry will be introduced in a progressive manner. The basic philosophy behind steady alignment model is also valid in the unsteady state except for the consideration of time-dependent variables. Consider a propeller subject to a non-axisymmetric inflow $\vec{U}_{in}(x,y,z,t)$, which consists of the effective wake, $\vec{U}_{eff}(x,r,\theta)$, and the angular velocity of the rotating propeller. $\vec{U}_{eff}(x,r,\theta)$ includes the effect of propeller-hull interaction and the vorticity due to the propeller. When the propeller rotates at a constant angular velocity, $\vec{\omega}$, the inflow velocity relative to the propeller becomes

$$\vec{U}_{in}(x,y,z,t) = \vec{U}_{eff}(x,r,\theta - \omega t) + \vec{\omega} \times \vec{x}(x,y,z) \tag{1}$$

where $r = \sqrt{(y^2 + z^2)}$ and $\theta = tan^{-1}\left(\frac{z}{y}\right)$.

The modeled propeller geometry in the Cartesian coordinate system is shown in Figure 1. Although the conical bulb and the cylindrical tip vortex (S_T) are presented in the figure, those tip vortex geometries are not considered in this paper. The main emphasis of this paper is on the effect of wake alignment scheme on the cavity patterns and on the propeller induced hull pressures, rather than on the effect of the developed tip vortex cavity on the propeller performance.

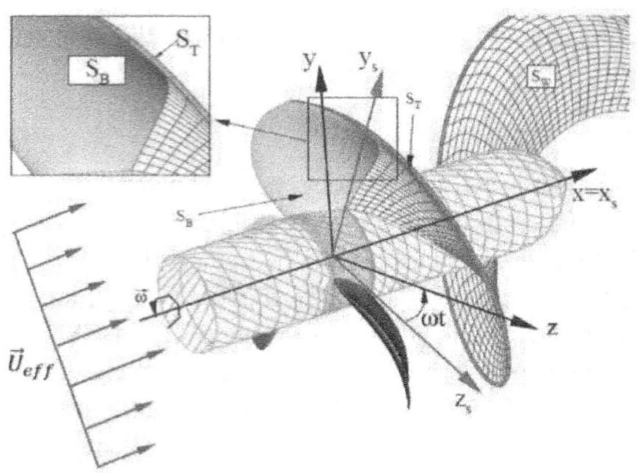

Figure 1. Propeller subject to a general inflow, assumed to be effective wake. (x,y,z) and (x_s,y_s,z_s) denote the propeller fixed and the ship fixed coordinates respectively; from (Lee 2002).

The potential at an arbitrary point on the discretized propeller geometry at time t can be expressed by the Green's third identity, that is, the superposition of the potentials due to source and dipole distributions on the propeller, and due to dipoles on the wake. The unknown strength of the singularities is determined by enforcing the kinematic and dynamic boundary conditions at control points. Once the velocity potentials are determined, the perturbation velocity on the wake panels at time step t can be calculated by taking the gradient on the Green's formula.

$$\begin{aligned}\vec{u}_i(t) = &\frac{1}{4\pi}\iint_{S_B}\left[\phi_q(t)\nabla\frac{\partial}{\partial n_q(t)}\left(\frac{1}{R(p;q)}\right) - \frac{\partial \phi_q(t)}{\partial n_q(t)}\nabla\left(\frac{1}{R(p;q)}\right)\right]dS \\ &+\frac{1}{4\pi}\iint_{S_W}\Delta\phi_w(r_q,\theta_q,t)\nabla\frac{\partial}{\partial n_q(t)}\left(\frac{1}{R(p;q)}\right)dS\end{aligned} \tag{2}$$

Different from steady state, the dipoles on the wake surface are convected along the wake surface with angular velocity $\vec{\omega}$ to satisfy the force free condition. In steady state, the trailing wake strength $\Delta\phi_w$ is constant along the circumferential direction and time invariant. In unsteady state, however, $\Delta\phi_w$ behaves as a function of the r and θ directions of the blade trailing edge in the cylindrical coordinates. Note that the dipole strength on each strip on wake surface is time and space dependent due to the convection of wake strength in Equation (2).

The inflow velocity, $\vec{U}_{in}(x, y, z, t)$, and the perturbation velocity, $\vec{u}_i(x, y, z, t)$, are evaluated at each node on the discretized wake surface and constitute the total velocity, $\vec{V}_{total}(x, y, z, t)$, with which the unsteady wake alignment is implemented in a progressive manner. At $t = 0$, steady alignment mode is firstly implemented followed by unsteady alignment mode and then fully unsteady mode ($t \geq 0$). Detailed description of each mode is presented in (Kim 2017). With the fully aligned wake from unsteady calculation, unsteady cavity calculation is performed at each time step to predict the cavity patterns on the blade surface.

Underwater Acoustic Model

To predict the propeller-induced pressure pulses on the ship hull, the boundary element method can, again, be used. First, the total unsteady pressure P_T is decomposed to a known radiated pressure P_R and an unknown diffraction pressure P_D which is governed by the Helmholtz equation.

$$\nabla^2 P_D = \frac{1}{c^2}\frac{\partial^2 P_D}{\partial t^2} \tag{3}$$

For near-field underwater noise, the sound wavelength is much longer than the dimension of the domain we are interested in. Therefore, an infinite sound speed can be assumed. As a result, equation (3) turns into a Laplace equation, as shown in equation (4), and can be solved with a potential flow BEM solver. Details on the evaluation of the diffraction potential due to the propeller flow field and on the boundary condition on the hull are given in (Su et al 2017a).

$$\nabla^2 P_D = 0 \tag{4}$$

Numerical Results

The geometrical arrangement in the experiment and fully aligned unsteady wake from BEM are presented in Figures 2 and 3, respectively. Although the BEM does not contain the hull geometry as opposed to the experiment, BEM/RANS takes the hull geometry into account when evaluating the effective wake (Su et al 2017b). As shown in Figure 3, the direction of the aligned wake is inclined toward hull surface since the incoming flow has relatively faster velocity around the ship bottom. Although the hub geometry in the experiments has a finite length in the axial direction, BEM assumes infinite hub length to prevent wake sheets from being distributed around hub center where the effective wake needs to be extrapolated. In this paper, only the time-averaged propeller/rudder interaction is included.

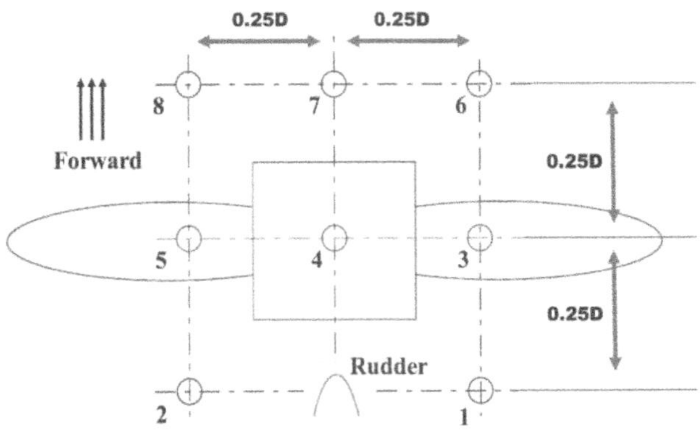

Figure 2. Experimental arrangement of SPSS P2772 propeller (left) and pressure transducer with numbering (right) (Li et al, 2015).

A cavitating case is investigated via BEM to validate the predicted cavity patterns. SPSS P2772 propeller is operated with $J_S = 0.8082$ and cavity number $\sigma = 2.914$ as indicated in the experiments. The predicted cavity patters, as well as the pressure fluctuation amplitude of the first blade harmonic on the ship hull are presented in Figure 4. Also, photographs taken during the experiments are shown for a comparison with the results from BEM at several angular positions, in Figure 5. As shown, both BEM and the experiments detect a small amount of sheet cavity.

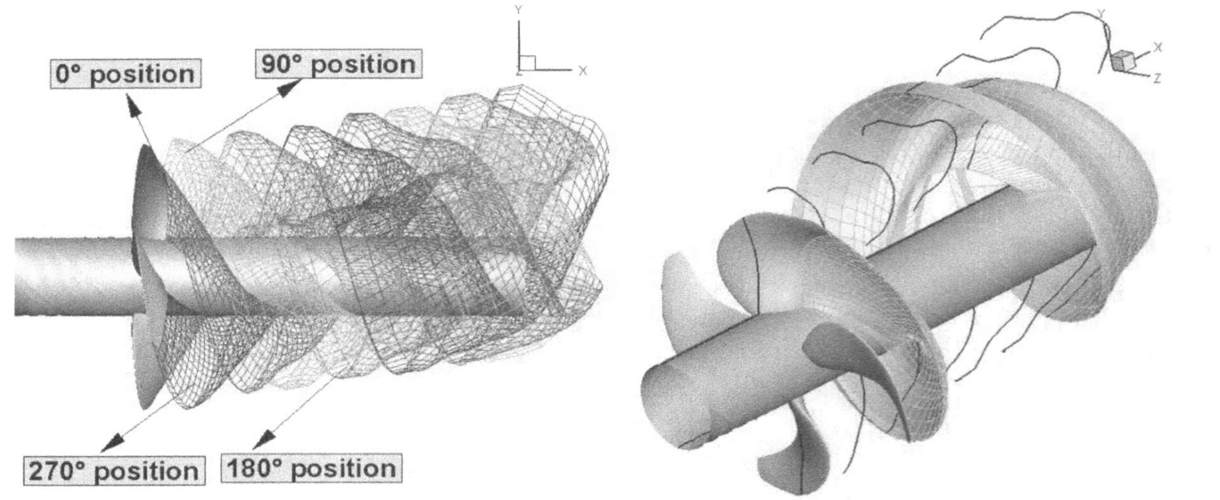

Figure 3. The projected view of aligned wake geometries for SPSS P2772 propeller with infinite hub, and the wake shapes at center-plane (right)

Although the experiment captures blade sheet and developed tip vortex cavity on blade, only the former can be detected by BEM, given that the developed tip vortex cavity model is not activated. The complete description of the developed tip vortex cavity is provided in (Lee 2002), (Lee & Kinnas, 2004). At angular positions from 0° to 30°, the comparison of the predicted sheet cavity from both the experiment and BEM is in good agreement with each other, but beyond that, the sheet cavity in the experiment is replaced by a developed tip vortex cavity.

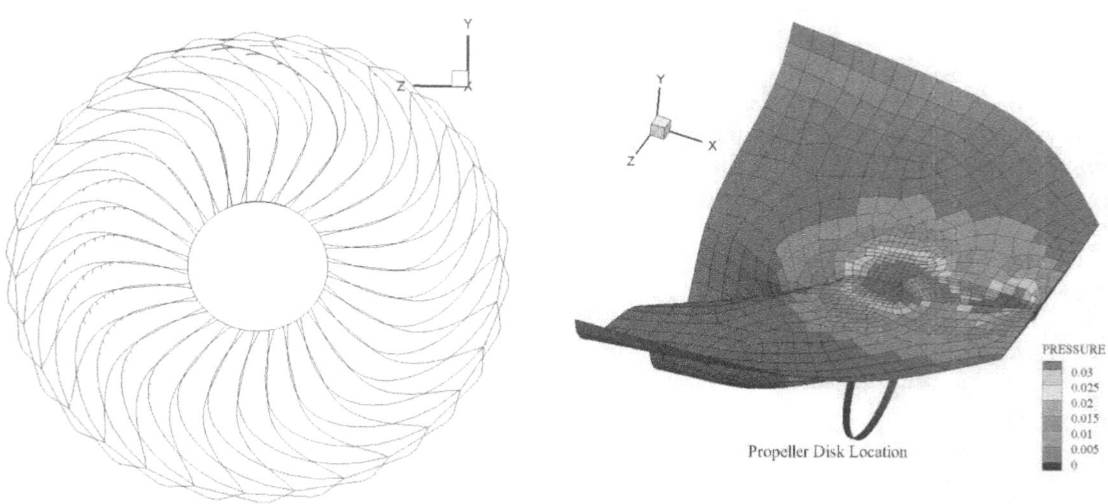

Figure 4. Pressure fluctuation amplitude on the ship hull (right) and the projected blade geometry with predicted cavity patterns at various angular positions (left). Blade at 0° position is colored blue for clarity with positive y axis pointing free surface. Note the variation of cavity patterns with blade angle is also due to the changes in the hydrostatic pressure.

The predicted hull pressure fluctuation is also compared with the experimental data. In the experiment, eight pressure transducers are placed on the hull surface above the propeller. The arrangement of the transducers is shown in Figure 2 (right). The pressure histories are recorded under the same load condition as mentioned above. Figure 6 (right) shows the pressure fluctuation amplitude of the first blade harmonic on the ship hull. Figure 6 compares the pressure history predicted by the acoustic solver with that from the measurement. As shown in Figure 6, the current method can predict the peak amplitude on most of the transducer locations. However, the experiment contains a high-frequency component which the current model cannot capture. This might be due to the lack of the developed

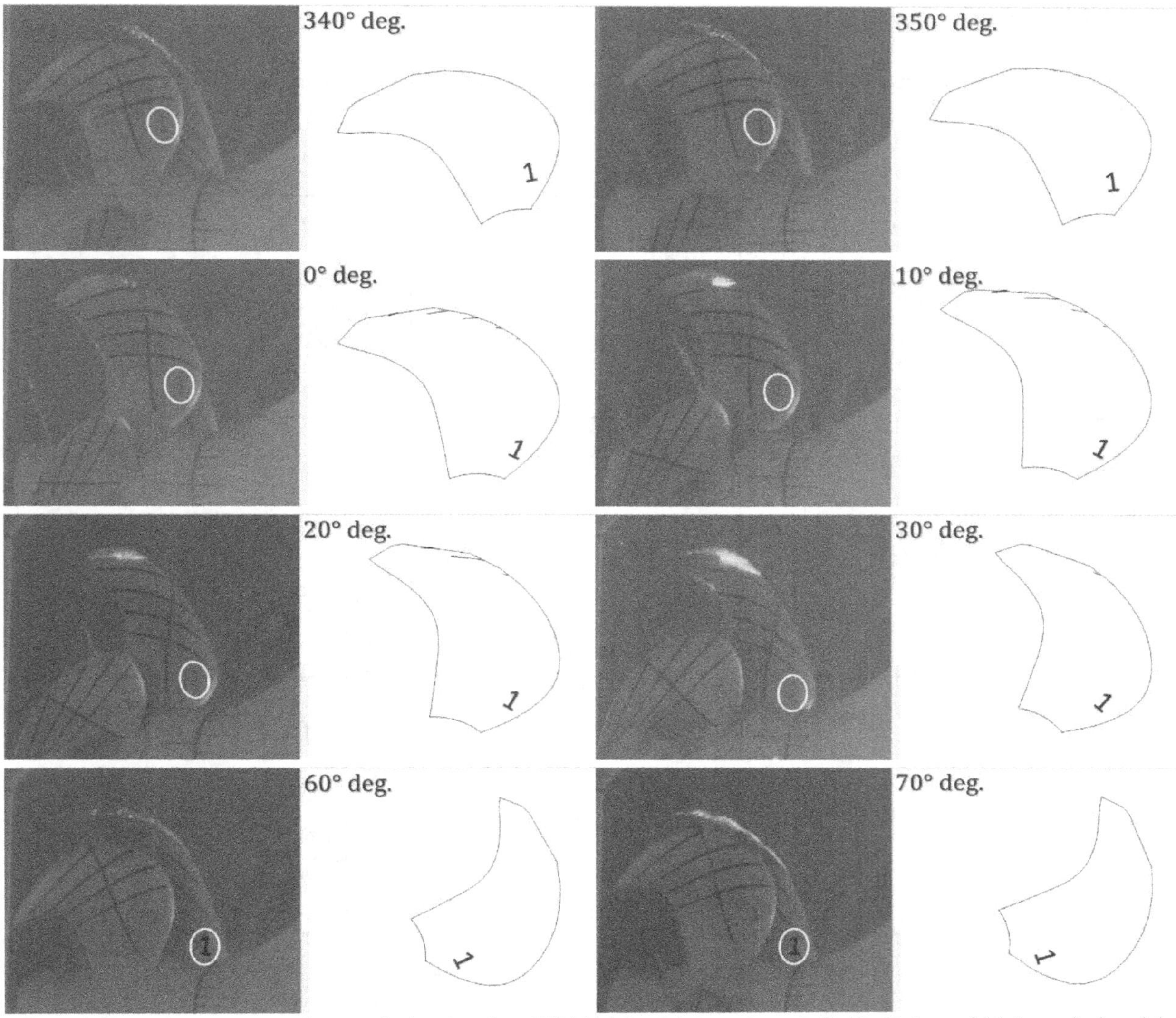

Figure 5. The cavity patterns at various angular locations from BEM (right) and the experiment (left). The blade on which the cavity is mainly detected is marked with the number 1 to indicate the key blade.

tip cavity model in the current solver. On transducers #3, #6, and #7, different types of wake alignment models do not affect the predicted hull pressure as much. This is because these transducers are located upstream of the blade (right-handed). For the same reason, since transducer #1, #2, #4, #5, and #8 are closer to the downstream wake, a better wake alignment scheme can significantly improve the numerical result. Also, it should be noted that the predicted hull pressures in cavitating flow show better correlation with those measured in the experiment than those in fully-wetted flow. The blade sheet cavity therefore results into a major change in the peak values of the predicted hull pressure fluctuations, and improves the correlation with those measured in the experiment.

Conclusions

The propeller-induced hull pressure is predicted for a cavitating condition. The unsteady wake alignment is further improved to better capture the cavity patterns in unsteady state. As a result, the predicted cavity from BEM is qualitatively well compared with that of the experiment. The predicted hull pressure fluctuations are also compared with the experimental data. Unsteady wake alignment and the inclusion of cavitation significantly improves the numerical results downstream of the propeller. In the future, the inclusion of the developed tip vortex cavity model (Lee & Kinnas, 2004) in the current model is expected to improve the prediction of higher harmonics.

Acknowledgement

Support for this research was provided by the U.S. Office of Naval Research (Grant Number N00014-14-1-0303 and N00014-18-1-2276; Dr. Ki-Han Kim) partly by Phases VII and VIII of the Consortium on Cavitation Performance of High Speed Propulsors.

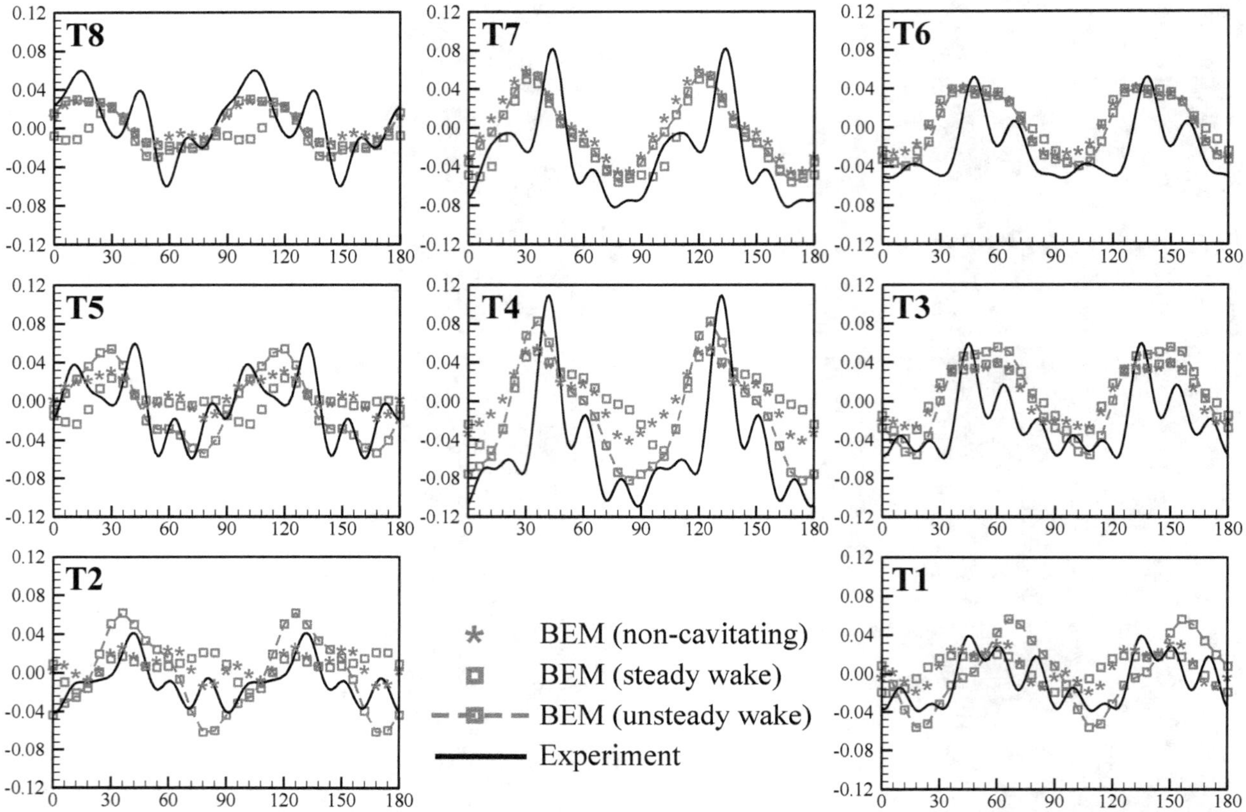

Figure 6. Comparison of hull pressures predicted by the pressure-BEM solver with different treatment of the effective wake field. Hull pressure (coefficient) fluctuation with respect to the blade angle (only 0° to 180° is plotted due to periodicity).

References

[1] Lee. H. (2002). *Modeling of Unsteady Wake Alignment and Developed Tip Vortex Cavitation.* Ph.D. diss., Ocean Engineering Group, CAEE, UT Austin.

[2] Tian, Y., Kinnas. S. A. (2012). *A wake model for the prediction of propeller performance at low advance ratios.* International Journal of Rotating Machinery, Volume 2012, Article ID 372364.

[3] Kim. S. (2017). *An Improved Full Wake Alignment Scheme for the Prediction of Open/Ducted Propeller Performance in Steady and Unsteady Flow*. M.S. thesis., Ocean Engineering Group, CAEE, UT Austin, August.

[4] Su. Y., Kim, S., Du, W., Kinnas, S.A., Grekula, M., Hallander, J., and Li, D.-Q. (2017a). *Prediction of the Propeller-induced Hull Pressure Fluctuation via a Potential-based Method: Study of the Rudder Effect and the Effect from Different Wake Alignment Methods.* Fifth International Symposium on Maritime Propulsion, *smp'17*, Espoo, Finland.

[5] Li, D.-Q., Hallander, J., Johansson, T., & Karlsson, R. (2015). 'Cavitation dynamics and underwater radiated noise signature of a ship with cavitating propeller'. 6th Inter. Conference on Computational Methods in Marine Eng/ring, MARINE2015, Rome Italy.

[6] Greely, D. S., and Kerwin, J. E. (1982). *Numerical methods for propeller design and analysis in steady flow*. Trans. SNAME.

[7] Su. Y. and Kinnas. S. A. (2017b). *A generalized potential/RANS interactive method for the prediction of propulsor performance*. Journal of Ship Research, 61(4), 214-229.

[8] Lee. H. and Kinnas S. A. (2004). *Application of boundary element method in the prediction of unsteady blade sheet and developed tip vortex cavitation on marine propellers*. Journal of Ship Research, 48(1):15-30.

© 2018 ASME

Probe into internal flow structures of a ventilated supercavity

[1]Yue Wu; [1,2]Yun Liu; [1,3]Siyao Shao; [1,3]Jiarong Hong*;

[1]Saint Anthony Falls Laboratory, University of Minnesota, Minneapolis, MN, USA; [2]Department of Mechanical and Civil Engineering, Purdue University Northwest, Hammond, IN, USA; [3]Department of Mechanical Engineering, University of Minnesota, Minneapolis, MN, USA

Abstract

This study presents the first systematic experimental investigation of the internal flow structures of a ventilated supercavity using flow visualization and particle image velocimetry (PIV). Experiments are conducted in the high-speed water tunnel at Saint Anthony Falls Laboratory. A backward-facing cavitator mounted on a hydrofoil is used to generate the cavity. Fog particles are introduced into the cavity through the ventilation line, and then illuminated by a laser sheet to visualize the internal flow. A high-speed camera is applied to record internal flow images both for visualization and PIV purposes. The visualization images capture the shape of the supercavity and reveal the patterns of particle movements inside the cavity. The PIV images are processed to obtain the instantaneous and averaged 2D velocity fields. The measurements are performed for the supercavities of two closure modes, i.e. the re-entrant jet (RJ) and the twin-vortex (TV). The flow visualization shows some common features of internal flow across RJ and TV supercavities, including the recirculating flow region near the cavitator, the internal boundary layer extended from the interface as well as the reverse flow in the center portion of the cavity. Furthermore, time-averaged internal velocity field from PIV provides a quantification of streamwise extent of the recirculating flow region, the variation of the thickness of the internal boundary layer along the interface and the magnitude of reverse flow. The analysis of the PIV flow field indicates the TV supercavity yields a longer streamwise extent of the recirculation region, a thinner internal boundary layer and a stronger reverse flow.

Keywords: ventilated supercavitation; optical visualization; particle image velocimetry

Introduction

Ventilated supercavitation is a technique to create a gas bubble enclosing a moving object by artificially injecting gas into the liquid. This technique allows significant reduction of the skin friction on the object, which has attracted considerable research attentions as a promising approach to achieve high-speed underwater transport **[1]**. Ventilated supercavitation has been conventionally studied using the following non-dimensional parameters, including cavitation number, $\sigma_C = 2(P_\infty - P_C)/(\rho_W U^2)$, Froude number, $Fr = U/\sqrt{gd_C}$, and air entrainment coefficient, $C_Q = Q/Ud_C^2$, where P_∞ and P_C are the test-section pressure upstream of the cavitator and the cavity pressure, respectively, ρ_W is the liquid density, U refers to the free-stream velocity, g is the gravitational acceleration, d_C denotes the cavitator diameter, and Q is the volumetric ventilation rate under standard condition. Despite of decades of research focusing on deriving the semi-empirical relations among these dimensionless parameters, there is still a significant lack of fundamental understanding of detailed physical processes involved in ventilated supercavitation. Recently, a number of studies have suggested the flow structures inside a ventilated supercavity (referred to as internal flow hereafter) play a crucial role in determining its gas entrainment/leakage mechanism and closure modes. Specifically, the gas leakage from the cavity is a key process that dictates the ventilation demand for sustaining a supercavity. The majority of prior studies which estimate the ventilation demand qualitatively assume that the gas is entrained in the bulk flow and directly exits from the vortex tubes formed at the rear portion of the supercavity due to gravity **[2-4]**. However, for ventilated supercavitation at high Fr, Spurk hypothesized that the gas is mainly carried by the internal boundary layer formed at the gas-liquid interface **[5]**. Recently, using numerical simulation, Kinzel et al. refined Spurk's entrainment theory by considering the difference between the amount of air entrained by the internal boundary near the closure region and the ventilation input, which causes a reverse or a forward flow in the center portion of the supercavity **[6]**. These internal flow phenomena can also significantly impact the flow topology in the closure region of a supercavity **[7-8]**. In particular, based on the hypothesis of a non-uniform pressure field and pressure-driven internal flow, Karn et al. reconciled the variation of different modes across a broad range of ventilation and flow conditions from different flow test facilities **[7]**. Due to immense technical challenges, there are only very small number of studies that have examined internal flows of a supercavity, primarily relying on theoretical analysis **[9]** and numerical simulations **[10]**. Specifically, based on the analogy of the cavitation and the wake flow behind a cylinder,

*Corresponding Author, Jiarong Hong: jhong@umn.edu

Wu postulated the presence of counter rotating circulation of gas flows inside the cavity **[9]**. Using a delayed –detached eddy simulation, Kinzel et al investigated the disturbances of the internal flow structures due to the presence of a downstream facing ventilation jet inside the supercavity **[10]**. Only very recently, experimental studies of the internal flows were conducted using flow visualization **[11]** or quantitative optical diagnostic techniques such as particle image velocimetry (PIV) **[12]**. However, generally speaking, these studies only presented a rude picture of internal flows with relatively low resolution, and did not perform in-depth analysis to elucidate the connection between internal flows and the key processes involved in ventilated supercavitation. Hence, in the current study, we present the first systematic experimental investigation of internal flows using flow visualization and PIV with a goal of examining the impact of such internal structures on the air entrainment mechanisms.

Methodology

The experiments are conducted using the high-speed water tunnel at Saint Anthony Falls Laboratory (SAFL). The tunnel has a horizontal test section of 1.20 m in length and 0.19 m × 0.19 m cross section with the bottom and two side windows made of Plexiglas to allow optical access. It has been used for a number of supercavitation experiments in recent years which provide detailed information of the flow facility **[13-15]**. As shown in figure 1(a), a backward facing cavitator (with detailed information provided in **[13]**) of 30 mm-diameter is employed to generate a free standing ventilated supercavity for the experiments. The cavitator is mounted at the trailing of a hydrofoil strut (connect through an 18 mm extension tube) of 5 mm maximum thickness to minimize the influence from the wake of the mounting structure on the supercavity.

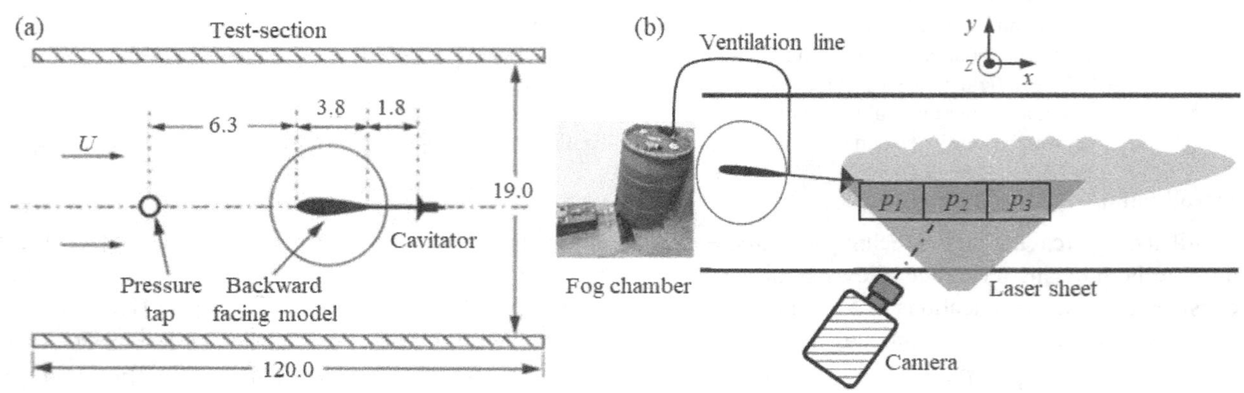

Figure 1. Installation of the backward facing model (a) and schematic experiment setup (b)

Flow visualization and planar PIV are performed using fog tracer particles for qualitative and quantitative analysis of internal flows. Fog particles are generated using a Rosco1700 fog machine which yields a particle mean diameter of 28 μm and a standard deviation of 13 μm. As shown in figure 1(b), the fog machine feeds fog particle directly into a 42 gallon fog chamber, which is then connected to a ventilation line that goes through the hydrofoil strut and the cavitator to introduce the fog particles into the supercavity. The large fog chamber, maintained at atmospheric pressure, is used to ensure consistent fog supply for optical measurements. A laser sheet of ~1 mm thickness, generated using a Photonics DM30-527 Nd/YLF high-speed laser (30 mJ/pulse at 3 kHz repetition rate), is employed to illuminate the particles on the central streamwise-vertical plane (i.e. x-y plane with $z = 0$ as defined figure 1(b)) of the supercavity. During the experiments, the expansion angle and the center position of the light sheet are adjusted for measurements observing different regions of the supercavity with different magnifications. To capture the images of internal flow fields, a Photron APX-RS high speed camera (3000 fps with 1024 × 1024 pixels size on the imaging sensor), equipped with a Nikon macro 105 mm/f5.6 lens, is positioned perpendicular to the laser sheet. During the experiments, the laser and the camera are synchronized at 3 kHz to record flow visualization and PIV images with two different magnifications (corresponding to a field of view of 34.5 mm × 34.5 mm with a pixel resolution of 0.03 mm/pixel, and 409.5 mm × 409.5 mm with 0.4 mm/pixel, respectively). The recorded particle images are post-processed by a modified histogram equalization program to enhance the contrast. The enhanced images are then processed using a multi-pass cross-correlation method by LaVision DaVis7.2 to obtain instantaneous velocity vector fields. The method applies a minimal interrogation window size of 32 × 32 pixels with 75% overlap, resulting in a vector spacing of 0.27 mm for the velocity vector fields. The velocity vector fields are further improved using the ray tracing mapping method presented in **[16]** to eliminate the distortion caused by the gas-liquid interface of the supercavity. In addition to the optical measurements, we also conduct pressure measurements using the hypodermic tube at the backside of the cavitator (for P_C) and the pressure tap located upstream of the airfoil strut (for P_∞). These measurements are performed separately from the optical measurements, under the same flow and ventilation conditions in order to obtain cavitation number σ for each test case. It is worth noting that during the experiments the angle of attack of the airfoil strut is set at $5 \pm 0.5°$ to ensure a relatively smooth bottom interface of the supercavity for the laser sheet to remain undisturbed as it penetrates the interface. The measurements are conducted for the supercavities of different closure modes, i.e. the re-entrant jet (RJ) closure and the twin-vortex (TV) closure, with the corresponding flow and ventilation conditions listed in Table 1. For each closure mode, the PIV images are recorded in three adjacent regions (i.e. P1, P2 and P3 as shown in figure 1(b)). At each region, in total of 4 sequences of internal flow images are recorded. Each sequence contains 2048 images and corresponds to a sampling period of 0.68 s.

U (m/s)	Fr	P_∞ (kPa)	P_C (kPa)	Q (SLPM)	C_Q	σ_C	Closure	Cavity length L (m)
3.2	5.9	96.5	94.7	10.0	0.0550	0.347	RJ	0.272
		90	88.2	13.5	0.0752	0.349	TV	0.319

Table 1. Experimental conditions for the supercavities with RJ and TV closure modes

Results

Flow visualization based on fog patterns provides a qualitative view of detailed gas flows inside a ventilated supercavity. As shown in figure 2(a), for the ventilated supercavity with RJ closure, the internal flow exhibits the following three main features: (i) a recirculating flow region observed behind the cavitator; (ii) a flow region of high shear extended from the air-water interface (referred to as internal boundary layer); (iii) a reverse flow region occurring in the center of the supercavity. Specifically, in the recirculating flow region, the air is entrained from the internal boundary layer to the reverse flow region, and circulated back to the boundary layer. The recirculating flow region has a vertical extent close to d_C and a streamwise extent (i.e., reattachment length) in the scale of d_C. The internal boundary layer is formed when the ventilated air is entrained by the moving air-water interface shortly after it exits the ventilation holes on the cavitator. The boundary layer develops similar to the one over a moving foil that moves at the same velocity as the local water flow. In the recirculating flow region, the internal boundary layer is strongly disturbed by the reverse flow and the impingement of gas jets from the ventilation holes. These disturbances also cause the boundary layer to transition directly to turbulent boundary layer. Downstream of the recirculating region, the development of internal boundary layer with appreciable air entrainment towards the interface can be observed. The boundary layer gradually grows thicker till a location slightly downstream of the location of maximum supercavity diameter, and then starts to thin as the supercavity reduces its diameter. The presence of this internal boundary layer and its important role on air entrainment is consistent with the theory from Spurk **[5]**, and the general trend on the change of boundary layer thickness according to the supercavity geometry observed here was described by Kinzel et al **[6]**. However, the prior studies did not specifically comment on the impact of the recirculation region on the general characteristics and the development of the boundary layer. The reverse flow region is observed throughout the entire streamwise span of the flow visualization (i.e., till ~80% L downstream of the cavitator) in the center portion of supercavity. Downstream of the recirculating region, the reverse flow covers majority of the supercavity and travels with almost uniform streamwise velocity, which is much slower than that of the moving air-water interface. Besides, highly coherent spanwise vortical structures are convected upstream with no visible deformation in the reverse flow region, indicating the flow in this region is more laminar compared to that of the internal boundary layer. When the reverse flow approaches the location approximately $0.5d_C$ downstream of the cavitator, a clear deceleration is observed. It is worth noting that both Spurk **[5]** and Kinzel et al **[6]** did not indicate the presence of the reverse flow for the supercavity with RJ closure even though Kinzel et al noted such flow could occur for TV type supercavities. In addition, some subtle flow features can be observed from our flow visualization. For example, in the recirculating region, groups of tiny vortices with both clockwise and counter-clockwise rotation appear behind the cavitator and right above the air-water interface, as shown from the close-up view in figure 2(a). These vortical structures are produced by ventilation jets and are strongly dependent on the configuration of the ventilation schemes. Moreover, at the outer extent of the internal boundary layer, a series of hairpin-like vortical structures emerge and are entrained into the reverse flow region. Note that the rotational direction of these coherent structures is opposite to that of the hairpin vortices observed in the wall-bounded flows, possibly due to the reversed wall shear stress associated with the moving air-water interface as well as the reverse flow above the boundary layer.

As shown in figure 2(b), the supercavity with TV closure displays similar main flow features as those in the RJ supercavity. However, some apparent discrepancies on the detailed characteristics of these features are also observed. Specifically, the recirculating region appears to extend much farther downstream. The reverse flow shows a wider extent, almost fulfilling the entire lateral span of the supercavity, and moves faster in the streamwise direction. Moreover, there seems to be an increasing frequency of coherent vortical structures being rolled-up from the boundary layer and entrained into reverse flow region. Additionally, as shown in the close up view in figure 2(b), more tiny vortices with enlarged size are produced in the vicinity of the location where ventilation jet impinges the air-water interface, due to the increase of flow rate of ventilation. Moreover, the close up view of the supercavity with TV closure presents a series of coherent vortices rolled-up along the shear interface between the reverse flow and the boundary layer. These vortices may be induced by enhanced Kelvin-Helmholtz (KH) instability, which is caused by the increasing velocity shear in the recirculating region.

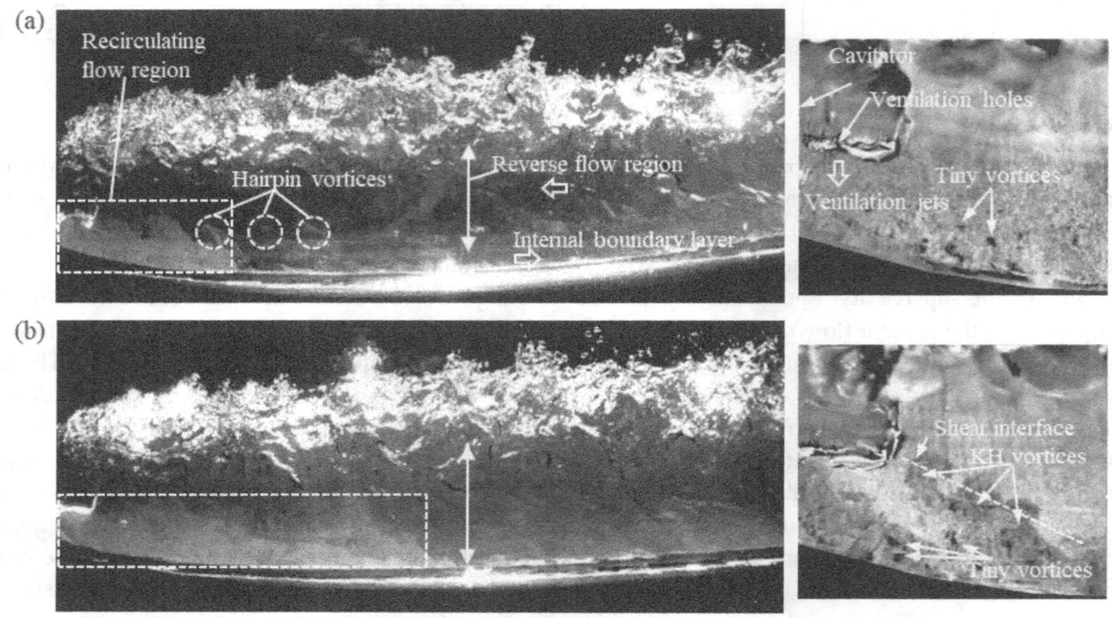

Figure 2. Internal flow patterns inside the supercavity with RJ closure (a) and TV closure (b).

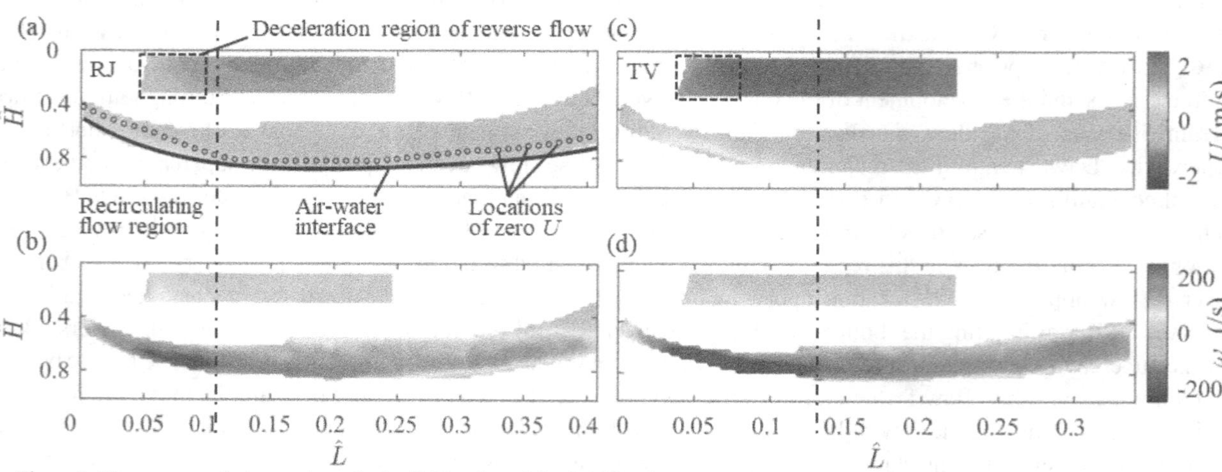

Figure 3. Time-averaged streamwise velocity field and vorticity field for the supercavity with RJ closure (a and b), and TV closure (c and d)

To provide quantitative characterization of the main internal flow features observed from the flow visualization, the time-averaged internal flow fields are examined for supercavities of both RJ and TV closure modes. As shown in figure 3(a), for the RJ closure supercavity, the mean streamwise velocity field is obtained by interpolating the mean velocity fields at P1, P2 and P3. The streamwise location (l) and vertical location (h) of mean velocity field are normalized using the cavity length (L) and the cavitator diameter (d_C), respectively (i.e., $\hat{L} = l/L$, $\hat{H} = h/d_C$). Due to the optical noise generated from laser scattering at the interface and the disturbance from the upper cavity surface, a large portion of the sampling areas for flow visualization cannot be used to obtain reliable information of the velocity field, which are blanked out in figure 3. Nevertheless, the remaining portion of the imaging area can still provide some quantitative assessment of the main flow features discussed above. Specifically, the mean flow field shows the presence of recirculation region behind the cavitator, marked by the region of high velocity gradient in figure 3(a), and also evidenced by the region of strong vorticity (i.e., ω_z) in figure 3(b). Moreover, based on the distribution of ω_z, the streamwise extent of the recirculation region is determined to be $1d_c$ for the supercavity with RJ closure. In comparison, as shown in figure 3(c) and 3(d), the TV supercavity presents an enlarged recirculating flow region with an elongated streamwise extent of $1.5d_c$. This phenomenon is analogous to the elongation of reattachment length of flow behind a backward-facing step associated with increasing turbulent level of the incoming flow **[17]**.

Based on the mean velocity field (figure 3(a)), the variation of the thickness of the internal boundary layer after the recirculation zone can be quantified. At each streamwise location, the thickness is defined as the distance between the air-water interface and the position where the internal flow has zero streamwise velocity. As shown in figure 3(a), the interface (marked with solid curve) is extracted using a fourth-order polynomial fit, and the points of zero streamwise velocity (marked with circles) are located by spatial interpolating the mean velocity fields. The variation of the boundary layer thickness along the interface (represented by the streamwise location) is further examined quantitatively in figure 4. Starting from $0.11L$ where the measured thickness ($\delta_{\text{exp,RJ}}$) has a minima of 0.8 mm, it gradually increases, and reaches the maxima of 1.8 mm at about $0.3L$. Then the thickness starts to decrease slowly and drops to 1.6 mm at $0.41L$, possibly due to the increase of adverse pressure gradient in the streamwise direction as the closure is approached. Note that the measured thickness shows some jitters along the interface which may be caused by the uncertainties involved in determining the interface and the convergence of the mean flow field. Therefore, to yield clearer trend, the boundary layer is approximated by utilizing the semi-empirical expression from turbulent boundary layer over a flat plate **[18]** with the effect of adverse pressure gradient as well as the speed and curvature of the moving interface incorporated, which yields the following form:

$$\delta_{\text{cal}} = \begin{cases} 0.14\nu C_1 \text{Re}_l/[U\ln(\text{Re}_l)] & 0 < l \le l_{\text{cr}} \\ C_2(\nu/U)^{0.2}\,(Re_{L-l})^{0.8} & l_{\text{cr}} < l \le L \end{cases}$$

where ν is the kinematic viscosity of the air (1.5×10^{-5} m^2/s), l_{cr} the critical location where the thickness starts to drop ($0.32L$), Reynolds numbers used in the expression are defined using streamwise length scale and the streamwise velocity of the moving air-water interface (equivalent to water tunnel velocity U), and C_1 (1.5) and C_2 (0.0694) are empirical constants determined by least fitting of the measurement thickness. The semi-empirical curve is also obtained for TV supercavity with slightly different C_1 (1.2) and C_2 (0.0687). Comparing the fitting curves for RJ and TV supercavities, the boundary layer of TV is thinner than that of RJ for majority portion of the boundary layer.

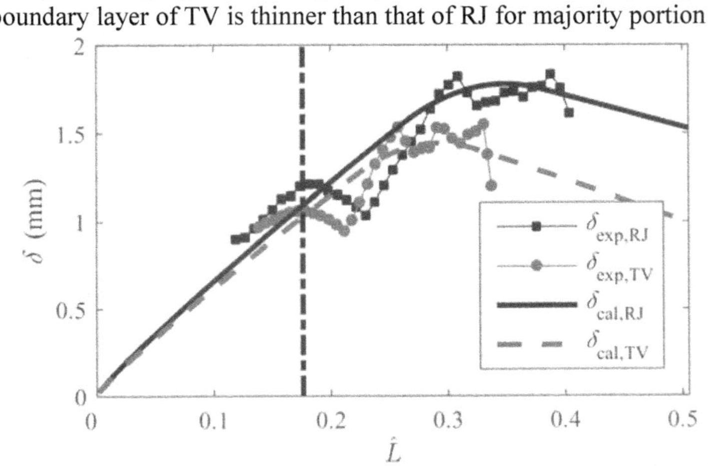

Figure 4. Comparison of the internal boundary layer thickness measurements and semi-empirical calculations with respect to streamwise locations

Like other main flow features, the reverse flow can also be quantitatively evaluated using the mean flow field in figure 3. Particularly, for the RJ closure supercavity, the mean velocity of the reverse flow can be estimated by spatially averaging the velocities in the upper rectangular window with streamwise range of $[0.05L\ 0.25L]$ and vertical range of $[0\ 0.3d_c]$, as shown in figure 3(a). The magnitude of the resulting mean velocity is 1.03 m/s, approximately one third of the free stream velocity. As observed from figure 3(b), unlike the recirculating flow and the internal boundary layer, the reverse flow does not possess significant vorticity. Traveling from $0.25L$ to $0.1L$, the reverse flow accelerates slightly due to the converging supercavity. From $0.1L$ to $0.05L$, marked as the deceleration region in figure 3(a), the flow gradually decelerates with approximately constant velocity gradient, primarily due to the blockage of the cavitator. In comparison, the reverse flow of the TV supercavity shows the similar trend of velocity variation in the streamwise direction but yields a significantly higher reverse flow of 1.42 m/s. Such discrepancy can be attributed to the thinner boundary layer for TV closure supercavity, which indicates more amount of air is entrained by the reverse flow, resulting in an increase of reverse flow speed.

Conclusion

This work presents the systematic experimental investigation of the internal flow structures of a ventilated supercavity using flow visualization and PIV. The flow visualization shows some common features of internal flow across RJ and TV supercavities, including the recirculating flow region near the cavitator, the internal boundary layer extended from the interface and the reverse flow in the center portion of the cavity. Moreover, time-averaged internal velocity field from PIV provides a quantitatively study of streamwise extent of the recirculating flow region, the variation of the thickness of internal boundary layer along the interface and the magnitude of reverse flow. The analysis of the PIV flow field indicates the TV supercavity yields an elongated streamwise extent of recirculation region, a thinner internal boundary layer and an enhanced reverse flow. Further investigation will be primarily on the influence of internal flow on the supercavity leakage and closure types.

References

[1] Nesteruk, I. (2012). Supercavitation: Advances and Perspectives. A collection dedicated to the 70th jubilee of Yu. N. Savchenko. Springer Science & Business Media.

[2] Cox, R. N., Clayden W. A. (1955). Air entrainment at the rear of a steady cavity. In *Proceedings of the Symposium on Cavitation in Hydrodynamics, London.*

[3] Epshtein, L. A. (1973). The characteristics of ventilated cavities and some scale effects. In *Proceedings of the International Symposium IUTAM, Unsteady Water Flow with High Velocities, Moscow*, pp. 173-185.

[4] Logvinovich, G. V. (1972). Hydrodynamics of free-boundary flows. Halsted Press.

[5] Spurk, J. H. (2002). On the gas loss from ventilated supercavities. Acta mechanica. 155(3), 125-135.

[6] Kinzel, M. P., Lindau, J. W., Kunz, R. (2009). Air entrainment mechanisms from artificial supercavities: insight based on numerical simulations. In *Proceedings of the 7th International Symposium on Cavitation, CAV 2009, Ann Arbor*, vol. 136.

[7] Semenenko, V. N. (2001). Artificial supercavitation: physics and calculation. Ukrainian academy of sciences, Kiev Inst of hydromechanics.

[8] Karn, A., Arndt, R. E. A., Hong, J. (2016). An experimental investigation into supercavity closure mechanisms. Journal of Fluid Mechanics. 789, 259-284.

[9] Wu, T. (1972). Cavity and wake flows. Annual Review of Fluid Mechanics, 4(1), 243-284.

[10] Kinzel, M. P., Krane, M. H., Kirschner, I. N., Moeny, M. J. (2017). A numerical assessment of the interaction of a supercavitating flow with a gas jet. Ocean Engineering, 136, 304-313.

[11] Savchenko, Y. N., Savchenko, G. Y. (2012). Gas Flows in Ventilated Supercavities. In Supercavitation (pp. 115-126). Springer Berlin Heidelberg.

[12] Wang, Z., Huang, B., Wang, G., Zhang, M., Wang, F. (2015). Experimental and numerical investigation of ventilated cavitating flow with special emphasis on gas leakage behavior and re-entrant jet dynamics. Ocean Engineering, 108, 191-201.

[13] Karn, A., Arndt, R. E., Hong, J. (2015). Dependence of supercavity closure upon flow unsteadiness. Experimental Thermal and Fluid Science, 68, 493-498.

[14] Karn, A., Shao, S., Arndt, R. E., Hong, J. (2016). Bubble coalescence and breakup in turbulent bubbly wake of a ventilated hydrofoil. Experimental Thermal and Fluid Science, 70, 397-407.

[15] Shao, S., Karn, A., Ahn, B. K., Arndt, R. E., Hong, J. (2017). A comparative study of natural and ventilated supercavitation across two closed-wall water tunnel facilities. Experimental Thermal and Fluid Science, 88, 519-529.

[16] Kumar, S. S., Sun, Y., Zou, S., Hong, J. (2016). 3D Holographic Observatory for Long-term Monitoring of Complex Behaviors in Drosophila. Scientific reports, 6.

[17] Adams, E. W., Johnston, J. P. (1988). Effects of the separating shear layer on the reattachment flow structure part 2: Reattachment length and wall shear stress. Experiments in Fluids, 6(7), 493-499.

[18] Schlichting, H., Gersten, K., Krause, E., Oertel, H. (1955). Boundary-layer theory (Vol. 7). McGraw-hill.

© 2018 ASME

10th International Symposium on Cavitation - CAV2018
Baltimore, Maryland, USA, May 14 – 16, 2018 CAV18-05178

Propeller Cavitation Noise Radiated from Single and Twin-Screw Cargo Liners: CFD Prediction and Full Scale Validation

[1]Nobuaki Sakamoto*; [1]Hikaru Kamiirisa

[1]National Maritime Research Institute(NMRI), Mitaka, 181-0004, Tokyo, JAPAN

Abstract

Present study is dedicated to estimate sound pressure level (SPL) radiated from cavitating marine propellers equipped on two different cargo liners by means of viscous computational fluid dynamics (CFD). Since limited kind of model tests were carried out for the target ships, flow configurations (e.g. full scale ship wake, thrust coefficient and cavitation number) were estimated via CFD and full scale sea trial data. Estimated configurations are utilized to carry out CFD simulations of cavitating propellers. The estimated SPL were validated against full scale data which were measured according to ISO standard. It is encouraging that estimated SPL by the present CFD agree well to the full scale measurements up to 900Hz. This evidences that 1) the flow configurations estimated by the present CFD are adequate enough to reproduce cavitation condition, and 2) assumption of "near field acoustic pressure can be approximated by near field hydrodynamic pressure" is likely to be appropriate.

Keywords: underwater ship radiated noise; viscous CFD; full scale validation

1. Introduction

Underwater ship radiated noise (USRN) has been a large interest among maritime industries in terms of mariners' comfort as well as environmental protection. The latest response to this trend includes the recommendation made by Convention of Biological Diversity Subsidiary Body on Scientific, Technical and Technological Advice (CBD-SBSTTA) on 2016 that the ratifiers should prepare toolkits for assessment of USRN[1]. One of the major acoustic sources of USRN is propeller cavitation. The present study is dedicated to estimate and validate the computational procedure for predicting propeller cavitation noise via viscous computational fluid dynamics (CFD).

2. Computational Method

2.1 Viscous CFD solver

The computational results are obtained using general-purpose commercial CFD package STAR-CCM+ ver11.04 (double precision version). The governing equations are the continuity and unsteady RaNS equations with dimensional forms, and are solved by finite volume method. Rotational motion of a propeller is modeled by dynamic overset technique. Among several turbulence models on the solver, k-omega Shear Stress Transport (k-ω SST) based Detached Eddy Simulation (DES) is selected based on the past studies[2]. The boundary layer is assumed being "fully developed", in other words, no transition model has been introduced. The turbulent boundary layer is fully resolved without wall function. As the two-phase flow modeling, Volume of Fluid (VoF) method is utilized in conjunction with the cavitation model by Schnerr and Sauer. Divergence free condition is satisfied by solving pressure Poisson equation iteratively using Semi-Implicit Method for Pressure-Linked Equation (SIMPLE)-type algorithm. The order of accuracy is second in both space and time. The code is parallelized by Message Passing Interface (MPI) and is ran on distributed memory type machines with 152 CPUs. Physical time inclement is set in such a way that the propeller rotates 1.0deg with 20 sub-iterations per physical time advancement.

2.2 Computational grid

The coordinate system is right-handed Cartesian, and its center is coincident with the boss center of a propeller. The flow direction is x negative to x positive, and the propeller rotates clockwise around x-axis observing from downstream. Figure 1 presents the computational grid in the blade vicinity for the two propellers utilized in the present study. The computational domain consists of two cylindrical volume parts, e.g. rotator and stator, and necessary local refinements. The propeller is placed inside the rotator, and the flux between the rotator and the stator is exchanged through donor and receptor cells updated every time step. To increase spacial resolution, local refinement regions are set at r/R~0.5 toward the tip. Trimmed-cell is utilized for grid generation, and nondimensional viscous length $y+$ is close to 1 in order to meet the requirement for the present turbulence model. Total number of cells for cavitation simulations is

*Corresponding Author, Nobuaki Sakamoto: sakamoto@nmri.go.jp

approximately 11M. From the past studies[2], the distance from the coordinate center to the inflow boundary, far-field boundary and outflow boundary are $3D_p$ (uniform inflow to estimate propeller open water characteristics (POC)) / $0.7D_p$ (non-uniform inflow to estimate propeller cavitation), $5D_p$ and $5D_p$, respectively, where D_p(m) is the diameter of propellers.

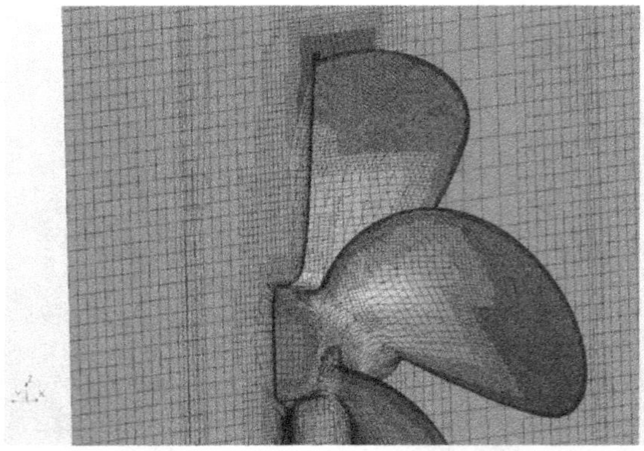

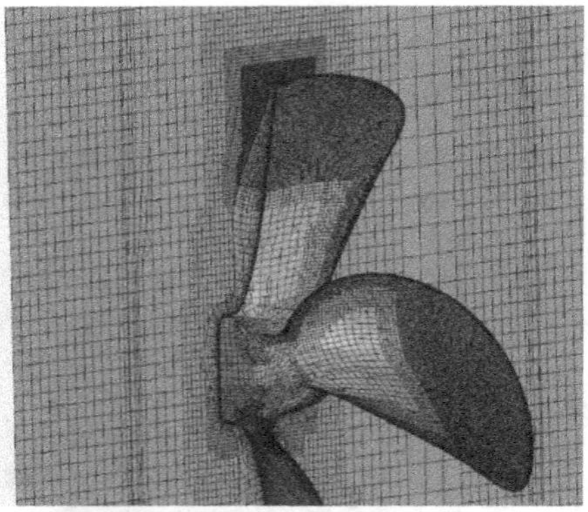

Figure 1 Computational grid in the propeller vicinity, Ship A (left) and Ship B (right)

2.3 Target geometry

Table 1 summarizes the major dimensions of two target hulls and their propellers. These are inland cargo liners operating at domestic remote islands of Japan. Ship A is the single screw vessel while the Ship B is the twin screw vessel with split stern. In Table 1, L_{pp}(m) is the length between perpendiculars, B(m) is the breadth, d(m) is the draught at design load, C_B is the block coefficient, P/D_p is the pitch ratio, B.R. is the boss ratio, E.A.R. is the expanded area ratio, Z is the number of blades, respectively.

Table 1 Major dimensions of the target ships and their propellers

	Ship A	Ship B
L_{pp}/B	5.667	4.750
B/d	2.813	3.529
C_B	0.544	0.552
D_p/L_{pp}	0.045	0.044
P/D_p	0.835	0.938
B.R.	0.187	0.172
E.A.R	0.867	0.623
Z	5	4
Rotating dir.	right	Right/left

2.4 Approximate resolution of the pressure wave in space and time

According to the past studies[2], it is recommended that approximately 20 cells should be prepared to resolve one wave length of pressure wave. For the present simulation, approximate maximum frequency f_{max_s}(Hz) in space to be resolved in model scale is

$$f_{\max\_s} = \frac{c_0}{20\Delta} \sim 12kHz \tag{1}$$

where c_0=1438(m/s) is the sound speed in water and Δ=6.0e-03(m) is the cubic cell size prepared between blade tip and sampling point of the pressure. According to the scaling criterion discussed in section 4.2 Eq. (11), this f_{max_s} corresponds to 3kHz in full scale. The size of time step Δt is set to 1.388E-04(s) for cavitation simulation which yields approximate maximum frequency f_{max_t}(Hz) in time taking Nyquist frequency into consideration as

$$f_{\max\_t} = \frac{1}{2\Delta t} \sim 3.6kHz \tag{2}$$

According to the scaling criterion discussed in section 4.2 Eq. (11), this f_{max_t} corresponds to approximately 800 to 900Hz in full scale.

3. Simulation Design

3.1 Test cases

Table 2 summarizes test cases simulated in the present study. Tests #1, #2 and #3 are carried out as a pre-process for cavitation simulation. The estimated POC through #1 is utilized as "calibration curve" to adjust thrust coefficient K_T defined as

$$K_T = \frac{T}{\rho n^2 D_p^4} \quad (3)$$

where T(N) is the propeller thrust, ρ(kg/m^3) is the fluid density and n(rps) is the propeller rotation speed. The cavitation number σ_n is defined as

$$\sigma_n = \frac{P_A + \rho g(I-h) - P_v}{0.5\rho(nD_p)^2} \quad (4)$$

where P_A(Pa) is the atmospheric pressure, g(m/s^2) is the gravitational acceleration, I is the immersion depth of the propeller, h(m) is the datum height, and P_v(Pa) is the vapor pressure of fresh water. When ships are running, the I could be different to that of still water configuration due to changes of running attitude as well as to the wave elevation at the propeller plane, therefore, #2 is necessary to set realistic I and resultant σ_n. Full scale ship wake is simulated by #3 and the result (e.g. axial component) is utilized as the inflow boundary condition (BC) to the cavitation simulation. Test #4 is carried out to obtain propeller cavitation and resultant noise. Tests #1 and #2 can be replaced by model testing, however, specifically for the present study, very limited model testing results are available and that is the reason to simulate #1 and #2 configurations by CFD.

Table 2 Test cases

Test #	Geometry	Cavitation	Inflow BC	Model/Full	Purpose
1	Propeller	NO	uniform	model	Obtain POC to tune K_T
2	Ship	NO	uniform	model	Obtain trim, sinkage and wave elevation at propeller plane to estimate σ_n
3	Ship	NO	uniform	full	Obtain full scale ship wake for propeller cavitation simulation
4	Propeller	YES	non-uniform	model	Obtain propeller cavitation pattern and pressure fluctuation at a sampling point

3.2 Conditions for cavitation simulation

To determine target K_T for cavitation simulations, full scale sea trial data are utilized. In the sea trial, brake power P_B(kW) was measured at certain ship speeds V_s(m/s). Given the transmission efficiency η_T, T(N) in Eq. (3) can be estimated as

$$T = 1000 \cdot \frac{P_B \eta_T}{V_s} \quad (5)$$

The σ_n for cavitation simulation is estimated with/without ships' running attitude in two different engine outputs (e.g. Half and Navigation-full). According to our computational results, both Ship A and Ship B run bow-up configuration, in other words, I in Eq. (4) becomes larger when ships are running than those in rest. In consequence, more severe configurations in cavitation (e.g. ships in still water with smaller σ_n) are adopted for the present study. CFD simulations of propeller cavitation are carried out in model scale for which the Kempf's Reynolds number RnK is defined as

$$RnK = \frac{c_{0.7R}\sqrt{V_a^2 + (0.7\pi n D_p)^2}}{\nu} \quad (6)$$

where $c_{0.7R}$(m) is the chord length of the blade section at 70% of propeller radius (termed r/R=0.7), V_a(m/s) is the advance speed, ν(m^2/s) is fluid kinematic viscosity. Table 3 summarizes the simulation conditions for the present propeller cavitation.

Table 3 Conditions for cavitation simulation

	Engine output	K_T	σ_n	RnK*
Ship A	Half	0.103	2.824	1.10E+06
Ship B		0.185	2.966	1.07E+06
Ship A	Navi-full	0.251	1.732	9.80E+05
Ship B		0.199	1.909	9.76E+05

*: Based on D_p=0.25m with constant n(rps) in model scale

4. Computational Results

4.1 Cavitation pattern and extent in time

Figure 2 summarizes instantaneous (e.g. key blade position θ=40deg.) cavitation patters depicted by volume fraction α of 0.1 iso surface at two different engine outputs. Definition of θ is counter-clockwise positive observing from fore with 0deg at 12 o'clock position. Animated version of Fig. 4 will be shown on site. For Ship A "Half", sheet cavitation becomes apparent at leading edge of r/R=0.8 at θ=345deg. Its extent becomes maximum at θ=13deg and disappears at θ=53deg. For Ship B "Half", sheet cavitation becomes apparent at leading edge of r/R=0.5 at θ=320deg. Its extent becomes maximum at θ=7deg and disappears at θ=56deg. For "Half" engine output, tip vortex cavitation (TVC) cannot be resolved for both Ship A and Ship B. Different to "Half" engine output, the sheet cavitation is apparent in all blades for both Ship A and Ship B in "Navi-full" engine output. TVC is resolved by the present CFD at 48deg.≤θ≤56deg. and 54deg.≤θ≤77deg. for Ship A and Ship B, respectively. Figure 3 presents the area of cavitation extent $A_c(m^2)$ on the key blade in one propeller rotation relative to the blade area $A_D(m^2)$. Ac is computed as

$$A_c(t) = \sum_{i=1}^{N_f} s_i\beta_i(t) \ where \ \beta_i(t) = \begin{cases} 1 \ if \ \alpha_j(t) \geq 0.1 \\ 0 \ otherwise \end{cases} \quad (7)$$

where N_f is the number of cell faces on the key blade, $s_i(m^2)$ is the area vector of cell face i, β_i is the flag to determine the extent of cavitation on cell face i and α_j is the void fraction of cell j which contains face i. The maximum value of Ac/A_D will be the input variable to Brown's formula which estimates the upper bound of broadband noise of cavitating marine propeller[3] as

$$SPL = K + 10\log\left(ZD_p^4 n^3 f^{-2}\right) + 10\log(A_c/A_D) \quad (8)$$

where SPL(dB rel.1μPa/√Hz@1m) is the source spectral density, K=153 is the tuned empirical constant (originally K=163) and f(Hz) is the target frequency.

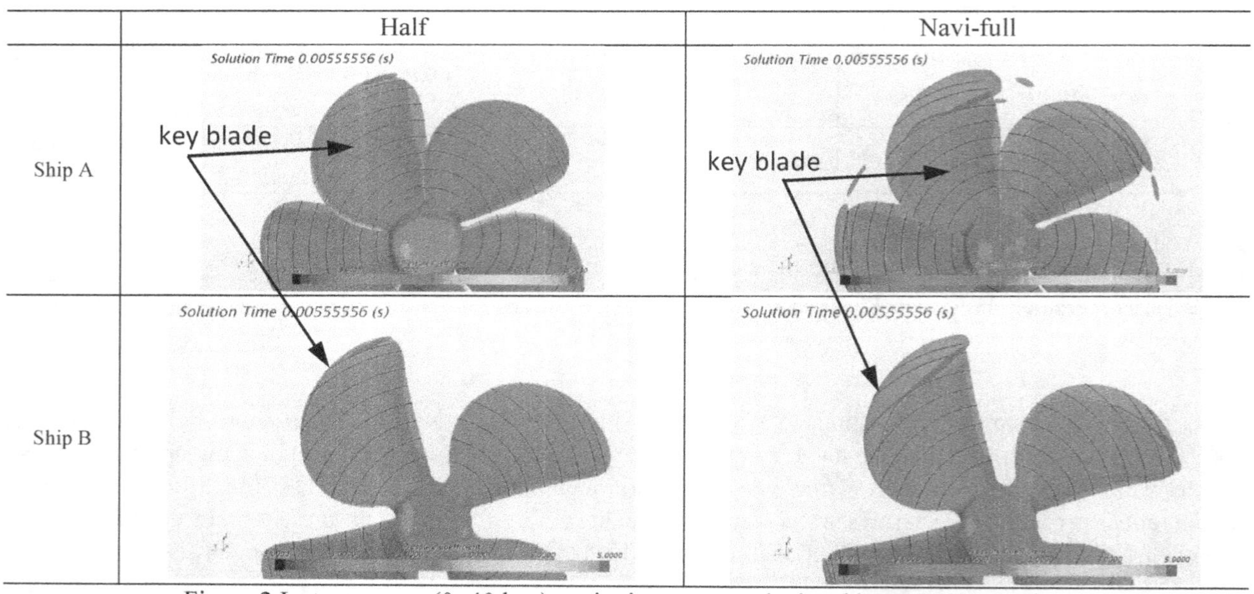

Figure 2 Instantaneous (θ=40deg.) cavitation patterns depicted by α=0.1 iso surface

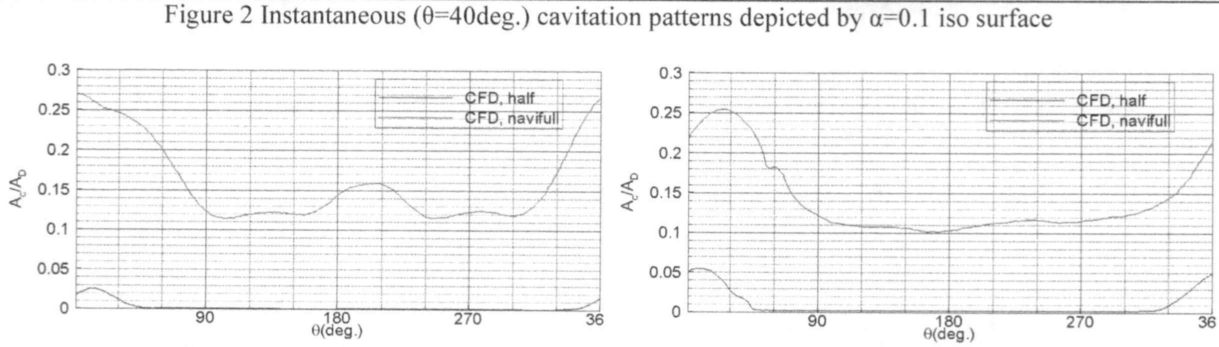

Figure 3 Cavitation extent on the key blade in time: Ship A(left) and Ship B(right)

Figure 4 presents the fluctuation of pressure coefficient $K_P(=p/\rho n^2 D_p^2)$ in time logged at one point, e.g. r/R=0.7 of propeller generator line in x, shifted 5%D_p towards port from propeller boss center in y and tip clearance coordinate in z. Such sampling point is usually adopted in the model tests in a cavitation tunnel with wire-mesh screen[4]. Volumetric cavitation extent $V_c(m^3)$ in the computational domain is estimated in one propeller rotation as

$$V_c(t) = \sum_{j=1}^{N_c} v_j \alpha_j(t) \ where \ \alpha_j(t) = \begin{cases} \alpha_j(t) \ if \ \alpha_j(t) \geq 0.1 \\ 0 \ otherwise \end{cases} \quad (9)$$

where N_c is the number of cells in the computational domain and $v_j(m^3)$ is the volume of jth cell. In Fig. 4 the profile of second time derivative of V_c is presented as d^2V_c/dt^2. K_p and d^2V_c/dt^2 are in-phase each other in several orders of blade frequencies which supports the fact that the pressure fluctuations can be well approximated by d^2V_c/dt^2 of single spherical bubble[5]. Volumetric change of cavitation extent in time can be considered as source/sink. Since the acoustic originated from periodic source/sink is considered as monopole source, acoustic interpretation of in-phase relationship between K_p and d^2V_c/dt^2 is that the monopole source of propeller cavitation noise is the volumetric change of cavitation extent in time.

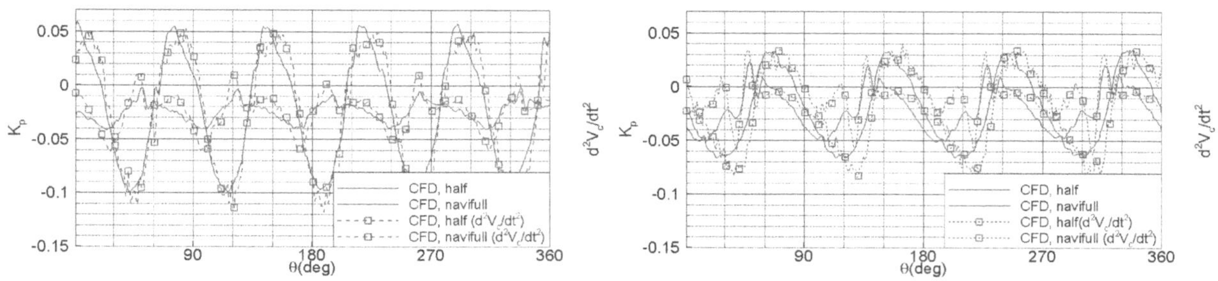

Figure 4 Change of cavitation volume in time: Ship A(left) and Ship B(right)

4.2 Propeller cavitation noise

Figure 5 summarizes the validation results of SPL(dB.rel.1μPa@1m) of USRN for Ship A and Ship B in two different engine outputs. Horizontal axis f_c(Hz) is the 1/3 Octave band central frequency. Notice that the Lloyd-mirror (LM) correction is applied for the results of "Navi-full" to take the effect of acoustic wave reflection due to the presence of free surface into consideration. The measurements at the sea trial were carried out according to the ISO/DIS 16554.3. Detail procedure for the present measurements can be found in the reference[2]. The pressure fluctuation obtained from model scale CFD is directly utilized to estimate model scale SPL, in other words, the present study assumes that the near field acoustic pressure is approximated by the hydrodynamic pressure. Scale effect correction suggested by the International Towing Tank Conference (ITTC)[6] is applied to the CFD results in order to make direct comparison with the sea trial data in full scale. The ITTC correction formula is

$$\lambda = D_{ps}/D_{pm} \quad (10)$$

$$f_s/f_m = n_s/n_m \quad (11)$$

$$\Delta SPL = 20\log_{10}\left[\lambda^Z (r_m/r_s)^X (\sigma_{ns}/\sigma_{nm})^{/2} (\lambda n_s/n_m)^{/2} (\rho_s/\rho_m)^{/2}\right] \quad (12)$$

where the subscripts "s" and "m" describes "full scale" and "model scale", respectively, r_m is the distance from blade tip to the hydrophone location in model scale CFD, r_s is set to 1(m) so that the SPL becomes source level, ρ_s=1025(kg/m^3), ρ_m=998(kg/m^3), n_s(rps) is the measured data at sea trial, and n_m(rps) is the value used for the present CFD simulations. Since the present CFD holds the law of similarity in σ_n, σ_{ns}/σ_{nm}=1. Exponents (X, Y and Z) appeared in the right hand side (RHS) of Eq. (12) are all 1 for the present study. Specifically for Ship B, 3dB is added to the RHS of Eq. (12) to take the effect of twin screw into consideration, e.g. acoustic energy originated from the propeller is doubled for twin screw configuration (assuming in-phase rotation between port and starboard) in comparison to the single screw configuration. As shown in Fig. 5, the increase of maximum SPL due to the change of engine output is 17dB and 6dB for Ship A and Ship B, respectively, at the sea trial. The present CFD quantitatively captures this trend well. Brown's formula appropriately estimates upper bound of broadband noise for both Ship A and Ship B. Two implications from these results are 1) present estimations of σ_n, K_T, full scale ship wake through viscous CFD and sea trial data are acceptable to reproduce cavitation extent for the two propellers, and 2) it is practically adequate to assume that the near field acoustic pressure can be approximated by the hydrodynamic pressure.

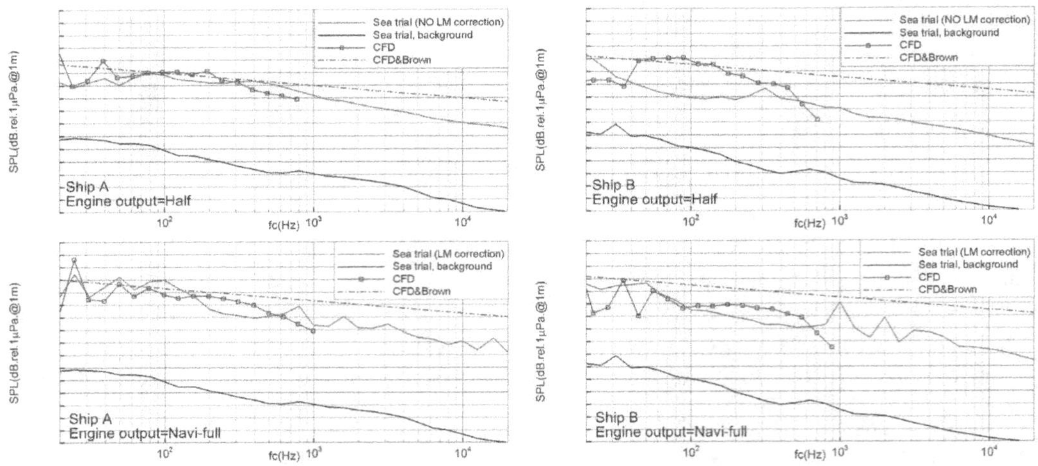

Figure 5 Validation of SPL of USRN: Ship A(left) and Ship B(right), 10dB interval for the vertical axis.

5. Conclusion

Practical methodology, for which its flow diagram is summarized in Fig. 6, has been developed to estimate propeller cavitation noise by viscous CFD and available experimental/sea trial data, and its validity has been proofed.

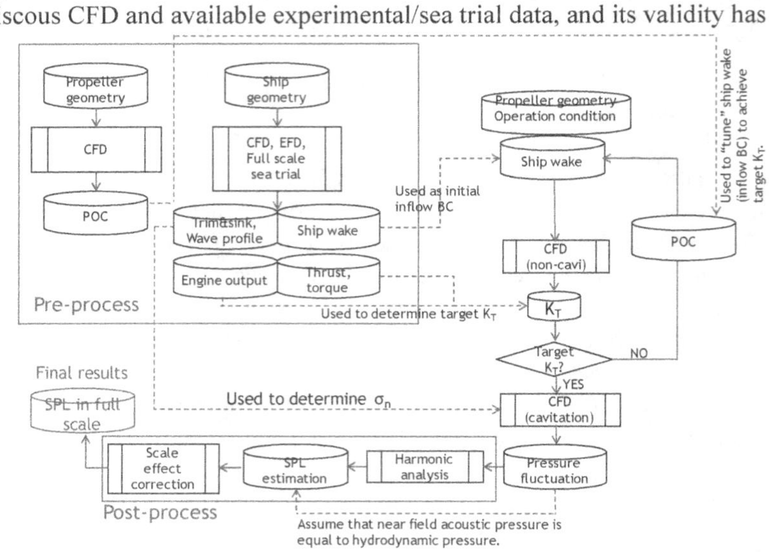

Figure 6 Procedure to estimate propeller cavitation noise by viscous CFD

Acknowledgements

This study is funded by Japan Ship Technology Research Association with a financially support by the Nippon Foundation.

References

[1] Recommendations adopted by the SBSTTA. (2016). *UNEP/CBD/SBSTTA, UNEP/CBD/COP/13/5*. pp. 56-61.

[2] Kamiirisa et al. (2016). *Exploratory study for computation/measurement of underwater ship radiated noise toward marine environmental protection.* NMRI Technical Report. 16(1). pp. 65-87.

[3] Brown N.A. (1976). *Cavitation noise problem and solutions*, Proc. International Symposium on Shipboard Acoustics.

[4] Ukon. Y. (1991). *Study on experimental prediction on ship hull vibration induced by propeller and cavitation.* SRI Technical Report. 28(4). pp. 19-52.

[5] Hoshino et al. (2010). *Full scale cavitation observations and pressure fluctuation measurements by high-speed camera system and correlation with model test.* Proc. IPS'10.

[6] International Towing Tank Conference. (1987). *Report of the cavitation committee*, Proc. ITTC87, pp. 159-219.

 © 2018 ASME

Pulsating Supercavities: Occurrence and Behavior

[1]Samuel Hansford, [1]Timothy Brungart, [1]Jules Lindau*, [1]Michael Moeny

[1]*Penn State Applied Research Lab, PO Box 30, State College, PA, USA,*

Abstract

Low Froude number ventilated cavities developed behind a disk cavitator, in a horizontal flow, under the influence of a vertical gravity field are considered. Pulsation is shown to occur in two different regions on the $C_Q - \sigma$ phase map, the hysteresis region and at vent rates greater than those necessary to maintain twin vortex cavities. Two conditions are necessary to initiate pulsation. One condition is a ventilation rate too great for either re-entrant jet or twin vortex closure regimes. The other condition is a finite amplitude pressure wave needs to force the cavity into pulsation.

Keywords: Supercavitation, Pulsation

Introduction

There are three main cavity closure regimes: (1) re-entrant jet, (2) twin vortex, and (3) pulsation. While the first two regimes have been studied at length, less attention has been given to the third regime and it remains less understood. Another regime is the over-ventilated cavity, foamy cavity, or free jet. This type is not a contiguous gas-filled cavity, but a bubbly gas-liquid mixture. The distinguishing feature of over-ventilated cavities is mixing occurs immediately behind the cavitator. Pulsation is a resonance phenomenon, with the cavity gas acting as a spring, resulting in gas ejection through periodic events. Pulsating cavities may have characteristics of both re-entrant jet and twin vortex closure but they are distinguished by their overall regular, periodically pulsating size and shape [1,2]. Ventilated supercavity pulsation has been experimentally observed by a number of researchers and under somewhat differing conditions. Some researchers have noted that pulsation occurs in what is typically referred to as the *hysteresis region* [3,4] while other researchers have observed pulsation at roughly the maximum sustainable cavity ventilation flow rate prior to cavity breakdown [5,6]. Thus, on a ventilated cavity phase map, pulsation has been observed in two distinct regions. It does not appear that any single previously published study has reported pulsation in both regions.

Methods

Experiments were conducted in the 0.305 m diameter water tunnel at the Penn State University Applied Research Laboratory. The test setup is shown in Figure 1. Gas was supplied from 6 gas bottles that vented out through the vent ports to form a cavity behind a 34.29 mm diameter disk. The freestream speed was set to 1.7 m/s so the Froude number was 3.0. The tunnel pressure was 103 kPa and to facilitate pulsation, the tunnel test section was filled just until there was no free surface visible. A Measurement Specialties XPM5 pressure sensor measured the cavity interior pressure while a Benthowave BII-7071 hydrophone measured the radiated sound pressure. The sampling rate and sample time was 25 kHz and 10 seconds, respectively. An in-situ calibration was used to remove the effects of the tunnel walls and transfer the signal to the equivalent free-field condition. Gas ventilation rates were measured by a Sierra Instruments FlatTrak 780S thermal gas flow meter and manufacturer supplied conversion factors were used to compensate for different ventilation gases.

The computational portion of this work was performed with StarCCM+. Previous work has shown that an approach using StarCCM+ captures cavity pulsation [7]. It was found that time dependent finite volume discretization with a compressible gas and incompressible liquid without a turbulence model was sufficient to capture pulsation. The same scheme employed previously was used in this study to computationally capture pulsation at a Froude number of 24.6 because pulsation could not be easily captured computationally at such a low Froude number (3.0). A 34.29 mm diameter floating disk without tunnel walls was studied.

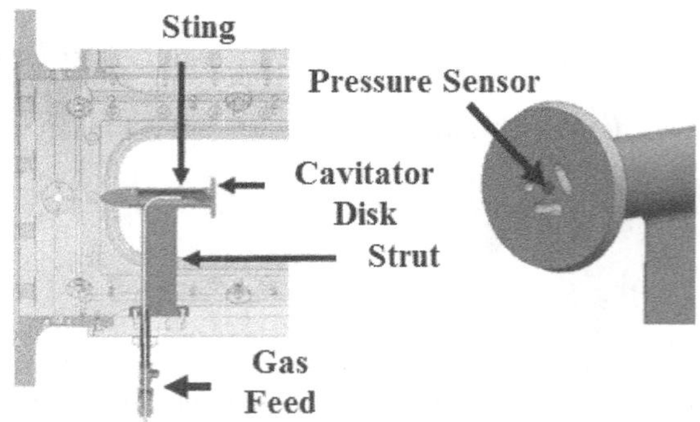

Figure 1: Experimental setup in the 0.305 m diameter water tunnel.

Results

There are two heuristic conditions necessary to induce pulsation: (1) the cavity must be transitioning between closure regimes and (2) a finite amplitude pressure wave must initiate pulsation. The first pulsation region is found at the transition between re-entrant jet and twin vortex closure regimes, commonly called the hysteresis region. Pulsation in this region will be called hysteresis pulsation. The second pulsation region is at the transition between twin vortex and over-ventilated cavities. Pulsation in this region will be called high-vent pulsation. Pulsating cavities can occur in both regions but the nomenclature is adopted to distinguish between the two. In Figure 2a, from CFD simulations, the cavitation number history for a one second duration of Fr=24.6 cavity pulsation is given. Pulsation of both the hysteresis and high-vent kind are presented. The hysteresis pulsation has a higher frequency at approximately 17 Hz while the high-vent case pulses at roughly 14 Hz. Per the Song model [8], the higher ventilation rate results in a larger cavity while not significantly increasing the pressure so the lower frequency is expected. In Figure 2b, water tunnel experimental results are presented. The high-vent pulsation (green trace in Figure 2a and horizontal line in Figure 2b) falls within the experimentally observed pulsation regions. The pulsation regions are denoted by P1 and P2 and closure regimes are separated by dashed blue lines. The hysteresis pulsation (black trace in Figure 2a and horizontal line in Figure 2b) falls within the range that is believed to be the hysteresis region, given by RJ/HYS. However, for the water tunnel data, the noise floor of the flow meter is higher than the ventilation rate for the hysteresis region. The red triangle close to the horizontal axis with p_c approximately equal to 104 kPa is a re-entrant jet cavity while the next lowest vent rate (3) is an RJ/TV transitioning cavity.

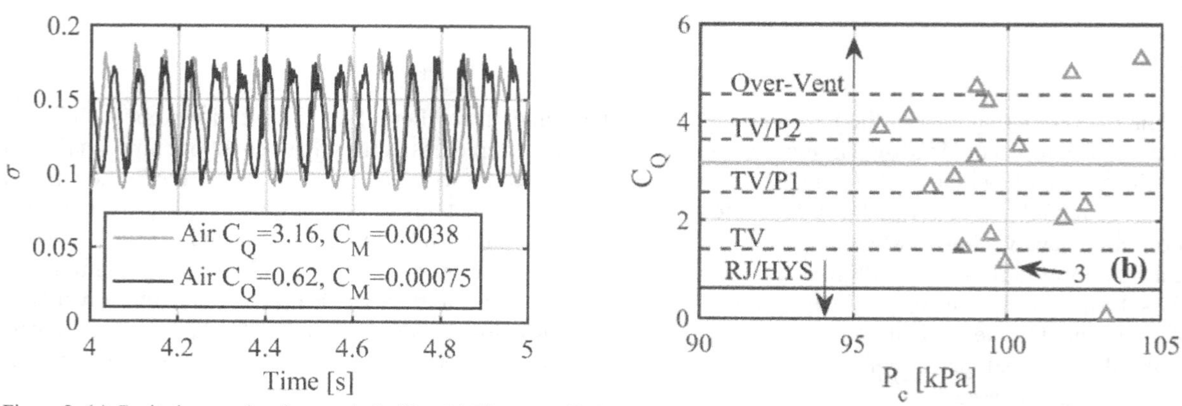

Figure 2: (a) Cavitation number for a period of 1 s. (b) Those ventilation rates correspond to both pulsation regions where the red triangles are experimentally measured cavity pressures.

While it was not possible (in the water tunnel) to visually ascertain pulsation in the hysteresis region, at the second lowest ventilation rate in Figure 2b, denoted by (3), there are strong indications of pulsation. The cavity interior and +25 dB shifted radiated pressure spectra for this cavity are given in Figure 3. Around 30 Hz is a tone that is related to pulsation and is pointed out by the blue arrow. High-vent pulsation had a tone at 33 Hz with an amplitude at 190 dB. While the frequency difference between the two pulsation experimental regions is similar to the CFD results, the pulsation amplitudes are nearly the same in the CFD results between the two pulsation regions. The cavity in Figure 3 has not fully transitioned to pulsation nor been initiated. Nonetheless, these results indicate there are two distinct pulsation regions and they occur when the ventilation rate is too great for either a re-entrant jet or twin vortex cavity. In this condition, pulsation is an efficient mechanism for gas removal from a cavity. While the ventilation condition is necessary, it is not sufficient for pulsation; there must be some kind of trigger to force the cavity into pulsation in regions susceptible to pulsation.

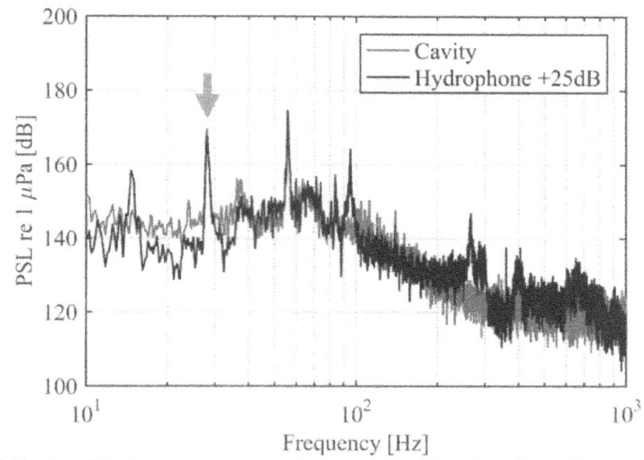

Figure 3: Cavity interior and +25 dB shifted radiated pressure spectra for an air ventilated cavity with C_M=0.0014/C_Q=1.15.

There are two methods, documented during water tunnel testing, to initiate pulsation. One method is through a pressure impulse and the other is by a high supply pressure. In Figure 4a, a pulsating cavity generated using the high pressure method, supply pressure set to 689.5 kPa, is shown. There is no cavity when a ball valve is opened and gas begins to vent into the tunnel. This method uses a high enough supply pressure to perturb the system into immediate pulsation without the need for an existing cavity. In Figure 4b, the impulse method was used to generate the cavity shown. This case is from work presented in reference [6] and was generated with a lower supply pressure (~69 kPa). To generate this pulsating cavity, a steady cavity is first formed, then the gas is shut off. When the cavity transitions to a re-entrant jet cavity, the ball valve is quickly opened, and the pressure wave generated from the unsteady action of quickly opening the ball valve initiates pulsation. The relatively low supply pressure demonstrates that high pressure is not needed to maintain pulsation. Thus, maintaining pulsation appears to be a linear process.

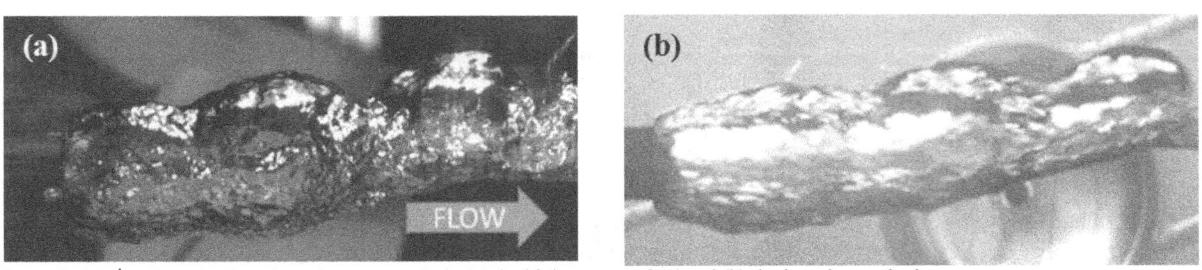

Figure 4: A 2nd order pulsating air cavity generated via (a) the high pressure feed and (b) the impulse method.

The high pressure feed method was verified using SF_6 as a ventilation gas. Since the density of SF_6, at fixed conditions, is greater than that of air, the pressure drop in the ventilation system is greater. Therefore, a higher supply pressure is needed to initiate pulsation. To trigger pulsation with SF_6, the needed supply was roughly 1379 kPa. In Figure 5a, the resulting cavity is shown. A twin vortex cavity, in Figure 5b, was generated with a supply pressure of 690 kPa. Undulations on the lower and upper interface indicate pulsation, seen in (a). The C_Q and pressure values are 0.67 and 95.6kPa (a) and 0.66 and 104.4kPa (b). The pressure reduction is also an indication of pulsation versus twin vortex. At the same ventilation rate, doubling the supply pressure caused the SF_6 supplied cavity to transition from twin vortex to pulsation; thus, altering the supply pressure for a non-choked flow leads to changes in cavity dynamics. This indicates a feedback loop between the ventilation system and cavity. Initiating cavity pulsation is a nonlinear process and requires a finite amplitude perturbation.

Skidmore *et al.* [9] have shown, both analytically and experimentally, that without altering the mean ventilation rate, pulsation can be disrupted with small amplitude modulation of the ventilation rate. By modulating the gas supply flow, waves on the cavity interface, evidence of which is given in the pressure spectra in Figure 3 and shown on the cavity interface in Figure 5a, are disrupted and pulsation is prevented. The amplitude dependence of this method of pulsation disruption has not been thoroughly investigated.

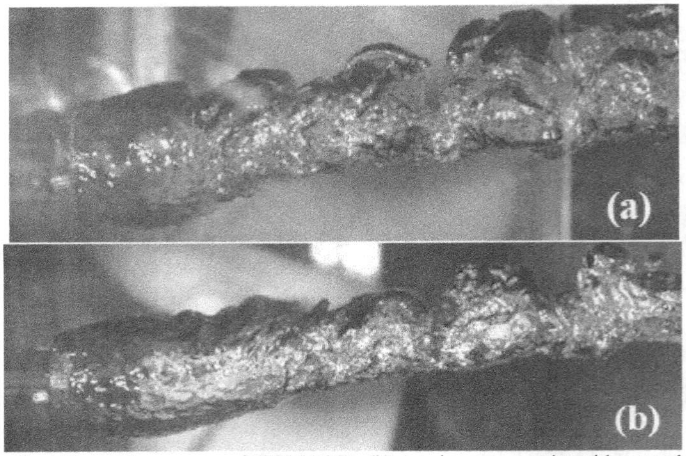

Figure 5: (a) A pulsating SF_6 cavity with supply pressure of 1378.95 kPa. (b) A twin vortex cavity with a supply pressure of 689.5 kPa

Conclusions

Two conditions are necessary for pulsation. One is that the cavity must be transitioning between closure regimes and the other is that an initializing, finite amplitude perturbation is required. Thus, unlike twin vortex or re-entrant jet cavities, pulsation is never history independent; twin vortex and re-entrant jet cavities can exist at vent rates where pulsation is not possible and other regimes are not attainable. When cavity closure regimes cannot efficiently remove gas, a cavity can pulsate, which is, for those cases, the most efficient method of gas removal, but causes strong fluctuations on the cavity interface. Mitigating pulsation occurs through disruption of the waves on the cavity interface that force pressure (and volume) fluctuations of the gas.

Acknowledgments

This work was sponsored by the Office of Naval Research under Grant N00014-11-10322 and monitored by Dr. David Drumheller.

References

[1] Michel J M 1984 Some Features of Water Flows With Ventilated Cavities *J. Fluids Eng.* **106** 319
[2] Paryshev E V. 2006 Approximate mathematical models in high-speed hydrodynamics *J. Eng. Math.* **55** 41–64
[3] Semenenko V N 2002 Artificial Supercavitation: Physics and Calculation *DTIC ADP012080* 33
[4] Silberman E and Song C C S 1961 Instability of Ventilated Cavities *Jour. Sh. Res.* **Vol. 5** 13–33
[5] Karn A, Arndt R E A and Hong J 2016 An experimental investigation into supercavity closure mechanisms *J. Fluid Mech.* **789** 259–84
[6] Skidmore G M, Brungart T A, Lindau J W and Moeny M J 2015 Noise generated by ventilated supercavities *Noise Control Eng. J.* **63** 94–101
[7] Skidmore G M, Lindau J W, Brungart T A, Moeny M J and Kinzel M P 2017 Finite Volume , CFD Based Investigation of Supercavity Pulsations *J. Fluids Eng.* **139** 91301
[8] Song C S 1962 Pulsation of Ventilated Cavities *J. Sh. Res.* **5** 8–20
[9] Skidmore G M, Brungart T A, Lindau J W and Moeny M J 2016 The control of ventilated supercavity pulsation and noise *Int. J. Multiph. Flow* **85** 14–22

© 2018 ASME

Re-entrant Gas Flows of High-speed Ventilated Supercavity

Wang Zou*; Lei-Ping Xue; Ben-Long Wang; Xin-Tao Xiang;

MOE Key Laboratory of Hydrodynamics, School of Naval Architecture, Ocean and Civil Engineering, Shanghai Jiao Tong University, Shanghai, China

Abstract

Re-entrant jet closure mechanism usually occurs for ventilated supercavitating flows at high speed in unbounded flows. The tail closure pattern is closely related to gas-leakage mechanism. The multi-fluid model, which involves interphase momentum and mass interactions, is applied to simulate the flows. Re-entrant flow mechanism is analyzed with more detail, and gas velocity distribution laws are presented in the re-entrant flow region. It is also found that maximum gas velocity occurs in the region. The maximum value is much greater than the incoming flow velocity in the opposite direction, and the relationship between the two is illustrated. This study contributes to deeply understand the gas entrainment and closure mechanism of high-speed supercavity.

Keywords: ventilated supercavitating flows; re-entrant flows; velocity distribution

Introduction

Supercavitation drag reduction has a revolutionary impact on the research and development of high-speed underwater vehicle and receives worldwide attentions. A better understanding of supercavitating flow mechanism leads to apply this method and technology quite flexibly. The flows involve many problems, such as phase change, re-entrant jets, gas loss, and these phenomena are interrelated and intricate. As an important part of flow mechanisms, flow structure inside supercavity has been an emergent task to study.

Since ventilated supercavitation was discovered, researchers have made endeavours in this area and some achievements in the formation, hydrodynamics and stability of supercavity[1-5], but inner flow structures remain little focused. Unlike natural case, where vapor-water mixture interacts[6], there are gas vortex ring structures inside ventilated supercavity[7,8]. It is inextricably linked with gas loss[9,10], which not only determines supercavity tail closure[11], but also is crucial to the flow stability and control[12,13].

The multi-fluid model, involving momentum interactions and mass transport between gas, vapor and water, is applied to reveal re-entrant gas flows in the tail region for high-speed supercavitating flows. The re-entrant mechanism is demonstrated by analyzing flow structure in more detail. The velocity distribution laws are presented in the re-entrant regions at different Reynolds and cavitation numbers, and the maximum gas velocity is found and the relationship between the velocity and cavitation number is obtained.

Multiphase flow model

Considering momentum interactions in different phases, the gas-water two-fluid model is qualified for ventilated supercavitating flows[14-16]. However, some unsteady factors or processes can induce natural cavitation in high-speed states. Further considering the vapor-water mass transport, the gas-vapor-water multi-fluid model was developed in our recent work[8] and is applied here. In the model, governing equations consist of the continuity and momentum equations of each phase and the volume conservation and pressure constraint equations. SST model (shear stress transport) based on $k-\omega$ [17] is used to close Reynolds equations. The cavitation model deduced from Rayleigh-Plesset equation[18], which represents the vaporization and condensation processes, is added in vapor and water continuity equations.

Results and discussion

Based on the multi-fluid model, supercavitating flows are simulated to reveal gas flows in the re-entrant flow region defined from sections S_{t1} to S_{t4} in Figure 1, where S_{t1} is the cavity section where re-entrant flows start to

*Corresponding Author, Wang Zou: hopingzou@sjtu.edu.cn

occur; S_{t4} is the section where the pressure difference between S_{t3} and S_{t4} does not exceed 5%; $\overline{V}$ is the non-dimensional gas velocity, $\overline{V} = V_g/V_\infty$, V_g and γ_g are the gas velocity and volume fraction; V_∞ is the inflow velocity. It can be clearly shown that gas escapes along cavity wall and that most gas flows back into cavity interior. There is a strong adverse pressure gradient, as shown in Figure 2, where $\overline{x}$, $\overline{r}$ and $\overline{p}$ are the non-dimensional axial and radial coordinates and pressure, $\overline{x} = x/L_c$, $\overline{r} = r/r_m$ and $\overline{p} = p/p_f$, L_c is the supercavity length, r_m is the cavity radius, p_f is the ambient pressure. Because the pressure gradient ($-\partial p/\partial x$) is large along the re-entrant flow direction in the beginning, pressure decreases and velocity continues to increase from sections S_{t1} to S_{t3}. Gas speed achieves its maximum in section S_{t3} and is about 1.86 times as large as the inflow speed. Then the pressure gradient is greatly reduced, while the effect of viscous force is significant to decrease the speed. The radial pressure distribution in the cavity section is almost uniform in Figure 2(c), which is consistent with the expansion principle of cavity section[19,20].

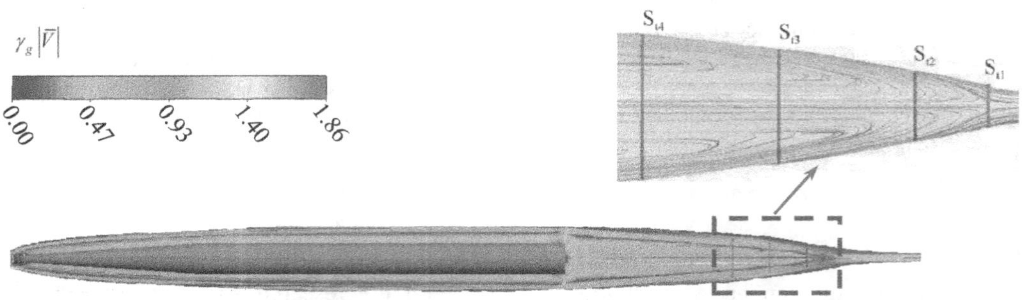

Figure 1 Ventilated supercavitating flows ($Re = 5.20\times10^5$, $\sigma_c = 0.0233$)

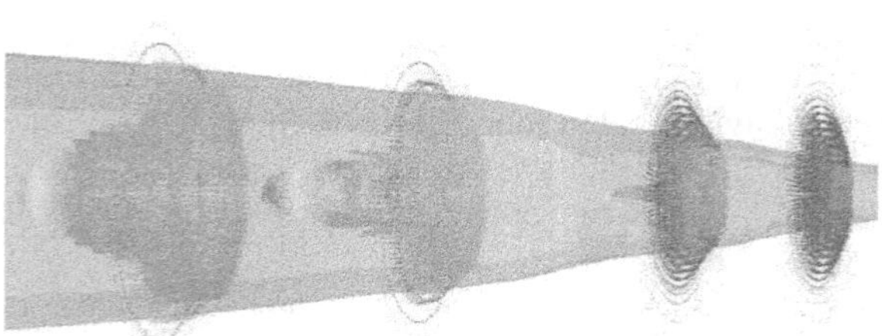

Figure 2(a) Gas velocity distribution

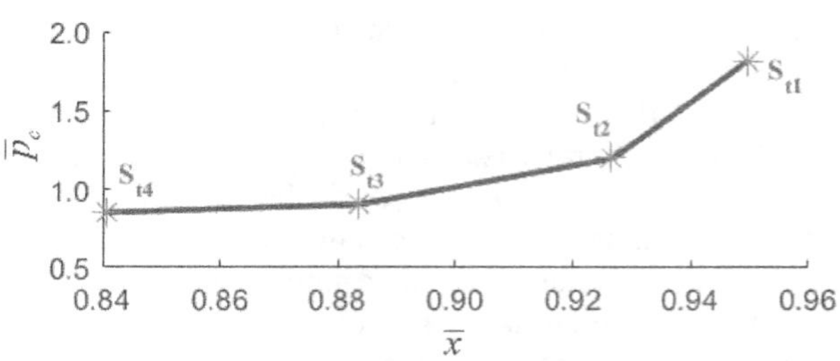

Figure 2(b) Axial pressure distribution

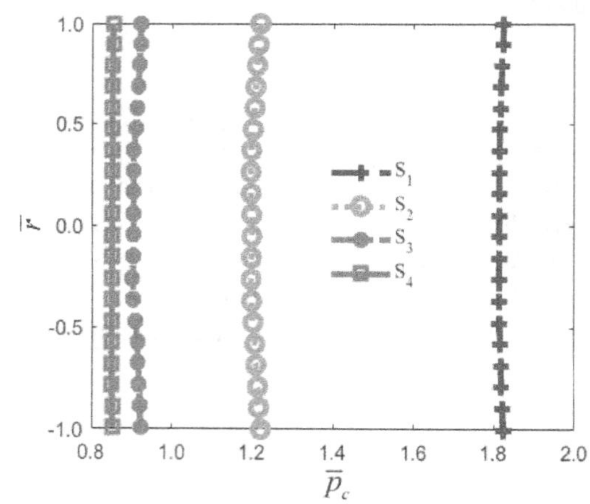

Figure 2(c) Radial pressure distribution

Figure 2 Re-entrant flows under adverse pressure gradient ($Re = 5.20\times10^5$, $\sigma_c = 0.0233$)

Flows simulations are carried out to at different Reynolds and cavitation numbers, and the gas velocity distribution laws are obtained by the analyses of flows in the characteristic sections S_{t1} and S_{t3}. In Figure 3(a), gravity effects are obvious to float supercavity tail, where a group of Froude numbers, $Fr = [258.29, 413.26, 628.12]$, corresponds to different Reynolds numbers; the smaller $\sigma_c F_r$, the greater gravity effect and the more velocity symmetry axis moves up; Cavitation number decreases to increase the pressure in the section S_{t1} so that gas in the central area of the section is entrained away more difficultly and the velocity distribution curves become smoother. The maximum velocity $\overline{V}_m$, $\overline{V}_m = V_m/V_\infty$, always occurs in the re-entrant flow region, as shown in Figure 3(b). It can be seen from Figure 4 that there is a multiple relationship between the maximum and inflow speeds. The smaller cavitation number, the greater the maximum value.

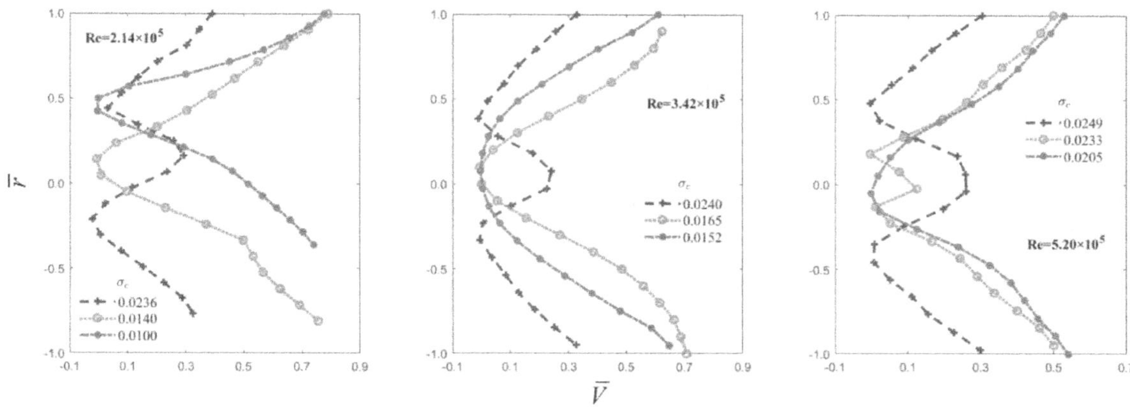

Figure 3(a) Gas velocity in the initial re-entrant section S_{t1}

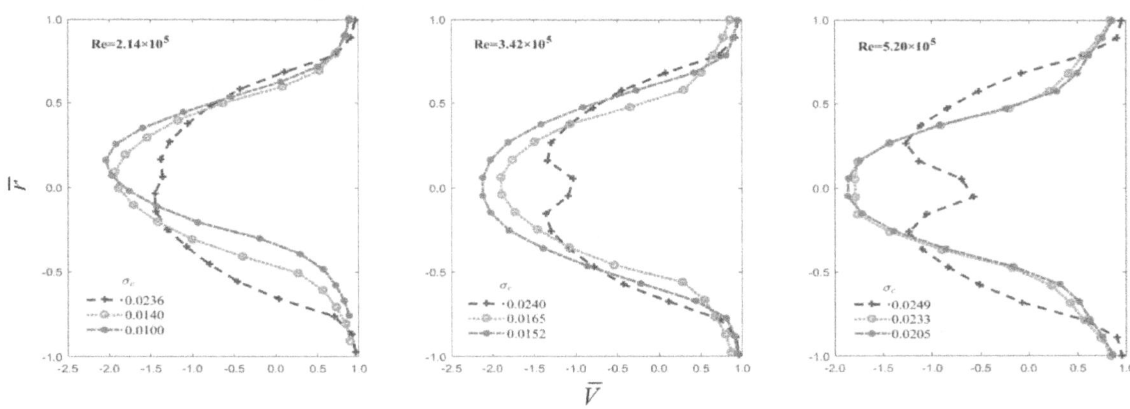

Figure 3(b) Gas velocity in the maximum velocity section S_{t3}

Figure 3 Gas velocity distributions in the characteristic sections

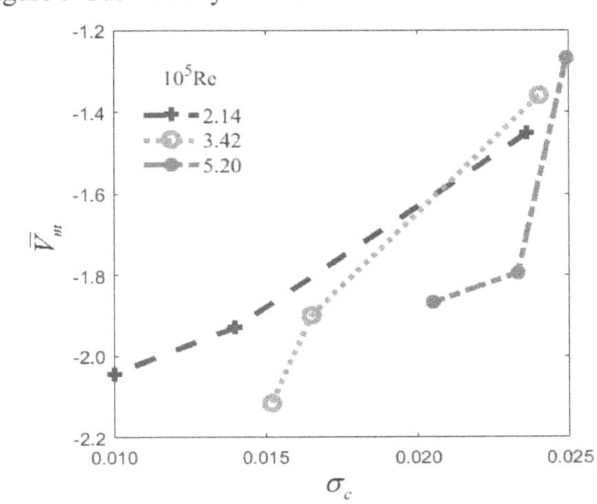

Figure 4 Maximum gas velocities inside supercavities

Conclusion

The developed multi-fluid model is applied to reveal flow structures in the re-entrant flow region for ventilated supercavitating flows at high speed. Re-entrant gas flows occur under the adverse pressure gradient, and the velocity distribution characteristics are presented at different Reynolds and cavitation numbers. Due to the multiple relationship between the maximum gas and inflow velocities, gas compressibility needs to be considered for high-speed ventilated supercavitating flows.

References

[1] Stinebring, D. R., Billet, M. L., Lindau, J. W., et al. (2002). *Developed cavitation-cavity dynamics.* In: RTO AVT Lecture Series on "Supercavitating Flows", Brussels, 2001-01-12~16. Canada: St. Joseph Ottawa/Hull.

[2] Kirschner, I. N., Fine, N. E., Uhlman, J. S., et al. (2001). *Supercavitation research and development.* Undersea Defense Technologies, Hawaii, Waikiki, HI.

[3] Lindau, J. W., Kunz, R. F. (2004). *Advancement and application of multiphase CFD modeling to high speed supercavitating flows.* DTIC document, Pennsylvania State University Park Applied Research Lab, Pennsylvania.

[4] Lu, C. J., He, Y. S., Chen. X., et al. (2007). *Numerical and experimental research on cavitating flows.* In: Proceedings of the Fifth International Conference on Fluid Mechanics, Shanghai, China.

[5] Wei, Y. J., Cao, W., Wang, C., et al. (2007). *Experiment research on character of ventilated supercavity.* In: Proceedings of the Fifth International Conference on Fluid Mechanics, Shanghai, China.

[6] Li, X. B., Wang, G. Y., Zhang, M. D., et al. (2008). *Structures of supercavitating multiphase flows.* International Journal of Thermal Sciences. 47.

[7] Savchenko, Y. N., Savchenko, G. Y. (2012). *Gas flows in ventilated supercavities.* Supercavitation, Springer, Berlin.

[8] Zou, W., Xue, L. P., Jin, W. W., et al. (2017). *Investigation into internal flow velocity distribution and gas loss of high-speed supercavitating flows.* Proceedings of ASME 2017 Fluids Engineering Division Summer Meeting, Waikoloa, Hawaii, USA.

[9] Spurk, J. H. (2002). *On the gas loss from ventilated supercavities.* Acta Mechanica. 155.

[10] Kinzel, P. K., Lindau, J. W., Kunz, R. F. (2009). *Air entrainment mechanisms from artificial supercavities: insight based on numerical simulations.* In: Proceedings of the Seventh International Symposium on Cavitation, Ann Arbor, Michigan, USA.

[11] Paryshev, E. V. (2006). *Approximate mathematical models in high-speed hydrodynamics.* Journal of Engineering Mathematics. 55.

[12] Paryshev, E. V. (2003). *Mathematical modeling of unsteady cavity flows.* In: Proceedings of the Fifth International Symposium on Cavitation, Osaka, Japan.

[13] Zou, W., Liu, H. (2015). *Control of the ventilated supercavity on the maneuvering trajectory.* Ocean Engineering. 101.

[14] Xiang, M., Cheung, S. C. P., Tu, J. Y., et al. (2011). *Numerical research on drag reduction by ventilated partial cavity based on two-fluid model.* Ocean Engineering. 38.

[15] Yu, K. P., Zhou, J. J., Min, J. X., et al. (2010). *A contribution to study on the lift of ventilated supercavitating vehicle with low Froude number.* Journal of Fluids Engineering. 132.

[16] Kunz, R. F., Gibeling, H. J., Maxey, M. R., et al. (2007). *Validation of two-fluid Eulerian CFD modeling for microbubble drag reduction across a wide range of Reynolds numbers.* Journal of Fluids Engineering. 129(1).

[17] Menter, F. R. (1994). *Two-equation eddy-viscosity turbulence models for engineering applications.* AIAA Journal. 32(8).

[18] Bakir, F., Rey, R., Gerber, A.G., et al. (2004). *Numerical and experimental investigations of the cavitating behavior of an inducer.* International Journal of Rotating Machinery, 10.

[19] Logvinovich, G. V. (1969). *Hydrodynamics of flows with free boundaries.* Naukova Dumka, Kiev, Ukraine (in Russian).

[20] Serebryakov, V. V. (2009). *Physical-mathematical bases of the principle of independence of cavity expansion.* In: Proceedings of the Seventh International Symposium on Cavitation, Ann Arbor, Michigan, USA.

© 2018 ASME

Regime Classification Methods for a Gas Jet in a Liquid Co-Flow

[1]Melissa Fronzeo*, [1]Samuel Hansford, [1]Zachary Berger, [1]Michael Kinzel

[1]*The Pennsylvania State University, University Park, PA, USA*

Abstract

The present effort evaluates various methods to classify flow regimes from computational fluid dynamics (CFD) simulations of various multiphase flows. The present effort specifically aims to understand a gas jetted into a liquid co-flow, with parametric variations of the freestream liquid velocity and jet mass flow rate. Previous CFD indicated several flow regimes occur in this interaction. However, the method of regime classification relied solely on visual inspection of simulation videos, as well as time-averaged solutions. This work compares several methods to classify gas-liquid interactions, and assesses the regime identification capabilities of each method. The methods evaluated are as follows: Proper Orthogonal Decomposition (POD), Fast Fourier Transformation (FFT), and Mixture of Gaussian (MOG) methods. The results indicate that POD may be the most efficient, effective method to classify the flow regime.

Introduction

The classification of multiphase flows can be a complicated process. Specifically, it relies on qualitative evaluations of the flow that are prone to bias. For this reason, we want to develop quantitative methods for flow regimes specifications. With such methods, we anticipate it will help to develop flow-regime maps based on quantitative descriptions as opposed to an opinion-based assignment.

In this effort, we investigate gas jets in a liquid co-flow in a pipe, which is a flow indicated in Figure 1. Such a flow is being examined to understand gas mixing within a liquid co-flow, which is a process that is dominated by the initial cavity formation and its dynamics. Unlike single-phase jets, such multiphase jets are poorly understood. This is, despite, widespread industrial use for gas jets in a quiescent liquid, such as chemical mixing, nuclear reactor maintenance, and underwater cutting [1-3]; difficulties in experimental visualization cause the full jet formations to be understudied. One factor that makes these jets so interesting is that despite the gas jets having high speeds/momentums, they remain weak with respect to the liquid. Such a feature leads to highly dynamic flow phenomena and multiple flow regimes. CFD is employed to better understand these multiphase jets, which have previously been applied, using two-dimensional, axisymmetric simulations to examine submerged multiphase jets. While varying the freestream velocity (V_∞) and the mass flow rate of the gas jet ($\dot{m}_{jet}$), it became clear that five distinct regimes occur. However, classification of these various regimes has only been conducted through visual inspection, using a combination of videos extracted from the simulations and time-averaged images. The combination of these two visual methods resulted in the classification of five regimes for a gas (air) jet submerged in a liquid (water) co-flow. Therefore, the focus of this paper is to investigate various additional tools to aid in the classification of these multiphase jet regimes.

The following work details the findings of three different methods: POD, FFT, and MOG. The three methods will be compared in their ability to differentiate between the flow regimes, as well as ease of use.

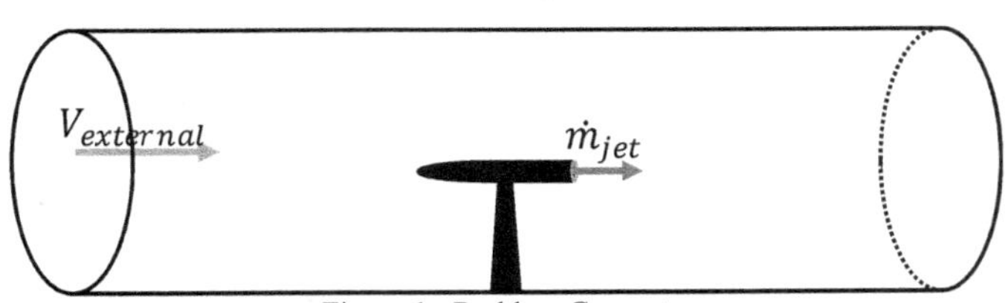

Figure 1: Problem Geometry

Methods

The results discussed in this work focus on different classification methods for multiphase jet flow regimes. The simulations used were conducted using Star-CCM+ [4]. The physics modeling was validated against experimental data from Weiland's PhD dissertation [1], and the validation process is described in the same work which identifies the flow regimes [5,6]. The flow regimes identified in the previous work detail the interactions between the gas jet (using air) and the liquid freestream (using water),a) and are as follows:

*Corresponding Author, Melissa Fronzeo: maf5524@psu.edu

- Shedding Jet: Large bubble ejections; gas is attached to the nozzle
- Toroidal Cavity: Small gaseous cavity; liquid enters cavity in constant stream.
- Pulsating Cavity: Medium gaseous cavity; large interface waves; liquid enters cavity
- Stable Cavity: Large gaseous cavity; little to no liquid enters cavity; small interface waves
- Over-Ventilated Cavity: Small gaseous cavity; large bubble ejections from end of cavity.

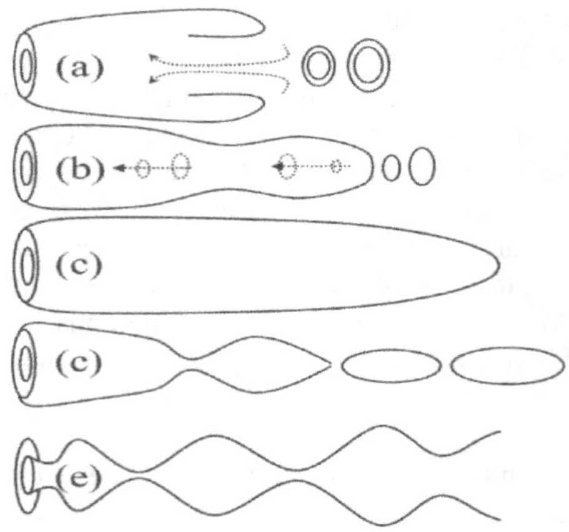

Figure 2: Cartoons of each mode; (a) Toroidal Cavity, (b) Pulsating Cavity, (c) Stable Cavity, (d) Over-Ventilated Cavity, (e) Shedding Jet.

Examples of each of these regimes are shown in Figure 2 (cartoons) and Figure 3 (simulation). The grayscale lines are volume fraction, which outline the gaseous cavity (where regions outside the black are pure liquid and white indicates gas). The color contours are Mach, with dark blue being low Mach, and the brighter blue/cyan colors showing the jet. The top images are of the time-averaged simulation, while the bottom images are a snapshot of the unsteady simulation.

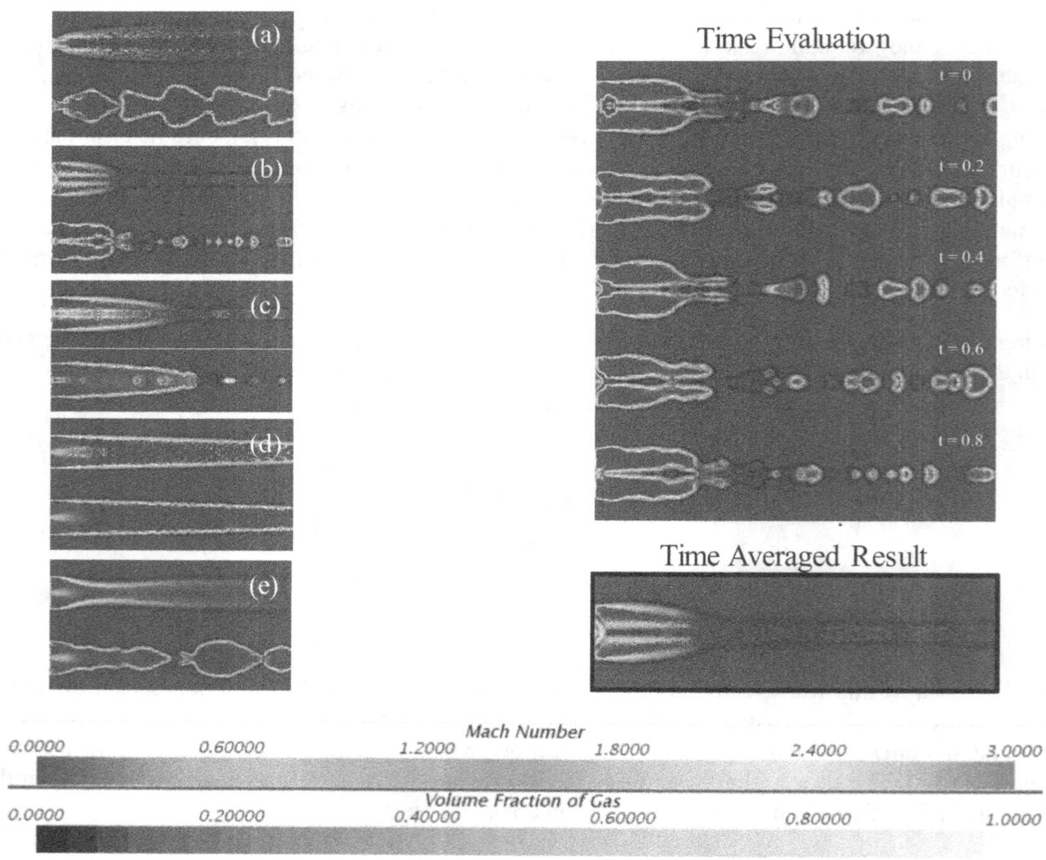

Figure 3: (a) Shedding Jet, (b) Toroidal Cavity, (c) Pulsating Cavity, (d) Stable Cavity, (e) Over-Ventilated Cavity

Figure 4: Time Sequence (top) and time-averaged solution (bottom) of a Toroidal Cavity

Previously, the regimes were classified through visual inspection, comparing videos created from the unsteady simulations and the time averaged solutions. This provided the insight to pick out the properties described above, and subsequently classify regimes. A sample of this process is provided for a Toroidal Cavity case in Figure 4.

The new methods of regime analysis to be investigated (POD, FFT, and MOG) all require the full simulation data. The POD and FFT methods require a video, or series of snapshots, extracted from the simulation while it was running. MOG can be conducted on mean data collected at the end of the simulation.

The first tool of interest is the use of POD. The POD approach offers several advantages(1) it is statistically based, (2) it has clear analytical foundations for understanding its limitations and capabilities, and (3) it allows for the extraction of spatial and temporal structures from a turbulent field (Berkooz, Holmes, and Lumley) [7]. Applying POD to a set of multiphase jet cases, simulated through CFD, isolates large-scale structures within the regimes, such as waves on the interface. The exact method that will be applied in this paper is a singular value decomposition (SVD). The approach uses a series of images, with a known time delay between images, and computes the spatial eigenmodes, or POD modes. The number of modes is dependent on the number of images input into the code. Each mode has a given amount of energy, with lower number modes being the most energetic, and thus they cause the more pronounced interfacial motion on the jet regimes. The original images can then be reconstructed using a set energy cut off, which allows one to determine the dominant modes and large scale motions in the overall movement. This in turn aids in the identification of regimes based on significant interfacial characteristics. In regards to this method, the input parameters, such as initial image quality and the number of input images, must also be investigated. Using very large or high-quality images can become computationally expensive, with large amounts of computer memory required to process the data. This method provides additional visual information for study in the spatial modes, and the amount of energy required to reconstruct an image provides some quantitative data.

The second method is employing an FFT of the temporal behavior linked to the POD modes. Thus, this FFT data is collected and output within the same code as the POD analysis. This gives the frequency characteristics of the corresponding POD modes from the latter analysis. However, this is highly dependent on the number of images input, as well as their temporal spacing, as the method is incapable of extracting frequency data below the temporal cutoff. However, it is worthwhile to investigate the frequency peaks which correspond to the different spatial POD modes, looking for a method of classification

The final method investigated with respect regime classification is the MOG method [8]. This method takes a histogram of the resulting field within each condition and using the distribution to map to a specific regime. More specifically, the resultant peaks in data (such as internal pressure, volume fraction of gas in a defined region, internal velocity, or turbulent kinetic energy within the system) can then be compared from regime to regime, to see if they have any discerning power in regime classification. These peaks can be statistically analyzed, and the overall system of peaks can be reconstructed using a number of Gaussian peaks. This method has previously been applied to cavitation and supercavitation to identify regions and physical phenomenon, and it will be interesting to apply the method to regime classification.

Results

This section details some of the results of each method, along with a description of the regime discerning power with regards to the original visual method.

POD is similar to the time-averaged solution images, but the coherent structures of the waves are easier to identify. Such a process allows for better discriminatory power of the flow regimes, though it still requires visual inspection and comparison. A case for each regime (classified by the original visual method) is shown below (Figure 5) The bottom half of each image is the 1st spatial eigenmode, which is the mode with the largest energy. The black and white patches distinguish wave formations, giving an idea of frequency and distance over which they occur. The modes have been placed under a slice of the original simulation (converted to black and white to highlight the gas-liquid interface), at a single time step.

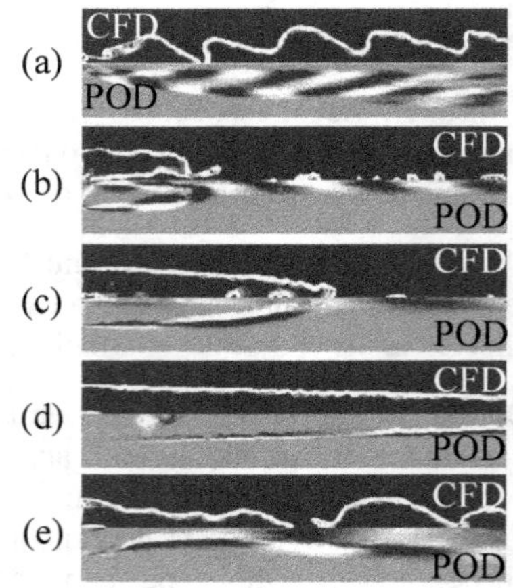

Figure 5: POD of Shedding Jet (a), Toroidal Cavity (b), Pulsating Cavity (c), Stable Cavity (d) and Over-Ventilated Cavity (e)

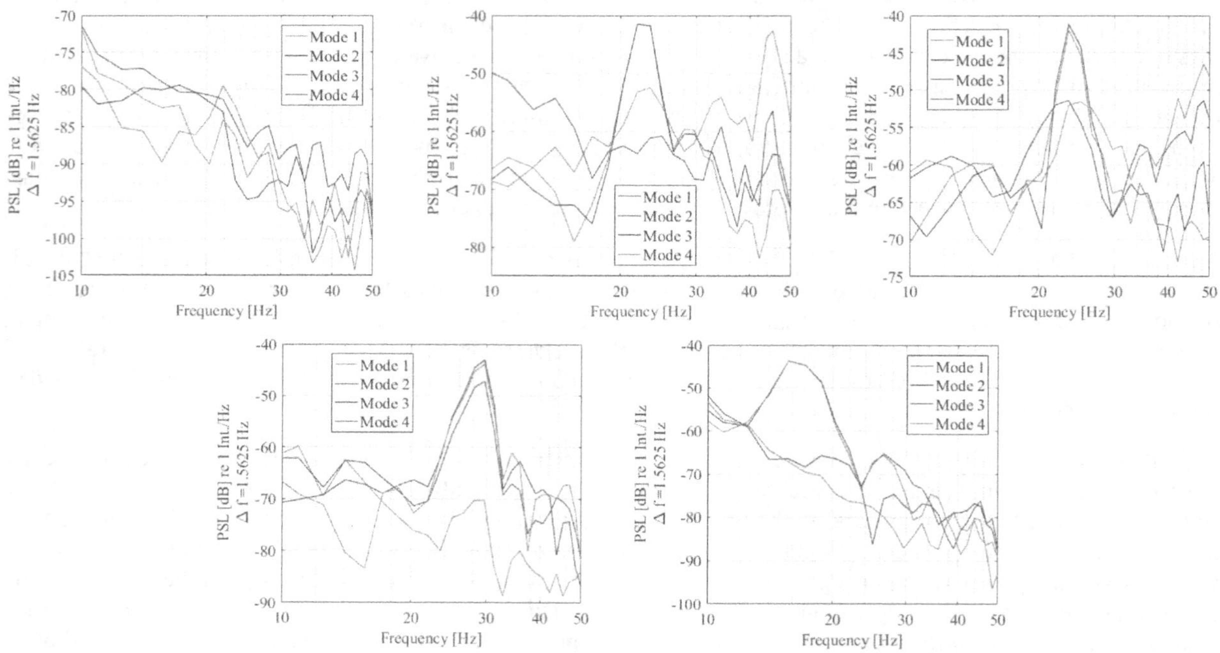

Figure 6: FFT of Shedding Jet (top left), Toroidal Cavity (top middle), Pulsating Cavity (top right), Stable Cavity (bottom left) and Over-Ventilated Cavity (bottom right)

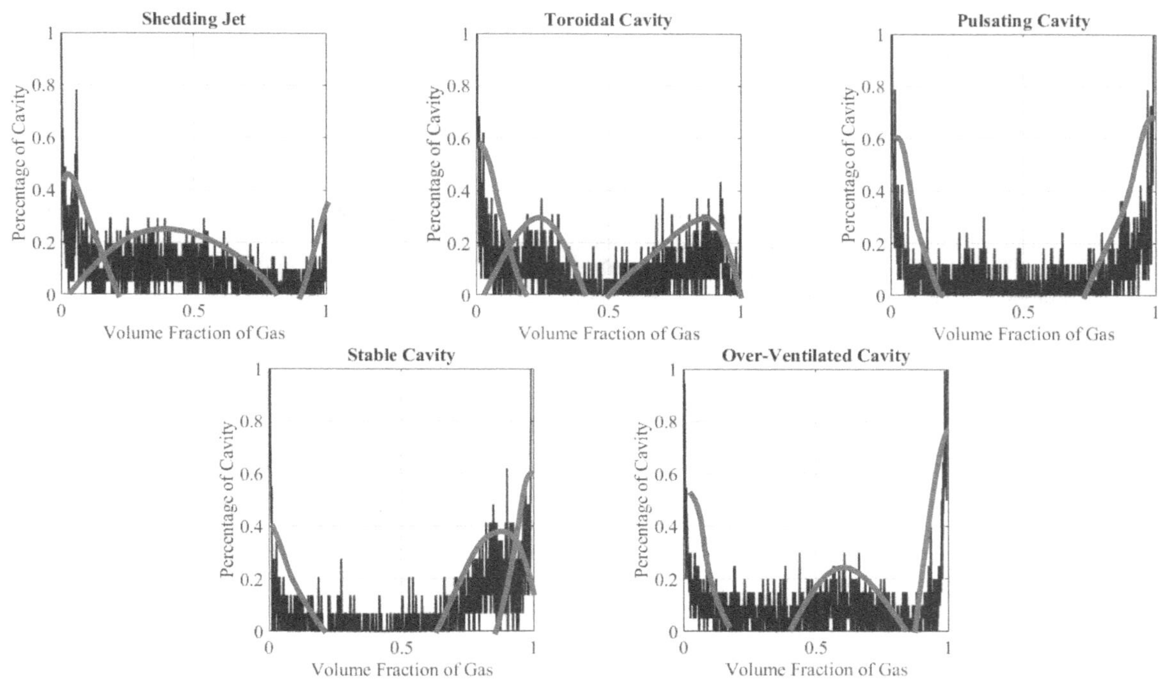

Figure 7: Volume Fraction PDFs of a Shedding Jet (left) and Pulsating Cavity (right)

FFTs are limited by image quality and number of images that can be processed at once, which is overall limited by processing power. The number of images input in these cases limits the overall frequency that can be extracted from the simulations. Due to this, only the lower frequency waves on the gas-liquid interface can be expressed as peaks. This is apparent in the toroidal and pulsating cavities (Figure 6), where the 1st and 2nd modes result in lower frequency peaks. However, there is no discriminatory power, currently limited by computational inputs, for the higher frequency regimes such as the shedding jet, stable cavity, and over-ventilated cavity.

Initial data from the MOG method (currently displayed using thee mean volume fraction of gas as the underlying PDFs) indicates there is some discerning capability to this method. The plots are displayed in Figure 7. Through comparison of plots for multiple cases from each regime, there are several trends that stand out. The shedding jet regime typically has the majority of the cavity at lower volume fractions of gas (0-0.2), and can typically be identified by this trait. Similarly, the stable cavity regime is heavily weighted towards high volume fractions of gas (0.7-1.0). The over-ventilated case has roughly three peaks (towards 0, 1.0, and one in the 0.4-0.8 range). The toroidal and pulsating cavities are the hardest cases to distinguish using this method, both being weighted at the 0 and 1.0 ends of volume fraction of gas. The possible difference may be that the center area (0.4-0.6) of the toroidal cases is very low (less than 0.2%), but more cases may need to be evaluated.

Conclusion

All three methods (POD, FFT, and MOG) still rely on some form of visual inspection. However, POD does provide more information than the original two methods (video and time average images). The ability of FFT to distinguish regimes is highly dependent time step and number of images fed through the process. The MOG method has some discerning ability, though many cases must be compared to classify the unique differences between each regime from the PDF plots. It seems that it is not necessarily peak location/size, but more statistical ranges in the jet's downstream volume fraction of gas. Additional methods of regime classification will still be of assistance in better differentiating intermediate regimes, or regimes that are not easily distinguished from only video and time averaged solutions.

References

[1] Weiland, C. J., 2009, "Characteristics of the high speed gas-liquid interface," Ph.D. thesis, Virginia Tech, Blacksburg

[2] Shi, H., Wang, B., and Dai, Z., 2010, "Research on the mechanics of underwater supersonic gas jets," Sci. China. Ser. G: Phys. Mech. Astron., **53**(3), pp. 527–535.

[3] Shi, H.-H., Guo, Q., Wang, C., Dong, R.-L., Zhang, L.-T., Jia, H.-X., Wang, X.-G., and Wang, B.-Y., 2010, "Oscillation flow induced by underwater supersonic gas jets," Shock Waves, **20**(4), pp. 347–352.

[4] Siemens, "StarCCM+," 2017.

[5] Fronzeo, Melissa, Michael Kinzel, and Jules Lindau "A Flow Regime Map for Gas Jets in a Bubbly Flow" Proceedings of the 91st Fluid Engineering Division Summer Meeting FEDSM July 2017.

[6] Fronzeo, M. A., 2017, "A Computational Study of Gaseous Jets Submerged in a Liquid Co-Flow", M.S. thesis, The Pennsylvania State Univeristy, State College

[7] Berkooz, Gal, Philip Holmes, and John L. Lumley. "The proper orthogonal decomposition in the analysis of turbulent flows." *Annual review of fluid mechanics* 25.1 (1993): 539-575.

[7] Esplin, J. J., Kim, B. J., Kinzel, M. P., and Culver, R. L., 2016, "Bulk cavitation extent modeling: An energy-based approach," J. Acoust. Soc. Am., 140(4), pp. 3121–3121.

© 2018 ASME

Sensitivity Analysis of Zwart-Gerber-Belamri Model Parameters on the Numerical Simulation of Francis Runner Cavitation

[1]Xavier Escaler*; [1]Rafel Roig; [2]Víctor Hidalgo

[1]*Universitat Politècnica de Catalunya, Barcelona, Spain;* [2]*Escuela Politécnica Nacional, Quito, Ecuador*

Abstract

A series of numerical simulations of the 3D cavitation development in the GAMM Francis turbine runner were carried out with the Zwart-Gerber-Belamri (ZGB) two-phase model available in ANSYS® CFX v16.2. The aim of the investigation was to evaluate the influence of the ZGB parameters on the location, size and shape of the simulated cavitation. To begin, a sector of the entire fluid domain comprising a single blade was created taking profit from the rotational symmetry of the geometry and a stage simulation was set to simulate the flow field without cavitation. A grid independence analysis was carried out using the shear-stress transport (SST) turbulence model. Results such as the pressure drop and the torque at the best efficiency point (BEP) compared reasonably well with the experimental values, thus proving the model validity. The cavitation inception and development was simulated at BEP using the ZGB default parameters by progressively decreasing the sigma value. As expected, the onset of cavitation took place at the blade suction side close to the junction with the band, and the length and area of the blade cavity increased for decreasing sigma values. The cavity shape also showed a good agreement with the observations at the GAMM turbine model. Finally, the values corresponding to the mean nucleation site diameter, the nucleation site volume fraction and the empirical coefficients of evaporation and condensation were modified individually while keeping the rest constant in order to observe the changes in cavitation behavior. As a result, it was concluded that all the parameters have a significant effect in the cavity length, the pressure distribution and the torque. Moreover, extreme values leading to unrealistic results were also found.

Keywords: Francis runner, GAMM turbine, stage simulation, Zwart cavitation model, empirical constants, nucleation diameter

Introduction

Cavitation in hydraulic turbines is well known to be harmful because it can reduce the performance and provoke vibrations and erosion of the solid components. The turbine design combined with particular operating conditions such as the actual setting level of the unit and the deviation from the best efficiency point (BEP), among others, are the main factors that will determine the risk of cavitation and its consequences. Moreover, the cavitation phenomenon inside a turbine occurs under unsteady flows with 3D effects. For that, numerical and experimental investigations are still required to understand the physics of the phenomenon [1]. The computational fluid dynamics(CFD) software ANSYS® CFX v16.2 incorporates the unsymmetrical cavitation model proposed by Zwart-Gerber-Belamri (ZGB) [2], which is based on the interphase mass transfer, $\dot{m}$, derived from the Rayleigh-Plesset equation. The ZGB model is expressed by (1):

$$\dot{m} = \begin{cases} -F_v \frac{3r_{nuc}(1-\alpha)\rho_v}{R_B}\sqrt{\frac{2}{3}\frac{P_v-P}{\rho_l}} & if\ P < P_v \\ F_c \frac{3\alpha\rho_v}{R_B}\sqrt{\frac{2}{3}\frac{P-P_v}{\rho_l}} & if\ P > P_v \end{cases}, \tag{1}$$

where F_v and F_c are the empirical calibration coefficients for vaporization and condensation respectively, r_{nuc} is the nucleation site volume fraction and R_B is the typical bubble radius of the nucleation site in water [3]. More specifically, the model default parameters are $F_v = 50$, $F_c = 0.01$, $r_{nuc} = 5\cdot10^{-4}$ and $R_B = 2\cdot10^{-6}$. The rest of variables are the vapor volume fraction, α, the liquid density, ρ_l, the vapor density, ρ_v, the time averaged pressure, P, and the liquid saturation pressure, P_v. The ZGB model has already been used and validated in several investigations for the numerical prediction of sheet cavitation around hydrofoils [4, 5], centrifugal pumps [6], marine propellers and Kaplan turbines [7] with fairly good results. Nevertheless, in all the cases it has been necessary to tune the model empirical coefficients to ensure the accuracy of the results.

*Corresponding Author, Xavier Escaler: xavier.escaler@upc.edu

In order to evaluate the influence of the ZGB model parameters on the numerical prediction of cavitation on the blades of a Francis runner, the GAMM Francis model geometry and performance [8, 9] were taken as a study case. More specifically, cavitation inception and development were simulated at the best efficiency point (BEP) which is characterized by a flow rate Q = 0,376 m3/s, a head H = 5,957 m, a rotating speed ω= 52,36 rad/s, and a torque T= 388 Nm.

The computational domain did not consider the distributor stay and guide vanes. The runner fluid domain was bounded upstream by a cylindrical surface located in the middle of the distributor channel and extended downstream up to a circular surface in the draft tube channel as indicated in Figure 1a. To reduce computational effort, a sector comprising only one inter-blade channel was created taking profit from the rotational symmetry of the runner and the existence of 13 blades as indicated in Figure 1b. The simulations were carried out using the Rotating Frames of Reference (RFR) and the Shear-Stress Transport (SST) turbulence models. Initially, a grid independency analysis was done with no cavitation model activated. The simulated torque and head values showed a good agreement with the experimental ones at BEP. As a result, an accurate mesh and the correct boundary conditions necessary to predict the expected behavior were established.

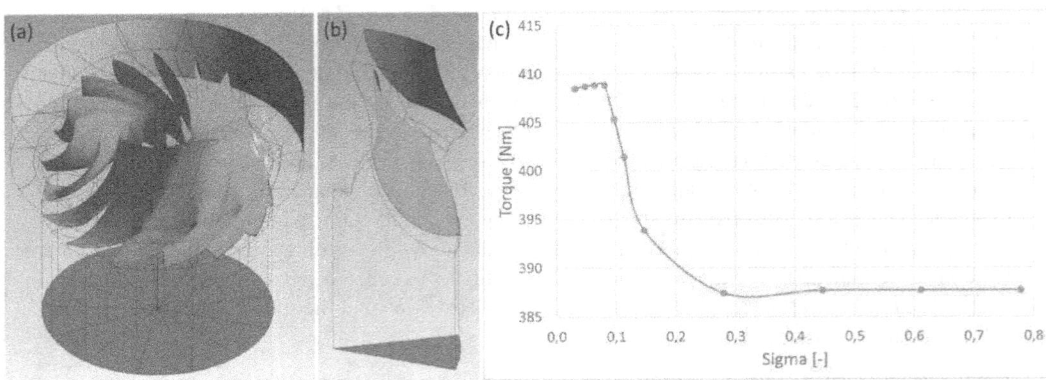

Figure 1. (a) Runner geometry; (b) sector computational domain; (c) simulated runner torque evolution.

Cavitation simulations

Initially, the cavitation development was simulated at BEP for different values of the outlet pressure, P_{out}, and using the ZGB default parameters. As a result, the typical torque rise for decreasing Sigma, σ, was correctly simulated (see Figure 1c). In our case, σ is calculated with (2) where g is the gravity:

$$\sigma = \frac{P_{out}-P_v}{\rho_l g H}. \qquad (2)$$

In Figure 2, the location, size and shape of the simulated leading edge cavity has been visualized with an isosurface of water vapor volume fraction equal to 0.5. As σ decreases, cavitation develops on the blades. A good similarity with the visual observations in the reduced scale GAMM turbine model at σ = 0.14 [6] is found with the numerical results at σ = 0.15, which is also in accordance with the works of Susan-Resiga *et al.* [7-8].

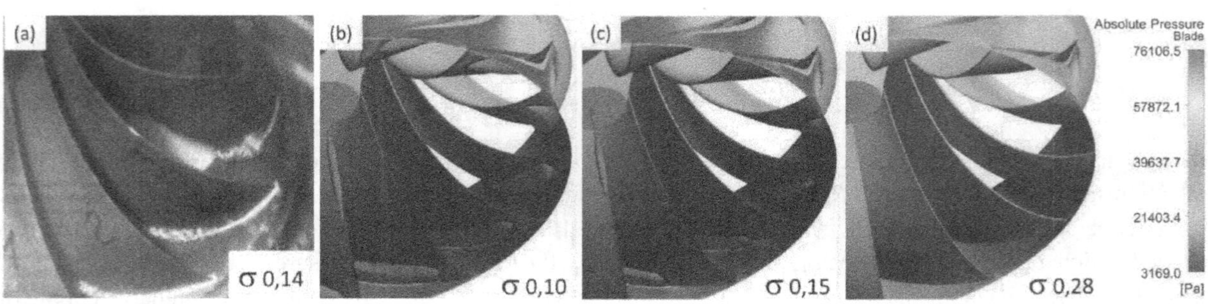

Figure 2. (a) Photography of the inlet edge cavitation in GAMM Francis turbine at BEP for σ= 0.14 [6]; contours of absolute pressure and isosurfaces of water vapor volume fraction 0.5 for σ = 0.10 (b), 0.15 (c) and 0.28 (d) using the ZGB default parameters.

	Min		**Default**		Max
F_v	$5\cdot10^{-1}$	$5\cdot10^{-0}$	$\mathbf{5\cdot10^{1}}$	$5\cdot10^{2}$	$5\cdot10^{3}$
F_c	$1\cdot10^{-4}$	$1\cdot10^{-3}$	$\mathbf{1\cdot10^{-2}}$	$1\cdot10^{-1}$	$1\cdot10^{-0}$
r_{nuc}	$5\cdot10^{-6}$	$5\cdot10^{-5}$	$\mathbf{5\cdot10^{-4}}$	$5\cdot10^{-3}$	$5\cdot10^{-2}$
R_B	$2\cdot10^{-8}$	$2\cdot10^{-7}$	$\mathbf{2\cdot10^{-6}}$	$2\cdot10^{-5}$	$2\cdot10^{-4}$

Table 1. Tested values of the ZGB parameters for $\sigma = 0.15$.

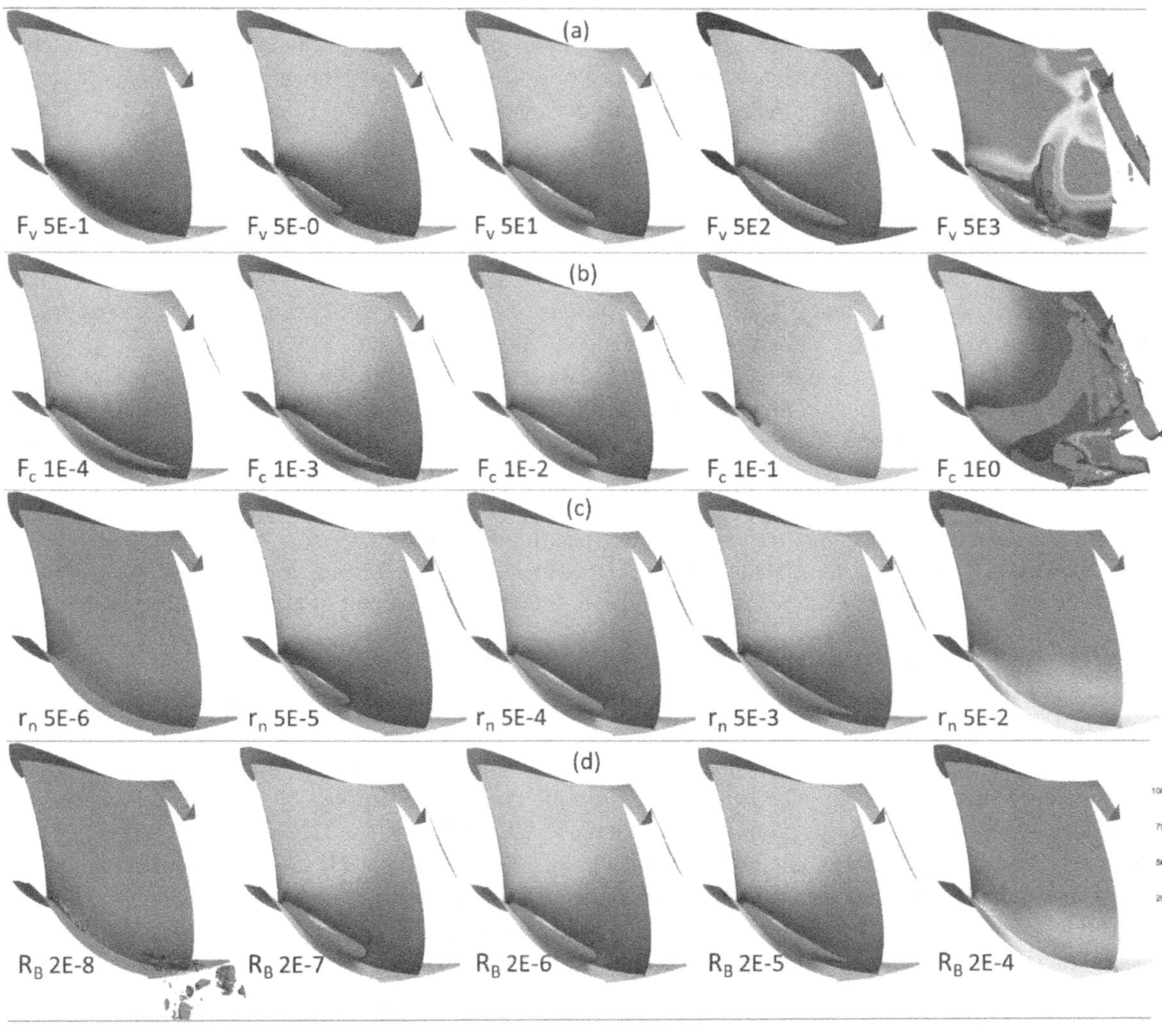

Figure 3. Contours of absolute pressure on walls and isosurface of water vapor volume fraction 0.5 at $\sigma = 0.15$ for different values of F_v (a), F_c (b), r_{nuc} (c) and R_B (d).

To study the influence of the ZGB parameters on the numerical results at $\sigma = 0.15$, each of the values indicated in Table 1 were considered individually while keeping the rest of them at their default values. The plots shown in Figure 3 correspond to the contours of absolute wall pressures and the 0.5 water vapor volume fraction isosurfaces for the different F_v (a), F_c (b), r_{nuc} (c) and R_B (d) values, from top to bottom rows respectively.

As it can be observed, unrealistic results were found for extreme values two orders of magnitude higher or lower than the default ones. In such cases, the torque can deviate more than 10% from the nominal value because the pressure distribution around the blade and the water vapor phase appear significantly altered. In particular, the possible range of values for each parameter that keep the expected runner efficiency are [0.5 – 500] for F_v, [0.0001 - 0.1] for F_c, [$5\cdot10^{-5}$ - $5\cdot10^{-3}$] for r_{nuc} and [$2\cdot10^{-7}$ - $2\cdot10^{-5}$] for R_B.

Regarding the cavitation development on the blade, all the parameters have a main influence on its length along the junction between the blade and the runner band. More specifically, F_v determines the location of both the cavitation inception and of the closure region, meanwhile F_c only determines the location of the cavitation closure. On the other hand, r_{nuc} shows a similar effect to F_v, and R_B shows a similar effect to the reciprocal of F_c.

Conclusion

The default values of the Zwart-Gerber-Belamri cavitation model available in ANSYS® CFX v16.2 appear to provide results that are in accordance with the cavitation observations in the GAMM Francis turbine runner at BEP. Nevertheless, the sensitivity of the inception and closure locations of the inlet edge cavitation to these parameters is found to be significant when they are increased or decreased by at least one order of magnitude. Moreover, unrealistic results are obtained if extreme values of the parameters are considered.

References

[1] Ji, B., Long, Y., Long, X., Qian, Z., and Zhou, J. (2017) *Large eddy simulation of turbulent attached cavitating flow with special emphasis on large scale structures of the hydrofoil wake and turbulence-cavitation interactions.* Journal of Hydrodynamics, Ser. B, 29(1):27-39.

[2] Zwart, P., Gerber, A. and Belamri, T. (2004) *A Two-Phase Flow Model for Predicting Cavitation Dynamics.* Proceedings of the ICMF 2004 Int. Conf. on Multiphase Flow, Yokohama, Japan.

[3] Hidalgo, V. H., Luo, X. W., Escaler, X., Ji, J. and Aguinaga, A. (2014). *Numerical investigation of unsteady cavitation around a NACA 66 hydrofoil using OpenFOAM.* IOP Conf. Ser.: Earth Environ. Sci. 22 052013.

[4] Hidalgo, V., Luo, X. W., Escaler, X., Bin, J. I. and Aguinaga, A. (2015). *Implicit large eddy simulation of unsteady cloud cavitation around a plane-convex hydrofoil.* Journal of Hydrodynamics, Ser. B, 27(6), 815-823.

[5] Morgut, M., Nobile, E. and Biluš, I. (2011) *Comparison of mass transfer models for the numerical prediction of sheet cavitation around a hydrofoil.* International Journal of Multiphase Flow, 37(6):620–626.

[6] Liu, H-l., Wang, J., Wang, Y., Zhang, H. and Huang, H. (2014). *Influence of the empirical coefficients of cavitation model on predicting cavitating flow in the centrifugal pump.* International Journal of Naval Architecture and Ocean Engineering, 6(1):119–131.

[7] Morgut, M., Jošt, D., Škerlavaj, A., Nobile, E. and Contento, G. (2017) *Numerical Predictions of Cavitating Flow Around a Marine Propeller and Kaplan Turbine Runner with Calibrated Cavitation Models.* Strojniški vestnik – Journal of Mechanical Engineering. Article in Press.

[8] Avellan, F., Dupont, P., Farhat, M., Gindroz, B., Henry, P., Hussain, M., Parkinson, E. and Santal, O. (1990) *Flow Survey and blade pressure measurements in a Francis turbine model.* Proceedings of 15th IAHR Symposium on modern technology in hydraulic energy production, September 1990, Belgrade, vol. 2, Session I, Paper 14, 14 pages.

[9] Avellan F., Dupont P., Farhat M., Gindroz B., Henry P. and Hussain M. (1993) *Experimental flow study of the GAMM turbine model.* Proceedings of the GAMM Workshop "3D-computation of incompressible internal flow" Ed. Sottas G. and Ryhming I.L., NNFM 39, Vieweg Verlag, Braunschweig, pp. 33-53.

[10] Susan-Resiga, R., Muntean, S. and Anton, I. (2002) *Numerical Analysis of Cavitation Inception in Francis Turbine.* Proceedings of the Hydraulic Machinery and Systems 21st IAHR Symposium, Lausanne, Switzerland.

[11] Susan-Resiga, R., Muntean, S., Anton, I. and Bernard, S. (2003) *Numerical Investigation of 3D Cavitating Flow in Francis Turbines.* Conference proceedings / Conference on Modelling Fluid Flow (CMFF '03) The 12th Int. Conf. on Fluid Flow Technologies, Budapest, Hungary.

© 2018 ASME

Shape Oscillation of an Encapsulated Bubble in Electric Fields

[1]Yunqiao Liu*; [2]Dongdong He; [1]Xiaobo Gong; [3]Huaxiong Huang

[1]MOE Key Laboratory of Hydrodynamics, Department of Engineering Mechanics, Shanghai Jiao Tong University, Shanghai 200240, China; [2] School of Science and Engineering, The Chinese University of Hong Kong, Shenzhen 518172, China; [3]Department of Mathematics and Statistics, York University, Toronto, Ontario M3J 1P3, Canada

Abstract

We investigate the dynamics of an encapsulated bubble in electric fields theoretically, based on the leaky dielectric model. On the bubble surface, the applied electric field generates a Maxwell stress, in addition to hydrodynamic traction and membrane mechanical stress. Our model also includes the effect of interfacial charge due to the jump of the current and the stretching of the interface. We focus on the axisymmetric deformation of the encapsulated bubble induced by the electric field and carry out our analysis using the Legendre polynomials. In our first example, the encapsulating membrane is modelled as an incompressible interface with bending rigidity. Under the steady uniform electric field, the encapsulated bubble resumes an elongated equilibrium shape, dominated by the second and fourth order shape modes. The deformed shape agrees well with experimental observations reported in the literature. For our second example, we considered a bubble encapsulated with a hyperelastic membrane with bending rigidity, subject to an oscillatory electric field. We show that the bubble can modulate its oscillating frequency and reach a stable shape oscillation at an appreciable amplitude.

Keywords: encapsulated bubble; electric field; shape oscillation

Introduction

Encapsulated bubbles are widely used in various biomedical applications. It is important to understand the dynamics of encapsulated bubbles subject to applied physical forces and a key set of parameters are their natural frequencies. The natural frequencies of shape modes are crucial for controlling membrane stability. An approach capable of controlling the shape oscillation and obtaining the natural frequencies of shape mode directly is needed. In this work, we show that applying an oscillatory electric field offers a noninvasive way for estimating natural frequencies of shape modes when an encapsulated bubble undergoes oscillations.

Problem formulation

We consider an encapsulated bubble suspended between two conducting plates parallel to each other. Applying temporally constant or varying voltage yields steady or oscillatory electric field. The flow field and electric field are assumed to be axisymmetric with the symmetric axis along the direction of the applied electric field z and through the center of the bubble.

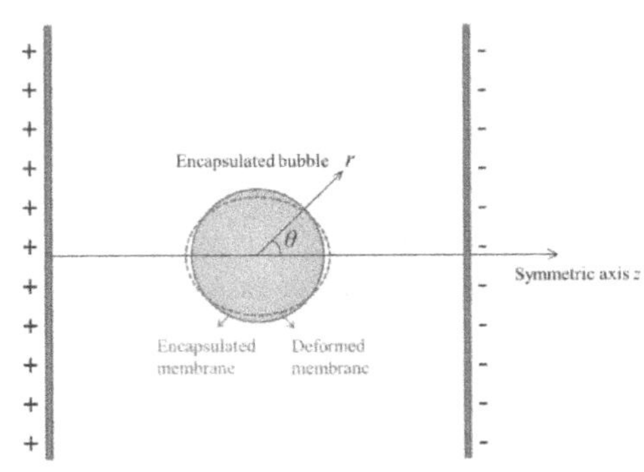

Figure 1. The coordinate of the encapsulated bubble in a spatially uniform electric field.

We establish a polar coordinate system based at the center of the bubble and consider a perturbation to a spherical shape. The positions of the material points at the bubble interface are thereby determined by the radial and zenith directions and expanded in terms of the (associated) Legendre polynomials as:

*Corresponding Author, Yunqiao Liu: yunqiaoliu@sjtu.edu.cn

$$r(\theta,t)=R(t)+\sum_{k=2}^{\infty}a_k(t)P_k(\cos\theta),\quad \Theta(\theta,t)=\theta+\frac{1}{R(t)}\sum_{k=1}^{\infty}b_k(t)P_k^1(\cos\theta). \tag{1}$$

We assume that the flow field inside the bubble is negligible and consider only the flow field outside the bubble by the incompressible Navier-Stokes equations:

$$\nabla\cdot\mathbf{u}=0,\quad \rho\frac{\partial\mathbf{u}}{\partial t}+\rho[(\mathbf{u}-w\mathbf{e}_z)\cdot\nabla]\mathbf{u}=-\nabla p+\mu\nabla\cdot(\nabla\mathbf{u}+\nabla\mathbf{u}^{\mathrm{T}}), \tag{2}$$

where $\mathbf{u}$ and p are the velocity and pressure, respectively, and w is the translation velocity, ρ and μ are the liquid density and viscosity, respectively. The velocity and pressure are decomposed into potential part (with subscript p) and viscous pressure correction (with subscript v) [1]

$$\mathbf{u}=\mathbf{u}_p+\mathbf{u}_v,\qquad p=p_p+p_v. \tag{3}$$

We assume that there is no free charge in the bulk, and it only exists on the bubble interface. According to Gauss's law and Faraday's law without induction, the external and internal electric fields $\mathbf{E}_{\mathrm{ex}}$ and $\mathbf{E}_{\mathrm{in}}$ are irrotational and solenoidal. Therefore, the electric fields can be expressed by the electric potentials as

$$\mathbf{E}_{ex}=-\nabla\psi_{ex},\qquad \mathbf{E}_{in}=-\nabla\psi_{in}, \tag{4}$$

and the electric potentials ψ_{ex} and ψ_{in} satisfy the Laplace equations:

$$\nabla^2\psi_{ex}=0,\quad \nabla^2\psi_{in}=0. \tag{5}$$

Due to the spatially uniform distribution of electric field as $r\to\infty$ and the requirement of regularity at $r=0$, the solutions to the Laplace equations have the forms of

$$\psi_{ex}(r,\theta,t)=\sum_{k=0}^{\infty}\frac{A_k(t)}{r^{k+1}}P_k(\cos\theta),\qquad \psi_{in}(r,\theta,t)=\sum_{k=0}^{\infty}B_k(t)r^kP_k(\cos\theta), \tag{6}$$

where the coefficients A_k and B_k are determined by the boundary conditions [2,3], including the continuity of the tangential components of the electric fields

$$\mathbf{t}\cdot\nabla\psi_{ex}=\mathbf{t}\cdot\nabla\psi_{in}, \tag{7}$$

and the jump condition for the normal components of the electric displacement vectors

$$-\varepsilon_{ex}\mathbf{n}\cdot\nabla\psi_{ex}+\varepsilon_{in}\mathbf{n}\cdot\nabla\psi_{in}=q/\varepsilon_0, \tag{8}$$

where $\mathbf{t}$ and $\mathbf{n}$ are the unit tangential and normal vectors, respectively. ε_0 is the permittivity of vacuum, $\varepsilon_{\mathrm{ex}}$ and $\varepsilon_{\mathrm{in}}$ are the relative permittivity of the liquid and gas phase, respectively. q is the interfacial charge density, satisfying the transport equation

$$-\kappa_{ex}\mathbf{n}\cdot\nabla\psi_{ex}+\kappa_{in}\mathbf{n}\cdot\nabla\psi_{in}=\frac{\mathrm{d}q}{\mathrm{d}t}+q\nabla_s\cdot\mathbf{u}. \tag{9}$$

The stress balance at the interface includes the pressure difference, the jump of the viscous and the Maxwell stresses, and the membrane stress (for encapsulated bubbles only) or surface tension (for gas bubbles only), in the normal and tangential directions

$$-p_{ex}+\mathbf{n}\cdot\mathbf{T}_{ex}\cdot\mathbf{n}+\mathbf{n}\cdot\mathbf{M}_{ex}\cdot\mathbf{n}=-p_{in}+\mathbf{n}\cdot\mathbf{M}_{in}\cdot\mathbf{n}+F_n, \tag{10}$$

$$\mathbf{n}\cdot\mathbf{T}_{ex}\cdot\mathbf{t}+\mathbf{n}\cdot\mathbf{M}_{ex}\cdot\mathbf{t}=\mathbf{n}\cdot\mathbf{M}_{in}\cdot\mathbf{t}+F_t, \tag{11}$$

where p_{ex} and p_{in} are the pressure outside the inside the bubble, respectively. $\mathbf{T}_{ex}=\mu(\nabla\mathbf{u}+\nabla\mathbf{u}^T)$ is the viscous stress tensor at the liquid side, while the viscous stress in the bubble is neglected due to the small viscosity of gas. $\mathbf{M}$ (the subscript is omitted temporarily for brevity) is the Maxwell stress

$$\mathbf{M}=\varepsilon\varepsilon_0\left(\mathbf{EE}-\frac{1}{2}|\mathbf{E}|^2\mathbf{I}\right). \tag{12}$$

The membrane stress is calculated by [4,5]

$$\mathbf{F}=F_n\mathbf{e}_n+F_t\mathbf{e}_t=-(\mathbf{P}\cdot\nabla)\cdot(\boldsymbol{\tau}+\mathbf{Qn}), \tag{13}$$

where $\mathbf{P}$ is the tangential projection operator. $\boldsymbol{\tau}$ is the in-plane stress. $\mathbf{Q}$ is the transverse shear tension, associated with the bending moment as

$$\mathbf{Q}=[(\mathbf{P}\cdot\nabla)\cdot\mathbf{m}]\cdot\mathbf{P}. \tag{14}$$

We adopt the neo-Hookean law to connect the in-plane stress and strain, the Love law to connect the bending moment and curvature. Expanding (10) and (11) in terms of the Legendre polynomials, we obtain the dynamic equations for radial oscillation, translation and shape oscillation. The derived equation is referred to [6].

Results

To validate our model about the coupling of electric field and the dynamics of an encapsulated bubble, we focus on the effects of electric stress and membrane stress on the bubble shape. Due to the lack of experimental results for an encapsulated bubble, we compare with the results by Kummrow& Helfrich [7] which are for a vesicle. The vesicle is coated with lipid bilayer and only bears bending rigidity. Accordingly, we modify our membrane model to exlclude hyperelastic stress and only consider the bending moments. The in-plane strain is prescribed to be 1, implying the membrane is tensionless. For this case, the compressibility of gas plays little role. Therefore, the results of a vesicle and an encapsulated bubble can be compared, although the inside of a vesicle is not gas and the compositions of encapsulating membranes are different. According to Kummrow & Helfrich[7], we choose a bubble with initial radius $R_0 = 23$ μm and bending modulus $G_b = 2.47 \times 10^{-20}$ Nm, and apply a steady electric field $E_0 = 10^4$ Vm^{-1}. The external fluid is assumed to be water, with density ρ = 1000 kg m^{-3}, viscosity μ = 0.001 kg(m s)$^{-1}$, relative permittivities $\varepsilon_{ex} = 81$ and conductivities $\kappa_{ex} = 5 \times 10^{-4}$ S m^{-1}. The internal fluid is considered as gas, with relative permittivities $\varepsilon_{in} = 1$ and conductivities $\kappa_{in} = 3 \times 10^{-15}$ S m^{-1}. The ambient pressure is set as $p_\infty = 10^5$ Pa. Figure 2(a) suggests the second and fourth order shape modes develop under the electric field and become stable to be a new shape within 0.2 s. This time scale is close to Kummrow and Helfrich's result. Plotting the shape at the stable state based on the magnitudes of different shape modes and comparing with Kummrow and Helfrich's observation, we find that the shapes agree well with each other (Fig. 2(b)).

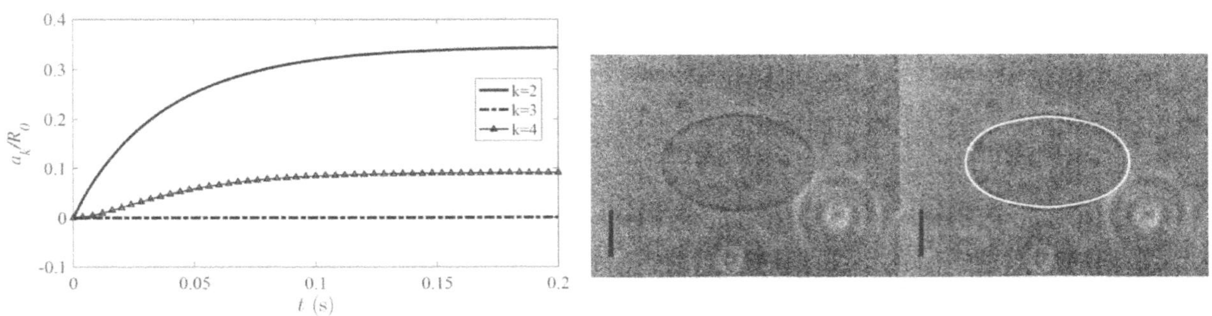

(a) (b)

Figure 2. (a) Time evolutions of shape modes for k = 2, 3 and 4 for an encapsulated bubble. (b) Comparison of the stable shape with the experimental observation by Kummrow & Helfrich[7]. The bar represents 20 μm.

For a general membrane with hyperelasticity and bending rigidity, we add the hyperelastic stress to the above bubble by setting elastic modulus $G_s = 0.15$ N m^{-1}. The deformation of the bubble is much smaller than the bubble whose membrane only has bending rigidity (Fig. 3(a)). It is not easy to obtain the characteristic parameters such as elastic modulus, natural frequency, etc. using steady electric field. Even we increase the electric field, e.g. applying $E_0 = 10^6$ Vm^{-1}, the deformation at the stable state is an order of magnitude smaller than the bubble only with bending rigidity (Fig. 3(b)).

The results above suggest that the effect of a steady electric field on encapsulated bubble dynamics is small. To explore the effect of a temporally oscillatory electric field, we apply $E_0 = E_a \sin(2\pi f_e t)$, where E_a is the amplitude of the oscillatory electric field and f_e is the oscillatory frequency. Since the second order shape mode has a quadratic relation with the electric field[6], if we set the electric frequency equal to half of the natural frequency, the shape oscillation is resonant and thus the amplitude of the shape mode is enlarged. The natural frequency of shape modes for the encapsulated bubble is calculated to be $f_2 = 45.2$ kHz[4]. Accordingly, we set $f_e = 22.6$ kHz and find that the second order shape mode oscillates continuously with an amplitude twofold larger than the stable magnitude under a steady electric field (Fig. 4).

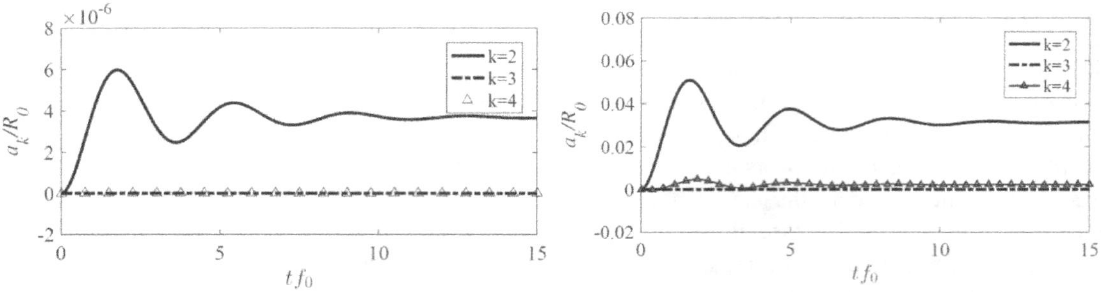

(a) (b)

Figure 3. Time evolutions of shape modes under a steady electric field with $E_0 = 10^4$ Vm^{-1} (a) and $E_0 = 10^6$ Vm^{-1} (b).

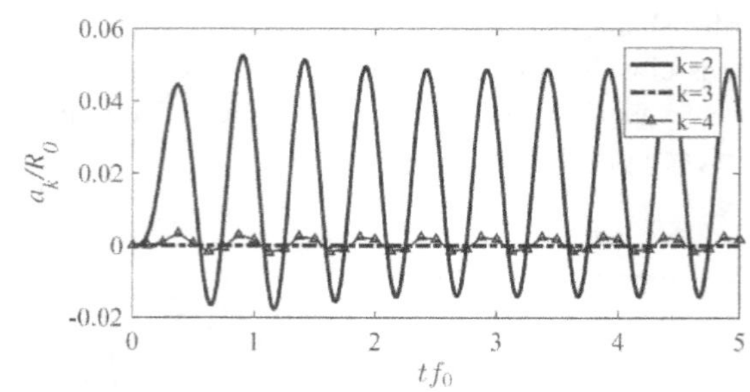

Figure 4. Time evolutions of shape modes under an oscillatory electric field.

To obtain the natural frequency of shape mode for an encapsulated bubble, we can modulate the electric frequency to induce resonance. As an example, we take a gas bubble whose natural frequency of shape mode is calculated to be 42.7 kHz and assign F_n in (10) as the surface tension and F_t in (11) to be zero for a shear-free condition. Sweeping the electric frequency and recording the maximal amplitude of the second order shape mode after the transient state, we find that the maximum of amplitude appears at around 21 kHz (Fig. 5(a)), which is half of its natural frequency. In the same way, we find that the maximum for the encapsulated bubble is at around 20 kHz (Fig. 5(b)), a little smaller than half of the natural frequency we calculated (22.6 kHz). The smaller value is attributed to the significant effect of viscosity for an encapsulated bubble, which makes the resonant frequency smaller than the natural frequency without considering the effect of viscosity. Nevertheless, frequency modulation of an oscillatory electric field is a feasible way to estimate the natural frequencies of shape modes. In addition, the maximum a_2 at resonance is more than twice of the stable a_2 under steady electric field, making it easier to detect bubble deformation.

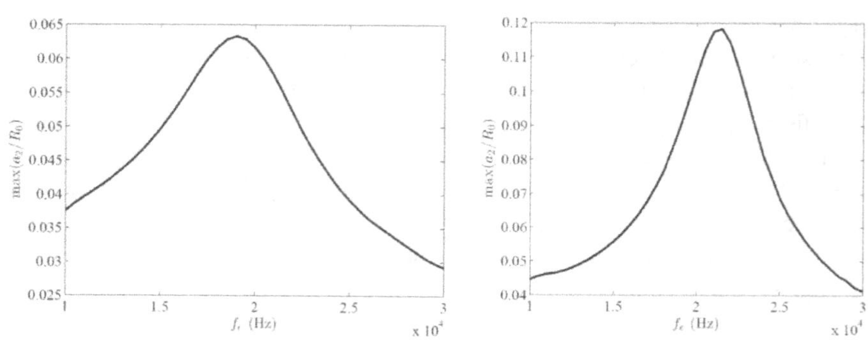

(a) (b)

Figure 5. The maximal amplitudes of the second order shape mode under various electric frequencies for a gas bubble (a) and an encapsulated bubble (b).

Conclusion

In this paper, we have proposed a leaky dielectric model for investigating the dynamics of an encapsulated bubble in an electric field. The applied electric field is spatially uniform and temporally constant or oscillatory. Outside the bubble, we solve a potential flow field with a viscous pressure correction. The electric field and flow field are coupled through force balance on the bubble surface, including electric, hydrodynamic traction and membrane stresses. Our results for a bubble coated by a membrane with bending rigidity show that it undergoes elongated deformation subject to a steady electric field, whose shape agrees well with the observation by Kummrow & Helfrich[7] with the same parameters. For a membrane both with elasticity and bending rigidity, our results show that the deformation of an encapsulated bubble are not as visible under a steady applied electric field. By modulating the frequency of an oscillatory electric field, on the other hand, we can induce shape mode resonance. Since the deformation is significant, natural frequencies of the shape model can be estimated directly with less error.

Acknowledgement

This work was supported by the National Natural Science Foundation of China (11772196, 11302128, 11402174).

References

[1] Prosperetti, A. (1977). *Viscous effects on perturbed spherical flows*. Q. Appl. Maths. 34(4).

[2] Lee, S. M., Kang, I. S. (1999). *Three-dimensional analysis of the steady-state shape and small-amplitude oscillation of a bubble in uniform and non-uniform electric fields*. J. Fluid. Mech. 384.

[3] Saville, D. A. (1997). *Electrohydrodynamics: the Taylor-Melcher leaky dielectric model.* Annu. Rev. Fluid Mech. 29(1).

[4] Liu, Y., Sugiyama, K., Takagi, S., Matsumoto, Y. (2012). *Surface instability of an encapsulated bubble induced by an ultrasonic pressure wave*. J. Fluid Mech. 691.

[5] Liu, Y., Sugiyama, K., Takagi, S. (2016). *On the interaction of two encapsulated bubbles in an ultrasound field*. J. Fluid Mech. 804.

[6] Liu, Y., He, D., Gong, X., Huang, H. (2017). *Deformation of an encapsulated bubble in steady and oscillatory electric fields*. J. Fluid Mech. in revision.

[7] Kummrow, M., Helfrich, W. (1991). *Deformation of giant lipid vesicles by electric fields*. Phys. Rev. A. 44(12).

© 2018 ASME

Simulating the effect of in-nozzle cavitation on liquid atomisation using a three-phase model

M.G. Mithun*[,1], P. Koukouvnis[1], I. K. Karathanassis[1], M. Gavaises[1]

[1]*City, University of London, United Kingdom*

Abstract

The aim of this article is to present a fully compressible three-phase (liquid, vapor, air, and mixture) cavitation model and its application to the simulation of in-nozzle cavitation effects on liquid atomization. The model employs a combination of barotropic cavitation model with an implicit sharp interface capturing Volume of Fluid (VoF) approximation. The results from the simulation are compared against the experimental results obtained by (1) for injection of water into the air from a stepped nozzle. Large Eddy Simulation (LES) model is utilized for resolving turbulence. Simulations are performed for a condition where developing cavitation is observed. Model validation is achieved by qualitative comparison against the available images for the cavitation, spray pattern. The model predictions suggest that the experimentally observed void inside the nozzle is not purely vapor, but a mixture of both vapor and back-flowing air. The simulation also identified periodic air entrainment that occurs at developing cavitation condition which further improves primary atomization.

Keywords: three-phase flow; cavitation; Volume of Fluid; Large Eddy Simulation; atomization

Introduction

Many researchers in the past have reported that the in-nozzle flow has a significant influence on the spray formation. Among geometric and other in-nozzle flow features, cavitation has the significant effect on the spray. Despite knowing this fact, there are only very few numerical studies available in literature where the interaction between cavitation and spray is considered (2,3). Most of the studies either consider only the in-nozzle flow with/without cavitation or the primary/secondary atomization without cavitation.

In the past, models based on approaches such as the Eulerian-Lagrangian (4,5) and Eulerian-Eulerian (6,7) or their combination (8,9) were widely utilized for modeling spray atomization. Their choice purely depends on the physics sought, Lagrangian based models are best suited for lean spray modeling, and the fully Eulerian models are proved to be better for dense spray predictions. Another popular approach is the interface tracking methods such as the VoF (10,11) which is useful when many topological changes such as interface pinching and merging, and the interface motion is to be tracked accurately.

For modeling cavitation, heterogeneous models which consider the non-equilibrium between the phases and the homogeneous mixture models (12) where the slip between the phases is neglected are widely utilized. Models that solve transport equation for the secondary phase with appropriate mass transfer terms (13) and the single-fluid approach which uses an equation of state (EoS), to relates density with pressure and temperature or pressure alone (barotropic) are the most popular of homogeneous models. The barotropic models, based on the assumption of pressure equilibrium and infinite mass transfer between the phases are best suited for complex simulations owing to their simplicity and numerical stability (14).

To numerically study the effect of in-nozzle flow on primary atomization, a model that can handle the transport and interaction between the three phases present, namely liquid, vapor and air, is required. There are only very few studies available in the literature which deal with such problem, and they are typically an extension of an existing cavitation model (15,16). An alternative approach for modeling the co-existence of three-phases is by employing the Volume of Fluid (VoF). To the author's knowledge, there are five studies available in the literature, that attempted to link a two-phase VoF model with a cavitation model for studying the in-nozzle effects on atomization (2,3,17–19). These models differ in the way cavitation, and compressibility is treated. A linear barotropic model with VoF was used for modeling atomization in a gasoline injector by (2). A comparative study of two transport-based cavitation models with the application of VoF can be found in (19) for atomization occurring from a single solid cone injector. Further studies that assume the phases to be incompressible can be found in (3,17).

In this study, we present a compressible three-phase model which considers compressibility of both the mixture and pure phases, using non-linear isentropic relations. Such a consideration in the compressibility is essential to capture the nonlinear effects of the flow even when the phase change is not dominant. All the three-phase models available in

*Corresponding Author, Mithun MG: mithun.murali-girija@city.ac.uk

literature and presented above, consider the phases to be incompressible or assume linear compressibility. To the best of author's knowledge, this is the first work to consider a non-linear compressible model in conjunction with VoF and LES for studying primary atomization.

1. Three-Phase Model

The cavitation model used in this study is a piecewise barotropic function employing three different equations corresponding to liquid, liquid-vapor mixture and vapor phases. The Tait equation of state is used for modelling liquid ($\rho \geq \rho_l$); the pure vapor phase ($\rho < \rho_v$) is modeled using the isentropic gas equation, and the equation for the mixture phase ($\rho_v \leq \rho \leq \rho_l$) is derived by integrating Eq- 1.1with respect to mixture density for an isentropic process, using the Wallis speed of sound; the reader can refer to (14) for the detailed derivation:

$$c^2 = \left(\frac{\partial p}{\partial \rho}\right)_s \qquad \text{Eq- 1.1}$$

where c is the speed of sound, and s represents isentropic flow process.

Combination of the individual equations of state with the assumption of the homogeneous equilibrium results in Eq- 1.2 for a two-phase mixture:

$$p = \begin{cases} B\left[\left(\dfrac{\rho}{\rho_l(T_l)}\right)^N - 1\right] + p_{sat,l} & \rho \geq \rho_l \\ \dfrac{c_v^2 c_l^2 \rho_l \rho_v (\rho_v - \rho_l)}{c_v^2 \rho_v^2 - c_l^2 \rho_l^2} \ln\left(\dfrac{\rho}{c_l^2 \rho_l (\rho_l - \rho) + c_v^2 \rho_v (\rho - \rho_v)}\right) + p_{ref} & \rho_v \leq \rho \leq \rho_l \\ C_{vap} \rho^{\kappa} & \rho < \rho_v \end{cases} \qquad \text{Eq- 1.2}$$

In Eq- 1.2, B is the bulk modulus, ρ_l, $p_{sat,l}$ and N are the saturation density, saturation pressure and the stiffness of the liquid, respectively. The parameter p_{ref} in the mixture equation is tuned to ensure continuous variation of density between the liquid and mixture phases. C_{vap} is the constant of the isentropic process and κ is the heat capacity ratio for the vapor phase. The subscript m, v, and l correspond to mixture, vapor, and liquid phases, respectively.

The third phase, i.e., the non-condensable gas (air) is modeled using isentropic gas equation Eq- 1.3, and it is assumed to be immiscible with the barotropic fluid.

$$p = C_{gas} \rho^{\gamma} \qquad \text{Eq- 1.3}$$

where, C_{gas} is the constant of the isentropic process for air and γ is the heat capacity ratio for air. The thermodynamic properties of water, vapor and gas (air) along with the constants used in Eq- 1.2 and Eq- 1.3 are listed in Table 1.

The barotropic fluid and the non-condensable gas equations (Eq- 1.2 and Eq- 1.3) are then combined using an implicit VoF model to closure the interaction of the three-phase system. The discretization of the phase volume fraction is performed using the compressive scheme, similar to the CICSAM (Compressive interface capturing for arbitrary meshes) scheme of (20). The discretization of the governing equations used in this study is based on the finite volume approach as implemented in ANSYS Fluent v17.1. In the results that follow, the variables are made non-dimensional based on based on the mean velocity inside the nozzle ($V_n = 18.3\ m/s$) and the length of the bottom wall ($l_{ref} = 9mm$).

2. Test case and simulation setup

The experimental geometry and the computational domain used for the simulation are shown in Figure 1. Pressurized tap water at 293K is injected through the nozzle to ambient air at 1 bar and then gravitated to a buffer tank. The computational domain is extended to model the ambient air as shown. A block-structured mesh with appropriate refinement near the walls is used to ensure $y^+ < 1$. The initial estimate of the mesh resolution for LES is calculated based on the Kolmogorov and Taylor length scales (~0.84 μm and ~48 μm computed for the maximum injection pressure condition). At the inlet, the absolute pressure is set to 3 $bars$ which corresponds to a developing cavitation condition inside the nozzle. The outlet pressure is set to ambient (1 $bars$).

3. Results and Discussion

The results are presented for an injection pressure ($p_{inj} = 3\ bars$) at which developing cavitation occurs inside the nozzle. The comparison of the in-nozzle cavitation and the near-exit spray formation between the experimental result

(1) and the present simulation is shown in Figure 2. The asymmetry in geometry initiates cavitation inside the nozzle with more cavitation occurring from the lower wall compared to the upper wall. The results presented in Figure 2 confirms that the liquid jet atomizes faster on the cavitating side of the step-nozzle as more ligaments and droplets are formed on this side. Compared to the experiments, the numerical simulation shows the entrainment of ambient air moving back inside the orifice and the total void seen inside the nozzle is a mixture of vapor and gas as highlighted in Figure 2. This could be the case in the experiment as well, however differentiating between the gaseous and vaporous cavitation during experiments is still an open question. The comparison of the void fraction and spray pattern shows a good match between the experimental result and simulation.

The evolution of the spray at developing cavitation condition is shown in Figure 3. The formation of the mushroom head during the early stages of the spray development due to the aerodynamic instabilities and the initial droplet formation from the periphery of the mushroom can be seen from Figure 3(a, b). The relative velocity difference between the water and ambient air initiates the Kelvin-Helmholtz instabilities at the interface. However, their effects on jet disintegration are limited due to the high density ratio between the two fluids in the present study (ratio of liquid over air density ~ 1000) in line with the observations of (21). As the time progress, the flow inside the nozzle becomes turbulent, and the interaction of the turbulent structures with the liquid-air interface initiates the disintegration of the jet core, Figure 3d. At the same time, the sheet cavity formed at the inlet edge of the bottom wall quickly transforms into small vortices that are transported downstream by the flow. The variation in vortex transport velocity causes these vortices to merge to form vapor clouds. The presence of the low-pressure cloud at the exit produces a pressure gradient across the interface that pulls the ambient air into the nozzle. Strong primary atomization and a sharp increase in spray cone angle are observed during this event, Figure 3(e, f). The entrained gas pushes the liquid away from the wall causing a restriction to the flow, a reduction in spray width and an increase in the jet velocity is associated with this. An effect of which can be seen in Figure 3i towards the end of the domain. Subsequently, the pressure build-up caused by the flow restriction pushes the entrained gas back, recovering the flow area. An increase in atomization and spray width is again observed during this event. The entrainment of the air in this study is a cyclic process, and its frequency depends on the dynamics of the cavitation.

4. Conclusions

This paper presents a fully compressible three-phase model developed by utilizing a barotropic cavitation model along with a sharp interface VoF approach with the application of LES for resolving turbulence. The simulations result is compared with the experimental result from (1) and further utilized to enhance the understanding of the complex interaction between the in-nozzle flow and primary atomization. It is identified that the main in-nozzle effects that influence the primary atomization are the cavitation, air entrainment, and the liquid turbulence. In addition to that, aerodynamic forces further assist the atomization process to a smaller extent. The capability of the model to distinguish between the vapor and air has revealed that the voids reported in the experiment are not entirely vapor, but a mixture of vapor and entrained air. A periodic phenomenon of air-entrainment is observed at developing cavitation condition which is found to enhance the atomization process. From the results obtained, it can be stated that developing cavitation is a favorable condition for better primary atomization. However, a detailed study of the erosion occurring at this condition is required before making a conclusion, as high erosive potential vapor cloud generation due to vortex merging is observed at this condition.

Table 1: Thermodynamic properties for water, vapor and gas at 20°C.

Liquid Properties			Vapor properties			Gas Properties		
B	3.07	GPa	C_{vap}	27234.7	Pa/(kg/m^3)	C_{gas}	75267.84	Pa/(kg/m^3)
N	7.15	--	κ	1.327	--	γ	1.4	--
ρ_l	998.16	Kg/m^3	ρ_v	0.0173	Kg/m^3			
c_l	1483.26	m/s	c_v	97.9	m/s			
$P_{sat,l}$	4664.4	Pa	$P_{sat,v}$	125	Pa			
μ_l	1.02	mPa s	μ_v	9.75	μPa s	μ_g	0.178	μPa s
surface tension	*0.0728*	*N/m*						

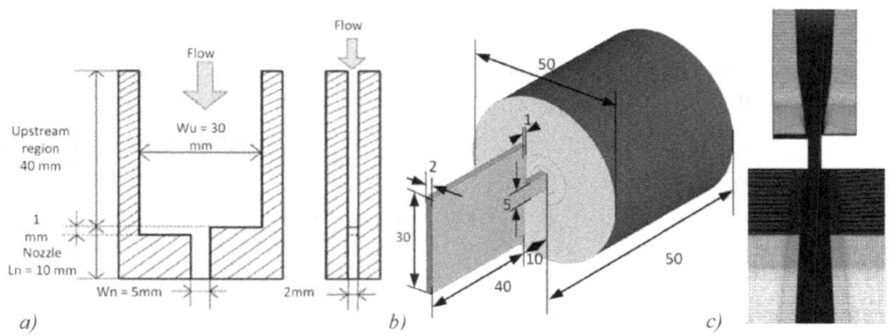

Figure 1: a) Step- Nozzle geometry as reported in (1) *b) Computational domain c) the numerical grid used for LES.*

Figure 2: Comparison of in-nozzle cavitation and near-exit spray formation between experimental results from (1) *and current numerical study at 3 bar inlet pressure. Iso-surfaces of 50% vapor (white) and 95% gas(blue) are shown.*

Figure 3: Evolution of in-nozzle cavitation and liquid jet at P_{inj}=3 bar. Iso-surfaces of 50% vapor (white) and 95% gas (blue) volume fraction shown. The instances are chosen randomly over time to highlight the major events.

References

1. Abderrezzak B, Huang Y. A contribution to the understanding of cavitation effects on droplet formation through a quantitative observation on breakup of liquid jet. Int J Hydrogen Energy. 2016;41(35):15821–8.
2. Ishimoto J, Sato F, Sato G. Computational Prediction of the Effect of Microcavitation on an Atomization Mechanism in a Gasoline Injector Nozzle. J Eng Gas Turbines Power. 2010;132(8):082801.
3. Edelbauer W. Numerical simulation of cavitating injector flow and liquid spray break-up by combination of Eulerian-Eulerian and Volume-of-Fluid methods. Comput Fluids. 2017;144:19–33.
4. Lippert AM, Chang S, Are S, Schmidt DP. Mesh Independence and Adaptive Mesh Refinement For Advanced Engine Spray Simulations. In: SAE Technical Paper. SAE International; 2005. p. 2005-01–0207.
5. Ning W, Reitz RD, Lippert AM, Diwakar R. Development of a Next-generation Spray and Atomization Model Using an Eulerian- Lagrangian Methodology. Int Multidimens Engine Model User's Gr Meet. 2007;(April).
6. Behzadi A, Issa RI, Rusche H. Modelling of dispersed bubble and droplet flow at high phase fractions. Chem Eng Sci. 2004;59(4):759–70.
7. Vujanović M, Petranović Z, Edelbauer W, Baleta J, Duić N. Numerical modelling of diesel spray using the Eulerian multiphase approach. Energy Convers Manag. 2015;104:160–9.
8. Vallet A, Burluka AA, Borghi R. Development of a eulerian model for the "atomization" of a liquid jet. At Sprays. 2001;11(6):24.
9. AVL. Fire Manual. 2013.
10. Berlemont A, Bouali Z, Cousin J, Desjonqueres P, Doring M, Menard T, et al. Simulation of liquid/gas interface break-up with a coupled Level Set/VOF/Ghost Fluid method. In: ICCFD7. Big Island, Hawaii; 2012.
11. Arienti M, Li X, Soteriou MC, Eckett CA, Sussman M, Jensen RJ. Coupled Level-Set/Volume-of-Fluid Method for Simulation of Injector Atomization. J Propuls Power. 2013 Jan 12;29(1):147–57.
12. Lauer E, Hu XY, Hickel S, Adams NA. Numerical investigation of collapsing cavity arrays. Phys Fluids. 2012 May 1;24(5):052104.
13. Schnerr GH, Sauer J. Physical and Numerical Modeling of Unsteady Cavitation Dynamics. In: Fourth International Conference on Multiphase Flow. 2001.
14. Koukouvinis P, Naseri H, Gavaises M. Performance of turbulence and cavitation models in prediction of incipient and developed cavitation. Int J Engine Res. 2017 Apr 28;18(4):333–50.
15. Wang Y, Qiu L, Reitz RD, Diwakar R. Simulating cavitating liquid jets using a compressible and equilibrium two-phase flow solver. Int J Multiph Flow. 2014;63:52–67.
16. Örley F, Trummler T, Hickel S, Mihatsch MS, Schmidt SJ, Adams NA. Large-eddy simulation of cavitating nozzle flow and primary jet break-up. Phys Fluids. 2015 Aug;27(8):086101.
17. Marcer R, Le Cottier P, Chaves H, Argueyrolles B, Habchi C, Barbeau B, et al. A Validated Numerical Simulation of Diesel Injector Flow Using a VOF Method. In: SAE Technical Paper. SAE International; 2000. p. 2000-01-2932.
18. Ghiji M, Goldsworthy L, Brandner PA, Garaniya V, Hield P. Analysis of diesel spray dynamics using a compressible Eulerian/VOF/LES model and microscopic shadowgraphy. Fuel. 2017;188:352–66.
19. Yu H, Goldsworthy L, Brandner PA, Garaniya V. Development of a compressible multiphase cavitation approach for diesel spray modelling. Appl Math Model. 2017;45:705–27.
20. Ubbink O. Numerical Prediction of Two Fluid Systems With Sharp Interfaces. Imperial College of Science, Technology and Medicine; 1997.
21. Wu P-K, Faeth GM, Arbor A. Aerodynamic Effects on Primary Breakup of Turbulent Liquids. At Sprays. 1993;3(3:265):289.

© 2018 ASME

Simulation of an Internal Nozzle Flow Using an Euler-Lagrange Method

[1]Andreas Peters*; [1]Udo Lantermann; [1]Ould el Moctar

[1]*University of Duisburg-Essen, Duisburg, Germany*

Abstract

In this work, an Euler-Lagrange method to simulate cavitation and predict cavitation erosion is presented. The method considers bubble motions as well as growth and collapse of a discrete number of single, spherical bubbles. A two-way coupling approach is implemented so that both liquid phase and vapour bubbles interact with each other. In contrast to Euler-Euler methods, this allows a more detailed simulation of bubble transport and dynamics. We used the developed method to simulate the internal cavitating flow in a nozzle.

Keywords: Euler-Lagrange, cavitation, erosion, particle tracking, bubble dynamics

Introduction

Hydrodynamic cavitation is a process of evaporation caused by liquid pressures below the equilibrium vapour pressure at constant temperature. Depending on water quality, flows of liquid water contain a certain amount of free gas bubbles. When these bubbles reach regions of low pressure in a flow they grow due to evaporation processes which start on the bubble walls. This growth continues until they reach regions of higher pressure again. The condensation processes start on the bubble walls causing the bubbles to rapidly collapse. As these collapses take place, high pressures are generated in the bubbles' centres and pressure waves are radiated through the fluid. These pressure waves are attenuated as they travel further through the fluid so that the pressure amplitudes become smaller. Though, if a bubble is located in the vicinity of a solid surface, the bubble pressure after the collapse or the formation of a liquid water jet, the so-called *microjet*, may lead to material damage. First, this erosion process causes plastic deformation of the material during the so-called *incubation time*. Second, after a longer exposure time, the material will break and mass loss will occur.

The Euler-Lagrange approach considers a continuous liquid phase and a discrete number of spherical vapour bubbles. In contrast to an Euler-Euler method where both phases are continuous and share equal streamlines, the motions of the single bubbles can differ from the streamlines of the liquid phase depending on multiple forces acting on the bubbles. The dynamics of each single bubble are calculated based on the forces acting on the bubble wall using the Rayleigh-Plesset equation. This equation can be solved iteratively to obtain the bubble wall velocity and acceleration and therefore calculate a new bubble radius. The phases are two-way coupled. Thus, both phases influence each other.

In technical applications, flows of high flow gradients are common. Here, the bubbles' traces may differ significantly from the streamlines of the liquid flow. These deviations are large for vortical flows, flows around sharp edges near the tips of propeller blades or flows in rudder gaps. In addition, a better understanding of bubble collapses near surfaces and erosion mechanisms can be gained from analysing the behaviour of single bubbles. Euler-Lagrange methods were used to investigate details of cavitation in flows around propellers, hydrofoils and other technical applications [1-3]. Especially for vortical flows [1] and cloud collapses [4] the Euler-Lagrange method was used to provide more detailed insights into the regarded physical processes. In this work, we used the Euler-Lagrange method to simulate the internal cavitating flow in an axisymmetric nozzle and to predict cavitation erosion from the information about flow properties.

Numerical Method

We used the open source CFD framework OpenFOAM to combine Lagrangian particles with a Reynolds-averaged Navier-Stokes solver. For the multiphase flow on an Eulerian grid, we solved the equations of mass and momentum conservation. During one Eulerian time step, multiple time steps of the Lagrangian transport of the single bubbles may be calculated depending on bubble velocity and growth related to the grid size. The Lagrangian equation of motion reads as [1]

*Corresponding Author, Andreas Peters: andreas.peters@uni-due.de

$$m\frac{d\boldsymbol{u}_b}{dt} = \sum_i \boldsymbol{F}_i \,.$$

$m = m_b + m_a$ is the effective bubble mass. $m_b = \rho_b V_b$ is the bubble mass with the bubble density ρ_b and bubble volume V_b. $m_a = \frac{1}{2}\rho_l V_b$ is the added mass with ρ_l as the liquid density. $\boldsymbol{u}_b$ is the bubble velocity and t is the time. $\boldsymbol{F}_i$ are forces acting on a bubble due to pressure gradient, virtual mass, drag, volume variation, lift and gravity. The bubble position $\boldsymbol{x}_b$ can then be obtained by integrating its velocity over time:

$$\frac{d\boldsymbol{x}_b}{dt} = \boldsymbol{u}_b \,.$$

After the Lagrangian equation is solved and the new bubble position is defined the bubble dynamics are calculated. The dynamics of a cavitation bubble take place in even shorter time scales than the bubble motions and are, therefore, integrated in an internal time loop. The time steps are adjustable and become smallest during the collapse phase of a bubble. The equation of bubble dynamics is based on the Rayleigh-Plesset equation. Following the work of Brennen [5] and Chahine [3] it reads as

$$\frac{p_b - p}{p} - \frac{2\sigma}{\rho_l R} - \frac{4\nu}{R}\dot{R} + \frac{(\boldsymbol{u} - \boldsymbol{u}_b)^2}{4} = R\ddot{R} + \frac{3}{2}\dot{R}^2 \,.$$

$p_b = p_v + p_g$ is the pressure inside of the bubble where p_v is the vapour pressure and p_g the pressure of non-condensable gas. p is the pressure in the liquid carrier phase at the position of the bubble. σ and ν are the surface tension and kinematic viscosity of liquid water, respectively. $\boldsymbol{u}$ is the velocity of the carrier phase flow. $R, \dot{R}$ and $\ddot{R}$ are the bubble radius, bubble wall velocity and bubble wall acceleration. Using this equation we can determine if a bubble is going to grow or shrink.

Dynamics and transport of the bubbles are influenced by forces coming from the carrier phase. The bubbles will influence the carrier phase because of their own motion which is transferred to the momentum equation and because they change mixture density $\rho = \alpha_v\rho_v + (1 - \alpha_v)\rho_l$ and viscosity $\mu = \alpha_v\mu_v + (1 - \alpha_v)\mu_l$, where the indices l and v stand for liquid and vapour phase, respectively. The vapour volume fraction $\alpha_v = V_v/V$ is the volume of vapour V_v in a regarded cell of volume V. The volume of vapour in a control volume is obtained based on the volume of accumulated bubbles in the control volume. Furthermore, the Poisson equation which is used to calculate the pressure in the flow contains source terms depending on the vapour volume fraction. We chose the approach by Sauer and Schnerr [6] to calculate these source terms.

To predict erosion, we used the erosion model by Peters et al. [9], based on the microjet hypothesis by Dular and Coutier-Delgosha [11]. We hypothesise that a bubble will collapse asymmetrically if it is positioned close to a surface. This asymmetric collapse causes the formation of a high velocity microjet, which is almost always directed towards the surface. Depending on the liquid pressure at the bubble wall, the bubble is collapsing more or less rapidly and creates a microjet which may damage the surface. The dimensionless coefficient c_{Def} was introduced [9] to predict erosion sensitive areas taking into account the amount of impacts and the impacts' intensities in an area. The erosion model uses information from the flow solution including the vapour volume fraction which, in this work, was calculated based on the bubble distribution in the domain.

Results

The cavitating flow through an axisymmetric nozzle has been investigated experimentally [7-8] and numerically [9-10] before. A sketch of the nozzle cross section is shown in Figure 1. Water flowed from top to bottom through an inlet cylinder and radially outwards through a radial divergent part. The two parts were connected by an 1 mm radius at which unsteady cavitation was generated. In the experiments, the bottom surface of the radial divergent part functioned as a probe on which erosion occurred in a defined radial distance from the rotation axis of the geometry. We generated a numerical simulation domain from 0° to 22.5° of the full 360° geometry. The domain was discretised with a cylindrical mesh using hexahedral control volumes.

Figure 2 shows the shedding and collapse processes of different bubble clouds (grey) from the side. The pressure is depicted on a plane behind the cloud with low pressures shown in blue and high pressures shown in red. Two clouds were shed from the radius and travelled further downstream. The upper cloud touches the top of the domain. We can see that bubbles follow this cloud from the cavitation region at the connecting radius. The lower cloud has been separated from other cavitation regions and touches the bottom surface while encountering a region of higher pressure again. The lower cloud was collapsing shortly after the depicted moment when it encountered higher liquid pressures further downstream. The collapse of this cloud generated high pressures traveling through the domain and causing the upper cloud to collapse as well. While both clouds started to collapse, another cloud was generated at the connecting radius.

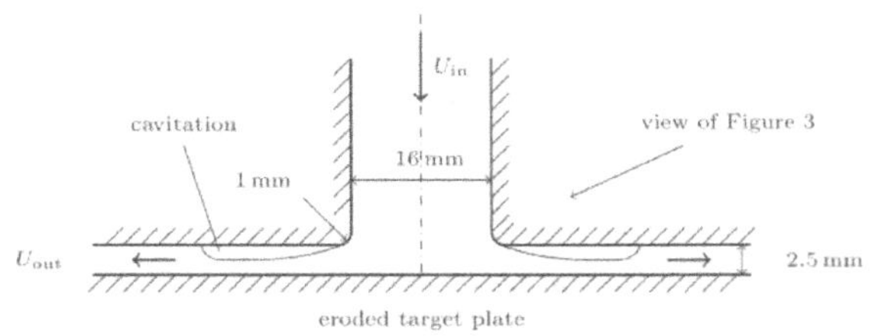

Figure 1: sketch of nozzle cross section taken from [9]

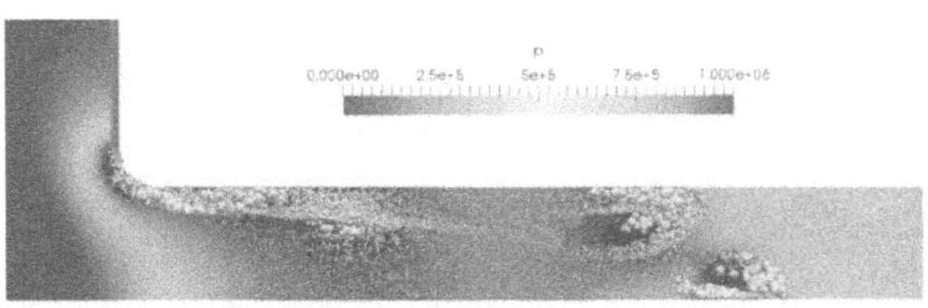

Figure 2: side view of cavitation bubbles collapsing near the bottom surface

Figure 3: perspective view of cavitation bubbles and erosion prediction into axisymmetric nozzle

Figure 2 shows the shedding and collapse processes of different bubble clouds (grey) from a side view into the nozzle. The pressure is depicted on a plane behind the cloud with low pressures shown in blue and high pressures shown in red. Two clouds were shed from the radius and travelled further downstream. The upper cloud touches the top of the domain. We can see that bubbles follow this cloud from the cavitation region at the connecting radius. The lower cloud has been separated from other cavitation regions and touches the bottom surface while encountering a region of higher pressure again. The lower cloud was collapsing shortly after the depicted moment when it encountered higher liquid pressures further downstream. The collapse of this cloud generated high pressures traveling through the domain and causing the upper cloud to collapse as well. While both clouds started to collapse, another cloud was generated at the connecting radius.

Figure 3 shows a perspective view into the nozzle as denoted in the sketch of Figure 1. The same time instant is shown as in Figure 2. The outer wall of the inlet cylinder and the connecting radius are shown in grey. The bottom of the radial outlet part is shown in grey as well whereas the top is made transparent. The growth of the cavitation bubbles (blue) starts at the radius. They are accumulated and form bubble clouds which travel further downstream and collapse in regions of higher liquid pressure. The black area on the bottom surface shows the erosion prediction by the microjet erosion model. Most clouds collapse in the vicinity of the area predicted to be eroded.

The cloud collapse and cloud shedding processes reoccurred periodically and with different intensities. The vapour volume in the simulation domain changed as new clouds were generated and cavitation regions collapsed. Furthermore, high forces were exerted on the bottom surface of the nozzle when cavitation clouds collapsed near the surface.

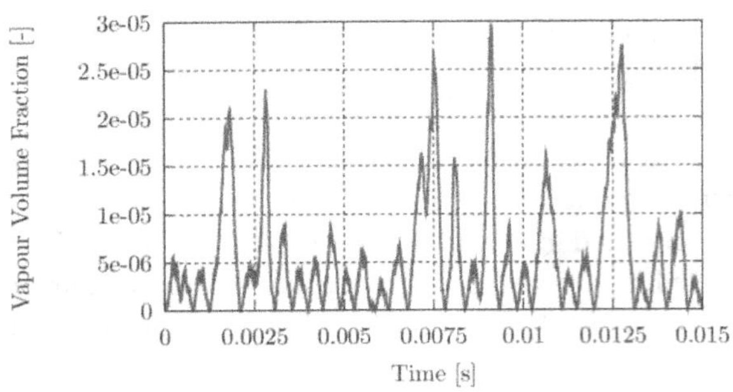

Figure 1: weight averaged vapour volume fraction

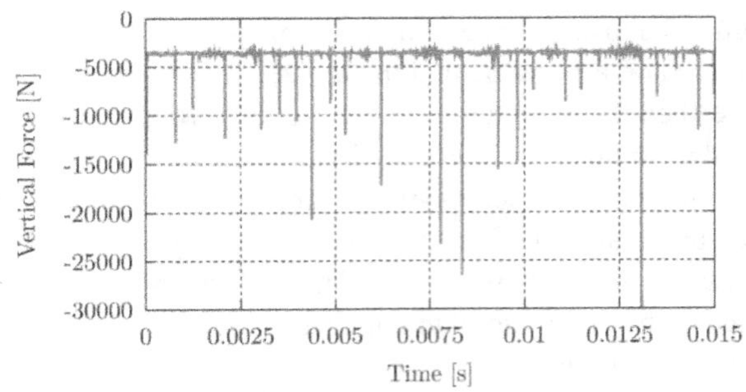

Figure 2: vertical force on the bottom surface

Figure 4 shows the temporal progress of the vapour volume fraction in the simulation. The volume fraction was weight averaged taking the cell volumes into account. The vapour volume oscillated periodically with different frequencies. Oscillations of higher frequencies were of lower amplitudes whereas oscillations of lower frequencies were of higher amplitudes. Figure 5 shows the time history of the vertical force on the bottom boundary. The average magnitude of the vertical force was about 3600 N. Peaks of large negative forces appeared harmonically. The peaks exceeded this value by a factor of 2 to 10. Analysing the frequencies of the time history of the force with a fast Fourier transform, we obtained a characteristic frequency of about 2000 Hz. This frequency corresponded to the negative peaks in the vertical force as well as to the high frequency oscillations in vapour volume. We concluded that these high magnitude force peaks were caused by collapsing clouds of cavitation bubbles. A fast Fourier transform of the time history of the volume fraction did not yield a clear characteristic frequency, probably because of the fact that the simulation should have been run for a longer time.

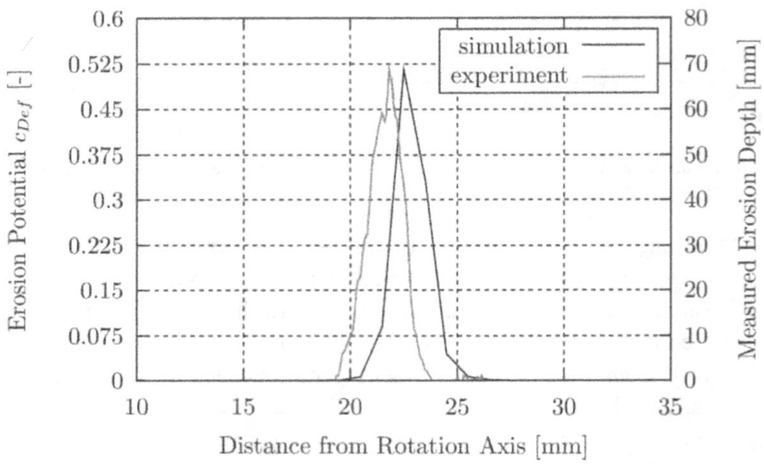

Figure 3: numerical erosion prediction (red) and experimentally measured erosion depth (black) [7] on the bottom surface

Figure 6 shows radial distribution of the numerical prediction and the experimental measurement of erosion [7] on the bottom surface of the nozzle. In the experiments, eroded areas were summed up over the circumference of the nozzle for different radial intervals. This approach was also used for the distribution of the numerical coefficient of erosion potential c_{Def} using intervals of 1 mm. The maximum of measured erosion depth was found to be at a distance of 21.8 mm from the central. The maximum erosion potential was calculated close to the measured maximum within an interval of 22 mm to 23 mm. The overall shape of the predicted erosion distribution agreed well with the experiment. Deviations in the maximum of the distributions may be accounted for by the nuclei distribution in the numerical simulation because it has a large influence on the inception and extent of cavitation.

Conclusions and Outlook

We have shown the ability of the Euler-Lagrange approach to simulate hydrodynamic cavitation and enable the prediction of cavitation induced erosion. The consideration of single bubbles ensures a more accurate simulation of the bubbles' behaviour. First simulations have shown that cavitation behaviour as well as erosion prediction will differ with variation of the nuclei distribution in the domain which will be investigated in further work. In the future, the present approach enables the possibility to quantitatively predict erosion by correlating bubble collapse rates with pitting patterns from experimental tests.

Acknowledgement

The present work was funded by the German Research Foundation [grant number EL 611/2-1].

References

[1] Abdel-Maksoud, M., Hänel, D., Lantermann, U. (2010). *Modeling and computation of cavitation in vertical flow*. International Journal of Heat and Fluid Flow. 31(6).
[2] Yakubov, S., Cankurt, B., Maquil, T., Schiller, P., Abdel-Maksoud, M., Rung, T. (2011). *Euler-Euler and Euler-Lagrange Approaches to Cavitation Modelling in Marine Applications*. International Conference on Computational Methods in Marine Engineering (MARINE), Lisbon, Portugal.
[3] Chahine, G. L. (2008). *Numerical Simulation of Bubble Flow Interactions*. Cavitation: Turbo-machinery & Medical Applications, WIMRC Forum, Warwick, United Kingdom.
[4] Ma, J., Hsiao, C.-T., Chahine, G. L. (2015). *Euler-Lagrange Simulations of Bubble Cloud Dynamics Near a Wall*. Journal of Fluids Engineering. 137.
[5] Brennen. C. E. (2005). *Fundamentals of Multiphase Flows*. Cambridge University Press. ISBN 0521 848040.
[6] Sauer, J., Schnerr, G. H. (2000). *Unsteady Cavitating Flow – a New Cavitation Model Based on Modified Front Capturing Method an Bubble Dynamics*. In Proceedings of FEDSM, 4th Fluids Engineering Summer Conference.
[7] Franc, J.-P., Riondet, M. (2006). *Incubation Time and Cavitation Erosion Rate of Work-Hardening Materials*. In Proceedings of the 6th International Symposium on Cavitation, CAV2006, Wageningen, Netherlands.
[8] Franc, J.-P., Riondet, M., Karimi, A., Chahine, G. (2011). *Impact Load Measurements in an Erosive Cavitating Flow*. Journal of Fluids Engineering. 133(12).
[9] Peters, A., Sagar, H., Lantermann, U., el Moctar, O. (2015). *Numerical Modelling and Prediction of Cavitation Erosion*. Wear. 338-339.
[10] Peters, A., Lantermann, U., el Moctar, O. (2015). *Numerical Modelling and Prediction of Erosion Induced by Hydrodynamic Cavitation*. Journal of Physics: Conference Series. 656(012054).
[11] Dular, M., Coutier-Delgosha, O. (2009). *Numerical modelling of cavitation erosion*. International Journal for Numerical Methods in Fluids. 61.

© 2018 ASME

Single Oscillating Bubble from Radiofrequency Fiber Electrode

Fenfang Li*; George Sankin;

Duke University, Durham, NC, USA

Abstract

The radiofrequency (RF) power in the 0.3~30 MHz band is often used for coagulation, ablation, and modification of tissues in medicine while in the lower frequency band both shock waves and cavitation may occur, which has found applications in lithotripsy for stone fragmentation and water purification. In this study, we demonstrate that single oscillating bubble is produced at the electrode tip in cell culture medium, which avoids the random inception of cavitation in liquid. With high-speed image recordings and Fast Fourier Transform (FFT) spectrum analysis, the frequency of the bubble oscillations is found to be dependent on the driven radio frequency, electrode diameter and the supplied voltage. With microelectrodes that are incorporated into flexible fibers (2-6 Fr), sustained oscillation is generated in the frequency range of 5.7-17 kHz. The oscillation frequency decreases with the radio frequency and electrode diameter, while an optimum voltage exists to reach maximum frequency of bubble oscillation. The RF power supply also offers flexibility to control the bubble dynamics with different voltage waveforms, duty cycles and pulse repetition rate.

Keywords: self-oscillating bubble; microelectrode; radiofrequency ablation

Introduction

RF generators and probe electrodes are effective tools for the generation of heat, which has found applications in pain relief or functional modification in biological tissue (1). Electrode tip gauge, length of the electrode tip, frequency and treatment time all substantially affect heat lesion width and length. Coagulation necrosis in the central zone of RF lesion seems to be caused mainly by thermal injury while both thermal and electrical energy contributes to lesion formation at the border zone. However, it has been reported that "white cavitation" artifact in ultrasound imaging may occur during RF treatment indicating bubble formation around the electrode tip (2). These microbubbles can cause significant cell response (including cell fragmentation and permeabilization) and could serve as a useful ablation tool for applications that require submillimeter precision (3). However, little is known about the bubble dynamics and bioeffects when reducing the electrode size. It has been reported that oscillate boiling occurs on metal-coated surface submerged in liquid when illuminated by focused continuous wave (CW) laser (4) and on blackened optic fiber tip (5). In this study, the bubble dynamics from RF driven microelectrode is investigated in order to produce sustained cavitation in cell culture medium using electrical heating.

Materials and Methods

A sinusoidal RF power supply around 100-500 V_{PP}, 0.3-30 MHz with 100,000 cycles is applied to the microelectrode incorporated in flexible fibers (Lithotech Medical) by a function generator and a 55 dB RF amplifier (A-150). The function generator, oscilloscope and a delay generator is used to synchronize the high-speed camera (Phantom 7.3) to capture the bubble dynamics from the side view, which is illuminated by a fiber coupled Halogen lamp, see Fig. 1. Structures of the microelectrodes of different sizes are captured using x-rays (Nikon CT), which include a core wire in the center with insulator materials between the core wire and the cannula wall (ground). The microelectrode is mounted firmly and submerged in DMEM culture medium (approximate conductivity: $\sigma = 1.4$ S/m, and dielectric constant: $\varepsilon_r \approx 80$) (6) in a glass cuvette (dimension: 1cm x 2cm x 4cm, filled with 5mL DMEM) during experiment. High speed imaging is taken at an inter-frame time of 7 μs with an exposure time of 1 μs, see Fig. 2. Recorded images are processed with Matlab by thresholding to extract the shadow area during bubble oscillations, which includes both the electrode and the cavitation bubble. Thus the change of the shadow area reveals the bubble oscillation. The amplitude spectrum vs. the oscillation frequency are then calculated by FFT of the shadow area change, see Fig. 3. The center frequency is thus obtained over the whole bubble oscillation period (determined by the RF power supply duration).

*Corresponding Author

Three sets of experiments are performed. The effects of driven radio frequency (0.3, 1, 5, 30 MHz with 3.6 Fr electrode under 281 V_{PP} driven voltage), electrode probe size (2, 2.4, 2.7, 3, 3.3, 3.6, 6 Fr under driven voltage of 281 V_{PP} and radio frequency of 1 MHz) and applied voltage (112, 169, 225, 281, 337 V_{PP} with 2 Fr electrode and 1 MHz driven radio frequency) are investigated on the center frequency of bubble oscillation in DMEM medium, see Fig. 4.

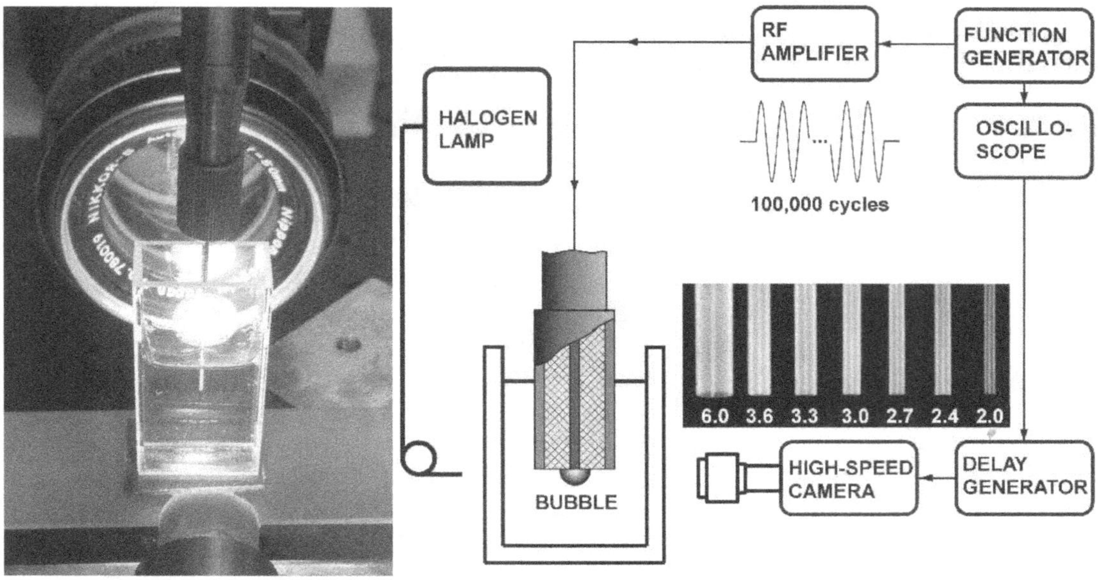

Fig.1 Experimental setup. *Left*: a fiber electrode submerged in DMEM medium in a glass cuvette with illumination and camera imaging from the side view. *Right:* a schematic of the experimental system. A high frequency RF power supply of 100,000 cycles with function generator and RF amplifier is applied to the microelectrode of 2-6 Fr (0.66-2 mm) incorporated in flexible fibers. Inset shows the x-ray images of the microelectrodes of different sizes. A Halogen lamp and high-speed camera are used to capture the bubble dynamics from the side view. The RF power supply and high speed camera are synchronized by the function generator, oscilloscope and delay generator.

Results

High-speed video recordings reveal that a single oscillating cavitation bubble is generated at the RF electrode tip. For example, with 3.6 Fr electrode under 281 V_{PP} driven at 1 MHz, the cavitation bubble gradually expands to the maximum diameter of 0.62±0.01 mm during 77 μs (11th frame from the left, Fig. 2b) then rapidly collapse during the next 28 μs before the next oscillation cycle starts. Small gas bubbles are extruded from the oscillating cavitation bubble to compensate rectified diffusion, as reported previously (4). Similar bubble dynamics are also observed for the driven frequency of 0.3 and 5 MHz (Fig. 2a and Fig. 2c). When driven by 30 MHz RF, the expansion of the cavitation bubble takes a longer time of 112 μs (16th frame from the left, Fig. 2d) before its collapse during the next 91 μs.

Bubble oscillations are processed with MatLab by extracting the shadow area in the high-speed images, as shown in Fig. 3a. The oscillation amplitude spectrum is then obtained by Fast Fourier Transform (window size 256 x 7 μs) of the shadow area change (Fig. 3b). The results reveal a consistent frequency of bubble oscillation during 100 ms (Fig. 3c). The center frequency f_{OB} is calculated by averaging over the whole observation period.

Parametric study reveals that f_{OB} decreases with the driven frequency (Fig. 4a), which can be attributed to the attenuation in the RF transmission line and transition from resistive heating (Joule heating, Fig. 2(a-c)) to dielectric heating (>10 MHz, Fig. 2d). With 3.6 Fr electrode under 281 V_{pp}, f_{OB} decreases from 9.5 kHz to 5.5 kHz when increasing the driven frequency from 0.3 MHz to 30 MHz. Similar to previous study (7), f_{OB} also decreases with the probe size (Fig. 4b). For example, f_{OB} decreases from 15.5 kHz to 6 kHz when increasing the electrode size from 2 Fr to 6 Fr (1 MHz RF, U_{RF}=281 V_{pp}). The bubble diameter at maximum expansion increases with voltage in the

112~337 V_{pp} range with maximum f_{OB} value around 17 kHz at 169 V_{pp}. Interestingly, similar trend is observed in (7) with direct current (DC) electrode.

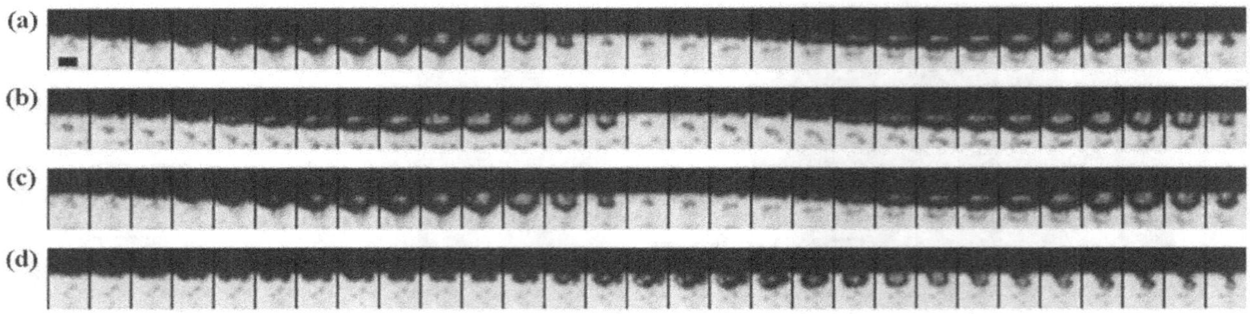

Fig.2 High-speed image recordings of bubble dynamics 1.75 ms after starting the AC power supply with electrode size of 3.6 Fr and driven AC voltage of 281V_{pp} under different RF of (a) 0.3 MHz; (b) 1.0 MHz; (c) 5.0 MHz; (d) 30 MHz. The interframe time is 7 μs with the scale bar denoting 200 μm.

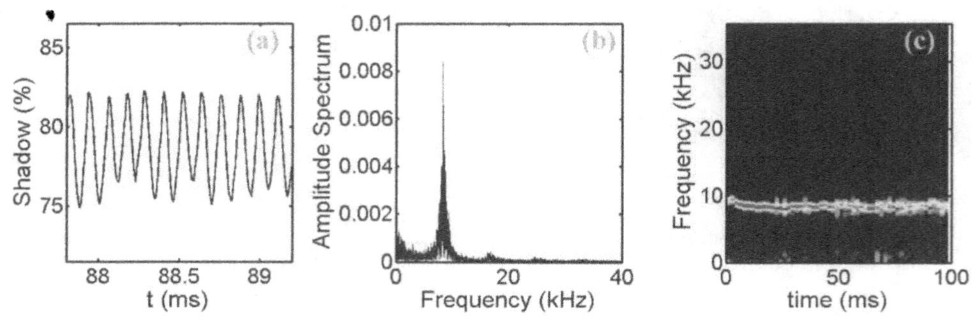

Fig.3 An exemplary case of 3.6 Fr electrode under 1MHz RF, 281V_{PP} for image processing and frequency analysis of the bubble oscillation. (a) Normalized shadow area including the electrode and bubbles in the high-speed images indicating bubble oscillations; (b) the amplitude spectrum averaged over 100 ms for the bubble in (a); (c) The detailed frequency spectrum in 0 to 100 ms time interval of the bubble from (a).

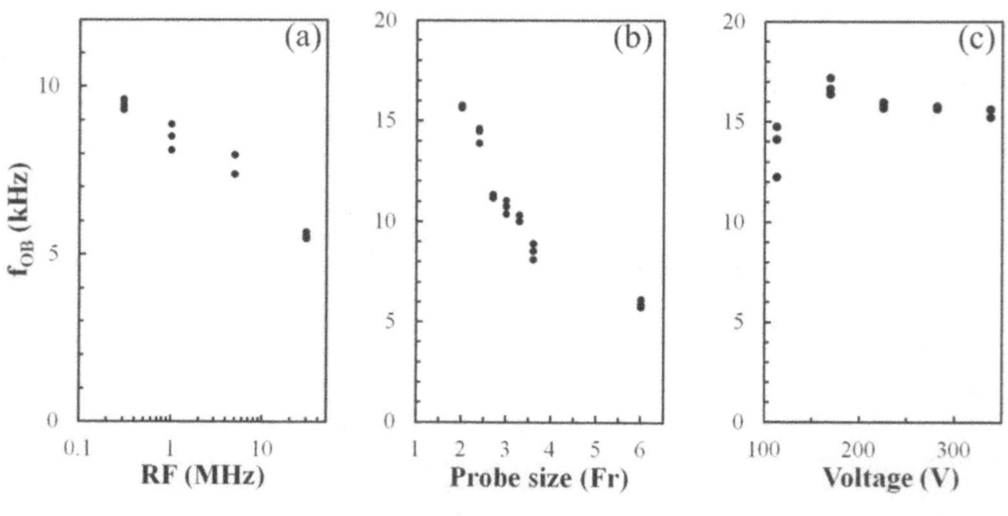

Fig.4 Effects of driven radio frequency, probe size and applied RF voltage (V_{pp}) on the center frequency of bubble oscillation in DMEM medium. (a) Center frequency of bubble oscillation vs. the driven radio frequency with the probe size of 3.6 Fr and driven RF voltage of 281V_{pp}; (b) Center

frequency of bubble oscillation vs. the probe size with driven RF voltage of 281 V_{pp} and driven frequency of 1 MHz; (c) Center frequency of the bubble oscillation vs. the driven RF voltage with 1 MHz driven frequency and 2 Fr probe size. Each condition is repeated three times.

Conclusion

Using high-speed photography, sustained single bubble oscillation is observed within 5.7~17.6 kHz frequency range as determined by well-controlled experimental conditions, such as applied voltage and the electrode diameter. The micron-sized electrode is incorporated into flexible fiber, which offers flexibility to control the bubble dynamics with variable amplitude, duty cycle and pulse repetition rate. This is in contrast to the random variation of cavitation inception in water by ultrasonic pulses and the inherent complexity of the multi-bubble systems used in previous studies. There exists a correlation between the electrode diameter and the frequency of bubble oscillation, which is likely to be generated by Joule heating (Fig. 2, a-c). Moreover, rapid expansion of the hemispherical bubble (0.36~0.8 mm in maximum diameter) with jet formation during collapse phase leads to the micro-streaming and intense heat exchange. Both consequences are of great importance for a diverse range of applications in therapeutic medicine, such as localized biological liquid boiling and macromolecule delivery into biological cells.

Technical support from Lithotech Medical, Israel on the use of the intracorporeal lithotripter probes is acknowledged.

References

[1] Cosman ER, Dolensky JR, Hoffman RA (2014) Factors that affect radiofrequency heat lesion size. Pain Medicine v15: 2020-2036.

[2] Viallon M, et al. (2010) Observation and correction of transient cavitation-induced PRFS thermometry artifacts during radiofrequency ablation, using simultaneous Ultrasound/MR imaging. Medical Physics v37: 1491-1506.

[3] Enbo Yang, et al. (2018) Cell Fragmentation and Permeabilization by a 1-ns Pulse Driven Triple Point Electrode. BioMed Research International. (In print)

[4] F Li, SR Gonzalez-Avila, DM Nguyen, CD Ohl (2017) Oscillate boiling from microheaters. Physical Review Fluids 2(1), 014007.

[5] Chudnovskii VM, et al. (2017) Laser-induced boiling of biological liquids in medical technologies. Quantum Electronics 47(4):361-370.

[6] Chen, Meng-Tse et al. (2009) Two-Dimensional Nanosecond Electric Field Mapping Based on Cell Electropermeabilization. PMC Biophysics 2: 9.

[7] Medvedev RN, Chernov AA, Karpov DI (2013) Calculations of pulsation period of a toroidal bubble during diaphragm discharge in electrolyte. International Journal of Heat and Mass Transfer v64:743-750.

© 2018 ASME

Size and concentration measurements of cavitation nuclei in the wake of a ship model

[1]M. Birvalski*, [1]M.X. van Rijsbergen

[1]*Maritime Research Institute of the Netherlands (MARIN), Wageningen, The Netherlands*

Abstract

Cavitation nuclei are added to the flow to achieve adequate scaling of cavitation phenomena in model-scale facilities with respect to full-scale conditions. During propeller cavitation studies in MARIN's Depressurized Wave Basin (DWB), cavitation nuclei are added to the flow using electrolysis. In the present study, Interferometric Particle Imaging (IPI) was applied to determine the size and the concentration of nuclei in the wake of a ship model. The dependence of the nuclei size and concentration spectra on the electrolysis current and the basin pressure showed good agreement with theory. One limitation of the present technique was identified: the minimum size of nuclei that can be resolved is approximately 40 μm. This also limits the basin pressure at which nuclei measurements can be performed to a maximum of approximately 100 mbar.

Keywords: cavitation nuclei; interferometric particle imaging; propeller cavitation; size and concentration measurement

Introduction

The pressure fluctuations and noise generated by a cavitating propeller are commonly assessed in model-scale facilities as a part of the design process for a new ship. Free-stream nuclei are added to the flow – either by injection of gas, supersaturated water or electrolysis – to minimize the scale effects on cavitation inception, the resulting cavitation extent and nuisances. In MARIN, the nuclei are generated by electrolysis using two metal strips – connected to an electric power supply – and fixed to the hull, several propeller diameters upstream of the propeller plane. A semi-empirical formula, based on model speed, basin pressure and the length of the electrolysis strips is used to set the nominal electric current. Although each facility has developed its optimum settings for the nuclei generation, little information is available on the produced nuclei size vs. concentration spectra. Moreover, to obtain a more fundamental understanding of cavitation inception mechanisms, in-situ measurements of the nuclei spectrum are needed. Previously [1], a technique based on Interferometric Particle Imaging (IPI) [2,3] was applied to a ship model towed in vacuum conditions in MARIN's Depressurized Wave Basin (DWB). Valuable measurement results were obtained using a simple laser pointer, but on average only approximately 65% of the detected bubbles could be sized and a reliable concentration calculation was not possible. Therefore, the system was upgraded with a laser with a higher power and beam quality, enabling a higher ratio of sizable bubbles. Furthermore, the technique was extended with a method for determining the nuclei concentration, similar to the one developed in [2,4,5].

Experimental setup

The measurements were performed in MARIN's DWB. A ship model – without propeller and rudder – was towed at a range of velocities and ambient pressures. The IPI technique, consisting of a laser light source (Laser Quantum, Torus 532 nm, 750 mW nominal power, 480 mW measured power), a camera (Photron FASTCAM Multi, 1280x1024 px^2 at 4800 fps, 12-bit with an Olympus M.Zuiko 45 mm f/1.8 lens) and a processing algorithm (developed in-house using MATLAB) was applied as shown in Figure 1a. Both the laser and the camera were placed in vacuum-proof housings inside the ship model. The laser produced a perpendicularly polarized beam of 1.7 mm diameter (at $1/e^2$ of the maximum intensity, according to specifications) that was shooting through a Perspex plate in the hull of the ship model. The camera was submerged in a water-filled box, observing the laser beam just in front of the propeller plane through a curved window in the hull.

A typical image recorded using this configuration is shown in Figure 1b. The size of the FOV was 89.2x55.8 mm^2; the scattering angle was θ=104.1° and the collecting angle was γ=2.93°, which resulted in the IPI conversion factor of κ=15.7 μm/fringe. The nuclei images were detected using a Hough transform, after which the number of fringes was determined using a zero-padded Fourier transform [1]. Finally, a parabolic function was fitted to the nuclei size vs. nuclei image intensity plot [1] for every measured case. The parabola represents the maximum intensity of a nucleus in a given particle size class; the nuclei that had an intensity higher than the value of this function were

*Corresponding Author: Miloš Birvalski, m.birvalski@marin.nl

discarded as incorrectly sized. The same parabola – along with the FOV height of 89.2 mm, a minimum detectable nuclei image intensity of 10 counts and a nuclei image saturation level of 4095 counts (12-bits) – was used to determine the nuclei size-dependent measurement volume [2,4-6], that finally resulted in nuclei concentrations.

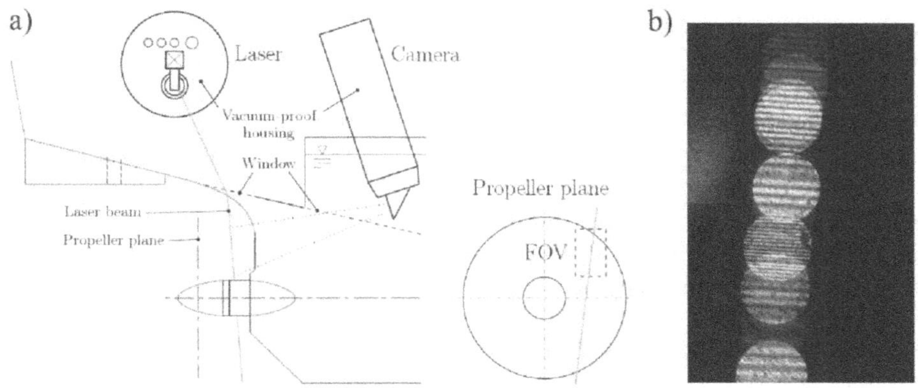

Figure 1. Disposition of the IPI measuring equipment in the ship model (a) and a typical image recorded in the measurements (b). The camera observes the nuclei upstream of the propeller plane in a FOV of 89.2x55.8 mm^2 (height x width).

Results

During the two-day measurement campaign, twenty three cases were measured, twelve of which are presented here (see Table 1). All the cases were measured using the same IPI system parameters: the exposure time was 250 μs, the framerate was 250 fps and the defocus level (expressed as nuclei image area) was the same for all but for case 2 (12565 px^2 for case 2, 30858 px^2 for the rest). The flow parameters were varied systematically: the electrolysis current from 0.125 A to 1 A, the model velocity from 0.45 m/s to 1.8 m/s and the basin pressure from 38 mbar to 550 mbar. The total number of nuclei was relatively large, and the convergence of the mean diameter and the standard deviation to within several percent was achieved for all but for the last case (12, with only 420 valid nuclei). Importantly, the proportion of valid nuclei compared to the total number of nuclei was high (on average 87%, varying between 80% and 92%), with a slight trend detected of a lower number corresponding to cases with a higher number of nuclei per image ratio. This is explained by the higher likelihood of overlapping nuclei images, which can be detected, but not always sized. However, since this is a random error and since the overall number of valid nuclei is high, it is concluded that the nuclei populations were sized correctly.

Table 1. A selection of twelve cases from the measurement campaign. The IPI system parameters are kept constant, while the flow parameters such as the electrolysis current, the model velocity and the basin pressure were systematically varied.

Case	Electrolysis current, i (A)	Model velocity, v (m/s)	Basin pressure, p (mbar)	Total number of images	Total number of nuclei	Tot. no. of valid nuclei	Mean diameter, d_p (μm)	St. dev. diameter (μm)	Mode diameter (μm)	Nuclei concentration, C (cm^{-3})
1	0.125	1.8	38	19480	2782	2402	207.3	141.3	60	0.3
2	0.255	1.8	38	19900	16025	14485	149.9	123.7	50	1.9
3	0.5	1.8	38	16960	30002	24952	117.7	61.0	80	3.7
4	0.75	1.8	38	18280	31943	26430	147.8	71.9	90	3.2
5	1	1.8	38	19050	32838	27134	166.2	80.4	100	3.0
6	0.5	0.45	38	34940	6848	6156	75.4	21.0	70	0.6
7	0.5	0.9	38	34940	10869	9815	105.1	75.9	60	0.9
8	0.5	1.35	38	26220	38535	33775	179.6	94.6	80	2.8
9	0.5	1.8	80	20930	25550	22365	96.8	53.7	60	3.1
10	0.5	1.8	98	19500	18521	16549	92.3	49.9	60	2.5
11	0.5	1.8	200	18570	5643	5214	82.7	58.8	50	0.9
12	0.5	1.8	550	20850	522	420	131.1	134.5	60	0.1

The largest detectable bubble diameter can be estimated from the nucleus image diameter, the Nyquist frequency and a safety factor of 2, resulting in approximately 780 μm for all but case 2. However, no bubbles larger than 500 μm were measured. The histograms of nuclei concentration (with 10 μm wide bins centred at d_p=0 μm, 10 μm, 20 μm, etc.) for cases with varying current, velocity and pressure are shown in Figure 2. In Figure 2a, five cases (1-5) with increasing current are presented. The first case, with half of the nominal current, has a very low nuclei concentration (0.3 cm^{-3}) and a distribution with two peaks (at approximately 60 μm and 180 μm). Case 2, with the nominal current, also has a distribution with two peaks, but with a significantly higher total concentration (1.9 cm^{-3}). Cases 3-5, with twice, three times and four times the nominal current, respectively, have distributions with only one peak, whose total concentration decreases slightly from the maximum of 3.7 cm^{-3}. Furthermore, the histograms in cases 1-5 have a monotonically increasing number of nuclei in the larger size bins, as indicated with an arrow in Figure 2a. Apparently, at the largest currents, larger bubbles are formed at the cost of a smaller concentration. However, the total volume of gas per unit of liquid volume (i.e. the gas void fraction) increases linearly with the electrolysis current, as expected from basic electrolysis theory.

In [1], the same ship model was used in similar conditions. There, the size distribution and the mode diameter stayed approximately the same with increasing electrolysis current; the number of bubbles initially increased, then stayed at the same level and then slightly decreased. The present data – in which the nuclei size increases with increasing current – supports the assertion made in [1] that a significant number of large nuclei has not been sized in those measurements. This was likely caused by either a relatively low camera defocus level (as stated in [1]) or a laser with a lower beam quality (e.g. short coherence length).

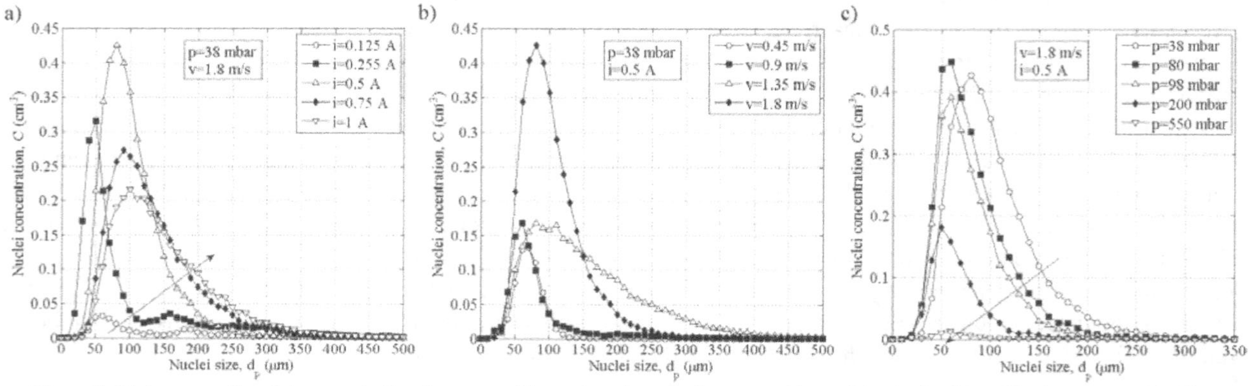

Figure 2. Histograms of nuclei concentration for cases with varying electrolysis current (a), model velocity (b) and basin pressure (c). The histograms have 10 μm wide bins, centered at 0 μm, 10 μm, 20 μm, etc. Figure (a) presents cases 1-5, figure (b) cases 6-8, 3 and figure (c) cases 3, 9-12.

Cases with increasing ship model velocity are presented in Figure 2b. The nuclei concentration increases monotonically, but not linearly, from 0.6 cm^{-3} to 3.7 cm^{-3}, and the modal diameter of the distributions stays approximately the same at 60-80 μm. The distributions, however, look quite different. The first two cases (6 and 7) are very similar, despite the increase in model velocity from 0.45 m/s to 0.9 m/s. Case 8 has a pronounced 'tail' with many nuclei between 100 μm and 300 μm. The fourth case (3), with the nominal (i.e. Froude-scaled) velocity of the model has a narrower distribution than case 8, with the standard deviation of the diameter of approximately 61 μm compared to 95 μm for case 8. A possible explanation of these results is offered by the mechanism of nuclei generation on the electrolysis strips. Namely, since the electrolysis current is the same for the four cases (0.5 A), the mass of produced gas should be the same. However, since the velocity is changing, two flow parameters are changing along with it: the shear at the electrolysis strips and the state of the boundary layer at the aft of the model. When the velocity is low, the shear is low, which tends to produce larger bubbles. Besides this, it is expected that at low velocity the boundary layer is laminar, which keeps the larger bubbles close to the hull. This would result in these nuclei not reaching the measurement volume (cases 6 and 7). With somewhat increased velocity, turbulence may stimulate a rather wide spectrum of nuclei to move away from the hull and reach the propeller plane (case 8). Finally, with further increasing velocity, many nuclei reach the measurement volume, but they are smaller since the shear is larger (case 3).

Cases measured at increasing basin pressure are shown in Figure 2c. As indicated by the arrow, the distributions are moving towards smaller sizes, as expected, since the higher pressure compresses the nuclei [1]. Furthermore, since the mass of produced gas is the same, it is expected from applying the ideal gas law that the product of basin pressure and the volume of nuclei (in a given liquid volume) is the same for the different cases. This is indeed the case for cases 3, 9 and 10, but not for cases 11 and 12, where this product is lower. This would suggest that the IPI technique, as it is currently applied, can resolve the nuclei spectrum adequately above approximately 40 μm. Consequently, the nuclei spectrum as generated by electrolysis can be measured properly at lower pressures (in cases 1-10), but not at pressures above 100 mbar (in cases 11 and 12).

A decrease in the total number of detected bubbles above basin pressures of 100 mbar was found also in [1]. Next to hydrodynamic reasons – such as the production rate of the bubbles and their trajectories – it was speculated in [1] that this was caused by the nuclei size-dependent measurement volume. With new insight gained from the present data, this reason appears less likely, since the same trend has been found when applying a size-dependent measurement volume. Another reason for the reduction of the number of nuclei mentioned in [1] is the inability of the measurement system to resolve the bubbles, i.e. the intensity of the small bubbles might be too low to reach above the detection threshold. This appears to be the case also in the present measurements. There are three variables that determine the intensity of the bubble image: laser light power, laser beam diameter and camera defocus level. For example, case 2 with a 60% smaller defocus level, shows significantly more nuclei in the 20 and 30 μm bins than all other cases. The camera sensor and lens characteristics also have an influence, but they are the same for all cases and the same as in the previous measurements [1]. When comparing with the measurements in [1], the light power has increased four times (although the beam diameter and therefore the light intensity per unit area in [1] is unknown) and the defocus level has been increased to capture larger bubbles. Apparently, this combination of effects in the present measurements resulted in approximately the same size limit as in [1]. In future measurements, the defocus level can be used to optimize the lower and higher size limits, since the sensor and the laser parameters are normally fixed.

The measured bubble sizes and concentrations are in the same order of magnitude as obtained previously with the IPI technique in cavitation tunnels [4–6]. A more detailed comparison with these studies is in most cases not meaningful, due to differences in the nuclei generation mechanism, the operating pressure and the velocity. In one case, however, nuclei were generated by the same electrolysis strips on a ship model [4,5]. The measurement volume was located approximately 100 mm downstream of the strips, very close to the hull. Since the current was not specified, the concentrations cannot be compared. The mode diameter at a velocity of 1 m/s and atmospheric pressure was 45 μm [5], which is similar to the values found in our measurements. However, the measured concentrations decreased with increasing velocity, whereas an increase with velocity was found in our study. This may be caused by the different relative position of the measurement volume.

The concentration of solid particles found in cavitation tunnels is often 10 to 20 times larger than the bubble concentration [4–6]. In the present measurements in the DWB, however, a negligible number of solid particles were found. This can be explained by the regular filtering of the water with 1 μm pore size filters, which would leave only the smallest solid particles in the flow. The amount of light these particles scatter is likely too small to be detected with the present system.

Conclusions

- Nuclei size and concentration measurements using IPI were performed successfully in depressurized conditions in MARIN's Depressurized Wave Basin. To obtain a high-quality interference pattern and therefore a high percentage of correctly sized nuclei, a good quality laser with sufficient power is necessary [cf. 1].
- With increasing electrolysis current, there is a shift towards producing larger nuclei. The total volume of produced gas increases linearly with the electrolysis current, as predicted by electrolysis theory.
- A complex relationship between the nuclei spectra and the model velocity was measured. It is proposed that this is the result of the combined influences of the shear at the electrolysis strips and the state of the boundary layer.

- The IPI technique, as it is currently applied, adequately measured the shrinking nuclei when increasing the basin pressure, but only up to p=100 mbar, above which an increasing number of nuclei fell below the detection limit.

Acknowledgements

This research is partly funded from the TKI-allowance of the Dutch Ministry of Economic Affairs.

References

[1] Birvalski, M., van Rijsbergen, M.X. (2017). *Exploratory measurements of cavitation nuclei in the wake of a ship model.* 5th International Conference on Advanced Model Measurement Technology for the Maritime Industry (AMT), Glasgow, UK.

[2] Albrecht, H.-E., Borys, M., Damaschke, N., Tropea, C. (2003). *Laser Doppler and Phase Doppler measurement techniques.* Springer-Verlag, ISBN: 3-540-67838-7.

[3] Tropea, C. (2011). *Optical Particle Characterization in Flows.* Annu. Rev. Fluid Mech., 43:399-426.

[4] Ebert, E., Kleinwächter, A., Kostbade, R., Damaschke, N. (2016). *HDNC - Nuclei Size and Number Concentration Estimation with Detection Volume Correction.* 31st Symposium on Naval Hydrodynamics, pp. 11–16.

[5] Ebert, E. (2017). *Optische Messtechnik zur Charakterisierung maritime Kavitationskeime.* Ph.D. Thesis, University of Rostock, Germany.

[6] Mées, L., Lebrun, D., Allano, D., Walle, F., Lecoffre, Y., Boucheron, R., Fréchou, D. (2010). *Development of interferometric techniques for nuclei size measurement in cavitation tunnel.* 28th Symposium on Naval Hydrodynamics, Pasadena, USA.

© 2018 ASME

Spatial Distribution of the Cavitation Aggressiveness in a High Speed Cavitation Tunnel

[1]Jan Hujer*; [1]Milos Muller

[1]*Technical University of Liberec, Department of Power Engineering Equipment, Liberec, Czech Republic*

Abstract

The cavitation aggressiveness was investigated by means of impact loads measurements. The cavitation impact load spectra were measured using PVDF film sensors placed at different locations in the eroded area of an aluminum disk. A high speed cavitation tunnel was used to generate a cavitating flow near to the sample surface. The eroded area was divided into seven regions and exposed to six cavitation regimes. The data was acquired by a high speed oscilloscope and automatically analyzed to identify the collapses peaks. Peak amplitude and peak count were obtained from the signal analysis. PVDF sensors were calibrated by a ball drop method. Each sensor had an individual calibration constant. The results represent relations between the cumulative peak rates, impact forces, sensor location positions and flow velocities.

Keywords: cavitation aggressiveness; PVDF film sensors; cumulative peak rate

Introduction

Collapses of vapor structures accompanied by microjets and pressure waves are responsible for the cavitation erosion of material in fluids engineering. Microjets and pressure waves form high intensity pulses of order of hundreds of MPa and of a short time duration (micro-nano second). Near surfaces are impacted and eroded. During the erosion incubation period, material surface gets plastically deformed and pits of order of tens micrometers occur. This deformation is caused by high pressure pulses applied on small areas. This initial period is in great interest of researchers who study cavitation erosion damage. The prediction of the cavitation erosion damage, i.e. the investigation of the aggressiveness of cavitating flow is conducted mostly by two ways, by pitting tests on polished samples of material at initial stage of cavitation erosion when pits are not overlapped and by measurements of impact loads on materials surfaces where suitable sensors are placed [1].

Momma and Lichtarowicz [2] correlated produced erosion on aluminium specimens A6063 with pit counting on polished soft aluminium discs and with pressure pulse data in a cavitating jet apparatus. Authors used a 110 μm thick piezoelectric pressure transducer made of polymer PVDF (polyvinylidene fluoride) mounted in gap on stainless steel target. The transducers was protected by two 70 μm layer of polyamide tape from water and the cavitation damage. Okada et al. [3] studied the relation between impact load by collapsing bubbles and the erosion damage. They developed a pressure detector consisting of a piezoelectric ceramic disc with one of test specimens made of aluminum, pure copper and stainless steel SUS304. The detector was used to measure the impact load, pits size and volume loss simultaneously.

The cavitation aggressiveness can be characterized by two ways, by conducting the pitting tests and by the direct measurement of impact loads. The aim of presented work is to investigate the spatial distribution of the cavitation aggressiveness in chosen cavitation testing device – in a high-speed cavitation tunnel. The cavitation aggressiveness is measured for several positions in the region of maximum damage and for different flow velocities at constant cavitation number. These results are expected to be compared later with pitting tests on polished aluminum samples Al 7075 at the same test conditions. The measurement of impact loads was conducted by a piezoelectric polyvinilidene fluoride (PVDF) film transducer. These transducers are characterized by high mechanical resistance, high resonance frequency, much clearer and higher signal than from a piezoelectric ceramic, low acoustic impedance and ease with which transducers can be made and mounted [2], [4].

Experiment

The investigation of spatial distribution of the cavitation aggressiveness was performed for the high-speed cavitation tunnel of the LEGI laboratory which test section is shown in figure 1. The cavitation tunnel is presented in more details in [5]. The tunnel axial inlet flow goes throw a nozzle and is bended to a radial outlet flow between two flat surfaces defined by the nozzle and tested sample. Behind the nozzle, a cavity of certain length occurs which finally collapses and leads to the erosion. The eroded area takes a form of a centred ring defined by two diameters, which positions depend on the cavitation number. Presented tests were carried out at a constant value $\sigma = 0.9$ of the cavitation number.

*Corresponding Author, Jan Hujer: jan.hujer@tul.cz

Based on a mass loss test on aluminum disc (Al 7075, 40 bar, 90 m/s, 10 hours) for which an eroded profile is between radiuses $R_{min} = 20\ mm$ and $R_{max} = 30{,}5\ mm$ an area of interest was chosen. The chosen area was divided into seven smaller regions. For that reason, seven aluminum discs with holes for transducer placement of the same geometry but at different radius were made. Position 1 is the closest to the centre and position 7 is the furthest from the centre. Commercial piezoelectric pressure film PVDF sensors (DTI-028K/L – Measurement Specialties, Inc.; TE Connectivity) were placed in these holes. The used PVDF transducers consist of 28 μm thick PVDF film of active area 12 mm x 30 mm given by both side silver ink metallizations, protecting urethane coatings over metallizations and lead wires with rivets. The transducers lengths were reduced from 30 mm to 1,5 mm, so the active areas of sensors were 12 mm x 1,5 mm. The PVDF transducers were mounted on aluminum discs as shown in detail in figure 1. The active area of PVDF film transducer was glued by a cyanoacrylate glue (Loctite Super Attak Power Flex Gel) on measured position. Holes for wires, gap for rivets and rest of space were filled by transparent epoxy glue (Bison Epoxy 5 minutes) which also creates a 1 mm thick protective layer of the transducer.

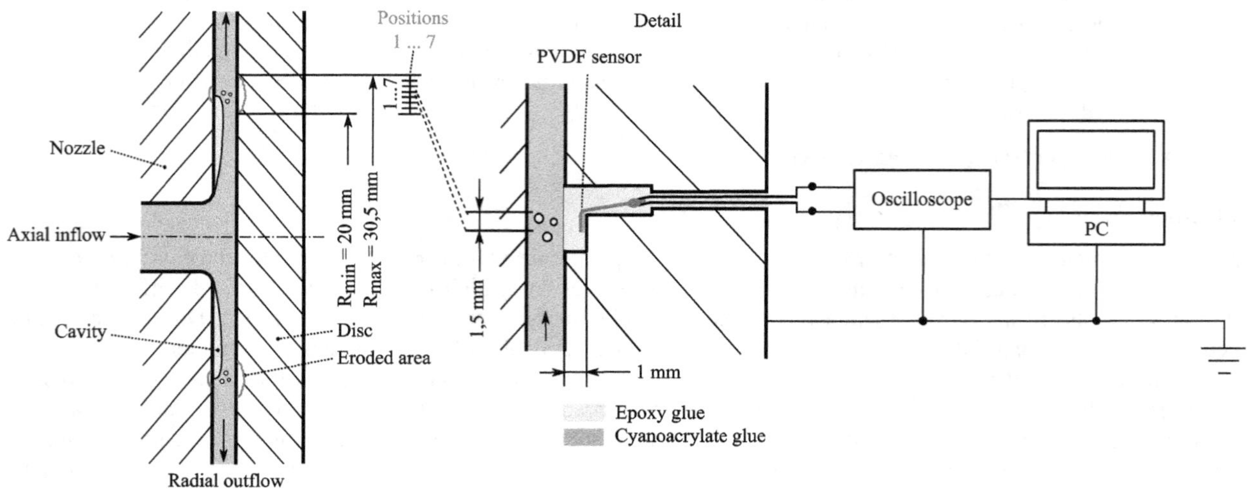

Figure 1. Measurement setup

Calibration of PVDF sensor

A ball drop method, pendulum-typed ball impact method and lead breaking method are used for the calibration of the PVDF film transducers [6]. The ball drop method was chosen to calibrate PVDF sensors in presented work. A stainless steel ball of certain weight m drops vertically from initial position of height h_1 above the calibrated PVDF sensor flush mounted on tested disc, hits the calibrated sensor and rebounds to position 2 of height h_2 above the calibrated sensor. A plastic tube of appropriate internal diameter with the ball diameter was used for the ball guiding. A camera and a rule are used to measure the heights h_1 and h_2. During the impact, a part of the initial ball energy is transformed in a deformation of transducer which generates an electric charge. Electric charge is measured by a digitizer NI-USB 5132 as a direct electric voltage output from which the maximum voltage U_{max} and the time duration of the impact τ are determined. Mean impact force is calculated from the following formula based on the impulse-momentum theorem

$$F_{mean} = \frac{1}{\tau}\int_{t_1}^{t_2} F(t)dt = \frac{m}{\tau}(v_1 + v_2). \tag{1}$$

Velocities before and after the impact are calculated from the following formula

$$v_{1,2} = \sqrt{2gh_{1,2}} \tag{2}$$

where $h_{1,2}$ are heights and g is gravity acceleration. Maximum force is double of the mean force, which results from the shape of the measured signal

$$F_{max} = 2 \cdot F_{mean}. \tag{3}$$

The calibration method results in a relation between the maximum impact force and the generated voltage. As example, calibrations of PVDF sensor at the sixth position before and after the measurement in the tunnel are shown in figure 2(a) for which the overall sensitivity was 0,0127 V/N.

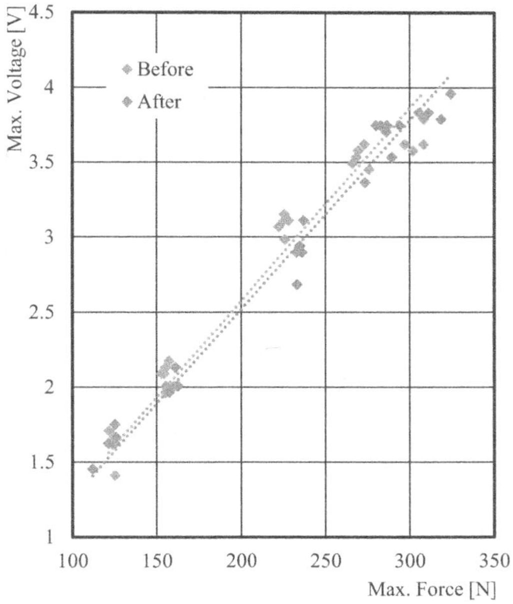

Figure 2(a). Calibration

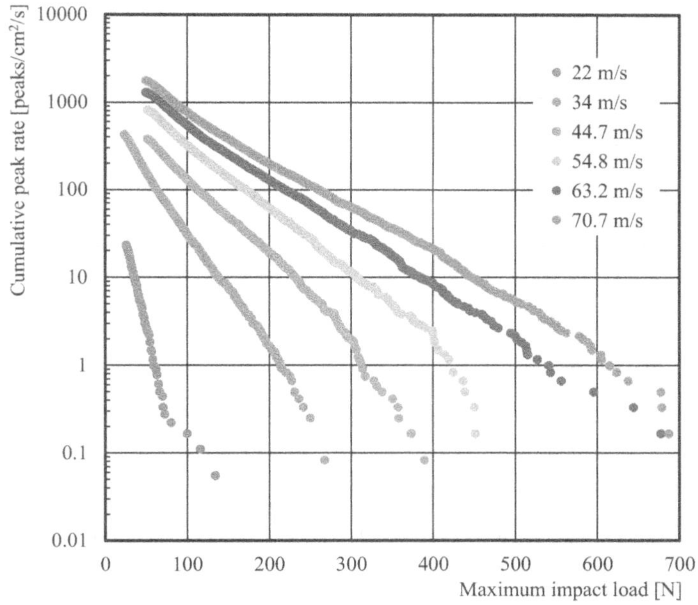

Figure 2(b). Impact load spectra at different flow velocities

Measurement and analysis of the data

The measurements of cavitation impact loads were done for several reference flow velocities on cavity in the test section of the cavitation tunnel. The measurement velocities and acquisition times (in brackets) were 22 m/s (15 s), 34 m/s (10 s), 44.7 m/s (10 s), 54.8 m/s (5 s), 63.2 m/s (5 s) and 70,7 m/s (5 s) m/s. Signal from PVDF sensor including impulses of cavitation impacts was measured by a 12-bit oscilloscope NI PXI 5105 at sampling frequency of 60 MS/s controlled by a LabVIEW virtual instrument. The measured data was analyzed in the Matlab. Firstly, a Savitzky-Golay filter was applied to a set of data for smoothing the data and then peak amplitudes were identified using the function findpeaks. Parameters of the used function findpeaks were minimum peak distance, minimum peak height and minimum peak prominence. The minimum peak distance of 10 μs was set as a locking time to avoid multiple counting of one impact. The minimum peak height was a threshold to exclude peaks corresponding to a noise. The minimum peak prominence selects only peaks that have a relative importance of at least set value. The minimum peak height and minimum peak prominence depend on noise of flow regime and are of the order of tenths of volts. The identified peaks maximum voltage was recalculated as the maximum force by means of calibration sensitivity.

Conclusion

The results are presented as cumulative histograms for one measured position at different flow velocities and for all positions at one chosen flow velocity. The relationship between cumulative peak rates [peaks/unit surface of the sensor/unit time] and maximum impact loads [N] for position 3 for different flow velocities is shown in figure 2(b). Both the peak rate and the maximum impact load increases as flow velocity increases. Parts of the spectra at low amplitudes are affected by set thresholds given by noise and data analysis method. Parts of the spectra at high amplitudes suggest a maximum impact load for each regime. Linear parts of the spectra correspond with an exponential law.

The relationship between cumulative peak rates and maximum impact loads for all measured positions for the flow velocity 63,2 m/s is shown in figure 3(a). Three regions are distinguishable in the histogram, which represents different flow aggressiveness at measured regions. The intensity profile is presented in figure 3(b). Values of cumulative peak rates at maximum impact loads 100 N, 150 N, 200 N and 250 N are displayed for all measured positions. The load levels are marked as colored vertical lines in figure 3(a). The highest intensity level is reached at positions 2 and 3. The intensity decreases away from these positions. For the larger impacts the results can be influenced by the size of the PVDF film area. This phenomena, discussed e.g. by Momma [2], predicts that the impacts on the boundary area of the PVDF film are included in the signal only partially.

The next step of the presented work is to correlate the measured data with pitting tests at the same regions and for the same flow velocities, i.e. to study the relation between impact load measurement and erosion damage at initial stage of cavitation erosion. The verification of the PVDF film active area influence on the measurement results is also desirable.

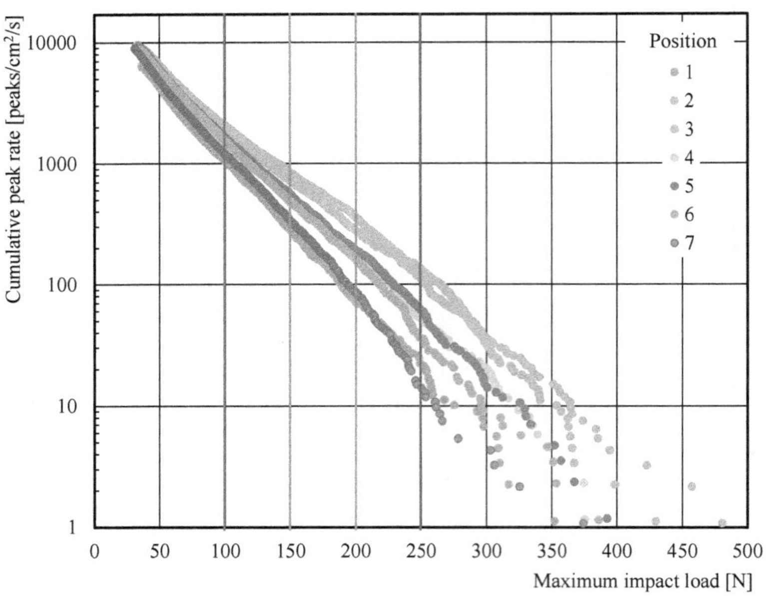

Figure 3(a). Impact load spectra at different positions

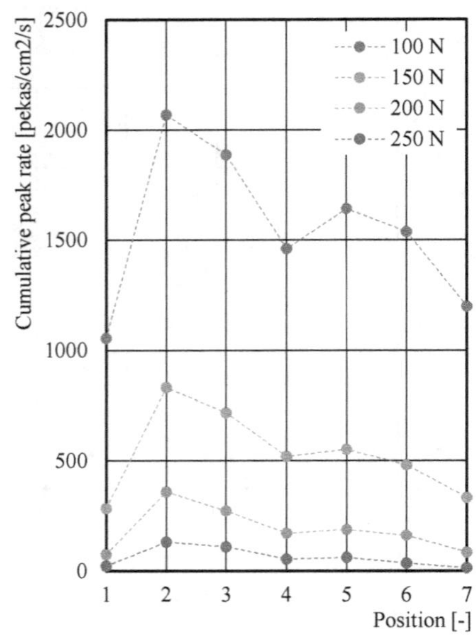

Figure 3(b). Intensity profile

Acknowledgment

This publication was written at the Technical University of Liberec as part of the project "SGS 21124" with the support of the Specific University Research Grant, as provided by the Ministry of Education, Youth and Sports of the Czech Republic in the year 2018. Authors are grateful to the LEGI laboratory for the access to the cavitation tunnel. Authors also would like to acknowledge the help provided by Jean-Pierre Franc and Michel Riondet from the LEGI laboratory.

References

[1] Franc, J.-P., Michel, J.-M. (2004). Fundamentals of Cavitation. Kluwer Academic Publishers.

[2] Momma, T., Lichtarowicz, A. (1995). *A study of pressures and erosion produced by collapsing cavitation.* Wear. 186-187.

[3] Okada, T., Iwai, Y., Hattori, S., Tanimura, N. (1995). *Relation between impact load and the damage produced by cavitation bubble collapse.* Wear. 184.

[4] Okada, T. Iwai, Y. (1989) *A study of cavitation bubble collapse pressures and erosion part 1: A method for measurement of collapse pressures.* Wear. 133(2).

[5] Franc, J.-P. (2009). *Incubation time and cavitation erosion rate of work-hardening materials.* Journal of Fluids Engineering. 131.

[6] Soyama, H., Lichtarowicz, A., Momma, T., Williams, E., J. (1998). *A New Calibration Method for Dynamically Loaded Transducers ans Its Application to Cavitation Impact Measurement.* Journal of Fluids Engineering. 120(4).

© 2018 ASME

10th International Symposium on Cavitation - CAV2018
Baltimore, Maryland, USA, May 14 – 16, 2018

Spectroscopic studies of nonequilibrium plasma generated by acoustic cavitation in aqueous solutions

Sergey I. Nikitenko; Rachel Pflieger

Institut de Chimie Séparative de Marcoule (ICSM), UMR 5257-CEA-CNRS-UM-ENSCM, Centre de Marcoule, Bat. 426, BP 17171, 30207 Bagnols sur Cèze, Cedex, France

Abstract

Molecular emissions in multibubble sonoluminescence (MBSL) spectra offer a powerful experimental tool to probe the extreme conditions inside the cavitation bubbles. Emission spectra of $OH(C^2\Sigma^+\text{-}A^2\Sigma^+)$, $OH(A^2\Sigma^+\text{-}X^2\Pi)$ and $NH(A^3\Pi\text{-}X^3\Sigma^-)$ systems revealed the formation of nonequilibrium plasma during multibubble cavitation in aqueous solutions saturated with noble gases. In nonequilibrium plasma, the electron temperature (T_e) exceeds the vibrational (T_v) and rotational (T_r) temperatures: $T_e > T_v > T_r \approx T_g$, where gas temperature (T_g) is usually considered equal to T_r. Spectroscopic analysis highlighted two kinds of plasmas produced by acoustic cavitation. Weakly excited plasma is observed at low ultrasonic frequency (20 kHz) in the presence of Ar and characterized by near-Boltzmann behavior of $OH(A^2\Sigma^+)$ and $NH(A^3\Pi)$ species. On the other hand, strongly excited plasma is formed at high frequency ultrasound (200-1000 kHz) and in the presence of Ar or Xe and even at 20 kHz in the presence of Xe. In a strongly excited plasma, the relative population distribution of $OH(A^2\Sigma^+)$ vibrational states deviates from Boltzmann equation. This type of nonequilibrium is created by vibration-vibration pumping (Treanor) mechanism. In general, spectroscopic study of MBSL allows to conclude that sonochemistry can be considered as a part of plasma chemistry.

Keywords: sonoluminescence; sonochemistry; nonequilibrium plasma, emission spectroscopy

Introduction

In the pioneering work on acoustic cavitation Rayleigh suggested that rapid bubble implosion should lead to adiabatic heating of the gas inside the cavity [1]. This idea was further developed by Noltingk and Neppiras [2] and later by Flynn [3]. Using Rayleigh-Plesset equation and presuming adiabatic conditions during bubble implosion they calculated the temperature and the pressure within the bubble at the moment of total collapse. For nitrogen-filled bubble in water their calculations provided 4200 K and 975 atm respectively. Despite numerous contradicting points, the adiabatic heating model based on simple physical principles was quite rapidly accepted by sonochemists since it allowed to interpret the sonochemical phenomena in terms of easily understandable flame-like processes. However, in 2005 Flannigan and Suslick observed emission from ionized oxygen O_2^+ during single-bubble sonoluminescence of concentrated H_2SO_4 pre-equilibrated with 50 mbar of Xe [4]. The formation of O_2^+ species is inconsistent with any thermal process. Indeed, the ionization energy of O_2 molecules is more than twice its bond dissociation energy. Therefore, the strong heat should lead to O_2 homolytic dissociation rather than to ionization. By contrast, the O_2^+ species can be formed by electron impact in plasma [5]. In the light of the above mentioned finding we performed a study of plasma formed during multibubble cavitation in aqueous solutions which is important for comprehensive understanding of numerous practically important sonochemical processes.

Body

The MBSL spectra of pure water pre-equilibrated with noble gases are composed of the emission lines of excited $OH^\bullet$ radicals in $A^2\Sigma^+$ and $C^2\Sigma^+$ states and broad continuum ranging from UV to near-infrared spectral range, which probably results from the superposition of several emissions: $H+OH^\bullet$ recombination, water molecule de-excitation, $OH(B^2\Sigma^+\text{-}A^2\Sigma)$ emission, and low-energy part of bremsstrahlung emission [6]. Figure 1 shows that not only the intensity but also the shape of MBSL spectra are strongly influenced by ultrasonic frequency and saturating noble gas. Significant broadening of $OH(A^2\Sigma^+\text{–}X^2\Pi_i)$ band with the increase of ultrasonic frequency was attributed to population of higher vibrational levels of $A^2\Sigma^+$ state [7]. Interesting that the relative intensity of $OH(C^2\Sigma^+\text{–}A^2\Sigma^+)$ emission band also increases with the increase of ultrasonic frequency or after the replacement of Ar by Xe [7,8]. The $OH(C^2\Sigma^+)$ state cannot be populated by any thermal process since the excited water molecule required for formation of such species must possess an excitation energy of at least 16.1 eV (i.e. >180000K). Most probably, its

*Corresponding Author, Sergey I. Nikitenko: serguei.nikitenko@cea.fr

formation involves the promotion of an electron from the $1b_2$ molecular orbital of the water molecule to a higher, vacant b_2 orbital upon collision with a sufficiently energetic electron [9]:

$$H_2O + e\ (\geq 16.1\ eV) \rightarrow H_2O\ (^1A_1) \rightarrow OH\ (C^2\Sigma^+) + H\ (^2S)$$

Therefore, the observation of $OH(C^2\Sigma^+-A^2\Sigma^+)$ emission clearly indicates the formation of nonthermal plasma during multibubble cavitation in water similar to that reported by Flannigan and Suslick [4] for single bubble in concentrated H_2SO_4.

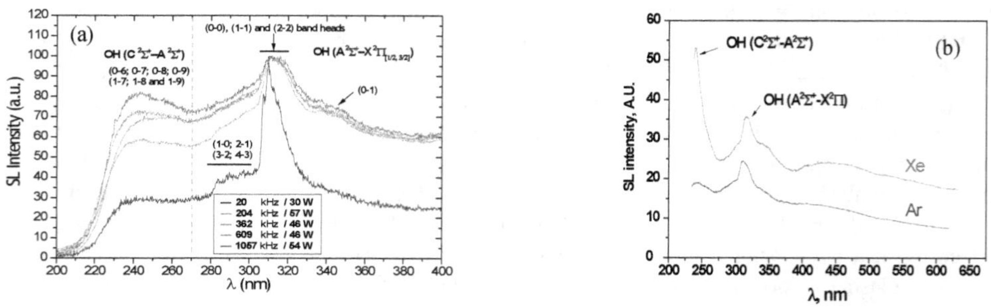

Figure 1. MBSL spectra in water: (a) effect of ultrasonic frequency in Ar at 10−11°C, intensity was normalized to OH(A-X) 0-0 transition [7]; (b) effect of noble gas at 204 kHz and 10-12°C, intensity in Ar has been multiplied by 3 to be comparable to that in Xe for the same exposition time [8]. Spectral resolution was ~2 nm for all spectra.

MBSL spectra of argon-saturated 0.1M $NH_3{\cdot}H_2O$ solution exhibit $OH(A^2\Sigma^+-X^2\Pi)$ and $NH(A^3\Pi-X^3\Sigma^-)$ emissions and a broad continuum as well (Fig. 2) [10]. At 20 kHz, the $NH(A^3\Pi-X^3\Sigma^-)$ system exhibits well-resolved (0-0)Q and (1-1)Q vibrational branches. By contrast, these branches strongly overlap at 359 kHz (Fig. 3). Vibrational (T_v) and rotational (T_r) temperatures of NH ($A^3\Pi$) and OH ($A^2\Sigma^+$) radicals were calculated as a function of ultrasonic frequency with Specair software widely used in spectroscopic studies of plasma [11]. The calculated values of rovibronic temperatures are summarized in Table 1.

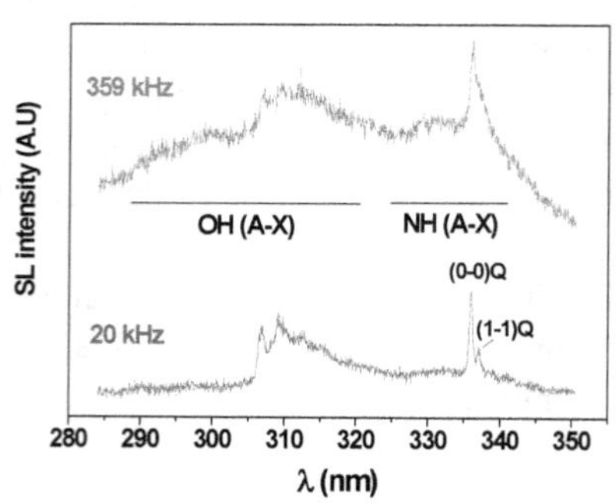

Figure 2. Typical SL spectra of 0.1 M $NH_3{\cdot}H_2O$ solutions obtained at 359 kHz (top spectrum) and 20 kHz (bottom spectrum) in the presence of Ar. Spectral resolution: ~0.3 nm.

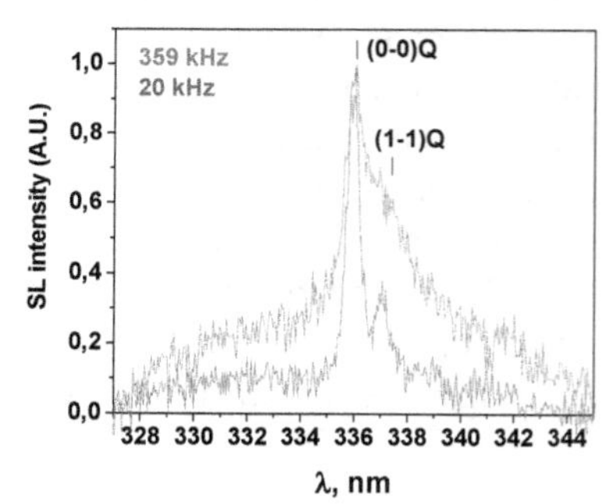

Figure 3. Comparison of NH ($A^3\Pi-X^3\Sigma^-$) emission for 20 kHz and 359 kHz of 0.1 M $NH_3{\cdot}H_2O$ in the presence of Ar. The spectra were normalized to (0-0)Q intensity after subtraction of a linear baseline.

In all four studied cases, vibrational temperature of radicals is higher than rotational or, in other words, gas temperature ($T_v > T_r \sim T_g$) indicating the absence of thermal equilibrium inside the cavitation bubble similar to pure water studied recently [8]. For both spectroscopic probes, increasing the US frequency results in significant differences in the shape of the emission spectra: "secondary" bands increase in intensity, reflecting an increase in vibrational temperature. Higher vibrational temperature highlights more drastic intrabubble conditions at higher ultrasonic frequency. This conclusion is in full agreement with MBSL study in water saturated with N_2-Ar gas mixture [12]. At 20 kHz, solely the molecular emission of $OH(A^2\Sigma^+-X^2\Pi_i)$ is observed in addition to a broad continuum typical for multibubble sonoluminescence. On the contrary, at high frequency (359 kHz) a second emission band is present around 336 nm which is assigned to the NH($A^3\Pi-X^3\Sigma^-$) system. The sonoluminescence of NH radicals in this system indicates more effective intrabubble dissociation of N_2 molecules at high ultrasonic frequency compared to low-frequency (20 kHz) ultrasound.

Spectroscopic transition	US frequency, kHz	T_v, K	T_r, K
NH ($A^3\Pi$-$X^3\Sigma^-$)	20	7000 ± 1000	4000 ± 500
	359	10000 ± 1000	2200 ± 500
OH ($A^2\Sigma^+$-$X^2\Pi$)	20	9000 ± 1000	5000 ± 500
	359	13000 ± 2000	6000 ± 1000

Table 1. T_v and T_r values calculated by fitting experimental SL spectra of NH ($A^3\Pi$-$X^3\Sigma^-$) and OH ($A^2\Sigma^+$-$X^2\Pi$) systems.

Replacing argon with xenon causes several dramatic changes in SL spectra as it can be seen from Figure 4. The much higher total MBSL intensity in the presence of Xe can be explained by its higher solubility in water compared to Ar, providing a larger concentration of cavitation bubbles. However, significant modifications of the spectral shape point out more drastic conditions inside the Xe-bubble. Thus, SL spectra obtained in $NH_3 \cdot H_2O$ solutions saturated with Xe exhibit stronger OH($C^2\Sigma^+$-$A^2\Sigma^+$) emission compared to Ar indicating higher electron temperature (higher ionization degree) of the intrabubble plasma which is most probably related to the lower ionization potential of Xe (12.13 eV for Xe, 15.76 eV for Ar). In general, vibrational excitation of molecules in plasmas increases with electron temperature [5]. This reflects in the intensity of (1-1)Q branch (around 337 nm) relative to (0-0)Q branch (at 336 nm) of NH($A^3\Pi$-$X^3\Sigma^-$) system obtained at 20 kHz that is higher with Xe (Fig. 4) than with Ar (Fig. 2 and 3). In addition, the SL spectra in Xe at 359 kHz exhibit NH($c^1\Pi$-$a^1\Delta$) emission at 324 nm which is not observable at 20 kHz ultrasound. Formation of NH($c^1\Pi$) state requires more energy (10.1 eV) than NH($A^3\Pi$) state (7.6 eV). Therefore, this observation confirms more drastic intrabubble conditions at high-frequency ultrasound.

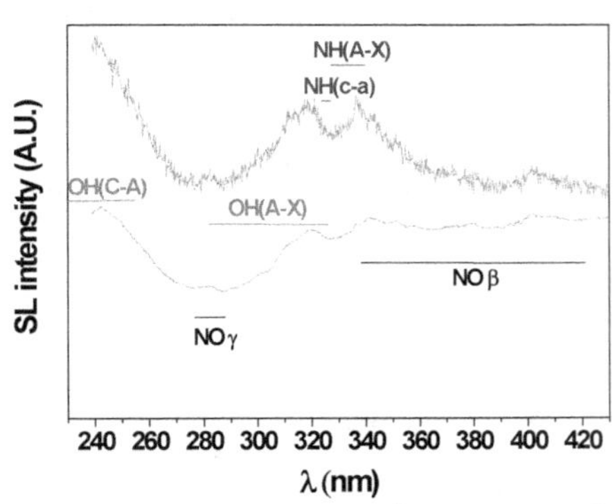

Figure 4. MBSL spectra of 0.1M $NH_3 \cdot H_2O$ solution in the presence of Xe at 20 kHz (green curve) and 359 kHz (red curve). Spectral resolution: ~2 nm.

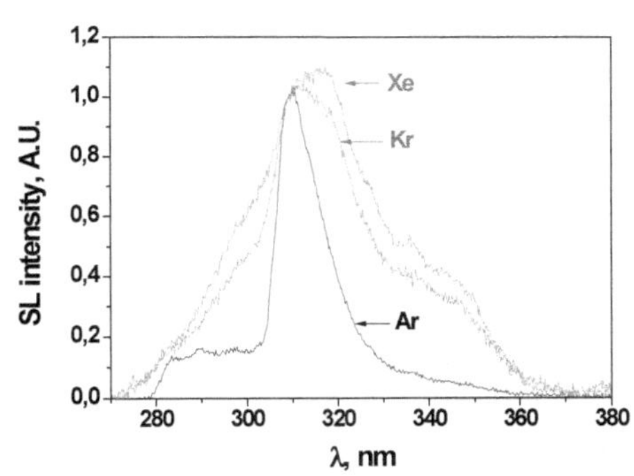

Figure 5. MBSL spectra of OH($A^2\Sigma^+-X^2\Pi_i$) system normalized to (0-0) transition in water saturated with Ar, Kr, and Xe at 20 kHz. The spectra were normalized to 0-0 transition after background subtraction. Spectral resolution: ~2 nm.

Emissions of the NH ($A^3\Pi$-$X^3\Sigma^-$) system appear significantly broadened in the presence of Xe compared to Ar (Fig. 2 and 4). This strong line broadening under Xe does not allow the use of Specair software for calculation of rovibronic temperatures and intrabubble pressure whatever the ultrasonic frequency. Other sources of line broadening than pressure and instrument response must be considered in a dense plasma such as Stark effect and non-Boltzmann overpopulation of higher vibrational levels. The latter phenomenon can be seen from the OH($A^2\Sigma^+-X^2\Pi_i$) mission band in water saturated with noble gases (Fig. 5) [8]. In Xe the intensity of (1-1) transition (314 nm) of the OH($A^2\Sigma^+-X^2\Pi_i$) system becomes comparable with that of (0-0) transition (309 nm). This type of nonequilibrium vibrational population distribution is created by vibration-vibration inelastic collisions between the excited molecules, in which quanta of vibrational energy are exchanged by the anharmonic vibrational modes of the collision partners.

Another striking feature of the SL spectra in Xe is the appearance of several weak emission bands in the wavelength interval 280-420 nm attributed to NO_γ and NO_β emission bands (Fig. 4). Formation of NO can be understood in terms of more efficient sonochemical splitting of NH_3 and H_2O in the presence of Xe compared to Ar followed by $NH^{\bullet}$ radicals oxidation with $OH^{\bullet}$ radicals:

$$NH_3 \text{ —))) } NH^{\bullet} + H_2$$

$$H_2O \text{ —))) } H + OH^{\bullet}$$

$$NH^{\bullet} + OH^{\bullet} \rightarrow NO + H_2$$

where the symbol ")))" indicates a process triggered by cavitation event.

Conclusion

The main new finding to emerge from recent MBSL spectroscopy studies of $OH(C^2\Sigma^+\text{-}A^2\Sigma^+)$, $OH(A^2\Sigma^+\text{–}X^2\Pi_i)$ and $NH(A^3\Pi\text{-}X^3\Sigma^-)$ systems is the absence of thermal equilibrium inside the cavitation bubbles in aqueous solutions. The most probable mechanism of $OH(C^2\Sigma^+)$ state formation involves electron impact indicating plasma formation during multibubble cavitation in water. The vibrational temperature of excited species increase with ultrasonic frequency indicating more drastic intrabubble conditions (higher ionization degree) at high-frequency ultrasound. The current state of research does not permit to establish the exact origin of this phenomenon. The MBSL spectra of water saturated with Xe unequivocally pointed out stronger vibrational excitation of $OH(A^2\Sigma^+)$ and $NH(A^3\Pi)$ states and higher yield of $OH(C^2\Sigma^+)$ state than in the presence of Ar. In terms of nonequilibrium plasma model this is the result of the lower ionization potential of Xe compared to Ar. A lower ionization potential provides a plasma with higher electron density and, consequently, higher vibrational temperature of radicals. Moreover, in Xe and at high-frequency ultrasound the relative population distribution of $OH(A^2\Sigma^+)$ vibrational states deviates from the Boltzmann distribution and follows V-V pumping anharmonic mechanism. The finding of nonequilibrium plasma formation during acoustic cavitation represents a paradigm shift for sonochemistry. Since a quasi-adiabatic heating model alone is not able to account for it, the sonochemical processes should be considered in terms of a more sophisticated plasma chemical approach that suggests the possibility to observe specific quantum effects in sonochemistry. Obviously, the average gas temperature around 5000 K usually reported to interpret the sonochemical activity presuming thermal equilibrium inside the bubble cannot be applied to plasma far from equilibrium. Instead, the mechanism of sonochemical reactions should be considered taking into account multiple different temperatures related to different plasma particles and different degrees of freedom.

References

[1] Rayleigh, L. (1917). *On the pressure developed in a liquid during the collapse of a spherical cavity*. Philos. Mag. 34(94).

[2] Noltingk, B.E., Neppiras, E.A. (1950). *Cavitation produced by ultrasonics*. Proc. Phys. Soc. 63B(674).

[3] Flynn, H. (1975). *Cavitation dynamics. Free pulsations and models for cavitation bubbles*. J. Acoust. Soc. Am. 58(1160).

[4] Flannigan, D.J., Suslick, K.S. (2005). *Plasma line-emission during single-bubble cavitation*. Phys. Rev. Lett. 95(N 044301).

[5] Fridman, A. (2008) *Plasma chemistry*. Cambridge University Press, Cambridge, USA.

[6] Nikitenko, S.I., Pflieger, R. (2017). *Toward a new paradigm for sonochemistry: Short review on nonequilibrium plasma observations by means of MBSL spectroscopy in aqueous solutions*. Ultrason. Sonochem. 35(623).

[7] Ndiaye, A.A., Pflieger, R., Siboulet, B., Molina, J., Dufreche, J.F., Nikitenko, S.I. (2012) *Nonequilibrium vibrational excitation of OH radicals generated during multibubble cavitation in water*. J. Phys. Chem. A 116(4860).

[8] Pflieger, R., Brau, H.P., Nikitenko, S.I. (2010). *Sonoluminescence from $OH(C^2\Sigma^+)$ and $OH(A^2\Sigma^+)$ radicals in water: Evidence for plasma formation during multibubble cavitation*. Chem. Eur. J. 16(11801).

[9] Quickenden, T.I., Irvin, J.A., Sangster, D.F. (1978). *Time resolved emission from $OH(C^2\Sigma^+)$ produced by pulse-radiolysis of water-vapor*. J. Chem. Phys. 69(4395).

[10] Pflieger, R., Ouerhani, T., Belmonte, T., Nikitenko, S. I. (2017). *Use of $NH(A^3\Pi\text{-}X^3\Sigma)$ sonoluminescence for diagnostics of nonequilibrium plasma produced by multibubble cavitation*. Phys. Chem. Chem. Phys. 19(26272).

[11] Laux, C.O., Spence, T.G., Kruger, C.H., Zare, R.N. (2003). *Optical diagnostics of atmospheric pressure air plasmas*. Plasma Sources Sci. Technol. 12(125).

[12] Ouerhani, T., Pflieger, R., Ben Messaoud, W., Nikitenko, S. I. (2015). *Spectroscopy of sonoluminescence and sonochemistry in water saturated with N_2-Ar mixtures*. J. Phys. Chem. B. 119(15885).

© 2018 ASME

10th International Symposium on Cavitation - CAV2018
Baltimore, Maryland, USA, May 14 – 16, 2018

CAV18-05191

Steel processed by water jet cavitation with ultrasonic irradiation

[1]Masataka Ijiri*; [1]Daisuke Nakagawa; [1]Kumiko Tanaka; [1]Toshihiko Yoshimura;

[1]Tokyo University of Science, Yamaguchi, JAPAN;

Abstract

Alloy steels for machine structural is used for industrial equipment. However, the operating environment is often harsh, requiring the development and application of various surface treatments. The authors focus on the use of water jet cavitation (WJC) with ultrasonication. WJC technology enables the generation of high pressure during cavitation collapse near the surface when a water jet ejected from a nozzle impacts a metal surface. This pressure causes a slight deformation in the impacted surface region and introduces a compressive residual stress due to the elastic constraints of the underlying and surrounding metal. If ultrasonic irradiation is applied to WJC bubbles with diameters of several hundred microns, the bubbles are subjected to alternating high and low sound pressures, which leads to a high-pressure and high-temperature reaction field. This technique is referred to as multifunction cavitation (MFC). The authors have reported several characteristics of specimens processed by MFC to date. For example, titanium oxide powder, which is widely used as a photocatalyst, can be fabricated in the form of nanoparticles by MFC. In addition, the indium tin oxide (ITO) film deposited on the transparent electrode mounted on liquid crystal displays can be melted by MFC treatment, leading to the release of indium, which has a low melting point. In present study, the compressive residual stress and corrosion resistance of Cr-Mo and Ni-Cr-Mo steels were improved by MFC treatment. Moreover, the authors compared conventional WJC technology to MFC technology. MFC was found to lead to higher compression residual stresses and higher corrosion resistances compared to conventional WJC. The corrosion resistance was revealed by the formation of an oxide film through selective oxidation and the concomitant reduction of surface defects. The oxide coating formed by a reaction between the dissolved oxygen in water with Cr on the metal surface during processing.

Keywords: Water jet cavitation; Multifunction cavitation; Low-alloy steel; Surface reforming

Introduction

Alloy steel for machine structural has been used for a wide range of machine parts mainly automobile parts. The improvement of fatigue characteristics isn't always satisfactory with the conventional surface modification at present, because of the use environment of this material is harsh year by year. To solve this problem, development of surface treatment and its practical application are expected. As one method, the authors take particular note of a multifunction cavitation (MFC) method with high-temperature and high-pressure microjet (MJ) which features both water jet cavitation (WJC) and ultrasonic cavitation (UC) [1]. In the WJC technology, high pressure occurs when the cavitation caused by high pressure water jetted from a nozzle collapses on the metal surface. This impact pressure results in slight plastic deformation of the surface layer and generates compressive residual stress by an elastic restraining force from the lower layer portion and the surroundings. When the compression force is converted to compression deformation, the deformation returns to its original state after the cavitation collapse; however, if a small amount of plastic deformation occurs, then compressive residual stress is generated after cavitation collapse. When WJC is applied to a welded part where tensile residual stress exists or a surface finished by surface grinding, the tensile residual stress on the surface is reduced and a compressive residual stress is produced. As a result, fatigue strength [2] and stress corrosion cracking resistance [3] are improved. Other than WJC technology, there are vibration cavitation called UC and laser induced cavitation. In the vibration cavitation, there are low sound pressure and high sound pressure in the liquid, because ultrasonic vibration generates strong and dense longitudinal waves in the liquid by applying strong ultrasonic waves in the liquid. Bubbles under low sonic pressure are expanded and then quickly shrink under high pressure. After repeating isothermal expansion or adiabatic compression, the bubbles collapse. UC is typically applied in chemistry, biotechnology, and medicine, using several microsize bubbles with a high temperature reaction field [4] generated by repetition of the expansion or contraction processes. Recently, we have developed a cavitation method with high-temperature and high-pressure microjets (MJs) that features both WJC and UC. This cavitation technique is so-called the mechanical and electrochemical cavitation method because it has both of mechanical action by MJ with high pressure of WJC and electrochemical action [5] by MJ with high temperature reaction field of UC. Moreover, it can be called MFC because it has new functions given to various materials. The

*Corresponding Author, Masataka Ijiri: ijiri@rs.tusy.ac.jp

authors have reported several characteristics of specimens processed with MFC so far. For example, firstly, titanium oxide powder is widely used as a photocatalytic material. It has been reported that the titanium oxide powder can be microfabricated to nano level when processed with MFC [6]. Secondly, the ITO film is a material applied to the transparent electrode mounted on the liquid crystal panel. It has also been reported that melting and releasing of indium having a low melting point occurs when an ITO (Tin-doped Indium Oxide) film coated on a transparent electrode mounted on a liquid crystal panel is subjected to MFC treatment [1]. In present study, the compressive residual stress and corrosion resistance of Cr-Mo and Ni-Cr-Mo steels were improved by MFC treatment. Moreover, the authors compared conventional WJC technology to MFC technology.

Experimental methods

The material used for these tests was Cr-Mo steel (SCM435) and Ni-Cr-Mo steel (SNCM630), both structural steels, the chemical compositions of which are shown in Table 1.

Table 1. Chemical composition of the low-alloy steels used in this work (mass%).

	C	Si	Mn	P	S	Ni	Cr	Mo	Cu	Fe
SCM435	0.37	0.32	0.81	0.014	0.012	0.012	0.95	0.15	0.14	Bal.
SNCM630	0.29	0.24	0.44	0.11	0.018	2.93	3.01	0.56	0.03	Bal.

Figure 1 is a schematic diagram of the equipment used for MFC processing. This equipment was similar to a conventional WJC apparatus, in that a jet of room temperature tap water was discharged from a nozzle at 35 MPa. The nozzle diameter was 0.8 mm. The distance between the nozzle and the specimen was 65 mm. The size of the water tank was $60\times45\times37$ mm^3, and it was filled with tap water. In MFC treatment, an ultrasonic transducer is placed next to the water jet nozzle for conventional WJC, and ultrasonic waves are propagated to the water jet flowing from the nozzle. The ultrasonic wave had an output of 225 W and frequency of 28 kHz. As shown in Figure 2, bubbles from WJC isothermally expand when the sonic pressure due to ultrasonic irradiation exceeds the Blake threshold, and after the bubbles expand to a certain size, Rayleigh shrinkage occurs rapidly (adiabatic compression). The flow cavitation, including the hot spot, becomes MFC by the repetition of isothermal expansion and adiabatic compression. At the same time that the MFC begins to collapse, it approaches the specimen surface. Figure 3 shows the bubble shape change from photographs taken by a high-speed camera during bubble collapse into an aspheric shape [7]. As the volume of the bubble decreases, a MJ with a columnar shape is formed and impinges on the specimen surface. This phenomenon is referred to as micro-forging because the MJ provides high-temperature and high-pressure processing in a microscopic area. Conventional WJC produces large high-pressure bubbles (diameter of several hundred micrometers at approximately 1,000 MPa), whereas bubbles produced with conventional UC are small (several micrometers) with high temperature (several thousand degrees Celsius). In contrast, the MJ that occurs in a bubble during MFC is a deforming liquid-phase body similar to a liquid jet (column) at the terminal stage of bubble collapse, and the interior temperature and pressure of the bubble become high (several thousand degrees Celsius and approximately 1,000 MPa, respectively). The processing time for all three methods was 2 min. Residual stress was measured by the full-width at the half-value breadth method from the (211) lattice spacing strain using Cr Kα X-ray diffraction (MSF-3M, Rigaku Corporation). Micro Vickers hardness tests were performed at room temperature under a load of 1.96 N for 10 s. The hardness values were averaged from eight measurements after the minimum and maximum of ten measurements were discarded. In addition, hardness measurements near the surface after processing were performed after slight mechanical polishing to remove the oxide film and peening marks formed during processing. Microstructure observations were conducted using optical microscopy (OM) and scanning electron microscopy (SEM; S-4800, Hitachi, Ltd.). Before observation, the specimens were etched in 5 vol% Nital. The specimens used for all analyses after processing were cut to 1×1 cm^2. For dissolved oxygen (DO) measurement, a portable type DO analyzer (OM–71, Horiba, Ltd.) was used. This measurement was carried out by appropriately collecting treated water in which cavitation occurred within a reactor made of SUS304 for a processing time of 30 min.

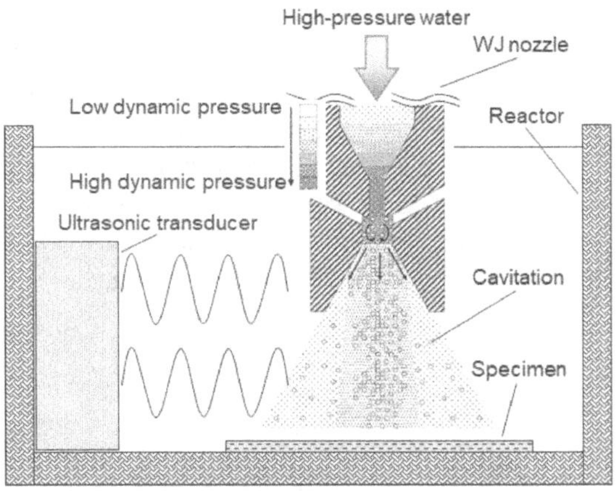

Figure 1. Equipment employed for surface machining by water jet cavitation under ultrasonication.

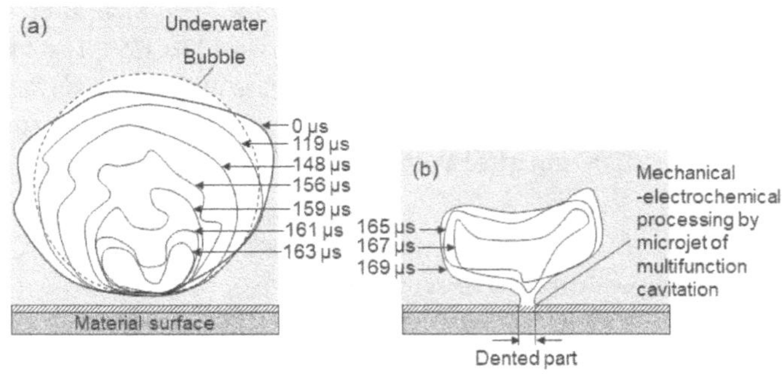

Figure 3. Images showing the aspherical collapse of a bubble at time intervals of (a) 0-163 μs and (b) 165-169 μs.

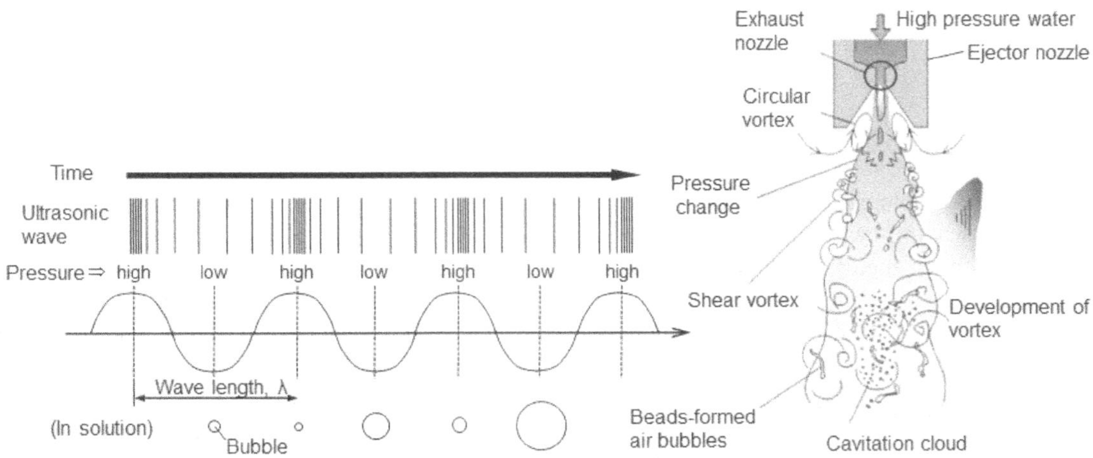

Figure 2. Multifunction cavitation mechanism.

Results and discussion

Table 2 shows the results of residual stress measurements. To further increase the compressive residual stress generated after WJC and MFC processing, the specimen surface was first subjected to hard grinding, which imparted tensile residual stress by a shakedown effect. These measurements were conducted in a state without rust on the specimen surface just after each processing and were performed in a direction parallel to the grinding direction. It was confirmed that grinding imparted a tensile residual stress in the grinding direction of the unprocessed specimens, while a compressive residual stress was imparted in the vertical direction. In WJC and MFC processing, compressive residual stress is generated when the specimen surface is compressed by the pressure from cavitation because the MJ applies force in the lateral direction, leading to plastic deformation around the region that is elastically constrained by its surroundings. Figure 4 shows the change in the micro Vickers hardness for each processed surface in the depth direction, but does not include the oxide film and peening marks. In the hardness of the near-surface region in the unprocessed specimen, SCM435 was 246 HV, and SNCM630 was 256 HV. The hardness distribution of the unprocessed specimen remained unchanged in the depth direction. It is considered that the unprocessed specimen is not significantly influenced by the strong surface grinding that was performed to impart tensile residual stress. The hardness in the near-surface region of the WJC-processed SCM435 was 271 HV. The hardness decreased at a depth of ca. 0.75 mm from the surface layer. On the other hand, the hardness in the near-surface region of the MFC-processed specimen was 276 HV. The hardness decreased at a depth of ca. 0.5 mm from the surface layer. This hardness change depends on the value (compressive stress field) where compressive residual stress was added in the depth direction. The compressive stress field was deeper in WJC than MFC. The compressive stress field occurred at a depth of ca. 0.5 mm after WJC processing, and it occurred at a depth of ca. 0.25 mm after MFC processing. About the depth of ca.

0.50 mm of WJC processed specimen, it is considered that the hardness decreased compared with the abrasive material due to voids and cracks. The reason why the hardness decreased from the depth of ca. 0.5 mm in the specimen after WJC processing is considered to be the influence of voids and cracks inside the specimen. Recently, the authors report that voids and cracks tend to form inside the specimen after WJC processing [8]. The hardness in the near-surface region of the WJC-processed SNCM630 was 270 HV. The hardness decreased at a depth of ca. 1.00 mm from the surface layer. On the other hand, the hardness in the near-surface region of the MFC-processed specimen was 264 HV. The hardness decreased at a depth of ca. 1.00 mm from the surface layer. The compressive stress field occurred at a depth of ca. 0.75 mm after WJC processing, and it occurred at a depth of ca. 0.50 mm after MFC processing.

Table 2. Compressive residual stress of specimens before and after various treatments (MPa).

		SCM435	SNCM630
As received after grinding	Vertical direction	-173	-238
	Parallel direction	+202	+155
UC		-216	-196
WJC		-361	-450
MFC		-293	-481

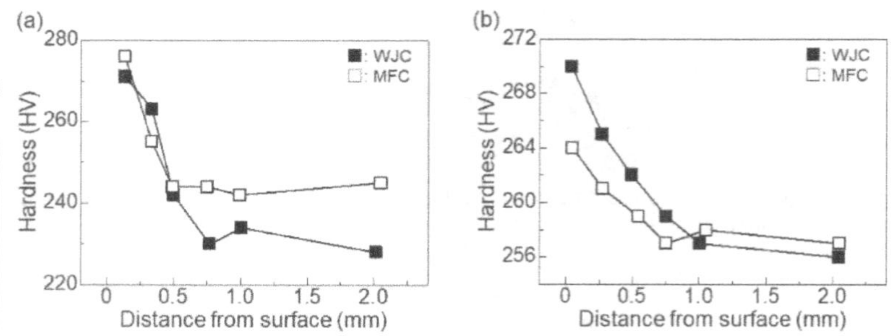

Figure 4. Hardness depth profile of the specimens after WJC and MFC treatment: (a) SCM435 and (b) SNCM630.

Figure 5 shows surface photographs of SCM435 (a, b) and SNCM630 (c, d) specimens after processing. The specimen surface was painted with oil-based ink prior to processing to identify the peening positions. The condition of the ink coating showed that the diameter of the peened portion of SCM435 was 40.8 mm after WJC processing and 42.2 mm after MFC processing. The peened diameter of SNCM630 was 41.3 mm after WJC processing and 42.7 mm after MFC processing. Therefore, MFC produced a slightly larger area than WJC produced. In addition, when the processed specimens were kept at room temperature for several months, corrosion developed after WJC processing, whereas there was almost no corrosion after MFC processing. Perhaps corrosion of the SNCM630 steel was prevented due to the formation of a dense oxide film on the specimen by increased temperature on the surface. Figure 6 shows OM micrographs of the SCM435 specimens after (a) WJC and (b) MFC. The particle size in the depth direction from the surface after each process was not significantly affected. In the near-surface region, no significant corrosion was observed after WJC processing, although it occurred easily after MFC processing. Corrosion was not observed in the ferrite phase but in the pearlite phase. The rust formed to a significant extent from the surface to a depth of approximately 1 mm.

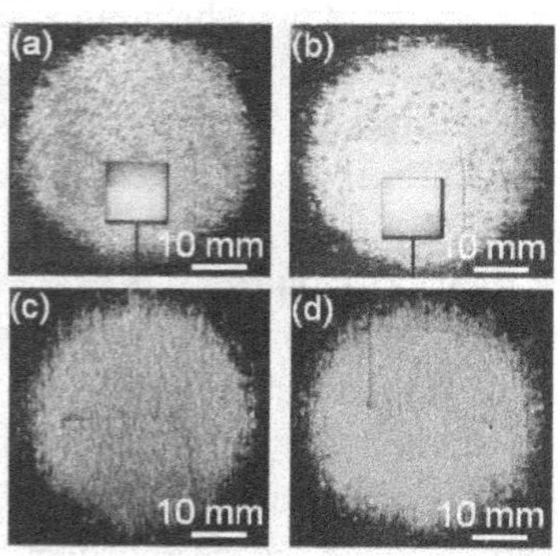

Figure 5. Surface change after treatment: of (a, c) WJC and (b, d) MFC. (a) and (b) are SCM435 and (c) and (d) are SNCM630.

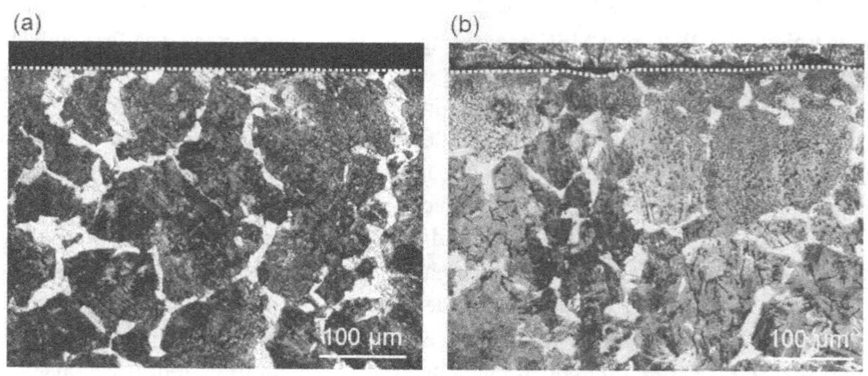

Figure 6. Optical micrographs of the SCM435 depth profile after (a) WJC and (b) MFC. A dotted line indicates a specimen surface.

Figure 7 shows SEM micrographs of the SNCM630 specimen depth profiles after WJC and MFC. The particle size in the depth direction from the surface after each process was not significantly affected. In the near-surface region, no significant corrosion was observed after WJC processing, although it occurred easily after MFC processing. Thus, selective oxidation is considered to have occurred in the bulk of the MFC-processed specimen. The element that is easy to oxidize in this specimen is Cr, which migrates to the topmost surface and forms a Cr-poor region beneath the surface during heating by MFC processing. As a result, rust can be easily generated because the amount of Cr decreased in the near-surface region. The corrosion resistance of the specimen surfaces was assessed by measuring surface potentials, and the results are summarized in Table 3.

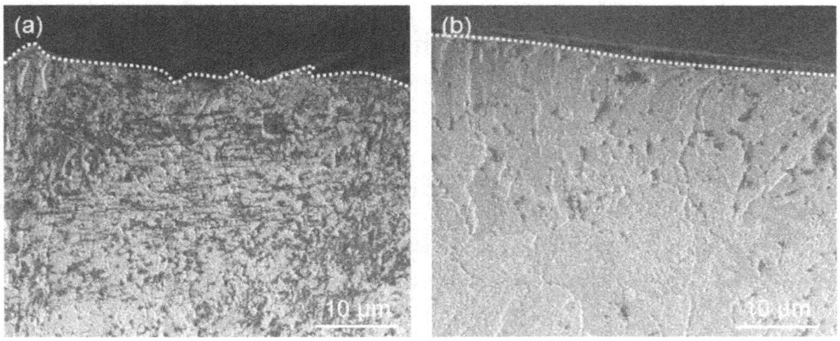

Figure 7. SEM images of the SNCM630 depth profile after (a) WJC and (b) MFC. A dotted line indicates a specimen surface.

Table 3. Surface potential of specimens after various treatments (mV).

	SCM435	SNCM630
As received	234 ± 14	359 ± 25
UC	194 ± 13	342 ± 18
WJC	382 ± 26	329 ± 18
MFC	544 ± 17	445 ± 16

The work function corresponding to the surface potential decreases as the surface roughness increases. The cavitation at the material surface during WJC generates impact pressures greater than several thousand megapascals. In addition, because some thermal energy is transferred to the material surface, a passive (corrosion-resistant) layer is formed. Soyama et al. [9] have also reported the formation of a passive layer as part of the mechanism by which the corrosion resistance of carbon steel is improved following the application of a cavitation jet. Regarding SNCM630, the surface potential of the WJC-processed specimen was lower than the abrasive material. It is probably influenced by the increase in surface defects compared with SCM435. During MFC, the bubbles generated by WJC (which are larger than UC bubbles) are irradiated with ultrasonic waves, and thus possess both high temperature (several thousand degrees Celsius) and high pressure (approximately 1,000 MPa). As these bubbles collide with the surface, a more stable passive layer is formed than that obtained with WJC. The reason for the selective oxidation inside the specimen during MFC processing was investigated by monitoring the water temperature, and Figure 8 summarizes the results. While UC did not increase the water temperature, both WJC and MFC did raise the temperature in the reactor. This occurred as a result of the pressurization of the water by the high-pressure pump, because the kinetic energy of the water was converted to thermal energy. From the viewpoint of bubble energy, it is thought that both WJC and MFC generate larger bubble collapse energies than UC processing, which also increases the water temperature. The slightly higher temperature obtained during MFC compared to WJC processing is due to the application of ultrasonic energy, reflecting the hot spots generated in the bubbles. Figure 9 plots the DO in the reactor over time for each processing method. The DO concentration was almost unchanged during UC, whereas it decreased during both WJC and MFC processing. In WJC processing, the pressure applied to the specimen surface increases due to the shock wave generated by the MJ. A part of the foam is deformed or bubbles become large again over a very short time span, and so both the water temperature and the temperature of the processed surface increase.

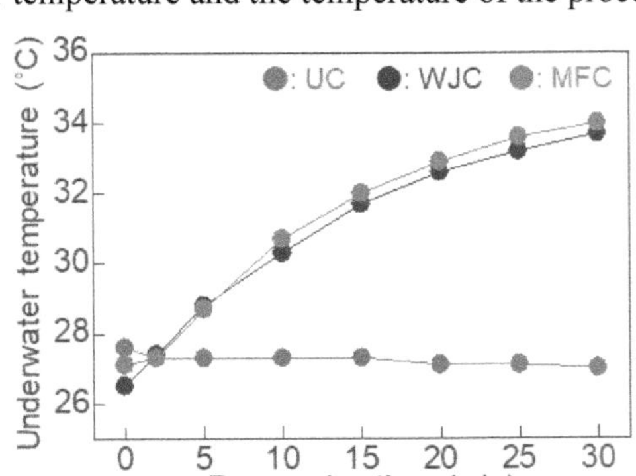

Figure 8. Relationship between processing time and underwater temperature.

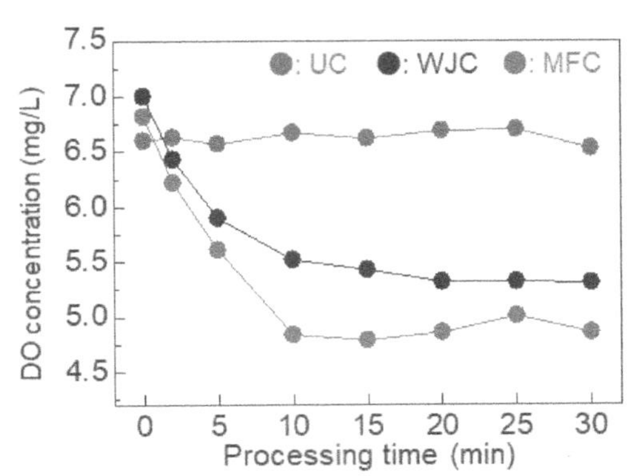

Figure 9. Relationship between processing time and DO concentration.

As a result, it is believed that DO combines with Cr to form oxides on the processed specimen surface, causing DO in the water to decrease. During MFC, the bubbles (which have a high internal temperature and pressure) increase the sample surface temperature to a greater extent than during WJC processing due to the hot-spot phenomenon inside the bubbles. Thus, the DO concentration is reduced to a greater degree as a passive layer is generated. As described above, in the MFC technique, bubbles having characteristically high temperature and high pressure collide with the specimen surface as the result of irradiating large bubbles generated by WJC with ultrasonic waves. This method effectively improves the surface residual stress via surface modification and also increases the specimen strength and corrosion resistance.

Conclusion

Improvements in residual stress and surface modification leading to high strength and corrosion resistance were found to occur at the surfaces of specimens processed with MFC. Corrosion resistance was improved via the formation of an oxide film by selective oxidation, as well as the reduction of surface defects. We concluded that this oxide film was formed by dissolved oxygen in the water reacting with Cr on the metal surface during the MFC treatment.

Acknowledgment

This work was supported by the Innovative Science and Technology Initiative for Security Program of the Acquisition, Technology & Logistics Agency (ATLA) of Japan.

References

[1] T. Yoshimura, K. Tanaka, N. Yoshinaga (2003). *Development of mechanical-electrochemical cavitation technology, Journal of Flow Engineering*. 20.
[2] D. Odhiambo, H. Soyama (2003). *Cavitation shotless peening for improvement of fatigue strength of carbonized steel*, J. Fatifue. 25.
[3] N. Saitou, K. Enomoto, K. Kurosawa, R. Morinaka, E. Hatashi, T. Ishikawa, T. Yoshimura (2003). *Development of water jet peening technique for reactor internal components nuclear power plant*. J. Jet Flow Eng 20.
[4] B. Gompf, R. Gunther, G. Nick, R. Pecha, W. Eisenmenger (1997). *Resolving sonoluminescence pulse width with time-correlated single photon counting*. Phys. Rev. Lett., 79.
[5] JP Lorimer, TJ Mason (1999). *Applications of Ultrasound in Electroplating in Electroplating, Electrochemistry*., 67.
[6] K. Tanaka, T. Yoshimura and N. Yoshinaga (2016). *Study on photocatalyst properties improvement of titanium oxide by the mechanical-electrochemical cavitation*. J. Jet Flow Eng 32.
[7] C. L. Kling and F. G. Hammitt (1972). *A photographic study of spark-induced cavitation bubble collapse*. J. Basic Eng 94.
[8] M. Ijiri, D. Shimonishi, D. Nakagawa, T. Yoshimura (2017). *Evolution of microstructure from the surface to the interior of Cr-Mo steel by water jet peening*. Materials Sciences and Applications 8.
[9] H. Soyama and M. Asahara (1999). *Improvement of the corrosion resistance of carbon steel surface by a cavitating jet*. J. Mater. Sci. Lett. 18 (5).

© 2018 ASME

Stochastic processes to model impact events in a vibratory cavitation erosion apparatus

[1]Gabriel Taillon*; [1] Kaoruko Onishi; [1]Sumio Saito, [1]Kazuyoshi Miyagawa;
[1]Waseda University, Shinjuku, Tokyo, Japan

Abstract

Cavitation erosion is caused by bubble collapse causing micro jets and shockwave that impact the surfaces of materials: these impacts occur randomly over time. As such, Stochastic processes, specifically the Poisson process, can be used to model impacts as random events occurring over time. Peaks in pressure amplitude measurements made in time with a high-speed PVDF transducer in a vibratory cavitation erosion apparatus based on the ASTM G32 standard were defined as impact events. Using cumulative impact as a function of time, the rate parameter of the Poisson process was observed to vary periodically, which indicates that the process may be a non-homogeneous Poisson process (NHPP). Using a nonparametric estimation method, the NHPP's cumulative rate function was estimated, and then was fitted to a sinusoidal function with a frequency effectively equal to the vibratory apparatus' vibration and an average impact rate of 2.10×10^5 impacts per second.

Keywords: Impacts; Cavitation erosion; Vibratory apparatus; Poisson process; Stochastic process

Introduction

Cavitation occurrence in turbines, pumps and other turbomachines is a troublesome issue: cavitation bubble impacts are a major cause of erosion and mass loss itself is difficult to model accurately [1]. Significant discrepancies between modeled erosion rate and observed damage are still noted today [2]. One can realize the complexity and scope and the task undertaken: turbulent flow leading to nucleation, complex bubble-bubble interaction then violent collapse, forming shockwaves and micro-sized water jets, leading to deformation and erosion under high stress/strain rate [3].

Cavitation impact occurrence depends on parameters local to vapor bubbles that are difficult to measure [4,5], and cavitation bubble behavior is complex, a field of study on its own: bubble dynamics. It could prove useful to model such events as randomly occurring over time using stochastic processes, specifically Poisson point processes that can describe the occurrence of events in space and time [6-8]. From the point of view of the material, impacts also occur randomly on a 2D plane of generally flat geometry, and have other measurable characteristics: an amplitude, an area of effect, impact duration etc. that vary randomly. Here we will concentrate on modelling cavitation impacts events in time, and ignore other random variables. The confidence in the applicability of the Poisson process, the simplest of the point processes, to model cavitation impact events will be computed.

Body

The impact amplitude distribution was measured on a vibratory cavitation erosion apparatus inspired by the ASTM G32 standard [9]. The vibratory apparatus used had a vibration frequency of 19.5kHz and a peak-to-peak vibration amplitude that could be adjusted from 7μm to 20μm. A pump was used to recirculate water from the double-sided beaker to a water container: the tap water around the specimen was maintained at $25 \pm 5°C$. A schematic of the whole setup measurement setup is presented in Figure 1.

A Müller high-speed PVDF transducers model M60-1L-M3 was used: it has a rise time of 60ns, a sensitive diameter of 1mm, a sensitivity of -23.30 MPa/V and has a the following dynamic pressure range: -3MPa to 40MPa. A Tektronix MSO4054 oscilloscope was used to visualize and record the transducer output, with a sampling rate of at least 25MS/s, equivalent to a timestep of 40ns, close to the rise time. The event threshold was around 10mV equivalent to a minimum detected impact pressure of 200kPa, after filtering the data between 4kHz and 2MHz using Python.

The basic definition of a stochastic process is: a collection of random variables defined on a common probability space indexed by a set of numbers, generally time, which describes the system's evolution. In this context, the Poisson

*Corresponding Author, Gabriel Taillon: taillon.gabriel@asagi.waseda.jp

process is interpreted as a counting process [6-8]: we write $\{N(t), t \in T\}$, with T the continuous time $T = |0, \infty)$ and $N(t)$ the total number of impact events that occurred in the interval $[0, t]$. An impact event is defined as a local voltage/pressure maxima in the transducer data that exceeded the 10mV noise threshold.

The simplest Stochastic process that can be used to count events is the Poisson process [6]. It has several key properties, among which we cite: the number of events in a time interval Δt is completely independent from the number of events in all other intervals. It seems reasonable to assume that the impact events are independent, though some known physical mechanism demonstrate clustering behavior (a parent event leading to several children events). Bubble dynamic computations, confirmed by experiments, demonstrate that a single bubble collapse event can produce multiple predictable impact events because of multiple collapse-rebound cycles, under certain conditions [11]. Such repeated events can be taken into account by the so-called Cluster Poisson Process if they arise.

The Poisson process is generally defined using its probability density function (PDF), which outputs the probability P that the total number of events $N(t)$ in the interval $[0, t]$ is equal to i, noted $P[N(t) = i]$:

$$P[N(t) = i] = \frac{(\lambda t)^i}{i!} e^{-\lambda t}, \qquad t \geq 0 \tag{1}$$

With λ the single parameter of the Poisson process: the rate, or intensity. To compare the Poisson process with the recorded data, it is useful to learn how to how to estimate the PDF from observations, and how to draw samples using a PDF, or to simulate random variates. For the Poisson process, the expected number of points in a region is simply $E[N(t)] = \lambda t$. The strong law of large numbers leads to: $\lim_{t \to \infty} \left(\frac{N(t)}{t}\right) = \lambda$, which demonstrates that the best estimation method for the parameter of a Poisson process interpreted as a counting process is the total number of measured events divided by the total recording time.

As for sample generation, a simple method can be derived using the fact that the impact interarrival times are exponentially distributed with parameter λ: $P[\tau_i > t] = e^{-\lambda t}$. This function takes a time and outputs a probability: by isolating the time, we produce a function that outputs the time to the next event, as a function of a probability. Then, using a uniform random number generator a distribution of interarrival times can be obtained iteratively:

$$t_{next} = -\frac{\ln(u)}{\lambda} \tag{2}$$

with $u \sim U(0,1)$ the randomly generated probability. With this basic understanding of stochastic processes, we can start the analysis of the impact data. In Figure 2 (a), the raw pressure as a function of time is shown. There seems to be some periodicity: regions with high density of impacts repeat every 40-60μs at minimum, indicating it is the fundamental frequency of a periodic phenomenon, which implies a Poisson process may not be suitable to model such a phenomenon. By comparing the interarrival times of a Poisson process of rate equal to the total impacts divided by the record time, Figure 3 can be obtained. Cavitation impact events do not seem to follow a Poisson distribution: interarrival times are supposed to be exponentially distributed e.g. the number of long time periods with no impacts is too large. To take the apparent periodicity of the rate function into account, one can turn to the Non-Homogeneous Poisson process (NHPP)

The Poisson process described in Equation 1 indeed is homogeneous: the rate is constant in time. But one can obtain a NHPP by letting the rate function vary with time : $\lambda(t) \geq 0$. In the present case, the rate function is clearly periodic. The period of the pressure wave created by the apparatus is approximately 51.3μs, which provides a naïve explanation of this phenomenon: cavitation bubbles may increase in size in the low-pressure and collapse during the high-pressure period. Unfortunately, the bubble dynamics in a vibratory apparatus are complex [12] and outside the scope of the present research.

In any case, it is quite a bit more a bit more difficult to estimate the NHPP's rate function, but not at all impossible. The estimation of the cumulative rate function by the nonparametric method proposed by Leemis [13] was used. The cumulative rate function of a NHPP, which is its expectation function $\Lambda(t) \equiv E[N(t)]$, is defined [6-8]:

$$\Lambda(t) = \int_0^t \lambda(\tau) d\tau, \qquad t > 0 \tag{3}$$

The Leemis estimator gives an approximation of Λ, which is noted $\widehat{\Lambda}$, for a number k of realizations of the observation of the NHPP in a time window $(0, S]$. S can be, for example, all opening hours (e.g. 9AM to 5PM) of a restaurant that wishes to count the arrival of customers. The estimator is written:

$$\widehat{\Lambda}(t) = \frac{in}{(n+1)k} + \left[\frac{n(t-t_i)}{k(n+1)(t_{i+1}-t_i)}\right], \qquad t_i < t \leq t_{i+1};\ i = 0,1,2,\dots,n \tag{4}$$

$\widehat{\Lambda}$ is the estimation of Λ, n is the total number of impacts recorded in k realizations, t_i are the sorted arrival times of the n impacts and t represents an arbitrary large number of times between t_i and t_{i+1}. In this case, S was chosen to be long enough to include at least 10 periods: $S = 4ms$, for $k = 100$ realizations. The value of the estimator is linear between, t_i and t_{i+1}. If two or more impact have the same arrival time t_m, the estimator jumps to the value:

$$\widehat{\Lambda}(t_m) = \widehat{\Lambda}(t_{m+1}) = \frac{mn}{k(n+1)} \tag{5}$$

Using this estimator, the rate function could be fitted to a function of the form: $\lambda(t) = A(\sin(2\pi\nu t + \phi) + 1)$ (see Equation (3)). In the present case, for a distance of 1.4mm between the vibrating and head and the sensor and a vibration amplitude of 7μm, $A = 2.10 \times 10^5$ impacts per second. Also, the estimated period is 50.6μs, close to the expected 51.3μs. The frequency of the rate function (19.8kHz) is quite similar to the vibratory apparatus' (19.5kHz), but it is not possible to comment further without additional impact data, and studies on the bubble dynamics.

In Figure 4 are presented the results of the estimation $\widehat{\Lambda}$ compared with the fitted sinusoidal function, with 95% confidence intervals (CI). One must take heed as the CI in Figure 4 (a) are different from those in Figure 4 (b): in Figure 4 (a) the CI are on the expected number of events given the rate of a Poisson process (one can learn to compute such intervals using the Poisson distribution such as described in [16] p. 182 for example), in Figure 4 (b) the CI are on the estimation $\widehat{\Lambda}$ given the observed data, as described by Leemis [13]. Figure 4 (c) presents the error between the estimation $\widehat{\Lambda}$ and the fitted sinusoidal function, compared with the error between the estimation and its CI bounds. One expects the cumulative function to be contained in the CI bounds in 95% cases: the current fit is outside these bounds in 8.98% cases, which is high. If the current estimation $\widehat{\Lambda}$ is correct, 31.6% of all measured points are out of the confidence bounds of the NHPP, as can be observed on the data in Figure 4 (a). The authors feel the experimental setup may explain this unusually high variance. For example, local variation of temperature in the cavitation region may be due to the vibrating head acting as a heat source: temperature was observed to rise close to the sample even with temperature control using a double-sided beaker, because of the long time data recording necessitated (2min by realization). A higher variance could be explained by the rate function $\lambda(t)$ itself varying over time. Better control could provide clearer insight. Another explanation may be that some parameter in the rate function varies randomly:the local temperature may oscillate randomly between realizations and influence the rate. In such a case, the Mixed Poisson process could be used to model the impact events. If the rate function itself is a stochastic process, then the cavitation impacts events can be modelled using the doubly stochastic Poisson process (Cox process).

This analysis was proposed as the first step in the elaboration of a stochastic process that could capture the random nature of the cavitation erosion process. In subsequent experiments, the vibration frequency, vibration amplitude, water temperature among others should be changed to observe the effect on the rate function. Also, steps will be taken to model events in the vibratory apparatus using the mentioned other stochastic processes. Such stochastic processes could be used to model cavitation impact events in turbines and pump as well as cavitation erosion laboratory apparatuses such as the cavitation impact jet described in the ASTM G134 standard [17]. They could also prove useful to build a mass-loss prediction model, by combining them with deformation and failure under strain hardening and high strain-rate conditions, such as the Johnson-Cook dynamic failure model [10].

Figures & Tables:

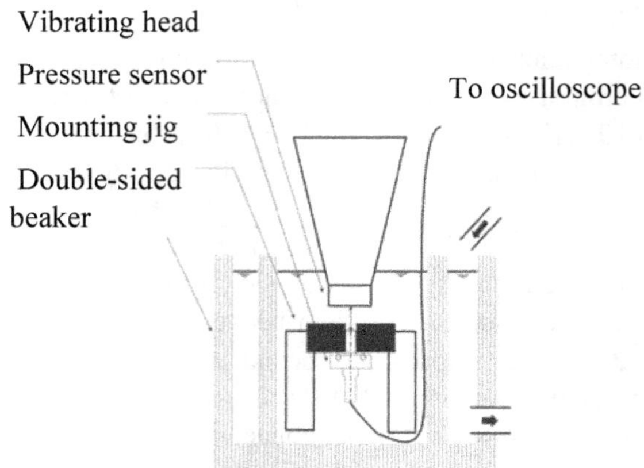

Figure 1: Vibratory apparatus setup to record impact pressure as a function of time

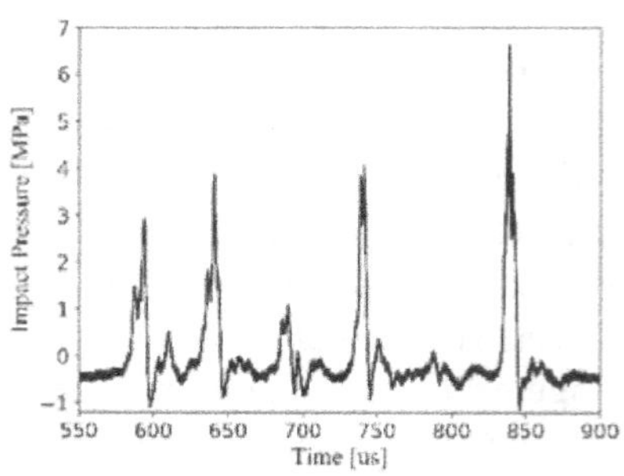

(a): Raw high pressure sensor output

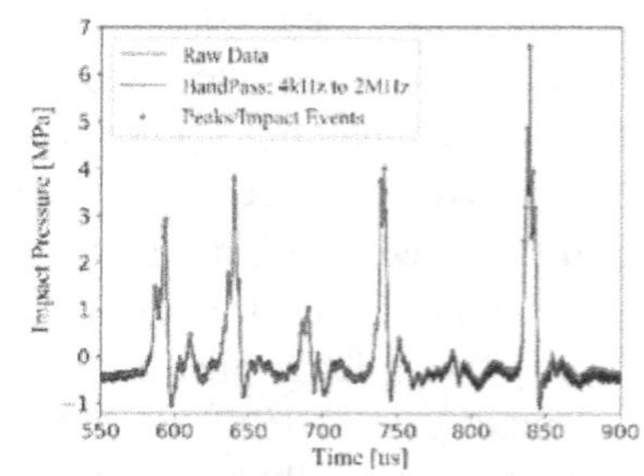

(b): Peaks were detected taking the first order difference (with peakutils, a python package)

Figure 2: Impact pressure as a function of time for 1.4mm distance and 7µm amplitude

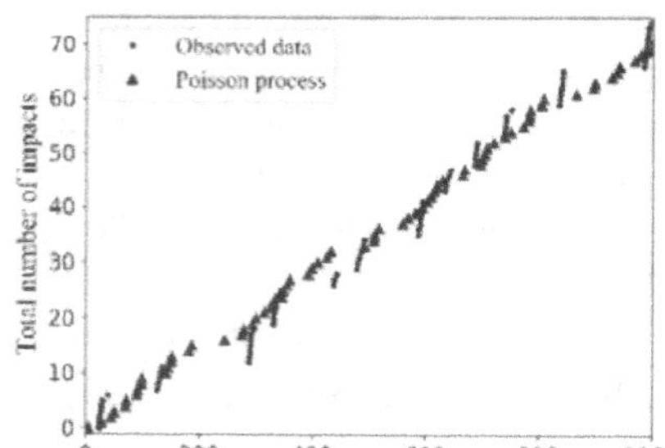

Figure 3: Cumulative events as a function of time: Comparison with Poisson process of equivalent rate

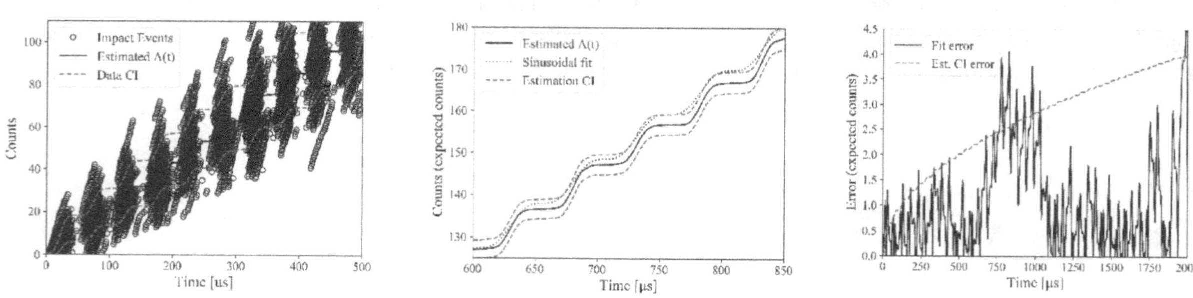

(a): Observed data for $k = 100$ realizations compared with $\widehat{\Lambda}$ estimation **(b):** Estimated rate function fitted with a sinusoidal rate function **(c):** Fitting error: 8,98% of points of the fit are outside the 95% CI

Figure 4: Rate function estimated using the nonparametric method in Leemis [13]

Conclusion

Cavitation erosion impacts occur randomly over time and as such can be modeled as a Stochastic process. The simplest is the Poisson process, for which the only parameter is the rate parameter. Using a high-speed PVDF pressure sensor to detect impacts in a vibratory apparatus, periodic spikes in the arrival indicated that the rate parameter varies over time: the Poisson process is non-homogeneous. Using a non-parametric method by Leemis [13], the cumulative rate function was observed to be sinusoidal, with a frequency close to the driving transducer: 19.8kHz compared to 19.5kHz. Cavitation impacts were recorded for 4ms, 100 times to produce said estimation. An average of 2.10×10^5 impacts per second were observed. Fitting the estimated cumulative rate function using a sinusoidal function proved relatively accurate, but 31.8% of all data points fall outside the 95% confidence intervals which is unusually high.

References

[1] Singh, R., Tiwari, S. K., & Mishra, S. K. (2012). Cavitation erosion in hydraulic turbine components and mitigation by coatings: Current status and future needs. *Journal of materials engineering and performance*, *21*(7), 1539-1551.
[2] Turbomachinery Society of Japan, TSJ Guideline; Guideline for Prediction and Evaluation of Cavitation Erosion in Pumps, TSJ G 001:2011, (2016), 93-96. (Partly translated from Japanese to English)
[3] Kim, K.-H., Chahine, G., Franc, J.-P., & Karimi, A. (2014). *Advanced Experimental and Numerical Techniques for Cavitation Erosion Prediction*. (R. Moreau & A. Thess, Eds.) (Vol. 106). Dordrecht: Springer Netherlands.
[4] Leigthon, T. G. (1994). *The Acoustic Bubble* (First ed.): Academic Press Limited.
[5] Brennen, C. E. (1995). *Cavitation and bubble dynamics*: Oxford University Press.
[6] Snyder, D. L., & Miller, M. I. (1991). *Random point processes in Time and Space* (Second ed.): Springer-Verlag.
[7] Nelson, R. (1995). *Probability, Stochastic Processes and queuing theory* (First ed.): Springer Science+Business Media, LLC.
[8] Resnick, S. I. (2002). *Adventures in Stochastic Processes* (3rd ed.): Springer Science+Business Media, LLC.
[9] ASTM G32-16, Standard Test Method for Cavitation Erosion Using Vibratory Apparatus, ASTM International, West Conshohocken, PA, 2016, www.astm.org
[10] Johnson, G. R., & Cook, W. H. (1983). A constitutive model and data for metals subjected to large strains, high strain rates and high temperatures. In *7th International Symposium on Ballistics* (pp. 541–547). http://doi.org/10.1038/nrm3209
[11] *Advanced Experimental and numerical Techniques for cavitation erosion prediction*. (2014). (K.-H. Kim, G. Chahine, J.-P. Franc, & A. Karimini Eds. Vol. 106): Springer.
[12] Hansson, I., & Mørch, K. A. (1980). The dynamics of cavity clusters in ultrasonic (vibratory) cavitation erosion. *Journal of Applied Physics*, *51*(9), 4651-4658.
[13] Leemis, L. M. (1991). Nonparametric estimation of the cumulative intensity function for a nonhomogeneous poisson process. *Management Science, 73*(7).
[14] Pasupathy, R. (2010). Generating homogeneous poisson processes. *Wiley Encyclopedia of Operations Research and Management Science*.
[15] Çinlar, E. (2013). *Introduction to stochastic processes*. Courier Corporation.
[16] Sachs, L. (2012). *Applied statistics: a handbook of techniques*. (Second ed.) Springer Science & Business Media.
[17] ASTM G134-17, Standard Test Method for Erosion of Solid Materials by Cavitating Liquid Jet, ASTM International, West Conshohocken, PA, 2016, www.astm.org

© 2018 ASME

10th International Symposium on Cavitation - CAV2018
Baltimore, Maryland, USA, May 14 – 16, 2018

CAV18-05193

Study on the Hybrid Method of CFD and Bubble Dynamics for Marine Propeller Cavitation Noise Prediction

[1] Satoko Ando *; [1] Koyu Kimura; [1] Kohei Segawa; [1] Koutaku Yamamoto;

[1] *AKISHIMA LABORATORIES(MITSUI ZOSEN)INC., Tokyo, JAPAN;*

Abstract

In this paper, the authors have been investigating the possibility of the prediction of propeller cavitation noise. This method is hybrid method of CFD analysis and bubble dynamics model. Obtaining a cavitation volume by using a commercial CFD code, we estimate the size distribution of the cavitation bubbles and calculate the propeller cavitation noise by the bubble dynamics model.

Keywords: marine propeller; cavitation noise; bubble dynamics

Introduction

In ship design, it is essential to improve propulsive performance and consequently fuel oil consumption, and high-efficiency propeller is always required. Furthermore, it is recognized that the reduction of both vibration and noise is important from the perspective of the structural strength, safety and habitability on ship. Especially, within the last decade there is concern about increase of the underwater ship radiated noise (USRN) from the perspective of the negative influence on marine environment. Thus, the demand for reduction and prediction of USRN at propeller design stage is growing rapidly. One of the main factors of USRN is the propeller cavitation noise. There are some methods for prediction of propeller cavitation noise. What is studied a lot recently is the hybrid method of CFD and Ffowcs Williams-Hawkings(FW-H) (Prins et al.[1], Hallander et al.[2]). However, this method is the one to analyze flow noise and does not represent the physical phenomenon of the radiated noise from cavitation bubbles. Though some prediction methods considering cavitation bubble were suggested (Matusiak[3],Kamiirisa et al.[4]), it was difficult to predict the cavitation volume and the size distribution of cavitation bubbles then. On the other hand, the authors have been investigating the possibility of the application of CFD to the propeller performance evaluation and cavitation. A good agreement with the experiment result was confirmed for the marine propeller cavitation (Kimura. et al.[5]). So the authors has been developed the hybrid method of CFD and bubble dynamics model for marine propeller cavitation noise prediction in order to resolve the above issue(Ando et al.[6]). In this paper, the authors improved this hybrid method in term of estimation of the size distribution of cavitation bubbles especially and described a possibility and problems of the hybrid method of CFD and bubble dynamics for propeller cavitation noise.

Methodology of marine propeller cavitation noise prediction

The calculation flow of the presented prediction method of a propeller cavitation noise was shown in Figure.1. At first, the unsteady cavitating simulation was computed using a RANS code for the marine propeller. And the volume of sheet cavitation and the intensity of tip vortex were estimated from the computed results.

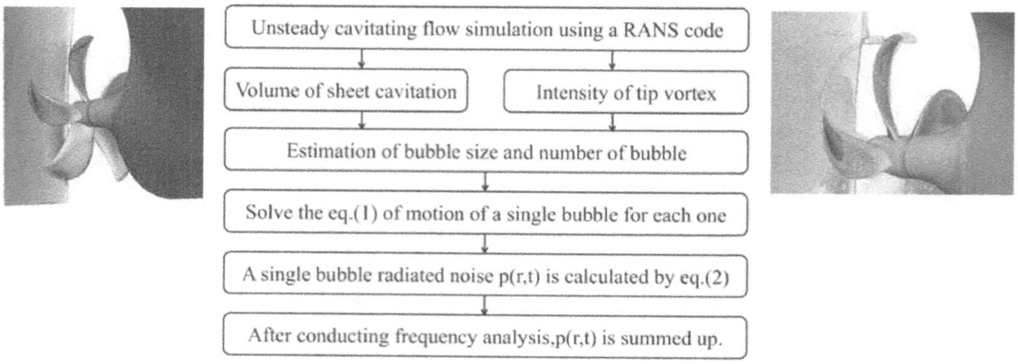

Figure1: Calculation flow of propeller cavitation noise

*Corresponding Author, Ando Satoko: ando@ak.mes.co.jp

And the reference bubble size was also estimated from the computed results. The size distribution of bubbles was assumed the normal distribution. Next, the equation of motion of the bubble with the second order correction of the liquid compressibility was calculated according with ref.[7].

$$R\ddot{R}\left[1-(1+\varepsilon)\frac{\dot{R}}{C_\infty}+\left(-1+\frac{7}{2}\varepsilon-\frac{\varepsilon^2}{5}\right)\frac{\dot{R}^2}{C_\infty^2}\right]$$
$$+\frac{3}{2}\dot{R}^2\left[\frac{4-\varepsilon}{3}-\frac{4\dot{R}}{3C_\infty}+\left(-\frac{4}{3}+4\varepsilon-\frac{43}{30}\varepsilon^2+\frac{\varepsilon^3}{6}\right)\frac{\dot{R}^2}{C_\infty^2}\right]$$
$$\left[\left(-2+\frac{7}{2}\varepsilon-\varepsilon^2\right)\dot{R}^2+\frac{3}{2}\cdot\frac{p_\infty-p_{1_{r=R}}}{p_\infty}\right]\}]=0 \quad (1)$$

The pressure in liquid at the bubble wall $p_{i_{r=R}}$ can be written as follows, and the subscript represents i-th approximation.

$$p_{1_{r=R}}=-\frac{2T}{R}+P_0\left(\frac{R_0}{R}\right)^{3\gamma}+\rho_g\varepsilon\dot{R}^2-\varepsilon\frac{4\mu\dot{R}}{R}$$

$$p_{2_{r=R}}=p_{1_{r=R}}+\frac{\varepsilon}{C_\infty^2}\left(\frac{4\mu}{3}+\zeta\right)\left\{\frac{3}{2}\varepsilon(1-\varepsilon)\frac{\dot{R}^3}{R}-\frac{\dot{P}_{1_{r=R}}}{\varepsilon\rho_\infty}+(1-\varepsilon)\frac{\left(P_\infty-P_{1_{r=R}}\right)\dot{R}}{\varepsilon\rho_\infty R}\right\}$$

$\varepsilon = 1-\rho_g/\rho_\infty$、$C_\infty$: Sound velocity in liquid、T: Surface tension、
ρ_g: Density of gas、μ: Viscosity of liquid、ζ: Bulk viscosity of liquid

Assuming that one bubble is acoustic monopole, the sound pressure is defined by the following equation (2)

$$p_m=\frac{\rho_0}{4\pi r}\frac{d^2V}{dt^2}\left(t-\frac{r}{c_0}\right) \quad (2)$$

$V = 4/3\,\pi R^3$, r = distance from center of buble, $\rho_o = density\ in\ liquid$, $c_o = velocity\ of\ sound$

After conducting frequency analysis, p_m is summed up.

Experimental setup

The experiments were carried out in HYKAT, the largest cavitation tunnel of the HSVA(Figure.2)[8]. The propeller cavitation noise was measured by the hydrophone located in the hydroacoustic chamber directly below the test section at a distance of ab.2.3m from the acoustic center of the propeller.

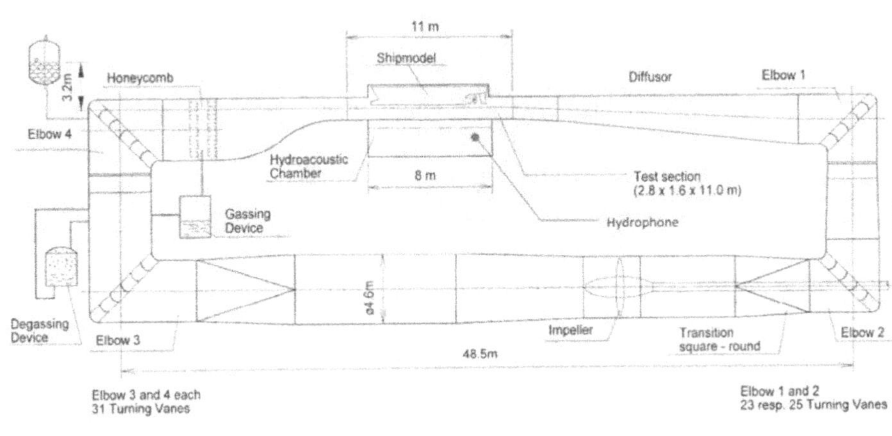

Figure2: Hydrodynamics and Cavitation Tunnel(HYKAT)

Table1 and Figure3 show principal particulars and lines of ship model. It is Japan Bulk Carrier(JBC) which was published by CFD Workshop 2015 Tokyo[9].On the other hand, the model propeller was designed based on the result of JBC's towing tank test. The model propeller's principal particulars and photo are shown in Table2 and Figure4.

Table1 : Principal particulars of JBC

JBC(Japan Bulk Carrier)		ship	model
LENGTH P.P.	(M)	280.000	7.778
LENGTH D.W.L.	(M)	285.000	7.917
BREADTH MOULDED	(M)	45.000	1.250
DEPTH MOULDED	(M)	25.000	0.694
DESIGHN DRAFT MOULD	(M)	16.500	0.458
PROPELLER DIAMETER	(M)	9.000	0.250
Scale	1/α	1/	36.0

Table2 : Principal particulars of the model propeller

MPNo.P628R		
Propeller Diameter	(M)	0.25
Number of Blades		5.00
Mean Pitch Ratio		0.80
Expanded Area Ratio		0.48
Boss Ratio		0.17
Turning Direction		Right

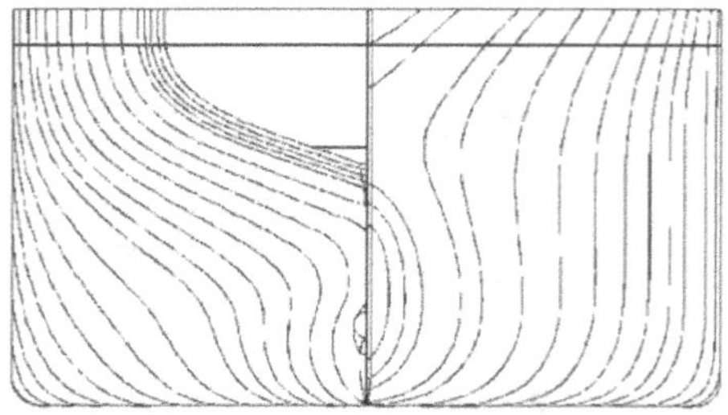

Figure3: Lines of JBC

Figure4: Model Propeller

Figure5 shows the experimental set up for cavitation tests. The tests consisted of cavitation observation, pressure pulse measurement and hydro acoustic measurement. The test condition was as follows.

$$\mathrm{KT} = \frac{T}{\rho n^2 D^4} = 0.28 \quad (3)$$

$$\sigma_n = \frac{P - P_v}{\frac{\rho}{2}(\pi n D)^2} = 0.37 \quad (4)$$

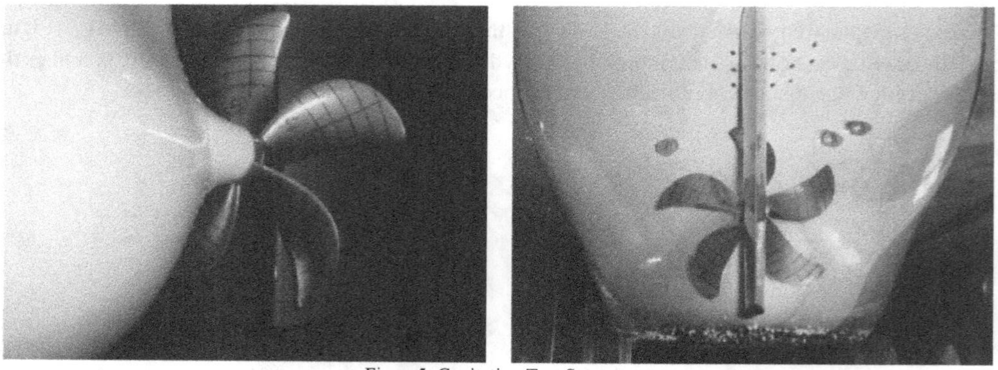

Figure5: Cavitation Test Set-up

CFD analysis

For estimation of a sheet cavitation noise, we need to assume the size distribution and number of the cavitation bubbles. We used a commercial CFD code to estimate of the sheet cavitation volume and the intensity of tip vortex. Figure6 shows CFD mesh to simulate of the experiment of HYKAT. The mesh was generated using Pointwise Ver18.0R2.The grid sizes were ab.0.6M and 1.3M cells for the sliding block and the other blocks, respectively. A commercial CFD software Fluent version 17.2 was used for the numerical simulations.Table3 showed summary of simulation methods.

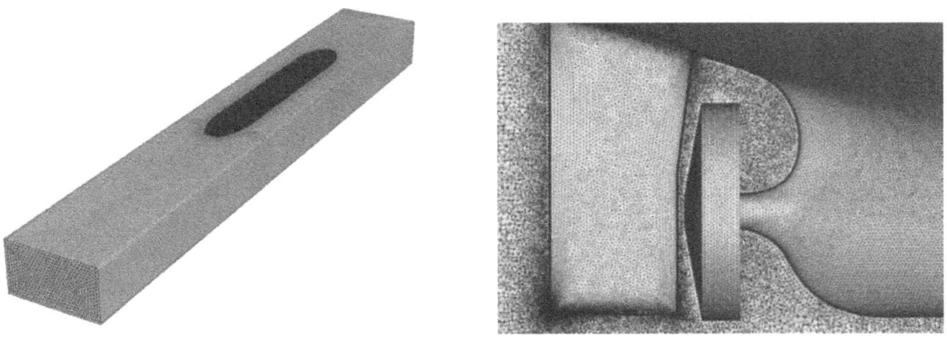

Figure6: CFD mesh

Table3: Summary of simulation methods

Govering equation	RANS
Turbulence model	k-ω SST
Cavitation model	Mixture Schnerr-Sauer
P-V coupling	Coupled
Pressure solver	PRESTO!
Momentum solver	QUICK
Transient Formulation	Second Order

Results and discussion

The comparison of cavitation pattern was shown in Figure7. This figure shows the iso-surface of vapor volume fraction of 50%. The sheet cavitation on the blade surface is well-reproduced compared with the experiment result. However, the occurrences of the tip vortex cavitation (TVC) were not sufficient compared with experiment result, because the resolution of CFD mesh around propeller tip was not fine. Figure8 shows the comparison of the pressure fluctuation on P1 between experiment and computed result.P1 is located at 0.125Dp upstream and starboard from the cross point hull center line and the propeller generator line on the hull surface. These results confirm that present simulation is capable of reproducing with sufficient accuracy the cavitation behavior due to practically high-skew propellers and the sheet cavitation volume.

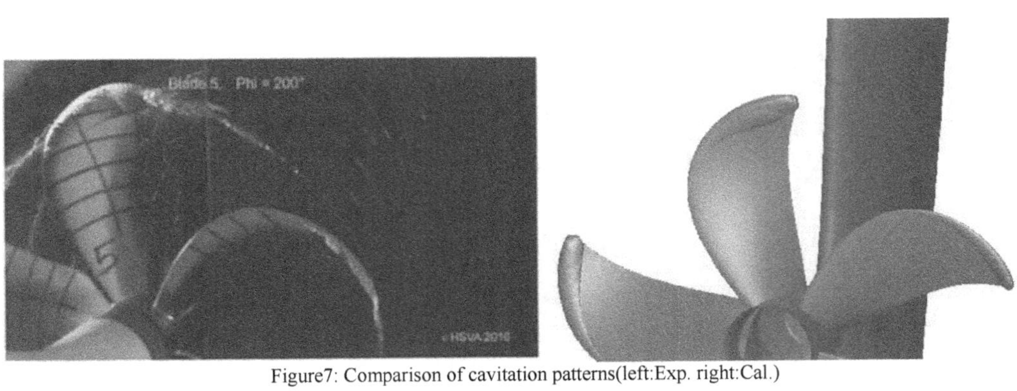

Figure7: Comparison of cavitation patterns(left:Exp. right:Cal.)

At the maximum blade angle position, the reference bubble diameter is assumed as shown in Figure9. As shown in Figure10, the bubble distribution was assumed to be a normal distribution, and the number of bubbles was determined such that the assumed total bubble volume was equal to the computed sheet cavitation volume. The noise level by sheet cavitation is shown in Figure11, but it is found that it is smaller by about 13 dB than the experiment result. This was attributed to not being evaluating noise by TVC, and TVC was also evaluated by the bubble dynamics as well as sheet cavitation. It has been difficult to reproduce TVC using CFD. However, from computed result, assuming that the reference bubble diameter of TVC is 2.5 mm too. And the number of bubbles was estimated by assuming the length of TVC to 1.5Dp from cavitation observation. As result, it was confirmed that the prediction method of cavitation noise with considering sheet cavitation and TVC have good agreement with the experimental results.

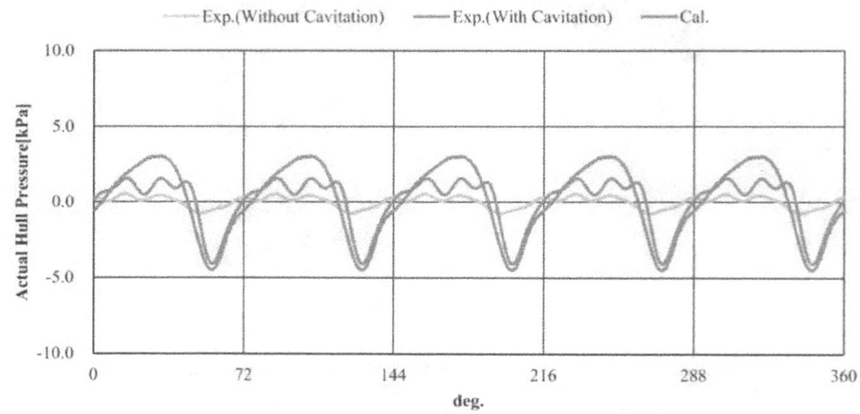

Figure8: Hull Pressure Time Signal on P1

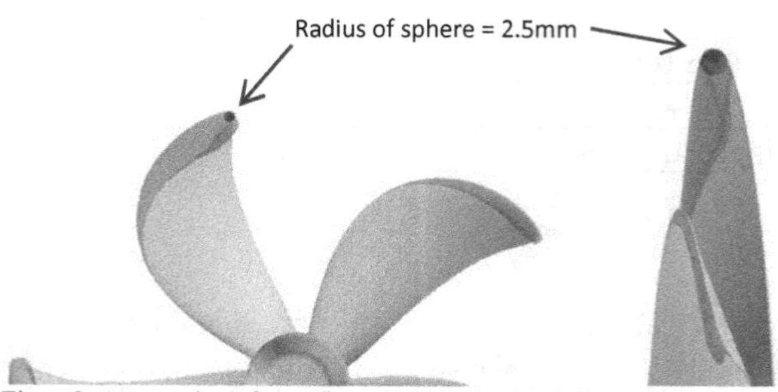

Figure9: Assumption of significant bubble size (left: front view, right: side view)

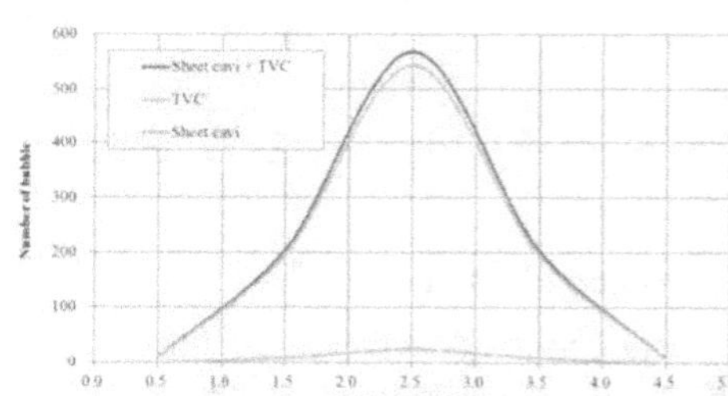

Figure10: The size distribution of cavitation bubbles

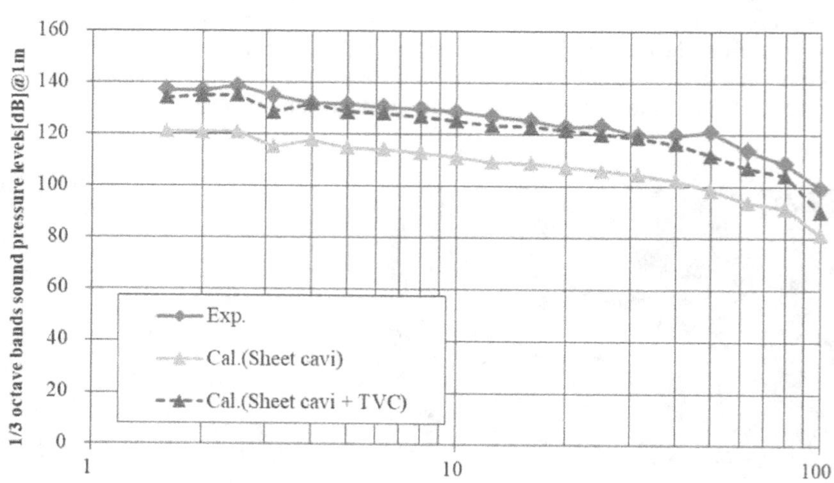

Figure11: Comparison of cavitation noise with Exp. and predictions

Conclusions

Authors were presented the hybrid method of CFD analysis and bubble dynamics model regarding the cavitation noise of the marine propeller. Also, it was confirmed that the predicted result by the present method agreed well with the experimental result. Therefore, we confirmed that the present method is effective as a practical method for the cavitation noise prediction of the marine propeller.

As future research, it is necessary to investigate the improvement of the hybrid prediction method by performing a comparison of the cavitation noise of the actual ship. And we aim to establish a method for estimating the propeller cavitation noise of the actual ship at the design stage.

References

[1] H.J.Prins et al. 2016, "Suppression of underwater noise induced by cvitation:SONIC "*Transportration Research Procedia 14*

[2] Li, D.Q., Hallander, J. and Karlsson R., 2015, "Progress in predicting pressure pulses and underwater radiated noise induced by propeller with pressure side cavitation", NuTTS'15, Cortona,Italy.

[3] Jeray Matusiak , 1992, "Pressure and noise induced by a cavitating marine screw propeller" VTTPUBLICATION 87,ESPOO

[4] Hikaru Kamiirisa et al. 2005, "Development of Prediction Method for Ship Underwater Noise by Bubble Dynamics" Mitsui Engineering & Shipbuilding Co Technical Review No.185

[5] Koyu Kimura. et al. 2009 , "Study on Unsteady Cavitating Flow Simulation around Marine Propeller using a RANS CFD code",CAV2009 , Michigan.

[6] Satoko Ando. et al. 2016,"Study on Cavitation and Noise Performance of Marine Propellers Using CFD Analysis, Symposium about cavitation, Japan.

[7] Yukio TOMITA,Akira SHIMA(1977),"On the Behavior of a Spherical Bubble and the Impulse, Bulletin of the JSME,Lol.20,No.149

[8] http://www.hsva.de/our-facilities/hykat.html

[9] http://www.t2015.nmri.go.jp/

© 2018 ASME

Surface modification of Cr-Mo steel by a new water jet cavitation technology

[1]Masataka Ijiri*;[1] Daichi Shimonishi; [1]Kumiko Tanaka; [1]Toshihiko Yoshimura;

[1] *Tokyo University of Science, Yamaguchi, JAPAN;*

Abstract

Low-alloy steels with improved corrosion resistance are obtained by alloying carbon steel with elements in amounts of several mass% or less or 1 mass% or less in various combinations. They cost slightly more than carbon steel, but have significantly improved resistance to specific types of corrosion. At present, the improvements obtained in fatigue characteristics and corrosion resistance by conventional surface modification techniques are not always satisfactory, because the operating environment of this material becomes increasingly harsh with time. Surface modification technologies such as water jet cavitation (WJC) have been applied to improve the stress corrosion cracking and fatigue strength of alloy steels. The tensile residual stress applied to the weld zone and the surface grinding face is improved to a compressive residual stress by the WJC technique, and the fatigue strength and stress corrosion cracking resistance are excellent. However, it is difficult to improve corrosion resistance. The authors have developed multifunction cavitation (MFC), a novel technique combining ultrasonication (UC) and WJC. It can utilize the same high-temperature and high-pressure microjet featured in both WJC and UC. In this study, the microstructure, hardness, and corrosion resistance on and just beneath the surface of Cr-Mo steel processed by MFC were investigated using scanning electron microscopy and a micro Vickers hardness meter. The dependence of the microstructure and hardness on the MFC processing time was also investigated, and the results were compared to those for WJC. The corrosion resistance of the specimen surface processed by MFC was improved owing to selective oxidation within the interior of the specimen. The residual stress at specimen surfaces treated by WJC or MFC was improved, changing from tensile to compressive. The compressive residual stress induced by WJC and MFC affected the surface hardness and microstructure. MFC processing caused the surface to harden, down to a depth of ca. 1 mm. As for the microstructure of the WJC-processed specimen, surface protrusions were observed and the cementite in the pearlite grains on the surface disappeared. In addition, voids and cracks were found at depths of 0.5–1 mm from the surface. No voids or cracks were observed at depths of 2–3 mm from the surface. Microstructural observation indicated that the cementite within the pearlite grains at the surface of the MFC-processed specimen became spherical and that this spheroidization of cementite was observed at depths of 0.5–1 mm beneath the surface. No voids or cracks were evident in the specimen interior.

Keywords: water jet cavitation; multifunction cavitation; SCM435 steel; Surface reforming;

Introduction

Low-alloy steel is based on carbon steel combined with several alloying elements, such as Ni, Cr, and Mo, in an amount of several mass% or less (in many cases, 1 mass% or less). Different types of low-alloy steels are made for various applications: heat-resistant steel, low-temperature steel, weather-resistant steel, and high tensile-strength steel. This steel costs slightly more than carbon steel but has better corrosion resistance. However, the corrosion resistance of low-alloy steel is inferior to that of stainless steel. In low-alloy steel, rust formation is prevented by two methods. Firstly, the surface can be covered by paint or plating. Secondly, the alloy contains Cu, Cr, P, and Ni, which are corrosion-resistant alloying elements. In appropriate amounts, these alloys encourage the formation of a dense rust on the surface as the steel is exposed to alternating moderate dry and wet conditions in the atmosphere over time. This dense rust protects the underlying steel material, and the formation of deeper rust is gradually suppressed. However, elements effective for improving corrosion resistance are harmful to the excellent mechanical properties and good weldability Low-alloy steel is required to have. As a result, there is a limit to the amount added, and a limit to the corrosion resistance that can be "built in" to the alloy mixture. As one method, the authors take particular note of a multifunction cavitation (MFC) method with high-temperature and high-pressure microjet (MJ) which features both water jet cavitation (WJC) and ultrasonic cavitation (UC) [1]. In the WJC technology, high pressure occurs when the cavitation caused by high pressure water jetted from a nozzle collapses on the metal surface. This impact pressure results in slight plastic deformation of the surface layer and generates compressive residual stress by an elastic restraining force from the lower layer portion and the surroundings. When the compression force is converted to

*Corresponding Author, Masataka Ijiri: ijiri@rs.tusy.ac.jp

compression deformation, the deformation returns to its original state after the cavitation collapse; however, if a small amount of plastic deformation occurs, then compressive residual stress is generated after cavitation collapse. When WJC is applied to a welded part where tensile residual stress exists or a surface finished by surface grinding, the tensile residual stress on the surface is reduced and a compressive residual stress is produced. As a result, fatigue strength [2] and stress corrosion cracking resistance [3] are improved. Other than WJC technology, there are vibration cavitation called UC and laser induced cavitation. In the vibration cavitation, there are low sound pressure and sound high pressure in the liquid, because ultrasonic vibration generates strong and dense longitudinal waves in the liquid by applying strong ultrasonic waves in the liquid. Bubbles under low sonic pressure are expanded and then quickly shrink under high pressure. After repeating isothermal expansion or adiabatic compression, the bubbles collapse. UC is typically applied in chemistry, biotechnology, and medicine, using several microsize bubbles with a high temperature reaction field [4] generated by repetition of the expansion or contraction processes. Recently, we have developed a cavitation method with high-temperature and high-pressure microjets (MJs) that features both WJC and UC. This cavitation technique is so-called the mechanical and electrochemical cavitation method because it has both of mechanical action by MJ with high pressure of WJC and electrochemical action [5] by MJ with high temperature reaction field of UC. Moreover, it can be called MFC because it has new functions given to various materials. The authors have reported several characteristics of specimens processed with MFC so far. For example, firstly, titanium oxide powder is widely used as a photocatalytic material. It has been reported that the titanium oxide powder can be microfabricated to nano level when processed with MFC [6]. Secondly, the ITO film is a material applied to the transparent electrode mounted on the liquid crystal panel. It has also been reported that melting and releasing of indium having a low melting point occurs when an ITO (Tin-doped Indium Oxide) film coated on a transparent electrode mounted on a liquid crystal panel is subjected to MFC treatment [1]. In the present study, the microstructure and micro Vickers hardness at and below the surface of MFC-treated Cr-Mo steel (SCM435) were investigated using scanning electron microscopy and micro Vickers hardness measurements, respectively. Moreover, we also compared the results to those for WJC.

Experimental methods

The material used for these tests was Cr-Mo steel (SCM435), the chemical composition of which is shown in Table 1. Round bar (rod) specimens were heated at 860 °C as a solution treatment, followed by quenching. Tempering was performed at 600 °C. The specimens were subsequently cut into rectangular specimens with dimensions of 100 mm×100 mm×3 mm. Figure 1 shows a schematic diagram of the equipment used for MFC processing. In conventional WJC apparatus, a nozzle is fixed in water at room temperature and water jets are pumped at a discharge pressure of 35 MPa. The nozzle diameter was 0.8 mm. The distance between the nozzle and the specimen was set to be 65 mm. In MFC treatment, an ultrasonic transducer is set to the vertical direction of the water jet nozzle for conventional WJC, and ultrasonic waves are irradiated to the water jet flow from the water jet nozzle. As shown in Figure 2, WJC isothermally expands when the sonic pressure due to ultrasonic irradiation exceeds the break threshold, and Rayleigh shrinkage then occurs rapidly (adiabatic compression) after the cavitation expands to a certain size. The flow cavitation including the hot spot becomes MFC by the repetition of isothermal expansion and adiabatic compression. At the same time that the MFC begins to collapse, it approaches the specimen surface. Moreover, as shown in Figure 3, the volume in the bubble decreases, and a MJ with a shape like sticking the specimen surface is formed [7]. This phenomenon is referred to as micro forging because the MJ provides high-temperature and high-pressure processing in a microscale area. Conventional WJC produces large high-pressure (several 100 μm and approximately 1000 MPa, respectively) bubbles, whereas those produced with conventional UC are small (several micrometers) with high temperature (several 1000 °C). On the other hand, the MJ of MFC is a large size bubble (several 100 μm) according to the above principle, and the interior of the bubble becomes high temperature (several 1000 °C) and high pressure (approximately 1000 MPa). The processing time of WJC and MFC was 2 min, respectively. Residual stress was measured by the full width at half maximum method and the peak top method from the (211) lattice spacing strain using Cr Kα X-ray diffraction (XRD; MSF-3M, Rigaku Corporation). Microstructure observations were conducted using optical microscopy (OM) and scanning electron microscopy (SEM; S-4800, Hitachi). The specimen for observation was corroded in 5 vol% Nital etchant.

Table 1. Chemical composition of the Cr-Mo steel used in this work (mass%).

C	Si	Mn	P	Ni	Cr	Mo	Cu	Fe
0.37	0.32	0.81	0.014	0.012	0.95	0.15	0.14	Bal.

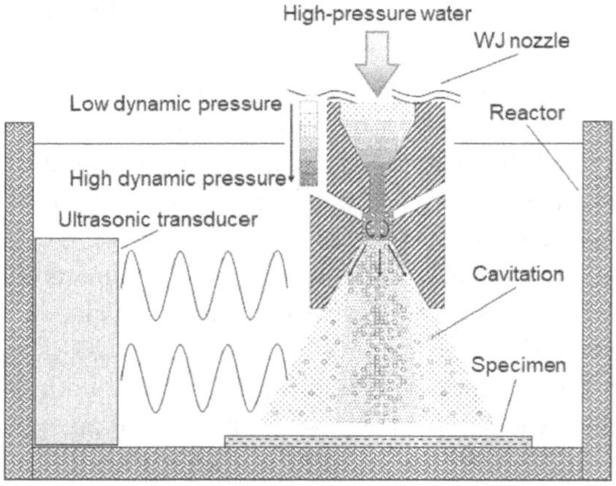

Figure 1. Equipment employed for surface machining by water jet cavitation under ultrasonication.

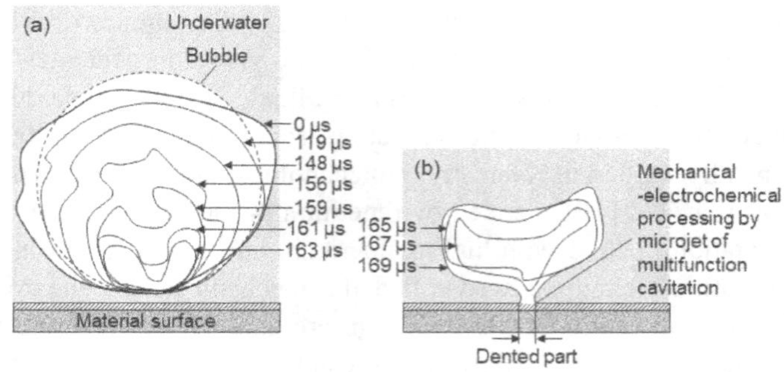

Figure 3. Images showing the aspherical collapse of a bubble at time intervals of (a) 0-163 μs and (b) 165-169 μs.

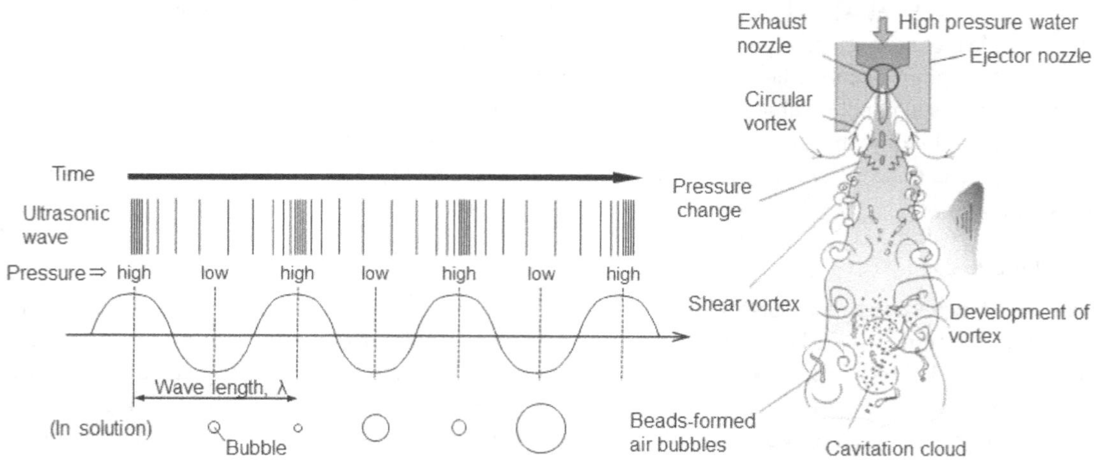

Figure 2. Multifunction cavitation mechanism.

Results and discussion

Figure 4 shows surface photographs of specimens processed for 2 min. The specimen surface was painted with oil-based magic ink prior to processing to identify the peening position. Observation of the oil-based magic ink marking after processing indicated the diameter of the WJC processed region was 40.8 mm, while that with MFC processing was 42.2 mm; therefore, the MFC processing area was slightly larger than the WJC processing area. In addition, when the processed specimen was kept at room temperature for several months, corrosion proceeded after WJC processing, where there was almost no corrosion after MFC processing. Perhaps, a minute oxide film is formed on the specimen surface after MFC processing, and it is considered to have prevented the occurrence of rust.

Table 2 shows the results of residual stress measurements. To further increase the compressive residual stress generated after WJC and MFC processing, the specimen surface was first subjecting to hard grinding, which imparted tensile residual stress. Compared to the case of no tensile residual stress applied, the compressive residual stress is considered to increase after processing by a shakedown effect. These measurements were conducted in a state without rust on the specimen surface after each processing, and were performed from the grinding direction in the as-received specimen (parallel direction). It was confirmed that the tensile residual stress imparted in the grinding direction of the WJC and MFC unprocessed specimen, while the compressive residual stress was imparted in the vertical direction. This tensile residual stress meant that the compressive residual stress was improved under various cavitation processing conditions. Compressive residual stress is generated when the specimen surface is compressed by the cavitation collapse pressure, because MJ pushes out in the lateral direction, which leads to plastic deformation around the area that is elastically constrained from the surroundings. It is considered that the specimen surface hardens considerably due to this plastic deformation. Compressive residual stress imparted by processing decreased with an increase of each processing time. In contrast, it increased with of the MFC processing time.

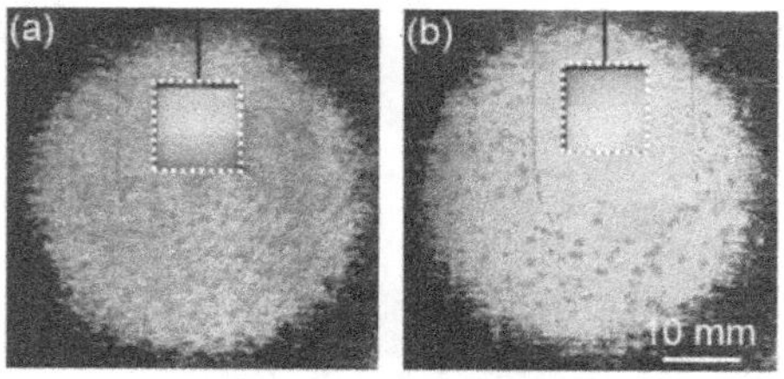

Figure 4. Change in surface morphology by (a) WJC and (b) MFC treatment.

Table 2. Compressive residual stress of specimens after various treatments.

		Half value breadth method (MPa)	Peak top method (MPa)
As received	Vertical direction	-173	-158
	Parallel direction	+202	+179
WJC		-361	-394
MFC		-293	-254

Figure 5 shows the change in the micro Vickers hardness for each processed surface in the depth direction, but does not include the oxide film and peening marks. The hardness of the near-surface region in the unprocessed specimen was 246 HV. The hardness distribution of the unprocessed specimen remained unchanged in the depth direction. It is considered that the unprocessed specimen is not significantly influenced by the strong surface grinding that was performed to impart tensile residual stress. The hardness in the near-surface region of the WJC-processed specimen and MFC-processed specimen were 272 HV and 276 HV, respectively. The hardness decreased at a depth of ca. 0.5 mm from the surface layer. Thus, it is considered that the compressive stress field reaches a depth of ca. 0.5 mm from the surface.

Figure 6 shows an SEM image of the specimen surface after WJC processing for 2 min. This SEM observation was performed after removing a sufficient amount of the surface oxide layer and peening marks. The microstructure of the unprocessed specimen was simple with only pearlite and ferrite observed. Pearlite and ferrite were also observed in the various processed specimens, as well as the unprocessed specimen. The shape of the ferrite grains did not change, but the lamellar structure of the pearlite was bent and kinked bands were formed, which was similar to the cold-rolled texture.

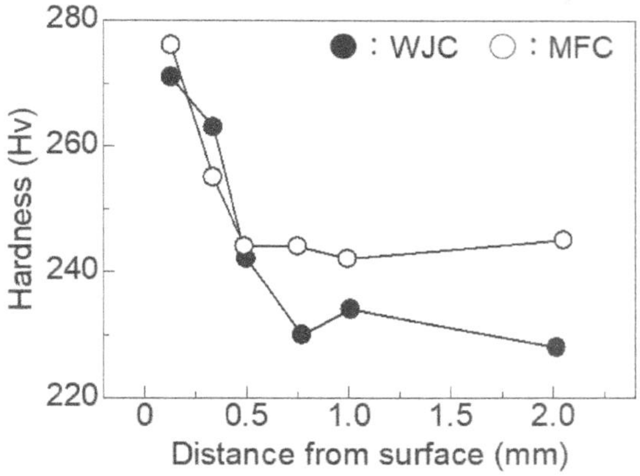

Figure 5. Hardness change of the specimen after each processing from surface to interior.

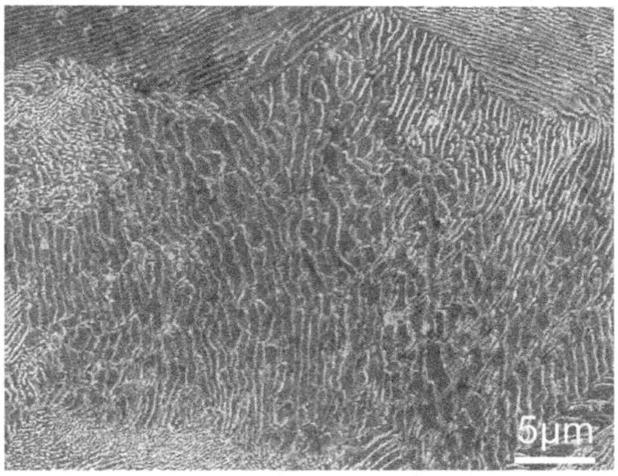

Figure 6. SEM image of specimen surface after WJC treatment for 2 min.

Apart from the microstructure shown in Figure 6, a circular product as shown in Figure 7 was confirmed. (a) shows that this product is distributed in large numbers on the specimen surface, and the size and shape are different. The surroundings are flaked like (b) and the product where the interior is raised and the product whose the inside was broken like (c) were observed.

Figure 8 shows an SEM image of a specimen side after WJC processing for 2 min. This observation was performed at a depth of approximately 0.5–1 mm from the surface. Deformation of the lamellar structure within the pearlite was observed in inside as well as the surface. Overall, the lamellae layer of pearlite had a collapsed shape, and the ferrite grains were elongated.

An enlarged view of a broken line shown in Figure 8 is shown in Figure 9. The void formation was observed in around the ferrite. It is considered that it is not compressive stress on the surface but was formed by an increase in internal tensile stress. In addition, voids are mainly formed at grain boundaries where ferrite and pearlite are in contact. Only local observation was possible in SEM, but was distributed a lot over other points.

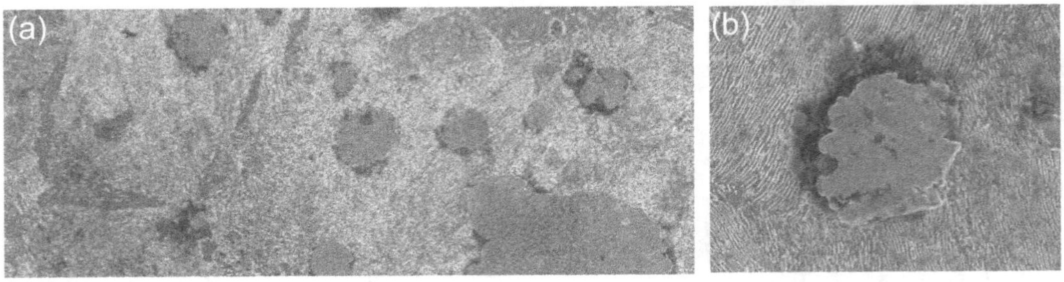

Figure 7. (a) SEM images of the specimen surface after WJC treatment for 2 min. (b) and (c) are one enlarged images of the products in (a).

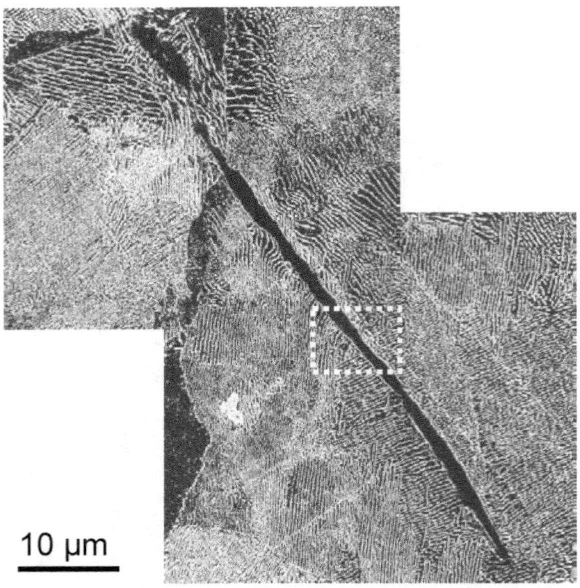

Figure 8. SEM image of the specimen side after WJC treatment for 2 min.

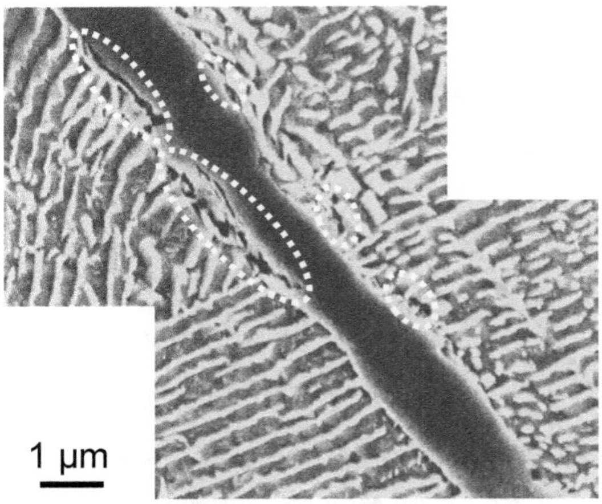

Figure 9. Enlargement of the dashed box region in Figure. 8. The dashed line indicates the voids.

The void increased compared with the specimen processed at 2 min, but in some cases voids were connected and cracks were observed. No voids or cracks were observed at depths of approximately 2–3 mm from the surface. Thus, WJC processing imparts compressive residual stress without changing the properties of the material surface. However, it was revealed that voids and cracks are likely to be formed at grain boundaries where ferrite and pearlite grains are in contact within the specimen bulk after WJC processing.

Figure 10 shows SEM images of the specimen side after MFC processing for 2 min. This observation was performed at a depth of approximately 0.5–1 mm depth from the surface. The plastically deformed ferrite is evident in Figure 10 (a), in the same manner as that for WJC processing. The lamellar structure in the pearlite was disordered and granular products were observed. In Figure 10 (b), at a different location from that shown in Figure 10 (a), even more granular products were observed. This is considered to be spherical cementite. When pearlite is heated to a temperature along the A1 point (727 °C), layered cementite breaks and gradual spheroidization of cementite occurs. However, this reaction proceeds slowly and it is difficult to obtain a completely spheroidized cementite structure by mere heat treatment. On the contrary, when cold-worked pearlite is annealed, the spheroidization of cementite is significantly promoted. Shape change of pearlite like WJC was observed in MFC processed surface and side. In MFC, spherical cementite was formed from MJ has significant compressive deformation from the surface to the interior and the specimen surface, and specimen surface heated in bubbles of high temperature and high pressure. Perhaps, in MFC, it is considered that compression deformation by MJ and heating by high-temperature and high-pressure bubbles occur on the specimen surface, and spheroidization of cementite was formed to the specimen side. Finally, voids and cracks observed in the specimen bulk after WJC processing were not observed in the specimen subjected to MFC processing. This is considered to be due to a decrease in the internal strain during processing when the surface temperature increased.

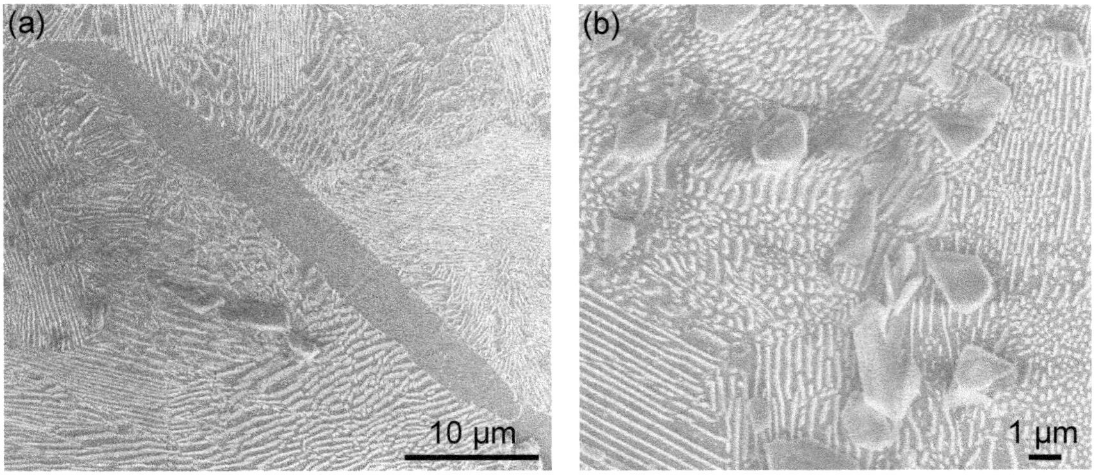

Figure 10. (a) SEM image of the specimen side after MFC treatment for 2 min and (b) enlarged SEM image of a different region.

Conclusion

The specimen surface after MFC processing improved corrosion resistance to rust than it after WJC processing. It is judged that this cause is due to formation of oxide film by the selective oxidation. The tensile residual stress of the ground specimen was improved to compressive residual stress by WJC processing and MFC processing. In the microstructure after WJC processing, voids were observed in a depth of 0.5–1mm. This defect was likely to occur at grain boundaries where ferrite and pearlite are in contact. On the other hand, in the microstructure after MFC processing, spheroidization of cementite was confirmed at a depth of 0.5–1 mm. Compared with WJC processing, voids and cracks didn't occur the inside of specimen. Finally, voids didn't exist in the depth direction of 2–3 mm in each processing.

Acknowledgment

This work was supported by the Innovative Science and Technology Initiative for Security Program of the Acquisition, Technology & Logistics Agency (ATLA) of Japan.

References

[1] T. Yoshimura, K. Tanaka, N. Yoshinaga (2003). *Development of mechanical-electrochemical cavitation technology*, J. Jet Flow Eng 20.
[2] D. Odhiambo, H. Soyama (2003). *Cavitation shotless peening for improvement of fatigue strength of carbonized steel*, J. Fatifue., 25.
[3] N. Saitou, K. Enomoto, K. Kurosawa, R. Morinaka, E. Hatashi, T. Ishikawa, T. Yoshimura (2003). *Development of water jet peening technique for reactor internal components nuclear power plant*. J. Jet Flow Eng 20.
[4] B. Gompf, R. Gunther, G. Nick, R. Pecha, W. Eisenmenger (1997). *Resolving sonoluminescence pulse width with time-correlated single photon counting*, Phys. Rev. Lett 79.
[5] JP Lorimer, TJ Mason (1999). *Applications of Ultrasound in Electroplating in Electroplating*. Electrochemistry 67.
[6] K. Tanaka, T. Yoshimura and N. Yoshinaga (2016). *Study on photocatalyst properties improvement of titanium oxide by the mechanical-electrochemical cavitation*. J. Jet Flow Eng 32.
[7] C. L. Kling and F. G. Hammitt (1972). *A photographic study of spark-induced cavitation bubble collapse*. J. Basic Eng 94.

© 2018 ASME

Synchronized Measurement of Cloud Cavitating Flow around a 3D Twisted Hydrofoil

[1,2]Y. T. Cao*, [1,2]X. X. Peng, [1,2]L. H. Xu, [1,2]M. T. Song

[1]*National Key Laboratory on ship Vibration & Noise, China Ship Scientific Research Center, Wuxi, 214082, China;*

[2]*Jiangsu Key Laboratory of Green Ship, China Ship Scientific Research Center, Wuxi, 214082, China*

Abstract

Cloud cavitation shedding from sheet cavitation usually can be seen on propellers, rudders and shaft brackets of fast running ships. The unsteady and instable behavior of these cavitation structures generally can produce severe impingement on the neighboring wall, which is one of the main reasons for vibration, noise and erosion induced by cavitation. Therefore, it is significant to know the relation between the cavitation structure and its reduced pressure signals, which can help to recognize the generation mechanism of cavitation impact. In this paper, cavitation behavior, pulse pressure and noise were synchronized measured to illustrate the dynamic characteristics of cavitation structures around a 3D twisted hydrofoil. The pressure fluctuation and noise were recorded under different cavitation numbers, including cases of flow without cavitation and with cavitation at the same incoming velocity. Then cavitation behavior and relevant pressure signals were compared both in time domain and in frequency domain to investigate the relation between the unsteady cavitation structures and the corresponding pressure response.

Keywords: synchronized measurement; cloud cavitating flow

Introduction

Cavitation is a unique phenomenon in liquid due to the local pressure dropping to the saturated vapor pressure. It is a source of vibration, noise and erosion for ships under fast speed. In general, cavitation can be divided into different types according to its appearance, such as bubble cavitation, sheet cavitation, cloud cavitation and vortex cavitation. Among these types, cloud cavitation shedding from sheet cavitation is usually considered to be one of the most harmful one since its evolution can produce strong dynamic pressure. Extensive investigations have been focused on the collapse of a single isolated bubble to illustrate the generation mechanism of violent radiated pressure waves. However, cloud cavitating flow is a complicated process coupled with phase change, vertical structures and micro bubble clusters, the understanding on the dynamic process is still not enough at this moment. Recent study (Cao et al. 2015, Peng et al. 2016) has shown that there exist distinct U-shaped vortex structures in large scale cloud cavitating flows under various flow configurations. And the overall contribution of different cavitation structures to cavitation erosion was evaluated qualitatively using painting method (Cao et al. 2017). To further understand the reason for the test result of paint loss regions, the pressure fluctuations and acoustic signals were recorded simultaneously with cavitation behavior.

Cavitation on spanwise uniform (2D) hydrofoil sections has been investigated extensively (such as Kjeldsen et al. 2000, Arndt et al. 2000, Ganesh et al. 2016). However, the shedding location on this kind of models in spanwise direction is random, which may bring more uncertainty and complexity in the association of instantaneous cavitation structure to corresponding dynamics. Foeth (2008) introduced a 3D twisted hydrofoil that can produce periodic shedding cloud cavitation resulted from sheet cavitation in his work. As a result, we choose to use this kind of configure to estimate the relationship between the cavitation structures and its reduced dynamic signals.

Test facility and experimental setup

Cavitation tests were conducted in the high speed cavitation tunnel (as shown in Figure 1) located in China Ship Scientific Research Center (CSSRC). The test section is composed of 8 perspex window which permits to observe cavitation structures from different sides. The length of test section is 1600mm×225mm×225mm. The maximum flow velocity can reach 25m/s and the background pressure can be adjusted from 5kPa to 500kPa, which allows the cavitation number defined as formula 1 to be easily changed. In addition, this tunnel has a fast degassing system allowing to degass the water in limited time.

*Corresponding Author, Y.T. Cao: caoyantao@126.com
This work was supported by the National Natural Science Foundation of China (Project No. 11332009 & 11072223).

$$\sigma = \frac{p_\infty - p_v}{0.5\rho U^2} \tag{1}$$

The test model used in this test is a NACA16012 3D twisted hydrofoil (as shown in Figure 2) with a chord length of 100mm and a spanwise length of 225mm. It is a symmetric hydrofoil with spanwise varied attack angle and the maximum angle of 11 degree in the middle. The AOA (angle of attack) was set to be 0 degree based on the end of straight part in spanwise. The cavitation structures were observed synchronously by two high speed cameras from the bottom and backside of test section (as shown in Figure 3), which consequently can get both the top view and backside view of cavitation structures around the hydrofoil. Two LED lights provide light for high speed cameras. The images were recorded at a frame rate of 6000 FPS during the experiment. Figure 4 shows the location of pressure transducers and hydrophone. The range of pressure transducers is 300 kPa with an accuracy of 0.3%. In this paper the result of P2 was considered to be more convincible because signals of P1 contain some high frequency electrical noise. The hydrophone used in this test is B&K 8103 installed inside a water filled cylindrical box which is put outside of side window. The sampling rate of transducers was set to be 195652.18Hz. However, it should be noticed that the response frequency of pressure transducers could not reach so high. As a result, in this test the pressure transducers can measure relatively low frequency (within 1000Hz) dynamic signals accurately while the hydrophone can measure signals with both low and high frequency components.

Figure 1 The high speed cavitation tunnel in CSSRC.

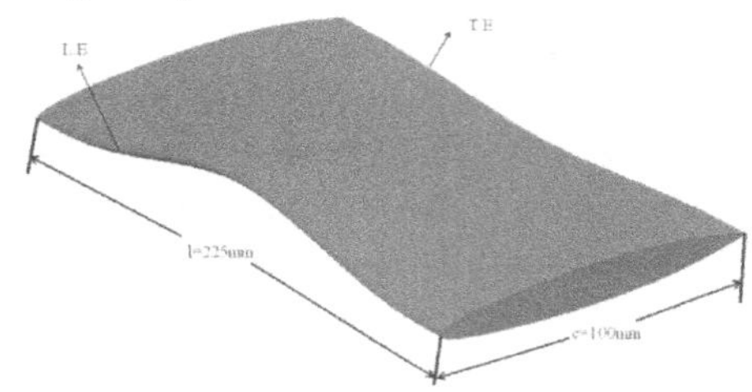

Figure 2 A sketch of NACA16012 3D twisted hydrofoil.

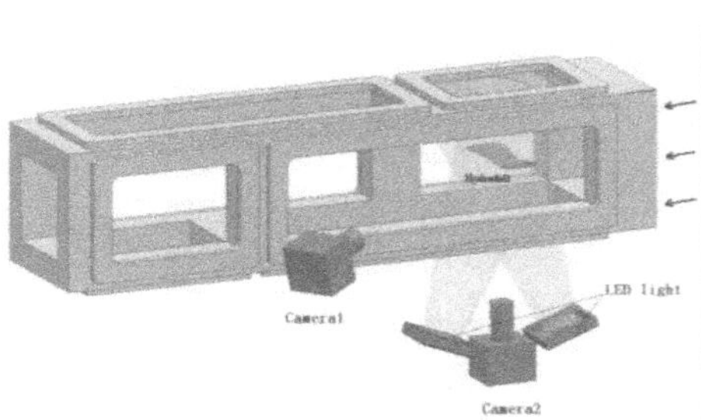

Figure 3 Setup of high speed cameras and lights.

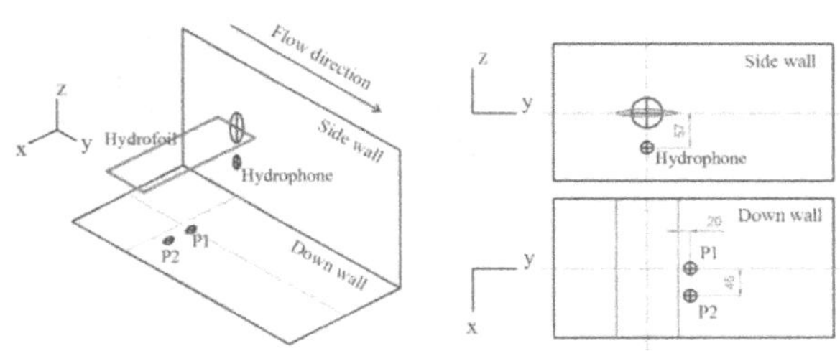

Figure 4 Location of pressure transducer and hydrophone.

Results and discussion

A serious of cases with different cavitation numbers were conducted during the experiment. The incoming flow velocity was set to be 7m/s and the AOA was set to be 0 degree which allows the maximum angle in the middle to be 11 degree. In this paper, we give the results of cavitation behavior only under the condition of $\sigma = 1.0$. Pressure and acoustic spectrum of different cavitation numbers under the condition 7m/s were given for all test conditions. The signal spectrums were expressed in SPL (Sound Pressure Level) and were averaged using 10 group data at the same condition.

The evolution of cavitation behavior

The evolution of cavitation behavior was observed by high speed cameras. The results show that the shedding process of large scale cloud cavitation around the 3D twisted hydrofoil is qusi-periodic. Figure 5 shows the behavior in a classical period and Figure 6 is the corresponding processed noise and pressure signals. The interval between two images is 2.83 ms (17 frames taken with frame rate 6000FPS). In this paper the beginning of a shedding cycle is considered to be the incident when sheet cavity reaches the maximum length (Shown as T1). At this moment, re-entrant flows beneath the sheet reach the leading edge of the hydrofoil and then impinge on the surface of sheet cavity. Thus results in the primary shedding that almost the whole part of cavity was detached from the leading edge (Shown as T2). Then the shed part moved downward and ultimately developed into a U-shaped vortex structure (Shown as T10). Simultaneously a reborn sheet grew up from the leading edge (Shown as T3). During the sheet growing process (From T4 to T6), re-entrant flow continually formed at the two side rear part of the new sheet and induced the local shedding or secondary shedding (Shown as T6). When the reborn cavity approached to the maximum length, re-entrant flow at both side of the rear part moved forward to the center of hydrofoil and gathered together in the spanwise symmetric plane. That's the typical behavior of large scale cloud cavitating flow around a twisted hydrofoil. It can't be denied that the cavitation structure doesn't evolve exactly the same in different cycles, but in most cases it develops in this way or tend to be in this way. The Strouhal number of the main shedding expressed with maximum sheet length is 0.23.

Cao et al. 2017 showed that the unsteady rear part of sheet cavitation is the most erosive structure, and the route where the U-shaped vortex structure evolved from cavitation cloud passes can also be erosion risk region. It can be seen from Figure 6 that the maximum pressure peak in a cycle is around T3, when the cavitation cloud was just formed from the breakup of sheet cavity. It can be interpreted that the generation process of cloud cavitation from sheet cavity could produce relatively high magnitude pressure fluctuation, which was mainly due to the change of cavitation volume. This provides a proof for why the most erosive region was there. From the noise signal showed in Figure 6, it can be seen that there exists some high magnitude pulse signals in time domain which could not be seen from the pressure signals. Considering the response frequency of transducers, it can be deduced that the pulse signal is produced in limited short time. The HSV results (Figure 7, sequential images taken at 6000FPS, between T7 and T8) showed that this pulse in noise signal might be related to the collapse and rebound behavior of part of cloud structure. The data in Figure 6 was normalized with half of the difference between the maximum and minimum value.

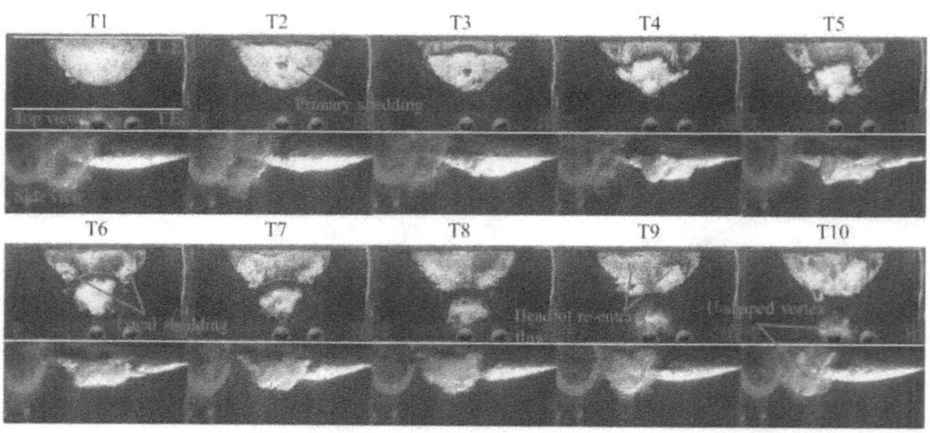

Figure 5 Cavitation behavior in a period (U=7m/s, $\sigma = 1.0$)

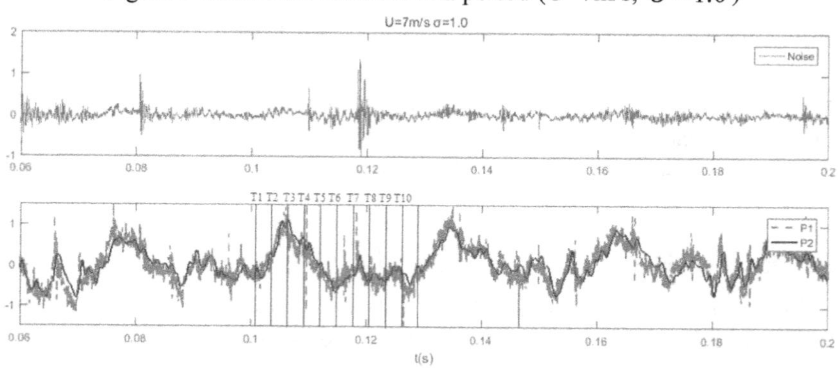

Figure 6 Dynamic signals in a period (U=7m/s, $\sigma = 1.0$)

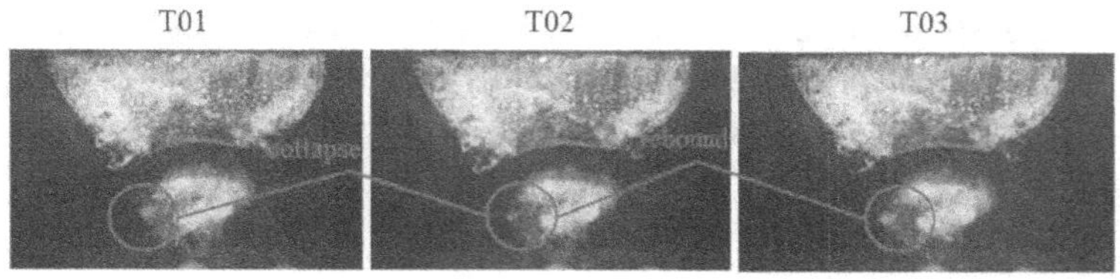

Figure 7 Collapse and rebound of part of cavitation cloud (U=7m/s, $\sigma = 1.0$)

Spectrum of dynamic signals

The spectrum of noise and pressure signals is shown in Figure 8 and Figure 9 respectively. It can be seen that when cavitation appears, the noise magnitude rises rapidly all over the frequency domain, especially in high frequency segment, whereas the pressure magnitude rises mainly in the low frequency part. Combining the time domain results described above, it can be speculated that the high magnitude fluctuations in low frequency part seems to be mainly related to the sheet and high frequency signals mainly related to the shed cloud. The peak around 800-900Hz was the vibration induced by the interaction between the flow and the foil structure, which can be heard obviously during the test.

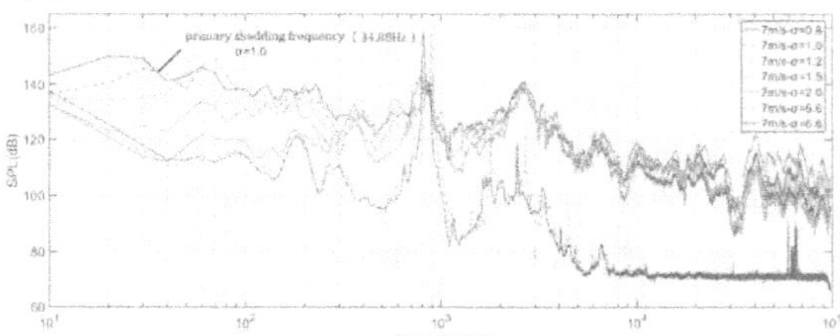

Figure 8 Spectrum of noise signals

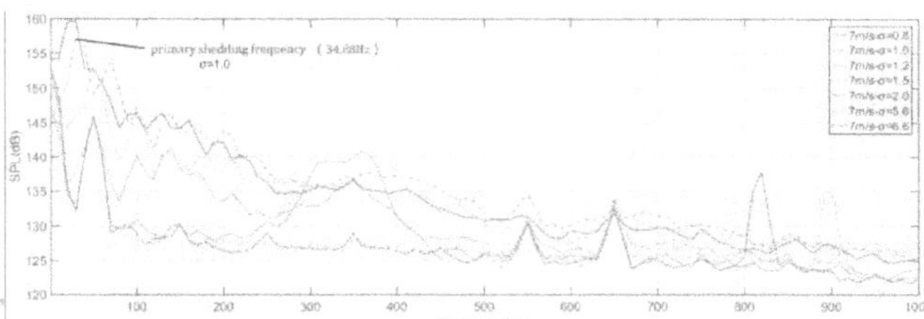

Figure 9 Spectrum of pressure signals measured by P2

Conclusions

This paper introduced the recent experimental work of measurement of cloud cavitation dynamics around a 3D twisted hydrofoil in CSSRC. Combining high speed cameras, pressure transducers and hydrophone, the relationship between cavitation structure and corresponding pressure signals was compared. It is revealed that the maximum magnitude of cavitation fluctuation pressure was related to the initial stage of cloud cavitation evolution, when sheet cavity just breakup into cloud, whereas the high magnitude pulse noise signal seemed to be related to the collapse behavior of cloud.

References

[1] Cao, Y.T., Peng, X.X., Xu, L.H. & Hong, F.W. (2015). *U-shaped Vortex Structures in Large Scale Cloud Cavitation.* 9TH International Symposium on Cavitation, Lausanne, Switzerland, Journal of Physics: Conference Series 656 (2015) 012165.

[2] Peng, X.X., Ji, B, Cao, Y.T., Xu, L.H., Zhang, G.P., Luo, X.W.& Long X.P. (2016) *Combined experimental observation and numerical simulation of the cloud cavitation with U-type flow structures on hydrofoils*. International Journal of Multiphase Flow 79, pp.10-22

[3] Cao, Y.T., Peng, X.X., Yan, K., Xu, L.H., Shu, L.W. (2017). *A Qualitative Study on the Relationship Between Cavitation Structure and Erosion Region around a 3D Twisted Hydrofoil by Painting Method.* Fifth International Symposium on Marine Propulsors, smp'17, Espoo, Finland

[4] Kjeldsen, M., Arndt, R.E. and Effertz, M., *Spectral characteristics of sheet/cloud cavitation.* Journal of Fluids Engineering, Vol. 122, No. 3, Mar. 2000, pp. 481-487

[5] Arndt, R.E., Song, C.C.S., Kjeldsen, M., He, J. and Keller, A. *Instability of partial cavitation: a numerical/experimental approach.* Proc. Twenty-Third Symposium on Naval Hydrodynamics, Val de Reuil, France, 2000, pp. 1-16

[6] Ganesh, H., Wu, J., Ceccio, S.L., (2016) *Investigation of Cavity Shedding Dynamics on a NACA0015 Hydrofoil Using Time Resolved X-ray Densitometry.* 31st Symposium on Naval Hydrodynamics, Monterey, CA, USA

[7] Foeth, E. J. (2008). *The structure of three-dimensional sheet cavitation.* Mechanical Maritime & Materials Engineering.

© 2018 ASME

Synchronized X-ray Densitometry and Surface Pressure Measurements on a Cavitating NACA0015 Hydrofoil to Estimate Bubbly Shock Properties

Juliana Wu; Anubhav Bhatt; Harish Ganesh*; Steven L. Ceccio;

[1]*University of Michigan, Ann Arbor, MI, USA;*

Abstract

Measuring properties of the shock waves and their sensitivity to change in flow conditions can help in the understanding and prediction of the role of propagating bubbly shock waves in the multi-modal cavitation dynamics of NACA0015 hydrofoil. In the current study, we obtain time-resolved X-ray densitometry measurements on a larger model (compared to [3]) for better spatial resolution cavity shedding dynamics due to propagating bubbly shocks. Due to lack of availability of surface pressure measurements at the moment, observed shedding dynamics are discussed. Surface pressure measurements and shock properties will be compared in the presentation.

Keywords: cloud cavitation, multi-modal shedding, NACA 0015 hydrofoil

Introduction

Cavitation dynamics on a NACA0015 hydrofoil has been studied extensively. It has been observed that the dynamics change abruptly with conditions such as attack angle or cavitation number [1] and [2].

Time resolved X-ray densitometry was used to understand the physics behind abrupt change in dynamics of a NACA0015 hydrofoil [3]. The study found that shedding dynamics was well behaved at higher cavitation numbers (Type 1) and multi-modal at lower cavitation numbers, designated as Type-2 cavitation [1] and [2]. A typical Type 2 shedding can have intermediary steps caused by the interference of the cloud collapse on cavity growth, along with propagating bubbly shocks [4]. Lack of surface pressure measurements and modest spatial resolution of X-ray measurements made it difficult to further understand the change in dynamics caused by these flow features.

Present study further explores cavitation dynamics on a NACA0015 hydrofoil, using a larger model than [3] to facilitate the use of unsteady surface pressure transducers and have better spatial resolution to resolve flow features. By synchronizing X-ray densitometry measurements with dynamic pressure measurements on the foil surface, an explanation for the observed mechanism is sought. In the current paper, observations obtained from X-ray densitometry are discussed.

Experimental Set-up

Cavitation experiments were carried out at the University of Michigan re-circulating Water Tunnel. A 6:1 round contraction transitions into a test section with a diameter of 9", which then transitions to 8.25" by 8.25" square cross-section with chamfered corners. Static pressure, p_0, and flow velocity, U_0, in the tunnel test section can be varied from near vacuum to 200 kPa and 0 to 20 m/s respectively. Cavitation number at the inlet (σ_0),

$$\sigma_0 = \frac{(p_0 - p_v)}{\frac{1}{2}\rho U_0^2} \tag{1}$$

Here, p_v : vapor pressure, and ρ: liquid density. σ_0 in the present study was multiplied by a factor of λ_1^2/λ_2^2, where λ_1, and λ_2 are blockage from present study and [3] respectively. Test model consisted of a NACA0015 hydrofoil with a span of 8.25" and chord of 6.52", compared to 3" span and 50 mm chord used in [3]. The foil was mounted to the side windows at a constant attack angle of α = 10 degrees. U_0 was maintained at 8m/s, while inlet pressure was varied from 48-120kpa. Observation was made through top view using a Phantom V710 high-speed video camera. Variations

*Corresponding Author, Harish Ganesh: gharish@umich.edu

in span-wise averaged void fraction fields were measured using time-resolved X-ray densitometry. A detailed description of the method can be found in [5]. Figure 1 shows the schematic of the hydrofoil in the test-section.

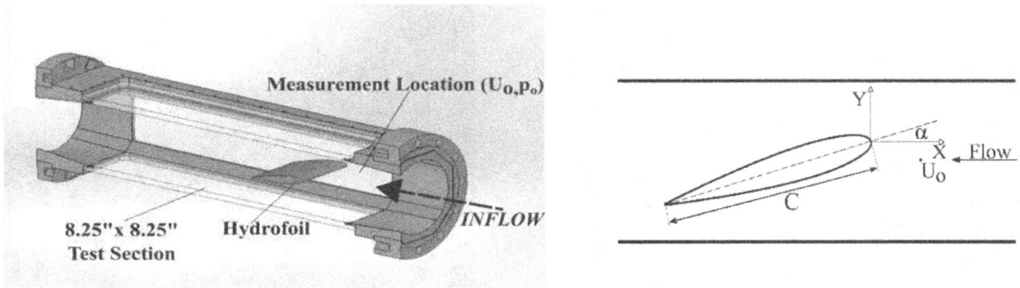

Figure 1: Experimental set-up.

Cavitation Behavior:

High-speed cinematography and time resolved X-ray densitometry were used to record flow features during cavitation. Figure 2 shows the variation of non-dimensional cavity length (L_c/C) with cavitation parameter ($\sigma/2\alpha$) for larger planform model used in current study and smaller planform model used in [3].

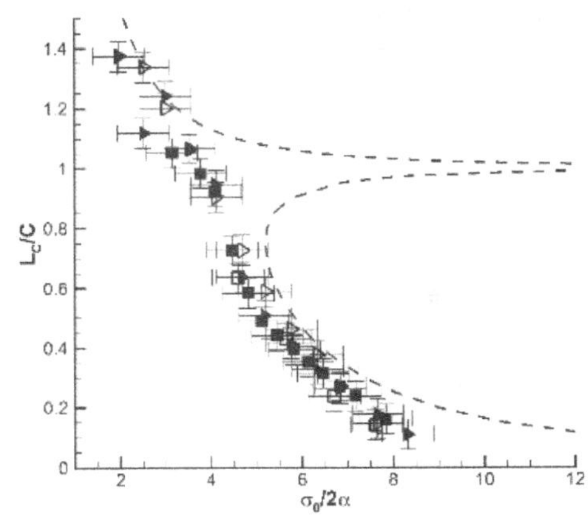

Figure 2: Comparison of Lc/C for current model from HS(■) and XR(•) measurements for cavitation parameter ($\sigma_0/2\alpha$). (■) larger model used in current study. Dashed line represents analytical solution (Watanabe et. al.). Smaller planform model [3] is shown as (▷) for XR and (▶) for HS.

From figure 2, it can be observed that variations follow the same trend for both models, and the effect of cavitation induced unsteadiness is seen in the "bulged" portion of the curve. Cavitation dynamics varied significantly with changes in inlet cavitation. Stable cavities of nearly constant cavity length were observed at higher cavitation numbers, while shedding occurred at lower cavitation numbers. Figure 3(a) shows a typical stable cavity and 3(b) shows a snapshot during a shedding cycle.

The mechanism of shedding observed in the present study is significantly different from that observed in [3]. Observed shedding is summarized as follows: (1) The collapse of the shed cloud did not completely arrest cavity growth as observed in [3]. (2) At lower cavitation numbers, X-ray densitometry measurement revealed the presence of a re-entrant flow with a stationary front (sometime causing shedding) in addition to propagating bubbly shock waves. It is observed that the adverse pressure gradient is not strong enough for the re-entrant flow to cause pinch-off. On the other hand, lower cavitation number resulted in stronger condensation shocks. Figure 4 shows a typical shedding cycle with the two mechanisms co-existing.

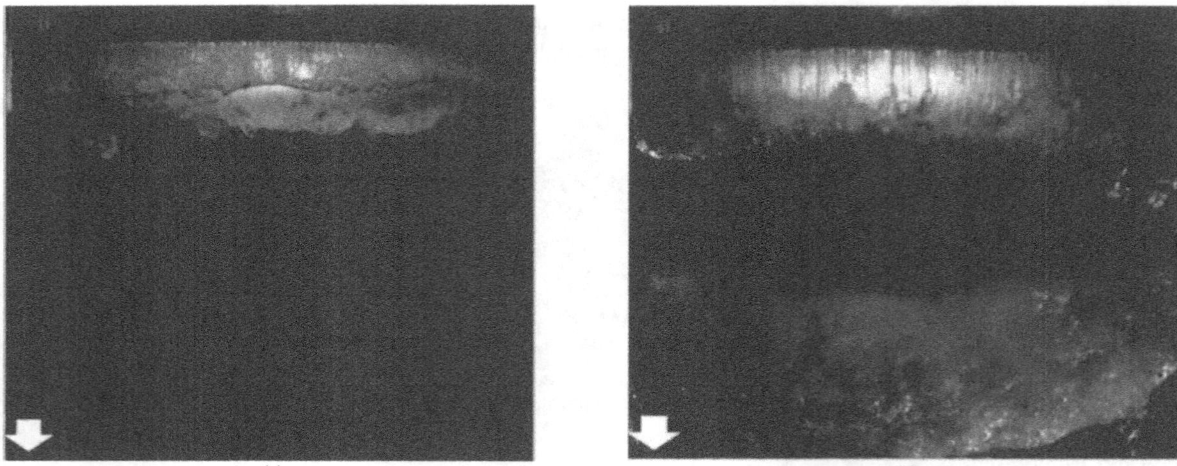

(a) (b)

Figure 3: (a) Instantaneous high-speed video snapshot of a stable incipient cavity. The cavity length remained nominally stable with occasional pinch-off from the rear. (b) Instantaneous high-speed video snapshot of a shedding cavity. The shedding was from the leading edge, and cavity collapse did not influence cavity growth.

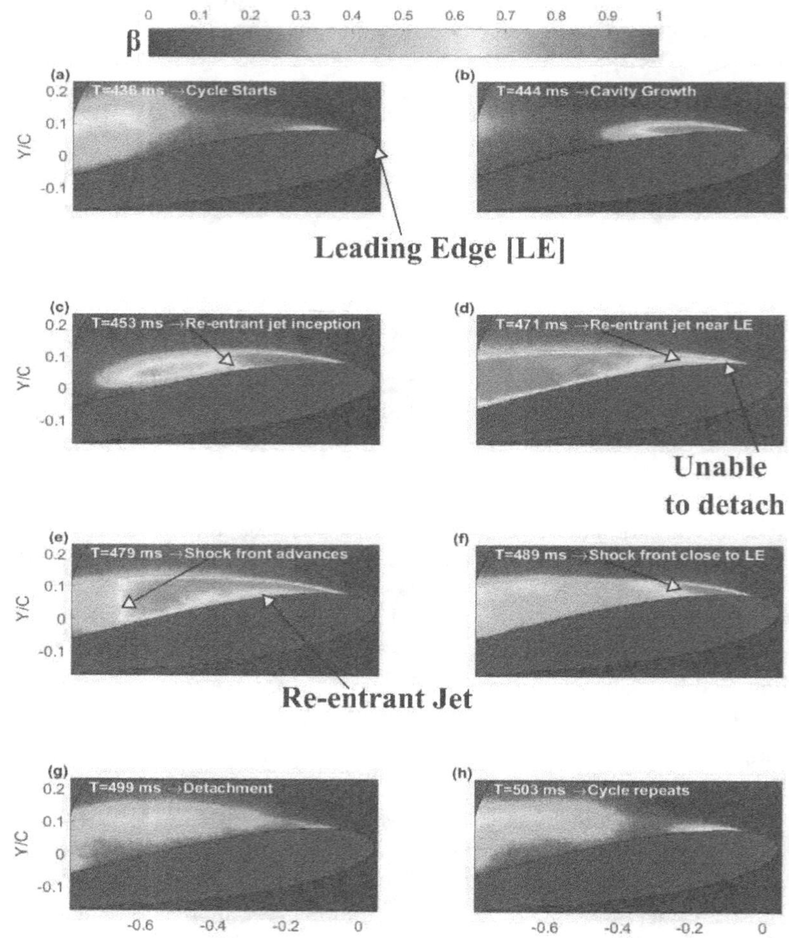

Figure 4: Instantaneous void fraction flow fields based on X-ray densitometry depicting a shedding cycle with both re-entrant liquid flow and propagating bubbly shock waves. Propagating bubbly shocks were the dominant mechanism for shedding, causing pinch-off from the leading edge of the hydrofoil ($\sigma_0 = 0.96$)

Modal analysis of shedding was performed by probing a control area to estimate the time variation of spatially averaged void fraction in the vicinity of cavity closure, and near the leading edge. Figure 5(a) and 5(b) show spectral content of the time-varying void fraction signal for different cavitation numbers. Strouhal number (St), is expressed using the chord (C), frequency (f), and inlet speed (U_0), and k the blockage factor:

$$St = \frac{fC}{k * U_0} \tag{2}$$

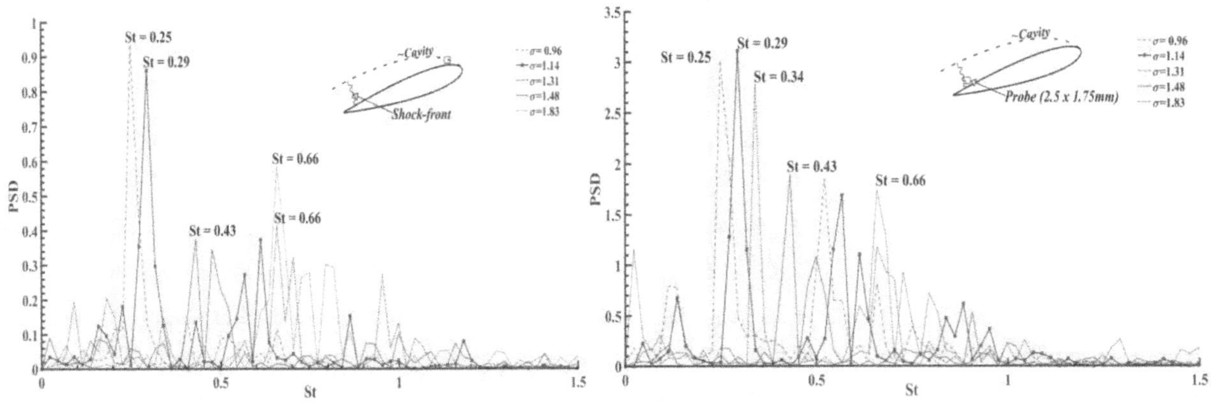

(a) (b)

Figure 5: Power Spectral density analysis of lower cavitation numbers reveals multi-modal nature of the system. The spatially averaged void fraction intensity at each time frame for entirety of capture (787 frames at 1000fps) forms the time varying signal here.

The variation in first peaks (depicted in figure 5) follow an increasing trend as reported by [1]. Presence of first and second peaks can be clearly viewed with the probe at trailing end (figure 5(b)). This affirms the observation of both re-entrant flow front and propagating bubbly shocks affecting flow dynamics.

Conclusion

Cavitation dynamics on a larger planform hydrofoil compared to [3] was studied using both high-speed cinematography and time-resolved X-ray densitometry. From the observations it was found that mechanisms responsible for shedding in the larger model are different from those observed in [3]. Use of X-ray densitometry revealed the presence of both re-entrant liquid flow and propagating bubbly shocks in a shedding cycle, with the bubbly shocks being the dominant mechanism of shedding.

References

[1] Kjeldsen, Morten, Roger EA Arndt, and Mark Effertz. Spectral characteristics of sheet/cloud cavitation. Journal of Fluids Engineering 122.3 (2000): 481-487.

[2] Kravtsova, A. Y., Markovich, D. M., Pervunin, K. S., Timoshevskiy, M. V., & Hanjalić, K. (2014). High-speed visualization and PIV measurements of cavitating flows around a semi-circular leading-edge flat plate and NACA0015 hydrofoil. International Journal of Multiphase Flow, 60, 119-134.

[3] Ganesh, H., Wu, J., & Ceccio, S. L., 2016, "Investigation of cavity shedding dynamics on a NACA0015 hydrofoil using time-resolved X-ray densitometry", Proceedings of the 31st Symposium on Naval Hydrodynamics, Monterey, USA, August 2016

[4] Ganesh, H., Mäkiharju, S.A., & Ceccio, S. L., (2015). Bubbly shock propagation as a mechanism of sheet-to-cloud transition of partial cavities. Journal of Fluid Mechanics, under review

[5] Makiharju, S.A., Chang, N., Gabillet, C., Paik, B.-G., Perlin, M. & Ceccio, S.L. (2013) Time Resolved Two Dimensional X-Ray Densitometry of a Two Phase Flow Downstream of a Ventilated Cavity. Experiments in Fluids, 54

© 2018 ASME

10th International Symposium on Cavitation - CAV2018
Baltimore, Maryland, USA, May 14 – 16, 2018 CAV18-05197

Temperature and pressure effects on cavitation erosion in diesel-like fuels

[1,2]Markus Hosbach*, [1]Tobias Sander, [2]Uwe Leuteritz, [1]Michael Pfitzner

[1]*Bundeswehr University Munich, Germany;* [2]*Contintental Automotive GmbH, Regensburg, Germany*

Abstract

In the present work hydrodynamic cavitation erosion in a square channel made of Al99.0 is experimentally investigated under static flow conditions. It was found that the normalized maximum erosion rate $(nRoE)_{max}$ increases nonlinearly with the rail pressure between 100 bar and 500 bar. The presence and magnitude of back pressure determines the axial position of erosion in the test section. The flow temperature influence on cavitation erosion in calibration fluid ISO 4113 is investigated between the nondimensional temperatures $\theta = 0.13$ and $\theta = 0.40$. The highest cavitation aggressiveness is detected at about $\theta = 0.29$. To explain this result, numerical calculations based on the Rayleigh-Plesset equation with respect to the thermodynamic effect of cavitation are conducted. For the single-component fuel n-decane the same overall trend and a temperature dependent maximum erosion rate of $\theta = 0.29$ was calculated.

Keywords: cavitation erosion; diesel-like fuel; Rayleigh-Plesset equation; thermodynamic effect

Introduction

If cavitation bubbles in liquid filled with vapor collapse close to a solid wall, cavitation damage occurs. This erosive wear can occur in high-pressure components of gasoline and diesel fuel injection systems like pumps and injectors. Cavitation erosion can have negative effects on the component durability and consequently on the mixture generation. This in turn can affect the performance and exhaust emissions of modern engines. Several studies reveal the influence of temperature and flow velocity on hydrodynamic cavitation erosion in water. Pitting tests on aluminum are carried out by Franc et al. [1] using a nozzle within a closed flow loop (water pressure between 10 bar and 40 bar) to evoke cavitation erosion on a wall close to the nozzle exit. The pitting rate rises with the fifth power of the flow velocity. Further studies confirm the power law exponent being within a range of 5-8 [2, 3, 4].

Fluid temperature has an important effect on cavitation erosion. Most recently hydrodynamic cavitation erosion in water at temperatures between 30 °C and 100 °C is investigated by Dular [5]. The maximum erosion rate at about 50 °C - 60 °C is reproduced and confirmed by numerical means using the Rayleigh-Plesset equation.

In contrast to the working fluid in the literature mentioned above, the present work aims at a deeper investigation of two main influence parameters of diesel-like fuels on hydrodynamic cavitation erosion: the flow velocity respectively the ratio of rail pressure to outlet pressure and the fluid temperature.

Experiments

To investigate cavitation erosion under diesel-like conditions, a cavitation test bench was built up at the Institute for Thermodynamics at the University of the Bundeswehr Munich, Germany. It consists of a high pressure generator for diesel-like fuels, which supplies the test section (see figure 1) with fuel at a pressure up to 2200 bar and a temperature range between 15 °C and 150 °C. Pressure and temperature sensors are located upstream and downstream of the test section, a mass flow sensor is located in the return low pressure region of the test bench.

*Corresponding author, Markus Hosbach: markus.hosbach@unibw.de

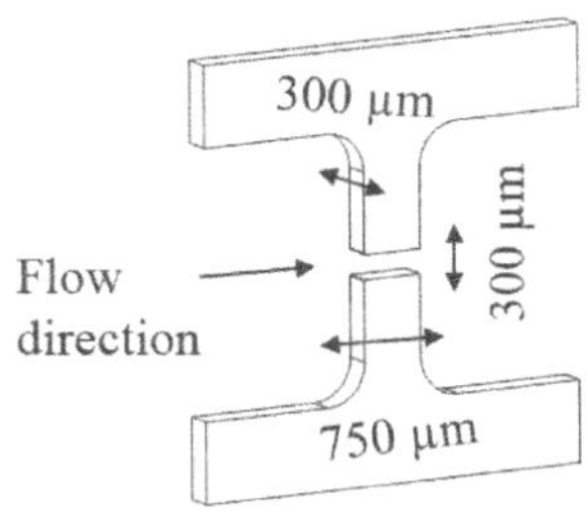

Figure 1: Test section

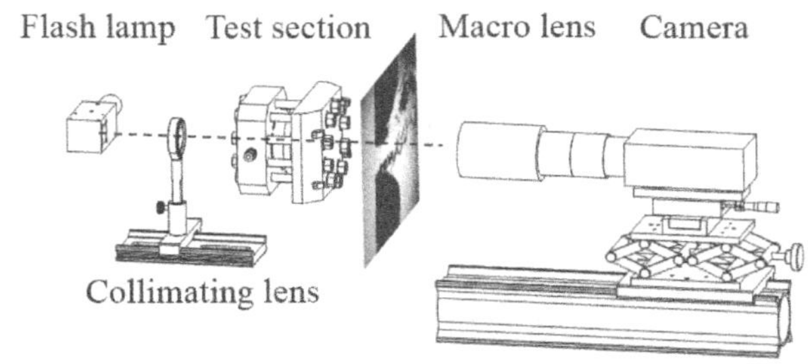

Figure 2: Optical setup

In the present work, hydrodynamic cavitation erosion is experimentally investigated, varying the nondimensional fluid temperature between $\theta = 0.13$ and $\theta = 0.40$ and the rail pressure between 100 bar and 500 bar. T_{in} is the inlet temperature, $T_{t,decane}$ it the tripple point temperature of n-decane and $T_{c,decane}$ is the critical point temperature of n-decane (see equation 1).

$$\theta = \frac{T_{in} - T_{t,decane}}{T_{c,decane} - T_{t,decane}} \tag{1}$$

To track the erosion progression of 2D-throttle plates, which are made of Al 99.0, the well known shadowgraph technique is applied. To ensure sharp images, the throttle geometry is back-illuminated for 39 ns by a flash lamp, while a 4MP CMOS camera captures the resulting shadows of the cavitating flow at a frequency of 7 Hz.

For each measuring interval, which is 5 to 20 minutes, the actual throttle plate surface, the mean cavitation probability and the corresponding standard deviation is calculated from a series of 100 frames (see figures 3 and 4). During all measurements, the cavitation number σ, defined in equation 2, is held constant.

$$\sigma = \sqrt{\frac{p_{rail} - p_{out}}{p_{rail}}} \tag{2}$$

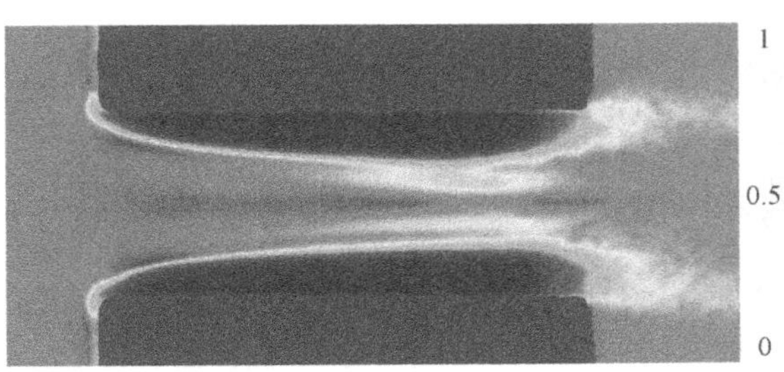

Figure 3: Mean cavitation probability, $p_{rail} = 300\,bar$, $\sigma = 0.81$, $\theta = 0.19$

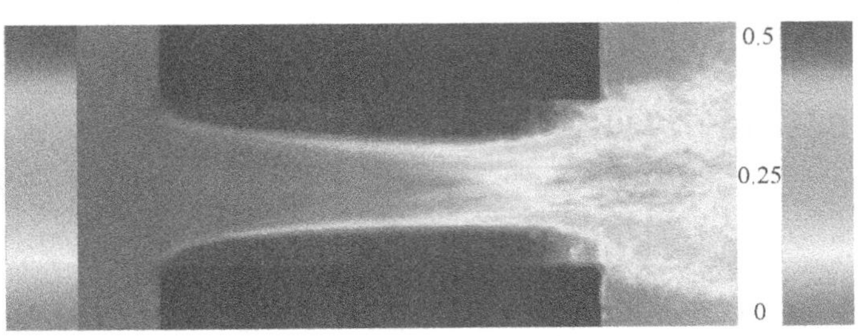

Figure 4: Standard deviation, $p_{rail} = 300\,bar$, $\sigma = 0.81$, $\theta = 0.19$

In figure 5 a typical erosion progression is shown, which was captured at a rail pressure of 300 bar, a cavitation number of 0.81 and an inlet temperature $\theta = 0.19$. The known erosion phases (also see e.g. [6]): incubation period, acceleration and deceleration period and steady state period can be easily recognized. The erosion aggressiveness of each test case is determined by evaluating the normalized maximum erosion rate $(nRoE)_{max}$ of the erosion progression. It is obtained by dividing the derivative with respect of time of the throttle area A (length x height) at the time, where the maximum erosion rate during the acceleration period occurs by the initial throttle area A_0 of 750 µm x 300 µm (see equation 3).

$$(nRoE)_{max} = \frac{d}{dt}\left(\frac{A}{A_0}\right)\bigg|_{max} \tag{3}$$

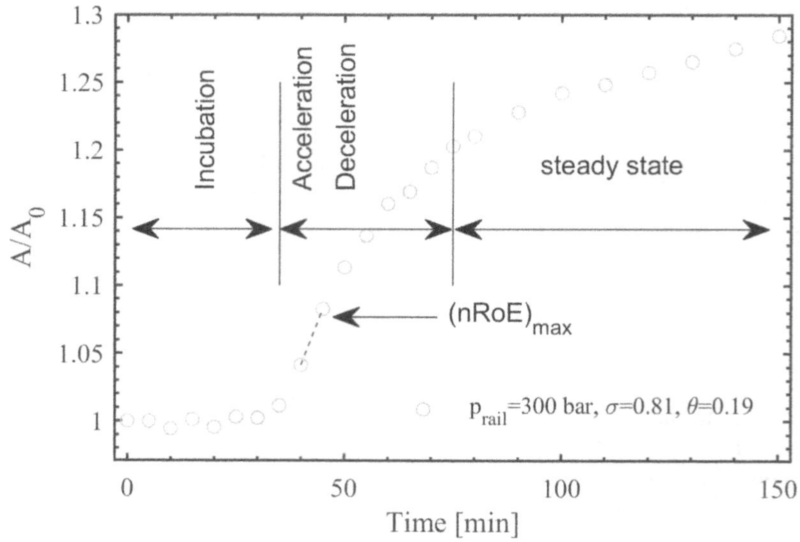

Figure 5: Erosion progression

Results - pressure influence

To investigate the pressure influence on hydrodynamic cavitation erosion in diesel-like fuels, the rail pressure was varied between 100 bar and 500 bar for diesel and calibration fluid (CF). It was found, that the maximum erosion rate increases nonlinearly with the rail pressure (see figure 6). Increasing the rail pressure leads to a higher shedding frequency, i.e. the detachment frequency of cavitation structures, drifting in high-pressure regions [7, 8]. With increasing the pressure also the mass flow augments. Hence the number of bubbles, collapsing close to the solid wall rises, leading to a more severe erosive damage. Furthermore the static pressure in the area of erosion augments when keeping the cavitation number constant and the bubble collapses get more intense hence the erosion aggressiveness rises. The resulting erosion rates for calibration fluid and diesel fuel can be well fit by power laws (see equation 4), with n in the range of $3-4$ and a coefficient of determination $R^2 > 0.83$.

$$(nRoE)_{max} \approx p_{rail}^n \tag{4}$$

Based on the measured mass flow, the mean flow velocity at the vena contracta v_{vc} is calculated. With

$$(nRoE)_{max} \approx v_{vc}^m \tag{5}$$

one receives an exponent $m_{Diesel,\theta=0.19}$ in the range of $5-6$ and $R^2_{Diesel,\theta=0.19} = 0.88$ for diesel fuel at an inlet temperature of $\theta = 0.19$. This result fits well with former studies of cavitation erosion in H_2O, where the exponent is in the range of $5-8$ [1, 2, 3, 4].
It was found, that both liquids, calibration fluid and diesel fuel show similar erosive behavior regarding the rail pressure influence.

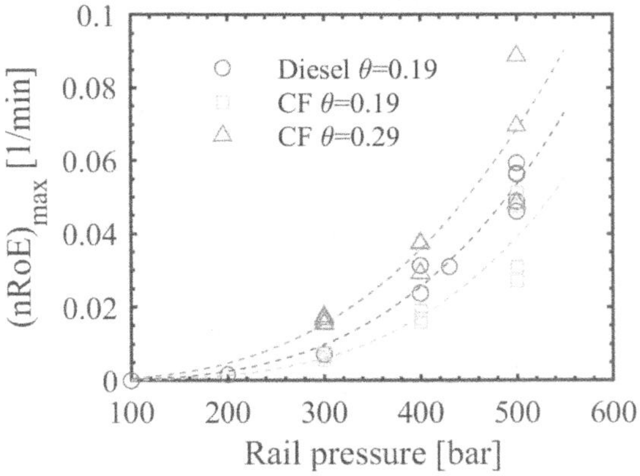

Figure 6: $(nRoE)_{max}$ for diesel-like fuels

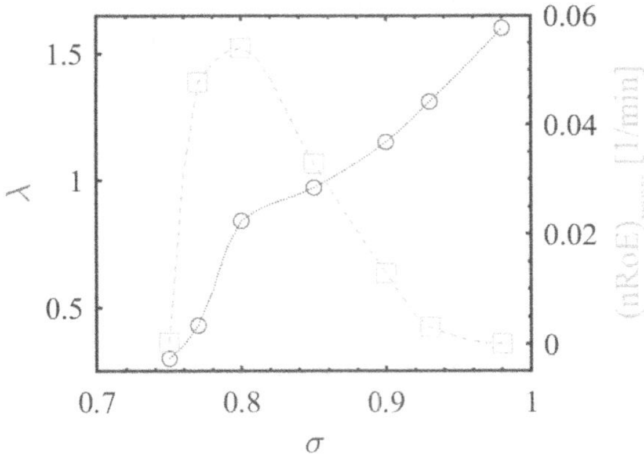

Figure 7: Normalized cavitation length and maximum erosion rate against cavitation number

In a second step, in the experiments with diesel fuel, the rail pressure is held constant at $p_{rail} = 500\,bar$ whereas the back pressure is altered from measurement to measurement. The temperature is set to $\theta = 0.19$. Consequently the cavitation number ranges between 0.75 and 0.98. In figure 7 one can see the effect on the normalized cavitation length λ and the normalized maximum erosion rate $(nRoE)_{max}$ plotted against the cavitation number σ. λ is the ratio of the measured cavitation length to the throttle length. It was found that λ increases as the back pressure is lowered , i.e. the attached cavitation structures expands downstream. Erosion occurs slightly downstream of the attached cavitation. This axial position correlates well to the cavitation length λ (see figure 8). For $\sigma \leq 0.75$ no or little cavitation is visible. Even after several hours no erosive damage could be observed. Rising the cavitation number from 0.75 to 0.8, the measured mass flow and according to Dular et al. [9] the shedding frequency increases. Consequently the erosive aggressiveness rises (Compare figure 8). As soon as the choked flow conditions are reached ($\sigma > 0.8$), the maximum mass flow is achieved. But in this case the cavitation length exceeds the throttle length, so the erosion decreases as the area of bubble collapse moves farther downstream of the channel exit. Erosion could be detected far downstream of the test section.

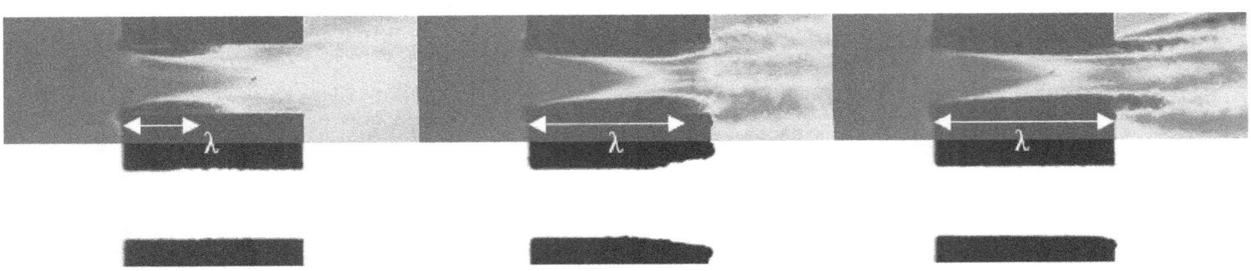

Figure 8: First row: standard deviation and cavitation length λ for $p_{rail} = 500\,bar$, $\theta = 0.19$, left: $\sigma = 0.77$, middle: $\sigma = 0.80$ and right: $\sigma = 0.93$.
second row: erosive damage after 2 min.

Results - temperature influence

To study the effect of the fluid temperature on cavitation erosion, a further measurement campaign with calibration fluid is conducted. Here, the fluid temperature is stepwise altered between $\theta = 0.13$ and $\theta = 0.40$ from measurement to measurement, the rail pressure p_{rail} is set to 300 bar respectively 500 bar and the cavitation number $\sigma = 0.81$. Rising the inlet temperature from $\theta = 0.13$ to $\theta = 0.29$ leads to a higher cavitation aggressiveness (see figure 9). In fact the $(nRoE)_{max}$ quadruples. Going beyond this temperature, the maximum erosion rate drops again. As one can see the rail pressure apparently has no significant effect on the curve progression.

For a deeper understanding of the temperature influence on cavitation erosion, a computational model was developed. It is based on the assumption of a spherical cavitation bubble collapse, i. e. the measured erosion rate $(nRoE)_{max}$ corresponds to the bubble collapse pressure p_{sc}. For H_2O this approach was already suggested by Dular [5]. The collapse pressure (see equation 8) can be derived from the Rayleigh-Plesset equation, which describes the dynamics of a single bubble in liquid (see equation 6). Here R is the bubble radius, t is the time, p_g is the pressure of non condensable gas, p_v is the vapor pressure, $p_\infty(t)$ is the liquid pressure, ϱ_l is the liquid density, S is the surface tension and μ is the viscosity. The driving force for the bubble oscillation was chosen $p_v - p_\infty(t) = 1.4 \cdot p_0 \cdot \sin(2\pi f t) - p_0$, with $f = 25\,kHz$ and $p_0 = 1\,bar$. The initial bubble radius is $R_0 = 2\,\mu m$.

$$R\ddot{R} + \frac{3}{2}\dot{R}^2 + \Sigma \dot{R}\sqrt{t} = \frac{p_g + p_v - p_\infty(t)}{\varrho_l} - \frac{2S + 4\mu\dot{R}}{R\varrho_l} \tag{6}$$

To consider the effect of local temperature decrease through latent heat on bubble growth, the thermodynamic parameter Σ, introduced by Brennen [10], is implemented (see equation 7). Here ϱ_v is the vapor density, L is the evaporate latent heat, c_{pl} is the isobaric mass heat capacity, T_∞ the ambient fluid temperature and α is the thermal diffusivity of the liquid. This thermodynamic parameter Σ is based on the assumption that a thermal boundary layer surrounding the bubble develops if the fluid temperature rises, inhibiting the bubble growth. This effect gains more and more influence with rising fluid temperature, so that the bubble collapse pressure p_{sc} diminishes.

$$\Sigma = \frac{(\varrho_v L)^2}{\varrho_l^2 c_{pl} T_\infty \sqrt{\alpha_l}} \tag{7}$$

As fluid for the calculation n-decane was chosen, because its temperature dependent maximum collapse pressure $p_{sc,max}$ fits best to the measurement. If one compares the measured maximum erosion rate of the cavitating liquid to the calculated collapse pressure of a single bubble in liquid, it appears that both show the same overall trend regarding the temperature influence (see figure 9).

$$p_{sc}(r,t) = p_\infty(t) + \varrho_l\left(\ddot{R}\frac{R^2}{r} + 2\dot{R}^2\left(\frac{R}{r} - \frac{R^4}{r^4}\right)\right) \tag{8}$$

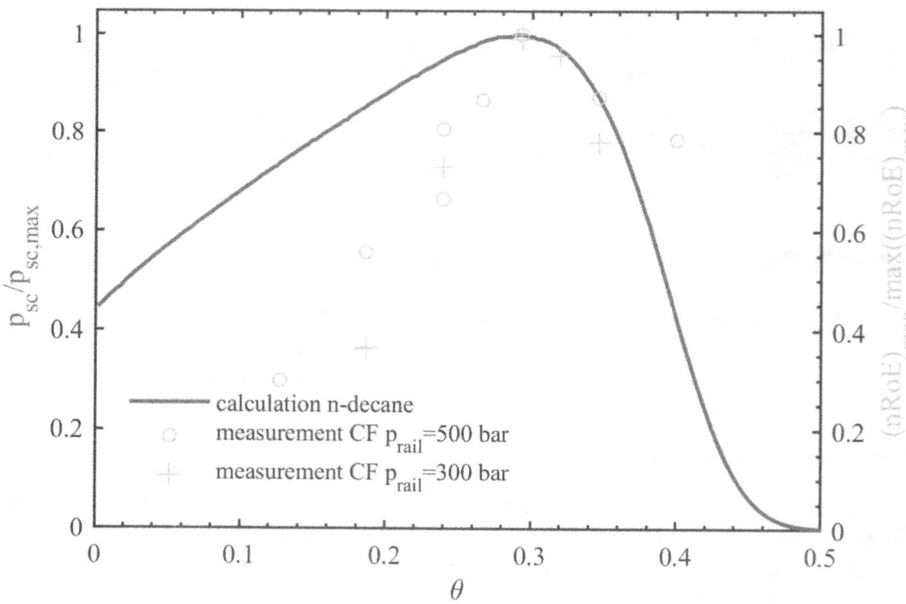

Figure 9: Temperature influence on cavitation aggressiveness

Conclusion

Hydrodynamic cavitation erosion of a square channel is experimentally investigated. Pressure has an important effect on erosion, because $(nRoE)_{max}$ increases nonlinearly with p_{rail}. The back pressure determines the axial position of erosion within the test section which is closely related to the cavitation length λ. Investigating the temperature effect on erosion in calibration fluid, the highest cavitation aggressiveness is detected at about $\theta = 0.29$. Numerical calculations based on the Rayleigh-Plesset equation for the single-component fuel n-decane conclude the same overall trend and a temperature dependent maximum erosion rate of $\theta = 0.29$. Future work will further investigate the temperature influence on cavitation erosion.

References

[1] Jean-Pierre Franc et al. "Material and velocity effects on cavitation erosion pitting". In: *Wear* 274-275 (2012), pp. 248–259. ISSN: 00431648. DOI: 10.1016/j.wear.2011.09.006.

[2] Matevž Dular, Bernd Stoffel, and Brane Širok. "Development of a cavitation erosion model". In: *Wear* 261.5-6 (2006), pp. 642–655. ISSN: 00431648. DOI: 10.1016/j.wear.2006.01.020.

[3] Jiu-Gen He and Frederick G. Hammitt. "Velocity exponent and cavitation number for venturi cavitation erosion of 1100-O aluminium and 1018 carbon steel". In: *Wear* 80 (1982), pp. 43–58. ISSN: 00431648.

[4] József Varga and Gy Sebestyen. "Observations on cavitation velocity-damage exponent in a flowing system". In: *Periodica Polytechnica Mechanical Engineering* 8.3 (1964), pp. 343–352.

[5] Matevž Dular. "Hydrodynamic cavitation damage in water at elevated temperatures". In: *Wear* 346-347 (2016), pp. 78–86. ISSN: 00431648. DOI: 10.1016/j.wear.2015.11.007.

[6] Frederick G. Hammitt. "Cavitation Erosion: The State of the Art and Predicting Capability". In: *Applied Mechanics Review* 32.6 (1979), pp. 665–675.

[7] Ralph Böhm. "Erfassung und hydrodynamische Beeinflussung fortgeschrittener Kavitationszustände und ihrer erosiven Aggressivität". Dissertation. Darmstadt: Technische Universität Darmstadt, 1998.

[8] Matevž Dular. "Investigations into Dependence Between Cavitation Structures and Cavitation Erosion". In: *Advanced Experimental and Numerical Techniques for Cavitation Erosion Prediction*. Ed. by Ki-Han Kim et al. Vol. 106. Fluid mechanics and its applications. Dordrecht: Springer Netherlands, 2014, pp. 239–256. ISBN: 978-94-017-8538-9. DOI: 10.1007/978-94-017-8539-6_10.

[9] Matevž Dular and Rudolf Bachert. "The Issue of Strouhal Number Definition in Cavitating Flow". In: *Journal of Mechanical Engineering* (2009), pp. 666–674.

[10] Christopher Brennen. "The Dynamic Behavior and Compliance of a Stream of Cavitating Bubbles". In: *Journal of Fluids Engineering* 95.4 (1973), p. 533. ISSN: 0098-2202. DOI: 10.1115/1.3447067.

© 2018 ASME

The acoustic pressure generated by the non-spherical collapse of laser-induced cavitation bubbles near a rigid boundary

[1]Silvestre Roberto Gonzalez-Avila* and [1]Claus-Dieter Ohl

[1]*Nanyang Technological University, Singapore, 637371*

Abstract

In this paper we study the acoustic transients emitted during the collapse of a laser-induced cavitation bubble. A 150MHz bandwidth fiber optic probe hydrophone, resolves complex signals produced during the non-spherical collapse of the bubble near a rigid boundary. Simultaneous high-speed video recording up to 200,000 frames/s are correlated with the hydrophone trace. Additionally, we employ an intensified CCD-camera to record snapshots of the shock waves. The acoustic signal contains several pressure peaks with rise times as fast as 18 ns; they correspond to shock waves emitted as the bubble is reaching its minimum volume. For a limited range of standoff parameter values, $1.8 \leq \gamma \leq 5.0$, the strength and time of emission of the first and the second emitted shock waves are related. The maximum radii of the bubbles tested are between 560 μm and 760 μm. We obtain peak pressures of up to 1 kBar during the collapse phase for the bubbles collapsing closes to the boundary.

Keywords: Shock wave; laser-induced bubble; acoustic pressure

Introduction

Acoustic measurements are important in diverse fields of science and engineering; however, the measurements of transient acoustics is not trivial, especially in non-linear acoustic fields, where very fast rise times and high pressures are present. For instance, focused ultrasound waves as those used in lithotripsy peak pressures of a few hundred bars and rise times of less than 50 ns are typically encountered. For this challenge Staudenraus and Eisenmerger [1] developed a fiber optic probe hydrophone (FOPH) that uses the reflected light at the tip of a glass or polymer fiber to deduce the local pressure. Over time this device gained acceptance and became ISO standard to measure intense acoustic fields [2-6], yet it lacks sensitivity. Therefore, to increase the signal-to-noise-ratio averaging of 20 and up to 1000 waveforms has been reported [2]. However, important details varying between experiments which are real and would deserve a closer look may be removed. In this paper we present single acoustic pressure measurements with a FOPH and reveal how the emitted shock waves at the collapse of the bubble depend on γ. Also, we present the pressure developed by single cavitation bubbles that collapse near a solid boundary.

Experimental setup

A schematic diagram of the test section is portrayed in figure 1a. It consists of a modified acrylic cuvette inside which a pulsed laser beam (green Q-switched Nd:YAG laser, pulse duration 6 ns) is focused to create single cavitation bubbles at specific locations from a solid boundary. The optical arrangement is similar to a setup previously reported [7]; however, in the current setup we sample a small portion of the laser beam energy, ~4%, to accurately detect the time when the laser beam appears. The rising signal from the photo diode (see figure 1a) triggers the data acquisition by the oscilloscope (LeCroy HRO64Zi 2GHz 12-Bit) and the pressure generated during the cavitation event is recorded by a fiber optic probe hydrophone, FOPH (690 ONDA, 150 MHz bandwidth). The 125 μm tip of the fiber was inserted into the test section through a small hole of 0.6 mm in diameter, and positioned with a 3-axis stage at a distance *d* from the bubble inception location. A pulse delay generator (Berkley Scientific, BNC model 575) triggers and synchronizes the laser and the high-speed camera (Photron SAX-2) coupled to a 60 mm macro lens (Nikor) at full magnification (resolution 20 um per pixel). The scene is illuminated with mildly diffused light from a LED fiber optic lamp (REVOX SLG 150V). The shockwaves emitted during the bubble expansion or collapse are recorded with an intensified 12 bit CCD camera (PCO Dicam Pro). With this camera single images with a minimum exposure time down to 3ns can be captured. There we increase the spatial resolution to 2.2 μm/pixel by coupling the camera to a long distance microscope (Infinity CF-4). A circular aperture placed in front of a light emitting diode (Cree CBX 1512) was used for illumination and to record shadowgraphs of the emitted shock waves.

*Corresponding Author, S. Roberto Gonzalez-Avila: roberto_glez83@hotmail.com

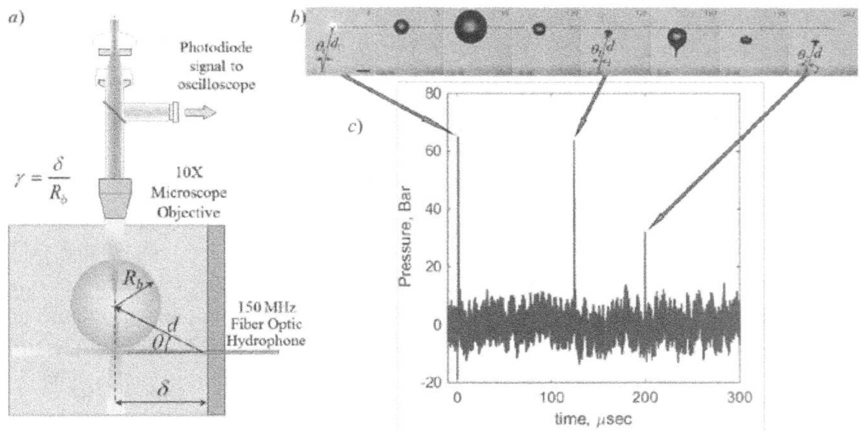

Figure 1. (a) Experimental setup and experimental parameters; (b) Laser-induced cavitation bubble created near a solid wall; the laser beam appears from right to left; the fiber optic can be seen in the lower left side of each frame, $\gamma = 2.7$; $R_b = 616$ μm. The length of the bar in the lower right corner of the first image is 500 μm; time is in μs. The dotted black is added as an aid to the eye; (c) the acoustic signal recorded by the FOPH; the three prominent peaks from left to right correspond to the bubble's inception, bubble's first collapse and the bubble's second collapse, respectively; $d_i = 1.7$ mm; $d_1 = 1.5$ mm; $d_2 = 1.2$ mm; $\theta_i = 13°$; $\theta_1 = 14°$; $\theta_2 = 17°$.

Figure 2b displays selected frames from a high-speed recording, $\gamma = 2.7$ and $R_b = 616$ μm; the images are rotated 90° clockwise. From left to right they depict the bubble dynamics from inception to the second collapse. The high-speed camera and the FOPH were triggered simultaneously, thus the images can be correlated to the acoustic trace. The dotted black line guides the eye to emphasis the translation of the cavitation bubble towards the boundary in its first and second collapse. The acoustic trace, figure 2c, shows three distinct pressure peaks; from left to right the first peak is due to the shockwave emitted during the nucleation of the cavitation bubble; the second and third peaks are due to the shock waves emitted during the first and second collapse of the bubble, respectively. During bubble expanding and before reaching its minimum volume, $5 \leq t \leq 120$ μs no pressure peaks or any other prominent features in the acoustic trace are observed. Also, after the bubble reaches its minimum volume no other peak is detected; only around the time the second collapse takes place a third peak is visible in the graph.

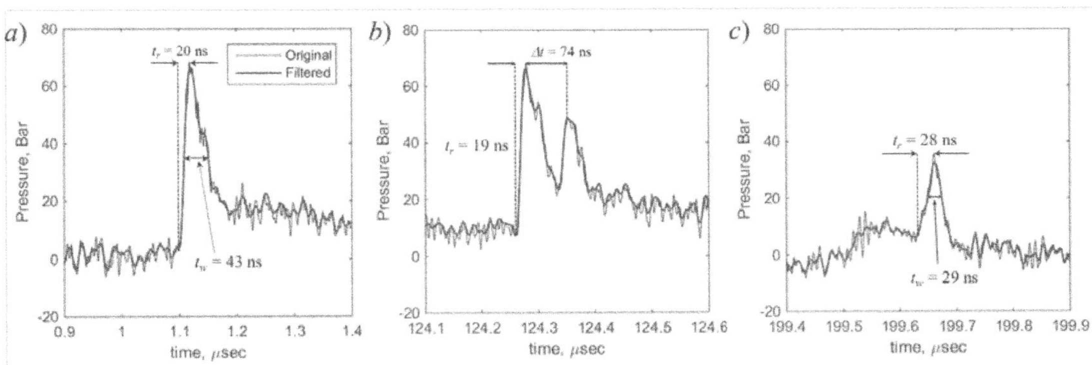

Figure 2. Close up of the acoustic pressure peaks portrayed in figure 1c. (a) inception; (b) bubble's first collapse; (c) bubble's second collapse.

The peaks from figure 1c are zoomed in, see figure 2a-c. Notice that these curves are essentially compression waves and do not show a tensile component which is different from previous measurements [5,6]. Interestingly, the rise time is only 20 ns and the width of the peak about 43 ns. Figure 2b shows an interesting feature which becomes visible at high temporal resolution during the first bubble collapse. The acoustic trace consists of two peaks emitted with a time delay of only 74 ns. We believe these two peaks are due to the non-spherical bubble collapse. Figure 2c also displays a short peak that is correlated to the time of the second collapse of the bubbles. These figures, 2a-c, present two traces each, the light color trace is the original acoustic signal recorded, while the darker trace is the signal after high frequency components (>70 MHz) have been removed.

Thus, to get a better evaluation of the peak pressure developed on the wall during cavitation bubble collapses, it is first necessary to evaluate the variation of the peak pressure measured as a function of the angular position θ and distance *d*. Figure 3a displays the change in the peak pressure detected with angular position θ. Each dot is a single event; in these tests both the maximum bubble size, R_b, and the distance between the bubble's inception location and the fiber tip, *d*, were kept constant (within experimental uncertainty); the red circles plot the peak pressure as a function of incidence angle of the shock wave with the fiber tip, $d = 1.68 \pm 0.02$ mm; the FOPH trace is recorded synchronously with the high-speed video images, thus we can clearly obtain the maximum size obtained by the bubble. For the lower lower data points (blue squares) in figure 3a the CCD camera was used. It was synchronized to capture the shock wave as it arrives at the fiber tip for the different γ values tested. The inset in the lower left corner of figure 3a depicts a sample image; the distance between the bubble's inception location and the fiber tip is $d = 1.00 \pm 0.01$ mm; the pressure is adjusted to a pressure of $d = 1.68$ mm. For these tests the time at which the first pressure peak is detected, T_c, is used to estimate the bubble size with $R_b = T_c(P_\infty/\rho)/1.83$; the average bubble size of the tests is 520 ± 6 μm. Figure 3b portrays the variation of maximum pressure with distance *d*; the parameter *d* was measured by locating the centroid of the plasma and the tip of the fiber for each test; snapshots of the plasma are shown,,too . In these tests a single cavitation bubble is created 2.6 mm away from the tip of the fiber optic then the fiber is moved closer to the inception location and both high-speed video and the acoustic signal are recorded. The figure portrays the peak pressure detected by the FOPH during this process. The angle formed by the axis of the fiber and the nucleation site, θ, was fixed at $15 \pm 1°$ together with the laser energy resulting in bubbles of $R_b = 695 + 16$ μm (average of 36 cavitation events). The distance between the fiber tip and the nucleation site, *d*, varied in the range $0.27 \text{ mm} \leq d \leq 2.6$ mm. In the linear regime an inversely proportional decay of the wave amplitude with distance (1/d decay) is expected [8] as confirmed by the data in figure 3b.

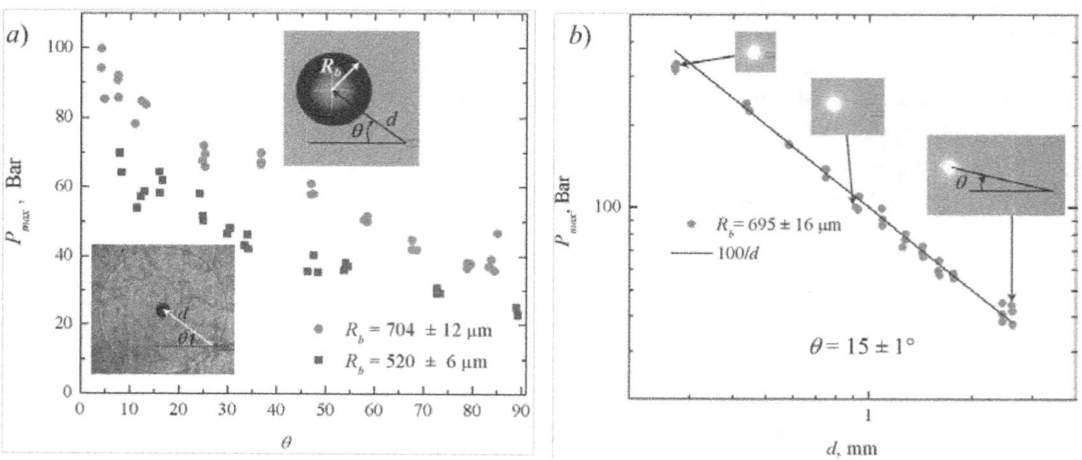

Figure 3. (a) Change of peak acoustic pressure with angular position; (b) Change of peak acoustic pressure with distance from the fiber tip

The prime motivation for this work was to study the pressure generated by the collapse of a single bubble near a solid wall. Here we find interesting acoustic traces for certain ranges of γ values. In the range $1.8 \leq \gamma \leq 5.0$ up to *three* distinct pressure peaks were detected during the bubble collapse. The third peak is usually weaker and more difficult to detect thus we focus here on the features of the first and second pressure peaks of the first bubble collapse. Samples (non-averaged) for increasing γ are shown in figures 4a-d. For $\gamma = 1.9$ the rise time of the first peak is comparably large, $t_r = 72$ ns. Yet, for larger stand-off distances the pressure rises within less than 30 ns. Additionally, the variation of the ratio of the first pressure peak, P_1, to the second pressure peak, P_2, with γ and the time delay between them, Δt changes as a function of γ. In figure 4a the magnitude of the first pressure peak is ~50% that of the second peak; both pressure peaks display approximately the same magnitude in figure 4b and in figure 4c the amplitude of the first peak is more than twice that of the second peak. From $\gamma = 4.9$ on a single peak is resolved, figure 4d which is consistent that Δt decreases with γ.

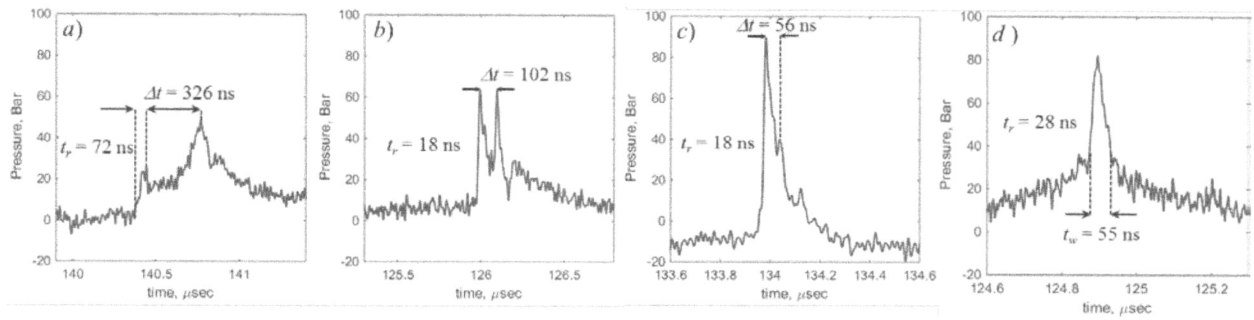

Figure 4. Pressure peaks detected at the first collapse of a cavitation bubble; a) $\gamma = 1.9$, $R_b = 677$ μm; b) $\gamma = 2.5$, $R_b = 620$ μm; c); $\gamma = 3.0$, $R_b = 679$ μm; e) $\gamma = 4.9$, $R_b = 701$ μm.

Unfortunately, the high-speed camera does not allow us to visualize the shock waves emitted either at the inception or at the collapse of the bubble we therefore use the single frame ICCD camera with a 3ns exposure time. Figure 5a displays two circles in the lower left corner of the image; the outer circle has a radius $R_o = 183$ μm, while the smaller circle has a radius $R_i = 108$ μm. Assuming constant speed of sound at 1480 m/s the estimated time delay between in the emission of the shock waves is $\Delta t = 50$ns which nicely fits with the time between the pressure peaks shown in figure 5b of the event, $\gamma \sim 2.3 \pm 0.2$. Vogel and Lauterborn (1988) and more recently by Supponen et al. (2107) have reported the emissions of multiple shock waves from a single bubble collapse have. Next, we want to demonstrate that the timing and strength emission of the first and the second shock waves are very well reproducible and how they depend on γ.

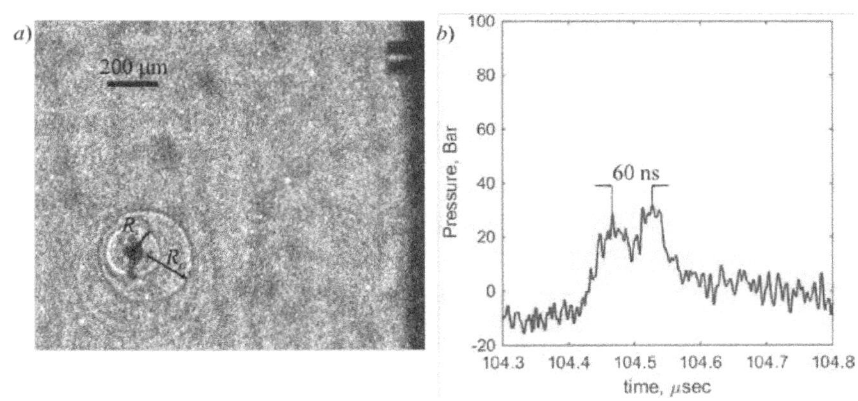

Figure 5. (a) Acoustic transients emitted at the collapse of the cavitation bubble, g ~ 2.3, exposure time is 3 ns; b) the acoustic signal recorded by the FOPH around the time reaches its minimum volume.

Therefore, we have extracted from many experiments the ratio of the first to the second pressure peak, and the time delay between these two peaks, Δt, and graphed them as a function of γ in figures 6a-b. In figure 6a this relationship between the P_{jet}/P_{tip} ratio with γ is clearly observed up to $\gamma \approx 1.8$, below the amplitudes were too weak to discern them. In the range $1.8 \leq \gamma \leq 3.0$, there is a increase of P_{jet}/P_{tip} with γ. The final stage of the collapse of a cavitation bubble is too fast to be resolved with our imaging system. However, a possible explanation for the change in the strength of the pressure peaks is as follows: The first pressure peak occurs due to the impact of the jet on the opposite wall of the bubble (see Ohl et al 1999). For the lower γ tests $1.8 \leq \gamma \leq 2.5$ the jet penetrates the bubble early in the collapse phase, thus the opposite wall has not gained enough velocity. As the jet impacts later in the collapse phase at slightly larger γ values the opposite wall of the bubble is moving faster and the water hammer pressure upon impact of the jet with the opposite bubble wall is stronger and the jet shock wave, P_{jet}, increases. Around $\gamma = 2.5 \pm 0.1$ the strength of the shocks are approximately the same and the ratio is maximum at $\gamma = 3.0 \pm 0.1$. As the jet develops later during in the collapse phase the collapse of the bubble occurs quasi-spherically and the shocks may not be discernible or (unlikely) as single shock being emitted. With respect to the time delay between the shock waves at the lower γ values the collapse of the bubble is more deformed, then not only the jet pierces sooner in the collapse phase but also the volume trapped between the tip of the jet and the opposite wall of the bubble continues to compress until a second shock wave, the tip bubble shock wave, P_{tip}, is emitted. For larger γ values the collapse becomes more symmetric thus the jet pierces at a later stage in the collapse of the bubble, and the small volume trapped cannot compress much further, hence the time delay between the first and the second shock wave is shorter than during collapse at lower γ [11].

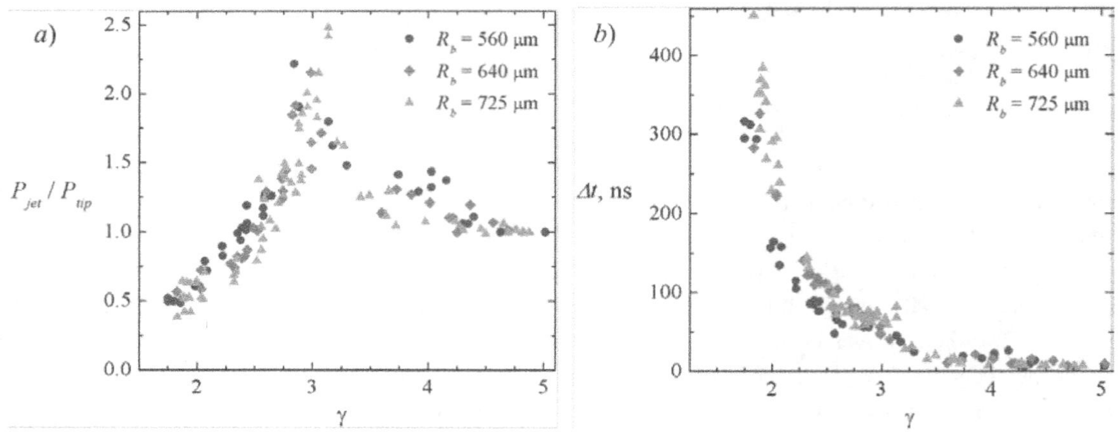

Figure 6. (a) The ratio of the peak pressures P_{jet}/P_{tip} from the cavitation bubble collapse with stand-off parameter; b) time delay, Δt, between the emission of the jet shock wave, P_{jet} and the tip bubble shock wave, P_{tip}.

Finally, a summary of the tests conducted near the solid boundary, $0.1 \leq \gamma \leq 3.0$, is portrayed in figure 7. In figure 7a the original peak pressure detected is shown; in figure 7b the pressure at the location where the bubble collapse takes place is portrayed; the original pressure values are corrected for distance and the angle of incidence, *θ*. The data in figure 7b shows more variation than the original data in figure 7a because the distance from the fiber tip to the bubble was different for different tests. In some cases the minimum volume detected corresponded to a torus while in other cases the minimum volume captured was quasi spherical. Already bubbles of sub-millimeter size can generate pressures on the boundary of up to 1 kBar for a few ns. Also, notice that for γ ~ 0.9 the sound emission decreases significantly; this is in good agreement with previous observations [13,14,10]. This is a consequence of the alteration of the bubble collapse due to the presence of the boundary and the formation of a vortex ring.

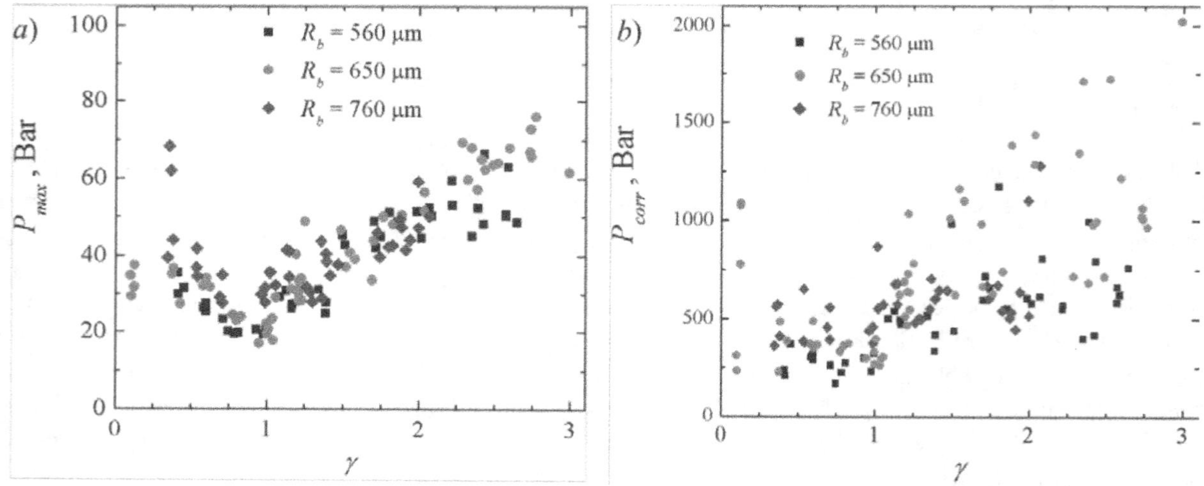

Figure 7. The acoustic pressure near the cavitation bubble at its first collapse; a) the peak pressure measured; b) the corrected peak pressure for distance and angular position at its first collapse, P_{corr}.

For larger γ values where the collapse is quasi spherical the involution of the bubble takes place approximately in a radial fashion thus the pressure peaks emitted are strong; however, when the bubble collapses close to a boundary, γ ~ 0.9, the jet penetrates through it before it has reached its minimum volume. As the jet impact on the wall it spreads radially transferring kinetic energy into circulation, forming a system of vortex rings. There the compression of the bubble content is less violent and sound emission diminished [15]. Figure 7a depicts how the acoustic emission decreases for γ = 0.9.

Conclusion

We have presented single acoustic pressure measurements of the acoustic transients emitted during the non-spherical collapse of laser-induced cavitation bubbles near a rigid boundary. The data is recorded for a range of stand-off parameters, far from the boundary, γ ~ 5, down to the collapse next to the boundary, γ = 0.1. The results show that strength of the shock waves emitted, namely the shock wave emitted due to the jet impact on the opposite wall of the bubble, P_{jet} and the tip bubble shock wave, P_{tip}, are a complex but comprehensible function in the range $1.8 \leq \gamma \leq 5.0$. This information we think is suitable for comparison with simulations to unravel the interaction leading to cavitation induced material damage.

Acknowledgements

We appreciate funding through ONR under grant number Grant No. N62909-16-1-2139.

References

[1] Stadenraus, J.,Eisenmenger, W. (1993). *Fiber-optic probe hydrophone for ultrasonic and shock wave measurements in water*, Ultrasonics, 31(4).
[2] Wang, Z. Q., Pecha, R., Gompf, B., Eisenmenger, W. (1999). *Single bubble sonoluminescence: Investigations of the emitted pressure wave with a fiber optic probe hydrophone*, Phys Rev-E, 59(2).
[3] Sankin, G. N., Simmons, W. N., Zhu, S. L., Zhong, P. (2005). *Shock Wave Interaction with Laser Generated Single Bubbles*. Phys Rev Lett, 95, 034501.
[4] Sankin, G. N., Zhong, P. (2006). *Interaction between shockwave and single inertial bubbles near an elastic boundary*. Phys Rev E, 74, 046304.
[5] Sankin, G. N., Zhou, Y., Zhong, P. (2008). *Focusing of shock waves induced by optical break down in water*. J. Acoust. Soc. Am. 123 (6).
[6] Zijlstra, A., Ohl, C-D. 2007. *On fiber optic hydrophone measurements in a cavitating liquid*. J Acoust Soc Am., 123, 29-32.
[7] Gonzalez-Avila, S. R., OHL, C-D. (2106). *Fragmentation of acoustically levitating droplets by laser-induced cavitation bubbles*. J. Fluid Mech. 805.
[8] Cole, R. H. (1948). *Underwater Explosions*. Princeton University Press.
[9] Ohl, C-D., Kurz, T., Geisler, R., Lindau, O., Lauterborn, W. (1999) *Bubble dynamics, shock waves and sonoluminescence*. Phil. Trans. R. Soc. Lond. A. 357.
[10] Vogel, A., Lauterborn, W. (1988). *Acoustic transient generation by laser-produced cavitation bubbles near solid boundaries*. J Acoust. Soc. Am.84 (2).
[11] Supponen, O., Obreschkow, D., Kobel, P., Tinguely, M., Dorsaz, N., Farhat, M. (2017). *Shock waves from nonspherical cavitation bubbles*, Phys Rev Fl, 2, 093601.
[12] Lindau, O., Lauterborn, W. (2003). *Cinematographic observation of the collapse and rebound of a laser-produced Cavitation bubble near a wall*. J Fluid Mech. 479.
[13] Naudé, C. F., Ellis, A. T., (1961). *On the Mechanism of Cavitation Damage by Nonhemispherical Cavities Collapsing in Contact With a Solid Boundary*, J Basic Eng. 83 (4).
[14] Tomita, Y., Shima, A. (1986). *Mechanisms of impulsive pressure generation and damage pit formation by bubble collapse*. J. Fluid Mech., 169.
[15] Chahine, G. L., Genoux, P. F., (1983). *Collapse of a Cavitaing Vortex Ring*. Trans. ASME J. Fl Eng, 105 (4).

© 2018 ASME

The cavitation flow in the jet pump cavitation reactor and its sterilization of *E. coli*

[1,2]Jiong Wang; [1,2]Xinping Long*; [3]Junqiang Zhang; [3]Longzhou Xiao; [1,2]Bin Ji;

[1] *Hubei Key Laboratory of Waterjet Theory and New Technology, Wuhan University, Wuhan 430072, China;*

[2] *School of Power and Mechanical Engineering, Wuhan University, Wuhan 430072, China;*

[3] *Wuhan Second Ship Design and Research Institute, Hubei, 430064 China*

Abstract

In the present paper, experiments are conducted to investigate the effects of a jet pump cavitation reactor (JPCR) on the sterilization of *e. coli*. The basic performance of the JPCR is tested firstly. And then the sterilization influences of cavitation state and initial bacterial concentration are investigated. The results show that the cavitation in the JPCR has satisfying effects on *e. coli* sterilization. With the decrease of pressure ratio, flow ratio increases and reaches a maximum value, which is so-called limited operation cavitation stage. In this stage, severe cavitation occurs, which is used to sterilize *e. coli* in this paper. Generally speaking, as the treating time increases, sterilization rate increases slowly at first, and then surges. This hysteresis effect indicates that *e. coil* shows resistance in a short time with the explosure to cavitation. The unsteady cavitation oscillating in throat can provide a better sterilization effect. Moreover, with the decrease of initial bacteria concentration, the sterilization rate increase. Our work provides a new method of sterilization with great potentiality.

Keywords: cavitation; jet pump cavitation reactor; *e. coli*; sterilization effect

1. Introduction

Cavitation is a phase transition between liquid and vapor occurring within the liquid inside or at the liquid-solid interface because of the power in the flow [1]. Cavitation nearly occurs in all kinds of fluid machinery and generates undesirable effects such as noise, vibration, material damage and performance reductions [2-4]. But in fact, cavitation is not entirely harmful. The local high temperature, high pressure, microjet, shock wave and the free radicals induced by the cavitation bubble can build special reaction site [5-7]. This favorable effect can be used in food engineering, bioengineering, mining industry and sewage treatment [8-10], etc.

There have many studies on cavitation applications. Ultrasonic cavitation is the most commonly used but it has high energy costs. Hydraulic cavitation has wide application prospect because of simple structure, high efficiency and good economy. The hydraulic cavitation reactors is various, such as orifice, venture, high pressure homogenizer and high speed homogenizer. Many studies have been conducted to investigate the application of cavitation reactors. Jyoti's [11] experiments shows that the hybrid technique combined hydrodynamic cavitation appears to be an attractive alternative to a single technique for the reduction in the heterotropic plate count bacteria. Balasundaram [12] carries out the disruption of *e. coli* by orifice plate for the purpose of preferentially releasing the intracellular proteins from organisms. Shivram [13] carries out the seawater disinfection using cavitation produced by vortex diode and have proved its effectiveness in killing of various types of zooplanktons in the seawater. These researches are satisfying and promising in sterilization application but further investigate is needed in hydraulic cavitation method. JPCR is a kind of cavitation reactor combining the feature of orifice and venture. Because of high velocities gradient and severe shear flows in JPCR, cavitation occurs easily, so it is of high potential in sterilization. However, relative investigations are rarely reported.

Therefore, experiments are conducted to obtain more understanding of cavitation flow in JPCR and its sterilization effect in the current paper. The basic performance of JPCRs is tested and the corresponding sterilization experiments are also conducted to study the influence of the cavitation state and initial bacterial concentration on sterilization effect.

2. Experiment Setup

2.1 *The structures of JPCR*

The structure of a typical JPCR is shown in **Fig. 1.** The main structure parameters are set as follows: inlet pipe diameter, D_{in}=25mm; outlet pipe diameter, D_{out}=50mm; throat length, L_{th}=96mm; contraction angle, α=45deg; and diffuser angle, β=12deg. Since JPCRs with the same area ratio (throat to the nozzle exit area ratio) have similar performance, three area ratios are obtained by only altering the size of the nozzle but fixing the sizes of the throat and the other parts for the convenience of machining. In this paper, the nozzle outlet diameters, D_n=8, 10, 12mm are employed to obtain three area ratios 4.00, 2.56, 1.78.

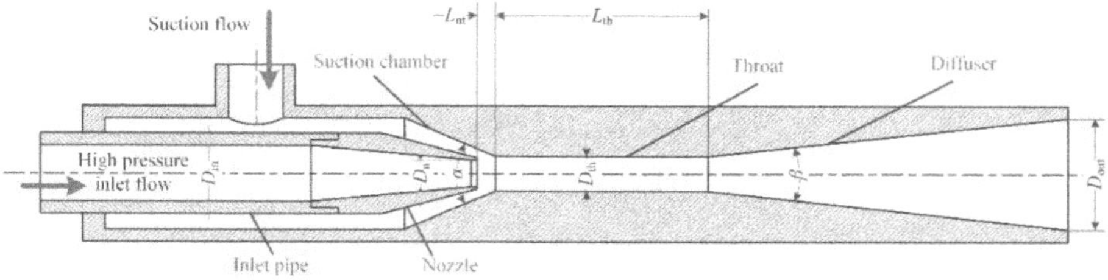

Fig. 1 Schematic drawing of JPCR

The basic performance of a JPCR can be characterized by the classical parameter such as area ratio (m), flow ratio (q) and pressure ratio (h). All these dimensionless parameters are defined as follows,

$$m = \frac{A_{\text{th}}}{A_n} \tag{1}$$

$$q = \frac{Q_s}{Q_{\text{in}}} \tag{2}$$

$$h = \frac{P_{\text{out}} - P_s}{P_{\text{in}} - P_s} = \frac{\left(\frac{p_{\text{out}}}{\rho_{\text{out}} g} + \frac{v_{\text{out}}^2}{2g} + z_{\text{out}}\right) - \left(\frac{p_s}{\rho_s g} + \frac{v_s^2}{2g} + z_s\right)}{\left(\frac{p_{\text{in}}}{\rho_{\text{in}} g} + \frac{v_{\text{in}}^2}{2g} + z_{\text{in}}\right) - \left(\frac{p_s}{\rho_s g} + \frac{v_s^2}{2g} + z_s\right)} \tag{3}$$

where A represents the cross sectional area, Q represents the flow rate, ρ represents the density, P represents the total pressure, p represents the hydrostatic pressure, z represents the elevation, v represents the average velocity of flow, respectively. Subscript in, th, n, s, out represent inlet fluid, throat inlet, nozzle exit, suction fluid, outlet fluid respectively.

2.2 *Experiment rig and procedure*

The experimental rig is illustrated in **Fig. 2.** The high pressure inlet flow (Q_{in}) supplied by a centrifugal pump entrains the suction flow (Q_s) from the water tank. The high pressure inlet flow and the suction flow mixes in the nozzle outlet and induces strong shear cavitation, the mixed flow run through the throat. The inlet and mixed flow rates are controlled by valves and measured by electromagnetic flow meters. Three pressure transducers are installed in the main water flow inlet, the suction pipe inlet and the diffuser outlet.

Because of high speed gradient and shear flows in throat, the cavitation develops violently and a large number of cavitation bubbles occurs, which collapse when passing through the diffuser due to high back pressure. In addition, different cavitation state can be obtained by adjusting the outlet valve 11.

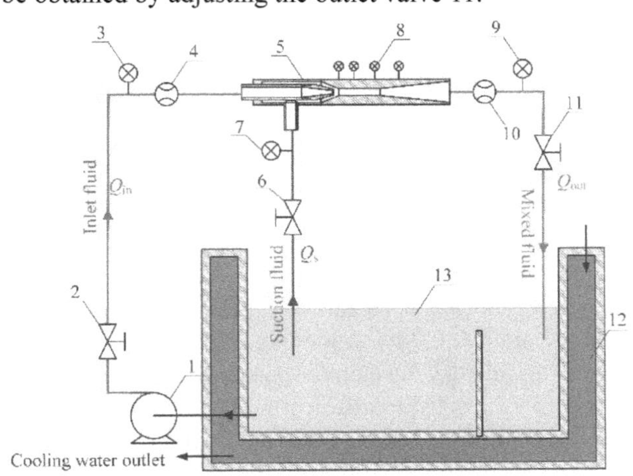

1-centrifugal pump; 2,6,11-valves; 4,10-electromagnetic flowmeter;
3,7,8,9-pressure sensor; 5-JPCR; 12-cooling water tank; 13-water tank

Fig. 2 Sketch of the experimental rig

The treatment of bacteria solution with different initial bacterial concentration was carried out by JPCRs at different cavitation state. For all the tested experiments, fixed treating fluid volume of 60L and a circulation time through the device of 3-4 hours was maintained. In all the treating conditions, the cooling water tank was open to keep the water temperature constant (20-30℃). Samples were taken at regular intervals and the concentration of bacterium was measured by plate count method. The disinfection rate (γ) was calculated as follow,

$$\gamma = \left(1 - \frac{N_1}{N_0}\right) \times 100\% \tag{4}$$

where N_0, N_1 represents the concentration of live bacteria before and after treatment.

3. Result and discussion

Before the sterilization experiment, some check experiment was carried out. The results shows that the concentration of live bacteria didn't change under no cavitation stage and tested water temperature. So the death of bacteria is caused by cavitation.

3.1 *Basic performance of JPCR*

Fig. 3 shows the basic performance of JPCRs. Under different pressure ratios, the cavitation state changes significantly. With the decrease of pressure ratio (h), flow ratio (q) increases gradually. The maximum q is then the so-called cavitating flow ratio, q_c. The corresponding flow condition is defined as the limited operation stage. Severe cavitation occurs during this stage with massive cavity clouds completely filling the throat and even expanding into the diffuser when h is very low. The cavitation flow chokes the flow at the throat and the velocity may reach the sound velocity, so q remains almost constant with further decrease of p_{out}. In current paper, the limited cavitation stages are used for sterilization. Below the cavitating flow ratio, the h-q curves overlap for the same area ratio but different inlet pressures. Smaller area ratios result in steeper performance curves. However, during the limited operation stage, the curves separate and become vertical with different q_c for different p_{in} and a smaller area ratio leading to a smaller q_c, which means that JPCRs with small area ratio can more easily reach the limiting condition. As the flow ratio decreases from the maximum, the cavitation weakens and finally disappears.

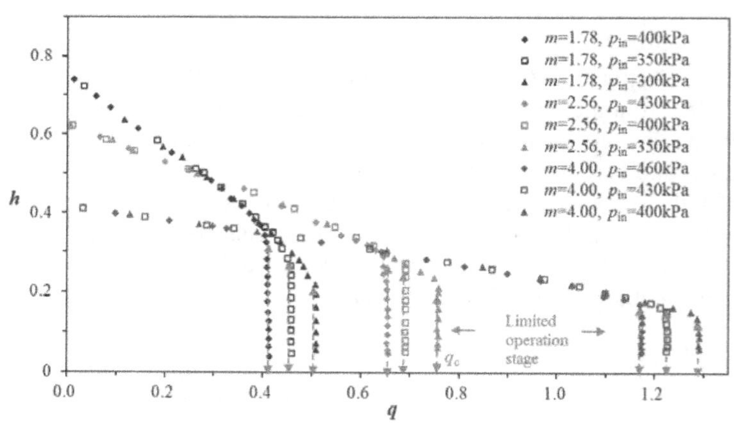

Fig. 3 The basic performance of JPCR

3.2 *The sterilization effect under different cavitation state*

Fig. 4 shows the sterilization effect under different cavitation state in limited cavitation stage. It can be seen that the sterilization rate gradually increase with increasing treating time. The sterilization rate is lower within the former 30 min. However, the sterilization rate significantly increased at 60 min, and then it slows down and gets close to 100% at 150 min. It indicated that the cavitation of JPCR has satisfying effect on *e. coli*, but there exists hysteresis effect. *E. coli* cannot be killed immediately unless damage accumulates to a certain degree, at this time the sterilization rate increases significantly. But the sterilization speed is different in different treatment conditions although the change rules are consistent. It can be seen that the sterilization rate in stable limited operation cavitation stage (the collapse position is in the diffuser) is much lower than that in unstable limited operation cavitation stage (the collapse position is in the throat). With the increase of pressure ratio, the back pressure increases and cavitation area shrinks. The cavitation clouds switch from stable development to unstable pulsation in throat. As a result, the corresponding sterilization rate increase. That is to say, the bactericidal effect is not determined solely by the number of cavitation bubbles, but also depends on the back pressure and pulsating pressures of the environment. In unstable limited operation cavitation stage, the cavitation cloud chokes the flow in the throat with greater pressure pulsation than stable limited operation cavitation stage. And the *e. coli* was killed quickly under high frequency pulsating pressures.

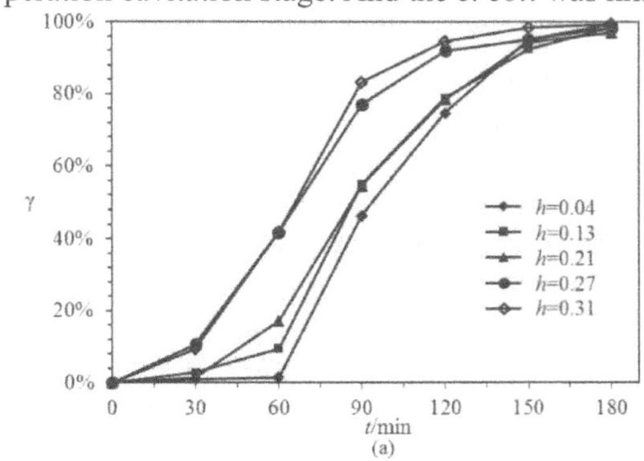

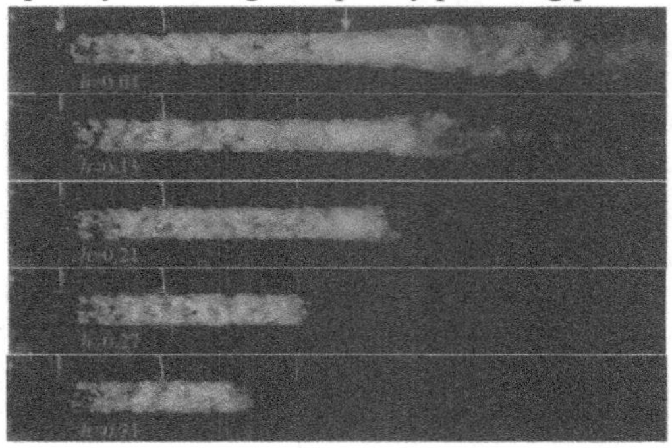

Fig. 4 The sterilization effect under different cavitation state (m=1.78, p_{in}=400kPa)

3.3 *The sterilization effect under different initial bacterial concentration*

Fig. 5 shows the sterilization effect under different initial bacterial concentration. It can be seen that the change rule of sterilization rate is the same for different initial bacterial concentration. But with the decrease of initial bacterial concentration, the sterilization rate increase significantly. The lower the initial bacterial concentration is, the faster the sterilization speed is. In the lower initial bacterial concentration, the number of bacteria treated with cavitation is less and the cumulative time of damage is shorter, so the corresponding effects of sterilization rate and treating time is much greater.

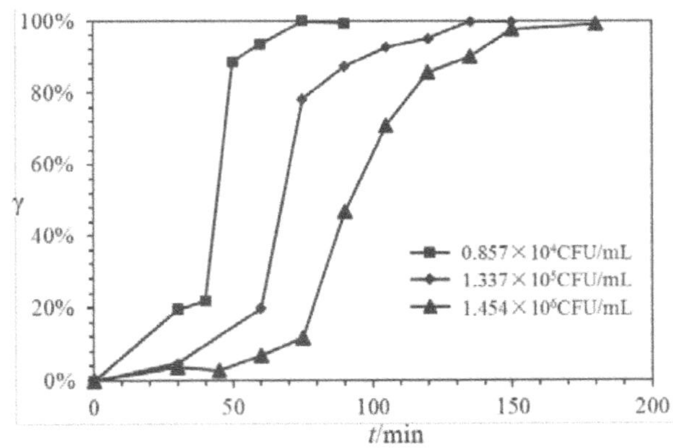

Fig. 5 The sterilization effect under different initial bacterial concentration (m=2.56, h=0.20).

4. Conclusions

In this paper, the basic performances of JPCRs are investigated and the corresponding sterilization experiments are also conducted to study the influence of the cavitation state and initial bacterial concentration on sterilization effect. The main conclusions can be draw as follows,

(1) As pressure ratio decreases, flow ratio reaches a maximum value and cavitation gets severer, the cavitation in JPCR reaches the limited cavitation stage with massive cavitation bubbles when flow ratio increases to the maximum value. And the severe cavitation is the applied sterilization conditions;

(2) The cavitation of JPCR has satisfying effect on *e. coli*, but there exists hysteresis effect. *E. coli* cannot be killed immediately unless damage accumulates to a certain degree for long time;

(3) The sterilization effect can be improved by adjusting pressure ratio to make the cavitation state switch from stable limited cavitation stage to unstable limited cavitation stage. The cavitation effects in the throat is stronger than that in the diffusor. Moreover, with the increase of concentration of initial bacteria, the sterilization rate decreases.

Acknowledgements

This work was financially supported by the National Natural Science Foundation of China (Project Nos. 11472197, 51679169 and 51576143).

References

[1] Pan, S., Peng, X. (2013). *Physical Mechanism of Cavitation*. National Defence Industry Press.

[2] Korkut, E., Atlar, M. (2012). *An experimental investigation of the effect of foul release coating application on performance, noise and cavitation characteristics of marine propellers*. Ocean Engineering, 41(1).

[3] Wittekind, D., Schuster, M., (2016). *Propeller cavitation noise and background noise in the sea*. Ocean Engineering. 120.

[4] Shi, W., Atlar, M., Rosli, R., et al., (2016). *Cavitation observations and noise measurements of horizontal axis tidal turbines with biomimetic blade leading-edge designs*. Ocean Engineering. 121.

[5] Suslick, K.S., (1989). *The Chemical Effects of Ultrasound.* Scientific American. 260.

[6] Mason, T.J., Lorimer, J.P., (2002). *Applied Sonochemistry: Uses of Power Ultrasound in Chemistry and Processing.*

[7] Badve, M., Gogate, P., Pandit, A.B., et al., (2013). *Hydrodynamic cavitation as a novel approach for wastewater treatment in wood finishing industry*, Separation & Purification Technology, 106(6).

[8] Gogate, P.R., Kabadi, A.M., (2009). *A review of applications of cavitation in biochemical engineering/biotechnology*. Biochemical Engineering Journal. 44(1).

[9] Gogate, P.R. (2010). *Hydrodynamic Cavitation for Food and Water Processing*. Food and Bioprocess Technology, 4(6).

[10] Save, S.S., Pandit, A.B., Joshi,J.B., (1997) *Use of Hydrodynamic Cavitation for Large Scale Microbial Cell Disruption*. Food and Bioproducts Processing. 75(1).

[11] Jyoti, K.K., Pandit, A.B., (2003). *Hybrid cavitation methods for water disinfection*. Ultrasonics sonochemistry. 14(1).

[12] Balasundaram, B., Harrison, S.T., (2006). *Study of physical and biological factors involved in the disruption of E. coli by hydrodynamic cavitation*. Biotechnology progress. 22(3).

[13] Sawant, S.S., Anil, A.C., Krishnamurthy, V., et al., (2008). *Effect of hydrodynamic cavitation on zooplankton:A tool for disinfection*. Biochemical Engineering Journal, 42(3).

© 2018 ASME

The effect of gas content on cavitation shedding and test facility dynamics

[1,2] Ingrid K. Vilberg *; [1] Morten Kjeldsen; [3] Roger E.A. Arndt; [2] Torbjørn K. Nielsen

[1]Flow Design Bureau AS, Stavanger, Norway; [2]Norwegian University of Science and Technology, Trondheim, Norway, [3]University of Minnesota, Minneapolis, MN, USA

Abstract

This study investigates experimentally the coupling between cavitation shedding, considered as the source of tunnel dynamics, and the system response for various degrees of tunnel gas content. The experimental work took place in the high-speed water tunnel at the St. Anthony Falls Laboratory at the University of Minnesota. The test object was a c=81 mm NACA 0015 hydrofoil spanning the whole test section, 190mm by 190mm, at speeds around 10 m/s. Gas content was controlled by either comprehensive degassing, by allowing different gas saturations and eventually by continuously injecting air directly into the flow downstream of the test object. Most of the gas was removed by a gas collector, unique to the facility, upstream the test section. By combining simultaneous readings of pressure dynamics at three locations in the tunnel loop, together with measured shedding dynamic loads in parallel with high frequency impacts using the same externally mounted accelerometers, the research team aimed at concluding the effect of gas content on several cavitation characteristics. The results presented, and upon which discussion and conclusions are made, include excitation strength as a function of test section gas saturation level and the frequency response between the various pressure transducers and observed shedding strength.

Keywords: cavitation; hydrofoil; dynamics; cavitation tunnels

Introduction

The main benefit of using laboratory scale cavitation experiments is the relative ease of observing and documenting the underlying physics, and the ability of controlling all factors affecting the same physics. The International Towing Tank Conference (ITTC) demonstrated the lack of complete control for cavitation inception experiments, and observations on the same test-body at several facilities showed a wide scatter in results [1], [2]. The culprit of the observed discrepancy is often taken to be water quality, and most notably the tensile strength of the water in the test facility [3].

In 1997 one of the authors (MK) joined professor Arndt at St. Anthony Falls Laboratory (SAFL) at the University of Minnesota (UMN) to study the global dynamics of sheet and cloud cavitation, and the main findings were reported in Kjeldsen et al [4] and Kjeldsen and Arndt [5]. Based on observations in those studies the question was raised about the influence of the test facility on such studies. Following the SAFL studies and addressing the influence of test facility, Arndt repeated with co-workers many of the SAFL studies both at the Technical University of Munich /Obernach and University of Osaka, see Arndt [6] for a summary. Qualitatively the dynamics of cavitation shedding appears similar, as would be expected from the underlying cavitation physics, but with differences in measured unsteady lift amplitudes. Such differences can be attributed the response of the test facility, which is a line of investigation that has been addressed by e.g. Duttweiler et al. [7] and Kjeldsen et al. [8]. Other inspiration for the current work, but not necessarily including the role of system dynamics, stem from research on the role of water quality and gas content on cavitation intensity [9], as well as investigations into the shedding similarity between gaseous and natural cavitation [10].This study tries to bring these efforts to a conclusion by modifying the one parameter that easily can be changed in the test facility: the sonic speed. This was accomplished by injecting gas downstream the cavitating hydrofoil, and thus modifying acoustic properties within most of the tunnel loop. The experimental program took place in 2015 at the SAFL high speed water tunnel.

The interaction between cavitation and general system dynamics is not limited to laboratory testing. Large scale flow systems, such as hydropower plants, experience a wide variety of dynamic oscillations that can affect and interact with local cavitation phenomena and possibly induce extra loads on structures.

Experimental test facility

An overview of the high-speed water tunnel at SAFL is shown in Figure 1. The same figure illustrates the position of the absolute pressure transducers used to capture facility loop dynamics. The permanent instrumentation of the tunnel provided pressure and velocity measurements. Extracting water at the elbow downstream the test section and through a DO meter vessel, established the dissolved oxygen content. The air injection port, positioned shortly downstream the foil, allowed for continuous injection controlled and measured by an airflow controller. All experiments use the NACA 0015 hydrofoil, c=81mm and span = 190mm, built in brass and polished. The foil being a cantilever construction attached to a rigid support base. On the latter three accelerometers (BK 4397) are mounted as shown in Figure 2, all being able to sense both low frequency lift and drag oscillations and high frequency cavitation intensity.

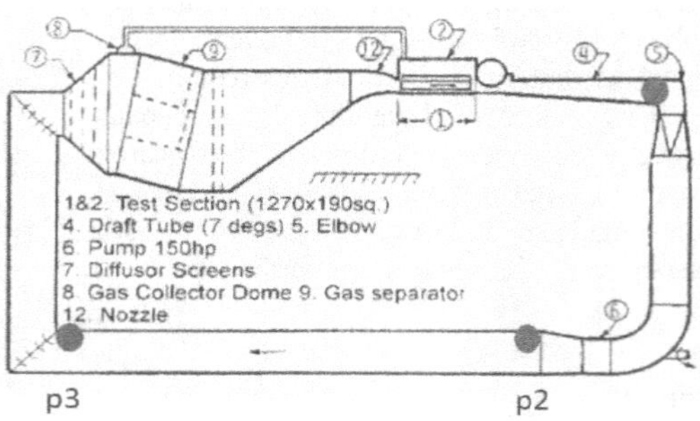

Figure 1: The cavitation tunnel at SAFL with indication for locations of dynamic pressure transducers.

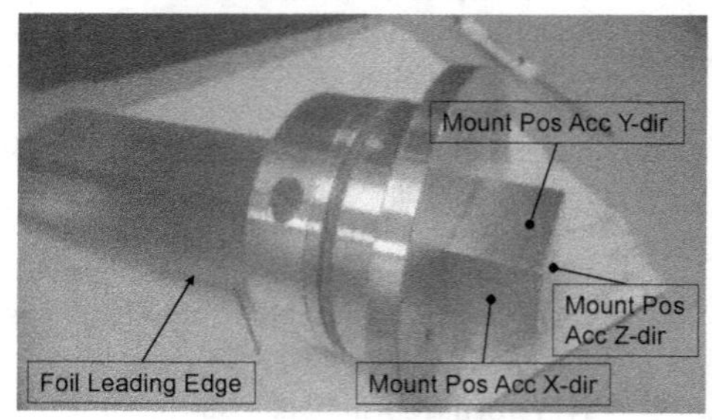

Figure 2: Test object, NACA0015, mounted on supporting base and with indications for location of high-frequency accelerometers.

Experimental observations

The measurements cited herein did not include direct measurements of lift. Instead the accelerometers attached to the base provide this information. There will be a frequency dependent transfer function associated to the determination of oscillating shedding loads using the accelerometers, but the experiment does not include a controlled frequency response experiment. Figure 3 compares the current experiment with results from a similar experiment carried out at Obernach in 1999, where the lift was measured directly. The obvious difference is the apparent higher amplitudes of oscillations for higher values of $\sigma/2\alpha$ when measured by the accelerometers (SAFL 2015), than by direct dynamic lift (Obernach 1999). The latter reading corresponds however qualitatively with other methods used for quantifying lift dynamics at SAFL.

Given the intention of the current study, focus is given the $\sigma/2\alpha$ range between 5.5 and 6.5, and for investigating the medium to high frequency range of 50-90 Hz. All SAFL 2015 studies are made at 8 degrees angle of attack. In addition, the $\sigma/2\alpha$ range was explored using velocities, U, in the range 7-10.5 m/s, and four different tunnel water qualities: 1) Degassed water with a low measured dissolved oxygen content, 2) high measured dissolved oxygen content, 3) continuous air injection of 10 slpm (standard liters per minute) and 4) continuous air injection of 20 slpm. The air injection point was at the tunnel test section wall approximately two chord lengths downstream the foil trailing edge.

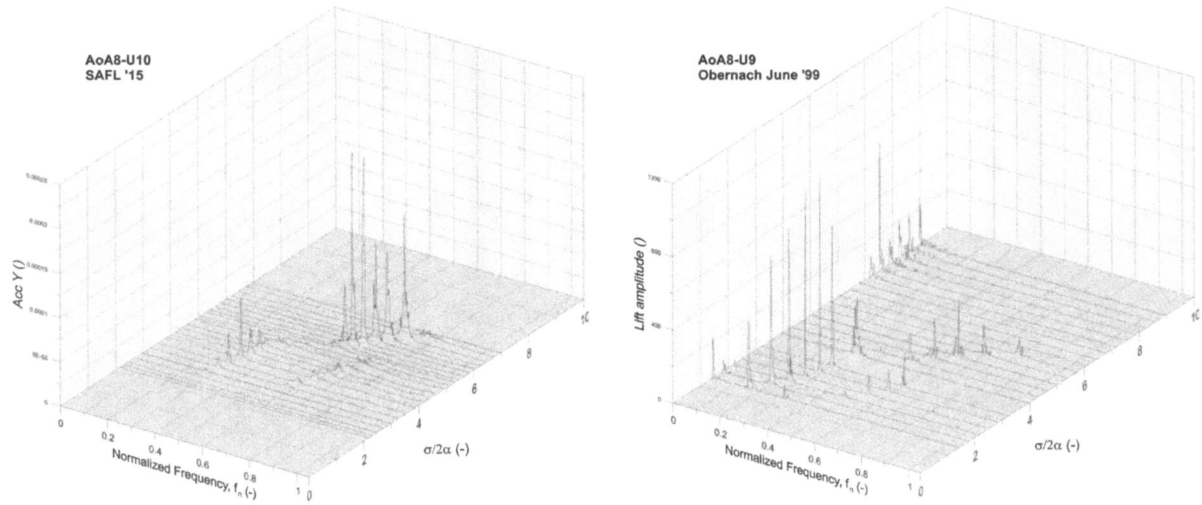

Figure 3: Comparison between SAFL2015 (left) and Obernach1999 (right) experiments, using accelerometers and direct dynamic load cells respectively, for the determination of lift oscillations on a cavitating NACA0015 hydrofoil.

Determination of excitation loads

One of the main objectives of the current study relates to the interaction between cavitation shedding physics and the systems. From the system perspective, the cavitation shedding defines the excitation and is characterized by its amplitude and frequency. Figure 4 shows the observed amplitudes and frequencies captured by the lift-direction mounted accelerometer. It appears that the normalized amplitude can be expressed as a function of frequency alone, both for different test section velocities and tunnel water qualities. The amplitude level has been calculated as the sum of spectral components, $A = \sum_{i=n}^{m} a_i$ in a manually determined frequency range $\langle f_n, f_m \rangle$ containing the cavitation shedding information. The shedding frequency is now calculated as $f_{cen} = A^{-1} \sum_{i=n}^{m} f_i \cdot a_i$.

Characteristics of excitation dynamics

Kjeldsen and Arndt describe the shedding frequency using the composite parameter *fl/U* [4]. Here *l* is the observed maximum cavity length and is a function of σ/2α. Figure 5 shows the results from a visual determination of cavity length. A linear relation between cavity length is said to exist, and by using a type of expression as shown by Eq.1. Furthermore, *fl/U* is said to be constant within given σ/2α ranges. This allows the estimation of the *l/c* relationship based on measured shedding frequencies using e.g. the method of least squares on Eq.2 with data as given in Figure 4. Table 1 shows the results of this exercise, and Figure 5 shows the resulting *l/c* estimates compared with visually observed *l/c* data. The visual observation made my one of the authors (MK) in 1998.

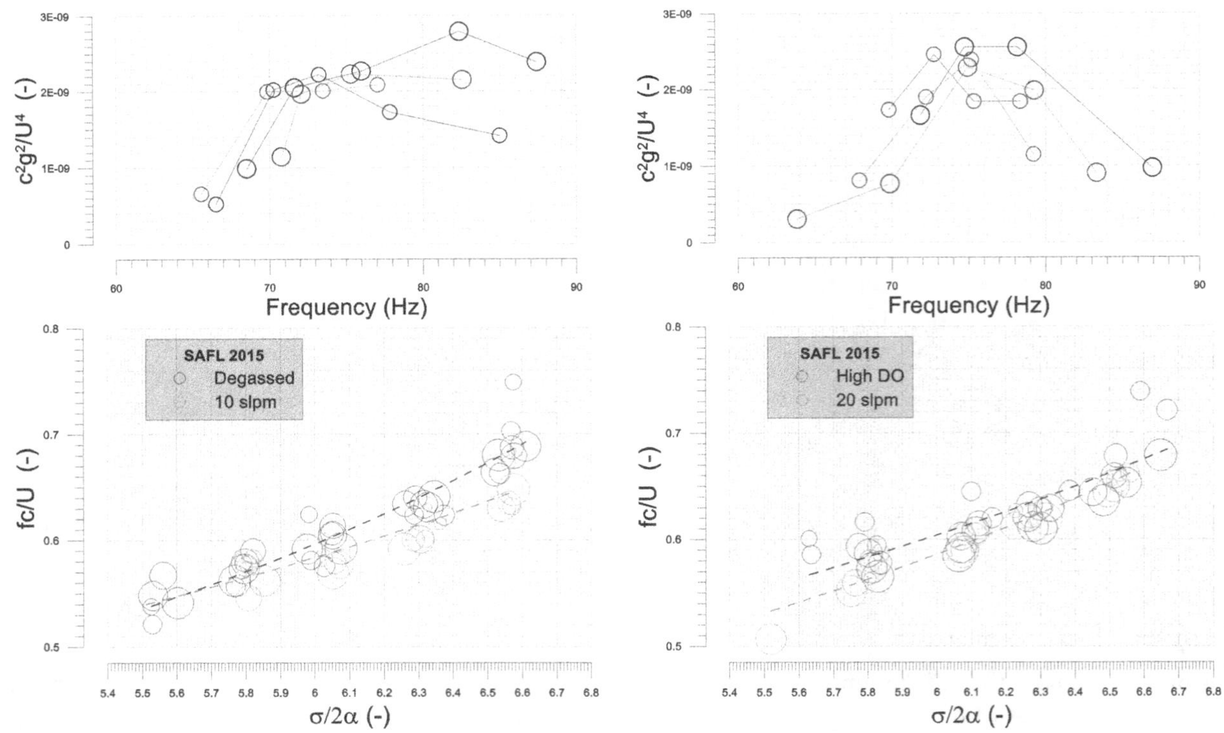

Figure 4: Cavitation shedding amplitudes (above) and frequencies (below). Amplitudes shown for two speeds, U=10 and 10.5 m/s the latter using slightly larger symbols, while the size of the symbols indicate velocities from 7 to 10.5m/s using increasingly sized symbols. Dashed line indicates best estimates for *fl/U=constant*.

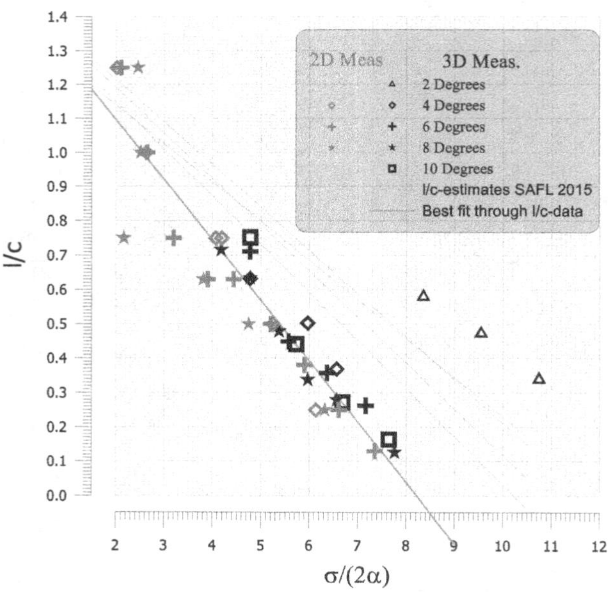

Figure 5: Maximum cavity extension length on a NACA0015 hydrofoil.

$$\frac{l}{c} = C_1 - C_2 \frac{\sigma}{2\alpha} \quad \text{Eq. 1}$$

$$\frac{fl}{U} = C_0 \Leftrightarrow \frac{fc}{U} = C_0 \left(C_1 - C_2 \frac{\sigma}{2\alpha} \right)^{-1} \quad \text{Eq.2}$$

Quality	C_0	C_1	C_2
Degassed	0.367	1.429	0.1396
High DO	0.431	1.466	0.1253
10 slpm	0.400	1.367	0.1137
20 slpm	0.383	1.459	0.1341

Table 1: Estimates of cavity length relation based on observed shedding dynamics.

Observed system dynamics

Figure 6 shows cross-power spectra that indicate that the frequency content associated with shedding is present at the position of the dynamic pressure sensors. The cross-power spectra originate from the same set of data used for the spectrum of Figure 3. These plots reveal a more pronounced correlation at higher values of $\sigma/2\alpha$ and hence for higher frequencies. Based on this observation the SAFL 2015 study focuses on the cavitation shedding dynamics in the $\sigma/2\alpha$ range 5.5 - 6.5. The absolute frequency levels also allow more than one whole acoustic wave to be contained in the tunnel: $f=a/\lambda$ gives f=33 and 40 Hz for a=1000 and 1200 m/s, while assuming the wave length (λ) to be similar the tunnel loop length of approximately 30 m.

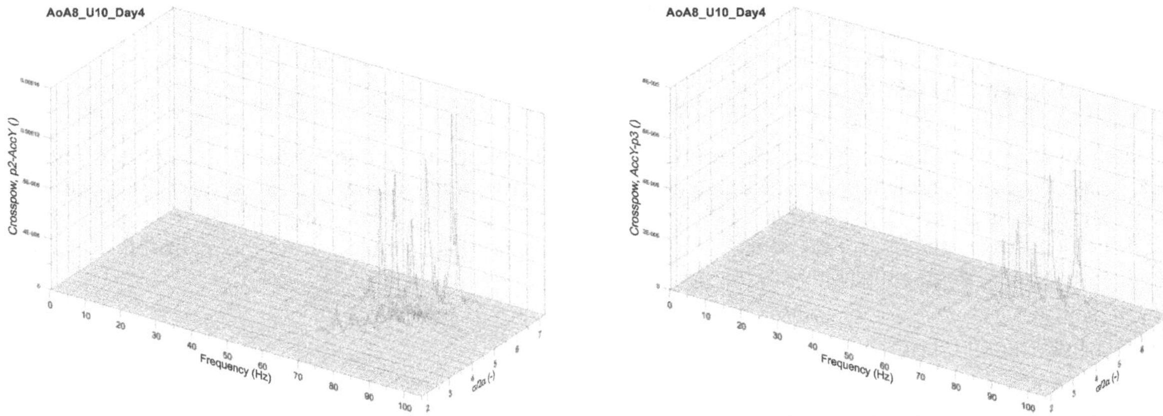

Figure 6: Cross power spectrum between lift-direction accelerometer and the two sensors at the lower leg of the tunnel (p2 and p3).

System modeling

The effect of the tunnel system dynamics on cavitation was evaluated using the time-domain solver, FloMASTER from Mentor Graphics. The software solves 1D transient flow problems using the Method of Characteristics (MOC). A time step of dt=10^{-4} seconds was used for all calculations, allowing to resolve the smaller pipe segments, L=0.6 m, using MOC. Figure 7 shows the FloMASTER network used to model the SAFL cavitation tunnel. In addition to system layout, the required input includes sonic speed in the different pipe segments and modeling of the cavitation shedding excitation. These inputs are described more detailed in the following section.

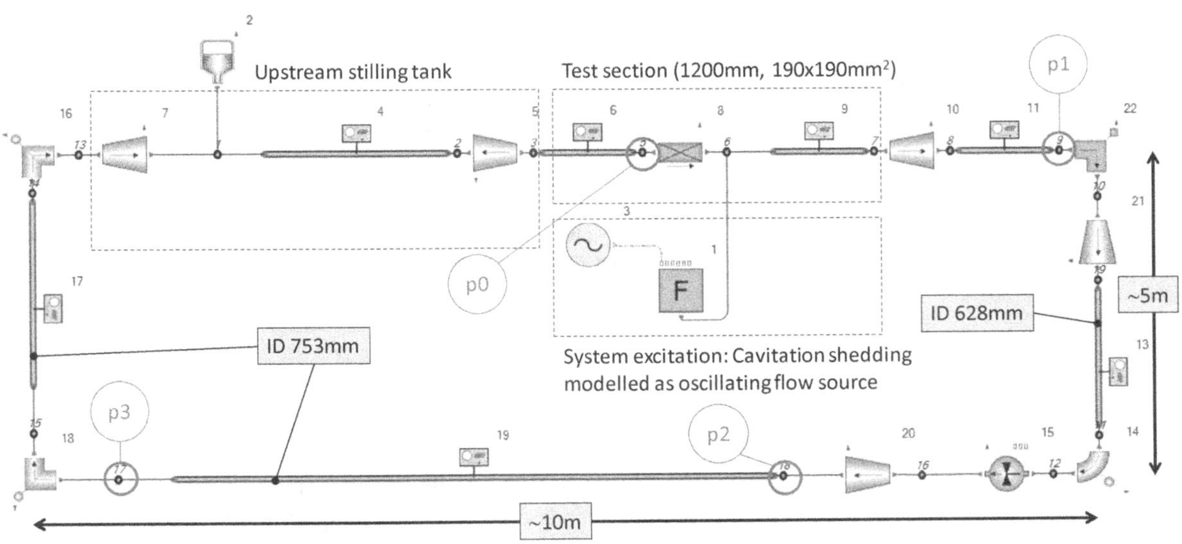

Figure 7: SAFL cavitation tunnel modelled in the software FloMASTER from Mentor Graphics. Pressures measured experimentally at locations p1, p2 and p3. Results from calculations include pressure response at position upstream the hydrofoil (p0).

Hydrofoil-dynamics

The hydrofoil cavitation dynamics can, in a first approach, be described using continuity and momentum equations as stated below.

1. Continuity over the hydrofoil: $U_2 A_{CS} = U_1 A_{CS} - \frac{dV_{Cav}}{dt}$, where the cavity volume (V_{Cav}) is a function of σ/2α. A_{CS} is the cross-sectional area of the test section.
2. Momentum: Pressure loss over the element: $\Delta(p_1 - p_2)\cdot A \approx \frac{\partial F}{\partial C_D}\cdot \Delta C_D + \frac{\partial F}{\partial U_1}\cdot \Delta U_1$, i.e. the pressure difference is dependent on the drag coefficient C_D, which is a function of σ/2α, and the speed of flow upstream the hydrofoil.

For system analysis, the cavitation shedding affects the cavity volume. In the system modeling this effect is included as an oscillating flow source, element 1. The drag coefficient, as given in the momentum equation, is included as a stationary resistance in element 8. The effect of the system is now determined by investigating the response in the pressure directly upstream the hydrofoil, node 5. A variation in this pressure should impact the hydrofoil cavitation shedding, and hence the degree of steady-oscillatory behavior.

Sonic speed of gas/liquid mixture pipes

The sonic speed of gas-water mixture can be calculated using Eq.3. Figure 8 shows the dependency of the mixture sonic speed on the gas void fraction (γ). For the flow loop the gas void fraction within the different pipes is defined as: $\gamma = Q_{Gas}/Q_{Tot}$. The gas volumetric flow rate also depends on the absolute pressures of the pipe, that is $Q_{Gas,Pipe} = Q_{Gas,0}(p_0/p_{pipe})$. For the current analysis $Q_{Gas,0}$ refers to the measured air injection flow rates which is taken at $p_0 = p_{atm}$. The pipe pressure is taken as the average between inlet and outlet. Table 2 gives an indication in the variation in sonic speeds for the lower leg as a function of operating conditions (U and $\sigma/2\alpha$).

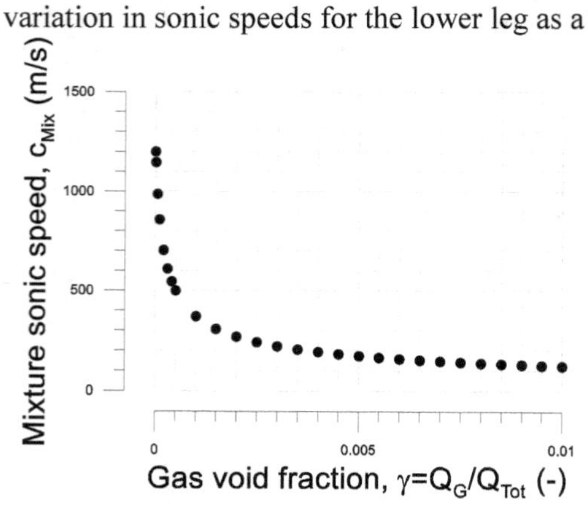

Figure 8: Mixture sonic speed.

$$\frac{1}{c_{mix}^2} \approx \frac{(1-\gamma)^2}{c_l^2} + \frac{\gamma^2}{c_G^2} + \gamma(1-\gamma)\frac{\rho_L}{\rho_G c_G^2} \qquad \text{Eq.3}$$

U , σ/2α	10slpm	20slpm
10.5, 5.5	667	512
10.5, 6.5	685	529
7, 5.5	465	342
7.5,6.5	477	351

Table 2: Calculated sonic speeds, c_{Mix}, in lower leg pipe for various air injection rates ($Q_{Gas,0}$) and operating conditions, c_L=1200 m/s

Results from modeling

Figure 9 shows the results for the pressure frequency response on the oscillating flow source. The latter is used to mimic the cavitation shedding action. The modeling has been used for one flow speed series, i.e. that of 10.5 m/s, and with various sonic speeds. Also included is the expected sonic speeds when injecting air at 20 slpm. Figure 9 also includes the measured pressure gain using the shedding load as the expected excitation. The calculations also include the pressure gain at the position of the hydrofoil, denoted p0.

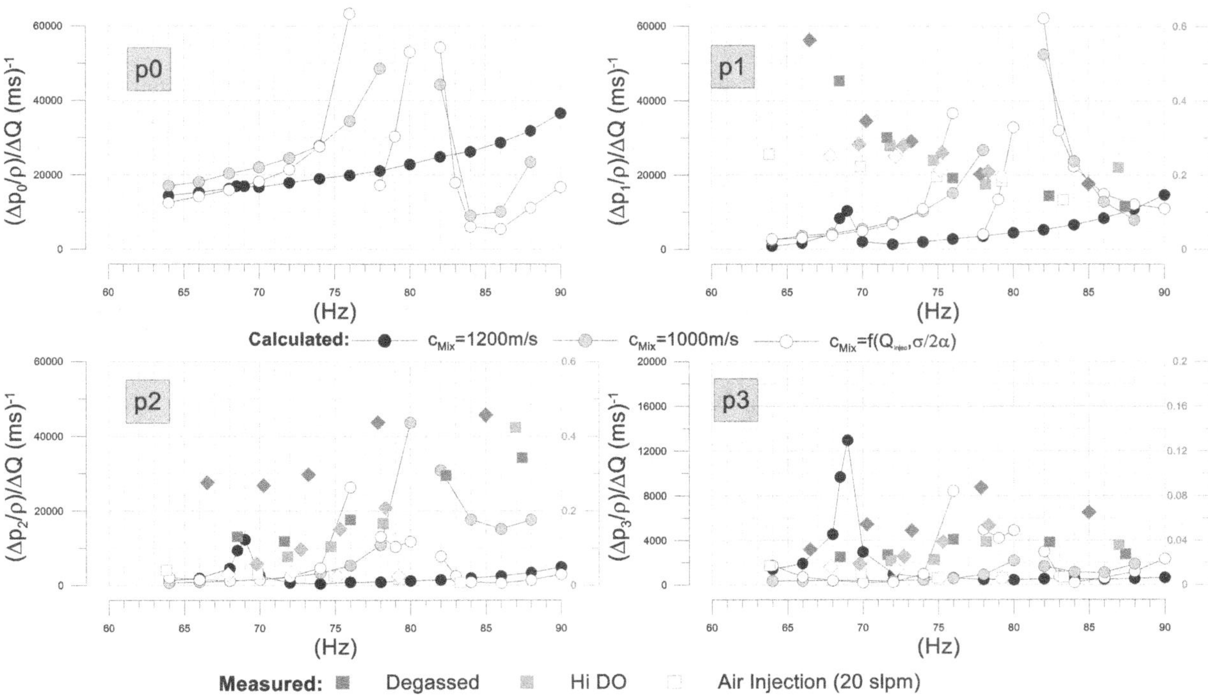

Figure 9: Calculated frequency response using the flow source as excitation, and measured pressure-dynamics gain relative to measured shedding load at shedding frequency. Calculations made for 10.5m/s and experiments presented for a test section velocity of 10.5 m/s (squares), and 10 m/s (diamonds)

Discussion

No direct modeling of the tunnel sonic speed exists, and thus the actual sonic speeds used during calculations remain uncertain. The use of continuous air injection should on the other hand affect the tunnel sonic speed to such a degree that any difference in facility dynamics due to sonic speed should be detectable.

Figure 9 compares calculated and measured response. The calculated response shows the presence of resonance peaks in the frequency range investigated. The measured response appears consistent, except for the offset in amplitude response for the two tunnel velocities at the p2 location and for the degassed tunnel quality. The observed response does not show similar tendencies of resonance conditions as the calculated response. The difference between calculated and measured or observed frequency response can be attributed to:

- A system amplified pressure oscillation at the position of the hydrofoil prevents a steady oscillatory condition since this requires $\sigma/2\alpha$, and hence the pressure, to be constant. Therefore, the true response for a given frequency will not be observed.
- For various reasons, free air can be trapped along the pipe loop. This will in effect lead to a multi-accumulator effect significantly altering system dynamics.
- As already mentioned the frequency dependency of the signal path between the hydrodynamic loads on the hydrofoil and the acceleration at the base is unknown, i.e. there is a chance that this unknown can obscure the readings in the frequency range tested.

While observing the frequency response for continuous air injection the dynamic response at the p2 and p3 locations is highly attenuated, while constant for the frequency range investigated in the p1 location. This suggests that the system acoustics is close to non-existing when injecting air into the tunnel. This observation supports the trapped gas argument.

Conclusion

The work presented suggests that cavitation shedding dynamics, quantified by shedding amplitudes and frequencies, is independent of tunnel water quality. The observed excitation loads as presented in Figure 4, support this conclusion. This study also indicates the presence of a systems response, both observed in Figure 6 and modelled and presented in Figure 9. Although the apparent negligible impact on excitation, the system response needs to be controlled and/or observed.

It is observed that continuous air injection attenuate system dynamics. At this point it must be emphasized the unique and efficient gas removal capabilities in the stilling tank of the SAFL water tunnel. I.e. the excitation, or hydrofoil cavitation shedding, is not affected by free air/ bubbles injected into the tunnel downstream the hydrofoil. Still, a conclusion from this work is the need for control and observation of pressure variations in the test section, this to cover the p0 dynamics depicted in Figure 9.

From a modeling perspective, an improvement can be made by modeling the interaction between the system and the hydrofoil cavitation shedding. For the case presented herein it can be achieved by coupling the transient 1D calculation with the 3D calculation of the hydrofoil cavitation. Also from an engineering perspective and for industrial application the interaction between the flow system and cavitation shedding interaction needs to be considered to avoid unfortunate combinations.

References

[1] Lindgren, H. and Johnsson, C. A. (1966) "*Cavitation inception on headforms, ITTC comparitive experiments*," Proceedings of the 11th Towing Tank Conference, Tokyo, pp. 219-232

[2] Johnsson, C. A. (1969) "*Cavitation inception on headforms, further tests*," Proc. 12 Int. Towing Tank Conf. Rome, pp. 381-392

[3] Keller, A. P. (1974) "*Investigations concerning scale effects of the inception of cavitation*," Proc. I. Mech. E. Conf. on Cavitation, pp. 109-117

[4] Kjeldsen, M., Arndt, R. E. A., and Effertz, M. (2000) "*Spectral Characteristics of Sheet/Cloud Cavitation*," Journal of Fluids Engineering, vol. 122, no. 3, pp. 481-487

[5] Kjeldsen, M. and Arndt, R. E. A. (2001) "*Joint Time Frequency Analysis Techniques: A study of Transitional Dynamics in Sheet/Cloud Cavitation* " in Fourth International Symposium on Cavitation (CAV 2001), California Institute of Technology, Pasadena, USA.

[6] Arndt, R. E. A. (2012) "*Cavitation research from an international perspective*," IOP Conference Series: Earth and Environmental Science, vol. 15, no. 1,

[7] Duttweiler, M. E., Schell, S. E., and Brennen, C. E. (2000) "*The effect of experimental facility dynamics on a cavitation instability*," in ASME 2000 Fluids Engineering Division Summer Meeting, Boston, Massachusetts, USA.

[8] Kjeldsen, M., Vennatrø, R., Arndt, R. E. A., and Keller, A. P. (1999) "*Discussion on cyclic cavitation in closed water tunnels and the influence from the dynamic response of the tunnel*," in IAHR Work group on the behavior of hydraulic machinery under steady oscillatory conditions, Brno, Czeck Republic.

[9] Ekanger, J. V., Kjeldsen, M., Escaler, X., Kawakami, E., and Arndt, R. E. A. (2012) "*Cavitation Intensity Measured on a NACA 0015 Hydrofoil with Various Gas Contents*," in 8th International Symposium of Cavitation, Singapore.

[10] Kjeldsen, M. and Arndt, R. E. A. (2009) "*Blade load dynamics in cavitating and two phase flows*," in 7th International Symposium on Cavitation, Ann Arbor, Michigan, USA.

© 2018 ASME

The effects of air content on unsteady tip vortex cavitation

Xiaoxing Peng, Lianghao Xu, Yuwen Liu, Yantao Cao

National Key Laboratory on Ship Vibration & Noise, China Ship Scientific Research Center, Wuxi, 214082, China

Abstract.
The phenomena of tip vortex cavitation (TVC) in unsteady incoming flows were studied experimentally in this paper. An elliptic hydrofoil with section NACA 66_2-415 was adopted as test model and high-speed video camera was employed to observe the tip vortex cavitation inception in the fixed foil with oscillation incoming flows. The effect of air content was considered in this investigation. The results show that the cavitation inception is delayed by the decrease of the non-dimensional frequency. The effect of air content in water is still very important in the case of unsteady tip vortex cavitation. The results indicate that the unsteady effect should be further considered in the scale effect study of TVC inception.
Keywords: tip vortex, cavitation, unsteady, experiment

Introduction

For the designer of naval and merchant ships cavitation is always a major concern because it induces hydrodynamic noise and vibration of marine propellers, as well as the potential of cavitation erosion. Among various kinds of cavitation tip vortex cavitation (TVC) begins in tip vortices and appears at lower ship speed than other forms of cavitation in most cases. The prediction of tip vortex cavitation inception is very important to develop high performance and quiet propellers. In practice marine propellers generally operate under unsteady conditions. The propeller blade always experiences varying pressure and velocity conditions due to the revolution of propeller, the wake of the hull and the heave movement of ship. Therefore, it is important to know the influence of unsteadiness on the inception and the development of cavitation.

Although tip vortex cavitation was extensively studied for many years, but most of the works were for tip vortex cavitation for steady situation. A very few researches related to tip vortex cavitation under unsteady conditions could be found. Hart et al. (1991[1], 1992[2]) and McKenney and Hart (1993[3]) studied a oscillating hydrofoil of rectangular planform with NACA64A309 cross section experimentally and showed an very complicated vortex structures due to the interaction between the tip vortex and the spanwise vortices shed from the trailing edge in the case of oscillating hydrofoil. Boulon, Franc and Michel (1997[4]) conducted experiments on a oscillating hydrofoil with symmetrical NACA 16-020 cross-section and an elliptical planform to study the influence of nuclei content and oscillation frequency on the inception of tip vortex cavitation. They found that the inception of tip vortex cavitation on an oscillating hydrofoil strongly depends upon the oscillation frequency and the nuclei concentration. Difference from the previous work in this paper another unsteady condition, fixed foil with oscillation incoming flows, is studied experimentally. The focus will be on the TVC inception under the effect of unsteady incoming flow and the influence of air content of water.

Experimental setup

Experiments were conducted in cavitation mechanism tunnel of China Ship Scientific Research Center (CSSRC). The tunnel has two exchanged test sections of circular and square respectively. There is a large degassing tank installed in the downstream of test section to control the air content of water and the tunnel also equipped a nuclei seeding system. In present study the test section with a square cross section of 225×225 mm was used that the maximum incoming velocity of this section can be up to 25m/s.

[1] Correspondence author: Dr. Xiaoxing Peng, henrypxx@163.com.
* The work supported by National Natural Science Foundation of China (No. 11332009 & No. 11772305)

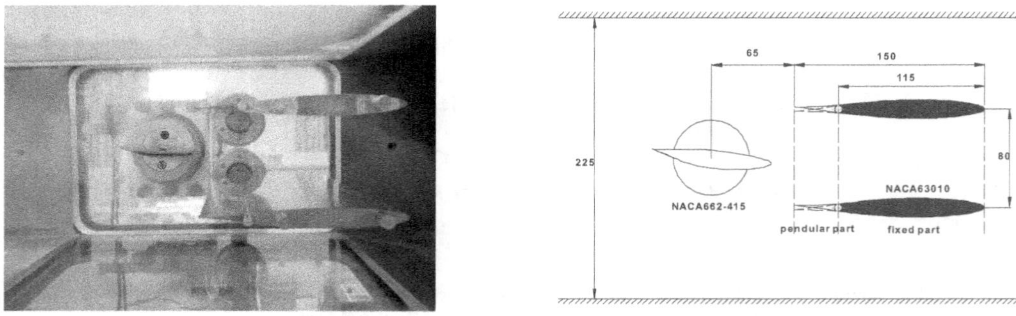

Figure 1. The arrangement of test model and swing system.

Test model is an elliptic hydrofoil with cross section NACA 66_2-415. The model is 94.2mm and 112.5mm in maximum chord and half span respectively and installed in the horizontal center of test section with the tip in the centreline of test section. The model was supported by a device which could adjust the attack angle of the foil. The tunnel installs an oscillating system in the front of test model to produce the oscillated incoming flows. The device has two hydrofoils installed in the tunnel, each hydrofoil includes a swing foil in its trailing part, as shown in Fig.1. A crank-link mechanism is set outside the tunnel and drives the two swing foils oscillating synchronously. A high-speed video camera of type of Photron APX and a LED lamp with 3000fps was used to observe the tip vortex cavitation inception and visualize the development of tip vortex cavitation. Air content of water was measured by by dissolved oxygen meter and expressed as *DO*, which is the relative dissolved oxygen to the saturation case.

Usually there are two ways to define the TVC inception criteria. One is the occurrence or disappearing of intermittent cavitation along the vortex path. The other is the occurrence of TVC core attached or detached to the tip of the foil. In present test TVC inception was distinguished by observation the images obtained by high speed video camera. In one oscillation circle TVC often experiences stretch out and draw back from the foil tip so it hardly to tell the cavitation number of attaching or detaching. Therefore, the TVC inception was defined as disappearing of intermittent cavitation along the vortex path in this paper.

Experimental results

In the experiments test model was fixed with 7°attack angle and the incoming flow oscillated with a swing mechanism in front of test foil. During the tests the oscillation amplitudes of swing foils were $\Delta\alpha$= 5.7°, 4.3°and 3.2°respectively, the oscillation frequencies were f=1.8, 3.6, 5.3 and 7.1Hz, mean incoming velocities were V= 5, 10 and 15m/s, air contents were DO=0.30, 0.62, and 1.01.

To represent the effect of oscillating incoming flows a non-dimensional frequency f^*, similar to the reference [4], was adopted to reduce the results, defined as:

$$f^* = \frac{fc}{V}$$

where f is oscillation frequency, V mean incoming velocity and c maximum chord of test foil.

1.1. Effect of Renolds number

The effect of Reynolds number on tip vortex cavitation inception was investigated by varying the mean incoming flow velocity. A series of cavitation inception tests were carried out in the air content DO=1.01 and oscillating amplitude $\Delta\alpha$= 5.7° by changing the incoming velocity and oscillation frequency. The relation between incipient cavitation number and non-dimensional frequency was shown in Fig.2. The results demonstrate the cavitation inception is delay by the decrease of the non-dimensional frequency f^*. For each Reynolds number the TVC inception tends to become independent of the oscillation frequency at higher frequencies. But it should be pointed out that the incipient cavitation numbers in steady incoming flows (f=0) were 1.99, 2.97 and 7.16 for V= 15m/s ,10m/s and 5m/s respectively, which were much higher than these in the unsteady case, especially in lower Reynolds number. It indicates that the unsteady incoming flows will delay the cavitation inception.

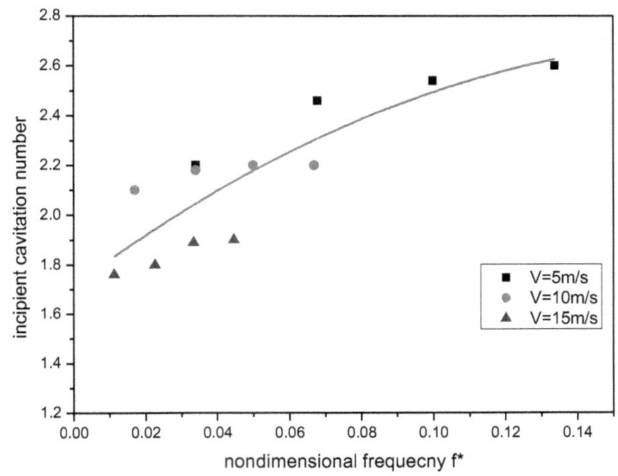

Figure 2. Incipient cavitation number vs non-dimensional frequency for different Reynolds number ($\Delta\alpha$ = 5.7°, *DO*=1.01)

1.2. Effect of oscillation amplitude

The test was conducted in the condition of mean incoming velocity V=10m/s, and air content DO=0.30 with three oscillating amplitudes, $\Delta\alpha$= 5.7°, 4.3°and 3.2°. The result is showed Fig. 3. It clearly demonstrates that the cavitation inception is delayed when the oscillating amplitude of incoming is increase. Also for each oscillation amplitude the TVC inception number tends to become independent of the oscillation frequency at higher frequencies.

1.3. Effect of air content

Three air contents, DO=0.30, 0.62, and 1.01, were tested in the condition of mean incoming velocity V=10m/s, and oscillating amplitude, $\Delta\alpha$= 5.7°. The result is shown in Fig. 4. It clearly shows the similar tendency that the cavitation inception is delayed with the decrease of the oscillation frequency for different air content of water. But the cavitation inception is heavily delayed in the degassing water where the air content is much lower.

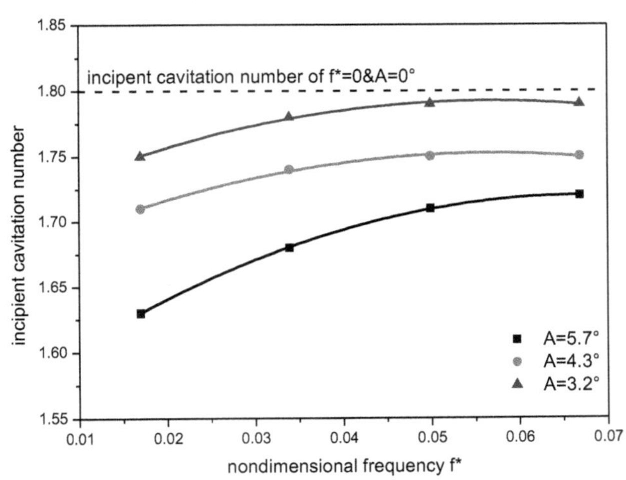

Figure 3. Incipient cavitation number vs non-dimensional frequency for different oscillation amplitudes

(V=10m/s, DO=0.30)

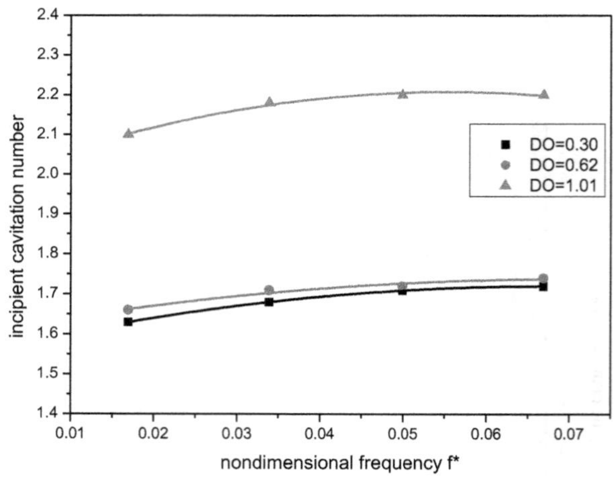

Figure 4. Incipient cavitation number vs non-dimensional frequency for different air contents ($\Delta\alpha$ = 5.7° , V= 10 m/s)

Conclusions

The tip vortex cavitation inception in unsteady incoming flows, formed by the swing foils in front of test model, was investigated experimentally in this paper. The effects of oscillation frequency, Reynolds number, oscillation amplitude, and air content of water were studied. The results show that the cavitation inception is delay by the decrease of the non-dimensional frequency f^*. But this delay tends to become independent of the oscillation frequency at higher frequencies. The decrease of air content and increase of oscillation amplitude will delay TVC inception.

The experimental results also show that the oscillating incoming flow will heavily delay the TVC inception compared to the steady case. It indicated that the unsteady effect should be further considered in the scale effect study of TVC inception.

Acknowledgments

Authors wish to acknowledge the support from the National Natural Science Foundation of China (Project No. 11332009 & 11772305).

References

[1] Hart D. P., 1991, "Cavitation Inception in the Tip Vortex Region of an Oscillating Hydrofoil," 1991 ASME Cavitation and Multiphase Flow Forum, FED-Vol, 109, pp. 35-41.

[2] Hart D. P., Acosta A., and Leonard A., 1992, "Observation of Cavitation and Wake Structure of Unsteady Tip Vortex Flows," International STG Symposium on Propulsors and Cavitation, June 22-24, Hamburg, Germany.

[3] McKenney E. A. and Hart D. P., 1993, "Experimental Determination of Bound Circulation and Shed Vorticity Induced by an Oscillating Hydrofoil," 1993 ASME Cavitation and Multiphase Flow Forum, FED-Vol. 153, pp. 87-91

[4] Boulon O., Franc J. P. and Michel J. M., "Tip Vortex Cavitation on an Oscillating Hydrofoil", Transactions of the ASME, Journal of Fluids Engineering, Vol. 119, pp. 752-758

© 2018 ASME

The Growth and Collapse of a Bubble between Parallel Flat Free Surfaces

[1]Toshiyuki Ogasawara*; [1]Seisuke Ito; [1]Hiroyuki Takahira

[1]Osaka Prefecture University, 1-1 Gakuen-cho, Naka-ku, Sakai-shi, Osaka, JAPAN

Abstract

The growth and collapse of a bubble between two parallel flat free surfaces and the corresponding water column formation on the free surfaces have been investigated experimentally and numerically. In the experiment, a laser-induced bubble generated inside a plane water jet and the dynamical motion of free surfaces were observed with a high-speed video camera. The results are characterized by two dimensionless parameters: the ratio of the maximum bubble radius to the water jet width, R_{max}^*, and the ratio of the initial bubble offset from the center line of the jet to the water jet width, ε^*. When a bubble is generated at the center ($\varepsilon^* = 0$), water columns are formed on both free surfaces during the growth phase of a bubble, following the formation of crown-like shaped water columns during the second growth and collapse of a bubble. Since the bubble does not translates, it grows and collapses at the center of the jet. However, a water column due to the bubble growth when $\varepsilon^* \neq 0$ tends to be formed only on the nearer free surface to the bubble during its growth phase. Then in the collapse phase the bubble translates toward the other free surface far from the bubble accompanying with a liquid-jet formation toward it. The liquid-jet and the rebound shockwave from the bubble cause the water column formation on both sides of free surfaces. Numerical simulations are conducted for the growth and collapse phases using the boundary element method (BEM) and for the collapse phase using the ghost fluid method (GFM). The water column formation during the growth phase, a toroidal bubble deformation during the collapse phase, and the translation of the collapsing bubble are well simulated by BEM. The result of GFM shows shockwave emission from the rebounding motion of the bubble which causes the water column formation when the shockwave impacts on the free surface with large curvature.

Keywords: bubble collapse; free surfaces; laser-induced bubble; boundary element method; ghost fluid method

Introduction

As a next generation nuclear spallation neutron source, liquid mercury jets are proposed as a novel target for various high-power nuclear and high-energy physics applications [1]. The cavitation bubbles within the jet which occur due to high-intensity pressure wave caused by the shock heating of the liquid-jet, however, raise target fragmentations and disturb a stable output of secondary particles. To predict the performance of those jets, the knowledge of the cavitation bubble behavior near free surfaces and/or inside those jets is necessary. The experiments and numerical analyses for the behavior of the bubble generated beneath a flat free surface were conducted by Chahine [2] and Blake and Gibson [3]. They showed that the bubble translates toward the direction away from the free surface during the bubble collapse phase. In addition, Chahine [2] found a limiting value above which the interaction between a bubble and a free surface is negligible. Afterwards, Robinson and Blake [4] and Tomita and Kodama [5] investigated the dynamics of single- and two-cavitation bubbles generated by a laser beam beneath a free surface. They concluded that the motion of two bubbles depends on the bubble size ratio and the distance between a bubble and a free surface. Zhang et al. [6] conducted the experiment for the behavior of the bubble generated beneath a free surface. They identified six distinctive patterns of free surface deformation accompanying the growth and collapse of the bubble. They also discussed the period of bubble oscillations, jet tip velocities, etc. Robert et al. [7] studied the growth and collapse of a laser-induced bubble inside an axisymmetric columnar water jet. They found that the shape and the timing of the ejection from columnar jet depend on the initial bubble position and the ratio of the bubble size to the jet diameter.

In the present study, we investigate the bubble collapse between two parallel free surfaces experimentally and numerically, and the influences of a bubble initial position and a maximum bubble size on its collapsing behavior and water column formations on the free surfaces are analyzed. In the experiment, a laser-induced bubble generated inside of a plain jet is observed using a high-speed video camera. The growth and collapse phases are simulated by

*Corresponding Author, Toshiyuki Ogasawara: oga@me.osakafu-u.ac.jp

the boundary element method (BEM) [8, 9] and the collapse phase is also simulated by the ghost fluid method [10-12].

Experimental study

Figure 1 shows schematic diagrams of an experimental setup and a nozzle geometry to form a plane jet, respectively. A nozzle outlet has a rectangular cross-section whose size is 3.4 mm x 26.0 mm and flat glasses are put on both ends in spanwise direction for the optical observation. To form a two-dimensional water jet in the streamwise direction, the shaded area on the glasses in Fig. 1(b) is coated so as to be hydrophobic. Nd: YAG laser (wavelength: 1064 nm, pulse energy: 325.0 mJ, beam diameter: 8 mm, pulse duration: 5-8 ns) is used to generate a laser-induced bubble in the plane water jet. The laser beam is collimated and focused by the achromatic lens whose focal length is 75 mm. Flow rate is 4.0 l/min and its corresponding average jet velocity is 0.75 m/s. Time scale of the bubble collapse is about 100 μs and it is so short that the effects of flow on the bubble behavior and the water column formation of the free surfaces are negligible. The bubble is generated about 8 mm below the nozzle and observed with a high-speed video camera. The frame rate and the exposure time are set to be 200 kfps and 0.4 μs, respectively. The shadow image of the bubble is visualized by the back illumination with a strobe light. The bubble size is controlled by an iris diaphragm and an initial bubble position is controlled by moving the nozzle in the perpendicular direction to the free surfaces using a linear stage. We introduce dimensionless time t^* which is a ratio of time t to the characteristic time of bubble collapse $t_0 = R_{max}(p_0/\rho)^{-0.5}$, here p_0 and ρ are the atmosphere pressure and the water density, respectively, and $t = 0$ s corresponds to the time of the laser emission. The results are characterized by two dimensionless parameters: the ratio of a maximum bubble radius to a water jet width, $R_{max}^* = R_{max}/w$, and the ratio of an initial bubble offset from the center line of the jet to a water jet width, $\varepsilon^* = \varepsilon/w$.

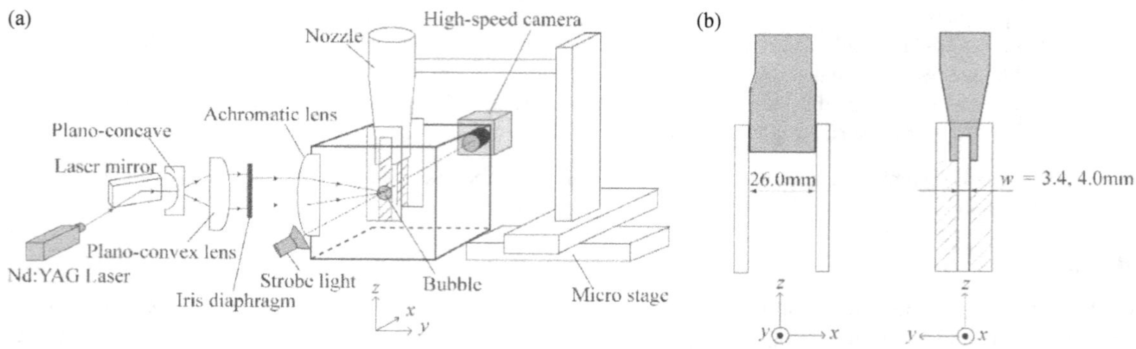

Figure 1 (a) Experimental setup and (b) the nozzle geometry.

The successive behaviors of a bubble and free surfaces are shown in Fig. 2 when the bubble is generated at the center of a plane water jet ($\varepsilon^* = 0$). Figures 2(i) and (ii) correspond to the cases of $R_{max}^* = 0.46$ and 0.61, respectively. The bubbles take the maximum volumes at the fourth top images (d). Initially, the bubble grows with horizontally long ellipsoidal shape. The bubble in Fig. 2(i) gradually becomes spherical shape during the growth phase. However, the bubble in Fig. 2(ii) keeps the horizontally long ellipsoidal shape at its maximum expansion. After the bubble volume becomes maximum, the bubble begins to shrink from the both sides near the free surfaces and collapses with vertically long ellipsoidal shape. In these cases with $\varepsilon^* = 0$, since the bubble position is symmetry to the plane water jet, the bubble centroid does not translate during the growth and collapse phases. The water columns are formed on both free surfaces. From the beginning of the bubble growth, the free surfaces gradually deform to be convex shape as the bubble grows. These deformations developing after the maximum expansion lead to water column formation normal to the free surfaces. Another type of water column formation is observed after the first rebound of the bubble. Crown-like shaped water columns surrounding the former water columns normal to the free surfaces develop afterwards. Both former and latter water columns grow more significantly when $R_{max}^* = 0.61$: the larger bubble case (Fig. 2(ii)).

Figure 3 shows the results in the case of $R_{max}^* \sim 0.5$ when the bubble initially offsets from the center line of the plane jet ($\varepsilon^* = 0.10, 0.18$). The bubble takes the maximum volume in (c). In the case of $\varepsilon^* = 0.10$, the collapsing behavior of the bubble and the free surface deformation are similar to those in the symmetric case in Fig. 2(i). The difference between these is that the bubble translates toward the farther (left) free surface slightly during the collapse phase because of a slightly stronger effect of the nearer (right) free surface. In the case of $\varepsilon^* = 0.18$, however, a water column due to the bubble growth is formed only on the right free surface and grows with splash form. During the collapse phase, the bubble surface starts shrinking from a right side near the free surface, following a liquid-jet formation and penetration toward the left free surface. As a result, the bubble shape becomes toroidal and collapses at the center of the plane water jet. The liquid-jet and the rebound shockwave from the bubble cause the water column formation on the left free surface after the bubble collapses and rebounds.

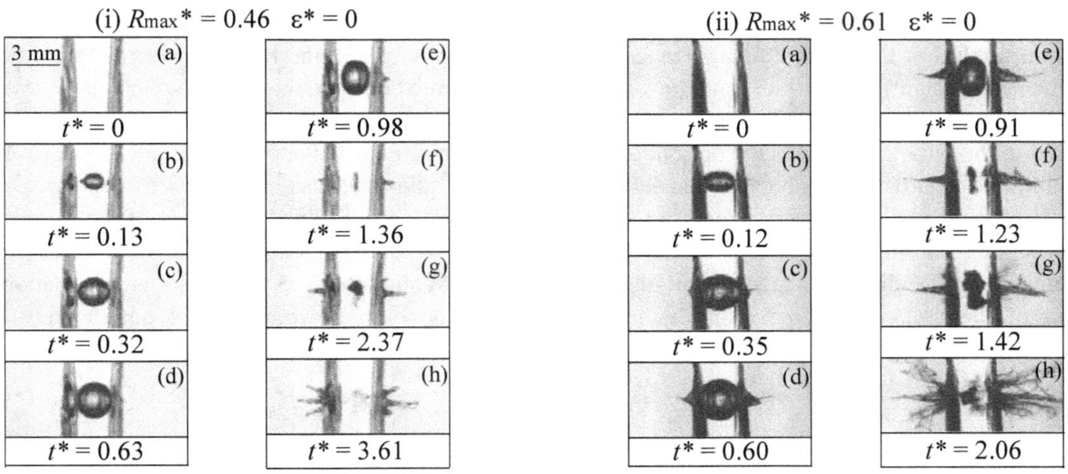

Figure 2 Snapshots of growth and collapse of a laser-induced bubble between free surfaces for $\varepsilon^* = 0$.

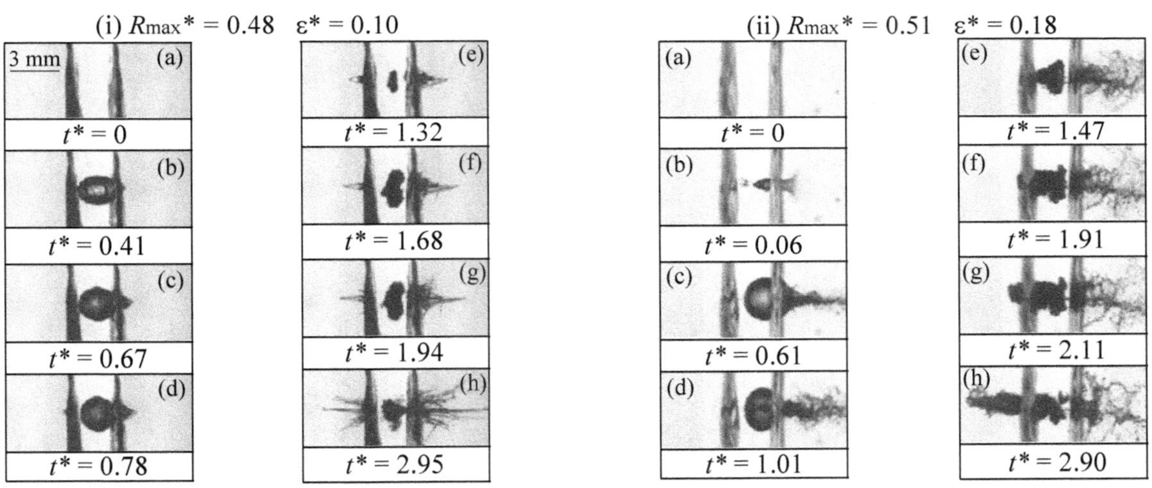

Figure 3 Snapshots of growth and collapse of a laser-induced bubble between free surfaces for asymmetric arrangement ($R_{max}^* \sim 0.5$).

Numerical studies

The above experiments are simulated using the boundary element method and the ghost fluid method. In the simulations, the axisymmetric situation in which the symmetry axis from the bubble center directs normally to the free surface is assumed.

The present boundary element analysis deals with the topological change in bubble shape using the method proposed by Best [13] and can simulate the toroidal bubble after a liquid-jet penetration. The initial ambient pressure in the liquid surrounding the bubble p_∞ is 10^5 Pa and the liquid density ρ_0 is 998.2 kg/m^3. The initial gas pressure inside the bubble p_{g0} is calculated from the integral form of Rayleigh-Plesset equation so that R_{max}^* takes the value of 0.47 (R_{max} = 1.88 mm). The mesh is divided non-uniformly into 405 on each free surface so as to arrange the finer meshes near the bubble, and the minimum size is $0.025R_{max}$. On the bubble surface, the boundary is divided into 32 meshes and the initial mesh size is $0.0048R_{max}$. The characteristic time is defined as $t_0 = R_{max}/(p_\infty/\rho)^{-0.5}$. Figure 4 shows the numerical results for symmetrical arrangement ((a) $\varepsilon^* = 0$) and asymmetrical one when the bubble initially offsets from the center line between the free surfaces ((b) $\varepsilon^* = 0.25$). The bubble grows and takes the maximum volume at the time in Fig. 4(a) (iii) to become a horizontally long ellipsoidal shape. After that, the bubble shrink begins from the both sides near the free surface. Development of the bubble surface instability leads to a biconcave disc shape during the collapse. Finally, the bubble is penetrated at the center of the disc shape and becomes toroidal shape. This toroidal bubble collapses accompanying with the liquid-jet formation from the inside of the torus to outside of the torus in Fig. 4(a) (vi-viii). On the other hand, the free surface begins to deform as the bubble grows and the water columns develop toward each outward normal direction. This development continues in the collapse phase which is consistent with the experimental result. For the asymmetrical arrangement (Fig. 4(b) $\varepsilon^* = 0.25$), the bubble grows with the translation towards the nearer (right) free surface, and the water column is formed only on this nearer (right) free surface. After the bubble takes the maximum volume in Fig. 4(b) (iii), the liquid-jet towards the farther (left) free surface develops and impact on the bubble surface. The toroidal bubble takes the minimum volume at the moment in Fig. 4(b) (vii), and rebounds afterwards as the bubble translates towards the liquid-jet direction (Fig. 4(b) (vii-x)). Under this condition, we could successfully simulate after the rebounding motion until another impact of the toroidal bubble surface occurs in Fig. 4(b) (x). In this case, not only the water column which develops during the growth and collapse phases on the right free surface, but also the other water column due to the liquid-jet penetration and the translation of the collapsing bubble on the left free surface is formed, which agrees qualitatively with the experimental results.

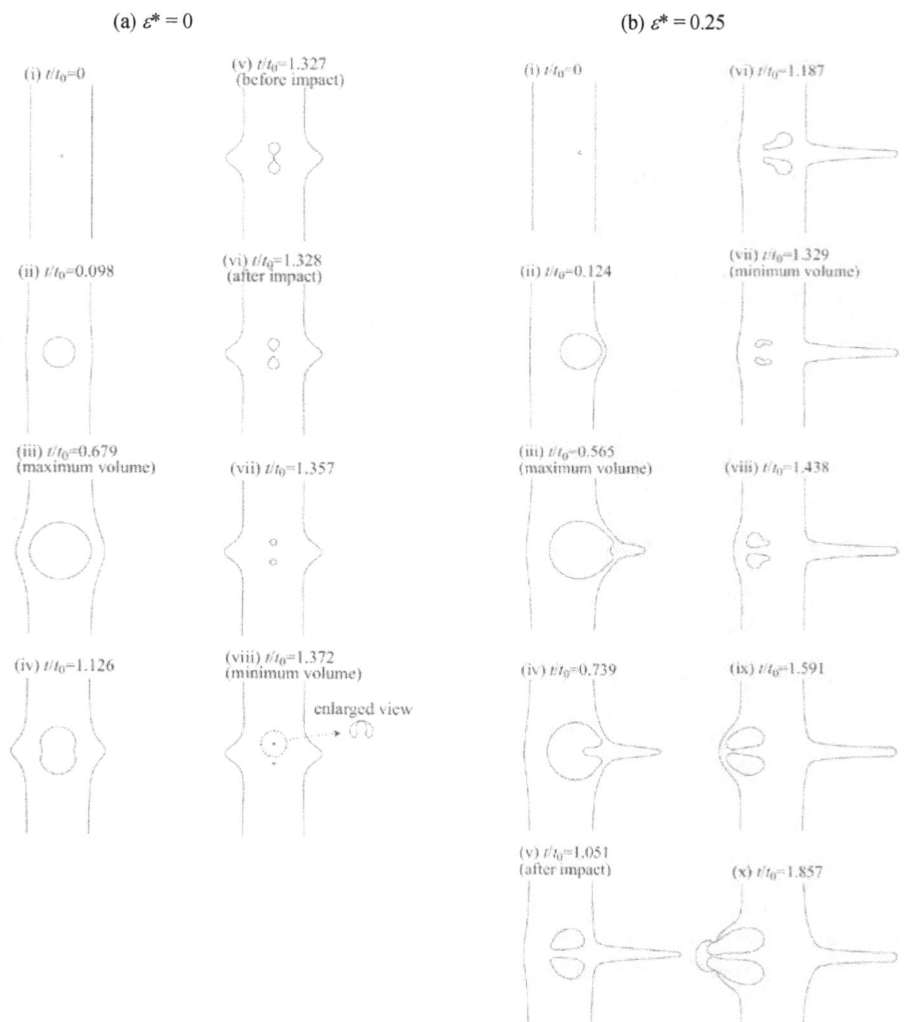

Figure 4 Successive bubble and free surface shapes simulated with the boundary element method ($R_{max}^* = 0.47$).

Since the liquid is assumed to be incompressible in the boundary element analysis, the physics of the shockwave cannot be treated. Therefore, we simulated the bubble collapse by solving Euler equations and a stiffened-gas equation of state [14] using the ghost fluid method [10-12] (GFM) with the level-set method [15, 16]. Initial conditions are given at the maximum expansion of a bubble: the initial shapes of the bubble and free surfaces are given from the corresponding numerical results of the boundary element analysis. The other initial conditions are as follows; $p_\infty = 10^8$ Pa, $\rho_0 = 1034$ kg/m^3, and $p_{g0}/p_\infty = 30$. Note that $t = 0$ in only this simulation indicates the time when the bubble takes the maximum volume. Figure 5 shows the successive Schlieren images in the symmetric arrangement ($\varepsilon^* = 0$) at $R_{max}^* = 0.46$ and 0.57. Under these conditions, the skirts of the water columns deform to be a concave shape during the bubble collapse. Then, shockwaves are emitted from the bubble due to the liquid-jet impact on the toroidal bubble surface and the rebounding motion. After that, the shockwaves propagate and reach the concaved free surface, which leads to the formation of ring-shaped inclined water columns shown in the last images in Fig. 5. This causes the crown-like shaped water column which is observed in the experiment after the first bubble collapse.

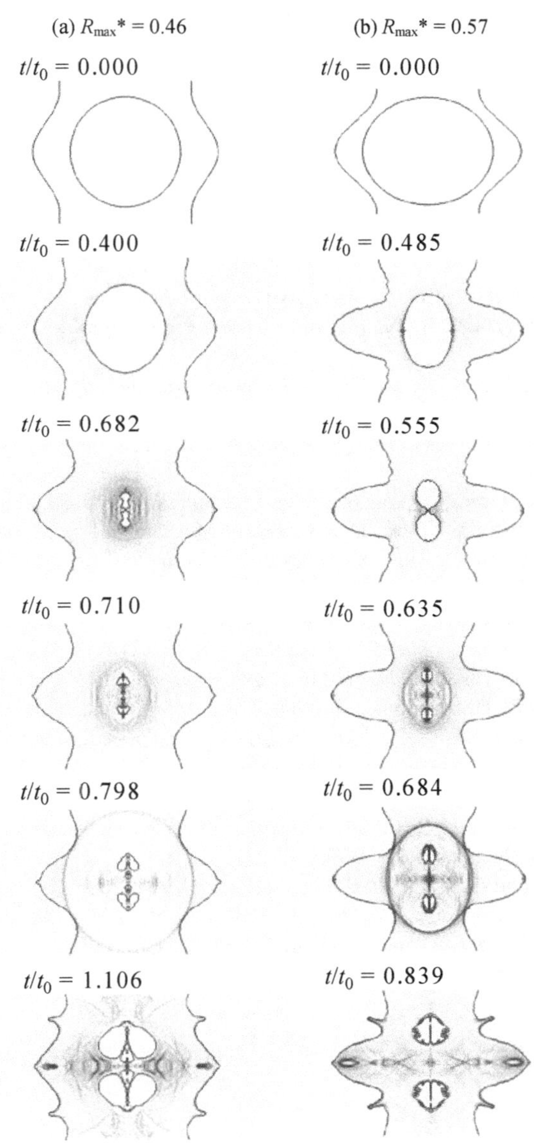

Figure 5 Successive Schlieren images with bubble and free surface shapes simulated with the ghost fluid method ($\varepsilon^* = 0$).

Conclusion

The growth and collapse of a bubble between two parallel free surfaces were investigated experimentally and numerically. In the experiment, the water column formations on the free surfaces are observed: one is the convex water column developing on the both free surfaces during the bubble growth, and the other is the crown-like shaped water column during the growth and collapse of a bubble collapse. The former water column tends to be formed only on the nearer free surface when the initial bubble position offsets from the center of a plane jet. The latter water column is formed by the liquid-jet and rebound shockwave from the collapsing bubble. When a bubble is generated at the center of the plane water jet, the bubble translation does not arise because of the symmetric arrangement of free surfaces to the bubble. The violent crown-like shaped water columns are developed on the closer free surfaces to the bubble. The bubble and free surface shapes in the simulation are qualitatively in good agreements with the corresponding experiments: the bubble is elongated toward the free surface at the maximum bubble volume, and a water column grows on each free surface during the first growth and collapse phases. It is shown that a shockwave is emitted from the bubble when the bubble becomes toroidal during the collapse. Also, another shockwave is emitted when the ring liquid-jet developed from the inner surface of the toroidal bubble impacts on the outer surface of the bubble. These shockwaves cause the generation of inclined crown-like shaped water columns on the free surfaces.

References

[1] Hassanein, A. (2000). *Liquid-metal targets for high-power applications: Pulsed heating and shock hydrodynamics*. Laser Part. Beams 18, 611–622.
[2] Chahine, G. L. (1977). *Interaction between an oscillating bubble and a free surface*. J. Fluid Eng. 99, 709–716.
[3] Blake, J. R. and Gibson, D. C. (1981). *Growth and collapse of a vapour cavity near a free surface*. J. Fluid Mech. 111, 123–140.
[4] Robinson, P. B. and Blake, J. R. (2001). *Interaction of cavitation bubbles with a free surface*. J. Appl. Phys. 89, 8225–8237.
[5] Tomita, Y. and Kodama, T. (2003). *Interaction of laser-induced cavitation bubbles with composite surfaces*. J. Appl. Phys. 94, 2809–2816.
[6] Zhang, S., Wang, S. P. and Zhang, A. M. (2016). *Experimental study on the interaction between bubble and free surface using a high-voltage spark generator*. Phys. Fluids. 28, 032109.
[7] Robert, E., Letty, J., Farhat, M., Monkewitz, P. A. and Avellan, F. (2007). *Cavitation bubble behavior inside a liquid jet*. Phys. Fluids. 19, 067106.
[8] Takahira, H., Murakami, K., Omori, H. Tanaka, S. and Kamiirisa, H. (2003). *Boundary element analysis of bubble collapse under a floating body*. Transactions of the Japan Society of Mechanical Engineers, B 69, 755–763 (in Japanese).
[9] Yasuda, A. and Takahira, H. (2003). *Numerical analysis of the dynamics of toroidal bubbles considering the heat transfer of internal gas*. JSME Int. J., Ser. B 46, 600–609.
[10] Fedkiw, R., Aslam, T., Merriman, B. and Osher, S. (1999). *A non-oscillatory Eulerian approach to interfaces in multimaterial flows (the ghost fluid method)*. J. Comput. Phys. 152(2), 457–492.
[11] Kobayashi, K., Jinbo, Y. and Takahira, H. (2011). *Application of multigrid ghost fluid method to the interaction of shock waves with bubbles in liquids*, Transactions of the Japan Society of Mechanical Engineers, B 77, 723–738 (in Japanese).
[12] Jinbo, Y., Ogasawara, T. and Takahira, H. (2015). *Influence of the nonequilibrium phase transition on the collapse of inertia nonspherical bubbles in a compressible liquid*. Exp. Therm. Fluid Sci. 60, 374–384.
[13] Best, J. P. (1993). *The formation of toroidal bubbles upon the collapse of transient cavities*. J. Fluid Mech. 251, 79–107.
[14] Saurel, R., Petitpas, F. and Abgrall, R. (2008). *Modelling phase transition in metastable liquids. Application to cavitation and flashing flows*. J. Fluid Mech. 607, 313–350.
[15] Sussman, M., Smereka, P. and Osher, S. (1994). *A level set approach for computing solutions to incompressible two-phase flow*. J. Comput. Phys. 114. 146–159.
[16] Enright, D., Fedkiw, R., Ferziger, J. and Mitchell, I. (2002). *A hybrid particle level set method for improved interface capturing*. J. Comput. Phys. 183. 83–116.

© 2018 ASME

The influence of fluid-structure interaction on cloud cavitation about a hydrofoil

[1]Samuel M Smith*; [1]James A Venning; [1]Paul A Brandner; [1]Bryce W Pearce; [1]Dean R Giosio; [2]Yin Lu Young

[1]*Cavitation Research Laboratory, Australian Maritime College, Launceston, Tasmania, Australia*

[2]*Department of Naval Architecture and Marine Engineering, University of Michigan, Ann Arbour, MI, USA*

Abstract

The dynamics of cloud cavitation about rigid and flexible 3D hydrofoils is investigated in a cavitation tunnel. The two hydrofoils have identical undeformed geometry of tapered planform, NACA-0009 section and cantilevered setup at the hydrofoil root. The rigid model is made of stainless steel and the flexible model of carbon and glass-fibre reinforced epoxy resin with an effectively quasi-isotropic lay-up without material bend-twist coupling. Tests were conducted at a fixed incidence of 6°, a chord-based Reynolds number of 0.7×10^6 and a cavitation number ranging from 1.0 to 0.2. Unsteady force measurements were made simultaneously with high-speed imaging to enable correlation of forces and with cavity dynamics. High-resolution force spectra at discrete cavitation numbers and separate pressure sweeps were taken to acquire spectrograms of frequency response as a function of cavitation number. Three shedding modes, designated as types 1, 2 and 3, are apparent for both rigid and flexible hydrofoils although significant differences in peak amplitudes were observed. Types 2 and 3 shedding occur at high cavitation numbers where frequency varied with cavitation number and high-speed imaging showed the dominant shedding mechanism to be due to re-entrant jet formation. The type 1 shedding that developed with reduction in cavitation number, once cavity lengths grew to about full-chord, occurred at a nominally constant frequency. In this case, the imaging showed the dominant mechanism to be shockwave formation. This behaviour has been reported upon extensively in literature although there are some new features apparent from the data. The flexibility of the composite hydrofoil was found to increase the magnitude of the force fluctuations for the low frequency type 1 mode compared to the rigid hydrofoil. However, hydrofoil flexibility was seen to dampen the fluctuating magnitude of the high-frequency type 2 and 3 modes, despite being close to the hydrofoil's natural frequency.

Keywords: Fluid-structure interaction; Cloud cavitation; Hydrofoil

Introduction

The adverse effects of cavitation on hydrofoils, such as unsteady loading and induced vibration due to the shedding of cloud cavitation, can be delayed and mitigated through utilizing passive control of the 3-D morphology. Recent research on these self-adaptive properties of lifting surfaces has been carried out on composite propellers [1,2] and active control surfaces [3]. Geometric aspects of hydrofoils, such as skew and pitch, can be passively tailored allowing for the suppression or delay of cavitation on propellers operating in unsteady inflows [4].

The effects of unsteady cloud cavitation on the hydroelastic response of hydrofoils has previously been investigated [5] with Akcabay et al. [6] showing that greater spanwise flexibility of isotropic hydrofoils with the centre of pressure upstream of the shear centre causes increased cavity length, reduced shedding frequency and broadening of the induced vibration frequencies. Depending on the flow conditions, several possible mechanisms have been identified as the primary instability causing periodic shedding. These include re-entrant jet formation [7-10], shockwave propagation [9-14] and growth of interfacial instabilities such as Kelvin-Helmholtz waves [15]. In a recent study on cloud cavitation about a sphere, all three mechanisms have been observed occurring either under varying flow conditions or as a complex coupled mechanism [13]. Each of these instabilities have certain flow conditions in which they become the critical driver of shedding in cloud cavitation, resulting in the formation of two distinct modes [16]. The low-frequency (type 1) mode, typically defined as transitional cavity oscillation, occurs at relatively low cavitation number to incidence ratios ($\sigma/2\alpha$) where the long cavity is periodically shed due to shockwave propagation generated from the collapse of a previously shed cavity. The shedding frequency of the type 1 mode is typically independent of $\sigma/2\alpha$, occurring at chord-based Strouhal numbers, *St*, between 0.15-0.3. The type 2 mode, typically defined as partial cavity instability, occurs at higher $\sigma/2\alpha$, where a re-entrant jet is the cause of periodic shedding with its frequency changing with σ due to cavity length dependence on σ.

Fluid-structure interaction (FSI) between cloud cavitation and hydrofoils is an intricate phenomenon due to complex interactions between turbulent flow structures, phase-change dynamics, and the structural response of the hydrofoil (Figure 1). To investigate this phenomena, we present force measurements with synchronised high-speed photography, providing insight into the spectral content of unsteady force oscillations. The data obtained also sheds light on the disparity in some previous findings as discussed in [17]. The aim of this study is to further the understanding of cloud cavitation about a 3D hydrofoil and how FSI can influence its behaviour by assessing the correlation between the cavity dynamics and forces.

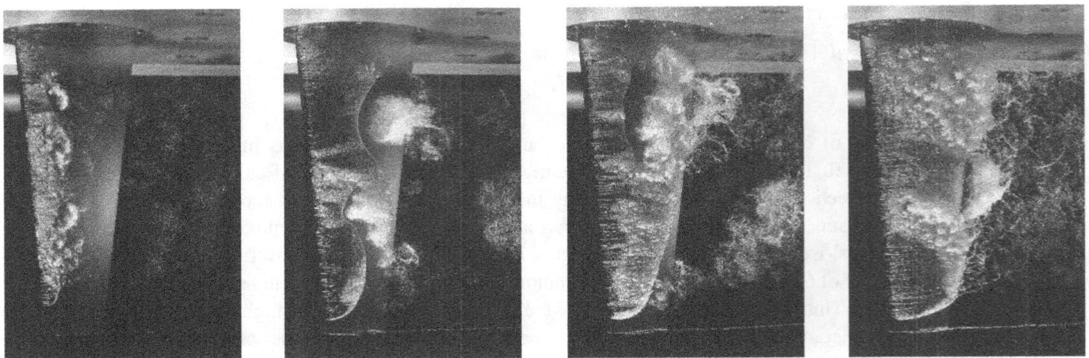

(a) $\sigma = 1.0$ (b) $\sigma = 0.8$ (c) $\sigma = 0.6$ (d) $\sigma = 0.4$

Figure 1 Photographs of the cloud cavitation about a stainless steel NACA-0009 hydrofoil with variations in σ at $\alpha = 6°$ and $Re = 0.7 \times 10^6$.

Experimental overview

The geometry and the mechanical properties of the hydrofoil models were selected based on modelling the static and dynamic fluid-structure interaction representative of propellers and control surfaces. Measurements were conducted using a flexible composite hydrofoil and a nominally rigid steel hydrofoil. The chosen geometry was a symmetric trapezoidal planform of 300 mm span (s) with a chord of 60 mm at the tip and 120 mm at the root, providing a geometric aspect ratio of 3.33 without considering the double body effect provided by the fixed well at the root. The chord length was chosen to be compatible with the mounting to the water tunnel test section and to achieve a chord-based Reynolds number, $Re_c = U_\infty c/\nu = 0.7\times10^6$ with $c = 90$ mm as the mean chord, which is representative of full scale applications. The geometry, in conjunction with a span-wise alignment of the fibre orientation, was intentionally chosen to principally consider bending deformation only of the flexible hydrofoil. A modified NACA-0009 section profile with a thicker trailing edge was selected for improved manufacture of the flexible composite model (see [18] for further details). Deflection measurements [18] showed that the maximum tip bending deflection of the flexible hydrofoil was 15% of the mean chord and negligible twist deformation, while the stainless steel hydrofoil was nominally rigid. The force data for the stainless steel model was found to be nominally invariant with Reynolds number for $\alpha \leq 6°$.

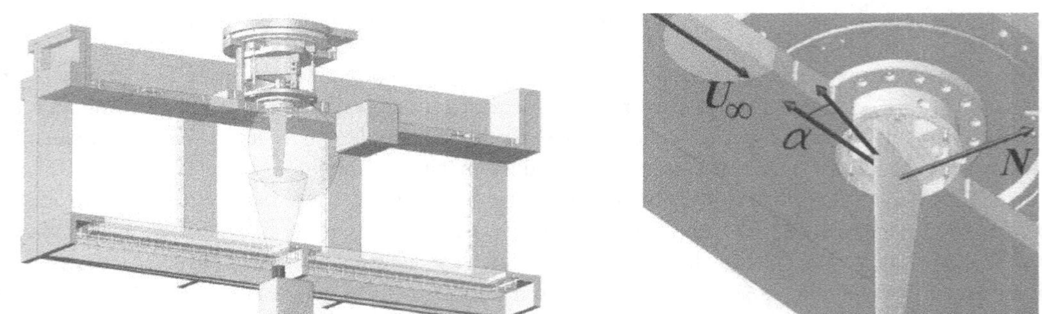

Figure 2 (Left) Section view of the experimental setup where the model hydrofoils are mounted via a 6-component force balance while the cavitation behaviour and tip deflection are imaged using two high-speed cameras. Tip deflection data is presented in [8]. (Right) A close-up of the hydrofoil showing the coordinate system used.

The flexible (composite) model was manufactured as a carbon/glass-epoxy hybrid structure with the lay-up sequence that yield quasi-isotropic response. The construction procedure is detailed in [18]. The nominally rigid (stainless steel) model was machined from a 316 grade stainless steel billet. First mode natural frequencies were obtained in air at 96 Hz and 112 Hz, and in water at 54 and 40 Hz for the stainless steel (rigid) and composite (flexible) models, respectively, determined from impact tests and hydrofoil loading spectra [18].

Measurements were carried out in the Cavitation Research Laboratory water tunnel at the Australian Maritime College. The tunnel test section is 0.6 m square by 2.6 m long, in which the operating velocity and pressure ranges are 2 to 12 m/s and 4 to 400 kPa absolute, respectively. The tunnel volume is 365 m^3 of demineralised water. A detailed description of the facility is given in [19]. Two profiled plates were used to clamp the model within a housing that was attached to a 6-component force balance (Figure 2) with estimated precision of less than 0.1% on all components. Further description of the mounting arrangement may be found in [8]. Data was obtained for cavitation numbers from 1.0 to 0.2 and at a Reynolds number of 0.7×10^6 and an incidence of 6°. The hydrofoil tip cavitation number is defined as $\sigma = 2(p_\infty - p_v)/\rho U^2_\infty$, where p_∞ is the static pressure at the test section centreline, p_v is the vapour pressure, ρ is the water density and U_∞ is the test section velocity.

The cavitation behaviour was recorded using high-speed photography with a HighSpeedStar8 (LaVision, Germany) camera set up as shown in Figure 2. A Nikkor f/1.4 50 mm lens was used resulting in a magnification factor of 3.28 px/mm. High-speed images were recorded with a spatial resolution of 1024×1024 for 1 s at 7,000 Hz for the rigid hydrofoil while the flexible hydrofoil was recorded for 5 s at 1,000 Hz (due to synchronisation compatibility of the two cameras) to allow tip deflection measurements for data presented in [8]. The high-speed photography was synchronized with the force measurement acquisition by simultaneous triggering. Force and tunnel flow data were sampled at 7,000 Hz and 1,000 Hz, respectively.

Results and Discussion

High-resolution spectra of the hydrofoil normal force (perpendicular to the planform as shown in Figure 2) are presented in Figure 3 for all σ. Spectra were obtained using the Power Spectral Density (PSD) derived with the Welch estimate of the PSD [20] utilizing a window size of 4096 samples (4.1 s) and overlap of 256 samples (0.26 s). At each σ, both the rigid and flexible hydrofoils exhibited similar peak frequencies. As σ decreases, the frequency at which these peaks occur reduces due to the increasing cavity length requiring greater time for the cavity to grow and detach. It was also observed that there is an increase in the peak frequency amplitude by approximately one order of magnitude for every 0.2 reduction in σ (note the changing vertical scale in Figure 3).

Comparing the force spectra from both models, it is evident that the amplitude of the rigid hydrofoil is approximately double that of the flexible at σ = 0.8 and 0.6 for the peak frequencies. However, at σ = 0.4, the amplitude of the flexible is more than 3 times that of the rigid. The small differences at σ = 1.0, noting the scale of the PSD, is negligible. The horizontal axis is the cavity shedding frequency (f) Strouhal number defined using the mean chord, $St = fc/U_\infty$. Inconsistencies between rigid and flexible hydrofoil St is attributed to varying flow velocity between runs to maintain a constant Re_c at different temperatures. The 135 Hz (St = 1.43) peaks are associated with the structural response of the force balance based on previous experiments [21] and is evident in that the peaks occur at the same frequency for all σ.

At σ = 0.8, the common 37 Hz (St = 0.39) peak is linked to the periodic shedding of cloud cavitation from the upper half of the hydrofoil as seen in Figure 1(b). The secondary peaks at 42 and 49 Hz (St = 0.44 and 0.52) for the flexible and rigid hydrofoil, respectively, are attributable to the presence of another shedding event on the lower half of the span evident in the power distribution maps (Figure 5) discussed below.

Through measuring the normal force on the hydrofoils while the pressure is gradually increased, a joint time-frequency analysis (JTFA) can be conducted to investigate spectral characteristics at fine σ increments. The JTFA spectrograms shown in Figure 4 are obtained from 256 individual PSD spectra based on time series with 2048 samples (2 s) each and overlap of 1024 (1 s). Overlaying the high-resolution PSDs from Figure 3, good correspondence between the pressure sweep and steady state condition tests is observed showing the transient nature of the JTFA does not influence the resulting spectra. The pressure sweep was conducted both by increasing and decreasing σ and revealed no evidence of hysteresis.

Three trends are evident in the spectrograms corresponding to differing cloud cavitation shedding mechanisms discussed in the introduction. A shockwave-driven (type 1) mode occurs at $St \approx 0.1$ and within a range of σ from 0.3 – 0.55 for both hydrofoils. A slight dependence on σ is observed in both cases with the frequency increasing slightly with σ. A similar trait observed in [17] attributing it to low dissolved oxygen levels ($\approx$ 7 ppm) where current results are for a constant dissolved oxygen level of about 3 ppm. Comparing the hydrofoils, the flexible model exhibits significantly greater amplitude than the rigid for virtually the entire type 1 mode region. As σ increases from 0.2, the flexible hydrofoil reaches its maximum at $\sigma \approx 0.38$, compared to $\sigma \approx 0.45$ for the rigid. As

σ increases to approximately 0.5, the cavity becomes smaller, closing further upstream from the hydrofoil trailing edge. This reduction in cavity size causes a change in the pressure distribution and the resulting unsteady forces acting on the hydrofoil. With the cavity closing on the hydrofoil, a re-entrant flow may form (evident in Figure 1), becoming the primary shedding instability mechanism as σ increases.

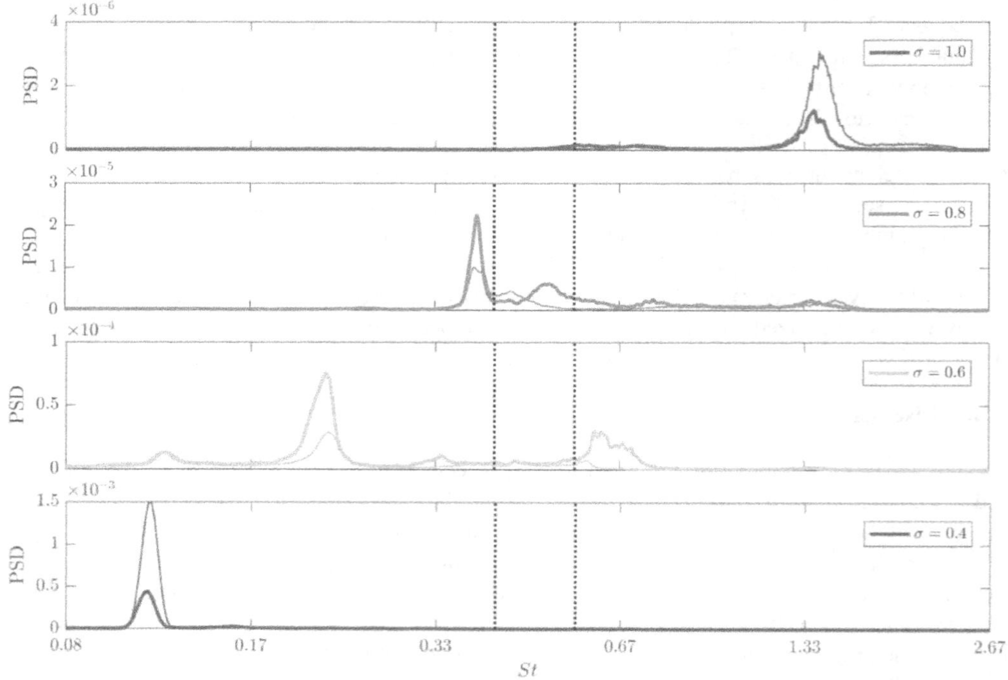

Figure 3 High-resolution non-dimensional normal-force spectra for the rigid (thick line) and flexible (thin line) hydrofoils for various σ. The vertical dashed lines show the flexible and rigid hydrofoils wetted natural frequencies. Note the ordinate axis limits change for each σ.

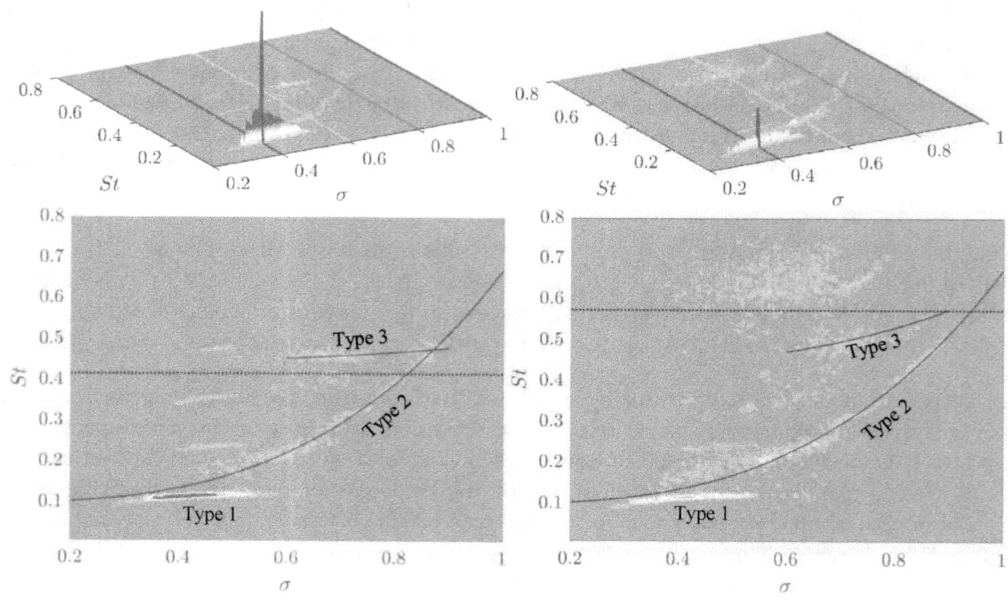

Figure 4 (Top) Isometric overlay of high-resolution spectra and low resolution spectrograms. (Bottom) Top view of the spectrograms with identified shedding modes and hydrofoil wetted natural frequency (dotted line). The left and right columns are data for the flexible and rigid hydrofoils, respectively. The results show convergence of type 2 into type 3 shedding modes with decreasing σ. Type 2 and 3 mode shedding frequencies varies as a power law with σ where $St_2 \approx 0.57\sigma^3 + St_1$, $St_{3F} \approx 0.05\sigma^3+0.44$ and $St_{3R} \approx 0.2\sigma^3+0.43$, where the subscripts are the shedding modes.

The type 2 mode, driven by re-entrant flow, follows an almost identical trend for both hydrofoils, showing strong dependence on σ and diminishing by $\sigma \approx 0.9$. Figure 4 shows the type 2 mode varies with σ as a power law with an exponent of 3 where $St_2 \approx 0.57\sigma^3+St_1$, where the subscripts are the shedding modes. From approximately $\sigma =$

0.5 to 0.6, types 1 and 2 modes co-exist, indicating that shedding-induced unsteady forces may be due to either re-entrant flow, shockwave induced cavity collapse or both in this region. This is reinforced by the high-resolution PSD in Figure 3 where at $\sigma = 0.6$, the spectra exhibit peaks at both 11.7 Hz ($St = 0.12$, shockwave-driven) and 21.5 Hz ($St = 0.23$, re-entrant jet-driven) with visual evidence observed in high-speed imaging for both hydrofoils.

The third mode (type 3) is evident in the spectrograms at $0.6 < \sigma < 0.9$, co-existing with type 2 for both hydrofoils. The type 3 mode is also seen to vary with σ as a power lower with an exponent of 3 but the trend differs between hydrofoils where $St_{3F} \approx 0.05\sigma^3+0.44$ and $St_{3R} \approx 0.2\sigma^3+0.43$ for the flexible and rigid, respectively. Identification of the tertiary mode is made evident in the power distribution maps below.

To gain spatial insight into the shedding behaviour, power distribution maps of identified shedding modes are shown in Figure 5. The maps are obtained by performing a FFT on the time series of each pixel and extracting the frequency specific power, then plotting the values to form a plot of distribution of power through the image. At $\sigma = 0.8$, both hydrofoils exhibit two shedding modes (type 2 and 3) of which the type 2 shedding occurs at similar frequencies of 37.7 and 37.4 Hz ($St = 0.392$ and 0.397, also shown in Figure 3) for the flexible and rigid hydrofoils, respectively. Power distributions reveal that this fundamental frequency is driven by shedding between $s/6$ and $s/2$ from the root for both hydrofoils.

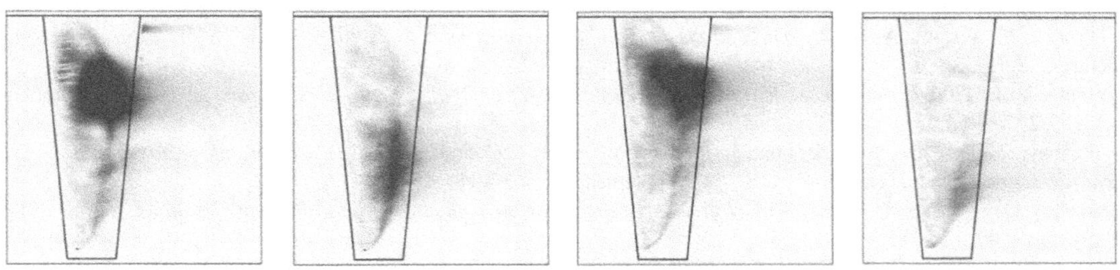

(a) Flexible: 37.7 Hz $St = 0.392$ (b) Flexible: 44.4 Hz $St = 0.462$ (c) Rigid: 37.4 Hz $St = 0.397$ (d) Rigid: 49.9 Hz $St = 0.53$

Figure 5 Power distribution of shedding events for the (a & b) flexible and (c & d) rigid hydrofoil with type 2 mode shown in (a) and (c) while a type 3 mode is shown in (b) and (d). All data is for $\sigma = 0.8$.

The tertiary mode (type 3) of each hydrofoil occurs at different frequencies coinciding with those observed in the normal force PSD (Figure 3) at 44.4 Hz ($St = 0.462$) and 49.9 Hz ($St = 0.53$) for the flexible and rigid hydrofoil, respectively. The power concentration of the tertiary mode is localized to the lower half of the span in contrast to the type 2 mode in the top half. Due to the principle spanwise bending deformations of the flexible hydrofoil, the type 3 mode power distribution extends further along the span than that of the rigid hydrofoil. High-speed imaging show that both the type 2 and 3 shedding modes are re-entrant jet driven.

With both hydrofoils exhibiting two shedding modes and power distributions showing they cover separate regions of the planform indicates the formation of two shedding events on the hydrofoil for a particular σ. The occurrence of the differing shedding events along the span is supported by the span-wise space-time plots in [8] and high-speed imaging. The formation of multiple span-wise shedding events is not observed at σ other than 0.8 as the cavity length, both stream-wise and span-wise, are incompatible with the hydrofoil span and, instead, result in an irregular break-off mechanism [22]. In the present study additional variations in spanwise geometry, hydrostatic pressure and hydroelastic response may also attribute to this behaviour.

Conclusions

Fluid-structure interaction is seen to have an effect on the cavity dynamics and induced normal force fluctuations experienced by a hydrofoil operating under cloud cavitation conditions. JTFA identified three shedding modes, type 1, 2 and 3, where the type 2 and 3 modes are seen to vary as a power law with σ. Power in the peak frequencies is seen to drop by approximately one order of magnitude with every 0.2 reduction in σ. Furthermore, it is shown that the rigid hydrofoil exhibits double the power of the flexible hydrofoil at the fundamental peak frequencies for σ between 0.6 and 0.8. However, the flexibility of the composite hydrofoil resulted in 3 times the power of that of the rigid at σ equal to 0.4, where the shedding mode has transitioned from type 2 to 1. The fundamental peak frequencies at each σ were not seen to vary significantly between hydrofoils, suggesting flexibility has more effect on the severity of cloud cavitation than its shedding frequency. Power distribution maps show the existence of two shedding modes (type 2 and 3) along the span of the hydrofoils that match those in the PSDs. The common type 2 mode frequency (37 Hz, St = 0.39) is seen to be due to shedding cavities on the upper half of the hydrofoil. The flexibility of the composite hydrofoil is seen to drop the type 3 mode frequency from 49 Hz (St = 0.52) to 42 Hz (St = 0.44) as well as change the spatial distribution of the secondary shedding across the bottom half of the span. The complex behaviour found under cavitation conditions indicates that any simple design assumptions applied to flexible hydrofoils may be not sufficient and a detailed analysis of these structures is required.

Acknowledgements

This project was supported by the University of Tasmania, the Defence Science and Technology Group, US Office of Naval Research (Dr. Ki-Han Kim, Program Officer) and ONR Global (Dr. Woei-Min Lin) through NICOP S&T Grant no. N62909-11-1-7013. The authors would like to acknowledge the assistance of Mr Steven Kent and Mr Robert Wrigley from the Australian Maritime College for their essential help with setting up and carrying out the experiments.

References

[1] Young, Y. L. (2008). *Fluid–structure interaction analysis of flexible composite marine propellers*. Journal of Fluids and Structures. 24(6):799–818.

[2] Motley, M., Liu, Z., & Young, Y. (2009). *Utilizing fluid–structure interactions to improve energy efficiency of composite marine propellers in spatially varying wake.* Composite Structures. 90(3), 304-313.

[3] Turnock, S., Wright, A. (2000). *Directly coupled fluid structural model of a ship rudder behind a propeller.* Marine Structures. 13(1), 53-72.

[4] Young, Y. L. (2007). *Time-dependent hydroelastic analysis of cavitating propulsors*. Journal of Fluids and Structures, 23(2), 269-295.

[5] Akcabay, D. T., Chae, E. J., Young, Y. L., Ducoin, A., & Astolfi, J. A. (2014). *Cavity induced vibration of flexible hydrofoils.* Journal of Fluids and Structures, 49 (Supplement C), 463-484

[6] Akcabay, D. T., & Young, Y. L. (2014). *Influence of cavitation on the hydroelastic stability of hydrofoils.* Journal of Fluids and Structures, 49, 170-185.

[7] Callenaere, M., Franc, J.-P., Michel, J.-M., & Riondet, M. (2001). *The cavitation instability induced by the development of a re-entrant jet.* Journal of Fluid Mechanics, 444, 223-256.

[8] Smith, S., Venning, J., Brandner, P., Pearce, B., Giosio, D., & Young, Y. L. (2017). *Cloud cavitation behaviour on a hydrofoil due to fluid-structure interaction.* 17th International Symposium on Transport Phenomena and Dynamics of Rotating Machinery, Maui, Hawaii.

[9] Ganesh, H., Wu, J., & Ceccio, S. (2016). *Investigation of cavity shedding dynamics on a NACA0015 hydrofoil using time resolved x-ray densitometry.* 31st Symposium on Naval Hydrodynamics, ONR.

[10] Ganesh, H., Mäkiharju, S. A., & Ceccio, S. L. (2016). *Bubbly shock propagation as a mechanism for sheet-to-cloud transition of partial cavities.* Journal of Fluid Mechanics, 802, 37-78.

[11] Reisman, G., Wang, Y.-C., & Brennen, C. E. (1998). *Observations of shock waves in cloud cavitation.* Journal of Fluid Mechanics, 355, 255-283.

[12] Venning, J., Giosio, D., Smith, S., Pearce, B. & Brandner, P. (2018). *The influence of nucleation on the spectral content of cloud cavitation about a hydrofoil.* The 10th International Symposium on Cavitation, Baltimore, Maryland, accepted paper.

[13] De Graaf, K. L., Brandner, P. A., & Pearce, B. W. (2017). *Spectral content of cloud cavitation about a sphere*. Journal of Fluid Mechanics, 812, R1.

[14] Venning, J., Smith, S., Brandner, P., Giosio, D., & Pearce, B. (2017). *The influence of nuclei content on cloud cavitation about a hydrofoil.* 17th International Symposium on Transport Phenomena and Dynamics of Rotating Machinery, Maui, Hawaii.

[15] Avellan, F., Dupont, P., & Ryhming, I. L. (1988). *Generation mechanism and dynamics of cavitation vortices downstream of a fixed leading edge cavity.* 17th Symposium on Naval Hydrodynamics, Berkeley, California.

[16] Kjeldsen, M., Arndt, R. E., & Effertz, M. (2000). *Spectral characteristics of sheet/cloud cavitation.* Journal of Fluids Engineering, 122(3), 481-487.

[17] Kawakami, D. T., Fuji, A., Tsujimoto, Y., & Arndt, R. (2008). *An assessment of the influence of environmental factors on cavitation instabilities*. Journal of Fluids Engineering, 130(3).

[18] Zarruk, G. A., Brandner, P. A., Pearce, B. W., & Phillips, A. W. (2014). *Experimental study of the steady fluid–structure interaction of flexible hydrofoils.* Journal of Fluids and Structures, 51, 326-343.

[19] Brandner, P., Lecoffre, Y., & Walker, G. (2007). *Design considerations in the development of a modern cavitation tunnel.* 16th Australasian Fluid Mechanics Conference, Gold Coast, Australia.

[20] Welch, P. (1967). *The use of fast Fourier transform for the estimation of power spectra: a method based on time averaging over short, modified periodograms.* IEEE Transactions on audio and electroacoustics, 15(2), 70-73.

[21] Smith, S., Pearce, B., Brandner, P., Clarke, D., Moreau, D., & Xue, Y. (2016). *Steady and unsteady loads acting on a hydrofoil immersed in a turbulent boundary layer.* 20th Australasian Fluid Mechanics Conference, Perth, Australia.

[22] Kawanami, Y., Kato, H., & Yamaguchi, H. (1998). *Three-dimensional characteristics of the cavities formed on a two-dimensional hydrofoil.* Third International Symposium on Cavitation. Grenoble, France.

© 2018 ASME

The Influence of Liquid Rocket Engine Inducer Compliance and Mass Flow Gain Factor on Cavitation Surge Frequencies

[1]David E. Jackson; [1]John A. Schwille*; [1]James M. Gariffo; [2]Zoltan Spakovszky; [2]Claudio Lettieri; [2]Vincent Wang;

[1]The Aerospace Corporation, El Segundo, CA, USA; [2]Massachusetts Institute of Technology, Cambridge, MA, USA

Abstract

A primary driver in cryogenic pump transfer functions is the presence of cavitation and cavitation instabilities in the pump's inducer. Fluid mode frequencies in a water test loop are measured and confirmed to be planar wave surge (cavitation surge) through analysis of an array of dynamic pressure transducers. In addition, steady computational fluid dynamics (CFD) simulations are conducted and used to estimate the cavitation compliance and mass flow gain factor at both room temperature and elevated temperature where thermal suppression effects are dominant. Excellent agreement is found between measured fluid mode frequencies and predictions using compliance estimated from steady CFD and the inlet feedline inertance.

Keywords: inducer; cavitation instabilities

Introduction

Cavitation surge, or long column mode (LCM), is the presence of planar waves that can be present in rocket turbomachinery feed systems. These instabilities are heavily influenced by the transfer functions of the pump [1], which in turn are driven by the cavitation behavior of the inducer. These transfer functions have typically been written as

$$[TP]=\begin{bmatrix} 1+\left.\dfrac{\partial\left(\Delta P^T\right)}{\partial P_1^T}\right|_{m_1} & \left.\dfrac{\partial\left(\Delta P^T\right)}{\partial m_1}\right|_{P_1^T} \\ j\omega\rho_L\left.\dfrac{\partial V_c}{\partial P_1^T}\right|_{m_1} & 1+j\omega\rho_L\left.\dfrac{\partial V_c}{\partial m_1}\right|_{P_1^T} \end{bmatrix} \tag{1}$$

Of particular importance in terms TP_{21} and TP_{22} are the cavitation compliance (K) and the mass flow gain factor (M), which are defined as

$$K=\left.\frac{\partial(V_c)}{\partial P^T}\right|_{m_1} \tag{2}$$

$$M=\left.\frac{\partial(V_c)}{\partial m_1}\right|_{P^T} \tag{3}$$

The cavitation compliance describes the sensitivity of cavitation vapor perturbation to inlet total pressure fluctuations. The mass flow gain factor describes the proportionality between cavitation vapor fluctuations and mass flow rate fluctuations. Experiments to measure these parameters require inducing fluctuations into a feed system and measuring the unsteady response in mass flow. Since these measurements require specific hardware, there are few measurements of actual inducer transfer functions [2-5]. This study aims to increase the understanding of several aspects of cavitation instabilities. First, measurements are made of the LCM frequency in a water test rig using an inducer. Second, a series of computational fluid dynamics (CFD) simulations are used to estimate a test inducer's compliance and mass flow gain factor, which are then combined with calculated inertance to predict the LCM frequencies in the test loop. Third, a stability model is used to demonstrate instability onset and termination similar to those observed during experiments. These efforts expand the data set available to engineers and provide modeling tools that illustrate the underlying physics of cavitation instability.

*Corresponding Author, John Schwille: Jay.Schwille@aero.org
Approved for public release. OTR 2018-00168.

Methodology

Experiments were conducted at the Aerospace Cavitation Test Facility to observe system and inducer behavior over a range of inlet pressures. This closed loop water test facility's primary purpose is to measure steady and unsteady performance data for inducers. A photograph of the test loop is shown in Fig. 1a. The facility is sized for a nominal 76.2 mm (3 inch) test article, and is designed for speeds up to 6000 rpm. The test section is instrumented with circumferential arrays of static and/or dynamic pressure transducer ports at multiple planes to allow detailed measurement of the cavitating flow field. The facility allows investigation of thermally-limited cavitation by using heaters, allowing for operation at water temperatures up to 400 K. To verify that the identified signatures are due to cavitation surge, a circumferential array of 4 dynamic pressure transducers at the inducer inlet tip plane was used to characterize the spatial structure of the cavitation disturbance. The current test article, shown in Fig. 1b, is a four-bladed inducer that was designed to represent a modern rocket turbopump inducer.

To estimate the cavitation compliance and mass flow gain factor, the commercial code ANSYS-CFX™ is used to solve the conservation equations. The default submodel for cavitation lacks the thermal terms necessary to account for the effect of thermodynamic suppression. To capture thermal effects, that model is replaced with one based on work by Tsuda et al. [6], which modifies the vapor pressure used in the cavitation submodel without calculating the temperature field by solution of the energy equation. This formulation tracks the vapor as a second Eulerian phase, with all the individual bubble or cloud growth relegated to the cavitation submodel. The results from a series of steady state runs are used to estimate the response of the vapor with respect to changes in pressure or mass flow in order to calculate the cavitation compliance and mass flow gain factor.

Results

Images from the optical test section are shown in Fig. 2. The eight images show a transient progression of the vapor cloud. In the first four images, the vapor cloud grows into the blade passages, almost reaching the trailing edge. The subsequent images depict the other half of the surge cycle when the cloud reverses this travel. The surge frequencies decrease from 19 Hz to ~5 Hz as cavitation number decreases.

As shown in Fig. 3a and 3b, the estimates of cavitation compliance and mass flow gain factor calculated from steady CFD simulations compare well to previous measurements on a variety of inducers [2-5]. It is interesting to note that preliminary results show similar values of cavitation compliance between room temperature water and elevated temperature water. Since room temperature water exhibits inertially limited cavitation and elevated temperature water exhibits thermally limited cavitation, the similar results are surprising and will be further investigated.

The natural frequency, ω, of the feedline and inducer system can be calculated from

$$\omega = \frac{1}{2\pi(LK)^{1/2}} \tag{4}$$

where K is the inducer cavitation compliance and L is the combined inertance of the feedline and inducer. The cavitation compliance shown in Fig. 3 is used to tailor Eq. 2 to the current test article.

The result of this frequency calculation is compared to measured surge frequencies in Fig. 4. Both the calculated and the measured frequencies decrease with decreasing cavitation number. The agreement between the measured frequencies and calculated frequencies is quite good. Although a number of challenges to calculating the behavior of cavitating turbomachinery remain, the CFD study conducted here shows continuing progress toward that goal.

One key observation from the current data is that the LCM is amplified in a range of cavitation numbers, rather than just increasing in magnitude as cavitation number decreases. Fig. 5 shows this behavior in the experimental apparatus. In these figures, cavitation number is decreasing with time as inlet pressure is lowered during the test. Once the cavitation number reaches 0.03, the amplitude of the LCM frequency abruptly increases. The amplitude remains at that level until the cavitation number reaches 0.02, at which point the amplitude dramatically decreases.

In order to explore this behavior of the LCM, a stability model of the experimental test loop was constructed. The model is primarily based on a framework previously published [7], which uses transfer matrices to represent the elements of the water loop. The formulation of the transfer matrices for the general flow elements like the pipes, flow straightener, and contraction is given by that reference. In the present work, the system of transfer matrices is augmented by the addition of a pump model that uses the pump's compliance and the mass flow gain factor. Again, in this study, the CFD-calculated compliance and mass flow gain factor provide specific values that tailor the instability model to this specific inducer.

Figure 6 shows the results of the stability model. When the system damping is negative, the system instability that manifests results in the increased amplitude observed in Fig. 5. When the parameters in the model are calculated with the CFD results, Fig. 6a shows that the system stability margin decreases with decreasing cavitation number, just as observed during experiment. In this case, the threshold for system stability is too high, occurring at a cavitation number of 0.06. However, it is useful to explore the stability model by modifying the mass flow gain factor. Although a large reduction in the CFD-calculated mass flow gain factor of approximately 47% is necessary, it is possible to match the threshold for instability shown in the experiments at a cavitation number of 0.03. The interesting result that occurs with the reduced mass flow gain factor is that the system regains stability at a cavitation number of 0.02. This matches the experimental behavior and shows promise in the stability model in understanding the complex phenomena of instability onset and termination. Of course, the discrepancy between the CFD calculations and the value that prompts the onset and termination behavior motivates further work in understanding whether the steady state CFD approach fails to capture all the physics present in the dynamic behavior of the system.

Figures & Tables:

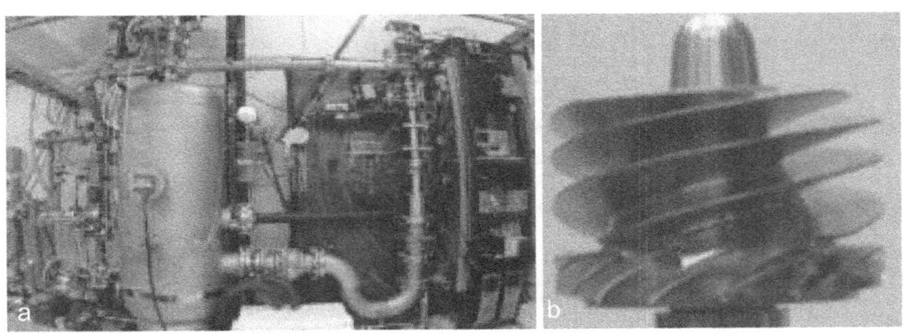

Figure 1. The Aerospace Cavitation Test Facility (a) and the test inducer (b).

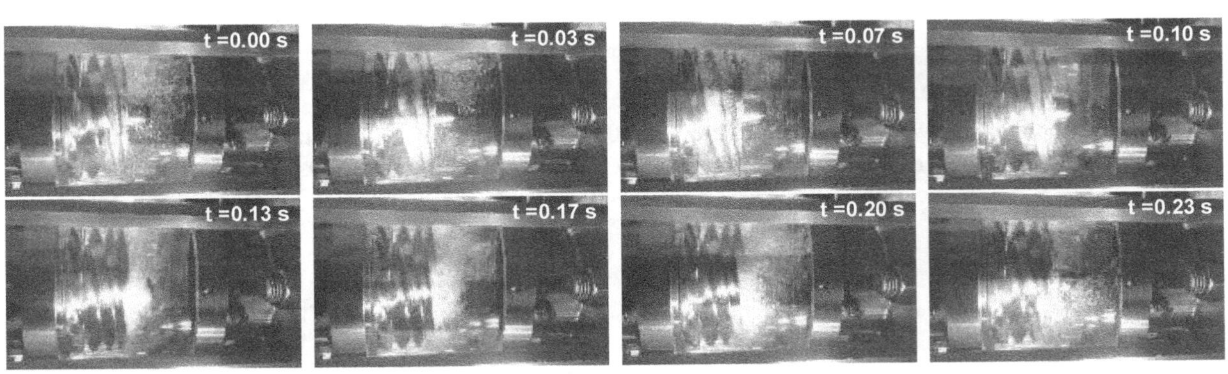

Figure 2. Progression of cavitation surge in the current inducer at $\phi=0.06$.

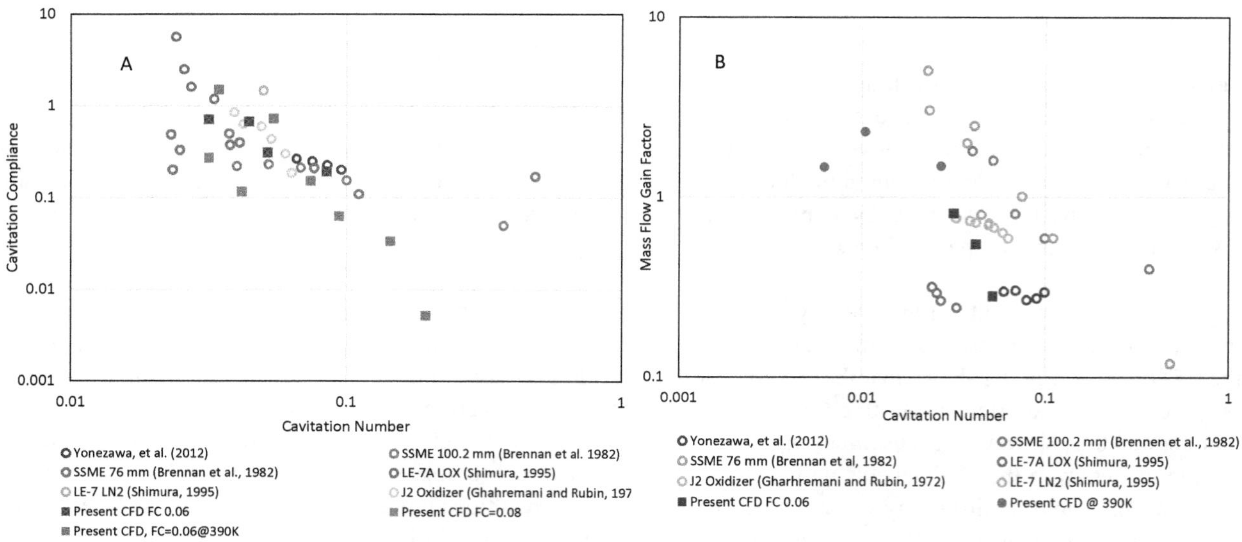

Figure 3. Pump parameters (a) compliance and (b) mass flow gain factor calculated from CFD compared with experimental data from a variety of hardware [2-5].

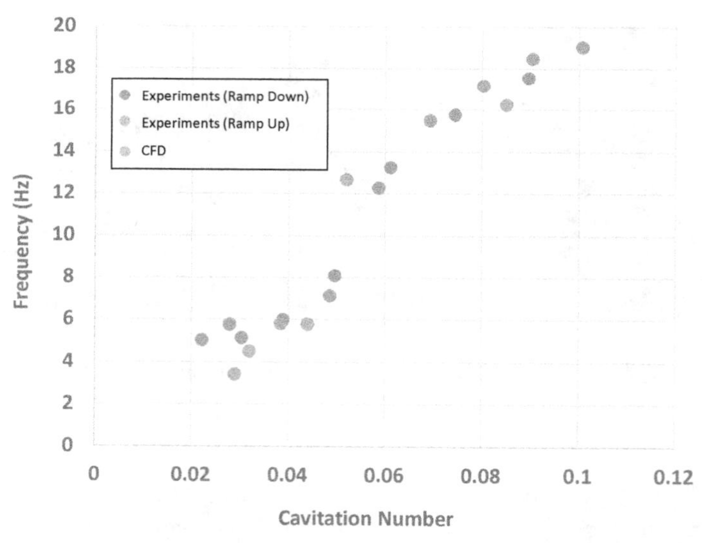

Figure 4. Comparison between surge frequencies at ϕ=0.06 in the Aerospace Cavitation Test Facility and those estimated from parameters derived from steady state CFD.

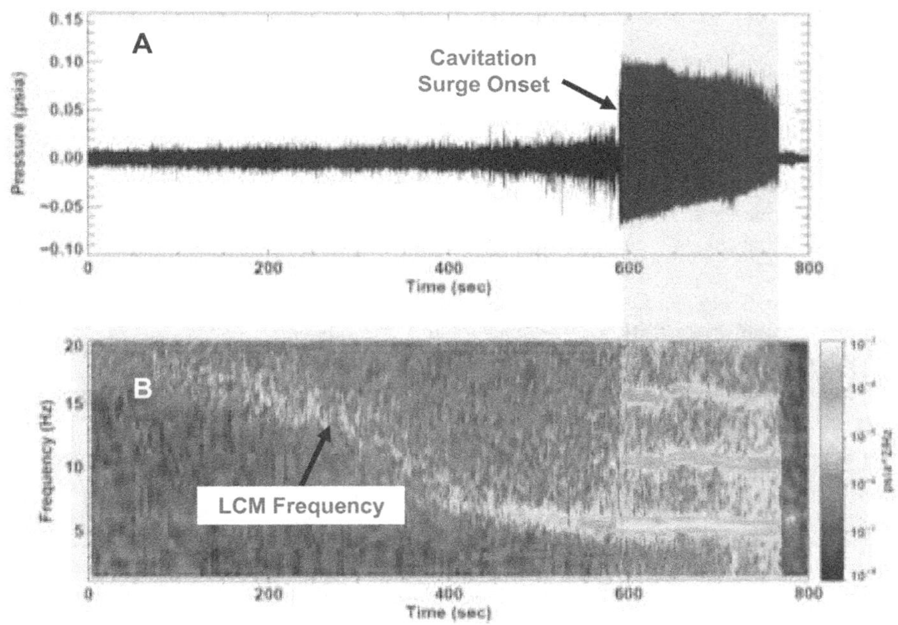

Figure 5. Cavitation instability in the inducer as demonstrated by (a) system pressure and (b) a contour plot of spectral density.

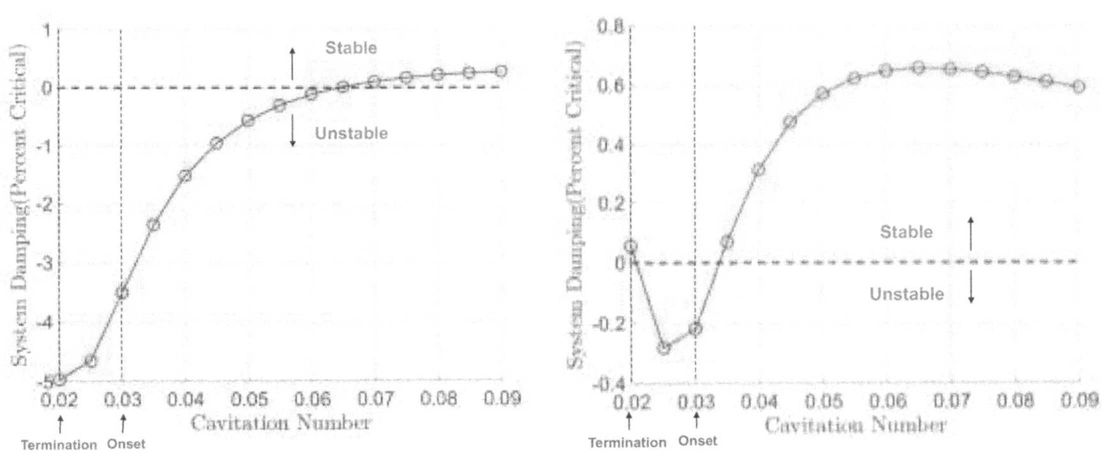

Figure 6. System damping showing onset and termination of high amplitude LCM with CFD calculated compliance. Mass flow gain factor is (a) calculated from CFD and (b) adjusted to match the observed onset and termination of instability.

Conclusion

The combination of experiments, CFD, and instability modeling shows significant progress into understanding the behavior and impact of cavitation in rocket propellant feed systems. The study shows two main conclusions. First, the compliance and the mass flow gain factor calculated from CFD can be combined with characteristics of the inlet feedline to provide an accurate prediction of the LCM frequency. Second, the sensitivity of the stability model to mass flow gain factor shows the critical nature of that parameter. This sensitivity will be the focus of future research.

References

[1] Brennan, C. E., Hydrodynamics of Pumps, Concepts ETI, Vermont, 1994, pp. 218.
[2] Yonezawa, K., Aono, J., Kang, D., Horiguchi, H., Kawata, Y., & Tsujimoto, Y. (2012). *Numerical evaluation of dynamic transfer matrix and unsteady cavitation characteristics of an inducer*. International Journal of Fluid Machinery and Systems, 5(3), 126-133.
[3] Brennen, C. E., Meissner, C., Lo, E. Y., & Hoffman, G. S. (1982). *Scale effects in the dynamic transfer functions for cavitating inducers*. Journal of Fluids Engineering, 104(4).
[4] Shimura, T. (1995). *Geometry effects in the dynamic response of cavitating LE-7 liquid oxygen pump*. Journal of Propulsion and Power, 11(2)
[5] Ghahremani, F. G., & Rubin, S. (1972). *Empirical evaluation of pump inlet compliance*. NASA-CR-123963.
[6] Tsuda, S., Tani, N., and Yamanishi, N. (2012), *Development and Validation of a Reduced Critical Radius Model for Cryogenic Cavitation*, J Fluid Eng, 134 (5).
[7] Greitzer, E., Tan, C., Graf, M. (2004). *Internal Flows: Concepts and Applications*. Cambridge, NY: Cambridge University Press.

© 2018 ASME

The influence of nucleation on the spectral content of cloud cavitation about a hydrofoil

[1]James Venning*; [1]Dean Giosio; [1]Samuel Smith; [1]Bryce Pearce; [1]Paul Brandner;

[1]*Cavitation Research Laboratory, Australian Maritime College, University of Tasmania, Australia;*

Abstract

The influence of nucleation on the spectral content of cloud cavitation is investigated using unsteady force measurements and simultaneous high-speed imaging of a rectangular planform NACA-0015 section hydrofoil, with and without upstream nuclei injection. Experiments were conducted for cavitation numbers over the range of 0.60 to 0.45, at a chord-based test Reynolds number of 1.4×10^6, and at an incidence of 6°. The introduction of freestream nuclei was found to significantly alter the cavity topology and spectral content of the force response. Three shedding modes were observed in the case without nuclei injection, while two modes were present with the injection of abundant freestream nuclei. In the nuclei deplete case the dominant shedding frequency is associated with upstream shock propagation and was found to decrease with test cavitation number from a Strouhal number of 0.311 to 0.243, while the sub-harmonic and first harmonic were associated with localized tip and root activity. With the introduction of nuclei the dominant of two distinct shedding frequencies was found to be the lower frequency mode at approximately half that of the corresponding nuclei deplete case, with Strouhal number decreasing from 0.159 to 0.148. The higher frequency mode was shown to be due the propagation of an initial shockwave that only partially condenses the cavity prior to the passage of a second shockwave that leads to full span-wise condensation.

Keywords: cavitation; nucleation; spectral content; cavity dynamics

Introduction

The formation, separation and collapse of sheet cavities (i.e. partial cavitation) is generally described as cloud cavitation. For sufficiently thick cavities the re-entrant jet, associated with the cavity closure, has sufficient intensity to propagate upstream and penetrate the cavity interface causing a portion of the cavity to detach. This detached cavity undergoes condensation, resulting in a bubbly cloud that is then advected away by the bulk flow. This phenomenon has been widely studied, both experimentally and numerically, in various contexts including hydrofoils [1-3], rotating machinery [4], nozzles/venturies [5], wall-mounted cavitators [6] and bluff bodies [7, 8]. Recently it has been shown that the formation of condensation shockwaves produce another shedding mechanism, either independently or coupled with a re-entrant jet, leading to cavity de-stabilisation [9, 10]. Three-dimensional flow features have been shown to significantly alter the re-entrant flow behaviour and the resulting shedding physics [11], as was also observed in the present study. The inception of hydrodynamic cavitation in practical flows is invariably due to heterogeneous nucleation. These discrete sites of weakness, termed nuclei, are typically either free microbubbles within the liquid volume, or volumes of gas trapped in or on adjacent surfaces. Nuclei have been shown to control not only inception [12] but also the dynamics of cavitation [13]. In the current work the effect of nuclei injection and cavitation number on the spectral content of a cavitating finite span hydrofoil is presented.

Experimental Overview

Experiments were performed in the variable pressure Cavitation Research Laboratory (CRL) water tunnel at the Australian Maritime College (AMC). The 365 m^3 capacity tunnel (demineralized water, conductivity of order 1μS/cm) is fitted with ancillary systems for strict control of dissolved oxygen content and continuous removal of nuclei and incondensable gas volumes. The free stream velocity and operating pressure in the 0.6 m square test section can be independently controlled within the ranges of 2 to 12 ms^{-1} and 4 to 400 kPa absolute, respectively. The test section velocity is spatially uniform to within ±0.5%, with temporal variations of less than 0.2%, while the free stream turbulent intensity is approximately 0.5%. Upstream of the test section a geometrically configurable array of microbubble generating devices allow for controlled nuclei injection. A schematic of the experimental facility is given in Figure 1, while a more detailed description of the facility is given in [14].

*Corresponding Author, James Venning: James.Venning@utas.edu.au

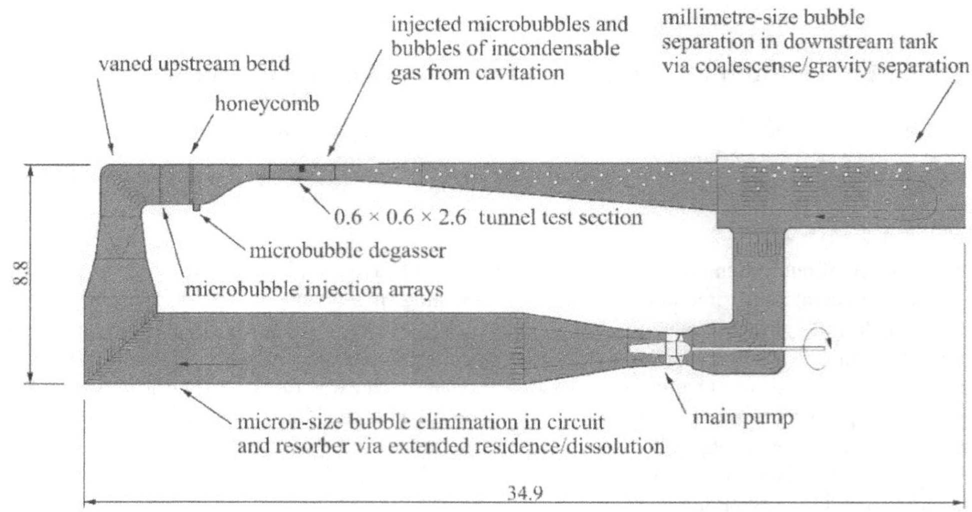

Figure 1: Schematic of the CRL water tunnel circuit. The facility is designed for continuous injection and separation of incondensable gas and cavitation nuclei via processes of coalescence/gravity separation of bubbles greater than about 100 μm in a downstream tank and dissolution of smaller sized bubbles via extended residence in the lower limb, or resorber. All dimensions are given in metres.

An anodized aluminium NACA-0015 section hydrofoil was vertically mounted in the centre of the test section ceiling on a 6-component force balance for acquisition of dynamic force data. The rectangular planform hydrofoil had a chord length, c, of 0.15 m and a span-wise dimension, s, of 0.30 m with a faired tip.

Measurements were made at a chord-based Reynolds number of 1.4×10^6 over a range of cavitation numbers from 0.45 to 0.60 with the hydrofoil at a fixed incidence of 6°. The Reynolds number is defined as $Re = U_0c/\nu$, where U_0 is the test section velocity, c is the chord, and ν is the kinematic viscosity of water; the cavitation number is defined as $\sigma = (p_0 - p_v)/\frac{1}{2}\rho U_0^2$, where p_0 is the static pressure at the hydrofoil tip (test section centerline), p_v is the vapour pressure, and ρ is the water density. Simultaneous measurements of hydrofoil normal force and high-speed photography of the cavitation were acquired at 7 kHz for a period of 3 s at each test condition for investigating the phasing of the shedding mechanism. High-speed images were acquired using a LaVision HighSpeedStar8 12-bit camera at a resolution of 1024 x 1024 pixels with a Nikkor f/1.4 50 mm lens. Additional force data for obtaining high-resolution spectra were recorded at 1 kHz for 240 s giving about 5,000 cycles of the dominant frequency. The Power Spectral Density (PSD) were derived using the Welch estimate with a window size of 4096 points (4.1 s) and overlap of 256 points (0.26 s).

Two conditions were investigated where the freestream is either deplete of nuclei, or with abundant injected nuclei, at a dissolved oxygen content of ~3 ppm for both cases. The nuclei deplete case is devoid of any active nuclei [15] while for the abundant case homogeneous nuclei injection was achieved using a 3×10 array of poly-disperse microbubble generators producing a dominant size of 10 to 15 μm at a concentration of ~100 cm^3 [16].

Results

For all cases the cavity shedding was found to be highly periodic as shown in the PSD of the hydrofoil body-normal force, presented below in Figure 2. Continuous wavelet transforms of force data (not presented here) were performed to discriminate between and identify those peak frequencies associated with the cavitating flow physics from those being artefacts of the PSD processing. Additionally, fast Fourier transform (FFT) analysis of pixel intensities from the high-speed photography was undertaken to highlight regions of maximum variation, or energy, within the flow. Amplitudes of the identified peak frequencies are presented in Figure 3 to assist in visualising the shedding mechanisms associated with each frequency, while space-time plots at various hydrofoil locations are presented later in Figure 4.

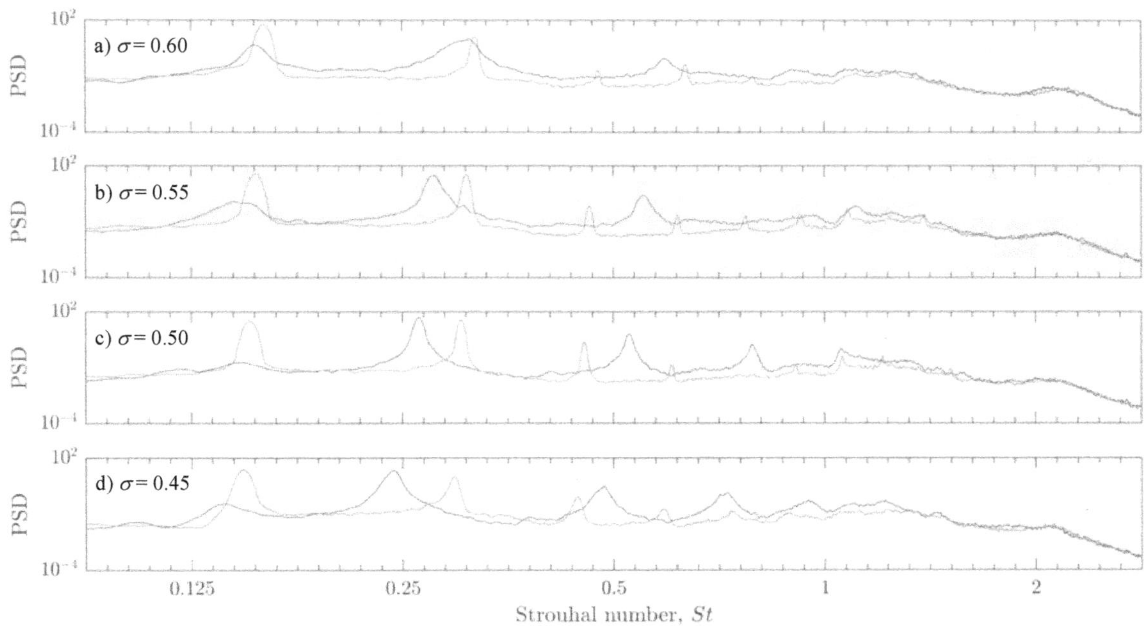

Figure 2: Comparison of the hydrofoil body-normal force spectra with abundant nuclei injection (red), and deplete of nuclei (blue).

For the nuclei deplete case three peak frequencies were observed over the range of tested σ values, $0.60 \geq \sigma \geq 0.45$, the dominant frequency decreasing with cavitation number from a Strouhal number, $St = fc/U$, of 0.311 to 0.243 (Figure 2(a) – (d)), where f is the shedding frequency. Visual examination and FFT analysis of the high-speed imaging reveals this dominant frequency to be associated with large-scale cavity growth and subsequent condensation/collapse occurring over a significant portion of the hydrofoil. This is evident in Figure 3(c) showing a reconstruction of the image frame at $\sigma = 0.55$ expressed as the FFT amplitude of each pixel a Strouhal number of 0.277. The high amplitude region is seen to extend over a distance of approximately $0.5s$, centred at $\sim 0.35s$. The first frequency peak, occurring at half the dominant frequency for all σ values, is associated with local cavity shedding near the hydrofoil tip (Figure 3(b)) such that there are two distinct, but coupled shedding regions, while the third peak at around twice the main shedding frequency is associated with both tip and root activity (Figure 3(d)). This spectral behavior is further illustrated in the space-time plots given in Figure 4. By inspection of the force time-series, and analysis of the space-time plots, it was shown that the relative strength between the first harmonic and the sub-harmonic shedding mode tended to alternate irregularly. Furthermore, it can be seen from the PSD data that the sub-harmonic peak becomes a minimum at $\sigma = 0.50$ while, conversely, the higher frequency peak reaches its maximum at this condition.

The large-scale shedding occurs over the mid-half to two-thirds of the span as it is only in this region that the cavity undergoes full chord-wise collapse. As such, during the growth phase the cavity exhibits a crescent-shaped closure region as the cavity regrows from the hydrofoil leading edge at mid-span (Figure 3(a)). A re-entrant jet is formed, the upstream extent of which remains almost stationary as the growing cavity reaches the hydrofoil trailing edge. Although the re-entrant jet may occasionally break through the surface of the growing cavity it does not initiate a shockwave as has been observed for flows about bodies with much greater adverse pressure gradients [10]. As the cavity reaches the trailing edge region shockwaves are triggered, initially at the root and tip which then propagate and focus at the mid-span, leading to the condensation of the cavity in this region as the wave travels upstream and the cycle begins anew.

For the nuclei abundant case the shedding physics are altered due to the high concentration of continuously activated nuclei. The fundamental frequency of the bulk shedding is at $St \approx 0.15$ corresponding in this case to the first peak in the force PSD (Figure 3) and shows only a slight decrease with cavitation number. This primary shedding is approximately two-dimensional, occurring over the full hydrofoil span (Figure 3(f)).

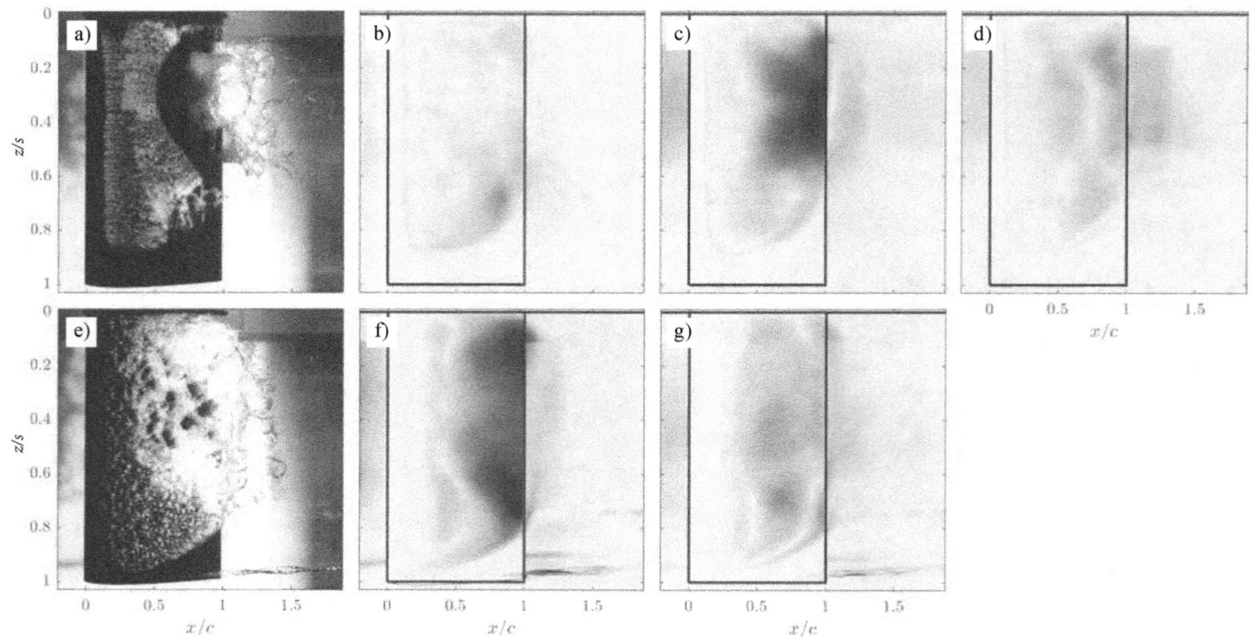

Figure 3: Photographs of the NACA-0015 hydrofoil (a) without, and (e) with nuclei injection, showing the alteration of cavity topology with the addition of nuclei. FFT analysis of pixel intensity from high-speed images are presented for peak frequencies without (top row) and with nuclei injection (bottom row), where (b) $St = 0.144$, (c) $St = 0.277$, (d) $St = 0.553$, (f) $St = 0.155$, and (g) $St = 0.308$. Both images and FFT analysis are taken at a test condition of $Re = 1.4 \times 10^6$ and $\sigma = 0.55$.

The second peak in the nuclei abundant cases is consistently twice that of the fundamental and can be shown to be associated with the propagation of the first of two shockwaves each cycle. The first shockwave forms once the growing cavity reaches the trailing edge. This shockwave propagates upstream but loses strength and speed, stalling before reaching the cavity leading edge resulting in small-scale localized shedding and condensation. Shortly after this another shockwave forms that travels upstream with greater strength causing large scale condensation with subsequent near- two-dimensional re-growth along the hydrofoil span [17]. Due to the continuous supply of activated nuclei there is no clear cavity detachment line but a discontinuous cavity leading edge such that the cavity is not condensed completely each shedding cycle as was the case for the nuclei deplete case. Nuclei activated at the hydrofoil leading edge grow as they are advected downstream merging to form a contiguous cavity volume. During the growth phase, advected bubbles reach the cavity trailing edge and condense simultaneous with the formation of re-entrant flow. Cavity growth is due to the volume of activated nuclei being greater than the condensing volume at the cavity trailing edge. As described above, once the slow growing cavity reaches the trailing edge, the first shockwave is initiated. As with the nuclei-deplete case the re-entrant jet apparently has no direct influence on the shedding. The peaks at higher frequencies can be shown to be harmonic artefacts of the FFT process.

The space-time plots presented below provide further evidence of the multi-modal nature of the shedding in both nuclei deplete (Figure 4), and nuclei abundant cases (Figure 5). Figure 4(a) presents an ensemble of a single column of pixels extracted from sequential image frames across the length of the span near the trailing edge ($0.8c$), while Figures 4(b) & 4(c) are chord-wise extractions taken at $0.4s$ and $0.7s$, respectively. The large-scale shedding is clearly identified in Figure 4(a) with a cycle composing of the growth phase with re-entrant jet flow, cavity (high intensity pixels) and the passage of a dark vertical band indicating the absence of a cavity. The tip shedding that occurs every alternate cycle is indicated in Figure 4(a) & 4(c), while the consequent alternating strength of the large-scale shedding, initiated at the trailing edge, is evident in Figure 4(b) as shown by the increased speed (solid line) of the shock travelling towards the leading edge following a tip shedding event (as compared with the dashed line).

The two-dimensional nature of the full-span shedding with injected nuclei is evident in the span-wise space-time plot given in Figure 5(a), while the higher frequency small-scale shedding can be seen at mid-span in Figure 5(a) and by the short upstream travelling shock-front indicated in Figure 5(b).

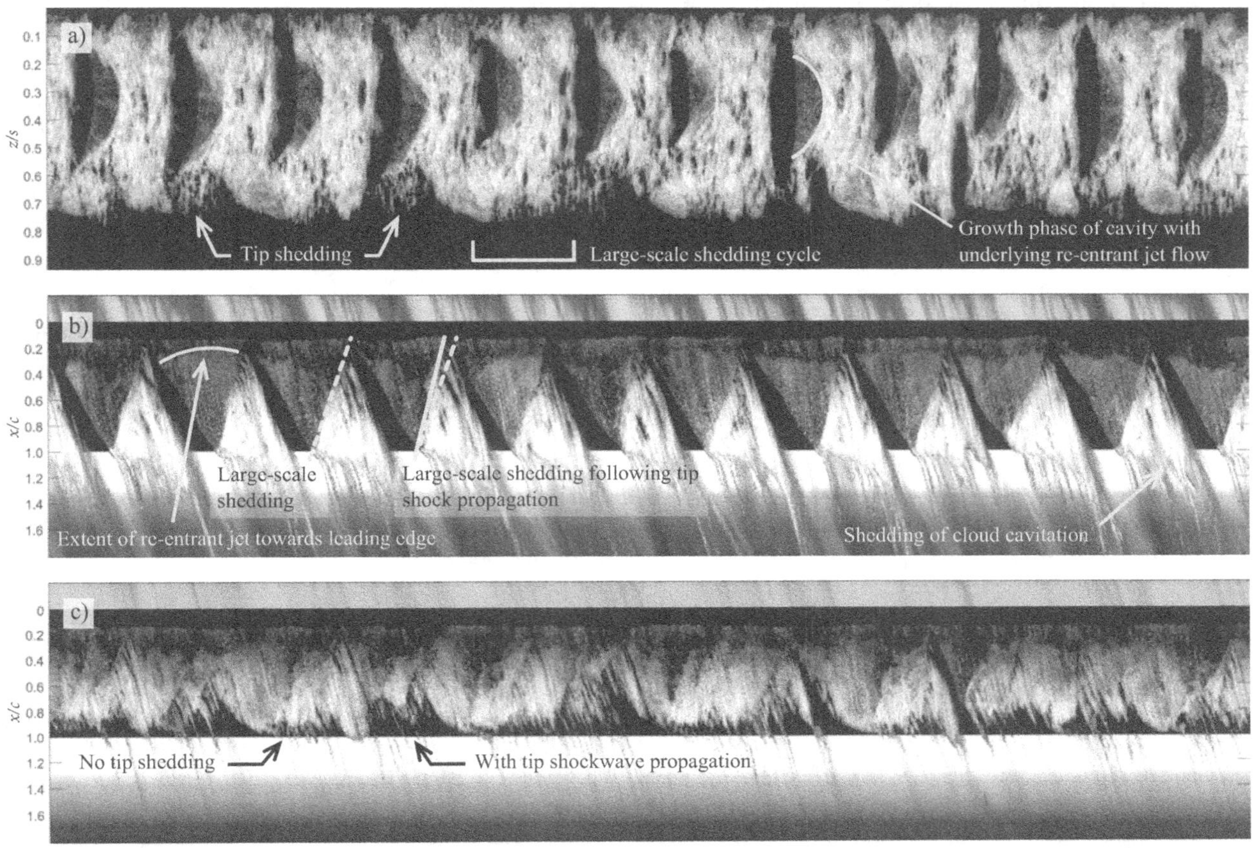

Figure 4: Space-time plots for the nuclei deplete case at $\sigma = 0.55$. Time progresses left-to-right, while position is the vertical dimension for the respective orientation: a) span-wise plot at $0.8c$, b) chord-wise plot at $0.4s$, and c) chord-wise plot at $0.7s$.

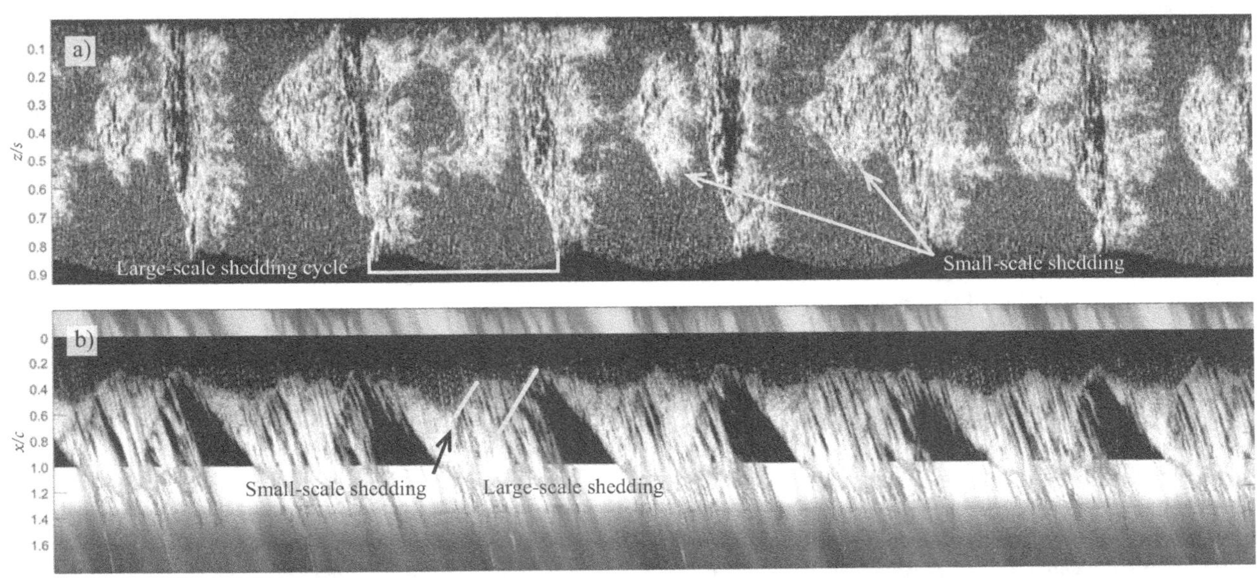

Figure 5: Space-time plots for the nuclei abundant case at $\sigma = 0.55$. Time progresses left-to-right, while position is the vertical dimension for the respective orientation: a) span-wise plot at $0.4c$, and b) chord-wise plot at $0.4s$.

Conclusion

The spectral content of the body-normal force and the shedding dynamics of a cavitating NACA-0015 section hydrofoil has been studied to examine the effect of both cavitation number and freestream nuclei content. Tests were conducted at a constant Reynolds number over a range of cavitation numbers for test conditions both deplete of freestream nuclei, and with the injection of a high concentration of nuclei. Three distinct, yet coupled frequencies were observed in the nuclei deplete cases – the dominant frequency associated with large-scale shockwave induced cavity shedding at mid-span, the sub-harmonic with the alternating strength of localized tip shedding, and the first harmonic at twice the dominant mode frequency was associated with small-scale shedding activity at both root and tip. The frequencies were found to decrease with cavitation number over the narrow range tested. The presence of abundant freestream nuclei was found to alter the hydrofoil response and shedding characteristics. Two distinct frequencies were observed in the latter case, the dominant low-frequency mode associated with large-scale full-span cavity shedding. This dominant mode was observed at approximately half the frequency of the corresponding nuclei deplete mode in all cases. The higher frequency mode was due to an initial, weaker shock propagation centred at mid-span that resulted in only localized cavity condensation in between the passage of the large-scale shedding events.

Acknowledgements

This project was supported by the Defence Science and Technology Group (Mr. Brendon Anderson and Dr. David Clarke), the University of Tasmania, and the US Office of Naval Research (Dr. Ki-Han Kim, Program Officer) and ONR Global (Dr. Woei-Min Lin) through NICOP S&T Grant no. N62909-15-1-2019.

References

[1] Kawakami, D. T., Fuji, A., Tsujimoto, Y., & Arndt, R. E. (2008). *An assessment of the influence of environmental factors on cavitation instabilities*. Journal of Fluids Engineering, 130(3).

[2] Kjeldsen, M., Arndt, R. E., & Effertz, M. (2000). *Spectral characteristics of sheet/cloud cavitation.* Journal of Fluids Engineering, 122(3).

[3] Ganesh, H., Wu, J., & Ceccio, S. L. (2016). *Investigation of cavity shedding dynamics on a NACA0015 hydrofoil using time resolved x-ray densitometry.* In 31st Symposium on Naval Hydrodynamics.

[4] Tan, D., Li, Y., Miorini, R., Vagnoni, E., Wilkes, I., & Katz, J. (2014). *Role of large scale cavitating vortical structures in the rotor passage of an axial waterjet pump in performance breakdown.* In 30th Symposium on Naval Hydrodynamics.

[5] Gnanaskandan, A., & Mahesh, K. (2016). *Large Eddy Simulation of the transition from sheet to cloud cavitation over a wedge*. International Journal of Multiphase Flow, 83.

[6] Barbaca, L., Pearce, B. W., & Brandner, P. A. (2017). *Experimental study of ventilated cavity flow over a 3-D wall-mounted fence*. International Journal of Multiphase Flow, 97.

[7] Ceccio, S., & Brennen, C. (1992). *Dynamics of attached cavities on bodies of revolution.* Journal of Fluids Engineering, 114(1).

[8] Brandner, P. A., Walker, G. J., Niekamp, P. N., & Anderson, B. (2010). *An experimental investigation of cloud cavitation about a sphere*. Journal of Fluid Mechanics, 656.

[9] Ganesh, H., Mäkiharju, S. A., & Ceccio, S. L. (2016). *Bubbly shock propagation as a mechanism for sheet-to-cloud transition of partial cavities*. Journal of Fluid Mechanics, 802.

[10] de Graaf, K. L., Brandner, P. A., & Pearce, B. W. (2017). *Spectral content of cloud cavitation about a sphere*. Journal of Fluid Mechanics, 812.

[11] Laberteaux, K. R., & Ceccio, S. L. (2001). *Partial cavity flows. Part 2. Cavities forming on test objects with spanwise variation.* Journal of Fluid Mechanics, 431.

[12] Gindroz, B., & Billet, M. L. (1998). *Influence of the nuclei on the cavitation inception for different types of cavitation on ship propellers*. Journal of fluids engineering, 120(1).

[13] de Graaf, K., Pearce, B., & Brandner, P. (2016). *The influence of nucleation on cloud cavitation about a sphere*. In 16th International Symposium on Transport Phenomena and Dynamics of Rotating Machinery (ISROMAC 2016).

[14] Brandner, P. A., Lecoffre, Y., & Walker, G. J. (2007). *Design considerations in the development of a modern cavitation tunnel*. In 16th Australasian Fluid Mechanics Conference.

[15] Venning, J. A., Khoo, M. T., Pearce, B. W., & Brandner, P. A. (2018). *Background nuclei measurements and implications for cavitation inception in hydrodynamic test facilities*. Under consideration for publishing in Experiments in Fluids.

[16] Giosio, D., Pearce, B. W., & Brandner, P. A. (2016). *Influence of Pressure on Microbubble Production Rate in a Confined Turbulent Jet*. In 20th Australasian Fluid Mechanics Conference.

[17] Venning, J., Smith, S., Brandner, P., Giosio, D., & Pearce, B. (2017). *The influence of nuclei content on cloud cavitation about a hydrofoil*. In 17th International Symposium on Transport Phenomena and Dynamics of Rotating Machinery (ISROMAC 2017).

© 2018 ASME

The numerical scheme for the history force integrals in hydrodynamics

[1,2]Vladimir Vanovskiy

[1]*Moscow Institute of Physics and Technology, Dolgoprudny, Russia;*
[2]*Institute for Problems in Mechanics of the Russian Academy of Sciences, Moscow, Russia*

Abstract

A high-order precision numerical scheme was proposed for the Basset type kernel integrals and the Basset type integro-differential equations that often arise in hydrodynamics. The scheme was tested on the model integrals and on the problems of particle and bubble focusing in a standing acoustic wave in liquid. The scheme showed around 2 orders higher speed of integral estimation compared with the existing analogs. The obtained results show that the integral can be approximated with high order of precision and the real problem is simulated well by the proposed scheme. A variable step technique was used to increase the precision of integro-differential equation simulation even more by eliminating the discrepancy at the very start of the simulation. The proposed numerical scheme may find its applications in many biological and medical problems related to acoustophoresis or to particle and bubble focusing by ultrasound.

Keywords: Basset integral; Basset force; memory forces; integrodifferential equations; acoustophoresis; fractional derivative;

Introduction

The numerical simulation of the integro-differential equations describing "history" forces or physical systems with memory is complicated because at any step of the numerical integration all the previous steps should be considered and the time of execution may become quadratic in the number of simulation steps. In case of the integrals with the singular kernels the situation may become even worse as the usual numerical integration schemes of the Newton-Cotes family are useless. The example of such an integral is the Basset kernel integral described by the integral operator

$$\hat{I}(f) = \int_{\tau=0}^{t} \frac{f(\tau)d\tau}{\sqrt{\pi(t-\tau)}} \tag{1}$$

This operator is often named a semiderivative of function f and is studied well in mathematical applications. The kernel is singular at $\tau \rightarrow t$ and its fast computation is a big challenge in all the problems where such a kernel appears such as Basset force consideration in hydrodynamics or fractional derivative calculation. Many approaches of such an integral approximation were developed in the last years. A comprehensive review that contains most of them was recently published [1]. We try to focus on the exact integral estimation without changing the kernel as it was done in [2]. However the proposed here approach is much simpler and the integral is calculated more than 100 times quicker. Also the proposed approach gives the possibility for the exact estimation of the obtained error of integration for the functions with limited corresponding order derivative value.

Body

The integrand function in (1) is assumed to be $k-1$ times continuously differentiable over the closed interval $[0, t]$ and to have the k-th derivative in the opened interval $(0, t)$ limited by some constant. The integral is assumed to be taken over the uniform mesh containing n points with step h, $t = nh$. It was shown that under such assumptions there exists a numerical scheme providing the integration error limited by $Mh^k\sqrt{t}$ where M stays for some constant. The numerical scheme coefficients were calculated for $k = 3$. The integral was rewritten as follows

$$I = I(f) = 2\sqrt{h/\pi}\sum_{i=1}^{n} I_i;\ \mathrm{I}_i = -\int_{i-1}^{i} f(hx)d\sqrt{n-x} \tag{2}$$

*Corresponding Author, Vladimir Vanovskiy: vladimir.vanovsky@gmail.com

With the help of decomposition into Taylor series with the Lagrange form residual the following approximation is obtained together with the error estimation. The error estimated here proves the declared whole integration error where M is the upper estimation of the function f third derivative on the all interval multiplied by some constant.

$$\begin{aligned} & I_i \approx \tilde{I}_i \approx \left(\sqrt{n-i+1}-\sqrt{n-i}\right)\left(\alpha_i f_{i-2}+\beta_i f_{i-1}+\gamma_i f_i\right) \\ & \frac{|I_i-\tilde{I}_i|}{\sqrt{\varepsilon_i}} < \frac{h^3}{6} f'''_{i\max}\left(\frac{11}{12}+O\left(\frac{1}{n-i}\right)\right) \\ & \alpha_i = -1/12+\varepsilon_i^2/60 \\ & \beta_i = 2/3-\varepsilon_i/6-\varepsilon_i^2/30 \\ & \gamma_i = 5/12+\varepsilon_i/6+\varepsilon_i^2/60 \\ & \varepsilon_i = (\sqrt{n-i+1}-\sqrt{n-i})^2 \\ & f'''_{i\max} = \max(f'''(\xi)),\ \xi \in ((i-2)h, ih) \end{aligned} \quad (3)$$

A big advantage of the scheme proposed is that the coefficients α, β, γ in the scheme depend only on n−i and may be calculated in advance once for all the n in order to boost the simulation of the integro-differential equation when the integral is calculated on every step. Also a variable step technique is used to increase the precision of integro-differential equation simulation even more by eliminating the discrepancy at the very start of the simulation.

The numerical schemes of the orders 1–4 were evaluated on a model function $f(t) = \cos t$ integrated over interval $[0,50]$. The answer can be expressed analytically in terms of Fresnel integral as $\sqrt{2}C(10) \approx 0.70696351315804088$. The function and interval were chosen the same as in [1] in order to compare the obtained results. All the calculations were made in Matlab environment with a modern i7 CPU. Matlab is considered to be far from the fastest computing environment and some speed increase from 3 to 5 times may appear if the simulations were done using Fortran as in [1]. However in figure 1 one may observe that the precision of the schemes corresponds to declared and calculation speed is more than two orders higher than of the third order Daitche scheme and much better than any other scheme investigated in [1].

The Basset-Boussinesq-Oseen equation (BBO equation) is used to describe the motion of a small spherical particle in unsteady flow with velocity field $v(x, t)$ at low Reynolds numbers. The modeling of the BBO equation in oscillatory flow is a challenging problem because of the big number of periods to model and an integrodifferential part with a Basset kernel. We have used the numerical scheme proposed to model the movement of a small stiff particle in the stationary acoustic wave. In figures 2a and 2b by points, solid and dashed line are denoted the results of the BBO equation modeling without the Basset force consideration, with the Basset force consideration and the asymptotic particle trajectory obtained analogically to [5]. One may observe that our approach works well for such a problem and gives one the possibility for the precise modeling of the BBO equation in hydrodynamics extending the limits of integration much further than in [5].

In figures 2(a) and 2(b) by q is denoted the dimensionless particle coordinate (2π equals to the standing wave wavelength) and by τ_1 is denoted the “slow” dimensionless time (defined in [6]). Figure 2(a) may be considered as proof of the Basset force consideration importance in the acoustophoresis problem as the particle trajectory and focusing time changes drastically when one starts considering the Basset force. The figure 2(b) demonstrates that the Basset force consideration may even change the particle focusing point from standing wave pressure node to antinode. The difference between the exact calculation results (solid line) and the asymptotic particle trajectory (dashed line) appeared because of very long integration time (around 10000 wave periods, 1.2 million steps, 8 hours integration time for figure 2(a)) and not very small acoustic wave magnitude. When the wave magnitude tends to zero the precise simulation results practically coincide with the asymptotics as follows from the Krylov-Bogolyubov averaging technique theory [7]. However the time to simulate the BBO equation until particle focuses grows as $1/\varepsilon^4$ where ε stands for the standing wave magnitude so from the consideration of realistic time to computation the magnitude cannot be decreased to much smaller values and many real case situations can be modeled only by the asymptotical solution.

Figures & Tables:

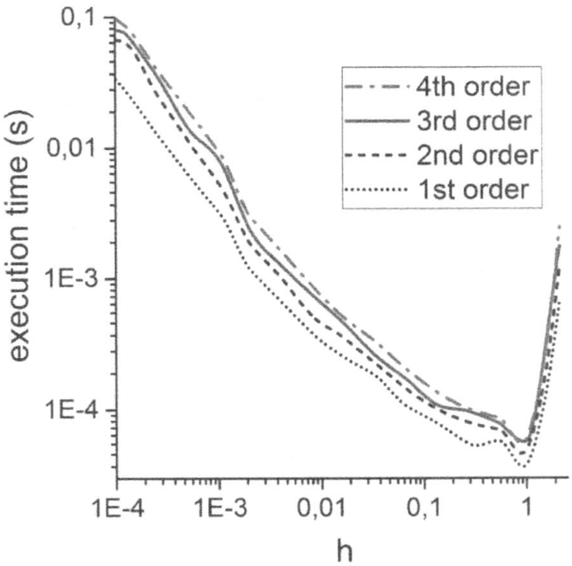

Figure 1(a). The time values vs step size plotted for the different order of accuracy methods. Dotted, dashed, solid and dash-dot lines stay for the 1st-4th order numerical schemes correspondingly.

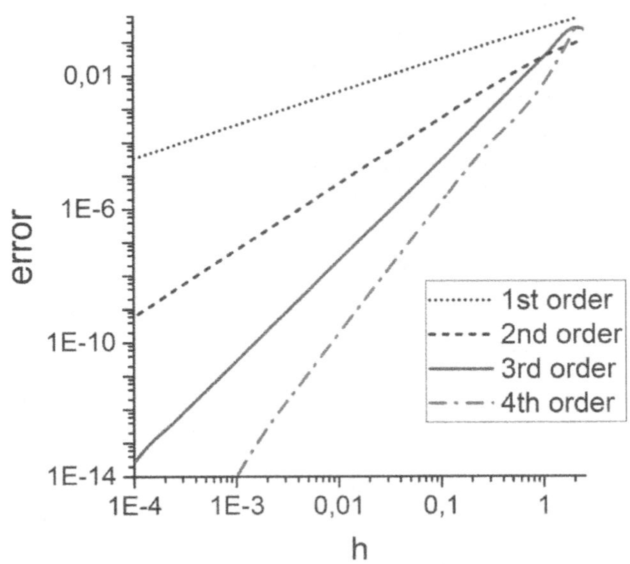

Figure 1(b); The error values vs step size plotted for the different order of accuracy methods. Dotted, dashed, solid and dash-dot lines stay for the 1st-4th order numerical schemes correspondingly.

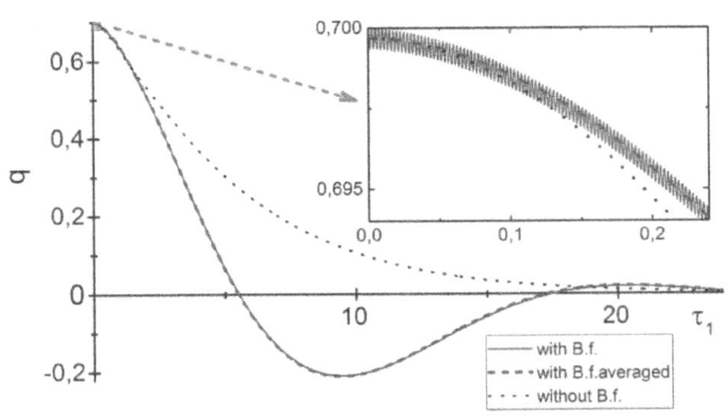

Figure 2(a). The simulation of particle trajectory for the dense particle (particle to fluid density ratio 1.5) moving in fluid in standing acoustic wave. The asymptotic and the exact numerical solutions of the integro-differential BBO equation (light solid line and dashed line) and the numerical solution of the same equation without the Basset force (dotted line) are plotted. The inset shows the magnified part of the dependence at the process start. Around 10000 wave periods or 1.2 million steps were simulated. The simulation took around 8 hours. The asymptotical solution was simulated by the same numerical technique in 0.17 sec.

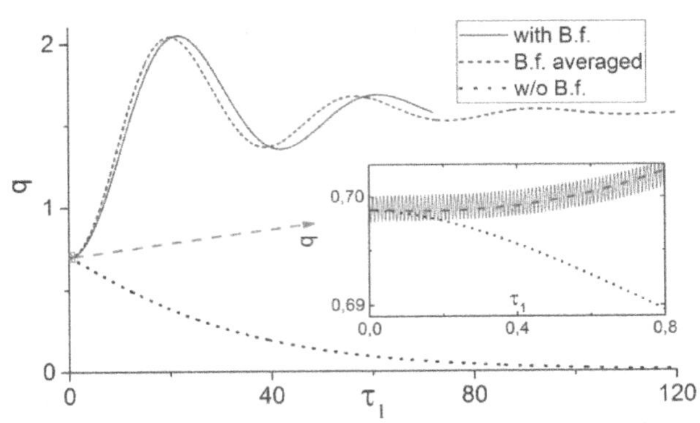

Figure 2(b). The simulation of particle trajectory for the light particle (particle to fluid density ratio 0.6) moving in fluid in standing acoustic wave. The asymptotic and the exact numerical solutions of the integro-differential BBO equation (light solid line and dashed line) and the numerical solution of the same equation without the Basset force (dotted line) are plotted. The inset shows the magnified part of the dependence at the process start. Around 7500 wave periods or 0.9 million steps were simulated. The simulation took around 6 hours. The asymptotical solution was simulated by the same numerical technique in 0.05 sec.

Conclusion

The proposed numerical scheme for the Basset integral approximation proves to be much simpler and faster than the analogs. It also allows some optimizations such as the precomputed coefficients for the integro-differential equation solution and the smaller step size in the beginning of the grid. The work of the scheme with the 3rd order of the integral approximation was demonstrated on the example of the model integral and in a model problem of particle focusing in the standing acoustic wave. The proposed scheme allows the solution to be extended to much bigger interval than in the previous works on this problem up to the moment of the particle focusing in a standing wave pressure node. Also with the help of the numerical scheme proposed it was demonstrated that the particle may even change its focusing point in the standing wave from the pressure node to antinode. The proposed numerical scheme may find its applications in many biological and medical problems related to acoustophoresis or to particle and bubble focusing by ultrasound.

References

[1] Moreno-Casas, P.A., Bombardelli, F.A. (2016). *Computation of the Basset force: recent advances and environmental flow applications.* Environmental Fluid Mechanics. 16(1). pp. 193–208.

[2] Daitche, A. (2013). *Advection of inertial particles in the presence of the history force: Higher order numerical schemes.* Journal of Computational Physics. 254. pp. 93–106.

[3] Michaelides, E.E. (1992). A novel *way of computing the Basset term in unsteady multiphase flow computations.* Physics of Fluids A: Fluid Dynamics. 4(7). pp. 1579–1582.

[4] Maxey, M.R., Riley, J.J. (1983). *Equation of motion for a small rigid sphere in a nonuniform flow.* Physics of Fluids. 26(4). pp. 883–889.

[5] Aksenov, A.V., Petrov, A.G., Shunderyuk, M.M. (2011). Motion of Solid Particles in Fluid in a Nonlinear Standing Ultrasonic Wave. Doklady Physics. 439(1). pp. 37–41.

[6] Vanovskiy V., Petrov A. (2018). The Numerical Scheme for the Basset Type Integro-Differential Equation in Hydrodynamics. In Applied Physics, System Science and Computers. APSAC 2017. Lecture Notes in Electrical Engineering. V. 428. pp. 69–75.

[7] Bogoliubov, N. N. and Y. A. Mitropolsky. (1961). Asymptotic *methods in the theory of nonlinear oscillations.* Gordon and Breach, New York: 298 pages.

© 2018 ASME

The role of diffusion-driven nucleation in hydrodynamic cavitation

[1]Tim F. Groß; [1]Alexander Terwort; [1]Gerhard Ludwig; [1]Peter F. Pelz*;

[1]*Chair of Fluid Systems, Technische Universität Darmstadt, Otto-Berndt-Straße 2, 64287 Darmstadt, Germany*

Abstract

Most of the time it is argued surface nuclei only plays a minor role when it comes to the onset of cavitating flows. Experimental findings indicate diffusion-driven nucleation from surface nuclei is an important mechanism that needs to be considered in this regard as well. A physical model being experimentally validated is introduced to describe diffusion-driven nucleation in a cavitating flow.

Keywords: hydrodynamic cavitation, diffusion-driven nucleation; physical model

Introduction

Cavitation in hydraulic machines is usually accompanied by negative aspects. It has a negative impact on the efficiency, causes vibration, noise and damage of structural elements [1] [2]. Yet, cavitation has its useful applications. It is used to clean surfaces [3], fragment kidney stones in the human body [4] and enhance the efficiency of mixing processes [5].

Regardless of whether cavitation should be avoided or is used for beneficial purposes all cavitating flows have one fundamental similarity. The smallest elements of cavitating flows are cavitation nuclei. In cavitation research a distinction is made between bubbles and particles floating freely in the liquid and nuclei being attached to walls. Bubbles and particles in the bulk of the liquid are referred to as free stream nuclei while nuclei present in the walls bounding the flow are referred to as surface nuclei [1]. Harvey et al. were the first to postulate the existence of nuclei in hydrophobic crevices [6]. Atchley & Prosperetti later introduced the crevice model as the standard model for acoustic cavitation today [7]. A widespread opinion is that the onset of cavitation is mainly associated with free stream nuclei and surface nuclei are of minor importance. This assumption is usually based on the estimation of diffusion processes being much slower than typical cavitation timescales [1]. This is usually argued based in the findings of Epstein & Plesset who investigated the growth and shrinkage of bubbles in supersaturated and undersaturated liquids [8]. However, Epstein & Plesset conducted their experiments in quiescent liquids. Wijngaarden and Parkin & Kermeen reported that forced convection in a fluid flow strongly enhances mass diffusion and thus the growth of a gas bubbles that might act as cavitation nuclei [9] [10]. Peters and Honza designed an experiment to investigate bubble formation at small crevices in a laminar flow and were able to observe nucleation rates of 0.1 Hz to 1 Hz [11]. They found out that the nucleation rate depends on the supersaturation of the liquid and the shear rate at the wall but did not provide any functional dependency. Groß and Pelz improved the experimental setup and reached nucleation rates of up to 1000 Hz [12]. Based on these findings a physical model is developed to describe diffusion-driven nucleation which is presented in this paper. Figure 1 provides a schematic overview of the nuclei and nucleation theories currently available. It distinguishes between equilibrium and non-equilibrium with respect to the gas content of the liquid. A more detailed discussion of the topic can be found in [13].

Diffusion-driven nucleation

Figure 2 shows a surface nucleus located in a crevice. Supersaturated liquid passes from left to right. Due to the supersaturation of the liquid with non-condensable gas, a diffusion mass flux from the liquid into the nucleus is initiated. The nucleus grows and after reaching a critical size a bubble detaches. Usually some gas is left behind inside the crevice. This is a self-exciting process with bubble detachment as a trigger continuing as long as the liquid is supersaturated. Fick's first law of diffusion states the mass flux $\dot{m}$ depends on the concentration gradient at the liquid-gas interface [14]

$$\dot{m} = -M\mathcal{D} \int_A \nabla c \cdot \vec{n} dA. \tag{1}$$

*Corresponding Author, Peter F. Pelz: peter.pelz@fst.tu-darmstadt.de

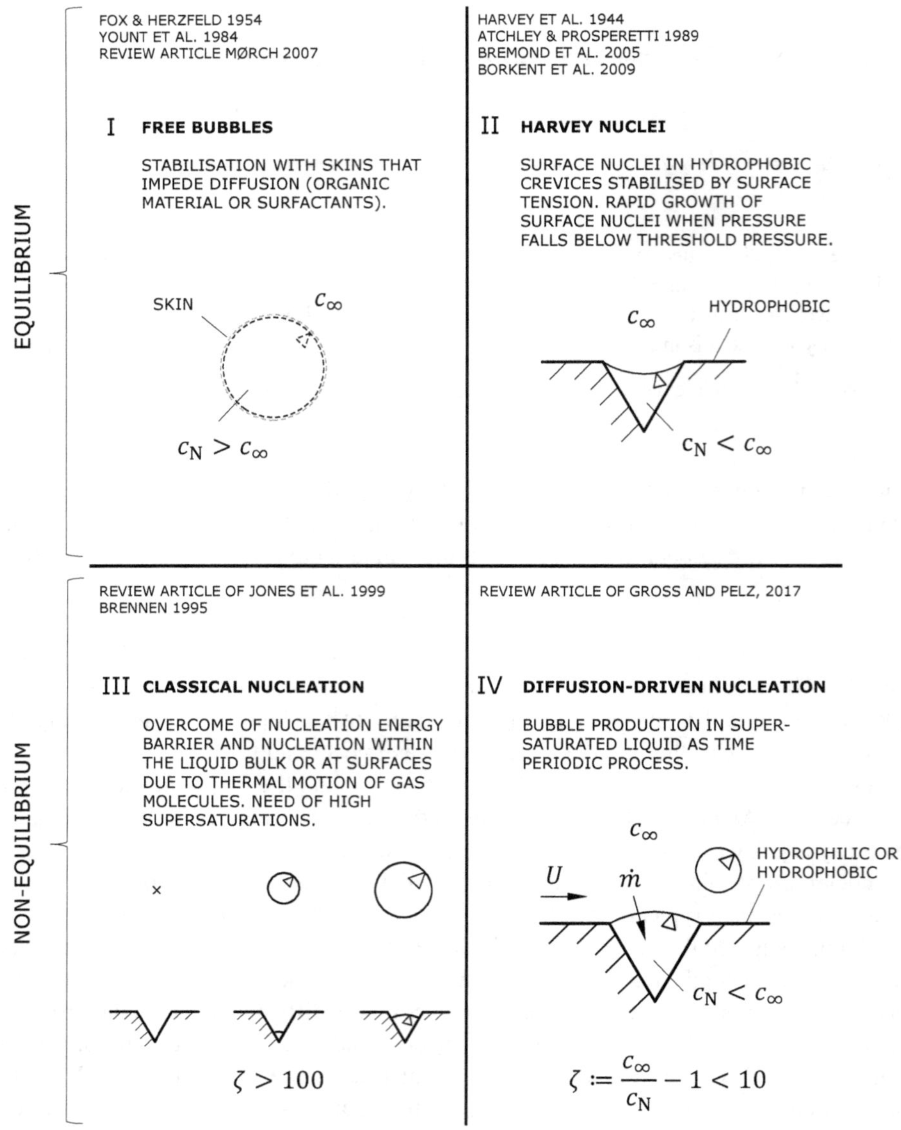

Figure 1: Schematic overview of nuclei and nucleation in liquids. The theories are distinguished between equilibrium and non-equilibrium with respect to the gas content. Adapted from [13].

M is the molar mass, $\mathcal{D}$ the diffusion coefficient, c the concentration and A the surface normal to $\vec{n}$. In an equilibrium state the gas concentration is given by Henry's law

$$c_s = p_s \mathcal{H}, \tag{2}$$

with the saturation pressure p_s and the Henry coefficient $\mathcal{H}$ [15]. Once the pressure of the liquid decreases, the liquid becomes supersaturated. The supersaturation is defined as

$$\zeta := \frac{c_\infty}{c_N} - 1. \tag{3}$$

c_∞ is the gas concentration in the liquid and c_N the concentration at the liquid-gas interface of the nucleus. Gas diffuses from the liquid into the nucleus – the process shown in Figure 2 is initialized. While the bubble growth is relatively slow, bubble detachment occurs rapidly after reaching a critical size. The nucleation rate f is then given by

$$f = \frac{\dot{m}}{m_B} \tag{4}$$

with the diffusion mass flux $\dot{m}$ and the mass of the bubble m_B.

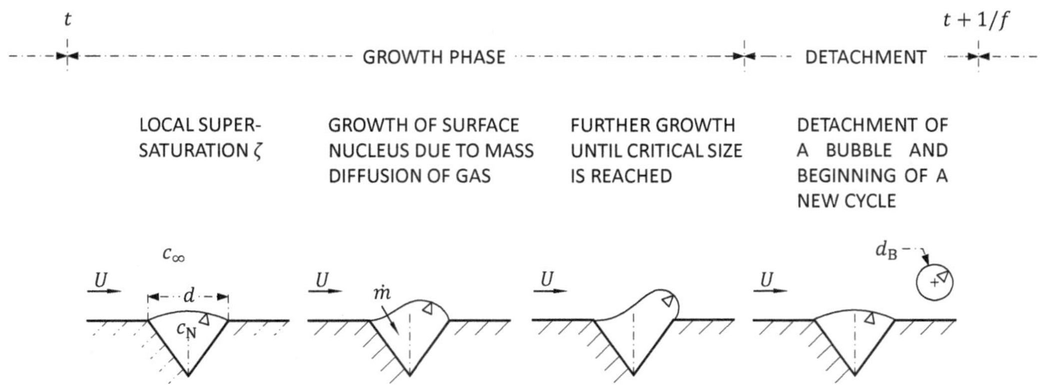

Figure 2: Principle sketch of the growth of a surface nucleus and detachment of a bubble. Adapted from [13].

The concentration field can be described by the advection-diffusion equation

$$\frac{\partial c}{\partial t} + \vec{u} \cdot \nabla c = \nabla \cdot (\mathcal{D} \nabla c), \tag{5}$$

where $\vec{u}$ is the velocity field of the flow [13]. From this approach the solution

$$Sh \approx 0.66 Pe^{1/3} \tag{6}$$

is derived, with the Sherwood number and Péclet number defined as

$$Sh := \frac{\dot{m}}{c_N \zeta M \mathcal{D} d}, \quad Pe := \frac{\dot{\gamma} d^2}{\mathcal{D}}. \tag{7}$$

A detailed discussion can be found in [13]. In the present case, d is the diameter of the crevice and $\dot{\gamma}$ the shear rate at the liquid-gas interface. When it comes to bubble detachment, three different cases are distinguished. In the first case, the size of the detaching bubble is of the same order of magnitude as the surface nucleus and the characteristic length of the device is of the same order of magnitude as the crevice. In the second case, the detaching bubble is of the same size as the nucleus, but the characteristic length of the device is much larger than the length of the crevice. In the third case, the detaching bubbles are much smaller than the crevice and the characteristic length of the device. From this, three different solutions for the dimensionless size of the gas bubble δ^3 are derived

$$\delta^3(We) = \begin{cases} (\epsilon_I We)^{-1/3} & \text{for case I,} \quad (8) \\ (\epsilon_{II} We)^{-3/4} & \text{for case II, and} \quad (9) \\ (\epsilon_{III} We)^{-1} & \text{for case III} \quad (10) \end{cases}$$

with

$$\delta = \frac{d_B}{d}, \quad We := \frac{\varrho \dot{\gamma}^2 d^3}{S}. \tag{11}$$

We is the Weber number and ϵ a dimensionless constant. Groß and Pelz also derive the Strouhal number Sr as dimensionless nucleation frequency given by

$$Sr \approx 1.26\, \Lambda \zeta Pe^{1/3} \delta^{-3}. \tag{12}$$

Experimental results

The experimental setup consists of two tanks with a volume of 50 litres each. They are connected by a pipe with a transparent test section. Inside the test section there is a silicon wafer with holes of a diameter between 25 µm and 100 µm acting as nucleation sites. The experiments were conducted by filling one tank with a working fluid. The second tank was left empty at atmospheric pressure. The first tank was pressurized by compressed air. Due to the pressure difference the liquid streamed from one tank into the other one, passing the test section. The bubbles emerging from the surface nuclei were recorded by a high speed camera and saved for later processing. The test section is a transparent rectangular channel, so the images can be recorded either from the top view or from the side view. A more detailed description of the setup can be found in [13]. To estimate the size of the bubble the images just before bubble detachment were used. It has to be noted that the calculation of the bubble volume has a relatively large measurement uncertainty. This is due to the bubble deformation just before detachment as a result of high shear rates, the calculation of the bubble volume based on a two dimensional image and the limited resolution of the camera.

The investigated bubble detachment describes a case II detachment mechanism, since the crevice diameter is much smaller than the characteristic length of the channel and the bubble is of the same size as the nucleus. Figure 3 pictures the results obtained in the experiments as dimensionless bubble diameter over the Weber number. The grey symbols show the real measurement data. Each symbol represents a different hole diameter. For comparison, the black dots represent the data calculated by the physical model. As expected the bubble volume decreases with increasing shear rate and thus increasing Weber number. Furthermore, a linear dependency on the supersaturation and a non-linear dependency on the shear rate is derived. As Figure 3 indicates, the model is a reasonable approach to describe diffusion-driven nucleation from surface nuclei.

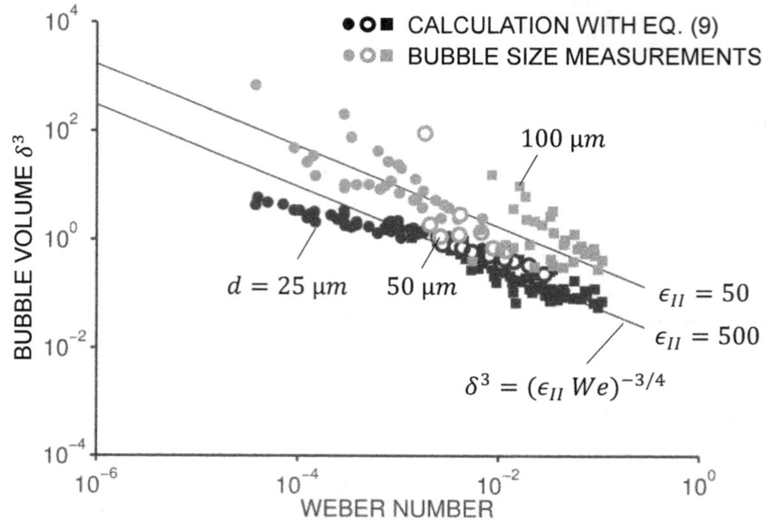

Figure 3: Dimensionless bubble volume δ^3 vs. Weber number We. The solid lines are calculated with equation 9. Grey symbols are based on measurement of the size of the bubbles. Black symbols are the calculated bubble volumes with equation 12. The symbols mark different crevice diameters: 25 μm filled circle, 50 μm open circle, 100 μm square. All measurements were performed with tap water. Adapted from [13].

The dimensionless parameter ϵ_{II} yields to 5×10^2 in case of the model, while $\epsilon_{II} = 5 \times 10^1$ seems to be more feasible regarding the experiments. There are two main reasons for this difference in the data sets. First, as stated before the systematic measurement uncertainty of the bubble size is relatively large. Second, the mass flux model contains uncertainties as well. In reality, the surface of the surface nucleus is larger, so the mass flux is much greater. Furthermore, the influence of the protruding surface nucleus on the thickness of the concentration boundary layer is neglected. The underestimation leads to a smaller calculated bubble diameter and thus to a higher value of ϵ_{II}. Therefore, the derived mass flux can be seen as a lower bound.

In Figure 4 another bubble detachment mechanism is shown that might be relevant in some technical applications. In this case, the working fluid was replaced by silicon oil with a kinematic viscosity of $20 \times 10^{-6}\, m/s$. Blind holes with a diameter of 0.8 mm worked as nucleation sites. The experiments were conducted in the same test section as the experiments described before. The crevice diameter is of the same order of magnitude as the characteristic length of the channel, but the detaching bubbles are much smaller. Hence, a case III detachment mechanism occurs. The images in Figure 4 illustrate the detachment based on a Plateau-Rayleigh instability [16]. A surface nucleus grows over the edge of its crevice and forms a cylindrical body with a diameter d_c. Small perturbations with different wavelength deform the cylindrical body due to the surrounding fluid in motion. If the wavelength is smaller than a critical wavelength, the surface increases, higher wavelength cause a shrinkage of the surface. Surface increasing perturbations stabilize the nucleus, while surface decreasing perturbations are not damped but increased. The critical wavelength of a cylindrical body is approximately its circumference πd_c. When the growing nucleus reaches the critical length it starts oscillating up and down. Since the perturbations are not attenuated, the cylindrical part deforms so strong it constricts and a bubble detaches [16]. While the shape of the bubble and its expansion over the edge of the bore result from acting forces, the Plateau-Rayleigh instability is a geometrical criterion.

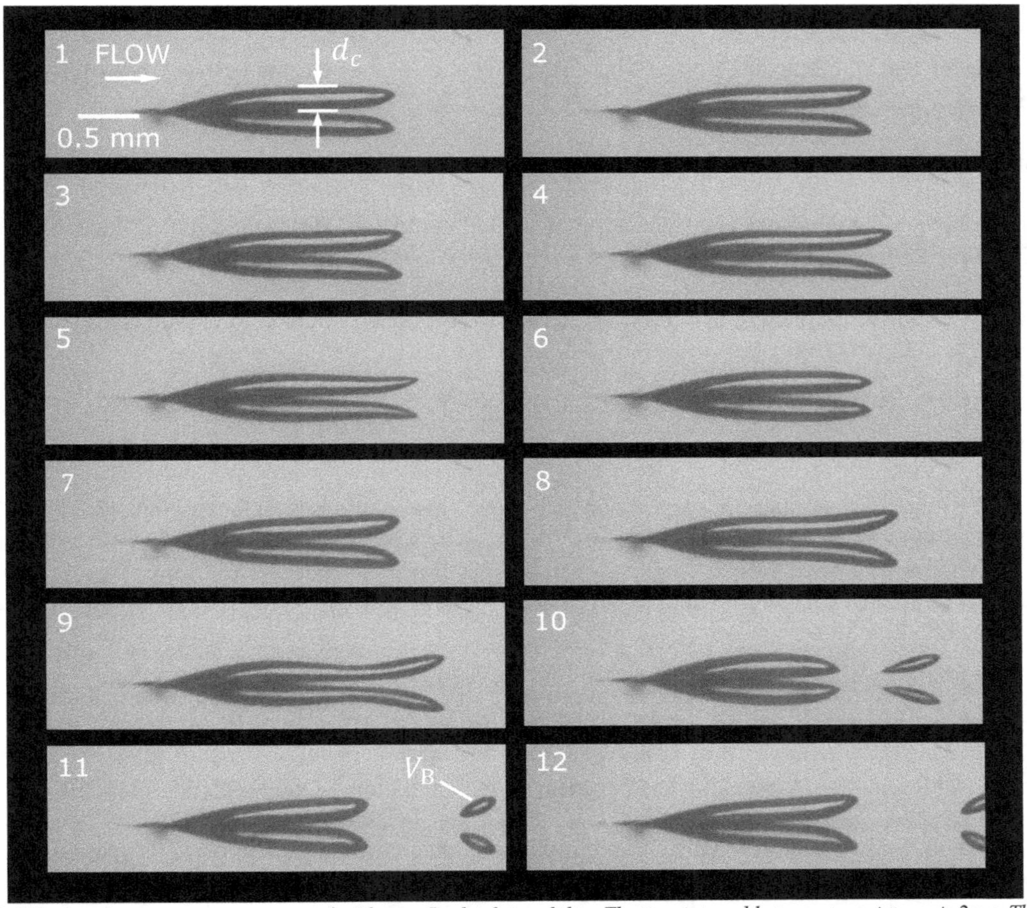

Figure 4: Bubble detachment as a consequence of a Plateau-Rayleigh instability. The time interval between two pictures is 2 ms. The flow goes from left to right. Adapted from [13].

Conclusion

It is widely accepted that free stream nuclei play an important role for the onset of cavitating flows while surface nuclei are only of minor importance. The findings presented in this paper indicate that the influence diffusion is often underestimated. The physical models serve well for the analysis of streak cavitation and capture the physics of nucleation. Furthermore, the models can be used to evaluate the plausibility of nucleation rates and bubble diameters used in Euler-Lagrange simulations of cavitating flows. The transfer to other forms of cavitating flows.

References

[1] C. Brennen, Cavitation and Bubble Dynamics, New York: Oxford University Press, 1995.

[2] P. Pelz, T. Keil and G. Ludwig, "On the Kinematics of Sheet and Cloud Cavitation and Related Erosion" in *Advanced Experimental and Numerical Techniques for Cavitation Erosion Prediction*, Springer Netherlands, 2014, chapter 9.

[3] B. Verhaagen and D. Fernandez Rivas, "Measuring cavitation and its cleaning effect," *Ultrasonics Sonochemistry* 29, 619-628, 2015.

[4] C. Brennen, "Cavitation in medicine," *Interface Focus,* 2015.

[5] E. Spiridonov, "Characteristics and calculation of cavitation mixers," *Procedia Engineering* 129, 446-450, 2015.

[6] E. Harvey, D. M. W. Barnes, A. Whiteley, D. Pease and K. Cooper, "Bubble formation in aminals," *J. Cell. Physiol.* 24, 1-22, 1944.

[7] A. Atchley and A. Prosperetti, "The crevice model of bubble nucleation," *J. Acoust. Soc. Am.* 86, 1065-1084, 1989.

[8] P. P. M. Epstein, "On the stability of gas bubbles in liquid-gas solutions," *J. Chem. Phys.* 18, 1505-1509, 1950.

[9] L. van Wijngaarden, "On the growth of small cavitation bubbles by convective diffusion," *Int. J. Heat. Mass. Trans.* 10, 127-134, 1967.

[10] B. Parkin and R. Kermeen, "The roles of convective air diffusion and liquid tensile stresses during cavitation inception," in *Proc. IAHR Symp. on Cav. and Hyd. Mach.*, 1963.

[11] F. Peters and R. Honza, "A benchmark experiment on gas cavitation," *Exp. Fluids* 55, 1786, 2014.

[12] T. F. Gross, G. Ludwig and P. F. Pelz, "Experimental and theoretical investigation of nucleation from wall-bounded nuclei in a laminar flow," in *Proceedings of the 16th International Symposium on Transport Phenomena and Dynamics of Rotating Machinery*, Honolulu, USA, 2016.

[13] A. Fick, "Über Diffusion," *Ann. Phys.* 170, 59-86, 1855.

[14] T. P. P. Groß, "Diffusion-driven nucleation from surface nuclei in hydrodynamic cavitation," *J Fluid Mech* 830, 138-164, 2017.

[15] W. Henry, "Experiments on the quantity of gases absorbed by water at different temperatures and under different pressures," *Philos. Trans. R. Soc. London* 93, 29-274.

[16] L. Rayleigh, "On the capillary phenomena of jets," *Proc. R. Soc. Lond.,* 71-97, 1879.

[17] S. Bankoff, "Entrapment of gas in the spreading of a liquid over a rough surface," *AIChE Journal* 4, 24-26, 1958.

[18] G. Liger-Belair, "The physics and chemistry behind the bubbling properties of champagne and sparkling wines: A state-of-the-art review," *J. Agric. Food Chem.* 53, 2788-2802.

© 2018 ASME

Thermal Effects in Ultrasonic Cavitation of Ionic Liquids

[1]Ross M. Elder; [1]Michael L. Calvisi*

[1]*University of Colorado, Colorado Springs, CO, USA*

Abstract

Ionic liquids have favorable properties for sonochemistry applications in which the high temperatures and pressures achieved by cavitation bubbles are important drivers of chemical processes. A model was developed to simulate ultrasonic cavitation in ionic liquids. The model uses the finite element method (FEM) and accounts for spatial variations in pressure and temperature in the interior and exterior of the bubble. This model provides insight into heat transfer across the bubble surface and thermal penetration into the liquid. Parametric studies are presented for sonochemistry applications involving ionic liquids as a solvent, examining a range of realistic ionic liquid properties to determine their effect on temperature and pressure inside the bubble and in the surrounding liquid. Results are presented for parametric variations including viscosity, thermal conductivity, and acoustic frequency. An additional study examines thermal penetration into the surrounding ionic liquid during bubble collapse. Among the most significant findings are that liquid viscosity and acoustic frequency have a strong effect on altering bulk and surface temperatures and pressures of the bubble. In all cases, the amount of thermal penetration into the exterior liquid is very small. Overall, the results suggest the prospect of tuning ionic liquid properties for specific applications.

Keywords: ultrasonic cavitation; sonochemistry; ionic liquids; heat transfer; finite element method

Introduction

Ionic liquids (ILs), salts that are liquid at or near room temperature, have favorable properties for chemical applications such as high thermal stability and negligible vapor pressure. ILs have been utilized for a variety of applications, including as catalysts and solvents for chemical reactions and ultrasonic cavitation [1]. In ultrasonic cavitation, diffuse acoustic energy is focused through the generation and oscillation of bubbles in a liquid, causing an oscillatory pressure and subsequent temperature rise in the liquid that can promote certain chemical reactions. Violent collapses, due to the nonlinearity of the bubble dynamics, can produce temperatures on the order of several thousand Kelvin and pressures on the order of several thousand atmospheres in the interior of the bubble [2]. While the massive temperatures tend to exist in a localized hot spot in the bulk of the bubble interior, the surrounding liquid may heat up significantly enough to undergo chemical reactions under the appropriate conditions.

ILs are comprised entirely of cations and anions, and the chemical structure has a strong influence on the thermophysical properties. As shown by Smiglak et al. [3], a variety of ILs can be combustible under significant heating due to their positive heats of formation, oxygen content, and decomposition products. While no research to our knowledge has stated the primary factors in producing a combustible IL, several ILs have been ignited, including protonated imidazolium nitrates, protonated C-nitro-substituted imidazolium nitrates and picrates, and 1-butyl-3 methylimidazolium azolates [3]. Others have been identified, synthesized, and characterized for propellant usage [4]. The combustibility of ILs presents a safety concern when performing acoustically-driven cavitation in ILs, due to the very large temperatures that can be exhibited in the interior of the bubbles. However, due to the competition between the time scales of motion and heat diffusion, the surrounding liquid may potentially experience little to no temperature change.

With over one trillion different IL possibilities [5], the potential for design freedom exists to tailor the cation and anion selection to yield favorable IL properties for a given application. To effectively utilize ILs in ultrasonic cavitation applications, it is important to understand how the liquid properties and system parameters affect the bubble dynamics, particularly the temperatures and pressures generated during collapse. To better understand this relationship, a numerical model of a spherical bubble was developed based on the finite element method (FEM). The state of the bubble interior during collapse is strongly influenced by several processes, including compressibility, surface tension, liquid viscosity, and heat transfer [2]. The numerical model accounts for all of these effects, in addition to spatial gradients in pressure and temperature throughout the flow domain. The model is used to conduct parametric studies on the effects of variations in acoustic frequency, IL viscosity, and IL thermal conductivity on the temperature and

*Corresponding Author, Michael L. Calvisi: mcalvisi@uccs.edu

pressures achieved in the bubble interior. This model is further used to study the effects of heat transfer through the bubble interior and into the surrounding IL, which provides insight into the feasibility of IL combustion.

Model Development

The object of interest is a single, spherical bubble surrounded by a Newtonian liquid driven by an acoustic standing wave. Because the bubble is spherical, the flow field is purely radial and therefore vorticity in the bubble wall region is considered negligible [6]. The bubble interior is initially composed of a noncondensable gas, such as argon, and potentially water vapor. The noncondensable gas is regarded as insoluble to the surrounding liquid based on the short time scale of the acoustic forcing, thus neglecting rectified diffusion. The outward flux of the noncondensable gas into the liquid has been shown by Storey and Szeri to have negligible effects on the bubble dynamics, unless at low ambient pressures [7]. Therefore, the number of moles of noncondensable gas present in the bubble interior is fixed throughout the entire oscillation. Additionally, the noncondensable gas and vapor are assumed to be inviscid. The model includes the effects of liquid viscosity, surface tension, and compressibility. The FEM model accounts for heat transfer throughout the bubble interior and exterior in the form of conduction and convection, but neglects mass transfer of water vapor at the bubble wall, which is beyond the scope of this paper. The physical processes of radiative heat transfer, light emission, chemical reactions, and buoyancy are neglected, as they are likely to have an insignificant effect on the results. The effects of shock waves are also not included in this study, for the diffusive transport in noble gas bubbles inhibits the steepening of wavy disturbances, such that shocks are not likely to occur inside argon bubbles [8, 9]. At cavitation collapse, though, the speed of the bubble wall can approach or surpass the speed of sound in the liquid. However, the total time the bubble wall is supersonic is a small fraction of the oscillation cycle; thus, the modeling error from neglecting shock waves in the liquid does not substantially affect the bubble dynamics [6].

COMSOL®, a commercial finite element software package, was used to develop the model. The program models a single, spherical bubble that undergoes a time and 1-D space-dependent change of volume in a compressible, Newtonian fluid at rest at an infinite distance relative to the initial bubble radius. The main bubble dynamics are described by laminar, two-phase compressible flow physics of immiscible fluids separated by a moving interface. A compressible formulation of the Navier-Stokes equations is implemented in both the interior and exterior domains. The velocity field, pressure field, and mesh deformation are solved for and the location of the interface is tracked by a deformed mesh. Upon post-processing, the Reynolds number of the simulations were shown to range from approximately 0.015 to 110, well within the laminar limit. The mesh nodes conform to the moving interface and other boundaries in the model. Regarding boundary conditions, a normal stress balance is applied at the gas-liquid interface to couple the separate Navier-Stokes equations used for the inner and outer domains. In addition, no-slip and kinematic boundary conditions are applied at the bubble surface so that it advects with the surrounding fluid. The outer liquid boundary is considered a pressure inlet to model the acoustic forcing. A sinusoidal pressure, $P_a = -A\sin(\omega t)$, is applied here through a Dirichlet boundary condition. The outer boundary wall is placed at 100 times the radius of the bubble in order to minimize numerical artifacts at the bubble surface due to the applied pressure at the outer boundary.

To account for heat transfer in the bubble interior, surrounding liquid, and fluid-fluid interface, transient conduction with pressure work, originating from the conservation of energy, was coupled to the Navier-Stokes equations. The pressure work term accounts for the heating of the fluids under adiabatic compression and thermoacoustic effects.

The FEM model above was validated against results in the literature, including those of Putterman and Weninger [10], and found to have good agreement. Due to the limited scope of this paper, further details are not included here.

Results

A baseline for the simulation parameters was determined. The IL 1-butyl-3-methylimidazolium tetrafluoroborate ([BMIM][BF4]) is one of the most commonly studied, commercially available ionic liquids and has been used experimentally in cavitation studies [11, 12]. Unlike many ILs, fluid property information for [BMIM][BF4] is readily available. Thus, the baseline fluid properties are chosen as those of [BMIM][BF4] at standard temperature and pressure. The bubble interior is assumed to consist of only the noncondensable gas argon. Water vapor can be present in the bubble interior if water is present or mixed with [BMIM][BF4], but consideration of the effects of water vapor is beyond the scope of the present study. Due to the low IL vapor pressure, the presence of IL vapor in the bubble interior can be neglected. The values selected for liquid and gas properties, and additional parameters are as follows:

ρ_l = 1201.2 kg/m^3, μ_l = 0.099200 Pa-s, σ_l = 0.045330 N/m, c_l = 1572.3 m/s, k_l = 0.18600 W/(m-K), $c_{p,l}$ = 1577.6 J/(kg-K), $T_{0,l}$ = 293.15 K, γ_g = 1.6700, k_g = 0.017574 W/(m-K), M_g = 0.039900 kg/mol, $c_{p,g}$ = 520.00 J/(kg-K), $T_{0,g}$ = 293.15 K, R_0 = 2.5-5.0 microns, P_∞ = 1.00 atm, T_∞ = 293.15 K, A = 1.10 atm, and f = 30.0 kHz, where ρ is density, μ is viscosity, σ is surface tension, c is speed of sound, k is thermal conductivity, c_p is specific heat, T is temperature, γ is ratio of specific heats, M is molecular weight, R is bubble radius, P is pressure, $f = \omega/2\pi$ is the acoustic forcing frequency, and the subscripts l, g, 0, and ∞ refer to the liquid, gas, initial state, and far-field, respectively. The initial bubble pressure is set so that it is in static equilibrium with the surrounding liquid.

Variation of Acoustic Frequency: To understand the effects of the wide range of frequencies that are commonly used for cavitation applications, the acoustic forcing frequency is varied in the range of 10-1000 kHz using the baseline parameters outlined above. For radii of 2.5, 5.0, and 15 μm, little change in temperature or pressure occurs with frequency. The temperature and pressure changes for the larger radii of 30 and 50 μm are much higher, and increase as frequency decreases. At lower frequencies, the forcing period is longer, which allows the bubble to expand to larger maximum radii and collapse to smaller minimum radii, thus increasing the compression ratio and, consequently, the maximum bulk temperatures and pressures achieved inside the bubble. All pressures approach the initial bubble pressure at the higher frequencies.

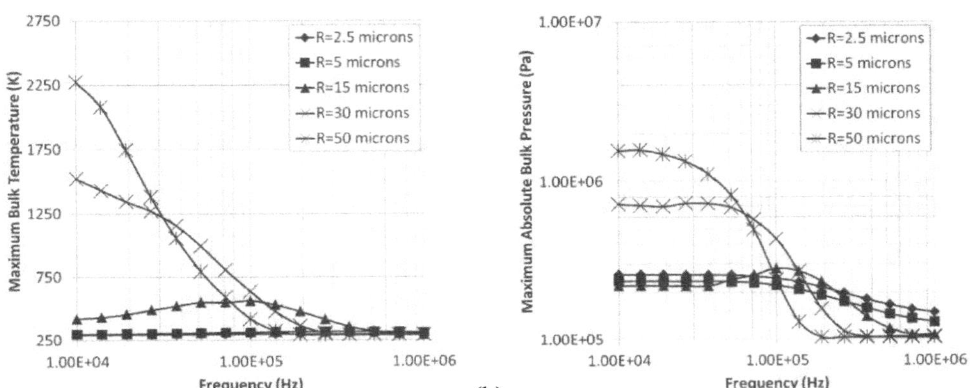

(a) (b)

Figure 1: (a) Maximum bulk temperature vs. frequency, and (b) maximum bulk pressure vs. frequency.

The surface temperature and pressure across the frequency range are shown in Figure 2. Despite the high temperatures at the center of the bubble, the bubble wall experiences significantly lower temperatures. The bubble wall also exhibits generally slightly higher maximum pressures than the bubble interior, typically occurring just after the point of minimum volume.

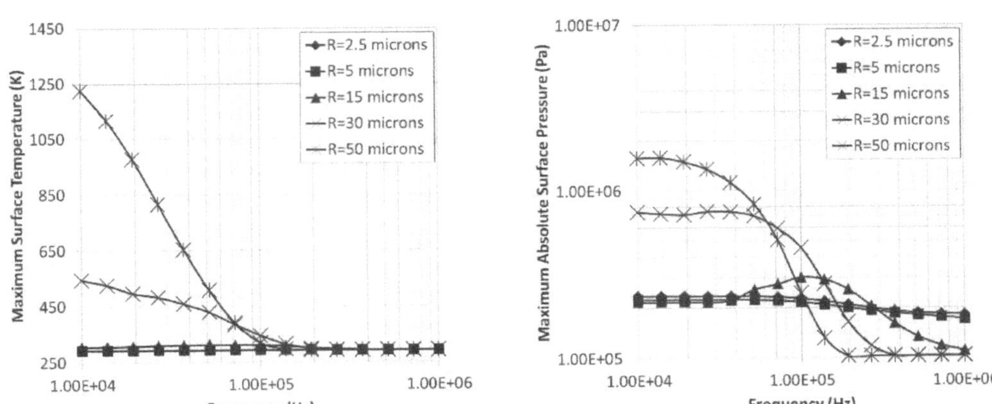

(a) (b)

Figure 2: (a) Maximum surface temperature vs. frequency, and (b) maximum surface pressure vs. frequency.

Variation of Viscosity: According to Endres et al. [13], ILs tend to exhibit viscosities within the range of 0.01-0.5 Pa-s compared to water, which has a viscosity of approximately 0.001 Pa-s at standard temperature and pressure. A larger range for ILs was stated by Wilkes [14] of 0.01-1 Pa-s. In order to understand the effect of viscosity, the viscosity was varied from 0.001-1 Pa-s. Although 0.001 Pa-s is not currently a feasible viscosity for an IL, further research may yield viscosities approaching this order of magnitude. The maximum bulk and surface temperature results are shown in Figure 3. Generally, as viscosity increases, the maximum temperatures decrease. Reducing viscosity yields a very significant increase in the temperatures achieved. As with lowering frequency, as the liquid viscosity decreases, the bubble achieves larger maximum radii and smaller minimum radii, thus increasing the compression ratio and, consequently, the maximum bulk temperatures achieved. Interestingly, one can see by comparing the bulk and surface temperature plots that for smaller initial radii, bubbles experience the largest temperature gradient from the center of the bubble to the wall. The time scales of collapse for the smaller bubbles are significantly shorter than that of the larger bubbles. This may potentially "trap" a larger portion of the heat at the center of the bubble, not allowing enough time for the heat to significantly diffuse through the bubble interior and through the bubble wall. This yields a larger temperature gradient for the smaller bubbles from the center to the liquid interface.

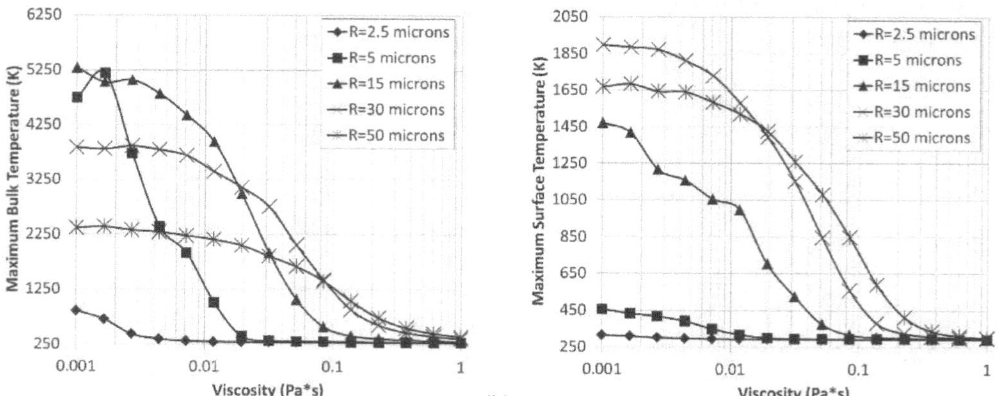

(a) (b)

Figure 3: (a) Maximum bulk temperature vs. viscosity, and (b) maximum surface temperature vs. viscosity.

The maximum bulk pressure vs. viscosity is shown in Figure 4. Along with large bulk temperatures, the simulations demonstrate very high bulk pressures, some on the order of 1000 atm, which occurs at the deepest depths of the ocean. On a logarithmic plot, the surface pressures are not noticeably different from the bulk pressures and, hence, are not shown. The majority of these scenarios exhibit rather uniform pressures throughout the bubble interior. However, in some scenarios, there is as much as a 67% difference in pressure between the bubble center and wall.

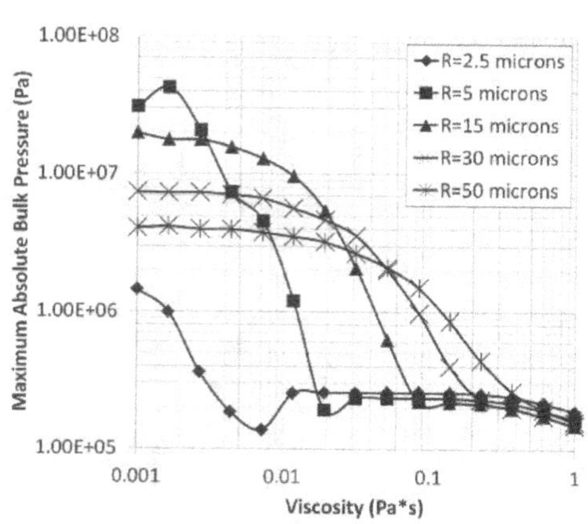

Figure 4: Maximum bulk pressure vs. viscosity.

Variation of Thermal Conductivity: While the results are not shown here for the sake of brevity, the simulations show that variations in IL thermal conductivity over a typical range of 0.1-0.5 W/(m-K) have a negligible effect on the bulk temperature. However, increasing thermal conductivity can significantly reduce the bubble surface temperature for initial radii equal to or greater than 30 microns. The larger thermal conductivity in the liquid promotes heat conduction away from the bubble, thereby lowering surface temperatures by as much as ~25% in these cases.

Thermal Penetration: As the above results show, the temperatures can decrease significantly from the bubble center to the surface. Several simulations were run, with the same baseline parameters as above, to determine the temperature gradient in the liquid and its penetration depth for varying parameters, including acoustic frequency, IL viscosity, and IL thermal conductivity. In the first study, the frequency of the acoustic forcing was varied from 10-1000 kHz. To compare the results of each scenario, the dimensionless volume of liquid heated as a function of the minimum heated dimensionless temperature (within that volume) is plotted in Figure 5 for initial radii of 5 and 30 microns, respectively, at the instant of minimum bubble volume. For an initial radius of 5 microns, decreasing the frequency to 10 kHz increases the bulk temperature and surface temperature greatly – as shown above – yet the liquid experiences much less of a temperature increase. The heat is transferred further into the liquid at the time of minimum volume, yielding higher volumes of heated liquid at the lower temperatures than other frequencies, but results in a much lower volume of higher temperature liquid. Furthermore, despite the high temperatures developed in the bulk, only a small liquid region experiences significant temperature change, with a maximum temperature increase of 83.2% for a 30 micron bubble at 10 kHz driving frequency.

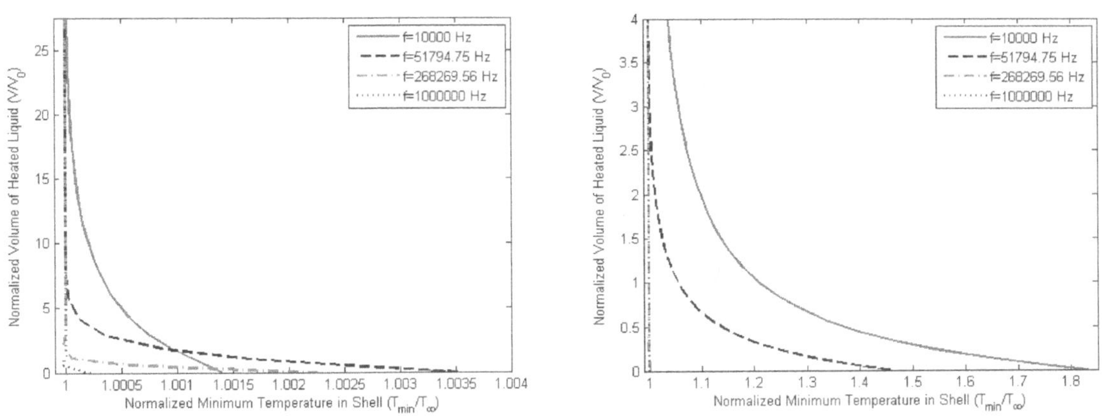

Figure 5: Dimensionless liquid volume heated vs. minimum dimensionless temperature at instant of minimum bubble volume with varying frequency. Left: $R_0 = 5$ microns. Right: $R_0 = 30$ microns.

Additional thermal penetration studies were performed by varying the IL viscosity from 0.001-0.5 Pa-s and the IL thermal conductivity over a typical range of 0.1-0.5 W/(m-K). Similar results were found; namely, the temperature increase in the liquid is relatively small – considering the large bulk temperatures achieved – and is confined to only a thin region around the bubble.

Conclusion

A model was developed to simulate ultrasonic cavitation in ionic liquids using the finite element method (FEM). Parametric studies were performed for sonochemistry applications involving ionic liquids as a solvent, examining a range of realistic ionic liquid properties to determine their effect on temperature and pressure inside the bubble and in the surrounding liquid. Results were presented for parametric variations including viscosity, thermal conductivity, and acoustic frequency. An additional study examined thermal penetration into the surrounding ionic liquid during bubble collapse. The results show that liquid viscosity and acoustic frequency have a strong effect on altering bulk and surface temperatures and pressures of the bubble. In all cases, the amount of thermal penetration into the exterior liquid is small relative to the initial bubble radius. Overall, the results suggest the prospect of tuning ionic liquid properties for specific applications.

References

[1] Calvisi, M. L., Lindau, O., Blake, J. R., Szeri, A. J. (2007). *Shape stability and violent collapse of microbubbles in acoustic traveling waves*. Phys. Fluids 19(047101).

[2] Storey, B. D., Szeri, A. J. (2001). *A reduced model of cavitation physics for use in sonochemistry*. Proc. R. Soc. Lond. 457.

[3] Smiglak, M., Reichert, W. M., Holbrey, J. D., Wilkes, J. S., Sun, L., Thrasher, J. S., Kirichenko, K., Singh, S., Katritzky, A. R., Rogers, R. D. (2006). *Combustible ionic liquids by design: Is laboratory safety another ionic liquid myth?* Chem. Commun.

[4] Sims, J., Forton, M., Allan, B., Rogers, R., Shamshina, J., Koelfgen, S. *Ionic liquids: Unlocking the gate to replacing hydrazine*. NASA Tech Briefs DRC-010-041.

[5] Forsyth, S. A., Pringle, J. M., MacFarlane, D. R. (2004). *Ionic liquids - An overview*, Aust. J. Chem. 57.

[6] Brenner, M. P., Hilgenfeldt, S., Lohse, D. (2002). *Single-bubble sonoluminescence*. Rev. of Modern Phys. 74.

[7] Storey, B. D., Szeri, A. J. (2000). *Water vapour, sonoluminescence and sonochemistry*. Proc. Math. Phys. Eng. Sci. 456(1999).

[8] Vuong, V. Q., Szeri, A. J. (1996). *Analytical integration and a numerical method for 1-D compressible flow with diffusive transport*. Computers and Fluids 25(1).

[9] Vuong, V. Q., Szeri, A. J. (1996). *Sonoluminescence and diffusive transport*. Phys. Fluids 8(9).

[10] Putterman, S. J., Weninger, K. R. (2000). *Sonoluminescence: How bubbles turn sound into light*, Annu. Rev. of Fluid Mech. 32(1).

[11] Flannigan, D. J., Hopkins, S. D., Suslick, K. S. (2005). *Sonochemistry and sonoluminescence in ionic liquids, molten salts, and concentrated electrolyte solutions*. J. Organometallic Chem. 10.

[12] Oxley, J. D., Prozorov, T., Suslick, K. S. (2002). *Sonochemistry and sonoluminescence of room-temperature ionic liquids*. J. Am. Chem. Soc. 125(37).

[13] Endres, F., Zein El Abedin, S. (2006). *Air and water stable ionic liquids in physical chemistry*, Phys. Chem. Chem. Phys. 8.

[14] Wilkes, J. S. (2004). *Properties of ionic liquid solvents for catalysis*. J. Molecular Catalysis A: Chem. 214.

© 2018 ASME

Thermocavitation in a microchannel with a low power light source

Loreto Oyarte Gálvez and David Fernández Rivas

Mesoscale Chemical Systems, Universiteit Twente, Post Office Box 217, 7500AE Enschede, The Netherlands

Abstract

We have studied the formation of vapour bubbles in a microchannel, a confined quasi-two dimensional system where one dimension is much smaller than the other two. The bubble is created by thermocavitation, i.e. by overheating a small region of the fluid, using a continuous wave laser diode. To the best of our knowledge, this is the first time that thermocavitation is achieved with such low power light sources ($\sim$ 100mW). Our fast-imaging experimental results are compared with theoretical models and show good agreement with the predicted bubble growth. Considering the bubble as a piston able to transfer kinetic energy to the fluid, we show that this microfluidic system is a low-cost micro-engine able to convert thermal energy into work.

Keywords: Thermocavitation, CW-laser, Confined fluid

Introduction

Cavitation induced by lasers has been ultimately used for development of technolgies, such as needle free injection [1, 2, 3] and ink-jet printing [4, 5]. Pulsed-laser-based systems produce very fast jets, however they are expensive, noisy, heavy, among others [1, 6, 7]. Recently, it has been demonstrated that continuous wave lasers -which are cheaper, lighter, more portable and with lower safety risks than pulsed lasers- are able to induce jets by cavitation with speeds up to $\sim$100m/s [1, 8], becoming a more efficient alternative for the mentioned technologies.

In this work, we analize the development of a vapour bubble in a microcell created by thermocavitation. A low power light source, particularly a continuous wave laser diode, is used to inject energy into a liquid column and induce the nucleation.

Experimental setup description & Results

The microfluidic cell consists of a shallow glass container, as described by Berrospe *et al.* [1]. The microfluidic cell is partially filled with colored water (Direct Red 81, Sigma Aldrich, $wt = 5\%$ [9]) and a blue laser diode laser, $\lambda = 450$nm and $P \simeq 150$mW, is focused at the bottom of the cell.

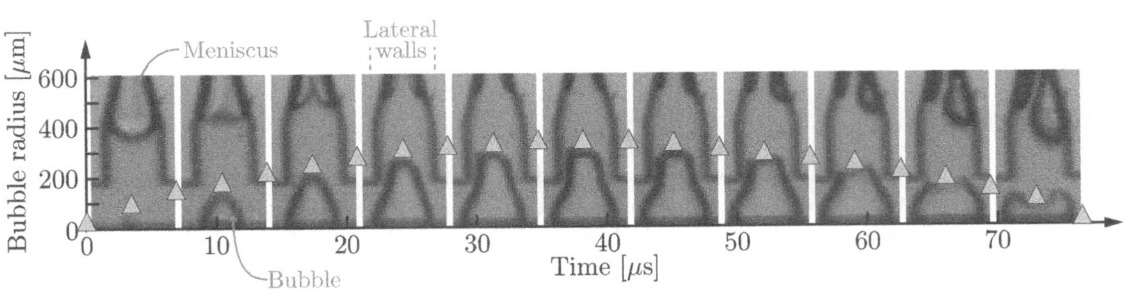

Figure 1: The bubble radius as a function of time is plotted with the respective image sequence.

A bubble grows radially always from the entry point of the laser beam, until the diameter of the bubble d reaches the lateral walls. From then, the bubble grows further in the axial direction with an elongated pancake shape along the walls of the cell [10, 11, 12, 13, 14] and a constant cross-section area, as shown in Figure 1. The growth and collapse of the bubble depend on the initial amount of liquid filling the channel, i.e. the initial meniscus position H with respect to the bottom of the microcell [7, 15].

Depending on the total volume and energy input, two scenarios have been observed: the first one occurs where all the liquid is vaporised, and a second one where the collapse phase exhibits complex after-bounces, and secondary cavitation close to the bottom of the cell (Figure 1).

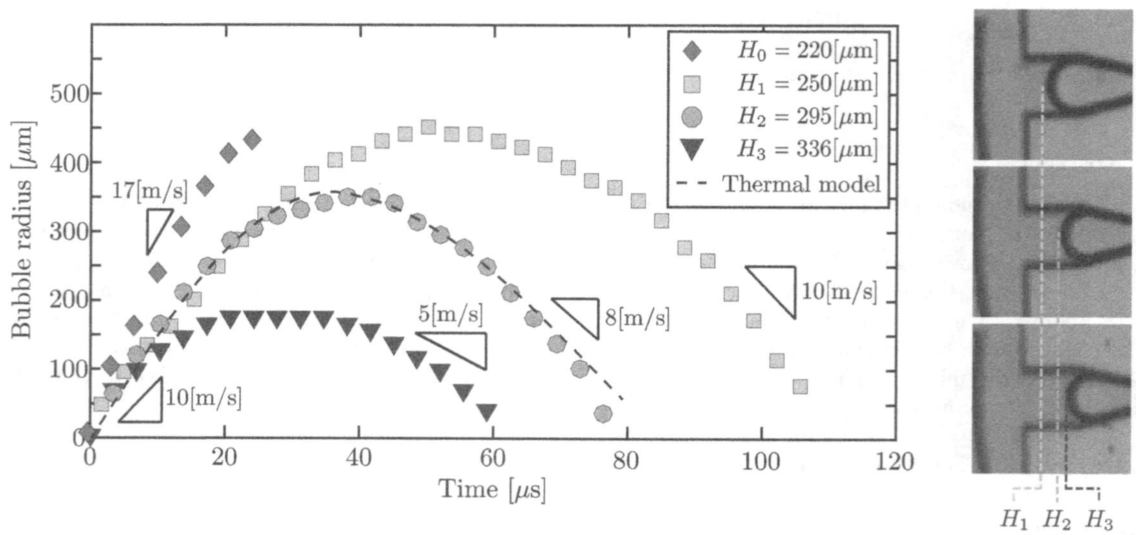

Figure 2: The bubble radius in time is compared for four meniscus positions, $H = 220, 250, 295$ & 336 μm. The dashed line is the fit of the bubble radius using the thermal model, for the case $H = 295\mu$m.

In the second scenario, the bubble growth is always faster than the bubble collapse, as was observed and described by Zwaan *et al.* [9] in non-confined conditions, and by Sun *et al.* [16] in microtubes. The influence of the meniscus position H, i.e. the liquid volume, is shown in Figure 2.

The bubble radius is compared with the theoretical thermal model developed by Sun *et al*, which predicts the growth and collapse of a vapour bubble in a microtube [16]. A good agreement, between theory and experiments is observed, as shown in Figure 2.

Conclusions & Outlook

We studied the formation of a vapour bubble, expansion and collapse, in a confined quasi-two dimensional geometry, induced by a continuous wave laser diode. More specifically, we obtained the radius of the bubble during time and compared with a theoretical thermal model showing a good agreement.

We are currently studying further confinement effects of bubble nucleation, growth and collapse and the induced jet. We change the geometry of the microcell in order to improve the energy transfer efficiency from the laser diode to the liquid column.

Acknowledgements

We thank C. Berrospe Rodríguez -INAOE, Cholula, Mexico- F. Segerink, H. Offerhaus, S. Vázquez Córdova -OS, University of Twente- V.S. Arza and D. vd Meer -POF, University of Twente- R. Sanders -MSS, University of Twente- S. Schlautmann -MCS, University of Twente- for stimulating disscusions and contribution to this work.

References

[1] C. Berrospe-Rodríguez, C.W. Visser, S. Schlautmann, D. Fernandez-Rivas, and R. Ramos-García. Toward jet injection by continuous-wave laser cavitation. *Journal of Biomedical Optics*, 22:22 – 22 – 9, 2017.

[2] S. R. Gonzalez Avila, C. Song, and C. D. Ohl. Fast transient microjets induced by hemispherical cavitation bubbles. *Journal of Fluid Mechanics*, 767:31Ð51, 2015.

[3] Y. Tagawa, A. E. Oudalov, N.and Ghalbzouri, C. Sun, and D. Lohse. Needle-free injection into skin and soft matter with highly focused microjets. *Lab Chip*, 13:1357–1363, 2013.

[4] P. Delrot, M. A. Modestino, D. Psaltis, and C. Moser. Laser-assisted inkjet printing of highly viscous fluids with sub-nozzle resolution, 2016.

[5] P. Delrot, M. A. Modestino, F. Gallaire, D. Psaltis, and C. Moser. Inkjet printing of viscous monodisperse microdroplets by laser-induced flow focusing. *Phys. Rev. Applied*, 6:024003, Aug 2016.

[6] J. P. Padilla-Martinez, C. Berrospe-Rodríguez, G. Aguilar, J. C. Ramirez-San-Juan, and R. Ramos-García. Optic cavitation with CW lasers: A review. *Physics of Fluids*, 26(12):122007, 2014.

[7] I. R. Peters, Y. Tagawa, N. Oudalov, C. Sun, A. Prosperetti, D. Lohse, and D. van der Meer. Highly focused supersonic microjets: numerical simulations. *Journal of Fluid Mechanics*, 719:587Ð605, 2013.

[8] C. Berrospe-Rodríguez, C.W. Visser, D. Schlautmann, S. Ramos-García, R., and Fernandez-Rivas. Continuous-wave laser generated jets for needle free applications. *Biomicrofluidics*, 10(1):014104, 2016.

[9] E. Zwaan, S. Le Gac, K. Tsuji, and C. D. Ohl. Controlled cavitation in microfluidic systems. *Phys. Rev. Lett.*, 98:254501, Jun 2007.

[10] S. W. Ohl and C. D. Ohl. Acoustic Cavitation in a Microchannel. In *Handbook of Ultrasonics and Sonochemistry*, pages 99–135. Springer Singapore, 2016.

[11] S. Xiong, L. K. Chin, K. Ando, T. Tandiono, A. Q. Liu, and C. D. Ohl. Droplet generation via a single bubble transformation in a nanofluidic channel. *Lab on a Chip*, 15(6):1451–1457, 2015.

[12] B. Shen, M. Leman, M. Reyssat, and P. Tabeling. Dynamics of a small number of droplets in microfluidic hele-shaw cells. *Experiments in Fluids*, 55(5):1728, 2014.

[13] L. Zhu and F. Gallaire. A pancake droplet translating in a Hele-Shaw cell: lubrication film and flow field. *Journal of Fluid Mechanics*, 798:955–969, 2016.

[14] E. Brujan. Cavitation in non-newtonian fluids. pages XI, 269. Springer-Verlag Berlin Heidelberg, 2011.

[15] Y. Tagawa, N. Oudalov, C. W. Visser, I. R. Peters, D. van der Meer, C. Sun, A. Prosperetti, and D. Lohse. Highly focused supersonic microjets. *Phys. Rev. X*, 2:031002, Jul 2012.

[16] C. Sun, E. Can, R. Dijkink, D. Lohse, and A. Prosperetti. Growth and collapse of a vapour bubble in a microtube: the role of thermal effects. *Journal of Fluid Mechanics*, 632:5Ð16, 2009.

© 2018 ASME

Towards development of surface coating for improved cavitation resistance

[1]Ravikant Kamble; [1]Sureka P; [1,3]C. Syamsundar; [1]Dhiman Chatterjee*; [2]M. Kamaraj, [1]Oswald J. Lobo

[1]Department of Mechanical Engineering IIT Madras Chennai India; [2] Department of Metallurgical and Materials Engineering IIT Madras Chennai India; CMR Engineering College Hyderabad India

Abstract

Hydrodynamic cavitation is known to have many harmful effects and of these effects, cavitation-induced erosion and surface pitting are detrimental to the life of hydraulic machines. One approach to reduce material damage could be through development of new materials and coatings with improved cavitation resistance. Though a large body of work exists on surface treatment methods, especially those involving surface hardening processes, use of non-metallic materials such as polymers as coatings on steel substrate has not received adequate attention. Present study addresses this gap through systematic study of different coatings involving vibratory cavitation (ASTM G-32 protocol) as well as studying the dynamics of single bubble near a material surface with appropriate coatings. Mass loss for different coatings under vibratory cavitation as well as the scanning electron microscopy study of these specimens has been reported. Pressure measurement as well as high speed imaging of single bubble collapse near the coated surface is also highlighted to bring out the efficacy of this system.

Keywords: cavitation; erosion; ASTM G-32; high speed imaging; pressure measurement

Introduction

Hydrodynamic cavitation is known to have many harmful effects. So different attempts were made in the past to suppress or postpone cavitation through improved design, air injection [1] or use of ultrasound [2]. Among different harmful effects, cavitation-induced erosion and surface pitting are detrimental to the life of a hydraulic machine like a pump or a hydraulic turbine. Air-injection is often used to retard the collapse of cavitating bubbles near material surface. However, for real-life applications, what is more desirable is to develop materials or coatings with improved cavitation resistance. Though a large body of work exists on surface treatment methods especially involving surface hardening processes [3], the use of non-metallic materials, such as polymers, rubber, epoxy resin and synthetic elastomeric coatings on metallic structures to mitigate cavitation-induced erosion have not received adequate attention [4-8]. Choi and Chahine [5] had used three different polyurethane coating thicknesses and demonstrated that the thinner coatings are superior to thicker coatings. Zhang et al. [7] used different non-metallic coatings and concluded that polyurethane provided the best anti-cavitation resistance. Hattori and Itoh [8] showed that impact loads on thermoplastics and thermosetting plastics are lower than on metals. Based on these studies, it was decided to use polyurethane coatings filled with different concentrations of hard nanoparticles like boron carbide (B_4C) or silicon carbide (SiC).

Vibratory cavitation can produce accelerated damage tests of specimens as per ASTM G-32 protocol. The same was followed in this work to bring out the effect of nanoparticle concentration on cavitation resistance. Vibratory cavitation produces a stream of cavitating bubbles in the vicinity of the sample and hence it is difficult to understand the interaction between any cavitating bubble and the material specimen. This interaction is better captured by studying the interaction of a single collapsing bubble near a wall. It has been shown by Tomita and Shima [9] and Shima [10] that the mechanism of cavitation damage depends on the separating distance between the bubble and the material surface. Based on Shima's study, it can be concluded that microjet formation is the dominant mechanism if $0.6 \le \frac{H}{R_{\max}} \le 0.8$ and shock-wave is dominant for $0.3 < \frac{H}{R_{\max}}$ and $\frac{H}{R_{\max}} > 1.5$. In these expressions, H is the separating distance between bubble centre (electrode location in their experiments) and R_{max} is the maximum radius of the bubble. Both the mechanisms play an important role in the intervening parameter space. Hence, in this work, single bubble experiments were performed near the material surface at varying distance from the wall and simultaneous high speed imaging and pressure measurements were carried out and reported in this work.

*Corresponding Author, Dhiman Chatterjee: dhiman@iitm.ac.in

Next section deals with the experimental methodology and that is followed by the presentation of results and finally the major conclusions are presented.

Experimental Methodology

Experimental methodology can be broadly divided into three category:

i) Preparation of sample: Base material used for testing of cavitation resistance is made of 16 Cr-5 Ni martensitic stainless steel (MSS) — a material typically used as materials for blades of hydraulic turbines. This material was chosen because of its good mechanical properties and high toughness [11]. Irathane 155, a Polyurethane (PU) material and Irabond UU55/252A adhesive was used for preparing coatings on the substrate. Further, in some of the coatings, hard nanoparticles of boron carbide (B_4C) and silicon carbide (SiC) at varying weight percentages were used. Further details about preparation and characterization of PU-coatings can be found in the work of Syamsundar et al. [12].

ii) Experiments with vibratory cavitation: Specimens, thus prepared following step (i) above, was characterized for cavitation resistance in the vibratory cavitation test facility (figure 1). A Roop Telsonic-make ultrasound transducer/horn (SE26/20) and a generator producing acoustic excitation was used to produce an acoustic pressure field at 20 ± 0.5 kHz with constant peak-to-peak displacement of 50 μm ± 5% in a constant temperature (25 ± 2 °C) deionized water. A specimen button of proper mass was attached to the transducer horn by threading it into the tip of the horn. The test specimen was weighed accurately before testing and also during periodic interruptions of the test in order to obtain the time history of the mass loss of the specimen. Total duration to which each specimen was subjected to cavitation was 10 hours. The test protocol followed was ASTM G-32.

iii) Single bubble experiment: Single bubble can be produced near a material surface either using laser or through underwater spark-discharge. In case of spark discharge method, typically two electrodes are separated by a small gap and a very high (2-20 kV) voltage is passed. However, in this work, low voltage (typically less than 100 V) is applied to create the spark through electrodes which are in contact with each other [13-14]. This method, besides being safer to handle, employs smaller electrode sizes (100-300 μm) in comparison to that (≈ 1 mm) in high voltage method. Thus, it is expected to minimize the influence of electrodes on bubble dynamics. The disadvantage of this method is the need to replace electrodes after each spark. Figure 2(a) shows the circuit used for the production of low (at 60 V) voltage spark based on the descriptions given in prior work [13-14]. Figure 2(b) shows the backlight LED lamp used to capture high speed images at 40,000 fps using Photron SA-4 camera. Along with photographic observation of bubble dynamics, pressure pulse was captured using miniature pressure transducer (PCB112A22) mounted below the surface of the test specimen in a way similar to that of Chahine et al. [15]. Voltage output from pressure sensor was acquired using data acquisition system (National Instruments) and stored in computer for post processing.

Results and discussion

The results obtained using the experimental technique described in the previous section is presented now. For the ease of description these are presented as:
i) Results of vibratory cavitation, and,
ii) Dynamics of a single bubble near a material surface

i) Results of vibratory cavitation

One of the surest tests of characterizing the resistance of a given material subjected to cavitation is to analyze the mass loss information over an extended period of time and figures 3 and 4 show these results. It can be seen from figure 3 that the mass loss reduces in case of PU and PU-reinforced with nanoparticles (B_4C and SiC) from that in case of base material (MSS substrate). It is well-known that cavitation erosion of materials is not uniform and can be divided into three prominent zones of incubation, growth and saturation. This is brought out in figure 4.

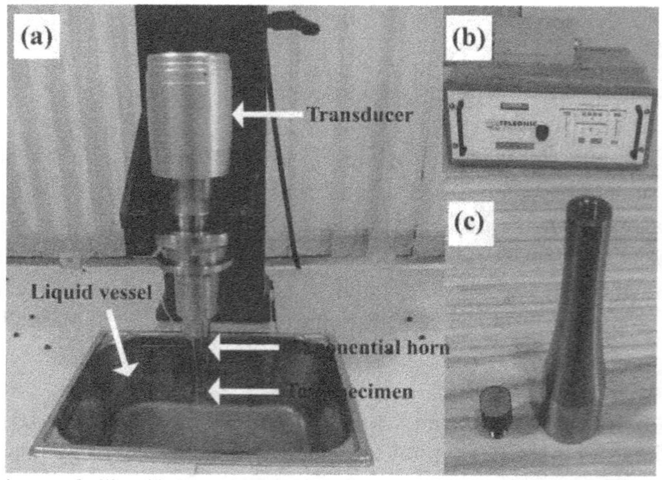

Figure 1: Vibratory cavitation erosion test facility. (a) Overall assembly, (b) power amplifier, and, (c) exponential horn with test specimen (Photo courtesy: BHEL, Hyderabad, India)

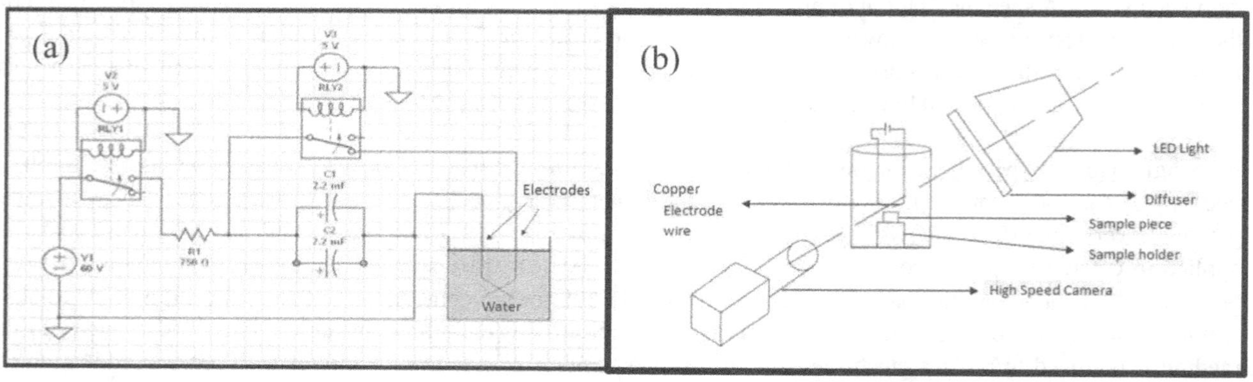

Figure 2: (a) Circuit diagram used to produce low-voltage spark, (b) schematic of a set up used for studying dynamics of a single bubble near a surface

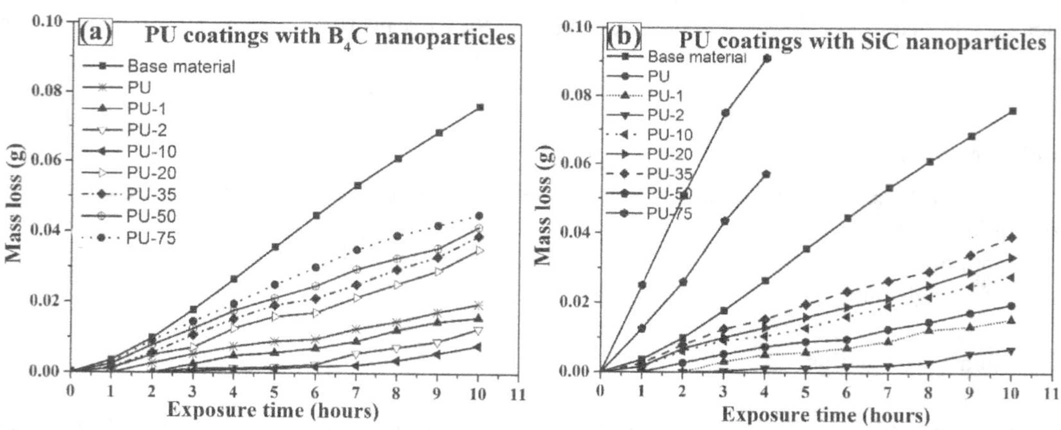

Figure 3: Plot of mass loss of PU coatings as a function of cavitation exposure time in hours. (a) refers to the reinforcement of B_4C nanoparticles in PU coatings; (b) refers to the reinforcement of SiC nanoparticles in PU coatings.

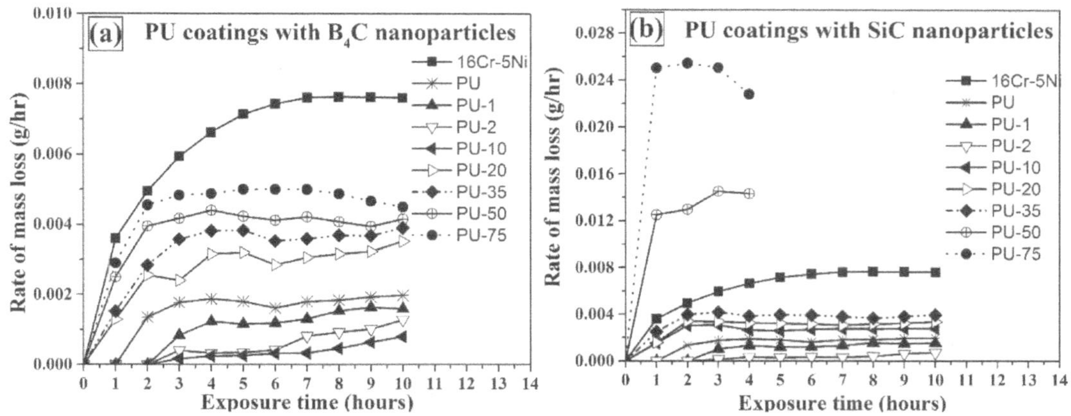

Figure 4: Rate of CE as a function of exposure time of PU coatings exposed to cavitation for 10 hours duration, (a) refers to the reinforcement of B_4C nanoparticles in PU coatings; (b) refers to the reinforcement of SiC nanoparticles in PU coatings

Both figures 3 and 4 indicate that there is an optimum concentration of nanoparticles which produces the minimum mass loss. For B_4C, it is 10% weight while for SiC it is 2%. Further it is clear from figure 4 that mere addition of PU-layer on top of MSS substrate increases the incubation period to 1 hour before any significant mass loss is noted. This incubation period gradually increases with increasing concentration and at 10% B_4C (or, 2% SiC) it reaches 2 hours. This could be because the PU-based coatings can absorb impact loading much better than the bare substrate. SEM studies were carried out for these material specimens and figure 5 presents SEM observations for MSS, pure PU coating, 10 wt% B_4C and 2 wt% SiC. Beyond 10 wt% B4C (or, 2 wt% SiC), the performance deteriorates.

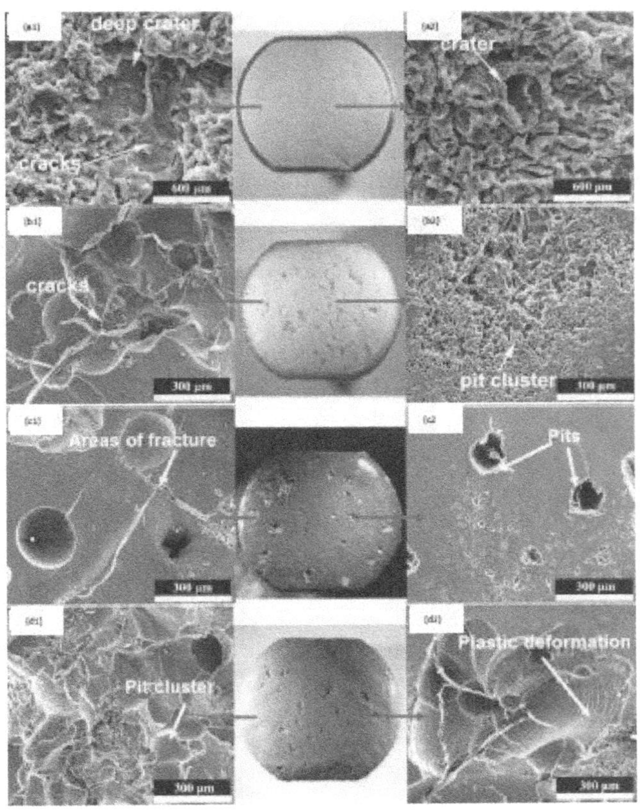

Figure 5: SEM images of surface damage after 10 hours of cavitation study. (a) MSS, (b) pure PU, (c) PU reinforced with 10 wt% B4C, and, (d) PU reinforced with 2 wt% SiC. Left column shows cavitation damage near the periphery of the damage zone while right one corresponds to the central region of cavitation damage for each of the specimens

ii) Dynamics of a single bubble near a material surface

Dynamics of single cavitating bubble was studied for varying distance. Figure 6 shows a typical variation for a single bubble near the MSS specimen surface is shown as an example. The microjet formation is clear from figure 6 (n-p). Pressure measurements were carried out and typical pressure sensor output is shown in figure 7. It is clear that both B_4C sample performs better than the bare MSS substrate. The reason behind this different behavior is understandable from the study of vertical sections of different specimens. It was shown by Syamsundar [16] that the depth of penetration of erosion zone is restricted in case of nanoparticle-reinforced PU coatings.

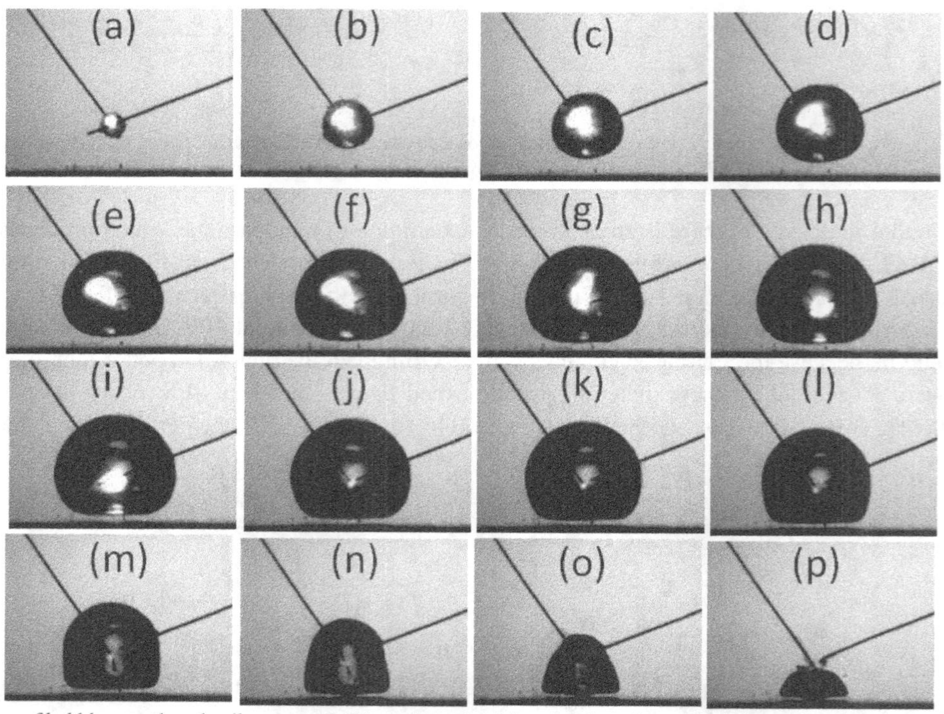

Figure 6: Sequence of bubble growth and collapse near MSS specimen (visible as black surface at the bottom of each image). Time gap between each of these images is 75 μs. Microjet formation is clear for (n) to (p). Normalized distance from the surface ($\frac{H}{R_{max}}$) is 0.84.

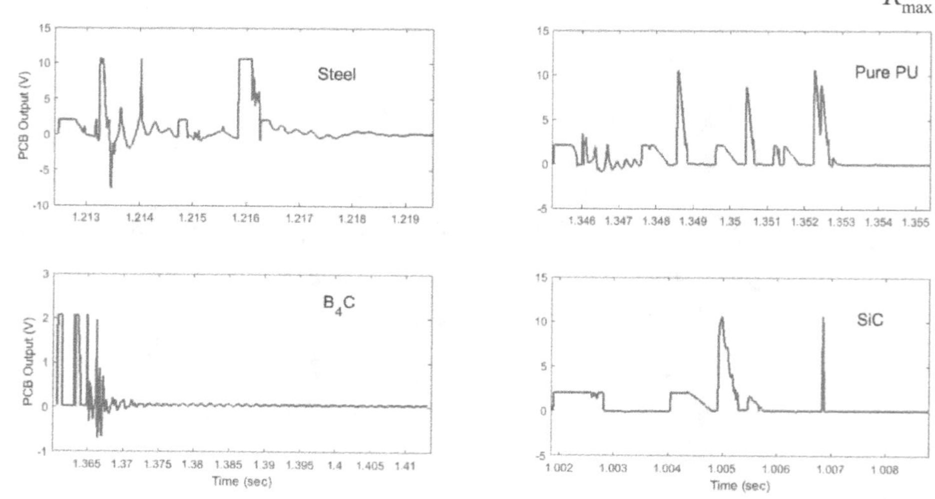

Figure 7: Pressure sensor output for different specimens. Separating distance ($\frac{H}{R_{max}}$) is about 0.4 in each case.

Conclusion

This paper brought out cavitation resistance of different PU-based coatings over MSS substrate. Vibratory cavitation showed that B_4C- and SiC- nanoparticle reinforced PU is very effective in containing cavitation damage. SEM images and pressure measurement for collapse of single bubble near the specimen surface show that the extent of damage is less in these coatings because the coatings absorb greater energy and this results in restriction of zone of erosion in PU-based coatings.

References

[1] Arndt, R.E.A., Ellis, C.R., Paul, S. (1995). *Preliminary investigation of the use of air injection to mitigate cavitation erosion* J. Fluids Engg 117.

[2] Chatterjee, D., Arakeri, V.H. (2004). *Some investigations on the use of ultrasonics in travelling bubble cavitation control*. J. Fluid Mech. 504.

[3] Wu, Y., Hong, S., Zhang, J., et al. (2012). *Microstructure and cavitation erosion behavior of WC-Co-Cr coating on 1Cr18Ni9Ti stainless steel by HVOF thermal spraying*. Intl. J. Refractory Metals Hard Matls 32.

[4] Correa, C.E., García, G.L., García, A.N., Bejarano, W., Guzmán, A.A., Toro, A. (2011). *Wear mechanisms of epoxy-based composite coatings submitted to cavitation*. Wear 271.

[5] Choi, J. K., Chahine, G. L. (2015). *Experimental and numerical study of cavitation erosion resistance of a polyurea coating layer*. Fourth Intl Symp Marine Propulsors, USA.

[6] Deplancke, T., Lame, O., Cavaille, J-Y., Fivel, M., Riondet, M., Franc, J.P. (2015). *Outstanding cavitation erosion resistance of ultra high molecular weight polyethylene (UHMWPE) coatings*. Wear 328-329.

[7] Zhang, J., Richardson, M.O.W., Wilcox, G.D., et al. (1996). *Assessment of resistance of non-metallic coatings to silt abrasion and cavitation erosion in a rotating disk test rig*. Wear 194.

[8] Hattori, S., Itoh, T. (2011). *Cavitation erosion resistance of plastics*. Wear 271.

[9] Tomita Y., Shima A. (1990). *High-speed photographic observations of laser-induced cavitation bubbles in water*. Acustica 71.

[10] Shima, A. (1997). Studies on bubble dynamics. Shock Waves 7.

[11] Syamsundar, C., Chatterjee, D., Kamaraj, M. (2014). *Experimental characterization of silt erosion of 16Cr–5Ni steels and prediction using artificial neural network*. Trans. Indian Inst. Met. 68(4).

[12] Syamsundar, C., Chatterjee, D., Kamaraj, M., Maiti, A. K. (2015). *Erosion characteristics of nanoparticle-reinforced polyurethane coatings on stainless steel substrate*. J. Mater. Eng. Perform. 24(4).

[13] Goh, B.H.T., Oh, Y.D.A., Klaseboer, E., et al. (2012). A low-voltage spark-discharge method for generation of consistent oscillating bubbles. Rev. Sc. Instruments 84.

[14] Müller, M., Unger, J., Budarek, T., et al., (2012). *Collapse of a cavitation bubble generated by low voltage discharge in water*. EPJ Web of Conference 25.

[15] Chahine, G.L., Franc, J-P., Karimi, A. (2014). *Cavitation impulsive pressures*. Adv. Expt Num. Tech Cavitation Erosion Prediction (Kim, K.H. et al. (Ed)), Springer.

[16] Syamsundar, C. (2015). *Experimental investigation of silt and cavitation induced erosion of blade material used in hydro turbines*. Ph.D. Thesis, IIT Madras.

© 2018 ASME

Towards real-time optical measurement of microbubble content in hydrodynamic test facilities

[1]Patrick S. Russell*; [1]Dean R. Giosio; [1]James A. Venning; [1]Bryce W. Pearce; [1]Paul A. Brandner
V. Aumelas[2] and G. Maj[2]

[1]*Cavitation Research Laboratory, University of Tasmania, Launceston, Tasmania, Australia*
[2]*YLEC Consultants, Grenoble, France*

Abstract

Real-time measurement of microbubble concentrations is desirable in order to inform experimental results, particularly in studies of cavitation physics. To develop these capabilities a controlled experiment using a micro-fluidic T-junction to produce mono-disperse microbubbles was devised with the size and frequency of microbubbles measured using a line-scan camera capable of acquiring 45k images per second. Measurements were able to be obtained and reported in under 3 seconds from the triggering time. Tests were carried out in quiescent water and implementation in non-stationary environments would extend the operational range. The principal operating mode produced microbubbles on the order of 80 to 130 μm in size at frequencies ranging from 750 to 3200 bubbles per second across the range of air and water pressures tested.

Keywords: microbubbles; cavitation

Introduction

Even small concentrations of microbubbles present within a fluid can greatly impact upon its mechanical properties. In particular the inception of cavitation inside facilities and on board vessels is often controlled by the largest microbubble which will pass through a region of low pressure. Consequently, real-time measurement of these concentrations is desirable in order to inform predicted performance and experimental results.

In hydrodynamic flows of practical interest cavitation nucleation is invariably heterogeneous where microbubbles entrained in the flow or ejected from surface hydrophobic crevices provide sites of weakness, or nuclei, for the initiation of phase change from liquid to vapour [1, 2]. The equilibrium of a microbubble becomes unstable below a critical pressure, depending on its diameter, after which it will grow explosively. Once activated bubbles fill with vapour and interact with the surrounding stationary or flowing liquid developing into macroscopic cavitation phenomena. Rigorous experimental modelling of the inception and dynamics of hydrodynamic cavitation in water tunnels thus requires control and measurement of the microbubble population. The variable pressure water (cavitation) tunnel within the Cavitation Research Laboratory (CRL) at the Australian Maritime College has been developed with ancillary systems for continuous artificial seeding and removal of microbubbles to provide controlled nuclei populations in the test flow [3, 4]. To date this capability has been developed using direct or dilute injection of poly-disperse microbubble populations generated through the rapid de-pressurisation and cavitation of supersaturated water [5]. Whilst poly-disperse populations (typically 10 to 100 μm in diameter) are always required to model real flows the use of mono-disperse nuclei provides several advantages for basic research and for comparative experimental and computational work.

Microfluidic or lab-on-chip devices have been developed for mono-disperse generation of micro, or nano-bubble populations for sono-fluidic or sono-chemical processes such as contrast agents or drug delivery vectors in medical applications [6]. These devices typically generate smaller bubbles than those suitable for nuclei and may involve the use of surfactants [7]. Commercial devices using common materials and simple experimental set-ups generating microbubbles of order 10 to 100 μm at rates of order 10^3 to 10^4 have been developed by YLEC Consultants, France. The present work is a collaboration between the CRL and YLEC Consultants to investigate the operational range and use of these devices for mono-disperse cavitation nuclei seeding.

*Corresponding Author, Patrick Russell: Patrick.Russell@utas.edu.au

To characterise such devices, or measure populations within the CRL hydrodynamic test facility long range microscopy-shadowgraphy is typically used. With the use of PIV cameras and dual pulse lasers bubble velocities may also be measured. However, the processing of these high resolution image pairs is demanding such that results can not, at present, be obtained in real-time. To address this problem a line-scan camera with a single row of densely spaced pixels was used. The reduced number of pixels allowed the acquisition speed of the camera to be high and the spatial resolution to be increased from standard PIV cameras in one dimension at the sacrifice of the second spatial dimension. When stacked these line measurements produce a space-time plot from which size and frequency information is readily extracted.

Experimental overview

The L10 device from Ylec Consultants was mounted within a quiescent tank at atmospheric pressure and supplied by independently pressurised air and water lines. The size and frequency of bubbles produced was dependent on the balance of these pressures. The device was deemed to be in its principal operating mode when the standard deviation of the bubble size was 10% of the measured mean, with a single production frequency.

The T-junction device was tested in a 0.05 m square acrylic tank filled with distilled water. The junction was fitted through the base of the tank on a mount to which was connected the 4 mm pressurised air and water supply lines. Air pressure was regulated then conditioned through a 1.0 μm filter before it reached the device. Water was supplied from a reservoir of distilled water pressurised by air connected to a second regulator. Water and air pressures ranging between 1.5 and 7 bar absolute in 0.5 bar increments were tested, measured using two Siemens Sitrans P DS III, Range: 0-500 kPa, Model number: 7MF4333-1GA02-2AB1 absolute pressure transducers. These were connected to a National Instruments 6366 USB-DAQ which sampled the pressures for 1 second during image acquisition at a rate of 1000 Hz. The mean and standard deviation of these measurements were recorded. The standard deviation of all pressure measurements was below 0.01 bar and the mean within 0.05 bar of the nominal pressure.

Bubbles were illuminated using a Constellation 120 W white 5600K LED light positioned directly behind the acrylic tank. Images were captured using a Teledyne DALSA Linea Mono linescan camera with 8192×1 pixel resolution. The single line of pixels was sampled 10000 times at a rate of 45 kHz. These rows of pixels when stacked produced a space-time plot which was recorded on board the camera and then transmitted as a single frame to MATLAB for processing. The camera was coupled to a Questar QM100 long-range microscope using a 1.5× Barlow lens then bellows giving a field of view of approx 900 μm with a spatial resolution of 0.121 μm/pixel. A schematic of the overall experimental set-up is shown in figure 1. For each condition 4 ensemble images were collected and processed separately. All data is presented as the mean of these measurements with error bars of two standard deviations applied where appropriate.

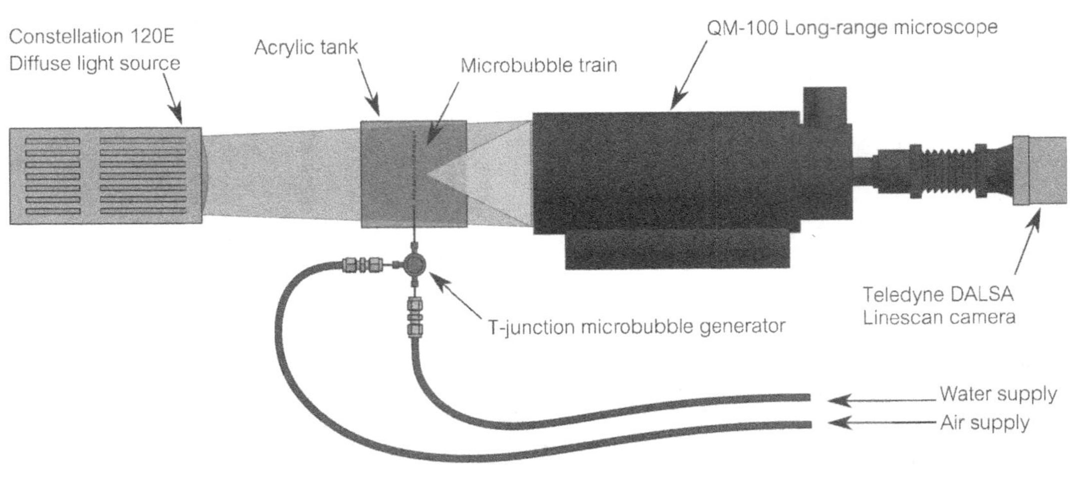

Figure 1: Schematic of experimental setup for the measurement of microbubble size and production rate using shadowgraphy

Processing method

A small portion of an image frame is shown in figure 2. Images were acquired using the MATLAB image acquisition toolbox. Once in MATLAB each row of pixels was summed together to create a time series of integral intensity. Peaks extracted from the series correspond to the passage of a bubble through the measurement area. The single row of pixels corresponding to the maximum of each peak was extracted and the diameter of the bubble calculated via thresholding of the smoothed pixel intensity. Two methods were used to collect bubble production frequency. The first was extracted from the fast Fourier transform (FFT) of the intensity time series. In periodic production regimes this was sufficient to assess production frequency. As the device reached the limits of its operating envelope the size and frequency of the bubble produced fluctuated. The second method simply estimated the integral production frequency by dividing the number of bubbles observed in the intensity time series by the length of the sample. Comparison of this frequency estimate with the peak frequency observed in the FFT of the series allowed immediate determination of when the device was operating in its intended manner. Table 1 shows a summary of the differences between these frequencies. Large numbers signalled that the device was no longer producing mono-disperse bubbles.

Frequency Difference	Δf (hz)	p_{air} (bar)											
		1.5	**2.0**	**2.5**	**3.0**	**3.5**	**4.0**	**4.5**	**5.0**	**5.5**	**6.0**	**6.5**	**7.0**
p_{water} (bar)	**1.5**	-	-	-	-	-	-	-	-	-	-	-	-
	2.0	-	-	-	-	-	-	-	-	-	-	-	-
	2.5	-	-	-	-	-	-	-	-	-	-	-	-
	3.0	-	-	-	-	1	2	3	1	3	776	-	-
	3.5	-	-	77	1	1	0	1	1	1083	68	-	-
	4.0	-	-	116	2	2	1	3	2	3	378	-	-
	4.5	-	26	3	13	2	1	1	1	2	2	2	-
	5.0	-	-	-	1	0	1	3	1	373	17	594	-
	5.5	-	-	-	633	1	1	0	2	37	1723	1334	-
	6.0	-	-	-	-	-	-	1	5	15	1	21	897

Table 1: For each condition the table shows the absolute difference (in Hz) between two measures of the production frequency. The first measure uses the dominant bubble production frequency from largest peak in the FFT of pixel row intensity. The second measure of frequency is gathered by dividing the total number of bubbles observed over the acquisition time. Large values in this table indicates that the device was operating outside the intended principal operating mode. Conditions whose frequency measures are in agreement, $\Delta f < 50\ Hz$, have been highlighted in blue. A difference in frequency greater than 50 Hz is marked in red. A dash indicates that for this pressure combination the device either did not produce bubbles or was not tested.

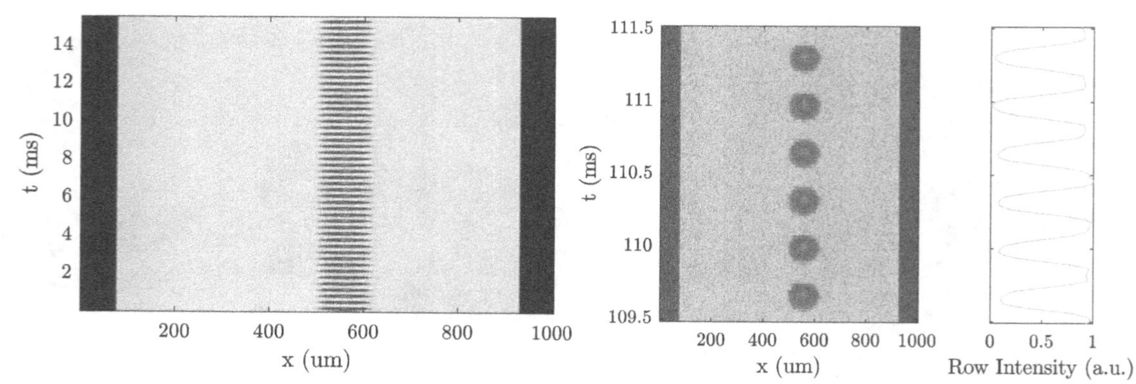

Figure 2: A sample image (left) - $p_{air} = 5$ bar, $p_{water} = 4.5$ bar - showing 700 lines from the line-scan camera. Here the vertical axis is time, with spatial coordinates on the horizontal axis to produce a space time plot. A portion of the regular stream of similarly sized bubbles observed in the image has been enlarged and stretched (centre). Each row of pixels in this centre image has been summed and non-dimensionalised by the minimum and maximum value of the overall series to produce the plotted row intensity time series (right). This time series is used to detect bubbles and find the centre location of each bubble for sizing.

Results

The range of the principal operating mode can be established from examination of Table 1. For convenience the pressure ratio is defined with the air pressure as the numerator. It was observed that a pressure ratio near unity is required to remain inside the principal operating mode. The device remained stable over the largest range when an air pressure of 3-5 bar(abs) was used. At low pressure ratios there was a rapid cessation of production with the device simply ejecting water upon exceeding the critical ratio. In contrast, the breakdown of consistent production at high ratios was more subtle. The device would continue to operate but sporadically shift in and out of periodic production into a chaotic production mode. Qualitatively it was observed that this transition is often due to the disturbances created by coalescence further up in the bubble train, however this is not revealed in the data presented here.

With the conditions of interest established, the relevant data was plotted in terms of the size and frequency in figure 3. Naturally, lower air pressures produced smaller bubbles. Lower air pressures also produced bubbles at lower frequencies. As water pressure increased the bubble size reduced and the production frequency increased. This indicates bubble pinch-off occurred quicker inside the junction at these pressures. The largest bubbles of consistent size were 133 µm in diameter, produced at an air pressure of 5 bar and water pressure of 4 bar. While the smallest bubble size was approximately 70 µm, the two conditions where this size was observed were very close to device breakdown ($p_{Air} = 3$ bar, $p_{water} = 5.5$ bar & $p_{air} = 5$ bar, $p_{water} = 6$ bar). The smallest stable condition then was 83 µm at an air pressure of 3 bar and water pressure of 5 bar.

Bubble production rates were of order 10^3 to 10^4 bubbles per second. In contrast to bubble diameter production frequency increased with water pressure but also with air pressure. The lowest production frequency was 770 Hz ($p_{air} = 3$ bar, $p_{water} = 3.5$ bar), while the highest consistent frequency was 3200 Hz ($p_{air} = 5$ bar, $p_{water} = 5.5$ bar). Higher, stable, mono-disperse production frequencies were achieved, but they were very close to device breakdown and could not be repeated when the combination was later re-tested. Further extreme cases were found that produced bubbles outside of these size and frequency bounds but they were either inconsistent or likely to disappear during testing.

A plane was fitted through the size data as a function of water and air pressure.

$$Size(p_{air}, p_{water}) = 110 + 13.4\ p_{air} - 13.83\ p_{water} \quad (\mu m) \tag{1}$$

In equation form it is clear that the mean bubble size produced was 110 µm and that variation of water and air have similar influence but opposite effects. The size reduced or increased by approximately 13.5 µm as the difference between these two pressures increases by 1 bar.

The adjusted R-squared residual of this fit was 0.76. Consequently approximately 24% variation in the size was left unexplained. This result is not as strong as would be desired. To improve this result, due to the speed with which measurements may be collected, electronically controlled pressure regulators could be used to conduct a more detailed sweep with increased repetition of tests. However, before testing it was assumed that the device was operating independent of its supply pressure history. Observations during testing indicate that this may not be true across short time periods at all conditions. Acquisition of samples were slowed to give time for these effects to decay - approximately 2 minutes for each pressure combination - but this effect may have contributed to errors observed here. The cause for this effect may lie in the pressure supply system. Detailed tracking of the evolving production characteristics following large pressure changes is to be conducted and compared to the supply pressure measurements in time.

Conclusion

Bubble sizes were measured using a linescan camera to collect size and frequency measurements in near real time. Tests were carried out in quiescent water and implementation in non-stationary environments would extend the operational range. However, in this environment the principal operating mode produced microbubbles on the order of 80–130 µm in size at frequencies ranging from 750–3200 bubbles per second were produced across the range of air and water pressures tested. Bubbles near these sizes could be used as seeded cavitation nuclei within hydrodynamic test facilities through dilute or direct injection.

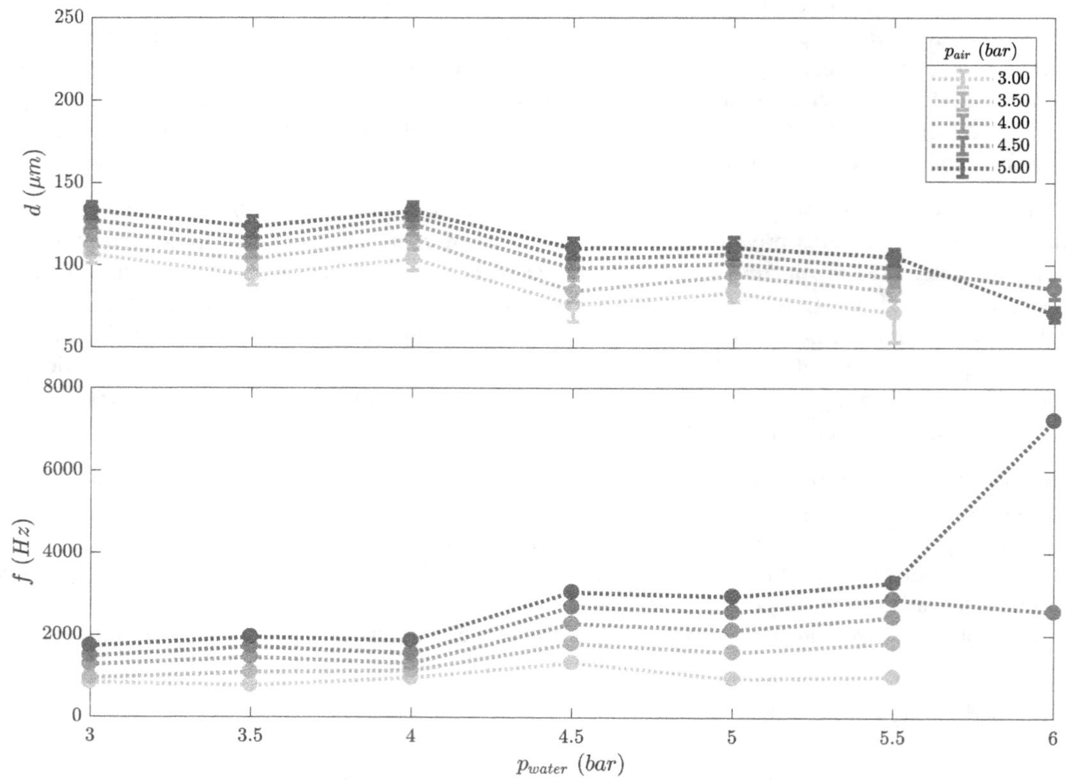

Figure 3: The diameter of bubbles produced (top) decreases as water pressure increases for a constant air supply pressure. Diameters increase with air pressure. Error bars denote two standard deviations. (bottom) Bubble production frequency increases both with air and water pressure.

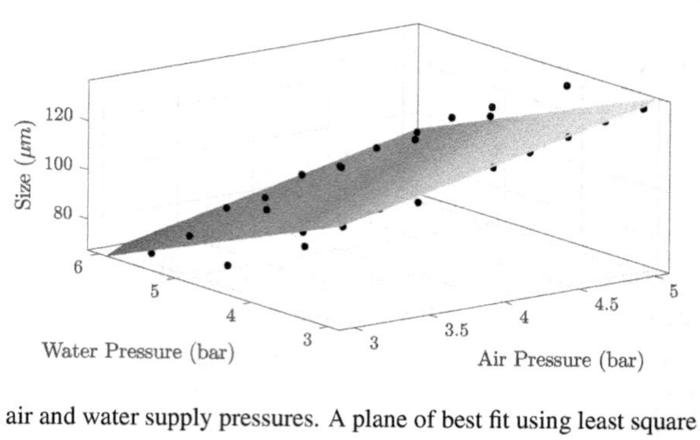

Figure 4: Size is plotted against both air and water supply pressures. A plane of best fit using least square residuals is created though the data.

Acknowledgements:

This project was supported by the Defence Science and Technology Group (Mr. Brendon Anderson and Dr. David Clarke), the University of Tasmania, and the US Office of Naval Research (Dr. Ki-Han Kim, Program Officer) and ONR Global (Dr. Woei-Min Lin) through NICOP S&T Grant no. N62909-15-1-2019.

References

[1] Christopher E. Brennen. *Cavitation and bubble dynamics*. Cambridge University Press, 2013.
[2] Jean-Pierre Franc and Jean-Marie Michel. *Fundamentals of Cavitation*. Springer Science & Business Media, 2006.
[3] PA Brandner, Y Lecoffre, and GJ Walker. *Development of an Australian National Facility for Cavitation Research*. In Sixth International Symposium on Cavitation, 2006.
[4] P. A. Brandner, Y. Lecoffre, and G. J. Walker. *Design Considerations in the Development of a Modern Cavitation Tunnel*. In 16th Australasian Fluid Mechanics Conference, 2007.
[5] D Giosio, BW Pearce, and PA Brandner. Influence of pressure on microbubble production rate in a confined turbulent jet. In *20th Australasian Fluid Mechanics Conference*, 2016.
[6] Joshua Owen, Paul Rademeyer, Daniel Chung, Qian Cheng, David Holroyd, Constantin Coussios, Peter Friend, Quentin A Pankhurst, and Eleanor Stride. *Magnetic targeting of microbubbles against physiologically relevant flow conditions*. Interface focus, (5), 2015.
[7] Shelley Lynn Anna. *Droplets and Bubbles in Microfluidic Devices*. Annual Review of Fluid Mechanics, 48(1), 2016.

© 2018 ASME

Transient Cavitation by Quick Closing Pincers

[1]Francisco A. Godínez*; [1]Margarita Navarrete; [2]Oscar Chávez; [1]Enrique Guzmán

[1]*Universidad Nacional Autónoma de México, Instituto de Ingeniería, Cd Mx, México,* [2]*Instituto Tecnológico de Chihuahua, Chihuahua, México*

Abstract

A mechanical device that mimics the snapping-claw mechanism of alpheid shrimps was developed to study experimentally transient cavitation. The apparatus consists of a pair of clamps (jaws), each with an inner contour that conforms to a half-Venturi geometry. One of the jaws is rigidly attached between two optically transparent parallel plates (forming a socket) and the other one works on a pivot. The clamps are mounted in such a way that when they close a Venturi tube-like with open and closed ends is formed. For the experiments, the device was immerse a few centimeters under tap water at lab conditions. The pivoting jaw was locked into an open position; then, its rapid closing was triggered mechanically and was driven by the contraction of rubber bands. High-speed video was used to capture the characteristics of the flow induced by the sudden relative motion between clamps. A water jet formed along with a variety of cavitation structures, such as cloud shedding, vapor lobes, travelling vortex rings, and bubble clouds were revealed. The developed technique could be an inexpensive alternative to test the performance of different Venturi geometries in cavitation regimes without the need for pipes, valves, tanks, and pumps typically used in experiments with hydraulic circuits.

Keywords: transient cavitation; Venturi tube; snapping-claw; cavitation structures

1. Introduction

Snapping shrimps (Family Alpheidae) possess a claw that can grow to be half the size of their entire body [1]. When this claw is closed rapidly, a cavitation front forms and collapses to produce a loud snap [2] with energy over a broad range of frequencies ranging from a few hundred hertz to 200 kHz [3]. The majority of studies to characterize this snap sound have been conducted with live animals and in their natural habitat. Nevertheless, greater control, repeatability and scalability may be attained through the use of artificial devices to emulate the claw geometry and its fast closing mechanism, in order to understand and clarify the phenomena associated with the emission of the snap sound. In this sense, Hess et al. [4] fabricated and tested an enlarged transparent model (which basically consists of a pivoting plunger and a socket) to reproduce the closure of the snapper claw and the flow generated by this process. Experiments were performed using both high-speed particle image velocimetry (HS-PIV) and flow visualization. During claw closure a channel-like cavity was formed between the plunger and the socket featuring a nozzle-type contour at the orifice. Hess's measurements revealed maximum flow velocities of 1 m/s in the nozzle with the formation of a vortex ring that grows in time. However, these velocities are too low to produce pressure drops below the vapor pressure of the used liquid (water-glycerine mixture) at lab conditions, and thus no cavitation process was observed. The aim of this paper was to design and test a mechanical device to produce transient cavitation in a repetitive and controlled way by mimicking the snapping-claw mechanism of alpheid shrimps. The proposed design closes the pivoting jaw in less than 9×10^{-3} s and is powerful enough to drive the formation of cavitating structures, as revealed by high-speed videography and image processing.

2. Mechanical device

The CAD model used for experiments is shown in figure (1a). It is worthy of note that the pair of walls (4) serve to shape the socket (3) and are made of transparent PMMA to facilitate flow visualization; additionally, they serve as support for the lower clamp (2), pivot (5), attachments (7), and mechanical trigger (8). Each one of the clamps have an inner contour that conforms to a half-Venturi geometry. The lower clamp (2) is rigidly attached between the two optically transparent parallel plates (4) (which in turn are fixed to corbel type supports), while the upper clamp (1) works on the pivot (5). When closing the pincers a Venturi tube-like with open and closed ends is formed. The violent closure of the clamps is achieved by the contraction of the set of rubber bands (6). Figure (1b) depicts the size of the Venturi tube formed between the socket walls and the clamps (1) and (2) when they are closed. It is important to note that the used Venturi geometry is not typical; in the present design, the cross section of the tube is rectangular as that used in some scrubbers [5] (see figure 1c). Besides, the converging section is larger than the diverging one and the

*Corresponding Author, Francisco Godínez: fgodinezr@gmail.com

throat is as large as the converging section. Despite these differences with traditional Venturi geometries, the present configuration allows the formation of a strong jet that moves beyond the outlet of the divergent section; the direction of this flow is indicated in figure (1b) with an arrow.

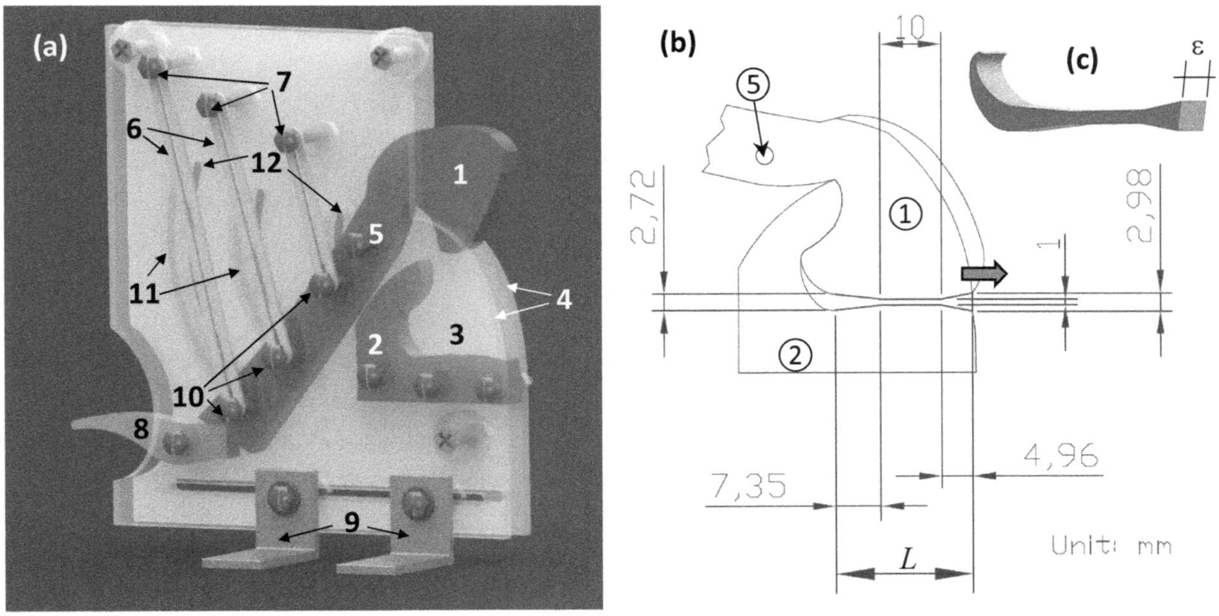

Figure 1 a) 3D CAD-model of the mechanical device to generate transient cavitation (open); 1) pivoting upper clamp, 2) fixed lower clamp, 3) socket, 4) walls, 5) pivot, 6) set of rubber bands, 7) set of band attachments, 8) mechanical trigger, 9) corbel type supports, 10) sliding bolts, and 11) curved sliding guides. b) Venturi tube-like geometry formed when clamps are totally closed. c) Three-dimensional scheme of the tube with rectangular cross-section formed when pincers are fully closed. The total length of the Venturi tube is $L = 22.31$ mm, and the distance between the walls (4) is $\varepsilon = 5.8$mm.

3. Experiments with high-speed video camera

3.1 Experimental setup

For the experiments, a rectangular tank (580 mm long, 500 mm wide and 270 mm deep) with transparent walls and filled with tap water at lab conditions (20° C, 77.86 kPa) was used. The upper clamp was locked into an open position (as seen in figure (1a)). The device was immersed in the tank 150 mm below the water level. The upper clamp was released by the mechanical trigger and its motion was recorded with a Phantom® v1212 high speed camera. To make the shots, the region of interest was illuminated with a 36,000 lumens lamp.

3.2 Dynamics and energetics of the closing process

Image processing and video analysis were applied to obtain the graph of the angular position of the upper clamp versus time (figure (3a)). The displacement ranges from the fully open position, when the clamp is mechanically released by the trigger (8), to the final position where the clamps are fully closed and the Venturi is formed. Figure (3b) represents the angular velocity developed by the upper clamp during the closing process. This curve was obtained by differentiating the curve of angular position of figure (3a). An acceleration region (i) is observed at the beginning of the closing process ($0 \le t/T \le 0.42$); region (ii) shows an almost constant rotation motion ($0.42 < t/T \le 0.7$); then a deceleration process is observed in the region (iii) ($0.7 < t/T \le 1$), which is slightly shorter than the region (i); finally, after the total closure, a series of bounces are visible in the region (iv) ($1 < t/T \le 1.29$). Such bounces are due to impact forces between the sliding bolts (10) and the uppermost part (12) of the curved sliding guides (11); see figure (1a).

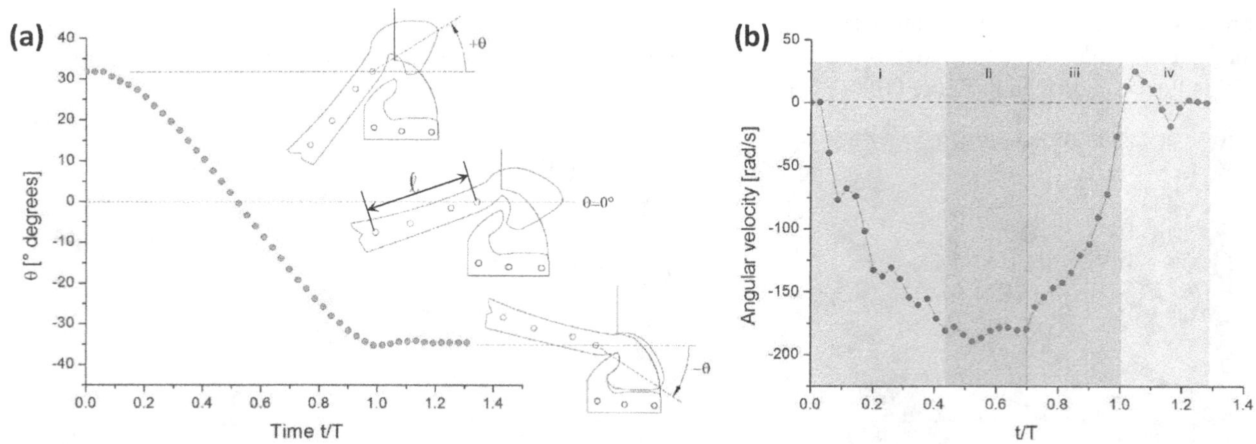

Figure 3 Time evolution of a) the angular position (data are averages of three experiments with SD of 0.6 [°degrees] or less), and b) the angular velocity of the upper clamp. In both graphs the horizontal axis is normalized with the total closure time, which is $T \approx 9$ ms.

Figure (4a) shows the curve of the torque (as a function of the angular position) that must be applied to open the upper clamp to its initial position; in terms of energy this represents around 3 J. By combining the angular velocity and torque data from figures (3b) and (4a), respectively, a power delivery curve is constructed. The average value of the power delivered is around 347 W (0.465 hp).

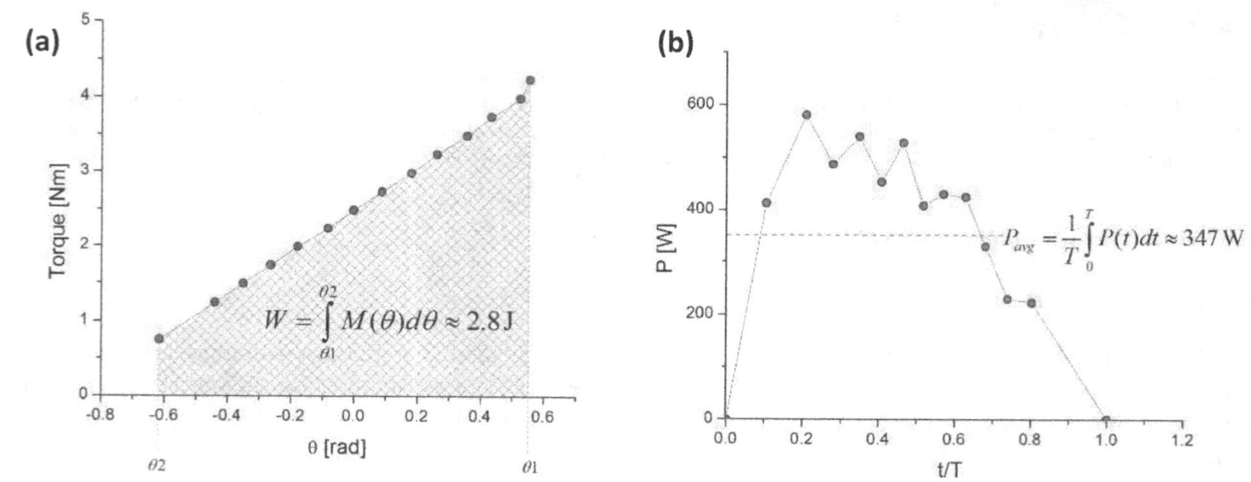

Figure 4 a) Torque (M) vs. angular position (θ); the points (solid symbols) were measured with a dynamometer considering a lever arm of length $\ell = 53mm$ (see figure (3a)). b) Power delivered by the pincers during the closing process.

3.3 Cavitation patterns and flow regimes

The high-speed photographic sequences of different cavitation patters and flow regimes were extracted from videos taken with a sampling frequency of 40,000 fps, and exposure time 10 μs. These were ordered chronologically from the moment in which the upper clamp is released, until a few instants after it closes completely.

3.3.1 No cavitation regime

Starting from t=0 s, when the upper clamp starts moving, and up to t ≈ 6.1 ms, no cavitating structures are observed.

3.3.2 Cavitation cloud shedding and vortex cavities

The motion of the upper clamp generates an incipient cavitating flow which is visible at t ≈ 6.125 ms, as indicated in the snapshot 1 of figure 5. This flow evolves and forms a cavitation cloud (snapshots 4, 5), and a series of vortices whose sizes increase as the upper clamp closes (snapshots 3, 6, 7). Bubble clouds originate in the leading-edge of the upper clamp (indicated in snapshot 2); from these clouds, a series of cavitating vortices are detached (snapshots 3, 6). This flow resembles in some sense the cavitating flow observed in hydrofoils and convergent divergent nozzles [6]. In the present case, however, the profile moves in relation to the static fluid and the flow is not periodic. Furthermore, an incipient cloud shedding is observed in the lower clamp (snapshot 5), a few moments later, tiny vortices are also formed and expelled into the free fluid (snapshots 7, 8).

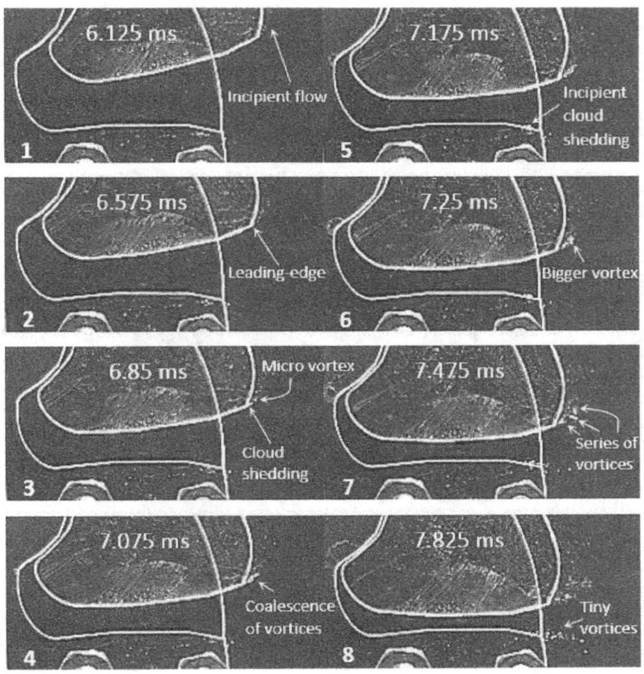

Figure 5 High-speed Photographic sequence of the cavitating flow induced by the pincers closing during the lapse $6.125 \leq t \leq 7.825$ ms.

3.3.3 Vapor lobes, counter-rotating travelling vortex rings and cavitating sheet

Cavitation inception is observed at the lower edge of the throat, tiny vapor loves appear in both the upper and lower clamps (see snapshot 10 in fig.6). As time progresses, these lobes increase in size surrounding a water channel in the middle (snapshots 11, 12), something similar has been reported in [7]. Around the same time, some other cavitation structures appear. Photos 9 to 14 show the progressive growth of two counter-rotating travelling vortices in the free fluid; an upper vortex ring (indicated as V1) of small bubbles and a lower cavitating vortex ring (indicated as V2). A similar flow of two counter-rotating vortices (with no cavitation) at the exit of the diverging nozzle is experimentally observed in [4]. The inception and growth of a cavitating sheet in the upper wall of the throat is also observed from snapshot 13 onwards.

3.3.4 Vapor bubbles, cavitating vortex ring, bubble clouds and reverse flow

Photo 13 shows how the tips of the vapor lobes leave the divergent nozzle and come together to form a single lobe, as seen in photo 14. At the same time, millimetric vapor bubbles become visible at the upper clamp in the closed-convergent section. Their, growth, collapse (which release pressure waves), and rebound are visible from photo 14 onwards. Snapshots 15 to 17 show the process of formation of a leading vortex ring (indicated as V3). It is noticed that part of it begins to detach from the main lobe at 8.7 ms. At 8.75 ms, the vortex ring has practically detached from the single lobe, and moves with an average velocity of 14 m/s through the free fluid while it begins to collapse. Two clouds of bubbles (indicated as CB1 and CB2) are observed at 8.9 ms; these are the product of the collapse and expansion of the vortex ring V3. Later on, the clouds CB1 and CB2 collapse and bounce to give rise to the single cloud CB3 observed at 9.35 ms; this cloud keeps moving while it disperses. Finally, at 9.775 ms, a cluster of tiny bubbles is observed, which still shows the basic structure of vortices V1, V2 and V3.

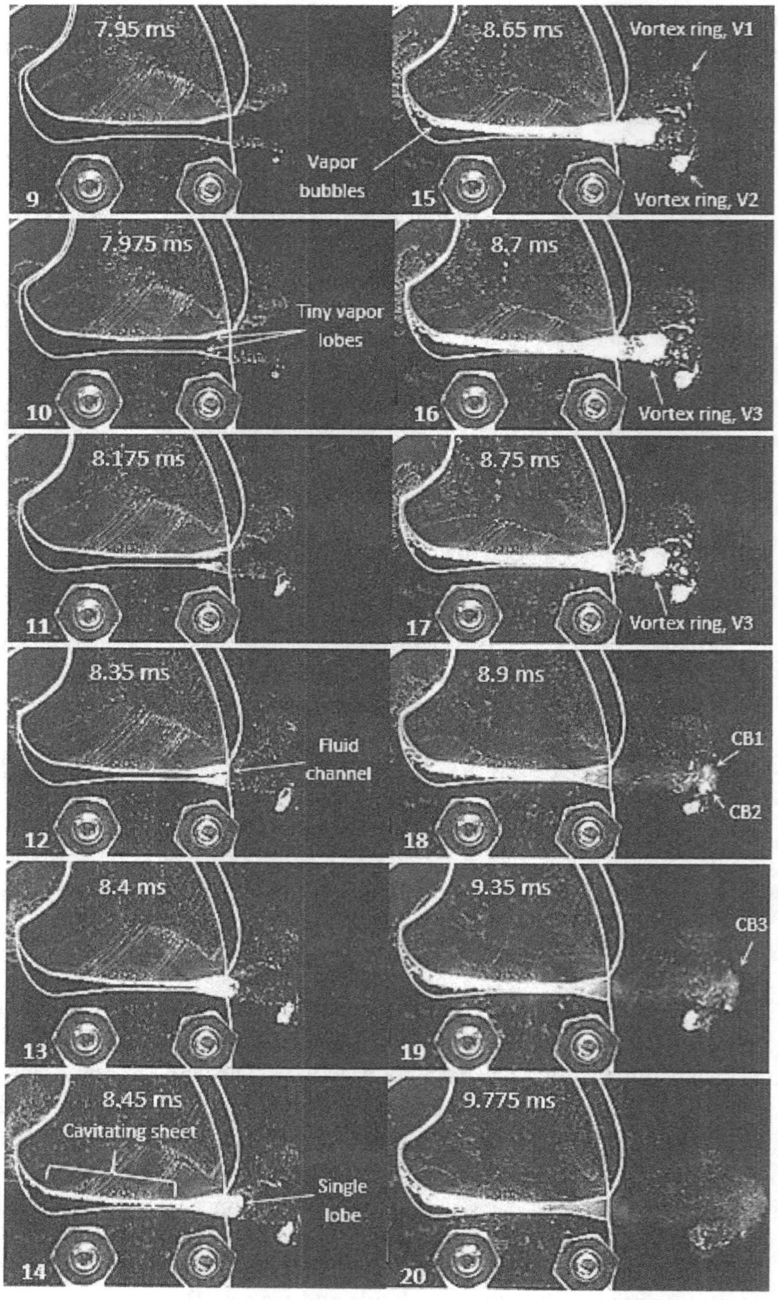

Figure 6 High-speed photographic sequence showing the cavitating flow induced by the motion of the pincers.

4. Dimensionless numbers of the induced flow

Table 1 compares the values of some dimensionless parameters that characterize the flow induced by the rapid closure of the shrimp claw and the pincers device. The average speed of the flow was measured from the exit of the throat to a few millimeters beyond the exit of the divergent section. The procedure consisted in following the motion of the tip of the upper cavitation lobe originated at the exit of the throat. Then the analysis from snapshots 9 to 14 (figure (6)) yields a value of $U \approx 14.5$ m/s. However, the flow velocity developed in the throat when the Venturi is formed (snapshots 9 to 11 in figure (6)) is greater than U. Respectively, the Cavitation, Reynolds and Strouhal numbers, were estimated with [4]: $Ca = \frac{p - p_v}{1/2 \rho U^2}$, $\mathrm{Re} = \frac{U \cdot L}{\nu}$, $Sr = \frac{L}{T \cdot U}$. It is remarked that a very similar value of the Ca parameter was calculated for the shrimp claw and the pincers device; this value is less than 1, thus indicating the occurrence of cavitation [8] in both systems, a fact corroborated by our experiments.

	Alpheid shrimp	***Pincers***	***Scaling ratio***
Geometry	$L_0 \cong 1.41mm$	$L \cong 22.31mm$	$L / L_0 \cong 16$
Medium properties	$\nu_0 \cong 1cm^2 s^{-1}$	$\nu \cong 1cm^2 s^{-1}$	$\nu / \nu_0 \cong 1$
	$\rho_0 \cong 1g cm^3$	$\rho \cong 1g cm^3$	$\rho / \rho_0 \cong 1$
	$p_{v0} \cong 2333 Pa$	$p_v \cong 2333 Pa$	$p_v / p_{v0} \cong 1$
Closure time	$T_0 \cong 500 \mu s$	$T_0 \cong 9ms$	$T / T_0 \cong 18$
Flow velocity	$U_0 \cong 17ms^{-1}$	$U \cong 14.5ms^{-1}$	$U / U_0 \cong 0.85$
Cavitation Nr.	$Ca_0 \cong 0.685$	$Ca \cong 0.732$	$Ca / Ca_0 \cong 1$
Reynolds Nr.	$\mathrm{Re}_0 \cong 24000$	$\mathrm{Re} \cong 323495$	$\mathrm{Re}/ \mathrm{Re}_0 \cong 13.5$
Strouhal Nr.	$Sr_0 \cong 0.17$	$Sr \cong 0.17$	$Sr / Sr_0 \cong 1$

Table 1 Comparison between parameters of the shrimp claw and the pincers device, the data in the first column were taken from [4].

5. Conclusions

A mechanical device that emulates the snapping-claw mechanism of the pistol shrimp was designed and experimentally tested. The apparatus reproduces a variety of transient bubbly flow-regimes and cavitation-structures in a Venturi tube, and in the fluid near the open end of the tube. High-speed video recording proved to be insufficient to accurately establish the velocity of the water jet formed in the Venturi. Techniques such as PIV will be used in the future to precisely measure the velocity field. Furthermore, a photodiode and a hydrophone will be used to detect shock waves and luminescence, respectively. Both phenomena presumably result from the observed collapse and rebound of bubble clouds. The developed technique is simple, inexpensive, and shows great potential for testing the performance of different Venturi geometries. It proves particularly adequate to study cavitation regimes without the need for pipes, valves, tanks and pumps typically used in experiments with hydraulic circuits.

Acnowledgment: This research is funded by Programa de Apoyo a Proyectos de Investigación e Innovación Tecnológica PAPIIT de la UNAM. IN105117-2: Erosión por cavitación óptica, ultrasónica e hidrodinámica.

References

[1] Bohnenstiehl, D.R., Lillis, A., Eggleston, D.B. (2016). *The Curious Acoustic Behavior of Estuarine Snapping Shrimp: Temporal Patterns of Snapping Shrimp Sound in Sub-Tidal Oyster Reef Habitat.* PLoS ONE. 11(1).

[2] Versluis, M., Schmitz, B., Heydt von der, A., Lohse, D. (2000). *How snapping shrimp snap: Through cavitating bubbles.* Science. American Association for the Advancement of Science. 289.

[3] Cato, D.H., Bell, M.J. (1992) *Ultrasonic ambient noise in Australian shallow waters at frequencies up to 200 kHz*. Materials Research Laboratory, Australia; 1992 p. 27. Report No.: MRL-TR-91-23.

[4] Hess, D., Brücker, C., Hegner, F., Balmert, A., Bleckmann, H. (2013). *Vortex Formation with a Snapping Shrimp Claw.* PLoS ONE. 8(11).

[5] Overcamp, T. J. &. Bowen, S. R. (1983) *Effect of Throat Length and Diffuser Angle on Pressure Loss Across a Venturi Scrubber*, Journal of the Air Pollution Control Association. 33(6).

[6] Sato, K., Wada, Y., Noto, Y., Sugimoto, Y. (2010) *Reentrant Motion in Cloud Cavitation due to Cloud Collapse and Pressure Wave Propagation.* ASME. Fluids Engineering Division Summer Meeting, ASME 2010 3rd Joint US-European Fluids Engineering Summer Meeting: Volume 2, August 1–5, Montreal, Canada.

[7] Abdulaziz, A.M. (2014) *Performance and image analysis of a cavitating process in a small type venture.* Experimental Thermal and Fluid Science, 53.

[8] Gogate, P. R. & Pandit, A. B. (2000) *Engineering Design Methods for Cavitation Reactors II: Hydrodynamic Cavitation.* AIChE Journal. 46(8).

© 2018 ASME

Transient Cavitation in Pre-Filled Syringes During Autoinjector Actuation

Jean-Christophe Veilleux*; Kazuki Maeda ; Tim Colonius ; Joseph E. Shepherd

California Institute of Technology, Pasadena, CA, USA

Abstract

Cavitation has been observed in the cone of a syringe actuated by an autoinjector device. Numerical simulations were used to determine if the cone can enhance the collapse of a bubble. We found the collapse of a bubble in a cone can be more violent than the collapse of a bubble close to a flat wall or in a free space due to the reflected wave focusing on the axis of symmetry.

Keywords: transient cavitation; tension waves; shock focusing; autoinjector; syringe

Introduction

Autoinjectors are used for emergency purposes and for long-term treatments requiring the frequent injection of biopharmaceuticals [1]. Even if the specific design of each autoinjector device may differ, in most devices currently available on the market, the mechanism is spring actuated [2, 3]. Activation of the device may result in mechanical impacts between the moving components, and large accelerations/decelerations of the moving components may occur.

Figure 1 is a schematic of the internal components and actuation sequence of a SureClick®[1] autoinjector. In panel A the device is in its initial state just before actuation; the device bottom features are in contact with the patient's skin. In panel B the device has been activated; the spring-actuated driving rod is moving forward and impacts on the plunger-stopper. The impact of the driving rod on the plunger-stopper forces the plunger-stopper to start moving. This compresses and pressurizes the air gap (panel C) and sends a train of compression waves in the liquid from top to bottom. Because there is friction between the plunger-stopper and the syringe, the impact of the driving rod on the plunger-stopper also accelerates the syringe forward (panel C). The abrupt acceleration of the syringe creates a train of tension waves traveling from bottom to top in the syringe. In panel D, the syringe reaches its travel limit and is decelerated to a complete stop. The deceleration of the syringe creates a train of compression waves traveling from bottom to top in the syringe. In panel D, the force exerted by the spring-actuated driving rod on the plunger-stopper pressurizes the syringe and the medicament is extruded through the needle.

Experimental measurements by Veilleux and Shepherd [4, 5, 6] have shown the driving rod impacts on the plunger-stopper at about 10 m/s, and the acceleration/deceleration of the syringe during device actuation can be as large as

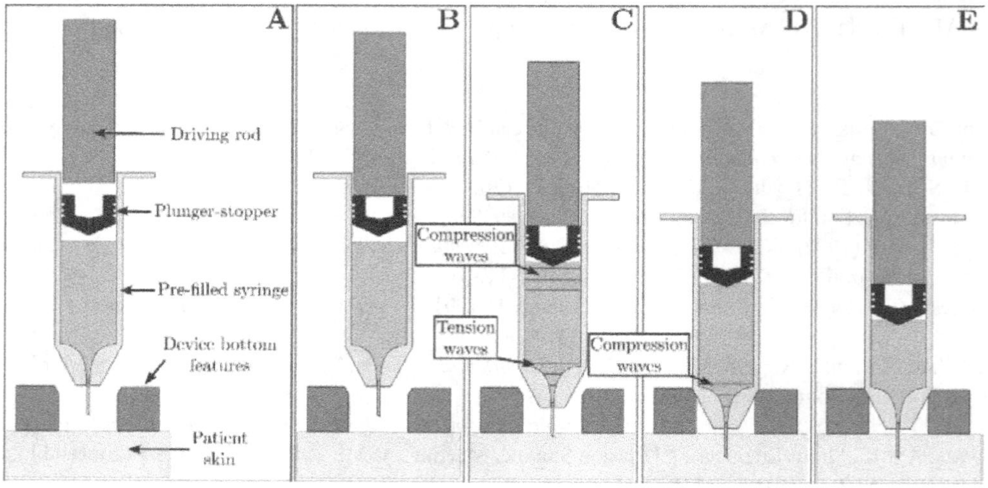

Figure 1. Actuation sequence of a Sureclick® autoinjector.

[1] SureClick® is a registered trademark of Amgen.

*Corresponding Author, Jean-Christophe Veilleux: jc.veilleux@caltech.edu

10,000-35,000 m/s^2. Such large accelerations of the syringe create tension waves with magnitudes large enough to reduce the liquid pressure to vapor pressure, and this results in cavitation in the syringe [4]. In particular, transient cavitation is consistently observed at the bottom of the syringe in the cone area.

Tension waves are also generated when compression waves reflect on the air gap located between the plunger-stopper and the liquid. Details on the complex interaction between the pressure waves, the air gap and the plunger-stopper are available in [7].

An example of the liquid pressure in the syringe during actuation is shown in figure 2. Events A through E are clearly indicated. It is important to note that event E (extrusion of the drug) is only partially represented in figure 2; the extrusion continues beyond 6 ms for several seconds. The multiple pressure peaks shown in figure 2 between 0.8 ms and 2.0 ms are caused by the collapse of cavitation bubbles. In figure 3 it is possible to see the growth and collapse of a bubble within the cone upon acceleration of the syringe, and evidence of cavitation away from the cone. The time window over which frames 1 through 18 of figures 3 have been obtained is indicated in figure 2 (grey area — frame 1 corresponds to event A).

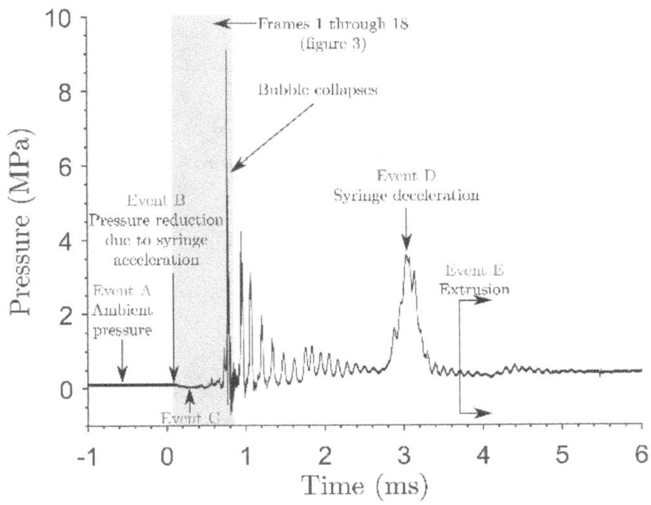

Figure 2. Liquid pressure during the transient events following actuation of a SureClick® autoinjector.

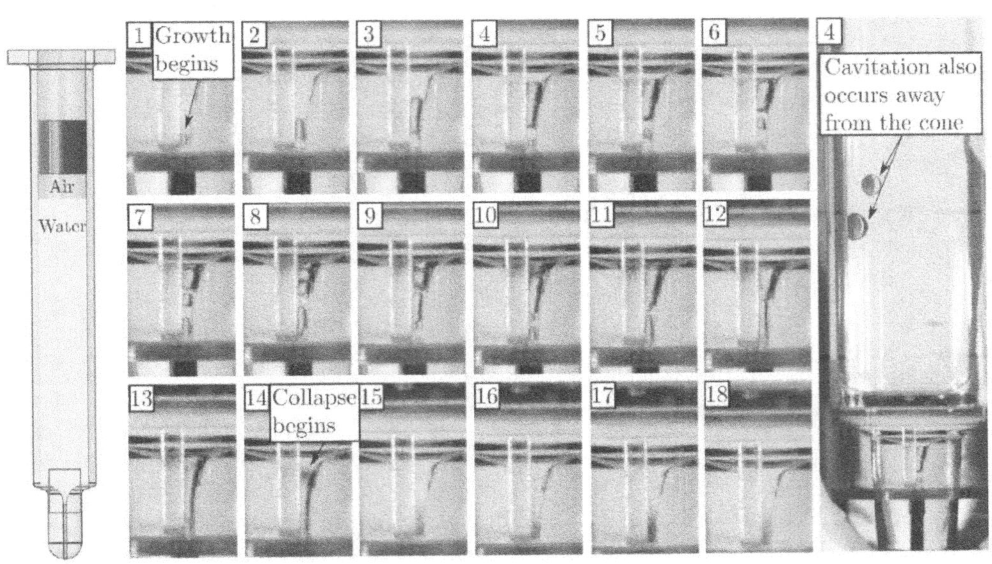

Figure 3. Sequence of images showing the growth and collapse of a bubble in the cone area of a syringe mounted inside a SureClick® autoinjector during acceleration of the syringe (30 µs between each frame). Frame 4 also shows evidence of cavitation away from the cone.

Studying cavitation within the cone of a syringe mounted in an autoinjector is complicated by the small size of the syringe. Experiments on a scaled-up model autoinjector and numerical simulations have been used to gain insights. In this paper we report on the numerical simulations used to study the collapse of a cavitation bubble within the cone of the syringe. The main objective is to determine if and how the cone can enhance the collapse of a bubble.

Numerical Setup

The simulations have been performed with an interface-capturing method for compressible multi-component flows that employs an Eulerian finite-volume, weighted essentially non-oscillatory (WENO) scheme. The details about the numerical methods and verification/validation of the code are presented elsewhere [8, 9].

A first step in understanding the collapse of a bubble in the cone of a syringe is to study the shock-induced collapse of a gas bubble in an otherwise quiescent media, depicted in figure 4. This configuration avoids having to address the randomness in nucleation events. Initially an air bubble of radius R_0 is located at a distance H_0 from the apex of a straight, water-filled cone of half-angle α and depth L. An incident weak shock is traveling down the cone. The pressure and density ahead of the shock are respectively P_0 and ρ_0. The pressure behind the shock is P_s (other properties can be calculated with the shock-jump conditions). Inside the air bubble the initial pressure and density are respectively $P_{b,0}$ and $\rho_{b,0}$. At a later time, the weak shock has entered the cone (right-hand-side of figure 4); a reflected wave and a diffracted wave form. The pressure increases in the cone and the bubble is forced to collapse.

All numerical simulations reported in this paper were performed in 2D axisymmetric coordinates. A uniform grid is used in the shock-bubble interaction region, with a resolution of 108 cells per original bubble diameter. All quantities are normalized using R_0, ρ_0 and a_0. A stiffened gas equation of state is used for both fluids:

$$P = (\gamma - 1)\rho e - \gamma\pi_\infty$$

where P is the pressure, e is the internal energy, ρ is the density. The heat capacity ratio γ and the normalized stiffness π_∞ are respectively 1.4 and 0 for air, and 6.68 and 195.25 for water.

Shock-induced collapses have been simulated with $P_s/P_0 = 500$, $\rho_{b,0}/\rho_0 = 1.2\text{E-}3$, $L/R_0 = 10$, and $P_{b,0} = P_0$. For comparison purposes, we also considered a Rayleigh collapse in free space where the bubble is initially set to a low pressure, $P_{b,0} = 0.002\ P_0$ and the shock is not present ($P_s = P_0$). Numerical simulations have been performed for α ranging from 25 to 60 degrees, and H_0/R_0 ranging from 1 to 10.

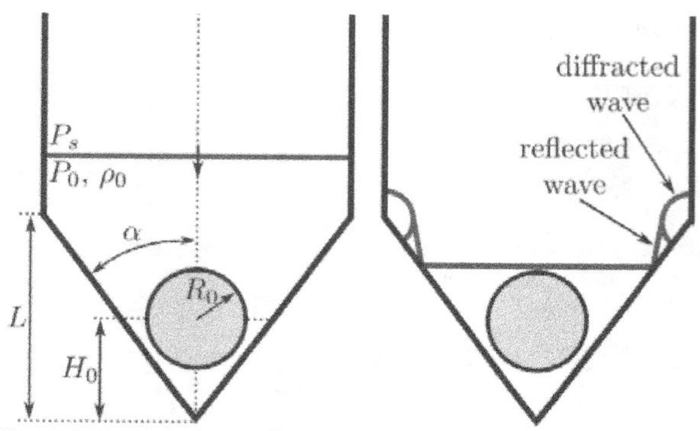

Figure 4. Numerical setup for the simulations reported in this paper.

Results & Discussion

The non-dimensional collapse time $\tau = a_0 t_{collapse}/R_0$ is shown in figure 5 as a function of the offset distance for half-angles of 30, 45, 60 and 90 degrees. The latter corresponds to the collapse of a bubble close to a flat wall. The collapse time for the free-space collapse (i.e., shock induced collapse of a bubble in a free space) and the Rayleigh collapse are also shown. As expected, the collapse time tends toward the free-space collapse time for large offset distances. This means the bubble does not feel the cone or the flat wall if the offset distance is large. When the offset distance is such that the bubble collapse is affected by the cone or the wall, the collapse time is generally reduced. This suggests the collapse is enhanced by the presence of the cone. The collapse time can even be less than the Rayleigh collapse. The minimum collapse time is achieved at an offset distance that decreases when the half-angle is decreased. Furthermore, the collapse of a bubble in a cone appears to be generally more violent than the collapse of a bubble close to a flat wall.

The peak pressure on the axis of symmetry is shown in figure 6 for the same cases shown in figure 5. There are also 4 cases labeled as *no bubble*: these cases correspond to the situation where there is no bubble in the cone, and the peak pressure on the axis results from shock focusing alone.

The peak pressures on the axis appear to always be larger than for a free-space collapse, even when the bubble is far from the apex of the cone. The peak pressure on the axis is due to the combined effect of the collapsing bubble and shock focusing. Unless the offset distance is very small (i.e., $H_0/R_0 \sim 1$), the peak pressure on the axis is larger when the bubble is collapsing within a cone than when it collapses close to a flat wall. The collapse of a bubble within the cone gives large peak pressures for intermediate offset distances ranging between 3 and 6. However, the peak axis pressure is not always larger than for the no bubble case. The same conclusions apply to the peak pressure on the wall (not shown in this paper) instead of the peak pressure on the axis.

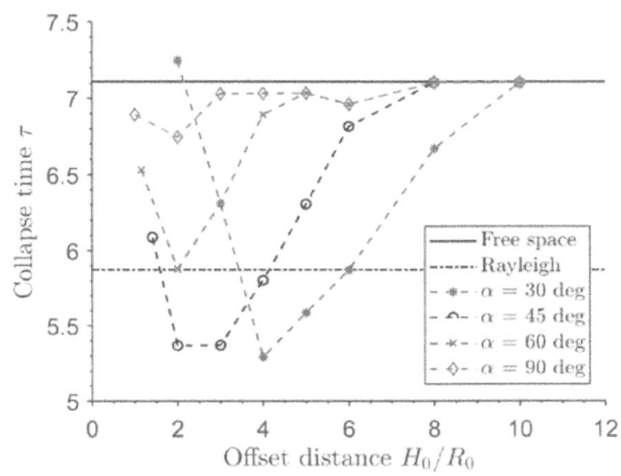

Figure 5. Collapse time of the bubble for various offset distances and half-angles.

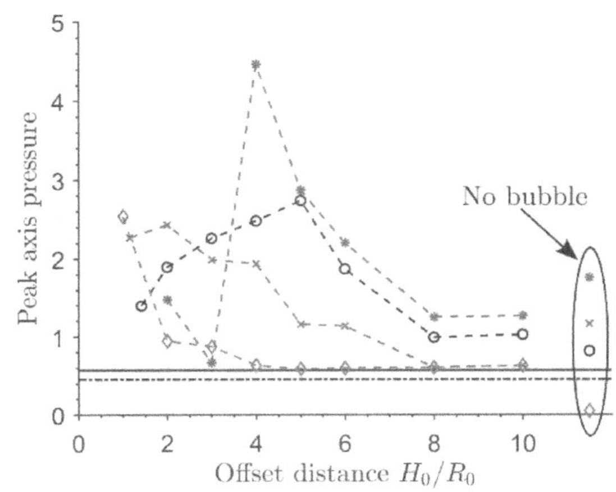

Figure 6. Peak pressure on the axis of symmetry for various offset distances and half-angles.

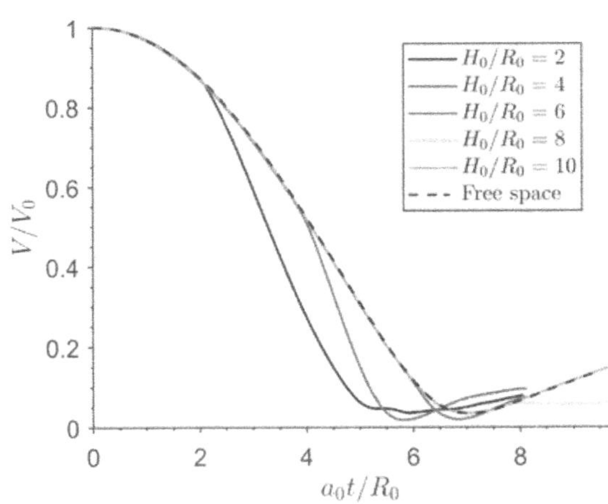

Figure 7. Bubble voulme history for various offset distances at a half-angle of 45 degrees.

The results above suggest the collapse of a bubble in a cone is generally more violent (i.e., smaller collapse time and larger peak pressures) than the collapse of a bubble close to a flat wall or a Rayleigh collapse. In order to gain insights about the mechanism of enhancement, we consider the bubble volume histories shown in figure 7. The traces shown are for a half-angle of 45 degrees, and for a free-field collapse. Time 0 corresponds to the arrival of the incident wave on the proximal side of the bubble.

For all offset distances, the initial phase of the collapse is almost identical to the free-space collapse, before the cone influences the collapsing bubble. The effect of the cone begins at a time a_0T/R_0 which increases with the offset distance. This time a_0T/R_0 is well approximated by:

$$\frac{a_0 T}{R_0} \approx 2\left(\frac{H_0}{R_0}\right)(sin \propto)^2$$

For $a_0\, t/R_0 > a_0\, T/R_0$ the volume history of a bubble collapsing in a cone differs greatly from that of the free-space collapse; the collapse is enhanced and happens more rapidly. The reflected wave is responsible for the enhancement of the collapse. In fact, if one assumes the reflected wave travels at sound speed and the bubble radius does not change between the arrival of the incident wave and the arrival of the reflected wave, it is possible to show that $2(H_0/R_0)(\sin \propto)^2$ corresponds exactly to the non-dimensional time it takes for the reflected wave to reach the bubble.

More details are provided in figure 8 for an offset distance of 4 and a half-angle of 45 degrees. The pressure field (top of each frame) and the numerical schlieren field (bottom of each frame) are shown. The bubble volume history is also shown along with the corresponding times for frames 1 through 7 (red, discontinuous lines).

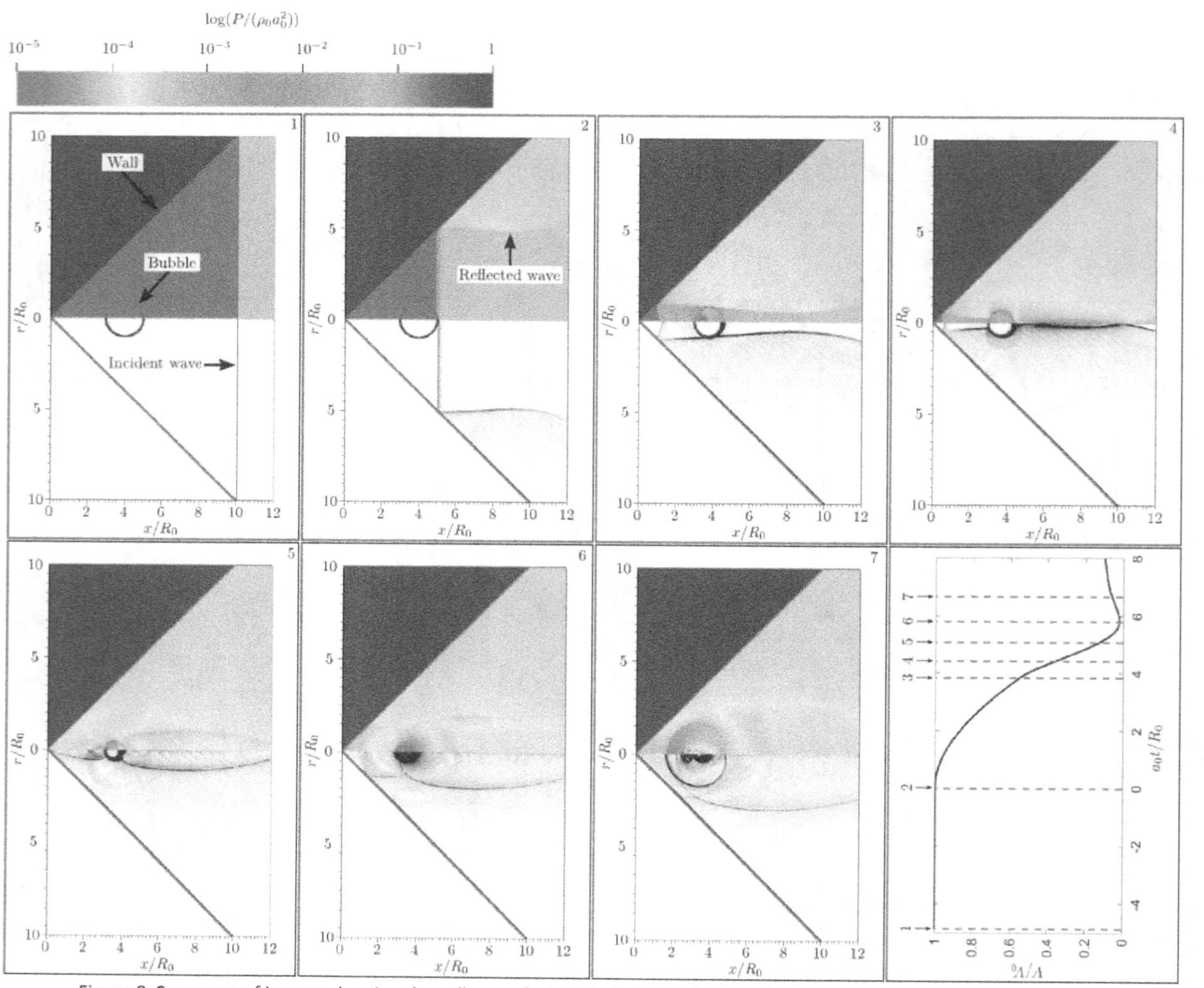

Figure 8. Sequence of images showing the collapse of a bubble in a cone for a half-angle of 45 degrees and an offset of 4.

The incident and the reflected waves are indicated in frames 1 and 2 of figure 8. Because the reflected wave converges toward the axis of symmetry the pressure rise it incurs is substantial. More interestingly, the enhancement of the collapse begins precisely when the reflected wave begins to interact with the bubble (frame 3). It turns out this is not a coincidence and the same observation can be made for all half-angles and offset distances.

If the non-dimensional collapse time τ is less than $a_0 T/R_0$ — see, for example, offset distances of 8 and 10 in figure 7 — the reflected wave does not affect the collapsing bubble; the bubble collapses before the arrival of the reflected wave. This means the bubble does not feel the presence of the cone until after it has collapsed.

Conclusion

We experimentally observed the growth and collapse of cavitation bubbles within the cone of a syringe mounted inside an autoinjector device. Using numerical simulations, we were able to show the shock-induced collapse of a bubble in a cone can be more violent than the shock-induced collapse of a bubble located close to a flat wall or in a free-field, thus increasing the potential for wall damage.

The reflected wave which forms as the incident wave enters the cone is responsible for the enhancement of the collapse. The reflected wave converges on the axis of symmetry and creates a substantial pressure rise. The collapse is enhanced when the reflected wave begins to process the bubble. If the collapsing bubble is too far from the apex of the cone, the reflected wave will only reach the bubble after it has collapsed. Therefore, the collapsing bubble is affected by the conical section only if the non-dimensional collapse time $\tau > 2(H_0/R_0)(\sin \propto)^2$.

References

[1] Akers, M.J. (2010). *Sterile Drug Products: Formulation, Packaging, Manufacturing and Quality.* Series: *Drugs and the Pharmaceutical Sciences* (Book 208). CRC Press.

[2] Thompson, I., Lange, J. (2013). *Pen and Autoinjector Drug Delivery Devices.* In Kohle, P. Shah, M. Rathore (Eds.) *Sterile Product Development: Formulation, Process, Quality and Regulatory Considerations.* Series: *AAPS Advances in the Pharmaceutical Sciences Series* (Volume 6). Springer-Verlag New York. DOI: 10.1007/978-1-4614-7978-9.

[3] French, D., Collins, J. (2010). *Advances in Parenteral Injection Devices and Aids.* In Nema, S. Ludwid, J. (Eds.) *Pharmaceutical Dosage Forms: Parenteral Medications* (Volume 3, pp. 71-85), 3rd edition. CRC Press.

[4] Veilleux, J.-C., Shepherd, J.E. (2017). *Dampers and Methods for Performing Measurements in an Autoinjector.* U.S. Patent Application 15649329. Filed July 2017. Available online January 2018.

[5] Veilleux, J.-C., Shepherd, J.E. *Understanding the Pressure and Stress Transients in Autoinjector Devices for Improved Designs and Reliability.* Paper in preparation.

[6] Veilleux, J.C., Jazayeri, J., Eu, B., Shepherd, J.E. (2017). *Primary Packaging & Delivery Systems Interaction.* In 10th Workshop on Monoclonal Antibodies of the Parenteral Drug Association, Berlin, Germany. DOI: 10.13140/RG.2.2.10549.37601.

[7] Veilleux, J.-C., Shepherd, J.E. (2017). *Impulsively-Generated Pressure Transients and Strains in a Cylindrical Fluid-Filled Tube Terminated by a Converging Section.* In Pressure Vessels and Piping Conference, Volume 4: Fluid-Structure Interaction. Waikoloa, Hawaii, USA. Paper PVP2017-65471. DOI: 10.1115/PVP2017-65471.

[8] Coralic, V., Colonius, T. (2014). Finite-volume WENO scheme for various compressible multicomponent flows. *Journal of Computational Physics*, 274, pp. 95-121. DOI: 10.1016/j.jcp.2014.06.003.

[9] Meng, J.C. (2016). *Numerical Simulations of Droplet Aerobreakup.* PhD Thesis, Caltech.

© 2018 ASME

Ultrasonic cavitation structures in a conical vessel

[1]Margarita Navarrete*; [2]Jorge Naude; [1]Francisco Godínez; [1]Olivia Zurita; [2]Federico Méndez

[1]Insitituto de Ingeniería, Universidad Nacional Autónoma de México, Ciudad de México, CDMX, México; [2]Facultad de Ingeniería, Universidad Nacional Autónoma de México, Ciudad de México, CDMX, México

Abstract

During the acoustic cavitation of non-Newtonian liquids, the bubbles that emerge upon "isothermal rupture" of a liquid do not disperse homogenously. The ultrasonic method provides well-defined pressure fields, and its combination of interfacial motion, vortices, and liquid streams generate self-organized structure ensembles in the form of clusters, filamentary patterns extending to webs of bubbles, and dense clouds in the bulk liquid. This behavior is not clearly identified by a simple inspection of the dynamic models. Therefore, it is essential to obtain insight into the collective and bulk behavior of these fluid structures from the experiments. In particular, we observe the geometrical characteristics of the recurrent structures in a cavitating flow and analyze their fluctuations as a function of driving frequency from 20 kHz to 200 kHz, using a viscoelastic liquid inside a conical glass vessel. We implemented a variable ultrasonic frequency generator with a bandwidth of 2 MHz and 50 W power that replaced the amplifier of a commercial acoustic horn system, and used as an ultrasonic source. This, along with a high-definition digital camera and acoustic emission measurements were used to characterize the behavior of the cavitation structures as a function of driving frequency. This system induces cavitation in the viscoelastic liquid confined in an axisymmetric conical glass vessel with a volume of 1000 mL. We used degassed 1,2-propanediol with an initial temperature between 5–9°C. Experimental runs showed unexpected self-organization as the frequency increased, with bubble accumulations, an acoustic pressure field, and liquid viscosity. We observed that outside of the strong acoustic field, small increases in both the frequency and liquid temperature produced a significant change in the type of fluidic structure observed, with a bubble cloud transforming into multiple bubble rings.

Keywords: acoustic cavitation; bubble motion; high-speed recording; driving frequency

Introduction

In order to properly use sonochemical reactors in industrial and medical application, the density and distribution of the bubbles that occur within the affected volumetric space must be known. This is an arduous task since it is necessary to consider the space-time dynamics of acoustic multi-bubble systems that span several time and space scales [1]. One theoretical approach considers an N-body problem comprising individual bubbles that move under the influence of the acoustic forces [2]. However, fast translational motion of bubbles relative to the liquid is observed in most acoustic cavitation experiments, along with their self-organization into complicated structures. These structures are reviewed in [3] and some are explained from results derived from single bubble physics [3-5]. Despite this, few data are available in the literature describing the sizes, velocities, and densities of bubbles in cavitation fields, since bubbles only become visible after expanding to a few tens of micrometers. R. Mettin [6] proposed a classification of acoustic cavitation structures based on visual aspects, their acoustic environment and using water as prototype liquid. His study included optical and high-speed imaging, a lower ultrasonic range (20 to 50) kHz and a rectangular cuvette with transparent walls. Mettin also categorized these structures as being associated standing waves, traveling waves, and surface-bound figures. Structures resulting from standing waves include streamer filaments, bows, rings, and layer structures (jellyfish and starfish). Several studies [e.g., 7, 8] focus on fluidic structures that form beneath of output tip of the sonotrode (the surface that irradiates the acoustic energy), and compare different types of sonotrodes, concluding that cone-type bubble structure originate in a thin layer of bubbles attached to the surface of the sonotrode tip.

Lixin Bai *et al.* [9] studied the generation and control of the five typical acoustic cavitation structures, discussing the source of nuclei and the relationship and mutual transformation among them. In addition, to produce Acoustic Lichtenberg Figures, AFL, they suggested that the field of the standing waves could be controlled by adjusting the reflective surfaces by using container geometries. They also describe a method to adjust the depth of the sonotrode tip (node or antinode) inside the liquid. Rahimi *et al.* [10] and Klíma *et al.* [11] proposed the optimization of the sonoreactor geometry based on a numerical simulation of local ultrasonic intensity and its qualitative comparison with experimental results. Both groups predicted an acoustic streaming field distribution inside the sonoreactor, with the first group studying the induced flow pattern with a moving boundary condition, and the second paying attention to the intensity distribution of the ultrasonic energy, driving frequency, dip size of the sonotrode tip, proposing a "horn up" optimized at 20 kHz. They also attempted to extend Mettin's classification. Tzanakis *et al.* [12] characterized the cavitation development and acoustic spectrum in three optically transparent liquids (water, ethanol, and glycerin) and molten aluminum using a 40-mm diameter sonotrode with acoustic power from (90 to 230) W and 20 kHz frequency. They studied the effects of liquid properties on the cavitation zone, the behavior of the cavitation cloud, and the corresponding acoustic emissions, and found varying behavior due the physical parameters used. An excellent review of fluidic structures using acoustic fields can be found in [3-9]. Thus, the distribution of bubbles is an important factor for any sonochemical or sonomechanical device where bubbles are involved, and detailed knowledge of bubble density in space is desirable. Understanding and predicting cavitation structures is, however, of a quite complex. Newtonian and non-Newtonian fluids drastically change their rheological properties under ultrasonic cavitation, with high-amplitude acoustic waves providing well-defined pressure fields. If low-molecular-weight liquid polymers are used, the actual triggers for cavitation and structure formation can be identified due to their unique properties such as viscosity, low vapor pressure and low thermal conductivity.

Methods and materials

Experimental setup

Many commercially available acoustic cavitation devices operate in the range 20–50 kHz. In these devices, the active element responsible for converting electrical into mechanical energy is known as a sonotrode, which is an assembly of a piezoelectric ceramic transducer (PCT) confined by a metal structure (mechanical amplifier) which increases the intensity of the vibration generated by the transducer. The entire ultrasonic assembly is tuned to the same frequency to transfer maximum energy. If the resonant frequencies of the components of the ultrasonic assembly differ, vibrations of different frequencies would be induced, leading to energy loss and possible fatigue damage to certain components. Thus, key issues involved in sonotrode design are the material, shape, frequency, resonant length, and gain used. We implemented an ultrasonic variable frequency generator with a bandwidth of 2 MHz and 50 W power by replacing the amplifier of a commercial acoustic horn system (VC505-500 Sonics, 20 kHz fixed frequency, 13-mm-tip face diameter) which is an acoustic source. The PCT was encapsulated in a Ti-6Al-4V titanium alloy housing, which was highly resistant to traction, corrosion, and erosion. The characterization of the sonotrode as a function of the reactance, *Xc,* is shown in Fig 1.

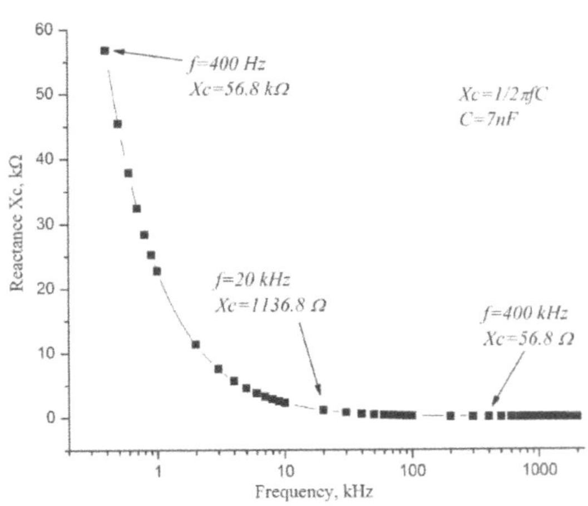

Fig. 1. Variation in reactance, *X*c, as a function of frequency.

The power stage (amplifier) was designed so that the ultrasonic generator supplied maximum power at high frequencies, taking into account that the increased frequency caused a decreased *Xc*, in turn implying an increased current passing through the PCT. The power stage used a Power Operational Amplifier (AOP, PA107DP from Cirrus Logic) to control the amplitude of the applied voltage, which also managed any type of signal with a bandwidth and current gain appropriate for this application. The power stage had a gain voltage of 50, an output 140 V peak-to-peak, a maximum current of 1 A, and a maximum output power of 37 W/cm^2. Unlike other experimental arrangements, where both frequency and power are fixed [12], our electronic arrangement allowed for sweeping the frequency and the system to absorb the needed power. A diagram of the experimental configuration is shown in Fig 2.

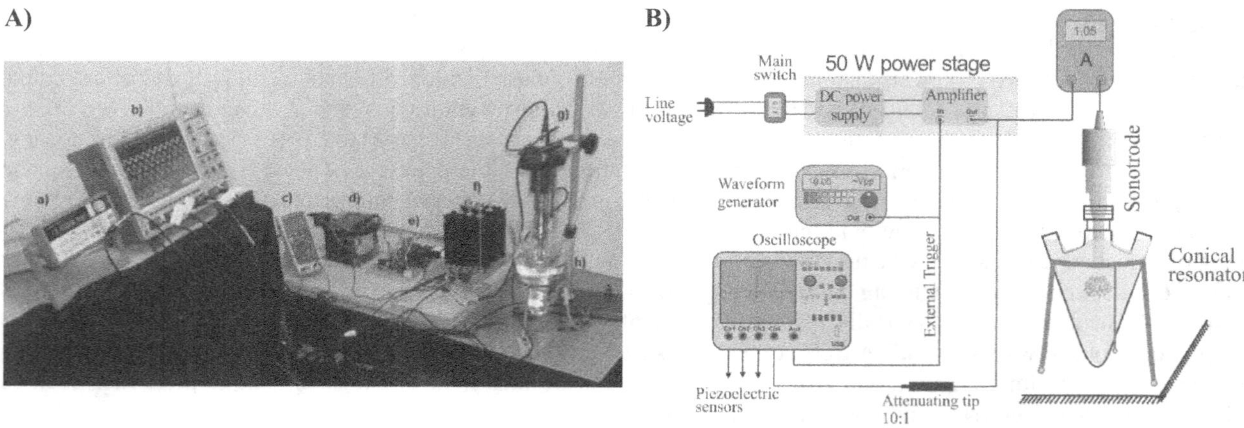

Fig. 2. A) System for acoustic cavitation: a) waveform generator, b) oscilloscope, c) digital multimeter, d)-f) electronic power amplifier, g) sonotrode, and h) conical resonator; B) Connection diagram among instrumentation, amplification power stage, sonotrode, data acquisition and conical resonator.

Vessel characteristics and liquid properties

The conical vessel was a 1000 mL four-neck tapered sonochemical flask (Ace Glass, Inc.). The sonotrode was placed in the central neck and its tip was submerged up to 15 mm in the liquid. The other three necks allowed for monitoring the effects of ultrasonic wave propagation inside the liquid by means of piezoelectric sensors and changes in temperature. Degassed 1,2-propanediol was used as a viscoelastic liquid sample in all experiments; its physical properties are displayed in Table 1. Its viscosity curve as a function of temperature is shown in Fig 3.

1,2-propanediol	
Chemical formula	$C_3H_8O_2$
Molar mass	76.10 g·mol^{-1}
Density	1.036 g/cm^3
Melting point	-59 °C
Boiling point	188.2 °C
Solubility in water	Miscible
Dynamic viscosity	
15°C	48.6 [mPa·s]
25°C	40.4 [mPa·s]
60°C	8.42 [mPa·s]
Vapor pressure at 25°C	(9.3-26.7) [Pa]
Liquid density at 15°C	1032 [Kg/m^3]
Vapor density at 100 °C	2062 [Kg/m^3]
Vapor density at 20°C	1006 [Kg/m^3]

Table 1. Physical properties of the 1,2-propanediol.

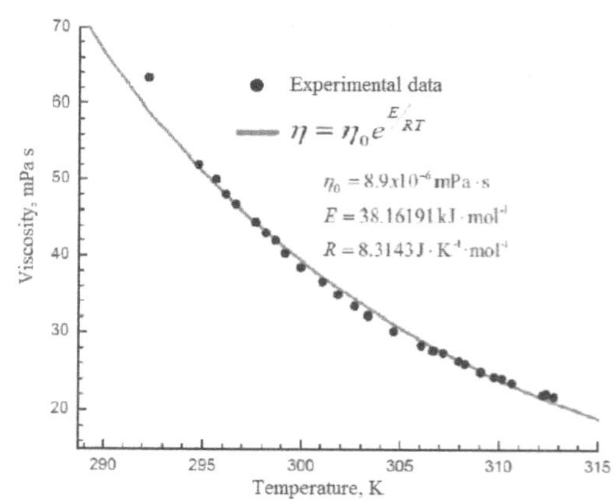

Fig. 3. Dynamic viscosity of 1,2-propanediol for temperatures between 17–40 °C.

Data acquisition0

The auxiliary instrumentation consisted of piezoelectric transducers centered on 1 MHz and 5 MHz (Olympus), whose signals were stored in a 1 GHz digital oscilloscope (Lecroy LC584AM). We fixed a PVDF piezoelectric film on the external face of the conical vessel for the acquisition of acoustic signals there. We apply a Hanning window to the raw voltage signals, which were then transformed into the frequency domain through a Fast Fourier Transform (FFT). Data acquisition was performed when local steady-state was reached. The experimental run was carried out three times to ensure reproducibility of results.

Results and discussion

Fig 4a shows the evolution profile of the different fluidic structures produced inside the conical resonator when the frequency varied between 10 kHz and 170 kHz, observe in the upper bar, as the liquid temperature increased. Fig 4b shows the gain system with and without load (bubble cavitation). Considering the zone near to the sonotrode tip, the cavitation pattern in 1,2-propanediol was apparently much more confined than in water.

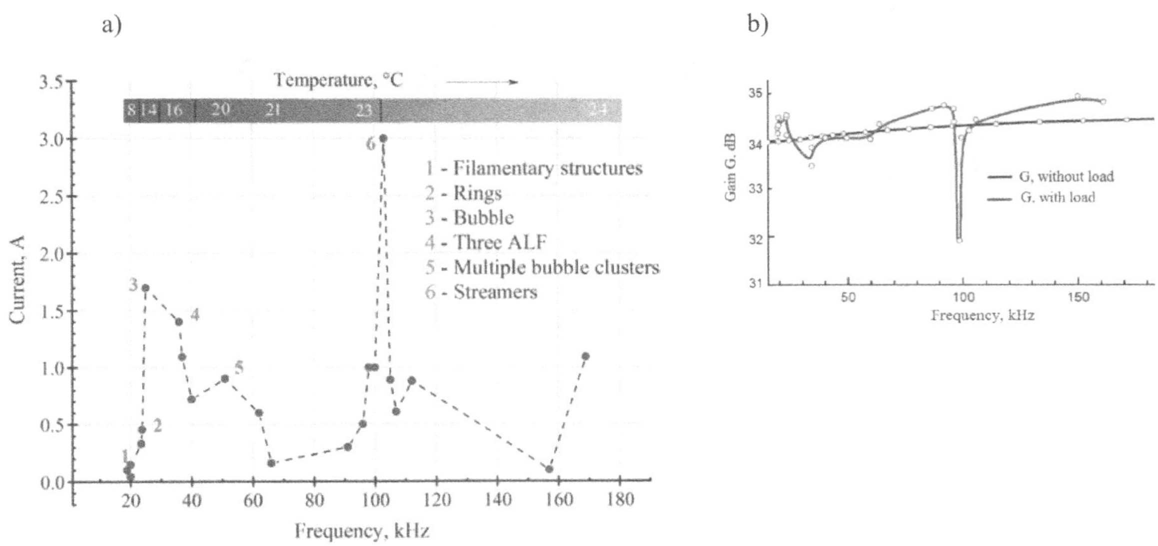

Fig. 4. a) Evolution of fluidic structures as a function of applied frequency in a conical resonator filled with 1,2-propanediol and initial temperature 7 °C; b) fluctuation of the amplifier gain (dB) as a function of driving frequency with and without load.

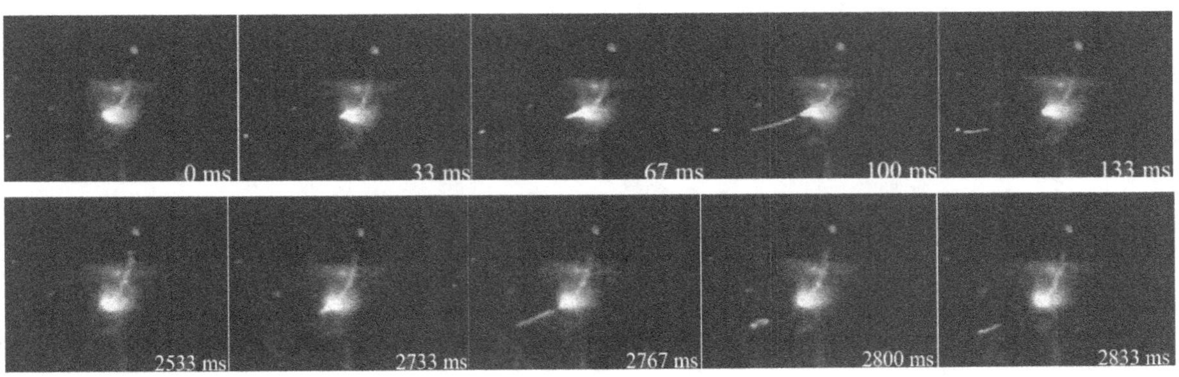

Fig. 5. Bubble ejection mechanism towards the solid wall, with 1,2-propanediol at 7°C.

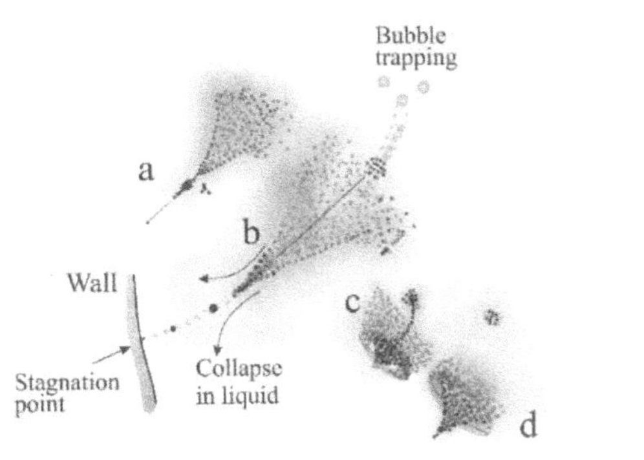

Fig. 6. Fluid structure ejecting bubbles towards to a solid wall.

Fig 5 shows a series of images where a cloud of mushroom-shaped bubbles captured and dragged bubbles along a "central filament", then ejecting them synchronously towards the reactor wall. The ejection mechanism consisted of the contraction and expansion of a tip of the bubble cloud. This mechanism depended on the temperature, since a change of a few degrees decreased or stopped the ejection frequency. Sketches of the mechanism are depicted in Fig 6: a) conical bubble cloud, b) expansion tip and bubble ejection, c) contraction tip, and d) new expansion tip. Fig 7 shows a series of images of various experimental runs showing unexpected self-organization as the frequency increased.

Each image is accompanied by its frequency spectrum and spectral noise obtained from the signals acquired by the piezoelectric sensors. Those sensors captured the behavior of the acoustic field inside and outside the conical resonator and its perturbation by the cavitation in the fluidic structures.

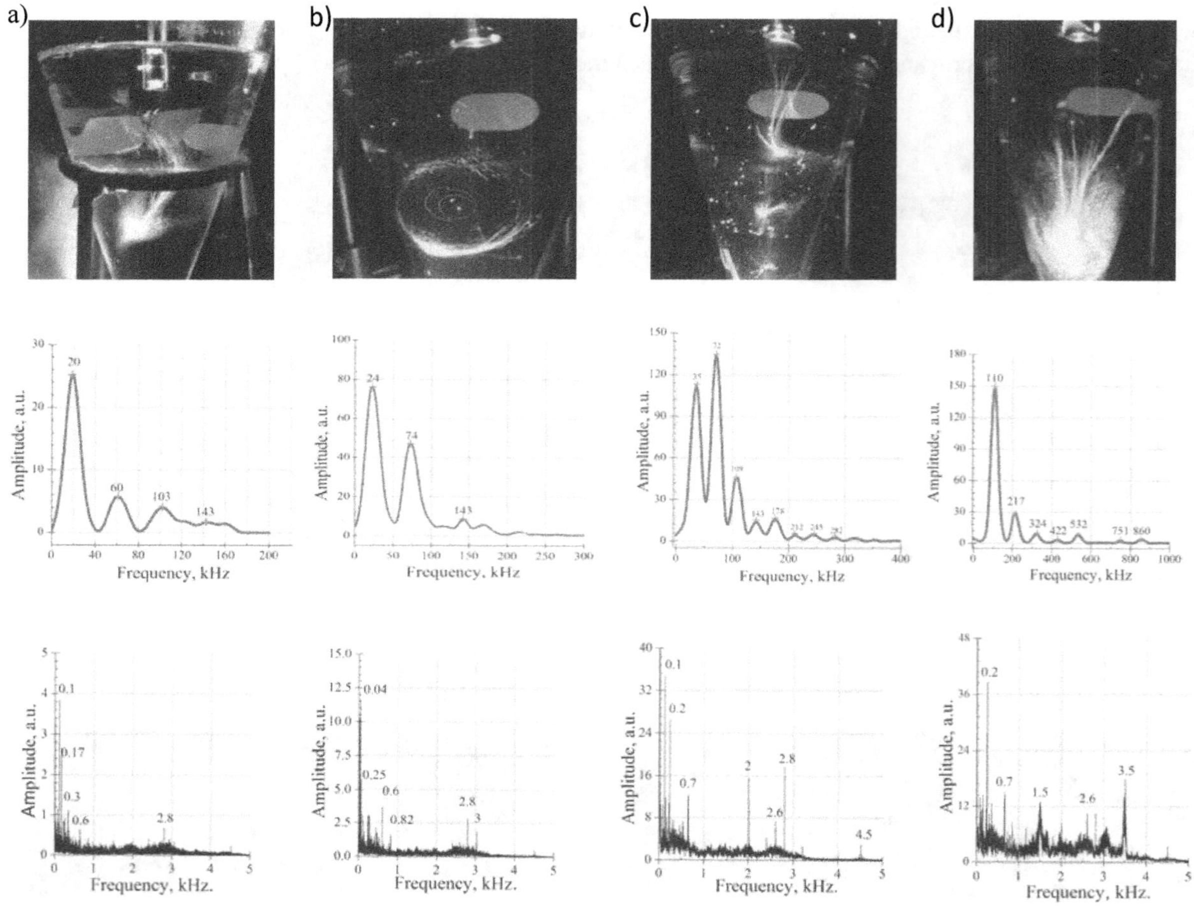

Fig. 7. Experimental runs showing unexpected self-organization as frequency increased. a-d) Images of the fluidic structures, below which its frequency spectrum is displayed, the first peak corresponds to the driving frequency. In the third row, the noise spectrum generated by each fluidic structure is shown. These included sharp peaks, sub-harmonics, and continuous broadband components.

Fig 8 shows the transformation process from the conical bubble structure in (a) to a ring structure (b), were only the frequency varied and the energy absorption was left free. The mutual transformation was very stable at low liquid temperatures.

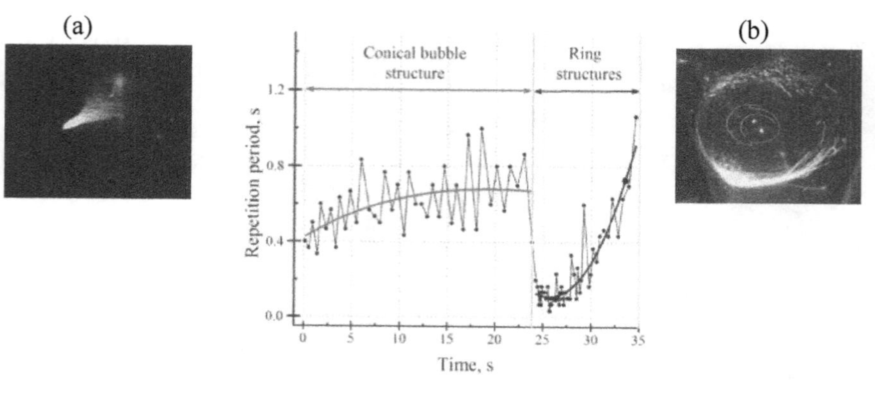

Fig. 8. Transformation process from a conical bubble to a ring structure.

In general, we note three luminescent spatial distributions using 1,2-propanediol at low temperatures when only the driving frequency was modified: a) around 20 kHz, a cluster of long vertical streamers emerged from tip of the sonotrode with the same curvature changes and accumulated in a large arrow-shaped cloud, at a 2/3 the height of the conical glass vessel; b) near 24 kHz, an array of filaments formed a hoop were kept in a vertical plane in the middle of the vessel; c) close 33 kHz, three-bubble cumuli were observed, with almost equal shift, developing bubble networks. In addition, in a particularly intriguing phenomenon, a nearly spherical bubble cloud formed in the center of vessel, periodically changing its configuration to a retractable conical nozzle, and expelling bubble streamers that condensed into a large bubble that collided and rebounded on the vessel wall, as shown in Fig 5. As well, we observed three stability conditions for trapping bubbles in a standing ultrasonic field, positional, spherical, and diffusional as mentioned [11-13].

Conclusion

Cavitation experiments with 1,2-propanediol using acoustics and a conical vessel demonstrated many fluidic phenomena. Liquid density, viscosity, surface tension, vapor pressure, temperature and concentration of dissolved gas, as well as the geometry of the vessel, influenced the bubble resonance frequency and thus were key in supporting translational motion under an oscillatory pressure field. These cavitation structures showed unexpected self-amplification effects over time, with bubble accumulation and pressure field attenuation. We obtained a self-organization using a large volume of liquid, which established fluid structure according to its temperature and local acoustic wave velocity. The stability of these patterns depends mainly on three main parameters: the driving frequency temperature of liquid, and geometry of the glass container.

Acknowledgments

The authors are grateful to the Programa de Apoyo a Proyectos de Investigación e Innovación Tecnológica de la Universidad Nacional Autónoma de México, PAPIIT-UNAM, for the financial assistance for this research within the project IN105117-2. The authors also thank Eng. J. Roberto Alvarado M. for the assistance in the instrumentation.

References

[1] I. Akhtov, U. Parlitz, W. Lauterborn (1996). *Towards a theory of self-organization phenomena in bubble-liquid mixtures.* Physical Review E 54(5) 4990.

[2] R. Mettin, C.-D. Ohl, W. Lauterborn (1999). *Particle approach to structure formation in acoustic cavitation,* in: L. A. Crum et al. (Eds.), *Sonochemistry and Sonoluminescence*, (Proceedings of the NATO Advanced Study Institute, Leavenworth (WA), USA, 18-29 Aug. 1997), pp. 138-144, Kluwer Academic Publishers, Dordrecht.

[3] R. Merttin, S. Luther, C.-D Ohl, W. Lauterborn (1999). *Acoustic cavitation structures and simulations by a particle model.* Ultrasonic Sonochemistry 6, 25.

[4] U. Parlitz, R. Mettin, S. Luther, I. Akatov, M. Voss, W. Lauterborn (1999). *Spatio-Temporal dynamics of acoustic cavitation bubble clouds.* Phil. Trans. R. Soc. Lond. A 357, 313.

[5] R. Merttin, (2007). *From a single bubble to bubble structures in acoustic cavitation*, in: T. Kurz, U. Parlitz, U. Kaatze (Eds.) Oscillations, Waves, and Interactions, 171-198. Universitatsverlag Gottingen.

[6] R. Merttin (2005). *Bubble structures in acoustic cavitation*, in: A. A. Doinikov (Ed.): Bubble and Particle Dynamics in Acoustic Fields: Modern Trends and Applications, pp. 1-36. Research Signpost, Kerala (India).

[7] A. Moussatov, C. Granger, B. Dubus (2003). *Cone-like bubble formation in ultrasonic cavitation field.* Ultrasonic Sonochemisty 10, 191.

[8] B. Dubus, C. Vanhille, C. Campos-Pozuelo, C. Granger (2010). *On the physical origin of conical bubble structure under an ultrasonic horn.* Ulltrasonic Sonochemistry 17, 810.

[9] L. Bai, W. Xu, J. Deng, C. Li, D. Xu, Y. Gao (2014). *Generation and control of acoustic cavitation structure.* Ultrasonic Sonochemistry 21, 1696.

[10] M. Rahimi, S. Movahedirad, S. Shahhosseini (2017). *CFD study of the flow pattern in an ultrasonic horn reactor: Introducing a realistic vibrating boundary condition.* Ultrasonic Sonochemistry 35, 359.

[11] J Klíma, A. Frias, J. Gonzalez, J. Ludvik, V. Sáez, J. Iniesta (2007). *Optimization of 20 kHz sonoreactor geometry on the basis of numerical simulation of local ultrasonic intensity and qualitative comparison with experimental results.* Ultrasonic Sonochemistry 14, 19.

[12] I. Tzanakis, G. S. B. Lebon, D. G. Eskin, K. A. Pericleous (2017). *Characterizing the cavitation development and acoustic spectrum in various liquids.* Ultrasonic Sonochemistry 34, 651.

[13] R. W. Time, A. H. Rabenjafimanantsoa (2011). *Cavitation bubble regimes in polymers and viscous fluids.* Annual Transactions of the Nordic Rheology Society, 19, 1-12.

© 2018 ASME

Ultrasound and lipid-coated microbubbles effect on proliferation and osteogenic differentiation of mesenchymal stem cells in 3D printed tissue scaffold

[1]Jenna Osborn; [1]Mitra Aliabouzar; [1]Xuan Zhou; [1]Raj Rao; [1]Lijie Grace Zhang; [1]Kausik Sarkar*

[1]*George Washington University, Washington, DC, USA*

Abstract

With increasing incidence of bone disorders in a rapidly aging, sedentary, and overweight population, engineered bone tissues promise to be a better alternative for conventional bone grafts. Human mesenchymal stem cells (hMSCs) extracted from a patient's bone marrow can be cultured to create bone tissue via osteogenic differentiation of the stem cells. The engineered bone tissue may then be implanted into the injury site. However, bone tissue engineering currently suffers due to the inability to form mechanically strong, porous structures that can be grown quickly and efficiently. Low intensity pulsed ultrasound (LIPUS) stimulation has proven to enhance bone healing and recovery rates. In this study, the beneficial effects of LIPUS stimulation are harnessed to improve bone tissue engineering methods and are further improved with the addition of in-house prepared, lipid-coated microbubbles (MBs). Bone marrow hMSCs were seeded on 3D printed poly(lactic acid) (PLA) porous scaffolds and cultured to study the effects of LIPUS in the presence of MBs on proliferation and osteogenic differentiation. For proliferation studies, the samples were treated with LIPUS in the presence of 0.5% (v/v) for 3 minutes a day at 30 mW/cm2 with a frequency of 1.5 MHz and a duty cycle of 20%. A significant increase in cell number was observed after 1, 3, and 5 days of culture as compared to control. For osteogenic differentiation studies, the samples were treated with LIPUS in the presence of 0.5% (v/v) MBs for 3 days for 3 minutes a day with the same parameters as the proliferation studies. The samples were then cultured for 1, 2, and 3 weeks for osteogenic differentiation to occur. Total protein, alkaline phosphatase (ALP) activity, and total calcium content were found to increase with LIPUS with and without the presence of MBs. Integrating LIPUS and MB appears to be a promising strategy for bone tissue engineering and regeneration therapies.

Keywords: ultrasound; microbubbles; tissue engineering

Introduction

Osteoporosis and osteoarthritis are major health problems, affecting over 6 million people per year, and that number is expected to grow. With increasing incidence of these disorders, more options for therapies are desired. Tissue Engineers have been exploring the opportunity of using human mesenchymal stem cells (hMSCs) as a potential source of healthy cells that can help expedite the healing process of bone injuries and diseases. The advantage of the hMSCs is their ability to differentiate into different cells types, such as osteoblasts, chondrocytes, or adipocytes when exposed to the proper environmental factors.

3D printing technologies for tissue engineering applications have gained attention for their ability to have patient injury-specific scaffolds for replacements and grafts [1] [2]. The geometry of the patient's injury site can be obtained from a MRI or CT image and can be 3D printed from a porous, biodegradable material. These scaffolds can be used for stem cells to grow and differentiate into the desired tissue. This can then be implanted into the injury site, and as the scaffold material degrades, the tissue fills in. To induce the differentiation of the stem cells on the scaffolds, hMSCs can be exposed to osteogenic growth factors and mechanical stimuli, such as ultrasound [1].

Low intensity pulse ultrasound (LIPUS) has been shown to be effective for helping fracture healing at every phase of healing [2]. Previous studies have also shown improvements of proliferation and differentiation of hMSCs [1]. Microbubbles (MBs) have been used as ultrasound contrast agents in imaging domains. Their compressible gas core gives rise to highly echoic behavior. The compressibility also generates bubble oscillations and complex fluid patterns around the bubbles. The fluid motion can give rise to extra shear stress on the cell membranes and can lead to additional mechanical stimuli to the cells. In this study, the effects of LIPUS and LIPUS in the presence of MBs on hMSC proliferation and osteogenic differentiation were evaluated.

Methods

A. Cell Culture

*Corresponding Author, Kausik Sarkar: sarkar@gwu.edu

Under an IRB-approved protocol with written, informed consent, human bone marrow mesenchymal stem cells (hMSCs) were obtained from healthy individuals at Tulane University. The cells were characterized at the Institute for Regenerative Medicine at Texas A&M Health Science Center with a fully executed Material Transfer Agreement. The cells were cultured in α-Minimum Essential Medium Eagle (α-MEM) supplemented with 16.5% (v/v) fetal bovine serum (FBS), 1% (v/v) L-glutamine, and 1% (v/v) penicillin and streptomycin. The hMSCs were incubated in a 37°C, 5% CO_2, and 95% relative humidity environment. For the studies requiring osteogenic differentiation of the hMSCs, the cells were cultured in Dulbecco's Modified Eagle Medium (D-MEM) supplemented with 10% (v/v) FBS, 50 µg/mL of L-ascorbate acid, 10 mmol/L of β-glycerophosphate, 10 nmol/L of dexamethasone, and 1% (v/v) penicillin and streptomycin. An image of the hMSCs can be seen in figure 1(a).

B. Synthesis of Microbubbles

The MBs used in this experiment were prepared locally using a lipid shell and perfluorocarbon gas. For the shell of the MBs, 0.75 mg/mL of 1,2-dipalmitoyl-sn-glycero-3-phosphocholine (DPPC), 1.5 mg/mL of 1,2-dipalmitoyl-sn-glycero-3-phosphoethanolamine-N-[methoxy(polyethylene glycol)-2000] (DPPE-PEG2000), and 3.0 mg/mL of 1,2-dipalmitoyl-3-trimethylmmonium-propane (TAP) were dissolved in heated (above 45°C) propylene glycol for 30 minutes. Then glycerol and phosphate buffer solution (PBS) were added and mixed while heated for 40 minutes. The solution was kept at 4°C overnight. The following day, 1.5 mL of the lipid solution was placed in a 2 mL vial, crimped closed, underwent a gas exchange with perflurobutane.

The bubbles were generated by mechanical agitation of the vial for 45 seconds using a Vial Mixer (Bristol Myers Squibb). The resulting bubbles are highly polydisperse in size. To narrow the range of diameters, the bubbles were diluted to 50 mL in PBS and centrifuged at 40 relative centrifugal force (RCF) for 4 minutes. The bubble size distribution was measured using qNano (Izon Science™, MA) with nanopore membrane NP2000 (1000 nm- 5000nm). The average size of the particles was roughly 1.1-1.2 µm and an average concentration of about 10^8 - 10^9 particles/mL as seen in figure 1(b).

C. Production and Characterization of Scaffolds

The 3D printing technique used to fabricate the scaffolds was fused deposition modeling (FDM) (Solidoodle, NY). The scaffolds were fabricated using poly(lactic)-acid (PLA). First, a 3D computer-aided design (CAD) model of the scaffolds was made to dictate the dimensions of the scaffolds. Then, the infill density, scaffold pore geometry, and the speed of the printer nozzle were set using Slic3r computer numerical control conversion software. We chose a square pore geometry which was previously shown to be optimal for the hMSC growth [1]. To assess the pore size, the scaffolds were imaged using scanning electron microscopy (SEM) (Zeiss Nvision) as seen in figure 1(c). The average pore dimensions of the scaffolds were 700 µm and 1000 µm.

D. Ultrasound Parameters

The ultrasound parameters for hMSC parameters were optimized previously and the setup described by Aliabouzar et al [1] was used. Briefly, a flat ultrasound transducer (Olympus NDT, Waltham, MA) was used to expose the cells from above with a frequency of 1.5 MHz, intensity of 30 mW/cm^2, and a duty cycle of 20% for 3 minutes per day. The MB concentration was also optimized (results not shown) to be 0.5% (v/v). The experimental setup can be seen in figure 1(d).

E. Proliferation of hMSCs

The effect of LIPUS with and without the presence of microbubbles on proliferation was evaluated. The hMSCs were cultures in 24 well plates with 2.5 x 10^4 cells per well overnight. The samples were divided into three different groups, control (no LIPUS or MB), LIPUS (no MBs), and LIPUS + MBs. The control cells had the LIPUS transducer dipped into the wells without any excitation. The cells were tested after 1 day, 3 days, and 5 days of treatment using Alamar Blue Assay. The cells were incubated for 2 hours at 37°C and then tested for absorbance at 570 nm and 600 nm. The absorbance measurements were converted to percent reduction of the assay using the manufacturer's instructions. The percent reduction of the assay is proportional to cell number. The cells were rinsed with PBS three times and replenished with fresh medium.

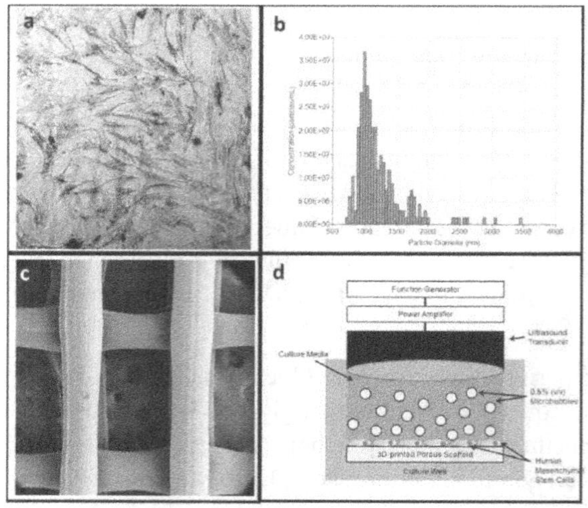

Figure 1. (a) The hMSCs in culture, (b) The size distribution as measured by qNano of the centrifuged microbubbles (c) An SEM image of the 3D printed PLA scaffolds, (d) the experimental setup of the ultrasound exposure of the hMSCs

F. Osteogenic Differentiation of hMSCs

For osteogenic differentiation studies, the hMSCs were cultured in 24 well plates with 1 x 10^5 cells per well overnight. The cells were divided into the same three groups and were then immersed in osteogenic media and exposed to ultrasound with the parameters mentioned above for 3 days. After that, the cells were incubated in osteogenic media for 1, 2, or 3 weeks. Alkaline phosphatase (ALP) activity and extracellular calcium deposition are two biomarkers that are indicative of osteogenic differentiation of hMSCs. ALP activity was tested using a standard assay kit (Bioquest, Sunnydale, CA) according to the manufacturer instructions. Calcium deposition was also tested using a standard assay kit (Pointe Scientific, Canton, MI) according to the manufacturer instructions.

Results and Discussion

For the effects of LIPUS and LIPUS in the presence of 0.5% (v/v) MBs on hMSC proliferation, the percent reduction of Alamar Blue assay was measured after 1, 3, and 5 days of treatment. The results can be seen in figure 2 for all three groups (control, LIPUS, and LIPUS+MBs). After one day, there was 8% increase in percent reduction when the cells were exposed to LIPUS in the presence of microbubbles as compared to control ($p < 0.001$). With only LIPUS stimulation, there was a 4% percent increase as compared to control ($p < 0.05$). After three days the cells with LIPUS stimulation in the presence of microbubbles had a 11% increase in percent reduction as compared to control, while LIPUS alone had a 5% increase ($p < 0.001$). After five days, a 12% increase in percent reduction was observed after 5 days in the LIPUS in the presence of MBs when compared to control and only 7% with LIPUS alone ($p < 0.05$). The addition of the LIPUS in the presence of microbubbles improved cells number significantly.

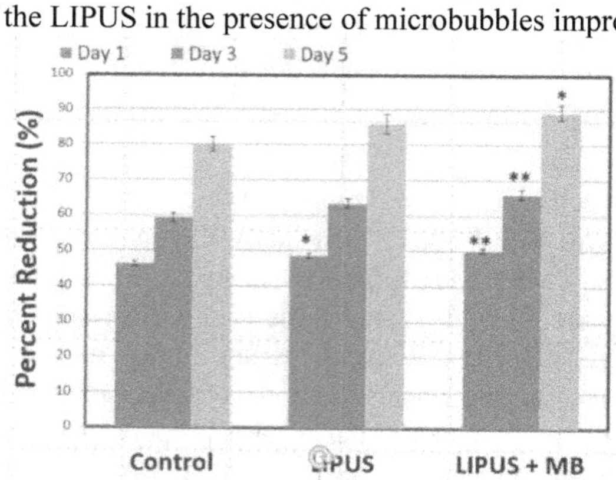

Figure 2. The percent reduction of the Alamar Blue assay after 1, 3, and 5 days of treatment. (*p<0.05, **p<0.001)

For osteogenic differentiation, the two biomarkers that were tested were ALP Activity (figure 3(a)) and calcium deposition (figure 3(b)). ALP is an enzyme that is important in the formation and mineralization of the bone extracellular matrix (ECM) [1] . The ALP activity is an early stage indicator of osteogenic differentiation [1]. The addition of LIPUS in the presence of MBs improved the ALP activity within the samples. After 2 weeks, there was 74% increase as compared to control, while LIPUS alone there was only a 7% increase. After 3 weeks, there was a 20% increase as compared to control and LIPUS alone had no statistical significant increase in ALP activity.

Extracellular calcium deposition is another biomarker for osteogenic differentiation of hMSCs. However, calcium deposition is a late-stage differentiation marker. Calcium has shown to be important for bone strength and stability [5]. After 3 weeks, there was a 7% increase in calcium concentration in samples that were exposed to LIPUS in the presence of MBs as compared to control. While LIPUS alone only had a 4% increase. Overall, LIPUS in the presence of MBs has shown to enhance proliferation and differentiation potential of the hMSCs. The mechanism behind LIPUS stimulated cell behavior is not fully understood. There are several pathways that LIPUS may stimulate, such as RhoA-GTP, p-ERk1/2, p-FAK, p-MEK1/2, p-p38, p-Akt, p-IKKα/β, and NK-kB [1]. The mentioned pathways are involved with bone regeneration and repair.

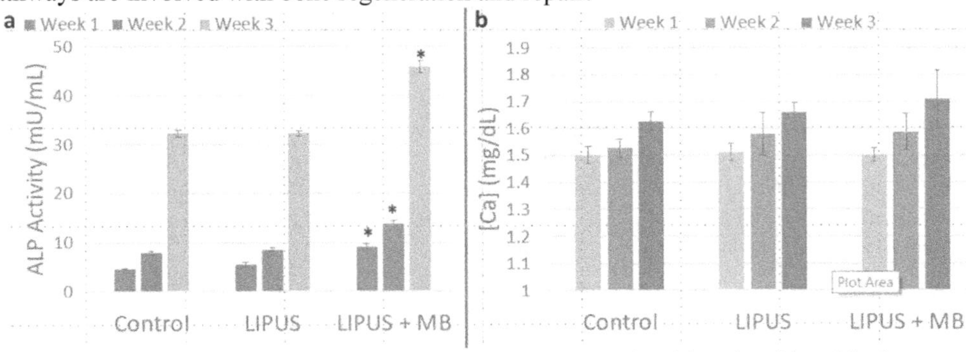

Figure 3. (a) The ALP Activity of the three groups after 1, 2, and 3 weeks (*$p<0.05$), (b) the calcium deposition of the three groups after 1, 2, and 3 weeks.

Conclusion

Overall, LIPUS in the presence of MBs has shown to enhance proliferation and differentiation potential of the hMSCs as shown by ALP Activity and Calcium Deposition. After 5 days of culture, there was a 12% increase in percent reduction of Alamar Blue, which is proportional to cell number, as compared to control. While LIPUS alone, only had a 7% increase after 5 days. For osteogenic differentiation of the hMSCs, two biomarkers were quantified, ALP Activity and extracellular calcium deposition. The ALP Activity increased by 25% after 3 weeks of culture in the samples exposed to LIPUS in the presence of MBs, while LIPUS did not have a statistically significant effect. For calcium deposition, there was a 7% increase after 3 weeks in samples exposed to LIPUS in the presence of microbubbles and only a 4% increase in LIPUS alone. The addition of LIPUS in the presence of MBs shows to be a promising enhancement of the osteogenic differentiation of hMSCs and could be a promising strategy for future bone tissue engineering applications.

References

[1] Zhou, X., Zhu, W., Nowicki, M., Miao, S., Cui, H., Holmes, B., Glazer, R. I., Zhang, G. L. (2016). *3D Bioprinting a Cell-Laden Bone Matrix for Breast Cancer Metastasis Study*, ACS Applied Materials & Interfaces, 8(44).

[2] Zhou, X., Nowicki, M., Cui, Zhu, W., Fang, X., Miao, S., Lee, S.-J., Keidar, M., Zhang, L. G. (2017) *3D Bioprinted graphene oxide-incorporated matrix for promoting chondrogenic differentiation of human bone marrow mesenchymal stem cells,* Carbon. 116

[3] Zhou, X., Castro, N. J., Zhu, W., Cui, H., Aliabouzar, M., Sarkar, K., Zhang, L. G. (2016). *Improved Human Bone Marrow Mesenchymal Stem Cell Osteogenesis in 3D Bioprinted Tissue Scaffolds with Low Intensity Pulsed Ultrasound.* Scientific Reports. 6(32876).

[4] Mundi, R., Petis, S., Kaloty, R., Shetty, V., Bhandari, M. (2009). *Low-intensity pulsed ultrasound: Fracture healing*. Indian Journal of Orthopaedics. 43(2).

[5] Aliabouzar, M., Zhang, L. G., Sarkar, K. (2016). *Lipid Coated Microbubbles and Low Intensity Pulsed Ultrasound Enhance Chondrogenesis of Human Mesenchymal Stem Cells in 3D Printed Scaffolds*. Scientific Reports. 6(37728).

[6] Marom, R., Shur, I., Solomon, R., Benayahu, D. (2004). *Characterization of adhesion and differentiation markers of osteogenic marrow stromal cells.* Journal of Cellular Physiology. 202(1).

[7] McCulloch, R. G., Bailey, D. A., Houston S., Dodd, B. L.. (1990). *Effects of physical activity, dietary calcium intake and selected lifestyle factors on bone density in young women.* Canadian Medical Association Journal. 142(3).

© 2018 ASME

Unsteady cavitating flow around axisymmetric headform at small angles of attack

[1]Hua Liu*; [1]Yousheng He; [2]Churui Wan

[1]Shanghai Jiao Tong University, Shanghai, China; [2]Marine Design and Research Institute of China, Shanghai, China

Abstract

Cavity shedding of cavitating flows around an axisymmetric headform belongs to one of the unsteady cavitating flows in the condition of steady incoming current. The periodic characteristics of unsteady cavitating flows around an axisymmetric body at small angles of attack are investigated experimentally and numerically. The evolution and shedding process of the three dimensional sheet cavitation are computed numerically by the Reynolds Averaged Navier-Stokes equations and the RNG k-eposilon model. The eddy viscosity coefficient modification approach for in the transition area of the two-phase flow is adopted to reproduce the shedding process of cavitating flows. The computed frequency of the cavity shedding coincides with the experimental data for the cases of unsteady cavitating flows around axisymmetric bodies with four headforms. Given the cavitation number, the shedding process of the cavitating flow depends on the headform of the axisymmetric body significantly. If the angle of attack of the axisymmetric body is greater than a critical value, the violent shedding of the sheet cavitation could not appear.

Keywords: unsteady cavitating flow; axisymmetric headform; frequency

Introduction

Unsteady behavior of cavitation on two dimensional hydrofoils in the case of steady incoming flows has been received intensive investigations. The Strouhal number for the lift and the drag fluctuation, defined by the chord length, was obtained experimentally as 0.14 for an Eppler foil[1]. Using the high speed photography, the evolution and abrupt cavity separation of unsteady cavitation on a hydrofoil NACA4412 were analyzed and the Strouhal number, corresponding to the violent deformation of the cavity, was proposed as 0.158[2]. Through numerical simulation, it was found that, for the unsteady cavitating flow on two dimensional hydrofoils NACA0015, the computed value of the Strouhal number is equal to 0.1[3].

Regarding the cavitation pattern on an axisymmetric body with different headforms, it can be divided into cavitation inception, developed cavitation and supercavitaion. There are several experimental investigations on the mechanism of unsteady cavitating flows through analyzing the cavity profile measured by high speed photography. It turns out that the reentrant jet moving through the cavity would be the major factor causing the cavity to detach from the front near the cavitator leading edge[4]. Such periodic fluctuation of the cavity profile affects hydrodynamic lift and drag on the headform[5].

In order to model the fluctuating hydrodynamic loads, this article focuses on the frequency of unsteady cavitating flows around the axisymmetric headforms at small angles of attack.

Results and discussion

An experimental investigation on unsteady cavitating flows on an axisymmetric headform was carried out at SSRI K15 Cavitation Tunnel. The characteristics of frequency of cavity evolution and hydrodynamic loads were discussed based on the high speed movies and the wavelet analysis of the measured time series of forces. The length of the test section is 2.6m. The cross section of it is a square of 0.6m by 0.6m. The maximum flow speed reaches 12m/s and the minimum static pressure can be controlled from near vacuum to 200kPa. A de-aeration system was used to control the dissolved air content of the water. The relative dissolved air content is around 0.3 in the experiments measured by a Van Slyke equipment. The cavity profiles were recorded by the high speed photograph HIGHSPEED10-16. The recording rate was 6000 frame per second. A three component balance was adopted to measure the lift, drag and moment acting on a slender model with different headforms. We selected four headforms in the experiment, including an elliptic headform, a 1/4-caliber ogive headform, a 600-conical headform and a hemispheric headform.

*Corresponding Author, Hua Liu: hliu@sjtu.edu.cn

The heaform diameter is 62.9mm. The axisymmetric model consisted of a headform and a parallel cylinder in 657mm length. The strut-mounted slender model was attached to the wall of the test section of the cavitation tunnel.

The cavitation number σ is defined as

$$\sigma = \frac{p_\infty - p_0}{0.5\rho V_\infty^2} \tag{1}$$

in which, p_∞ and V_∞ are the reference velocity and pressure, respectively, p_v is the vapor pressure at the bulk temperature of the water, and ρis the water density. The Strouhal number St is defined as

$$St = \frac{fD}{V_\infty} \tag{2}$$

where f stands for the fluctuating frequency of unsteady cavitating flows and D is the diameter of the headform.

Analyzing the time series of the drag acting on the headforms by FFT and the images of cavity by digitalizing high speed records. The averaged value of the Strouhal number obtained by analyzing the measured drag on the models is equal to 0.107±0.014.

We carried out the numerical simulation of unsteady cavitating flows around the axisymmetric body with four headforms by the computational code Fluent. The RANS equations and the RNG k-epsilon model are used to simulate the cavitating turbulent flows[6]. The cavitation model developed by Singhal et al. was adopted in the present study[7]. The grid convergence was checked carefully in terms of the pressure distribution on the axisymmetric body. The grid number of the meshes for the numerical simulations is 1.67M and the corresponding y+=30-60. In order to reproduce the shedding process of the cavitating flow around an axisymmetric body at small angles of attack, the modification approach of decreasing the vortex viscosity of two phase flow inside the cavitation region is implemented in the numerical simulations[8]. The computational domain and grids are shown in Fig. 1.

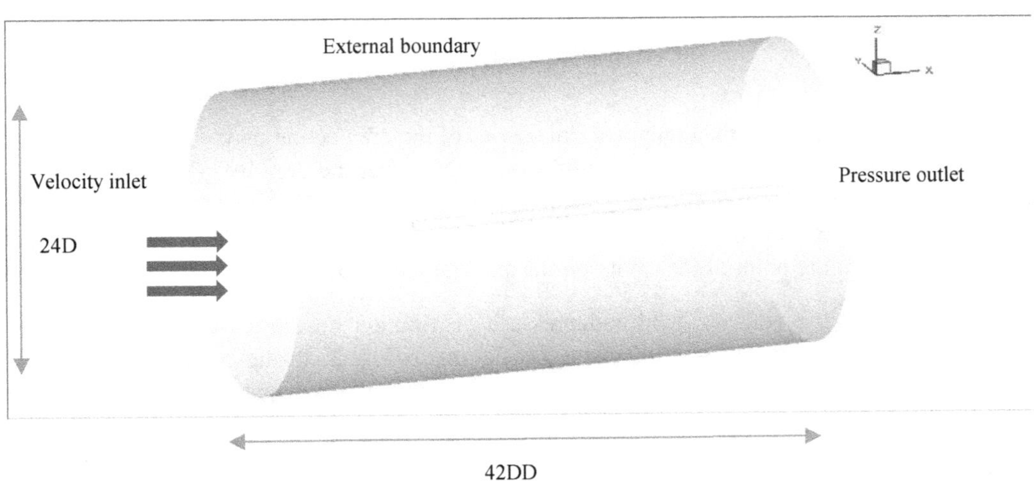

(a) Computational domain.

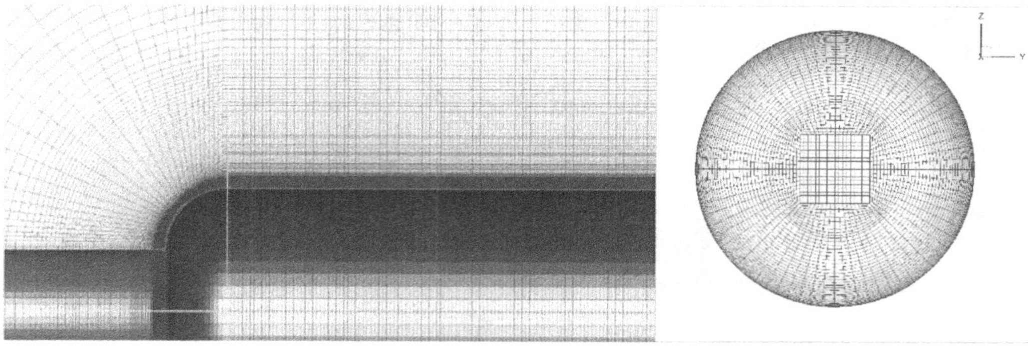

(b) Grid system

Figuer 1 Computational domain and grid system.

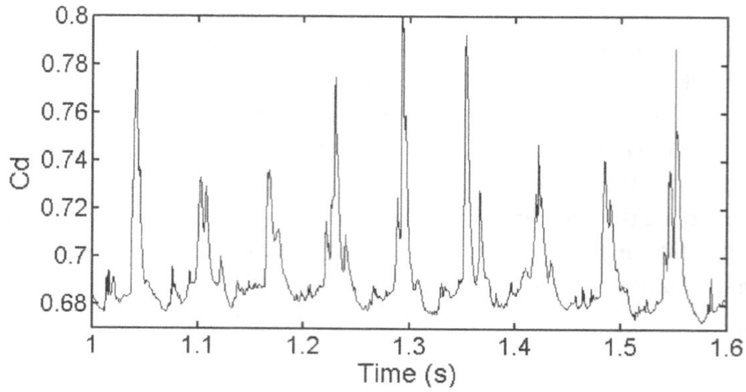

Figure 2 Time series of drag force on cavitating a 1/4-caliber ogive headform.

For the 1/4-caliber ogive headform, the computed time series of the drag acting on the axisymmetric body is plotted in Fig. 2 for the case of $\sigma = 0.38$ and zero angle of attack. Analyzing the drag by FFT, the peak frequency of the spectra is equal to 15.6Hz and the corresponding Strouhal number is 0.124. The snapshots of the shedding process of unsteady cavitating flows are shown in Fig. 3 for the case of a small angle of attack. It can be seen that the violent detachment near the separating point of the cavity on the axisymmetric body appear with three dimensional cavities.

Fig. 4 and Fig. 5 present the velocity field and pressure distribution on the surface of the axisymmetric body respectively in which the panels corresponds to the subfigures of Fig. 3. In the computation, the velocity of the incoming flow is 7.5m/s, the background pressure is 13200Pa, and the vapor pressure is 2368Pa. A high pressure distribution appears at the region where the detached sheet cavitation transfers to the cloud cavitation with violent changes of the cavity shape. During the process of shedding of the sheet cavitation and developing of the cloud cavitation, the maximum pressure could be up to 25000Pa, which is around twice of the background pressure, in the high pressure region on the surface of the body at a small angle of attack.

Comparing the computed drag for the cases of smaller angle of attack, it turns out that the fluctuation peaks appear at almost the same time and, however, for the case of the angle of attack is equal to 8^0, the computed drag is significantly different with those at smaller angles of attack. As the angle of attack is higher than a critical value, the violent shedding of the sheet cavitation seems to be depressed.

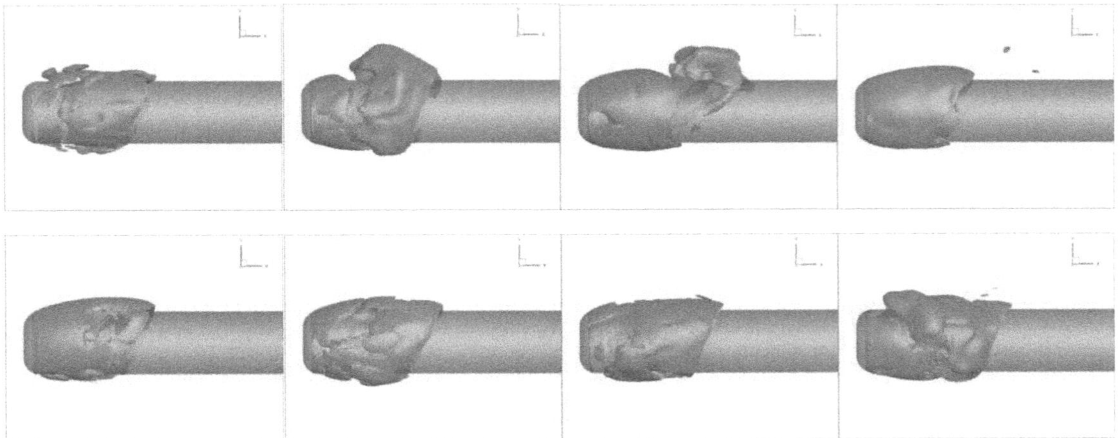

Figure 3 Snapshots of cavity shedding process for 1/4-caliber ogive headform at 2^0 AOA. The fluctuation period T=0.074s . Time interval is 1/8 the fluctuation period.

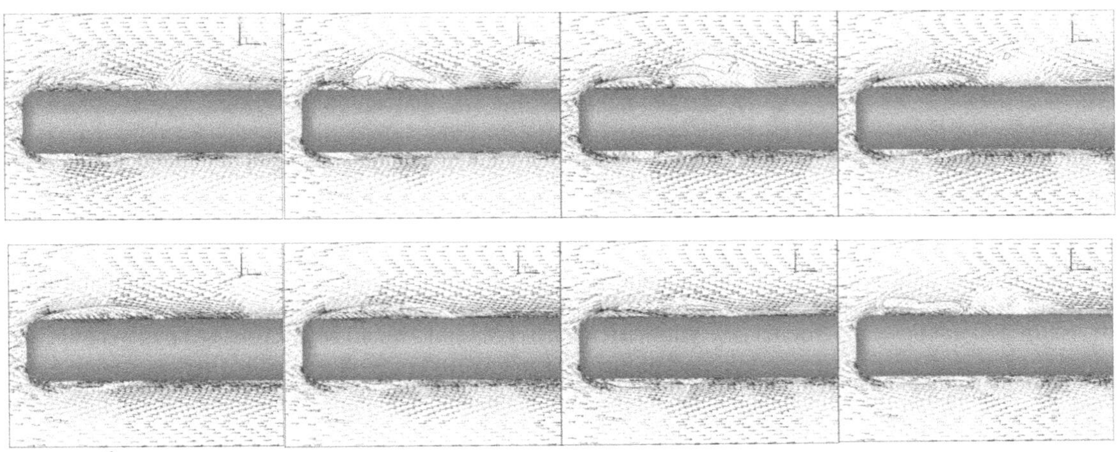

Figure 4 Snapshots of velocity fields for 1/4-caliber ogive headform at 2^0 AOA. The panels correspond to those of Fig. 2.

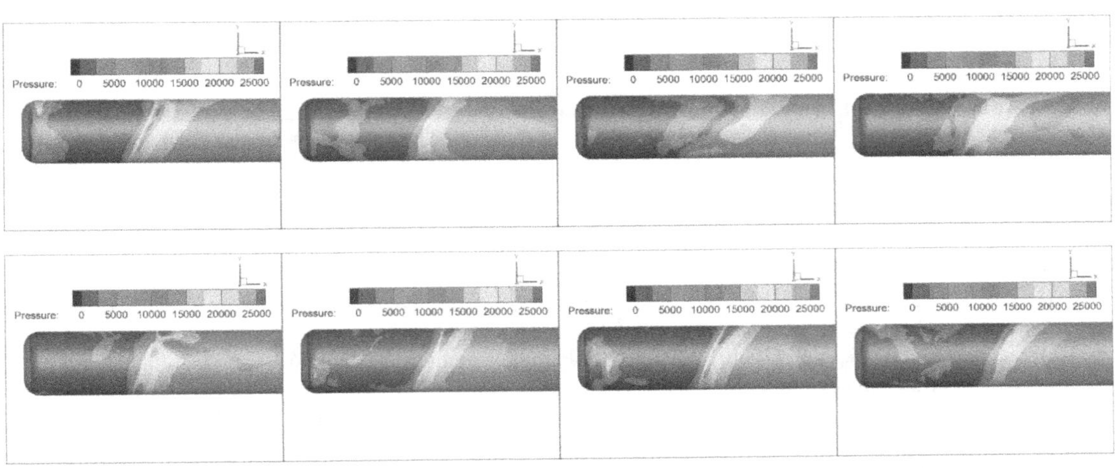

Figure 5 Pressure distributions on the surface of a cavitating axisymmetric body. The panels correspond to those of Fig. 2.

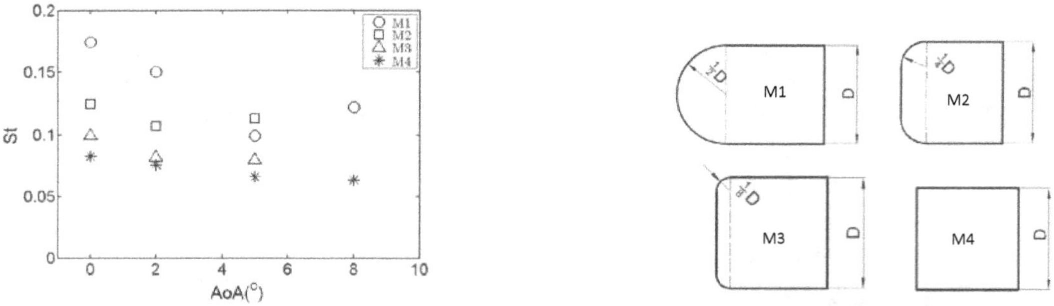

(a) Computed Strouhal number (b) Headforms

Figure 6 The relationship between Strouhal number and the angle of attack of an axisymmetric body with different headforms.

Carrying out a series of numerical experiments, we can obtain the relationship between the frequency characteristics and the angle of attack for the axisymmetric body with four headforms. As shown in Fig. 6, it is interesting to note that, for a given headform, the Strouhal number of the shedding sheet cavity decreases as the angle of attack increases. For the case of nonzero angle of attach, the sheet cavitation becomes to be asymmetry, in which the length of the sheet cavitation on the leeside of the axisymmetric body is greater than that on the upwind side. Compared with the re-entrant jet for the case of zero angle of attack, the re-entrant jet becomes weak due to the asymmetry of the sheet cavitation for the case of nonzero angle of attach and, then, the period of the transition from sheet to cloud cavitation becomes greater. For a given angle of attack, the Strouhal number becomes greater from the blunt headform to the hemispheric headform due to the variation of the length of the sheet cavitation on different headforms. The length of the sheet cavitation on the blunt headform and is the largest one among those of four headforms while the length of the sheet cavitation on the hemispheric headform is the smallest one.

Concluding remarks

The shedding process of unsteady cavitating flows and the corresponding hydrodynamic loads are studied experimentally and numerically for an axisymmetric body with four blunt headforms at small angles of attack. Based on the measured time series of drag and high speed image of cavity, the Strouhal number of low frequency fluctuation of the violent shedding sheet cavitation can be estimated by St=0.107±0.014 for different headforms. It turns out that, for a given headform, the Strouhal number of the violent shedding of the unsteady cavitating flows around an axisymmetric body decreases as the angle of attack increases. The violent shedding will not appear as the angle of attack is greater than a critical value. More effort should be made to understand the rapid changes of the pressure distribution on the surface of a cavitating axisymmetric body owing to the violent shedding of cavitaing flows.

Acknowledgments

The authors acknowledge the financial support from National Natural Science Foundation of China (10972138 and 11632012).

References

[1] Sakagami, K. et al. (1990). Behavior of the lift and the drag fluctuations on a foil in cavitation violent vibration. *Proc. of Japanese Society of Mechanical Engineers* (B), 56 (525):1223-1229.

[2] Liu, H., He, Y. S. (2012). Frequency characteristics of unsteady cavitating flows around axisymmetric headforms. *Proc. of the 8th International Symposium on Cavitation*, CAV2012, Singapore.

[3] Arndt, R. E. A., Song, C. C. S. et al. (2000). Instability of Partial cavitation: A Numerical/Experimental Approach. *The 23rd Symposium on Naval Hydrodynamics*, Rouen, France.

[4] Stinebring, D. R., Billet, M. L., et al. (2001). Developed Cavitation-Cavity Dynamics[R]. RTO EN-010, the von Kármán Institute (VKI) in Brussels, Belgium.

[5] Gnanaskandan, A., Mahesh, K. (2017). Numerical investigation of near-wake characteristics of cavitating flow over a circular cylinder. *Journal of Fluid Mechanics*, 790:453-491.

[6]Coutier-Delgosha, O., Fortes-Patella, R., Reboud, J. L. (2003). Evaluation of the turbulence model influence on the numerical simulations of unsteady cavitation. *Journal of Fluids Engineering*, 125(1):38-45.
[7]Singhal, A. K., Athavale, M. M., Li, H. et al. (2002). Mathematical basis and validation of the full cavitation model[J]. *Journal of Fluids Engineering*, 124(3):617-624.
[8] Wan, C. R., Liu, H. (2017). Shedding frequency of sheet cavitation around axisymmetric body at small angles of attack. *Journal of Hydrodynamics*, 29(2):175-182.

© 2018 ASME

VISVE, a Vorticity Based Model Applied to Cavitating Flow around a 2-D Hydrofoil

[1]Lu Xing*; [1]Chunlin Wu; [1]Spyros A. Kinnas

[1] *Ocean Engineering Group, The University of Texas at Austin, Austin, TX, USA*

Abstract

In this study, a method solving the viscous vorticity equation (VISVE) and the mixture model, which introduces an additional transport equation for the vapor volume fraction, are coupled to predict the partial cavitating flow around a two-dimensional hydrofoil. The VISVE method is designed to be both spatially compact and numerically efficient in comparison with the commonly used RANS models. The cavitation is modeled by the Schnerr-Saucer or the Zwart-Gerber-Belamri models. The predicted vorticity, velocity, pressures and the vapor volume fraction are compared with those from a commercial RANS solver.

Keywords: vortex method; mixture model; cavitation; hydrofoil

Introduction

Cavitation is a complex physical phenomenon of rapid phase change between liquid and vapor, resulting in very large and steep variation of density and viscosity. In many engineering applications – particularly hydrofoils, liquid pumps and marine propellers – cavitation, as a subject of research has attracted much attention mainly due to the detrimental effects (e.g. structural failure, corrosion, and noise radiation) of collapsing bubbles [4]. In particular, a type of cavitation that is very common on marine propellers is sheet cavitation, in which a distinct thin vapor bubble is attached to the blade surface. Other types of cavitation which also occur include cloud and bubble cavitation. However, this work addresses the computational analysis of sheet hydrofoil cavitation, though the method by nature is eligible to predict cloud and bubble cavitation.

VISVE method has been proposed and implemented in our previous work by Tian and Kinnas [1]. The VISVE method is designed to be both spatially compact and numerically efficient in comparison with the commonly used RANS models. Especially, after parallelization the efficiency can be improved significantly. The method was recently extended in the case of 2-D and 3-D hydrofoil in backing conditions by Wu et al [5] and cylinder in alternating flow by Li and Kinnas [6]. Reliable results were obtained in both cases. Mixture Model is a homogeneous fluid approach based on solving the transport equation for vapor volume fraction, where mass transfer rate due to cavitation is modeled by a mass transfer model. The mass transfer between two phases is controlled by including appropriate source terms in the volume fraction equation. There are several models [2][3][7] to formulate the simple rational expression of these source terms. In our work, we applied the model proposed by Schnerr and Sauer [2] and the model by Zwart-Gerber- Belamri [3].

The goal of this study is to incorporate in VISVE the mixture model to predict the cavitating flow around a hydrofoil. To verify the coupled scheme, the results are compared with those from the commercial RANS solver *ANSYS/Fluent*.

Numerical Methods and Models

1. Governing system

The cavitation mixture model assumes that the bubble-liquid flow under investigation is a single fluid with the homogeneous mixture of two phases (liquid and vapor). Therefore, only one set of equations is needed to simulate the cavitating flows:

$$\frac{\partial \rho_m}{\partial t} + \vec{\nabla} \cdot (\rho_m \vec{u}) = 0 \tag{1}$$

$$\frac{\partial \omega}{\partial t} + \vec{\nabla} \cdot (\omega \vec{q}) = \nabla^2(\nu\omega) + \vec{\nabla} \cdot \left[\nu \frac{\overrightarrow{\nabla\rho}}{\rho} \omega \right] \tag{2}$$

Phases are considered to be incompressible and to share the same velocity field [3]. Equation (1) represents the continuity equation and Equation (2) is the simplified vorticity equation under the assumption that the two-dimensional hydrofoil is placed along the x-axis, with inflow also along the x-axis, such that $\frac{\partial}{\partial x} \ll \frac{\partial}{\partial z}$, especially within the narrow region close to the hydrofoil and along the wake where $\omega \neq 0$. The complete formulation of vorticity

*Corresponding Author, Lu Xing: xinglu@utexas.edu

equation (2) can be found in Kinnas [9]. The above equations are closed with the constitutive relations for the density and dynamic viscosity:

$$\rho_m = \alpha\rho_v + (1-\alpha)\rho_l;\ \mu_m = \alpha\mu_v + (1-\alpha)\mu_l \tag{3}$$

Where, α represents the volume vapor fraction within the mixture. Detailed implementation schemes will be presented in Xing [8].

2. Pressure Calculation

Pressure is not a direct solution out of vorticity equation. It can then be calculated through post-processing. The momentum equation in the direction normal to the hydrofoil surface is given by the following equation [9]:

$$\frac{\partial \boldsymbol{q_n}}{\partial t} + \frac{\partial}{\partial n}\left(\frac{q^2}{2}\right) = -\frac{\partial P}{\partial n}\cdot\frac{1}{\rho} + \boldsymbol{q}\times\omega - \upsilon\nabla\times\omega - \frac{(\boldsymbol{\nabla}\cdot\boldsymbol{q})}{\rho}\cdot\frac{\partial\mu}{\partial n} \tag{4}$$

Where t is time, P is pressure, ρ is density, υ is the kinematic viscosity, $\boldsymbol{q}$ is the total velocity, ω is vorticity and μ is dynamic viscosity. Under the assumption that the normal velocity $\boldsymbol{q_n}$ near the hydrofoil surface can be neglected, equation (4) then be simplified as [9]:

$$\frac{\partial P}{\partial n}\cdot\frac{1}{\rho} = -\frac{\partial}{\partial n}\left(\frac{q^2}{2}\right) - \upsilon\boldsymbol{\nabla}\times\boldsymbol{\omega} - \upsilon\boldsymbol{\nabla}\times\boldsymbol{\omega} - \frac{(\nabla\cdot\boldsymbol{q})}{\rho}\cdot\frac{\partial\mu}{\partial n} \tag{5}$$

Since in VISVE the grid is generated normal to the hydrofoil, a line integral can be conducted along the direction normal to the hydrofoil from the outer boundary to any point (X) within the domain to calculate the pressure at that point. Therefore, as long as the pressure on the outer boundary of the domain is given, the pressure within the whole domain can be obtained.

$$P_X = \int_{outer\ boundary}^{X} \frac{\partial P}{\partial n}\cdot d\boldsymbol{n} \tag{6}$$

3. Cavitation model

With Equation (3), the continuity equation (1) can then be split into the following system:

$$\frac{\partial(\alpha\rho_v)}{\partial t} + \boldsymbol{\nabla}\cdot(\alpha\rho_v\boldsymbol{q}) = R \tag{7}$$

$$\boldsymbol{\nabla}\cdot\boldsymbol{q} = R\left(\frac{1}{\rho_v} - \frac{1}{\rho_l}\right) \tag{8}$$

Where, R is the phase change rate: $R = R_c$ or $R = -R_e$, with R_c, R_e representing condensation rate and evaporation rate, respectively, which can be obtained by the following two models. Equation (7) is the vapor transport equation, while Equation (8) models the "compressibility" of the mixture due to the generation and collapse of the cavity bubbles.

3.1 Schnerr and Sauer's model

In this model, the vapor volume fraction is defined as the ratio of the volume of vapor over the cell volume and then reformulated [2]:

$$\alpha = \frac{V_{vapor}}{V_{cell}} = \frac{N_{bubbles}\cdot\frac{4}{3}\pi\Re_B{}^3}{V_{vapor}+V_{liquid}} = \frac{n_0V_{liquid}\cdot\frac{4}{3}\pi\Re_B{}^3}{n_0V_{liquid}\cdot\frac{4}{3}\pi\Re_B{}^3 + V_{liquid}} = \frac{n_0\cdot\frac{4}{3}\pi\Re_B{}^3}{1+n_0\cdot\frac{4}{3}\pi\Re_B{}^3} \tag{9}$$

Where V_{vapor}, V_{liquid}, V_{cell} represent the volume of vapor, liquid and cell respectively; $\Re_B$ is the radius of the vapor bubble; n_0 is the bubble concentration per unit volume of pure liquid and is set to be $10^{13}/m^3$. The evaporation and condensation phase change rate is then modeled from Rayleigh-Plesset equation as follows:

$$R_e = \frac{\rho_l\rho_v}{\rho_m}\alpha(1-\alpha)\frac{3}{\Re_B}\sqrt{\frac{2}{3}\frac{(P_V-P)}{\rho_l}} \tag{10}$$

$$R_C = \frac{\rho_l\rho_v}{\rho_m}\alpha(1-\alpha)\frac{3}{\Re_B}\sqrt{\frac{2}{3}\frac{(P-P_V)}{\rho_l}} \tag{11}$$

$$\Re_B = \left(\frac{\alpha}{1-\alpha}\frac{3}{4\pi}\frac{1}{n_0}\right)^{\frac{1}{3}} \tag{12}$$

Here, P_V is the vapor pressure.

3.2 Zwart- Gerber- Belamri's model

This model assumes constant bubble radius and simulate the bubble growth and collapse as [3]:

$$R_e = F_{evap} \frac{3\alpha_{nuc}(1-\alpha)\rho_v}{\Re_B} \sqrt{\frac{2}{3} \frac{(P_V - P)}{\rho_l}} \tag{13}$$

$$R_C = F_{cond} \frac{3\alpha\rho_v}{\Re_B} \sqrt{\frac{2}{3} \frac{(P - P_V)}{\rho_l}} \tag{14}$$

In the above equations, α_{nuc} is the nucleation site volume fraction, F_{evap} and F_{cond} are two empirical calibration coefficients for the evaporation and condensation process, respectively. In VISVE, the above parameters are set as follows: $\alpha_{nuc} = 5.0 \times 10^{-4}, \Re_B = 1.0 \times 10^{-6} m,\ F_{evap} = 50,\ F_{cond} = 0.01$.

4. Boundary treatment

The boundary treatment scheme is shown in Figure 1. After solving the above coupled system, we could obtain the vorticity field. However, velocities calculated from vorticity field using Biot-Savart law will not satisfy the non-slip boundary conditions on the wall. A vorticity creation scheme based on the Boundary Element Method (BEM) is designed to eliminate tangential and normal velocity, denoted as $\boldsymbol{q^n}, \boldsymbol{q^s}$ on the wall. After assigning the newly created vorticity into cells in the first layer, the non-slip boundary condition will then be satisfied.

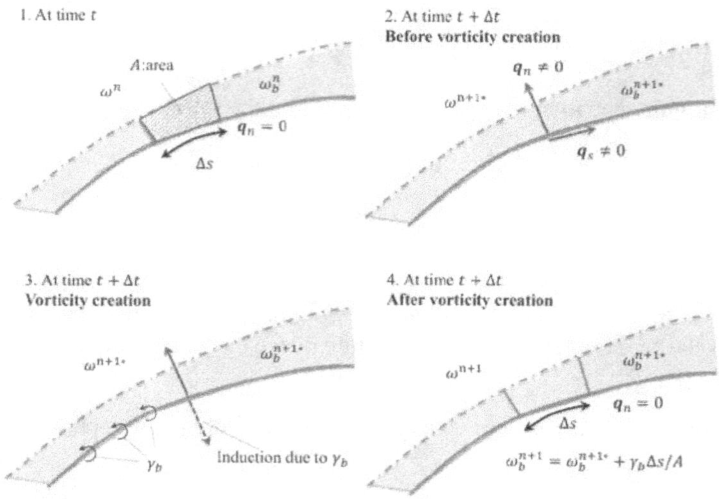

Figure 1: Schematic figure for the vorticity creation algorithm [1]

Results and comparison with RANS

To compare the computation time, VISVE and RANS are run on the same Stampede2 node of Texas Advanced Computing Center using 16 CPUs, and the parameters are shown in Table 1. VISVE is proven to be compact and efficient. Time step is set as $\Delta t = 0.0001\ s$ in both VISVE and RANS, and they are run for 10000 steps, so we get results for the first second. CPU time is calculated simply by taking the average of several cases.

1. Grid Configurations and numerical setup

The computation domains were generated around a NACA66 hydrofoil with thickness to chord ratio 4% ($\frac{t}{L} = 4\%$), camber to chord ratio1% ($\frac{c}{L} = 1\%$). The grid configurations in VISVE and RANS [1] are shown in Figure 2 and Figure 3, respectively. It can be noticed that VISVE enjoys a much more compact domain than RANS model. The total number of cells for the two models are shown in Table 1.

[1] All the RANS simulations shown in this paper have been performed by using *ANSYS/Fluent*.

Table 1: CPU time comparison

	Total Cell Number	CPU Time
VISVE	20,000	55min
RANS	50,000	4.5 h

In the RANS model, the cavitation is modeled as the steady flow. Second order discretization is used for density, momentum and turbulent kinetic energy. PRESTO!, QUICK and Least-square-cell-based schemes are applied to pressure, vapor phase transport equation and gradient calculation respectively. The simulations were run at Reynolds number Re=4,800, Angle of Attack ($AOA = 4°$)and the laminar model was adopted in VISVE and RANS. The boundary condition in RANS simulation is shown in Figure 3. The computation domain in VISVE is also open.

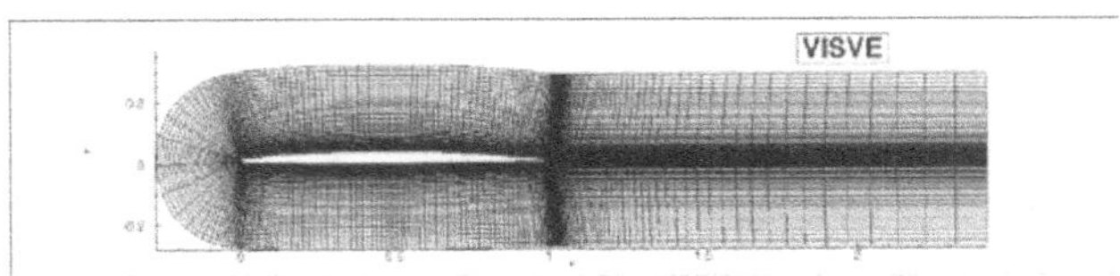

Figure 2: VISVE computation domain and grid configuration

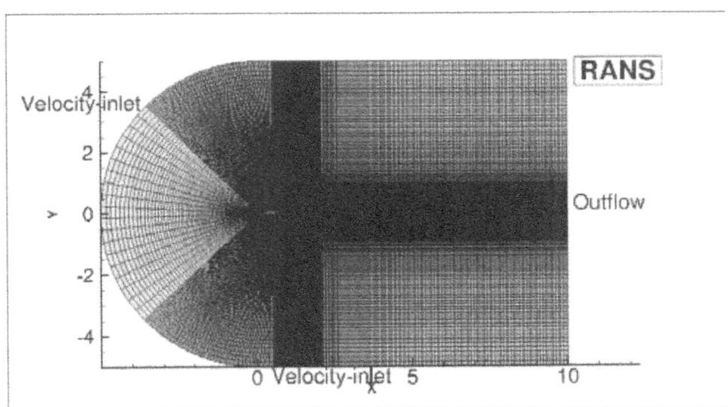

Figure 3: RANS computation domain and grid configuration

For the computational results, we use the following non-dimensional parameters:

$$\text{Cavitation Number } \sigma = \frac{P-P_v}{\frac{1}{2}\rho_l u_\infty{}^2} \tag{12}$$

$$\text{Reynolds Number } Re = \frac{u_\infty L}{\nu_l} \tag{13}$$

With chord length L= 1m, the free stream velocity u_∞ is set to be 2.44m/s. The results of a wetted case $\sigma = 1.6$ and cavitation cases of $\sigma = 1.2$, $\sigma = 1.0$ and $\sigma = 0.8$ are presented next.

2. Wetted case ($\sigma > 1.6$)

Before investigating the cavitating flow, the study of velocity and vorticity field was first carried out in fully wetted case. The pressure coefficient on the foil as well as the profiles of velocity and vorticity predicted by VISVE and RANS are compared in Figure 4 and Figure 5, respectively. The two models yield very close profiles to each other. The good correspondence verifies the reliability of VISVE model in fully wetted cases.

3. Steady Cavitation

The cavitating results shown below was simulated using the Zwart- Gerber- Belamri's model. However, with the runtime option given, users can easily choose their preferable model. More results are shown in Xing [8]. The pressure coefficients and the cavity shape at $\sigma = 1.2$, $\sigma = 1.0$ and $\sigma = 0.8$ are shown in Figure 6-11. It can be seen from Figures 6-9 that when the cavity is moderate ($\sigma = 1.2$, $\sigma = 1.0$), the pressure coefficient and cavity shape predicted by VISVE show good agreement to those simulated by RANS. However, for smaller cavity numbers ($\sigma = 0.8$), as shown in Figures 10 and 11, VISVE yields thicker and more diffused cavity shape, compared to that from RANS.

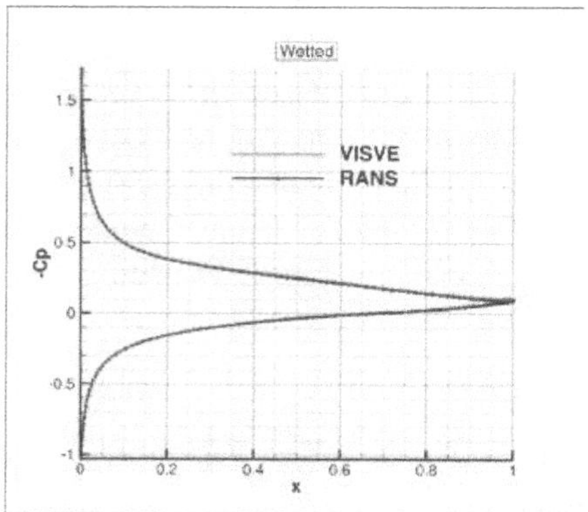

Figure 4: Pressure coefficient comparison between VISVE and RANS in wetted case.

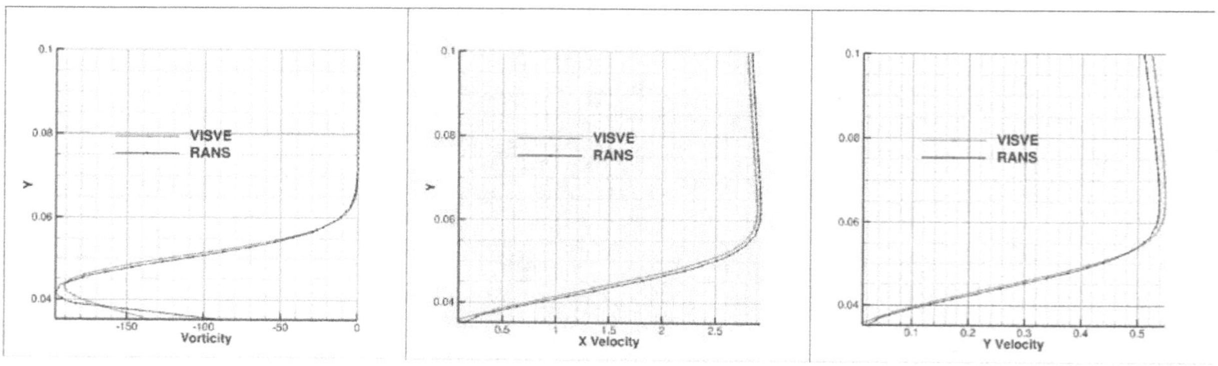

Figure 5 Vorticity, x-velocity, y-velocity profile comparison at 10% chord at Re= 4,800

Conclusions

A new numerical model which couples VIScous Vorticity Equation (VISVE) model with mixture model to simulate cavitating flow is proposed and the predicted velocity, vorticity, pressure and cavity patterns are compared with those from Reynold Averaged Navier-Stokes (RANS) simulations. In fully wetted conditions, the VISVE model predicts vorticity and velocity profiles, and pressure distributions which are very close to those predicted from RANS. In the cases of moderate cavitation number ($\sigma = 1.2$, $\sigma = 1.0$), VISVE predicts pressures and cavity shapes which are comparable to those from RANS. With lower cavitation number, $\sigma = 0.8$, VISVE predicts larger cavity than that from RANS. Further efforts will be made in the near future to resolve the differences in the cavity size predicted by the present method and by RANS.

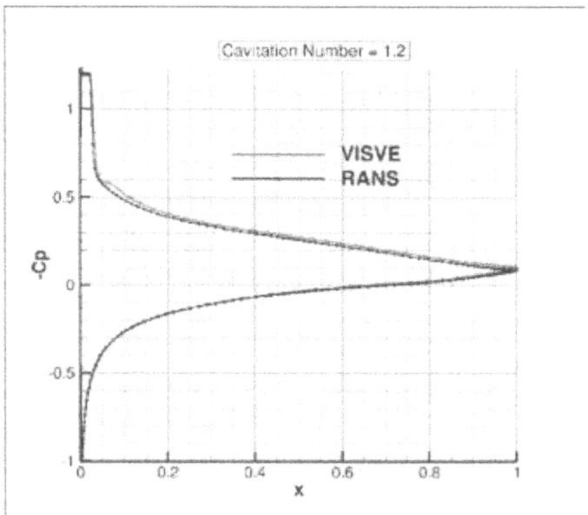

Figure 6: Pressure coefficient comparison between VISVE and RANS at Re= 4,800 $\sigma = 1.2$

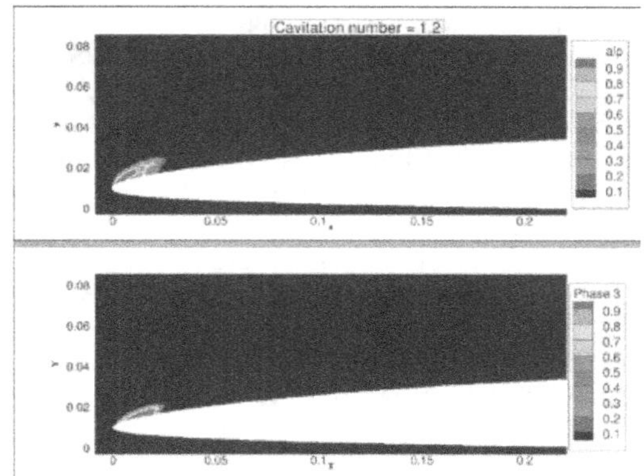

Figure 7: Vapor fraction contour comparison between VISVE and RANS at Re= 4,800 $\sigma = 1.2$

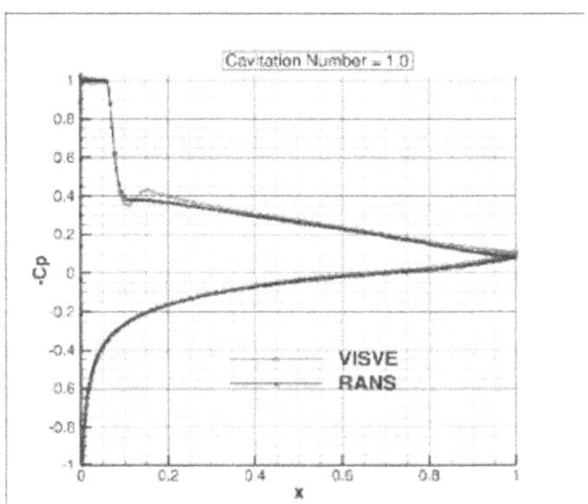

Figure 8: Pressure coefficient comparison between VISVE and RANS at Re= 4,800 $\sigma = 1.0$

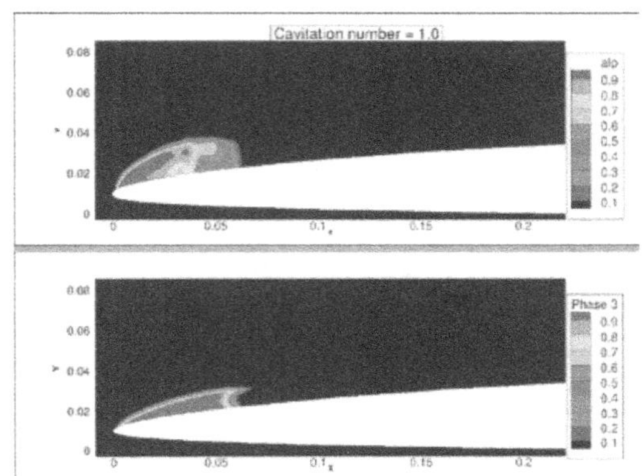

Figure 9: Vapor fraction contour comparison between VISVE and RANS at Re= 4,800 $\sigma = 1.0$

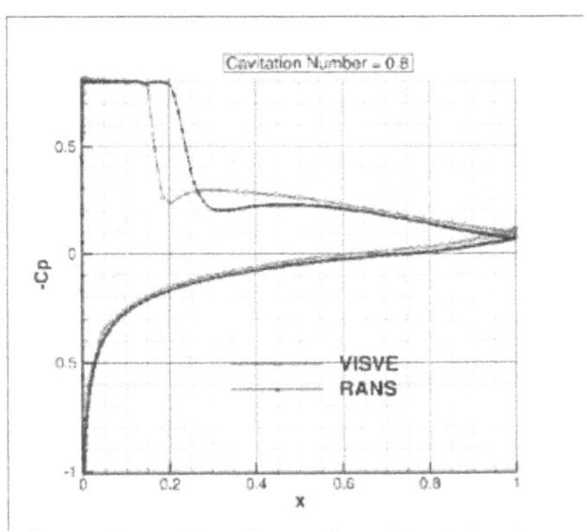

Figure 10: Pressure coefficient comparison between VISVE and RANS at Re= 4,800 $\sigma = 0.8$

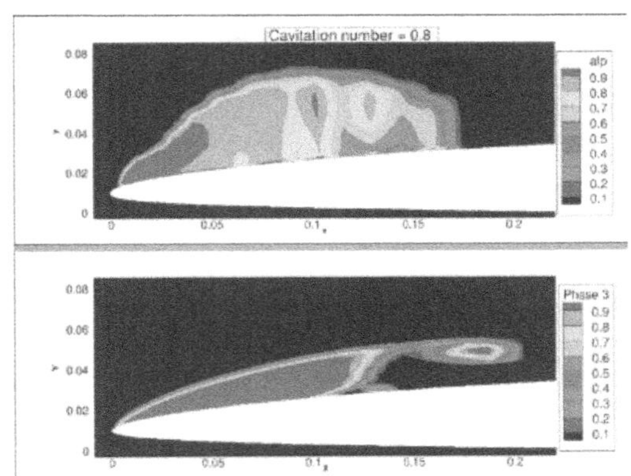

Figure 11: Vapor fraction contour comparison between VISVE and RANS at Re= 4,800 $\sigma = 0.8$

Acknowledgement

Support for this research was provided by the U.S. Office of Naval Research (Grant Nos. N00014-14-1-0303 and N00014-18-1-2276; Dr. Ki-Han Kim) partly by Phases VII and VIII of the "Consortium on Cavitation Performance of High Speed Propulsors".

References

[1] Tian, Y. and Kinnas, S.A., "A Viscous Vorticity Method for Propeller Tip Flows and Leading Edge Vortex", 4th International Symposium on Marine Propulsors, smp'15, Austin, Texas, USA, May 31-June 3, 2015.

[2] Schnerr, G.H. and Sauer, J., "Physical and Numerical Modeling of Unsteady Cavitation Dynamics", In Fourth International Conference on Multiphase Flow, New Orleans, USA. 2001.
[3] Zwart, P.J, Gerber, A.G., and Belamri, T., "A Two-Phase Flow Model for Predicting Cavitation Dynamics", In Fifth International Conference on Multiphase Flow, Yokohama, Japan. 2004.
[4] Kinnas, S.A. and Fine, N.E., "A Numerical Nonlinear Analysis of the Flow around Two-and Three-Dimensional Partially Cavitating Hydrofoils", Journal of Fluid Mechanics 254: 151-181, 1993.
[5] Wu, C., Xing, L., Kinnas, S.A., "A Viscous Vorticity Equation (VISVE) Method applied to 2-D and 3-D hydrofoils in both forward and backing conditions", 23rd SNAME Offshore Symposium, Texas, Houston, February 14, 2018.
[6] Li, Z. and Kinnas, S.A., "VISVE, a Vorticity Based Model Applied to Unidirectional and Alternating Flow around a Cylinder," 22nd SNAME Offshore Symposium, Houston, Texas, February 2, 2017.
[7] Singhal, A.K., Li, H.Y., Athavale, M.M., and Jiang Y.,"Mathematical Basis and Validation of the Full Cavitation Model", ASME FEDSM'01. New Orleans, Louisiana 2001.
[8] Xing, L., "VISVE, a Vorticity Based Model Applied to 2-D Hydrofoils in Backing and Cavitating Conditions", MS thesis, Ocean Engineering Group, The University of Texas at Austin, May 2018.
[9] Kinnas, S.A., "Vorticity Equation for Turbulent Flows with Variable Density and Viscosity" (submitted for publication), July 2018.

© 2018 ASME

Wall shear rates induced by a single cavitation bubble collapse

[1,2]Fabian Reuter*; [2]Hemant Sagar; [2]Bettar el Moctar; [1]Robert Mettin

[1]*Georg-August-Universität Göttingen, Drittes Physikalisches Institut, Christian Doppler Laboratory for Cavitation and Microerosion, Germany*

[2]*Universität Duisburg-Essen, Institute of Ship Technology, Ocean Engineering and Transport Systems; Germany*

Abstract

When a cavitation bubble collapses in vicinity to a solid surface, high flow velocities are induced. They involve remarkably high but unsteady wall shear rates. Even though they are crucial in ultrasonic surface cleaning or cavitation erosion, and their knowledge is needed for validation of numerical methods, they have not been measured so far due to experimental difficulties. Here, a wall shear rate raster microscope was developed. It bases on an electrochemical principle and involves a model to solve the appropriate convection-diffusion equation. As wall shear rate sensor, a microelectrode was flush-mount into a solid surface. With this method, wall shear rates on micrometer and microsecond scales can be resolved. The wall shear rates produced during the collapse of a single, laser generated bubble (maximum radius about 400 μm) were measured in planes. Via the synchronously performed high speed imaging, the wall shear rates can be clearly related to details and the different stages of the bubble dynamics. This way the respective impacts of the flow phenomena involved on the generation of wall shear stress were evaluated. For example, the jet that accompanies the bubble collapse was resolved in terms of wall shear rates during its impact on the wall and its subsequent spreading in radial direction. The above experimental data are compared with results of numerical simulations of the collapse of a single bubble obtained with a compressible two-phase flow solver that uses barotropic equations of state.

Keywords: bubble collapse, wall shear rates, cavitation

Introduction

Collapsing cavitation bubbles may induce significant mechanical effects on nearby solid surfaces. This is the case for example in ultrasonic cleaning [1][2], cavitation erosion [3][4][5] or cell manipulation [6][7][8]. Most of the mechanical interaction from a cavitation bubble is mediated by the shock wave that is emitted during the bubble collapse [9] and via the intense microflows generated by the bubble dynamics. Because a cavitation bubble can generate its intense microflows directly in the near-wall liquid layer at a solid boundary, the involved flows possess extraordinarily high wall shear rates. However, because the transient bubble dynamics occur are fast and local, the wall shear flows are short-lived and local. Consequently, there are hardly experimental techniques available for their measurement. Thus, the wall shear rates exhibited by cavitation bubbles are largely unknown. Here, we present measurements on the time-resolved wall shear rates produced by a single cavitation bubble. An electrochemical wall shear rate microscope was developed and used to spatially scan the wall shear rates produced by a single laser-induced cavitation bubble. The experimental results are compared to numerical computations performed using a Navier Stokes equations solver implemented in the open source toolbox OpenFOAM [10].

Experimental methods

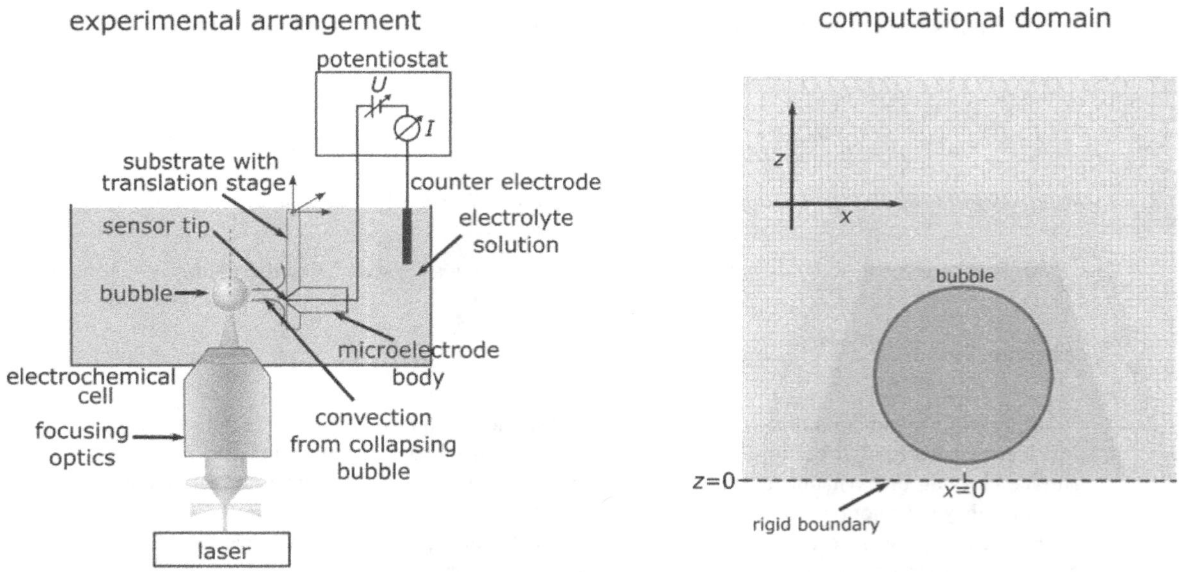

Figure 1 *Left*: Experimental arrangement. *Right*: Discretization of the computational domain with a refinement around the bubble.

To measure the wall shear rates of a single collapsing cavitation bubble, an electrochemical wall shear rate raster microscope was developed, a sketch is given in Figure 1 (left). As sensor, it employs an electrode that is flush-mounted into a solid substrate. The substrate is connected to a two axis precision translation stage (12 nm resolution) to allow for a rastering of the wall shear rate in a plane (450 μm x 450 μm). At each measurement position, the complete time series of wall shear of a collapsing bubble (radius at maximum expansion: $R_{\max} \cong 425$ μm) is recorded. Thus, one bubble needs to be generated for each measurement position. The preferred means for the repeatable production of spherical cavitation bubbles of defined properties at a confined position is by the focusing of a laser pulse. In the laser focus, the power density is large enough to induce optical breakdown, i.e. a plasma is generated. Around the plasma, a cavitation bubble forms, a mechanism termed optical cavitation [11][12]. This mechanism is used here: a laser pulse (5 ns pulse duration, 532 nm, ~ 1 mJ) is focused close to the substrate (PMMA slide of 1 mm thickness) and the electrode (platinum electrode $d_E = 25$ μm diameter) into an aqueous solution in a glass cuvette of 125 ml volume. To avoid any spurious effects from the intense laser light, the laser beam is aligned vertical to the electrode surface (see [13]). With a high speed video camera, the bubble shape is imaged synchronized to the wall shear rate measurements allowing to relate the wall shear rate to the different stages of bubble dynamics. The wall shear rate is derived from a chronoamperometric measurement performed with a potentiostat of type Gamry Reference 600 that works at its maximum sampling frequency of 300 kHz. The sensor electrode is operated as working electrode, for counter and reference electrode a platinum wire is submerged close to the substrate. As buffer electrolyte KNO_3 is used. As analyte, a Faradaic electrolyte is dissolved in the deionized water ($Ru(NH_3)_6^{3+}$, concentration $c_0 = 0.03$ M). An electrical potential of $U_W = -0.7$ V is maintained between working electrode and counter electrode. It allows for the reversible reduction of the Ru-complex at the surface of the working electrode, in turn giving rise to a reduction current of:

$$I = \frac{FA_E \partial c(z,t)}{\partial z}\bigg|_{z=0}, \tag{1}$$

where F is the Faraday constant, A_E the electrode surface, c the concentration, and z the coordinate perpendicular to the electrode surface whereas $z = 0$ refers to the electrode surface [14]. The reaction kinetics of the Ru-complex are very fast, so a molecule of the Ru-complex can be considered to be reduced instantaneously, once it is found at the electrode surface. Therefore, the reaction current is limited only by the rate at which molecules of the Ru-complex reach the electrode. This rate is limited here only by diffusion and convection. The equation describing the concentration change is the convection diffusion equation. It reads:

$$\frac{\partial c(\vec{x},t)}{\partial t} = D\Delta c(\vec{x},t) - \vec{u}(\vec{x},t)\nabla c(\vec{x},t), \tag{2}$$

where D is the diffusion constant, c the concentration, and $\vec{u}$ the velocity. Here, by convection, the cavitation bubble increases the transfer rate of the electrolyte concentration to the electrode. This effect is exploited here. For the present situation, Equation (2) can be simplified: The electrode is operated as a macro electrode. This is achieved by the application of the working potential in sufficiently short intervals. Then, the flux to the electrode is planar, allowing for a one-dimensional treatment. Furthermore, the diffusion layer thickness δ is smaller than d_{E}. Within the small diffusion layer at the solid ($\delta < 25$ µm), the flow can be assumed to be parallel to the boundary and the flow profile is assumed to be linear. With these assumptions one arrives at the following parabolic differential equation for the wall shear rate $G(t) = \left.\frac{\partial u(t)}{\partial z}\right|_{z=0}$, where u is the x-component of the flow velocity:

$$\frac{\partial c(z,t)}{\partial t} = D\frac{\partial^2 c(z,t)}{\partial z^2} + G(t)\frac{z}{d_{\mathrm{E}}}(c_0 - c(z,t)). \quad (3)$$

In Equation (3), the wall shear rate appears as a source term. Its determination is an implicit problem. We solve it here with an optimization approach. For different candidates of G, the concentration profile was simulated using the toolbox Matlab (pdepe function). For each concentration profile, the resulting electrical currents were calculated. Then G was selected as the value at which the simulated current equals the measured current I. More details on this procedure and on the wall shear rate microscope can be found in [15]. From the wall shear rate, the wall shear stress s can be calculated by: $s = \mu G$, where μ is the viscosity of water.

Numerical methods

The bubble collapse is simulated using the density based CavitatingFoam solver of OpenFOAM. It is a fully compressible two-phase flow solver that applies barotropic equations of state and an Euler-Euler approach to solve the Navier-Stokes equations on a fixed Eulerian grid, considering gas and liquid phases as homogeneous mixture. Here, the liquid phase is water, the gas phase is water vapor. The mass fraction of the vapor in the mixture α is assumed to be a continuous function of space and time. It is determined as:

$$\alpha = \frac{\rho_{\mathrm{m}} - \rho_{\mathrm{w,sat}}}{\rho_{\mathrm{v,sat}} - \rho_{\mathrm{w,sat}}}, \quad (4)$$

where ρ_{m} is the mixture mass density, $\rho_{\mathrm{w,sat}}$ is the density of water at saturation and $\rho_{\mathrm{v,sat}}$ is vapor saturation density. Drop in the saturated liquid density below vapor saturation density indicates the presence of cavitation. $\alpha = 0$ means the presence of only the liquid phase, whereas $\alpha = 1$ indicates cavitation zones. The mixture compressibility φ_{m} is modelled using Wallis linear model [16] given by:

$$\varphi_{\mathrm{m}} = \alpha\varphi_{\mathrm{v}} + (1-\alpha)\varphi_{\mathrm{w}}, \quad (5)$$

where φ_{v} and φ_{w} are the liquid and the vapor compressibility. Further, the mixture viscosity μ_{m} is computed as:

$$\mu_{\mathrm{m}} = \alpha\mu_{\mathrm{v}} + (1-\alpha)\mu_{\mathrm{w}}, \quad (6)$$

where μ_{w} and μ_{v} are the viscosity of water and vapor respectively. The relation between pressure and density is defined as:

$$\frac{D\rho_{\mathrm{m}}}{Dt} = \varphi\frac{DP}{Dt}. \quad (7)$$

Above equation can be used directly in the continuity equation to formulate a pressure equation or integrated to obtain the pressure as a function of the density.

Simulations are carried out in a three dimensional domain without the assumptions of any symmetries in order to also simulate the bubble dynamics after the first collapse, when many bubble related phenomena become inherently three dimensional. The size of the computational domain is 5 x 5 x 5 mm^3. Figure 1 (right) illustrates the plane section of the finite volume discretized domain. The flow is assumed to be laminar. The maximum acoustic Courant number is limited to 0.5 by choosing appropriate time steps. To reduce computational costs, the grid is refined to a minimum cell size of 6.25 μm only in region of interest, i.e. the collapse region. Furthermore, the bubble is initialized at its maximum expansion, defining the initial internal pressure and vapor volume fraction. Thus, the initial expansion from plasma seeding to maximum expansion is not simulated.

Results

An overview over the bubble dynamics is given in Figure 2 through the evolution of the bubble shape. The bubble collapses in vicinity to the substrate with the integrated wall shear rate sensor. The bubble was seeded at $t = 0$ μs. The bubble life time T_L is defined as the time from bubble generation (plasma seeding) to bubble collapse. Here, times are given in normalized form $\tau = t/T_L$, where $T_L \cong 89$ μs.

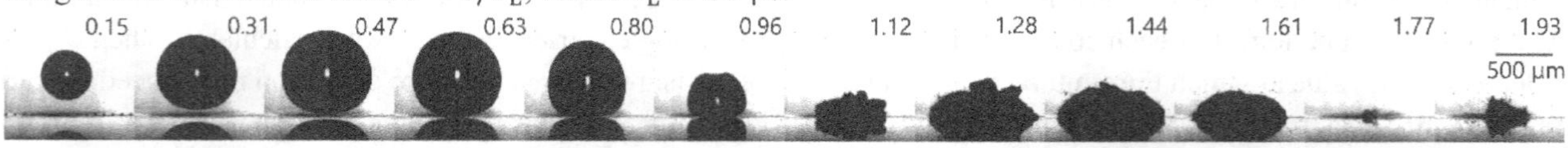

Figure 2 Photographic image series of the time evolution of the bubble shape. Normalized time τ are given in the top right corner of each image. The rigid boundary is depicted in the first frame. On the solid surface a mirrored image of the bubble shape is visible.

From $\tau = 0$ to $\tau = 0.5$ the bubble expands radially. After $\tau = 0.5$, the bubble collapses, initially in a merely spherical manner, towards the end of the collapse phase, increasingly aspherically. At $\tau = 0.96$, a small indentation on the bubble north pole is visible, a consequence of the jet that pierces the bubble and impacts perpendicular on the substrate [12]. After the collapse, at $\tau = 1.00$, the bubble is in the rebound phase up to $\tau \cong 1.35$: it takes a flat shape and spreads directly over the boundary. The bubble now has a toroidal shape [17][18], which is however not visible in this side projection. Then, the bubble collapses (toroidally) at $\tau \cong 1.75$. After the second collapse, main parts of the bubble have disintegrated and only a small bubble in the center shows some afterbounces.

In Figure 3 the respective shear rates are shown for four instances of time, in the top column experimental data is shown, in the bottom column data obtained from the numerical simulation. At the time instance $\tau = 0.72$, the bubble only undergoes a radial oscillation (sometimes called volume pulsation), i.e. it remains merely spherical. During this type of dynamics, wall shear rates are very low. Below the bubble center they are minimum because a stagnation area is formed. At the next time instance, $\tau = 1.03$, the jet has pierced the bubble and impacts on the substrate. From the jet impact, wall shear rates are now about 25 times larger. At the axis of symmetry, again a stagnation area, this time from the jet, is formed. At $\tau = 1.23$, the wall shear rates are still large as the bubble rebounds at the substrate. Thereby liquid is accelerated annularly outwards. The ring of high wall shear consequently expands radially (not shown here). The figure at $\tau = 1.73$ presents an intriguing situation. The (toroidal) bubble is about to collapse. Now, the highest wall shear rates are measured $G_{\text{max}} = 5.88\ 10^6\ \text{s}^{-1}$ during the situation in which flows directed inwards from the outer torus wall and the flows from the inner torus wall directed outwards collide.

After this collision of flows, no significant shear forces are anymore actuated on the solid surface. This is consequence of the formation of a ring vortex that is ejected from the surface. Thereby, the moving liquid is convected remote from the solid surface into the bulk. Figure 4 illustrates the flow field, obtained from the numerical simulation. This behavior was experimentally observed and quantitatively investigated before [18].

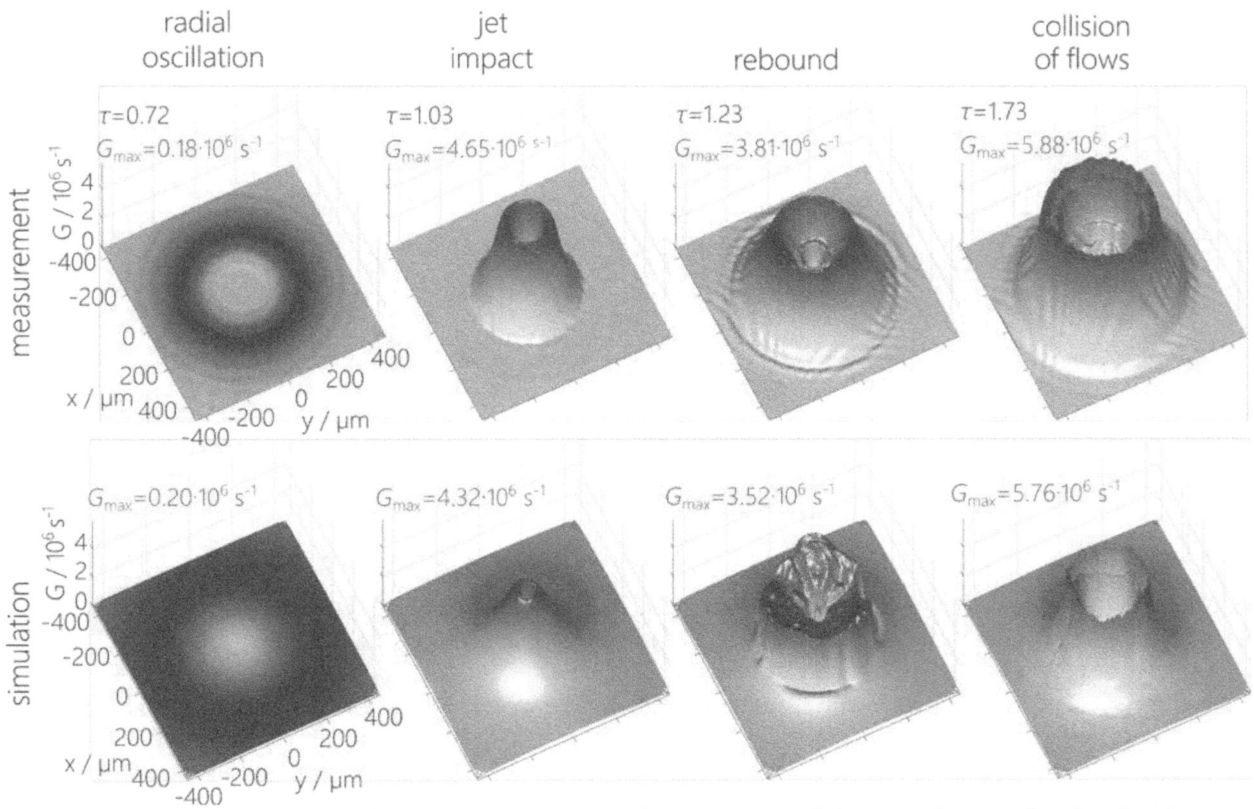

Figure 3 Comparison between measured wall shear rates (top column) and computed wall shear rates (bottom column) on the substrate surface. Four different times are chosen ($\tau = 0.72, \tau = 1.03, \tau = 1.23, \tau = 1.73$) that reflect different stages of the bubble dynamics (radial oscillation, jet impact, rebound, collision of flows). The data is height and color coded. The height coding is the same for every image (vertical axis wall shear rate $G = 0..6\ 10^6\ \mathrm{s}^{-1}$), the respective color levels are scaled to each maximum wall shear rate G_{max}.

Both datasets shown in Figure 3 match well. They show the same features and almost the same maximum wall shear rates. This can be considered a confirmation for the numerical as well as the experimental method. However, differences between experiments and computations can be observed. In general, the experimental data shows a broadened structure, while according to the numerical computation, the wall shear rates seem to be locally more confined. A main reason for this appearance can be found in the experimental rastering methodology, in which the wall shear rates are measured from many bubbles. Even though we took measures to take into account only same-sized, spherical bubbles, still some jitter on the bubble life time (and consequently the geometric extension of the bubble at a given instance of time) is unavoidable. This jitter, is expected to geometrically smooth out the experimental data, and can be especially seen on the volcano-like structure that decays much smoother outwards in the experimental case (see $\tau = 1.72$). The accuracy of the numerical simulation could be increased by a finer spatial and temporal mesh size. Also, the spherical initialization of the bubble at its maximum expansion lets the simulated situation differ from the experimental one, where the bubble as it grows out of its plasma spot deforms due to the influence of the boundary. Also, the determination of the bubble stand-off distance to the wall is subject to some uncertainty. This is important because the stand-off distance is known to critically influence the bubble dynamics and consequently the generation of wall shear forces.

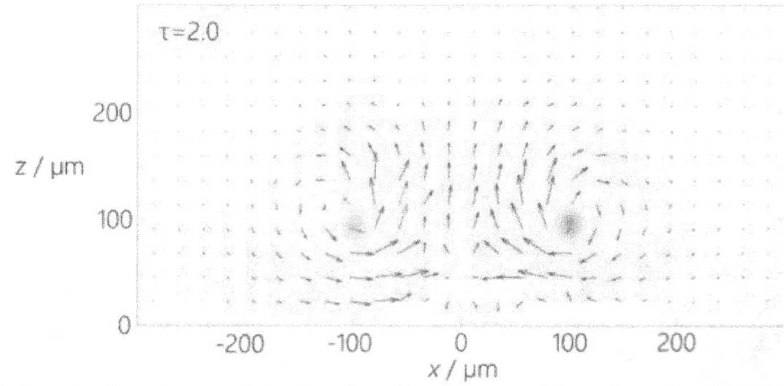

Figure 4 Flow field (simulated) after the detachment of the flow from the collision of flows has taken place and a ring vortex is formed, seen here in side projection. The boundary is located at $z = 0$. The gas phase is overlaid in gray.

Conclusion

A cavitation bubble generates intense microconvection directly at solid surfaces. A significant force transmission and potential surface alteration is induced by the wall shear flow. So far, experimental techniques for the measurement of the respective wall shear rate were limited. Here, the wall shear rates of a single collapsing cavitation bubble were measured in high spatial and temporal resolution. The wall shear rates were related to the bubble dynamics via synchronized high speed imaging. In addition, numerical simulations were carried out. The experimental results are consistent with numerical simulations. The wall shear rates during jet impact and toroidal bubble collapse could be quantified. They are shown to be the most significant ones generated by the bubble dynamics.

Acknowledgements

We thank Prof. Werner Lauterborn for constant commitment and scientific advice. F.R. and R.M. thank for the financial support by Lam Research AG, Austria, and by the Austrian Federal Ministry of Science, Research and Economy and the Austrian National Foundation for Research, Technology and Development.

References

[1] H. F. Okorn-Schmidt *et al.*, "Particle Cleaning Technologies to Meet Advanced Semiconductor Device Process Requirements," *ECS J. Solid State Sci. Technol.*, vol. 3, no. 1, pp. N3069–N3080, Dec. 2013.

[2] F. Reuter and R. Mettin, "Mechanisms of single bubble cleaning," *Ultrason. Sonochem.*, vol. 29, pp. 550–562, Mar. 2016.

[3] C. F. Naudé and A. T. Ellis, "On the Mechanism of Cavitation Damage by Nonhemispherical Cavities Collapsing in Contact With a Solid Boundary," *J. Basic Eng.*, vol. 83, no. 4, pp. 648–656, 1961.

[4] Y. Tomita and A. Shima, "Mechanisms of impulsive pressure generation and damage pit formation by bubble collapse," *J. Fluid Mech.*, vol. 169, pp. 535–564, Apr. 1986.

[5] W. Lauterborn and T. Kurz, "Physics of bubble oscillations," *Rep. Prog. Phys.*, vol. 73, no. 10, p. 106501, Oct. 2010.

[6] K. R. Rau, A. Guerra, A. Vogel, and V. Venugopalan, "Investigation of laser-induced cell lysis using time-resolved imaging," *Appl. Phys. Lett.*, vol. 84, no. 15, p. 2940, 2004.

[7] A. Vogel, P. Schweiger, A. Frieser, and M. N. Asiyo, "Intraocular Nd : YAG Laser Surgery : Light-Tissue Interaction , Damage Range , and Reduction of Collateral Effects," *IEEE J. Quantum Electron.*, vol. 26, no. 12, pp. 2240–2260, 1990.

[8] J. L. Compton, A. N. Hellman, and V. Venugopalan, "Hydrodynamic determinants of cell necrosis and molecular delivery produced by pulsed laser microbeam irradiation of adherent cells," *Biophys. J.*, vol. 105, no. 9, pp. 2221–2231, 2013.

[9] C. F. Delale, Ed., *Bubble Dynamics and Shock Waves*. Springer, 2013.

[10] "OpenFOAM 2.3.x, OpenFOAM Foundation, www.openfoam.org." 2015.

[11] W. Lauterborn, "Kavitation durch Laserlicht ('Cavitation by laser light')," *Acustica*, vol. 31, no. 2, pp. 51–78, 1974.

[12] W. Lauterborn and H. Bolle, "Experimental investigations of cavitation-bubble collapse in the neighbourhood of a solid

boundary," *J. Fluid Mech.*, vol. 72, no. 2, pp. 391–399, Nov. 1975.
[13] F. Reuter, C. Cairós, and R. Mettin, "Vortex dynamics of collapsing bubbles: Impact on the boundary layer measured by chronoamperometry," *Ultrason. Sonochem.*, vol. 33, pp. 170–181, Nov. 2016.
[14] J. Wang, *Analytical Electrochemistry*, Third edn. Wiley-VCH, 2006.
[15] F. Reuter and R. Mettin, "Electrochemical wall shear rate microscopy of collapsing bubbles," *Submitt. to Phys. Rev. Fluids*, 2017.
[16] G. B. Wallis, *One-dimensional two-phase flow*. New York, NY: McGraw-Hill Publishing Co., 1969.
[17] O. Lindau and W. Lauterborn, "Cinematographic observation of the collapse and rebound of a laser-produced cavitation bubble near a wall," *J. Fluid Mech.*, vol. 479, pp. 327–348, 2003.
[18] F. Reuter, S. R. Gonzalez-Avila, R. Mettin, and C.-D. Ohl, "Flow fields and vortex dynamics of bubbles collapsing near a solid boundary," *Phys. Rev. Fluids*, vol. 2, no. 6, p. 64202, Jun. 2017.

© 2018 ASME

X-ray computed tomography of cavitating flow in a converging-diverging nozzle

[1]Saad Jahangir*; [2]Evert C. Wagner; [2]Robert F. Mudde; [1]Christian Poelma

[1]Department of Process & Energy (Faculty 3mE), Delft University of Technology, Leeghwaterstraat 21, 2628 CA Delft, The Netherlands; [2]Department of Chemical Engineering (Faculty of Applied Sciences), Delft University of Technology, Julianalaan 136, 2628 BL Delft, The Netherlands

Abstract

Cavitation is a complex multiphase phenomenon, where the production of vapor bubbles leads to opaqueness of the flow. While it is nearly impossible to visualize the interior of the cavitation region with visible light, we show that with X-ray computed tomography it is possible to obtain the time-averaged void fraction distribution in an axisymmetric converging-diverging nozzle (venturi). This technique is based on the amount of energy absorbed by the material, based on its density and thickness. Time-averaged 3D reconstruction of the X-ray images is used (i) to distinguish between vapor and liquid phase, (ii) to get radial geometric features of the flow, and (iii) to quantify the local void fraction. The results show the presence of intense cavitation at the walls of the venturi, and the vapor fraction decreases downstream of the venturi with the vapor cloud.

Keywords: venturi; X-ray computed tomography; cloud cavitation

1. Introduction

Cavitation in a flow occurs when the static pressure at a given location in the flow drops below the vapor pressure of the liquid. If a cavitation bubble or cloud collapses close enough to the solid wall, it will induce a high-speed micro-jet and shock waves, which can result in erosion [1]. Understanding the correct cavitation physics is important because then the adverse consequences such as erosion can be reduced.

Among the studies on cavitation, high-speed visualization is the most popular technique to investigate the cavitation. However, quantitative information regarding the vapor fraction is difficult to obtain from the high-speed imaging, because the cavitation bubbles scatter light and make the flow opaque. Recently, X-ray densitometry has been demonstrated as a valuable technique to quantify the void fractions in various cavitation related studies [2, 3, 4]. Void fractions are of high importance in the understanding of shedding behavior in the periodic cavitation. Recently Ganesh et al. [5] found that under particular conditions a condensation shock can be the dominant mechanism for periodic cavitation shedding, instead of the re-entrant jet. Time-resolved X-ray densitometry was used to investigate the local void fractions in the flow field. They found that void fractions increase with an increase in cavitation intensity. These experiments were performed on a 2D wedge. Although cavitation can also be observed in other test geometries, a converging-diverging nozzle (venturi) is used in this study. Due to its high contraction ratio, wider cavitation dynamic range can be achieved.

X-ray computed tomography is extensively used in medical imaging. It uses the relation between the material properties and the attenuation coefficient of X-rays. Images are created of the attenuation along the beam paths. This capability will be utilized in this study to examine the vapor fractions, and radial geometric characteristics in the flow.

2. Experimental details

2.1. Flow facility

A schematic overview of the experimental setup is given in Figure 1(a). The flow in the closed loop system is driven by a centrifugal pump, and a flowmeter is used to measure the volumetric flow rate. The measurements from the upstream pressure transducer, downstream pressure transducer, and the temperature sensor are used to determine the cavitation number σ. A water column is present at an angle due to space restrictions to collect the air bubbles entrained in the flow during degasification, and to vary the global static pressure of the system. A vacuum pump is used to control the global static pressure below ambient pressure.

*Corresponding Author, Saad Jahangir: S.Jahangir@tudelft.nl

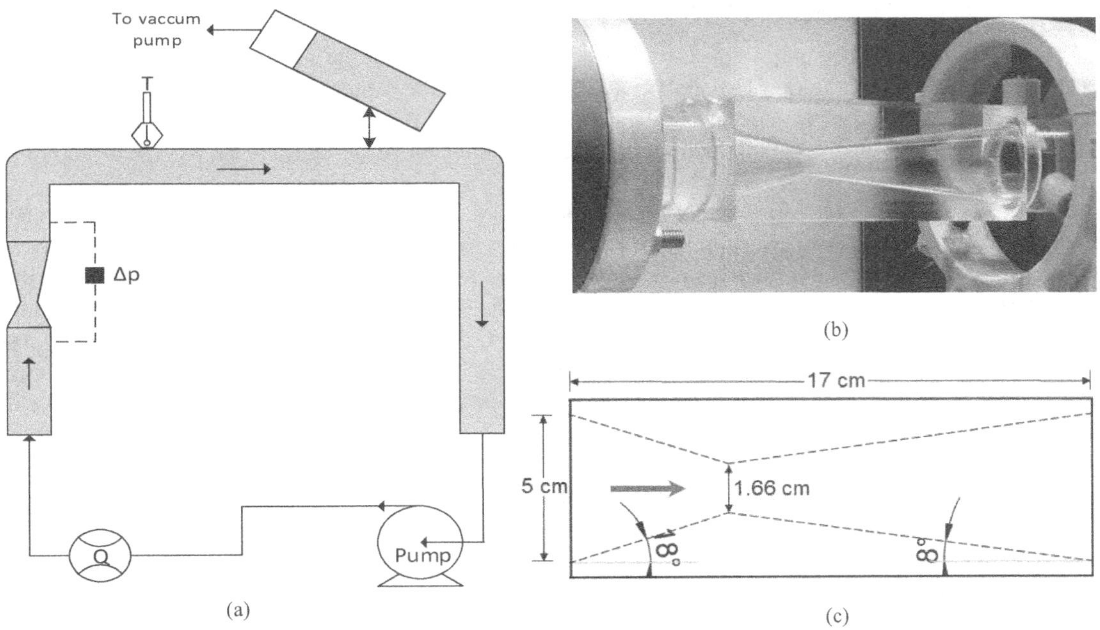

Figure. 1: (a) Schematic overview of the experimental setup. (b) Picture of the converging-diverging part (venturi) in the experimental setup. (c) The geometry and relevant dimensions of the venturi.

In Figure 1(b) and (c), a picture of the venturi can be seen with its geometrical parameters. The flow direction is from left to right in the figure, upward in reality. The convergence and divergence angles are 36° and 16°, respectively (inspired by previous studies: [6, 7, 8]). An area ratio of 1:9 (area of the throat versus exit area) is chosen. An oxygen sensor is used to determine the gas content in the system.

2.2. X-ray imaging

Figure 2(a) and (b) show a photograph of the measurement section in the X-ray setup and schematic overview of the method, respectively. A source-detector pair is used to measure the attenuation of the X-rays through the cavitating venturi. The source is operated at 120 keV and 5 mA in order to achieve a high contrast between the liquid and vapor phases within the venturi. The flat detector, Xineos-3131 CMOS model, consists of a 307×302 mm sensitive area. For the experiments, the field of view of 1548×660 pixels is chosen. Each pixel has a size of 198×198 μm with 14 bits of pixel depth. The images are recorded at 60 Hz during 1 min. As the typical shedding frequency is 40 Hz at $\sigma = 0.46$, this ensures that the statistics are based on sufficient shedding cycles.

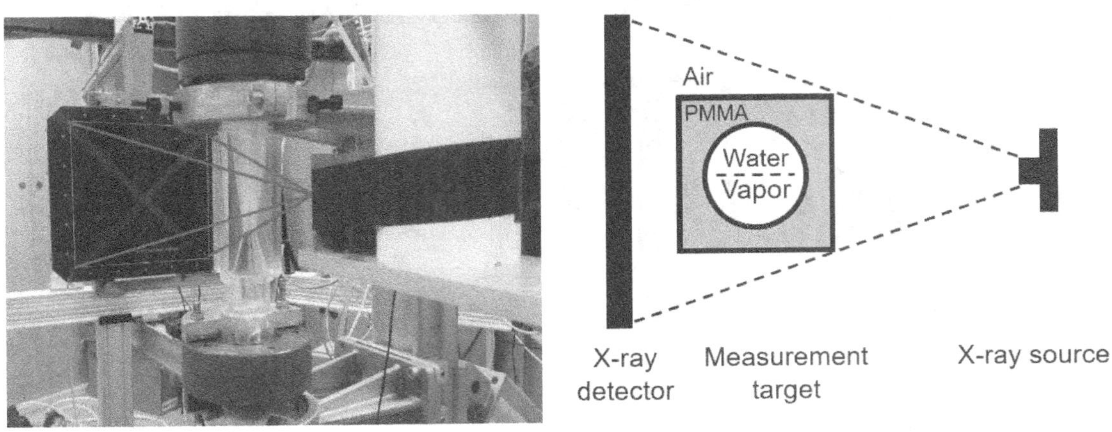

(a) (b)

Figure. 2: (a) Test-rig inside the X-ray setup. (b) Sketch of the X-rays imaging method, the viewing angle in the experiments is smaller than the depicted angle.

2.3. Experimental procedure and post-processing

The water is degasified and the setup is operated for a few minutes before the measurements are started in order to obtain a uniform water temperature. The global static pressure of the system is set to a fixed value and the measurements are started when the pressure measurements are constant. All the sensor values (pressure, flow rate, and temperature) and the X-ray images (explained in more detail in the later paragraph) are saved concurrently by using a data acquisition system.

A background subtraction operation is performed for the raw X-ray images, for which background images with only the liquid phase without flow are captured. In order to improve the contrast, an image adjustment operation is performed on images. This process involves rescaling the grayscale intensities in order to have 1% of the data being saturated at low and high intensities. 3600 instantaneous X-ray images are averaged to get a representative time-averaged image, averaging also reduces the noise considerably. Calibration for the void fraction is done by inserting 5 empty thin-walled plastic cylinders of known, different diameters into the venturi filled with water, and X-ray images are obtained. The wall thickness of the plastic cylinders is very small, so it is neglected. The obtained calibration curve is used to calculate the void fractions for the reconstructed images (explained in the results).

Filtered back projection is applied to the cone-beam 2D time-averaged X-ray image (Figure 3) using the ASTRA Toolbox [9]. It is a stabilized and discretized version of the inverse Radon transform. Density slices of the object perpendicular to the centerline axis are created. As the measurement section is axisymmetric, we assume axisymmetry in the flow. The 3D reconstruction is performed with a time-averaged image obtained from the projections recorded at a single viewing angle.

3. Results

Figure 3(a) shows the time-averaged X-ray intensity data. This panel is before tomographic reconstruction, i.e. a projection along the line between source and detector. Figure 3(b) shows reconstructed slices at different axial positions, showing growth of the cavitation cloud. Figure 3(a) and (b) show that most of the vapor is attached to the nozzle wall and persists until 4 diameters downstream of the throat. This is the point where the cavity detaches during the periodic shedding, this is also confirmed from the high-speed images [10]. After detaching, the vapor cloud moves towards the center of the venturi and diffuses with the liquid phase.

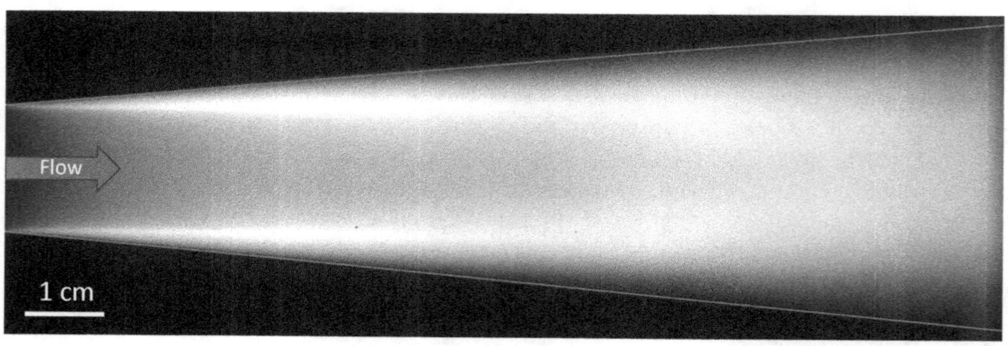

(a)

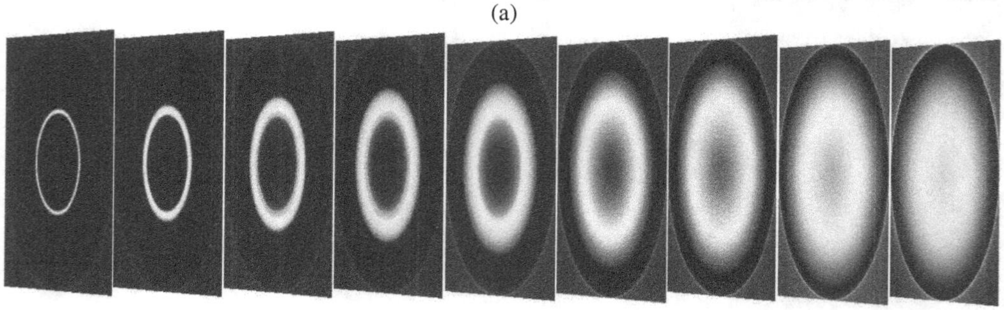

(b)

Figure. 3: (a) Time-averaged X-ray image of the cavitating venturi at σ = 0.46 (vapor is light gray, liquid is black). (b) Cross-sectional CT images at different axial positions showing growth of the cavitation cloud. The contrast in the images is adjusted for better understanding.

Figure 4 represents two slices from two different positions along the venturi length, where the vapor and liquid regions are shown in red and blue color, respectively. According to the calibration process discussed above and the corresponding color bar, the vapor fraction can be obtained easily for every pixel of the slice. It can be seen from Figure 4(a) and (b), the cavitation changes from sheet cavitation to cloud cavitation with the interface diffusing downstream of the venturi. The void fraction decreases as the cavitation structures expand in the liquid core.

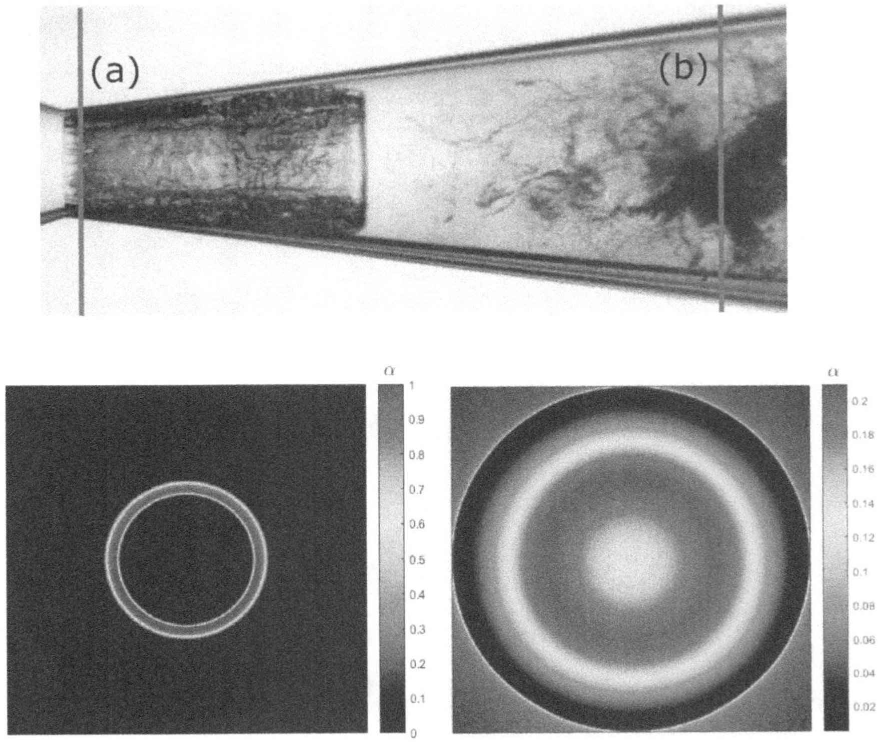

(a) (b)

Figure. 4: (Top) A high-speed snapshot showing cavitation in the venturi. In (a) and (b), the quantitative measurements of vapor fractions at two different locations along the venturi are shown. The red regions indicate the presence of vapor and the blue regions indicate the presence of liquid. Note the difference in color scales.

4. Conclusions and Outlook

In this study, the phenomenon of cavitation was examined by CT measurements of the flow through a venturi. The results look encouraging in terms of time-averaged vapor fractions. More detailed information about the cavitation development is extracted using the cross-sectional CT-measurements as compared to the high-speed imaging. We can now quantify the radial geometric features of this complex two-phase flow. This data will be essential to validate our assumptions regarding the physical mechanisms. Furthermore, it will be helpful for the validation of numerical studies.

Acknowledgments

SJ has received funding from the European Union Horizon 2020 Research and Innovation programme, Grant Agreement No 642536. CP has received funding from ERC Consolidator Grant "OpaqueFlows".

References

[1] C. Peng, S. Tian, and G. Li, "Joint experiments of cavitation jet: High-speed visualization and erosion test," *Ocean Engineering*, vol. 149, pp. 1–13, 2018.

[2] I. Khlifa, A. Vabre, M. Hočevar, K. Fezzaa, S. Fuzier, O. Roussette, and O. Coutier-Delgosha, "Fast x-ray imaging of cavitating flows," *Experiments in Fluids*, vol. 58, no. 11, p. 157, 2017.

[3] N. Mitroglou, M. Lorenzi, M. Santini, M. Gavaises, and D. Assanis, "Application of cone-beam micro-CT on high-speed Diesel flows and quantitative cavitation measurements," in *Journal of Physics: Conference Series*, vol. 656, p. 012094, 2015.

[4] D. Bauer, H. Chaves, and C. Arcoumanis, "Measurements of void fraction distribution in cavitating pipe flow using x-ray CT," *Measurement Science and Technology*, vol. 23, no. 5, p. 055302, 2012.

[5] H. Ganesh, S. A. Mäkiharju, and S. L. Ceccio, "Bubbly shock propagation as a mechanism for sheet-to-cloud transition of partial cavities," *Journal of Fluid Mechanics*, vol. 802, pp. 37–78, 2016.

[6] X. Long, J. Zhang, J. Wang, M. Xu, Q. Lyu, and B. Ji, "Experimental investigation of the global cavitation dynamic behavior in a venturi tube with special emphasis on the cavity length variation," *International Journal of Multiphase Flow*, vol. 89, pp. 290–298, 2017.

[7] S. Hayashi and K. Sato, "Unsteady Behavior of Cavitating Waterjet in an Axisymmetric Convergent-Divergent Nozzle: High Speed Observation and Image Analysis Based on Frame Difference Method," *Journal of Flow Control, Measurement & Visualization*, vol. 2, pp. 94–104, 2014.

[8] P. Tomov, S. Khelladi, F. Ravelet, C. Sarraf, F. Bakir, and P. Vertenoeuil, "Experimental study of aerated cavitation in a horizontal venturi nozzle," *Experimental Thermal and Fluid Science*, vol. 70, pp. 85–95, 2016.

[9] W. van Aarle, W. J. Palenstijn, J. Cant, E. Janssens, F. Bleichrodt, A. Dabravolski, J. De Beenhouwer, K. J. Batenburg, and J. Sijbers, "Fast and flexible X-ray tomography using the ASTRA toolbox," *Optics express*, vol. 24, no. 22, pp. 25129–25147, 2016.

[10] W. J. Hogendoorn, "Cavitation: Experimental investigation of cavitation regimes in a coverging-diverging nozzle," *Master's thesis, Delft University of Technology*, 2017. Available online at uuid:823a18f0-66a8-4ffd-a688-c3dadf62c4da.

© 2018 ASME

www.ingramcontent.com/pod-product-compliance
Lightning Source LLC
LaVergne TN
LVHW080059011026
849160LV00013B/994